Encyclopedia of Genetics, Genomics, Proteomics, and Informatics

George P. Rédei

Encyclopedia of Genetics, Genomics, Proteomics, and Informatics

3rd Edition

Volume 2
M–Z

With 1914 figures and 94 tables

Author:
George P. Rédei
Professor Emeritus, University of Missouri, Columbia
3005 Woodbine Ct. Columbia, MO 65203-0906
USA
redeia@mchsi.com
redeig@missouri.edu
www.missouri.edu/~redeig

A C.I.P. Catalog record for this book is available from the Library of Congress

ISBN: 978-1-4020-6753-2
This publication is available also as:
Electronic publication under ISBN 978-1-4020-6754-9 and
Print and electronic bundle under ISBN 978-1-4020-6755-6

Springer is part of Springer Science+Business Media

springer.com

Printed on acid-free paper SPIN: 12121977 2109 — 5 4 3 2 1 0

To Paige, Grace and Anne

Acknowledgements

I thank my wife Magdi for her patience during writing and for critical reading of the text. My daughter Mari introduced me into word processing. My son-in-law, Kirk has been very supportive. Granddaughters, Grace, Paige and Anne are most inspirational. I am grateful to countless numbers of colleagues on whose work this material is based and to whom I could not refer because of limitations of space. I am indebted to colleagues, especially to Dr. Csaba Koncz, Max-Planck-Institut, Cologne, Germany for many useful discussions. My students at the University of Missouri, Columbia, MO, and students and colleagues, particularly Dr. András Fodor, at the Eötvös Lóránd University of Basic Sciences, Budapest, provided purpose for undertaking this project. I am also indebted to Jane D Phillips, Director of Development for Life Sciences at the University of Missouri for her interest and encouragement.

Some of the chemical formulas are based on the Merck Index, on the Aldrich Catalog and Fluka Catalog.

During the preparation of the third edition, Mark Jarvis has been most helpful in resolving a variety of computer problems.

I appreciated the comments from the readers on the first and second editions. The first e-mail from Dr. SLC said: "Thank you for assembling such concise explanations of all genetic concepts in a single volume". Similar letters came from many others, which for reasons of space, I cannot quote here.

I am thankful to the public reviews for the constructive comments and suggestions.

Author is much indebted to Anil Chandy for expert advice, friendship, cooperation and understanding during all phases of the production of this book.

Cudweed by Konrad von Gessner (1597), the famous Swiss savant and zoological and botanical illustrator. Von Gessner's work has been borrowed by many, among them the German Joachim Camerarius (1500–1576) of Tübingen, a great authority on classics, religion and science and whose descendant Rudolph Jacob Camerarius (1665–1721) would become the first experimental geneticist, and discover the love life of plants.

Acknowledgements

Preface to the Third Edition

"When I acquired the right words…, it was easier for me to understand what I was thinking: it is difficult to reason through a problem if you cannot articulate what the problem is."

(John F Bruzzi 2006 *New England J. Med.* 354:665).

This volume is essentially an improved and enlarged new version of the previous, much-acclaimed book. The third edition has over 1,000 pages more of new material than the second. The latest progress in current hot topics such as stem cells, gene therapy, small RNAs, transcription factories, chromosomal territories, networks, genetic networks, ENCODE project, epigenetics, histone and protein biology, prions, hereditary diseases, and even patents are covered. The number of illustrations increased to nearly 2,000 and several hundreds are in four-color. The old entries have been revised, updated and expanded. Cross-references among entries have been increased. Retractions and corrigenda are pointed out. Nearly 1,800 database and web server addresses, about 14,000 journal paper references and more than 3,000 current book titles are included. Interesting historical vignettes lend some insight into the lighter sides of biology. I hope the reader finds the absence of laboratory jargon refreshing.

The encyclopedia will equip its reader to prepare journal papers, write or review research proposals or help organize a new course or update a current course. Students will find the topics useful for preparing for exams or for writing term papers. It may also appeal to basic and applied biologists and to practitioners of many other fields. The readers of previous editions appreciated the clarity of the basics in this book. A sample of what professional reviewers wrote about the book is carried for the reader's reference. In addition, about the current publications in general, the Editor of Science had the following comment. "Each specialty has focused in to a point at which even the occupants of neighboring fields have trouble understanding each others' papers"… "The language used in Reports and Research Articles is sufficiently technical and arcane that they are hard to understand, even for those in related disciplines." (Kennedy, D 2007 Science 318:715).

One of the most fascinating features of science is its continuous evolution. My goal was to emphasize the principles, provide numerical data and guide to resources that are more difficult to access from other publications. The book can assist the reader to make better use of the Internet but the Internet and textbooks are no substitutes for this book. Unlike the majority of books, this Encyclopedia will not be outdated because the continuously renewed and updated databases and web sites listed assure its usefulness even in the distant future. Describing individual genes in different organisms—with some exceptions—is no longer practical in a single book or even with the aid of an excellent resource such as PubMed due to the multitude of abstracts (more than 15,000,000 entries from more than 19,000 journals). The web sites—listed in this book—can however, provide great help in identifying many genes, their synonyms, functions and interacting partners as well as critical references to them. Although basic statistical concepts are explained in simple terms, most of the theoretical mathematical models or detailed laboratory procedures are not included because of the difficulties of describing the techniques within the limited space available in a single book. The abundance of references can lead the reader in the right direction.

Selection of papers for inclusion is also a continuous challenge. During 1992 to 2001 4,061 journals published more than 3.47 million peer-reviewed articles in health-related areas alone (Paraje, G et al. 2005 *Science* 308:959). Although this book is quite complex, integrated, and referenced, it does not include everything but it might be a useful guide to almost everything one needs in biology. For a proof of principle, I suggest that you look up any concept what you know or what you are uncertain or have doubts about.

I welcome all readers and I promise to respond to questions or comments.

GPR
redeia@mchsi.com

First to Read for the Third Edition

The organization of the expanded and revised third edition is slightly different from that of the previous ones. The material is still in alphabetical order but in a somewhat different style. Numbers involved with the entries do not affect the order. Entries beginning with Greek letters are sorted as if they would be in Roman or the Greek letter follows the term. Words followed by a comma and another word precede entry words without comma, e.g., "antibody, secondary" is followed by "antibody detection". Hyphenated entries are sorted as single words. The spelling of some terms vary because in the scientific literature some technical terms are written either with a *c* or *k* or with an *e* or *ae*. Some abbreviations are used in the literature as e.g., PGD or P.G.D. and both may not be used in this book. Here the most common usage is favored. Some entries are qualified by another word added after and in others the qualifier comes first. Most entries are followed by cross-references that guide to relevant topics. In particular cross-references qualifiers are not separated by comma even when they are with comma in the main list. In case you make electronic searches it is frequently more practical to use only part of the words because in the alphabetical list their ending may be different, e.g., maximum or maximal. An attempt has been made to guide the reader to the desired entry when necessary. In rare instances the reader may need to search for synonyms or related terms to find the desired entry. Thank you for your patience.

This book contains a large number of Internet addresses under the heading of "databases" and even more after many entries. Every effort has been made to keep the web addresses current. Unfortunately, they are altered frequently and it is likely that some will cease to exist or will change by the time the book gets to your hands. In some instances Word alone may not open the URL site directly, but Internet Explorer or Safari or even the search engines Google, Yahoo, AltaVista or others can be used to access the sites.

It must be remembered that data are not knowledge. The data must be integrated into science and the aim of the author has been the facilitation of such integration. The contents of the entries are based on the best information available in the literature at the time of the completion of the writing. As sources of information like most human products, are not always perfect, the author and publishers cannot claim perfection or assume legal responsibility for errors beyond their control.

"Knowledge…built on opinion only, will not stand". Linnaeus, 1735

Preface to the First Edition

The primary goal of this Manual is the facilitation of communication and understanding across the wide range of biology that is now called genetics. The emphasis is on recent theoretical advances, new concepts, terms and their applications. The book includes about 18 thousand concepts and over 650 illustrations (graphs, tables, equations and formulas). Most of the computational procedures are illustrated by worked-out examples. A list of about 900, mainly recent books, is provided at the end of the volume, and additional references are located at many entries and illustrations. The most relevant databases are also listed. The cross-references following the entries connect to a network within the book, so this is not just a dictionary or glossary. By a sequential search, comprehensive, integrated information can be obtained as you prepare for exams, or lectures, or develop or update a course, or need to review a manuscript, or just wish to clarify some problems. In contrast to standard encyclopedias, I have used relatively short but greater variety of entries in order to facilitate rapid access to specific topics. This Manual was designed for students, teachers, scientists, physicians, reviewers, environmentalists, lawyers, administrators, and to all educated persons who are interested in modern biology. Concise technical information is available here on a broad range of topics without a need for browsing an entire library. This volume can always be at your fingertips without leaving the workbench or desk. Despite the brevity of the entries, the contents are clear even for the beginner. Herbert Macgregor made the remarkable statement that in 1992 about 7,000 articles related just to chromosomes were scattered among 627 journals. Since then, the situation has become worse. Many publications—beyond a person's specialization—are almost unreadable because of the multitude of unfamiliar acronyms and undefined terms. Students and colleagues have encouraged me to undertake this effort to facilitate reading of scientific and popular articles and summarize briefly the current status of important topics. According to Robert Graves (a good poem) "makes complete sense and says all that it has to say memorably and economically". I hope you will appreciate the sense and economy of this Manual. I will be much indebted for any comment, suggestion and correction.

GPR
3005 Woodbine Ct.
Columbia, MO 65203–0906, USA
Telephone: (573) 442–7435,
e-mail: redeig@missouri.edu or redeia@mchsi.com

"I almost forgot to say that genetics will disappear as a separate science because, in the 21st century, everything in biology will become gene-based, and every biologist will be a geneticist." Sydney Brenner, 1993

Preface to the Second Edition

The majority of the users of the first edition considered this book as an encyclopedia because the cross-references tied the short entries into comprehensive reviews of the topics. In contrast to "big encyclopedias" in this work only a few entries exceed a couple of thousand words and that make it much faster to find the specific concept or term of interest. Unlike multi-author works this is practically free of redundancy and it is compact in size but not in depth of information. One of the reviewers pointed out that many of the topics covered could not be found in any other single book, including encyclopedias, dictionaries or glossaries. Another reviewer appreciated it as a broad resource of information that may take a lengthy search to uncover without it.

Since the publication of the 1st edition, I have steadily updated and improved on the topics. I have added many new concepts, illustrations, books and database addresses. (The database addresses are in an unfortunate flux and some may be out existence by the time you wish to log in; therefore I provided several to minimize the problem beyond my control.) This second edition contains about 50% more information and more than twice as many illustrations than the 1st edition. A new feature is the predominantly current, over 7,000 text references to journal articles. Their bibliographies may help to locate additional key and classical papers. The General References at the end include about 2,000 books. For additional medical genetics references I suggest the use of OMIM at The National Center for Biotechnology Information (http://www.ncbi.nlm.nih.gov/, see also Grivell, L 2002. *EMBO Rep*. 3:200). I have greatly expanded the cross-references among the entries because the users found this feature especially useful. Color plates were added to the end of the book. At the end of the files there are some historical vignettes.

Since the publication of the first edition the need for such a book became even more evident. In the literature unexplained concepts, terms and acronyms are on the increase and even a name DAS (*dreaded abbreviation syndrome*) has been coined for the malaise (*Science*, 283:1118). The users of the first edition agreed with the Nobel-laureate geneticist, HJ Muller who posed and answered the still current problem: "Must we geneticists become bacteriologists, physiological chemists and physicists, simultaneously with being zoologists and botanists? Let us hope so." (*Amer. Nat*. 1922, *56*:32).

The vision of genetics today is not less than the complete understanding how cells and organisms are built, how they function metabolically and developmentally, and how they have evolved. This requires the integration of previously separate disciplines, based on diverse concepts and tongues. Whatever is your specialization or interest, I hope you will find this single volume helpful and affordable.

Although I had the aim of comprehensiveness beyond all the available compendiums, there were hard decisions of what to include and what to pass. The same science may appear different depending on who, how and when looks at it (up or down) as Gerald H Fisher's art above (man or woman or both) illustrates the point.*

Thank you for the appreciation of the first edition. I will be much indebted for your comments and suggestions.

GPR
redeia@mchsi.com

*By permission of Perception and Psychophysics 4:189 (1968).

Preface to the Second Edition

M

M: Abbreviation for mitosis or in statistics for the mean. ►mitosis, ►mean

M9 Bacterial Minimal Medium: 5 concentrated salt solution g/L H_2O, Na_2HPO_4. 7 H_2O 64, KH_2PO_4 15, NaCl 2.5, NH_4Cl 5 → 200 μL, 1M $MgSO_4$ 2 mL, glucose 4 g, fill up to 1 L after any supplement (e.g., amino acids) added.

M Component: Same as paraprotein.

M Cytotype. ►hybrid dysgenesis

M13 Phage: ►bacteriophages, ►DNA sequencing

M Phase: Part of the cell cycle when the structure, movement and separation of chromosomes or chromatids are visible and the nuclear divisions are completed. ►mitosis and ►meiosis

M Phase Histone-1 Kinase: ►growth-associated kinase

M Phase Promoting Factor (MPF): Same as maturation promoting factor.

M RNA: ►Cowpea mosaic virus (CPMV)

M Virus: ►killer strains

Ma (*mille mille annus*): A million years. ►My

MAb: Monoclonal antibody.

Mab: Mouse antibody.

MAC: Mammalian artificial chromosome. ►HAC, ►HAEC, ►YAC, ►BAC, ►PAC, ►SATAC; Grimes B, Cooke H 1998 Hum Mol Genet 7:1635.

MAC-1: The oligonucleotide-binding heparin-like integrin on the surface of neutrophils, macrophages and killer cells (NK). ►integrin, ►heparin, ►killer cell, ►blood

Macaca: Rhesus Old World monkeys. ►cynomolgus, ►Rhesus, ►*Cercopithecidae*

Macadamia Nut (*Macadamia* spp): A delicious nut, 2n = 2x = 28.

Macaloid: A clay colloid capable of adsorbing ribonuclease during disruption of cells and can be centrifugally removed from the RNA preparation after extraction by phenol. ►RNA extraction

Macaroni Wheat (hard wheat): *Triticum durum* (2n = 4x = 28) varieties containing the AB genome and taxonomically classified as a subgroup of *T. turgidum* (see Fig. M1). The endosperm of the commercially used varieties has more β-carotenes and therefore, the milled products show yellowish color, desirable for pastas. It is milled to a grainier product (semolina) that favors cooking quality. The durum wheats have higher protein and mineral content than the soft wheats. The turgidum wheats, except the durums, are not used for food. ►*Triticum*

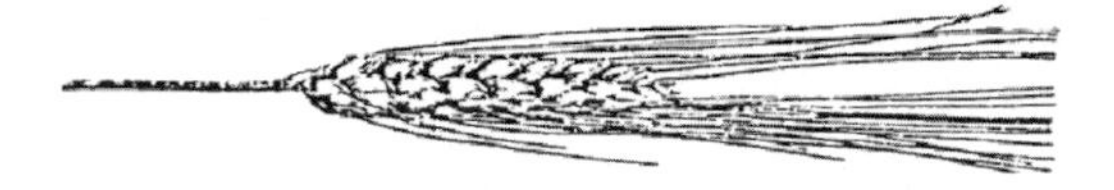

Figure M1. Macaroni wheat (*Triticum durum*)

Macerozyme: A pectinase enzyme of fungal origin; it is used for the preparation of plant protoplasts in connection with a cellulase. ►cellulase, ►protoplast

MACH (caspase 8): ►caspases

Machado-Joseph Syndrome (Azorean neurologic disease, spinocerebellar ataxia Type 3, MJD): It is a 14q32.1 chromosome dominant defect of the central nervous system involving ataxia, and other anomalies of motor control. In the gene locus, abnormally increased number of CAG repeats (from 13–36 in normal→61–84 in disease) occurs. If the long CAG repeats are translated into polyglutamine, cell death may result. The neurodegeneration may suppressed by the chaperone HSP70. ►ataxia, ►muscular atrophy, ►fragile sites, ►trinucleotide repeats, ►spinocerebellar ataxia, ►RNAi, ►Josephin domain

Machine, Biological: A set of interacting components performing functional specialized role. ►network

Machine Learning: The study of computer algorithms in the interest of improving the study of scientific data. A machine can learn from experience or extract knowledge from a database. ►support vector machine; Mjolsness E, DeCoste D 2001 Science 293:2051.

Machine Reading: A new form of computer, searching the scientific literature is under development. Its eventual use, if all publishers consent and either the user or the owner(s) of the publications consent it may revolutionize data mining. It may link various concepts across the field of interest, e.g., disease and molecule(s) in an extremely efficient way. See for example: http://arrowsmith.psych.uic.edu/arrowsmith_uiuc/index.html.

Macroarray Analysis: Mutations of unknown function are grown in the presence of compounds that make them sensitive if they are defective, e.g., cell wall-binding calcofluor dyes (fluorescent brightener 28/Tinopal) detects a group of yeast mutations defective in cell wall biogenesis. Mutations in several different gene loci involved in microtubule functions are sensitive to benomyl/benlate ($C_{14}H_{18}N_4O_3$, a fungicide and ascaricide). Genetic defects in sugar utilization may not grow on the non-fermentable glycerol if the mutation affected a certain domain of the protein product. Thus, this type of analysis can find NORFs involved in certain pathways. ►genomics, ►NORF, ►microarray hybridization, ►synexpression; http://strc.herts.ac.uk/bio/BioArray/doc/BioArray.htm.

Macrocephaly: A common feature of several mental retardations; it is classified as such when the circumference of the head of an individual exceeds by more than 2 standard deviations the mean of the population. ►microcephaly

Macroconidium: A large multinucleate conidium (see Fig. M2). ►conidia, ►microconidia, ►fungal life cycle

Figure M2. Macroconidium

Macrochromatin: ►lyonization

Macrocycles: These are cyclic macromolecules or a portion of macromolecules or low-molecular weight synthetic compounds. A DNA library can be translated into small macrocycles. These can be targeted to proteins and subjected to selection and used for probing biological functions (Gartner ZJ et al 2004 Science 305:1601).

Macroencapsulation: ►microencapsulation

Macroevolution: A major genomic alteration, which produced the taxonomic categories above the species level. ►evolution, ►microevolution

Macrogametophyte: A megaspore. ►gametogenesis in plants

Macroglobulinemia (Waldenström syndrome): HLA-linked immunodeficiency resulting in increase in IgM in the blood, thrombosis, skin, nose and gastrointestinal bleeding. ►immunodeficiency, ►immunoglobulins, ►multiple myeloma

Macroglossia: An abnormally large tongue; occurs in Down syndrome, Beckwith-Wiedemann syndrome and other hereditary disorders.

Macrolesion: Visible alterations in the chromosomes in genetic toxicology. ►Gene-Tox

Macrolide: Type of antibiotic (including more than one keto and hydroxyl groups such as erythromycin, troleandomycin) associated with glycoses. *Streptomyces* bacteria produce these compounds that inhibit the 70S ribosomes by binding to the large (50S) subunit and interfering with peptidyl transferase. ►antibiotics, ►maytansinoids, ►glycoses, ►*Streptomyces*, ►protein synthesis, ►ribosome, ►hysteresis; Retsema J, Fu W 2001 Int J Antimicrob Agents 18 Suppl 1:3; Berisio R et al 2003 Nature Struct Biol 10:366.

Macromere: A large blastomere. ►blastomeres

Macromelic Dwarfism (Desbuquois syndrome): A skeletal and digital anomaly.

Macromolecule: Molecules with molecular weight of several thousands to several millions such as DNA, RNA, protein and other polymers; macromolecular structure database: http://www.ebi.ac.uk/msd.

Macromutation: A genetic alteration involving large, discrete phenotypic change.

Macronucleus: A larger type of polyploid nucleus in Protozoa. While inheritance is mediated through the small micronucleus, the macronucleus is directing metabolic functions by being transcriptionally active and the latter is responsible for the phenotype of the cells. After the internally eliminated sequences (IES) are removed from the micronuclear DNA, the leftover tracts are joined into macronuclear-destined sequences (MDS) to form the macronucleus. The macronuclear genome is derived at sexual reorganization during conjugation. This involves various chromosomal rearrangements. Also during conjugation, the germline sequences are transcribed to produce double-stranded RNA, which can guide specific DNA deletions in the macronuclei. This mechanism—resembling RNAi—provides a means for elimination of transposons and other invaders (Yao M-C et al 2003 Science 300:1581; Yao M-C, Chao J-L 2005 Annu Rev Genet 39:537). Macronuclear differentiation includes site-specific fragmentation of the micronuclear chromosomes. A 15-bp site that is required for fragmentation marks each fragmentation site. The two ends of the fragments form telomeres and thus become *autonomously replicating pieces* (ARP). Although normally inheritance is mediated by the micronucleus or the mitochondria, in some instance, macronuclear inheritance is detected. In this case, apparently the old macronucleus determines some rearrangements (deletions) in the new macronucleus. The participation of a short scanning RNA (sRNA) has been suggested. The macronucleus contains about 45 copies of the ~300 subchromosomal fragments except the rDNA, which is amplified by ca. 10,000 times. The macronucleus divides by fission rather than by mitosis yet recombination between and within genes occurs. In the absence of recombination, groups of genes may stay together and this phenomenon is called *coassortment*. The genes of these polyploid macronuclei may sort out to pure form of the initial heterozygous state by a process of *phenotypic assortment*. On the basis of the assortments the genes may be mapped genetically. The macronuclei are formed by the fusion of two diploid micronuclei of the zygotes (see Fig. M3). ►*Paramecium*, ►*Tetrahymena*, ►chromosome diminution, ►protozoa, ►IES, ►RNAi; Katz LA 2001 Int J Syst Evol Microbiol 51(pt4): 1587.

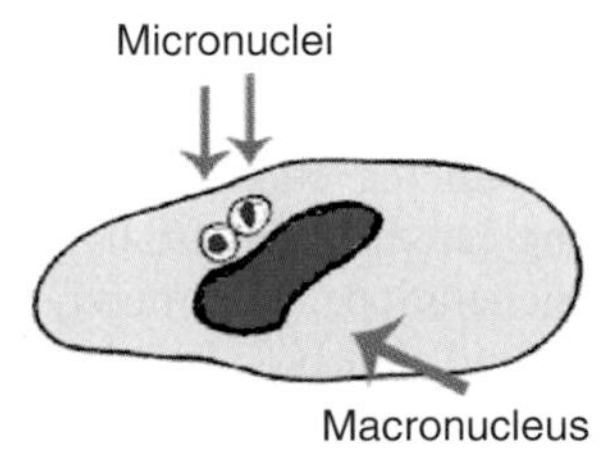

Figure M3. Paramecium

Macronutrient: Required in relatively large quantities by cells.

Macroorchidism: A condition of larger than normal testes. ►fragile X

Macropexophagy: The reduction of the number of peroxisomes in yeast grown on methanol after exposure to excessive amounts of glucose or ethanol. ►micropexophagy, ►pexophagy

Macrophage (Mφ): Phagocytic cell of mammals with accessory role in immunity. Macrophage cells are produced by the bone marrow stem cells as monocytes, which enter the blood stream, and within two days of circulation they enter the various tissues of the body and develop into macrophages. The macrophages may differ in shape but generally, they have single large round or indented nuclei, extended Golgi apparatus, lysosomes and vacuoles for the ample storage of digestive enzymes. The cell membrane is forming microvilli of various sizes, suited for phagocytosis (engulfing particles) and pinocytosis (uptake of fluid droplets) and thus are important as part of the cellular defense mechanism against foreign antigens, including even tumor antigens. The macrophages have a wide array of receptors, including the Fc, complement, carbohydrate, chemotactic peptide and extracellular matrix receptors. Some macrophages phagocytize apoptotic cells (see Fig. M4). Tumor necrosis factor (TNFα) moved from endosomes to the surface of the cells by VAMP facilitates its function in immunity (Murray RZ et al 2005 Science 310:1492). Macrophages can mediate angiogenesis, cell migration and metastasis of cancer. Because they are genetically quite stable, they may be good targets of therapeutic intervention in some cancers (Condeelis J, Pollards JW 2006 Cell 124:263). ►granulocytes, ►immune system, ►antibody, ►apoptosis, ►phagocytosis, ►TNF, ►VAMP, ►angiogenesis, ►metastasis

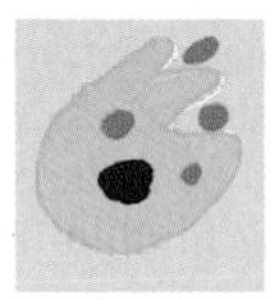

Figure M4. Macrophage cell ingesting bacteria.

Macrophage Activity Factor: MAF is a lymphokine, identical to IFN-gamma. ►lymphokine, ►interferon

Macrophage Colony Stimulating Factor (M-CSF, 5q33.2-q333): M-CSF's receptor is the KIT oncogene. The receptor MCSFR is a protein tyrosine kinase. It promotes B cell proliferation with the assistance of IL-7. ►FMS, ►KIT oncogene, ►colony stimulating factor, ►RAS, ►signal transduction, ►GM-CSF, ►G-CSF, ►IL-7, ►B cell, ►osteoclast; ►cherubism, Csar XF et al 2001 J Biol Chem 276:26211.

Macrophage-Stimulating Protein (MSP): MSP is the 80-kDa serum protein stimulating responsiveness to chemoattractants in mice, that induces ingestion of complement-coated erythrocytes and inhibits nitric oxide synthase in endotoxin- or cytokine-stimulated macrophages. MSP is structurally homologous to hepatocyte growth factor, scatter factor (HGF-SF) but their targets are different. MSP has been located to human chromosome 3p21, a site frequently deleted in lung and renal carcinomas. ►HGF-SF, ►Met, ►Ron, ►macrophage, ►scatter factor; Stella MC et al 2001 Mol Biol Cell 12:1341.

Macrothrompbocytopathy: ►thrombocytopenia, ►giant platelet syndrome

Macrospore: ►megaspore

Macula: A spot or thicker area, particularly on the retina, often colored. The macula of the eye contains photoreceptor cone cells, critical for color and other aspects of vision. ►retina, ►choroidoretinal degeneration, ►macular degeneration, ►macular dystrophy, ►foveal dystrophy

Macular Corneal Dystrophy (MCD, 16q22): MCD involves progressive recessive opacity of the cornea. Two types may be distinguished on the bases of absence (MCDI) and presence (MCDII) of keratan sulphate in the serum. A third type (CHST6) has a defect involving corneal N-acetylglucoseamine-6-sulphotransferase, which has overlapping mutations within the two other types. ►Stargardt disease, ►hypotrichosis

Macular Degeneration: Dominant-acting 6q25 deletion causing degeneration of the vitelline layer of the eye. The dominant 2p16 located disease Malattia Leventinese and Doyne honeycomb retinal dystrophy are similar, to some extent, to macular degeneration (see Fig. M5) and show drusen (bright speckles on the retina) due to mutation in the epidermal growth factor (EGF)—containing fibrillin-like extracellular matrix protein 1 (EFEMP1).

The Stargardt like STGD3 locus is in human chromosome 6q14. This gene is responsible for ELOVAL4 (elongation of very long chain fatty acid) and autosomal dominant macular dystrophy (see Fig. M6). These diseases manifest with some variations in expression and onset. Age-related macular degeneration (ARMD) causing yellow deposits (drusen) in the outer layer of the retinal epithelium is a very common cause of blindness in the West and it is controlled by several independent as well as by cooperative loci (Majewski J et al 2003 Am J Hum Genet 73:540). One major ARMD locus is at 15q21 (Iyengar SK et al 2004 Am J Hum Genet 74:20). Single amino acid substitution at site 402 (tyrosine→histidine) of the complement H gene accounts for the increased risk factor of ARMD from 4.6 (heterozygotes) to 7.4 (homozygotes) in the families (Klein RJ et al 2005 Science 308:385; Edwards AO et al 2005 Science 308:421). Age-related macular degeneration risk loci were confirmed in complement factor H (*CFH*, 1q32), and functionally related genes in chromosome 10q26.13 (Loc387715) and in genes C2 and BF in chromosome 6q21.3 (Maller J et al 2006 Nature Genet 38:1055; Li M et al Nature Genet 38:1049). Deletion of *CFHR1* and *CFHR3* is associated with lower risk of age-related macular degeneration (Hughes AE et al 2006 Nature Genet 38:1173; see corrigendum in Nature Genet. 39:567). Polymorphism of the HTRA1 promoter (10q26) greatly increases the risk of developing ARMD (DeWan A et al 2006 Science 314:989).

Nornicotine, a metabolic product of cigarette smoke catalyzes retinal isomerization and is a promoting factor of ARMD (Brognan AP et al 2005 Proc Natl Acad Sci USA 102:10433). Oxidative stress appears to play a role in age-related macular degeneration and $Sod1^{-/-}$ mouse appears to be workable model for studies (Imamura Y et al 2006 Proc Natl Acad Sci USA 103:11282). The ApoE E4 (apolipoprotein) allele increases susceptibility to ARMD as well as to Alzheimer disease (Malek G et al 2005 Proc Natl Acad Sci USA 102:11900).

Although several approaches are being suggested to control ARMD, only laser irradiation has limited success to curtail the choroidal neovascularization, the underlying cause of the visual impairment of wet macular degeneration. The newly formed blood vessels leak fluid and blood under the retina, form scar tissues and gradually destroy vision. One potential cure involves injecting antibodies against vascular endothelial growth factor A (VEGF-A) into the eye, a procedure under clinical trial. The ciliary margin of the retina contains active stem cells capable of self-renewal and provides possibilities for the cure of retinal disease (Coles B L K et al 2004 Proc Natl Acad Sci USA 101:15772). Currently RNAi technology is under clinical evaluation. The light-induced blindness, affecting the early microscopists such as August Weismann (1834–1914), has a molecular cause different from macular degeneration, and it is caused by degradation of rhodopsin (Lee S-J, Montell C 2004 Current Biol 14:2076). In June 2006, the US Federal Drug Administration approved the use of two new drugs for treatment. The (48 kDa) is a recombinant humanized monoclonal IgG1 kappa-isotype antibody fragment produced in an *E. coli* expression system. It is not glycosylated and is used for intraocular injection. Bevacizumab (149 kDa) is also a recombinant humanized monoclonal IgG1 antibody

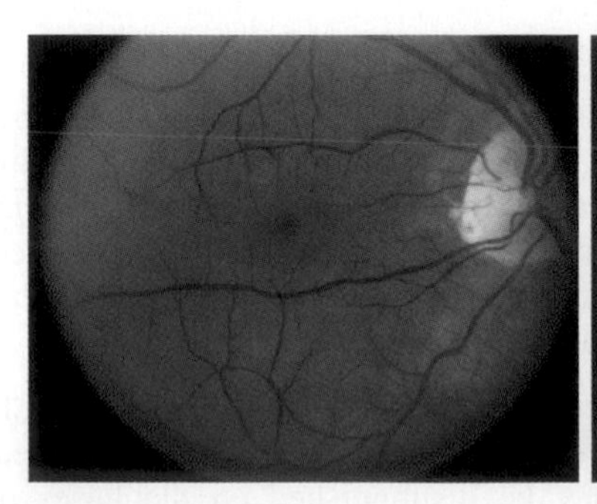
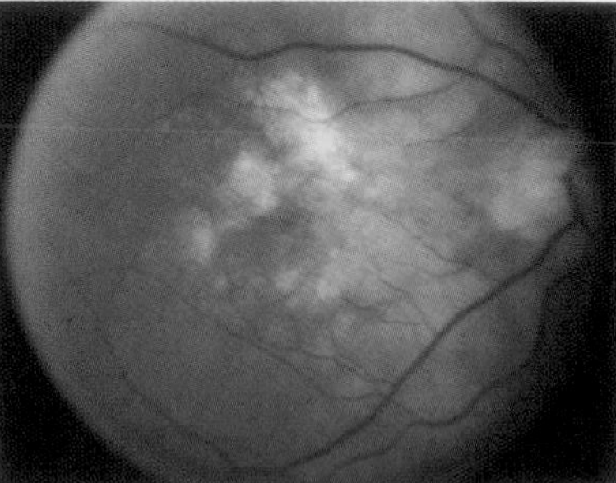
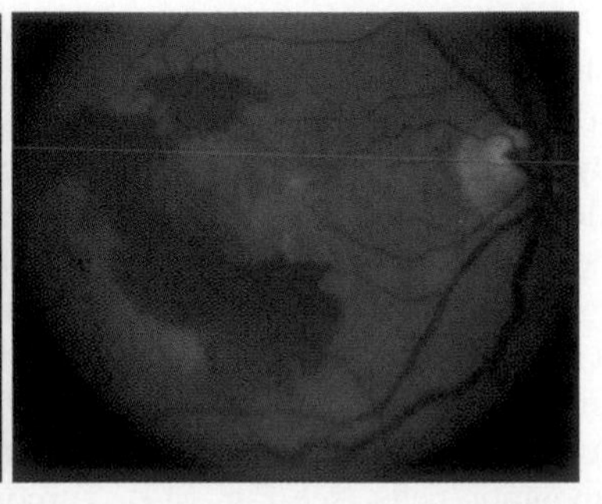

Figure M5. Macular degeneration. Left to right: Normal eye, age-related dry macular regeneration, macular degeneration with hemorrhage and fluid leakage. (The photographs are the courtesy of Dr. Timothy Holecamp and Ms. Jackie Bowman)

Figure M6. Peripheral vision in wet macular degeneration

produced in a Chinese-hamster-ovary mammalian-cell expression system, is glycosylated and used for intravenous infusion. Both the antibody fragment and the full-length antibody bind to, and inhibit all the biologically active forms of vascular endothelial growth factor (VEGF) A and are derived from the same mouse monoclonal antibody. Both drugs are promising, require repeated administration and are expensive (Steinbrook R 2006 New England J Med 355:1409). ►macula, ►eye diseases, ►RNAi, ►stem cells, ►vascular endothelial growth factor, ►angiogenesis, ►Stargardt disease, ►complement, ►monoclonal antibody, ►immunoglobulins, see Fig. I66; Marx J 2006 Science 311:1704; ARMD: de Jong PTVM 2006 N Engl J Med 355:1474; review: Rattner A, Nathans J 2006 Nature Rev Neurosci 7:860; Stone EM 2007 Annu Rev Med 58:477.

Macular Dystrophy: Autosomal recessive (8q24) and X-linked forms with symptoms similar to macular degeneration. The 1.4 Mb vitelline macular dystrophy (VMD2) gene (human chromosome 11q11-q13.1) encodes a 585 amino acid protein (bestrophin), named after the other name of the condition, Best's disease (autosomal dominant at 11q13). Its prevalence in the USA is about 3×10^{-5}. ►macula, ►eye diseases, ►Stargardt disease, ►retinal dystrophy

MAD: Multiwavelength anomalous dispersion analysis is a physical method for the determination of crystal structure of molecules.

Mad (*Mothers against decapentaplegic*): *Drosophila* gene controling several developmental events in the fly. Its human homolog is called SMAD1 (Sma in *Caenorhabditis*). SMAD is a TGF (transforming growth factor)/BPM (bone-morphogenetic protein) cytokine family-regulated transcription factor involving serine/threonine kinase receptors. The Mad family of proteins can be found in a wide range of organisms and upon dimerization with MAX they function as transcriptional repressors. The repressor function is correlated with the recruitment of the Sin3 protein binding to the Sin3-Mad interaction domain (Mad-Sid). SMAD is also called hMAD; MADR is its receptor. Mad is a component of the N-CoR/Sin3/RPD complex mediates the repression of certain classes of genes. ►tumor growth factor, ►cytokines, ►*dpp*, ►serine/threonine kinase, ►MYC, ►RPD, ►SMAD, ►Dpp, ►histone deacetylase, ►*decapentaplegic*

mad (*many abnormal discs*; 3–78.6): A homozygous disc and cell autonomous lethal, affecting several morphogenetic functions in *Drosophila.*

MAD2, MAD1 (mitotic arrest deficient, MAD2L1, 4q27; MAD2L2, 1p36): These are anaphase-regulating proteins, interacting with estrogen receptor-β. These are essential for normal mitosis checkpoints and their defect may lead to chromosome instability, cancer and embryonic lethality. Mad1 and Mad2 regulate the bipolar orientation of the chromosomes of yeast (Lee MS, Spencer FA 2004 Proc Natl Acad Sci USA 101:10655). The kinetochore localization of MADs requires the expression of Hec1/Ndc80p. The human Hec is highly expressed in cancer. Mad2 interacts with the kinetochores (Vink M et al 2006 Current Biol 16:755). ►anaphase, ►cell cycle, ►spindle, ►BUB; Skoufias DA et al 2001 Proc Natl Acad Sci USA 98:4492; Martin-Lluesma S et al 2002 Science 297:2267.

MADM: ►Mosaic Analysis with Double Marker

Mad/Max: Sequence-specific heterodimeric transcriptional repressors.

Mad Cow Disease: ►encephalopathy

MADS Box: A conserved motif of 56 residues in a DNA-binding protein of transcription factors involved in the regulation of *MCM1* (yeast mating type), *AG* (agamous homoeotic gene and a root morphogenesis gene of *Arabidopsis*) *ARG80* (arginine regulator in yeast), *DEF A* (deficient-flower) mutation of *Antirrhinum* and SRF (serum response factor in mammals) regulating the expression of the c-fos protooncogene. The MADS box has amino terminal sequence specificity and carboxyl dimerization domains. In addition, the SRF recruits accessory proteins such as ELK1, SAP-1 and MCM1 and relies on MATα1 and MATα2, STE12 and SFF. MADS box genes control vernalization responses in cereals. (See separate entries, ►mating type determination in yeast, ►vernalization, ►MEF; Jack T 2001 Plant Mol Biol 46:515).

M

Madumnal Allele: Derived from the female. ►Padumnal allele

MAF (macrophage activity factor): A lymphokine, an oncogene and a regulator of NF-E2 transcription factor. Heterodimers of Maf protooncogene family members promote the association with and the expression of NF-E2, whereas the homodimers are inhibitors of it. ►macrophage, ►lymphokine 4, ►oncogenes, ►NF-E2, ►homodimer, transcription factors; Swamy N et al 2001 J Cell Biochem 81:535.

MAF: Minor allele frequency.

MAFA (12p12–13): Inhibitory immune receptors on myeloid, mast and natural killer cells.

MAFFT-5: ►CLUSTAL W, ►interalign

Magainin: ►antimicrobial peptides

MAGE: The melanoma antigens encoded by several genes and expressed in different tumors. In normal tissue, MAGE is limited to the testes and wound healing. In normal tissues, the MART-1/Melan-A differentiation antigen represents the melanoma lineage. The immunogenic epitopes for MAGE-1 are EADPTGHSY and SAYGEPRKL and for MAGE-3: EVDPIGHLY, FLYGPRALV, MEVDPIGHLY. Each has different HLA specificity. The MART-1 peptide epitopes are: AAGIGILTV, ILTVILGVL, and GIGILTVL. ►antigen, ►melanoma, ►tumor antigen, ►amino acid symbols in protein sequences, ►Cancer-testis antigen; Otte M et al 2001 Cancer Res 61:6682.

M

MAGE (microarray and gene expression): ►microarray hybridization; http://www.mged.org/Workgroups/MAGE/mage.html.

MAGI-2 (membrane-associated guanylate kinase inverted-2): A scaffold protein enhancer of PTEN. ►PTEN; Vazqez F et al 2001 J Biol Chem 276:48627; Wu X et al 2000 Proc Natl Acad Sci USA 97:4233.

Magic Bullet: A specific monoclonal antibody is supposed to recognize only one type of cell surface antigen (e.g., one on a cancer cell) or growth factor receptors or differentiation antigens, and carries either the *Pseudomonas* exotoxin, or the diphteria toxin, or ricin or maytansinoids (an extract of tropical shrubs or trees) or a radioactive element (such as Y^{90}) capable of selectively unloading these harmful agents at the target cell and thus killing the cancer cell without much harm to any other cell. One such approach is to develop a special organ-homing peptide motif and combined with pro-apoptotic peptide domain. The homing peptide allows internalization only into the tumor cells where it onloads and disrupts the mitochondrial membranes and causes apoptosis (Ellerby HM et al 1999 Nature Med 5:1032). Some of these new drugs (Rituximab) carry an anti-CD20 monoclonal antibody and radionuclides are effective against B-cell non-Hodgkin lymphoma (Milenic DE et al 2004 Nature Rev Drug Discov 3:488). CD22 and CD33 proteins can be effectively targeted by monoclonal antibodies in lymphomas with the associated CalichDMH (*N*-acetyl-γ-chalicheamicin dimethyl hydrazid), which is a potent DNA-binding antibiotic (DiJoseph JF et al 2004 Blood 103:1807). There may be some problems with monoclonal antibodies because the same tumor tissue may express different antigens (Scanlan MJ et al 2002 Immunol Revs 188:22). ►ADEPT, ►vascular targeting, ►hybridoma, ►monoclonal antibody, ►receptor, ►antigen, ►diphteria toxin, ►ricin, ►immunotoxin, ►maytansinoid, ►isotopes, ►CD20, ►CD22, ►CD33; Frankel AE et al 2000 Clinical Cancer Rev 6:326.

Magic Number: The 64 triplet (4^3) combinations of the four nucleotides specify the 20 natural amino acids in all organisms. (See Gamov G, Yčas M 1955 Proc Natl Acad Sci USA 41:1011).

Magic Spots: These are pppGpp and ppGpp nucleotides that serve as effectors of the stringent control. RelA has apparently two catalytic sites, one for the synthesis of (p)ppGpp and another for hydrolysis. (p)ppGpp controls the elongation of DNA replication in response to the nutritional status of the cell (Wang JD et al 2007 Cell 128:865). ►stringent control, ►Rel oncogene; Schattenkerk C et al 1985 Nucleic Acids Res 13:3635.

Magnaporthe grisea: Rice blast pathogenic fungus with sequenced genome of 37,878,070 bp and with predicted 8,868 protein coding sequences in 7 chromosomes (See Rice, Dean RA et al 2005 Nature [Lond] 434:980). For efficient initiation of infection of rice, the formation of an appressorium, the conidial cells must die (Veneault-Fourray C et al 2006 Science 312:580). Insertion mutagenesis using *Agrobacterium* vector revealed 202 new pathogenicity loci in the fungus (Jeon J et al 2007 Nature Genet 39:561). ►rice, ►appressorium

Magnesium Transport: The level of Mg^{2+} is important for many cellular processes such as ATP-dependent reactions and stabilization of membranes and ribosomes. A two-component system acting at the 5′-untranslated sequences regulates magnesium transporter (MgtA) protein synthesis. In *Salmonella* in low-Mg^{2+} environment, the PhoQ senses magnesium and promotes phosphorylation of the DNA-binding PhoP of the two-component regulatory system. Phosphorylated PhoP binds to the single stem-loop structure in the promoter of *MgtA* and stimulates Mg^{2+} increase into the cytoplasm. When the level of Mg^{2+} reaches a higher level a two-loop structure forms in the 5′UTR and shuts off the transcription of the transporter (Cromie MJ et al 2006 Cell 125:71). ►two-component system

Magnetic Relaxation Switches: The nanometer-size colloidal metal particles may be coupled to affinity ligands and used as chemical sensors. Highly uniform magnetic nanoparticles can be covalently and stoichiometrically attached to oligonucleotides, nucleic acids, peptides, proteins, receptor ligands and antibodies. These nanomagnetic probes can be assembled also into larger units. These superparamagnetic scale particles can efficiently dephase the spins of the surrounding water protons and enhance spin-spin relaxation times. This technology can thus detect at miniaturized scale DNA–DNA, Protein–Protein, Protein–Small Ligand and enzyme reactions. Thus, e.g., non-sense DNA sequences, green fluorescent mRNA can be identified even in cell lysates. (See Perez JM et al 2002 Nature Biotechnol. 20:816).

Magnetic Resonance: ►nuclear magnetic resonance spectroscopy

Magnetic Targeting: Magnetic particles (ferrofluids) bound to mitoxanthrone (a cytostatic anthraquinone derivative) when injected into arteries can be concentrated in an external magnetic field aimed at a tumor (e.g., squamous cell carcinoma) can effectively cause remission without toxicity (Alexious C et al 2000 Cancer Res 60:6641). Similar procedure using superparamagnetic nanoparticles permit the tracking of progenitor cells (Lewin M et al 2000 Nature Biotechnol 18:410) or interaction of biomolecules within cells (Won J et al 2005 Science 309:121). ►magnetic relaxation switches

Magnetoreception: The response of organisms to the magnetic field of the Earth for behavior and orientation.

Magnetosome: Intracellular organelles in microbes that contain mineral crystals within a lipid membrane (see Fig. M7). The aquatic magnetotactic bacteria can follow the magnetic field lines of the earth. The magnetosomes may appear in rows. (See Bazylinski DA, Frankel RB 2004 Nature Rev Microbiol 2:217; Komelli A 2007 Annu Rev Biochem 76:351).

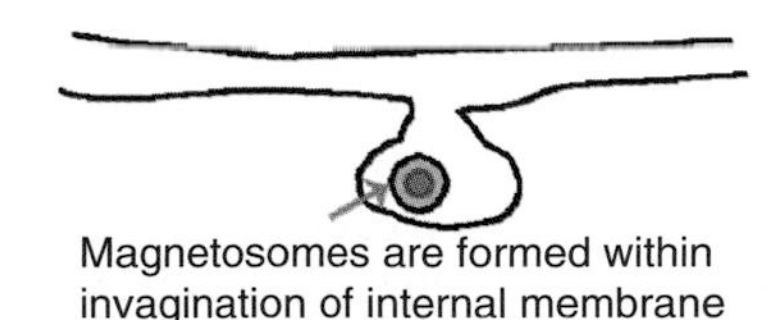

Magnetosomes are formed within
invagination of internal membrane

Figure M7. Magnetosome

Magnification: Increase in the units of ribosomal genes, hypothesized to occur by extra rounds of limited replication or unequal sister-chromatid exchange. At the *bobbed* (*bb,* 1–66.0) locus present in both sex-chromosomes of *Drosophila* in about 225 copies organized as large tandem arrays, separated by non-transcribed spacers. The copy numbers of *bb* in wild population Y-chromosomes may vary 6-fold. ►ribosomes, ►microscopy

MAGUKs (membrane-associated guanylate kinases): Mediate nuclear translocation and transcription. Membrane-associated guanylate kinases (PSD post-synaptic density-MAGUKs) mediate synaptic targeting, with remarkable functional redundancy within this protein family. PSD-95 and PSD-93 independently mediate AMPA-Receptor targeting at mature synapses. The loss of either PSD-95 or PSD-93, silences largely nonoverlapping populations of excitatory synapses. In adult PSD-95 and PSD-93 double knockout animals, SAP-102 (synaptic protein) is upregulated and compensates for the loss of synaptic AMPA-Rs (Elias GM et al 2006 Neuron 52:307). ►zona occludens, ►ELL, ►AMPA; McMahon L et al 2001 J Cell Sci 114 (pt 12):2265.

Maintenance Methylase: Keeps up methylation through cell divisions. ►methylation of DNA

Maize (*Zea mays*): It belongs to the family of *Gramineae* (grasses) and the tribe *Maydeae* along with teosinte (*Euchlena mexicana*) and the genus *Tripsacum* (with a large number of species). The male inflorescence of maize is more similar to that of teosinte than the female inflorescence. The ear or maize carries generally 8 to 24 rows of kernels whereas teosinte has only two rows. The teosinte ear is fragile the maize ear is not. The seed as is commonly named is really a karyopse (caryopse), a single-seed fruit (kernel). The basic chromosome number x = 10. *Tripsacum* resembles more some members of the *Andropogonaceae* family than to these two closer relatives and its basic chromosome number x = 18. Teosinte can be crossed readily with maize and the offspring is fertile whereas *Tripsacum* is more or less strongly isolated sexually and their hybrids are not fully fertile. Teosinte genes display the same chromosomal and gene arrangements as maize in contrast to *Tripsacum* that is more dissimilar. The evolution of modern maize has been traced to teosinte (Jaenicke-Després V et al 2003 Science 302:1206). The seed of teosinte is covered by a tough glume, which does not make it suitable for human consumption. Mutation in a single gene (*teosinte glume architecture, tga1*, maize chromosome 4) rendered maize grains naked during evolution and valuable for domestication (Wang H et al 2005 Nature [Lond] 436:714). All three genera have evolved in the

Western Hemisphere. Maize is one of the most thoroughly studied plants in genetics. It is a monoecious species; under natural conditions, it is allogamous. Approximately 500 kernels may be fixed on the cob for easy Mendelian analysis for a good number of endosperm, pericarp and embryo and seedling characters. Mendelian genes have been identified at about 1000 loci and more than 500 have been mapped. RFLP maps are also available. The gene number in maize is estimated to be ~33,000 to 54,000 (Fu Y et al 2005 Proc Natl Acad Sci USA 102:12282). The characteristic pachytene chromosomes facilitate cytogenetic analyses. Several genes controlling meiosis have been identified. Cytogenetic studies with maize contributed significantly to the understanding of chromosomal rearrangements. Transposable genetic elements were first recognized in maize (see Fig. M8).

Figure M8. Segregation in a maize ear

Transformation is possible but requires either electroporation or the biolistic technique. The discovery and the commercial production of hybrid corn (heterosis) made an unprecedented increase in the food and feed supply. ►biological control, ►domestication, ►non-coding DNA, for detailed chromosomal map of 1736 loci see Davis GL et al 1999 Genetics 152: 1137; Doebley J 2001 Genetics 158:487; Liu K et al 2003 Genetics 165:2117; evolution of maize: Doebley J 2004 Annu Rev Genet 38:37; genetics and genomics: http://www.maizegdb.org; http://maize.tigr.org/; molecular and functional diversity: http://www.panzea.org/; maize database: http://www.maizegdb.org.

Majewski Syndrome: The autosomal recessive phenotype has many similarities with oral-facial-digital syndromes, particularly with the Mohr syndrome. It is distinguished from the latter on the basis of laryngeal (throat) anomalies and polysyndactyly of the feet. ►Oro-facial-digital syndromes, ►polydactily

Major Facilitator Superfamily (MFS): These are ubiquitous transporters (more than 1000 members) in all biological groups. They transport ions, sugars, amino acids, peptides, nucleosides, drugs, neurotransmitters, etc. Their mutations may result in various diseases (seizures, diabetes, etc), may be responsible for antibiotic resistance, cancer chemotherapy (controlling drug efflux). ►membrane transport; Huang Y et al 2003 Science 301:616.

Major Gene: Determines clear, qualitative phenotypic trait(s). ►minor gene

Major Groove: The DNA helix as it turns displays two grooves in a 3.46 nm pitch; the wider one (almost 2/3 of the pitch) is the major groove ↗ (see Fig. M9). ►DNA, ►hydrogen pairing, ►Watson and Crick model

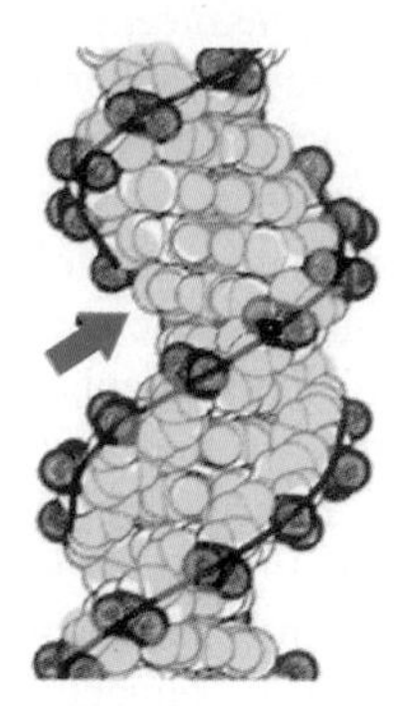

Figure M9. Major Groove

Major Histocompatibility Complex (MHC in humans, H-2 in mice): Triggers the defense reactions of the cells against foreign proteins (invaders). Their existence was first recognized by incompatibilities of tissue grafts. The MHC molecules are transmembrane glycoproteins encoded by the HLA complex in humans and by H-2 in mouse. The class I molecules have three extracellular domains at the NH_2 end and the COOH terminus reaches into the cytosol. The extracellular domains are associated with β_2 microglobulin. The class II MHC molecules are formed from α and β chains without the microglobulin. The class I heavy chain and the β_2 microglobulin are translocated to the endoplasmic reticulum before their translation is completed and their assembly takes place there with the assistance of the chaperones, BiP (a heatshock protein) and calnexin (glycoprotein). The MHC I molecules then associate with TAP and is ready for picking up a foreign antigenic peptides. After a peptide is bound, the system is released and after passing through the Golgi where their attached carbohydrates may be modified, exocytosis moves them to the surface of the cell membrane and this provides a chance for the $CD8^+$ T cells (CTL) to react to them. At this stage, the cells that recognize self-antigens are eliminated by apoptosis. The MHC molecules resemble immunoglobulins, and the class II molecules, especially the β chains are highly polymorphic (The HLA-DRB 1 locus has more than 100 alleles.). The MHC molecules bind foreign antigens and present them to the lymphocytes. The murine CD1 proteins bear similarities to the MHC molecules but they present to the T cells lipids and glycolipid antigens. MHC Class I proteins accumulate the cut pieces of peptides from inside the cells whereas Class II MHC proteins are attached to the pieces of antigens from outside the cell. Before a Class II protein is loaded with an invader peptide, it carries a neutral (dummy) peptide called CLIP which is then replaced by foreign pieces regulated by acidic conditions in the presence of DA. The Class I molecules are expressed on practically all cells that T cells recognize. Class II molecules are found on $CD4^+$ B cells and other antigen presenting cells. MHC class I peptides (CTL epitopes) stimulate the $CD8^+$ cytotoxic T cells and the MHC class II peptides are recognized by the $CD4^+$ T cells. The T_H cells destroy any cell that present the antigen with MHC II molecules. The helper T cells do not directly attack the invaders but stimulate the action of macrophages. The MHC genes rely on transactivation by CIITA (non-DNA-binding) and RFX5, NF-X, NF-Y and other DNA-binding transcription factor proteins. There is a great diversity among MHC molecules. Gene conversion may create variations in 10^{-4} range in sperms. The histocompatibility system of rabbits is called RLA, in rats RT1, in guinea pigs GPLA. MHC I homologs of herpesviruses may disarm natural killer T cells. ►HLA, ►Ii, ►immune system, ►blood cells, ►T cells, ►αβ T cells, ►CTL, ►microglobulin, ►DA, ►proteasome, ►TAP, ►antigen presenting cell, ►antigen processing and presentation, ►Bare lymphocyte syndrome, ►apoptosis, ►self-antigen, ►RFX; Beck S, Trowsdale J 2000 Annu Rev Genomics Hum Genet 1:117; Horton R et al 2004 Nature Rev Genet 5:889; http://www.meddb.info/index.php.en?cat=6&subcat=104; DNA and clinical resources: http://www.ncbi.nlm.nih.gov/mhc/MHC.cgi?cmd=init.

Majority Class Spores: ►polarized recombination

Mal de Maleda (MDM, 8q24.3): A rare, recessive keratoderma, redness on the face, brachydactyly and nail abnormalities. The basic defect is in a secreted Ly-6/uPAR domain related protein (defined by disulfide bonding pattern between 8 or 10 cysteine residues). ►keratoma, ►brachydactyly; Fischer J et al 2001 Hum Mol Genet 10:875.

Malaria: An infectious disease affecting 300–500 million people annually (Snow RW et al 2005 Nature [Lond] 434:214), and causing 1.5–2.7 million deaths among them. It is caused by one or another species of the *Plasmodium* protozoa (see Fig. M10). Transmission is mediated by *Anopheles* mosquito bites but transplacental infection or transfusion with contaminated blood may transmit it. The infection relies on a Duffy-like cysteine-rich module of the parasites that recognize host cell surface receptors. The crystal structure of this domain has been revealed

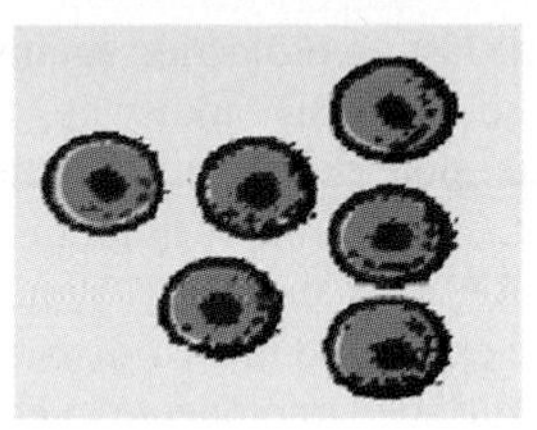

Figure M10. *Plasmodium* merozoites

(Kumar Singh S et al 2006 Nature [Lond] 439:741). This disease is prevalent in the subtropical and tropical areas of the world. The symptoms are chills, fever, sweating, anemia, etc. The attacks are recurring according to the major reproductive cycles of the parasite. Heterozygotes of sickle cell anemia are somewhat resistant to the disease and that explains the higher than expected frequency of this otherwise deleterious gene. Plasmodium degrades hemoglobin by two plasmalepsin proteases. Hemoglobin C may provide some protection by abnormal display of *P. falcipaum* erythrocyte membrane protein 1 (PfEMP1) on the erythrocytes and by reducing their sequestration in the microvasculature (Fairhurst RM et al 2005 Nature [Lond] 435:1117). Two minor and one major quantitative trait loci have been identified in *Anopheles gambiae* that inhibit the development of *Plasmodium cynomolgi* B in the midgut of the mosquito. The human body defends itself by the cytotoxic T cells. The human immune response cannot prevent the infection because the malaria-infected erythrocytes attach and slow down the maturation of the antigen-presenting dendritic cells and thus fail to activate properly the T lymphocytes. In some populations HLA and other ligands may cooperatively down-regulate this immune response. During the development of the parasite, the surface antigens are altered providing immune evasion and chances of reinfection after a period of apparent curing. Some success was obtained by vaccination before the parasite reaches the blood infection stage. Viral vector-delivered multigenic DNA vaccines encoding different epitopes and applied before the parasite would reach the blood stage may overcome the problems caused by the antigenic variations. The PEMP1 protein is responsible for most of the antigenic variation in the parasite. More than 50 var genes encode this multimodular (DBL, CIDR) adhesion protein. Various vaccines are under development either to prevent the multiplication of the parasite (sterilizing immunity) or by anti-disease approach that would prevent the development of malaria-associated pathology of the brain or the placenta in the infected individuals. Antibodies can be generated against many different, known antigens of the parasite but because of mutation in single proteins may eliminate lasting protection against single antigen. Transcriptome analysis indicates that about 197 proteins can potentially be targeted for vaccine development. The thrombospondin-related adhesive protein (TRAP) and the apical membrane antigen 1 (AMA-1) are expressed in the *Plasmodium falciparum* sporozoite surface microneme vehicles and later proteolytic cleavage generates soluble fragments. These microneme proteins are critical for the invasion of blood and hepatocytes. A subset of serine protease inhibitors can prevent proteolytic cleavage, essential for infection and are good targets for vaccine development (Silvie O et al 2004 J Biol Chem 279:9490; Waters A 2006 Cell 124:689). Application of attenuated sporozoite vaccines have been considered but were plagued by practical, technical difficulties.

One possible means of protection appears to immunize (rodents) by *uis3* deficient sporozoites. The *UIS3* gene (upregulated in infective sporozoites 3) is essential for the early liver-stage development and in its absence, the *Plasmodium berghei* will not establish the blood-stage infection (Mueller A-K et al 2005 Nature [Lond] 433:164). Transgenic mosquitos may fail to transmit the parasites (Ito J et al 2002 Nature [Lond] 417:452). First time gravid humans are more susceptible to malaria because the parasite attaches to the chondroitin sulfate A on the placenta. By subsequent pregnancies a partially protective anti-adhesion response develops. The size of the *Plasmodium falciparum* genome is ~30 Mb (~82% A = T) located in 14 chromosomes. It has two organelle genomes: the mitochondrial is 5.9-kb (tandemly repeated) and ~35-kb in the apicoplast, a plastid-like organelle. The technology of molecular biology greatly facilitates the identification of the biology of the protozoa and the development of more effective control measures. (Le Roch KG et al 2003 Science 301:1503). *P. falciparum*, the most fatal species is rapidly developing resistance to the drug chloroquine and to several newer drugs. More recently high hopes were placed in the protective effect by the old Chinese herbal plant extract artemisinin of *Artemisia* (see Fig. M11). Now even that treatment may result in the development of resistance in *Plasmodium falciparum* (Duffy PE, Sibley CH 2005 Lancet 366:1908). *Plasmodium* has 400 putative genes involved in targeting erythrocytes of which 225 encode virulence proteins and 160 apparently remodel host erythrocytes. Switching to positions conducive to transcription and heterochromatin remodeling regulate virulence genes (*var*), which control antigenic variation (Freitas-Junior LH et al 2005 Cell 121:25). These genes are potential targets of antimalarial drugs (Mart M et al 2004 Science 306:1930). New drugs (fosmidomycin) blocking the non-mevalonate pathway of isoprenoid biosynthesis in the apicoplast appear promising alternatives. Oil-based suspension sprays containing *Beauveria bassiana* and *Metarhizium anisopliae* directed against both *Plasmodium* and *Anopheles* mosquitos was up to ~90% effective control of malaria transmission (Blanford S et al 2005 Science 308 1638; Sholte E-J et al 2005 Science 308:1641).

Figure M11. *Artemisia*

The cerebral form of malaria attacks the brain and causes mental disorders, hyperthermia and about half of the infections is lethal. Malaria may mitigate Hepatitis B viral infections.

P. falciparum is the major cause of lethal malaria in humans but in chimpanzees, only moderate parazitization and no severe infection occurs. *P. reichenowi* is very similar to P. *falciparum* but affects only chimpanzees and other apes but not humans. This remarkable difference is based on the lack of N-glycolylneuraminic acid on human erythrocytes in contrast to other primates. *P. reichenowi* erythrocyte-binding antigene-175 requires this sialic acid (Martin MJ et al 2005 Proc Natl Acad Sci USA 102:12819; note figures scale bars are correctly 100 μm). ►hemoglobin, ►thrombospondin, ►microneme, ►sickle cell anemia, ►*Plasmodium*, ►merosome, ►CD36, ►QTL, ►glucose-6-phosphate dehydrogenase, ►cytotoxic T cell, ►HLA, ►immunization genetic, ►apicoplast, ►Oct-1, ►chloroquine, ►refractory genes, ►neuraminic acid, ►serpine, ►Duffy blood group; de Koning-Ward TF et al 2000 Annu Rev Microbiol 54:157; Kappe SHI et al 2001 Proc Natl Acad Sci USA 98:9895; Nature [Lond] 2002 Insight 415: 670–710; *Anopheles* genome: Science 298, 4 Oct. 2002; malaria issue: Nature [Lond] 430:925–944, 2004; http://www.malaria.org.

Malate Dehydrogenase: The product of the MDH1 (human chromosome 2p23) is cytosolic whereas the MDH2 enzyme (encoded in 7p13) is located in the mitochondria. ►Krebs-Szentgyörgyi cycle, ►mitochondria

Malattia Leventinese: ►macular degeneration

MALD (mapping by admixture linkage disequilibrium): The mapping on the basis of linkage disequilibrium in a population. Complex diseases, based on more than a single gene are difficult or impossible to map on the basis of conventional linkage. Association mapping is more effective but requires too much genotyping. The efforts can be reduced if neighboring markers of the haplotypes are considered. Unfortunately, even this approach requires 300,000 to 1,000,000 single nucleotide polymorphic markers (SNP) to be successful, and its cost is too high for single disease. When populations are mixed by gene flow (e.g., European, Asian and African ethnic groups), temporarily large haplotypes blocks form linkage disequilibrium. The association of these blocks (disequilibrium) can be statistically analyzed though the centuries and the association of disease with haplotypes markers can be inferred. Such a study requires 200–500 fewer markers than whole genome haplotype mapping. MALD is carried out in five steps. 1. A group of people (cohort) is selected from the admixed population that have high incidence of the disease. 2. Cases and controls are genotyped using markers of high frequency and informative for ancestry. 3. The patchwork of the ancestry is determined for individuals. 4. Chromosomal regions with elevated frequency of the ancestry markers and disease are identified. 5. The candidate genes are located then by the use of SNP-association. This procedure is best suited for diseases, which differ substantially in frequency between the ancestral groups. Environmental differences may confound and frustrate the analysis. It is necessary that the admixture would be higher than 10%. The range of admixture generally varies a great deal geographically. Continued introgression also influences the outcome. The required population size depends on the allele frequencies and extent of the admixture. Genes displaying similar frequencies in the ancestral populations are not informative. It is also important that the target genes could be safely identified (few false positives). The analysis requires ethical considerations because of racial sensitivities. ►linkage disequilibrium; Smith MW, O'Brian SJ 2005 Nature Rev Genet 6:623; Collins-Schramm HE et al 2002 Am J Hum Genet 70:737.

MALDI/TOF/MS: Matrix-assisted laser desorption ionization/time of flight/mass spectrometry can detect minute differences in masses of large DNA molecules (~2,000 nucleotide) in ~femtomole (10^{-15}) size samples. This technology is used for characterization and identification of proteins and protein fragments. It detects post-translational modifications and any alteration in protein mass during development and modulation of function.

For proteomics, generally tryptic peptide mixture is used. Trypsin cuts proteins at arginine and lysine residues and on that basis, unknown proteins may be identified on the basis of matches with known proteins in the databases. For post-translational modification of proteins, the tryptic peptides can be analyzed in two steps. First, the mass spectrum (MS) of the intact peptide fragments can be determined. Second, (MS/MS) peptide ions are fragmented in a gas phase to identify protein sequence and modifications. For these analyses MALDi or electrospray ionization (ESI) are used. By this "bottom-up" approach, the full spectrum of the fragmentation ladder generally cannot be obtained. Alternatively, the "top-down" approach can be used where intact protein ions are introduced into the gas phase of ESI and obtaining molecular mass of the proteins and protein ion fragmentation ladders (see Fig. M12). This latter process permits now fragmentation of

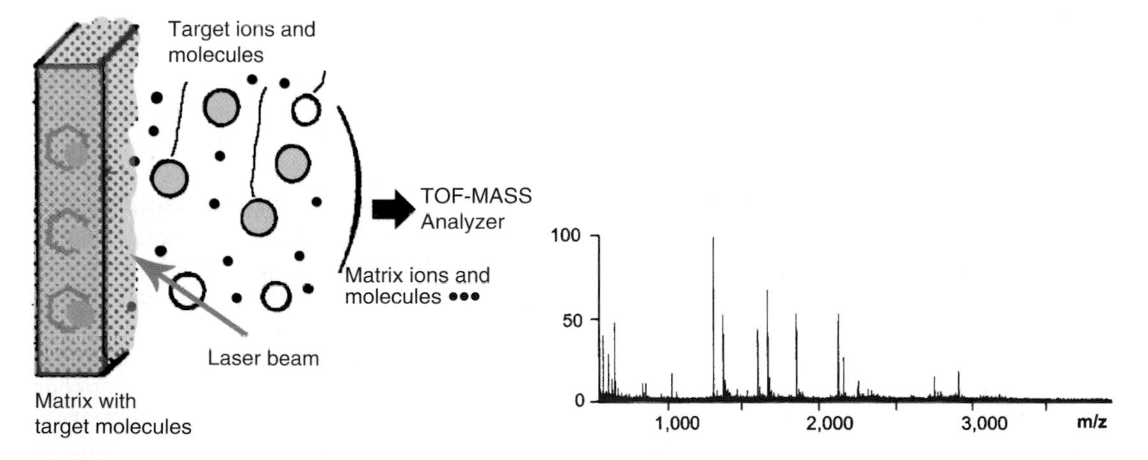

Figure M12. Diagram of the principle of MALDI/TOF/MS at left; MALDI/TOF/MS protein profile (for m/z see Mass Spectrum) Maldi/tof/ms protein profile (for m/z see Mass spectrum)

molecular masses up to 200 kDa (Chait BT 2006 Science 314:65; Han X et al 2006 Science 314:109).

The analytical procedure requires that the target molecules be co-precipitated with an excess matrix consisting of either α-cyano-4-hydroxycinnamic acid or dihydrobenzoic acid (DBH). The material is then dried on a metal surface and irradiated by a flanking nanosecond nitrogen laser at a wavelength of 337 nm to generate ionization. MALDI is used for relatively simple mixtures whereas ESI is integrated with liquids chromatography and is suitable for more complex mixtures. The sample is analyzed by a mass spectrometer (see Fig. M12). Some recent developments make these types of analyses very useful for biomedical applications (proteomics, pathology, tissue imaging, diagnostics, cancer monitoring, metabolomics, identification of biological agents, drug analysis, pharmaceutical process monitoring, in forensics and national security). ►DNA chips, ►proteomics, ►protein chips, ►electrophoresis, ►mass spectrometer, ►mass spectrum, ►fragmentation ladder, ►DESI, ►SIMS genomics, ►CID, ►SNIPS, ►quadrupole, ►electrospray MS, ►ESI, ►mole, ►PMF; Li L et al 2000 Trends Biotechn 28:151; Mann M et al 2001 Annu Rev Biochem 70:437; Bennett KL et al 2002 J Mass Spectrom 37:179; Cooks RG et al 2006 Science 311:1566, http://prowl.rockefeller.edu/, OMSSA.

Male Contraceptive: The oldest form is a mechanical barrier to the sperm (condom) to prevent its penetration of the reproductive tract of the female. Primarily females practice physiological fertility control in animals and humans. Adjudin (1-[2,4-dichlorobenzyl]-1H-indazole-3-carbohydrazide) is capable of inducing germ cell loss from the Sertoli cell epithelium. Oral dose of Adjudin (50 mg per kg body weight for 29 d) resulted, however, in liver inflammation and muscle atrophy. If Adjudin is specifically targeted to the testis by conjugating it to a recombinant follicle-stimulating hormone mutant as its vector infertility was induced in adult rats when 0.5 mμg Adjudin per kg body weight was injected intraperitonally with the same effectiveness as at the high oral dose (Mruk DD et al 2006 Nature Med 12:1323). There may eventually be a chance for human applications. ►contraceptive

Male-Driven Evolution (α): It is indicated by the higher rate of mutation in males than females. The extent of the ratio is not universally agreed upon. (See Makova KD, Li WH 2002 Nature [Lond] 416:624).

Male Gametocide: A chemical that destroys male gametes and thus may facilitate cross-pollination in plants or may serve as a human birth-control agent. ►crossing, ►birth control

Male Gametophyte: ►gametophyte

Male Recombination: It is normally absent in *Drosophila* (except when transposable elements are present; up to 1%) and in the heterogametic sex (females) of the silk worm. In case of the presence of a P element, most of the recombination is site-specific within a 4-kb tract near the element. Recombination then involves either adjacent duplication and deletion of a few to > 100-bp. Genetic markers can be mapped relative to known P element sites. Mitotic recombination may occur, however, even in the absence of the meiotic one. In maize and *Arabidopsis,* certain cases indicate reduced frequency of recombination in the megasporocytes compared to the microsporocytes. In rainbow trout the average female: male recombination ratio is 3.25:1 but shows significant variation according to chromosomal regions and families. Generally, in most species the recombination frequency in the heterogametic sex is slightly lower. ►recombination frequency, ►*MR*, ►hybrid dysgenesis, ►HEI, ►achiasmate; Meneely PM et al 2002 Genetics 162:1169; Hellig R et al 2003 Nature [Lond] 421:601.

Male-Specific Phage: It infects only those bacterial cells that carry a conjugative plasmid.

Male Sterility: Caused by various chromosomal aberrations, such as inversions, translocations, deficiencies, duplications, aneuploidy, polyploidy, etc., and by cytoplasmic factors. Generally in plants, male sterility is more common than female sterility because the pollen, the male gametophyte, has a more independent life phase than the megaspore and cannot rely well on support by sporophytic tissues. In plants, the male sterility may be caused by a mutation in the male gametophyte or it may be the result of the abnormal development of the anthers or other parts of the flower, e.g., failure to release the normally developed pollen. Common cause of male sterility is an alteration of the mitochondrial DNA, cytoplasmic male sterility (*cms*). Incompatibility between nuclear genes and certain cytoplasms as well as viral infection may also cause male sterility. Certain chemicals (mutagens), chromosome-breaking agents (maleic hydrazide) may have gametocidic effects. Male sterility may result in transgenic plants when a tapetum-specific promoter drives a ribonuclease gene within the anthers and thus destroy the pollen (see Fig. M13). The fertility can be restored if the sterile plants are employed as female in crossing with a male carrying the transgene *barstar* inhibiting the ribonuclease. Other self-destroying–restoring combinations have also been considered. A healthy human male ejaculates 25–40 million sperms each time, and if for any reason this number is reduced to 20,000 or below, male infertility/sterility may result although only a single sperm functions in fertilization. Polymorphy for hybrid male sterility is a factor in speciation (Reed LK, Markow TA 2004 Proc Natl Acad Sci 101:9009). ►cytoplasmic male sterility, ►chromosomal defects, ►gametocides, ►KIT oncogene, ►azoospermia, ►oligospermia, ►asthenospermia, ►speciation, ►hybrid sterility, ►*msl*; Hackstein JHP et al 2000 Trends Genet 16:565.

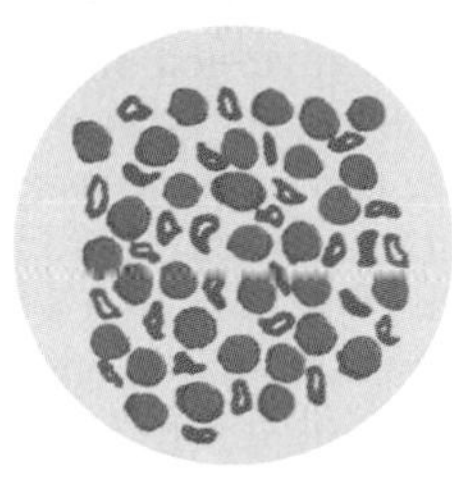

Figure M13. The male sterile pollen does not stain and is shrunken

Malécot Equation: Employed for estimation of marker frequency probability in linkage disequilibrium in populations: $(1 - L)Me^{-\varepsilon d} + L$, where L is an asymptote of the association for long distance and the intercept M represents association in the population founders, ε is a constant. ►linkage disequilibrium; Lonjou C et al 2003 Proc Natl Acad Sci USA 100:6069.

Male-Stuffing: In some social insects the workers may force (and kill) males into empty cells in order to keep them from using the scarce food and thus making it available to the larvae. The relatedness among workers in single-mated queen colonies is 75% whereas to males it is only 25%. Thus, this behavior is a form of kin selection. ►kin selection

Malformation: The abnormal formation of body parts. ►dysplasia

Malignant Growth: A defect in the regulation of cell division that may lead to cancer and may cause the spreading of the abnormal cells. ►metastasis, ►cancer, ►oncogenes, ►cell cycle

Malondialdehyde (MDA): A mutagenic carbonyl compound, generated by lipid peroxidation. ►adduct, ►pyrimidopurinone, ►lipid peroxidation

Malonic Aciduria (16q24): A recessive deficiency in malonyl-CoA decarboxylase and causes abnormally large amounts of malonic acid in the urine. It involves developmental retardation, constipation, abdominal pain and infantile death.

Malpighian Tubules: Excretory channels of arthropods emptying into the hindgut performing kidney-like function.

Malpractice: Improper practice due to negligence or misconduct or inadequate training or lack of adequate experience and causing physical, mental or financial harm in medicine or any other business activity.

Malsegregation: When the two homologous chromosomes are not recovered in 2:2 proportion at the end of meiosis. ►segregation distorter, ►meiotic drive, ►nondisjunction

MALT (mucosa-associated lymphoid tissue lymphoma): It is most common in the gastrointestinal tract and in non-Hodgkin lymphoma. MALT is frequently associated with translocations t(1;14)p22;p32 and to the apoptotic signaling gene BCL10. MALT1 gene is involved in the activation of NF-κ (Talwalkar SS et al 2006 Mod Pathol 19:1402). ►lymphomas, ►Hodgkin disease, ►BCL, ►apoptosis, ►fungal diseases

Malt: The specific activator of the bacterial maltose operon. ►maltose

Malthusian Parameter: It assumes that the age distribution in a population remains constant from one generation to the next. In case the age-specific birth and death rates would continue without any hindrance, the population growth would ultimately reach that level, frequently denoted by *r*. ►age-specific birth and death rate, ►population growth; Fisher RA 1958 The Genetical Theory of Natural Selection. Dover, New York; Demetrius L 1975 Genetics 79:535.

Malthusian Theorem: Postulates that population growth exceeds the rate of food production and thus jeopardizes the survival of humans. This nineteenth century idea was contradicted by the fact that between 1961 and 1983 the available food calories per capita has increased from 2320 to 2660 (over 7% in two decades). Unfortunately, this growth is slowing down again, and from 1984 to 1990, the increase was only about 1%. For about 2,000 years, the global human population grew by an annual rate of about 0.04%. From 1965 to 1970, the rate increased to 2.1% and in 1995 it was about 1.6% per year. ►human population growth; Black JA 1997 BMJ 315:1686; Lee RD 1987 Demography 24:443.

Maltose (glucopyranosyl glucose): A disaccharide. The Snf1 protein kinase controls the Mig1 transcriptional repressor of *SUC2* (sucrose), *GAL* (galactose) and *MAL* (maltose) genes. Induction of *MAL* requires transcriptional activator. Maltose utilization is controlled by several operons. A common activator of these is a 103-kDa activator MalT. Glucose and the global repressor *Mlc* repress *MalT*. The maltose transporter system is an integral part of maltose utilization. A maltose transporter is essential for starch degradation and carbohydrate export from the site of photosynthesis (Niittylä T et al 2004 Science 303:87). ►glucose effect, ►catabolite repression, ►Snf; Boos W, Böhm A 2000 Trend Genet 16:404.

MAMA (monoallelic mutation analysis): The mutations are screened in somatic cell hybrids of lymphocytes and hamster cells with aid of DNA techniques. ►mutation detection; Jinneman KC, Hill WE 2001 Curr Microbiol 43(2):129.

Mammalian Comparative Mapping Database: A part of the Mouse Genome Database URL: http://www.informatics.jax.org; questions, problems, etc. can be addressed to Mouse Genome Informatics Project: mgi-help@informatics.jax.org; tel: (207) 288–3371 ext. 1900, fax (207) 288–2516.

Mammalian Genome: A monthly publication of the Mammalian Genome Society. Contact V.M. Chapman, Dept. Molecular and Cellular Biology, Roswell Park Cancer Institute, Elm & Carlton, Buffalo, NY 1463, USA. ►databases for additional addresses.

Mammalian Species: See http://www.science.smith.edu/departments/Biology/VHAYSSEN/msi/.

Mammaprint: A government approved, high-cost 70-marker microarray test for the prediction of status and prospect of breast cancer at high accuracy (Glas AM et al 2006 BMC Genomics 7:278). (See information for physicians and patients: http://www.agendia.com/us/).

Mammary Tumor Virus of the Mouse (MTV): It causes an estrogen-stimulated adeno-carcinoma. The virus is transmitted through the milk. ►glucocorticoid response element; Kang CJ, Peterson DO 2001 Biochem Biophys Res Commun 287:402.

Mammoth (*Mammothus primigenius*): The best known of these now extinct Elephantine species of the Pleistocene era in Siberia. These animals were somewhat comparable to the Asian elephants (*Elephas maximus*) in size but they were well adapted to the harsh climate by having long wool and thick layers of fat under skin all over the body. The nuclear DNA is not well preserved even in the arid caves or permafrost region. The nuclear DNA fragments (average length 84 bp) analyzed matched (to 95% identity) the available sequences of the African elephant (*Loxodonta Africana*) in 30% but only in 0.4% to human DNA. The estimated divergence between the two animals is 5–6 million years. Better specimens of DNA (~89 bp in average length) could be prepared from bone mitochondria and could be amplified by PCR and the sequences matched in 95,93% to that of the African elephant (Poinar HN et al 2006 Science 311:392). By multiplex amplification, the entire mitochondrial DNA has been reconstructed and mapped (Krause J et al 2006 Nature [Lond] 439:724). The mastodon (*Mammut americanum*) is a North American relative (diverged 24–28 million years ago [mya] from the Elephanitine lineage); its mitochondrial DNA has been sequenced from fossil Alaskan tooth. The ancestors of African elephants diverged from the lineage leading to mammoths and Asian elephants approximately 7.6 mya, and mammoths (see Fig. M14) and Asian elephants diverged approximately 6.7 mya. (Rohland N et al 2007 PloS Biol 5(8):e207). ►ancient DNA, ►elephants, ►ancient organisms; Rogaev EI et al 2006 PLoS Biol 4(3):e73.

Figure M14. Mammoth in cave art of Roufignac, France

Manatee (*Trichechus manatus latirostris*): Large herbivorous aquatic mammal, 2n = 48.

MANET: The evolution of protein structure in metabolic networks: Kim HS et al 2006 BMC Bioinformatics 7:351, http://www.manet.uiuc.edu/.

Mango (*Mangifera*): A tropical fruit tree, 2n = 2x = 40.

Mania: A pathological neuronal disorder of mental and/or physical hyperactivity.

Mannans: The yeast cell wall polymers of mannose. ►mannose

Manic Depression (MAFD1): A psychological condition characterized by recurrent periods of excessive anguish (unipolar) or by manic depression (bipolar). The latter form is accompanied, in addition, by hyperactivity, obsessive preoccupation with certain things or events. Depression and other affective disorders may involve 2 to 6% of the human populations although the incidence of unipolar depression may be as high 21% among females and 13% among males during the entire lifetime. The concordance among monozygotic twins may be as high as 80% whereas among dizygotic twins, it is about 8%. It appears that the recurrence in the families is higher with the early onset types. Also, the bipolar types appear to have higher hereditary components. The genetic control of depression is unclear, X-linked recessive, autosomal dominant (chromosome 11p) genes have been implicated but the majority of assignments were not well reproducible. These may be major genes but other genes are also involved. Susceptibility loci were assigned to 0 12q23-q24, 18p and 18q, 5q, 8p, 21p22, and others. MFAD2 (major affective disorder 2 or bipolar affective disorder; BPAD) was assigned to Xq28. The physiological bases may also vary from defects in neurotransmitters to electrolyte abnormalities, etc. Commonly recommended therapy involves monoamine oxidase inhibitors, tranquilizers (prozac), lithium, etc. ►affective, ►disorders, ►schizophrenia, ►bipolar mood disorder, ►tyrosine hydroxylase, ►lithium; Molecular genetics of bipolar disorder and depression: Kato T 2007 Psychiatry Clin Neurosci 61:3.

Mannopine: N2-(1′-deoxy-D-mannitol-1′-yl)- L-glutamine. ►opines, ►Ti plasmid

Mannose: An aldohexose (see Fig. M15).

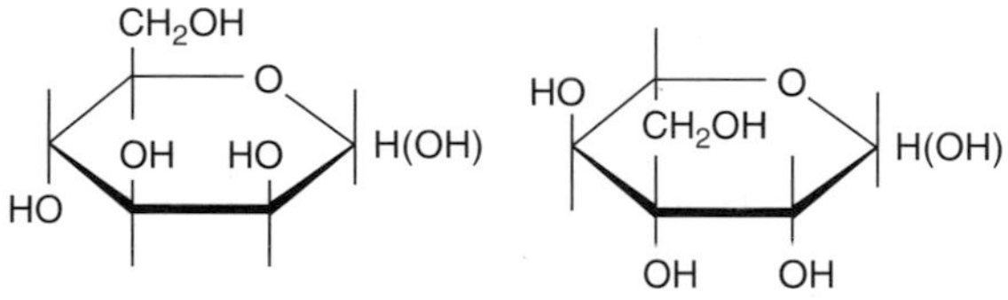

Figure M15. D Mannose and L Mannose

Mannosephosphate Isomerase (MPI): A Zn^{2+} monomeric enzyme converts mannose-6-phosphate into fructose-6-phosphate. It is coded in human chromosome 15q22.

Mannose-6-Phosphate Receptor (MPR): Same as insulin-like growth factor; it plays a role in signal transduction, growth and lysosomal targeting. LOH mutation of this receptor (human chromosome 6q26-q27) causes liver carcinoma. MPR is also a death receptor for granzyme B during CTL-induced apoptosis. ►insulin-like growth factor, ►LOH, ►death receptor, ►granzyme, ►CTL, ►apoptosis, ►RAB, ►TIP47, ►GGA

Mannosidosis: The recessive (α-mannosidase B) deficiency has been located to human chromosome 19cen-q12 (MAN2B1) and involves large increase of mannose in the liver causing susceptibility to infection, vomiting, facial malformations, etc. Another mannosidase defect at another autosome (4q22-q25, MANBA) caused excessive mannosyl-1 4-N-acetylglucosamine and heparan in the urine, apparently involving glycoprotein abnormalities and a variety of physical and mental defects.

Mannosyltransferases: Defects in several genes involve mental retardation, psychomotor defects, hypomyelination, coloboma, etc.

Mann-Whitney Test: A powerful non-parametric method for determining the significance of difference between two normal-distributed populations. This method is useful for evaluating scores of samples even if their size is not identical. The procedure is illustrated by small samples, however samples of n > 20 are preferred. The scores are ranked (T) in Table M1 (in case of ties the average ranks are assigned to the two):

Table M1. A hypothetical example for the use of the Mann-Whitney test

Populations	I	II	II	I	I	II	II	Sum of I = T_I = 10
Scores	1	4	5	7	8	9	10	Sum of II = T_{II} = 18
Rank	1	2	3	4	5	6	7	n_I = 3, n_{II} = 4

The null hypothesis to be tested is that the distribution of the two populations (I and II) is identical. Then the U is determined for sample I: $U_I = n_I n_{II} + \{[n_I(n_I + 1)]/2\} - T_I = 3 \times 4 + (12/2) - 10 = (12 + 6) - 10 = 8$. In case the resulting U value is larger than $(n_I n_{II})/2$ then calculate $U' = n_I n_{II} - U_I$. For large populations, the sampling distribution for U is approximately normal and $E(U) = (n_I n_{II})/2$ and the variance is determined $\sigma^2 = [n_I n_{II}(n_I + n_{II} + 1)]/12$; hence the z value (the standard normal variate) = $[U - E(U)]/\sigma_U$ and the probabilities corresponding to z can be read from statistical tables of the cumulative normal probabilities and a few commonly used corresponding values are as follows for z = 1.65, 1.96, 2.58 and 3.29, the P values are 0.90, 0.95. 0.99 and 0.999, respectively. (See Wilcoxon's signed rank test, Standard deviation, Probability, Null hypothesis).

Mantel Test: Estimates the common odds ratio in two-by-two contingency tables that result from different populations. The formula for the estimation is: $\sum_{i=1}^{k} a_i d_i / \sum_{i=1}^{k} c_i b_i$, where k is the number of the two-by-two tables and a_i, b_i, c_i and d_i are the counts in the tables. ►association test

Mantle Cell Lymphoma (MCL, human translocation [11;14][q13;q32] or mutation at 11q22-q23): A non-Hodgkin type lymphoma. Derived from naïve CD5$^+$ B cells of the primary follicles or of the mantle zones of the secondary follicles. The segment deleted from chromosome 11q22-q23 includes the ataxia telangiectasia locus. The translocation juxtaposes cyclin D1 (CCND1) and the transcriptional control element of the immunoglobulin G gene although other factors (c-Myc) are also required for the development of the lymphoma. ►anaplastic lymphoma, ►ataxia telangiectasia

Manx Cat: Tailless, because of fusion, asymmetry and reduction in size of one one or more caudal vertebrae (see Fig. M16). A dominant gene that is lethal in homozygotes causes this phenotype. The Manx protein is essential also for the development of the notochord of lower animals. ►brachyury, ►notochord

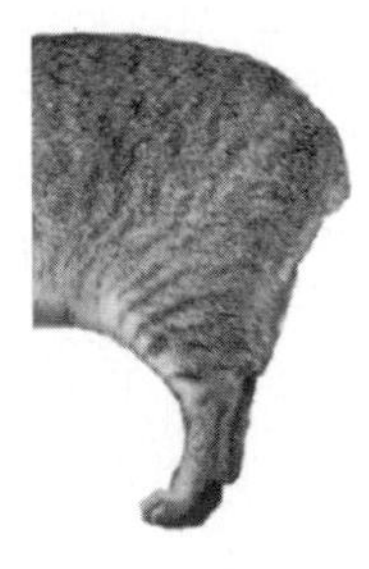

Figure M16. Caudal end of Manx cat

MAO (monoamine oxidase): This enzyme is involved in the biosynthetic path of neurotransmitters from amino acids. Mutation in the 8th exon (amino acid position 936) converted the glutamine codon CAG to TAG (chain termination codon) and resulted in mild mental retardation, continued and impulsive aggression, arson, attempted rape, and exhibitionism in human males with this X-chromosomal recessive defect. The block of MAOA resulted in accumulation of normetanephrine (a derivative of the adrenal hormone epinephrine), and tyramine (an adrenergic decarboxylation product of tyrosine) and a decrease in 5-hydroxyindole-3-acetone. The heterozygous women were not affected behaviorally or metabolically; monoamine oxidase B level remained normal. Both enzymes are in human chromosome Xp11.2-p11.4 and the enzymes are located in the mitochondrial membrane. The enzymes may affect various psychiatric disorders such as the Gilles de la Tourette syndrome, panic disorder, alcoholism, etc. ►mitochondria, ►Norrie disease, ►Tourette syndrome

MAP: see Microtubule associated proteins. ►*ASE1*, ►microtubule, ►centrosome, ►map genetic

MAP (multi-use affinity probe): A biarsenic tag of amino acids in proteins facilitating easy monitoring interactions (Cao H et al 2007 J Amer Chem Soc 129: 12123).

MAP-1 (modulator of apoptosis): A mitochondria-associated protein and activates BAX when it is translocated into the mitochondria from the cytosol upon apoptotic signals. RNAi regulates MAP-1. ►apoptosis, ►BAX, ►RNAi; Tan KO et al 2005 Proc Natl Acad Sci USA 102:14623.

MAP-Based Cloning (positional cloning): Isolation of gene(s) on the basis of chromosome walking and propagation usually by YAC and/or cosmid clones. After a genome sequence has been completed, positional cloning will no longer be needed as long as the function of the gene is also known. Positional cloning may be complicated by the fact that some phenotypes are affected by more than a single locus. ►chromosome walking, ►chromosome landing, ►position effect; Lukowitz W et al 2000 Plant Physiol 123:795; Tanksley SD 1995 Trends Genet 11:63.

MAP Distance: Indicates how far syntenic genes are located from each other in the chromosome as estimated by their frequency of recombination; 1 map unit = 1% recombination = 1 centi Morgan. The greater the distance between two genes, the higher is the chance that they are separated by recombination. A single recombination between two genes in a meiocyte produces maximally 50% recombination that is 50 map units. The distance between syntenic genes may exceed 50 map units several times; these longer distances are determined then in a staggered manner, proceeding step-wise from left to right and right to left. In prokaryotes, using conjugational transfer and recombination, map distances are measured in minutes of transfer. ►conjugational mapping, ►recombination frequency, ►radiation hybrids, ►mapping function

MAP Expansion: The distance between two distant markers exceeds the sum of the distances of markers in between; it is commonly observed in gene conversion. ►gene conversion; Holliday R 1968 p 157. In: Replication and Recombination of Genetic Material. Peacock WJ, Brock RD (Eds.) Australian Acad Sci, Canberra.

Map, Genetic: The order of genes (markers) in chromosomes determined on the basis of recombination frequencies. ►mapping, ►recombination frequencies, ►physical map, ►radiation hybrids, ►RFLP, ►RAPD, ►mapping function; Human genes: OMIM; http://linkage.rockefeller.edu/; genetic & physical of various organisms: http://www.ncbi.nlm.nih.gov/Genomes/index.html.

Map Genetic Versus Physical Map: ►coefficient of crossing-over, ►gene number; Ashburner M et al 1999 Genetics 153:179.

Map Kinase (MPK): A family of serine/threonine protein kinases associated with mitogen activation (growth) and stress responses. Three groups exist: ERK, cJun N-terminal kinase (JNK) and p38. These paths of responses interact at various levels. They have a key role in signal transduction pathways. The p42 and p44 MPKs are also called ERK2 and ERK1, respectively. The MAP kinase family is activated by STE20, RAS, Raf protein serine/threonine kinases. In normal T cells, T cell receptor and CD28 synergistically activate p38. An alternative activation pathway involves Tyr323 by antigen-stimulated T cells but not B cells (Salvador JM et al Nature Immunol 6:390). MAPK81P1 (11p11.2-p12) may be a transactivator of SLC2A2 gene encoding GLUT2 glucose transporter and one of the factors responsible for MODY diabetes. The MAP kinases follow the signaling path: MAPKKK → MAPKK → MAPK. ►cell cycle, ►signal transduction, ►MAPK, ►JNK/SAPK, ►ERK, ►p38, ►STE diabetes, ►MODY, ►GLUT; Cobb MH 1999 Progr Biophys Mol Biol 71:479; English J et al 1999 Exp Cell Res 253:255; Chang L, Karin M 2001 Nature [Lond] 410:37; Dong C et al 2002 Annu Rev Immunol 20:55; Park S-H et al 2003 Science 299:1061; Schwartz MA, Madhani HD 2004 Annu Rev Genet 38:725.

MAP Kinase Kinase (MAPKK): ►Ste 7

MAP Kinase Kinase Kinase (MAPKKK): ►Ste 11

MAP Kinase Kinase Kinase Kinase (MAPKKKK): ►Ste 20

MAP KINASE PHOSPHATASE (MPK): MPK-3 dephosphorylates phosphotyrosine and phosphothreonine and inactivates the MAP kinase family proteins. Binding activates it by its non-catalytic C-end to ERK2 without a need for phosphorylation. The homologous MPK-4 is also activated by ERK2 but also by JNK/SAPK and p38. ►signal transduction and other proteins under separate entries; Zhang T et al 2001 Gene 273:71.

Map Manager v 2.5: The software for storing, organizing genetic recombination data and data base for RI strains of mouse. Information: K.F. Manly, Roswell Park Cancer Institute, Elm & Carlton, Buffalo, NY 14263, USA. Phone: 716–845–3372. Fax: 716–845–8169. kmanly@mcbio.med.buffalo.edu.

Map, Metric: An on-scale ordered FISH map where cosmid clones can orderly be positioned. ►FISH, ►cosmid

Map, Physical: ►physical map

Map, Self-Organizing: Constructed on the basis of mathematical cluster analysis; recognizes and classifies complex multidimensional data such as information of an array of genes involved in differentiation or other complex pathways. ►cluster analysis

Map Unit: 1% recombination = 1 map unit (m.u. or 1 centiMorgan, c.m. or cM). In approximate kilobase pairs equivalent to one centiMorgans in a few species: *Arabidopsis* ≈140; tomato ≈ 510; human ≈ 1,108 (in human chromosome 10 the recombination frequency is 1.32 cM/Mb^{-1}); maize ≈ 2,140. The *Salamanders* have the largest known genetic map 7291 cM. ►recombination frequency

Map Viewer: Provides genomic information by chromosomal location of a variety of eukaryotic organisms from animals to plants, fungi and protozoa. It can be queried by gene name, sequence alignment (BLAST), etc. ►genome, ►BLAST; http://www.ncbi.nlm.nih.gov/mapview/.

MAPCS (multipotent adult progenitor cells): These are "universal stem cells" present in small numbers in some adult tissues and have the capacity to produce other types of cells in a manner similar to embryonic stem cells. ►stem cells; Jiang Y et al 2002 Nature [Lond] 418:41; Schwartz RE et al 2002 J Clin Invest 109:1291.

M

MAPK (mitogen-activated protein kinase): Distinct kinases responding to different environmental cues and sets into motion different development/physiological pathways. The family includes KSS1 (filamentous growth), HOG-1 (hypertonic stress), FUS3 (mating), Mpk1 (cell wall remodeling), SLT-2, sapk-1 (stress activated protein kinase), FRS (FOS regulating kinase), erk-1 (extracellular signal regulated kinase), Smk1 (sporulation), etc. In yeast association with the Ste12 transcriptional activator regulates the specificity. Ste12 in combination with other proteins may bind to the pheromone response element (PRE). Ste12 protein associates with the filamentation and invasion response element (FRES), including A Ste12 protein binding site (TGAAACA) and a neighboring CATTCY sequences specific for the Tec1 (non-receptor tyrosine kinase) transcription factor. The mating and filamentous growth pathways are initiated in a similar way but for mating the FUS kinase is activated rather than the KSS. In the absence of phosphorylation KSS inhibits the Ste12-Tec1 complex and filamentous growth. The Dig proteins appear to be cofactors of the inhibitory path. In the inhibitory MAPKs (KSS), there is the MKI (MAP kinase insertion) site, which is remodeled upon phosphorylation and thus conversion into an activator. The specificity of the MAPKs is secured also by the complexes of recruited proteins, and each of these selects the appropriate MAPK. ►signal transduction, ►MAPKK, ►MEK, ►arrestin, ►Ask, ►JNJ FOS, ►Ste, ►anthrax; Ito M et al 1999 Mol Cell Biol 190:7539; Roberts CJ et al 2000 Science 287:873; Pouysségur J 2000 Science 290:1515; Barsyte-Lovejoy D et al 2002 J Biol Chem 277:9896.

MAPKK (mitogen-activated [MAP] protein kinase kinase): This protein mediates signal transduction pathways by phosphorylating RAS, Src, Raf and MOS oncogenes. When such an active kinase was introduced into mammalian cells, the AP-1 transcription factor was activated and the cells formed cancerous foci and became highly tumorigenic in nude mice, indicating that MAPKK is sufficient for tumorigenesis. ►signal transduction, ►tumor, ►anthrax, ►NPK

MAPKKK: Mitogen activated protein kinase kinase kinase. ►TAK1

Maple (*Acer* spp.): Hardwood trees; sugar maple is used for collecting syrup, 2n = 26.

Maple Syrup Urine Disease: ►isoleucine-valine biosynthetic pathway

Mapmaker 3.0: A software for constructing linkage maps using multipoint analysis in testcross and F_2. MAPMAKER/QTL is for quantitative trait loci. Available for Sun (Unix), PC (DOS) and Macintosh. Contact: Eric Lander, Whitehead Institute, 9 Cambridge Center, Cambridge, MA 02142, USA. Fax: 617–258–6505. INTERNET: mapmaker@genome.wi.mit.edu.

Mapping: Establishes the sequential location of genes, restriction fragments or PCR products. Molecular mapping of individual genomic DNA molecules labeled with fluorescent dyes at specific sequence motifs by the action of nicking endonuclease followed by the incorporation of dye terminators with DNA polymerase can be used. The labeled DNA molecules are then stretched into linear form on a modified glass surface and imaged using total internal reflection fluorescence (TIRF) microscopy. By determining the positions of the fluorescent labels with respect to the DNA backbone, the distribution of the sequence motif recognized by the nicking endonuclease can be established with good accuracy in a manner similar to reading a barcode (Xiao M et al 2007 Nucleic Acids Res 35(3):e16). ►genome projects, ►physical maps, ►recombination, ►restriction endonuclease, ►PCR, ►SNIPs, ►radiation hybrids, ►databases; http://linkage.rockefeller.edu/.

Mapping by Admixture Linkage Disequilibrium: ►MALD

Mapping by Dosage Effect: If the activity of enzymes is proportional to their dosage and disomics can be distinguished from critical trisomics, genes (for the enzymes) located in a specific trisome can be identified and assigned to that specific chromosome or in the case of telotrisomic to a particular chromosome arm. Theoretically, the enzyme activity is expected as follows if the locus is situated in the long arm of a particular chromosome (see Fig. M17).

In human trisomy 21 several genes show increased expression ranging from 1.21 to 1.61 relative to the normal disomic condition. Thus, in practice, the dosage effect may not be perfectly additive, yet it may be clear enough for classification. ►trisomics; Carlson PS 1972 Mol Gen Genet 114:273.

Mapping, Genetic: The mapping of chromosomes can be carried out on the basis of recombination frequencies of chromosomal markers, either genes or DNA markers such as RFLPs, RAPDs, etc., or molecular methods used in physical mapping (chromosome walking). As a hypothetical example assuming that genes *a, b, c, d* and *e* are syntenic, the recombination frequencies between them was found to be:

a 0.06 ***b*** 0.04 ***c*** 0.06 ***d*** 0.16 ***e*** 0.18 ***g***

The sum of the recombination frequencies is 0.06 + 0.04 + 0.06 + 0.16 + 0.18 = 0.50 indicating that the segregation between *a* and *g* is independent. The results shown could be obtained between two genes at a time but such a two-point cross would not have permitted the determination of the order of the genes relative to each other. For the determination of genes at "left" and genes at "right" a three-point cross is required as a minimum, and multipoint crosses are even more helpful ►recombination frequencies. For the results of a hypothetical three-point testcross see Table M2.

According to the data, the number of recombinants in interval I was 10 + 2 and in interval II 20 + 2 (the double recombinants had recombination in both interval I and II and their number must be added to the numbers observed). Thus the frequency of

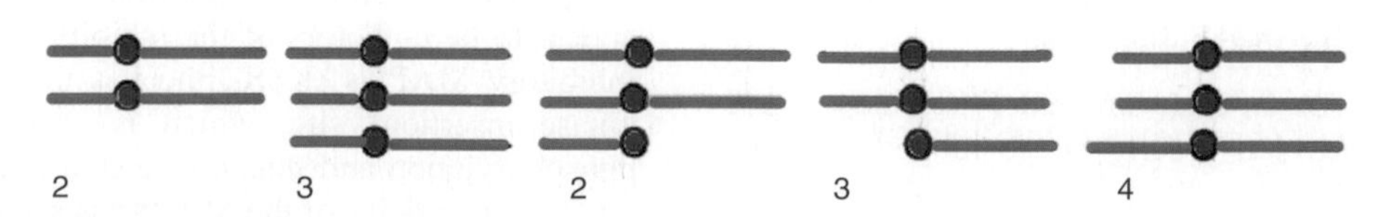

Figure M17. Mapping by dosage effects

Table M2. A hypothetical three-point testcross

Phenotypic classes →	ABD	abd	Abd	aBD	ABd	abD	AbD	aBd
Number of individuals	34	34	5	5	10	10	1	1
	Parental		Recombinants interval I		Recombinants interval II		Recombinants intervals I + II	
Total of 100	68		10		20		2	

recombination between A and B is 12/100 – 0.12 and between B and D 22/100 = 0.22. The number of recombinations between A and D is 10 + 20 + 4 = 34 (the 4 is the double of the number of recombinants in intervals I and II because these represented double recombination events).

Thus the relative map positions are $A - B - D$. Had it been found that the combined parental numbers were 68 but the recombinants *Abd* plus *aBD* 10 and *ABd* plus *abD* 2, and *AbD* plus *aBd* 20, it had to be concluded that the gene order was $A - D - B$, because the lowest frequency class ($0.10 \times 0.20 = 0.02$) must have been the double recombinants, and thus the gene order would have been $A - D - B$. The observed recombination frequencies may have to be corrected by mapping functions because not all double-crossovers might have been detected (see Mapping functions). The recombination frequencies may be biased also by interference when the frequency of double-crossovers are either higher or lower than expected on the basis of the product of the two single-crossovers (see ►Coincidence, Interference). The (corrected) recombination frequencies can be converted to map units by multplication with 100 and 1 map unit (m.u. or centiMorgan [cM]) is 0.01 frequency of recombination. Recombination frequencies can be estimated also in F_2 by using the product ratio method (see ►Product ratio method). In the latter case recombination frequencies can be calculated only between pairs of loci yet from the data of two pairs involving 3 loci, the gene order can be determined. On the basis of SNIPs human linkage maps of 3.9 cM resolution have been generated (Matise TC et al 2003 Am J Hum Genet 73:271). Availability of complete nucleotide sequences of the genomes does not obviate the calculations of recombination frequencies because the physical maps and the genetic map lengths are not necessarily identical. ►Recombination frequencies, ►crossing over, ►QTL, ►deletion mapping, ►chromosome walking, ►physical maps, ►mapmaker, ►joinmap, ►maximum likelihood method applied to recombination, ►mapping functions, ►radiation hybrids, ►F_2 linkage estimation, ►genomic screening, ►skeletal map, ►comparative map, ►unified genetic map, ►integrated map, ►consensus, ►SNIPs, ►linkage, ►coefficient of crossing over, ►exclusion mapping; Multilocus mapping algorithm for humans: Lander ES, Green P 1987 Proc Natl Acad Sci USA 84:2363; Ott J, Hoh J 2000 Am J Hum Genet 67:289; Grupe A et al 2001 Science 292:1915; survey of newer multipoint algorithms: Gudbjartsson DF et al 2005 Nature Genet 37:1015.

Mapping Functions: Correct map-distance estimates from recombination frequencies when the recombination frequency in an interval exceeds 15–20% and double crossing overs are undetectable because of the lack of more densely positioned markers. ►Haldane's mapping function, ►Kosambi's mapping function, ►Carter-Falconer mapping function, ►mapping, ►recombination frequency, ►coefficient of coincidence, ►stationary renewal process, ►count-location models; Zhao H, Speed TP 1996 Genetics 142:1369.

Mapping in Silico: Feeding phenotypic information to a computer program facilitates fast information gathering on the regulation and chromosomal locations of the multiple factors involved in a polygenic disease. ►polygenic; Grupe A et al 2001 Science 292:1814.

Mapping Panels: These are DNA sequences with known chromosomal location and can be used to locate unknown sequences to chromosomes. ►radiation hybrid panels

Mapping Sets: ►genomic screening

MapSearch: Locates regions of a genomic restriction map that resemble best a local restriction the so-called probe.

MapShow: A computer program displaying MapSearch alignments and draws Probe-to-Map alignments in Sun Workstations. ►mapping

MAR (matrix attachment region): Attaches chromatin loops to the nuclear matrix. The attachment region has a consensus of so-called A box (AATAAATCAA) or a T box (TTA/TAA/TTTA/TTT). The MARs are about 100 to 1,000 bp long and frequently include replicational origins and transcription factor binding sites. ►chromatin, ►loop domains mode, ►scaffold; Stratling WH, Yu F 1999 Crit Rev Eukaryot Gene Expr

9(3–4):311; Pemov A et al 1998 Proc Natl Acad Sci USA 95:14757; http://www.futuresoft.org/MAR-Wiz.

Maranhar: ►killer plasmids, ►*Neurospora*

Marburg Virus: A negative-sense single-stranded RNA virus containing seven genes within a lipid envelope (see Fig. M18). Besides transcribing its full genome, it produces subgenomic RNAs without encapsidation and these can also be translated into viral protein. The name comes from the German city Marburg where it was discovered in 1967 in monkeys shipped from Uganda. The virus causes hemorrhagic fever and very high mortality. Current efforts are directed to the development of efficient vaccine and RNAi based defense systems. ►Ebola virus, ►RNAi, ►vaccines; Fowler T et al 2005 J Gen Virol 86:1181.

Figure M18. Marburg virus

MARCKS (myristoylated alanine-rich C kinase substrate): Protein substrates of the protein kinase C (PKC), involved in differentiation; they bind actin filaments. (See Spizz G, Blackshear PJ 2001 J Biol Chem 276:32264).

mardel10: A supernumerary human chromosome 10 with a large deletion at the regular centromere. The deletion, however, activates, a functional neocentromere at 10q25, which lacks α-satellite and CENP-B centromeric protein although it shows some other centromere proteins. ►centromere, ►neocentromere, ►human artificial chromosome; Voullaire LE et al 1993 Am J Hum Genet 52:1153; Choo KH A 1997 Am J Hum Genet 61:1225.

Marek's Disease: A lymphoproliferative viral chicken disease. The growth hormone, GH1 conveys resistance. ►herpes, ►Epstein-Barr virus; Liu H-C et al 2001 Proc Natl acad Sci USA 98:9203; Levy AM et al 2005 Proc Natl Acad Sci USA 102:14831.

Marfan Syndrome (MFS, FBN1): Common symptoms include tall thin stature, long limbs and fingers, chest deformations. The three most consistent defects are skeletal, heart-vein (cardiovascular) and eye (ectopia lentis) abnormalities. The disease may affect the development of the fetus and may be recognized in early development and the life expectancy in serious cases may not much exceed 30. The cause of death is generally heart failure but the defect may be surgically corrected in some cases. The penetrance appears very good but the expressivity is highly variable. The symptoms frequently overlap with other anomalies, particularly with those of the Ehlers-Dunlop syndrome. The latter involves a defect in collagen. The primary defect in MFS involves the elastic fiber system glycoprotein, fibrillin. This protein of the connective tissue contains repeats resembling sequences in the epidermal growth factor (EGF) where the lesion observed leads to the identification of the basic molecular cause. Formerly a collagen defect was suspected.

Several investigators confirmed a transversion mutation at codon 293 leading to CGC (Arg) → CCC (Pro) replacement. Similar molecular defects have been identified in *Drosophila, Caenorhabditis* and cattle. This dominant gene has been assigned to human chromosome 15q21.1. Interestingly, mosaicism for trisomy 8 causes similar symptoms. The prevalence of MFS/FBN1 is about 1×10^{-4} but this figure may not be entirely reliable because of the wide range of manifestation of the symptoms (see Fig. M19). The recurrence risk is about 50%; 15–30% of the cases may be due to new mutation that is the cause of the most severe cases, whereas the familial incidence generally entails milder symptoms. The estimated mutation rate is $4 - 5 \times 10^{-6}$. FBN1 may be dominant negative. The mutation in fibrillin-1 affects the regulation of transforming growth factor (TGF-β) resulting in apoptosis of the alveolar cells of the lung. Excessive signaling of TGF-β family cytokines can cause aortic aneurysm in mice and can be blocked by angiotensin II type1 receptor blocker drug, Losartan (Habashi JP et al 2006 Science 312:117).

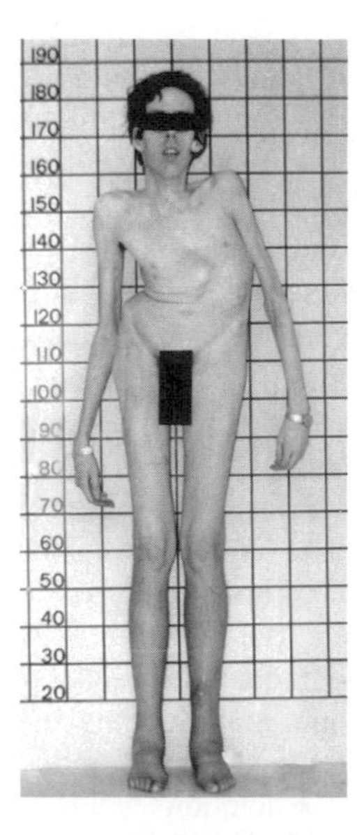

Figure M19. Marfan syndrome. (Courtesy of Dr. D.L. Rimoin, Los Angeles)

A higher probability of ectopia lentis was found for patients with a missense mutation substituting or producing a cysteine, when compared with other missense mutations. Patients with an FBN1 premature termination codon had a more severe skeletal and skin phenotype than did patients with an inframe mutation. Mutations in exons 24–32 were associated with a more severe and complete phenotype, including younger age at diagnosis of type I fibrillinopathy and higher probability of developing ectopia lentis, ascending aortic dilatation, aortic surgery, mitral valve abnormalities, scoliosis, and shorter survival (Faivre L et al 2007 Am J Hum Genet 81:454).

It has been suggested that President Abraham Lincoln, the famous musician Niccolo Paganini, renowned composer Sergey Rachmaninof and Pharaoh Akhenatan were afflicted with this type of anomaly. ►Marfanoid syndromes, ►coronary heart disease, ►cardiovascular disease, ►connective tissue disorders, ►penetrance, ►expressivity, ►transversion mutation, ►Ehlers-Dunlop syndrome, ►fibrillin, ►angiotensin, ►aneurysm, ►dominant negative, ►TGF, ►apoptosis, ►arachnodactyly, ►inbreeding depression, ►frameshift mutation; Dietz HC, Pyeritz RE 1995 Hum Mol Genet 4:1799; Neptune ER et al 2003 Nature Genet 33:407.

Marfanoid Syndromes: It may resemble the Marfan syndrome but one of the forms has no ectopia lentis (displacement of the crystalline lens of the eye). Another form does not involve cardiovascular defects. The marfanoid-craniosynostosis is called Shprintzen-Goldberg syndrome and the anomaly is caused by mutation in the fibrillin-1 gene. ►Marfan syndrome, ►eye diseases, ►Craniosynostosis syndromes, ►fibrillin, ►Lujan syndrome

Marijuana: ►cannabinoids, ►*Cannabis*

Mariner: Probably the smallest transposable element in eukaryotes (1,286 bp). It has not been observed in *Drosophila melanogaster* but has been detected in African species of the *D. melanogaster subgroup*, *D. sechellia* (1–2 copies), *D. simulans* (usually 2 copies), *D. yakuba* (about 4 copies), *D. teissieri* (10 copies), and *D. mauritiana* (20 to 30 copies). Mariner-like element has been detected also in soybeans.

Mariner contains 28-bp inverted terminal repeats and a single open reading frame (1,038-bp) beginning with an ATG codon at position 172 and termination with an ochre (TAA). Overlapping AATAA bases may serve as polyadenylation signal (see Fig. M20).

The target in the untranslated leader of the w^{pch} is 5′- TGGCGTA↓TAAACCG-3′. The arrow marks the insertion and the TATA indicate probably the target site duplication. *Mariner* is different from other transposable elements inasmuch as inducing a high frequency of somatic sectors (4×10^{-3}) at the w^{pch} (white *peach*) locus. Germline mutation is about 2 to 4×10^{-3}, with about twice as high in the males than females (no sex difference in somatic mutation).

Before the transposable element was recognized, the somatic instability was attributed to the factor named *Mos* in chromosome 3. This element also causes dysgenesis but it does not dis-play the reciprocal difference observed in the $P - M$ and $I - R$ systems, however, *mariner* trans-mitted through the egg shows higher rates of somatic excisions. *Mariner* homologs occur in other species too, including humans but the (*human*) sequences are pseudogenic although in-crease unequal crossing over in human chromosome 17p11.2-p12. The mariner type DNA-based transposons are most common in the human genome, representing ~1.6% of it. The *mariner* sequences are also expressed in *Leishmania* and *Caenorhabditis*. ►hybrid dysgenesis, ►transposable elements, ►Charcot-Marie-Tooth syndrome, ►neuropathy, ►HNPP, ►unequal crossingover, ►MLE, ►MITE, ►*Leishmania*, ►sleeping beauty; Hartl DL et al 1997 Annu Rev Genet 31:337; Zhang L et al 2001 Nucleic Acids Res 29:3566; Feschotte C, Wessler SR 2002 Proc Natl Acad Sci USA 99:280.

Marinesco-Sjögren Syndrome (MSS, 5q31): The major symptoms of the recessive hereditary disease are cerebellar ataxia, cataracts, retarded developmental/mental maturation. Hypergonadotropic hypogonadism is also a common feature of the condition. Chylomicron retention deficiency is coded in the same chromosomal region but physiologically not linked to this disease. The SIL1 nucleotide exchange factor for the Hsp70 chaperone BiP mutations are the primary cause of MSS. Affecting the endoplasmic reticulum may explain the multiple consequences of the mutations. ►hypergonadotropic hypogonadism, ►chylomicron, ►nucleotide exchange factor, ►endoplasmic reticulum, ►Hsp70; Senderek J et al 2005 Nature Genet 37:1312.

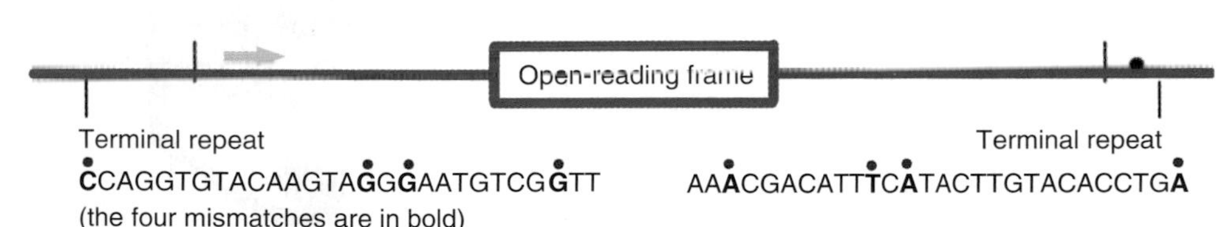

Figure M20. *Mariner* transposable element

Marker: Any gene or detectable physical alteration in a chromosome (e.g., knobs, microsatellites) or cytoplasmic organelle, used as a special label for that chromosome or chromosomal area. Molecular marker is a macromolecule (nucleic acid or protein) of known size and electrophoretic mobility to be used as a reference point in estimating the size of unknown fragments and molecules. In genomics Type I markers are the protein-coding genes, Type II markers are the highly polymorphic microsatellites and Type IIIs are SNPs. ►RFLP, ►RAPD, ►AFLP, ►ladder, ►microsatellite, ►minisatellite, ►SNIPs, ►linkage, ►association test; Schlötterer C 2004 Nature Rev Genet 5:63; plant markers: http://markers.btk.fi/.

Marker-Assisted Selection: Used for animal and plant breeding once linkage has been established between physical markers (RFLP, microsatellite loci) and others, economically desirable traits (e.g., disease resistance, productivity) that have low expressivity and/or poor penetrance because of the substantial environmental influences can be better identified. The physical markers should not have ambiguity of expression and thus may facilitate much faster progress in breeding. ►RFLP, ►microsatellite, ►expressivity, ►penetrance, ►QTL; Hospital EF 2001 Genetics 158:1363; Davierwala AP et al 2001 Biochem Genet 39(7–8):261; Hayes B, Goddard ME 2001 Genet Sel Evol 33(3):209.

Marker Effect: Theoretically, by generalized transduction, any bacterial gene should be transferred by the transducing phage. In fact some genes are transduced 1,000 fold better than others. The differences have been attributed to the distribution of *pac* sites in the bacterial chromosome. Also recombination by transduction may vary along the length of the bacterial chromosome. Marker effects have been observed also in eukaryotic recombination. ►generalized transduction, ►*pac* site

Marker Exchange Mutagenesis: ►targeting genes

Marker Exclusion: Occurs upon joint infection of bacteria by phages T4 and T2. In the progeny, T2 genes are recovered in about 30% rather than 50% as expected. The apparent basis of the bias is that T4 harbors 13 sequence-specific endonucleases that selectively eliminate particular tracts from the other phage. ►homing endonucleases; Edgell DR 2002 Current Biol 12:R276.

Marker Panels: DNA probes or genes that cover reasonably well portions of the genome regarding linkage. ►probe, ►linkage

Marker Rescue: The integration of markers into normal DNA phage from mutagen-treated (irradiated) phage during mixed infection of the host; it is similar to cross reactivation where from defective phages by recombination normal phages can be obtained. Alternatively, a wild type DNA fragment can be inserted by recombination into a mutant one and the mutation can thus be mapped. Marker rescue is the most commonly used method of mapping in phages. ►reactivation, ►multiplicity reactivation, ►Weigle reactivation; Barricelli NA, Doermann AH 1961 Virology 13:460; Thompson CL, Condit RC 1986 Virology 150:10.

Marker Transfer: Gene replacement by recombination.

Markov Chain Monte Carlo Algorithm: It applies Bayesian inference for evaluation of the posterior distribution. It is used also in classical likelihood calculation. (See Gilks WR et al (Eds.) 1996 Markov Chain Monte Carlo in Practice. Chapman and Hall, London, UK; Larget B, Simon DL 1999 Mol Biol Evol 16:750; ►Monte Carlo method, ►Bayes' theorem).

Markov Chain Statistics: A sequence x_1, x_2...of mutually dependent random variables constitutes a Markov chain if there is any prediction about x_{n+1}. Knowing $x_1 \ldots x_n$ may without loss be based on x_n alone. Among others, it is used in physical mapping of genomes, for ascertaining frequency distributions in populations. The *hidden Markov model* (HMM) type analysis—seeks the probabilities of an event occurring also prior to and after another event—permits the identification of protein domains in peptide chains or in nucleotide sequences and frequency distribution through time. ►TB-parse, ►alignment; Stephens DA, Fisch RD 198 Biometrics 54:1334.

Marmosets: New World Monkeys. In marmoset (*Callithrix kuhlii*) twins all somatic tissue types sampled were found to be chimeric by the use of microsatellite DNA (see Fig. M21). Chimerism was present in hematopoietic tissues and also in germ-line tissues, an event never before documented as naturally occurring in a primate. Chimeric marmosets often transmit sibling alleles acquired in utero to their own offspring. Thus, an individual that contributes gametes to an offspring is not necessarily the genetic parent of that offspring. Chorions of the twins'

Figure M21. Marmoset

placentas begin to fuse on day 19 and the process is complete by day 29, forming a single chorion with anastomoses connecting the pre-somate stage embryos. The fusion of the chorions and a delay in embryonic development at this stage allows the exchange of embryonic stem cells via blood flow between the twins. As a result, the infants are genetic chimeras with tissues derived from self and sibling embryonic cells (Ross CN et al 2007 Proc Natl Acad Sci USA 104:6278). ►*Callithrichicidae*, ►Chimera

Marmota (groundhog, woodchuck): Squirrel-like rodents; *Marmota marmota* 2n = 38; *Marmota monax* 2n = 38 (see Fig. M22).

Figure M22. Marmota

Maroteaux-Lamy Syndrome: ►mucopolysaccharidosis

Marquio Syndrome: ►mucopolysaccharidosis

Marriage: In every state of the United States and many other countries marriages between parent and child, grandparent-grandchild, aunt-nephew, uncle-niece and brother-sister are illegal. In about half of the states first cousin and half sibling unions are also prohibited. In some societies, consanguineous marriages are legal. ►consanguinity, ►miscegenation

Marry-in: A member of a pedigree without his parents being members of the pedigree. ►pedigree

Mars Model: Developed to interpret the fate of mitochondria during the development of an organism. The acronym is derived from (i) accumulation of defective **m**itochondria, (ii) accumulation of **a**berrant proteins, (iii) effect of oxygen-free **r**adicals and antioxidant enzymes, (iv) turnover of proteolytic **s**cavengers. ►mitochondria, ►aging

Mars Shield: A symbol of male; in pedigrees usually squares are used (see Fig. M23). ►venus mirror

Figure M23. Mars shield

Marshall Syndrome: ►Stickler syndrome

Marsupial: A mammalian group of animals that carry their undeveloped offspring in a pouch, e.g., the kangaroos and other Australian species, also the North-American opossum. The organization of the genetic material of marsupials differs in several ways from other placental mammals. Ohno's law is not entirely complied with inasmuch as genes in the short arm of the eutherian X chromosomes are dispersed in three autosomes. Marsupial Y chromosomes are extremely small yet they contain testis-determining and other sequences homologous to other mammals although the Y chromosome is not critical for sex determination. Their sex chromosomes also carry a pseudoautosomal region at their tip but do not form synaptonemal complex and do not recombine. ►Eutheria, ►Monotrene, ►Ohno's law, ►SRY, ►pseudoautosomal, ►vomeronasal organ; Marshall Graves JA 1996 Annu Rev Genet 30:233; Zenger KR et al 2002 Genetics 162:321.

MART-1 (Melan-A): ►MAG

Marten (*Martes americana*): 2n = 38.

Martin-Bell Syndrome: ►Fragile Xq27.3

Martsolf Syndrome: An apparently autosomal recessive cataract-mental retardation-hypogonadism. ►cataracts, ►hypogonadism, ►mental retardation, ►Cerebro-oculo-facio-skeletal syndrome; Hennekam RC et al 1988 Eur J Pediatr 147:539.

MaRX: A method of isolation of mammalian cells of a certain type. A DNA library is introduced into a packaging cell line (linX cells) with the aid of retroviral vectors. The virus-infected cells are subjected then to selection for the desired phenotype. Proviruses are recovered and used for the production of virus and further screening. ►DNA library, ►retroviral vectors, ►provirus, ►packaging cell line; Hannon GJ et al 1999 Science 283:1129.

MAS: ►Marker-assisted selection

MAS Oncogene: MAS1 was assigned to human chromosome 6q27-q27; it encodes a trans-membrane protein. ►transmembrane protein, ►oncogenes

MASA Syndrome (Xq28): Involves mental retardation, aphasia, shuffling gate, adducted and/or clasped thumb, etc. The basic defect is in L1 CAM cell adhesion molecule of 143 molecular mass of 1,256 amino acids. MASA is allelic to X-linked hydrocephalus (prevalence: $\sim 3 \times 10^{-4}$). ►aphasia, ►hydrocephalus, ►CAM

Masculinization: ►sex reversal

MASDA (multiplex allele-specific diagnostic assay): A mutation detection test capable of simultaneous detection of up to 100 nucleotide changes in multiple genes concerned with disease. ►single strand conformation polymorphism, ►peptide nucleic acids; Shuber AP et al 1997 Hum Mol Genet 6:337.

MASH2: A mammalian helix-loop-helix transcription factor controling extraembryonic trophoblast development but not that of the mouse embryo. ►MYF-3, ►DNA-binding protein domains

Mask: A computer program that produces ambiguous files in order to protect confidential DNA sequences in EcoSeq programs. ►EcoSeq

Masked mRNA: It is present in eukaryotic cells in such a form and cannot be translated until a special condition is met. ►mRNA; Spirin AS 1996 p 319. In: Hershey JWB et al (Eds.) Translational Control, Cold Spring Harbor Lab. Press, Cold Spring Harbor, NY.

Masked Sequences: Masked sequences of nucleic acids are associated with either proteins or other molecules to protect them from degradation.

Masking DNA Sequences: It is replacing nucleotide regions with certain properties with 'N' characters, or by converting the nucleotides within the region to lower-case letters. The program called RepeatMasker most frequently used to masks repeats: http://www.repeatmasker.org/. Repeats around SNPs may cause failure of assay or give mixed signals from different genomic regions. Variations may cause biased signal due to allele-specific binding of primers. Automatic masking tool: http://bioinfo.ebc.ee/snpmasker/.

Maspin: A protease inhibitor of the serpin family; maspin sequences are frequently lost from advanced cancer cells. Maspin is an inhibitor of angiogenesis. ►serpin, ►angiogenesis; Maass N et al 2000 Acta Oncol 39:931; Futscher BW et al 2002 Nature Genet 31:175.

Mas6p: ►mitochondrial import

Mass-Coded Abundance Tagging (MCAT): A method of protein characterization for proteomic studies. It is based on differential guanidination of the C-terminal lysine residues in tryptic peptides and followed by capillary liquid chromatography–electrospray tandem mass spectrometry. ►capillary electrophoresis, ►electrospray MS, ►proteomics, ►trypsin; Cagney G, Emili A 2002 Nature Biotechnol 20:163.

Mass Spectrum: When in the mass spectrometer molecules are exposed to energetic electrons they are ionized and fragmented. Each ion has a characteristic mass to charge ratio, m/e or m/z. The m/e values are characteristic for particular compounds and provide the mass spectrum for chemical analysis. Quantitative mass spectrometry uses the incorporation of a stable isotope (N^{15}) derivative that shifts the mass of the peptides in a known extent. The ratio between the derivatized and underivatized target can be measured. Shotgun mass spectrometry identifies components of a protein mixture by first identifying the mass of peptide fragments. ►laser desorption mass spectrum, ►electrospray mass spectrum, ►affinity-directed mass spectrometry, ►genomics, ►proteomics, ►MALDI, ►MS/MS, ►quadrupole, ►electrospray; Oda Y et al 1999 Proc Natl Acad Sci USA 96:6591; Mann M et al 2001 Annu Rev Biochem 70:437; Figeys D et al 2001 Methods 24:230; Cohen SL 2001 Annu Rev Biophys Biomol Struct 30:67; Aebersold R, Mann M 2003 Nature [Lond] 422:198; OMSSA, for protein structure: Sharon M, Robinson CV 2007 Annu Rev Biochem 76:167; virtual lab exercises: http://mass-spec.chem.cmu.edu/VMSL; Ms/Ms peptide spectra search: http://pubchem.ncbi.nlm.nih.gov/omssa.

Massively Parallel Signature Sequencing (MPSS): The method permits large-scale analysis of transcription templates of an entire genome without physical separation of the DNA fragments. The 16–20 base fragments were generated by repeated digestion with restriction endonucleases and ligated by appropriate adaptors. Complex DNA mixtures are cloned in vitro onto glycidyl methacrylate microbeds in quantities sufficient for biochemical and enzymatic analyses using fluorescent probes. The template-containing microbeads are assembled in a flow cell in a close planar, fixed array while the sequencing reagents are pumped through. The sequencing is monitored with the aid of the fluorescent signals. The procedures facilitate the early recognition of gene products involved in the development of disease and may permit intervention. The improved system is capable of sequencing 25 million bases within four hours in picoliter size wells 1.6 million on a 6.4-cm^2 slide, using a modified pyrophosphate-based method (pyrosequencing). It is about two-order magnitude faster than the standard Sanger method and has an accuracy of >99%. It does not require subcloning of the DNA in bacteria. Although it seems very convenient for relatively small (microbial) genomes, it is not yet practical with large (mammalian) systems (Margulies M et al 2005 Nature [Lond] 437:376). ►gene expression, ►microfluidics, ►small RNA, ►DNA sequencing, ►pyrosequencing; Brenner S et al 2000 Nature Biotechnol 18:630; Reinartz J et al 2002 Brief Funct Genomic Proteomic 1:95; Hood L et al 2004 Science 206:640; http://www.massivelyparallel.com/.

Mass-to-Charge (m/e, m/z): ►mass spectrum

Mast Cells: They reside in the connective or hemopoietic tissues and play an important role in natural and acquired immunity. They release TNF-α (tumor necrosis factor), histamines and attract eosinophils (special white blood cells) and destroy invading microbes especially if they have IgE or IgG (immunoglobulins) on their surface. They are responsible for the inflammation reactions in allergies and susceptibility to bee and snake venoms by facilitating vascular permeability. On the other hand, they release carboxypeptidase A and probably other proteases, which can degrade the venoms (Metz M et al 2006 Science 313:526). Mast cells are essential for regulatory T-cell controlled immune tolerance. ►IL-10, ►IL-9, ►immune system, ►TNF, ►immunoglobulins, ►carboxypeptidase, ►venome, ►T cell regulatory, ►immune tolerance

Mast Syndrome: A recessive (15q22.31) brain developmental disorder (paraplegia, dementia) caused by premature termination of transcription of the maspardin protein gene due to single nucleotide pair insertion. The protein is situated in the endosomal/trans-Golgi transport vesicles. ►paraplegia, ►endosome, ►Golgi; Simpson MA et al 2003 Am J Hum Genet 73:1147.

Master Chromosome: The large circular genome within the mitochondria. ►mtDNA

Master Genes: They have major role in a range of functions. (See Prior HM, Walert MA 1996 Mol Med 2:405; Silver LM 1994 Mamm Genome 5:S291).

Master molecule: Regulates series of reactions in differentiation, involving several genes. ►morphogenesis in *Drosophila*, ►neuron-restrictive silencer factor

Master-Slave Hypothesis. The hypothesis interpreted redundancy in the genomes by multiple copying 'the slaves' from the original 'master' sequences based on the structure of the loops of the lampbrush chromosomes of the newt *Triturus*. ►redundancy, ►Lampbrush chromosomes; Callan HG 1967 J Cell Sci 2:1.

MAT1: A RING finger protein subunit stabilizing cyclin H-CDK7 complex or CAK. ►RING finger, ►cyclin, ►CDK, ►CAK, ►four-hybrid system; Devault A et al 1995 EMBO J 14:5027.

MAT Cassette: ►mating type determination in yeast

***MAT* Locus**: In yeast, it is involved in the determination of mating type in yeast. ►mating type determination in yeast

Matched Pairs t-Test: Checks the equality of the means of paired observations. The difference between the matched pairs is tested by the formula: $t = \frac{\bar{d}}{s_d/\sqrt{n}}$ $\bar{d}$ = the mean of the differences, s_d = standard deviation and t is determined at n – 1 degrees of freedom from a t table. The null hypothesis is that the means of the paired observations are true. ►mean, ►standard deviation, ►Student's t table, ►Null hypothesis

Mate Killer: mu particle. ►symbionts hereditary

Mate Pairs: Randomly sequenced DNA fragments are fit together by their matching mate pair ends into a continuous sequence. DNA sequences read from opposite ends of fragments are the mate pairs. ►WGS, ►Human genome

Maternal Behavior: A very complex trait, it is regulated by hormones such as estradiol, progesterone, prolactin, oxytocin and β-endorphin. The Mest/Peg1 (mouse chromosome 6, human chromosome 7q32) is expressed only from the paternally transmitted allele and its defect leads to altered maternal behavior. ►hormones under separate entries, ►imprinting

Maternal Contamination: During amniocentesis, maternal cells may contaminate the sample withdrawn and may become a cause of genetic diagnostic error. ►amniocentesis

Maternal Coordinate Genes: Expressed during oogenesis and determine positional information in the egg.

Maternal Effect Genes: Display delayed inheritance because only the offspring of the homozygous or heterozygous dominant females is affected; these females themselves may appear normal. Also, genes with products (RNA, protein) in the follicle and nurse cells that may diffuse into the oocytes and the embryo, and thus not just the zygotic genes, affect early development. After the initial phases, the maternal transcripts and proteins are destabilized. (See Fig. M24 of ►morphogenesis, ►delayed inheritance, ►cadherin, ►indirect epistasis, ►transgenerational effect, ►imprinting; Evans MMS, Kermicle JL 2001 Genetics 159:303; Tadros W et al 2003 Genetics 164:989.

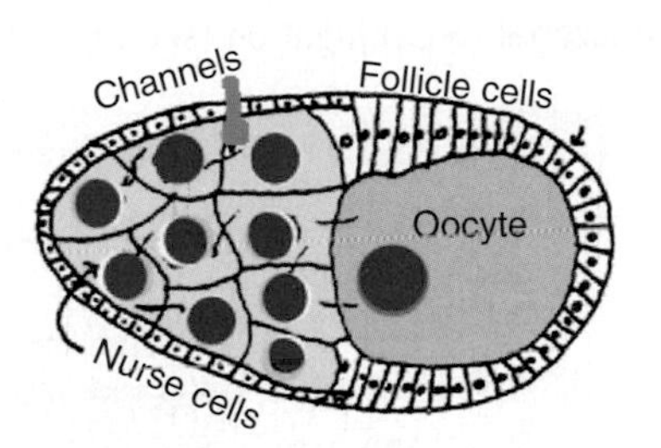

Figure M24. The Drosophila egg chamber cysts are surrounded by follicle cells and inside the diploid primary oocyte is shown with the often polyploid nurse cells. These are connected by cytoplasmic bridges. The maternal genes affect oogenesis and the early development of the embryo

Maternal Embryo: Develops from an unfertilized egg. ►apomixis, ►parthenogenesis

Maternal Genes: ►maternal effect genes, ►*Limnaea*

Maternal Inheritance: Genetic elements (generally extranuclear) are transmitted only through the female. ►mtDNA, ►mitochondrial genetics, ►doubly uniparental inheritance, ►chloroplast genetics

Maternal Performance: A complex of physiological and behavioral traits of the mother that affects the wellbeing and survival of the offspring. It includes such maternal attributes as nursing, grooming, nest building, etc. The heritability of such traits is variable and generally low yet it contributes measurably to fitness. ►maternal behavior; Peripato AC et al 2002 Genetics 162:1341.

Maternal Tolerance: The embryo expresses antigens that is supposed to be foreign to the maternal tissues yet usually the embryo does not suffer immunological rejection. The cause of the tolerance appears to be due to the secretion of corticotropin-releasing hormone (CRH) and the activation of the pro-apoptotic Fas ligand (FasL) resulting in killing of activated T lymphocytes. Some of the female infertility may be caused by lack of adequate level of CRH in the endometrium. ►incompatibility, ►corticotropin releasing factor, ►FAS, ►apoptosis, ►T cell; Makrigiannakis A et al 2001 Nature Immunol 2:1018.

Maternity Verification: Rarely needed as the Romans held "mater certa" (mother is certain); in case of legal disputes, the methods of forensic genetics are available. ►forensic genetics, ►paternity testing

MATH (meprin and TRAF homology): The protein domain shared by metalloendopeptidases and TRAF proteins regulating the folding to activable forms of the molecules. ►meprin, ►TRAF

Mating Assortative: ►assortative mating

Mating Bacterial: ►conjugation (see Fig. M25).

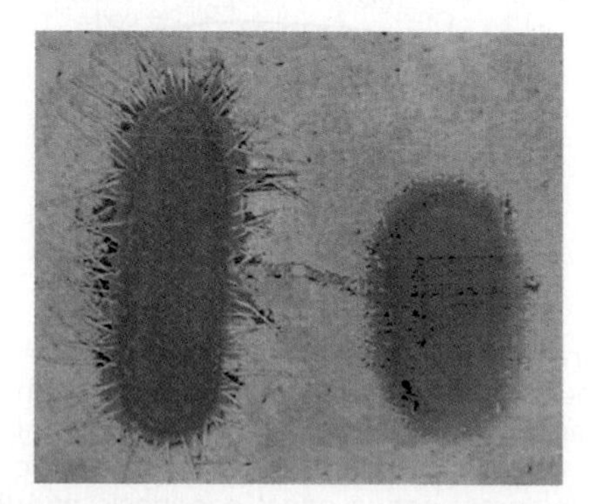

Figure M25. Mating bacteria

Mating Controlled: ►controlled mating

Mating Interrupted: ►interrupted mating, ►conjugation mapping

Mating Nonrandom: ►inbreeding, ►assortative mating, ►autogamy (self-fertilization), ►controlled mating

Mating, Physiological Consequences: In some species the male seminal fluid may contain substances toxic to the female. In contrast, the seminal fluid of the male cricket (*Gryllus lineaticeps*) increases the life expectancy of the female by about a third and multiple matings (common in this species) almost doubles the fertility of the female (Wagner WE Jr et al 2001 Evolution 55:994).

Mating Plug: A gelatinous material deposited at vulva after mating of hermaphroditic *Caenorhabditis*. During mating between some hermaphroditic males, plugs may be visible at the head. ►vaginal plug

Mating Random: ►random mating, ►mating systems

Mating Success (K): Expressed as

$$\frac{\text{No. of females mated by mutants/No. of mutant male}}{\text{No. of females mated by wild types/No. of wild type males}}$$

Increasing success in mating in insect species may reduce the level of phenoloxidases in both sexes. Thus, a major humoral immune component is weakened and the individuals become more susceptible to infection. Therefore, copulation may adversely influence fitness. (See Rolff J, Siva-Jothy MT 2002 Proc Natl Acad Sci USA 99:9916).

Mating Systems: It can be random mating, self fertilization, inbreeding, assortative mating or a combinations of these. ►Hardy-Weinberg theorem

Mating Type: The designation of individuals with plus or *a* ("male") and minus or *α* ("female") labels when sex like in higher eukaryotes cannot be recognized yet genetically two types exist that do not "mate" within group but in between groups, and the diploid zygote subsequently undergoes meiosis and reproduces the mating types in 2:2 proportion. (See Ferris PJ et al 2002 Genetics 160:181).

Mating Type Determination in Yeast: *Saccharomyces cerevisiae* yeast can exist in homothallic and heterothallic forms. The heterothallic yeast cells are haploids and are either of ***a*** or ***α*** mating type. The homothallic yeast cells are diploid and heterozygous for the mating type genes (***a/α***). The diploid cells arise by fusion of haploid cells. Recognition of the opposite mating type cells is mediated by pheromones, the ***α*** factor (13 amino acids) and the ***a*** factor (12 amino acids) peptide hormones, respectively. The two types of cells are equipped also by surface receptors for the opposite mating type pheromones. The diploid cells lack these factors and receptors and do not fuse but may undergo meiosis and sporulate by releasing both α and a haploid cells.

The two mating types are coded by genes in chromosome 3 on the left and right sides of the centromere, respectively. These genes are clustered within the so-called mating type cassettes and are silent at their positions named *HML**α*** and *HMR**a*** locations. They are expressed when transposed to the mating-type site, *MAT*. At *MAT*, either the left *HMLα* or the right *HMR**a*** cassette can be expressed within a particular homologous chromosome. The *MAT* site is approximately 2-kb and it is about 187 kb from *HML**α*** and at about 93 kb from *HMR**a***.

(See Fig. M26 for the overall structure of the regions). In the outline the *MAT* site is shown empty but in reality either the *HML**α*** or by the *HMR***a** cassette occupies it. When one or the other silent complex is unidirectionally transposed to the *MAT* site, the expressed *MAT**α*** and *MAT**a***, respectively, are generated. At the original (left and right) locations the *Y**α*** (747 nucleotides) and the ***Ya*** (642 nucleotides) are kept in place and silent by the product of the *SIR1–4* gene (**S**ilent **In**formation **R**egulator). Actually four Sir proteins exist, 1, 2, 3 and 4. Repression is mediated also by the autonomous replication sequences (ARS) and the pertinent enhancer (E) elements. The transposition activity of a *MAT* cassette requires also the presence of nuclease hypersensitive sites in the flanking regions. The transposition at the *MAT* sites is initiated when the HO endonuclease makes a staggered cut at the *Y Z1* junction-generating strands with four base overhangs:

The *HO* gene is expressed only in the haploid cells but not in the diploids, and only at the end of the G1 phase of the cell cycle and both new cells have the same mating type and can transpose their mating type gene only after the completion of the next cell cycle. The product of gene SIN1–5 represses the function of HO and the expression of gene *SW1–5* is required for the expression of *HO*. Mother cells selectively transcribe *HO*. In daughter cells, Ash1p suppressor of *HO* may accumulate as a result of asymmetric mRNA distribution. For the actual mating the function of other (sterility, *STE*) genes are also needed. *STE2* and *STE3* are cell type-specific receptors of G proteins. *GPA1* encodes the Gα subunit, *STE4* the Gβ subunit and *STE18* the Gγ subunit. The GβGγ subunits, after activation, regulate further downstream units of the pheromone-signaling pathway, including the kinases STE20, a series of MAP kinases (encoded by *STE11*, *STE7* and *FUS/KSS* genes). The Ste5p protein (encoded by *STE5*) is presumed to be a scaffold for organizing all the kinases.

Transposition involves pairing with the homologous *Z* sequences, followed by an invasion of a double-stranded receiving *Y* site and degradation of these *Y* sequences. Subsequently the *X* site is invaded. New DNA synthesis then takes place using as template the sequences of the invader molecule that is replacing the old DNA tract. Integration is mediated in the pattern of a gene conversion mechanism as suggested by the Holliday model of recombination.

The α mating type gene has two elements ***α***1 (transcribed from right to left) that induces the ***α***mating functions and ***α***2 (transcribed in left to right similarly to a) that keeps in check the expression of the *a* mating type expression. The simultaneous expression of *a*1 and *α*2 represses also *SIR* and other haploid-specific genes. These processes recruit also proteins PRTF (Pheromone Receptor Transcription Factor) and GRM (General Regulator of Mating type) that recognize specific nucleotide sequences within the haploids and diploids. The ***α***2 protein also interacts the MCM1 DNA binding and the MADS box proteins. The binding of MAT***α***2 with MCM1 represses ***a***-specific genes in haploids but in diploid cells MAT***α***2 heterodimerizes with MAT***a***1 and thus haploid-specific genes are repressed (see Nature [Lond] 391:660 for the structural interactions of these proteins). In *Candida albicans* (and some other fungi) an apparently evolutionarily more ancient regulation operates the basically identical mating type determination system (Tsong AE et al 2006 Nature [Lond] 443:415).

Whether *HML* or *HMR* switching occurs depends on the surrounding sequences and it is not intrinsic to the elements. *Mat a* cells recombine with *HML* almost an order of magnitude more frequently than

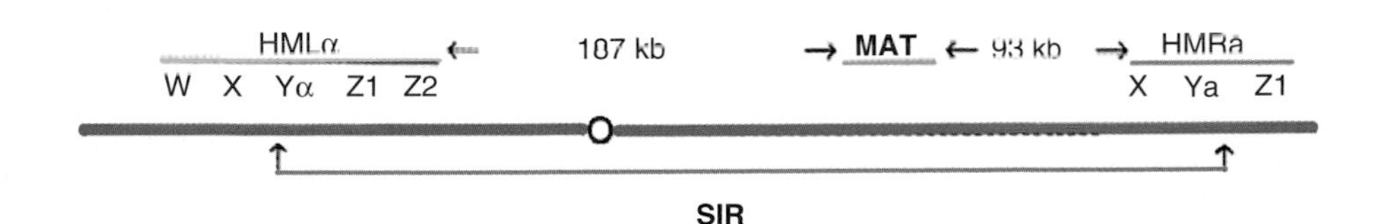

Figure M26. Mating type region in yeast

5′-GCTTT↓CGGCAACAGTATA-3′ *MATa*
3′-CGAAAGGCG↑TTGTCATAT-5′

5′-ACTTCGCGC AACA↓GTATA-3′ *MATα*
3′-TGAAGCGCG↑TTGTCATAT-5′

Figure M27. Nucleotide sequences in the *MAT* loci

with *HMR* (see Fig. M27). *MATα* cells recombine in 80–90% of the cases with *HMR*. The switching is controlled by 700-kb element 17-kb proximal to *HML* by a recombinational enhancer.

The expression of the mating type genes involves a complex cascade of events. The mating type protein factors interact with G protein-like receptors situated in the cell membrane. These G proteins then transduce the mating signals through a series of phosphorylation reactions to transcription factors that control the turning on/off genes mediating the cell cycle, cell fusion and conjugation. The industrial strains of yeast are frequently diploid or polyploid and may be heterozygous for mating type. In such a case, they fail to mate. If they are plated on solid medium, they may sporulate and the chromosome number will be reduced. ►Holliday model, ►signal transduction, ►cell cycle, ►silencer, ►HML and HMR sex determination, ►*Candida albicans*, ►pheromones, ►MCM1, ►MADS box, ►homothallic, ►heterothallic, ►rare-mating, ►regulation of gene activity, ►Ty, ►*Schizosaccharomyces pombe*, ►ORC, ►RAP1, ►Abf; Haber JE 1998 Annu Rev Genet 32:561; Dohlman HG, Thorner JW 2001 Annu Rev Biochem 70:703; Rusche LN et al 2003 Annu Rev Biochem 72:481; Tsong AE et al 2003 Cell 115:389.

Matrilineal: Descended from the same maternal ancestor, e.g., the mtDNA. ►mtDNA

Matrix: Solutes in cells, organelles or chromosomes, etc.; the *extracellular matrix* fills the space among the animal cells and it is composed of a meshwork of proteins and polysaccharides, secreted by the cells. The viral matrix connects the genomic core with the envelope. ►nuclear matrix

Matrix Algebra: It deals with elements that are arranged as shown in the figure. A matrix with *r* rows and *c* columns is of *order r x c* or an *r x c matrix*. If r = c, it is a *square matrix* (see Fig. M28). A matrix with 1 row is *row vector* and a matrix with only 1 column is a *column vector*. A *weight matrix* can be generated for a particular motif by the use of Bayes' theorem. Each column in the matrix represents one position of the motif. Each row of the matrix corresponds to the probability that a corresponding, e.g., nucleotide, occurs at a position in the motif. Matrix algebra can be used for correlations, in numerical taxonomy, evolution, comparative genomics, etc. ►motif,

$$\begin{bmatrix} 1 & 4 & 7 \\ 2 & 5 & 8 \\ 3 & 6 & 9 \end{bmatrix} \begin{bmatrix} 2 & 1 \\ 4 & 6 \end{bmatrix}$$

Figure M28. Matrix arrangements

►Bayes's theorem; Hays WL, Winkler RL 1970 Statistics: Probability, inference and decision. Holt, Rinehart and Winston, New York.

Matrix Attachment Region: ►MAR

Matrix-Assisted Laser Desorption Ionization/Time of Flight/Mass Spectrometry (MALDI-TOF): A procedure for separation of DNA fragments mixed with a carrier that is painted subsequently on the surface of a solid face target. Laser desorbs and ionizes the fragments and acceleration in a mass spectrometer is used to determine fragment length. The MALDI-TOF procedure may detect as small as a nucleotide change in the length of a DNA sequence (ca. 100 perhaps 1,000 bp) or changes in the microsatellite numbers. This technique can be utilized in DNA sequencing, discrimination among mutations of a gene, for identification of STS. It may eventually be used also for clinical and diagnostic procedures. ►laser, ►mass spectrum, ►laser desorption mass spectrum, ►microsatellite, ►SNIPs, ►proteomics, ►STS, ►TOFMS, ►MALDI-TOF; Shahgholi M et al 2001 Nucleic Acids Res 29:E91; Chu J et al 2001 Clin Chim Acta 311:95; Nordhoff E et al 2001 Electrophoresis 22:2844.

Matrix Diseases: These diseases periodically/seasonally reoccur, frequently in a somewhat different form such as the annual influenza epidemics or the occasional pandemics.

Matroclinous: The offspring resembles the mother because it developed either from an unfertilized egg or from an egg that underwent nondisjunction and carries two X chromosomes or from a female with attached X-chromosomes or by imprinting or dauermodification or due to sets of dominant genes or the failure of transmission of a particular chromosome through the sperm or caused by non-nuclear (mitochondrial, plastid) genes. ►*Rosa canina*, ►chloroplast genes, ►mtDNA, ►mitochondrial genetics

Matthiola (garden stock, wallflower): A cruciferous ornamental (2n = 14). Lethal factor (*l*), is tightly linked to the simple flower character (S), has been of special interest (see Fig. M29). This causes the appearance of the "ever-segregating "full-flower" trait (s) in 1:1 proportions rather than in 3:1. By sophisticated breeding techniques and seed selection in trisomic offspring, the commercially available seed germinates and develops into nearly 100% full-flower/double flower plants that are, however, completely sterile due to the recessive lethal factor. Thus, the seed supply is entirely dependent on commercial sources. (See Kappert H 1937 Ztschr Ind Abst- Vererb-lehre 73:233; Roeder AHK, Yanofsky MF 2001 Dev Cell 1:4).

Figure M29. Matthiola

Maturases: Proteins that mediate a conformational change in the pre-mRNA transcript and cooperate in the splicing reactions. ►introns, ►mitochondrial genetics

Maturation Divisions: Same as meiosis.

Maturation of DNA: Phage proteins cut the linear, continuous DNA into pieces that can be accommodated by the phage capsids.

Maturation Promoting Factor: ►MPF

Maurice of Battenberg: A hemophiliac grandson of Queen Victoria of England; son of carrier daughter Beatrice. ►hemophilias

Mauriceville Plasmid: ►*Neurospora* mitochondrial plasmids

MAVS: ►RIG oncogene

MAX: A b/HLH/LZ (basic helix-loop-helix/leucine zipper) protein hetero-oligomerizes with the MYC oncoproteins, and this state is required for malignant transformation by c-MYC. MAX alone lacks the transactivator domain of MYC. Its DNA recognition site is CACGTG. MAX may be orchestrating the biological activities of b/HLH/LZ transcription factors. The basic α helices follow the major groove of the DNA in a *scissors grip*. ►MYC, ►helix-loop-helix, ►RFX, ►leucine zipper, ►Mxi/Max, ►major groove

Maxam-Gilbert Method: ►DNA sequencing

Maxicells: Bacterial cells that lost most or their entire chromosomal DNA because of heavy irradiation by UV light. Therefore, they do not replicate their DNA. The plasmid they contain may have escaped the irradiation, and represents an appropriate replicon, and can carry on replication of that plasmid and direct the synthesis of plasmid-coded proteins. This makes such cells ideal for the expression of the plasmid-born protein without a background of cellular proteins. Especially useful are those maxi cells containing lambda vectors, which have sufficient expression of the phage repressor and thus do not permit λ protein expression. ►mindless, ►lambda phage, ►replicon, ►plasmids; Jemiolo DK et al 1988 Methods Enzymol 164:691.

Maxicircle: The large mitochondrial genome. ►mtDNA; Carpenter LR, Englund PT 1995 Mol Cell Biol 15 (12):6794.

Maximal Equational Segregation: It takes place when a gene is segregating independently from the syntenic centromere. It has particular significance in polyploids because it facilitates an increase of double (or multiple) recessive gametes and thus affects segregation ratios as a function of the map distance between gene and centromere. ►autopolyploids, ►trisomic analysis, ►synteny, ►polyploidy; Mather K 1935 J Genet 30:53; Rédei GP 1982 Genetics. Macmillan, New York.

Maximal Parsimony: ►evolutionary tree

Maximal Permissive Dose: ►radiation hazard assessment

Maximization of Gene Expression: It can be achieved by the selection or modification of optimal promoters or in prokaryotes by varying the bases immediately after the Shine-Dalgarno sequence or manipulation of the triplet preceding the first methionine codon. ►regulation of gene expression, ►regulation of protein synthesis, ►Shine-Dalgarno

Maximum Likelihood Method Applied to Recombination Frequencies: The justification for the use of the maximum likelihood principle in estimating recombination frequencies is that the value obtained has the smallest variance among all procedures. The estimation is based on the maximization of:

$$\frac{n!}{a_1!a_2!\ldots a_t!}(\mathrm{m}_1)^{\mathrm{a}_1}(\mathrm{m}_2)^{\mathrm{a}_2}\ldots.(\mathrm{m}_t)^{\mathrm{a}_t}$$

where n is the population size, $a_1 \ldots a_t$ stand for the number of individuals in the different phenotypic or genotypic classes, $m_1 \ldots m_t$ represent the expected proportions of individuals in classes $1 \ldots t$. After maximizing the logarithm of the likelihood (L)

expression with respect to the recombination fraction (p), we have:

$$L = C + a_1 \log m_1 + a_2 \log m_2 + \ldots a_t \log m_t,$$

where C is a constant of the maximum likelihood that is eliminated upon differentiation:

$$\frac{dL}{dp} = a_1 \frac{d \log m_1}{dp} + a_2 \frac{d \log m_2}{dp} + a_t \frac{d \log m_2}{dp} = 0$$

For the coupling experiment (Table M3) below:

$$L = 4032 \log\left(\frac{1}{2} - \frac{1}{2}p\right) + 149 \log\left(\frac{1}{2}\,p\right) + 152 \log\left(\frac{1}{2}\,p\right) + 4035 \log\left(\frac{1}{2} - \frac{1}{2}\,p\right)$$

After maximization and differentiation:

$$\frac{dL}{dp} = \frac{4032}{1-p} + \frac{149}{p} + \frac{152}{p} - \frac{4035}{1-p} = 0 \text{ and}$$

$$\mathrm{p} = \frac{149 + 152}{8368} \cong 0.03597$$

The standard error s_p is calculated:

$$-\frac{1}{V_p} = S\left(mn \frac{d^2 \log m}{dp^2}\right)$$

Since

$$a \frac{d \log m}{dp}$$

was defined earlier, after a second differentiation and substitution (mn) for (a), we obtain:

$$-\frac{1}{V_p} = -\frac{n}{2} \times \frac{1}{1-p} + \frac{1}{p} + \frac{1}{p} + \frac{1}{1-p} = \frac{n}{p(1-p)} \cong \frac{8368}{0.034676} \cong 241,319$$

and hence

$$V_p = 0.000004143$$

and

$$s_p = \sqrt{V_p} \cong 0.00204$$

or by the general formula

$$s_p = \sqrt{\frac{p[1-p]}{n}}$$

Recombination in F_2 can also be estimated with the aid of the maximum likelihood principle and a coupling phase progeny will exemplify it (see Table M4):

Table M3. Hypothetical test cross examples

	Parental	Recombinant	Recombinant	Parental	
Gametic genotypes →	AB	Ab	aB	ab	Σ
Observed in **coupling**	4032	149	152	4035	8368
Expected coupling	$\frac{1}{2}n(1 - p)$	$\frac{1}{2}n(p)$	$\frac{1}{2}n(p)$	$\frac{1}{2}n(1 - p)$	n
	Recombinant	**Parental**	**Parental**	**Recombinant**	
Gametic genotypes →	AB	Ab	aB	ab	Σ
Observed **repulsion**	638	21,379	21,096	672	43,785
Expected repulsion	$\frac{1}{2}n(p)$	$\frac{1}{2}n(1 - p)$	$\frac{1}{2}n(1 - p)$	$\frac{1}{2}n(p)$	n

Table M4. Recombination in F_2 coupling

	Parental	Recombinant	Recombinant	Parental		
Phenotypic classes	AB	aB	Ab	ab	Σ	
Expectation	$\frac{n}{4}(2+P)$	$\frac{n}{4}(1+P)$	$\frac{n}{4}(1-P)$	$\frac{n}{4}P$	n	**(1)**
Observed	663	36	40	196	935	

$$L = 663 \log\left(\frac{1}{2}+\frac{1}{4}P\right)+36 \log\left(\frac{1}{4}-\frac{1}{4}P\right) + 40 \log\left(\frac{1}{4}-\frac{1}{4}P\right)+196 \log\left(\frac{1}{4}P\right) \quad (2)$$

Upon maximization:

$$\frac{dL}{dP} = \frac{663}{2+P} - \frac{36}{1-P} - \frac{40}{1-P} + \frac{196}{P} = 0 \quad (3)$$

This can be reduced:

$$\frac{663(1-P(P)}{2P-P^2-P^3} - \frac{76(2+P)(P)}{2P-P^2-P^3} + \frac{196(2+P-2P-P^2)}{2P-P^2-P^3} \quad (4)$$

Common denominator omitted and multiply

$$663(P-P^2) - 76(2P+P^2) + 196(2+P-2P-P^2)(5)$$

Multiplication completed:

$$663P - 663P^2 - 152P - 76P^2 + 392 + 196P - 392P - 196P^2 \quad (6)$$

Terms summed up: $392 + 315P - 935P^2 = 0.0001585$ (close to zero)

(Designate terms) c b a (7)

The right side of eq. (7) can be determined only after solving the quadratic equation below:

$$P = -b \pm \frac{\sqrt{b^2-4ac}}{2a} = 315 \pm \frac{\sqrt{99225+1466080}}{1870} = -315 \pm \sqrt{\frac{1565305}{1870}} = -315 \pm \frac{1251.1215}{1870} = \frac{-1566.1215}{1870}$$

$= -0.837498$; after changing sign, $P = 0.837498$

Thus $P = 0.837498$, and $\sqrt{P} = 0.9151492 = 1-p$, and hence the recombination fraction $p = 1 - 0.9151492 = 0.0848508$.

The variance of P,

$$V_P = \frac{2P[1-P][2+P]}{n[1+2P]} = 0.0004495,$$

where n (= Σ) = 935 and the variance of p,

$$V_p = \frac{VP}{4P} = 0.0001342$$

and the standard error

$$s_p = \sqrt{Vp} = \sqrt{0.0001342}, = 0.01158$$

Thus, the frequency of recombination between the two genes is ~0.085 ± 0.012. Data may be entered at step (6) to expedite routine calculations. ►maximum likelihood principle, ►recombination frequency, ►F_2 linkage estimation, ►information; Mather K 1957 The Measurement of Linkage in Heredity. Methuen, London, UK; Wu R, Ma C-X 2002 Theor Population Biol 61:349.

Maximum Likelihood Principle: It provides a statistical method for estimating the optimal parameters from experimental data. The best statistics for the computations is that it provides the smallest variance., E.g., the variance of the median of a sample is $\frac{\pi\sigma^2}{2n}$ which is $\frac{\pi}{2} = 1.57$ times the size of the variance of the mean ($\bar{x}$). Therefore the mean is a much better characteristic of the population than the median. The binomial probability is expressed as: $\binom{n}{r}p^r(1-p)^{n-r}$ giving the probabilty (p) that (r) events occur in a sample of (n).

The relative probability of r/n events for different values of (p) is called the *likelihood*. The procedure that facilitates finding a population parameter (θ) that maximizes the likelihood of a particular observation is a *maximum likelihood procedure.* If the dispersion of a population follows the normal distribution, the variance $V = \sigma^2$, is a maximum likelihood estimator of the distribution of that population. All other methods need to be compared with and tested against this method before their results can be accepted and used.

Naturally, all statistics provide only predictions and not direct proof regarding the biological mechanism concerned. Therefore, careful collection of data, replications, sufficient sample sizes, etc., are indispensable for accuracy and predictability. The maximum likelihood mandates that the choice of the parameter (θ) makes the likelihood, $L(X_1, X_2 \ldots X_n|\theta)$ the largest value. Example: a random sample of 20 is obtained and among them, say 12 belongs to a particular class. We can hypothesize that the true frequency of this class is either (I): p = 0.6 or (II): 0.5 or (III): 0.7. According to the normal distribution then:

$$(\text{I})\binom{20}{12}(0.6)^{12}(0.4)^8 = \frac{20!}{12!8!}(0.6)^{12}(0.4)^8 = 125.970 \times 0.002176782 \times 0.00065536 \cong 0.17971$$

$$(\text{II})\binom{20}{12}(0.5)^{12}(0.5)^8 = \frac{20!}{12!8!}(0.5)^{12}(0.5)^8 = 125.970 \times 0.000244140 \times 0.00390625 \cong 0.12013$$

M

$$(\mathrm{III})\binom{20}{12}(0.7)^{12}(0.3)^{8} = \frac{20!}{12!8!}(0.7)^{12}(0.3)^{8}$$
$$= 125.970 \times 0.013841287 \times 0.00006561$$
$$\cong 0.11440$$

Obviously, hypothesis (I) has the maximum likelihood to be applicable to this case. After this simple demonstration, we can generalize the likelihood function as:

$$L(X_1, X_2 \ldots\ldots X_N|p) = \binom{N}{r}p^r(1-p)^{N-r}$$

where X are the samples, N = population size, p = probability, and r = 0, 1,...N. The maximized likelihood is conveniently expressed by the logarithm of the likelihood function:

$$\log L = \log\binom{N}{r} + (r)\log(p) + (N-r)\log(1-p)$$

After differentiation to (p) and equating it to zero:

$$\frac{d}{dp}\log L = \frac{r}{p} - \frac{N-r}{1-p} = 0.$$

After bringing it to the common denominator:

$$\frac{r(1-p)-(N-r)p}{p(1-p)} = 0$$

The denominator omitted:

$r(1-p)-(N-r)p = 0 = r - rp - Np + rp$, and hence $p = r/N$ and this is the *maximum likelihood estimator* of *p*. Similarly, it can be shown that for a population in normal distribution the arithmetic mean of the sample $\left(\frac{\sum x_i}{N}\right)$ is the maximum likelihood estimator of the μ. The probability P for a multinomial distribution is: $\frac{N!}{X!Y!Z!\ldots}p^X q^X r^X$ where p, q, r... are the probabilities of X,Y,Z...classes. Although we may not know these probabilities but we may have experimentally observed the classes (genotypes, alleles, etc.), and we can derive the likelihood function which permits the estimation of the parameters of p, q, etc. If in a random mating population the proportion of *A* is p^2, that of *B* is 2pq and that of *C* is q^2, we can write the likelihood function as:

$$L = \frac{N!}{A!B!C!}(p^2)^A(2pq)^B(q2)^C$$

from which after logarithmic conversion and differentiation we can obtain the value of $p = \frac{2A+B}{2N}$ and $q = \frac{2C+B}{2N}$ and the variance $V_p = \frac{pq}{2N}$.

For an in-depth treatment of maximum likelihood, mathematical statistics monographs should be consulted. The maximum likelihood method is widely used in decision-making theory. In genetics, it is most commonly used for the estimation of recombination and allelic frequencies. (See maximum likelihood method applied to recombination frequencies, probability, information)

Maximum Parsimony: same as maximal parsimony. ►evolutionary tree

Maximum Tolerated Dose (MTD): It does not cause more than 10% weight loss, does not cause clinical toxicity, death or disease that would shorten life span.

Maxizyme: A dimeric ribozyme

May-Hegglin Anomaly (Dohle leukocyte inclusions with giant platelets): An asymptomatic dominant granulocyte and platelet disorder resulting often in thromobocytopenia located to chromosome 22q12.3-q13.1. The locus is about 0.7 megabase DNA. In the granulocytes, spindle-shaped cytoplasmic inclusions have been observed that appear to be the depolarization relics of ribosomes. The Fechtner syndrome and the Sebastian syndrome share the major characteristics and the chromosomal location (22q13.3-q13.2). The non-muscle myosin heavy chain IIA (MYH9) mutations appear to account for the three diseases. The Alport syndrome shares some of the symptoms also although that is a different disease. ►hemostasis, ►platelet anomalies, ►giant platelet; Martignertti JA et al 2000 Amer J Hum Genet 66:1449; Kelley MJ et al 2000 Nature Genet 26:106.

Maytansinoids: The extract of tropical trees or shrubs with an LDLo of 190 μg/kg as intravenous dose for humans (see Fig. M30). Related compounds are rifamycin, streptovaricin, macrolides, etc. ►magic bullet, ►LDLo

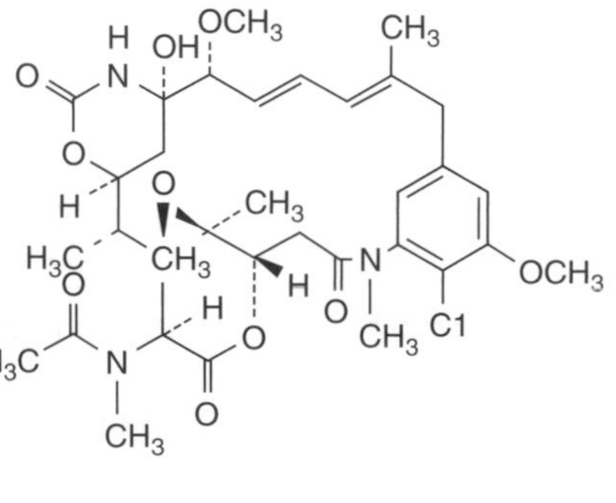

Figure M30. Maytansin

MBD (methyl-binding domain): MBD proteins are members of the histone deacetylase complex and bind to DNA and remodel the chromatin resulting in gene silencing. *Mbd2*$^{-}$ homo or heterozygous mice displayed fewer intestinal tumors than the wild type for the gene. ►histone deacetylase, ►5-azacytidine; Sansom OJ et al 2003 Nature Genet 34:145.

MBF: ►Mbp1, ►Swi

MBP: Maltose binding protein, encoded by gene *malE* (91 min) of *E. coli*.

Mbp: Megabase pair, 1 million base pair.

Mbp1 (mitotic binding protein): The components of the MBF (microtubule-binding factor) with Swi6, mediate S phase expression of the cell cycle. ►cell cycle, ►SBF, ►Swi; Iyer VR et al 2001 Nature [Lond] 409:533.

μC (microcurie): 3.7 × 10^4 dps [disintegration/second]. ►Curie, ►isotopes

MCA (metabolic control analysis): The study of complex enzyme systems in response to any changes in substrate(s). It may facilitate the detection of thresholds potentially leading to disease. It may permit the classification of genes and protein regarding their role in metabolic networks and thus may help drug discovery. ►proteomics; Cascasnte M et al 2002 Nature Biotechnol 20:243.

McArdle's Disease: ►glycogen storage disease type V

McCune-Albright Syndrome (MAS, GNAS1, 20q13.2): A pituitary neoplasia resulting from excessive secretion of growth hormone, caused by mutation and constitutive expression of the GTP-binding subunit (G_s) of a G-protein. It is characteristically a polyostotic fibrous dysplasia, café-au-lait skin lesion, and gonadotropin-independent gonadal precocious activation. MAS is generally not inherited, probably because germline G_s-activating mutations are lethal. It is believed that the somatic mutation in MAS patients occurs early in development, and therefore the clinical spectrum in each individual is determined by the tissue distribution of mutant-bearing cells (Rey RA et al 2006 Hum Mol Genet 15:3538). ►pituitary gland, ►pituitary tumor, ►G-proteins, ►securin, ►Albright hereditary osteodystrophy, ►pseudohypoparathyroidism, ►gonadotropin, ►café-aut-lait

McDonald-Kreitman Hypothesis: The excess of replacement substitutions in the amino acids in proteins indicates that they are the consequences of selectively advantageous mutations. Reduction in non-synonymous mutations indicates that they are selected against. ►mutation neutral, ►amino acid replacement, ►Ka/Ks, ►selection inferred from DNA sequences; McDonald JH, Kreitman M 1991 Nature [Lond] 354:114.

MCF Oncogene (synonymous with DBL, ROS): The human mammary carcinoma proto-oncogene was assigned to human chromosome Xq27. It encodes a serine-phosphoprotein (p66). ►oncogenes, ►ROS

MCH (melanin-concentrating hormone): MCH reduces appetite and increases metabolic rate. It encodes a neuropeptide precursor at 12q23 and PMCHL1 at 5p14 and PMCHL2 at 5q13 are truncated versions of MCH. ►obesity, ►leptin, ►melanin

Mch: ICE-related proteases. ►ICE, ►apoptosis

MCK: A muscle-specific kinase. ►MyoD

McKusick-Kaufman Syndrome (MKKS, 20p12): A recessive developmental anomaly including accumulation of fluids in the uterus and vaginal area (hydrometrocolpos), extra finger at the area of the little finger (postaxial polydactyly), heart disease, etc. Several other genes map apparently to the same chromosomal location, e.g., that of the Bardet-Biedl (BBS) syndrome. The distinction between MKKS and BBS is by the three criteria named above. Some of the symptoms are shared also the Ellis-van Creveld syndrome that is at another chromosomal location. The critical protein appears to be a chaperonin. ►Bardet-Biedl syndrome, ►Ellis-van Creveld syndrome; David A et al 1999 J Med Genet 36:599; Slavotinek AM et al 2000 Nature Genet 26:15.

McLeod Syndrome (XK): A recessive human Xp21 region deficiency of the Kx blood antigen precursor. The symptoms vary because of overlapping defect with closely linked genes, particularly CGD (chronic granulomatous disease). It may be associated with acanthocytosis, characteristic for abetaliproteinemia. ►abetalipoproteinemia, ►granulomatous disease chronic, ►Kell-Cellano blood group, ►contiguous gene syndrome

MCM1 (licensing complex): A yeast DNA-binding protein, product of the minichromosome maintenance gene also involved in the regulation of mating type; it controls the entry into mitosis. All eukaryotes have apparently at least six different MCMs assisting in DNA replication. MCM 2–7—a putative helicase—appears to become activated in telophase before the ext cell cycle (Dimitrova DS et al 2002 J Cell Sci 115:51). Some have helicase and DNA-dependent ATPase function. Mcm10 is a component of the replication fork and apparently is required for the function of DNA polymerase-α (Ricke RM, Bielinsky A-K 2004 Mol Cell 16:173). The licensing actually means that the chromatin must be subject to quality control to qualify for replication after mitosis. MCM also prevents the re-entry into S phase. ►mating type determination in yeast, ►MCM3, ►ARS, ►Cdc45/Cdc46/Mcm5, ►cell cycle, ►CDC19, ►CDC21,

M

►Cdc6, ►Cdc18, ►Cdt1, ►geminin, ►reinitiation of replication, ►DNA polymerases, ►sex hormones; Tye BK 1999 Annu Rev Biochem 68:649; Lee J-K, Hurwitz J 2000 Proc Natl Acad Sci USA 98:54; Nishitani H et al 2001 J Biol Chem 276:44905.

MCM3: It is apparently the same as the replicational licensing factor (RLF), that appears in tight binding to DNA during interphase but released during S phase. This factor assures that within a cell just one cycle of DNA replication occurs. This protein belongs to the family of MCM1 to MCM5 factors detected in yeast. ►MCM1, ►replication licensing factor, ►cell cycle

MCP (membrane cofactor protein, CD46): MCP regulates (along with other proteins such as DAF, factor H and C-4 binding protein) complement functions and protects the cells from attacks by their own defense system. MCP and DAF control also reproductive functions (spermatozoa, extrafetal tissues) besides infectious diseases and xenografts. These functions are encoded in human chromosome 1q3.2 region. MCP (14 exons) has four isoforms generated by alternative splicing of a single transcript. MCP shares a 34 amino acid signal peptide with DAF (11 exons). The signal sequence is followed by *complement control protein repeats* (CCPR) where C3b and C4b complement components bind. Glycosylated and serines, threonines, prolines (STP) residues follow the CCPR sequences. The other regions of MCP and DAF are different. ►complement, ►decay accelerating factor, ►xenotransplantation, ►signal sequence, ►atherosclerosis; Kemper C et al 2001 Clin Exp Immunol 124(2):180.

MCP-1 (monocyte chemoattractant protein): It controls (along with IL-8) adhesion of monocytes to the vascular epithelium. It is inducible by the platelet-derived growth factor (PDGF). Mice lacking these receptors are more prone to atherosclerosis. ►atherosclerosis, ►chemokines, ►monocytes; Yamamoto T et al 2001 Eur J Immunol 31:2936.

MCR (mutation cluster region): A segment of a gene where mutations occur at high frequency.

mcr: ►methylation of DNA

MCS: Multiple cloning sites. ►polylinker

M-CSF: ►macrophage colony stimulating factor, ►macrophage

MDA (multiple displacement amplification): MDA can be carried out from crude whole blood or tissue culture cell. As small as 1–10 copies of human genomic DNA can be amplified to 20–30 microgram. It has been used for analysis of the entire genome of single microbial cells. The product is suitable for various genetic analyses such as SNP, Southern blots, and various diagnostic purposes. ►amplification, ►SNP, ►Southern blot; Dean FB et al 2002 Proc Natl Acad Sci USA 99:5261.

Mda-7 (IL24): Encodes a cytokine, which selectively inhibits human melanoma cancer cells without conspicuous side effects. ►IL24; Fisher PB et al 2003 Cancer Biol Ther 2 Suppl 1:S23.

MDC1: A mediator of DNA checkpoint control. ►checkpoint, ►histone variants; Lou Z et al 2003 Nature [Lond] 421:957.

mdg: ►copia

mDIP (methylated DNA immunoprecipitation): It uses antibodies specific for 5-methylcytosine residues and the procedure can distinguish between some cancerous and normal tissues (Keshet I et al 2006 Nature Genet 38:149). ►landmark genomic scanning

Mdl1: A mitochondrial export protein of the AAA transporter family. ►AAA proteins

MDM2 (murine double-minute homolog, MDM2 is at 12q14.3-q15): A cellular oncoprotein that can bind and downregulate p53 tumor suppressor, attach to the retinoblastoma suppressor, and can stimulate transcription factors E2F1 and DP1, and thus may promote tumorigenesis. MDM2 is an E3 ubuiquitin ligase that assists in the degradation of p53. The cis-imidazoline analogs (nutlins) are antagonists of MDM2 and can activate the p53 pathway, leading to cell cycle arrest, apoptosis (see Fig. M31) and inhibition of tumor growth (Vassilev LT et al 2004 Science 303:844). Nutlin prevents the interaction of MDM2 with p53, and the growth of half of the cancers by the normal tumor suppressor function of p53 is restored (Harris C 2006 Proc Natl Acad Sci USA 103:1659; Tovar C et al 2006 Proc Natl Acad Sci USA 103:1888).

MDM2 can suppress TGF-β effect also without the inactivation of p53. MDM2 is regulated by RASA through the Raf/Mek/Map kinase pathway. Raf also activates $p19^{ARF}$, an inhibitor of MDM2. MDM2 involves TGF resistance in various tumor cell lines. MDM2 prevents transcriptional activation by Sp1 but the retinoblastoma protein displaces Sp1 from MDM2 and restores Sp1 transcriptional activity. The RING

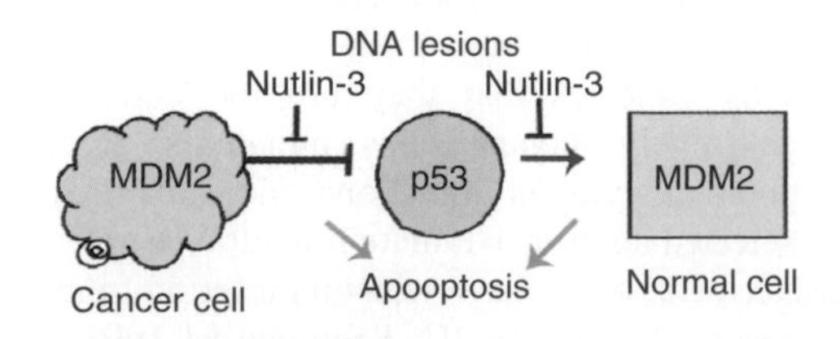

Figure M31. MDM2 functions

domain of MDM2 may stimulate ubiquitins (Minsky N, Oren M 2004 Mol Cell 16:631). Loss of MDM2 or MDM4 leads to lethality in p53-dependent manner in the mouse. Mice lacking MDM2 in the central nervous system develops hydranencephaly after 14.5 days of pregnancy whereas lacking MDM4 results in porencephaly (cerebrospinal fluid-filled cavities of the brain) after 17.5 days of embryonic development. Both mutations can be rescued by inhibition of p53 (Xiong S et al 2006 Proc Natl Acad Sci USA 103:3226). Similar diseases are known also in humans that are encoded in different chromosomes. ►oncogenes, ►transcription factors, ►retinoblastoma, ►E2F, ►DP1, ►p53, ►TGF, ►cyclin G, ►ubiquitins, ►ARF, ►apoptosis, ►anencephaly, ►endocytosis, ►Zinc finger; Johnson-Pais T et al 2001 Proc Natl Acad Sci USA 98:2211.

MDR: ►multidrug resistance

MDS (macronuclear destined sequences): From the germline DNA during vegetative development of ciliates internal sequences are eliminated by the process of chromosome diminution and only the MDS is retained in the macronucleus. ►chromosome diminution, ►*Paramecium*, ►macronucleus; Prescott DM 1997 Curr Opin Genet Dev 7:807.

Meal Worm (*Tenebrio molitor*): A larger X and a smaller Y chromosome in this insect determine sex. ►antifreeze protein

Mealybug: A member of the coccidian taxonomic group of animals with the name reflecting the "mealy" appearance of the wax coat of the body of the insects and the mealy appearance of their colonies on the surface of plants (see Fig. M32). They received attention by the peculiarity of their chromosome behavior. During the cleavage divisions immediately after fertilization, all the chromosomes are euchromatic. After blastula one-half of the chromosomes (2n = 10) becomes heterochromatic in the embryos which develop into males. At interphase these heterochromatic chromosomes clump into a chromocenter. By metaphase, the heterochromatic and euchromatic sets are no longer distinguishable. In the males, the first meiotic division is equational and during the second division the two types of chromosomes go to opposite poles. Two of the four nuclei are heterochromatic and two euchromatic. The heterochromatic nuclei then disintegrate and the euchromatic cells proceed to spermiogenesis. The euchromatic set of the fathers becomes later the heterochromatic chromosomes of the sons. This was verified by X-raying the females and males. Only 3% of daughters of males irradiated by 16,000 R survived but the sons were unaffected even after 30,000 R. Some sons survived even after 90,000 R exposure of the fathers. Thus, sex appears to be determined developmentally in these insects. ►chromosomal sex determination; Nur U 1967 Genetics 56:375; Palotta D 1972 Can J Genet Cytol 15:809.

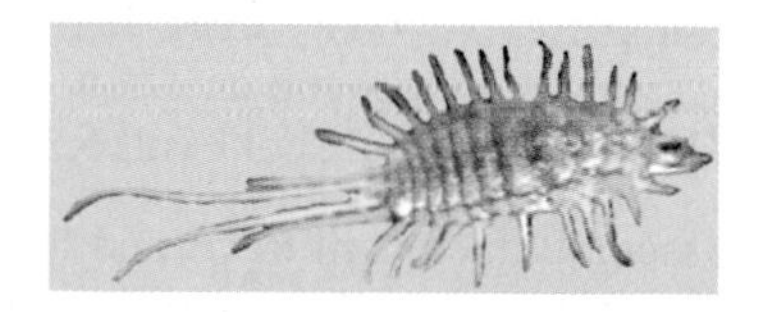

Figure M32. Mealybug

Mean: The *arithmetic mean* $\overline{x}$ is equal to the sum (Σ) of all measurements (x) divided by the number (n) of all measurements, or $\overline{x} = \frac{\sum x}{n}$. The *geometric mean* (G) is the nth root of the product of all measurements: $G = \sqrt[n]{x_1.x_2...x_n}$. The *harmonic mean* (H) is the inverse average of the reciprocals of the measurements $H = \frac{n}{\Sigma[1/x]}$.

Examples:

$$\overline{x} = \frac{2+8}{2} = 5, G = \sqrt[3]{2 \times 8 \times 4} = \sqrt[3]{64} = 4,$$

$$H = \frac{2}{[1/2] + (1/8)} = 3.2.$$

The *weighted mean* is the calculated mean multiplied by the pertinent frequency of the groups in a population. (See variance)

Mean Lethal Dose: Mutagens or toxic agents are denoted by LD_{50}. ►LD_{50}, ►LDLo, ►LC50

Mean Squares: The average of the squared deviations from the mean; it is obtained by dividing the sum of the squared deviations by the pertinent degrees of freedom. Basically, this is the estimated variance. ►variance, ►variance analysis, ►intraclass correlation

Meander: When two consecutive β-sheets of a protein are adjacent and antiparallel. ►protein structure

Measles: An infectious viral disease, practically eliminated in the USA since compulsory vaccination was introduced in the 1960s. Sporadically—after importation from a foreign country—local measles outbreaks have however occurred (Parker AA et al 2006 N Engl J Med 355:447).

Measurement Units: Length: 10 ångström (Å) = 1 nanometer (nm), 1000 nm = 1 micrometer (μm), 1000 μm = 1 millimeter (mm), 10 mm = 1 centimeter (cm), 100 cm = 1 m.

Volume: 1000 microliter (μL or λ = 1 milliliter (mL), 1000 mL = 1 liter (L);

Weight: 1000 picogram (pg) = 1 nanogram (ng), 1000 ng = 1 microgram (μg), 1000 μg = 1 milligram (mg), 1000 mg = 1 gram (g), 10 g = 1 dekagram (dg), 100 dg = 1 kilogram (kg).

M

Generally: milli = 10^{-3}, micro = 10^{-6}, nano = 10^{-9}, pico (p) = 10^{-12}, fempto (f) = 10^{-15}, atto (a) = 10^{-18} and kilo (k) = 10^3, mega (M) = 10^6, giga (G) = 10^9 and tera (T) = 10^{12}. ►M_r, ►dalton, ►agricultural measures; Unit Converter: http:// mypage.bluewin.ch/berthod/vuc/.

MEC: ►degenerin, ►ion channels

MEC1: A kinase locus of yeast (member of the PIK family); its phosphorylates RAD53 and RAD9, signal transducers of DNA damage. Mec3 protein seems to regulate telomere length. Mec1 and Rad53 are also involved in the G1, S and G2 checkpoint control. Tel1 can carry out the Mec1 function also. The homologs are *SAD3, ESR1,* and the human gene is homologous to AT, responsible for ataxia telangiectasia. ►ataxia, ►RAD, ►signal transduction, ►DNA replication, ►cell cycle, ►checkpoint, ►double-strand break, ►PIK, ►telomeric silencing; Tercero JA, Diffley JFX 2001 Nature [Lond] 412:553; Lopes M et al 2001 Nature [Lond] 412:557.

Mechanism-based Inhibition: ►regulation of enzyme activity

Mechanosensory Genes: These are involved in the neurobiological control of proprioceptory sensations. The proprioceptory nerve terminals are located in the muscles, joints, tendons and the ears and perceive the information about movements, touch, balance and hearing. The process converts mechanical forces into electrical signals through special ion channels. ►zona pellucida

Mecillinam: A β-lactam type antibiotic, which targets the penicillin-binding protein 2 (PBP2) required for the elongation of the bacterial cell wall. ►β-lactamase

Meckel Syndrome (MKS): A rare complex recessive syndrome (MKS1, 17q22-q23) with most specific characteristics are the brain defects, cystic kidneys and polydactyly. MKS1 is also called the Meckel-Gruber syndrome. In Finnish populations, it may occur however in the near 10^{-4} range. The mouse homolog is at chromosome 11; similar genes occur in the majority of organisms and appear to have ciliary functions (Kyttälä M et al 2006 Nature Genet 38:155). MKS genes were located also at 11q13, and at 8q21.13-q22.1. The latter encodes a 995-amino acid seven-transmembrane receptor protein (meckelin) of unknown function (Smith UM et al 2006 Nature Genet 38:191). ►neural tube defects, ►polydactily, ►kidney diseases, ►cilia, ►Joubert syndrome

MeCP1 (methyl-CpG binding protein, MBD1): It binds methylated CpG sequences in the DNA and is part of a transcriptional repression complex along with histone deacetylase. It is encoded at human chromosome 18q21. ►CpG, ►islands, ►histone deacetylase, ►methylation of DNA

MeCP2: methyl-CpG binding protein encoded at Xq28 and it is involved with the Rett syndrome autism. It seems to be involved in brahma-containing SWI/SNF chromatin remodeling system (Harikrishnan KN et al Nature Genet. 38:964). ►autism, ►brahma, ►Rett syndrome; Chen WG et al 2003 Science 302:885.

MED-1: ►null promoter

Medaka (*Oryzias latipes*): A small, fertile fish (see Fig. M33) well suited for developmental studies because it is genetically pigment-free (due to four recessive genes). Transparent stock is available that permits the direct visualization through a stereoscopic microscope, the major internal organs (heart, spleen, blood vessels, liver, gut, gonads, kidney, brain, spinal cord, eye lens, air bladder, etc.) in live animals. Medaka draft genome(700 megabases), is less than half of the zebrafish genome and 20,141 genes are predicted (Kasahara M et al 2007 Nature [Lond] 447:714). ►zebrafish; Wakamatsu Y et al 2001 Proc Natl Acad Sci USA 98:10046; ►pufferfish; http://medaka.utgenome.org.

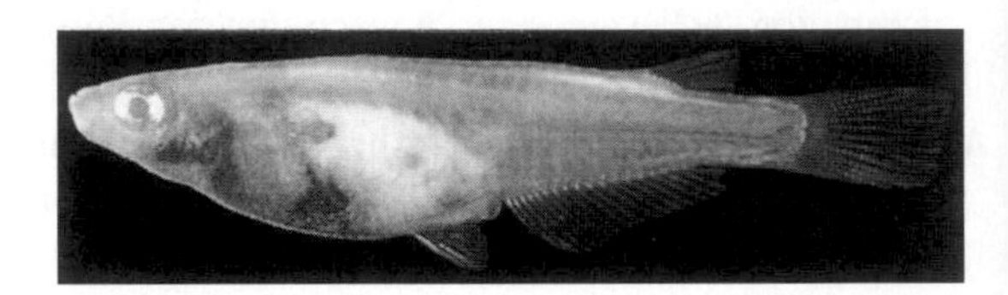

Figure M33. Medaka female (courtesy of Dr. Yuko Wakamatsu)

Medea Factor (maternal effect dominant embryonic arrest): A lethality factor transmitted through the egg cytoplasm, which kills the offspring, in, e.g., *Tribolium*, unless it inherits from either parent a rescuing M factor. ►killer genes, ►*Tribolium*; Beeman RW, Friese KS 1999 Heredity 82:529; Grossniklaus U et al 1998 Science 280:446.

Median: A statistical concept that indicates that equal numbers of (variates) observations are on its sides at both minus and plus directions. ►mean, ►mode

Median-Joining Networks: A statistical procedure for reconstructing phylogenies from intra-specific data. The difficulties are in the large required sample sizes and small genetic differences and a network of potential evolutionary paths overcomes these obstacles to construct simple evolutionary trees. It proved useful for the analysis of mitochondrial DNA phylogenies (Bandelt HJ et al 1999 Mol Biol Evol16:37).

Mediator: Assembly factors of RecA and like recombinases and single-strand DNA binding proteins. The mediator of the GAL gene of yeast is associated with the upstream activator sequences and not with the core promoter or with TBP or TFIID (Kuras L et al 2003 Proc Natl Acad Sci USA 100:13887). ►RecA, ►core promoter, ►UAS, ►TBP, ►transcription, ►factors; Gasior SL et al 2001 Proc Natl Acad Sci USA 98:8411.

Mediator Complex (Meds): A group of ~20 or more proteins involved in the facilitation of transcription by RNA polymerase II in yeast and other eukaryotes (see Bjorklund S et al 1999 Cell 96:759). The mediator complex is a co-activator, co-repressor and a general transcription factor. These proteins share subunits and participate in a large variety of different complexes, which have different functions in gene expression and developmental control. They may mediate chromatin remodeling, interact with various proteins (activators), general transcription factors and directly or indirectly with RNA polymerase II. One form of RNA polymerase II includes an additional polypeptide, Gdown1 and it is called Pol II(G). The latter type of polymerase definitely requires the Mediator for efficient transcription (Hu X et al 2006 Proc Natl Acad Sci USA 103:9506). The yeast Mediator proteins interacting with Med17(Srb4) as a 223 kDa complex is called a *mediator head module*. It interacts with an RNA polymerase II-TFIIF complex, but not with the polymerase or TFIIF alone. This interaction is lost in the presence of a DNA template and associated RNA transcript, recapitulating the release of Mediator that occurs upon the initiation of transcription (Takagi Y et al 2006 Mol Cell 23:355). MSA specifically restricts PIC (preinitiation complex) function in the absence of the Mediator. This function is fully restored in the presence of the Mediator, indicating that Mediator dependency in the metazoan cell is imparted, at least in part, through factors that negatively modulate unregulated PIC function. That one such activity resides in a complex containing hSpt5 and hSpt4, which were previously identified as components of the transcription elongation factor DRB sensitivity-inducing factor DSIF (Malik S et al 2007 Proc Natl Acad Sci USA 104:6182). ►Srb, ►DRIP, ►NAT, ►chromatin remodeling, ►activator proteins, ►transcription factors, ►TBP, ►TAF, ►co-activator, ►re-initiation, ►elongator, ►preinitiation complex, ►DRB, ►DSIF, ►PIC; Svejstrup JQ et al 1997 Proc Natl Acad Sci USA 94:6075; Gustafsson CM et al 1998 J Biol Chem 273:30851; Myers LC, Kornberg RD 2000 Annu Rev Genet 69:729; Gustafson CM et al 2001 Mol Microbiol 41:1; Boube M et al 2002 Cell 110:143; Conaway RC et al 2005 Trends Biochem Sci 30:250.

Medicago: ►alfalfa

Medical Error: It kills 44,000 to 98,000 people in US hospitals annually (Hayward RA, Hofer TP 2001 J Am Med Assoc 286:415).

Medical Genetics: Genetics applied to medical problems. ►clinical genetics, ►human genetics, ►genomic medicine; http://research.marshfieldclinic.org/genetics/.

Medical Terminology (UMLS): http://umlsks.nlm.nih.gov, acronyms: http://invention.swmed.edu/argh/; http://medstract.med.tufts.edu/acro1.1/index.htm; abbreviations: http://www.hpl.hp.com/research/idl/projects/abbrev.html.

Medicine: ►diseases in humans, ►preventive medicine; history and encyclopedia of medicine: http://www.mic.ki.se/History.html.

Medicinal Chemistry: It is involved with drug design and development.

Mediterranean Fever, Familial (FMF): A human chromosome-16p13 recessive disease with recurrent spells of fever, pain in the abdomen, chest and joints and red skin spots (erythema). It is a type of amyloidosis. The basic defect involves a 781-amino acid protein, pyrin. In some populations the prevalence, gene frequency and carrier frequency may be 0.00034, 0.019, and 0.038, respectively. ►amyloidosis; Schaner P et al 2001 Nature Genet 27:318.

Medline: A medical bibliographic system of the National Library of Medicine USA. It can be reached on-line as part of the MEDLARS database, http://www.ncbi.nlm.nih.gov/entrez/query.fcgi?DB=pubmed. ►BITOLA

MedMiner: It extracts and organizes relevant sentences in the scientific literature based on a gene, gene-gene or gene drug query. (See http://discover.nci.nih.gov/textmining/main.jsp).

medRNA: mini-exon-dependent RNA. ►*Trypanosoma brucei*

Medulla: The inner part of organs, the basal part of the brain connecting with the spinal chord. ►brain human

Medullary Bone: An ephemeral tissue inside the long bones of female birds. It is formed during ovulation by the increased level of estrogen. Its role is to accumulate calcium that is mobilized for the building of eggshells.

Medulloblastoma (17p13.1-p12, 10q25.3-q26.1, 1p32): A brain cancer of childhood. Its frequency in adenomatous polyposis increases by about two orders of magnitude. Dysregulation of the sonic hedgehog signaling may lead to the disease. ►Gardner syndrome, ►nevoid basal cell carcinoma, ►squamous, ►sonic hedgehog

Meekrin-Ehlers-Danlos Syndrome: A connective tissue disorder.

MEF: A series of myocyte enhancer binding factors that specifically potentiate the transcription of muscle genes and thus differentiation of various types of muscles and myoblasts. The MEF group belongs to the family of MADS domain protein. MEF2 is a Ca^{2+}- regulated transcription factor and it is actively transcribed during development of the central nervous system. MEF2 is activated when it dissociates from histone deacetylase in response MAPK signals. Nur77 and Nor1 steroid receptors mediate apoptosis of T cell receptors controled by MEF2. Mutation in MEF-2 may lead to coronary heart disease and myocardial infarction (Wang L et al 2003 Science 302:1578). The MEF2C is involved in inflammation responses and it is stimulated by lipopolysaccharides of Gram-negative bacteria. MEF2C transactivation is regulated through phosphorylation by p38. ►MADS box, ►MAPK, ►MyoD, ►MYF5, ►MRF4, ►myocyte, ►p38, ►Gram-negative, ►Nur77, ►Nor1, ►apoptosis, ►coronary heart disease, ►myocardial infarction, ►T cell; Mora S et al 2001 Endocinology 142:1999; Han A et al 2003 Nature [Lond] 422:730; regulation of excitatory synapses: Flavell SW et al 2006 Science 311:1008.

Megabase (Mb): 1,000,000 nucleic acid bases.

Megablast: ►BLAST; http://genopole.toulouse.inra.fr/blast/megablast.html.

Megabyte (MB): 1 megabyte is 1024 K (1 K= 1024 bytes [2^{10}]; 1 byte = 8 bits [binary digits]). Conversion of pages into MB may be affected substantially by font size, formatting, size of the window, illustrations included, etc.

Megadalton (Mda): Mda is 10^6 dalton; 1 da (or Da) = 1.661×10^{-24} g.

Megalencecephalic Leukoencephalopathy (MLC, 22qtel): The recessive enlargement of the brain, defective motor functions (ataxia, spasms) and mental deterioration caused by defects in a transmembrane protein. Onset is within a year of birth and a slow progressive realization of the symptoms follows.

Megaevolution: The process and facts of descent of higher taxonomic categories. ►evolution, ►macroevolution

Megagametophyte: One of the four, the functional, haploid products of female meiosis (megaspores) in plants, it develops into embryo sac. Its origin and most prevalent developmental paths are outlined in a figure at gametophyte. ►gametophyte [female]

Megakaryocytes: Large cells in the bone marrow with large lobed, polyploid nuclei; their cytoplasm produces the platelets. Megakaryocyte formation from stem cells is regulated by surviving, the cytokine receptor cMpl and its ligand the megakaryocyte lineage-specific growth factor (meg-CSF), which is homologous to erythropoietin and has both meg-CSF and thrombopoietin-like activities. ►erythropoietin, ►thrombopoietin, ►platelet, ►survivin

Megalin: A cell surface apolipoprotein-B receptor. Its deficiency may lead to holoprosencephaly. There is some disagreement concerning megalin-mediated pathway of sex steroid movement (Rosner W 2006 Cell 124:455; Willnow TE, Nykjaer A 2006 Cell 124:456). ►holoprosencephaly, ►cholesterol, ►apolipoprotein; Barth JL, Argraves WS 2001 Trends Cardiovasc Med 1:26.

Megaloblast: Large, nucleated, immature cells giving rise to abnormal red blood cells.

Megaloblastic Anemia: The autosomal dominant (human chromosome 5q11.2-q13.2) deficiency of dehydrofolate reductase (involved in the biosynthetic path of purines and pyrimidines) resulting in hematological and neurological anomalies. Magaloblastic anemia 1 (MGA1, 14q32) is a vitamin B12 absorption defect. The symptoms may be alleviated by 5-formyltetrahydrofolic acid. A rare recessive type (10p12.1) is caused by intestinal malabsorption of vitamin B_{12}, due to defects in cubilin, the intrinsic factor (IF)-B_{12} receptor. In the Imerslund-Grasbeck syndrome (11q12) the vitamin 12 deficiency is not corrected by cubilin alone but with co-administration of B12 it is effective (Tanner SM et al 2005 Proc Natl Acad Sci USA 102:4130).

Another form (TRMA, Rogers syndrome, 1q23.2-q23.3) responds favorably to thiamin. In chromosome 5p15.3-p15.2 methyl-cobalamine and methionine synthase reductase have been located. Megaloblastic anemia may be found in several other syndromes too. ►phosphatase [ACP1], ►megaloblast, ►folic acid, ►thiamin, ►anemia, ►transcobalamine deficiency; Tanner SM et al 2003 Nature Genet 33:426.

Megalocytosis: Induction of gradually increasing cell size.

Megaspore: ►megagametophyte, ►gametogenesis, ►megasporocyte

Megaspore Competition: It determines which of the four products of meiosis (megaspore) in the female (plant)

becomes functional (see Fig. M34). It occurs only in a few species such as *Oenotheras*. ►certation, ►pollen, ►gametogenesis

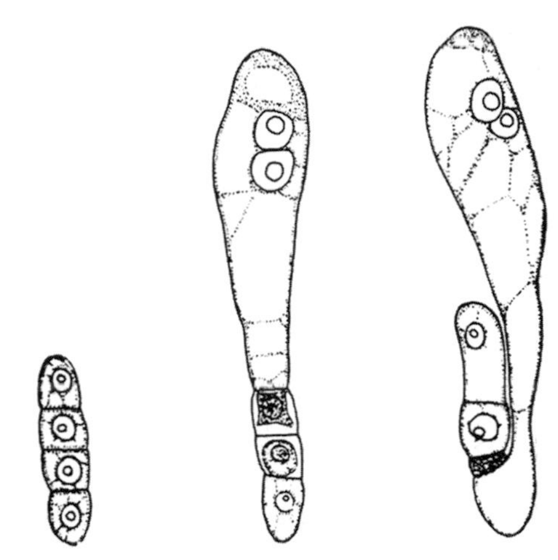

Figure M34. Megaspore competition in the *Oenotheras* where normally the top spore of the tetrad is functional but in case there is a deleterious gene in the top spore, the basal spore may compete with it successfully. (After Renner O. from Goldschmidt R 1928 Einführung in die Verebungswissenschaft. Springer-Vlg. Berlin, Germany)

Megaspore Mother Cell: A diploid cell that produces the haploid megaspores through meiosis in the female plant (see Fig. M35). ►gametophyte [female]

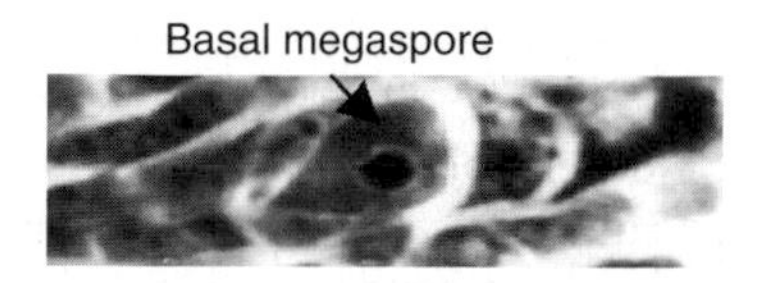

Figure M35. Megaspore mother cell

Megasporocyte: The same as megaspore mothercell. ►gametogenesis, ►gametophyte

MEI41: A 270-kDa *Drosophila* phosphatidylinositol kinase. When inactivated meiotic recombination is reduced. ►PIK

Meiocyte: The cell that undergoes meiosis. ►meiosis

Meiosis: The two step-nuclear divisions, which reduce somatic chromosome number (2n) to half (n) and is usually followed by gamete formation. (See Fig. M36). Meiosis is genetically the most important step in the life cycle of eukaryotes. Meiosis proceeds from the 4C sporocyte (in diploids) and includes one numerically reductional and one numerically equational chromosome divisions.

Synapsis takes place at prophase I to metaphase I. Chiasmata may be visible by the light microscope during prophase I. Centromeres do not split at anaphase I and the sister-chromatids are held together at the centromeres during the separation of the bivalents at anaphase I. At the completion of meiosis I the chromosome number is reduced to half (2C) and by the end of meiosis II, the C-value of each of the 4 haploid daughter cells is 1. The major stages of meiosis are shown in the diagram. These stages are rather transitional than absolutely distinct. The nucleolus is not shown. The nuclear membrane is generally not discernible by the light microscope from metaphase to anaphase but reappears at telophase. Dashed and solid thin lines represent the spindle fibers. The genetic consequences of the meiotic behavior of the chromosomes are best detected in ascomycete fungi with linear tetrads.

The duration of meiosis generally much exceeds that of mitosis and the longest is the prophase stage. In the majority of plants, the completion of mitosis requires 1–3 hours whereas meiosis may need from 1–8 days. In yeast, meiosis takes place in about 7 hours. The stages most revealing for the cytogeneticist, pachytene (2– 8), diplotene (0.5–1), metaphase I (1.5–2), and anaphase I (0.5–1) require generally the number of hours indicated in parenthesis (see Fig. M37). In human females meiosis begins at the early embryonic stage (in mice ovaries it is microscopically detectable by day 12.5) and it stalls at the late prophase stage, at dictyotene, and the subsequent divisions take place only before the onset of ovulations (and following fertilization), a period repeated approximately 13 times annually during about 40 years. In males, meiosis is initiated in the testes only after birth. In both ovaries and testes, the onset of meiosis is under the control of gene *Stra8* (*stimulated by retinoic acid gene 8*). Retinoic acid is key regulator of the initiation of meiosis in both sexes (Koubova J et al 2006 Proc Natl Acad Sci USA 103:2474).

The activation of meiosis requires C_{29} sterols in both male and female. In *Schizosaccharomyces pombe* starvation initiates the switch from mitosis to meiosis. Starvation stimulates the expression of the *ste11* gene through cAMP-dependent protein kinase. Ste11 protein activates many genes, including *mei2*, which produces an RNA-binding protein. The meiRNA is transcribed from gene *sme2*. Mutant ste11 cannot support meiosis. The Mmi1 RNA-binding protein is required for the elimination of RNA conducive to meiosis. Actually, a larger number of mRNAs must be blocked for meiosis to proceed

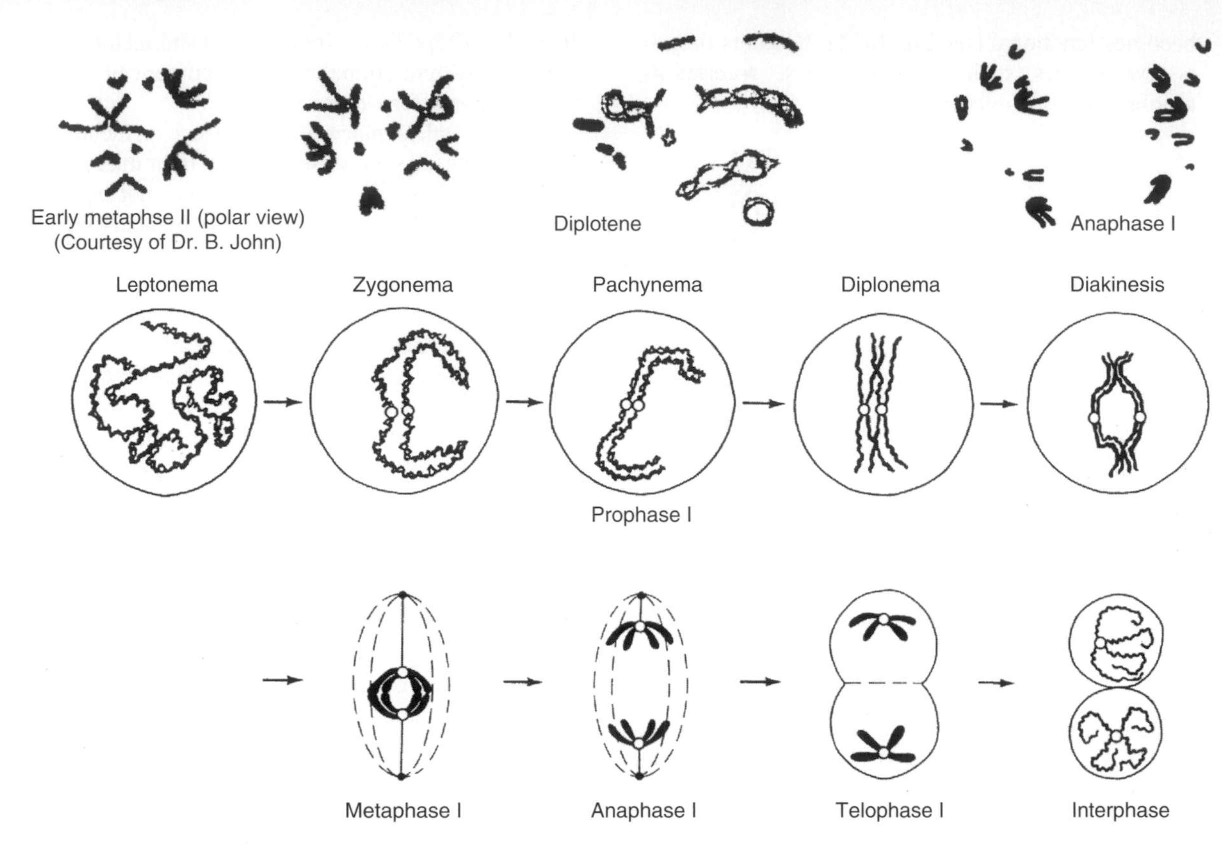

Figure M36. A generalized course of meiosis represented by only one pair of chromosomes with characteristic photomicrographs of the male grasshopper *Chortippus paralellus*

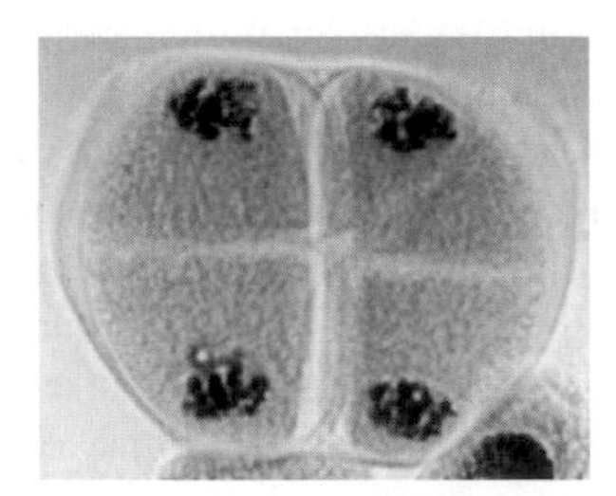

Figure M37. The four male meiotic products in plants

(Harigaya Y et al 2006 Nature [Lond] 442:45). Under such conditions protein kinase Pat1/Ran1 is inactivated by the expression of the *mei3*$^{+}$gene. The substrate of this kinase is the RNA-binding protein Mei2. Dephosphorylation of Mei2 protein also causes a change from mitosis into meiosis. Mei2 is required for premeiotic DNA synthesis and the polyadenylated meiRNA promotes the first nuclear division. Mei2 is localized in the cytoplasm in mitotic cells but during meiotic prophase, it has been visualized in the microtubule-organizing center. Meiosis in budding yeast is controlled by about 150 genes but meiosis affects the expression of over ten times more loci.

Maternal mRNA is transmitted and stored in the egg and it is selectively translated after fertilization during the early stages of embryogenesis. Testicular mRNA is present in the sperm and it is translated postmeiotically. The translation is developmentally regulated, depending on the nature of the genes concerned (Iguchi N et al 2006 Proc Natl Acad Sci USA 103:7712). ►leptotene, ►zygotene, ►pachytene, ►diplotene, ►diakinesis, ►metaphase, ►anaphase, ►cohesin, ►separin, ►shugoshin, ►interphase, ►tetrad analysis, ►mitosis, ►dictyotene, ►nucleolus, ►cell cycle, ►gametophyte, ►C amount of DNA, ►Ran1, ►Pat1, ►Cdc28, ►sister chromatid cohesion, ►monopolin, ►recombination; Zickler D, Kleckner N 1998 Annu Rev Genet 32:619; Zickler D, Kleckner N 1999 Annu Rev Genet 33:603; Rabitsch KP et al 2001 Curr Biol 11:1001; Davis L, Smith GR 2001 Proc Natl Acad Sci USA 98:8395; Forsburg S 2002 Mol Cell 9:703; Nakagawa T, Kolodner RD 2002 J Biol Chem 277:28019; Nagakawa T, Kolodner RD 2002 J Biol Chem 277:28019; Lemke J et al 2002 Am J Hum Genet 71:1051, general and species-specific features of meiosis: Gerton JL, Hawley RS 2005 Nature Rev Genet 6:477.

Meiosis I: It is the first stage of meiosis when through reduction of the chromosome number the 4C amount of DNA in the meiocyte takes place and each of the two daughter nuclei has 2C amounts of DNA. During meiosis I, the sisterchromatids are held together at the centromere by the protective action of the shugoshin proteins until meiosis II. Shugoshins are localized to the centromere by the kinase Bub1 (Vaur S et al 2005 Curr Biol 15:2263). The cohesin protein complex mediates the cohesion of the sisterchromatids. This complex includes protein Scc1. During meiosis, Rec8 replaces Scc1 and Rec8 persists until metaphase II. Anaphase promoting complex (APC) mediates the degradation of securin, which releases Esp1 endopeptidase and that in turn cleaves Scc1 and leads to the termination of the cohesion (Kitajima TS et al 2004 Nature [Lond] 427:510). ►meiosis, ►C amount of DNA, ►cohesin, ►securin, ►cohesin, ►separin, ►shugoshin, ►Esp1, ►Bub1, ►Scc, Fig. M36.

Meiosis II: It follows meiosis I, and is basically an equational division of the chromosomes resulting in four daughter nuclei of the meiocyte that each have only 1C amount of DNA. Mes1 protein is essential for the completion of meiosis II but it is non-essential for mitosis. ►meiosis, ►cohesin, ►separin, ►shugoshin, ►C amount of DNA, ►CDC13

Meiotic Drive: It results in unequal proportions of two alleles of a heterozygote among the gametes in a population because certain meiotic products are not or less functional and consequently the proportion of other gametes increases. In spite of the preferential transmission of the segregation distorters, the various populations display fewer carriers of the distorter than expected. Meiotic drive usually requires the *drive locus* (with driving and non-driving alleles) and the *target locus* (with sensitive and resistant alleles). These two loci are usually tightly linked in repulsion. Coupling the drive and sensitive alleles is expected to eliminate the system. These loci are usually within inversions and they involve the sex chromosomes more commonly than the autosomes probably because the two sex chromosomes (X and Y) in *Drosophila* do not recombine, except in pseudoautosomal region. Meiotic drive may become a microevolutionary factor because it may alter gene frequencies. Meiotic drive may favor selectively disadvantageous gene combinations and thus may contribute to the genetic load of a population. Meiotic drive may be subject to genetically determined modification. Meiotic drive operates in the males or in the females but not in both. In the mouse, cytoplasmic factors may affect meiotic drive and it has been attributed also to the effect of the paternal allele of the *Om (ovum)* locus by inducing the maternal allele to go preferentially to the polar body after fertilization. Meiotic drive may be caused by deletions from the pericentromeric heterochromatin of the X chromosome and Y chromosome—autosome translocations. X-linked insertions containing the 240-bp rRNA intergenic spacer or the rRNA genes may restore the normal conditions. The failure of pairing in the pseudoautosomal region causes meiotic drive and XY nondisjunction. Meiotic drive may be detected by sperm typing. ►segregation distorter, ►preferential segregation, ►spacer DNA, ►rRNA, ►pseudoautosomal, ►polarized segregation, ►certation, ►megaspore competition, ►genetic load, ►transmission, ►polycystic ovarian disease, ►symbionts hereditary, ►killer strains, ►brachyury, ►sexual dimorphism, ►sexual selection, ►sex-ratio, ►sperm typing, ►mitotic drive; Hurst GD, Werren JH 2001 Nature Rev Genet 2:597; Jaenike J 2001 Annu Rev Ecol Syst 32:25.

meiRNA: The polyadenylated meiotic RNA cooperates with meiotic non-phosphorylated protein Mei2 to promote premeiotic DNA synthesis and nuclear division. ►meiosis, ►Pat1, ►Ran; Ohno M, Mattaj IW 1999 Curr Biol 9:R66.

Meisetz: A meiosis-induced factor containing a PR/SET and a zinc-finger domain is a histone methyltransferase required for the progression of early meiotic prophase. The PR/SET domain is the catalytic domain of histone methyltransferases. It only trimethylates lysine 4 of histone H3. It is important for the formation of sex body, fertility, homologous chromosome pairing and double-strand DNA repair. ►histone meththyltransferases, ►sex body; Hayashi K et al 2005 Nature [Lond] 438:374.

MEIV (multilocus exchange with interference and viability): A statistical model of recombination and achiasmate segregation in tetrads. (See Zwick ME et al 1999 Genetics 152:1615).

MEK: A member of the extracellular signal-regulated kinase (ERK) family. About 30% of human cancers have activated MEK1 and MEK2 tyrosine/threonine protein kinase activity. The similar structures have a pocket that binds inhibitors, causes conformational changes in the unphosphorylated proteins and inhibition (Ohren JF et al 2004 Nature Struct Mol Biol 11:1192). ►signal transduction, ►BRAF, ►MAPK, ►MP1, ►Cranio-facio-cutaneous syndrome; Widmann C et al 1999 Physiol Rev 79:143.

MEKK (196-kDa protein serine/threonine kinase): MEK kinase (i.e., a kinase kinase). ►MEK, ►signal transduction; Yujiri T et al 1998 Science 282:1911.

MEL Oncogene: Isolated from human melanoma although its role in melanoma is unclear; it was assigned to the broad area of human chromosome 19p13.2-q13.2). ►melanoma, ►oncogenes, ►p16^{INK4}

Melancholy: Severe depression (seeing the world bitter dark; melano: dark, chole: bile). Increased level of norepinephrin appears in the cerebrospinal fluid and increases the chances for heart failure (Gold PW et al 2005 Proc Natl Acad Sci USA 102:8303). ►depression, ►animal hormones

Melandrium (synonymous with *Lychnis, Silene*): A dioecious plant (see Fig. M38) (2n = 22 + XX or 22 + XY). *Caryophyllaceae*. ►intersex, ►*Lychnis*; Lardon A et al 1999 Genetics 151:1173; Lebel-Hardenack S et al 2002 Genetics 160:717.

Figure M38. *Melandrium*

Melanin: ►albinism, ►pigmentation of animals, ►piebaldism, ►agouti, ►hair color, ►melanosome

Melanie: A computer software package that can match two-dimensional protein gel data to information in database. ►databases, ►annotation; http://www.expasy.org/melanie/.

Melanism: Increased production of the dark melanin pigment. ►melanin, ►industrial melanism, ►*Biston betularia*

Melanocortin: Synthesized as a complex pre-pro-opiomelanocortin, which by processing contributes to the formation of the adrenocorticotropin hormone (ACTH), melanocyte-stimulating hormone (MSH) and β-endorphin. UV-induction of POMC/MSH under the control of p53 regulates the response of the cells to ultraviolet light by tanning (Cui R et al 2007 Cell 128:853). MSH regulates the brain melanocortin-4-receptor (MC4R) and thereby leptin. Haplo-insufficient, morbid autosomal (18q29) mutation in MC4R leads to obesity in the carriers. The agouti and related gene products are natural antagonists of the MC4R ligand MSH. In many fair-skinned individuals melanocortin 1 receptor (MC1R) is not functional and therefore lack protection against UV light that would be secured by tanning. In this case melanocyte-stimulating hormone fails to stimulate pigmentation in the keratinocytes (see Fig. M39). In red/blonde mice forskolin application can restore pigmentation even in the absence of UV and functional MC1R. This information indicates the

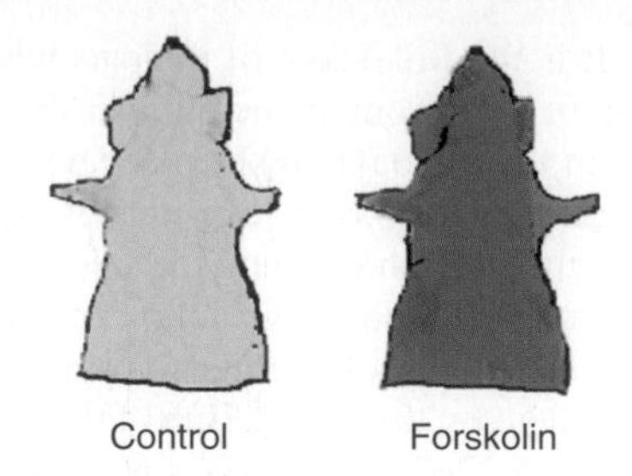

Figure M39. Mice pigmentation

topical application of this chemical can protect against skin damage and cancer (D'Orazio JA et al 2006 Nature [Lond] 2006 443:340). ►opiocortin, ►ACTH, ►melanocyte stimulating hormone, ►POMC, ►p53, ►leptin, ►endorphin, ►agouti, ►pigmentation of animals, ►nociceptor, ►haplo-insufficient, ►hyperphagia, ►melanoma, ►forskolin, ►keratin, ►tanning, ►ultraviolet light; Huszar D et al 1997 Cell 88:131; Kistler-Heer V et al 1998 J Neuroendocrinol 10:133; Chen AS et al 2000 Nature Genet 26:97.

Melanocyte: Produces melanin. ►melanin

Melanocyte Stimulating Hormone (MSH): MSH exists in forms α-MSH, β-MSH and γ-MSH. These adenocortical hormones regulate melanization and also energy utilization. ►opiocortin, ►agouti, ►pigmentation of animals, ►leptin, ►syndecan; Haskell-Luevano C et al 2001 J Med Chem 44:2247.

Melanoma (CMM1, 1p36; CMM2, 9p21): Forms of cancer arising in the melanocytes or other tissues. The most prevalent form appears as a mole of radial growth of reddish, brown and pink color with irregular edges that penetrate, as they progress, into deeper layers. It may originate in dark freckles on the head or other parts of the body. Excessive exposure to sunlight may condition its development although autosomal dominant genes determine susceptibility to melanoma. Three alleles of the melanocortin receptor (MC1R, 16q24.3) double the cutaneous malignant melanoma in MC1R variants (CMM1) risk for red-haired individuals. Mutations in BRAF strongly increase the incidence of melanoma among Caucasians who are not exposed chronically to sun-induced skin damage (Landi MT et al 2006 Science 313:521). Melanoma may be one of the most aggressively metastatic cancers. The development of melanoma is regulated by the melanoma mitogenic polypeptide, encoded by GRO human gene in chromosome 4q21. The melanoma-associated antigen, ME49 appears in the early stages of this cancer, and it is coded by an autosomal dominant locus (MLA1) in human chromosome 12q12-q14. Melanoma-associated antigen, MZ2-E, is coded for by another autosomal dominant locus. The melanoma-associated antigen p97, a member of the iron-binding transferrin protein family, is coded by an autosomal dominant gene, MAP97 (MF12) at human chromosome 3q28-q29. An autosomal dominant gene in human chromosome 15 encodes the melanoma-specific chondroitin sulfate proteoglycan, expressed in melanoma cells. In cultured melanoma and metastatic tissues a mutant CDK4/CDKN2A (9p21 protein was found that was unable to bind the $p16^{INK4a}$ protein and thereby interferes with normal regulation of the cell cycle inhibitor. The ERK growth factor is hyperactivated in up to 90% of melanomas. RAS-ERK and phosphoinositide-3-OH kinase signal to its development. The microphthalmia-associated transcription factor (MITF, 3p14.1-p12.3) is a regulator of melanomas. Therapies may involve several drugs targeted to various kinases in the melanoma development pathways (Gray-Schopfer V et al 2007 Nature [Lond] 445:851).

In human chromosome 11 several presumptive melanoma suppressors have been found. Many of the hereditary melanomas are attributed to G→34T mutation at this locus. Because of this mutation an AUG initiation codon is created at site –35 and the resulting 4-kDa translation-product shows no homology to CDKN2A/p16. Deletions of p16 may frequently be responsible for malignant melanoma. Catenin seems to be a transcriptional coactivator of Tcf and Lef and seems to affect melanoma progression. Approximately 5 to 10% of the melanoma patients have at least one afflicted family member. Non-melanoma skin cancer in fair skinned redheads may be associated with the melanocortin-1 receptor (MCR1). The *fat-1* mutation in mice can convert n-6 fatty acids into n-3 and it reduced melanoma development in melanoma transgenic animals and establish a favorable balance between the two groups of fatty acids. In *fat-1* mice prostaglandin E3 and PTEN were upregulated (Xia S et al 2006 Proc Natl Acad Sci USA 103: 12499). Genetic hybrids between the species of the platyfishes (*Xiphophorus*) are prone to develop melanoma. Melanoma is frequently an aggressive cancer and is refractory to anticancer drugs. Cisplatin is sequestered into subcellular organelles of melanosomes and thus hindered from accessing the nucleus (Chen KG et al 2006 Proc Natl Acad Sci USA 103:9903). ►MEL oncogene, ►p16, ►melanocyte, ►cancer, ►catenins, ►Tcf, ►Lef, ►MAGE, ►metastasis, ►freckles, ►cisplatin, ►Apaf, ►survivin, ►CDKN2A, ►skin cancer, ►mole, ►BRAF, ►fatty acids, ►obesity, ►prostaglandin, ►PTEN; Mellado M et al 2001 Curr Biol 11:691; van der Velden P et al 2001 Am J Hum Genet 69:774; melanoma statistics database: http://www.pcabc.upmc.edu/main.cfm?dis=statmel.

Melanoma Growth-Stimulatory Factor (GROα): An interleukin-related protein like SDF. ►IL-8, ►SDF, ►CXCR, ►chemokines; Wang et al 2000 Oncogene 19:4647.

Melanopsin: A sensory retinal photopigment in mammals and other species (Bellingham J et al 2006 PLoS Biol 4(8):e254).

Melanosomes: Specialized pigment-containing organelles of animals produced by the melanocytes. They include tyrosinase, and related proteins, GP100, etc. and bear resemblance in function to lysozymes. ►melanocyte, ►albinism, ►pigmentation of animals, ►lysozymes, ►skin color, ►amyloids; Kishimoto T et al 2001 Proc Natl Acad Sci USA 98:10698; Marks MS et al 2001 Nature Rev Mol Cell Biol 2:738.

Melas Syndrome: ►mitochondrial disease in humans

Melatonin: Hormone synthesized in the pineal gland and is controling reactions to light, diurnal changes, seasonal adjustment in fur color, aging, sleep, and reproduction. It is a scavanger of oxidative radicals and it protects from ionizing radiation without substantial side effects. It is synthesized from serotonin by serotonin N-acetyltransferase and regulated by cAMP. ►serotonin, ►sulfhydryl, ►radioprotectors, ►circadian rhythm; Reiter RJ et al 2007 World Rev Nutr Diet 97:211.

MELD: Overlapping, Merged DNA fragments.

$MELD_{10}$ (mouse equivalent lethal dose): ►LD50

Melibiose (galactopyranosyl-glucose): A fermentable disaccharide (see Fig. M40).

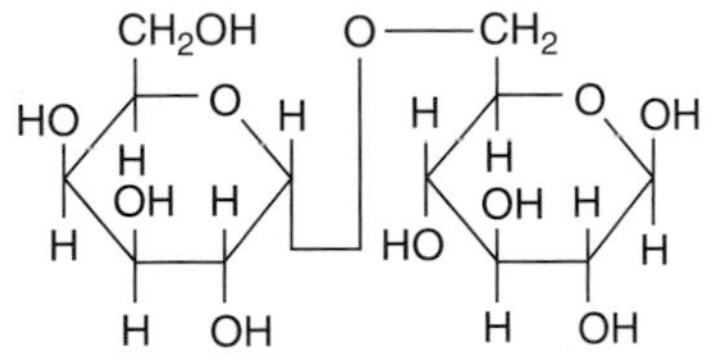

Figure M40. Melibiose

MELK (multi-epitope-ligand-'Kartographie'): Technology capable of analyzing the pattern, topological arrangement and interaction of proteins within single cells rather than cell homogenates. It thus detects single combinatorial protein patterns (s-CCP) and combinatorial protein pattern motifs. The pattern may provide specific signatures of cell types, cell states in health and disease and response to drugs. It provides information in parallel also on carbohydrates nucleic acids and lipids and reveals their interaction networks the toponome. ►genetic networks; Schubert W 2003 Adv Biochem Engin/Biotechnol 83:189.

Melnick-Needles Syndrome (xq28): A hereditary bone defect due to mutation in filamin A. ►filamin

Melt-and-Slide Model: The dual functions of DNA polymerase I are carried out by pol I occupying the duplex primer-template site for the polymerase action, and for the editing reaction the DNA melts (strands separate), unwinds and the single-strand DNA is transferred to the exonuclease site of the polymerase enzyme. ►DNA repair, ►DNA replication

Meltdown, Mutational: Can occur when the mutation rate is high, the genetically effective population size is small and genetic drift is high. It may lead to extinction. ►effective population size, ►genetic drift, ►extinction; Zeyl C et al 2001 Evolution 55:909.

Melting: The breakdown of the hydrogen bonds between paired nucleic acid strands (►Denatur-Ation, ►breathing of DNA, ►Watson and Crick model).

Melting Curve of DNA: Higher temperatures cause progressively higher disruption of the hydrogen bonds between DNA strands; it is affected also by the base composition of the DNA because there are three hydrogen bonds between G≡C and two between A=T. ►re-naturation, ►C_0t curve, ►hydrogen pairing, ►melting temperature

M

Melting Temperature: The temperature where 50% of the molecules is denatured (T_m); the DNA strands may be separated (depending on the origin of the DNA, G≡C content, solvent, and homology of the strands) generally above 80° C and melting may be completed below 100°. ►C_0t curve, ►hyperchromicity, ►melting curve of DNA

Meltrins: Metalloproteinase proteins mediating the fusion of myoblasts into myotubes. ►myotubes, ►metalloproteinase, ►ADAM; Inoue D et al 1998 J Biol Chem 273:4180.

Memapsins: Cloned and sequenced aspartic proteases with β secretase activity. ►secretase

Membrane: Lipid protein complexes surrounding cells and cellular organelles and forming intracellular vesicles. Synthetic membranes can now be engineered on polymer surfaces and their functions studied along with associated proteins using semiconductor technology. The function of membranes is important for many physiological mechanisms, including screening for pathogens and drugs. ►cell membranes, ►semiconductor; Tanaka M, Sackmann E 2005 Nature [Lond] 437:656.

Membrane Attack Complex (MAC): C5b can be converted to C5b6 and then to C5b67 by binding to C6 and C7 complement components and then with C8 and C9 resulting in the formation of MAC. This complex can protect the cells against certain foreign, intruder cells but it must be regulated to protect the cells own membrane system. The so-called homologous restriction factor (HRF) is a 65-kDa glycoprotein bound to cell membranes through glycosylphosphatidylinositol (GPI) can perform this task. A smaller (20-kDa) immunoglobulin G (IgG1) called also HRF20 (CD59), and MIRL (membrane inhibitor of reactive lysis) function in a similar manner. In paroxysmal nocturnal hemoglobinuria is a mutational defect in GPI anchoring of HRF20 to the hematopoietic membrane. ►complement, ►paroxysmal nocturnal hemoglobinuria, ►phosphoinositides, ►immunoglobulins, Linton S 2001 Mol Biotechnol 18:135.

Membrane Channels: These permit passive passing of ions and small molecules through membranes. ►cell membrane, ►ion channel

Membrane Filters: Used for clarification of biological or other liquids, trapping macromolecules, for exclusion of contaminating microbes, for Southern and Northern blotting, etc. The filters may be cellulose, fiberglass, nylon and might have been specially treated to best fit for the purpose.

Membrane Fusion: Maintains sub cellular compartments. The process requires ATPases (NSF), accessory proteins (SNAP), integral membrane receptors of SNAP (SNARE) and GTPases (RAB) and additional proteins. The SNARE function may be transient. Vacuoles Ca^{2+}/calmodulin regulates membrane bilayer mixing in the final steps (see Fig. M41). Protein phosphatase 1 (PP1) has essential role in membrane mixing. Infection by viral pathogens—penetration of the egg by the sperm, vesicular transport, etc.—involves fusion of membranes.

►ATPase, ►RAB, ►SNAP, ►SNARE, ►protein phosphatases; Eckert DM, Kim PS 2001 Annu Rev Biochem 70:777; Jahn R, Grubmüller H 2002 Current Opin Cell Biol 14:488; Ostrowski SG et al 2004 Science 305:71.

Membrane Potential: Eelectromotive force difference across cell membranes. In an average animal cell inside it is 60 mV relative to the outside milieu. It is caused by the positive and negative ion differences between the two compartments.

Membrane Proteins: These may be *integral* parts of the membrane structure and cannot be released. The *transmembrane* proteins are single amino acid chains folded into (seven) helices spanning across the membrane containing lipid layers. The latter have a hydrophobic tract that passes through the lipid double layer of the membrane and their two tails, one pointing outward from the membrane and the other

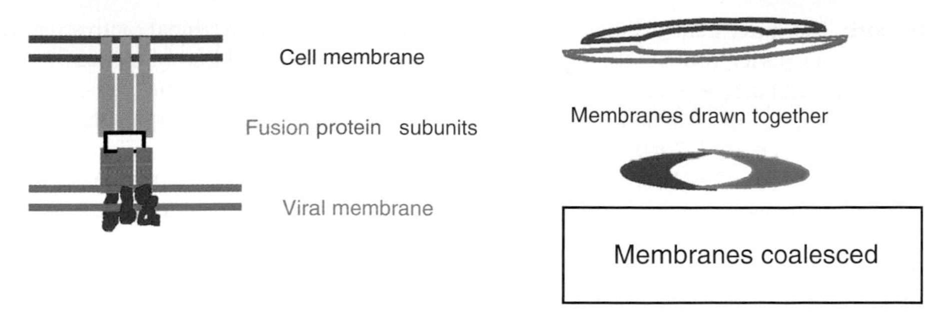

Figure M41. Fusion of viral membranes with the cell membrane during infection. The several subunits of the fusion protein complex draws the membranes together. This is an oversimplified representation. (See Modis Y et al 2004 Nature [Lond] 427:313; Gibbons DL et al 2004 Nature [Lond] 327:320)

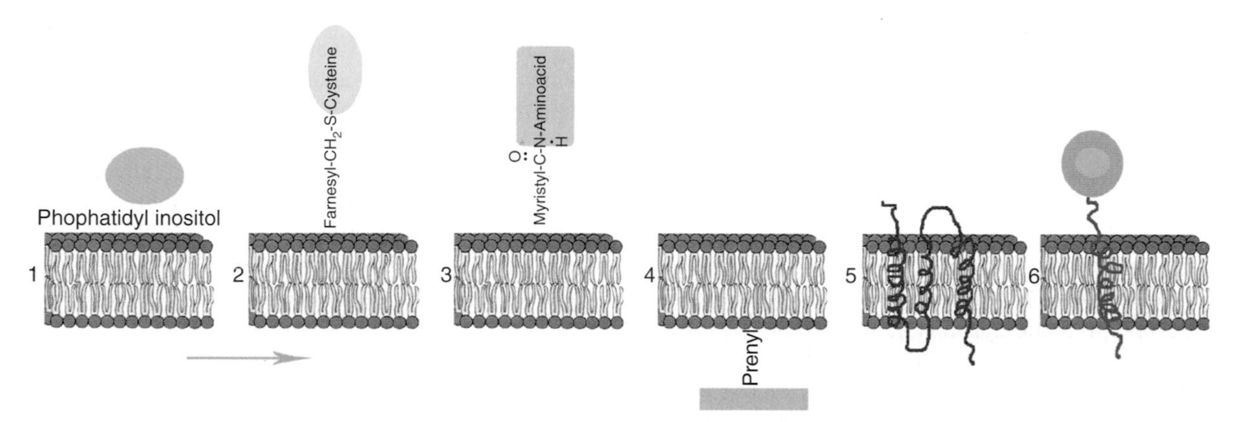

Figure M42. 1. Globular protein is attached to the outer surface by a glycosyl-phosphatidyl inositol anchor. 2. The protein is attached to the outer layer of the lipid bilayer by a thioether linkage between the sulfur of cysteine and a farnesyl molecule. 3. Amino acid and myristil anchor join a protein to the outer surface of the membrane. 4. Prenyl residue anchors the protein to the inner part of the membrane. 5. Transmembrane protein chain passes through the membrane three or seven times. 6. Transmembrane protein anchors another protein without covalent linkage

reaching into the cytosol, are hydrophilic. Some of the membrane proteins are attached only the outer or to the inner layer of the membrane lipids and they are called *peripheral membrane proteins*. The topology of >50,000 prokaryotic and ~15,000 eukaryotic membrane proteins have been determined and there is a remarkable similarity between these two groups (Kim H et al 2006 Proc Natl Acad Sci USA 103:11142). The membrane proteins regulate not just the cell membranes but cell morphology by anchoring to cytoskeleton, pH, ion channels and general physiology of the cell. The structure of membranes can now be analyzed with the aid of membrane mutants, available in several organisms. Method has been developed for the computational design of peptides that target (TM) transmembrane helices in a sequence-specific manner. The designed peptides specifically recognized the TM helices of two closely related integrins in micelles, bacterial membranes, and mammalian cells (Yin H et al 2007 Science 315:1817). (See Fig. M42, ►prenylation, ►farnesyl pyrophosphate, ►myristic acid, ►cell membranes, ►cytoskeleton, ►integrins; Dalbey RE, Kuhn A 2000 Annu Rev Cell Dev Biol 16:51; Sachs JN, Engelman DM 2006 Annu Rev Biochem 75:707; membrane protein structure: Elofsson A, von Heijne G 2007 Annu Rev Biochem 76:125.

Membrane Segment: Antigen receptor immunoglobulins posses a membrane-bound segment at the C-end of their heavy chains. The transmembrane part is composed of 25–26 highly conserved amino residues and the intracellular portion is of 25-amino acids in IgG and 14 in IgE and IgA in mouse. Membrane bound IgG, IgE and IgA are involved in the stimulated B cell receptors whereas IgM and IgD are parts of the naive (immature) B cell receptors. ►immunoglobulins

Membrane-Spanning Helices: ►membrane proteins, ►seven membrane proteins

Membrane Transport: The movement of polar solutes with the aid of a transporter protein through cell membranes. ►cell membrane, ►ABC transporters, ►major facilitator superfamily; Kaback HR et al 2001 Nature Rev Mol Cell Biol 2:610, membrane traffic: Pfeiffer SR 2007 Annu Rev Biochem 76:629; membrane transport proteins: http://www.membranetransport.org/.

MEME: A unit of cultural transmission (imitation), which bears some similarities to the gene that it is "transmitted" horizontally and not vertically from generation to generation as cultural inheritance and not as a biological entity. (See Boyd R, Richerson PJ 2000 Sci Amer 283(4):70).

MEME: Replicator functions of organisms that also control co-evolution of genes and organisms. (See Bull L et al 2000 Artif Life 6(3):227; motif alignment tool: http://meme.nbcr.net/meme/intro.html).

Memory: Information storage in the brain or in a computer. The mammalian brain deals with synaptic strength as memory. If synapses are used repeatedly, the strength is improved and long-term potentiation (LTP) takes place. Arc/Arg3.1 regulates endophilin 3 and dynamin 2, two components of the endocytosis machinery. Genetic ablation of *Arc/Arg3.1* in mice or overexpression in culture suggests that Arc/Arg3.1 regulates AMPA receptor trafficking and synaptic plasticity (See Neuron 52(3):445–474 2006). The opposite of LTP is LTD (long-term depression). The latter may erase the effect of LTP. Acetylcholine, dopamine and norepinephrine mediate different memory signals acting in the prefrontal cortex of the brain. Both of these mechanism are triggered by the inflow of Ca^{2+}, regulated by an ion channel (NMDAR), glutamate-activated N-methyl-D-aspartate receptor channel. Multiple protein kinases (adenylate cyclase, CREB) appear to be involved in extremely complex manners. It is not entirely clear how discrimination between LTP and LTD is accomplished. Nerve growth factor (NGF) gene transfer to basal forebrain of rats resulted in recovery from age-related memory loss. There are some indications that memory is controled by positive and negative signals. LTP seems to require protein synthesis but long-term memory (LTM) may not. Long-term memory seems to require synthesis of new proteins in the dendrites (Sutton MA, Schuman EM 2006 Cell 127:49). Modern neurobiology uses a variety of techniques for exploring the mechanisms of memory and learning, including mutations, knock-outs, transgenes, tomography, magnetic resonance imaging, etc. Unwanted memories are eliminated by increased dorsolateral prefrontal activation and reduced hippocampal activation in the brain (Anderson MC et al 2004 Science 303:232). *Explicit* (declarative) *memory:* the remembrance of facts, resides in the hippocampal region of the brain. *Implicit* (non-declarative) *memory* involves perceptual and motor skills (based on basal ganglia) that may be more widely distributed; memory of conditional fear appears to be in the corpus amygdaloideum (an almond-shaped part in the temporal lobe of the brain with connections to the hippocampus, thalamus and hypothalamus). The *working memory* functions during the period when a mental task is performed. *Social memory* (parental care, etc.) in mice is based on olfactory functions controled by oxytocin in the brain. *Drosophila* mutants *dunce* (encoding cAMP phosphodiesterase) and *rutabaga* (encoding adenylate cyclase) appear to have an ability to learn yet display short memory. The mutant *amnesiac* is defective in a pituitary peptide required for activation of adenylate cyclase, and mutation *linotte* is affected in a helicase-like function. In *Aplysia* and mice CREB affects long-term memory but not the short-term memory. All these pieces of information point to the role of cAMP in memory. Ca^{2+}/calmodulin dependent protein kinase (CaMKII) is an important signaling molecule for memory. Dynamin, a microtubule-associated GTPase is important for synaptic vesicle recycling. Protein phosphatase 1 seems to be a suppressor of learning and memory (Genoux D et al 2002 Nature [Lond] 418:970). Single amino acid substitutions in brain derived neutrophilic factor (BDNF) in rats seem to effect long-term potentiation of memory (Egan MF et al 2003 Cell 112:257). Recalling olfactory memory in *Drosophila* requires the function of the mushroom bodies although this organelle is not necessary for learning or storage of information. In *Drosophila,* the most central part of the brain—in the so-called fan-shaped body (FB)—stores short-term memory of visual pattern recognition. Memories of different shape patterns are localized in groups of neurons in different, parallel, horizontal branches of FB. In the flies the rutabaga protein (an adenylyl cyclase regulated by Ca^{2+}/calmodulin and G_{α} protein) is required for conditioned and unconditioned memory. For the experiment *rut*$^{-}$ mutants were employed. Then the investigators looked for neurons where transgenic *rut*$^{+}$ restored learning. Using sophisticated genetic regulatory system and well-controlled environment providing behavioral (and memory) control permitted mapping the visual memory pattern in the brain (Liu G et al 2006 Nature [Lond] 439:551).

Diacylphosphatidyl ethanol (present in fish oil) in the diet may significantly increase learning capacity of young rats but not the old ones (Barceló-Coblijn G et al 2003 Proc Natl Acad Sci USA 100:11321). Age in humans generally reduces memory and may result in Alzheimer disease. In mice, which express human amyloid-β protein precursor (APP), a variant of 56-kDa amyloid-β plaques accumulate in the neurons by aging and cause memory loss (Lesné S et al 2006 Nature [Lond] 440:352). ►brain, ►human, ►mushroom body, ►ARC, ►AMPA, ►dynamin, ►endophilin, ►synaps, ►ion channels, ►long-term potentiation, ►NMDAR, ►flyn, ►hebbian mechanism, ►synapse, ►adenylate cyclase, ►oxytocin, ►dopamine, ►acetylcholine, ►norepinephrine, ►C/EBP, ►CREB, ►long-term potentition, ►knockout, ►transgene, ►tomography, ►nuclear magnetic resonance spectroscopy, ►Alzheimer disease; Silva AZ et al 1997 Annu Rev Genet 31:527; Zars T et al 2000 Science 288:672; Sweatt JD 2001 Curr Biol 11: R391; Waddell S, Quinn WG 2001 Science 293:1271; Wadell S, Quinn WG 2001 Annu Rev Neurosci 24:1283; maintenance of long-term memory: Bailey CH et al 2004 Neuron 44:49.

Memory, Epigenetic: ►epigenetic memory

Memory, Immunological: Immunological memory rests with the lymphocytes and this is the basis of vaccination. There are three phases in its development: (i) activation and expansion of the CD4 and CD8 T cells, (ii) apoptosis, and (iii) stability (memory). During phase (i), lasting for about a week the antigen selects the appropriate cells and an up to 5,000-fold expansion of the specific cells takes place. They also differentiate into effector cells. Between a week and a month's time, as the antigen level subsides, the T cells die and effector function declines. This is called activation-induced cell death (AICD). THE AICD is a defense against possible autoreactive function. The memory stage may then last for many years and they respond to low exposure of the reintroduced antigen and respond very rapidly. A certain level of antigen maintenance seems to be required to keep memory cell at high level but it appears that memory cells can be saved even in the absence of the antigen (Maruyama M et al 2000 Nature [Lond] 407:636). The conditions for CD4 and CD8 cell maintenance appear different. Memory B and T cells do not provide immediate protection against infection but after infection they start rapid formation of effectors. The duration of the memory depends on the strength of immunization. The protection against peripheral re-infection is antigen-dependent. The long-term success of immunization depends on the presence of memory cells. Therefore there is much significance in selecting T cells that may become memory cells. The division of $CD8^+$ memory T cells is slow and it can be increased by IL-15 but IL-2 has the opposite effect. T-bet and eomesodermin transcription factors facilitate cytokine-dependent memory T cell and natural killer cell formation. These two transcription factors enhance the expression of CD122, which mediates interleukin-15 (IL-15) responsiveness (Intlekofer AM et al 2005 Nature Imunol 6:1236). Eomesodermin (Eomes) is a T-box transcription factor homologous to T-bet. Thus these memory T cells are under the control of interleukin balance. The 'central memory cells' have $CCR7^+$ receptors whereas the 'effector memory T cells' are $CCR7^-$. ►immune system, ►apoptosis, ►CD4, ►CD8, ►lymphocytes, ►germinal center, ►T cell, ►B cell, ►T-bet, ►eomesodermin, ►killer cell, ►immunization, ►vaccines, ►immunotherapy active specific, ►and affinity maturation, ►IL-15, ►IL-12, ►CCR, ►effector, ►CD8; Mackay CR, von Andrian UH 2001 Science 291:2323; Sprent J, Tough DF 2001 Science 293:245; Fearon DT et al 2001 Science 293:248; Sprent J, Surh CD 2002 Annu Rev Immunol 20:551.

Memory, Molecular: Heritable specific pattern of gene expression.

Memory, Transcriptional: Maintains transcriptional state and transmits it epigenetically to cell progeny. In yeast the ATP-dependent SWI/SNF chromatin-remodeling system is essential for the GAL1 gene and it antagonizes the ISWI system (Kundu S et al 2007 Genes & Dev 21:997). ►ISWI, ►SWI/SNF, ►Gal

MEN (multiple endocrine neoplasia): MEN1 (dominant) was located to human chromosome 11q13, responsible for the production of a 610 amino acid protein, Menin, encoded by 10 exons predisposing to pancreatic islet cell tumors. Menin is both a tumor suppressor and a cofactor of mixed lineage leukemia (MLL) oncoprotein (Yokoyama A et al 2005 Cell 123:207). MEN1 homozygosity is lethal. MEN2 is encoded at 10q11.2 and puts individuals at risk of thyroid cancer. The RET tyrosine kinase is defective. ►endocrine neoplasia, ►p27, ►ganglioneuromatosis, ►RET oncogene; Crabtree JS et al 2001 Proc Natl Acad Sci USA 98:1118; Krapp A et al 2004 Current Biol 14:R722.

MEN (mitotic exit network): Includes cyclin-dependent kinase (CDK) inactivators, CDC15, CDC14, CDC5, DBF, TEM1, etc. (See factors named, ►mitotic exit, ►FEAR; Asakawa K et al 2001 Genetics 157:1437).

Menaquinone (vitamin K_2): Synthesized by bacteria and is an electron carrier. ►vitamin K

Menarche: The first menstruation event, followed by a period of about three years of no ovulations before the regular menstruation begins and continues until menopause. ►menstruation, ►menopause

Mendelian Laws: The term was first used by Carl Correns (1900), one of the rediscoverers of these principles, which he named (1) Uniformitäts- und Reziprozitätsgesetz, (2) Spaltungsgesetz, (3) unabhängige Kombination). *First law*: uniformity of the F_1 (if the parents are homozygous) and the reciprocal hybrids are identical (in the absence of cytoplasmic differences). *Second Law*: independent segregation of the genes in F_2 (in absence of linkage). *Third law*: independent assortment of alleles in the gametes of diploids. Thomas Hunt Morgan (1919) also recognized three laws of heredity: (1) free assortment of the alleles in the formation of gametes, (2) independent segregation of the determinants for different characters, (3) linkage-recombination. In some modern textbooks only two Mendelian laws are recognized but this is against the tradition of genetics that the first used nomenclature is upheld. Mendel himself never claimed any rules as such to his credit. He did not observe any linkage among the 7 factors he studied in peas although he had less than 1% chance for all factors segregating independently. This was called "Mendel's luck". If he had found the linkage it would have in all probability been recorded in his notes. Unfortunately after his death, his successor at the abbey, Anselm Rambousek, disposed most of the records. After he experimented with *Hieracium*, an apomict (unknown that time), he developed some doubt about the general validity of his discoveries. Yet, before ending his career he stated: "My scientific work has brought me a great deal of satisfaction, and I am convinced that I will be appreciated before long by the whole world". That appreciation began in 1900, 16 years after his death and continues since. In 1936, the famous statistician and geneticist Ronald Fisher (Annals of Sci. 1:115) questioned the authenticity of Mendel's observations on the basis that the data were 'too good to be true'. Fisher's criticism has been widely touted by many without thoroughly studying the original paper of Mendel or that of Fisher. The statistician F. Weiling (1966 Züchter 36:359) pointed out that Fisher misunderstood some of the experiments and erred in his analysis too. More recently Hartl & Fairbanks (2007 Genetics 175:975) concluded, "Fisher's allegation of deliberate falsification can finally be put to rest, because on closer analysis it has proved to be unsupported by convincing evidence." The experimental data accumulated during the one and a half century since the publication of Mendel's paper convincingly prove that Mendel was right and honest (Rédei GP 2002 p 1. In: Quantitative Genetics, Genomics and Plant Breeding. Kang MS (Eds.) CABI Publishing, New York). ►Mendelian segregation, ►epistasis, ►modified Mendelian ratios

Mendelian Population: A collection of individuals, which can share alleles through interbreeding. ►population, ►genetics

Mendelian Segregation: Mendelian segregation for independent loci can be predicted on the basis of the Table M5. Mendelian segregation ratios may show only apparent deviations in case of epistasis (see Fig. M43). Reduced penetrance or expressivity may also confuse the segregation patterns and in such cases it may be necessary to determine the difference between male and female transmission. The wrinkled/shrunken seeds shown in Fig. M43 accumulate water-soluble sugars at the expense of complex starch and when the seed dries it shows the wrinkled phenotype. ►epistasis, ►penetrance, ►expressivity, ►segregation distorter, ►certation, ►modified Mendelian ratios

Table M5. Mendelian expectations

Number of different allelic pairs	1	2	3	4	n
Kinds of gametes and number of phenotypes in case of dominance	2	4	8	16	2^n
Number of phenotypes (in case of no dominance) and number of genotypes	3	9	27	81	3^n
Number of gametic combinations	4	16	64	256	4^n

Figure M43. Segregation for smooth and wrinkled within a pea pod

Mendelizing: Segregation corresponds to the expectations by Mendelian laws. ►Mendelian laws, ►Mendelian segregation

Ménière Disease (COCH, 14q12-q13): A late-onset, dominant non-syndromic deafness although in some cases vertigo (dizziness-like sensation of whirling of the body or the surroundings) and tinnitus (ringing inside the ears) may occur periodically. In the majority of cases mutation at nucleotide position C-T^{208} results in proline^{51}→serine substitution in the COCH protein. ►deafness

Menin1: A tumor suppressor protein (binding JunD) encoded at 11q13 by the gene responsible for multiple endocrine neoplasia. ►endocrine neoplasia, ►Jun; Guru SC et al 2001 Gene 263:31.

Meninges: The three membranes (pia, arachnoid, dura maters) surrounding the brain and the spinal cord.

Meningioma: A slow proliferating brain neoplasias classified into different groups on the basis of anatomical features. Generally meningiomas involve the loss of human chromosome 22 (hemizygosity) or part of its long arm or some lesions at 22q12.3-q13 where the SIS oncogene, responsible for a deficit of the platelet derived growth factor (PDGF) is located. Chromosomes 1p, 14q, and 17 have also been implicated. ►SIS, ►cancer, ►neurofibromatosis, ►meninges, ►ERM

Meningocele: ►spina bifida

Menkes Syndrome (MNK, kinky hair disease): The gene is situated in the centromeric area of the human X-chromosome (Xq12-q13). The phenotype involves hair abnormalities, mental retardation, low pigmentation, hypothermia and short life span. Apparently, the defect is in the malabsorption of copper through the intestines resulting in copper deficiency of the serum. The prevalence is in the 10^{-5} range or less. It is detectable prenatally and the heterozygotes can be identified although its inheritance is apparently recessive. Lysyl oxidase and other copper-dependent enzyme levels (tyrosinase, monoamine oxidase, cytochrome c oxidase, ascorbate oxidase) are reduced in the afflicted individuals. In the mouse homolog, Mottled-Bridled, the non-exported copper is tied up by metallothionein and the afflicted individual dies within a few weeks after birth. The Occipital Horn Syndrome may be an allelic variation of MNK and of cutis laxa. ►mental retardation, ►Wilson disease, ►acrodermatitis, ►hemochromatosis, ►Ehlers-Danlos syndrome, ►Cutis laxa, ►collagen, ►metallothionein

Menopause: The end of the periodic ovulation (menstruation) and fertility around age 50 in human females. Animals in the wild usually stay fertile in old age. The evolutionary cause of menopause is not known but it has been hypothesized it is a protection against the increase of chromosomal aberrations in old egg cells. An alternative hypothesis assumes that life beyond menopause may aid the rearing of the last offspring or grandchildren and this conveys fitness by kin selection. ►menstruation, ►menarche, ►andropause, ►age-specific birth and death rates, ►kin selection, ►dictyotene stage, ►porin

Menses: Same as menstruation

Menstruation: The monthly discharge of blood from the human (primate) uterus in the absence of pregnancy. If the egg is not fertilized, it dies and the endometrial tissue of the uterus is removed amids the bleeding. Fertilization takes place within the oviduct. About three days are required for the egg to reach the uterus through the oviduct where it is implanted within a day or two, and about a week after being fertilized. Fertilization may occur if the coitus takes place in period about two weeks after the beginning of the last monthly menstruation. The calendar rhythm method of birth control relies on knowledge of this receptive period. Unfortunately, its effectiveness is not very high. ►hormone receptors, ►sex hormones, ►ovulation, ►menarche

Mental Retardation: A collection of human disabilities caused by direct or indirect genetic defects and acquired factors such as diverse types of infections (syphilis, toxoplasma coccidian protozoa), viruses (rubella, human immunodeficiency virus, cytomegalovirus, herpes simplex, coxsackie viruses), bacteria (*Haemophilus influenzae*, meningococci, pneumococci, mechanical injuries to the brain pre-, peri- and postnatally, exposure to lead, mercury, addictive drugs, alcoholism or deprivation of oxygen during birth, severe malnutrition, deficiency of thyroid activity, social and psychological stress, etc. An estimated 2 to 3% of the population is suffering from mild (IQ 50–70%) or more or less severe (IQ below 50%) forms of it. Special education programs can help an estimated 90% of the cases. Approximately 10% of the human hereditary disorders have some mental-psychological debilitating effects.

Autosomal dominant type hemoglobin H disease associated mental retardation due to a lesion in the α-globin gene cluster with chromosomal deletion and without it have been observed. Other cases of mental retardation were also observed involving autosomal dominant inheritance caused by breakage in several chromosomes. Autosomal recessive inheritance was involved in mental retardation associated with head, face, eye, and lip abnormalities, hypogonadism, diabetes, epilepsy, heart and kidney malformations, phenylketonuria. X-chromosome linked mental retardation was observed as part of the syndromes involving the development of large heads, intestinal

M

defects, including anal obstructions, seizures, short statures, weakness of muscles, obesity, marfanoid appearance, etc. In some cases the "kinky hair" syndrome (Menkes syndrome), caused apparently by abnormal metabolism of copper and zinc, also involved mental retardation. A fragile site in the X chromosome (Xq27-q28) apparently based on a deficiency of thymidine monophosphate caused by insufficient folate supply is associated with testicular enlargement (macroorchidism), big head, large ears, etc. A defect in the IL-1 receptor accessory protein, encoded at Xp22.1–21.3, affects learning ability and memory. The transmission of the fragile X sites (FRAX) is generally through normal males. The carrier daughters are not mentally retarded and generally do not show fragile sites. In the following generation, about a third of the heterozygous females display fragile sites and become mentally retarded. This unusual genetic pattern was called the Sherman paradox and it is interpreted by some type of a pre-mutational lesion. The pre-mutation ends up in a genuine mutation only after being transmitted by a female, which already had a microscopically undetectable rearrangment.

The risk of the sons was estimated to be 50% from mentally retarded heterozygous females, 38% from normal heterozygous mothers, and 0% from normal transmitting fathers. The probabilty of these sons being a mentally sound carrier was estimated as 12, 0 and 0%, respectively. The risk of the daughters of the same mothers to become a mentally affected carrier was calculated to be 28, 16 and 0%, and being a mentally normal carrier was estimated 22, 34 and 1%, respectively. The chance of mental retardation for the brother of a proband whose mother has no detectable fragile X site, may vary from 9–27% and among first cousins this is reduced to 1–5%. It was proposed (Laird 1987 Genetics 117:587) that the expression of the fragile X syndrome is mediated by chromosomal imprinting. The imprinting can, however, be erased by transmission through the parent of the other sex. The fragile X syndrome is apparently caused by localized breakage and methylation of CpG islands at the site (Bardoni B, Mandel J-L 2002 Current Op Genet Dev 12:284). Currently, the most reliable diagnosis of this condition is based on DNA probing. Submicroscopic deletions (1.5–2.9 Mb) associated with mental retardation can be detected by microarray hybridization (Vissers LELM et al 2003 Am J Hum Genet 73:1261). In addition, the fragile X syndrome autosomal and sex-chromosomal trisomy and chromosome breakage associated with translocations may be contributing factors of mental retardation. Human chromosome 17q21.31 microdeletions accompanied mental retardation and face morphology as detected by fluorescence in situ hybridization (Koolen DA et al 2006 Nature Genet 38:999). Further, mutations causing metabolic disorders (phenylketonuria, homocystinuria), defects in the branched-chain amino acid pathway (maple syrup urine disease) anomalies in amino acid uptake (Hartnup disease), defects involving mucopolysaccharids (Hunter, Hurler and Sanfilippo syndromes), gangliosidoses and sphingolipidoses (most notably the Tay-Sachs disease, Farber's disease, Gaucher's disease, Niemann-Pick disease, etc.), galactosemias, failure of removal of fucose residues from carbohydrates (fucosidosis), defects in acetyl-glucosamine phosphotransferase (I-cell disease), defects in HGPRT (Lesch-Nyhan syndrome), hypothyroidism, a variety of defects of the central nervous system, and other genetically determined conditions may be responsible for mental retardation. The incidence, establishment of genetic risks and possible therapies are as variable as the underlying causes.

Mental retardation is defined as borderline: IQ ≈70–85, mild in case of IQ ≈50–70, moderate: IQ ≈35–50, severe: IQ ≈25–35 and profound: IQ ≤ 20.

In utero radiation exposure may cause mental retardation; 140 rad during the first 8–15 weeks may result in such damage in 75% of the fetuses and in 46% of the irradiated at any stage. (For more specific details see ►Huntington's disease, ►biotidinase deficiency, ►myotonic dystrophy, ►muscular dystrophy, ►hydrocephalus, ►cranofacial dysostosis, ►spina bifida, ►tuberous sclerosis, ►neurosfibromatosis, ►Menke's syndrome, ►Smith-Lemli-Opitz syndrome, ►Smith-Magenis syndrome, ►Seckel's dwarfism, ►Laurence-Moon syndrome, ►Noonan syndrome, ►Lowe's syndrome, ►mental retardation X linked, ►Apert syndrome or Apert-Crouzon disease, ►Prader-Willi syndrome, ►Rubinstein syndrome, ►cerebral gigantism, ►Langer-Giedion syndrome, ►Miller-Dieker syndrome, ►Walker-Wagner syndrome, ►Wilms tumor, ►Roberts syndrome, ►Russel-Silver syndrome, ►Opitz-Kaveggia syndrome, ►De Lange syndrome, ►Bardet-Biedl syndrome, ►focal dermal hypoplasia, ►ceroid lipofuscinosis, ►autism, ►dyslexia, ►human intelligence [IQ], ►psychoses, ►aspartoacylase deficiency, ►glutamate formiminotransferase deficiency, ►CADASIL, ►Cohen syndrome, ►Coffin-Lowry syndrome, ►Juberg-Marsidi syndrome, ►fragile sites, ►FMR1 mutation, ►trinucleotide repeats, ►human intelligence, ►human behavior, ►head/face/brain defects, ►craniosynostosis, ►double cortex, ►periventricular heterotopia, ►oligophrenin, ►heritability, ►QTL, ►PAK, ►neurodegenerative diseases, ►neurotrypsin, ►IL-1, ►serpines, ►tetraspanin, ►microdeletion; Shea SE 2006 Semin Pediatr Neurol 13:(4):262).

Mental Retardation X-Linked (MRX): These hereditary defects occur in a variety of forms: (i) MRXS with diplegia (bilateral paralysis), (ii) associated with psoriasis (skin lesions), (iii) with lip deformities, obesity and hypogonadism, (iv) Renpenning type with short stature and microcephaly, (v) with seizures (EFMR), (vi) with Marfan syndrome-like habitus, (vii) with fragile X-chromosome sites among others. Altogether more than a dozen different types have been characterized. Translocation involving chromosome 7 results in nonsyndromic cognition deficit due to a defect in a zinc-finger protein (ZNF41). Several of the MRX disorders map in the human Xq28 region and in this general area apparently 12 genes are located. At Xp22.1–21.3 there is the IL1RAPL (IL-1 receptor accessory protein) gene, expressed in the hippocampus. Translocations involving Xq26 and 21p11 (ARHGEF6) involve defects in a guanine exchange factor for Rho GTPases. Some of the sequences observed at Xp22 are found also at Yq11.2 where infertility and azoospermia factors are situated. ►Marfan syndrome, ►fragile X-chromosome, ►Juberg-Marsidi syndrome, ►Lowe's syndrome, ►Coffin-Lowry syndrome, ►Rett syndrome, ►West syndrome, ►Partington syndrome, ►mental retardation, ►oligophrenin, ►tetraspanin, ►non-syndromic; Fukami M et al 2000 Am J Hum Genet 67:563; Chelly J, Mandel J-L 2001 Nature Rev Genet 2:669; Shoichet SA et al 2003 Am J Hum Genet 73:1341; review: Skuse DH 2005 Human Mol Genet 14 (1):R27.

Mentalizing: The ability to read the mental states of others and self in the process. It engages many neural processes (Frith CD, Frith U 2006 Neuron 50:531).

Menthas: A group of dicotyledonous species of plants of various (frequently aneuploid) chromosome numbers. *M. arvensis*: 2n = 12, 54, 60, 64, 72, 92, *M. sylvestris*: 2n = 24, 48; *M. piperita*: 2n = 34, 64. They were the source of menthol (peppermint camphor) and other oils used in cough drops, nasal medication, anti-itching ointments, candy, liquors, etc. Menthol appeared non-carcinogenic although doses above 1 g/kg may cause 50% death in laboratory rodents when administered subcutaneously or orally.

Mentor Pollen Effect: The simultaneous application of dead or radiation damaged compatible (mentor) pollen with incompatible pollen may in some instances help to overcome the incompatibility of the latter and fertilization may result. ►incompatibility alleles; Stettler RF 1968 Nature [Lond] 219:746.

Menu: In a computer, menu lists the various functions to choose. The menu bar on top of the screen of the monitor displays the titles of the menus available.

Meprin: A metalloendoproteinase with α and β subunits. The β subunit is a kinase-splitting membrane protease. The α subunit stays in the endoplasmic reticulum until its transmembrane and cytoplasmic domains are cleaved off and then it moves out. (See Ishmael FT et al 2001 J Biol Chem 276:23207).

MEPS (minimal efficient processing segment): The identical nucleotide tract length required for efficient initiation of recombination. ►recombination mechanisms in eukaryotes

MER (from the Greek μεροσ, part): It is used as, e.g., octamer, indicating it is built of 8 units (octomer would be a Latin-Greek hybrid to be avoided even by geneticists).

MER: Medium reiteration frequency sequence; ~35 copies/human genome. ►redundancy

Mer: Human T cell protooncogene encoded receptor tyrosine kinase. It mediates phagocytosis and apoptosis of thymocytes. ►TCR, ►thymocytes, ►phagocytosis, ►apoptosis, ►receptor tyrosine kinases

Mer$^-$ Phenotype: Mammalian cell defective in methylguanine-O^6-methyltransferase.

Mercaptoethanol: Keeps SH groups in reduced state and disrupts disulphide bonds while proteins are manipulated in vitro. ►DTT, ►thiol

Mercaptopurine: A purine analog inhibiting DNA synthesis and is therefore cytotoxic. The drug 6-mercaptopurine (6-MP) is effective against acute lymphoblastic leukemia (see Fig. M44). The enzyme thiopurine-*S*-methyltransferase (TPMT, 245 amino acids, encoded at human chromosome 6p22.3) catalyzes *S*-methylation of 6-MP and inactivation. The dominant allele (frequency ~0.94) determines high activity whereas the recessive (frequency ~0.06) conveys no detectable activity in the homozygotes and in the heterozygotes a low intermediate level occurs. At low or no TPMT activity 6-MP treatment may be fatal to the patients (about 1% of the population). The testing for TPMT may help to adjust the dosage yet some physicians do not favor it because delaying the treatment may also be risky. ►leukemia, ►TPMT

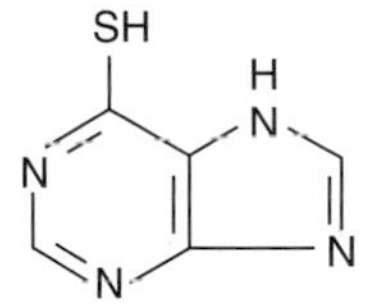

Figure M44. Mercaptopurine

Merge: Merged images.

Merged Sequence Contig: The overlapping initial sequence contigs are merged. ►initial sequence contig, ►contig

Mericlinal Chimera: The surface cell layers are different from the ones underneath just like in the periclinal chimeras but the difference is that the different surface layer does not cover the entire structure but only a segment of it (see Fig. M45). ►chimera, ►periclinal chimera; Jørgensen CA, Crane MB 1927 J Genet 18:247.

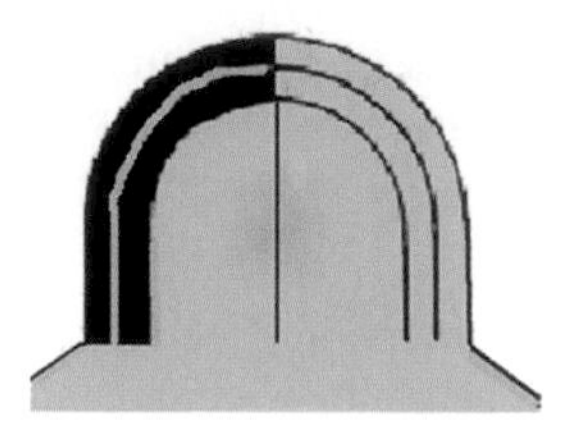

Figure M45. Mericlinal chimera

Meristem: Undifferentiated plant cells capable of production of various differentiated cells and tissues, functionally similar to the *stem cells* of animals. (See Fig. M46, ►stem, ►cells, ►flower differentiation, ►Seed germination, ►*Arabidopsis* mutagen assay; Weigel D, Jürgens G 2002 Nature 415:751; Nakajima K, Benfey PN 2002 Plant Cell 14:S265; axial patterning of shoot meristem: Grigg SP et al 2005 Nature [Lond] 437:1022.

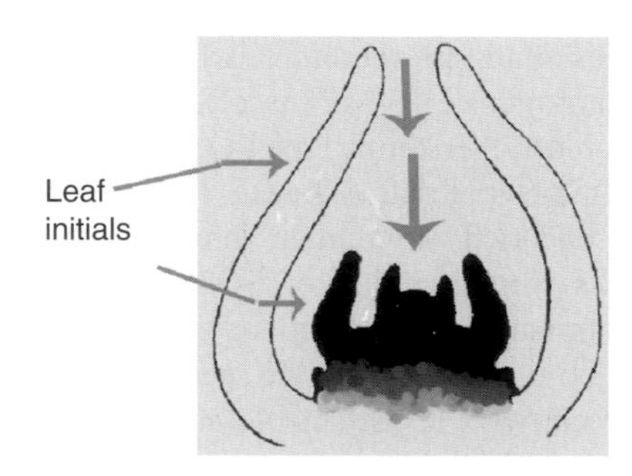

Figure M46. Apical meristem

Meristemoid: For example, stomatal precursor cell. ►stoma

Meristic Traits: Quantitative traits that can be represented only by integers, e.g., the number of kernels in a wheat ear or the number of bristles on a *Drosophila* body. ►quantitative traits

7mer-9mer: Seven or nine base long conserved sequences in the vicinity of the V-D-J (variable, diversity, junction) segments of immunoglobulin genes in the germline DNA. ►immunoglobulins

Merit, Additive Genetic: The same as the breeding value of an individual.

MERLIN (moesin, ezrin, radixcin-like protein): The same as schwannomin. ►neurofibromatosis, ►ERM; Bretscher A et al 2000 Annu Rev Cell Dev Biol 16:113.

MeRNA: Metal-binding RNA is required for proper folding of the molecule: http://merna.lbl.gov/.

Merodiploid: ►merozygous

Merogenote: ►merozygous

Merogone: A fragment of an egg.

Meromelia: ►limb defects in humans

Merosin: ►muscular dystrophy

Merosome: Parasite-filled transport vehicles of malaria that mediate migration in the bloodstream of the host and thus evade immunity. ►malaria; Sturm A et al 2006 Science 313:1287.

Merotelic Attachment: The capture of single kinetochores by microtubules from both centrosomes, i.e., the kinetochore is attached to both spindle poles. It may cause aneuploidy. ►kinetochore, ►centrosome, ►aneuploidy, ►spindle pole; Stear JH, Roth MB 2002 Genes & Development 16:1498.

Merozoite: ►*Plasmodium*

Merozygous: A prokaryote, diploid for part of its genome (merogenote). Prokaryotes are functionally haploid but transduction or plasmid may add another gene copy into the cell. ►transduction, ►conjugation; Wollman E L et al 1956 Cold Spring Harbor Symp Quant Biol 21:141.

MERRF: ►mitochondrial diseases in humans

MERRY: ►M genes causing developmental defects of the germ cells in the offspring

MES (maternal effect sterility): MES occurs due to recessive genes causing developmental defects of the germ cells in the offspring.

Meselson-Radding Model Of Recombination (1975 Proc Natl Acad Sci USA 72:358): Explains gene conversion (occurring by asymmetric heteroduplex, symmetric heteroduplex DNA) and crossing over occurring from one initiation event as indicated by the data of *Ascobolus* spore octads. In yeast the aberrant conversion tetrads arise mainly from asymmetric heteroduplexes as suggested by the Holliday model ►Holliday model). Symmetric heteroduplex covers the same region of two chromatids whereas asymmetric heteroduplex means that the heteroduplex DNA is present in only one chromatid. The heteroduplexes can be genetically detected very easily in asci containing spore octads. In the absence of heteroduplexes, the adjacent (haploid) spores are identical genetically. If the heteroduplex area carries different alleles the two neighboring spores may become different after post-meiotic mitosis. Actually heteroduplexes may be detectable also in yeast (that forms only 4 ascospores) by sectorial colonies arising from single spores. Branch migration indicates that the exchange points between two DNA molecules can move and eventually they can reassociate in an exchanged manner in both DNA double helices involved in the recombination event. Rotary diffusion indicates that the joining between single strands can take place by movement of the juncture in either direction, thus making the heteroduplex shorter or longer. See Fig. M47, ►recombination models, ►recombination molecular mechanisms

Meselson-Stahl Model: Proved that in bacteria DNA replication is semi-conservative. See Fig. M48 ►semi-conservative replication; Meselson M, Stahl FW 1958 Proc Natl Acad Sci USA 44:671; Hanawalt PC 2004 Proc Natl Acad Sci USA 101:17894.

Mesenchyma: Unspecialized early connective tissue of animals that may give rise also to blood and lymphatic vessels. The epithelial–mesenchymal transition is mediated by transcription factors that repress E-cadherin expression. As a consequence embryonic morphogenesis can proceed. The transition also plays a part in the initiation of metastasis.

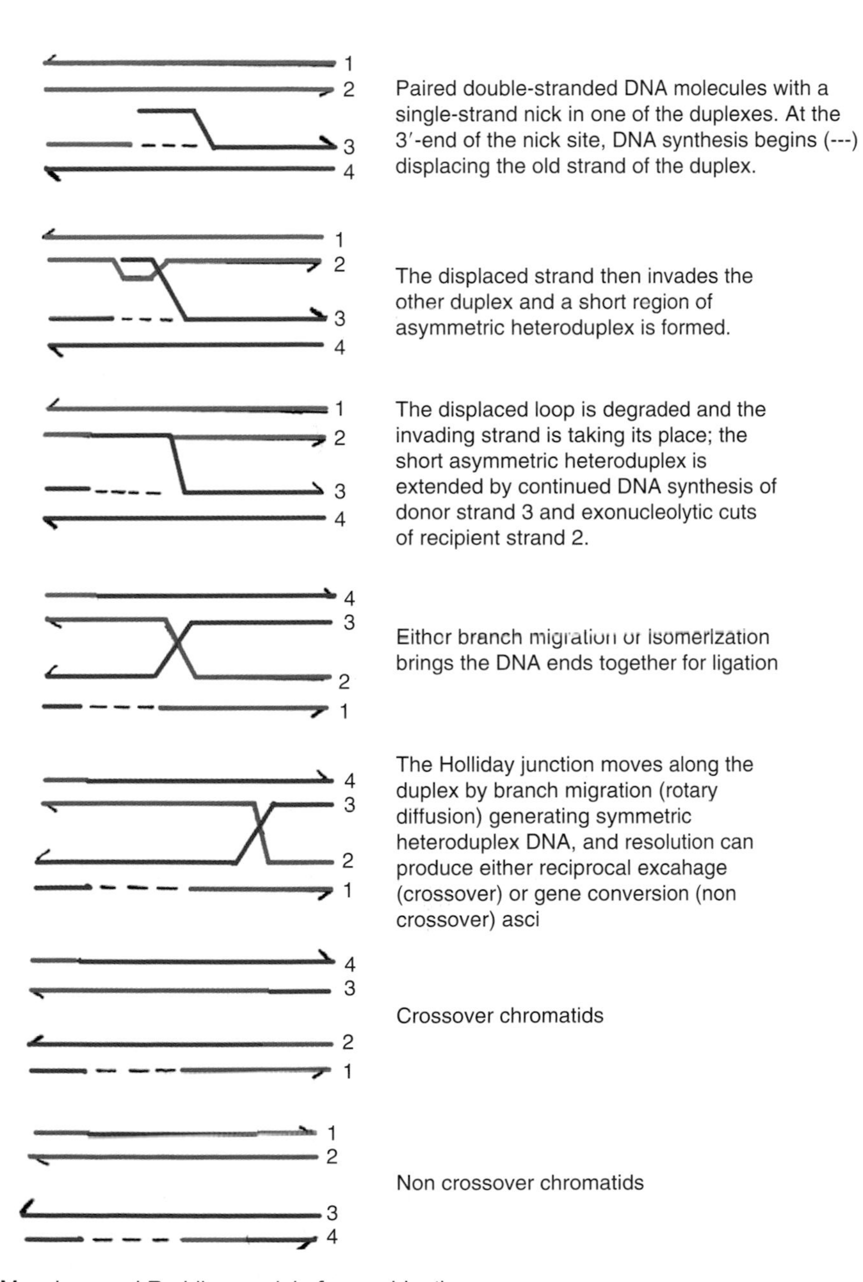

1
2
3
4
Paired double-stranded DNA molecules with a single-strand nick in one of the duplexes. At the 3′-end of the nick site, DNA synthesis begins (---) displacing the old strand of the duplex.
1
2
3
4
The displaced strand then invades the other duplex and a short region of asymmetric heteroduplex is formed.
1
2
3
4
The displaced loop is degraded and the invading strand is taking its place; the short asymmetric heteroduplex is extended by continued DNA synthesis of donor strand 3 and exonucleolytic cuts of recipient strand 2.
4
3
2
1
Either branch migration or isomerization brings the DNA ends together for ligation
4
3
2
1
The Holliday junction moves along the duplex by branch migration (rotary diffusion) generating symmetric heteroduplex DNA, and resolution can produce either reciprocal excahage (crossover) or gene conversion (non crossover) asci
4
3
2
1
Crossover chromatids
1
2
3
4
Non crossover chromatids

Figure M47. Meselson and Radding model of recombination

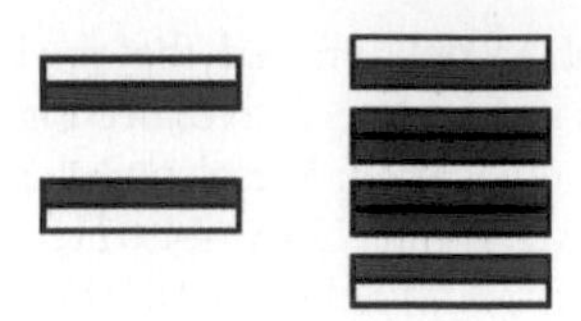

Figure M48. Density labeling was used (open box light DNA single strands), colored box (heavy strands). At left: the original double-stranded DNA molecule, in the middle: the first generation after one replication, at right: the second generation daughter molecules. This experiment thus indicated semi-conservative replication rather than conservative one in the absence of recombination

Mesenchymal stem cells are pluripotent and a balance between Runx2 and PPARγ determine whether osteoblasts or adipocytes develop from the stem cells. The transcriptional activator with PDZ-binding domain (TAZ), a 14–3–3-binding protein coactivates Runx-dependent gene transcription and represses PPAR and functions like a rheostat (Hong J-H et al 2005 Science 309:1074). ►osteoblast, ►adipocyte, ►RUNX, ►PPAR, ►PDZ, ►mesoderm, ►metastasis, ►cadherins, Kang Y, Massagué J 2004 Cell 118:277.

MeSH: Database for information retrieval using certain MEDLINE terminologies. (http://www.ncbi.nlm.nih.gov/entrez/query.fcgi?db=mesh&cmddearchand term=).

Mesocarp: The middle part of the fruit wall. ►exocarp, ►endocarp

Mesoderm: The middle cell layer of the embryo developing into connective tissue, muscles, cartilage, bone, lymphoid tissues, blood vessels, blood, notochord, lung, heart, abdominal tissues, kidney and gonads. Members of the transforming growth factor (TGFβ) family of proteins regulate mesoderm development. ►morphogenesis, ►TGF, ►eomesodermin, ►endomesoderm, Furlong EEM et al 2001 Science 293:1629; Kimelman D 2006 Nature Rev Genet 7:360.

Mesogen (liquid crystal): Some compounds may exhibit transitions between crystalline and liquid forms and can be manipulated by rotation or external electric fields causing rotation. They have various applications (in cellular phones, pocket computers, etc) and because of these properties they may also be used for detection of binding of ligands to specific molecules. Heat responding mesogens may generate optical anisotropy and thus may transduce optical signals and thus facilitate molecular diagnostics without invasive procedures.

Mesokaryote: Organism(s) that occupy some kind of a middle position between prokaryotes and eukaryotes. They are endowed with cytoplasmic organelles like plant cells but their nuclear structure reminds of prokaryotes. The amount of chromosomal basic proteins in the nucleus is low, the chromosomes are attached to the membrane yet they develop a nuclear spindle apparatus. Several microtubules pass through the nuclear membranes and in the majority of the species then pull the chromosomes to the poles with the membrane and without being attached to the chromosomes. In the dinoflagellate *Cryptecodinium cohnii* 37% of the thymidylate is replaced by 5-hydroxymethyluracil. More than half of the DNA is repetitious. The vegetative cells appear to be haploid and thus mutations can be readily detected. Both homo- and heterothallic species are known. ►prokaryote, ►eukaryote, Hamkalo BA, Rattner JB 1977 Chromosoma 60:39.

Mesolithic Age: Period about 12,000 years ago when domestication of animals and agriculture started. ►paleolithic, ►neolithic

Mesomere: A blastomere of about medium size. ►blastomeres

Mesonephros: Secretory tissue of the embryo along the spinal axis supporting the development of the gonads.

Mesophyll: The parenchyma layers of the leaf blade (see Fig. M49) (below cuticle and epidermis).

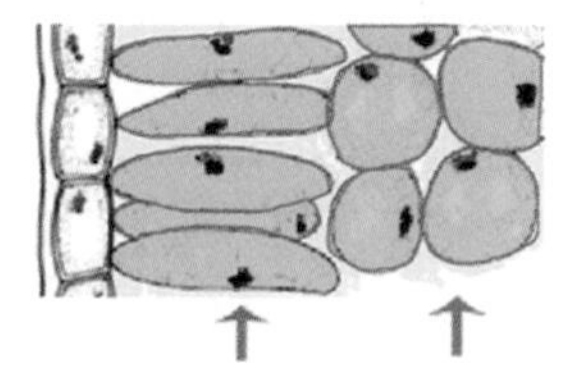

Figure M49. Mesophyll

Mesophyte: Plants avoiding extreme environments such as wet, dry, cold or warm.

Mesosome: An invaginated membrane within a bacterial cell.

Mesothelin (MSLN/MPF, human chromosome 16): A 33kDa protein expressed in the lung and other organs. Various monoclonal antibodies have been developed that react with ovarian cancer (Chang F, Pastan I 1996 Proc Natl Acad Sci USA 93:136). ►ovarian cancer, ►monoclonal antibody, ►humanized antibody

Mesothorax: The middle thoracic segment of insects, bearing legs and possibly wings (see Fig. M50). ►*Drosophila*

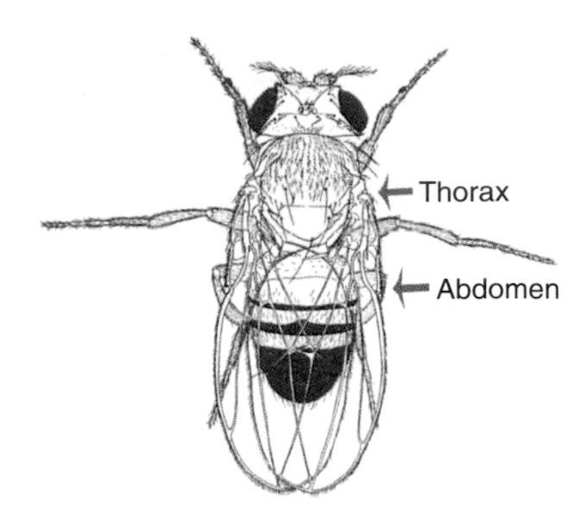

Figure M50. Mesothorax

Mesozoic: Geological period in the range of 225 to 65 million years ago; age of the life and extinction of dinosaurs and several other reptiles. ►geological time periods

Messenger Polypeptides: Extracellular signaling molecules that can pass the cell membrane, enter the cell nucleus, recognize in the DNA a special sequence motif and activate then transcription. These messengers are different from hormones or other signaling molecules because they do not need membrane receptors or special ligands or series of adaptors or phosphorylation to be activated and becoming nuclear co-activators of gene expression. ►lactoferrin

Messenger RNA: ►mRNA

Mestizo: Offspring of Hispanic and American Indian parentage. ►miscegenation, ►mulatto

MET Oncogene: Hepatocytic growth factor receptor gene in human chromosome 7q31. Its α subunit is extracellular and its β subunit is an extra- and transmembrane protein. It is a tyrosine kinase as well as a subject for tyrosine phosphorylation. The receptor of HGF-SF (hepatocyte growth factor/scatter factor) is the product of Met. Expression of the MET oncogene in the liver leads to hepatocarcinogenesis after blood coagulation and internal hemorrhages. The pathogenesis involves the upregulation of plasminogen activator inhibitor type 1 (PAI-1) and cyclooxygenase (COX-2) genes (Boccaccio C et al 2005 Nature [Lond] 434:396). An engineered decoy for MET inhibits its binding to the hepatocyte growth factor receptor; interferes with tumor proliferation, angiogenesis, suppresses or prevents metastasis synergizes with radiotherapy without affecting the physiological functions in mice (Michielli P et al 2004 Cancer Cell 6:61). ►tyrosine kinase, ►oncogenes, ►hepatocyte growth factor, ►papillary renal cancer, ►cyclooxygenase, ►plasminogen activator; Birchmeier C et al 2003 Nature Rev Mol Cell Biol 4915.

$$\chi^2_{2m} = -2\sum_{i=1}^{m} \ln(P_i)$$

Meta-Analysis of Linkage: Used when the information in an experiment is inadequate for drawing definite conclusions. In such cases, data from other comparable experiments are pooled the best possible way. Usually the formula of R.A. Fisher (Statistical Analysis for Research Workers) is used where the P = the probability of obtaining as or more extreme information under the assumption that there is no linkage, m = the data sets. The χ^2 is calculated by 2 degrees of freedom. Meta-analysis may be used for other synthetic purposes when the information available is inconclusive. Its conclusions, therefore, may have to be regarded with some reservations. Meta-analysis may reveal substantial variation/heterogeneity in the genetic association of disease symptoms. ►linkage, ►chi square, ►association test, ►GSMA; Allison DB, Heo M 1998 Genetics 148:859; Wise LH et al 1999 Ann Hum Genet 63 (Pt 3):263; Ioannidis JPA et al 2001 Nature Genet 29:306.

Metabolic Block: A non-functional enzyme (due to mutation in a gene, or to an inhibitor) prevents the normal flow of metabolites through a biochemical pathway.

Metabolic Footprinting: Monitors extracellular metabolites in the culture medium during different physiological states of wild type and mutant cells with the aid electrospray ionization-mass spectrometry. (See Electrospray MS, Allen J et al 2003 Nature Biotechn 21:692).

Metabolic Pathway: A series of sequential biochemical reactions mediated by enzymes under the control of genes. Genetic studies greatly contributed to their understanding along with the use of radioactive tracers. ►radioactive tracer, ►auxotrophy, ►MCA; http://expasy.ch; http://ecocyc.PangeaSystems.com/ecocyc; http://path-a.cs.ualberta.ca; metabolic pathway annotation tool: http://kobas.cbi.pku.edu.cn/; pathway reconstruction tool: http://bioinformatics.leeds.ac.uk/shark/; pathway networking tools for downloading: http://biocyc.org/download.shtml.

Metabolic Syndrome (MetS): The syndrome develops through the interplay of obesity and metabolic susceptibility. It provides information and guidance on insulin signaling, intensity of drug therapy for elevated cholesterol, aspirin prophylaxis, blood pressure and glucose control in cardiovascular disease (Grundy SM 2007 J Clin Endocrinol Metab 92:399; Petersen KK et al 2007 Proc Natl Acad Sci USA 104:12587). ►diabetes, ►heart disease, ►cholesterol

Metabolism: Enzyme-mediated anabolic and catabolic reactions in cells. ►BRENDA, ►MetaCyc, ►EX-Prot, ►MCA; compound/enzymes/reactions: http://www.genome.ad.jp/ligand/.

Metabolite: The product of metabolism. ►metabolism; http://metacyc.org/.

Metabolite Connectivity: The number of reactions of a metabolite. ►pathway tools

Metabolite Engineering: The transformation of certain genes into a new host may lead to production of substances that the organisms never produced before. It can also synthesize larger quantities or different qualities of certain proteins. Examples include: production of indigo or human insulin in *E. coli* or novel antibiotics or expressing antigens of mammalian pathogens in plants or changing the pathway of diacetyl formation in yeast to acetoin production and thus shortening the lagering process in brewing beers, etc. ►genetic engineering, ►protein engineering; Kholodenko BN et al 2000 Metabol Eng 2(1):1.

Metabolome: A complete set of (low-molecular weight) intermediates in a cell or tissue metabolism. Plants produce about 200,000 metabolites. ►gene function; Tweeddale H et al 1998 J Bacteriol 180:5109; Fiehn O 2002 Plant Mol Biol 48:155; human metabolome: http://www.hmdb.ca.

Metabolon: A supramolecular-associated complex of sequential metabolic enzymes. (See Reithmeier RA 2001 Blood Cells Mol Dis 27:85).

Metabonomics: The study of the metabolic status in the biofluids of animals (urine, serum, tissues) by high-resolution nuclear magnetic resonance spectroscopy and appropriate statistics to reveal possible toxic effects of chemicals used for drug development. Metabonomic studies may make possible personalized drug treatment if predose phenotyping variations are considered before administration of drugs to which individual differences in response can be predicted. Besides the metabonomic state a number of other factor nutritional state, gut microbiota, age, disease and co- or pre-administration of other drugs may have great influence (Clayton TA et al 2006 Nature [Lond] 440:1073). ►nuclear magnetic resonance spectroscopy; Robosky LC et al 2002 Comb Chem High Throughput Screen 5:651.

Metabotropic Receptor: The binding of a ligand initiates intracellular metabolic events. Many have transmembrane regions and may respond to second messengers such as cAMP, cGMP or inositol triphosphate. ►ionotropic receptor; Ramaekers A et al 2001 J Comp Neurol 438(2):213.

Metacarpus: The area of the hand between the wrist and fingers (see Fig. M51). ►metatarsus

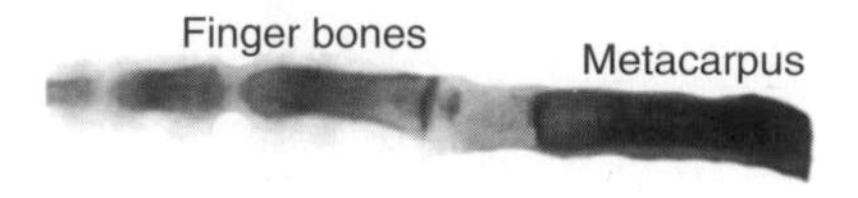

Figure M51. Metacarpus

Metacentric Chromosome: Chromosome whose two arms are nearly equal in length (see Fig. M52). ►chromosome morphology

Figure M52. Metacentric chromosome

Metachondromatosis: Autosomal dominant multiple exostoses particularly on hands, feet, knees and limb bones, without deforming the bones or joints and which disappears with time. ►exostosis, ►fibrodysplasia

Metachromasia: When the same stain colors different tissues in different hues.

Metachromatic Leukodystrophy (MLD): A sulfatide lipidosis. Two distinct forms have been identified that are due to two recessive alleles of a gene, mutations in *A* and *I*, in human chromosome 22q13.31-qter. Due to the deficiency arylsulfatase A, cerebroside sulfate accumulates in the lysosomes. The accumulation of galactoside-sulfate-cerebrosides in the plasma membrane and particularly in the neural tissues (myelin) causes a progressive and fatal degeneration in the peripheral nerves, liver and kidneys. As a consequence failure of muscular coordination (ataxia), involuntary partial paralysis, hearing and visual defects as well as lack of normal brain function arise after 18 to 24 months of age and usually causes death in early childhood. In the juvenile form of the disease the symptoms appear between age 4 and 10 years. There is also an adult type of the disease with an onset after age 16 and involves schizophrenic symptoms. The mutations involve either a substitution of tryptophan at amino acid residue 193, or at threonine 391 by serine or a defect at the splice donor site at the border of exon 2. MLD has been observed also in animals. The reduced activity of the enzyme can be identified also in cultured skin fibroblasts of heterozygotes and prenatally in cultured amniotic fluid cells of fetuses. ►arylsulfates, ►sphingolipidoses, ►sphingolipids, ►Krabbe's leukodystrophy, ►prenatal diagnosis, ►lysosomal storage diseases, ►saposin

Metacline Hybrids: In *Oenotheras,* called the exceptional progeny that occurred only in reciprocal crosses due to the difference in transmission through egg and sperm of the different complex translocations. ►complex heterozygotes, ►certation, ►megaspore competition

MetaCyc: A database of more than 445 metabolic pathways involving more than 1,115 enzymes in >300 organisms. ►pathway tools; http://metacyc.org/; http://www.pubmedcentral.nih.gov/articlerender.fcgi?artid=99148; http://bioinformatics.ai.sri.com/ptools/.

Metacyclic *Trypanosoma*: Parasitic eukaryote that lives in the salivary gland of insects, which spread sleeping sickness. ►*Trypanosoma*

Metacyclogenesis: The differentiation of the promastigote of *Leishmania* into a highly infective form in the sandfly. ►*Leishmania*

Metadata: Data about the data, i.e., closer information regarding the methods of collection, analysis, etc., of data available in databases or in reviews or in other publications.

Metafemale: Having more than the usual dose of female determiners; it may be XXX (see Fig. M53). ►*Drosophila*, ►triplo-X, ►euploid

Figure M53. Euploid metafemale *Drosophila*

Metagene: Metagene has several meanings.

1. A gene-predicting commercial software.
2. Network of genes mediating certain physiological functions as revealed by joint expression in microarrays and which display similarities across evolutionary boundaries (Stuart JM et al 2003 Science 302:249).
3. A mystical, hypothetical RNA formed under exceptional conditions imparting superhuman qualities to individuals, which form it.

Metagenesis: Sexual and asexual generations alternate.

Metagenomics: The analysis of the genomes of groups of organisms in samples from the environment, including special environmental locations, such as deep sea, hospital waste, acid mine water, and archeological remains. Generally nucleotide sequences and functions are used. Such studies yield valuable information on evolution, new genes, new functions and new antibiotics. The human intestinal microbiota is composed of 10^{13} to 10^{14} microorganisms and their collective genome contains at least 100 times as many genes as the human genome (Gill SG et al 2006 Science 312:1355). ►genomics, ►microbiome; Riesenfeld CS et al 2004 Annu Rev Genet 38:525; Green Tringe S, Rubin EM 2005 Nature Rev Genet 6:805.

Metal Metabolism And Disease: ►Wilson syndrome, ►Menkes syndrome, ►Occipital horn syndrome, ►hemachromatosis, ►acrodermatitis enteropathica, ►hyperzincemia, ►selenium-binding protein, ►glutathione peroxidase, ►aceruloplasminemia, ►metalloproteinases, ►metallothionein, ►protoporphyria erythropoietic, ►porphyria, ►coproporphyria; Thompson KH, Orvig C 2003 Science 300:936.

Metallo-Base Pairing: The nucleobases (e.g., hydroxypyridone) are modified and paired not by hydrogen bonds but by an interstrand metal such as copper. The metal-containing bases are incorporated into the DNA by phosphoramidite chemistry or by automated DNA synthesis. Such a duplex may have advantage for some applications because of higher thermal stability or for the construction of molecular magnets. ►hydrogen pairing, ►phosphoramidates, ►DNA chemical synthesis; Tanaka K et al 2003 Science 299:1212.

Metalloprotein: The prosthetic group of the protein is a metal, e.g., hemoglobin. ►hemoglobin, ►zinc-finger; Annotations and Structure: Shi W et al 2005 Structure 13:1473; http://metallo.scripps.edu.

Metalloproteinases (metalloproteases): Cell surface endopeptidase enzymes mediating the degradation of the extracellular matrix, cartilage formation, the release of tumor necrosis factor α, a cytokine involved in inflammatory reactions, in embryogenesis, and cell migration. Membrane-bound metalloproteinase (matrix metalloproteinase, MMP3, stromelysin 1, 11q13) deficiencies cause cranio-facial anomalies, arthritis, dwarfism and other defects due to collagen and connective tissue problems. MMP11 (stromelysin III, 22q11.2) is overexpressed in metastatic breast cancers. Cancer tissues are associated with increased protease activities and it is supposed that these activities facilitate the bursting out of the tumor from the normal cell milieu and increase angiogenesis. MMP-9 (gelatinase, 20q11.2-q13.1) is a contributor to skin carcinogenesis. MMP8 I is collagenase I (11q21-q22), MMP2 is collagenase type IV (16q13) mediating endometrial breakdown during menstruation and may facilitate carcinogenesis by promoting angiogenesis. MMP1 is also a collagenase (11q22-q23). MMP26 (matrilysin 2,

11p15) is involved in tissue healing and remodeling. MMP7 (11q21-q22) is a relatively short uterine matrilysin protein. MMP15 (16q13-q21) and MMP16 (8q21) are membrane enzymes with roles in normal physiological as well as pathogenic processes at special organs. Metalloproteinase inhibitors are explored for cancer therapy. ►tace, ►meltrin, ►stromelysin, ►collagenase, ►bone morphogenetic protein, ►night blindness [Sorsby syndrome], ►ADAM, ►arthritis, ►disintegrin, ►extracellular matrix; Nagase H et al 1992 Matrix Suppl 1:421; Bode W et al 1999 Cell Mol Biol Life Sci 55:639; Sternlicht MD, Werb Z 2001 Annu Rev Cell Dev Biol 17:463; Coussens LM et al 2002 Science 295:2387; Egeblad M, Werb Z 2002 Nature Rev Cancer 2:163; Saghatelian A et al Proc Natl Acad Sci USA 101:10000.

Metallothionein: It is an SH-rich metal-binding protein in mammals. Its main function is detoxification of heavy metals. The promoter is activated by the same heavy metal the protein product binds. This promoter is very useful for experimental purposes because structural genes attached to it can be turned on and off by regulating the amount of heavy metal in the drinking water of the transgenic animals. It is encoded in human chromosome 16q13. ►transgenic, ►Menke's disease, ►metals; Vasák M, Hasler DW 2000 Curr Opin Chem Biol 4:177.

Metals: They may contribute oxidative damage to the DNA and may hinder repair and enhance radiation damage. Beryllium, chromium and lead salts may enhance radiation damage. ►metallothionein

Metamale: ►supermale, ►*Drosophila*

Metamerism: Anterior–posterior segmentation of the body of annelids and arthropods (see Fig. M54). In chemistry it is rarely used for a type of structural isomerism when different radicals of the same type are attached to the same polyvalent element and give rise to compounds possessing identical formulas.

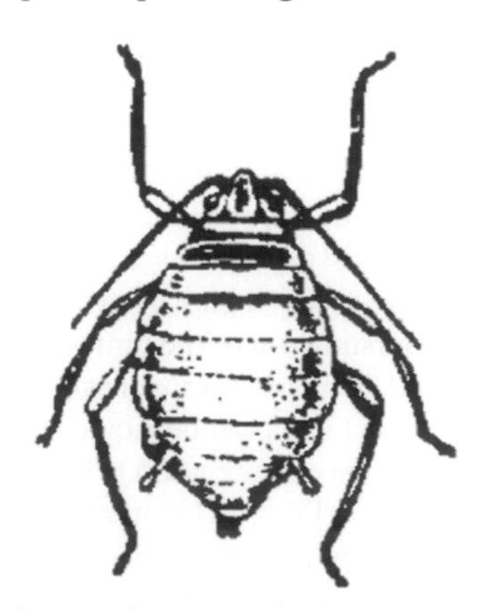

Figure M54. Metameric insect

Metamorphosis: The distinct change from one developmental stage to another, such as from larva to adult or from tadpole to toad. ►*Drosophila*

Metanomics: The use of gas chromatography and mass spectrometry to trace metabolic changes upon mutation or environmental changes. ►gas-liquid chromatography, ►mass spectrum

Metaphase: A stage in mitosis and meiosis when the eukaryotic chromosomes have reached maximal condensation and spread out on the equator of the cell (metaphase plane) and their arm ratios and some other morphological features can be well recognized. In meiosis I the bivalent chromosomes may be associated at their ends if chiasmata had taken place during pro-phase. The ring bivalents indicate crossing over between both arms whereas rod bivalents are visible when crossing over was limited to only one of the two arms. Figure M55 shows a HeLa cell at metaphase stained with DAPI, anti-tubulin (green) and human anti-centromere autoantibody (red), chromatin (blue). Courtesy of Kevin F. Sullivan, Department of Cell Biology, The Scripps Research Institute, San Diego, California and Don W. Cleveland, Ludwig Institute for Cancer Research, University of California, San Diego, California. ►meiosis, ►mitosis, ►chromosome rosette, ►ring bivalent

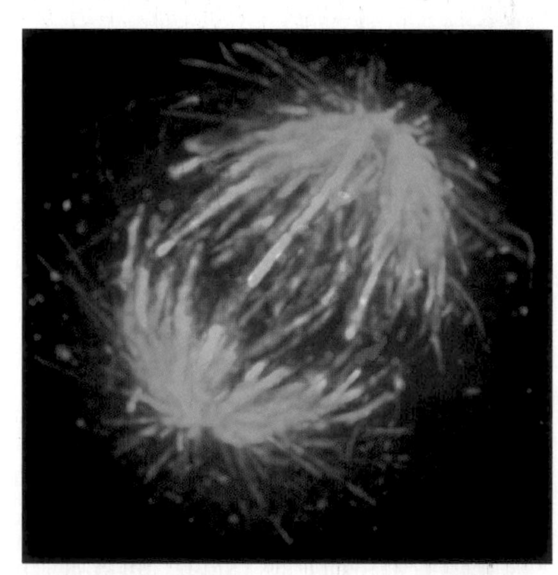

Figure M55. A fluorochrome-stained metaphase of HeLa human cells

Metaphase Arrest: Can occur if toxic agents or two separate kinetochores joined by translocation block the nuclear division. Homologous recombination does not lead to metaphase arrest indicating kinetochore tension may be responsible for the event. ►meiosis

Metaphase Plane: The central region of the cell where the chromosomes are located during metaphase. Often incorrectly called metaphase plate (but no plate is involved). ►mitosis, ►meiosis

Metamphetamine (ecstasy): A psychoactive, addictive drug. In the presence of white light and riboflavin it is photodegradable to the non-psychoactive enantiomer via a type I photooxidation process by monoclonal antibody derived from mouse (Xu Y et al 2007 Proc Natl Acad Sci USA 104:3681).

Metaplasia: One cell type gives rise to another and in the tissue the two may be adjacent. Normally these two cell types are not expected to occur together. The phenomenon is attributed to different activation of a particular cell(s). ►stem cells; Tosh D, Slack JMW 2002 Nature Rev Mol Cell Biol 3:187.

Metapopulation: A large population composed of smaller, local populations. (See Coalescent F, Hanski I et al (Eds.) 1997 Metapopulation Biology: Ecology, Genetics, and Evolution. Academic Press, San Diego).

Metastable: A potentially transitory state; it can change to more or less stable form.

Metastasis: The spread of cancer cells through the blood stream and thus establishing new foci of malignancy in any part of the body although, e.g., breast cancer cells frequently metastase to the lung or to the liver. Metastasis has been attributed to epigenetic changes and/or mutation during cancer progression. Several genes have been identified that predispose lung cancer to metastase specifically or mainly to lung (Minn AJ et al 2005 Nature [Lond] 436:518). Three membrane-anchored proteases, type-1, type-2, and type-3 metalloproteinases, independently confer cancer cells with the ability to proteolytically open the basement membrane scaffolding, initiate the assembly of invasive pseudopodia, and propagate transmigration (Hotary K et al 2006 Genes Dev 20:2673). VEGFR1-positive hematopoietic bone marrow progenitor cells scout out suitable niches where metastatic cells can attach and support the incoming lung tumor cells. The VEGFR1 (vascular endothelial growth factor receptor) cells also express integrin α4β1, which upregulate fibronectin, an integrin ligand and have essential role in establishment of cancerous growth of metastized cells (Kaplan RN et al 2005 Nature [Lond] 438:820).

The invasiveness of cancer cells requires an active state of the integrin system, cell surface gelatinases (collagenases, proteases) so they could penetrate the extracellular matrix of the target cells. On the cell surface actually precursors of the gelatinases are found that are proteolytically activated by metalloproteinases. Heatshock protein 90α seems to an activator of the surface metalloproteinase 2 (Eustace BK et al 2004 Nature Cell Biol 6:507). The plasminogen activator proteases include urokinase and tissue type activators. Plasminogen activator inhibitor (PAI1) is also required. IL-18 and TNF-α may promote cell adhesion and metastases by upregulating vascular cell adhesion molecule (VCAM-1, 1p32-p31) synthesis. Also ICE may facilitate metastasis after processing the precursors of IL-1β and IL-18 proinflammatory cytokines. The CCR gene product appears to suppress metastasis by apoptosis of small cell lung carcinoma and melanoma cells. CXCR4 and CCR7 are highly expressed in breast cancer cells and metastasis.

Migrastatin, a macrolide antibiotic produced by *Streptomyces* and their synthetic analogs inhibit 91–99% of migration of some tumor cells of mouse and human tumor cells but did not inhibit cell proliferation at relatively low toxicity. These compounds were highly selective in their effect on tumor cells compared to normal cells (see Fig. M56). (Shan D et al 2005 Proc Natl Acad Sci USA 102:3772).

It seems that these chemokine receptors and chemokines play important role in metastasis. The direction of invasiveness may depend on the organs/tissues where these receptors are expressed. TIP30 kinase, which appears to be the same as CC3, upregulates some apoptotic genes by phosphorylating the C-terminal domain of the largest subunit of DNA-dependent RNA polymerase II. Some interfering RNAs

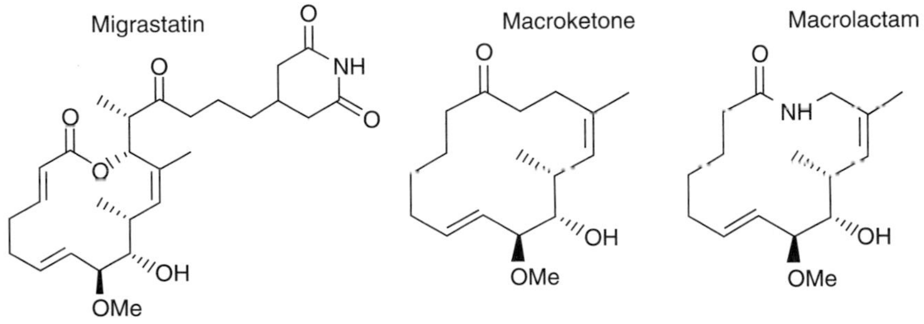

Migrastatin
Macroketone
Macrolactam

Figure M56. Migrastatin, Macroketone, Macrolactam

(RNAi) suppressed metastasis of ovarian cancer cell line (SKOVB-3) by affecting a cyclin-dependent kinase (CDK7), a dual-specificity tyrosine phosphorylation-regulated kinase (DYRK1B), mitogen-activated protein kinase (MAP4K4) and serpin (SCCA-1). Such a procedure may have therapeutic potential for cancer (Collins CS et al 2006 Proc Natl Acad Sci USA 103:3775).

Mucin-type glycoprotein precursor of the core 3 structure (GlcNAcβ1–3GalNAcα1-Ser/Thr) is synthesized by the enzyme β1,3-*N*-acetylglucosaminyltransferase 6. Core 3 structure is markedly reduced in gastric and colorectal carcinomas in comparison to the Core 1 structure (Galβ1,3GalNAcα1-Ser/Thr) and core 2 structure and metastasis occurs. When the Core 3 synthase activity is restored by transfection of the active gene, metastasis is suppressed (Iwai T et al 2005 Proc Natl Acad Sci USA 102:4572). Platelet-derived lysophosphatidic acid promotes the progression of breast and ovarian cancer metastasis to bone. Silencing its receptor by siRNA provided effective protection (Boucharaba A et al 2006 Proc Natl Acad Sci USA 103:9643).

Lysyl oxidase (LOX) expression is regulated by hypoxia-inducible factor (HIF) and the high level of LOX presents poor prognosis for breast cancer and other tumors in mice (Erler JT et al 2006 Nature [Lond] 440:1222).

In *Drosophila* cooperation between Ras^{V12} and one of the cell polarity genes results in metastasis of tumor cells (Pagliarini RA, Xu T 2003 Science 302:1227). Techniques have been designed for the identification of genes, which control/suppress metastasis. In a forward mutation test of mice, disulfide isomerases (thiol isomerases), which catalyze disulfide bond formation, reduction and isomerization, were found to mediate metastasis. Overexpression of ERp5 promotes both in vitro migration and invasion and in vivo metastasis of breast cancer cells. These effects were shown to involve activation of ErbB2 and phosphoinositide 3-kinase (PI3K) pathways through dimerization of ErbB2. Activation of ErbB2 and PI3K subsequently stimulates RhoA and β-catenin, which mediate the migration and invasion of tumor cells (Gumireddy K et al 2005 Proc Natl Acad Sci USA 104:6696).

Hammerhead ribozyme proteins are hooked to RNA helicases. Such a system may destroy the transcripts of metastasis suppressor genes and thus enhance cell migration and can be assayed by an in vitro system (Suyama E et al 2003 Proc Natl Acad Sci USA 100:5616). Metastasis of melanoma cells may be initiated by an increase in the expression of fibronectin, RhoC and thymosin β4 as visualized by microarray hybridization. Microarray hybridization of primary tumor tissue transcripts may permit the prognostication of tumor progression and patient survival. ►CD44, ►cancer, ►oncogenes, ►malignant growth, ►organizer, ►contact inhibition, ►saturation density, ►collagen, ►extracellular matrix, ►VEGF, ►fibronectin, ►Rho, ►thymosin, ►metalloprotein, ►DAP kinase, ►intravasation, ►KISS, ►urokinase, ►plasminogen activator, ►TNF, ►ICE, ►IL-1, ►IL-18, ►apoptosis, ►PRL-3, ►anoikis, ►RAGE, ►CCR, ►CXCR, ►ACIS, ►ADM, ►FAST, ►macrolide, ►wound-healing assay, ►lysophosphatidic acid, ►ERBB1, ►PIK, ►catenins, Al-Mehdi AB et al 2000 Nature Genet 6:100; Müller A et al 2001 Nature [Lond] 410:50; Liotta LA, Kohn EC 2001 Nature [Lond] 411:375; Trusolino L, Comoglio PM 2002 Nature Rev Cancer 2:289; Chambers AF et al 2002 Nature Rev Cancer 2:563; Dudley ME et al 2002 Science 298:850; Ramaswamy S et al 2003 Nature Genet 33:49; Steeg PS 2003 Nature Rev Cancer 3:55; Christofori G 2006 Nature [Lond] 441:444; Nguyen DG, Massagué J 2007 Nature Rev Genet 8:341.

Metatarsus: The middle bones beyond the ankle but preceding the toes in a human foot. In insects the basal part of the foreleg distal to the tibia but proximal to the tarsal segments and the claw (see Fig. M57). It carries the sexcombs in the male *Drosophila*. ►*Drosophila*, ►sex comb, ►metacarpus

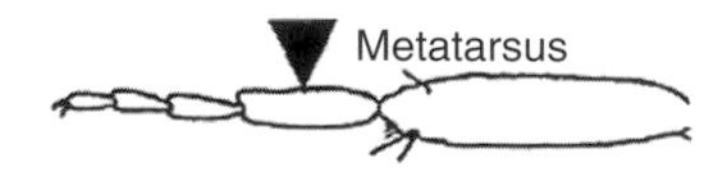

Figure M57. Metatarsus

Metaxenia: A physiological modification of maternal tissues of the fruit in plants by the genetically different embryo, e.g., in green hybrid apples the exocarp (skin) may become reddish if the pollen carries a dominant gene for red color. ►xenia, ►fruit

Metazoa: All animals with differentiated tissues; thus protozoa are excluded.

Methacrylateaciduria: β-hydroxyl-isobutyryl CoA deacylase deficiency—involving the catabolism of valine—leads to urinary excretion of cysteine and cysteinamine conjugates of methacrylic acid, and to teratogenic effects. Methacrylic acid [CH_2= C(CH_3) COOH] is a de-gradation product of isobutyric acid [$(CH_3)_2$CHCOOH] and the amino acid valine [$(CH_3)_2$CH (NH_2)CHCOOH]. ►isoleucine-valine biosynthetic pathway

Methanococcus jannaschii: Bacterium with a genome of 1.67×10^6 bp DNA and 1,738 ORF. *Methanococcus thermoautotrophicum's* genome is 1.75×10^6 and has 1,855 ORF. ►ORF, ►missing genes; Bult CJ et al 1996 Science 273:1058.

Methanogenesis: It takes place by decay of various types of organic material brought about by Archaebacteria and other microorganisms (see Fig. M58). They liberate methane (CH_4) and carbondioxide (CO_2), which contribute to greenhouse gas levels. Burning fossil fuels, wetlands, landfill fermentation, livestock manure, termites, etc., also produce these gases. More recently a very substantial role in methane production is attributed to the vegetation (Nature [Lond] 2006 442:730). Archaebacteria of the oceans also consume methane (reverse methanogenesis) and reduce the level of this environmentally undesirable gas (Hallam SJ et al 2004 Science 305:1457). Sphagnum mosses in peat bogs cover 10–15% of their C need from the microbially liberated CO_2 (Raghoebarsing AA et al 2005 Nature [Lond] 436:1153). (See sources of variations of methane in the atmosphere: Bousquet P et al 2006 Nature [Lond] 443:439).

Figure M58. Sphagnum moss

Methemoglobin (ferrihemoglobin, hemiglobin): A hemoglobin with the ferroheme oxidized to ferriheme and has impaired reversible oxygen binding ability. Mutation and certain chemicals (ferricyanide, methylene blue, nitrites) may increase the normally slow oxidation of hemoglobin. Reduction of methemoglobin may be accomplished either through the Embden-Meyerhof pathway or through the oxidative glycolytic pathway (pentose phosphate pathway). The principal methemoglobin-reducing enzyme (22q13.31-qter) is NADH-methemoglobin reductase that may be adversely affected by autosomal recessive genetic defects. Methemoglobinemia may also be caused by a deficiency of cytochrome b5 (18q23). As a consequence, lifelong cyanosis (bluish discoloration of the skin and mucous membranes) results following the accumulation of reduced hemoglobin in the blood. Males are apparently more affected. This anomaly may be corrected with the reducing agents, methylene blue or ascorbic acid. NADH-methemoglobin reductase is often called diaphorase. ►hemoglobin, ►NADH, ►Embden-Meyerhof pathway, ►pentose phosphate pathway, ►hexose monophosphate shunt

Methionine Adenosyltransferase Deficiency (MAT, 10q22, 2p11.2): Homocystinuria and tyrosinemia may cause hypermethioninemia. The most direct cause of hypermethioninemia is an autosomal recessive defect in methionine adenosyltransferase. Besides these genetically determined causes methionine accumulation may be due to prematurity at birth or the overactivity of cystathionase when cow milk (rich in methionine) is fed. Hypermethioninemia itself may not have very serious consequences. Methionine adenosyl transferase catalyzes the biosynthesis of *S*-adenosylmethionine a major methyl donor in the cell. ►methionine biosynthesis, ►AdoMet, ►homocystinuria, ►tyrosinemia, ►amino acid metabolism

Methionine Biosynthesis: Methionine is biosynthesized from homocysteine by methionine synthase using N^5-methyltetrahydrofolate as a methyl donor:

$$\underset{\text{homocysteine}}{HSCH_2CH_2CH(NH_2)COOH} \rightarrow \underset{\text{methionine}}{CH_3SCH_2CH_2CH(NH_2)COOH}$$

Methionine then can be used by methionine adenosyltransferase to generate S-adenosylmeth-ionine and subsequently S-adenosylhomocysteine which after hydrolyzing off adenosine yields again homocysteine. ►homocystinuria, ►amino acid metabolism, ►methionine, ►adenosyltransferase deficiency, ►AdoMet

Methionine Malabsorbtion: Methionine malabsorbtion is under the control of an autosomal recessive gene, and results in the excretion of α-hydroxybutyric acid (with its characteristic odor) in the urine. Mental retardation, convulsions, diarrhea, respiratory problems and white eye accompany the condition. ►methionine biosynthesis

Methionine Synthase Reductase (MTRR, 5p15.3-p15.2): The deficiency may lead to megaloblastic anemia, hyperhomocyteinemia and hypomethioninemia and meiotic nondisjunction of human chromosomes. ►megaloblastic anemia, ►hyperhomocyteinemia hypomethioninemia; Leclerc D et al 1998 Proc Natl Acad Sci USA 95:3059.

Methionyl tRNA Synthetase (MARS): MARS charges $tRNA^{Met}$ by the amino acid methionine. The MARS gene is in human chromosome 12. ►aminoacyl-tRNA synthetase

Methotrexate (amethopterin): A folic acid antagonist and an inhibitor of dihydrofolate re-ductase. It is an inhibitor of thymidylate synthase, glycinamide ribonucleotide transformylase and 5-aminoimidazole-4-carboxamide ribonucleotide formyltransferase. It is used as antineoplastic and autoimmunity drug. Because of its multiple metabolic effects, response to it may vary according to the genetic constitution of the patient. It is used also as selectable agent in genetic transformation. A mutant dehydrofolate reductase may improve the resistance to methotrexate and may be advantageous for cancer chemotherapy. ►folic acid, ►transformation, ►selectable, ►nanotechnology, see formula at ►amethopterin, ►DB [doubleminute] chromosome; Goodsell DS 1999 Stem Cells 17:314.

Methylacetoaceticaciduria (11q22.3–23.1): A 3-oxothiolase deficiency in the degradation of isoleucine resulting in 2-methyl-hydroxybutyric acid, 2-methylacetoacetic acid, tiglylglycine and butanone in the urine. Tiglic acid [$CH_3CH:C(CH_3)CHO$] is trans-2-methyl-2-butenoic acid. ►isoleucine-valine biosynthetic pathway

Methylase: Enzymes in bacteria protect the cell's own DNA from type II restriction endonucleases by transferring methyl groups from *S*-adenosyl methionine to specific cytosine or adenine sites within the endonuclease recognition sequence (cognate methylases). When eukaryotic DNAs are transfered to *E. coli* cells by transformation for cloning, their methylation pattern may be lost because the methylation system of the prokaryotic cell is different from that of the eukaryote. *E. coli* does not methylate the C in a 5′-CG-3′ but the dam methyltransferase methylates A in the 5′-GATC-3′ sequence and the dcm methyl transferase methylates the boxed C in the 5′C[C]$\frac{A}{T}$GG-3′ group. Such methylations may change the restriction pattern of cloned DNAs depending whether the particular restriction enzyme can or cannot digest methylated DNA. ►methylation of DNA, ►methylation-specific PCR; Cheng X, Roberts RJ 2001 Nucleic Acids Res 29:3784.

Methylation Filtration: The procedures that separate methylated and unmethylated DNA sequenced. Since in, e.g., maize ~95% of the exons are unmethylated, active, functional genes and active transposons can be separated from the bulk DNA. In higher plants, generally 60–80% of the DNA is repetitive and methylated; only ~7% of the repetitive DNA is unmethylated. In *E. coli* strain JM107MA2 the McrA and McrC modification-restriction systems are defective and do not permit cloning of methylated DNA. Thus cloning enriches the unmethylated genic and retroposon sequences and this "filtration" can reduce the difficulties of sequencing of genes of large eukaryotic genomes (Palmer LE et al 2003 Science 302:2115; Whitelaw CA et al 2003 Science 302:2118; Rabinowicz PD et al 1999 Nature Genet 23:305). ►methylation of DNA, ►cot filtration

Methylation Interference Assay: Detects whether a binding protein can attach to the specific DNA sites and thus provides information on the binding site and on the protein. The analysis is carried out by combining DNA and binding proteins and followed by treatment with methylating enzyme. If the protein binds to a specific guanine site(s), that base will not be methylated. Piperidine breaks DNA at bases modified by methylation, and sites protected from methylation by bound protein are not cleaved by peperidine. ►methylation of DNA; Shaw PE, Stewart AF 2001 Methods Mol Biol 148:221.

Methylation of DNA (DNMT): In many eukaryotes 1–6% or more of the bases in DNA is methylcytosine. In T2, T4 and T6 bacteriophage DNA 5-hydroxymethylcytosine occurs in place of cytosine. Methylation of other bases thymine (= 5-methyluracil), adenine and guanine may also occur in prokaryotes. Adenoviral DNA is not methylated but adenoviral as well as other foreign DNA integration into the mammalian genome may be followed by methylation of cytosine (Orend G et al 1995 J Virol69:1226). The extent of methylation of the same 5′-CG-3′ nucleotides varies in different tissues. Some of the alkylations of DNA bases lead to mutations by base substitutions. Methylation protects DNA from most of the restriction endonucleases (►restriction endonuclease types). In the majority of *E. coli* strains two enzymes are responsible for DNA methylation, *dam* and *dcm* methylase; *dam* methylates adenine at the N^6 position within the sequence 5′-GATC- 3′. This sequence occurs at the recognition sites of a number of frequently employed restriction enzymes (Pvu I, Bam HI, Bcl I, Bgl II, Xho II, Mbo I, Sau 3AI, etc). Mbo I (↓GATC) and HpaII (C↓CGG) are sensitive to methylation but Sau 3AI and MspI, respectively, are not, and their recognition sites are identical (isoschizomers), therefore when the DNA is methylated, the latter ones still can be used. For several restriction enzymes to work, the DNA must be cloned in bacterial strains that do not have the *dam* methylase. Mammalian DNA is not methylated at the N^6 position of adenine, therefore Mbo I is always supposed to work, as well as Sau 3AI. The DNAs of eukaryotes are most commonly methylated on C nucleotides, in $\frac{CG}{GC}$ sequences. The dcm methylase methylates the internal C positions in the sequences 5′-CCAGG-3′ and 5′-CCTGG-3′; this methylation interferes with cutting by EcoRII [↓CC(A/T)GG] but not by BstN I, although at another position [CC↓(A/T)GG] of the same sequence. *E. coli* strain K also has methylation-dependent restriction systems that recognize only methylated DNA: *mrr* (6-methyladenine), *mcrA* [5-methyl-C(G)], *mcrB* [(A/G)5-methyl C]. Mammalian DNA with extensive methylation at 5-methyl C(G) is, for, e.g., restricted by *mcrA*. Once the DNA is methylated, this feature may be transmitted to the following cell generation(s) by an enzyme, *maintenance methylase* although methylation is usually lost through the meiotic cycle. In bacteria, the expression of methylated genes may be reduced by a factor of 1000 but in mammals, the reduction may be of six orders of magnitude. Methylation in the promoter region usually prevents transcription initiation but not RNA chain elongation of that gene in mammals.

In the mouse, *Dnmt3a* and *Dnmt3b* are necessary for embryonic survival (*b*) or development after birth (*a*). The double mutants (*a, b*) cannot develop beyond gastrulation (Li E et al 1992 Cell 69:915). *Dnmt1* is incapable for de novo methylation; its role is maintenance methylation of the unmethylated strand generated by replication across the methylated strand in the double helix. Dnmt 1 and PCNA (proliferating cell nuclear antigen) are attracted to damaged DNA sites and this assures the maintenance of the methylated state after replication and repair (Mortusewicz O et al 2005 Proc Natl Acad Sci USA 102:8905). Inactivation of DNMT1 does not affect the status of methylation of CpG doublets in human cancer cells that is involved in the silencing of tumor susceptibility genes in several types of cancers (see Fig. M59). In melanoma and several solid tumors about 40% of the promoter of the p16^{INK4a} gene is hypermethylated. In endometrial, colon and gastric cancer cells the microsatellite instability is accompanied by up to 70–90% methylation. By base specific cleavage of DNA and the use MALDI/TOF/MS using high-throughput technology methylation pattern can be identified (Ehrich M et al 2005 Proc Natl Acad Sci USA 102:15785). The DAP kinase gene in Burkitt's lymphoma may be completely methylated. Mutation in the human DNMT3B gene is responsible for the rare immunodeficiency, ICF. ICF (20q11.2) involves immunodeficiency, pericentromeric hypomethylation and instability in lymphocytes and facial malformations. In some fungi (*Ascobolus, Neurospora*) peptide chain elongation may also be inhibited. The inhibition of initiation is attributed to the reduced binding of transcription factors to methylated DNA. Also, methylation-dependent DNA-binding proteins (MDBP) may suppress transcription. Methylation by Dnmt1 may affect also histone deacetylation and chromatin remodeling.

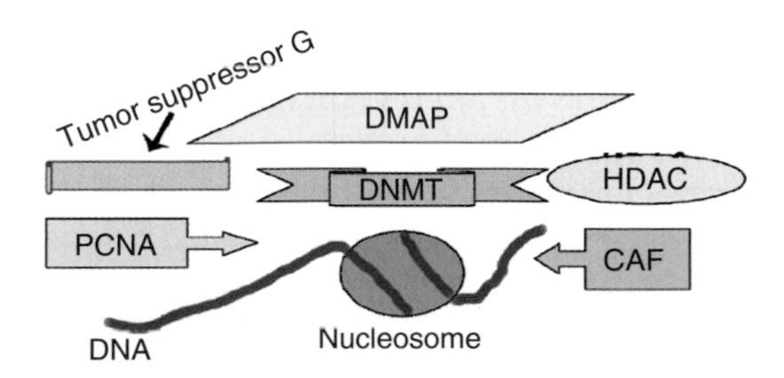

Figure M59. Some of the major factors controlling gene silencing. Proliferating Cell Nuclear Antigen (PCNA) and Chromatin Assembly Factor (CAF) cooperate in building the nucleosomal structure. DNA Methyltransferase (DNMT) is assited by DNA Methytransferase Associated Protein (DMAP) and the Tumor Suppressor Gene 101 methylates primarily cytosine residues and silences gene. Histone Deacetylase (HDAC) joins the system

Proteins MeCP1 and MeCP2 mediate silencing of gene expression due to methylation of CpG. MeCP1 effect on silencing depends on the density of methylation near the promoter. MeCP2 binds only single methylated CpG pairs and can act at a distance from the transcription factor binding sites. The effect of MeCP2 is apparently not gene-specific, it is rather global. Genomic imprinting is caused by differential methylation. In mice demethylation or lack of MeCP2 may prevent the normal completion of the embryonic development. In *Arabidopsis* plants demethylation to about 1/3 of the normal level caused either by a DNA (*ddm1*) demethylation mutation or by introducing by transformation an antisense RNA of the cytosine methyltransferase (MET1) caused alterations in the morphogenesis and developmental time of the plants. Demethylation of the *XIST* leads to its expression but that inactivates both X-chromosomes in the female and the single X in the male, an obviously lethal condition. The gene silencing by methylation in *Arabidopsis* is controled by the *MOM* gene. This protein contains a sequence with similarity to the ATPase region of SWI2/SNF2 protein family, which is involved in chromatin remodeling. Inactivating *MOM* by antisense technology releases heavily methylated genes from the silenced state. In *Arabidopsis* the *mom1* mutation releases silencing independently from DNA demethylation (Mittelsten Scheid O et al 2002 Proc Natl Acad Sci USA 99:13659). Genome-wide methylation of *Arabidopsis* DNA has been mapped at 35-bp resolution. Pericentromeric heterochromatin, repetitive sequences and small interfering RNA regions are heavily methylated. About one-third of the expressed genes contain methylated bases within transcribed areas contrary to expectations. These genes are constitutively and highly expressed. Only 5% of the promoters display methylation and frequently show tissue-specific expression. Methylation epigenetically controls hundreds of genes and intergenic noncoding RNAs (Zhang X et al 2006 Cell 126:1189). Moderately transcribed genes of *Arabidopsis* are most likely to be methylated, whereas genes at either extreme are least likely. In turn, transcription is influenced by methylation: short methylated genes are poorly expressed, and loss of methylation in the body of a gene leads to enhanced transcription (Zilberman D et al 2007 Nature Genet 39:61).

Demethylation of tumor suppressor genes (by antisense technology) restores their suppressor function whereas methylation of the same genes silences them (Robert M-F et al 2003 Nature Genet 33:61). In some organisms with small genomes (*Caenorhabditis*) methylation of the DNA is very low, thus methylation may not have a general developmental regulatory role. In *Drosophila* DNA 5-methylcytosine is detectable only at very low frequency and mainly in the embryo but interestingly the genome encodes two proteins that resemble cytosine DNA methyltransferase and methyl-CpG-binding-domain proteins. RNAi and microRNA in several organisms are also epigenetic silencing factors.

If 5-methylcytosine is deaminated, thymine results and a C≡G pair may suffer a transition mutation to T = A. It seems that the genome of higher eukaryotes, including humans, includes 35% or more active or silent transposable elements. It had been suggested that methylation of infective (inserted) DNA is part of the eukaryotic defense system. Actually most of the methylated cytosines in mammals are in the parasitic transposable elements. This methylation suppresses their transcription and the C→T mutation leads to the formation of pseudogenes. In the small invertebrate chordate *Ciona intestinalis* non-methylated transposons and normally methylated genomic sequences were detected (Simmen et al 1999 Science 283:1164). Not uncommonly the DNA of cancer cells is undermethylated at C residues indicating the demethylation of their parasitic sequences (SINE, LINE, Alu, etc.) and leading to the destabilization of the genome. The DNA methylase (methyltransferase) enzymes therefore have been supposed to be the means to defend the genome against the deleterious effects of the infective transposable/viral elements. In repetitive DNA CpG methylation is 37% higher than in non-repetitive sequences (Meunier J et al 2005 Proc Natl Acad Sci USA 1023:5471). Although this hypothesis is in agreement with many observations, it does not seem to be of general validity, particularly for the methylation of plant transposable elements. The epigenetic state of methylation can be transferred in the ascomycete, *Ascobolus* by a mechanism resembling or related to recombination. After fertilization, the methyl moieties are generally removed from the CpGs and an unmethylated state is maintained through blastula stage. Some of the genes involved in tumorigenesis display an increased methylation on aging. Housekeeping genes stay unmethylated whereas the methylation of tissue-specific genes varies by tissues. Reduced methylation causes developmental anomalies in plants and animals. The maternal genomes in haploid and diploid gynogenetic one-cell mammalian embryos are always methylated. The polar bodies are always methylated. The paternal genomes in haploid or diploid androgenetic embryos are de-methylated. Triploid digynic embryos show two methylated maternal and one de-methylated paternal chromosome set. In the diandric triploid embryos the methylation pattern is the opposite (Barton SC et al 2001 Hum Mol Genet 10:2983). In the mammals, the active X chromosome displays more than two times as much allele-specific methylation as the inactive X. This methylation is concentrated at gene bodies, affecting multiple neighboring CpGs. Before X inactivation, all of these active X gene body–methylated sites are biallelically methylated. A methylation-demethylation program results in active X-specific hypomethylation at gene promoters and hypermethylation at gene bodies (Hellman A, Chess A 2007 Science 315:1141). Methyltransferase enzymes comprising enhanced zinc-finger arrays coupled to methyltransferase mutants are functionally dominated by their zinc-finger component. Both in vitro plasmid methylation studies and a novel bacterial assay reveal a high degree of target-specific methylation by these enzymes (Smith AE, Ford KG 2007 Nucleic Acids Res 35:740).

Changes in methylation are apparently not required for the regulation of development of zebrafish. Methylation of the normally barely methylated *Drosophila* DNA reduces viability. In the embryonic tissues of mice CpA and CpT are also methylated to some extent not just CpG. The methylation pattern of cancer cell DNA is usually altered (the promoter of tumor suppressor genes is heavily methylated) but the overall extent of methylation is lower. Methylation of the promoter may interfere with the attachment of the transcription factors. The silencing effect of methylation may be associated with the simultaneous deacetylation of the nucleosomes. Selective methylation of specific DNA sequences may be an important goal of regulating gene expression. The DNA adduct of an oligonucleotide–quinon methide may mediate the transfer of the methide to a complementary base and thus may assure selective targeting (Zhu Q, Rokita SE 2003 Proc Natl Acad Sci USA 100:15452). Methylation may strongly suppress recombination between repeated elements and thus preventing deleterious rearrangements of the genome (Maloisel L, Rossignol JL 1998 Genes Dev 12:1381). Genomic methylation is important factor in epigenetic modification of DNA and development. The program HDFINDER can detect CpG methylation in the human genome with ~86% accuracy (Das R et al 2006 Proc Natl Acad Sci USA 103:10713). ►transposition, ►transposable elements, ►LINE, ►SINE, ►Alu, ►silencing, ►cross linking, ►chemical mutagens, ►alkylation, ►mDIP, ►paramutation, ►imprinting, ►regulation of gene activity, ►lyonization, ►cancer, ►Sp1, ►methylation resistance, ►demethylation, ►hypermethylation, ►hypomethylation of DNA, ►ascomycete, ►5-azacytidine, ►HMBA, ►methylation-specific PCR, ►methyltransferase, ►PCNA, ►CpG islands, ►MeCP, ►histone deacetylase, ►base flipping, ►RIP, ►MIP, ►miRNAP, ►RISC, ►integration, ►chromatin remodeling, ►histone deacetylase, ►histone methyltransferases, ►methylation of RNA, ►housekeeping genes, ►carcinogenesis, ►RLSG, ►immunodeficiency, ►methylome, ►PAD4, ►epigenesis, ►DAM, ►digyny, ►diandry, ►RNAi,

►microRNA, ►RNAS-directed DNA methylation, ►methylation filtration, ►bisulfite reaction, ►MALD/TOF/MS, ►MS-RDA, ►AP-PCR, ►DMH, ►Rett syndrome, ►ATRX, ►fragile X chromosome, ►choline, ►chromomethylase; Bird AP, Wolffe AP 1999 Cell 99:451; Robertson KD, Wolffe AP 2000 Nature Revs Genet 1:11; Baylin SB et al 2001 Hum Mol Genet 10:687; Reik W et al 2001 Science 293:1089; Aoki A et al 2001 Nucleic Acids Res 29:3506; Bender J 2001 Cell 106:129; Cervoni N et al 2002 J Biol Chem 277:25026; DNA methylation and cancer: Laird PW 2003 Nature Rev Cancer 3:253; DNA methylation in *Arabidopsis*: Chan S W-L et al 2005 Nature Rev Genet 6:351; Goll MG, Bestor TH 2005 Annu Rev Biochem 74:481, http://www.methdb.net.

Methylation of DNA in Human Disease: ►imprinting, ►Albright hereditary osteodystrophy, ►Angelman syndrome, ►Beckwith-Wiedemann syndrome, ►cancer, ►diabetes mellitus, ►muscular dystrophy, ►fragile X syndrome, ►Prader-Willi syndrome, ►Rett syndrome; Robertson KD 2005 Nature Rev Genet 6:597.

Methylation of Proteins: Reversible post-translational alteration involved in regulation most frequently at arginine and lysine sites. ►histone methyltransferases; Server for methylation prediction: http://www.bioinfo.tsinghua.edu.cn/~tigerchen/memo.html.

Methylation of RNA: 2′-O-methyladenosine, 2′-O-methylcytidine, 2′-O-methylguanosine, 2′-O-methyluridine, 2′-O-methylpseudouridine are minor nucleosides in RNA. The cap of mRNA is a 7-methyl guanine, and a N^6-methyladenosine occurs near the poly-adenylated tract. DNA methyltransferase homolog Dnmt2 does not methylate DNA rather it methylates $tRNA^{Asp}$ at cytosine 38 in mouse, *Drosophila* and *Arabidopsis* (Goll MG et al 2006 Science 311:395). ►capping enzyme; Santoro R, Grummt I 2001 Mol Cell 8:719.

Methylation Resistance: ►MNNG

Methylation-Specific PCR (MSP): Methylation of DNA is usually detected by the inability of the majority of restriction endonucleases to cleave methylated sites. MSP detects methylated CpG sites in minute amounts of DNA. The DNA is treated with sodium bisulfite to convert all non-methylated cytosines to uracil. Two different primers designed to represent original CpG rich sequences then amplify the methylated and unmethylated DNAs. The amplified DNA can then be sequenced to compare the differences between the two samples. Alternatively, the DNA is digested by restriction enzyme HpaII, which does not cut at sites of methylated DNA (CC^mGG) but its isoschizomer MspI does. The DNA is exposed also RsaI (GT'AC) to generate smaller fragments. The fragments are amplified by PCR in the presence of α ^{32}P-dCTP and separated by gel electrophoresis. If a fragment is present in both the RsaI and (RsaI + HpaII) digest but not in the (RsaI + MspI) samples, it is considered to be methylated. Other isoschizomer pairs with one methylation-sensitive restriction enzyme can be used similarly. ►methylation of DNA, ►polymerase chain reaction, ►bisulfite reaction; Velinov M et al 2001 Genet Test 5(2):153; Oshimo Y et al 2004 Int J Cancer 110:212.

Methylated DNA-Binding Proteins: They bind methylated DNA and recruit additional proteins and the methylome silences transcription of DNA. ►methylation of DNA; Nan X et al 1998 Nature 393:311; Ng HH et al 1999 Nature Genet 23:58.

Methylator: It simultaneously methylates several sites in different cancer suppressor genes. ►CIMP, ►tumor suppressor, ►methylation of DNA

3-Methylcrotonyl Glycinemia: It is caused by the deficiency of a mitochondric enzyme involved in the degradation of leucine, β-methylcrotonyl-CoA-carboxylase (MCCA, 19 exons, 3q25-q27). The β subunit is encoded at 5q12-q13 (MCCB, 17 exons). As a consequence of this autosomal recessive condition, muscle defects, and in some cases urinary overexcretion of 3-methylcrotonyl glycine and 3-hydroxyisovaleric acids occurs. Some patients respond favorably to biotin because this vitamin is a cofactor of the enzyme. ►isoleucine-valine metabolic pathway, ►isovaleric acidemia, ►3-methylglutaconaciduria, ►3-hydroxy-3-methyl-glutaricaciduria, ►amino acid metabolism; Gallardo ME et al 2001 Am J Hum Genet 68:334.

5-Methylcytosine (5mC): Common in tRNA and DNA of eukaryotes. Methylcytidylic acid is frequently called the fifth nucleotide. Deamination of 5-methylcytosine at CpG sites into thymidine is one of the most common causes of disease and accounts for 20% of all human point mutations (see Fig. M60). The methyl-CpG-binding domain protein MBD4 enzymatically removes T or U from the mismatched sites and protects against tumorigenesis (Millar CB et al 2002 Science 297:403). Thymine DNA glycosylase (TDG) has similar repair function. Sodium bisulfite converts cytosine but not 5mC into uracil. On this basis the location of 5mC in the nucleotide sequence can be determined. ►5-hydroxymethyl cytosine, ►5-azacytidine, ►bisulfite reaction, ►methylation of DNA; Clark SJ et al 1994 Nucleic Acids Res 22:2990.

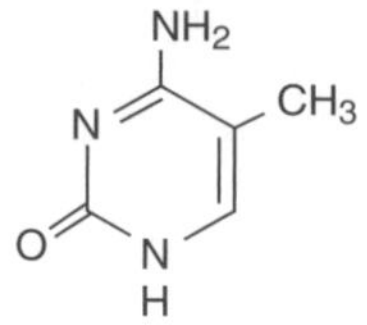

Figure M60. 5-Methylcytosine

Methyl-Directed Repair: ►mismatch repair

Methylene Blue: Aniline dye, for microscopic specimens, an indicator of oxidation-reduction, an antiseptic, an antidote for cyanide and nitrate poisoning.

5,10-Methylenetetrahydrofolate Dehydrogenase (MTHFD1): An enzyme in folate and purine biosynthesis is encoded in human chromosome 14q24. ►folic acid

Methylenetetrahydrofolate Reductase (MTHFR, 1p36.3): MTHFR deficiency is responsible for homocystinuria and may affect human chromosomal nondisjunction. ►homocystinuria, ►nondisjunction

Methylglutaconicaciduria: The autosomal recessive condition is caused by a deficiency of 3-methylglutaconyl-CoA hydratase, an enzyme mediating one of the steps in the degradation of leucine. The patients may develop nerve disorders such as partial paralysis, in-voluntary movements, eye defects, etc. In some cases there is a marked increase of methylglutaric and methylglutaconic acid (an unsaturated dicarbonic acid) in the body fluids. Leucine administration may exagerate the symptoms. 3-methylglutaconic aciduria (MGA) type III is encoded at 19q13.2–q13.3. ►isoleucine-valine metabolic pathway, ►isovalericacidemia, ►3-hydroxy-3-methylglutaricaciduria, ►amino acid metabolism, ►endocardial fibroelestosis; Anikster Y et al 2001 Am J Hum Genet 69:1218; Ijlst L et al 2002 Am J Hum Genet 71:1463.

Methylgreen-Pyronin: A histological stain; coloring DNA blue-green and RNA red. ►stains; Brachet J 1953 Quart J Microscop Sci 94:1.

Methylguanine-O^6-Methyltransferase: An enzyme that reverses the alkylation of this base, and it is thus antimutagenic and anticarcinogenic. Cells defective in the enzyme (Mer$^-$, Mex$^-$) are extremely sensitive to DNA-alkylating agents. Mutant enzymes with increased methylguanine-O^6-methyltransferase activity may decrease the cells's sensitivity to alkylating mutagens. ►methylation of DNA, ►antimutator; Lips J, Kaina B 2001 Mutation Res 487:59; Zhou Z-Q et al 2001 Proc Natl Acad Sci USA 98:12566.

2-Methyl-3-Hydroxybutyryl-Coa Dehydrogenase Deficiency (HADH2, Xp11.2): This deficiency results in motor function disorder, blindness and epilepsy due to inborn error of isoleucine metabolism. ►isoleucine-valine biosynthetic pathway; Ofman R et al 2003 Am J Med Genet 72:1300.

Methylimidazole: ►pyrrole

Methyljasmonate: The fragrance of jasmine and rosemary plants; it is a proteinase inhibitor.

Methylmalonicaciduria: There are several forms of the metabolic disorder, methylmalonic-CoA mutase deficiency (MUT), in human chromosome 6p21, and another is caused by a defect in the synthesis of adenosyl-cobalamin (cblA, vitamin B12) a necessary co-factor in the biosynthesis of succinyl-CoA from L-methylmalonyl-CoA by MUT. A third type of methylmalonic aciduria is due to a defect in the enzyme epimerase (racemase) that converts D-methylmalonyl-CoA to the L form. This pathway is represented in Fig. M61. In these disorders methylmalonic acid and glycine may accumulate in the body fluids, and the affected individuals may show serious (growth and mental retardation, acidosis [keto acids in the blood]) or almost no adverse effects. High protein diet (valine, isoleucine) may aggravate the condition. Administration of vitamin B_{12} may alleviate the problem in some cases. ►methylcrotonylglycemia, ►amino acid metabolism, ►vitamin B_{12} defects, ►ketoaciduria

Methylmercuric Hydroxide (MMH): MMH may be added to the electrophoretic agarose running gel of RNA, and when it is stained with ethidium bromide in 0.1 M ammonium acetate the color of the RNA is enhanced. It is also used to treat mRNA for preventing the formation of secondary structure during the synthesis of the first strand of cDNA. Note that MMH is an extremely toxic volatile compound. ►cDNA, ►electrophoresis

Methylmethanesulfonate: A powerful alkylating agent and mutagen/carcinogen. ►ethylmethane sulfonate

Methylome: The methylated part of the genome, the factors involved in methylation of DNA. ►methylation of DNA, ►methylated DNA binding proteins;

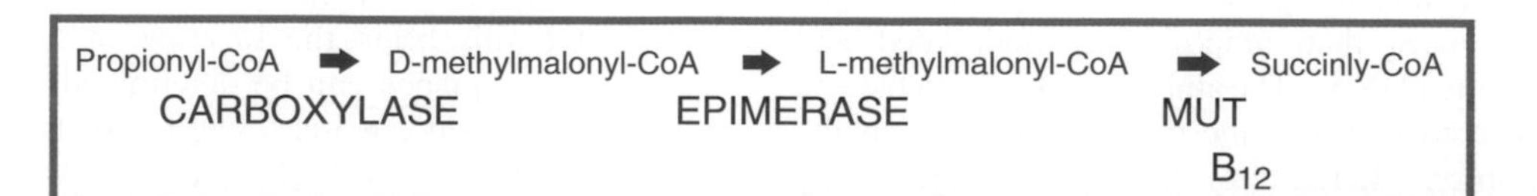

Figure M61. Methylmalonyl pathway

methylome of the *Arabidopsis* genome: Silberman D et al 2007 Nature Genet 39:61.

Methylphosphonates: Oligonucleotide analogs used for antisense operations. They are readily soluble in water and resistant to nucleases. The oligonucleoside methylphosphonates form stable complexes with both RNA and single- and double-stranded DNA. They have both antiviral and anticarcinogenic effects (see Fig. M62). ►antisense technologies, ►mixed backbone oligonucleotide; Schweitzer M, Engels JW 1999 J Biomol Struct Dyn 16(6):1177.

$$CH_3P(O)(OH)_2$$

Figure M62. Methylphosponic acid

Methyltetrahydrofolate Cyclohydrolase Deficiency: Affects folate and purine metabolism. ►phosphoribosylglycinamide formyltransferase, ►folic acid

Methylthioadenosine Phosphorylase (MTAP); MTAP is encoded in human chromosome 9p21 area, and its defect or deletion is characteristic for many malignant tumors, and it may be associated (linked) to a tumor suppressor activity of CDK-4. Methylthioadenosine is abundant in some human tissues and it is an important donor for methylation. ►CDK, ►methylation of DNA; Hori Y et al 1998 Int J Cancer 75:51.

Methyltransferase, DNA (dnmt1, dnmt1-b, 19p13.3-p13.2): Responsible for the methylation of CpG sites. These are encoded in humans by the same gene locus but the transcript is spliced alternatively. The activity of these enzymes is increased during the initiation and progression of carcinogenesis but the methylation of DNA in tumors is altered and usually reduced. The 1,620-amino acid mammalian dnmt is essential for embryonic development of the mouse. Methyltransferases have important role in regulation of gene activity, restriction-modification in bacteria, mutagenesis, DNA repair, cancer, imprinting, chromatin organization and lyonization. DNA methylating enzymes are scarce in some insects (*Drosophila*), which do not have methylation in the genetic material. ►cancer, ►methylation of DNA, ►methylguanine-O^6-methyltransferase, ►immunodeficiency, ►restriction-modification, ►set motif, ►histone tail, ►histone methyltransferases; Adams RLP 1995 Bioassays 17:139; Bestor TH 2000 Hum Molec Gen 9:2395; Kiss A 2001 et al 2001 Nucleic Acids Res 29:3188; Verdine GL, Norman DPG 2003 Annu Rev Biochem 72:337; Goll MG, Bestor TH 2005 Annu Rev Biochem 74:481.

Methylviolet: An aniline dye for bacterial microscopic examination.

Metree: A computer program package for inferring and testing minimum evolutionary trees; designed by the Institute of Molecular Evolution, Pennsylvania State University, Philadephia, PA, USA. ►evolutionary tree

METRO (message transport organizer): A mechanism and center for sorting out molecules within the developing oocyte. ►morphogenesis, ►RNA localization

Metronidazole (2-methyl-5-nitro-1-imidazole ethanol): A radiosensitizing agent and mutagen for chloroplast DNA, and a suspected carcinogen (see Fig. M63). ►chloroplast genetics, ►formula

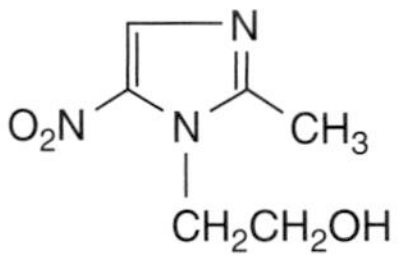

N
O_2N N CH_3
CH_2CH_2OH

Figure M63. Metronidazole

MeV (mega electron volt): A million electron volt. ►electron volt

Mevalonicaciduria (MVK, 12q24): A recessive (human chromosome 12q24) defect with huge increase of mevalonic acid in the urine, caused by a defect in mevalonic acid kinase. ►hyperimmunoglobulinemia; Houten SM et al 2002 Hum Mol Genet 11:3115.

Mex$^-$ (methylation excision minus): Deficient in methyltransferase DNA repair.

Mex67: Protein carrying leucine-rich nuclear localization signal and which serves as an export adaptor. ►export adaptor, ►nuclear localization sequence

MF1: A 5′ to 3′ exonuclease. ►DNA replication eukaryotes

Mfd: Prokaryotic protein, which repairs preferentially the template strand of a transcriptional unit. ►backtracking; Stanley LK, Savery NJ 2003 Arch Microbiol 179:381.

MFG: A Moloney murine leukemia retrovirus-based vector.

MFISH (multicolor in situ hybridization): Detects probes in 27 fluorescent colors. ►FISHI, ►in situ hybridization

MGD: ►Mouse genome database

Mge1: A 26-kDa GrpE homolog in *Saccharomyces* mitochondria. ►Grp.)

MGMT: ►Methylguanine-O^6-methyltransferase

MGSA (melanoma growth stimulating activity): ►KC, ►N51, ►gro

MGT: The MGMT group of enzymes; they are encoded in human chromosome 10q and protect the cells against the genotoxic, recombinogenic and apoptotic effects of O^6-methylguanine with the aid of mismatch repair. When MGMT is expressed at high level the formation of O^6-methylguanine is substantially reduced. ►MGMT, ►Mer; Kaina B et al 2001 Progr Nucleic Acid Res Mol Biol 68:41.

MHC: Major histocompatibility complex is involved in immunological reactions; it is controled by linked multigene families (HLA) determining cell surface antigens and thus cellular recognition. The acronym is used also to the myosin heavy chain. ►HLA, ►Immune system, ►major histocompatibility complex; MHC binding peptides: http://www-bs.informatik.uni-tuebingen.de/SVMHC/.

MHC Restriction: ►immune system

MIAME (minimum information about a microarray experiment): A required standard for acceptance of microarray data by the major journals since 2002. (See Nature Genet 32:333; Science 298:539; Gene Expression Omnibus; http://www.mged.org).

MIC: Minimal inhibitory concentration. ►micRNA

MICA/B Proteins: Intracellular proteins, which mark (cancer) cells for destruction by natural killer cells. Association of MICA with endoplasmic reticulum protein 5 (similar to disulphide isomerase) potentiates shedding of tumor-associated killer cell ligand NKG2D and promotes the immune evasion of tumor cells (Kaiser BK et al 2007 Nature [Lond] 447:482). ►killer cells, ►disulphide bridge, ►immune evasion

Micelle: A round body of (protein) substances surrounded by lipids.

MICER (mutagenic insertion and chromosome engineering resource): A collection of 93,960 insertional targeting vectors; 5,925 of them inactivate genes with 28% efficiency (Adams DJ et al 2004 Nature Genet 36:867). ►targeting genes, ►insertional mutation, ►chromosome engineering

Michaelis–Menten Equation: It measures enzyme kinetics in a process:

$$(E) + (S) \overset{k1}{\Leftrightarrow} (E) \overset{k3}{\rightarrow} \text{Product(s)} + (E),$$

where k_1, k_2, k_3 are constants of the reactions, (E) = enzyme, (S) = substrate concentration and $v = \frac{V(S)}{K_m + (S)}$ or $K_m = (S)\left[\frac{V}{v} - 1\right]$ where v is the velocity of the reaction when half of the substrate molecules is combined with the enzyme, V = the maximum velocity, and $K_m = \frac{[(E)-(ES)](S)}{(S)}$ is the Michaelis – Menten constant. also ►Linweaver-Burk plot).

Michel Syndrome (oculopalatoskeletal syndrome): An autosomal recessive multiple defect involving the eyelid, opacity of the cornea, cleft lip and palate, defects of the inner ear and spine column, etc. It causes complete deafness. ►deafness, ►eye diseases

micRNA (messenger RNA-interfering RNA): This RNA is transcribed on short sections of the complementary strand DNA; prevents gene expression. ►antisense RNA, ►RNAi, ►microRNA; Mizuno T et al 1984 Proc Natl Acad Sci USA 81:1966.

Microarray: ►Microarray hybridization, ►DNA chips, ►small molecule microarray, ►protein array, ►protein chips; microarray data repository: http://www.ebi.ac.uk/arrayexpress-old/.

Microarray Hybridization: A microtiter tray with wells containing DNA to which fluorochrome-labeled-RNA can be hybridized. After incubation and scanning the amounts of the mRNAs derived from the same tissue can then be quantitated (by fluorescence intensity). A microarray tray looks similar to the pattern shown in the Fig. M64. Its size may be as small as 18 × 18 mm. The thousands of genes expressed on a single tray may

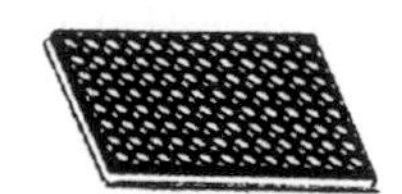

CLONED DNA PLACED IN THE WELLS OF MICROTITER TRAY. AFTER AMPLIFICATION BY PCR AND PURIFICATION, THE SAMPLES (~5 nL) ARE SPOTTED BY A ROBOT PIPETTER ONTO SPECIFICALLY COATED MICROSCOPE SLIDES.

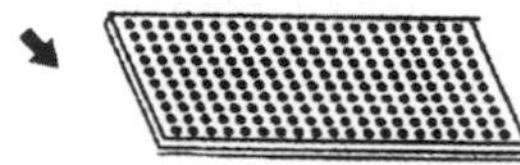

CELLULAR RNA TRANSCRIPTS OF SAMPLE AND AN APPROPRIATE REFERENCE ARE LABELED BY TWO DIFFERENT FLUOROCHROMES AND HYBRIDIZED UNDER STRINGENT CONDITIONS TO THE CLONES.
THE SLIDES ARE IRRADIATED BY LASER AND THE EMISSION IS MEASURED BY CONFOCAL LASER MICROSCOPE. THE MONOCHROME IMAGES ARE TRANSFERED TO A COMPUTER SOFTWARE WHICH GENERATES PSEUDO-COLOR IMAGES. THE COLOR INTENSITIES AS WELL AS INTENSITY RATIOS OF SAMPLE AND REFERENCE (THE NORMALIZED RATIO) ARE DETERMINED, AND THE DIFFERENCE IN EXPRESSION RATIO REVEALS HOW THE GENES ARE EXPRESSED UNDER THE TWO DIFFERENT CONDITIONS.

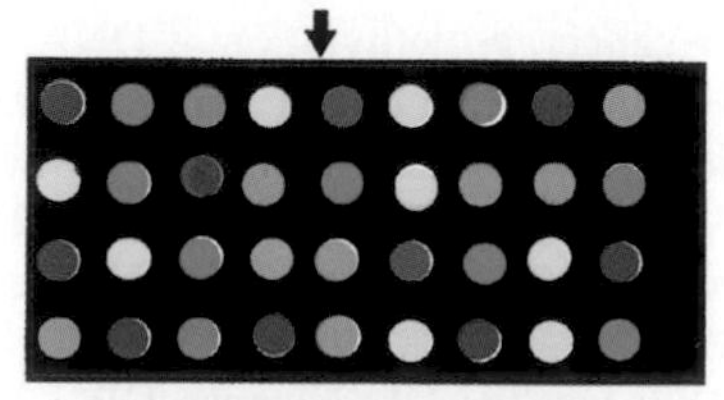

THE SIMULATED COLORS REPRESENT DIFFERENT LIGHT INTENSITIES

Figure M64. Microarray Hybridization

display different fluorescent colors according to the fluorochrome labeling of the probe in a spot test. Such an analysis may identify the simultaneous expression of large sets of genes at a particular developmental or disease/health stage and thus, permits a functional study at the entire genome level. Most commonly, the colors shown in the publications represent the intensity of hybridization (red the highest).

If the tissues are, for, e.g., from healthy and diseased sources, the genetic cause of the disease can be inferred on the basis of the level of expression. Depending on the type of fluorescent label used, the RNA sample may vary from 10 μg to 50 ng and the number of cells needed to extract this much material may vary from >1 to 10^9. The analysis may begin with cDNA clones or PCR-amplified samples transferred to glass plates or nylon filters where appropriate probes are hybridized to the arrays. For the glass slides, usually fluorochrome labeling is used whereas with the nylon support phosphor imager (P^{33}, half-life ~25 days) is employed. The data is first evaluated by image processing after the information is fixed in JPEG or GIF formats. Then statistical and biological methods are applied (JPEG, TIFF and GIF are Adobe PhotoShop computer formats). The information reveals the function of many genes in a context in different tissues, developmental or pathological states, and may be used for simultaneous characterization of many tumor biopsies. By sophisticated computer-based analysis of the biological meaning of the information, *data mining* permits the identification of genes, which are either expressed simultaneously or in certain tissues; in a specific state of a disease or diseases; which appear epistatic or respond to a particular condition.

Microarray interrogated with short synthetic RNA oligomers permit the analysis of polymorphism at ~2% accuracy in up to 40,0000 genes per microarray, which is more than enough for a genome the size of yeast. By the use of hierarchial clustering algorithms, parameterization or profiling methods, gene expression can be "mapped" not unlike the methods of genetic or physical mapping. However, clustered functions are independent from the physical location of the genes but may indicate the co-transcribed compartments and interaction of gene products. The pattern of expression reflects the dynamic network of timing, the physiological and developmental processes. The microarrays may be used to study the co-regulation of genes and the exploration of the global or particular effects of mutation or repression of single genes. Furthermore, the co-expression patterns of pathogens and host cells can also be revealed.

The *spotting method* outlined above uses a single clone for the analysis of each mRNA. The GeneChip Expression (Affymetrix, 3380 Central Expressway, Santa Clara, CA 95051) employs about 16 pairs of specific oligonucleotide probes to interrogate each transcript. (The description given is based on information and images provided by Affymetrix and Gene Microarray Shared Resource, Ohio State University.) The latter procedure is better suited for reducing and identifying non-specific hybridization and background effects and therefore it is more sensitive and more accurate. Basic principles can be outlined as follows. A *target sequence* of ~600 nucleotides of gene is selected from a public database. About 16 to 20 *probes* are generated to match the sequences of the database. A pair of 25-base probe is of two types.

The *perfect match* (PM) is entirely identical to a tract of the DNA. The *mismatch probe* (MM) is the same as the PM except for a single mismatch in the middle. The rationale for using a MM probe is to have a control for non-specific hybridization. These probes are synthesized on a GeneChip, a small glass plate. The synthesis is a light-directed process (photolitographic fabrication) yielding large quantity of accurate probes at an economical manner (see Fig. M65).

From the biological sample, biotinylated mRNA is prepared, fragmented by heat and applied to a *probe array*, which is 1.28 × 1.28 cm glass surface held on a small tray. Hybridization is allowed for about 16 hours. The extent of hybridization is ascertained by the fluorescence intensity as detected by a laser scanner.

From a small segment of image, the intensity of the hybridization (i.e., the identity of the target and the probe pair) may have the alternative matches shown below. White corresponds to high hybridization intensity, black no measurable hybridization signal. Intermediate shades or colors correspond to intermediate signals. When the Perfect Match and the Mismatch signals are inconsistent, the hybridization is not detectable (see Fig. M66).

The Microarray Suite (MAS) software manages the Affymetrix GeneChip experiments. The relative expression of a transcript is determined by the difference between each probe pair (PM minus MM) and averaging the difference over the whole probe set (*Avg Diff*). The term *Abs Call* for a probe set is a qualitative measure based on three different determinations collected by MAS 4.0. The Absolute Call for a probe set can be "A" for non-detectable, "M" for marginal and "P" for present. *Diff Call*, difference call is the qualitative call for a probe set representing the outcome of one array set compared to another. There are five possibilities:

"I" increased, "MI" moderately increased, "NC" no change, "MD" moderately decreased and "D" decreased call.

Microarray hybridization is an essential tool in proteomics, for the study of genetic networks that

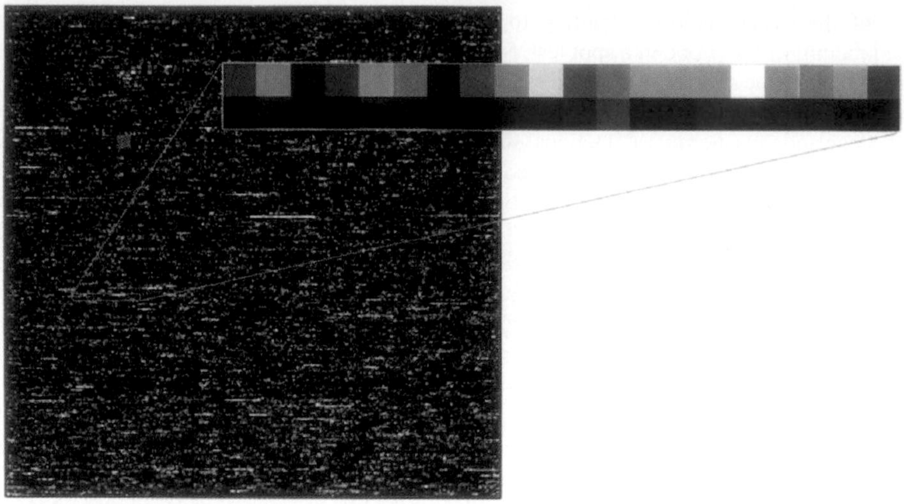

Figure M65. DNA chips. Different 25-mer oligonucleotide probes in an area of 1.28 × 1.28 cm. The array contains probe sets for more than 6,800 human genes, and the image was obtained after overnight hybridization of an amplified and labeled human mRNA sample. Image by courtesy of Affymetrix Inc. and from Fields S et al 1999 Proc Natl Acad Sci USA 96:8825; [Copyright by the National Academy of Sciences, U.S.A. 1999]

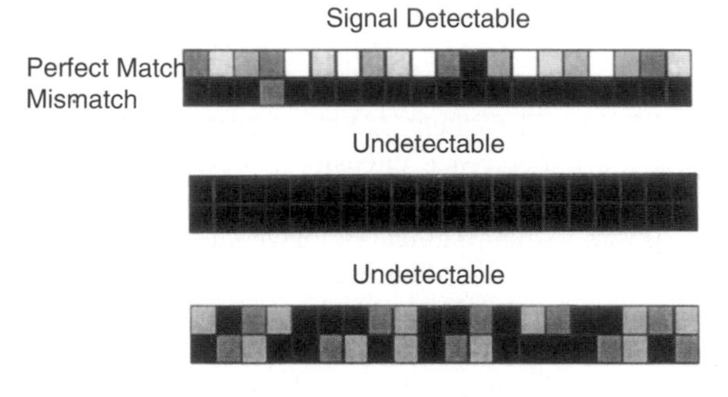

Figure M66. Hybridization signal evaluation

play an important role in development and in studies of the reaction to drugs. The impact of stress or cancer on gene expression may become amenable to interventions. The simultaneous effect of various drugs on a family of genes can be assessed. The availability of these methods is completely revolutionizing information gathering on the expression of genes, and may be exploited for the study of the simultaneous expression of many genes involved in a pathway or related pathways. A number of factors may affect the efficiency of hybridization, such as temperature, base composition (A=T versus G≡C) and even at the same composition, the actual sequences of the bases, secondary structures, sequence length, distribution of mismatches, etc. Unexplained variations have also been seen with the present-day—steadily improving—technologies. An alternative approach for monitoring gene expression pattern optically measures light emission of transcriptional gene fusions of luciferase structural genes to promoters of operons or regulons. Gene expression (mRNA), and protein levels can be analyzed under a variety of conditions and an integrated picture can be derived of metabolic networks. The same basic principle has also been applied to an array of animal cells, which are transfected with a variety of cDNAs and the expression of the transgenes that affect cellular physiology is monitored. ►SAGE, ►laser desorption MS, ►electrospray MS, ►two-hybrid method, ►DNA chips, ►epistasis, ►synexpression, ►operon, ►regulon, ►luciferase, ►transcriptional gene fusion, ►base-calling, ►IMAGE clones, ►genetic network, ►global single cell reverse transcription-polymerase chain reaction, ►macroarray analysis, ►genomics, ►interrogation genetic, ►cluster analysis, ►support vector machine, ►activity-based protein profiling, ►TOGA, ►protein arrays, ►lymphochips, ►SOM, ►tissue microarray, ►light directed parallel synthesis, ►MAGE, ►MELANIE, ►MIAME, ►Atlas™ human cDNA, ►linkage; Ermolaeva O et al 1998 Nature Genet 20:19; Nature Genet 21(1) Supplement Nature Biotechn 17:974; Scherf U et al 2000 Nature Genet 24:236; Van Dyk TK et al 2001 Proc Natl Acad Sci USA 98:2555; Ideker T et al 2001 Science 292:929; Ziauddin J, Sabatini DM 2001 Nature [Lond] 411:107; Stoughton RB 2005 Annu Rev Biochem 74:53; computations: Quackenbush J 2001 Nature Rev Genet 2:418; Zhao LP et al 2001 Proc Natl Acad Sci USA 98:56321; Schulze A, Downward J 2001 Nature Cell Biol 3:E190; Yang YH, Speed T 2002 Nature Rev Genet 3:579; Nature Genet Suppl 2002 32:461–552; transcript concentration analysis: Held GA et al 2003 Proc Natl Acad Sci

USA 100:7575; statistical analysis of gene expression during development in microarrays: Storey JD et al 2005 Proc Natl Acad Sci USA 102:12837; data analysis: Allison DB, Page GP 2006 Nature Rev Genet 7:55; [note Fig. 1 y-axis should have been $-Log_{10} \times (\frac{[p-value]}{1-[p-value]})$]; extensions of the classical microarrays: Hohenheisel JD 2006 Nature Rev Genet 7:200; GE OMNIBUS: http://www.ncbi.nlm.nih.gov/geo; http://www.ebi.ac.uk/microarray/; http://www.ebi.ac.uk/arrayexpress; http://www.biochipnet.de/; http://www.genomethods.org/caged; microarray standards: http://www.mged.org/; comparison of published microarray data to experimenter's results and can help in interpreting the biological and medical meaning of the results: http://depts.washington.edu/l2l; gene expression pattern analysis tool: http://gepas.bioinfo.cipf.es/; gene expression pattern comparison software: http://ihome.cuhk.edu.hk/~b400559/arraysoft_image.html; Affymetrix and Applied Biosystem microarray analysis tool: https://carmaweb.genome.tugraz.at/; prokaryotic microarray–gene ontology tool: http://www.jprogo.de; tools and software: http://smd.stanford.edu/; copy number variations: aCGH; microarray expression pathway: http://bioinfoserver.rsbs.anu.edu.au/utils/PathExpress/.

Microarray Image Analysis: It performs gridding (location of spots on the slide), segmentation (differentiates the pixels between spots and background), information extraction (calculation of fluorescence intensity from the segmentation information). ►microarray hybridization; http://ihome.cuhk.edu.hk//~b400559/arraysoft_image.html.

Microbe: Small organisms like the eukaryotic fungi, algae and protozoa and the prokaryotic blue-green algae, bacteria and viruses. The 2–3 billion microbial species exhibit enormous genetic diversity and are largely unknown. They are frequently associated with disease. The identification of particular microbes may be quite difficult by morphology or culture methods. Recently, broad-range polymerase chain reaction, representational difference analysis and expression library screening for specific sequences became feasible and greatly improved identification by molecular means. ►RDA, ►PCR broad-base, ►microarray hybridization, ►microbiome; http://pbil.univ-lyon1.fr/emglib/emglib.html; microbial genomes: http://microbialgenome.org; http://www.biocrawler.com/encyclopedia/Microorganism; microbial orthologous groups: http://mbgd.genome.ad.jp/.

Microbial Safety Index: The logarithm of the reciprocal number of microbes that have survived after a procedure of sterilization. Sterilization, in principle, is the destruction of all infective agents but, in practice, a small fraction of $\sim 10^{-6}$ may survive the treatment.

Microbiome: The collective genomes of the very large number of microbes that inhabit the human or animal body or a particular space. Their populations of 500–1000 species were estimated to exceed that of the number of somatic cells of the body. The human gut may include 10^{11} foreign cells per mL of proximal colonic content. There are 2 to 4 million genes encode metabolic capacities absent from the human proteome and thus play important roles in human nutrition (Xu J et al 2003 Science 299:2074). One study found 128 phylotypes (belonging to a taxonomic phylum) of microbes in the human stomach and the microbial community was different than that in the mouth or in the esophagus (Bik EM et al 2006 Proc Natl Acad Sci USA 103:732). The human oral cavity may harbor ~500 different microbial species (Paster BJ et al 2001 J Bacteriol 183:3770). The human intestinal flora of bacteria is transmitted during the birth process (vaginal delivery) from the mother and maintained in the offspring. It may be different, however, between marital partners although some changes may occur depending on toilet use, antibiotic treatment, shift in diet or infection with different parasites. Some lateral gene transfer among the various bacteria within the gut does occur yet it does not entirely homogenize the populations. The microbiome inhabiting different mammalian species is different although some similarities may be established by environmental factors and selection. The host immune system may be somewhat responsive to these microbes yet does not eliminate them. When gut microbiome was reciprocally transplanted between germ-free mouse and zebrafish the transplanted community resembled its origin but in the lineage, it resembled the abundance of the recipient host indicating a selective pressure of the host gut habitat (Rawls GF et al 2006 Cell 127:423).

According to a new estimate there appears to be 10^{16} prokaryotes in a ton of soil, much more than the estimated number of stars in the galaxy (10^{11}). These estimates may not be entirely accurate because they are based on some data and hypothesis and mathematics (Curtis TP, Sloan WT 2005 Science 309:1331). The secretion through/by the roots—that varies according to species—determines the microbe communities in the rhizosphere (Mark GL et al 2005 Proc Natl Acad Sci USA 102:17454). ►gut, ►metagenomics, ►oral bacterial film, ►vagina, ►inclusive fitness, ►Paneth cell; gastrointestinal microbial flora in metabolism and health: Nicholson JK et al 2005 Nature Rev Microbiol 3:431; plant pathogenic RNA viruses in human feces: Zhang T et al 2006 PLoS Biol 4(1):e3; review of human intestinal microflora: Ley RE et al 2006 Cell 124:837; human microbiome: http://bioinformatics.forsyth.org:7070//homd/; human oral microbiome: http://bioinformatics.forsyth.org/old_homd/.

Microbodies: ►peroxisomes (see Fig. M67)

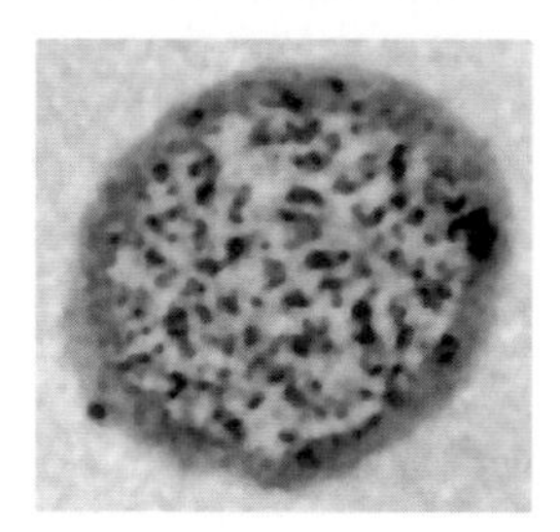

Figure M67. Microbody

Microcalorimetry: Detects the minute change in heat energy resulting from molecular interactions, associations, and dissociations. Two procedures are used: *differential scanning calorimetry* (DSC) to record the changes in temperature while proteins are unfolding and *isothermal titration calorimetry* (ITC) to record the changes in temperature while the solutions are mixed. ►immunoprecipitation, ►surface plasmon resonance

Microcell: A micronucleus, a piece of chromatin, a chromosome or a few chromosomes surrounded by a membrane. Microcells are suitable vehicles for the transfer of blocks of genetic material or entire chromosomes between organisms. ►transchromosomic, ►micronucleus

Microcell Hybrid: Contains a single (e.g., human) chromosome in a complete other (e.g., mouse) genome. The hybrid, using deletions, may permit the identification of specific functions associated with segments of the critical chromosome. ►somatic cell genetics, ►transchromosomic; Cao Q et al 2001 Cancer Genet Cytogenet 129(2):131.

Microcephaly (19.13.1, 15q15, 9q33, 1q31, 13q12.2): Abnormal smallness of the head (brain volume ~400 cm^2); generally involves mental retardation. It is a condition due to various genetic and environmental causes (e.g., X-ray, exposure of the fetus to heat [febrile]). The incidence of the autosomal recessive form is about 2.5×10^{-5}. A deletion of chromosomal segment 1q25-q32 or mutation in 1q31–32 may be the cause of severe cases. The 8p23 locus encodes the microcephalin protein regulating brain size and it continuously undergoes adaptive positive selection for larger brain (Evans PD et al 2005 Science 309:1717). It has been suggested that the microcephalin gene was acquired by introgression into modern human genomes 37,000 years ago and was an important factor of evolution of brain size (Evans PD et al 2006 Proc Natl Acad Sci USA 103:18178). Some of the suggestions concerning correlations among genes for microcephalin and ASPM (abnormal spindle-like, microcephaly-associated gene; human locus 1q31) and evolution of intelligence are controversial (Currat M et al 2006 Science 313:172).

Recurrence risk among sibs was estimated to be 0.19 but the risks might vary depending on the cause of the defect. It is generally accompanied by other abnormalities. Microcephaly with normal intelligence characterizes the autosomal recessive Nijmegen breakage syndrome. The latter is associated with chromosomal instability, immunodeficiency and radiation sensitivity. Microcephaly with immunodeficiency is caused by mutation in (Cernunnos) NHEJ protein factor (Buck D et al 2006 Cell 124:287). ►mental retardation, ►hydrocephalus, ►brain, ►craniofacial dysostosis (Crouzon syndrome), ►cerebral gigantism, ►NHEJ, ►macrocephaly; Pattison L et al 2000 Am J Hum Genet 67:1578; Bond J et al 2005 Nature Genet 37:353.

Microchaetae (pl.): Hairs of insects. ►chaetae for illustration

Microchannel Plate Detector: Analytical equipment for the detection of radioactive labels in proteins separated by 2-dimensional gel electrophoresis. ►two-dimensional gel electrophoresis; Richards P, Lees J 2002 Proteomics 2:256).

Microchimerism: The survival of donor cells in the recipient after transplantation of foreign tissues or organs. Preliminary evidence indicates the rate of survival of pig cells in humans appears to be as low as 10^{-5}. ►xenograft, ►xenotransplantation, ►PERV; Johnson KL et al 2001 Arthritis Rheum 44:2107.

Microchip: Integrated computer circuitry. ►DNA chip

Microchromosomes: Very small chromosomes of the avian genome uniform in size and rich in GC content. They replicate ahead of the larger (macro) chromosomes. Their density of genes appears very high probably because their introns are short. ►intron, ►human artificial chromosome; McQueen HA et al 1996 Nature Genet 12:321.

Microcin: An enterobacterial heptapeptide (Acetyl-Met-Arg-Thr-Gly-Asn-Ala-Asp-X) inhibiting protein synthesis where the first amino acid is acetylated and X is an acid labile group. Microcin J25 obstructs the nucleotide uptake channel of the RNA polymerase (Mukhopadhyay J et al 2004 Mol Cell 14:739; Adelman K et al 2004 Mol Cell 14:753). It is encoded by 21 bps and thus appears to be one of the smallest translated genes. ►gene size

Microcinematography: A time-lapse motion photorecording of living material. Successive frames delayed in real time, are projected at normal speed giving the sensation as if the events, movements of cells or chromosomes, etc., would have taken place at an accelerated time sequence. Thus, for instance the progress of mitosis that requires 1–2 hours can be seen in motion, in minutes. ►phase contrast microscopy; Nomarski, Matter A 1979 Immunology 36(2):179.

Micrococcal Nuclease: Prepared from *Staphylococcus aureus* degrades DNA (with preference for heat-denatured molecules) and RNA and generates mono- and oligonucleotides with 3′-phosphate termini.

Microconidia: Small uninucleate conidia. ►macroconidium, ►conidia, ►fungal life cycles

Microcytotoxicity Assay: Detects allelic variants of MHC proteins. Cells are usually labeled by the green fluorescein diacetate and exposed to two different antibodies. The immunologically reacting cells are lysed and their nuclei are stained red by propidium iodide. Thus, the two alleles are distinguished in vitro. (See Wahlberg BJ et al 2001 J Immunol Methods 253:6).

Microdeletion: The loss of a segment too short to be visualized by light microscopy. Comparative genomic hybridization can detect submicroscopic alteration of 100 kb in the chromosome. Conventional karyotyping can detect changes only of 5–10 million bases. In 10% of mentally retarded individuals, 540 kb to 12 Mb de novo alterations were identified (Bert B et al 2005 Am J Hum Genet 77:606). ►karyotype, ►comparative genomic hybridization

Microdose: Generally 1% of a therapeutic dose of a drug, labeled by a radioactive tracer to facilitate imaging (by positron emission tomography or accelerator mass spectrometry) without toxicity to the treated individual. The microdose permits detection of absorption, distribution, metabolism and excretion of the drug (Lappin G, Garner RC 2003 Nature Rev Drug Discovery 2:233). ►tomography, ►mass spectrum

Microdot DNA: A molecular version of steganography (concealing text) used to communicate secret spy messages. The messages are represented in nucleotide sequences generated by PCR using 20-base forward and reverse primers. The encrypt code may be represented by base triplets, e.g., CGA for the letter A, CCA for B, etc. The sequences within the PCR products can be hidden within the total human DNA or a mixture of DNAs of different organisms from where it can be fished out despite the enormous complexity of the mix. This total mixture, including the secret PCR message in ~10 ng would not occupy a much larger microdot than about a full stop (.) on a filter paper. Primers may permit amplification of the message, and it can be sub-cloned and sequenced. If the cipher is known, the message is readable in English (or any other language) by those who are familiar with the primers but for those who are not it is virtually impossible to decipher it. (See C. Taylor C et al 1999 Nature [Lond] 399:533, ►PCR).

Microelectrodes: Chemical sensors with micrometer or smaller dimension. The potentiometric units can detect changes across chemically selective membranes. The voltametric units are microscopic in size and detect substances on the basis of their oxidation or reduction and can detect charge changes involved in neurotransmission and in other biological systems (Wightman RM 2006 Science 311:1570). ►electrode

Microencapsulation: To envelop cells with a thin, generally spherical, semipermeable polymer film. He capsule protects cell viability, permits diffusion, and may release its content. Chemically and mechanically fragile structure may be used for cell therapy. Macroencapsulation involves the use of a hollow cylinder of selectively permeable, thicker membrane filled with cells suspended in a matrix and sealed. Its utility is similar to microcapsules. It may damage the tissues during manipulations.

Microencephaly: Abnormal smallness of the brain caused by developmental genetic blocks or degenerative diseases. ►microcephaly

Microevolution: A minor variation within species that may lead to speciation. ►evolution, ►macroevolution; Hendry AP, Kinnison MT 2001 Genetica 112:1.

Microfilaments: Actin and myosin containing fibers in the cells serving as part of the cytoskeleton and mediatecell contraction, amoeboid movements, etc. ►cytoskeleton

Microfluidic Digital PCR: It can analyze single cells by multiplex PCR. ►PCR multiplex; Ottesen EA et al 2006 Science 314:1464.

Microfluidics: These systems reduced to micrometer scale behave differently from what we usually see in the physical world. Diffusion, surface tension and viscosity affect the movement of fluids (10^{-9} to 10^{-18} liters) in a different way such as in cells and other microscopic systems. Devices that integrate diverse cellular parameters within sub-nanoliter volumes for the characterization of systems biology and such a medicine is now opening up for experimental approaches. The major merits of microfluidics are the small volumes used, small quantities of reagents are needed, high resolution and high sensitivity in analysis is possible by taking advantage of new developments in microelectronics. Microfluidics has potential for many different applications such as separations coupled to mass spectrometry, high throughput screening and synthesis of drugs, examination of single cells or single molecules and synthesis of labeled compounds for positron emission tomography, development of new medical diagnostic tools, etc. A microfluidic genetic analysis system performs nucleic acid purification through solid-phase extraction, followed by target sequence amplification by PCR and microchip electrophoretic amplicon separation and detection is completed in <30 min. The presence of *Bacillus anthracis* (anthrax) in 750 nl of whole blood from living asymptomatic infected mice and of *Bordatella pertussis* in 1 μl of nasal aspirate from a patient suspected of having whooping cough could be confirmed by the resultant genetic profile (Easley CJ et al 2006 Proc Natl Acad Sci USA 103:19277). ►systems biology, ►massive parallel signature sequencing, ►dielectrophoresis, ►tissue engineering, ►protein synthesis, ►anthrax, ►whooping cough; Werdich AA et al 2004 Lab Chip 4:357; Shaikh KA et al 2005 Proc Natl Acad Sci USA 102:9745; Atencia J, Beebe DJ 2005 Nature [Lond] 437:648; Whitesides GM 2006 Nature [Lond] 442:368.

Microfusion: The fusion of protoplast fragments with intact protoplasts to generate cybrids (generally by electroporation). ►cell fusion, ►somatic hybridization, ►cybrid

Microglia: The mesodermal cells supporting the central nervous system and constitute about 10% of the central nervous system. Microglia are a class of monocytes capable of phagocytosis. They can bind, through scavanger receptors, β-amyloid fibrils (present in Alzheimer plaques) resulting in the production of reactive oxygen species and leading to cell immobilization, and cytotoxicity toward neurons. ►monocyte, ►Alzheimer disease; Giulian D 1999 Am J Hum Genet 65:13; Nakajima K, Kohsaka S 2001 J Biochem 130:169; Fetler L, Amigorena S 2005 Science 309:392.

Microglobulin β_2: The class I MHC α chain is non-covalently associated with this polypeptide, which is not encoded by the HLA complex. The α_3 domain and the microglobulin are similar to immunoglobulins and they are rather well conserved, in contrast to the α_1 and α_2 domains, which are highly variable. Its defect results in renal amyloidosis. ►TAP, ►HLA, ►amyloidosis; Hamilton-Williams EE et al 2001 Proc Natl Acad Sci USA 98:11533.

Micrognathia: Abnormally small jaws.

Microgonotropens (MGT): These are inhibitors of transcription factor—DNA interactions. Transcription factors usually bind to the major groove of the DNA, whereas MGTs associate primarily with the minor groove but to some extent also to the major groove. Efficient MGTs may regulate gene expression and may be of therapeutic value. ►transcription factors, ►DNA grooves; Wemmer DE, Dervan PB 1997 Curr Opin Struct Biol 7:355; White CM et al 2001 Proc Natl Acad Sci USA 98:10590.

Micrograph (photomicrograph, electronmicrograph): The photograph taken through a microscope (light- or electronmicroscope). ►microscopy

Microinjection: A method of delivery of transforming DNA or other molecules into animal or other cells by a microsyringe. This procedure is not considered as a highly efficient method of transformation in plants. An advantage is, however, that the delivery can be targeted to cells but not into chromosomal location unless gene targeting is used. ►transformation animals, ►caged compounds, ►targeting genes, ►gene transfer by microinjection

Microinversions: Short inverted DNA sequenced in the genomes microinversions that occur across all species at a frequency of about once per megabase per 66 million years of evolution. Microinversions have low homoplasy and thus provide ample characters for phylogenetic studies. Algorithms are available for their analysis (Chaisson MJ et al 2006 Proc Natl Acad Sci USA 103:19824). ►inversion, ►evolution of the karyotype, ►homoplasy

Microlesion: In genetic toxicology it denotes microscopically undetectable change in the genetic material.

Micromanipulation of the Oocyte: The penetration of some types of disadvantaged sperms can be facilitated by mechanically opening an entry point through the zona pellucida (a non-cellular envelope of the oocyte) and thus facilitating the penetration of the sperm (PZD). Alternatively, with the aid of a microneedle the sperm can directly be deposited under the zona pellucida (SZI) into the space before the vitellus (egg yolk). ►ART, ►preimplantation genetics, ►in vitro fertilization, ►oocyte donation

Micromanipulator: A mechanical device usually employing glass needles or microsyringes to carry out dissections or injections of cells, while viewed under the microscope.

Micromere (small micromeres): Small cells in the vegetal pole arising from the 8-cell blastomere and giving rise to the coelom. ►vegetal pole, ►blastomere, ►coelom

Micromirror: A spatial light modulator (SLM) that provides excellent resolution, high contrast, brightness, true colors and fast response. Due to these qualities, it is used in imaging nucleic acid microarrays and for many other purposes. It is a type of a liquid crystal display that has been used in computer monitors, high-definition TV, watches and various types of optical devices.

Microneme: The tubular organelle of apicomplexan protozoa that develop into rhoptry at the apex of the organism. ►apicoplast, ►*Plasmodium*, ►malaria, ►rhoptry

Micronucleus: The reproductive nucleus of *Infusoria*, as distinguished from their vegetative macronucleus; also a small additional nucleus containing only one or a few chromosomes in other taxonomic groups. In some organisms, broken chromosomes may be visible as micro-nuclei. ►macronucleus, ►*Paramecium*, ►*Tetrahymena*, ►microcell, ►micronucleus formation as a bioassay, ►protozoa

Micronucleus Formation as a Bioassay: Micronuclei are formed when broken chromosomes, chromosomal fragments fail to be incorporated into the daughter nuclei during cell division. Also, the damage to the spindle apparatus may result in the appearance of micro-nuclei. These phenomena have been exploited for testing mutagenic agents specifically causing these types of genetic damage to animal and plant cells. Such assays can be done in cultured cells but in vivo assay of meiotic plant cells or mammalian bone marrow polychromatic erythrocytes have also been used. ►bioassays in genetic toxicology, ►*Tradescantia*, ►protozoa; Riccio ES et al 2001 Environ Mol Mutagen 38:69, (see Fig. M68).

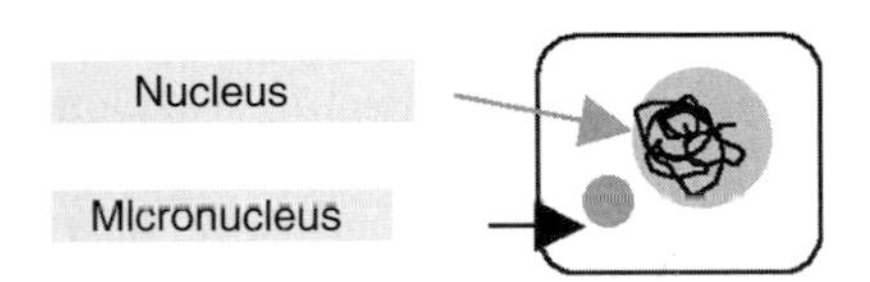

Figure M68. Micronucleus

Micronutrients: They are required for nutrition in small or trace amounts. This class of approximately 40 different compounds includes vitamins, minerals, etc. The deficiency of folate, Vitamins B12, B6, niacin, C and E, etc., have been suggested to be responsible for chromosomal damage and certain types of tumors (See Ames BN 2001 Mutation Res 475:7, and ff. articles in the same issue). Some of the micronutrient deficiencies also cause mitochondrial decay with oxidant leakage and cellular aging and are associated with late onset diseases such as cancer (Ames BN 2006 Proc Natl Acad Sci USA 103:17589).

Microorganisms: *prokaryotic* → bacteria, *eukaryotic* → protozoa, fungi, algae.

Micropenis: It is much reduced in length but the testes are normal in size (see Fig. M69). This condition occurs in several developmental anomalies as part of the syndrome.

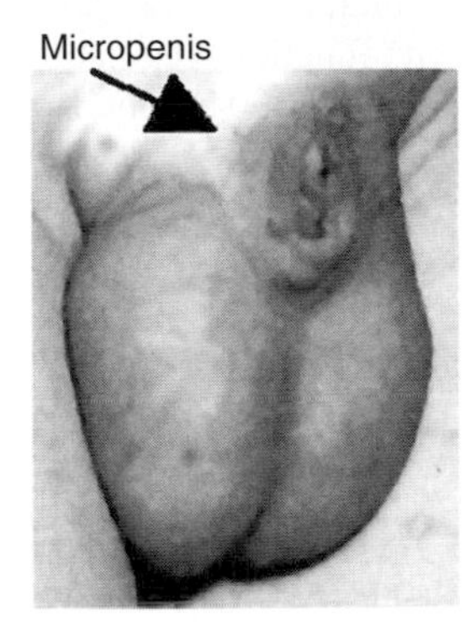

Figure M69. Micropenis

Micropexophagy: The destruction of excess peroxisomes when the need for them is reduced. ►peroxisome, ►pexophagy

Microphthalmia: ►microphthalmos

Microphthalmos (nanophthalmos): Genetically determined (dominant and recessive) forms involve (extreme) reduction of the eye(s) (see Fig. M70). In some instances, it does not involve additional defects. Its frequency in the general population is low, about 0.004%, and the incidence among Caucasian sibs is about 12–14% in the recessive form. Transformation of mice with diphteria toxin genes attached to eye-specific (γ-crystalline, a globulin of the lens) or the pancreas-specific Elastase I, a collagen-digesting enzyme, promoted this developmental condition (ablation). The microphthalmia-associated transcription factor (MITF) mutations transform fibroblasts into melanocytes. Some of the *Mitf* mutations in mice involve partial or entire-body albinism.

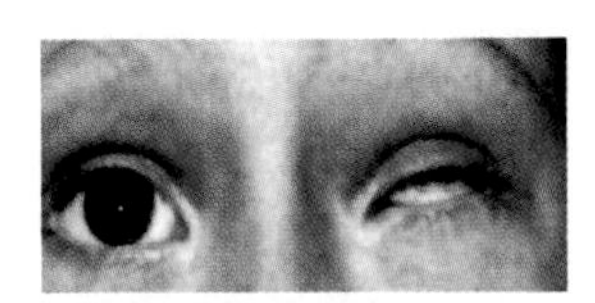

Figure M70. Reduction of eye in microphthalmos

Stimulation of melanoma cells by Steel Factor (*Sl*) activates a MAP kinase that phosphorylates the MITF resulting in the transactivation of a tyrosinase pigmentation promoter (see Fig. M71). In humans, ~5 loci control the condition. Single amino acid replacement mutations at 14q24.3 of the retinal homeobox gene CHX10 have been identified (Heilig R et al 2003 Nature [Lond] 421:601). Dominant colobomatous micropthlamia was assigned to 15q12-q15. ►eye diseases, ►anophthalmos, ►Waardenburg syndrome, ►ablation, ►Kit oncogene, ►Steel factor, ►coloboma; Planque N et al 2001 J Biol Chem 276:29330; Mitf function, regulation and signaling: Steingrímsson E et al 2004 Annu Rev Genet 38:365.

Figure M71. Dominant eyeless *Drosophila*

Micropia: ►copia

Microplasmid (miniplasmid): πVX, ►recombinational probe

Microprocessor: A multiprotein complex, also containing Drosha ribonuclease and it generates microRNAs. ►microRNA, ►Drosha; Gregory RI et al 2004 Nature [Lond] 432:235.

Microprojectile: ►biolistic transformation

Micropropagation: Regeneration plants from somatic, usually apical meristem cells, by in vitro techniques. It may be useful for rapid propagation of rare plants or non-segregating hybrids or secure virus free stocks. ►synthetic seed, ►embryo culture; Evans DA et al (eds) 1983 Handbook of Plant Cell Culture. Macmillan, New York.

Micropyle: A pore of the plant ovule, between the ends of the integuments, through which the pollen tube (sperms) reaches the embryosac. The diagram shows the plant micropyle at the eight-nucleate stage of the embryosac (see Fig. M72). The pores on the ovules of arthropods (and some other invertebrates) serve for the penetration of the sperm. ►embryosac

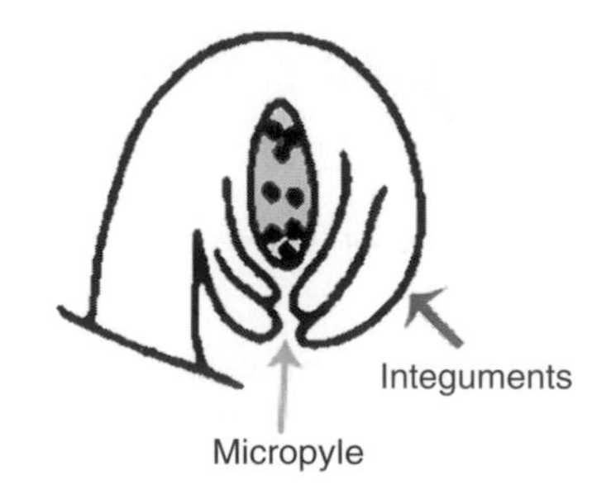

Figure M72. Plant ovule with micropyle

microRNA (micRNA/miRNA): About 18–25 nucleotide long regulatory molecules of diverse sequence. miRNAs have an imperfect match to the 3′ region of the mRNA. Experimental data suggest that association with any position on a target mRNA is mechanistically sufficient for a microribonucleoprotein to exert repression of translation at some step downstream of initiation (Lytle JR et al 2007 Proc Natl Acad Sci USA 104:9667). In the human genome there are > 500 miRNA genes, which target thousands of genes. Apparently, the loss of a single miRNA is not much consequence because there seems to be enough redundancy among miRNAs and in *Caenorhabditis,* three miRNAs (mir-46, mir-84 and mir-241) jointly affect a developmental stage (Abbott AL et al 2005 Dev Cell 9:403). Perfect pairing with 6- to 8-base is not a generally reliable predictor for an interaction of a miRNA. Rather, it can interact functionally with its target site only in specific 3′ UTR contexts (Didiano D, Hobert O 2006 Nature Struct Mol Biol 13:849).

Exportin, a RAN-GTP dependent cargo transporter protein that carries pre-miRNAs (ca. 70 nucleotides) is transported to the cytoplasm. The first Drosha, then Dicer RNase III cut them from one arm of endogenous hairpins and they regulate translation/transcription. For the proper function of Dicer-1 there is a requirement for a double-strand RNA-binding domain protein (DGCR), Loquacious in *Drosophila* or TRBP in humans (Förstemann K et al 2005 PLoS Biol 3(7):e236). Some of the miRNAs apparently do not cause the degradation of their target(s). The let-7 microribonucleoproteins or the Argonaute protein may interfere with recognition of the cap of human mRNAs and inhibit initiation of translation (Pillai RS et al 2005 Science 309:1573). MicroRNA interference can provide viral immunity in *Drosophila* (Wang X-H et al 2006 Science 312:452). Human RISC (RNA-induced silencing complex) associates with a multiprotein complex that contains MOV10 – a translational repressor – and proteins of the 60S ribosomal subunit. This complex contains the anti-association factor eIF6 (also called ITGB4BP or p27BBP), a ribosome inhibitory protein known to prevent productive assembly of the 80S ribosome. Depletion of eIF6 in human cells specifically abrogates miRNA-mediated regulation of target protein and mRNA levels (Chendrimada TP et al 2007 Nature [Lond] 447:823).

Several methods are available for the detection of microRNAs (Jiang J et al 2005 Nucleic Acids Res 33:5394). In plants, it may not be a Drosha but one of the multiple Dicers that assumes this role after moving to the nucleus. Uridine (less frequently adenosine) is added to the 3′ or 5′ ends of the miRNA that enhance the decay of the RNA (Shen B, Goodman BM 2004 Science 306:997). The so-called *seed sequence* (2–7 nucleotides) at the 3′ end of the miRNA initiates the binding to the 5′ end of the mRNA (Bartel DP 2004 Cell 116:181; Brennecke J et al 2005 PloS 3: e85) although other elements may affect their effectiveness. Presence of Argonaute protein(s) is required for initiating the cleavage in the target; the Piwi domain of this protein binds the 5′ end and the PAZ domain attaches to the 3′ end of miRNA. The Piwi domain is a slicer structural homolog of ribonuclease H (Liu J et al 2004 Science 305:1437). miRNA affects translation by attaching to the cap-binding protein eIF-4E or it may accelerate decay of mRNA through the polyA tail (Humphries DT et al 2005 Proc Natl Acad Sci USA 102:16961). After the cut, the RNA is released to the cytoplasm where exosome and Xrn1 exonuclease destroy the fragments (Orban TI, Izaurralde E 2005 RNA 11:459). miRNA and siRNA bound to Argonaute2 may move to the cytoplasm to the site of P bodies, which can destroy them.

A mouse chromosome 12 (human chromosome 14q32) encodes two micRNAs (*mir-136, mir-127*) expressed in the maternally inherited chromosome and act as antisense to retroposon-like gene (*Rtl1*) expressed only from the paternal allele (imprinting, Seitz H et al 2003 Nature Genet 34:261). The Kaposi sarcoma-associated herpes virus encodes 11 distinct micRNAs; some of them are expressed in up to 2,200 copies per cell and facilitate the maintenance of the infected state by suppressing mammalian gene expression (Cai X et al 2005 Proc Natl Acad Sci USA 102:5570). Cellular miRNA provides defense against the primate retrovirus (PFV-1), a foamy virus in humans (Lecellier C-H et al 205 Science 308:557). In *Caenorhabditis* there are about 130 miRNAs and ~800 (regulating 5,300 genes) are expected in humans; among them 53 are unique to primates (Bentwich I et al 2005 Nature Genet 37:766). The Epstein-Barr virus also encodes miRNAs to control both host and viral gene expression (Pfeffer S et al 2004 Science 304:734). Human cytomegalovirus produces miRNAs, which target the major histocompatibility complex class I–related chain B (MICB) gene. MICB is a stress-induced ligand of the natural killer (NK) cell activating receptor NKG2D and is critical for the NK cell killing of virus-infected cells and tumor cells and thus constitutes an immune evasion means (Stern-Ginossar N et al 2007 Science 317:376).

The small RNAs play an important role in the development of animals and plants by inhibiting or degrading specific mRNAs, for, e.g., the mRNAs coding for the TCP transcription factors. miRNAs can affect differentiation by acting on chromatin structure through methylation of the DNAs (Bao N et al 2004 Dev Cell 7:653). Brain morphogenesis in zebrafish requires miRNAs (Giraldez AJ 2005 Science 308:833). Methylation is important for the biogenesis of plant micRNA through the action of the HEN1 protein, which can bind glutathione-S-transferase (Yu B et al 2005 Science 307:932). All small RNAs of plants are apparently modified by a proteinase at the 2′ hydroxyl of the terminal ribose (Ebhardt HA et al 2005 Proc Natl Acad Sci USA 102:13398). Application of the advantages of mRNAs may be enhanced by the use of vectors that can provide regulatable expression of miRNAs. A lentiviral system (pSLIK) permits tetracycline-regulated expression of mRNA-like short hairpin RNAs in any organism, including mouse (Shin K-J et al 2006 Proc Natl Acad Sci USA 103:13759).

miRNAs regulate leaf polarity in dicots and monocot of plants (Juarez MT et al 2004 Nature [Lond] 428:84). Both endogenous and synthetic miRNAs can be effectively targeted to genes controlling morphogenesis of the stem of *Arabidopsis* as well as in tomato and tobacco (Alvarez JR et al 2006 Plant Cell 18:1134; Schwab R et al 2006 Plant Cell 18:1121). About 30% of the human genes appear to be conserved targets for miRNA (Lewis BP et al 2005 Cell 120:15). The various small non-coding RNAs such as the miRNA, RNAi, siRNA are basically identical. In plants, generation of siRNAs most often (but not always) requires two cleavage events (Axtell MJ et al 2006 Cell 127:565). Some miRNAs down regulate a large number of mRNAs and proteins (Lim LP et al 2005 Nature [Lond] 433:769) and coordinately regulate cell-specific target genes. The mir-17–92 cistron (13q31) modulates the expression of several cancer genes (He L et al 2005 Nature [Lond] 435: 828). The core promoter region contains two functional E2F transcription factor-binding sites. miR-17–92 promotes cell proliferation by shifting the E2F transcriptional balance away from the pro-apoptotic E2F1 and toward the proliferative E2F3 transcriptional network (Woods K et al 2007 J Biol Chem 282:2130). Two miRNAs negatively regulate the E2F1 transcription, and the c-Myc proto-oncogene activates the expression of six clustered miRNAs in human chromosome 13 (O'Donell KA et al 2005 Nature [Lond] 435:839). MicroRNA arrays are generally quite different in a variety of cancer cells compared to normal cells. The level of most of the miRNAs increases in solid tumors but others display reduction and they present a signature for cancer. In ovarian cancer 37.1%, in breast cancer 72.8% and in melanoma 85.9% of the miRNA genes displayed copy number changes and presumably functional differences (Zhang L et al 2006 Proc Natl Acad Sci USA 103:9136). Several miRNAs are shared by different tumors (Volinia S et al 2006 Proc Natl Acad Sci USA 103:2257). miRNA-372 and miRNA-373 down-regulate RAS and p53 tumor suppressors and contribute to the formation of germ cell tumors (Voorhoeve PM et al 2006 Cell 124:1169). Statistically significant association was found between the chromosomal location of miRNAs and those of mouse cancer susceptibility loci that influence the development of solid tumors (Cevignani C et al 2007 Proc Natl Acad Sci USA 104: 8017).

Hepatitis C virus replication is apparently aided by human miRNA-122 but it did not affect mRNA translation or stability (Jopling CL et al 2005 Science 309:1577). Several good computational methods exist for the identification and prediction of miRNA targets in different organisms (Krek A et al 2005 Nature Genet 37:495; Rajewsky N 2006 Nature Genet 38 Suppl 1: S8; Miranda KC et al 2006 Cell 126:1203). Several human miRNAS are most abundant in the cells where their mRNA targets are expressed yet several highly abundant mRNAs in the same cells do not have miRNA recognition sites (Sood P et al 2006 Proc Natl Acad Sci USA 103:2746). In human hepatocarcinoma cells, the cationic amino acid transporter 1 (CAT-1) mRNA and reporters bearing its 3′-UTR can be relieved from the miRNA miR-122-induced inhibition under stress conditions (Bhattacharyya SN et al 2006 Cell 125:1111). ►RNAi, ►shRNA, ►stem cells, ►antagomir, ►tncRNA, ►small RNA, ►imprinting, ►RNA non-coding, ►TCP, ►PAZ domain, ►P body, ►eIF-2, ►eIF-4E, ►imprinting, ►Dicer, ►Drosha, ►exosome, ►Xrn1, ►Kaposi sarcoma, ►DiGeorge syndrome, ►microprocessor, ►Simian Virus 40, ►Invader, ►longevity, ►polyadenylation signal; Ambros V 2001 Cell 107:823; Hutvágner G, Zamore PD 2002 Science 297:2056; Carrington JC, Ambros V 2003 Science 301:336; Palatnik JF et al 2003 Nature (Lond] 425:257; Lee Y et al 2003 Nature [Lond] 425:415; MicroRNA Registry: Griffith-Jones S 2004 Nucleic Acids Res 32; Database issue D109, He L, Hannon GJ 2004 Nature Rev Genet 5:522; review: Zamore PD, Haley B 2005 Science 309:1519; genomics: Kim NV, Nam JW 2006 Trends Genet 22:165; review of applications to neurology: Kosik KS, Krichevsky AM 2005 Neuron 47:779; microRNA and disease: Chang T-C, Mendell TC 2007 Annu Rev Genomics Hum Genet 8 (Sept) doi: 101146; review of miRNA and development: Plasterk RHA 2006 Cell 124:877; MicroRNA base: http://www.sanger.ac.uk/Software/Rfam/mirna; MicroRNA Database: http://microrna.sanger.ac.uk/sequences/; http://mirnamap.mbc.nctu.edu.tw/; miRNA target server: http://www-ab.informatik.uni-tuebingen.de/software/welcome.html; miRNA target prediction: http://bibiserv.techfak.uni-bielefeld.de/rnahybrid/; miRNA target sites in QTLs: http://compbio.utmem.edu/miRSNP/; miRNA targets in viruses: http://vita.mbc.nctu.edu.tw/; human miRNA analysis and prediction: http://cbit.snu.ac.kr/~ProMiR2/; animal miRNA organization; co-transcription and targeting: http://www.diana.pcbi.upenn.edu/miRGen; true pre-miRNA identification: http://www.bioinf.seu.edu.cn/miRNA/.

Microsatellite: The mono-, tetra- or hexanucleotide repeats distributed at random (10 – 50 copies) in the eukaryotic nuclear and mitochondrial chromosome, and can be used for constructing high-density physical maps, for the rapid screening for genetic instability, evolutionary relationships, detection of cancer in bladder cells found in the urine, etc. It was initially assumed that the alterations at the microsatellites follow a step-wise pattern and are therefore, very useful for following changes within populations or between populations. However, these expectations were not entirely realized because the variability depends on the length of the arrays such as possible size constraints to the expansion, possible increase in the flanking regions due to mutations, etc. The microsatellite region may expand (most commonly) or contract and thus, result in genetic instabilities. Microsatellite loci may mutate at a frequency of ~0.8% per gamete, more frequently than in somatic cells. Some estimates for eukaryotes are in the 10^{-4} to 10^{-5} range and make them useful for linkage analysis using PIC. The rate of mutation varies according to organisms, the microsatellite length and site, etc. In bacteria, the mutation rate in repeats appears two orders of magnitude higher. In general, longer repeats display higher mutation rates. The length of the microsatellite repeats appears longer in vertebrates than in *Drosophila*. In the majority of the cases, the expansion of the loci does not change the linkage phase of the flanking markers and therefore gene conversion is implicated. In humans they estimated one (≥4 bp) microsatellite per 6 kbp genomic DNA. Three quarters of the repeats are A, AC, AAAN or AG. The CA and TG repeats are distributed at random in the genome and used most commonly for linkage studies. In the human X chromosome, 3 and 4 base repeats occur in every 300–500 bp. The most common poly(A)/ poly(T) repeats are not well suited for genetic studies because of instability at PCR. Large-scale survey indicates that most of the microsatellites were generated as a 3′ extension of retrotranscripts and may serve as pilots to direct integration of retrotransposons. Defective mismatch repair may be caused by mutations in the human gene MBD4/MED1 at 3q21-q22 leading to carcinomas with microsatellite instability. The *perfect microsatellite* represents one repeated motif without any insertion of a different base. The *imperfect microsatellite* has repeats with one base different from the main type. *Interrupted microsatellite* contains several bases repeated that are different from the pattern of the rest of the repeat structure. *Compound microsatellite* includes two or more different types of repeats in adjacent microsatellites. In *Saccharomyces* the poly(AT) sequences are common. Prokaryotes display relatively few repeats and they are mainly poly(A)/poly(T). About 30% of the human microsatellites are conserved in murines. The microsatellites appear to have regulatory roles but some may be transcribed and translated as coding trinucleotides (e.g., the CAG polyglutamine tracts). Some of the untranslated upstream repeat elements are conserved among related species and may serve as enhancers by binding transcription factors. The enhancer activity may be only moderate but may increase or decrease by high copy number. Discrete repeat lengths may carry specific adaptive value in some populations. Diversity in microsatellite repeats may be generated by replicational slippage, unequal crossing over (although the reciprocal products are not always present), by gene conversion, additions and deletions. The microsatellites show a tendency of association with the non-repetitive portion of the genome. Microsatellite repeat length in the vicinity of the vasopressin 1a receptor gene of voles (*Microtus*) is correlated with behavioral traits, such as grooming the females and offspring care by males. Since vasopressin is supposed to affect social behavior, microsatellite length appears regulatory to gene expression (Hammock EAD, Young LJ 2005 Science 308:1630). ►hereditary non-polyposis colorectal cancer, ►mismatch repair, ►trinucleotide repeats, ►minisatellite, ►VNTR, ►PIC, ►DNA fingerprinting, ►PCR, ►cryptically simple sequences, ►unequal crossing over, ►slippage, ►enhancer, ►stutter bands, ►slip-strand mispairing, ►IAM, ►SMM, ►TPM, ►KAM, ►behavior genetics, ►oxytocin; Graham J et al 2000 Genetics 155:1973; Tóth G et al 2000 Genome Res 10:967; Morgante M et al 2002 Nature Genet 30:194; free software for small repeats: http://www.bio.net/hypermail/methods/2000-November/086109.html; microsatellite markers in Taiwanese human populations: http://tpmd.nhri.org.tw; http://tpmd.nhri.org.tw/php-bin/index_en.php; insect microsatellite database: http://210.212.212.8/PHP/INSATDB/home.php.

Microsatellite Mutator (MMP): MMP increases frameshift and other mutations in G-rich microsatellites. ►microsatellite, ►minisatellite

Microsatellite Typing: In case of linkage disequilibrium the inheritance of microsatellite markers can be followed on PCR amplified DNA. Example the CAAT repeat in the human tyrosine hydroxylase genes (chromosome 11p15.5) using the Z alleles displayed the pat-tern of segregation (see Table M6) in the ethidium bromide stained non-denaturing gel. (After Hearne CM, Ghosh S et al 1992 Trends Genet 8:288). ►DNA fingerprinting, ►paternity testing, ►microsatellite, ►PIC; Calbrese PP et al 2001 Genetics 159:839.

Microscopic Polyangiitis: An autoimmune blood vessel inflammation: http://vasculitis.med.jhu.edu/typesof/polyangiitis.html.

Microscopy: The use of a special optical device for viewing objects that are not discernible clearly by the naked eye. ►light microscopy, ►stereomicroscopy, ►resolution, ►fluorescence microscopy, ►multiphoton microscopy, ►second-harmonic imaging microscopy, ►dark-field microscopy, ►phase-contrast microscopy, ►Nomarski, ►confocal microscopy, ►electronmicroscopy, ►atomic force microscopy, ►atom microscopy, ►scanning electronmicroscopy, ►scanning tunneling, ►photoactivated localization microscopy, ►two-photon microscopy, ►ADM, ►FAST, ►Cryo-EM; Sharpe J et al 2000 Science 296:541; Jain RK et al 2002 Nature Rev Cancer 2:266; Stephens DJ, Allan VJ 2003 Science 300:82;

Table M6. Microsatellite typing

Father	Mother	Sib-1	Sib-2	Sib-3	Paternal grandfather	Paternal grandmother	Alleles
■■■	■■■		□□	■■■	■■■		Z
					■■■		4
	■■■	■■■				■■■	12
■■■		■■■		■■■		■■■	16

The ■ symbolize gel bands and the □ stand for homozygosity of the Z allele

Hadjantonakis A-K et al 2003 Nature Rev Genet4:613; http://micro.magnet.fsu.edu/.

Microsomes: The membrane fragments with ribosomes and enzymes obtained after grinding eukaryotic cells and separating the cellular fractions by centrifugation. After about 9,000 x g force (10 min) the microsomes (S9) float, while other cellular particulates sediment. For the Ames genotoxicity bioassay, generally Sprague-Dawley rat livers are used. The rats are previously fed the drinking water with polychlorinated biphenyl (PCB), Araoclor 1254, a highly carcinogenic substance (requires special caution of disposal!) in order to induce the formation of the P-450 monooxygenase activating enzyme system associated with the endoplasmic reticulum. ►Ames test, ►activation of mutagens, ►carcinogen, ►centrifuge

Microspectrophotometer: A spectrophotometer, which can measure monochromatic light absorption of microscopical objects. ►spectrophotometer

Microsporangium: The sac that contains the microsopore tetrad of plants and the micro- spores of fungi and some protozoa. ►microspore

Microspore (small spore): The immature male spore of plants that develops into a pollen grain. ►microspore mother cell, ►microspores, ►microsporocyte, ►meiosis, ►gametogenesis

Microspore Culture: The in vitro method for the production of haploid plants by direct or indirect androgenesis. (See Fig. M73).

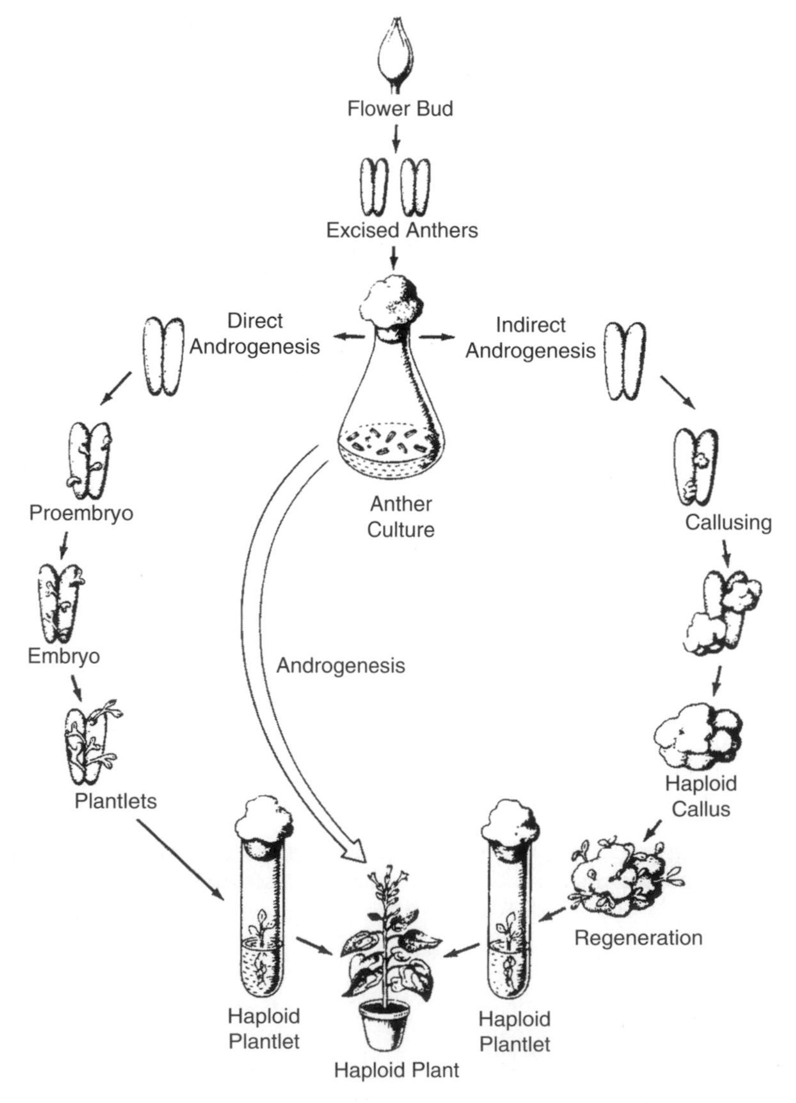

Figure M73. Microspore culture (After Reinert J and Bajaj YPS 1977 Plant Cell, Tissue and Organ Culture. Springer, New York)

Microspore Dyad: The microspore mother cell after the end of the first meiotic division is divided into two haploid cells (the microspore dyad). ►meiosis

Microsopore Mother Cell: ►microsporocyte

Microspores: Haploid products of male meiosis in plants that develop into pollen grains, and the smaller generative cells of lower organisms. ►microsporocyte, ►megaspore

Microsporidia: (Archezoa): It represents an evolutionary branch of anaerobic protists, which diverged from the main line of Eukarya before the acquisition of mitochondria. ►evolution

Microsporocyte: The cell within which meiosis takes place and the microspores develop. ►microspore, ►gametogenesis

Microsoporogenesis: see ►Meiosis in plants, Ma H 2005 Annu Rev Plant Biol 56:393.

Microsporophyll: A leaf on which microsporangia develops in lower plants.

Microsurgery: Dissection or other surgical operations carried out under the microscope, generally with the aid of a micromanipulator. ►micromanipulator

Microtechnique: The procedures used for the preparation of biological specimens for microscopic examination (involving fixation, staining, sectioning, squashing, etc.).

Microtiter Plates: It most commonly contains 96 wells but 384 or even 9600-well (etched glass or silicon wafers) are also used for storage of samples. ►microarray hybridization

Microtome: An instrument that, by means of various (sliding or rotating or rocking) motions, cuts serial thin (usually within the range of 1–20 μm) sections of the embedded or frozen specimens to be examined by light or electronmicroscopy. The cutting edge may be steel or glass. Electronmicroscopy also requires sectioning of the specimens, usually employing a diamond knife. ►embedding, ►sectioning, ►stains, ►smear

Microtubule: Various types of long cylindrical filaments of about 25 nm in diameter within cells built by polymerization of α and β tubulin and actin proteins. Microtubules are hollow tubulin filaments of the spindle apparatus of the dividing nuclei, elements of the cytoskeleton, cilia, flagella, etc. The energy for polymerization is provided by hydrolysis of GTP to GDP. The beginning of the polymerization of the microtubules is called nucleation and in animal cells that begins at the centrosomes. Microtubule elongation is a polarized process. Microtubules move around chromosomes and various protein complexes according to the blueprints of differentiation. Kinetochore-mediated and polar ejection forces (PEF) move chromosomes toward the spindle equator. Kid and KIF-4 proteins are behind PEF (Brouhard GJ, Hunt AJ et al 2005 Proc Natl Acad Sci USA 102:13903). The nerve axon microtubules are oriented with the plus end (their growing point) away from the cell body whereas in epithelial cells they point toward the basement membrane. In fibroblasts and macrophage cells, the microtubules originate in the center of the cell and the plus end face the outer regions. The transport function requires the cooperation of motor proteins (kinesins, dynein) that push the protein complex organelles on the microtubule trails. Microtubules are somewhat unstable molecules and antimitotic drugs such as colchicine or colcemid may block their growth. Taxol (an anticancer extract of yew plants) stabilizes the microtubules and arrests the cell cycle in mitosis. After polymerization the microtubules undergo modification, e.g., a particular lysine of α-tubulin may be acylated and tyrosine residues removed from the carboxyl end. Microtubule-associated proteins (MAP) mediate the "maturation" of microtubules. MAPs aid the differentiation of nerve axons and dendrites that are loaded with microtubules. Microtubules move various organelles such as the chromosomes during nuclear divisions in the cells with assistance of the proteins kinesin and dynein. The cortical protein Kar9 and the Bim1/EB1 microtubule stabilizing protein control the positioning of the microtubules. The EB1 is also a suppressor of the *adenomatous polyposis coli* tumors. The cilia and flagella involved in movements are built of bundles of microtubules. The microtubule protein complexes bind ATP at their "head" and associate with organelles by their "tail". The "plus" end of microtubules where tubulin subunits are added rapidly and the "minus" end where the addition is more slow. The microtubules usually are in a dynamic instability, i.e., they alternate between extension and shortening but they may remain also in equilibrium. There are several protein factors that move the tubulins and associated structures. In prophase, the duplicated centrosomes are separated by the bimC, KAR3 and cytoplasmic dyneins (450–550-kDa, motility 75 μm/minute). The bimC family members (120–135-kDa and with 1–2 μm/min motility) have different names in the different species (KLP61F and KRP_{130} in *Drosophila*, Cin8 in *Saccharomyces*, cut7 in *Schizosaccharomyces*). The KAR3 family (65–85-kDa, motility 1–15 μm/min) is involved also in spindle stabilization during metaphase. Cdc14 mediates the localization of the microtubule-stabilizing proteins for successful anaphase separation (Higuchi T, Uhlmann F 2005 Nature [Lond] 433:171). During prometaphase, the

M

microtubules are captured by the kinetochore with the assistance of the MCAK (mitotic centromere associated kinesin). The chromosomes congregating at the metaphase plane are moved by KIF4 (140-kDa) and related proteins that are also involved in chromosome alignment during metaphase. During anaphase, CENP-E and the KAR3 family of proteins propel the movement of the chromosomes toward the poles. The MKLP, bimC and cytoplasmic dynein proteins mediate the elongation of the spindle fibers. ►actin, ►tubulin, ►Cdc14, ►kinesins, ►Pac-Man model, ►dynein, ►mitosis, ►meiosis, ►centrosome, ►spindle, ►centromere, ►tau, ►katanin, ►filaments, ►dynamic instability, ►treadmilling, ►polyposis adenomatous intestinal, ►colchicine, ►taxol, ►vinblastin; Nogales E 2000 Annu Rev Biochem 69:277; Downing KH 2000 Annu Rev Cell Dev Biol 16:89; Popov AV et al 2001 EMBO J 20:397; Schuyler SC, Pellman D 2001 Cell 105:421; Lloyd C, Hussey P 2001 Nature Rev Mol Cell Biol 2:40; McIntosh JR et al 2002 Annu Rev Cell Dev Biol 18:193; Howard J, Hyman AA 2003 Nature [Lond] 422:753; Westermann S, Weber K 2003 Nature Rev Mol Cell Biol 4:938; microtubule capture by kinetochore: Kotwaliwale C, Biggins S 2006 Cell 127:1105.

Microtubule Associated Proteins (MAP): It controls stability and organization of microtubules. ►microtubule, ►kinesin, ►dynein

Microtubule Organizing Center: The areas in the eukaryotic cells from where the microtubules emanate and grow such as the mitotic centers (poles) that give rise to the mitotic spindle (see Fig. M74). The microtubule-binding protein TPX2 binds Aurora A kinase and targets it to the mitotic spindle and this stabilizes microtubules connecting to the spindle pole (Özlü N 2005 Dev Cell 9:237). ►spindle, ►mitosis, ►MTOC, ►centrosome, ►Aurora

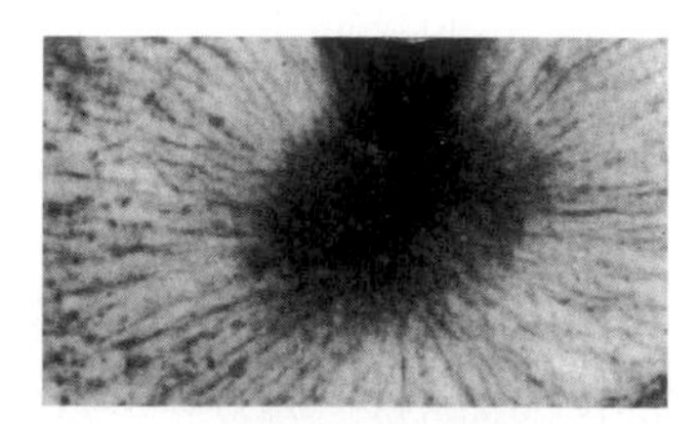

Figure M74. Microtubule organizing center

Microvilli: These are small emergences on the surface of various cells and increase the surface of the cells. The microvilli of the chorion may be sampled for amniocentesis in prenatal genetic examinations. ►amniocentesis

Microwave Radiation: Electromagnetic radiation in the $\sim 10^9$ to $\sim 10^{11}$ nm range. Genetic effects are difficult to separate from the heat effects. In the 2–3 Ghz range ambiguous mutagenic effects were reported and the mutagenicity of the radiations could not be confirmed. Microwave radiation can damage cell membranes.

MIDA1: A helix-loop-helix associated protein of mammals. It is inactive in erythroleukemic cells. ►eryhtroleukemia, ►zuotin, ►DNA-binding protein domains, ►helix-loop-helix

Midbody (midzone): After division, microtubule fragments (midbodies) may be detected in animal cells connecting the two daughter cells by a refringent structure (see Fig. M75). PRC1 is a midzone-associated microtubule bundling protein, a substrate of Cdk. Cdk-mediated phosphorylation of PRC1 keeps it in inactive, in monomeric state. During metaphase–anaphase transition, PRC1 is dephosphorylated and that promotes polymerization of PCR1 and lead to the formation of the midbody (Zhu C et al 2006 Proc Natl Acad Sci USA 103:6196). The proteins in the midbody are essential for chromosome segregation and cytokinesis. MudPIT analysis revealed the presence of 577 different proteins. Of the 172 proteins disrupted by RNAi analysis, 58% interfered with cytokinesis (Skop AR et al 2004 Science 305:610. ►microtubules; MudPIT, Piel M et al 2001 Science 291:1499; photomicrograph is the courtesy of Drs. Kevin Sullivan and D. W. Cleveland; red color indicates the centromeres of the blue HeLa chromosomes; green tubulins.

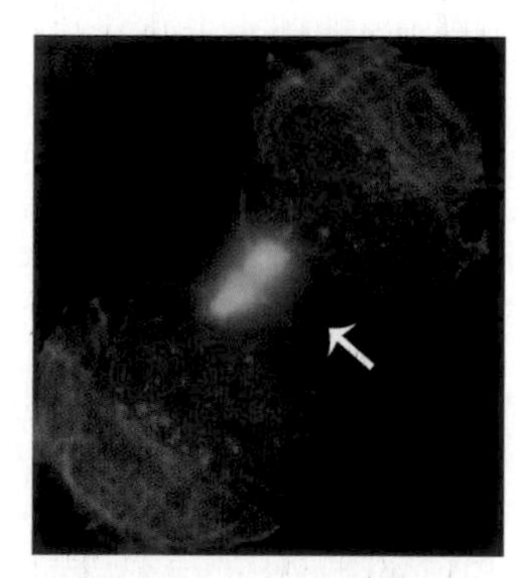

Figure M75. Midbody

Midbrain: The middle part of the brain. Degeneration of the motor neurons in this region may be responsible for Parkinson disease. ►brain human, ►Parkinson disease

Middle Lamella: The material (pectin mainly) that fills the intercellular space in plants.

Middle Repetitive DNA: It is made up of relatively short repeats dispersed throughout the genome of (higher) eukaryotes. ►SINE, ►LINE, ►redundancy

Midgut: The middle portion of the alimentary tract of insects and other invertebrates.

Midkine: ►pleiotrophin

Midparent Value: ►breeding value [midpoint]

Midpoint: ►breeding value

Midzone: A network of antiparallel, interdigitating, nonkinetochore microtubules, which initiate and complete cytokinesis. The PRC1 cyclin-dependent protein kinase is moved to the midzone by the Kif4 motor protein and both are essential for cytokinesis (Zhu C, Jiang W 2005 Proc Natl Acad Sci USA 102:343). ►cytokinesis, ►midbody, ►PRC1, ►Kif

MIF (macrophage inhibitory factor): A pro-inflammatory pituitary factor; it may override glucocorticoid-mediated inhibition of cytokine secretion. MIF binds to cytoplasmic protein Jab1, which induces the phosphorylation of c-Jun and through it, the activity of AP-1. Jab1 also helps the degradation $p27^{Kip1}$. ►cytokines, ►glucocorticoid, ►septic shock, ►AP1, ►Jun, ►p27; Froidevaux C et al 2001 Crit Care Med 29 Suppl 7:S13.

Mifepristone (11-[44-(dimethylamino)phenyl-17-hydroxy-17-(1-propynyl)-(11_{β},17_{β})-estra-4,9-dien-3-one): An antiprogestin, targeted at the progesterone receptors. This receptor family includes the glucocorticoid, mineralocorticoid, androgen, estrogen and vitamin D receptors. It is used as a birth-control drug. ►RU486 and the others mentioned, ►hormone receptors, ►breast cancer; Ho PC 2001 Expert Opin Pharmacother 2:1383; Zalányi S 2001 Eur J Obstet Gynecol Reprod Biol 98(2):152.

Mighty Mouse: ►insulin-like growth factors

Migraine: A neurological anomaly causing recurrent attacks of headaches, nausea, light and sound-avoidance. It may or may not be preceded by an aura (subjective sensation) and it appears to be controlled at chromosome 4q21 (Björnsson Á et al 2004 Am J Hum Genet 73:986). Migraine may be triggered by dietary factors (monosodium glutamate, tyramine and phenylethylamine occurring in chocolates, citrus fruits or certain cheeses) in persons with low levels of phenolsulphotransferase activity. (This enzyme catalyzes the conjugation of sulfate of catecholamines and phenolic drugs.) Its duration may be from minutes to days. The sexual migraine pops up during sexual activity, commonly at or near orgasm. It may affect females (18–24%) more frequently than males (6–12%) to a very variable degree. Migraine is relatively rare in children (~4%) and its onset is usually begins after age 30. In males, it usually ceases after age 45 but in women it may continue well beyond menopause. It is generally attributed to multiple genes. Familial hemiplegic migraine (FHM) is however, a rare autosomal dominant condition coded in human chromosome 19p13 as is episodic ataxia type 2. CADASIL has been localized to the same chromosomal area and involves migraine. It appears that the basic defect is in a brain-specific Ca^{2+} ion channel α-1 subunit translated from 47 exons of the CACN1A4 gene. Spinocerebellar ataxia type 6 and episodic ataxia are allelic to the latter gene. Some migraines are associated with mtDNA-encoded MELAS syndrome. Another migraine susceptibility locus appears to be at Xq24. ►ion channels, ►spinocerebellar ataxia, ►mitochondrial diseases in humans, ►CADASIL, ►CACN1A4, ►serotonin; Guida S et al 2001 Am J Hum Genet 68:759; Wessman M et al 2002 Am J Hum Genet 70:652.

Migration: ►gene flow

Migration of DNA in Gels: It is affected by molecular size, configuration of the macro-molecule, concentration of the support medium (agarose, polyacrylamide), voltage, changing direction in the electric field, base composition, presence of intercalating dyes, buffer, etc.

Migration Inhibition Factor: MIF is a lymphokine. ►lymphokines, ►lymphocytes, ►immune system

Migration in Populations: ►Gene flow

Migration, Nuclear: Mediates polarization of cell division and provides direction for cell growth. (See Bloom K 2001 Curr Biol 11:R326)

Mik1: A mitotic kinase and an inhibitor of Cdc2. Mik1 accumulates in S phase and may mediate the transition from S phase to mitosis. ►cell cycle, ►Cdc2, ►Wee1

MIL Oncogene: The avian representative of the RAF murine oncogene, a protein serine kinase; it is also related to the murine leukemia virus (MOS). ►oncogenes, ►RAF, ►MOS

MILC (maximum identity length contrast): A statistical method for genetic analysis of multifactorial diseases. It looks for an excess of identity of parental haplotypes transmitted to the affected offspring as compared to non-transmitted haplotypes. ►haplotype; Bourgain C et al 2002 Ann Hum Genet 66:99.

Miliary: Minute lesions (resembling millet seeds)

Miller Units: ►β-galactosidase (activity measurement)

M

Miller-Dieker Syndrome: Characterized by smooth brain (lissencephaly I, LIS1, AFAH1B1), more like in the early fetus, defects in other internal and external organs, mental retardation and death before age 20. Apparently, deletions in area 17p13.3 are responsible for the recessive phenotype. It is a cell-autonomous disease inhibiting neuronal migration. The LIS1 protein contains a C-terminal seven-blade-β-propeller domain whereas the N-terminal fragment includes the Lis homology (LisH) motif, which is widespread in more than 100 eukaryotic genes yet its function is not known (Kim MH et al 2004 Structure 12:987). The deletion always involves protein 14–3–3ε and when two sites are simultaneously affected, the severity of the disease increases. (Toyo-oka K et al 2003 Nature Genet 34:274). Other genes also involving hydrocephaly and severe brain lesions may cause Lissencephaly II (DCX, Xq22.3-q23-linked doublecortin). The gene PAF (lipid platelet activating factor) encodes a subunit of brain platelet-activating acetylhydrolase. ►deletion, ►head/face/brain defects, ►malformations, ►Walker-Warburg syndrome, ►Walker-Wagner syndrome, ►lissencephaly, ►HIC; Cardoso C et al 2003 Am J Hum Genet 72:918.

Millets (*Eleusine, Pennisetum*): Arid climate grain crops (see Fig. M76). The cultivated *E. coracana* is 2n = 4x = 36, tetraploid. *P. americanum* is 2n = 2x = 14 diploid. The common millet (*Panicum miliaceum*) is an old grain crop; 2n = 4x = 36. The foxtail millet (*Setaria italica*) is 2n = 2x = 18 is mainly a hay crop.

Figure M76. Millet

Milroy Disease: ►lymphedema hereditary

MIM: ►mitochondrial import

MIME (Multipart Internet Mail Extensions): This was used before the Extensible Markup Language (XML) was developed ►XML

Mimicry: The process and result of protective change in the appearance of an organism that makes it resemble the immediate environment for better hiding or imitating the features of other organisms that are distasteful or threatening to the common predators. The blister beetle *Meloe franciscanus*, which parasitize nests of the solitary bee *Habropoda pallida*, cooperate to exploit the sexual communication system of their hosts by producing a chemical cue that mimics the sex pheromone of the female bee. Male bees are lured to larval aggregations, and upon contact (pseudocopulation), the beetle larvae attach to the male bees. The larvae transfer to female bees during mating and subsequently are transported to the nests of their hosts. To mimic the chemical and visual signals of female bees effectively, the parasite larvae must cooperate, emphasizing the adaptive value of cooperation between larvae (Saul-Gershenz LS, Millar JG 2006 Proc Natl Acad Sci USA 103:14039). It has been observed that bats quickly learned to avoid the noxious tiger moths first offered to them and associated the warning sounds with bad taste. They then avoided a second sound-producing species regardless of whether it was chemically protected or not. A subset of the red bats subsequently discovered however, the palatability of the Batesian mimic (Barber JR, Conner WE 2007 Proc Natl Acad Sci USA 104:9331). ►Batesian mimicry, ►*Biston betularia*, ►industrial melanism, ►Müllerian mimicry, ►molecular mimics, ►pheromone; Mallet J 2001 Proc Natl Acad Sci USA 98:8928.

Mimicry, Macromolecular: Some proteins may mimic nucleic acids in shape, structure and, even to some extent, function. Such mimicry may protect a phage from the host restriction endonucleases. (See Nissen P et al 2000 EMBO J 19:489; Walkinshaw MD et al 2002 Mol Cell 9:187).

Mimicry, Structural: Pathogenic microorganisms produce virulence factors that are molecular mimics of host proteins. This way they may evade the host defense system. ►molecular mimics; Stebbins CE, Galán JE 2001 Nature [Lond] 412:701.

Mimics: Individuals that develop mimicry as a form of adaptation and evasion of predators. Also genes, which control practically the same phenotype yet they are not allelic.

Mimiotope (mimeotope): ►mimotope

Mimivirus: A very large double-stranded DNA virus (1,181,404 bp) in a 400-nanometer icosahedral particle. It contains 1,262 putative open reading frames; 10% of which encode proteins of known functions. Many of the genes are involved in protein synthesis-processing machinery. ►icosahedral, ►open reading frame, ►protein synthesis; Raoult D et al 2004 Science 306:1344; Suzan-Monti M et al 2005 Virus Res 117:145.

Mimotope: A conformational mimic of an epitope without great similarity in amino acid residues or as a response to microbial anti-DNA antibodies. ►epitope; Wun HL et al 2001 Int Immunol 13(9):1099; Mullaney BP et al 2002 Comp Funct Genomics 3:254.

***Mimulus* sp**: (2n = 28, 2n = 16) are hermaphroditic/outcrossing species (~160) of plants with about two-months generation time and high (~2000) seed output (see Fig. M77). Its genome size is ~500 Mb; total map length ~2,000 cM in *M. guttatus* (2n = 28). (See http://www.biology.duke.edu/mimulus/).

Figure M77. *Mimulus luteus*

Mineral Corticoid Syndrome (ame): It causes hypertension without overproduction of aldosterone. This syndrome is activated by cortisol. 11β-hydroxysteroid dehydrogenase converts cortisol to cortisone and thus, activates the mineralcorticoid receptor. The patients are deficient in this enzyme, which is inhibited by glycyrrhetinic acid (enoxolone [$C_{30}H_{46}O_4$], present in licorice). The mineral corticoid receptor and other steroid receptors regulate the activity of many genes. ►nuclear receptor, ►hypertension, ►aldosteronism, ►Liddle syndrome; Pearce D et al 2002 J Biol Chem 277:1451.

Mineral Requirements of Plants: The 9 macro elements (H, C, O, N, Ca, Mg, P, S) and 7 micro elements (Cl, B, Fe, Mn, Zn, Cu, Mo). Under some circumstances, other elements may also be beneficial. Using Inductively Couple Plasma (ICP) spectroscopy, 18 essential and nonessential macro- and microelements in plant tissues could be quantitatively determined. Mutations were also identified at about four dozen loci of *Arabidopsis* that control the metabolic role of these elements. It has been estimated that 2–4% of the genome of this plant involves the control of nutrient and trace element control, the "ionome" (Lahner B et al 2003 Nature Biotechnol 21:1215). ►embryo culture

Mini Mu: The deletion variants of phage Mu cloning vehicles that still carry the phage ends, a selectable marker and replicational origin. ►Mu bacteriophage

Miniaturization: The development of analytical tools and methods for the study of small samples, single cells, proteins or nucleotides, etc. (See Sauer S et al 2005 Nature Rev Genet 6:465, sequencing 25 million bases in four-hour run with 99% or better accuracy: Margulies M et al 2005 Nature 437:326).

Minicell: DNA-deficient bodies surrounded by cell wall (in bacteria). Since they have no DNA, they cannot incorporate labeled precursors either into RNA or protein. In case the mini-cells descended from parents with plasmids, they may contain DNA and can thus, make RNA and direct protein synthesis depending on the nature of the DNA they carry. ►maxicells

Minichromosome: In eukaryotic viruses (SV40, polyoma virus) it is the histone-containing, small nucleosome-like structure of genetic material; also in eukaryotes, an extra chromosome with extensive deletions. Such minichromosomes can be generated by the insertion of human telomeric sequences $(TTAGGG)_n$ between the centromere and the natural telomere and by eliminating the sequences distal to the insertion point. Human mini Y chromosomes have about 32.5 to 4 Mb compared with the normal Y chromosome of 50–75 Mb. Minichromosomes may also be generated by removal of non-essential distal genes from each arm. Neocentromeres may serve as well as regular centromeres for human artificial chromosomes. These minichromosomes permit the analysis of the role of different sections of the chromosomes and possibly, it may become feasible to extend the analysis of mammalian chromosomes in yeast cells. Artificial chromosomes may be exploited as large capacity cloning vectors. ►human artificial chromosome, ►neocentromere, ►MCM1; Saffery R et al 2001 Proc Natl Acad Sci USA 98:5705.

Minichromosome Maintenance Factor: ►MCM

Mini-F: The basic replicon of the bacterial F plasmid. ►F plasmid, ►replicon

Minigels: Minigels are used for the separation of small quantities (10 to 100 ng DNA), in small fragments (<3-kb) on about 5 × 7.5 cm slides (10 to 12 mL agarose), in a small gel box (ca. 6 × 12 cm) for 30 to 60 min at 5 to 20 V/cm. ►electrophoresis

Minigene: Some of its internal sequences are deleted in vitro before transfection. ►transfection

Minihelix of tRNA: It consists only of the TΨC arm and the amino acid acceptor arm and may be aminoacylated by some aminoacyl-tRNA synthetases according to the rule of the operational RNA code. The minihelix is considered to be the most ancestral part of the tRNA. ►operational RNA code, ►transfer

RNA, ►tRNA, ►aminoacyl-tRNA synthetase; Nordin BE, Schimmel P 1999 J Biol Chem 274:6835.

Minimal Genome Size: The smallest number of genes required for survival (replication) in a specific milieu for a free-living organism. Based on transposon inactivation experiments of the genome *Mycoplasma genitalium*, the estimate was ~265–350 bp. Some viruses have only 4 genes but they are genetic parasites. Among free-living microorganisms, *Plagibacter ubique*, an abundant, heterotrophic marine proteobacterium has the smallest genome size of 1,308,799 bp (Giovannoni SJ et al 2005 Science 309:1242). ►genome, ►gene number, ►*Mycoplasma*

Minimal Medium: Provides only the minimal (basal) menu of nutrients required for maintanence and growth of the wild type of the species. ►complete medium

Minimal Promoter: It includes the most essential sequences to facilitate transcription of genes (basal promoter) without some other regulatory elements such as enhancer, transcriptional activator. ►promoter, ►basal promoter, ►enhancer, ►transcriptional activators

Minimal Residual Disease (MRD): Even after apparently successful chemotherapy of cancer, some residual cancerous cells may persist. Their detection is important to develop treatment for the prevention of relapse. Molecular techniques (PCR) are applicable for this goal. ►PCR; Kim YJ et al 2002 Eur J Hematol 68:272.

Minimal Tiling Path: A tightly overlapping set of bacterial vector clones, suitable for sequencing eukaryotic genomes. ►tiling

Minimization: ►crossing over

Minimum Description Length (MDL): MDL is the principle frequently used in characterizing macromolecular sequence information. The best principle is that uses the least number of bits for the theory and the data. L(T) indicates the complexity of the theory by the number of bits that encode the theory. L(D/T) is the number of bits required to define the data in connection with the theory and reveals the consistency of the data with the theory. ►bit

Minimum Evolution Methods: These methods use an estimate of a branch length of an evolutionary tree construct on the basis of pair-wise distance data calculated by various mathematical algorithms. The most plausible tree should be that which provides the smallest sum of total branch length. ►evolutionary distance, ►evolutionary tree, ►least square methods, ►four-cluster analysis, ►neighbor joining method, ►algorithm

Miniorgan (neo-organ) **Therapy**: The *ex vivo* genetically modified group of cells capable of synthesis of immunotoxins, angiogenesis inhibitors, hormones, ligands or other proteins and enzymes delivered to target cells/tissues in vivo with the purpose of correcting acquired or genetic disorders. In case the promoters can be regulated, the supply of the gene product(s) can be adjusted according to the need. ►retroviral vectors, ►immunotoxin, ►angiogenesis, ►promoter, ►gene therapy, ►cancer gene therapy, ►*ex vivo*, Bohl D, Heard JM 1997 Hum Gene Ther 8:195; Rosenthal FM, Kohler G 1997 Anticancer Res 17(2A):1179.

Miniplasmid: ►πVX microplasmid, ►recombinational probe

Miniprep: A small-scale quick preparation of DNA from plasmids or from other sources. (See Ferrus MA et al 1999 Int Microbiol 2(2):115).

Minireplicon: A vector consisting of a pBR322 replicon, a eukaryotic viral replicational origin (SV40, Polyoma) and a transcriptional unit. These vectors can be shuttled between *E. coli* and permissive mammalian cells. Also, deficient replicons containing only the replicational origin. ►replicon, ►shuttle vector; Roberts RC, Helinski DR 1992 J Bacteriol 174:8119.

Minisatellite: In eukaryotic genomes short (14–100-bp) tandem, highly polymorphic repeats occur at many locations with repeat arrays of 0.5–30 kb. In forensic work, the minisatellites are used for DNA fingerprinting. They are supposed to be products of replicational errors (slippage) and localized amplifications. Their high variability is probably due to frequent unequal crossing over and duplication-deficiency events. Gene conversion may also expand or contract minisatellites. These sequences are highly variable. The variations are associated with diabetes mellitus and various types of cancers. The minisatellite mutation rate in the human germline was estimated as ~5.2% although it may vary at different loci and may be different in the two sexes. They are used as RFLP or PCR probes in physical mapping or for characterization of populations. In contrast to humans mutations in general, minisatellite intra-allelic mutation rate is quite low at 5×10^{-6} per sperm (Bois PRJ et al 2002 Genomics 80:2). ►microsatellite, ►SINE, ►VNTR, ►DNA fingerprinting, ►RFLP, ►PCR, ►small-pool PCR, ►MVR, ►unequal crossing over, ►trinucleotide repeats, ►MVR, ►IAM, ►SMM, ►TPM, ►gene conversion10:899; Vergnaud G, Denoeud F 2000 Genome Res 10:899.

Minisegregant: Bud-like extrusions of animal cells with pinched-off DNA.

MINK: MAP kinase kinase kinase; also represented as MAPKKK or Ste 11. ►Ste

MinK: A 15 K protein which in association with other proteins forms potassium ion channels.

Minocycline: A tetracycline type antibiotic capable of passing the blood-brain barrier and it is a neuroprotective by inhibition of caspases. It may delay the progression of neurodegenerative diseases (Zhu S et al 2002 Nature [Lond] 417:74).

Minor Allele Frequencies (MAF): These frequencies have a role in the susceptibility to a particular disease. Generally, the frequencies are less than 0.1 and because of their relatively small effect, the odds ratio is less than 1.3. Therefore, their identification—at an acceptable level of statistical significance—requires large populations ($>10^4$).

Minor Grove of DNA Double Helix: It is marked by an arrow in the Fig. M78.

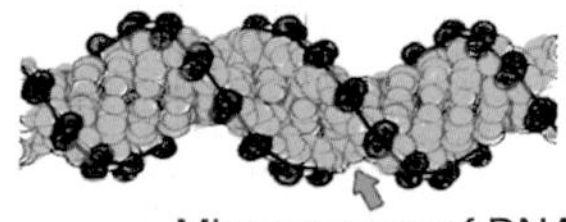

Figure M78. Minor groove of DNA double helix

Minor Histocompatibility Antigen: It has some role in immune reactions but it is not coded in the HLA region. ►HLA, ►MHC, Dazzi F et al 2001 Nature Med 7:769.

Minos: A 1,775-basepair transposable element of *Drosophila hydei* with 255-bp inverted terminal repeats and with two non-overlapping open reading frames. ►transposable elements animals; Zagoraiou L et al 2001 Proc Natl Acad Sci USA 98:11474.

Minus END: of microtubules or actin filaments is less liable for elongation. ►plus end

Minus Position of Nucleotides: It indicates the upstream distance from first translated triplet of the transcript of a gene. ►triplet code, ►RNA polymerase

Minutes: Approximately 60 dominant mutations in *Drosophila* that slow down the development of heterozygotes and is lethal when homozygous. The phenotypes vary but generally the bristles are reduced (see Fig. M79). Some mutants increase somatic crossing over. Minutes have several chromosomal locations. They are defective in ribosomal proteins. (See Lambertson A 1998 Adv Genet 38:69; Marygold SJ et al 2005 Genetics 169:683).

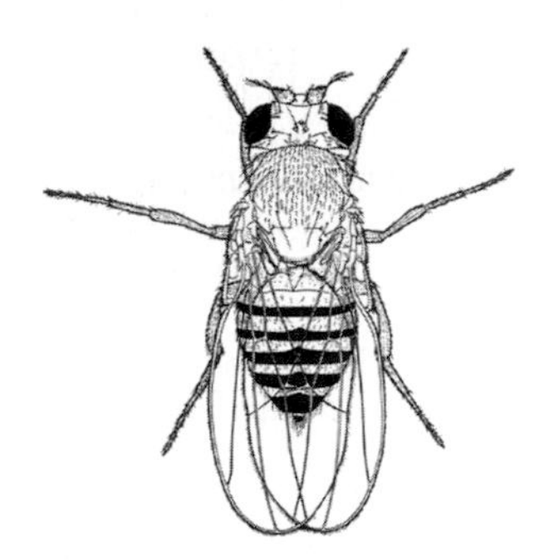

Figure M79. Minute (M1)n fly

MIP (methylation induced premeiotically): The mechanism of gene silencing by causing apparently a stall of the RNA polymerase before completing the transcription of fungal genes. ►RIP

MIP-1α (macrophage inflammatory protein; CCL3L1/CC chemokine ligand 3-like): A chemokine mediating virus and other microbial induced inflammation protein related to RANTES. It belongs to the family of FK506-binding proteins. ►RANTES, ►blood cells, ►FK506, ►peptidyl-prolyl isomerases, ►acquired immunodeficiency syndrome; Matzer SP et al 2001 J Immunol 167:4635.

MIPS (Munich Information Center for Protein Analysis): A database of functional genomes and proteomes. (See Mewes HW et al 2002 Nucleic Acids Res 30:31; http://mips.gsf.de).

MIR: Mammalian-wide interspersed repeat, ~12–30 × 10^4 copies/primate genome. They may regulate the expression of genes, alternative splicing, polyadenylation sites and evolution. ►redundancy; Matassi G et al 1998 FEBS Lett 439:63.

MIR: ►ILT

Mirabilis jalapa (four-o'clock): Ornamental plant and early object of inheritance, 2n = 58 (see Fig. M80).

Figure M80. *Mirabilis*

miRE: MicroRNA responsive element ►microRNA

miRNA: ►MicroRNA

miRNP: A complex of ribonuclease containing miRNA and several proteins are involved in processing of interfering RNAs. ►microRNA, ►RNAi, ►Dicer, ►RISC; Morelatos Z et al 2002 Genes Dev 16:720.

Mirtron (pre-microRNA intron): Debranched introns, which mimic the structural features of pre-miRNAs to enter the miRNA-processing pathway without Drosha-mediated cleavage. ►microRNA, ►Drosha, ►intron; Ruby JG et al 2007 Nature [Lond] 448:83.

MIS (Müllerian inhibitory substance): ►Müllerian duct, ►gonads

Mis12/Mtw1: A protein controlling kinetochore orientation in yeast. ►kinetochore

Miscall: ►base-calling

Miscarriage: The loss of pregnancy. It occurs spontaneously in about 15–20% of the pregnancies at least once during the life of human female. Repeated miscarriage before 20 weeks of pregnancy occurs in 0.5–2% of the women and it may be caused by environmental, extrinsic factors (infections) or by uterine anomalies, hormonal problems, chromosomal aberrations, autoimmune reactions or other immune problems (Rh), metabolic dysfunction (folic acid deficiency, hyperhomocystinemia, defects in nitric oxide synthase, etc.). Although most commonly genetic (chromosomal) anomalies of the fetus is responsible for the miscarriage, recent observations point to an essential role of mutation or deletion in the HLA-G (Ober C et al 2003 Am J Hum Genet 72:1425). (See terms under separate entries, ►abortion spontaneous; Tempfer C et al 2001 Hum Reprod 16:1644).

Miscegenation: Sexual relations between partners of different human races. In the majority of human tribes, marriage was generally limited within the tribes, however, marriage by "capture" existed in ancient societies where the conquerors in war abducted females. Miscegenation was applied particularly to marriage between whites and blacks in the United States and in some South-American societies. There is no genetic justification against interracial marriage. Although marriage between Blacks and Orientals was not prohibited by any law, marriage between Caucasians and Blacks was unlawful in about 15 states of the U.S. until 1967 when the Federal Court ruled that the choice to marry resides with the individual. Racial and social (cast) discrimination in marriage still exists in many underdeveloped countries and in backward communities. ►mulatto, ►mestizo, ►racism, ►marriage; Hulse FS 1969 J Biosoc Sci Suppl 1:31.

Mischarged tRNA: It is linked to wrong amino acids. ►aminoacyl-tRNA synthetase, ►protein synthesis

Miscoding: ►ambiguity in translation

Misconduct, Scientific: The fabrication, falsification or embellishing of data, the use of inadequate statistics or techniques, making unjustified conclusions, plagiarism, omitting relevant facts concerning the experimental data or the literature, misrepresenting previous or competing publications, claiming undue credit or any other unethical behavior. A survey reported in 2005 (Martinson BC et al Nature [Lond] 435:737) indicates that 33% of the (several, thousands) scientists supported by National Institutes of Health USA anonymously admitted bad behavior although only 0.3% confessed falsifying data but 27.5% felt guilty of inadequate record keeping. The corrosive behavior was substantially (15.5%) promoted by inadequate institutional policies. ►ethics, ►publication ethics, ►scientific misconduct

Misdivision of the Centromere: A vertical rather than longitudinal division; generates two telochromosomes from one bi-armed chromosome. The telochromosomes may open up to isochromosomes and may undergo misdivision again forming telochromosome (see Fig. M81). ►centromere, ►heterochromatin; Darlington CD 1939 J Genet 37:341; photo at telochromosome.

Misexpression of a Gene: ►ectopic expression

Misincorporation: During DNA replication a normal nucleotide or an analog is placed into the growing strand, at a site where correct pairing (e.g., A = T, G≡C) is not available. Such an event may lead to base substitution in the following step of replication, e.g.,

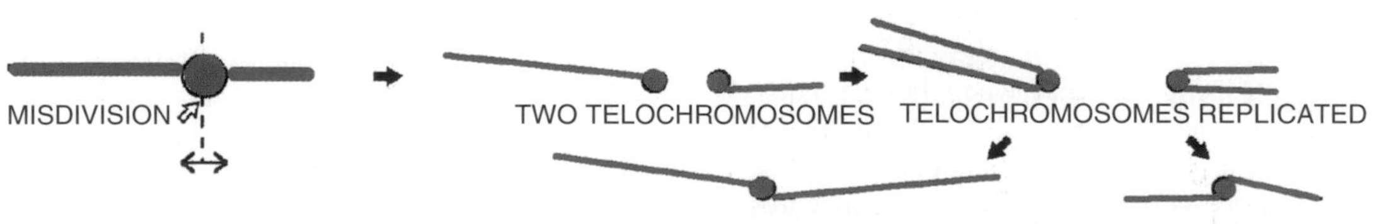

Figure M81. Misdivisional cycles

A + T→ A:**C**→ **G≡C**, resulting in mutation unless mismatch repair corrects it. Misincorporation can also occur during transcription of the DNA into RNA. ►base substitution, ►mismatch repair, ►DNA repair, ►error in transcription; Freese E 1963 p 207. In: Molecular Genetics. Taylor JH (Ed.) Academic Press, New York.

Misinsertion: DNA polymerase binds to a correctly matched 3′ end of a primer where right and wrong dNTP substrates compete for insertion and occasionally the wrong may succeed resulting in misinsertion. ►DNA editing, ►dNTP

Mis-Localization: Gene products are normally localized according to a specific pattern within the tissues, cells or at subcellular sites. Deviations from the normal pattern may lead to disease. (See Sutherland HGE et al 2001 Hum Mol Genet 10:1995).

Mismatch: One or more wrong (non-complementary) base(s) in the paired nucleic acid strands. Mismatches are generally identified (localized) by the use of flurochrome labeled probes or they can be labeled by nanoparticle gold and analyzed on the basis of reduction silver (in the presence of hydroquinine at pH3.8) by ordinary flatbed scanners with high sensitivity. Chemical cleavage of mismatch (CCM) is a hydroxylamine/osmium tetroxide-based (HOT) analysis of sequence variability in DNA. This procedure has been improved by adapting fluorescence techniques.

A bulky rhodium (a platinum type metal) complex can bind to mismatched and matched sites in the DNA of the oligonucleotide 5′-(dCGG*A*AATT***C***CCG) 2–3′. At the AC mismatch site, the structure reveals ligand insertion from the minor groove with ejection of both mismatched bases and destabilized mispairs in DNA may be recognized. This unique binding mode contrasts with major groove intercalation, observed at a matched site, where doubling of the base pair rise accommodates stacking of the intercalator. Mass spectral analysis reveals different photocleavage products associated with the two binding modes in the crystal, with only products characteristic of mismatch binding in solution. This structure, illustrating two clearly distinct binding modes for a molecule with DNA, provides a rationale for the interrogation and detection of mismatches (see Fig. M82). DNA polymerases generate mismatches at the rate of 10^{-4} to 10^{-5} per base pair at the nucleotide insertion step. These mistakes are typically reduced to 10^{-7} per base pair per replication by exonucleases associated with the DNA polymerase and are further reduced to 50- to 1,000-fold by the mismatch repair machinery. Deficiencies in mismatch repair increase the rate of mutation and subsequently the risk of developing cancer (Pierre VC et al 2007 Proc Natl Acad Sci USA 104:429). ►transition mismatch, ►transversion mismatch, ►mismatch repair, ►DNA grooves; Ellis TP et al 1998 Hum Mutat 11:345.

ATCG A GTCA
TAGC C CAGT
Mismatch

Figure M82. Mismatch

Mismatch Extension: DNA polymerase binds to either a matched or mismatched primer end, and the mismatch is extended in replication. ►mismatch

Mismatch of Bases: Non-complementary bases in a (hetero)duplex DNA. ►mismatch

Mismatch Repair (MMR): An excision repair that removes unpaired or mispaired bases and replaces them through unscheduled DNA synthesis with correct base pairs. The long-patch mismatch repair may have to correct tracts of hundreds of nucleotides. Mutations in *E. coli* gene *mutL* (homodimer, subunits ~68 kDa) and *mutS* (~95 kDa subunits) increase genetic instabilities and in yeast, defects in *PMS1, MLH1* and *MLH2* may increase genetic instability 100 to 700-fold because of deficient mismatch repair. In yeast Msh6 (S1036P) the mutant complex, which binds the mispaired base is defective in ATP-induced sliding clamp formation and assembly of the ternary complex with Mlh1–Pms1 and prevents the access of other repair systems to access to the defect. In another mutation of Msh6 (G1142D), which binds mispaired sequences and is defective in the ATP-induced sliding clamp formation but permits the assembly of the Mlh1–Pms1 complex yet either prevents the access of other repair system or hinders Mlh1–Pms1 to activate the other repair systems (Hess MT et al 2006 Proc Natl Acad Sci USA 103:558). mutS may also inhibit both homologous and homeologous recombination (Pinto AV et al 2005 Mol Cell 17:113). Mismatch repair deficiency may enhance mutation rate by 1–2 orders of magnitude, depending on the background (Ji HP, King M-C 2001 Hum Mol Genet 10:2737). The consequences of deletions of RTH1, encoding a 3′→5′ exonuclease are similar. The G/T binding protein (GTBP, 100K) and hMSH2 (160 K, homolog of bacterial mismatch-binding protein) are essential for mismatch recognition in human cells. In fission yeast the mismatch repair enzyme, exonuclease I, reduces mutation rate. Defects in the bacterial methyl-directed mismatch repair system may also enhance mutability (Burdett V et al 2001 Proc Natl Acad Sci USA 98:6765). The repair system is directed to DNA strands by methylation of adenine in the d(GATC) sequences. Since newly synthesized strands are not methylated, this is the criterion for recognition by the repair system (MutH, MutL, MutS, ATP). The mismatch may be located kilobases away from the d(GATC) tract. Interaction of this complex with heteroduplexes activates the MutH-associated endonuclease that responds to an initial sequence discontinuity or nick. Excision from the 5′ side of the mismatch requires the RecJ exonuclease or exonuclease VII and digestion from the 3′-end needs exonuclease I and also a helicase to unwind the strands because these proteins hydrolyze only single strands. RecJ and exo VII require that the unmethylated GATC would be downstream from the defect. Exo I usually does not have this precondition. The replacement synthesis requires DNA polymerase III. Other polymerases do not work in this system. In the eukaryotic systems, the MutL homologs are PMS1 in yeast and PMS1, PMS2 in mammals. The yeast PMS1 system corrects G-T, A-C, G-G, A-A, T-T and T-C mismatches but C-C or long insertion/deletion sequences were barely repaired if at all. The human repair system fixes 8–12 pair mismatches and C-C. An MutS homolog in mammals is called GTBP (G-T binding protein). Defects in PMS1, MLH1, MSH2 and [MSH3] enhance mutation rate in the yeast mitochondria, increases somatic mutability up to three orders of magnitude at certain loci (e.g., canavanine resistance), destabilizes $(GT)_n$ sequences. Mutations in the yeast Rth1 5′→3′ exonuclease also increases mutability, particularly of plasmid-encoded genes. MutS function is required for the normal progression of meiosis and the maintenance of the female gonads. The yeast homologs of the bacterial MutS and MutL not only correct errors in replication but block recombination between not entirely homologous (homoeologous) sequences. Heterocomplexes of the mismatch repair genes have important role in meiotic recombination (Snowden T et al 2004 Mol Cell 15:437). Defects in mismatch repair result in increased frequency of recombination, mutation and indirectly on evolution. PCNA and related proteins mediating DNA replication are essential for repair synthesis. Cadmium is a potent inhibitor of mismatch repair and greatly increases mutation (Jin YH et al 2003 Nature Genet 34:326). Double-strand DNA breaks may be repaired by gene conversion. Mismatch repair deficiency may increase the susceptibility to methylating drugs but not for others (e.g., cisplatin) or ionizing radiation. ►unscheduled DNA synthesis, ►DNA repair, ►glycosylase, ►PCNA, ►MRD, ►incongruence, ►gene conversion, ►double-strand break, ►Walker boxes, ►mutator genes, ►microsatellite, ►homeologous recombination, ►hereditary non-polyposis colorectal cancer, ►Lynch cancer families, ►Muir-Torre syndrome, ►gonads, ►excision repair, ►short patch repair, ►slippage, ►Huntington's chorea, ►cisplatin, ►morphogenic; Nakagawa T et al 1999 Proc Natl Acad Sci USA 96:14186; Oblomova G et al 2000 Nature [Lond] 407:703; Lamers MH et al 2000 *ibid* 701; Buermeyer A et al 1999 Annu Rev Genet 33:533; Harfe BD, Jinks-Robertson S 2000 Annu Rev Genet 34:359; Evans E, Alani E 2000 Mol Cell Biol 20:7839; Junop MS et al 2001 Molecular Cell 7:1; Wang H, Hays JB 2002 J Biol Chem277:26136, 26143; recombination: Sugawara N et al 2004 Proc Natl Acad Sci USA 101:9315; Kunkel TA, Erie DA 2005 Annu Rev Biochem 74:681.

Misoprosto (Misoprostol): A synthetic prostaglandin E_1 analog that inhibits secretion of gastric acid (as an antacid pill) but it also induces uterine contractions and can cause abortion. It is used also for non-surgical termination of pregnancy. ►prostaglandins, ►RU486

Mispairing: Ooccurs when the homology between paired nucleotide sequences is imperfect and illicit binding takes place between nucleotides.

Misreading: A triplet is translated into an amino acid different from its standard coding role. ►ambiguity in translation

Misrepair: ►DNA repair, ►SOS repair

Misreplication: ►error in replication

Missense Codon: Inserts an amino acid different from that encoded by the wild type codon at the site in the polypeptide chain. ►nonsense codon

Missense Mutation: A change in DNA sequence that results in an amino acid substitution in contrast to mutations in synonymous codons where the change involves only the DNA but not the protein. ►nonsense mutation

Missense Suppressor: Eenables the original amino acid to be inserted at a site of the peptide in the presence of a missense mutation. ►suppressor tRNA; Benzer S, Weisblum B 1961 Proc Natl Acad Sci USA 47:1149.

Missing Genes: In the archaebacterium, *Methanococcus janaschii*, four of the 20 aminoacyl-tRNA synthetases were not detected in the completely sequenced genome. It has been hypothesized that glutamine and asparagine are incorporated into polypeptides as transamidated derivatives of glutamate and aspartate. Furthermore, cysteine was assumed to be inserted as a trans-sulfurated serine and $tRNA^{Lys}$ synthetase function was replaced by a protein quite dissimilar to known aminoacyl-tRNA synthetases. ►aminoacyl-tRNA synthetase; Bult CJ et al 1996 Science 273:1058.

Missing Link: The lack of transitional forms in evolution in between two species, of which one was/is supposed to have descended from the other. In 2006, a 385–359 million-year old fossil of a fish that lived in the Late Devonian period in Canada was reported to have been found. It had bony structures in the fins and other morphological features indicating a missing link to tetrapods (Daeschler EB et al 2006 Nature [Lond] 440:757). ►tetrapod, ►geological–►evolutionary time periods

Missing Self Hypothesis: One of the functions of natural killer lymphocytes (NK) is to recognize and eliminate cells that do not express class I MHC (major histocompatibility complex) molecules. ►killer cells, ►MHC; Ljunggren HG, Karre K 1990 Immunology Today 11:237.

Missplicing: Incorrect splicing. ►splicing

Mistranslation: ►misreading, ►ambiguity in translation

***MIT*.**: The general designation of mitochondrial point mutations.

Mitchurin: A nineteenth-twentieth century Russian plant breeder who contributed a large number of improved varieties, mainly fruits, to the Russian and then to the Soviet agriculture. His career is often compared to that of the American Burbank. There was a very important difference, however, Burbank produced over 100 varieties to the USA agriculture without governmental pretensions that he made novel basic scientific discoveries. Mitchurin, on the other hand, published undigested and misunderstood theories and became the official forefather of lysenkoism, a state-supported charlatanism. He drew sweeping conclusions on the basis of tenuous experimental technology. ►lysenkoism, ►Burbank, ►acquired characters inheritance, ►graft hybridization

MITE (mariner insect transposon-like element, miniature inverted-repeat transposable element): A 1,457-bp DNA sequence in the vicinity of MLE containing a 24-bp tract with homology to Mos1 and supposedly responsible for the high frequency of unequal crossing over resulting in a duplication appearing as the Marie-Charcot-Tooth disease, and in the complementary deletion appearing as HNPP. Similar MITE elements occur also in rice, maize, sugarcane and other plants and control transcription. ►Marie-Charcot-Tooth disease, ►HNPP, ►unequal crossing over, ►hot spot, ►Mos1, ►mariner, ►heartbreaker, ►MLE; Casa AM et al 2000 Proc Natl Acad Sci USA 97:10083; Jiang N et al 2003 Nature (Lond) 421:163.

Mitochip: The mitochondria-specific oligonucleotide microarray that has potential to define mechanisms of disease progression and drug toxicities involved in mitochondrial dysfunction (Desai VG, Fuscoe JC 2007 Mutation Res 616:210). ►microarray; Zhou S et al 2006 J Mol Diagn 8:476.

Mitochondria: Cellular organelles 1–10 μm long and 0.5–1 μm wide, surrounded by double membranes. The outer membrane has a diameter of 50–75 and the inner 75–100 Å. The latter forms the structures, called cristae. Mitochondria are associated frequently with the cytoskeleton and the latter may control the distribution of this organelle within the cell and their transmission during cytokinesis.

The outer membrane is associated with monoamine oxidase and a NADH-cytochrome c reductase. Cardiolipin, phosphatidyl inositol, and cholesterol are major compounds associated with the inner membrane. Mitochondria have an important role in generation of ATP and in electron transport. ATP synthesis requires oxidative phosphorylation. The mitochondria generally encode cytochrome b, cytochrome oxidase (COX) subunits, ATPase and NADH dehydrogenase (see Fig. M83). Three protozoa (*Giardia, Trichomonas, Vairimorpha*) and *Entamoeba*s are the only animals without mitochondria. All human cells, except the erythrocytes contain mitochondria. In human mitochondria 13 out of the 78 polypeptides involved in the electron transport are coded for by mitochondrial DNA.

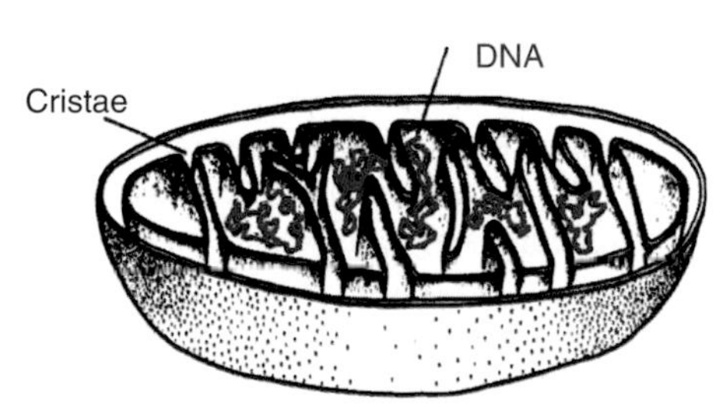

Figure M83. Diagram of a longitudinal section of a mitochondrion showing the cristae and the DNA strings in between them. according to electronmicroscopic view the cristae are tubular rather than lamellar structures

Mitochondria are the major source of reactive oxygen (ROS) in the body. In humans, about 300 nuclear genes control mitochondrial functions. The larger mitochondrial genomes of fungi and plants have larger coding capacity. The mitochondrial DNA and the prokaryotic type ribosomes in this organelle are capable of independent protein synthesis although they may not have all their necessary tRNA genes within the mt genome. The majority of the proteins within the mitochondria are synthesized in the cytosol and imported. This import depends also on the function of cytosolic chaperones such as Hsp70 and Hsp60. In yeast and some other organisms, under some circumstances, a single giant mitochondrion is formed that spreads all over the cell and can break up into smaller organelles (see Fig. M84). Mitochondria are usually transmitted only through the eggs. Exceptions exist, however. In a human cell, about 100–1000 mitochondria may be found but a single mouse oocyte has $\sim 1 \times 10^5$. Each may contain 2–10 or more DNA molecules and about 1000 peptides of which only 13 are coded by mtDNA. In addition, the mtDNA genome (16.5×10^3 bp in higher animals) encodes two rRNAs and 22 tRNAs. (In other organisms, the number of genes in the mitochondria may vary between 5 and 60.) Mitochondria are involved in the production of ~80% of the cellular ATP, carry out respiration, synthesize amino acids, nucleotides, lipids, heme, regulate inorganic ion channels and apoptosis.

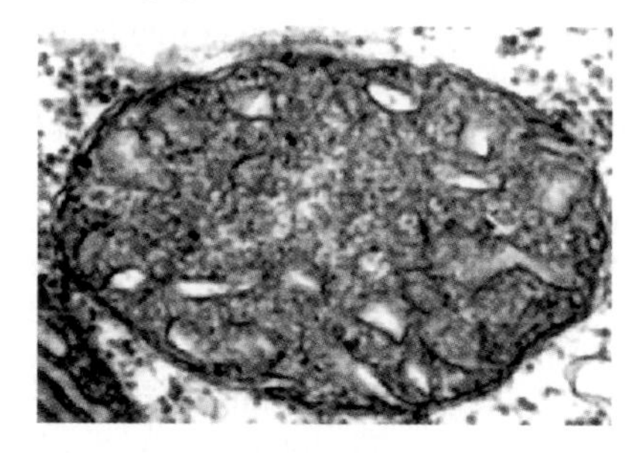

Figure M84. Plant mitochondrion

During mammalian oogenesis, the number of mitochondria may be 4,000–200,000 but their number may be reduced substantially during early development of the embryo. The number of mitochondria seems to be regulated by mtTFA and mtTFA is regulated by a nuclear regulatory factor (NRF-1). The large number of mitochondria within mammalian and other higher eukaryotic cells was probably necessitated by the limited capacity of DNA repair within the mitochondria. The large number can compensate for the defective organelles. There is apparently no nucleotide excision repair (NER) in the mammalian mitochondria but base excision repair (BER), by the use of glycosylases, has been detected. Mismatch repair (MMR) also exists in mitochondria. It is not clear whether recombinational repair has a role in mitochondria. The evolutionary origin of the mitochondria has been interpreted by syntrophic hypotheses assuming the fusion of a primitive amitochondrial eukaryote with a *Clostridium*-like eubacterium or a *Sulfolobus*-like archaebcterium. Genome signatures, the selective advantage of expanded energy metabolism, and the possibility that *Clostridium* provided an opportunity for the development of the nucleus and the cytoplasm of the eukaryotic cell support the latter hypothesis. Mitochondrial DNA analysis (Finnilä S et al 2001 Am J Hum Genet 68:1475) is a very important tool for evolutionary and demographic studies. Some anaerobic or near-anaerobic unicellular eukaryotes do not have typical mitochondria but hydrogenosomes. An integral membrane protein, mitofusin, mediates fusion of mitochondria (Koshiba T et al 2004 Science 305:858). ►mtDNA, ►paternal leakage, ►doubly uniparental inheritance, ►Eve foremother of mitochondrial DNA, ►CSGE, ►compatibility of organelles with the nuclear genome, ►respiration, ►photorespiration, ►promitochondria, ►mitochondrial genetics, ►petite colony mutations, ►organelle sequence transfer, ►aspartate aminotransferase [glutamate oxaloacetate transaminase, ►GOT2], ►mitochondrial diseases in humans, ►cytoplasmic male sterility, ►mitochondrial abnormalities in plants, ►mitochondrial genetics, ►mitochondrial plasmids, ►mitochondrial import, ►polar granules, ►*Neurospora*, ►mitochondrial plasmids, ►mtTFA, ►sorting out, ►endosymbiont theory, ►Rickettsia, ►evolution of organelles, ►Tid50, ►Hsp70, ►Hsp60, ►Ydj, ►rho factor, ►hydrogenosome, ►apoptosis, ►aging, ►ROS, ►signature genomic, ►membrane proteins, ►amitochondriate, mitosome, evolution of mitochondria: Gray MW et al 2004 Annu Rev Genet 38:477; MAVS, Chen XJ, Butow RA 2005 Nature Rev Genet 6:815; mitochondrial fusion and fission: Okamoto K, Shaw JM 2005 Annu Rev Genet 39:503; mitochondrial origins and evolution: Embley TM, Martin W 2006 Nature [Lond] 440:623; Mt DNA in human oocyte: Shoubridge EA, Wai T 2007 Curr Top Dev Biol 77:87; human mitochondrial database: http://www.genpat.uu.se/mtDB/; mitochondrial proteome: http://www.mitop2.de; human mitochondria: http://www.mitomap.org; *Arabidopsis* mitochondrial proteins: http://www.ampdb.bcs.uwa.edu.au/.

Mitochondria and Cancer: Dysfunction of apoptosis may lead to abnormal cell proliferation and thus to cancer. The apoptosis blocking Bcl-2 and other proteins are localized in the mitochondrial membrane. The oxidative processes in the mitochondria may cause mutagenic lesions leading to tumor initiation and progression. ►apoptosis, ►cancer

Mitochondria, Cryptic: Double-strand membrane-enclosed structures in microsporidia that were supposed to lack mitochondria. Microsporidia are intracellular parasites of some animals and protists. The cryptic mitochondria are about 1/10 in size of mitochondria of animals and contain only a few proteins. The majority of the mitochondrial functions were apparently lost in these parasites as a consequence of retrograde evolution. ►retrograde evolution, ►hydrogenosomes; Roger AJ, Silberman JD 2002 Nature [Lond] 418:827.

Mitochondrial Abnormalities in Plants: The cytoplasmic male sterility, widespread in many species of plants and the *non-chromosomal striped* mutations of maize have proven mitochondrial defects. The latter have deletions either in the cytochrome oxidase (*cox2)* or in NADH-dehydrogenase (*ndh2*) or in ribosomal protein (*rps3*) mitochondrial genes. The mtDNA defects apparently originate by recombination between repeats (6–36-bp) at different locations within the genome. Recombinations involving larger repeats may also give rise to normal mtDNA as well as duplication-deficiency molecules. The phenotype (green and bleached stripes on the leaves, stunted growth, etc.) might suggest as if the plastids would have been involved. A similar mt-coded enzyme causes mitochondrial defects in other plants as well as in animals. ►cytoplasmic male sterility, ►killer plasmids, ►senescence, ►mitochondrial diseases in humans, ►mitochondrial genetics, ►mitochondrial plasmids, ►mitochondria; Newton KJ et al 1996 Dev Genet 19:277.

Mitochondrial Complementation: When mitochondria with long deletions (4696 bp), including the site of cytochrome oxidase c (COX) is lost, mice did not show any disease when COX^+ mitochondria were introduced. Within single cells, apparently only normal mitochondrial function could then be detected, indicating genetic complementation. The results promise the possibility of gene therapy by mtDNA. ►mitochondrial heterosis; Nakada K et al 2001 Nature Med 7:934.

Mitochondrial Control-Regions: They encompass the areas where replication of the heavy strand (control region I, bp 16040–16400 in humans) and light strand (control region II, bp 39–380 in humans) takes place. The control regions display greater variation than the rest of the molecule and therefore, these regions are frequently studied for base variations in evolutionary and population studies. ►DNA replication mitochondria, ►D loop, ►mtDNA, ►DNA 7 S, ►mitochondrial genetics

Mitochondrial DNA Depletion Syndromes (MDDS): MDDS involve tissue-specific decrease of mtDNA copy number and organ failure. Mitochondrial DNA synthesis depends on import of deoxyribonucleotide phosphates from the cytosol or salvage from DNA inside mitochondria. Human deoxyguanosine kinase (dGK, 2p13) and thymidine kinase-2 (TK2, 16q12) are expressed in the mitochondria. These two enzymes phosphorylate purines (guanine, adenine) and pyrimidines (thymidine, cytidine) respectively. Mutations in TK2 can reduce its activity by 14–45% and involve devastating myopathy (Saada A et al 2001 Nature Genet 29:342). Similarly, mutation in the last step of the salvage pathway (ADP forming succinyl-CoA ligase) resulted in mtDNA depletion and encephalomyopathy (Ellpeleg O et al 2005 Am J Hum Genet 76:1081). A hepatocerebral mtDNA depletion locus (MPV17, 2p21-p23) affect the inner mitochondrial membrane and causes defects in (oxidative phosphorylation) OXPHOS (Spinazzola A et al 2006 Nature Genet 38:570). ►Alpers progressive poliodystrophy, ►oxidative phosphorylation myopathy, ►mitochondrial diseases in humans, ►hepatocerebral, ►mtDNA

Mitochondrial Diseases in Humans: Up to the 1970s no human disease was attributed to mtDNA. Approximately 0.001 fraction of the newborns have a mitochondrial disease. The role of mitochondrial DNA was definitely proven only after molecular analysis became practical. The inheritance pattern is not always clear because of heteroplasmy. *Chloramphenicol* resistance in human cells is caused by several mutations in the mitochondrial DNA-encoded 16S ribosomal RNA genes. Leber hereditary optic atrophy (LHON), a progressive disease (caused most commonly by single base pair substitution [G(A(Arg→His]) resulting in wasting away of the eyes (optic atrophy), abnormal heart beat accompanied by neurological anomalies characterizes this disease (involving 5 NADH dehydrogenases at different sites in the mtDNA) that it appears a human X-chromosomal factor has a role in determining susceptibility to the mitochondrial defect (see Fig. M85). In mice, three nuclear quantitative gene loci appear to affect mitochondrial sorting out (Battersby BJ et al 2003 Nature Genet 33:183). In LHON models of mice and dogs adeno-associated virus vector carrying the gene for the retinal pigment epithelium photoreceptor successfully restored visual function by this gene therapy (Acland GM et al 2001 Nature Genet 28:92). *Kearns-Sayre syndrome* (KSS/ PEO) a progressive external ophtalmoplegia (paralysis of the eye muscles) is characterized by pigmentary degeneration of the retina and cardiomyopathy (inflammation of the heart muscles) and other less specific symptoms are caused either by deletion or base substitution mutation in nucleotide 8,993 in the mtDNA. Deletions—in 30–50% of the afflicted—are flanked by the 5′-ACCTCCCTCACCA direct repeats and show the loss of nucleotides 8,468–13,446 and a less frequent deletion between the repeats 5′-CATCAACAACCG at positions 8,468–16,084. The larger deletions may affect simultaneously also other mitochondrial genes. The KSS involves most commonly heteroplasmy and its occurrence is frequently sporadic. The basic defect is either in a protein phage T7 gene-4-like pimase/helicase of DNA (Spellbrink JN et al 2001 Nature Genet 28:223) or in DNA polymerase γ, located at 15q22-q26 (Van Goethem G et al 2001 Nature Genet 28:211). In both cases, multiple

M

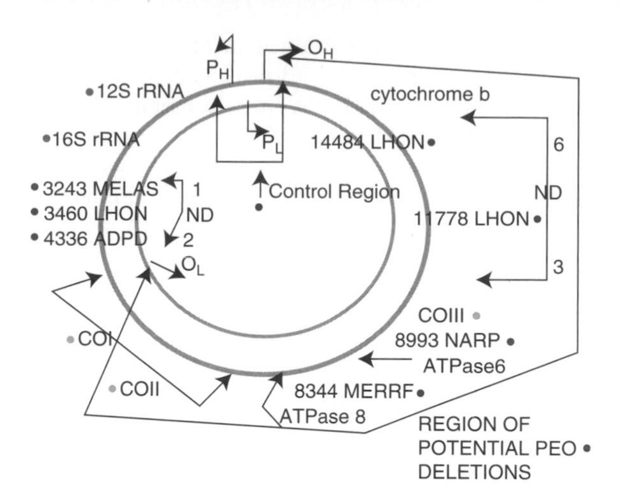

Figure M85. Incomplete Human mtDNA Map. O_H:Origin of heavy-chain replication, P_H: promoter of heavy chain replication, P_L: light strand promoter, ND: NADH-coenzyme Q oxidoreductase subunits, ADPD: Alzheimer disease late onset, CO:cytochrome oxidase subunits, NARP: neuropathy, ataxia, retinitis pigmentosa syndrome, MELAS: mitochondrial myopathy, encephalopathy, lactice acidosis, stroke-like syndrome (associated mainly with mutation in the $tRNA^{Leu}$), LHON: Leder's hereditary optic neuropathy, MERRF:myoclonic epilepsy and ragged-red fibers, PEO: progressive external ophthalmoplegia (most commonly deletions)

mutation/deletions occur in the mtDNA. One form of *myoclonic epilepsy*, MERRF (shock-like convulsions and ragged red fibers [when stained with Gömöri's modified trichrome stain]) is caused by a point mutation in the mtDNA-coded $tRNA^{Lys}$. *Mitochondrial myopathy* (muscle/eye degeneration), *encephalopathy* (brain degeneration), *encephalomyopathy* (MTTL2) mutation in $tRNA^{Leu}$, *lactic acidosis* (accumulation of lactic acid in the blood), and *stroke-like episodes* (MELAS syndrome, *mitochondrial encephalopathy with lactic acidosis and stroke-like symptoms*) are due to a mtDNA encoded $tRNA^{Leu}$ base substitution (A→G) at nucleotide 3,243 and at other sites and by mutation in $tRNA^{Cys}$ and $tRNA^{Val}$. The same $tRNA^{Leu}$ mutation occurs in about 20% of the patients with the recessive autosomal *Progressive external ophtalmoplegia* (PEO) and in some cases of diabetes mellitus ($tRNA^{Leu}$). Mutation in $tRNA^{Leu\ A3243G}$ leads to gradual pancreatic β-cell dysfunction upon aging and diabetes (*Mitochondrial diabetes*, Maassen JA et al 2004 Diabetes 53:S103). The *Pearson marrow-pancreas syndrome* is caused by mtDNA deletions affecting subunit 4 of NADH dehydrogenase, subunit 1 of cytochrome oxidase and subunit 1 of ATPase. This rare disease, usually heteroplasmic, is frequently fatal at infancy. *Oncocytoma*, responsible primarily for benign solid kidney tumors, loaded densely with mitochondria, has deletions in subunit 1 of cytochrome oxidase.

In general, human males have more, or more are affected by mitochondrial mutations. About 85% of the *Leber's optic dystrophy* cases are found in males. Also, Alzheimer disease ($tRNA^{Gln}$), and Parkinson disease are associated with defective mitochondrial energy metabolism, reduced sperm motility and fertility. *Leigh syndrome* (an ATP synthase defect at nucleotide 8527–9207) is a progressive encephalopathy of children and it is accompanied by over 90% mutant mitochondria in the blood, muscle and nerve cells, although the inheritance appears autosomal. Several myopathies are associated with mutations also in various tRNAs of the mitochondria ($tRNA^{Phe}$, $tRNA^{Ile}$ [cardiomyopathy], $tRNA^{Glu}$ [cardiomyopathy], $tRNA^{Met}$, $tRNA^{Ala}$, tRN^{Asn}, $tRNA^{Cys}$ [ophthalmoplegia], $tRNA^{Tyr}$, $tRNA^{Ser}$, $tRNA^{Asp}$, $tRNA^{Lys}$ [MERRF syndrome], $tRNA^{Gly}$, $tRNA^{Arg}$ [LHON], $tRNA^{Ser}2$, $tRNA^{Leu(CUN)}$ [skeletal myopathy], $tRNA^{Glu}$, $tRNA^{Thr}$ [cytochrome b subunits]. U → C transition mutation adjacent to the anticodon GAU of $tRNA^{Lys}$ may involve hypertension, dyslipidema and atherosclerosis (Wilson FH et al 2004 Science 306:1190). In MELAS Leu^{UUR} tRNA five mutations (sites A3243G, G3244A, T3258C, T3271C and T3291C) lacked the normal taurine-containing modification (5-taurinomethyluridine) at the anticodon wobble position. Other mutations in tRNA Leu^{UUR} (G3242A, T3250C, C3254T and A3280G) display mitochondrial disease but not the MELAS symptoms (Kirino Y et al 2005 Proc Natl Acad Sci USA 102:7127).

Aminoglycoside-sensitivity (modest doses of streptomycin) may lead to hearing defects due to mutation in the 12S rRNA gene (Li R et al 2004 Am J Med Genet 124A:113; Li X et al 2004 Nucleic Acids Res 32:867). About 5×10^{-5} fraction of the human population is hypersensitive to chloramphenicol and may become anemic from the drug. Deletion of the COX (cytochrome oxidase) gene may result in *myoglobinuria*.

The human mitochondria harbors about 1,500 proteins and but most of them are encoded by nuclear genes. By computational technology, 1,080 gene products were allocated to mitochondria with an estimated false positive rate of 10%. The predicted number includes 8 genes implicated in disease (Calvo S et al 2006 Nature Genet 328:576). A decrease in the number of mitochondria, caused by nuclear genetic factors or drugs may lead also to disease.

The autosomal dominant progressive external ophthalmoplegia (adPEO, CPEO), a muscle weakness affecting the eyes primarily, is caused by a mutant gene in human chromosome 10q23.3-q24.3 that causes multiple deletions.

Hereditary spastic paraplegia is encoded by chromosome 16q but the 795 amino acid paraplegin protein is located in the mitochondria. Some of the late onset neurodegenerative diseases apparently have mitochondrial components. Human *colorectal tumors* frequently display purine transition mutations and appear homoplasmic. *Hereditary paraganglioma* is caused by mutation at 11q23 in the SHDS gene encoding a small subunit of cytochrome b involved in succinate-ubiquinone oxidoreductase (sybS). This disease shows vascularized benign tumors in the head and neck due to defects in the carotid body (the main artery to the head), which senses oxygen levels in the blood (Vanharanta S et al 2004 Am J Hum Genet 74:153).

The mtDNA contains 37 genes, most of them transcribed from the heavy chain, but 9 are read in opposite direction from the light chain of the DNA and they are thus somewhat overlapping. The human mitochondria encode only 13 respiratory chain proteins although it may harbor about ~1,500 proteins (see Fig. M86). Some human diseases or syndromes (hypotonia [reduced muscle tension], ptosis [dropping down eyelids], ophthalmoplegia [eye muscle paralysis], high level of lactate in the blood serum, liver defects) may be associated with a reduced level of mtDNA. Genetic counseling with mitochondrial disease is difficult because the transmission of the heteroplasmic conditions is irregular. Treatment of mitochondrial diseases so far appeared rather elusive. The mitochondrial respiratory chain involves the multisubunit complexes (I) NADH-UQ oxidoreductase, (II) succinate dehydrogenase, (III) UQ-cytochrome c oxidoreductase, (IV) cytochrome c oxidase, and (V) ATP synthase. Blocking complex I by 1-methyl-4-phenylpyridinium (MPP^+) mimicks Parkinson disease and LHON. Inhibition of complex II by 3-NPA (3-nitropropionic acid) causes the symptoms of Huntington's chorea. Cyanide and azide inhibition of complex IV resembles the expression of Alzheimer disease. The effect of oligomycin on complex V involves neuropathy, ataxia and retinitis pigmentosa-like symptoms. There are a number of diseases, which are not encoded by the mitochondrial DNA and neither are the proteins localized in the mitochondria yet they affect mitochondrial functions. Mitochondrially determined disease due to ATP deficiency could be alleviated (Manfredi G et al 2002 Nature Genet 30:394) by introduction into the nucleus the functional mitochondrial gene. Also, the transplantation of nuclei from mitochondrial mutants into normal cytoplasm may restore normal health if the mutational defect was limited only to mitochondria (Sato A et al 2005 Proc Natl Acad Sci ISA 102:16765). Dominant optic atrophy (Kjer type, 3q28-q29) causes serious vision defects in childhood and adulthood due to degeneration a mitochondrial intermembrane dynamin-like OPA1 protein (Kim JY et al 2005 Neurology 64:966). ►mitochondria, ►mitochondrial DNA depletion syndromes, ►mtDNA, ►mitochondrial genetics, ►protoporphyria, ►aging, ►senescence, ►chondrome, ►optic atrophy, ►epilepsy, ►spastic paraplegia, ►antibiotics, ►Leigh's encephalopathy, ►diabetes mellitus, ►ophthalmoplegia, ►myoglobin, ►Wolfram syndrome, ►Friedreich ataxia, ►Wilson disease, ►Alzheimer disease, ►Parkinson disease, ►Charcot-Marie-Tooth disease, ►spastic paraplegia, ►MNGIE, ►amyotrophic lateral sclerosis, ►mitochondrial gene therapy, ►homoplasmy, ►heteroplasmy, ►colorectal cancer, ►apoptosis, ►mtPTP, ►mitochondrial abnormalities in plants, ►atresia, ►pleiotropy, ►oxidative DNA damage, ►transmitochondrial, ►NARP syndrome, ►myoneurogastrointestinal encephalopathy, ►dimethylglycine dehydrogenase, ►taurine, ►wobble, ►Gomori's stain; Wallace DC 1992 Annu Rev Biochem 61:1175; Smeitink J, van den Heuvel B 1999 Am J Hum Genet 64:1505; Acland GM et al 2001 Nature Genet 28:92; Thornburn DR, Dahl HH 2001 Am J Med Genet 106:102; Steinmetz LM et al 2002 Nature Genet 31:400; Taylor RW, Turnbull DM 2005 Nature Rev Genet 6:389; mtDNA – tRNA human disease: Florentz C et al 2003 Cell Mol Life Sci 60:1356; wobble site modification in tRNA leads to disease specificity: Kirino Y et al 2005 Proc Natl Acad Sci USA 102:7127; Wallace DC 2005 Annu Rev Genet 39:359; mitochondrial proteome: http://www.mitop.de/; human mitochondrial genome: http://www.genpat.uu.se/mtDB/; human mitochondrial variation database: http://www.genomic.unimelb.edu.au/mdi/dblist/mito.html.

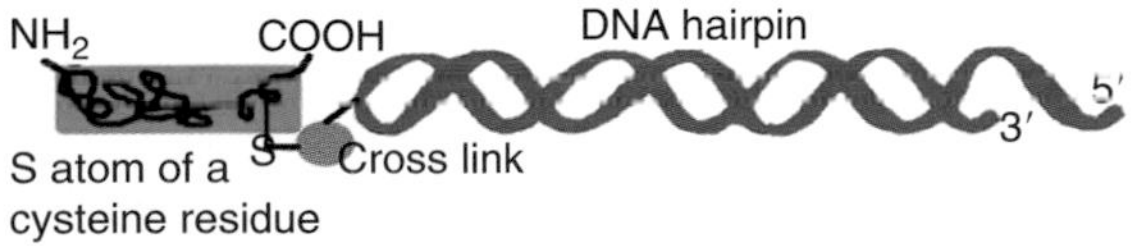

Figure M86. Targeting a transit peptide to DNA (Modified after Seibel P et al. 1998 in Mitochondrial DNA in Aging, Disease and Cancer. Singh KK ed.Springer, New York)

Mitochondrial DNA: ►mtDNA

Mitochondrial Export: ►polar granules

Mitochondrial Gene Therapy: Several mitochondrial diseases became ascertained and studied since the 1970s. They may include (i) nuclearly encoded gene products imported to the mitochondria, (ii) strictly mitochondrially encoded defects, and (iii) gene products under dual control. Correcting the defect of proteins encoded by the nucleus and translated in the cytoplasm has about the same requisites as gene therapy in general. There are more hurdles to overcome when the therapeutic DNA sequences are intended for transport into the mitochondria with the aid of cytoplasm-located vectors. The first requirement is the construction of appropriate signal peptide, which would mediate the import. There is experimental evidence for the feasibility to targeting a hairpin-shape DNA into the mitochondrial matrix with the aid of an appropriate transit/signal peptide (Fig. M86). The transforming vector must use the coding system particular for the mitochondria that may be different in different organisms. The transcript and the translation product must be suitable for processing within the mitochondria by its own protein-synthesizing machinery. The mitochondrial ribosomes are also different from the cytosolic ones. The promoter must respond to the regulatory system of the mitochondria and the expression of the gene would be optimally modulated in that environment. The maintenance replication of the introduced transgene must be secured for both somatic divisions and for the maternal germline. Problems may arise if the defective, mutant polypeptide interferes with the assembly of the functional protein complex even in the presence of a correct product of the transgene. Since the majority of the mtDNA mutations involve deletions, duplications and rearrangement of the mitochondrial genome, some diseases may not be amenable to exogenously delivered corrections. The mitochondrial genome in higher organisms is present in "polyploid" forms and the rules of the mitochondrial segregation within the cells are not entirely known. There is evidence that in the relatively very simple yeast system the mitochondrial ATPase 8 defect could be corrected by a genetic vector product in the cytoplasm. It has been reported that the β subunit of the mitochondrial ATP synthase fused to the bacterial *CAT* (chloramphenicol acetyltransferase) and driven by the cauliflower mosaic virus CaMV 35 S promoter and introduced into tobacco (*Nicotiana plumbaginifolia*) by an agrobacterial vector, was expressed mainly in the mitochondria but not in the chloroplasts. The introduction of transgenes by the biolistic methods to both mitochondria and chloroplasts has been achieved repeatedly. There may be ways to target therapeutic agents to the mitochondria by coupling, e.g., antioxidants to alkytriphenylphosphonium cations (Smith RAJ et al 2003 Proc Natl Acad Sci USA 100:5407).

Defects of the mtDNA frequently result in faster replication of the mutation and this type of anomaly may be corrected by antisense technologies. Antisense technology may also silence a mitochondrial gene with harmful effect. Eventually gene replacement techniques may also become practical. ►transit peptide, ►signal peptide, ►genetic code, ►biolistic transformation, ►ATPase, ►transformation [plants], ►antisense technologies, ►gene replacement, ►mitochondrial genetics, ►mitochondrial diseases in humans, ►mitochondrial anomalies in plants, ►gene therapy, ►mitochondrial complementation acid, ►peptide nucleic acid; Murphy MP, Smith RA 2000 Adv Drug Deliv Rev 41:235.

Mitochondrial Genetics: The mitochondrial DNA is generally transmitted only by the egg cytoplasm. The failure of transmission of the mtDNA by the sperm in the majority of the species appears to be due to selective degradation. It appears that during spermatogenesis the mitochondrial nucleoid number is reduced and after fertilization the mtDNA is rapidly digested (Nishimura Y et al 2006 Proc Natl Acad Sci USA 103:1382). The wild type yeast mtDNA is inherited biparentally. The hypersuppressive petite mutants of yeast are transmitted, however, only uniparentally. In the bivalve molluscs a specific F (female) mitochondrial DNA is transmitted to both males and females whereas the M (male) mtDNA is transmitted only to the sons of males. These two types of mtDNAs recombine in the somatic tissues but the recombinant molecules are rarely transmitted (Passamonti M et al 2003 Genetics 164:603). In some species a low frequency of male transmission (paternal leakage) exists (Fig. M87).

Segregation of mitochondrial genes is followed by cell phenotype because individual mitochondria are not amenable to direct genetic analysis (see Fig. M88). (Molecular analysis can identify, however, differences among mitochondria.) Some of the genes are specific for the mitochondria and others, e.g., the cytochrome oxidase complex share subunit coding with the nucleus. This sharing may vary in different evolutionary groups, which seems to indicate interchanges between the nuclear, mitochondrial and chloroplast genomes. Despite the large number of mitochondria (>100/cell) and mtDNA molecules (4–6 per nucleoid), homoplasmic condition may be obtained at a much faster rate than expected on the basis of random sorting out. In mouse, it has been estimated that the number of segregating mitochondrial units is about 200 and the rapid sorting out was attributed to random genetic drift. (Some estimate the number of sorting mammalian mtDNAs >50,000.) The sorting out of the mitochondria—according to recent studies of yeast mutants of the MDM (mitochondrial distribution and morphology) group— is controlled by proteins of the cytoskeleton and integral proteins of the outer mitochondrial membrane, such as a dynamin-like protein. In fission yeast, mitochondrial distribution is mediated by microtubules tethered to the spindle pole bodies (Yaffe MP 11424). The ubiquitin-protein ligase suppressor, Rsp5p and the Ptc1p (serine/threonine phosphatase) seem also be involved in the regulation of mitochondrial transmission. In the mouse oocytes

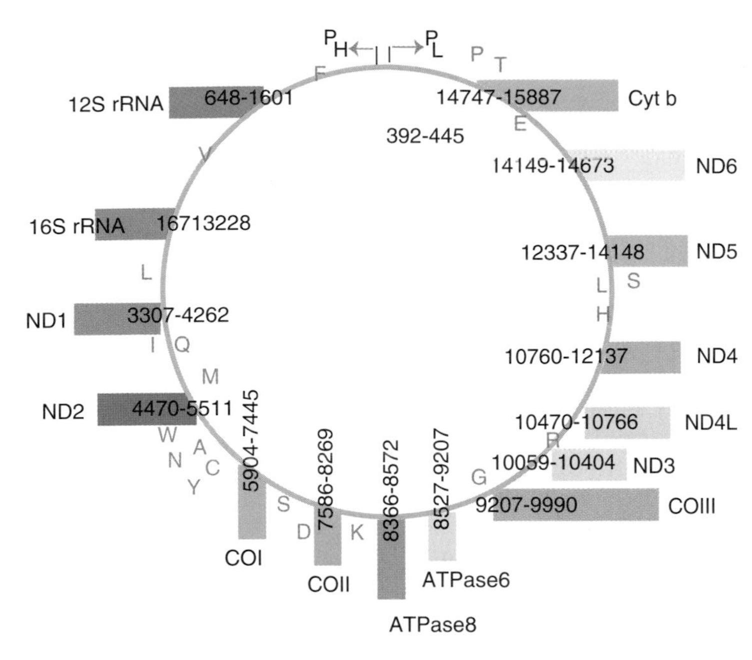

Figure M87. An incomplete map. P: promoters, Cyt b:cytochrome b, ND: nicotineamide-adenine dinucleotide reduced, CO: cytochrome oxidase, rRNA: ribosomal RNA, blue letters are amino acid tRNAs. For more information see www.mitomap.org. See also mitochondrial diseases in humans. Numbers stand for nucleotide positions

$$P_{1-5} = \frac{\begin{bmatrix} 10-2i \\ 5-j \end{bmatrix} \begin{bmatrix} 2i \\ j \end{bmatrix}}{\begin{bmatrix} 10 \\ 5 \end{bmatrix}}$$

Figure M88. The probability (P) of segregation of mtDNA molecules under the assumption that each of the cells carry 10 rings of mtDNA, and among them the number of mutants may be 0, i (1) to j. The bracketed expressions represent the binomial probability functions. (From Kirkwood, TBL & Kowald, A 1998, p.131 in Mitochondrial DNA Mutations in Aging, Disease and Cancer. Singh, KK, ed., Springer-Verlag, New York)

$\sim 10^5$ mtDNA molecules may occur yet it appears that a relatively small number of mitochondria (20? to 200?) is transmitted by the egg cytoplasm to the zygote and this bottleneck facilitates a relatively rapid sorting out. Generally, 20 cell generations of the zygotes are used for appropriate mitochondrial typing followed by the use of selective media. In mammals $\sim$100 mitochondria may be transmitted by the spermatozoa, these are, however, destroyed within less than a day in the egg cytoplasm. If wild type and mutant (MERRF) mitochondria (coexisting since the mutational event) were transferred into cells lacking mtDNA (rho$^-$), the mutant mitochondria were complemented by the wild type. When, however two independent, non-allelic chloramphenicol-resistant mutations were introduced into chloramphenicol sensitive cells both mutant types were maintained separately for prolonged periods of time (Yoneda M et al 1994 Mol Cell Biol 14:2699).

In natural populations of higher eukaryotes, mitochondrial recombination is considered to be rare because mitochondria are usually inherited through the egg although paternal leakage may occur. Homologous recombination enzymes have been detected in mammalian somatic cells (Thyagarajan B et al 1996 J Biol Chem 271:27563). The highest frequency of recombination in heteroplasmic cells is usually 20–25% because recombination between identical molecules also takes place and multiple "rounds of matings" occur within longer intervals. In humans and chimpanzees, the decline of linkage disequilibrium of mitochondrial markers was interpreted as evidence for recombination. In rare cases when paternal transmission of mtDNA occurs in humans (Schwartz M, Vissing J 2002 N Engl J Med 347:576) recombination may be detected by molecular analysis (Kraytsberg Y et al 2004 Science 304:981). Mitochondrial recombination in human skeletal muscle cells has been observed (Zsurka G et al 2005 Nature Genet 37:873). In *Schizosaccharomyces* the distribution of the mitochondria is mediated by microtubules. It has been suggested that mitochondria move along the cytoskeleton with a combination of motor proteins. The Apf2 protein seems to be involved in the control of segregation and recombination of mtDNA in budding yeast. In the mouse

mtDNA seems to be involved in neuron development and affect axonal and synaptic activity as well as cognitive functions of the brain (Roubertoux PL et al 2003 Nature Genet 35:65).

Recombinational mapping is generally useful within short distances (≈1 kb). Genetic maps of mtDNA are generally constructed by physical mapping procedures. Also, the relative position of the genes may be ascertained by co-deletion–co-retention analyses. These methods resemble the deletion mapping of nuclear genes. Genes simultaneously lost by deletion, or retained in case of deletion of other sequences, must be linked. Mitochondrial genes have been mapped also by polarity. It was assumed that the yeast mt DNA carried some sex-factor-like elements, ω^+ and ω^-, respectively. The omega plus (ω^+) cells appeared to be preferentially recovered in polar crosses ([ω^+] x [ω^-]). In these so called *heterosexual crosses* the order of certain genes, single and double recombination and negative interference were observed. The exact mechanism of the recombination was not understood. The ω^+ factor turned out to be an intron within the 21S rRNA gene with a transposase-like function. Non-polar crosses have also been used to determine allelism. In the latter case all progeny was parental type and recombination was an indication of non-allelism. Linkage and recombination could also be detected in non-polar crosses but mapping was impractical. Complementation tests can be carried out in respiration deficient mutations on the basis that in "non-allelic" crosses respiration is almost completely restored within 5–8 hours whereas recombination may produce wild types only in 15–29 hours. Different loci are complementary and appear unlinked. Some loci may not complement each other. Mitochondrial fusion and complementation of mitochondrial genes in human cells has also been reported (Ono T et al 2001 Nature Genet 28:272; Westermann B 2002 EMBO Rep 3:527). Mitochondrial fusion requires the sequential interaction of the outer and inner mitochondrial membranes (Pfanner N et al 2004 Science 305:1723). The fusion of the outer membranes needs low levels of GTP (mediated by GTPase Fzo1) but the inner membrane fusion requires high GTP levels as well as inner-membrane electrical potential. Inner membrane fusion as well as the maintenance of the structure of the crystae requires the dynamin-related GTPase Mgm1 (Meeusen S et al 2006 Cell 127:383). Mutations in the genes controlling mitochondrial fusion (mitofusin, MFN2. 1p36.2) may account for the 2A type neuropathy of the Charcot–Marie–Tooth disease (Zuchner S et al 2004 Nature Genet 36:449) and optic atrophy (Zuchner S et al 2006 Ann Neurol 59:276). Primarily, dynamin-like proteins control mitochondrial fission. Mitochondrial fission entails mitochondrial disintegration during apoptosis (Parone PA, Martinou JC 2006 Biochim Biophys Acta 1763:522). This protein is involved also in the control of cytokinesis (Chanez AL et al 2006 J Cell Science 119:2968).

Homologous recombination has been extensively studied in yeast by the Flp/FRT system. Recombination of mtDNA in vertebrate cybrids does not readily occur. In case of fusion of haploid cells, gene conversion may take place.

Transformation of mitochondria requires the biolistic method or by implantation of cell hybrids with mutant mitochondria. Unfortunately, only a fraction of the mutant mitochondria are transmitted and only through the female. Usually co-transformation by nuclear and mitochondrial DNA sequences is employed. The nuclear gene (*URA3*) is used for selection of transformant cells. The ρ^- cells cannot be selected because they are defective in respiration. They are crossed then with mt^- cells that carry a lesion in the target gene and among the recombinants, the transformants may be recovered. It is possible that the wild type *URA3* gene is transferred from the mitochondrion into the *ura3* mutant nucleus whereas the cytochrome oxidase gene (*cox2*) is maintained in the nucleoid.

The large subunit of the mitochondrial ribosomal RNA may house a mobile element, ω^+. This element (a Group I intron) may show polar transmission and can be used in a limited extent for recombination analysis as mentioned above. It encodes 235 amino acids and represents *Sce*I endonuclease. A *Chlamydomonas smithii* intron (from the *cytb* mitochondrial gene) can move into the same but intronless gene in *C. reinhardtii*. Several other Group I introns have similar mobility and the introns usually share the peptide LAGLI-DADG, a consensus conserved also in some maturases. Other Group I introns display the GIY-10/11aa-YIG pattern. The VAR-1 yeast gene (ca. 90% A + T) encoding the small mt ribosomal subunit is also an insertion element. Other mobile elements with characteristic G + C clusters are common features of these insertion elements. Group II introns present in various fungi are also mobile. Splicing/respiration deficient mutants become revertible if the defective intron is removed. Similar mobile elements can be found in the cpDNA of *Chlamydomonas* algae. Mitochondrial plasmids may be circular or linear. Some of the circular mt plasmids display structural similarities to Group I introns and their transcripts are reminiscent of reverse transcriptases (resembling protein encoded by Group II introns).

In yeast, nuclear genes *MDM1* (involved in the formation of non-tubulin cytoplasmic filaments), *MDM2* (encodes Δ9 fatty acid desaturase) and *MDM10* (determines mitochondrial budding) control the transmission of mitochondria to progeny cells.

The subunit(s) of the mitochondrial transcriptase and the mtTFA transcription factor are coded in the yeast nucleus.

Nuclear and mitochondrial promoters share cis elements. It seems that mitochondrial signals regulate nuclear genes controlling mitochondrial functions. Partially reduced intermediates of NADH dehydrogenase and coenzyme Q (ubiquinone) are held responsible for the production of ROS. In human diseases such as cancer, ischemic heart diseases, Parkinson disease, Alzheimer disease, and diabetes ROS products have been implicated. The mitochondrial ROS activates mammalian transcription factors NF-κB, AP and GLUT glucose transporters.

Majority of the prokaryotic mutagens are apparently effective in causing mutations in the mtDNA. The ROS molecules generated within the mitochondria may be responsible for a variety of alterations in the mtDNA.

Deletions and duplication, besides point mutations, may also occur spontaneously during the processes of aging. Mutation rates per mitochondrial D loop has been estimated to be 1/50 between mothers–offspring but other estimates used for evolutionary calculations are 1/300 per generation or 3.5×10^{-8} per site per year although the DNA polymerase γ appears to have low error rates. Some of the (recessive?) mitochondrial mutations are apparently non-detectable without molecular analysis because functionally the mitochondria are "polyploid" and their numbers in the cells are commonly very high. Due to the large number of cells, the large number of mitochondria per cell and the multiple copies of the mtDNA per mitochondria, chances for mutation are high within a multicellular organism. Also, within a single individual different mitochondrial mutations may occur and their spectrum may vary in the different tissues. By 1998, nearly 1000 mitochondrial point mutations were known and 65 were involved in human disease. Mutations in the mtDNA (5×10^{-7}/base/human generation) are much higher than in the nuclear DNA (8×10^{-8}). In the control region of human mtDNA mutation rate of 1/bp/10^6 years was reported (Howell N et al 2003 Am J Hum Genet 72:659). The estimates of mitochondrial mutations may be biased by genetic drift, selection, recurrent mutations, and hot spots in some hypervariable sequences.

The mutation rate/bp in both mtDNA and nuclear DNA varies substantially from region to region. About 40% of the mtDNA deletions involve 7000–9000 bp and 20% encompass 4000–5000. Their distribution is biased in as much as 95% of the deletions affect the region between bp 3300 to 12000.

Mitochondrial DNA frequently classified into macrohaplogroups including subhaplogroups on the basis of the mitochondrial protein sequence and geographic origin. The African haplogroups (subhaplogroups) are L1, L2 and L3. The M and N haplogroups originated on L3 background in eastern Africa but migrated into Eurasia and also to the Americas from Siberia where additional variations, A, B, C, D and G accumulated (Mishmar D et al 2003 Proc Natl Acad Sci USA 100:171). Several additional haplogroups are distinguishable in Europe, Asia and the Americas facilitating the tracing of human migration (Pakendorf B, Stoneking M 2005 Annu Rev Genomics Hum Genet 6:165). High mutation rates, especially in the hypervariable sequences, may be an obstacle to definite conclusions regarding the populations' origin. A further difficulty is that mtDNA sheds light only on maternal lineages. Another complication is the transfer of genes and gene fragments between organelles where their rate of mutation adjusts to that of the recipient organelle.

In plants, recombination repeats occur that may generate additional variation by interchanges. In somatic cell hybrids of plants, the mitochondria may recombine and non-parental sequences and duplications of genes may be generated. The pseudogenic sequences are also attributed to recombination. In maize, extensive intramolecular recombination occurs between/among the main mtDNA molecule or the subgenomic, smaller mtDNAs (Fauron C et al 1995 Trends Genet 11:228). These recombinational events may lead to variegation and other morphological alterations (Sakamoto W et al 1996 Plant Cell 8:1377). Early studies indicated an apparent lack of genetic repair systems in the mitochondria. Recent studies detected, however, several proteins (glycosylases, excinucleases, mismatch repair enzymes, Rec A like proteins, etc.) that may mediate genetic repair in the mitochondria. The *CHM* locus of *Arabidopsis* apparently encodes a mismatch repair protein, similar to the prokaryotic MutS. The protein is mitochondrially located and its mutations lead to variegation (Abdelnoor RV et al 2003 Proc Natl Acad Sci USA 100:5968). ►physical mapping, ►mapping function, ►petite colony mutants, ►deletion mapping, ►rounds of matings, ►interference, ►mutations in cellular organelles, ►homoplasmy, ►sorting out, ►biolistic transformation, ►transformation of organelles, ►chloroplast, ►genetics, ►mitochondria, ►doubly uniparental inheritance, ►transmitochondrial, ►dynamin, ►paternal leakage, ►mtDNA, ►mitochondrial import, ►introns, ►maturase, ►mitochondrial plasmids, ►mitochondrial diseases in humans, ►killer plasmids, ►senescence, ►MSS, ►mismatch repair, ►Parkinson disease, ►Alzheimer disease, ►ischemic, ►NF-κB, ►AP, ►GLUT, ►ROS, ►mitochondrial plasmids, ►mt, ►bottleneck effects, ►heteroplasmy, ►mitochondrial recombination, ►Romanovs, ►sorting out, ►RNA editing,

►endosymbiont theory, ►RU maize, ►conplastic, ►mitochondrial mutation, ►mitochondrial suppressor, ►recombination repeat, ►DNA repair, ►binomial probability, ►mitochondrial gene therapy, ►replicative segregation, ►spindle pole body, for an abbreviated human mtDNA map see ►mitochondrial diseases in humans, ►atresia, ►mutation spontaneous, ►Eve foremother, ►forensic genetics, ►out-of-Africa; Mutation Res 1999 Vol 434 issue 3; Elson JL et al 2001 Am J Hum Genet 68:145; Tully LA, Levin BC 2000 Biotechnol Genet Eng Rev 17:147; Pakendorf B, Stoneking M 2005 Annu Rev Genomics Hum Genet 6:165; division and fusion: Hoppins S et al 2007 Annu Rev Biochem 76; 751; http://megasun.bch.umontreal.ca/ogmp/projects/projects.html; http://bighost.area.ba.cnr.it/mitochondriome; human mitochondrial genome: http://www.genpat.uu.se/mtDB/.

Mitochondrial Heterosis: It was claimed that the presence of different mitochondria might lead to complementation and increased vigor. ►hybrid vigor, ►mitochondrial complementation; Sarkissian IV, Srivastava HK 1973 Basic Life Sci 2:53.

Mitochondrial Import: Transport into the mitochondria must pass through two cooperating membrane layers. The outer membrane carries four receptors (Tom37, -70, -20, and -22). The transmembrane import channel is built of at least six proteins; the transmembrane translocation system uses at least three proteins (Tom40, -6 -7). The Tom5 forms the link (a relay) between the outer and inner channel proteins. The inner membrane import channel proteins (Tim) are Mas6p (MIM23), Sms1p (MIM17), and Mpi1p (MIM44/ISP45). The mitochondrial heatshock 70 protein (Hsp70), the MIM44 complex, and ATP play a central role in import. It appears that MIM44 first binds the incoming unfolded polypeptide chain as it is passing the entry site, and it is then transferred to Hsp70 as ATP dissociates the complex. After this the polypeptide moves further and binds again to the complex and eventually traverses also the inner membrane. Tim10p and Tim12p mediate the import of the multispanning carriers into the inner membrane. Tim12 is bound to Tim22. Tim23 passes proteins through the inner membrane. Although the mitochondria contain their own protein-synthetic machinery (ribosomes, tRNAs), the majority of the mitochondrial proteins are encoded by the nucleus and translated in, or imported from, the cytosol. Imported cytosolic tRNAs may correct mutations of mitochondrial tRNAs. The import system is also nuclearly encoded. The general import factors include the Hsp proteins, cyclophilin 20, ADP/ATP carrier (AAC), and proteases. More specialized is the role of the imported assembly facilitator proteins. Yeast genes *SCO1, PET 117, PET191, PET100, OXA1, COX14, COX11,* and *COX10* are such facilitators. The latter two are actually involved in heme biosynthesis. The COX10 product farnesylates protoheme b. Rescue of cells with defective mitochondia by transfer of normal mitochondria from adult stem cells or somatic cells has been reported without clarification of the mechanism involved (Spees JL et al 2006 Proc Natl Acad Sci USA 103:1283). ►mitochondria, ►mitochondrial genetics, ►mtDNA, ►chloroplast import, ►cyclophilin, ►heat-shock proteins, ►Hsp70, ►Ydj, ►Hsp, ►MSF, ►farnesyl, ►heme, ►Brownian ratchet, ►mitochondrial gene therapy; Annu Rev Biochem 66:863, Bauer MF et al 2000 Trends Cell Biol 10:25; Schneider A, Maréchal-Drouard L 2000 Trends Cell Biol 10:509; Wiedemann N et al 2001 EMBO J 20:951; Rehling P et al 2001 Crit Rev Biochem Mol Biol 36(3):291; Kovermann P et al 2002 Mol Cell 9:363; Pfanner N, Wiedemann N 2002 Current Opin Cell Biol 14:400; Neupert W, Brunner M 2002 Nature Rev Mol Cell Biol 3:555; Wiedemann N et al 2003 Nature [Lond] 424:565; Chacinska A et al 2005 Cell 120:817; Wilcox AJ et al 2005 Proc Natl Acad Sci USA 102:15435; protein import review: Dolezal P et al 2006 Science 313:314; Ryan MT, Hoogenraad NJ 2007 Annu Rev Biochem 76:701; ADP/ATP carriers: Nury H et al 2006 Annu Rev Biochem 75:713.

Mitochondrial Mapping: ►mitochondrial genetics

Mitochondrial Mutations: Mitochondrial mutations occur about ten times or more frequently than in the nuclear genes (Marcelino LA, Thilly WG 1999 Mutation Res 434:177). This may be due to inadequate DNA repair. The mtDNA is "naked" (free of histones) and there is relative abundance of free oxygen radicals in this organelle. Large deletions are common in the human mitochondrial DNA encompassing the "5-kb deletion" ($mtDNA^{4977}$) between nucleotide positions 8470–8482 and 13447–13459, respectively. Some deletions may encompass even longer segments of the mtDNA and cover the position of more than a single mitochondrial gene. These deletions are most common in muscle tissues and the brain and frequently occur by aging. The average mutation rate in the elderly appeared to be 2×10^{-4}/bp in mtDNA (Lin MT et al 2002 Hum Mol Genet 11:133). Other deletions usually occur between direct repeats of 13 to 5 nucleotides. Some of the deletions may involve only single bases. Single base mutations Thymidine 7512Cytosine and Guanine 7497Adenine in the $tRNA^{Ser(UCN)}$ reduced the amount of this tRNA below 10% and reduced protein synthesis by 45%.

Aminoacylation was not affected so it is supposed that a posttranslational structural alteration might be involved in the pathogenic condition (Möllers M et al 2005 Nucleic Acids Res 33:5647). The deletion mutants and their relative extent are characterized by various PCR procedures. The data on increased frequency of point mutations during aging are somewhat ambiguous. Apparently deletions and duplications, however, accumulate by aging. Sometimes it may be difficult to ascertain whether a particular mutation occurs in the nuclear or mitochondrial DNA. In cases of ambiguity, the problem may be resolved by transferring a new nucleus (karyoplast) into an enucleated cytoplasm (cytoplast). If the expression of the mutation is limited to the donated cytoplast, the mitochondrial origin of the mutation can be proved in animal cells. In plant cells, the plastids may complicate the identification. In cancer cells, the mtDNA mutations occurred 19 to 229 times as abundantly as in the nuclear p53 gene. Interestingly, the cancer cells were largely homoplasmic for the mtDNA mutations indicating that the mutation had selective advantage. With the techniques available, the complete sequence of the mtDNA can be determined in single cells. Deficiency for uracil-DNA glycosylase in yeast results in mitochondrial mutator phenotype. Although fidelity of DNA polymerase γ, responsible for the replication of mtDNA, is not poor ($<10^{-5}$/base substitution) and the enzyme has 3′ exonuclease function to correct defects, the relative mutation rate is elevated. One possible cause of the increased mutation frequency is the disproportionally increased deoxygunanidine nucleotide pool in the mitochondria of rats (Song S et al 2005 Proc Natl Acad Sci USA 102:4990). The Twinkle helicase and the accessory B subunit of DNA polymerase are absolutely essential for mtDNA replication pol γ. The duplex DNA binding activity of the B subunit is needed for coordination of POLγ holoenzyme and Twinkle helicase at the mtDNA replication fork. Mutations in Twinkle and the catalytic A and accessory B subunits of the POLγ holoenzyme may result in autosomal dominant progressive external ophthalmoplegia, which is associated with deletions in mtDNA (Farge G et al 2007 Nucleic Acids Res 35:902). ►oxidative DNA damage, ►oxygen effect, ►aging, ►mutation rate, ►mitochondrial diseases in humans, ►petite colony mutants, ►PCR, ►somatic cell hybrids, ►homoplasmic, ►p53; Inoue K et al 2000 Nature Genet 26:176; Taylor RW et al 2001 Nucleic Acids Res 29 (15):e74; Jacobs HT 2001 Trends Genet 17:653; Chatterjee A, Singh KK 2001 Nucleic Acids Res 29:4935; Taylor RW, Turnbull DM 2005 Nature Rev Genet 6:389; http://www.mitomap.org.

Mitochondrial Myopathy: ►mitochondrial disease in human

Mitochondrial Plasmids: Plasmids are circular or linear and occur in the mitochondria of some cytoplasmically male sterile lines of maize plants and relatives. Their sizes vary between 1.4 to 7.4 kb. The main types are S, R, and D. S2, R2, and D2 are the same and R1 and D1 are apparently also identical with each other. S1 appears to have emerged as a recombinant between R1 and R2. The *cms*-S plants carry the S1 and S2 plasmids. During the formation of S1, a terminal part of R1 (R*) was lost and is inserted at two sites in the mtDNA. The S elements can be either integrated or free mitochondrial episomes. The S2 element encodes (*URF1*), a protein somewhat homologous to a viral RNA polymerase, whereas another (*URF3*) appears to be homologous to a DNA polymerase. The *cms*-C and *cms*-T nucleoids are free from these plasmids, whereas in the N-nucleoids the R1 and R2 sequences (from RU) are integrated. Another 2.3-kb plasmid (or a 2.15-kb derivative) is homologous to the $tRNA^{Trp}$ and $tRNA^{Pro}$ in the cpDNA, and also represents the only functional $tRNA^{Trp}$ in the mitochondrion. ►cytoplasmic male sterility, ►mitochondrial genetics, ►cpDNA, ►mitochondria, ►tRNA, ►killer plasmids, ►senescence, ►episome, ►cytoplasmic male sterility, ►*Neurospora* mitochondrial plasmids, ►endosymbiont theory, ►RU maize; Bok JW, Griffith A 2000 Plasmid 43:176.

Mitochondrial Proteins: Most of the mitochondrial proteins are imported from the cytosol, but the ~17 kbp mammalian mitochondrial DNA transcribes 13 polypeptides, which are then translated by the organelle. The ~337 kbp *Arabidopsis* mtDNA encodes 32 proteins. In plants, many mitochondrial genes display multiple promoters (Tracy RL, Stern DB 1995 Current Genet 28:205; Kühn K et al 2005 Nucleic Acids Res 33:337) although two nuclear genes encode the transcriptase. The total number of proteins located in mitochondria may be about 1,500. The human heart mitochondria include more than 600 distinct proteins (Taylor SW et al 2003 Nature Biotechn 21:281). The proteome of the yeast mitochondria includes >750 proteins (Sickmann A et al 2003 Proc Natl Acad Sci USA 100:13207). ►mitochondrial diseases, ►mitochondrial import; Kenmochi N et al 2001 Genomics 77:65; Taylor SW et al 2003 Trends Biotechn 21:82; mitochondrial import of proteins: Neupert W, Hermann JM 2007 Annu Rev Biochem 76:723; human, yeast, mouse mitochondrial proteome: http://www.mitop.de/.

Mitochondrial Recombination: Mitochondrial recombination is generally inferred from observations of linkage disequilibrium. However, the problem is somewhat controversial. Paternal leakage is rare and reduces the chance for recombination. Furthermore, the paternal mitochondria may be eliminated after fertilization even if they are transmitted. In different human tissue samples from the A8344G/A16182C (MERRF syndrome) and A3243G/G16428A (MELAS syndrome) double-heteroplasmic family, all four allelic combinations of the two heteroplasmic mutations were present in the family, although, as expected, the distribution showed significant differences between individuals. Two family members carried the double-mutant (and possibly recombinant) allelic combination; and high amounts of the possibly recombinant genotype, along with all other possible allelic combinations, were present in the fibroblast sample from one individual. The data indicate that recombinant mtDNA molecules can be inherited (Zsurka G et al 2007 Am J Hum Genet 80:298). ►mitochondrial genetics, ►paternal leakage, ►mitochondrial diseases in humans; Wiuf C 2001 Genetics 159:749; Innan H, Nordborg M 2002 Mol Biol Evol 19:1122.

Mitochondrial Suppressor: Enzyme complexes involved in oxidative phosphorylation are encoded by cooperation among nuclear (*PET*) and mitochondrial genes (*oli/ATP9*). Mutation in, e.g., in *AEP2* nuclear gene regulating *oli1* mRNA stability may prevent the formation of a functional subunit 9 of ATP synthase. Mutation in the 5′-untranslated region of *oli1* may however suppress the mutation in *aep2*. ►mitochondrial genetics; Chiang CS, Liaw GJ 2000 Nucleic Acids Res 28:1542; Alfonzo JD et al 1999 EMBO J 18:7056; Bennoun P, Delosme M 1999 Mol Gen Genet 262:85; Chen W et al 1999 Genetics 151:1315.

Mitochondrion: The singular form of mitochondria. ►mitochondria

Mitochondriopathies: ►mitochondrial diseases in humans

Mitogen: Collective name of substances stimulating mitosis and thus cell proliferation (such as growth factors).

Mitogen-Activated Protein Kinases: ►MAPK

Mitogenesis: Processes leading to cell proliferation.

Mitokinesis: A hypothetical mechanism of orderly distribution of mitochondria between two cells during cell division possibly using the cytoskeleton as a vehicle.

Mitomap: (human mitochondrial genome database): http://www.mitomap.org.

Mito-Mice: Mice models of mitochondrial diseases. They display deletions in their mitochondrial DNA. (See Sato A et al 2005 Proc Natl Acad Sci USA 102:16765).

Mitomycin C ($C_{15}H_{18}N_4O_5$): Antibiotic, antineoplastic agent that after activation causes DNA crosslinks, enhances mitotic recombination, inhibits DNA synthesis, etc.

MitoNuc: A database of nuclear-encoded mitochondrial proteins. ►mitochondrial genetics; http://bighost.area.ba.cnr.it/mitochondriome

MitoPark: Conditional knockout mice with Parkinsonism caused by mitochondrial malfunction due to dopamine neuron deficiency. ►knockout, ►Parkinson disease, ►dopamine

Mitosenes: DNA cross-linking compounds (such as mitomycin), frequently with antitumor activity. ►mitomycin

Mitosis: Nuclear division leading to identical sets of chromosomes in the daughter cells. Mitosis assures the genetic continuity of the ancestral cells in the daughter cells of the body. It involves one fully equational division preceded by a DNA synthetic phase. There is normally no (synapsis) pairing of the chromosomes. The centromeres split at metaphase separate at anaphase and the chromosomes relax after moving to the poles in telophase. The centromere–spindle association is regulated by a number of kinases (Mad1, Bub1, Msp1). Msp1 is essential for meiosis II but dispensable for mitosis. Before anaphase takes place, a number of proteins (Pds1, Scc1, Ase1) regulating sister chromatid cohesion are degraded by proteasomes under the control of the cyclosome (APC). The critical features of mitosis are diagrammed and comparable photomicrographs are also shown. Mitosis is thus, different from meiosis, in which there is one reductional and one numerically equational division. During meiotic prophase, the homologous chromosomes (bivalents) are synapsed and chiasmata may be observed. At the first meiotic division the centromeres do not split, and the undivided centromeres separate at anaphase I. Meiosis reduces the chromosome number to half, in contrast to mitosis, which preserves the number of chromosomes in the daughter cells. A comparison between mitosis and meiosis is also diagrammed. Mitosis and meiosis are the genetically most important processes of the eukaryotic cells (organisms).

Mitosis maintains genetic continuity in the development of an organism from conception to the end of its life. Somatic mutation may bring about changes in the exact continuity but this usually affects only a small fraction of the genes. The subsequent mitotic divisions provide the precise mechanism for the maintenance of such mutations. The mitotic and the meiotic cell cycles show some differences in the molecular mechanisms. In mitosis before the S phase, the two key cyclins, Clb5 and Clb6, are kept inactive by the *cyclin-dependent kinase inhibitor protein* Sic1. Two cyclins (Cln1 and 2) along with Cdc28/Cdc2 degrade Sic. In case Clb5 and Clb6 are non-functional, the initiation of the S phase is delayed until cyclins CLB1–4 are activated at a subsequent stage. In meiosis, Cdc28/Cdc2 is not required and its role is taken over by a similar protein, SPF (S phase promoting factor) encoded by the yeast gene *IME2*. Note that only a single S phase is required for meiosis I, and meiosis II does not require the synthesis of new DNA because it involves only the segregation of the sister chromatids after reduction. The pre-replication complex in yeasts includes Cdc18/CDC6 and the minichromosome maintenance proteins (MCM). For meiosis in fission yeast, these proteins are apparently not mandated. Also the vegetative replication checkpoint genes are not active when there are problems during the S-phase events. Cdc2 kinase and Cdc10 transcription factor and Cdc22 functions were also needed for meiosis. Mutation in DNA polymerases α and ε and ligase (Cdc17) functions delayed meiotic S phase. Mutation in DNA polymerase-δ (Cdc6) and in GEF (Cdc24), similarly to mitosis, reduced the number of divisions completed. ►meiosis, ►cell cycle, ►nucleolus, ►nucleolar organizer, ►nucleolar reorganization, ►CENP, ►cyclosome, ►proteosome, ►condensin, ►sister chromatid cohesion, ►mitotic exit, ►lamins, ►Cdc18, ►CDC6, ►MCM, ►cdc10, ►cdc22, ►Cdc2, ►APC, ►spindle fibers, ►centromere, ►kinetochore, ►centrosome, ►sister chromatid cohesion, ►separin, ►interphase, ►CDC13, see Figs. M89 and M90; Forsburg SL, Hodson JA 2000 Nature Genet 25:263; Tóth A et al 2000 Cell 103:1155; Nasmyth K 2001 Annu Rev Genet 35:673; Georgi AB

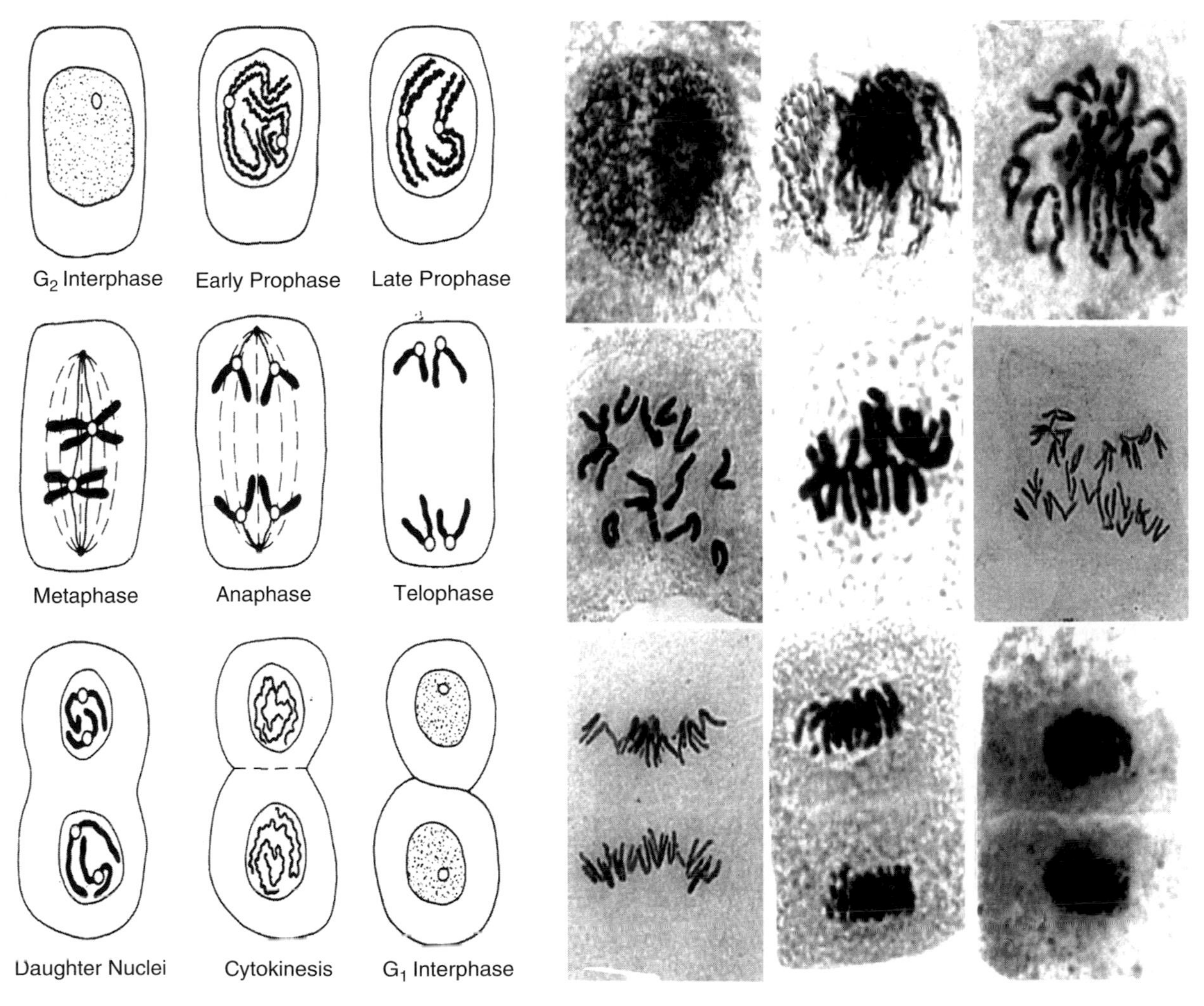

Figure M89. Mitosis. Left: The major steps of mitosis diagramed by only two chromosomes. Right: Photomicrographs of mitosis in barley 2n = 14 (Courtesy of T. Tsuchiya). Top Right to Left: Interphase, early prophase, late prophase, middle; early metaphase, metaphase, early anaphase, Bottom: Early telophase, late telophase, two daughter nuclei in the two progeny cells

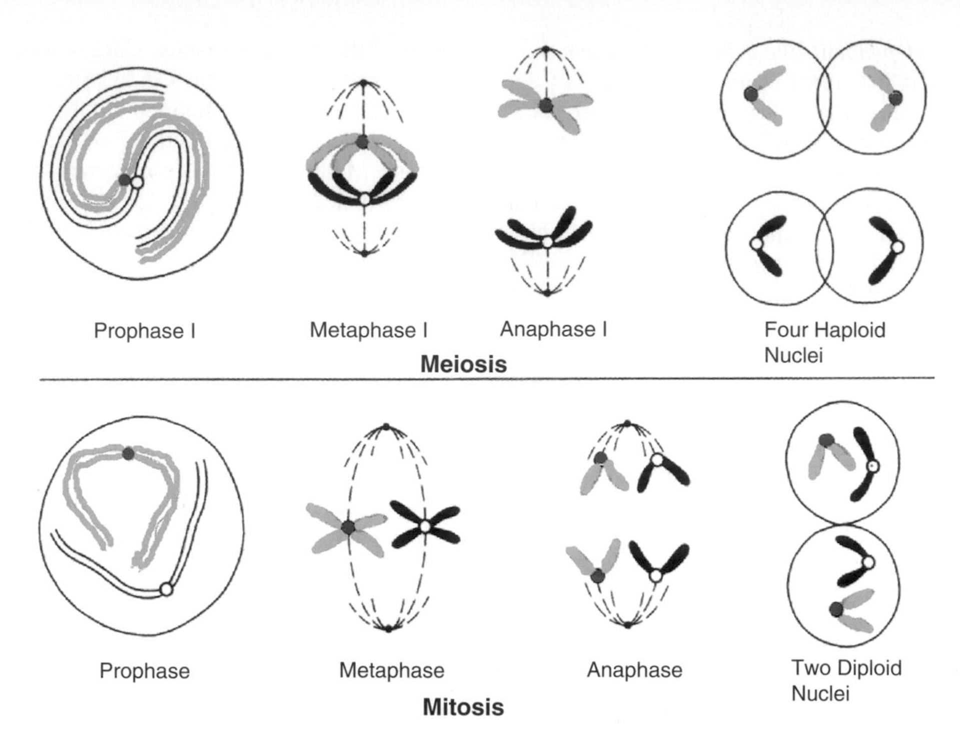

Figure M90. Meiosis and mitosis compared by cytological behavior. Meiosis: One numerical reduction & one numerically equational division synapsis & chiasma at prophase, centromeres do not split at metaphase and separate undivided at anaphase i chromosome number reduced. Mitosis: One fully equational division, no pairing & no chiasma, centromeres split in metaphase and separate at anaphase, chromosome number maintained in daughter cells

et al 2002 Curr Biol 12:105; Rieder CL, Kodjakov A 2003 Science 300:91; review of segregation of chromosomes by microtubules: Kline-Smith SL et al 2005 Current Opin Cell Biol 17:35.

Mitosome: A substantially reduced form of the mitochondrion. Mitosomes do not generate ATP but assemble iron-sulfur clusters, which are required for making ATP. Apparently, these organelles lost some of the mitochondrial functions during evolution (Tovar J et al 2003 Nature [Lond] 426:172), yet mitosomes seem to have some of the basic mitochondrial functions such as protein targeting (Dolezal P et al 2005 Proc Natl Acad Sci USA 102:10924). ►mitochondria, ►hydrogenosome, ►protein targeting, ►*Giardia*, ►*Trichomonas*

Mitospore: Meiotic product of fungi ready for mitotic divisions.

Mitostatic: Stopping or blocking the mitotic process. Many anti-cancer agents are mito-static.

Mitotic Apparatus: Subcellular organelles involved in nuclear divisions. Mitotic apparatus include organelles such as the spindle (microtubules), centromere (kinetochore), centriole, poles, and CENP. In yeast cells, the mitotic apparatus is within the nucleus, whereas in plants and animals the system is cytoplasmic and the nuclear membrane disappears, while the chromosomes are distributed to the poles and the membrane is reformed after completion of the process. In some organisms, the nuclear membrane does not disintegrate entirely during mitosis. ►nuclear membrane

Mitotic Catastrophe: Mitotic catastrophe is caused by DNA damage resulting in chromosomal abnormalities, polyploidy, and formation of micronuclei. CDH1 and the slow breakdown of cyclin B1 interfere with the metaphase promoting complex. ►CDH1, ►cyclin B1, ►micronucleus; Huang X et al 2005 Proc Natl Acad Sci USA 102:1065.

Mitotic Center: ►centrosome

Mitotic Chromosomes: Chromosomes undergoing mitosis.

Mitotic Crossing Over: Recombination in somatic cells. Recombination is generally a meiotic event (►crossing over) but in some organisms, the chromosomes may associate also during mitosis and this may be followed by genetic exchange of the linked markers. Although this may take place spontaneously, several agents capable of chromosome breakage (radiation,

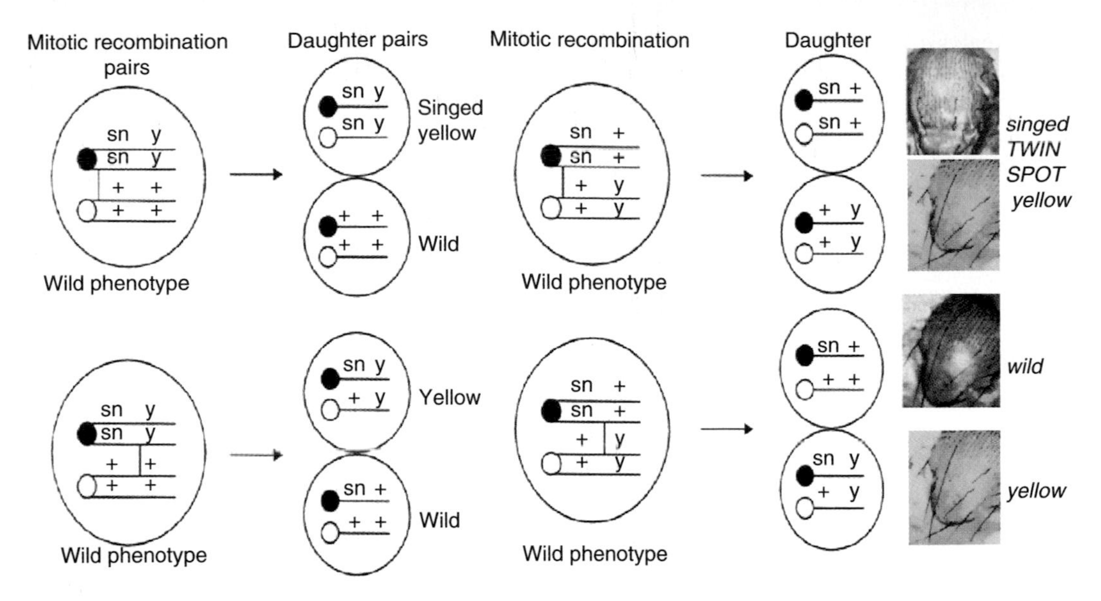

Figure M91. Mitotic crossing over (at sign |) with markers in coupling and repulsion and the consequences of recombination. (On the basis of experiments of Curt Stern,1936.Genetics 21:625.)

chemicals) may enhance its frequency. For the detection of somatic recombination in higher eukaryotes, the markers must be cell and tissue autonomous, i.e., they should form sectors (spots) at the locations where such an event has taken place. In the first experiments with *Drosophila*, the chromosomal construct diagrammed here, was used. If an exchange has taken place between *sn* and *y*, then only a yellow sector (homozygous for *y*) was formed. *Twin spots* were observed only when in repulsion, the exchange took place between the proximal locus and the centromere (see Fig. M91).

Somatic recombination takes place in the *Drosophila* male too, although this usually does not happen in meiosis. The characteristics of somatic recombination are that exchange is between two chromatids at the four-strand stage, but instead of reductional division, as in meiosis, the centromeres are distributed equationally. Mitotic recombination is a relatively rare event in higher eukaryotes. Somatic recombination can be studied also in plants when the chromosomes are appropriately marked, and generative progeny can be isolated from the sectors, either from the sectorial branches or by tissue culture techniques. In fungi (lower eukaryotes), in which a longer diploid phase exists (*Aspergillus*), mitotic recombination can be used even for chromosomal mapping, because some of the parental and crossover products can selectively be isolated. The meiotic and mitotic maps are collinear, but the relative recombination frequencies in the different intervals vary.

Homologous mitotic recombination in mouse can easily be detected in a transgenic construct. Similar techniques are available also for plants. Two copies, of a differently truncated vector cassette, containing the enhanced yellow fluorescent protein (EYFP), are introduced into the animal cells. The fluorescence is expressed only when the recombination, gene conversion, or repair of collapsed replication fork takes place within the coding sequence, shown in the diagram. Spontaneous recombination in primary fibroblasts from adult ear tissue occurred at 1.3 ± 0.1 per 10^6 per cell division. In embryonic fibroblasts, or if Mitomycin C was used, the rate was found to increase by an order of magnitude.

Homology of the chromosome is required in both meiotic and mitotic recombination (Shao C et al 2001 Nature Genet 28:169). A single nucleotide difference may reduce recombination frequency to a detectable extent. Larger sequence differences may reduce recombination by two-three orders of magnitude or eliminate it completely.

Some researchers assume that the mechanisms of the meiotic and mitotic recombination are the same. Several facts, however, cast some doubt about this view. Many proteins are involved in both meiotic and mitotic recombination. In yeast, e.g., both Rad51 and Dmc1 are required for meiotic recombination, but Dmc1 is not specific for the mitotic event (Shinohara M et al 2003 Genetics 164:855). In general, caution must be taken in interpreting genetic phenomena in somatic cells, unless classical progeny tests or molecular information can confirm the assumptions. (See Figs. M92 and M93, ►mitotic mapping, ►parasexual mechanisms, ►mitotic recombination as a bioassay in genetic toxicology, ►twin spot, ►recombination, ►GUS; Pontecorvo G 1958 Trends in Genetic Analysis. Columbia University Press, New York; McKim KS et al 2002 Annu Rev Genet 36:205.

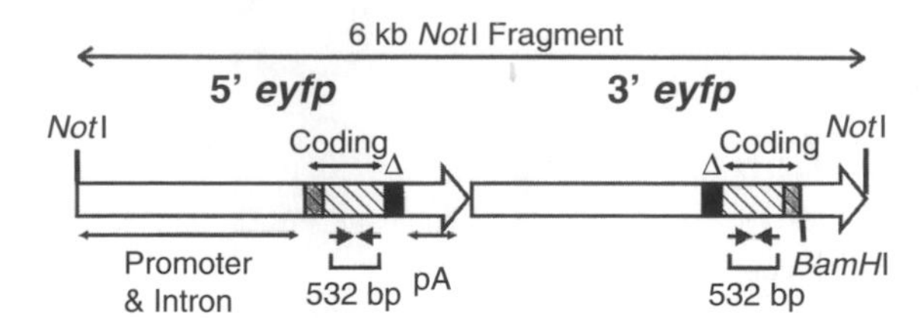

Figure M92. The tandem arrangements of the two deletion (Δ) constructs are represented by the arrows. The coding sequences are hatched. The chicken β-actin promoter and a cytomegalovirus enhancer drive the expression. NotI, BamHI are restriction enzymes, pA is polyadenylation signal. (By permission from CA Hendricks, KA Almeida, MS Stitt, VS Jonnalagedda, RE Rugo, GF Kerrison & BP Engeward 2003 Proc Natl Acad Sci USA 100:6325. Copyright 2003 National Academy of Sciences USA)

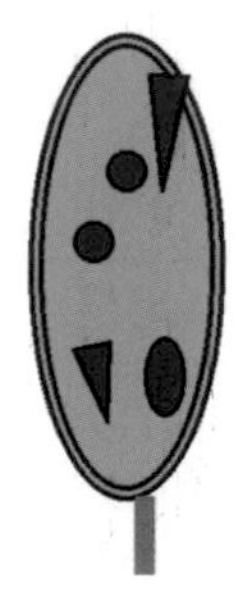

Figure M93. Plants heterozygous for transgenic constructs of the *uidA* (β-glucuronidase, GUS) alleles, which have not-overlapping defects upon spontaneous or induced (by DNA-damaging agents) recombination within the locus produce cells with restored enzyme function. By providing the appropriate substrates, the staining of the spots reveal the recombinational event at a frequency of 10^{-6} to 10^{-7} per cell. The size and shape of the spots or sectors are determined by the pattern of differentiation of the tissue. The recombination can be further verified by polymerase chain reaction at the site as well as by the presence of molecular markers flanking the gene. Schemating drawing based on the report by Swoboda P et al 1994 EMBO J 13:484)

Mitotic Drive: Stepwise expansion of trinucleotide repeats. ►trinucleotide repeats, ►anticipation, ►meiotic drive; Khajavi M et al 2001 Hum Mol Genet 10:855.

Mitotic Exit: In a mitotic exit, after metaphase the mitotic spindle moves the chromosomes during anaphase to the poles of the cell and eventually two daughter cells are formed. PDS and Clb5 protein components of the anaphase inhibitory complex inhibit the formation of the spindle apparatus. Both PDS and Clb5 need to be degraded by the activation of Cdc20 to prepare the path for exit from mitosis. PDS degradation activates the Cdc14 phosphatase. Clb5 kinase degradation facilitates the dephosphorylation of Sic1, Cdh1, and various cyclin kinases by Cdc14, and thus permits the transition from anaphase into G1 interphase. ►FEAR, ►PDS, ►Clb5, ►Cdc20, ►Cdc14, ►MEN, ►Sic1, ►Cdh1, ►mitosis; Bardin AJ, Amon A 2001 Nature Rev Mol Cell Biol 2:815; role of CDC14: Stegmeier F, Amon A 2004 Annu Rev Genet 38:203.

Mitotic Index: The fraction of cells involved in the process of mitosis at a particular time.

Mitotic Mapping: Genetic recombination during mitosis is generally a very rare event in the majority of organisms. In some diploid fungi, its frequency may reach 1 to 10% that of meiosis. As a consequence of ionizing irradiation and certain chemicals the frequency of mitotic exchange may increase and on the basis of mitotic recombination genetic maps can be constructed (see Fig. M94). Although the order of genes is the same in mitotic and meiotic maps, the recombination frequencies may be quite different. ►mitotic crossing over, ►parasexual mechanisms, ►mitotic recombination as a bioassay in genetic toxicology

Mitotic Non-conformity: Genetic change in fungi due to chromosome translocations.

Mitotic Recombination: ►mitotic crossing over, ►mitotic mapping

Mitotic Recombination as a Bioassay in Genetic Toxicology: Mitotic recombination has been used in

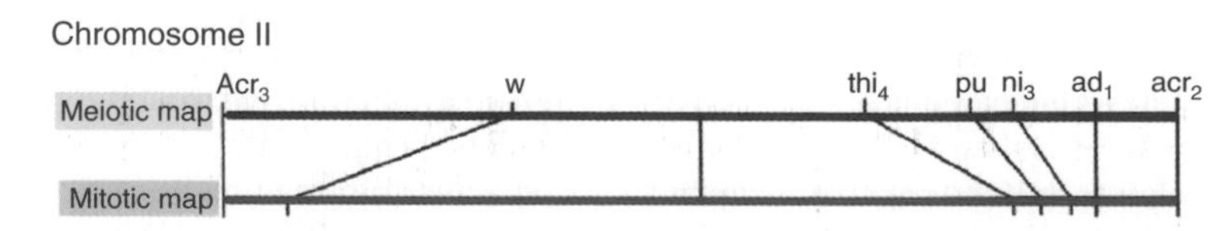

Figure M94. Comparison of a segment of the meiotic and mitotic maps of the fungus *Aspergillus nidulans* at identical scales. (Redrawn after Pritchard, RH 1963. Methodology in Basic Genetics, Burdette, WJ, ed., p. 228. Holden-Day, San Francisco.)

Saccharomyces cerevisiae yeast. In a diploid strain (D5) heterozygous for *ade2–40* and *ade2–119* alleles, twin spots are visually detectable in case of mitotic crossing over. Homozygosity for *ade2–40* has an absolute requirement for adenine and involves formation of red color; it is often called red adenine mutation. Homozygosity for *ade2–119*, a leaky mutation, results in pink coloration. Since the two genes are complementary, the heterozygous cells do not require adenine and the colonies are white. Any environmental or chemical factor that promotes mitotic recombination is thus detectable. This procedure has not been extensively used in genetic toxicology programs. ►mitotic crossing over, ►bioassays in genetic toxicology

Mitotic Spindle: ►spindle, ►spindle fibers

Mitoxantrone ($C_{12}H_{28}N_4O_6$): Antineoplastic drug with the ability to break DNA double strands.

Mitral Cells: Mitral cells can make junctions in a beveled fashion.

Mitral Prolapse: A buckling of the heart atrial leaflets as the heart contracts, resulting in backward flow (regurgitation) of the blood. It is caused by an autosomal (11p15.4) dominant allele and it is a common congenital heart anomaly affecting 4–8% of young adults, particularly females. It may accompany various other syndromes such as ►Marfan syndrome, ►Klinefelter syndrome, ►osteogenesis imperfecta, ►Ehlers-Danlos syndrome, ►fragile X syndrome, and ►muscular dystrophies. ►Non-syndromic mitral valve prolapse has been mapped to 16p11.2-p12.1 and a similar disease is cased by mutation at Xq28. In an autosomal recessive form, ophtalmoplegia (paralysis of the eye muscles) is also involved. (See ►heart disease and the other conditions named at separate entries, Towbin JA 1999 Amer J Hum Genet 65:1238; Freed LA et al 2003 Am J Hum Genet 72:1551).

Mixed Backbone Oligonucleotide (MBO): MBO is composed of a phosphorothionate backbone and 2′-O-methyloligoribonucleotides or methylphosphonate oligodeoxyribonucleotides. Such an antisense construct is resistant to nucleases and forms stable duplexes with RNA. ►antisense technologies

Mixed Lymphocyte Reaction (MLR): MLR is used for the detection of histoincompatibility in vitro. The principle of the test is that lymphocytes of two individuals are mixed. One (donor) serves as *responder* and the other as *stimulator*. The proliferation of the antigen-presenting cells is blocked by irradiation or by mitomycin C. The CD4 T cells proliferate as they recognize foreign MHC II molecules, and the measure of their proliferation is assessed by the incorporation of H^3-thymidine. The cytotoxic CD8 T cells recognize differences primarily in MHC I molecules and the extent of the killing reaction is detected by using Cr^{51} (sodium dichromate)-labeled cells. In seven days, the reaction may detect histoincompatibility between two persons. ►MHC, ►lymphocytes, ►HLA, ►microcytotoxicity test

Mixed-Function Oxidases: Generally, flavoenzymes that oxidize NADH and NADPH, and in the process, may activate promutagens and procarcinogens. ►promutagen, ►procarcinogen, ►P-450, ►microsomes, ►monooxygenases

Mixoploid: In a mixoploid, the chromosome numbers in the different cells of the same organism vary. (Bielanska M et al 2002 Fertility & Sterility 78:1248).

Miyoshi Myopathy (MM, 2p13.3-p13.1): ►dysferlin

MKDOM: A protein domain analysis program (Gouzy J et al 1999 Comput Chem 23:333).

MKI: MAP kinase insertion factor. ►MAPK

MKK: A homolog of MEK. ►MEK

MKP-1: MKP-1 dephosphorylates Thr^{183} and Tyr^{185} residues, and thus regulates mitogen-activated MAP protein kinase involved in signal transduction. ►MAP, ►PAC

Mle (*maleless*): *mle* is located in *Drosophila* chromosome 2–55.2, *mle3* in *Drosophila* chromosome 3.25.8, and similar genes are also present in the X-chromosomes. The common characteristics are that homozygous females are viable but the homozygous males die. The underdeveloped imaginal discs of *mle3* may develop normally, if transplanted into wild type larvae. MLE protein appears to be an RNA helicase and its amino and carboxyl termini may bind double-stranded RNA. Ribonuclease releases MLE from the chromosomes without affecting MSL-1 and -2 and RNA. MLE and MSL proteins appear to control dosage compensation. ►*Msl*, ►dosage compensation

MLE (mariner-like element): A transposable element in *Drosophila* that is present also in the human genome where it is responsible for a recombinational hot spot in chromosome 17. ►mariner, ►Charcot-Marie-Tooth disease, ►HNPP, ►MITE

MLH (muscle enhancer factor, MLH2A): A MADS box protein inducing muscle cell development in cooperation with basic helix-loop-helix proteins. ►MyoD, ►MADS box, ►bHLH

MLH2: ►hereditary non-polyposis colorectal cancer

MLH3: A mismatch repair gene that occurs with frequent variation in hereditary nonpolyposis cancer.

M

▶mismatch repair, ▶hereditary non-polyposis colorectal cancer; Wu Y et al 2001 Nature Genet 29:137.

MLINK: A computer program for linkage analysis.

MLK (mixed lineage group kinases): Members of the MAP kinase family of phosphorylases mediating signal transduction. ▶MAP kinase, ▶signal transduction; Gallo KA, Johnson GL 2002 Nature Rev Mol Cell Biol 3:663.

MLL (mixed lineage leukemia, All-1, Htrx): The protein encoded in human chromosome 11q23 has four homology domains, the A-T hook region, DNA methyltransferase homology, Zinc fingers, and the *Drosophila trithorax* (*trx*, 3–54.2) homology. Overexpression of the MLL5 protein, encoded at 7q22, inhibits the cell cycle. In its vicinity at 7q36, MLL3 is situated. Loss of the long arm 7 leads to myeloid malignancies. Other MLL genes were located to chromosomes 12q12 and 19q13. ▶leukemia, ▶cancer stem cell, ▶methyltransferase, ▶SET, ▶COMPASS; Deng L-W et al 2004 Proc Natl Acad Sci USA 101:757.

MLP1, MLP2: Proteins associated with the nuclear pore complex that suppress the export of unspliced mRNA (Galy V et al 2004 Cell 116:63). ▶nuclear pores, ▶PML39

MLS (maximum likelihood lod score, multipoint lod score): A method for the analysis of linkage (in human families). The MLS increases when the proportion of relatives carrying an allele identical by descent is higher than expected on the basis of the degree of relatedness and independent segregation. ▶relatedness degree, ▶maximum likelihood, ▶lod score, ▶maximum likelihood method applied to recombination

MLV: ▶Moloney mouse leukemia virus

MLVA (multi-locus variable-length repeat analysis): Basically, the same as VNTR, used for the identification of related but different bacterial strains. ▶VNTR

mm: Prefix for mouse (*Mus musculus*) protein or DNA, e.g., mmDNA.

MMP (microsatellite mutator phenotype): ▶microsatellite mutator

MMP: Matrix metalloprotease. ▶metalloproteinases

MMR: ▶mismatch repair

MMTV: Murine mammary tumor virus.

M-MuLV: Moloney murine leukemia virus.

MMTV: ▶mouse mammary tumor virus (MTV)

MN Blood Group (MN): The human chromosomal location is 4q28-q31. α-sialoglycoprotein (glycophorin A) is responsible for the M blood type, while the δ peptide (glycophorin B) is responsible for the N type (see Fig. M95). Glycophorin deficient erythrocytes are resistant to *Plasmodium falciparum* (malaria). The En(a-) blood group variants also lack glycophorin A. The M and N alleles are closely linked to the S/s alleles and the complex is often mentioned as a MNS blood group. The frequencies of the allelic combinations in England were MS (0.247172), Ms (0.283131, NS (0.08028), and Ns (0.389489). ▶blood groups

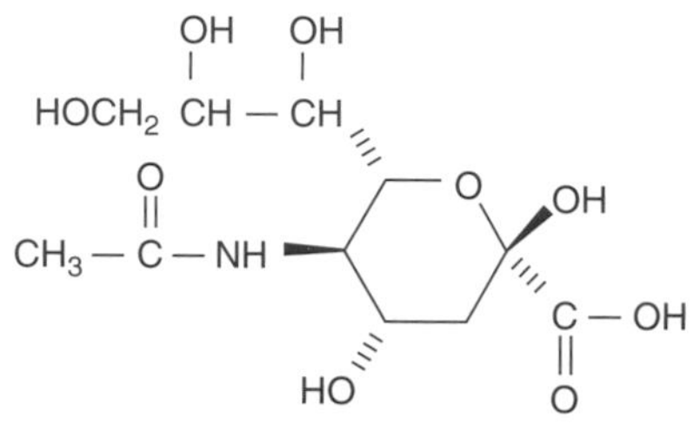

Figure M95. Sialic acid

MND2: An antagonist of the anaphase-promoting complex during meiotic prophase. ▶anaphase-promoting complex; Penkner AM et al 2005 Cell 120:789.

MNEMONS: L. Cuénot's historical (early 1900s) term for genes.

MNGIE (mitochondrial neurogastrointestinal encephalopathy): An autosomal recessive disease involving myopathy with ragged-red fibers, reduced activity of the respiratory chain enzymes, etc. The basic defect is attributed to the deficiency of TK2 (a thymidine kinase), which phosphorylates normally also the deoxynucleosides of other pyrimidines. The reduction of the activity of the enzyme apparently affects the maintenance of mtDNA. The TK2 gene has been located to human chromosome 16q22 (earlier to chr. 17), and more recently MNGIE was assigned to 22q13.32-qter. ▶mtDNA, ▶mitochondrial diseases in humans

MNNG: *N*-methyl-*N'*-nitro-*N*-nitrosoguanidine is a monofunctional alkylating agent and a very potent mutagen and carcinogen (see Fig. M96) (rapidly decomposing in light). It methylates the O^6-position of guanine. Cell lines exist that are highly resistant to the cytotoxic effects of MNNG, but they are even more sensitive to the mutagenic effects. ▶mutagens, ▶alkylating agents

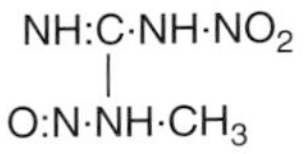

NH:C·NH·NO$_2$
|
O:N·NH·CH$_3$

Figure M96. Methylnitrosoguanidine

MNU (*N*-methyl-*N*-nitrosourea): A monofunctional mutagen and carcinogen forming O^6-methylguanine (see Fig. M97). ►mutagens, ►alkylating agents, ►monofunctional

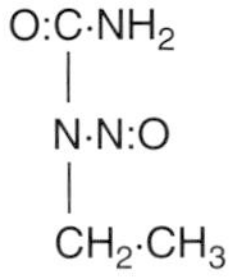

O:C·NH$_2$
|
N·N:O
|
CH$_2$·CH$_3$

Figure M97. Ethylnitrosourea

MØ: ►macrophage

MO15: A CDK-activating kinase, related to CAK. Association of MO15 with cyclin H greatly increases its kinase activity toward Cdk2. ►cell cycle, ►cyclin, ►CDK, ►CAK, ►Cdk2

MoAb: ►monoclonal antibody

mob: The *mob* bacterial gene facilitates the transfer of bacterial chromosome or plasmids into the recipient cell. In order to transfer the plasmids, there is a need for the cis-acting *nick* site and a *bom* site. At the former the plasmid is opened up (nicked) and the *bacterial origin of mobilization* (*bom*) makes possible the conjugative transfer. ►Hfr, ►conjugation

Mobile Genetic Elements: Mobile genetic elements occur in practically all organisms, represent different types of mechanisms, and serve diverse purposes. They share some common features. They are capable of integration and excision from the genome (like the temperate bacteriophages), or movement within the genome (like the insertion and transposable elements). These elements may fulfill general regulatory functions in normal cells (such as the switching of the mating type elements in yeast). Phase variation in *Salmonella*, antigenic variation as a defense system in bacteria (*Borrelia*) and protozoa (*Trypanosomas*), parasitizing plant genomes by agrobacteria, etc., may depend on such elements. They have a role in the generation of antibody diversity in vertebrates by transposition of immunoglobulin genes. They have played an important role in the evolution of the genomes (Kazazian HH Jr 2004 Science 303:1626). There is an apparent paucity of DNA transposon mobilization in mammals, and in amniotes in general in comparison with other organisms. (e.g., plants) This paucity may result from the relative difficulty in horizontal transfer into animals' germ lines (Han K et al 2007 Science 316:238), although about 50% of the genome of simians consists of transposable elements. (See items under separate entries, ►transposable elements, ►SDR, ►SINE, ►LINE, ►Alu, ►organelle sequence transfers, ►transposons, ►retroposons, ►retrotransposons, ►pathogenecity islands; classification: http://aclame.ulb.ac.be/.

Mobility Shift Assay: When two molecules (DNA-protein, protein-protein) bind, upon electrophoretic separation the mobility is retarded. ►gel retardation assay; Filee P et al 2001 Biotechniques 30:1944.

Mobilization: In mobilization, the binding of the ribosome to a mRNA initiates polysome formation. Mobilization also refers to the process of conjugative transfer; and to the release of a compound in the body for circulation.

Mobilization of Plasmids: The transfer of conjugative plasmids to another cell. ►conjugation

Mobilome: Potentially mobile part of the genome, such as episomes, transposons, integrons, and genomic islands.

Möbius Syndrome: In this condition, congenital partial or full paralysis and dysfunction of a cranial nerve, face and limb malformations, and mental retardation also may occur. A dominant locus was found at 3q21-q22 and another dominant on the long arm of chromosome 10. Recessive and X-linked inheritance was also suspected.

Mod Score: A maximized lod score over recombination fractions (θ) and genetic models (φ).►lod score, ►model genetic

Modal: Adjective of mode. ►mode

Mode: The value of the variates (class of measurements) of a population that occurs at the highest frequency. ►mean, ►median

Model: A model represents the essential features of a concept with minimum detail. It assists in making predictions about an operation under different conditions and sheds light on the contribution of the components of the system. In the *procedural system* of modeling, the focus is on the steps from the data to the conclusion. An example: looking for binding sites of transcription factors in the promoter regions of genes to shed light on expression by searching for over-represented elements in the promoters of each gene cluster. Or we can test the genes with similar binding sites to determine whether they are co-expressed. In the *declarative approach*, the modeling extends to both binding sites in the promoter and to expression levels. In the next step, the validity of the model is tested by the maximum likelihood estimation regarding its fit to the data available. The testing of the sequences may be evaluated by the hidden Markov chain procedure. ►maximum likelihood, ►hidden Markov chain model, ►probabilistic graphical models of cellular networks, ►small-world networks, ►modeling; Friedman N 2004 Science 303:799.

Model, Genetic: In human genetic analysis, the genetic model is specified by the type of inheritance, e.g., dominant/recessive, autosomal/sex-linked, penetrance/expressivity, frequency of phenocopies, mutations, allelic frequencies, pattern of onset, etc.

Model, Molecular: Three-dimensional physical representation of molecules. (See molecular modeling for beginners: http://www.usm.maine.edu/~rhodes/index.html; http://www.natsci.org/Science/Compchem/feature14b.html).

Model Organisms: All organisms share basic molecular mechanisms. Relatively simple biological systems of common evolutionary descent facilitate rigorous experimental studies because of ease of manipulation, short lifecycle, small genomes, etc. "Model organisms" are frequently of no economic interest beyond making research efficient, fast and less demanding in labor and funds yet they make possible the understanding basic genetic and molecular mechanisms common in other biological systems. About 75% of the human disease genes have some counterparts in *Drosophila* and even particular genes of the plant *Arabidopsis* display about 40% similarity to a human disease gene, e.g., responsible for breast cancer. Such models are *E. coli* for prokaryotes, *Saccharomyces cerevisiae* for fungi and other eukaryotes, *Caenorhabditis elegans, Drosophila melanogaster,* zebrafish for lower animals and humans, mouse and rat for higher animals and humans, *Arabidopsis thaliana* for higher plants and can provide useful information for more complex systems. No single "model organism" is suitable for the study of all biological problems. Obviously dogs can be exploited better for the study of behavior than yeast. For the study of chromosomal mechanisms broad bean (*Vicia faba*) is better than *Arabidopsis*. The use of several types of research methods is immoral or otherwise objectionable in humans (e.g., controlled mating, testing of new drugs and mutagens, genetic engineering) and mouse models are indispensable and invaluable. Since DNA sequence and proteome information is becoming available generalizations on functions, development and evolution is much facilitated. Model organisms have provided useful means to study human disease, regulatory networks, development, behavior, comparative analysis of gene function and other basic and applied biological problems. The animal disease models usually do not fully represent the human conditions. (See Reinke V, White KP 2002 Annu Rev Genomics Hum Genet 3:153, survey of recent genetic technologies: Nagy A et al 2003 Nature Genet 33 (Suppl): 276; Davis RH 2004 Nature Rev Genet 5:69; Neff MW, Rine J 2006 Cell 124:229; ►Homophila; human disease *Drosophila* gene database: http://superfly.ucsd.edu/homophila; http://www.nih.gov/science/models/).

Model-Free Analysis: A term from human genetics indicating no need for a genetic model because the persons and ancestors have already been genotyped. ►genotyping, ►model genetic

Modeling: Physical, mathematical, and/or hypothetical construction for the exploration of the reality or mechanism of theoretical concepts. In modeling a protein structure, it is assumed that that the amino acid sequence specifies the three-dimensional structure. The normal structure represents most probably the global free-energy maxima. ►simulation, ►Monte Carlo method, ►model; protein structure prediction: http://predictioncenter.org; protein modeling principles and tools: Schueler-Furman O et al 2005 Science 310:638; annotated mathematical models of biological systems: http://www.ebi.ac.uk/biomodels/.

Modem (modulator/demodulator): The modem either links the computer to another computer through a telephone line, e.g., fax modem (sends printed and graphic information), or a modem sends and retrieves information from a mainframe computer to other computer operators through information networks such as BITNET, INTERNET, and various online services.

Modification: Most commonly in molecular biology, methylation. ►methylation of DNA, ►methylase

Modified Bases: Modified bases occur primarily in tRNA and are formed mainly by post-transcriptional alterations. More than 35 different modified nucleosides were known by the 1970s (Hall RH 1971 The Modified Nucleosides in Nucleic Acids. Columbia Univ. Press, New York), and over 100 became known later (http://medlib.med.utah.edu/RNAmods/). The most common modified nucleosides are ribothymidine, thiouridine, pseudouridine, isopentenyl adenosine, threonyl-carbamoyl adenosine, dihydrouridine, 7-methylguanidine, 3-methylcytidine, 5-methylcytidine, 6-methyladenosine, inosine, etc. These modified bases have important roles in the function of tRNA, e.g., 1-methyladenosine in the TψC loop at position 58 facilitates the translation from the $tRNA_1^{Met}$ initiation codon of yeast. ►transfer RNA, ►ribosome, ►initiation codon, ►thiouracyl; Anderson J et al 2000 Proc Natl Acad Sci USA 97:5173; http://genesilico.pl/modomics/index2.pt; rRNA small subunit: http://medlib.med.utah.edu/SSUmods/.

Modifier Gene: A modifier gene affects the expression of another gene. ►epistasis; Nadeau J 2001 Nature Rev Genet 2:165; Burghes AHM et al 2001 Science 293:2213.

Modified Mendelian Ratios: These ratios are observed when the product(s) of genes interact. Such situations in case of two loci can be best represented by modified checkerboards using the zygotic rather than the gametic constitutions at the top and at the left side of the checkerboards.

At the top, a standard constitution is shown where for each locus the genotypes are 1*AA*, 2*Aa*, 1*aa*, and 1*BB*, 2*Bb*, 1*bb*, respectively, etc. Although in the boxes only the relative numbers of the phenotypes (genotypes) are shown (as a modification of the 9:3:3:1 Mendelian digenic ratio), their genetic constitution can be readily determined from the top left checkerboard. These schemes assume complete dominance. Additional variation in the phenotypic classes occurs in case of semidominance or codominance and the involvement or more than two allelic pairs. In common usage these modifications are mentioned as gene interactions but actually the products of the genes do interact.

Sometimes it is not quite easy to distinguish between two segregation ratios within small populations and statistical analysis may be required. Example: let us assume that we wish to ascertain whether the segregation is 3:1 or 9:7. With 3:1 the standard error of the recessives is $\sqrt{\frac{3n}{16}}$ and with 9:7 it is $\sqrt{\frac{63n}{256}}$. Since deviations from the expectation can be either + or −, the statistically acceptable misclassification of 0.025 can rely on the deviate of 0.05. From a table of the normal deviates (►normal deviate), the expected number of recessives must be or exceed 1.9599 times the standard error. Thus, for 9:7 (7/16n - recessives) $= 1.9599\sqrt{\frac{63n}{256}}$ and for the 3:1 segregation the expected is $r - 1/4(n) = 1.9599\sqrt{\frac{3n}{16}}$.

By adding up, we get:

$$n([7/16] - [1/4] = 1.9599\sqrt{n}\left(\sqrt{\frac{63}{256}}+\sqrt{\frac{3}{16}}\right.$$

$$\text{and } \sqrt{n} = \frac{16}{3}\left\{1.9599\left[\left(\frac{7.9373}{16}\right)+\left(\frac{1.7321}{4}\right)\right]\right\}$$

$$= 937121.$$

Since $n = 9.7121^2 - 94.32$, about 95 individuals permit a distinction between the two hypothetical segregations at a level of 0.025 probability. In a similar way, other segregation ratios can also be tested. ►Mendelian segregation, ►Punnett square, ►semidominance, ►codominance, ►phenotype, ►genotype, ►normal deviate, see Fig. M98.

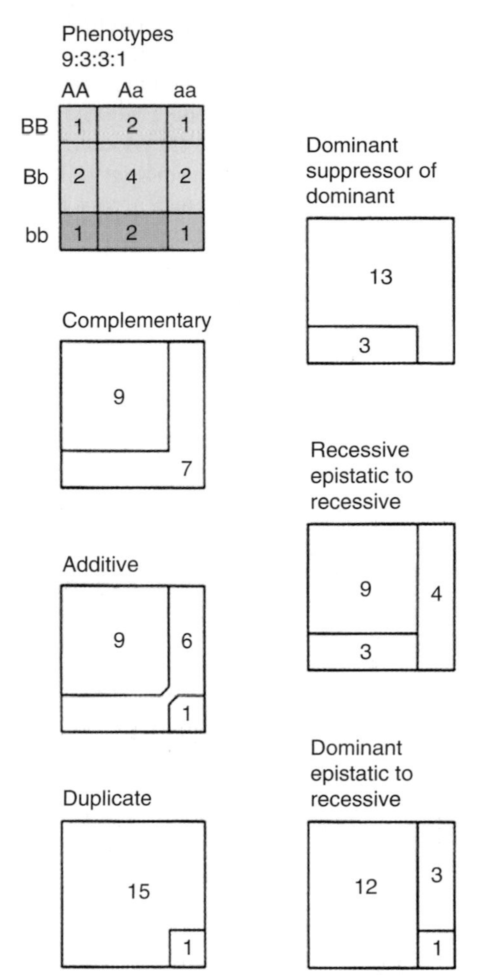

Figure M98. Modified Mendelian ratios

Modularity: Separability of a pattern into independently functionable units of a complex system. Developmental domains may be formed by specific transcription factors. These modules may subsequently affect other sets of transcription factors in a series of cascades of gene expression until the final differentiated structures arise. (See evolution of modularity: Kashtan N, Alon U 2005 Proc Natl Acad Sci USA 102:13773).

Modulation: Reversible alteration of a cellular function in response to intra- or extracellular factors.

Module: Single or multiple motif containing structural and functional units of macromolecules (proteins, DNA, RNA) and possibly small molecules. Their function is not a simple sum of that of the isolated components. Modules may be combined in alternative patterns and increase the functionality of the cell. Modules may be spatially isolated in the cell, e.g., as a ribosome, the complex site of polypeptide synthesis, or may only functionally be separated as, e.g., a particular signal transduction pathway. Modules have the ability, however, for hierarchial interaction. One evolutionary advantage of the modular organization is that a mutation in a modular component may alter the function of that module without an upset in the whole system of the organism. Module components involved in a central role of the biology of the cell, e.g., histones, must be well conserved across phylogenetic ranges. Gene modules are coexpressed groups of genes within a genetic network. ►motif, ►genetic network, ►regulated gene, ►Williams syndrome; Bar-Joseph Z et al 2003 Nature Biotechnol 21:1337.

Modulon: ►origon

Modulins: Protein factors modulating bacterial infection. (See Neff L et al 2003 J Biol Chem 278:27721).

MODY (maturity-onset diabetes of the young): Apparently dominant, monogenic forms of familial diabetes with an onset at or after puberty but generally before age 25. Human chromosomal location is 20q11.2 (MODY1), 7p13 (MODY2), or 12q24 (MODY3), and it is a heterogeneous disease. MODYs are responsible for about 2–5% of non-insulin-dependent cases of diabetes. MODY1 is responsible for the coding of HNF-4α, a hepatocyte nuclear transcription factor, a member of the steroid/thyroid hormone receptor family and an upstream regulator of HNF-1α. MODY2 encodes the glycolytic glycokinase, which generates the signal for insulin secretion. MODY3 displays defects in the hepatocyte nuclear factor HNF-1a, a transcription factor that normally transactivates an insulin gene. SHP/NRPB2 (1p36.1) modulates the expression of HNF-4α and contributes to obesity. ►diabetes mellitus, ►diabetes insipidus, ►GLUT; Barrio R et al 2002 J Clin Endocrin Metab 87:P2532.

Moebius Syndrome: The major characteristics of the moebius syndrome are the facial paralysis of the sixth and seventh cranial nerves and often limb deformities and mental retardation. The dominant disorder was located to human chromosome 13q12.2-q13. The recurrence rate is below 1/50. ►neuro-muscular diseases, ►mental retardation, ►periodic paralysis, ►limb defects

Mohr Syndrome: ►orofacial digital syndrome II

Mohr-Tranebjaerg Syndrome (ddp): A recessive deafness encoded in xq21.3-xq22, involving poor muscle coordination, mental deterioration, but not blindness. In the same chromosomal region there is a transcribed region, DFN-1, which shows symptoms similar to DDP, but alsoblindness. ►deafness

MOI: ►multiplicity of infection

Molar Solution: 1 gram molecular weight compound dissolved in a final volume of 1 L.

Mole: *Talpa europaea*, 2n = 34, an insect-eating small underground mammal (see Fig. M99).

Figure M99. Mole

Mole: ►gram molecular weight; ►fleshy (placental) neoplasia

Mole: An abnormal mass of tissue in the uterus, developed from a degenerated oocyte. Alternatively, it may be of neoplastic nature. Hydatidiform moles may occur when one or two spermatozoa fertilize an enucleated egg. Moles frequently occur in the melanocytes of the skin and are generally benign and limited in growth. Mutations in the protein kinase BRAF, a downstream effector of the RAS oncogene, occasionally progress toward melanoma. Generally, the mutation induces cell cycle arrest and the induction of $p16^{INK4a}$, a tumor suppressor. In addition, the cells express a senescence-associated acidic β-galactosidase (SA- β-gal). The $p16^{INK4a}$ and other factors apparently prevent proliferation and the cells and the mole remains senescent. This is an evidence of RAS oncogene induced senescence (Michaloglou C et al 2005 Nature [Lond] 436:720). ►nevus, ►tumor suppressor, ►galactosidase, ►senescence, ►RAS, ►melanoma, ►hydatidiform mole, ►$p16^{INK4a}$

Molecular Ancestry Network: http://www.manet.uiuc.edu/.

Molecular Beacon: A molecular tool for allele discrimination; it is a short hairpin oligonucleotide probe that binds to a specific oligonucleotide sequence and produces fluorescence signal. ►beacon molecular; Tyagi S et al 1998 Nature Biotechnol 16: 19; Rizzo J et al 2002 Mol Cell Probes 16(4):277.

Molecular Biology: The study of biological problems with physical and chemical techniques and interpretation of the functional phenomena on macromolecular bases. (See database: http://nar.oupjournals.org).

Molecular Breeding: The goal of molecular breeding is to develop highly efficient new vectors for genetic engineering. The procedure includes the reshuffling the coding regions of viral glycoprotein genes, which determine tropism of the proteins. This is followed by selection of vectors with exquisite specificity, improved gene transfer, etc. Molecular breeding may also mean application of molecular biology methods for the development of new crops or animal stocks. ►molecular pharming, ►plantibody, ►plant vaccine; Stöger E et al 2002 Mol Breeding 9(3):149, http://www.phytome.org.

Molecular Chaperone Heterocomplex (MCH): MCH includes heatshock proteins Hsp90 and Hsp70, chaperone interacting proteins Hop, Hip and p23, and peptidyl-prolyl isomerases, FKBp51, 52, and Cyp-40. ►chaperones, ►chaperonins, ►heat-shock proteins, ►cyclophilins, ►FK506, ►FKB; Bharadwaj S et al 1999 Mol Cell Biol 19:8033.

Molecular Clock: ►evolutionary clock

Molecular Cloning: Reproduction of multiple copies of DNA with the aid of a vector(s). (See Sambrook J, MacCallum P 2006 Molecular Cloning. Cold Spring Harbor Lab, Press).

Molecular Combing: Spreading and aligning purified and extended DNA molecules on a glass plate and appropriate probes and optical tools reveal quantitatively at high resolution the amplification or losses of the genome. (See Allemand JF et al 1997 Biophys J 73:2064; Gueroui Z et al 2002 Proc Natl Acad Sci USA 99:6005).

Molecular Computation: Molecular computation could be called "reversed mathematics" because it uses DNA or protein sequence information in an attempt to solve complex mathematical problems. ►DNA computer

Molecular Disease: The term "molecular disease" implies that the molecules involved in a disease have been identified, e.g., in sickle cell anemia, a valine replaces a glutamic acid residue in the hemoglobin β-chain. Since this first example, numerous diseases have been explained in molecular terms.

Molecular Drive: Copies of redundant DNA sequences are rather well conserved in the genomes although one would have expected divergence by repeated mutations. The force behind this tendency for uniformity within species has been named molecular drive or concerted evolution. The current literature uses mainly the latter term, which has priority. ►concerted evolution

Molecular Evolution: Studies the relationship of the structure and function of macromolecules (DNA, RNA, protein) among taxonomic groups. ►evolutionary distance, ►evolutionary tree, ►evolutionary clock, ►evolution of the genetic code, ►polymerase chain reaction. ►K_A/K_S, ►RNA world, ►origin of life, ►phylogenomics; Nei M 1996 Annu Rev Genet 30:371; MANET, Wrenn SJ, Harbury PB 2007 Annu Rev Biochem 76:331; Phylemon: http://phylemon.bioinfo.cipf.es/.

Molecular Farming: The use of transgenic animals or plants for the production of substances needed for the pharmaceutical industry or for other economic activities. ►transgenic, ►pharming

Molecular Genetics: The application of molecular biology to genetics. It is a somewhat unwarranted distinction because genetics as a basic science must always use all the best integrated approaches.

Molecular Hybridization: Annealing two different but complementary macromolecules (DNA with DNA, DNA with RNA, etc.). ►c_0t curve, ►probe

Molecular Imaging: A non-invasive imaging of targeted molecules in living organisms. The purpose is to identify a specific [altered] biological process with the aid of a molecular probe that is subject to alteration by the process and the alteration is detectable by light or near-infrared emission or by radioisotopes. An example is shown by the Fig. M100.

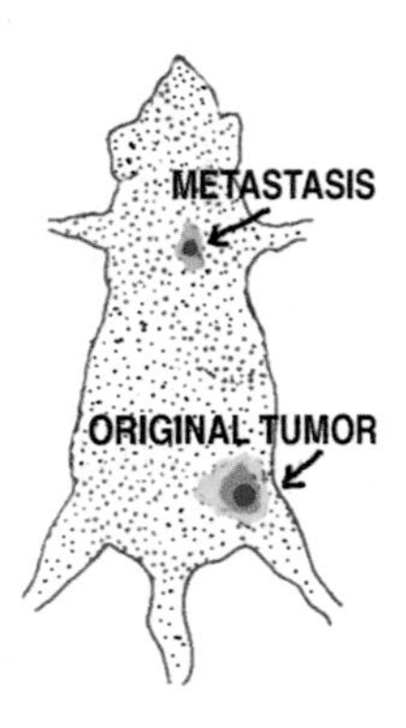

Figure M100. Using an appropriate vector firefly luciferase driven by a tissue-specific promoter is injected at the original tumor site. When metastasis takes place a new tumor site is detectable

Another example is that the dopamine receptor gene (D2R) in an adenovirus vector is injected into the tail of a mouse, and two days later the positron-labeled D2R ligand, (^{18}F) FESP (the dopamine analog 3-(2-(^{18}F) fluoroethylpiperone, is injected into the mouse's bloodstream. MicroPET tomography then reveals the concentration of the molecular target by direct labeling. There are a wide variety of techniques for imaging other molecules. Direct-binding probes reveal the quantity of the target whereas indirect probes shed information on the activity status of the target. For the detection, one of the tomography or magnetic resonance imaging techniques or optical devices for luciferase activity are used. ►tomography, ►nuclear magnetic resonance spectroscopy, ►luciferase, ►GUS; Herschman H R 2003 Science 302:605.

Molecular Markers. ►RFLP, ►RAPD, ►VNTR, ►molecular weight, ►ladder

Molecular Medicine: The integration of molecular, clinical and genomic information for developing improved therapy. (See Bellazzi R, Zupan B 2007 Int J Med Inform 40:787).

M

Molecular Mimics: Molecular mimics may trigger an autoimmune reaction because they have sequence homology to bacterial or viral pathogens. It is assumed that viral infection may lead to the synthesis of antigens structurally of functionally resembling self-proteins. Naïve T cells may be then activated by this antigen and may also mount an immune response to self-proteins. Some proteins appear to mimic the structure of nucleotides, e.g., domain 4 of EF2 peptide elongation factor mimicks the anticodon stem and loop of tRNA. Similarly, the eukaryotic release factors mimic tRNA. ►autoimmune diseases, ►bystander activation, ►immune response, ►immune tolerance, ►tRNA, ►structural mimics; Putnam WC et al 2001 Nucleic Acids Res 29:2199; Quin W et al 2007 Mol Immunol 44:2355.

Molecular Modeling: Aims to determine the three-dimensional atomic structure of macromolecules. Database and software information is available at http://www.ncbi.nlm.nih.gov/Structure/; http://www.rcsb.org/pdb/home/home.do.

Molecular Motor: ►motor protein

Molecular Pharming: ►pharming

Molecular Plant Breeding: This technique uses DNA markers to map agronomically desirable quantitative traits by employing the techniques of RFLP, RAPD, DAF, SCAR, SSCP, etc. The purpose is to incorporate advantageous traits into crops. (See separate entries, ►QTL)

Molecular Plumbing: The generation of blood vessels by tissue culture techniques for the purpose of replacing defective organs.

Molecular Structure Database Tool: http://bip.weizmann.ac.il/oca-bin/ocamain.

Molecular Weight: The relative masses of the atoms of the elements are the atomic weights. The sum of the atomic weights of all atoms in a molecule determines the molecular weight of that molecule (MW). In the determination of the relative mass the mass of C^{12} is used, which is approximately 12.01 (earlier the mass of the hydrogen was used, 1.08 but the calculations with it were more difficult). The relative molecular weight is abbreviated usually as M_r. The molecular weight of macromolecules is generally expressed in daltons, 1 Da = 1.661×10^{-24} gram. Molecular weights are determined by a variety of physicochemical techniques. In gel electrophoresis of DNA fragments are compared with sequenced (known base number) restriction fragments. For λ phage the fragments can be generated by *HinD* III [125 to 23,130 bp], by *HinD* III - *Eco* RI double digests [12 to 21,226 bp], or *Eco* RI [3,530 to 21,226 bp] or pUC18 plasmid cleaved by *Sau* 3AI [36 to 955 bp] or ΦX174 digested by *Hae* III [72 to 1,353 bp]. Alternatively, commercially available synthetic *ladders* containing 100-bp incremental increases from 100 to 1,600 bp or several other sizes are most frequently used. For the large chromosome-size DNAs studied by pulsed-field gel electrophoresis, T7 (40-kb), T2 (166-kb), phage G (758-kb), or even larger constructs obtained by ligation are used. For protein electrophoresis, bovine serum albumin (67,000 M_r), gamma globulin (53,000 and 25,000 M_r), ovalbumin (45,000 M_r), cytochrome C (12,400 M_r), and others can be employed. ►ladder

Molecule: Atoms covalently bound into a unit.

Moloney Mouse Leukemia Oncogene (MLV): MLV integrates at several locations into the mouse genome; an integration site has been mapped to human chromosome 5p14. The Moloney leukemia virus-34 (Mov34) integration causes recessive lethal mutations in the mouse. Homolog to this locus is found in human chromosome 16q23-q24. The Mos oncogene encodes a protein serine/threonine kinase. It is a component of the cytostatic factor CSF and regulates MAPK. ►oncogenes, ►CSF, ►MAPK

Moloney Mouse Sarcoma Virus (MSV, MOS): The c-oncogene maps to the vicinity of the centromere of mouse chromosome 4. The human homolog MOS was assigned to chro-mosome 8q11-q12 in the vicinity of oncogene MYC (8–24). Break-point at human 8;21 trans-locations have been found to be associated with myeloblastic leukemia. It has been suspected that band 21q22, critical for the development of the Down syndrome, is responsible for the leukemia that frequently affects trisomic individuals for chromosome 21. The cellular mos protein (serine/threonine kinase) is also required for the meiotic maturation of frog oocytes. If the c-mos transcript is not polyadenylated maturation is prevented. Overexpression of c-mos causes precocious maturation, and after fertilization, cleavage is prevented but disruption of the gene may lead to parthenogenetic development of mouse eggs. This protein is active primarily in the germline, may cause ovarian cysts and teratomas, and oncogenic transformation also in somatic cells. ►cell cycle, ►teratoma, ►polyadenylation, ►leukemia, ►Down syndrome, ►c-oncogene

Moloney Murine Leukemia Virus: ►MoMuLV

Molscript: A computer program for plotting protein crystal structure. ►protein structure; Kraulis PJ 1991 J Appl Crystallogr 24:946; Spock program manual: http://quorum.tamu.edu/Manual/Manual/Manual.html.

Molting: Shedding the exoskeleton (shell) by insects during metamorphosis (developmental transition stages). ►*Drosophila*

Molybdenum: A cofactor of several enzymes in prokaryotes and eukaryotes. In mammals, it is required by xanthine dehydrogenase, aldehyde, and sulphite oxidase. An autosomal recessive molybdenum cofactor (molybdopterin) deficiency/sulphite oxidase deficiency involves neonatal seizures, neuronal damage, and is a rare lethal condition that manifests between ages two to si two open reading frames in human chromosome 6p encode the protein. ►nitrate reductase, ►gephyrin

Moments: The expectations of different powers of a variable or its deviations from the mean, e.g., the first moment is E(X) = the mean, the second moment is $E(X^2)$, the third $E(X^3)$. The first *moment about the mean:* $E(X - E[X])^2$, the second $E(X - E[X])^3$. The third moment about the mean (if it is not 0) indicates skewness, the fourth moment reveals kurtosis. The moment of the joint distribution of X and Y is the covariance. ►skewness, ►kurtosis, ►covariance, ►correlation

MoMuLV (Moloney murine leukemia virus): A retrovirus with two copies of the genomic RNA in the capsid. The surface of the virus is decorated with trimeric transmembrane glycoprotein (Env). During infection, Env spikes mediate binding of the virus to the cell surface receptor, and structural rearrangements of Env enable the fusion with the cell membrane and entry of the virus into the cell. Cryo-electron tomography was found to reveal the structural basis of this process. The viral membrane is represented by purple and the green structure represents Env (see Fig. M101). The ~30 kDa receptor binding domain fits into the Env.

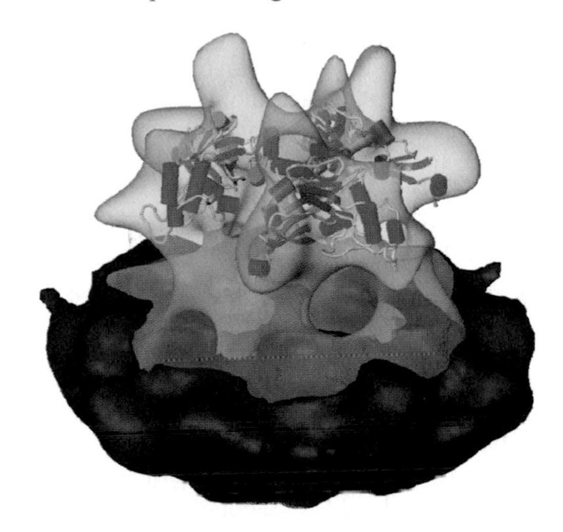

Figure M101. Envelope structure complex of MoMuLV *in situ* as mentioned in the text. For details see Föster F, Medalia O, Zauberman N, Baumeister W & Fass D 2005 Proc. Natl. Acad. Sci. USA 102:4729. (Coutesy of Dr. Friedrich Föster)

MoMuLV is frequently used for vector construction. It is potentially useful for transformation of both rodent and human cells. Integration is very efficient because the viral genome is retained in the reverse-transcribed DNA form of the virus. It can be targeted to rapidly dividing and not to post-mitotic cells. The *gag*, *pol*, and *env* protein genes are dispensable and also ensure that the virus is incapable of autonomous replication, but the ψ packaging signal is required. If the target cells do not have the ψ, the virus does not spread to other tissues. This is a particular advantage when tumor cells are targeted but normal cells are spared. For example, herpes simplex virus thymidine kinase (*HSV-tk*)-containing viral vectors thus convey ganciclovir-sensitivity only to tumor cells (which divide), and 10–70% may then be selectively killed by this cytotoxic drug. Similar vector designs can be applied to some degenerative diseases. (Parkinson disease, Huntington's chorea, Alzheimer disease and other neurodegenerative diseases). The protective gene can be transfected into cultured cells and then can be grafted back onto the target tissue of the same patient. ►ganciclovir, ►retroviral vectors, ►gene therapy; D'Souza V, Summers MF 2004 Nature [Lond] 431:586; Förster F et al 2005 Proc Natl Acad Sci USA 102:4729.

MONA (Al Aqueel-Sewairi syndrome, 16q12-q21): Multicentric osteolysis with nodulosis and arthritis due tom deficiency of matrix metalloproteinases.

Monarch Butterfly: ►navigation, ►*Bacillus thüringiensis*

Monastrol: A protein causing monopolar (rather than the normal bipolar) spindle in dividing cells. It blocks kinesin-related proteins (such as those in the bimC family, e.g., Eg5). ►kinesin, ►bimC; Kapoor TM et al 2000 J Cell Biol 150:975.

Mondrian: A type of representation of microarray of RNA transcripts resembling the "Broadway boogie-woogie" by the Dutch painter Piet Mondrian (1872–1944). It identifies open reading frames within the sequenced genomic DNA. (See Penn SG et al 2000 Nature Genet 26:315; art in science).

Mongolian Spot: A transient or long-lasting bluish birthmarks most commonly found upon the buttocks of infants. Its prevalence among Asians and East Africans 95–100%, Native Americans 85–90%, Hispanics 5–70%, and Caucasians 1–10%. It is caused by the entrapment of melanocytes during their migration from the neural crest into the epidermis during fetal development. (After Dr. Numabe, H., Tokyo University).

Mongolism (mongoloid idiocy): Now rejected name of Down's syndrome, human trisomy 21. Besides the epicanthal eyefold other—non-mongoloid—features are more characteristics for the condition (see Fig. M102). ►Down syndrome, ►trisomy

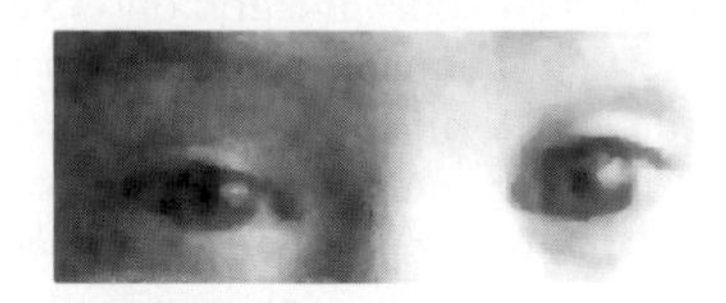

Figure M102. Mongoloid eyefold

Monilethrix: Autosomal dominant (human chromosome 14q) and possibly autosomal recessive baldness (alopecia) due to defects of the keratin filaments of the hair. ►filaments, ►baldness, ►hair; Schweizer J 2006 J Invest Dermatol 126:1216.

Monitor: A video monitor receives and displays information directly received by a computer, while a television monitor accepts broadcast signals. Monitoring: keeping track of something.

Monoallelic Expression: In a diploid (or polysomic) cell or individual only one of the alleles is expressed such as the genes situated in one of the mammalian X chromosomes or as is the case of maternal or paternal imprinting. Besides X-chromosomal genes, imprinted autosomal genes display monoallelic expression. In addition several other autosomal genes (odorant receptors, pheromone receptors, immunoglobulins, T cell receptors, interleukins) are transcribed in random monoallelic manner. These alleles also show asynchronous replication, which may be responsible for their differences in expression. The asynchrony within a particular chromosome appears coordinated (Singh N et al 2003 Nature Genet 33:339). Monoallelically expressed autosomal genes of mammals display a substantially higher frequency of LINE1 sequences (Allen E et al 2003 Proc Natl Acad Sci USA 100:9940). Actually some lower organisms (*Plasmodium, Trypanosoma*) also express only one allele of some genes of the diploid. ►lyonization, ►imprinting, ►allelic exclusion, ►*Plasmodium*, ►*Trypanosoma*

Monoamine Oxidase: ►MAO, ►Norrie disease

Monobrachial Chromosome: A monobrachial chromosome has only one arm (see Fig. M103). ►telocentric, ►chromosome morphology

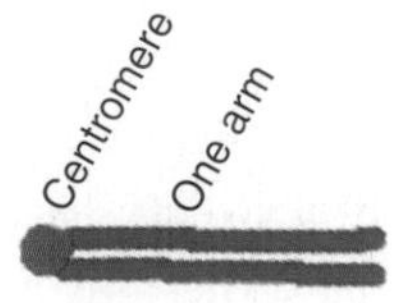

Figure M103. Monobrachial chromosome

Monocentric Chromosome: A monocentric chromosome has a single centromere, as is most common.

Monochromatic Light: Literally, light of a single color, practically light emission with a single peak within a very narrow wavelength band.

Monocistronic mRNA: A monocistronic mRNA codes for a single type of polypeptide. It is transcribed from a single separate cistron (not from an operon). Eukaryotic genes are usually monocistronic. The class I genes, 18, 5.8 and 28 S rRNAs, including spacers, are transcribed as a single unit containing one of each gene's pre-rRNAs and are cleaved subsequently into mature rRNA. The 5 S RNA genes are transcribed from separate promoters by pol III. ►ribosome, ►cistron

Monoclonal Antibody (MAb): The MAb is of a single type, produced by the descendants of one cell and specific for a single type of antigen. In case the epitome domains are highly conserved, the specificity of the monoclonal antibody is reduced. MAbs are generated by injecting into mice, purified antigens (immunization), than by isolating spleen cells (splenocytes, B lymphocytes) from the immunized animals and fusing these cells, in the presence of polyethylene glycol (a fusing facilitator), with bone marrow cancer cells (myelomas) that are deficient in thymidine kinase (TK), or HGPRT. This process assures that the non-fused splenocytes rapidly senesce in culture and die. The unfused myeloma cells are also eliminated on a HAT medium because they cannot synthesize nucleic acids, either by the de novo or by the salvage pathway. The selected myelomas do not secrete their own immunoglobulins and thus, hybrid immunoglobulins are not produced. Single hybridoma cells are then cultured in multi-well culture vessels and screened for the production of specific MAbs by the use of radioimmunoassays (RIA) or by enzyme-linked immunosorbent assay (ELISA). Most of the hybridomas will produce many types of cells and are of limited use; a few, however, may be more specific and these are re-screened for the specific type needed. E.g., animals immunized with melanoma cells produce HLA-DR (Human Leukocyte Antigen D-related) antibodies or melanotransferrin (a 95-kDa glycoprotein) or melanoma-associated chondroitin proteoglycan (heteropolysaccharides, glucosaminoglycans attached to extracellular proteins of the cartilage). Monoclonal antibodies have been used effectively for identification of tumor types in sera and histological assays. A human trial using humanized mouse monoclonal antibody in the hope to find an effective treatment for autoimmune conditions had a tragic outcome. Six healthy individuals in a London hospital were injected with anti-CD28 mAB that rather than suppressing regulatory T cells—as expected on the basis of experiment with mice and monkeys—turned 'superagonist' and activated all types of T cells, and not just the regulatory T cells. The individuals in the trial developed serious organ failures and one fell into a coma for three weeks. Apparently, the mAB of TGN1312 caused an unexpected 'cytokine storm' indicating that there is still a lot to be learned before antibodies can be generally used for therapeutic purposes (Vitetta ESD, Ghetie VF 2006 Science 2006 313:308).

Some MAbs have been successfully applied for direct tumoricidal effect. Radioactively labeled monoclonal antibodies of Melanoma Associated Antigens (MAA) have been used for imaging (immunoscintigraphy) and detecting melanoma cells in the body when other clinical and laboratory methods failed. The "magic bullet" approach of combining specific monoclonal antibodies (prepared from individual cancer patients) against a specific cancer tissue with Yttrium^{90} isotope (emitting β rays of 2.24 MeV energy and 65 h half-life), or ricin (LD50 for mice by intravenous administration 3 ng/kg) or the deadly diphtheria toxin, or tumor necrosis protein (TNF) is expected to home in on cancer cells and destroy them. This attractive scheme has not been proven successful in clinical trials so far. (►abzymes). More recently, several monoclonal antibody drugs targeting tyrosine kinase (Gleevec, Trastuzumab) and antiangiogenesis Avastin have been approved for therapeutic use alone or in combination with other treatments. A newer approach is to clone separately cDNAs of a variety of light and heavy chains of the antibody and allowing them to combine in all possible ways, thus producing a combinatorial library of antibodies against all present and possible, and emerging and future epitopes. These antibodies can then be transformed into bacteria by using λ phage or filamentous phage such as M13. The phage plaques can then be screened with radioactive epitopes. The advantage of using filamentous phages is that they display the antibodies on their surface and permit screening in a liquid medium that enhances the efficiency by orders of magnitude. Although monoclonal antibodies did not completely fulfill all the (naïve) therapeutic expectations, they still remain a power-tool of biology and currently they represent more than 20% of the biopharmaceuticals being evaluated in clinical trials. Monoclonal antibodies cannot be produced against self-antigens or against less than about 1-kDa antigens. In the latter case, high molecular weight carrier proteins such bovine serum albumin or limpet hemocyanin are used. Typical yield of MAb from hybridoma cells is 10–100 μg/10^6 cells/day. Currently, mammalian cells are used mainly to produce IgG and obtain post-translational modifications (e.g., glycosylation), which are essential for the maintenance of the conformation of the immunoglobulin effector function. For transient expression of antibody genes, most commonly COS cells are employed. For stable-expression myeloma cell lines, SP2/0, NS0, and Chinese hamster ovary (CHO) cells are used. Antibodies can be produced in transgenic mammals, insects, and plants. Mouse adenocarcinoma-associated antigen (EpCAM [GA73302]) is a highly expressed target of mouse monoclonal antibody (mAb CO17–1 A) and it can be produced in transgenic *Nicotiana tabacum* plants after transformation by *Agrobacterial* vectors. Although the glycosylation pattern of the plant monoclonal antibody differs from that of the animals, it inhibited colon cancer cells in nude mice similarly to its mammalian counterpart (Ko K et al 2005 Proc Natl Acad Sci USA 102:7026). ►antibody, ►immunoglobulins, ►immune system, ►T cells, ►B cells, ►lymphocytes, ►somatic cell hybrids, ►hybridoma, ►MAb, ►TK, ►HGPRT, ►HAT medium, ►senescence, ►RIA, ►ELISA, ►hybridoma, ►heterohybridoma, ►melanoma, ►ricin, ►LD50, ►diphteria toxin, ►epitope, ►combinatorial library, ►phage display, ►quasi-monoclonal, ►bispecific monoclonal antibody, ►keyhole limpet hemocyanin, ►humanized antibody, ►CD28, ►immunization genetic, ►phage display, ►HAMA, ►COS, ►transformation transient, ►CHO, ►plantibody, ►genetic engineering, ►antibody polyclonal, ►receptin, ►aptamer, ►genetic medicine; Ritter MA, Ladyman HM (Eds.) 1995 Monoclonal Antibodies. Production, Engineering and Clinical Application, Cambridge University Press, New York; Alkan SS 2004 Nature Rev Immunol 4:153; http://www.antibodyresource.com/; http://bioresearch.ac.uk/browse/mesh/D000911.html.

Monoclonal Antibody Therapies: Monoclonal antibody therapies open a variety of applications in medicine. Although at the moment they do not offer perfect solutions, they do have a lot of potential. Reocclusion of blood vessels after angioplasty—caused by unwanted aggregation of the platelets—may be alleviated by the 7E3 antibody fragments of the chimeric mouse-human Fab. Single- or two-chain fragments of the antibody variable region and the urokinase fusion protein (scFv-uPA) or tissue plasminogen activator (tPA) conjugated to anti-fibrin antibody have been designed for thrombolysis. Monoclonal antibodies may be generated against interleukins (IL-1), TNF, and cell adhesion molecules (selectins, integrins) to block inflammatory responses after wounding or autoimmune diseases (rheumatoid arthritis, septic shock), T cells, T cell receptors, etc. Monoclonal antibody-mediated cytotoxicity responses are activated or inhibited through the Fcγ receptors of the antibody. By computational design and engineering, the affinity of these receptors can be increased by two orders of magnitude and can broaden the therapeutic applicability to cancer (Lazar GA et al 2006

M

Proc Natl Acad Sci USA 103:4005). ►magic bullet, ►ADEPT, ►vascular targeting, ►immunotoxins, ►antiviral antibodies, ►tissue plasminogen activator, ►thrombin, ►platelet, ►fibrin, ►selectins, ►integrin, ►rheumatic fever, ►septic shock, ►autoimmune diseases, ►T cell, ►T cell receptor, ►plantibody, ►antibody, ►antibodies intracellular, ►abzymes, ►gene therapy, ►cancer gene therapy, ►genetic medicine, ►biomarkers; Baselga J 2001 Eur J Cancer 37 Suppl 4:16.

Monocotyledones: Plants that form only one cotyledon such as the grasses (cereals).

Monocytes: Mononuclear leukocytes that become macrophages when transported by the blood stream to the lung and the liver. ►microglia, ►atherosclerosis, ►MCP-1, ►leukocyte, ►macrophage

Monoecious: Plants with separate male and female flowers on the same individual (see Fig. M104). ►autogamous, ►outcrossing, ►protandry, ►protogyny

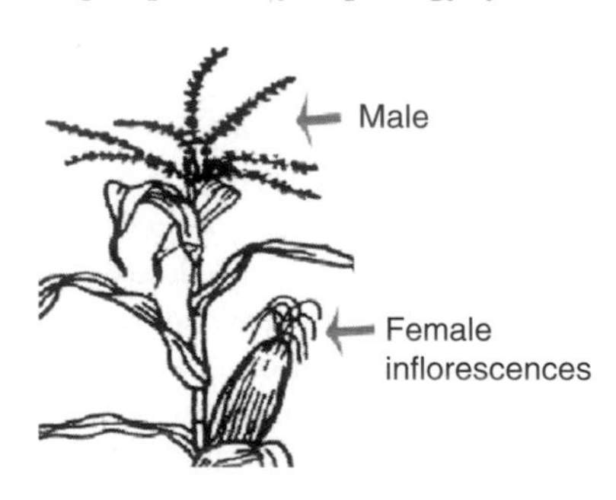

Figure M104. Monoecious plant

Monofactorial Inheritance: A single (dominant or recessive) gene (factor) determines the inheritance of a particular trait. Although this term has some value for classification of inheritance, in most of the case the phenotype is the function of a number of genes with different alleles that respond to a variety of internal and external factors at the same time. Thus, strictly monogenic inheritance may not exist. The so-called monogenic human diseases generally are expressed as syndromes. As of February 2006, 1,822 "monogenic" human allelic variants were identified by OMIM. Sequencing of the human disease genes and encoded proteins often reveals more than a single alteration even within the single gene. Also, environmental factors and the genetic background may modify the phenotypes substantially. ►Mendelian laws, ►reaction norm, ►syndrome, ►pleiotropy, ►SNIPs, ►epistasis, ►digenic diseases, ►QTL; Nadeau JH 2001 Nature Rev Genet 3:165.

Monofunctional Alkylating Agent:: A monofunctional alkylating agent has only a single reactive group.

Monogenic Heterosis: Same as overdominance, superdominance.

Monogenic Inheritance: Same as monofactorial inheritance.

Monogerm Seed: A monogerm seed contains only a single embryo. The fruit of some plants (e.g., sugar beet) is frequently used for propagation and usually it contains multiple seeds. This fruit may, however, be genetically modified to contain a single seed or mechanically fragmented to become monogerm. The agronomic advantage of the monogerm seed is that the emerging seedlings are not crowded and the labor-consuming thinning may be avoided or is at least facilitated.

Monogyne: Social insects with single functional female (queen) in the colony. ►polygyne

Monohybrid: A monohybrid is heterozygous for only one pair of alleles. ►monofactorial inheritance

Monoisodisomic: In wheat, 20" + i1", 2n = 42, ["=disomic, i=isosomic].

Monoisosomic: In wheat, 20"+ i', 2n = 41, ["=disomic, '=monosomic, i=isochromosome].

Monoisotrisomic: In wheat, 20"+ i2"', 2n = 43, ["=disomic, i=isosomic, "'=trisomic].

Monokine: Lymphokine produced by monocytes and macrophages. ►lymphokine

Monolayer: Non-cancerous animal cell cultures grow in a single layer in contact with a solid surface, i.e, in a monolayer. It also designates a single layer of lipid molecules. ►tissue culture, ►cell culture

Monolithic Subtrate: Polymer or silicon substrate with microchannels or other functional elements.

Monomer: One unit of a molecule (which frequently has several in a complex); a subunit of a polymer ►protein

Monomorphic Locus: In the population, a monomorphic locus is represented by one type of allele.

Monomorphic Trait: A monomorphic trait is represented by one phenotype in the population.

Mononucleosis: Caused by cytomegalovirus; infectious mononucleosis is the result of activation of the Epstein-Barr virus. ►cytomegalovirus, ►Epstein-Barr virus

Monooxygenases: Monooxygenases introduce one atom of oxygen into a hydrogen donor, e.g., P450 cytochromes, and have varied functions in cells in normal development, as detoxificants, and in converting promutagens and procarcinogens into active compounds. ►mixed-function oxidases, ►P450

Monoparous: Species, which generally produce a single offspring at a time. ►multiparous

Monophyletyic: Monophyletic organisms have evolved from a single line of ancestry. ►polyphyletic

Monoploid: A monoploid has only a single basic set of chromosomes. ►haploid

Monopolar Spindle: A monopolar spindle pulls all chromosomes to one pole (see Fig. M105).

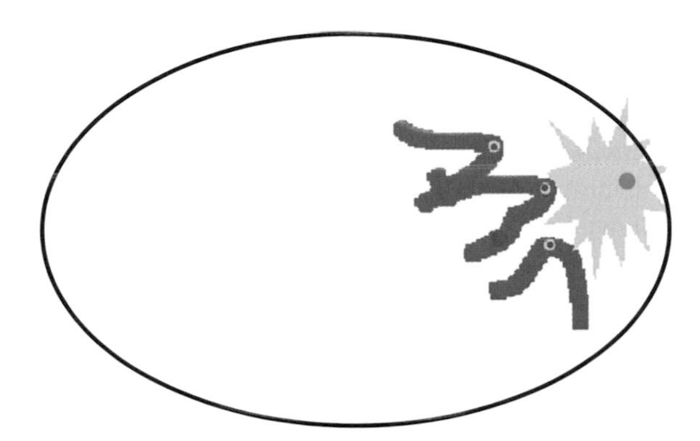

Figure M105. Monopolar spindle

Monopolin (Mam1): A kinetochore associated protein that assures that during meiosis I the homologous chromatids (held together by cohesin) are pulled to the same pole. The Csm1/Lrs4 proteins reside in the nucleolus and shortly before meiosis I they leave that location to associate with Mam1. ►meiosis, ►cohesin, ►co-orientation; Toth A et al 2000 Cell 103:1155; Rabitsch KP et al 2003 Dev Cell 4:535.

Monoprotic Acid: A monoprotic acid has only a single dissociable proton.

Monosaccharide: A carbohydrate that is only one sugar of basic units $C_nH_{2n}O_n$ and thus can be diose, triose, pentose, hexose, etc., depending on the number of C atoms; the sugars can also be also aldose (e.g., glucose, ribose) or ketose sugars (e.g., fructose, deoxyribose).

Monosaccharide Malabsorption (22q13.1): Monosaccharide malabsorption is controlled by a sodium/glucose transporter at the intestinal brush border. Glucose/galactose malabsorption may be remedied by fructose.

Monosodium Glutamate (HOOCCH(NH2)CH2CH2 COONa.H2O): Monosodium glutamate is used as a flavor enhancer (0.2–0.9%) in salted food or feed, for repressing the bitter taste of certain drugs, or as a medication for hepatic coma. ►Chinese restaurant syndrome

Monosomic: In a monosmic, the homologous chromosome(s) is represented only once in a cell or all cells of an individual. The gametes of diploids are monosomic for all chromosomes. A monosomic individual of an allopolyploid has 2n − 1 chromosomes. About 80% of the human 45 + X monosomics involve the loss of a paternal sex chromosome. Medical cytogeneticists frequently refer to deletions as "partial monosomy", but this is a misnomer and should be avoided because such a condition is hemizygosity for a particular locus or chromosomal region. Monosomics can produce both monosomic and nullisomic gametes. ►nullisomics, ►chromosome substitution, ►monosomic analysis

Monosomic Analysis: Monosomic analysis is very efficient in Mendelian analysis of allopolyploids, such as hexaploid wheats. Monosomic individuals can produce monosomic and nullisomic gametes and their proportions are different in males and in females (see Fig. M106). The proportion also depends on the individuality of the particular chromosomes. On the average, the gametic output and the zygotic proportion of selfed monosomics of wheat is as shown in the body of the table. Very few nullisomic sperms are functional, whereas the majority of the eggs are nullisomic because most of the monosomes (in the absence of a partner) remain and get lost in the meta-phase plane and the eggs receive only one representative from each chromosome that has paired during prophase I. The monosomics can be used to assign genes to chromosomes and if the proper genetic constitution is used cytological test may not be required. On the basis of the Fig. M106 and the Table M7 it is obvious that 75% of the F_1 will be of recessive phenotype if the genes are in the chromosomeof the monosomic female. In case the recessive

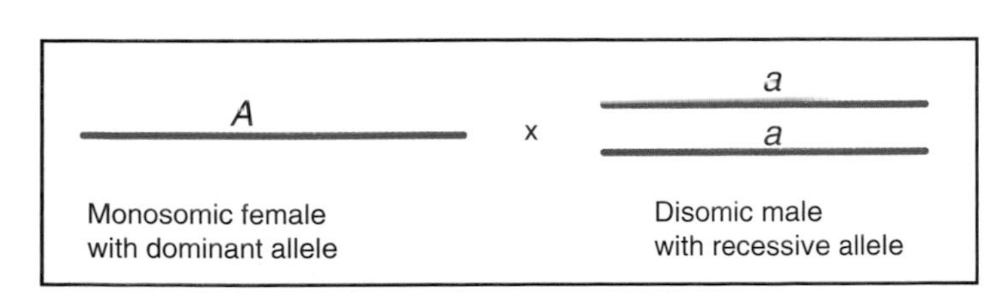

Figure M106. Gene localization to chromosome with monosomics

Table M7. Monosomic analysis

Eggs ⬇	Sperms: Monosomic (0.96)	Sperms: Nullisomic (0.04)
	Zygotes	
Monosomic 0.25	Disomic ≈ 24% (0.25 x 0.96)	Monosomic ≈ 1% (0.25 x 0.04)
Nullisomic 0.75	Monosomic ≈ 72% (0.75 x 0.96)	Nullisomic ≈ 3% (0.75 x 0.04)

is in the monosomic female, only 3% will be recessive in the F_2 offspring. ►nullisomy; Morris R, Sears ER 1967 In Wheat and wheat improvement; Quisenberry KS, Reitz LP (Eds.) Am Soc Agron Madison WI Sears E R 1966 Hereditas Suppl 2:370; Tsujimoto H 2001 J Hered 92:254.

Monospermy: Fertilization by a single sperm.

Monotelodisomic: In wheat, 20" + t2"', 2n = 43, ["=disomic, t = telosomic, "'=trisomic].

Monotelomonoisosomic: In wheat, 20"+ t'+ i', 2n= 42, ["=disomic, '=monosomic, t = telosomic, i = isosomic].

Monotelo-Monoisotrisomic: In wheat, 20"+ (t+ i)1"', 2n = 43, ["=disomic, "'=trisomic, t = telosomic, i = isosomic].

Monotelosomic: In wheat, 20" + t', 2n = 41, ["=disomic, '=monosomic, t = telosomic].

Monotocous Species: A monocotous species produces a single offspring by each gestation.

Monotrene: An animal belonging to the taxonomic order Monotremata, a primitive small mammalian group, including the spiny anteater and the duck-billed platypus (see Fig. M107). They lay eggs and nurse their offspring within a small pouch through a nipple-less mammary gland. They show other organizational similarities to marsupials. ►marsupials, ►sex determination

Figure M107. Platypus

Monotypic: A taxonomic category represented by one subgroup, e.g., a monotypic genus includes a single species.

Monoubiquitin: In a monoubiquitin, the surface of the ubiquitin may be ordained for separate multiple functions, e.g., for degradation or endocytosis. The various monoubiquitins are specialized to one function only although the function itself may vary, e.g., it may involve regulation of histones H2A and H2B, endocytosis, or budding of the retroviral Gag polyprotein. Monoubiquitins can be attached to separate sites, or *polyubiquitin* chains are attached at a single lysine[48]. Polyubiquitin attachments at Lys^{63}activates several proteins but this complex is not targeted to the proteasome. ►ubiquitin, ►sumo, ►histones, ►endocytosis, ►retrovirus, ►proteasome, ►deubiquitinating enzymes; Hicke L 2001 Nature Rev Mol Cell Biol 2:195.

Monozygotic Twins: Monozygotic twins develop from a single egg fertilized by a single sperm, therefore they should be genetically identical, except mutations that occur subsequent to the separation of the zygote into two blastocytes after implantation into the wall of the uterus. Monozygous female twins may be phenotypically discordant in case of heterozygosity of X-linked genes. The difference may be caused by the inactivation of one or the other X chromosome. Although MZ twins are generally identical genetically, their birthweight may be different because of differences in intrauteral nutrition due the path of blood circulation. Developmental malformations may affect only one of the MZ twins. The concordance in susceptibility to infectious diseases has been investigated but the information is not entirely unequivocal. Multifactorial diseases are more concordant in MZ twins than among dizygotic twins. ►twinning, ►dizygotic twins, ►heritability estimation in humans, ►zygosis, ►concordance, ►discordance, ►co-twin, ►lyonization

Monte Carlo Method: A computer-assisted randomization of large sets of tabulated numbers that can be used for testing against experimentally obtained data to determine whether their distribution is random or not. It is used for simulation when the analysis is intractable or too complex. ►modeling, ►simulation, ►Markov chain; Roederer M et al 2001 Cytometry 45:47; Roederer M et al *ibid* 37.

Moolgavkar-Venzon Model: A revised version of Knudson's two-mutation model of carcinogenesis. It considers the possibility that after the first mutation not all the cells survive, and allows for differential growth of the cells before the occurrence of malignant transformation. ►Knudson's two-mutation hypothesis; Armitage-Doll model, Holt PD 1997 Int J Radiat Res 71(2): 203.

Moore-Federman Syndrome: A dwarfism with stiff joints and eye anomaly. ►dwarfism

Mooring Sequence: An 11 nucleotide element anchored downstream to the base **C** to be RNA edited (**C**AAUUUUGAUCAGUAUA). ►RNA editing

Moose, North American: (*Alces alces*), 2n = 70.

Morality: Socially accepted principles and guidelines for the distinction of right from wrong in human behavior, usually based on customs and generally required to be abided by, by society. Moral standards are categorically binding principles independent of one's wishes or desires and therefore transcend all other kinds of potentially conflicting norms. Some moral rules are present in all human societies; these rules may have, however, distinct variations. ►behavior genetics, ►ethics, ►ethology, ►informed consent

Morbidity: The diseased condition or the fraction of diseased individuals in a population. A wide range of biomarkers, reflecting activity in a number of biological systems (e.g., neuroendocrine, immune, cardiovascular, and metabolic), has been found to prospectively predict disability, morbidity, and mortality outcomes in older adult populations. Levels of these biomarkers, singly or in combination, may serve as an early warning system of risk for future adverse health outcomes. Markers for neuroendocrine functioning (epinephrine, norepinephrine, cortisol, and dehydroepiandrosterone), for immune activity (C-reactive protein [one of the acute phase proteins that increase during systemic inflammation], fibrinogen, IL-6, and albumin) is appropriate. For cardiovascular functioning (systolic and diastolic blood pressure), and for metabolic activity [high-density lipoprotein (HDL) cholesterol, total to HDL cholesterol ratio are being used, Glycosylated hemoglobin] over a 12-year period in a sample of men and women (*n* = 1,189) 70–79 years of age was informative cardiovascular problems. Almost all these markers entered into one or more high-risk pathways, although combinations of neuroendocrine and immune markers appeared frequently in high-risk male pathways, and systolic blood pressure was present in combination with other biomarkers in all high-risk female pathways (Gruenewald TL et al 2006 Proc Natl Acad Sci USA 103:14158). ►mortality; Petronis A 2001 Trends Genet 17:142.

Morgan: 100 units of recombination, 100 map units, usually the centiMorgan, 0.01 map unit is used. ►map unit

Morphactins: Various plant growth regulators.

Morphallaxis: ►regeneration in animals, ►epimorphosis

Morpheeins: Proteins in which one functional homo-oligomer can dissociate, change conformation, and reassociate into a different oligomer. ►conformation

Morphine: An opium-like analgesic and addictive alkaloid; methylation converts it to the lower potency codein. Noradregenic signaling mediates opiate reward (Olson VG et al 2006 Science 311:1017). ►alkaloids, ►opiate, ►animal hormones

Morphoallele: Gene involved in morphogenesis. ►morphogenesis

Morphogen: A compound that can affect differentiation and/or development, and can correct the morphogenetic pattern of a mutant that cannot produce it if it is supplied by an extract from the wild type. In some sea algae (*Ulva, Enteromorpha*), marine bacteria (Cytophaga-Flavobacterium-Bacteroides), on the surface of the algae, control leaf-like morphology. The bacteria-produced thallusin (see Fig. M108) determines foliaceus morphology and spore germination (Matsuo Y et al 2005 Science 307:1598). In the higher plant *Cardamine hirsuta* (see Fig. M109), the KNOX (KNOTTED1-like homeobox) regulated by the asymmetric leaf 1 protein makes the leaves dissected, whereas in another crucifer, *Arabidopsis* leaves, KNOX is excluded making the leaves simple (Hay A, Tsiantis M 2006 Nature Genet 38:942). In tomato, the wildtype plant has compound leaves composed of several leaflets. The semidominant *lanceolate* (*LA*) mutation in chromosome 7 alters the leaf shape and structure. *LA* encodes the TCP transcription factor

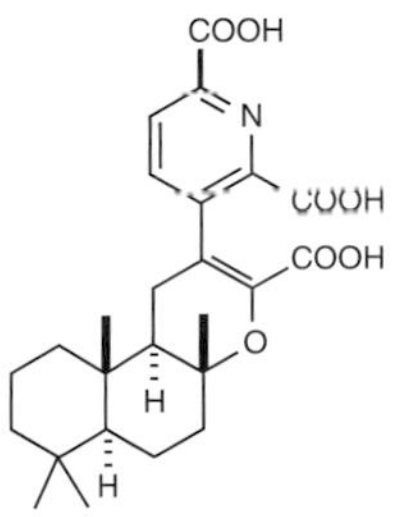

COOH
N
COOH
COOH
O
H
H

Figure M108. Thallusin

with miR319 (microRNA) binding site. Base substitution mutations in the binding site reduce or eliminate TCP inhibition required for the development of the wild type compound leaf (see Fig. M110).

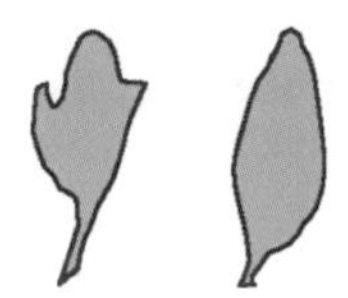

Figure M109. *C. hirsuta A. thaliana*

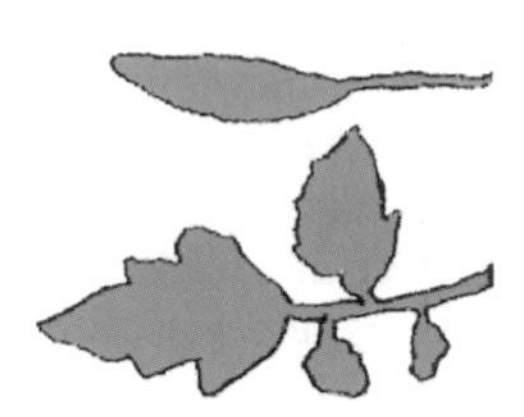

Figure M110. Wild type leaflet of tomato (Redrawn after Ori N et al. 2007 Nature Genet. 39:787)

The exact chemical nature of many morphogens is unknown, however, some hormones and proteins have these properties. There is some evidence that some morphogens (produced by Wingless and Hedgehog) are moved in the cell by association with glycophosphatidylinositol (Panáková D et al 2005 Nature [Lond] 435:58). The morphogen is either a transcription factor or a type of transcriptional regulator. The slope of the *Decapentaplegic* gene expression acts as a morphogen on wing development and the expression gradient regulates growth (Rogulja D, Irvine KD 2005 Cell 123:449). Control by morphogens involves concentration gradients over a threshold level. Generally, the initial morphogen signal recruits other similar signals and signal receptors (encoded by genes) which act positive or negative manners and determine cell fates and bring about a differentiational event(s). Generally the production rate, the effective diffusion coefficient, the degradation rate, and the immobile fraction of the morphogen contribute to the pattern of the morphogenic differentiation (see Fig. M111) (Kicheva A et al 2007 Science 315:521). The activin signal (a member of the transforming growth factor-β family) spreads in a passive gradient about 300 μm or 10 cells diameter within a few hours in the vegetal cells of animals. ►morphogenesis, ►vegetal pole, ►activin, ►cross-talk, ►signal transduction, ►selector genes, ►cue, ►wingless, ►hedgehog, ►TCP, ►microRNA; Pagès F, Kerridge S 2000 Trends Genet 16:40; Tabata T 2001 Nature Rev Genet 2:620; Gurdon JB, Bourillot P-Y 2001 Nature [Lond] 413:797; Nüslein-Volhard CN 2004 Cell 116:1; Lander AD 2007 Cell 128:245.

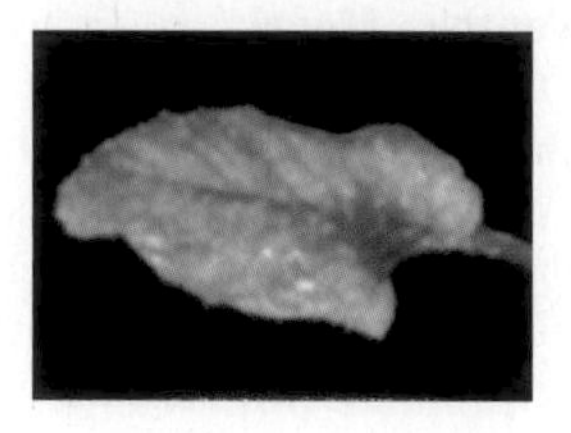

Figure M111. *Arabidopsis thaliana*: asymmetric leaf

Morphogenics: Morphogenics employs a dominant negative mismatch repair gene to create genetic diversity within defined cellular systems and results in a wide range of phenotypes. It has been used in immunology and drug development (Nicolaides NC et al 2005 Ann NY Acad Sci 1059:86). ►mismatch, ►mismatch repair

Morphogenesis: The process of development of form and structure of cells or tissues and eventually the entire body of an organism beginning from the zygote to embryonic and adult shape. Morphogenesis is mediated through morphogens in response to inner and outer factors. Morphogenesis takes place in three phases: determination, differentiation, and development. Determination is a molecular change preparing the cell (or virus) to competence for differentiation. Differentiation is the realization of molecular and morphological structures that determines the differences among cells that are endowed with identical genetic potentials. The events of differentiation are coordinated in sequences of development. These steps generally occur in this order, yet may run in overlapping courses for different aspects of morphogenesis. The morphogenetic events vary in different organisms because this is the basis of their identity, yet some basic principles are common to all. The start point of morphogenesis is difficult to pinpoint because these events run in cycles of generations (what comes first: the egg or the hen?). The life of an individual begins with fertilization of the egg by the sperm and the formation of the zygote. The zygote has all the genes that it will ever have (*totipotency*) yet many of these genes are not expressed at this stage. (New genes may be acquired during life cycle by mutational conversion of existing ones or by transformation and transduction). Also, there are *maternal effect* genes that control oogenesis and affect the zygote from outside without being expressed at this stage within the embryo's own gene repertory. Morphogenesis cannot be explained by one general set of theory, e.g., *gradients of morphogens* or the *signal relay* system. Apparently, evidence for either can be found in different morphogenetic pathways. Many structures in higher eukaryotes are under the control of partially redundant gene families. In different tissues different sets of transcription factors or transcriptional activators and enhancers may orchestrate in a combinatorial manner the expression of different cell types, adapted to a particular organ. The cytoskeleton and cell adhesion may determine cell shape. Signals originating outside or inside the cell may act as morphogens and by their concentration gradients along their distribution path may trigger morphogenetic changes. Alternatively, it has been suggested that signals induce a change in one type of cell that then through a relay of series of cells influences other cells to various types of changes. For example, signals transmitted to metalloproteinases, collagenase-1, and stromelysin-1 mediated by the RAS family (Rac) GTP-binding proteins are also involved in cell morphogenesis. ►*Drosophila*, ►homeotic genes, ►morphogens, ►signal transduction, ►clonal analysis, ►developmental genetics, ►signal transduction, ►RNA localization, ►RAS, ►Rac, ►metalloproteinases, ►cytoskeleton, ►CAM, ►imaginal disks, ►oncogenes, ►morphogenesis in *Drosophila*, ►pattern formation, ►RNA localization, ►founder cells, ►segregation asymmetric, ►organizer, ►anchor cell, ►Hensen's node, ►morphogenetic furrow, ►development, ►mRNA targeting, ►compartmentalization, ►primitive streak, ►cue, ►Notch, ►left-right asymmetry, ►selector genes; Madden K, Snider M 1998 Annu Rev Microbiol 52:687; Chase A 2001 Bioessays 23:972; high resolution X-ray computed tomography of animal morphology: http://www.digimorph.org/.

Morphogenesis in *Drosophila*: Morphogenesis in *Drosophila* has been the most extensively studied with genetic techniques (see Fig. M112). Oogenesis requires four cell divisions within the oocyst (see illustration of maternal effect genes) resulting in the formation of 16 cells, one oocyte, and 15 nurse cells. The latter become polyploid and are surrounded by a single layer of somatic (diploid) follicle cells. The nurse cells are in communication with the egg through cytoplasmic channels. Both the nurse cell genes (somatic maternal genes) and the egg (germ-line maternal genes) influence the fate of the zygote through morphogens. The oocyte itself is transcriptionally not active. These maternal effect genes determine the polarity of the zygote (see Fig. M113). The anterior–posterior gradients of the morphogens account for the future position of the head and tail, respectively. Genes *gurken* (*grk*, 2–30) and *torpedo* (2–10) play an important role in anterior–posterior polarity and later during development also of dorso-ventral determination. The dorso-ventral determination is responsible for the sites of the back and belly, respectively. The medio-lateral polarities are involved in the determination of the left and right sides of the body. The larvae and the adults develop from 12 compartments, one for the head, three for the thorax, and nine for the abdomen, formed already during the blastoderm stage of the embryo.

Mutation of the maternal genes mentioned in the box below usually cause recessive embryo lethality although the homozygous mothers are generally normal. The males are usually normal and fertile. The molecular basis of their actions is known in a few cases. E.g., the amino terminal of the DL protein is homologous to the C-REL (avian reticuloendothelial viral oncogene homolog) protooncogene, which is present in human chromosome 2p13–2cen, and in mouse chromosome 11. The N terminus of *dorsal* is homologous to the product of gene *en* (*engrailed,*

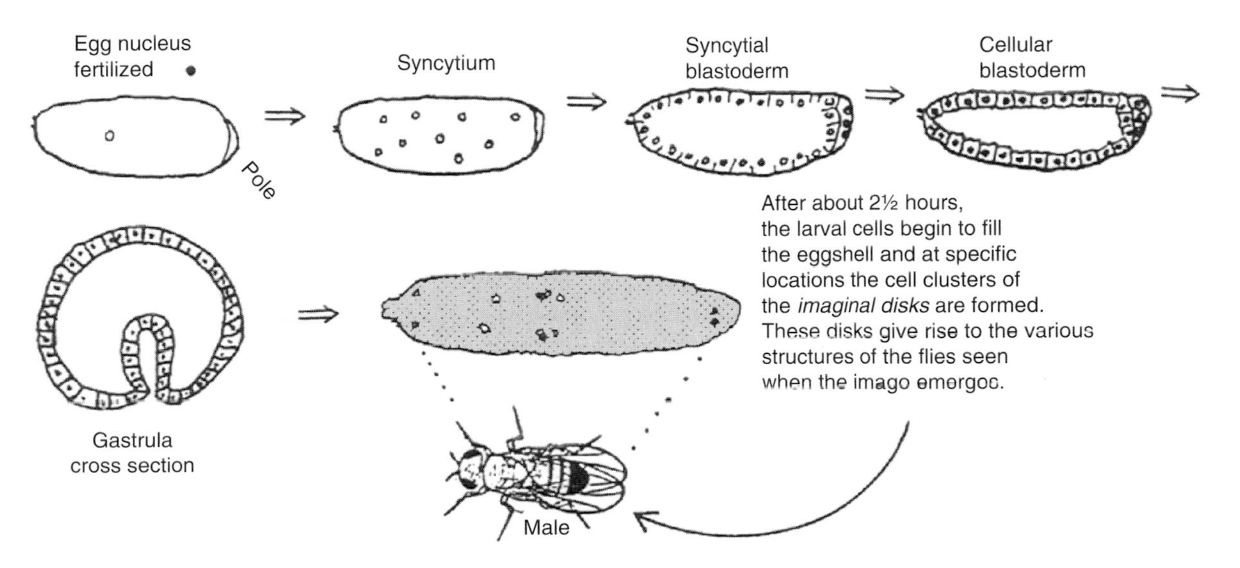

Figure M112. Embryogenesis in *Drosophilia*

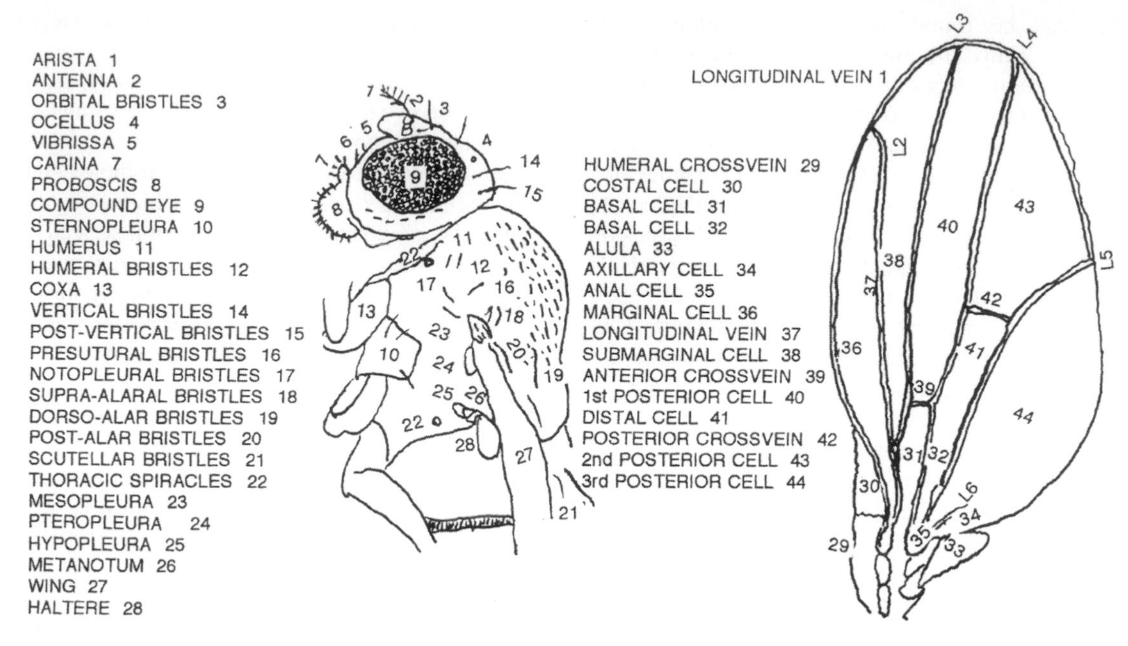

Figure M113. Head and wing landmarks of *Drosophilia*

2–62) that is also expressed during gastrulation in stripe formation of the embryo. The carboxyl terminus of the *sna* gene product appears to contain five Zinc-finger motifs indicating a DNA binding mechanisms of transcription factors.

Somatic maternal effect lethal genes: in *Drosophila* {1} *tsl* (*torsolike*, chromosome 3–71) controls the anterior-posteriormost body structures (labrum, telson). Genes {2} *pip* (*pipe*, 3–47), {3} *dl* (*dorsal*, 2–52.9) eliminate ventral and lateral body elements, and their product is homologous to the *c-rel* proto-oncogene, a transcription factor homolog, NF-κB. {4} *ndl* (*nudel*, 3–17) exaggerates dorsal elements of the embryos, {5} *wbl* (*windbeutel*, 2–86) controls dorsal epidermis, and the {6} *sna* (*snail*, 2–51) strong alleles eliminate most mesodermal tissues. {7} The *gs* (*grandchildless*, 1–21), *gs(2)M* (chromosome 2) cause blockage of the embryos of normal-looking homozygous females before the blastoderm stage at temperature above 28.5 C°.

Maternal and germline genes (A very incomplete list. Numbers in parenthesis indicate map locations):

{8} *ANTC* (*Antennapedia Complex*, 3–47.5): This contains elements *lab* (*labial*), *pb* (*proboscipedia*) *Dfd* (*Deformed*), *Scr* (*Sex combs reduced*), and *Ant* affecting head structures in anterior-posterior relations. The *lab* and *pb* elements share cuticle protein genes. Elements *(ftz) fushi tarazu,* (*zen*) *zerknüllt*, and *z* (*zen-2* [*zpr*]) affect segment numbers and (*bcd*) *bicoid* functions as a maternal effect gene eliminating anterior structures (head) and duplicating posterior elements (telsons). The product of *bcd* also binds RNA and acts as a translational suppressor of the *caudal* (*cad*) protein product. The *bcd* and *zen* genes are included in a 50-kb transcription unit called *Ama* (*Amalgam*). The latter four do not have homeotic functions. *Scr* regulates the segmental identity of the anterior thorax and the posterior part of the head is under the control of (*Scr, Dfd, pb* and *lab*). The entire complex encompasses 355-kb genomic DNA with multiple exons. Eight of the gene products are transcription factors and *Ant, Scr, Dfd, pb,* and *lab* have homeotic functions. The *Antp* (*Antennapedia*) gene has both lethal and viable loss-of-function and gain-of-function recessive and dominant alleles controlling structures anterior to the thorax and the thoracic segments. The first mutations observed converted the antennae of the adult into mesothoracic legs. The gene has two promoters and four transcripts that control differently the spatial expression, relying also on alternate splicing. The *bcd* gene is situated within *zen*, and *Ama* is a maternal lethal that affects head and thorax development. The strong alleles in the females replace the head and thorax of the embryos with duplicated telsons.

Injection of bcd^{+} cytoplasm into the embryo (partially) remedies the topical alterations brought about by mutant alleles. The RNA transcript sticks to the anterior pole of the embryo and forms a steeply decreasing gradient in the posterior direction. This *bcd* gradient is regulated by genes *exu, swa*, and *stau*, (see them below) and the gradient may be eliminated by mutations in these genes. In bcd^{-} embryos, the anterior activity of *hb* is eliminated and replaced by mirror image posterior *hb* stripes. The four exons are transcribed in either a long, complete RNA (2.6-kb) or a short one (1.6-kb) with exons 2 and 3 spliced out. The protein contains homologous tracts to the non-maternal effect genes *prd* (*paired*, 2–45, involved in the control of segmentation) and *opa* (*odd paired*, 3–48, that deletes alternate metasegments). Exon 3 contains the homeodomain with only 40% homology to other homeoboxes. The C-termini of the bic protein appears to be involved in transcription activation by binding to five high-affinity upstream sites of *hb* (TCTAATCCC). *Dfd* (*Deformed*) is a weak homeotic gene with recessive and dominant lethal alleles affecting the anterior ventral structures of the head; occasionally thoracic bristles may also appear on the dorsal part of the head. It is composed of five exons coding for a 586-residue protein. Gene *ftz* (*fushi tarazu* [segment deficient in Japanese]) has both recessive late embryo lethal and dominant and viable regulatory alleles affecting genes in the *BXC* (*Bithorax complex*), *Ubx* (*Ultrabithorax*, 3–58.8), involved in the control of the posterior thorax and abdominal segments. The general characteristics of *ftz* are the pair-rule feature, i.e., in the mutants the even numbered abdominal and nerve cord segments are deleted (or fused). The striped pattern of the abdomen is controlled within a 1-kb tract upstream of the beginning of transcription, whereas a more distal upstream element regulates the central nervous system and an even more distal tract is required for the maintenance of the striped pattern. The homeobox is within the second of the two exons of this gene. The *ftz Rpl* mutations may transform the posterior halteres into posterior wing, while the *ftz Ual* mutations convert patches of the first adult abdominal segment into a third abdominal segment-like structure. (The latter two types are not embryonic lethal as the others). The *lab* (*labial*) mutations are embryonic lethal because of the failure of head structures. The protein product of the gene contains *opa* (*odd paired*, 3–48; deletes alternate metasegments) as well as a homeodomain although it does not display homeotic transformations. Gene *pb* (*proboscipedia*) may convert labial ("lip") portions into prothoracic leg structures or antennae. From the nine exons, #4 and #5 contain the homeobox and in exon 8 there are again *opa* sequences. The gene products (RNA and protein) are localized to the general area affected by the mutations. Null mutations in gene *Scr* (*Sex combs reduced*) are embryonic lethals. Homeotic transformations involve the labial and thoracic areas. Dominant mutation reduces the number of sex comb teeth. Gene *z2* (*zen-2*) has no detectable effects upon development. Gene *zen* (*zerknüllt*) mutations may involve embryo lethality and the products may be required for post-embryonic development.

{9} *arm* (*armadillo*, 1–1.2): Cell lethal at the imaginal disc stage because an anterior denticle belt replaces the posterior part of each segment. Transcripts have been found in all parts of the larva.

{10} *bcd* (*bicoid*): See *AntC*

{11} *bic* and *Bic* (*bicaudal*, 2–67, 2–52, 2–52.91): Genes affect the anterior poles of the embryo by replacing these segments with posterior ones in opposite orientation. *Bic* apparently encodes a protein homologous to actin, a part of the cytoskeletal system.

{12} *btd* (*buttonhead*, 1–31): Mutations fail to differentiate the head.

{13} *BXC* (*Bithorax complex*, 3–58): *BXC* is a cluster of genes that determine the morphogenetic fate of many of the thoracic and abdominal segments of the body. The second thoracic segment, which develops the second pair of legs and a pair of wings, is the most basic part of the complex. The genetic map appears as follows:

abx bx Cbx **Ubx** bxd pbx iab2 **abd-A** Hab iab3 iab4 Mcp iab5 iab6 iab7 **Abd-B** iab8 iab9

The entire complex is organized into three main integrated regions: *Ubx* (*Ultrabithorax*) is responsible for parasegments PS 5–6, *abd-A* (*abdominal-A*) defines the identity of PS7–13, and *Abd-B* (*Abdominal-B*) is expressed in PS10–14. In the *Ubx* domain, *anterobithorax-bithorax* (*abx-bx*) region specifies PS5 and *bithorax-postbithorax* (*bxd-pbx*) defines PS6. In the *abd-A* region are *iab2, iab3,* and *iab4,* and in *Abd-B* are *iab5* to *iab9* elements. Mutations in *iab* (*infra-abdominal*) tracts cause the homeotic transformation of an anterior segment to a more anterior abdominal (A) segment (e.g., A2→A1 or A3 [or more posterior ones]→A2, etc.) Mutation *abx* (*anterobithorax*) causes changes in thoracic (T) and abdominal (A) segments: T3 → T2, *bx* (*bithorax*): T3 → T2, *Cbx* (*Contrabithorax*): T2 → T3, *Ubx* (*Ultrabithorax*): A1+ T3 → T2 and T2+ T3→T1, *bxd* (*bithoraxoid*): A1 → T3, *pbx* (*postbithorax*): T3 → T2, *abd-A* (*abdominal-A*): A2 to A8 → A1, *Hab* (*Hyperabdominal*): A1+ T3 →A2, *Mcp* (*Miscadastral pigmentation*): A4 & A5 to an intermediate between A4 & A5, *Abd-B* (*Abdominal-B*): A5, A6, A7, which may be weakly transformed into anterior forms. Most of these changes involve only some structures in the segments but additional alterations may also occur.

{14} *cact* (*cactus*, 2–52) mutations reduce the dorsal elements and enhances ventral structures. This gene encodes a homolog of the Iκ-B protein, which forms a complex with product of *dl*, a NF-κB homolog transcription factor, which is released from the complex upon phosphorylation of the *cact* product. After entering the cell nucleus, it may participate in the activation of its cognate genes.

M

{15} *capu* (*cappucino*, 2–8) mutations may be lethal. It causes somewhat similar alterations as *stau* in addition to making pointed appendages on the head.

{16} ci^D (*cubitus interruptus*$^-$, Dominant, 4.0): The wing vein 4 is twice interrupted proximal and distal to anterior crossvein. In homozygotes, the anterior portions of the denticle belts are duplicated in a mirror image manner in place of the posterior parts, and they are lethal.

{17} *dpp* (*decapentaplegic*, 2–4.0, [old name was *ho*]) is a complex locus with multiple developmental functions (it is homologous to BMP, TGF-β). The haploid-insufficiency *Hin/+* condition is dominant embryonic lethal because of defects in gastrulation. The *Hin-Df* (deficiency) and *hin-r* (recessive) are also lethal. The *hin-emb* is an embryonic lethal mutation. It is complementary to the *shv* (*shortvein*) and *disk* (*imaginal disk*) region genes in the same complex. The *shv-lc* recessive larva-lethal mutants complement all *disk* alleles but mutant *Hin* and *hin-r*. Mutants *shv-lnc* do not complement *disk*, *Hin* or *hin-r*, and *shv-p* alleles are viable and complement the disk mutants but not *Hin* or *hin-r*. Another mutation in the *shv* region, *Tg* (*Tegula*) causes the roof-like appearance of the wings. This mutation is complementary to all *dpp* genes. The *disk* group of mutations is either viable or lethal and may affect the eyes, wings, haltere, genitalia, head, imaginal disks, etc. The *dpp* gene apparently acts as regulator of mesodermal genes and it encodes a secreted protein of the TGF-β family transforming growth factor, involved in mammalian cancerous growth.

{18} *ea* (*easter*, 3–57) involves maternal effect lethal with loss- or gain-of-function effects on the dorsal, mesodermal, or lateral structures, depending on the alleles. The mutants may be rescued by injection of normal cytoplasm.

{19} *ems* (*empty spiracles*, 3–53): The interior of the breathing orifices are partially missing and it is embryo lethal.

{20} *en* (*engrailed*, 2–62): some point mutations and chromosome breaks are viable. Others display "pair rule" defects, adjacent thoracic and abdominal segments fuse.

{21} *eo* (*extra organs*, 1-[66]) in the homozygous lethal embryos causes head defects and ventral hole.

{22} *eve* (*even skipped*, 2–58), lethal segmentation and head defects, "pair rule" effects. Its expression is reduced in *h* mutants and *en* segments do not appear. The expression of *Ubx* protein is high in the odd-numbered parasegments 7 to 13 rather than in every segment from 6 to 12; the *ftz* segments are disrupted.

{23} *exu* (*exuperentia*, 2–93) replaces the anterior part of the head with an inverted posterior mid-gut and anal pit (proctodeum).

{24} *fs(1)K10* (*female sterile*, 1–0.5) and a whole series of other *fs* genes in chromosome 1 may have both specific and overlapping effects. Expression may depend on cues from the oocyte. Eggs of homozygous females are rarely fertilized but if they are, the gastrulation is abnormal, the anterior ends are dorsalized.

{25} *fu* (*fused*, 1–59.5): Veins L3 and L4 are fused beyond the anterior crossvein with the elimination of the latter. Heterozygous daughters of homozygous mothers have a temperature-sensitive segmentation problem that is not observed in reciprocal crosses.

{26} *ftz* (*fushi tarazu*): See *AntC* above (also in a separate entry).

{27} *gd* (*gastrulation defective*, 1–36.78) causes dorsal and ventral furrowing of the gastrula stage embryos.

{28} *gsb* (*goosberry*, 2–107.6) is homozygous lethal because the posterior part of segments are deleted and the anterior parts duplicated in mirror image fashion.

{29} *gt* (*giant*, 1–0.9): increased size larvae, pupae and imagos based on increased cell size. DNA metabolism is abnormal and both viable and lethal alleles are known affecting many ways the entire embryo. The protein product appears similar to that of *opa*.

{30} *h* (*hairy*, 3–26.5) displays extra microchaetae along the wing veins, membranes, scutellum, and head. It is also a "pair rule" gene and affects the expression of *ftz*. It also regulates the expression of genes in the ASC (*achaete-scute* complex, 1.-0.0) involved in the control of hairs and bristles. The major gene products are located in the posterior and adjacent anterior parts of segment primordia.

{31} *hb* (*hunchback*, 3–48) alleles have different effects; the class I alleles among other effects lack thoracic and labial segments, class II mutants retain the prothoracic segment, class III al-leles retain also the labial parts, class IV mutations prevent the formation of the mesothoracic segments, class V alleles also cause various gaps as well as segment transformations. These alleles are transcribed from two different promoters and produce up to five different transcripts. The products of this locus interact with those of *kr* and *ftz*. The product of *nos* activates some of the *hb* alleles. The dMi-2 proteins in cooperation with *hb* repress *Polycomb* (see ►*Polycomb* under separate entry at P).

{32} *hh* (*hedgehog*, 3–81): in homozygous embryos a posterior-ventral portion of each segment is removed and the anterior denticle belt is substituted for in a mirror image. The embryos may not have demarcated segments. The gene has two activity peaks: during the first 3–6 hr and at 4–7 days of development (see more under separate entry).

{33} *kni* (*knirps*) [3–46]): these are lethal zygotic gap mutants. Its shorter transcript is expressed only until the blastoderm stage but the longer one is expressed even after gastrulation. The NH_2 end of the protein is homologous to one of the vertebrate nuclear hormone receptors. The *kni* box, a Zn-finger domain, is homologous to parts of the products of genes *knrl* (*knirps-related*, 3-[46]) and *egon* (*embryonic gonad*, 3-[47]).

{34} *Kr* (*Krüppel*, 2–107.6) mutants show gaps in thoracic and abdominal segments and other anomalies, and are lethal when homozygous. The protein product has similarity to the Zn-finger domain of transcription factor TFIIIA. It interacts with transcription factors TFIIB and TFIIEβ. In monomeric form it is an activator, in the dimeric form at high concentration it represses transcription. This protein binds to the AAGGGGTTAA motif upstream of *hb*. It also affects other maternal effect genes. The GLI oncogene is a homologue.

{35} *nkd* (*naked cuticle*, 3–47.3): Denticle bands are partially missing; germ bands are shortened and thus lethal.

{36} *nos* (*nanos*, 3–66.2) is active in the pole cells, and transport of its product in anterior direction is required for the normal abdominal pattern, and it is essential for the normal development of the germline. It is a maternal lethal gene. Its product represses that of *hb* in the posterior part of the embryo. Deficiency of both *hb* and *nos* is conducive, however, to normal development.

{37} *oc* (*ocelliless*, 1–23.1): Some alleles are viable although the ocelli are eliminated; others (*otd*) involve lethality because of neuronal defects.

{38} *odd* (*odd skipped*, 2–8): Embryonic lethal because posterior part of the denticle bands are replaced by the anterior parts in mirror image fashion in T2, A1, A1, A3, A5, and A7.

{39} *opa* (*odd paired*, 3–48): Alternate metasegments are genetically ablated. Denticle bands of T2, A1, A3, A5, A7, and naked cuticle of T3, A2, A4, and A6 are absent. The product of *en* is lost but that of *Ubx* increases in even-numbered parasegments.

{40} *osk* (*oskar*, 3–48): Homozygous females and males are fertile but the embryos produced by homozygous females are defective in the pole cells and consequently also in abdominal segments; IT affects also *BicD* and *hb* expression.

{41} *phl* (*pole hole*, 1–0.5) blocks the formation of anterior–posterior end structures as well as the entire 8th abdominal segment. The phenotype is similar to that caused by *tso*. It is the *raf* oncogene in *Drosophila*.

{42} *pll* (*pelle*, 3–92) gene causes maternal embryo lethality by preventing the formation of ventral and lateral structural elements.

{43} *prd* (*paired*, 2–45): In strong mutants the anterior parts of T1, T3, A2, A4, A6, and A8 and posterior parts of T2, AS1, A3, A5, and A7 are absent.

{44} *run* (*runt*, 1-[65.8]; syn. *legless* [*leg*]): A "pair rule" embryo lethal; eliminates the central mesothoracic and the uneven numbered abdominal denticle belts. The deletions are accompanied by duplication of the more anterior structures. The wild type allele appears to regulate the expression of *eve* and *ftz*.

{45} *slp* (*sloppy paired*, 2.8): Parts of the naked cuticle are missing from T2, A1, A3, A5, and A7 in an irregular way and it causes lethality.

{46} *sna* (*snail*, 2–51): Embryonic lethals causes the dorsalization and reducing or eliminating of most of the mesodermal tissues. The C-terminus of the polypetide encoded has five Zn-finger motifs.

{47} *snk* (*snake*, 3–52.1) is a maternal lethal gene with dorsalizing effects. The encoded polypeptide contains elements homologous to serine proteases.

{48} *spi* (*spitz*, [*spire*], 2–54) is embryonic lethal blocking the development of anterior mesodermal tissues.

{49} *spz* (*spätzle*, 3–92): Maternal lethal alleles accentuate dorsal structures.

{50} *stau* (*staufen*, 2–83.5) ablates pole cells and other anterior structures and some abdominal segments; causes the *grandchildless-knirps* syndrome. The Staufen protein binds RNA, and along with the Prospero RNA, is asymmetrically partitioned into the ganglion mother cells (GMC) and thus, they determine neuroblast cell fate. It functions also nonsense-mediated decay.

{51} *swa* (*swallow*, 1–15.9) is expressed in the nurse cells and the product is transported to the oocyte and into the blastoderm until gastrulation, leading to problems with nuclear divisions, head, and abdominal defects. In *swa* homozygotes the bcd^+ products are disrupted.

{52} *tl* (*Toll*, 3–91): Females heterozygous for the dominant or homozygous for the recessive *tl* mutant alleles produce lethal embryos that are defective in gastrulation and dorsal or ventral structures. The gene product is an integral membrane protein. Toll-like proteins are receptors of interleukin signals in the innate immunity system of mammals.

{53} *tll* (*tailless*, 3–102) deletes several posterior structures (Malpighian tubules, hindgut, telson), but brain and other anterior structures are also missing. Its expression is required for the manifestation of "pair rule" genes, *h* and *ftz*, in the 7th abdominal segment and for site-specificities of *cad* (*caudal*, 2-[55] involved with head, thorax, and abdominal structures and regulation of *ftz*), *hb* (see above), and *fkh* (*fork head*, 3–95, involved with homeosis in both anterior and posterior structures of this non-maternal embryonic lethal). Gene *tll* has negative effects on genes *kni*, *Kr*, *ftz* and *tor*.

M

{54} *tor* (*torso*, 2-[57]) locus has both loss-of-function and gain-of-function alleles. The former type of alleles eliminate anterior-most head structures and segments posterior to the 7th abdominal segment. The latter type alleles are responsible for defects in the middle segments and enlargement of the most posterior parts of the body. The gene is expressed in the nurse cells, oocytes, and early embryos. The expression of *ftz* is reduced in the gain-of-function *tor* mutants, whereas *phl* mutations are epistatic to *tor*. With the exception of the NH_2 terminus, the protein is homologous to the growth factor receptor kinases of other organisms. It is concentrated in both pole cells and at the surface cells. Apparently, the product of this gene receives and transmits maternal information into the interior of the embryo.

{55} *trk* (*trunk*, 2–36) mutants lack anterior head structures as well as segments posterior to the 7th abdominal band.

{56} *Tub* (*tubulin* multigene families, scattered in chromosome 2 and 3). The *αTub* genes are responsible for the production of α-tubulin and are apparently active in the nurse cells; the transcripts accumulate in the early embryo and the ovaries and control mitotic and meiotic spindle and cytoskeleton. The β-tubulin genes are expressed in the nurse cell, early embryos, and various different structures and organs.

{57} *tud* (*tudor*, 2-[97]): The germline autonomous mutants display the "grandchildless-knirps" phenotype and lack pole cells, yet about 30% of the embryos survive into sterile adults.

{58} *tuf* (*tufted*, 2–59): Segment boundaries are duplicated in a mirror image and other parts deleted and neuronal pattern altered.

{59} *twi* (*twist*, 2–100): Mutants are embryo lethal with defects in mesodermal differentiation. The embryos are twisted in the egg case. Mutations at the *dl, ea, pll,* and *Tl* loci prevent the expression of *twi*. The *twi* polypeptides are homologous to DNA-binding myc proteins.

{60} *Ubx* (*Ultrabithorax*, 3–58.8): See *BXC*

{61} *vas* (*vasa*, 1-[64]) affects the pole region and segmentation (*grandchildless-knirp* syndrome). The protein product is homologous to murine peptide chain translation factor eIF-4A.

{62} *vls* (*valois*, 2–53): The phenotype is very similar to that caused by *vas*.

{63} *wg* (*wingless*, 2–30): Visible, viable, and lethal alleles control segmentation pattern and imaginal disk pattern (wings and halteres). Its protein product is homologous to the mouse mammary oncogene *int-1* (INT1/Wnt). It appears that *frizzled* (*fz*) is the receptor of *wg*. The signal then may be transmitted to *Dsh* and *Arm* (β-catenin) and transcription of *En* is turned on with the assistance of other factors. (See more under ►wingless and ►wnt entries).

{64} *zen* (*zerknüllt*) is a segment of *ANTC* {8} affecting segmentation and dorsal structures of the early embryo.

The maternal germline mutations are genetically identical in the unfertilized egg and their gene products generally cause lethality in the egg or in the homozygous embryos. The heterozygous mothers are semi-sterile and the males are usually fairly normal in appearance and in function. The molecular bases of a few such lethal genes are known. E.g., the *phl* gene appears to be homologous to the *v-raf* protooncogene and *phl* gene is the *Drosophila raf* gene that encodes a serine–threonine kinase protein also in humans and mice. The *snk* locus encodes a protein that appears to have a calcium-binding site at the NH_2-end and with homology to several serine proteases at its C terminus. The product of the *ea* gene has some homology to an extracellular trypsin-like serine proteases. Specific cytoplasm extracted from some (e.g., *osk, tor, ea, bcd*) unfertilized normal eggs or from normal embryos when injected into the mutant cytoplasm end may rescue the embryos that develop into sterile adults. The 923-amino acid protein encoded by *tor* has no homology to other known proteins in the NH2-end but the rest of it is similar to a growth factor receptor tyrosine kinases and a hydrophobic segment appears to be associated with the cell surface membrane. The product of the *Tl* locus is an integral membrane protein with both cytoplasmic and extracytoplasmic domains containing 15 repeats of leucine-rich residues resembling yeast and human membrane proteins. The cytoplasmic domain is homologous to the interleukin-1 receptor (IL1R), the heterodimeric platelet glycoprotein 1b, and coded in human chromosome 2q12 and mouse chromosome 1 near the centromere.

Although all of the mutants assigned to different chromosomal locations have different molecular functions, the phenotypic manifestation may not necessarily distinguish that. The majority of the morphogenetic-developmental mutations is pleiotropic (e.g., affects the pole cells and causes the *grandchildless-fushi tarazu* syndrome). Many display epistasis and indicate the complex interactions of the regulatory processes involved (e.g., the expression of *twi* may be prevented by mutations at *dl, ea, pll, and Tl*). The genes may be expressed at a particular position but their products form a diminishing gradient (e.g., *bcd*, *nos*). The DNA-binding protein encoded by *bcd*, depending on its quantitative level, then regulates qualitatively the transcription of, e.g., *hb*. Some of the genes display the so-called "gap" effect. They eliminate particular body segments, e.g., *kni, Kr,* and *hb*. The so-called "pair rule" genes may eliminate certain body segments and replace them with others (e.g., *ftz, eve*) or eliminate half of the segments and fuse them together in pairs (e.g., *en*). Several of the genes, particularly those in the huge *BTC* and *ATC* clusters, may display homeotic effects. Although a great deal of information has been gathered on these genes primarily during the last decade, more specific knowledge is required for understanding the precise functions (especially the interacting circuits) of the morphogenetic processes. With the aid of microarray hybridization, information on the interacting systems will greatly expand. It appears that these genes are frequently preferentially expressed at a particular position. Their mRNA or protein product is then spread in a gradient to the sites of the required action. In some instances, only the RNA is spread and the protein is made locally. The position of morphogenetic function is controlled by a hierarchy of signals. These genes are expressed differently in different time frames and in coordinated sequences. The coordination is provided by the interaction of gene products. The same protein may turn on a set of genes and turn off others, depending also on the local concentration of the products.

According to their **main** effects, some of these genes may be classified into groups, others are more difficult to place in any of the groups because they all affect several stages and different structures. (The horizontal line stands for the embryo axis and the arrows or gaps illustrate the typical sites of action; the best representative genes are bracketed):

The *Ax* (*Abruptex*) homozygotes reduce the length of longitudinal vein L5 and commonly L4, L2, and sometimes L3. The various *Ax* alleles are either positive or negative regulators of *Notch.* The Notch signals are modulated by *fringe* and this gene determines the dorsal ventral boundaries. *Co* (*Confluens*) causes thickening of the veins. The *fa* alleles affect the eye facets and are non-complementary to the *spl* gene that also causes rough eyes and bristle anomalies. Gene *nd* (*notchoid facet*) is homozygous viable and displays some of the characteristics of the other genes within the complex. *E(spl)* (*Enhancer of split*, 3–89) mutation became known for exaggerating the expression of *spl* (enhancer in this context does not correspond to the term enhancer as used in molecular biology). This gene produces 11 similar transcripts. They share homologies with *c-myc* oncogene (a helix-loop-helix) protein and also with the β-transducin G protein subunit, known to involve signal transduction. *Dl* (*Delta,* 3–66.2) causes thickening of the veins (and a number of other developmental defects) and is responsible for a protein that has an extracellular element with nine repeats resembling the EGF, an apparent transmembrane and an intracellular domain with apparently five glycosylation residues. *Egfr* (*Epidermal growth factor receptor* [synonyms *top {Torpedo}, Elp {Ellipse}, fbl {faint little ball*}, 2–100) genes cause embryonic lethality and a number of other developmental effects including extra wing veins, eyes, etc.

The genes *wg* (see above {63} and *dpp* (see {17}) are involved in anteriorposterior specifications of the

MATERNAL ANTERIOR GENES:→____ 1, 10 see [8] 23, 51,

MATERNAL POSTERIOR GENES:_____← 6, 11, [36], 40, 50, 57, 61, 62

MATERNAL END SEGMENT:—— ←1, 24, [54], 55,

MATERNAL DORSO-VENTRAL:—↓↑—2, [3] 4, 5, [14], 17, 18, 42, 46, 49, 52, 59, 63

ZYGOTIC GAP:__ ___ ____ __ ____ _ 29, [31], [33], [34], 53

ZYGOTIC PAIR RULE:—_—_—_— [22], 26, [30], 38, 39, [14], 45

ZYGOTIC SEGMENT POLARITY:__→←__→←__ 9, 16, [20], 25, 28, 32, 35, 43, [63]

HOMEOTIC GENES: genes within the *Antennapedia Complex* (8), and the *Bithorax Complex* (13)

The development of wing veins is controlled by several genes. Locus *vn* (*vein*, 3–16.2, *Vein*, 3–19.6) disrupts longitudinal vein L4, posterior crossvein and sometimes L3, *ri* (*radius-incompletus* 3–46.8) interrupts L2, and mutations in *px* (*plexus,* 2–100.5) produce extra veins. *N* (*Notch*, 1–3.0) complex (*Ax, Co, fa, l(1)N, N, nd, spl*), with a very large number of (dominant and recessive) alleles, are homo- and hemizygous lethal and remove small portions of the ends of the wings and affect also hair and embryo morphogenesis, thickening of veins L3 and L5, and hypertrophied nervous system. *Notch* gene homologs are also present in vertebrates. The N complex codes for a protein with EGF-like repeats.

embryo and thus, also in wing formation. Several other genes also affect wing and vein differentiation. Molecular evidence permits the assumptions that Dsh (dishevelled), a cytoplasmic protein is one of the receptors of the Wg protein signal. Dsh binds to N (Notch, a transmembrane protein) and inhibits its activity. When N binds to Delta it activates Su(H), the suppressor of hairless (H). Su(H) then moves to the cell nucleus and operates as a transcription factor. The Wg signal can also activate the Shaggy-Zeste white (Sgg-Za3) serine- threonine kinase and the phosphorylation of Armadillo may lead the Wg-dependent gene expression. The level of Armadillo may be elevated also by binding Wg to a member of

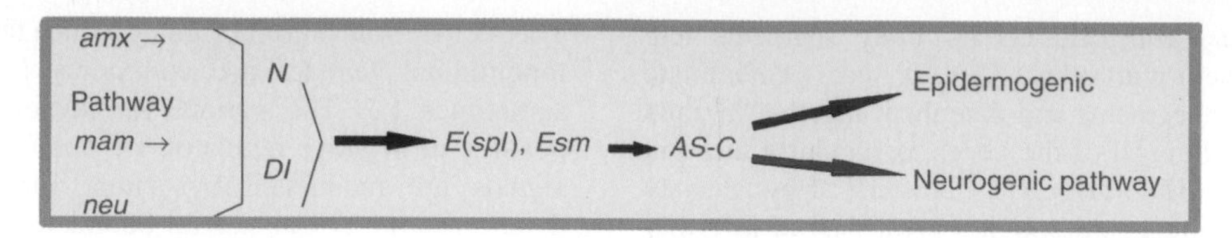

Figure M114. Epidermogenic and neurogenic pathways in *Drosophila*

the frizzled family (Dfz2). This *frizzled* (*fz*, 3–41.7) appears to be another receptor of Wg. Diffusible protein product of *optomotor-blind* (*omb*, 1-{7.5}), which was initially identified on the basis locomotor activity, is also required for the development of the distal parts of the wing within the *wg-dpp* system.

A general picture of neurogenesis is also emerging (see Fig. M114). (Modified after Campos-Ortega, Knust 1992 p. 347. In: Development. Russo VEA et al (Eds.) Springer Vlg., New York):

amx (*almondex*, 1–27.7) is a locus with multiple functions including eyes [some alleles complement the *lozenge* mutants]. Most relevantly, in the mutants there is hyperplasia of the central - peripheral nervous system and a concommitant reduction in epidermogenesis.

mam (*master mind*, 2–70.3) also affects the eyes but mutations lead to neural hyperplasia and epidermal hypoplasia.

neu (*neural*, 3–50) mutations also cause neural hyperplasia and epidermal hypoplasia.

N and Dl have been briefly described above. Both have protein products with epidermal growth factors (EGF)-like repeats. EGF has growth promoting signals and has ability to bind appropriate ligands.

E(spl): The wildtype alleles encode a protein with helix-loop-helix motifs, characteristic for transcription factors and displays similarities also to one subunit of the trimeric G proteins, which have key roles in several signal transduction pathways.

ASC (*achaete-scute complex*, 1–0.0) controls sensilla and micro- and macrochaetae that are sensory organs of the flies and correspond to the peripheral nervous system. The ca. 100-kb region contains four major distinguishable areas *ac* (*achaeta*), *sc* (*scute*), *l(1) s* (*lethal scute*) and *ase* (*asense*). All four reduce and alter the pattern of the sensory organs. The dominant components, the *Hw* (*Hairy wing*) mutations, increase hairiness. Another regulatory mutation has been named *sis-b* (*sisterless b*). The complex includes nine transcription units. Four of them appear to be transcription factors because of the helix-loop-helix motifs.

svr (*silver*), elav (*embryonic lethal abnormal vision*), and vnd (*ventral nervous system defective*) all are affected in the nervous system and located at 0 position of the X chromosome just like *ASC*. A large number of other genes at various locations are also involved in the nervous system. One must remember that in even a relatively simple organism, such as *Drosophila* ~ 14,000 genes exist and their functions are too complex to be represented by simple models. ►*Drosophila*, ►homeotic genes, ►morphogens, ►gap genes, ►clonal analysis, ►developmental genetics, ►signal transduction, ►imaginal disks, ►oncogenes, ►pattern formation, ►selector genes, ►RNA localization, some genes are described with more details under separate entries, ►selector genes; Morisato D, Anderson KV 1995 Annu Rev Genet 29:371; Mann RS, Morata G 2000 Annu Rev Cell Dev Biol 16:243.

Morphogenetic Field: An embryonal compartment capable of self-regulation.

Morphogenetic Furrow: An embryonic tissue indentation marking the front line of the differentiation wave. Signal molecules mediate its progression. ►morphogenesis, ►morphogenesis in *Drosophila*, ►daughter of sevenless

Morpholinos: Antisense oligomers that block cell proliferation, interfere with normal splicing of pre-mRNAs, and generate aberrant splicing (see Fig. M115). They are highly specific and immune to nuclease (RNase H). They are suitable for the inactivation (knock- down) of targeted genes. (See Summerton J 1999 Biochim Biophys Acta 1489:141).

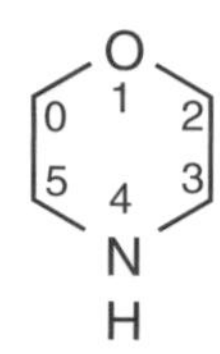

Figure M115. Morpholine

Morphology: Study of structure and forms.

Morphometry: Quantitative study of shape and size of a body or organ. ►QTL

Morphosis: A morphological alteration, a phenocopy rather than being a mutation.
►phenocopy, ►epigenetic

Morquio Disease: ►gangliosidosis type I

Morsier Syndrome: ►septooptic dysplasia

Morsus Diaboli (literally devil's bite): The fringe end of the oviduct (Fallopian tube) at the ovary. ►uterus

Mortality: The condition of being mortal, i.e., subject to death. The rate of mortality is computed as the average number of deaths per a particular (mid-year) population. An important factor of (particularly) infant death rate is the coefficient of inbreeding. First cousin marriages (inbreeding coefficient 1/16) approximately double infant mortality. The rate of human mortality calculated as the number of deaths per 1000 population may vary substantially in different parts of the world. In the years 1955–59, in the USA it was 9.4, in England 11.6, in Japan 7.8. In some other parts of the world it was double or higher. *Extrinsic mortality* is an age and condition-independent concept. Infant mortality rate is highly correlated with the level of income in the world's 231 countries. In the 48 high-income countries, it is 0.8%, whereas in the lowest-income 67 countries, it was 79.2% by latest reports in 1999 (United Nations Demographic Yearbook, New York). In the same group of countries, birth was 7.9 and 55.8%, respectively. These figures indicate high birth rates and high infant and child mortality in the low-income countries. In the high-income countries the female literacy rate was 96, whereas in the low-income countries it was 68 and the health expenditure in dollars of purchasing power parity (PPP) was 2435 and 59 per capita, respectively. In the AIDS pandemic South Africa and Botswana, in the age group 15–49, mortality was 20.1 and 38.8%, respectively between 1997–1999. The median age in years in England, a high-economy country, was 40 and in the lower-resource-but-transitional country of Iran it was 20. More than 50/1000 births in developed countries are affected by genetically determined anomalies and almost 30% of them involve early mortality. Congenital heart disease involves more than 40% mortality and congenital malformation of the central nervous system is the second greatest ause of mortality at about 26%. In Western Europe, about 50% of the pregnancies with severe genetic/developmental defects are medically terminated. ►inbreeding coefficient, ►age-specific birth and death rates, ►longevity, ►aging, ►juvenile mortality, ►morbidity, ►alcoholism substance abuse, ►addiction, ►smoking, ►counseling genetic, ►family planning, ►prenatal diagnosis; Christianson A, Modell B 2004 Annu Rev Genomics Hum Genet 5:219; www.mortality.org.

Morula: A mass of blastomeres, containing 8–16 cells in the mouse (see Fig. M116) (the number of cells in humans is about twice or more) before implantation of the zygote onto the uterus by transfer the morula from the oviduct. ►blastomeres, ►blastocyst

Figure M116. Morula

Mos: ►*mariner*

MOS (*mos*): ►Moloney mouse sarcoma virus oncogene, ►MITE

Mosaic: Mixture of genetically different cells or tissues. Mosaicism is generally the result of somatic mutation, change in chromosome number, deletion or duplication of chromosomal segments, mitotic recombination, nondisjunction, sister chromatid exchange, sorting out of mitochondrial or plastid genetic elements, infectious heredity, intragenomic reorganization by the movement of transposons or insertion elements, lyonization, gene conversion, lyonization etc.

A mosaic hybrid may be the result of co-dominance of the parental alleles (see Fig. M117). Some breeds, however, display homozygosity for mosaicism. Introduction of foreign DNA into the cells by transformation or gene therapy may also be the cause of mosaic tissues. Somatic mutation may be of medical importance and may lead to oncogenic transformation and the expression of cancer. Somatic reversion of recessive genes causing disease, e.g., reversion of adenosine deaminase (ADA) or the tyrosinemia (FAH) genes, followed by selective proliferation of the normalized sector, may alleviate the disease. Loss of the extra chromosome in trisomics or nondisjunction in monosomics may restore the normal chromosome number. When a mutation is induced only in a single strand of the DNA, its expression may be delayed, but in mice fur patches may occur in the heterozygotes

Figure M117. "Mosaic hybrid" rooster

and these animals are called *masked mosaics*. ►sex-chromosomal anomalies in humans, ►variegation, ►lyonization, ►allophenic, ►mosaic variegated aneuploidy; Extavour C, Garcia-Bellido A 2001 Proc Natl Acad Sci USA 98:11341; Yousouffian H, Pyeritz RE 2002 Nature Rev Genet 3:748.

Mosaic Analysis with Double Marker (MADM): Two reciprocally chimeric genes, each containing the N terminus of one marker and the C terminus of the other marker interrupted by a loxP-containing intron, are knocked in at identical locations on homologous chromosomes (see Fig. M118). Functional expression of markers requires Cre-mediated interchromosomal recombination. Such a system can generate loss of heterozygosity in tumor suppressor genes and their fate in cancer development can be followed (Zong H et al 2005 Cell 121:479; Mazumdar MD et al 2007 Proc Natl Acad Sci USA 104:4495). ►Cre/loxP, ►LOS, ►tumor suppressor gene

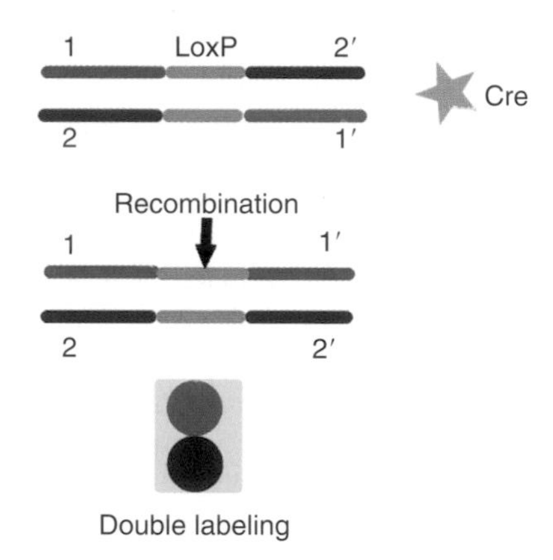

Figure M118. MADM

Mosaic genes: Mosaic genes contain exons and introns. Also some nuclear genes borrow from or loan exons to organelles or vice versa. Transposition can generate mosaics. Exon shuffling also generates mosaic genes. ►introns, ►exons, ►exon shuffling, ►organelle sequence transfer, ►transposition; mosaic genes in human chromosome 22: Bailey JA et al 2002 Am J Hum Genet 70:83.

Mosaic theory: ►suicide vectors

Mosaic Variegated Aneuploidy (MVA, BUB1B/BUBR1, 15q15): MVA is characterized by somatic aneuploidy (trisomy, monosomy) of different chromosomes and in different tissues, apparently due to defects in the spindle checkpoints and nondisjunction. The premature termination of the transcript leads to the formation of a truncated protein. The recessive disorder causes reduced intrauterine growth, microcephaly, eye anomalies, and other morphological defects (Hanks S et al 2004 Nature Genet 36:1159). The mutation leads to carcinogenesis, and the condition is frequently associated with rhabdomyosarcoma, Wilms tumors, leukemia, colorectal cancer, etc. The same defect occurs in yeast and various vertebrates. ►cancer, ►aneuploidy, ►Wilms tumor, ►leukemia, ►colorectal cancer

Mosolov model: In the Mosolov model, an extremely long DNA fiber is tightly packed into a very compact eukaryotic chromosome (see Fig. M119). One of the many existing models is shown here. Each line represents an elementary chromosome fiber, including the nucleosomal structure at various stages of folding. E conceptualizes the chromomeres. The elementary fiber is about 25 Å in diameter and forms a tubular coil (solenoid). D signifies the tightly coiled coils of the metaphase chromosomes. ►packing ratio, ►nucleosome, ►chromosome structure, ►chromosome coiling, ►SMC, ►condensin

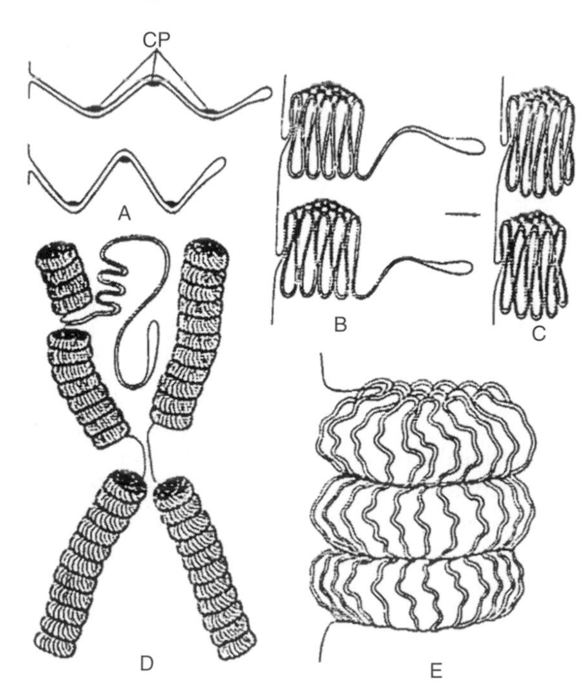

Figure M119. Mosolov model of chromosome structure. (Diagram from Kushev VV 1974 Mechanisms of Genetic Recombination. By permission of the Consultants Bureau, New York)

Mosquito: *Culex pipiens*, 2n = 6. ►malaria, ►*Anopheles gambiae*, ►*Aedes aegypti*; Atkinson P W, Michel K 2002 Genesis 32:42; http://www.tigr.org/tdb/tgi/.

MOST: is web tool for extraction of motifs in DNA sequences. (See Pizzi C et al 2005 Nucleic Acids Res. 33 (15):e135; http://telethon.bio.unipd.it/bioinfo/MOST/).

Mossy Fibers: Microtubules on nerve axons (see Fig. M120).

Figure M120. Mossy fiber

Most Recent Common Ancestor: ►MRCA, ►coalescent

Motheaten: Mutation in the SH2 domain of a tyrosine phosphatase of mice leading to autoimmune anomalies. ►tyrosine phosphatase, ►autoimmune disease, ►SH2; White ED et al 2001 J Leukoc Biol 69:825; Hsu HC et al 2001 J Immunol 166:772.

Mother-Fetus Compatibility: ►immune tolerance, ►Rh blood groups, ►incompatibility, ►mylotarg

Motif: A small, non-random assembly of structural domain or sequence of amino acids or nucleotides present in different macromolecules. Motifs are generally preserved during evolution and usually convey some functionality. The probability of a motif is matrix of probabilities. The column number in the matrix is the length (N) of the motif. The number of rows is four in DNA (the four kinds of nucleotides) and it is 20 in protein (the 20 common amino acids). Row number *i* and column number *j* specify the probability for finding a special appearance (instance) of a motif. ►protein domains, ►module, ►MP-score, ►matrix algebra, ►MOST; http://www.bioinf.man.ac.uk/dbbrowser/PRINTS/; motifs in different sets of genomic data: http://fraenkel.mit.edu/webmotifs/; new motif discovery: http://genie.dartmouth.edu/scope/.

Motif Ten EL6:47 PM (MTF): MTF promotes transcription by DNA-dependent RNA polymerase II when it is located in the core promoter at + 18 to + 22 from the initiator element (A+ 1). ►core promoter; Lim CY et al 2004 Genes Dev 18:1606.

Motif-Trap Technology: In motif-trap technology, random fragments of DNA fused to fluorescent protein are screened and their location in the cells is monitored. The procedure may facilitate the identification of the function of the unknown sequences. ►aequorin; Cutler SR et al 2000 Proc Natl Acad Sci USA 97:3718.

Motilin: A 22-amino acid, well-conserved, peptide hormone regulating motility of the intestinal tract. (See Coulie B et al 2001 J Biol Chem 276:35518).

Motor Proteins: Motor proteins can move along microtubules and actin filaments or macromolecules by deriving energy from the hydrolysis of energy-rich phosphates such as ATP. They transport various molecules and vesicles within the cell. Motor proteins mediate some of the processes involved in establishing body plans, such as left-right, dorso-ventral, and anterior-posterior differentiation. Representatives include myosin, kinesin, dynein, helicases, bimC, etc. The MyoVa motor is involved in melanosome transport and organization of the endoplasmic reticulum in Purkinje cells. Motor protein defects are involved in several human diseases. ►microtubule, ►anaphase, ►duty ratio/duty cycle, ►myosins, ►actin, ►kinesin, ►dynein, ►Purkinje cells, ►albinism, ►cytoskeleton, ►caveolae; Karcher RL et al 2001 Science 293;1317; Schliwa M, Woehlke G 2003 Nature [Lond] 422:759; review: Mallik R, Gross SP 2004 Current Biol 14:R971; motor proteins in nanotechnology: van den Heuvel MGL, Dekker C 2007 Science 317:333.

Mountjack (*Muntiacus muntjack*): The male is 2n = 7 and the female is 2n = 6 but the *Muntiacus reevesi* is 2n = 46 (see Fig. M121).

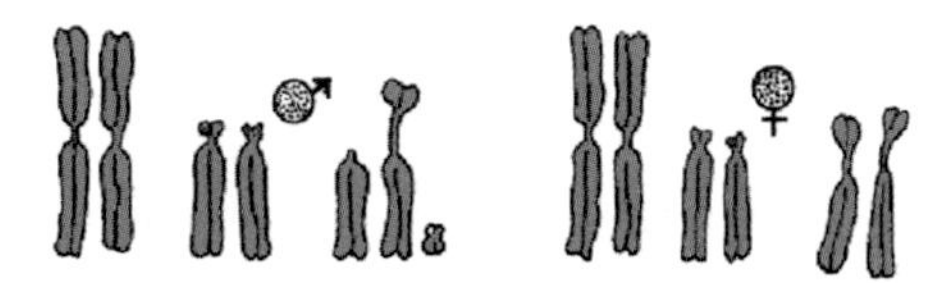

Figure M121. Mountjack chromosomes after Wurster & Bernischke

Mouse: The mouse of a computer is a pointer device.

Mouse (*Mus musculus*, 2n = 40): Rodent belonging to the subfamily *Murinae*, including about 300 species of mice and rats (see Fig. M122). They are extensively used in genetics and physiological studies because of their small size (25–40 g), short lifecycle (10 weeks), life span of about two years, gestation ~19 days, having 5–10 pups/litter, and their practically continuous breeding. The genome is $\sim 1.8 \times 10^6$ kDa in 2n = 40 chromosomes. The mouse genome is ~14% smaller than the human genome, yet about 40% of their genomes can be aligned. In one m^2 laboratory space up to 3,000 individuals can be studied annually. From embryonic stem cells, viable, fertile adults can be differentiated (Eggan K, Jaenisch R 2003 Methods Enzymol 365:25). Very detailed linkage information is available. According to the 1996 map the genetic length, based on 7,377 genetic markers including RFLP and other markers, is 1,360.9 units. The average spacing between markers then was 400 kb. A nucleotide sequence draft became available in 2002 and the final annotated sequence was expected by 2006. Transformation, gene targeting and other modern techniques of molecular genetic manipulations are well worked out. It is also very important

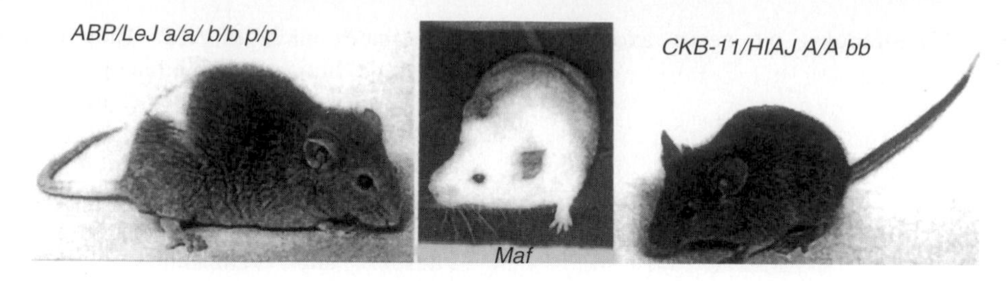

Figure M122. Mouse has a very large number of spontaneous and induced variations involving morphological, physiological, biochemical, and behavioral traits. Left: *pink-eyed dilution, A/a Tyrp1^b/Tyrp1^b bt/bt p/p*, chromosome 7–28.0. Middle: Kit ligand, *Mgf$^{S1\text{-}pan}$/Mgf$^{S1\text{-}pan}$*, chromosome 7 10–57.0. Right: tyrosine related, A/A Tyrp1^b/ Tyrp1^b, chromosome 4–38. (Courtesy of Dr. Paul Szauter, http://www.informatics.jax.org/mgihome/other/citation.shtml).

that the mouse be used as a human genetic model for immunological, cancer, and other human diseases. About >85% of the autosomal gene repertory of the mouse displays conserved synteny with that in the human genome. The X chromosomal genetic structure is practically identical in man and mouse. *Mus spretus*, "a non-laboratory" relative, is similar in appearance to the laboratory mouse but it harbors many genetic differences and can form fertile female hybrids with the laboratory strains, which are inbred. Both of these species differ from the house mouse, *Mus domesticus*. In mice, Robertsonian translocations are common, and therefore the chromosome number may vary in different populations and in different individuals: http://www.immunologylink.com/transgen.htm. ►databases, ►animal models, ►ENU, ►MICER, ►IGTC, ►Encyclopedia of the Mouse Genome, ►Mouse Genome Database, ►Mouse Genome Informatics Group, ►Portable Dictionary of the Mouse Genome; Mapping information in *Nature* (Lond.) 380:149, knockout, radiation hybrid map: Nature Genet 1999 22:383; YAC-based physical map: Nature Genet 1999 22:388; Festing MFW 1979 Inbred Strains in Biomedical Research. Oxford Univ. Press, New York; origin of outbred strains, including genealogy of Swiss mice: Chia R et al 2005 Nature Genet 37:1181; large-scale gene expression information about mouse tissues and cells:Siddiqi AS et al 2005 Proc Natl Acad Sci USA 102:18485; Fox JG et al (Eds.) 2006 The Mouse in Biomedical Research. Academic Press, San Diego, California; http://cgap.nci.nih.gov/SAGE#mouse; electronic information sources: http://www.informatics.jax.org; http://www.genome.wi.mit.edu or by "help" to mailto:genome_database@genome.wi.mit.edu; http://genome.rtc.riken.go.jp BodyMap, nude mouse; http://www.rodentia.com/wmc/index.html; book: http://www.informatics.jax.org/silver; genetrap: http://www.genetrap.org/; transgenic and knockout: http://www.immunologylink.com/transgen.htm; Mouse Atlas: http://genex.hgu.mrc.ac.uk; genetic map: Dietrich WF et al 1996 Nature [Lond] 3880:149; Mouse Sequence Consortium: http://www.ensembl.org; http://www.ncbi.nlm.nih.gov/genome/seq/MmProgress.shtml; transcript view: http://www.ensembl.org/Mus_musculus/transcriptsnpview; functional annotation; cDNA clones: http://fantom.gsc.riken.go.jp/db/; mouse gene expression pattern database: www.genepaint.org/; functional annotation: http://mips.gsf.de/genre/proj/mfungd/; embryo development: http://genex.hgu.mrc.ac.uk/Emage/database/emageIntro.html; Center for Rodent Genetics: http://www.niehs.nih.gov/crg/cprc.htm; gene expression database: http://www.informatics.jax.org/menus/expression_menu.shtml; mouse phenotypes: http://www.jax.org/phenome; GXD, Adv Genet 35:155; Joyner AL (Ed.) 2000 Gene targeting. A practical approach. Oxford University Press, Oxford, UK; mouse chromosome 16 versus human genome: Mural RJ et al 2002 Science 296:1661; physical map: Gregory SG et al 2002 Nature [Lond] 418:743; mouse genome: Nature [Lond] 420:509 2002, 15,000 cDNA sequences: Strausberg RL et al 2002 Proc Natl Acad Sci USA 99:16899; Paigen K 2003 Genetics 163:1; book with cDNA samples: Kawai J, Hayashizaki Y (Eds.) 2003 RIKEN Mouse Genome Encyclopedia DNA Book, RIKEN Genomic Sciences Center, Waco, Japan, genetic screens: Kile BT, Hilton DJ 2005 Nature Rev Genet 6:557; genome engineering: Glaser S et al 2005 Nature Genet 37:1187; genetic resources: Peters LL et al 2007 Nature Rev Genet 8:58.

Mouse Genome: A newsletter of Journal Subscriptions Department, Oxford University Press, Walton Str., Oxford, OX2 6DP, UK.

Mouse Genome Database (MGD): MGD integrates various types of information, mapping, molecular, phenotypes, etc. Contact: Mouse Genome Informatics, The Jackson Laboratory, 600 Main Str., Bar Harbor, ME 04609, USA, Phone: 207-288-3371, ext. 1900. Fax: 207-288-2516. INTERNET: mgi-help@infor matics.jax.org; for transgenic and knockout: http://www.immunologylink.com/transgen.htm.

Mouse Genome Informatics Group: An electronic bulletin board. Information: Mouse Genome Informatics User Support, Jackson Laboratory, 600 Main Str., Bar Harbor, ME 04609, USA. Phone: 207-288-3371, ext. 1900. Fax: 207-288-2516. INTERNET: mgi-help@informatics.jax.org.

Mouse Lymphoma Test for Genotoxicity (MLA): MLA employs thymidine kinase heterozygous ($TK^{+/-}$) and homozygous ($TK^{-/-}$) mouse lymphoma cells (L5178Y) and classifies the treated cultures for chromosomal aberrations and colony morphology. ►bioassays in genetic toxicology; Hozier J et al 1981 Mutation Res 81:169; Clements J 2000 Mutation Res 455:97.

Mouse Mammary Tumor Virus (MMTV): MMTV causes mammary adenocarcinomas (see Fig. M123). The virus is transmitted to the offspring by breast-feeding. If the virus is transposed within 10-kb distance to the *Wnt-1* oncogene (homolog of the *Drosophila* locus *wingless* [*wg*]), the insertion may activate the oncogene because of the very strong enhancer in the viral terminal repeat. APOBEC3 proteins are packaged into virions and inhibit retroviral replication in newly infected cells, at least in part by deaminating cytidines on the negative strand DNA intermediates. A3 provides partial protection to mice against infection with MMTV (Okeoma CM et al 2007 Nature [Lond] 445:927). Retroviral insertion mutagenesis revealed 33 insertion sites as potential sources of tumor (Theodorou V et al 2007 Nature Genet 39:759). ►DNA-PK, ►pattern formation, ►retroviruses, ►APOBEC, ►breast cancer

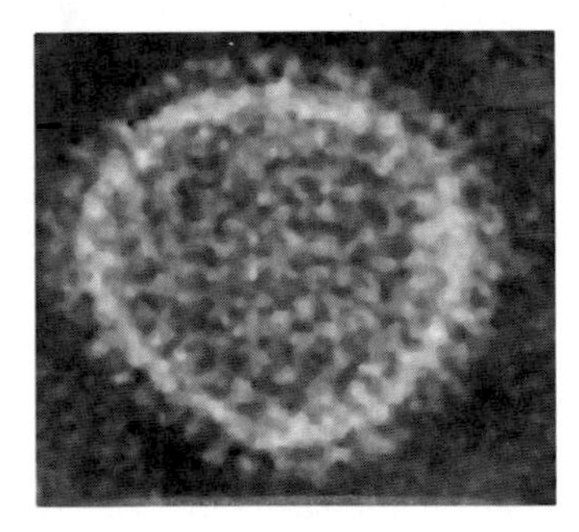

Figure M123. MMTV

Mouse Tumors, Spontaneous: http://www.informatics.jax.org.

Mov34: ►Moloney mouse leukemia virus oncogene

Movable Genetic Elements: ►mobile genetic elements, ►transposons, ►transposable elements

Movement Proteins: The synthesis of moveent proteins is directed by plant viruses in order to spread the infectious particles through the plasmodesmata with the aid of microtubules. ►microtubules, ►plasmodesma, ►motor proteins

MOWSE (new name Gene Service): A peptide mass and molecular database http://www.geneservice.co.uk/home/.

MOZ (monocytic leukemia zinc finger domain): ►CREB, ►leukemia

Mozart, Wolfgang Amadeus (1756–1791): Probably the greatest musical genius of all time, starting a musical career at age four and composition at six. His father was also a gifted violinist. Among his seven sibs, the only other survivor, Maria Anna, was also a very talented performer, although less renowned (see Fig. M124). ►musical talent, ►genius

Figure M124. Mozart family portrait. Maria Anna (Nannerl), Amadeus, mother Anna Maria, father Leopold. (Coutesy of Steve Boerner)

M6P (mannose 6-phosphate): The M6P proteins are transmembrane proteins in the trans Golgi network. ►endocytosis, ►Golgi apparatus

MP1 (MEK partner 1): MP1 is involved in the activation of MEK and ERK in the signal transduction pathway. ►signal transduction, ►MEK, ►ERK; Schaeffer HJ et al 1998 Science 281:1668.

Mpl: A regulator of megakaryocyte formation. ►megakaryocyte

MPD: The maximal permitted dose. ►radiation threshold, ►radiation hazards, ►radiation effects

Mpd: A 36-kDa yeast protein in the endoplasmic reticulum with disulfide isomerase activity. ►PDI, ►Eug

MPF (maturation protein factor/M-phase promoting factor): The MPF contains two subunits; a protein kinase (coded for by the $p34^{cdc2}$ gene) and a B cyclin. Activation takes place (probably by phosphorylation at threonine 161 of the $p34^{cdc2}$ protein) during M phase and deactivation is mediated by degradation of cyclin subunit during the rest of the cell cycle. MPF is deactivated also by phosphorylation at Thr-14 and Tyr-15 amino acid residues. These sites are dephosphorylated probably by the product of gene $p80^{cdc25}$ or a homolog before the onset of mitosis. ►protein kinases, ►cyclin, ►signal transduction, ►$p34^{cdc,2}$, ►cell cycle, ►APC; Frank-Vaillant M et al 2001 Dev Biol 231:279; Taieb FF et al 2001 Curr Biol 11:508.

MPF: Mating pair formation proteins of bacteria.

MPI (minimal protein identifier): The MPI serves as an index of protein identity on the basis of proteomics characteristics. ►proteomics

MPK2: ►p38

Mp1p: ►See mitochondrial import

MPR: Mannose-6-phosphate receptor.

Mps: A kinetochore-associated kinase required for maintaining the anaphase checkpoint until the microtubules are attached to the kinetochore. It acts before the anaphase promoting complex. Mps1 participates also in the duplication of the centrosome in cooperation with Cdk2. ►kinetochore, ►checkpoint, ►anaphase, ►APC, ►Cdk2, ►centrosome; Abrieu A et al 2001 Cell 106:83; Fisk HA, Winey M 2001 Cell 106:95.

MP-Score: The mean sum of pairs score for a column in an alignment of motifs. In other words, the value of the SP-score divided by the total number of, e.g., amino acids. The SP-score is the sum of pair scores for a column in a multiple alignment. ►motif

MPSS (multiple parallel signature sequencing): A system biology approach determining libraries of cDNAs from normal and diseased tissues. ►cDNA, ►massive parallel signature sequencing, ►transcriptome; Brenner S et al 2000 Nature Biotechnol 18:630.

MPT (mitochondrial permeability transition): MPT indicates protein release from mitochondria and apoptosis and may be detected by ICAT. ►apoptosis, ►ICAT; Bruno S et al 2002 Carcinogenesis 23:447.

MQM: Marker-QTL-marker. ►QTL

M_r (relative molecular weight): Relative molecular mass of a molecule compared to that of the mass of a C^{12} carbon atom. This is different from the gram molecular weight, traditionally used in chemistry. ►dalton

MR *(Male recombination)* factor of *Drosophila* (map location 2–54): Apparently, a defective P element (►hybrid dysgenesis) that cannot move, yet it can facilitate the movement of other P elements. Besides causing recombination in the male, it induces many of the symptoms of hybrid dysgenesis, including chromosomal aberrations, high mutation rate, and mitotic exchange. ►male recombination, ►hybrid dysgenesis

MRCA (most recent common ancestor): An evolutionary concept for divergence or coalescence. If MRCA is old, many mutations could occur around a particular locus. However, if MRCA is young, the genetic background to the locus most likely did not change much during evolution. Also, if a mutation is young, the frequency of the allele may be low, because there was not yet enough time for it to diffuse (unless the selection coefficient was very large and positive). This would mean that rare mutations are young and young mutations are rare. On the bases of these three parameters, statistical procedures can be used to estimate the age of MRCAs (Patterson NJ 2005 Genetics 169:1093). ►evolutionary tree, ►coalescence, ►diffusion genetic

MRD (mismatch repair detection): In MRD, DNA fragments are cloned and inserted into bacterial plasmids, and bacteria are transformed by the heteroduplex constructs. If there is a mismatch, the growing bacterial colonies become white, whereas in case of no mismatch the colonies are blue. The reporter gene is *LacZα*. The template for repair is a hemimethylated double-stranded DNA. Mismatches activate repair and the unmethylated strand is degraded, whereas the methylated strand becomes the template for repair. The method can scan up to 10-kb fragments for mismatches that are below five nucleotides in length. A revised procedure more suitable for high throughput analysis uses the Cre/lox recombinase, a tetracycline resistance (Tet^R), and a streptomycin sensitive (Str^S) marker. Two vectors are constructed and one of them carries a 5-bp deletion for the *Cre* gene. In other respects, the two vectors are identical. DNA fragments cloned in the active Cre^+ vector are propagated in a bacterial strain, which is dam methylase free. The two clones are made only to be used as standard panels for testing human DNA samples. Human DNA fragments from each individual to be tested are first amplified and then pooled. Linearized methylated DNAs from the Cre-deletion and the Cre^+ unmethylated DNA are combined in one vessel and single stranded PCR-amplified DNA is added. Subsequently, Taq ligase is added to generate hemimethylated, closed heteroduplex circles. The remaining linear strands are removed by exonuclease III digestion. The heteroduplexes are cloned into an *E. coli* strain (carrying Str^R in its chromosome) carrying an F' plasmid with Tet^R and Str^S cassettes, each, flanked by two lox sites. The DNA strands, without mismatch, replicate normally in the bacterium and both types of plasmids survive. The Cre^+ protein mediates recombination between the two lox sites and as a consequence, the Tet^R as well as the Str^S genes are removed. The bacterial cell thus becomes tetracycline-sensitive but because of the presence of the chromosomal Str^R gene, it will thrive on streptomycin but not on tetracycline media. The mismatch in the heteroduplex will be repaired and the unmethylated Cre^+ strand will be degraded. The inactive Cre strand will, however, be used as a template for replication and thus, the intact Tet^R and Str^S cassettes will stay functional and cells carrying them will remain tetracycline-resistant and streptomycin-sensitive. The tetracycline and streptomycin resistant cells are propagated on Petri plates and their restricted DNA content can be assayed by gel electrophoresis. ►mismatch repair, ►lactose, ►Cre/lox, ►tetracycline, ►streptomycin, ►dam methylase, ►PCR, ►electrophoresis, ►restriction enzymes; Faham M, Cox DR 1995 Genome Res 5:474; Faham M et al 2001 Hum Mol Genet 10:1657.

MRE11: MRE11 mediates double-strand break repair in somatic cells in cooperation with Ku. In meiosis, it is expressed without Ku and apparently mediates repair of the recombined DNA strands. Its defects may cause symptoms of the Nijmegen breakage syndrome and ataxia telangiectasia. Mre11, Rad50, and Xrs2 complexes are important for DNA damage control, maintenance of telomere length, cell cycle checkpoint control, and meiotic recombination (Ghosal G, Muniyappa K 2005 Nucleic Acids Res 33:4692). ►Ku, ►DNA repair, ►PIK, ►Nijmegen breakage syndrome, ►ataxia telangiectasia, ►double-strand break, ►telomerase, ►MRE11, ►RAD50, ►XRS; Petrini JH 2000 Curr Opin Cell Biol 12:293; Costanzo V et al 2001 Mol Cell 8:137; Goldberg M et al 2003 Nature [Lond] 421:952.

MRF: A member of the family of muscle proteins. ►MEF, ►myogenin, ►MYF5, ►MYOD

MRI: Magnetic resonance imaging. ►magnetic resonance

MRN: A protein complex of Mre11, Rad50, and Nbs1/Xrs2, involved in chromosome breakage and telomeric fusion. ►MRE11, ►Rad50, ►Nijmegen breakage syndrome; Ciapponi L et al Current Biol 14:1360.

mRNA: Messenger RNA carries genetic information from DNA for the sequence of amino acids in protein. Its half-life in prokaryotes is about 2 min, in eukaryotes 6–24 hr, or it may survive for decades in trees. In individual cells, the life of the mRNA may vary by an order of magnitude (Grunberg-Manago M 1999 Annu Rev Genet 33:193). Unstable mRNAs even in eukaryotes may last only for a few minutes. The mRNA is produced by a DNA-dependent RNA polymerase enzyme, using the sense strand of the DNA as template, and it is complementary to the template. The mRNA is derived from the primary transcript by processing, including the removal of introns (in eukaryotes). The size of the mRNA molecules varies a great deal because the size of the genes encoding the polypeptides is quite variable. Upstream of the coding sequences of the mRNA there are several regulatory sequences (G box, CAAT or AGGA box, etc.), which are important for transcription but are not included in the mRNA. Transcription begins by the recognition and attachment of the transcriptase to a TATA box and the assembly of the transcription complex. The eukaryotic mRNA is capped with a methylated guanylic acid after transcription. Preceding the first amino acid codon (Met in eukaryotes and fMet prokaryotes), there is an untranslated leader sequence that helps in the recognition of the ribosome. In prokaryotes, the leader includes the Shine-Dalgarno sequence which assures a complementary sequence on the small ribosomal unit. In eukaryotes, such a sequence is not known; however, usually there is a AG—CC**AUG**G preferred box around the first codon and the ribosomal attachment is relegated to "scanning" for it in the leader. The structural genes then follow that in eukaryotes have been earlier spliced together from a highly variable number of exons. In the eukaryotic mRNA, there are untranslated sequences also at the 3′ end that include a polyadenylation signal (most commonly AAUAAA) to improve stability of the mRNA by post-transcriptional addition of over hundred adenylic residues. This signal is used also for discontinuing transcription. In prokaryotes, either a rho protein-dependent palindrome or a rho-dependent GC-rich palindrome or a polyU sequence serves the same purpose. The number of mRNA molecules per cell varies according to the gene and the environment. In yeast cells, under good growing condition, 7 mRNA molecules were detected with a half-life of about 11 minutes, indicating that the formation of one mRNA molecule required about 140 seconds. The maximal transcription initiation rate for mRNA in yeast was found to be 6–8 second. The level of mRNA is frequently used for the assessment of proteins. The estimate so obtained may be biased, however, because proteins may be in dynamic state subject maturation and degradation, and many proteins are regulated by alternative splicing of the mRNA and also by post-translational processing. The cellular distribution of mRNAs can be monitored by nuclease resistant molecular beacons used to monitor the path of the mRNA within the nucleus and its passing though the nuclear pores (Bratu DP et al 2003 Proc Natl Acad Sci USA 100:13308); (Bratu DP et al 2003 Proc Natl Acad Sci USA 100:13308). The movement of the mRNA and attached proteins may be stalled within the chromatin and to switch to the mobile phase requires ATP (Vargas D Y et al 2005 Proc Natl Acad Sci USA 102:17008). ►introns, ►RNA polymerase II, ►transcription factors, ►transcription complex, ►open promoter complex, ►Hogness box, ►Pribnow box, ►UAS, ►up promoter, ►G-box, ►GC box, ►enhancer, ►leader sequence, ►cap, ►Shine-Dalgarno box, ►mRNA tail, ►transcription termination in eukaryotes, ►decapping, ►transcription termination in prokaryotes, ►regulation of gene activity, ►mono-cistronic mRNA, ►operon, ►regulon, ►mRNA degradation, ►aminoacyl-tRNA synthetase, ►mRNA targeting, ►RNA non-coding, ►RNA surveillance, ►ribonuclease E, ►ribonuclease III, ►polynucleotide phosphorylase, ►degradosome, ►polyA polymerase, ►polyadenylation signal, ►beacon molecular, ►Mlp1, ►PML39, ►half-life, ►PMAGE; Brenner S et al 1961 Nature [Lond] 190:576; Maquat LE, Carmichael GG 2001 Cell 108:173; Moore MJ 2005 Science 309:1514; mRNA and small RNA of some plants: http://mpss.udel.edu/, coding potential of RNA transcripts: http://cpc.cbi.pku.edu.cn/.

mRNA CAP: ►cap

mRNA Circularization: The 5′ and 3′ ends of messenger RNA may be joined through the transcription initiation polypeptides eIF-4G, eIF-4E, and PABp, and this structure favorably modulates translation initiation and possibly prevents the truncation of the message and the polypeptide. The rotavirus mRNA, which has a GUGACC rather than a poly(A) tail, uses a special binding protein NSP3 and probably employs this structure to suppress host protein synthesis. ►PABp, ►polyadenylation signal; Mazumder B et al 2001 Mol Cell Biol 21:6440.

mRNA Degradation: mRNA degradation is not an incidental random process but it is under precise genetic control. The decay in eukaryotes may begin by the shortening or removal of the poly(A) tail, that is followed by removal of the Cap and digestion by 5′→3′ exoribonucleases (encoded by *XRN1*, *HKE1*). The decay by 3′→5′ exonuclease is of minor importance. The poly(A) tail is removed by the PAN 3′→5′ exonuclease but requires for activity the PABP (poly(A)-binding protein) and other proteins. The decapping is mediated by pyrophosphatases. The degradation is predominantly cytoplasmic although it may take place in the nucleus before the transfer to the cytoplasm. Most frequently, those mRNAs are attacked that have a termination codon 50–55 nucleotide upstream of the last exon–exon junction. AU-rich elements (AURE) and other factors in the downstream regions regulate the decay process. Decay regulatory purine-rich elements (180–320 base) reside also within the coding region. The histone mRNAs lack poly(A) tail and their stability depends on 6-base double stranded stem and a 4-base loop, and their stability depends also on the 50-kDa SLBP (stem-loop-binding protein) and other similar proteins. The destabilization of these mRNAs depends on the process of translation, and probably on the association of the stem-and-loop proteins with the ribosomes, and it appears to be auto-regulated also by histone(s). The degradation of the mRNA is initiated by defects in the process of the translation or by encountering nonsense codons, wrong splicing, upstream open reading frames (uORFs), and transacting protein factors in the cytoplasm and the nucleus. *RNA surveillance* (non-sense-mediated decay) disposes defective mRNAs. Similarly, if the mRNA is elongated beyond the normal stop signal, *non-stop decay* destroys it. If eukaryotic mRNA stalls during the process of translation, endonucleolytic cleavage takes place leading to *no-go decay*. The latter requires the presence of Dom34p and Hbs1p protein factors that resemble eRF1 and eRF3 translation termination factors (Doma MK, Parker R 2006 Nature [Lond] 440:561). Steroid hormones, growth factors, cytokines, calcium, and iron also affect mRNA stability. Cytokine and proto-oncogene mRNAs are degraded after binding the AUF1 protein to the 3′ untranslated AU-rich sequences. This AU factor forms complexes with heatshock proteins, translation initiation factor eIF4G, and a polyA-binding protein. When the AUF1 is dissociated from eIF4G, AU-rich mRNA is degraded. Exo- and endoribonucleases, regulated by various factors, may also be involved in processing and also in protecting mRNA. Viral infections may rapidly destabilize host mRNAs without affecting the rRNAs and tRNAs. Premature termination of translation may account for Fanconi anemia Duchenne muscular dystrophy, Gardner syndrome, ataxia telangiectasia, breast cancer, polycystic kidney disease, desmoid disease, etc. A method exists for assessing the dynamics of the steady-state level and degradation of mRNA by the use of the cellular pyrimidine salvage pathway and affinity-chromatographic isolation of thiolated mRNA (Kenzelmann M et al 2007 Proc Natl

Acad Sci USA 104:6164). ►mRNA surveillance, ►non-stop decay, ►AMD, ►NMD, ►mRNA, ►polyA mRNA, ►exonuclease, ►endonuclease, ►nonsense codon, ►histones, ►Cap, ►transcription termination, ►mRNA tail, ►URS, ►transcription factors, ►eIF4G, ►heat-shock proteins, ►polyA mRNA, ►ubiquitin, ►RNAi, ►exosome; Ross J 1995 Microbiol Rev 59:423; Wilson GM et al 2001 J Biol Chem 276:8695; Wilusz CJ et al 2001 Nature Rev Mol Cell Biol 2:237; Chen C-Y et al 2001 Cell 107:451; van Hoof A, Parker R 2002 Current Biol 12:R285, degradation pathway in yeast: Kuai L et al 2005 Proc Natl Acad Sci USA 102:13962.

mRNA Display: An entirely in vitro technique for the selection of peptide aptamers to protein targets. The binding of the aptamers does not require disulphide bridges or special scaffold (e.g., antibody), yet the affinities are comparable to those of monoclonal antibody–antigen complexes. The polypeptides are linked to their mRNA in vitro, while stalling the translation on the ribosomes with the aid of puromycin, which attaches to the 3′-end of the mRNA. The mRNA–peptide fusions are then purified and selected in vitro. The procedure is highly efficient and permits selection in libraries of about 10^{13} peptides. It obviates the need for transformation, a disadvantage of phage display. ►aptamer, ►phage display, ►puromycin, ►RNA display; Wilson DS et al 2001 Proc Natl Acad Sci USA 98:3750.

mRNA Migration: After synthesis and release from the nucleus, the mRNA is not translated until it reaches its subcellular destination for translation. The localization is mediated by binding the onco-fetal protein ZBP1 (Zipcode binding protein 1) to a conserved 54-nucleotide element (Zip code) in the 3′-untranslated region of the β-actin mRNA. ZBP1 promotes translocation of actin to the periphery of the cell. Src protein kinase promotes translation by phosphorylating a key tyrosine residue (Tyr^{396}) required for ZBP1 binding to mRNA (Hüttelmaier S et al 2005 Nature [Lond] 438:512). ►actin

mRNA Leader: ►leader sequence

mRNA Surveillance (nonsense-mediated decay, NMD): RNA transcripts containing premature stop codons or other defects are more liable to destruction by specific exonucleolytic proteins than normal ones. In yeast, the decay begins at the cap-binding protein or at the eIF4E translation initiation factor (Gao O et al 2005 Proc Natl Acad Sci USA 102:4258). The decay is generally initiated at the 5′ or at the 3′ end, but in *Drosophila* it starts in the vicinity of the nonsense codon (Gatfield D, Izaurralde E 2004 Nature [Lond] 429:575). Prematurely terminated mRNA molecules may cause about 1/3 of the human hereditary diseases. In human cells, three proteins, hUpf1 (cytoplasmic ATP-dependent RNA helicase), hUpf2 (perinuclear), and hUpf3b (nuclear export-import signals shuttle between nucleus and cytoplasm), are involved. The NMD complex is formed apparently already in the nucleus at the exon-exon junctions, and triggers destruction of mRNA in the cytoplasm beginning downstream of the translation termination site. In *Caenorhabditis*, seven genes (*Smg*) are involved in NMD. Drugs (e.g., gentamycin) in cultured cells of the patients can inhibit the NMD pathway and the nonsense transcripts can be stabilized. The drug-induced changes are than revealed in normal and test systems with the aid of microarray hybridization. By identifying NMD inhibition, human disease genes can be recognized. This way, map location and biological function of the disease genes can be determined even when no *a priori* information is available. Some mutations causing NMD can be corrected by the application of appropriate drugs and the shortened polypeptide chain may be sufficient to alleviate defects normally causing hereditary disease. ►mRNA, ►mRNA degradation, ►alternative splicing, ►microarray hybridization, ►toeprinting, ►eIF-4E; Hilleren P, Parker R 1999 Annu Rev Genet 33:229; Lykke-Andersen J et al 2000 Cell 103:1121; Noensie EN, Dietz HC 2001 Nature Biotechnol 19:413; Amrani N et al 2004 Nature [Lond] 432:112; suppression of NMD for trapping mouse stem cells: Shigeoka T et al 2005 Nucleic Acids Res. 33(2):e20.

mRNA Tail: ca. 200 adenylate residues generally tails mature mRNAs of eukaryotes. This is not the end of the primary transcript; transcription may continue by a thousand or more nucleotides beyond the end of the gene. Polyadenylation requires that the transcript be cut by an endonuclease and then a poly-A RNA polymerase attaches to this poly-A tail, which is probably required for stabilization of the mRNA. Several of the histone protein genes do not have, however, a poly-A tail. Other histone mRNAs, which are not involved with the mammalian cell cycle and histone mRNAs of yeast and *Tetrahymena,* are polyadenylated. The common post-transcriptional polyadenylation is signaled generally by the presence near the 3′-end an AATAAA sequence, which is followed two dozen bases downstream by a short GT-rich element. Polyadenylation takes place within the tract bound by these two elements. Within most gene tracts, AATAAA occur at more upstream locations but they are not used for poly-A tailing. Several genes may have alternative polyadenylation sites, however, and thus can be used for the translation of different molecules, e.g., for membrane-bound or secreted immunoglobulin, respectively. The polyadenylation signal may also have a role in signaling the termination of transcription, no matter how much further downstream that takes place. In the nonpolyadenylated histone genes, there is a 6-base pair palindrome that forms a stem for a 4 base loop near the 3′-end of the mRNA, and it is followed further downstream by a short polypurine sequence. The latter may pair with a U7 snRNP that facilitates termination. The U RNA transcripts of eukaryotic RNA polymerase II are not polyadenylated, either. The formation of an appropriate 3′-tail requires that it would be transcribed from a proper U RNA promoter and the transcript would have the 5′ trimethyl guanine cap. The 3′-end of U1 and U2 RNAs is formed by the signal sequence GTTN$_{0-3}$AAAPUAGA [PU any purine, PY any pyrimidine] near the end. ►transcription termination, ►decapping, ►polyadeylation signal, ►polyA polymerase; Hilleren P et al 2001 Nature [Lond]:413:538.

mRNA Targeting: In order that proteins required for differentiation and morphogenesis would be available at the location needed, the mRNA generally carries a relatively long 3′-UTR (untranslated region). These long sequences appear to provide means for binding multiple proteins while being transported to the target sites. Additional 'zip codes' may be located in the 5′-untranslated sequences. During the process, the mRNA is bound and stabilized at the "ordained" sites from where the translated product may diffuse (in a gradient). ►morphogenesis, ►compartmentalization, ►differentiation; Jansen R-P 2001 Nature Rev Mol Cell Biol 2:247.

mRNA, Transgenic: The result of trans-splicing of the transcripts. ►trans-splicing; Finta C, Zaphiropoulos PG 2002 J Biol Chem 277:5882.

mRNA Turnover: ►mRNA degradation

mRNP: The messenger ribonucleoprotein is a repressed mRNA. It is found in the eukaryotic cytoplasm and stored until later translation. In the female gametocyte of *Plasmodium*, this translational repression is mediated by an RNA helicase (DOZI, development of zygote inhibited) as a means of sexual development. Mutation in DOZI inhibits the formation of mRNAP and 370 transcripts are degraded. ►mRNA, ►gametocyte, ►*Plasmodium*; Mair GR et al 2006 Science 313:667.

MRP (mitochondrial RNA processing): An RNase, which cleaves the RNA transcribed on the H strand of mtDNA at the CSB elements. MRP seems essential for normal exit from cell cycle at the end of mitosis (Cai T et al 2002 Genetics 161:1029). MRP designates frequently also mitochondrial ribosome proteins. ►CSB, ►DNA replication mitochondria, ►mtDNA, ►ribosomal RNA, ►ribosomal protein, ►ribonuclease P, ►Rex, ►chondrodysplasia McKusick type; Ridanpaa M et al 2001 Cell 104:195.

MRP (matrix representation with parsimony): A method for constructing composite phylogenetic trees base on published data. The source phylogenetic information is encoded as a series of binary characters that represent the branching pattern of the original trees. The data matrix of the phylogenies is evaluated by parsimony and integrated into a composite tree. ►evolutionary tree, ►maximum parsimony; Bininda-Emonds OR 2000 Mol Phylogenet Evol 16:113.

mrr: ►methylation of DNA

MRX: ►mental retardation X-linked

MRX: A DNA double-strand break repair and telomerase complex. ►DNA repair, ►double-strand break, ►telomerase

MS: ►multiple sclerosis

MS2 Phage: A mainly single-stranded icosahedral RNA bacteriophage of about 3.6 kb with 4 genes, completely sequenced. The structure of phage RNA genetic material is represented in Fig. M125, M126.

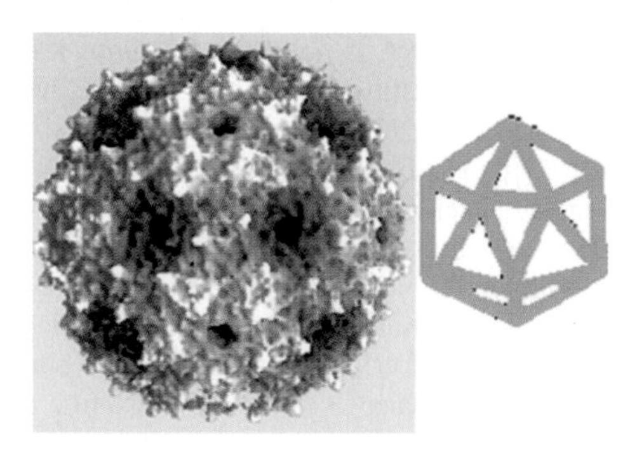

Figure M125. MS2 phage

The sequence is composed with permission on the basis of information provided by Min Jou W Ysebaert M et al 1972 Nature [Lond] 237:82; Fiers W 1975 In: Phages RNA, Zinder ND (ed) Cold Spring Harbor Laboratory, Cold Spring Harbor, N., pp353–396; Fiers W, Contreras R, et al 1976 Nature [Lond] 260:500. Gene 4 is not shown (see Fig. M126). ►structure, ►bacteriophages; Bollback JP, Huelsenbeck JP 2001 J Mol Evol 52:117.

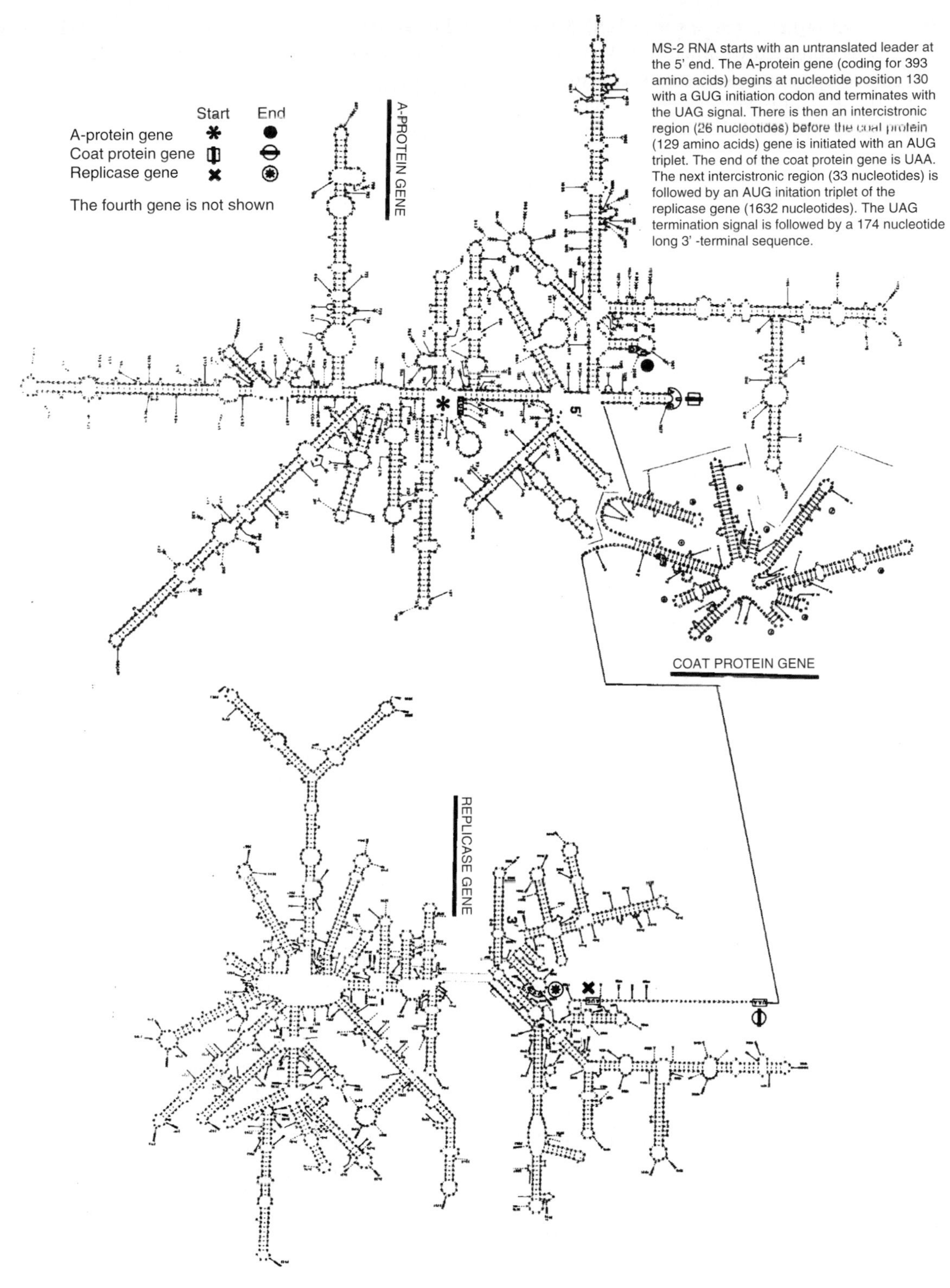

MS-2 RNA starts with an untranslated leader at the 5' end. The A-protein gene (coding for 393 amino acids) begins at nucleotide position 130 with a GUG initiation codon and terminates with the UAG signal. There is then an intercistronic region (26 nucleotides) before the coat protein (129 amino acids) gene is initiated with an AUG triplet. The end of the coat protein gene is UAA. The next intercistronic region (33 nucleotides) is followed by an AUG initation triplet of the replicase gene (1632 nucleotides). The UAG termination signal is followed by a 174 nucleotide long 3' -terminal sequence.
Start
End
A-protein gene
Coat protein gene
Replicase gene
The fourth gene is not shown
A-PROTEIN GENE
COAT PROTEIN GENE
REPLICASE GENE
5'
3'

Figure M126. The complete nucleotide sequence of the MS2 RNA phage.Composed according to Min Jou, W. et al. 1972 Nature 237:82; Fiers, W.1975 in RNA Phages (Zinder ND ed), Cold Spring Harbor Lab.p.353; Fiers, W. et al. 1976 Nature 260:500.

MSAFP (maternal serum α-fetoprotein): MSAFP analysis may detect prenatally chromosomal aneuploidy, (trisomy), and open neural-tube defects between the first 15–20 weeks of pregnancy. Oxidation products of estradiol, estrone (estriol), and chorionic gonadotropin generally accompany the high level of this protein. ►fetoprotein, ►gonadotropin, ►trisomy, ►Turner syndrome, ►Down syndrome; Ochshorn Y et al 2001 Prenat Diagn 21:658; Miller R et al 2001 Fetal Diagn 16(2):120.

MSCI (meiotic sex chromosome inactivation): Silences unpaired (unsynapsed) chromosome regions in yeast and also in mouse during both male and female meiosis. The tumor suppressor protein BRCA1 is implicated in this silencing, mirroring its role in the meiotic silencing of the X and Y chromosomes in normal male meiosis. These findings impact the interpretation of the relationship between synaptic errors and sterility in mammals (Turner JM et al 2005 Nature Genet 37:11; Turner JM et al 2006 Dev Cell 10:521; MSUC).

MSCRAMM (microbial surface components recognizing adhesive matrix molecule): Microbial protein(s) recognizing host fibronectin-, fibrinogen-, collagen-, and heparin-related polysaccharides, important for host invasion, colonization, and as virulence factors. ►heparin, ►receptin; Patti JM et al 1994 Annu Rev Microbiol 48:585.

msDNA (multicopy single-stranded DNA): msDNA is formed in bacteria, which contain a reverse transcriptase and is associated with this enzyme. This DNA has a length of 162 to 86 bases and it may be repeated a few hundred times. The 5′ end of the msDNA is covalently linked by a 2′-5′-phosphodiester bond to an internal G residue and thus forms a branched DNA-RNA copolymer of stem-and-loop structure. In some cases, because of processing, the msdRNA does not form a branched structure and exists only as a single-stranded DNA. The 5′-region of the msDNA is part of internal repeats within the copolymer. The system is transcribed from a single promoter and thus constitutes an operon. ►retron, ►reverse transcription; Lampson B et al 2001 Progr Nucleic Acid Res Mol Biol 67:69.

msdRNA: ►msDNA; Lima TM, Lim D 1997 Plasmid 38:25.

MSF (mitochondrial import stimulating factor): The MSF selectively binds mitochondrial precursor proteins and causes the hydrolysis of ATP. The MSF-bound precursor is made up of at least four proteins: Mas-20p, -22p, -70p, -37p. ►mitochondria, ►mitochondrial import; Hachiya N et al 1993 EMBO J 12:1579.

MSH: ►melanocyte stimulating hormone, ►hereditary non-polyposis colorectal cancer

MSH2 (2p22-p21): A homolog of the bacterial MutS mismatch repair gene (Mazur DJ et al 2006 Mol Cell 22:39). ►mismatch repair

MSH5/MSH4: MSH5/MSH4 are encoded in the central MHC class III region, and their obligate heterodimerization have a critical role in regulating meiotic homologous recombination and antibody gene switching (Sekine H et al 2007 Proc Natl Acad Sci USA 104:7193).

MSI: Microsatellite instability. ►microsatellite

mSin: mSin proteins are transcriptional co-repressors. ►co-repressor, ►nuclear receptors; Ayer DE et al 1996 Mol Biol Cell 16:5772.

MSK: An enzyme of the MAPK family, closely related to RSK, and involved in phosphorylation of Histone 3 and thereby, in chromatin remodeling. ►chromatin remodeling, ►RSK

Msl (male-specific lethal): At least five of the Msl genes exist in *Drosophila* that along with *Mle* (maleless) assure that a single X-chromosome in the male carries out all the functions at approximately the same level (dosage compensation) as two X-chromosomes in the female *Drosoophila*. *Msl-1*, (2–53.3) and *Msl-2* (2–9.0) are located in the 2nd chromosome but similar genes are found also along the X-chromosome. Normally, the male chromatin is highly enriched in a histone 4 monoacetylated at lysine-16 (H4Ac16). Mutation at this lysine alters the transcription of several genes. Mutation in *Msl* genes prevents the accumulation of H4Ac16 in the male X-chromosome. Msl-2 protein by containing a Zinc-binding RING finger motif may specifically recognize X-chromosomal sequences and this way distinguishes between X and autosomes. The *Msl* complexes contain at least two RNAs on the X: roX1 (3.7 kb) and roX2 (0.6 kb). These RNAs, similarly to *Xist* RNAs in mammals, spread along over 1 Mbp of the chromosome although not as broadly as *Xist*, which may span over >100 Mbp. ►dosage compensation, ►*Mle*, ►ring finger, ►*Xist*, ►histone acetyltransferase; Larsson J et al 2001 Proc Natl Acad Sci USA 98:6273; Smith ER et al 2001 J Biol Chem 276:31483; Park Y et al 2002 Science 298:1620.

MS/MS: Tandem mass spectrometer. It can be used as an automatable tool for the diagnosis of amino acid sequence in proteins, etc. ►mass spectrometer, ►proteomics

***Msp* I**: Restriction endonuclease with recognition site C↓CGG.

MS-RDA (methylation-sensitive representational difference analysis): A subtractive hybridization method designed for the detection of differences in methylation patterns among genes. ►methylation of DNA, ►subtractive cloning, ►differential hybridization mapping, ►RDA; Ushijima T et al 1997 Proc Natl Acad Sci USA 94:2284.

Mss (mammalian suppressor of Sec4): A guanosine-nucleotide exchange factor, regulating RAS GTPases. ►Ypt1, ►Rab, ►GTPase, ►RAS

MSUC (meiotic silencing of unsynapsed chromosomes): In MSUC, any chromosome region unsynapsed during pachytene of male and female mouse meiosis is subject to transcriptional silencing (Turner JM et al 2006 Dev Cell 10:521). Silencing seems to be due to extensive replacement of nucleosomes within unsynapsed chromatin and results in the exclusive incorporation of the histone H3.3 variant, which appears to be associated with transcriptional activity (van der Heijden GW et al 2007 Nature Genet 39:251). ►MSCI, ►silencer

MSUD (meiotic silencing by unpaired DNA): In MSUD, DNA unpaired in meiosis silences all homologous sequences including genes that are paired. MSUD requires the function of an RNA-dependent RNA polymerase. The mRNA is degraded after transcription. ►co-suppression, ►silencer, ►MSCI, ►quelling; Shiu PKT et al 2001 Cell 107:905; Shiu PKT et al 2006 Proc Natl Acad Sci USA 103:2243.

MSV (*msv*): ►Moloney mouse sarcoma virus

MSX (muscle segment homeobox, *msh*): Pleiotropic loci in humans (MSX1, 4p16.1) and MSX2 (5q34-q35) and homologous genes in mice control cranofacial (skull, teeth, etc.) development. ►pleiotropy, ►craniosynostosis, ►tooth-and-nail dysplasia; Milan M et al 2001 Development 128:3263; Cornell RA, Ohlen TV 2000 Curr Opin Neurobiol 10:63.

MTA: 5′-methylthioadenosine, one of the methyl donors in biological methylation.

MTA α: Yeast mating type α gene. ►mating type determination in yeast

MTA *a*: Yeast mating type *a* gene. ►mating type determination in yeast

MTD (maximum tolerated dose): The maximum tolerated dose of a treatment is that, which does not affect the longevity of the animal (except by cancer if the agent is a carcinogen) or does not decrease its weight under long exposure by more than 10%. MTD has been used in classification of carcinogens; in some instances at such high doses certain chemicals appear carcinogenic, however, under the conditions of normal use they may not pose a risk. Another potential problem is that genetic differences among test animals and humans may affect the sensitivity or susceptibility to the agents. ►bioassays in genetic toxicology; Leung DH, Wang YG 2002 Stat Med 21:51.

mtDNA (mitochondrial DNA): mtDNA in mammals consists of generally 5–6 small 16.5×10^3 bp mtDNA rings per organelle, but there are hundreds or thousands of mitochondria per cell. The yeast mtDNA genome is circular and 17–101-kb. In *Paramecium*, the mtDNA is linear and 40 kb. In about 1/3 of the yeasts and many other species (protozoa, fungi, algae), mtDNA may be linear. In plants, it varies in the range of 200 to 2,500-kbp, and occurs in variable size of mainly circular molecules, although smaller linear mtDNAs also exist and show inverted terminal repeats (Ward BL et al 1981 Cell 25:793). The mtDNA genome of *Plasmodium falciparum* is only 6-kbp, that of the plant *Arabidopsis* contains 366, 924-bp, and hexaploid wheat is 452,528 bp. More than 80% of the plant mtDNA is non-coding, whereas in protists only ~10% is non-coding. The human mtDNA genome contains 16,596 base pairs, is similar to that of other mammals, encodes 13 proteins, and transcribes two ribosomal genes and a minimal (22) set of tRNAs. The 13 proteins are: ATP synthetase subunits 6 and 8, cytochrome oxidase subunits I, II, III, apocytochrome b, and NADH dehydrogenase subunits 1–6 and 4L. A data set of 827 carefully selected sequences shows that modern humans contain extremely low levels of divergence from the mitochondrial consensus sequence, differing by a mere 21.6 nt sites on average. Fully 84.1% of the human mitochondrial genome was found to be invariant (Carter RW 2007 Nucleic Acids Res 35:3039).

Among the metazoan mtDNAs, some variations exist in gene number and the base composition of the coding strands also vary. *Reclinomonas* protozoons encode 97 genes by the mtDNA. Before the small (~16 kbp) metazoan mtDNAs have evolved, the unicellular protists had larger and genetically more complex mitochondria (Burger G et al 2003 Proc Natl Acad Sci USA 100:892).

Arabidopsis borrows two tRNAs from the chloroplasts. The wheat mtDNA includes 55 genes with exons, including 35 coding for protein, 3 rRNA, and 17 tRNA. Three percent of the wheat mtDNA genes are of chloroplast origin, ~16% of the total mtDNA is genic, and intramolecular recombination is evident (Ogihara Y et al 2005 Nucleic Acids Res 33:6235).

The number of mitochondria in the eukaryotic cells may run into hundreds to thousands. The buoyant density of the mtDNA is surprisingly uniform among many species; it varies between 1.705 to 1.707 g/mL.

In the marine mussels (*Mytilus*), there is an M (transmitted by the male to the sons) and an F (transmitted by the female to sons and daughters) mitochondrial DNA resulting in high degree of heteroplasmy among the males.

The replication of the mouse mtDNA proceeds generally from two origins of replication, thus forming D loops, a tripartite structure of two DNA, and a nascent RNA strand. Replication of the heavy strand begins at the single origin (O_H) and continues until the origin of the light chain (O_L) is reached. Then the "lagging" strand replication is initiated in the opposite direction (for an overview of the human mtDNA map see p. 1,261). Another major D-loop originates at position 57 in several human cell lines. This replication type is responsible for the maintenance of mtDNA under steady-state conditions, whereas the other D-loop origin is more important for the recovery of mtDNA after depletion or for accelerating mtDNA replication in response to physiological demands (Fish J et al 2004 Science 306:2098). Restriction fragments including the replication forks when studied by 2-dimensional gels reveal structures that are called "Y arc". In vertebrates, only a single DNA polymerase (polymerase γ) is involved in replication (Iyengar B et al 2002 Proc Natl Acad Sci USA 99:4483). This enzyme has two subunits;the larger (~125 to ~140-kDa) is involved in polymerization and the smaller (35–40-kDa) subunit may have a role in the recognition of the DNA replication primer and may function as a processivity clamp. The smaller, the accessory subunit, shares structural homology with the aminoacyl-tRNA synthetases. The large subunit also has a 3′→5′-exonuclease activity, assuring the high fidelity of DNA synthesis. Replication also in mitochondria requires a number of accessory proteins. The origin of replication of the light chain is at some distance from that of the heavy chain. The replication is regulated by the mtTFA protein, which also assures proper maintenance of the DNA and regulates transcription. The mutation rate of the mtDNA, despite the exonuclease function of polymerase γ, is an order of magnitude higher than that of the nuclear DNA. This higher rate has been attributed to oxygen, generated by oxidative phosphorylation. DNA repair may take place by nucleotide excision, mismatch repair, and alkyltransferase, but some aflatoxin B1, bleomycin, and cisplatin damage as well as pyrimidine dimers cannot be repaired efficiently.

In animals the heavy strand codes for 2 rRNAs, for 14 tRNAs, and for 12 polypeptides of the respiratory chain (ATP synthase, cytochrome b, cytochrome oxidase, 7 subunits of NADH dehydrogenase). In human mtDNA, the rRNA termination factor (mTERF) binds—by forming a loop—to both the termination and initiation sites of the two rRNA tracts and also drives mt rRNA synthesis (see Fig. M127) (Martin M et al 2005 Cell 123:1227; see diagram drawn after the electronmicrographs of Martin et al).

The plant mtDNA may encode three rRNAs and 15–20 tRNAs. The heavy chain of the mammalian mtDNA may be transcribed in a single polycistronic unit and subsequently cleaved into smaller functional units. There are also abundant shorter transcripts of the heavy chain of the mtDNA. The light DNA strands of vertebrates are transcribed into eight tRNAs and into one NADH dehydrogenase subunit. The tRNA genes are scattered over the genome. The small mammalian mtDNA is almost entirely functional, and only a few (3–25) bases lie between the genes without introns, and some genes overlap. The promoter sequences are very short and are embedded in non-coding regions. The human mtDNA employs two forms of transcription factors that bear structural similarities to rRNA modifying enzymes (see Fig M128). These two factors (TFBM1 and TFBM2, the latter being much less efficient) bind to the core RNA polymerase (PolRMT), which can recognize the two promoter sites (LSP and HSP). TFAM is the protein that binds upstream to both promoters, unwinds the DNA template, and facilitates bidirectional transcription (Shoubridge EA 2002 Nature Genet 31:227). The upstream leader sequence is minimal and the typical eukaryotic cap is absent. In addition to the AUG initiator codon, AUA, AUU, and AUC may start translation as methionine codons. The genetic code dictionary of mammalian and fungal (yeast) mitochondria further differs somewhat from the "universal code". The UGA stop codon means tryptophan, the AUA isoleucine codon represents methionine in both groups, whereas the CUA leucine

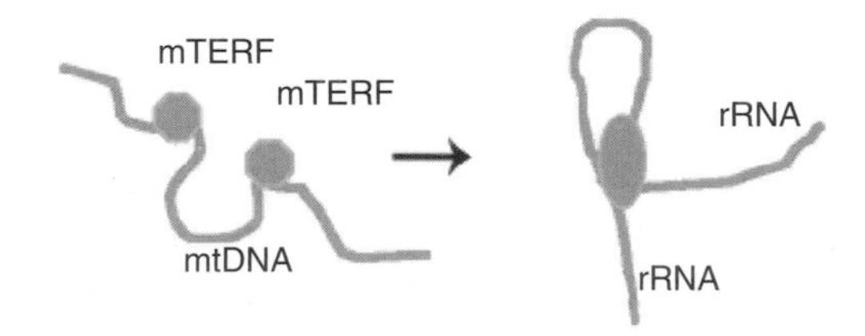

Figure M127. rRNA termination factor in mitochondria

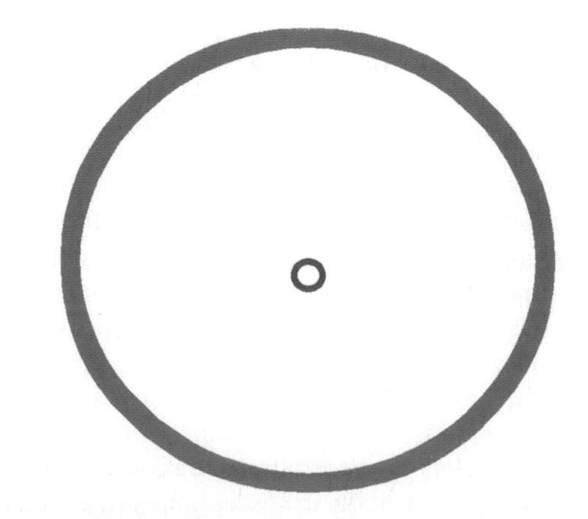

Figure M128. Arabidopsis mtDNA (outer circle) relative to mammalian mtDNA (inner circle)

codon in yeast mtDNA spells threonine (but in *Neurospora* mtDNA it is still leucine), and the AGA and AGG arginine codons are stop codons in mammalian mtDNA. Other coding differences may still occur in other species. Some mtDNA genes have no stop codons at the end of the reading frame but a U or UA terminate the transcripts after processing that may become, by polyadenylation, a UAA stop signal. Since mitochondria have only 22 tRNAs, the anticodons must use an unusual wobbling mechanism. The two-codon-recognizing tRNAs can also form a G*U pairing, and the four-amino-acid codon sets are base-paired either by two nucleotides only, or the 5′ terminal U of the anticodon is compatible with any other of the 3′ bases of the codon. Nuclear mRNA may not be translated in the mitochondria because of the differences in coding. Mitochondrial translation does not use the Shine-Dalgarno sequence for ribosome recognition in contrast to prokaryotes. Rather, it depends on translational activator proteins that connect the untranslated leader to the small ribosomal subunit. Plant mitochondria utilize the universal code.

The fission yeast and *Drosophila* mtDNAs are just slightly larger (19 kb) than that of mammals. The mtDNA of the budding yeast is about 80 kb, yet its coding capacity is about the same as that of the mammalian mtDNA. The large mitochondrial DNA is rich in AT sequences and contain introns. Some yeast strains have the same gene in the long form (with introns) while other strains in the shorter form (without intron). The introns may have maturase functions in processing the transcripts of the yeast genes that harbor them (cytochrome b) or they may be active in processing the transcripts of other genes (e.g., cytochrome oxidase). The 1.1-kb intron of the 21S rDNA gene contains coding sequences for a site-specific endonuclease that facilitates its insertion into some genes lacking this intron as long as they contain the specific target site (5′-GATAACAG-3′). Thus, this intron is also an insertion element.

For transcription of the vertebrate mtDNA the transcription factor mtTFA (a high-mobility group protein) must bind upstream (−12 to −39 bp) to the promoter to unwind the DNA for the mtRNA polymerase, which binds downstream. The distance between these two binding sites is important.

The *Saccharomyces* mtDNA genes are replicated from several (7 or more) scattered replicational origins, each containing 3 GC-rich segments separated by much longer AT tracts. Transcription is mediated by a mitochondrial RNA polymerase (similar to T-odd phage polymerases) that is coded, however, within the nuclear DNA. The conserved 5′-TTATAAGTAPuTA-3′ promoter is positioned within nine bases upstream from the transcription initiation site. Unlike the much smaller mammalian mtDNA genes, yeast genes have more or less usual upstream and 3′ sequences. The much more compact mammalian or fission yeast genes do not use upstream regulatory sequences (UTRs). In the mtDNA, a 13-residue sequence embedded in the $tRNA^{Leu(UUR)}$, a 34-kDa protein, (mTERM) is bound and the complex is required for the termination of transcription. Transcription initiation also requires—besides sc-mtTFA—the protein sc-mtTFB. It seems that the primers for the heavy chain transcription are synthesized on the light chain of the mtDNA. Mammalian light chains contain three conserved sequence blocks (CSBs) serving apparently as primers for the transcription of the heavy chain. These seem to be analogous to the GC-rich segments of yeast. In yeast for the transcription of mtDNA heavy chain, a R loop is also formed. Near the origin of the heavy chain there is an RNase, specific (RNase MRP) for processing the 3′-OH ends of the origin-containing RNAs.

The rRNA genes, unlike in the nuclear genomes, are separated by other coding sequences. Yeast mtDNA encodes at least one ribosomal protein. In mammals, all mitochondrial ribosomal proteins are coded in the nucleus and imported into this organelle. The majority of the other proteins are also imported. Nuclear proteins mediate the translational control locus by specific or global manners.

The transport of proteins through the mitochondrial membrane takes place in several steps. The NH_2 terminus passes into the inner membrane through the protein import channel. The mitochondrial heatshock protein 70 (mtHsp70) stabilizes the translocation intermediate in an ATP-dependent process. The traffic may be in two ways, and it may be a passive transport.

The mtDNA of plants is much larger than that of other organisms, varying between 208 to 2,500 kbp. The mtDNA of *Arabidopsis* is 366,924 bp and contains 57 genes (Unseld M et al 1997 Nature Genet 15:57). The mtDNA of plants frequently exists in multiple size groups. This DNA is interspersed with large (several thousand kb) or smaller (200–300 kb) repeated sequences. Recombination between these direct repeats of a "master circle" may generate in stoichiometric proportion smaller, subgenomic DNA rings. In species without these repeats (e.g., *Brassica hirta*), subgenomic recombination does not occur. The very small mtDNA molecules are called plasmids. Plasmids, *S-1,* and *S-2* share 1.4-kb termini. These plasmids may integrate into the main mtDNA.

The variety and the number of repeats may vary in the different species and some species (*Brassica hirta, Marchantia*) may be free of recombination repeats. The recombination repeats may contain one or more genes. The repeats display recombination within and between the mtDNA molecules. These events then generate chimeric sequences, deficiencies, duplications, and a variety of rearrangements.

Recombination between direct repeats tosses out DNA sequences between them and generates smaller circular and possibly linear molecules. The 570-kb maize mtDNA “master circles” may have six sets of repeats, whereas in the T cytoplasm the repeats are quite complex. The size of the repeats may be 1 to 10-kb. Not all of them promote recombination and even the recombination repeats may vary in different species. Plant mitochondria may also contain single and double-stranded RNA plasmids. The latter may be up to 18-kb size. The single-stranded RNA plasmids may be replication intermediates of the double-stranded ones. The base composition of some is different from that of the main mtDNA and appears to be of foreign (viral) origin.

It has been reported that passage of cells thorough in vitro culturing (tissue culture) may incite rearrangements in the mtDNA genome. This phenomenon may be the result, however, of different amplification of preexisting alterations. Formation of cybrids may also result in new combinations of the mtDNAs (►somaclonal variation). The mitochondrial protein complexes may be organized by chaperone-like mitochondrial proteases.

Some of the DNA sequences in the plant mitochondria are classified as “promiscuous” because they occur in the nuclei and in the chloroplasts too. These promiscuous DNA sequences may contain tRNA ($tRNA^{Pro}$, $tRNA^{Trp}$) and 16S rRNA genes. Into the human nuclear DNA, mtDNA has been inserted several times during evolution and is detectable mostly as pseudogenes. It has been estimated that mtDNA sequences are translocated into the nuclear chromosomes at the high frequency of 1×10^{-5} per cell generation. With the aid of biolistic transformation any type of DNA sequence can be incorporated into organellar genetic material, including mtDNA. The plant mitochondrial tRNAs may be encoded by mtDNA, nuclear DNA. or plastid DNA directly, or by plastid DNA inserted into the mtDNA. Plant mitochondria may code for some mitochondrial ribosome proteins. These ribosomal proteins may be different in different plant species. The liverwort mtDNA genome is very large and contains at least 94 genes.

Human mtDNA was also used for evolutionary studies to trace the origin of the present-day human population to a common female ancestral line, the so-called phylogenetic Eve. Restriction fragment length polymorphism carried out on the mtDNA of 147 people, representing African, Asian, Australian, Caucasian, and New Guinean populations indicated an evolutionary tree by the maximal parsimony method. Accordingly, it appeared that Eve lived about 200,000 years ago in Africa because that population was the most homogeneous in modern times and other populations shared the most common mtDNA sequences with these samples. The advantage of the mtDNA for these studies was that mitochondria are transmitted through the egg and thus recombination with males would not alter its sequences. Later studies have shown, however, that in mice (and probably in humans) a small number of mitochondria are transmitted also through the sperm. Therefore some of the conclusions of the original study regarding the time and population scales may require revision, although the basic ideas about the human origin may be correct. The unique descendence of males can be traced through the Y chromosome. The original human mtDNA sequence (‘Cambridge reference’ Anderson S et al 1981 Nature [Lond] 290:457) has numerous errors because of variations particularly in the control region. These errors may affect demographic/population genetics conclusions and may lead to serious bias in forensic evaluations (Forster PM et al 2003 Ann Hum Genet 67:2).

In vitro in heteroplasmic cells, the mutant mtDNA replication could be inhibited by peptide nucleic acid, complementary to the mutant sequences. Mitochondria depend, either on the salvage pathway for deoxyribonucleotides or on special transporters to provide these DNA precursors. Human mitochondria have only two deoxyribonucleoside kinases. The thymidine kinase phosphorylates both deoxythymidine and deoxycytidine. The deoxyguanosine kinase phosphorylates both deoxypurine nucleosides.

Mutations can cause mtDNA depletion in humans and include: TP, which is associated with mitochondrial neurogastrointestinal encephalopathy (MNGIE, 22q13.32-pter); TK2 (16q22), encoding thymidine kinase, which causes mitochondrial myopathy2; DGUOK (2p13) that codes for a deoxyguanidine kinase, associated with liver failure; POLG (5q25) that is responsible for Alpers progressive infantile poliodystrophy; MPV17 (2p23-p21) that can cause neurohepatopathy; and SUCLA2 (succinate-coenzyme-A ligase,13q12.2-q13), which causes encephalomyopathy (Ostergaard E et al 2007 Am J Hum Genet 81:383).

►mitochondria, ►introns, ►intron homing, ►spacers, ►kinetoplast, ►DNA replication mitochondrial, ►D loop, ►pan editing, ►RNA editing, ►mitochondrial import, ►numts, ►cryptogene, ►cytoplasmic male sterility, ►mitochondrial abnormalities in plants, ►RFLP, ►evolution by base substitution, ►DNA fingerprinting, ►chondrome, ►plastid male transmission, ►organelle sequence transfers, ►paternal leakage, ►*Plasmodium*, ►Eve foremother, ►mitochondrial genetics, ►petite colony mutations, ►maximal parsimony, ►MSF, ►ancient DNA, ►heat-shock proteins, ►chaperone, ►passive transport, ►heteroplasmy, ►peptide nucleic acid, ►mismatch repair, ►mtTFA, ►bottleneck effect,

►RU maize, ►mitochondrial disease in humans, ►sublimon, ►R loop, ►DNA 7S, ►RNA 7S, ►DNA repair, ►aflatoxin, ►bleomycin, ►cisplatin, ►methyltransferase, ►Y chromosome, ►polymerase chain reaction, ►mitochondrial-control regions, ►median-joining networks; Lang BF et al 1999 Annu Rev Genet 33:351; Richards M et al 2000 Am J Hum Genet 67:1251; Holt IJ et al 2000 Cell 100:515; Moraes CT 2001 Trends Genet 17:199; replication of mammalian mtDNA: Falkenberg M et al 2007 Annu Rev Biochem 76:679; for the evolutionary sequences of the mtDNA: http://megasun.bch.umontreal.ca/gobase/.

MTF (mouse tissue factor): A membrane protein initiating blood clotting.

mtFAM (mitochondrial transcription factor, Tfa, TCF6): A high mobility group nuclear gene product controlling transcription and replication in the mammalian mitochondria and possibly affecting transcription also in the nucleus. It is also indispensable for embryogenesis. Heart-specific local mutation (deletion) in mtFAM in mice leads the expression of heart anomalies appearing in Kearns-Sayre syndrome in humans. ►mtDNA, ►mitochondrial genetics, ►high mobility group proteins, ►mitochondrial diseases in humans

MTHF: ►photolyase

Mtj1: A transmembrane murine chaperone with homology to DnaJ and Sec63. ►DnaJ, ►Sec63

MTOC: microtubule organizing centers are formed early in the 16-cell stage within the cyst that gives rise to the primary oocyte and nurse cells. ►microtubule organizing center, ►centromere, ►spindle, ►spindle pole body, ►centrosome, ►maternal effect genes, ►oocyte; Rieder CL et al Trends Cell Biol 11:413.

MtODE: A mitochondrial AP lyase that cleaves double-stranded DNA at 8-oxoguanine sites and repair oxidative damage. ►oxidative DNA damage, ►AP lyase; Croteau DL et al 1997 J Biol Chem 272:27338.

mtPTP (mitochondrial transition pore): An opening on the mitochondrial inner membrane through which mitochondrial proteins can be released to the cytosol and may initiate, e.g., apoptosis. ►apoptosis, ►AIF

M-Tropic (e.g., virus): Homes on macrophages. ►tropic

MTS1 (multiple tumor suppressor): The *MTS1* gene encodes an inhibitor (p16) of the cyclin-dependent kinase 4 protein. When it is missing or inactivated by mutation, cell division may be out of control. *MTS1* is implicated in melanoma and pancreatic adenocarcinoma. ►p16, ►melanoma, ►tumor suppressor, ►pancreatic adenocarcinoma

MTT Dye Reduction Assay: The MTT dye reduction assay measures cell viability by spectrophotometry.

mtTF: The mitochondrial transcription factor. ►transcription factors, ►transcription complex, ►mtDNA

MtTGENDO: mtGENDO refers to oxidative damage-specific mitochondrial AP lyases. It cleaves TG mismatches in double-stranded DNA. It does not recognize 8-oxodeoxyguanine or uracil sites. ►oxidative DNA damage, ►MtODE, ►AP lyase; Stierum RH et al 1999 J Biol Chem 274:7128.

Mu *(Mutator)*: A transposable element system of maize increases the frequency of mutation of various loci by more than an order of magnitude (10^{-3}–10^{-5}). About 90% of the mutations induced carry a *Mu* element. Their copy number in the genome may be 10–100. The element comes in various sizes but the longest are less than 2-kb. The shorter elements appear to have originated from the longer by internal deletions. There are relatively long (0.2-kb) inverted repeats at the termini, and 9-bp direct repeats are adjacent to them. The inverted terminal repeats are conserved among the at least five different classes of *Mu* elements, although the sequences in between them may be quite different. The *Mu* elements appear to transpose (in contrast to other maize transposable elements) by a replicative type of mechanism. The two best-studied forms, *Mu1* (1.4-kb) and *Mu1.7* (1.745-kb), were identified also in circular extrachromosomal states. When *Mu* is completely methylated, the mutations caused by it become stable. Less than complete methylation of the element and some other sequences (e.g., histone DNA) are associated with mutability. *Mu* is regulated by MuDR regulatory transposon (4.9-kb) carrying the *mudrA* and *mudrB* genes with transcripts of 2.8-kb and 1.0- kb, respectively. The MuDR element frequently suffers deletions limited to the MURB region. Mutator-like elements (MULEs) with substantial variations in structure and size have been discovered also in *Arabidopsis*. These 22 mutator elements are normally dormant in Arabidopsis but a mutation that decreases DNA methylation (DDM1) activates them. ►transposable elements, ►insertional mutation, ►controlling elements, ►hybrid dysgenesis, ►methylation of DNA; Singer T et al 2001 Genes Dev 15:591; Miura A et al 2001 Nature [Lond] 411:212; Yin Z, Harsdhey RM 2005 Proc Natl Acad Sci USA 102:18884.

Mu Bacteriophage: A 37-kbp temperate bacteriophage. Its linear double-strand DNA is flanked by 5-bp direct repeats. Transcription of the phage genome during the lytic phase requires the *E. coli* RNA polymerase holoenzyme. During replication, it may integrate into different target genes of the bacterial chromosome and thus, may cause mutations (as the name indicates). Mu may integrate in more than one copy and if the orientation of the prophage is the same, it may cause deletion, or if the orientation of the two Mu DNAs is in reverse, inversion may take place by recombination between the transposons. The phage carries three terminal elements at both left and right ends where recombination with the host DNA takes place. There is a transpositional enhancer (internal activation sequence) of about 100-bp at 950 bp from the left end. This left-enhancer-right complex is called LER. The 75-kDa transposase binds to the two ends and to the enhancer. A complex of nucleoproteins, the transposome, mediates the transposition. The transposome includes four subunits of the MuA transposase and each contains a 22-base pair recognition site. These recognition sites—in case there is a shortage of Mu DNA—may recruit non-Mu DNA and then can transpose it too. In the presence of bacterial binding proteins (HU and IHF), a stable synaptic complex (SSC) is formed. Then at the 3′-ends, Mu is cleaved by the transposase and they form the Cleaved Donor Complex (CDC). Transesterification at the 3′-OH places Mu into the host DNA. In the Strand Transfer Complex (STC), the 5′-ends are still attached to the old flanking DNA but the 3′-ends are joined to the new target sequences and a cointegrate is generated by replication or nucleolytic cleavage separates Mu from the old flanks and then the gaps are repaired and the transposition is completed. ►mini Mu, ►transposons, ►mutator phage, ►cointegrate, ►transposition site; Bukhari AI 1976 Annu Rev Genet 10:389; Abbes C et al 2001 Can J Microbiol 47 (8):722; Goldhaber-Gordon I et al 2002 J Biol Chem 277:7694; Pathania S et al 2002 Cell 109:425.

Mucins: Glycosylated protein components of the mucosa that lubricate the intestinal tract. Mutation in the genes involved may lead to cancer. ►D-amino acids; Velcich A et al 2002 Science 295:1726; Hollingsworth MA, Swanson BJ 2004 Nature Rev Cancer 4:45.

Muckle-Wells Syndrome: Dominant (1q14) inflammatory disease accompanied by fever, abdominal pain, and urticaria (red or pale skin eruptions). Progressive nerve deafness and amyloidosis may follow. This, and two other diseases, involve mutations in the CIAS1 gene (cold autoinflammatory response) at the NALP locus. Cryopyrin (Nod family protein), along with CAIS1, regulates the inflammatory response (Dowds TA et al 2003 Biochem Biophys Res Commun 302:575; Sutterwala FS et al 2006 Immunity 24:317). ►amyloidosis, ►deafness, ►urticaria familial cold, ►cold hypersensitivity, ►hydatiform mole, ►inflammasome

Mucolipidoses (ML): ML include a variety of recessive diseases connected to defects in lysomal enzymes. ML I is a neuroaminidase deficiency and ML II is an N-acetylglucosamine-1-phosphotransferase deficiency (human chromosome 4q21-q23). This enzyme affects the targeting of several enzymes to the lysosomes. Congenital hip defects, chest abnormalities, hernia, and overgrown gums, but no excessive excretion of mucopolysaccharides are observed. ML III is apparently allelic to ML II although the mutant sites seem to be different. MP III (pseudo-Hurler polydystrophy) also has similarity to the Hurler syndrome (►mucopolysaccharidosis), although the basic defects are not identical. MP III A is a heparan sulfate sulfatase deficiency (►mucopolysaccharidosis, ►Sanfilippo syndrome A), whereas MP IIIB is basically the Sanfilippo syndrome B (►mucopolysaccharidosis). MP IV (19p13.2-p13.3) is a form of sialolipidosis with typical lamellar body inclusions in the endothelial cells that permit prenatal identification of this disease that may cause early death. ►mucopolysaccharidosis, ►Hurler syndrome

Mucopolysaccharidosis (MPS): Hurler syndrome (*MPS I*) recessive 22pter-q11 de-ficiency of α-L-iduronidase (IDUA) resulting in stiff joints, regurgitation in the aorta, clouding of the cornea, etc. The latest chromosomal assignment of MPS1 is 4p16.3. The IDUA protein (about 74 kDa) includes a 26-aminoacid signal-peptide. The *MPS II* (*Hunter syndrome,* I-cell disease) is very similar to *MSP I*, but it is located in human chromosome Xq27-q28. The symptoms in this iduronate sulfatase deficiency are somewhat milder and the clouding of the cornea is lacking. Dwarfism, distorted face, enlargement of the liver and spleen, deafness and excretion of chondroitin, and heparitin sulfate in the urine are additional characteristics. The following MPSs are controlled by autosomal recessive genes and the deficiencies involve: *MPS IIIA* (Sanfilippo syndrome A; human chromosome 17q 25.3) heparan sulfate sulfatase, in *MPS IIIB* (Sanfilippo syndrome B, in human chromosome 17q21) N-acetyl-α-D-glucosaminidase, in *MPS IIIC* acetylCoA:α-glucosaminide-N-acetyl-transferase, *MPS IIID* (Sanfilippo syndrome D, 12q14) N-acetylglucosamine-6-sulfate sulfatase, in *MPS IVA* (Morquio syndrome A, 16q24.3) galactosamine-6-sulfatase, in *MPS IVB* (Morquio syndrome B) β-galactosidase, *in MPS VI* (Maroteaux-Lamy syndrome, 5q11-q13) N-acetylgalactosamine-4-sulfatase (arysulfatase B), in *MPS VII* (Sly syndrome, 7q21.11) β-glucuronidase, and in *MPS VIII* (DiFerrante syndrome) glucosamine-3–6-sulfate sulfatase. The Hurler syndrome and Sly disease may be treated by implantation of neo-organs, tissues of skin fibroblasts secreting β-glucuronidase, or α-L-iduronidase, respectively. ►mucolipidoses, ►iduronic acid, ►heparan sulfate, ►glucuronic acid, ►arylsulfates, ►hyaluronidase deficiency, ►coronary heart disease, ►eye diseases, ►deafness, ►miniorgan, ►geleophysic dysplasia, ►enzyme replacement therapy

Mucosal Immunity: The mucosal membranes (in the gastrointestinal tract, nasal, respiratory passages, and other surfaces that have lymphocytes) are supposed to trap 70% of the infectious agents. The mucosal cells rely on immunoglobulin A (IgA) rather than IgG used by the serum. In the gastrointestinal system are located the Peyer's patches that trap infectious agents and pass them to the antigen-presenting cells, T and B cells. The B cells generate the IgA anti-bodies. Concommitantly, the released antigen may stimulate the formation of IgG too. Providing attenuated forms of the pathogen (e.g., *Vibrio cholerae*) orally can trigger the mucosal immunity system. Another approach is to deliver the antigens by bacteria that more effectively stimulate simultaneously both IgA and IgG production. Other delivery systems may use a biodegradable poly (DL-lactide-co-glycolide), PLG, or liposomes. The oral polio vaccine has been quite successful because it provides lasting protection but most of oral vaccines are short in this respect. ►Peyer's patch, ►antigen-presenting cell, ►T cell, ►B cell, ►immunoglobulins, ►antigen, ►immune system, ►vaccines, ►plantibody; Simmons CP et al 2001 Semin Immunol 13(3):201.

Mucosulfatidosis: A multiple sulfatase deficiency with excessive amounts of mucopolysaccharides and sulfatides in the urine. Multiple sulfatase deficiency encoded by SUMF, causing a lysosomal storage disease, results from the lack or co- or post-translational replacement of cysteine by 2-amino-3-oxopropionicacid in several sulfatases (Schmidt B et al 1995 Cell 82:2712; Landgrebe J et al 2003 Gene 316:47). The numerous human sulfatases share common domains. The defect results in abnormal development and early childhood death because of abnormal signaling by gene WNT (Dhoot GK et al 2001 Science 293:1663). Another sulfatase (RsulfFP1) normally mediates proteoglycan signaling by desulfation (Ohto T et al 2002 Genes Cells 7:521). Some of the symptoms of sulfatase deficiency are shared by leukodystropy, Maroteaux-Lamy, Hunter, Sanfilippo A, and Morquio syndromes, as well as by ichthyosis. Several sulfatase genes occur in bacteria to animals but few in plants. Heterologous bone transplantation, somatic cell gene transfer, and enzyme replacement offer therapeutic potentials in several sulfatase deficiency diseases. (See diseases mentioned in separate entries, ►genetic engineering; Diez-Roux G, Ballabio A 2005 Annu Rev Genomics Hum Genet 6:355).

Mud: A transposon modified (derived) from bacteriophage Mu. ►Mu bacteriophage

MudPIT (multidimensional protein identification technology): A method for the study of complex protein mixtures. Multidimensional liquid chromatography is followed by mass spectrometric analysis by digesting the cell lysate with the help of endoproteinase lysC, and then further purification. The databases are searched by SEQUEST algorithm. The peptide sequences can be used for the identification of proteins in a mixture. ►mass spectrum, ►MALDI-TOF, ►proteomics, ►SEQUEST; Washburn MP et al 2001 Nature Biotechnol 19:242; Smith RD et al 2002 Proteomics 2:513.

Muenke Syndrome (nonsyndromic coronal craniosynostosis): Primarily a skull bone fusion disorder that sometimes affects finger and toe development. It is generally caused by mutation affecting the proline^{250} site in the fibroblast growth factor receptor 3 (FGFR3) gene at 4p16.3. Its incidence is higher in women than in men. ►craniosynostosis syndromes

Muir-Torre Syndrome: A familial autosomal dominant disease involving skin neoplasias, apparently due to hereditary defects in the genetic (mismatch) repair system. ►mismatch repair, ►colorectal cancer, ►Gardner syndrome, ►polyposis hamartomatous, ►hereditary nonpolyposis colorectal cancer

Mulatto: Offspring of white and black parentage. ►miscegenation

Mulberry (*Morus* spp): fruit tree; x = 14. *M. alba* and *M. rubra* are 2n = 28 whereas the Asian *M. nigra* is 2n = 38.

Mulibrey Dwarfism: ►dwarfism [Mulibrey nanism]

Mule: ►hinny

MULE (Mcl ubiquitin ligase E3): Mc*l*-1 is an anti-apoptotic member of the Bc*l* protein family controlling DNA-damage-induced apoptosis (Zhong Q et al 2005 Cell 121:1085). ►apoptosis, ►BCL

Mulibrey Nanism (MUL, 17q22-q23): An autosomal anomaly causing small size. It involves all or some of theSE characteristics: low birth weight, small liver, brain, eye, growth, triangular face, yellow eye dots, etc. The basic defect is in a Zinc-finger protein (RBCC), containing a ring finger, a B box, and a coiled-coil region. It appears to be a peroxisomal disorder. ►stature in humans, ►Zinc finger, ►B box, ►ring finger, ►coiled coil, ►peroxisome; Kallijärvi J et al 2002 Am J Hum Genet 70:1215.

Muller 5 Technique: ►*Basc*

Muller's Ratchet: Genetic drift can lead to the accumulation of deleterious mutations, particularly in asexual populations. By each mutation the ratchet (a toothed wheel) may click by one notch. The expected time of losses of individuals with successive minimal number of mutations depends on the absolute number of individuals with minimal number of mutations: $N_m = q_m$ where q is their expected frequency and N = effective population size. Muller's ratchet may operate during the evolution of transposons and retroviruses and fix shorter sequences than the initial elements and establish elements that would depend on trans-acting elements for transposition. In case the number of non-deletrious mutations is small (n_0), the equilibrium value $n_0 = Ne^{-\mu/s}$ where N = effective population size, μ = expected number of deleterious mutations per genome, s = selection coefficient, e = base of natural logarithm. Conversely, the rapid mutations of retroposons may eliminate elements that would lose their transposase and thus can escape the Muller's ratchet. Deleterious mutations may be eliminated by reversions, by compensating new mutations, or by genetic recombination. The original hypothesis (Muller H J 1932 Amer Nature 66:118) suggested the advantage of the evolution of sex and recombination as a means to purify the population from deleterious mutations. ►Y chromosome, ►genetic drift, ►transposable elements, ►retrovirus; Gabriel W, Bürger R 2000 Evolution Int J Org Evolution 54:1116; Gordo I, Charlesworth B 2000 Genetics 156:2137.

Müllerian Ducts: Gonadal cells begin to develop before the mouse embryo is two-weeks old. From the unspecialized primordial cells in the male, the Wolffian ducts develop, and from those in the female, the Müllerian ducts. Initially, however, both sexes form both of these structures and appear bisexual but later, according to sex—one or the other—degenerates. The Müllerian inhibiting substance (MIS, Jost factor) causes the degeneration of the Müllerian ducts. This is a member of the TGF-β (transforming growth factor family of proteins) or is the AMH (anti-Müllerian hormone) and it is encoded in human chromosome 19p13-p13.2. Defect in the AMH receptor causes pseudohermaphroditism with uterine and oviductal tissues in males observed in PMDS (persistence of Müllerian duct syndrome). MIS regulates NFκB signaling and breast cancer cell growth in vitro and can induce apoptosis in ovarian and cervical cancer cell lines (Renaud EJ et al 2005 Proc Natl Acad Sci USA 102:111). In mammals, the Wnt-4 signaling is required for ovarian morphogenesis. ►gonad, ►Wolffian ducts, ►SRY, ►SF-1, ►sexual dimorphism, ►TGF, ►pseudohermaphroditism; Segev DL et al 2001 J Biol Chem 276:26799; Bédécarrats G et al 2003 Proc Natl Acad Sci USA 100:9348.

Müllerian Inhibitory Substance (Jost factor): ►Müllerian duct

Müllerian Mimicry: In Mullerian mimicry, two monomorphic species share morphological similarities signaling a defense (e.g., distastefulness, poisonousness) to predators, and both benefit from the trait. Similar mimicry occurs in various species. A single gene locus may be shared among the different *Heliconius* species (see Fig. M129). ►Batesian mimicry; Kapan DD 2001 Nature [Lond] 409:338; Joron M et al 2006 PLoS Biol 4(10:e303).

Figure M129. *Heliconius* species of different geographical areas display Müllerian mimicry. From top to bottom: southern Ecuador, southern Brazil, northern Ecuador, western Brazil, Peru. (Coutesy of Fred Nijhout, Duke University)

Multibreed: A population including purebred and crossbred groups. ►pure-breeding, ►cross breeding

Multicase (multiple computer automated structure evaluation): Multicase relies on information of molecular fragments as descriptors of potential biological (carcinogenic) activity of chemicals. It compares the structure–activity relationship among many compounds in order to find commonality. ►CASE, ►SAR, ►biophore; Cunningham AR et al 1998 Mutation Res 405:9.

Multicellular Organisms: In a multicellular organism, the cells of an individual are coordinated for different function(s) and situated closely enough to ensure interaction. A bacterial colony is formed from many cells but these cells are not coordinated even when they display patterned growth. The multicellular condition apparently evolved repeatedly during evolution. The structural condition for evolution of a multicellular system was apparently the synthesis of an adhesive polymer matrix needed for cell alignment (Vellicer GJ, Yu Y-t N 2003 Nature [Lond] 425:75). Evolution of multicellularity involved signal transmission, reception, and localized expression of specialized genes. Their main advantages are in feeding and dispersion. ►quorum-sensing; Kaiser D 2001 Annu Rev Genet 35:103.

Multicompartment Virus: In a multicompartment virus, each individual viral particle may carry only part of the total genome and complementation between the particles can provide the full function. Such viral strategy permits a combinatorial advantage to the virus.

Multicomponent Virus: The genome of a multicomponent virus is segmented, i.e., its genetic material is in several pieces similarly to the chromosomes in eukaryotic nuclei. ►bacteriophages

Multicopy Plasmids: Multicopy plasmids have several copies per cell. ►plasmid types

Multidrug Resistance (MDR/ABCB1): Multidrug resistance is mediated by the multidrug transporter (MDT, 1,280 amino acids) phosphoglycoprotein that regulates the elimination (or uptake) of chemically quite different drugs from mammalian (cancer) cells in an ATP-dependent manner. MDR is generally carried by extrachromosomal elements in bacteria such as plasmids, transposons, and integrons. In cancer cells, the MDR gene is generally amplified as episomes and double-minutes. Intercellular transfer of phosphoglycoproteins can mediate acquired drug resistance (Levchenko A et al 2005 Proc Natl Acad Sci USA 102:1933). The MDR gene may be used for gene therapy by protecting the bone marrow from the effects of cytotoxic cancer drugs. The MDT gene controls a very broad base drug-resistance. The MRP (multidrug resistance associated protein) is a glutathione conjugate protein that belongs to the same family as MDR and may be expressed in cell lines where MDR function is usually limited. The bacterial multidrug resistance gene, *LmrA*, is very similar to the human gene and it is expressed in human cells when introduced by transformation. Besides gene mutation, MDR can be acquired also by rearrangement of the genome (aneuploidy) in cancer cells.

Whole-genome sequencing identified steps in the evolution of multidrug resistance in isogenic *S. aureus* isolates recovered periodically from the bloodstream of a patient undergoing chemotherapy with vancomycin and other antibiotics. After extensive therapy, the bacterium developed resistance, and treatment failed. Sequencing the first vancomycin susceptible isolate and the last vancomycin nonsusceptible isolate identified genome wide 35 point mutations in 31 loci (Mwangi MM et al 2007 Proc Natl Acad Sci USA 104:9451). ►amplification, ►multiple drug resistance, ►ABC transporters, ►Emre, ►episomes, ►integron, ►double-minute, ►ABC transporters, ►P-glycoprotein, ►SOD, ►MGMT, ►DHFR, ►aldehyde dehydrogenase, ►glutathione-*S*-transferase, ►chemosensitivity, ►Crohn disease; Hipfner DR et al 1999 Biochim Biophys Acta 1461:359; Rosenberg MF et al 2001 J Biol Chem 276:16076; Duesberg P et al 2001 Proc Natl Acad Sci USA 98:11283; Yu EW et al 2003 Science 300:976; crystal structure of multidrug transporter: Murakami S et al 2006 Nature [Lond] 443:173; review: Alekshun MN, Levy SB 2007 Cell 128:1037.

Multifactorial Cross: In a multifactorial cross, the mating is between parents, which differ at multiple gene loci.

Multifactorial Disease: Multifactorial disease is caused or controlled by more than single genes. Generally, the number of factors involved cannot be determined in a straight forward manner and environmental influences may be significant. ►QTL, ►sporadic

Multiforked Chromosomes: In bacteria, replication of the DNA may start again before the preceding cycles of DNA synthesis have been completed and thus display multiple replication forks (see Fig. M130). ►replication fork, ►replication bidirectional, ►DNA replication; diagram from Sueoka N 1975 Stadler Symp 7:71.

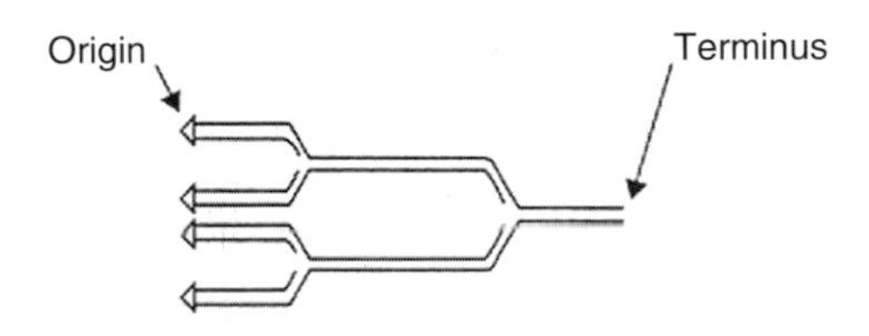

Figure M130. Multiforked chromosome of bacteria

Multifunctional Proteins: ►one gene–one enzyme

Multigene Family: In a multigene family, clusters of similar genes evolved through duplications and mutations and display structural and functional homologies. Some members of the families may be at different locations in the genome and some may be pseudogenes. ►pseudogenes

Multigenic: ►polygenic inheritance

Multi-Locus Probe (MLP): In a multi-locus probe, in DNA fingerprinting alleles of more than one locus are examined simultaneously. ►DNA fingerprinting

Multi-Locus Sequence Typing: Multi-locus sequence typing detects allelic variations (mutations and recombination) within about 450 bp internal sequences of seven bacterial housekeeping genes. Since the integrity of housekeeping gene function is generally vital, most of the alterations detected are neutral. ►DNA typing, ►housekeeping genes; Feil EJ et al 2000 Genetics 154:1439.

Multimap: A computer program for linkage analysis using lod scores. ►lod score

Multimeric Proteins: Multimeric proteins have more than two polypeptide subunits.

Multinomial Distribution: Multinomial distribution may be needed to predict the probability of proportions: $P = \frac{n!}{r_1!r_2!...r_z!}(a)_1^r(b)_2^r...(z)_z^r$ where P is the probability r1, r2…rz stand for the expected numbers in case of a, b…z theoretical proportions and n = the total numbers. Example: in case of codominant inheritance and expected 0.25 AA, 0.50 AB, and 0.25 BB, the probability that in an AB x AB mating we would find in F2 among five progeny 2AA, 2AB, and 1 BB is: $P = \frac{5!}{2!2!1!}(0.25)^2(0.50)^2(0.25)^1 = \frac{120}{4}(0.0625)(0.25)(0.25) = 30(0.00391) \approx 0.117$. binomial distribu tion, distribution)

Multiparental Hybrid: A multiparental hybrid can be generated by fusion of two or more different embryos of different parentage at the (generally) 4 - 8 cell stage and then reimplantion at (generally) the blastocyst stage into the uterus of pseudo-pregnant foster mouse. Such hybrids may appear chimeric if the parents were genetically different and can be used advantageously for the study of development (see Fig. M131). ►allophenic, ►pseudo-pregnant, ►tri-parental human embryo

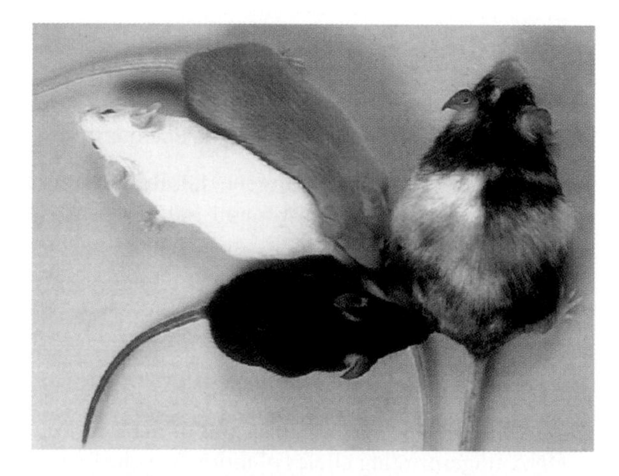

Figure M131. Hexaparental mother mouse displaying yellow, black and white sectors (at right). It was obtained by aggregation of three strains in vitro that were implanted into a pseudopregnant female. Her offspring are at left. (Coutesy of RM Petters & CL Markert)

Multiparous: An animal species that usually gives birth to multiple offspring by each delivery; a human female that gives birth to twins a least twice.

Multipaternate Litter: A multipaternate litter is produced if the receptive multiparous female mates during estrus with several males. ►estrus, ►superfetation, ►last-male sperm preference

Multiphoton Microscopy: A non-invasive fluorescence microscopy for various tissues. Molecular excitation by two simultaneous absorptions of two photons provides intrinsic three-dimensional resolution of laser scanning fluorescence microscopy (Denk W et al 1900 Science 248:73). ►microscopy, ►laser scanning cytometry; Zipfel WR et al 2003 Nature Biotechnol 21:1369.

Multiple Alleles: More than two different alleles at a gene locus. The number of different combinations at a particular genetic locus can be determined by the formula [n(n + 1)]/2 where the number of different alleles at the locus is *n*. ►allelic combinations

Multiple Birth: ►twinning

Multiple Cloning Sites (MCS): ►polylinker

Multiple Cross Mapping (MCM): A procedure for chromosomal localization of quantitative trait and complex loci using several inbred strains of mice. The technique can identify linked modifiers. ►QTL, ►modifier gene; Hitzemann R et al 2002 Genes Brain & Behavior 1(4):214; Park Y-G et al 2003 Genome Res 13:118.

Multiple Crossovers: Multiple crossovers are genetically detectable only when the chromosomes are densely marked. In the absence of interference, the frequency of multiple crossovers is expected to be equal to the products of single crossovers. ►coincidence, ►mapping, ►mapping function

Multiple Drug Resistance (MDR): MDR is based on several mechanisms such as active detoxification system, improved DNA repair, altered target for the drugs, decreased uptake, increased efflux, and inhibition of apoptosis. The MDR genes control multiple drug resistance in mammals. The 27 exon MDR1 locus (human chromosome 7q21.1) encodes a phosphoglycoprotein and controls drug transport and drug removal ('hydrophobic vacuum cleaner' effect). A repressor binding to a −100 to −120 GC-rich sequence regulates the MDR1 gene. At −70 to −80 is a Y box (binding site of the NF-Y transcription factor). This area overlaps with the binding site of Sp1. There is also a 13-bp sequence (IN) surrounding the transcription initiation site and promoting the accuracy of transcription. If tumors have high expression of this gene, the prognosis for chemotherapy is not good. Upon drug treatment, the activity of the MDR protein may increase several-fold. The MDR activity can be suppressed by some calcium channel blockers, antibiotics, steroids, detergents, antisense oligonucleotides, anti-MDR ribozymes, etc. Non-P-glycoprotein-mediated MDR has also been restricted by MRP (a multiple-drug-resistance-related glycoprotein), encoded in human chromosome 16p13.1. The MDR gene targeted to special tissues of transgenic animals by tissue-specific promoters may protect these tissues (e.g., bone marrow) during cancer chemotherapy. Also, the MDR transgenic cells may display selective advantage and thus be advantageous during the curing process. In mouse tumors like in human tumors, the response to drugs of individual tumors varies, but eventually they all become resistant to the maximum tolerable dose of doxorubicin or docetaxel. The tumors also respond well to cisplatin but do not become resistant, even after multiple treatments in which tumors appear to regrow from a small fraction of surviving cells. Classical biochemical resistance mechanisms, such as up-regulated drug transporters, appear to be responsible for doxorubicin resistance, rather than alterations in drug-damage effector pathways (Rottenberg S et al 2007 Proc Natl Acad Sci USA 104:12117). ►ABC transporter, ►cancer therapy, ►chemotherapy, ►multidrug resistance, ►drug intersection, ►transcription factors, ►Sp1, ►promoter, ►epistasis; Gottesman MM et al 1995 Annu Rev Genet 29:607; Litman T et al 2001 Cell Mol Life Sci 58(7):931; review: Higgins CF 2007 Nature [Lond] 446:749.

Multiple Endocrine Neoplasia (MEN): ►endocrine neoplasia

Multiple Epiphyseal Dysplasia (MED): An osteochondrodysplasia causing short stature and early onset osteoarthritis. Mutations in a cartilage oligomeric-matrix protein (a pentameric 524-kDa glycoprotein, COMP, 19p13.1), and in collagen IX (COL9A2, 1p33-p32.2, COL9A3, 20q13.3) cause MED. ►osteochondromatosis, ►osteoarthritis, ►collagen, ►Ehlers-Danlos syndrome, ►pseudoachondroplasia, ►COMP, ►thrombospondin; Briggs MD et al 1995 Nature Genet 10:330.

Multiple Hamartoma Syndrome (MHAM, PTEN, MMAC1): Hamartomas are groups of proliferating, somewhat disorganized, mature cells—occurring under autosomal dominant control—on the skin, breast cancer, thyroid cancer, mucous membranes, and gum, but may be found also as polyps in the colon or other intestines. The lesions may become malignant. The symptoms may be associated with a number of other defects. The Cowden disease was assigned to human chromosome 10q22-q23. The Bannayan-Riley-Zonona syndrome resembles Cowden disease and maps to the same location. The molecular basis is a dual function phosphatase with similarity of tensin. Another form involving megalocephaly and epilepsy symptoms is the Lhermitte-Duclos disease. ►cancer, ►breast cancer, ►tensin, ►PTEN, ►Bannayan-Riley-Zonona syndrome; Backman SA et al 2001 Nature Genet 29:396; Kwon C-H et al *ibid.* pp 404.

Multiple Hit: In a multiple hit, the mutagen causes mutation at more than one genomic site. ►kinetics

Multiple Myeloma: A bone marrow tumor-causing anemia and decrease in immunoglobulin production, frequently accompanied by the secretion of the Bence-Jones proteins. Multiple myeloma may be genetically determined as it may appear in a familial manner. In some forms that have independent origin, within the same family or even within the same person, the protein may show some variations. It can be also induced in isogenic mice by injection of paraffin oil into the peritoneal cavity. Such animals may produce large quantities of the light chain immunoglobulin that can be subjected to molecular analysis. Such globulins are not necessarily monoclonal although from single transformation essentially similar molecules are expected. Multiple myeloma is also called amyloidosis encoded by a gene in human chromosome 11q13. Translocation of the IgH (14q32) immunoglobulin H gene to the amyloidosis locus or to the fibroblast growth factor receptor 3 (4p16.3) may also evoke multiple myeloma. The lymphocyte-specific interferon regulatory factor 4 (6p25-p23) may also be involved. The cyclin D1 protein mediating cell cycle events and the expression of oncogenes also occurs at the 11q13 location. ►Bence-Jones proteins, ►immunoglobulin, ►monoclonal antibody, ►macroglobulinemia, ►cyclin D, ►interferon; Kuehl WM, Bergsagel PL 2002 Nature Rev Cancer 2:175.

Multiple Regression: A continuous variable, say y regresses to other variables such as x_1, $x_2 \ldots x_n$. The expected value is: $E(y) = b_0 + b_1x_1 + \ldots + b_nx_n$ where the b regression coefficients are generally estimated by least squares. ►correlation, ►least squares, ►regression

Multiple Sclerosis (MS): a disease caused by loss of myelin from the nerve sheath or even defects of the gray matter. It is called multiple because it frequently is a relapsing type of condition involving incoordination, weakness, and abnormal touch sensation (paresthesia) expressed as a feeling of burning or prickling without an adequate cause. Myelin oligodendrocyte glycoprotein (MOG) is expressed in the central nervous system oligodendrocytes and in the outermost myelin lamellae. Anti-MOG antibodies have proven effective in destroying multiple sclerosis (demyelination) in some animal models. There is now

M

evidence that IgG binding to MOG can detect the disease at the preclinical early inflammatory stage in humans and marmoset monkeys (Lalive PH et al 2006 Proc Natl Acad Sci USA 103:2280).

The exact genetic basis is not clear. It may be determined polygenically or it may be recessive; in both cases with much reduced penetrance. It is associated with defects of the HLA-DQ and HLA-DR system in white as well as in American black populations (Oksenberg JR et al 2004 Am J Hum Genet 74:100). There is molecular evidence that in some forms α-B-crystalline, a small heatshock protein, is formed as an autoantigen. Viral initiation has also been considered. MS is basically an autoimmune disease. Encephalitogenic antigens and adjuvants induce *experimental allergic encephalomyelitis* in mice and the condition significantly modified autimmune susceptibility in both females and males by the Y chromosome derived from the father and grandfather (Teuscher C et al 2006 Proc Natl Acad Sci USA 103:8024). In a rodent model, blocking cytosolic phospholipase A_2 ($cPLA_2$) may alleviate the condition if the disease has not progressed too far. The controls may reside in the thymus where the T lymphocytes develop. Also damage in the white matter of the central nervous system may be responsible. It appears that MS is induced by damage to the blood-brain barrier (BBB), caused by inflammatory cytokines (TFN, IFN-γ). Among close relatives the incidence may be 20 times higher than in the general population. The concordance between monozygotic twins is about 30%. HLA DR2 carriers have a four-fold increased relative risk. Females have almost twice the risk than males in the relapsing type but in the progressive MS male affliction is slightly more common. The prevalence in the USA is about 1/1,000 but it varies by geographical areas. The onset is between ages 20 to 40. Some of the symptoms may be shared with other diseases, and conclusive identification requires laboratory analysis of myelin. The synonym for MS is disseminated sclerosis. Several procedures have been suggested for the suppression or mitigation of the autoimmune reaction. Anti integrin $\alpha_4\beta_1$ agents (cognate antibodies, synthetic antagonists) may delay demyelination. In a mouse model, isolated neural precursor cells from the mouse brain was injected into the blood or the spinal fluid. These cells express the α4 integrin protein, which may assist their movement from the blood to the brain where they differentiate into oligodendroglial cells and make myelin or may become neurons. As a consequence, the autoimmunological damage may be repaired (Pluchino S et al 2003 Nature [Lond] 422:688). Copaxone (Cop-1), and a synthetic copolymer of tyrosine, glutamate, alanine, and lysine may protect motor neurons against acute and chronic degeneration (Kipnis J, Schwartz M 2002 Trends Mol Med 8:319).

MS is under the control of several genes, yet linkage to 5p14-p12 with a lod score of 3.4 was found. The Pelizaeus-Merzbacher disease, a late onset multiple sclerosis-like disorder with an autosomal dominant expression, has been frequently confused with MS. Two principal regions in human chromosomes 17q22 and 6p21 (in the HLA area) control epistatically the susceptibility to MS. Besides these loci, chromosome 1 – centromere region also displays linkage with MS (lod score 4.9) or even stronger based on large African-European admixture populations (Reich D et al 2005 Nature Genet 37:1113). Genomewide association studies indicated linkage with the HLA region (6p21) and with interleukin recptor loci IL2RA (CD25, 10p15) and IL7RA (CD127, 5p13) and to a lesser degree with KIAA (multiple intereacting proteins, 16p13) and CD58 (lymphocyte-associated antigen, 1p13) among other SNPs Ramagopalan SV et al 2007 New England J. Med. 357:2199. ►myelin, ►oligo1, ►epilepsy, ►neuromuscular diseases, ►autoimmune disease, ►MHC, ►HLA, ►heatshock protein, ►Pelizaeus-Merzbacher syndrome, ►Addison-Schilder disease, ►neurological diseases, ►leukodystrophy, ►HLA, ►concordance, ►twinning, ►lod score, ►TFN, ►IFN, ►BBB, ►CD45; Klein L et al 2000 Nature Med 6:56; Smith T et al 2000 Nature Med 6:62; Barcellos LF et al 2001 Nature Genet 29:23; Schmidt S et al 2002 Am J Hum Genet 70:708; Keegan BM, Noseworthy JH 2002 Annu Rev Med 53:285.

Multiple Translocations: ►translocation complex

Multiple Tumor Suppressor: ►MTS1

Multiplex Amplifiable Probe Hybridization: Multiplex amplifiable hybridization uses PCR and electrophoresis for the detection of small deletions and duplications within genes. (See White S et al 2002 Am J Hum Genet 71:365).

Multiplexing: In multiplexing, several pooled DNA (or other) samples are sequenced/processed/analyzed simultaneously to expedite the process. ►genome project

Multiplication Colony: Multiplication colony is an expansion by random mating of the foundation stock of inbred rodents for experimental use. ►foundation stock, ►inbred

Multiplication Rule: ►joint probability

Multiplicative Effect: As per the multiplicative effect, alleles at more than one gene locus together have higher than simply additive contribution to the phenotype. In genetically determined disease, the relative risk with two alleles is the square of the relative risk with only one allele. ►additive effects

Multiplicity of Infection: A single bacterial cell is infected by more than one phage particle. ►double infection

Multiplicity Reactivation: As per multiplicity reactivation, when bacterial cells are infected with more than one phage particle, each, that have been inactivated by heavy doses of DNA-damaging mutagens, the progeny may contain viable viruses because replication and/or recombination has restored functional DNA sequences. ►Weigle reactivation

Multipoint Cross: More than two genes are involved in a multipoint cross. (See http://www.broad.mit.edu/ftp/distribution/software/).

Multipolar Spindle: During mitosis, more than two poles exist under exceptional conditions, e.g., aneuploidy caused by fertilization by more than one sperm. Such a condition may be the result of centrosome defects in animals (see Fig. M132). Multiple centrosomes may be formed in case the tumor suppressor gene p53 is not functioning. ►mitosis, ►centrosome, ►p53, ►spindle

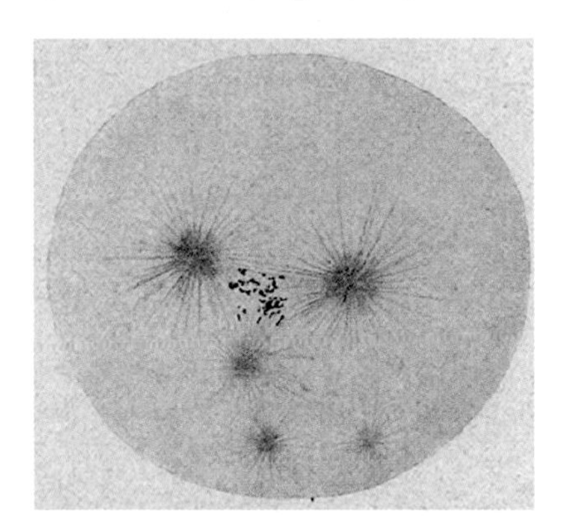

Figure M132. Multipolar spindle in a starfish egg

Multipotent: Multipotency is an ability to develop into more than one type of cell or tissue. ►pluripotency, ►totipotency, ►regeneration

Multi-Regional Origin: ►out-of-Africa hypothesis

Multisite Gateway Technology: Invitrogen Gateway technology exploits the integrase/*att* site-specific recombination system for directional cloning of PCR products and the subsequent subcloning into destination vectors. One or three DNA segments can be cloned using Gateway or MultiSite Gateway, respectively (Magnani E et al 2006 BMC Mol Biol 7:46; http://www.invitrogen.com/content.cfm?pageid=8012&sku=12537023).

Multisite Mutations: Multisite mutations occur generally when an excessively large dose of a mutagen(s) is applied. Frequently, deletions or other chromosomal aberrations are included. ►mutation, ►chromosomal aberration, ►deletion

Multistranded Chromosomes: Multistranded chromosomes contain more than two chromatids, such as in the polytenic chromosomes produced by repeated replication without separation of the newly formed strands. In the early period of electronmicroscopic studies of ordinary chromosomes, apparent multiple strands were observed, and it was assumed that each chromosome has many parallel strands of DNA (see Fig. M133). This assumption could not be validated by subsequent investigations. Each chromatid has only a single DNA double helix that is folded to assure proper packaging. ►packing ratio, ►polytenic chromosomes, ►Mosolov model

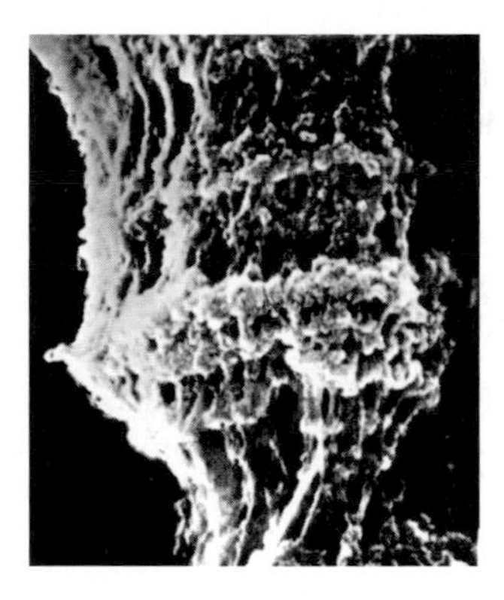

Figure M133. Scanning electronmicrograph of a segement of polytenic salivary gland chromosome. (Courtesy of Dr. Tom Brady)

Multitasking: The multitasking computer program performs more than a single operation simultaneously.

Multivalent: Multivalent refers to the association of more than two chromosomes in meiosis I. ►meiosis

Multivariate Analysis: Multivariate analysis has many uses in genetics when a decision is needed to classify syndromes with overlapping symptoms and when the diagnosis requires quantitation. It may be also used to classify populations with similar traits. The assumption is that the variates $X_1 \ldots X_k$ are distributed according to a multivariate normal distribution. The variance of X_i, σ_{ii} and the covariance of X_i and X_j are presumed to be identical in the two populations, but σ_{ii} or σ_{ij} are not the same from one variate to another or one pair of variates to another pair, respectively. The difference between the two means for X_i is $\delta_I = \mu_{2i} - \mu_{1i}$. Hence the *linear discrimination function* $\Sigma L_i X_i$ provides the lowest probability of incorrect classification. The L_i values will be determined according to the procedure shown. The δ/σ is maximalized for minimization of the chance of misclassification. The *generalized squared distance* is $\Delta^2 = (\sum L_i \delta_i)^2 / \sum\sum L_i L_j \sigma_{ij}$ and L_i is obtained by a set of equations: $\sigma_{11}L_1 + \sigma_{12}L_2 \ldots + \sigma_{1k}L_k = \delta_1$, $\sigma_{k1}L_1 + \sigma_{k2}L_2 + \ldots + \sigma_{kk}L_k = \delta_k$. ►discriminant function, ►covariance [look up at correlation], ►Euclidean distance)

Multivariate Normal Distribution:

$$f(x_1, x_2 \ldots x_q) = (2\pi)^{-q/2} |\sum|^{1/2} \exp - 1/2(x - \mu)' \sum^{-1}(x - \mu)$$

Multivesicular Body (MSB): Proteins or peptides covalently tagged with ubiquitin(s) are sorted through the MVB pathway and delivered to lysosome for breakdown. ►ubiquitin

MUM (maximum unique match): Indicates the homology of the nucleotide sequences between two DNA single strands.

MUMmer: A genome sequence alignment algorithm. ►algorithm; Delcher AL et al 2002 Nucleic Acids Res 30:2478.

Mummies: Dried animal or human bodies, preserved by chemicals and/or desiccation. They may contain proteins (blood antigens), and DNA sequences to carry out limited molecular analysis using PCR technology. ►ice man, ►ancient DNA, ►ancient organisms; Buckley SA et al 2004 Nature [Lond] 431:294.

Mumps: An infectious disease caused by paramyxovirus. It is spread through contact, sneezing, saliva, or through other body fluids, and generally affects children under 15. Swelling of the parotid (salivary) glands situated in front and below the ears is its primary target, but it may affect other organs such the ovaries, testes, pancreas, and brain. In the majority of cases, clinical treatment is not required. Vaccination is available.

MUNC18: A mammalian homolog of the Unc18 protein of *Caenorhabditis,* and binds syntaxin. Sec1/Munc18 activate SNARE fusion and are required for neurotransmitter release (Shen J et al 2007 Cell 128:183). ►syntaxin, ►snare, ►neurotransmitter

Mung Bean Nuclease: A single-strand specific endonuclease.

Munchausen Syndrome: Medically, a psychological disorder of simulating disease in order to get attention. It was so named by Batshaw ML et al (1985 in New England J Med 312:1437) for some patients simulating Torsion Dystonia (9q34), a neurological disorder involving involuntary motions. Freiherr/Baron von Münchausen, a German raconteur published his fancy, mendacious adventures in a *Vademecum for Happy People* in 1760, which appeared in different editions and languages and made him a legend of tall stories. ►dystonia

Muprinting: Analysis of the pattern of integration sites of Mu phage in large populations of bacteria. ►Mu bacteriophage, ►insertional mutation

Murashige & Skoog Medium (MS1): MS1 for plant tissue culture is suitable for growing callus and different plant organs. Composition mg/L: NH_4NO_3 1650, KNO_3 1900, $CaCl_2.2H_2O$ 440, $MgSO_4.7H_2O$ 370, KH_2PO_4 170, KI 0.83, H_3BO_3 6.2, $MnSO_4.4H_2O$ 22.3, $ZnSO_4.7H_2O$ 8.6, $Na_2MoO_4.2H_2O$ 0.25, $CuSO_4.5H_2O$ 0.02, $CoCl_2.6H_2O$ 0.025, Ferric-EDTA 43, sucrose 3%, pH 5.7, inositol 100, nicotinic acid 0.5, pyridoxine.HCl 0.5, thiamin.HCl 0.4, indoleacetic acid (IAA) 1 – 30, and kinetin 0.04 – 10. Microelements, vitamins, and hormones may be prepared in a stock solution and added before use. For kinetin, other cytokinins may be substituted such as 6-benzylamino purine (BAP) or isopentenyl adenine (or its nucleoside), for IAA, naphthalene acetic acid (NAA), or 2,4-D (dichlorophenoxy acetic acid) may be substituted or a combination of the hormones may be used in concentrations that are best suited for the plant and the purpose of the culture. For solid media, use agar or gellan gum. Heat labile components are sterilized by filtering through 0.45 μm syringe filters. Variations of this medium are commercially available as a dry powder ready to dissolve but the pH needs to be adjusted. ►agar, ►gellan gum, ►syringe filter, ►Gamborg medium; Murashige T, Skoog F 1962 Physiol Plant 15:473.

Murine: Pertaining to mice (or rats).

Murine Leukemia Virus (MuLV): ►viral vectors

Muscarinic Acetylcholine Receptors: Muscarinic acetylcholine receptors are activated by the fungal alkaloid, muscarine, a highly toxic substance, causing excessive salivation (ptyalism, sialorrhea), lacrimation (shedding tears), nausea, vomiting, diarrhea, lower than 60 pulse rate, convulsions, etc. Deficiency of the M3 receptor decreases appetite in mice and the animals stay lean. Antidote: atropine sulfate. ►signal transduction

Muscular Atrophy: Peroneal muscular atrophy (of the fibula) may be classified as (I) demyelinating form and sensory neuropathy, (ii) axonal motor and sensory neuropthy and (iii) distal hereditary neuropathy or distal spinal muscular dystrophy. ►Werdnig-Hoffmann disease, ►Kennedy disease, ►Charcot-Marie-Tooth disease, ►olivopontocerebellar atrophy, ►dentatorubral-pallidoluysian atrophy, ►Machado-Joseph syndrome, ►spinal muscular atrophy, ►trinucleotide repeats

Muscular Dystrophy: A collection of anomalies involving primarily the muscle, and controlled by X and autosomal recessive or dominant genes. The most severe form is the (DMD) *Duchenne muscular dystrophy* in human chromosome Xp21.2 (12q21). Its transmission is through the females because the males affected do not reach the reproductive stage. Similar defects were observed also in animals. The prevalence is about 3×10^{-4}. The frequency of female carriers is about twice as high as the affliction of males. The onset is around age 3 and within a few years, the affected persons fail to walk and usually die by age 20. Mental retardation is common in this disease. The earliest diagnosis may be made by an abnormally high serum creatine phosphokinase (CPK) level at birth. Prenatal diagnosis is feasible and successful in about 90% of the cases. The dystrophin gene is one of the largest human genes involving more than two megabases (about half of the size of the entire genome of *E. coli).* This gene includes many introns with an average size of about 16 kb, whereas in the about 79 exons average size is only about 50 kb. In a tissue specific manner, it may chose among eight promoters located at different positions. The full-length dystrophin is ~427-kDa but the R dystrophin is 260, the B 140, the S 114, and the G 71-kDa. The pertinent promoter used 5′ downstream to determine the length of the protein. All have the same C-terminal domain, except the 40-kDa apo-dystrophin-3, which has a truncated C-terminal synthrophin-binding domain. The gene in the muscles appears to use a promoter other than in the brain. The majority of the cases are deletions and other chromosomal defects, yet gene mutations (predominantly frameshifts) are common too because of the large size of the gene. Intragenic recombination frequency was estimated to be 0.12. Dystrophin connects F-actin through its N-terminus, and its C-terminus binds to a complex of glycoproteins and through them to the extracellular matrix and the sarcolemma (Bassett DI et al 2003 Development 130:5851). In DMD, the dystrophin protein may be entirely missing, whereas in the milder form of the diseases, the *Becker type* (BMD), the dystrophin protein is just shorter. With antisense technology (in a mouse model, using 2′-O methyl oligoribonucleotides), skipping of exon 23 induction altered the reading frame and the mRNA was translated into a protein resembling that of the Becker type dystrophin. Such a gene therapy significantly mitigated the symptoms of the disease (Mann CJ et al 2001 Proc Natl Acad Sci 98:42; Goyenvalle A et al 2004 Science 306:1796). Adeno-associated virus vector can deliver an antisense sequence, through the tail vein to rodents had beneficial therapeutic effects (Denti MA et al 2006 Proc Natl Acad Sci USA 103:3758). The dog model of DMD responded favorable to intra-arterial delivery of blood vessel-associated stem cells making the wild type protein (Sampaolesi M et al 2006 Nature [Lond] 444:574). Combining the effects of nitric oxide with nonsteroidal anti-inflammatory activity by using HCT 1026, (a nitric oxide-releasing derivative of flurbiprofen) (see Fig. M134) in murine models for limb girdle and Duchenne muscular dystrophies, significantly ameliorated the morphological, biochemical, and functional phenotype in the absence of secondary effects and slowed down disease progression. In addition, HCT 1026 enhanced the therapeutic efficacy of arterially delivered donor stem cells, by increasing 4-fold their ability to migrate and reconstitute muscle fibers (Brunelli S et al 2007 Proc Natl Acad Sci USA 104:264).

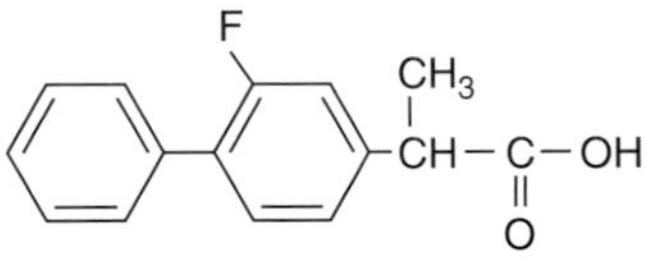

Figure M134. Flurbiprofen

The frequency of BMD is about 0.1 of that of DMD, and the patients may live until 35 and thus may have children. Apparently, allelic genes code for BMD and DMD. Deletions in the DMD gene may affect genes proximal to the centromere: CGD (chronic granulomatous disease), or the XK (McLeod syndrome) or the XP6 (retinitis pigmentosa) and distally GKD (glycerol kinase) or AHC (adrenal hyperplasia) may be affected. The two forms of the *Emery-Dreifuss dystrophy* affecting the shoulder muscles (scapulohumeral dystrophy) are autosomal (1q21-q23, encoding lamins A and C) and X-linked (Xq28, encoding emerin). In the *limb-girdle muscular dystrophy* (LGMD, 3 dominant, 8 recessive), the defects are in the sarcoglycan subunits; α-sarcoglycan is encoded in chromosome 17q12-q21, the β subunit is coded in chromosome 4q12, whereas the γ subunit was localized to 13q12. Mouse defective in α-sarcoglycan can be successfully treated by arterial delivery of mesangioblasts (stem cells) transduced by lentiviral vectors expressing the sarcoglygan gene (Sampaolesi M et al 2003 Science 3001:487). The human chromosome 4q35 region harbors the dominant genes, which are misregulated in FSHD (*facioscapulohumoral muscular dystrophy*). In FSHD, deletion is found in a 4q35 duplicated region (D4ZA) containing a transcriptional silencer. Overexpression of FRG1 (but not the FRG2) gene in the region leads to the FSHD disease because of inappropriate alternative splicing of the pre-mRNA of mice (Gabellini D et al 2006 Nature 439:973). LGMD1A (human chromosome 5q) and LGMD2 (human chromosome 1q11-q21) are both the dominant forms. LGMDC (3q25) involves a defect in the plasma membrane protein caveolin-3. Chromosomes 2p13, 5q22.3–31.3, and 15q15 also encode LGMD subunits. The recessive (chromosome 9q31) *Fukuyama congenital muscular dystrophy* (FCMD) has a prevalence of $\sim 10^{-4}$ in Japan. It involves polymicrogyria (numerous small convolutions of the brain) due to defects of migration of the neurons and involves mental retardation. The mutation is caused by a ~ 3-kbp insertion of a retrotransposal tandem repeat at the 3′-untranslated sequences of a gene encoding a secreted protein, fukutin. The recessive *Miyoshi myopathy* (2q13.2) has an early adulthood onset due to defect in the 6.9-kb cDNA encoding the dysferlin protein (named after dystrophy and fertility in *Caenorhabditis*). The *oculopharyngeal muscular dystrophy* (OPMD), a late adulthood onset disease of swallowing, dropping eye lids, and limb weakness is caused by 8–13 expansion of a (GCG)6 tract specifying polyalanine encoded by the poly(a) binding protein 2 gene (PABP2) at human chromosome 14q11-q13. The symptoms of the muscular dystrophies display considerable variation in onset and severity of expression. The face and shoulder (fascioscapulohumeral) dystrophy is limited to the named body parts. The gene was assigned to human chromosome 4q35. The congenital dystrophy gene causing deficiency in the laminin α2 chain (merosin) around the muscle fibers was located to human chromosome 6q22-q23. Laminin-α2-deficiency can be compensated for by agrin in mouse. Adeno-associated virus vector—introduced into multiple muscles—mediated overexpression of agrin and restored the structural integrity of the basal lamina of myofibers, inhibited interstitial fibrosis, and ameliorated the pathological symptoms by such somatic gene therapy in mice (Qiao C et al 2005 Proc Natl Acad Sci USA 102:11999). For improving plasmid-mediated integration of dystrophin, phage ΦC31 integrase was used in mouse model of gene therapy for Duchenne muscular dystrophy. Using luciferase expression vector, the integration can be modified in non-invasive manner (Bertoni C et al 2006 Proc Natl Acad Sci USA 103:419). The *Ullrich* (scleroatonic) *muscular dystrophy* is a recessive defect in collagen 6 (COL6A1, COL6A2, 21q22.3, COL6A3, 2q37). Mitochondria in cells from Ullrich patients, unlike those in myoblasts from healthy donors, depolarized upon addition of oligomycin and displayed ultrastructural alterations that were worsened by treatment with oligomycin. The increased apoptosis, ultrastructural defects, and the anomalous response to oligomycin could be normalized by Ca^{2+} chelators, by plating cells on collagen VI, by treatment with cyclosporin A, or with the specific cyclophilin inhibitor methylAla3ethylVal4-cyclosporin, which does not affect calcineurin activity (Angelin A et al 2007 Proc Natl Acad Sci USA 104:991). Other types of muscular dystrophies may accompany symptoms of other human diseases. Dystrophin is a rod-like cytoskeletal protein, normally localized at the inner surface of the sarcolemma (the muscle fiber envelope). Dystrophin is attached within the cytoplasm to actin filaments and also to the membrane-passing β subunit of the dystroglycan (DG) protein while the α subunit of DG joins the basal lamina through the laminin protein (encoded in chromosome 6q22-q23). In the membrane, DG is associated with the β and γ subunits of sarcoglycan (SG); the α sarcoglycan is extracellularly attached to the other two subunits. In *Caenorhabditis*, the acetylcholine transporter SNF-6 mediates synaptic activity at the neuromuscular junctions. Improper clearing of acetylcholine in the mutants seems to contribute to muscular dystrophy pathogenesis (Kim H et al 2004 Nature [Lond] 430:891). In some LGMD patients, there is also a defect in the muscle-specific protease calpain-3, encoded in chromosome 15q15. [The LGMD disease is also called severe childhood autosomal recessive muscular dystrophy or SCARMD]. In some of the SCARMDs, the defect was associated with adhalin, a 50-kDa sarcolemma dystrophin-associated glycoprotein, encoded in human chromosome 17q. Bone marrow cells, which normally are precursors of cartilage, bone, and some parenchyma cells may differentiate into myotubes and have been considered as genetic therapeutic agents for the treatment of muscular dystrophy. Carriers of dystrophies can be *biochemically normalized* when from over-expressed positive cells the protein is diffused to the nearby negative cells. In case of *genetic normalization*, degenerated myonuclei are replaced by nuclei from dystrophin-positive *muscle satellite cells* (muscle stem cells). Gene therapy is possible by the injection of integrating vectors carrying the normal genes into muscle satellite cells that may fuse with the muscle fibers. Adenovirus vectors, however, do not integrate into the host nuclei and replicate only in extrachromosomal form. Therefore, after a variable period of time they may be lost. Retroviral vectors integrate and are stably expressed in the chromosomes. They can be introduced in vivo or *ex vivo*. Direct injection of supercoiled plasmid DNA may also be successful for transformation although such DNAs are not integrated into the host genome and frequently display variable expression. Another therapeutic approach is the injection of fusogenic, in vitro cultured myoblast cells that may cure in a mosaic pattern the host's defect.

Upregulation of the utrophin gene (6q24, its encoded protein is 65–80% homologous to dystrophin) has also been considered as a remedy. The utrophin gene is normally active only during embryonal development and then it falls dormant. ►myotonic dystrophy, ►Charcot-Marie-Tooth disease, ►neuromuscular diseases, ►muscular atrophy, ►neuropathy, ►lamins, ►agrin, ►RFLP, ►apoptosis, ►emerin, ►myotubes, ►gene therapy, ►adenovirus, ►viral vectors, ►retroviral vectors, ►myoblast, ►sarcolemma, ►dsytrophin, ►caveolin, ►dysferlin, ►mitochondrial diseases in humans, ►contiguous gene syndrome, ►antisense technologies, ►readthrough, ►Walker-Wagner syndrome, ►collagen; Koenig M et al 1989 Am J Hum Genet 45:498; Arahata K 2000 Neuropathology Dystrophin S34-S41; Burton EA, Davies KE 2002 Cell 108:5; Durbeej M, Campbell KP 2002 Current Op Genet Dev 12:349; Chamberlain JS 2002 Hum Mol Genet 11:2355; Hauser MA et al 2002 Am J Hum Genet 71:1428, potential gene therapies: van Deutekom JCT, van Ommen G-JB 2003 Nature Rev Genet 4:774.

Mushroom: The fruit body of basidiomycete fungi. The mushroom is generally a genetic mosaic arising from mycelial fusion of two to several genetically compatible colonies. ►basidiomycetes, ►fungal life cycles

Mushroom Body: A central nerve (neuropil) complex in the insect brain and the pairs of mushroom bodies are implicated in olfactory memory recall and elementary cognitive functions. Mushroom bodies regulate sleep in *Drosophila* (Joiner WJ et al 2006 Nature [Lond] 441:757). ►memory, ►olfactogenetics; McGuire SE et al 2001 Science 293:1330.

Music of Macromolecules: http://www.geneticmusic.com/dnamusic; music of the genome: http://www.pandora.com.

Musical Talent: Pitch perception appears to be associated with an associative auditory area of the brain (asymmetry of the planum temporale). The immediate and correct recognition of the musical pitch, an auditory tone without an external reference, is rare autosomal dominant factor in the human populations and it is called the perfect or absolute pitch (AP). The heritability of musical ability appears high (70–80). The development of this ability is favored by early musical training. Whales and several species of birds use music somewhat similarly to humans (Gray PM et al 2001 Science 291:52). The neurobiological bases of human music perception and performance is discussed by Tramo MJ 2001 Science 291:54. ►Bach, ►Beethoven, ►Mozart, ►Strauss, ►dysmelodia, ►amusia, ►prosody, ►pitch; Baharloo S et al 1998 Am J Hum Genet 62:224 ibid. 67:755 2000; Peretz I et al 2002 Neuron 33:185; reviews on music faculty: 2003 Nature Neurosci 6:663–695.

Mustard Gas (dichloroethyl sulfide, $[ClCH_2CH_2]_2S$): A radiomimetic, vesicant poisonous gas, used for warfare in World War I. ►nitrogen mustards, ►radiomimetic

Mustards (family of cruciferous plants): The taxonomy of this family is not entirely clear. The white mustard (*Sinapis alba*) is 2n = 24. The black mustard (*Brassica nigra*) is 2n = 2x = 16 and supposedly the donor of its genome to the Ethiopian mustard (*B. carinata*) 2n = 34, (2 × [8 + 9]), an amphidiploid that received 9 chromosomes from *B. oleracea* (2n = 18). The brown mustard (*B. juncea*) 2n = 36 (2 × [8 + 10]) has one genome (x = 8) of the black mustard and another genome (x = 10) from *B. campestris* (2n = 20). The 38 (2 × (9 + 10) chromosomes of the rapes and swedes descended from *B. oleracea* (2n = 18) and *B. campestris* (n = 20). Sometimes *Arabidopsis* (2n = 10) is also called mustard since mustard also means crucifer. ►*Brassica*

Muta^{tm} Mouse: Commercially available transgenic strain with the *LacZ* insertion useful for testing mutagens/carcinogens. The transgene is extracted from mice and introduced into λ phage and subsequently into *E. coli* bacteria. In the presence of Xgal substrate, the mutant bacterial colonies appear colorless on a blue background because of the inactive galactosidase. Actually, mutations inside the mouse are ascertained outside the animals for convenience of detectability. ►Big Blue, ►Lac operon, ►β-galactosidase, ►Xgal; Nohmi T et al 2000 Mutation Res 455:191.

Mutability: Mutability indicates how prone is a gene to mutate; it indicates a genetic instability. This may be an intrinsic property of the gene or due to the presence of transposable elements or it may depend on the functionality of the genetic repair systems.

Mutable Gene: A mutable gene has higher than usual rate of mutation. ►mutator genes, ►transposable elements, ►mismatch repair, ►DNA repair, ►mutator

Mutagen: Physical, chemical, or biological agent capable of inducing mutation. ►distal mutagen, ►promutagen, ►proximal mutagen, ►ultimate mutagen, ►activation of mutagen, ►physical mutagens, ►chemical mutagens, ►biological mutagens, ►effective mutagen, ►efficient mutagen, ►supermutagen, ►triple helix formation, ►mutagen direct, ►mutagen indirect, ►environmental mutagens

Mutagen Assays: ►bioassays in genetic toxicology, ►Ames test, ►*ClB*, ►autosomal dominant assays, ►autosomal recessive assays, ►*Basc*

Mutagen Direct: A direct mutagen acts by modifying DNA (or RNA bases) by, e.g., alkylation or deamination, cross-linking DNA strands, and breaking the nucleic acid backbone.

Mutagen Indirect: An indirect mutagen damages DNA by its metabolic products (activation of promutagens), increase of reactive free radicals (ROS), affects (reduce) apoptosis, increases recombination, mutates oncogenic suppressors, etc.

Mutagen Information Center: ►databases

Mutagen Sensitivity: Mutagen sensitivity is determined by DNA repair and the metabolic enzymes degrading or activating mutagens/promutagens. ►DNA repair, ►promutagen; Tuimula J et al 2002 Carcinogenesis 23:1003.

Mutagenesis: Induction and procuring of mutations either by mutagenic agents in vivo or by the use of reversed genetics. Mutagenic agents usually inflict damage to the genetic material. The cell attempts to repair the lesions and uses transcription factors to meet this goal. By binding 30 damage-related transcription factors to the DNA, the DNA damage response pathway can be mapped (Workman CT et al 2006 Science 312:1054). ►mutation induction, ►cassette mutagenesis, ►transposable elements, ►insertion elements, ►insertional mutation, ►REMI, ►localized mutagenesis, ►directed mutation, ►oligonucleotide-directed mutagenesis, ►sequence saturation mutagenesis, ►RID, ►reversed genetics, ►genome-wide functional analysis, ►synthetic lethals, ►DNA repair; Jackson IJ 2001 Nature Genet 28:198.

Mutagenesis, Genome-Wide: ►insertional mutation; Spradling AC et al 1995 Proc Natl Acad Sci USA 92:10824; Szabados L et al 2002 Plant J 32:233.

Mutagenesis, Site-Selected: ►localized mutagenesis, ►directed mutation

Mutagenesis, Site-Specific: ►localized mutagenesis, ►directed mutation, ►targeting genes, ►site-specific mutation, ►RID, ►mutation locus-specific in humans

Mutagenic Potency: Mutagenic potency is difficult to determine for many agents that have low mutagenicity (see Table M8). Therefore, the genetic and carcinogenic hazard of many compounds is unknown, and possibly harmless agents may have been found mutagenic in some studies while other studies do not verify their harmful effect. There are other agents that may be definitely mutagenic and/or carcinogenic but may have escaped attention. Also, there is no perfect way to quantitate mutagenic effectiveness, particularly for human hazards because of the different quantities humans may be exposed to. As an example, the mutagenicity in the Ames *Salmonella* assay of a few compounds is listed in the table. It is interesting to note that ethylmethane sulfonate, the probably most widely used mutagen, is only the ninth on this list and nitrofuran, an erstwhile food preservative, exceeds its effectiveness as a mutagen on molar basis more than three thousand fold. ►environmental mutagens, ►Ames test, ►mutagen assays, ►bioassays in genetic toxicology

Mutagenic Specificity: As per muagen specificity, base analogs (5-bromouracil, 2-aminopurine) affect primarily the corresponding natural bases. The target of hydroxylamine is cytosine. Alkylating agents preferentially affect guanine. There are no simple chemicals, however, that would selectively recognize a particular gene or genes, yet the frequency of mutation may vary according to the gene locus and the mutagen used. With the techniques of molecular biology, specifically altered genes can be produced by synthesis and introduced into the genetic material by transformation. Gene replacement by double crossing-over can substitute one allele for another (see Table M9).

Table M8. Mutagenic potency

Compound	Mutations/nmole	Compound	Mutations/nmole
Caffeine	0.002	Ethidium bromide	80
EDTA	0.002	Sodium azide	150
Sodium nitrite	0.010	Acridine ICR-170	260
Ethylmethane sulfonate	0.160	Nitrosoguanidine	1375
Captan (fungicide)	25.000	Aflatoxin B-1	7057
Proflavine	38.000	Nitrofuran (AF-2)	20800

(After McCann, J. et al. 1975. *Proc. Natl. Acad. Sci. USA* 72:5135.)

Table M9. Mutations/10^{-6} bacterial cells induced by three different agents at 9 loci. (M. Demerec 1955 Amer. Nat. 89:5)

$MnCl_2$		UV		X-rays	
Gene	Mutations	Gene	Mutations	Gene	Mutations
phe-1	11	hi-1	22	leu-2	12
leu-2	24	phe-1	100	hi-1	34
ar-3	63	ar-2	440	ar-2	54
hi-1	121	leu-2	1,200	try-3	113
try-2	448	try-3	1,800	ar-3	468
leu-3	1,050	try-5	3,110	try-2	1,160
ar-2	1,720	ar-3	4,600	leu-3	1,380
try-3	10,200	leu-3	6,300	try-5	1,563
try-5	14,000	try-2	10,700	phe-1	2,460

►synthetic genes, ►frameshift, ►gene replacement, ►transformation, ►localized mutagenesis, ►TAB mutagenesis, ►Cre/Lox, ►knockout, ►knock in, ►targeting genes, ►chimeraplasty, ►triplex, ►homolog-scanning mutagenesis, ►TFO, ►Zinc-finger nuclease

Mutagenicity and Active Genes: Experimental data indicate that replicating or actively transcribed genes are preferred targets of mutagens, probably because of the decondensation of the genetic material. Spontaneous deleterious mutation does occur during the stationary phase of *E. coli* when replication is very low (Loewe L et al 2003 Science 302:1558).

Mutagenicity of Electric and Magnetic Fields: A large body of the published positive results lack rigorous experimental verification and reproducibility. Some newer data do not apparently suffer from these shortcomings. An electric field has electric charges at specific points and in the magnetic field magnetic force prevails. (See McCann J et al 1998 Mutation Res 411:45), ►mutagenicity of active genes

Mutagens-Carcinogens: A very large fraction of mutagens (genotoxic agents) are also carcinogens but not all carcinogens are mutagenic. The difference is based on the biology of the two events. Mutation is a single step alteration in the DNA/RNA and carcinogenesis is a multistep process. ►bioassays in genetic toxicology, ►environmental mutagens and carcinogens; Zeiger E 2001 Mutation Res 492:29; http://potency.berkeley.edu/cpdb.html.

Mutant: An individual with mutation.

Mutant Enrichment: ►screening, ►mutation detection, ►filtration enrichment

Mutant Frequency: The frequency of mutant individuals in a population, disregarding the time or event that produced the mutation. It is thus a concept different from mutation frequency. ►mutation rate, ►jackpot mutation

Mutant Hunt: Inducing/collecting/isolating particular mutants for a purpose.

Mutant Isolation: See Figs. M135 and M136 for the isolation of mutations. Basically similar procedures can be used in various microorganisms. The isolation of mutations is greatly facilitated if selective techniques are available.

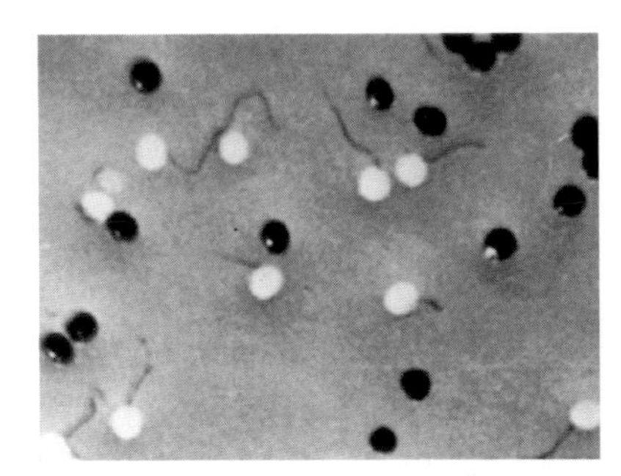

Figure M135. Allylalcohol selection of alcohoil dehydrogenase mutant maize pollen. The black is dead the germinating white pollen is resistant. (Courtesy of M. Freeling, photo by D.S.K. Cheng, see also Nature [Lond] 267:154)

Revertants of auxotrophs can grow on minimal media whereas the resistants (to antibiotics, heavy metals, metabolite analogs, etc.) can be selectively isolated in the presence of the compound to what resistance is sought. Herbicide-resistant plants can be

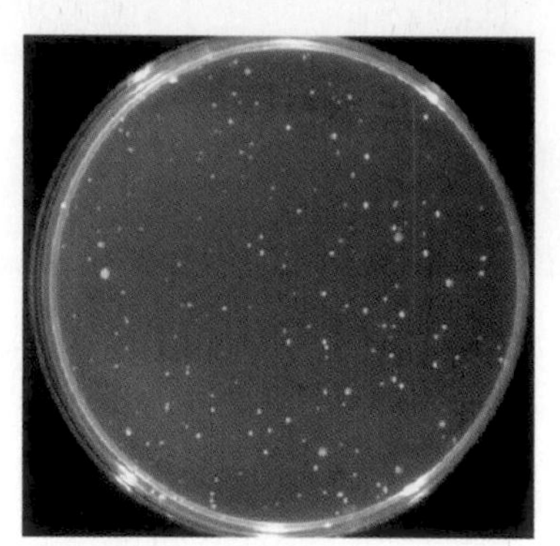

Salmonella revertants
(G.P. Rédei, unpublished photos)

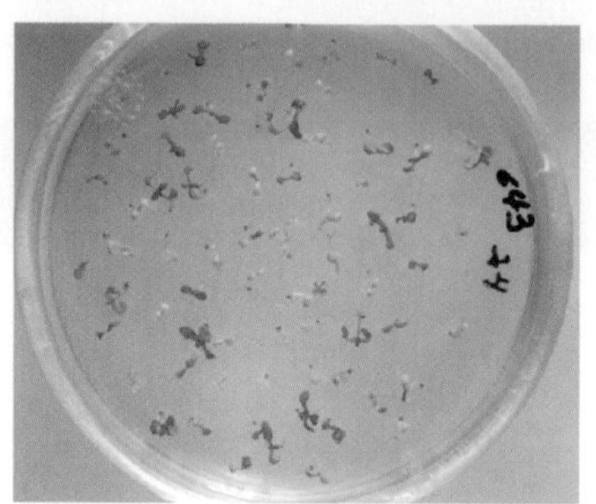

Kanamycin resistant *Arabidopsis* seedlings

Thiamine-revertant *Arabidopsis*

Figure M136. Selective isolation of mutants of bacteria and plants

selected in the presence of the herbicide. (►mutation detection)

Auxotrophic animal cells could be isolated in the presence of 5-bromodeoxyuridine because the wild type cells that incorporated the nucleoside analog after exposure to visible light—because of breakage of the analog-containing DNA—are inactivated. The non-growing mutant cells fail to incorporate the analogs and thus stay alive and after transferring them to supplemented media they may resume growth. Feeding the cells allyl alcohol can isolate alcohol dehydrogenase mutations. The wild type cells convert this substance to the very toxic acrylaldehyde and are killed, whereas the alcohol dehydrogenase inactive mutants (microorganisms, plants) cannot metabolize allyl alcohol and selectively survive. ►replica plating, ►fluctuation test, ►filtration enrichment, ►penicillin screen, ►selective medium, ►Ames test, ►reverse mutation, see Fig. M137.

Mutarotation: A change in optical rotation of anomers (isomeric forms) until equilibrium is reached between the α and β forms.

Mutase: An enzyme mediating the transposition of functional groups.

Mutasome: The enzyme complex (Rec A, UmuC-UmuD′, pol III/pol V) involved in the error-prone DNA replication in translesion. In *E. coli,* the repair polymerase V and to some extent RecA are required. ►DNA repair, ►DNA polymerases, ►RecA, ►translesion pathway; Schlacher K et al 2005 Mol Cell 17:561.

Mutasynthesis: In mutasynthesis, a biochemically altered mutant of an microorganism may produce modified metabolites either directly or from a precursor analog; the new product (e.g., antibiotic) may be more useful as a drug.

Mutation: Heritable change in the genetic material; the process of genetic alteration (see Figs. M138 and M139). ►mutagen, ►spontaneous mutation, ►mutagenic potency, ►mutagenic specificity, ►mutagenicity in active genes, ►mutant isolation, ►mutagen assays, ►mutation detection, ►mutant frequency, ►mutation rate, ►mutation in multicellular germline, ►somatic mutation, ►mutation in cellular organelles, ►mutator genes, ►mutation chromosomal, ►transposable elements, ►mutation beneficial, ►mutation neutral, ►mutation pressure, ►equilibrium mutations, ►mutation useful, ►mutation in human populations, ►genetic load, ►somatic hypermutation, ►auxotrophy, ►forward mutation, ►reverse mutation, ►polymorphism

Mutation, Adaptive: ►mutation beneficial

Mutation, Age of: The age of a mutation may be inferred on the basis of breaking up the linkage disequilibrium of the ancestral haplotype by additional mutations and recombination. The task links the present haplotype to the haplotype of the coalescence. The identification of the ancestral haplotype is, however, problematic generally. There is no way to determine with certainty the constitution of the original haplotype. One approximation is the study of closely linked markers. The probability that the haplotype stays ancestral during the period elapsed since the mutation occurred can be statistically inferred. If we consider that p is the probability of the ancestral haplotype in a very large population in which all lineages are basically independent, then $G - - l_n(p)/r$ where G is the generations, and r is the frequency of mutations and recombination. ►haplotype, ►coalescent, ►linkage disequilibrium, ►genealogy; Reich DE, Goldstein DB p. 129 in Microsatellites, Goldstein DB, Schlötterer C (Eds.) 1999 Oxford University Press, Oxford, UK; Slatkin M, Rannala B 2000 Annu Rev Genomics Hum Genet 1:225.

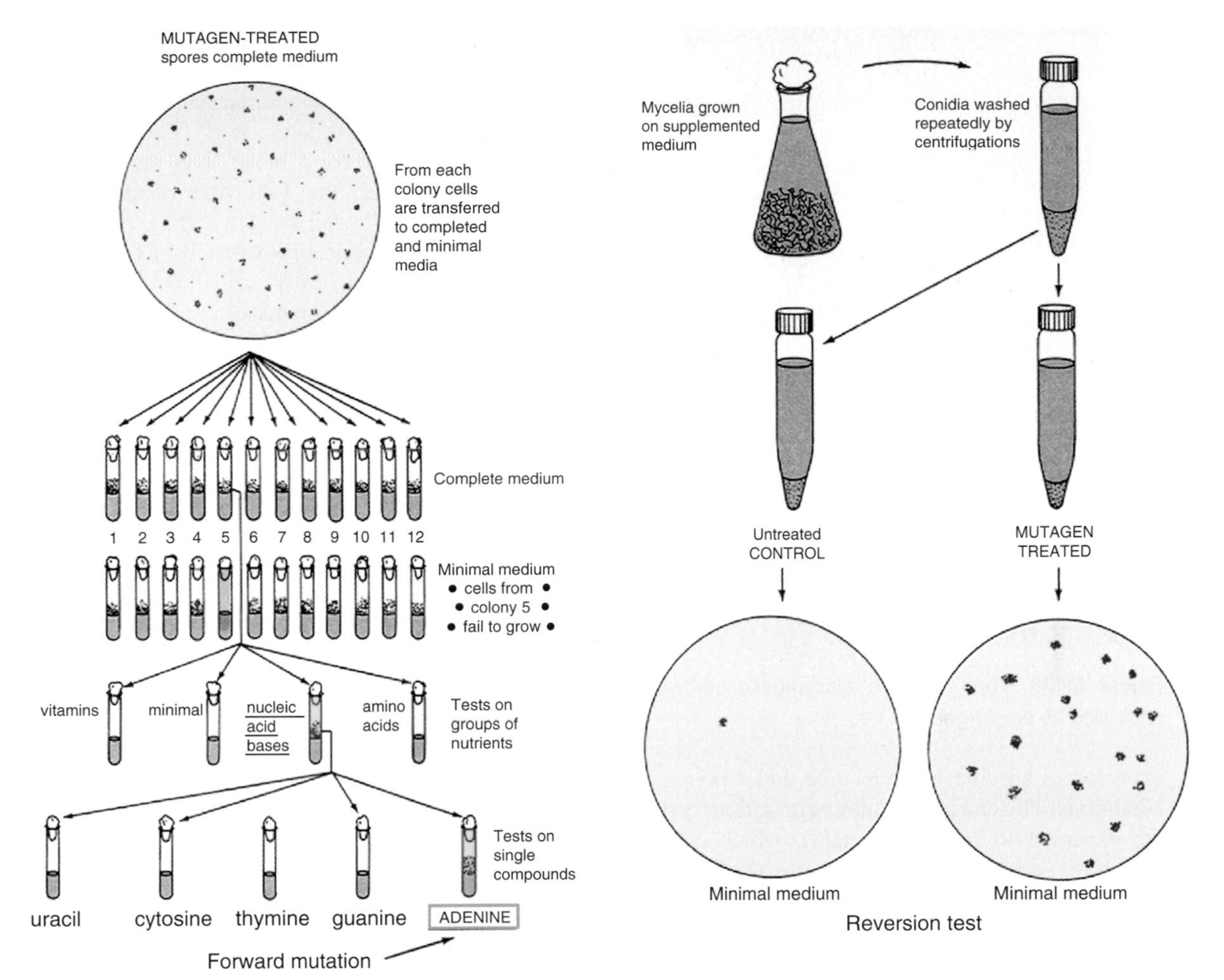

Figure M137. General procedures for the isolation of forward and reverse mutants in fungi or other haploid organisms. The chart (left) illustrates nonselective screening of forward mutants. Mutagen-treated haploid cells are propagated on a complete medium (containing vitamins, aminoacids, nucleic acid bases, etc.). Each colony on the Petri plate is the progeny of a single cell. Inocula are transferred both to minimal (without organic supplements) and to complete media to test the nutritional requirements of the colonies.Wild-type cells are expected to grow on both minimal and complete media. Auxotropic mutants will grow on the complete medium (as shown in test tube 5) but not on the minimal. Then cells from the complete medium culture 5 are tested further for growth on groups of nutrients (vitamins, nucleic acid bases, amino acids, etc.). The mutant will grow only on the medium which contains the compound that cannot be synthesized because of the genetic defect. Then the response of the mutant is tested on media containing single compounds. An adenine mutant will grow only on adenine. For reverse mutant isolation (right chart), cells from an auxotroph are washed repeatedly to remove the needed compound. Then the cells are exposed to a mutagen or only to the solvent of the mutagen (control), and after washing they are spread onto minimal media. Only the revertants are expected to grow. The control plate provides information on spontaneous rate of reversion. The treated cells show the effect of the potential mutagen.

Mutation and Allelic Frequencies: ►equilibrium mutations

Mutation and DNA Replication: Most commonly, mutation is caused by base substitution as replicational error. In multicellular eukaryotes, the detection of the time of mutations is technically difficult or impossible although dormant seeds exposed to ionizing radiation regularly display mutations in the progeny. These mutations are due, however, to chromosomal breakage or to the production of reactive chemicals that may act during subsequent cell divisions. In prokaryotes, tests can be devised for the detection of mutations occurring during the stationary phase. Using a conditional lethal system for selection, the cells may survive under both restrictive and permissive conditions. Mutations arising under restrictive growth conditions are not

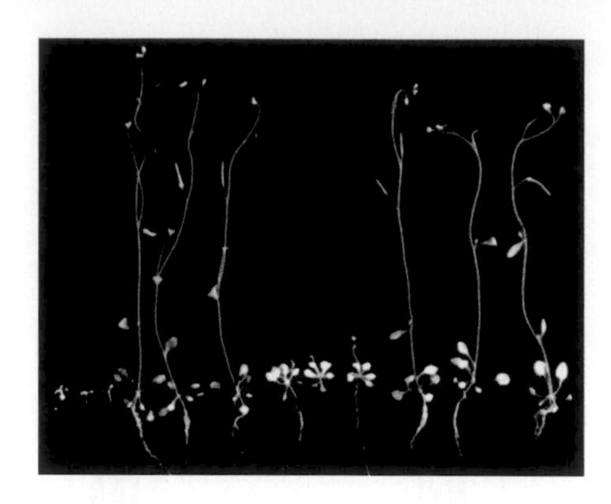

Figure M138. The thiazole-requiring mutants of *Arabidopsis* fail to grow on basal or pyrimidine media but display normal growth on thiazole or thiamine. (From left to right: Basal, Thiazole, Pyrimidine, Thiamine media)

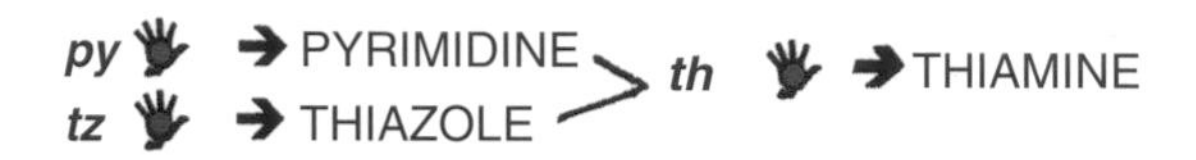

Figure M139. The Thiamine biosynthetic pathway is controlled by single genes

expected to survive but may live under permissive conditions. Thus bacterial cultures in stationary phase are expected to lose all mutants, which occurred during the preceding exponential growth phase. Thus when the culture is shifted to permissive conditions, all the immediately detectable mutations must have their origin in mutations without replication. ►base substitution, ►mutagenicity in active genes, ►growth curve, ►conditional lethal, ►permissive condition; Freese E 1963 p 207. In: Molecular Genetics Pt. I. Taylor JH (Ed.) Acad. Press, New York.

Mutation Asymmetry: As per mutation asymmetry, the base substitution in the transcribed and non-transcribed strands is not equally frequent. It may be caused by transcription-coupled repair. ►DNA repair; Majewski J 2003 Am J Hum Genet 73:688.

Mutation Avoidance: Removal of potential mutations by DNA repair. ►DNA repair

Mutation, Beneficial: when a new mutant appears in the population with a reproductive success of 1.01 (selective advantage 0.01), the odds against its survival in the first generation, $e^{-1.01} \cong 0.364$. Its chances for elimination by the 127th generation will be reduced to 0.973 compared to the probability to a neutral mutation that has a 0.985 chance of extinction. Ultimately, at a selective advantage of 0.01, its chance of survival will be 0.0197. According to the mathematical argument, even mutations with twice as great fitness than the prevailing wild type, have a high (<13%) probability of being lost, $e^{-2} \cong 0.1353$, during the first generation. Under normal conditions, the mutants' selective advantage (s) is generally very small and the probability of its ultimate survival is $(y) = 2s$. The chance of its extinction is $(l) = 1 - 2s$. In order that a mutant to have a better than random chance (0.5) for survival, the following conditions must prevail: $(1 - 2s)^n < 0.5$ or $(1 - 2s)^n > 2$, hence $-n \log_e (1-2s) > \log_e 2$ or approximately $-n(-2s) > \log_e 2$, and therefore $n > \log_e 2/(2s)$, i.e., $0.6931/(2s)$.

If $s = 0.01$ and n = number of mutations, *n* must be larger than $0.6931/(2 \times 0.01) = 34.655$.

In simple words, approximately 35 times must a mutation, with 0.01 selection coefficient, occur to be ultimately accepted and fixed. If the spontaneous rate of mutation at a locus is 1×10^{-6}, a population of 35 million is required for providing an adequate chance for the survival of a mutant of that type. Since the proportion of advantageous mutations is probably no more than 1 per 10,000 mutations, very few new mutations have a real chance of survival. The cause of this may be attributed to the fact that during the long history of evolution of the organisms, the majority of the possible mutations have been tried and the good ones were adopted by the species. The chances of new mutations is favored, however, when environmental conditions, such as agricultural practice, change in pests, pathogens, and predators, etc. take place for which historical adaptation does not exist.

A new study suggests that 1 in 150 newly arising mutations in bacteria is beneficial and that 1 in 10 fitness-affecting mutations increases the fitness of the individual carrying it. This high beneficial mutation rate is contingent of the relatively small size of the bacterial population ($\sim 10^4$) where clonal interference/competition among mutations would not eliminate many beneficial mutations. Hence, an enterobacterium has an enormous potential for adaptation and may help explain how antibiotic resistance and virulence evolve so quickly (Perfeito L et al 2007 Science 317:873). ►mutation neutral, ►fitness, ►selective advantage, ►selective sweep, ►mutation rate, ►adaptive evolution, ►genetic load, ►deleterious mutation

Mutation Bias: As per mutation bias, microsatellite sequences undergo frequent unequal recombination or replicational errors but the changes increase when the size of the tracts is short and decrease when it is longer than a particular size (~200) because the longer tracts are likely to include imperfections. This bias may be maintained in the population by selective forces. In various organisms, one type of base pair substitutions are more frequent than the others, presumable caused by differences of genetic repair. ►microsatellite, ►unequal crossing over; Udupa SM, Baum M 2001 Mol Genet Genomics 265:1097; Green P et al 2003 Nature Genet 33:514.

Mutation, Chromosomal: Partial chromosomal losses (deletion, deficiency), duplications, and rearrangements of the genetic material (inversion, translocation, transposition) are considered chromosomal mutations. The various types of aneuploids (nullisomy, monosomy, trisomy, polysomy, etc.) or increase in the number of genomic sets (polyploidy) can also be classified as chromosomal mutations. (See the specific entries for more details).

Mutation Clearance: The removal of disadvantageous mutations from a population. The emergence of sex and recombination facilitate the process.

Mutation, Compensatory: Compensatory mutation restores stability and/or functionality of another mutation.

Mutation, Cost of Production: The cost of production of mutations varies a great deal from organism to organism and also depends on the developmental stage when the mutagen is applied. Mutation induction at the gametic stage may be less expensive than at the multicellular (diploid) germline stage. In higher plants such as *Arabidopsis*, mutagens are generally applied to the mature seed and thus two generations are required for the isolation of mutants to be used for further experimental studies. The cost of raising these two generations may not be equal and both contribute to the final cost. Although in large M_2 families there is an increased probability for recovering the mutations induced, from the viewpoint of cost effectiveness large number of families with minimal size each are desirable. In case of recessive mutations, depending on the (n) number of individuals in M_2 families, the probability of recovery (P) of a mutant is at n = 1 (P = 0.25), n = 2 (P = 0.437), n = 4 (P = 0.683), n = 8 (P = 0.899), n = 16 (P = 0.989), and n = 24 (P = 0.998) in case of heterozygosity of the M_1 generation, derived from a single cell. It is thus obvious that increasing the M_2 size from 1 to 24 involves the increase of probability of recovery only from 0.25 to 0.998. In the final cost, the cost of both generations must be included along with the effectiveness of recovery of the mutations induced. Table M10 indicates that M_2 family sizes larger than four individuals generally increase the labor and the cost. The recovery of mutations is most effective if selective techniques are applicable. Unfortunately, in some cases (morphological mutants) this may not be possible. (See Rédei GP et al 1984 In: Mutation, Cancer and Malformation, Chu EHY, Generoso WM (Eds.) Plenum, New York, pp 295)

Mutation, Dating of Origin: G (number of generations since the emergence of the mutation) can be determined in a population using the information of linkage disequilibrium (δ) concerning closely linked genetic or molecular markers and their known recombination frequencies (θ):

$$G = \frac{\log[1 - Q/(1 - pN)]}{\log(1 - \theta)} \times \frac{\log \delta}{\log(1 - \theta)}$$

where δ = (pD − pN)/(1 − pN) and pD and pN represent the two different linkage phases of the homologous chromosomes. Q (probability) = $(1 - [1-\theta]^G)$ (1-pN). (See Guo SW, Xiong M 1997 Hum Hered 47(6):315; Colombo R 2000 Genomics 69:131).

Mutation Detection: Detection of mutation depends on the general nature of the organisms. In haploid organisms (bacteria, algae, some fungi), in haploid cells (microspores, pollen, sperm), in hemizygous cells (e.g., Chinese hamster ovary cell cultures), or in heterozygotes for easily visible somatic markers, the mutations are readily detectable. In diploid or polyploid cells, only the dominant mutations can immediately be

Table M10. Calculation of the cost of mutation

M_2 family size ➡	1	2	4	8	24
$M_1 + M_2$ size ➡	24 + 24	12 + 24	6 + 24	3 + 24	1 + 24
Mutant Expected ➡	6	5.244	4.098	2.697	1
Cost M_1:M_2 ⬇	Cost of 1 Mutation in Arbitrary Units Under the Conditions Shown Above and in the Left-Side Column of This Tabulation				
1:1	8	**6.845**	7.321	10.011	25
1:2	12	**11.442**	13.177	18.910	49
1.3	16	16.018	19.034	27.809	73
2:1	12	9.153	**8.785**	11.123	26
3:1	16	11.442	**10.249**	12.236	27
4:1	20	13.783	**11.713**	13.384	28

observed and for the detection of recessive mutations a more elaborate procedure is required (►mutation in the multicellular germline). The most effective methods involve selective screening. In diploid cell cultures, recessive mutation can be detected if mitotic recombination generates homozygosity at the mutant locus (Guo G et al 2004 Nature [Lond] 429:891; Yusa K et al ibid pp 896) or if the dominant wild type allele is deleted. Molecular methods are also available but these are generally not suitable for large-scale screening. When single-stranded DNA is subjected to electrophoresis in non-denaturing gel, changed pattern is detectable by SSCP. Heteroduplexes can be resolved by instability in non-denaturing gradient gels (DGGE). Also, heteroduplexes move differently in non-denaturing gels. Cleavage of heteroduplexes by chemicals or enzymes may detect mutant sites. Polymorphism at a single locus can be detected by automated analysis using PCR and fluorescent dyes coupled with a quencher (see more under ►polymorphism). Mutation may be detected by chemical modification of the mismatches. Carbodiimide generates electrophoretically slow moving DNA containing mismatched deoxyguanylate or deoxythymidylates. Hydroxylamine modifies deoxycytidylate and osmium tetroxide modifies deoxycytidylate and deoxythymidylate in such a way that the DNA at these residues becomes liable to cleavage by strong bases. Ribonuclease A cleaves at mismatches (depending on the context) in RNA-DNA hybrids. The Mut Y glycosylase of *E. coli* excises adenine from A-G and, with somewhat less efficiency, from A-C mispairs. In *E. coli*, methyl-directed mismatch system is working primarily up to 3-nucleotide insertions or deletions but longer sequences barely respond, if corrected at all, and C-C pairs are ignored. The MutH/L/S multi-enzyme complex misses only 1% of the G-T mispair-induced cuts at nearby GATC sequences. The repair is identified by PCR, which also may be a source of replicational error. The latter errors can be estimated as $f = 2lna$, where f is the estimated fraction of mutations within the sequence, l = the length of the amplified sequence, n = the PCR cycles, and a = the polymerase-specific error rate/nucleotides; (a for Taq polymerase is within the range of 10^{-6} to 10^{-7}). Direct sequencing may provide the most precise information at the highest investment of labor. ►selective screening, ►replica plating, ►sex-linked lethal tests, ►specific-locus tests, ►mutation in human populations, ►Ames test, ►host-mediated assays, ►EMC, ►SSCP, ►DGGE, ►DNA sequencing, ►gel electrophoresis, ►polymorphism, ►PCR, ►footprinting genetic, ►mutant isolation, ►bioassays in genetic toxicology, ►substitution mutation, ►padlock probe, ►SNIPS, ►Comet assay, ►heteroduplex analysis, ►allele-specific probe for mutation, ►ARMS, ►alanine-scanning mutagenesis, ►hemiclonal, ►cancer [INTER-SS PCR]

Mutation Equilibrium: ►equilibrium mutations

Mutation Frequency: The same as mutation rate; see also ►mutant frequency (that is different from mutation frequency).

Mutation, Implicit: A replication error rather than base substitution, insertion, deletion, or recombination. This term is used in genetic algorithms. ►algorithm genetic

Mutation in Cancer Cells: ►cancer, ►Knudson's two-mutation theory

Mutation in Cellular Organelles: Cellular organelles mitochondria and plastids are generally present in multiple copies per cell, and within individual organelles generally several copies of DNA molecules exist. Therefore, it generally is not expected that the mutations would be immediately revealed. For their visible manifestation they have to "sort out", i.e., the mutations may not be visible until single organelles become "homogeneous" regarding the mutation, and single cells would be "homoplasmonic" regarding that particular organelle. Although this process is frequently claimed to be stochastic, the direct observation does not seem to support the assumption. Organelles divide by fission, and the daughter organelles are expected to stay in the vicinity of the parental organelle within the viscous cytosol. Thus it is not mere chance where they are located, and the progeny organelles tend to remain clustered unless the plane of the cell division separates them. Therefore, sorting out may be a relatively fast process. Mutations in chloroplast gene generally do not affect the morphology of the plants, however, inactivation of the *clpP1* (caseinolytic protease P1) gene interferes with shoot development in tobacco (Kuroda H, Maliga P 2003 Nature [Lond] 425:86). Mutation in $tRNA^{Leu}$ (MTTL1) in lung carcinoma cybrid cells may be higher than 95% and suppressor mutations in mitochondrial DNA have been identified in only about 10% of the total mtDNA. ►mitochondrial genetics, ►chloroplast genetics, ►mtDNA, ►chloroplasts, ►mitochondrial diseases in humans, ►suppressor gene, ►suppressor tRNA, ►sorting out

Mutation in Human Populations: Mutation in human populations is difficult to study directly because the random or assorted mating systems do not favor the identification of recessive mutations. Therefore, human geneticists generally rely on the relative increase of "sentinel phenotypes" such as new dominant mutations, hemophilia, muscular dystrophy, and cancer that may be used as "epidemiological indicators" for an increase of mutation. Also, sister chromatid exchange, chromosomal aberrations, and sperm motility assays may reveal mutations.

Molecular methods, such as RFLP, RAPD, and PCR have become more available recently and have changed the previously held view that was based on the fact that no transmitted genetic effects of radiation had been clearly detected by the traditional methods in human populations. Mutational hazard for humans is generally inferred from mouse data, which include induced mutations. The immature mouse oocytes are insensitive to radiation-induced mutation but quite likely to be killed by 60 rads of neutrons and 400 R of X-rays or γ rays. Maturing or mature mouse oocytes on the other hand are very susceptible to mutation by acute radiation, although the sensitivity to low-dose rate irradiation is 1/20 or less. In contrast, the immature human oocytes are not susceptible to killing but their susceptibility to mutation is not amenable to testing. The estimated mutational hazard of mouse oocytes at various stages ranged from 0.17 to 0.44 times that in spermatogonia, indicating lower hazard to females than to males. In industrialized societies, chemical mutagens constitute the major hazard, and because of their variety and potency, their effects are difficult to assess, especially at low levels of exposure. Mutation rates can be estimated also on the basis of base substitutions in synonymous and non-synonymous codons. It may be assumed that the majority of non-synonymous substitutions are more or less deleterious, whereas the synonymous substitutions are neutral. It has been estimated that a total of 100 new mutations occur in the genome of each human individual and that the rate of deleterious mutations per generation per diploid human genome is ~1.6. (See more under individual entries, ►mutation rate [undetected mutation], ►doubling dose, ►bioassays in genetic toxicology, ►specific locus mutations test, ►RBE, ►atomic radiations, ►base substitution mutation, ►synonymous codon, ►mutation neutral, ►genetic load, ►diversity, ►human mutation assays, ►mutation locus-specific in humans; http://mutview.dmb.med.keio.ac.jp; http://archive.uwcm.ac.uk/uwcm/mg/hgmd0.html; http://www.hgmd.cf.ac.uk/ac/index.php; human gene mutation database: http://www.hgmd.org/.

Mutation in Multicellular Germline: Mutation in the multicellular germline reveals the numbers of cells in that germline at the time the mutation occurred (see Fig. M140). E.g., if a plant apical meristem is treated with a mutagen at the genetically effective 2-cell stage, in the second generation of this chimeric individual the segregation for a recessive allele will not be 3:1 but 7:1. This is so because one of the cells will produce four homozygous wild type individuals, the other cell will yield one homozygous wild type + 2 heterozygotes (=3 dominant phenotypes) and one homozygous recessive mutant ([4 + 3]:1). In case the germline contains four cells, the segregation ratio is expected to be 15:1 (1/16). In case it consists of eight cells, the proportions are 31:1. ►mutation rate, ►GECN; Rédei GP et al 1984 In: Mutation, Cancer and Malformation, Chu EHY, Generoso WM (Eds.) Plenum, New York, pp 295.

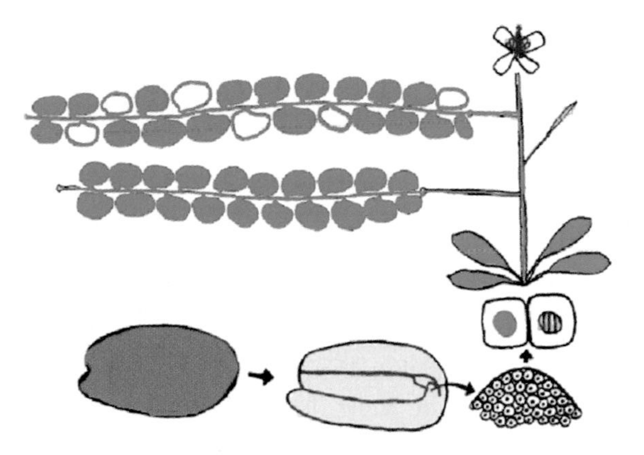

Figure M140. Mutation in multicellular diploid germline

Mutation Induction: See mutagen, directed mutation, localized mutagenesis, cassette mutagenesis, chemical mutagens, physical mutagens, environmental mutagens.

Mutation Load: ►genetic load

Mutation, Locus-Specific in Humans: Database of mutations specific for human genes: http://archive.uwcm.ac.uk/uwcm/mg/docs/oth_mut.html.

Mutation, Neutral: A neutral mutation is neither advantageous nor deleterious to the individual homo- or heterozygous for it. Its chance of immediate loss is determined by the first term of the Poisson series, e^{-1}: ≅ 0.368. (►evolution non-Darwinian). Consequently, its chance for survival is 1 − 0.362. The chance for its extinction during the second generation is $e^{-0.632}$ ≅ 0.532, and by the third generation it is $e^{-(1-0.532)}$ ≅ 0.626 (e = 2.71828). In general terms, the extinction of any neutral mutation is e^{x-1} where x is the probability of its loss in the preceding generation. According to R.A Fisher, by the 127th generation the odds against the survival of a neutral mutation is 0.985 and may eventually reach 100%. Several evolutionists, notably M. Kimura, have argued statistically against the conclusion of Fisher (and the Darwinian theory of the necessity of selective advantage). Accordingly, if the rate of mutation for a locus is μ, the population size is N, and the organism is diploid, the number of new mutations occurring per generation per gene are 2 Nμ. The chance for random fixation is supposed to be (1/2)N because the new allele is represented only once among the total of 2N. The probability of fixation, accordingly, is expected to be $(2N\mu) \times [(1/2)N] = \mu$, indicating that the rate of incorporation of a new neutral mutation into the population is equal to the mutation rate. The average number of generations required for fixation of a neutral mutation (according to Kimura and Ohta) is expected to be $4N_e$ where N_e = the effective population size. Thus, if the effective population size = 1,000 individuals, it takes 4,000 generations for a neutral mutation to be fixed in the population.

Sequencing 419 genes from 24 lines of *Drosophila melanogaster* and some related species, nearly 30% of the amino acid substitutions appeared to be adaptive (Shapiro JA et al 2007 Proc Natl Acad Sci USA 104:2271). In *Caenorhabditis elegans*, 65–80% of the mutations do not have a visible phenotype and thus do not seem to have much effect on fitness. In some human families, a 28-kb, apparently neutral, deletion was detected in chromosome 15q11-q13 area involved in imprinting (Buiting K et al 1999 Am J Hum Genet 65:1588). On the basis of nucleotide sequences, the neutrality can be statistically estimated by the HKA formula (Hudson RR et al 1987 Genetics 116:153) at the constant rate neutral model under specified conditions:

$$\chi^2 \sum_{i=1}^{L} (S_i^A - \hat{E}[S_i^A])^2 / \hat{V}ar[S_i^A] + \sum_{i=1}^{L} (S_i^B - \hat{E}[S_i^B])^2 /$$

$$\hat{V}ar[S_i^B + \sum_{i=1}^{L} (D_i - \hat{E}[D_i])^2 / \hat{V}ar[D_i]$$

where L indicates the locus (loci, l) in species A and B sequenced. S_i^A and S_i^B stand for the number nucleotide sites that are polymorphic at locus i in the number of gametes of species A and B, respectively. D_i indicates the number of differences at locus i between a random sample of gametes of species A and B, respectively. $\hat{E}$ and $\hat{V}ar$ stand for the estimates of expectation and variance, respectively. In the past, sequence variation in mitochondrial DNA has been considered as neutral but this assumption may not be generally valid. It is generally assumed the synonymous substitutions in the codons are neutral because these do not alter the amino acid sequence and composition of the proteins. This assumption may not be entirely valid because the codon usage is not stochastic and synonymous base substitutions may still affect RNA splicing and stability (Chamary JV et al 2006 Nature Rev Genet 7:98). The basic rate of neutral mutation can be detected in non-functional pseudogenes. ►effective population size, ►non-Darwinian evolution, ►adaptive evolution, ►neutral space, ►mutation beneficial, ►ultraconserved DNA elements, ►mutation rate, ►fitness, ►imprinting, ►radical amino acid substitution, ►McDonald-Kreitman hypothesis, ►Tajima's method, ►codon usage; Hudson RR et al 1987 Genetics 116:153; Kimura M, Ohta T 1977 J Mol Evol 9 (4):367; Gerber AS et al 2001 Annu Rev Genet 35:539; Clark AG et al 2003 Science 302:1960.

Mutation Pressure: Repeated occurrence of mutations in a population.

Mutation Pressure Opposed by Selection: Mutations are frequently prevented from fixation by chance alone (see mutation neutral, mutation beneficial, non-Darwinian evolution). An equilibrium of selection pressure and mutation pressure may be required for the new mutations to have a chance for survival. Mathematically, the frequency of homozygous mutants is $q^2 = \mu/s$ where μ is the mutation rate and s is the coefficient of selection. ►allelic fixation

Mutation-Preventive Concentration: A high concentration of a drug that does not permit the survival of even resistant mutant cells.

Mutation Rate: Frequency of mutation per locus (genome) per generation. This calculation is relatively simple if the cells population is haploid (prokaryotes, most of the fungi, or when the gametes are mutagenized). Mutation rate can be expressed also as alteration per nucleotide or as mutation per replicational cycle. If mutation takes place in the multicellular germline of higher eukaryotes, only indirect procedures can be used. First, the genetically effective cell number (GECN) in the germline must be determined (see genetically effective cell number). One must know also the level of ploidy. Thus mutation rate in these germline cells (R) is:

$$R = \frac{\text{number of independent mutational events}}{\text{survivors} \times \text{GECN} \times \text{ploidy}}$$

The standard error of mutation rates (s_m) can be computed as $\sqrt{\mu[1-\mu]/n}$ where μ = the mutation rate observed, and n = the size of the population. The calculation of the mutation rate on the basis of survivors may pose problems if two different agents are compared. An example: if we use 10,000 haploid cells as a concurrent control and find 10 mutations, the spontaneous mutation rate would be 10/10,000 = 0.001. If we expose another 10,000 cells to a mutagen that is lethal to 5,000 cells but again we obtain 10 mutations, the calculated apparent induced mutation rate is 10/5,000 = 0.002, when actually the treatment may not have induced any mutation, it may have only reduced survival. In case the mutagen is very potent and in each genome or family multiple mutations occur, it may become very difficult to distinguish the multiple mutations from the single ones without further genetic analysis. Some information indicates that in bacteriophages, in case of high mutation rate (induced by mutator factors) more mutants contain more than a single (clustered) mutation as expected by random distribution (Drake JW et al 2005 Proc Natl Acad Sci USA 102:12849). Since the distribution of mutations follows the Poisson distribution, in such a case the average mutation rate may be better determined on the basis of the size of the fraction of the population that shows no mutation at all. This is the zero class of the Poisson series, $e^{-\mu}$. If, e.g., the fraction of mutations of the population is 0.3, then zero class is $1–0.3 = 0.7 = e^{-\mu}$. Hence $-\mu = ln\ 0.7 = -0.3566$ and $\cong$ 0.36. In yeast for the determination of spontaneous backmutation rate, the formula of von Borstel may be used. Accordingly, $M = e^{(N_0/N)} - m_b/2C$, where N = the total number of compartments, N_0 = number of compartments without reversions, m_b = the average number of mutants in the inoculum, and C = the average number of cells in the compartment after the stoppage of growth. If no mutations are found at all in an experimental population, that does not necessarily mean that the mutation rate is zero. An approximation to the possible mutation frequency may be made. If we assume perfect penetrance and "normal distribution" of undetected mutations, we may further assume that in a population of n genomes the frequency of mutations is $(1 - q)$ at a probability of P. In order to obtain an estimate of q, we have to solve the equation: $(1 - q)^n = (1 - P)/2$, hence $\hat{q} = 1 - \times \sqrt[n]{[1 - P]/2}$. As an example, after arbitrary substitution of 10,000 for n and 0.99 for P, $\hat{q} \approx 1 - \sqrt[10000]{0.01/2} \approx 1 - 0.9995 \approx 5.3 \times 10^{-4}$. In simple words, this means that if we observed no mutations at all, there is a high probability that actually more than five may have occurred under these conditions but we have missed them by chance. It is also possible—of course—that the rate of mutation in this case is much below the 10^{-4} range or it may even be zero.

Mutation rate in human populations in case of dominance may be estimated:

$$\frac{\text{number of sporadic cases}}{2\,x\ \text{number of individuals}} = \mu.$$

The number of individuals stands for the number of total population studied and 2 is used to account for diploidy. Frequently, it may be necessary to use a correction factor for penetrance or viability. Recessive X-linked mutations can be estimated on the basis of the afflicted males and the formula becomes $\mu = (1/3)s(n/N)$, where n = the number of new mutations, N = the size of the population examined and s = the relative selective value and/or penetrance. (The 1/3 multiplier is used because the females have 2 and the males have 1 X chromosome). Estimation of the rate of recessive mutations is very difficult in human populations because the detection of homozygotes would require controlled mating (inbreeding). Mutation rate in mammals varies a great deal (Ellegren H et al 2003 Curr Opin Genet Dev 13:562). The majority of the spontaneous mutations among humans occur in the males (achondroplasia, acrodysostosis, Marfan syndrome, oculodentaldigital syndrome, Pfeiffer syndrome) because more cell divisions take place in the male germline than in that of the female. Some analyses indicate four times higher mutation rates in males than in females and lower base substitutions in the human X-chromosomes than in the autosomes. Some other studies claim much smaller excess of mutations in the human males (Crow JF 2000 Trends Genet 16:525). On the basis of sequencing the human genome the male:female substitution rate $\alpha_m:\alpha_f = 2.1$. Mutability increases with paternal age. In some instances, the increased male mutation rate was attributed to a paradoxical selective advantage of the spermatogonia carrying the mutation (Goriely A et al 2003 Science 301:643). The number of indel mutations in the X and Y chromosomes is practically the same but in the Z chromosome of birds about twice as many indels were detected than in the W chromosome (Sundström H et al 2003 Genetics 164:259). In some human diseases, (e.g., Duchenne muscular dystrophy, neurofibromatosis) there is no large sex bias, however. Actually, deletions may be more frequent in females than in males (presumably because of the long dictyotene stage). In point mutation there is a reverse tendency. Human geneticists may also have great difficulty in identifying mutations that have symptoms overlapping several syndromes. For single cases, multivariate statistics may be used but for many this procedure may be arduous.

Precise calculation of mutation rate is almost intractable by classical techniques but may be solved by molecular methods. DNA sequencing indicates genetic variations in the range of about 1 per kb. Many of these are synonymous at the protein level or do not alter the visible phenotype. The rate of synonymous mutations is higher than that of the non-synonymous (Li WH et al 1985 Mol Biol Evol 2:150). Calculating mutation rate may not be accurate if the incidence of the mutation is considered at a particular age but potential loss of the mutations by abortus or neonate death is not considered. The mutation rate on the basis of single nucleotide polymorphism in a population for the Y chromosomes and autosomes has been estimated within the range 1.9×10^{-9} to 5.4×10^{-9}, and for mitochondria 3.5×10^{-8} per site per year. Other estimates for average nucleotide changes/human genome are higher (2.5×10^{-8}, ~175/diploid genomes/generation). Single nucleotide replacements appear an order of magnitude higher than "length mutation". Both transitions and transversions are most common at CpG sites. The average deleterious mutation rate (U) in humans appears 3 or less. In the HIV1 retrovirus, mutation rate per nucleotide per generation was estimated to be as high as 1 per 10^{-5} to 10^{-4}.

Mutation rate in bacteria has been estimated by various means.

Rate = *ln* 2(M2 − M1)/(N2 − N1), alternatively $R = 2ln2\left(\frac{M2}{N2} - \frac{M1}{N1}\right)/g$ where M1 and M2 are the number of mutant colonies at time 1 and 2, respectively; N1 and N2 are the corresponding bacterial counts, *ln* = natural logarithm, g = number of generations. In order to obtain reliable estimates on the rates, the culture must be started by large inocula to avoid bias due to mutation at the early generations when the population is still small, and the experiments must be maintained over several generations under conditions of exponential growth. The majority of the *Caenorhabditis* genes do not mutate to visible, lethal, or sterile phenotypes. Some double mutants may display, however, mutant phenotype. In *Caenorhabditis*, by sequencing PCR-amplified genomes the spontaneous mutation rate per nucleotide per generation was found to be 2.1×10^{-8}, about an order higher than previous estimates for the organism (Denver DR et al 2004 Nature [Lond] 430:679) and for some of the figures given above for humans. More than half of the mutations were insertions or deletions. The rate of mitochondrial mutations was also higher (1.6×10^{-7}) than the neutral mutation rate in bacteria (Ochman H 2003 Mol Biol Evol 20:2091) or in human mitochondria as shown. The true mutation rate can be determined only by sequencing of the DNA. ►mutation spontaneous, ►fluctuation test, ►multivariate analysis, ►discriminant function, ►F_{ST}, ►polymorphism, ►diversity, ►chromosome replication, ►synthetic lethal, ►band-morph variants, ►hemophilia, ►SNIP, ►substitution mutation, ►base substitution, ►error in replication, ►DNA repair, ►mitochondrial mutation, ►indel, ►mutator gene, ►hot spot; Crow JF 2000 Nature Rev Genet 1:40; Nachman MW et al 2000 Genetics 156:297; Kumar S, Subramanian S 2002 Proc Natl Acad Sci USA 99:803.

Mutation Rate, Effective (w): The product of the mutation rate and the variance of mutational changes.

Mutation Rate, Evolution of: The evolution of the mutation rate is determined by the presence of mutator and antimutator factors. ►mutator, ►antimutator; Baer CF et al 2007 Nature Rev Genet 8:619.

Mutation Rate, Induced: Inducing N-ethyl-N-nitrosourea in mice at specific loci, 1.5×10^{-3} mutations per locus were observed. This rate may, however, be an overestimate for other eukaryotes, and the range appears to be about 10^{-4} or less. The induced mutation rate per cGy per locus in mouse was estimated to be 2.2×10^{-7} and in *Drosophila* $1.5–8 \times 10^{-8}$. It appeared also that in mouse the females had lower rates than the males although the frequencies are much affected by the developmental stages. ►mutation rate spontaneous, ►supermutagens, ►Gy, ►MNU, ►error in replication

Mutation Rate in Fitness: In *Drosophila*, about 0.3/genome/generation was estimated as the frequency of deleterious mutations. In *E. coli*, the rate of deleterious mutations per cell was estimated to be 0.0002. This is larger than three orders of magnitude difference. The *Drosophila* genome is about 20 fold larger than that of *E. coli* and during a generation approximately 25 divisional cycles take place compared to one in the bacterium. If we take the liberty of making adjustments for these differences, $0.0002 \times 20 \times 25 = 0.1$, the deleterious mutation rate in the eukaryote and the prokaryote falls within close range. Mutation rate in fitness is very difficult to estimate and in *Caenorhabditis* it appears three order of magnitude lower than in *Drosophila*. (See Zeyl C et al 2001 Evolution Int J Org Evolution 55:909).

Mutation Scan: A system of mutation detection.

Mutation Screening: Selective isolation of mutation(s).

Mutation Shower: Chronocoordinate multiple mutations spanning multiple kilobases (Wang J et al 2007 Proc Natl Acad Sci USA 104:8403).

Mutation Spectrum: The mutation spectrum indicates the range of mutations observed in a population under natural conditions or after exposure to different types of mutagens (see Fig. M141). The theoretical expectations would be that the common genetic material would mutate at the same rate at identical nucleotides in different organisms or under different conditions. This expectation is not met because some mutagenic agents have specificities for certain bases. Others act by breaking the chromosomes, depending on their organization and the physical and biological factors present. Also, genes present in multiple copies per genome may suffer mutational alterations but this

Figure M141. Sample of the spectrum of morphological mutations in *Arabidopsis* expressed at the rosette stage grown under 9 hours daily illumination. Columbia wild type is at the top right corner

may not be observed by classical genetic analysis although molecular methods may detect them. In general, obligate auxotrophic mutations are very limited in higher plants when the screening uses selective culture media. A range of morphological mutants of ►*Arabidopsis* (Figure M141); Li SL et al 1967 Mol Gen Genet 100:77; Reich DE, Lander ES 2001 Trends Genet 17:502.

Mutation, Spontaneous: Spontaneous mutation occurs at low frequency and its specific cause is unknown (Figure M142). The rate of spontaneous mutation per

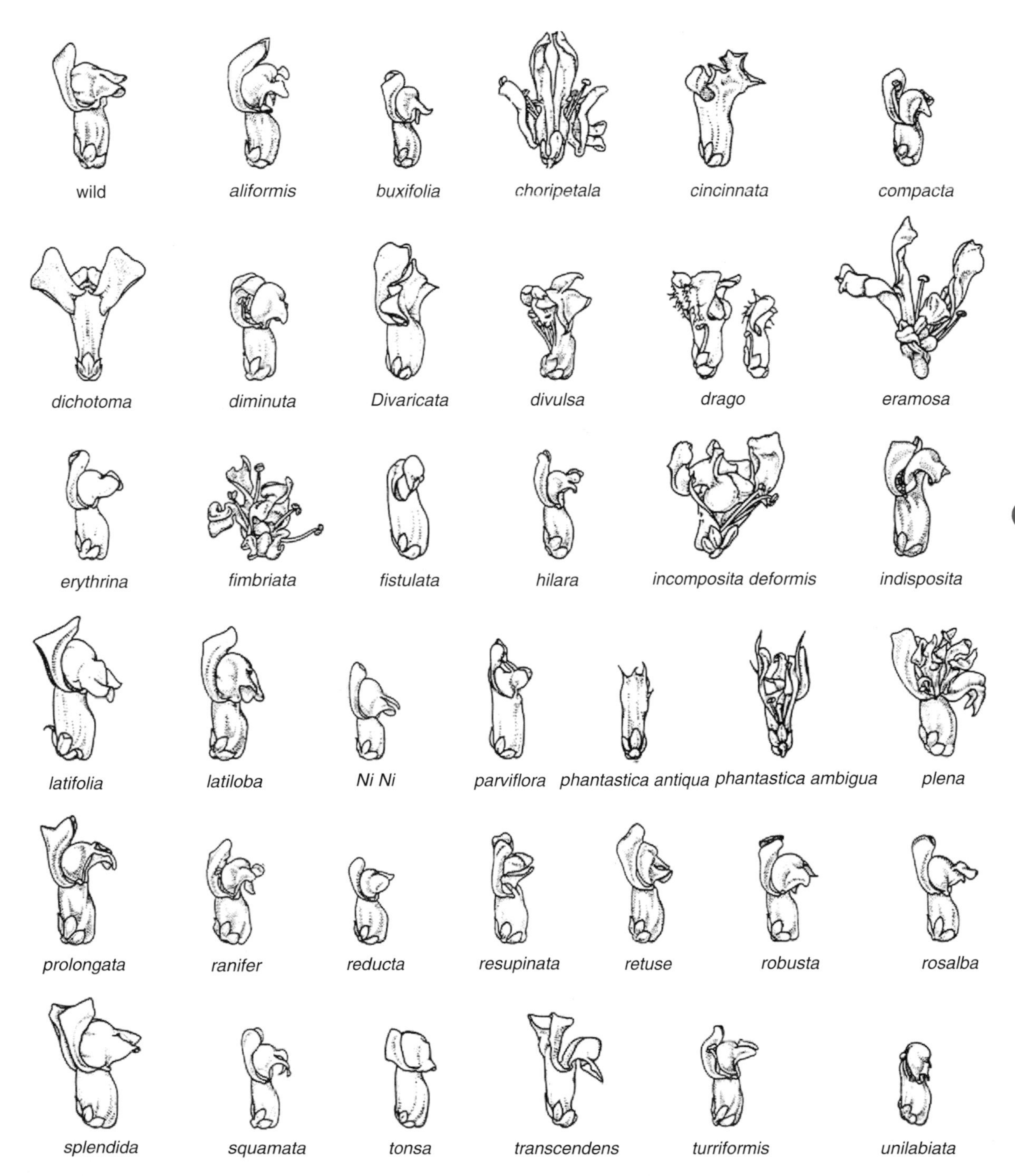

Figure M142. Mutations affecting the shape of the flower of snapdragon. The standard wild type is in the upper left corner. (Composed on the basis of H. Stubbe 1966 Genetik and Zytologie von Antirrhinum L. Sect. Antirrhinum. Fischer, Jena, Germany)

genome may vary. In bacteriophage DNA, it is 7×10^{-5} to 1×10^{-11}, in bacteria 2×10^{-6} to 4×10^{-10}, in fungi 2×10^{-4} to 3×10^{-9}, in plants 1×10^{-5} to 1×10^{-6}, in *Drosophila* 1×10^{-4} to 2×10^{-5}, in mice for seven standard loci is 6.6×10^{-6} per locus, in humans, 1×10^{-5} to 2×10^{-6} rates have been reported. The total number of detectable spontaneous mutations per live birth had been estimated as 10–3% but neither of these % estimates is very accurate because of the uncertainties of identifying very low levels of anomalies. In RNA viruses, the very high rates of 1×10^{-3} to 1×10^{-6} per base per replication have been estimated. The RNA viruses lack repair system. Amino acid substitutions in proteins may occur at the rate of 10^{-4}/residue. The error rate of the reverse transcriptase is extremely high because the enzyme does not have editing function. The induced rate of mutation may be three orders of magnitude higher but it varies a great deal depending on the locus, the nature of the mutagenic agent, the dose, etc. Mutation rates per mitochondrial D loop have been estimated to be 1/50 between mothers and offspring but other estimates used for evolutionary calculations are 1/300 per generation. Mutation rate in mitochondria on the basis of phylogenetic studies was estimated ~0.17 ±0.15–2.2/site/Myr to 1.35 ± 0.72–1.98/site/Myr at the 95% confidence level. Mutation rate per base in the human mitochondria has been estimated as 5×10^{-7}. ►mutation rate induced, ►mutation rate, ►confidence interval, ►Myr; for human diseases see Sankaranarayanan K 1998 Mutation Res 411:129; Maki H 2002 Annu Rev Genet 36:279.

Mutation, Useful: Although the majority of the new mutations have reduced fitness and are rarely useful for agricultural or industrial applications, some have obvious economic value.

Spontaneous and induced mutations have been incorporated into commercially grown crop varieties by improving disease and stress resistance, correcting amino acid composition of proteins, and eliminating deleterious chemical components (erucic acid, alkaloids, etc.).

Most of the natural variation in the species arose by mutation during their evolutionary history, and many have been added to the gene pool of animal herds and cultivated plants. Many of the floricultural novelties are induced mutants. Some of the animal stocks have accumulated single mutations, others such as the platinum fox are based on a single dominant mutation of rather recent occurrence (see Fig. M143).

Genetic alterations in industrial microorganisms have contributed very significantly to the production of antibiotics. (See Demain AL 1971 Adv Biochem Eng 1:113; Sakaguchi K, Okanishi M (Eds.) 1980 Molecular Breeding and Genetics of Applied Microorganisms, Academic Press, New York; Quesada V et al 2000 Genetics 154:421).

Figure M143. Dark platinum fox mutant (From Mohr O, Tuff P 1939 J Heredity 30:227)

Mutational Bias: Microsatellites tend to change to longer repeats sequences, thus the bias is in favor of longer repeat tracts although long repeats may not recombine if mismatches are included that impede chromosome pairing. ►microsatellite

Mutational Decay: Losing DNA sequences during evolution when their presence is no longer indispensable.

Mutational Delay: As per mutational delay, there is a time lag between the actual mutational event and the phenotypic expression because recessive mutations may show up only when they become homozygous. In organelles, the mutations must sort out before becoming visible, etc. ►sorting out, ►pre-mutation

Mutational Dissection: Analysis of a biochemical, physiological, or developmental process with the aid of mutations in the system.

Mutational Distance: The number of amino acid or nucleotide substitutions between (among) macromolecules that may indicate their time of divergence on the basis of a molecular clock. ►evolutionary distance, ►evolutionary clock

Mutational Envelope Scanning: As per mutational envelope scanning, functional similarities between proteins may be hidden because only a limited number of residues are critical for function, e.g., for ligand binding. Systematic mutations induced (e.g., by alanine scanning mutagenesis) at the critical side chains may reveal the hidden functional similarities. These hidden features may not be revealed by common biochemical, crystallographic, or nuclear magnetic resonance studies. ►homologue-scanning mutagenesis; Christ D, Winter G 2003 Proc Natl Acad Sci USA 100:13202.

Mutational Load: ►genetic load

Mutational Spectrum: The array and frequency of the different mutations that have been observed under certain conditions or in specific populations. The spectrum of auxotrophic mutations is much lower in most eukaryotes (except yeast and other fungi) than in lower and higher photoautotrophic organisms. ►mutation spectrum

Mutations Undetected: ►mutations rate

Mutator Genes: Mutator genes may be functioning on the basis of abnormal level of errors in DNA synthesis (error-prone DNA polymerase III) or abnormally low level of genetic repair due to defect of proofreading exo-nuclease in bacteria. The actual repair DNA polymerase in bacteria is pol I that also has 5′→3′ polymerase and double and single-strand 3′→5′ exonuclease capability. MutS protein recognizes DNA mismatches while MutL protein scans for nicks in the DNA and then through exonuclease action slices back the defective strand beyond the mismatch site and facilitates the replacement of the erroneous sequences with new and correct sequences. In bacteria, MutH protein recognizes mismatches not by the proofreading system of the exonucleases but by the distortion of the DNA molecules newly made. The recognition of the new strands depends on the not yet methylated A or C sites within GATC sequences in the new strands. Once the distortion is found, the mismatched base can be selectively excised. If either of these repairs fail, mutator action is observed.

Any defects in the bacterial gene *dam* (DNA methylase) may also result in high mutation rates because the correct base may be excised and replaced erroneously (see Fig. M144).

The active sites of DNA polymerases are highly mutable and the mutant enzymes may act as mutators of other genes (Patel PH, Loeb LA 2000 Proc Natl Acad Sci USA 97:5095).

Introducing point mutations in three structural domains of bacterial Pol I DNA polymerase resulted in ~8×10^4 increase of transition mutations at random sites but with preference for plasmids DNA. The desired mutations can be selectively isolated and used for synthetic chemistry, gene therapy, and molecular biology (Camps M et al 2003 Proc Natl Acad Sci USA 100:9727).

Some of the so called mutator genes of the past were actually insertion elements that moved around in the genome and caused mutation by inactivation of genes through disrupting the coding or promoter sequences or altering specific bases in the gene(s). The mutator genes *mutA* and *mutC* of *E. coli* cause the A • T→T • A and G • C→T • A transversions in the anticodon of two different copies of tRNA genes, which normally recognize the GGU and GGC codons. As a consequence, Asp replaces Gly at a rate of 1 to 2%. Defective methyl-directed mismatch repair (gene MutS) occurs in 1–3% of *E. coli* and *Salmonella* raising the mutation rate to antibiotic resistance (Rif [rifampicin], Spc [spectinomycin] and Nal [nalidixic acid]) up to hundreds of fold, depending on the strain

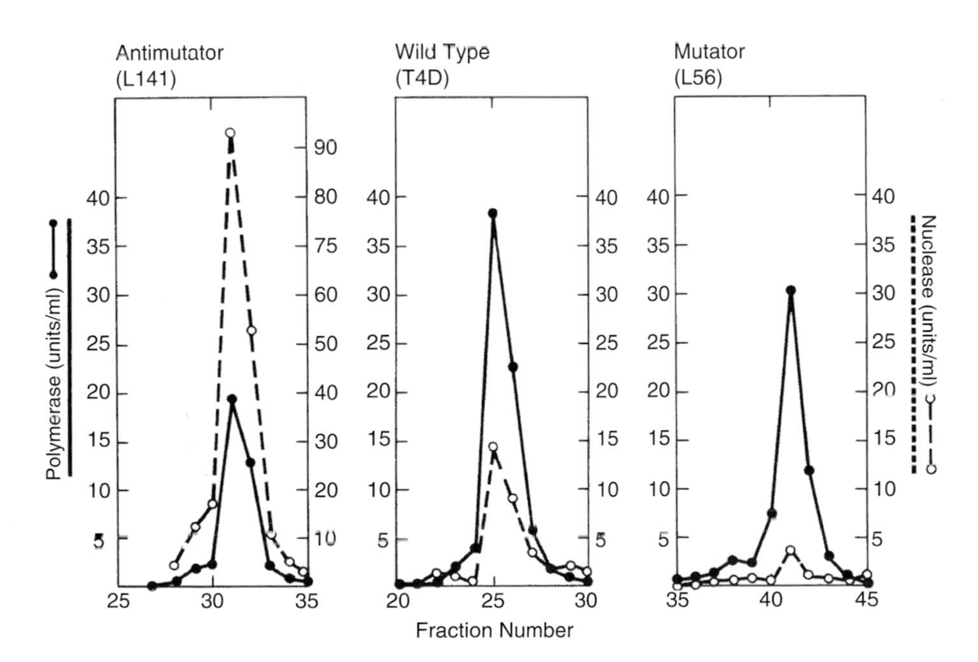

Figure M144. Mutator and antimutator activity may depend on the balance between DNA polymerase and nuclease functions. (From Muzyczka N et al. 1972 J. Biol. Chem, 247:7116)

M

of bacteria. Besides polymerase and repair defects, increase in mutation rate may be caused by anomalies of the replication accessory proteins such as RPA/RAF, PCNA, and RAD27/Rthp/Fen-1. Single-nucleotide substitutions in normal and neoplastic human tissues is two-hundred-fold different. In normal tissues, the frequency of spontaneous random mutation was exceedingly low, less than 1 × 10–8 per base pair. In contrast, tumors from the same individuals exhibited an average frequency of 210 × 10–8 per base (Bielas JH et al 2006 Proc Natl Acad Sci USA 103:18238).

High mutator activity may increase evolutionary adaptation in bacteria. Once the particular strain is well established, the high mutator activity is expected to decrease in that habitat. High mutator activity of a nuclear gene of *Arabidopsis*, *chm*, has been traced to mutation in mutS-like mismatch repair gene. The protein localized to the mitochondria and the mutations are manifested most obviously in the plastids (Abdelnoor RV et al 2003 Proc Natl Acad Sci USA 100:5968). The defect in *chm* apparently fails to correct the defects in substoichiometric shifting in the mtDNA. ►insertional mutation, ►transposable elements, ►DNA replication prokaryotes, ►DNA repair, ►antimutators, ►mutations in cellular organelles, ►mismatch repair, ►mutation detection, ►RAD, ►DNA polymerases, ►anticodon, ►tRNA, ►substoichiometric shift, ►transversion, ►amino acids, ►RPA, ►PCNA, ►RAD27, ►Mu, ►proofreading, ►cancer, ►mutation rate, ►hot spot; Giraud A et al 2001 Science 291:2606; Shah AM et al 2001 J Biol Chem 276:10824; Shinkai A, Loeb LA 2001 J Biol Chem 276:46759; Shaver AC et al 2002 Genetics 162:557; Rédei GP 1973 Mutation Res 18:149.

Mutator Phage: The best characterized mutator phage is the Mu bacteriophage. The Mu particle includes a 60 nm isosahedral head and a 100 nm tail (in the extended form) containing also base plates, spikes, and fibers. The phage may infect Enterobacteria (*E. coli, Citrobacterium freundii, Erwinia, Salmonella typhimurium*). Its DNA genetic material consists of a 33-kb (α) and a 1.7-kb (β) double-stranded sequences separated by a 3-kb, essentially single-stranded, G-loop (specifies host range). It also has variable length (1.7-kb) single-strand "split ends" (SE). Besides the coat protein genes, the *c* gene is its repressor (prevents lysis). The other genes are *ner* (negative regulator of transcription), *A* (transposase), *B* (replicator), *cim* (controls superinfection [immunity]), *kil* (killer of host in the absence of replication), *gam* (protein protects its DNA from exonuclease V), *sot* (stimulates transfection), *arm* (amplifies replication), *lig* (ligase), *C* (positive regulator of the morphogenetic genes) and *lys* (lysis). Upon lysis, 50 to 100 page particles are liberated. The Mu chromosome may exist in linear and circular forms. Mu can integrate at about 60 locations in the host chromosome with some preference. At the position of integration, 5-bp target site duplications take place. The integration events cause insertional mutation in the host. Mu causes host chromosome deletions, duplications, inversions, and transpositions. These functions require gene *A*, the intact termini of the phage and replication of the phage DNA. A related other phage, D108, has several DNA regions that are non-homologous. Its host range is the same as that of Mu. ►bacteriophages, ►temperate phage, ►Mu bacteriophage, ►insertion elements; Nakai H et al 2001 Proc Natl Acad Sci USA 98:8247.

Mutein: Mutant protein.

Muton: An outdated term meaning the smallest unit of mutation. For three decades it is known that single nucleotides or nucleotide pairs are basic units of mutation. Benzer S 1957 In: The Chemical Basis of Heredity. McElroy WD, Glas B (Eds.) Johns Hopkins University Press, Baltimore, MD, pp 70).

MutS: ►mismatch repair

Mutual-Best Blast Matches: Mutual-best BLAST matches use BLAST computer program for determining orthologous sequences in different species. ►BLAST, ►orthologous loci

Mutual Exclusiveness: An example of mutual exclusiveness is that alternative alleles at particular genetic sites in a haploid cannot exist simultaneously.

Mutualism: A mutually beneficial or alternatively a selective situation for increased exploitative association of organisms. ►symbionts, ►commensalism, ►sociogenomics; Curie CR 2001 Annu Rev Microbiol 55:357.

MVR (minisatellite variant repeat): Within different-length minisatellite allele pairs of the same size different alterations may occur at different sites and generate isoalleles (see Fig. M145).

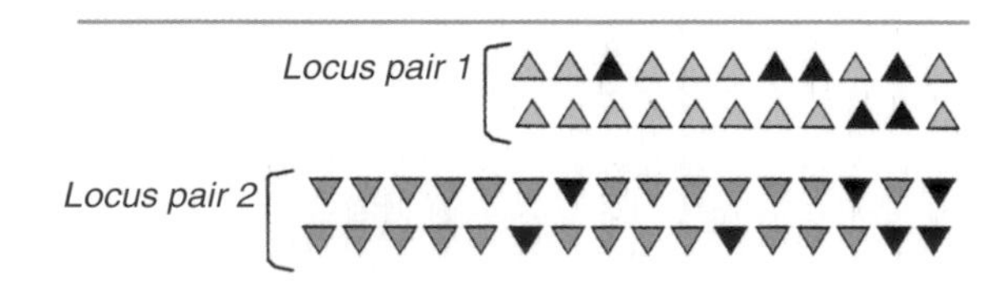

Figure M145. Minisatellite alleles

The hypervariable alterations (represented by the triangles) may be identified with aid of restriction enzyme digestion and electrophoresis and can yield profile of an individual. The majority of the mutations show polarity, i.e., the alterations are most common at the end of the minisatellite site. Many of the mutations generate alleles of different lengths as a result of recombination of the parental alleles. Mutation rate is variable depending on alleles but apparently it is independent from the length of the repeat as long as homologies are not much violated. ►minisatellite, ►DNA fingerprinting; Junge A 2001 Forensic Sci Int 119:11.

Mx Proteins: 70–80-kDa interferon-inducible GTPases of the dynamin family that interfere with the replication and transcription of negative-strand RNA viruses. Mx1: myxovirus (influenza) resistance; it is located at human chromosome 21q22.3. ►dynamin, ►replicase; Regad T et al 2001 EMBO J 20:3495.

Mxi1: A tumor suppressor gene at human chromosome 10q24–26. ►prostate cancer, ►MYC; Wang DY et al 2000 Pathol Int 50(5):373.

Mxi/Max: Heterodimeric specific transcriptional repressors. ►Max, ►Mad/Max, ►E box; Billin AN et al 2000 Mol Cell Biol 20:9945.

My: A million years during the course of evolution; MYA means million years ago. ►KYA

Myasis (myiasis): The infestation of a live body with fly larvae (maggots) such as occurs as a consequence of screw worm, a common pest of southern livestock (see Fig. M146). ►genetic sterilization

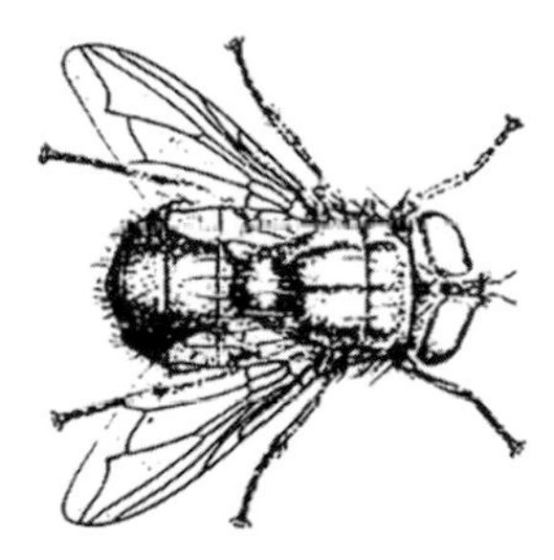

Figure M146. Adult screwworm fly. It has an average life span of three weeks and during the period of time it may travel 100 to 200 kilometers but it can survive only during the warm winters such as exist in Mexico or in the southern USA. (From Stefferud A (Ed.) 1952 Insects. Yearbook of Agriculture, USDA, Washington, DC)

Myasthenia (gravis, MG, 17p13): Generally involves muscular (eye, face, tongue, throat, neck) weakness, shortness of breath, fatigue, and the development of antibodies against the acetylcholine receptors. The genetic determination is ambiguous, generally it appears autosomal recessive. The infantile form lacks the autoimmune feature, although it also includes a defect in the acetylcholine receptor. Anticholinesterase therapy and immunosuppressive drugs may be helpful. Newborns of afflicted mothers may be temporarily affected through placental transfer. In persons with "limb girdle," pattern of muscle weakness may have small neuromuscular junction but normal acetylcholine receptor and acetylcholinesterase action (Beeson D et al 2006 Science 313:1975). ►neuromuscular diseases, ►epistasis indirect, ►IVIG, ►rapsyn, ►acetylcholine

Myb Oncogene: An avian myeloid leukemia (myeloblastosis) oncogene; its human homolog was assigned to chromosome 6q21-q23. Its product is a transcription factor. The same gene is also called AMV (*v-amv*). The MYB gene product is translated into ~75-kDa protein in immature myeloid cells and its activity is substantially reduced as differentiation proceeds.

The Myb protein (see Fig. M147), with a nuclear protein-binding leucine zipper in the regulatory domain, recognizes the 5′-PyAAC(G/Py)G-3′ core sequence and in case the C- or N- terminus or both are deficient it becomes a potent activator of leukemia, although overexpression of the entire Myb may also be oncogenic. The Myb oncogene regulates hematopoiesis. The Myb A protein is required for normal spermatogenesis and mammary gland development. Using antisense oligodeoxynucleotides for codons 2–7 may reduce its transcription. The same treatment may be effective against chronic myelogenous leukemia. Recent data indicates that the Myc/Max/Mad proteins may regulate ~15% of all the genes (Orian A et al 2003 Genes Dev 17:1101). Myb homology domains occur in multiple copies in both monocotyledonous and dicotyledonous plants and are involved in various regulatory functions, such as in transcription factors. ►E box, ►oncogenes, ►leukemias, ►AMV, ►hematopoiesis, ►antisense technologies, ►incompatibility alleles

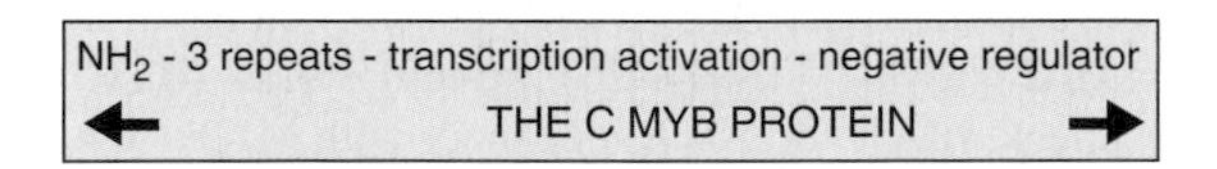

Figure M147. The C MYB protein

MYC (8q24.12-q24.13): An oncogene named after the myelocytomatosis retrovirus of birds from which its protein product was first isolated. It is a widely present DNA-binding nuclear helix-loop-helix phosphoprotein in eukaryotes. c-MYC is expressed only in dividing cells and it is an inhibitor of terminal differentiation. C-MYC induces the expression of the cell cycle activating kinase, CDK4. The MH-2 virus carries, besides the *v-myc* gene, also the *v-mil* gene, which encodes serine/threonine kinase activity and invariably causes monocytic leukemia. Most retroviruses in this group cause only the formation of non-immortalized macrophages requiring growth factors for proliferation. The cellular forms of the oncogenes are specially named after the type of cells in which, they are found. NMYC occurs in neuroblastomas (and retinoblastomas). Homologues of this gene were assigned to human chromosome 2p24 and to mouse chromosomes 12 and 5. The LMYC genes are involved in human lung cancer where one was located to chromosome 1p32 and two to the mouse chromosomes 4 and 12. Alternative splicing and polyadenylation produce several distinct mRNAs from a single gene. Metastasis is favored by the presence of a 6-kb restriction fragment of the DNA. In human chromosome 7, another form, the MYC-like one gene was detected with a 28-bp near perfect homology to the avian virus. In both man and mouse, translocation sequences were identified between the Myc gene (human chromosome 8) and several immunoglobulin genes encoding the heavy, κ and λ chains. cMyc overexpression causes telomeric association of interphase chromosomes, which may result in breakage-bridge-fusion cycles and various types of chromosomal rearrangements (Louis SF et al 2005 Proc Natl Acad Sci USA 102:9613).

(PVT1) has an activation role in Burkitt's lymphomas (caused by the Epstein-Barr virus) and plasmacytomas (neoplasia). Myc/MAX heterodimer binds to the CDC25 gene and activates transcription after binding to the DNA sequence CACGTG. This binding site is recognized by a number of MYC-regulated genes. Frequent targets of activated Myc are the cycline genes as well as ornithine decarboxylase, lactate dehydrogenase and thymosine β4. When cellular growth factors are depleted, Myc can induce apoptosis with the cooperation of CDC25. MYC in cooperation with RAS mediate also the progression of the cell cycle from G1 to the S phase through induction of the accumulation of active cyclin-dependent kinase and transcription factor E2F. MAD (Mxi1; 10q24–26) protein holds MYC in check. MAD forms a heterodimer with MAX (MAD-MAX), and this successfully competes with MYC-MAX heterodimers for transcription and thus can control malignancy. Down-regulation of the human ferritin gene by MYC leads to cellular proliferation. Myc expression may mediate Fas-FasL (Fas-ligand) interaction and an apoptosis pathway through FADD. The BIN1 protein interacts with MYC and serves as a tumor suppressor. Myc apparently affects the expression of thousands of genes and its effect can be detected by microarray hybridization and correlation analysis (Remondini D et al 2005 Proc Natl Acad Sci USA 102:6902). ►oncogenes, ►hepatoma, ►cancer, ►Burkitt's lymphoma, ►Gardner syndrome, ►PVT, ►CDC25, ►MAD, ►BIN1, ►apoptosis, ►Fas, ►FADD, ►RAS, ►CDK, ►E2F, ►CDF, ►MAX, ►ferritin, ►immunoglobulins, ►E box, ►Id protein, ►breakage-fusion-bridge cycles; Henriksson M, Lüscher B 1996 Adv Cancer Res 68:109.

Mycelium: a mass of fungal hyphae. ►hypha

Mycobacteria: are Gram-positive bacteria and cause tuberculosis (killing 3 million people annually) and responsible for leprosy (*Mycobacterium leprae,* 3.27 Mb, 1604 genes), respectively. *M. leprae* is somewhat difficult to study because it does not grow in axenic cultures. The bacteria can be propagated to sufficient quantities for analysis in the nine-banded armadillo (*Dasypus novemcinctus*). About half of its genome is inactivated as pseudogenes. The populations have much lower variability than other bacteria. The bacterium probably originated in Eastern Africa and than spread to other parts of the world, primarily to India and Brazil and Western Africa (Monot M et al 2005 Science 3008:1040). Their cell wall has low permeability and therefore they are rather resistant to therapeutic agents.

Mycobacterium tuberculosis (see Fig. M148) H37Rv genome is 4,411,529-bp and includes about 4,000 genes. (See Nature [Lond] 393:537 1998, see also corrections in Nature 396:190 1998). Through *mycobacterial protein fragment complementation* (in analogy to the yeast two-hybrid system) protein interaction network could be revealed that sheds light on virulence pathway of the tuberculosis bacterium (Singh A et al 2006 Proc Natl Acad Sci USA 103:11346). Envelope metalloprotease enzymes regulate the composition and virulence of the tuberculosis bacteria (Makinoshima H, Glickman MS 2005 Nature [Lond] 436:406). Worldwide about 2 billion people could be infected and 9 million have the active disease. The tuberculosis bacteria reside in the macrophages during latent infection and kept in check there by nitric oxide and other nitrogen intermediates (see Fig. M149). When the mycobacterial proteasome cleaves the inducible nitric oxide synthase, susceptibility to the disease increases

Figure M148. *M. tuberculosis*

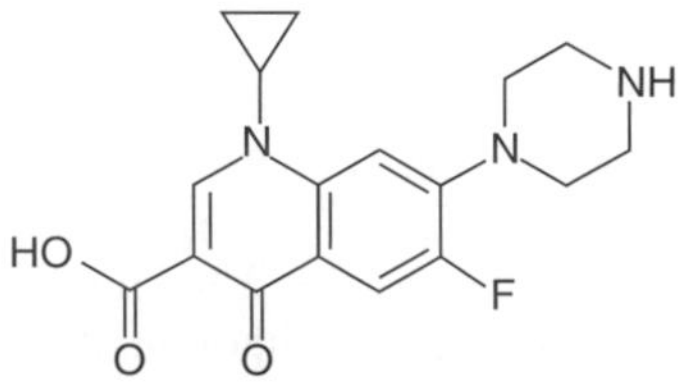

Figure M149. Cyprofloxacin (a fluoroquinolone)

(Darwin KH et al 2003 Science 302:1963). Mouse chromosome 1 harbors a gene (*sst*) for super susceptibility to tuberculosis and within the same locus there is a gene, *Intracellular pathogen resistance* (*Ipr1*). The closest human homolog to *Ipr1* is SP110 and there are three variants of it in West Africa (Gambia, Guinea-Bissau and Guinea) that convey resistance to tuberculosis. These variants lie within a 31-kb block and are in strong linkage disequilibrium (Tosh K et al 2006 Proc Natl Acad Sci USA 103:10364). The same gene conveys resistance also *Listeria monocytogenes* (Pan H et al 2005 Nature [Lond] 434:767). Infection by human immunodeficiency virus (AIDS) greatly increases the susceptibility to tuberculosis and antiretroviral drugs may provide protection thus against both diseases (Williams BG, Dye C 2003 Science 301:1535). Tuberculosis of cattle is caused by *M. bovis* and it can be transmitted also to humans and other animals. The transmission to humans is usually by not or improperly Pasteurized milk. *Mycobacterium avium* K-10 (paratuberculosis) genome has 4,829,781 bp and 4,350 open reading frames and more than 3,000 genes are homologous with the human tuberculosis bacterium but 161 sequences are unique to the bird bacterium (Li L et al 2005 Proc Natl Acad Sci USA 102:12344).

Tuberculosis is most commonly treated by isoniazid (INH) and as a result saturated hexacosanoic acid (C26:0) accumulates on a 12-kDa acyl protein carrier (see Fig. M150). This complex associates also with β-ketoacyl carrier protein synthase (KasA). Apparently INH acts by inhibiting KasA. When mutations occur in the latter, INH resistance develops. Apparently INH acts by inhibiting KasA. Nitroimidazopyran drugs are also promising against tuberculosis. The cure of tuberculosis generally requires strict multidrug therapy to avoid the emergence of drug-resistance, which is a very serious threat nowadays.

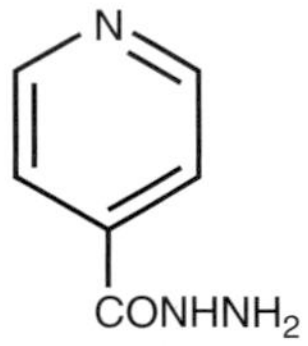

N
$CONHNH_2$

Figure M150. Isoniazid

Diarylquinoline (see Fig. M151), which apparently active on the proton pump of ATP synthase, at 0.06 μg/mL in vitro, exceeded significantly the efficacy of isoniazid and rifampin and does not appear to have serious side effects (Andries K et al 2005 Science 307:223).

Fluoroquinolones (e.g., moxifloxacin and gatifloxacin) interact with DNA gyrase and DNA

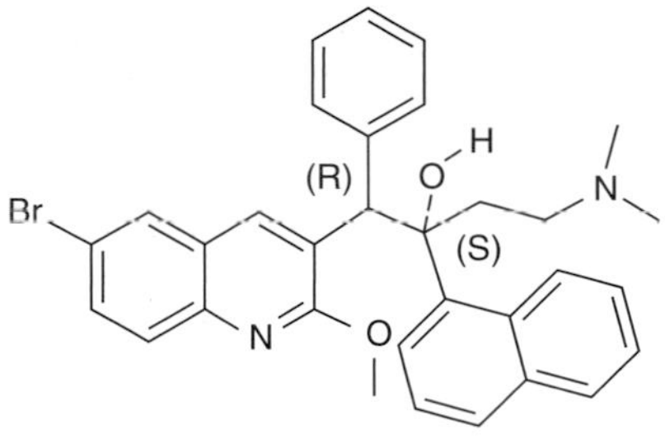

Br
(R)
O
H
N
(S)
N
O

Figure M151. Diaryquinoline

topoisomerase are effective antibacterial agents yet the MfpA bacterial plasmid-encoded protein, which contains as every fifth amino acid either leucine or phenylalanine can convey resistance even against these newer fluoroquinolones. The MfpA protein forms a right-handed quadrilateral β helix (a DNA mimic), binds to DNA gyrase and inhibits its activity and conveys resistance to fluoroquinolone in vivo (Hegde SS et al 2005 Science 308:1480).

The mouse strain DBA/2 is very susceptible to tuberculosis whereas C57BL/5 is more resis-tant. The resistance is controled by a QTL in chromosome 19 and 1. In humans natural resistance factors (controling bacterial proliferation) have been found in chromosomes 2q, 15q and Xq (Mitsos L-M et al 2003 Proc Natl Acad Sci USA 100:6610). *Mycobacterium vaccae* and *Bacillus Calmette-Guerin* (BCG, also a mycobacterium) produce cross-reactive material against *M.l.*, and *M.v.* is suited for vaccine production against leprosis. *M.l.* cannot efficiently be propagated in vitro for vaccine production and only the nine-banded armadillo would be a suitable animal host. The virulence of mycobacteria seems to be controled by the *erp* (exported repetitive protein) gene. Some major (2q32-q35, 15q11-q13) and apparently several minor genetic factors determine the susceptibility to tuberculosis and other diseases. Gene UBE3A, a ubiquitin ligase is located in the imprinted region of chromosome 15. Mycobacterial infection susceptibility genes encode the interferon-γ receptor (INFGR-1, 6q23-q24; INFGR-2, 6q23-q24), interleukin-12 β-1 chain receptor (9p13) and STAT1 (2q32). Some IFNGR-2 mutations gained new N-glycosylation sites (Vogt G et al 2005 Nature Genet 37:692).

The use of the BCG (bacillus Calmette0-Guérin) vaccine is substantially effective for childhood treatment but much less efficient for adult infections. BCG primes T_h1 (helper T cells) and promotes the release of interferon (IFN-γ) by the macrophages. Newer vaccines incorporating antigen 85A of the bacterium with some adjuvants appear promising (Young D, Dye C 2006 Cell 124:683). ►vaccine, ►*Bacillus Calmette-Guérin*, ►leprosy, ►AIDS, ►glycosylation, ►PGRS, ►interferon, ►T_h1;

Abel L, Casanova J-L 2000 Am J Hum Genet 67:274; Glickman MS, Jacobs WR Jr 2001 Cell 104:477; Russel DG 2001 Nature Rev Mol Cell Biol 2:569; Cervino ACL et al 2002 Hum Mol Genet 11:1599; *Mycobacterium bovis* genome sequence: Gernier T et al 2003 Proc Natl Acad Sci USA 100:7877; susceptibility: Fortin A et al 2007 Annu Rev Genomics Hum Genet 8:163; http://genolist.pasteur.fr/.

Mycology: the discipline of studying the entire range of mushrooms (fungi).

Mycophenolic Acid (MPA): ►IMPDH

Mycoplasma: Parasitic and/or pathogenic bacteria. They may also contaminate animal cell cultures. *Mycoplasma genitalium* was the first free-living organism to be completely sequenced by 1995, containing 580,070-bp DNA, 480 open reading frames, and 37 RNA transcribing genes; 31.8% of its known genes (127) is involved in translation.

The intact genomic DNA from *Mycoplasma mycoides* large colony (LC), virtually free of protein, was transplanted into *Mycoplasma capricolum* cells by polyethylene glycol-mediated transformation. Cells were selected for the tetracycline resistance marker of the donor, and proved that they contained the complete donor genome and were free of detectable recipient genomic sequences (Lartigue C et al 2007 Science 317:632).

Mycoplasma pneumoniae has ~0.82×10^6 bp DNA genome and 679 genes. *Mycoplasma penetrans* sequenced circular genome is 2,358,633 bp and includes 1038 predicted coding sequences, one set of rRNA genes, and 30 tRNA genes (Sasaki Y et al 2002 Nucleic Acids Res 30:5293). ►phytoplasmas, ►gene number; Glass JI et al 2000 Nature [Lond] 407:757; Flynn JL, Chan J 2001 Annu Rev Immunol 19:93; http://genolist.pasteur.fr/.

Mycorrhiza: A symbiotic association between plant roots and certain fungi. Some of the mycorrizal fungi are genetically quite peculiar in as much they are haploid but heterokaryotic with multiple nuclei and substantial differences among the nuclei (Hijri M, Sanders IR 2005 Nature [Lond] 433:160). ►nitrogen fixation, ►symbionts; Smith KP, Goodman RM 1999 Annu Rev Phytopath 37:473, nitrogen path: Govindarajulu M et al 2005 Nature [Lond] 435:819; arbuscular mycorrhiza: http://www.ffp.csiro.au/research/mycorrhiza/intro.html.

Mycosis: Infection or disease by fungi. The infection may be superficial, subcutaneous, and systemic by inhalation of the spores. Most commonly, fungi cause skin diseases. Immune-compromised individuals are particularly susceptible to mycoses.

Mycotoxins: Fungal metabolites that damage DNA. The best known are: aflatoxins, sterigmacystin, ochratoxin, zearalenone, and some penicillin toxins. ►aflatoxin

MyD88: (myeloid [bone marrow like] differentiation primary response factor, encoded in human chromosome 3p22-p21.3): An adaptor protein in the Toll — NF-κB signaling pathway. ►NF-κB, ►Toll/TRAF6, ►TIRAP

Myedema (hyperthyroidism): Myedema is also attributed to hypothyroidism caused by autosomal recessive/dominant conditions. It involves dry skin with wax-like deposits. Some HLA genes and autoimmune causes have been implicated. ►Graves disease

Myelencephalon: The part of the brain of the embryo that develops into the medulla oblongata (the connective part between the pons [brain stem]) and the spinal cord. ►brain

Myelin: A lipoprotein forming an insulating sheath around nerve tissues. ►Pelizaeus-Merzbacher disease, ►Marie-Charcot-Tooth disease, ►multiple sclerosis; Greer JM, Lees MB 2002 Int J Biochem & Cell Biol 34:211.

Myeloblast: A precursor (formed primarily in the bone marrow) of a promyelocyte and eventually a granular leukocyte. It contains usually multiple nucleoli. ►leukocyte, ►Wegener granulomatosis

Myeloblastin: A serine protease, specific for myeloblasts controlling growth and differentiation and myelogenous leukemia. ►leukemia

Myelocyte: A precursor of the granulocytes, neutrophils, basophils and eosinophils.

Myelocytoma: ►myeloma

Myelodysplasia (MDS): A recessive form of leukemia, generally associated with monosomy of human chromosome 7q. The critical deletion apparently involves chromosomal segment 7q22-q34. Peptide mass fingerprinting and quadrupole TOF MS identified two differential proteins: CXC chemokine ligands 4 (CXCL4) and 7 (CXCL7), both of which had significantly decreased serum levels in MDS, as confirmed with independent antibody assays (Aivado M et al 2007 Proc Natl Acad Sci USA 104:1307). ►leukemia, ►nucleophosmin, ►chemokine, ►CXCR, ►mass spectrum, ►proteomics, ►MALDI, ►quadrupole, ►TOFMS

Myeloma: Bone marrow cancer. CREB seems to promote myeloid transformation (Shankar DB et al 2005 Cancer Cell 7:351). Myeloma cells are used to produce hybridomas. ►Bence Jones proteins, ►monoclonal, ►antibody, ►hybridoma, ►CREB

Myelomeningocele: ►spina bifida

Myelopathy: Diseases involving the spinal chord and myelination of the nervous system. About 15 different human anomalies are involved. ►Dejerine-Sottas neuropathy, ►Charcot-Marie-Tooth disease

Myelopoiesis: The formation of granulocytic and monocytic cell lineages. ►granulocyte, ►monocyte

Myelosuppression: Myelosuppression interferes with bone marrow function and blood cell formation. It is the most common target of cancer chemotherapy drugs. These drugs are aimed at highly proliferative cells and the bone marrow produces 4×10^{11} cells daily for long periods of time.

MYF-3 (myogenic factor): MYF-3 was located to human chromosome 11-p14 and its mouse homolog, *MyoD* is in chromosome 7. MYF-3 controls muscle development, may be subject to imprinting, and may possibly be involved in the formation of embryonic tumors. The protein is a helix-loop-helix transcription factor (MASH2) and it is a mammalian member of the *achaete-scute* complex of *Drosophila*. ►imprinting, ►Mash-2, ►morphogenesis in *Drosophila*.

Myhre Syndrome: ►dwarfism

Myleran (1,4-di[methanesulphonoxy] butane): A clastogenic (causing chromosome breakage) alkylating mutagen. Also called busulfan

Mylotarg: Monoclonal antibody used for targeting acute myeloid leukemia with the toxin calicheamycin. ►leukemia, ►monoclonal antibody, ►magic bullet

Myoadenylate Deaminase: ►adenosine monophosphate deaminase

Myoblast: Muscle cell precursor. Myoblasts can be well cultured in vitro and reintroduced into animals where they fuse into myofibers. They can be used also as gene or drug delivery vehicles. The paired-box proteins, Pax3 and Pax7, are essential for the formation of skeletal muscles (Relaix F et al 2005 Nature [Lond] 435:948). ►vectors, ►dermomyotome

Myocardial Infarction: Obstruction of blood circulation to the heart; it may be accompanied by tissue damage. Infracted (dead/decaying) heart muscle cells may be replaced by transplantation of regenerating bone marrow cells. Arachidonat 5-lipoxygenase activating protein (encoded at 13q12-q13) confers increased risk of myocardial infarction and stroke. This gene overlaps the ALOX5AP lipoxygenase activating-protein (FLAP) coding region and increases the risk for stroke almost two-fold in Icelandic populations. The effect is due to the over production of leukotrienes B and inflammation of the arterial walls (Helgadottir A et al 2004 Nature Genet 36:233). Human haplotype HapK, spanning this region, confers an about three-fold greater risk of heart disease to Afro-Americans than to European-Americans (Helgadottir A et al 2006 Nature Genet 38:68). Presence of variants of the proprotein convertase subtilisin/kexin type 9 serine protease gene (PCSK9) involved substantial reduction cholesterol. The low density lipoprotein cholesterol and myocardial infarction decreased by 88% among US blacks. In whites 15% reduction in cholesterol and 47% reduction in the heart disease risks were observed among several thousands of individuals examined during a period of 15 years (Cohen JC et al 2006 N Engl J Med 354:1264). Complement-reactive protein (CRP) increases myocardial and cerebral infarct size. 1,6-bis(phosphocholine)-hexane can abrogate the infarct size and protects from cardiac dysfunction by blocking CRP action (Pepys MB et al 2006 Nature [Lond] 440:1217). Delivery to the myocardium myocytes and biotinylated insulin-like growth factor, a cardiomyocyte growth and differentiation factor in biotinylated nanofibers, improved systolic function significantly. The delivery activated Akt, decreased caspase-3 activation, and improved the expression of troponin I (Davis ME et al 2006 Proc Natl Acad Sci USA 103:8155). ►coronary heart disease, ►heart disease, ►MEF, ►lipoxygenase, ►leukotriene, ►ethnicity, ►haplotype, ►LDL, ►cholesterol coronary heart disease atherosclerosis, ►insulin-like growth factor, ►caspases, ►troponin, ►systole

Myocardium: The heavy muscles of the heart. ►heart disease; Buckingham M et al 2005 Nature Rev Genet 6:826.

Myoclonic Epilepsy: Myoclonic epilepsy occurs in different forms. The recessive juvenile EJM gene is situated within the boundary of the HLA gene complex in the short arm of chromosome 6p21 (Janz syndrome). It is characterized by generalized epilepsy with onset in early adolescence. Mutations in a subunit of the sodium channel SCN1A gene (2q24) cause slow growth from the second year of life and often become ataxic and later suffer from speech problems. Myoclonous epilepsy associated with ragged-red fibers is characterized also by epileptic convulsions, ataxia, and myopathy (enlarged mitochondria with defects in respiration). The defect was attributed to an adenine → guanine transition mutation at position

8344 in the human mitochondrial DNA involving the $tRNA^{Lys}$ gene. The recessive (EPM2A, 6q24; EPM2B, 6p22.3) myoclonous epilepsy (LaFora disease) shows up at about age 15 and results in death within ten years that after. The EPM2A disease involves defects in a protein tyrosine phosphatase (laforin protein). The EPM2B/NHLRC1 disease is due to deletions or various types of mutations in the 395 amino acid protein malin, which appears to be a single-subunit E3 ubiquitin ligase. Normally, malin ubiquitinates laforin but mutation in the gene leads to LaFora disease (Gentry MS et al 2005 Proc Natl Acad Sci USA 102:8501). Both of these proteins are associated with the endoplasmic reticulum and are normally involved in clearance of polyglucosans in dendrites and the defect disturbs neuronal synapsis (Chan EM et al 2003 Nature Genet 35:125).

The myoclonus epilepsy Unverricht and Lundborg (EPM1) is a 21q22.3 chromosome recessive with onset between ages six to 13, beginning with convulsions turning within a few years into shock-like seizures (myoclonus). The latter disease is caused by mutation in the cystatin B (cysteine protease inhibitor) gene. The EPM1 disease involves a large insertion (600–900 bp) consisting of 12-bp repeats (CCCCGCCCCGCG). The cases appeared to be initiated by pre-mutational changes of 12–13 repeats and usually by maternal transmission the "alleles" appeared stable. In the majority of cases of paternal transmission, the repeats increased. In some cases, a 18mer repeat was observed at the 5′ region of the promoter and a 15mer repeat at the 3′ region. The 18 and 15mer repeats included also six and four T and A bases, respectively. Benign familial adult epilepsy maps to 8q23.3-q24.1. Some human myoclonus diseases are apparently under dominant control. A *myoclonus* disease in cattle has an apparent defect in glycine-strychnine receptors. In mouse, the spastic mutation in chromosome 3 affects the α-1 glycine receptors, whereas the α-2 receptor defect is X-linked. The incidence of epilepsy, especially in some breeds, is more common in dogs than in humans (Lohl H et al 2005 Science 307:81). ►epilepsy, ►trinucleotide repeats, ►mitochondrial disease in humans, ►pre-mutation, ►ion channels

Myoclonus: Sudden, involuntary contractions of the muscle(s) like in seizures of epilepsy or sometimes as a normal event during sleep. Myoclonus-dystonia, a dominant disease, is located to 7q21 and the gene encodes ε-sarcoglycan (Müller B et al 2002 Am J Hum Genet 71:1303). ►myoclonic epilepsy, ►dystonia, ►sarcoglycan

MyoD: Muscle-specific basic helix-loop-helix protein that regulates muscle differentiation and the cessation of the cell cycle. Its DNA binding consensus is CANNTG in the promoter or enhancer of the genes controlled. MyoD may activate p21, and p16 and promotes muscle-specific gene expression. When MyoD is phosphorylated by cyclinD1 kinase (Cdk), it may fail to transactivate muscle-specific genes. ►p21, ►p16, ►cell cycle, ►helix-loop-helix, ►Myf3, ►cyclin, ►Cdk, ►myogenin, ►MEF, ►enhancer; Tedesco D, Vesco C 2001 Exp Cell Res 269:301.

Myofibril: Muscle fiber made of actin, myosin and other proteins (bundle of myofilaments). ►satellite cells

Myofibromatosis, Juvenile: Multiple fibroblastic tumors on the skin, muscles bones, and viscera in young infants due to autosomal recessive or dominant mutation. Prognosis for survival in severe cases is low.

Myogenesis: Muscle development.

Myogenin: A protein involved in muscle development. MyoD activates myogenin and overcomes the inhibitor of DNA-binding (helix-loop-helix) gene *Id*, and other muscle gene transcription and differentiation is set on course. ►MyoD, ►MYF5, ►MRF, ►MEF; Sumariwalla VM, Klein VH 2001 Genesis 30(4):239.

Myoglobin: A single polypeptide chain of 153 amino acids, attached to a heme group and functions in the muscle cells to transport oxygen for oxidation in the mitochondria. Recurrent myoglobinuria may result by a microdeletion in the mitochondrially-encoded cytochrome C oxidase (COX) subunit III. Some Antarctic fishes do not have myoglobin, and homozygous myoglobin knockout mice function in a practically normal manner. ►hemoglobin, ►mitochondrial disease in humans

Myo-Inositol: An active form of inositols. They are widely used as phospholipid head-groups (see Fig. M152). They are essential for signaling, membrane trafficking, etc. ►inositol, ►phosphoinositides, ►embryogenesis somatic, ►phytic acid

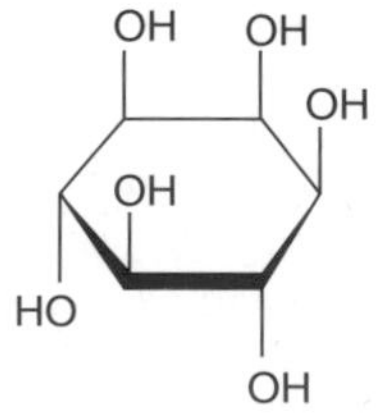

Figure M152. Myoinositol

Myokymia: Hereditary spasmic disorder of the muscles, caused by potassium channel defects. ►ion channels

Myoneurogastrointestinal Encephalopathy (mitochondrial myoneurogastrointestinal encephalopathy, MNGIE, 22q13.32-qter): A recessive defect in a nuclear gene, however, mitochondrial DNA aberrations my also accompany it. The onset usually sets in after the second decade of life and the clinical symptoms involve atrophy of the muscles (ophthalmoplegia), multiple neural disorders, lactic acidosis, etc. ►ophthalmoplegia, ►mitochondrial diseases in humans

Myopathy: A collection of diseases affecting muscle function, controlled by autosomal dominant, autosomal recessive, X-linked, and mitochondrial DNA. Besides the weak skeletal muscles, ophthalmoplegia (paralysis of the eye muscles) usually accompanies it. The recessive homozygotes for myotonic myopathy are also dwarf and have cartilage defects and myopia. Two types are known with carnitine palmitoyle transferase deficiency (CPT1; mutation at 1pter-q12 (CPT2) involves myoglobinuria, especially after exercise or fasting. Another recessive myopathy is based on succinate dehydrogenase and aconitase. Phosphoglycerate mutase deficiencies (human chromosome 10q25 and 7p13-p12) also involve myoglobinuria. X-linked forms display autophagy (cytoplasmic material is sequestered into lysosome associated vacuoles) or slow maturing of the muscle fibers, or swelling and hypertrophy in the quadriceps muscles, respectively. Either insertion, deletion, or frameshift mutations in the ITGA7 integrin gene cause congenital myopathy (human chromosome 12q13). The Xq28-linked tubular myopathy is caused by defects in the myotubularin tyrosine phosphatase. The centronuclear myopathy is due to missense mutation in dynamin 2 and causes defect in centrosome function (19p13.2; Bitoun M et al 2006 Nature Genet 37:1207) and another similar myopathy due to anomalies of the myogenic factors (MYF5, MYF6) is encoded at 12q21 (Cupelli L et al 1996 Cytogenet Cell Genet 72:250). Dominant and recessive base substitution mutations in the skeletal muscle actin, ACTA1 gene (1q42), involve thin muscle fibers and severe, hypotonia, muscle weakness, feeding, and breathing difficulties and most commonly death during the first few months after birth. Occasionally, some afflicted individuals may survive to adulthood. ►mitochondrial diseases in humans, ►cartilage, ►myopia, ►carnitine, ►Batten-Turner syndrome, ►actin, ►desmin, ►integrin, ►dynamin, ►Sbf1, ►troponin, ►inclusion body myopathy, ►myositis, ►muscular dystrophy, ►cardiomyopathies, ►adenosine monophosphate deaminase

Myopia (nearsightedness): Myopia is caused by the increased length of the eye lens in the front-to-back dimension, and focusses the refracted light in front of the retina (nearsightedness). Most forms are under polygenic control with rather high heritability (around 0.6). Autosomal dominant, autosomal recessive (infantile), and X-linked forms have also been suggested. Some studies indicate significant positive correlation between myopia and intelligence. Myopia may be concommitant with various syndromes. ►eye diseases, ►farsighted, ►human intelligence

Myosins: Contractile proteins that form thick filaments in the cells, hydrolyze ATP, bind to actin, and account for the mechanics of muscle function, organelle movement, phagocytosis, pinocytosis, cell movement, RNA transport, phototransduction, signal transduction, etc. Myosins are activated by kinases (MLCK) and deactivated by phosphatases. MLCK is inhibited by PAK. Members of the 11 families of myosin are represented in amoebas, insects, mammals, and plants. The majority of myosin motors move in one direction toward the plus (+) end of actins. A conformational change at the actin-binding site (converter) of myosin class VI may facilitate movement of the lever in the minus (−) direction. This class of myosin has a 50-amino acid insertion at the converter region. The three-dimensional structure of myosin 5 has been determined (Liu J et al 2006 Nature [Lond] 442:209; Thirumurugan K et al 2006 Nature [Lond] 442:212). ►filament, ►microfilament, ►myofibril, ►phagocytosis, ►pinocytosis, ►Usher syndrome, ►PAK, ►RAC, ►motor proteins, ►kinesin, ►deafness, ►spindle; Homma K et al 2001 Nature [Lond] 412:831; crystal structure of myosin VI motor: Ménétrey J et al 2005 Nature [Lond] 435:779; http://www.mrc-lmb.cam.ac.uk/myosin/myosin.html.

Myositis: The name of some myopathies. ►myopathy

Myostatin Gene (human chromosome 2q32.1 recessive, GDF8): The inactivation (partial deletion) of the myostatin gene substantially increases muscle development in mammals. This mutation is found in the Belgian Blue and Piedmontese cattle breeds. Transforming growth factor-β is involved. The condition is also named 'double muscling' and 'growth/differentiation factor 8'. Inhibition of myostatin leads to robust muscle regeneration and this fact may mean therapeutic value for muscular dystrophy (Wagner KR et al 2005 Proc Natl Acad Sci USA 102: 2519). ►insulin-like growth factors; Lee SJ, McPherron AC

2001 Proc Natl Acad Sci USA 98:93067; Schuelke M et al 2004 N Engl J Med 2004 350:2682.

Myotome: The area of the somites that forms involuntary muscles. ►somites; Gros J et al 2004 Dev Cell 6:875.

Myotonia: The recessive myotonia congenita (chloride channel, CLCN1) is apparently in human chromosome 7q35 (Becker disease) and there is also a dominant form (Thomsen disease, 7q35). Both involve difficulties in relaxing the muscles. The various forms (M. fluctuans, M. permanens, paramyotonia congenita, K^+-activated myotonia) may involve Na^+ ion channel malfunctions. ►periodic paralysis, ►ion channels

Myotonic Dystrophy (DM): A human disorder expressed as wasting of head and neck muscles, eye lens defects, testicular dystrophy, speech defects, frontal balding, frequently heart problems, and behavioral effects. It is controlled by a dominant gene (DMPK) at the centromere of human chromosome 19q13.2-q13.3 and it is caused by $(CTG)_n$ expansion at the 3′ untranslated region (Steinert disease). DM2 is a $(CTG)_n$ expansion at 3q21 (Bachinski LL et al 2003 Am J Hum Genet 73:835). The DMPK related kinase, DK serine/threonine kinase, is at 1q41-q42. The prevalence is highly variable. It seems that DM may be a contiguous gene syndrome. In some isolated populations, it may occur in 1/500 to 1/600 proportion while in other populations the occurrence is about 1/25,000. The manifestation of the symptoms is enhanced in subsequent generations, and more are found in children of affected mothers than those of affected fathers. This observation may be due, however, to anticipation. The condition is dominant and it is quite polymorphic. Molecular evidence indicates that the DNA sequences concerned are unstable in the 3′-untranslated tracts downstream of the last exon of the protein kinase gene. The difficulties involving nucleosome assembly in DNA containing CTG triplet repeats in the 3′-untranslated region of a serine/threonine kinase gene seem to be concerned. It appears that the CUG repeats of the RNA transcript bind proteins (CUG-BP), which interfere with the proper splicing of the transcript. This toxic RNA can be silenced and DM can be normalized (Mahadevan MS et al 2006 Nature Genet 38:1066). The crystal structure of CUG repeats is known (Mooers BHM et al 2005 Proc Natl Acad Sci USA 102:16626). A troponin protein has been implicated with the binding. On the basis of linkage with chromosome 19 centromeric genes, prenatal tests are feasible. Mutation frequency was estimated within the $1.1–0.8 \times 10^{-5}$ range. Ribozyme-mediated trans-splicing of the trinucleotide repeats have been tried to reduce the expansion. Mutation in the DM1 gene may result in the excessive binding of transcription factors to mRNA and mRNA is sequestered in the nucleus and thus the DM protein kinase gene is not translated sufficiently (Ebralidze A et al 2004 Science 303:383). A second myotonic dystrophy gene (DM2) was mapped to a 10-cM region of human chromosome 3q21 (ZNF9). The latter does not display CTG expansion (~15 kb) seen in DM1 but large expansion (~44 kb) of the tetranucleotide repeat (CCTG). Although the amplification involves untranslated sequences, neighboring genes (chloride channel subunit 1, insulin receptor, insulin intolerance) are spliced incorrectly in mammals (Houseley JM et al 2005 Hum Mol Genet 14:873). The expanded mRNAs are not distributed normally to the cytoplasm and are not translated into the needed protein in sufficient amounts nor do they interfere with the function of nuclear RNA-binding proteins. In the brain, hyperphosphorylated short tau protein may occur. ►muscular dystrophy, ►Pompe's disease, ►Werdnig-Hoffmann disease, ►Duchenne periodic paralysis, ►Schwartz-Jampel syndrome, ►Steinert disease, ►anticipation, ►prenatal diagnosis, ►mental retardation, ►imprinting, ►fragile sites, ►trinucleotide repeats, ►troponin, ►transsplicing, ►contiguous gene syndrome, ►tau; Liquori CL et al 2001 Science 293:864; Sergeant N et al 2001 Hum Mol Genet 10:2143.

Myotubes: Aggregate of myoblasts into a multinucleated muscle cell. ►cadherin, ►integrin, ►fertilin, ►meltrin, ►acetylcholine

Myristic Acid (tetradecanoic acid, $CH_3(CH_2)_{12}COOH$): A natural 14-carbon fatty acid without double bonds. Myristoylation are common in oncoproteins, protein serine/threonine and tyrosine kinases, and protein phosphatases in the α-subunit of heterotrimeric G proteins, transport proteins, etc. By myristoylation of nascent proteins, cell membranes can be targeted with the aid of myristoyl-CoA:protein N-myristoyltransferase enzyme. ►fatty acids, ►kinase, ►G protein; Farazi TA et al 2001 J Biol Chem 276:39501.

Myt1: ►Cdc2

Myxobacteria: Slime-secreting bacteria, which may form cysts that contain lots of cells.

Myxococcus xanthus: The genome of myxococcus xanthus is a single circular chromosome with

9,139,763 bases of GC- rich DNA (69%) predicted to encode 7,331 coding sequences. ►Myxobacteria; http://www.xanthusbase.org.

Myxoviruses: A group of RNA viruses. ►RNA viruses

m/z (m/e): Mass-to-charge. ►mass spectrum

Historical vignettes

HJ Muller was the second geneticist recipient of the Nobel Prize, on 31 October 1946, for his research on the influence of X-rays on genes and chromosomes. At the Cold Spring Harbor Symposia on Quantitative Biology (9:163, in 1941) he stated:

"We are not presenting…negative results as an argument that mutations cannot be induced by chemical treatment."

"…it is not expected that chemicals drastically affecting the mutation process while leaving the cell viable will readily be found by our rather hit-and-miss methods. But the search for such agents, as well as the study of the milder, 'physiological' influences that may affect the mutation process, must continue, in the expectation that it still has great possibilities before it for the furtherance both of our understanding and our control over the events within the gene".

Timothy Taylor (archeologist) in 2001 in Nature (Lond) 411:419.

"Perhaps I have no business commenting on genetics, but what I am really interested in is an explanatory imperialism, which threatens to subsume even archeology."

MZ: Monozygotic twin. ►twinning

MZEF: A free, exon detection program based on quadratic discriminant function for multivariate statistical pattern recognition. ►GENESCAN, ►Genie, ►FGENE; Zhang MQ 1997 Proc Natl Acad Sci USA 94:5495.

N

n: The gametic chromosome number, 2n = the zygotic chromosome number. ►x, ►genome, ►polyploid

N: ►Newton

N_2, N_3...N_n: Designate backcross generations. ►B

N51: A serum-inducible gene, identical to KC, MGSA and *gro*.

N50 Length: The largest length such that 50% of all base pairs are contained in contigs of this length or larger. In the sequenced human genome N50 appeared to be 82 kb by the HGP and 86 by the Celera map. ►contig

N Nucleotides: ►immunoglobulins

N Value Paradox: A concept similar to the C value paradox indicating that the complexity of an organism is not directly proportional to the number of its genes because the same genome may code for a much larger proteome and the complexity may depend also on genetic networks. ►C value paradox, ►proteome; Claverie JM 2001 Science 291:1255.

NAA (α-naphthalene acetic acid): A synthetic auxin. ►plant hormones, ►somatic embryogenesis, see formula given under ►naphthylacetic acid

NAADP (nicotinic acid adenine dinucleotide phosphate): The NAADP may coordinate agonist-induced Ca^{2+} signaling and other metabolic processes (see Fig. N1). ►agonist

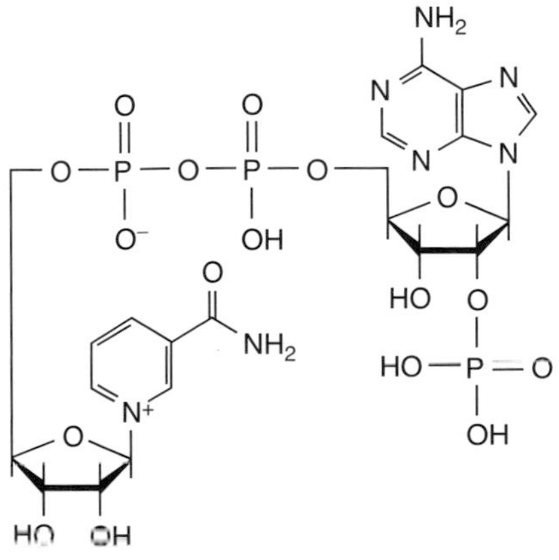

Figure N1. NAADP

NAC (nascent-polypeptide associated complex): ►protein synthesis

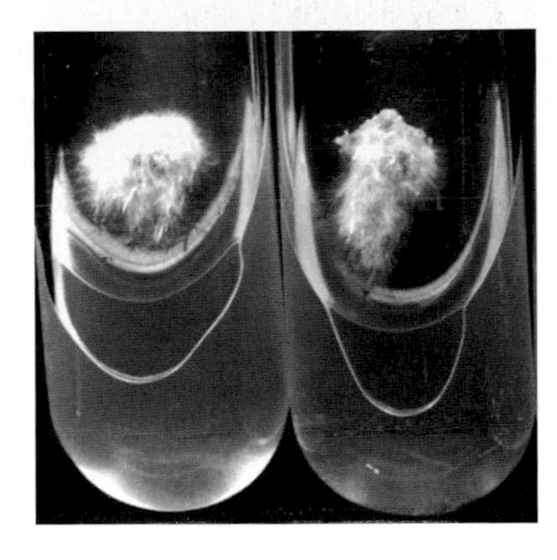

Figure N2. Plant callus transferred to naphthaleneacetic acid medium develops abundant roots

NACHT: NACHT are ATPases involved in apoptosis and MHC gene activation. ►NTPase, ►apoptosis, ►MHC; Koonin EV, Aravind L 2000 Trends Biochem Sci 25(5):223.

NAD (β-nicotinamide-adenine dinucleotide [diaphorase]): A cofactor of dehydrogenation reactions and an important electron carrier in oxidative phosphorylation. DIA1 (cytochrome b5 reductase) is in human chromosome 22q13-qter, DIA2 in chromosome 7, and DIA4 in 16q22.1.

NADH: The reduced form of NAD. ►NAD

$NADP^+$ (nicotine adenine dinucleotide phosphate): An important coenzyme in many biosynthetic reactions. The enzyme reduced nicotinamide-adenine dinucleotide oxidase is an important producer of reactive oxygen species in the cell. Its deficiency may lead to chronic granulomatous disease. ►NAD, ►NADH, ►granulomatous disease, ►Rossmann fold

NAE Genes: Non-annotated expressed genes. ►annotation, ►AE genes, ►ANE genes

Nail-Patella Syndrome (Turner-Kieser syndrome, Fong disease): Characterized by malformation or absence of nails, poorly developed patella (a bone of the knee), defective elbows and other bone defects, kidney anomalies, collagen defects, etc. The dominant locus was assigned to human chromosome 9q34.1 encoding a LIM domain protein. ►collagen, ►LIM domain, ►patella aplasia-hypoplasia

NAIP (neuronal apoptosis inhibitory protein, 5q12.2-q13.3): The partial deletion of the NAIP leads to spinal muscular atrophy. ►apoptosis, ►spinal muscular atrophy; Crocker SJ et al 2001 Eur J Neurosci 14:391.

NAIS (nucleotide analog interference suppression): A short nucleotide sequence is ligated to a special RNA site to reveal the functional consequences of this modification, e.g., for self-splicing of introns.

(See Ryder SP Strobel SA 1999 J Mol Biol 291:295; Szewczak AA et al 1998 Nature Struct Biol 5:1037).

Naïve: The word naïve means unaffected. For example, a B lymphocyte that has not yet been exposed to a foreign antigen.

NAK (NF-κB activating kinase): An IKK kinase that mediates NF-κB activation in response to growth factors and phorbol ester tumor promoters that stimulate protein kinase C-ε. ►IKK; Tojima Y et al 2000 Nature [Lond] 404:778.

Naked DNA: Naked DNA is not embedded in protein; it is pure DNA. It may be used as a low-efficiency, low-immunogenicity, and short-lived DNA vector.

Nalidixic Acid (1-ethyl-1,4-dihydro-7-methyl-4-oxo-1,8-naphthydrine-3-carboxylic acid): An antibacterial agent that inhibits DNA synthesis. It is also used in veterinary medicine against kidney infections.

NALP: A member of the NLR family of proteins forming a caspase-activating complex. ►caspase

Nance-Horan Syndrome: The gene encompasses ~650-kb genomic DNA at Xp22.13 and encodes a 1,630 amino acid nuclear protein, which regulates eye, tooth, brain, and craniofacial development (Burdon KP et al 2003 Am J Hum Genet 73:1120).

Nanism: ►Mulbrey nanism, ►dwarfism

Nanobacteria (nannobacteria): These are ~30–100 nm mineralized particles (see Fig. N3) isolated from fetal bovine serum (FBS), human serum, and kidney and dental pulp. Their volume is about 1/1000 that of the majority of common bacteria. The living nature of these particles and their nucleic acid content could not be unambiguously confirmed. (See Cisar JO 2000 Proc Natl Acad Sci USA 97:11511).

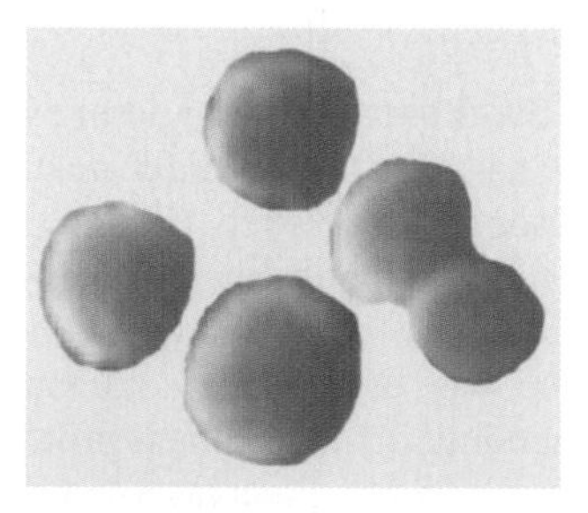

Figure N3. Nanobacteria

Nanocrystals, Semiconductor: Semiconductor nanocrystals may be used in the same way as fluorochromes. They can be excited at any wavelength shorter than their emission peak. Many sizes of nanocrystals can be excited by one wavelength resulting in the simultaneous production of different colors. Nanocrystals can be readily synthesized with different properties and shapes. ►fluorochromes, ►microfluidics, ►FISH, ►quantum dot; Holmes JD et al 2001 J Am Chem Soc 123:3743; Hamad-Schifferli K et al 2002 Nature [Lond] 415:152; Yin Y Alivisatos P 2005 Nature [Lond] 437:664.

Nanoelectrospray Mass Spectrometry: ►electrospray

NANOG: A human chromosome 12p region encoded homeodomain protein directing the infinite propagation, and sustaining pluripotency, in embryonic stem cells. The mouse homolog is quite similar. Primarily in human cells, the C-terminus domain is involved in gene transactivation whereas in mice the N-terminal is more active (Oh JH et al 2005 Exp Mol Biol 37:250). ►homeodomain, ►transactivator, ►pluripotency, ►stem cells

Nanoparticle-based Bio-Bar Code (MNP probe, barcode amplification, BCA): This bio-bar code can be applied for the ultra-sensitive detection of proteins at attomolar (10^{-18}) concentration. Magnetic microparticle probes are equipped with monoclonal antibodies and with DNA unique to the protein of interest. The outline of the procedure is diagrammed (see Fig. N4). If a polymerase chain reaction follows the separation, the sensitivity may be as low as 3 attomolar. This procedure can be six orders of magnitude more sensitive than conventional assays.

Nanoparticles: Usually less than 1 μm colloidal capsules of biodegradable or non-biodegradable vehicles for the protected delivery of molecules into the cells of an organism. They are usually internalized by endocytosis and are transported to the lysosomes where the contents may or may not be saved from degradation. This technology may facilitate the detection of DNA with extremely high selectivity (Park S-J et al 2002 Science 295:1503; Martin CR, Kohli P 2003 Nature Rev Drug Discovery 2:29). Hepatitis B virus envelope L particles to form hollow nanoparticles. Such vectors home in on liver cells and deliver genes and drugs to hepatocytes, specifically. They pose no virus risk. (Yamada T et al 2003 Nature Biotechnol 21:885). Single-walled carbon nanotubes can deliver cargoes across cellular membranes without harming the cells. Nucleotides can be delivered to cell nuclei. Near-infrared laser pulses can selectively destroy cells by heat after sensitization of the nanotubes with folate receptor tumor markers (Shi Kam NW et al 2005 Proc Natl Acad Sci USA 102:11600). Technology is under development for engineering different atomic and molecular structures (Barth JV et al 2005 Nature [Lond] 437:671).

Silica nanoparticles, encapsulating fluorescent dyes and covered by cationic-amino groups, can efficiently complex with DNA. This also protects DNA from digestion by DNase I. Such a vector

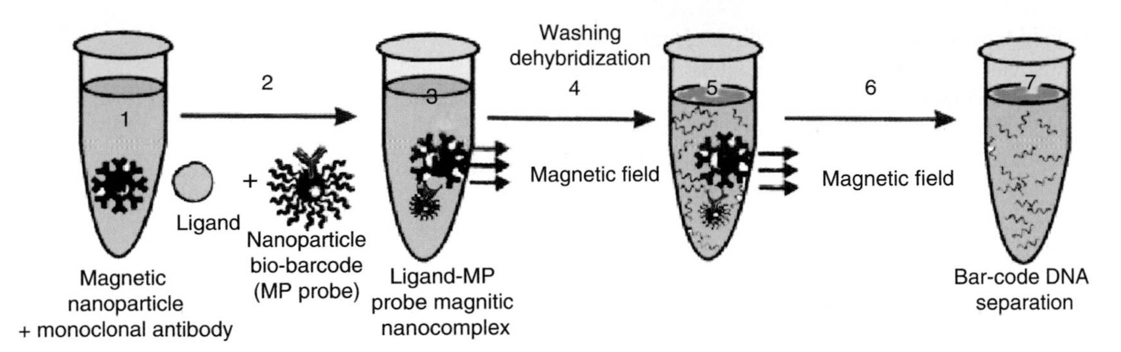

Figure N4. Outline of nanoparticle and bio-bar code technology for protein detection. 1. Magnetic microparticle with monoclonal antibody (Y). (2). Ligand or target protein and gold particle equipped with bar-code DNA and protein plus polyclonal antibody (Y). 3. The MMP probe and NP probe are combined in a centrifuge tube (a magnified image of this step is shown at left). 4. Purification and exposed to a magnetic field. 5. The NP probe is dehybridized (heat) and in a magnetic field (6) the bar-code DNA is seperated and collected at step 7. The DNA that anchored the protein is analyzed in a chip system for protein identification. The system has orders of magnitude greater sensitivity than conventional tests. (Modified after Nam J-M *et al.* 2003 Science 301: 1884; Geoganopoulou DG et al 2005 Proc. Natl. Acad. Sci. USA 102:2273)

system permits monitoring of the fate of the particles, which are incorporated into the nucleus. It is promising for targeted therapy and real time monitoring of drug action (Roy I et al 2005 Proc Natl Acad Sci USA 102:279). ►biomimics, chart in diagram (see Fig. N4); nanoparticles environmental–health hazard paper summaries: http://icon.rice.edu/research.cfm.

Nanophthalmos: A rare human developmental anomaly involving recessive or weak dominant mutation in the MFRP gene at 11q23.3 resulting in two small eyes that are functional without few major structural changes. The distance between lens and retina is very short, causing the focal point of the eye to be well behind the retina resulting in extreme farsightedness (hyperopia). Secondary complications occur and glaucoma may develop. The protein encoded is homologous to Tolloid proteases and the Wnt-binding domain of the Frizzled transmembrane receptors. ►microphthalmos, ►Tolloid, ►Wnt, ►farsighted, ►frizzled, ►glaucoma; Sundin OH et al 2005 Proc Natl Acad Sci USA 102:9553.

Nanopore Technology: An upcoming development for extremely rapid detection of single-stranded DNA, RNA, and nucleotides. It may be applicable to sequencing these macromolecules. (See Deamer DW, Akeson M 2000 Trend Biotechn 18:147; see also articles in Science 290:1524 ff).

Nanotechnology: This technology generally uses atomic force microscopy or scanning tunneling microscopy to manipulate objects at the level of individual molecules with the aid of special tweezers, pipettes, and nanolithography dip pens. This technology may be exploited for the manipulation of gene chips for construction of branched DNA structures and other novel DNA motifs, including those that appear during crossing over, etc. Using polymers, metal oxides, semiconductor material like cadmium selenide and carbon nanotubes with special coating can facilitate tissue/cell targeting, drug delivery, imaging capabilities and present medical applications of 25 to 70-nm size particles. Using dendrimers with folate can facilitate targeting folate-receptor-rich cancer cells with nanoparticles containing methotrexate, without poisoning normal cells. Oligonucleotide-modified gold nanoparticles have higher affinity to complementary nucleic acids than unmodified oligonucleotides. They are resistant to nucleases and cells take them up easily and can be used for intracellular gene regulation (Rosi NL et al 2006 Science 312:1027; Famulok M, Mayer G 2006 Nature [Lond] 239:666.) reviewed the ingenious applications of nanotechnology. Gold nanoparticles of 10–100 nm are "cameleon-like" molecules inasmuch as the separated molecules are red in color, but when they aggregate they become blue. Gold nanoparticles can be prepared with attached short DNA strands. Then a DNA aptamer, which can bind adenosine, is added and the aptamer links two or more gold nanoparticles by virtue of complementarily to the DNA strands, which are attached to the gold particle. The aggregation turns the color of gold from red to blue-purple. The addition of adenosine to the aptamer-gold aggregates brings about a conformational change resulting in the dissociation of the nanoparticle network and, as a result, the gold particles change back to red color. This change is also an indication of the presence of adenosine. The aptamer still associated with a gold particle forms a fold for adenosine binding and its three-dimensional structure can be analyzed by nuclear magnetic resonance spectroscopy. The technology is also applicable for the detection of 50–500 micromole/liter cocaine or other molecules. A modification of the system uses quantum dots, which produce fluorescent probes for biodetection, say nucleic acids. Microbeads with quantum dots can be constructed for exhibiting specific fluorescence for DNA sequences. Fluorescence resonance energy transfer (FRET) can alter the emission spectrum and the color produced, and manage different specificities. These new developments are very promising for various applications. A self-assembling nanofiber scaffold permits the regeneration of injured nerve axons and the knitting together of the injured brain tissue in hamsters and probably in other species (Ellis-Behnke RG et al 2006 Proc Natl Acad Sci USA 103:5054).

Recently, nanotechnology increasingly came under attack because of the fear of the unknown consequences of its applications. The bases of the charges were not well defined (grey goo, green goo), although nanoparticles have some potentially harmful effects when they land at undesirable locations in the body. The actual risks of this rapidly developing industry—expected impact of $1 trillion by 2015—requires closer study (Service RF 2005 Science 310:1609; Maynard AD et al 2006 Nature [Lond] 444:267). ►atomic force microscopy, ►scanning tunneling microscopy, ►DNA chips, ►extracellular matrix, ►amphiphile, ►dendrimer, ►methotrexate, ►PAMAM, ►Z DNA, ►labeling, ►quantum dot, ►FRET; Seeman NC 1998 Annu Rev Biophys Biomol Struct 27:225; Zandonella C 2003 Nature [Lond] 423:10; for the status of nanobiotechnology: Nature Biotechnol. 2003 Vol 21. issue 10; DNA nanotechnology: Condon A 2006 Nature Rev Genet 7:565; Toxic potentials of nanopartricles: Nel A et al 2006 Science 311:622); http://www.pa.msu.edu/cmp/csc/nanotube.html; http://www.nano.gov/; http://www.nanohub.org.

Nanowires: Can be produced from silicon in between specific electrodes in a diameter of 1/5,000 of the human hair. Such electrodes may permit direct electrical detection of small molecules without any labeling. This technology may serve drug discovery. (See Wang WU et al 2005 Proc Natl Acad Sci USA 102:3208).

Naphthaleneacetic Acid (naphthylacetic acid): ►NAA

Naphthylacetic Acid: The same as NAA (see Fig. N5).

Narcissism: A psychological anomaly of excessive self-appreciation or erotic self-interest.

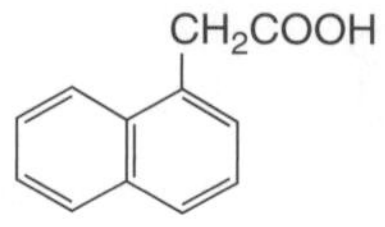

CH_2COOH

Figure N5. 1-Naphthylacetic acid

Narcolepsy (17q21): A pathological frequent sleep, commonly associated with hallucinations, loss of muscle tone, and paralysis. It may be associated with other syndromes such as cataplexy. Cataplexy may occur in the Prader-Willi syndrome, in the Niemann-Pick disease Type C, and in the Norrie disease. Its sporadic incidence is about 2×10^{-3}. Among first degree relatives, narcolepsy is 20–40 times higher than in the general population. Deficiency of hypocretin (orexin, hypothalmus-specific neuroexcitatory peptides) in the cerebrospinal fluid is the suspected cause. The prepro-orexin gene was assigned to 17q21. Susceptibility to narcolepsy is associated with the HLA-DQB1 region yet there is no firm evidence of an autoimmune cause for this anomaly. ►autoimmune diseases, ►orexin, ►apnea, ►sleep, ►HLA, ►cataplexy; Mignot E et al 2001 Am J Hum Genet 68:686; Chabas D et al 2003 Annu Rev Genomics Hum Genet 4:459.

Naringenin: A trihydroxyflavonone; with naringin it causes the bitter flavor in grapefruit.

Narp Syndrome: A complex disease involving neuronal defects, ataxia, seizures, feeble-mindedness, and developmental retardation caused by mutation in subunit 6 of mitochondrial ATPase. ►mitochondrial diseases in humans

Narrow-Sense Heritability: ►heritability

NARS: Nutritional Adequacy Ratios.

NarX: An *E. coli* kinase affecting nitrate reduction by regulator protein NarL. ►nitrate reductase, ►nitrogen fixation; Wei Z et al 2000 Mol Plant Microbe Interact 13:1251.

NAS (nonsense-associated altered splicing): NAS may be executed by skipping an exon, choosing, alternate splice sites, or insertion of an intron as a consequence of stop codons. ►nonsense codon, ►splicing, ►intron, ►ESE, ►NMD; Liu H-X et al 2001 Nature Genet 27:55; Li CM et al 2001 Eur J Hum Genet 9(9):685.

NASBA (nucleic acid sequence based amplification): An RNA strand in the presence of a primer is amplified by reverse transcriptase. Then, the RNA strand is removed by RNase H. The resulting cDNA, using RNA primers, is amplified by (T7) RNA polymerase into sufficient quantities of RNA. ►RT-PCR, ►PCR, ►LCR; Borst A et al 2001 Diagn Microbiol Infect Dis 39(3):155.

Nascent: Nascent means just being born or synthesized, e.g., the mRNA still associated with the ribosomes. The molecule is not yet combined with any other molecule and may be highly reactive.

Nasopharyngeal Carcinoma: ►Epstein-Barr virus

Nastic Movements: Nastic movements in plants do not follow the direction of the causative source, e.g., the folding of leaves in response to night or cold. ►tropism, ►circumnutation

NAT (negative regulator of activated transcription): A 20-polypeptide (approx.) complex in human cells including homologs of yeast proteins (Srb, Med, RGR, CDK8) negatively regulating transcription of RNA polymerase II. A defect or deficiency of the NAT2 gene may increase cancer risk. ►Srb, ►mediator, ►RGR; Gadbois EL et al 1997 Proc Natl Acad Sci USA 94:3145.

Native: Original to a particular region, a natural form of a substance.

Native Americans: People who mainly speak the Amerind languages (Eskimo-Aleuts and Na-Dene are different) and whose migration to the American continent across the Bering land bridge (when during the ice age the water level receded) apparently started 11,000 years ago and was followed by two additional waves 9,000 and 4,000 years ago. This idea of populating the American continent is somewhat tenuous and better information is provided by the four mitochondrial DNA haplotypes (A, B, C, D) that are distributed widely in the Western Hemisphere. In addition, a fifth, somewhat debated original haplotype, X, also occurs at much limited frequencies. The same four mitochondrial haplotypes are also present in Chinese, Mongolian, and Tibetan populations, and the A, C and D haplotypes occur in Siberian populations. The mitochondrial haplotypes may occur up to 91% in some populations of North Asia (see Fig. N6). The mitochondrial haplotypes were identified in ancient DNAs extracted from pre-Columbus skeletal remains. Two Y chromosomal haplotypes C (~5%) and Q (~76%) are also characteristic of Native Americans. The Y haplotypes are uncommon elsewhere, except in northern Asia where the frequencies are 28% and 18%, respectively. The limited genetic evidence suggests that perhaps only a single wave of migration from Asia occurred, about 15,000 to 20,000 years ago. The molecular information might have been biased by severe population reductions after migration and after contact with European immigrants due to wars,

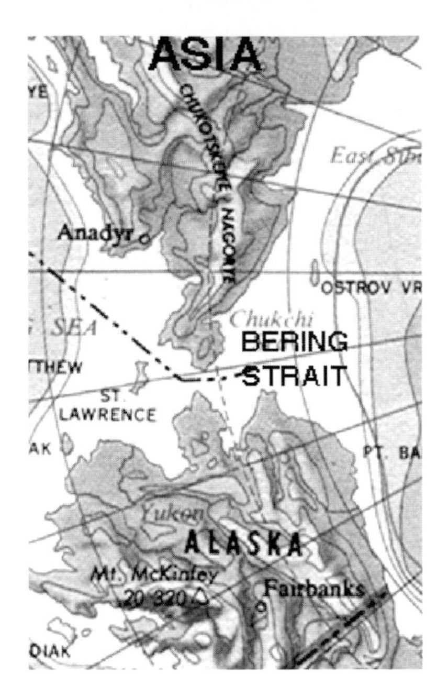

Figure N6. Northern Asia and Alaska separated by the Bering Strait

genocide, and new diseases. Interestingly, genetic variability is lower in Native Americans that in other populations of other continents. This reduced variation, however, does not seem to be lower than that existing before the arrival of the Europeans. Alcoholism, diabetes, obesity, and heart disease are relatively common among Native Americans, but these conditions have rather complex hereditary bases and their origin is hard to trace. In addition, life style and environmental factors such as poverty may have confounding effects. Other diseases might have been caused by recent acquisition of the mutations through admixture because the R1162X mutation of cystic fibrosis in the Zuni tribe is also common among Italians and Spaniards. Native Americans developed unique architecture, industry, and art (see Fig. N7) hundreds of years before the beginning of the Columbian period. The great sixteenth-century European painter, Albrecht Dürer, remarked: "...never in my born days have I seen anything that warmed my heart as much as these things" [art objects]. It is a great loss to human history that this remarkable culture declined and disappeared during the centuries. (For a comprehensive review see: Mulligan CJ et al 2004 Annu Rev Genomics Hum Genet 5:295).

Native Conformation: The natural active structural form of a molecule.

Natural Antibody: Produced "spontaneously" without deliberate immunization. Immunoglobulins M, G, and A may be involved. A natural antibody may assist

Figure N7. Pre-Columbian silver ring

in antigen uptake and antigen presentation by B lymphocytes through complement. ►antibody, ►innate immunity, ►immmunoglobulins, ►immune reaction, ►complement

Natural Compounds: Biological products used for self-defense, and biomedical and industrial purposes: http://bioinformatics.charite.de/supernatural/.

Natural Immunity: ►immunity

Natural Killer Cell (NK): ►killer cell, ►T cell

Natural Selection: The actions of forces in nature that maintain or choose the genetically fittest organisms in a habitat. The *fundamental theorem of natural selection* is "The rate of increase in fitness of any organism at any time is equal to its genetic variance in fitness at that time". It is also mathematically expressed by $\Sigma\alpha\ dp = dt\Sigma\Sigma'(2pa\alpha) = Wdt$; where α = the average effect on fitness of introducing a gene, a = the excess over the average of any selected group, W = fitness, and $dt\Sigma'(2pa\alpha)$ the sum of increase of average fitness due to the progress of all alleles considered (according to RA Fisher 1929). An analysis of 3,377 potentially informative human gene loci indicated positive selection, i.e., in 304 (9%) there appeared rapid amino acid substitution within the species. In contrast, in 6,033 potentially informative loci of humans and chimpanzees, 813 (13.5%) indicated a paucity of amino acid substitutions, i.e., weak negative selection/balancing selection. Positive and negative selection varied greatly among the various biological functions. Transcription factors evolved rapidly whereas cytoskeletal proteins displayed little amino acid divergence between humans and chimpanzees (Bustamante CD et al 2005 Nature [Lond] 437:1153). A comparison of the human and chimpanzee genomes facilitates the assessment and timing of natural selection. Long haplotypes indicate more recent origin; lactase alleles in European populations (where the frequency is ~77%) are surrounded by a much longer haplotype than in African populations indicating rapid and recent positive selection for lactose tolerance. ►selection, ►fitness, ►cost of evolution, ►mutation neutral; Fischer RA 1958 The Genetical Theory of Natural Selection, Dover, New York; Edwards AWF 2002 Theor Population Biol. 61:335; generalization to biochemistry, chemical kinetics: Vlad MO et al 2005 Proc Natl Acad Sci USA 102:9843; molecular signature of natural selection: Nielsen R 2005 Annu Rev Genet 39:197; positive selection in the human lineage: Sabeti PC et al 2006 Science 312:1614.

Nature and Nurture: Those hereditary (nature) and environmental factors (nurture), which cooperatively mold the actual appearance of an organism. Francis Galton first used this term in science in 1874. (Although W. Shakespeare in *The Tempest* (1612) speaks of "…a born Devil, on whose nature nurture can never stick.") According to the more modern concept of R. Woltereck (1909) the genes assure a "reaction norm" upon which nutrition, climate, education, etc., act and form the phenotype. The relative impact of nature and nurture varies depending on the attributes. Some traits and faculties are under almost complete genic determination; others are more influenced by the environment in the broad sense. The environmental component of the response to pharmaceuticals (e.g., effect of food on calcium channel regulator proteins) may be critical for optimal dose in effective, individual medication. ►heritability, ►genetic networks, ►QTL, ►ion channels

Navajo Neuropathy: An apparently rare disease involving loss of pain and temperature sensation in the limbs due to poor myelination of the nerve fibers.

Navel Oranges: These oranges are commonly seedless but being "navel" may not rule out seed development since some navel citruses produce normal number of seeds. The Washington navel orange and the Satsuma mandarin are completely pollen sterile and that is the cause of the seedlessness when foreign pollen is excluded. The navel oranges (and grapefruits) have two or three whorls of carpels, resulting in a fruit-in-fruit appearance. This abnormal carpel formation may also interfere with pollination. ►seedless fruits, ►orange

Navigation: An ability of animals to migrate to their destination by using different cues. Birds, in order to find their wintering or breeding sites, rely on either the stars or the sun or the earth's magnetic field in a combinatorial manner. They may rely on geographical landmarks and smell too. Homing pigeons (*Columbia livia*) use compass orientation and landmark guidance in a system of simultaneous or oscillating dual control, although older birds can rely on landmark alone (Biro D et al 2007 Proc Natl Acad

Sci USA 104:7471). The Monarch butterflies (*Danaus plexippus*) (see Fig. N8) rely on the polarizing light in the UV spectrum of sunshine during their seasonal travel for up to 4,000 km. The opsin protein expressed in their eyes senses the light. A cryptochrome pigment connects the neural pathway to the circadian rhythm (Sauman I et al 2005 Neuron 46:457); two cryptochrome genes have already been identified in the Monarch genome. ►opsins, ►cryptochrome, ►circadian rhythm; Reppertr SM 2006 Cell 124:233.

Figure N8. Monarch

B-Back Test: The B-back test of working memory is to ascertain neurocognitive functioning of the frontal lobe by testing the ability to recall an earlier event or stimulus after a following event or stimulus.

NBM Paper: ►diazotized paper

NBT: Nitroblue tetrazolium used as a chromogen (0.5 g in 10 mL, 70% dimethyl formamide).

NC (negative cofactors): Negative cofactors interfere with the TFIIB transcription factor binding to the preinitiation complex. NC2 plays dual roles and also stimulates activator-dependent transcription through contact with TBP. ►transcription factors, ►TBP, ►PIC, ►preinitiation complex, ►DSTF; Cang Y, Prelich G 2002 Proc Natl Acad Sci USA 99:12727.

NcOR: ►sirtuin

NCp7: A viral nucleocapsid protein, which serves as a chaperone for the folding of the RNA. It also enhances recombination of two single-stranded RNAs within the capsid. (See Takahashi K et al 2001 J Biol Chem 276:31274).

N-CAM (neural cell adhesion molecule): A Ca^{2+}-independent immunoglobulin-like protein, which binds together cells by homophilic means (i.e., by being present on neighboring cells). NCAM may activate NF-κB. NCAM1 is encoded at 11q23.1, NCAM2 at 21q21, and L1CAM at Xq28. The latter mutation involves symptoms similar to the Kallmann syndrome and the MASA syndrome: a mental retardation, aphasia (a speech and writing defect), walking anomaly, and abnormal thumb position. Other cell adhesion defects may be coded at other locations. ►ICAM, ►neurogenesis, ►NF-κB, ►Kallmann syndrome

NCAM (N-CAM): Neural cell adhesion molecule. ►N-CAM; Thomaidou D et al 2001 J Neurochem 78:767.

NCBI (National Center for Biotechnology Information): Maintains nucleotide sequence information on genes and clones. ►GenBank, ►GSDB, ►databases

NCoA: ►signal transduction

N-CoR (nuclear receptor corepressor): A protein of M_r 270K binding to the ligand-binding domain of thyroid hormone and retinoic acid receptors by means of its carboxy-terminal. The binding of this protein to these hormone receptors mediates a ligand-independent transcriptional repression. These hormone-receptors are transcriptional repressors for their target genes without their cognate ligands. N-CoR seems to be associated with SIN3 and its binding proteins. ►Cor-box, ►Sin3, ►SMRT, ►signal transduction, ►histone deacetylase, ►SANT, ►chromatin remodeling; Guenther MG et al 2001 Mol Cell Biol 21:6091.

NCR (natural cytotoxic receptors): ►killer cells

ncRNA: ►RNA noncoding

N-Degron: The degradation signal of a protein, the NH_2-terminal amino acids, and an internal lysine residue of the substrate protein. At that lysine, several ubiquitin molecules form a multi-ubiquitin chain. The degradation may be mediated by a G-protein. ►degron, ►ubiquitin, ►N-end rule, ►G proteins, ►PEST; Suzuki T, Varshavsky A 1999 EMBO J 18:6017.

NDF: ►heregulin

nDNA: Nuclear DNA.

NDUFV (NADH-ubiquinone-oxidoreductase flavoprotein): A part of the mitochondrial oxidoreductase system and responsible for some of the symptoms in Alexander's and Leigh's diseases. ►Alexander's disease, ►Leigh's encephalopathy, ►mitochondrial disease in humans

Neanderthal (Neandertal) People: Hominids who lived about 30,000 to 200,000 years ago in France and Southern Germany and the Middle-East. Recent archaeological evidence argues for the chronological and potential demographic and cultural interactions with modern (Aurignacian, upper Paleolithic era ~35,000–20,000 B.C.E.) human populations at the Chatelperronian-type site in east-central France (Gravina B et al 2005 Nature [Lond] 438:51). They had large jaws and heavy bones and may have developed later (about 30,000 years ago) into the Cro-Magnon men that showed more similarity to present-day humans. Their precise relationship to *Homo sapiens* is not known but DNA evidence indicates a dead-end detour from human evolution. Based on an mtDNA (extracted from fossil bones) analysis revealed 22—36-bp difference from modern humans whereas the same mtDNA among pairs of humans varies within the range of 1–24. There is no evidence for existing Neanderthal sequences in the modern human mtDNA (Serre D et al 2004 PloS Biol 2:313) although, on the basis of currently available evidence, some low-level admixture cannot be ruled out. On the basis of stratigraphic, radiocarbon, and archaeological evidence, coexistence and interactions between Neanderthal and anatomically modern populations in Western Europe is inferred (Mellars P et al 2007 Proc Natl Acad Sci USA 1004:3657). Other archaeologists still question the view of the overlapping existence of Neandertals and the Aurignacians (early human populations in the Near-East and Europe 33,000–23,000 years ago). Early modern Europeans reflect both their predominant African, early modern human ancestry and a substantial degree of admixture of indigenous Neandertals. Due to limitations inherent in ancient DNA, this process is largely invisible at the molecular level. This seems, however, apparent in the bone morphological record (Trinkaus E 2007 Proc Natl Acad Sci USA 104:7367). The difference from chimpanzees and gorillas is much greater than from modern humans. The common ancestors of the Neanderthals and modern humans date back an estimated 550,000 to 690,000 years. The one million base pairs of DNA fragments extracted from ~38,000-year-old bones have been sequenced by a new version of pyrosequencing (Margulies M et al 2005 Nature [Lond] 437:376; note corrigendum Nature [Lond] 441:120) in the laboratory of Svante Pääbo (Green RE et al 2006 Nature [Lond] 444:330; Stiller M et al 2006 Proc Natl Acad Sci USA 103:13578; Noonan JP et al 2006 Science 314:1113). The Neanderthal sequences that were 30 base or longer were aligned on the sequences of each human chromosome with an average of 3.61 bases per 10,000 bases. The particular sample extracted from the Neanderthal bone fragment covered the X chromosome by 2.18 bases per 10,000 whereas by 1.62 bases/10,000 on the Y chromosome. This indicated that the bone fragment belonged to a Neanderthal male. The average divergence of the human linage from the Neanderthal genome is ~7.9% and from that, at 95% confidence interval, the divergence could be assumed to take place about 569,000 years ago. The ancestral effective population size of the Neanderthals might have been more similar to that of present-day humans than to that of other apes, which is larger. When sufficient information becomes available on the entire Neanderthal genome it may become possible to decipher their biology. Perhaps, by 2008, the draft of the entire Neanderthal genome will be within reach.

If enough DNA fragments can be extracted from the ancient bones and hooked up in sequence according to the modern human tract, the Neanderthal genome could be reconstituted. Such a genome could be transferred to an enucleated human egg, inserted into a human female "incubator" and carried to term. Such a series of procedures could recreate the extinct species if moral objections would not prevent such a project. Sequencing the Neanderthal genome may at least provide insight into human evolution (Dalton R 2006 Nature [Lond] 442:238); ►hominidae, ►ancient DNA, ►Neanderthal, ►B.C.E.; Ovchinnikov IV et al 2000 Nature [Lond] 404:490; Scholz M et al 2000 Am J Hum Genet 66:1927; Krings M et al 2000 Nature Genet 26:144; Klein RG 2003 Science 299:1525; Harvati K et al 2004 Proc Natl Acad Sci USA 101:1147; http://www.krapina.com/; http://www.biocrawler.com/encyclopedia/Neanderthal.

Nearest Neighbor Analysis: A technique to determine base sequences in oligonucleotides; a procedure of historical interest for DNA sequencing which gained new usefulness in designing DNA binding ligands for antisense therapy. The DNA is synthesized with 5′-P^{32}-labeled nucleotides. The sequence is then digested with micrococcal endonuclease and by spleen phosphodiesterase. In the digest, specific 3′-P^{32} mononucleotides and oligonucleotides are found thus indicating which base was nearest to the original radioactively labeled nucleotide (e.g., adenylic acid). Nearest-neighbor analysis of protein complexes may reveal the patterns of subunit association (Yi T-M, Lander ES 1993 J Mol Biol 232:1117). ►antisense RNA, ►endonuclease, ►phosphodiesterase; Josse J et al 1961 J Biol Chem 236:284; Crevel G et al 2001 Nucleic Acids Res 29:4834.

NEB: ►copia

Nebenkern (paranucleus): A cellular body; generally a mitochondrial aggregate resembling the cell nucleus by microscopical appearance situated in the flagellum of the spermatozoon.

Nebularine (9-β-D-ribofuranosyl-9H-purine): A natural product of some fungi and Streptomyces with an antineoplastic effect (see Fig. N9). Its subcutaneous LD50 for rodents varies from 220 mg/kg in rat, 100 mg/kg in mouse and 15 mg/kg in guinea pig. ►LD50

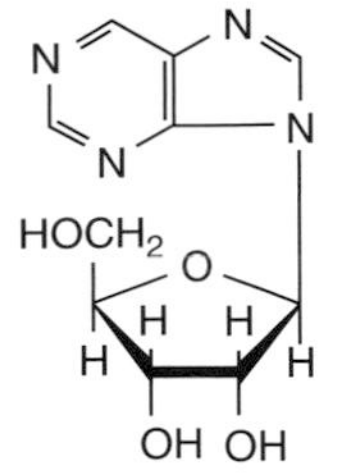

Figure N9. Nebularine

Nebulin: An actin-associated large protein in the skeletal muscles, built of repeating 35-residue units. ►titin

Necrophagous: An adjective for organisms that thrive by eating dead tissues.

Necropsy (autopsy): The examination of the body after death.

Necrosis: The death of an isolated group of cells or part of a tissue or tumor. It generally involves inflammation in animals because of the release of cell components toxic to other cells. Necrosis is different from apoptosis, which involves elimination of cells no longer needed, and it is a normally programmed cell death. ►apoptosis, ►hypersensitive reaction

Necrotic: The adjective for necrosis. ►necrosis

Necrotroph: The pathogen that kills the host and feeds on its dead tissue. ►biotroph

Nectar: A sweet plant exudate, fed on by insects and small birds.

Nectins: Nectins in four isoforms represent an immunoglobulin-like cell-cell adhesion system, which organizes adherens junctions cooperatively with the cadherin-catenin system. ►adherens junctions; Mizoguchi A et al 2002 J Cell Biol 156:555.

NEDD: ICE-related proteases. NEDD4–1 is a proto-oncogenic ubiquitin ligase that negatively regulates PTEN (Wang X et al 2007 Cell 128:129). ►ICE, ►apoptosis, ►cancer classification, ►PTEN; Murillas R et al 2002 J Biol Chem 277:2897.

Neddylation: To degrade protein(s) with the aid of NEDD, a ubiquitin-like molecule. ►COP, ►NEDD; Stickle NH et al 2004 Mol Cell Biol 24:3251; Bornstein G et al 2006 Proc Natl AcadSci USA 103:11515.

Need for Eugenic Reform: According to R. A. Fisher (The Genetical Theory of Natural Selection, 1929): "The various theories, which have thought to discover in wealth a cause of infertility, have missed the point that infertility is an important cause of wealth". ►eugenics

Needleman-Wunsch Algorithm: The Needleman-Wunsch algorithm finds optimal global alignment of macromolecule-building blocks. ►algorithm; Laiter S et al 1995 Protein Sci 4:1633.

Negative Binomial: $(q - p)^{-k}$ after expansion $P_x = [(k + (x - 1)] \frac{R^x}{q^k}$ where p = m/k, m = mean number of events, q = 1 + p, px = probability for each class, R = p/q = m/(k + m), x = number of events/class, and *k* must be determined by iterations using z_i scores and approximate *k* values until the z_i becomes practically zero. The detailed procedure cannot be shown here (see Rédei GP, Koncz C 1992 p 16. Methods in Arabidopsis Research, In: Koncz C et al (Eds.) World Scientific, Singapore). The negative binomial distribution resembles that of the Poisson series. It may provide superior fit to data that are subject to more than one factor, each affecting the outcome according to the Poisson distribution, e.g., mutations occur according to the Poisson series but their recovery is following an independent Poisson distribution, etc. ►Poisson distribution, ►distributions

Negative Complementation: An intraallelic complementation when the polypeptide chain translated on one cistron of a locus interferes with the function of the normal polypeptide subunit(s) encoded by the same or another cistron of the multicistronic gene. Thus, there is, in fact, an interference with the expression of the locus concerned. ►allelic complementation, ►complementation, ►hemizygous ineffective, ►cistron; Garen A, Garen S 1963 J Mol Biol 7:13.

Negative Control: The prevention of gene activity by a repressor molecule. The gene can be turned on only if a ligand molecule (frequently the substrate of the enzyme, which the gene encodes) binds to the repressor resulting in moving the repressor from the operator site or from an equivalent position. ►*lac* operon, ►*tryptophan* operon, ►lambda phage, ►DNA binding proteins, ►DNA-binding domains

Negative Cooperativity: The binding of a ligand or substrate to one subunit of a multimeric protein precludes the binding to another. (See Horovitz A et al 2001 J Struct Biol 135(2):104).

Negative Dominant: ►dominant negative, ►negative complementation (See Müller-Hill B et al 1968 Proc Natl Acad Sci USA 59:1259).

Negative Feedback: ►feedback inhibition

Negative Interference: ►coefficient of coincidence, ►interference, ►rounds of matings

Negative Numbers in Nucleotide Sequences: These numbers indicate the position of bases upstream of the position (+1) where translation begins. ►upstream

Negative Regulator: A negative regulator suppresses or reduces transcription or translation in a direct or indirect manner. Such systems may use antisense technology, genetic suppressors, and SETGAP (Singhi AD et al 2004 Proc Natl Acad Sci USA 101:9327). ►*Lac* operon, ►lambda phage, ►NAT, ►SETGAP; Maira SM et al 2001 Science 294:374.

Negative Selection of Lymphocytes: ►positive selection of lymphocytes

Negative Staining: Negative staining is used for the study of macromolecules. The electron microscopic grid with the specimen on it is exposed to uranyl acetate or phosphotungstic acid that produces a thin film over it except where the macromolecule is situated. The electron beam "illuminates" the non-covered macromolecules while the metal-stained parts appear denser. This then gives a negative image of viruses, ribosomes, or other complex molecular structures. ►stains, ►electron microscopy

Negative-Strand Virus: Their genomic and replicative intermediates (plus strand) exist as viral ribonucleoprotein (RNP). For replication they require viral RNA polymerase and precise 5′ and 3′ termini for both replication and packaging. ►replicase, ►positive strand virus; Pekosz A et al 1999 Proc Natl Acad Sci USA 96:8804.

N

Negative Supercoil: A double-stranded DNA molecule twisted in the opposite direction as the turn of the normal, right-handed double helix (e.g., in B DNA). ►supercoiled DNA, ►Z DNA

Neighbor Joining Method: A relatively simple procedure for inferring bifurcations of an evolutionary tree. Nucleotide sequence comparisons are made pair-wise and the nearest neighbors are expected to display the smallest sum of branch lengths. ►evolutionary distance, ►evolutionary tree, ►least square methods, ►four-cluster analysis, ►unrooted evolutionary trees, ►transformed distance, ►Fitch-Margoliash test, ►DNA likelihood method, ►protein-likelihood method, ►gene tree, ►population tree; Saitou N, Nei M 1987 Mol Biol Evol 4:406; Romano MN, Weigend S 2001 Poult Sci 80:1057.

Neighborliness: A statistical method for the estimation of the position of two taxonomic entities in an evolutionary tree. ►evolutionary tree, ►transformed distance, ►Fitch-Margoliash test; Charleston MA et al 1994 J Comput Biol 1(2):133.

Neisseria gonorrhoeae, N. meningitidis: Gram-negative bacteria. Strain B (2,272,351 bp) and Strain A (2,184406 bp) of *N. meningitidis* are responsible for inflammation of the brain membranes (meninges); blood poisoning (septicemia) has been completely sequenced. This bacterium—on the average—has three genomes per cell (Tobiason DM, Seifert HS 2006 PLoS Biol 4(6):e185). Normally, *N. meningitides* is a harmless commensal organism, inhabiting the nose and the pharynx (the back part of the oral cavity). In a small fraction of humans, the bacteria invade the blood stream and cross the blood-brain barrier. These invasive cells contain a filamentous prophage. The bacteria secrete the phage through pili, which can invade other cells and can cause a deadly disease within hours (Bille E et al 2005 J Exp Med 201:1905). Serogroup A causes large epidemics in sub-Saharan Africa whereas B and C are responsible for sporadic outbreaks worldwide. *Neisserias* display great abilities for antigenic variation making vaccine development difficult. The sequenced genomes may facilitate defense measures against these pathogens. ►antigenic variation, ►vaccines, ►prophage, ►BBB, ►commensalisms, ►pilus; Merz AJ, So M 2000 Annu Rev Cell Dev Biol 16:423; Tettelin H et al 2000 Science 287:1809; Parkhill J et al 2000 Nature [Lond] 404:502.

NELF: A negative protein factor of mRNA elongation by polymerase II. ►transcript elongation, ►DSIF, ►TEFb, ►DRB; Ping YH, Rana TM 2001 J Biol Chem 276:12951.

Nemaline Myopathy: A collection of muscle fiber gene defects. (Nemaline = thread or rod-like.) In nemaline myopathy, the majority of the muscles may be affected and severity may vary from intrauterine death to relatively mild anomalies. The heart muscles are usually not affected. The dominant tropomyosin-3 is encoded at 1q22-q23. The recessive alpha skeletal muscle myofilament disease gene (actin myopathy ACTA1) is at 1q42.1. The recessive nebulin gene (2q22) encodes the filaments in the sarcomeres. ►heart diseases; North KN et al 1997 J Med Genet 34:705.

Nematodes: The very large number and different species of nematodes are both model organisms in biological research and important parasites of agricultural and horticultural crops. ►*Caenorhabditis*, ►*Pristionchus pacificus*, ►*Photorhabdus*, ►*Xenorhabdus*; nematode genomics: Mitreva M et al 2005 Trends Genet 21:573; nematode resistance of plants: Williamson VM, Kumar A 2006 Trends Genet 22:396.

NEMO (NF-κB essential modulator, Xp28): The γ subunit of IKK. NEMO links innate immunity to chronic intestinal inflammation (Nenci A et al 2007 Nature [Lond] 446:557). ►IKK, ►NF-κB, ►incontinentia pigmenti, ►Crohn disease; Courtois G et al 2001 Trend Mol Med 7:427; Aradhya S et al 2001 Hum Mol Genet 10:2557.

N-End Rule: The half-life of a protein is determined by the amino acids at the NH_2 end. Three N-terminal amino acids—aspartate, glutamate, and cysteine—are arginylated by arginine transferase. The oxidation of the N-terminal cysteine is essential for arginylation and nitric oxide is required for the process. In eukaryotes, E3 ligase (N-recognin) recognizes the N-end substrate and mediates the degradation by proteasome. In *E. coli* chaperone ClpA in complex with ClpP peptidase and ClpS adaptor are required for proteolysis (Erbse A et al 2006 Nature [Lond] 439:753). ►N-degron, ►protein degradation within cells, ►destruction box, ►proteasome, ►Johanson-Blizzard syndrome; Rao H et al 2001 Nature [Lond] 410:955; Huy R-G et al 2005 Nature [Lond] 437:981.

Neobiogenesis: The idea that living organisms arose from organic and inorganic material. It is also used to describe the formation of new organelles. ►spontaneous generation

Neocarzinostatin (NCS): A naturally occurring enediyne antibiotic. The NCS chromophore attacks, specifically, a single residue in a two-base DNA bulge. It attacks HIV type I RNA and other viruses. ►enediyne, ►bulge, ►HIV; Maeda H et al (eds) 1997 Neocarzinostatin: The Past, Present, and Future of an Anticancer Drug. Springer, New York.

Neocentric: ►neocentromere

Neocentromere: An extra spindle-fiber attachment site in the chromosomes in eukaryotes of certain genotypes, different from the regular centromere position yet containing 100–200-bp tandem repeats of satellite DNA in plants (see Fig. N10). In barley, the centromeric repeat is not required for centromeric function (Nasuda S et al 2005 Proc Natl Acad Sci USA 102:9842). The human neocentromeres are not repetitive and do not contain the 171-bp α-satellites present in the regular human centromeres. The human neocentromeric DNA sequences do not share commonality and, therefore, they may be the result of epigenetic alterations rather than basic sequence specificity. The proteins and functions associated with human neocentromeres are shared with the same of the regular centromeres. Neocentromeres occur at low frequencies. Some genetic constitutions (e.g., in maize and rye) promote neocentromere formation. The neocentromeres are not surrounded by heterochromatin. From the maize neocentromeres centromeric protein CENP-C is absent. CENP-A, a histone-like centromeric protein, can be mislocalized in *Drosophila* and can nucleate the formation of functional kinetochores at ectopic sites, i.e., neocentromeres. This results in anomalous segregation, aneuploidy, and growth defects (Heun P et al 2006 Developmental Cell 10:303). ►centromere, ►kinetochore, ►CENP, ►centromere activation, ►centromere silencing, ►breakage-fusion-bridge cycles, ►holocentric, ►preferential segregation, ►knob, ►human artificial chromosomes; Warburton P et al 2000 Am J Hum Genet 66:1794; Hiatt EN et al 2002 Plant Cell 14:407; Amor DJ, Choo KHA 2002 Am J Hum Genet 71:695.

Neocortex: The neocortex is supposed to control motor, sensory, and cognitive functions of the brain. ►brain

Neo-Darwinian Evolution: Evolution is attributed to small random mutations accumulated by the force of natural selection and characters acquired as a direct adaptive response to external factors have no role in

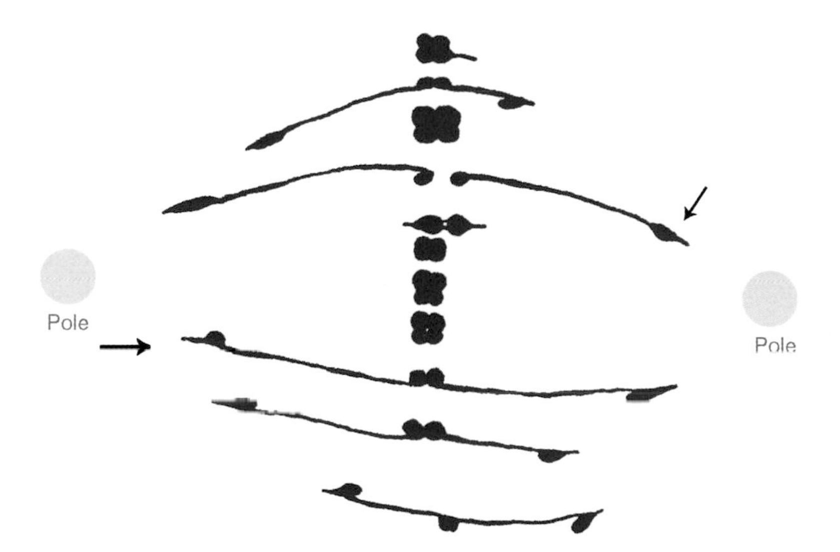

Figure N10. Precocious neocentromeres (near terminal knobs) at metaphase I in 5 chromosome pairs. (After MM Rhoades)

evolution. ►mutation beneficial, ►mutation neutral, ►selection, ►fitness, ►Darwinism, ►directed mutation; Matsuda H, Ishii K 2001 Genes Genet Syst 76(3):149.

Neofuntionalization: The acquisition of a new function, generally by duplication and modification of a gene. ►subfunctionalization

Neo-Lamarckism: ►Lamarckism

Neolithic Age: About 7,000 years ago when humans turned to agricultural activity from hunting and gathering. ►Paleolithic, ►Mesolithic

Neomorphic: A mutation that displays a new phenotype of any structural or other change that evolved recently. (See Muller HJ 1932 Proc 6th Int Congr Genet 1:213).

Neomycin (neo, $C_{23}H_{46}N_6O_{13}$): A group of aminoglycoside antibiotics (see Fig. N11). ►antibiotics, ►neomycin phosphotransferase, ►kanamycin, ►geneticin

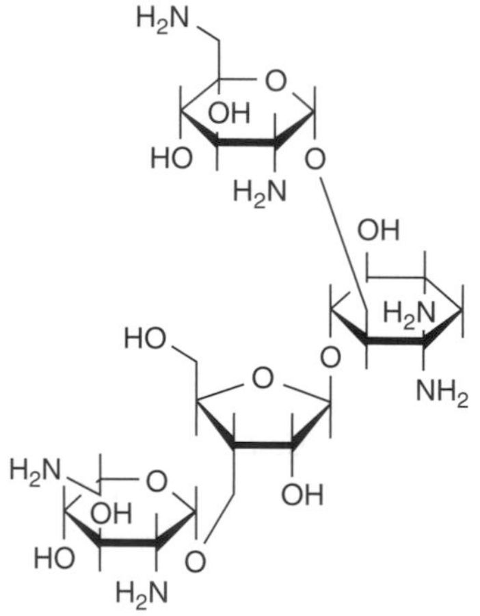

Figure N11. Neomycin

Neomycin Phosphotransferase (NPTII): ►kanamycin resistance, ►aminoglycoside phosphotransferase, ►aph(3′)II, ►*neo*^{r}

Neonatal Screening: ►genetic screening

Neonatal Tolerance: ►immune tolerance

Neo-organ: ►mini-organ

Neoplasia: A newly formed abnormal tissue growth such as a tumor; it may be benign or cancerous. ►cancer, ►carcinogens, ►oncogenes

neo^r: A neomycin resistance gene encoding the APH(3′) II enzyme. ►APH(3′)II, ►aminoglyco-side phosphotransferases, ►neomycin

Neo-sex-chromosome: A translocation involving an autosome and the X or Y chromosome; some have binding sites for the MSL (male-specific lethal) proteins. ►MSL; Steinemann M, Steinemann S 1998 Genetica 102–103:409.

Neoteny: The retention of some juvenile (or earlier stage) function in a more advanced stage.

Neo-X-chromosome: ►neo-sex-chromosome

Neo-Y-chromosome: ►neo-sex-chromosome

Nephritis, Familial (nephropathy): Familial nephritis is autosomal dominant without deafness or eye defects. It closely resembles the Alport syndrome. Elevated blood pressure, proteinuria, and only microscopically detectable blood cells in the urine precede kidney problems. In one form, immunoglobulin G accumulates in the serum and the expression of the disease is promoted by dietary conditions (e.g., high gluten). Several gene loci may be responsible for nephropathy (Chung KW et al 2003 Am J Hum Genet 73:420). ►Alport's disease, ►hypertension, ►kidney disease

Nephrolithiasis (kidney stones): Nephrolithiasis may be caused by the CLCN5 (Xp11.22) voltage-gated chloride channel and by Dent's disease, which maps to the same cytological position and causes a similar disorder. Human chromosomes 12q12-q14, 1q23-q24, 10q21-q22, and 20q13.1-q13,3 may harbor additional genes affecting the disease. Kidney stones (calcium oxalate, calcium phosphate, uric acid, ammonium-magnesium sulfate [struvite] and cystine) may afflict about 10% of Western populations with substantial variations according to genetic factors, diet, and climatic conditions. A nephrocystin-interacting protein may be involved in Leber congenital amaurosis and Joubert syndrome (Arts HH et al 2007 Nature Genet 39:882). ►kidney diseases, ►Dent's disease; Ombra MN et al 2001 Am J Hum Genet 68:1119; Gianfrancesco F et al 2003 Am J Hum Genet 72:1479; ►Joubert syndrome, ►amaurosis

Nephron: The functional tubular units of the kidney.

Nephronophthisis: An autosomal (3q22) recessive kidney disease characterized by anemia, passing of large amounts of urine (polyuria), excessive thirst (polydipsia), wasting of kidney tissues, etc. It is the most common renal failure in children. The GLIS2 family Zn-finger transcription factor mutation fails to mediate the transition from epithelia to mesenchyma (Attanasio M et al 2007 Nature Genet 39:1018). ►diabetes, ►cilia; Otto E et al 2002 Am J Hum Genet 71:1161; Otto EA et al 2003 Nature Genet 34:413; Simons M et al 2005 Nature Genet 37:537.

Nephropathy, Juvenile Hyperuricemic (16p21.2): A dominant hyperuricemia (excessive amounts of uric acid in the blood); an elevated serum creatinine but reduced uric acid excretion, gout, and renal malfunction. ►nephritis familial, ►kidney disease

Nephrosialidosis: An autosomal recessive kidney inflammation caused by oligosaccharidosis. ►lysosomal storage diseases, ►Hurler syndrome

Nephrosis, Congenital: An autosomal recessive inflammation of the kidney. It can be detected prenatally by the accumulation of α-fetoprotein in the amniotic fluid. The basic defect is in the basement membrane structure of the glomerular membranes. Its incidence may be as high as 1.25×10^{-4} in populations of Finns or of Finnish descent. ►kidney diseases, ►fetoprotein, ►prenatal diagnosis, ►basement membrane, ►glomerulonephrotis

Nephrotic Syndrome, Steroid Resistant (SRN1, 1q25-q31): A recessive fetal kidney disease due to a defect in the NPHS2 gene encoding podocin (42 kDa), an integral membrane protein. In mice, the expression is most evident in the kidneys, brain, and pancreas. The concomitant massive proteinuria cause neonatal death. ►kidney diseases; Boute N et al 2000 Nature Genet 25:125; Putaala H et al 2001 Hum Mol Genet 10:1.

Nephrotome: A part of the mesoderm that contributes to the formation of the urogenital tissues and organs.

Nepotism in Selection: Some disadvantageous individuals or groups may be favored by selection in case they promote the fitness of the reproducing groups that share genes with them or the ability to favorably recognize kin. ►selection, ►inclusive fitness, ►kin selection; Mateo JM, Johnston RE 2000 Proc R Soc Lond B Biol Sci 267:695.

Neprilysin: A 94 kDa metalloendopeptidase that normally degrades amyloid β peptides but not in Alzheimer disease and then the brain plaques develop. This enzyme may become a target for modification in order to degrade amyloid plaques. ►Alzheimer disease; Iwata N et al 2001 Science 292:1550.

NER (nucleotide exchange repair): ►DNA repair, ►excision repair

Nernest Equation: $\left(\frac{RT}{zF}\ln\frac{C_o}{C_i} = V\right)$ expresses the relation of electric potential across bio-membranes to ionic concentration at both sides; V = equilibrium potential in volts, C_o and C_i are outside and inside ionic concentrations, R = 2 cal $mol^{-1}°K^{-1}$, T = temperature in K (Kelvin), $F = 2.3 \times 10^4$ cal V^{-1} mol^{-1}, z = valence, *ln* = natural logarithm. ►chemiosmosis

Nerve Cell: ►neuron

Nerve Function: The nervous system includes the central (brain and spinal cord) and the peripheral system. The central nervous system in humans contains an immensely large number of cells (~30 billion in the cerebellum). The elements of this complex system are well coordinated. The neurons communicate with each other through the synapses of the dendrites. The signals are transmitted through the neurotransmitters and received by the receptors. The information is passed along the axons. The impulses may be forwarded in two opposite directions electrically by changes in the polarity of the cell membranes. The membranes are equipped with ion channels. The nervous system begins to develop at the embryo stage and continuously expands up to the adult stage and beyond. Mutation in the genes, rearrangement of the chromosomes, infectious diseases, drugs, malnutrition, and changes in the metabolism due to various causes may affect this development. Sleeping, learning, emotional events, pain, and all types of normal changes in the external or internal events may bring about fluctuations and alterations in the function of the system. Differences may be based on the genetic constitution of the individual, age, sex, etc. Generally, monozygotic twins respond very similarly and their narrow sense heritability may be quite high (22–88%). Modern neurobiology can monitor and map in vivo the activity in different parts of the brain or in the peripheral system by the use of electroencephalography, positron emission tomography, functional magnetic resonance imaging and other methods. ►individual entries, ►nanotechnology; Kennedy MB 2000 Science 290:750; Kandel ER, Squire LR 2000 Science 290:1113.

Nerve Growth Factor (NGF, neurotrophin): A stimulatory factor of growth and differentiation of nerve cells. NGF promotes neuronal survival partly with the aid of CREB transcription factor activation of anti-apoptotic gene (*Bcl*-2) activity. Structurally it is different from the ciliary neurotrophic factor. It is present in small amounts in various body fluids. NGF is a hexamer composed of α, β and γ subunits but only the β subunit is active for the ganglions. The active component of the mouse submaxillary factor has a dimeric structure of two 118-amino acid residues (MW 13 kDa). The mouse gene for α is in chromosome 7, β is in chromosome 3. The human gene for β NGF is in chromosome 1p13. It acts primarily on a tyrosine kinase receptor. The NGF receptors (NGFR) share similarities to the TFNRs (tumor necrosis factor receptors). ►signal transduction, ►growth factors, ►neurogenesis, ►neuropathy, ►ciliary neurotrophic factor, ►trk, ►TNFR, ►CREB, ►*Bcl*-2, ►apoptosis, ►protein 4.1N, ►ERK; Saltis J, Rush RA 1995 Int J Dev Neurosci 13:577.

NES (nuclear export signal): ►RNA export, ►chromosome maintenance region 1, ►CRM1

Nested Genes: Nested genes are partially overlapping genes, sharing a common promoter; the structural genes may be read in different registers. This is a very economical solution for extremely small genomes (ϕX174) for multiple utilization of the same DNA sequences within different reading frames. ►overlapping genes, ►recoding, ►contiguous gene syndrome, ►knockout; Turner SA et al 2001 J Bacteriol 183:5535.

Nested Models (hierarchical models): A series of statistical models, each with a different hypothesis. Each subsequent model adds or deletes one more factor than the preceding one. Such models are suitable for testing interactions between more relevant factors.

Nested Primers: The product of the first PCR amplification houses internally a second primer in order to minimize amplification of products by chance. ►PCR, ►primers; Menschikowski M et al 2001 Anal Cell Pathol 22(3):151.

Net1: An inhibitor of CDC14; it moves CDC14 into the nucleolus, and removes inhibitory phosphates from APC. It anchors Sir2 in the nucleolus. ►CDC14, ►APC, ►Sir; Shou W et al 2001 Mol Cell 8:45.

Netherton Syndrome (NS, 5q32): A recessive ichthyosis; its characteristic features include hair shaft defect, skin allergy (atopy), hayfever, and high serum IgE. The gene co-localizes with LEKTI (lymphoepithelial Kazal-type inhibitor), a serine protease inhibitor (SPINK5). ►ichthyosis; Walley AJ et al 2001 Nature Genet 29:175; Lauber T et al 2001 Protein Exp Purif 22:108; Descargues P et al 2005 Nature Genet 37:56.

Netrins: Netrins are secreted by neuronal target cells and assist homing in the proper nerve axons under the guidance of Ca^{2+}. Netrins regulate many developmental processes including angiogenesis (Wilson BD et al 2006 Science 313:640). Netrins also serve as a maintenance factor. Netrin-1 controls colorectal tumorigenesis by regulating apoptosis (Mazelin L et al 2004 Nature [Lond] 431:80). ►UNC-6, ►semaphorin, ►collapsin, ►axon guidance, ►tenascin, ►axon, ►colorectal cancer [DCC], ►*frazzled*; Kennedy TE et al 1994 Cell 78:425; Arakawa H 2004 Nature Rev Cancer 4:978.

Netropsin: Organic molecules that recognize A-T base pairs in the minor groove of DNA. ►lexitropsin; Wemmer DE 2000 Annu Rev Biophys Biomol Struct 29:439; Wang L et al 2001 Biochemistry 40:2511.

Network Motifs: Network motifs are circuits of interactions from which the networks are built. For design and tests for transcriptional motifs see Alon U 2007 Nature Rev Genet 8:450.

Networks: In the past, the goal of the biologists was to study the phenomena in isolation hoping that without the confounding factors their nature could be better revealed (reductionism). The cellular components, organelles, gene products, and small and large molecules are not independent but most likely interact. In recent years, molecular and bioinformatics tools have become available for the simultaneous study of the expression of thousands of genes by the use of microarray hybridization. The discovery the two-hybrid and similar systems permitted the analysis of the interaction of proteins pair-wise, and subsequently by high-throughput systems. If the function of one protein is known, the function of its associated partner can also generally be inferred. On this basis, a network of interacting proteins can be determined (see Fig. N12).

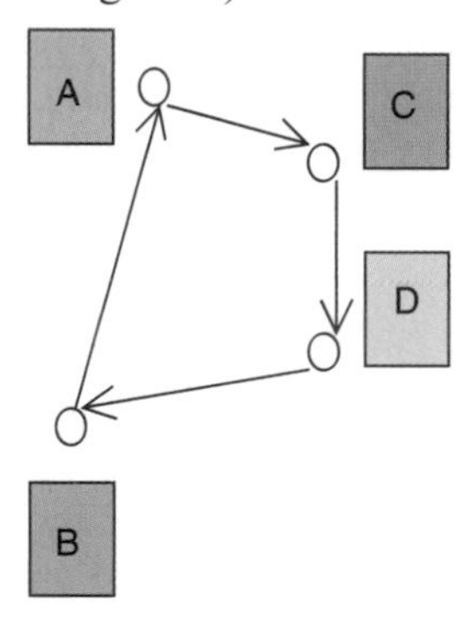

Figure N12. Network

The networks include *nodes* where incoming relations meet or from where they originate. The *node* (also called vertex) is characterized by its *degree of connectivity*, represented by k_{in} and k_{out}, respectively. In the diagram (see Fig. N13) the red node

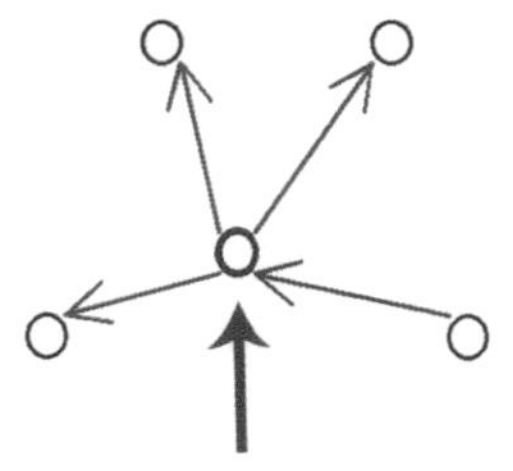

Figure N13. Node

shows three green arrows, corresponding to $k_{out} = 3$, which connect to three other (black) nodes, and one incoming blue arrow indicates $k_{in} = 1$. The number of nodes with *k* links divided by the number of nodes determines the *degree of distribution*. The highly connected nodes are called *hubs*. Most of the biological networks are called *scale-free networks* because their degree of distribution approximates a power law, i.e., $P(k) \sim k^{-\gamma}$. The small γ increases the role of the hub. Larger than 3 γ indicates irrelevance of the hub, whereas when it is between 2 and 3 there is a hierarchy of hubs and the most connected hub is linked with a small fraction of all nodes. When $\gamma = 2$ a *hub-and-spoke* network appears and the largest hub is linked to a large fraction of the nodes. Biological systems interact in a nonrandom, scale-free manner reflecting the existence of special functional hubs. The general transcription factors regulate the transcription of many genes whereas the special transcription factors may affect the expression of only a few. In practice, many protein interaction, gene regulation and metabolic networks are only subsets of some larger networks and may not qualify for scale-free classification (Stumpf MPH et al 2005 Proc Natl Acad Sci USA 102:4221).

The *path length* indicates the number of links between two nodes. The smallest numbers of paths that are traveled to arrive to a certain node measures the shortest path between nodes. The scheme at Figure N14 is called a *directed network* because from a node (e.g., B) both in and out paths occur. Such is the situation in metabolic networks when substrate and product has a direction of flow. Proteins usually form *undirected networks* because they have mutual interactions. From B we can arrive to A by a single path whereas from A to B three paths must be taken (A →C→D→B). The path length is measured by the average length (l), i.e., the average over the shortest paths between all pairs of nodes. This indicates then the network overall *navigability*. When one node is connected to several others then the *clustering coefficient* can be determined. The clusters may form different types of hierarchal structures. In *stochastic networks* the connections (edges) are probabilistic whereas in the *deterministic networks* the connections are either present or absent. Many of the biological networks have probabilistic components (Jiang R et al 2006 Proc Natl Acad Sci USA 103:9404).

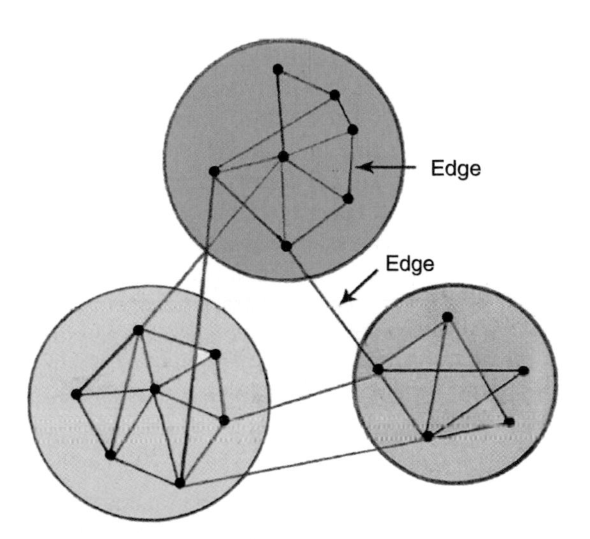

Figure N14. Sets of vertices ahown in different colors form modules. Numerous edges within modules and fewer among modules

Functional modules carry out the majority of biological activities, i.e., by coordinated nodes. Certain patterns of mutual connection are displayed more commonly in the network than expected at random and thus constitute a *network motif.* In metabolic networks, typically 80% of the nodes are connected within their own nodules. Metabolites participating in relatively few reactions and connected to different modules are more conserved evolutionarily than hubs, which display links mainly within a single nodule (Guimerà R, Amaral N 2005 Nature 433:895). A structural kinetic modeling procedure of metabolic networks is presented by Steuer R et al (2006 Proc Natl Acad Sci USA 103:11868). Comparative genomics and proteomics indicate the evolutionary origin and conservation of the basic functional modules. The modules form hierarchal networks (see Fig. N14). The different networking procedures may, however, indicate somewhat different hierarchal arrangements of the same system because the enormous complexity within a cell may be differently interpreted by alterations in some parameters. Another complicating although exciting feature of the networks is that the expression of genes and the activities of proteins, signal transducers and external factors are in dynamic flux in time and space. An insight into the networks provides possibilities for intervention for a cure. It may also forewarn against the possible side effects.

The biological systems themselves are generally characterized by *robustness*. This means, that despite their complexities, the inactivation of some nodes by mutation or other perturbation does not necessarily lead to total breakdown and death. Perturbation of major hubs may have more severe consequences. In reality, the modular organization of the cell and cellular systems may convey homeostatic advantages. An algorithm exists for the identification of "reporter metabolites," which indicate the most significant transcriptional changes upon perturbation of the system when expression changes occur in specific parts of the metabolism (Patil KR, Nielsen J 2005 Proc Natl Acad Sci USA 102:2685).

Network systems also pervade the social system and display some unexpected overlaps with a variety of systems (Palla G et al 2005 Nature [Lond] 435:814). In a hippocampal neuron function, 545 components/nodes and 1,259 interactions—identified by graph theory—represented signaling pathways and molecular machines. Such a system shed information on the regulatory interactions, accounting for cellular choices between homeostasis and plasticity (Ma'ayan A et al 2005 Science 309:1078). Networks are generally constructed on the basis of the yeast two-hybrid method. This procedure may be burdened by substantial noise. A new procedure based on desolvation/solvation of the target molecules is supposed to provide more reliable information of interactions (Deeds EJ et al 2006 Proc Natl Acad Sci USA 103:311). ►genetic network, ►probabilistic graphical models of cellular networks, ►cell models, ►small-world networks, ►systems biology, ►regulation of gene activity, ►microarray hybridization, ►two-hybrid system, ►hub, ►fractals, ►origons, ►machines biological, ►brain human, ►homeostasis, ►plasticity, ►graph theory; Schwikowski B et al 2000 Nature Biotechnol 18:1257; Barabási A-L, Oltvai ZN 2004 Nature Rev Genet 5:101; de novo designed network: Ashkenasy G et al 2004 Proc Natl Acad Sci USA 101:10872; Middendorf M et al 2005 Proc Natl Acad Sci USA 102:3192; systems biology, evolution of scale-free networks: Zhu H, Lipowsky R 2005 Proc Natl Acad Sci USA 102:10052; likelihood approach to biological network data: Wiuf C et al 2006 Proc Natl Acad Sci USA 103:7566; network modules in evolution: Spirin V et al 2006 Proc Natl Acad Sci USA 103:8774; Biomolecular Interaction Network Database: http://www.unleashedinformatics.com/index.php?pg=products&refer=bind; http://wwwmgs.bionet.nsc.ru/mgs/gnw/genenet; molecular interaction gateway: http://pstiing.licr.org/; http://www.thebiogrid.org.

Neu: Synonymous with ERBB2. ►ERBB1

Neu-Laxova Syndrome: Fetal growth retardation, microcephaly, abnormal extremities and genitalia, edema, skin lesions, etc. It is probably due to an autosomal recessive condition.

Neural Code: A neural code represents the various stimuli as sensory experiences such as visual, olfactory, gustatory, mechanical, auditory qualities as well as intensities, frequencies, and spatial relations. (See Fairhall AL et al 2001 Nature [Lond] 412:787).

Neural Crest: Ectodermal cells along the neural tube cleaved off from the ectoderm. These cells migrate through the mesoderm to form the peripheral nervous system, pigment cells and possibly other tissues such as thyroid and adrenal gland, connective tissues, heart, eye, etc. The neural crest anchors the head onto the anterior lining of the shoulder girdle and the Hox-gene-controlled mesoderm links the trunk muscles to the posterior neck and shoulder skeleton (Matsuoka T et al 2005 Nature [Lond] 436:347). Neural crest development is critical for the craniofacial skeleton and the peripheral ganglia. Injury to the neural crest may lead to human cleft palate. In chickens, neural crest specification occurs early, during development, and it requires Pax7 (Basch ML et al 2006 Nature [Lond] 441:218). ►germ layer, ►cleft palate, ►PAX

Neural Plate: A thickened notochordal overlay of nerve cells that develops into neural tubes.

Neural Tube: The central nervous system of the embryo, derived from epithelial cells of the neural plate. ►floorplate

Neural Tube Defects: Neural tube defects occur at a frequency of 1–2/1,000 birth. It results from the failure of closing (normally by the fourth week of a human pregnancy) of the neural tube. The recurrence risk is usually 3–5%. Folic acid administration may prevent it in about 70% of the cases. At the molecular level, the defect may be caused by mutation in the Slug Zn-finger transcription factor. Protein *Bcl*10 seems important for neural tube closure and lymphocyte activation. Mutation in the low-density lipoprotein receptor-like protein 6, a co-receptor in Wnt signaling may cause it according to a mouse model (Carter M et al 2005 Proc Natl Acad Sci USA 102: 12843). ►anencephaly, ►spina bifida, ►Meckel syndrome, ►microcephaly, ►Wnt, ►hydrocephalus; such a condition may occur as part of a number syndromes including trisomy 18 and 13, ►*Bcl*

Neuraminic Acid: A pyruvate and mannosamin-derived 9-carbon aminosugar. Its derivatives—such as sialic acid—are biologically important. ►sialic acid (where formula is shown), ►sialidoses, ►sialiduria, ►sialidase deficiency, ►neuraminidase deficiency

Neuraminidase Deficiency (sialidosis): A recessive autosomal lysosomal storage disease with multiple and variable characteristics. The basic common defect in the various forms is a deficiency of the sialidase enzyme. Sialidase (6p21.3) cleaves the linkage between sialic acid (*N*-acetyl-neuraminic acid) and a hexose or hexosamine of glycoproteins, glycolipids, or proteoglycans. The uncleaved molecules are then excreted in the urine. In sialurias, a rather free sialic acid is excreted. For the normal expression of the 76-kDa enzyme, the integrity of the structural genes in human chromosome 10pter-q23, and a 32-kDa glycoprotein coded by chromosome 20q13.1, are required. The deficiency is characterized by cherry red muscle spots, progressive myoclonous (involuntary contraction of some muscles), and loss of vision, but generally normal intelligence. ►lysosomal storage diseases, ►neuraminic acid, ►influenza virus; Lukong KE et al 2001 J Biol Chem 276:17286.

Neuregulins (NDF [neuron differentiation factor], GGF [glial growth factor], ARIA [acetylcholine receptor inducing activity]): Human chromosomes 8p22-p11 (NRG1) and 10q22 (NRG3) encode protein signals of the epidermal growth factor family (EGF) that activate acetylcholine receptor genes in synaptic nuclei. They also have multiple effects on various processes of differentiation. Some neuregulins belong to the tyrosine kinase transmembrane receptor family (NRG2, 5q23-q33). Neuregulin-1 controls the thickness of the myelin sheath on the axons and determines conduction velocity (Michailov GV et al 2004 Science 304:700). ►heregulin, ►acetylcholine, ►agrin, ►EGF, ►synaps, ►schizophrenia; Frenzel KE, Falls DL 2001 J Neurochem 77:1.

Neurexins: Neurexins are a great variety of nerve surface proteins generated by alternative splicing of three genes. They may function is cell-to-cell recognition and to interact with synaptotagmin, latrotoxin, and neuroligins. ►splicing, ►synaptotagmin, ►latrotoxin, ►neurexophilin, ►neuroligin; Sugita S et al 1999 Neuron 22:489.

Neurexophilin: A small neurexin-α binding molecule with probable signal function. ►neurexin; Missler M et al 1998 J Biol Chem 273:34716.

Neurite: An extension from any type of neuron (axon and dendrite).

Neuritis: A nerve inflammation, increased sensitiveness or numbness, paralysis or reduced reflexes caused by one or more (polyneuritis) defects of nerves. ►neuropathy

Neuroantibody: An antibody secreted by the nerve cells. Although antibody secretion is the duty of the plasma cells many other types of cells of animals (and plants) are capable of antibody production when the proper immunoglobulin genes are present and expressed. The degree of secretion, however, varies a great deal. ►immunoglobulins, ►antibody, ►plasma cell, ►lymphocytes; Ruberti F et al 1993 Cell Mol Biol 13:559.

Neuroaxonal Dystrophy, Late, Infantile (Hallervorden-Spatz disease, NBI1, 20p13-p12.3): A recessive nerve-degenerative disease with onset generally between ages 10 and 20 and death before age 30. It involves involuntary movements, speech defects, difficulties with swallowing, and progressive mental deterioration. Brown discoloration is visible in the brain (substantia nigra, globus pallidus) after autopsy. Both J. Hallervorden and H. Spatz discredited themselves during the era of the Third Reich by being involved in murderous human experimentation. Therefore, it has been suggested that the eponymous designation of the disease be changed to NBI1. Another similar but separate disease is also called infantile neuroaxonal dystrophy. ►neurodegenerative diseases

Neuroblast: A cell that develops into a neuron. ►neurogenesis

Neuroblastoma: A nerve cell tumor located most frequently in the adrenal medulla (in the kidney). A MYC oncogene seems to be responsible for its development. Deletions of the short arm of human chromosome 1 (NB, 1p36.3-p36.2) inactivate the relevant tumor suppressor gene. Histone deacetylase inhibitors (HDACI) make Ku70 release BAX, which then translocates to the mitochondria and triggers cytochrome c release leading to caspase-dependent apoptosis. Thus, HDACI is antineoplastic (Subramainian C et al 2005 Proc Natl Acad Sci USA 102:4842). Microarray analysis can predict the prognosis for neuroblastomas (Ohira M et al 2005 Cancer Cell 7:337). ►cancer, ►MYC, ►ERBB1, ►tumor suppressor gene, ►Ku70, ►histone deacetylase, ►BAX, ►apoptosis, ►cytochromes, ►microarray hybridization, ►caspase

Neurod: A helix-loop-helix regulatory protein of pancreas development. Its defect may lead to diabetes mellitus II and defects of the sensory neurons. ►diabetes mellitus, ►helix-loop-helix; Kim WY et al 2001 Development 128:417.

Neurodegenerative Diseases: For several of these diseases *Drosophila* models are available (Bilen J, Bonini NM 2005 Annu Rev Genet 39:153). ►Alzheimer's disease, ►Parkinson's disease, ►Huntington's chorea, ►Wilson's disease, ►frontotemporal dementia, ►schizophrenia, ►Friedreich ataxia, ►amyotrophic lateral sclerosis, ►Guam disease, ►prions, ►encephalopathies, ►fatal familial insomnia, ►tau, ►FTDP-17, ►Pick's disease, ►dementia, ►trinucleotide repeats, ►Lewy body, ►synuclein, ►mental retardation, ►affective disorders, ►neuroaxonal dystrophy, ►trinucleotide repeats, ►neuromuscular diseases, ►endophenotype, ►nanotechnology; Diagnostic and Statistical Manual of Mental Disorders, American Psychiatric Association, Washington DC 2000 (See 2006 reviews: Nature [Lond] 443:767–810).

Neuroectoderm: A neuroectoderm contributes to the formation of the nervous system. ►neurogenesis

Neuroendocrine Cancer: A neuroendocrine cancer occurs in the neuroectoderrmal tissue or in the endoderm-derived epithelia (pheochromocytoma and small cell lung carcinoma, respectively). Generally, it is detected after metastasis. Microarray hybridization and mass spectrophotometric analysis may provide early identification (Ippolito JE et al 2005 Proc Natl Acad Sci USA 102:9901). ►pheochromocytoma, ►small cell lung carcinoma, ►microarray hybridization, ►metastasis, ►MALDI

Neuroendocrine Immunology: Studies the interactions among the central-peripheral nervous system, endocrine hormones, and the immune reactions.

N

Neuroepithelioma: ►Ewing sarcoma

Neurofibromatosis: Autosomal dominant (NF-1, incidence ~$3x10^{-4}$), near the centromere of human chromosome 17q11.2, and another gene (NF-2) in the long arm of chromosome 22q12.2. They affect the developmental changes in the nervous system, bones and skin and cause light brown spots and soft tumors (associated with pigmentation) over the body (see Fig. N15), and NF-2 affects particularly the Schwann cells of the myelin sheath of neurons and involves mental retardation, etc. The NF-2 protein is called schwannomin or merlin and it is practically absent from schwannomas, meningiomas, and ependymomas (nerve neoplasias). Merlin-deficient cells lose contact inhibition. Merlin is a tumor suppressor and it is activated by dephosphorylation at serine 518 by myosin phosphatase, MYPT-1–PP1δ. The phosphatase is inhibited by a 17-kDa-protein kinase (CPI-17). Tumor suppression can be hindered either by mutation in NF-2 or by upregulation of CPI-17 (Jin H et al 2006 Nature [Lond] 442:576). In some instances, schwannomin is broken down by calpain. Schwannomin, a tumor suppressor, interacts with the actin-binding site of fodrin and thus with the cytoskeleton. The literature distinguishes several forms of this syndrome. The loss of the NF-1 protein activates the RAS signaling pathway through the granulocyte macrophage colony stimulating factor and makes the cells prone to develop juvenile chronic myelogenous leukemia. NF-1 mutations also inactivate p53. NF-1 also activates an adenylyl cyclase coupled to a G protein. It is deeply involved in the regulation of the overall growth of *Drosophila*. Neurofibromin (a product of NF-1) regulates longevity and stress resistance in *Drosophila* through cAMP regulation of mitochondrial respiration and ROS production, and NF-1 may be treatable using catalytic antioxidants (Tong JJ et al 2007 Nature Genet 39:476).

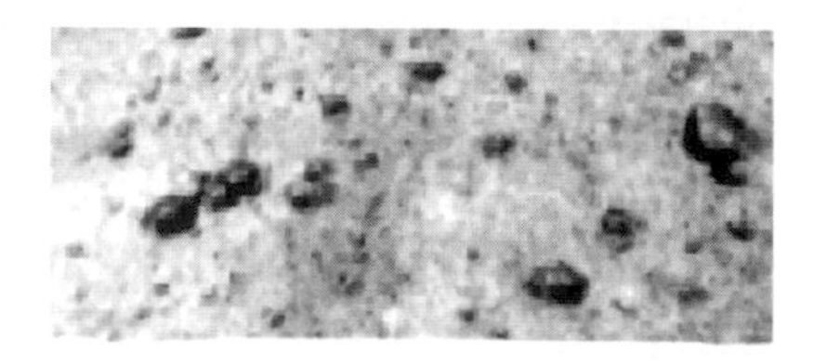

Figure N15. Neurofibromatosis. (Photo modified from Dr. Curt Stern, originally from Dr. V McKusick)

A pseudogene (NF1P1) was located in human chromosome 15. Estimated mutation rate 1–0.5 × 10^{-4}. The neurofibromin protein (guanosine triphosphatase activating, GAP) is encoded by NF-1 in a 350-kb DNA tract with 59 exons. About 5–10% of the affected individuals have ~1.5-kb microdeletions due to unequal crossing over occurring mainly during female meiosis. Radiation and cyclophosphamid therapy of NF-1 heterozygotes increase the chances of secondary carcinogenesis (Chao RC et al 2005 Cancer Cell 8:337). ►epiloia, ►mental retardation, ►ataxia, ►hypertension, ►eye disease, ►Cushing syndrome, ►leukemia, ►RAS, ►GCSF, ►café-au-lait spot, ►Lisch nodule, ►von Recklinghausen disease, ►cancer, ►ependymoma, ►schwannoma, ►merlin, ►meningioma, ►fodrin, ►actin, ►cytoskeleton, ►tumor suppressor, ►calpains, ►p53, ►GAP, ►contact inhibition; Ars E et al 2000 Hum Mol Genet 9:237; Nguyen R et al 2001 J Biol Chem 276:7621; Gutman DH et al 2001 Hum Mol Genet 10:1519.

Neurogastrointestinal Ecephalomyopathy (MNGIE): A human chromosome 22q13-qter external ophthalmoplegia with drooping eyelids (ptosis), abnormality of the intestines, defects of the nervous system and the muscles, lactic acidosis (accumulation of lactic acid), and slender body. The mitochondria in the skeletal muscles are structurally defective, the mtDNA show deletions. The activity of the respiratory enzymes is reduced. The basic defect appears to involve nuclearly encoded thymidine phosphorylase controlling maintenance of mtDNA. ►ophthalmoplegia, ►mitochondrial diseases in humans, ►mtDNA, ►encephalopathies

Neurogenesis: The nervous system has a great deal of similarity among all animals. The *neuroblasts* (the cells which generate the nerve cells) develop from the ectoderm. The *neural tube* (originating by ectodermal invagination) gives rise to the *central nervous system*, composed of *neurons* and the supportive *glial* cells. The differentiated neural cells do not divide again. The *neural crest* produces cells that eventually migrate all over the body and forms the *peripheral nervous system*. A neuron contains a dense cell body from which emanates the *dendrites* (reminding to the root system of plants) (see Fig. N16). From the neurons extremely long *axons* may emanate which at

Figure N16. Neurons and axon connections

the highly branched termini make contact with the target cells by *synapses*. The dendrites and axons are also called by the broader term *neurites*.

Initially, the various components of the system develop at the points of migration and are subsequently connected into a delicate network. The point of growth is named *growth cone,* which manages a fast expansion, also generating some web-like structures (*microspikes or filo- podia* organized into *lamellipodia*). The neurites are frequently found in *fascicles*, indicating that several growth cones travel the same track across tissues. This motion is mediated by cell adhesion molecules (N-CAM, cadherin, integrin). When the migrating growth cones reach their destination, they compete for the limited amount of *neutrophic factor* (NGF) released by the target cell and about a half of them starve to death.

The nerve cells function at the target by their inherent *neuronal specificity* rather than by their positional status. When the branches of several axons populate the same territory of control, they are trimmed back and a process of activity-dependent synapse elimination disposes of some. Due to this process, the synaptic function becomes more specific. The migration of the motor neurons from the central nervous system toward the musculature begins already in the tenth hour of embryo development in *Drosophila*. It has been found that the membrane-spanning receptor tyrosine phosphatases, DLAR, DPTP99A, DPT69D and others determine the "choice point" when the neuron heads toward the muscle fiber bundle. Apparently, after receiving the appropriate extracellular signal, they dephosphorylate the relevant messenger molecules. Deleting these surface molecules, the axons lose their guidance system. When the neurons break out of the nerve fascicles, they are homing on to the muscles by the attraction of the fasciclin III proteins in the muscle membranes. At this stage, the muscles also secrete the chemorepellent semaphorin II and thus the neurons settle down to form eventual synapses. The late bloomer protein (LBP) mediates the slowing down of the growth cone and the formation of the synapse. The neurotransmitters are then released by the terminal arbors. The agrin protein secreted by the nerve cells binds to heparin and α-dystroglycan and causes the clustering of the acetylcholine receptors. Other molecules, still not identified, are likely to be involved. The genetic study of neurogenesis is best defined in *Drosophila* and *Caenorhabditis*. In the latter, the nervous system contains only about 300 cells and a large array of mutant genes (many cloned) are available. In *Drosophila*, the large *achaete-scute complex* (*AS-C,* in chromosome 1–0.0) and several other genes cooperate in neurogenesis. The differentiation and function of the systems appear to be operated by cell-to-cell contacts rather than by diffusing molecules. In most of the mutants this communication is disrupted. For a long period of time regeneration of neuronal tissues was unsuccessful. Recently, experiments with EAK, RGD, and RAD peptides (the letters stand for the amino acid symbols in protein sequences) promoted the formation of microscopic fibers on nerve cells giving some hope that, eventually, problems with regeneration can be resolved for therapeutic purposes. Neural differentiation from embryonic stem cells can be accomplished by different culture procedures. Using stromal cells (SDIA) as feeders, neural differentiation occurs in more than 90% within five days. By the use of sonic hedgehog, bone morphogenetic protein and retinoic acid along with SDIA, various types of neural cell differentiation can be secured. Alternatively, human noncellular amniotic membrane matrix extracts exhibit potent SDIA-like inducing ability. The cells obtained by induced differentiation appeared effective in the monkey Parkinson's disease model in improving motor function (Ueno M et al 2006 Proc Natl Acad Sci USA 103:9554). ►morphogenesis, ►neuron, ►nerve growth factor, ►GSK, ►N-CAM, ►cadherin, ►integrin, ►signal transduction, ►neuron-restrictive silencer factor, ►ciliary neurotrophic factor, ►stem cells, ►signal transduction, ►amino acid symbols in protein sequences; Reh TA 2002 Nature Neurosci 5:392.

N

Neurogenetics: The study of the genetic bases of nerve development, behavior and hereditary anomalies of the nervous system.

Neurogenic Ectoderm: A set of cells that separates from the epithelium and forms the neurons by moving into the interior of the developing embryo. ►gastrulation

Neurogenin: A basic loop helix transcription factor promoting the expression of neuronal genes. Its overexpression may inhibit glial differentiation. ►neuron, ►glial cell; Sun Y et al 2001 Cell 104:365.

Neurohormone: A neurohormone is secreted by neurons such as vasopressin and gastrin. ►vasopressin, ►gastrin, ►gonadotropin releasing factor; Maestroni GJ 2000 An NY Acad Sci 917:29.

Neuroleptic: The changes effected by the administration of antipsychotic drugs.

Neuroligins: Neuroligins are nerve cell-surface proteins in the brain, encoded by five genes: NKGN1 (3q26), NLGN2 (17p13), NLGN3 (Xq13), NLGN4 (Xp22.3), and NLGN4Y (Yq11.2). They bind to neurexins. Genes 3 and 4 are associated with autism. ►autism, ►neurexins; Bolliger MF et al 2001 Biochem J 356(pt 2):581.

Neurological Disorders: ►neuropathy, ►hypomyelination, ►lysosomal storage disease, ►gangliosidoses, ►neuromuscular diseases, ►mucopolysaccharidoses, ►trinucleotide repeats, ►Alzheimer's disease, ►Parkinson's disease, ►Huntington's chorea, ►amyotrophic lateral sclerosis, ►encephalopathy, ►phenylketonuria, ►epilepsy, ►Down's syndrome, ►neurofibromatosis, ►affective disorders, ►mental retardation, ►multiple sclerosis. Many of these diseases have animal models which may not completely represent these complex human diseases. (See Watase K, Zoghbi HY 2003 Nature Rev Genet 4:296).

Neuromedin: A peptide widely available in the central nervous system and the gut. It stimulates smooth muscles, blood pressure, controls blood flow, regulates adrenocortical functions, and appears to control feeding. (See Kojima M et al 2000 Biochem Biophys Res Commun 276:435).

Neuromodulin (GAP-43): A protein abundant at the nerve ends and which may be involved in the release of neurotransmitters. It is a negative regulator of secretion at low levels of Ca^{2+}. ►neurotransmitter, ►synaptotagmin; Slemmon JR et al 2000 Mol Neurobiol 22:99.

N

Neuromuscular Diseases: These diseases include a variety of hereditary ailments with the common symptoms of muscle weakness and the inability to normally control movements. The specific and critical identification is often quite difficult because of the overlapping symptoms. In *Caenorhabditis*, 185 genes were identified in neuromuscular function on the basis of decreased acetylcholine secretion upon bacterial RNAi feeding. Mutations that decreased acetylcholine secretion became resistant to the pesticide Aldicarb (Sieburth D et al 2005 Nature [Lond] 436:510). ►lipidoses, ►gangliosidoses, ►sphingolipidoses, ►aminoacidurias, ►glutaric aciduria, ►muscular dystrophy, ►Werdnig-Hoffmann syndrome, ►myotonic dystrophy, ►chromosome defects [Down's syndrome, ►trisomy 18, ►Cri-du-chat and Prader Willi syndrome], ►glycogen storage diseases, ►brain malformations, ►dystrophy, ►amyotrophic lateral sclerosis, ►Parkinson's disease, ►cerebral palsy, ►atrophy, ►multiple sclerosis, ►Kugelberg-Welander syndrome, ►Charcot-Marie-Tooth disease, ►abetalipoproteinemia, ►Zellweger syndrome, ►Refsum diseases, ►diastematomyelia, ►Leigh's encephalopathy, ►Moebius syndrome, ►ataxia, ►dystonia, ►palsy, ►myasthenia, ►aspartoacylase deficiency, ►carnosinemia, ►spinal muscular atrophy, ►Kennedy's disease, ►Brody disease, ►muscular dystrophy; ►www.neuro.wustl.edu/.

Neuron: A nerve cell capable of receiving and transmitting impulses (see Fig. N17). In an adult

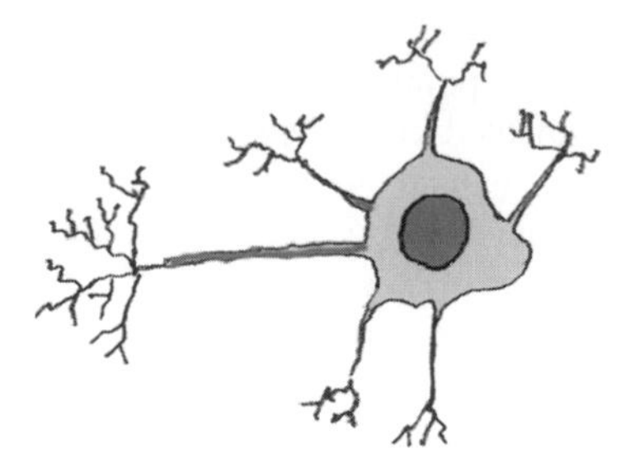

Figure N17. Neuron

human body there are about 10^{12} neurons connected by a very complex network. According to some estimates, there may be 10,000 different types of neurons. The enormous diversity of neurons is difficult to define and many different molecular mechanisms can account for the diversity; DNA recombination, like in the immunoglobulin genes, is apparently not involved. Aneuploidy has been detected in neurons and it seems to be a major cause of diversity. LINE elements can also cause structural modifications in the neural genomes. Use of alternative promoters, alternative splicing, RNA editing, and epigenetic modification can contribute to the diversity of neuron function (Moutri AR, Gage FH 2006 Nature [Lond] 441:1087).

For a long time regeneration of neural cells was not expected. Recently, pluripotent neural stem cells have been identified that can give rise to neurons, myelinating and other types of cells. In non-neuronal cells, neuron genes are repressed by carboxyl-terminal domain phosphatase targeted to class-C RNA polymerase II (Yeo M et al 2005 Science 3007:596). ►neurogenesis, ►aneuploid, ►LINE; Neuron 2003 40(2) for reviews.

Neuronopathy: ►muscular atrophy

Neuron-restrictive Silencer Factor (NRSF/REST/XBR): A protein that can bind to the neuron-restrictive silencer element (NRSE) by virtue of eight non-canonical zinc fingers. NRSE sequences were detected in at least 17 genes expressed only in the nervous system. It is a repressor of some neural-specific genes in neural and non-neural tissues but in case of its derepression it may induce the expression in non-neural cells neural-specific genes without converting these cells into neural cells. ►master molecule, ►silencer, ►Zn-finger; Kuwahara K et al 2001 Mol Cell Biol 21:2085.

Neuropathy: A noninflammatory disease of the peripheral nervous system. Sensory neuropathies may be associated with deficiency of α-methylacyl-CoA racemase (AMACR). This peroxisomal enzyme mediates the conversion of pristanoyl-CoA and C27-bile acyl-CoA to their (s)-stereoisomers (see Fig. N18). This is, however, a heterogeneous disease based on different enzymes encoded at several locations (Kok C et al 2003 Am J Hum Genet 73:632). One locus at Xq21.32-q24 encodes PRPS1 (phosphoribosyl pyrophosphate synthetase 1), an isoform of the PRPS gene family ubiquitously expressed in human tissues, including cochlea. It affects hereditary peripheral hearing loss and optic neuropathy (CMTX5). The enzyme mediates the biochemical step critical for purine metabolism and nucleotide biosynthesis (Kim H-J et al 2007 Am J Hum Genet 81:552). ►Charcot-Marie-tooth disease, ►Riley-Day syndrome, ►giant axonal neuropathy, ►hypomyelination, ►Egr, ►Krox, ►pain-insensitivity, ►sensory neuropathy 1, ►HNPP, ►Navajo neuropathy; Verhoeven K et al 2006 Curr Opin Neurol 19:474; mitochondrial origin: Finsterer J 2006 Acta Neurol Scand 114:217.

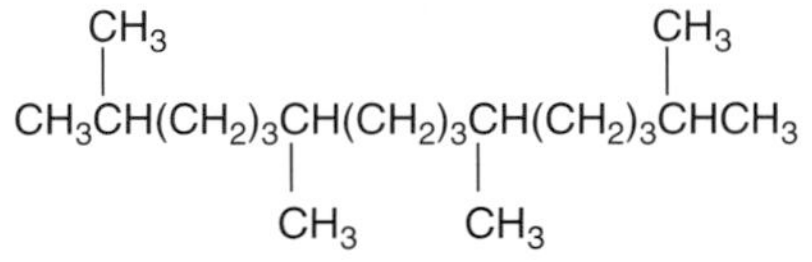

$$CH_3CH(CH_3)(CH_2)_3CH(CH_3)(CH_2)_3CH(CH_3)(CH_2)_3CH(CH_3)CH_3$$

Figure N18. Pristane

Neuropeptide: A signaling molecule (peptide) secreted by nerve cells. (See Hansel DE et al 2001 J Neurosci Res 66:1).

Neuropeptide Y (NPY, encoded in mouse chromosome 4): A 36-amino acid neurotransmitter that is supposed to modulate mood, cerebrocortical excitability, hypothalamic-pituitary signaling, cardiovascular physiology, sympathetic nerve function and increased feeding behavior. NPY-receptor-deficient mice become very sensitive to various pain signals. The latter claim is not be fully supported by the evidence obtained with NPY mutant mice; however, several NPY receptors exist. The ablation of the NPY peptide and agouti-related protein by the diphteria toxin is well tolerated in neonatal mice but it leads to serious starvation in adult animals (Luquet S et al 2005 Science 310:683). ►CART, ►ghrelin, ►PYY_{3-36}, ►leptin, ►obesity; Niimi M et al 2001 Endocrine 14(2):269.

Neurophysin: A group of soluble carrier proteins (M_r about 10,000) of vasopressin and oxytocin and related hormones secreted by the hypothalmus. ►oxytocin, ►vasopressin, ►diabetes insipidus, ►hypothalmus, ►brain human; Assinder SJ et al 2000 Biol Reprod 63:448.

Neuropil (neuropile): A bunch of dendrites, axons in cells, and neuroglia in the gray matter of the brain.

Neuropilin-1 (NRP): A 130-kDa transmembrane receptor that directs axonal guidance in cooperation with semaphoring. It may be coupled with VEGF and aids angiogenesis. ►VEGF, ►KDR, ►angiogenesis, ►axon, ►semaphorin; Marin O et al 2001 Science 293:872.

Neurospora crassa (red bread mold): An ascomycete (n = 7, DNA 4 × 10^7 bp) was introduced into genetic research more than half a century ago for the exact study of recombination in ordered tetrads. A high-quality draft sequence of the genome has been completed (Galagan JE et al 2003 Nature [Lond] 422:489) and 10,082 protein-coding genes, almost twice as many as *Saccharomyces cerevisiae* (ca. 6300), were found. The number of tRNA genes is 424, and 5S rRNA genes is 74. The average gene size is 1,673 bp. About 1.5% of the cytosines are methylated by two methyltransferases (dim-2, rid). Its original home is in tropical and subtropical vegetation. The asexual life cycle is about one week whereas the sexual cycle requires about three weeks. The two mating types (*A* [5 kb] and *a* [3 kb]) are determined by a single locus. It was the first fungus to be used for centromere mapping on the basis of the frequency of second division segregation in the linear asci of eight spores. The first auxotrophic mutants were induced in *Neurospora* and these experiments led to the formulation of the one-gene-one enzyme hypothesis that formed the cornerstone of biochemical genetics. Although it is a haploid organism, the availability of heterokaryons permits the study of dominance and allelic complementation. This was the first eukaryote where genetic transformation with DNA became feasible. Currently, about 10^4 to 10^5 transformants can be obtained per microgram, DNA. *Neurospora* (and other ascomycetes) may display RIP (repeat induced pointmutation) observed at high frequency when duplicated elements are introduced into the nuclei (by transformation) resulting in GC→AT transitions and eviction of the duplications, resulting also in chromosomal rearrangements. The genetic maps of *Neurospora* are generally shown in relative distances only because of the different *rec* genes present in the various mapping strains that may alter recombination frequencies by an order of magnitude. Mitochondrial mutants are known. The respiration-deficient *poky* mutants are somewhat similar to the petite colony mutants of budding yeast. The *stopper* mutants are also 350–5,000-bp deletions and unable to make protoperithecia. Related species used for genetic studies are *N. sitophila* and *N. tetrasperma*. (See Fig. N19; ►tetrad analysis, ►channeling, ►RIP, ►argonaute, ►quelling,

N

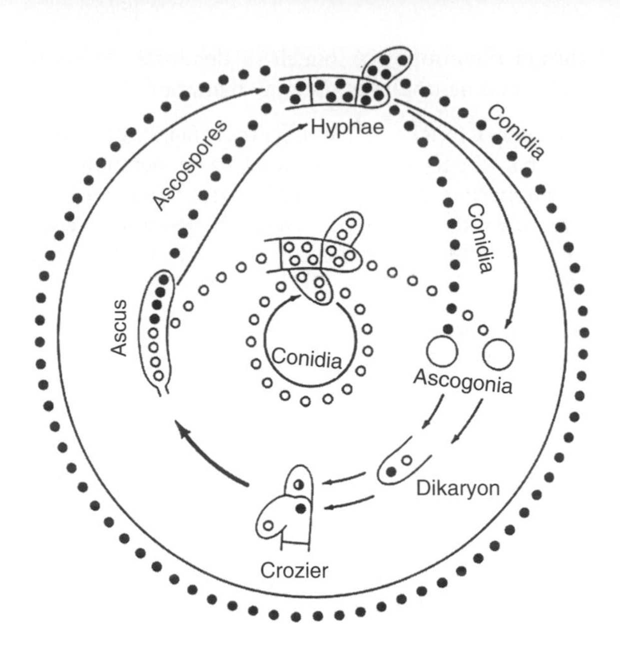

Figure N19. Life cycle of *Neurospora*

►*Neurospora* mitochondrial plasmids, ►fungal life cycles; Perkins DD et al 2000 The Neurospora Compendium, Academic Press, San Diego, California; http://www.fgsc.net/outlink.html; http://mips.gsf.de/genre/proj/ncrassa/.

Neurospora Mitochondrial Plasmids: *Neurospora crassa* and *N. intermedia* strains may harbor 3.6–3.7-kb mitochondrial plasmids with a single open reading frame encoding an 81-kDA reverse transcriptase protein. The proteins appear similar to the class II intron-encoded reverse transcriptase-like elements. The Mauriceville and Varkud plasmid DNAs, apparently, use as primers a 3′-tRNA like structure. The latter two plasmids integrate into the mitochondrial DNA at the 5′-end of the major plasmid transcript. *N. intermedia* also contains a smaller "Varkud satellite plasmid" (VSP) that may be either linear or circular single-stranded RNA. The VSP plasmid bears resemblance to class I introns. ►mitochondrial plasmids, ►introns; Kevei F et al 1999 Acta Microbiol Hung 46(2–3):279.

Neurotactin: A membrane bound chemokine encoded in human chromosome 16q. It is overexpressed during inflammation of brain tissues. ►chemokine; Pan Y et al 1997 Nature [Lond] 387:611.

Neuroticism: Emotional instability such as extreme anxiety, depression, low self-esteem, and diffidence. The additive variance is ~27–31% and the non-additive is 14–17%. Expressivity is generally higher in females (59.3%) versus 40.7% in males. Chromosomes 1q, 4q, 7p, 12q, and 13q are likely to be involved. ►variance, ►expressivity; Fullerton J et al 2003 Am J Hum Genet 72:879.

Neurotoxin: Destroys nerve cells. ►toxins

Neurotransmitter: Neurotransmitters include several types of signaling molecules (acetylcholine, noradrenaline, glutamate, glycine, serotonin, γ-aminobutyric acid, catecholamines, neuropeptides, nitric oxide, carbon monoxide, adenosine, etc.) used for information transmission of neuronal signals. These signals can be excitatory (e.g., glutamate, acetylcholine, serotonin) or inhibitory such as gamma-aminobutyric acid and glycine and may open Cl^- channels. The secretion is activated by electric nerve impulses in response to extracellular signals. The signals are transmitted to other cells by chemical synapses across the synaptic cleft by attaching to the transmitter-gated ion channels. Although it was earlier believed that individual neurons could release only a single type of neurotransmitter, recent information indicates that both glycine and GABA can be simultaneously released by spinal interneurons, such as the neurons between the sensory- and motor neurons. ►ion channels, ►AMPA, ►transmitter-gated ion channel, ►voltage-gated ion channel, ►signal transduction, ►synaps, ►calmodulin, ►GABA, ►synapse, ►synaptic cleft; see also other separate entries, ►dopamine, ►DARPP, ►complexin, ►epinephrine, ►Munc18; Matthews G 1996 Annu Rev Neurosci 19:219; Lin RC, Scheller RH 2000 Annu Rev Cell Dev Biol 16:19; historical: Snyder SH 2006 Cell 125:13.

Neurotrophins (NTs): Neurotrophins are growth and nutritive factors for the nerves but may also potentiate nerve necrosis. Their receptor is either a glycoprotein (LNGFR) or trk. Neurturin, a structurally NT-related protein, also activates the MAP kinase signaling pathway and promotes neural survival. ►NGF, ►MAP, ►trk, ►LNGFR, ►FOS oncogene, ►aging; Huang EJ, Reichardt LF 2001 Annu Rev Neurosci 24:677; McAllister AK et al 1999 Annu Rev Neurosci 22:295; He X-I, Garcia C 2004 Science 304:870.

Neurotrypsin: An 875-amino acid secreted trypsin-like serine protease encoded at 4q24-q25. Its mutation may lead to nonsyndromic mental retardation (Molinari F et al 2002 Science 298:1779). ►mental retardation

Neurovirulence: A viral infection that attacks the nerves. ►virulence

Neurturin: ►neurotrophins, ►GDNF

Neurula: An embryonic stage when the development of the central nervous system begins; about 19–26 days after fertilization in humans, or ~19 h in the frog (see Fig. N20).

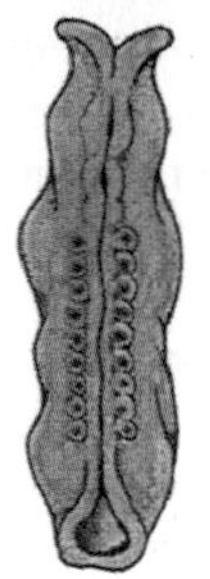

Figure N20. Neurula (not on scale)

Neurulation: The formation of the neural tube (the element that gives rise to the spinal chord and the brain) from the ectoderm during gastrulation. The process depends on multiple genes. A nucleosome assembly (NAP)-like protein, Nap1/2 (Nap1 like 2), plays a specific role. ►gastrula, ►ectoderm, ►neural tube defects; Colas JF, Schoenwolf GC 2001 Dev Dyn 221(2):117; Copp AJ et al 2003 Nature Rev Genet 4:784.

Neutered: An ovariectomized (ovaries surgically removed) female. ►spaying, ►castration, ►oophorectomy

Neutral-loss Scan: Ions can decompose into both charged and uncharged (neutral) fragments. The products can be analyzed by mass spectrometry. The loss of the charged part affects the peaks in the mass spectrum of, e.g., proteins and peptides. ►CID; Vrabanac JJ et al 1992 Biol Mass Spectrom 21(10):517.

Neutral Mutation. ►mutation neutral

Neutral Petite: ►petite colony mutants

Neutral Space: Where alternative configurations can solve the same biological need. ►mutation neutral, ►robustness

Neutral Substitution: Amino acid changes in the proteins that have no effect on function.

Neutrality, Conditional: When the difference among QTLs is distinguishable conditionally.

Neutrality Index in Proteins: (see Fig. N21)

Neutralizing Antibody: A neutralizing antibody covers the viral surface and prevents binding of the virus to cell surface receptors thereby interfering with infection. The IgG1 subtype protects against mucosal infection and infection by HIV1 and SIV. ►antibodies, ►acquired immunodeficiency, ►SIV; Grossberg SE et al 2001 J Interferon Cytokine Res 21(9):743.

Neutron: ►physical mutagens

Neutron Flux Detection: Personal monitoring may be based on microscopic examination of proton recoil tracks on a film or dielectric materials (i.e., which transmit by induction rather than by conduction) such as plastic or glass. Precise measurement of neutron flux is more complex. ►radiation measurement

Neutropenia, Chronic: An autosomal dominant or recessive defect of the neutrophils. ►neutropenia cyclic, ►neutrophil

Neutropenia, Cyclic: A rare blood disease in humans and dogs. In humans it appears to be autosomal dominant (19p13.3) while in dogs it is recessive. In dogs, the adaptor protein complex 3 β-subunit directing trans-Golgi export of transmembrane cargo proteins to lysosomes is the cause of the disease (Benson KF et al 2003 Nature Genet 35:90). In 21-day cycles in humans (12 in dogs), fever, anemia, a decrease in neutrophils, granulocytes, eosinophils, lymphocytes, platelets, and monocytes recurs and enhances the chances of infection. In some instances this childhood disease is outgrown by adulthood. A granulocyte-colony-stimulating factor may alleviate the condition. Neutrophil elastase (ELA2, 19p13.3; 240 amino acids), a serine protease mutation may account for the condition. Recessive mutation in the HAX1 mitochondrial protein (encoded at 1q21.3) is critical for maintaining the inner mitochondrial membrane potential and protecting against apoptosis in myeloid cells (Klein C et al 2007 Nature Genet 39:86). ►endocardial fibroblastosis, ►granulocyte-colony-stimulating factor, ►elastase; Person RE et al 2003 Nature Genet 34:308.

Neutrophil: One of the specialized white blood cells (leukocyte). Generally, they are distinguished by their irregular-shaped, large nuclei, and granulous internal structure (polymorphonuclear granulocyte) (see Fig. N22). They contain lysosomes and secretory vesicles enabling them to engulf and destroy invading bodies (bacteria) and antigen-antibody complexes. They produce reactive oxygen species (superoxide, singlet oxygen, hydroxyl radical, N chloramines, etc.)

$$\frac{\text{Number of polymorphic replacement sites/number of fixed replacement sites}}{\text{Number of polymorphic synonymous sites/number of fixed synonymous sites}}$$

Figure N21. Neutrality index of proteins

and cationic antimicrobial proteins (defensins). Neutrophils expressing matrix metalloprotease type 9 can activate early stage of tumorigenesis by activating angiogenic pathway (Nozawa H et al 2006 Proc Natl Acad Sci USA 103:12493). ►macrophage, ►eosinophils, ►basophils, ►granulocytes, ►angio-genesis, ►metalloproteinase; Reeves EP et al 2002 Nature [Lond] 416:291.

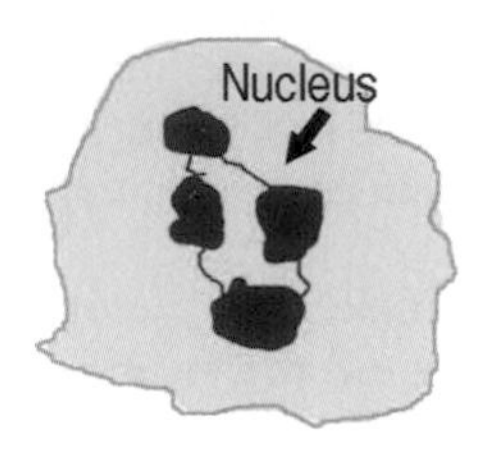

Figure N22. Neutrophil

Nevirapine: A non-nucleoside inhibitor of HIV-1 reverse transcriptase. It is much less expensive than AZT. Resistance may develop in 23% of the women, who are treated to prevent the transmission of the virus to fetus (Palmer S et al 2006 Proc Natl Acad Sci USA 103:7094). ►TIBO, ►acquired immunodeficiency syndrome, ►AZT; De Clercq E 2001 Curr Med Chem 8:1529.

Nevoid Basal Cell Carcinoma (NBCCS, Gorlin-Goltz syndrome): NBCCS is quite frequently caused by a new dominant mutation in human chromosomes 9q31, 9q22.3, or 1p32. The patients generally have bifid ribs, bossing on the head, tooth and cranofacial defects, poly- or syndactyly and reddish birthmarks that may become neoplastic. These cutanious neoplasias frequently become horny but are, in the majority of cases, sporadic and benign basal cell carcinomas (BCCs) although they may become metastatic (see Fig. N23). The skin is very sensitive to ionizing radiation and ultraviolet spectrum of the sunshine and in response develops numerous new spots (nevi). The more serious form is the basal cell nevoid syndrome (BCNS). This is the most common type of cancer in the USA and about 750,000 cases occur annually. It turned out that the major culprit in this disease is a homolog of the *Drosophila* gene *patch* (*ptc* or *tuf* [*tufted*] in chromosome 2–59). *Ptc* encodes the receptor for the product of the *sonic hedgehog* (*Shh*) gene and a mutational event prevents *Ptc* from inhibiting SMO (smoothened), a seven-span transmembrane protein, a factor in carcinogenesis. The Gli1 transcription factor signals the expression of BCC in the basal cells. This segment polarity locus encodes a transmembrane protein and the wild type allele represses the transcription of members of the TGF-β (transforming growth factor). *Drosophila* gene *hh* (*hedgehog*, 3–81) has the opposite effect, i.e., it promotes the transcription of the TGF proteins. Homologs of these genes exist in other animals, including mouse. The human homolog has now been cloned by the use of a mouse probe of *ptc*. Nucleotide sequencing revealed that in one of the families, a 9-bp (3-amino acid) duplication occurred in the afflicted member of the family; in another kindred, an 11-bp deletion was associated with the expression of the BCNS. Other developmental genes may also contribute to the expression of these genes. ►cancer, ►squamous, ►TGF-β, ►morphogenesis in *Drosophila*, ►Gli, ►*Sonic hedgehog*, ►*hedgehog*; Bale AE, Yu K-p 2001 Hum Mol Genet 10:757.

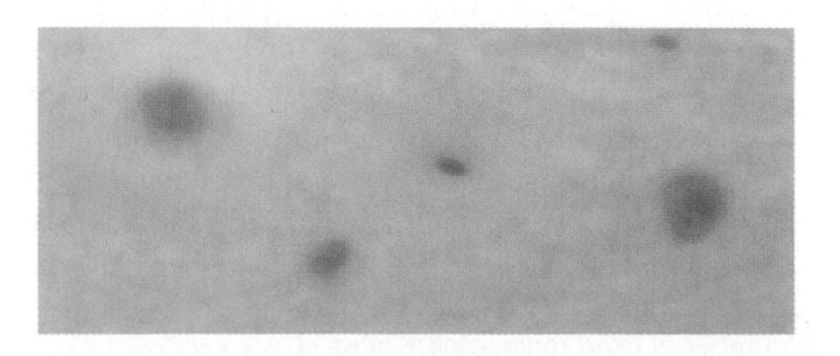

Figure N23. Papules of different size and surface may appear at several parts of the body and frequently become cancerous

Nevus (mole): An autosomal dominant red birthmark on infants that usually disappears in a few months or years. The autosomal dominant *basal cell nevus syndrome* is a carcinoma. It also causes bone anomalies of diverse types, eye defects, sensitivity to X-radiation, etc. ►vitilego, ►skin diseases, ►carcinoma, ►cancer, ►mole

Newcastle Disease: An avian influenza that may be transmitted to humans. The 15,186-nucleotide viral RNA genome has been completely sequenced. Newcastle virus (NDV) containing the hemagglutinin-neuraminidase gene of avian influenza virus provided protection to chickens against both viruses (Veits J et al 2006 Proc Natl Acad Sci USA 103:8197). When animals immunized with NDV expressing S glycoprotein were challenged with a high dose of SARS-Corona Virus (SARS-CoV), direct viral assay of lung tissues taken by necropsy at the peak of viral replication demonstrated a 236- or 1,102-fold (depending on the NDV vector construct) mean reduction in pulmonary SARS-CoV titer compared with control animals (DiNapoli JM et al 2007 Proc Natl Acad Sci USA 104:9788). ►oncolytic viruses, ►SARS

Newfoundland: A rare blood group.

Newt (*Triturus viridescens*): A salamander; 2n = 2x = 22. Its lampbrush chromosomes have been extensively studied (see Fig. N24). ►lampbrush chromosome, ►regeneration

Figure N24. Newt

Newton (N): A physical unit of force; 1 N = 1 kg • m • sec^{-2}. In molecular kinetics, usually pN (picoN, 10^{-12} N) is used.

Newton, Isaac (1642–1727): A famous mathematician, physicist, and natural philosopher. Newton was afflicted by gout. ►gout

Nexin: A protein connecting the microtubules with a cilium or flagellum. ►PN-1

Nexus: A junction.

Nezelof Syndrome: An autosomal recessive T lymphocyte deficiency immune disease. The affected individuals lack cellular immunity while displaying humoral immunity. The development of the thymus is abnormal. ►immunodeficiency

NF: Nuclear factors required for replication. NF-1 (similar to transcription factor CTF) binds to nucleotides 19–39 of adenovirus DNA and stimulates replication initiation in the presence of DBP. NF-2 is required for the elongation of replicating intermediates. NF-3 (similar to OTF-1 transcription factor) binds nucleotides 39–50 adjacent to the NF-1 binding site and stimulates replication initiation. The NFs bind to the same DNA sequence and yet they display somewhat different functions in the regulation of development. ►CTF, ►OTF, ►DBP

NF-1: Human neurofibromatosis; it involves proteins homologous to GAP and IRA. NF is a CCAAT binding protein. ►GAP, ►IRA1, ►IRA2, ►CAAT box, ►neurofibromatosis

NF-AT (nuclear factor of activated T cells): A family of transcription factors (NF-AT_c and F-AT_p) that activate the interleukin (IL-2) promoter or the dominant negative forms block IL-2 activation in the lymphocytes. The CD3 and CD2 antigens activate NF-AT jointly but not separately and they are suppressed by cyclosporin (CsA). For the nuclear import of the cytoplasmic NF-AT, it is dephosphorylated by calcineurin at the nuclear import site, the nuclear localization signal (NLS). When the level of calcineurin is lowered, NF-AT is rephosphorylated and exported from the nucleus and the transcription regulated by it may cease. The Crm1 export receptor mediates the rephosphorylation process at the nuclear export site (NES). Calcineurin binding to NF-AT may prevent Crm1 binding to NF-AT. The different members of the NF-AT family have sequence-differences at their NLS and NES regions and may be acted upon by different kinases. These transcription factors, AP1 and FOS-JUN, cooperatively bind to DNA. The NF-AT_c transcription factor plays an important role in the differentiation of heart valves and septum. Its nuclear localization is blocked by FK506, an inhibitor of calcineurin. ►lymphocytes, ►interleukin, ►immune system, ►T cell, ►transcription factors, ►NF-κB, ►cyclosporin, ►cyclophyllin, ►AP, ►JUN, ►FOS oncogene, ►calcineurin, ►FK506, ►Noonan syndrome; Rao A et al 1997 Annu Rev Immunol 15:707.

NFAT (nuclear factor activated T cell): A family of phosphoprotein transcription factors that mediate the production of cell surface receptors and cytokines and regulate the immune response. Its nuclear localization is facilitated by phospholipase C (PLC-γ) through signals by a cell membrane receptor. It is then that the phosphoinositide pathway is activated and NFAT moves into the nucleus after dephosphorylation. ►T cell, ►T cell regulatory, ►phosphoinositides, ►rel, ►NF-AT, ►Noonan syndrome; Graef IA et al 2001 Curr Opin Genet Dev 11:505; Macian F et al 2001 Oncogene 20:2476; Crabtree GR et al 2002 Cell 109:S67; McCullagh KJA et al 2004 Proc Natl Acad Sci USA 101:10590.

NF-E2: A transcription factor with a basic leucine zipper domain; it regulates different specific genes. Its binding site in erythroid-specific genes is GCTGAGTCA and in β-globin genes the NF-E2-AP1 motif (a subset of an antioxidant response element) is GCTGAGT-CATGATGAGTCA. ►Maf, ►NRF, ►LCR, ►DNA-binding protein domains; Francastel C et al 2001 Proc Natl Acad Sci USA 98:12120.

NF-κB (nuclear factor kappa binding, a *Drosophila* homolog is *Dorsal*): A transcription factor family dimeric with a REL oncoprotein. They are specific for the IκB (immunoglobulin kappa B lymphocytes, *Drosophila* homolog Cactus) proteins. NF-κB is involved—among its many other roles—in inflammation and the development and progression of cancer. NF-κB is activated by the interleukin 1 (IL-1R1) signaling cascade with the aid of TNF, hypoxia, and viral proteins. Genotoxic stress activates NF-κB and promotes the survival of cancer cells targeted by chemicals. PIDD (p53-inducible death domain protein) activates caspase-2 and apoptosis. In stress, PIDD complexes with a kinase and NEMO is formed but sumoylation and ubiquitination of NEMO follows. Thus, PIDD serves as a switch between cell survival and death (Jannsens S et al 2005 Cell

123:1079). NF-κB normally stays inactivated in the cytoplasm by being associated with inhibitor IκB. The activation dissociates the two proteins and IκB is degraded before NF-κB moves to the nucleus where it binds to the GGGACTTTCC consensus and activates the transcription of *Twist*, which is involved in the establishment of embryonal germ layers and may inhibit *bone morphogenetic protein*-4 (*Drosophila* homolog *decapentaplegic*). The target genes are involved either with cellular defense or with differentiation such as IL-2, IL-2 receptor, phytohemagglutinin (PHA), and phorbol ester (PMA) synthesis. NF-κB suppresses caspase-8 and thus apoptosis is mediated by the tumor necrosis factor (TNF-α) and its receptor (TNFR). Na-salicylate, aspirin, glucocorticoids, immunosuppressants (cyclosporin, rapamycin), nitric oxide, etc., inhibit NF-κB involved with inflammation and infection processes. Down-regulation of NF-κB by de-ubiquitinating of the ubiquitin ligase domain of the inhibitor protein A20 protects against inflammation (Wertz II et al 2004 Nature [Lond] 430:694). In arthritis, lupus erythematosus, Alzheimer's disease, HIV and influenza infection, and in several types of cancer, NF-κB levels are elevated. The function REL subunit is necessary for the activation of TNF and TNF-α dependent genes and for the NF-κB protection against apoptosis. Inhibition of NF-κB translocation to the nucleus enhances apoptotic killing by ionizing radiation, some cancer drugs, and TNF. Thus, managing NF-κB may assist in drug therapy of cancer. ►REL, ►morphogenesis in *Drosophila* [**{3}** 14, 17], ►interleukins, ►TNF, ►IκB, ►IKK, ►NIK, ►T cell, ►immunosuppressants, ►nitric oxide, ►glucocorticoid, ►autoimmune diseases, ►Alzheimer's disease, ►HIV, ►apoptosis, ►phorbol-12-myristate-13-acetate, ►bone morphogenetic protein, ►TFD, ►NFKB, ►NF-AT, ►signal transduction, ►inflammation, ►NEMO, ►SUMO, ►ubiquitin; Baldwin AS Jr 1996 Annu Rev Immunol 14:649; Ghosh S et al 1998 Annu Rev Imunol 16:225; Martin AG et al 2001 J Biol Chem 276:15840; De Smaele E et al 2001 Nature [Lond] 413:308; Tang G et al 2001 *ibid*. 313; Ghosh S et al 2002 Cell 109:S81; Larin M et al 2002 Nature Rev Cancer 2:301; Karin M 2006 Nature [Lond] 441:431.

NFKB (nuclear factor kappa B, NF-kB, NF-κB): NFKB are transcription factors. NFKB1 is in human chromosome 4q23 and NFKB2 in human chromosome 10q24. The former codes for a protein p105 and the latter for p49. Both are regulators of viral and cellular genes. IKKβ is essential for IκB phosphorylation and its subsequent degradation, leading to the activation of NFKB1. IKKα seems to be involved in the activation of NFKB2 through phosphorylation of histone H3 (Anest V et al 2003 Nature [Lond] 423:659). They also have homology to the REL retroviral oncogene and the product of the *Drosophila* maternal gene *dl* and some regulatory proteins of plants. NF-κB is selectively activated by the nerve growth factor using the neurotrophin p75 receptor ($p75^{NTR}$), a member of the tyrosine kinase family, located in the Schwann cells. NF-κB usually exists within the cell, in association with the inhibitory molecule IκB. Their dissociation (induced by TNF, IL-1, etc.) permits NF-κB to move into the nucleus. NF-κB regulates the expression of cytokine genes, acquired immunodeficiency, cancer metastasis, rheumatoid arthritis, inflammation, and other processes. Protein kinase Θ mediates NF-κB activation via a T cell receptor and CD28. ►oncogenes, ►cancer, ►angiogenesis, ►morphogenesis in *Drosophila* **{3}**, ►Schwann cell, ►Akt, ►IκB, ►IKK, ►NF-κB; Senftleben U et al 2001 Science 293:1495; Claudio E et al 2002 Nature Immunol 3:958.

NF-X, NF-Y: Both are CAAT-binding transcription factors. ►major histocompatibility complex, ►RFX; Stroumbakis ND et al 1996 Mol Cell Biol 16:192; Linhoff MW et al 1997 Mol Cell Biol 17:4589.

NG (nitrosoguanidine): ►MNNG

Ng: The effective number of genes per organelle locus; in a diploid population it is about 0.25 of the nuclear genes (Palumbi SR et al 2001 Evol Int J Org Evol 55:859). ►effective population size

NGF: ►nerve growth factor

NGFI-A: A mitogen-induced transcription factor (probably he same as egr-1, zif/268, Krox-24, and TIS8, Slade JP, Carter DA 2000 J Neuroendocrinol 12(7):671).

NGFI-B: ►nur77, ►TIS1

NHEJ (nonhomologous end-joining): A repair process for limited homology chromosomes, broken in both chromatid strands by mutagenic agents. Proteins mediating the NHEJ repair pathway include XRCC4, DNA-PK, Ku, and DNA ligase IV. Several of the NHEJ complexes are telomere bound and participate in the maintenance of telomere length. ►XRCC, ►DNA-PK, ►ligase DNA, ►Ku, ►DNA repair, ►nonhomologous end-joining, ►V(D)J, ►gene evolution; Haber JE 2000 Trends Genet 16:259; Barnes DE 2001 Curr Biol 11: R455; Ma Y et al 2004 Mol Cell 16:701.

NHR: A nuclear hormone receptor. ►hormone receptors, ►nuclear receptors

Niacinamide (nicotinamide): A vitamin (see Fig. N25). ►nicotinic acid

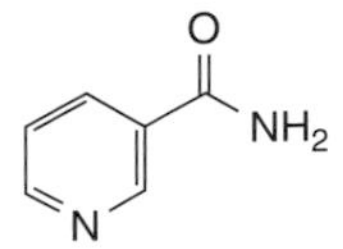

Figure N25. Niacinamide

Nibrin: ►p95

Nicastrin (encoded at human chromosome 1): A 709-amino acid transmembrane glycoprotein, which associates with presenilins and acylpeptide hydrolase (APH, 3p21) and participates in the generation of the amyloid-β-peptide fragment (Aβ) contributing to Alzheimer's disease. ►presenilins, ►Alzheimer's disease, ►secretase; Fagan R et al 2001 Trends Biochem Sci 26(4):213; Kopan R, Goate A 2002 Neuron 33:321.

Niche: A small depression on a surface or a special area of nature, favorable for the species.

Niche: A sheltered environment that sequesters stem cells from stimuli to differentiate. The niche(s) is/are involved in transdetermination and self-renewal of the stem cells. The niche is not necessarily a topological concept because the hematopoietic cells are circulatory; thus, the niche has physiological determinants too. Several different paracrine signals, metabolic, physical and neural factors contribute to the definition of the niche. The Jak-Stat signaling pathway mediates in *Drosophila* spermatogenesis from stem cells. ►stem cells, ►transdetermination, ►signal transduction, ►paracrine; Spradling A et al 2001 Nature [Lond] 414:98; Kiger AA et al 2001 Science 294:2542; Trulina N, Matunis E 2001 Science 294:2546; Scadden DT 2006 Nature [Lond] 441:1075.

Nicholas II: Nicholas II was the czar of Russia and the distressed father of a hemophiliac son, Alexis. Some historians have suggested that this monogenic disease (transmitted by mother Alexandra, granddaughter of Queen Victoria of England) had been one important factor in the disability of Nicholas II to deal with the social problems of his country. Major historical upheavals lead to his murder and to communist rule for about three-fourths of a century. Therefore, a single gene altered the history of about one-fourth of the population of the world. ►hemophilia, ►antihemophilic factors, ►Romanov

Nick: The disruption of the phosphodiester bond in one of the chains of a double-stranded nucleic acid by an endonuclease. It reduces the twist of the double helix. ►phosphodiester bond, ►endonuclease, ►DNA twist

Nick Translation: A restriction enzyme cuts DNA. DNA polymerase I enzyme of *E. coli can* attach to nicks and add labeled nucleotides to the 3′ end while slicing off nucleotides from the 5′ end, thus moving (translating) the nicks through this nucleotide labeling. This is a process of replacement replication. ►DNA replication, ►Klenow fragment; Rigby PWJ et al 1977 J Mol Biol 113:237; see Fig. N26.

Nicked Circle: A circular DNA with nicks. ►Nick

Nicking Enzyme: ►Nick

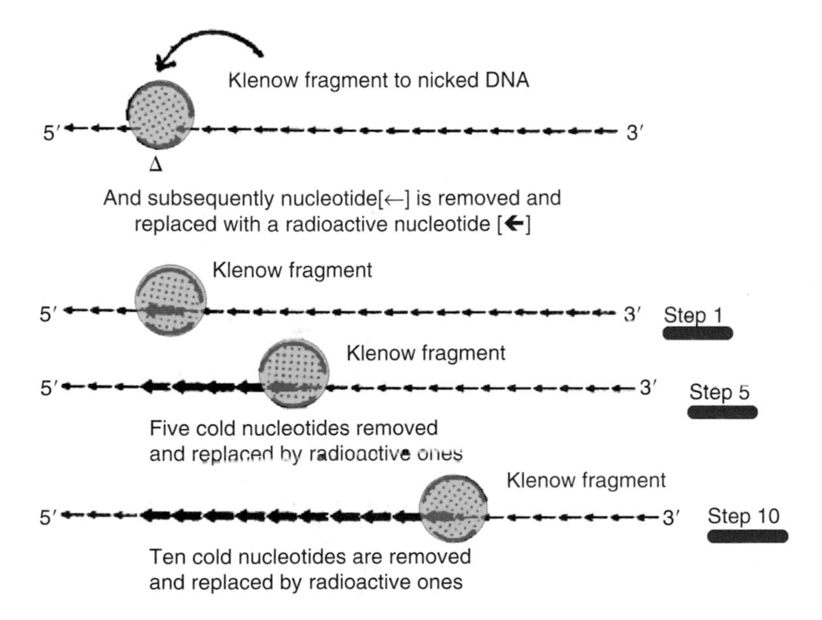
Klenow fragment to nicked DNA
5′
3′
Δ
And subsequently nucleotide[←] is removed and
replaced with a radioactive nucleotide [🡐]
Klenow fragment
5′
3′
Step 1
Klenow fragment
5′
3′
Step 5
Five cold nucleotides removed
and replaced by radioactive ones
Klenow fragment
5′
3′
Step 10
Ten cold nucleotides are removed
and replaced by radioactive ones

Figure N26. Nick translation

Nicotiana (x = 12): Genus in the *Solanaceae* family; it includes over 60 species. The majority is of new world origin although some are native to Australia and the South Pacific islands. About 10 species have been used for smoking and alkaloid production. Some tobaccos are ornamental. Economically the most important is *N. tabacum* (2n = 48) (see Fig. N27). The latter is an amphidiploid, presumably

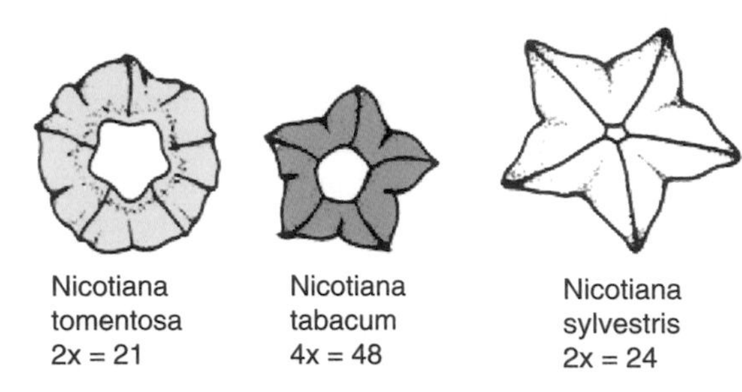

Figure N27. *Nicotiana*. (Figures of flower shape after Goodspeed TH 1954. The Genus Nicotiana. Chronica Bot. Waltham, MA)

of *N. syvestris* (2n = 24) and either *N. tomentosiformis* (2n = 24) or *N. otophora* (2n = 24). Synthetic amphidiploids with the latter resemble less *N. tabacum* than *N. tomentosiformis.*

The fertility of *sylvestris* x *otophora* amphiploids is better than that of *sylvestris x tomentosiformis*. *N. tabacum* has been widely used for cytogenetic analysis; monosomic and trisomic lines are available. This species has been extensively studied by in vitro culture techniques, including regeneration of fertile plants from single cells, because it is very easy to establish cell and protoplast cultures and carry out transformation by agrobacterial vectors. This was also the first higher plant species where Nina Fedoroff could express foreign transposable elements (*Ac* of maize) in Jeff Schell's laboratory (Max-Planck Institut fur Zuchtungsforschung. Köln-Vogelsang. Germany). For manipulations involving isolated cells, the SR1 (streptomycin-resistant chloroplast mutation of *P. Maliga*) is most commonly used. This stock originated from the variety Petite Havana that is relatively easy to handle because of its small size.

Mendelian experiments are difficult to conduct with *N. tabacum* because of its allotetraploid nature. Single gene mutations or chromosomal markers are generally not available. Antibiotic resistant chloroplast mutations (streptomycin, lincomycin) have been produced in *Nicotiana* and served as tools to demonstrate recombination of the plastid genome.

Transformation of chloroplast genes has been accomplished both by the biolistic methods and polyethylene glycol treatment of protoplasts. *N. plumbaginifolia,* one of the diploid species offers some advantages for Mendelian analysis, but it is less suitable for cell cultures and regeneration. ►tobacco

Nicotine (β-pyridyl-α-*N*-methylpyrrolidine): A contact poison insecticide and an indispensable metabolite for nerve and other functions. Nicotine is considered to be an addictive substance. Nicotine is metabolized to cotinine (an antidepressant) by the wild type enzyme CYP2A6. Individuals with the inactive alleles CYPA6*2 and CYPA6*3 are less likely to become smokers, or if they smoke, are will consume less tobacco products. The carriers or homozygotes for the null alleles are also less likely to be affected by diseases common among smokers (cancer, Alzheimer's disease, Tourette syndrome, ulcerative colitis, etc.) Therapeutic treatment targeted to the wild type enzyme may be a means of reducing the smoking addiction. The nature of the α4* nicotinic acetylcholine receptors may control smoking pleasure, tolerance, and sensitization (Tapper AR et al 2004 Science 306:1029). Ornithine aminotransferase and GABA levels quantitatively modify *Drosophila* survival time upon chronic nicotine exposure (Passador-Gurgel G et al 2007 Nature Genet 39:264). (►smoking, ►alkaloids, ►insecticide resistance, ►ornithine aminotransferase deficiency, ►GABA; Pianezza MI et al 1998 Nature [Lond] 393:750; neonicotinoids: Tomizawa M, Casida JE 2003 Annu Rev Entomol 48:339; a model of nicotine addiction: Gutkin BS et al 2006 Proc Natl Acad Sci USA 103:1106).

Nicotine Adenine Dinucleotide: ►NAD^+

Nicotine Adenine Dinucleotide Phosphate: ►$NADP^+$

Nicotine Acetylcholine Receptors: Nicotine acetylcholine receptors are ion-channel linked receptors in the skeletal muscles and presynaptic neurons. Thus nicotine and/or acetylcholine activate the nicotine-acetylcholine-regulated calcium channels resulting in the release of glutamate to its receptor in the post-synaptic cells of the hippocampus. This modulation of synaptic transmission may play an important role in learning, memory, arousal, attention, information processing. In Alzheimer's disease the nicotinic cholinergetic transmission degenerates. After smoking a cigarette, about 0.5 μM nicotine may be delivered within 10 s to the brain and lungs. ►signal transduction, ►nicotine, ►smoking; Itier V, Bertrand D 2001 FEBS Lett 504(3):118.

Nicotinic Acid (niacin): Vitamin PP, coenzyme component (see Fig. N28). ►niacinamide

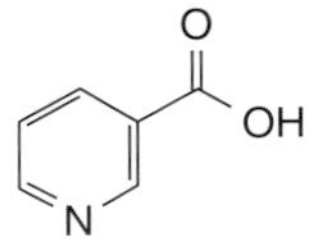

Figure N28. Nicotinic acid

NIDDM: Non-insulin-dependent diabetes mellitus. ►diabetes mellitus

Nidus (Latin: nest): The origin of a process, well, pit, or shallow depression.

Niemann-Pick Disease (NP, sphingomyelin lipidosis): Types C (18p11-q12), A, B, and E (11p15.4-p15.1) are known as autosomal recessive hereditary disorders of sphingomyelinase deficiency. The differences among these types involve the level of activity of the enzyme and the onset of the symptoms. The most common form, Type A, is identified as a severe enlargement of the liver, degeneration of the nervous system, and generally death by age four. In the lysosomes, phosphorylcholine ceramide (sphingomyelin) accumulates. Type B is also deficient in sphingomyelinase and since the nervous system is not affected, patients may live to adulthood. Type C is a milder form of Type A with low activity of the enzyme; the afflicted persons may survive up to age 20. In C1, the accumulation of LDL-derived cholesterol is observed as a consequence of a defect in the NPC-1 protein (Heilig R et al 2003 Nature [Lond] 421:601), a permease with similarity to 3-hydroxy-3-methyl-glutaryl coenzyme A reductase. This protein also displays homology to the Patched morphogen receptor of *Drosophila*. The C1-like protein is the direct target of the anti-cholesterol drug Ezetimibe (Garcia-Calvo M et al 2005 Proc Natl Acad Sci USA 102:8132). The NPC-2 type of the disease is due to deficiency in HE1 (14q24.3), a lysosomal protein. In Type D, the symptoms are similar to those in Type C and sphingomyelin is accumulated, and yet the activity of the enzyme appears close to normal. The latter type apparently involves cholesterol transport from the lysosomes. In the Type E disease the nervous system remains normal and sphingomyelin accumulation is limited to some organs. Type E may not be directly determined genetically. A human acid sphingomyelinase gene has been mapped to chromosome 11p15.1-p15.4. Prevalence of NP is $\sim 7 \times 10^{-6}$. Transferred human acid sphigomyelinase to a mouse model of NP corrects neuropathological and motor deficits (Dodge JC et al 2005 Proc Natl Acad Sci USA 102:17822). ►sphingolipidoses, ►sphingolipids, ►epilepsy, ►lysosomal diseases, ►LDL, ►hedgehog, ►sonic hedgehog, ►SCAP, ►SREBP, ►cholesterol, ►Ezetimibe, ►gene therapy; Sun X et al 2001 Am J Hum Genet 68:1361; Millat G et al 2001 Am J Hum Genet 69:1013; Friedland N et al 2003 Proc Natl Acad Sci USA 100:2512.

Nieuwkoop Center: A position of the early embryo for dorsal and bilateral differentiation. ►Wnt, ►chordin, ►noggin

nif: ►nitrogen fixation

Nigericin (antibiotic of *Streptomyces*): Nigericin facilitates K^+ and H^+ transport through membranes and activates ATPase in the presence of a lipopolysaccharide. It promotes the processing of interleukin-1β precursors. ►ionophore, ►interleukins, ►ATPase; Cascales E et al 2000 Mol Microbiol 38:904.

Night Blindness (nyctalopia): Defective vision in dim light is caused by several choroid and retinal defects. The X-linked phenotype (Xp11.3) is distinguished from the autosomal type by frequent association with myopia (nearsightedness). The autosomal dominant Sorsby syndrome involves retinal degeneration caused by mutation in the tissue inhibitor metalloproteinase-3 gene (TIMP3). Mutation in the leucine-rich proteoglycan, nyctalopin (481 amino acids), encoded by a gene (NYX, Xp11.4) apparently disrupts retinal interconnections. A mutation in rhodopsin may cause autosomal dominant congenital night blindness. ►color blindness, ►day blindness, ►optic atrophy, ►Oguchi's disease, ►Stargardt's disease, ►retinitis pigmentosa; Pusch CM et al 2000 Nature Genet 26:324; Jin S et al 2003 Nature Neurosci 6:731.

Nightshade (*Atropa belladonna*): A solanaceus alkaloid-producing poisonous plant (see Fig. N29), 2n = 72. ►burdo

Figure N29. Nightshade

NIH: National Institutes of Health, Bethesda, MD, USA.

Nijmegen Breakage Syndrome (NBS): A rare 8q21 recessive phenotype closely resembling the characteristics of ataxia telangiectasia with the exception of the absence of ataxia, telangiectasia and elevated levels of α-fetoprotein. It involves chromosomal instability, increased susceptibility to radiation damage, microcephaly, retarded growth, immunodeficiency,

N

and increased chances for (breast) cancer. The 50-kb gene encodes a 754- amino acid protein—nibrin (Nbn) in mouse and NBS in humans. The absence of NBS, class-witch recombination of immunoglobulin G is reduced in B cells (Reina-San-Martin B et al 2005 Proc Natl Acad Sci USA 102:1590). The ataxia telangiectasia kinase phosphorylates the NBS protein. The NBS protein may be translated in an alternative manner resulting in a somewhat modified phenotype. ►ataxia telangiectasia, ►fetoprotein, ►p95, ►Mre11, ►PIK; Zhao S et al 2000 Nature 405:473; Wu X et al 2000 Nature 405:477; Moser RS et al 2001 Nature Genet 27:417; Williams BR et al 2002 Curr Biol 12:648.

NIK (NF-κB inducing kinase): ►NF-κB, ►NκB, ►IKK, ►NEMO

NIL (near isogenic line): ►inbred

NIMA: A mitosis-specific protein kinase. ►parvulins, ►cell cycle, ►histone phosphorylation

Nina: *Drosophila* cyclophilin mediating rhodopsin maturation in the endoplasmic reticulum. ►cyclophilin, ►rhodopsin

Ninhydrin: A triketohydrindene hydrate, a reagent for ammonia. Amino acids upon heating liberate ammonia that reduces ninhydrin and produces a blue color. This reagent is used for very sensitive colorimetric estimation of amino acids or as a spot test on paper or thin-layer chromatograms. ►paper chromatography, ►thin-layer chromatography

NIR (nucleotide incision repair): NIR is carried out by an AP-endonuclease enzyme, incising DNA 5′ of various oxidatively damaged bases. NIR generates both 3′ and 5′ ends and permits repair of the dangling modified nucleotide independent from DNA-glycosylase activity. It is similar to BER in action; the latter however removes the abnormal base and deoxyribose, leaving an abasic site or a single-strand break in the DNA (Ishchenko AA et al 2006 Proc Natl Acad Sci USA 103:2564). ►DNA repair, ►BER, ►glycosylase

NIS (sodium [Na]-iodide symporter): A transmembrane carrier protein also responsible for transporting radioactive iodine into the thyroid and breast epithelial cells and causing cancer. NIS activity is correlated with pregnancy and the onset of lactation. Iodine is essential for normal development. Radioactive iodine has been one means of treatment of Basedow/Grave's disease. Regulating NIS activity may be medically useful. ►goiter, ►isotopes, ►atomic radiation

NISH: Non-isotopic in situ hybridization. ►FISH, ►in situ hybridization

Nishimine Factor: A special hemostatic factor required for the generation of thromboplastic (blood clot forming) activity. ►antihemophilic factors

Nisin (lanthionin-containing peptide antibiotic): Nisin is produced by some strains of the bacterium *Lactococcus lactis*. (Lanthionine is bis[2-amino-2-carboxyethyl] sulfide.) It makes pores in the membranes of Gram-positive bacteria. It is highly effective and until 1999 resistance against it was not found. ►antimicrobial peptides, ►lantibiotics; Breukink E et al 2000 Biochemistry 39:10247.

Nitrate Reductase: Nitrate reductase is used in prokaryotes to reduce nitrate to nitrite under the conditions of anaerobic respiration. The enzyme is associated with molybdenum and is essential for normal function. Nitrate reductase is generally also associated with cytochrome b. Plants and fungi reduce nitrate before it can enter into the amino acid synthetic path. Nitrate reductase mutations are generally selected on chlorate media. ►chlorate, ►nitrogenase, ►molybdenum, ►nitrogen fixation; Heath-Pagliuso S et al 1984 Plant Physiol 76:353; Lejay L et al 1999 Plant J 18:509.

Nitric Oxide (NO): Endothelial cells may make and release NO in response to the liberation of acetylcholine by the nerves (NOS1, 12q24.2) in the blood vessel walls. The hepatic expression gene, NOS2A, is at 17cen-q11.2. The endothelial enzyme, NOS3, is encoded at 7q36. NOS4 is a chondrocyte enzyme. Nitric oxide synthase isozymes (NOS) produce NO from L-arginine. NO activates potassium channels through a cGMP-dependent protein kinase. It is then that the smooth muscles of the veins become relaxed and blood flow is boosted. This mechanism is the basis of penal erection and is the target of the action of the anti-impotence drug Viagra (Bivalacqua TJ et al 2000 Trends Pharmacol Sci 21:484). Penile erection is mediated by NO produced by alternatively spliced endothelial and neuronal NOS (Hurt KJ et al 2006 Proc Natl Acad Sci USA 103:3440). *In vivo* gene transfer of the endothelial isoform of nitric oxide synthase may also alleviate erectile dysfunction in animal models. NO is contributing toward the activation of macrophages and neutrophils in the body's defense reaction. In *Bacillus subtilis*, NO transiently suppresses cysteine and activates catalase, which is suppressed by cysteine (Gusarov I, Nudler E et al 2005 Proc Natl Acad Sci USA 102:13855). NO may also cause cytostasis during differentiation. NO is also called a physiological messenger of the cardiovascular, immune and nervous system, memory, and learning. NOS proteins interact with more than 20 other proteins regulating a wide array of metabolic systems (Zimmermann K et al 2002

Proc Natl Acad Sci USA 99:17167). Deficiency of NO may cause pyloric stenosis, depression, hypotension, inflammation, aggressive behavior, recurrent miscarriage, etc. NO may contribute to the formation of reactive nitrogen species (RNS) and reacting with $O^{2}\bullet^{-}$ it may generate $ONOO^{-}$ (peroxynitrite) radicals and may contribute to the formation of peroxides. NO may have a disinfectant effect against fungi and viruses. $ONOO^{-}$ can oxidize sulfhydryl groups, peroxidize lipids and nitrate, or deaminate DNA bases that may cause cytotoxicity and mutation. NO also regulates angiogenesis and thus may have both promoting and suppressive effects on tumor growth. The inducible nitric oxide synthase II gene introduced into several tumors by adenoviral vectors displayed concentration-dependent antitumor and antimetastasis activity despite the fact that it up-regulated angiogenesis factors (Le X et al 2005 Proc Natl Acad Sci USA 102:8758).

Atherosclerosis, pulmonary hypertension, pyloric stenosis, stroke, multiple sclerosis, etc., may be related to NO metabolism. NOs expression/activity is controlled by steroid sex hormones. NO also controls the metabolism of metals. NO gene therapy has been considered to relieve hyperoxic lung injury and pulmonary hypertension. Unfortunately, the technology is not yet satisfactory because NO is inhibitory to the adenoviral vectors used (Champion HC et al 1999 Circ Res 84:1422). NO also regulates plant growth and hormonal signaling (Guo F-Q et al 2003 Science 302:100). Inducible nitric oxide synthase can be visualized in situ by PIF (pyrimidine imidazole FITC) probes (Panda K et al 2005 Proc Natl Acad Sci USA 102:10117). The 1998 Nobel Prize in medicine recognized the discovery of the physiological role of nitric oxide. ►depression, ►hypotension, ►pyloric stenosis, ►aggression, ►circadian rhythm, ►cGMP, ►ion channels, ►olfactogenetics, ►hypersensitivity reaction, ►Akt oncogene, ►chemical mutagens, ►luciferase, ►peroxynitrite, ►ROS, ►phosphodiesterase, ►priapism, ►FITC, ►cyclooxygenase; Stuehr D et al 2001 J Biol Chem 276:14533; Ganster RW et al 2001 Proc Natl Acad Sci USA 98:8638.

Nitrification: The conversion of ammonium into nitrate. (See Hipkin CR et al 2004 Nature [Lond] 430:98).

Nitroblue Tetrazolium ($C_{40}H_{30}Cl_2N_{10}O_6$): ►NBT, ►BCIP

Nitrocellulose Filter: A nitrocellulose filter can be pure nitrocellulose or of cellulose acetate and cellulose nitrate mixtures. It can be used for the immobilization of RNA in Northern blots, for Southern blotting of DNA, for replica plating and storage of λ phage or cosmid libraries, bacterial colonies, for Western blotting, for antibody purification, etc. Nucleic acids generally have superior binding to nylon filters. (See entries separately, Thomas PS 1980 Proc Natl Acad Sci USA 77:5201).

Nitrogen (N_2 = 28.02 MW): An odor and colorless gas that becomes liquid at –195.8°C and solidifies at –210°C; four-fifth of the volume of air is nitrogen.

Nitrogen Cycle: ►nitrogen fixation

Nitrogen Fixation: The mechanisms of incorporating atmospheric nitrogen (N_2) into organic molecules with the aid of the nitrogenase enzyme complex, present in a few microorganisms (*Clostridia, Klebsiella, Cyanobacteria, Azotobacters, Rhizobia, etc.*). The first step in the process is the reduction of N_2 to NH_3 or NH_4. It has been estimated that microbes reduce, annually, about 120 million tonnes of atmospheric nitrogen. Atmospheric nitrogen can also be converted into ammonia industrially at high temperature and high atmospheric pressure (for the production of fertilizers and explosives). Many organisms are capable of using ammonia and can oxidize it into nitrite (NO_2^-) and nitrate (NO_3^-) by the process called *nitrification*.

Bacterial nitrogen fixation relies primarily on two enzymes: the tetrameric *dinitrogenase* (M_r 240,000), containing 2 molybdenum, 32 ferrum, and 30 sulfur per tetramer and the dimeric *di-nitrogenase reductase* (M_r 60,000) with a single Fe_4-S_4 redox center and two ATP binding sites. Nitrogen fixation begins by the reduced first enzyme. The reduction is the job of this second enzyme that hydrolyzes ATP. The role of ATP is to donate chemical and binding energy. The nitrogenase complex uses different means to protect itself from air (oxygen) that inactivates it. The crystal structure of nitrogenase complexes with multiple docking sites has been identified (Tezcan FA et al 2005 Science 309:1377). In the symbiosis between *Rhizobia* and legumes, leghemoglobin, an oxygen-binding protein, is formed in the root nodules for protecting the nitrogen fixation system. In *Rhizobium leguminosarum* and related species, the genes required for nitrogen fixation and nodulation are in the large conjugative plasmids. The large plasmid of the *Rhizobium* species NGR234 contains 536,165 bp, including 416 open reading frames; 139 of these genes seem to be unique in any organism. In *R. meliloti* the corresponding genes are within the bacterial chromosome. The fast growing species (colony formation in four to five days), *Bradyrhizobium*, are common in tropical areas while in moderate climates the various *Rhizobia* species are found on legumes. These two main groups cannot utilize N_2 in culture but only in the nodules. The *Azorhizobia* can use N_2 without a symbiotic relation. The nodule formation requires a special interaction between the

N

bacteria and the host. A series of nodulation factors have been identified—including the *ENOD40*-encoded 10-amino acid oligopeptide, a member of the TNF-R family, among others. A homolog of this gene has been found in non-legumes too. The *Rhizobium leguminosarum* (*Vicia*) nodulation genetic system may be represented as:

T N M L EF **D** *A B C I J*
→ ←←← ← ←← ← →→→→→

The *nod* (nodulation) genes of the bacteria are located in a plasmid within the bacterial cell. The *nodD* expression is induced by root exudates (the flavonoids eryodyctyol, genistein). The rhizobial signals (lipochitin-oligosaccharide; ►chitin) are perceived by the legume roots Nod factor; one of them, NFR5, is a transmembrane serine/threonine receptor-like kinase (Madsen EB et al 2003 Nature [Lond] 425:637). Two symbiotic, photosynthetic, *Bradyrhizobium* strains, BTAi1 and ORS278, do not have the canonical *nodABC* genes and typical lipochito-oligosaccharidic Nod factors. Mutational analyses indicates that these unique rhizobia use an alternative pathway to initiate symbioses, where a purine derivative may play a key role in triggering nodule formation (Giraud E et al 2007 Science 316:1307).

This is similar to the infection process of plants by *Agrobacteria* (related to *Rhizobia*) where the *virulence* gene cascade is induced by flavonoids (acetosyringone). As a consequence of turning on the *nod* system, signals are transmitted to the root hairs to curl up and develop into nodules inhabited then by bacteria. The different legume species produce different inducers and the bacteria may have different genes to respond to, and this interaction specifies host and bacterial functions. Once within a root hair the infection spreads to neighboring cells. The bacteria are enveloped by plant membranes and form *bacteroids*. The nodule formation assures the right supply of oxygen to the bacteroids and at the same time protects the nitrogenase system from oxygen that is detrimental to it.

Nitrogen fixation is controlled by the *nif* and *fix* genes, encoding the FixLJ proteins. The FixL protein is a transmembrane kinase that is activated by low oxygen tension. This kinase then phosphorylates the FixJ protein that in turn switches on *FixK* and *nifA* genes. The product of the latter interacts with an upstream element to promote the transcription of *fix* and *nif* operons. Nodule formation and nitrogen fixation then depend on complex circuits between bacterial and (and over 20) plant genes and environmental stimuli. There is a rather high degree of host-specificity for nodulation. *R. leguminosarum bv. viciae* hosts include pea, vetch, sweet pea, and lentil. *R. meliloti* nodulates sweet clover, alfalfa, soybean, and *Trigonella*. *Bradyrhizobium japonicum* uses soybean and cowpea (*Vigna sinensis*) (see Fig. N30) Some other *Rhizobia* may have a larger

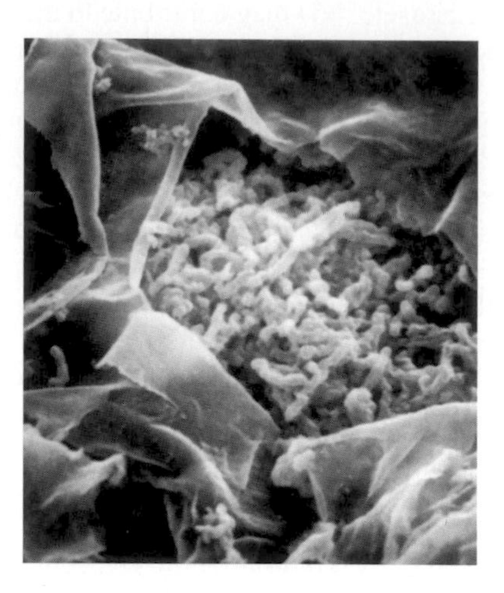

Figure N30. *Rhizobium* bacteria within a soybean root nodule. (Courtesy of Dr. WJ Brill)

range of hosts. The relationship between the host and bacteria is mutually advantageous and the symbiotic system also has enormous economic value in maintaining soil fertility and crop productivity. It would be desirable to transfer the genes required for nitrogen fixation into crop plants. Some of the homologs of the nodulin genes are already present in plants (leghaemoglobin). The *nif* genes of *Klebsiella* had been transferred to plants but they failed to express. To overcome this problem, transformation of the (prokaryote-like) chloroplasts was considered but it was not possible to protect the O_2-sensitive nitrogenase from the oxygen evolution concomitant with photosynthesis. Plants reduce nitrate to nitrite by nitrate reductase (NR). Then nitrite reductase (NiR) converts nitrite to ammonium. Glutamine synthetase (GS) and glutamine-2-oxoglutarate aminotransferase (GOGAT) incorporate the nitrogen into organic compounds (glutamate, asparaginate, etc.). The *Mesorhizobium loti* chromosome (~7 Mb) and its two plasmids (~0.35 Mb and ~0.208 Mb) and the *R. meliloti* (*Sinorhizoboium meliloti*) chromosome (3.65 Mb), and two of its plasmids (1.35 Mb and 1.68 Mb) have been completely sequenced (Kaneko T et al 2000 DNA Res 7:331; Galibert F et al 2001 Science 293:668; resp.). ►*Agrobacteria*, ►bacteria, ►symbionts, ►symbiosome, ►autoregulation, ►glutamine synthetase, ►nodule; Freiberg C et al 1997 Nature [Lond.] 387:394; van de Sande K, Bisseling T In: Essays in Biochemistry. Bowles DJ (ed) Portland Press, London, pp 127; Spaink HP, Kondorosi Á, Hooykas PJJ (eds) 1998 The Rhizobiaceae. Kluwer, Dordrecht; Christiansen J et al 2001 Annu Rev Plant

Physiol Mol Biol 52:269; Proc Natl Acad Sci USA 2001 Vol. 98, pp. 9877–9894; Endre G et al 2002 Nature [Lond] 417:962; Geurts R, Bisseling T 2002 Plant Cell 14:S239; Radutolu S et al 2003 Nature [Lond] 425:585; Kaló P et al 2005 Science 308:1786.

Nitrogen Mustards: Nitrogen mustards are (radiomimetic) alkylating agents. See Fig. N31 for the general formula where Al means alkyl groups; in this form it is trifunctional and has three chlorinated alkyl groups. When it is monofunctional it has one chlorinated alkyl group; the bifunctional has two such groups. The nitrogen mustards family of antineoplastic and immunosuppressive compounds includes cyclophosphamide, uracil mustard, melphalan, chlorambucil, etc. ►sulfur mustards, ►mustards; Ross WCFJ 1962 Biological Alkylating Agents. Butterworth, London, UK.

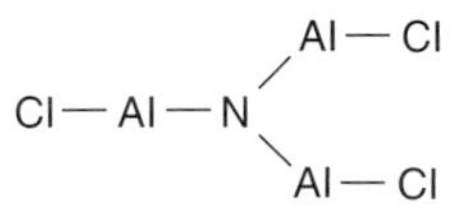

Al—Cl
Cl—Al—N
Al—Cl

Figure N31. Nitrogen mustard

Nitrogenase: ►nitrogen fixation

Nitrogenous Base: A purine or pyrimidine of nucleic acids.

Nitrosamines: Nitrosamines are highly toxic carcinogens and mutagens (see Fig. N32). Lethal Dose Low (LDLo) of nitrosodimethylamine for mouse, orally, is 370 mg/kg, and intraperitonally 7 mg/kg; for dogs, orally, it is 20 mg/kg. ►ENU, ►chemical mutagens

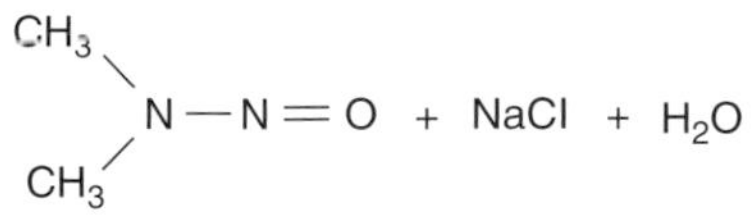

$(CH_3)_2N—N=O + NaCl + H_2O$

Figure N32. *N*-Nitrosodimethylamine

Nitrosomethyl Guanidine (NNMG): A very potent alkylating agent, mutagen and carcinogen (see Fig. N33). It is light sensitive and potentially explosive. ►chemical mutagens

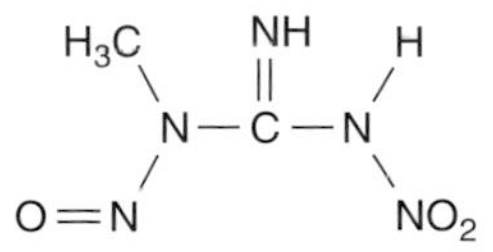

H_3C, NH, H, N—C—N, O=N, NO_2

Figure N33. Nitrosoguanidine

Nitrosourea: ►ENU

Nitrous Acid Mutagenesis (HNO_2, MW 47.02): ►chemical mutagens (ii)

NK: ►killer cell

NKG2 (KLRK, 12p13): Regulatory lectin-like immune receptors of NK and T cells. The NKG2D ligands are upregulated in non-tumor cell lines by DNA damage. Subsequently, either DNA repair or apoptosis is stimulated. (See Gasser S et al 2005 Nature [Lond] 436:1186).

NKG2C, NKG2D, NKG2E: Activating receptors of CTLs. HLA-E is the ligand. ►CTL, ►HLA

NKT Cells (natural killer T lymphocytes): NKT cells are similar in function to NK cells and share receptors. They also play a regulatory function for lymphocytes by releasing, promptly, large amounts of IFN-γ and IL-4. ►killer cells, ►interferon, ►interleukin; Seino K-i et al 2001 Proc Natl Acad Sci USA 98:2577.

NLI: A nuclear LIMB domain-binding factor. ►Lim domain, ►LDB

NLP: A shared relationship analysis according to co-occurrence in the scientific literature (Wren JD, Garner HR 2004 Bioinformatics 20:191).

NLS: ►nuclear localization sequence

NMD (nonsense-mediated mRNA decay): ►mRNA surveillance, ►NAS

NMDA (*N*-methyl-D-aspartate) Receptor: The NMDA receptor is located in the postsynaptic membranes of the excitatory synapses of the brain. Activated NMDA receptors control the Ca^{2+} influx and excitatory nerve transmission. Tyrosine kinases and phosphatases regulate NMDA. NMDA receptors may cause high sensitivity to acute injury to brain oligodendrocytes that can be the cause of cerebral palsy, spinal cord injury, multiple sclerosis, and stroke (Káradóttir R et al 2005 Nature [Lond] 438:1162; Salter MG, Fern R 2005 Nature [Lond] 438:1167). ►synapse, ►PTK, ►PTP, ►DARPP, ►CaMK, ►nociceptor, ►PDZ domain, ►lupus erythematosus, ►Eph, ►AMPA; Klein RC, Castellino FJ 2001 Curr Drug Targets 2:323; crystal structure of NMDA: Furukaqwa H et al 2005 Nature [Lond] 438:185.

NMDAR: ►NMDA receptor

***N*-Methyl-N′-Nitro-*N*-Nitrosoguanidine**: ►MMNG

NMR: ►nuclear magnetic resonance spectroscopy

NMYC: ►MYC

Noah's Ark Hypothesis: The low level of divergence among the different human populations indicates recent origin of modern humans and the relatively fast replacement of antecedent populations. ►Eve foremother, ►Y chromosome; Brookfield JFY 1994 Curr Biol 4:651.

Nociceptin (orphanin): ►opiate

Nociceptor: A pain receptor involved in injuries. It is regulated by the NMDA-receptor and substance P neurotransmitters. The $P2X_3$ ATP receptor mediates ATP-gated cation channels and pain sensation. The melanocortin-1 receptor gene mediates κ-opioid analgesia only in female mice and humans. Fair-skinned, red-haired women seem to display greater analgesia to the opioid pentazocine (Mogil FS et al 2003 Proc Natl Acad Sci USA 100:4867). ►opiate, ►analgesic, ►NMDA, ►capsaicin; Zubieta J-K et al 2001 Science 293:311; Julius D, Basbaum AI 2001 Nature [Lond] 413:203; mutants: Tracey WD et al 2003 Cell 113:261.

Nocturnal Enuresis (bedwetting): The dominant gene has been assigned to human chromosome 13q13-q14.3. The condition generally afflicts children under age 7 and then gradually improves and usually disappears later. Its cause is the improper regulation of the antidiuretic hormone and social maladjustment. ►antidiuretic hormone

NOD: A kinesin-like protein on the surface of the oocyte chromosomes of *Drosophila* that stabilizes the chromosomes to stay on the prometaphase spindle. A similar protein, Xklp1, is found in *Xenopus* oocytes. ►kinesin, ►oocyte, ►spindle, ►nondisjunction; Clark IE et al 1997 Development 124:461.

NOD (nucleotide-binding oligomerization protein): Similar to the Toll-like receptor transmembrane proteins, NODs are important for the innate immunity of humans, animals, and plants. The different NOD proteins have variable amino-terminal effector-binding domains, a central nucleotide-binding oligomerization domain and carboxyl-terminal ligand recognition domain, and leucine-rich repeats. The effector domain includes a caspase-recruitment domain (CARD) and other sites to bind proteins mediating apoptosis. NODs recognize peptidoglycans of invading bacteria. NODs are also activated in many human diseases. The host responds by generation of reactive oxygen species and chemokines for defense. ►glycoseaminoglycan, ►peptidoglycan, ►Toll, ►leucine-rich repeat, ►caspases, ►apoptosis, ►immunity innate; Inohara N et al 2005 Annu Rev Biochem 74:355.

NOD Mouse: A NOD mouse is non-obese diabetic. ►diabetes; Zucchelli S et al 2005 Immunity 22:385.

Nodal: A vertebral embryonic signaling protein resembling TGF. It is involved in the control of mesoderm and endoderm differentiation, anterior-posterior determination, specification of the left-right axis, etc. ►TGF, ►DSTF, ►left-right asymmetry, ►mesoderm, ►endoderm, ►organizer; Iratni R et al 2002 Science 298: 1996.

Node: A knot- or swelling-like structure; the widened part of the plant stem from where leaves, buds, or branches emerge; also a crossover site in the DNA (see Fig. N34). ►internode, ►junction of cellular network

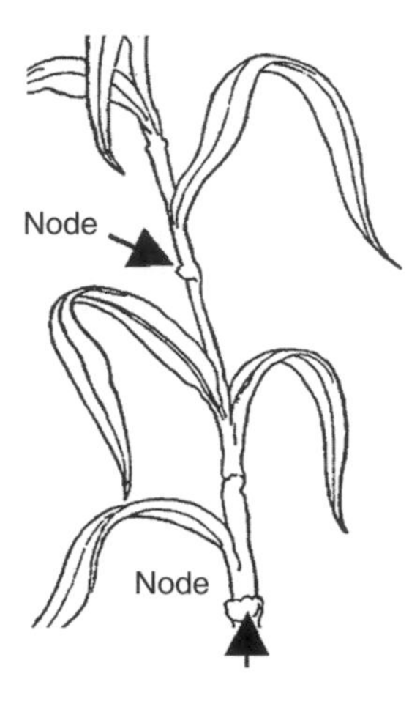

Figure N34. Node

Nodule: A small roundish structure. The root nodule of legumes harbors nitrogen-fixing bacteria. The rhizobial nodulation genes determine the formation of lipo-chitooligosaccharide signals that specify the Nod (nodulation) factors (NF). Perception of Nod leads to Ca^{2+} oscillations by protein kinase Ca^{2+} /Calmodulin-dependent protein kinase (CCaMK). CCaMK has an autoinhibitory domain and its removal autoactivates nodulation even in the absence of rhizobia, but in the presence of GRAS family transcriptional regulators. Some non-leguminous plants have somewhat homologous proteins to CCaMK; thus there might be a possibility to engineer nitrogen fixation into non-legumes (Gleason C et al 2006 Nature [Lond] 441:1149; Tirichine L et al 2006 Nature [Lond] 441:1153). Nodulation in legumes is cytokinin-dependent and histidine kinase is the cytokinin receptor. Specific mutation in the histidine kinase may lead to spontaneous nodulation in the absence of nitrogen fixing bacteria (Tirichine L et al 2007 Science 315:104). Other mutation in the receptor may eliminate nodule formation even when colonized by Rhizobia (Murray JD et al 2007 Science 315:101) The plant cells in the nodule replicate DNA without cytokinesis and consequently become highly polyploid (endoploidy). Interestingly, the bacteroid also replicate their DNA excessively and their cells elongate as a result. These polyploidy bacteria are metabolically functional but lose their ability to resume growth as free-living bacteria do. The polyploidization of the bacteria in the nodule is controlled by the host plant and differs accordingly (Mergaert P et al 2006 Proc Natl Acad Sci USA 103:5230). ►nitrogen fixation, ►recombination nodule, ►bacteroid; Spaink HP 2000 Annu Rev Microbiol 54:257.

Nodule DNA: Nodule DNA is formed when two H-DNAs (a purine-rich *H-DNA-PU3′ and pyrimidine-rich H-DNA-PY3′ are combined into a structure as shown in Fig. N35. ►H-DNA, ►recombination nodule

Figure N35. Nodule DNA

Nodulin: A plant protein in the root nodules of leguminous plants. ►nitrogen fixation

NodV: *Rhizobium* kinase affecting nodulation regulator protein NodW. ►nitrogen fixation; Loh J et al 1997 J Bacteriol 179:3013; Stacey G 1995 FEMS Microbiol Lett 127:1.

Noggin: A bone/chondrocyte morphogenetic protein encoded at 14q22-q23. Noggin mutations at the 17q22 gene cause brachydactyly, deficiencies of interphalangeal (finger and toe) joints, and bone fusion (multiple synostoses). ►organizer, ►bone morphogenetic protein; Marcelino J et al 2001 Proc Natl Acad Sci USA 98:11353; Brown DJ et al 2002 Am J Hum Genet 71:618; Groppe J et al 2002 Nature [Lond] 420:636.

Nogo: A myelin-associated group of neurite growth inhibitors that prevent repair of damage and regeneration in the central nervous system and the spinal cord. (See Grandpre T, Strittmatter SM 2001 Neuroscientist 7(5):377).

No-Go-Decay: ►mRNA degradation

Noise: Random variations in a system, either biological or mechanical or electronic, or disturbance and interference in any of these systems. Noise in gene expression causes phenotypic heterogeneity in isogenic populations (e.g., identical twins or clones). Increased variability in gene expression, affected by the sequence of the TATA box, can be beneficial after an acute change in environmental conditions. Mutations within the TATA region of an engineered *Saccharomyces cerevisiae GAL1* promoter can be characterized as being either highly variable and rapid or steady and slow. A stable transcription scaffold can result in "bursts" of gene expression, enabling rapid individual cell responses in the transient and increased cell-cell variability at a steady state. Increased cell-cell variability enabled by TATA-containing promoters confers a clear benefit in the face of an acute environmental stress (Blake WJ et al 2006 Mol Cell 24:853). Signal amplification and damping may cause fluctuations and phosphorylation may counteract damping. The noise can be intrinsic due to rare and few molecules. Extrinsic noise is caused by extracellular events due to fluctuations in transcription factors and it is manifested differently among different cells. In yeast, the proteins are present in 50 to million copies and mRNA in 0.001 to 100 copies per cell. The stochastic expression of the genes is also affected by their position along the chromosomes (Becskei A et al 2005 Nature Genet 37:937). Genetic networks are affected by noise because chemical reactions are probabilistic and many genes, RNAs, and proteins occur in low numbers (Paulsson J 2004 Nature [Lond] 427:415). Noise in a single-cell proteome level can be monitored by GFP fusion to the carboxy-terminals of proteins and flow cytometric analysis (Newman JRS et al 2006 Nature [Lond] 411:840). ►transcriptional noise, ►plasticity, ►genetic networks; Ozbudak EM et al 2002 Nature Genet 31:69; Elowitz MB et al 2002 Science 297:1183; Rao C et al 2002 Nature [Lond] 420:231; Raser JM, O'Shea EK 2005 Science 309:2010; delay-induced oscillations in gene regulation: Bratsun D et al 2005 Proc Natl Acad Sci USA 102:14593; possible role *Wnt* in noise filtering in development: Arias AM, Hayward P 2006 Nature Rev Genet 7:34; noise in gene function: Maheshri N, O'shea EK 2007 Annu Rev Biophys Biomol Struct 36:413.

Nomad: Specially constructed vectors built of fragment modules in a combinatorially rearranged manner. These vectors allow sequential and directional insertion of any number of modules in an arbitrary or predetermined way. They are useful for studying promoters, replicational origins, RNA processing signals, construction of chimeric proteins, etc. (See also Rebatchuk D et al 1996 Proc Natl Acad Sci USA 93:10891).

Nomadic Genes: Nomadic genes are dispersed repetitive chromosomal elements with a high degree of transposition. ►copia, ►insertion elements, ►transposon, ►retroposon, ►retrotransposon

Nomarski Differential Interference Contrast Microscopy: A phase shift is introduced artificially in unstained objects to cause a field contrast that is due to interference with diffracted light. As a result either dark field (phase contrast) or bright field contrast (Nomarski technique) can be obtained depending on the direction of the light by a quarter wave phase shift in special microscopic equipment. In colored microscopic specimens the phase shift (and contrast of the image) is brought about by the differential staining by microtechnical dyes. ►resolution optical, ►fluorescence microscopy, ►phase-contrast microscopy, ►confocal scanning, ►electron microscopy

Nomenclature: ►gene symbols, ►gene nomenclature assistance, ►databases. (See Turnpenny P, Smith R 2003 Nature Rev Genet 4:152).

Nomothetics: The study of laws of nature and law-like generalizations that are applicable for classes of individuals as found in the evolutionary process.

Nonallelic Genes: Nonallelic genes belong to different gene loci and are complementary. ►allelism

Nonallelic Non-complementation: A relatively rare phenomenon when recessive alleles of different gene loci fail to be complementary in the heterozygote despite the presence of a dominant allele at both loci. Such a situation occurs when the two gene loci encode physically interacting products. The mechanism may involve a particular *dosage* problem. Reduced dosage at one locus still supports the wild phenotype but simultaneous reduction of dosage at another gene may not permit wild expression of the wild phenotype. Such may be the case when a ligand and its receptor mutate the same time. An alternative mechanism may be based on poisoning the expression of one mutation by another, e.g., one of the locus controls polymerization of another product. The first mutation is innocuous as long as the second locus is fully expressed, but decrease of the level of the product of the second gene may have deleterious consequences. In fact, physical interaction of the products of the two loci is not a requisite for the phenomenon although, most commonly, physical interaction occurs between the two proteins. ►allelism test, ►allelic complementation; Yook KJ et al 2001 Genetics 158:209.

Nonautonomous Controlling Element: A transposable element that has lost the transposase function and can move only when this function is provided by a helper element. A nonautonomous controlling element is *Ds* or *Spm* of maize whereas the helper element may be *Ac* and *Spm*, respectively (see Fig. N36). ►transposable elements, ►*Ac-Ds*, ►*Spm*; McClintock B 1955 Brookhaven Symp Biol 8:58; McClintock B 1965 Brookhaven Symp Biol 18:162.

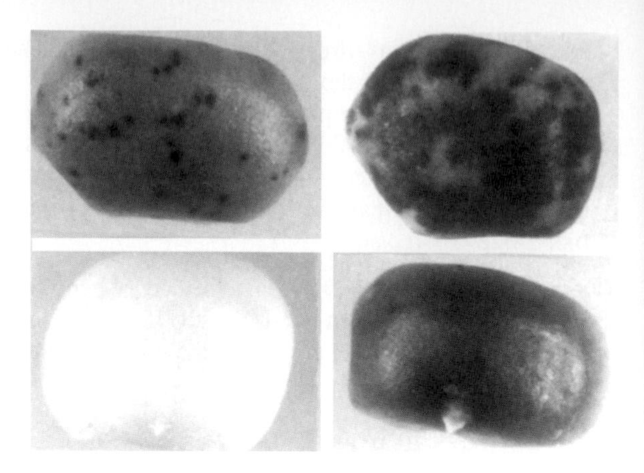

Figure N36. Active Spm transposase (top maize kernels) and inactive transposase (bottom endosperms) in either solid deeply colored (A) or inactivated (a-m[1]) states. (Courtesy of Barbara McClintock)

Nonchromosomal Genes: Genes that are not located in the cell nucleus. It is more logical to call them extranuclear genes, like the genes in mitochondria and plastids.

Noncoding DNA: Noncoding DNA is not translated into protein, e.g., some of the sequences of introns and intergenic regions. These sequences were considered of no or minimal evolutionary significance. Some of these sequences—in *Drosophila*—seem to be subject to positive selection for maintenance and may be of significance for adaptive evolution (Andolfatto P 2005 Nature [Lond] 437:1149) and their preservation is not due to mutational cold spots (Drake JA et al 2006 Nature Genet 38:223). Conserved noncoding (CNE/CNS) elements of transposon origin represent~3.5% of the human genome (Xie X et al 2006 Proc Natl Acad Sci USA 103:11649). The mean CNS length in the plant *Arabidopsis* is 31 bp, ranging from 15 to 285 bp. There are 1.7 CNSs associated with a typical gene, and *Arabidopsis* CNSs are found in all areas around exons, most frequently in the 5′ upstream region. Gene ontology classifications related to transcription, regulation, or to external or endogenous stimuli, especially hormones, tend to be significantly overrepresented among genes containing a large number of CNSs, whereas protein localization, transport, and metabolism are common among genes with no CNSs (Thomas BC et al 2007 Proc Natl Acad Sci USA 104:3348). ►junk DNA, ►conserved noncoding elements, ►ENCODE

Noncoding RNA (ncRNA): The DNA is transcribed into RNA but the RNA does not specify oligo- or polypeptides; examples are the ribosomal rRNAs, transfer tRNAs, small nuclear snRNAs, small nucleolar snoRNAs, microRNAs, RNAi, as well as small RNA components of RNase P and other protein complexes. Actually, about half of the RNA transcripts of eukaryotes are noncoding. According to some estimates, about 1.2% of the human genome codes for protein, but 60–70% of the mammalian genome is transcribed into RNA on one or both strands (Mattick JS, Makunin IV 2006 Human Mol Genet 15:R17). More recent estimates indicate that more than 90% of the genomic DNA is transcribed (ENCODE Project 2007). Many of the noncoding RNAs have regulatory functions such as microRNA and RNAi and they are conserved across species boundaries. Some of the ncRNAs are much longer (~2 kb) than the small RNAs. NRON (noncoding repressor of NFAT) dramatically affects the function of transcription factor NFAT and other proteins involved in numerous cellular activities such as nucleocytoplasmic transport (importins), translation initiation, E3 ubiquitin ligase, DEAD-box proteins, calmodulin-binding protein, protein phosphatase 2, and JNK-associated leucine zipper proteins. In maize, large stretches of noncoding DNA affects the expression of downstream quantitative genes (Clark RM et al 2006 Nature Genet 38:594). (See mentioned terms under separate entries, ►ENCODE, ►noncoding sequences, ►heterochromatin, ►gene number, ►Xist, ►DHFR, ►ribozymes, ►snoRNA, ►tRNA, ►gRNA, ►U RNA, ►tiling, ►TUF, ►snoRNA, ►RNAi, ►microRNA, ►chimpanzee; Mattick JS 2005 Science 309:1527; Claverie J-M 2005 Science 309:1529; Hüttenhofer A, Schattner P 2006 Nature Rev Genet 7:475; review of non-coding RNA in gene silencing: Zaratiegui M et al 2007 Cell 128: 763; regulatory ncRNAs: Prasanth KV, Spector DL 2007 Genes & Development 21:11; ncRNA database: http://biobases.ibch.poznan.pl/ncRNA/; http://research.imb.uq.edu.au/RNAdb; functional non-coding RNA: http://www.ncrna.org/.

Noncoding Sequences: Noncoding sequences do not specify an amino acid sequence or a tRNA or rRNA. About 98 to 99% of the human genome is noncoding. They are generally clustered (e.g., in human chromosome 5q31) in noncoding intergenic regions (~45%), in introns, in the nontranslated upstream (promoter) and downstream sequences, and sometimes overlap with coding sequences. Conserved noncoding sequences contained 233 motifs matching a total of 60,000 conserved instances across the human genome. These motifs include known regulatory elements, such as the histone 3′-UTR motif and the neuron-restrictive silencer element, as well as striking examples of novel functional elements. The most highly enriched motif corresponds to the X-box motif. Their numbers may run to hundreds of thousands in the mammalian genomes and may be well conserved in some areas among species. Nearly 15,000 conserved sites serve as likely insulators, and nearby genes separated by predicted CTCF sites show markedly reduced correlation in gene expression. These sites may thus partition the human genome into domains of expression (Xie X et al 2007 Proc Natl Acad Sci USA 104:7145).

They regulate various types of interleukins (IL-4, -5, -13) and are well conserved between mouse and humans. The regulation has some cell-specificity (T_H2) and may selectively affect genes at a distance over 120 kb. About 0.3 to 1% of the human genome contains highly conserved nongenic sequences and these must play important roles as judged from their conservation across evolutionary ranges (Dermitzakis ET et al 2003 Science 302:1033). Noncoding DNA is apparently subject to adaptive evolution in *Drosophila* too (Andolfatto P 2005 Nature [Lond] 437:1195). ►coding sequence, ►CTCF, ►insulator, ►X-box, ►UTR, ►small RNA, ►T cell, ►interleukins, ►RNA regulatory; Loots GG et al 2000 Science 288:136; Cliften PF et al 2001 Genome Res 11:1175.

Noncompetitive Inhibition: In noncompetitive inhibition, increasing the concentration of the substrate does not relieve enzyme inhibition. ►inhibitor, ►repression, ►regulation of gene activity

Nonconjugative Plasmids: Nonconjugative plasmids do not have transfer factors (*tra*) and are not transferred by conjugation. ►plasmid mobilization, ►plasmid conjugative, ►conjugation

Noncovalent Bond: A noncovalent bond is not based on shared electrons; individually these are weak bonds, but many of them may result in substantial interactions. ►covalent bond, ►hydrogen bond

Non-crossover Recombinant: Recombination most commonly takes place by crossing over at the four-strand stage of meiosis, or rarely at mitosis. About half of the gene conversion events do not involve crossing over of outside markers, yet the syntenic markers are recombined. Similarly, repair of double-strand breaks may produce recombination without crossing over. ►recombination, ►crossover, ►crossing over, ►mitotic crossing over, ►gene conversion, ►DNA repair; Allers T, Lichten M 2001 Cell 106:45.

Noncyclic Electron Flow: A light induced electron flow from water to $NADP^+$ as in photosystem I and II of photosynthesis. ►photosynthesis, ►Z scheme, ►drift genetic

Non-Darwinian Evolution: Non-Darwinian evolution is evolution by a random process rather than based on selective value. ►evolution non-darwinian, ►mutation neutral, ►Darwinian evolution; Matsuda H, Ishii K 2001 Genes Genet Syst 76(3):149.

Non-directiveness: Non-directiveness is the principle of genetic counseling that only the facts are disclosed but the decision is left to the persons seeking information. Non-directiveness means disclosing all the facts in an unbiased manner without influencing the patient's decision on what course of action to take, but rather support whatever decisions he/she makes, even if the counselor personally disagrees with it. Non-directiveness is supposed to help to work out decisions that are best for patients, in view of their own values and helping people to adjust and cope with genetic conditions and understand their options, with the present state of medical knowledge, so that they can make their own informed decisions. The counselor should not tell the clients what he/she would do in their situation. ►counseling genetic

Nondisjunction: Nondisjunction is when one pair of chromosomes goes to the same pole in meiosis and the other pole then will have neither of them; in mitosis, the movement of both sister chromatids is to the same pole (see Fig. N37). In meiosis, nondisjunction can occur at meiosis I when the two non-disjoining chromosomes are the homologues (the bivalent) or at meiosis II when the non-disjoining elements are sister chromatids.

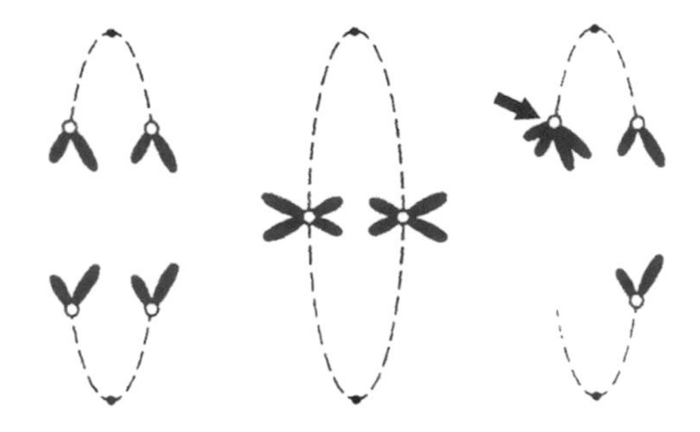

Figure N37. Nondisjunction. Left: normal disjunction, Center: normal metaphase, Right: mitotic nondisjunction. (After Lewis KR, John B 1963 Chromosome Marker, Little Brown, Boston)

Nondisjunction in a normal euploid cell is called primary nondisjunction and when the event takes place in a trisomic cell it is called secondary nondisjunction. Normal disjunction of meiotic chromosomes seems to have the prerequisite of chiasma (chiasmata). In the absence of chiasma the likelihood of nondisjunction increases. In cases when nondisjunction occurs after chiasma, the chiasma seems to be localized to distal position of that bivalent.

The human X chromosome fails to disjoin in the achiasmatic case whereas the autosomes may be nondisjunctional even when distal exchanges take place. Meiosis II nondisjunction occurs when chiasmata take place in the near-centromeric region. In achiasmatic meiosis orderly disjunction depends on the centromeric heterochromatin. It is interesting to note that in human females nondisjunction increases with advancing age although the recombinational event determining chiasmata takes place prenatally. These facts require the assumption that the proteins (spindle and motor proteins) regulating normal chromosome segregation become less efficient during later phases of development. Nondisjunction occurs more frequently in the *Drosophila* and in human females than in the males. In natural populations of the flies, frequencies for the X chromosomes varied from 0.006 to 0.241. Defects in maternal methylenetetrahydrofolate reductase (MTHFR) may create an increased risk for human chromosome 21 nondisjunction (James SJ et al 1999 Am J Clin Nutr 70:495). Both MTHFR and methionine synthase reductase (MTRR) may affect nondisjuntion of the human sex chromosomes and some autosomes. ►aneuploidy, ►trisomy, ►monosomy, ►nullisomy, ►Down's syndrome, ►him, ►chromokinesin, ►nod, ►methylene tetrahydrofolate reductase, ►methionine synthase reductase; Bridges CB 1916 Genetics 1:1, 107; Hassold TJ et al 2001 Am J Hum Genet 69:434; X chromosome nondisjunction in *Drosophila*: Xiang Y, Hawley RS 2006 Genetics 174:67.

Nondivisible Zones: Nondivisible zones flank the prokaryotic terminus of replication and inversions within the zones are either deleterious or impossible.

Nonessential Amino Acids: Nonessential amino acids are not required in the diet of vertebrates because they can synthesize them from precursors (alanine, asparagine, aspartate, cysteine, glutamate, glutamine, glycine, proline, serine, tyrosine). ►amino acids, ►essential amino acids

Non-genic DNA: About 0.3 to 1% of the human genome is highly conserved (CNG) across evolutionary boundaries, although these sequences are not transcribed or translated into protein. Their conservation is significantly higher than protein-coding genes or noncoding RNA genes. Although their functions are unknown, their profiles resemble that of protein-binding coding sequences. There are some indications that certain genetic anomalies are caused by alterations in these CNGs (Dermitzakis ET et al 2005 Nature Rev Genet 6:151). ►RNA noncoding; Dermitzakis E et al 2003 Science 302:1033.

Nonheme Iron Proteins: Nonheme iron proteins contain iron but no heme group.

Non-histone Proteins: Non-histone proteins are proteins in the eukaryotic chromosomes. In contrast to histones which are rich in basic amino acids, the

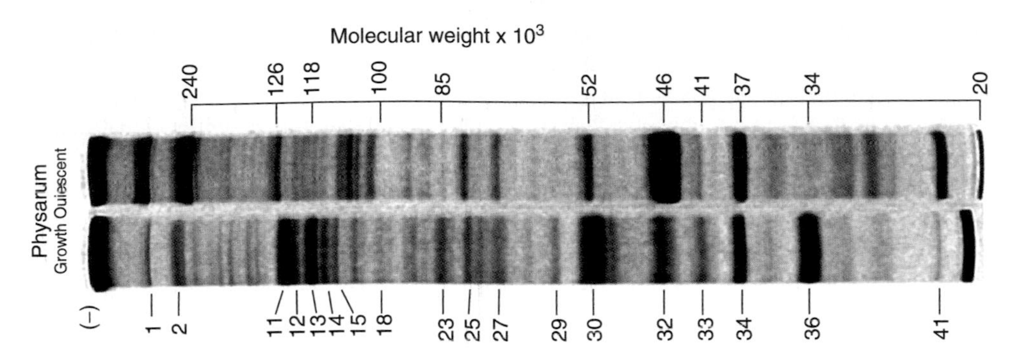

Figure N38. Phenol-soluble acidic protein profiles, separated by electrophoresis from *Physarum* at two developmental stages. (From Lestourgeon A et al 1975 Arch Biochem Biophys 159:861)

non-histone proteins are acidic and are frequently phosphorylated. These proteins include enzymes, DNA- and histone-binding proteins, and transcription factors. Since they play regulatory role(s), they are a heterogeneous group varying in tissue-specific and developmental stage-specific manner. Along with DNA and histones they are part of the chromatin (see Fig. N38). ►histones, ►high-mobility group proteins, ►chromatin; Wilhelm FX et al 1974 Nucleic Acids Res 1:1043); Earnshaw WC, Mackay AM 1994 FASEB J 8:947; Perez-Martin J 1999 FEMS Microbiol Rev 23:503.

Non-Hodgkin Lymphoma: ►anaplastic lymphoma, ►lymphoma

Nonhomologous: When chromosomes do not pair in meiosis and their nucleotide sequences do not show substantial sequence similarities, except at the telomeric regions. (See Mefford HC 2001 Hum Mol Genet 10:2363).

Nonhomologous End-Joining (NHEJ): A mechanism of V(J)D recombination of immunoglobulin genes. It is used for repairing double-strand breaks in not completely homologous chromosomes as occuring in illegitimate recombination and chromosomal rearrangements; it is catalyzed by Ku. KU70/Ku80 recruit DNA-PK, RAD50 and other proteins to the broken ends. NHEJ repair frequently involves insertion of extraneous DNA into the double-strand break in yeast, plant, and mammalian cells. ►immunoglobulins, ►NHEJ, ►end-joining, ►V(D) J, ►DNA repair, ►Ku, ►DNA-PK, ►DNA50, ►XLF/Cernunnos, ►double-strand breaks, ►polynucleotide kinase, ►histone deacetylase, ►integration, ►illegitimate recombination; Essers J et al 2000 EMBO J 19:1703; Pospiech H et al 2001 Nucleic Acids Res 29:3277; Lin Y, Waldman AS 2001 Nucleic Acids Res 29:3975; Kysela B et al 2005 Proc Natl Acad Sci USA 102:1877; Daley JM et al 2005 Annu Rev Genet 39:431.

Nonhomologous Recombination: The genetic exchange between chromosomes with less than the usual similarity in base sequences. The genetic repair system of the cell may work against such recombination. ►illegitimate recombination, ►Ku, ►DNA ligases, ►heterochromosomal recombination, ►mismatch repair; Derbyshire MK et al 1994 Mol Cell Biol 14:156.

Nonidet : A nonionic detergent used in electrophoresis.

Non-isotopic Labeling: ►non-radioactive labeling, ►biotinylation, ►fluorochromes, ►FISH, ►nanocrystal semiconductor, ►quantum dot

Nonlinear Tetrad: In a nonlinear tetrad the ascus contains the spores in a random group of four. ►tetrad analysis, ►unordered tetrad; see Fig. N39.

Figure N39. Nonlinear tetrad

Non-Mendelian Inheritance: Non-Mendelian inheritance is when extranuclear genetic elements are involved or gene conversion alters the allelic proportions. A novel type of non-Mendelian inheritance was observed in *Arabidopsis*. Several independent mutations homozygous for the recessive *hth* (hothead) allele, involving fusion of the flower base, reverted to the dominant wild type at a high frequency (in the 10^{-2} range). The most unusual feature of the event was that the parental line as well as its first progeny did not display such events but the reversion occurred only in the third generation. The revertants contained the exact nucleotide sequence of the wild type allele. Gene conversion—contamination by seed or pollen—gene mutation, and transposons were ruled out. The hypothesis is that that the nuclei contained an RNA template cache from an earlier generation and it facilitated the retrotranscription into the correct wild type DNA in an atavistic fashion (Lolle SJ et al 2005 Nature [Lond] 434:505) (Unfortunately this observation and interpretation appear incorrect. see paramutation). ►extranuclear genes, ►gene conversion, ►paramutation, ►mitochondrial genetics, ►chloroplast genetics, ►imprinting, ►cortical inheritance, ►meiotic drive, ►atavism

Non-MHC-restricted Cell: ►killer cells

Nonnatural Amino Acids: ►unnatural amino acids, ►amino acids

Non-orthologous Gene Displacement: Non-orthologous gene displacement, in evolutionarily related species genes, encodes a particular function with dissimilar nucleotide sequence. ►orthologous, ►paralogous

Nonparametric Linkage Test: ►affected-sib-pair method, ►GSMA

Nonparametric Tests: Nonparametric tests do not deal with parameters of the populations (such as means). They estimate percentiles of a distribution without defining the shape of the distribution by the means of parameters. Nonparametric tests are simple and, often, they are the only ways of statistical estimations. The Wilcoxon's signed-rank test and the Mann-Whitney test substitute for a t-test by using paired samples. The Spearman rank-correlation test determines correlation between two variables. The chi-square goodness of fit test, the association test and many other frequently used tests in genetics are nonparametric. ►statistics, ►parametric methods in statistics, ►robustness, see also above-mentioned tests for description and use, ►distribution-free, ►Kruskal-Wallis test; Kruglyak L et al 1996 Am J Hum Genet 58:1347.

Nonparental Ditype (NPD): A tetrad or octad of spores in an ascus that display only two types of recombinant spores. ►tetrad analysis

Nonpermissive Condition: A regime where a conditional mutant may not thrive; the growth of the cell is not favored under these circumstances. ►temperature-sensitive mutation, ►conditional lethal

Nonpermissive Host: A cell that does not favor autonomous replication of a virus; it may not prevent, however, the integration of the virus into the chromosome where the virus may initiate neoplastic transformation.

Non-permuted Redundancy: At the ends of T-even phage DNAs repeated sequences occur in the same manner in the entire population, e.g., 1234....1234 1234....1234. ►permuted redundancy

Non-photosynthetic Quenching: Non-photosynthetic quenching is required in plants when the quantity of absorbed photosynthetic light exceeds the level that the plant can utilize under a particular culture or environmental condition. The protein concerned is a component of the chlorophyll-binding photosystem II complex. It protects the cells against reactive oxygen damage. ►photosystem, ►Z scheme

Non-plasmid Conjugation: Some transposons (Tn1545, Tn916) are capable of conjugal transfer at frequencies about 10^{-6} to 10^{-8}. These larger transposons are very promiscuous and transfer DNA by a non-replicative process to different bacterial species. ►transposable elements, ►conjugation bacterial recombination, ►plasmids, ►restriction-modification; Macrina FL et al 1981 J Antimicrob Chemother 8 Suppl D:77.

Nonpolar Crosses (homosexual crosses): In mitochondrial crosses some genes appeared to segregate in a polar manner and others apparently failed to display such polarity. The polarity was attributed to ω^+ and ω^- condition in yeast mitochondria. Actually, ω is an intron within 21S rRNA genes. ►mtDNA, ►mitochondrial genetics; Linnane AW et al 1976 Proc Natl Acad Sci USA 73:2082.

Nonpolar Molecule: A nonpolar molecule lacks dipole features, repels water and therefore has poor solubility (if any) in water-based solvents.

Nonpolyposis Colorectal Cancer: ►hereditary nonpolyposis colorectal cancer

Nonproductive Infection: In a nonproductive infection the virus DNA is inserted into the chromosome of the eukaryotic host and replicated as the host genes but no new virus particles are released. In such a situation the viral oncogene may initiate the development of cancer rather than the destruction of the cell as in a productive infection. ►temperate phage, ►oncogenes; Butler SL et al 2001 Nature Med 7:631.

Non-progressor: A non-progressor does not follow the expected course, e.g., a person infected by a virus does not develop all the symptoms of the disease.

Nonradioactive Labels: For DNA, biotinylated probes are used most frequently. Biotin 16-ddUTP (dideoxy-uridine tri-phosphate) contains a linker between biotin and the 5′-position of deoxyuridine triphosphate is used for 3′end labeling of oligonucleotides with the aid of nucleotidyl terminal transferase (see Fig. N40). Biotin-16-dUTP (deoxy-uridine-5-triphosphate) can replace thymidylic acid in nick translation. Then, a signal-generating complex is added, containing a streptavidin (protein with strong affinity for biotin) complex and peroxidase. After addition of peroxide and diaminobenzidine tetrahydrochloride substrates a dark precipitate results if the probe is present. Some procedures use photobiotin (*N*-biotinyl-6-aminocaproic hydrazide) or digoxigenin-conjugated compounds (toxic!) or other chemoluminiscents. ►fluorochromes, ►aequorin, ►FISH, ►chromosome painting, ►immunolabeling

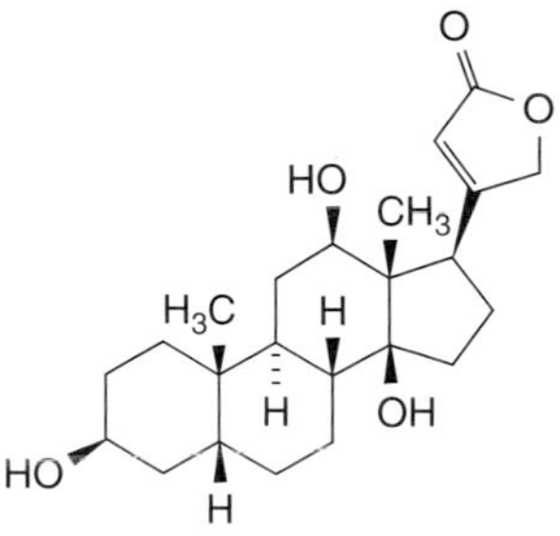

Figure N40. Digoxigenin

Nonrandom Mating: In nonrandom mating mate selection is by choice, either by the organism or by the experimenter. ►assortative mating, ►inbreeding, ►autogamy, ►mating systems

Nonreciprocal Recombination: ►gene conversion

Nonrecurrent Parent: A nonrecurrent parent is not used for back-crossing of the progeny. ►recurrent parent, ►backcross

Non-repetitive DNA: Non-repetitive DNA consists of unique sequences such as the euchromatin of the chromosomes that codes for the majority of genes and displays slow reassociation kinetics in annealing experiments. It does not involve many redundancies such as SINE, LINE. ►unique DNA, ►euchromatin, ►SINE, ►LINE, ►c_0t curve, ►c_0t value

Non-ribosomal Peptides: Built of amino acids by non-ribosomal peptide synthetases and are natural products such as penicillins, cyclosporin, etc.. Iterated modules in these megasynthetases determine the identity and sequence of the amino acids. Each module activates a specific amino acid by means of coupled domains. An adenylation domain generates an aminoacyl-*O*-adenosine monophosphate, which is then joined to a phosphopantetheinyl group by thioesther linkage of a peptidyl carrier. The chain elongates with the aid of acyl-S-enzymes. Polyketide synthases (PKS) and non-ribosomal peptide synthetases (NRPS) synthesize some bacterial and fungal secondary metabolites such as erythromycin (PKS) and vancomycin (NRPS), bleomycin antitumor agent, and rapamycin immunosuppressant (PKS/NRPS) hybrid enzyme. Chimeric (with swapped elements) and often nonfunctional NRPSs can be activated by directed evolution. Using rounds of mutagenesis (repeated cycles) coupled with in vivo screens for NRP production, rapidly isolated variants of two different chimeric NRPSs with ~10-fold improvements in enzyme activity and product yield, including one that produces new derivatives of the potent NRP/polyketide antibiotic andrimid (Fischbach MA et al 2007 Proc Natl Acad Sci USA 104:11951); ►aminoacyl-tRNA synthetases [in ribosomal systems], ►protein synthesis, ►transcriptome, ►polyketides, ►erythromycin, ►vancomycin, ►bleomycin, ►rapamycin; Silakowski B et al 2001 Gene 275:233; Patel HM, Walsh CT 2001 Biochemistry 40:9023; Lai JR et al 2006 Proc Natl Acad Sci USA 103:5314.

Non-synonymous Codon: A non-synonymous codon codes for a different amino acid than the synonymous nucleotide replacement. ►synonymous codons

Non-secretor: ►secretor, ►ABH antigens

Nonselective Medium: Any viable cell, irrespective of its genetic constitution, can use a nonselective medium in contrast to a selective medium containing, e.g., a specific antibiotic, in which only the cells resistant to the antibiotic can survive. ►selective medium

Nonsense Codon: A nonsense codon is a stop signal for translation. About 1/3 of the mutations involved in human disease involve new nonsense codons. ►nonsense mutation, ►code genetic, ►readthrough

Nonsense-Associated Altered Splicing: ►NAS

Nonsense-Mediated Decay: A mRNA surveillance mechanism that may destroy mRNAs with a premature translation termination signal. ►translation termination, ►mRNA degradation, ►mRNA surveillance; Kim VN et al 2001 Science 293:1832.

Nonsense Mutation: Nonsense mutation converts an amino acid codon into a peptide-chain-terminator signal (UAA, ochre; UAG, amber; UGA, opal). ►genetic code, ►code genetic

Nonsense Supressor: A nonsense suppressor allows the insertion of an amino acid at a position of the peptide in spite of the presence of a nonsense codon at the collinear mRNA site. Chemically-acylated suppressor tRNA can insert particular amino acids in response to a stop codon (Noren CJ et al 1989 Science 244:182). ►suppressor tRNA

Non-small-cell Lung Carcinoma Suppressor (NSCLCS): By deletion analysis NSCLCS has been located to a ~700-kb region in 11q23. Non-small-cell carcinoma is the most common type (~80%) of the disease. In 50% of the latter (in adenocarcinomas) there is a mutation in the epidermal growth factor receptor (EGFR), which is sensitive to kinase inhibitors. In these cases the application of the tyrosine kinase inhibitor drug Gefitinib (Iressa) evokes a favorable response to this targeted therapy. Small inversion within chromosome 2p results in a fusion gene of the echinoderm microtubule-associated protein-like 4 (EML4) gene and the anaplastic lymphoma kinase (ALK) gene in non-small-cell lung cancer (NSCLC) cells. The EML4–ALK fusion transcript was detected in 6.7% (5 out of 75) of NSCLC patients (Soda M et al 2007 Nature [Lond] 448:561). ►small cell lung carcinoma, ►lung cancer; Minna JD et al 2004 Science 304:1458.

Nonspecific Binding: Nonspecific binding involves adhesion to a surface. ►cross-reaction

Nonspecific Pairing: ►illegitimate pairing, ►chromosome pairing

Nonstop Decay: If mRNA lacks a stop codon, it may be destroyed by the exosome to prevent the synthesis of an abnormal protein (van Hoof A et al 2002 Science 295:2262). ►exosome, ►RNA surveillance, ►mRNA degradation

Non-Syndromic: Non-syndromic is when few or no other symptoms than the main characteristic of a condition appear in the diseases. ►syndrome

Non-synonymous Mutation: Non-synonymous mutation results in amino acid replacement in the protein.

Nontranscribed Spacer (NTS): Nucleotide sequences in-between genes, which are not transcribed into RNA, e.g., sequences between ribosomal RNA genes.

Nonviral Retroelements: There is a class of nonviral retrotransposable elements in *Drosophila* and other organisms (*Ty* in yeast, *Cin* in maize, *Ta* and *Tag* in *Arabidopsis, Tnt* in tobacco). These probably transpose by RNA intermediates but lack terminal repeats although they encode reverse transcription type functions. They contain AT-rich sequences at the 3′-termini and are frequently truncated at the 5′-end. Some of these elements are no longer capable of movement because of their extensive (pseudogenic) modification. The *Drosophila*: "*D*" is a variable length element with up to 100 copies and variable number of target site duplications. "*F*" is of variable length in 50 copies with 8–22-bp target site replication. The longest element of the *F* family is about 4.7 kb. The "*Fw*" element (3,542 bp) contains two long open reading frames (ORF). The longer of these bears substantial similarity to the polymerase domains of retroviral reverse transcriptases. The other transcript (ORF1) codes for a protein that has similarities to the DNA-binding sections of retroviral gag polypeptides. The *FB* family of transposons of *Drosophila* is not related to the *F* family mentioned here ►hybrid dysgenesis. The "*G*" transposable element is of variable length of up to 4-kb in 10 to 20 copies with 9-bp target site duplication. These elements are similar to the *F* family inasmuch as they have polyadenylation signals and poly-A sequences at the 3′-end. The *G* elements insert primarily into centromeric DNA. "*Doc*" is of variable length (up to 5 kb) with 6 to 13-bp target site duplication. The element has a variable 5′-terminus but a conserved 3′-terminus. "*Jockey*" is of variable length of up to 5 kb, in 50 copies. It has incomplete forms named *sancho* 1, *sancho* 2, and *wallaby.* "*I*" is of variable length, up to 5.4 kb, with 0 to 10 complete copies plus about 30 incomplete elements with variable number of target site duplications (usually 12 bp). ►copia, ►hybrid dysgenesis; Higashiyama T et al 1997 EMBO J 16:3715; Schmidt T et al 1995 Chromosome Res 3(6):335.

Nonviral Vector: Nonviral vectors are laboratory constructs for the purpose of introducing selected genes into another organism. Because they lack essential viral elements, the accidental reconstruction of a pathogenic virus in the target organism is prevented. ►viral vectors, ►packaging cell lines, ►vectors

Noonan Syndrome (male Turner syndrome, female pseudo-Turner syndrome, 12q24.1): Noonan syndrome is apparently caused by an autosomal dominant gene with an incidence of $1–2.5 \times 10^{-3}$. It is about twice as common among male than female offspring. Noonan mutations were assigned to human chromosomes 17q, 22q, and 12q224.1 but the identification is hampered by the variable penetrance. The SHP2/PTPN11 gain of function mutations enhance calcium oscillations and impair NFAT signaling (Uhlén P et al 2006 Proc Natl Acad Sci USA 103:2160). PTN11, KRAS, and SOS1 are responsible for 60% of Noonan cases. The symptoms are variable and complex yet there is some resemblance to females with Turner syndrome; but Noonan also affects males. Short stature, a webbed neck, mental retardation, heart and lung defects (stenosis), cryptorchidism, etc.—without visible chromosomal abnormalities—frequently accompany it. Cystic hygroma (lymphangioma) or nuchal lucency (shiny outgrowth on the back of the neck) may be detected by fetal ultrasonography. Missense mutation in the PTPN11 gene encoding non-receptor protein tyrosine phosphatase (SHP2) is responsible for over 50% of the cases. SHP2 is required for RAS-ERK MAP kinase cascade activation, and Noonan syndrome mutants enhance ERK activation *ex vivo* and in mice. Missense mutations in SOS1 (2p22-p21), which encodes an essential RAS guanine nucleotide-exchange factor (RAS-GEF), is found in 20% of cases of Noonan syndrome without PTPN11 mutation (Roberts AE et al 2007 Nature Genet 39:70; Tartaglia M et al 2007 Nature Genet 39:75; see correction in Nature Genet 39:276). The LEOPARD syndrome bears similarity to Noonan syndrome and they appear to be allelic. Both conditions can be caused by single nucleotide replacements. The mutation affects signal transduction (SHP-2). The large majority of the PTPN11 mutations are of paternal origin. The age of fathers of Noonan progeny was higher than average. Also gain-of-function mutations in RAS/RAF signaling, involving several loci, seem to be involved in the hypertrophic cardiomyopathy in both Noonan and LEOPARD syndromes (Pandit B et al 2007 Nature Genet 39:1007; Razzaque MR et al 2007 Nature Genet 39:1013). ►turner syndrome, ►stenosis, ►heart disease, ►face/heart defects, ►cryptorchidism, ►MAPK, ►SOS, ►GEF, ►LEOPARD syndrome, ►signal transduction, ►cardio-facial-cutaneous syndrome, ►NF-AT, ►*Pterydium*, ►*Popliteal pterydium*; Tartaglia M et al 2001 Nature Genet 29:465; Tartaglia M Gelb BD 2005 Annu Rev Genomics Hum Genet 6:45.

Nopaline: A dicarboxyethyl derivative of arginine is produced in plants infected by *Agrobacterium tumefaciens* (strain C58) (see Fig. N41). The nopaline synthase (*nos*) gene is located within the T-DNA of the Ti plasmid. The bacteria also have a gene for the catabolism of nopaline (*noc*). This and other opines serve bacteria with a carbon and nitrogen source. The *nos* promoter and tailing have been extensively used in the construction of plant transformation vectors. ►*Agrobactrium*, ►opines, ►octopine, ►T-DNA

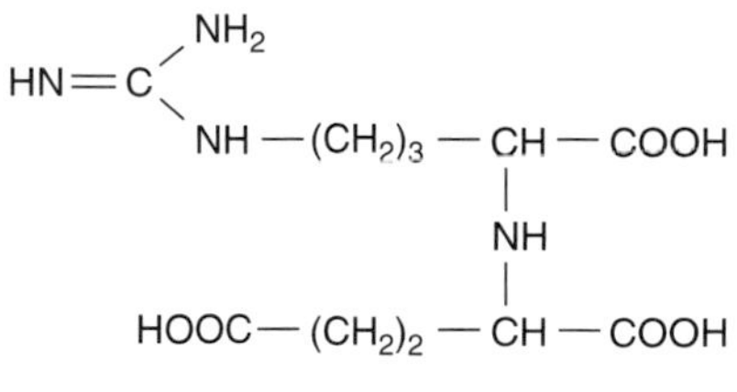

NH_2
HN=C
$NH-(CH_2)_3-CH-COOH$
NH
$HOOC-(CH_2)_2-CH-COOH$

Figure N41. Nopaline

NOR: ►nucleolar organizer

NOR1: A steroid receptor. ►Nur

Noradrenaline: ►norepinephrine

Norepinephrine (noradrenaline): ►animal hormones

NORF (not-annotated open reading frame): ►annotation, ►ORF

Normal Deviate: A normal deviate is the difference between two estimates divided by the standard deviation. The normal deviates have been tabulated by RA Fisher (see Fisher RA, Yates F 1963 Statistical Tables. Hafner, New York) from where the probabilities can conveniently be read. An example: the difference between two means = 4.5. The standard deviation of the difference of the means is $s_d^2 = s_1^2 + s_2^2 = 3^2 + 1^2 = 9 + 1 = 10$, hence $s_d = \sqrt{10} = 3.16$. Thus, 4.5/3.16 = 1.42 = the normal deviate.

In order to find the probability sought, the value nearest to the normal deviate calculated is to be located in the body of the Table N1. The number on the second line under 0.05 (in the heading) is 1.4395, not too far from our estimated normal deviate of 1.42. Therefore, the probability will be at least 0.05 + 0.1 = 0.15. Thus the difference between the two means of this example is statistically not significant. Had we had a normal deviate of 1.96 or higher (on the first line) the probability would have exceeded 0.05 + 0.0 = 0.05, a minimal value of significance by statistical conventions. Interpolations for the tables may be required. ►normal distribution, ►Z distribution

Normal Distribution: Normal distribution is derived from the binomial and not from experimental data. The normal distribution is continuous yet it can be represented by the binomial distribution $(p + q)^k$ where k is an infinitely large number. The normal distribution takes the shape of a bell curve, however a variety of bell curves may exist depending on the two critical parameters μ (mean), and σ (standard deviation). The mean corresponds to the center of the bell curve, and σ measures the spread of the variates. The normal distribution probability density function is represented by the formula:

$f(x) = Z = \frac{1}{\sqrt{2\pi\sigma^2}} e^{-(Y-\mu)^2/2\sigma^2}$ fμ = 0 and σ = 1

then $f(z) = \frac{1}{\sqrt{2\pi}} e^{-z^2/2}$

where Z indicates the height of the ordinate of the curve, π = 3.14159, e = 2.71828, Y = variable. Although in absolute sense perfect normal distribution is only a mathematical concept, experimental

Table N1 The probability of each entry is found by adding the column heading to the value of the far-left column at the appropriate line. This table is incomplete and is used only for illustrating the procedure of the calculation

Probability →	0.00	0.001	0.002	0.01	0.02	0.05	0.08	0.09
0.0	∞	3.2905	3.0902	2.5758	2.3263	1.9599	1.7507	1.9954
0.1	∞			1.6449	1.5982	1.4395	1.3408	1.3106

data may fit in a practically acceptable manner and can be used for the characterization of the available set of data. (See Fig. N42; ►mean, ►standard deviation).

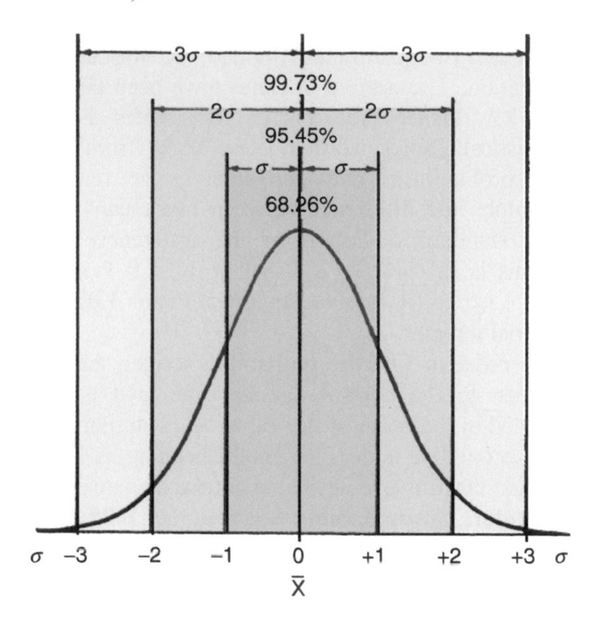

Figure N42. The normal distribution is represented by a normal curve, characterized by the standard deviation σ and within ±1, ±2 and ±3σ, 68.26, 95.45, and 99.73% of the population is expected around the mean, e.g., if the population is 100, the mean is 30, and the standard deviation = 5, approximately 68 individuals will be within the range of 30 ± 5

Normal Solution: In a normal solution each liter (L) contains gram molecular weight equivalent quantity (quantities) of the solute; thus it can be 1N or 2N... 5N, etc. In other words, in 1 L of the solvent there is a solute equivalent in gram(s) atom of hydrogen. Thus, 1 normal HCl (MW = 36.465), analytical grade hydrochloric acid requires 36.465 g HCl in 1 L of water. (The commercially available HCl contains about 38% HCl [specific gravity about 1.19], thus the 1 N solution should have approximately 80.62 mL of the reagent in a total volume of 1 liter). The molecular weight of sulfuric acid (H_2SO_4) is 98.076 and thus requires, for the 1 N solution, 98.076/2 = 49.038 gram/L because H_2SO_4 has 2 hydrogen atom equivalence. The solutions are generally prepared in volumetric flasks. ►mole

Normalization: Normalization is a process in gene discovery by which the mutant gene is used to identify the wild-type (normal) sequences. Normalization selectively reduces the level of representation of multicopy gene products and thus facilitates the discovery of specific gene functions among the large pool of mRNAs. Before normalization, a typical cell expresses 1,000 to 2,000 different abundant messages at the level of >500 copies per cell. Middle abundant messages before normalization may appear in 15–500 copies per cell. The ca. 15,000 rare messages may appear in less than 5 copies per cell. Normalization is carried out by the generation of cDNA from a single tester RNA and a single cDNA with the aid of poly (dT) primer. The use of the poly(dT) primer results in long poly(dA/dT) sequences that become tangled and causes the loss of many templates during normalization and subtraction. The alleviation of this problem is expected by the use of anchored oligo(dT) primers. Microarray data can be evaluated by RMA (Irizarry RA et al 2003 Biostatistics 4:249) or by PDNN (Zhang L et al 2003 Nature Biotechnol 21:818) or by other procedures. There is no general consensus regarding their relative merits. ►gene discovery, ►subtractive cloning, ►transcriptome; Wang SM et al 2000 Proc Natl Acad Sci USA 97:4162.

Normalome: A standard (reference) protein pattern in two-dimensional gel electrophoresis. ►two-dimensional gel electrophoresis

Normoxia: Normal oxygen tension. ►hypoxia

Norrie Disease: An X-linked (Xp11.2-p11.4) retinal neoplastic disease (a pseudoglioma) often complicated by other diverse symptoms, cataracts, microcephalus, etc. The same region also includes two monoamine oxidase (MAOA and MAOB) genes. The deletion of this site may result in neurodegeneration, blindness, hearing loss, and mental retardation. MAO defects have been associated with several psychiatric disorders. ►eye diseases, ►contiguous gene syndrome, ►MAO

Northern Blotting: Electrophoresed RNA is transfered from gel to especially impregnated paper to which it binds. It is then hybridized to labeled (radioactive or biotinylated) DNA probes, followed by autoradiography or streptavidin-bound fluorochrome reaction for identification. This procedure is also suitable for the study of the expression profile of a large number of open reading frames under different conditions. ►autoradiography, ►biotinylation, ►amino-benzyloxymethyl paper, ►North-Western blotting, ►Southern blotting, ►Western blotting; Brown AJP et al 2001 EMBO J 20:3177.

Norum Disease: ►lecithin:cholesterol acyltransferase deficiency

Norvaline (DL-2 aminopentanoic acid, $CH_3(CH_2)_2CH(NH_2)COOH$): An arginase inhibitor, a methionine analog. It can be incorporated into a protein and increases the activation of methionyl-tRNA synthetase.

It can replace leucine in proteins if its concentration is high.

Norwalk Virus: A single-stranded, positive-sense, ~7.7-kb RNA virus within a shell of 180 copies of a single ~5.6-kDa protein organized in an icosahedron. It is responsible for 96% of the nonbacterial gastroenteritis cases in the USA. Its cultivation is difficult. ►icosahedron; Belliot GM 2001 J Virol Methods 19(2):119.

Nosocomial: A hospital originated condition, e.g., an infection acquired in a hospital. ►iatrogenic, ►*Acinetobacter*, ►*Clostridium difficile*; Webb GF et al 2005 Proc Natl Acad Sci USA 102:13343.

Nosography: The description of a disease.

Nosology: The subject area of disease classification.

NOT (CDC39): A global regulator complex of yeast. ►CDC39; Maillet L, Collart MA 2002 J Biol Chem 277:2835.

Notch (*N*): *Drosophila* gene locus; map location 1–3.00. Homozygotes and hemizygotes are lethal; the wing tips of heterozygotes are "notched" (gapped), and thoracic microchaetae are irregular. The expression is greatly variable, however, in the independently obtained mutants. Homozygotes may be kept alive in the presence of a duplication containing the normal DNA sequences. All the mutants, whether homozygotes or hemizygotes, display aberrant differentiation in the ventral and anterior embryonic ectoderm. The nervous system is abnormal. The genetic bases of the mutations are deletions, rearrangements, or insertions. Homologs of the *Notch* gene are also found in vertebrates (including humans) and the transmembrane gene product is involved in cell fate determination. The 300-kDa Notch protein is a single-pass membrane receptor for several ligands in various developmental pathways. The extracellular domain includes 36 tandem epidermal growth factor (EGF)-like repeats and three cysteine-rich repeats. The intracellular part includes six tandem ankyrin repeats, a glutamine-rich domain, and a PEST sequence. *N* activation requires proteolytic cleavage of the intracellular domain of the encoded protein product and its association with the protein complex called CSL (Cbf1, Su[H], Lag-1). The transcription of several genes may ensue. In *Drosophila* the two most important proteins, which interact with the extracellular domain of N, are Delta and Serrate, both single-pass transmembrane proteins (the equivalents of the latter in vertebrates are Delta and Jagged and LAG-2 and APX-1 in *Caenorhabditis*). Fringe proteins display fucose-specific β1,3 N-acetylglucosaminyl-transferase activity. They elongate the fucose residues on the epidermal growth factor-like sequence repeats of Notch and modulate signaling. The primary target of N signaling is the gene *Enhancer of split* (*E[spl]*) in *Drosophila* where it encodes a basic helix-loop-helix nuclear protein. The transcription factor Suppressor of hairless (Su[H]) is a major effector (in mammals CBF1/RJBk and in the nematode LAG-1 perform the same task). Several extracellular and intranuclear proteins are also involved in the regulation of Notch. Notch orchestrates the up and down modulation of numerous proteins. Notch signaling initiates the metamerism of the somites. The Mesp2 basic helix-loop-helix protein contributes to the rostral (anterior) specificity whereas presenilin is required for the caudal differentiation. Both mediate the action of the Delta ligands of Notch. Notch affects practically all types of morphogenetic/developmental processes such as the central and peripheral nerve systems, eyes, spermatogenesis, oogenesis, muscle development, heart, imaginal disc, apoptosis, proliferation, etc. ►*Drosophila*, ►morphogenesis, ►knirps, ►CADASIL, ►helix-loop-helix, ►presenilins, ►EGF, ►ankyrin, ►Furin, ►PEST, ►ADAM; a picture at CSL (see Fig. N43); Baonza A, Garcia-Bellido A 2000 Proc Natl Acad Sci USA 100:2609; Allman D et al 2002 Cell 109:S1; EGF-like domain structures: Hambleton S et al 2004 Structure 12:2173; Notch signaling: Wilson A, Ratke F 2006 FEBS Lett 580:2860; cell fate decisions: Ehebauer M et al 2006 Science 314:1414.

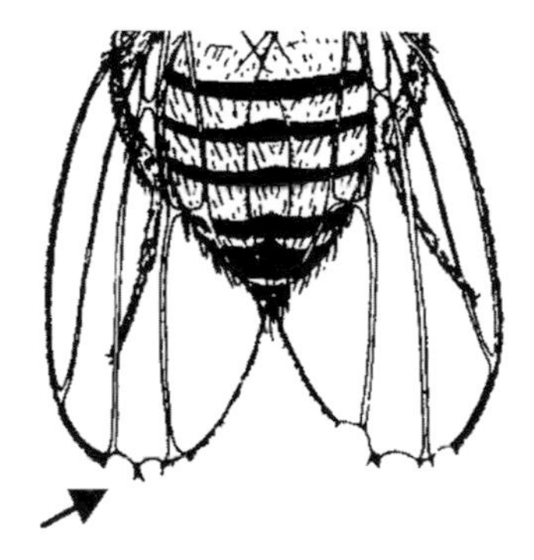

Figure N43. Notch

Notochord: A rod-shape cell aggregate that defines the axis of development of the animal embryo giving rise to the somites that develop into the vertebral column in higher animals. ►organizer, ►somite, ►chordoma, ►epiboly

Notum: The dorsal part of the body segments in arthropods. ►dorsal

NPC (nuclear pore complex): The NPC is involved in the import of proteins into the nucleus. It is a 125-Mda complex, embedded into the nuclear envelope. It is associated with cytoplasmic proteins, importin α and β, Ran (small guanosine triphosphatase), Nup214/CAN, and nuclear transport factor 2 (NTF2). ►nuclear pore; Walther TC et al 2001 EMBO J 20:5703; Komeili A, O'Shea EK 2001 Annu Rev Genet 35:341; Walther TC et al 2003 Nature [Lond] 424:689.

NPH (nucleoprotein helicase): NPH disrupts the interaction between U1A and the spliceosome, and unwinds double-stranded RNA and protein associations. ►snRNA, ►DExH, ►spliceosome; Jankowsky E et al 2001 Science 291:121.

Npi: A peptidyl-prolyl isomerase recognizing the nuclear localization sequences of the FKBP family. ►PPI, ►nuclear localization sequence, ►FKBP

NPK: A plant MAPKKK required for the activation of the MAPK path involved in the regulation of auxin responsive transcription. ►MAPK, ►MAPKKK

NPP-1: ►PP-1

NPTII: ►aph(3′)II, ►neomycin phosphotransferase, ►amino glycoside phosphotransferase

NPXY: A protein amino acid sequence: Asn-Pro-X-Tyr. ►amino acid symbols in protein

nr: The prefix for rat (*Rattus norvegicus*) protein or DNA, e.g., nrDNA.

NRE1: ►DNA-P

NRF (NF-E related factor): ►NF-E

NRON: A noncoding repressor of NFAT transcription factor. The gene has three exons transcribed into 0.8 to 3.7-kb RNAs. Its expression is highest in the embryo and thymus of mouse and humans. ►NFAT, ►noncoding RNA; Willingham AT et al 2005 Science 309:1570.

NRSF: ►neuron-restrictive silencer factor

NRY: The non-recombinant part of the Y chromosome. ►Y chromosome

NSF: National Science Foundation, USA. An agency which supports basic research and science.

NSF (*N*-ethylmaleimide-sensitive factor): A vesicle transport ATPase component of SNAP (soluble NSF attachment protein) in the fusion complex of vesicles (see Fig. N44). This then binds to the membrane receptor SNARE or tSNARE (target membrane SNARE). ATPase NSF and SNAP can also disrupt SNARE, the synaptobrevin and syntaxin ternary complex, anchored to the lipid bilayer of the membrane. These play a role in the transport between the endoplasmic reticulum and the Golgi apparatus and neurotransmitter release. SNAPs are also involved in the regulation of transcription of snRNA by RNA polymerases II and III. ►SNARE, ►RAB, ►synaptic vessel, ►snRNA, ►membrane fusion, ►caveolae, ►endocytosis; Yu RC et al 1999 Mol Cell:4:97; May AP et al 2001 J Biol Chem 276:21991.

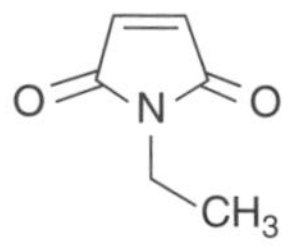

Figure N44. *N*-Ethylmaleimide

nsL-TP: Non-specific lipid transfer proteins.

NTE (neuropathy target esterase): The primary target of the nerve defect syndromes of organophosphate insecticides. ►cholinesterase; Winrow CJ et al 2003 Nature Genet 33:477.

N-Terminus (amino end): The end of a polypetide chain where translation started.

N-TEF: ►TFIIS

NTP: NTP stands for nucleotidetriphosphates.

NTPase (nucleotide triphosphatase, 14q14): NTPase are divalent cation-dependent transmembrane enzymes.

NtrB: A kinase in the bacterial nitrogen fixation system affecting NtrC. ►nitrogen fixation

NtrC: A protein which regulates the bacterial nitrogen assimilation signaling path by phosphorylation. ►nitrogen fixation, ►signal transduction

NTS: ►nontranscribed spacer

Nu Body: The unit of a chromosome fiber containing 8 + 1 molecules of histones and about 240 nucleotide pairs of DNA. ►nucleosome

Nu End: The Nu end of the DNA first enters the phage capsid. ►packaging of DNA

NuA3, NuA4: Nucleosomal acetyltransferase of histone 3/histone 4, respectively. ►histone acetyltransferase, ►double-strand breaks; John S et al 2000 Genes Develop 14:1196.

Nuage: Large cytoplasmic inclusions in the cells, destined to become the germline. Observations suggest that the nuage—encoded by several genes—functions as a specialized center that protects the genome in the germ-line cells by gene regulation mediated via repeat-associated small interfering RNAs (Lim AK, Kai T 2007 Proc Natl Acad Sci USA 104:6714). (See Ikenishi K 1998 Dev Growth Differ 40:1; ►chromatoid body).

Nucellar Embryo: Develops from a diploid maternal nucellus. ►nucellus, ►apomixes

Nucellus: The maternal tissue of the ovule surrounding the embryo sac of plants. ►embryo sac, ►megagametophyte development; see Fig. N45.

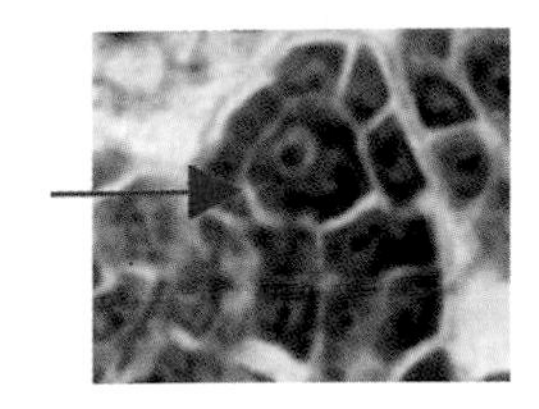

Figure N45. Nucellar cell layer surrounding the prominent archespore

Nuclear Bodies: Nuclear bodies are aggregated proteins—without membrane enclosure—recruited by protein-protein interaction. Their role is apparently storage or enzyme activity.

Nuclear Dimorphism: There is the genetically active micronucleus and the much larger macronucleus in *Ciliates*, responsible for the metabolic functions of the cell but not for its inheritance. ►paramecium

Nuclear Envelope: ►nuclear membrane

Nuclear Export Sequences (NES): Leucine-rich tracts of proteins and the Ran GTPase complex without GTPase activity. ►Ran, ►GTPase, ►nuclear localization sequences, ►nuclear pore, ►CRM1; Ossareh-Nazari B et al 2001 Traffic 2(10):684.

Nuclear Family: Consists of parents and children living in the same household. The extended family includes other relatives of the household. These are actually social terms. ►family

Nuclear Fission: The splitting of heavy atoms to elements—e.g., uranium, and may thus produce barium, krypton, etc.—resulting in the liberation of a great amount of thermal energy causing the split nuclei to fly apart at great velocity. ►nuclear fusion, ►isotopes, ►nuclear reactor

Nuclear Fusion: Atomic energy can be liberated either by nuclear fission or by fusing lighter atomic nuclei (e.g., hydrogen) into heavier ones and, in the process of thermonuclear reaction, generate huge amount of energy. If this reaction could be made slower (rather than explosive), humanity could get access to vast amounts of inexpensive energy. ►nuclear fission

Nuclear Import: ►nuclear pore, ►nuclear export

Nuclear Inclusions: In diseases caused by trinucleotide repeats, polyglutamine proteins may accumulate in the neurons causing degeneration. ►trinucleotide repeats; Ross CA 1997 Neuron 19:1147; Chai Y et al 2001 J Biol Chem 276:44889.

Nuclear Lamina: In a nuclear lamina there are three polypeptides which form a fibrous mesh within the cell nucleus attached to the inner nuclear membrane and participate in the formation of the nuclear pores. They anchor chromosomes to the membrane and control the dissolution of the nuclear membrane during mitosis. ►mitosis, ►nuclear pores; Guillemin K et al 2001 Nature Cell Biol 3(9):848.

Nuclear Localization Sequences (NLS): Nuclear localization sequences direct the movement of proteins imported into the nucleus through ATP-gated pores. The targeting proteins (also called nuclear localization factor, NLF) usually contain one or more basic amino acid clusters. The SV40 T antigen has a ^{126}PKKKRKV^{132} at the location shown by the raised numbers. The nucleoplasmin has two bipartite clusters KRPAAIKKAGQAKKKK. Proteins so equipped are transported to the nucleus by karyopherin/importin proteins. ►nuclear proteins, ►ion channels, ►importin, ►transportin, ►karyopherin, ►nuclear pore, ►Ran, ►footprinting, ►second cycle mutation, ►nuclear export sequences, ►nuclear pore, ►cofilin, ►T antigen, ►NF-κB, ►export signals, ►export adaptors, ►nucleoplasmin; Post JN et al 2001 FEBS Lett 502:41.

Nuclear Magnetic Resonance Spectroscopy (NMR): A physical method for studying three-dimensional molecular structures. Electromagnetic radiation is pulsed at small 15–20-kDa proteins in a strong magnetic field. This results in a change in the orientation of the magnetic dipole of the atomic nuclei. When the number of protons and neutrons in an atomic nucleus is not equal they display a spin angular momentum. In the strong magnetic field the spin is aligned but it becomes misaligned in an excited state as a consequence of the radio frequencies of the electromagnetic radiation. When they return to the aligned state, electromagnetic radiation is emitted. This emission displays characteristics dependent on the neighbors of the atomic nuclei. It is thus feasible to estimate, from the emission spectrum, the relative position of hydrogen nuclei in different amino acids of the protein. If the primary structure of the polypetide is known from amino acid sequence data, the three-dimensional arrangement of the molecules can be determined. The NMR technology—unlike X-ray crystallography—does not require that the material be in a crystalline state. NMR has found its use for the analysis of medical specimens and also in plant development and infection of tissue by pathogens. Relatively thick

(500 μm) tissue slices can be studied using this nondestructive method. With the aid of color video processors the distribution of water content can be followed and photographed. Magnetic resonance imaging (MRI) may permit viewing targets deep in the body and visualization of enzymes within a living organism (tadpole) without destruction. Single micrometer-sized iron oxide particles can be detected in cultured cells and, in mouse embryos, treated at a single cell stage after many cell divisions, up to the 11.5 day of embryonic development (Shapiro EM et al 2004 Proc Natl Acad Sci USA 101:10901). NMR applied to intact mouse is capable of assessing mitochondrial function (Padfield KE et al 2005 Proc Natl Acad Sci USA 102:5368). NMR is becoming a tool of proteomics. ►X-ray crystallography, ►electromagnetic radiation, ►circular dichroism, ►Raman spectroscopy, ►imaging, ►proteomics, ►metabonomics; Yee A et al 2002 Proc Natl Acad Sci USA 99:1825; Walter G et al 2000 Proc Natl Acad Sci USA 97:5151; Ratcliffe RG, Shachar-Hill Y 2001 Annu Rev Plant Physiol Plant Mol Biol 52:499; Palmer AG III 2001 Annu Rev Biophys Biomol Struct 30:129; Yee A et al 2002 Proc Natl Acad Sci USA 99:1825; stereo-array isotope labeling [SAIL]: Kainosho M et al 2006 Nature [Lond] 440:52; new tools: Mittermaier A, Kay LE 2006 Science 312:2124; http://www.bmrb.wisc.edu.

N

Nuclear Matrix: An actin-containing scaffold extended over the internal space within the nucleus participating in and supporting various functions of the DNA, including replication, transcription, processing transcripts, receiving external signals, chromatin structure, etc. Loops of chromatin fibers are associated with non-histone proteins. ►scaffold, ►actin; Lepock JR et al 2001 Cell Stress Chaperones 6(2):136; http://www.rostlab.org/db/NMPdb/.

Nuclear-Matrix Associated Bodies: A number of different proteins involved in controlling cellular proliferation and oncogensesis. ►PML; Zuber M et al 1995 Biol Cell 85:77; Carvalho T et al 1995 J Cell Biol 131:45.

Nuclear Membrane (nuclear envelope): The nuclear membrane surrounds the cell nucleus of eukaryotes and the pores of this membrane facilitate the export and import of molecules. The nuclear membrane seems to disappear during the period of prophase-metaphase and is reformed again during anaphase and telophase in the majority of eukaryotic cells. In the ascomycete fungus *Aspergillus nidulans* the nuclear membrane does not entirely disintegrate during mitosis (De Souza CP et al 2004 Current Biol 14:1973). The assembly of the envelope is induced by Ran and enhanced by RCC1. Dynein and Dynactin proteins mediate the breakdown of the membrane. The Brr6 integral membrane protein surrounds the nuclear pore and regulates the transport through the pore. Apparently, a large number of proteins are involved in the formation of the membrane and their defects are associated with a large number of diverse diseases (Schirmer EC et al 2003 Science 301:1380). The integral inner nuclear membrane protein elements are targeted to the inner membrane by karyopherins and RAN GTPase (King MC et al 2006 Nature [Lond] 442:1003). On the inner face of the nuclear envelope, active genes localize to nuclear-pore structures whereas silent chromatin localizes to non-pore sites. Nuclear-pore components seem to not only recruit the RNA-processing and RNA-export machinery, but also contribute a level of regulation that might enhance gene expression in a heritable manner (Ahtar A, Gasser SM 2007 Nature Rev Genet 8:507). ►mitosis, ►dynactin, ►dynein, ►lamins, ►Ran, ►RCC1, ►nuclear pore, ►karyopherin, ►chromosome positioning, ►chromosome territory, ►laminopathies; de Bruyn Kops A, Guthrie C 2001 EMBO J 20:4183; Aitchison JD, Rout MP 2002 Cell 108:301; Burke B, Stewart CL 2002 Nature Rev Mol Cell Biol 3:575; dynamics of pore complexes: Tran EJ, Wente SR 2006 Cell 125:1041.

Nuclear Pores: Perforations in the nuclear membrane, defined on the inner side by the nuclear pore complex consisting of eight large protein granules in an octagonal pattern and it is also associated with nuclear lamina (see Fig. N46). The pores selectively

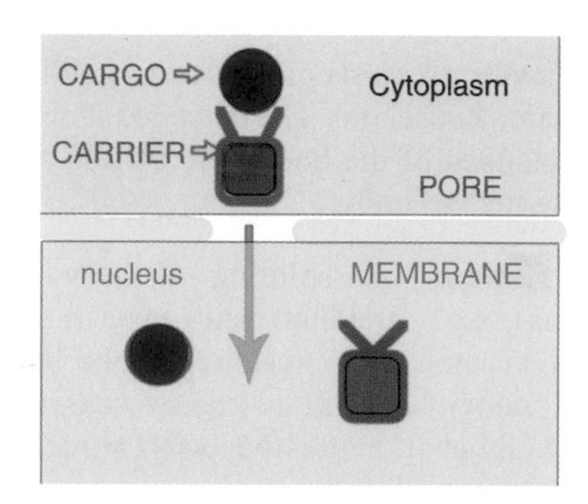

Figure N46. Nuclear import

control the in and out traffic through their about 100-nm channel. An active mammalian cell nucleus may have 3,000–4,000 nuclear pore complexes. The nuclear pore may have a mass of 125-MDa and is built of about 50–100 polypeptides (some of them in replicates). Small molecules may have a passive passage but larger ones (up to 25 nm in diameter) require active transport. In a single minute, about 100 ribosomal proteins and three ribosomal subunits may pass through a pore. About 1,000 molecules can be transferred per second, several of them simultaneously (Yang W et al 2004 Proc Natl Acad Sci USA 101:12887). The transport requires appropriate recognition sites without clear consensus sequences. The major import factors are *importin* α and β; α picks up the molecule to be imported and β eases them through the pore by Ran GTPase; protein *pp15* provide the energy. The translocation may require effectors and other energy sources too. After delivery, the import complex disengages and the two importin subunits quickly return to the cytoplasm. The export of RNA—generally following trimming and splicing—requires binding to special nuclear export sequences of export proteins. This process may be mediated by the Rev factor that has binding sequences to the HIV-1 transcripts and may also allow through unspliced RNA.

Human cells have hRip (human Rev interacting protein) or Rab. Viral RNAs, mRNA, tRNA, rRNA, and snRNA may each have somewhat different export system variants. The general transcription factor TFIII has some features that may qualify it for the viral Rev functions.

The capped (by m^1GppN-5′) RNA ends are apparently joined in a cap-binding CBP80 and CBP20 protein complex (CBC) for export. The M9 region of the heterologous nuclear ribonucleoprotein (hnRNAP) also appears to be involved in RNA trafficking through the nuclear pores. The Ca^{2+} content of the nucleus may regulate the traffic of intermediate size molecules (>70-kDa) across the membrane.

Some of the molecules exported from the nucleus to the cytoplasm are very large (see Fig. N47), e.g.,

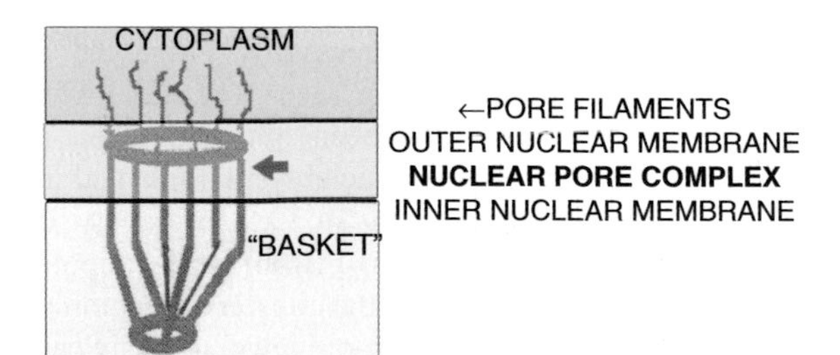

Figure N47. Schematic details of the nuclear pore. (Modified after Science 279:1129)

the ribosomal nucleoprotein components. These molecules can pass through the pore only in extended forms. Mutational analysis, particularly in yeast, revealed a great deal about the protein components of the nuclear traffic. Some small molecules (proteins and RNA, < 25–40 kDa) may traffic by a passive diffusion. The yeast proteins Mlp1 and Mlp2, localized to the basket filaments, play a role in the active transport. These two proteins as well as the associated Ku70 protein, nuclear pore protein 145 (Nup145), and Rap1 also regulate the transcription of the genes situated in the telomeric region. ►nucleus, ►cap, ►Rev, ►RCC1, ►transcription factors, ►hnRNA, ►Ran, ►RCC, ►CRM1, ►CIITA, ►RNA transport, ►NPC, ►NUP, ►nuclear localization sequence [NLS], ►export signal, ►nucleoporin, ►karyopherin, ►transportin, ►importin, ►BR RNP, ►Rap, ►Mlp1, ►PML39; http://www3.shinbiro.com/~virbio/index.htm; ►RNA export, ►nucleocytoplasmic interactions; Fabre E, Hurst E 1997 Annu Rev Genet 31:277; Mattaj IW, Englmeier L 1998 Annu Rev Biochem 67:265; Nakielny S, Dreyfuss G 1999 Cell 99:677; Route MP et al 2000 J Cell Biol 148:635; Wente SR 2000 Science 288:1374; Ribbeck K, Görlich D 2001 EMBO J 20:1320; Rout MP, Aitchison JD 2001 J Biol Chem 276:16593; Cyert MS 2001 J Biol Chem 276:20805; Enninga J et al 2002 Science 295:1523; Nemergut ME et al 2002 J Biol Chem 277:17385; Ishii K et al 2002 Cell 109:551; pore complex modular architecture: Devos D et al 2006 Proc Natl Acad Sci USA 103:2172.

Nuclear Proteins: These proteins function within the nucleus. ►nuclear localization signals

Nuclear Reactor (atomic reactor): A nuclear reactor splits Uranium or Plutonium nuclei, and the neutrons in chain reactions split additional atomic nuclei. Cadmium and boron rods that can absorb neutrons regulate the process. For moderation (slowing down the neutron) heavy water and pure graphite is used. This makes the splitting more efficient. Only the $Uranium^{235}$ is fissionable but more than 99% of the naturally occurring Uranium is U^{238} that makes enrichment mandatory. In some reactors U^{238} and $Thorium^{232}$, by capturing one neutron and releasing two electrons, yield fissionable $Plutonium^{239}$ and U^{233}. The liberated thermal energy can thus be used to drive water vapor turbines to generate electric energy for peaceful purposes.

In some countries (France, Hungary) very substantial parts of the electric energy are provided by atomic power plants. Atomic reactors also produce radioactive tracers used in basic research and medicine. By combining U^{235} and Plutonium (Pu^{239}) atomic bombs may be produced. In the hydrogen bombs Uranium or Plutonium ignites a fusion process between Deuterium (heavy hydrogen isotope [D] of atomic weight 2.0141) and Tritium (H^3) thereby generating helium in the process and millions of degrees of heat. All nuclear reactions generate some fallout of radioactive isotopes that may pose serious genetic danger to living organisms. The use of nuclear energy for peaceful purposes, under carefully shielded and monitored conditions, may be justified as a compromise between environmental protection and sustainable industrial society. The use of thermonuclear weapons is not justifiable. One of the serious problems of the utilization of remaining nuclear energy is the safe disposal of the radioactive waste. ►radiation effects, ►radiation hazard assessment, ►radiation protection, ►fallout, ►atomic radiations, ►cosmic radiation, ►isotopes, ►plutonium

Nuclear Receptors: Nuclear receptors provide links between signaling molecules and the transcriptional system. They use several domains with different functions. The A/B domains are receptive to transactivation. The conserved C domains with Zn-fingers bind DNA (usually after dimerization). The D domain is a flexible hinge. The E domain binds ligands (hormones, vitamin D, etc.), dimerizes, and controls transcription. The role of the F domain is not quite clear. Fusing to them specific transactivator and receptor domains enhances the efficiency of the nuclear receptors. The nuclear receptor superfamily includes more than 150 transcription factors. *Drosophila* has only 18 nuclear receptor genes, far fewer than other genetic model organisms; yet these represent all six classes of vertebrate receptors (King-Jones K, Thummel CS 2005 Nature Rev Genet 6:311). There are ~200 nuclear receptor co-activators (Lonard DM, O'Malley BW 2006 Cell 125:411). Co-activators are also considered to be integrators of various activities such as metabolism, growth, and signaling. The co-activators recruit co-coactivators such as protein remodelers, ubiquitin-conjugating enzymes, methylases, acetylases, kinases and other proteins and RNAs to the transcription machinery. Histone methyltransferase dependent inhibitory histone code requires specific histone demethylases such as LSD1 to permit ligand- and signal-dependent activation of genes (Garcia-Bassets I et al 2007 Cell 128:505). LSD1 is a flavin-dependent amine oxidase that catalyzes the specific removal of methyl groups from mono- and dimethylated Lys4 of histone H3. ►orphan receptors, ►transactivation, ►ligand, ►receptor, ►hormone-response elements, ►estrogen receptor, ►glucocorticoid response elements, ►retinoic acid, ►co-activators, ►co-repressor, ►NHR; Weatherman RV et al 1999 Annu Rev Biochem 68:559; Anke C et al 2001 Annu Rev Biophys Biomol Struct 30:329; Rosenfeld MG, Glass CK 2001 J Biol Chem 276:36865; Chawla A et al 2001 Science 294:1866; Xu W et al 2001 Science 294: 2507; Xu HE et al 2002 Nature [Lond] 415:813; see Fig. N48; Nuclear Receptor Signaling Atlas: http://www.nursa.org/.

Nuclear RNA: ►hnRNA, ►transcript

Nuclear Shape: A nuclear shape is influenced by the organization of the cytoskeleton and nuclear matrix proteins. These factors also impact gene expression. ►cytoskeleton; Thomas CH et al 2002 Proc Natl Acad Sci USA 99:1972.

Nuclear Size and Radiation Sensitivity: Both are directly proportional except in polyploids where multiple copies of the genes may greatly delay the manifestation of the damage although the larger nucleus is a more vulnerable target for damaging agents; the multiple copies of the same gene may compensate for each other's function. ►physical mutagens, ►radiation effects, ►radiation vs. nuclear size; Sparrow AH et al 1961 Radiat Bot 1:10.

Nuclear Targeting: ►nuclear localization sequences, ►receptor-mediated gene transfer

Nuclear Traffic: ►nuclear pore, ►RNA transport, ►nuclear localization sequences

Nuclear Transfer: ►nuclear transplantation

Nuclear Transplantation: The nuclei of plant and animal cells can be inactivated by UV or X-radiation and then the nucleus can be replaced with another. Alternatively, the nucleus of one cell is evicted (enucleation, leading to the formation of a cytoplast) by cytochalasin (a fungal toxin) and the nucleus (karyoplast) is saved. Another cell is subjected to a similar procedure and the enucleated cytoplast saved (its karyoplast is discarded). Then, by reconstitution, the nucleus of cell #1 is introduced into the cytoplast of cell #2. In successful cases the transplanted nucleus is functional and can direct the normal development of the cell into an organism (frog), indicating totipotency of nuclei harvested from different cell types. Similar transplantation procedures have been successfully employed for various mammals, including sheep and cattle where viable progeny has also been obtained when the oocytes with transplanted nuclei were implanted into ewes or cows after several in vitro passages. The donor nuclei were obtained from still totipotent embryonic tissues. In 1997, the transfer of a mammary cell nucleus of a six-year-old ewe to enucleated metaphase II stage eggs resulted in a normal lamb (Dolly). Dolly died in 2003 due to lung tumors and arthritis but left behind six healthy offspring, the results of natural procreation. Before transfer, the mammary cells were cultured at low concentration (0.5%) serum to force the cells to exit from the cell cycle into G_0 stage. This step was required to assure compatibility of the donor nucleus (reprogramming) with the egg and avoiding DNA replication that may result in polyploidy and other types of chromosomal anomalies. It was assumed that the success of

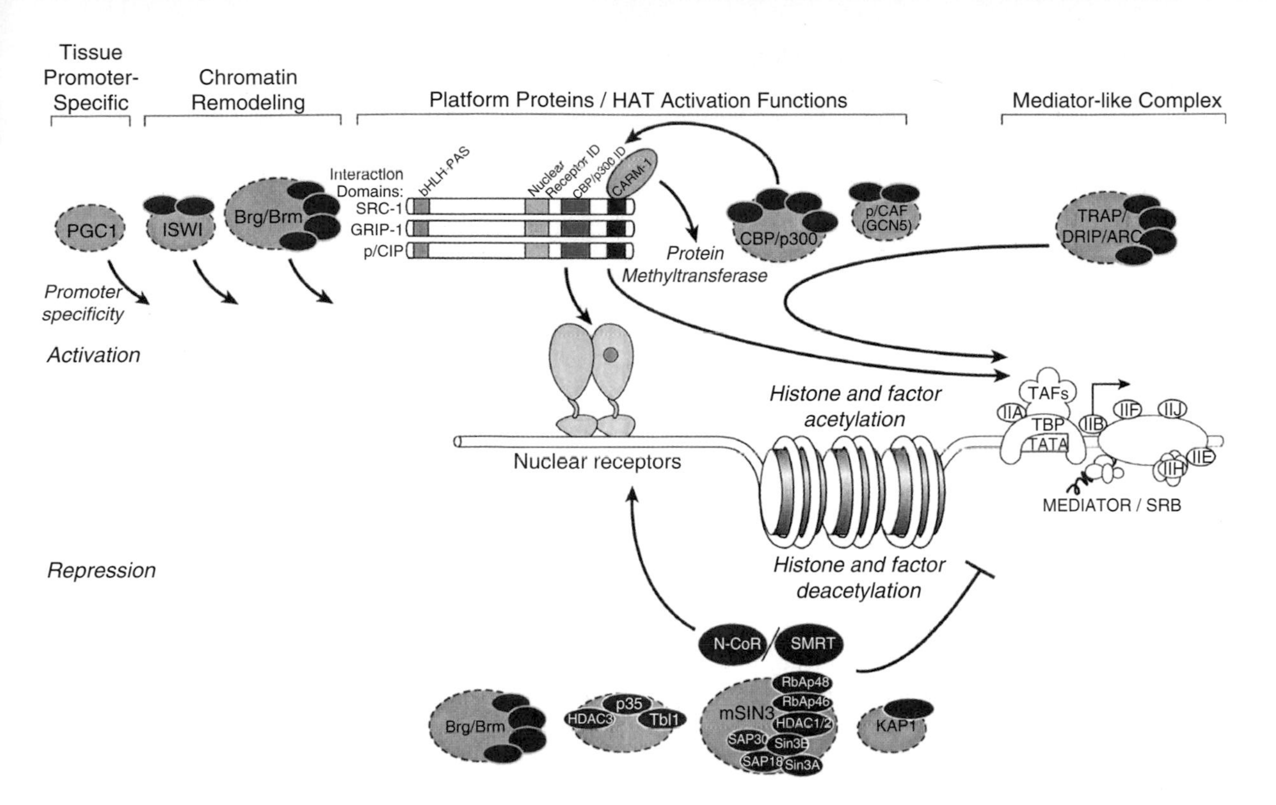

Figure N48. The activation and repression circuits tied to the nuclear receptor. PGC1: A peroxisome proliferator-activated receptor-γ co-activator; ISWI: The ATPase subunit of NURF, a chromatin remodeling protein; Brg/Brm: DNA-dependent human ATPase, active in SWI/SNF-like manner in chromatin remodeling; SRC-1: A hormone receptor co-activator and it enhances the stability of the transcription complex controlled by the progesteron receptor; GRIP: A glutamate receptor interacting protein, which contains seven PDZ domains and interacts with C end and links AMPA (a glutamate receptor targeting protein) to other proteins; CARM-1: Mediates transcription when recruited to the steroid receptor co-activator (SRC-1) and cofactors p300 and P/CAF; p/CIP: A signal-transducer and co-activator protein, interacting with p300 (CREB). It regulates transcription and somatic growth of mammals; GCN5: A yeast transcriptional co-activator; TRAP/DRIP/ARC: A multiprotein complex regulating transcription; TAFs: TATA box associated transcription factors; TBP: TATA box-binding protein; IIA, IIB, IIE, IIF, IIJ, IIH: General transcription factors; PolII: A DNA-dependent RNA polymerase; MEDIATOR is a complex of several proteins facilitating transcription; SRB: Stabilizes the polymerase and its association with transcription factors; N-CoR: Proteins that are negative regulators of transcription; SMRT: Silencing mediator of retinoic a thyroid hormone receptors; mSIN: Proteins are transcriptional co-repressors; RbAp: Proteins of the WD-40 family of widely present regulators of chromatin, transcription and cell division; HDAC: Histone deacetylase proteins; Sin 3: Repressor proteins; SAP: Stress-activated regulatory proteins; p35: a CDK5-protein; Tbi1: A thyroxin-binding regulator. (See more under separate entries) The illustration is the courtesy of Dr. Michael G. Rosenfeld. By permission of the American Society for Biochemistry and Molecular Biology.

reprogramming or remodeling of the donor nucleus was due to the presence of appropriate transcription factors and DNA-binding proteins. Nucleosomal ATPase ISWI may affect the course remodeling of the nucleus and the TATA box-binding protein is released from the nuclear matrix. Electrical pulses facilitated the uptake of the nucleus. The identity of the donor and recipient genotypes was verified using microsatellite markers. The possibility of transplantation of nuclei from mature tissues and raising viable offspring after reintroducing the reconstituted egg into incubator females was the first example of cloning of an adult mammal. Mitochondrial DNA evidence was not included in the original report. In cattle, actively dividing embryonal fibroblasts were the successful nuclear donors. In a similar manner, human cord fibroblast nuclei were successfully transferred into enucleated bovine oocytes, which developed up to the blastocyst stage (Chang KH et al 2004 Fertil Steril 82:960); chicken somatic nuclei worked similarly (Kim TM et al 2004 Fertil Steril 82:957) using this interspecies somatic cell nuclear transfer technique (ISCNT). Essentially, similar transplantation experiments supposedly produced viable offspring in rhesus monkeys (Meng L et al 1997 Biol Reprod 57:454). When nuclei from skin cells of an adult rhesus monkey (*Macaca mulatta*) were introduced into enucleated monkey eggs, some were developed into blastocysts as suitable sources for the production of embryonic stem cells in a primate (Baker M 2007 Nature [Lond] 447:891) indicating the feasibility of the procedure, which earlier had encountered pessimism for primates. Microinjected nuclei from cloned embryonic stem cells also resulted in normal mice showing that, from a single cell, several identical offspring could be obtained. A primordial germ cell at the stage of 8.5 to 9.5 days post coitum can be successfully used to develop rather normal offspring when inserted into oocytes; after 11.5 days the nuclei were not suitable to support full development of the fetus (Yamazaki Y et al 2005 Proc Natl Acad Sci USA 102:11361).

The sheep experiment indicated unexpected shortening of the telomeres of the cloned lamb, similar to what happens during aging. In cattle, this nuclear senescence did not occur in the cloned calves. Actually, older donor nuclei returned to a juvenile state and appeared normal. The cells of the clones displayed an extended life span (Lanza RP et al 2000 Science 288:665). Recently, a high frequency of pregnancy loss was observed during the early stages of the fetuses and again during the perinatal stages in sheep and other animals. Some of the defects resembled the cases known in the human Beckwith-Wiedeman syndrome, the Simpson-Golabi-Behmel syndrome, the Alagille syndrome, pulmonary hypertension, etc. (Rhind SM et al 2003 Nature Biotechn 21:744). Mice cloned from somatic cells have a shorter life because of pneumonia and hepatic failure (Oganuki N et al 2002 Nature Genet 30:253). The obesity of cloned mice is not inherited in the progeny (Tamashiro KLK et al 2002 Nature Med 8:262).

N

When nuclear transplantation takes advantage of genetically modified donors, special genes can be targeted for recombination and thus insertion or knockout. One must keep in mind that "perfect" cloning of males may not be feasible because the nuclear transfer takes place into an egg that may have different mitochondrial DNA. Also, if the eggs are collected from different females, even if the nuclei are obtained from a single individual, the offspring may not be entirely identical. In cattle, a low degree of mitochondrial heteroplasmy had been generated by nuclear transplantation. The propagation of embryos by splitting the cleavage stage blastomeres produces, presumably, entirely identical embryos because in this case not only the cell nucleus but also the cytoplasm is cloned. These experiments foreshadowed the possibility of eventual cloning of all other mammals, including humans. On 27 December 2002, the *New York Times* reported the announcement of the birth of the "first cloned" (unconfirmed!) human baby girl. Therefore, new political and ethical concerns surfaced regarding the potential manipulation of the human race. In addition, there exist biological difficulties with the cloning of primates. Experimental evidence indicates that after somatic nuclear transfer into enucleated eggs a matrix protein, which is required for spindle pole assembly (NuMA), and a kinesin motor protein (HSET) are not detectable in the cells of primates. Consequently, mitosis is abnormal and the fertilized egg fails to develop into a normal, viable fetus (Simerly C et al 2003 Science 3000:297). It is noteworthy that germ cells are inadequate as nuclear donors because the somatic methylation pattern is erased as they are reprogrammed for either male or female sex-specific manner. This (in mice) interferes with normal embryonic development and causes death during gestation (Yamazaki Y et al 2003 Proc Natl Acad Sci USA 100:12207).

Genetically the clones would not be expected to differ from monozygotic twins. The manipulations, however, may cause abnormal development and—according to some observations—premature aging unless the procedure is perfected. In cloned bovine embryos, the various developmental anomalies were attributed to abnormal methylation, i.e., the methylation pattern of the donor nucleus persisted (Kang YK et al 2001 Nature Genet 28:173). Frequently respiratory and circulatory problems arise and the fetus and the newborns may display abnormally large size. Using the nuclei of embryonic stem cells involves fewer developmental problems than the use of nuclei from adult tissues. More recent information is contrary to this view: mouse hematopoietic cells, when tested at different differentiation stages—hematopoietic stem cells, progenitor cells and granulocytes—indicated increased cloning efficiency over the differentiation hierarchy; also, terminally differentiated postmitotic granulocytes yield cloned pups with the greatest cloning efficiency (Sung L-Y et al 2006 Nature Genet 38:1323). Some observations indicate subtle abnormalities even when the animals so cloned are apparently normal. The expression of genes varies in nuclear transplantation (and even after artificial insemination) from that after normal fertilization. In one global study of fertilization in cattle, from the 50 genes studied by microarray hybridization, the expression of 25 genes was different after artificial insemination from nuclear transplantation; 17 genes were specific to artificial insemination; and eight were specific for nuclear transplantation. The expression of several other genes varied according to the type of the generation of the embryos by these methods. The differences in reprogramming the embryos at the blastocyst stage explains why there is so much difference during development (Smith SL et al 2005 Proc Natl Acad Sci USA 102:17582). When nuclei from hair follicle stem cells and other skin keratinocytes were used as donors, 19 live-born mice were obtained, nine of which survived to adulthood. Embryonic keratinocytes and cumulus cells also gave rise to cloned mice. Although cloning efficiencies were similar (<6% per transferred blastocyst), success rates were consistently higher for males than for females (Li J et al 2007 Proc Natl Acad Sci USA 104:2738). The cloning of animals may contribute to the development of improved livestock and may facilitate medical research. By cloning, embryonic stem cells obtained with the aid of nuclear transplantation from an individual may facilitate tissue and organ replacement for the same individual without the danger of immunological rejection. For xenotransplantation of cloned pig tissues or organs into humans, the removal or blocking of the porcine α-1,3-galactosyl transferase cell surface antigen may prevent immunological rejection by humans. In the past, the homozygosity of animal herds and flocks was pursued by inbreeding. This could lead to reduced vigor (inbreeding depression) which is avoided by cloning in the short range; in the long range, however, appropriate mating schemes are required for sexual reproduction. Nuclear transplantation may rescue endangered species. Nuclei prepared from dead female mouflons (*Ovis orientalis musimon*), injected into enucleated sheep (*Ovis aries*) oocytes, and then the transfer of the blastocyst stage embryos to domesticated sheep foster mothers produced apparently normal mouflon. ►totipotency, ►differentiation, ►reprogramming, ►methylation of DNA, ►aging, ►cytoplast, ►cytochalasin, ►cloning, ►transplantation of organelles, ►allopheny, ►cell cycle, ►G^0, ►inbreeding coefficient, ►mating systems, ►stem cells, ►*Acetabularia*, ►ethics, ►knockout, ►rejection, ►large embryo/offspring syndrome, ►hinny, ►clonote, ►epigenetic memory, ►somatic nuclear transfer, ►Beckwith-Wiedeman syndrome, ►Simpson-Golabi-Behmel syndrome, ►Alagille syndrome, ►pulmonary hypertension, ►GMO; Poljaeva IA et al 2000 Nature [Lond] 407:86; Humpherys D et al 2001 Science 293:95; Loi P et al 2001 Nature Biotechnol 19:962; Lanza RP et al 2002 Science 294:1893; Wilmut I 2002 Nature Med 8:215; Gurdon JB, Byrne JA 2003 Proc Natl Acad Sci USA 100:8048; Eggan K et al 2004 Nature [Lond] 428:44.

Nucleases: ►endonuclease, ►exonuclease, ►restriction enzymes, ►ribonucleases, ►DNase

Nuclease-Hypersensitivity: ►nuclease sensitive sites

Nuclease-sensitive Sites: In the transcriptionally active chromatin nuclease enzymes more easily digest some short DNA tracts than the rest of the chromatin. These hypersensitive regions indicate that in order to be accessible to RNA polymerase and transcription factors, the DNA must be in a particular and more open conformation. A well-conserved, 5′-CCGGNN-3′ repeat sequence seems to exclude nucleosomes. The longer the repeat the more effective is the nucleosome exclusion. Nuclease hypersensitive sites seem to be absent in regions where there is no potential transcription. ►nucleosomes, ►regulation of gene activity, ►chromatin remodeling, ►FAIRE; Li G et al 1998 Genes Cells 3(7):415.

Nucleation: In general usage it means the formation of an initial critical core in a process. During polymerization, first a smaller number of monomers assemble in the proper manner, which can then be followed by a rapid extension of the polymer.

Nucleic Acid Bases: In DNA, adenine (A), guanine (G), thymine (T), and cytosine (C), in RNA uracil (U) is comparable to (T). The common bases in DNA are adenine (A), guanine (G), thymine (T) and cytosine (C) and in RNA uracil (U) has a role comparable to (T) in DNA. Besides these main bases, hydroxymethyl cytosine (in bacteriophages) and methyl cytosine (in eukaryotes) are regular components of the DNA. To a lesser extent, other methylated bases may also occur in both DNA and RNA. In the transfer RNAs additional minor purine and pyrimidine bases may be found such as dimethyladenine, hypoxanthine, isopentenyl adenine, kinetin, zeatin, pseudouracil, 4-thiouracil, etc. The minor bases are modified after biosynthesis of the regular nitrogenous bases (see Fig. N49).

N

Nucleic Acid Chain Growth: Nucleic acid chain growth is secured by adding 5′-triphosphates of nucleosides to the replicating system. The additions take place at the 3′-OH end of the ribose (deoxyribose) by joining only the α phosphate; the β and γ phosphate groups (pyrophosphate) are split off. Nucleic acid chains can grow only at the 3′-OH terminus. The first nucleotide in the chain may retain all three phosphates. The single-stranded DNA chain elongates in the same manner as the RNA strand. The only difference between the two is that in the DNA (solid arrow) there is 2′ deoxyribose, whereas in the RNA, at the same position, there is OH.

Direct evidence shows that the RNA polymerase protein (RNAP) moves a single nucleotide at a time (base-pair stepping) as a nucleotide is added to the elongating RNA (Abbondanzieri EA et al 2005 Nature [Lond] 438:460). ►DNA replication in

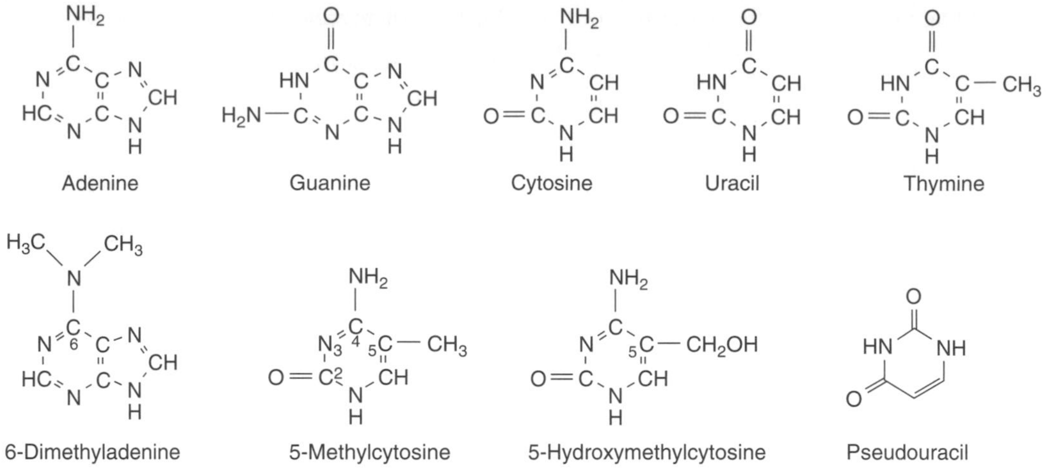

Figure N49. Major nucleic acid bases

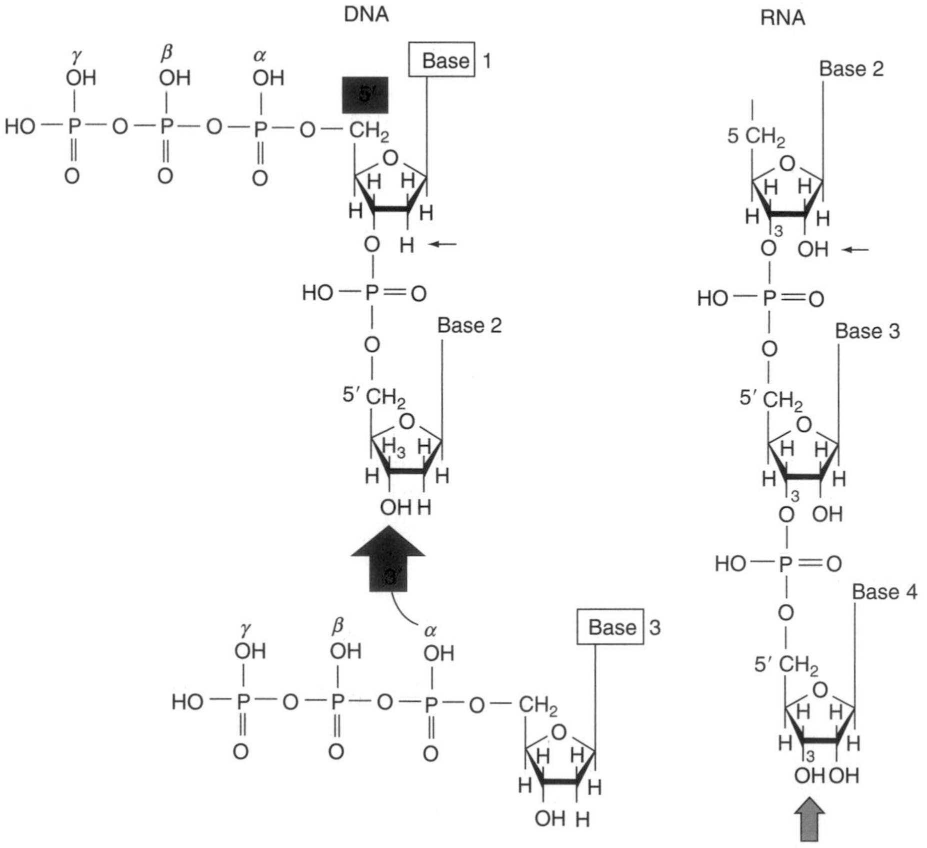

Figure N50. Nucleic acid chain growth

prokaryotes, ►replication fork, ►transcript elongation, ►DNA sequencing; see Fig. N50.

Nucleic Acid Homology: Nucleic acid homology is based on the complementarity of the bases in the nucleic acids. Complementary bases can anneal by the formation of hydrogen bonds, e.g., in two single-stranded nucleic acid chains such as A-T-**A**-**G**-**C**-T-G and G-C-**T**-**C**-**G**-T-A; only the bases represented by bold letters are com-plementary. ►hydrogen bond; nucleic acids structure alignment server: http://correlogo.abcc.ncifcrf.gov/.

Nucleic Acid Hybridization: Annealing single strands of complementary DNAs or DNA with RNA. ►nucleic homology, ►DNA hybridization, ►c_0t curve, ►Southern blotting, ►in situ hybridization; Marmur J, Lane D 1960 Proc Natl Acad Sci USA 46:453; Doty P et al 1960 Proc Natl Acad Sci USA 46:461; Hall BD, Spiegelman S 1961 Proc Natl Acad Sci USA 47:137.

Nucleic Acid Probe: ►probe

Nucleic Acids: Deoxyribonucleic acid (DNA) and ribonucleic acid (RNA) are nucleic acids; they are polymers of nucleotides. ►DNA, ►RNA; http://ndbserver.rutgers.edu.

Nuclein: A crude nucleic acid-containing preparation first obtained by Friedrich Miescher in 1868. He identified it as a common constituent of the nuclei in pus ENVELOPE cells and other tissue cells and yeast. ►nucleic acid

Nucleobindin (calnuc): A Ca^{2+} -binding protein in the cytosol and the Golgi apparatus. It interacts with the α-5 helix of the $G_{ai}3$ mediating signal transduction. ►Golgi apparatus, ►signal transduction, ►G proteins; Kubota T et al 2001 Immunol Lett 75(2):111.

Nucleocapsid: The viral genome, including an inner protein layer that is not part of the viral envelope protein (see Fig. N51).

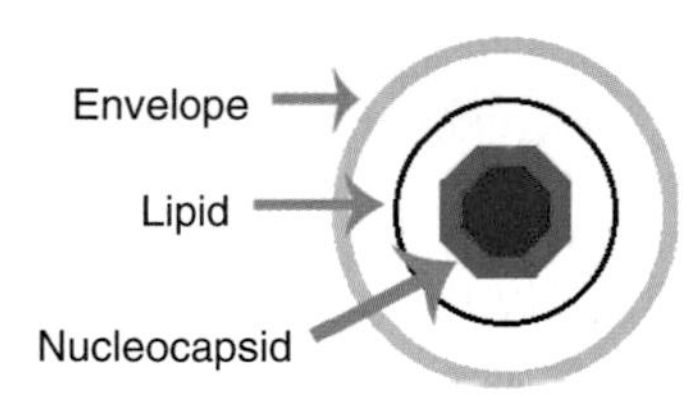

Figure N51. Viral nucleocapsid

Nucleo-cytoplasmic Interactions: In nucleo-cytoplasmic interactions the products of chromosomal genes may affect the expression of organellar genes and organellar gene products may influence the expression of nuclear genes. These interactions are facilitated through the complex structure of the nuclear pore. The communication is bidirectional and nuclear localization signals dictate the fate of a protein population rather than that of the individual molecules that bear it, which remain free to shuttle back and forth (Kopito RB, Elbaum M 2007 Proc Natl Acad Sci USA 104:12743). ►nuclear pore, ►cytoplasmic male sterility, ►restorer genes, ►chloroplast genetics, ►nuclear localization sequences, ►mitochondrial genetics; Adam SA 1999 Curr Opin Cell Biol 11:402; Simons KJ et al 2003 Genetics 165:2129; Cook A et al 2007 Annu Rev Biochem 76:647.

Nucleocytoplasmic Factory: There are thousands of about 50-nm diameter centers in the eukaryotic nucleus where several DNA-dependent RNA polymerases are actively turned on in different transcription units. The factory forms a cloud of RNA loops. (See Cook PR 2002 Nature Genet 32:347).

Nucleoid: A region at the prokaryotic cell membrane where the cell's DNA is condensed. This DNA is not surrounded, however, by a membrane, as is the case with the eukaryotic nucleus. The DNA in the mitochondria and chloroplasts is organized in a similar manner without a special envelope; however, the organellar genetic material is present in multiple copies. In *E. coli* bacteria, the volume of the nucleoid is ~0.2 μ^3 within the ~1 μ^3 cell. The bacterial chromosome is compacted by ~10^3- to 10^4-fold. The DNA occupies about 5% of the nucleoid. Various proteins maintain the structure of the chromosome in the nucleoid (Dame RT et al 2006 Nature [Lond] 444:387) such as the histone-like HU, HNS (a nucleoid-structuring protein family), IHF (integration host factor, histone-like protein) and SMC (a structure and function regulating group of proteins with homologs in eukaryotes), and leucine-responsive regulatory protein (Lrp). The Dna proteins, helicases, and AAA ATPases are involved in various steps of replication. The SeqA proteins prevent premature reinitiation by sequestering the replicational origin. Topoisomerases are used for separating the two DNA strands and gyrase removes positive supercoiling and separate catenated molecules. Recombination of the replicating chromosomes (occurring in every six generations in *E. coli*) results in dimeric chromosomes that are resolved to monomers by tyrosine recombinases XerC and XerD acting at the replication termination sites *dif*. The Fts protein complex regulates chromosome segregation. ►prokaryote, ►AAA protein, ►catenated, ►topoisomerase, ►gyrase, ►DNA replication prokaryotes, ►reinitiation of replication, ►mtDNA, ►replication fork, ►chloroplast, ►differentiation of plastid nucleoids, ►chondriolite; Sherratt DJ 2003 Science 301:780.

Nucleolar Chromosome: A nucleolar chromosome has a nucleolar-organizing region where the ribosomal genes are located. ►nucleolus, ►ribosomal RNA, ►nucleolar organizer, ►nucleolar reorganization

Nucleolar Dominance: In allopolyploid species usually only one of the parental sets of ribosomal RNA genes is expressed. In most cases the nucleolar dominance is characteristic of the components of the allopolyploid. In some *Arabidopsis thaliana* x *Cardaminopsis arenosa* hybrids both the *Cardaminopsis* and *Arabidopsis* nucleolar organizers are expressed (Pontes O et al 2003 Proc Natl Acad Sci USA 100:11418). ►nucleolus; Pikaard CS 2000 Trends Genet 16:495.

Nucleolar Localization Signals: Nucleolar localization signals target proteins to the nucleolus. Usually they contain basic amino acid motifs. ►nuclear localization sequences; nucleolar proteome: http://lamondlab.com/NOPdb/.

Nucleolar Organizer: The location of the highly repeated genes responsible for coding ribosomal RNA. It also assembles the products of their transcription until utilization. Morphologically, the nucleolar-organizing regions (NOR) are identified by secondary constrictions. The number of NORs per genome varies from one to several (see Fig. N52). The human genome has nucleolar organizers in five chromosomes (13, 14, 15, 21 22). ►nucleolus, ►ribosome, ►rRNA, ►satellited chromosome; Hourcade D et al 1973 Cold Spring Harbor Symp Quant Biol 38:537.

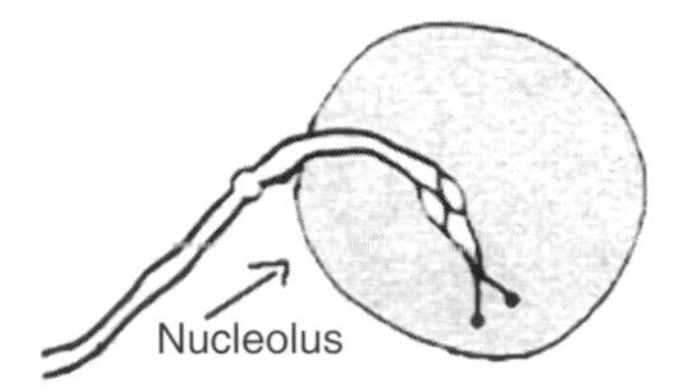

Figure N52. Nucleolar organizer region of maize chromosone 6

Nucleolar Reorganization: The nucleolus becomes dispersed beginning with the prophase. It is invisible by the light microscope by metaphase and anaphase and it starts reorganization by telophase; it finishes the reorganization by the end of the G1 phase. ►nucleolus, ►mitosis

Nucleolin: A 110-kDa nucleolar protein, coating the ribosomal transcripts within the nucleolus. Nucleolin accumulates in the cells, at the chromosomes, and is phosphorylated before mitosis. It has numerous functions: chromatin decondensation, transcription and processing of mRNA, cell proliferation, differentiation, apoptosis, shuttling between the nucleus and the cytoplasm, binding lipoproteins, laminin, growth factors, the complement inhibitor J, etc. ►nucleolus, ►p53; Westmark CJ, Malter JS 2001 J Biol Chem 276:1119.

Nucleolomics: The proteomic inventory of the nucleolus. Dundr M Misteli T 2002 Mol Cell 9:5.

Nucleolus: A body (1–5 μm) associated with the nucleolus-organizer region of one or more eukaryotic chromosomes at the coding region of ribosomal RNAs. Besides RNA, the nucleolus contains various proteins and it is the site of the production of the ribosomal subunits. In the human nucleolus, at least 271 proteins occur (Andersen JS et al 2002 Curr Biol 12:1). Actually, the flux of 489 endogenous proteins has been detected more recently (Andersen JS et al 2005 Nature [Lond] 433:77). It has a larger variety of proteins than those incorporated into the ribosomes. By electron microscopy in the nucleolar organizer region, paler fibrillar regions are distinguished where the DNA is not transcribed. The darker fibrillar elements indicate the rRNAs positions, and a granular background contains the ribosomal precursors (see Fig. N53). Although in some species there are several nucleolar organizer regions, the number of nucleoli may be smaller because of fusion. During nuclear division the nucleoli gradually disappear from view (as rRNA transcription tapers off) to be reformed again, beginning with the telophase. Apparently, during the stage of no-show, the ribosomal components are not destroyed but only dispersed to the surface of the chromatin bundle. ►cell structure, ►ribosome, ►rRNA, ►nucleolin, ►nucleolar organizer, ►nucleolar chromosome, ►nucleolar reorganization, ►satellited chromosome, ►coiled body, ►CDC14, ►paraspeckles, ►transcription factories; Pederson T 1998 J Cell Biol 143:279; Nomura M 1999 J Bacteriol 181:6857; Filipowicz W Pogacic V 2002 Curr Opin Cell Biol 14:319; review: Raska I et al 2006 Curr Opin Cell Biol 18:325.

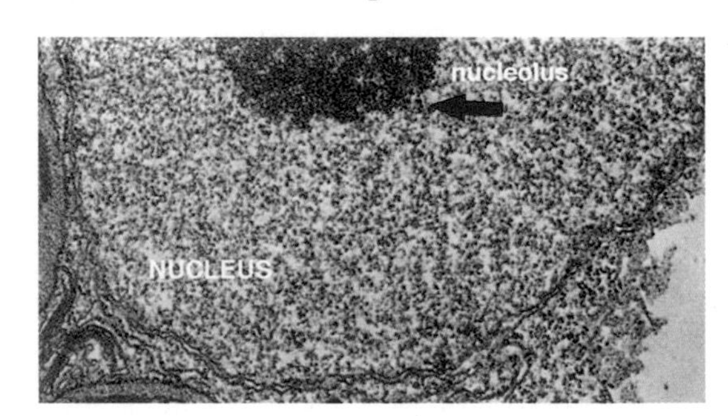

Figure N53. Electron micrograph of nucleus and nucleolus.

Nucleolytic: The function of nucleases cutting phosphodiester bonds.

Nucleomorph: A vestigial nucleus-like structure in some algae, enclosed within the chloroplast endoplasmic reticulum. The nucleomorph of the chlorachniophytes is only 380 kb, the smallest among eukaryotes. It contains three linear chromosomes, with subtelomeric rRNA genes at both ends. Two protein genes are co-transcribed. The 12 introns are very short (18–20 bp), and the spacers are too. Most of the genes are involved with the maintenance of the nucleomorph. In the cryptomonads, the nucleomorph is somewhat larger (550–600 kb), yet it also has only three small chromosomes. In some organisms there are three or four layers of membranes indicating that they absorbed the nucleomorph(s) of the symbiont(s) and retained it as a plastid(s) enclosed by the chloroplast. ►chloroplast; Douglas S et al 2001 Nature [Lond] 410:1091; Gilson PR et al 2002 Genetica 115:13.

Nucleophile: An electron-rich group that donates electrons to electron-deficient (electrophile) carbon or phosphorus atoms. ►electrophile

Nucleophilic Attack: The reaction between a nucleophile and an electrophile. ►nucleophile, ►electrophile

Nucleophosmin (NPM, encoded at 5q35): Its cDNA encodes a 294-amino acid protein required for the maintenance of genomic stability. In heterozygous or homozygous loss leads anomalous replication of the centrosome and myelodysplasia and anaplastic lymphoma (leukemia). (See Grisendi S et al 2005 Nature [Lond] 437:147).

Nucleoplasm: A solute within the eukaryotic cell nucleus. (see Fig. N53)

Nucleoplasmin: An acid-soluble protein which mediates the assembly of histones and DNA into chromatin. It is present in the nucleoplasm and in the nucleoli and also carries out chaperonin functions. It removes sperm-specific basic proteins from the pronucleus after fertilization and replaces them with H2A and H2B histones. Tyrosine-124-dephosphorylated nucleoplasmin mediates chromosome condensation but not the fragmentation of chromosomes during apoptosis (Lu Z et al 2005 Proc Natl Acad Sci USA 102:2778). ►chromatin, ►histones, ►apoptosis; Andrade R et al 2001 Chromosoma 109(8):545; Dutta S et al 2001 Mol Cell 8:841; Burns KH et al 2003 Science 300:633; No38 structure: Namboodiri VMH et al 2004 Structure 12:2149.

Nucleoporin: The structural elements of the nuclear pore. The Nup98 and Nup96 and the Nup98-Nup96 complex are translated from an alternately spliced mRNA. These two proteins are post-translationally processed by auto-proteolysis (without a need for proteases). The nuclear pore complex of budding yeast is built of 30 nucleoporins whereas the mammalian complex has about 60 nucleoporins. The nuclear pore plays an important role in activation/suppression of gene expression and it recruits DNA sequences responding to RAP 1 and GCR co-activators (Menon BB et al 2005 Proc Natl Acad Sci USA 102:5749). ►nuclear pore, ►RAP1A, ►karyopherin, ►APC (anaphase-promoting complex); Allen NP et al 2001 J Biol Chem 276: 29268; Patel SS et al 2007 Cell 129:83.

Nucleoprotein: Nucleic acids associated with protein.

Nucleoside: A purine or pyrimidine covalently linked to ribose or deoxy-ribose or to another pentose.

Nucleoside Diphosphate Kinase: Mediates the transfer of the terminal phosphate of a nucleoside 5′-triphosphate to a nucleoside 5′-diphosphate.

Nucleoside Monophosphate Kinase: Transfers the terminal phosphate of ATP to a nucleoside 5′-monophosphate.

Nucleoside Phosphorylase Deficiency: A dominant human chromosome 14q22 anomaly involving T cell immunodeficiency and neurological problems. ►T cell

Nucleosome: A histone octamer is wrapped around by about 1 and 3/4 times/particle and these nu bodies are connected with a 40–60-nucleotide long DNA linker with either histone 1 or histone 5 (►histones). The total size of the nucleosomes is different in different species and may vary between 146 (147) to 250-nucleotide pairs. In the centromeric region of *Drosophila* interphase chromosomes the nucleosomes are tetrameric in vivo, with one copy of CenH3 (the centromere-specific H3 histone variant), H2A, H2B, and H4 each, wrapped one full turn of DNA indicating that these nucleosome particles are only half as high as the common nucleosomes (see Fig. N54) (Dalal Y et al 2007 PloS Biol 5(8):e218).

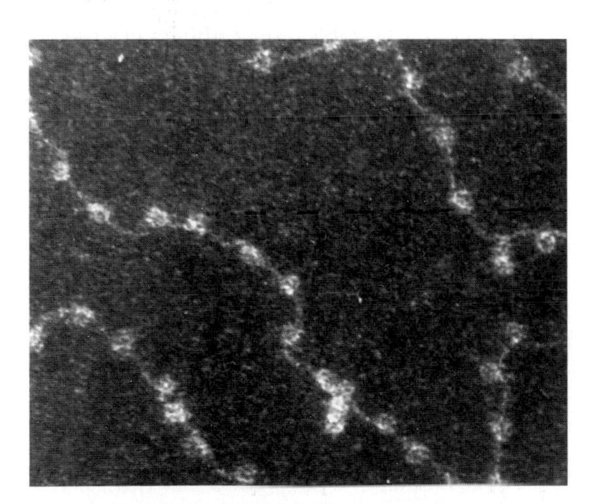

Figure N54. A dark field electron micrograph (260,000 X) of a nucleosome string of chicken. (Courtesy of Drs. Olins AL and Olins DE.)

Organization similar to the nucleosome core is found in the dTAF$_{II}$ proteins associated with the TFIID transcription factors (see Fig. N55). Before transcription, the nucleosomal structure is remodeled.

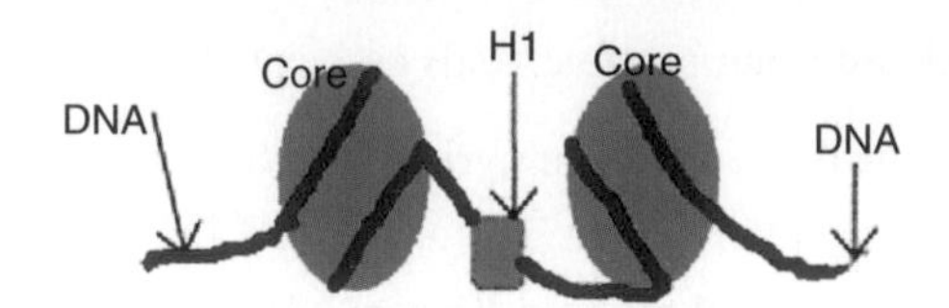

Figure N55. Model of a nucleosome string

The 2000-kDa SWI/SNF complex generates DNase hypersensitive sites by loosening the association of the DNA with histones and permitting the entry of the transcription factors (see Fig. N56). This function of SWI/SNF is transient and requires ATP. There is another nucleosome remodeling factor (NURF, an ATPase) that remains associated with the nucleosomes.

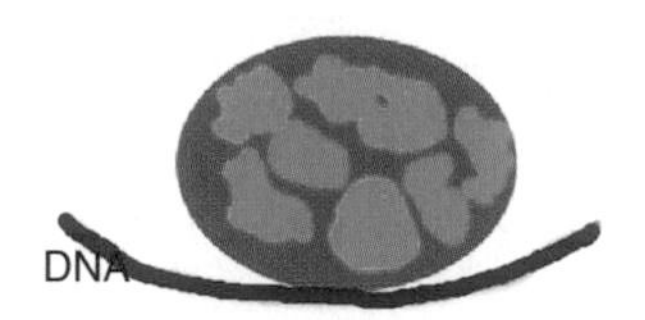

Figure N56. Open nucleosome permits access of protein factors to DNA during transcription. (See Tomchik M et al. 2005 proc. Natl. Acad. Sci. USA 102: 3278)

Nucleosomes are not distributed at random along the DNA sequences, but high and low affinity tracts are known. Nucleosome occupancy appears to be highest over the centromeric regions. Nucleosomal organization is unstable over highly expressed genes. The highly expressed ribosomal and transfer RNA genes seem to have low nucleosome occupancy. Low occupancy facilitates chromatin remodeling, a requisite with gene expression due to more facile access of transcription factors. Methods are available for mapping nucleosomal positions in the genome (Dennis JH et al 2007 Genome Res 17:928).

About 10-bp AA/TT/TA dinucleotides oscillate in phase with each other and out of phase with ~10-bp GC dinucleotides that apparently facilitate the DNA helical repeat in different DNAs and facilitate the sharp bending of DNA around the nucleosomes. These genome-wide observations indicate a code for nucleosome positioning (Segal E et al 2006 Nature [Lond] 442:772). Comparative genomics facilitated genome-wide mapping of nucleosome positioning sequences (NPSs) in the vicinity of all *Saccharomyces cerevisiae* genes. The underlying DNA sequence provides a very good predictor of nucleosome locations that have been experimentally mapped to a small fraction of the genome. Notably, distinct classes of genes possess characteristic arrangements of NPSs that may be important for their regulation. DNA sequences that favor nucleosome formation are enriched with AA dinucleotides spaced 10-bp apart, resulting in a deficiency of TT dinucleotides at the same location. Five to six nucleotides in either direction, where the complementary strand faces the histone core, the trend is reversed (TT enrichment and a deficit of AA). Genes that are positively regulated by nucleosomes tend to be TATA-less, whereas nucleosome-inhibited genes tend to have TATA boxes. TATA-less promoters comprise 80% of the yeast genome, whereas TATA-containing promoters comprise 20% (Ioshikhes IP et al 2006 Nature Genet 38:1210).

The HIV-1 viral integrase is unusual; it prefers DNA bent around the nucleosomal structure rather than the naked DNA (Pryciak PM et al 1992 Cell 69:769). The compact yeast nucleosomes apparently lack histone 1. H1 occurs in subtypes *a* to *e* and these may be differentially involved in the regulation of mammalian gene expression (Alami RT et al 2003 Proc Natl Acad Sci USA 100:5920). Recent studies indicate that H1 may be attached not only to the linker portion of the DNA but may be situated within the coiled section. Also, the role of the histones, considered earlier only to block transcription, in some cases may be actually slightly stimulative to transcription. At high salt concentrations the nucleosomal assembly assumes a zigzag format, and at low physiological salt concentration it appears as "beads-on-a-string" (see Fig. N57) (Sun J et al 2005 Proc Natl Acad Sci USA 102:8180). The intact nucleosomal structure in place prevents the function of the DNA-dependent RNA polymerase. Therefore, the histone octamer is temporarily displaced and reestablished after the polymerase has passed through (Svejstrup JQ 2003 Science 301:1053). The DNA-dependent RNA polymerase II is preceded by a about 200-base pair nucleosome-free region in front of the start codon. The majority of the transcription factor-binding sites was devoid of nucleosomes, suggesting their role in the access of the transcription machinery (Yuan G-C et al 2005 Science 309:626). In the active euchromatin the available lysine sites of H4 histone are acetylated while in heterochromatin the acetylation is minimal. The histone acetylation and methylation map reveals the structural modification of the chromatin associated with transcription in yeast (Pokholok DK et al 2005 Cell 122:517).

The chromatin-accessibility factor (CHRAC) facilitates access to the chromatin and also in the assembly of the nucleosomal structure. CHRAC is made of five subunits including the ATPases of NURF (topoisomerase II ?), ISWtI, and Acf1, an accessory factor (Eberharter A et al 2001 EMBO J 20:3781). For the assembly of the nucleosomal H4-H3 structure by the Chromatin Assembly Factor1 (CAF1), lysine

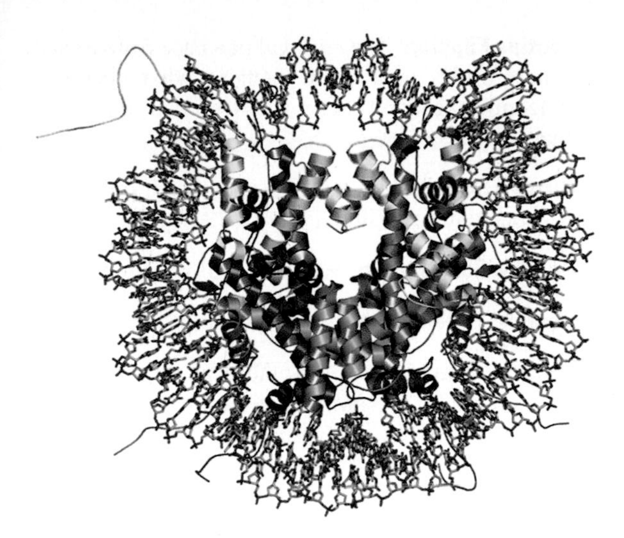

Figure N57. The closed nucleosome core particle containing four main types of eight histone molecules wrapped around by 146 DNA nucleotides. The DNA atoms are colored as carbon = green, oxygen = red, nitrogen = blue, phosphorus = purple. The histones are displayed in ribbon format. H3: Two shades of gold (yellow shades are hard to differentiate). H4: Two shades of blue (the darker is H4–1, associated with H3–1). H2A: Two shades of lavender (the darker is H2A-1). H2B: Two shades of green (the darker is H2B-1; H2B-2 is almost completely hidden behind H2B-1). The picture is courtesy Dr. Gerard J Bunick, Oak Ridge Natl. Lab

residues 5, 8 or 12 of H4 require acetylation. CAF1 is a complex of p150, p60, and p48 proteins in humans. This complex may, however, vary. In several organisms, the Nucleosome Assembly Protein1 (NAP1) joins histones 2A and 2B and H3-H4 (McBryant SJ et al 2003 J Biol Chem 278:44574). The function of the histone acetyl transferase (HAT) enzyme may be a requisite for the initiation of transcription. Histone deacetylase (HDA) proteins on the other hand may prevent gene expression. In *Drosophila* and yeast heterochromatin—which is transcriptionally inert—only lysine 12 is acetylated. It appears thus that acetylation at Lys12 controls the silencing of genes. Heterochromatin packaging in yeast is mediated by the RAP1 (repressor/activator protein). The various SIR (silencing information regulator) proteins mediate silencing. The effect of histones is not necessarily global in silencing of genes. Depletion of histone 4 in yeast may actually reduce the expression of some telomere-proximal genes and has little influence on many others. In yeast, the deletion of the N-terminal 4–23 residues of histone 4 may reduce GAL1 transcription 20-fold, indicating that this region facilitates the unfolding the nucleosome structure for the initiation of transcription (Durrin LK et al 1991 Cell 65:1023). For the binding of HNF3, acetylation of the nucleosome is not required (Cirillo LA, Zaret KS 1999 Mol Cell 4:961). When only lysine 16 is acetylated, nucleosome assembly does not take place. The nucleosomal structure is assembled at the replication fork as an initial step of the maturation of chromatin. The 30-nm chromatin fiber nucleosomal structure is further compacted and assumes an organization called two-start helix, which stacks two adjacent nucleosome cores by connecting with straight linker DNA. This model is different from the one-start class where nucleosomes were assumed to form a solenoid of six to eight units around a central cavity. The histone H4 tail appears crucial for the compaction (Dorigo B et al 2004 Science 306:1571). ►transcription, ►transcription factors, ►DNase hypersensitive site, ►histones, ►chromatosome, ►altosome, ►SWI, ►NURF, ►CHRAC, ►RSF, ►ACF, ►RAP, ►regulation of gene activity, ►chromatin remodeling, ►BRG1, ►*Polycomb*, ►histone acetyl transferase, ►histone deacetylase, ►histone phosphorylation, ►heterochromatin, ►silencer, ►SRC-1, ►signal transduction, ►high-mobility group of proteins, ►rhabdomyosarcoma, ►histone fold, ►RCAF, ►ASF1, ►ORC, ►solenoid; Workman JL, Kingston RE 1998 Annu Rev Biochem 67:545; Wolffe AP, Kurumizaka H 1998 Progr Nucleic Acids Res Mol Biol 61:379; Lomvardas S, Thanos D 2001 Cell 106:685; Lucchini R et al 2001 EMBO J 20:7294; Jacobs SA, Khorasanizadeh S 2002 Science 295:2080; Ray-Gallet D et al 2002 Mol Cell 9:1091; Becker PB, Hörz W 2002 Annu Rev Biochem 71:247; Ahmad K, Henikoff S 2002 Cell 111:281; Richmond TJ, Davey CA 2003 Nature [Lond] 423:145; NAP-1 crystal structure: Park Y-J, Luger K 2006 Proc Natl Acad Sci USA 103:1248; nucleosome/histone modification: Kouzarides T 2007 Cell 128:693; chromatin and transcription: Li B et al 2007 Cell 128:707; nucleosome region positioning database: http://srs6.bionet.nsc.ru/srs6/; nucleosome-free DNA tract detection: http://www.sfu.ca/~ibajic/NXSensor/.

Nucleosome Phasing: Nucleosome positions are not entirely random along the length of the DNA, but small variations exist and serve as controls of transcription and packaging of eukaryotic DNA. ►nucleosome; Sykorova E et al 2001 Chromosome Res. 9(4):309; Kiyama R, Trifonov EN 2002 FEBS Lett 523:7.

Nucleosome Remodeling: ►chromatin remodeling

Nucleostemin: A nucleolar protein that controls cell cycle progression in the stem cells of the central nervous system and also in some cancers. Its level rapidly decreases upon differentiation. Its overexpression leads to reduced proliferation of stem cells and cancer cells. Mutants that lack the N-terminal GTP domain of nucleostemin do not enter mitosis and undergo apoptosis in a p53-dependent manner. ►stem cells, ►cell cycle, ►p53; Tsai RY, McKay RD 2002 Genes Dev 16:2991.

Nucleotide: A purine or pyrimidine nucleoside with 1 to 3 phosphate groups attached.

Nucleotide Analog Interference Mapping (NAIM): NAIM identifies atoms, chemical groups, ligand binding, and active sites in RNA. (See Ryder SP, Strobel SA 2002 Nucleic Acids Res 30:1287).

Nucleotide Biosynthesis: *Purine biosynthetic path* → PHOSPHORIBOSYLAMINE → GLYCINAMIDE RIBONUCLEOTIDE → FORMYLGLYCINAMIDE RIBONUCLEOTIDE → FORMYLGLYCINAMIDINE RIBONUCLEOTIDE → 5-AMINOAIMIDAZOLE RIBONUCLEOTIDE → 5-AMINOIMIDAZOLE-4-CARBOXYLATE RIBONUCLEOTIDE → 5-AMINOIMIDAZOLE-4-N-SUCCINOCARBOXAMIDE RIBONUCLEOTIDE → 5-AMINOIMIDAZOLE-4-CARBOXAMIDE RIBONUCLEOTIDE → 5-FORMAMIDOIMIDAZOLE-4-CARBOXAMIDE RIBONUCLEOTIDE → INOSINATE. From the latter ADENYLATE is formed through adenylosuccinate, and GUANYLATE is synthesized through xanthylate. Free purine bases may be formed by the hydrolytic degradation of nucleic acids and nucleotides.

The pyrimidine biosynthetic pathway → N-CARBAMOYLASPARTATE → DIHYDROOROTATE → OROTIDYLATE → URIDYLATE → CYTIDYLATE. From uridylate THYMIDYLATE is made by methylation. ►DNA replication, ►salvage pathway, ►orotic acid

Nucleotide Chain Growth: At the initial position the 5′ phosphates are retained; at subsequent sites nucleotide monophosphates are attached to the 3′ position of the preceding ribose or deoxyribose after two phosphates of the triphosphonucleotides have been removed. The nucleotide chain always grows in the 5′→3′ direction. ►nucleic acid chain growth

Nucleotide Diversity: τ = the average number of nucleotide differences between two sequences randomly chosen. It also permits the estimation of polymorphism within a species and divergence among species. In humans, the average τ was estimated to be 0.063% ±0.036%; these values are about an order of magnitude lower than those determined for *Drosophila* from large populations. ►evolutionary distance, ►diversity

Nucleotide Exchange Factor: ►GEF, ►GRP

Nucleotide Excision Repair: ►NER, ►DNA repair

Nucleotide Flipping: The removal of a base from a stack of nucleotides in a chain such as what occurs in nucleotide exchange repair. ►DNA repair, ►excision repair

Nucleotide Sequencing: ►DNA sequencing; nucleotide sequence database: http://www.ebi.ac.uk/embl/.

Nucleotide Substitution: ►base substitution

Nucleotide Triplet Repeat: ►trinucleotide repeat

Nucleotidyl Transferase: Transfers nucleotides from one substance to another. ►terminal nucleotidyl transferase, ►DNA polymerase, ►RNA polymerase, ►polyA polymerase; Ranjith-Kumar CT et al 2001 J Virol 75:8615.

Nucleus: The genetically most important organelle (5–30 μm) in the eukaryotic cell surrounded by a double-layer membrane (ca. 25 nm) that encloses the chromosomes, proteins and RNA, besides other solutes. The nuclear membrane is equipped with well-organized pores for transport of macromolecules in both directions. ►nuclear pore, ►nucleolus, ►chromosomes, ►nucleomorph, ►chromosome territories; Pederson T 2002 Nature Cell Biol 4(12):E287; organization of functional structures within the nucleus: Taddei A et al 2005 Annu Rev Genet 38:305.

Nuclides: Atoms with characterized atomic number, mass, and quantum. There are almost 1,000 nuclear species and about 40 are natural radioactive nuclides. By bombardment with radioactive energetic particles many additional ones have been generated in the laboratory. ►radionuclides, ►mass spectrometry

Nude Mouse: A genetically hairless mouse; it lacks thymus and thymic lymphocytes (see Fig. N58). It is commonly used in immunogenetic research. Due to the weakened immune reaction these animals do not reject xenografts. Also, they are very useful for testing carcinogens because of the lack of immune surveillance that may eliminate the transformed cells. They are particularly advantageous for testing skin carcinogens because of their hairless skin. Cervical thymus grafts into nude mouse improved lymphocyte production. ►lymphocytes, ►immune reaction, ►HLA, ►alopecia, ►mouse, ►immunological surveillance, ►xenograft, ►thymus, ►plantibody

Figure N58. Nude mouse

Nuisance Parameter: A generally unknown but mostly needed parameter in a model that has no scientific interest. We would like to know the mean of a normal distribution but the variance is unknown. The likelihood of the mean involves the variance and different variances lead to different likelihoods. The problems may be overcome if parameter estimates (conditional likelihood) are used that do not involve unwanted parameters. In a genetic association test, allele frequency may create a nuisance parameter. ►association test

Null Allele: A non-expressed allele; it is commonly a deletion. ►deletion

Null Hypothesis: Assumes that the difference between the actually observed data and the theoretically expected data is null. Statistical methods are then used to test the probability of this hypothesis. Obviously, if the data observed does not fit to the null hypothesis considered, it may be false to conclude that the null hypothesis is not valid. The right procedure is to determine what is the probability that the data might comply with the expectation. ►t-test, ►maximum likelihood, ►probability, ►significance level

Null Mutation: Eliminates the function of a gene entirely; it may be a deletion.

Null Promoter: A null promoter lacks a TATA box and Initiator element and the transcription may begin at multiple start site sequences (MED-1). ►promoter, ►core promoter, ►base promoter

Nullichiasmate: A nullichiasmate does not show crossing over or recombination.

Nulliplex: A polyploid or polysomic individual that at a particular locus has only recessive alleles. ►simplex, ►duplex, ►triplex, ►quadruplex

Nullisomic: A condition in which a cell or individual lacks both representatives of a pair of homologous chromosomes. Nullisomy is viable only in allopolyploids where the homoeologous chromosomes can compensate for the loss.

In the photo (see Fig. N59), A, B and D denote the genomes, and the numbers indicate the particular chromosome within the three series of 7. The bottom right ear represents the normal hexaploid. (Photo courtesy Professor E.R. Sears).

Nullisomy may come about by selfing monosomics or by nondisjunction at meiosis I or II. In diploids, however, it results in lethal gametes. Nullisomy is a normal condition for the Y chromosome in females (XX) whereas nullisomy for the X chromosome is lethal. In allohexaploid wheat, nullisomy has on an average only 4% transmission through the male

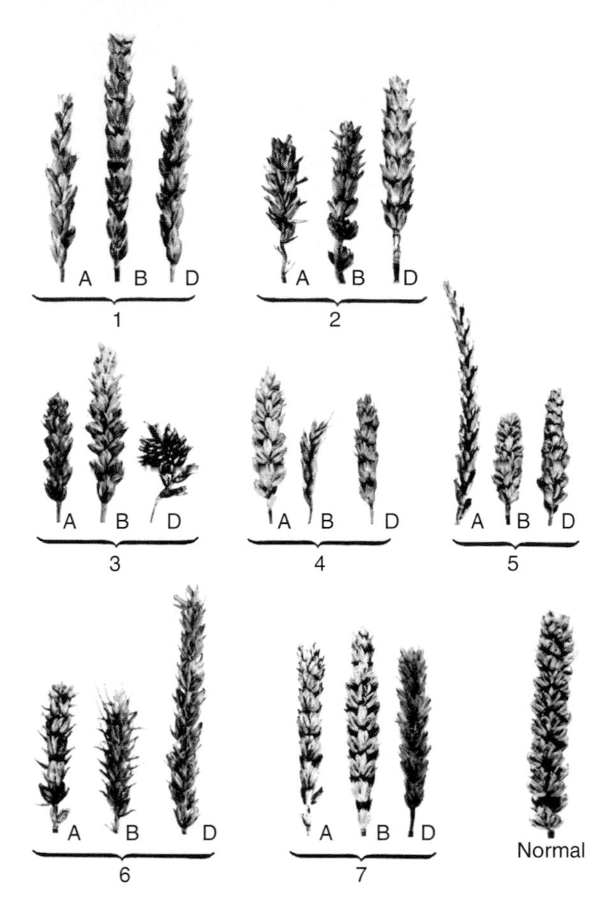

Figure N59. The complete set of the 21 nullisomics of hexaploid wheat, Chinese spring

whereas about 75% the eggs of monosomics are nullisomic. The cause of the high frequency of nullisomic eggs is that during meiosis I the univalent chromosome (of monosomics) fails to go to the pole and is thus lost. ►allopolyploid, ►monosomic, ►monosomic analysis, ►sex determination, ►nullisomic compensation, ►genome, ►*Triticum*; Sears ER 1959 p 164. In: Handbuch der Pflanzenzüchtung, vol 2., Kappert H, Rudorf W (Eds.) Parey, Berlin, Germany.

Nullisomic Compensation: Allopolyploids can survive as nullisomics but it is a deleterious condition. If, however, they are made tetrasomic for another homoeologous chromosome, their condition is ameliorated because of some degree of restoration of the genic balance (see Fig. N60).

If, however, they are madetetrasomic for another (non-homeologous) chromosome, their condition is further aggravated. The response to the added chromosome varies according to the specific chromosome. The compensation may occur spontaneously by occasional nondisjunction. If such

Figure N60. Nullisomic compensation. **Top row**: nullisomic 3A, nulli 3A-tetra 3B, nulli 3A-tetra 3D, nulliisomic 3B, 3A, nulli 3B-tetra 3A, nulli 3B-tetra 3D, nullisomic 3D, nulli 3D-tetra 3A, nulli 3D-tetra 3B. **Bottom row**: normal hexaploid (N), nulli 2B-tetra 4D, nulli 4B-tetra 5A, nulli 5D-tetra 4A, nulli 6D-tetra 1A, nulli 7A-tetra 1B, nulli 7A-tetra 4D, nulli 7A-tetra 6B, nulli 3A-trisomic 4A. Obviously the corresponding homoeologous chromosomes compensated for the entire loss of that chromosome but the non-homoeologous addition even aggravated the condition. (Courtesy of Professor ER Sears)

a nondisjunction takes place in the germline, the tissue receiving the compensating homoeologous chromosome will be at an advantage in producing gametes and there is also a higher chance for improved fertility. ►nullisomics, ►monosomic, ►homoeologous, ►tetrasomic, ►dosage effect; Sears ER 1954 Res Bull 572 Missouri Agric Exp. Sta, Columbia, MO.

Nullizygous: A loss of both alleles in a diploid. ►nullisomic

Nullosomic: Same as nullisomic (so used mainly by some human cytogeneticists).

NuMA (nuclear mitotic apparatus, centrophilin, ~11q13): A non-histone protein of about 250 kDa. It is present in the interphase nucleus and accumulates at the poles of the mitotic spindle until anaphase. Together with dynein and dynactin, NuMA tethers microtubules in the spindle pole and they assure the assembly and stabilization of the spindle pole. Mitotic interaction between Rae1 (messenger RNA export factor) and NuMA balances these two proteins for bipolar spindle formation (Wong RW et al 2006 Proc Natl Acad Sci USA 103:19783). ►mitosis, ►non-histone proteins, ►spindle, ►spindle fibers, ►dynein; Gobert GN et al 2001 Histochem Cell Biol 115(5):381; Gordon MB et al 2001 J Cell Biol 152(3):425.

Numerator: Genetic element(s) that "count" the number of X chromosomes in sex determination and for dosage compensation. ►sex determination, ►dosage compensation

Numerical Aperture (NA): The numerical aperture of a microscope lens determines the efficiency of the objective. The optical resolution of a dry lens with 0.75 NA at green light is about 0.5 μm and the depth of focus is about 1.3 μm. An oil immersion lens of 2-mm focal length and 1.3 NA has a resolution of 0.29 μm and a focal depth of 0.4 μm. ►resolution optical

Numerical Taxonomy: The classification of organisms into larger, distinct categories on the basis of quantitative measurements.

Numts (nuclearly located mitochondrial DNA sequences): These tracts may be only 100 nucleotides in length or up to 270 kb. In *Arabidopsis* plants, 75% of the mtDNA is present in the nucleus. In humans the 300–400 numts may correspond to 0.5 to 88% of the mtDNA. An analysis of numts confirmed that in the oral polio vaccines (produced with the aid of chimpanzee kidney tissue cultures) macaque mtDNA sequences occur, but not chimpanzee mtDNAs. Thus, the hypothesis that the AIDS was initiated from simian virus, SIV by the use of polio vaccines does not seem likely ►mtDNA, ►organelle sequence transfer, ►acquired immunodeficiency, ►proteobacteria; Vartanian J-P, Wain-Hobson S 2002 Proc Natl Acad Sci USA 99:7566.

Nup214: ►nuclear pore

Nup475: A transcription factor similar to TIS11, but it differs in amino sequence at the NH_2 and COOH termini. ►TIS, ►transcription factors

NUPT (nuclear integrant of plastid DNA): ►organelle sequence transfer

nur77 (TR3): A ligand-binding transcription factor, including steroid and thyroid hormone receptors (similar to NGFIB and TIS1). Its level is high in apoptotic lymphocytes but not in growing T cells. It permeabilizes the mitochondrial membrane when it migrates from the nucleus to the mitochondria. ►apoptosis; Langlois M et al 2001 Neuroscience

106:117; Sohn YC et al 2001 J Biol Chem 276:43734.

NuRD: A histone deacetylase complex of ~2 MDa containing at least seven subunits. ►histone deacetylase, ►Sin3; Ahringer J 2000 Trends Genet 16:351.

NURF: A four-protein complex mediating the sliding of the nucleosomes using ATP hydrolysis for energy. ►ISWI, ►nuclear receptors, ►nucleosomes; Xiao H et al 2001 Mol Cell 8:531.

Nurse Cells: In insect ovaries 15 (generally polyploid) nurse cells surround the oocyte within the follicles. Their gene products affect, and play a morphogenetic role in, the differentiation of the embryo at the early stages of development. ►morphogenesis, ►dumping, ►oocyte

Nurture: Nutritional factors (and also other environmental factors) that affect the manifestation of the hereditary properties (nature). In human genetics, for the separation of the two components of the phenotype twin studies are used. The differences between identical twins permit the quantitation of the extent of the influence of nurture. ►twinning

nusA, nusB: Lambda bacteriophage genes involved in the regulation of RNA chain elongation. ►lambda bacteriophage, ►DSIF, ►transcript elongation, ►TRAP

Nutlin: ►MDM2

NUTs: *NUTs* are negative regulatory elements of transcription of the Mediator family. ►mediator

Nutmeg (*Myristica fragrans*): An evergreen, dioecious spice plant; 2n = 6x = 42.

Nutrigenetics (nutritional genomics): The study of diets that are best for the genetic constitution of the individual to secure good health and long life. The goals include prevention of nutritional deficiencies and harmful metabolic responses to food according to the special need of the genotype. The special responses are not easy to assess because of the complexities of the diet; the other interacting environmental factors also play a role. Principal component analysis, cluster analysis and other statistical methods are needed for obtaining reliable information. Biomarkers are essential for the scientific assessment of food and health correlations in relation to genetic makeup, age, sex, lifestyle, and environment. Reliable genotyping procedures and chemical, molecular technologies are indispensable. It may make a difference whether the conclusions are based on single genes, QTLs, or haplotypes. Phenylketonuria, galactosemia, fructose intolerance, lactose intolerance, and celiac disease are relatively simple examples of reactions to food according to genetic constitution. The frequencies of these mutations and the particular gene or allele involved vary in different populations and different geographical areas. The sensitivity to the reaction to the diet may also be age-dependent. Although celiac disease is a reaction to cereals, the species and genotype of the grain crops can make a difference in addition to the HLA allele of the individual. Familial hypercholesterolemia can be caused by mutations at several gene loci and the effected individuals (5×10^{-2} in the USA but with much lower incidence among Orientals in China and Japan) are at risk because of a high cholesterol producing diet (LDL) and are susceptible to coronary heart disease due to advancing age, especially with certain lifestyles. Several other inborn errors of metabolism such as cystinuria, cystinosis, cystathionuria, vitamin B_{12} deficiency, hyperhomocysteinemia, hyperuricemia, hypervalinemia, hemochromatosis, and other relatively simple genetic defects can be modified by diet. Obesity, dyslipidemia, diabetes, ketotic glycinemia, kwashiorkor, lysine intolerance, and other diseases can be alleviated by appropriate diet. Apolipoproteins are involved in several health problems including aging, Alzheimer's disease, atherosclerosis and others. Cancer prevention may be favored by high-fiber, low-sugar, low-fat, low-salt diet, and moderate use of alcohol. Alcohol may also interfere with folate absorption, a factor in cancer susceptibility. Consumption of fruits and vegetable, rich in β-carotene, may be beneficial if free of toxic insecticides or fungicides. Overcooked proteins may reduce the level of enzymes (cytochrome P450 and N-acetyltransferase) involved in detoxifying naturally ingested carcinogens and overcooking of meats may generate benzo(a)pyrene and other compounds that are carcinogens/mutagens. Plant products contaminated by *Aspergillus flavus* may contain aflatoxin, a potent carcinogen. High glutathione-S-transferase content can detoxify carcinogens/mutagens. Several prescription and nonprescription drugs have side effects of different consequences, depending on the genotype of the individuals. (See disease mentioned in the alphabetical entries; ►environmental mutagens, ►carcinogens, ►Recon; Ordovas JM, Corella D 2004 Annu Rev Genomics Hum Genet 5:71).

Nutritional Mutant: ►auxotroph, ►mutation

Nutritional Therapy: Humans cannot synthesize the essential amino acids and depend on their diet for a steady supply. Similarly, there may be a dependence on an exogenous (dietary or medicinal) supply of vitamin C or other vitamins or minerals, etc. Some epileptics may benefit from the administration of pyridoxine. Various hereditary defects are known in folic acid metabolism. Hereditary fructose intolerance, galactosemia, and lactose intolerance can be kept in check by limiting the supply of these carbohydrates in the diet or by infant formulas. Phenylketonurics must avoid phenylalanine consumption. ►epilepsy, ►fructose intolerance, ►galactosemia, ►disaccharide intolerance, ►phenylketonuria

nvCJD (new variant of Creutzfeldt-Jakob disease): ►Creutzfeldt-Jakob disease

Nyctalopia: ►night blindness

Nymph (nympha): A sexually immature stage between larvae and adults of some arthropods such as ticks (*Ixodes*). ►ixodiodia

Nypmha of Krause: The same as clitoris. ►clitoris

Nymphomania: An excessive sexual drive (abnormally long estrus) in the mammalian female based on hormonal disorders and usually accompanied by reduced fertility in mares and cows. The condition may have a clear hereditary component. ►estrus

Nystagmus: An involuntary eye movement (displayed by some albinos). This condition may be controlled by autosomal recessive, dominant, or X-linked inheritance and may be associated as parts of some syndromes. Idiopathic congenital nystagmus has been located to Xq26–q27. Its manifestation varies and its incidence is about 1/1,000 (Tarpey P et al 2006 Nature Genet 38:1242). ►eye diseases, ►achromatopsia; Gottlob I 2001 Curr Opin Ophthamol 12(5):378.

Nystagmus-Myoclonous: Nystagmus accompanied by involuntary movement of other parts of the body. It is a rare congenital anomaly. ►nystagmus, ►myoclonus

Nystatin: An antibiotic produced by a *Streptomyces* bacterium. It is effective against fungal infections; it is also used as a selective agent in mutant isolation of yeast and other fungi. It kills, primarily, the growing cells. (See Arikan S, Rex JH 2001 Curr Opin Investig Drugs 2(4):488).

NZB Mouse: A non-inbred strain; used mainly for autoimmunity research.

NZCYM Bacterial Medium: H_2O 959 mL, casein hydrolysate (enzymatic, NZ amine) 10 g, NaCl 5 g, bacto yeast extract 5 g, casamino acids 1 g, $MgSO_4.7 H_2O$ 2 g, pH adjusted to 7 with 5 N NaOH and filled up to 1 L. ►casamino acids, ►bacto yeast extract

NZM Medium: The same as NCZYM but without casamino acids.

N

Historical vignette

Hermann J Muller, recipient of the Nobel Prize in 1946, commented in 1931:

"It is too late to protest that the choice of our own genes was determined by the sheer caprices of a generation now dead. but it is not too late for us to make sacrifices to the end that the children of tomorrow will start life with the best equipment of genes that can be gathered for them…but it must also be remembered that a prime condition for an intelligent and moral choice of genes is an intelligent and moral organization of society."

(Quoted by G Pontecorvo in 1968, Annu Rev Genet 2:1)

O

O: Refers to replicational origin. ►replication, ►replication fork, ►bidirectional replication

Ω (Omega): This is the insertion element present in 0 to 1 copy per mitochondrion in yeast. ►mitochondria, ►mtDNA, ►insertion elements

O Antigen: ►ABO blood group, see also O-type lipopolysaccharide-protein antigen of gram negative bacterial capsids

O Blood Group: ►ABO blood group

ω-Agatoxin: Refers to the ion channel blocking proteins present in the *Agelenopsis* spider.

Oak (*Quercus* ssp.): This is a forest as well as an ornamental tree, which has many morphological varieties, 2n = 24. The delineation of some of the numerous species may be difficult because of the not uncommon spontaneous hybridization.

Oats (*Avena* ssp.): A major cereal crop with somewhat reduced acreage since farm mechanization has led to a decrease in the number of horses used in agriculture (see Fig. O1). The cultivated species (*A. sativa*) is an allohexaploid but diploid and tetraploid forms are also well known. Basic chromosome number, x = 7. (See http://plants.usda.gov/java/profile?symbol=AVFA; genes: http://wheat.pw.usda.gov/GG2/index.shtml).

Figure O1. Avena

Oaz (ornithine decarboxylase antizyme): A multi-Zn finger protein affecting both the BMP-Smad and olfactory Olf signaling pathways. ►BMP, ►Smad, ►Zinc finger, ►olfactogenetics; Hata A et al 2000 Cell 100:229.

OBA (Office of Biotechnology Activities): Information about regulations can obtained from http://www4.od.nih.gov/oba.

Obesity: This condition is characterized by the accumulation of excessive body weight (primarily fat) beyond the physiologically normal range. Differences in predisposition to obesity have long been recognized by animal breeders and the different breeds of swine have large (over 100%) differences in fat content per body weight. Obesity is a health problem in humans because diabetes mellitus, hypertension, hyperlipidemia, heart disease and certain types of cancer appear to be associated with obesity. In mice obesity is regulated by a gene *ob* in chromosome 6, sequenced in 1994. It has been suggested that the 167 amino acid protein product synthesized in the adipose (fat) tissues is secreted into the bloodstream and regulates food intake by signaling to the hypothalamus. Reduction in this gene product or specific lesions in the ventromedial (basal-central) region of the hypothalamus stimulates food consumption and reduces the expenditure of energy. Mice lacking Nocturnin (a circadian deadenylase) remain lean on high fat diets, with lower body weight and reduced visceral fat. However, unlike lean lipodystrophic models, these mice do not have fatty livers and do not exhibit increased activity or reduced food intake. Data on gene expression have indicated that *Nocturnin* knockout mice have deficits in lipid metabolism or uptake, in addition to changes in glucose and insulin sensitivity (Green CB et al 2007 Poc Natl Acad Sci USA 104:9888).

It has been observed that in obese and normal mice the gut has different species of microbes (Ley RE et al 2005 Proc Natl Acad Sci USA 102:11070). Some experimental data point to the *db* (*diabetes*) gene (mouse chromosome 4) as a receptor for the *ob*-encoded factor, leptin. The *tubby* gene (mouse (chromosome 7) also causes maturity-onset obesity, insulin resistance, vision and hearing deficit. Tubby is activated by signal transduction from G protein-linked receptors. Phospholipase C (PLC) releases Tubby from the phosphatidylinositol-4,5-bisphosphate of the plasma membrane and then triggers its movement to the cell nucleus where it functions as a transcription factor. The *fat* mutation in mice has a later onset than *ob*. In *ob/ob* mice the serum insulin level decreases with an increase of blood glucose level. In the *fat/fat* mice exogenously supplied insulin reduces the serum glucose level. Fat mice store 70% of their insulin as proinsulin. Apparently, the *fat* gene causes deficiency in carboxypeptidase, an enzyme that normally processes proinsulin. The protein tyrosine phosphatase-1B gene (PTP-1B) of mice is a negative regulator of insulin signaling. Mutational loss of PTP-1B activity results in decreased phosphorylation of the insulin receptor. Insulin receptor knockout protects against obesity (Blücher M et al 2002 Dev Cell 3:25). Consequently, the mutant animals gain much less weight than those with the wild type allele. Phosphorylation of the insulin receptor apparently promotes glucose uptake and weight gain. In humans, obesity has been attributed to both dominant and recessive genetic factors with environmental (diet) factors accounting for about 40% of the variation in obesity. *Mahagony* (*mg*) locus has wide-ranging pleiotropic effects by suppressing obesity of the *agouti-lethal-yellow* locus. The *mg* gene encodes a transmembrane signaling receptor of 1,428 amino acids with homology to the human attractin protein produced by T lymphocytes and cross-talking between the immune system and melanocortin. Major genes in humans seem to determine 40% of the variation in body and fat mass. Heritability of increased body weight in humans may reach > 90% (Stunkard AJ 1991 Res Publ Assoc Nerv Ment Dis 69:205). There is a strong linkage between the growth hormone secretagogue receptor (ghrelin receptor, 3q26) and obesity (Baessler A et al 2005 Diabetes 54:259). Vaccination of rats with various forms of ghrelin slowed down weight gain and reduced obesity (Zorilla EP et al 2006 Proc Natl Acad Sci USA 103:13226). In humans, there is some indication of greater effect of either maternal or paternal body weight on the obesity of the progeny. The geneticist faces problems in measuring such a complex trait as obesity, e.g., in terms of body mass, fat mass, visceral adipose tissue amounts, metabolic rate, respiratory quotient and insulin sensitivity. Many human obesity factors have been implicated mainly on the basis of mice models and putative linkage of quantitative trait loci (QTL). A major susceptibility locus was detected in the short arm of human chromosome 10 (MLS 5.28) and minor quantitative factors appeared in chromosomes 2 (lod score 2.68) and 5 (lod score 2.93). Neuropeptide Y (NPY) appears to be a stimulant of food intake and an activator of a hypothalamic feeding receptor (Y5). A cAMP-dependent protein kinase (PKA) also plays an important role in obesity. This holoenzyme is a tetramer, containing two regulatory (R) and two catalytic (C) subunits. The catalytic function is phosphorylation of serine/threonine and the regulatory units slow down the enzyme when the level of cAMP is low. A knockout of the RIIβ regulatory subunit leads to the stimulation of energy expenditures in mice and they remain lean even on a diet that is normally conducive to obesity. In the inner membrane of the mammalian mitochondria, body heat is generated by uncoupling oxidative phosphorylation. Uncoupling proteins UCP1 (4q23), UCP2 (11q13) and UCP3 (11q13) also regulate obesity to some extent. UCP2 is associated with a slightly reduced tendency for obesity (Esterbauer H et al 2001 Nature Genet 28:178). These proteins regulate energy balance and cold tolerance. Mice mutants in the uncoupling protein (ICP) have increased food intake but because of the increase in the rate of metabolism they do not become obese. Melanocyte regulatory factors (POMC, α-MSH, MC3-R, MC4-R) and bombesin receptor-3 (BRS-3) are modulators of energy balance and thus obesity and associated diseases. MC4-R regulates food intake and possibly energy use; MC3-R affects the efficiency of the feed and the storage of fat. Mice mutants for both these hormones eat excessively (because of MC4-R deficiency) and store fat excessively (because of MC3-R deficiency) and become quite obese. Systemic impairment of oxidative phosphorylation (OXPHOS) due to *Cre*-mediated targeted disruption, and unexpected ubiquitous reduction of mitochondrial frataxin protein expression lead to a significant reduction in total energy expenditure paralleled by increased expression of ATP citrate lyase, a rate-limiting step in de novo synthesis of fatty acids and triglycerides contributes to obesity (Pamplun D et al 2007 Proc Natl Acad Sci USA 104:6377). Perilipin (an adipocyte protein) modulates hormone-sensitive lipase (HSL) activity. HSL hydrolyses triacylglycerol, which stores energy in the cell. Deficiency of perilipin protects against obesity. The peptide motif CKGGRAKDC, which associates with prohibitin, a multifunctional membrane protein, targets angiogenesis in the adipose tissue and causes apoptosis. This results in resorption of white fat cells and normalizes metabolism without adverse effects (Kolonin MG et al 2004 Nature Med 10:625). Lymphocytic leakage – due to mutation at the mouse locus *Prox1* – also controls adult-onset obesity and lymphatic vascular disease (Harvey NL et al 2005 Nature Genet 37:1072). Close to 30 genetic sites are known to be involved in human obesity. Further, 16–22% of adults who are homozygous for FTO (fat mass and obesity, human chromosome 16q22.2 [mouse chromosome 8]) risk allele weighed about 3 kilograms more and had 1.67-fold increased odds of obesity when compared with those who did not inherit a risk allele (Frayling TM et al 2007 Science 316:889; Dina C et al 2007 Nature Genet 39:724).

Obesity may be controlled by appropriate exercise regimes as well as restricted food intake. Anti-obesity drugs may target appetite, intestinal fat absorption, increase energy expenditures, stimulate fat mobilization or decrease triglyceride synthesis. Some of the drugs of the dexfenfluoramine family (Fen Phen, inhibit serotonin re-uptake and stimulate its release) may reduce obesity by 10% but may have life-threatening side effects in a few cases. The newer drug Sibutramine does not stimulate serotonin release and is considered safe. Current research has focused on the mechanism of action of leptin (response of the hypothalamus to it) and the level of leptin biosynthesis or degradation. Cholecystokinin hormone receptor stimulants, glucagon-like peptide may reduce food intake. Fatty acid synthase (FAS) inhibitors may reduce food intake by inhibiting the removal of FAS from the cells and thus ensuring that the level of malonyl coenzyme A remains high. Malonyl-CoA (an appetite inhibitor) is generated from acetyl-CoA with the aid of acetyl-CoA carboxylase. During fasting lipid stores are mobilized from the adipose tissue in part by phosphorylation and inactivation of acetyl-coenzyme A carboxylase (ACC) involved in fatty acid synthesis. The pseudokinase Tribbles 3 (TRB3) also promotes lipolysis during fasting by degrading ACC by recruiting COP1 (ubiquitin ligase), a photomorphogenesis protein. Lipid metabolism and obesity may be controlled by both these pathways besides blocking or eliminating insulin signaling (Qi L et al 2006 Science 312:1763). Obesity in mice and humans may be promoted by an increase of Bacteriodetes and Firmicutes bacteria in the gut. These bacteria facilitate greater utilization of food energy resulting in increased body fat production (Turnbaugh PJ et al 2006 Nature [Lond] 444L1027). Germ-free animals – in contrast to mice with a gut microbiota – are protected against obesity that is a consequence of consuming a Western-style, high-fat, sugar-rich diet. Their persistently lean phenotype is associated with increased skeletal muscle and phosphorylated AMP-activated protein kinase (AMPK) in the liver and its downstream targets involved in fatty acid oxidation (acetyl-CoA carboxylase and carnitine-palmitoyltransferase). Moreover, germ-free knockout mice lacking fasting-induced adipose factor (Fiaf), a circulating lipoprotein lipase inhibitor whose expression is normally selectively suppressed in the gut epithelium by the microbiota, are not protected from diet-induced obesity (Bäckhed F et al 2007 Proc Natl Acad Sci USA 104:979).

Insects have developed a mechanism to avoid obesity. *Plutella xylostella* caterpillars reared for multiple generations on carbohydrate-rich foods (either a chemically defined artificial diet or a high-starch *Arabidopsis* mutant) progressively developed the ability to eat excess carbohydrate without storing it as fat, providing strong evidence that excess fat storage has a fitness cost. In contrast, caterpillars reared in carbohydrate-scarce environments (a chemically defined artificial diet or a low-starch *Arabidopsis* mutant) had a greater propensity to store ingested carbohydrate as fat. Moreover, insects reared on the low-starch *Arabidopsis* mutant evolved a preference for laying their eggs on this plant, whereas those reared on the high-starch *Arabidopsis* mutant showed no such preference. These observations provide an experimental example of metabolic adaptation in the face of changes in the nutritional environment and suggest that changes in plant macronutrient profiles may promote host-associated population divergence (Warbrick J et al 2006 Proc Natl Acad Sci USA 103:14045).

In the twenty-first century obesity has become a serious human biological/social problem because of diabetes, heart disease and other ill effects on health. In the past centuries some degree of corpulence was regarded as a status symbol or opulence or beauty as revealed by many Renaissance Italian paintings and other masterpieces of art (see Fig. O2). In recent years there is an increased awareness and need for controlling obesity by all available means. Although the government is not supposed to intrude on personal free choice, it is becoming evident that some legislative controls are required. Precedents have been set by controlling the use of alcohol, tobacco and narcotics. Further legislative controls of food for schoolchildren, advertising high sugar- and fat-containing food for children seem to be warranted to curb the obesity epidemics. Also, schools should promote physical education and exercise for children (Mello MM et al 2006 N Engl J Med 354:2601). ►leptin, ►resistin, ►histamine, ►PYY_{3-36}, ►neuropeptide Y, ►orexin, ►bombesin, ►cholecystokinin, ►frataxin, ►melanocyte stimulating hormone, ►melanocortin, ►melanoma, ►body mass index, ►diabetes, ►IL-6, ►insulin, ►insulin receptor, ►secretagogue, ►ghrelin, ►obestatin, ►adiponectin, ►paternal transmission, ►hypertension, ►Prader–Willi syndrome, ►Alström syndrome,

Figure O2. Tintoretto's *Suzanna with the Elders*; 16th century oil. By permission of the Kunsthis torisches Museum, Wien, Austria

►Bardet–Biedl syndrome, ►hypogonadism, ►QTL, ►triaglycerols, ►serotonin, ►glucagon, ►bulimia, ►anorexia, ►lipodystrophy, ►lod score, ►MLS, ►MCH, ►ZAG, ►attractin, ►ciliary neurotrophic factor, ►GATA, ►muscarinic acetylcholine receptors, ►phosphoinositides, ►G proteins, ►cachexia, ►photomorphogenesis, ►COP1, ►acetyl-CoA carboxylase deficiency, ►fatty acids, ►JNK, ►microbiome; Barsh GS et al 2000 Nature [Lond] 404:644; Robinson SW et al 2000 Annu Rev Genet 34:255; Spiegelman BM, Flier JS 2001 Cell 104:531; Brockmann GA, Bevova MR 2002 Trends Genet 18:367; Unger RH 2002 Annu Rev Med 53:319; Czech MP 2002 Mol Cell 9:695; Phan J, Reue K 2005 Cell Metab 1:73; Bell CG et al 2005 Nature Rev Genet 6:221; Horvath TL 2005 Nature Neurosci 8:561; Morton GJ et al 2006 Nature [Lond] 443:289; Muoio DM, Newgard CB 2006 Annu Rev Biochem 75:367; Murphy KG, Bloom SR 2006 Nature [Lond] 444:854; Coll AP et al 2007 Cell 129:251; http://obesitygene.pbrc.edu/.

Obestatin: This is a peptide hormone transcribed from the ghrelin gene but its effects are opposite to those of ghrelin; it suppresses appetite, food intake and weight gain (Zhang JV et al 2005 Science 310:996). ►ghrelin, ►obesity

OBF (oct-binding factor; synonyms BOB.1, OCA-B): Regulates the lymphocyte-specific oct sequence in the promoter of the transcription of immunoglobulin genes; it is required for the development of the germinal centers. ►oct, ►immunoglobulins, ►germinal center

Objective Lens: Refers to the microscope lens next to the object to be studied. ►light microscopy

Obligate: This means restricted to a condition or necessarily of a type, e.g., obligate parasite, obligate anaerobe. The latter can thrive only in the absence of air.

Oblique Crossing Over: In the case of adjacent (tandem) duplications in homologous chromosomes pairing may take place in more than one register and crossing over may yield unequal products (see Fig. O3). ►unequal crossing over

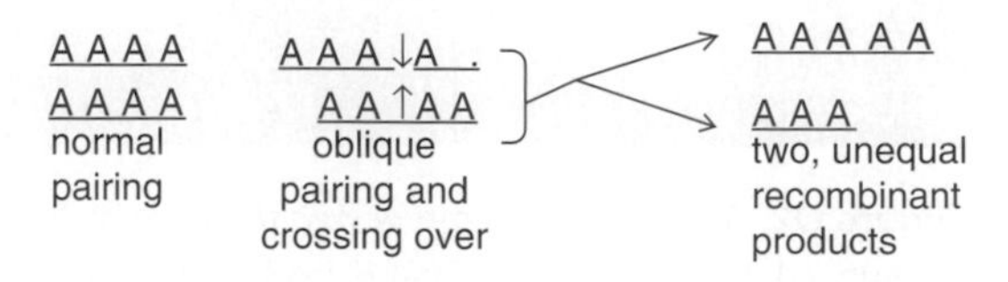

Figure O3. Oblique crossing over

Obsessive-Compulsive Disorder (OCD): This is a type of schizophrenic behavior bearing some resemblance to the Tourette syndrome (see Fig. O4).

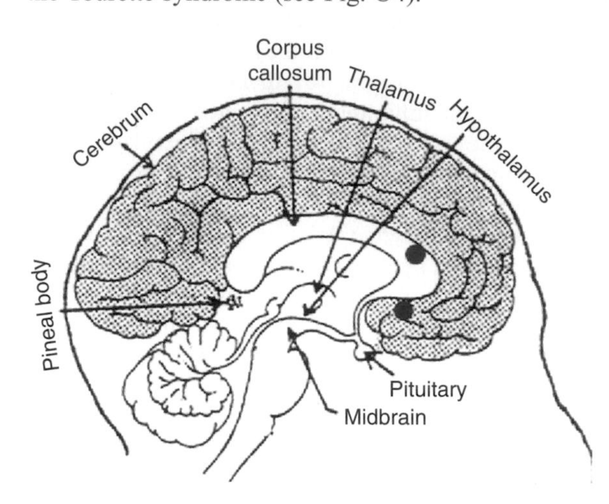

Figure O4. The red marked spots of the brain indicate some of the approximate areas of successful therapeutic stimulation. Other points (not shown here) were also used for the same purpose

A major dominant gene with minor modifiers determines this condition without difference in the two sexes. In some psychotherapy and electroconvulsive therapy resistant patients, who exhibit overactivity in the subgenual cingulated region of the brain (Brodmann area 25), electrical stimulation applied deep into the region was remarkably successful in reversing the condition in a limited trial (Mayberg HS et al 2005 Neuron 454:6512). ►schizophrenia, ►panic disorder, ►Tourette syndrome; Hanna GL et al 2002 Am J Med Genet 114:541.

OCA-B: This is the same as OBF or Bob. ►OBF

Occam's Razor (Ockham's razor): This is the philosophical precept of William Ockham [1280–1349], a rebellious clergyman and a venerabilis inceptor [= reverend innovator]: *"pluralites non est ponenda sine necessitate"*, meaning that multiple alternatives should not be offered in logical argumentation but the simplest yet adequate explanation should be chosen. ►maximal parsimony

Occipital Horn Syndrome (cutis laxa): ►Menkes syndrome, ►cutis laxa

Occluded Virus Particle: The virus is surrounded by proteinacious material that protects it from the adverse environment, e.g., when the insect host dies and decomposes. When the insect eats plant material the alkaline gut fluid dissolves the occlusion and the infectious (baculovirus) particles are released. The lipoprotein viral envelope fuses with the gut cell walls and the nucleocapsids are transmitted to the cytoplasm and eventually to the cell nucleus. ►baculoviruses; Hu Z et al 1999 J Gen Virol 80(pt 4):1045; Braunagel SC et al 2003 Proc Natl Acad Sci USA 100:9797.

Occlusion: Transcription from one promoter reduces transcription from a downstream promoter. ►downstream, ►promoter, ►transcription

Occupational Hazard: Refers to the presence of genotoxic (carcinogenic) agents at the workplace. Monitoring includes urine analysis for chemicals, sisterchromatid exchange, abnormal sperm count or deformed sperm or SNIPS, etc. ►mutation detection, ►epidemiology

Occupational and Safety and Health Administration: ►OSHA

Occurrence Risk: Refers to the chance that an offspring of a particular couple will express or become a carrier of a gene. ►genetic risk, ►recurrence risk

Ocellus (plural ocelli): This is a simple light sensor (eyelet, eyespot) on the top of the head of insects (see tiny arrows in Fig. O5), behind the compound eyes (see Fig. O5). ►compound eyes, ►ommatidium, ►rhabdomere, ►morphogenesis in *Drosophila*

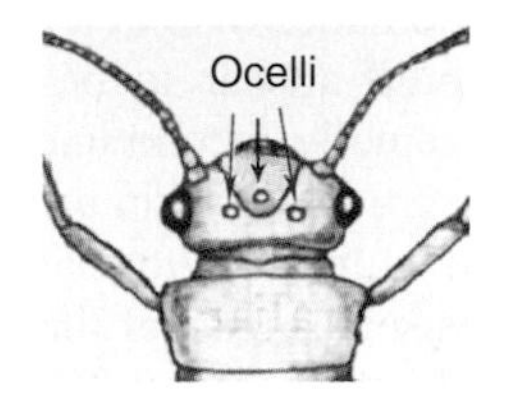

Figure O5. Ocelli in insects

Ochre: Denotes chain-terminator codon (UAA).

Ochre Suppressor: Refers to mutation in the anticodon of tRNA that permits the insertion of an amino acid at the position of a normally chain terminating UAA RNA codon; ochre suppressors frequently suppress amber (UAG) mutations. ►code genetic, ►nonsense codon, ►suppressor tRNA

Ochronosis: This is the blue pigmentation in alkaptonuria. ►alkaptonuria

ω-Conotoxins: These are *Conus* snail inhibitors of calcium ion channels. ►ion channels

Oct: This is a mammalian gene regulatory protein (helix-turn-helix transcription factors) with octa recognition sequence: ATTTGCAT. Oct-1 and Oct-2 regulate B cell differentiation. Oct3/4 mediates differentiation and dedifferentiation of embryonic stem cells. The Oct-6 transcription factor regulates Schwann cell differentiation. The presence of Oct-1 allele may lead to a fourfold increase in susceptibility to the cerebral form of malaria. ►Schwann cell, ►lymphocytes, ►OBF, ►B cell, ►immunoglobulins, ►Oct-2, ►Octa, ►malaria, ►octamer; Pesce M, Schöler HR 2001 Stem Cells 19:271.

Oct-2: A lymphoid transcription factor that is similar to Oct-1; both respond to BOB.1/OBF.1 activators. ►OBF, ►OCT

Octa: This is an 8-base sequence (ATTTGCAT) in the promoter of H2B histone gene and some other genes. Several slightly different octa sequences are found in the promoter regions of various genes. (Octo in Latin, οκτασ in Greek: number 8, Oct-1). ►OCT

Octad: This means comprising eight elements, e.g., the spores in an ascus if meiosis is followed by an immediate mitotic step as in *Neurospora*, *Ascobolus*, etc. ►tetrad analysis

Octadecanoic Acid (stearic acid): This is an inducible plant defense molecule against insects. ►sphingolipids

Octamers: These are conserved key elements in the promoter of immunoglobulin genes where several transcription factors bind. ►octa, ►OCT, ►immunoglobulins; Matthias P 1998 Semin Immunol 10:155.

Octaploid: This is a cell nucleus carrying eight genomes, 8x. ►polyploid

Octopine: This derivative of arginine is synthesized by a Ti plasmid gene (*ocs*) in *Agrobacterium* 5 (see Fig. O6). ►opines, ►*Agrobacterium*

```
        NH2
       /
HN=C
       \
        NH-(CH2)3-CH-COOH
                  |
                  NH
                  |
               H3C-CH-COOH
```

Figure O6. Octopine

Ocular: This refers to the microscope lens next to the viewer's eye. ►objective

Ocular Albinism: ►albinism

Ocular Cicatricial Pemphigoid: This is an autosomal dominant autoimmune disease of the eye and possibly of other mucous membranes. It may be associated with defects in the HLA system. ►eye diseases, ►autoimmune disease, ►HLA

Oculocutaneous Albinism: Several forms of the recessive disease are known. Type I (11q14-q21) is tyrosinase negative recessive whereas types II (15q11.2-q121) and III (9p23) are tyrosinase positive (brown). There is another brown type at 15q. The most prevalent is type II, its incidence is ~1/1,100 among the Ibos of Nigeria, ~1/10,000 among US blacks and ~1/36,000 among the general population of the US. ►albinism

Oculodentodigital Dysplasia: An autosomal dominant disorder characterized by defects in the eyes (microphthalmos), small teeth and polydactyly or syndactyly. Its mutation rate reveals a large paternal effect. The basic defect may be in the connexin 43 gene (6q22-q23) that may cause disturbance in the gap junctions that are required for the exchange of small ions and signaling molecules. ►eye diseases, ►microphthalmos, ►polydactyly, ►syndactyly, ►connexins, ►gap junction; Paznekas WA et al 2003 Am J Hum Genet 72:408.

O.D.: Optical density indicates the absorption of light at a particular wavelength by a compound in a spectrophotometer. O.D. can be used to characterize a molecule, e.g., pure nucleic acids have maximal absorption at 260 nm but contamination and the solvent may alter the absorption pattern. ►DNA measurements, ►extinction

Odds Ratio (OR): A comparison of the effect of a treatment or exposure on a particular group (of a certain genotype) versus the same treatment or exposure on another different group. The OR may reveal the response of the organism of a certain genotype to the particular exposure. The OR can be estimated statistically by a two-by-two contingency table and as ad/bc. Greater differences from 1 (+/−) indicate stronger association. ►association test, ►lod score

Odontoblast: Refers to the connective tissue cell that forms dentin and dental pulp of the teeth.

Odor-Sensing: ►olfactogenetics, ►pheromones; Hurst JL et al 2001 Nature [Lond] 414:631; Jacob S et al 2002 Nature Genet 30:175; Hallem EA, Carlson JR 2006 Cell 125:143.

ODP (origin decision point): This is a checkpoint for the initiation of replication. It precedes the restriction point. ►R point; Wu JR, Gilbert DM 2000 FEBS Lett 484:108.

Oedipus Complex: According to Freudian theory, during adolescence a child may become more attached to the parent of the opposite sex, and if the condition persists it may lead to neurotic behavior.

Oenocytes: These insect organs accumulate lipid droplets during starvation and cooperate with fat bodies in regulating lipid metabolism (Gutierrez E et al 2007 Nature [Lond] 445:275). ►fat body

Oenothera (*Onagraceae*, x = 7): Several species, diploid and polyploid have been used for the cytological study of multiple translocations and the nature and inheritance of plastid genes. ►gaudens, ►translocation, ►complex heterozygote, ►megaspore competition, ►zygotic lethal; Cleland RE 1972 Oenothera: Cytogenetics and Evolution, Acad. Press, New York; Hupfer H et al 2000 Mol Gen Genet 263:581; Mracek J et al 2006 Genomics 2006 88:372.

Oestrogen (estrogen, estradiol): This is a steroid hormone. ►animal hormones, ►estradiol

Oestrus (estrus): Refers to the periodically recurrent sexual receptivity, concomitant with sexual urge (heat) in mammals, except humans. In mice it lasts for ~3–4 days, depending on crowding, exposure to male pheromones or hormone treatment and the daily light/dark cycles. ►uterus

OFAGE (orthogonal-field alternation gel electrophoresis): This device is used to isolate small chromosomal size DNA of lower eukaryotes. ►pulsed field electrophoresis

Offermann Hypothesis: Purported to provide a mechanism for recombination within a short chromosomal region that would appear to be intragenic. It was assumed that actually recombination separated two genes, which were involved in position effect. These loci were not supposed to have any detectable phenotype themselves, except the position effect on the neighbor. This idea emerged in 1935, years before pseudoallelism has been discovered in 1940 by C.P. Oliver. ►pseudoallelism; Carlson EA 1966 The Gene: A Critical History. Saunders, Philadelphia, PA.

Offspring-Parent Regression: ►correlation

O-GlcNAc Transferase (OGT): This is an indispensable cellular enzyme mediating post-transcriptional glycosylation of many different proteins involved in regulatory functions. (See Hanover JA 2001 FASEB J 15:1865)

OGOD (one gene - one disorder): A hypothesis based on the analogy of the one gene - one poly-peptide (one enzyme) theorem. The majority of diseases in humans (and animals) cannot be reconciled with a single gene mutation, and most of the symptoms (syndromes) are under multigenic control although particular genes may have a major effect. ►behavior in humans, ►one gene—one enzyme theorem; Plomin R et al 1994 Science 264:1733.

Oguchi Disease: This condition is characterized by a recessive human chromosome-2q mutation or deletion in the arrestin protein modulating light signal transduction to the eye or by defects in the arrestin and the rhodopsin kinase genes. Night vision is impaired but otherwise vision is normal. ►night blindness, ►arrestin, ►retinal dystrophy, ►eye diseases

Ohm (Ω = 1V/A): A unit of resistance of a circuit in which 1 volt electric potential difference produces a current of 1 ampere.

OH˙ (hydroxyl radical): This is responsible for the oxidative damage of superoxide and hydrogen peroxide. ►superoxide, ►hydrogen peroxide, ►Fenton reaction, ►ROS

Ohno's Law: The gene content of the X chromosome is basically the same in all mammals. Some exceptions are seen in humans, marsupials and in the monotreme, *Platypus* where genes in the short arm of the X chromosome may be of autosomal origin. There are other exceptions such as the chloride channel gene (*Cln4*) is autosomal in the mouse *Mus musculus* but it is X-linked in *Mus spretus*. Similarly, the human steroid sulfatase (STS) gene is near the pseudo-autosomal region whereas in lower primates it is autosomal. The rationale of Ohno's Law is that translocations between autosomes and X chromosome would upset the sex determination gene balance. In the X chromosomes of various mammals sequence homologies are conserved. FISH probes have, however, revealed some rearrangements. ►sex determination, ►FISH; Ohno S 1993 Curr Opin Genet Dev 3:911; Palmer S et al 1995 Nature Genet 10:472.

Oidia: Asexual fungal spores produced by fragmentation of hyphae into single spores.

Oil Immersion Lens: This is the highest power objective lens of the light microscope. It is used with a special non-drying immersion oil, available at different viscosity with a refractive index of about 1.5150 for D line at 23°C, and it increases light-gathering power and improves resolution. ►light microscope, ►objective, ►resolution optical, ►numerical aperture

Oil Spills: It is known that around 22 bacterial genera have genetically determined ability to degrade petroleum hydrocarbons. Although a procedure invented by A.M. Chakrabarty did not involve molecular genetic engineering but only the introduction of two different plasmids into *Pseudomonas* (*P. aeruginosa* and *P. putida*) to degrade several harmful products, it was the first US patent (#4,259,444) issued in 1981 for unique microorganisms. ►patent, ►biodegradation; Díaz MP et al 2002 Biotechnol Bioeng 79:145.

OK Blood Group: This is encoded in human chromosome 19pter-p13. This antigen is present in the red cells of chimpanzees and gorillas but not in rhesus monkeys, baboons or marmosets. ►blood groups

Okadaic Acid ($C_{44}H_{68}O_{13}$): An inhibitor of PP1 and PP2a protein phosphatases. ►PP-1

Okayama & Berg PROCEDURE: This procedure permits cloning of full length mRNA genes. The mRNA is extracted from post-polysomal supernatant of reticulocyte lysate of rabbits, made anemic by phenylhydrazine injection. The globin mRNA is recovered in the alcohol-precipitate of phenol extract or with the aid of a guanidinium thiocyanate method. The poly-A tailed mRNA is annealed to plasmid pBR322 that is equipped with a poly-T attached to a SV40 fragment inserted into the vector. In the next step an oligo-G linker is constructed, separated, and purified by agarose gel electrophoresis. Now cloning of the mRNA can begin. The poly-A tail is annealed to the poly-T end of the vector. Using reverse transcriptase, a DNA strand, complementary to the mRNA strand, already in the plasmid, is generated. Using terminal transferase enzyme poly-C tails are added to one strand of the plasmid vector as well as to the DNA strand of the RNA-DNA double strand. Now the oligo-G linker is added to the oligo-C ends and the plasmid is made circular by DNA ligase. Then the mRNA strand is removed by RNase H and replaced by a complementary DNA strand, generated by DNA polymerase I and the construction is completed by ligation into a circular cloning vector. This new vector, containing the full length cDNA, is transformed into *E. coli* cells for propagation. ►cloning vectors, ►cDNA, ►ribonuclease H, ►linker, ►reverse transcriptase, ►terminal transferase, ►guanidinium thiocyanate, ►phenylhydrazine, ►RNA extraction; Okayama H Berg P 1982 Mol Cell Biol 2:161.

Okazaki Fragments: These are short (generally less than 1 kilobase in eukaryotes and about 2 kb in prokaryotes) DNA sequences formed during replication (of the lagging strand) and subsequently ligated into a continuous strand. Okazaki fragments are needed because nucleic acid chains can grow only by adding nucleotide to the 3′ end and the lagging strand template would not allow continuous chain elongation like the leading strand of DNA. Six steps are involved in the generation of Okazaki fragments: (i) polymerase α and primase synthesize an RNA primer for 3′ ← ← ← growth on the lagging strand template, (ii) RFC assists in binding the primer to the DNA and in the displacement of polymerase α, (iii) PCNA promotes the assembly of the replicator DNA polymerase δ complex, (iv) RNase H, Fen1 and Dna2 endonuclease digest off the RNA primer under the control of replication protein RPA as DNA synthesis proceeds, (v) the gaps between the Okazaki fragments are filled by the DNA, and (vi) DNA ligase joins the fragment into a continuous strand of DNA. In eukaryotes the process is more complex than in prokaryotes. ►DNA replication, ►replication fork, ►primosome, ►RCF, ►PCNA, ►Rad27/Fen1, ►ribonuclease H, ►DNA ligase, ►alpha accessory factor, ►polymerase switching, ►processivity; Bae S-H, Seo Y-S 2000 J Biol Chem 275:38022; Jin YH et al 2001 Proc Natl Acad Sci USA 98:5122; Bae S-H et al 2001 Nature [Lond] 412:456; Jin H Y et al 2003 J Biol Chem 278:1626.

Okihiro Syndrome: ►Duane retraction syndrome

Okra (*Abelmoschus esculenta*): An annual vegetable of the Malvaceae with nearly 30–40 species having variable chromosome numbers, generally higher than n = 34–36 as revealed by reports.

OKT3: A monoclonal antibody capable of blocking interleukin production. ►Oct

Oleorupein: This complex substance is found in olive tree leaf extracts and it is used as a herbal medicine for various ailments.

Olig1: This is a basic helix-loop-helix transcription factor required for the repair of demyelinated lesions of the central nervous system. It is also needed for repairing the myelin sheath in multiple sclerosis. ►multiple sclerosis, ►helix-loop-helix; Arnett HA et al 2004 Science 306:2111.

2′-5′-Oligoadenylate: This is synthesized by 2–5A synthetases (2′5′AS), which are induced by interferons and they activate RNase L in defense of viral infection. ►interferon, ►ribonuclease L; Bonnevie-Nielsen V et al 2005 Am J Hum Genet 76:623.

Oligo-Capping: Refers to the replacement of the original cap of the mRNA by a short synthetic oligoribonucleotide. The removal is due to tobacco acid pyrophospatase and the ligation by T4 RNA ligase. In uncapped mRNAs the 5′ phosphate is first removed by alkaline phosphatase. The purpose is labeling prior to first-strand DNA synthesis. The 5′ end of the mRNA is identified by reverse transcription-polymerase chain reaction. ►cap, ►RNA ligase, ►RT-PCR; Maruyama K, Sugano S 1994 Gene 138:171.

Oligohydramnios: ►renal tubular dysgenesis

Oligonucleotide: A short nucleotide tract (about 15 to 30 units). See http://basic.northwestern.edu/biotools/OligoCalc.html.

Oligopeptides: The natural oligopeptides are generally not more than 50 amino acids long and frequently have regulatory functions. (See database: http://erop.inbi.ras.ru/).

Oligostickiness: This is a measure of the binding affinity of 12-base (dodeca-) oligonucleotides to the genome of a species. The affinity is characteristic of genomes as well as of chromosomes of a species and hence facilitates identification. It is also known as chromosome texture. (See Nishigaki K and Saito A 2002 Bioinformatics 28:1153).

OL(1)p53: A phosphorothioate oligonucleotide sequence (5′-d[CCCTGCTCCCCCCTGGCTCC]-3′) used as antisense DNA to suppress p53 function to pass the checkpoint into the S phase of the cell cycle as a potential treatment for acute myeloblastic leukemia. ►leukemia, ►Myb oncogene, ►antisense DNA, ►phosphorothioate, ►p53; Bishop MR et al 1997 J Hematother 6(5):441.

Oleuropein: A phenolic secoiridoid glycoside in the leaves of privet (*Ligustrum*) and when activated by herbivores becomes a protein cross-linking, lysine-decreasing glutaraldehyde-like structure, an α,β-unsaturated aldehyde as a means of self-protection (see Fig. O7). ►plant defense, ►host–pathogen relations; Konno K et al 1999 Proc Natl Acad Sci USA 96:9159.

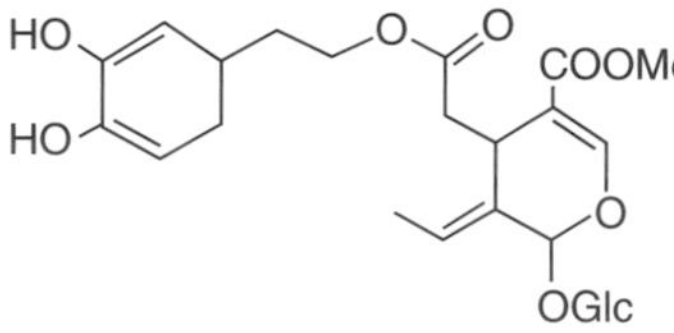

Figure O7. Oleuropein. The left side of the molecule represented here is a catechol and the right side is the secoiridoid moiety

Olfactogenetics: This is concerned with the genetically determined differences in (body) smell and the ability to recognize it. In the human brain the olfactory bulbs situated under the frontal lobes interpret the olfactory signals (see Fig. O8). The mouse olfactory bulbs contain ~two glomeruli for each olfactory receptor. The different olfactory receptors are segregated into groups of glomeruli. Scent is influenced by the chemical nature of secretions and to a large extent by the diet and the microflora of the body. Human polymorphism in olfactory responses is of concern for the cosmetics (perfume) industry. It has been claimed that the ability to distinguish between various odors is determined by the *H2* locus of mice (an analog of the human *HLA* complex). Some people are incapable of smelling, anosmic for isobutyric acid or cyanide or urinary excretes of asparagus metabolites. The regulation of the olfactory responses involves cAMP or phosphoinositide (IP_3)-regulated ion channels and G protein ($G_\alpha olf$) coupled receptor kinases. The olfactory memory in sheep is triggered by nitric oxide that potentiates the release of glutamate and GABA neurotransmitters leading to an increase of cGMP in the mitral cells of the nose, the site of perception of smell. The olfactory memory formation has been localized in the mushroom body of the brain. In *Drosophila* cAMP phosphodiesterase, encoded by gene *dunce* (*dcn*, chromosome 1–3.9), calcium—calmodulin-dependent adenylyl cyclase (encoded by *rutabaga* [*rut*, 1-{46}]), the catalytic subunit of the cyclicAMP-dependent protein kinase A, and the α-integrin subunit encoded by *Volado* (*vol*, located in the X and the 2nd chromosomes) are involved in the control of olfactory memory. There are many other unidentified genes and proteins involved in different olfactory functions. It appears that each neuron of the olfactory epithelium expresses only one olfactory receptor. In the human genome ~900 olfactory receptor (OR) genes (encoding seven-transmembrane proteins) or pseudogenes (60%) have been identified. In humans, the non-functional, pseudogenic olfactogenes are twice as high as in non-human primates and in mice the proportion of pseudogenes of olfaction is only 20% (Gilad Y et al 2003 Proc Natl Acad Sci USA 100:3324). They are common in chromosomes 7 and 17 but are found in most other chromosomes as well, generally clustered as 6–138 genes. Of 856 olfactory receptors in the human genome, 40% is located in chromosome 11 in 28 single- and multi-gene clusters (Taylor TD et al 2006 Nature [Lond] 440:497). In each olfactory sensory neuron only one odorant receptor (MOR in mouse, HOR in humans) gene is expressed (Serizawa S et al 2003 Science 302:2088). A single transacting enhancer element may allow the stochastic activation of only one olfactory receptor allele in the olfactory sensory neurons of mice (Lomvardas S et al 2006 Cell 126:403). No olfactory receptor genes seem to be coded in human chromosomes 20 and Y. Different olfactory sensory neurons in the nose express a different complement of the ORs and transmit the information through their axons to the olfactory bulbs in the brain. In humans, mice and fishes there is another class of chemosensory receptors, TAARs (trace and amine-associated receptors). One group of TAARs recognizes volatile amines in the urine of mouse linked to stress and two groups are involved in sensing compounds enriched in the male urine (Liberles SD, Buck LB 2006 Nature [Lond] 442:645).

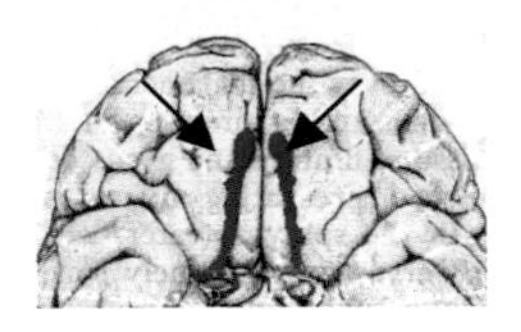

Figure O8. Olfactory bulbs in the human brain situated under the frontal lobes

In fishes approximately 100 genes are involved in olfactory functions whereas in rodents there are about 1,300 and humans have around 1,000. In humans many of the olfactory genes are not functional. In mammals only a single odor receptor gene is expressed per cell. Dogs and rats have a superior ability to smell. In dogs 1,094 and in rats 1,493 olfactory receptor genes were identified from shutgun sequences but nearly 20% were pseudogenes (Quignon P et al 2005 Genome Biol 6(10):R83). The *Drosophila* in general expresses two odor receptors per cell whereas the *Caenorhabditis* may express nine chemoreceptors per cell. The *Caenorhabditis* has around 1,500 G-protein-coupled odor receptors (GPCR). *Drosophila melanogaster* has about 60 genes for odorant receptors and an equal number but different gustatory receptor genes. These low numbers are characteristic of other insects as well. A combinatorial use of the olfactory receptors permits the distinction of an almost indefinite variety of odors. The ligands of the receptors also contribute to sensory specificity. Different glomeruli respond qualitatively and quantitatively to the types and intensities of the odors. According to some estimates olfactory receptors may represent 1% of all genes. It has been reported that the major histocompatibility complex is a main source of unique individual odors in animals and women can detect differences among male odor donors with different MHC genotypes. This ability is dependent on the HLA allele inherited by the human female from her father but not from her mother (Jacob S et al 2002 Nature Genet 30:175). In the vomeronasal organ—sensing pheromones—there are ~35 V1R and ~150 V2R receptor family members. The number of human OR genes exceed 1,000 and many of them are pseudogenes. In animals the major histocompatibility complex is a source of olfactory recognition of mating preferences and various other behavioral traits. The rodent ORs rarely, if at all, are pseudogenic. Natural odors are often a blend of several components present in specific ratios. Therefore, an enormous number of odor (fragrance) signals may exist. Several OR genes have been cloned. Pheromones are perceived by >240 proteins of the vomeronasal system including special 7-transmembrane proteins. The removal of the vomeronasal organ interferes with the pheromone response but not with other odor perceptions. The olfactory pathway also mediates pheromone responses (Shepherd GM 2006 Nature [Lond] 439:149). The Ras-MAPK signal transduction pathway mediates odor perception and transmission of sensory signals to the *Caenorhabditis* olfactory neurons. Deficiencies in the olfactory system of the *Caenorhabditis* seem to prolong life. In 2004, Richard Axel and Linda Buck received the Nobel Prize for their work on olfaction. ►bisexual, ►cAMP, ►cGMP, ►neurotransmitter, ►nitric oxide, ►phosphoinositide, ►IP_3, ►MHC, ►signal transduction, ►*Asparagus officinalis*, ►pheromones, ►fragrances, ►vomeronasal organ, ►taste, ►brain human, ►Kallmann syndrome, ►Bruce effect, ►mushroom body, ►odor-sensing; Pilpel Y et al 1999 In: Higgins SJ (ed) Molecular Biology of the Brain, Princeton Univ. Press, Princeton, NJ, pp 93; Buck LB 2000 Cell 100:611; Glusman G et al 2001 Genome Res 11:685; Firestein S 2001 Nature [Lond] 413:211; Mombaerts P 2001 Annu Rev Genomics Hum Genet 2:493; Young JM et al 2002 Hum Mol Genet 11:535; Nakagawa T et al 2005 Science 307:1638); Bargmann C 2006 Nature [Lond] 444:295; Shepherd GM 2006 Nature [Lond] 444:316; Jefferis GSXE et al 2007 Cell 128:1187; Lin H-H et al 2007 Cell 128:1205; http://senselab.med.yale.edu/senselab/ordb/; www.leffingwell.com.

Olfactory: This is related to the sense of smell. ►olfactogenetics

Oligodendrocyte: Refers to the non-neural cells that form the myelin sheath (neuroglia) of the central nervous system and they coil around the axons. They may become the stem cell of the central nervous system. ►neurogenesis; Lu QR et al 2002 Cell 109:75.

Oligodeoxyribonucleotide Gated Channel: Composed of a 45-kDa protein, this is involved in Ca^{2+}-dependent uptake of oligonucleotides through an approximately 5-μm pore. Protein kinase C and a few organic molecules are inhibitors. ►protein kinases, ►liposomes, ►DNA uptake sequences; Salman H et al 2001 Proc Natl Acad Sci USA 98:7247.

Oligodontia: The lack of development of six or more permanent teeth is based on frameshift mutation in the PAX9 gene in human chromosome 14. ►tooth agenesis, ►hypodontia, ►PAX; Heilig R et al 2003 Nature [Lond] 421:601.

Oligodynamic Action: Refers to the antimicrobial effect of trace amounts of heavy metals. ►sterilization.

Oligogenes: This is a small group of genes responsible for a particular trait; usually one has a major role. ►breast cancer, ►prostate cancer, ►polygenes, ►QTL

Oligo-Labeling Probes: These are short (10–20 nucleotides), commonly radioactively or of fluorochrome labeled, synthetic probes for the identification of genes for isolation, labeling in gel retardation assays, screening of DNA libraries, etc. ►label, ►probe, ►gel retardation assay, ►variant detector assay

Oligomer: A polymer of relatively few units (amino acids, nucleotides, sugars, etc.). An oligomeric protein has a quaternary structure associated with non-covalent bonds.

Oligonucleotide-Directed Mutagenesis (Kunkel mutagenesis): Synthetic oligonucleotide with the mutation sought is annealed to a single-stranded M13 phage DNA template with a number of uracil residues in place of thymine in a strain *dut*$^-$ (dUTPase) and *ung*$^-$ (uracil-DNA-glycosidase). The *E. coli* transformation medium should contain the 4 deoxynucleotide triphosphates, T4 DNA polymerase and T4 DNA ligase to generate double-stranded circular DNA and M13 phages are selected with the mutation desired. The heteroduplex is introduced into a wild type *dut* and *ung* strain to maintain the mutation. ►site-specific mutagenesis; Kunkel TA et al 1987 Methods Enzymol 154:367.

Oligophrenia Phenylpyruvica: Refers to mental retardation due to phenylketonuria. ►phenylketonuria

Oligophrenin: This RAS-like GTPase protein (91-kDa) encoded in human chromosome Xq12 is responsible for cognitive impairment. Similar mental retardation genes are scattered in the genome and afflict about 0.15–0.3% of males. Its level is higher in several cancerous tissues. ►mental retardation, ►RAS; Pinheiro N A 2001 Cancer Lett 172:67.

Oligoribonucleotide Synthesis: ►silyl-phosphite chemistry

Oligosaccharides: These consist of sugar residues, such as glycans, and they may be present in many metabolically, immunologically and structurally important molecules. Oligosaccharides may carry information for the folding of proteins in the endoplasmic reticulum. Their sequence can be determined by exoglycosidase mediated digestion or by sequencing the amino acids of the protein. In both procedures electrophoretic (MALDI) analysis may be used. ►glycan, ►electrophoresis, ►protein folding, ►folding, ►endoplasmic reticulum; Billuart P et al 1998 Nature [Lond] 392:923; Lehrman MA 2001 J Biol Chem 276:8623.

Oligospermia: This condition is characterized by low sperm content in the semen. Chromosomal rearrangement (translocations, inversions, ring chromosome, etc.) or aneuploidy may be responsible for this condition. ►sperm, ►semen, ►azoospermia, ►cytoplasmic male sterility

Oligozyme: Refers to nuclease resistant RNA oligomers (29–36 residues) that can cleave specific RNA sequences. ►ribozyme; Kitano M et al 2001 Nucleosides Nucleotides Nucleic Acids 20:719.

Olive (*Olea europea*): There are about 30 species of this oil-producing tree. The cultivated forms are $2n = 2x = 46$ although aneuploids have been identified. Because of its oleocanthal content freshly pressed oil is anti-inflammatory as it lowers cyclooxygenase enzymes COX-1 and COX-2 but has no substantial effect on lipoxygenase. ►cyclooxygenase, ►lipoxygenase; Beauchamp GK et al 2005 Nature [Lond] 437:45.

Olivopontocerebellar Atrophy (OPCAI): The autosomal dominant or recessive, variable types of expressions involve ataxia, paralysis, incoordination, speech defects, and brain and spine degeneration. In some forms eye defects and other anomalies have been observed. In several cases a linkage has been observed to the HLA complex in human chromosome 6p21.3-p21.2, but in others such a linkage was not evident. Patients with this disease exhibit 50% or less glutamate dehydrogenase activity. ►ataxia, ►palsy, ►glutamate dehydrogenase

Omega-3-Fatty Acids: This is a collective name for polyunsaturated fatty acids. The ω name is derived from the first double bond position. The early synthetic products with 1, 2 and 3 double bonds are shown. Mammals lack the enzymes (desaturases) to add double bonds beyond the 9th carbon atom of the fatty acid chain. Therefore, they cannot synthesize these *essential*, unsaturated fatty acids and depend on plants. Fish oil – because of feeding on algae – may also be rich in ω-3 fatty acids (see Fig. O9). In the absence of these fatty acids skin lesions, defects in the lipid membranes, kidney disease, reduced fertility, prostaglandin deficiency, etc. occur. Arachidonic acid, a ω-6-fatty acid, is the precursor of prostaglandins mediating anti-inflammatory responses. Omega-3 fatty acids are known to lower risks of cardiovascular diseases, to reduce depression, bipolar mental afflictions and other nerve disorders. On the other hand, these fatty acids may increase hemorrhages, lower immune responses, etc. According to the Federal Drug Administration, the intake of these fatty acids through food and supplements should be limited to 3mg/day. Some nutritionists have recommended higher doses. Since mammals lack the desaturase enzyme that can convert ω-6 fatty acids to ω-3, it is of interest to introduce the gene for this enzyme from another animal such as *Caenorhabditis elegans*. Following the transfer of the *fat-1* gene to mice (Kang JX et al 2004 Nature [Lond] 427:504) and pigs (Lai L et al 2006 Nature Biotechnol 24:435) it was possible to produce a high ratio of n-6/n-3 of 0.7 in the muscles of transgenic mice compared to 49.0 in the wild type. In 8 transgenic pigs this ratio was variable but the average was 1.69 versus 8.52 in the wild type controls. For the experiments the humanized version of the pCAGGS-fat-1 expression vector was used with a cytomegalovirus enhancer and driven by the chicken β-actin promoter. The vector containing G418 selective marker was electroporated into fetal porcine fibroblasts. The transgenic cells were then used for nuclear transplantation into pig ova and cloned. A total of 1,633 engineered embryos was transferred to 14 females in estrus which resulted in pregnancies. Five of them were carried to term resulting in 10 live and two dead piglets of which six contained the *fat-1* transgene. Initially all the offspring were completely normal, however three piglets developed heart failure by the age of 3 weeks. This defect appeared to be due to the nuclear transfer procedure rather than to the *fat-1* gene. Although such transgenic pigs are not yet available for commercial meat production, the method appears very promising from the viewpoint of potential improvements in human diet and health. ►fatty acids, ►acetyl-CoA, ►syntaxin, ►erucic acid, ►prostaglandins, ►arachidonic acid, ►desaturase, ►cardiovascular disease, ►enhancer, ►promoter, ►cytomegalovirus, ►G418, ►electroporation, ►nuclear transplantation

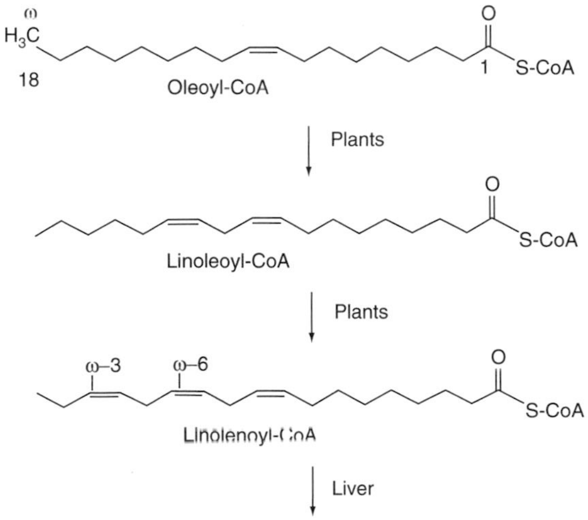

Figure O9. The ω numbering begins at the C of the methyl end; the n-3 or ω-3 indicate the double bond at the 3rd or 6th carbon from the methyl end. Chemists start the numbering (1) at the other end

Omega Sequence: This is the viral nucleotide sequence in the mRNA 5′-region of eukaryotic genes that enhances translation.

Omenn Syndrome: ►reticulosis familial histiocytic

OMIA (online Mendelian inheritance animals): This is a database for animal (excluding man and mouse) genes and hereditary diseases and disorders. (http://omia.angis.org.au).

Omics: Refers to genomics, transcriptomics, proteomics, etc.; the global information gathering and its tools in biology.

OMIM (online Mendelian inheritance man): This is an up-to-date catalog of autosomal dominant, autosomal recessive, X-linked and mitochondrial genes of humans available through the Internet (http://www.ncbi.nlm.nih.gov/entrez/query.fcgi?db=OMIM). The database provides information on relevant literature on ('morbid') genes and connections to other databases. As of May 1 2006 OMIM identified 1,871 genes in which at least one disease related mutation has been described and the number of phenotypes in these genes was 3,112 (McKusick V 2006 Annu Rev Hum Genet Genomics 7:1). By 2007 the l number of entries totaled 17,370 (McKusick V 2007 Amer J Hum Genet 80:588). It is also available in book form: McKusick, Victor 1997 *Mendelian Inheritance in Man*. The Johns Hopkins University Press, Baltimore, MD. OMIM proteins are annotated at http://www.hprd.org/. ►BITOLA, ►OMIA; hereditary disease frequency: http://www.findbase.org.

Ommatidium (Plural ommatidia): A self-sufficient element (facet) of the compound eye of arthropods, such as *Drosophila* (see Fig. O10). E- and N-cadherins regulate the shape of the cells but the shapes are also controlled by the physical requirement to minimize surface (Hayashi T, Carthew RW 2004 Nature [Lond] 431:647). ►compound eye, ►rhabdomere, ►cadherins; Wernet MF et al Nature [Lond] 440:174.

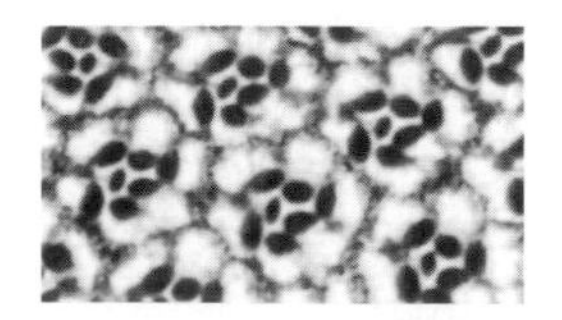

Figure O10. *Drosophila ommatidia* with eight photoreceptors inside each. There are about 800 ommatidia per eye

Ommochromes: These are insect eye pigments synthesized from tryptophan and the formation and condensation of hydroxykinurenine into xanthommatin and complexing with other components into brown pigment granules. ►pteridines, ►*w* locus, ►pigmentation in animals

Omnibank (OST): A collection of 80–700-nucleotide long rapid amplification cDNA 3′ ends by PCR (RACE), generated from mouse embryonic stem cells for the purpose of identification of sequence-tagged mutations. ►stem cell, ►PCR, ►RACE, ►sequenced-tagged sites; Zambrowicz BP et al 1998 Nature [Lond] 392:608.

Omnipotent: ►totipotent

Omp A: Refers to the outer membrane protein A of the bacterial cell. ►Rickettsia

OmpC, OmpF: Bacterial outer membrane proteins are regulated by kinase EnvZ in response to osmolarity (at low osmolarity by C and at high by F). OmpR is an outer membrane regulatory protein. OmpT is a serine protease which cleaves cyclin A. ►cell membrane, ►membrane proteins

Omphalocele: This refers to umbilical hernia. It is probably autosomal recessive with a prevalence of 1 to 2×10^{-4}. ►gastroschisis

OMSSA: The search engine for identifying MS/MS peptide spectra. ►mass spectrum; http://pubchem.ncbi.nlm.nih.gov/omssa; http://pubchem.ncbi.nlm.nih.gov/omssa/download.htm.

Oncocytoma: ►mitochondrial disease in humans, ►kidney diseases

Oncogene Antagonism Therapy: This therapy leads to transformation by vector constructs carrying tumor suppressor genes, dominant negative genes, suicide genes, antisense nucleotides, toxin genes that may prevent tumor formation or disable tumor cells. ►gene therapy, ►cancer gene therapy, ►suicide vectors, ►antisense technologies

Oncogene Collaboration: Some oncogenes do not transform cell cultures when applied single, e.g., rat embryo fibroblast requires the simultaneous presence of both RAS and MYC for complete transformation. ►oncogenes, ►transformation oncogenic

Oncogene-Induced Senescence: Certain oncogenes accelerate derepression of the *CDKN2a* locus by cooperation with the Polycomb group proteins. ►CDKN2, ►Polycomb; Collado M Serrano M 2006 Nature Rev Cancer 6:472.

Oncogenes: These genes of RNA viruses (v-oncogene) or similar genes in animal cells (c-oncogene) are responsible for the initiation of cancer. The c-oncogenes (proto-oncogenes) perform a normal function in animal cells but may cause abnormal proliferation by activation or amplification or promoter/enhancer fusion (translocation), mutation, deletion or inactivation. The transforming genes of DNA viruses do not have cellular counterparts and they induce tumors by interacting with tumor suppressor genes. An allelic form of an oncogene represents such a gain-of-function that favors cancerous transformation. The primary target of the majority of oncogenes is the cell cycle and they deregulate the function of genes that normally control the initiation or progression of the cell cycle. The human genome contains about 30 recessive and over 100 known oncogenes. For details see ►ABL, ►AKT1, ►AMV, ►ARAF, ►ARG, ►BLYM, ►BMYC, ►CBL, ►DBL, ►ELK, ►EPH, ►ERBB, ►ERG, ►ETS, ►EVI, ►FES, ►FGR, ►FLT, ►FMS, ►FOS, ►GLI, ►HIS, ►HKR, ►HLM, ►HST, ►INT, ►JUN, ►KIT, ►LCA, ►LCK, ►LYT, ►MAS, ►MCF, ►MEL, ►MET, ►MIL, ►MYB, ►MYC, ►NGL, ►NMYC, ►OVC, ►PIM, ►PKS, ►PVT, ►RAF, ►RAS, ►REL, ►RET, ►RHO, ►RIG, ►ROS, ►SPI1, ►SEA, ►SIS, ►SK, ►SNO, ►SRC, ►TRK, ►VAV, ►YES, ►YUASA. Many of the oncogenic transformations are caused by cis-activation of proto-oncogenes by non-oncogenic viruses. ►cancer, ►carcinogens, ►retroviruses, ►non-productive infection, ►proto-oncogene, ►CATR1, ►tumor suppressor, ►amplification, ►gene fusion, ►protein tyrosine kinases; Kung HJ et al 1991 Curr Top Microbiol Immunol 171:1; Liu D, Wang LH 1994 J Biomed Sci 1:65; Dua K et al 2001 Proteomics 1:1191; http://cgap.nci.nih.gov/; http://oncodb.hcc.ibms.sinica.edu.tw.

Oncogenic Transformation: This refers to the development of a cancerous state. It may begin by the loss or suppression of tumor suppressor genes. ►oncogenes, ►oncoproteins, ►oncogenic viruses; Di Croce L et al 2002 Science 295:1079.

Oncogenic Viruses: These can integrate into mammalian cells and rather than destroying the host, they can induce cancerous proliferation of the target tissues by inhibiting the cellular tumor suppressor genes. Oncogenic viruses may have double-stranded DNA genetic material such as adenoviruses (genome size ca. 37-kbp), Epstein-Barr virus (ca. 160-kbp), human papilloma virus (ca. 8-kbp), polyoma virus (ca. 5–6-kbp) and the single-strand RNA viruses (retroviruses, 6–9-kb). ►adenoviruses, ►SV40, ►papilloma virus, ►Epstein-Barr Virus, ►papova viruses, ►hepatitis B, ►retroviruses, ►acquired immunodeficiency, ►tumor viruses, ►Kaposi sarcoma, ►Moloney, ►Kirsten, ►Rous sarcoma, ►avian

Oncolytic Viruses: Herpes simplex-1, Newcastle disease virus, reovirus and adenovirus may selectively replicate and destroy tumor cells without the causing disease itself. Selectively targeting tumor cells may be accomplished by fusing antibody fragments or erythropoietin or heregulin to the viral envelope protein. Erythropoietin recognizes its receptor on erythroid precursor cells. Heregulin is a nerve growth factor required specifically for breast cancer and fibrosarcoma cells. Engineering tumor-specific promoter to the viral genes may enhance tumor-specific viral gene expression. The herpes simplex (HSV) or adenovirus may directly lyse the tumor cells. Cyclophosphamide enhances glioma therapy in rats by HSV by inhibiting immune responses (Fulci G et al 2006 Proc Natl Acad Sci USA 103:12873). Newcastle disease virus may increase sensitivity to tumor necrosis factor (TNF). The parvovirus may promote apoptosis of the cancer cells. The viral antigens bound to cellular MHC class I proteins may become targets for cytotoxic T lymphocytes (CTL). ►cancer gene therapy, ►herpes, ►Newcastle disease, ►parvovirus, ►reovirus, ►adenovirus, ►heregulin, ►TNF, ►apoptosis, ►CTL, ►ONYX-015, ►cyclophosphamide; Wildner O 2001 Annals Med 33 (5):291; Smith ER et al 2000 J Neuro-Onc 46(3):268.

Oncomine: This is a cancer microarray database and integrated data mining platform. (See http://www.oncomine.org/).

Oncomouse: This is the trade name for a mouse strain prone to breast cancer and suitable for this type of research. ►breast cancer; Kerbel RS 1998–99 Cancer Metastasis Rev 17(2):301.

Onconase: This is an anti-tumor protein with ribonuclease activity to tRNA. ►tRNA, ►ribonuclease; Notomista E et al 2001 Biochemistry 40:9097.

Oncoprotein: A product of an oncogene which is responsible for the initiation and/or maintenance of hyperplasia and malignant cell proliferation. ►oncogenes

Onco-Retroviral Vectors: These are engineered from onco-retroviruses such as murine leukemia virus and Rous sarcoma virus. The viral protein genes required for disease are removed and replaced by transgenes. Infectious particles are generated through packaging cells. ►retroviruses, ►viral vectors, ►transgene, ►packaging cell lines

Oncostatin M: ►APRF; Radtke S et al 2002 J Biol Chem 277:11297.

Ondex: Facilitates data acquisition from linked, integrated and visualized bases by graph analysis techniques. It can handle many hundreds of thousands of data and examine their potential relationships by identifying equivalent concepts and filtering out unattached nodes. (See http://ondex.sourceforge.net/).

One Gene-One Enzyme Theorem: This theorem recognizes that one gene is generally responsible for one particular biosynthetic step, mediated through an enzyme. More precisely stated is the one gene (cistron)-one polypeptide rule because some enzymatically active protein aggregates may be encoded by more than a single gene. There are some other apparent exceptions, e.g., one mutation blocking the synthesis of homoserine may also prevent the synthesis of threonine and methionine because homoserine is a common precursor of these amino acids (see Fig. O11). Further, in the branched-chain amino acid (isoleucine-valine) pathway, ketoacid decarboxylase and ketoacid transaminase enzymes control the pathways leading to both isoleucine and valine. As data on genomics accumulate it becomes increasingly evident that the one nucleotide sequence (depending on the multiple promoters and processing of the transcript) may carry out different (pleiotropic) functions. Sequencing and proteomic information of the human genome indicates that the same gene (DNA tract) is spliced in three alternate ways and this may be construed as one gene—three functions. ►homoserine, ►isoleucine-valine biosynthetic pathway, ►bifunctional enzymes, ►overlapping genes, ►contiguous gene syndrome, ►pleiotropy, ►genetic network; Beadle GW 1945 Chem Rev 37:15; Boguski MS 1999 Science 286:543; Venter JC et al 2001 Science 291:1304; Chen J et al 2002 J Biol Chem 277:22053; Kondrashov FA 2005 Nature Genet 37:9.

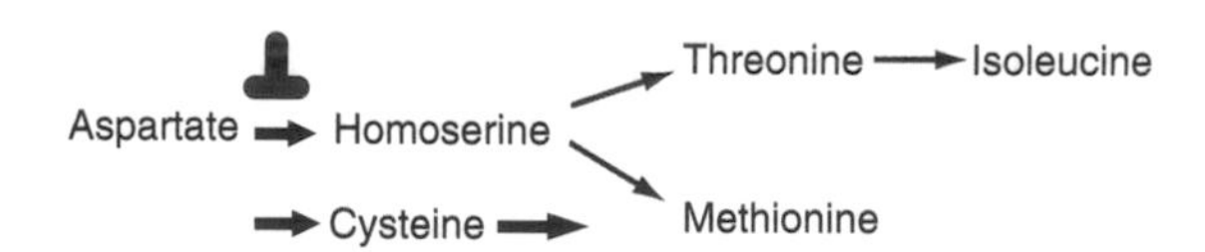

Figure O11. Block in homoserine biosynthesis generates requirements for threomine and methiomine

One Gene-One Polypeptide: ►one gene-one enzyme theorem

One-Hybrid Binding Assay: This is basically a gene fusion assay where a transcriptional activation domain is attached to a particular gene. The function of such a "hybrid" may be assessed by the expression of an easily monitored reporter gene (e.g., luciferase), depending on the signal received from the particular gene. ►two-hybrid assay, ►gene fusion, ►luciferase, ►split-hybrid system, ►three-hybrid system; Wilhelm JE, Vale RD 1996 Genes Cells 1(3):317; Murakami A et al 2001 Nucleic Acids Res 29:3347.

One-Step Growth: Bacteriophages multiply within the bacterial cell and in one step, within less than 10 min, during the "rise" period, all the particles are released. The temperate phages may have a longer period preceding the rise after infection because the phage DNA may be integrated into the bacterial chromosome and then replicates synchronously with the bacterial genes (see Fig. O12). Upon induction the phage may switch to a lytic cycle, which begins with autonomous replication followed by liberation of the phage particles. The number of phage particles released is called the burst size. See Fig O12, ►bacteriophages, ►phage lifecycles; Hayes W 1965 The Genetics of Bacteria and Their Viruses. Wiley, New York.

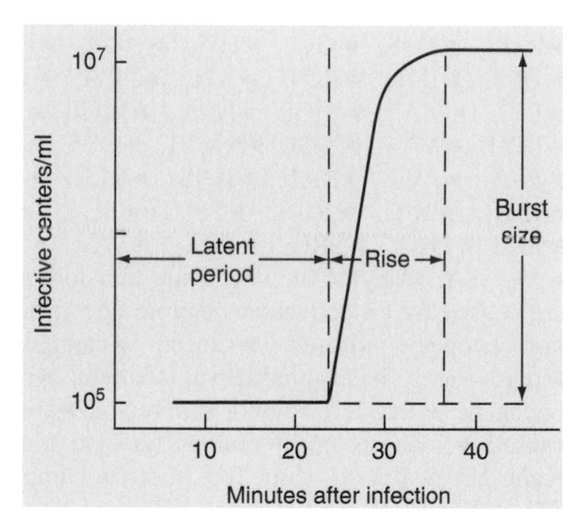

Figure O12. One-step growth

Onion (*Allium* spp.): The *Alloideae* subfamily of lilies has 600 species. *A. cepa*, 2n = 2x = 16. Most of the North American species have x = 8 (see Fig. O13). Polyploid species occur in different parts of the world with somatic chromosome numbers 24, 32, 40, 48. The eye-irritating lachrymatory factor (propanal S-oxide) is generated from 1-propenyl-L-cysteine sulphoxide by the enzyme lacrymatory factor synthase. It may be possible to develop a variety of onion that retains its full flavor without the pungent characteristics (Imai S et al 2002 Nature [Lond] 419:685). ►garlic

Figure O13. Allium x = 8

One-Tailed Test: The deviation is measured in one direction of the null hypothesis.

Onozuka R-10: ►cellulase

Onset: Denotes stage (time) of expression of a genetically determined trait.

Onto Tools: Function analysis of protein-coding genes from microarray data. http://vortex.cs.wayne.edu/Projects.html.

Ontogeny: Describes the developmental course of an organism. ►phylogeny

Ontology, Genetic: This seeks out the function and meaning of genes and genetic elements; e.g., the gene is a functional unit that can be either a DNA or a RNA but it is not a protein or other macromolecule. ►gene ontology; http://bcl.med.harvard.edu/proj/gopart.

Ontos: Refers to *E. coli* computer data sets organized into object-oriented database management system.

Onyx-015: This is a genetically modified adenoviral vector which lacks E1B-55K. It was earlier believed that the absence inactivates p53 tumor suppressor protein and facilitates oncolytic viral replication and selectively destroys cancer cells. However, it now known that the loss of E1B-55K leads to the induction rather than the inactivation of p53 by selective RNA export function (O'Shea CC et al 2004 Cancer Cell 6:611). ►adenovirus, ►oncolytic viruses, ►p53; Galanis E et al 2001 Crit Rev Oncol Hematol 38 (3):177.

Oocyte Donation: This is a means to overcome infertility in women who for some reason (older age, genetic risks, etc.) do not want or cannot conceive in the normal manner but can serve as a recipient of either their own ovum obtained earlier and preserved or of an ovum from a donor. These women can thus carry to term a normal baby. From a genetic viewpoint it is important that the ovum implanted is carefully analyzed to ensure that there is no risk involved. ►artificial insemination, ►surrogate mother, ►ART, ►micromanipulation of the oocyte; Noyes N et al 2001 Fertil Steril 76:92.

Oocyte, Primary: This has the same chromosome number as other common body cells (2n, 4C) but upon meiotic division each gives rise to two haploid *secondary oocytes* (n, 2C). The smaller of the two is called the 1st polar body. By another division the egg and three polar bodies (n, 1C) are formed. The egg may become fertilized but the polar bodies do not contribute to the progeny and fade away. In human females meiosis begins in the four month old fetus and proceeds to the diakinesis (dictyotene) stage until sexual maturity. After puberty in each four-week cycle one oocyte reaches the stage of the secondary oocyte and after completing the equational phase of meiosis (meiosis II), an egg is released during ovulation. Each of the two human ovaries contains about 200,000 primary oocytes (see Fig. O14). On an average around 400 eggs are produced during the entire fertile period of the human female, spanning a period of 30 to 40 years, i.e., beginning at puberty and terminating with the onset of menopause. Apoptosis leads to the loss of oocytes. The exhaustion of oocytes nutrients and the inability to generate NADPH may be contributing factors. It has been observed that pentose-phosphate-mediated inhibition of cell death is due to inhibitory phosphorylation of caspase 2 by calcium-dependent protein kinase II (Nutt LK et al 2005 Cell 123:89). In vitro differentiation of oocytes from embryonic stem cells can be studied experimentally. According to some reports, bone marrow and peripheral blood stem cells can repopulate the ovaries and can reinitiate the ovulation of oocytes after the loss of this capacity due to radiation injury or disease (Johnson J et al 2005 Cell 122:303). These findings have significance for postmenopausal women or those who suffered from cured ovarial cancer. A recent study did not confirm these observations and concluded that parabiosis, allowing efficient circulation of humoral and cellular factors, did not contribute to ovulated oocytes in mice but only to previously committed blood cells (Eggan K et al 2006 Nature [Lond] 441:1109).

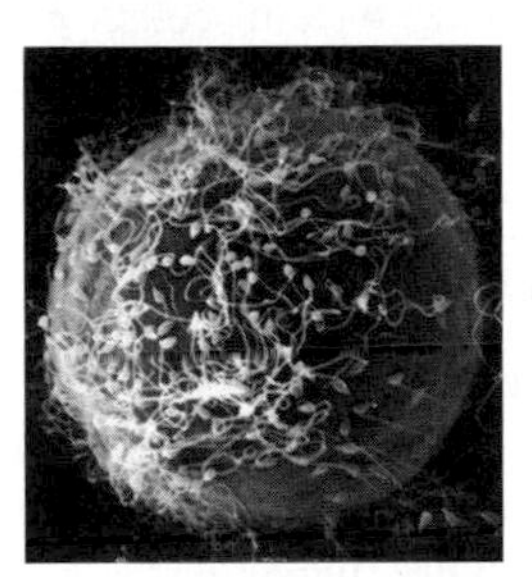

Figure O14. Human oocyte with about 1,500 spermatozoa (Courtesy of Dr. Mia Tegner)

In the human metaphase II oocytes, compared with reference samples from other tissues, 5,331 transcripts were significantly upregulated and 7,074 transcripts significantly down regulated. Of the oocyte upregulated probe sets, 1,430 had unknown function. A core group of 66 transcripts was identified by intersecting significantly upregulated genes of the human oocyte with those from the mouse oocyte and from human and mouse embryonic stem cells. GeneChip array results were validated using RT-PCR in a selected set of oocyte-specific genes. Within the upregulated probe sets, the top overrepresented categories were related to RNA and protein metabolism, followed by DNA metabolism and chromatin modification (Kocabas AM et al 2006 Proc Natl Acad Sci USA 103:14027).

O

Mouse embryonic stem cells in culture can develop into oogonia that enter meiosis, recruit adjacent cells to form follicle-like structures, and later develop into blastocysts (Hübner K et al 2003 Science 300:1251). Stem cells isolated from the skin of porcine fetuses have the intrinsic ability to differentiate into oocyte-like cells. When differentiation was induced, a subpopulation of these cells expressed markers such as Oct4, growth differentiation factor 9b (GDF9b), the Deleted in Azoospermia-like (*DAZL*) gene and Vasa (a germ line-specific protein) — all consistent with germ cell formation. On further differentiation, these cells formed follicle-like aggregates that secreted estradiol and progesterone and responded to gonadotropin stimulation. Some of these aggregates extruded large oocyte-like cells that expressed oocyte markers, such as zona pellucida, and the meiosis marker, synaptonemal complex protein 3 (SCP3). Some of these oocyte-like cells spontaneously developed into parthenogenetic embryo-like structures (Dyce PW et al 2006 Nature Cell Biol 8:384). These developments may eventually obviate the need for embryonic stem cells in therapeutic cloning. ►egg, ►spermatocyte, ►meiosis, ►C amount of DNA, ►menopause, ►gametogenesis, ►*Xenopus* oocyte culture, ►oogonium, ►oospore, ►oocyte translation system, ►nurse cell, ►fertilization, ►spermatozoon, ►stem cells, ►NADP, ►protein kinase, ►caspase, ►parabiosis, ►azoospermia, ►therapeutic cloning; Johnstone O Lasko P 2001 Annu Rev Genet 35:365.

Oocyte, Secondary: ►oocyte primary

Oocyte Translation System: mRNAs injected into the amphibian oocyte (*Xenopus*) nucleus (germinal vesicle) may be transcribed, and in the cytoplasm these exogenous messengers are translated, the proteins may be correctly processed, assembled, glycosylated,, phosphorylated, and delivered (targeted) to the proper location, etc. Similarly, injections of foreign DNA into the fertilized embryos may be replicated and inserted into the chromosomes. ►translation in vitro; Skerrett IM et al 2001 Methods Mol Biol 154:225.

Oogamy: Refers to the fertilization of a (generally) larger egg with a (smaller) sperm. ►oocyte.

Oogenesis: Refers to the formation of the egg. Meiosis in the mammalian oocyte begins much before fertilization but it is arrested at late prophase (dictyotene stage). The G_s protein-linked receptor (GPR3) maintains the pause (Mehlmann LM et al 2004 Science 306:1947). Shortly before ovulation meiosis is resumed. In vertebrates, the luteinizing hormone of the pituitary acts on the surrounding somatic cells and the continuation of meiosis is mediated by CDK1. ►gametogenesis, ►dictyotene stage, ►G_s protein, ►luteinizing hormone, ►CDK1; Navarro C et al 2001 Curr Biol 11:R162; Matzuk MM et al 2002 Science 296:2178; Schmitt A, Nebreda AR 2002 J Cell Sci 115:2457.

Oogonium: This is the female sex organ of fungi fertilized by the male gametes. It is also called a zygote.

Oogonium in Animals: Refers to the primordial female germ cell that is enclosed in a follicle by the term of birth of the individual and becomes the oocyte. ►gametangium, ►gametogenesis

Ookinete: This is the protozoan zygote at a motile stage within the malaria host mosquito.

Oophorectomy: The surgical removal of the ovary is also known ovariectomy, neutering, or spaying. ►castration

Ooplasm: This is an egg cytoplasm. The transfer of 5–15% of ooplasm from a donor may facilitate pregnancy in certain cases of in vitro fertilization. The transfer involves mitochondria and mRNA. ►ART; Brenner CA et al 2000 FertilSteril 74:573.

Oospore: A fertilized egg in (fungi) which is either dikaryotic or diploid and is frequently covered by a thick wall (see Fig. O15). ►oogonium

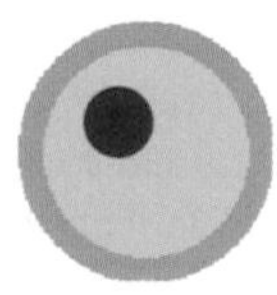

Figure O15. Oospore

Oozing: The binding of a protein at one site facilitates the adjacent binding of additional proteins until the resulting multimeric protein complex extends to the site of transcription initiation (Talbert PB, Henikoff S 2006 Nature Rev Genet 7:793).

Opal: This is the chain-terminator codon (UGA). ►code genetic

Opaque (*o*): Refers to genes in maize (several loci); *o-2* in particular attracted attention because its presence reduces the levels of prolamine and zein type proteins and increases the lysine content of the kernels. This improves the nutritional value significantly which is of concern to people in some parts of the world where corn is the main staple food. ►*floury*, ►kwashiorkor

Open-Label Trial: In contrast to the double-blind tests both subjects and experimenters are aware of the prospective medicine tested and the dose used. ►double-blend trial

Open Promoter Complex: Refers to a partially unwound promoter (the DNA strands separated) to facilitate the operation of the RNA polymerase. This separation is thought to be the result of the attachment of the transcriptase to the promoter. The TATA box of the promoter is a logical place for the attachment of the pol enzyme because there are only two hydrogen bonds between A and T in contrast to the three bonds between G and C and thus the separation of the double helix is easier. This is followed by initiation of transcription. After the attachment of the RNA elongation proteins, the σ subunit of the bacterial pol enzyme is evicted and transcription proceeds. Transcription factor TFIIB has a 7-bp recognition element immediately upstream of the TATA box and TFIIB and TBP are required for the formation of the preinitiation complex for RNA polymerase II. Several protein elements are important for forming an open complex. ►closed promoter complex, ►promoter melting, ►PIC, ►pol, ►Pribnow box, ►Hogness box, ►TBP, ►TAF, ►transcription factors, ►RAD25, ►regulation of gene activity, ►nucleosome, ►reinitiation, ►PSE, ►IHF, ►FIS, ►CAP; Uptain SM et al 1997 Annu Rev Biochem 66:117; Ranish JA et al 1999 Genes Dev 13:49; Davis C A et al 2005 Proc Natl Acad Sci USA 102:285.

Open Reading Frame (ORF): This is a nucleotide sequence between an initiation and a terminator codon. In higher organisms one of the two DNA strands is usually transcribed into functional products although there are open reading frames in both strands. ►initiation codon, ►nonsense codon, ►coding sequence, human and mouse ORF database: http://orf.invitrogen.com/.

Open System: This system exchanges material and energy within its environment.

Operand: Refers to what is supposed to be operated (worked) on.

Operational Concepts: These concepts were frequently used in genetics for providing an explanation when the underlying mechanism was not fully understood but from the manifest behavior a conceptualization was possible in agreement with what was known. For example: T.H. Morgan defined the gene as the unit of function, mutation and recombination before the nature of the genetic material was discovered. F.H.C. Crick concluded on the basis of frameshift mutagenesis—before the genetic code was experimentally determined—that the genetic code probably used nucleotide triplets.

Operational RNA Code: This is a sequence/structure-dependent aminoacylation of RNA oligonucleotides that are devoid of an anticodon. The specificity and efficiency is determined by a few nucleotides near the amino acid acceptor arm. ►aminoacylation, ►aminoacyl-tRNA synthetase, ►transfer RNA, ►mini-helix of tRNA; Schimmel P et al 1993 Proc Natl Acad Sci USA 90:8763; de Pouplana LR, Schimmel P 2001 J Biol Chem 276:6881.

Operational Taxonomic Unit (OTU): ►character matrix

Operator: This is the recognition site of the regulatory protein in an operon or possibly in other systems such as suggested for controlling elements (transposable elements) of maize. ►transposable elements, ►operon, ►*Lac* operon, ►*Ara* operon

Operon: This is a functionally coordinated group of genes producing polycistronic transcripts (co-transcription). Operons have been discovered in prokaryotes but are exceptional in eukaryotes. Similar organization occurs in the homeotic gene complexes of eukaryotic organisms, such as *ANTP-C* and *BX-C* of *Drosophila*. Many genes of the nematode (ca. 15%), *Caenorhabditis* seem to be coordinately regulated and transcribed into polycistronic RNA that is processed into monocistronic mRNA by transsplicing (Blumenthal T et al 2002 Nature [Lond] 417:851). ►*Lac* operon, ►*Arabinose* operon, ►*Tryptophan* operon, ►*His* operon, ►dual-gene operons, ►supraoperon, ►morphogenesis in *Drosophila*, ►homeotic genes, ►coordinate regulation, ►überoperon, ►transsplicing, ►SL1, ►SL2; Hodgman TC 2000 Bioinformatics 16:10; Blumenthal T, Seggerson Gleason K 2003 Nature Rev Genet 4:119; Ben-Shahar Y et al 2007 Proc Natl Acad Sci USA 104:222; http://regulondb.ccg.unam.mx:80/index.html; operon database: http://odb.kuicr.kyoto-u.ac.jp/.

Operon, Selfish: Clustered genes are more likely to be transmitted as a group. This is a selfish evolutionary feature because the clustering itself may not have any physiological benefit for the host, except the joint transmission. ►lateral transmission; Lawrence JG, Roth JR 1996 Genetics 143:1843; Lawrence J 1999 Curr Opin Genet Dev 9:642.

Ophthalmoplegia: Autosomal dominant phenotypes (incidence $\sim 1 \times 10^{-5}$, encoded at 10q23.3-q24.3, 3p14.1-p21.2, 4q34-q35) involve defects in moving the eyes and the head as well as some other variable symptoms. The 4q locus encodes a tissue-specific adenine nucleotide (ADP/ATP) translocator (ANT) and also controls mtDNA integrity. The autosomal recessive ophthalmoplegic sphingomyelin lipidosis appears to be allelic to the Niemann–Pick syndrome gene and it is associated with mitochondrial DNA

O

mutations. ►mitochondrial diseases in humans, ►neurogastrointestinal encephalomyopathy, ►Kearns–Sayre syndrome, ►Niemann–Pick syndrome, ►eye disease, ►myopathy, ►inclusion body myopathy, ►horizontal gaze palsy

Opiate: An opium-like substance which regulates pain perception and pain signaling pathways and mood (see Fig. O16). *Endogenous opiates*, enkephalins and endorphins, were isolated from the brain and the pituitary gland, respectively. They contain a common 4-amino acid sequence and bind to the same cell surface receptors as morphine (and similar alkaloids). Nociceptins (orphanin) are 17 amino acid antagonists of the opioid receptor-like receptor. Opioids are opiate-like, but they are not derived from opium. Nocistatin with an evolutionarily conserved C-terminal hexapeptide blocks pain transmission in mammals. Opioids activate the expression of FAS which upon binding its ligand (FasL) promotes apoptosis of lymphocytes and thereby the immune system. Opioids may affect the immune system by suppressing cytokine synthesis. Opioids modulate stress responses, learning and memory, metabolism, and may lead to addiction, etc. A region of chromosome 14q with a non-parametric lod 3.3 is responsible for opioid addiction (Lachman HM et al 2007 Human Mol Genet 16:1327). The major types of opiums are plant alkaloids. ►enkephalin, ►endorphin, ►dynorphin, ►morphine, ►FAS, ►apoptosis, ►immune system, ►cytokines, ►nociceptor; formula after Massotte D, Kieffer BL 1998 Essays Biochem 33:65.

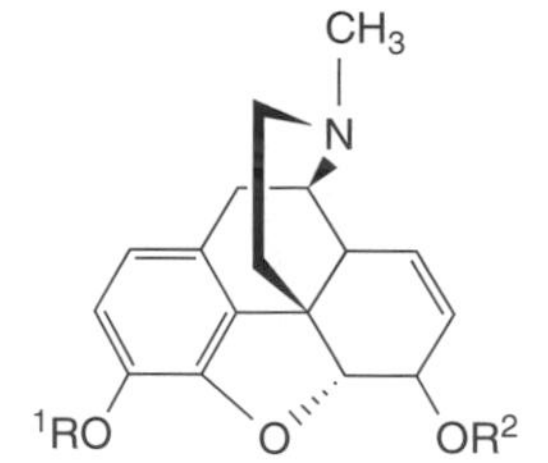

CH_3

N

^1RO O OR^2

Morphine: $R^1 = R^2 = H$

Codeine: $R^1 = CH_3$, $R^2 = H$

Heroin: $R^1 = R^2 = CH_3 - CO$

Figure O16. Opiates

Opines: These are synthesized in crown-gall tumors of dicotyledonous plants under the direction of agrobacterial plasmid genes. The bacteria use these opines as carbon and nitrogen sources. The octopine family of opines comprises octopine, lysopine, histopine, methiopine and octopinic acid. The nopaline group includes nopaline and nopalinic acid. Agropines are agropine, agropinic acid, mannopine, mannopinic acid and agrocinopines. ►*Agrobacteria*, ►T-DNA octopine, ►nopaline; Petit A et al 1970 Physiol Vég 8:205.

Opiocortin: This is a prohormone (pro-opiocortin) translated as a precursor of several corticoid hormones and cut by proteases and processed into corticotropin, β-lipotropin, γ-lipotropin, α-MSH (melanin-stimulating hormone), β-MSH and β-endorphin as shown here (see Fig. O17) ►animal hormones, ►POMC, ►individual peptide hormones under separate entries, ►pigmentation of animals; Lowry PJ 1984 Biosci Rep 4(6):467; De Wied D, Jolles J 1982 Physiol Rev 62:976; Challis BG et al 2002 Hum Mol Genet 11:1997.

Opisthotonus: A motor protein (myosin) dysfunction resulting in spasms and backward pulling of the head and heels while the body seems to move forward. ►Usher syndrome

Opitz Syndrome (G syndrome, BBB syndrome, 22q11.2): An apparently autosomal dominant anomaly with complex features such as hypertelorism (the distance between paired organs is abnormally increased), defects in the esophagus (the passageway from the throat to the stomach), hypospadias (the urinary channel opens in the underside of the penis in the vicinity of the scrotum), etc. In an autosomal recessive form it is characterized by polydactyly, heart anomaly, triangular head, failure of the testes to descend into the scrotum and suspected deficiency of the mineralocorticoid receptor. An Xp22 gene encodes the 667-amino acid Mid1 (midline) Ring finger protein and its mutant forms may interfere with microtubule function. ►corticosteroid, ►head/face/brain defects, ►Opitz-Kaveggia syndrome; Liu J et al 2001 Proc Natl Acad Sci USA 98:6650.

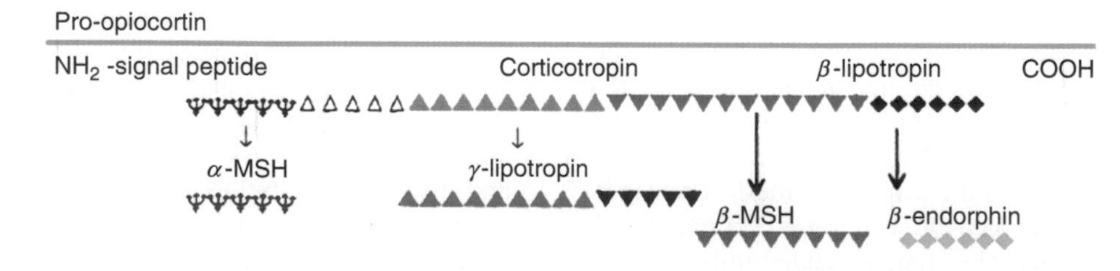

Figure O17. Pro-opiocortin

Opitz–Kaveggia Syndrome: This is a Xp22-linked phenotype characterized by a large head, short stature, imperforate anus, heart defect, muscle weakness, defect in the white matter of the brain (corpus callosum) and mental retardation. A recent report noted that *MED12* (Xq13) encodes a subunit of the macromolecular complex of Mediator, which is required for thyroid hormone-dependent activation and repression of transcription by RNA polymerase II. The MED12 protein is not part of the Mediator core complex but is a member of a Mediator module that may act as an adaptor for specific transcription factors and a defect in it leads to the syndrome (Risheg H et al 2007 Nature Genet 39:451). ►stature in humans, ►heart disease, ►mental retardation, ►head/face/brain defects, ►Opitz syndrome

Opossum (Metatherians/Marsupials): *Caluromys derbianus* 2n = 14, *Chironectes panamensis* 2n = 22, *Monodelphis domestica* 2n = 18. The latter is a South American species, closely related to the North American Virginia opossum. A draft sequence of its genome is available which contains 3,475 Mb comprising about 18,000 to 20,000 protein-coding genes. The autosomes of opossum are extremely large whereas the X chromosome is much smaller than in most eutherians. The average rate of autosomal recombination is lower (0.2–0.3 cM Mb^{-1}) compared to other amniotes (0.5– > 3 cM Mb^{-1}) but in the X chromosome it is larger than in the autosomes. The G + C content is about 2% lower than in other amniotes. Segmental duplications are much shorter and lower in frequency than the sequenced amniotes and shorter sequences separate them. The stability of the genome is greater. Conserved non-coding elements in opossum evolved from transposons. The eutherian protein-coding sequences are largely conserved in this species (Mikkelsen TS et al 2007 Nature [Lond] 447:167; Goodstadt L et al 2007 Genome Res 17:969; Belov K et al 2007 Genome Res 17:982; Gentles AJ et al 2007 Genome Res 17:992).

Opsins: These are photoreceptor proteins of the retina but non-visual photoinduction opsin is found in the pineal gland. The chromophore of opsins is either 11-cis-retinal or 3-dehydroretinal. Red-green color vision depends on these molecules. The red- and green-sensitive pigments differ mainly in amino acids at sites 180, 277 and 28 respectively. In red sensitives alanine and phenylalanine, and in green sensitives serine, tyrosine and threonine, respectively are most common. In hominids and Old World monkeys two X chromosomal genes encode these pigments. In all mammals (with the exception of the New World monkeys) there is only one locus encoding either red or green opsins. In humans, additional sites have minor effects. In pigeons five retinal opsin genes have been distinguished. ►pineal gland, ►rhodopsin, ►retinoic acid, ►color vision, ►color blindness, ►navigation; Yokoyama S, Radlwimmer FB 1999 Genetics 153:919.

Opsonins: These trigger phagocytosis by the scavenger macrophages and neutrophils. These substances bind to antigens associated with immunoglobulins IgG and IgM and facilitate the recognition of the antigen-antibody complexes by the defensive scavenger cells. Also, they may bind to the activated complement of the antibody and assist in the recognition of the cell surface antigens and thus mediate their destruction. ►antibody, ►complement, ►immunoglobulins, ►macrophage, ►neutrophil; Moghimi SM, Patel HM 1988 FEBS Lett 233:143.

Opsonization: Foreign invaders of the cells are coated by opsonins to facilitate their destruction by phagocytosis. ►opsonin, ►phagocytosis; Mevorach D 2000 Ann NY Acad Sci 926:226.

Optic Atrophy: This is determined either by autosomal dominant, recessive (early onset types) or X-linked or mitochondrial defects of the eye, the ear and other peripheral nerve anomalies. OPA1 gene in 28 exons encodes a dominant optic atrophy at human chromosome 3q28-q29. Its product is a 960-amino acid dynamin-like protein localized in the mitochondria. The prevalence of OPA1 is 5×10^{-4}. ►Behr's syndrome, ►night blindness, ►Leber optic atrophy, ►Kearn–Sayre syndrome, ►Wolfram syndrome, ►mitochondrial disease in humans, ►myoclonic dystrophy, ►eye diseases, ►color blindness, ►dynamin, ►methylglutaconic aciduria; Pesch UEA et al 2001 Hum Mol Genet 10:1359.

Optical Density: ►O.D.

Optical Mapping: This may be used for ordering restriction fragments of single DNA molecules. The fragments are stained by fluorochrome(s) and the restriction enzyme-generated gaps can be visualized. The contigs are assembled automatically by using the Gent algorithm. Most of the optical mapping procedures are useful only for mapping small genomes. An adaptation of these principles for the assembly of very large genomes is also available (Valuev A et al 2006 Proc Natl Acad Sci USA 103:15770). ►physical mapping, ►mapping genetic, ►FISH, ►Gent algorithm; Lin J et al 1999 Science 285:1558; Aston C et al 1999 Trends Biotechnol 17:297.

Optical Rotatory Dispersion: Denotes a variation of optical rotation by the wavelength of polarized light; it depends on the difference in refractive index between left-handed and right-handed polarized light. The rotation is measured as an angle. It is similar to circular dichroism. ►base stacking, ►circular dichroism

Optical Scanner: A device that generates signals from texts, diagrams, pictures, electrophoretic patterns, autoradiograms, etc. that can be then read or printed out with the assistance of a computer.

Optical Tweezer: This consists of a special laser beam linked to a microscope system. It facilitates the manipulation of cell membranes, protein folding, DNA condensation, structure of chromosomes, etc. ►scanning force spectroscopy; Hayes JJ, Hansen JC 2002 Proc Natl Acad Sci USA 99:1752; Grier DG 2003 Nature [Lond] 424:810.

Optimon: This DNA unit avoids recombination and is preserved as such during evolution.

Opus: ►copia

OR: ►odds ratio, ►lod score

Oral Bacterial Films: The human oral cavity and the gut may be inhabited by more than 500 different taxa. These microorganisms cause tooth decay, gingivial bleeding and other health problems. ►microbiome, ►gut, ►*Helicobacter*, Kolenbrander PE 2000 Annu Rev Microbiol 54:413; Hooper LV, Gordon J I 2001 Science 292:1115.

Oral-Facial-Digital Syndrome: ►orofacial–digital syndrome

Orange (*Citrus aurantium*, 2n = 18): This is a common fruit tree. Botanically, the peel of the fruit is the pericarp, containing at the lower face the fragrant oil glands. The juice sacs are enclosed by the carpels, containing the seeds. ►navel orange

ORC (origin recognition complex): This six subunit complex (including Rap1 and Abf1 silencers) is required before DNA replication can begin in the eukaryotic cell. The N-terminal domain of Orc1 protein (the largest subunit) interacts with Sir1 and Sir1 mediates the recruitment of the other Sir proteins. Sir2 is a NAD-dependent histone deacetylase and Sir3 and Sir4 play structural roles. Sir3 and Sir4 interact with Rap1 and bind the N-terminal ends of histones H3 and H4 and control the epigenetic mechanism of sex determination in yeast and chromatin remodeling (Hsu H-C et al 2005 Proc Natl Acad Sci USA 102:8519). In yeast the ORC binds to the autonomously replicating sequence (ARS). During a cell cycle replication can be initiated only once but hundreds of sites of replicational origin exist. The ORC may place the nucleosomes at the DNA replication initiation site to facilitate the process. The ORC also mediates sister chromatid cohesion in yeast (Shimada K, Gasser SM 2007 Cell 128:85). In budding yeast the ORC complex and Cdc6 in the presence of a tandem array of GAL4 binding site in a plasmid is sufficient for the initiation of replication from this artificial construct (Takeda DY et al 2005 Genes Dev 19:2827).

The replication cannot begin before the *origin of licensing* is created in the M phase with the participation of the MCM protein(s). This is followed by the *origin of activation* in the S phase. Both these steps are controlled by the ORC. Protein subunit ORC2 (Orp2 in fission yeast) apparently interacts with Cdc2, Cdc6 and Cdc18 proteins that regulate replication. The ORC is required for silencing the *HMRa* and *HMLα* loci of yeast, involved in mating type determination. The ORC homolog of *Drosophila* is DmORC2. Budding yeast genome appears to have ~429 replication origins. The replication origins in *Schizosaccharomyces pombe* are quite different from the same functional elements in budding yeast. They are mainly in intergenic regions, rich in AT (~70%) sequences (Dai J et al 2005 Proc Natl Acad Sci USA 102:337). ►replication, ►Cdc2, ►Cdc18, ►mating type determination in yeast, ►HML, ►HMR, ►ARS, ►MCM, ►replication licensing factor, ►ARS, ►cell cycle, ►Rap1, ►Abf1, ►histone deacetylase, ►nucleosomes, ►cohesin, ►sister chromatid cohesion; Lipford JR, Bell SP 2001 Mol Cell 7:21; Vashee S et al 2001 J Biol Chem 276: 26666; Dhar SK et al 2001 J Biol Chem 276:29067; Gilbert DM 2001 Science 294:96; Wyrick JJ et al 2001 Science 294:2357; Fujita M et al 2002 J Biol Chem 277:10345.

Orchard Grass (*Dactylis glomerata*): A shade and drought-tolerant forage crop, 2n = 28.

Orchids (*Orchideaceae*, 2n = 20, 22, 34, 40): These are monocotyledonous tropical ornamentals.

Ord: 55-kDa chromosomal protein with a role in chromatid cohesion. ►sister chromatid cohesion; Bickel SE et al 1998 Genetics 150:1467.

Order: A taxonomic category above *family* and below *class*, e.g., order of primates within the class of mammals.

Ordered Tetrad: The spores in the ascus represent the first and second meiotic divisions in a linear sequence such as Ⓐ Ⓐ ⓐ ⓐ. ►tetrad analysis

Figure O18. Ordered tetrad

Orestes: Contraction of the words open reading frames + EST (expressed sequence tags). The procedure is aimed at sequencing mid-portions of the genes in contrast to ESTs which deal with either 5′ or 3′ ends. ORESTES may help in the annotation of the genomes. By 2001 700,000 ORF tags were available for the definition of the human proteome. ►ORF,

►EST, ►proteome; de Souza SJ et al 2000 Proc Natl Acad Sci USA 97:12690; Camargo AA et al 2001 Proc Natl Acad Sci USA 98:12103.

Orexin A and B (hypocretin): Appetite-boosting polypetides synthesized in the lateral hypothalmus area of the brain. Their aberrant splicing (caused by insertional mutation) may lead to narcolepsy. Their G-protein coupled receptors are Hcrtr-1 and -2. Hcrtr-2 was assigned to human chromosome 17q21. Orexins also regulate the sleep-wake cycles in neurodegenerative diseases, and drug (e.g., cocaine) and food reward. ►leptin, ►obesity, ►narcolepsy; Scammell TE 2001 Curr Biol 11:R769; Petersen A et al 2005 Hum Mol Genet 14:39; Harris GC et al 2005 Nature [Lond] 437:556.

ORF: Denotes open reading frame, i.e., the nucleotide sequences between the translation initiator and the translation terminator codons, e.g., AUG →→→→→→→ UAG. ►cORF, ►hORF, ►kORF, ►qORF, ►sORF, ►tORF, ►OST, ►Gateway cloning; Snyder M Gerstein M 2003 Science 300:248.

Orfome (ORFeome): The collection of all defined open reading frames of an organism (Harrison PM et al 2002 Nucleic Acids Res 30:1083). ►Gateway cloning

Organ: A body structure destined to a function.

Organ Culture: Refers to growing organs in vitro to gain insight into function, differentiation and development. Plant organ cultures, such as propagating roots, stem tips and embryos under axenic conditions on synthetic media are known for decades. More recently, research has focused on generating human tissues and organs as replacements in case of injury or disease. The organs generated from the patients' own tissues avoid some of the problems of graft rejection. Since 1997 the US Food and Drug Administration has approved the clinical use of cultured cartilage. Laboratory production of blood vessels, bladder, cardiovascular tissues and even kidneys and livers is expected. Usually, researchers employ biodegradable polymer scaffolds (polyglycolic acid, polylactide) to permit the development of thicker cell layers permeable to nutrients. ►tissue culture, ►grafting in medicine, ►stem cells

Organelle: Refers to intracellular bodies, such as the nucleus, mitochondria and plastids. The division of plastids requires – among other proteins – FtsZ and DRP. The division of the mitochondria does not require an ancestral prokaryotic system but relies on DRP and other proteins. Both these organelles universally require dynamin-related guanosine triphosphatases for division (Osteryoung KW, Nunnari J 2003 Science 302:1698).

The term organelle is also used for specialized protein complexes. Earlier no bacterial organelles were detected. Recently, membrane-enclosed acidocalcisomes have been identified in *Agrobacterium tumefaciens* (Seufferheld M et al 2003 J Biol Chem 278:29971) that are similar to the acidic calcium storage compartments of some unicellular eukaryotes (trypanosomas, apicomplexan parasites, algae and slime molds). In *Arabidopsis* the number of proteins localized in different organisms was as follows: endoplasmic reticulum 182, Golgi apparatus 89, plasma membrane 92, vacuoles 24, mitochondria and plastids 140 and 162 to unknown sites (Dunkley TPJ et al 2006 Proc Natl Acad Sci USA 103:6518). These numbers are probably much lower than the actually figures. A map of the mouse liver reveals 10 subcellular locations of 1,404 proteins (Foster LJ et al 2006 Cell 125:187). ►nucleus, ►mitochondria, ►chloroplast, ►apicoplast, ►FtsZ, ►DRP, ►dynamin, ►*Agrobacterium*, ► *Trypanosoma*, ►slime mold, ►evolution of organelles, ►tissue-specificity; origin of organelles: Dyall SD et al 2004 Science 304:253; genomes: http://megasun.bch.umontreal.ca/gobase/; protein database: http://organelledb.lsi.umich.edu; protein map in organelles: http://proteome.biochem.mpg.de/ormd.htm; mitochondrial and chloroplast genome database: http://gobase.bcm.umontreal.ca/.

Organelle Genetics: ►mitochondrial genetics, ►chloroplast genetics, ►sorting out, ►Golgi, ►gobase

Organelle Sequence Transfers: During evolution sequences homologous among the major organelles were transferred in the direction shown (see Fig. O19). In budding yeast mtDNA sequences may be regularly transferred to the nucleus during double-strand break repair. In the 2nd chromosome of *Arabidopsis* 135 genes appear to be of chloroplast origin. In the centromeric region of the same chromosome of one *Arabidopsis* ecotype ~618-kb appears identical to part of the mitochondrial genome (Stupar RM et al 2001 Proc Natl Acad Sci USA 98:5099). The organelle genomes (mitochondrial, plastidic) are adopted by initial symbiosis. The complete genomes of the originally free-living organisms were not retained during evolution, some were lost and others were redistributed among the organelles. Some of the genes were apparently sequestered into the organelles in order to assure a homeostatic balance in the redox potential. Approximately 18% of the nuclear protein-coding genes of *Arabidopsis* appear to be acquired from the cyanobacterial ancestors of the plastids (Martin W et al 2002 Proc Natl Acad Sci USA 99:12246). Enhanced production of reactive oxygen—by metabolic accident—may kill the sensitive cells unless the damage is readily corrected at the origin. A survey

277 genera of angiosperm plants indicated that the ribosomal protein gene, *rps10*, has been transferred from the mitochondrion to the nucleus at a very high rate during evolution and this probably still continues in plants but not in animals. The transfer of functional genes from the mitochondria to the nucleus seems to be more common in selfing or clonal plants than in outcrossing plants. It has been suggested that selfing and vegetative reproduction conserve adaptive mitochondrial – nuclear gene combinations, allowing functional transfer, whereas outcrossing prevents transfer by breaking up these combinations (Brandwain Y et al 2007 Science 315:1685).

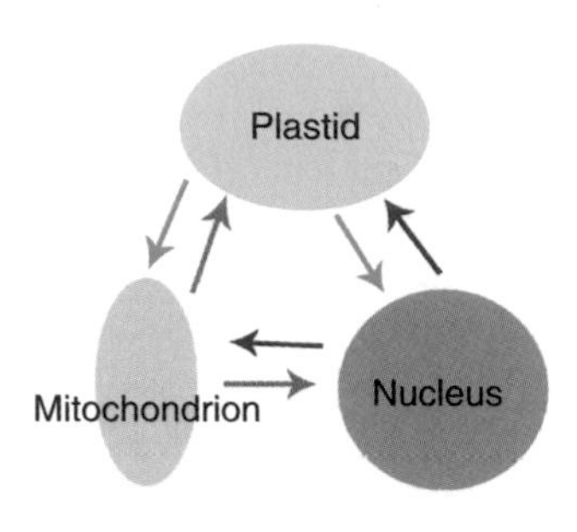

Figure O19. Organelle sequence transfer

Nuclear mutations in yeast control the escape of mitochondrial DNA into the nucleus (Thorsness PE Fox TD 1993 Genetics 134:21). The process is mediated by zinc-dependent mitochondrial protease (Weber ER et al 1996 Mol Biol Cell 7:307). There are interactions between organellar and nuclear functions without the actual transfer of genetic material (Traven A et al 2001 J Biol Chem 276:4020). Microarray hybridization revealed that in the petite strains of yeast (devoid of mitochondrial DNA), the expression of several nuclear genes (citrate synthase, lactate dehydrogenase, etc.) is altered. The petite strains exhibited increased resistance to heatshock and pleiotropic drug resistance. The rate of transfer of genes from the mitochondria to the nucleus appears to be within the range of 2×10^{-5} and from the plastid to the nucleus is about 6×10^{-5} per cell generations. The potential expression of organellar genes within the nucleus depends of the nature of the promoters. Prokaryotic type organelle promoters may not function at nuclear locations. Some of the aminoacyl-tRNA synthases (24 known) can be shared between mitochondria and chloroplasts (~15) and 5 are shared between the cytosol and mitochondria and one also present in the chloroplasts (Duchène A-M et al 2005 Proc Natl Acad Sci USA 102:16484). The phototrophic unicellular ciliate protozoon *Myrionecta rubra/Mesodinium rubrum* can intake chloroplasts, mitochondria and nuclei from the cryptomonad alga *Geminigera cryophila* and these organelles can survive and function up to 30 days within the predator (Johnson MD et al 2007 Nature [Lond] 455:426). ►mobile genetic elements, ►chloroplasts, ►mtDNA, ►double-strand break, ►nupt, ►numts; Race HL et al 1999 Trends Genet 15:364; Adams KL et al 2000 Nature [Lond] 408:354; Blanchard JL, Lynch M 2000 Trends Genet 16:315; Hedtke B et al 1999 Plant J 19:635; Adams KL et al 2002 Plant Cell 14:931; Huang CY et al 2003 Nature [Lond] 422:72; Martin W 2003 Proc Natl Acad Sci USA 100:8612; Timmis JN et al 2004 Nature Rev Genet 5:123; http://megasun.bch.umontreal.ca/gobase/gobase.html.

Organelles: These are membrane-enclosed cytoplasmic bodies such as the nucleus, mitochondrion, plastid, Golgi and lysosome. See under separate entries, ►organelle, ►organelle sequence transfer; organelle protein database of several eukaryotes: http://organelledb.lsi.umich.edu/.

Organic: A carbon containing compound or something associated with a metabolic function.

Organic Evolution: Refers to the historical development of living beings in the past and present times. ►geological time periods

Organismal Genetics: Studies inheritance in complete animals and plants by biological means and does not employ molecular methods or the tools of reversed genetics. ►reversed genetics, ►inter-organismal genetics

Organizer (Spemann organizer): The dorsal lip of the blastopore (an invagination that encircles the vegetal pole of the embryo) becomes a signaling center for differentiation (see Fig. O20). The formation of the organizer is preceded by induction of the mesoderm cell layer, resulting in the expression of organizer-specific homeobox genes and transcription of genes coding for signal molecules. This is followed by recruitment of the neighboring cells into the axial mesoderm and neural tissues. Several of the nuclear genes responsible for the component of the organizer have been identified by genetic and molecular means. The organization is controlled by proteins such as Wnt that determine the body axis and anterior posterior development. Activin and fibroblast growth factor (FGF) involve signal reception to organize the formation of the mesoderm or by noggin that controls dorsal/ventral differentiation. Chordin affects the development of the notochord, and follistatin is an antagonist of activin, etc. Chordin and noggin are also required for the development of the forebrain. These proteins apparently bind to and antagonize the BMP protein of vertebrates (homologous to decapentaplegic of *Drosophila*). The organizer may show some

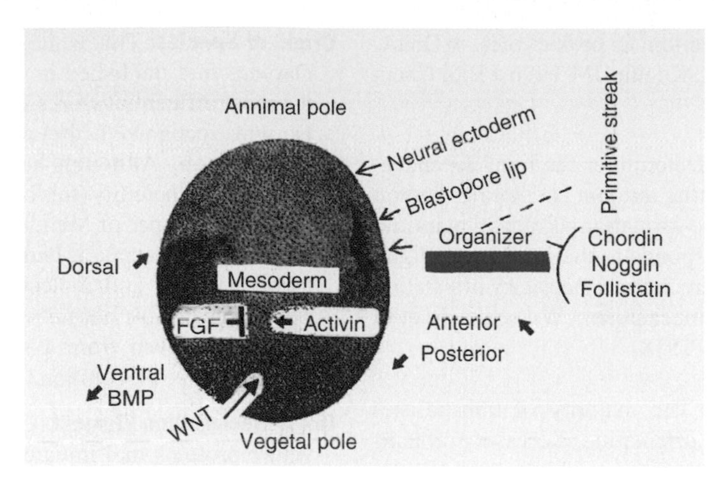

Figure O20. Organizer

additional variations in different species. It is diagrammatically not possible to represent correctly the complexity of the embryonal differentiation, which involves thousands of genes directly and indirectly. The discovery of the organizer by Hans Spemann and co-workers in the 1920s is considered a major milestone in experimental embryology and in recognition of his work Spemann was awarded the Nobel Prize in 1935. This line of research initiated a new series of inquiries in "chemical embryology" and became the focus of contemporary molecular biology. Many of the genes involved in embryonic induction, regulation of differentiation and development have now been cloned. The discovery of new techniques based on genetics, immunology, radioactive tracers, fluorochromes, microarray hybridization, scanning electron microscopy, etc., coupled with genetic transformation will provide the answer to the most basic problems of development and the nature of disease, cancer, etc. The *Goosecoid* homeobox transcription factor of the organizer promotes tumor cell malignancy and indicates that other conserved organizer genes may function similarly in metastasis of human cancer (Hartwell KA et al 2006 Proc Natl Acad Sci USA 103:18969). ►vegetal pole, ►animal pole, ►morphogenesis, ►gastrula, ►epiblast, ►homeobox, ►homeotic genes, ►signal transduction, ►LIM, ►BMP, ►FGF, ►decapentaplegic, ►activin, ►follistatin, ►bone morphogenetic protein, ►noggin, ►pattern formation, ►induction, ►signal transduction, ►primitive streak, ►left-right asymmetry, ►Spemann organizer, ►self-regulation, ►metastasis; Harland R Gerhart J 1997 Annu Rev Cell Dev Biol 13:611; De Robertis EM et al 2000 Nature Revs Genet 1:171; Shilo B-Z 2001 Cell 106:17; Lee HX et al 2006 Cell 124:147; Yu J-K et al 2007 Nature [Lond[445:613.

Organogenesis: The development of organs from cells differentiated for special purposes.

Organophosphates: These are common ingredients of insecticides and biological warfare agents such as sarin, soman and paraoxon. Their acute toxicity is due to irreversible inhibition of acetylcholinesterase, an enzyme that inactivates the neurotransmitter acetylcholine. Galanthamine (see Fig. O21), a reversible inhibitor, when administered in combination with atropin to guinea pigs provides effective protection soon after exposure to organophosphates (Albuquerque EX et al 2006 Proc Natl Acad Sci USA 103:13220). Goats transgenic for recombinant butyrylcholinesterase produced active rBChE in milk, sufficient for prophylaxis of humans at risk for exposure to organophosphate agents (Huang Y-J et al 2007 Proc Natl Acad Sci USA 104:13603). ►acetylcholinesterase, ►acetylcholine, ►sarin, ►insecticide resistance

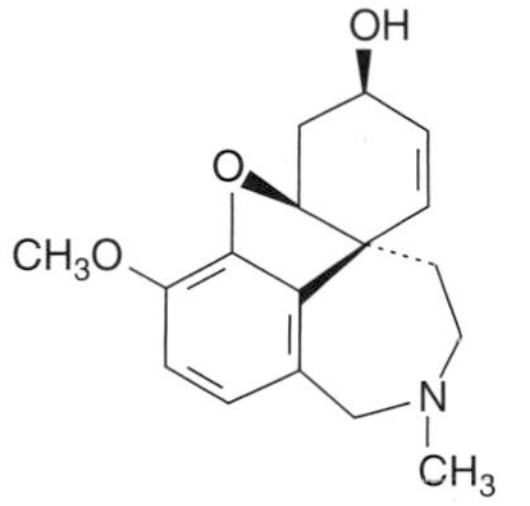

OH
O
CH_3O
N
CH_3

Figure O21. Galanthamine

ori: Refers to origin of replication. ►DNA replication, ►replication, ►plasmid ori_T, ►ori_V

oriC: This gene controls the origin of replication in *E. coli* and binds the replication proteins DnaA and

DnaB. ►DNA replication in prokaryotes, ►DnaA, ►DnaB; Margulies C, Kaguni JM 1996 J Biol Chem 271:17035.

Orientation, Magnetic: Determines the light-dependent migration of birds during autumn and spring. Photon absorption leads to the formation of radical pairs and the avian compass responds to the intensity of light. However, the fixed direction in migratory orientation depends on another mechanism (Wiltschko R et al 2005 Current Biol 15:1518)

Orientation Selectivity: The majority of transposases work well only in one orientation. Accessory proteins associated with the transposases choose this orientation. It is synonymous with the directionality of transposons. ►transposable elements, ►transposons

Origin of Life: The prebiotic evolution laid down the foundations for the origin of life ►evolution prebiotic, ►evolution of the genetic code. The following steps may have been involved in this process: 1. generation of organic molecules, 2. polymerization of RNA, capable of self-replication and heterocatalysis (cf. ribozymes), 3. peptide synthesis with the assistance of RNA, 4. the evolution of translation and polypetide assisted replication and transcription, 5. reverse transcription of RNA into DNA, 6. the development of DNA-RNA-Protein auto- and heterocatalytic systems, 7. the sequestration of this complex into organic micellae formed by fatty acid - protein membrane-like structures, 8. the appearance of the first cellular organisms about 3–4 billion years ago, and 9. the development of photosynthesis and autonomous metabolism. According to some estimates, the starter functions of life may have been carried out by as few as 20–100 proteins. The extensive duplications in all genomes may be an indication that the early enzymes did not have strong substrate specificity. The minimal cellular genome size may have been comparable to that of *Mycoplasma genitalium*, 580-kb long and coding 482 genes. The time required to develop from the 100-kb genome to a primitive heterotroph cyanobacterium with 7,000 genes may have taken about 7 million years. Life originated on earth about 3.2 billion years ago (Nofke N et al 2006 Geology 34:253). ►spontaneous generation unique or repeated, ►exobiology, ►geological time periods, ►evolution of the genetic code; Nisbet EG, Sleep NH 2001 Nature [Lond] 409:1083; Rotschild LJ, Mancinelli RL 2001 ibid:1092; Carroll SB 2001 ibid. 1102; Hanczyc MM et al 2003 Science 302:618.

Origin of Replication: This is the starting point of replication during cell proliferation. ►replication

Origin of Species: This is the title of a book by Charles Darwin, first published in 1859. The book is one of the most influential works on human cultural history. Darwin recognized the role of natural selection in evolution. Although he did not understand the principles of heredity (this book was published 7 years prior to the paper of Mendel). With the development of modern concepts of heredity, cytogenetics, population genetics and molecular biology, the seminal role of this book became generally recognized and appreciated even from a sesquicentennial distance. ►Darwinism, ►evolution

Origin Recognition Element (ORE): This is the DNA site where proteins that initiate replication bind. ►replication fork

Origons: These are the processors of environmental perturbations in transcriptional networks. Origons are sub-networks originating in a single transcription regulatory network. A pleiotropic regulator controls all nodes in *modulons*. *Simulons* include all nodes affected by an environmental signal and are composed of all origons rooted in transcription factors sensitive to the signal. ►networks, ►genetic networks, ►modulon; Balázsi G et al 2005 Proc Natl Acad Sci USA 102:7841.

ori_T: This refers to the origin of transfer of the bacterial plasmid. Its A=T content is higher than the surrounding DNA. It has recognition sites for a number of conjugation proteins. It contains promoters for the *tra* genes that are localized in such a way that only after the complete transfer of the plasmid can all of them be transferred. The transfer of the single strand proceeds in 5′→3′ direction in all known cases, including the transfer of T-DNA to plant chromosomes. The 3′ end can accept added nucleotides. After the transfer has terminated, the plasmid re-circularizes. All these processes require specific proteins. ►bom, ►conjugation, ►T-DNA

Ori_V: Refers to the origin of vegetative replication that is used during cell proliferation of bacteria.

Ornithine ($NH_2[CH_2]_3CH[NH_2]COOH$): A non-essential amino acid for mammals. It is synthesized through the urea cycle. ►urea cycle

Ornithine Aminotransferase Deficiency (hyperornithinemia, OAT): Ornithine is derived either from arginine or *N*-acetyl glutamic semialdehyde or glutamic semialdehyde, and carbamoylphosphate synthetase converts it to citrulline. The decarboxylation of ornithine (a pyridoxalphosphate [vitamin B_6] requiring reaction) yields polyamines such as spermine and spermidine. OAT deficiency causes ornithinemia, 10–20 times increase of ornithine in blood plasma, urine and other body fluids. It also leads to the degeneration of eye tissue and tunnel vision and night blindness by late childhood. Restricted ornithine diet may prevent eye defects due to deficiency of the γ-chain. The OAT locus (21-kb, 11 exons) is in human chromosome 10q23qter and its mutation blocks the metabolic path between pyrroline-5-carboxylate and ornithine, Pseudogenes are at Xp11.3-p11.23 and at Xp11.22-p11.21. ►amino acid metabolism, ►urea cycle, ►ornithine transcarbamylase deficiency, ►ornithine decarboxylase, ►ornithine transcarbamylase deficiency, ►pseudogene, ►hyperornithinemia, ►nicotine

Ornithine Decarboxylase (ODC): The dominant allele (human chromosome 2p25, mouse chromosome 12, in *E. coli* 63 min) controls the ornithine→putrescine reaction and its activity is very sensitive to hormone levels. There is a second ODC locus in human chromosome 7 but the latter has a reduced function. Elevated levels of ODC may be indicative of skin tumorigenesis. ►amino acid metabolism, ►ornithine aminotransferase deficiency, ►antizyme

Ornithine Transcarbamylase Deficiency (OTC): This X-linked (Xp21.1) enzyme normally expressed primarily in the liver mitochondria, catalyzes the transfer of a carbamoyl group from carbamoyl phosphate to citrulline and ornithine is made while inorganic phosphate is released (see Fig. O22). The defect may lead to an accumulation of ammonia (hyperammonemia) because carbamoyl phosphate is generated from NH^{4+} and HCO_{3-} (in the presence of 2 ATP). The high level of ammonia may cause emotional problems, irritability, lethargy, periodic vomiting, protein avoidance and other anomalies. Na-benzoate, Na-phenylacetate and arginine may medicate the plasma ammonium level. The incidence of OTC deficiency in Japan was found to be about 1.3×10^{-3}. In mice an OTC deficiency mutation is responsible for the *spf* (sparse fur) phenotype. This single gene metabolic defect may be corrected by somatic gene therapy. Unfortunately, one of the initial attempts using a very high dose of adenoviral vector (6×10^{14} particles/kg) ended in an unexpected fatal immune reaction in Jesse Gelsinger in 1999. ►amino acid metabolism, ►ornithine aminotransferase, ►ornithine decarboxylase, ►channeling, ►hyperammonemia

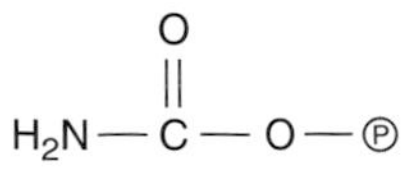

$$H_2N-C(=O)-O-Ⓟ$$

Figure O22. Carbamoyl phosphate

Ornithine Transporter: ►urea cycle

Orofacial–Digital Syndrome (OFD): OFD I is a dominant Xp22-linked or a recessive autosomal syndrome. The dominant form is lethal in males. OFD I is characterized by loss of hearing and polydactyly of the great toe, mental retardation, face and skull malformation, cleft palate, brachydactyly (short fingers and toes), defective kidney cysts, etc. (Ferrante MI et al 2006 Nature Genet 38:112). OFD II (Mohr syndrome) is characterized by polysyndactyly, brachydactyly [short digits] and lobate tongue. OFD III and OFD IV include mental retardation, eye defects, teeth anomalies, incomplete cleft palate, hexadactyly, hunchback features, etc. OFD IV is distinguished on the basis of tibial dysplasia (shinbone defects). OFD V (Váradi–Papp syndrome) also includes a nodule on the tongue and neural defects, etc. ►polydactyly, ►limb defects, ►mental retardation, ►Majewski syndrome; Ferrante MI et al 2001 Am J Hum Genet 58:569.

Orosomucoid: There are two orosomucoid (serum glycoprotein) genes in humans, ORM1 and ORM2 in 9q31-q32. (See Yuasa I et al 2001 J Hum Genet 46(10):572).

Orotic Acid:

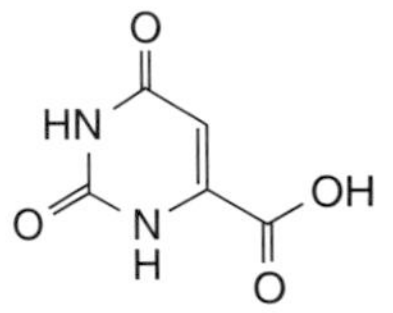

Figure O23. Orotic acid monohydrate

Oroticaciduria: This is a recessive deficiency of the enzyme orotidylate decarboxylase, encoded at human chromosome 3q13 or a deficiency of orotate phosphoribosyl transferase. Rare human diseases have their counterparts in other higher eukaryotes. The clinical symptoms include anemia with large immature erythrocytes and urinary excretion of orotic acid (see Fig. O24). The administration of uridylate and cytidylate alleviates the symptoms. Homologous mutations have been detected in cattle. ►orotic acid

Precursor in the de novo pyrimidine synthesis: N-CARBAMOYLASPARTATE $\xrightarrow{1}$ L-DIHYDROORORATE $\xrightarrow{2}$ OROTATE $\xrightarrow{3}$ OROTIDYLATE $\xrightarrow{4}$ URIDYLATE $\xrightarrow{5}$ URIDYLATE (UTP) $\xrightarrow{6}$ CYTIDYLATE (CTP). The enzymes mediating the reactions numbered are: (1) ASPARTATE TRANSCARBAMOYLASE, (2) DIHYDROOROTASE, (3) DIHYDROOROTATE DEHYDROGENASE, (4) OROTATE PHOSPHORIBOSYLTRANSFERASE, (5) OROTIDYLATE DECARBOXYLASE, (6) LINASES, (7) CYTIDYLATE SYNTHETASE.
► oroticaciduria

Figure O24. Orotate in the pyrimidine pathway

Orphan Genes: These are open reading frames in yeast (about 1/3 of the ORFs) that cannot be associated with known functions (FUN genes) and do not seem to have homologs in other organisms. With advancing information their number is expected to decline. ►ORF, ►*Saccharomyces*, ►fun genes; Enmark E Gustafsson JA 1996 Mol Endocrinol 10:1293; Brenner S 1999 Trends Genet 15:132.

Orphan Receptor: This has either no known ligand or it has the ligand-binding domain but lacks the conserved DNA binding site. ►nuclear receptors, ►receptor; Xie W Evans RM 2001 J Biol Chem 276:37739.

Orphanin: ►opiate

Orphon: This is a former member of a multigene family, but at a site separate from the cluster, and contains one coding region or they are pseudogenes. ►gene family, ►pseudogene

Orthogenesis: Evolution follows a straight line toward a goal or proceeds along a predetermined path rather than being influenced by Darwinian selection.

Orthogonal-Field Alternation Gel Electrophoresis: ►OFAGE

Orthogonal Functions: These can be employed for comparisons between observed data which have not affected each other.

Orthogonal mRNAs: The suppressor tRNA is not a substrate for endogenous aminoacyl-tRNA synthetases, and the mutant aminoacyl-tRNA synthetase aminoacylates only the suppressor tRNA with an unnatural amino acid but no other tRNA. Such a system facilitates the incorporation of unnatural amino acids into peptides. ►unnatural amino acids

Orthologous Loci: These genes are in the direct line of evolutionary descent from an ancestral locus, e.g., 1,000 genes in *Bacillus subtilis* are similar to those in *Escherichia coli*, indicating the common ancestry. Among the eukaryotes, *Saccharomyces cerevisiae* and *Caenorhabditis elegans* ~57% of the protein pairs of the two organisms are represented by just one from each of these organisms. Intermediary metabolism (in 28%), DNA and RNA synthesis and function (in 18%), protein folding and degradation (in 13%), transport and secretion (in 11%) and signal transduction (in 11%) are carried out by genes preserved through evolution from a common origin in these two eukaryotes. ►paralogous loci, ►duplications, ►evolution of proteins, ►non-orthologous gene displacement, ►xenology, ►homolog, ►orthologous proteins, ►genome projects, ►comparative genomics, ►constrained elements; Koonin EV 2005 Annu Rev Genet 39:309; http://www.ncbi.nlm.nih.gov/COG; orthologous groups of 17 eukaryotes: http://inparanoid.cgb.ki.se/; human–mouse orthologous miRNA gene homologies ftp://ftp.ncbi.nih.gov/pub/HomoloGene/; http://www.informatics.jax.org/searches/homology_form.shtml; orthologs in 9 eukaryotes: http://www.sanger.ac.uk/PostGenomics/S_pombe/YOGY; ortholog server: http://oxytricha.princeton.edu/BlastO/.

Orthology: Refers to common ancestry in evolution. ►orthologous loci, ►paralogy

Orthostitchies: ►phyllotaxis

Orthotopic: This means in normal position. ►ectopic

Oscillator: This molecular mechanism generates the circadian rhythm, heart, neuronal and other cellular functions. The bursts of cell divisions oscillate between pauses and transcription of genes within the entire genome displays temporal clusters. In baker's yeast cell suspensions with billions of cells show synchronized metabolic oscillations in NADH fluorescence with a period of duration around one half minute (see Fig. O25). Acetaldehyde and glucose (if the glucose transporters are open) are oscillators. The oscillation indicates that the cells are in communication (Danø S et al 2007 Proc Natl Acad Sci USA 104:12732). ►circadian rhythm, ►quantized, ►repressilator, ►entrainment, ►gene-switch cassette, ►gene circuits; Goldbeter A 2002 Nature [Lond] 420:238; Klevecz RR et al 2004 Proc Natl Acad Sci USA 101:1200, Farré EM et al 2005 Current Biol 15:47; Fung E et al 2005 Nature [Lond] 435:118.

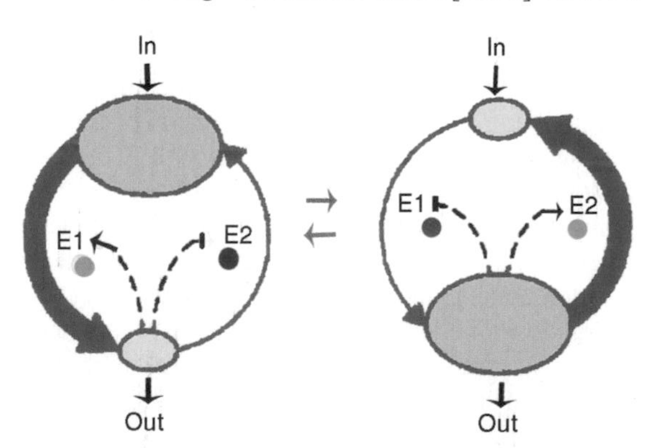

Figure O25. Schematic representation of metabolic oscillations. Enzymes E1 and E2, respectively turned on (arrow) or off (bar). The substrates and products are represented by shaded ovals

Oscillin: This is a protein formed in the egg after penetration by the sperm, and it is involved with the oscillation of the level of Ca^{2+}. It is apparently involved in triggering the early development of the embryo. (See Nakamura Y et al 2000 Genomics 68(2):179).

Osha: This is the acronym for Occupational Safety and Health Administration which is based in the USA. It provides information about and protects against potential hazards involved in laboratory and industrial operations.

Oskar (*osk*, 3–48): *Drosophila* homozygous females and males are normally viable and fertile. The embryos produced by the homozygous females lack pole plasm, the abdominal segments do not form and the embryos die (see Fig. O26). The *osk* mRNA appears in the posterior of the embryo. Kinesin I facilitates the transport of *oskar* and *Staufen* mRNAs along the microtubules and the Mago Nashi (Kataoka N et al 2001 EMBO J 20:6424) and Y14 proteins (Hachet O, Ephrussi A 2001 Curr Biol 11:1666) are involved in the splicing of the pre-mRNA (Palacios IM 2002 Curr Biol 12:R50). Translation initiation factor eIF4E-binding protein Cup interacts with silencing protein Bruno and inhibits translation. The large (50S-80S) silencing particle involving the oligomerized mRNA cannot be accessed by the ribosomes although it may be advantageous for mRNA transport (Chekuleava M et al 2006 Cell 124:521). ►eIF-4E

Figure O26. Oskar

Osler-Rendu-Weber Syndrome: ►telangiectasia hereditary hemorrhagic

Osmiophilic: This readily stains by osmium or osmium tetroxide.

Osmium Tetroxide (OsO_4): This is a fixative for electron microscopic specimens. ►fixatives, ►microscopy

Osmolarity: Refers to the osmotic pressure of a solution as the molarity of dissociated particles.

Osmosis: This is a process by which water passes through a semipermeable membrane toward another compartment (cell) where the concentration of solutes is higher. The osmo-sensing signal seems to be mediated by the Jnk protein kinase, a member of the mitogen-activated large family of proteins in mammals. In bacteria, a histidine kinase sensor (EnvZ) and a transcriptional regulator (OmpR) are involved. In yeast a similar mechanism is implicated, also involving HOG-1. ►HOG-1, ►histidine kinase, ►HSP

Osmotic Pressure: In two compartments separated by a semipermeable membrane the solvent molecules pass toward the higher concentration of solutes. This flow may be prevented by applying high pressure to the compartment toward which the flow is directed. Osmotic stress may regulate the expression of genes as a homeostatic control. (See Xiong L et al 2002 J Biol Chem 277:8588)

Osmotin: Refers to protein in plants that may accumulate at high concentrations of salts. It is a homolog of mammalian adiponectin, which regulates lipid and sugar metabolism. ►adiponectin; Narasimhan M L et al 2005 Mol Cell 17:171.

O-Some: A nucleosome-like structure of lambda phage involved in DNA replication. It is formed from dimeric λO (lambda origin) protein-coding sequences and four inverted repeats. During the beginning of replication the dimer of λP (λ promoter) and a hexamer of DnaB forms $ori\lambda{\sim}\lambda O{\sim}(\lambda P{\sim}DnaB)_2$ which gives rise to the pre-primosomal complex. ►lambda phage, ►nucleosome, ►DNA replication prokaryotes, ►primosome; Zylicz M et al 1998 Proc Natl Acad Sci USA 95:15259.

OSP: ►allele-specific probe for mutation

OSP94: ►HSP

Ossification of the Longitudinal Filaments of the Spine: This common disorder in humans (over 3% above the age of 50) is caused by ectopic bony development of the spinal muscles. In mice, defects in the nucleotide pyrophosphatase (*Npps*, chromosome 10), which regulates soft tissue calcification and bone mineralization by producing inorganic phosphate, seem to be a major factor.

OST: This is an open reading frame tagged by EST. ►EST, ►Gateway cloning

Osteoarthritis: A common human affliction involving degeneration of the cartilage, caused by the replacement of the [α1(II)] collagen by αII chains with reduced glycosylation. Several gene loci at 4q26, 7p15, 2q12, 11q and others are involved. More than 70% of the US population over 65 years of age may be afflicted to varying degrees. GDF5 (growth and differentiation factor) or CDMP1 (cartilage-derived morphogenetic protein 1) is a member of the transforming growth factor (TGF) superfamily and is closely related to bone morphogenetic proteins (BMPs). GDF5 is expressed in the regions of future joints during early development, and mutations in both mouse and human *GDF5* cause abnormal joint development and is a factor in the development of osteoarthritis (Myamoto Y et al 2007 Nature Genet 39:529). ►collagen, ►arthritis; Stefánson SE et al 2003 Am J Hum Genet 72:1448.

Osteoblast: This is a cell involved in bone production from mesenchymal progenitor and is positively controlled in the mouse mainly by genes CBFA-1. Gene Hoxa-2 and leptin are negative regulators. Osteoblast and skeletal differentiation are mediated by Runx2 and ATF4 transcription factors, controlled by SATB2 (Dobreva G et al 2006 Cell 125: 971). ►osteoclast, ►ATF, ►RUNX, ►mesenchyma, ►ITAM; Olsen BR et al 2000 Annu Rev Cell Dev Biol 16:191; Science 2000 Vol. 289:1501–1514.

Osteocalcin (1q25-q31, three γ-carboxyglutamic acid, 5.9 kDa): This is an abundant, non-collagen type protein in the bones dependent on Ca^{2+}. It is involved in blood coagulation. In Neanderthal relics osteocalcin residues have been found which are similar to that observed in modern humans and modern primates (Nielsen-Marsh CM et al 2005 Proc Natl Acad Sci USA 102:4409).

Osteochondromatosis (dyschondroplasia, osteochondrodysplasia): An inhomogeneous group of autosomal dominant bone and cartilage defects encoded at 8q24 and 11p11-p12 and possibly at other sites. ►spondyloepimetaphyseal dysplasia, ►spondyloepiphyseal dysplasia, ►Dyggve-Melchior-Clausen dysplasia

Osteochondroplasias (osteochondrodysplasias): These cartilage disorders are caused by defects in different collagen genes. ►collagen, ►chondrodysplasias

Osteoclastogenesis Inhibitory Factor (OCIF): ►osteoprotegrin

Osteoclasts: These multinucleate cells mediate bone resorption. The osteoclast differentiation factor (ODF) receptor is TRANCE/RANK. The ITAM motif plus RANK ligand and macrophage colony stimulating factor are required for maintenance between bone formation and resorption (Koga T et al 2004 Nature [Lond] 428:758). ►Paget disease, ►TRANCE, ►osteoblast, ►bone morphogenetic protein, ►RANK, ►ITAM, ►macrophage colony stimulating factor, ►osteoporosis; Karsenty G 1999 Genes & Development 13:3037; Väänänen KH et al 2000 J Cell Sci 113:377; Motickova G et al 2001 Proc Natl Acad Sci USA 98:5798; Feder ME, Mitchell-Olds T 2003 Nature Rev Genet 4:649.

Osteogenesis Imperfecta (OI, 11q12-q13, [neonatal lethal] 17q21.3, [perinatal] 7q22.1): These are rare (prevalence 1/5,000 – 1/10,000) autosomal dominant or recessive disorders of collagen, resulting in abnormal bone formation due to mutation at numerous loci (ca. 17) that replace, e.g., a glycine residue by a cysteine. Such a substitution disrupts the Gly - X - Pro tripeptide repeats in collagen that assures the helical structure of the molecules. Type I (chromosome 17) is characterized by postnatal bone fragility, blue coloring of the eye white (sclera) and ear problems. Type II (Vrolik type) generally involves dominant negative lethality at about birth time due to a variety of bone defects. Most of the cases are considered new dominant mutations. Type III may lead to prenatal bone deformities and the survivors are crippled. The symptoms may also include hearing loss. This may be either recessive or dominant. Type IV is characterized by a milder expression (17q21.3) of the similar symptoms than the other classes and may not prevent survival. The four types have defects in collagen genes COL1A1 or COL1A2. OI types V – VII have no mutations in collagen; the *fragilitas osseum* mutation of mouse leads to severe osteogenesis and dentinogenesis imperfecta because of deletion of sphingomyelin phosphodiesterase 3 (Aubin I et al 2005 Nature Genet 37:803). Cartilage associated protein (CRTAP) is required for prolyl 3-hydroxylation and its mutation causes OI (Morello R et al 2006 Cell 127:291). ►collagen, ►sphingomyelin, ►dentinogenesis imperfecta, ►hydrocephalus, ►mitral prolapse, ►connective tissue disorders, ►dominant negative; Millington-Ward S et al 2002 Hum Mol Genet 11:2201.

Osteolysis: This group of diseases is characterized by bone resorption and osteoporosis controlled by several different loci. ►bone diseases

Osteolysis, Familial Expansile: ►Paget disease

Osteomalacia: ►hypophosphatemia; Feng JQ et al 2006 Nature Genet 38:1310.

Osteonectin (5q31.3-q32): This is a glycine-rich protein involved in bone proliferation, formation of the extracellular cell matrix, repair, morphogenesis, etc.

Osteopenia: This condition is characterized by reduced bone mass. It may be due to defects in zinc transporter, ZNT5 resulting in defective maturation of osteoblasts to osteocytes (Inoue K et al 2002 Hum Mol Genet 11:1775)

Osteopetrosis: This condition is manifested in different forms and is controlled by autosomal dominant or recessive genes in humans. The recessive form at 11q13 may be due to defects in a 3-subunit osteoclast-specific proton pump encoded by ATP6i. Osteopetrosis (increased bone mass) with renal tubular acidosis is based on deficiencies in the carbonic anhydrase B or CA2 enzyme. The disease makes the calcified bones brittle. It generally involves mental retardation, visual problems, reduced growth, elevated serum acid phosphatase levels, and higher pH of the urine because of the excretion of bicarbonates and less acids in the urine. In the recessive Albers-Schönberg disease (16p13.3, 1p21), osteopetrosis causes reduction in the size of the head, deafness, blindness, increased liver size and anemia. Defects in

O

a colony stimulating factor (CSF) have been inferred from mouse experiments. A mild form of osteopetrosis is also known. A severe lethal form affects the fetus by the 24th week and may cause stillbirth. In mice and humans the loss of chloride ion channel (CIC-7, 16p13.3) leads to dominant osteopetrosis, i.e., Albers-Schönberg disease. ►carbonic anhydrase, ►osteoporosis, ►TRANCE, ►colony stimulating factor, ►ion channels, ►cathepsins; Lazner F et al 1999 Hum Mol Genet 8:1839; Kornak U et al 2001 Cell 104:205; Bénichou O et al 2001 Am J Hum Genet 69:647; Sobachi C et al 2001 Hum Mol Genet 10:1767; Lange PF et al 2006 Nature [Lond] 440:220.

Osteopoikilosis (BOS, Buschke–Ollendorf syndrome, 12q14): Refers to dominant, spotty (sclerotic) areas at the ends of bones and skin, as well as connective tissue lesions. (See Hellemans J et al 2004 Nature Genet 36:1213)

Osteopontin: This glycoprotein is produced by osteoblasts and is encoded in human chromosome 4q21-q25 at the SPP1 locus. Its expression is repressed by Hoxc-8 but Smad1 prevents Hoxc-8 from binding to the osteopontin promoter and facilitates its transcription. Osteopontin is also called Eta-1 (early T lymphocyte activation factor 1) because it is a ligand of CD44 and plays a role in cell-mediated immunity. Its deficiency severely impairs the immunity of mice to Herpes simplex virus 1 and bacterial (*Listeria monocytogenes*) infection. Osteopontin deficiency interacting with CD44 reduces the expression of IL-10. A phosphorylation-dependent interaction of Eta-1 and its integrin receptor stimulates IL-12. Osteopontin normally affects cell migration, calcification, immunity and tumor cell phenotype. Milk production in cattle is affected by six quantitative trait locus (QTL) scattered in the genome. BTA6 QTL controls the milk protein level. The osteopontin gene aids in the characterization of the neighboring QTL (Schnabel RD et al 2005 Proc Natl Acad Sci USA 102:6896). ►osteoblast, ►homeotic genes, ►Smad, ►integrin, ►CD44, ►L-10, ►IL-12, ►Herpes, ►QTL; Agnihotri R et al 2001 J Biol Chem 276:28261.

Osteoporosis (17q21.31-q22.05, 7q22.1, 7q21.3): This condition is characterized by abnormal thinning of the bone structure because of reduced activity of the osteoblasts to manufacture bone matrix from calcium and phosphates. The frequency of fracture of the vertebrae may increase up to ~30 times in postmenopausal women. Steroids and other hormones and vitamin D can regulate this condition. The autosomal recessive juvenile form (IJO) may be caused by defects in bone formation and the symptoms may be alleviated spontaneously by adolescence or may be treated successfully with steroid hormones. Another autosomal recessive form in infancy involves eye defects (pseudoglioma) and possibly problems of the nervous system. The collagen genes COLIA1 and COLIA2 have been implicated. An extracellular non-collagen proteoglycan (biglycan) deficiency in mice may lead to an osteoporosis-like phenotype. A vegetable rich diet may retard bone resorption. Osteoporosis is caused by bone resorption mediated through the osteoclasts. Inactivation of the cannabinoid type 1 receptors results in increased bone mass and protects ovariectomized rodents from bone loss. Also, cannabinoid receptor antagonists protect against bone loss after ovariectomy (Idris AI et al 2005 Nature Med 11:774). Osteoclasts mediating bone resorption in osteoporosis when inhibited by the antibiotic reveromycin A (see Fig. O27) caused apoptosis of these cells and thus attenuated osteoporosis (Woo J-T et al 2006 Proc Natl Acad Sci USA 103:4729). ►aging, ►LDL, ►hormone-receptor elements, ►estradiol, ►Cbl, ►Src, ►collagen, ►ABL oncogene, ►osteoblast, ►osteoclasts, ►apoptosis, ►ovariectomy, ►TNF, ►Paget disease

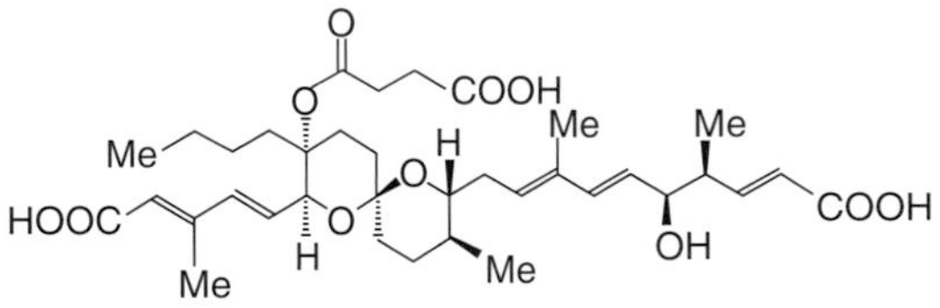

Figure O27. Reveromycin A (from Woo, JT et al 2006)

Osteoprotegrin (osteoclastogenesis inhibitory factor, OPG/OCIF): This is a receptor of the TNF family of proteins. The ~29-kb human gene encodes an inhibitor of TRAIL and other TNF ligands. It regulates bone formation and the mutant form may lead to the early onset of osteoporosis and arterial calcification, i.e., calcium deposits in the veins. ►TNF, ►TRAIL, ►osteoporosis; Bucay N et al 1998 Genes Dev 12:1260.

Osteosarcoma: This autosomal recessive, usually malignant bone tumor has an onset in early adulthood. It is normally part of a complex disease. People affected by bilateral (but not those afflicted by unilateral) retinoblastoma are at a high risk to develop bone cancer. Circulating monocytes mediate the inflammation in osteosarcoma. These cells can be used as reporters for the disease by the genetic markers carried (Patino WD et al 2005 Proc Natl Acad Sci USA 102:3423). ►retinoblastoma, ►sarcoma

Osteosclerosis: This condition is characterized by abnormal hardening of the bones. ►cathepsins, ►osteopetrosis

Ostiole (ostium): This is a small (mouth-like) opening on various structures such as fungal fruiting bodies or internal organs.

Ostrich (*Struthio camelus*): The zoological name reflects the ancient belief that this huge bird descended from a misalliance between the sparrow and the camel. It differs from other birds because it has only two toes. Also, while other birds copulate by bringing together their cloacas (the combined opening for urine, faeces and reproductive cells), the male ostrich everts a penis-like structure from the cloaca during the process.

OTC: The expanded form of this acronym is over-the-counter drug, i.e., available without prescription.

OTF-1, OTF-2: These binding proteins recognize the consensus octamer 5′-ATGCAAAT-3′ and facilitate transcription in several eukaryotic genes. OTF-1 is identical to NF-3 replication factor of adenovirus. ►NF, ►binding proteins

Otopalatodigital Syndrome (OPD): This condition involves human chromosome Xq28, semidominant, variable expression deafness, cleft palate, broad thumbs and big toes. ►cranioorodigital syndrome, ►deafness, ►cleft palate

Otosclerosis: This is an autosomal dominant hardening of the bony labyrinth of the ear resulting in lack of mobility of the structures and thus conductive hearing defect. It may be a progressive disease starting in childhood and fully expressed in adults. The incidence of otosclerosis leading to loss of hearing is about 0.003 among US whites and among blacks its frequency is of a lower magnitude. OTSC1 is at 15q25-q26 and OTSC2 at 7q34-q36. ►deafness

Otter: *Amblonyx cinerea*, 2n = 38; *Enhydra lutris*, 2n = 38 (see Fig. O28).

Figure O28. Otter

Otto: This is a gene predictor program based on integrated multiple evidence such as homology, EST, mRNA, and RefSeq. ►EST, ►mRNA, ►RefSeq

OTU (operational taxonomic unit): Refers to an entity such as a particular population or a species which is used in evolutionary tree construction. ►evolutionary tree, ►character matrix

Ötzi: ►ice man (named after the Ötzthal where the mummy was found)

Ouabain (3[{6-deoxy-α-L-mannopyranosyl}oxy]1,5,11α, 11,19-pentahydroxy-card-20(22)-enolide): A cardiotonic steroid capable of blocking potassium and sodium transport through the cell membranes; a selective agent for animal cell cultures. Ouabain regulates cell adhesion, differentiation, migration and metastasis through hormone-like action (Larre I et al 2006 Proc Natl Acad Sci USA 103:10911). ►steroids; Aizman O et al 2001 Proc Natl Acad Sci USA 98:13420.

Ouchterlony Assay: ►antibody detection

Outbreeding: This is the opposite of inbreeding; mating is between unrelated individuals (crossbreeding, allogamy). ►allogamy, ►autogamy, ►inbreeding, ►protogyny, ►protandry

Outbreeding Depression: Co-adapted and different subpopulations can no longer interbreed successfully because of divergence. ►speciation, ►co-adapted genes

Outcome Space: Denotes all possible genotypic constitutions in a specific pedigree compatible with Mendelian inheritance and parental genotypes.

Outcrossing: Refers to pollination of an autogamous plant by a different individual or strain or mating between animals of different genetic constitution. ►protandry, ►protogyny, ►incross

Outgroup: This means that at least two species can be used to distinguish an ancestral from a derived species and for the rooting of phylogenetic trees. Characters (physical or molecular) shared between the ingroup and the outgroup are considered to be ancestral. ►evolutionary tree, ►PAUP

Outlier: Refers to a gene or genes which is/are more divergent from the typical members of a gene family than expected. These represent highly specialized function(s). In statistical treatment of data, the extreme values that appear inconsistent with the bulk of the data are called outliers. ►gene family

Out-of-Africa: A hypothesis proposes that the origins of the human race are in Africa. The human race diverged from chimpanzee-like ancestors 4.5 million years ago according to DNA information (Takahata N, Satta Y 1977 Proc Natl Acad Sci USA 94:4811). The first human fossil records date back to ~3.6 million years and 2.5 million years ago humans were found only in Africa (Stringer C 2003 Philos Trans Roy Soc London

B 357:563). The African origin thesis is supported by mtDNA composition showing that more differences exist in this respect in Africa than other parts of the world, supposedly because a few founders (highly conserved mtDNA) emigrated, spread and evolved relatively recently into the majority of the existing human racial groups. However, insufficient number and quality of archeological and paleontological relics plague the current concepts of hominine dispersion. It is also difficult to distinguish between the various early species such as *Homo erectus* (*Pithecanthropus erectus*) and *Homo ergaster* and the many other primitive taxonomic categories (Dennell R, Roebroeks W 2005 Nature [Lond] 438:1099).

At the niche of origin longer evolutionary period permitted greater divergence. This hypothesis has been further substantiated by linkage disequilibrium between an Alu deletion at the CD4 locus in chromosome 12 and short tandem repeat polymorphisms (STRP) used as nuclear chromosomal markers. The two markers are separated by only 9.8 kb. The mapping data indicate that 1 cM of the human genome corresponds to about 800 kb. The Alu deletion was mainly associated with a single STRP in Northeast Africa and non-African populations sampled from 1,600 individuals from European, Asian, Pacific and the Amerindian groups. In contrast in the sub-Saharan Africa a wide range of STRP markers were with the Alu deletion. These data also indicate that migration from Africa took place relatively recently, i.e., an estimated 102,000 to 313,000 years ago. This estimate is in relatively close agreement with the age of the first human fossil records in the Middle East (90,000–120,000 years old). Other information has indicated that dispersion from Africa began about 1.7 million years ago (Gabunia L et al 2000 Science 288:1019). Recent information based primarily on mitochondrial DNA evidence of the oldest isolated human populations, e.g., the Andaman people in the island between India and Burma and the Orang Asli people of Malaysia, has revealed other possible routes of migration as shown on the map (see Forster P Matasumura S 2005 Science 308:965). *Homo sapiens* diverged from *H. erectus* around 800,000 or more years ago. The European continent was settled from the Levant nearly 40,000 years ago in several waves during the paleolithic, mesolithic and neolithic eras. In 2005, Paleolithic flint artifacts dated to about 750,000 (700 kyr) years ago were found in the Suffolk area of England, indicating very old human activity in northern Europe (Parfitt SA et al 2005 Nature [Lond] 438:1008). An analysis of the mtDNA of Paleolithic central European human remains (7500 years old) revealed that mtDNA sequences that were present in 25% of the ancient populations were present in only 0.2% of the present populations (see Fig. O29). This finding suggests that the Paleolithic people made a minimal contribution to the current gene pool (Haak W et al 2005 Science 310:1016). More recent analysis of mtDNA suggests that the Levant area was populated by a migration from Southeast Asia toward the Levant and from the

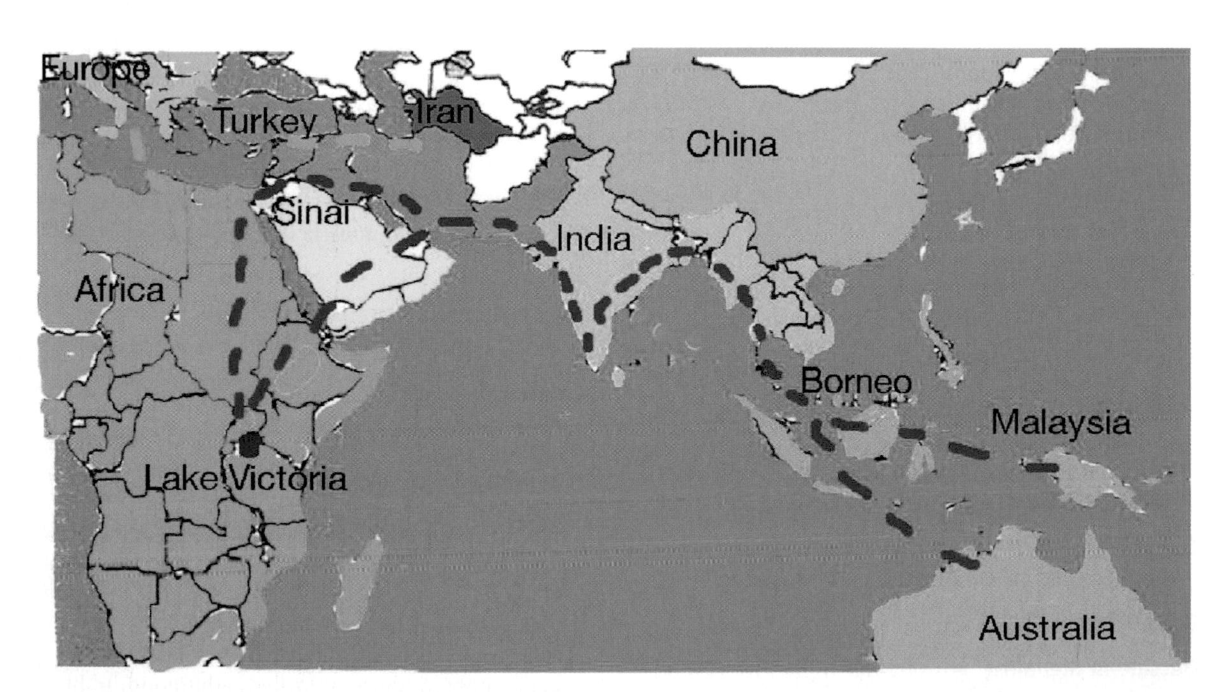

Figure O29. One of the most plausible routes (red path) of human migration from Africa about 85,000 years ago, based on mtDNA and archeological evidence. From Asia Minor and North Africa migration to Europe (green path) also took place

Levant they re-entered North Africa (Olivieri A et al 2006 Science 314:1767). Near the Don River of Russia 45,000–42,000 years old human remains were unearthed (Anikovich MV et al 2007 Science 315:223). In South Africa 36,000–33,000 years old human skulls were found (Grine FE et al 2007 Science 315:226).

The subsequent waves of migration were expected to replace the preceding ones (*replacement theory*). Some anthropologists, however, do not accept this theory of human evolution and have suggested *multiregional origin* of modern humans. An analysis of the human Y chromosome lends support not only to the African origin thesis of male descent, but also to the same general region as proposed by the Eve foremother theory in the case of females. Some genetic variations exist among the major continents and –to a smaller extent – within continents. If appropriate and sufficiently large number of loci are selected, the majority of human groups exhibit differences, especially in regions of limited migration. Even such small differences may be of significance from a medical point of view and for social policy (e.g., lactose intolerance and school lunch programs). Some of the European populations migrated to that continent from Asia as the gradient of gene flow from the Mid-East toward Northern and Western Europe decreased. Patterns of gene differences among the groups indicate population expansion and demography (Harpending HC et al 1998 Proc Natl Acad Sci USA 95:1961). The effective population sizes of males appeared to be lower than those of females indicating polygyny and elimination of males but not females during wars and conquests (Dupaloup I et al 2003 J Mol Evol 57:85). Such studies mandate appropriate case controls in stratification. ►mtDNA, ►Eve foremother of mitochondrial DNA, ►Y chromosome, ►founder principle, ►Garden of Eden, ►CD4, ►Alu, ►hominids, ►ancient DNA, ►demography, ►case control, ►stratification, ►ethnicity, ►radiocarbon dating; Quintana-Murci L et al 1999 Nature Genet 23:437; Harpending H Rogers A 2000 Annu Rev Genomics Hum Genet 1:3611; Barbujani G Bertorelle G 2001 Proc Natl Acad Sci USA 98:22; Adcock GJ et al 2001 Proc Natl Acad Sci USA 98:537; Templeton AR 2002 Nature [Lond] 416:45; Tishkoff SA Williams SM 2002 Nature Rev Genet 3:611; Barbujani G Goldstein DB 2004 Annu Rev Genomics Hum Genet 5:119.

Outparalog: This means evolved through ancient duplications. ►paralogous loci, ►inparalog

Output Trait: ►input trait

Outside-in Signaling: ►inside-out signaling

Ovalbumin: This is a nutritive protein of the egg white of chickens ($M_r \approx 4{,}500$); each molecule carries a carbohydrate chain. The gene is split by 7 introns into 8 exons and its transcription is induced by estrogen or progesterone. In the presence of the effector hormone the mRNA has a half-life of about a day but in the absence of it, its half-life may be reduced to 20%. The gene is part of a family of 3. ►oviduct, ►intron

Ovalocytosis: ►elliptocytosis, ►acanthocytosis

Ovarian Cancer: This is probably due to mutation of the OVC oncogene (9p24). In 2004 it accounted for approximately 26,000 new cases and around 16,000 deaths in the USA. The expression of several other genes is altered. The same genes affect a large variety of other cancers too. Ovarian cancer may be associated with breast cancer and may increase following the administration of tamoxifen medication. Mutation in the so-called ovarian cancer cluster region (OCCR, nucleotides 3059–4075 and 6503–6629) of the BRCA2 increases the relative risk of ovarian cancer by 0.46–0.84. Ovarian germ cell cancer involves the germ cells in the ovaries and ~5% of the cases are due to genetic causes. OPCML (opioid binding protein/cell adhesion-like molecule, encoded at 11q25) appears to be a tumor suppressor of ovarian cancer (Sellar GC et al 2003 Nature Genet 34:337). Hormone (estradiol family) replacement therapy of menopausal women may increase the risk. It has been suggested that the use of oral contraceptives may lower the risk. Apparently microarray hybridization profile provides an early diagnosis. Bombesin, somatostatin and luteinizing hormone releasing hormone coupled with 2-pyrrolino-doxorubicin can be targeted to the receptors of these substances in the ovary and may block ovarian cancer (Buchholz S et al 2006 Proc Natl Acad Sci USA 103:1043). ►breast cancer, ►tamoxifen, ►sex hormones, ►mesothelin, ►bombesin, ►somatostatin LRF, ►doxorubicin; Welsh JB et al 2001 Proc Natl Acad Sci USA 98:1176; Thompson D, Easton D (Breast Cancer Linkage Consortium) 2001 Am J Hum Genet 68:410; Cannistra SA 2004 New Engl J Med 351:2519.

Ovariectomy (oophorectomy): Surgical removal of the ovary. ►castration, ►spaying, ►neutering

Ovariole: ►ovary

Ovary (ovarium): This contains the ovules of plants, the egg-producing organ of animals, and a female gonad. The ovaries of *Drosophila* contain *ovarioles*, each with 2–3 germ line stem cells at the tip and associated with the *terminal filament cells* with the duty of somatic signaling (see Fig. O30). In the *Caenorhabditis* asymmetric differentiation of the germ cell lineage is not observed rather the distal mitotic stem cells undergo meiosis through adulthood. In humans the pre-implantation development is mediated though the brain-derived neurotrophic factors (BDNF) through TrkB and some low-affinity receptors.

BDNF enhances exclusion of the first polar body and oocytes development (Kawamura K et al 2005 Proc Natl Acad Sci USA 102:9206). Ovary transplantation (in the case of developmental failure) from one monozygotic twin to the other can be successful and can lead to the restoration of fertility (Silber SJ et al 2005 N Engl J Med 353:58). ►gonad, ►uterus, ►gametogenesis, ►germarium, ►pole cells, ►germ line, ►cytoblast, ►ovary genes, ►functions, ►mutations, ►TRK, ►BDNF, ►polar body, ►twinning; biomedical relevance: http://ovary.stanford.edu.

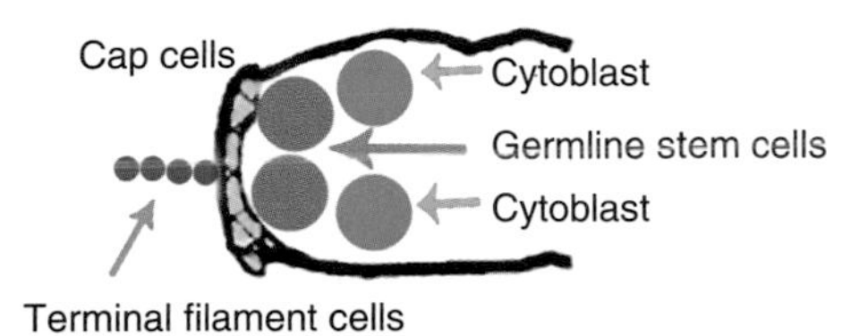

Figure O30. Ovary of Drosophila

OVC Oncogene: This was discovered in an ovarian cancer in a fusion of human chromosomes 8–9. ►oncogenes

Overdispersion: The empirically observed variance of the data exceeds expectation for a certain model, e.g., the variation is not compatible with the binomial distribution or with the Poisson distribution because of the numbers of outliers encountered or specifications used. ►binomial distribution, ►Poisson distribution, ►outlier

Overdominance: The *Aa* heterozygote surpasses both *AA* and *aa* homozygotes. Deleterious recessive genes may occur in heterozygotes at a higher frequency (at linkage disequilibrium) than expected because of the association with neutral genes; this is called *associative overdominance* (see Fig. O31).

Figure O31. Allelic complementation (= overdominance) of temperature-sensitive pyrimidine mutants of *Arabidopsis*. Top and bottom rows are homozygous mutants, in the middle row: the F_1 hybrids. (From Li, S.L. & Rédei, G.P)

Pseudo-overdominance occurs when two genes are closely linked and the aB and Ab combinations simulate overdominance whereas the increased performance is actually due to dominance. Polar overdominance characterizes the non-Mendelian inheritance of the callipyge genes in sheep and cattle. Callipyge (*CLPG1*) is expressed as a muscular hypertrophy in the buttock due to a single nucleotide substitution (A→G) in the heterozygotes but not in the homozygotes in the intergenic region of the imprinted genes *Delta-like* (*DLK1*) and the maternally *Expressed Gene 3* (*MEG3*). These two genes are at the distal end of bovine chromosome 18. The CLPG gene is expressed biallelically during the prenatal stage but only monoallelically in adults. During prenatal development CpG sites flanking the mutation is normal but postnatally it increases, except when two alleles are present. In the CLPG sheep DLK1 expression is enhanced because of the reduction in chromatin condensation at the location. Imprinting is not affected by calliypege (Murphy SK et al 2006 Genome Res 16:3430; Takeda H et al 2006 Proc Natl Acad Sci USA 103:8119). ►hybrid vigor, ►superdominance, ►heterosis, ►overdominance and fitness, ►polar overdominance

Overdominance and Fitness: It is assumed that in the case of overdominance the selection against the two homozygotes at a locus is 0.05. Thus, instead of the Hardy-Weinberg proportions, the population will be *AA* 95, *Aa* 200 and *aa* 95. The proportion of the surviving zygotes will be 390/400 = 0.975. This means 2.5% reduction in the size of the population in a single reproductive phase. This may not be very serious in the case of a single locus and large populations, but in the case of 10, 100 and 500 loci it would be $0.975^{10} \cong 0.776$, $0.975^{100} \cong 0.0795$ and $0.975^{500} \cong 0.0000032$, respectively.

It is likely that such a large reduction in population size would not be tolerated. Thus, overdominance may be advantageous only at 1 or very small number of loci. ►heterosis, ►fitness, ►hybrid vigor, ►allelic complementation, ►monogenic heterosis; Rédei GP 1982 Cereal Res Commun 10:5.

Overdrive: A consensus of 5′-TAAPuTPyNCTGTPuTNTGTTTTGTTTG-3′ in the vicinity of the T-DNA right border of some octopine plasmids facilitating the transfer of the T-DNA into the plant chromosomes. The VirC1 protein binds this region. In the *overdrive mode* of the RNA polymerase the system does not respond to protein signals of transcription termination or pause. ►*Agrobacterium*, ►T-DNA, ►virulence genes of *Agrobacterium*, ►antitermination, ►terminator, ►transcription factors, ►RNA polymerase; DeVos G Zambryski P 1989 Mol Plant-Microbe Interact 2:43.

O

Overepistasis: This is an archaic term for overdominance.

Overgrowth: Denotes increase of cell size and/or number in the neighboring tissues.

Overhang: A double-stranded nucleic acid has a protruding single strand end (see Fig. O32)

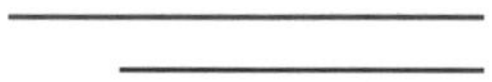

Figure O32. Single-strand overhang

Overlapping Code: This is a historical idea about how neighboring codons may share nucleotides. ►overlapping genes; Yčas M 1969 The Biological Code. North-Holland, Amsterdam.

Overlapping Genes: These occur in the small genomes of viruses, eukaryotic mitochondria and rarely in eukaryotic nuclear genes. In *Drosophila* 3,152 (~22%), in humans more than 1,000 protein coding (~1,600) and in the mouse 2,481 protein coding and non-protein coding gene pairs appeared overlapping in either a parallel or an antiparallel fashion or in a nested arrangement (Boi S et al 2004 Curr Genomics 5:509). In human chromosome 11 one-quarter of the protein coding genes overlaps other genes (Taylor TD et al 2006 Nature [Lond] 440:497). *In Arabidopsis* 1,340 potential overlapping (cis-natural) gene sequences have been identified and evidence has been obtained for 957 cis-nat-pair transcription (Wang X-J et al 2005 Genome Biol 6(4):r30). Earlier, a multitude of sense and antisense transcripts was detected in rice (Osato N 2003 Genome Biol 5:R5). Sequences of the genetic material may be read in different registers; thus the same sequences may represent two or more genes in a complete or partial overlap because they do not need equal amounts of proteins from the overlapping genes. The means for achieving this goal is either stop codon read-through or translational frame shifting on the ribosome. The murine leukemia virus (MLV) and the feline leukemia virus (FeLV) use the first alternative. In the HIV virus the RNA transcript makes a small loop and in the sequence 5′ UUU[UUUA]GGGAAGAU-LOOP-GGAU a one base slippage (framed) takes place. This allows the normally UUA recognizing $tRNA^{Leu}$ to pair with UUA (leucine) and in place of GGG (glycine) an AGG (arginine) was inserted, and thus the stop codon was thrown out of frame which permitted the synthesis of a fusion protein (*gag* + *pol*, i.e., envelop + reverse transcriptase). This frameshift translation is generally stimulated by a pseudoknot located 5–9 nucleotides (3′) from the shift site. After the translation the pseudoknot unfolds and the RNA returns to a linear shape. Such a ribosomal frameshifting occurred only in a fraction (11%) of the ribosomes but as a consequence instead of a 10–20 envelope:1 reverse transcriptase protein, the proportion changed to 8 envelope:1 reverse transcriptase (*gag* + *pol*) production. After the completion of sequencing of the genome, of the 879 nested genes found in *Drosophila* 574 were transcribed from the opposite polarity strands (Misra S et al 2002 Genome Biol 3(12):res0083). Within a ~3-Mb region of the long arm of chromosome 2 of *Drosophila* 12/17 were transcribed from the antiparallel strand. Deletions encompassing overlapping genes usually result in sterile or lethal phenotypes. In compact eukaryotes such as cryptomonads (e.g., microsporidia, *Guillardia*) single RNA molecules encode more than one gene but the phenomenon is not due to operons (Williams BAP et al 2005 Proc Natl Acad Sci USA 102:10936). ►gene, ►transcription, ►frameshift, ►decoding, ►wobble, ►recoding, ►tRNA, ►deletion, ►knockout, ►contiguous gene syndrome, ►translational hopping, ►φX174, ►pseudoknot, ►fuzzy logic; Pedersen AM, Jensen JL 2001 Mol Biol Evol 18:763; Besemer J et al 2001 Nucleic Acids Res 29:2607; Barrette IH et al 2001 Gene 270:181.

Overlapping Inversions: A part of one chromosomal inversion is included in another inversion (see Fig. O33). ►inversion, ►figure, ►salivary gland chromosomes

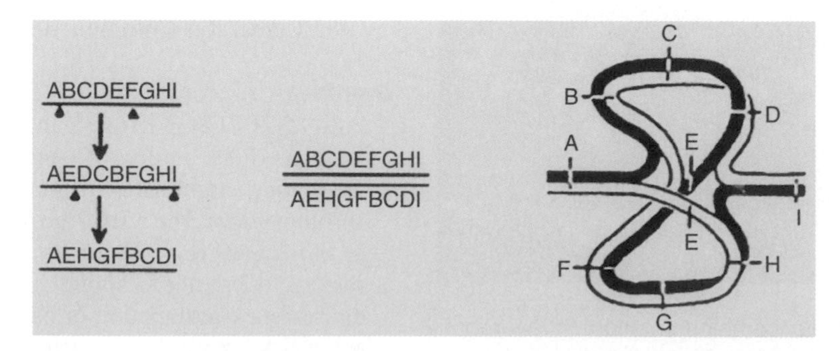

Figure O33. Two overlapping inversions in a heterozygons salivary gland cell of *Drosophila*, The salivary gland chromosomes display somatic pairing. (From Dobzhansky T and Sturtevant AH 1938. Genetics 23:28)

Overlay Binding Assay: This is used to determine interactions between protein subunits (on a membrane) after labeling of the SDS-PAGE separated proteins still retaining the critical conformation. ►SDS-polyacrylamide gels ; Kumar R et al 2001 J Biosci 26(3):325.

Overproduction Inhibition: A highly active promoter of a transposase may actually reduce the rate of transposition of a transposable element. ►transposable elements

Overwinding: This generates supercoiling in the DNA. ►supercoil

Oviduct: This is the passageway through which externally laid eggs are released (e.g., in birds), the channel through which the egg travels from the ovary to the uterus (e.g., mammals). In mammals fertilization takes place in the oviduct. ►ovary, ►uterus, ►ovalbumin, ►morula

Ovis aries (sheep, 2n = 54): This forms a fertile hybrid with the wild mouflons but the sheep x goat (*Capra hircus*, 2n = 60) embryos rarely develop in a normal way although some hybrid animals have been produced. ►nuclear transplantation

Ovist: ►preformation

Ovotestis: Refers to a gonad that abnormally contains both testicular and ovarian functions. ►hermaphrodite; Sato A et al 2002 Genetics 162:1791.

Ovulation: This involves the release of the secondary oocyte from the follicle of the ovary that is eventually followed by the formation of the egg and the second polar body. Bone morphogenetic protein 15 (BMP15, Xp11.2-p11.4), a member of the TGFβ family of proteins, is expressed specifically in the oocytes and is essential for normal follicular development and ovulation. Heterozygosity for a mutation (in sheep) in the gene may lead to twinning but homozygosity for the mutation impairs follicular growth beyond the primary stage and results in female sterility. The number of oocytes released in a cycle is called the ovulation quota that varies among multiparous and monoparous species. ►gametogenesis, ►twinning, ►fertility, ►luteinizing hormone-releasing factor, ►multiparous; Richards JS et al 2002 Annu Rev Physiol 64:69.

Ovule: This is the megasporangium of plants in which a seed develops; the animal egg enclosed within the Graafian follicle. ►Graafian follicle, ►nucellus

Ovulin: A seminal fluid protein which stimulates egg laying and the release of oocytes.

Ovum (plural ova): This is an egg(s) ready for fertilization under normal conditions.

Ox (plural oxen): Refers to a castrated male cattle. ►cattle

Ox40: This is a member of the tumor necrosis factor receptor family. ►TNF; Rogers PR et al 2001 Immunity 15:445.

Oxalosis (hyperoxaluria): Autosomal recessive forms involve the excretion of large amounts of oxalate and/or glycolate through the urine caused either by a deficiency of 2-oxoglutarate (α-ketoglutarate): glyoxylate carboligase or a failure of alanine: glyoxylate aminotransferase or serine:pyruvate aminotransferase. The accumulated crystals may cause disease of the kidney and liver. The glycerate dehydrogenase defects also increase urinary oxalates and hydroxy-pyruvate. Hydroxypyruvate accumulates because the dehydrogenase does not convert it to phosphoenolpyruvate. It is also called peroxisomal alanine:glyoxylate aminotransferase deficiency (AGXT) traced to human chromosome 2q36-q37. Hyperoxaluria II (9cen) is a deficiency of glyoxylate reductase/hydroxypyruvate reductase. ►glycolysis, ►peroxisome

oxi3: Denotes the mitochondrial DNA gene responsible for cytochrome oxidase.

Oxidation: Refers to loss of electrons from a molecule.

Oxidation–Reduction: This is a reaction transferring electrons from a donor to a recipient.

Oxidative Burst: In plants resistance to microbial infection may be mediated by the production of reactive oxygen species (ROS) such as superoxide anion ($O^{\bullet-}$), OH^-, H_2O_2. The oxidative burst may induce the transcription of defense genes in the plant tissue infected by even avirulent or non-pathogenic agents. The reactive oxygen may kill the cells along with the invader and may modify the cell wall proteins, which then reinforce the barrier to the infectious microbes. Several genes encoding proteins with functional similarity to RAC, a RAS family of proteins involved in signal transduction, mediate these reactions. ►host–pathogen relations, ►RAC, ►ROS; Martinez C et al 2001 Plant Physiol 127:334; Siddiqi M et al 2001 Cytometry 46(4):243.

Oxidative Deamination: For example, O replaces a NH_2 group of cytosine (C) resulting in the conversion of C→ Uracil. ►nitrous acid mutagenesis, ►chemical mutagens II

Oxidative Decarboxylation: ►pyruvate decarboxylation complex

Oxidative DNA Damages: These occur when ionizing radiation or oxidative compounds hit cells. The major class of damages includes the formation of thymine glycol (isomers of 5,6-dihydroxy-5,6-dihydrothymine), 5-hydroxymethyluracil, 5,6-hydrated cytosine, 8-oxo-7,8-di-hydroguanine, 2,2-diamino-4-[(2-deoxy-β-D-erythropentofuranosyl)amino]5(2H)-oxazolone and its precursor, cross-linking between the DNA and protein, elimination of the ribose. 8-oxoguanine represents about 10^{-5} of the mammalian guanine in the DNA. A minor oxidative damage product is 2-hydroxyadenine, which can also pair with guanine. These two purines are regularly generated within the cells and cause mutation or death unless repaired by glycosylases (Ushijima Y et al 2005 Nucleic Acids Res 33:672). 8-oxoguanine can be incorporated and paired in the DNA with A and C residues. DNA polymerases λ and η with the assistance of proliferating cell nuclear antigen (PCNA) and replication protein-A (RP-A) permit the correct incorporation of dCTP (compared to the incorrect dATP) opposite an 8-oxo-G template 1,200-fold and 68-fold better, respectively (Maga G et al 2007 Nature [Lond] 447:606). If 8-oxoguanine is not repaired A= T→G≡C transitions or G≡C→T= A transversions occur. The oxidative damages are usually repaired by exonucleases, endonucleases, glycosylases and other repair enzymes in prokaryotes and eukaryotes (Fromme JC et al 2004 Nature [Lond] 427:652). 8-oxoguanine-DNA glycosylase OGG1 excises 8-oxo-guanine from the DNA. Defects in OGG1 have been associated with cancer in humans; enhanced mutagenesis and accelerated senescence have been observed in a mouse strain with thermolabile OGG1 (Kuznetsov NA et al 2007 J Biol Chem 282:1029). OGG1 initiates age-dependent CAG trinucleotide expansions in somatic cells resulting in Huntington's chorea (Kovtun IV et al 2007 Nature [Lond] 447:447). In mice mutation of the MMH (MutM homolog) repair enzyme results in a threefold to sevenfold increase in 8-hydroxyguanine. In prokaryotes MutY and in humans hMYH glycosylases normally remove oxoguanine from the DNA. MutY oxidation mediated by guanine radicals with the aid of charge transport of the DNA duplex is the first step in signaling to DNA repair upon oxidative damage (Yavin E et al 2005 Proc Natl Acad Sci USA 102:3546). Oxidation of mRNA leads to premature termination of the translation process and the proteolytic degradation of the modified full length polypeptide resulting from translation errors induced by oxidation (Tanaka M et al 2007 Proc Natl Acad Sci USA 104:66).

A method has been developed for the genome-wide identification of functional polymorphisms in the antioxidant response element (ARE), a *cis*-acting enhancer sequence found in the promoter region of many genes that encode antioxidant and detoxification enzymes/proteins. In response to oxidative stress, transcription factor NRF2 (nuclear factor erythroid-derived 2-like 2) binds to AREs, mediating transcriptional activation of its responsive genes and modulating in vivo defense mechanisms against oxidative damage (Wang X et al 2007 Hum Mol Genet 16:1188). ►DNA repair, ►oxidative deamination, ►ROS, ►abasic site, ►photosensitizer, ►transition mutation, ►transversion mutation, ►MtODE, ►MtGendo, ►glycosylases, ►deoxyoxoguanine; Nunez ME et al 2001 Biochemistry 40:12465; Kaneko T et al 2001 Mutat Res 487:19; Vance JR Wilson TE 2001 Mol Cell Biol 21:7191; Aitken RJ, Krausz C 2001 Reproduction 122(4):497; Sartori AA et al 2004 Nucleic Acids Res 32:6531; Banerjee A et al 2005 Nature [Lond] 434:612; David SS et al 2007 Nature [Lond] 447:941.

Oxidative Phosphorylation (OXPHOS): ATP is formed from ADP as electrons are transferred from NADH or $FADH_2$ to O_2 by a series of electron carriers. Such processes take place in the inner membrane of the mitochondria with the assistance of cytochromes as electron carriers. This is the main mechanism of ATP formation. OXPHOS minus yeast mutants are petite and fail to grow on a non-fermentable carbon source and cannot convert into a red pigment, an intermediate of the adenine (*ade*$^-$) biosynthetic path. Nearly 20 different mutations in nuclear genes are now known, which control various functions of the mtDNA. Also, 82 structural subunits of mitochondrial genes are encoded in the nucleus versus the 13 coded by the mtDNA. mtDNA haplotypes from NIH3T3 and NZB/B1NJ mice induce significantly different OXPHOS performance as compared with those from C57BL/6J, BALB/cJ and CBA/J mice. In addition, a single nucleotide polymorphism (SNP) in the *mt-Tr* gene (mitochondrial $tRNA^{Arg}$) is the sequence variation responsible for the phenotypes observed. The differences in OXPHOS performance are masked by a specific upregulation in mitochondrial biogenesis, triggered by an increase in the generation of reactive oxygen species (ROS) in cells carrying NIH3T3 or NZB/B1NJ mtDNA (Moreno-Loshuertos R et al 2006 Nature Genet 38:1261). ►mitochondria, ►mitochondrial DNA depletion syndromes, ►ATP, ►NADH, ►FAD, ►petite colony mutants, ►mitochondrial diseases in humans; ROS, Saraste M 1999 Science 283:1428; Shoubridge EA 2001 Hum Mol Genet 10:2277.

Oxidative Stress: Such stress is exerted by free radicals (reactive oxygen species, ROS, singlet oxygen, nitric oxide) and peroxides and hydroxyl species through the action of enzymes involved in mixed-function oxidation and auto-oxidation (P450 cytochrome complex, xanthine oxidase, phospholipase A_2). These reactions play an important role in mutagenesis, aging, mitochondrial functions, signal transduction, etc., and are thought to affect neuronal degeneration in Alzheimer's and Parkinson's diseases and in amyotrophic lateral sclerosis. The transcription of protooncogenes may be regulated by redox systems. Glutathione and thioredoxin-dependent enzymes provide some protection. See diseases mentioned under separate entries, ►ROS, ►superoxide dismutase, ►catalase, ►redox reaction, ►glutathione, ►thioredoxins, ►peroxidase, ►Yap1, ►vitellogenin, ►FOX; Carmel-Harel O, Storz G 2000 Annu Rev Microbiol 54:439; Rabilloud T et al 2002 J Biol Chem 277:19396.

Oxidizing Agent: Refers to an acceptor of electrons.

Oxindoles: These are protein tyrosine kinase inhibitors. They inhibit the fibroblast growth factor and other receptor tyrosine kinases. For example, chemically they are 3-[4-(formylpiperazine-4-yl)benzyl-idenyl]-2-indolinone and 3-[(3-(carboxyethyl)-4-methylpyrrol-2-yl)methylene]-2-indolinone. (See Cane A et al 2000 Biochem Biophys Res Commun 276:379).

8-Oxodeoxyguanine: ►oxidative DNA damage, ►abasic site

Oxoguanine: ►8-oxodeoxyguanine (see Fig. O34).

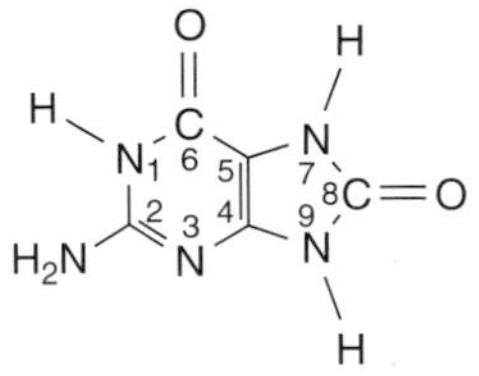

Figure O34. 8-Oxoguanine

Oxoprolinuria: ►glutathione synthetase deficiency

Oxphos (oxidative phosphorylation): A mitochondrial process by which molecular oxygen is combined with the electron carriers NADH and $FADH_2$ by the enzymes of the respiratory chain and mediate the ADP + P_i → ATP conversion. ►oxidative phosphorylation

Oxygen Effect: In the presence of air or oxygen the frequency of chromosomal aberrations induced by ionizing radiation increases in comparison to conditions of anoxia (lack of air in the atmosphere). Bacterial irradiation experiments have indicated that oxygen must be present before or at the radiation pulse for reducing survival but after the pulse its effect rapidly decreases. Glutathione level in the cells protects against oxygen effect. Exogenous thiols (cysteine, cysteinamine, WR2721, Amifostil, [2-(3-amino propyl)aminoethyl-phosphorothioic acid]) also have a protective role. Nitroaromatic compounds (e.g., antiprotozoan metronizadol) and 5′-halogen-substituted pyrimidines (bromouracil) are sensitizers under anoxia. The higher mutability of the mtDNA relative to that of the nucleus is attributed to the presence of reactive oxygen in this organelle. The ambient atmospheric oxygen level has almost doubled during the last 205 million years (beginning in early Jurassic and Eocene periods) and has contributed to the evolution of large placental mammals (Falkowski PG et al 2005 Science 309:2202). Deprivation of oxygen may lead to human death within seconds. ►physical mutagens, ►radiation effects, ►mtDNA; Thoday JM, Read JM 1947 Nature [Lond] 160:608; Schulte-Frohlinde D 1986 Adv Space Res 6(11):89; Hsieh MM et al 2001 Nucleic Acids Res 29:3116; Ragu S et al 2007 Proc Natl Acad Sci USA 104:9747.

Oxygenases: ►mixed function oxidases

Oxyntic Cells: These cells secrete HCl acid in the lining of the stomach.

Oxyntomodulin: The small intestine secretes oxyntomodulin which reduces appetite and body weight; it suppresses ghrelin. ►ghrelin, ►obesity

Oxysterols: These are derivatives of cholesterol with extra hydroxyl or keto groups, usually at the 7 position on the B ring or at the 24, 25, or 27 positions on the side chain. Oxysterols are synthesized in various tissues by specific hydroxylases and play a role in the export of excess cholesterol from the brain, lung and other organs. They are intermediates in bile acid synthesis. Oxysterols are potent feedback regulators of cholesterol homeostasis despite the fact that their concentration is lower by several orders of magnitude (Rhadhakrishnan A et al 2007 Proc Natl Acad Sci USA 104:6511). ►sterol, ►SREBP, ►cholesterol, ►statins

Oxytocin: This is an octapeptide stored in the pituitary that regulates uterine contraction and lactation. During delivery in rats, oxytocin triggers an inhibitory switch in GBA signaling in the brain (Tyzio R et al 2006 Science 324:1788). It is synthesized in the hypothalamus with other associated proteins. They are assembled in the neurosecretory vesicles and transported to the nerve ends in the neurohypophysis (posterior lobe of the pituitary) and may eventually be secreted into the bloodstream. Vasopressin and oxytocin (12-kb between them) and neurophysin genes are linked within the arm of human chromosome 20pter-p12.21. They are transcribed from the opposite strands of the DNA. It appears that pre-proarginine - vasopressin - neurophysin are transcribed jointly and post-translationally separated by proteolysis. Oxytocin favorably affects the social behavior of animals and humans (Kosfeld M et al 2005 Nature [Lond] 435:673). ►vasopressin [antidiuretic hormone], ►neurophysin, ►animal hormones, ►memory, ►behavior genetics, ►microsatellite, ►GABA; Breton C et al 2001 J Biol Chem 276:26931.

Oxytricha: This is a group of ciliate protozoa with a germ line micronucleus and a somatic macronucleus (see Fig. O35). During conversion of the micronucleus to macronucleus about 150,000 internal DNA segments are routinely eliminated by recombinational events. The processes permit shuffling of the ca. 30,000 genes at reconstitution and high rates of mutation. ►protozoa; Prescott DM 1999 Nucleic Acids Res 27:1243; http://oxytricha.princeton.edu/dimorphism/database.htm.

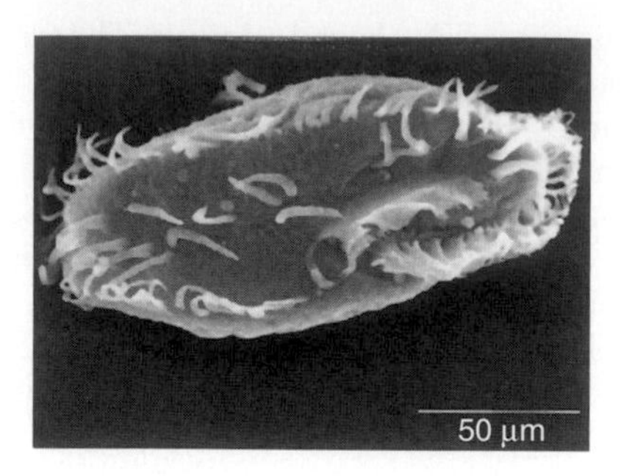

Figure O35. *Oxytricha*. (Courtesy of National Human Genome Research Institute)

Ozone (O_3): This is a bluish, highly reactive form of oxygen. It is a disinfectant and its presence in the atmosphere acts as a screen protecting the earth from excessive ultraviolet radiation coming from the sun.

Historical vignettes

"First your opponents say it can't be true. Next they say if it is true it can't be very important. Finally they say well, we've known it all along."

Jeffrey Kluger (2004) citing Jonas Salk in his book *Splendid Solution. Jonas Salk and the Conquest Polio.* Putnam, New York.

On July 9, 1909 (more than two decades earlier than the Neurospora work), FA Janssens, Professor at the University of Louvain, presented his theory of chiasmatypy in the journal *La Cellule* (25:389–411):

"In the spermatocytes II, we have in the nuclei chromosomes which show one segment of two clearly parallel filaments, whereas the two distal parts diverge… The first division is therefore reductional for segment A and a and it is equational for segment B and b … The 4 spermatids contain chromosomes 1st AB, 2nd Ab, 3rd ab, and 4th aB. The four gametes of a tetrad will thus be different…

"The reason behind the two divisions of maturation is thus explained… The field is opened up for a much wider application of cytology to the theory of Mendel."

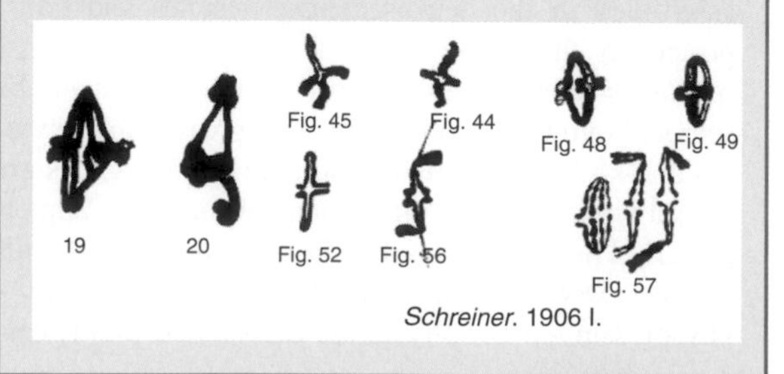

P

P: Parental generation.

P: ►probability

P (Polyoma): A regulatory DNA element in the viral basal promoter. ►Simian virus 40, ►polyoma

^{32}P: Phosphorus isotope. ►isotopes

P1: Double-stranded DNA, temperate *E. coli* phage. It is also a vector used DNA sequencing with a load capacity of ~80 kb.

P_1, P_2: Designations of the parents, homozygous for different alleles, at the critical locus (loci) in a Mendelian cross. ►Mendelian laws, ►gametic arrays, ►genotypic segregation, ►allelic combinations

p (petit): Short arm of chromosomes, also denote frequencies. ►q

π: ►diversity

π: 22/7 ≈ 3.132857, Ludolf number, indicates the ratio of the circumference to the diameter of a circle.

ϕ (phi): The symbol of some phages.

ψ (psi): ►pseudouridine (see Fig. P1), formula. Also, packaging signal for virions. The packaging signal is located at the 5′ LTR repeat and reaches over into the upstream end of the gag gene. It is not translated. ►retroviruses

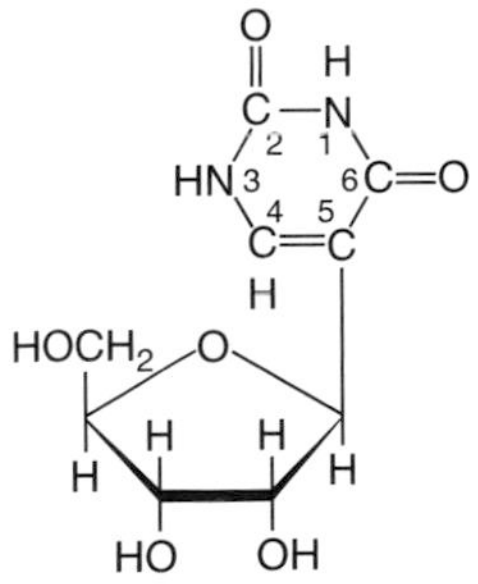

Figure P1. Pseudouridine

p13suc1: The yeast cell cycle activating enzyme binding to the cyclosome. ►cyclosome, ►cell cycle; Simeoni F et al 2001 Biochemistry 40:8030.

p14: ►ARF

p15^{INK4B}: An inhibitor of CDK4 (encoded in human chromosome 9p21) appears to be an effector of TGF-β, a protein known to control the progression from G1 phase of the cell cycle to S phase. ►cell cycle, ►TGF, ►p16^{INK4}, ►p18, ►p19, ►cancer; Seoane J et al 2001 Nature Cell Biol 3:400.

p16 (*MTSI* [multiple tumor suppressor], CDKN2): A cell cycle gene (in human chromosome 9p21) that has a major role in tumorigenesis. In 50% of the melanoma cells, it is deleted and in 25%, it is mutated. Over 70% of the bladder cancer cases are associated with deletions of 9p21 in both homologous chromosomes, in head and neck tumors 33%, in renal and other cells usually this tumor suppressor gene is lost. It normally restrains CDK4 and CDK6. ►melanoma, ►pancreatic adenocarcinoma, ►CDK4, ►CDK6, ►cell cycle, ►ARF; Serrano M et al 1993 Nature [Lond] 366:704.

p16: A weak ATPase and a packaging protein of φ29 bacteriophage. It interacts with the viral packaging RNA (pRNA) and the phage portal protein and assists in pumping the double-stranded DNA into the phage head. p16 and other packaging components are not found within the head. ►packaging of DNA; Ibarra B et al 2001 Nucleic Acids Res 29:4264.

p16^{INK4} (CDKN2A): A protein inhibitory to CDK4/CDK6 and thus appears to be a tumor suppressor because it inhibits the progression of the cell cycle from G^1 to S. Thus, it may also direct the cell toward senescence rather than neoplasia. p16^{INK4a} induces age-dependent decline in cell regenerative capacity (Krishnamurty J et al 2006 Nature [Lond] 443:453). Absence of p16^{INK4a} may reduce aging of hematopoietic stem cells and acts similarly to ARF in tumor suppression. The two proteins share coding sequences but ARF does not seem to affect stem cell aging (Janzen V et al 2006 Nature [Lond] 443:421). Both p16^{INK4} and p19^{INK4} are transcribed from the same 9p21 locus but from alternative alleles. ►CDK4, ►tumor suppressor, ►PHO81, ►cell cycle, ►cancer, ►p18, ►p19, ►Ets oncogene, ►Id protein, ►senescence, ►mole, ►centrosome; Quelle DE et al 1995 Cell 83:993; Wang W et al 2001 J Biol Chem 276:48655; Reinolds PA et al 2006 J Biol Chem 281:24790.

p18^{INKC}: Cell cycle inhibitors that block cell cycle kinases CDK4 and CDK6. ►cell cycle, ►cancer, ►p15, ►p16, ►p19; Blais A et al 2002 J Biol Chem 277:31679.

p19^{Arf}: ►Arf, ►p16^{INK4}

p19^{INK4d}: Cell cycle inhibitor of CDKs. ►p15, ►p16, ►p18, ►cell cycle, ►cancer, ►ARF

p21: A transforming protein of the Harvey murine sarcoma virus. A Ras-gene encoded 21 kDa protein binds GDP/GTP and hydrolyzes bound nucleotides and inorganic phosphate. This protein is involved in signal transduction, cell proliferation and differentiation and p21 controls the cyclin-dependent kinases (Cdk4, Cdk6, Cdk2), and binds to DNA polymerase δ processivity factor and it inhibits in vitro PCNA-dependent DNA replication but not DNA repair. In the absence of p21, cells with damaged DNA are arrested temporarily at the G2 phase and that is followed by S phases without mitoses. Consequently hyperploidy arises and apoptosis follows. Gene *p21* is under the control of p53 protein and the retinoblastoma tumor suppressor gene RB. MyoD regulates expression of the p21 Cdk inhibitors (p21Cip/WAF1, 6p21.2) during differentiation of muscle cells and non-muscle cells, and then withdrawal from the cell cycle does not require the participation of p53. Experimental evidence indicates that telomere dysfunction induces p21-dependent checkpoints in vivo that can limit longevity at the organismal level (Choudhury AR et al 2007 Nature Genet 39:99). RAF and RHO activate protein $p21^{Cip/Waf}$, leading to inhibition of the transition of the cell cycle into the S phase. Actually, RAS activates the serine/threonine kinase Raf that may facilitate the transition to the S phase but upon excessive stimulation by RAF and RHO, $p21^{Cip/Waf}$ has the opposite effect. Also, p53 is an independent activator of $p21^{Cip/Waf}$. After mitosis, p21 is expressed at the onset of differentiation but it may again be down-regulated at later stages of differentiation due to proteasome activity. In addition, p21 may not have an absolute requirement for induction by MyoD. The N terminal domains of p27 and p57 provide other antimitogenic signals. $P21^{SNFT}$ is a 21 kDa nuclear factor of T lymphocytes. Its overexpression represses transcription from the interleukin-2 and AP1-driven promoters. ►cell cycle, ►Cdk, ►mitosis, ►cancer, ►hyperploidy, ►apoptosis, ►p53, ►PCNA, ►p27, ►p57, ►MyoD, ►cell cycle, ►RAS, ►RASA, ►proteasome, ►interleukin, ►AP1, ►NF-AT, ►epigenesis; Prall OWJ et al 2001 J Biol Chem 276:45433; Wu Q et al 2002 J Biol Chem 277:36329; Bower KE et al 2002 J Biol Chem 277:34967.

p23: A component of the steroid receptor complex with Hsp90. ►Hsp, ►steroid hormones, ►molecular chaperone interacting complex, ►telomerase; Knoblauch R, Garabedian MJ 1999 Mol Cell Biol 19:3748; Munoz MJ et al 1999 Genetics 153:1561.

p24: A family of evolutionarily conserved small integral membrane proteins, which form parts of the COP transport vesicles or regulate the entry of cargo into the vesicles. It also mediates viral infection. ►protein sorting, ►COP transport vesicles; Blum R et al 1999 J Cell Sci 112(pt 4):537; Hernandez M et al 2001 Biochem Biophys Res Commun 282:1.

p27 ($p27^{Kip1}$): A haplo insufficient tumor suppressor protein, which inactivates cyclin-dependent protein kinase 2. Its mutation (homo- or heterozygous) results in increased body and organ size and neoplasia in mouse. p27 in cooperation with RAS controls metastasis (Bessom A et al 2004 Genes Dev 18:862). The wild type allele as a transgene may retard cancerous proliferation. Germ-line mutations in p27*Kip1* can predispose to the development of multiple endocrine tumors in both rats and humans (Pellegata NS et al 2006 Proc Natl Acad Sci USA 103:15558). Inactivation of $p27^{Kip1}$ is triggered by phosphorylation and mediated by the proteasome complex. Non-receptor tyrosine kinases phosphorylate $p27^{Kip1}$ and decrease its stability leading to entry into the cell, cycle (Kaldis P 2007 Cell 128:241). ►Cdk2, ►p21, ►cell cycle, ►cancer, ►KIP, ►Knudson's two-mutation theory of cancer, ►cancer gene therapy, ►metastasis, ►RAS, ►transgene, ►proteasome, ►p38, ►SKP, ►dyskeratosis, ►MEN; Mohapatra S et al 2001 J Biol Chem 276:21976; Malek NP et al 2001 Nature [Lond] 413:323.

$p27^{BBP}$: ►eIF-6

$p34^{cdc2-2}$: The gene coding for the catalytic subunit of MPF in *Schizosaccharomyces pombe* (counterparts, *CDC28* in *Saccharomyces cerevisiae*, *CDCHs* in humans have 63% identity with *cdc2*; these genes are present in all eukaryotes). The gene product is a serine/tyrosine kinase and its function is required for the entry into M phase of the cell cycle. If prematurely activated, it may cause apoptosis. ►cell cycle, ►MPF; Shimada M et al 2001 Biol Reprod 65:442; Nigg EA 2001 Nature Rev Mol Cell Biol 2:21.

p35 (Cdk5 regulatory subunit): A homolog of the baculoviral (survival) protein and cyclin-dependent regulator of neural migration and growth and which blocks apoptosis in a variety of eukaryotic cells. It colocalizes in the cells with RAC and Pac-1. Truncation of p35 results in a very stable p25 protein that is present in Alzheimer disease tangles. ►baculoviruses, ►RAC, ►Pac-1, ►Alzheimer disease, ►CDK; Tarricone C et al 2001 Mol Cell 8:657; Lin G et al 2001 In Vitro Cell Dev Biol Anim 37(5):293.

p38 (MPK2/CSBP/HOG1): A stress-activated protein kinase of the MEK family. It accelerates the degradation of $p27^{Kip1}$, and it phosphorylates H3 histone. p38 is one of the factors required for the initiation of G2/M checkpoint after UV radiation. The down-regulation of E-cadherin during mouse gastrulation requires p38 and a p38-interacting protein (Zohn IE et al 2006 Cell 125:957). The p38-activated protein (PRAK) is essential for RAS-induced senescence and tumor suppression (Sun P et al 2007 Cell 128:295). The mitogen-activated protein kinase

(MAPK) p38 controls inflammatory responses and cell proliferation. Using mice carrying conditional *Mapk14* (also known as *p38α*) alleles, when specifically deleted in the mouse embryo, fetuses developed to term but died shortly after birth, probably owing to lung dysfunction. Fetal hematopoietic cells and embryonic fibroblasts deficient in p38 showed increased proliferation resulting from sustained activation of the c-Jun N-terminal kinase (JNK) pathway (Hui L et al 2007 Nat Genet 39:741). p38 is a lung tumor suppressor (Ventura JJ et al 2007 Nat Genet 39:750). ►MEK, ►MEF, ►MAP kinase, ►JNK, ►cadherins, ►gastrula, ►checkpoint, ►UV; Schrantz N et al 2001 Mol Biol Cell 12:3139; Bulavin DV et al 2002 Curr Opin Genet Dev 12:92.

p40: A tumor suppressor encoded in human chromosome 3q, produced by alternative splicing of p51. Also an L1 RNA transcript (of ORF II) binding protein required of retrotransposon movement. ►p51, ►p53, ►L1, ►retrotransposon, ►ORF; Hess SD et al 2001 Cancer Gene Ther 8(5):371; Henning D, Valdez BC 2001 Biochem Biophys Res Commun 283:430.

p42: ►MAPK

p50: The N-terminus of the p105 light chain of NF-κB encoded at 4q23-q24. ►NF-κB; Yamada H et al 2001 Infect Immun 69:7100.

p51: A cell proliferation inhibitor protein related to p73 and encoded at 3q28. p51A is 50.9 kDa and p51B is 71.9 kDa. ►p73, ►p40, ►p53; Guttieri MC, Buran JP 2001 Virus Genes 23:17.

p52^{SHC}: A RAS G-protein regulator protein; it is regulated through CTLA-4–SYP associated phosphatase. ►CTLA-4, ►SYP, ►RAS; Joyce D et al 2001 Cytokine Growth Factor Rev 12:73.

p53 (TP3, 17p13.1): A tumor suppressor gene when the wild type allele is present but single base substitutions may eliminate suppressor activity and the tumorigenesis process may be initiated.

p53 is a tetramer with separate domains for DNA binding, transactivation and tetramerization (see Fig. P2).

The transcriptional coactivator p300 binds to and mediates the transcriptional functions of the tetrameric tumor suppressor p53. In the four domains of the complex between tetrameric p53 and p300, the latter wraps around the four transactivation domains of p53 (see Fig. P3) (Teufel DP et al 2007 Proc Natl Acad Sci USA 104:7009). Protein p53 as tetramers of tetramers binds to different DNA sites and in case of stress activates the expression of genes involved in apoptosis. Wild type alleles of oncogenes and DNA damage, both activate tumor suppressor activities of p53 although through separate metabolic routes (Efeyan A et al 2006 Nature [Lond] 443:159; Christophorou MA et al 2006 Nature [Lond] 443:214). New information is available for the structural framework in interpreting mechanisms of specificity, affinity and cooperativity of DNA binding as well as regulation by regions outside the sequence-specific DNA-binding domain (Kitayner M et al 2006 Mol Cell 22:741). Protein p53 recognizes specific DNA sequences, activates transcription from promoters with p53 protein binding sites and represses transcription from promoters lacking p53-binding sites; p53 regulates more than 160 genes. Recent data based on chromatin immunoprecipitation and paired end ditag sequencing identified 542 binding sites for p53 (Wei CL et al 2006 Cell 124:207). After DNA damage, the level of p53 increases by new translation and increased half-life of its mRNA. Ribosomal protein L26 binds to the 5′-untranslated region of mRNA and enhances its translation. Protein nucleolin binds to the same region and decreases translation (Takagi M et al 2005 Cell 123:49). Tumor suppressors p53, p63 and p73 express multiple splice variants and can use different promoters, thereby determine tissue-specificity of their expression (Bourdon J-C et al 2005 Genes Dev 9:2122; Murray-Zmijewski F et al 2006 Cell Death Differ 13:962). It promotes annealing of DNAs, inhibits replication, controls G1 and G2 phase checkpoints, leads to apoptosis or just blocks cytokinesis if the DNA is damaged, interferes with tumorous growth, maintains genetic stability, reduces radiation hazards by its regulatory role in the cell cycle. For the maintenance of G2 arrest after DNA damage it also requires the presence of p21. Protein p53 binds to a somewhat conserved consensus and it is phosphorylated at serine 315 residues by CDK proteins during S, G2 and M phases of the cell cycle but not at G1 although p53 controls an important G1

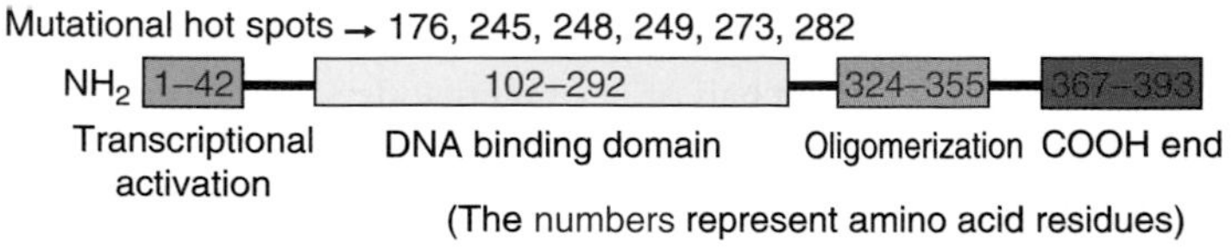

Figure P2. Structure of the p53 protein

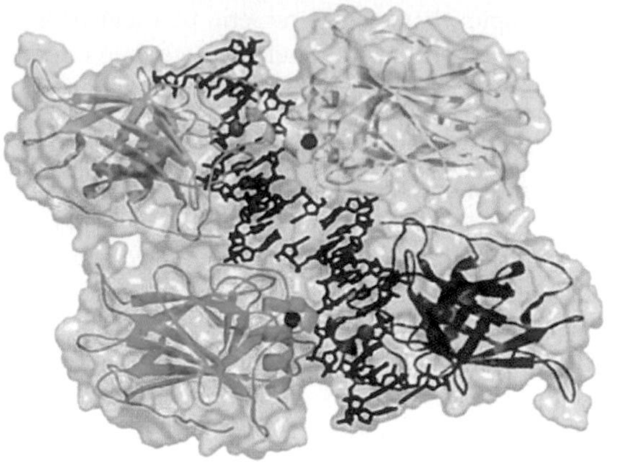

Figure P3. Ribbon diagram of the four core domains of p53 (light blue, green, yellow, maroon) interacting with DNA (dark blue). The Van der Waals surface is shown in gray; the four Zn irons are represented by mangenta spheres. (Courtesy of Professor Z. Shakked, Weizmann Institute of Science)

checkpoint. Binding subunits may have ubiquitin-conjugating role. Histone H3, methylated at lysine 79, targets 53BPP to DNA double-strand breaks (Huyen Y et al 2004 Nature [Lond] 432:406). p53 also controls proteins p21, p27 and p57. The p53 protein binds to the four copies of its consensus in DNA (5′-PuPuPuGA/T-3′). The C-terminal domain controls tetramerization and the N-terminal domain is responsible for transcriptional activation and for the regulation of down-stream genes. Small-angle x-ray scattering information in solution defined its shape, and NMR identified the core domain interfaces and showed that the folded domains had the same structure in the intact protein as in fragments. The combined solution data with electron microscopy on immobilized samples provided medium resolution 3D maps (Tidow H et al 2007 Proc Natl Acad Sci USA 104:12324).

One study indicated that at least 34 different transcripts were induced by p53 more than ten-fold although there was heterogeneity in the response. Co-activators TAFII40, TAFII 60 and other TATA box binding factors mediate its transcriptional activation. When the first six exons of the gene are deleted, the mRNA is still translated into a C-terminal protein fragment. Such a mutation enhances tumor suppression but leads to premature aging in mice (Tyner SD et al 2002 Nature [Lond] 415:45). p53 is encoded in human chromosome 17p13.105-p12; in about half of the human tumors the normal allele is altered. p53 regulates mitochondrial respiration through cytochrome oxidase c (SCO2) and may account for glycolysis in cancer cells and aging versus the respiratory pathway in normal cells (Matoba S et al 2006 Science 312:1650). Protein 53 plays a central role in cellular metabolism and its expression is induced by many factors such as oncogene expression, chemotherapy, oxidative stress, hypoxia, etc. Topoisomerase I is a p53-dependent protein. A p53-induced apoptosis may require transcriptional activation or it may occur in the absence of RNA or protein synthesis. The activation of p53 may be followed by FAS transport from Golgi intracellular stores without a need for synthesis of FAS. Tumor suppressor p53 product in the nucleus regulates pro-apoptotic genes such as *FAS, BAX, Bid, Noxa* and *PUMA*. In the cytoplasm, the p53 protein activates Bcl-2 and facilitates the permeability of the mitochondria for the release of pro-apoptotic molecules such as cytochrome c. In the cytoplasm, p53 is dislodged from Bcl-2 by PUMA to act on the mitochondrial permeability and apoptosis; thus PUMA interconnects the nuclear and cytoplasmic functions of p53 (Chipuk JE et al 2005 Science 309:1732).

The activity of p53 is also regulated by methylation of lysine 4 residues of histone-3 by Set9 methyltransferase and thereby the protein is better stabilized (Chuikov S et al 2004 Nature [Lond] 432:353). The Smyd2 protein mediates methylation of Lys370 in p53 and represses its activity; Set9 mediated methylation at Lys372 reduces methylation at Lys370 (Huang J et al 2006 Nature [Lond] 444:629). Acetylation of lysines 373 and 382 increases its DNA binding (Luo J et al 2004 Proc Natl Acad Sci 101:2259). Single nucleotide polymorphism in the promoter of MDM2 increases the affinity of the Sp21 transcriptional activator and attenuation of the p53 resulting in accelerated tumorigenesis (Bond GL et al 2004 Cell 119:591).

The ASPP family member regulatory proteins bind to the proline-rich region of p53, which contains the most common p53 polymorphism at codon 72. iASPP (inhibitory ASPP) binds to and regulates the activity of p53Pro72 more efficiently than the alternative amino codon p53Arg72. Hence, escape from negative regulation by iASPP is a newly identified mechanism by which p53Arg72 activates apoptosis more efficiently than p53Pro72 (Bergamaschi D et al 2006 Nature Genet 38:1133).

The calcium-binding proteins S100B and S100A4 bind preferentially to the tetramerization domain at lower oligomerization states, disrupt tetramerization and control intracellular movement of the p53 protein (Fernandez-Fernandez, M.R et al. 2005 Proc Natl Acad Sci USA 102:4735).

Protein p73 has functions similar to those of p53. p33^{INGI}, encoded at human chromosome 13q34, cooperates with p53 by protein-protein interaction in repressing cellular proliferation and the promotion of apoptosis.

The DNA-dependent protein kinase (DNA-PK) activates p53 in case the DNA is damaged. When amino acid site 376 is dephosphorylated by ionizing radiation protein 14-3-3 binds to p53 and increases its ability for DNA binding (see Fig. P4). Still other proteins such as IRF may be involved with p53 and other cooperating proteins. p53 activity may also be affected by oncoproteins RAS and MYC. Chemotherapeutic agents, UV light and protein kinase inhibitors may also activate p53. p53 is reversibly blocked by pifithrin-α (2[2-imino-4,5,6,7-tetrahydrobenzothiazol-3-yl]-1-polyethanone) and may protect from the undesirable side effect of anticancer therapy without causing new tumors in the absence of p53 function. When 33,615 human unique genes were tested by cDNA microarrays, 1,501 genes responded one way or another to p53 (Wang L et al 2001 J Biol Chem 276:43604).

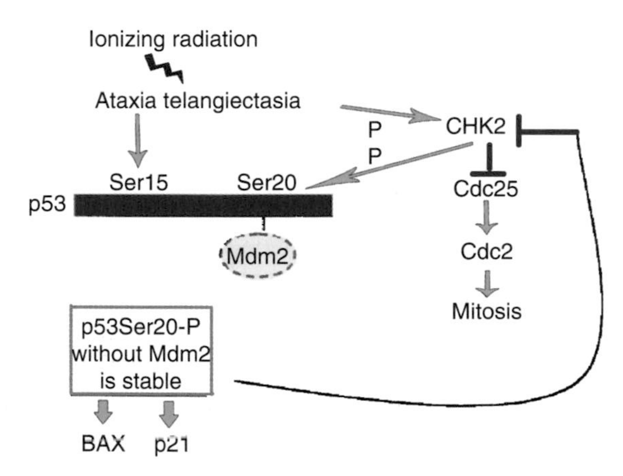

Figure P4. Some of the circuits of p53. Ionizing radiation damage to the ataxia telangiectasia protein results in the phosphorylation of the serine 15 residue of p53 and checkpoint 2 (CHK2), which in turn phosphorylates residue 20. The latter point is normally occupied by Mdm2/HDM2 a negative regulator of p53 that is now dislodged by the phosphorylation of site 20. The p53 ser20-p protein is now stabilized, despite the radiation and can induce BAX (a porin with anti-apoptosis function) and p21, a suppressor of mitosis. The blocking of CHK2 than prevents its inhibition of cell division cycle proteins (Cdc25 and Cdc2). The process to mitosis is then facilitated

Pharmacological compounds (CP-31398, CP-257042, etc.) have been selected by large-scale screening that could stabilize the DNA-binding domain of mutant p53 and activate its transcription as well as to slow tumor development in mice. In some cancers, p53 may increase sensitivity to chemotherapeutic agents but in others, it does not affect them. NADH quinone oxidoreductase may stabilize the p53 protein. Synthetic siRNA with single base difference may suppress the expression of mutation in p53 and may selectively block tumorigenesis by restoring wild type function (Martinez LA et al 2002 Proc Natl Acad Sci USA 99:14849). Using RNAi technology p53 function can be reactivated in murine liver carcinomas by triggering cellular senescence and innate immunity and consequently leading to tumor clearance (Xue W et al 2007 Nature [Lond] 445:656). Blocking p53 expression by introducing a stop cassette into the gene or removing it by the use of the Cre/loxP system, lymphomas and sarcomas of mice could be induced and then regressed after restoration of p53 function (Ventura A et al 2007 Nature [Lond] 455:661). The antiviral and anticancer effects of p53 is mediated by interferons (α and β), which boost the response to stress signals and thus promote apoptosis (Takaoka A et al 2003 Nature [Lond] 424:516).

Germline mutations occur in about 1×10^3 of the Caucasian populations but in about half of the sporadic cancers, it is somatically mutated. (Science magazine declared p53 the molecule of year 1993). ►tumor suppressor gene, ►annealing, ►apoptosis, ►PUMA, ►BCL, ►TAF, ►TBP, ►transactivator, ►nucleolin, ►cancer, ►Sp1, ►cell cycle, ►p21, ►p27, ►p40, ►p51, ►p57, ►p63, ►p73 substitution mutation, ►DAP kinase, ►MDM2, ►lactacystin, ►ARF, ►IRF, ►papilloma virus, ►DNA-PK, ►protein 14-3-3, ►E2F, ►ribonucleotide reductase, ►ataxia telangiectasia, ►CHK, ►Cdc, ►porin, ►p21, ►GADD45, ►RNAi, ►interferon, ►paired-end diTAG; Vogelstein B et al 2000 Nature [Lond] 408:307; Vousden KH 2000 Cell 103:691; Asher G et al 2001 Proc Natl Acad Sci USA 98:1183; Johnson RA et al 2001 J Biol Chem 276:27716; Olivier M et al 2002 Hum Mut 19:607; minireview: Kastan MB 2007 Cell 128:837; p53 mutation database: http://www-p53.iarc.fr/index.html; http://p53.free.fr.

p55 (TNFR1): A tumor necrosis factor receptor of Fas. ►Fas, ►TNF; Dybedal I et al 2001 Blood 98:1782; Longley MJ et al 2001 J Biol Chem 276:38555.

p56chk1: A protein kinase and a checkpoint for mitotic arrest after mutagenic damage inflicted by UV, ionizing radiation or alkylating agents. The DNA damage results then in the phosphorylation of this protein in yeasts. The phosphorylation may prevent the mitotic arrest yet the cells may die later; p56 is not involved in DNA repair. Phosphorylation is required so that other checkpoint genes become/stay functional. ►cell cycle, ►DNA repair; Feigelson SW et al 2001 J Biol Chem 276:13891.

p57 ($p57^{Kip2}$): An antimitogenic protein; its carboxyl end assures nuclear localization and the amino end is involved in the inhibition of CDK proteins. ►CDK, ►p21, ►p27, ►cell cycle, ►KIP; Thomas M et al 2001 Exp Cell Res 266:103.

p58^{IPK}: A heatshock protein 40 family member that inhibits interferon-induced, double-stranded RNA-activated eukaryotic translation initiation factor eIF2α protein kinase, PERK. Stress in the endoplasmic reticulum (ER) caused by unfolded proteins activates the translation of the gene and thereby reduces protein overload in the ER. ►heat-shock proteins, ►eIF2; Yan W et al 2002 Proc Natl Acad Sci USA 99:15920.

p60: Binds to Hsp70 and Hsp90 and chaperones the assembly of the progesterone complex. ►Hsp70, ►Hsp, ►progesterone, ►animal hormones, ►chaperone; Mukhopadhyay A et al 2001 J Biol Chem 276:31906.

p63: A member of the p53 tumor suppressor gene family, encoded at 3q27-q29. The gene expresses at least 6 transcripts, involved with transactivation of p53 and p73, DNA binding and oligomerization. It controls ectodermal (limb, craniofacial and epithelial) differentiation. Protein p63 regulates also the commitment to prostate cell lineage development (Signoretti S et al 2005 Proc Natl Acad Sci USA 102:11355). The first direct target of p63 appears to be the Perp protein localized in the desmosomes. Absence of Perp leads to post-natal death in mice (Ihrie RA et al 2005 Cell 120:843). Although p63 belongs to the p53 tumor suppressor family, its function seems to be different. Mouse heterozygotes for p63 were not prone chemically induced tumorigenesis (Keyes WM et al 2006 Proc Natl Acad Sci USA 103:8435). p63 protects female germ line—by apoptosis—during meiotic arrest (Suh E-K et al 2006 Nature [Lond] 444:624). Epithelial stem cells require p63 for proliferation (Senoo M et al 2007 Cell 129:523). ►p53, ►p73, ►desmosome, ►EEC syndrome, ►Hay-Wells syndrome; van Bokhoven H et al 2001 Am J Hum Genet 69:481; van Bokhoven, Brunner HG 2002 Am J Hum Genet 71:1.

p65: A component of the NF-κB complex. It can be exploited advantageously for gene activation in a chimeric construct with a mutant progesterone-receptor-ligand binding domain of gene GAL4. ►NF-κB, ►GAL4, ►gene-switch, Burcin MM et al 1999 Proc Natl Acad Sci USA 97:355.

p70/p86: The Ku autoantigen. ►DNA-PK, ►Ku

p70^{s6k}: Phosphorylates S6 ribosomal protein at serine/threonine residues before translation. Also called S6 kinase. ►translation initiation, ►p85^{s6k}, ►S6 kinase, ►signaling to translation; Harada H et al 2001 Proc Natl Acad Sci USA 98:9666.

p73: It has homology in amino acid sequence to p53 protein and is encoded in human chromosome 1p36.3. Similarly to p53, it regulates apoptosis and anti-tumor activity (upon E2F1 induction), hippocampal dysgenesis, hydrocephalus, immune reactions and pheromone sensory pathways. It affects proliferation, although in a somewhat different manner, but its loss does not lead to tumorigenesis in mice. p73 may compete with p53. This protein may have antiapoptotic effect in neurons. ►p53, ►apoptosis, ►p63, ►E2F; Sasaki Y et al 2001 Gene Ther 8:1401; Stiewe T, Putzer BM 2001 Apoptosis 6:447; Melino G et al 2002 Nature Rev Cancer 2:605.

p75: A non-tyrosine kinase receptor protein, TNFR 2 (tumor necrosis factor receptor 2). It is a Fas receptor. ►Fas, ►TNF; Hutson LD, Bothwell M 2001 J Neurobiol 49(2):79; Wang X et al 2001 J Biol Chem 276:33812.

p80^{sdc25}: A protein phosphatase that activates $p34^{cdc2}$-cyclin protein kinase complex by dephosphorylating Thr^{14} and Tyr^{15}. ►cell cycle, ►Ku; McNally KP et al 2000 J Cell Sci 113[pt 9]:1623.

p85^{s6k}: Phosphorylates S6 ribosomal protein before translation at serine/threonine sites; also called S6 kinase. p85 protein is also involved in a p53-dependent apoptotic response to oxidative damage and activation of natural killer. p85 phosphoinositide 3-kinase also mediates developmental and metabolic functions. ►translation initiation, ►p70^{s6k}, ►S6 kinase, ►phosphoinositides; Fruman DA et al 2000 Nat Genet 26:379.

p95: is Fas and is involved in apoptosis; its mutation leads to the Nijmegen breakage syndrome and other double breakage of the chromosomes. ►acrosomal process, ►Fas, ►APO, ►Nijmegen breakage syndrome, ►apoptosis

p97 (Cdc48): About M_r 600 ATPase and mediates membrane fusion. ►endoplasmic reticulum-associated degradation; Hirabayashi M et al 2001 Cell Death Differ 8(10):977.

p105: ►p50

p107: A retinoblastoma protein-like regulator of the G1 restriction point of the cell cycle. ►restriction point, ►tumor suppressor, ►retinoblastoma, ►cell cycle, ►pocket; Charles A et al 2001 J Cell Biochem 83:414.

p110α: The catalytic subunit of PIK. It has a critical role in insulin signaling and with TOR it controls the development of gliomas. ►PIK/PI(3)K, ►insulin, ►glioma

p110^{Rb}: The protein encoded by the retinoblastoma (Rb) gene. When not fully phosphorylated it interferes with the G_0 and G_1 phases of the cell cycle by inhibition of the E2F transcription factor. ►retinoblastoma, ►cell cycle, ►tumor suppressor, ►E2F, ►killer cells; DeCaprio JA et al 1988 Cell 54:275.

p115: A monomeric GTPase with a specific guanine exchange factor (GEF) for RHO (p115 Rho GEF). ►GTPase, ►RHO, ►GEF; Wells CD et al 2001 J Biol Chem 276:28897.

p125^{FAK} (focal adhesion kinase): A non-receptor tyrosine kinase. ►CAM; Yurko MA et al 2001 J Cell Physiol 188:24.

p130: A retinoblastoma protein-like regulator of the G1 restriction point of the cell cycle. p130^{Cas} is involved in the organization of myofibrils, actin fibers, anchorage-dependence of cultured cells. ►restriction point, ►tumor suppressor, ►retinoblastoma, ►CAS, ►pocket; Tanaka N et al 2001 Cancer 92:2117.

p160: A family of transcriptional co-activators such as SRC-1, GRIP1/TIF2 and p/CIP. They modify chromatin structure by methylating some of the histones. ►chromatin remodeling, ►nuclear receptors, ►p/CIP; Mak HY 2001 Mol Cell Biol 21:4379.

p300 (CBP): A cellular adaptor protein preventing the G_0/G_1 transition of the cell cycle, it may activate some enhancers and stimulate differentiation. It is also a target of the adenoviral E1A oncoprotein. Its amino acid sequences are related to CBP, a CREB-binding protein. Nuclear hormone-receptors interact with CBP/p300 and participate in gene transactivation. PCAF is a p300/CBP-associated factor in mammals, and it is the equivalent of the yeast Gcn5p (general controlled nonrepressed protein), an acetyltransferase working on histones 3, 4 (HAT A) and thus regulating gene expression. p300 functions also as a co-activator of NF-κB. p300 also binds PCNA. In several human cancers, p300 mutations were identified indicating that the protein is a tumor suppressor. p300 may show ubiquitin ligase activity for p53. ►adenovirus, ►CREB, ►NF-κB, ►PCNA, ►histone acetyl-transferase, ►E1A, ►bromodomain, ►chromatin remodeling, ►histone methyltransferases, ►p53; Lin CH et al 2001 Mol Cell 8:581; ►CARM

p350: A DNA-dependent kinase, it is a likely basic factor in severe combined immunodeficiency and it may also be responsible for DNA double-strand repair, radiosensitivity and the immunoglobulin V(D)J rearrangements. In association with the KU protein, it forms a DNA-dependent protein kinase. ►severe combined immunodeficiency, ►kinase, ►KU, ►DNA-de-pendent protein kinase, ►immunoglobulins; Chan DW 1996 Biochem Cell Biol 74:67.

P450 (CYP): A family of genes coding for cytochrome enzymes involved in oxidative metabolism. They are widely present in eukaryotes and scattered around several chromosomes. All mammalian species have at least eight subfamilies. The homologies among the subfamilies are over 30% whereas the homologies among members of a subfamily may approach 70%. These cytochromes possess monooxygenase, oxidative deaminase, hydroxylation, sulfoxide forming, etc., activities. *Aspergillus oryzae* has ~149 cytochrome P450 genes in multiple copies. The proteins are generally attached to the microsomal components of homogenized cells (endoplasmic reticulum [fragments]), often called S9 fraction. Some of these enzymes (subfamily IIB) are inducible by phenobarbital. Their expression may be tissue-specific, predominant in the liver, kidney or intestinal cells. Mammalian P450 cytochrome fraction is generally added to the *Salmonella* assay media of the Ames test in order to activate promutagens. The pregnane X receptor (PXR) is activated by a variety of compounds and is thus responsible for the activation of different drugs involved in mutation, cancer and interaction with other drugs. One member of the P450 series is involved in the regulation of the synthesis of the 6th class of plant hormones, brassinosteroids. P450 enzymes require the cofactor NAD or NADPH and their activity is favored by the presence of peroxides as oxygen donors. By mutagenesis, industrially more useful P450 variants are being produced. The P-450 (CYP1A1) dioxin and aromatic compound-inducible P450 maps to human chromosome 15q22-qter. CYP1A2 is phenacetin O-deethylase. Phenacetin is an analgesic and antipyretic carcinogen. CYP2D (22q13.1) is a debrisoquin 4-hydroxylase. Debrisoquine is a toxic anti-hypertensive drug. CYP51 (7q21.2-q21.3) is lanosterol 14-α-demethylase is a sterol biosynthetic protein. ►Ames test, ►cytochromes, ►hypoaldosteronism, ►steroid hormones, ►brassinosteroids, ►peroxide, ►NAD, ►analgesic, ►antipyretic, ►cyclophilin; Fujita K, Kamataki T 2001 Mutat Res 483:35; Ingelman-Sundberg M 2001 Mutat Res 482:11; crystal structure of p450 3A4: Williams PA et al 2004 Science 305:683.

P Blood Group: Controlled by two non-allelic loci. The non-polymorphic P blood group is located in human chromosome 6 and it is encoding globoside whereas the polymorphic P1 locus in human chromosome 22 encodes paragloboside. The frequency of the P gene in Sweden was found to be 0.5401 and that of P1 0.4599. According to other studies, the frequency of

P among caucasoids is about 0.75. The P1 blood type facilitates bacterial attachment to the epithelial cells of the urinary tract and kidney. Therefore, infections are more common. Some P alleles raise the risk of abortions, and others may increase the chances of stomach carcinomas. Some of the literature calls P as P1 and P1 as P2. ►blood groups, ►globoside; Stroud MR 1998 Biochemistry 37:17420.

P Body (processing bodies, cytoplasmic body): A small number of specific sites in the cytoplasm involved in the decapping of mRNA after deadenylation of the polyA tail. At this location, mRNAs occur at various stages of degradation mediated by several proteins. Argonaute 2 of the RISC complex of RNAi is also localized in the P bodies (Sen GL, Blau HM 2005 Nat Cell Biol 7:633). The mammalian protein elongation factor eIF4E, its transporter (eIF4E-T) as well as the DEAD-box helicase rck/p54 are also located at these cytoplasmic sites (Andrei MA et al 2005 RNA 11:717). P bodies can recycle to the polysomes and when conditions are favorable can be translated (Brengues M et al 2005 Science 310:486). ►decapping, ►mRNA, ►RNAi, ►microRNA, ►eIF-4E, ►DEAD-box, ►RNA surveillance; Sheth U, Parker R 2003 Science 300:805; review: Parker R, Sheth U 2007 Mol Cell 25:635.

P1 Cloning Vectors: They have a carrying capacity up 100 kbp DNA; thus they fall between Lambda and YAC vectors. ►vectors, Park K, Chattoraj DK 2001 J Mol Biol 310:69; Grez M, Melchner H 1998 Stem Cells 16(Suppl. 1):235.

P Cytotype: ►hybrid dysgenesis

P Element: ►hybrid dysgenesis

P Element Vector: Constructed from the 2.9 kb transposable element P of *Drosophila* equipped with 31 bp inverted terminal repeats. The gene to be transferred is inserted into the element but in order to generate stable transformants the transposase function located in the terminal repeats is disabled. Functional transposase is provided in a separate helper plasmid (pπ25.7wc). Such a binary system permits the separation of the two plasmids and the screening of the permanent transgenes if a selectable marker is included. Both plasmids are mixed in an injection buffer and delivered into pre-blastoderm embryos. The various P vectors have been widely used in for gene tagging, induction of insertional mutation and for exploration of functional genetic elements in *Drosophila* and in some other insects. A newer type of vector, Pacman contains the P transposase and the phage φC31 integration site. The φC31 integrase mediates recombination between the engineered phage *attP* in the *Drosophila* genome and a bacterial *attB* site in an injected plasmid. Such a system permits integration of large tracts of DNA (up to 133 kb) at specific sites. Such a targeted transgenesis can rescue much larger lethal mutations than it would be possible with only P element (Venken KJT et al 2006 Science 314:1744). ►hybrid dysgenesis, ►transposon vector, ►*att* sites; Sullivan W et al 2000 *Drosophila* Protocols, Cold Spring Harbor Laboratory Press, Cold Spring Harbor, New York.

P Granule: Serologically definable elements in the cytoplasm of animal cells at fertilization that segregate to the posterior part of the embryo where stem cell determination takes place. During embryogenesis, the P granules (RNA) may segregate asymmetrically into the blastomeres that produce the germline. (See Harris AN, Macdonald PM 2001 Development 128:2823).

P Nucleotides: ►immunoglobulins

P1 Phage: An *E. coli* transducing phage and vector with near 100 kb carrying capacity. (See Lehnherr H et al 2001 J Bacteriol 183:4105).

P22 Phage: The temperate bacteriophage of *Salmonella typhimurium*; its genome is about 41,800 bp. (See Vander Byl C, Kropinski AM 2000 J Bacteriol 182:6472).

P1 Plasmid: A cloning vector with a carrying capacity of about 100 kb. ►vectors, ►P1 phage; Bogan JA et al 2001 Plasmid 45(3):200.

P Region of GTP-Binding Proteins: Shares the G-X-X-X-X-G-K-(S/T) motif (►amino acid symbols) and is suspected to involve the hydrolytic process of GTP-binding and several nucleotide triphosphate-utilizing proteins. ►GTP binding protein superfamily

P Site: The peptidyl site on the ribosome where the first aminoacylated tRNA moves before the second charged tRNA lands at the A site as the translation moves on. The binding of the tRNA to the 30S ribosomal subunit appears to be controlled by guanine residues at the 966, 1401 and 926 positions in the 16S rRNA. ►A site, ►protein synthesis, ►ribosome; Feinberg JS, Joseph S 2001 Proc Natl Acad Sci USA 98:11120; Schäfer MA et al 2002 J Biol Chem 277:19095.

PABp: The poly(A) binding protein (~72 kDa) is the major protein that binds to the poly A tail of eukaryotic mRNA and converts it to mRNAP. Pab1p connects the mRNA end to the eIF-4H subunit of the eukaryotic peptide initiation factors eIF-4G and eIF-4F. It contains four RRM motifs. PABp also interacts with PAIP a translational co-activator protein in mammals. ►binding proteins, ►mRNAP, ►mRNA tail, ►polyadenylation signal, ►mRNA decay, ►eIF-4F, ►eIF-4G, ►Xrn1p, ►ribosome scanning, ►translation initiation, ►translational termination, ►RRM, ►mRNA circularization; Kozlov G et al 2001 Proc Natl Acad Sci USA 98:4409.

PAC (phage artificial chromosome): P1 phage PAC carries about 100–300 kb DNA segments. Most PAC vectors lack selectable markers suitable for mammalian cell selection but can be retrofitted by employing the Cre/loxP site-specific recombination system. ►BAC, ►YAC; Poorkaj P et al 2000 Genomics 68:106.

pac: A site in the phage genome where terminases bind and cut during maturation of the DNA before packing it into the capsid. ►terminase, ►packaging of the DNA

PAC-1: Dephosphorylates Thr^{183} and Tyr^{185} residues and thus regulates mitogen-activated MAP protein kinase involved in signal transduction. The Pac1 nuclease removes the 3′ external transcribed spacers from the nascent rRNAs in cooperation with RAC. PAC1 is activated by p53 protein during apoptosis and suppresses carcinogenesis. ►MAP, ►MKP-1, ►apoptosis, ►p53; Boschert U et al 1997 Neuroreport 8:3077; Spasov K et al 2002 Mol Cell 9:433.

PACAP (pituitary adenylyl cyclase-activating polypeptide-like neuropeptide): A neurotransmitter at the body-wall neuromuscular junction of *Drosophila* larvae. It mediates the cAMP-RAS signal transduction path. ►signal transduction, ►RAS, ►RAF; Kopp MD et al 2001 J Neurochem 79:161.

Pacemaker: Maintains rhythmic balance like the pulse of the heart or circadian rhythm.

Pachynema: Literally "thick thread" of chromosomes at early meiosis when the double-stranded structure of the chromosomes is not distinguishable by light microscopy because the chromatids are tightly appositioned (see Fig. P5). Also, the two homologous chromosomes are closely associated, unless structural differences prevent perfect synapsis. If a pair of chromosome is not completely synapsed by pachytene, they will not pair later either. In pachytene the chromosomal knobs and chromomeric structure is visible and can be used for identification of individual chromosomes. After pachytene, the synaptonemal complex is dismounted and the chromosomes progressively condense. In case the chromosomes are defective at this stage the Red1 (required for chromosome segregation) and Mek1 proteins serve as checkpoint control by preventing further progress of meiosis. Normally MEK kinase phosphorylates Red. Phosphatase Glc7 dephosphorylates Red. More than two dozens of other proteins (named differently in different organisms) are also involved in pachytene controls. ►meiosis, ►pachytene analysis, ►synapsis, ►chiasma, ►chromomere, ►MEK; Bailis JM, Roeder GS 2000 Cell 101:211; Roeder GS, Bailis JM 2000 Trends Genet 16:395.

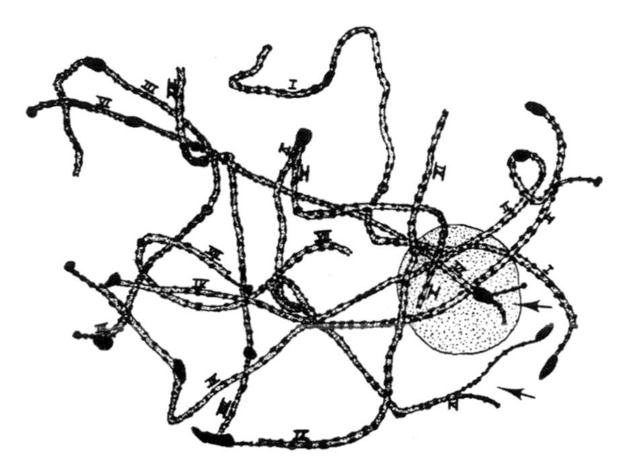

Figure P5. Naturalistic drawing of the 10 pachytene chromosome pair of a teosinte x maize hybrid. Note (←) unpaired ends of chromosomes V, VII. And some terminal and near-terminal knobs (Courtesy of Dr. A. E. Longley, see also 1937 J Agric Res 54:835)

Pachyonychia: A rare autosomal dominant keratosis of the nails and skin. ►keratosis

Pachytene Analysis: The study of meiotic chromosomes at the pachynema stage when cytological landmarks, chromomeres, and knobs are distinguishable by the light microscope, and chromosomal aberrations (deletions, duplications, inversions, translocations, etc.) can cytologically be identified and correlated with genetic segregation information. The pachytene analysis of plants is analogous to the study of giant chromosomes in dipteran flies and other lower animals. The bands of the (somatic) salivary chromosomes are tightly appositioned chromomeres in these endomitotic chromosomes. ►meiosis, ►salivary gland, ►chromomere, ►endomitosis, ►recombination nodule; McClintock B 1931 Missouri Agric Exp Sta Bull 163; Carlson WR 1988 In: Corn and Corn Improvement, Agricultural Monograph 18, ASA-CS-SSA, Madison, Wisconsin, p 259.

Pachytene Stage: The chromosomes form pachynema. pachynema.

Packaging Cell Lines (For Retroviral Vectors): For the replication of the vector the viral proteins gag, pol, env are required but these are deleted from the vectors to prevent the production of disease-causing virions. The packaging signal ψ is however retained in the vector. Another solution is to insert these viral genes into host chromosomes or remove from the helper virus the packaging signal (Ψ [psi]) and delete the 3′-LTR. In neither case could the production (by two recombinations) of replication-competent virions be completely eliminated. Thus, the nucleic acid (with the transgene) can be packaged although the virions are defective. An improved construct removed LTRs from the structural genes and replaced them with heterologous promoters and polyadenylation signals. The *gag* and *pol* genes are placed on a plasmid different from the one that carries the *env* gene. Thus, in the packaging cell lines, these two are inserted at different chromosomal sites. Also, if the number of cell divisions is limited, the chance of recombination between vector and helper is reduced. In an improved packaging system, a stop codon is engineered into *gag* reading frame to prevent the assembly of a fully competent virus. In the packaging cell lines, the appropriate envelope protein for the intended target (ecotropic or amphotropic) should be present in the helper virus (pseudotyping) to insure optimal transfection. The envelope protein may need modification in order to ensure the proper targeting to the intended types of cells. Antibodies, specific for certain cell surface antigens or against particular receptors may be employed. Although some of these procedures appear very attractive, they may not always be equally efficient. These technical problems are obviously attracting serious research efforts. ►retroviral vectors, ►ecotropic retrovirus, ►amphotropic retrovirus, ►pseudotyping, ►viral vectors; Thaler S, Schnierle BS 2001 Mol Ther 4(3):273.

P

Packaging of Phage DNA: λ phage gene A recognizes the cos sites, gene D assists in filling the head (capsid) and genes W, F, V ILK and GMH assemble the phage from prefabricated elements and act in the processes shown diagrammatically in Fig. P6.

The DNA that first enters the phage capsid has the Nu end and the opposite end (the last) is the R end. The organization of the DNA in the phage head is not random; the geometry of the arrangement is determined by writhe of the DNA (Arsuaga J et al 2005 Proc Natl Acad Sci USA 102:9165). ►lambda phage, ►p16, ►development, ►writhing number, ►heedful rule; Smith DE et al 2001 Nature [Lond] 413:748, Kindt J et al 2001 Proc Natl Acad Sci USA 98:13671.

Packaging Signal (ψ): Allows the stuffing of the viral genome into the viral capsid.

Packing Ratio: The DNA molecule is much-much longer than the most extended chromosomes fibers. The packing ratio was defined as the proportion of the DNA double helix and the length of the chromosome fibers. In the human chromosome complement, the packing ratio was estimated to be more than 100:1 at metaphase. The length of the *Drosophila* genome at meiotic metaphase was estimated to be 7.8 μm and the length of a chain of 3000 nucleotides is approximately 1 μm. The *Drosophila* genome contains about 9×10^7 bp, hence the total length of DNA within the *Drosophila* genome is about 30,000 μm and that would indicate a packing ratio of 3846:1. The packing ratio indicates some of the problems the eukaryotic chromosomes encounter in condensing an enormous length of DNA to a small space and still replicating, transcribing and recombining it in an orderly manner. To illustrate the problems in a trivial way: many eukaryotes have the same packing problem as folding a 2.5 km (1.6 mi) long thread into a 2.5 cm (1″) skein. Prokaryotic type DNA—such as without nucleosomal structure—the excessive amount of plasmid DNA forms liquid crystalline molecular supercoils. ►Mosolov model, ►supercoiled DNA; see photo of bacterial chromosome at lysis, p. 1,150; DuPraw EJ 1970 DNA and Chromosomes. Holt, Rinehart and Winston, New York; Holmes VF, Cozzarelli NR 2000 Proc Natl Acad Sci USA 97:1322; Cook PR 2002 Nature Genet 32:347.

Pack-MULEs: The abundant (3000/rice genome) transposable elements in different plant species that carry fragments of cellular genes (~1000 in rice) derived from all chromosomes. These fragments can have multiple chromosomal origins, and can be functional. During millions of years, the fragments could be rearranged, amplified and contributed to evolution of plant genes. ►transposable elements plants; Jiang N et al 2004 Nature [Lond] 431:569.

Paclitaxel: ►taxol

Pac-Man Model: Kinetochores induce depolymerization of the microtubules of the kinetochore at their plus end and that allows the sister-chromatids to move toward the poles during mitosis by, so to say, chewing

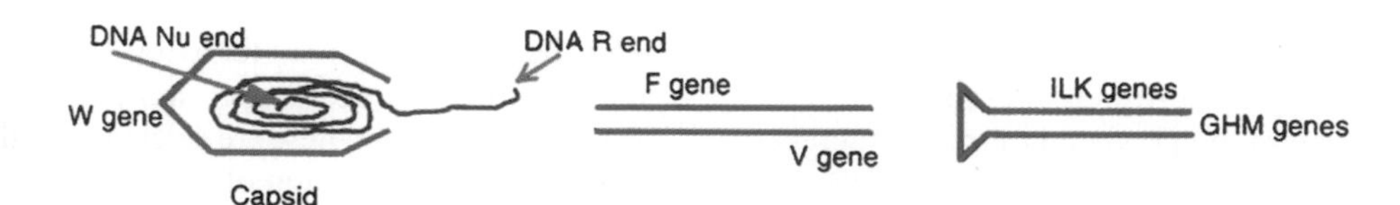

Figure P6. Packaging of phage DNA

up spindle fiber tracks. ►anaphase, ►spindle fibers, ►microtubules, ►kinetochore; Rogers GC et al 2004 Nature [Lond] 327:364; Liu J, Onuchic JN 2006 Proc Natl Acad Sci USA 103:18432.

PACT (p53 associated cellular protein): A negative regulator of p53. ►p53

Pactamycin: An inhibitor of eukaryotic peptide chain initiation (see Fig. P7).

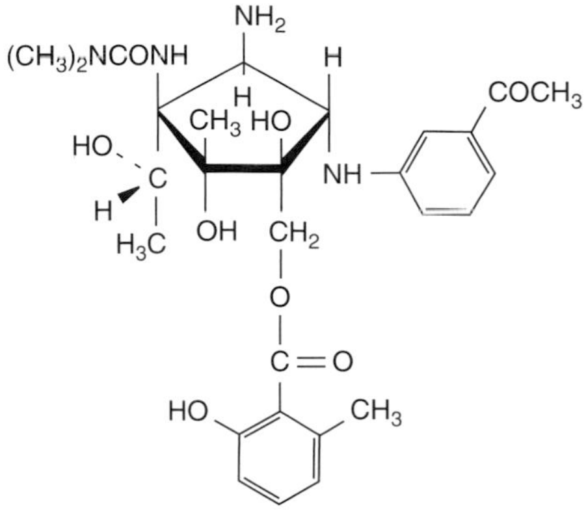

Figure P7. Pactamycin

PAD4: peptidylarginine deiminase. It converts methylarginine to citrulline and releases methylamine. It targets also multiple sites in histones H3 and H4. ►arginine, ►citrulline, ►methylation of DNA, ►histones, ►epigenesis; Wang Y et al 2004 Science 306:279.

Padlock Probe: Contains two target-complementary segments connected by linker sequences (see Fig. P8). Hybridization to target sequences brings the two ends close to each other and can be covalently ligated. The so circularized probes are thus catenated to the DNA (≈) like a padlock (OO). Such probes permit high-specificity detection and distinction among similar target sequences and can be manipulated without alterations or loss. By using circularizable or circularized allele-specific probe, primers and rolling circle, amplification can detect mutations in short genomic sequences (see Fig. P9). The principle of the procedure is shown modified after Lizardi PM et al 1998 Nature Genet 19:225.

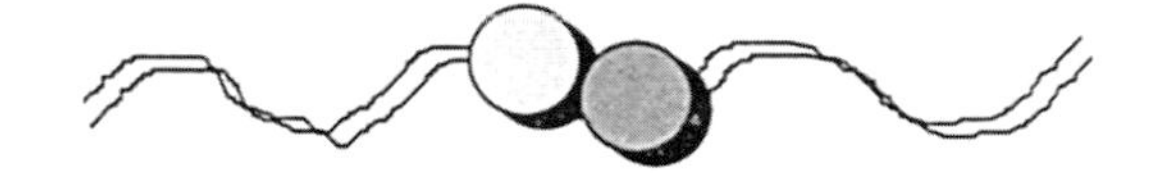

Figure P8. Padlock

Alternatively, for the rolling circle amplification two primers were used (see Fig. P10). After the first primer (P1) initiated the replication the second (P2) primer is bound to the tandem repeats and both primers generate repeats in opposite directions ⮌ or ⮎ using either the (+) or (−) strands, respectively. In 90 min, at least 10^9 copies of the circles are generated making it possible to detect very rare somatic mutations. Rolling circle amplification can detect gene copy number single base mutations and can quantify the transcribed mRNA (Christian AT et al

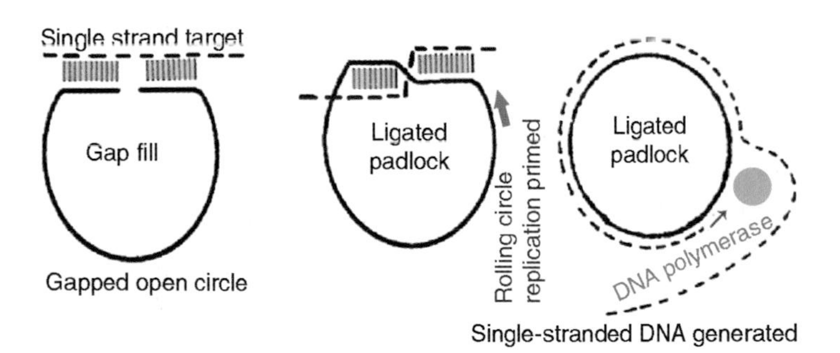

Figure P9. The probe at left, interrupted by a 6- to 10-base gap, hybridizes to the target DNA. The gap is filled with an allele-specific or DNA polymerase-generated sequence. After ligation, it generate a closed duplex padlock. A complementary (18-base) primer was then employed with a DNA polymerase. The original target DNA is not shown

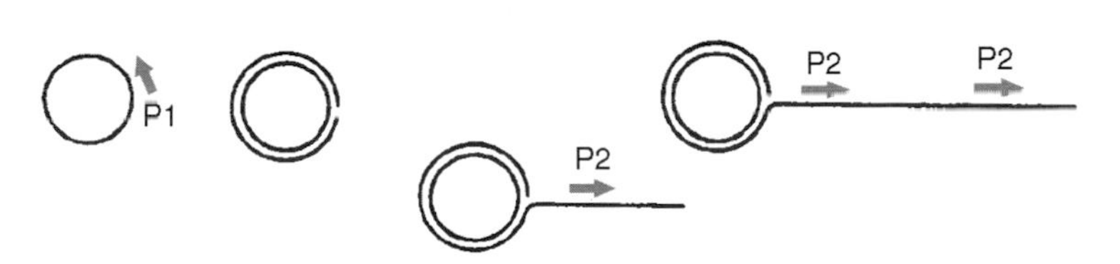

Figure P10. Padlock primers

2001 Proc Natl Acad Sci USA 98:14238). The procedure generated replication products, which were hybridized to either fluorescein- or Cy3-labeled deoxyribonucleoprotein-oligonucleotide (DNP) tags, respectively. The tag was anti-DNP immunoglobulin M (IgM). This process of condensation of amplification circles after hybridization of encoding tags is called CACHET. The procedure permits also the identification of single-copy genes by epifluorescence microscopy. ►probe, ►rolling circle, ►fluorochromes, ►immuno-globulins, ►DNA polymerases, ►microscopy, ►mutation detection; Baner J et al 2001 Curr Opin Biotechnol 12:11; Roulon T et al 2002 Nucleic Acids Res 30 (3):e12.

Padumnal Allele: Derived from the male. ►madumnal allele

PAF (population attributable fraction): PAF represents the fraction of the disease that would be eliminated if the risk factor were removed. High risk alleles generally show PAF > 50% whereas in rare alleles in common diseases it is generally <10%. The common modest-risk alleles account for greater PAF in common diseases than the rare high-risk alleles. This hypothesis is called the common disease/common variant (CDCV). ►complex disease, ►QTL, ►correlation, ►association, ►HapMap; Carlson CS et al 2004 Nature [Lond] 429:446.

PAF: A positive and negative regulator of RNA polymerase II mediated transcription (Shi X et al 1996 Mol Cell Biol 16:669). The Paf complex has a role also in polyadenylation of mRNA.

PAF-AH: platelet activating factor acetylhydrolase coupled with dynein affects neural migration in lissencephaly. ►lissencephaly, ►dynein; Tarricone C et al 2004 Neuron 44:809.

Page: An acronym for polyacrylamide gel electrophoresis. ►gel electrophoresis

Paget Disease: Two autosomal dominant forms have been described involving cancer of the bones or of the anogenital region (the region of the anus and genitalia) or the breast. The disease is an anomaly of osteoclastogenesis. BDB1 gene was located to 6p21.3 and PDB2 (also called familial expansile osteolysis, FEO) to 18q21-q22. Mutations in the tumor necrosis factor receptor, TNFR seem to be involved and affect the signaling by NF-κB (RANK, receptor activator of nuclear factor κB). Additional loci mapped to 5q35-qter (PDB3), to 5q31 (PDB4), to 2q36 (PDB5) and 10p13 (PDB6). ►osteoclast, ►osteoporosis, ►NF-κB, ►TNFR; Laurin N et al 2001 Am J Hum Genet 69:528.

PAH (polyaromatic hydrocarbon): The majority of PAH are carcinogenic (see Fig. P11).

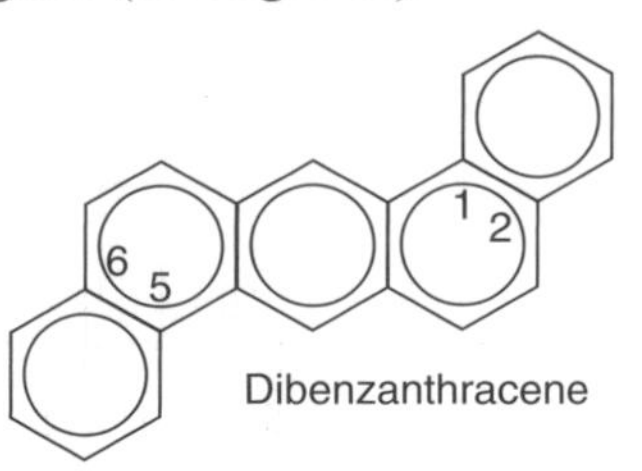

Figure P11. PAH

PAH (paired amphipathic helix motif): It may mediate protein-protein interactions in regulating enzyme functions. ►amphipathic

PAI: plasminogen activation inhibitor. ►plasminogen activator; Eilers AL et al 1999 J Biol Chem 274:32750.

Pain-Insensitivity: Controlled by defects causing hereditary sensory neuropathies. In the dominant form, the dorsal ganglia are degenerated. In the recessive neuropathy, the loss of myelinated A-fibers cause touch insensitivity. The congenital pain insensitivity with anhidrosis (CIPA, 1q21-q22) involves a defect of the nerve growth factor receptor (TRKA), and in the congenital insensitivity to pain without anhidrosis, the small myelinated A-delta fibers are defective. Mutation in the Na-channel subunit (SCN9A, encoded at 2q24.3) leads to pain-insensitivity (Cox JJ et al 2006 Nature [Lond] 444:894). The apparent insensitivity to the self-torture of the fakirs (Hindu ascetics) may be based on such genetic condition. ►neuropathy, ►Riley-Day syndrome, ►TRK, ►sensory neuropathy 1, ►anhidrosis; Mardy S et al 2001 Hum Mol Genet 10:179; Cheng H-YM et al 2002 Cell 108:31.

Pain-Sensitivity: May be traditionally treated with analgesics. Gene therapy by introduction of genes producing analgesic substances (catecholamines, enkephalins) or antinociceptive peptides are potential molecular approaches. The capsaicin or vanilloid receptor (VR1) control heat-gated ion channel with response to low temperature (~43°C) stimuli whereas the VRL-1 receptor responds to about 52°C. The vanilloid channel receptor is induced by protein kinase C. The transcriptional repressor DREAM constitutively suppresses prodynorphin in the neurons of the spinal chord. When DREAM is knocked out, there is still sufficient expression of dynorphin but there is a strong reduction in pain-sensitivity. Single amino acid substitutions (val^{158}/met) in catechol-*O*-methyltransferase (COMT) may modulate pain sensitivity/insensitivity. Simultaneous, two synonymous and one non-synonymous, divergence in the human haplotype of the gene modulate COMT protein expression by altering mRNAs secondary structure (Nackley AG et al 2006 Science 314:1930). Prostaglandin E^2 is a mediator of inflammatory pain-sensitization via glycine receptor α3 (Harvey RJ et al 2004 Science 304:884). Expectation of pain reduced the subjective feeling of pain as well as activation of pair-related areas of the brain (Koyama T et al 2005 Proc Natl Acad Sci USA 102: 12950). Covalent modification of reactive cysteines within TRPA1 (Transient Receptor Potential family of ion channels) by noxious compounds causes channel activation, rapidly signaling potential tissue damage through the pain pathway (Macpherson LJ et al 2007 Nature [Lond] 445:541). ►analgesic, ►nociceptor, ►catecholamines, ►enkephalins, ►endorphin, ►dynorphin, ►DREAM, ►protein kinase, ►prostaglandins, ►allodynia, ►temperature-sensitive mutation; Samad TA et al 20001 Nature [Lond] 410:471; Costigan M, Woolf CJ 2002 Cell 108:297; Mantyh PW et al 2002 Nature Rev Cancer 2:201; Zubieta J-K et al 2003 Science 299:1240.

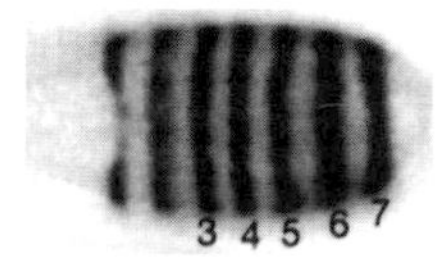

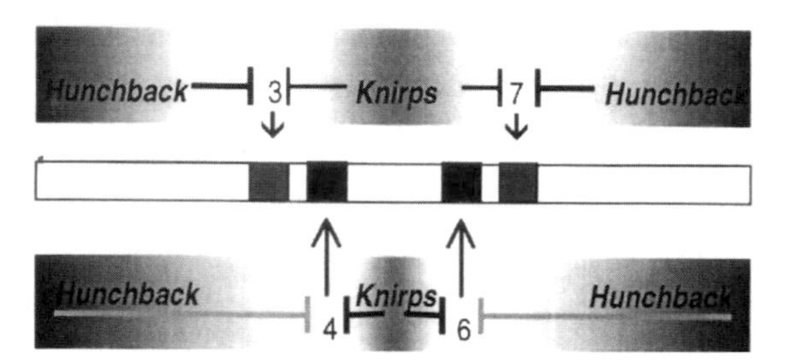

Figure P12. Pair rule genes. The banding pattern on the body of the Drosophila embryo is under the control of several regulatory genes. In the case of the *even-skipped* gene (chromosome 2.58) the seven stripes in the syncytial blastoderm are under the control of five enhancers. Three of them #1, #2, #3) drive the expression of single stripes and the remaining two control the expression of pairs of stripes (3 + 7 and 4 + 6). The stripes are formed at the boundary of interactions of the suppressor gradients of genes *Hunchback* (encoded at 3.48) and *Knirps* (encoded at 3.46) and the *even-skipped* enhancers. It is assumed that binding-site affinity and distribution on the enhancers determine the sensitivity to the repressors as illustrated on the diagram. The four bands in the middle represent body stripes (not chromosome bands!) brought about by the interactions of the suppressor gradients. (Modified after Clyde DE et al 2003 Nature [Lond] 426:849)

Pair Rule Genes: Determine the formation of alternating segments in the developing embryo as shown in Figure P12. Similar segment pattern, although with variations, occurs in other insects too. ►morphogenesis in *Drosophila*, ►metamerism, ►*fushi tarazu*, ►*knirps*, ►*engrailed*, ►*Runt*

Paired Box Genes: ►PAX

Paired t-Test: ►matched pairs test, ►Student's *t* distribution, ►*t* value

Paired-End diTAG (PET): PET uses chromatin immunoprecipitation to enrich DNA fragments for the mapping transcription factors across the genome. It separates signature sequences from the 5′ and 3′ ends of the fragments, concatenates them and maps them (Ng P et al 2005 Nat Methods 2:105). ►immunoprecipitation, ►transcriptome, ►transcription factor map

Paired-End Sequence: The product of the first sequencing of both ends of a cloned DNA tract. (See Zhao S et al 2000 Genomics 63:321).

Paired-End Sequence Method: The method used to identify structural alterations in the DNA. The standard genome sequence is compared with another genome represented by fosmid paired-end sequences (see Fig. P13). The procedure detects fine-scale variations in the genome that may be important for disease. ►fosmid; Tuzun E et al 2005 Nature Genet 37:727.

Pairing (synapsis): The intimate association of the meiotic chromosomes mediated by several protein factors. In prokaryotes, the RecA and the RecT protein have important role and in yeast and humans, the Rad52 protein carries out similar functions. ►meiosis, ►zygotene, ►pachytene analysis, ►somatic pairing, ►hydrogen pairing, ►base pair, ►tautomeric shift, ►synapsis,

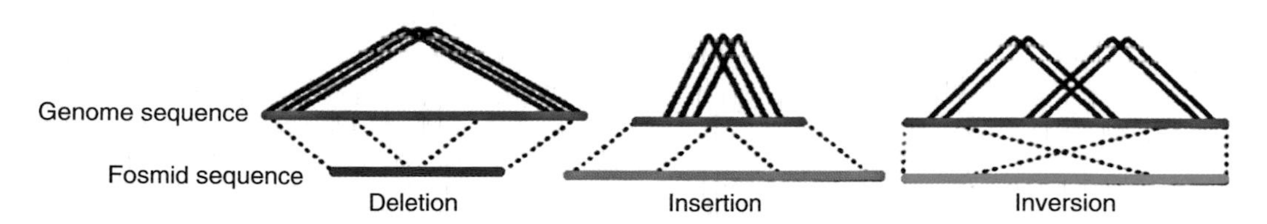

Figure P13. Paired-ed sequence method

P

►RecA, ►RecT, ►Rad, ►*Ph* gene; Kagawa W et al 2001 J Biol Chem 276:35201.

Pairing Alkylated Bases: ►alkylation

Pairing Centers: The cis-acting sites required for accurate segregation of homologous chromosomes during meiosis of *Caenorhabditis elegans* (MacQueen AJ et al 2005 Cell 123:1037). The *HIM-8* gene, encoding a zinc-finger protein, concentrates at the pairing centers on the X chromosome mediates chromosome-specific synapsis (Phillips CM et al 2005 Cell 123:1051).

Pairing-Sensitive Repression: Polycomb (PC) proteins bind to Polycomb-response elements (PREs) and thus cause repression. Repression is enhanced when two such elements are present. There are other similar elements like Mcp. ►Polycomb

Pair-wise Likelihood Score: Estimates the potential relationship between pairs of individuals on the basis of allele sharing. (See Smith BR et al 2001 Genetics 158:1329).

PAK (p21 activated kinase): The serine/threonine kinases activated by GTPases, Rac and Cdc42. Pak3 regulates Raf-1 by phosphorylating serine 338 in rats. Paks regulate the actin cytoskeleton, cell motility, neurogenesis, angiogenesis, signal transduction, apoptosis, metastasis, etc. A non-syndromic mental retardation (human chromosome Xq22, yeast homolog is *STE20*) prematurely terminates PAK3 transcription. ►GTPase, ►Rac, ►raf, ►Cdc42, ►p21, ►p35, ►PDK, ►actin, ►cytoskeleton, ►apoptosis, ►mental retardation, ►non-syndromic; Xia C et al 2001 Proc Natl Acad Sci USA 98:6174; Bokoch GM 2003 Annu Rev Biochem 72:743.

PAL: Phenylalanine ammonia lyase.

Palea: The inner, frequently translucent, bract around the grass flower (see Fig. P14).

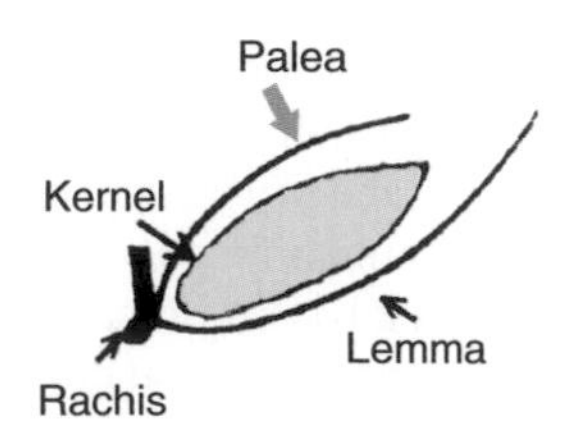

Figure P14. Palea

Paleogenomics: ►ancient DNA

Paleolithic Age (Old Stone Age): More than 20,000 years ago it marked the beginning of human tool formation and cave artistry by the Cro-Magnon humans. ►neolithic, ►mesolithic, ►geological time periods, ►Lascaux, ►Neanderthal

Paleologous Loci: Include ancient duplications.

Paleontology: Deals with the relics of past geological periods. Its methods and materials are used for the study of the evolution of biological forms. ►paleolithic age, ►geological time periods; Eurasian Miocene and Pleistocene land mammals and excavation sites: http://www.helsinki.fi/science/now/.

Paleozoic: The geological period between about 225 to 570 million years ago. During the later part of this period land plants, amphibians and reptile appeared. ►geological time periods

Palindrome: The region of a DNA strand where complementary bases are in opposite sequence, such as ATGCAC*GTGCAT (see Fig. P15). Palindromes may come about by inverted repeats of sections of the double-stranded DNA where these sequences of the opposite strands read the same forward and backward. Upon folding of these sequences in a single strand, they can assume structures with paired bases.

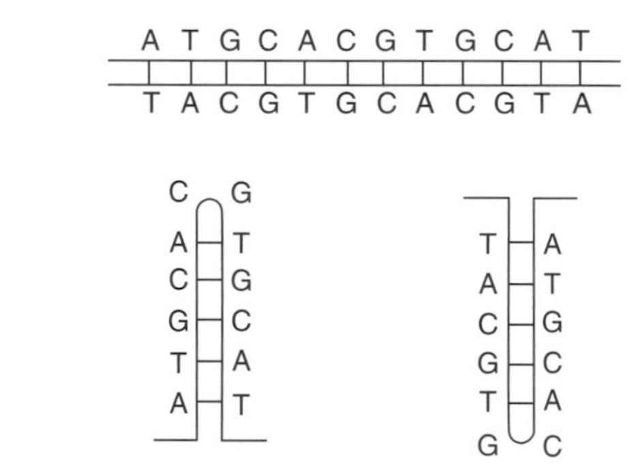

Figure P15. Palindromic DNA and possible pairing within single strands of it. (From Flavell RB, Smith DB 1975 Stadler Symp 7:47)

Palindromic sequences in the DNA reassociate very rapidly because of the complementary bases are in close vicinity. A simple palindromic word is MADAM, it reads the same from left to right or from right to left. Palindromic sequences are often unstable; recombination within palindromes results in deletions and duplications. The restriction enzyme recognition sites are palindromic. Palindromes are common in cancer cells and provide a platform for gene amplification and chromosomal aberrations (Tanaka H et al 2005 Nature Genet 37:320). ►stem and loop, ►inverted repeats, ►insertion elements, ►restriction enzyme, ►RecA independent recombination; Leach DR 1994 Bioessays 16:893; Nasar F et al 2000 Mol Cell Biol 20:3449; Zhu Z-H et al 2001 Proc Natl Acad Sci USA 98:8326; short palindromic repeat detection: http://crispr.u-psud.fr/Server/CRISPRfinder.php.

Palingenesis: The regeneration of lost organs and parts or the reappearance of evolutionarily ancestral traits during ontogeny. According to Ernst Haeckel (1834–1914) the ontogeny recapitulates the phylogeny. ►ontogeny, ►phylogeny

Palisade Cells: Oblong cells and arranged in a row; the large palisade parenchyma cells are below the upper epidermis of plant leaves and loaded with chloroplasts (see Fig. P16).

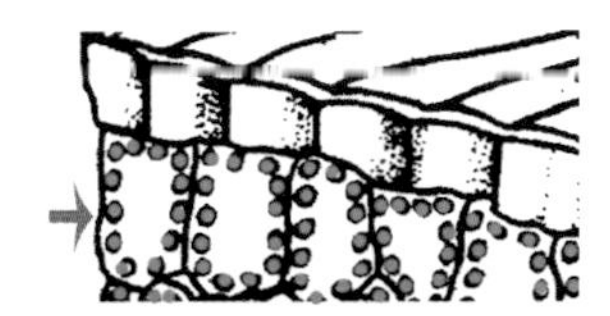

Figure P16. Palisade cells in a leaf

Pallister-Hall Syndrome: postaxial polysyndactyly (PAPA1), encoded in human chromosome 7q13, at the same location as the Greig syndrome. The gene is also called GLI3 and its product has a homology to the *Drosophila* Krüppel Zn-finger protein. The clinically distinct phenotypes of the two diseases are due to different allelic mutations of the same gene (Johnston JJ et al 2005 Am J Hum Genet 76:609). A similar syndactyly was recorded by French scientist, Maupertuis, in the Prussian royal court in Berlin in 1756. ►polydactyly, ►Greig cephalopolysyndactyly, ►DNA-binding protein domains, ►EEC syndrome, ►Smith-Lemli-Opitz syndrome

Palmitoylation: The covalent attachment of fatty acids (mainly palmitate) to cysteine residues to membrane proteins by creating a thioester link. This process tethers protein reversibly to cellular membrane surfaces. The enzymes involved belong to the protein acetyltransferase (PAT) family (e.g., Akr1, Erf2) and share a common domain of DHHC (aspartate-histidine-histidine-cysteine; see Mitchell DA et al 2006 J Lipid Res 47:1118). Palmitoylated proteins play key roles in cell signaling, membrane traffic, cancer and synaptic transmission, heterotrimeric G proteins and many non-receptor tyrosine kinases (Fyn, Lck, Yes) and the epithelial nitric oxide synthase are palmitoylated. By proteomic analysis, 35 new PAT proteins have been identified in budding yeast (Roth AF et al 2006 Cell 125:1003). ►fatty acids, ►G proteins, ►Fyn, ►Lck, ►Yes, ►nitric oxide

Palmprint: ►fingerprint, ►Down's syndrome, ►simian crease

Palomino: A horse with light tan color fur and flaxen mane and tail. The genetic constitution is *AAbbCCDd* (see Fig. P17).

Figure P17. Palomino

PALS: ►alternative splicing

Palsy (paralysis): Cerebral palsy may be caused by physical injuries or may be part of the symptoms of diverse genetic syndromes. ►syndrome, ►tau

PAM: ►evolutionary clock

PAMAM (polyamidoamine): The dendrimers can carry chemotherapy compounds (methotrexate) to the cancer cell, especially if connected with folate hooks that attach preferentially to the abundant folate receptors of cancer cells. ►dendrimer, ►methotrexate, ►nanotechnology; Najlah M et al 2007 Bioconjug Chem 18:937.

PAMP: pathogen-associated molecular pattern recognition is an important step in developing reaction in animals and plants against infection. ►vaccine

PAN: The genus of chimpanzees. ►primates, ►hominidae; Gagneux P 2002 Trends Genet 18:327.

PAN Editing: Adding of U residues to the primary transcripts of mtDNA and thus causing extensive post-transcriptional changes in RNA. ►kinetoplast, ►RNA editing

Pancreas: A large gland behind the stomach, between the spleen and the duodenum. It secretes insulin, glucagons, and protein-digesting enzymes. ►spleen, ►duodenum, ►insulin, ►glucagons, ►diabetes, ►Langerhans islets, diabetes: http://www.cbil.upenn.edu/EPConDB.

Pancreatic Adenocarcinoma: The cancer of the pancreas is frequently associated with loss or defect of DCC, or p53 or MTS1 oncogene suppressors. One study revealed an average of 15 annotated gene alterations (amplifications, deletions, tumor suppressors) in 24 adenocarcinoma lines (Aguirre AJ et al 2004 Proc Natl Acad Sci USA 101:9067). It was also attributed to mutations in codon 12 of the c-K-ras

P

(Kirsten RAS) gene encoded at human chromosome 12p12. A dominant susceptibility locus was assigned to 4q32-q34. Chromosomal aberrations, telomere shortening, methylation of CpG islands and point mutation may be the underlying cause. Silencing of cancer suppressors by hypermethylation at multiple genes may be involved. HER2/NEU overexpression is also a frequent cause. Tyrosine-kinase growth factor receptors may be involved. It may be associated with the Peutz-Jeghers syndrome (66%) and other syndromes such as hereditary pancreatitis (40%), BRCA2 (5–10%), etc. Its prevalence in the USA is $\sim 3 \times 10^{-5}$, and its prognosis is bad. ►DCC, ►p53, ►p16, ►Kirsten-Ras, ►p21, ►Peutz-Jeghers syndrome, ►HER2, ►NEU; Eberle MA et al 2002 Am J Hum Genet 70:1044; Hansel DE et al 2003 Annu Rev Genomics Hum Genet 4:237.

Pancreatitis, Hereditary: Autosomal dominant (7q35) gene (80% penetrance and variable expressivity) has an onset before the teen years, appearing as abdominal pain and other anomalies. The basic defect is in a cationic trypsinogen. ►trypsin, ►Johanson-Blizzard syndrome

Pancytopenia: Low blood cell number.

Panda (*Ailurus fulgens*): 2n = 36.

Pandemic: The infection by a microbe or virus spread over large areas (countries, continents).

Paneth's Cells: The secretory intestinal epithelial cells expressing defensin and other antimicrobial peptides. In the Paneth cells of mice, 149 transcripts expressed 2 to 45-fold by microbial colonization. Among them was very abundant (31-fold increase) a bactericidal lectin (RegIIIγ) that may represent a primitive evolutionary form of innate immunity (Cash HL et al 2006 Science 313:1126). ►defensin, ►microbiome; Ghosh D et al 2002 Nature Immunol 3:583.

Pangenesis: An ancient misconception about heredity that originated in the Aristotelian epoch and periodically revived during the centuries. Charles Darwin has also interpreted inheritance as pangenesis. Accordingly, all the information expressed during the life of the individuals is transported to the gametes from all parts of the body. Thus, pangenesis is the means of the inheritance of all, including the acquired characters. ►lysenkoism, ►acquired characters

Pangenome: The essential, shared sequences among all representatives of a species (Tettelin H et al 2005 Proc Natl Acad Sci USA 102:13950).

Panic Disorder (PD): Episodic panic attacks involving palpitations, sweating, shortness of breath, feeling of choking, chest pain, false touch sensations usually in the absence of physical contact (such as burning, prickling), nausea, etc. The heritability appeared 0.48. Panic disorder may be associated with a number of physical diseases. Chromosome 13q appears to harbor several genes responsible for PD and chromosome 22 may carry some susceptibility factors. ►psychoses, ►panic obsessive disorder, ►anxiety; Hamilton SP et al 2003 Proc Natl Acad Sci USA 100:2550.

Panicle: An inflorescence of a compound raceme structure such as of oats. ►raceme

Panmictic Index: ►fixation index, ►panmixis

Panmixis (panmixia): Random mating; in a population there is equal chance for each individual to mate with any other of the opposite sex. ►Hardy-Weinberg theorem

Panning: The use of antibody affinity chromatography or ELISA for the separation of specific molecules (in analogy to the gold-washing pans of the gold hunters). ►affinity chromatography, ►ELISA, ►phage display; Chen G et al 2001 Nature Biotechnol 19:537.

Panspermia: The theory claiming that life has originated at several places in the universe and spread to earth by meteorites or by other means. ►origin of life

Panther (*Panthera pardus*, leopard): 2n = 38: A feline species; in captivity may be crossed with lion but no mating is known to take place in the wild where conspecific sexual partners are available.

Panther: The database for functionally related proteins, signaling pathways. http://panther.applied biosystems.com.

Pantothenic Acid: A precursor of Coenzyme A (see Fig. P18).

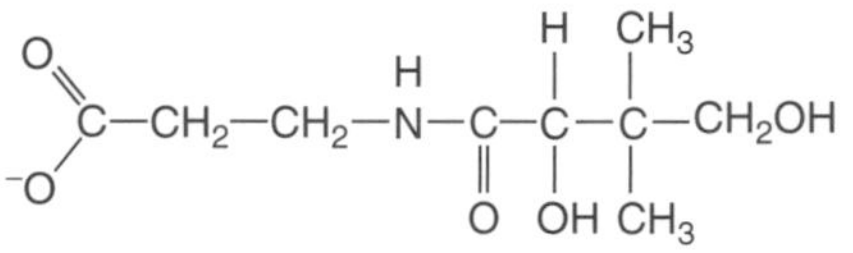

Figure P18. Panthotenate

Pantropic: Can affiliate with many different types of tissues.

PAP (purple acid phosphatases): Ubiquitous metallo-phosphoesterase proteins with phosphatase, exonuclease, 5′-nucleotidase, etc. functions. (See Li D et al 2002 J Biol Chem 277:27772).

Pap (Papanicolaou) Test: A cytological test for pre-malignant or malignant conditions (used primarily on smears obtained from the female urogenital tract). It detects also papilloma virus infections. ►malignant growth, ►papilloma virus

PAPA Syndrome (pyogenic sterile arthritis and acne and familial recurrent arthritis, 15p): It may involve also ulcerative skin lesions (pyoderma gangrenosum). It may be caused by defects in the 15-exon CD2-binding proteinl (CD2BP1). ►CD2; Wise CA et al 2002 Hum Mol Genet 11:961; Shoham NG et al 2003 Proc Natl Acad Sci USA 100:13501.

Papain: A member of a family of proteolytic enzymes with an imidazole group near the nucleophilic SH group, and the former plays a role as a proton donor to the cleaved-off part. Papain cleaves immunoglobulin G into three near equal size fragments and this helped in clarifying the structure of antibodies. ►proteolytic, ►calpain, ►immunoglobulins, ►antibody

Papanicolaou Test: ►PAP test

Papaver: ►poppy

Papaya (*Carica papaya*): A melon-like, edible fruit, latex-producing small tree with four genera and all 2n = 2x = 18; it is the source of the proteolytic enzyme papain. ►papain

PapD: Gram-negative bacterial chaperone (28.5 kDa) delivers the components of the pilus from the periplasm. ►periplasma, ►chaperones, ►pilus, ►Gram negative/Gram positive

Paper Chromatography: A technique for the separation of (organic) molecules in filter paper by applying the mixture in a spot or band at the bottom of the paper and allowing an appropriate solvent to be sucked up and thus carry the components at different speed (to different height) so they can be separated (see Fig. P19). The components become visible by their natural color or by the application of specific reagents. A large variety of different modifications were worked out in one or two dimensions, in ascending and descending ways. Nowadays paper chromatography is not used very much. ►chromatography, ►thin layer chromatography, ►Rf value, ►column chromatography, ►high performance liquid chromatography, ►affinity chromatography, ►ion-exchange chromatography

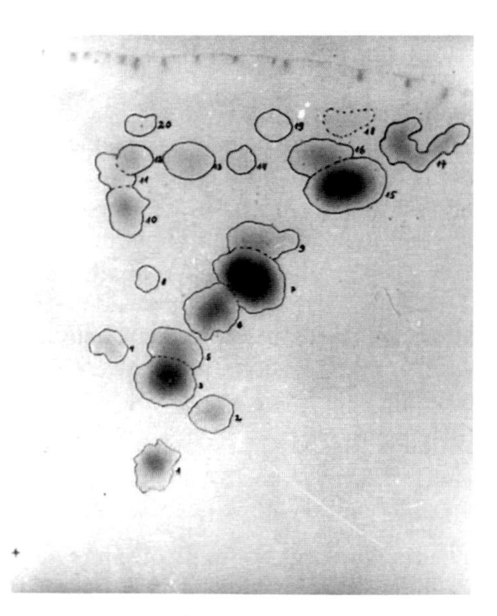

Figure P19. Two-dimensional separation of 20 amino acids in a plant extact

Papillary Renal Cancer, Hereditary: Based on the MET oncogene at human chromosome 7q31, encoding a hepatocyte growth factor receptor. ►MET, ►HGF, ►receptor tyrosine kinase

Papillary Thyroid Carcinoma: Caused by the RET oncogene. It accounts for about 80% of the thyroid cancers that have prevalence in the 10^{-5} range. Its incidence is higher in females than males. ►RET

Papillation, Bacterial: Secondary colonies develop on the colonies of bacteria.

Papilloma: Pre-malignant neoplasia displaying epithelial and dermal finger-like projections.

Papilloma Virus (HPV, human papilloma virus): A double-stranded DNA ($\approx 5.3 \times 10^6$ Da or ~8 kb) virus causing animal and human warts and squamous carcinomas in mice. The HPV-16 and 18 are frequently present in cervical cancer. The E6 viral protein of HPV-16—through a ubiquitin path—is prone to degrade the p53 tumor suppressor if at amino acid position 72 there is an arginine rather than a proline. The E7 protein of strain HPV-18 degrades another tumor suppressor protein RB. The E1 protein is a hexameric helicase (Enemark EJ, Joshua-Tor L 2006 Nature [Lond] 442:270). Apparently, highly effective vaccines have been developed against HPV strains 16 and 18 that most commonly cause viral cervical cancer. HPV is a critical factor in the majority of cases of cervical cancer that which allowed development of strategies to prevent this form of oncogenesis. It is important to note that several other cancers are also associated with HPV infection, including head and neck cancers. Cancer prevention will require the long-term observation of a large number of treated women and it is necessary in the meantime to monitor for unintended adverse consequences of vaccination (Baden LR et al 2007 N Engl J Med 356:1990). There is also an unresolved moral and ethical problem regarding the age of vaccination of adolescent, sexually not active girls. HPV has about 100 different strains and some pose cancer risk whereas some others do not seem to be carcinogenic. HPV is one of the most common causes of sexually transmitted disease and condom use does not offer perfect protection because the transmission is through

P

the skin. HPV has been used as a genetic vector. ►papova viruses, ►p53, ►retinoblastoma, ►tumor suppressor, ►cervical cancer, ►Pap test, ►condom, ►helicases; Wolf JK, Ramirez PT 2001 Cancer Invest 19:621.

Papillon-Lefèvre Syndrome: ►periodontitis

Papova Viruses: A large class of (oncogenic) animal viruses of double-stranded, circular DNA includes the polyoma viruses, the bovine papilloma virus and simian virus 40 (SV40), etc. that have been used as genetic vectors for transformation of animal cells. Also, they have been extensively studied by molecular techniques to gain information on structure and function. ►polyoma, ►Simian virus 40, ►papilloma virus; Soeda E, Maruyama T 1982 Adv Biophys 15:1.

PAPS: 3′-phosphoadenosine-5′-phosphosulfate is a sulfate donor in several biochemical reactions, involving cerebrosides, glycosaminoglycans and steroids. It is generated by the pathway: ATP + sulfate → adenosine-3′-phosphosulfate (APS) + pyrophosphate, APS + ATP → PAPS + ADP. Mutations at 10q23-q24 (spondyloepimetaphyseal dysplasia, SEMD) locus encoding the PAPSS2 cause short bowed limbs, large knee joints, brachydactyly, curved spinal column (kyphoscoliosis). ►bone diseases

PAR (pseudoautosomal region): Where recombination may take place between the X and Y chromosomes. The ends of both of the short arm (PAR1) and the long arm (PAR2) have pseudoautosomal regions. ►pseudoautosomal, ►X chromosome, ►Y chromosome, ►lyonization; Dupuis J, Van Eerdewegh P 2000 Am J Hum Genet 67:462.

PAR (protease-activated receptors): The seven-transmembrane G protein-coupled receptors that mediate thrombin-triggered phosphoinositide hydrolysis. Pars are involved in thrombosis, inflammation and vascular biology. Matrix metalloprotease (MMP-1) is an agonist of Par1 (Boire A et al 2005 Cell 120:303). Pars may also activate proteases and protect the airways. Par1 has a role in invasive and metastatic cancers. PAR is a cofactor of PAR4. Binding and phosphorylation of PAR4 by Akt is essential for the survival of cancer cells. Inhibition of the PI3K-Akt pathway leads Par-4—dependent apoptosis (Goswami A et al 2005 Mol Cell 20:33). Par2 is a trypsin receptor. ►thrombin, ►phosphoinositides, ►metalloproteinases, ►AKT, ►apoptosis; Kamath L et al 2001 Cancer Res 61:5933.

Parabiosis: Two animals joined together naturally such as Siamese twins or by surgical methods and can be used to study the interaction of hormones, transduction signals, etc., in-between two different individuals. Intrauterine parabiosis develops immune tolerance.

Paracellular Space: The intercellular space in the tissues.

Paracentric Inversion: ►inversion paracentric

Paracentrotus lividus: Sea urchin; extensively studied by embryologists. ►sea urchins

Paracrine Effect: A ligand (e.g., hormone) is released by a gland and affects neighboring cells.

Paracrine Stimulation: When one type of cell affects the function (such as proliferation) of another (nearby) cell. ►autocrine; Janowska-Wieczorek A et al 2001 Stem Cells 19:99.

Paracytosis: Passing bacteria through cell layers without disruption of the cells (See van Schilfgaarde M et al 1995 Infect Immun 63:4729).

Paradigm: A model or an example to be followed.

Paradox: A statement or phenomenon, which is apparently contradictory to current knowledge but may actually be true.

Paraesthesia (paresthesia): Peripheral nerve damage, disease-caused itching, burning and tickling sensation.

Paraganglioma: ►mitochondrial diseases in humans

Paraganglion: Cells originating from the nerve ectoderm flanking the adrenal medulla, and darkly stained by chromium salts. These cells may form a type of phaeochromocytoma tumors that secrete excessive amounts of epinephrine and norepinephrine. ►SHC oncogene

Paragenetic: phenotypic alterations not involving hereditary mutation.

Parahemophilia: Determined by homozygosity of semi-dominant autosomal genes. The symptoms involve bleeding similar to the conditions observed in hemophiliacs, bleeding from the uterus (menorrhagia) several days following childbirth. The physiological basis is a deficiency of proaccelerin, a protein factor (V) involved in the stimulation of the synthesis of prothrombin. The therapy requires blood or plasma. ►antihemophilia factors, ►hemophilia, ►pro-thrombin deficiency, ►hemostasis

Parahox Genes: The hox-like genes in clusters, separate from the hox genes, and has originated by duplication from an ancestral protohox gene. ►homeotic genes

Parainfluenza Viruses: A group of immunologically related but distinguishable pathogens responsible for some respiratory diseases. ►Sendai virus

Paralinin: ►karyolymph

Parallel Cascade Identification: Non-linear systems modeling approach. In biology, it can be used to predict long-term treatment response for cancer on the basis of small differences of gene expression levels. (See Korenberg MJ 2002 J Proteome Res 1:55).

Parallel Substitution: Various organismal lineages may display similar or different nucleotides at a number of sites. The chance of these substitutions at a site (*p*) in (*n*) lineages can be predicted on the basis of the binomial distribution, $(p + [1 - p])^n$ and upon expansion, e.g., for n = 5 it becomes $p^5 + 5p^4(1-p) + 10p^3(1-p)^2 + 10p^2(1-p)^3 + 5p(1-p)^4 + (1-p)^5$ and the same change per any two lines is p^2. ►evolution and base substitutions, ►evolutionary substitution rate, ►evolutionary tree, ►parallel variation

Parallel Synthesis: An approach commonly applied in drug development. Several similar compounds are generated simultaneously rather in a sequence, one after the other in order to speed up the process of discovery of effective drugs. ►combinatorial chemistry

Parallel Variation: Within taxonomically closely or even distantly related groups of organisms similar mutations may occur during evolution. Mutation in regulatory switches may be the basic cause of these alterations. ►parallel substitution; Vavilov NI 1922 J Genet 12:47; Pagel M 2000 Brief Bioinform 1(2):117.

Paraloci: They have the same properties as pseudoalleles. ►pseudoalleles

Paralogon: A pair of genes evolutionarily derived from common ancestral sequences.

Paralogous Loci: Originated by duplication that was followed by divergence. However, Paralogs may provide backup in case of defect or damage of one of the pairs (Kafri R et al 2005 Nature Genet 37:295). ►orthologous loci, ►isolocus, ►evolution of proteins, ►non-orthologous gene displacement, ►gene family, ►duplication, ►subfunctionalization, ►tetralogue, ►outparalog, ►inparalog; Yamamoto E, Knap HT 2001 Mol Biol Evol 18:1522.

Paralogy: Evolution by duplication of a locus. ►orthologous loci, ►orthology, ►homolog

Paramecium: Unicellular Protozoon. Normally reproduces by binary fission, i.e., a single individual splits into two. Each cell has two diploid micronuclei and a polyploid macronucleus. At fission, the micronuclei divide by mitosis while the macronucleus is simply halved. These animals also have sexual processes (*conjugation*). Two of the slipper-shaped cells of opposite mating type attach to each other and proceed with meiosis of the micronuclei (see Fig. P20). Only one of the four products of meiosis survives in each of the conjugants. Each of these haploid cells divides into four cells (gametes). One of these gametes (male) is passed on into the other conjugating partner through a *conjugation bridge* and fuses with a haploid gamete (female).

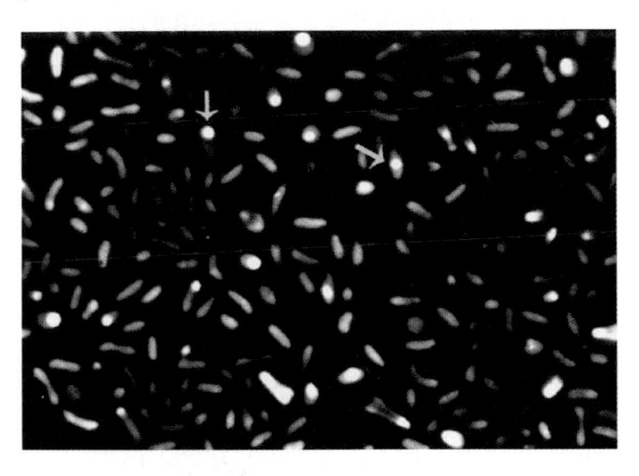

Figure P20. *Paramecium aurelia* (500 X) cells with bright and non-bright kappa particles symbionts (1,650X). The symbionts are bacteria. The bright particles contain the so-called R (refractive) bodies, which are bacteriophages. The non-bright kappa can give rise to bright indicating lysogeny. The kappa-free (*kk*) paramecia are sensitive to the toxin produced by the bright particles and may be killed. The K, killer stocks are immune to the toxin. (From Preer JR et al 1974 Bacteriol Rev 38:113 [Photo by C. Kung]; courtesy of Dr. J. R. Preer)

This is a reciprocal fertilization, resulting in diploid nuclei in the conjugants. Subsequently the pair separates into two *exconjugants*. The macronucleus disintegrates then in both. The diploid zygotic nuclei undergo two mitoses and form four diploid nuclei each. Two of the four nuclei function as separate micronuclei of the cells whereas the other two fuses into a macronucleus that become polyploid, and that is responsible for all metabolic functions and for the phenotype. Besides this sexual reproduction (conjugation), *Paramecia* may practice self-fertilization (*autogamy*).

Meiosis takes place and the one surviving product divides twice by mitoses. Two of these identical cells then fuse and form two diploid, isogenic micronuclei.

If the conjugation lasts longer, cytoplasmic particles may also be transfered through the conjugation bridge. Chromosome numbers may be 63–123; in the macronuclei there may be 800 or more chromosomes. *P. tetraurelia* genome includes 39,642 genes and 80% carry introns of mean 25 bp. The large gene number is apparently the result of whole genome duplication (Aury J-M et al 2006 Nature [Lond] 444:171). ►killer strains, ►symbionts hereditary, ►*Ascaris*, ►macronucleus, ►chromosome diminution, ►duplication, ►polyploidy, ►internally eliminated sequences, ►cortical inheritance, ►conjugation paramecia; Sonneborn TM 1974 In: King RC (Ed.) Handbook of Genetics, vol. 2, Plenum, New York, p. 469; Prescott DM 2000 Nature Rev Genet 1:191; *Paramecium tetraurelia* database: http://paramecium.cgm.cnrs-gif.fr.

Parameter: A quantity that specifies a hypothetical population in some respect or a variable to which a constant value is attributed for a specific purpose or process. Statistics usually denotes parameters by Greek letters and Latin letters indicates the computed values.

Parameter Alpha: ►alpha parameter

Parametric Methods in Statistics: Involves explicit assumptions about population distribution and parameters such as the mean, standard deviation of the normal distribution, the *p* parameter of the *Bernoulli process*, etc. ►Bernoulli process, ►normal distribution, ►non-parametric statistics, ►robustness

Parameter Space: In a dynamic system the values of the various parameters of a model are constrained within this limit.

Paramyxoviruses: Negative-sense retroviruses of 15–19 kb RNA containing 6–10 genes. They are infectious to a wide range of mammals. In humans, they cause influenza-like respiratory diseases. (See Gotoh B et al 2002 Rev Med Virol 12:337).

Paramutation: A *paramutable* allele becomes a *paramutant* (paramutated) in response to a *paramutagenic* allele if the two are in heterozygous condition. The alteration is similar but not identical to the paramutagenic allele. Both *paramutability* and paramutagenic functions are allele-specific. In contrast to gene conversion, paramutation may take place at low frequency and also in the absence of a paramutagenic allele. At the *R* locus of maize partial reversion of the paramutant may happen but this has not been observed at the *B* locus of maize. The paramutant phenotype at the *R* locus may vary but at the *B* locus, the phenotype appears to be uniform. The exact mechanism of this heritable alteration is not fully understood.

Apparently at the *R* locus of maize hypermethylation is involved, at the *B* locus involvement of methylation has not been detected. Distant upstream sequences play regulatory role in the expression and paramutation of the *B′* locus (Stam M et al 2002 Genetics 162:917). At the *pl* locus the paramutation seems to results in a genetic alteration of the chromatin structure, which affects the regulation of the expression of the gene during development. It appears that the level of transcription is reduced at the paramutant allele compared to that in the paramutable. one.

Although paramutation has been considered an endogenous mechanism, it appears that in the promoter region of the two *r* alleles in the homozygotes the *doppia* (CACTA) transposable elements are present within the 387 bp σ region that is intercalated between the two S elements in opposite orientation. (These elements are called S because they are responsible for anthocyanin coloration of the seed by this complex locus. The elements responsible for coloration of the plant were named P). Paramutation of the *b1* locus of maize depends on an RNA polymerase encoded by the *mop1* gene (*mediator of paramutation*). Paramutation at the *b1* locus involves the presence of non-coding tandem repeats of an

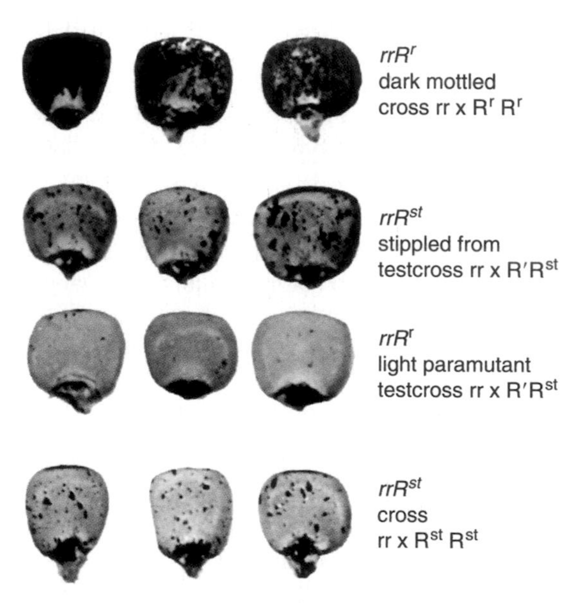

Figure P21. Paramutation results in reduced pigmentation in the triploid aleurone of maize. The R^r homozygotes are fully colored (when all other color-determining alleles are present). The r homozygotes are colorless. The rr^r genotype is responsible for the dark mottled aleurone. R^{st} causes paramutation (stippling) of the R^r paramutable allele that may be manifested in different grades. In the crosses the pistillate parents are shown first, left. (Courtesy of Brink RA see also 1956 Genetics 41:872)

853 bp sequence 100 kb upstream. The number of repeats may be 7 to 1. The strength of the paramutation is correlated with the number of these repeats and a single copy is not sufficient for paramutation. The RNA polymerase transcribes both strands of the repeats (Alleman M et al 2006 Nature [Lond] 442:295).

Paramutation is not a general property of all genes although similar phenomena have been observed at a few other genes in maize and other plants. This phenomenon seems to violate the Mendelian principle that alleles segregate during meiosis independently and during the process, no "contamination" takes place. Paramutation in the broad sense involves several types of gene silencing in various organisms. Paramutation-like phenomenon attributed to transmethylation was observed in mouse (Herman H et al 2004 Nature Genet 34:199). A new mechanism of paramutation was reported in mouse. Insertion of a 3 kilobase *LacZ-neomycin* cassette into the *Kit* gene downstream of initiator ATG site resulted in mutation involving white spots in the animals because inactivation of a tyrosine kinase receptor and defect in melanogenesis (►phenotype on the reconstructed image of an animal). Although the homozygotes were lethal, the heterozygotes expressed the phenotype as illustrated. The proven wild type individuals (lacking the insertion) among the progeny of heterozygotes displayed the white patches on the tail and feet as did the heterozygous mutants. This phenotype was transmitted by male and female to the offspring and there was a reduced level of *Kit* mRNA and an accumulation of abnormal size non-polyadenylated RNA molecules. The paramutant condition was transmitted though meiosis for several generations but eventually it was diluted out. Injection into fertilized eggs either the total RNA from $Kit^{tm1Alf/+}$ or *Kit*-specific microRNA also induced the white tail (see Fig. P22). The observations indicate a particular type of epigenetic inheritance of RNA molecules (Rassoulzadegan M et al 2006 Nature [Lond] 441:469). This phenomenon bears similarities to the case in the plant *Arabidopsis* claiming that a cache of RNA can be maintained in the nuclei and cause the reappearance of an atavistic trait (Lolle S et al 2005 Nature [Lond] 434:505). Newer information indicates, however, that the apparent "atavism" is due to contamination by pollen of the *Arabidopsis* mutant *HOTHEAD*, which has protruding stigma making unexpected cross-pollination easier (Pennisi E 2006 Science 313:1864). ►gene conversion, ►copy choice, ►directed mutation, ►localized mutagenesis, ►pangenesis, ►blending inheritance, ►presence-absence hypothesis, ►graft hybrid, ►co-suppression, ►RIP, ►epigenesis, ►position effect, ►transvection, ►tissue specificity, ►atavism; Brink RA 1960 Quart Rev Biol 35:120; Hagemann R, Berg W 1978 Theor Appl Genet 53:113; Chandler VL 2000 Plant Mol Biol 43 (2–3):121; Lisch D et al 2002 Proc Natl Acad Sci USA 99:6130; Chandler VL, Stam M 2004 Nature Rev Genet 5:532; Chandler VL 2007 Cell 128:641.

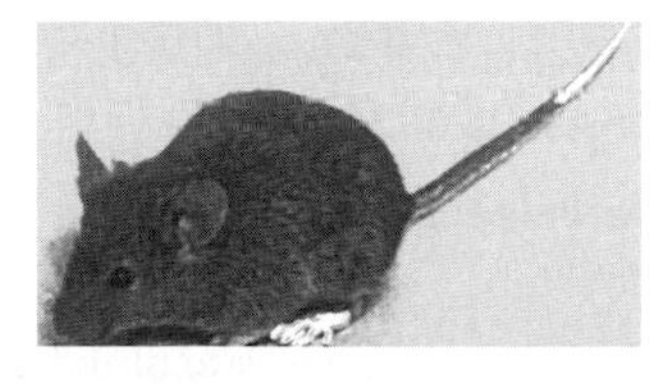

Figure P22. Kit^{tm1Alf}

Paramyotonia: Periodic paralysis (gene located in human chromosome 17). ►myotonia

Paramyxovirus: Single-stranded RNA viruses with a genome of 16–20 kb. Members of this group cause human mumps, respiratory diseases in human and other animals, including birds and reptiles. ►RNA viruses

Paranemic Coils: The two components of the coil can be separated from each other without any entanglement as one can easily pull apart two spirals that were pushed together after they were wound separately, i.e., they are not interlocked. ►plectonemic coils

Paraneoplastic Neurodegenerative Syndrome: ►autoimmune diseases

Paranoia: A psychological disorder in more (paranoia) or less severe (paranoid) state. The major characteristics are delusions of persecution (delusional jealousy, erotic delusions) or less frequently by feeling of grandiosity. It differs from schizophrenia in that, the rest of the personality and mental capacity may remain normal. Frequently, however, paranoid schizophrenia may occur. The precipitating factors are insecurity, frustration, physical illness, drug effects, etc. There is also an apparently undefined genetic component. ►schizophrenia, ►affective disorders

Paranome: Genes within gene families; the entire sets of duplicated genes. ►gene family

Paranormal: Beyond the normal biological expectation, e.g., extrasensory perception. It has been reported (Cha KY et al 2001 J Reprod Med) that prayer for success of in vitro fertilization approximately doubled the success of pregnancy in an international, randomized, double-blind clinical trial involving 219 women. More recently, experts questioned the outcome of this study and suspected inappropriate handling of the experiment (Nature [2004] 429:796). In the nineteenth century, Francis Galton, the father of biometrics, concluded that prayer has unlikely influence on worldly events because the number of shipwrecks was not effected by the fact that praying missionaries were aboard or royalties, for whom their subjects regularly prayed, did not live longer than other citizens. In some instances, when prayer provides comfort, beneficial psychological effects may result (Handzo G et al 2004 N Engl J Med 351:192; ►creationism

Paraoxonase (PON1, 7q21.3): May be associated with high-density lipoproptein in the blood plasma. It may protect against coronary heart disease by destroying oxidized lipids, responsible for inflammation. It may detoxify organophosphate pesticides (parathion, chloropyrifos [Dursban]). ►arylesterase, ►cholinesterase, ►pseudocholinesterase, ►HDL, ►Kupffer cell; Brophy VH et al 2001 Am J Hum Genet 68:1428.

Parapatric Speciation: Groups of organisms inhabiting an overlapping region become sexually isolated. ►allopatric, ►sympatric

Paraphyletic Group: Does not include all descendants of the latest common ancestor.

Paraplegia: The paralysis of the lower part of the body; it may be hereditary. ►Pelizaeus-Merzbacher disease, ►Silver syndrome, ►spastic paraplegia, ►Mast syndrome, ►ALS

Paraplegin: ►mitochondrial disease in humans

Paraptosis: A programmed neuronal cell death different from apoptosis in as much it is mediated by a caspase-9, which is independent of Apaf-1 and it does not respond to Bcl-X. ►apoptosis, ►Apaf-1, ►BCL; Sperandio S et al 2000 Proc Natl Acad Sci USA 97:14376.

Paraprotein: An abnormally secreted normal or abnormal protein, e.g., the Bence-Jones protein in myelogenous myeloma. It is also called M component. ►Bence-Jones protein

Paraquat: An artificial electron acceptor of photosystem I and a lung toxicant. It may produce oxidative stress by indirect production (through cellular diaphorases) of superoxide radicals. ►diquat, ►photosystem I, ►diaphorase, ►superoxide, ►ROS

Pararetrovirus: The genetic material is double-stranded DNA but it is replicated with the aid of an RNA molecule, e.g., in hepadnaviruses and caulimoviruses. They may occur in many copies in higher eukaryote genomes. ►animal viruses, ►plant viruses, ►retroviruses, ►hepatitis B virus, ►cauliflower mosaic virus; Richert-Poggeler KR, Shepherd RJ 1997 Virology 236:137; Gozuacik D et al 2001 Oncogene 20:6233.

Parascaris: A group of nematodes. ►*Ascaris*

Parasegment: The unit of a metameric complex consisting of the posterior part of one segment and the anterior part of another in insect larval and subsequent stages. ►morphogenesis

Paraselectivity: An apparent (but not real) selectivity in pollination among plants.

Parasexual Mechanism of Reproduction: The somatic-cell fusion and mitotic genetic recombination. The processes bear similarities to those common at sexual reproduction but do not involve sexual mechanisms. ►mitotic recombination, ►cell fusion, ►somatic cell genetics; Pontecorvo G 1956 Annu Rev Microbiol 10:393.

Parasitemia: The blood contains parasites, e.g., *Plasmodium*. ►thalassemia, ►*Plasmodium*

Parasitic: That which lives on and takes advantage of another live organism. ►biotrophic, ►parasitoid; http://www.ebi.ac.uk/parasites/parasite-genome.html; various parasites' database: http://fullmal.ims.utokyo.ac.jp.

Parasitic DNA: same as DNA selfish.

Parasitoid: Lives on another organism and eventually destroys it like some wasps and viruses. Some plants—upon attack and wounding by some insects—synthesize and emit host and parasite specific volatile compounds that attract parasitoid wasps that in turn may destroy the insects. The parasitoid wasp *Cotesia congregata* of the lepidopteran host *Manduca sexta* harbors Polydnavirus. The virus is injected into the host along with the parasitoid egg. The viral genome controls the host immune system and protects the wasp progeny development inside the host. The virus genome is 567,670 bp contained by 30 DNA circles of 5–40 kb, including 156 coding sequences of 66% AT; 69% of the viral genes has introns. The rest of the viral DNA is non-coding (Espagne E et al 2004 Science 306:286). ►parasitic, ►biological control, ►aphid

Paraspeckles: Formed from RNA-binding proteins in the cell nucleus within the interchromatin nucleoplasmic space, usually at the periphery of the nucleolus and in the vicinity of the nuclear speckles. ►speckles; Fox AH et al 2002 Curr Biol 12:13.

Parasterility: Caused by incompatibility between genotypes that may be fertile in other combinations. ►self-incompatibility alleles, ►Rh blood group

Parastichies: The imaginary helical line in phyllotaxis. ►phyllotaxis

Parathormone (parathyroid hormone, 11p15.3-p15.1): Produced by the parathyroid gland next to the thyroid gland. It is a regulator of calcium and phosphate metabolism (mediated by cAMP) primarily in the bones, kidneys and the digestive tract. A recessive hypoparathyroidism was mapped to Xp27. ►hypercalcemia-hypocalciuria, ►hyperpara-thyroidism, ►enchondromatosis; Healy KD et al 2005 Proc Natl Acad Sci USA 102:4724.

Parathyroid Hormone: Regulates Ca^{2+} level in animals. ►parathormone

Paratope: The epitope-binding site of the antibody Fab domain. ►antibody, ►epitope

P

Paratransgenic: An insect, which has transgenic symbionts inhabiting its gut. *Rhodnius prolixus* carries the actinomycete bacterium *Rhodococcus rhodnii* with which it has a symbiotic relationship. *R. prolixus* is a blood-sucking arthropod, vector of *Trypanosoma cruzi*, responsible for Chagas disease. When *R. rhodnii* is transformed by cecropin A, a 38-amino acid antimicrobial peptide derived from the moth *Hyalophora cecropia*, the peptide diffuses into the insect and lyses *Trypanosoma cruzi* within the insect without serious damage to *R. rhodnii* and thus effectively curtails the propagation of the protozoon. This is a more attractive defense than using chemical pesticides. ►transgenic, ►*Trypanosoma*, ►Chagas disease, ►CRUZIGARD; Beard CB et al 2001 Int J Parasitol 31:621.

Parcelation: The relative lack of pleiotropic effects between two sets of non-overlapping traits.

Parenchyma: In plant biology it means storage cells, either near isodiametric, *spongy parenchyma*, closer to the lower surface of the leaves or the *palisade parenchyma* consisting of one or two layers of columnar cells with their long axis perpendicular to the upper epidermis. Both types of tissues contain conspicuous intercellular space. In zoology, the parenchyma cells mean the functional units, rather than the network of an organ or tissue. ►palisade cells

Parens Patriae: The state or community right to intervene against individual rights or beliefs and protect the interest of a person against potentially serious or actually life-threatening conditions, e.g., compulsory immunization, genetic screening, prohibition of incest, etc.

Parental Ditype: ►tetrad analysis

Parental Histone Segregation: In front of the DNA replication fork the existing nucleosome structure is temporarily and reversibly disrupted to make nascent DNA readily accessible to the replication protein machinery (see Fig. P23).

Parenteral: The application of a substance by injection rather than by oral means.

Parent-of-Origin Effect: May be due to the differences in the cytoplasm, differential transmission of defective chromosomes through the two sexes, differences in trinucleotide-repeat expansions, endosperm: embryo chromosomal differences in the reciprocal crosses in case of polyploids and imprinting. ►imprinting, ►trinucleotide repeats, ►uniparental disomy, ►uniparental inheritance; Morrison IM, Reeve AE 1998 Hum Mol Genet 7:1599; Haghighi F, Hodge SE 2002 Am J Hum Genet 70:142.

Pareto Distribution: $f(x) = \frac{\gamma\alpha^{\gamma}}{x^{\gamma+1}}$ Applicable for the gene expression profiles. Some genes are expressed at very high level other transcripts occur once or even less per cell thus the distribution is highly skewed by the low-abundance transcripts (see Fig. P24). The Pareto probability distribution is: $\alpha \le x < \infty, \alpha > 0, \gamma > 0$, mean $= \gamma\alpha/(y-1, \gamma > 1$ and variance $= \gamma\alpha^2[(\gamma-1)^2(\gamma-2), \gamma > 2$ (See Kuznetsov VA et al 2002 Genetics 161:1321).

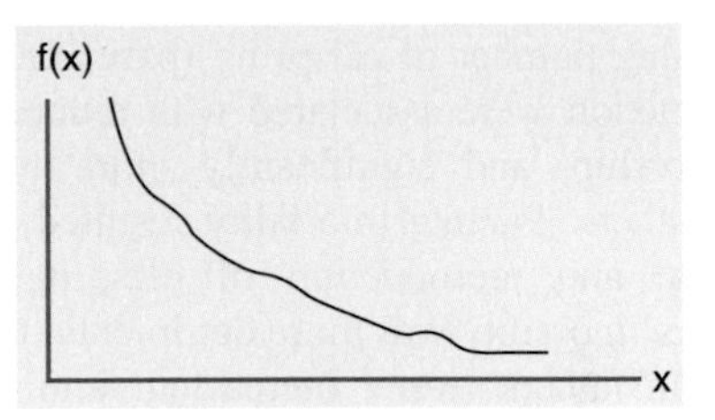

Figure P24. A hypothetical Pareto distribution

Parietal: Situated on the wall or attached to the wall of a hollow organ.

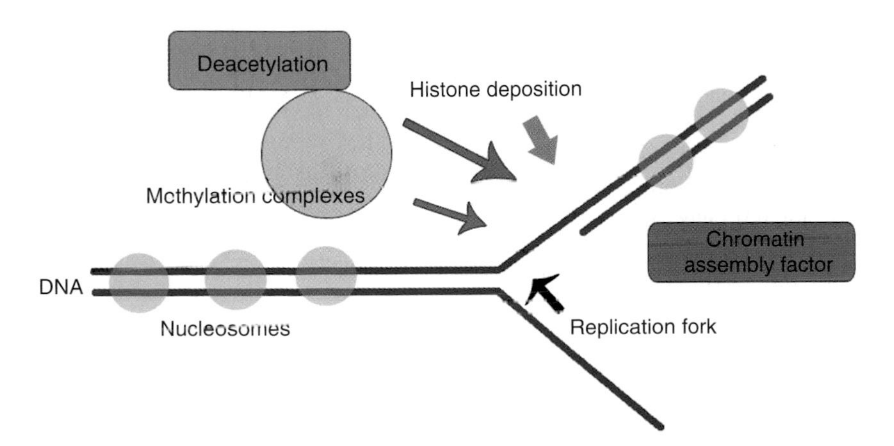

Figure P23. Nucleosome remodeling—Parental histone segregation

Periodontal Disease: Involves inflammation of the tissues surrounding teeth such as gingiva, ligaments, gum, etc. ►point-of-care test

Paris Classification: Paris classification of human chromosomes standardized (in 1971) the banding patterns and classified them by size groups; it is very similar to what is used today. Current maps show, however, only one of the two chromatids. ►human chromosomes, ►Denver classification, ►Chicago classification

Parity: Parity in gene conversion the process can go equally frequently in the direction of one or the other allele. ►gene conversion; Fogel S et al 1971 Stadler Symp 1–2:89.

Parity: In human biology, it is the condition that a woman had borne offspring. Natural selection does not necessarily favor maximal reproduction because reproduction imposes fitness costs, reducing parental survival, and offspring quality. Parents in a pre-industrial population in North America incurred fitness costs from reproduction, and women incurred greater costs than men. The survivorship and reproductive success (Darwinian fitness) of 21,684 couples, married between 1860 and 1895, identified in the Utah Population Database showed that increasing number of offspring (parity) and rates of reproduction were associated with reduced parental survivorship, and significantly more for mothers than fathers. Parental mortality resulted in reduced survival and reproduction of offspring, and the mothers' mortality was more detrimental to offspring than the fathers' were. Increasing family size was associated with lower offspring survival, primarily for later-born children, indicating a tradeoff between offspring quantity versus quality (Penn DJ, Smith KR 2007 Proc Natl Acad Sci USA 104:553). ►fitness, ►reproductive rate, ►fecundity, ►fertility

Parity Check: In a digital system it reveals whether the number of ones and zeros is odd or even.

Parkin: ►Parkinson's disease

Parking: A set of rules for seeding in which no seed overlaps with any other seedings. It is an iterative procedure that may be used at the early phase in genome sequencing. Each iteration sequences a new portion of non-overlapping piece of the DNA. Aside for non-overlaps, the sequences are chose at random. ►seeding, ►genome projects; Roach JC et al 2000 Genome Res 10:1020.

Parkinsonism: A secondary symptom caused either by drugs, or inflammation of the brain (encephalitis), or Alzheimer's disease, or Wilson's disease, or Huntington's chorea, etc. ►Parkinson's disease and the conditions named above, ►tau

Parkinson Disease (PD, PARK): A shaking palsy (bradykinesis) generally with late onset, however juvenile forms also exist. PD may include mental depression, dementia, reduced olfactory abilities and deficiency of several different substances, notably dopamine, from the nervous system. The genetic determination of the heterogeneous symptoms is unclear; autosomal dominant, recessive, X-linked, polygenic and apparently only environmentally caused phenotypes have been observed. The prevalence of PD is 0.001 and it may be 0.01 over age 50 but perhaps no more than 10% of the cases are familial.

Defects in the mitochondrial complex I resulting in oxidative stress favor the development of PD (Canet-Avilés RM et al 2004 Proc Natl Acad Sci 101:9103). Conditional knockout mice (termed MitoPark mice), with disruption of the gene for mitochondrial transcription factor A (*Tfam*) and in progressive degeneration of the nigrostriatal dopamine system (DA) neurons have reduced mtDNA expression and respiratory chain deficiency in midbrain DA neurons, which, in turn, leads to a parkinsonism phenotype with adult onset of slowly progressive impairment of motor function, accompanied by formation of intraneuronal inclusions and dopamine nerve cell death. Confocal and electron microscopy show that the inclusions contain both mitochondrial protein and membrane components demonstrating that respiratory chain dysfunction in DA neurons may be of pathophysiological importance in PD (Ektrand MI et al 2007 Proc Natl Acad Sci USA 104:1325). Endocannabinoids may have beneficial effects on motor deficits in PD and Huntington chorea (Kreitzer AC, Malenka RC 2007 Nature [Lond] 445:643).

A PTEN-induced putative kinase (*PINK1*) mutation in *Drosophila* that has high similarity to human PINK1 also encodes a mitochondrially located protein and it is complemented by parkin (Park J et al 2006 Nature [Lond] 441:1157; Clark IE et al 2006 Nature [Lond] 441:1162). Loss-of-function mutations in a previously uncharacterized, predominantly neuronal P-type ATPase gene, *ATP13A2*, underlying an autosomal recessive form of early-onset Parkinsonism with pyramidal degeneration and dementia (PARK9, Kufor-Rakeb syndrome) were observed. The pyramidal cells are excitatory neurons in cerebral cortex. The wild-type protein was located in the lysosome of transiently transfected cells; the unstable truncated mutants were retained in the endoplasmic reticulum and degraded by the proteasome (Ramirez A et al 2006 Nature Genet 38:1184).

The herbicide paraquat, the fungicide maneb, the insecticide rotenone and other environmental toxins may contribute to PD by inhibiting complex I (Dawson TM, Dawson VL 2003 Science 302:819).

Mitochondrially coded nicotinamide adenine dinucleotide dehydrogenase complex plays a role in reduced susceptibility (van der Walt JM et al 2003 Am J Hum Genet 72:804).

An early onset PD was located to human chromosome Xq28 and another (PARK6) to 1p35-p36. Mitochondrially localized PTEN-induced kinase 1 (PINK1, 1p35) mutations are involved in PARK6 (Valente EM et al 2004 Science 304:1158). PINK contains a serine-threonine protein kinase domain; it localized in mitochondria. An autosomal dominant form is in chromosome 22. Another locus encoding spheres of protofibrils of α-synuclein was assigned to human chromosome 4q21-q23. The α-synuclein gene is apparently responsible for only a minor fraction of Parkinsonism. A susceptibility gene (SNCA) has been located also to human chromosome 17q21. Multiple copies of SNCA may occur in a single nucleus. A low penetrance (40%), late onset (~60 years) gene is at 2p13. Autosomal dominant juvenile Parkinsonism gene (1395 bp ORF) encoding the 465-amino acid *parkin* protein was located to 6q25.2-q27. Parkin is an E3 ubiquitin protein ligase and ubiquitin-proteasome deficit favors the development of PD and its S-nitrosylation inhibits its normal protective action (Kung KKK et al 2004 Science 304:1328). Parkin disease is Parkinsonism without the formation of Lewy bodies. Several other loci are also involved with the developmental of this disease. Protein DJ-1 (1p36)—situated in the mitochondria—is a protein regulating mRNA stability and its defect can cause Parkinsonism.

The parkin gene appears to be a suppressor of ovarian cancer and adenocarcinoma (Cesari R et al 2003 Proc Natl Acad Sci USA 100:5956). In some cases a missense mutation in the carboxy terminal hydrolase L1 (UCH-L1, 4p14), component of the ubiquitin complex, localized in the Lewy bodies, is responsible for PD. Not all forms of PD shows Lewy bodies. Loss of dopaminergic neurons in the substantia nigra is usually associated with the disease. Glial cell line-derived neurotrophic factor (GDNF) has nutritive effects on the dopaminergic nigral neurons. In an autosomal recessive juvenile PD parkin-associated endothelin receptor-like (Pael-R) accumulates apparently because a misfolded Pael-R is not degraded if there is a defect in parkin. LRRK2 (PARK8, 12p12) is a leucine-rich repeat kinase gene involving Lewy body disease of advanced adult age.

Parkin has several other substrates and that explains the complicated etiology of the different forms of PD (Imai Y et al 2001 Cell 105:891; Shimura H et al 2001 Science 293:263).

Various tomography techniques facilitate detection of susceptibility to PD before the appearance of clinical symptoms. The imaging can measure the integrity of dopamine in the substantia nigra of the brain. The procedure is based on the conversion of 18F-DOPA into 18F-dopamine or by the use of DOPA transporter (DAT) ligands. The reduction in DAT ligand uptake correlates with the loss of dopamine in the corpus striatum (the striped gray substance in front and beside the thalamus in the brain) and this is characteristic for aging and particularly for incipient PD. Also, loss of fluorodeoxyglucose distinguishes PD from other neurodegenerative diseases. Dopamine dysfunction may be indicated also by olfactory impairment (see DeKosky ST, Marek K 2003 Science 302:830).

Gene therapy using lentivirus vector carrying the GDF gene, injected into the brain (striatum and substantia nigra) of old monkeys or young monkeys pretreated with a nigrostriatal degeneration inducing agent (1-methytl-4-phenyl-1,2,3,6-tetrahydropyridine [MPTP]) reversed the functional deficits and prevented degeneration, respectively. Anticholinergics (blocking choline) and dopamine, glial-cell-derived neurotrophic factor (GDNF) treatments may be somewhat beneficial. An alternative approach may be using gene therapy by expressing transfected tyrosine hydroxylase or aromatic amino acid decarboxylase in the striated muscle cells. These enzymes can produce dopamine yet the response is below expectation. New approaches appear more promising (▶gene therapy). Transplantation of fibroblasts equipped for secretion of BDNF and GDNF into the brain appears promising in animal models. Inhibition of cyclooxygenase (COX-2) seems to prevent the formation of the oxidant dopamine-kinone, which has been implicated in Parkinsonism (Teismann P et al 2003 Proc Natl Acad Sci USA 100:5473). Electric shocks localized to the globus pallidus (a medial part of the brain) had beneficial effects in some cases. Caspase-3 may be a conditioning factor in the apoptotic death of dopaminergic neurons in PD. Embryonic stem cells may develop into dopamine-producing neurons in the brain of the mouse and seem to be promising for cell-replacement therapy of PD (Kim J-H et al 2002 Nature [Lond] 418:50; Barberi T et al 2003 Nature Biotechnol 21:1200). Midbrain proteins Lmx1a and Msx1 mediate dopamine neuron differentiation of proneural protein NGN2 and seem important for cell replacement therapy in PD (Andersson E et al 2006 Cell 124:393). Activation of intracellular neurotrophic signaling pathways by vector transfer is a feasible approach to neuroprotection and restorative treatment of neurodegenerative disease. Adeno-associated virus 1 transduction with a gene encoding a myristoylated, constitutively active form of the oncoprotein Akt/PKB had pronounced trophic effects on dopamine neurons of adult and aged mice, including increases in neuron size, phenotypic markers, and sprouting. Transduction confers almost complete protection against apoptotic cell death in a highly destructive neurotoxin model (Ries V et al 2006 Proc Natl Acad Sci USA 103:18757). Nix, a pro-apoptotic BH3-only protein, promotes apoptosis of non-neuronal cells. Using a yeast two-hybrid screen with POSH (plenty of SH3 domains, a scaffold involved in activation of the apoptotic JNK/c-Jun pathway) as the bait, identified an interaction between POSH and Nix and contributed to cell death in a cellular model of Parkinson disease (Wilhelm M et al 2007 J Biol Chem 282:1288). The disease-linked processes are detectable in peripheral blood by 22 unique genes differentially expressed in patients with PD versus healthy individuals. Such an approach may provide biomarkers for early clinical detection of the disease (Scherzer CR et al 2007 Proc Natl Acad Sci USA 104:955). Multiple axon-guidance pathway genes may predispose to PD (Lesnick TG et al 2007 PloS Genet 3(6):e98). ►parkinsonism, ►neuromuscular diseases, ►dopamine, ►adeno-associated virus, ►Akt, ►tyrosine hydroxyls, ►Lewy body, ►Kufor-Rakeb syndrome, ►GDNF, ►BDNF, ►dopamine, ►mitochondrial disease in humans, ►subtantia nigra, ►synuclein, ►caspase, ►ubiquitin, ►tau, ►stem cells, ►brain human, ►tomography, ►PTEN, ►argyrophilic grains, ►BAK, ►cannabinoids, ►axon guidance; Dawson TM 2000 Cell 101:115; Kordower JH et al 2000 Science 290:767; Valente EM et al 2001 Am J Hum Genet 68:895; Vaughan JR et al 2001 Ann Hum Genet 65:111; Lansbury PT Jr, Brice A 2002 Curr Opin Genet Dev 12:299; Betarbet R et al 2002 BioEssays 24:308; Cookson MR 2005 Annu Rev Biochem 74:29; Farrer MJ 2006 Nature Rev Genet 7:306.

Paromomycin ($C_{23}H_{45}O_{14}N_5$): An aminoglycoside antibiotic. It may cause translational errors by increasing the initial binding affinity of tRNA. Oral LD^{50} in mice is 1625 mg/kg. ►phenotypic reversion, ►aminoglycoside antibiotics, ►LD50

Parotid Gland: The salivary gland; the proline-rich parotid glycoprotein is encoded in human chromosome 12p13.2.

Paroxysm: Recurring events such as convulsions (but most commonly normal conditions in between), sudden outbreak of disease.

Paroxysmal Nocturnal Hemoglobinuria: A dominant human chromosome 11p14-p13 susceptibility of the erythrocytes to destruction by the complement because of a deficiency in protectin (HRF20/CD59) and DAF. ►complement, ►membrane attack complex, ►DAF, ►angioneuritic edema, ►CD59, ►protectin, ►hemoglobin

PARP (poly[ADP-ribose] polymerase): An enzyme involved in surveillance and base excision repair of DNA and NAD^+-dependent chromatin remodeling (Kim MY et al 2004 Cell 119:803). PARP is involved in puff formation of *Drosophila* gene loci (Tulin A, Spradling A 2003 Science 299:560). It is cleaved by an ICE-like proteinase. Its deficiency increases the sensitivity to radiation damage, recombination and sister chromatid exchange. PARP is required also for the assembly and structure of the spindle (Chang P et al 2004 Nature [Lond] 432:645). PARP-deficient mice are viable and free of tumors and inhibitors of PARP/DNA repair activity can selectively kill BRCA2 defective tumor cells (Bryant HE et al 2005 Nature [Lond] 434:913). PARP deficiency prevents homologous recombination repair of damaged BRCA1 and BRCA2 eventually apoptosis eliminates the mutant cells (Farmer H et al 2005 Nature [Lond] 434:917). ►ICE, ►apoptosis, ►puff, ►tankyrase, ►telomeres, ►spindle, ►DNA repair, ►breast cancer, ►Kif; Bauer PI et al 2001 FEBS Lett 506(3):239; Lavrik OI et al 2001 J Biol Chem 276:25541.

PARS (poly[ADP-ribose] synthetase): PARS attaches ADP-ribose units to histones and to other nuclear proteins. It is activated when DNA is damaged by nitric oxide.

Parser: A software for reading flat files for further processing. ►flat file

Parsimony: ►maximal parsimony, ►evolutionary tree

Parsing: Resolve it to parts or components. ►exon parsing, ►pars means part in Latin

Parsley (*Petroselinum crispum*): The roots and leaves of parsley are used for flavoring; 2n = 2x = 22 (see Fig. P25).

Figure P25. Parsley

Parsnip (*Pastinaca sativa*): A root vegetable; 2n = 2x = 22 (see Fig. P26).

Figure P26. Parsnip

Parsonage–Turner Syndrome (Feinberg syndrome/Tinel syndrome/Kiloh-Nevin syndrome): The syndrome is most likely to be identical with hereditary amyotrophic neuralgia; it is also very similar to the Gillain-Barré syndrome. ►amyotrophy hereditary neuralgic, ►Guillain-Barré syndrome

Parthenocarpy: The development of fruit without fertilization. It may have horticultural application by producing seedless apple varieties such as Spencer Seedless or Wellington Bloomless. The gene responsible in these apples is homologous to *pistillata* of *Arabidopsis*. ►parthenogenesis, ►apomixia, ►seedless fruits, ►flower differentiation in *Arabidopsis*; Yao J et al 2001 Proc Natl Acad Sci USA 98:1306.

Parthenogenesis: Embryo production from an egg without fertilization. Parthenogenesis may be induced in sea urchins by hypotonic media or in some amphibia by mechanical or electric stimulation of the egg. In some fish, lizards and birds (turkey) it occurs spontaneously. Parthenogenesis in animals is most common among polyploid species. Parthenogenetic individuals produce only female offspring. On theoretical grounds, parthenogenesis may be disadvantageous because it deprives the species of elimination of disadvantageous mutations on account of the lack of recombination available in bisexual reproduction. Parthenogenesis may cause embryonic lethality in mouse if the imprinted paternal genes are not expressed. Parthenogenesis is not known to occur in humans (or generally in mammals but deletion of imprinting may yield viable parthenogenetic mouse), however it may exist as a chimera when after fertilization, the male pronucleus is displaced to one of the blastomeres and then the maternal chromosome set in the other blastomere is diploidized. The failure of parthenogenetic development of mammalian offspring is due to the requirement for imprinting (Kono T et al 2004 Nature [Lond] 428:860). Fusion of two oocyte nuclei can produce, however, viable, fertile mouse. Many plant species successfully survive by asexual reproduction as an evolutionary mechanism. Parthenogenesis in plants is called apomixis or apomixia. Asexually reproducing plant populations appear to be preponderant under conditions marginally suitable for the species is called *geographic parthenogenesis*. ►apomixia, ►gynogenesis, ►parthenocarpy, ►RSK, ►imprinting, ►oocyte; Mittwoch U 1978 J Med Genet 15:165; Cibelli JB et al 2002 Science 295:819; Krawetz SA 2005 Nature Rev Genet 6:633.

Parthenote: An egg stimulated to divide and develop to some extent in the absence of fertilization by sperm. Parthenotes may offer possibilities for stem cell production for therapeutic purposes when embryonic stem cells are prohibited. ►parthenogenesis, ►apomixia; Kiessling AA 2005 Nature [Lond] 434:145.

Partial Digest: The reaction is stopped before completion of nuclease action and thus the DNA is cut into various size fragments some of which may be relatively long because some of the recognition sites were not cleaved. ►restriction enzyme

Partial Diploid: ►merozygote

Partial Dominance: An incomplete dominance, semidominance.

Partial Linkage: The genes are less than 50 map units apart in the chromosome and can recombine at a frequency proportional to their distance. ►crossing over, ►recombination

Partial Trisomy: Only part of a chromosome is present in triplicate. ►trisomy

Particulate Inheritance: The modern genetic theory that inheritance is based on discrete particulate material (written in nucleic acid sequences) transmitted conservatively rather than according to the pre-Mendelian theory of pangenesis, which claimed that the hereditary material is a miscible liquid subject to continuous changes under environmental effects. According to the particulate theory of genetics, genes are discrete physical entities that are transmitted from parents to offspring without blending or any environmental influence, except when mutation or gene conversion or imprinting occurs. ►pangenesis, ►gene conversion, ►mutation, ►imprinting, ►blending inheritance

Particulate Radiation: ►physical mutagens

Partington Syndrome (Xp22.13): Primarily mild or moderate mental retardation accompanied by developmental anomalies caused by polyalanine expansion in the *aristaless-like* homeobox gene. (*al, aristaless* alleles were discovered early in *Drosophila* and mapped to chromosome 2.4; some involved cytologically detectable inversions. Mammalian homologs exist.) At this chromosomal area several other similar mental retardation genes occur. Some have recessive expression others are dominant. ►homeotic genes, ►mental retardation, ►mental retardation X-linked

Partitioning (segregation): The distribution of plasmids and/or the bacterial chromosome(s) into dividing bacterial cells. It may be a passive process mediated by the attachment of the DNAs to the cell membrane. The partition depends on ParA and ParB proteins by the P1 plasmid. ParB recognizes at the partition site a dimeric B box and at the opposite ends two A boxes with helix-turn-helix (HTH) domains (see Fig. P27). The HTH domains emanate from the dimerized DNA-binding modules composed of a six-stranded β-sheet coiled coil that binds the B boxes (Schumacher MA, Funnell BE 2005 Nature [Lond] 438:516). A bacterial tubulin-like protein FtsZ may carry out the separation. Plasmid and bacterial genes control the process. The bacterial chromosome or some of the plasmids seem to have a centromere-like protein that may bind to the cell poles and to 10 copies of a sequence situated along a 200 kb region near the replicational origin. The loss of the chromosome in the new cells is less frequent than 0.003. The segregation of plasmids present in multiple copies is more complex. Mechanisms ("addiction" system) exist to resolve plasmid dimers and to ensure that each cell would have at least one copy of a plasmid. The two-component addiction module include a stable toxin and a labile antitoxin. The plasmid-free cells may be eliminated (Engelberg-Kulka H, Glaser G 1999 Annu Rev Microbiol 53:43). ►segregation, ►cell division, ►addiction module, ►plasmid maintenance; Hiraga S 2000 Annu Rev Genet 34:21; Gordon GS, Wright A 2000 Annu Rev Microbiol 54:681; Draper GC, Gober JW 2002 Annu Rev Microbiol 56:567.

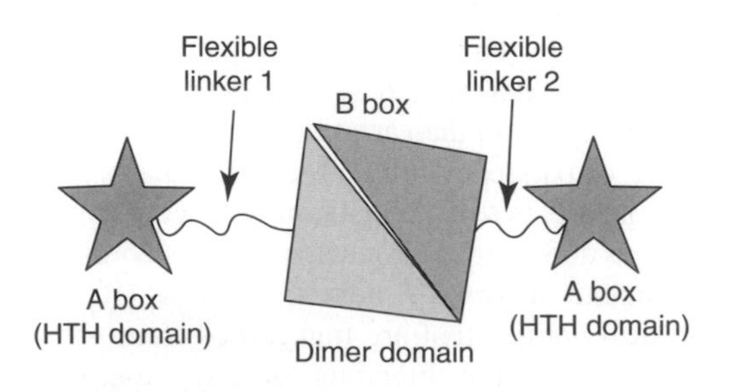

Figure P27. Partitioning

Partitioning: In statistics, breaking down the variances among the identifiable experimental components or in a compound chi square to reduce the quantity of the residual or error variance. ►analysis of variance

Parturition: The labor of child delivery.

Parvoviruses: Non-enveloped, icosahedral (18–25 nm), single-stranded DNA (~5.5 kb) viruses. The group includes the densoviruses of arthropods, the autonomous, lytic parvoviruses and the adeno-associated viruses. ►icosahedral, ►adeno-associated virus, ►autonomous parvovirus, ►oncolytic viruses; Lukashov VV, Goudsmit J 2001 J Virol 75:2729.

Parvulin: A very small monomeric 92-amino acid prolyl isomerase of *E. coli* involved in protein maturation. Similar proteins occur in yeast (Ess1, 19.2 kDa) and humans (Pin1, 18 kDa) and *Drosophila* (dodo, 18.3 kDa). Ess1 may not have isomerase activity, but Pin1 does. In the absence of Ess1 the nuclei fragment and growth ceases, dodo has apparently similar function as Ess1 and it is interchangeable. Pin regulates mitotic progression by interacting with CDC25 and NIMA. ►PPI, ►peptidyl-prolyl isomerases, ►CDC25, ►NIMA; Rulten S et al 1999 Biochem Biophys Res Commun 259:557.

PAS Domain (Per-Arnt-Sim): A shared motif of proteins involved in the regulation of the circadian rhythm of the majority of eukaryotes. PAS domain serine/threonine kinases also regulate several different signaling pathways including drug response. ►circadian rhythm; Rutter J et al 2001 Proc Natl Acad Sci USA 98:8991.

PASA: A special PCR procedure by which chosen allele(s) can be amplified if the primers match the end of that allele. ►PCR; Smith EJ, Cheng HH 1998 Microb Comp Genomics 3:13; Shitaye H et al 1999 Hum Immunol 60:1289.

Pascal Triangle: Represents the coefficients of individual terms of expanded binomials: $(p + q)^n$:

$$1p^n + \frac{n}{1!(n-1)!}p^{n-1}q + \frac{n!}{2!(n-2)!}p^{n-2}q^2 + \ldots + \frac{n!}{(n-1)!1!}p^{n-(n-1)}q^{n-1} + 1q^n$$

Since genetic segregation is expected to comply with the binomial distribution, the coefficients indicate the frequencies of the individual phenotypic (or in case of trinomial distribution) genotypic frequencies. ►binomial distribution, ►trinomial distribution, see Table P1.

Table P1. The Pascal triangle represents the coefficients of individual terms of expanded binomials. The exponent of the binomial is n. The figures display a symmetrical hierarchy. The frequency of a particular class can be readily calculated because of the sum of the coefficients is displayed at the bottom of the columns. Mendelian segregation follows the binomial distribution

n →	1	2	3	4	5	6	7	8	9	10
	1	1	1	1	1	1	1	1	1	1
	1	2	3	4	5	6	7	8	9	10
		1	3	6	10	15	21	28	36	45
			1	4	10	20	35	56	84	126
				1	5	15	35	70	126	210
					1	6	21	56	126	252
						1	7	20	84	210
							1	8	36	120
								1	9	45
									1	10
										1
SUMS	2	4	8	16	32	64	128	256	512	1024

n →	11	12	13	14	15	16	17	18	19	20
	1	1	1	1	1	1	1	1	1	1
	11	12	13	14	15	15	17	18	18	20
	55	66	78	91	105	120	136	153	171	190
	165	220	286	364	455	560	680	816	969	1140
	330	495	715	1001	1365	1820	2380	3060	3876	4845
	462	792	1287	2002	3003	4368	6188	8568	11682	15504
	462	904	1716	3003	5005	8008	12376	18564	27132	38760
	330	792	1716	3432	6435	11440	19448	31824	50388	77520
	165	495	1287	3003	6435	12870	24310	43758	75582	125970
	55	220	715	2002	5005	11440	24310	48620	92378	167960
	11	66	286	1001	3003	8008	19448	43758	92378	184756
	1	12	78	364	1365	4368	12376	31824	75582	167960
		1	13	91	455	1820	6188	18564	50388	125970
			1	14	105	560	2380	8568	27132	77520
				1	15	120	680	3060	11628	38760
					1	16	136	816	3876	15504
						1	17	153	969	4845
							1	18	171	1140
								1	19	190
									1	20
										1
SUMS	2048	4096	8192	16384	32768	65536	131072	262144	524288	1048576

Passage: The transfer of cells from one medium to another.

Passenger DNA: A DNA inserted into a genetic vector.

Passenger Proteins: Include the inner centromeric protein (INCENP), which is a substrate for Aurora, the Aurora B kinase, the TD-60 autoimmune antigen, the inhibitor-of-apoptosis protein Survivin/BIR-1. These proteins are situated at the centromeres and move the spindle at late metaphase and anaphase. ►centromere, ►mitosis, ►Aurora, ►Survivin; Bishop JD, Schumacher JM 2002 J Biol Chem 277:27577.

Passive Immunity: Acquired by the transfer of antibodies or lymphocytes

Passive Transport: Does not require special energy donor for the process. ►active transport

Pasteur Effect: The fast reduction of respiration (glycolysis) if O_2 is added to fermenting cells.

Pasteurella multocida: A pathogenic bacterium causing cholera in birds, bovine hemorrhagic septicemia, atrophic rhinitis (inflammation of the nasal mucosa) in pigs, and humans may be infected by it through cat or dog bites. The sequenced genome of 2,257,487 bp contains ~2014 coding sequences. ►*Yersinia*; May BJ et al 2001 Proc Natl Acad Sci USA 98:3460.

Pasteurization: Reducing (killing) the microbe population in a material by heating at a defined temperature for a specified period of time. ►aseptic, ►axenic, ►autoclaving

PAT1 (Ran1): A protein kinase required for the continuation of mitotic division in fission yeast. Its inactivation triggers the switch to meiosis. ►meiosis, ►Ran1

Patatin: A glycoprotein like storage protein in potato with lipid acid hydrolase and esterase activity. It inhibits pests' larvae. It constitutes ~40% of the soluble proteins in the tubers. ►potato; Hirschberg HJ et al 2001 Eur J Biochem 268:5037.

Pätau's Syndrome: Caused by trisomy for human chromosome 13. This is one of the few (X, Y, 8, 18, 21, 22) trisomies that can be carried to term but it generally leads to death within six months because of severe defects in growth, heart, kidney and brain failures. It is accompanied by face deformities (severe hare lip, cleft palate), polydactyly, clubfoot, defects of the genital systems, etc. (see Fig. P28). Definite identification is carried out by cytological analysis, including FISH with the available chromosome-13-specific probes. An old designation of trisomy 13 was trisomy D because chromosome 13 belonged to the D group of human chromosomes. ►trisomy, ►aneuploidy, ►polydactyly, ►hare lip, ►clubfoot

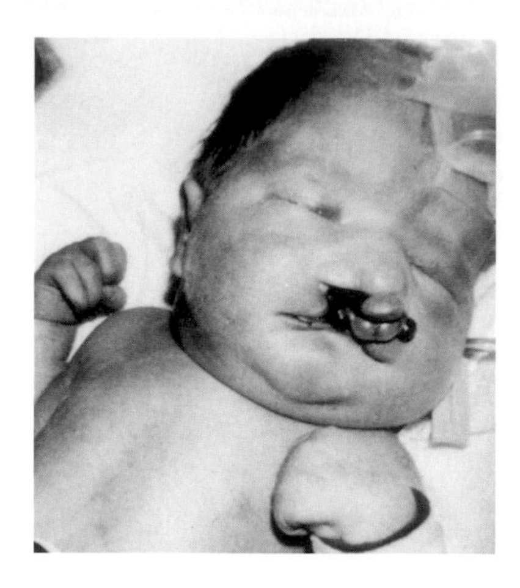

Figure P28. Pätau syndrome. (Courtesy of Dr. Judith Miles)

Patch (*Ptc*): ►sonic hedgehog, ►hedgehog

Patch (patched duplex): The resolution of a recombination intermediate (Holliday junction) without an exchange of the flanking markers (can be gene conversion). ►Holliday model L

Patch Clamp Technique: The method to measure the flow of current through a voltage gated ion channel by tightly pressing an electrode against the plasma membrane. It is used also for the sensitive in situ study of neurotransmitters. ►ion channels

Patch Mating: Actin patches of the cytoskeleton can be used in yeast to quantify the number of viable diploid cells in the presence of silencer genes, which regulate the expression of mating types. Actin patches are detectable for duration about 10 seconds at sites of polarized growth and then rapidly disappear. ►mating type determination in yeast; Smith MG et al 2001 J Cell Sci 114(pt 8):1505.

Patella Aplasia-Hypoplasia (PTLAH, 17q21-q22): The absence or reduction of the size of the knee cap. The symptoms occur also in various syndromes such as the Coffin-Siris syndrome, trisomy 8 syndrome. ►Coffin-Siris syndrome, ►nail-patella syndrome

Patent: The so-called gene patents do not protect the DNA (or ESTs) sequence itself, rather the process of manipulation is the object. The gene "ownership" only prevents the use or selling a particular sequence without permission. In general, according to US patent laws, the patent is protected for 17 years from date of issue. A requisite for patenting is that the subject of the patent application would be new and non-obvious and practically useful, e.g., a probe for a gene. Natural DNA sequences are not patentable, but purified or isolated recombinant molecules or parts of a vector are patentable. Legal patentability requirements: (i) usefulness, (ii) novelty, (iii) being non-obvious, and (iv) definiteness of description. A further requirement is "enablement," i.e., a trained person after reading the patent description can use the "invention" without further research. In October 1998, USA, the first patent was awarded to an EST. Once a patent is issued, even further, originally undisclosed applications are protected. Also, another person after isolating a full-length open reading frame using an STS may obtain a patent but not without the permission of the "inventor" of the patented STS. During the period of patent, the patent-holder can prevent anybody from using it, including those who invented the same independently or even those who improved on the procedure to such an extent that the second invention meets the requirements for patenting. However, the inventor is obligated by law to disclose the invention in sufficient technical detail so that anybody with proper expertise can use it. The fact that an invention was arrived at under federal financial support does not exclude patentability but the inventor must report the patentable invention to the sponsoring agency. The intention of the government is that the invention would be used at maximum benefit to the public that can be achieved most effectively by commercial private enterprises. Laboratory assays, reagents and procedures, including computer programs may also be patentable. By 2005, more than 4000 genes were patented and three-fourths of them by single individuals. Presenilin gene (PSN2) has 8 patent owners for 9 patents and breast cancer gene (BRCA1) has 12 owners for 14 patents. Of the 292 cancer genes reported by 2004, 131 are patented (Jensen K, Murray F 2005 Science 310:239).

The patenting of the outcome of genetic research may be harmful to science if the investigators keep the ongoing work a secret until it becomes patentable. Patenting basic research products (upstream inventions) is detrimental to society because it may prevent the development of new useful (commercial) products. The Bayh–Dole Act allows the "exemption for research" to facilitate the use of the results of basic research. Unfortunately, this exemption is difficult to define and is subject to controversy (Holman C 2006 Cell 125:629). The exemption can be applied to the patent (effectiveness, usefulness) itself but does not permit application of the patent (Kaye J et al 2007 Nature Biotechnol 25:739). If the discovery is published through proper means of scientific communications prior to the patent application, it is disqualified from patenting. It is generally easier to patent a product than a process. Natural products (e.g., proteins) are usually not patentable unless they are modified in some way and are different from the natural product in structure or function and these properties were not generally known. A DNA sequence, identified as the coding unit for a genetic disease or a genetic marker in its vicinity, may be patentable but a cloned gene that may be used for translating a protein may be not. DNA markers are patentable only if their direct use can be determined. The concept of patenting biological material raises several moral objections but it is defended by the biotechnology industry because it takes 100s of millions of dollars for the completion of such projects, and without the financial means, these investigations cannot be maintained. Between 1981 and 1995, a total of 1175 human DNA sequences were patented. If the subject of the patent has been published or in use for more than one year prior to the date of the patent application, it will not be approved. If another person can prove that he/she invented the object before the date of publication by others, the person may still be entitled to a patent. The patent laws vary in different countries, and new legislation may take place any time. The European Union is now approving patents for human genes and transgenic animals and plants. An alternative to patenting is Trade Secret Protection. One way of preventing another person from patenting an invention is public disclosure, e.g., publication in sufficient details (e.g., in a scientific journal). This ensures that another party would not be able to claim priority for the invention, which is one of the requisites for patenting. Publications may not necessarily provide an effective and lasting protection. Patent infringement usually does not entitle the patentee for more financial compensation than the reasonably calculated loss of royalty or profit caused by the infringement. It must also be verified that the original patent description do not include deceptive assertions.

The justification of an existing patent may be challenged administratively by *inter partes* re-examination request to the US Patent and Trademark Office (USPTO, Washington DC, USA). The claim must prove substantial new question of patentability based on a "prior art" document and requires a fee of $8800. This procedure has monetary advantages vis-à-vis litigation (Derzko NM, Behringer JW 2003 Nature Biotechnol 21:823). The patent regulations are subject to changes and it is advisable to seek consent from the owners of the patent even when the invention is used only for laboratory research. A recent analysis revealed many problems with patents granted by USPTO. The patents examined had problems with description (37.5%), enablement/utility (42.4%) novelty/non-obviousness (6.9%) and definiteness (13.1%). The conclusion faulted insufficient educational background of the examiners with most of the problems (Paradise J et al 2005 Science 307:1566). ►STS, ►EST, ►SNP, ►Herfindahl index, ►presenilin, ►breast cancer, ►cancer, ►Cohen-Boyer patent on recombinant DNA; Eisenberg RS 1992 p 226. In: Annas GJ, Elis S (Eds.) Gene Mapping, Oxford University Press, New York, DNA-based patents: Robertson D 2002 Nat Biotechnol 20:639; Arnold BE, Ogielska-Zei E 2002 Annu Rev Genomics Hum Genet 3:415; patented genetic sequence information retrieval: Dufresne G et al 2002 Nature Biotechnol 20:1269; ethical issues of DNA patenting: Resnik DB 2004 Owning the Genome: A Moral Analysis of DNA Patenting, State University New York Press, Albany, New York; Eisenberg RS 2003 Science 299:1018; Robertson JA 2003 Nature Rev Genet 4:162; property rights in plant breeding: Fleck B, Baldock C 2003 Nature Rev Genet 4:834; Paradise J, Janson C 2006 Nature Rev Genet 7:148; Van Overwalle G et al 2006 Nature Rev Genet 7:143; stem cell lines: Loring JF, Campbell C 2006 Science 311:1716; http://geneticmedicine.org or patents in general: http://www.uspto.gov; http://scientific.thomson.com/derwent; http://www.bioforge.org; http://www.bustPATENTS.COM/; gene and sequence patents for various organisms: http://www.patome.org/; ftp://ftp.ebi.ac.uk/pub/databases/embl/patent; ftp://ftp.wipo.int/pub/published_pct_sequences.

Patent Ductus Arteriosus (6p12): The ductus arteriosus connects the lung artery and the aorta and shunts away blood from the lung of the fetus. Normally it fades away after birth. In ~1/2000 cases, this does not happen and the duct stays open and causes heart defect. The disease may be caused by fetal rubella infection and apparently by autosomal dominant gene(s). The 6p12-p21 dominant *Char syndrome* involves patent ductus, facial anomalies and abnormal fifth digit of the hand (see Fig. P29).

Figure P29. Char syndrome. In the Char syndrome the middle phalanx of the fifth digit is missing

The basic problem is traced to TFAP2B neural crest-related transcription factor that does not bind properly to its target. Risk of recurrence in an affected family is about 1–2%. General incidence is less than 10% of that. ►risk, ►aneurysm; Zhao F et al 2001 Am J Hum Genet 69:695.

Paternal Leakage: The transmission of mitochondrial DNA through the males. Generally, mitochondria are not transmitted through the animal sperm because mitochondrial DNA of spermatozoon is destroyed by ubiquitination in the oocytes (Sutovsky P et al 2000 Biol Reprod 63:582). In mice, apparently the male transmission of mitochondria is within the range of 10^{-5}. In interspecific mouse crosses, paternal mitochondria are transmitted but they are eliminated during early embryogenesis or later during development (Kaneda H et al 1995 Proc Natl Acad Sci USA 92:4542). Heteroplasmy is rare. The role of transmission of mitochondria in humans is not clear. Some cytological observations may indicate the incorporation of the midpiece of the sperm (containing mitochondria) into the egg. Genetic evidence for human paternal transmission of mitochondria is rare (Schwartz M, Vissing J 2002 N Engl J Med 347:576).

In some molluscans (mussel), there is a strong biparental inheritance of mtDNA. In *Mytilus* the paternal and maternal mtDNA displays 10–20% nucleotide divergence. The females transmit just one type of mitochondria to sons and daughters whereas the males transmit a second type of mtDNA genome to the sons. Biparental transmission of mtDNA may also occur in interspecific crosses of *Drosophila*. In *Paramecia*, mitochondria may be transmitted through a cytoplasmic bridge. In fungi, the transfer is maternal although in some heterokaryonts cytoplasmic mixing may take place. In some slime molds, mtDNA transmission is also mating type dependent. In *Physarum polycephalum* different *matA* alleles regulate the mtDNA transmission but a plasmid gene may also be involved and recombination can take place between mtDNAs. In *Chlamydomonas* algae, several genes around the mating type factors were implicated. In the contact zone of hybridizing conifers (*Picea*) recombinant mtDNA was observed as apparent result of paternal leakage (Jaramillo-Correa JP, Bousquet J 2005 Genetics 171:1951). ►mtDNA, ►plastid male transmission, ►Eve foremother, ►mitochondrial disease in humans, ►mitochondrial genetics, ►plastid genetics, ►doubly uniparental inheritance, ►*Paramecium*, ►heteroplasmy; Eyre-Walker A 2000 Philos Trans R Soc Lond B Biol Sci 355:1573; Shitara H et al 1998 Genetics 148:851; Yang X, Griffith AJ 1993 Genetics 134:1055; Meusel MS, Moritz RF 1993 Curr Genet 24:539.

Paternal Transmission: Imprinting causes paternal transmission of certain genes. Some of the human insulin and the insulin-like growth factor alleles may be preferentially inherited through the paternal chromosome and cause early-onset obesity. ►imprinting, ►obesity, ►paternal leakage; Le Stunff C et al 2001 Nature Genet 29:96.

Paternity Exclosure: Based on genetic paternity tests. ►paternity testing, ►DNA fingerprinting, ►Y chromosome, ►alternate paternity

Paternity Testing: Frequently required in civil litigation suits, it might have significance for medical, population, immigration, archeological and other cases. The laboratory procedures are generally the same as used for DNA fingerprinting. Here, as in DNA fingerprinting in general, the exclusion of paternity is simple and straightforward. However, the determination of identity may pose more difficulties because in the multilocus tests more than 10% of the offspring may show one band difference and 1% may show two, due to mutation. Therefore, Penas and Chakraborty (Trends Genet 10:204 [1994]) recommended the formula shown in Figure P30.

$$PI = \frac{\binom{N}{U} \mu^{U} (1-\mu)^{N-U}}{\binom{n}{U} X^{n-U} (1-X)^{U}}$$

Figure P30. Penas-Chakraborty formula

PI = paternity index, μ = mutation rate, X = band-sharing parameter, N = total number of bands per individual, n = number of test bands, U = number of bands not present in the alleged father. In rare instances (mistakes at maternity wards) similar test may be necessary to test maternity. The biological father of a child—even if the paternity can be accurately proven—cannot assert paternal rights against the will of the mother if she was/is married to another man (Hill JL 1991 N Y Univ Law Rev 66:353). When "the child is born to a mother who is single or part of a lesbian couple, law does permit the biological father to assert his paternal rights, even if he clearly stated his intention prior to conception to have no relation" (Charo RA 1994 In: Frankel MS, Teich A (Eds.) The Genetic Frontier. Ethics, Law and Policy, American Association of Advance Science, Washington DC). ►Y chromosome, ►forensic genetics, ►DNA fingerprinting, ►forensic index, ►utility index, ►surrogate mother, ►microsatellite typing

Path Coefficient: This method of Sewall Wright was worked out for studying mathematically and by diagrams, the paths of genes in populations and genetic events determining multiple correlations. Here it is not possible to discuss meaningfully the mathematical foundations but one type of graphic application for determining some relations between offspring and parents can be found under *F* and inbreeding coefficient. ►inbreeding coefficient, ►correlation; Wright S 1923 Genetics 8:239; Wright S 1934 Ann Math Stat 5:161.

Pathogen: An organism (microorganism) capable of causing disease on another. (See pathogen database: http://www.nmpdr.org; ►vectors for pathogens, ►*Brucella*, ►*Rickettsia*, ►*Coxiella* and viruses: https://patric.vbi.vt.edu).

Pathogen Identification: The food industry may need rapid and highly sensitive methods for the detection of live pathogens in various products. In case of viable *E. coli* cells, this is feasible by infection with compatible bacteriophages carrying bacterial luciferase inserts. The genes in the phages are expressed only in live bacteria and if such are present, with a high-powered luminometer or by a microchannel plate enhanced image analyzer, even a single bacterial cell emitting light may be detected. Immunoassays and PCR are also useful for the detection of the presence of pathogens. An apparently very fast procedure is based on B lymphocyte sensors (CANARY: cellular analysis and notification of antigen risks and yields), which are engineered with the potential to express green fluorescent protein (GFP, aequorin) and membrane-bound antibodies specific for the pathogen of interest. GFP is a calcium-activated light emitter. When the antibody binds the pathogen, the intracellular calcium concentration is elevated within seconds and fluorescence is readily detectable. A bio-conjugated nanoparticle-based fluorescence immunoassay for in situ pathogen quantification detects single bacteria within 20 min (Zhao X et al 2004 Proc Natl Acad Sci USA 101:15027). For epidemic surveillance of respiratory pathogens, identification and strain typing can employ electrospray ionization mass spectrometry and polymerase chain reaction amplification from highly conserved genomic regions even from polymicrobial mixtures (Ecker DJ et al 2005 Proc Natl Acad Sci USA 102:8012). ►luciferase bacterial, ►immunological test, ►B lymphocytes, ►bioterrorism, ►PCR, ►aequorin, ►electrospray MS, ►polymerase chain reaction, ►quenched autoligation probe, ►quantum dot; Rider TH et al 2003 Science 301:213; GeneDB, bacterial, protozoa, fungal gene sequences: http://www.genedb.org/.

Pathogen-Derived Resistance: The protection of plants against certain pathogens by the transgenic expression of viral coat proteins, other proteins, antisense sequences, satellite and defective viral sequences. ►host–pathogen relations, ►plantibody

Pathogenesis Related Proteins (PR): A variety of acidic or basic proteins synthesized in plants upon infection with pathogens. The chitinases and glucanases apparently act by damaging the cell wall of fungi, insects or even bacteria. ►host–pathogen relations, ►SAR

Pathogenic: Capable of causing disease. (See Hill, A,V. S. 2001 Annu Rev Genome Hum Genet 2:373).

Pathogenicity Island (PAI): A group of genes in a pathogen involved in the determination and regulation of pathogenicity. In *Helicobacter pylori*, these islands are delineated by 31 bp direct repeats (DR), and indicate that horizontal transfer acquired these. Commonly the same genes are present between the ends of this large insert and the chromosomal genes in both pathogenic and non-pathogenic species of the same group of bacteria. Frequently, the insert is adjacent to a tRNA 3′ sequence or a codon for an unusual amino acid. The DNA inserts encode a rather specific secretory system (type III) and elements of transport and bacterial surface effectors located then next to host cell receptors. The PAI may carry insertion elements, integrases and transposases and their sequences may be unstable. Their location may change within the same bacterium. This organization is conducive to effective subversion the host defense system. The size of the pathogenicity islands may vary from 10 to 200 kb or more. The base composition of the islands and the codon usage may differ from that of the core DNA. In some bacteria only a single PAI occurs, in others there are several. Similar mechanisms operate in both animals and plants. ►transmission, ►cholera toxin, ►host–pathogen relations, ►integrase, ►transposase, ►secretion system, ►codon usage, ►*Helicobacter pylori*, ►symbionts; Hacker J, Kaper JB 2000 Annu Rev Microbiol 54:641; http://www.gem.re.kr/paidb.

Pathogenicity Islet: These are similar in some functions to pathogenicity islands but their size is much smaller, 1–10 kb.

Pathovar: Plant varieties or species, which share disease susceptibility/resistance genes. Alternatively, a plant pathogen that is specific for a taxonomic group of plants.

Pathway Tools (http://bioinformatics.ai.sri.com/ptools/): Software for determining metabolic and signaling pathways and database. ►EcoCyc, ►MetaCyc, ►BioCyc

Pathways: Guide to metabolic, molecular, immunological and many other processes and interactions: http://www.pathguide.org/; network tools: http://visant.bu.edu/.

Patrilocality: An anthropological term indicating that males more frequently bring in mates from outside their location than moving to the location of the females.

Patristic Distance: The sum of the length of all branches connecting two species in an evolutionary tree. ►evolutionary tree

Patroclinous: Inheritance through the male such as the Y chromosome, androgenesis, fertilization of a nullisomic female with a normal male, through nondisjunction the chromosome to be contributed by the female is eliminated, some of the gynandromorphs, sons of attached-X female *Drosophila*, etc. ►gynandromorph, ►nondisjunction, ►attached-X

Patronymic: A designation indicating the descent from a particular male ancestor, e.g., Johnson, son of John or O'Malley descendant of Malley. Common family names may assist in isolated populations to establish relationships. This analysis may be improved by studies of Y-chromosomal molecular markers. ►isonymy, ►Y chromosome

Pattern Formation during Development: Pattern formation specifies the arrangement of the cells in three dimensions. Developmental patterns may begin by intracellular differentiation (animal pole, vegetal pole, yolk), positional signals between cells and intracellular distribution of the receptors to various signals. Fibroblast growth factor and transforming growth factor-β have apparently major roles as epithelial and mesoderm induction signals. The *Drosophila* gene *fringe* (*fng*) is involved with mesoderm induction and in wing embryonal disc formation. Juxtaposition of *fng*-expressing and non-expressing genes is required for establishing the dorsal ventral boundary of wing discs. The gene *fng* is expressed in the dorsal half of the wing disc whereas *wingless* (*wg*, 2-30) is limited to the dorsal-ventral boundary. Gene *hedgehog* (*hh*, 3-81) is expressed in the posterior half, and *decapentaplegic* (*dpp*, 2-40) is detected in the anterior-posterior boundary. Anterior-posterior patterning in *Drosophila* is affected by the *trithorax* (*trx*), *Polycom* (*Pc-G*) family members such as *extra sex combs* (*esc*). The homolog of the latter, *eed* (embryonic ectoderm development) controls anterior-posterior differentiation in mouse. Homologs to these genes have been identified in other animals and humans as well. In *Xenopus*, it appears that the FNG protein is translated with a signal peptide (indicating that it is secreted), in the proFNG, peptide is terminated with a tetrabasic site for proteolytic cleavage and after this processing it is ready for the normal function. *Wingless* of *Drosophila* is homologous to *Wnt1* mouse mammary tumor gene. The branching pattern of trachea and lung, respectively in *Drosophila* and mammals is controlled primarily by the fibroblast growth factor signaling pathway. This is used reiteratively in repeated sequences of a branching. At each stage different feedback and other control signals provide the specifications (Metzger RJ, Krasnow MA 1999 Science 284:1635). In mice, mutation of the *Foxn1* gene causes follicular development to terminate just after it starts accumulating pigments. Then the immature follicles restart the developmental process and the skin color of the animal displays the striped pattern (Suzuki N et al 2003 Proc Natl Acad Sci USA 100:9680). Vascular mesenchymal cells can differentiate into specific embryonic structures and in adult diseases the process may be restarted again and bone osteoblasts may arise in the walls of the arteries or in the cardiac walls under the influence of morphogens. The process may be mathematically modeled (Garfinkel A et al 2004 Proc Natl Acad Sci USA 101:9247). In bacteria pattern formation can be programmed in a synthetic system by acyl-homoserine chemical signals emitted by labeled "sender" cells to "receiver" cells in Petri plate cultures (see Fig. P31).

Figure P31. Pattern formation induced by the sender cells (darker color) in the receiver cells (lighter color) lawn of bacteria. In the experiments the two types of cells were distinguished by flourescence. (Redrawn after Basu, S. et al. 2005 Nature [Lond] 434: 1130)

A newer idea posits that the mechanical control of form precedes pattern formation (Ingber DE 2005 Proc Natl Acad Sci USA 102:11571). Using microfabrication for controlling the organization of sheets of cells revealed the emergence of stable patterns of proliferative foci. Concentrated growth corresponded to regions of high tractional stress and could be measured by micromechanical force sensor. Inhibiting actomyosin-based tension or cadherin-mediated connections between cells disrupted spatial pattern of proliferation. The conclusion is that contraction of cells, an existence of pattern of mechanical forces, is due to multicellular organization and the tissue form is an active regulator of tissue growth (Nelson CM et al 2005 Proc Natl Acad Sci USA 102:11594).

Developmental pattern formation is under genetic control also in plants (Lee MM, Schiefelbein J 1999 Cell 99:473). The progress has been much slower, however, because the plant tissues and cells are more liable to dedifferentiation and redifferentiation. Mutants have been obtained with clear differences in morphogenesis, with the exception of flower differentiation and photomorphogenesis, much less is known about the molecular mechanisms involved. Down-regulation of the MYB transcription factor gene (*PHANTASTICA*) changes compound pinnate (left) leaves into palmate compound (right) leaves (see Fig. P32) (Kim M et al 2003 Nature [Lond] 424:438). ►morphogenesis, ►morphogenesis in *Drosophila*, ►RNA localization, ►flower differentiation, ►photomorphogenesis, ►MADS box, ►signal transduction, ►homeotic genes, ►cell lineages, ►fibroblast growth factor, ►actomyosin, ►cadherin; reaction–diffusion model of leaf venation as a mechanism of pattern formation: Dimitrov P, Zucker SW 2006 Proc Natl Acad Sci USA 103:9363; Malakinski G, Bryant P (Eds.) 1984 Pattern Formation: A Primer in developmental biology, Macmillan, New York; Comparative Pattern Formation in Plants and Animals: Willemsen V, Scheres B 2004 Annu Rev Genet 38:587.

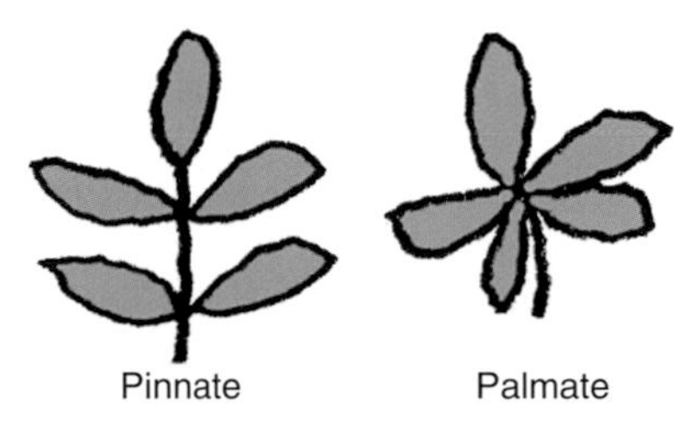

Figure P32. Pinnate (left); palmate (right)

Pattern Recognition Receptors (PRR): These are tools of the innate immunity system. They recognize pathogen-associated molecular patterns (PAMPs) that are essential for the survival of the pathogen and are therefore stable. PRRs are hereditary, are conserved across phylogenetic categories, expressed constitutively and do not require immunological memory. PRRs recognize pathogen-associated molecular patterns such as exist in lipopolysaccharides, proteoglycans or double-stranded RNA. (See Qkira S et al 2006 Cell 1214:763; innate immunity, Toll, host–pathogen relationship, downstream pathways: Lee MS, Kim Y-J 2007 Annu Rev Biochem 76:447.

PAU Genes: ►seripauperines

pauling: ►evolutionary clock

PAUP (phylogenetic analysis using parsimony): A computer program for the analysis of evolutionary descent on the basis of molecular data. ►evolutionary distance, ►evolutionary tree

Pause: RNA polymerase, I, II and III do not operate continuously at the same rate but due to various causes, their transcription may hesitate and then resume synthesis. A minimal functional element of PAUSE-1 is $TCTN_xAGAN_3T_4$ where x = 0, 2 or 4. Various elongation proteins such as ELL, Elongin, and transcription factor TFIIF may mediate pausing. The pause may facilitate the binding of regulatory factors. Pausing and backtracking allows the binding of the RfaH suppressor factor of early termination (Artsimovich I, Landick R 2002 Cell 109:193). Pausing may allow for proofreading and elimination of misincorporated bases (Shaevitz JW et al 2003 Nature [Lond] 426:684). Sequence-specific pause sites have been revealed (Herbert KM et al 2006 Cell 125:1083). ►attenuator region, ►Nus [►lambda phage], ►σ; Ogbourne SM, Antalis TM 2001 Nucleic Acids Res 29:3919.

Pausing, Transcriptional (hesitation): The discontinuity of the transcriptional process by all RNA polymerases. As a consequence there is a heterogeneity of the transcripts because of the differences in recognition of modulating factors such as attenuation, transcription factors TFIIF, ELL, silencers, nus (λ phage), antitermination signal, etc. Paucity of a nucleotide(s) or too high concentration of it may slow down transcription. RNA polymerase often pauses before a GTP is incorporated. Pause signals have been detected in both the template and the non-template DNA strands. Generally, hairpin structures (RNA base-pairing) favor pausing although secondary structure may not be the sole cause of it. DNA sequences 16–17 bp downstream of the pause may alter the conformation of the polymerase and the pause. Even the non-template strand may have an effect. ►arrest transcriptional, terms named under separate entries; Davenport RJ et al 2000 Science 287:2497.

PAX (paired box homeodomains): They are so called because they include two helix-turn-helix DNA-binding units. Several PAX proteins are known to be encoded in at least five different chromosomes and they mediate the development of the components of the eyes in insects (compound eyes) and humans, teeth, the central nervous system, the vertebrae, the pancreas and tumorigenesis. The 130-residue paired domain binds DNA and functions as a transcription factor for B cells, histones and thyroglobulin genes. The *Pax5* gene encodes the BSAP transcription factor. Mutation in *Pax5* arrests B cells at the pro-B stage. BSAP may also promote the expression of CD19 and indirectly IgE synthesis. BSAP may block the immunoglobulin heavy chain 3′-enhancer and isotype switching and the formation of the pentameric IgM antibody. The activator motifs of BSAP display about 20 times higher binding affinity to the DNA than the repressor motif yet the activator or repressor function depends primarily on the context of the motif. The level of BASP is high in the pre-B and immature B cell stages and after the antigen signal has arrived its level greatly diminished by signals from IL-2 and IL-5. Overexpression of BSAP results in its repressor activity. PAX6 (11p13) mutations cause absence of the iris of the eye (aniridia) without elimination of vision but other general neurodevelopmental problems. ►Waardenburg syndrome, ►DiGeorge syndrome, ►aniridia, ►Wilms tumor, ►renal-coloboma syndrome, ►hypodontia, ►rhabdosarcoma, ►animal models, ►B cell, ►immunoglobulins, ►histones, ►thyroglobulin, ►hox, ►homeotic genes, ►isotype switching, ►FKP, ►goiter familial, ►helix-turn-helix motif, ►integrin, ►PTIP, ►myoblast, ►neural crest; Balczarek KA et al 1997 Mol Biol Evol 14:829; Chi N, Epstein JA 2002 Trends Genet 18:41; Pichaud F, Desplan C 2002 Curr Opin Genet Dev 12:430; http://pax2.hgu.mrc.ac.uk/; http://pax6.hgu.mrc.ac.uk/.

PAZ Domain (Piwi/Argonaute/Zwille): In Argonaute 1 protein it consists of a left-handed, six-stranded β-barrel capped at one end by two α-helices and wrapped on one side by a distinctive appendage, which comprises a long β-hairpin and a short α-helix. The PAZ domain binds a 5-nucleotide RNA with 1:1 stoichiometry. It plays a role in RNAi function and it is present in both Argonaute and Dicer proteins. ►Argonaute, ►Dicer, ►RNAi; Yan KS et al 2003 Nature [Lond] 426:469; Lingel A et al 2003 Nature [Lond] 426:465; Ma J-B et al 2004 Nature [Lond] 429:318; PIWI domain structure: Parker JS et al 2005 Nature [Lond] 434:663; Ma J-B et al 2005 Nature [Lond] 434:666.

PBAF: An ATP-dependent chromatin remodeling complex. ►chromatin remodeling; structure of PBAF: Leschziner AE et al 2005 Structure 13:267.

Pbp74: ►Hsp70

pBR322: A non-conjugative plasmid (constructed by Bolivar & Rodriguez) of 4.3 kb, can be mobilized by helper plasmids because (although it lost its mobility gene) it retained the origin of conjugal transfer (see Fig. P33). It is one of the most versatile cloning vectors with completely known nucleotide sequence and over 30 cloning sites. It carries the selectable markers ampicillin resistance and tetracycline resistance. Insertion into these antibiotic resistance sites permit the detection of the success of insertion because of the inactivation of these target genes results in either ampicillin or tetracycline sensitivity. Although its direct use of this 20-year old plasmid has diminished during the last years, pBR322 components are present in many currently used vectors. ►plasmid, ►vectors, ►Amp, ►tetracycline; Bolivar, F., Rodriguez RL et al 1977 Gene 2:95.

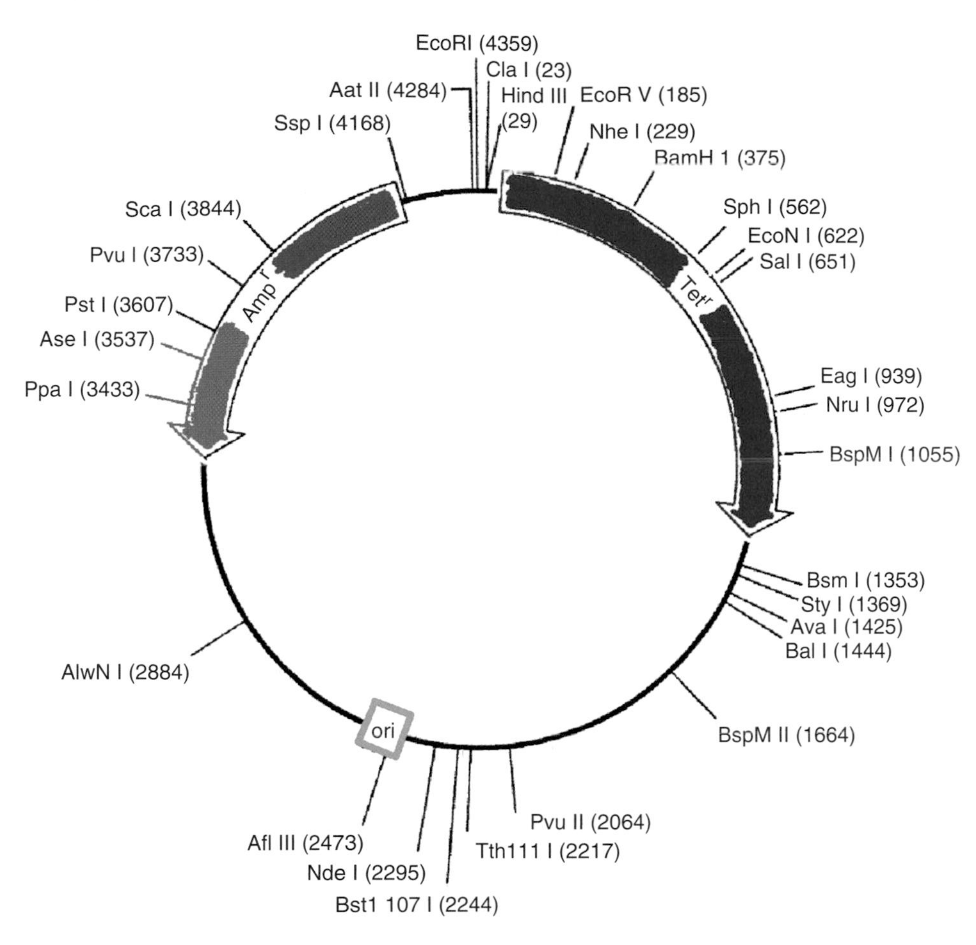

Figure P33. pBR vector (4361 bp), From Pharmacia Biotech Inc., by permission

PBS: Phosphate buffered saline. ►saline

PBSF (pre-B cell growth-stimulating factor): A ligand of CXCR controlling B cell development and vascularization of the gastrointestinal system. ►CXCR, ►lymphocytes; Egawa T et al 2001 Immunity 15:323.

PBX1, PBX2: Transcription factors involved in B cell leukemias, encoded in human chromosomes 1q23 and 3q222, respectively. ►leukemia

p. c.: ►post coitum

PC4 (positive co-activator of transcription): PC4 interacts with activator protein-1 (AP-2) and facilitates transcription, and may relieve the self-interference of AP-2. ►transcription, ►AP1; Zhong L et al 2003 Gene 320:155.

PCA (principal component analysis): A multivariate statistical method that separates the original variables into independent variables and associated variances. It may be useful for the interpretation of microarray data. (See Méndez MA et al 2002 FEBS Lett 522:24).

PCAF: A human acetyltransferase of histones 3 and 4. ►p300, ►$TAF_{II}230/250$, ►histone acetyltransferases, ►nucleosome, ►signal transduction, ►chromatin remodeling, ►bromodomain, ►INHAT; Blanco JC et al 1998 Genes Dev 12:1638.

PCB: Polychlorinated biphenyl is an obnoxious industrially employed carcinogen. A *Pseudomonas* enzyme may break it down. ►environmental carcinogens, ►sperm

PCD: ►apoptosis

pCIP: A co-activator protein, interacting with p300 (CREB) and regulates transcription and somatic growth of mammals. ►signal transduction, ►nuclear receptors; Wang Z et al 2000 Proc Natl Acad Sci USA 97:13549.

PCL: Putative cyclin. ►cyclin

PCNA (proliferating cell nuclear antigen): An auxiliary protein (a processivity factor) in pol δ and pol ε functions in eukaryotes. It has a similar role in DNA replication in general, and in the cell cycle and repair. Its function is similar to that of the β subunit of the prokaryotic pol III, it provides a "sliding clamp" on the DNA to be replicated. Binding to p21 may inhibit

PCNA replicative function. PCNA also binds other proteins such as cyclin D, FEN1/Rad27/MF1, DNA ligase 1, GADD, DNA methyltransferase, DNA repair proteins XPG, MLH and MSH. The protein interaction (PIP-box) provides a dock for interaction with replication and repair proteins (Bruning JB, Shamoo Y 2004 Structure 12:2209). The PCNA-interacting mutations in PCNA alter the conditions for nucleosomal assembly by interacting with CAF. Under such conditions, silencing by heterochromatin is reduced or lost. Some mutations in RFC may compensate for defects in PCNA. If the DNA is damaged, PCNA is modified by mono-ubiquitination at lysine residue 164 or a lysine 63-linked multi-ubiquitin chain to allow error-free or error-prone replication bypass of the damaged site. Ubiquitination of PCNA at lysine 164 specifically activates DNA polymerase η and Rev1, a deoxycytidyl-transferase in mutagenic replication of DNA (Garg P, Burgers PM 2005 Proc Natl Acad Sci USA 102:18361). In addition, SUMO-modified PCNA may recruit Srs2 helicase, which disrupts recombination by affecting Rad51 protein (Pfander B et al 2005 Nature [Lond] 436:428). ►DNA replication, ►eukaryotes, ►cell cycle, ►DNA polymerases, ►ligase DNA, ►p21, ►cyclins, ►ABC excinuclease, ►excision repair, ►mismatch repair, ►sliding-clamp, ►methylation of DNA, ►Rad27/Fen1, ►RFC, ►CAF, ►heterochromatin, ►ubiquitin, ►SUMO, ►Srs, ►Rad51; Karmakar P et al 2001 Mutagenesis 16:225; Ola A et al 2001 J Biol Chem 276:10168; López de Saro FJ, O'Donell M 2001 Proc Natl Acad Sci USA 98:8376; Lau PJ, Kolodner RD 2003 J Biol Chem 278:14; crystal structure of binding domains: Kontopidis G et al 2005 Proc Natl Acad Sci USA 102:1871; review: Moldovan G-L et al 2007 Cell 129:665.

PCR: ►polymerase chain reaction

PCR, Asymmetric: By using unequal amounts of amplification primers, an excess of single-strand copies of DNA can be obtained (Gyllensten UB, Erlich HA 1988 Proc Natl Acad Sci USA 85:7682). ►polymerase chain reaction, an improved procedure: Pierce KE et al 2005 Proc Natl Acad Sci USA 102:8609.

PCR, Allele-Specific: Used to screen a population for a particular allele-specific mutation, e.g., mutations responsible for MELAS in the aging human mitochondrial DNA. One of the most common mutations in this anomaly involves the transition A→G at site 3243. If the primer containing the complementary base C is used, the mutant sequence from appropriate tissues is successfully amplified and can be detected by gel electrophoresis. The same C primer does not generate substantial quantity of the fragment using the wild type template. Thus, by this procedure, the approximate frequency of this allele-specific mutation or recombination can be determined. ►polymerase chain reaction, ►mitochondrial disease in humans, ►transition mutation, ►hot-start PCR; Ugozzoli L, Wallace RB 1992 Genomics 12:670.

PCR, Broad-Based: Uses primers, which amplify a broad base of genes, e.g., the microbial rRNA genes or a group of viral genes common to the majority of related species in order to facilitate molecular identification of same pathogens. RDA and other procedures may supplement the analysis. ►RDA

PCR, Competitive: Used for quantifying DNA or RNA. The competitor nucleic acid fragment of known concentrations in serial dilutions is co-amplified with another (the experimental) nucleic acid of interest using a single set of primers. The beginning quantity of the experimental molecules is estimated from the ratio of the competitor and experimental amplicons obtained during the PCR procedure that are supposedly amplified equally. The quantity of the unknown DNA is determined by the equivalence-point (EQP) where the experimental and the competitor show the same signal intensity indicating that their amounts is the same. A simplified new version is described by Watzinger F et al 2001 Nucleic Acids Res 29(11):e52.

PCR, Discriminatory: A method to detect small mismatches or point mutations (see Fig. P34). It is a much easier method than sequencing larger sequences to distinguish, e.g., phylogenetic differences within taxonomic groups. (See Picard FJ et al 2004 J Clin Microbiol 42:3686).

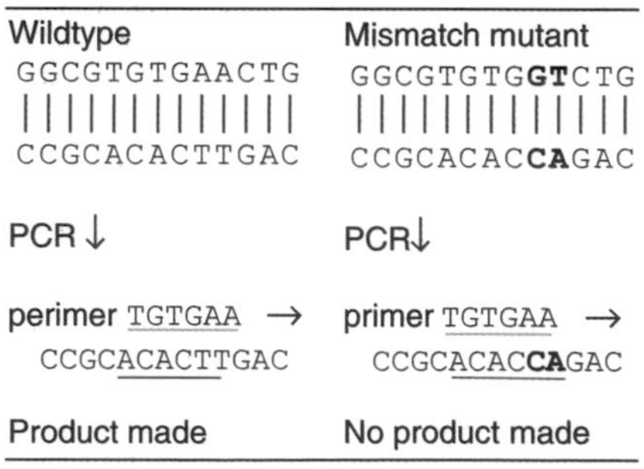

Wildtype	Mismatch mutant
GGCGTGTGAACTG	GGCGTGTG**GT**CTG
CCGCACACTTGAC	CCGCACAC**CA**GAC
PCR ↓	PCR↓
perimer TGTGAA →	primer TGTGAA →
CCGCACACTTGAC	CCGCACAC**CA**GAC
Product made	No product made

Figure P34. Discriminatory PCR

PCR, DOC (degenerate oligonucleotide-primed polymerase chain reaction and capillary electrophoresis of DNA): A random amplification technique combined with analysis on microchips. ►capillary electrophoresis, ►PCR, ►DOP-PCR; Cheng J et al 1998 Anal Biochem 257:101.

PCR, Electronic: ►electronic PCR

PCR, Methylation-Specific: ►methylation-specific PCR

PCR, Multiplex: Employs multiple sets of primers for amplification in a single reaction batch. (See Broude NE et al 2001 Proc Natl Acad Sci USA 98:206).

PCR, Nested: The use of two different internal primers to thus identify overlapping transcripts.

PCR, Overlapping: The use of two sets of primers; each has complementary sequences at the 5′-end. Two separate PCRs are carried out and then the products purified by gel to remove the unincorporated primers. A second PCR process uses only the outside primer pairs and the two primary products are joined. ►PCR

PCR, Quantitative: Determine gene expression quantitatively by optimized primers: http://primerdepot.nci.nih.gov/; http://mouseprimerdepot.nci.nih.gov/.

PCR, Real-Time Reverse Transcription: see Seeger K et al 2001 Cancer Res 61:2517.

PCR, Single Molecule: ►polony

PCR Targeting: ►targeting genes

PCR, Transcriptionally Active (TAP): 1. Specific primers amplify the gene of interest. 2. Mixtures of DNA fragments are equipped with promoter and terminator elements and then can be used for transfection in a suitable plasmid. They can be inserted into plasmids also by homologous recombination. TAP products can be used as DNA vaccines and generate antibodies against the encoded genes. The procedure is suitable for the generation of hundreds or thousands of transcriptionally active genes for genomic/proteomic studies. (Liang X et al 2002 J Biol Chem 277:3593).

PCR-Based Mutagenesis: Any base difference between the amplification primer will be incorporated in the future template through polymerase chain reaction. Actually only half of the new DNA molecules would contain the alteration present in the original amplification primer unless a device is used, e.g., the undesired strand would be made unsuitable for amplification and therefore lost from the reaction mixture. The method may include multiple point mutations, small insertions or deletions too. The amplification may also result in other nucleotide alterations as a result of the error-prone Taq polymerase. ►local mutagenesis, ►primer extension, ►polymerase chain reaction, ►DNA shuffling, ►VENT, ►small-pool PCR; Nelson RM, Long GL 1989 Anal Biochem 180:147.

PCR-LSA (polymerase chain reaction amplification): A method for the localization of SNIPS. ►SNIPS, ►RRS

PCR-Mediated Gene Replacement: The procedure replaces—by mitotic recombination—particular genes with an identifiable marker of a neutral phenotype. A 20 bp unique sequence tract tags each of these lines. On the basis of hybridization of the PCR products to a tag sequence, it is possible to quantitate the altered cell lines in a population. When one of the target genes is deleted in diploid yeast, the other is still expected to be functional. Some of them, however, due to haplo-insufficiency, will display a defective phenotype. These "heterozygotes" may also display increased sensitivity to drugs and may be used for pharmaceutical research. ►haplo-insufficiency; Giaever G et al 1999 Nature Genet 21:279.

PC-TP: phosphatidylcholine transfer protein mediating transfer of phospholipids between organelles within cell.

PD-1: An inhibitory immune receptor (2q37.3) on B, T lymphocytes and natural killer cells.

PDB (Protein Data Bank): http://www.rcsb.org/pdb/cgi/explore.cgi?pdbld.

PDECGF: Platelet-derived endothelial cell growth factor.

pDelta: $\gamma\delta$

PDF: An electronic publishing software readable with the aid of Adobe Acrobat reader. Frequently used by journals available through the Internet.

PDGF: ►platelet derived growth factor

PDGFR: Platelet derived growth factor receptor

PDI (protein disulfide isomerase): A co-factor of protein folding mediated by chaperones. ►chaperone, ►PPI, ►Eug, ►Mpd, ►Erp61

PDK (phosphoinositide-dependent kinase): A part of the MAPK, RSK signaling pathway. ►MAPK, ►RSK, ►phosphoinositides, ►PIK, ►Akt; Toker A, Newton AC 2000 Cell 103:185.

PDS: An anaphase inhibitory protein that must be degraded with the assistance of CDC20 component of APC before the cell cycle can exit from mitosis regulated also by the activity of the Cdc14 phosphatase. ►cell cycle, ►APC, ►Esp1, ►sister chromatid cohesion, ►checkpoint, ►mitotic exit, ►Cdc14, ►Cdc20; Salah SM, Nasmyth K 2000 Chromosoma 109:27.

pDUAL: $\gamma\delta$

PDZ Domains (post-synaptic density, disc-large, zo-1): Approximately 90 amino acid repeats involved in ion-channel and receptor clustering, and linking effectors and receptors. PDZ domain proteins are involved in the regulation of the Jun N-terminal kinase pathway, in the post-synaptic density (PSD) proteins at glutamatergic synapses, Rho-activated citron protein function, visual signaling, etc. The *Drosophila* gene, *scribble* (*scrib*) encodes a multi PDZ domain protein and in cooperation with a leucine-rich protein controls apical polarization of the embryo. ►ion channels, ►tight junction, ►protein folding, ►Van Gogh, ►receptor, ►effector, ►signal transduction, ►AMPA, ►HOMER, ►citron, ►Jun, ►Rho, ►NMDAR, ►mesenchyma, ►Fraser syndrome; Harris BZ, Lim WA 2001 J Cell Sci 114(pt 18):3219; Hung AY, Sheng M 2002 J Biol Chem 277:5699.

Pea (*Pisum* ssp): Several self-pollinating vegetable and feed crops: the Mendel's pea is *P. sativum*, and others are 2n = 2x = 14 (see Fig. P35). *Pisum*, photograph shows normal Mendelian segregation for smooth and wrinkled within a pod.

Figure P35. Pea

PEA: Death effector domain proteins.

Pea Comb: Comb characteristic of poultry of *rrP*(*P*/*p*) genetic constitution. ►walnut comb

Peach (*Prunus persica*): x = 7, the true peaches are diploid.

Peacock's Tail: An evolutionary paradigm when a clear disadvantage (like the awkward tail) turns into a mating advantage because of the females' preference for the fancy trait and thus increasing the fitness of the males that display it. ►fitness, ►selection, ►sexual selection

Peanut: ►groundnut

Pear (*Pyrus* spp): About 15 species; x = 17 and mainly diploid, triploid or tetraploids. It is very difficult to hybridize it with apples but can be crossed with some *Sorbus*. ►apple

Pearson Marrow Pancreas Syndrome: ►mitochondrial disease in humans

Pearson's Product Moment Correlation Coefficient: ►correlation

Pebble: ►scaffolds in genome sequencing

Pectin: The polygalacturonate sequences alternated by rhamnose and may contain galactose, arabinose, xylose and fucose side chains. Molecular weight varies from 20,000 to 400,000. Its role is intercellular cementing of plant cells. Acids and alkali may cause its depolymerization.

Pedicel: The stalk of flowers in an inflorescence. ►peduncle

PEDANT (protein extraction, description and analysis tool): See http://pedant.gsf.de/.

Pedigree Analysis: Generally carried out by examination of pedigree charts used in human and animal genetics where the family sizes are frequently too small to conduct meaningful direct segregation studies (see Fig. P36). The pedigree chart displays the lines of descent among close natural relatives. Females are represented by circles, males by squares and if the sex is unknown a diamond (◇) is used (see Fig. P37). The same but smaller symbols or by a vertical or slanted line indicates abortion or still birth over the symbol. For spontaneous abortions, triangles (q) may be used. Individuals expressing a particular trait are represented by a shaded or black symbol, and in case they are heterozygous for the trait, half of the symbol is shaded.

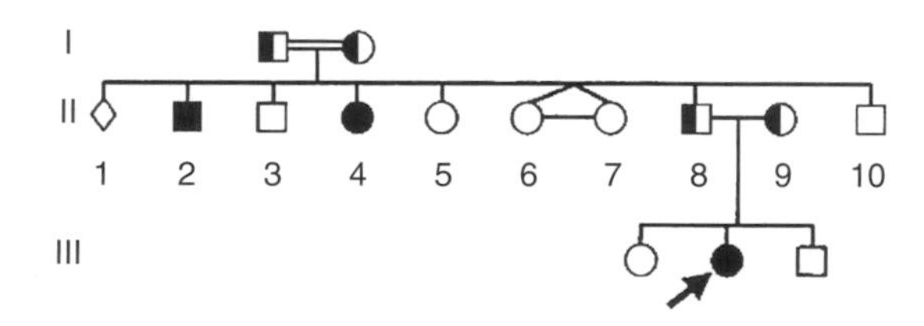

Figure P36. Pedigree chart

When an unaffected female is the carrier of a particular gene, there is a dot within the circle. In case segregation for traits needs to be illustrated in the pedigree, the individual displaying both traits may be marked by a horizontal and vertical line within the symbol or only by a horizontal or vertical line, respectively. Horizontal lines connect the parents and if the parents are close relatives, the line is doubled. The progeny is connected to the parental line with a vertical line and the subsequent generations are marked by Roman numerals at the left side of the chart, I (parents), II children, (III (grandchildren), and so on. Twins are connected to the same point of the generation line and if they are identical, a horizontal line connects them to each other. The order of birth of the offspring is from left to right and may be numbered accordingly below their symbol. An arrow to a particular symbol indicates the proband, the individual who first became known to the geneticist as expressing the trait. The appropriate symbols of

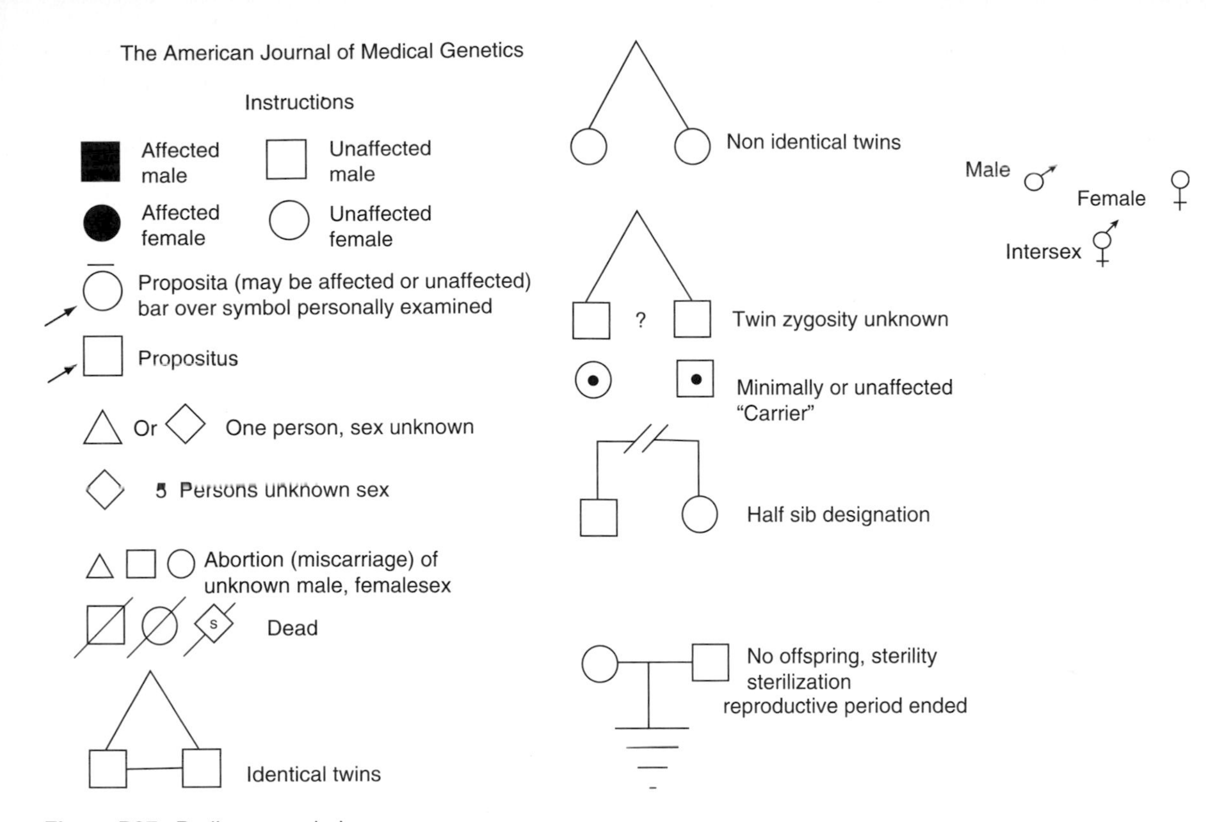

Figure P37. Pedigree symbols

adopted children may be bracketed. If a prospective offspring is considered at risk, broken lines draw the symbol. A horizontal line connected by a vertical line to the "parental" line may indicate lack of offspring by a couple.

Infertility may be represented by doubling a horizontal line under and connected to the male or female symbol, respectively. Egg or sperm donors (in case of assisted reproductive technologies [ART]) are indicated by a D and surrogate mothers by S within the symbols. *Mars shield* represents males and a *Venus mirror* represents females, and the sign in the box shown above at right indicates intersexes. ►ART; Bennett RL et al 1995 Am J Hum Genet 56:745; Am J Med Genet pedigree chart is reprinted by permission of John Wiley & Sons, Inc.

Pedogenesis: Egg production by immature individuals such as larvae.

PEDRo (Proteomics Experiment Data Repository): A model for the collection of information in proteomics. ►proteomics; Taylor CF et al 2003 Nat Biotechnol 21:247.

Peduncle: The stalk of single standing flowers (→); bundle of nerve cells (see Fig. P38). ►pedicel

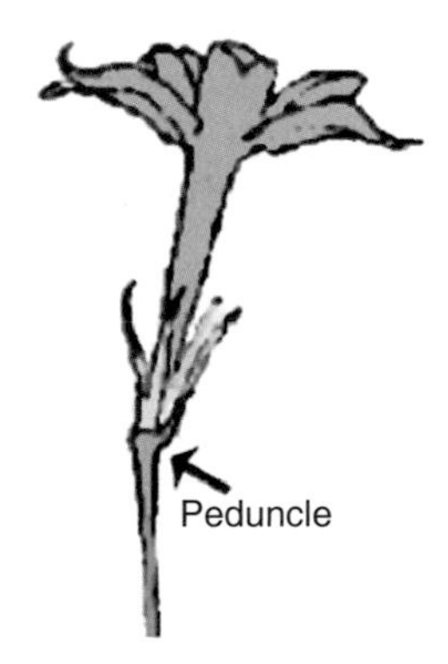

Figure P38. Peduncle

PEG (polyethylene glycol): May be liquid or solid and comes in a range of different viscosities (200, 400, 600, 1500, etc.). It facilitates fusion of protoplasts, uptake of organelles, precipitation of bacteriophages, plasmids and DNA, promoting end-labeling, ligation of linkers, reduction of immunogenicity when attached to humanized antibody, etc.

PEG-3 (progression elevated gene-3): In nude mice up-regulates carcinogenesis in progress via activation of VEGF. It can be blocked by antisense technology. ►VEGF, ►antisense technology

PEG (paternally expressed gene): ►imprinting

PEGylation: Attaches polyethylene glycol (PEG) to the polypeptide backbone of a protein drug, and renders it less liable to clearance from the body, and thus does not have to be supplied so frequently. PEGylated interferons have about 24 h half-life in the plasma whereas native interferons display only 4 h half-life. Interferon-α-2b (a recombinant product) when PEGylated is administered only once a week whereas without PEGylation it needs three dosing per week. ►interferon; Walsh G 2003 Nat Biotechnol 21:865.

PEK: ►HRI

Pelargonidine: ►anthocyanin

Pelargonium zonale (geranium): An ornamental plant. Some variegated forms transmit the non-nuclear genes also through the sperm whereas in the majority of plants the plastids are transmitted only through the egg. ►uniparental inheritance, ►chloroplasts, ►chloroplast genetics

Pelger(-Huet) Anomaly: An autosomal dominant condition in humans as well as in rabbits, cats, etc., characterized by fewer (1.1–1.6) than normal (2.8) nuclear lobes in the granulocytic leukocytes. Mutations may involve the lamin B receptor. It may be a mild anomaly but it may be associated with other more serious ailments. The prevalence varies from 1×10^{-3} to 4×10^{-4}. Similar phenotypes were described also as autosomal recessive or X-linked. ►laminopathies; Shultz LD et al 2003 Hum Mol Genet 12:61.

Pelizaeus-Merzbacher Disease: An Xq22 chromosomal recessive leukodystrophy that accumulates a proteolipoprotein (PLP, a 276-amino acid integral membrane protein) of the endoplasmic reticulum and the surface protein DM20 (26.5 kDa). The defect involves alternative splicing of the same mRNA. Duplications and deletion may be responsible for the disorder. The clinical symptoms are defective myelination (dysmyelination) of the nerves and defective interaction between oligodendrocytes and neurons, pathogenesis of the central nervous system and impaired motor development with an onset before age one. Mutation in the same gene encoding PLP is responsible also for X-linked spastic paraplegia type 2 (SPG-2) and the difference is in the degree of hypomyelination and motor dysfunction. Hereditary spastic paraplegia alleles were assigned to 8p, 16p, 15q and 3q27-q28. The corresponding defect in mouse displays *jimpy* and the myelin deficient (*msd*) phenotype. ►myelin, ►Charcot-Marie-Tooth disease, ►leukodystrophy, ►spastic paraplegia

Pelle: Serine/threonine kinase, involved in dorsal signal transduction. ►IRAK

Peloric: The circular symmetry of the flower in contrast to the bilateral symmetry of the wild type first described in *Linaria* by Linnaeus. Homologous mutations occur in *Antirrhinum* as shown in Figure P39. This variation of floral symmetry of *Linaria* is due to the different methylation of a gene, *Lcyc* (Cubas P et al 1999 Nature [Lond] 401:157). It is thus an epimutation. The first figure is the wild type *Antirrhinum* flower of bilateral symmetry. The second figure is the *cycloidea* mutant with radial symmetry (pelory). (Illustration is the courtesy of Professor Hans Stubbe; ►methylation of DNA, ►epigenesis, ►superman, ►snapdragon, ►cycloidea)

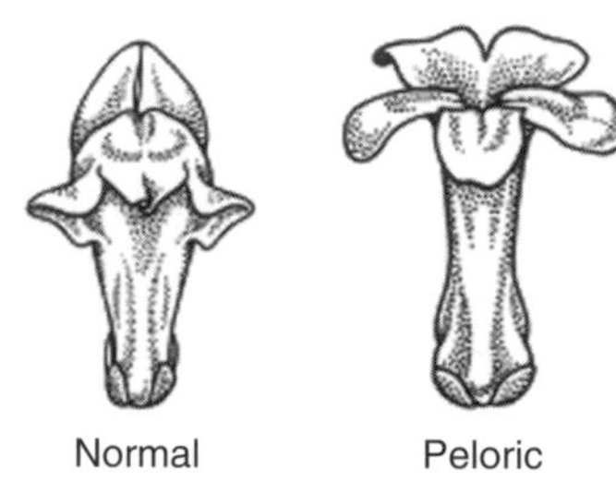

Figure P39. Normal and peloric snapdragon flowers

pelota: *Drosophila* gene involved in sperm function. ►azoospermia

Pemphigus: A collection of skin diseases with the general features of developing smaller or larger vesicles of the skin that may or may not heal and in extreme cases may result in death. The autosomal dominant familial pemphigus vulgaris is an autoimmune disease of the skin and mucous membranes. In the majority of cases, HLA-DR4 is involved. This anomaly is particularly common among Jews in Israel. In mice pemphigus vulgaris, inhibitor of MAPK (p38) can reduce the autoimmune blisters (Berkowitz P et al 2006 Proc Natl Acad Sci USA 103:12855). ►Hailey-Hailey disease, ►HLA, ►skin diseases, ►desmosome, ►autoimmune disease

Pena-Shokeir Syndrome: A fetal akinesis caused by brain malformations; X-chromosomal inheritance is suspected. ►akinesis

Pendred Syndrome: A recessive (7q31, PDS) thyroid anomaly and neurosensory deafness. The locus encodes *pendrin* an anion transporter, a presumed sulfate transporter localized in the cell membrane and a bicarbonate secretion in the kidney. Recent evidence indicates chloride and iodide transport too. This locus is responsible for about 1–10% of the genetically determined hearing loss. ►deafness, ►goiter; Royaux IE et al 2001 Proc Natl Acad Sci USA 98:4221.

Penetrance: The percentage of individuals in a family that express a trait determined by gene(s) they contain. The genetic basis of this phenomenon is poorly understood, and may cause serious problems in genetic counseling. ►expressivity

Penicillin: An antibiotic originally obtained from *Penicillium* fungi (see Fig. P40). The arrow points to the reactive bond of the β-lactam ring. ►antibiotics, ►*Penicillium*, ►lactam, ►β-lactamase

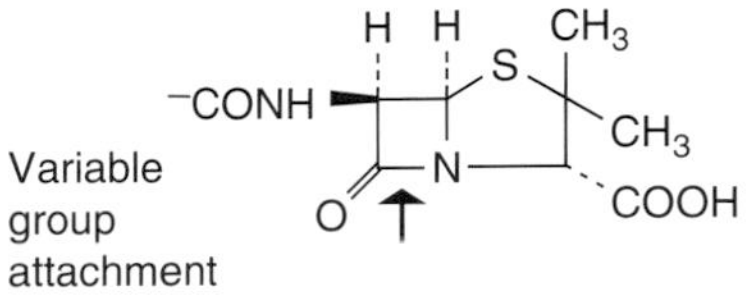

H H CH_3
S
−CONH
CH_3
Variable group attachment
N
O
COOH

Figure P40. Basic structure of penicillins

Penicillin Enrichment: ►penicillin screen

Penicillin Screen: Used for mass isolation of auxotrophic microbial mutations that failed to grow in basal media (in contrast to the wild type) and the presence of the antibiotic therefore did not lead to their death (in contrast to the wild type). After transfer to complete (or appropriately supplemented) media the auxotrophs grew and thus were selectively isolated. ►selective medium, ►replica plating, ►mutant isolation; Davis BD 1948 J Am Chem Soc 70:4267; Lederberg J, Zinder N 1948 J Am Chem Soc 70:4267.

Penicillinase: β-lactamase

Penicillin Binding Proteins: ►PBP

Penicillium notatum (fungus): x = 5 (see Fig. P41).

Figure P41. Penicillium conidiophore with conidia

Penis: The male organ of urinary excretion and insemination (homologous to the female clitoris). It contains the *corpus spongiosum* through which the urethra and sperm passes. Above that are the *corpora cavernosa* that become extended when erection takes place due to enhanced blood supply to this elastic tissue as a consequence of NO (nitrogen-monoxide) gas flows to the muscles of the blood vessel wall, initiated by acetylcholine. The release of acetylcholine is controlled by steroid hormones. Cyclic GMP-dependent kinase is an essential enzyme for the maintenance of the extended state of the *corpus cavernosum*. Injection of prostaglandin E1 blocks cGMP-degrading phosphodiesterase and facilitates erection. The penis of canines (and most other mammals including primates except humans and spider monkeys) contains a small bone (*baculum* or *os penis*). In all humans, the gulonolactone oxidase (EC 1.1.3.8) gene on chromosome 8p21 is defective and no baculum is formed. The lack of this enzyme also makes humans dependent on dietary ascorbic acid. (It has been suggested that God created Eve not from a rib but from the baculum of Adam.) The baculum of mammals assists quick penetration of the vagina. During sexual intercourse of dogs, the male first penetrates the vulva of the female and erection takes place only in a second phase. At the base of the dog's penis the oval *bulbus glandis* then expands and locks the penis in position and the mating pair cannot separate until the ejaculation is terminated. Some animals (snakes, lizards, crustaceans and insects) have two penises (*virgae*). In *Euborellia plebeja* (Dermatoptera) both are functional. ►animal hormones, ►acetylcholine, ►acetylcholine receptors, ►hypospadias, ►nitric oxide, ►cGMP, ►prostaglandin, ►baculum, ►erectile dysfunction, ►clitoris; Kamimura Y, Matsuo Y 2001 Naturwiss 88:447; insect penis evolution review: Palmer AR 2006 Nature [Lond] 444:689.

Pentaglycines: ►bacteria

Pentaploid: Its cell nucleus contains five genomes (5x). Pentaploids are obtained when hexaploids (6x) are crossed with tetraploids (4x). The pentaploids are generally sterile or semi-fertile because the gametes generally have unbalanced number of chromosomes. ►polyploids, ►*Rosa canina*

Pentatrico: A ~35 unit sequence of amino acids, generally repeated several times in some proteins. Pentatricopeptide repeat (PPR) proteins form one of the largest families in higher plants and are believed to be involved in the posttranscriptional processes of gene expression in plant organelles and RNA editing in the chloroplasts (Okuda K et al 2007 Proc Natl Acad Sci USA 104:8178). ►RNA editing, ►retrograde regulation

Penton: Capsomer with five neighbors in the viral capsid. ►capsomer, ►hexon; Zubicta C et al 2005 Mol Cell 17:121.

Pentose: A sugar with 5-carbon-atom backbone, such as ribose, deoxyribose, arabinose, xylose.

Pentose Phosphate Pathway: glucose-6-phosphate + 2 NADP + H_2O → ribose-5-phosphate + 2 NADPH + 2

H + + CO2, i.e., the conversion of hexoses to pentoses generates NADPH, a molecule that serves as a hydrogen and electron donor in reductive biosynthesis. ►Embden-Meyerhof pathway, ►Krebs-Szentgyörgyi cycle

Pentose Shunt: Same as pentose phosphate pathway.

Pentosuria: An autosomal recessive non-debilitating condition characterized by excretion of increased amounts of L-xylulose (1–4 g) in the urine because of a deficiency of the NADP-linked xylitol dehydrogenase enzyme. In Jewish and Lebanese populations, the frequency of the gene was about 0.013–0.03. ►gene frequency, ►allelic frequencies

PEPCK (phosphoenolpyruvate carboxykinase): A regulator of energy metabolism.

Pepper (*Capsicum* spp): It exists in a great variety of forms but all have 2n = 2x = 24 chromosomes. Some wild species are self-incompatible and the cultivated varieties yield better if they have a chance for xenogamy. ►self-incompatibility, ►xenogamy

Pepsin: An acid protease, formed from pepsinogens. It has preference for COOH side of phenylalanine and leucine amino acids.

Pepstatin ($C_{34}H_{63}N_5O_9$): A protease (pepsin, cathepsin D) inhibitor.

Peptamer: The exposed loop on the surface of a carrier protein; it is thus protected from degradation and its conformational stability is improved.

Peptidase (protease): Hydrolyzes peptide bonds. In humans, the peptidase gene PEPA is in chromosome 18q23, PEPB in 12q21, PEPC in 1q42, PEPD in 19cen-q13.11, PEPE in 17q23-qter, PEPS in 4p11-q12, and the tripeptidyl peptidase II (TPP2), a serine exopeptidase is in 13q32-q33. (See http://merops.sanger.ac.uk).

Peptide Bond: Amino acids are joined into peptides by their amino and carboxyl ends ↑ (and they lose one molecule of water) (see Fig. P42).

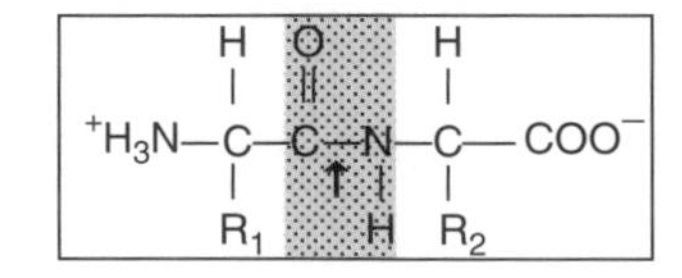

Figure P42. Peptide bond

Peptide Elongation: ►protein synthesis, ►aminoacylation, ►aminoacyl-tRNA synthetase, ►elongation factors (eIF), ►ribosome, ►tmRNA, ►cycloheximide

Peptide Initiation: ►protein synthesis, ►pactamycin

Peptide Mapping: The separation of (in)complete hydrolysates of proteins by two-dimensional paper chromatography or by two-dimensional gel electrophoresis for the purpose of characterization. The distribution pattern is the map or fingerprint, characteristic for each protein.

Peptide Mass Fingerprints: The protein is first cleaved by a sequence-specific protease such a trypsin and analyzed by MALDI-TOF and compared with protein sequences with similar lysyl or arginyl residues of the same mass. On this basis matching proteins even in a mixture can be identified. Modified proteins are detectable on the basis of peptide sequence with an incremental mass due to, e.g., a phosphogroup. ►proteomics, ►MALDI, ►trypsin; Mann M et al 2001 Annu Rev Biochem 70:437; Pratt JM et al 2002 Proteomics 2:157; Giddings MC et al 2003 Proc Natl Acad Sci USA 100:20; http://www.peptideatlas.org/.

Peptide Nucleic Acid (PNA): A nucleic acid base (generally thymine) is attached to the nitrogen of a glycine (or other amino acids) by a methylene carboxamide linkage in a backbone of aminoethylglycine units (see Fig. P43). Such a structure can displace one of the DNA strands and binds to the other strand.

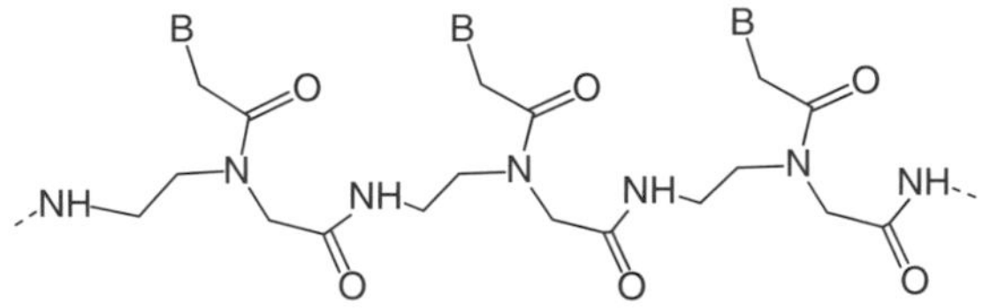

Figure P43. PNA backbone resembles that of DNA. The letter B stands for nucleic acid bases

They are DNA mimics. This highly stable complex has similar uses as the antisense RNA technology. PNA may be used to inhibit excessive telomerase activity in cancer cells. Homopyrimidine PNA may invade homopurine tracts in double-stranded DNA and may form triplex DNA and interfere with transcription. Peptide nucleic acid can target the polyguanine tract of HIV-1 and can arrest translation elongation (Boutiah-Hamoudi F et al 2007 Nucleic Acids Res 35:3907). PNA may also be used to screen for base mismatches and small deletions or base substitution mutations. Peptide nucleic acid complementary to mutant mtDNA selectively inhibits the replication of mutant mtDNA in vitro. PNA has been suggested to be the first pre-biotic genetic molecule rather than RNA. PNA may be useful for delivering genes to the mitochondria. PNA–DNA hybrids may be identified by binding of the dye 3,3′-diethylthiadicarbocyanine and used for the rapid detection of mutations of clinical importance. Site-specific recombination may be substantially enhanced by PNA. ►antisense RNA, ►antisense DNA, ►TFO, ►mtDNA, ►Hoogsteen pairing, ►RNA world, ►mitochondrial gene therapy, ►acquired immunodeficiency; Corey DR 1997 Trends Biotechnol 15:224; Chinnery PF et al 1999 Gene Ther 6:1909; Wilhelmsson LM et al 2002 Nucleic Acids Res 30(2): e3; Rogers FA et al 2002 Proc Natl Acad Sci USA 99:16695.

Peptide Processing: ►post-translational processing

Peptide Sequence Tag: ►electrospray MS

Peptide Transporters: ►TAP, ►ABC transporters

Peptide Vaccination: Synthetic polypeptides corresponding to CTL epitopes may result in cytotoxic T cell-mediated immunity but in some instances, it may enhance the elimination of anti-tumor CTL response. ►vaccination, ►CTL, ►epitope, ►cancer prevention, ►immunological surveillance; Vandenbark AA et al 2001 Neurochem Res 26:713.

Peptidoglycan: The heteropolysaccharides cross-linked with peptides constituting the bulk of the bacterial cell wall, especially in the Gram-positive strains. ►Gram negative; see chemical formula in Fig. P44.

Peptidomimetics: These are polymer analogs containing unnatural amino acids. The non-natural amino acids are incorporated by the translation machinery using suppressor amino acid-transfer RNAs or nonsuppressor tRNA in a modified system. Peptidomimetics may facilitate the study of translation, enable directed evolution of small molecules with desirable catalytic and pharmacological properties, and are potential blocking agents of carcinogenesis by promotion of apoptosis. ►translation, ►aminoacyl-tRNA synthetase, ►suppressor tRNA, ►genetic code, ►alloproteins; Forster AC et al 2003 Proc Natl Acad Sci USA 100:6353.

Peptidyl Site: ►P site, ►ribosome, ►protein synthesis

Peptidyl Transferase: Generates the peptide bond between the preceding amino acid carboxyl end (at the P ribosomal site) and the amino end of the incoming amino acid (at the A site of the ribosome). It is a ribozyme and the catalytic function resides in the 23S ribosomal RNA. Essential function is attributed to adenine 2451. ►ribosome, ►protein synthesis, ►macrolide

Peptidyl-Prolyl Isomerases (PPI): PPI mediate the interconversion of the cis and trans forms of peptide bonds preceding proline. PPI genes are in human chromosomes 4q31.3, 6p21.1, 7p13 and the mitochondrially located at 10q22-q23. This family includes cyclophilins, FKBs and parvulins. ►FK506, ►cyclophilins, ►parvulin; Shaw PE 2002 EMBO Rep 3:521; Wu X et al 2003 Genetics 165:1687.

Peptoid: A peptide-like molecule that results from the oligomeric assembly of N-substituted glycines. Peptoids have various biological applications.

per *(period)* locus: In *Drosophila* (map location 1-1.4, 3B1-2), *per* locus controls the circadian and ultradian rhythm, thus affecting eclosion, general locomotor activity, courtship, intercellular communication, etc. The mutations do not seem to affect the viability of the individuals involved, only the behavior is altered. When per^S (caused by a base substitution mutation in exon 5) in the brain is transplanted into per^{01} mutants (nonsense mutation in exon 4) causing short ultradian rhythm and multiple periods, some flies may be somewhat normalized. The locus has been cloned and sequenced and seems to code for a proteoglycan. Gene NONO of mammalian cells apparently positively modulates PER and gene WDR5 (involving a histone-methyltransferase subunit) controls methylation/expression of PER (Brown SA et al 2005 Science 308:693). The Per protein forms a heterodimeric complex with the Tim (*timeless* gene) protein

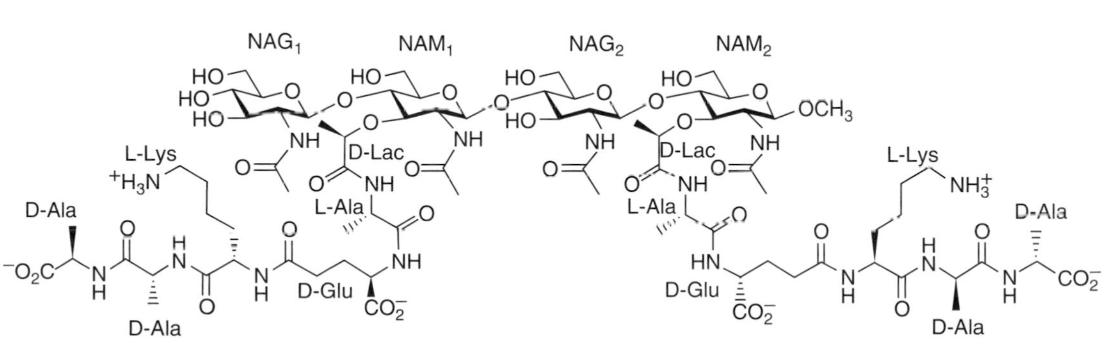

NAG1 NAM1 NAG2 NAM2 HO O OCH3 NH D-Lac L-Lys ^+H_3N NH_3^+ L-Ala D-Ala D-Glu CO_2^- ^-O_2C HN

Figure P44. Chemical structure of a segment of peptidoglycan. (NAG): N-acetylucosamine-N-acetyl muramic acid disaccharide (NAM) and attached pentapeptide. See Meroueh SO et al. 2006 Proc Natl. Acad. Sci. USA 103:4404

and jointly autoregulate transcription. Tim is degraded in the morning in response to light and that results in the disintegration of the complex that is reformed again in dark in a circadian oscillation. In mammals, three *mPer* loci have been identified controlling the circadian clock. ►circadian, ►ultradian, ►proteoglycan

Percent Identity Plot (PIP): A macromolecular sequence map displaying the percentage of identity between two sequences. (See http://bio.cse.psu.edu).

Percentile: The percentage of the distribution of variates. More than 50 percentile indicates a value higher than 50% of the variates.

Perdurance: The persistence and expression of the product of the wild type gene even after the gene itself is no longer there.

Perennial: Lasts through more than one year.

Perfect Flower: Has both male and female sexual organs, i.e., it is hermaphroditic (see Fig. P45). ►hermaphrodite, ►flower differentiation

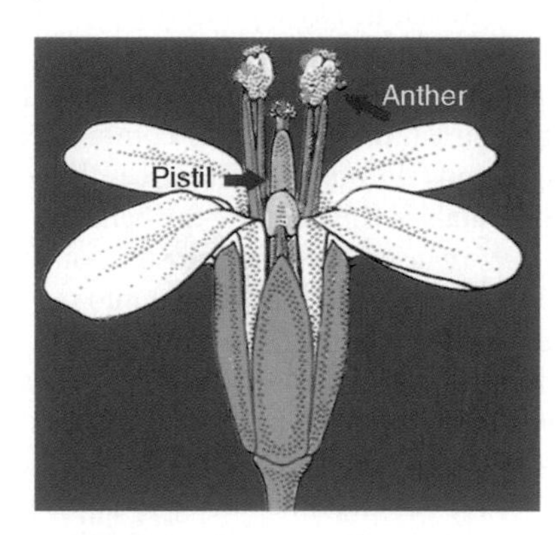

Figure P45. Perfect flower

Perforin (encoded at 10q22): Pore-forming protein (homologous to component 9 of the complement) that establishes transmembrane channels; it is stored in vesicles within the $CD8^+$ cytotoxic T cells (CTL). These vesicles contain also serine proteases. Perforin mediates apoptosis by permitting killer substances (granzyme) slowly enter the cell. Perforin may have a role in T cell-mediated destruction of pancreatic β-cell in diabetes mellitus type I. ►apoptosis, ►complement, ►T cells, ►fragmentin-2, ►granzymes, ►caspase, ►histiocytosis, ►diabetes mellitus; Keefe D et al 2005 Immunity 23:249.

Perfusion: Adding liquid to an organ through its internal vessels in vitro or in vivo.

Perianth: Designates both sepals and petals of the flowers. ►flower differentiation

Pericarp: The fruit wall (maternal tissue), developed from the ovary wall such as the pea pod, the outer layer of the wheat or maize kernels. The *Arabidopsis* silique, the peel of the citruses, and the skin of apple, the shell of the nuts, etc., are also similar but are exocarps. The outer layer of the common barley "seed" is not part of the fruit wall but it is a bract of the flower.

Pericentric Inversion: ►inversion pericentric

Pericentromeric Region: A highly redundant tract (<1 kb to ~85 kb), a transition zone, between the genic region of the chromosomes and the satellite heterochromatin. ►heterochromatin, ►centromere, ►satellite DNA; Horvath JE et al 2001 Hum Mol Genet 10:2215.

Perichromatin Fibers: Active genes occupy the surface of specific compartments in the interphase nucleus (chromosome territories) and represent the perichromatin fibers. ►SR motif, ►chromatin, ►chromosome territories; Cmarko D et al 1999 Mol Biol Cell 10:211.

Periclinal Chimera: Contains genetically different tissues in different cell layers. ►mericlinal chimera, ►chimera

Pericycle: The (root) tissue between the endodermis and phloem. ►root, ►endodermis, ►phloem

Peridium: The covering of the hymenium or the hard cover of the sporangium of some fungi. ►hymenium, ►sporangium

Perinatal: The period after 28 weeks of human gestation⇔four weeks after birth.

Perinuclear Space: A ~20 to 40 nm space between the two layers of the nuclear membrane. ►nucleus

Periodic Acid-Schiff Reagent (PAP): The tests for glycogen, polysaccharides, mucins, and glycoproteins. It breaks C-C bonds by oxidizing near hydroxyl groups and forms dialdehydes and generates red or purple color.

Periodic Paralysis (PP, KCNE): A group of autosomal dominant human diseases manifested in periodically recurring weakness accompanied by low blood potassium level (hypokalemic periodic paralysis, defect in the α subunits of Ca^{2+} channel) or in other forms with high blood potassium level (hyperkalemic periodic paralysis, paramyotonia). The latter types were attributed to base substitution mutations in a highly conserved region of the αsubunit of a transmembrane sodium channel protein. In another type of the disease, the blood potassium level appeared normal and the patients responded favorably to sodium chloride. The MiRP2 potassium channel defect is also associated with PP. ►ion channel, ►Moebius syndrome, ►myotonia, ►hyperkalemic periodic paralysis; Abbott GW et al 2001 Cell 104:217.

Periodicity: The number of base pairs per turn of the DNA or the number of amino acids per turn of an α-helix of a polypetide chain. ►protein structure, ►Watson and Crick model

Periodontitis: Several diseases involving inflammation of the gingiva, especially at the base of the teeth and the alveolae, the bone support of the teeth. It is usually associated with keratosis of the palms and soles. About 30% of the human population is affected by it. In the juvenile form (encoded at 4q11-q13) both milk and permanent teeth may be lost in early childhood. The disease is the result of bacterial infections (~500 different species may inhabit the human mouth). The Papillon-Lefèvre syndrome (11q14, prevalence 1–4 × 10^{-6}) is based on deficiency of cathepsin C, a dipeptidyl peptidase I. In the similar autosomal recessive periodontitis deficiency of IL-1 is suspected. Similar symptoms may occur also in the Ehlers-Danlos syndrome Type VIII. ►keratosis, ►cathepsin, ►IL-1, ►Ehlers-Danlos syndrome, ►dentinogenesis imperfecta; Travis J et al 2000 Adv Exp Med Biol 477:455.

Peripheral Nervous System: Resides outside the brain and the spinal chord.

Peripheral Proteins: These are bound to the membrane surface by hydrogen bonds or by electrostatic forces. ►membrane proteins

Peripherin (retinal degeneration slow protein): ►retinal dystrophy

Periplasma: The cell compartment between cell wall and cell membrane. In *E. coli* the Sec family of proteins mediate the translocation across the periplasmic and the outer membrane. The extracellular stress response factor σ^E regulates the assembly of the outer membrane. The two-component Cpx seems to be involved in the assembly of the pilus. ►Sec, ►two-component regulatory system, ►pilus; Danese PN, Silhavy TJ 1998 Annu Rev Genet 32:59; Raivio TL, Silhavy TJ 2001 Annu Rev Microbiol 55:591.

Peristalsis: The contraction of muscles of a tubular structures (e.g., intestines) propelling the content.

Peristome: A fringe of teeth at the opening of the sporangium of mosses or the buccal (mouth) area of ciliates.

Perithecium: A fungal fruiting body of disk or flask shape with an opening (ostiole) for releasing the spores (see Fig. P46). A perithecium of *Neurospora* contains about 200 asci. Its primordium is called protoperithecium. ►ascogonium, ►apothecium, ►cleistothecium, ►gymnothecium, ►*Neurospora*, ►ascus, ►tetrad analysis

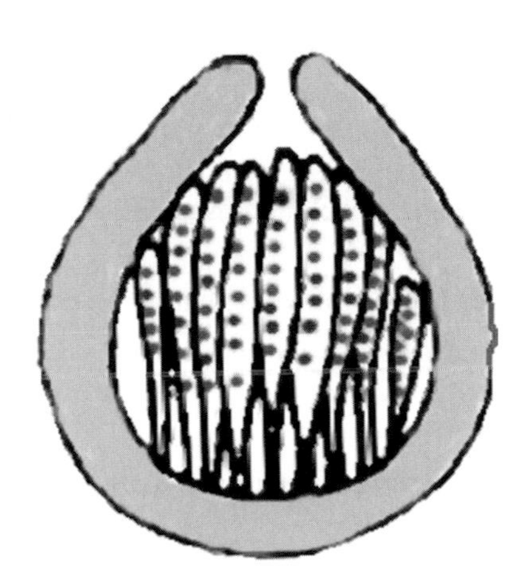

Figure P46. Perithecium

Periventricular Heterotopia: A human X-chromosomal mental retardation and seizures caused by anomalies of the brain cortex. The neurons destined for the cerebral cortex fail to migrate. The mutation involves the filamin gene (FLN1, Xq28) encoding an actin-cross-linking phosphoprotein. ►double cortex; Sheen VL et al 2001 Hum Mol Genet 10:1775.

PERK (PKR-like ER kinase): A phosphorylating enzyme in the endoplasmic reticulum, similar to the mammalian RNA-dependent protein kinase (PKR). It is interferon-inducible and activated by double-stranded RNA. PERK and the phosphorylation of eIF2α inhibit the initiation of translation. ►PKR, ►eIF2α, ►Ire, ►unfolded protein response, ►S6; Kumar R et al 2001 J Neurochem 77:1418.

Perl Script: Assembles and merges sequences from different DNA libraries. ►DNA library, ►PHRAP

Perlecan: Heparan sulfate proteoglycan that interacts with the extracellular matrix, growth factor receptors and affects signal transduction. ►proteoglycan, ►heparan sulfate, ►Schwartz-Jampel syndrome; Knox S et al 2002 J Biol Chem 277:14657.

Perlegen Sciences: Perlegen sciences have genotyped over 1.5 million unique genetic variants (SNPs), in 71 individuals of European American, African American, or Han Chinese ancestry. In total, more than 112 million individual genotypes were determined, with an average distance between adjacent SNPs of 1871 base pairs. The genotype browser permits accession to virtually all the SNP, linkage disequilibrium, and haplotype data reported in the study. ►SNIPs, ►linkage disequilibrium, ►haplotype;

P

expanded data: Hinds DA et al 2005 Science 307:1072; http://genome.perlegen.com/.

Permafrost: The soil layer in cold regions that remains permanently frozen even when the top may thaw.

Permanent Hybrid: ►complex heterozygote

Permease: The enzymes involved in the transport of substances through cell membranes. ►membrane transport, ►membrane channels, ►membrane potential, ►ion channels; Abramson J et al 2003 Science 301:610.

Permissible Dose: ►radiation hazard assessment

Permissive Condition: The condition at which a conditional mutant can survive or reproduce. ►conditional mutation

Permissive Host: A cell permits (viral) infection and/or development.

Permutation: Generating all possible orders of n numbers, and it can be obtained by the factorial: n!, e.g., the factorial of 4, $4! = 4 \times 3 \times 2 \times 1 = 24$. ►combination, ►variation

Permutation Test (randomization test): The test used to assess the association of QTLs with multiple (molecular) markers in a randomized array. ►QTL, Edington ES 1995 Randomization tests, Marcel Dekker, New York.

Permuted Redundancy: At the termini of phage DNA a collection of redundant sequences occur in the phage population that start and end with permuted sequences of the same nucleotide sequence, e.g., 1234...1234, 2341...2341, 3412...3412, etc. This arrangement is characteristic for T-uneven phages, e.g., T1, T3, T5, etc. ►non-permuted redundancy

P

Perodictus: ►*Lorisidae*

Peroxidase: Heme protein enzymes, which catalyze the oxidation of organic substances by peroxides. Glutathione peroxidase (and selenium) deficiency may cause hemolytic disease. Several peroxidase genes have been located in the human genome: GPX1 in 3q11-q12, GPX2 in 14q24.1, GPX3 in 5q32-q33, GPX4 (in testes) in chromosome 19, a rare eosinophil peroxidase (EPX) may compromise the immune system, a thyroid peroxidase deficiency (2p25) interferes with thyroid function. ►immune system, ►eosinophil, ►hemolytic disease, ►oxidative stress

Peroxidase and Phospholipid Deficiency: An autosomal recessive anomaly of the eosinophils involving the enzyme defects named. ►microbody, ►Refsum disease, ►Zellweger syndrome

Peroxides: They display the — O — O — linkage. Organic peroxides participate in activation and deactivation of promutagens, mutagens, procarcinogens and carcinogens and in many other physiological reactions. Peroxides are formed by the breakdown of amino acids and fatty material in the cell and may inflict serious damage. According to some views, spontaneous mutation may be caused to a great extent by these regular components of the diet. Therefore, eating rancy food may pose substantial risk. ►environmental mutagens, ►peroxidase, ►catalase peroxisomes, ►promutagen, ►ROS, ►P450

Peroxins: Peroxisome proteins. PEX1 (7q21-q22), PEX10, human chromosome 1) is a peroxisome biogenesis protein, PEX13 (2p15) in another peroxisome biogenesis protein. ►peroxisome, ►PEX, ►Zellweger syndrome, ►Refsum disease, ►microbodies; Walter C et al 2001 Am Hum Genet 69:35.

Peroxiredoxins: Antioxidant enzymes present in prokaryotes and eukaryotes and control signal transduction, apoptosis, tumor formation, HIV infection, etc. (See Wood ZA et al 2003 Science 300:650; Neumann CA et al 2003 Nature [Lond] 424:561).

Peroxisomal 3-Oxoacylcoenzyme A Thiolase Deficiency (pseudo-Zellweger syndrome): Autosomal recessive disease assigned to human chromosome 3p23-p22. ►microbody, ►Zellweger syndrome, ►adrenoleukodystrophy

Peroxisome: Peroxisome are ~0.15–0.5 μm diameter bodies, surrounded by single-layer membrane, in eukaryotic cells containing about 50 proteins, including oxidase and catalase enzymes. The peroxisomes synthesize ether phospholipids by dihydroxyacetonephosphate acetyltransferase (DHAPAT, human chromosome 1q42) and alkyl dihydroxyacetonephosphate synthase (ADHAPS, 2q31). In a human cell, the number of peroxisomes vary from less than 100 to more than 1000. In yeast either over expression or lacking the Inp 1 protein (inheritance of peroxisome protein 1) are detrimental for the right amount and the proper distribution of peroxisome. Inp binds several proteins involved in peroxisome division (Fagarasanu M et al 2005 J Cell Biol 169:765). The peroxisomes have indispensable roles in fatty acid β oxidation, phospholipid and cholesterol metabolism. Fatty acid granules are also named microbodies. The peroxisome biogenesis disorders (PBD) are recessive lethal diseases in variable forms. The extreme form is the Zellweger syndrome, the Refsum disease and the adrenoleukodystrophy are milder and the rhizomelic chondrodysplasia punctata (RCDP) involves bone defects. Peroxisome mutations have been identified also in yeast (PAS). Some rodent carcinogens increase the number of peroxisomes but in humans, these agents did not appear to be carcinogenic. Peroxisomal protein Pex2 controls photomorphogenesis in *Arabidopsis*. ►glyoxisome,

►microbodies, ►PPAR, ►peroxin, ►Zellweger syndrome, ►Refsum disease, ►adrenoleukodystrophy, ►chondrodysplasia, ►oxalosis, ►peroxidase and phospholipid deficiency, ►peroxisomal 3-oxo-acyl-coenzyme A thiolase deficiency, ►PPAR, ►micropexophagy, ►pexophagy, ►PEX; Gould SJ, Valle D 2000 Trends Genet 16:340; Sacksteder KA, Gould SJ 2000 Annu Rev Genet 34:623; Titorenko VI, Rachubinski RA 2001 Trends Cell Biol 11:22; Thai T-P et al 2001 Hum Mol Genet 10:127; Titorenko VI, Rachubinski RA 2001 Nat Rev Mol Cell Biol 2:357; Purdue PE, Lazarow PB 2001 Annu Rev Cell Dev Biol 17:701; Hu J et al 2002 Science 297:405; Matsumoto N et al 2003 Am J Hum Genet 73:233; Weller S et al 2003 Annu Rev Genomics Hum Genet 4:165; Wanders RJA, Waterham HR 2006 Annu Rev Biochem 75:295; http://www.peroxisomeDB.org.

Peroxynitrite ($ONOO^-$/ONOOH): The diffusion-limited product of nitric oxide with superoxide. It is strongly oxidizing and toxic. ►nitric oxide, ►superoxide, ►peroxides; García-Nogales P et al 2003 J Biol Chem 278:864.

Perp: ►p63

Persistence, Bacterial: A phenocopy-like phenomenon in bacteria. When the culture is exposed to strong stress (e.g., antibiotics), the majority of the cells die but a small fraction survives although without a heritable resistance. When they re grow the population become sensitive to the antibiotic. ►phenocopy; Balaban NQ et al 2004 Science 305:1622.

Personal Genomics: ►DNA fingerprinting, ►DNA sequencing

Personality: Can be characterized by five main groups of features: (1) *extraversion* (being outgoing) or the lack of it, ability to lead and sell their ideas versus reticent and avoiding company [heritability about 0.71]; (2) *neuroticism* (emotional versus stable) worrisome or self-assured, [heritability about 0.21]; (3) *conscientiousness* (well-organized versus impulsive) responsible or irresponsible, reliable or undependable, [heritability 0.38–0.32); (4) *agreeableness* (empathic or unfriendly) warm versus cold, cooperative versus quarrelsome, forgiving versus vindictive, [heritability about 0.49], and (5) *openness* (insightful or lacking intelligence) imaginative versus imitative, inquisitive or superficial). These heritability estimates vary a great deal, however, and may be very different in some populations. Based on twin studies, several investigators concluded that overall close to 50% of the variance could be attributed to additive or non-additive genetic determination. ►behavior in humans, ►behavior genetics, ►human intelligence, ►affective disorders, ►heritability in humans

Person/Year: Used as, for e.g., incidence of an event a (symptoms of a disease) per person per year.

Persyn: ►synuclein-γ

Perturbogen: Short peptides or protein fragments that can disrupt specific biochemical function in the cell.

Pertussis Toxin: Produced by the Gram-negative *Bordatella* bacteria, responsible for whooping cough. The toxin stimulates ADP-ribosylation of the $G\alpha_1$ subunit of a G-protein in the presence of ARF and thus GDP stays bound to the G-protein and adenylate cyclase is not inhibited and K^+ ion channels do not open. As a consequence, histamine hypersensitivity and reduction of blood glucose level follows. ►whooping cough, ►signal transduction, ►ARF, ►G-protein, ►ADP, ►GDP, ►cholera toxin, ►adenylate cyclase; Alonso S et al 2001 Infect Immun 69:6038.

PERV (porcine endogenous retrovirus): Exists in >50 copies/pig chromosome complement and has been feared to endanger humans with xenotransplantation of pig organs. So far, the limited information indicates minimal risk relative to the potential benefits. PERVs can be transferred to mice by xenotransplantation. ►xenograft, ►xenotransplantation, ►nuclear transplantation; Specke V et al 2001 Virology 285:177.

PEST (proline [P]-glutamate [E]-serine [S]-threonine [T]-rich motif): PEST in the carboxyl domain of IκB and other proteins (Ubc) is involved in the stimulation of proteolysis. ►IκB, ►NF-κB, ►proteasome, ►Ubc, ►ubiquitin

Pest Eradication by Genetic Means: ►genetic sterilization, ►*Bacillus thüringiensis*, ►host–pathogen relations (see Fig. P47).

Figure P47. *Bacillus thüringiensis* toxin transgene is lethal to worms (right) but the wild type plants (left) were destroyed. (Courtesy of Professor Marc Van Montagu, Rijksuniversiteit, Gent)

Pesticide Mutagens: ►environmental mutagens

Pesticin: The toxin of *Pasteurella* bacteria

Pestilence: An infectious epidemic of disease.

PET: ►tomography

PET: ►transcriptome, ►paired-end diTAG

Petaflop Computer: An extremely powerful supercomputer. "Peta" comes from the Latin word *peto* (I move forward) and in computer jargon, "flops" designate floating operations. This new hardware may be capable of performing one quadrillion flops/second, that are more than 10^6 times the efficiency of the best desktop computers.

Petals: Generally the second whorl of modified leaves from the bottom of the flower (see Fig. P48).

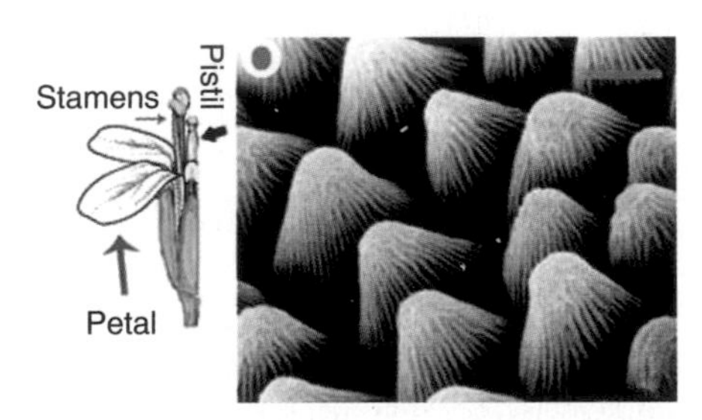

Figure P48. Petals of *Arabidopsis* are shown at left. At right: The adaxial ridge of the petal as viewed by scanning electronmicrography and reveal the beauty, which the naked eye cannot see. (From Bowman JL, Smyth DR 1994 In: Bowman JL (ed) Arabidopsis: An Atlas of Morphology and Development. By permission of Springer-Verlag, New York)

Frequently, they are quite showy because of their anthocyanin or flavonoid pigmentation. The petal number is a taxonomic characteristic, although petal number may be altered by homeotic mutations converting the anthers and/or pistils into petals and appearing as sterile double flowers of floricultural advantage. MicroRNA may have a role in the regulation of petal number (Baker CC et al 2005 Curr Biol 15:303). ►flower differentiation, ►flower pigments, ►homeotic mutants; Roeder AHK, Yanofsky MF 2001 Dev Cell 1:4.

PETCM (α-[trichloromethyl]-4-pyridineethanol): Stimulates caspase-3 activity and thereby apoptosis. ►caspase, ►apoptosis; Jiuang X et al 2003 Science 299:223.

Petiole: The stalk of a leaf (see Fig. P49).

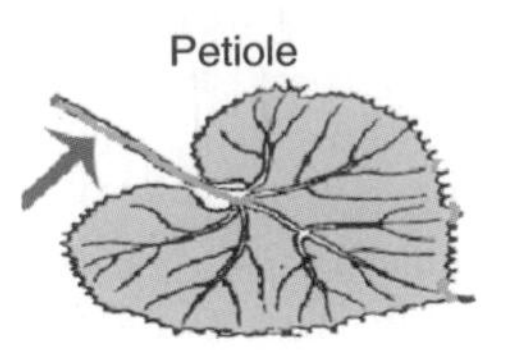

Figure P49. Petiole

Petite Colony Mutants: Petite colony mutants of yeast forms small colonies because they are deficient in respiration (OXPHOS minus) and lethal under aerobic conditions. The *vegetative petites* (ρ^-) are caused by (large) deletions in the mitochondrial DNA, the *segregational petites* are controlled by nuclear genes at over 200 loci. The mitochondrial mutations occur at high (0.1 to 10%) frequency and using ethidium bromide as a mutagen their frequency may become as high as 100%. The mitochondrial petites fail to transmit this character in crosses with the wild type except one special group the *suppressive petites* that may be transmitted at a low frequency in outcrosses with the wild type. In yeast, the A + T content of the normal mitochondrial DNA is about 83%, in some of the mitochondrial mutants the A + T content may reach 96% because the coding sequences were lost and only the redundant A + T sequences were retained and amplified so the mtDNA content is not reduced. The *hypersuppressive petite* mutants have short (400–900 bp) repeats that share 300 bp (*ori* and *rep*) sequences with the wild type, necessary for replication. *Neutral petites* produce wild type progeny when outcrossed to the wild type. Yeast cells that have normal mitochondrial function make large colonies and are called *grande*. Cells that can dispense with mitochondrial functions are sometimes called petite-positive whereas that absolutely need mitochondria are called petite-negative. Inactivation of an ATP and metal-dependent protease (Yme1p) associated with the inner membrane of the mitochondrium can convert the "positives" to "negatives" and the presence of the Yme1p function may have the opposite effect. Yme is not universally present in all yeasts. ►mitochondria, ►mtDNA, ►mitochondrial mutations, ►oxidative phosphorylation; Sager R 1972 Cytoplasmic Genes and Organelles, Academic Press, New York; MacAlpine DM et al 2001 EMBO J 20:1807; Chen XJ, Clark-Walker GD 2000 Int Rev Cytol 194:197.

Petri Plate: A (flat) glass or disposable plastic culture dish for microbes or eukaryotic cells (see Fig. P50).

Figure P50. Petri plate

Petunia hybrida (2n = 28): *Solanaceae*; predominantly self-pollinating but allogamy also occurs. It has been used extensively for cell, protoplast and embryo culture, intergeneric and inter-specific cell fusion, genetic transformation and the genetic control of pigment biosynthesis. It has a good number of related species.

Peutz-Jeghers Syndrome: ►polyposis hamartomatous

PEV (position effect variegation): ►position effect, ►heterochromatin, ►RPD3

PEX: Proteins import the peroxisome-targeting signals (PTS) to the peroxisomes. The various PEXs play roles in peroxisome biogenesis. ►peroxisome; Braverman N et al 1988 Hum Mol Genet 7:1195.

Pexophagy: Sequestration to and engulfing peroxisomes into vesicles and their destruction. ►peroxisome

Peyer's Patches: Aggregated lymphatic nodes. The Peyer's patches mediate the uptake of macromolecules, antigens and microorganisms through the epithelium of the gut. These plaques are instruments of mucosal immunity. B cells are required for the normal functions of the Peyer's patches. *Salmonella typhi* infection may cause perforation of the Peyer's patches. Tyrosine kinase receptor RET is a regulator of Peyer's patch formation (Veiga-Fernandes H et al 2007 Nature [Lond] 446:547). ►mucosal immunity, ►RET oncogene

Peyronie Disease: An apparently dominant autosomal disorder, it may be caused by a variety of acquired and genetic conditions. Its exact prevalence has not been determined but it may occur at ~1 to 8% of the human males. It involves fibrous, thickened collagen plaques on most commonly on the dorsal part of the penis and causes its curvature under painful conditions of erection. The condition may be transient. It occurs generally after age 40. Several drugs and in severe conditions surgical intervention have been used for treatment. (See Usta MF, Hellstrom WJG 2004 In: Seftel AD et al (Eds.) Male and Female Sexual Dysfunction, Mosby, St. Louis, Missouri, p 191).

PFAM: A database of over 3000 protein families and domains, multiple sequence alignments and profile hidden Markov models. ►alignment; Bateman A et al 2002 Nucleic Acids Res 30:276; http://pfam.wustl.edu; http://www.sanger.ac.uk/Software/Pfam/; http://pfam.jouy.inra.fr; http://pfam.cgb.ki.se/.

Pfeiffer Syndrome: The syndrome includes autosomal dominant bone malformation affecting the head, thumbs and toes (acrocephalosyndactyly) (see Fig. P51), the autosomal recessive head-bone (craniostenosis) and heart disease. The origin is primarily paternal. The latter type seems to co-segregate with fibroblast growth factor receptor 1 (FGFR1) in human chromosome 8p11.2-p11.1.

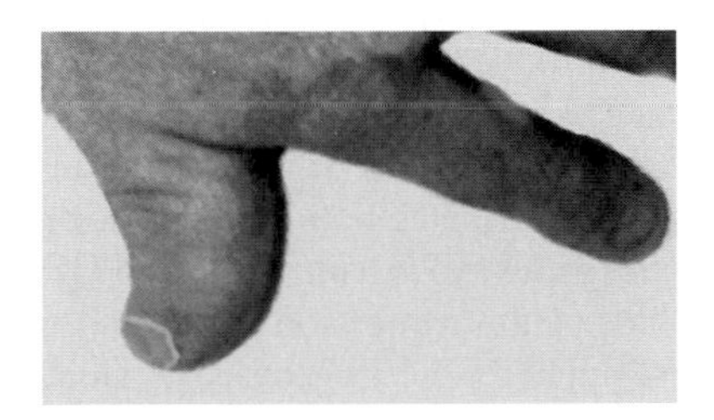

Figure P51. Short and broad thumb in Pfeiffer syndrome

Another locus in chromosome 10q26 represents also a fibroblast growth factor receptor, FGFR2. FGFR3 is located in chromosome 4p16.3 and its mutation is concerned with hypo-chondroplasia. Mutations in all three genes involve Pro→Arg replacements at identical sites, 253. This syndrome is allelic to the Crouzon and to the Jackson-Weiss syndromes. Some of the mutations represent gain-of-functions. ►fibroblast growth factor, ►Alpert's syndrome, ►Crouzon syndrome, ►Jackson Weiss syndrome, ►craniosynostosis syndromes, ►hypochondroplasia, ►achondroplasia, ►gain-of- function, ►receptor tyrosine kinase

PfEMP1: A group of *Plasmodium falciparum* protein ligands expressed on the surface of infected red blood cells and mediate cell adhesion (virulence factors) but may incite host immune reaction. PfEMP displays antigenic variation to evade this response. Other pathogenesis proteins of the parasite are rifins. ►Plasmodium, ►antigenic variation, ►rifin; Flick K et al 2001 Science 293:2009.

PFGE (pulsed field gel electrophoresis): PFGE separates very large nucleic acid fragments or even small chromosomes. The megabase size fragments can be used for physical mapping of large chromosomal domains (PFG mapping). ►pulsed field gel electrophoresis

pfu (p.f.u.): The plaque forming unit. The number of phage particles/mL that can invade a bacterial lawn and then after reaching about 10^7 particle numbers a clear spot appears on the Petri plate where the bacterial cells had been lysed. ►plaque, ►pu, ►CFU

PG: ►prostaglandins

PGA: Phosphoglyceric acid, a 3-carbon product of photosynthesis. ►photosynthesis, ►Calvin cycle, ►C3 plants

PGC (primordial germ cells): Gynogenetic/parthenogenetic cell, the primordial cell of female gonads. ►gonad, ►gynogenesis, ►parthenogenesis

PGC1 (PPAR-γ-coactivator): A regulator of transcription, body heat production, mitochondrial biogenesis and other processes. PGC-1β is a transcriptional coactivator for the production of cholesterol and triglycerides. PGC1α regulates both the gluconeogenetic and glycolytic pathways during fasting. PGC1α is also a regulator of the expression of several transcription factors required for the biogenesis of mitochondria and it is down regulated in Huntington disease (McGill JK, Beal MF 2006 Cell 127:465). Sirtuin interacts with and deacetylates PGC in a NAD-dependent manner as part of energy homeostasis, diabetes and lifespan (Rodgers JT et al 2005 Nature [Lond] 434:113). ►PPAR, ►nuclear receptor, ►sirtuin, ►gluconeogenesis, ►glycolysis, ►aging, ►resveratrol, ►Huntington's chorea; Tsukiyama-Kohara K et al 2001 Nature Med 7:1102.

PGD (preimplantation genetic diagnosis): PGD can be carried out for some human genetic disorders, for e.g., PCR or FISH (for fragile X, aneuploidy, etc.) examining polar bodies or blastomeres at the stage of a few cells in the embryo. The disorders that have been identified by PCR included cystic fibrosis, Tay-Sachs disease, Lesh-Nyhan syndrome, Huntington chorea, Marfan syndrome, ornithine transcarbamylase deficiency, Fanconi anemia, etc. The diagnosis may permit—by using in vitro fertilization—to develop a human offspring of a certain (disease-free) genetic constitution. Misdiagnosis is rare but varies somewhat in different laboratories. It seems likely that all cells in a single eight-cell embryo may not be identical. (See diseases mentioned under separate entries, ►PCR, ►FISH, ►ART; Bickerstaff H et al 2001 Hum Fert 4(1):24; Simpson JL 2001 Mol Cell Endocrinol 183(Suppl. 1):S69; Findlay I et al 2001 Mol Cell Endocrinol 183(Suppl. 1):S5; PGD tests carried out according to international survey: Sermon K et al 2005 Hum Reprod 20:19).

PGK-neo: A commonly used transformation cassette for gene knockout where the neomycinphos photransferase gene (*neo*) is fused to the phosphoglycerate kinase (PGK) promoter. ►knockout, ►vector cassette; Scacheri PC et al 2001 Genesis 30(4):259.

P-Glycoprotein: The 170 kDa product of the human multidrug resistance gene (MDR-1, 7q21.1) that exports different (mainly hydrophobic) toxic substances from the cells in an ATP-dependent manner. ►multidrug resistance

PGM: ►phosphoglucomutase

PGRS (polymorphic GC-rich repetitive sequences): Mycobacterium tuberculosis proteins (~70) with glycine-glycine doublets that have few charged amino acids and essentially contain no cysteine. These proteins are apparently involved in pathogenesis. ►*Mycobacteria*; Karlin S 2001 Trends Microbiol 9:335.

PgtB: Bacterial kinase that phosphorylates regulator protein PgtA. ►kinase, ►protein kinases

PH: Pleckstrin homology domain. ►pleckstrin

pH: $= -\log (H_3O^+)$, negative logarithm of the hydrogen ion concentration; pure water at 25°C contains 10^{-7} mole hydrogen ions; solutions of acids could contain 1 mole and solution of bases 10^{-13} moles per liter.

The pH meters measure the electrical property of solutions which is proportional to pH; pH 7 is neutral and below it is acidic above it is alkalic (basic) (see Fig. P52). The pH of body fluids and tissues is regulated by the function of the ion channels.

The pH of plant tissues is generally below 7 because of the presence of organic acids and the majority of plant tissues can be cultured best in media around pH 6. Most animal tissues display neutral pH (~7). In the human blood, the pH is normally within the narrow range of 7.3 to 7.5. If the blood pH approaches 7, acidosis may result causing coma and at about pH 7.8 alkalosis may cause tetany

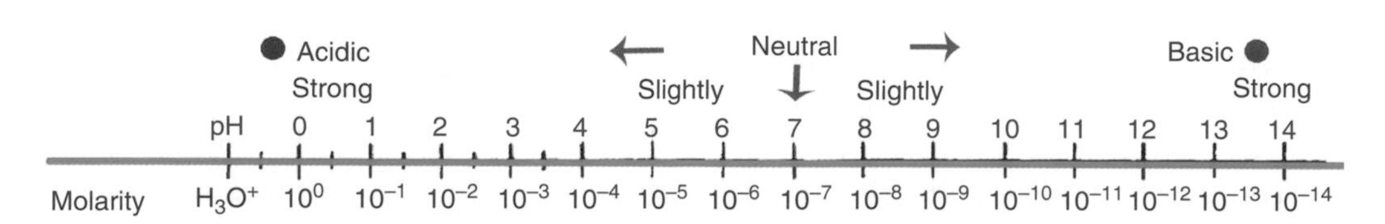

Figure P52. ph scale

(dangerous spasms) but the stomach secretions may be pH 0.9. The pH optima of enzymes vary a great deal but most commonly, it is in the range of 6 to 9. The preferred pH of microbes is also variable; the euryarchaeon bacterium, *Picrophilus torridus* thrives well at pH 0.7 and 60°C. (Fütterer O et al 2004 Proc Natl Acad Sci USA 101:9091). ►ion channels, ►buffer

Ph1 Chromosome: ►Philadelphia chromosome

***Ph*Gene** (pairing high): An approximately 700 Mb sequence that controls selective pairing in hexaploid wheat. In its presence, homoeologous chromosomes do not pair. It is in chromosome 5B. Plants nullisomic for this chromosome display multivalent associations in meiosis. A similar gene *Ph2* is in chromosome 3D. A *Ph* gene is present also in the A genome. Additional less powerful genes regulate chromosome pairing also. The *Ph* gene is absent from the genome of diploids but it is present in the B and G genomes in tetraploids. This observation indicates that *Ph* originated after polyploidization. ►*Triticum*, ►homoeologous, ►nullisomic; Sears ER 1969 Annu Rev Genet 3:451; Martinez-Perez E 2001 Nature 411:204; Griffith S et al 2006 Nature [Lond] 439749; Dvorak J et al 2006 Genetics 174:17.

PHA (phytohemagglutinin): A lectin of bean (*Phaseolus vulgaris*) plants; agglutinates erythrocytes and activates T lymphocytes. ►lectins, ►agglutination, ►erythrocyte, ►lymphocytes

Phaeochromocytoma (pheochromocytoma): A bladder-kidney carcinoma, over-producing adrenaline and noradrenaline. The disease may be caused by mutation in the von Hippel-Lindau gene or the neurofibromatosis 1 or the RET protooncogenes or by the multiple endocrine neoplasia gene MEN2. Mutations in the subunits of mitochondrial succinate dehydrogenase in the long arm of human chromosome 11 also may be involved. ►animal hormones, ►von Hippel-Lindau syndrome, ►neurofibromatosis, ►MEN, ►RET, ►succinate dehydrogenase; Astuti D et al 2001 Am Hum Genet 69:49; MaherER, Eng C 2002 Hum Mol Genet 11:2347.

Phaeomelanin: A mammalian pigment. ►pigmentation of animals

Phage (bacteriophage): A virus of bacteria. ►bacteriophages, ►development, ►phage life cycle

Phage Conversion: The acquisition of new properties by the bacterial cell after infection by a temperate phage. ►temperate phage

Phage Cross: ►rounds of matings

Phage Display: Filamentous bacteriophages (M13, fd) have a few copies (3–5) of protein III gene at the end of the particles (see Fig. P53). This protein controls phage assembly and adsorption to the bacterial pilus. When short DNA sequences are inserted into gene III (g3p), the protein encoded may be displayed on the surface of the particles. In case variable region fragments of antibody genes are inserted into the protein III coding sequences, specific antigens may be screened. The peptides can be separated with antibody affinity chromatography (panning). By repeated screening enormous arrays of recombinant libraries become available. The g3p product and the Fvs (fragments of variability) can be separated proteolytically or by inserting a stop codon between g3p and Fv. By the insertion of a large array of nucleotide sequences, a huge combinatorial library of soluble epitopes may be generated. The specificity of the antibodies can be further manipulated by mutation (random or targeted), by error-prone polymerase chain reactions, recombination, by chain shuffling, i.e., trying out various light and heavy chain combinations, synthetic CDR sequences, etc. Similarly, a variety of different antigens may be displayed on the surface of protein III and can be used to screen for cognate antibody. Phage display may be of applied significance for the pharmaceutical industry because extremely large number of variants (up to 10^8 to 10^{10}) of monoclonal antibodies can be selectively isolated and tested. For in vitro testing the two-hybrid method may be employed. The protein-protein interaction may then be studied in mammalian cells and screening techniques can be developed to isolate the cells that can neutralize the cytotoxic virus. The use of phagemid vectors may enhance the efficiency of the procedure. This procedure may facilitate the isolation of novel receptors, ligands, antibodies, anti-cancer reagents, transport proteins, signal transduction molecules, transcription factors, etc. Phage display technique may substitute for the construction of hybridomas (see Fig. P54). It can be used also for typing blood, for various diagnostic procedures, etc. A T7 phage display system permits the selection of RNA-binding regulatory proteins.

By screening large phage libraries for select tissue-specific organ antibodies, luminal endothelial cell

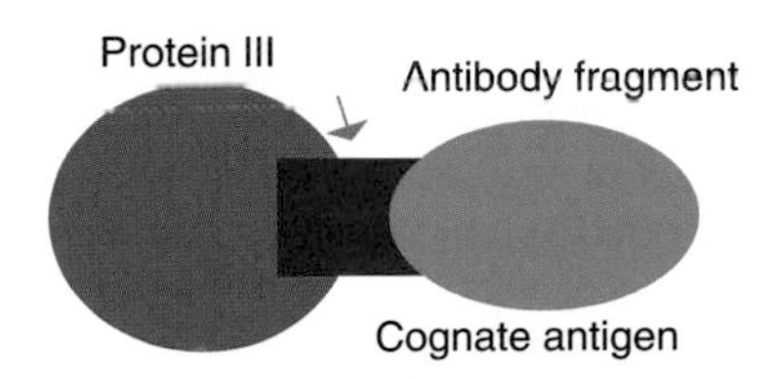

Figure P53. Protein III recognition

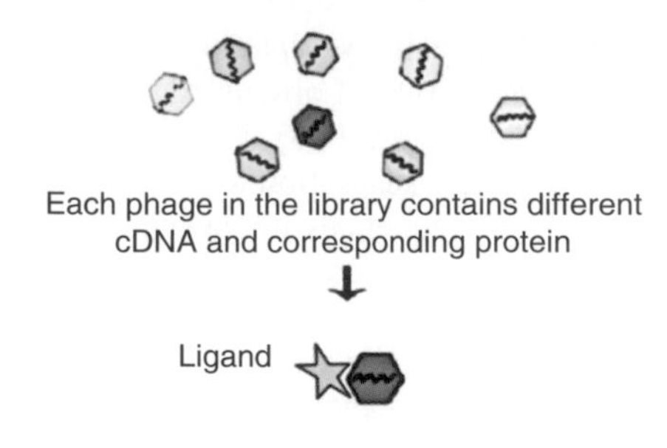
Each phage in the library contains different
cDNA and corresponding protein
Ligand
The desired type is bound to cognate ligand

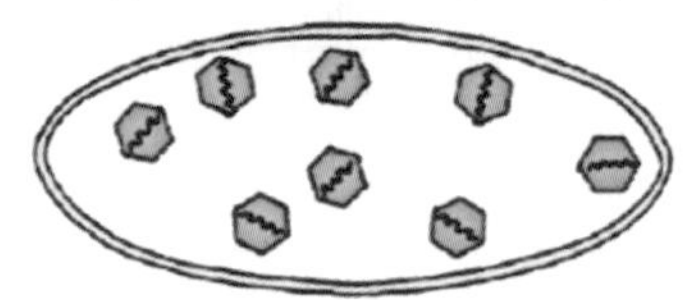
The selected type is amplified

Figure P54. Phage display. (Note that the filamentous phage is represented as 'globular' only for the convenience of drawing).

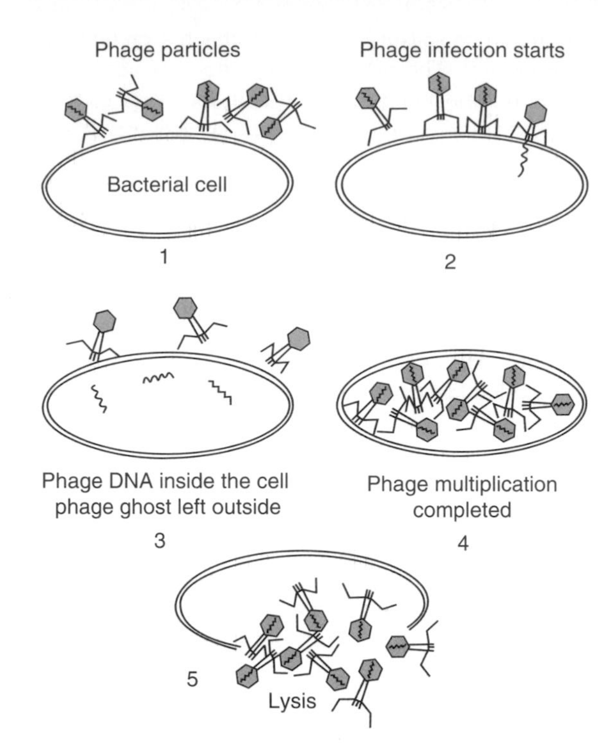
Phage particles
Phage infection starts
Bacterial cell
1
2
Phage DNA inside the cell
phage ghost left outside
3
Phage multiplication
completed
4
5
Lysis

Figure P55. Phage lifecycle. (Redrawn after the illustration provided by Drs. Simon LD, Anderson TF Institute of Cancer Research, Philadelphia, PA, USA

plasma membranes were enriched from the bloodstream. The phage-displayed antibodies were converted Fv-Fc fusion proteins and monitored target selection by whole-body γ-scintigraphic imaging. Mass spectrometry identified the antigen targets. The procedure permits monitoring the vascular route of specific substances (Valadon P et al 2006 Proc Natl Acad Sci USA 103:407). ►filamentous phages, ►affinity chromatography, ►epitope screening, ►combinatorial library, ►pilus, ►antibody engineering, ►CDR, ►monoclonal antibody, ►mRNA display, ►two-hybrid method, ►monoclonal antibody, ►hybridoma, ►phagemid, ►anchored periplasmic expression; Smith GP, Petrenko VA 1997 Chem Rev 97:391; Danner S, Belasco JG 2001 Proc Natl Acad Sci USA 98:12954; Arap W et al 2002 Nature Med 8:121.

Phage Ghost: The empty protein shell of the virus.

Phage Immunity: A lysogenic bacterium carrying a prophage cannot be infected by another phage of the same type. ►prophage, ►zygotic induction

Phage Induction: Stimulates the prophage to leave a site in the bacterial chromosome and become vegetative. Physical and chemical agents may be inducive (UV light, mutagens, zygotic induction).

Phage Lifecycle: See Fig. P55, Böhm J et al 2001 Curr Biol 11:1168.

Phage Morphogenesis: ►one-step growth, ►development, ►phage life cycle

Phage Mosaic: May be generated by phage display, expressing different molecular structures on the surface of a filamentous phage. ►phage display

Phage Therapy: The bacteriophages are bacteria-eating systems. D'Hérelle, the discoverer of phages, already attempted their therapeutic use in poultry as well as for humans with considerable success. With the discovery of antibiotics, the interest in phage therapy ebbed. Another cause for the decline of interest was the discovery of phage resistance in bacteria.

Also, bacteria encode restriction/modification systems of defense. The human body may also react immunologically against the phages. Recent studies, despite some technical problems indicate feasibility of this type of therapy. (See Summers WC 2001 Annu Rev Microbiol 55:437; Schuch B et al 2002 Nature [Lond] 418:884).

Phagemids: Genetic vectors that generally contain the ColE1 origin of replication and one or more selectable markers from a plasmid and a major intergenic copy of a filamentous phage (M13, fd1). When cells carrying such a combination are superinfected by a filamentous phage, it triggers a rolling circle type replication of the vector DNA. This single-stranded product is used then for sequencing by the Sanger type DNA sequencing system, for oligonucleotide-directed mutagenesis and as strand-specific probes. The phagemids can carry up to 10 kb passenger DNA. Their replication is fast (in the presence of a helper), and they can produce up to 10^{11} plaque-forming units (pfu)/mL bacterial culture. Their stability is comparable to conventional plasmids. They obviate subcloning the DNA fragments from plasmid to filamentous phage. The most widely used phagemids contain parts of phage M13 and pUC, πVX^c, and pBR322 vectors. ►phasmid, ►vectors, ►plasmovirus, ►vectors, ►pfu, ►pUC, ►DNA sequencing; Sambrook J et al 1989 Molecular Cloning, Cold Spring Harbor Laboratory Press, Cold Spring Harbor, New York; O'Connell D et al 2002 J Mol Biol 321:49.

Phagocytosis: A special cell (phagocyte) engulfs a foreign particle (microorganism, cell debris) and eventually exposes it to lysosomal enzymes for the purpose of destroying it. Dendritic cells and macrophages have important roles. In lower animals, this mechanism substitutes for the immune system. Phagocytosis pathway is controlled by a battery of *Ced* genes (and homologs, e.g., Dock180 in humans) during apoptosis. The CD14 human glycoprotein on the surface of macrophages recognizes and clears apoptotic cells (see Fig. P56). The major phagocyte receptors are CR3 (binds opsonized C3bi complement fraction) and the Fc gamma receptor, FcγR (binds immunoglobulin G). Both processes require the reorganization of the cytoskeleton under the control of RAC or RHO G proteins, respectively. ►pinocytosis, ►apoptosis, ►macrophage, ►complement, ►antibody, ►opsonins, ►RAC, ►RHO, ►lysosomes, ►cross presentation, ►cell fusion; Underhill DM, Ozinsky A 2002 Annu Rev Immunol 20:825; Stuart LM, Ezekowitz RAB 2005 Immunity 22:539.

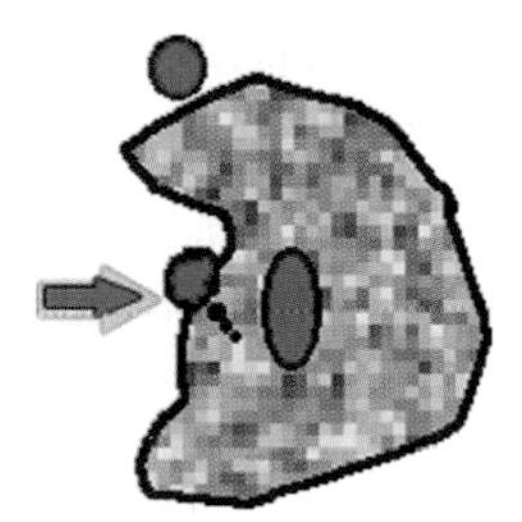

Figure P56. Phagocytosis

Phagosome: A body (vesicle) surrounded by plasma membrane of a phagocyte. They are fused with endosomes and lysosomal compartments to become degradative organelles (Touret N et al 2005 Cell 123:157). In *Drosophila* phagosomes 617 interactive proteins have been detected that are involved in immune reactions (Stuart LM et al 2007 Nature [Lond] 445:95). ►phagocytosis, ►endocytosis, ►endosome, ►lysosomes, ►macrophage

Phakomatoses (neurocutane syndromes): Hereditary and congenital diseases, which are of ectodermal origin and display spots on the body, such as neurofibromatosis, epiloia/tuberous sclerosis, FAP, von Hippel-Lindau syndrome, nevoid basal cell carcinoma, Cowden disease, Peutz-Jeghers syndrome, polyposis. (See separate entries; Tucker M et al 2000 J Natl Cancer Inst 92:530).

Phalange(s): The three bones in fingers and toes (at left) with the metacarpal bone (at right) (see Fig. P57).

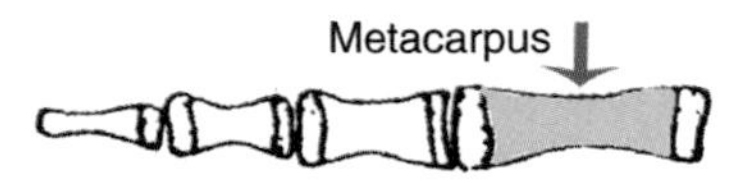

Figure P57. Phalange

Phalloidin: An amanotoxin, similar to, but faster in action than amanitin. When labeled with fluorescent coumarin phenyl isothiocyanate it is suitable to identify filamentous actin in the cells. It is extremely toxic. ►amatoxins, ►α-amanitin; Vetter J 1998 Toxicon 36:13.

Phallus: The penis, a symbol of generative power, also the fetal anlage of the penis and clitoris. ►penis, ►clitoris, ►anlage

Phantom Mutation: Artifacts of the DNA sequencing. They can be filtered out by statistical procedures. (Bandelt, H-J et al 2002 Am J Hum Genet 71:1150).

Pharate: The larva/adult emerging from the puparium.

Pharmaceuticals: The chemical agents used for medical purposes. Data collected on 352 marketed drugs (excluding anti-cancer agents, nucleosides, steroids and peptide-based formulations, which are known to affect DNA); 101 (28.7%) had at least one positive indication for genotoxicity. Four types of tests were used: bacterial mutagenesis, in vitro cytogenetics, in vivo cytogenetics and mouse lymphoma assay. One must keep in mind that carcinogenicity may involve routes that are not testable by these methods. Also, the laboratory assays are not 100% reliable. ►genotoxic chemicals, ►combinatorial chemistry, ►bioassays in genetic toxicology; Snyder RD, Green JW 2001 Mutat Res 488:151.

Pharmacogenetics: The study of the reaction of individuals of different genetic constitution to various drugs and medicines. Most of the differences are monogenic. Polymorphic genes frequently determine drug metabolism, drug transporters and drug responses of the body. Pharmacogenetics studies also study simultaneous drug responses by many genes. Based on these responses drugs with special, selective effect can be developed. Certain drugs have special side effects for individuals of particular genetic constitution ►cytochromes, ►SADR; Roses AD 2001 Hum Mol Genet 10:2261; Kuehl P et al 2001 Nature Genet 27:383; Roses AD 2002 Nature Rev Drug Discov 1:541; Goldstein DB et al 2003 Nature Rev Genet 4:937; Evans WE, Relling MV 2004 Nature [Lond] 429:464; problems and goals in pharmacogenetics/genomics: Need AC et al 2005 Nature Genet 37:671); variation in human genes and drug responses: http://www.PharmGKb.org; key therapeutic targets in proteins and nucleic acids: http://xin.cz3.nus.edu.sg/group/ttd/ttd.asp; pharmacogenetic substances [proteins, drugs]: http://bidd.cz3.nus.edu.sg/phg/.

Pharmacogenomics: The study of drug response of the entire genome of an organism. ►pharmacogenetics, ►SADR

Pharmacokinetics (pharmacodynamics): The study of absorption, tissue distribution, metabolism, and elimination (ADAME) as a function of time of biologically relevant molecules.

Pharmacoproteomics: The study of the proteins in sera or urine as a consequence of disease and/or drug therapy.

Pharmacodynamics: ►pharmacokinetics

Pharming: The production of pharmacologically useful compounds by transgenic organisms. ►transgenic, ►biopharming, ►molecular breeding, ►plantibody

PHAS-1: A heat stable protein ($M_r \approx 12,400$); when it is not phosphorylated it binds to peptide initiation factor eIF-4E and inhibits protein synthesis. Its Ser^{64} site is readily phosphorylated by MAP and then no longer binds to eIF-4E and protein synthesis may be stimulated. ►eIF-4E, ►MAP

Phase Diagram: A graphic representation of the equilibrium between/among components of a system. The phase is an identifiable part of a system. Phase diagrams are used in several scientific fields for the elucidation of the behavior of the phases of the components under dynamic conditions. In biology, phase diagrams can shed light on the mechanisms of interaction within a system, e.g., in a genetic network. Figure P58 represents three hypothetical internal (metabolic) components of the cell and two external factors (e.g., temperature and light) that determine in an interactive manner the node (red dot). ►networks, ►genetic networks; Park J, Barabási A-L 2007 Proc Natl Acad Sci 104:17916.

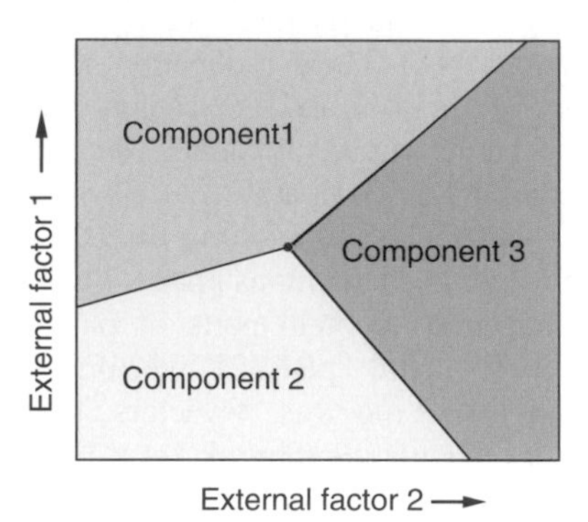

Figure P58. Phase diagram

Phase Variation: A programmed rearrangement in several genetic systems. The flagellin genes of the bacterium *Salmonella* display it at frequencies of 10^{-5} to 10^{-3}. The flagellar protein has two forms, H1 and H2. The *H1* gene is a passive element. When *H2* is expressed no H1 protein is made. When *H2* is switched off the H1 antigen is made. The expression of *H2* is regulated by the expression of the *rh1* repressor (repressing the synthesis of the H1 protein) and the promoter of *H2*. This promoter is about 100 bp upstream from the gene and it is liable to inversion, and then *H2* and *rh1* are turned off. Such an event then switches on the synthesis of H1 protein. Reversing the inversion flip switches back to H2. The Hin recombinase that is very similar to the invertases or recombinases of phage Mu or Cin from phage P1 catalyze the inversions. They can functionally substitute for each other. Hin binds to the *hixL* and *hixR* recombination sites. Additional genes are also involved in the fine-tuning. Defect in type III methyltransferase in restriction modification systems can cause phase variation of different genes in several bacterial species (Srikhanta YN et al 2005 Proc Natl Acad Sci USA 102:5547).

Somewhat similar mechanisms control the host-specificity genes of phage Mu and the mating type of budding yeast. ►cassette model, ►regulation of gene activity, ►antigenic variation, ►mating type determination in yeasts, ►*Trypanosoma*, ►flagellin, ►DNA uptake sequences, ►SSR; Hughes KT et al 1988 Genes Dev 2:937; Snyder LA et al 2001 Microbiology 147:2321.

Phase-Contrast Microscope: It alters the phase of light passing through and around the objects and this permits its visualization without fixation and/or staining. ►Nomarski, ►fluorescence microscopy, ►microscopy light, ►confocal microscopy, ►electronmicroscopy

Phaseolin ($C_{20}H_{18}O_4$): An antifungal globulin in bean (*Phaseolus*).

Phasing Codon: Initiates translation (such as AUG) and determines the reading frame. ►genetic code, ►reading frame

Phasmid (phage-plasmid): A plasmid vector equipped with the *att* site of the lambda phage and thus, enables the plasmid to participate in site-specific recombination with the λ genome resulting in incorporation of plasmid sequences into the phage (*lifting*). Because it contains both λ and plasmid origins of replication it may be replicated either as a plasmid or as λ. ►lambda phage, ►phagemid, ►vectors; Briani F et al 2001 Plasmid 45:1.

Phenacetin ($C_{10}H_{13}NO_2$): An analgesic and antipyretic drug and a carcinogen.

Phene: An observable trait that may or may not have direct genetic determination. ►gene

Phenetics: The taxonomic classification based on phenotypes.

Phenocopy: The phenotypic change that mimics the expression of a mutation. ►phenotype, ►epigenetic, ►epimutation, ►morphosis, ►genocopy

Phenodeviate: An individual of unknown genetic constitution displaying a phenotype attributed to various genic combinations within the population.

Phenogenetics: The attempts to correlate the function of genes with phenotypes.

Phenogram: ►character matrix

Phenolics: Compounds containing a phenol ring such as acetosyringone, hydroxyacetosyrin-gone, chalcone derivatives, phenylpropanoids, and some phytoalexins. The mentioned compounds may excite or suppress the *vir* gene cascade of *Agrobacterium* and may affect the response to plant pathogenic agents. Capsaicin, ginger, resveratrol may be anticarcinogens due to their antioxidative properties and may promote apoptosis. Phenylpropenes such as chavicol, *t*-anol, eugenol and isoeugenol repel animals and microorganism but may attract some pollinators (see Fig. P59). Humans use these compounds as spices, food preservatives and medicine. Coniferyl acetate and NADPH may form as a precursor for enzymatic synthesis of eugenols (Koeduka T et al 2006 Proc Natl Acad Sci USA 103:10128). ►*Agrobacterium*, ►virulence genes of *Agrobacterium*, ►phytoalexins, ►chalcones, ►acetosyringone, ►resveratrol, ►ginger, ►capsaicin, ►apoptosis, ►wound response; Nicholson RL, Hammerschmidt RE 1992 Annu Rev Plant Path 30:369.

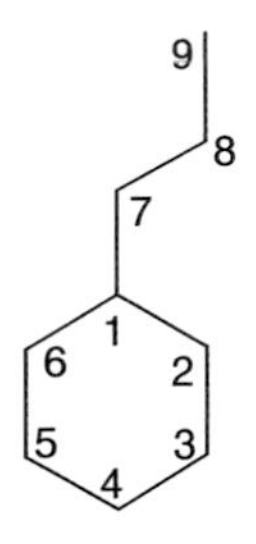

Figure P59. Phenylpropene backbone

Phenology: The study of the effects of the environment on live organisms.

Phenome: The collection of phenotypes; a group of organisms with shared phenotypes. (See Freimer N, Sabatti C 2003 Nature Genet 34:15; ►SNIPs; http://www.jax.org/phenome; http://www.phenomicdb.de.

Phenome Analysis: The attempts to determine the expression of genes at the RNA and protein level under different conditions, involving the environment and/or the prevailing genetic system (network). This may be an extremely difficult undertaking because gene expression may vary from complete silence to very variable and great complexity. ►phenotype, ►phenotype MicroArray; Bochner BR 2003 Nature Rev Genet 4:309.

Phenomenology: The description of facts as observed without metaphysical interpretation. The concept that behavior depends on how a person interprets reality rather than what is the objective reality.

Phenoptosis: An apoptosis-like phenomenon in unicellular organisms. ►apoptosis

Phenotype: The appearance of an organism that may or may not represent the genetic constitution. The proteins encoded by the DNA (the genotype) represent the phenotype. The 1014 human (disease) genes displayed 1429 distinguishable phenotypes (~141%). Microarray hybridization data can provide most comprehensive information on phenotype. In budding yeast, similar morphology indicates functional similarity of the coding genes. Triple-stained mutants facilitate the analysis of the morphology as a quantitative trait and can attribute function to many genes with previously unknown function (Ohya Y et al 2005 Proc Natl Acad Sci USA 102:19015). Mutation in the DNA is largely responsible for altered phenotypes. Alterations in transcription and translation also affect the phenotype and generally, as shown, there are more phenotypes than the number of genes (Bürger R et al 2006 Genetics 172:197). ►gene ontology, ►microarray hybridization, ►cell comparisons, ►endophenotype; Phenomic Data Base facilitates the identification of genes involved in a phenotype or gives the phenotype caused by genes in major organisms: http://www.phenomicDB.de; muse phenotypes and relevance to human disease: http://www.eumorphia.org/

Phenotype MicroArrays: An automated analysis of phenotypic expression of hundreds of genes on microplates and can be used to monitor the consequences of knockouts or other genetic alterations. ►microarray hybridization, ►knockout; Bochner BR et al 2001 Genome Res 11:1246.

Phenotypic Assortment: ►macronucleus

Phenotypic Knockout: It means somatic gene therapy that neutralizes intracellular harmful mechanisms. ►gene therapy, ►knockout

Phenotypic Lag: A period of time may be required for gene expression after transformation or mutagenic treatment. ►transformation genetic, ►premutation; Ryan FJ 1955 Am Nat 89:159.

Phenotypic Mixing: The mixed assembly of viral nucleic acids and proteins upon simultaneous infections by different types of viruses. Therefore, the coat protein properties of the virions do not match the viral genotype by serological or other tests. (See Hayes W 1965 The genetics of bacteria and their viruses. Wiley, New York).

P

Phenotypic Plasticity: An adaptive property of an organism enabling it to take advantage of local conditions without evolving a particular function and at the expense of another function. ►homeostasis genetic, ►canalization, ►plasticity

Phenotypic Reversion: An apparent restoration of the normal expression of a mutant gene; it is not, however, inherited. Aminoglycoside antibiotics (paromomycin, geneticin, etc.) may successfully compete with the translation termination factors in eukaryotes in cases when the mRNA carries a stop codon mutation. As a result, some of the polypeptide chains are not terminated/truncated but completed in the presence of the drug. Transposable elements may also alter gene function without causing mutation in the gene. Phenotypic reversion may be exploited for correcting the genetic defects in some diseases. ►phenocopy, ►amino-glycosides, ►G418, ►paromomycin, ►suppressor tRNA, ►translation termination; Gause M et al 1996 Mol Gen Genet 253:370; Franzoni MG, De Castro-Prado MA 2000 Biol Res 33:11; Biedler JL et al 1975 J Natl Cancer Inst 55:671.

Phenotypic Sex: It may not reflect the expectation based on the sex-chromosomal constitution. ►testicular feminization, ►hermaphrodite

Phenotypic Stability Factor: A measure of developmental homeostasis; it is calculated by the ratios of the quantitative expression of a parameter (gene) under two different environmental conditions. ►homeostasis, ►logarithmic stability factor; Lewis D 1954 Heredity 8:334.

Phenotypic Suppression: An apparently normal but non-hereditary phenotype brought about by translational error due to environmental effects and/or drugs. ►error in translation

Phenotypic Switch: Alters phenotypes more frequently than expected by point mutation and may be caused by epigenetic methylation or by protein folding. (See Lim HN, van Oudenaarden A 2007 Nature Genet 39:269).

Phenotypic Value: In quantitative genetics, it is defined as the mean value of a population regarding the trait under study and it is generally represented as P value. ►breeding value

Phenotypic Variance: ►genetic variance

Phenylalanine ($C_9H_{11}NO_2$): An essential water-soluble, aromatic amino acid (MW 165.19). Its biosynthetic path (*with enzymes involved in parenthesis*): Chorismate → (*chorismate mutase*) → Prephenate → (*prephenate dihydratase*) → Phenylpyruvate → (*aminotransferase*, glutamate NH_3 donor) → Phenylalanine. ►chorismate, ►tyrosine, ►phenylketonuria

Phenylalanine Ammonia Lyase (PAL): PAL deaminates phenylalanine into cinnamic acid and it is thus involved in the synthesis of plant phenolics. ►phenolics

Phenylalanine Hydroxylase (PAH, 12q24.1): The deficiency of PAH leads to phenylketonuria.

Phenylhydrazine ($C_6H_9ClN_2$): A hemolytic compound but it is also used as a reagent for sugars, aldehydes, ketones and a number of industrial purposes (stabilizing explosives, dyes, etc.). ►hemolysis

Phenylketonuria (PKU, PAH, Fölling disease): This gene was located to human chromosome 12q24.1. It is a recessive disorder that has a prevalence of about 1×10^{-4} (carrier frequency is about 0.02) in white populations. Thus, an affected person has about 0.01 chance to have an affected child in case of a random mate but the recurrence rate in a family where one of the partners is affected and the other is a carrier it is nearly 0.5 (see Fig. P60).

Figure P60. Mentally retarded heterozygous children of a phenylketonuric mother (indirect epistasis). Courtesy of Dr. C. Charlton Mabry 1963; by permission of the New England Journal of Medicine 269:1404

Its incidence is substantially lower among Asian and black people (one-third of that in whites). PKU is more frequent in European populations of Celtic origin than in the other Europeans. This has been interpreted as the result of natural selection because PKU heterozygosity conveys some tolerance to the mycotoxin, ochratoxin A, produced by *Aspergillus* and *Penicillium* fungi, common in humid northern regions. Before the nature of this disorder and the method of treatment were identified, about 0.5 to 1% of the patients in mental asylums were afflicted by PKU. A deficiency of the enzyme phenylalanine hydroxylase and consequently the accumulation of phenyl pyruvic acid and a deficiency of tyrosine cause the disease:

$$\text{PHENYL PYRUVIC ACID} \rightleftarrows \text{PHENYLALANINE} \Rightarrow \text{TYROSINE}$$

For the identification of the condition the Guthrie test has been used to cultures of *Bacillus subtilis* containing blood of the patients, β-2-thienylalanine was added. This phenylalanine analog is a competitive inhibitor of tyrosine synthesis. In the presence of excess amounts of phenylalanine, the bacterial growth does not stop. Since in the different families the genetically determined defect in the enzyme varies, so does the severity of the clinical symptoms. The accumulation of phenyl pyruvic acid is apparently responsible for the mental retardation and the musty odor of the urine of the patients.

The reduced amount of tyrosine prevents normal pigmentation (melanin) and thus results in pale color. The good aspect of this condition is that relative normalcy can be established if it is diagnosed early and dietary restrictions for phenylalanine are implemented. The restriction of phenylalanine must start as early as possible (before birth if feasible), and continue at least until age 10. Restriction should be observed before pregnancy, during pregnancy and during breast-feeding or during the entire life to avoid harm to the nervous system. Phenylketonuria of the mother may damage the nervous system of genetically normal fetus through placental transfer (indirect epistasis). Because of the multiple metabolic pathways involving phenylpyruvic acid, besides the deficiency of phenylalanine hydroxylase, other genes and conditions may cause similar clinical symptoms. Phenylalanine hydroxylase activity requires the availability of the reduced form of the co-factor 5,6,7,8-tetrahydrobiopterin that is made by the enzyme *dihydrobiopterin reductase* from 7,8-dihydrobiopterin. The dihydrobiopterin reductase enzyme is coded in human chromosome 4p15.1-p16.1. Defect in this enzyme also causes phenylketonuria symptoms but lowering the level phenylalanine in the diet does not alleviate the problems. Another form of phenylketonuria is based on a deficiency in dihydrobiopterin synthesis. Using the φBT1 integrating phage vector (containing a site-specific recombinase), equipped with the murine phenylalanine hydroxylase cDNA, PKU could be completely and persistently cured after three injections to the mouse liver (Chen L, Woo SLC et al 2005 Proc Natl Acad Sci USA 102:15581).

Prenatal diagnosis can be carried out by several methods. Mutations at various related metabolic sites in the mouse may serve as a model for studying phenylketonuria. ►epistasis, ►mental retardation, ►one gene-one-enzyme theorem, ►genetic screening, ►Guthrie test, ►tyrosinemia, ►alkaptonuria, ►phenylalanine, ►hyperphenylalaninemia, ►amino acid metabolism, ►prenatal diagnosis, ►enzyme replacement therapy, ►integrase, ►targeting genes; Ledley FD et al 1986 N Engl J Med 314:1276; Gjetting T et al 2001 Am J Hum Genet 68:1353.

Phenylpropanoid: ►phenolics, ►phytoalexins

Phenylthiocarbamide Tasting (PTC): A major incompletely dominant gene appears to be in human chromosome 7q35-q36 (Conneally PM et al 1976 Hum Hered 26(4):276). A major bitter testing locus is assigned to human chromosome 5p15 (Reed DR et al 1999 Am J Hum Genet 64:1478). The TAS2R10 PTC receptor appears to be in the short arm of chromosome 12. A single G protein coupled receptor with allelic variants may account for the taste perception (Bufe B et al 2005 Curr Biol 15:322). In humans and chimpanzees, two alleles at the TAS2R38 locus control bitter tasting but the human and chimpanzee alleles are different (Wooding S et al. 2006 Nature [Lond] 440:930). About 30% of North-American Whites and about 8–10% of Blacks cannot taste the bitterness of this compound. Persons affected by thyroid-deficiency (athyreotic) cretenism (mental deficiency) are non-tasters. Phenylthiocarbamide (syn. phenylthiourea) has been used for classroom demonstration of human diversity but it should be kept in mind that it is a toxic compound (LD_{50} oral dose for rats 3 mg/kg and for mice 10 mg/kg). ►taste, ►LD_{50}; Guo SW, Reed DR 2001 Ann Hum Biol 28(2):111; Kim U-K et al 2003 Science 299:1221.

Phenylthiourea: ►phenylthiocarbamide tasting (see Fig. P61).

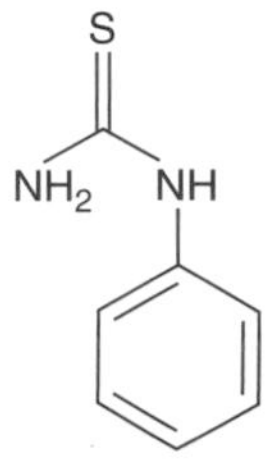

S
NH2 NH

Figure P61. Phenylthiourea

Pheochromocytoma (phaeochromocytoma): An adrenal tumor induced by the SHC oncogene. ►paraganglion, ►SHC, ►adenomatosis endocrine multiple, ►endocrine neoplasian neuroendocrine cancer

Pheresis (apheresis): The medical procedure of withdrawal of blood; after fractionation, some fraction(s) are reintroduced. Such a protocol may use stem cells, transfect them with a vector or apply to them chemotherapy, and eventually place them back in the body of the same individual.

P

Pheromones: Various chemical substances secreted by animals and cells for the purpose of signaling and generating certain responses by members of the species, such as sex-attractants, stimulants, territorial markers, or other behavioral signals and cues. The male pheromone of Asian elephants, frontalin (exists in two enantiomorphs) is secreted during musth (annual period of increased sexual activity and aggression) by the temporal gland on the face. The proportion of the two forms and the extent of secretion of frontalin are sensed primarily by the ovulating females (Greenwood DR et al 2005 Nature [Lond] 438:1097). About 100 genes control the pheromones and their receptors in rodents and the signals are transmitted through G protein-associated signal transduction pathways. The mouse sex pheromones in the male urine includes several compounds, among them the strongest response is evoked by (methylthio) methanethiol for females. Apparently, only a small number of cells at the olfactory bulb responded to this compound, which for humans had a garlic-like odor (Lin DY et al 2005 Nature [Lond] 434:470). The lacrimal glands of adult male mice secrete a 7 kDa peptide to which the vomeronasal receptors of the sensory neurons of the female animals respond (Kimoto H et al. 2005 Nature [Lond] 437:898). The role of pheromones in humans may not be generally agreed upon. The steroid compounds 4,16-androstadien-3-one (AND) present in the sweat of human males and estra-1,3,5(10)-tetraen-3-ol (EST) present in the female urine appear to have pheromone-like properties. Their smelling causes sex-differentiated activation of the anterior hypothalmus of the brain. Male homosexuals respond to AND but not to EST in contrast to heterosexuals (Savic I et al 2005 Proc Natl Acad Sci USA 102:7356). Heterosexual men were found to respond to AND; lesbian women processed AND (unlike heterosexual women), not by the anterior pituitary but by the olfactory network (Berglund H et al 2006 Proc Natl Acad Sci USA 103:8269).

The pheromone receptor genes in rodents include *V1r* and *V2r*; apparently there are no human homologs. The sexually deceptive orchid plant *Chiloglottis* attracts the *Neozeleboria* insect pollinator males by secreting a volatile compound (2-ethytl-5-propylcyclohexan-1,3-dione), which is chemically identical with the insect female sex pheromone (Schiestl FP et al 2003 Science 302:437). RNA interference technique revealed that in silkworm, for sex pheromone production pheromone activating neuropeptide receptor (*PBANR*), pheromone gland-specific (PG) fatty acyl reductase (*pgFAR*), PGZ11/Δ10,12 desaturase (*Bmpgdesat1*), PG acyl CoA-binding protein (*pgACBP*) genes are essential (Ohnishi A et al 2006 Proc Natl Acad Sci USA 103:4398). In *Drosophila*, the male-specific pheromone 1-*cis*-vaccenyl acetate (cVA) acts through olfactory receptor Or67d. The *fruitless* (*fru*) gene controls three classes of olfactory receptors, one of which is Or67d. This single receptor controls both male and female mating behavior. Mutant males display homosexual tendencies whereas mutant females are less receptive to courting indicating that cVA has opposite effects for the two sexes, i.e., inhibiting males' mating behavior but stimulating that in females (Kurtovic A et al 2007 Nature [Lond] 446:542). Male flies show less interest in females, which have mated before, apparently by sensing cVA (Ejima A et al 2007 Curr Biol 17:599). The trichoid sensilla on the antennae has three olfactory receptors; receptor T1 responds primarily to male odor in the cuticular extracts whereas to the odor of virgin females receptors T2 and T3 respond (van der Goes van Naters W, Carlson JR 2007 Curr Biol 17:606). In male *Drosophila*, the synthesis of octopamine (a norepinephrine related substance) mediates aggressive behavior toward other males. In flies defective for tyramine β-hydroxylase (the enzyme that converts octopamine from its precursor), courtship is observed (Cartel SJ et al 2007 Proc Natl Acad Sci USA 104:4706). Pheromone hydrocarbon chains are longer in females than in males. The transformer gene (tra) feminizes males and makes them produce longer (female type) hydrocarbon pheromones (Chertemps T et al 2007 Proc Natl Acad Sci USA 104:4273). ►mating type determination in yeast, ►sex determination, ►*fru*, ►olfactogenetics, ►vomeronasal organ, ►signal transduction, ►homosexual, ►RNAi, ►silkworm, ►kairomones, ►mimicry, ►Bruce effect; Kohl JV et al 2001 Neuroendocrinol Lett 22(5):309; Luo M et al 2003 Science 299:1196; Prestwich GD, Blomquist GJ 1987 Pheromone Biochemistry, Academic Press, Orlando, Florida; Wang Y, Dohlman HG 2004 Science 306:1508; Dulac C, Torello AT 2003 Nature Rev Neurosci 4:531; minireview: Stowers L, Marton TF 2005 Neuron 46:699; review of vertebrate pheromone communication: Brennanm PA, Zufall F 2006 Nature [Lond] 444:308; pheromone signaling circuits: Dulac C, Wagner S 2006 Annu Rev Genet 40:449; insect pheromones: http://www.pherobase.com.

Phialide: Fungal stem cells from which conidia are budded.

Philadelphia Chromosome: In the Philadelphia chromosome, the long arm (q34) of human chromosome 9, carrying the *c-abl* oncogene is translocated to the long arm (q11) of chromosome 22 carrying site *bcr* (break-point cluster region). The *bcr-abl* gene fusion is then responsible for 85% of myelogenous (Abelson) and acute leukemia as a consequence of the translocation and fusion. In the acute form, a 7.5 kb mRNA is translated into protein p190, and in the myelogenous form an 8.5 kb mRNA is translated into a chimeric protein p210. The fusion protein is a deregulated tyrosine kinase acting on hematopoietic cells and causes leukemia-like oncogenic transformation in mice. Synthetic antisense phosphorothioate oligonucleotides ([S]ODN) complementary to the 2nd exon of BCR or to the 3rd exon of the ABL of the fused genes block temporarily the proliferation of the chronic leukemic cells, without harming the normal cells, in a mouse model. The outcome of such a therapy could be improved by simultaneously targeting also the c-Myc oncogene with an antisense construct. Further effectiveness was observed by exposing the cells to a low concentration of mafosfamide, an antineoplastic drug that promotes apoptosis, or to cyclophosphamide. ►ABL, ►BCR, ►leukemia, ►hematopoiesis, ►cancer gene therapy, ►apoptosis, ►cyclophosphamide, ►antisense technologies, ►transresponder, ►Knudson's two mutation theory; Saglio G et al 2002 Proc Natl Acad Sci USA 99:9882; Goldman JM et al 2003 N Engl J Med 349:1451.

Phlebotomous: A blood-sucking (insect), or phlebotomy bloodletting surgical procedure.

Phloem: A plant tissue involved in the transport of nutrients; it contains sieve tubes and companion cells, phloem parenchyma, and fibers. ►sieve tube, ►parenchyma, ►xyleme, ►root, ►proteoglycan; Bonke M et al 2003 Nature [Lond] 426:181.

Phlorizin: A dichalcone in the bark of trees (*Rosaceae*); it blocks the reabsorption of glucose by the tubules of the kidney and causes glucosuria. ►disaccharide intolerance, ►chalcones

PHO81: A yeast CDK inhibitor homologous to $p16^{INK4}$. ►CDK, ►$p16^{INK4}$

PHO85: A cyclin-dependent kinase of *Saccharomyces cerevisiae*. ►CDK, ►KIN28, ►CDC28, ►PHO81

phoA: A gene for alkaline phosphatase.

Phobias: Phobias exist in different forms, all characterized by unreasonable avoidance of objects, events, or people. A phobia may amount to serious, morbid mental illness. Apparently, duplication in human chromosome 15q24-q26 is associated with one form and also with laxity of the joints. ►panic disorder, ►panic obsessive disorder, ►anxiety; Gratacós M et al 2001 Cell 106:367.

Phocomelia: The absence of some bones of the limbs proximal to the trunk. It may occur as a teratological effect of various recessive and dominant human genetic defects or as a consequence of teratogenic drugs, e.g., thalidomide use during human or primate pregnancy. ►limb defects in humans, ►Roberts syndrome, ►teratogen, ►thalidomide

PHOGE (pulsed homogeneous orthogonal field electrophoresis): A type of pulsed field gel electrophoresis, within the range of 50 kb to 1 Mb DNA, permitting straight tracks of large number of samples. ►pulsed field gel electrophoresis

PhoQ: *Salmonella* kinase, affecting regulator of virulence PhoP. ►virulence, ►*Salmonella*

PhoR: Phosphate assimilation regulated by PhoR kinase upon phosphorylation of regulator PhoB.

Phorbol Esters. Facilitators of tumorous growth that work by activating protein kinase C. ►TPA, ►protein kinases, ►procarcinogen, ►carcinogen, ►PMA, see formula at ►PMA

Phorbol 12-Myristate-13-Acetate (PMA): ►phorbol esters, ►PMA (see Fig. P62).

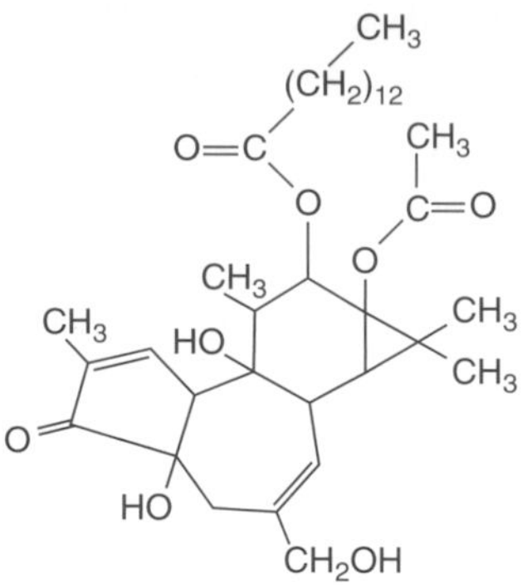

Figure P62. Myristoylphorbol acetate (ester)

Phosphatases: In animals, both acid and alkaline phosphatases are common, in plants acid phosphatases are found. Some of the phosphatases have high specificities and have indispensable role in energy release in the cells. A series of non-specific phosphatases carry out only digestive tasks. In humans, the erythrocyte and fibroblast expressed acid phosphatase (ACP1) isozymes are coded in chromosomes 2 and 4. It has been suggested that these enzymes split flavin mononucleotide phosphates. In megaloblastic anemia, ACP1 level is increased. The tartrate-resistant acid phosphatase type 5 (TR-AP) is an iron-glycoprotein of 34 kDa (human chromosome 15q22-q26), and it is increased in the spleen in case of Gaucher disease. Lysosomal acid phosphatase (ACP2) is located in human chromosome 11p12-p11. Alkaline phosphatase (ALPL) is present in the liver, bone, kidney, and fibroblasts, is often called the non-tissue-specific phosphatase (human chromosome 1p36-p34), and is deficient in hypophosphatasias. The alkaline phosphatase ALPP is located in the placenta (human chromosome 2q37) and several allelic forms have been identified. A similar alkaline phosphatase is present also in the testes and the thymus and the gene occurs at the same chromosomal location, but its expression is highly tissue-specific. Protein phosphatase 2A (PP2A) catalytic subunit, encoded at human chromosome 9q34, is involved in the control of many cellular processes, including the mitotic spindle (Sclaitz A-L et al 2007 Cell 128:115). The structure of the holoenzyme is known (Xu Y et al 2006 Cell 127:1239). ►serine/threonine and tyrosine protein phosphatases, ►hypophosphatasia, ►hypophosphatemia, ►dual-specificity phosphatase, ►megaloblastic anemia, ►Gaucher's disease

Phosphate Response of Plants: The inorganic phosphate level is frequently growth-limiting in plants. *Arabidopsis* is an extremely sensitive indicator of P is soils. The photograph illustrates *Arabidopsis* growth on a Missouri soil sample, without and with PO_4 addition (see Fig. P63) (Rédei GP 1966 unpublished).

Figure P63. *Arabidopsis* growth with and without PO_4 addition

The physical contact of the *Arabidopsis thaliana* primary root tip with low-phosphate medium arrests root growth. Loss-of-function mutations in *Low Phosphate Root1* (*LPR1*) and its close paralog *LPR2* strongly (encoding multicopper oxidases) reduce this inhibition (Svistoonoff S et al 2007 Nature Genet 39:792). The non-protein coding gene *IPS1* (Induced by phosphate starvation1) contains a motif with sequence complementarity to the phosphate (Pi) starvation-induced miRNA miR-399. When the pairing is interrupted by a mismatched loop at the expected miRNA cleavage site, IPS1 RNA is not cleaved but instead sequesters miR-399. Thus, IPS1 overexpression results in increased accumulation of the miR-399 target PHO2 mRNA (phosphate-starvation) and, concommitantly, in reduced shoot Pi content (Franco-Zorilla JM et al 2007 Nature Genet 39:1033).

Microarray hybridization of transcript abundance among 22,810 *Arabidopsis* genes indicated that 612 were coordinately induced, whereas 254 genes were suppressed by inorganic phosphate. These genes are involved with metabolic pathways, ion transport, signal transduction, transcriptional regulation and other cellular processes (Misson J et al 2005 Proc Natl Acad Sci USA 102:11).

Phosphatidate: A precursor of diaglycerol (see Fig. P64). ►diaglycerol, formula at right.

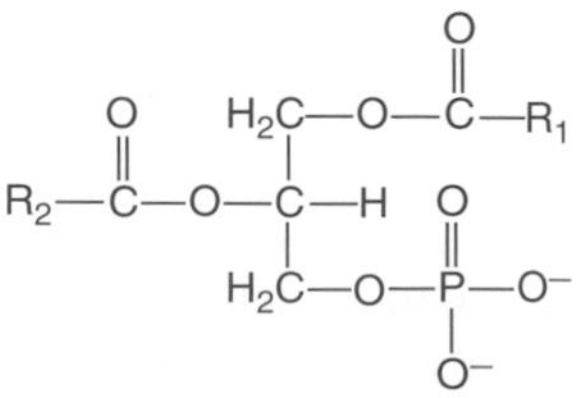

Figure P64. Phosphatidate

Phosphatidylinositol (1,2-diacyl-sn-glycero-3-phospho [1-o-myoinositol]): A cell membrane phospholipid. Phospha-tidylinositol-transfer protein is required for vesicle budding from the Golgi complex. Phosphatidylinositol-3,4,5-trisphosphate activates protein kinase B. Phosphatidylinositol-kinase-3 mutations are oncogenic (Kang S et al 2005 Proc Natl Acad Sci USA 102:802). ►PIK, ►pleckstrin domain, ►PKB, ►Golgi apparatus, ►PTEN, ►phosphoinositides, ►wortmannin; Bourette RP et al 1997 EMBO J 16:5880; Abel K et al 2001 J Cell Sci 114:2207.

3′-Phosphoadenosine-5′-Phosphosulfate: ►PAPS

Phosphodegron: SCFβ-TRCP promotes Chk1-dependent Cdc25A ubiquitination, and this involves serine 76, a known Chk1 phosphorylation site, but other sites of phosphorylated amino acids may make a protein liable to degradation by ubiquitination. ►SCF, ►Chk-1, ►CDC25, ►ubiquitin

Phosphodiester Bond: A phosphodiester bond attaches the nucleotides into a chain by hooking up the incoming 5′-phosphate ends to the 3′-hydroxy tail of the preceding nucleotide: R^1 and R^2 represent nucleosides, O: oxygen, H: hydrogen, P: phosphorus (see Fig. P65). ►Watson-Crick model, formula.

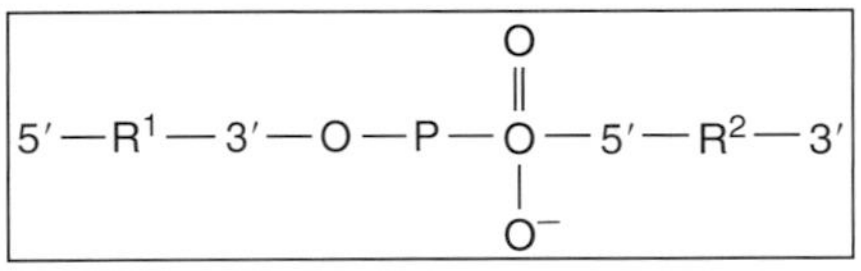

$$5'-R^1-3'-O-P(=O)(O^-)-O-5'-R^2-3'$$

Figure P65. Phosphodiester bond

Phosphodiesterases: Exonucleases. The snake venom phosphodiesterase starts at the 3′-OH ends of a nucleotide chain and splits off the nuclcoside-5′-phosphate. The 3′-phosphate terminus does not lend the nucleotide chain for its action. The spleen phosphodiesterase, on the other hand, generates nucleoside-3′-phosphate molecules by splitting on the other side of the nucleotides. Phosphodiesterase converts cyclic AMP into AMP or cGMP into GMP. Phosphodiesterase 5 inhibitors are therapeutics for erectile dysfunction and several other diseases. The sensitivity of RNA phosphodiesterases is affected by the secondary and tertiary structure of the RNA as well by the adjacent nucleotides. ►phosphodiester bond, ►nitric oxide, ►priapism, ►stroke; structure: Sung BJ et al 2003 Nature [Lond] 425:98; catalytic domains for different inhibitors: Card GL et al 2004 Structure12:2233.

Phosphoenolpyruvate: Phosphoenolpyruvate is efficient in transferring phosphate group to ADP to form ADP (see Fig. P66).

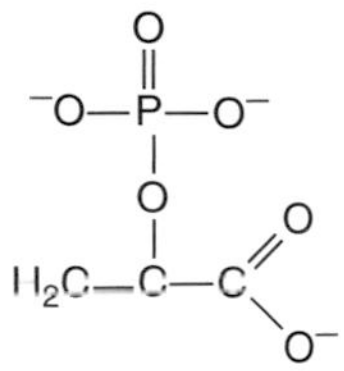

$$H_2C=C(O-PO_3^{2-})-COO^-$$

Figure P66. Phosphoenolpyruvate

Phosphofructokinase M (glycogen storage disease VII, 12q13.3): A phospho-fructokinase in the muscles (PFKM). It deficiency may cause muscle cramps and myoglobinuria. Lactate production is reduced and fatigue develops after exertion. ►glycogen storage diseases, ►myoglobin, ►fructose-2,6-bisphosphatase

Phosphofructokinase Platelet Type (PFKP, 10p15.3-p15.2): PFKP is expressed in the platelet but it displays 71% identity of amino acid sequence of the muscle type and 63% identity with the liver enzyme.

Phoshpofructokinase X (PFKX): A chromosome 2-encoded enzyme, which is expressed in the fibroblasts and the brain.

Phosphofructo-2-Kinase/Fructose-2,6-Bisphosphatase (PFKFB, PFRX, Xp11.21): A bifunctional enzyme encoded in the X chromosome of humans and rodents. The PFKFB3 locus is in 10p15-p14. The PFKFB4 enzyme is in 3p22-p21.

Phosphofructose Kinase 1 (PFK-1, PFKL, phosphofructokinase): PFK-1 enzyme catalyzes the formation of fructose-1-6-bisphosphate from fructose-6-phosphate in the presence of ATP and Mg^{2+}. PFKL (liver enzyme) is encoded in human chromosome 21q22.3. The tetrameric enzyme may exist in five different forms due to the random association of the products of two different loci.

Phosphofructose Kinase 2 (PFK-2): PFK-2 mediates the formation of fructose-2,6-bisphosphate from fructose-6-phosphate. It enhances the activity of fructosephosphate 1 enzyme by binding to it, and also inhibits fructose-2,6-bisphosphatase and therefore enhances glycolysis. ►glycolysis

Phosphoglucomutase (PGM): The enzyme that catalyzes the reaction:

$$\text{GLUCOSE}-1-\text{PHOSPHATE} \rightleftarrows \text{GLUCOSE}-6-\text{PHOSPHATE}$$

Phosphoglucomutase proteins are homologous in structure through the animal kingdom. In humans, there are several PGM enzymes and some with multiple allelic forms with characteristic patterns and are reasonably stable. Therefore, PGM is used in forensic genetics for personal identification on samples up to 6 months old. Their human chromosomal locations are: PGM1 (1p31, PGM2 (4p14-q12), PGM3 (6q12), and PGM5 (9p12-q12). ►forensic genetics

Phosphogluconate Oxidative Pathway: Same as pentose phosphate pathway.

3-Phoshoglycerate Dehydrogenase Deficiency (PHGDH, 1q12): PHGDH results in recessive serine biosynthetic defect, microcephaly, and neurological defects and seizures. ►serine

Phosphoglyceratemutase Deficiency: ►myopathy, ►glycerophospholipid

Phosphoglyceride: ►glycerophospholipid

Phosphohexose Isomerase (PHI): PHI catalyzes the glucose-6-phosphate⇔fructose-6-phosphate conversions. It is encoded in human chromosome 19cen-q12. Its defects result in dominant hemolytic anemia. ►anemia, ►hemolytic anemia

Phosphoinositide-3-Kinases: ►PIK, ►phosphoinositides

Phosphoinositides: Inositol-containing phospholipids. They play an important role as second messengers, and in phosphorylated/dephosphorylated forms they participate in the regulation of traffic through membranes, growth, differentiation, oncogenesis, neurotransmission, hormone action, cytoskeletal organization, platelet function, and sensory perception. The signals converge on phospholipase C (PLC, 20q12-q13.1). It hydrolyzes phosphatidylinositol-4,5-bisphosphate ($PtdInsP_2$) into inositol trisphosphate ($InsP_3$) and diacylglycerol (DAG). $PtdInsP_2$ segregation is mediated by PTEN, and CDC42 control apical morphogenesis (Martin-Belmonte F et al 2007 Cell 128:383). $InsP_3$ regulates Ca^{2+} household and DAG activates PLC. $InsP_3$ levels also regulate pronuclear migration, nuclear envelope breakdown, metaphase-anaphase transitions, and cytokinesis. Cytidine diphosphate-diacylglycerol synthase (CDS) is required for the regeneration of $PtdInsP_2$ from phosphatidic acid. CDS is a key regulator in the G-protein-coupled photo-transduction pathway. Pleckstrin homology domains selectively bind phosphoinositides. ►inositol, ►phospholipase, ►stoma, ►DAG, ►signal transduction, ►IP_2 [$InsP_2$ for formula], ►IP_3 [$InsP_3$ for formula], ►phosphatidylinositol, ►*myo*inositol, ►PIK, ►pleckstrin, ►PTEN, ►CDC42; Czech MP 2000 Cell 100:603; Vanhaesebroeck B et al 2001 Annu Rev Biochem 70:535; Sato TK et al 2001 Science 294:1881; De Matteis MA et al 2002 Curr Opin Cell Biol 14:434; review: Di Paolo G, Di Camilli P 2006 Nature [Lond] 443:651.

Phosphoinositide-Specific Phospholipase Cδ: Cδ signal transducers and generates the second messengers inositol-1,4,5-triphosphate and diaglycerol. ►signal transduction, ►second messenger

Phospholamban (phosphorylated pholamban, PLN, 6q22.1): A regulator of sarcoplasmic reticulum Ca^{2+}-ATPase (SERCA); it is a kinetic regulator of heart muscle function. Mutation in PLN leads to hereditary cardiomyopathy and premature death (Haghighi K et al 2006 Proc Natl Acad Sci USA 103:1388). ►cardiomyopathy

Phospholipase (PL) A, D, C: Each split specific bond in phospholipids. PLC-β generates diacylglycerol and phosphatidylinositol 2,4,5-triphosphate from phosphatidylinositol 4,5-bis-phosphate. These second messenger molecules play roles in signal transduction. PLC-γ is activated by receptor tyrosine kinases and one of its homologs is the SRC oncoprotein. PLC-γ1 with VEGF—through the FLT-1 receptor—controls the strength of the heartbeat by regulating calcium signaling in the myocytes (Rottbauer W et al 2005 Genes Dev 19:1624). PLA is present in mammalian inflammatory exudates. Form A2 is coded by human chromosome 12, the other PLA2B by chromosome 1. Phospholipase C is coded in human chromosomes at the following locations: PLCB3 (11q11), PLCB4 (20p12), and PLCG2 (16q24.1). ►serine/threonine phosphoprotein phosphatases, ►SRC, ►signal transduction, ►phosphoinositides, ►Ipk1, ►Ipk2, ►VEGF, ►stoma; Rhee SG 2001 Annu Rev Biochem 70:281; Wang X 2001 Annu Rev Plant Physiol Mol Biol 52:211.

Phospholipid: A lipid with phosphate group(s). ►liposome, ►lipids

Phosphomannomutase Deficiency: A rare defect of glycosylation displaying large differences in expressivity. It involves inverted nipples, fat pads, strabismus, hyporeflexia (sluggish responses), mental retardation, hypogonadism, and early death. (See Grünewald S et al 2001 Am J Hum Genet 68:347).

Phosphomannose Isomerase (MPI, 15q22-qter): The defects of MPI affect many glycosylation reactions in the cell. MPI is involved in the conversion of fructose-6-phosphate into mannose-6-phosphate. Clinically, it may cause diarrhea, enlarged liver, hypoglycemia with convulsions, coma, etc. The 5 kb gene includes 8 exons.

Phosphomonoesterase: A phosphatase digesting phosphomonoesters, such as nucleotide chains. ►phosphodiester bond

Phosphonitricin (Basta): ►herbicides

Phosphoramidates: Phosphoramidates are used in antisense technologies by the modification of the sugar-phosphate backbone of oligonucleotides (see Fig. P67). ►antisense technologies, ►trinucleotide-directed mutagenesis; Jin Y et al 2001 Bioorg Med Chem Lett 11:2057; Faria M et al 2001 Nature Biotechnol 19:40.

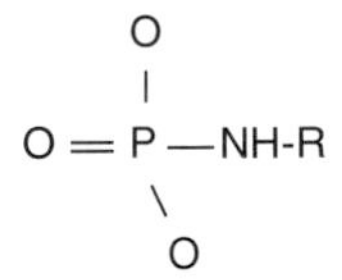

```
        O
        |
O = P — NH-R
        \
         O
```

Figure P67. Phosphoramidate

Phosphoramide/Ink Jotting: A rapid method of DNA analysis. (See Cooley P et al 2001 Methods Mol Biol 170:117).

Phosphorelay: ►two-component regulatory system

Phosphorescence: ►fluorescence, ►luminescence

Phosphoribosylglycinamide Formyltransferase: Phophoribosylglycinamide formyltransferase is human chromosome 21q22.1 dominant and it controls purine, pyrimidine biosynthesis, and folate metabolism.

Phosphoribosylpyrophosphate Synthetase (PRPS1, Xq22-q24): An enzyme of the purine/pyrimidine salvage pathway. Its deficiency may cause hyperuricemia (excessive amounts of uric acid in the urine), deafness, and neurological disorder.

Phosphorimaging: The detection of radioactive labels in tissues by phosphorescence.

Phosphorolysis: In phophorolysis, glycosidic linkage holding two sugars together is attacked by inorganic phosphate and the terminal glucose is removed (from glycogen) as α-D-glucose-1-phosphate.

Phosphorothioates: Analogs of oligodeoxynucleotides; they are used in antisense technology. Their attachment to the 3′-end inhibits the activity of nucleases that attacks RNA from that end. They can bind to proteins, but do not stimulate the activity of RNase H (phosphoro-dithioate modified heteroduplexes may stimulate RNase H), inhibit translation, and are relatively easily taken up by cells. Some of the truncated mRNAs can, however, be translated into truncated proteins (Hasselblatt P et al 2005 Nucleic Acids Res 33:114). Some of the effects of these molecules are not based on their anti-sense properties (e.g., binding to CD4, NF-κB, inhibition of cell adhesion, inhibition of receptors, etc.). Phosphorothioate-modified nucleotides (one of the oxygen attached to P is replaced by S) are used also in in vitro mutagenesis to protect the template strand from nucleases, while the strand to be modified is excised before re-synthesis in a mutant form. ►antisense technologies, ►antisense RNA, ►antisense DNA, ►OL(1)p53, ►ribonuclease H, ►CD4, ►NF-κB; Sazani P et al 2001 Nucleic Acids Res 29:3965; ►quenched autoligation probe

Phosphorylase b Kinase: An enzyme that phosphorylates two specific serine residues in *phosphorylase b*, thus converting it into *phosphorylase a* upon the action of cAMP-dependent protein kinase (synonym protein kinase A). Phosphorylase b kinase mediates glycogen breakdown. This enzyme is a tetramer and for activation the two regulatory subunits (R) must be separated from the two catalytic subunits to be able to function. The dissociation is mediated by cAMP through A-kinase. The δ subunit is calmodulin. ►epinephrine, ►cAMP-dependent protein kinase, ►cAMP, ►A-kinases, ►calmodulin; Brushia RJ, Walsh DA 1999 Front Biosci 4:D618.

Phosphorylases (kinases): ►serine/threonine kinases, ►tyrosine protein kinase, ►Jak kinase, ►phophorylase B, ►A-kinases, ►signal transduction, ►serine/threonine phosphoprotein phosphatases, ►phospholipase C, ►signal transduction, ►calmodulin, ►phosphorylase b kinase

Phosphorylation: Adding phosphate to a molecule. It may play an important role in signal transduction, and depending on which of the potentially several sites is phosphorylated, the function of some transcription factors may be altered. In the proteins serine and threonine, residues are frequently phosphorylated. If phosphoserine and phosphothreonine residues are replaced by the lysine analog aminoethylcyteine or β-methylaminoethylcysteine, the lysine-specific proteases cleave at these sites and thus these phosphorylation sites can be revealed (Knight ZA et al 2003 Nature Biotechnol 21:1047). Phosphorylation sites are detectable also by tandem mass spectrometry. The mouse genome contains more than 500 kinases. The nerve synapse system alone operates with 650 phosphorylation events involving 331 sites. Some proteins, like MAP1B, have 33 such sites. The bioinformatics information indicates that a small number of kinases phosphorylate many proteins and some substrates are phosphorylated by many kinases. These phosphorylations form elaborate interacting networks (Collins MO et al 2005 J Biol Chem 280:5972). ►oxidative phosphorylation, ►kinase, ►phosphorylases, ►tandem mass spectrometry, ►unstructured proteins; Whitmarsh AJ, Davis RJ 2000 Cell Mol Life Sci 57:1172; protein phosphorylation site prediction (DISPHOS): http://core.ist.temple.edu/pred/pred.html; PHOSIDA phosphorylation site database: http://www.phosida.com/; protein three-dimensional phosphorylation sites: http://cbm.bio.uniroma2.it/phospho3d.

Phosphorylation Potential (Δg_p): The change in free energy within the cell after hydrolysis of ATP.

Phosphoserinephosphatase: Phosphoserinephosphatase hydrolyzes O-phosphoserine into serine; it is encoded in human chromosome 7p15.1-p15.1.

Photoactivated Localization Microscopy: Photoactivated localization microscopy can detect activable fluorescent proteins within cells and cellular organelles at nanometer resolution (Betzig E et al 2006 Science 313:1642). ►microscopy

Photoaffinity Tagging: In photoaffinity tagging, the labels may be radioactive or fluorescent and bind to certain compounds by non-covalent bonds upon illumination. (See Knorre DG et al 1998 FEBS Lett 433:9).

Photoaging: In photoaging, skin collagens and elastin are damaged by the ultraviolet light induced metalloproteinases and this results in wrinkling of the skin similar to what occurs during aging. These enzymes are upregulated by AP-1 and NF-κB transcription factors. ►aging, ►collagen, ►elastin, ►AP-1, ►NF-κB

Photoallergy: Immunological response to a substance activated by light.

Photoautotroph: An organism that can synthesize in light all its required organic substances and energy from inorganic compounds. The majority of green plants are photoautotrophic. By introducing a glucose transporter gene into obligate photoautotrophic alga, the organism could be converted to light-independent growth on glucose. (See Zaslavskaia LA et al 2001 Science 292:2073).

Photochemical Reaction Center: The site of photon absorption and initiation of electron transfer in the photosynthetic system. ►photosynthesis

Photodynamic Effect: Photosensitivation, photodestruction. A dye or pigment absorbs light and converts the energy to a higher state and exerts specific effects. Photodynamic effects may have various therapeutic applications. Phenothiazines, phthalocyanines, porphyrines, and other molecules with photoactive properties have been successfully tested as photoinactivating agents against Gram-positive and Gram-negative bacteria. After absorption of light, singlet oxygen ($^1O_2$) may be generated and the oxidative damage to proteins and lipids may kill the bacteria even if they are resistant to antibiotics. ►ROS, ►singlet oxygen; Langmack K et al 2001 J Photochem Photobiol B 60:37; Maisch T et al 2007 Proc Natl Acad Sci USA 104:7223.

Photoelectric Effect: The photoelectric effect has very wide applications of modern technology (television, computers and other electronic instruments). Atoms may emit electrons when light hits a suitable target. When X-rays hit a target, very high energy photoelectrons may be generated.

Photogenes: Chloroplast DNA-encoded proteins involved in photosynthesis. One of the most studied is *photogene* 32 (*psbA*), which codes for a 32 kDa thylakoid protein involved in electron transport in photosystem II. Also, it binds the herbicide atrazine and by removing or altering this binding site, one can obtain plants resistant to the weed killer through molecular genetic manipulations. ►photosynthesis, ►herbicides; Rodermel SR, Bogorad L 1985 J Cell Biol 100:463.

Photography: Photography, in the laboratory, has special requirements depending on the objects. Cell cultures in Petri plates can be best photographed through macrolenses (for extreme close ups, use extension rings or teleconverter) and through using highly sensitive color films, such as Kodak Gold 400. To eliminate reflection, the blue photoflood lamps should be adjusted at an angle of about 45°. Agarose gels can be photographed with a polaroid camera mounted on a copying stand and using high speed (ASA 3000) films. Ultraviolet light sources of the longer wavelength are less likely to damage the DNA. The contrast can be enhanced by the use of orange filters on the camera (such as Kodak Wratten 22A). Note that ultraviolet light is dangerous to the skin and particularly to the eyes. Use gloves, goggles, and wear a long-sleeved shirt. For photomicrography, built-in automatic exposure meters are very advantageous, if frequently used. Otherwise, numerous exposures, at the proper color temperature, are necessary. For photocopying and editing, halftone image computers with (color) scanners can be used. The resolution now provided by digital cameras is satisfactory for most biological applications; they are very convenient and the high pixel (up to 8–10 megapixel [picture elements]) units are very powerful.

Photolabeling: Adding photoactivatable groups to proteins, membranes, or other cellular constituents in order to detect their reaction path. The labels are generally small molecules, stable in the dark and highly susceptible to light. They work without causing photolytic damage to the target and are stable enough to permit analytical manipulations of the sample. Synthetic peptides containing substances, such as 4′-(trifluoromethyl-diazirinyl)-phenylalanine or 4′-benzoyl-phenylalanine, etc., have been used to analyze biological structures (membranes, proteins, etc.). ►green fluorescent protein, ►luciferase

Photolithography: A modification of a more-than-a-century-old printing process. A solid plate is coated with a light-sensitive emulsion, overlaid by a photographic film, and then, illuminated. An image is formed after the plate is exposed to light. A similar principle has been adapted now to visualize DNA sequences for the purpose of large scale mapping, fingerprinting, and diagnostics. The process is also used for the synthesis of nucleotide probes. ►DNA chips, ►microarray hybridization; Barone AD et al 2001 Nucleosides Nucleotides Nucleic Acids 20 (4–7):525; review of techniques: Truskett VN, Watts MP 2006 Trends Biotechnol 24:312.

Photolyase: A flavoprotein repair enzyme (M_r 54,000) that splits cyclobutane pyrimidine dimers (Pyr < > Pyr) into monomers upon absorption of blue light. A photolyase-like 42-nucleotide deoxyribozyme is also capable of repairing thymine dimers optimally at 300 nm light (Chinnapen DJ-F, Sen D 2004 Proc Natl Acad Sci USA 101:65). In *E. coli*, two chromophores assist the process of photolyase action; 5,10-methenyltetrahydrofolate absorbs the photoreactivating light and 8-hydroxy-5-deazariboflavin, and the energy is then transferred to $FADH_2$, although the latter too absorbs some energy. The excited $FADH_2$* then transfers the energy to the dimer and while $FADH_2$ is regenerated, the dimer splits up, the recipient member of the dimer breaks down, and monomeric pyrimidines are formed. A second cofactor, 5,10-methenyl-tetrahydrofolylpolyglutamate (MTHF), may be the light harvester. It is interesting that the blue light photoreceptor cryptochromes of plants bear substantial similarities to the bacterial photolyase and its cofactors are also the same, yet the exact role of photolyases in plant DNA repair is unclear. Cyclobutane photolyase does not split the pyrimidine-pyrimidinone (6-4) photoproducts. The 6-4 photolyases are under the control of two different genes. Topical application of photolyase and light to sunburnt human skin may alleviate the symptoms by repair of the DNA damage. ►DNA repair, ►direct repair, ►photoreactivation, ►pyrimidine dimer, ►cyclobutane ring, ►cryptochrome, ►base flipping, ►pyrimidinone; Tanaka M et al 2001 Mutagenesis 16:1; Komori H et al. 2001 Proc Natl Acad Sci USA 98:13560; crystal structure: Mees A et al 2004 Science 306:1789; repair process: Kao Y-T et al 2005 Proc Natl Acad Sci USA 102:16128.

Photolysis: Degradation of chemicals or cells by light.

Photomixotrophic: An organism that can synthesize some of its organic requirements with the aid of light energy, while for others it depends on supplied organic substances.

Photomorphogenesis: Light-dependent morphogenesis. Light affects the growth and differentiation of plant meristems (photoperiodism), plastid differentiation, and directly or indirectly many processes of plant metabolism. Certain stages in photomorphogenesis can be reached at low intensity (fluence) illumination (or even in darkness), such as the formation of proplastids and etioplasts. Other steps such as the full differentiation of the thylakoid system and photosynthesis-dependent processes require high fluence rate and critical spectral regimes (red and blue). Several genes involved in the control of plastid development have been identified in *Arabidopsis* and other plants. The *lu* mutation is normal green at low light intensity but it is entirely bleached and dies at high light levels. Wild type plants can make etioplasts in the dark but the *deetiolated* (*det1*), *constitutive photomorphogenesis* (*cop1* and *cop9*) mutants develop chloroplasts in darkness. The Cop9 complex includes eight subunits, forming a signalosome in plants and a homolog is found also in animals. The *gun* (*genome uncoupled*) mutants grow normally in the dark but do not allow the development of etioplasts into chloroplasts. Various pale *hy* (*high-hypocotyl*) mutants, deficient in phytochrome, make light green plastids indicating that phytochrome is not a requisite for plastid differentiation to an advanced stage. The *blu* (*blue light uninhibited*) class of mutants is inhibited in hypocotyl elongation by far red light. The *HY4* locus of *Arabidopsis* encodes a protein, homologous to photolyases, and the recessive mutations are insensitive to blue light for hypocotyl elongation. Mutants were identified, some of which showed no response to blue light and others displayed very high blue light requirement for curvature. Most of these light responses appear to be mediated by signal transduction pathways. The chlorophyll-b free, yellow green mutants (*ch*) display chloroplast structure appearing almost normal by electronmicroscopy. Several mutations defective in fatty acid biosynthesis and/or photosynthesis are rather normal in photomorphogenesis. Some mutants are resistant to high CO_2 atmosphere, and normal chloroplast differentiation requires high CO2. Other mutants can be protected from bleaching only at 2% CO_2 atmosphere. The *Arabidopsis* nuclear mutants of the *im* (*immutans*) type display variegation under average greenhouse illumination, but they are almost normal green under low light intensity and short daily light cycles whereas at high intensity continuous illumination they are almost entirely free of leaf pigments. Under the latter condition, by continuous feeding of an inhibitor or repressor of the de novo pyrimidine pathway, the leaf pigment content may increase twenty-fold. In these variegated plants, the green cells have entirely normal chloroplasts whereas the white cells lack thylakoid structure. The azauracil-treated plants display fully functional, although morphologically altered thylakoids. An insertional mutation at the *ch-42 locus* (*cs*) identified a thylakoid protein, essential for normal greening of the plants without abolishing cell viability. The *PRF* (*pleiotropic regulatory factor*) locus, tagged by a T-DNA insertion, controls several loci involved in photomorphogenesis. The product of the gene is a subunit of the G-protein family. The *det2*, *cyp90*, *cop*, *fus*, *dim axr2*, and the *cbb* dwarf mutations develop their characteristic phenotypes because of defects in the brassinosteroid pathway. The nuclear gene *chm* (*chloroplast mutator*) induces a wide variety of plastid morphological changes, due to extranuclear mutation. ►photoperiodism, ►florigen, ►phototropism, ►phytochrome, ►circadian rhythm, ►signal transduction, ►brassinosteroids, ►COP, ►proteasome, ►dominance reversal; Wada M, Kadota A 1989 Annu Rev Plant Physiol Plant Mol Biol 40:169; von Arnim A, Deng X-W 1996 Annu Rev Plant Physiol Plant Mol Biol 47:215; Quail PH 2002 Nature Rev Mol Cell Biol 3:85.

Photon: A quantum of electromagnetic radiation, which has zero rest mass and an energy *h* times the frequency of the radiation. Photons are generated by collisions between atomic nuclei and electrons and other processes when electrically charged particles change momentum. ►measurement units

Photoperiodism: The response of some species of plants to the relative length of the daily light and dark periods.

Besides the length of these cycles, the spectral properties and the intensity of the light are also important. Responses of plants include, the onset of flowering, vegetative growth, elongation of the internodes, seed germination, leaf abscission, etc. *Short-day*, *long-day* and *day-neutral* plants are commonly distinguished on the basis of the critical daylength or, in the latter category, by the lack of it (see Fig. P68). The geographic distribution of plants is correlated with their photoperiodic response. In the near equatorial regions, short-day species predominate whereas in the regions extending toward the poles long-day plants are common. The onset of flowering of short-day plants is promoted by 15–16 h of dark periods whereas in long-day plants, the flowering is accelerated by continuous illumination or by longer light than dark daily cycles. The critical day-length is not an absolute term; it varies in different species. Usually, there is a minimum number of cycles to evoke the photoperiodic response. Mutants of *Arabidopsis* (*gi*, *co*, *ld*; Rédei GP 1962 Genetics 47:443) and others shed light on some of the basic mechanisms involved (Schultz TF, Kay SA 2003 Science 301:326). The most important photoreceptor chromoprotein is *phytochrome*. The effect of phytochrome is affected by different plant hormones. Typical long-day plants are henbane (*Hyoscyamus*), spinach, *Arabidopsis* [without a critical daylength], the majority of the grasses and cereal crops (wheat, barley, oats), lettuce, radish, etc. Typical short day plants are Biloxi soybean, cocklebur, aster, chrysanthemum, poinsettia, dahlia, etc. In the majority of species, the photoperiodic response is controlled by one or a few genes. Some processes in animals are also under photoperiodic control. In the Japanese quail, the gene encoding type 2 iodothyronine deiodinase, which catalyzes the conversion of the prohormone into the active 3,5,3′[-triiodothyronine is induced by light. The anatomical location of the response center is in the hypothalamus, while the target site is the differentiation of the gonads (Yoshimura T et al 2003 Nature [Lond] 426:178). ►phytochrome, ►florigen, ►cryptochromes, ►photomorphogenesis, ►circadian rhythm, ►phototropism, ►vernalization, ►flower evocation, ►floral induction, ►dominance reversal; Jackson SD, Prat S 1996 Plant Physiol 98:407; Amador V et al 2001 Cell 106:343; Quail PH 2002 Curr Opin Cell Biol 2002 14:180; Mockler T et al 2003 Proc Natl Acad Sci USA 100:2140S; Yanofsky MJ, Kay SA 2003 Nature Rev Mol Cell Biol 4:265; Chen M et al 2004 Annu Rev Genet 38:87.

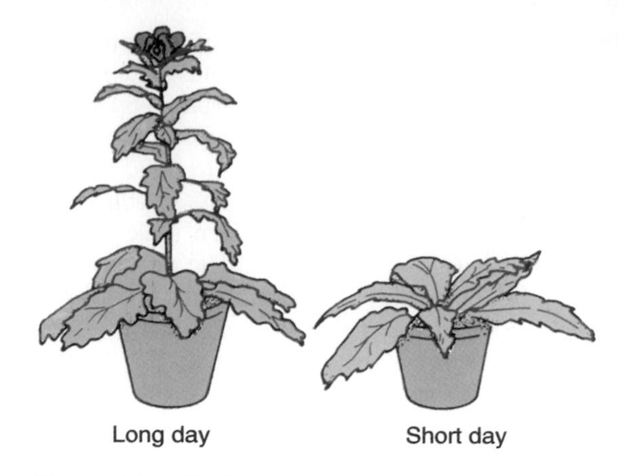

Figure P68. Henbane (*Hyoscyamus niger*) Long-day plants flower only under long daily light periods (after appropriate cold treatment). Courtesy of Professor G. Melchers

Photophosphorylation: ATP formation from ADP in photosynthetic cells.

Photoreactivation: Elimination of the harmful effects of ultraviolet irradiation by subsequent exposure to visible light (that activates enzymes splitting up the pyrimidine dimers in the DNA). With a few exceptions, e.g., *Haemophilus influenzae*, most organisms possess light-activated repair enzymes. The majority of mammals do not have efficient photoreactivation system, except the marsupials. ►light repair, ►photolyase dark repair, ►excision repair, ►glycosylases, ►error-prone repair, ►DNA repair; Kelner A 1949 J Bacteriol 48:5111; Tuteja N et al 2001 Crit Rev Biochem Mol Biol 36(4):337; Sancar GB 2000 Mutation Res 451:25.

Photoreceptors: Humans have, in the eye, the very sensitive rod cells, mediating black and white vision and the less sensitive cone cells for color vision. ►phytochrome, ►rhodopsin, ►CRX, ►metalloproteinases, ►phototropism, ►*sevenless*, ►S-cone disease; Calvert PD et al 2006 Trends Cell Biol 16:560.

Photoreduction: In photosynthetic cells, light induced reduction of an electron acceptor.

Photorespiration: Oxygen consumption in illuminated plants used primarily for the oxidation of the photosynthetic product phosphoglycolate; it also protects C3 plants from photooxidation. Step-wise nuclear transformation of *Arabidopsis* with five chloroplast-targeted bacterial genes encoding glycolate dehydrogenase, glyoxylate carboligase, and tartronic semialdehyde reductase converted chloroplast glycolate directly to glycerate. Transgenic plants grew faster, produced more shoot and root biomass, and contained more soluble sugars, reflecting reduced photorespiration and enhanced photosynthesis that correlated with an increased chloroplast CO_2 concentration (Kabeish R et al 2007 Nature Biotechnol 25:593). ►respiration, ►Calvin cycle, ►C3 plants; Wingler A et al 2000 Philos Trans R Soc Lond B Biol Sci 355:1517.

Photorhabdus luminescens: A gram-negative enterobacterium that maintains a mutualistic association with insect-feeding Heterorhabditis species of nematodes. When the nematodes invade the insects, the bacteria are released, kill the host with the help of the toxin, emit light, and make the cadaver luminescent. The toxins (tca and tcd) are potential insecticide, fungicide, and antibacterial agents, somewhat similarly to that of *Bacillus thüringiensis*. *Arabidopsis* plants, transgenic for the *TcdA* gene driven by the constitutive cassava vein mosaic virus promoter and equipped with the 5′ and 3′ untranslated sequences of the tobacco *osmotin* gene, were especially resistant to feeding insects (Liu D et al 2003 Nature Biotechnol 21:1038). The osmotin gene sequences increased the mRNA stability. The activity of the transgene was affected significantly by the position of the insertion site in the plant chromosome. Strain TT01 genome contains 5,688,987 bp and encodes presumably 4839 proteins (Duchaud E et al 2003 Nature Biotechnol 21:1307). ►*Bacillus thüringiensis*; Ehlers RU 2001 Appl Microbiol Biotechnol 56:623; Szállás E et al 1997 Int J Syst Bacteriol 47:402; Bowen D et al 1998 Science 280:2129.

Photosensitizers: Photosensitizers may increase the oxidative damage to DNA. Their action may involve initial electron or hydrogen transfer to the DNA by the excited photosensitizer, followed by the generation of free radicals. Alternatively, they generate singlet oxygen that interacts with the DNA and then produces peroxidic intermediates. Most commonly, guanine suffers lesions. ►oxidative DNA damages

Photosynthesis: Using light energy for the conversion of CO_2 into carbohydrates with the assistance of a reducing agent such as water. The photosynthetic system appears to have evolved from the core of the cyanobacterial genome (Mulkidjanian AY et al 2006 Proc Natl Acad Sci USA 103:13126). ►photosystems, ►Z scheme, ►chlorophyll binding proteins, ►thermotolerance, ►C3 plants, ►C4 plants, ►Calvin cycle; Matsuoka M et al 2001 Annu Rev Plant Physiol Mol Biol 52:297; Xiong J, Bauer CE 2002 Annu Rev Plant Biol 53:503.

Photosystems: In photosynthesis, photosystem I is excited by far red light (~700 nm) while photosystem II requires higher energy red light (~650–680 nm). In the thylakoids of the chloroplast of plants, the immunophilin FKB20-2, an FK-506 binding protein, is required for the assembly of the photosystem II complex (Lima A et al 2006 Proc Natl Acad Sci USA 103:12631). Photosynthesis in bacteria that does not evolve oxygen uses only photosystem I. Upon absorption of photons, photosystem I liberates electrons that are carried through a cascade of carriers to NADP+, which is reduced to NADPH. The departure of electrons generates a "void" in the P700 photoreaction center of photosystem I and that is filled then by electrons produced through splitting of water molecules in photosystem II. The overall reaction flow is:

$$2H_2O + 2\,NADP^+ + 8\,\text{photons} \rightarrow O_2 + 2NADPH + 2H^+$$

Mutants of *Chlamydomonas* alga lacking photosystem I survive as long as the actinic light (beyond violet) reaches 200 microeinsteins per m^2/second. The photosystem II of cyanobacteria (similar to that of plants and algae) is a complex of 20 proteins and 77 cofactors including 14 integrally bound lipids and their crystal structure has been determined at 3.0 Å resolution (Loll B et al 2005 Nature [Lond] 438:1040). Photosystem I has 17 protein subunits and the crystal structure of the supercomplex has been determined at 3.4 Å resolution (Amunts A et al 2007 Nature [Lond] 447:58). ►CAB, ►LHCP, ►antenna, ►chloroplast, ►thylakoid, ►Z scheme, ►immunophilins; Annu Rev

Genet 29:755, Guergova-Kuras M et al 2001 Proc Natl Acad Sci USA 98:4437; Jordan P et al 2001 Nature [Lond] 411:909; Chitnis PR 2001 Annu Rev Plant Physiol Plant Mol Biol 52:593; Szabó I et al 2001 J Biol Chem 276:13784; Rhe K-H 2001 Annu Rev Biophys Biomol Struct 30:307; Saenger W et al 2002 Curr Opin Struct Biol 12:244; Munekage Y et al 2004 Nature [Lond] 429:579; structure of photosystems: Nelson N, Yocum CE 2006 Annu Rev Plant Biol 57:521.

Phototaxis: A movement of organisms (plants, animals and microbes) in response to light.

Phototransduction: The transmission of light signals mediating gene expression. A scaffold protein (InaD in *Drosophila*) assembles the components of the light transduction pathway. ►signal transduction, ►rhodopsin, ►retinal dystrophy

Phototroph: An organism that uses light to generate energy and uses this energy to synthesize its nutrients from inorganic compounds.

Phototropin: Flavoprotein photoreceptors for plant phototropism. They have two flavin mononucleotide-binding domains (LOV1 and LOV2) and a serine-threonine kinase domain at the carboxyl end. Phototropins 1 and 2 are blue light-activated kinases for low and high intensity light. ►flavoprotein; Harper SM et al 2003 Science 301:1541.

Phototropism: The reaction of an organ or organism to light, involving apparently more than a single photoreceptor (see Fig. P69). In *Arabidopsis*, the phytochromes and two complementary cryptochrome mutations (*CRY1, CRY2*) have been identified. Inactivation of both is required to eliminate phototropic response. It was suggested that one of the receptors is a membrane protein with autophosphorylating ability. Additional genes (*NPH1, NPH2, NPH3, RPT, NPL1*) are required for processing the responses after perception of the signals. Phototropin (phot1) detects low fluence blue light. Phytochromes modulate phototropism by phytochrome A signaling components. Phytochrome kinase substrate proteins (Pks1, Pks2 and Pks4), in a complex with phot1 and NPH3 (non-phototropic hypocotyl), are involved in the signaling to phototropism (Lariguet P et al 2006 Proc Natl Acad Sci USA 103:10134). ►photoreceptors, ►gravitropism, ►phytochromes, ►cryptochromes, ►phototropin; Briggs WR, Liscum E 1997 Plant Cell Environ 20:768; Quail PH 2002 Curr Opin Cell Biol 2002 14:180; Chen M et al 2004 Annu Rev Genet 38:87.

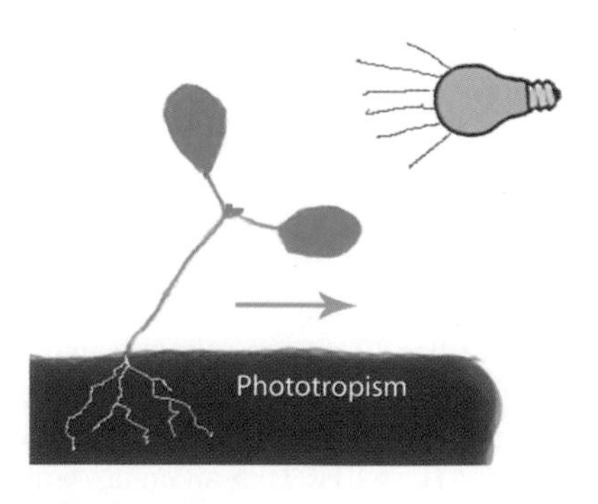

Figure P69. Phototropism

Phox: An oxidation subunit of proteins that is activated by phosphorylation. (Hoyal CR et al 2003 Proc Natl Acad Sci USA 100:5130).

Phragmoplast: A hollow-looking ring- or barrel-like structure formed near the end of mitosis in the middle plane of plant cells before the *cell plate* appears, separating the two daughter cells. ►mitosis; Gu X, Verma DP 1996 EMBO J 15:695; Zhang Z et al 2000 J Biol Chem 275:8779.

PHRAP: One of the frequently used DNA sequence alignment programs. A quality score of $10^{X/10} \approx 30$ corresponds to an accuracy of 99.9% regarding the base sequence. ►PHRED, ►CONSED, ►base-call; Harmsen D et al 2002 Nucleic Acids Res 30:416; http://www.phrap.org.

PHRED: An automated base-calling computer program. ►PHRAP, ►PolyPhred, ►base-call

Phycobilins: Highly fluorescent photoreceptor pigments in blue-green, red, and some other algae. They contain a linear tetrapyrrole prosthetic group for light harvesting. They also contain bile pigments and an apoprotein. This family of pigments includes the blue phycocyanins, the red phycoerythrins, and the pale blue allocyanins. These pigments may form phycobilisome, attached to the photosynthetic membrane. Phytochromes are also related pigments. ►light-harvesting protein, ►phytochrome; Wu SH et al 1997 J Biol Chem 272:25700.

Phycocyanin: The pigment of blue-green algae. ►phycobilins

Phycoerythrin: The red pigment of red algae. ►phycobilins, ►phycocyanin

Phycomycetes: Fungi with some algal characteristics. *Ph. blakesleeanus* is easy to grow with four-days-long asexual cycle and about two-months-long sexual cycle. It forms heterokaryons (n = 14) and can be subjected to formal genetic analyses, although the tetrads may be irregularly amplified. Transformation is feasible. It is well suited for physiological and developmental studies.

Phyletic Evolution: Gradual emergence of a species in a line of descent. The gaps in the fossil records are supposed to be due to accidents in the preservation of the intermediate forms.

Phyllody: Developmental anomaly of conversion of floral parts into leaves, generally after infection by pathogens.

Phylloquinone: Phylloquinone is composed of a p-naphthokinone and a phytol radical and it catalyzes oxydation-reduction reactions in plants. ►vitamin K

Phyllotaxy (phyllotaxis): In phyllotaxy, the consecutive leaves of plants do not occur above each other. Quite commonly, single leaves are at opposite positions (unless they occur in whorls) (see Fig. P70). This arrangement makes sense for the optimal utilization of light. In many plants, the leaves may not alternate in 180° but they may be arranged in any other determined pattern. This pattern is called phyllotaxy. If the leaves are opposite to each other, the phyllotaxy is 1/2. A common phyllotactic index is 2/5 (144°). This means that if the leaves are positioned by this index, leaf #1 will be followed by #2 at 144°, then #3 will take the place in a spiral at 288°, i.e., it will be above #1 (because 288:144 = 0.5 and 0.5 × 360 = 180), and so on. The arrangement of the fruits on the stem may also be caused by such an obliquity, following either clockwise or counterclockwise directions. The phyllotactic arrangement is determined by the flow of auxin, and it may be negatively regulated by cytokinins in the shoot meristem (Giulini A et al 2004 Nature [Lond] 430:1031). ►embryogenesis in plants, ►Fibonacci series, ►decussate; Hake S, Jackson D 1995 ASGSB Bull 8(2):29; Kuhlemeier C, Reinhardt D 2001 Trends Plant Sci 6:187; Reinhardt D et al 2003 Nature [Lond] 426:255; Jönsson H et al 2006 Proc Natl Acad Sci USA 103:1633; Smith RS et al 2006 Proc Natl Acad Sci USA 103:1301.

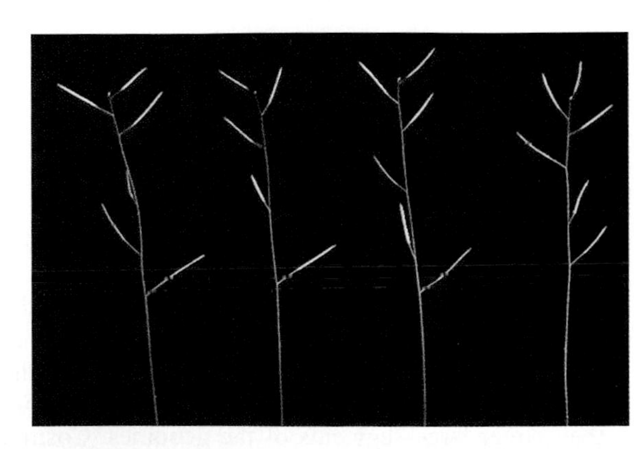

Figure P70. Phyllotaxy

Phylogenetic Analysis: Phylogenetic analysis in forensic science uses pathogen strain DNA comparisons for identifying the source of infection, e.g., the retroviral DNA in case of HIV. ►acquired immunodeficiency, ►DNA finger printing, ►forensic genetics

Phylogenetic Depth: The total number of genetic changes, which separate an organism from its ancestors.

Phylogenetic Profile Method: The phylogenetic profile method studies the correlations of inheritance of pairs of proteins among various species. These proteins are not necessarily homologous but they appear to be linked functionally. ►rosetta stone sequences; http://dip.doe-mbi.ucla.edu.

Phylogenetic Tree: The phylogenetic tree graphically represents the phylogeny of organisms. Trees have been constructed in the past on the basis of morphology, the sequences of single genes, or sequences of entire genomes. Similarity between two organisms can also be determined by dividing their total number of genes by the number of genes they have in common. Phylogenetic analysis based on molecular information greatly increases the precision of map construction. Although insights into the various genomes greatly facilitate the elucidation of phylogenetic relationships, none of the molecular methods are completely free of problems because duplications, deletions, horizontal gene transfer, and the evolution of new genes from various sequences may create problems in interpretation. ►evolutionary tree, ►BAMBE; Madsen O et al 2001 Nature [Lond] 409:610; Murphy WJ et al. 2001 Nature [Lond] 409:614; Kristian H et al 2007 Bioinformatics 23:793; http://www.treefam.org/.

Phylogenetic Weighting: As per phylogenetic weighting, DNA sequence information from various taxa is included in the phylogenetic tree in decreasing order of relationship. Thus, alignment from distant relatives should not precede alignment of closer relatives. This procedure prevents confounding similarity and descent. ►evolutionary tree, ►maximum parsimony, ►homology, ►DNA sequence alignment, ►homology; Robinson M et al 1998 Mol Biol Evol 15:1091.

Phylogenomics: Phylogenomics uses evolutionary information to infer function of genes or the reconstruction of phylogenetic history on the basis of genomes. (See Delsuc F et al 2005 Nature Rev Genet 6:361; metabolic networks from protein structure: Caetano-Anollés G et al 2007 Proc Natl Acad Sci USA 104: 9358; phylogeny of protein domains: http://www.bioinformatics.nl/tools/treedom/; Berkeley phylogenomics: http://phylogenomics.berkeley.edu; search several gene families simultaneously: http://www.cs.nuim.ie/distributed/multiphyl.php; distributed computing: http://distributed.cs.nuim.ie/multiphylOnlineManual.php; prokaryotic phylogenomics: http://genetrees.vbi.vt.edu).

Phylogeny: The evolutionary descent of a species or other taxonomic groups. ►evolution, ►ontogeny, ►speciation, ►genome conservation; Huelsenbeck JP et al 2001 Science 294:2310; information at the web site: http://beta.tolweb.org/tree/; http://mrbayes.csit.fsu.edu/.

Phylotype (phylogenetic type): A species representing a branch of a phylogenetic tree on the basis of shared similarity of nucleotide sequences. ►phylogenetic tree

Phylum: The first main category of the plant and animal, and other kingdoms.

Physarum polycephalum: A single-cell slime mold that displays physiological dioecy. The cell forms a plasmodium, i.e., the nuclei divide without cell division and thus, the cell becomes multinucleate. In the early embryos, only the S and M phases of the cell cycle are detectable.

Physcomitrella patens: A moss with a principal life phase as a haploid gametophyte. It can be used for the production of various mutants, for parasexual research, transformation, study of plant hormones on developmental processes, and various tropisms. Sequencing of the genome is nearly complete by 2005. The estimated genome size (n = 27) ~511 Mb (~0.53 pg). The chloroplast genome is 122890 bp encoding 83 proteins. (See Schaefer DG 2002 Annu Rev Plant Biol 53:477; Cove D 2005 Annu Rev Genet 39:339).

Physical Containment: ►containment

Physical Map: A map where the genome is ordered in DNA fragments or nucleotide sequences rather than in units of recombination. The first physical maps were constructed in bac-teriophages with small genomes. The DNA of phage P4 was cleaved completely by restriction endonuclease EcoRI into four fragments, which could be separated by electrophoresis according to size:

ν	ζ	ζ	ψ
A	C	B	D

After incomplete digestion for 5 min, larger fragments were also detected that contained fragments A + B + C, C + B, and the combined size of C + D appeared but no fragment appeared with the size B + D. The cause of the absence of B + D must have been that B and D were not adjacent in the circular DNA. Therefore, the sequence of the fragments in the chromosome could only have been: A − B − C − D.

The much larger polyoma genome was mapped by a different procedure. With a single EcoRI cut, the circular DNA was linearized and that cut was designated as the zero coordinate of the map. HinDIII cut the circle into two fragments: A = 55% and B = 45% (see Fig. P71). HpaII produced eight fragments: a = 27%, b = 21%, c = 17%, d = 13%, e = 8%, f = 7%, g = 5%, and h = 2% of the total genome. When EcoRI and HpaII cleaved the DNA, fragment b (21%) was not detected by electrophoresis, but instead, two new fragments of 1% and 20% were found. Obviously, the EcoRI cut was 1% from one end and 20% from the other end of fragment b. In the following step, the HindIII is shown to generate a fragment that was digested by HpaII. Fragments c, d, e, g, and h were found again (17 + 13 + 8 + 5 + 2 = 45) and two pieces of 3% and 7% were also obtained. When the HinDIII fragment of 45% length was exposed to HpaII fragment, f remained intact but two other fragments of 18% and 20% were recovered. Therefore, the fragments could be pieced together as follows:

HindIII A:	7%	-	45%	-	3%
	part of a				part of b
HindIII B:	18%	-	7%	-	20%
	part of b		f		part of a

Figure P71. Fitting the positions of hypothetical double digest fragments

Incomplete digestion of A by HpaII produced fragments: a + c, c + e, e + d, h + g, and g + b, therefore the polyoma DNA appeared as: b − f − a − c − e − d − h − g, with the zero coordinate in b and g near the 100 coordinate.

Larger genomes such as *E coli*, yeast, or those of higher eukaryotes are generally pieced together by a chromosome walking like procedure, using overlapping fragments generated by several restriction endonucleases, e.g.:

Fragments generated by enzyme A :

1	2	3
abcde	fghijklmn	oprstuvwz

Fragments generated by enzyme B :

4	5
cdefghi	jklmnoprst

will be tied into the order 1, 2, 3 on the basis of the hybridization of 4 with 1 and 2, and hybridization of 5 with 2 and 3, but not 5 with 1 or 4 with 3. In the initial steps, generally YAC clones are used because they cover large segments of the genomes. Cosmid clones usually follow this and eventually large continuities (contigs) are established without gaps. By the employment of anchors, fragments with genetically or functionally known sites, the physical map can be correlated with the genetic map determined by recombination frequencies, and thus *integrated maps* are generated. The individual fragments can then be sequenced and thus maps of ultimate physical resolution can be obtained. ►RFLP, ►chromosome walking, ►FISH, ►SAGE, ►integrated map, ►dynamic molecular combing, ►anchoring, ►contigs, ►cosmids, ►restriction enzymes, ►EcoRI, ►HindIII, ►HpaII, ►genomic screening, ►electronic PCR, ►PCR; Bhandarkar SM et al 2001 Genetics 157:1021.

Physical Mutagens: The most widely used forms of physical mutagens are *electromagnetic*, ionizing radiations such as X rays and γ rays emitted by radioisotopes. The most commonly used radiation sources for the induction of mutation by γ rays are cobalt^{60} (Co^{60}) and cesium^{137} (Cs^{137}). *Particulate radiations* such as produced by atomic fission are also ionizing. Ionization is the dislodging of orbital electrons of the atoms. The particulate (corpuscular) radiation source is uranium^{235}, which releases neutrons, uncharged particles (slightly heavier than that the hydrogen atom) with very high penetrating power and the ability to release about 15 times as much energy along their path as the hard X rays (of short wave length and high energy). The *fast neutrons* have energies between 0.5 and 2.0 MeV (million electron volt). The *thermal neutrons* have much lower level of energy (about 0.025 eV) because they have been "moderated" by carbon and hydrogen atoms. Radioactive isotopes emit also *β particles* (electrons). Their level of energy and penetrating power depend a great deal on the source; H^3 (tritium) has very short path (about 0.5 μm) and P^{32} is much more energetic (2600 μm). Beta emitters are rarely used for mutation induction. They can, however, be incorporated directly into the genetic material by using radioactively labeled precursors or building blocks of nucleic acids, and thus are capable of inducing localized damage, the degree of localization depends on the effective path length. Uranium238 emits *α particles* (helium nuclei) releasing thousands of times more energy per unit track than X rays. Because of the very low penetrating power, it can be stopped by a couple of sheets of cells in contrast to X rays and gamma rays which require heavy concrete or lead shielding. Alpha radiation, because of its high energy per short path, can very effectively destroy chromosomes. The most common genetic effect of all ionizing radiations is chromosome breakage and particularly deletions.

Another physical mutagen is *ultraviolet* (*UV*) radiation. The latter causes excitation, rather than ionization, in the biological material. Excitation may raise the orbital electrons to a higher level of energy, from which they return to the ground state very shortly. UV radiation sources are commonly mercury or cadmium lamps (black light, germicidal, and sun lamps). Natural light also includes UV radiation, especially in the clean air of the higher mountains. Near ultraviolet light, UV-B (290–400 nm) may be present in the emission of fluorescent light tubes and in the presence of sensitizers it may be genetically effective on a few layers of cells. The most common genetic effect of UV light is the production of pyrimidine dimers.

The effect of radiation on cells and organisms may be *direct*, i.e., the radiation actually hits the target molecules or it may be *indirect*, i.e., the radiation produces reactive molecules in the intra- or extra-cellular environment, and these in turn cause the genetic and/or physiological damage. Exposure to high temperature may enhance mutability. If radiation is received during DNA replication, damage is more likely than in the dormant state. Generally, hydrated cells and tissues are more sensitive to ionizing radiation than dry or non-metabolizing cells. ►X-rays, ►radioisotopes, ►radiation effects, ►ultraviolet light, ►chemical mutagens, ►maximal permissive dose, ►carcinogens, ►LET, ►chromosomal mutation, ►DNA repair, ►genetic sterilization, ►cosmic radiation, ►genomic subtraction, ►nuclear reactors, ►atomic radiations, ►electromagnetic radiation, ►pyrimidine dimer, ►cycloputane ring; Hollaender A (ed) 1954–56 Radiation Biology, McGraw-Hill, New York.

Physiology: The discipline dealing with the functions of living cells and organisms.

Phytanic Acid: A 20-carbon, branched chain fatty acid is formed from the phytol alcohol ester of chlorophylls and it degraded by β-oxidation into propionyl-, acetyl-, and isobutyryl-CoA. Deficiency of this oxidation leads to Refsum disease in humans. ►Refsum diseases, ►peroxisome

Phytic Acid (inositol hexaphosphoric acid, IP^6): Phytic acid combined with Ca^{2+} and Mg^{2+} salts are called phytins and are commonly present in plant tissues. Phytate also ties up iron in the plant tissues and limits its availability for human nutrition, unless it is degraded by phytase. ►*myo*inositol, ►phosphoinositides; engineered phytate-free seeds: Stevenson-Paulik J et al 2005 Proc Natl Acad Sci USA 102:12612.

Phytoalexins: Generally relatively low molecular weight, yet diverse, compounds synthesized through the phenylpropanoid pathway. They were attributed to defense systems against various plant pathogens. Currenty, they are considered to be mainly consequences of infection rather than active defense molecules. ►host-pathogen relation, ►phenolics; Hammerschmidt R 1999 Annu Rev Phytopath 37:285.

Phytochromes: Five regulatory proteins with alternating absorbance peaks in red and far-red light (see Fig. P72). Through their absorbance peaks (red [R] 660 nm and far-red [FR] 730), they control various photomorphogenic processes, such as short- and long-day onset of flowering, hypocotyl elongation, apical hooks, pigmentation, etc. These chromoproteins are homodimers of 124 kDa subunits and a tetrapyrrole complex, joined covalently through a cystein residue at about 1/3 distance from the NH_2 end. The molecule exists in two conformations corresponding to the R and FR absorption states. The interconversion between these states is mediated very rapidly by light of R and FR emission peaks. In etiolated plant tissue, the inactive P_r conformation may constitute up to 0.5% of the protein. The transition from the P_r conformation into the active P_{fr} form also entails the degradation of this receptor. The apoprotein, coded by different genes (*PHYA* and *PHYB*) in *Arabidopsis* may have only about 50% homology in amino acid sequences, although they bind the same chromophore. The specificity of PhyA (far-red) and PhyB (red) resides in the N-termini. The C-terminal domain of phytochrome B attenuates the transducing signals (Matsishita T et al 2003 Nature [Lond] 424:571). Phytochromes can induce and silence the expression of genes in a specific selective manner. The transcription of the phytochrome genes is also light regulated; R light reduces the transcription more effectively than FR. Phy-A perceives continuous FR, whereas phy-B responds to continuous red light.

Phytochrome B is also a photoreceptor in the circadian rhythm. Phytochrome A appears to be serine/threonine kinase. Phytochrome C is a light-stable molecule. SPA1 (suppressor of phy-A), a WD-protein with sequence similarity to protein kinases, mediates, among other factors, the photomorphogenic reactions. The phytochrome responses are under complex genetic regulatory systems involving light response elements, transcription factors, and components of the signal transduction circuits. PIF3 (phytochrome-inducing factor) is a basic helix-loop-helix protein that attaches to the non-photoactive C-terminus of phytochromes A and B and mediates their conversion into active forms. PIF3 also binds to a G-box in the promoter and thus regulates transcription. Nucleoside diphosphate kinase 2 (NDPK2) preferentially binds to the red light activated form of phytochrome and appears to play a role in eliciting light responses. In photomorphogenic responses, phytochromes interact with cryptochromes. Although phytochrome is known as a ubiquitous plant product, the yeast *Pichia* also synthesizes phytochromobilin (PΦB), a precursor of this plant chromophore. PΦB deficient plants can be complemented by the insertion of the algal phycocyanobilin gene (Kami C et al 2004 Proc Natl Acad Sci USA 101:1099). Also, a phytochrome-like protein (Ppr) has been identified in non-photosynthetic prokaryotes (*Deinococcus radiodurans, Pseudomonas aeruginosa*). In the *Rhodospirillum centenum*, a purple photosynthetic bacterium, a photoreactive yellow (PYP) pigment has been identified with a central domain resembling phytochromes.

In cyanobacteria, the circadian input kinase (CikA), a bacteriophytochrome, mediates the circadian oscillations. ►photoperiodism, ►photomorphogenesis, ►signal transduction, ►phycobilins, ►cryptochromes, ►brassinosteroids, ►WD-40, ►G box; Neff MM et al 2000 Genes Dev 14:257; Martinez-Garcia JF et al 2000 Science 288:859; Smith H 2000

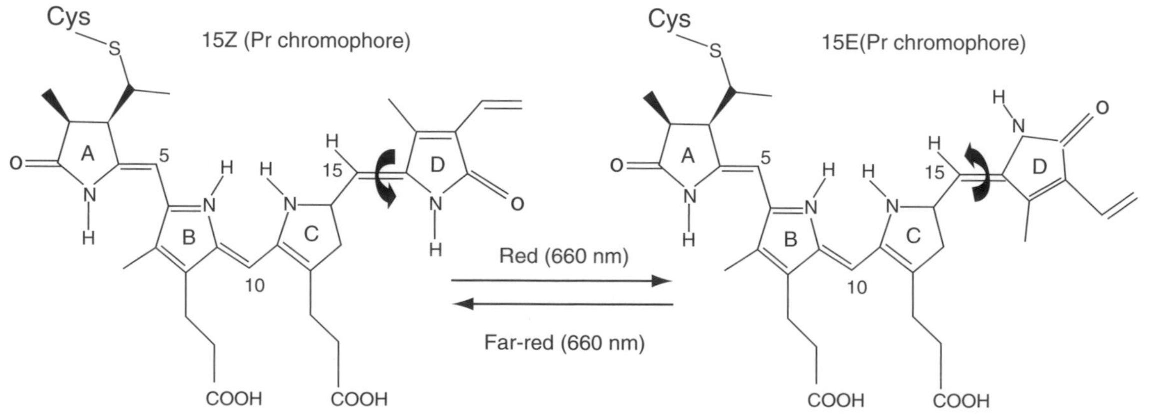

Figure P72. Phytochrome chromophores: The two isomers of phytochromobilin. (See Chen M et al 2004 Annu Rev Genet 38:87; courtesy of Dr. Meng Chen and Dr. Joanne Chory)

Nature [Lond] 407:585; Bhoo S-H. et al 2001 Nature [Lond] 414:776; Nagy F, Schäfer E 2002 Annu Rev Plant Biol 53:329.

Phytoestrogens: Estrogen-like plant products, such as the isoflavones (genistein, daidzein), and they can take advantage of the animal estrogen receptors and regulate gene expression similarly to other estrogens. Isoflavones can thus be used in hormone replacement therapies used to alleviate postmenopausal symptoms and for other purposes of selective modulation of estrogen receptors. ►estradiol, ►estrogen receptor, ►sterol, ►genistein; An J et al 2001 J Biol Chem 276:17808; Yellayi S et al 2002 Proc Natl Acad Sci USA 99:7616.

Phytoextraction: ►bioremediation

Phytohemagglutinin: ►PHA

Phytohormones: ►plant hormones

Phytophthora: A group of heterothallic plant pathogenic fungi. Each individual can produce both antheridia and oogonia. The fertilized oogonium develops oospores. The A1 mating type secretes α1 hormone (see Fig. P73), which induces oospore formation in A2 mating types, and A2 individuals secrete α2 hormone, which induces oospore formation in A1 types (See Qi J et al 2005 Science 309:1828). A draft of the genomes of *P. soyae* and *P. ramosa* is available, indicating 19,027 and 15,743 genes in the respective species and revealing evolutionary origin of related organisms (Tyler BM et al 2006 Science 313:1261). ►hormones, ►mating type, ►oöspore, ►oogonium, ►antheridium; genome: http://phytophthora.vbi.vt.edu/; functional genomics: http://www.pfgd.org/.

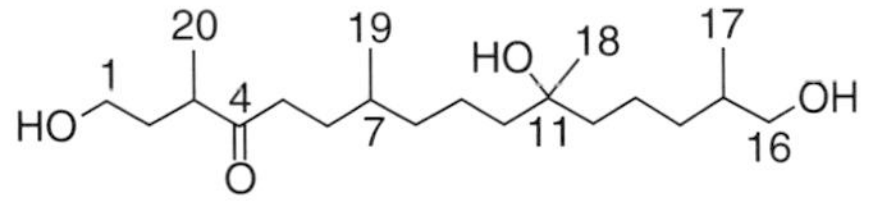

Figure P73. Alpha1 mating harmone

Phytoplankton: Aquatic, free-flowing plants. ►bacterioplankton

Phytoplasmas (Mollicutes, 530–1350 kbp circular DNA): Minute, round (200–800 μm or filamentous) bacteria without cell wall, infecting the phloem cells of plants and causing disease. The symptoms vary from yellowing to sterility, stunting, and heavy branching. Phytoplasmas resemble somewhat mycoplasmas of animals but cannot be cultured in cell-free media. They are propagated by sucking insects that cause economic loss in vegetables and trees. Phytoplasma infection may be exploited for gain by floriculture to obtain bushier Poinsettias (Lee I-M et al 1997 Nature Biotechnol 15:178). Phytoplasmas may be identified by DNA-DNA hybridization and serological means. ►mycoplasma, ►phyllody; Lee IM et al 2000 Annu Rev Microbiol 54:221.

Phytoremediation: ►bioremediation

Phytosulfokines (PSK): PSK-α, a sulfated pentapeptide, and PSK-β, a tetrapeptide, are cell proliferation promoting compounds of plants.

Phytotron: A plant growth chamber system with maximal physical regulation facilities.

Pi: Inorganic phosphate.

pI (pI_I). Isoelectric point. ►isoelectric focusing

PI 3 Kinase: ►phosphoinositide 3 kinase

PI Vector: The PI vector contains packaging site (*pac*) and allows about 115 kb to be packaged, and it infects *E. coli* at a pair of *lox P* recombination sites, at which the *Cre* recombinase circularizes DNA inside the host cell. ►vectors

Pibids: ►trichothiodystrophy

PIC (preinitiation complex): Proteins associated with RNA polymerase before transcription. During pre-initiation, the carboxyterminal domain (CTD) is hypophosphorylated but during initiation, the movement four kinases in a step-wise manner phosphorylate the RNA polymerase. Phosphorylation regulates the attachment of additional proteins. ►transcription factors, ►open promoter complex, ►TBP, ►transcript elongation, ►chromatin remodeling, ►mediator complex; He S, Weintraub SJ 1998 Mol Cell Biol 18:2876; Tsai FT, Sigler PB 2000 EMBO J 19:25; Soutoglou E, Talianidis I 2002 Science 295:1901; Wilcox CB et al 2004 Genetics 167:93; Chen H-T, Hahn S 2004 Cell 119:169.

PIC: ►polymorphic information content

PIC (SUMO): A ubiquitin-like protein associated with RanGAP. ►ubiquitin, ►UBL, ►sentrin, ►RanGAP, ►SUMO

Pick Disease (FTDP-17, frontotemporal dementia and parkinsonism): A chromosome 17q21.11 dominant behavioral, cognitive, and motor disease involving variable loss and atrophy of the frontal and temporal part of the brain, caused by defects in the splicing of the Tau microtubule-associated protein. The mutations responsible for the conditions occur in exon 10 of Tau or in its 5′-splicing site, resulting in duplications in Tau mRNA (14q24.3). Frontotemporal dementia (FTD) may be tau-negative in case of mutation/loss of progranulin, a 68.5 kDa regulatory protein encoded at 17q21.31 (Baker M et al 2006 Nature [Lond] 442:916; Cruts M et al 2006 Nature

[Lond] 442:920). ►dementia, ►Parkinsonism, ►tau, ►RNAi

Picornaviruses: The single-stranded RNA genomes of picornaviruses, measuring about 7.2 to 8.4 kb (ca. 2.5 to 2.9 × 10^6 Da), are transcribed into four major polypeptides. Their RNA transcript lacks the 5′ cap in the mRNA, characteristic for other eukaryotic viruses. A functional picornavirus IRES in a dicistronic mRNA may support the activity, not only of the downstream, but also of the upstream reporter gene at high salt concentrations in *cis*. Analysis of different experimental parameters influencing this effect shows that the enhanced availability of the initiation factor eIF4F provided by a functional picornavirus IRES on the same RNA molecule in *cis* causes this translation enhancement effect (Jünemann C et al 2007 J Biol Chem 282:132). They include *enteroviruses* (a group of mostly unsymptomatic intestinal viruses. The paralytic *poliovirus* may also belong to this group). *Cardioviruses* (responsible for myocarditis [causing inflammation of the heart muscles] and encephalomyelitis [inflammation of the brain and heart]), *rhinoviruses* (in over 100 variants responsible for the common cold and other respiratory problems in humans and animals), and *aphtoviruses* (causing foot-and-mouth disease in cattle, sheep and pigs and occasionally infecting also people) are other types of picornaviruses. The *hepatitis virus* may also be classified among the picornaviruses. ►papovaviruses, ►animal viruses, ►coxsackie virus, ►polio virus, ►IRES, ►eIF-4F; Knipe DM et al (Eds.) 2001 Fundamentals of Virology, Lippincott Williams & Wilkins, Philadelphia, Pennsylvania.

PIDD: A p53-inducible death domain protein, which promotes apoptosis. ►death domain, ►apoptosis, ►p53

PIE: Polyadenylation inhibition element. ►polyadenylation signal

Piebaldism: Piebaldism in animals is the result of hypomelanosis (low melanin), it is generally restricted to spots on the body; white spots occur on a black background. It may be a mutation of the KIT oncogene (4q12), or may be due to other factors. ►albinism, ►nevus, ►vitilego, ►melanin, ►Himalayan rabbit, ►mouse, ►pigmentation in animals, ►KIT oncogene, ►spotting, ►Hirschsprung disease, see Fig. P74.

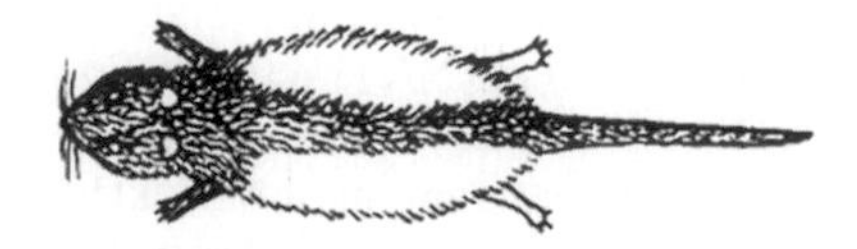

Figure P74. Piebald rat

Pierre-Robin Syndrome: An autosomal recessive defect, involving the tongue (glossoptosis), small jaws (micrognathia), and sometimes cleft palate and syndactyly of toes. In an autosomal dominant form, reduced digit number (oligodactyly) is also found. There is also an X-linked form involving clubfoot and heart defect. Another X-linked form shows and increase in the number of the bones in the digits (hyperphalangy). (See terms under separate entries).

Piezoelectric Mechanism: By piezoelectric mechanism, crystalline material, under pressure, may generate electricity. Also, expansion and contraction may take place in matter in response to alternative electric current mechanical stress. This latter property has been exploited for insertion of cell nuclei into eggs after the destruction of its original egg nucleus. This type of nuclear transplantation may help achieve cloning of higher animals. ►nuclear transplantation

PIF: Proteolysis inducing factor.

PIG (*Sus crofa*): 2n = 38. The domesticated breeds are the descendants of the crosses between the European wild boar and the Chinese pigs and they can still interbreed with the wild forms of similar chromosome number. The wild European pig is 2n = 36. The Caribbean pig-like peccaries (*Tayassuidae*) are 2n = 30. There are about 300 breeds of the domesticated pig. The various breeds of minipigs weigh generally less than 50 pounds as adults and are used for biomedical research. Sexual maturity sets in by about five to six months and the gestation period is about 114 days. It is a multiparous species with a litter size of 4–12. By adult somatic cell nuclear transplantation, live clones can be produced. ►animal genetics; Polejaeva IA et al 2000 Nature [Lond] 407:86; dispersal in Southeast Asia: Larson G et al 2007 Proc Natl Acad Sci USA 104: 4834; nuclear transplantation: http://www.toulouse.inra.fr/lgc/pig/hybrid.htm; http://www.animalgenome.org/QTLdb/; http://www.piggenome.org/; http://ascswine.rnet.missouri.edu/Description.html; http://www.piggis.org/; http://pig.genomics.org.cn/.

Pigeon: *Columbia livia*, 2n = 80. Great morphological variations among the various breeds of pigeons had already caught Darwin's attention, who made a few crosses between "pure races" and observed some "mendelian" patterns (see Fig. P75). Homing pigeons follow important landmarks such as railway tracks and highways as guided by their learned memory (Lipp H-P et al 2004 Curr Biol 14:1239). The "homing" ability, i.e., pigeons can return from great distances, is probably based on magnetoreception of the earth magnetic field facilitated by the upper beak area and may be aided also by olfactory nerves (Mora CV et al 2004 Nature [Lond] 432:508).

Figure P75. Variations in pigeons

PiggyBac: A cabbage moth (*Trichoplusia ni*) transposon-derived transformation vector of several different insect species. It is 2.5 kb with 13 bp inverted terminal repeats and contains a 2.1 kb open reading frame. Its specific target is TTAA. PiggyBac is particularly useful for large-scale and general disruption of *Drosophila* genes (Thibault S et al 2004 Nature Genet 36:283). PiggyBac efficiently transposes also in human and mouse cells (Ding S et al 2005 Cell 122:473). The high transposition activity of *piggyBac* and the flexibility for molecular modification of its transposase suggest the possibility of using it routinely for mammalian transgenesis (Wu SC-Y et al 2006 Proc Natl Acad Sci USA 103:15008). Frequently green fluorescent protein marker is used for its easy detection. ►transposon, ►open reading frame, ►transposon vector, ►sleeping beauty, ►GFP; Handler AM et al 1998 Proc Natl Acad Sci USA 95:7520; Horn C et al 2003 Genetics 163:647; inducible piggyBac: Cadiñanos J, Bradley A 2007 Nucleic Acids Res 35(12):e87.

Pigment Epithelium-Derived Factor (PEDF): A potent inhibitor of angiogenesis of the retina. Its defect leads to opacity of vision and blindness. ►angiostatin, ►endostatin, ►thrombospondin, ►angiogenesis

Pigmentation Defects: ►albinism, ►piebaldism, ►hypomelanosis, ►incontinentia pigmenti, ►pigmentation in animals, ►LEOPARD syndrome, ►Fanconi anemia, ►hematochromatosis, ►neurofibromatosis, ►tuberous sclerosis, ►Waardenburg syndrome, ►Hermansky-Pudlak syndrome, ►polyposis hamartomatous, ►Addison disease, ►focal dermal hypoplasia, ►erythermalgia, ►skin diseases

Pigmentation of Animals: In mammals, tyrosine is the primary precursor of the complex black pigment melanin. The enzyme tyrosinase (located in the melanosomes) hastens the oxidation of dihydroxyphenylalanine (DOPA) into dopaquinone, which is changed by non-enzymatic process into leukodopachrome. Leukodopachrome is an indole-derivative that is oxidized also by tyrosinase into an intermediate of 5,6-dihydroxyindole. After another step of oxidation, indole-5,6-quinone is formed. Coupling the latter to 5,6-dihydroxyindole is the first step in the addition of further dihydroxyindole units in the process of polymerization to melanin. When cysteine is combined with dopaquinone, through a series of steps, reddish pigments are formed in hair and feathers. The different pigments may have also other adducts at one or more positions to yield various colors. In the formation of the eye color of insects, tryptophan is a precursor to the formation of formylkynurenine → kynurenine → hydroxykynurenine → ommin, ommatin. The catabolic pathway of amino acids contributes to the formation of guanine and through the latter to pteridines that contribute to the coloration of insects, amphibians, and fishes, and serves also as a light receptor. The Xanthopterin and leucopterin account for the yellow and white pigmentation of butterflies, sepiapterin, is found in the eyes of *Drosophila* and biopterin is found in the urine and liver of mammals. The degradation of the heme group yields a linear tetrapyrrole from which the bile pigment biliverdin and ultimately bilirubin diglucuronide is synthesized. Bilirubin diglucuronide is secreted into the intestines and may accumulate in the eyes and other organs causing jaundice when the liver does not function normally. Oxidized derivatives of bilirubin, urobilin, and stercobilin color the urine. Mutations were detected already during the early years of genetics that block the biosynthetic paths of these pigments and thus contribute to understanding how genes affect the phenotype. The color of the skin in humans is determined by its melanin content. Phaeomelanin is a reddish pigment and eumelanin is black. The former is responsible for the light skin and red hair color and it also potentially generates free radicals and thus may make the individual susceptible to UV damage. Eumelanin provides protection against UV. The melanocyte-stimulating hormone (MSH) and its receptor (MC1R) regulate the relative proportion of these two melanins. A putative cation exchanger (SLC24A5, human chromosome 15q21) has modulatory effect on the formation of melanosomes and due to different single nucleotide polymorphisms has impact on the determination of pigmentation, depending also on the climatic regions and exposure to sunlight (Lamson RL et al 2005 Science 310:1782). In mice, about 100 genes are known that control pigmentation. Differences in the pigmentation of the human skin in various geographic areas of the world seem to be correlated with the degree of exposure to ultraviolet radiation. ►chorismate, ►tryptophan, ►tyrosine, ►phenylalanine, ►albinism, ►melanin, ►eye color in humans, ►Himalayan rabbit, ►Siamese cat, ►pigmentation in plants, ►agouti, ►melanocyte-stimulating hormone, ►hair color, ►tanning, ►opiocortin; hair and skin color: Rees JL 2003 Annu Rev Genet 37:67; Price T Borntrager A 2001 Curr Biol 11:R405; evolution, genetics, physiology [folate, vitamin D, UV light exposure] and variation in human skin color: Jablonski NG 2004 Annu Rev Anthropol 33:585; Sturm RA 2006 Trends Genet 22:464; Lin JY, Fisher DE 2007 Nature [Lond] 445:843.

pIgR (polymeric immunoglobulin receptor): ►antibody polymers

P_{II}: P_{II} proteins (are involved in bacterial glutamine synthesis) accelerate hydrolysis of NtrC in the presence of NtrB and ATP in limited N supply and low levels of 2-ketoglutarate. P_{II} uridylylation permits the increase of NtrC-phosphate level and increases transcription from the glnAp2 promoter. In excess N supply, PII is not altered resulting in no NtrC build-up and glnA2 activation ceases. ►*NtrB*, ►*NtrC*, ►*glnAp*

PI3K (PI(3)K): ►PIK

PIK/PI(3)K (phosphatidylinositol kinases): PI(3)K preferentially phosphorylates the 3 and 4 positions on the inositol ring. PIK-catalyzed reaction products (PtdIins) are second messengers. They participate in meiotic recombination, immunoglobulin V(D)J switches, chromosome maintenance and repair, progression of the cell cycle, etc. The mouse Pik3r1 regulatory gene encodes proteins p85α, p55α, and p50α. p55/p50 are essential for viability. Their defect may lead to immunological disorders and cancer. In ovarian cancer, increase of PIK3CA and increased PIK activity were detected. In different types of human cancers, mutations in the catalytic subunit is high (Samuels Y et al 2004 Science 304:554). PIK inactivation of its γ-subunit may lead to invasive colorectal cancer in mice. PI3Kγ may signal to phosphokinase B or to MAPK. PI3K is negatively controlled by PTEN. PIK related kinases are TOR, FRAP, TEL, MEI, and DNA-PK. Their inhibitor is wortmannin. ATM, ATR, DNA-P-related protein kinases, ATRIP, and Ku80 share a terminal amino acid sequence motif (734 AKEESLADDLFRYN-PYLKRRR) of the Nijmegen breakage syndrome (Nbs1) protein that recruits these kinases to the site of DNA damage and cell cycle checkpoint control and repair (Falck J et al 2005 Nature [Lond] 434:605). The nuclear GTPase PIKE enhances PIK activity and is regulated by protein 4.1N. ►phosphatidylinositol, ►second messenger, ►immunoglobulins, ►DNA repair, ►ATM, ►ATR, ►DNA-PK, ►Ku, ►ATRIP, ►Nijmegen breakage syndrome, ►cell cycle, ►wortmannin, ►MEC1, ►phosphoinositides, ►PTEN, ►protein 4.1N, ►colorectal cancer, ►chemotaxis, ►Langerhans islets; Kuruvilla FG, Schreiber SL 1999 Chem Biol 6:R129; Katso R et al 2001 Annu Rev Cell Dev Biol 17:615.

Pileus: The umbrella-shaped fleshy mushroom fruiting body. Also, it is a membrane that may be present on the head of newborns.

Pilin: The protein material of the pilus. ►pilus

Pilomatricoma: Usually benign, calcifying skin tumors, densely packed by basophilic cells and developing into hair follicle-like structures. Their origin is attributed to mutation in LEF/β-catenin. ►LEF, ►catenins, ►basophil, ►follicle

PILRα: Inhibitory receptor of myeloid cell encoded at human chromosome 7q22. ►ITIM

Pilus: A bacterial appendage, which may be converted into a conjugation tube through which the entire or part of the replicated chromosome is transferred from a donor to a recipient cell (see Fig. P76). It may also serve as protein conduit. In pathogenic enterobacteria (*Neisseria gonorrhoea*, *Vibrio cholerae*) and in some types of *E. coli*, the so-called pilus type IV may be formed. It facilitates bacterial aggregation (bundle-forming pilus, BFP, encoded by a 14-gene operon), and the expression of the LEE (enterocyte effacement) element enhances the association of the bacteria with the mucous intestinal membranes and triggers diarrhea. The pilin protein may undergo antigenic variation to escape host defenses. The pilus may also form an attachment to the invaded eukaryotic cell. ►conjugation, ►conjugation mapping, ►PapD, ►pilin, ►antigenic variation, ►mating bacterial, ►shoufflon, ►pseudopilus [ψ-pilus]; Jin Q, He S-Y 2001 Science 294:2556.

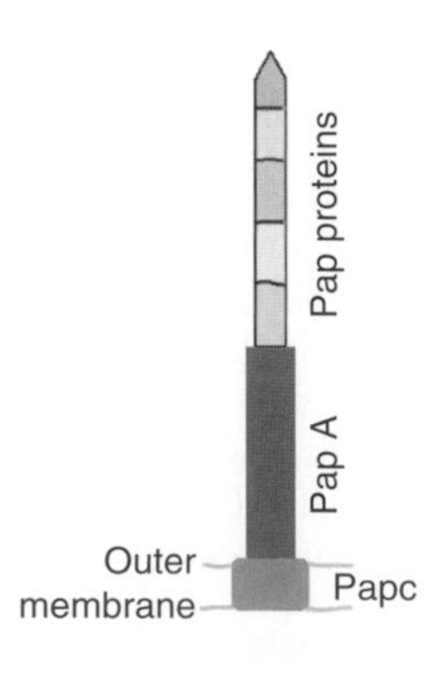

Figure P76. Pilus

PIM Oncogene: The PIM oncogene is located in human chromosome 6p21-p12 and in mouse chromosome 17. The gene is highly expressed in blood-forming (hematopoietic) cells and myeloid cells and overexpressed in myeloid malignancies and some leukemias. The human protein is a serine/threonine kinase. ►oncogenes, ►serine/threonine kinases

Pimento (*Pimento dioica*): Also called allspice. Tropical dioecious spice tree; 2n = 2x = 22.

PIN1: A peptidyl-prolyl cis/trans isomerase in human cells. It is important for protein folding assembly and/or transport. Its deficiency leads to mitotic arrest, while its overproduction may block the cell cycle in G2 phase. It interacts with NIMA kinase. PIN1 membrane protein of plants regulates auxin transport. ►cell cycle, ►NIMA, ►parvulin

PIN⁺: The prion form of the yeast protein Rnq1. ►prion; Bradley ME, Liebman SW 2003 Genetics 165:1675.

pIN: The promoter of the transposase gene of a transposon. There are two GATC sites involved in *dam* methylation within pIN. ►RNA-IN, ►*dam*

Pinch: A group of proteins with LIM and additional domain(s). ►CRP, ►LMO, ►LIM domain

Pineal Gland: The site of melatonin synthesis and photoreception in the brain. ►melatonin, ►opsins, ►Rabson-Mendenhall syndrome, ►brain

Pineapple (*Ananas comosus*): A monocotyledonous tropical or subtropical plant (2n = 50, 75, 100). The flowers and bracts sit on a central axis and form fleshy fruits. The lack of seeds is caused by self-incompatibility of the commercial varieties but they develop seeds if allowed to cross-pollinate with other varieties. ►seedless fruits

Pines (*Pinus* spp): Trees, all 94 species are 2n = 2x = 24. ►spruce

Ping-Pong Kinetics: The property of some dimeric or multimeric enzymes catalyzing two "half reactions." First, they release the first product and form an enzyme intermediate before binding of the second substrate. After the second reaction and release of the product, the enzyme returns to the initial state. (See Frank RAW et al 2004 Science 306:872).

Pinna: The ear lobe, the lobe of a compound leaf or frond. ►hairy ear

Pinning: Pinning uses a floating replication tool with about 100 or more pinheads to test yeast colonies on different culture media. ►replica plating

Pinocytosis: The formation of ingestion vesicles for fluids and solutes by the invagination of membranes of eukaryotic cells. ►phagocytosis, ►endocytosis

Pinosome: A small cytoplasmic vesicle originating by invagination of the cell membrane. ►endocytosis

PinPoint Assay: The PinPoint assay identifies single nucleotide polymorphism (SNIP). The polymorphic DNA site is extended by a single nucleotide with the aid of a primer annealed immediately upstream to the site. The extension products are analyzed by MALDI-TOF mass spectrophotometry. ►SNIP, ►primer extension, ►MALDI-TOF; Haff LA, Smirnov IP 1997 Genome Res 4:378.

PIN*POINT (protein position identification with a nuclease tail): An in vivo method to ascertain the position of the critical promoter-binding proteins involved in the LCR. Fusion proteins with an unspecific nuclease tail are studied for how the cleavage position affects the expression of the gene(s). ►LCR; Lee J-S et al. 1998 Proc Natl Acad USA 95:969.

PIP: Phosphatidylinositol phosphate. ►phosphoinositides, ►PIP2[PIP_2], ►PIP3

PIP2: Phosphatidylinositol (4,5)-bisphosphate is involved (with PIP3) in mediating the inositol phospholipid signaling pathway and in the activation of phospholipase C (PLC). PIP_2 also controls the ATP-regulated potassium ion channel (K_{ATP}) by binding to the intracellular C-domain of the channel protein and interfering with the binding of ATP. Since K_{ATP} channels affect pancreatic β cells and vascular and cardiac muscle tone, they may have relevance for human diseases, e.g., diabetes. Pleckstrin homology domains selectively bind phosphoinositides. ►phosphoinositides, ►PITP, ►ion channels, ►diabetes, ►InsP, ►pleckstrin, ►TIRAP; Martin TF 2001 Curr Opin Cell Biol 13:493.

PIP3 (phosphoinositol-3,4,5-trisphosphate): An intracellular messenger and stimulator of insulin, epidermal growth factor, etc., that works by adding another phosphate to PIP2 and activating PKB. ►PKB, ►PTEN, ►chemotaxis, ►InsP; Hinchliffe KA 2001 Curr Biol 11:R371.

Pipecolic Acid (homoproline): An intermediate in lysine catabolism (see Fig. P77). Increase of pipecolic acid (hyperpipecolathemia/hyperpipcolicacidemia) in the blood plasma and urine leads to increase in the size of the liver (hepatomegaly), resulting in growth retardation, vision defects, and demyelination of the nervous system.

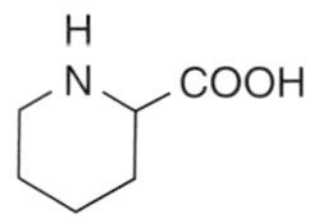

Figure P77. Pipecolic acid

Pipes (piperazine-*N,N'*-bis(2-ethanesulfonic acid): A buffer within the pH range of 6.2–7.3.

PIR (protein information resource): ►MIPS; http://pir.georgetown.edu/.

PIR-A, PIR-B: Immunoglobulin-like regulatory molecules (activator/inhibitor) on murine B cells, dendritic cells, and myeloid cells. A single gene encodes Pir-B whereas a multigene family encodes the six Pir-A proteins. ►ITIM; Dennis G Jr et al 1999 J Immunol 163:6371.

Piriformospora indica: Root endophytic fungus that may associate with both dicotyledonous and monocotyledonous plants and convey resistance to fungal disease, salt tolerance, improved nitrogen metabolism, and lead consequently to higher yield. ►symbiont, ►host–pathogen relationship, ►salt-tolerance; Waller F et al 2005 Proc Natl Acad Sci USA 102:13386.

piRNA (Piwi interacting RNA): 26-31-nucleotide-long RNA regulating germ and stem cell development when bound to Argonaute family proteins (Aubergine, Piwi, Ago3). In mouse, the MIWI/Piwi RNA associates with the polysomes and chromatoid body during spermatogenesis (Grivna ST et al 2006 Proc Natl Acad Sci USA 103: 13415). It is involved also in regulating transposons activity (Brennecke J et al 2007 Cell 128:1089). ►RNAi, ►Argonaute, ►chromatoid body, ►Slicer; Aravin A et al 2006 Nature [Lond] 442:203; Girard A et al 2006 Nature [Lond] 442:199; review: O'Donell KA, Boeke JD 2007 Cell 129:37.

PISA (protein in situ assay): In the PISA assay, PCR-generated DNA fragments are transcribed and translated in a cell-free protein expression system on a coated microtiter plate where the protein was immobilized. Single chain antibody fragments and luciferase have been successfully arrayed. ►PCR; He M, Taussig MJ 2001 Nucleic Acids Res 29(15):E73.

Pistil: A central structure of flowers (gynecium) consisting of the stigma, style, and ovary. ►gametophyte female, ►gametophyte male, ►flower differentiation

Pistillate: Flower or plants that carries the female sexual organs. A female parent in plants.

Pisum sativum (pea): A legume (2n = 14). It played an important role in establishing the Mendelian principles of heredity and contributed further information on genetics. Curiously, the famous "wrinkled" gene of Mendel turned out to be an insertional mutation. ►pea

Pit: An indentation. Also, the stony endocarp of some fruits, e.g., plums, apricot, cherry.

Pitalre (cdk9): ►acquired immunodeficiency, ►TEFb; Darbinian N et al 2001 J Neuroimmunol 121:3

Pitch: The length of a complete turn of a spiral (helix) and the translation per residue is the pitch divided by the number of the residues per turn. In a keratin alpha helix, it is 0.54 nm/3.6 = 0.15 nm. Also, a dark black residue after distillation. The auditory pitch is the physiological response of the ear to sound depending on the frequency of vibration of the air. Pitch-selective neurons are located in the auditory cortex of the brain in monkeys and humans (Bendor D, Wang X 2005 Nature [Lond] 436:1161). Perfect/absolute pitch is ability for recognizing musical notes by talented artists. ►musical talent, ►prosody

Pith: The parenchyma tissue in the core of plant stems, e.g., in elderberry (*Sambucus*).

Pithecia (saki monkey): ►Cebidae

PITP (phosphatidylinositol transfer proteins, 35 and 36 kDa): PITP is required by for the hydrolysis of PIP_2 (phosphatidyl-inositol bis-phosphate) by PLC (phospholipase C). In a GTP-dependent signal pathway, PITP is required also by epidermal growth factor (EGF) signaling. ►PIP, ►PIP_2, ►EGF, ►GTP, ►phosphoinositides; Cockcroft S 1999 Chem Phys Lipids 98:23.

PI-TR: Phosphatidylinositol transfer protein involved in transfer of lipids among organelles within cells.

PITSLRE: Members a cyclin-dependent protein kinase family involved in RNA transcription or processing. They are associated with ELL2, TFIIF, TFIIS, and FACT. ►ELL, ►transcription factors, ►TFIIS, ►protein 14-3-3; Trembley JH et al 2002 J Biol Chem 277:2589.

Pituitary (hypophysis): The hypophysis is located at the base of the brain and is connected also to the hypothalmus (a ventrical part of the brain). The anterior part secretes the pituitary hormones and the posterior part stores and releases them. ►brain human, ►gonads, ►septo-optic dysplasia; Fauquier T et al 2001 Proc Natl Acad Sci USA 98:8891; Scully KM, Rosenfeld MG 2002 Science 295:2231.

Pituitary Dwarfism: Pituitary dwarfism is due to recessive mutation, deletion, or unequal crossing over in the gene cluster containing somatotropin and homologs in human chromosome 17q22-q24. Administration of somatotropin may restore growth. The defect may also be in the hormone receptor (human chromosome 5p13.1-p12, mouse chromosome 15) and in these cases, the growth hormone level may be high (Laron types of dwarfisms). The level of somatomedin (insulin-like growth factors) may also be low. Somatomedin is a peptide facilitating the binding of proteins and in addition shows insulin-like activity. In either case, dwarfism may result. Dominant-negative mutations in IGHD2 (isolated growth hormone deficiency) are also known. ►dwarfism, ►GH, ►insulin-like growth factor, ►hormone receptor, ►binding protein, ►stature in humans, ►pituitary gland, ►growth hormone pituitary; Machinis K et al 2001 Am J Hum Genet 69:961.

Pituitary Hormone Deficiency, Combined Familial: The pituitary hormone deficiency fails to produce normally one or more of these hormones—growth hormone (HGH), prolactin, and thyroid-stimulating hormone (TSH)—because of mutation in the POU1F1 gene (3p11). Mutation in the PROP1 gene (5q), however, cannot produce luteinizing hormone (LH) and follicular stimulating hormone (FSH). Corticotropin deficiency is caused by mutation in the LHX4 gene. LHX3 is a homeobox gene with LIM repeats. ►animal hormones

Pituitary Tumor (GNAS1, 20q13.2): Pituitary tumor is caused by autosomal dominant mutations in the α chain of a G-protein (G_s). This protein is also called gsp (growth hormone secreting protein) oncoprotein. The human securin, mediating sister chromatid cohesion, has substantial sequence homology with the pituitary tumor-transforming gene. Securin may block sister chromatid separation and thereby can be responsible for chromosome loss or gain, common characteristics of tumors. ►G-protein, ►McCune-Albright syndrome, ►sister chromatin cohesion

Piwi: ►piRNA

Pixel: A picture element in the computer that represents a bit on the monitor screen or in the video memory. ►bit, ►byte

pK$_a$: The negative logarithm of the dissociation constant K_a; stronger acids have higher pK_a whereas weaker acids have lower. The dissociation of weaker acids is higher and that of stronger acids is lower. (See http://www.jenner.ac.uk/PPD/).

PKA: Protein kinase A (activated by cAMP). There are two types, PKA-I and PKA-II; they share a common catalytic subunit (C) but distinct regulatory subunits, RI and RII. RI/PKA-I controls positively cell proliferation and neoplastic growth. RII/PKA-II controls growth inhibition, differentiation, and cell maturation. RI is detectable in many types of cancers. Antisense methylphosphonate RNA of the RI_α subunit has been known to arrest proliferation of cancer cells without toxicity to normal cells. ►protein kinases, ►antisense technologies, ►cocaine, ►export adaptors

PKB (protein kinase B): A serine/threonine kinase, the same as Rac or Akt. It is activated by phosphatidylinositol-3,4,5-trisphosphate by binding to its pleckstrin homology domain. ►CaM-KK, ►protein kinases, ►phosphoinositides, ►pleckstrin domain

PKC: Protein kinase C. ►protein kinases

PKD: ►polycystic kidney disease

PKI (protein kinase I): A small protein, which attaches to the catalytic subunits of the heterotetrameric PKA and, with the aid of its nuclear localization sequence (NES), sends the complex to the nucleus. ►export adaptors, ►PKA, ►nuclear localization sequence

PKR: A double-stranded RNA-dependent serine-threonine protein kinase, involved in NF-κB signaling. One of the most important targets of PKR is the eIF-2A translation factor and thus, protein synthesis. It may control cell division, apoptosis, and may serve as tumor suppressor. Translation is required for viral infection of mammalian cells. Viral infection may trigger the activation PKR as a defense against infection through shutting off protein synthesis. PKR inhibits protein synthesis by autophosphorylation and phosphorylation of the Ser51 residue of eIF2α. Mutation at the Thr446 site prevents autophosphorylation at the catalytic domain activation segment and impairs phosphorylation of eIF2α and viral binding (Dey M et al 2005 Cell 122:901). The PKR active cell may succumb to apoptosis but the animal may survive. Several viruses (adenovirus, vaccinia virus, HIV-1, hepatitis C, poliovirus, SV40, etc.) use various mechanisms to inhibit activation of PKR by interfering either with its dimerization or RNA binding, or regulation of eIF-2A, etc. PKR preferentially binds mutant huntingtin protein in Huntigton disease. ►NFκB, ►oncolytic virus, ►reovirus, ►eIF-2A, ►PERK, ►interferon, ►apoptosis, ►Huntington's chorea; Kaufman RJ 1999 Proc Natl Acad Sci USA 96:11693.

PKS Oncogenes: PKS oncogenes are located in human chromosomes Xp11.4 and 7p11-q11.2. These genes display very high homology to oncogene RAF1 and apparently encode protein serine/threonine kinases. ►*raf*, ►oncogenes

PKU: ►phenylketonuria

PLAC: Plant artificial chromosome. ►artificial chromosome, ►YAC

Place Cells: Place cells in the brain are the locations for the firing of specific nerve cells.

Placebo: A presumably inactive but similar substance used in parallel to different individuals in order to serve as a concurrent (unnamed) control for testing the effect of a drug. In some instances the placebo has positive effects not by physical or chemical properties but by expectation-caused dopamine release, e.g., in Parkinson disease. ►concurrent control, ►double-blind test; de la Fuente-Fernández R et al 2001 Science 293:1164; Ramsay DS, Woods SC 2001 Science 294:785.

Placenta: The maternal tissue that is in most intimate contact with the fetus through the umbilical chord, found within the uterus of animals. Most commonly, the placenta is located on the side of the uterus; the placenta praevia is situated at the lower part of the uterus. The latter situation may be correlated with the age of the mother. Also, placenta refers to the wall of the plant ovary to which the ovules are attached. During pregnancy, the placenta of eutherian mammals includes both maternal and zygotic tissues in close association (feto-maternal interface). The interaction between these two types of tissues is essential for normal embryo development and viability of the conceptus. In normal pregnancy, the uterus is invaded by the cytotrophoblasts (the nutritive cells of the conceptus) but defects in the cell adhesion system may adversely affect the pregnancy and may lead to ecclampsia. In embryonic tissues of mouse, after 10.5 day hematopoietic stem cells develop to an extent comparable to the aorta-gonad-mesonephros (AGM) region (Gekkas C et al 2005 Dev Cell 8:365). The mesonephros is part of the embryonic kidney tissue. ►ecclampsia, ►imprinting, ►incompatibility; Zhou Y et al 1993 J Clin Invest 91:950; Georgiades P et al 2001 Proc Natl Acad Sci USA 98:4522; placenta of animal species: http://medicine.ucsd.edu/cpa/.

Placode: A heavy embryonal plate of the ectoderm from what organs may develop. ►ecto-derm, ►AER, ►ZPA, ►organizer, ►neural crest, ►germ-layer

PLADs (pro-ligand-binding assembly domains): aggregate the (death) receptors before binding the ligands. ►death receptors

Plagiary: ►publication ethics, ►ethics

Plague: The term plague has been used to loosely define widespread, devastating diseases. Strictly, the term applies today to infection by the *Pasteurella pestis* (*Yersinia pestis*) bacterium. The disease may occur in three main forms: *bubonic* plague (most important diagnostic features of is swelling lymph nodes, particularly in the groin area), *pneumonic* plague (attacking the respiratory system) and *septicemic* plague (causing general blood poisoning). Many of its symptoms overlap with those of other infectious diseases. A 1°C increase in spring temperature and wetter summers may increase the carrier gerbil (rodent) population and can result in >50% increase of the prevalence of plague in Central Asia (Stenseth NC 13110). It used to be known also as the "black death" on account of the dark spots, appearing in largely symmetrical necrotic tissue with coagulated blood. The bacilli spreads to human populations from rodents by fleas, but infections occur also through cough drops of persons afflicted by pneumonic plague. Various animal diseases are also called plague (pestis) but, except those in rodents, are caused by other bacteria or viruses. Pasteurellosis can be effectively treated with antibiotics although some strains become resistant to a particular type of antibiotics (streptomycin, chloramphenicol). Eradication of rodent pests is the best measure of prevention. During the great epidemics in the 14th century, the disease claimed an estimated 25 million victims. Sporadic occurrence is known even today in the underdeveloped areas of the world. ►zoonosis, ►*Yersinia*, ►plant vaccines, ►biological weapons; Parkhill J et al. 2001 Nature [Lond] 413:523.

Plakin: >200 kDa dimeric, coiled coil, actin-binding proteins forming molecular bridges between the cytoskeleton and other subcellular structures. They also bind microtubules. ►cytoskeleton, ►filaments, ►microtubule; Jefferson JJ et al 2007 J Mol Biol 366:244.

Plakoglobin: 83 kDa protein localized to the cytoplasmic side of the desmosomes. ►desmosome, ►adhesion, ►desmoplakin

Planar Cell Polarity (PCP): PCP is determined by several genes involved in embryonal development, neural tubes, cochlear sensory hair cells of the ear, etc.

Planarians (flatworms): Relatively simple carnivorous organisms inhabiting fresh waters. There are about 15,000 species, all with bilateral symmetry and elaborate digestive tract and nervous system. They are well suited for studies of regeneration. The tapeworms and flukes are serious human parasites. ►flatworm, ►regeneration; http://planaria.neuro.utah.edu/.

Planck Constant (h): A constant of energy of a quantum of radiation and the frequency of the oscillator that emitted the radiation. $E = h\nu$ where E = energy, ν = its frequency; numerically 6.624×10^{-27} erg^{-s}.

Plankton: The collective name of many minute free-floating water plants, animals, and prokaryotes. ►phytoplankton, ►bacterioplankton; microbial oceanography: DeLong EF, Karl DM 2005 Nature [Lond] 437:336; Giovannoni SJ, Stingl U 2005 Nature [Lond] 437:343; Arrigo KR 2005 Nature

P

[Lond] 437:349; viruses in the sea: Suttle CA 2005 Nature [Lond] 437:356; genetic and metabolic survey of microbial plankton from 10 m to 4000 m oceanic depth: DeLong EF et al 2006 Science 311:496.

Plant Breeding: An applied science involved in the development of high-yielding food, feed, and fiber plants. It is concerned also with the production of lumber, renewable resources of fuel, and many types of industrial raw products (such as latex, drugs, cosmetics, etc.). A major goal of plant breeding is to improve the nutritional value, safety, disease resistance, and palatability of the crops. Plant breeding and technological improvements in agriculture have resulted in a near 10-fold increase in maize production and doubled wheat yields in the twentieth and twenty-first century. Plant breeding is based on population and quantitative genetics, and biotechnology. (Mazur B et al 1999 Science 285:372).

Plant Defense: Plant defense against herbivores is mediated by the signaling peptide *systemin* activating a lipid cascade. Membrane linolenic acid is released by the damage and converted into phytodienoic and jasmonic acids, structural analogs to the prostaglandins of animals. As a consequence, tomato plants produce several systemic *wound response proteins*, similar to those elicited by oligosaccharides upon pathogenic infections. Mutation in the octadecanoic (fatty acid) pathway blocks these defense responses. ►host-pathogen relations, ►insect resistance in plants, ►jasmonic acid, ►prosta-glandins, ►fatty acids, ►systemin, ►oleuropein

Plant Disease Resistance: ►host-pathogen relation, ►plant defense

Plant Genomes: Plant genomes generally differ in size and organization from those in animals and pose new problems and answer unique questions of analysis. (See Peterson AH 2006 Nature Rev Genet 7:174).

Plant Genomics Database: http://sputnik.btk.fi; plant molecular markers: http://markers.btk.fi; see also crop plants and individual plant species in the alphabetical order.

Plant Hormones: Auxins, gibberellins, cytokinins, abscisic acid, brassinosteroids, jasmonate, and ethylene. Polypeptide hormones play roles in the defense systems of plants (Ryan CA et al 2002 Plant Cell 14:251). The natural *auxin* in plants is indole-3-acetic acid (IAA) but a series of synthetic auxins are also known such as dichlorophenoxy acetic acid (2,4-D), naphthalene acetic acid (NAA), indole-butyric acid (IBA), etc. Auxins are involved in cell elongation, root development, apical dominance, gravi- and phototropism, respiration, maintenance of membrane potential, cell wall synthesis, regulation of transcription, etc. The bulk (≈95%) of IAA in plants is conjugated through its carboxyl end to amino acids, peptides, and carbohydrates. The conjugate regulates how much IAA is available for metabolic needs, although some conjugates may be directly active as hormones. Enzymes have been identified that hydrolyze the conjugates. Over the developing tissues, auxins show concentration gradients, indicating its role in positional signaling similarly to animal morphogens. The conjugates may transport IAA within the plant. *Gibberellic acid* and gibberellins control stem elongation, germination, and a variety of metabolic processes. *Cytokinins* also occur in a wide variety of forms such as kinetin, benzylamino- purine (BAP), isopentenyl adenine (IPA), zeatin, etc. Their role is primarily in cell division but they regulate the activity of a series of enzymes. Regeneration of plants from dedifferentiated cells requires a balance of auxins and cytokinins. *Abscisic acid* and terpenoids control abscission of leaves and fruits, dormancy and germination of seeds and a series of metabolic pathways. *Ethylene* was recognized as a *bona fide* plant hormone more recently. It is involved in the control of fruit ripening, senescence, elongation, sex determination, etc. The hormone type action of *brassinosteroids* in controlling elongation and light responses has been recognized by genetic evidence only in 1996. *Jasmonic acid* is also a hormone like substance with role in parasite defense. Generally, the various plant hormones signal to each other and their dynamic cooperative effects are essential for plant responses (Schmelz EA et al 2003 Proc Natl Acad Sci USA 100:10552). A survey indicated that hormones affected 4666 genes of *Arabidopsis* but most commonly different hormones regulated distinct members of protein families (Nemhauser JL et al 2006 Cell 126:467). ►hormones, ►signal transduction, ►abscisic acid, ►ethylene, ►indole acetic acid, ►jasmonic acid, ►gibberellic acid, ►kinetin, ►zeatin, ►brassinosteroids, ►seed germination; Kende H 2001 Plant Physiol 125:81; Mok DWS, Mok MC 2001 Annu Rev Plant Physiol Mol Biol 52:89; http://www.ualr.edu/botany/hormimages.html.

Plant Pathogenesis: Plant pathogens pose risks for agricultural, horticultural, and forest plants and may damage natural habitats of different organisms, plants as well as animals. Several plant pathogens and saprophytes may pose human health hazards, especially for immunologically compromised individuals. (See Vidaver AK, Tolin S 2000 In: Fleming DO, Hunt DL (Eds.) Biological Safety, ASM, Washington DC, pp 27–33; ►host–pathogen relation; http://www.pathoplant.de.

Plant Vaccines: Transgenic plants may express immunogenic proteins, which by consuming the plant tissues by humans or animals, may protect against bacterial or viral diarrhea. Also, plant synthesized immunoglobulins may protects against *Streptomyces mutans*, responsible for dental caries and gum disease. Hepatitis B surface antigen (HBsAg), Norwalk virus capsid protein (NVCP), *E. coli* heat-labile enterotoxin B subunit (LT-B), cholera toxin B subunit (CT-B), and mouse glutamate decarboxylasc (GAD67) have been propagated in tobacco and potato tissues, respectively. Hepatitis B vaccine delivered by raw potatoes—when a sufficient quantity was consumed—increased the serum antiHB surface antigen titer in up to 62.5% of the volunteers (Thanavala Y et al 2005 Proc Natl Acad Sci USA 102:3378). So far, these edible vaccines have not shown clinical use. The S1 protein of the SARS corona virus propagated in tomato and tobacco plants displayed good immunogenicity in mice after both parenteral (injection) and oral administration. This result is similar to the early tests of gastroenteritis vaccine in swine and the infectious bronchitis virus vaccination of chickens by plant vaccines (Progrebnyak N et al 2005 Proc Natl Acad Sci USA 102:9062). Apparently, very effective vaccines can be produced by introducing into plants (tobacco) the F1 and V and the F1–V fusion antigens of *Yersinia*, the agent of plague (Santi L et al 2006 Proc Natl Acad Sci USA 103:861). Vaccinia virus antigenic domain B5 propagated in tobacco and collard plants when introduced orally in mice or the minipig (miniature pig) did not generate an anti-B5 immune response, but intranasal administration of soluble pB5 led to a rise of B5-specific immunoglobulins, and parenteral immunization led to a strong anti-B5 immune response in both mice and the minipig. Mice immunized i.m. (intramuscularly) with pB5 generated an antibody response that reduced smallpox virus spread in vitro and conferred protection from challenge with a lethal dose of vaccinia virus (Golovkin M et al 2007 Proc Natl Acad Sci USA 104:6864). ►vaccines, ►immunoglobulin, ►transformation genetic, ►plantibody, ►TMV, ►SARS, ►plague, ►*Yersinia*, ►bronchitis, ►gastroenteritis; Daniell H et al 2001 J Mol Biol 311:1001; Ruf S et al 2001 Nature Biotechnol 19:870; Sojikul P et al 2003 Proc Natl Acad Sci USA 100:2209.

Plant Viruses: Plant viruses vary a great deal in size, shape, genetic material, and host-specificity. The majority of them have single-stranded positive-strand RNA as genetic material and are either enveloped or not. The Reoviridae may have several double-stranded RNAs, and the Cryptovirus carries two double-stranded RNAs. The Cauliflower (Caulimo) virus has double-stranded DNA, whereas the Geminiviruses have single-stranded DNA genetic material. The size of their genome usually varies between 4 to 20 kb and their coding capacity is at least four proteins. The 5′-end may form methylguanine cap or it may have a small protein attached to it. The 3′-end may have a polyA tail or may resemble the OH end of the tRNA. Approximately, 600–700 plant viruses have been described. ►viruses, ►cap, ►polyA tail, ►tRNA, ►viroid, ►TMV, ►CaMV, ►geminivirus, ►viroid; Knipe DM et al (Eds.) 2001 Fundamental Virology, Lippincott Williams & Wilkins, Philadelphia, Pennsylvania; Harper G et al 2002 Annu Rev Phytopathol 40:119; Tepfer M 2002 Annu Rev Phytopathol 40:467; general virus database, including plant viruses: http://www.ncbi.nlm.nih.gov/ICTVdb/ictvdb.htm.

Plantibody (antibody synthesized by plants): A modified immunoglobulin produced in transgenic plants carrying the genetic sequences required for the recognition of the site of the viral coat or other proteins. The yield of the plantibody molecules is very high, up to 1% of the soluble plant proteins. The modification of the immunoglobulin involves usually the elimination of the constant region of the heavy chain while retaining the variable region. The plant antibodies are usually formed as single chains (ScFv). Other modifications for solubility and tissue-specific expression may be introduced. The plantibodies are modified also by intrinsic plant mechanisms (N-glycosylation) within the endoplasmic reticulum. Unfortunately, plant tissue lack β1,4-galactosyltransferase, which is required for the synthesis of mammalian-like glycans. By transformation, the gene of this enzyme has been transferred into tobacco plants and it functions normally. Retention and excretion of ScFv immunoglobulin molecules is increased if the KDEL amino acid sequence is present in the polypeptide chain. For some medical applications, the plantibodies may not be suitable because they may carry plant-specific β-1,2-xylose and α-1,3-fucose residues at the galactose-carrying N-glycans and cause allergic reactions in monoclonal antibodies. When, however, a hybrid enzyme called XylGalT that consists of the N-terminal domain of the *Arabidopsis* xylosyltransferase and the catalytic domain of human β-1,4- galactosyltransferases is used in tobacco plants, the core-bound xylose and fucose residues are sharply reduced. This type of monoclonal plantibody thus appears promising (Bakker H et al 2006 Proc Natl Acad Sci USA 103:7577). Single-chain variable fragment (scFv)-Fc (fragment crystalline) antibodies, with N-terminal signal sequence and C-terminal KDEL tag, can accumulate to very high levels as bivalent IgG-like antibodies in *Arabidopsis thaliana* seeds and illustrate that a plant-produced anti-hepatitis A virus scFv-Fc has similar antigen-binding and in vitro neutralizing activities as the corresponding full-length IgG. As expected, most scFv-Fc produced in seeds contained only oligomannose-type *N*-glycans, but, unexpectedly, 35–40% was never glycosylated. A portion of the scFv-Fc was found in endoplasmic reticulum (ER)-derived compartments delimited by ribosome-associated membranes. Additionally, consistent with the glycosylation data, large amounts of the recombinant protein were deposited in the periplasmic space, implying a direct transport from the ER to the periplasmic space between the plasma membrane and the cell wall. Aberrant localization of the ER chaperones calreticulin and binding protein (BiP) and the endogenous seed storage protein cruciferin in the periplasmic space suggests that overproduction of recombinant scFv-Fc disturbs normal ER retention and protein-sorting mechanisms in the secretory pathway (Van Droogenbroeck B et al 2007 Proc Natl Acad Sci USA 104:1430).

Monoclonal antibody against the non-protein Lewis Y oligosaccharide antigen is over-expressed in breast, lung, ovary, and colon cancers. Monoclonal antibody (mAb BR55-2) specific for LeY was expressed (30 mg/kg fresh weight of leaves) in low-alkaloid content in transgenic tobacco plants and bound specifically to SK-BR3 breast cancer and SW948 colorectal cancer cells. Its binding to the FcγRI receptor was the same as that derived from mammalian cells. The plantibody was effective in cytotoxicity assays as well as in grafting onto nude mice; thus, indicating its potential suitability for immunotherapy (Brodzik R et al 2006 Proc Natl Acad Sci USA 103:8804).

Plant-produced antibodies may find biomedical application in humans and animals. Transgenic plants may produce large quantities of IgA and IgG-IgA at low cost. In *Nicotiana benthamina* leaves, high-level expression of functional full-size monoclonal antibody (mAb) of the IgG class in plants has been ascertained. The process relies on synchronous coinfection and coreplication of two viral vectors, each expressing a separate antibody chain. The two vectors are derived from two different plant viruses that were found to be noncompeting. Unlike vectors derived from the same virus, noncompeting vectors effectively coexpress the heavy and light chains in the same cell throughout the plant body, resulting in yields of up to 0.5 g of assembled mAbs per kg of fresh-leaf biomass (Giritch A et al 2006 Proc Natl Acad Sci USA 103: 4701).

Also, other components of the immunization system may thus be synthesized with single plants after combining the genes through classical crossing procedures. By eating IgA secreting plant tissues, protection is expected through mucosal immunity or may protect against dental caries. If the antibody is expressed in seed tissues, it can be stored at room temperature (perhaps for years) without a loss of the variable region of the antibody and its antigen-binding ability. ►antibody, ►host-pathogen relations, ►ScFv, ►KDEL, ►immunization, ►mucosal immunity, ►monoclonal antibody, ►monoclonal antibody therapies, ►plant vaccine, ►molecular pharming, ►Lewis blood group, ►breast cancer, ►colorectal cancer, ►cancer therapy, ►cancer gene therapy, ►nude mouse, ►tobacco, ►BiP, ►calreticulin, ►periplasma; Bakker H et al 2001 Proc Natl Acad Sci USA 98:2899; Mayfield SP et al 2003 Proc Natl Acad Sci USA 100:438; Ma JK-C et al 2003 Nature Rev Genet 4:794.

PLAP Vector: ►axon guidance

Plaque: The clear area formed on a bacterial culture plate (heavily seeded with cells) as a consequence of lysis of the cells by virus; turbid plaques indicate incomplete lysis (see Fig. P78). ►lysis

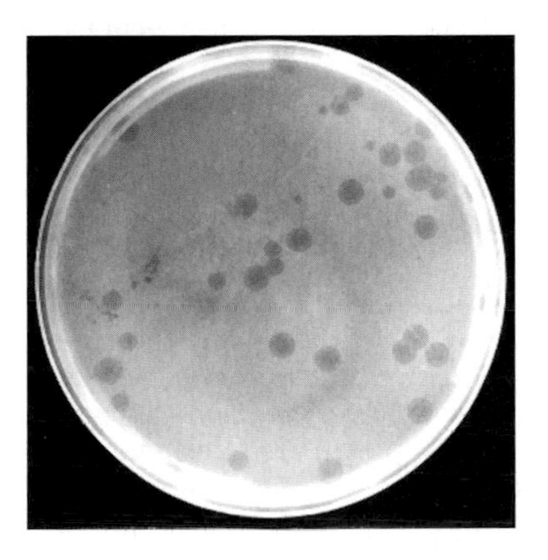

Figure P78. T3 bacteriophage plaques on petri plate heavily seeded by bacteria. (Courtesy of Dr. CS Gowans)

Plaque-Forming Unit: The number of plaques per mL bacterial culture.

Plaque Hybridization: ►Benton-Davis plaque hybridization

Plaque Lift: Plaque lifts on bacteriophage plates plaques are marked and overlaid by cellulose nitrate films. After denaturation and immobilization of the plaques on the filter, they are hybridized with probes to identify recombinants and return to the saved master plate for obtaining plugs of interest from the original plate. The procedure generally requires repetition in order to isolate unique single recombinants. ►colony hybridization; Frolich MW 2000 Biotechniques 29:30.

Plasma: The fluid component of the blood in which the particulate material is suspended. The blood plasma is free of blood cells but clotting is not allowed during its isolation, and it contains the platelets, which harbor animal cell growth factors. ►PDGF, ►platelets, ►serum, ►cytoplasm, ►cytosol

Plasma Cell (plasmacyte): B lymphocytes can differentiate into either memory cells or plasma cells and the latter secrete immunoglobulins. ►lymphocytes, ►immunoglobulins, ►immune system

Plasma Membrane: The plasma membrane envelops all cells. ►cell membranes

Plasma Nucleic Acid: During pregnancy, a small number of fetal cells can escape into the plasma and some can also shed their chromosomal DNA (see Fig. P79). Such plasma nucleic acids can be exploited for prenatal diagnosis without many intrusive procedures (Lo YMD et al 2007 Nat Rev Genet 8:71). ►prenatal diagnosis, ►DNA circulating

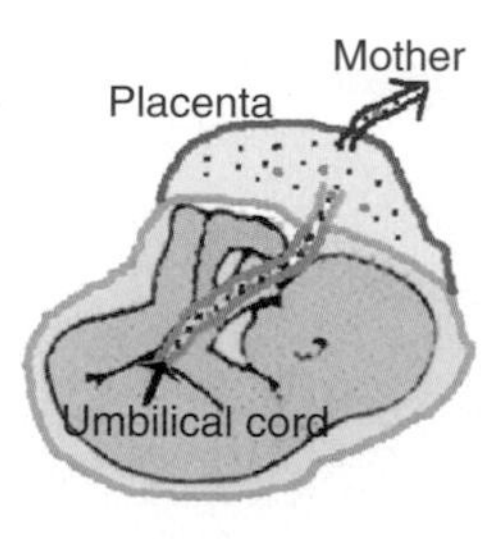

Figure P79. Contact is between fetus and mother through the umbilical cord and both normal (red) and different cells (green) are transferred to the maternal plasma

Plasma Proteins: Proteins in the blood plasma. The major components are serum albumin, globulins, fibrinogen, immunoglobulins, antihemophilic proteins, lipoproteins, α_1 antitrypsin, macroglobulin, haptoglobin, and transfer proteins, such as transferrin (iron), ceruloplasmin (copper), transcortin (steroid hormones), retinol-binding proteins (vitamin A), and cobalamin-binding proteins (vitamin B_{12}). The lipoproteins carry phospholipids, neutral lipids and cholesterol esters. In addition, there are a great variety of additional proteins present in the serum. In a small population of 96 healthy individuals, 76 structural variants were observed in 25 proteins by affinity-based mass spectrometric assays. This large variation predicts that analysis of plasma proteins may yield important biomarkers for medical purposes (Nedelkov D et al 2005 Proc Natl Acad Sci USA 102:10852)

Plasmablast: A precursor of plasmacyte or precursor cell of the lymphocytes.

Plasmacytoid Cell: Functionally, it is one type of dendritic cells with antigen-presenting properties (see Fig. P80). Plasmacytoid cells may not display dendritic morphology. They produce a large quantity of interferon α/β in response to bacterial or viral infection. They have roles in both innate and acquired immunity. ►dendritic cell, ►interferon, ►acquired immunity; McKenna K et al 2005 J Virology 79:17.

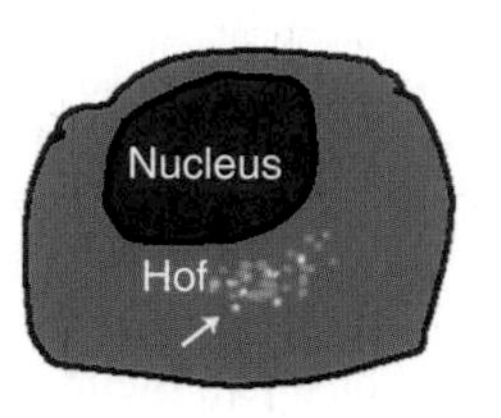

Figure P80. Plasmocytoid cell

Plasmacytoma: The cancer (myeloma) of antibody producing cells. Resistance against it is controlled mainly by different alleles of the complex FRAP. ►Fk506; Bliskovsky V et al 2003 Proc Natl Acad Sci USA 100:14982.

Plasmagene: Non-nuclear genes (mitochondrial, plastidic or plasmid). ►mitochondrial genetics, ►chloroplast genetics

Plasmalemma: The membrane around the cytoplasm or the envelope of the fertilized egg.

Plasmatocyte: Macrophage-like elements in the insect hemolymph. ►macrophage, ►hemolymph

Plasmid: The dispensable genetic element, which can propagate independently and can be maintained within the (bacterial) cell, and may be present in yeast and mitochondria of a number of organisms. The plasmids may be circular or linear double-stranded DNA. The conjugative plasmids possess mechanisms for transfer by conjugation from one cell to another. The non-conjugative plasmids lack this mechanism and are therefore preferred for genetic engineering because they can be easier confined to the laboratory. During evolution, some of the advantageous plasmid genes are assumed to have been incorporated into the chromosomes and the plasmids, lost. The persistence of the plasmids may be warranted by their ability to disperse genetic information horizontally. Plasmids occur also in the organelles of higher eukaryotes and lower eukaryotes. ►vectors, ►curing of plasmids, ►pBR322, ►pUC, ►transposon conjugative, ►cryptic plasmids, ►Ty; Summers DK 1996 The biology of plasmids, Blackwell; Thomas CM (Ed.) 2000 The Horizontal Gene Pool: Bacterial Plasmids and Gene Spread, Harwood Press, Durham, UK; http://plasmid.hms.harvard.edu.

Plasmid Addiction: The loss of certain plasmids from the bacterial cells may lead to an apoptosis-like cell death, called post-segregational killing or plasmid addiction. ►apoptosis

Plasmid, Chimeric: An engineered plasmid carrying foreign DNA.

Plasmid Incompatibility: Plasmids are compatible if they can coexist and replicate within the same bacterial cell. If the plasmids contain repressors effective for inhibiting the replication of other plasmids, they are incompatible. Generally, closely related plasmids are incompatible, and they thus belong to a different incompatibility group. The plasmids of enterobacteria belong to about two-dozen incompatibility groups. Plasmids may be classified also according to the immunological relatedness of the pili they induce to form (such as F, F-like, I, etc.). The replication system of the plasmids defines both the pili and the incompatibility groups. Cells with F plasmids may form F sex pili; the R1 plasmids belong to FII pili group, etc. ►pilus, ►F^+, ►F plasmid, ►R plasmids, ►enterobacteria, ►incompatibility plasmids

Plasmid Instability: Plasmid instability indicates difficulties in maintenance caused by defect(s) in transmission, internal rearrangements, and loss (deletion) of the DNA. ►cointegration

Plasmid, 2 μm: A 6.3 kbp circular DNA plasmid of yeasts, present in 50–100 copies per haploid nucleus. It carries two 599 bp inverted repeats separating 2774 and 2346 bp tracts. Re-combination between the repeats results in A and B type plasmids. Its recombination is controlled by gene *FLP* and its maintenance requires the presence of the *REP* genes. ►yeast; Scott-Drew S, Murray JA 1998 J Cell Sci 111:1779.

Plasmid Maintenance: Plasmid maintenance in prokaryotes is secured either by the high number of copies, or in low copy number plasmids, by a mechanism reminiscient to some extent to that of the centromere is mitosis of eukaryotes. The proteic plasmid maintenance system operates by the coordination of a toxin and an unstable antidote. When the labile antidote decays, the toxin kills the cells that do not have the plasmid. The antidote may be a labile antisense RNA that keeps in check the toxin gene. A plasmid-encoded restriction-modification may also be involved. When the modification system recedes beyond an effective level in the plasmid-free cells, the genetic material falls victim to the endonuclease. One of such systems in *E. coli* is the *hok* (host killing)-*sok* (suppressor of killing)-*mok* (modulation of killing) system of linked genes. ►partition, ►antisense RNA, ►restriction-modification, ►killer plasmids; Gerdes K et al 1997 Annu Rev Genet 31:1; Møller-Jensen J et al 2001 J Biol Chem 276:35707; Hayes F 2003 Science 301:1496.

Plasmid Mobilization: Plasmid mobilization may take place by bacterial conjugation. Plasmid vectors use the gene *mob* (mobilization) if they do not have their own genes for conjugal transfer. Some plasmids may rely on *ColK* (colicin K, affecting cell membranes) that nicks plasmid pBR322 at the *nic* site, close to *bom* (basis of mobility). Mobilization proceeds from the nicked site (base 2254 in pBR322). Plasmids lacking the *nic/bom* system, e.g., pUC, cannot be mobilized. (See Chan PT et al 1985 J Biol Chem 260:8925).

Plasmid Rescue: The plasmid rescue procedure was designed originally for transformation with linearized plasmids of *Bacillus subtilis* that normally does not transform these bacteria. The linearized plasmid could be rescued for transformation in the presence of the *RecE* gene if recombination could take place.

The linearized monomeric plasmid then could carry also any in vitro ligated passenger DNA into cells. If the host cells carry a larger number of plasmids (multimeric), special selection is necessary to find the needed one. Plasmid rescue has also been used for re-isolation of inserts (plasmids) from the genome of transformed cells of plants.

The re-isolation requires appropriate probes for (the termini) of the inserts to permit recognition, after which, ithe DNA is re-circularized and cloned in *E. coli* and they have at least one selectable marker and an origin of replication compatible with the bacterium. The cloned DNA insert or its fragments are inserted into the M13 phage for nucleotide sequencing. This permits the identification of any changes that may have taken place in the original transforming DNA and permits an analysis of the flanking sequences of the target sites as well. A number of different variations of the procedure have been adopted in prokaryotes, microbes, animals, and plants. ►T-DNA, ►DNA sequencing, ►Rec; Perucho M et al 1980 Nature [Lond] 285:207, see Fig. P81.

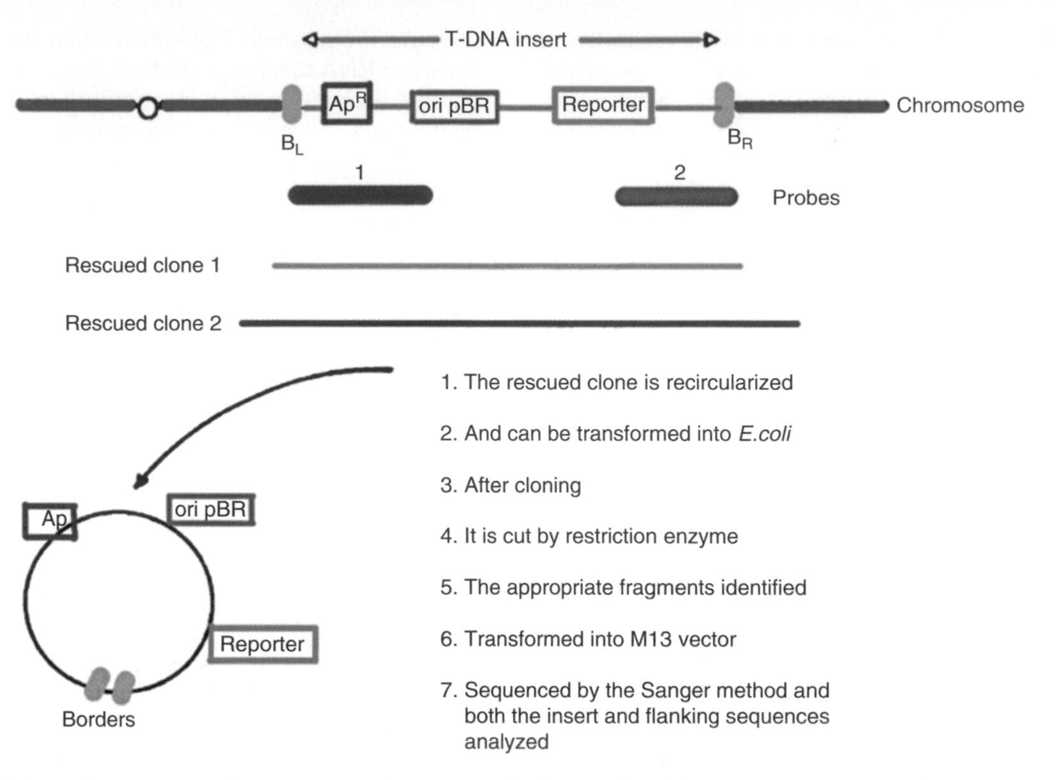

Figure P81. Outline of a plasmid rescue procedure exemplified by isolating T-DNA insert from *Arabidopsis*. Ap = ampicillin-resistance gene (Ap^R) of the pBR322, oripBR = origin of replication of the PBR322 plasmid present in the plant-transforming vector, reporter is hygromycin resistance, left (B_L) and | right (B_R) border sequence of the T-DNA. (After Koncz C, et al. 1989. Proc. Natl. Acad. Sci. USA 86:8467.)

Plasmid Segregation: In plasmid segregation, before cell division, plasmids are partitioned to ensure transmission to mother and daughter cells. The segregation is mediated by *par* (partitioning) loci of the plasmids. Type I pars encode Walker box ATPases and Type II *pars* encode actin-like ATPases. The actin-like filaments act in similarity to the microtubules in higher organisms. The *par* loci involve two proteins and a centromere-like cis-acting site. The presence of Type I par locus positions the plasmids in the center of the cell. Integration host factor (IHF) of the bacteria plays a role in plasmid segregation. The mechanisms of plasmid segregation vary among different plasmids. ►actin, ►ATPase, ►Walker box, ►IHF; Ebersbach G, Gerdes K 2005 Annu Rev Genet 39:453.

Plasmid Shuffling: The general procedure of plasmid shuffling in yeast first disrupts the particular gene in a diploid strain. After meiosis, the cells can be maintained only if the wild type allele is carried on a replicating plasmid (episome). Mutant copies of that particular gene are then introduced into the cell on a second episome and exchanged (shuffled) for the wild type allele. The phenotype of any of the mutant alleles can be studied in these cells that carry the disrupted (null) allele. (See Sikorski RS et al 1995 Gene 155:51; Zhao H, Arnold FH 1997 Nucleic Acids Res 25:1307).

Plasmid Telomere: Linear plasmids require exonuclease protection at the open ends. The problem may be resolved by capping with proteins or forming a lollipop type structure by fusing the ends of the single strands as shown in Figure P82. ►telomere

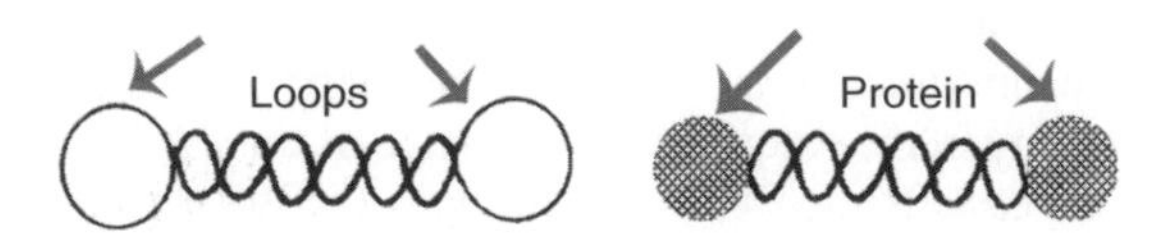

Figure P82. Plasmid telomeres

Plasmid Vehicle: A recombinant plasmid that can mediate the transfer of genes from one cell (organism) to another. ►vectors

Plasmids, Amplifiable: Amplifiable plasmids continue replication in the absence of protein synthesis (in the presence of protein synthesis inhibitor). ►amplification

Plasmids, Conjugative: Conjugative plasmids carry the *tra* gene, promoting bacterial conjugation and can be transferred to other cells by conjugation and can also mobilize the main genetic material of the bacterial cell. ►conjugation, ►F plasmid

Plasmids, Cryptic: Cryptic plasmids have no known phenotype.

Plasmids, Monomeric: Monomeric plasmids are present in a single copy per cell.

Plasmids, Multimeric: Multiple plasmids have multiple copies in a cell.

Plasmids, Non-Conjugative: Non-conjugative plasmids lack the *tra* gene required for conjugative transfer, but have the origin of replication and therefore when complemented by another plasmid for this function, they can be transferred. ►conjugation

Plasmids, Promiscuous: Promiscuous plasmids have conjugative transfer to more than one type of bacteria.

Plasmids, Recombinant: Recombinant plasmids are chimeric; they carry DNA sequences of more than one origin. (See Fig. P83).

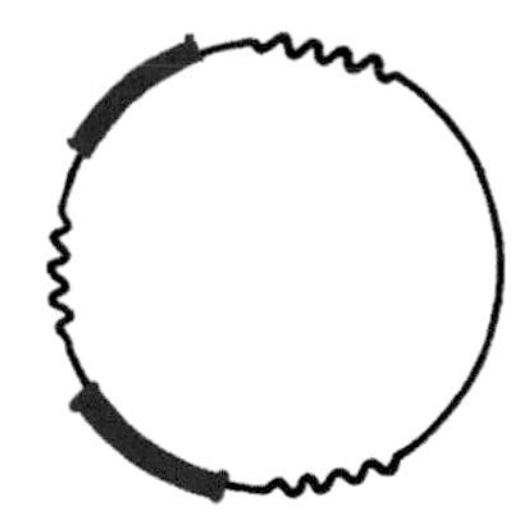

Figure P83. Recombinant plasmid

Plasdmids, Relaxed Replication: Relaxed replication plasmids may replicate to 1000 or more copies per cell.

Plasmids, Runaway Replication: In runaway replication plasmids, the replication is conditional, e.g., under permissive temperature regimes they may replicate almost out-of-control whereas under other conditions their number per cell may be quite limited.

Plasmids, Single Copy: Single copy plasmids may have single or very few copies per cell.

Plasmids, Stringent Multicopy: Stringent multicopy plasmids may grow to 10 to 20 copies in a cell.

Plasmin (fibrinolysin): Proteolytic protein (serine endopeptidase) with specificity of dissolving blood clots, fibrin, and other plasma proteins. For its activation, urokinases (tissue plasminogen activator) are required. Plasmin may be used for therapeutic purposes to remove obstructions in the blood vessels. ►urokinase, ►plasminogen, ►plasminogen activator, ►streptokinase, ►CAM; Lijnen HR 2001 Ann NY Acad Sci 936:226.

Plasmin Inhibitor Deficiency (PLI, AAP): Plasmin inhibitor deficiency is encoded in human chromosome 18p11-q11 as recessive gene, and is involved in the regulation of fibrinolysin. ►plasmin

Plasminogen: A precursor of plasmin. Human plasminogen markedly increases mortality of mice infected with streptococci due to bacterial expression of streptokinase. Streptokinase is highly specific for human plasminogen but not for other mammalian plasminogens (Hun H et al 2004 Science 305:1283). ►plasmin, ►plasminogen activator, ►angiostatin

Plasminogen Activator (PLAT): PLAT cleaves plasminogen into plasmin; it is encoded in human chromosome 8q11-p11. The plasmin activator inhibitor (PLANH1/PAI-1) is encoded in human chromosome 7q21-q22 and PLANH2 at 18q21.1-q22. The plasminogen activator receptor was localized to 19q13.1-q13.2. Tissue-specific plasminogen activator coupled to the surface of red blood cells can dissolve blood clots and prevent thrombosis (Murciano, J-C et al 2003 Nature Biotechnol 21:891). ►plasminogen, ►plasmin, ►urokinase, ►PN-1, ►streptokinase, ►thrombosis, ►MET oncogene

Plasmodesma (plural plasmodesmata): About 2μm or larger channels connecting neighboring plant cells, lined by extension of the endoplasmic reticulum. Functionally, they correspond to the gap junctions of animal cells. Various molecules, signals, including even viruses, may move through these intercellular communication channels. The plasmodesmata are subject to temporal and spatial regulation. ►gap junctions; Zambryski P, Crawford K 2000 Annu Rev Cell Dev Biol 16:393; Hake S 2001 Trends Genet 17:2; Haywood V et al 2002 Plant Cell 14:S303; Kim I et al 2005 Proc Natl Acad Sci USA 102:11945.

Plasmodium: A syncytium of the amoeboid stage of slime molds (such as in *Dictyostelium*).

Plasmodium: One of the several parasitic coccid protozoa causing malaria-like diseases in vertebrates, birds, and reptiles. A single *Plasmodium falciparum* (n = 14, ~23 Mb, ~5268 proteins) parasite transcribes simultaneously multiple *var* genes (at several chromosomal locations), encoding the erythrocyte-membrane protein (PfEMP-1) that binds to the vascular endothelium and red blood cells. Functionally related genes tend to be clustered in the subtelomeric regions of the chromosomes. A protein interaction network of *P. falciparum* expressed at the intra-erythrocyte-stage parasites involving >2000 fragments of 1295 genes has been constructed on the basis of the two-hybrid system. These networks can reveal potential drug targets (LaCount DJ et al 2005 Nature [Lond] 438:103). These interaction networks are quite different from those known in other organisms (Suthram S et al 2005 Nature [Lond] 438:108). Alignment of these *var* genes in heterologous chromosomes at the nuclear periphery may facilitate gene conversion and promotes diversity of antigenic determinants and adhesive phenotypes. Such a mechanism aids the evasion of the host immune system. The parasite invades the erythrocytes and destroys the host cells through the formation of merozoites (mitotic products) and spreading thus to other cells (see Fig. P84). The merozoites may develop into gametocytes (gamete forming cells) that infect blood-sucking mosquitos where they are transformed into sporozoites (the sexual generation) that are transmitted through insect bites to the higher animal host. The invaders first move to the liver where merozoites are formed and then return to the erythrocytes; thus the cycle continues. *Plasmodium falciparum* causes falciparum malaria. *P. malariae* is responsible for the *quartan*, or fourth day recurring malaria. The protozoon contains two double-stranded extranuclear DNA molecules; that of circular DNA resembles mitochondria whereas the second bears similarities to ctDNA, and contains 68 genes. Mutation in a single gene (*pfmdr1/PfEMP1*) encoding the P-glycoprotein homolog, Pgh1, may result in resistance to several antimalarial drugs of which some may or may not be chemically related. Transformation of the gene encoding the SM1 peptides into the *Anopheles* vector may render the insect resistant to *Plasmodium* infection. The rodent parasite *Plasmodium yoellii yoellii* genome (~23.1 Mb) is similar to that of *P. falciparum*. ►malaria, ►sex determination,

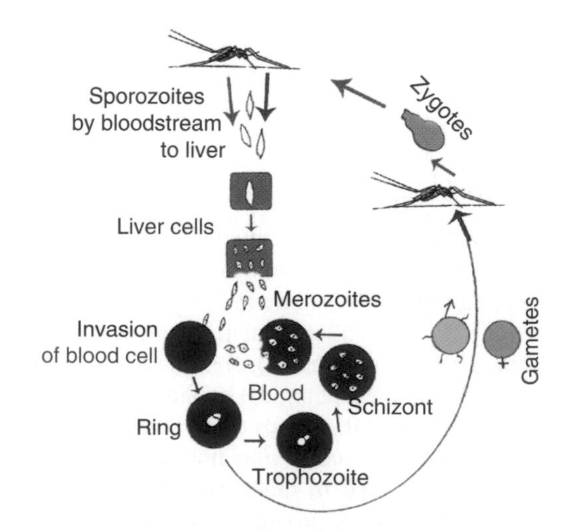

Figure P84. Life cycle of *Plasmodium* in humans and mosquitos

►thalassemia, ►antigenic variation, ►mRNAP, ►mtDNA, ►chloroplast, ►PfEMP1, ►rifins, ►gene conversion, ►serpine, ►antigenic variation, ►epigenetic memory; Fidock DA et al 2000 Mol Cell 6:861; Ito J et al 2002 Nature [Lond] 417:452; the sequenced *P. falciparum* genome: *Nature* 419, issue 6906, Oct 30, 2002; *Anopheles* genome: Science 298, 4 Oct 2002; Joy DA et al 2003 Science 3000:318; *P. falciparum* linkage and gene association: Su X et al 2007 Nature Rev Genet 8:497; comparative genome analysis of *P. berghei* and *P. chabaudi*: Hall N et al 2005 Science 307:82; invasion of the blood: Cowman AF, Crabb BS 2006 Cell 124:755; http://www.plasmodb.org/plasmo/home.jsp; http://www.tigr.org/tdb/tgi/.

Plasmogamy: fusion of the cytoplasm of two cells without fusion the two nuclei and thus resulting in dikaryosis. Plasmogamy is common in fungi but may occur in fused cultured cells of plants and animals. ►fungal life cycle, ►cell genetics

Plasmolemma (plasmalemma): Plant cell membrane; the ectoplasm of the fertilized egg of animals.

Plasmolysis: The shrinkage of the plant cytoplasm caused by high concentration of solutes (salt) outside the cell resulting in loss of water. The cytoplasm separates from the cell wall.

Plasmon: The sum of non-nuclear hereditary units such as exists in mitochondrial and plastid DNA. ►mtDNA, ►chloroplasts, ►plastome

Plasmon-Sensitive Gene: ►nucleo-cytoplasmic interaction

Plasmophoresis: A procedure to filter blood to allow the plasma proteins of the patient to be removed. At the same time, new donor plasma is replaced into the patient's blood, which is located in the plasmaphoresis machine. Subsequently, the blood is sent back to the patient. ►hemolysis

Plasmotomy: The fragmentation of multinucleate cells into smaller cells without nuclear division.

Plasmovirus: The plasmovirus bears some similarity to phagemids but in this case a retrovirus is combined with an independent vector cassette containing various elements. The envelope gene of the Moloney provirus is replaced by a transgene to prevent infective retroviral ability and it would not regain it by chance recombination with another retrovirus. Such a construct can express transgene(s), can multiply within the target cells, and provide a tool for cancer therapy. ►vectors, ►retrovirus, ►transgene, ►cancer therapy, ►viral vectors, ►phagemid; Morozov VA et al 1997 Cancer Gene Ther 4(5):286.

Plasticity: In general, the ability of a cell or organism to display different expressions (phenotype) dependent on the environment. The ability of cells to change from what they normally are enables them to perform tasks not normally found in differentiated cells. ►stem cells, ►MAPCs, ►adaptation, ►noise, ►reaction norm

Plastid: The cellular organelle of plants, containing DNA. It may differentiate into chloroplasts, etioplasts, amyloplasts, leucoplasts, or chromoplasts. In *Arabidopsis*, mechanosensitive ion channel proteins (localized in the plastid envelope) seem to control plastid size and shape (Haswell ES, Meyerowitz E M 2006 Curr Biol 16:1). ►under separate names, ►plastid number per cell, ►plastid male transmission, ►ctDNA, ►chloroplast, ►chloroplast genetics, ►apicoplast

Plastid Male Transmission: Generally, the genetic material in the plastids is transmitted only through the egg cytoplasm but in a few species of higher plants (*Pelargonium*, *Oenothera*, *Solanum*, *Antirrhinum*, *Phaseolus*, *Secale*, etc.) a variable degree of male transmission takes place. Biparental transmission of plastid genes (about 1%) may occur also in the alga *Chlamydomonas reinhardi*. The male transmission of these nucleoids is controlled by one or two nuclear genes. The nucleoids of the plastid and mitochondria of the male are usually degraded or if they are included in the generative cells of the male, most commonly fail to enter the sperm or are not transmitted to the egg cytoplasm. In contrast to the angiosperms, in conifers (pines, spruces, firs) the plastid DNA is usually transmitted through the males. In some interspecific hybrids, exclusively paternal, exclusively maternal, and biparental transmission were also observed. In other conifer crosses, the mtDNA is transmitted maternally. In redwoods, the transfer is paternal. One scientific claim posits that the destruction of the paternal ctDNA in the females is carried out by a restriction enzyme while the maternal ctDNA is protected by methylation. Others implicate a special nuclease C. ►chloroplast, ►genetics, ►ctDNA, ►mtDNA, ►paternal leakage; Diers L 1967 Mol Gen Genet 100:56; Avni A, Edelman N 1991 Mol Gen Genet 225:273; Sears B 1980 Plasmid 4:233.

Plastid Number Per Cell: In the giant cells of *Acetabularia* algae, there may be one million chloroplasts but in the alga *Chlamydomonas* there is only one per cell. In higher plants, the number of plastids vary according to the size of the cells, about 30–40 in the spongy parenchyma to about twice as many in the palisade parenchyma. ►plastid, ►ctDNA

Plastochrone: The pattern of organ differentiation in time and space is genetically controlled. (See Miyoshi K et al 2004 Proc Natl Acad Sci USA 101:875).

Plastocyanin: an electron carrier in photosynthesis between cytochromes and photosystem I. ►Z scheme; Ruffle SV et al 2002 J Biol Chem 277:25692.

Plastome: The sum of hereditary information in the plastids. ►ctDNA, ►chloroplast genetics

Plastome Mutation: Mutation in the plastid (chloroplast) DNA. ►chloroplast genetics, ►mutation in cellular organelles

Plastoquinone: An isoprenoid electron carrier during photosynthesis. ►isoprene

Plate: A Petri dish containing a nutrient medium for culturing microbial or plant cells. Cell plate divides the two daughter cells after mitosis.

Plate Incorporation Test: The most commonly used procedure for the Ames test when the *Salmonella* suspension (or other bacterial cultures), the S9 activating enzymes, and the mutagen/carcinogen to be tested are poured over the bacterial nutrient plate in a 2 mL soft agar. After incubation for two days at 37°C, the number of revertant colonies is counted. ►Ames test, ►spot test

Platelet Abnormalities: ►Glanzmann's disease, ►thrombopathic purpura, ►thrombopathia, ►giant platelet syndrome, ►Hermansky-Pudlak syndrome, ►May-Hegglin anomaly, ►platelets

Platelet Activating Factor (PAF): An inflammatory phospholipid. PAF acetylhydrolase may be a factor in atopy. ►platelets, ►atopy

Platelet-Derived Growth Factor (PDGF): A mitogen, secreted by the platelets, the 2–3 μm size elements in the mammalian blood, originated from the megakaryocytes of the bone marrow, and concerned with blood coagulation. PDGF controls the growth of fibroblasts, smooth muscle cells, blood vessel formation, nerve cells, cell migration in the oocytes, etc. This protein bears substantial homologies to the oncogenic product of the simian sarcoma virus, the product of the KIT oncogene, and the CSF1R (it activates also other oncogenes, such as c-fos). The PDGF is required for the healing of vascular injuries and in these cases the expression induced Egr-1 (early growth response gene product) may bind to the PDGF β chain promoter after displacing Sp1. PFGF- and insulin-dependent S6 kinase (pp70^{S6k}) is activated by phosphatidylinositol-3-OH kinase. Its receptor (PDGFR) is a tyrosine kinase. The detection of PDGF is facilitated by the construction of aptamers labeled with pyrene monomers at both ends. When an excimer is produced, the fluorescence emission is increased from 400 nm to 485 of PDGF bound excimer. The principle may be applicable to other molecules for facilitating biomedical analyses (Yang CJ et al 2005 Proc Natl Acad Sci USA

102:17278). ►oncogenes, ►growth factors, ►signal transduction, ►platelets, ►Sp1, ►S6 kinase, ►phosphatidyl inositol, ►aptamer, ►excimer, ►heart diseases; Betsholtz C et al 2001 Bioessays 23(6):494; Duchek P et al 2001 Cell 107:17.

Platelets: Platelets originate as cell fragments or "minicells" (without DNA) from the megakaryocytes of the bone marrow. Their function is in blood clotting and in the repair of blood vessels; they also secrete mitogen(s). Platelet abnormalities may cause stroke, myocardial infarction (damage of the heart muscles), and unstable angina (sporadic, spasmic chest pain). A balance between Bc*l*-x and Bak determines the life span of platelets. Inactive Bcl-x may cause thrombocytopenia (Mason KD et al 2007 Cell 128:1173). ►blood, ►megakaryocyte, ►platelet derived growth factor, ►blood serum, ►BAK, ►BCL, ►thrombocytopenia; Prescott SM et al 2000 Annu Rev Biochem 69:419.

Plating Efficiency: The percentage of cells or protoplasts placed on a Petri plate that grows. The relative plating efficiency compares the fraction of growing cells in a treated series to that of an appropriate control.

Platyfish (*Xiphophorus/Platypoecilus*): Tropical fishes with complex sex determination. WX, WY, and XX are females and the males are XY and YY. The pseudoautosomal region seems to be long. Their melanocytes frequently turn into melanoma. ►pseudoautosomal, ►sex determination, ►melanocyte, ►melanoma

Platykurtic: ►kurtosis

Platypus: ►monotrene

Platysome: The nucleosome core (when it was thought of as a flat structure). ►nucleosome

Playback: The number of non-repetitive sequences in a DNA can be determined by the saturation of single-strand DNA with RNA of unique sequences. The kinetics of saturation, R_0t (by analogy to C_0t), is then determined. The annealed fraction is generally a small percent of the eukaryotic DNA, which is highly redundant. To be sure that the RNA is hybridized to only the unique DNA sequences, in the DNA-RNA hybrid molecules the RNA is degraded enzymatically and the remaining DNA is subjected to a reassociation test to determine its C_0t curve. This "play-back" then reveals whether all the DNA so isolated, includes only genic DNA and is not redundant. Such studies may assist in estimating the number of housekeeping genes plus the genes that were transcribed when the RNA was collected. ►c_0t, ►housekeeping genes, ►gene number

PLC: Phospholipase C. ►phospholipase

Pleated Sheets: Relaxed β-configuration polypeptide chains hydrogen-bonded in a flat layer. ►protein structure

Pleckstrin Domain: The pleckstrein domain is approximately 100-amino acids in length and occurs in many different proteins such as serine/threonine kinases, tyrosine kinases, and the substrates of these kinases, phospholipase C, small GTPase regulators, and cytoskeletal proteins. Pleckstrin domains may participate in various signaling functions; they bind phosphatidylinosotol 4,5-bisphosphate. Pleckstrin is a substrate of protein kinase C in activated platelets. separate entries, ►PH, ►SHC, ►SH2, ►SH3, ►WW, ►PTB, ►adaptor proteins, ►phosphatidylinositol, ►platelets, ►desensitisation, ►phosphoinositides; Lemmon MA, Ferguson KM 1998 Curr Top Microbiol Immunol 228:39; Rebecchi MJ Scarlata S 1998 Annu Rev Biophys Biomol Struct 27:503.

Plectin: A 500 kDa keratin of the cytoskeleton encoded in human chromosome 8q24. ►epidermolysis ►[bullosa simplex], ►keratin

Plectonemic Coils: The DNA double helix represents plectonemic coils (see Fig. P85). Here, the two coils are wound together, therefore they can be separated only by unwinding rather than simple pulling apart, like in paranemic coils. ►paranemic coils

Figure P85. Plectonemic coil

Pleiomorphic: A pleiomorphic organism displays variable expression (without a genetic basis for the special changes).

Pleiomorphic Adenoma: A salivary gland tumor caused by human chromosome breakage points, primarily at 8q12, 3p21, and 12q13-15. The translocation t(3;8)(p21;q12) results in swapping the promoters of PLAG1, a Zn-finger protein encoded in chromosome 8 and β-catenin (CTNNB1), and activation of the oncogene. ►Zinc finger, ►β-catenin

Pleiotrophin (PTN): A 18 kDa heparin-binding cytokine, inducible by the platelet derived growth factor (PDGF). It is 50% identical with retinoic acid-inducible midkine growth factor, which like PTN is also a growth and differentiation factor. PTN reduces cell colony formation, interacts with receptor protein tyrosine phosphatase, and leads to tumor growth, angiogenesis, and metastasis. PTN regulates phosphorylation of serine 713 and 726 of β-adducin by activating protein kinase C and mediates its translocation to the nucleus. This phosphorylation contributes to uncoupling of the adducin–actin–spectrin complexes and stabilizes the cytoskeleton (Pariser H et al 2005 Proc Natl Acad Sci USA 102:12407). ►adducin, ►actin, ►protein kinase, ►spectrin, ►cytoskeleton; Meng K et al 2000 Proc Natl Acad Sci USA 97:2603; PTN in breast cancer: Chang Y et al 2007 Proc Natl Acad Sci USA 104:10888.

Pleiotropy: One gene affects more than one trait; mutation in various elements of the signal transduction pathways, in general transcription factors, or in ion channels may have pleiotropic effects. The existence of pleiotropy has been questioned with the emergence of the one gene–one enzyme theory. Earlier, it was inconceivable on the basis that one tract of DNA could code for more than a single function ("Pleiotropism non est… that is the dogma", p 161. In Genetics, 1959 Sutton EH (ed), Josiah Macey Found, New York). However, it has been since shown that mutation at different sites within single mitochondrial tRNA genes may lead to several different human diseases. Cytokines involved in signaling through different receptors in different pathways are pleiotropic molecules. The complete sequence of the *Drosophila* genome shows that ~13,601 genes encode ~14,113 transcripts indicating that a minimum of nearly 4% of the genes display pleiotropy. An analysis of 150,000 high-abundance human proteins derived from two-dimensional gels indicated an average of 10 isoforms per protein following the (MALDI-TOF) matrix-assisted laser desorption ionization/time of flight mass spectrometry (Humphery-Smith I 2004, p 2 In: Albala JS, Humphery-Smith I (Eds.) Protein Arrays, Biochips, and Proteomics, Marcel Dekker, New York). In yeast, pleiotropy is attributable to multiple consequences of single functions (He X, Zhang J 2006 Genetics 173:1885). *Antagonistic pleiotropy* claims that evolution does not work against variations, which adversely affect the individuals after the completion of the reproductive stage of life, and the alternative genotypes display opposite phenotypes. Actually, the genes displaying antagonistic pleiotropy can be silent in early life but are harmful later and contribute to aging; yet, they may have some selective advantage early and this assures their maintenance in populations.

The F1F0-ATP synthase is a ubiquitous mitochondrial enzyme that works as a rotary motor, harnessing the electrochemical proton gradients to carry out ATP synthesis from ADP and inorganic phosphate. It is composed of a membrane-embedded proton-translocating sector (F0), coupled to a soluble sector (F1) that contains catalytic sites for ATP synthesis/hydrolysis. Several of the most deleterious human mitochondrial diseases, such as the maternally inherited Leigh syndrome, neurogenic ataxia, retinitis pigmentosa, and some cases of Leber hereditary optic neuropathy are caused by point mutations in the mitochondrial *ATP6* gene that encodes subunit 6 of the ATP synthase F0 sector. The same single point mutation can produce either Leigh syndrome or retinitis pigmentosa, depending on the mtDNA mutation load. The mtDNA mutations most frequently associated with retinitis pigmentosa or Leigh syndrome are T8993G, T8993C, T9176G, and T9176C, which replace the conserved leucine residues at positions 156 or 217 of subunit 6 by arginine or proline, respectively. The primary molecular pathogenic mechanism of these deleterious human mitochondrial mutations is functional inhibition in a correctly assembled ATP synthase (Cortés-Hernández P et al 2007 J Biol Chem 282:1051). ►signal transduction, ►transcription factors, ►epistasis, ►two-hybrid method, ►mitochondrial diseases in humans, ►MALDI-TOF, see Fig. P86.

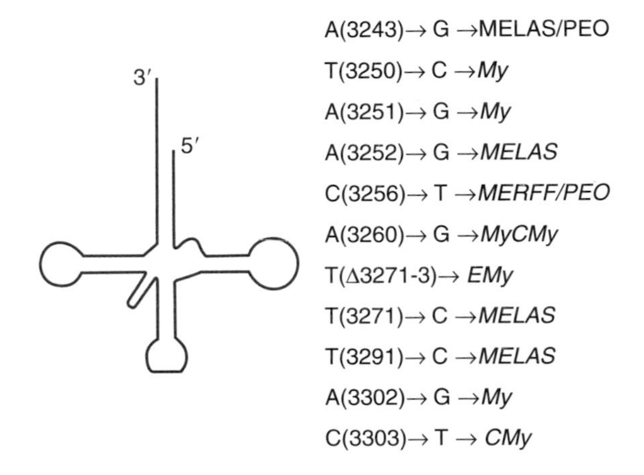

Figure P86. Pleiotropic mutations in the mtDNA Leu tRNA^{UUR} gene. The diagram displays the mutations in the human mitochondrial Leu tRNA^{UUR}. The first letter indicates the base that is changed, in parenthesis is the nucleotide number at the physical map, after the → the substituted base is given and after → the diseases described under mitochondrial diseases in humans are identified with abbreviations. Redrawn after Moraes CT 1998, p 167 In: Singh KK (Ed.) Mitochondrial DNA Mutations in Aging, Disease and Cancer, Springer, New York

Pleomorphism: Carl Wilhelm Nägeli's nineteenth century suggestion claiming lack of hard heredity in bacteria and that they simply exist in a variety of pliable forms. This idea held back the development of bacterial genetics, although physicians like Robert Koch and the taxonomist W. Migula sharply criticized it and stated that it ignored facts known by the 1880s. ►*Hieracium*

Plesiomorphic: A trait in its more primitive state among several evolutionarily related species. ►apomorphic, ►symplesiomorphic, ►synapomorphic

Pleura: Serous (moist) membrane lining the lung or insects' thoracic cavity.

Plexins: Receptors for semaphorins. Plexin-B1 is activates GTPase for RAS (Oinuma I et al 2004 Science 305:862). ►semaphorins, ►RAS, ►GTPase

PLGA: See Fig. P87, ►angiogenesis

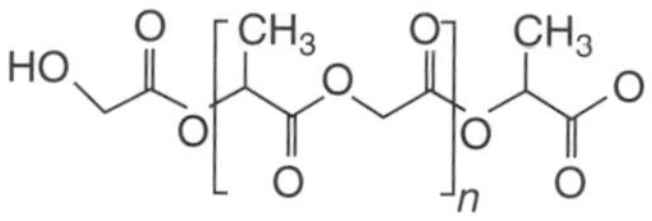

Figure P87. PLGA

Plk (polo-like kinase): Plk regulates the maturation of the centrosome, spindle assembly, the PICH checkpoint helicase, and the removal of cohesins, inactivates the anaphase promoting complex inhibitors, and controls mitotic exit and cytokinesis. ►polo, terms in alphabetical order; Baumann C et al 2007 Cell 118:101.

Ploidy: Ploidy represents the number of basic chromosome sets in a nucleus. The haploids have one set (x), the diploids two (xx), autotetraploids (xxxx), and so on. ►polyploidy

P-Loop: The ATP- and GTP-binding proteins have a phosphate-binding loop, the primary structure of which typically consists of a glycine-rich sequence followed by a conserved lysine and a serine or threonine. (See Saraste M et al 1990 Trends Biochem Sci 15:430).

PLTP (phospholipid transfer protein): The PLTP mediates the exchange of HDL cholesteryl esters with very low-density triglycerides and vice versa. ►HDL, ►cholesterol, ►CETP

Plug-In: A small circuit in a developmental function that can be present in several developmental networks.

Plum (*Prunus*): Basic chromosome number x = 7 but a variety of polyploid forms exist. (Bliss FA et al 2002 Genome 45:520).

Plumule: The embryonic plant shoot-initial.

Pluralism in Evolutionary Biology: Pluralism indicates sympatric speciation. ►sympatric

Pluripotency: A cell with pluripotency has the ability to develop into various, but not necessarily all types of tissues. Embryonic stem cells (from the inner mass of blastocysts), embryonic germ cells (primordial cells of the gonadal ridge), and the mesenchymal stem cells of the bone marrow possess pluripotency. Transcription factor Zfx controls the self-renewal of embryonic and hematopoietic stem cells (Galan-Caridad JM et al 2007 Cell 129:345). The good cultures may grow for more than 70 doublings ($2^{70} \geq 10^{20}$) and may be free of chromosomal defects. The ability of the embryonic stem cells to differentiate into many types of cells is regulated by MYC and Nanog proteins that regulate transcription factors, signal molecules, and suppress lineage specific cells. These two factors target a core set of 345 genes. The mouse and human MYC and Nanog target sites overlap in ~9 to 13% (Loh Y-H et al 2006 Nat Genet 38:431). Nanog proteins enable the reprogramming of somatic cells into pluripotent stem cells after fusion with embryonic stem cells of mouse (Silva J et al 2006 Nature [Lond] 441:997). Histone3 arginine26 methylation appears to be a crucial event in the formation of the pluripotent inner cell mass of the four-cell stage mouse embryos. CARM1 methyltransferase activity also upregulates Nanog and Sox2 proteins (Torres-Padilla A-E et al 2007 Nature [Lond] 445:214). ►totipotency, ►CARM1, ►MYC, ►Nanog, ►Sox, ►stem cells, ►ZFX; Donovan PJ, Gearhart J 2001 Nature [Lond] 414:92.

Plus and Minus Method (Sanger F et al 1975 J Mol Biol 94:441): The plus and minus method was an early version of DNA sequencing using dideoxy analogs of nucleosides (+ batch) during replication. After the analog was incorporated to a site, T4 exonuclease failed to continue degradation. In the minus (−) batch the synthesis stopped depending upon which single nucleotide was omitted (the precursor mixture containing only 3 deoxyribonucleotides). Thus nucleotide sequences of specific ends and length were generated and the fragments of different lengths were analyzed by electrophoresis. The Sanger et al (1977 Proc Natl Acad Sci USA 74:5463) method and its improvements replaced it. ►DNA sequencing

Figure P88. Plus and minus strands

Plus End: The preferential growing end of microtubules and actin filaments. ►minus end

Plus Strand: The plus strand of the single stranded DNA or RNA of a virus is represented in the mature virion whereas the minus strand serves as a template for the transcription (replication) of the plus strand and the mRNA (see Fig. P88). In most cases, the plus strands are synthesized far in excess to the minus strands. (►replicative form, ►RNA replication). The plus strand viral genomic RNA serves directly as mRNA.

Plutonium (Pu): A metallic fissile element (atomic number 94, atomic weight 242) produced by neutron bombardment of uranium (U^{238}) during the production of nuclear fuel and used for making nuclear weapons. Radioactive Pu powers some heart pacemakers. Thus, the wearers, as well as his/her family members and surgeons, will be exposed to some radiation, generally below 1.28 Sv per person per year, a little more than the average natural background (the doses are additive, however). If the highly toxic particles of Pu are inhaled (the most common type of ingestion), the element may affect the lung and may eventually be preferentially deposited in the skeletal system, causing bone cancer by the emission of X and γ rays. Pu^{238} has a half-life of 86.4 years. It propels some space vehicles. Pu^{239} has a half-life of 24.3×10^3 years and targets primarily the bone marrow. Other Pu isotopes have an even longer half-life. The level of Pu may be detected by radioactivity in the urine and by instruments placed on the body. Appropriate instruments can detect as low as 4 nCi (nanoCurie) values. ►atomic radiation, ►isotopes, ►radiation hazard assessment, ►Curie

Plx1: A kinase that phosphorylates the amino-terminal domain of Cdc25. ►Cdc25

Plymouth Rock: A recessive white-feathered breed of chickens with the genetic constitution of *iicc*. The dominant *I* gene is a color inhibitor and *C* symbolizes color. ►White Wyandotte, ►Leghorn White

PLZF: The zinc-finger protein encoded in human chromosome 11q23. It normally represses the promoter of cyclin A but a transposition of RARα (retinoic acid receptor) results in transactivation of the cyclin A gene, and may be involved in the initiation of cancer. ►cyclin A, ►RAR, ►transcriptional activator, ►transactivator, ►leukemia [acute promyelotic leukemia].

PMA: See ►phorbol 12-myristate-13-acetate (Fig. P89)

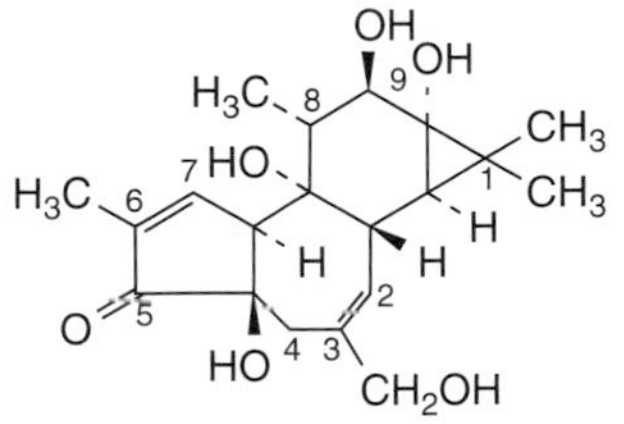

Figure P89. Phorbol

PMAGE (polony multiplex analysis of gene expression): PMAGE can detect single mRNA molecules in three cells (Kim JB et al 2007 Science 316:1481). ►polony

pMB1: ►ColE1

PMCA (protein misfolding cyclic amplification): ►prion

PMDS (persistence of Müllerian duct syndrome): ►Müllerian ducts

PME: Point mutation element where in the 3′-UTR regulatory proteins may bind and cause developmental switching.

PMF (peptide-mass fingerprinting): A method of rapid identification of proteins without sequencing but using mass spectrometry information. ►MALDI; Jonsson AP 2001 Cell Mol Life Sci 58:868.

PML (promyelotic leukemia): A putative Zinc finger protein, encoded in human chromosome 15q21. Formerly, this gene was called MYL. There are about 10–20 PML bodies of ~0.3–1 μm per mammalian nuclear matrix. In acute promyelotic leukemia, these bodies become disorganized as the PML-RARα oncogenic complex is formed. PML bodies are associated with caspase- and FAD-induced apoptosis. Casein kinase 2 (CK2) promotes PML ubiquitin-mediated degradation by phosphorylation at serine 517. In case of resistance to CK2, tumor-suppressor activity of PML is enhanced (Scaglioni PP et al 2006 Cell 126:269). Overexpressed PML also promotes apoptosis but without an enhanced caspase-3 activity. In the absence of PML ($PML^{-/-}$), the cells become resistant to ionizing radiation. ►leukemia, ►PLZF, ►nuclear matrix, ►RAR, ►apoptosis, ►POD, ►AKT; Lallemand-Breitenbach V et al 2001 J Exp Med 193:1361.

PML39: A yeast upstream effector regulating Mlp1/Mlp2 nucleopore-associated proteins suppressing nuclear export of un-spliced mRNA (Palancade B et al 2005 Mol Biol Cell 16:5258). ►nuclear pores

PMS1 (2q31-q33), **PMS2** (7q22) yeast homologs and colorectal cancer: Increased post-meiotic segregation in yeast and increased colorectal cancer (or Turcot syndrome) in humans due to mismatch repair deficiency. ►mismatch repair, ►colorectal cancer, ►Turcot syndrome

PN-1 (protease nexin): A 43 kDa inhibitor of serine proteases (thrombin, plasminogen activator). It is involved in the development of embryonic organs (cartilage, lung, skin, urogenital system, and nervous system). PN-1 is abundant in the seminal vesicle and its dysfunction leads to male infertility. ►thrombin, ►plasminogen activator, ►urokinase, ►nexin, ►infertility, ►claudin-11; Murer V et al 2001 Proc Natl Acad Sci USA 98:3029.

pN: ►Newton

PNA: ►peptide nucleic acid

Pneumococcus: ►*Diplococcus pneumoniae*

Pneumocystis carinii: A group of pathogenic ascomycetes with special susceptibility to immune-compromised individuals (e.g., AIDS patients) and rodents. It carries about 3740 genes in the about 8 Mb genome. It reproduces both asexually and sexually. ►acquired immunodeficiency, ►ascomycete; Kolls JK et al 1999 J Immunol 162:2890.

PNPase: ►polynucleotide phosphorylase

Pocket: The motif of the retinoblastoma (RB) tumor-suppressor protein family that binds to viral-DNA coded oncoproteins. Binding of RB to the E2F family of transcription factors blocks transcription, needed for the progression of the cell cycle. The pocket proteins share this retinoblastoma (RB) motif. ►E2F1, ►tumor suppressor, ►cell cycle, ►transcription factors, ►retinoblastoma, ►p107, ►p130; Botazzi ME et al 2001 Mol Cell Biol 21:7607.

POD (PML-oncogenic domain): ►PML

Podophyllotoxin (epipodophyllotoxin): Antimitotic plant product.

Podosomes (invadopodia): Actin-containing electron-dense adhesion structures on human primary macrophages, Src-transformed fibroblasts, and in some cancer cells, controlling cell motion migration and immune reactions. N-WASP WH2 nucleation promoting protein domains capture the barbed end (the protrusive attachment structure of actin) to the podosome (Co C et al 2007 Cell 128:901). ►Src, ►macrophage, ►actin, ►WASP, ►immune reaction; Linder S, Aepfelbacher M 2003 Trends Cell Biol 13:376.

P

Podospora anserina: n = 7, is a genetically well-studied ascomycete fungus.

Pof (Painting of fourth): *Drosophila* protein that binds only to the small 4th chromosome.

pogo: ►hybrid dysgenesis

Poikilocytosis: A hemolytic anemia with variable-shape red blood cells. The defect is due to the reduction of ankyrin binding sites or mutation in spectrin. ►ankyrin, ►spectrin, ►anemia

Poikiloderma Atrophicans (poikiloderma telangiectasia): ►Rothmund-Thompson syndrome

Poikiloploidy: In poikiloploidy, different cells of the body have different numbers of chromosomes.

Poikilothermy: In poikilothermy, the body temperature or the organism depends on the surrounding environmental conditions.

Point Mutation: Point mutations do not involve detectable structural alteration (loss or rearrangement of the chromosome), and are expected to involve base substitutions. The point mutation rate per locus in eukaryotes in about 10^{-5} and may vary from locus to locus and among various organisms. The rate per nucleotides of a locus is in the range of 10^{-8}. ►substitution mutation; Krawczak M et al 2000 Hum Mut 15:45.

Point-of-Care Technologies: Point-of-care technologies use small bench top analyzers (for example, saliva, blood gas, and electrolyte systems) and hand held, single use devices (such as urine albumin, blood glucose, and coagulation tests). Hand held devices have been developed using microfabrication techniques. They are outwardly simple but internally complex devices that perform several tasks for example, separate cells from plasma, add reagents, and read color or other end points (Price CP 2001 BMJ 322:1285). These are also called bedside technologies, because the samples do not have to be transferred for analysis to laboratories and therefore are much faster, especially when the newest microfluidic devices are used. ►microfluidics; Herr AE et al 2007 Proc Natl Acad Sci USA 104:5268.

Poise: ►viscosity, ►stoke

Poison Sequence: A poison sequence may be present in the genomes of some RNA viruses and thus even their cDNA cannot be cloned in full length in bacterial hosts. The problem may be overcome by propagating it in segments. (Brookes S et al 1986 Nucleic Acids Res 14:8231).

Poisson Distribution: Basically, an extreme form of the normal distribution, found when in large populations rare events occur at random, such as e.g., mutation. The general formula is e^{-m} $(m^i/i!)$, and expanded $e^{-m}(m^0/0!, m^1/m!, m^2/2! \ldots m^i/i!)$, where e = base of natural logarithm ($\cong$ 2.718), m = mean number of events, i = the number by which a particular m is represented at a given frequency, ! = factorial (e.g., 3! = 3 × 2 × 1, but 0! = 1). (See Fig. P90, ►negative binomial).

Pokemon: A repressor of the tumor suppressor ARF; thus, it represents a protooncogene. ►ARF, ►oncogenes; Maeda T et al 2005 Nature [Lond] 433:278.

poky (synonym: *mi-1*): A slow-growing and cyanide-sensitive respiration defective mitochondrial mutation in *Neurospora*. The basic defect appears to be a four-base deficiency of the 15 bp consensus at the 5′-end of the 19S rRNA of the mitochondria. Because of this defect, a further upstream promoter is used, making the transcript longer but during processing, shorter RNAs are made. It is analogous to the petite colony mutations of budding yeast. ►petite colony mutation, ►stoppers, ►mtDNA; Akins RA, Lambowitz AM 1984 Proc Natl Acad Sci USA 81:3791.

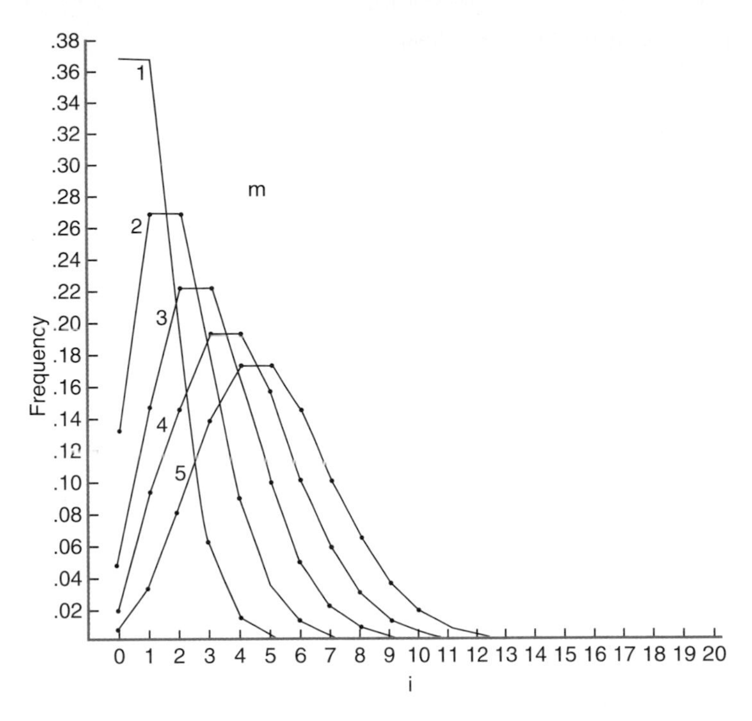

Figure P90. The Poisson distribution. Each curve corresponds to a numbered m value. The i classes represent the distribution of each mean value (m) with the ordinate indicating the frequencies

pol (bacterial RNA polymerase): pol synthesizes all the bacterial and viral RNAs in the bacterial cells. Its subunits are $\alpha\alpha\beta\beta'$ and σ. The σ subunit identifies the promoter sequences and is required for the initiation of transcription within the cell. After about a half dozen nucleotides are hooked up, it dissociates from the other subunits and further polymerization continues with the assistance of elongation protein factors. ►transcription, ►pol I, ►pol II, ►pol III eukaryotic RNA polymerases

pol I: Prokaryotic DNA polymerase, where the polymerase (Klenow fragment) and exonuclease functions are located about 30 Å distance apart in a subunit, and editing (removal of wrong bases) follows the melt and slide model. It plays a major role in prokaryotic repair and in the extension of the Okazaki fragments for joining them into a contiguous strand by ligase. It adds 10–20 nucleotides/second to the chain and so it is much slower than pol III. ►melt and slide model, ►DNA replication, ►replication fork, ►Klenow fragment, ►pol III, ►DNA ligase

pol I: RNA polymerase involved in the synthesis of ribosomal RNA (except 5S rRNA) in eukaryotes. By endonucleolytic cleavage, it generates the 3′ end of rRNA from longer transcripts. The upstream control regions for transcription initiation (binding proteins) vary from species to species. The human RNA pol I requires an activator UBF (upstream binding factor) and promoter selectivity factor SL1, including the TBF (TATA box binding protein) and associated subunits, TAF_I 110, TAF_I 63, and TAF_I 48. The former two keep contact with the promoter, whereas TAF_I 48 interacts with UBF and prevents RNA pol II from using this promoter site. ►ribosome; Reeder RH 1999 Progr Nucleic Acid Res Mol Biol 62:293; Grummt I 1999 Progr Nucleic Acid Res Mol Biol 62:109.

pol II: Prokaryotic DNA polymerase, the functions of which are not completely defined so far; it has known role in repair. ►DNA repair

pol II: The RNA polymerase transcribes messenger RNA and most of the snRNAs of eukaryotes with the assistance of different transcription factors. Nine or ten of its subunits are very similar to other polymerases; pol II has four to five smaller unique subunits. The two largest subunits are very similar in the three eukaryotic RNA polymerases and are similar also to prokaryotic subunits. It is most sensitive to α-amanitin inhibition (0.01 μg/mL). The site of sensitivity is in the largest 220 kDa polypeptide. This large subunit is activated by phosphorylation. At the carboxy terminal, there are 26 (yeast), 40 (*Drosophila*), or 52 (mouse) heptapeptide (Tyr-Ser-Pro-Thr-Ser-Pro-Ser) repeats. These repeats are essential for function. The Ser and Thr residues may be phosphorylated. Phosphorylation of the carboxy-terminal domain may affect the promoter-specificity of the enzyme. The C-terminus of (CTD) of the large subunit is instrumental also in the processing of the 3′-end of the transcript and the termination of transcription downstream of the polyA signal. CTD does not seem to affect initiation of transcription but it also mediates the response to enhancers. This enzyme is different from RNA pol I and RNA pol III inasmuch that it requires a hydrolyzable source of ATP for the initiation of transcription. RNA pol II is different from the other polymerases in its requirement for a large array of special transcription factors that modulate the transcription of the thousands of proteins. ►transcription factors, ►regulation of gene activity, ►α-amanitin, ►transcription factories, ►RNA polymerase; Cramer P et al 2001 Science 292:1863.

pol III: Prokaryotic DNA polymerase, where the α subunit carries out the replication function and the ε-subunit is involved in editing (exonuclease) activity. It plays a major role in the replication of the leading and lagging strands. The replication has a speed of ≈1 kb/sec. There are only about 10–20 copies of the 10-subunit holoenzyme/cell. ►DNA replication, ►replication fork, ►core polymerase, ►replisome

pol III: RNA polymerase involved in the synthesis of transfer RNA, 5S rRNA, 7S rRNA and U6 snRNA in eukaryotes. Transcription of pol III is higher during S and G2 phases of the cell cycle than during G1. Many neoplastic cells display high pol III activity indicating that protein synthesis is demanded for tumorous growth. The RET protein appears to be a suppressor of increased pol III activity. ►tRNA, ►ribosomal RNA, ►ribosomes, ►La; Geiduschek EP, Tocchini-Valentini GP 1988 Annu Rev Biochem 57:873; Huang Y, Maraia RJ 2001 Nucleic Acids Res 29:2675.

P

pol IV: A low-fidelity lesion bypass DNA polymerase belonging to the Y family. ►Y-family DNAS polymerases, ►RNA polymerase

polα: DNA polymerase (encoded in fission yeast by gene *pol1/swi7*), replicating the nuclear DNA (lagging strand) in cooperation with the primase of eukaryotes. Its mutation may result in mutator activity (Gutiérrez PJA, Wang TS-F 2003 Genetics 165:65). ►lagging strand, ►replication fork, ►DNA polymerases, ►primase

pol β: A eukaryotic DNA repair polymerase. ►DNA polymerases

pol δ: A eukaryotic DNA polymerase (replicating the leading strand) of the nuclear chromosomes. ►replication fork, ►DNA polymerases

pol δ_2: Synonymous with pol ε. ►DNA polymerases

pol ε: A eukaryotic DNA polymerase (*cdc20*) with repair role. ►DNA polymerases

pol γ: A DNA polymerase replicating eukaryotic organelle DNA. ►θ type replication, ►DNA polymerases

pol ζ: A eukaryotic DNA polymerase without exonuclease activity. It is a repair enzyme inasmuch that it can bypass pyrimidine dimers more efficiently than polα. It is insensitive to 200 μM aphidi-colin (and in this respect it is similar to polβ and polγ) and also insensitive to dideoxynucleotide triphosphates (which inhibit polβ and polγ). It is moderately sensitive to 10 μM butylphenyl-guanosine triphosphates. It is relatively inactive with salmon sperm DNA or primed homo-polymers. ►DNA polymerases

Poland Syndrome: An autosomal dominant defect with low penetrance. the teratogenic effects of diverse exogenous factors complicate the inheritance pattern. It is characterized by fusion of fingers (syndactily), short fingers, and anomalies of chest and, sometimes, other muscles. ►limb defects, ►syndactily, ►penetrance, ►teratogenesis

Polar: Hydrophilic, i.e., soluble in water; molecules with polarized bonds.

Polar Body: ►gametogenesis in animals

Polar Body Diagnosis: In polar body diagnosis, the genetic constitution of the polar body is tested by molecular techniques prenatally. ►prenatal diagnosis

Polar Bond: A polar bond is covalent, yet the electrons are more firmly tied to one of the two molecules and therefore the electric charge is polarized.

Polar Coordinate Model: The polar coordinate model of regeneration states that when cells are in non-adjacent positions, the process of growth restores all intermediate positions by the shortest numerical routes. The shortest intercalation mandates that small fragments may undergo duplication and large fragments may require regeneration. The position of each cell on a collapsed cone (the idealized primordium) is specified by the radial distance from a central point at the tip of the cone and the circumferential position on the circle defined by the radius of the base. ►distalization; Held LI 1995 Bioessays 17:721.

Polar Cytoplasm: Polar cytoplasm is situated in the posterior (hind) portion of the fertilized egg cell. ►pole cells

Polar Ejection Force (PEF): Microtubule

Polar Granules: The polar granules are present in the posterior pole region of insect eggs and have maternal effect and germ cell specification roles during embryogenesis. These granules are the mitochondrially coded 16S ribosomal RNA large subunits (mtRNA), exported from that organelle. ►animal pole, ►morphogenesis in *Drosophila*, ►RNA localization; Strom S, Lehmann R 2007 Science 316:392.

Polar Molecule: A polar molecule is generally soluble in water; the distribution of the positive and negative charges are not even, thus resulting in a polarized effect.

Polar Mutation: A polar mutation may be a base substitution (nonsense mutation), insertion, frame shift, or any chromosomal alteration that affects the expression of genes down-stream in the transcription–translation system. ►frame shift mutation; Jacob F, Monod J 1961 Cold Spring Harbor Symp Quant Biol 26:193.

Polar Nuclei: The polar nuclei occur in the embryosac of plants, and are formed at the third division of the megaspore. After they have fused (n + n) and have been fertilized by one sperm (n) they give rise to the triploid (3n) endosperm nucleus. ►megagametophyte, ►embryosac

Polar Overdominance: An unusual type of inheritance, i.e., mutants heterozygous for the dominant *callypige* gene of sheep (chromosome 18) display the (*CLPG*) allele only when inherited from the males but not from the females. The phenotype is a muscular hypertrophy resulting from the cis-regulation of four imprinted genes. ►imprinting, ►overdominance; Charlier C et al 2001 Nature Genet 27:367; Smit M et al 2003 Genetics 163:453.

Polar Transport: Certain metabolites move only in one direction in the plant body, e.g., the auxins under natural conditions are synthesized in the tissues over the ground and then move toward the roots.

Polarimeter: The polarimeter measures the rotation of the plane of polarized light.

Polarisome: A polarisome defines polarity within a cell with the aid of several proteins. (See Weiner OD 2002 Curr Opin Cell Biol 14:196).

Polarity, Embryonic: Embryonic polarity is required for differentiation and requires asymmetric cell divisions. In *Caenorhabditis*, the PAR proteins (serine/threonine kinase) control embryonic polarity and a non-muscle type myosin II heavy chain protein (NMY-2) is a cofactor of this polarity. Upon fertilization the Rho guanosine triphosphatase-activating protein CYK-4—enriched in the sperm—and the RhoA guanine exchange factor ECT-2 modulate myosin light chain activity and create an actomyosin gradient, which determines the anterior domain in the one-cell embryo in *Caenorhabditis* (Jenkins N et al 2006 Science 313:1298). In *Drosophila*, the major body axes, primarily the anterior-posterior polarity are controlled by the gurken-torpedo gene products, but other genes are also involved. Polarity may be achieved either by the asymmetric distribution of proteins or mRNA. ►morphogenesis in *Drosophila*, ►differentiation, ►polar cytoplasm, ►RNA localization, ►BUD, ►RHO, ►myosin, ►actomyosin; Drees BL et al 2001 J Cell Biol 154:549; Wodarz A 2002 Nature Cell Biol 4:E39; Frizzled pathway: Seifert JRK, Mlodzik M 2007 Nature Rev Genet 8:126.

Polarity of Hyphal Growth: See Fig. P91, Nelson WJ 2003 Nature [Lond] 422:766.

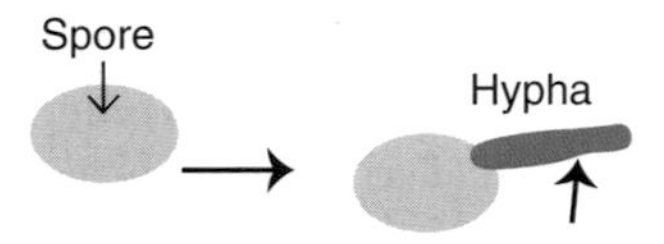

Figure P91. Polarity of hyphal growth

Polarity Mapping: ►mapping mitochondrial genes

Polarization: The distortion of the electron distribution in one molecule caused by another. ►bouquet of chromosomes

Polarized Differentiation: the basis of morphogenesis, chemotactic response, response to pheromones, etc. Polarized differentiation and growth is typical for neural and microtubule growth, for the pollen tubes, and for the roots of plants. Neuronal polarization requires the activity of SAD kinases in mammals (Kishi M et al 2005 Science 307:929). ►asymmetric cell division; Hepler PK et al 2001 Annu Rev Cell Dev Biol 17:159; Science [2002] 298:1941–1964.

Polarized Light: Polarized light exhibits different properties in different directions at right angles to the line of propagation. Specific rotation is the power of liquids to rotate the plane of polarization.

Polarized Recombination: ►polarized segregation

Polarized Segregation: Polarized segregation may be brought about by meiotic anomalies, e.g., in maize plants heterozygous for some knobbed chromosomes (and syntenic markers) are preferentially included into the basal megaspore. Polarized segregation has been observed as a result of gene conversion, e.g., in *Ascobolus immersus* alleles of the *pale* locus in the cross $\frac{188w^+}{188^+w}$ segregated in both cases, it is 6:2, but in the first case the results were (4*[188]* + 2 *[w$^+$]*): 2 (*w*) whereas in the cross $\frac{w137^+}{w^+137}$ the conversion asci were (4 *[w]* + 2 *[137]*): 2 (w^+). The genetic order of these alleles were *188 w 137*. Thus in the first cross *white* was in the minority class whereas in the second cross it was part of the majority class. ►gene conversion, ►meiotic drive, ►map expansion; Whitehouse HLK, Hastings PJ 1965 Genet Res 6:27.

Polarizing Microscope: The polarizing microscope uses a *polarizer* (a polaroid screen) in front of the light beam and an *analyzer* (permitting rotation) over the eyepiece. The anisotropic specimens (having difference in transmission or reflection depending on the angle of light) will display optical contrast. ►microscopy

Polarography: Electrochemical measurement of reducible elements.

Polaroid Camera: The Polaroid camera was developed during the 1940s and has found many uses in biological laboratories because it can provide almost immediate negative or positive images for recording observations such as those of electrophoretic gels. The combined developing and fixing solution is contained in between the exposed negative film and the receiving film or paper and when the storage "pod" bursts under pressure of pulling, the processing is carried out within the camera. For the majority of tasks, the digital cameras are even better suited for fast imaging.

Polaron: The part of a locus within which gene conversion (or recombination) is polarized. ►gene conversion, ►polarized segregation; Whitehouse HLK, Hastings PJ 1965 Genet Res 6:27.

Pole Cells: Pole cells localized in the posterior-most part of the cellularized embryo and give rise eventually to the germline. ►germline

Polintons: Typically, 15–20 kb long transposable elements in a wide range of lower and higher eukaryotes. They require a unique set of proteins for transposition such as protein-primed DNA polymerase B, retroviral integrase, cysteine protease, and ATPase. They show a 6 bp target site duplication and long inverted terminal repeats with 5′-AG and 3′-TC termini (Kapitonov VV, Jurka J 2006 Proc Natl Acad Sci USA 103:4540). ►transposable elements

Polioviruses: Icosahedral single-stranded RNA viruses with about 6.1 kb RNA in a total particle mass of about 6.8×10^6 Da; Type 1 was responsible for about 85% of the polyomyelitis (infantile paralysis) cases before successful vaccination (live oral, Sabin or inactivated, Salk) began to be widely used in the developed countries (see Fig. P92). These small RNA viruses are highly mutable because their genetic material lacks repair systems. The three serotypes produce a cell-surface receptor (PVR) by alternative

Figure P92. Apparent polio-stricken leg on a more than 3000 year old Egyptian hieroglyph

splicing of its transcript. Infectious poliovirus has been sythetized *de novo* without a natural template (Cello J et al 2002 Science 297:1016). Susceptibility to poliovirus was located to human chromosome 19q12-q13. Mice are very resistant to this virus because they lack the membrane receptor for the infection. ►picornaviruses, ►IRES, ►synthetic genes

Polled: A dominant/recessive gene (PIS) in goats and cattle, responsible for lack of horns/inter-sexuality. The forkhead transcription factor (FOXL2)—responsible also for blepharophimosis—may be involved. In goats, an 11.7 kb deletion at 1q43 (homologous to human band 3q23) normally encodes two mRNAs. The FOXL2 transcript is homologous with the human blepharophimosis syndrome gene. ►blepharophimosis; Crisponi L et al 2001 Nature Genet 27:159; Pailhoux E et al 2001 Nature Genet 29:453.

Pollen: The male gametophyte of plants developing from the microspores by two postmeiotic divisions (see Fig. P93). The first division results in the formation of a vegetative and a generative cell. The round vegetative cell directs the elongation of the pollen tube growing through the pistil toward the ovule. Pollen tube growth is guided by sporopohytic secretions (GABA, arabinogalactans and

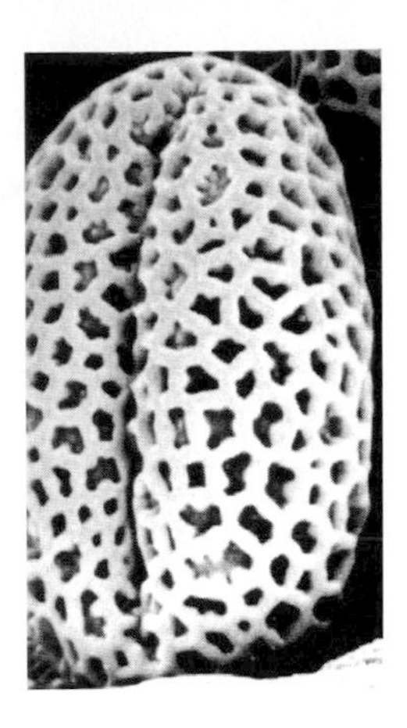

Figure P93. The sculptured surface of the mature pollen of *Arabidopsis*. (From Craig S, Chaudhury A 1994 In: Bowman JL (Ed.) *Arabidopsis*: An Atlas of Morphology and Development. Courtesy of Bowman JL By permission of Springer-Verlag, New York)

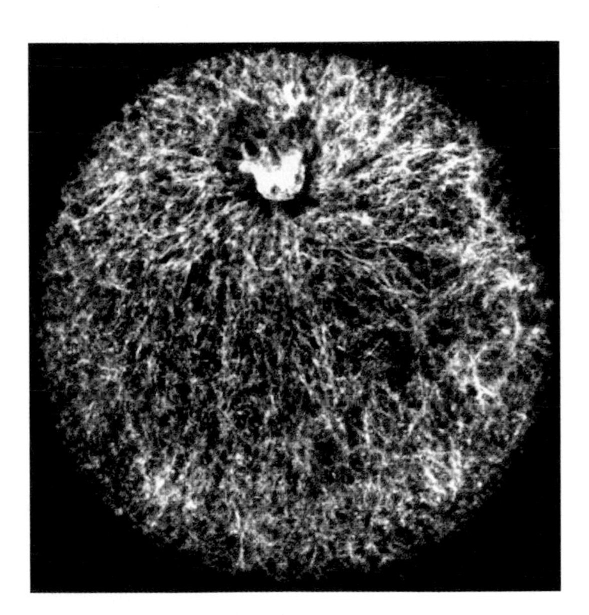

Figure P94. Fluorescent-phalloidin staining of the F-actin cytoskeleton in maize pollen. The bright spot on top is the pollen tube initial. (Courtesy of Dr. Chris Staiger. See Gibbon BC et al 1999 Plant Cell 11:2349)

proteins) (see Fig. P94). Synergids provide attraction by the 94-amino acid ZmEA1 protein in maize (Márton M et al 2005 Science 307:573).

The crescent-shaped generative cells may divide before or after the shedding of the pollen grains. One of them fertilizes the egg and thus, gives rise to the diploid embryo, the other fuses with the diploid polar cell in the embryosac and thus contributes to the formation of the endosperm. The pollen tube elongates quite rapidly; it may grow 15 cm in just 5 to 15 h. A protein that is glycosylated in that tissue regulates the pollen tube elongation. In allogamous species, a single individual may shed over 50,000,000 pollen grains whereas in autogamous species the number of pollen grains per anther may not exceed a couple of hundreds. Since the pollen grain is haploid and may be autonomous (gametophytic control), it may express its genetic constitution independently from the genotype of the anther tissues (e.g., waxy pollen, various color or sterility alleles), in some instances, however, the morphology of the pollen grain is under sporophytic control. Since the pollen is a more independent product than the megaspore, it is more likely to suffer from genetic defects for which the surrounding tissues cannot compensate, therefore pollen sterility is more common in plants than female sterility. However, pollen sterility may not necessarily affect the fertility of the individuals because of the abundance of functional pollen grains in case of heterozygosity for the defects. Under normal atmospheric conditions (high humidity) and high temperature, the viability of the pollen is maintained (depending on the species) for a few minutes or for several hours. In a refrigerator, at low humidity the viability of the pollen can be extended substantially. Freeze-dried and properly stored pollen of several species retains its ability to fertilize for years. Insects may carry viable pollen for long distances. According to a study, in rye populations cross-pollination (mediated by wind) may occur to 50% at 100 m distance and to 20% at about 400 m distance, but only to 3% at 600 to 700 m. Other studies reported in rye only 7% cross-pollination within a distance of 20 m. Creeping bentgrass (*Agrostis stolonifera*) pollen, transgenic for a resistance marker (5-enolpyruvylshikimate-3-phosphate synthase), was spread by the wind primarily within a distance of 2 km but some dispersal occurred up to 21 km (Watrud LS et al 2004 Proc Natl Acad Sci USA 101:14533). The prevailing environmental conditions (humidity, temperature, wind, etc.) and the quantity of the pollen influences the spread and viability. These problems gained new interest with the use of genetically engineered crops that are opposed by some environmentalists. The extracellular matrix of the pollen contains proteins, which recognize species-specificity and efficient pollination (Myfield JA et al 2001 Science 292:2482). These proteins are lipid-binding oleosins and lipases. ►microsporogenesis, ►gametogenesis, ►pollen tetrad, ►gametophyte, ►self-incompatibility, ►cross-pollination, ►GMO, ►autogamy, ►allogamy

Pollen Competition: ►certation

Pollen-Killer: Pollen-killer or spore-killer genes in wheat, tomato, and tobacco render the pollen incapable of functioning effectively in fertilization and may cause segregation distortion. ►segregation distorter, ►pollen tube competition, ►killer strains, ►killer plasmids, ►killer genes

Pollen Mother Cell: Microspore mother cell, microsporocyte. ►gametogenesis

Pollen Sterility: The inability of the pollen to function during fertilization. It can frequently be detected by the poor staining of the pollen grains with simple nuclear stains (acetocarmine, acetoorcein, etc.). Deletions, translocations, and inversion heterozygosity generally result in pollen sterility. Mitochondrial plasmids may also be responsible for some types of male sterility. ►pollen, ►certation, ►gametophyte, ►cytoplasmic male sterility, ►fertility restorer genes

Pollen Tetrad: The four products of a single male meiosis (see Fig. P95). The components of the pollen tetrad may not stick together and may shed in a scrambled state. In some instances (*Salpiglossis, Elodea*, some orchids), the tetrads remain together, however, in a way similar to the unordered tetrads of fungi. In *Arabidopsis*, induced mutations (*qrt1, qrt2, quartet*) cause the four pollen grains to stay together because of the alteration of the outer membrane of the pollen mother cell. Each tetrad may then fertilize four ovules. ►tetrad analysis

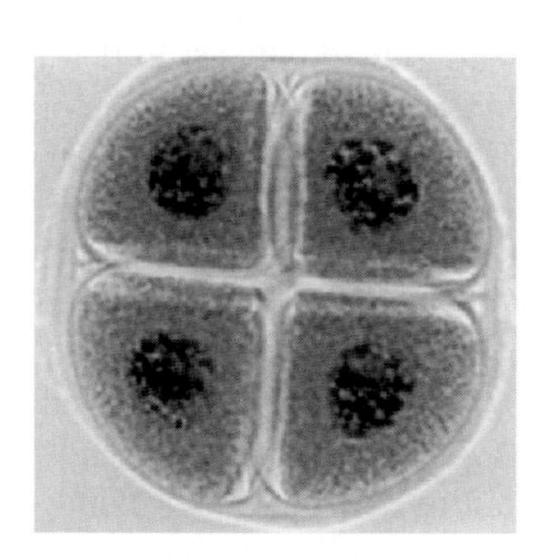

Figure P95. Pollen tetrad

Pollen Tube: In the majority of plants, the time between pollination and fertilization takes 24 to 48 h or less. In the alder tree (*Alnus*) the pollen tube travels slowly in the pistil (for about one month) because the ovary matures late and it arrives at fertilization in five successive steps. Also in some species, the ovaries may have multiple megaspore tetrads although, generally only one is fertilized (Sogo A, Tobe H 2005 Proc Natl Acad Sci USA 102:8770). ►pollen, elongating pollen tube in Figure P96, ►synergid, ►GABA, ►gametophyte, ►double fertilization; Palavinelu R, Preuss D 2000 Trends Cell Biol 10:517.

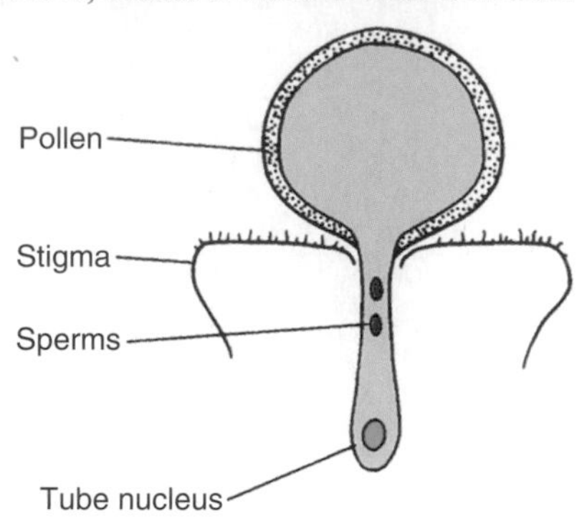

Figure P96. Pollen tube germination

Pollen-Tube Competition: ►certation

Pollination: The transfer of the male gametophyte to the stigmatic surface of the style (ovary). Obligate allogamous plant populations may suffer if the pollinator insect populations are reduced by adverse environmental conditions (Vamosi J C et al 2006 Proc Natl Acad Sci USA 103:956). ►gametophyte, ►autogamy, ►allogamy, ►fertilization, ►self-sterility

Pollinium: A mass of pollen sticking together and may be transported as such by the pollinator insects or birds.

Pollitt Syndrome: ►trichothiodystrophy

Pollution: Spoiling the environment by the release of unnatural, impure, toxic, mutagenic, carcinogenic, or any other undesirable and unaesthetic material, or to disturb nature by sound, odor, heat, and light. Pollution may cause mutation, cancer, and various other diseases. ►DNA-zyme; http://www.scorecard.org.

Polo: A 577-amino acid serine/threonine protein kinase of *Drosophila*, required for mitosis. It regulates centromeric cohesion protein MEI-S332 (Clarke AS et al 2005 Dev Cell 8:53). During interphase, it is predominantly cytoplasmic but at the end of prophase it associates with the chromosomes until telophase. Loss of polo kinase 4 allele(s) increases the probability of mitotic errors and cancer. In about 60% of the cells of mice, loss of heterozygosity takes place (Ko MA et al 2005 Nat Genet 37:883). ►CDC5, ►FEAR, ►cohesin; Llamazares S et al 1991 Genes Dev 5:2153; Alexandru G et al 2001 Cell 105:459.

Polony (polymerase colony): A small batch of DNA synthesized by PCR with the assistance of a primer to be used for sequencing. The PCR colonies on the glass microscope slide are the "polonies." On a single slide in the polyacrylamide film, as many as 5 million clones can be amplified. The amplified products stay at the vicinity of the linear DNA. If acrydite modification is used, the DNA is covalently attached to the polyacrylamide matrix and thus further enzymatic modifications are possible on all clones. The technology is well suited for genotyping and localizing of SNPs. ►PCR, ►SNIPs; Mitra RD, Church GM 1999 Nucleic Acids Res 27(24):e34; Mitra RD et al 2003 Proc Natl Acad Sci USA 1000:5926.

Poly I-G: A DNA strand containing more cytosine is called heavy chain of a DNA double helix because it binds more of the polyI-G (inosine-guanosine) sequences. Ultracentrifugation in CsCl separates these DNA heavy strands. ►ultracentrifuge, ►density gradient centrifugation, ►DNA heavy chain, ►inosine

PolyA⁺ Element: Transposons without long terminal repeats but poly-A sequences at the 3′-OH end. The RNA elements are usually mobilized via a DNA transcript with the aid of their encoded reverse transcriptase. Such elements are L1 (LINE) in mammals, the TART of *Drosophila*, the TRAS1 of silkworm, or the yeast Ty5 telomere-specific elements. Other polyA$^+$ elements (L1, I, and fungal and plant elements) can target a variety of other sites. ►TART, ►hybrid dysgenesis, ►transposable elements

polyA mRNA: Eukaryotic mRNAs post-transcriptionally polyadenylated at the 3′ tail before leaving the nucleus. Subsequently, in the cytoplasm, the tail may be reduced to 50–70 residues or further extended to hundreds. Polyadenylation improves the stability and efficiency of translation in cooperation with mRNA cap. The polyA tail and the mRNA cap seem to cooperate in the initiation of translation. PolyA tail is frequently added also to bacterial RNA. The addition of polyA tail accelerates the decay of RNA I of *E. coli*. All the data are consistent with polyadenylation being part of a quality control process targeting folded bacterial RNA fragments and nonfunctional RNA molecules to degradation. In *Escherichia coli*, polyadenylation may directly control the level of expression of a gene by modulating the stability of a functional transcript. Inactivation of poly(A)polymerase I causes overexpression of glucosamine-6-phosphate synthase (GlmS) and both the accumulation and stabilization of the *glmS* transcript (Joanny G et al 2007 Nucleic Acids Res 35:2494).

The majority of eukaryotic viruses (except areana- and reoviruses) also produce a poly A tail. In *Drosophila*, the length of the poly(A) tail may be correlated with the function in the differentiation of the mRNA. The regulatory mechanism of polyadenylation is interchangeable between mouse and *Xenopus*. Some genes use alternative polyadenylation sites (Edwalds-Gilbert G et al 1997 Nucleic Acids Res 25:2547). ►polyadenylation signal, ►mRNA tail, ►RNA I, ►PABP, ►mRNA degradation, ►capping enzymes, ►eIF; de Moor CH, Richter JD 2001 Int Rev Cytol 203:567; http://polya.umdnj.edu/.

polyA Polymerase (PAP): adds the polyA tail post-transcriptionally to the eukaryotic mRNA and anti-sense RNA transcripts (see Fig. P97). In yeast, at least two other genes *RNA14* and *RNA15* are involved in the processing of the 3′-end of the pre-mRNA.

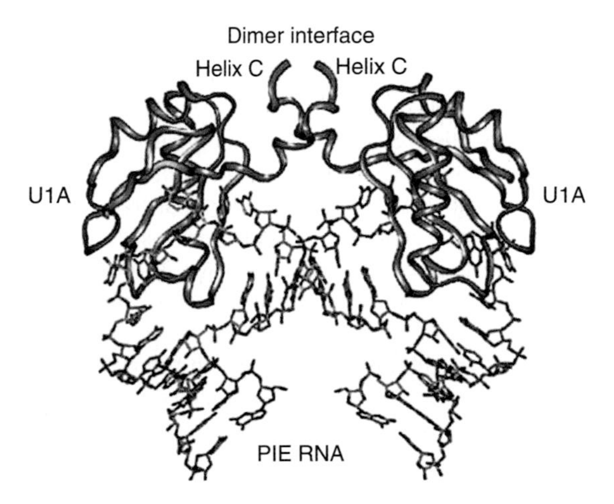

Figure P97. Poly(A) polymerase is regulated by inhibition (or stimulation) by the polyadenylation inhibition RNA element (PIE). PIE forms a trimolecular complex including the two U1A protein molecules. The U1A consists of a four-stranded β-sheet and two α-helices. The two C helices of the protein interface. The PIE RNA, which contains two asymmetric internal loops is separated by four Watson-Crick-paired nucleotides. (From Puglisi JD 2000 Nature Struct Biol 7:263)

E. coli also encodes at least two PAP enzymes. PolyA polymerase also facilitates the degradation of mRNA because it provides single-strand tails for polynucleotide phosphorylase. Bacterial PAP and tRNA nucleotidyl transferase are highly similar in structure but different in function inasmuch the latter catalyzes the addition of CCA to the 3′-end of tRNA. The C-terminal domain of nucleotidyl transferase restricts polymerization to these three nucleotides whereas a 27-aminoacid sequence determines whether the protein becomes a transferase or PAP. Both proteins have identical nucleotide recognition and incorporation domains (Betat H et al 2004 Mol Cell 15:389). ►mRNA tail, ►polyadenylation signal, ►polynucleotide phosphorylase; Dickson KS et al

2001 J Biol Chem 276:41810; Steinmetz EJ et al 2001 Nature [Lond] 413:327.

polyA Tail: ►polyadenylation signal, ►polyA mRNA

Polyacrylamide: See Fig. P98, ►electrophoresis, ►gel electrophoresis

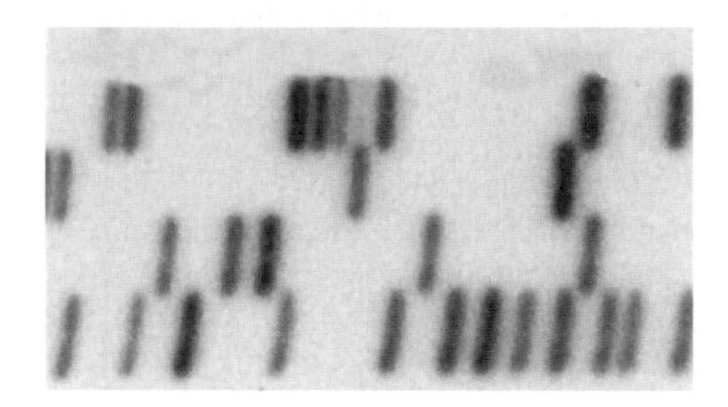

Figure P98. Polyacrylamide gel

Polyadenylation Signal: The endonucleolytic processing of the primary transcript of the majority of eukaryotic genes is followed by post-transcriptional addition of adenylic residues downstream of the structural gene. The consensus signal for the process is 5′-AAUAAA-3′ in animals and fungi, about half of the plants use the same signal, the rest rely on diverse signals. In humans, ~54%, and in mouse, ~32% of the genes have alternative polyadenylation sites and different cleavage sites resulting in heterogeneity of the transcripts, which may represent a type of regulation of gene expression (Tian B et al 2005 Nucleic Acids Res 33:201). In eukaryotes, the number of added A residues might vary from 50 to 250. The crystal structure of the 73 kDa subunit of the human polyadenylation-specificity endonuclease has been determined (Mandel CR et al 2006 Nature [Lond] 444:953).

Polyadenylation is under the control of several genes (see Fig. P99). The RNA transcript of eukaryotes besides the poly(A) signal contains a CA element (PyA in yeast) and a GU-rich downstream element.

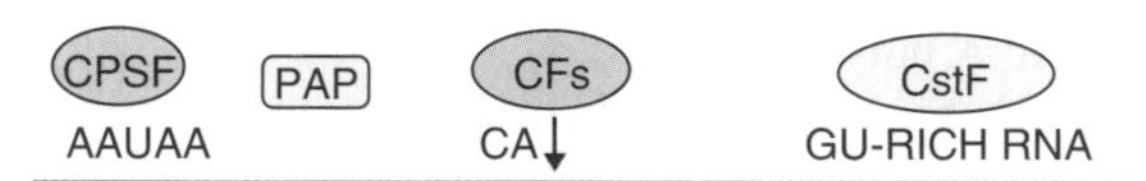

Figure P99. Some of the mechanisms in polyadenylation

To the AAUA AA *positioning element*, binds the *polyadenylation specificity factor* (CPSF) that is a tetrameric protein consisting of 160, 73, 100, and 33 kDa subunits. The *cleavage stimulating factor* (CstF), a trimeric protein of 64, 77, and 50 kDa subunits, binds to GU-rich element of the RNA. The *polyadenylation polymerase* protein (PAP) binds downstream of the CPSF binding sites. The *cleavage factors* (CFI) and (CFII) are positioned upstream of the GU-rich element and they terminate the mRNA. The polyadenylation complex of yeast is somewhat different. The PABP (poly-A-binding protein, 70 kDa) regulates mRNA stability, translation, and degradation. CPEB (cytoplasmic element binding protein), maskin, and cyclin B1 are regulators of polyadenylation and the transcription of some mRNAs. Poly(A)-specific ribonuclease (PARN) is a cap-interacting 3′ exonuclease. In cooperation with the cap, it mediates de-adenylation from cis position, or at low concentration it may be inhibitory to de-adenylation. From trans position, it inhibits de-adenylation if its concentration is high. The poly(A) tail, in synergy with the mRNA cap, stimulates the initiation of translation. The poly(A) binding protein (PABP) interacts with eIF4G of the eIF4F complex and eIF4E interacts with the cap and stabilizes mRNA. PABPs are located both in the nucleus and in the cytoplasm. In the human testes, a specific poly(A) binding protein occurs that is absent from other tissues (Féral C et al 2001 Nucleic Acids Res 29:1872). De-adenylation of the tail initiates mRNA decay and when less than ten A residue is left, an exonuclease attacks the RNA in 5′→3′ direction (Martînez J et al 2001 J Biol Chem 276:27923). In prokaryotes, rarely a few (14–60) adenine residues are also found at the mRNA 3′-terminus in 1 to 40% of the cases. In bacteria, *host factor q* (Hfq) plays a role similar to PABP. It stimulates the elongation of the polyA tail by poly(A) polymerase I (PAP) and protects against exoribonuclease attack. Some adenine sites are found in about 30% of both the early transcripts (transcribed by host polymerase) and late transcripts (transcribed by viral polymerase). Sometimes the poly-A sequence has interspersed other bases and may be located also within coding sequences. An interspersed long poly-(A) sequence was detected also in chloroplast RNA transcripts. In mitochondria, the poly-A tract (35–55 A residues) directly attaches to the termination codon without an untranslated sequence, after the endonucleolytic cleavage of the polycistronic transcript. In liver cancer, mitochondria tails of hundreds of As have been observed. In prokaryotes, two similar (36 and 35 kDa) poly-A polymerases with overlapping functions have been identified. With the exception of histone transcripts, all eukaryotic mRNAs appear polyadenylated, although some can be processed to become non-polyadenylated and the two types may coexist (bimorphic transcript). Polyadenylation of RNA in bacteria regulates plasmid replication and the degradation of RNAI. In yeast, the Trf4 complex recruits exosomes and the incorrectly folded polyadenylated RNA is degraded (Vaňácová S et al 2005 PLoS Biol 3(6):e189). (In *Archaea* short poly-A tracts exist). Cordycepin (3′-deoxyadenosine) is an inhibitor of polyadenylation.

In the *Drosophila melanogaster* genome, 17 polyadenylated sequences were detected without protein coding transcripts, yet many of these sequences were conserved in related species indicating some roles because of their conservation (Tupy JL et al 2005 Proc Natl Acad Sci USA 102:5495). Let-7 miRNPs, containing, Argonaute and GW182, dampen the synergistic enhancement of translation by the 5′-cap and 3′-poly(A) tail, resulting in translational repression (Wakyama M et al 2007 Genes Dev 21:1857). ►mRNA tail, ►U1 RNA, ►polyA polymerase, ►RNA I, ►cleavage stimulation factor, ►PABp, ►mRNA circularization, ►eIF4, ►TRAP, ►symplekin, ►microRNA, ►Argonaute, ►GW body; Hirose Y, Manley JL 1998 Nature [Lond] 395:93; Sarkar N 1997 Annu Rev Biochem 66:173; Beaudoing E et al 2000 Genome Res 10:1001; Mendez R Richter JD 2001 Nature Rev Mol Cell Biol 2:521; Wang L et al 2002 Nature [Lond] 419:312; http://polya.umdnj.edu/; http://polya.umdnj.edu/PolyA_DB2.

Poly(ADP-Ribose) Polymerase: A DNA-binding enzyme but it appears to have no indispensable function.

Polyamides: Polyamides containing *N*-methylimidazole and *N*-methylpyrrole amino acids have high affinity for specific DNA sequences and may regulate the transcription similarly to DNA binding proteins. ►binding proteins, ►inhibition of transcription, ►netropsin, ►lexitropsin; Maeshima K et al 2001 EMBO J 20:3218.

Polyamidoamine Dendrimers (PAMAM): Highly branched, soluble, non-toxic molecules with amino groups on their surface. They are suitable for attaching to this surface antibodies, various pharmaceuticals, and DNA. They are effective vehicles for transfection. (See Gebhart CL, Kabanov AV 2001 J Control Release 73:401).

Polyamines: Polyamines are various protein molecules derived in part from arginine and present in cells in millimolar concentrations, yet have important roles in RNA and DNA transactions, replication, supercoiling, bridging between strands, binding phosphate groups, biosynthesis, degradation, etc. Typical polyamines are spermine, spermidine, putrescine, etc. ►antizyme,►lexitropsins; Coffino P 2001 Nat Rev Mol Cell Biol 2:188; van Dam L et al 2002 Nucleic Acids Res 30:419.

Polyandry: A form of polygamy involving multiple males for one female. It may have the advantage of reducing the relatedness within colonies of social insects and thereby increasing fitness. In the live-bearing pseudoscorpions (*Cordylochernes scorpioides*), outbred embryos have beneficial effects on inbred half-siblings in mixed-paternity broods developing in the external, translucent brood sac and fed by nutrients of the maternal reproductive tract by an unclear mechanism (Zeh JA, Zeh DW 2006 Nature [Lond] 439:201). Honeybee queen matings with several drones enhances productivity and fitness of the colony (Mattila HR, Seeley TD 2007 Science 317:362). ►fitness; Tregenza T, Wedell N 2002 Nature [Lond] 415:71.

Polyaromatic Compounds: Polyaromatic compounds include various procarcinogens and promutagens, such as benzo(a)pyrene, dibenzanthracene, methylcholanthrene, etc. ►polycyclic hydrocarbons.

Polybrene (hexadimethrine bromide): A polycation used for introduction of plasmid DNA into animal cells; it is also an anti-heparin agent and an immobilizing agent in Edman degradation. Polybrene may have different toxicity to various cells. ►transformation genetic animal cells, ►heparin, ►Edman degradation

Polycentric Chromosome: ►neocentromeres

Polychlorinated Biphenyl (PCB): A highly carcinogenic compound and an inducer of the P-450 cytochrome group of monooxygenases. It had been used in electrical capacitors, transformers, fire retardants, hydraulic fluids, plasticizers, adhesives, pesticides, inks, copying papers, etc. *Pseudomonas* sp. KKS102 is capable of degrading PCB into tricarboxylic acid cycle intermediates and benzoic acid. ►microsomes, ►S-9, ►P-450, ►carcinogen; Ohtsubo Y et al 2001 J Biol Chem 276:36146.

Polychromatic: A polychromatic substance is stainable by different dyes or displays different shades when stained.

Polycistronic mRNA: A contiguous transcript of adjacent genes, such as exist in an operon but may also be formed in the short genes of eukaryotes, e.g., oxytocin. The *Trypanosomas* produce multicistronic transcripts. A gene (*mlpt*) of *Tribolium* involved in body segmentation also produces polycistronic mRNA. ►operon, ►oxytocin, ►*Trypanosoma*, ►*Caenorhabditis*, ►*Tribolium*

Polyclonal Antibodies: Polyclonal antibodies are produced by a population of lymphocytes in response to antigens. These are not homogeneous as are the monoclonal antibodies. ►monoclonal

Polycomb (*Pc*, chromosome 3-47.1): The *Drosophila* gene is a negative regulator of the *Bithorax* (*BXC*) and *Antennapedia* (*ANTC*) complexes (see Fig. P100). The homozygous mutants are lethal and the locus

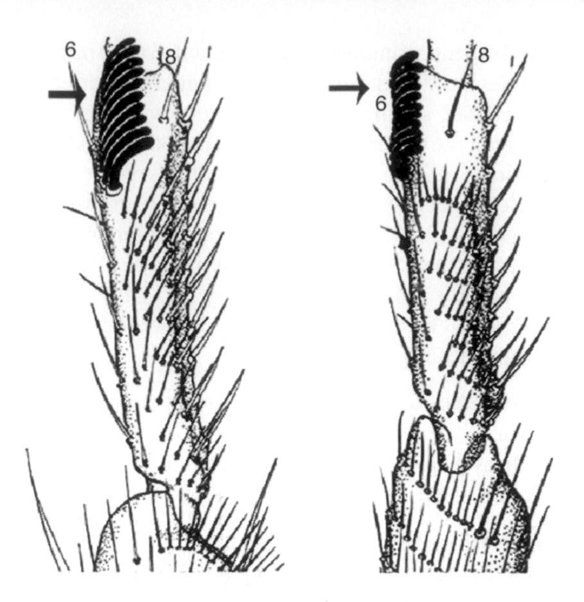

Figure P100. Sex combs on the second legs of male *Drosophila* in the *Sex comb extra* (*Scx*, 3.47, at left) and *Scx-Pc* (at right) homeotic mutants. Similar extra sex combs appear also on the third leg whereas sex combs in the wild type are limited to the first pair of legs. (After Hannah-Alava A 1958 Genetics 43:878)

(and its homologs in vertebrates [*M33* in mice]) is involved in the repression of homeotic genes, which control body segmentation. *Pc* is a member of a group (*Pc-G*) of repressors of homeotic genes (Cao R et al 2002 Science 298:1039). Although *Pc* is located in the euchromatin, it is involved in the silencing of genes by heterochromatin (Francis N et al 2004 Science 306:1574). Genes have been identified that are targeted for transcriptional repression in human embryonic stem (ES) cells by the Pc-G proteins suppressor of zeste 12 (*SUZ12*) and embryonic ectoderm development (*EED*), which form the Polycomb repressive complex 2 (*PRC2*) and which are associated with nucleosomes that are trimethylated at Lys27 of histone H3 (H3K27). Stem cell occupancy by *SUZ12* and *EED* and the trimethylation status of H3K27 for 77/177 genes showed evidence of cancer-associated DNA methylation when compared with matched normal colorectal mucosa. The observations suggest that the first predisposing steps towards malignancy may occur very early and are consistent with reports of field changes in histologically normal tissues adjacent to malignant tumors. These results provide a mechanistic basis for the predisposition of certain promoter CpG islands to cancer-associated DNA hypermethylation as an early epigenetic cancer marker (Widschwendter M et al 2007 Nat Genet 39:157).

Pc is involved in Histone2A ubiquitylation and the inactivation of mammalian X chromosome (de Napoles M et al 2004 Dev Cell 7:663). Insertion into the 5th exon of *M33* caused male→female sex-reversal. *Pc* is required for the activation of other silencing elements and its mutation may lead to derepression of these elements. The suppressive effect of *Pc* may be associated with chromatin remodeling and histone deacetylation. The Polycomb group of proteins forms a large complex and the TATA-box-binding proteins, Zeste and others, are associated with the general transcription machinery (Czermin B et al 2002 Cell 111:185). The SUZ12 subunit of the Polycomb Repressive Complex 2 (PRC2) extends over 200 genes encoding developmental regulators of human embryonic stem cells (such as Nanog, Oct4, Sox2, RNAp2 and SUZ12) (see Fig. P101). These genes are transcriptionally repressed because in the nucleosomes histone H3K27 is trimethylated. The PRC2 target genes are repressed in order to maintain pluripotency of these cells but they are activated during differentiation (Lee TI et al 2006 Cell 125:301). The PRC1 and PRC2 polycomb complexes co-occupy 512 genes and bind hundreds of others (►Venn diagram) coding for transcription factors during mouse embryonic development until differentiation (Boyer LA et al 2006 Nature [Lond] 441:349). ►morphogenesis in *Drosophila*, ►transdetermination, ►*SWI*, ►homeobox, ►homeotic genes, ►chromodomain, ►sex-reversal, ►*w* locus, ►*zeste*, ►*Antennapeadia*, ►*Bithorax*, ►*trithorax*, ►chromatin remodeling, ►nucleosomes, ►histone deacetylase, ►histones, ►TBP, ►transcription factors, ►Lyonization, ►ubiquitin, ►epigenesis, ►stem cells, ►genetic networks, ►pairing-sensitive repression; Breilling A et al 2001 Nature [Lond] 412:651; Simon JA, Tamkun JW 2002 Curr Opin Genet Dev 12:210; Polycomb repressor complex in epigenesis: Kuzmichev A et al 2005 Proc Natl Acad Sci USA 102:1859; review: Schwartz YB, Pirrotta V 2007 Nat Rev Genet 8:9; review: Schuttengruber B et al 2007 Cell 128:735.

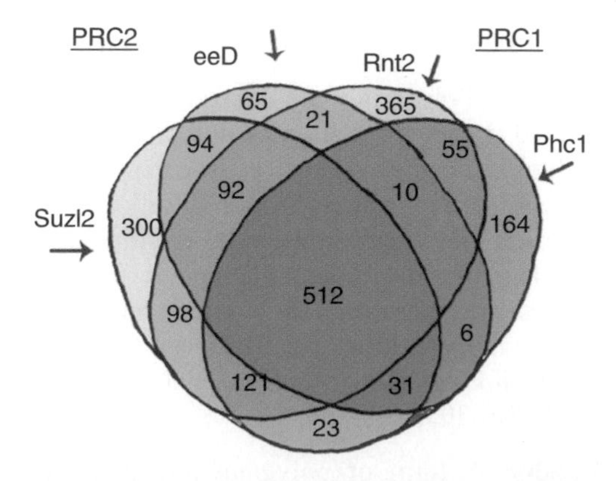

Figure P101. Polycomb group proteins overlap and repress many developmental regulators of mouse

Polycross: An intercross among several selected lines to produce a "synthetic variety" of a crop (See Tysdal HM et al 1942 Alfalfa Breeding, Nebr Agric Exp Sta Res Bull 124, Lincoln, Nebraska).

Polycyclic Aromatic Hydrocarbons (PAH): Generally, carcinogenic and mutagenic compounds. They become more active during the process of attempted detoxification by the microsomal enzyme complex. PAHs are the products of burning organic material (coal, charbroiling, smoking, etc.). Mice oocytes exposed to PAHs suffer apoptosis by activation of BAX. ►carcinogens, ►procarcinogens, ►mutagens, ►promutagens, ►benzo(a)pyrene, ►environmental mutagens, ►PAH, ►BAX, ►apoptosis; Matikainen T et al 2001 Nature Genet 28:355.

Polycystic Lipomemembraneous Osteodysplasia with Sclerosing Leukoencephaly (PLOSL, Nasu-Hakola disease, 19q13.1): A recessive psychosis turning into presenile dementia and bone cysts limited to the wrists and ankles. Prevalence in Finland is 2×10^{-6}. The basic problem is a loss of function of the TYROBP/DAP12 tyrosine kinase binding transmembrane protein, an activator of killer lymphocytes. ►killer cells

Polycystic Kidney Disease (PKD): PKD occurs in two main forms, and within each several form variations exist (see Fig. P102). The short arm of human chromosome 16p13.31-p13.12 apparently controls the adult type dominant (ADPKD), which involves fragility of the blood vessel walls. In the autosomal recessive ARPKD, the basic defect is in the Ca^{2+}-permeable non-selective cation channel. About 15% of the APKD cases are due to mutation in the gene (PKD2) encoding polycystin. Another gene (PKD1) is involved in the proliferation of the epithelial cells lining the cyst cavity, the thickening of the basement membrane, fluid secretion, and protein sorting (Bukanov NO et al 2002 Hum Mol Genet 11:923). ARPKD generally has an early onset. Both forms occur at frequencies of 0.0025 to 0.001. Even the late onset type may be detectable early by tomography. The symptoms vary and involve kidney disease, cerebral vein aneurism (sac like dilatation), underdeveloped lungs, liver fibrosis, and growth retardation, etc. The dominant type can be identified with high accuracy using chromosome 16p13 DNA probes but less than 10% of the cases are due to genes not in chromosome 16. The autosomal recessive form is at an unknown location and it can be identified after the third trimester by ultrasonic methods because the kidneys are enlarged. The genetic transmission of the dominant and recessive diseases is very efficient. One polycystic kidney (PKD1, 4300-amino acid integral membrane glycoprotein) locus was assigned to 6q21-p12, and sequences were also found in 2p25-p23 and 7q22-q31; these are homologous to polycystic kidney disease of the mouse. There is a PKD2 locus in 4q21-q23 and this is similar in function to PKD1. PKD2 interacts with PKD1 and PKD2 interacts also with the Hax-1 protein binding F-actin, suggesting that the system affects the cytoskeleton. Thus defect in PKD2 may be one of the causes of cyst formation in the kidney, liver, and pancreas. The

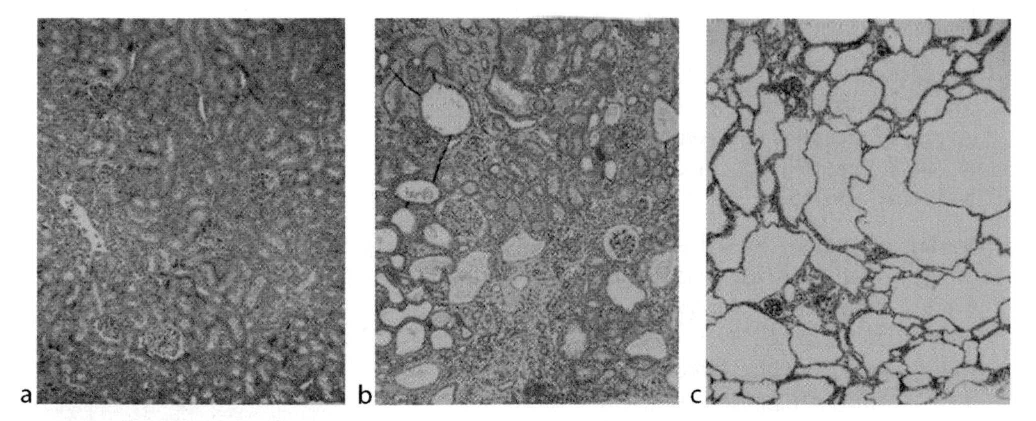

Figure P102. Polycystic kidney disease in RRRCHan:SPRD rat. a: wild type, b: heterozygote, and c: homozygous Pkdr1 mutant. The homozygous mutant rats die at age 3–4 weeks. Heterozygous males develop renal failures by about 6 months whereas the heterozygous females rarely progress to renal failure and death. Heterozygous males can be identified by blood urea nitrogen (BUN) level of the serum or plasma at age 9–10 weeks. PCR analysis and sequencing detected A for G substitution in exon 12 of the mutant gene. (The histological images are the courtesy of Professors Beth A. Bauer and Craig L. Franklin, Rat Resource and Research Center University of Missouri, Columbia, Missouri 6211; http://www.nrrrc.missouri.edu/Straininfo.asp?appn=46)

infantile type recessive PKD is also called Caroli disease. The ARPKD locus encodes a 968-amino acid protein, which forms six transmembrane spans with intracellular amino and carboxyl ends. It appears to be a voltage-activated Ca^{2+} (Na^{+}) channel protein. In a mouse model and in humans, TOR antagonist rapamycin protein may alleviate the dominant ADPKD disease (Schillingford JM et al 2006 Proc Natl Acad Sci USA 103:5466). The CDK inhibitor roscovitine ($C_{19}H_{26}N_6O$) appears to be an effective inhibitor of PKD in mouse (Bukanov NO et al 2006 Nature [Lond] 444:949). PKD1 may involve haplo-insufficiency. The PKD1 homolog in *Caenorhabditis* (*LOV-1*) controls sensory neurons required for male mating behavioral steps. The traditional Chinese drug, a diterpene triptolide (Lei Gong Teng), induces PC2-dependent calcium release and attenuates cyst formation (see Fig. P103) (Leuenroth SJ et al 2007 Proc Natl Acad Sci USA 104:4389). ►cardiovascular disease, ►Caroli disease, ►hypertension, ►genetic screening, ►ion channels, ►haplo-insufficient, ►rapamycin, ►CDK; Pei Y et al 2001 Am J Hum Genet 68:355; Lin F et al 2003 Proc Natl Acad Sci USA 100:5286; autosomal dominant polycystic kidney disease: http://pkdb.mayo.edu/.

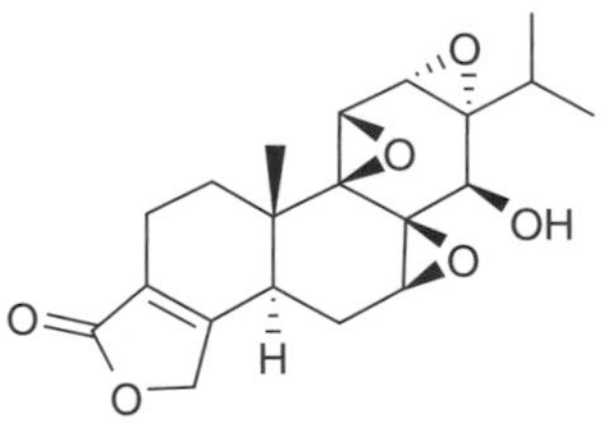

Figure P103. Triptolide

Polycystic Liver Disease; (PCLD, 19p13.2-p13.1): A dominant, often accompanying polycystic kidney disease. It involves fluid-filled cysts on the liver. The protein involved is hepatocystin. (See Drenth JPH et al 2003 Nat Genet 33:345).

Polycystic Ovarian Disease (Stein-Leventhal syndrome): Polycystic ovarian disease generally involves enlarged ovaries, hirsuteness, obesity, lack of or irregular menstruation, increased levels of testosterone high ratios of luteinizing hormone: follicle-stimulating hormone, and infertility. It appears to be due an autosomal factor, yet 96% and 82% of the daughters of affected mothers and carrier fathers, respectively, developed the symptoms indicating a meiotic drive-like phenomenon. Deficiency of 1-α-ketosteroid reductase/dehydrogenase (9q22) may cause polycystic ovarian disease as well as pseudohermaphroditism with gynecomastia in males. ►infertility, ►luteinization, ►Graafian follicle, ►meiotic drive, ►pseudohermaphroditism, ►gynecomastia

Polycystin: Polycystin proteins are supposed to regulate different functions, such as mating behavior, fertilization by the sperm, asymmetric gene expression, and mechanosensory transduction (Delmas P 2004 Cell 118:145).

Polycythemia (PFCP): An autosomal dominant proliferative disorder of the erythroid progenitor cells, resulting in an increase in the number of red blood cells and in vitro hypersensitivity to erythropoietin. Mutations in the von Hippel-Lindau protein are responsible for about half of the cases. Mutations of valine→phenylalanine at amino acid site 617 in Janus kinase 2 occurs in more than 80% of acquired polycytemic mice and leads to constitutive tyrosine phosphorylation and increased sensitivity to cytokinins (James C et al 2005 Nature [Lond] 434:1144) ►erythropoietin, ►Janus kinases, ►von Hippel-Lindau syndrome; Pastore Y et al 2003 Am J Hum Genet 73:412.

Polydactyly: The presence of extra fingers or toes. In *postaxial* polydactyly (the most common type), the extra finger is in the area of the "little finger" (see Fig. P104) and in *preaxial* cases, this malformation is on the opposite side of the axis (thumb) of the palm or foot. The various types of polydactyly may be determined by autosomal recessive or dominant gene(s) and their expression is usually part of other syndromes. Crossed polydactyly indicates coexistence of postaxial and preaxial types with discrepancy between hands and feet. Synpolydactyly is caused by an expansion of the normal 15 GCG trinucleotides to 22–29.

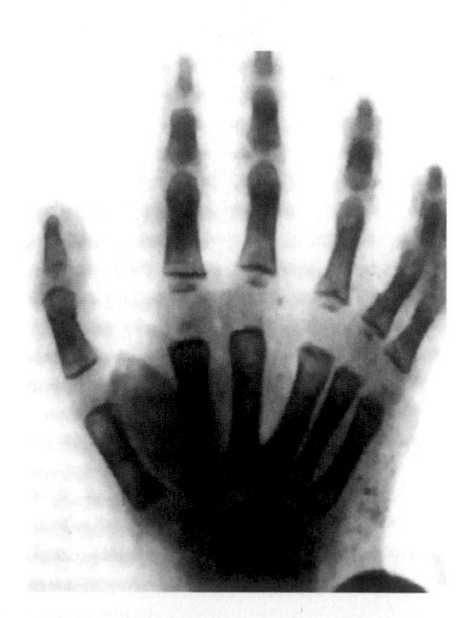

Figure P104. Polydactyly (From Bergsma D (ed) 1973 Birth Defects. Atlas and Compendium. By permission of the March of Dimes Foundation)

►Ellis-van Creveld syndrome, ►Opitz syndrome, ►Meckel syndrome, ►Majewski syndrome, ►orofacial-digital syndromes, ►Pätau's syndrome, ►diastrophic dysplasia, ►syndactyly, ►polysyndactyly, ►Greig's cephalopolysyndactyly syndrome, ►Rubinstein-Taybi syndrome, ►Pallister-Hall syndrome, ►focal dermal hypoplasia, ►ectrodactyly, ►adactyly, ►*hedgehog*

Polyelectrolytes: Polymers with attached anions and cations, respectively. Proteins and nucleic acids can be polyelectrolytes by carrying negatively and positively charged groups.

Polyembryony: In polyembrony, more than one cell of the embryo sac develops into an embryo in plants or in insects a single egg by clonal reproduction of hundreds of embryos. ►adventive embryos, ►embryo sac; Zhurov V et al 2004 Nature [Lond] 432:764.

Polydna Virus: ►parasitoid

Polyethylene Glycol (PEG): A viscous liquid or solid compound of low-toxicity, promoting fusion of all types of cells. PEG is widely used in textile, cosmetics, paint, and ceramics industry. ►PEG

Polyethyleneimine (PEI): A water-soluble polymer, which binds and precipitates DNA. It assists in the uptake of molecules including transforming DNA, especially in RGD-coated particles. Polyethyleneimine (25 kDa) contains N-acyl groups, which handicap its use for genetic transfection (see Fig. P105). Removal of these groups enhances its utility as an artificial vector. New linear PEIs synthesized by acid-catalyzed hydrolysis of poly(2-ethyl-2-oxazoline yielded products, which increased transfection efficiency up to 115-fold compared to deacetytlated commercial PEI. In addition, its efficiency for targeting lung cells increased 200-fold. Using this vector for RNAi delivery against the nucleocapsid protein gene of influenza virus dropped the virus titer in the lung of mice. A further advantage was the lower toxicity. Note: ethyleneimines are poisonous and mutagenic. ►RGD, ►vectors, ►gene therapy

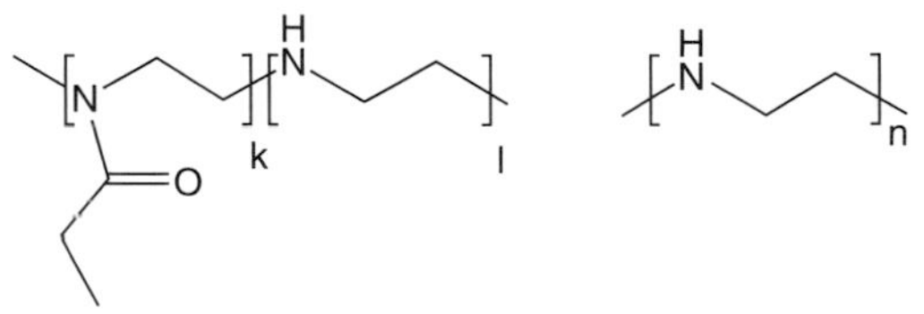

Figure P105. Left: PEI25 commercially available before hydrolysis. Right: Newly synthesized PEI after hydrolysis. After Thomas M et al. 2005 Proc. Natl. Acad. Sci. USA 102:5679

Polygalacturons: Complex carbohydrates in the plant cell wall.

Polygamy: Polygamy implies having more than one mating partner. In western human societies, it is illegal but in others, it is still acceptable for men to have more than one wife at the same time. Polyandry or polygyny is a common practice in animal breeding but it may be objectionable to humans on moral grounds. In the USA, polygamy laws are applied to all citizens, irrespective of religious affiliation or cultural tradition.

Polygenes: A number of genes involved in the control of quantitative traits. ►gene number in quantitative traits, ►QTL

Polygenic Inheritance: Polygenic inheritance is determined by a number of non-allelic genes, all involved in the expression of a single particular trait (such as height, weight, intelligence, etc.). Polygenic inheritance is characterized by counting and measurements and the segregating classes are not discrete but display continuous variation. ►quantitative genetics, ►QTL, ►complex inheritance, ►chaos, ►digenic diseases, ►selection long term, ►gain; Tanksley SD 1993 Annu Rev Genet 27:205; Klose J et al 2002 Nature Genet 30:385.

Polygenic Plasmids: are obtained when two plasmids carrying identical genes cointegrate. Such plasmids may have merit in genetic engineering if the genes show positive dosage effect for anthropocentrically useful traits. ►cointegration

Polygeny: In polygeny, one male has more than a single female mate. In *sororal polygeny*, the females are sisters. ►polygamy, ►effective population size

Polyglutamylase: The polyglutamylase enzyme adds several glutamic acids to the γ-carboxyl of a glutamate residue of proteins, such as tubulin and nucleosome assembly proteins (Janke C et al 2005 Science 308:1758).

Polyglutamine Diseases: ►trinucleotide repeats, ►resveratrol

Polygyne: Polygyne desribes social insect colonies with more than a single queen. ►monogyne

Polyhaploid: A polyhaploid has half the number of chromosomes of a polyploid. The gametes of polyploids are polyhaploid. ►polyploidy

Polyhedrosis Virus, Nuclear (BmNPV): An about 130 kbp DNA baculovirus of the silkworm (and other insects). It has been used (after size reduction) as a 30 kb cloning vector and it may propagate in a single silkworm larva about 50 μg DNA. ►baculoviruses,

►viral vectors, ►silkworm; Xia Q et al 2003 J Biol Chem 278:1094.

Polyhybrid: A polyhybrid is heterozygous for many gene loci.

Polyhydroxybutyrate (PHB): A bacterial polymer that can be manufactured by transgenic plants and is biodegradable.

Polyisoprenyl Phosphates: Intermediates in cholesterol biosynthesis; they play a role in signaling to the immune system. ►immune system, ►cholesterols

Polyketenes: Polymers of $CH_2 = C = O$ (ketene). Their biosynthesis is related to fatty acids. Several antibiotics (tetracycline, griseofulvin, etc.) contain ketenes. ►antibiotics

Polyketides: Various naturally occurring compounds, built from residues, which each usually contribute two carbon atoms to the assembly of a linear chain of which the β-carbon carries a keto group. These keto groups are frequently reduced to hydroxyls. The remaining keto groups at many of the alternate carbon atoms form the chains, which are called polyketides. Polyketide synthesis pathway resembles the fatty acid path. Flavonoids, mycotoxins, antibiotics, etc., occurring in plants from angiosperms to bacteria qualify for the polyketide collective name. Polyketide synthetases generate the precursors of erythromycin, rapamycin, and rifamycin antibiotics. ►lovastatin, ►epothilone; Khosla C et al 1999 Annu Rev Biochem 68:219; Walsh CT 2004 Science 303:1805.

Polykinetic Chromosome: A polykinetic chromosome has centromeric activity at multiple sites. ►neocentromeres

Polylinker: A DNA sequence with several restriction enzyme recognition sites (multiple cloning sites, MCS) used in construction of different cloning or transformation vehicles (plasmids). e.g., TTCTAGAATTCT sequence has an overlapping XbaI (TCTAGA) and an EcoRI recognition sites (GAATTC) and thus linking it to the DNA may generate both types of cloning sites. ►vectors, ►restriction enzymes, ►cloning sites, ►pUC

Poly(L-Lysine): A polycation that can form complex(es) with negatively charged DNA and mediate gene transfer using retroviral vector. In case the polycation has bound specific ligand(s), it can be targeted to special cell types. Without such a complex, the viral vector would have no target specificity. Some of the polycationic delivery systems are cytotoxic and/or may be subject to lysosomal degradation. ►transformation genetic; Putnam D et al 2001 Proc Natl Acad Sci USA 98:1200.

Poly-Marker Test: ►DNA fingerprinting

Polymer: A large molecule composed of a series of covalently linked subunits such as amino acids, nucleotides, fatty acids, carbohydrates, etc. ►DNA, ►protein; biopolymer motifs: http://bayesweb.wadsworth.org/gibbs/gibbs.html.

Polymerase: An enzyme that builds up large molecules from small units, such as the DNA and RNA polymerases generated from nucleotides DNA and RNA, respectively. ►pol

Polymerase Accessory Protein (RF-C): An essential part of the DNA replication unit in SV40. ►SV40

Polymerase Chain Reaction (PCR): A method of the rapid amplification of DNA fragments, employed when short flanking sequences of the fragments to be copied are known (see Fig. P106). The reaction begins by the denaturation of the target DNA, then primers are annealed to the complementary single strands. After adding a heat-stable DNA polymerase, such as Taq or Vent/Tli (originally the less thermostable Klenow fragment of polymerase I was used), chain elongation proceeds starting at the primers. The cycles are repeated 20–30 times, resulting in over a million fold (2^{20} = 1,048,576) replication of the target. The actual rate of replication may be less (80%) than that theoretically expected.

The DNA amplified can be subjected to molecular analysis such as preimplantation analysis, genetic

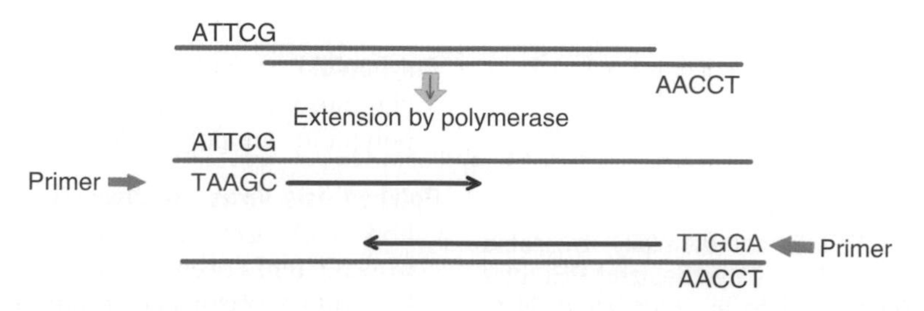

Figure P106. Polymerase chain reaction

screening, prenatal analysis, sperm typing, gene identification, etc. The error frequency for the Klenow fragment is about 8×10^{-5}, for Taq 10^{-5} to 10^{-4}, for Tli 2 to 3×10^{-5}. PCR amplification can be performed with a variety of mechanical devices, including chemical amplification on a microchip where the 20 cycles may be completed as fast as in 90 seconds. All types of technical information and references are available at http://apollo.co.uk/a/pcr. ►RAPDS, ►DNA fingerprinting, ►vectorette, ►sperm typing, ►genetic screening, ►prenatal analysis, ►preimplantation genetics, ►tissue typing, ►primer extension, ►ancient DNA, ►molecular evolution, ►RT-PCR, ►in situ PCR, ►recursive PCR, ►inverse PCR, ►capture PCR, ►PCR overlapping, ►tail-PCR, ►electronic PCR, ►PCR broad-base, ►PTPCR, ►methylation-specific PCR, ►AP-PCR, ►PCR asymmetric, ►PCR allele-specific, ►immuno-PCR, ►RNA-PCR, ►PCR-based mutagenesis, ►small-pool PCR, ►INTER-SS PCR, ►PRINS, ►reverse ligase-mediated polymerase chain reaction, ►thermal cycler, ►hot-start PCR, ►touch-down PCR, ►double PCR and digestion, ►PCR-LSA; Mullis KB, Faloona FA 1989, p 189 In: Wu R et al (Eds.) Recombinant DNA Methodology, Academic Press, San Diego, California; Innis M et al (Eds.) 1990 PCR Protocols: A Guide to Methods and Applications, Academic Press, San Diego, California; quantitative PCR primers: http://www.ncifcrf.gov/rtp/gel/primerdb/; http://medgen.ugent.be/rtprimerdb/; PCR primer design for mutation screening: http://bioinfo.bsd.uchicago.edu/MutScreener.html.

Polymerase Switching: In polymerase switching, DNA replication is initiated by the polymerase α/primase complex, but subsequently the chain elongation is continued by the eukaryotic polymerase δ. Polymerase ε may also have some role in the initiation and elongation. ►replication fork, ►DNA polymerases, ►primase, ►processivity

Polymery: In polymery, several genes cooperate in the expression of a trait. ►polygenes

Polymorphic: A trait that occurs in several forms within a population. The polymorphism may be balanced and genetically determined. ►polymorphism, ►balanced polymorphism, ►RFLP, ►SNP

Polymorphic Information Content (PIC): PIC is used to identify and locate a hard-to define marker locus. If the alleles of the marker locus are codominant, then PIC is the fraction of the progeny (the informative offspring) that cosegregates by phenotype with an index locus. The index locus (which is used for the detection of linkage with marker alleles) has two alternative alleles, a wild type and a dominant (mutant) allele. The marker locus is polymorphic for dominant (genetic or physical [nucleotide sequences]) alleles. Only those progenies are informative, where the index locus is homozygous in one of the parents and the other parent is heterozygous for the marker. The converse constitutions are not informative. In case both parents are heterozygous at the marker locus, only half of the offspring is informative.

$$PIC = 1 - \sum_{i=1}^{n} p_i^2 - \left(\sum_{i=1}^{n} p_i^2\right)^2 + \sum_{i=1}^{n} p_i^4$$

where p_i = frequency of the index allele and i and n are the number of different alleles. The PIC values may vary theoretically from 0 to 1. A hypothetical example: four A alleles occur in a population with frequencies A^1: 0.2, A^2: 0.1, A^3= 0.15, and A^4 = 0.55. After substitution, $PIC = 1 - (0.2^2 + 0.1^2 + 0.15^2 + 0.55^2) - (0.2^2 + 0.1^2 + 0.15^2 + 0.55^2)^2 + (0.2^4 + 0.1^4 + 0.15^4 + 0.55^4)$, thus $PIC = 1 - 0.375 - 0.140625 + 0.0937125 \approx 0.578$, and in this case almost 58% of the progeny is informative. Usually, PIC values of 0.7 or larger are required for showing good linkage. The larger the number of the marker alleles, the more informative is the PIC. ►microsatellite typing; Da Y et al 1999 Anim Biotechnol 10:25.

Polymorphism: In polymorphism, morphologically different chromosomes, or different alleles at a gene occur, or variable length restriction fragments are found within a population. Polymorphism can now be also detected through automated molecular techniques. During PCR amplification of a gene, one or more fluorescent reporter probes are attached to the 5′ end, and a quencher substance(s) added slightly downstream or at the 3′-end. During amplification, the quencher may be cleaved by the Taq polymerase if it hybridizes to an amplified segment. The cleavage of the quencher enhances the fluorescence of the reporter fluorochrome. The samples placed in a 96-well plate can be scanned at three wavelengths in about 5 min. The procedure may be sensitive enough to detect a single base difference. In the human DNA sequences, there is ca. one variation/500 bp. About 15% of the polymorphism involves insertions or deletions. At least 100 chromosomes are usually examined for base substitution before the alteration is considered as a polymorphism. The average estimated nucleotide polymorphism in human populations is $\sim 8 \times 10^{-4}$. The diversity is variable at different loci and affected by several factors. Among normal human individuals, on the average, 11 deletions and duplications of the average length of 465 kb have been observed (Sebat J et al 2004 Science 305:525). ►balanced polymorphism, ►mutation, ►diversity, ►mutation detection, ►fluorochromes, ►PCR, ►clone validation, ►RLP, ►SNP,

P

►microsatellite, ►blood groups, ►linkage disequilibrium, ►haplomap; Reich DE et al 2002 Nature Genet 32:135; polymorphism detection tool for large datasets: http://pda.uab.es/pda/; http://pda.uab.es/pda2/; mammalian: http://mampol.uab.es/.

Polymorphonuclear Leukocyte (PMN): ►granulocytes, ►leukocyte

Polyomyositis: The inflammation of muscle tissues, which may lead to rheumatoid arthritis, lupus erythematosus, scleroderma, Sjögren syndrome, and neoplasia. Polymyositis is caused by two autoantigens PMSCL1 and PMSCL2. Dermatomyosities is a form affecting the connective tissues. Polymyosities as such are not under direct genetic control. conditions under separate entries, ►IVIG; Wang HB, Zhang Y 2001 Nucleic Acids Res 29:2517.

Polyneme: The linear structure includes more than one strands, e.g., polytenic chromosomes (salivary gland chromosomes) may have 1024 (2^{10}) parallel strands.

Polynucleotide: A nucleotide polymer hooked up through phosphodiester bonds.

Polynucleotide Kinase (PK): PK phosphorylates 5′ positions of nucleotides in the presence of ATP, such as $ATP + XpYp \xrightarrow{PK} p - 5'XpYp + ADP$ (where X and Y are nucleotides), and can heal nucleic acid termini with ligase assistance. It functions in base excision repair and in non-homologous end-joining. ►ligase, ►DNA repair, ►non-homologous end-joining; crystal structure: Bernstein NK et al 2005 Mol Cell 17:657; Wang LK, Shuman S 2001 J Biol Chem 276:26868.

P

Polynucleotide Phosphorylase (PNPase): PNPase generates random RNA polymers [$(NMP)_n$]—without a template—from ribonucleoside diphosphates (NDP) and releases inorganic phosphate (P_i): $(NMP)_n + NDP \rightarrow\rightarrow\rightarrow (NMP)_{n+1} + P_i$. It degrades mRNA from the 3′-end.

Polynucleotide Phosphotransferase: Polynucleotide phosphotransferase transfers nucleotides to the ends of DNA or RNA sequences, such as in the polyadenylation of mRNA or nucleotidyl transferase of DNA. ►polyA polymerase, ►terminal deoxynucleotidyl transferase

Polynucleotide Vaccination: Inoculation by subcutaneous, intravenous, or particle bombardment-mediated transfer of specific viral or other nucleotides/nucleoproteins to develop an immune response. The immune reaction is generally low. ►immunization genetic

Polyoma: Neoplasia induced by one of the polyomaviruses. The globoid (icosahedral) mouse polyoma viruses (a papova virus of 23.6×10^6 Da) contain double-stranded, circular DNA (4.5 kb). The BK and the JC viruses infect humans. ►Papova viruses; Cole CN, Conzen SD 2001, p 985 In: Knipe DM, Howley PM (Eds.) Fundamental Virology, Lippincott Williams & Wilkins, Philadelphia, Pennsylvania.

Polyp: An outgrowth on mucous membranes such as may occur in the intestines, stomach, or nose. They may be benign, precancerous, or cancerous. Nasal polyps may occur in aspirin sensitivity and can be treated surgically or with nasal steroid drugs. ►PAP, ►polyposis adenomatous, ►Gardner syndrome

Polypeptide: A chain of amino acids hooked together by peptide bonds. ►protein synthesis, ►amino acids, ►peptide bond

Polyphenols: Polyphenols are catechol-related plant products causing the formation of melanin-like brown color. The polyphenols in tea (theaflavin, catechins) have apparently antimutagenic and anticarcinogenic effects. ►thea

Polypheny: In polyphony, the same gene(s) can determine alternative phenotypes in response to internal or external cues, e.g., the queens and workers in social insects. Polyphenism in insect coloration may be the result of mutation of juvenile hormone-regulatory pathway and temperature effects may reveal hidden genetic variations. The mechanism that regulates developmental hormones can mask genetic variations and can act as an evolutionary capacitor for facilitating novel adaptive changes by genetic accommodation (Suzuki Y, Nijhout HF 2006 Science 311:650). Some older dictionaries and glossaries equate it with pleiotropy but this does not conform to current usage. ►genetic accommodation

Polyphosphates: Linear polymers of orthophosphates (*n* up to 100 or more) present in all types of cells with roles similar to ATP in metal chelation, bacterial competence for transformation, mRNA processing, growth regulation, etc. (see Fig. P107). Polyphosphates can buffer cellular phosphate levels in case of limited external supply and affect phosphate uptake (Thomas MR, O'Shea EK 2005 Proc Natl Acad Sci USA 102:9565). ►competence of bacteria, ►chelation; Kulaev I, Kulakovskaya T 2000 Annu Rev Microbiol 54:709.

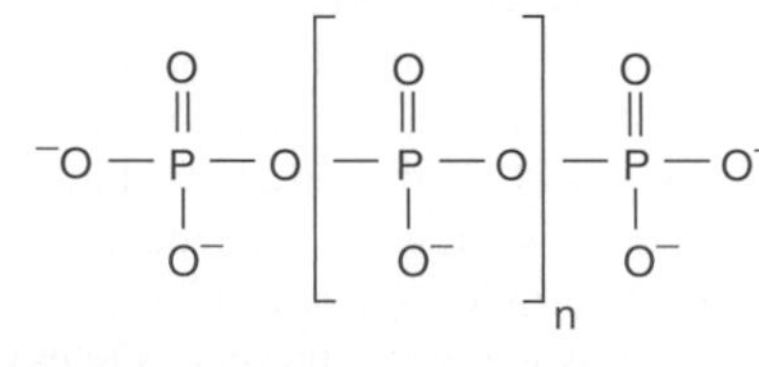

$$^{-}O-\overset{O}{\underset{O^{-}}{\overset{\|}{\underset{|}{P}}}}-O\left[-\overset{O}{\underset{O^{-}}{\overset{\|}{\underset{|}{P}}}}-O\right]_n-\overset{O}{\underset{O^{-}}{\overset{\|}{\underset{|}{P}}}}-O^{-}$$

Figure P107. Polyphosphate

PolyPhred: A computer program that automatically detects heterozygotes for single nucleotide substitutions by fluorescence-based sequencing of PCR products at high efficiency. It is integrated by the Phred, Phrap, and Consed programs. ►SNIP, ►PCR, ►Phred, ►Phrap, ►Consed; Nickerson DA et al 1997 Nucleic Acid Res 25:2745.

Polyphyletic: An organism (cell) that originated during evolution from more then one line of descent. A polyphyletic group may contain species that are classified into this group because of convergent evolution. ►convergence, ►divergence

Polyplexes: Polyplexes are employed for delivery of DNA to cells. They include DNA-binding and condensing molecules, cell-specific ligands, and other molecules necessary for protection and uptake.

Polyploid Crop Plants: The most important polyploid crop plants include alfalfa (4x), apple (3x), banana (3x), birdsfoot trefoil (4x), white clover (4x), coffee (4x, 6x, 8x), upland cotton (4x), red fescue (6x, 8x, 10x), johnsongrass (8x), cultivated oats (6x), peanut (4x), Euro-pean plum (6x), cultivated potatoes (4x), sugarcane (*x), common tobacco (4x), bread wheat (6x), and macaroni wheat (4x). Most of these are apparently allopolyploids. ►allopolyploid

Polyploidy: Having more than two genomes per cell. Definitive identification of polyploidy requires cytological analysis (chromosome counts), although many of the polyploid plants display broader leaves, larger stomata, larger flowers, etc. Polyploidy regulates the expression of individual genes in + or − manner. A yeast study (using microarray hybridization) found that the level of expression of some genes remained the same in haploid and tetraploid cells, whereas the expression of some cyclin genes decreased with tetraploidy. Additionally, a gene associated with cell adhesion was greatly over-expressed with tetraploidy (see Fig. P108).

Figure P108. Autotetraploid (top) and diploid (bottom) flowers of *Cardaminopsis petraea* (G.P. Rédei, unpublished)

Polyploidy may permit separate evolutionary paths for the additional gene copies. ►autopolyploid, ►endopolyploidy, ►inbreeding autopolyploids, ►chromosome segregation, ►duplication, ►maximal equational segregation, ►alpha parameter, ►allopolyploid, ►tetrasomic, ►trisomy, ►microarray hybridization; Otto SP, Whitton J 2000 Annu Rev Genet 34:401.

Polyploidy in Animals: Polyploidy in animals is rare and limited mainly to parthenogenetically reproducing species (e.g., lizards). It occurs also in bees, silkworm, and other species. Some cells in special tissues of the diploid body may have increased chromosome number as a normal characteristic. Among mammals, tetraploidy was found in the red visacha rat, *Tympanoctomys barrarae* (2n = 112). The rarity of polyploidy in animals is attributed to its incompatibility with sex determination and dosage compensation of the X chromosome. ►parthenogenesis, ►honey bee, ►silk worm; Zimmet J, Ravid K 2000 Exp Hematol 28:3; Wolfe KH 2001 Nature Rev Genet 2:333.

Polyploidy in Evolution: Polyploidy in evolution is common in the plant kingdom but the majority of polyploid species are allopolyploid. Some of the single copy genes of invertebrates are, however, detectable up to four copies in vertebrates. A survey of plants indicated only 38% of polyploid species in the Sahara region, 51% in Europe, 82% in the Peary Islands, and thus show an increasing trend towards the North. In yeast, after polyploidization, different genes were lost leading to speciation (Scannell DR et al 2006 Nature [Lond] 440:341). In parasites, haploidy is advantageous because selection favors organisms that express a narrow array of antigens and elicitors. In contrast, in the host mounting a defense response, selection favors a broader array of recognition molecules and thus diploids or polyploids (Nuismer SL, Otto SP 2004 Proc Natl Acad Sci USA 101: 11036). Polyploids are also less vulnerable to mutation despite the fact that the mutational target numbers are larger. ►allopolyploid, ►duplications; Otto SP, Whitton J 2000 Annu Rev Genet 34:401; Wu R et al 2001 Genetics 159:869; function of the duplicated genes: Kellog EA 2003 Proc Natl Acad Sci USA 100:4369; Adams KL et al 2003 Proc Natl Acad Sci USA 100:4649.

Polyposis Adenomatous, Intestinal (APC): APC is controlled by autosomal dominant genes responsible for intestinal, stomach (Gardner syndrome), or other types (kidney, thyroid, liver, nerve tissue, etc.) of benign or vicious cancerous tumors. The various forms are apparently controlled by mutations or deletions in the 5q21-q22 region of the human chromosome and represent allelic variations. Retinal lesions (CHRPE)

P

are associated with truncations between codons 463–1387; truncations between codons 1403–1528 involve extra-codonic effects, etc. In addition, it is conceivable that this is a *contiguous gene* region where adjacent mutations affect the expression of the polyposis. By the use of single strand conformation polymorphism technique, DNA analysis may permit the identification of aberrant alleles prenatally or during the presymptomatic phase of the condition. The situation is further complicated, however, by the possibilities of somatic mutations. The *Min* gene of mouse appears to be homologous to the human APC, thus, lending an animal model for molecular, physiological, and clinical studies. The expression of *Min* is regulated also by the phospholipase-encoding gene *Mom1*, indicating the involvement of lipids in the diet. Polyposis may affect a very large portion of the aging human populations, especially high is the risk for females. Certain forms of polyposis may affect the young (juvenile polyposis). Regular monitoring by colorectal examination is necessary for those at risk. Bloody diarrhea and general weakness are symptoms usually too late for successful medical intervention. Molecular genetic information suggests that vertebrates use the same pathway of signal transduction as identified by *Drosophila* genes: *porcupine* (*porc*, 1.59)→*wingless* (*wg*, 2-30.0)→*dishevelled* (*dsh*, 1-34.5)→*zeste white3* (z^{w3}, 1.1.0)→*armadillo* (*arm*, 1-1.2)→cell nucleus. The normal human APC gene appears to be either a negative regulator (tumor suppressor) or an effector, acting between z^{w} and the nucleus. When it mutates, it can either no longer carry out suppression or it may become an effector. The product of *dsh* also appears to be a negative regulator of z^{w}. When the *zeste* product, glycogen synthase kinase (GSK3β) is inactive, the *arm* product (catenin) is associated with the APC product and a signal for tumorigenesis is generated. Alternatively, when no signal is received, GSK phosphorylates and activates a second binding site on APC for catenin but that causes the degradation of catenin and thus no tumor signal is generated. The APC protein may act as a tumor gene also by docking at its COOH end with a human homolog of the *dlg1* (*disc large*, 1-34.82) of *Drosophila*). The *Dlg* product belongs to the *membrane associated guanylate kinase* protein family that is analogous to proteins in vertebrates sealing adjacent cell membranes (tight junction). *Dlg* is also considered to be a tumor gene. Although, the molecular information reveals a number of mechanisms of action, it is not clear which one is being used or if multiple pathways are involved in polyposis. APC/FAP has a prevalence of about 1×10^{-4}. The EB1 protein binds the APC protein, is situated on the microtubules of the mitotic spindle, and serves as a checkpoint for cell division. ►Gardner syndrome, ►Turcot syndrome, ►cancer, ►single-strand conformation, ►GSK3β, ►polymorphism, ►hereditary non-polyposis colorectal cancer, ►contiguous gene syndrome, ►animal models, ►tight junction, ►catenin, ►effector, ►polyposis hamartomatous, ►polyposis juvenile, ►spindle, ►cyclooxygenase, ►microtubule, ►PTEN

Polyposis Hamartomatous (Peutz-Jeghers syndrome, PJS): A chromosome-19p13.3 rare dominant overgrowth of mucous membranes (polyp), especially in the small intestine (jeujunum), but also in the esophagus (the canal from mouth to stomach), bladder, kidney, nose, etc. Melanin spots may develop on lips, inside the mouth, and fingers. Ovarian and testicular cancers were also observed. The susceptibility to this cancer is due to deletion in a serine/threonine protein kinase gene (LKB). LKB1 mediates glucose homeostasis in the liver (Alessi DR et al 2006 Annu Rev Biochem 75:137). This gene signals to VEGF and it is a player in the anterior-posterior axis formation as well as in epithelial polarity. ►pigmentation of the skin, ►cancer, ►Gardner syndrome, ►colorectal cancer Muir-Torre syndrome, ►polyposis adenomatous intestinal, ►multiple hamartomas, ►VEGF; Hemminki A et al 1998 Nature [Lond] 391:184; Sapkota GP et al 2001 J Biol Chem 276:19469; Ilikorkala A et al 2001 Science 293:1323; Bardeesy N et al 2002 Nature [Lond] 419:162.

Polyposis, Juvenile: An early onset polyposis frequently turning malignant, caused by a defect at the carboxyl terminal of the SMAD4/DPC4 (552 amino acids) protein, encoded in human chromosome 18q21.1. SMAD4 in a trimeric association is involved in TGF-β signaling pathway. Although some of the symptoms are similar to other hamartomas, the Cowden disease gene (PTEN, phosphatase and tensin homolog) is encoded in chromosome 10 and the Peutz-Jeghers syndrome is coded for in chromosome 19. ►multiple hamartomas, ►TGF, ►polyposis hamartomatous, ►SMAD, ►DCC, ►PTEN; Howe JR et al 2002 Am J Hum Genet 70:1357.

Polyprotein: A contiguously translated long chain polypeptide that is processed subsequently into more than one protein.

Polypurine: A stretch of purine residues in nucleic acids.

Polypyrimidine: A sequence of multiple pyrimidines (mainly Us) in nucleic acids adjacent to the 3′ splicing site. Py-tract-binding proteins (PTB) recognize these such as the essential splicing factor U2AF^{65}, the splicing regulator sex-lethal (*Sxl*), etc. ►splicing, ►introns, ►sex determination; Le Guinier C et al 2001 J Biol Chem 276:43677.

Polyribosome: Same as polysome (see Fig. P109). ►protein synthesis

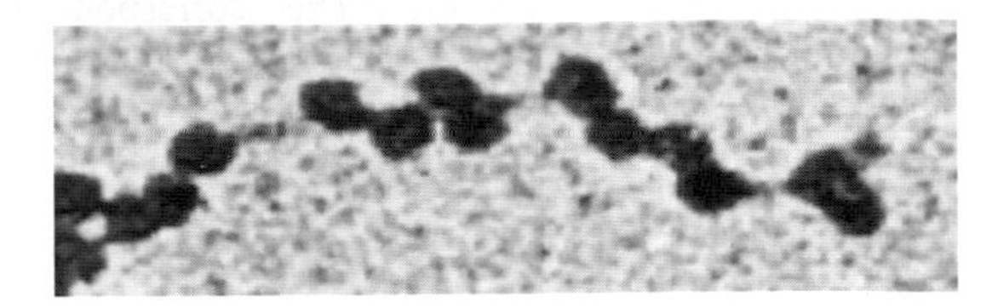

Figure P109. Polyribosome

Polysaccharide: Monosaccharides joined by glycosidic bonds (e.g., starch, glycogen, glycoprotein).

Polysome: In a polysome, the mRNA holds multiple ribosomes together. The ovalbumin polysomes comprise an average of 12 ribosomes and one peptide initiation takes place in every 6–7 s if all the required factors are functioning normally. The average polysome size for globin is ~5 ribosomes (1 ribosome/~90 nucleotides). Pactamycin may be an inhibitor of translation initiation and cycloheximide may interfere with peptide chain elongation. ►ribosome, ►mRNA, ►transcription, ►translation, ►pactamycin, ►cycloheximide

Polysome Display: In a polysome display, polysomes are isolated and screened by the affinity of the nascent peptides on an immobilized specific monoclonal antibody. The mRNA of the enriched pool of polysomes is reverse-transcribed into cDNA and amplified by PCR. The amplified template may be cloned and translated in vitro. The procedure is highly efficient for the screening of large, specific peptide pools. ►reverse transcription, ►cDNA, ►PCR, ►translation in vitro; Mattheakis LC et al 1994 Proc Natl Acad Sci USA 91:9022.

Polysomic Cell: In a polysomic cell, some chromosomes are present in more than the regular number of copies. The polyploids are polysomic for entire genomes. ►aneuploidy, ►polyploidy

Polysomy: In polysomy, some of the chromosomes in a cell are present in more than the normal numbers, examples of these cases in humans are 48,XXXX, 48, XXXY, 49,XXXXX or 49,XXXXY ►nondisjunction, ►polyploid, ►trisomy

Polyspeirism: In polyspeirism, one cell makes several types of related molecules, e.g., different chemokines. (See Montovani A 2000 Immunol Today 2(4):199).

Polyspermic Fertilization: In polyspermic fertilization, more than a single sperm enters the egg and, because each may provide a centriole, multipolar mitoses may take place resulting in aneuploidy and abnormal embryogenesis. ►fertilization

Polysyndactyly: Polysyndactyly is encoded by the HOXD13 gene at human chromosome 2q31-q32 (see Fig. P110). The amplification of the alanine codons (CCG, GCA, GCT, GGC) leads to an expanded (25 to 35) alanine residues in the protein. Some polysyndactyly is due to mutation in GLI3 gene at 7p13. ►trinucleotide repeats, ►syndactyly, ►Pallister-Hall syndrome

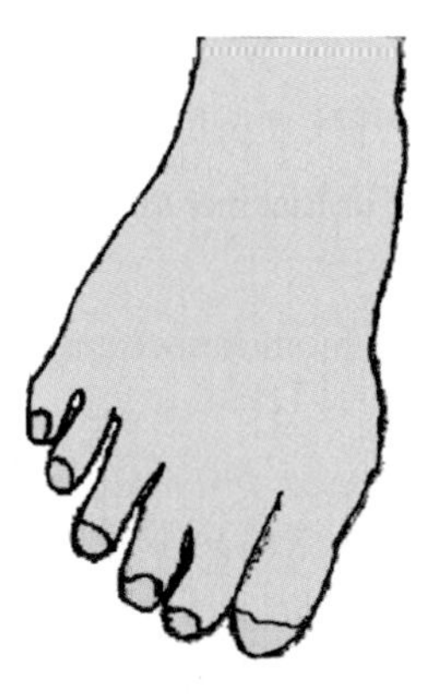

Figure P110. Polysyndactylic toes

Polytenic Chromosomes: Polytenic chromosomes are composed of many chromatids (e.g., in salivary-gland cell nuclei) because in such cases DNA replication was not followed by chromatid separation (see Fig. P111). The polytenic chromosomes in the salivary glands nuclei of diptera may have undergone ten cycles of replication ($2^{10} = 1024$) without division and may have over 1000 strands. Also, the polytenic chromosomes in the salivary glands are extremely long. A regular feature is the very close somatic pairing. Additionally, they all are attached at one point, at the chromocenter.

Polytenic chromosomes have been extensively exploited for analysis of deletions, duplications,

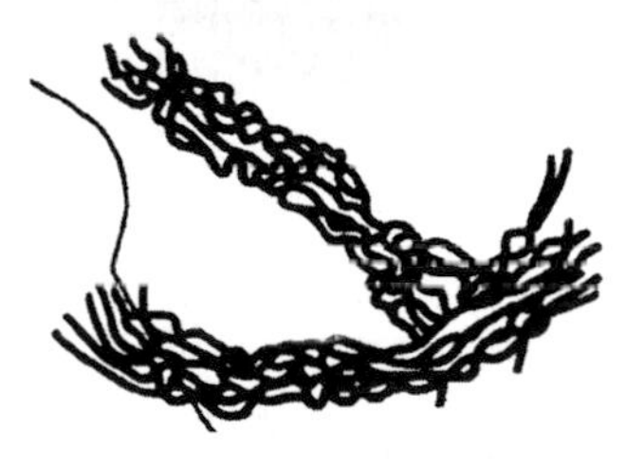

Figure P111. Polytenic chromosomes of *Allium ursinum*. Courtesy of G. Hasischka-Jenschke

inversions, and translocations. The characteristic banding pattern was used also as a cytological landmark for identification of the physical location of genes. Rarely, polyteny occurs in some specialized plant tissues (antipodals) too. ►salivary gland chromosomes, ►giant chromosomes, ►somatic pairing

Polytocous Species: Polytocous species produce multiple offspring by each gestation. ►monotocous

Polytomy: Multifurcating rather than bifurcating analysis of phylogenetic relations. ►evolutionary tree; Walsh HE et al 1999 Evolution 53:932.

Polytopic Protein (multispanning): A polytopic protein traverses the plasma membrane several times.

Polytropic Retrovirus: ►amphotropic retrovirus

Polytypic: A species that includes more than one variety or subtype.

POMC (pre-pro-opiomelanocortin): ►melanocortin, ►opiocortin, ►ACTH

Pomegranate (*Punica granatum*): A Mediterranean fruit tree, 2n = 2x = 16 or 18.

Pompe's Disease: ►glycogen storage diseases

Pongidae (anthropoid primates [hominoidea]): *Gorilla gorilla gorilla* 2n = 48 (see Fig. P112); *Hylobates con-color s* [gibbon] 2n = 52; *Hylobates lar* [gibbon] 2n = 44; *Pan paniscus* [pygmy chimpanzee] 2n = 48; *Pan troglodytes* [chimpanzee) 2n = 48; *Pongo pygmaeus* [orangoutan] 2n = 48; *Symphalangus brachytanites* 2n = 50. ►primates, ►chimpanzee

Figure P112. Gorilla

Pontin: ►chromatin remodeling

PO-PS Copolymers: Phosphorothioate-phosphodiester copolymers are used for antisense technologies. ►antisense RNA

POP′: POP′ symbolizes the ends of the temperate transducing phage genome integrating into the bacterial host chromosome. The corresponding bacterial integration sites are BOB′ and after integration (recombination) the sequence becomes: BOP′ and POB′, respectively ►attachment sites

Pop1p: A protein component of ribonuclease P and MRP. ►ribonuclease P, ►MRP

Poplar (*Populus* spp): 2n = 2x, 2n = 38 (see Fig. P113). Poplar includes cottonwood trees also. The genome of the black cottonwood (*Populus trichocarpa*, 485±10 Mb) has been sequenced and 45,000 putative protein-coding genes detected. Substantial portions of the nuclear and organellar (chloroplasts and mitochondria) genes the genome have been annotated in different tissues. About 8000 duplications were found (Tuskan GA et al 2006 Science 313:1596). (See Cervera M-T et al 2001 Genetics 158:787).

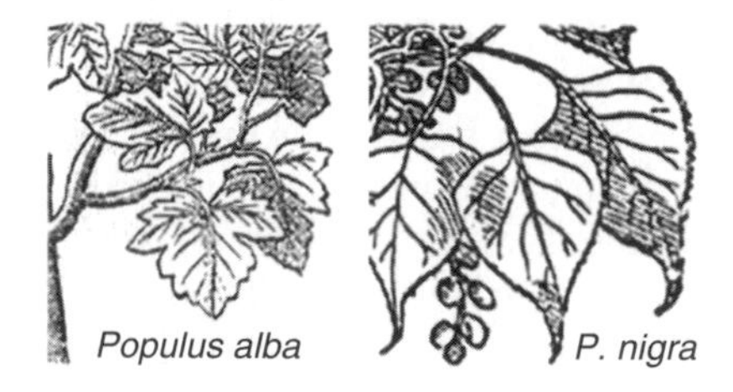

Figure P113. Poplars

Popliteal Pterygium Syndrome (PPS, 1q32-q41): PPS is allelic to the Van der Woude syndrome and it involves a defect in the interferon regulatory factor 6 (Ifr6). Clinical symptoms include cleft palate, harelip, and webbing of the skin. Pterygium is membrane or skin folding; popliteal indicates THE ligament behind the knee. ►Van der Woude syndrome, ►epithelial cell, ►Pterygium

Pop-Out, Chromosomal: Chromosomal pop-out originates due to the intrachromatid reciprocal exchange between direct repeats. It excises one of the repeats (the popout) but may retain the other member of the duplication. ►intrachromosomal recombination, ►sister chromatid exchange

Poppy (*Papaver somniferum*): The latex of poppy is a source of opium, codein, morphine, heroin, and other alkaloids. Their biosynthetic pathway, including a mutant blocked in the biosynthesis of the illicit drug (morphine and codeine) pathways (See Millgate AG et al 2004 Nature [Lond] 431:413). The plant is grown for its oil-rich seed as a food and also for pharmaceutical purposes (see Fig. P114). Basic chromosome number x = 11, diploid and tetraploid forms are known.

Figure P114. Poppy seed capsule

Population: A collection of individuals that may either interbreed and freely trade genes (Mendelian population, deme) or may be a closed population that is sexually isolated from other groups that share the same habitat. ►Hardy-Weinberg theorem, ►population equilibrium

Population Critical Size: ►critical population size

Population Density: The number of cells or individuals per unit volume or area.

Population Effective Size (N_e): The number of individuals in a group or within a defined area that actually transmit genes to the following reproductive cycles (offspring). Each breeding individual has 0.5 chance to contribute an allele to the next generation, and $0.5 \times 0.5 = 0.25$ is the probability to contribute two particular alleles. The probability that the same male contributes two alleles is $(1/N_m)$ 0.25 and for the same female it is $(1/N_f)$ 0.25 where N_m and N_f are the number of breeding males and females, respectively. The probability that any two alleles are derived from the same individual is $0.25N_m + 0.25N_f = 1/N_e$ and N_e is computed as $4N_mN_f /(N_m + N_f)$. ►founder principle, ►genetic drift, ►inbreeding and population size; Wright S 1931 Genetics 16:97.

Population Equilibrium: ►Hardy-Weinberg theorem

Population Genetics: Population genetics studies the factors involved in the fate of alleles in potentially interbreeding groups (see Fig. P115). The individuals within these groups (demes) may actually reproduce by random mating or selfing or by the combination of the two within this range. Population genetics can be entirely theoretical and developing mathematical formulas for predicting the allelic frequencies and the effect of various factors that affect these frequencies and the historical paths of the genes and factors as they emerge, become established or disappear, form equilibria or remain unstable during microevolutionary periods. Experimental population genetics conducts biological studies in the sense of the theoretical framework. Population genetics thus deals with the consequences of mutation, genetic drift, migration, selection and breeding systems and is also one of the most important approaches to experimental (micro) evolution. It provides also the theory for many human genetics, animal and plant breeding research efforts. The availability of molecular information greatly advanced the resolving power of population genetics. The availability of mitochondrial (maternally transmitted) and Y chromosomal (paternally transmitted) markers provide effective tools to study the dynamics and history of human populations. ►terms mentioned, and ►SNIPS, ►DNA chips, ►microsatellites, ►minisatellites, ►mtDNA, ►Y chromosome; population modeling software: http://www.trinitysoftware.com; population genetics tools and Internet resources: Excoffier L, Heckel G 2006 Nature Rev Genet 7:745; Arlequin, analysis of molecular genetic variations: Marjoram P, Tavaré S 2006 Nature Rev Genet 7:759.

Figure P115. Population genetics is concerned with the fate of genes in large collection of organism rather than in the descendants of single individuals

Population Growth, Human: $P_t = P_0(1 + r)^t$ where P_0 = the population at time 0, r = rate of growth and t = time. It can be calculated also by $P_t = P_0e^{rt}$ where e = the base of the natural logarithm. ►age-specific birth and death rates, ►human population growth, ►Malthusian parameter

Population Size, Ancestral: Ancestral population size can be estimated by different methods. The ancestral modern human population might have been about 10,000, whereas the common ancestral human and chimpanzee populations were of the order of 100,000. Newer estimate of the latter is only about 20,000 (Rannala B, Yang Z 2003 Genetics 164:1645).

Population Structure: Population structure is endemic by subpopulation groups. The dispersal of the subdivisions reflect adaptive genetic differences, gene flow and natural selection pressure, inbreeding, overlapping generations, effective population size, and sometimes genetic drift. Sometimes there are too few differences in some population and therefore it is difficult to trace the origin of possible demographic changes. Parasites, e.g., viruses may evolve much faster and from their dispersal one may get good information on the demography/distribution of the host in a region (Bick R et al 2006 Science 311:538). ►population genetics, ►endemic, ►natural selection, ►genetic drift, ►population effective size, ►stratification; Marth G et al 2003 Proc Natl Acad Sci USA 100:376.

Population Subdivisions: Smaller relatively separated breeding groups with restricted gene flow among them. ►gene flow, ►migration

Population Tree: The population tree is constructed on the basis of genes frequencies among populations indicating their evolutionary relationship. ►evolutionary tree, ►gene tree

Population Wave: Periodic changes in the effective population size. ►population size effective, ►random drift, ►founder principle, ►gene flow

Porcupine Man (ichthyosis histrix): Ichthyosis is a dominant form of hyperkeratosis. ►keratosis, ►ichthyosis

Porencephaly: Porencephaly is a generally rare dominant (13qter region) brain disease with cerebrospinal fluid-filled cavities or cysts, affecting primarily infants and young children. Few survivors are plagued by many other debilitating symptoms. In a mouse mutant, single-nucleotide alteration in collagen Col4a1 was the primary cause of the disease. ►collagen; Gould DB et al 2005 Science 308:1167.

Porin: Porin is a voltage-dependent anion channel. It is opened by Bax and Bak pro-apoptotic proteins and closed by the anti-apototic Bcl-x_L. Bax and Bak permit the exit of cytochrome c from the mitochondria and thus facilitate apoptosis by the activation of caspases. In case of IL-7 deficiency and increase in pH over 7.8 the conformation of Bax is altered and the protein moves from the cytoplasm to the mitochondria and facilitates apoptosis. The anti-apoptotic, 24 amino acid-peptide prevents the translocation of Bax to the mitochondria (Guo B et al 2003 Nature [Lond] 423:456). The Bcl-2 protein, localized to the mitochondrial membrane, normally suppresses the release of cytochrome c. Bax deficiency extends the ovarian life span into advanced age of mice. Normally the ovarian follicles fade by menopause in women and at similar developmental stages also in mice. Degradation of Bax by the proteasomes may protect against the apoptosis over-protective effect of Bcl-2 and reduce cancer cell survival. For drug therapy of epithelial cancer the state of BAX vs. Bcl-2 may be significant. ►Bak, ►ion channels, ►cytochrome c, ►apoptosis, ►hypersensitive reaction; Suzuki M et al 2000 Cell 103:645; Gogvadze V et al 2001 J Biol Chem 276:19066; Scorrano L et al 2003 Science 300:135.

Porphyria: Porphyria is a collective name for a variety of genetic defects involved in heme biosynthesis resulting in under- and/or over-production of metabolites in the porphyrin-heme biosynthetic pathway. These diseases may be controled by recessive or dominant mutations. The affected individuals may be suffering from abdominal pain, psychological problems and photosensitivity. The autosomal dominant acute *intermittent porphyria* (human chromosome 11q23-ter) is caused by a periodic 40–60% reduction in porphobilinogen deaminase enzyme resulting in in-sufficient supplies of the tetrapyrrole hydroxymethyl bilane that is normally further processed by non-enzymatic way into uroporphyrinogen I. It was speculated that the famous Dutch painter van Gogh was a victim of this rare disease. Prevalence is in the range of 10^{-4} to 10^{-5}. Exogenous effects such as barbiturate, sulfonamide, alkylating and many other drugs, alcohol consumption, poor diet, various infections and hormonal changes, generally elicit the periodic attacks. An *adult type* of (hepatocutaneous) porphyria, controled by another human gene locus (1p34), involves light-sensitivity and liver damage by the accumulation of, porphyrins caused by uroporphyrinogen decarboxylase deficiency. The general effect may be less severe than in the intermittent porphyria. The rare congenital *erythropoietic porphyria* (CEP) is the result of a defect in the enzyme uroporphyrinogen III co-synthetase controled by a recessive mutation in human chromosome 10q25.2-q26.3. The laboratory identification is generally based on urine analysis for intermediates in the heme pathway. Porphyrias affect also various mammals. Defects in the porphyrin pathways are involved in several types of pigment deficiency mutations of plants. The *variegate porphyria* is caused by a defect of protoporphyrinogen oxidase (PPOX) with symptoms basically similar to that of intermittent porphyria. This dominant disease has low penetrance. Its prevalence is very high (about 3×10^{-3}) in South-African populations of Dutch descent; it apparently represents founder effect. The mental problems of King George III of England (reigned during the US War of Independence) were also attributed to variegate porphyria.

ALAD porphyria also known as "Doss porphyria," is a very rare porphyric disorder linked to a profound lack of porphobilinogen synthase PBGS, also known as δ-aminolevulinate dehydratase (ALAD), is encoded by the *ALAD* gene (9q34). Human (PBGS) exists as an equilibrium of functionally distinct quaternary structure assemblies, known as morpheeins, in which one functional homo-oligomer can dissociate, change conformation, and reassociate into a different oligomer. In the case of human PBGS, the two assemblies are a high-activity octamer and a low-activity hexamer (Jaffe EK, Stith L 2007 Am J Hum Genet 80:329). ►porphyrin, ►heme, ►skin diseases, ►light-sensitivity diseases, ►founder effect, ►coproporphyria, ►aminolevulinic acid conformation

Porphyrin: Four special pyrroles joined into a ring; generally with a central metal, like iron in hemoglobin or in chlorophylls with magnesium (see Fig. P116). ►porphyria, ►coproporphyria, ►heme

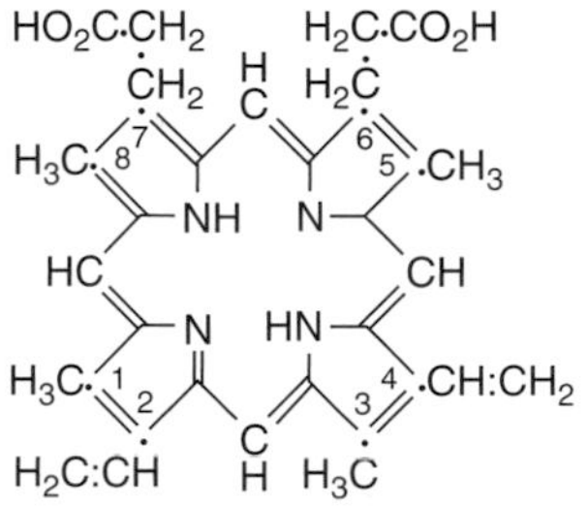

Figure P116. Protoporohyrin

Porphyrinuria: ►porphyria

Porpoise: *Lagenorhynchus obliquidens*, 2n = 44. ►dolphins

Portable Dictionary of the Mouse Genome: Data on ~12,000 genes and anonymous DNA loci of the mouse, homologs in other mammals, recombinant inbred strains, phenotypes, alleles, PCR primers, references, etc. The dictionary can be used on Macintosh, PC in FileMaker, Pro, Excel, and text formats, and is accessible through the Internet (WWW, Gopher, FTP), CD-ROM, or on floppy disk. Information: R.W. Williams, Center for Neuroscience, University of Tennessee, 875 Monroe Ave., Memphis, Tennessee 36163. Phone: 901-448-7018. Fax: 901-448-7266. e-mail: rwilliam@nb.utmem.edu.

Portable Promoter: An isolated DNA fragment, including a sufficient promoter that can be carried by transformation to other cells, and may function in promoting transcription. ►promoter, ►transformation, ►gene fusion

Portable Region of Homology: Insertion and transposon elements may represent homologous DNA sequences and can recombine. The recombination may then generate deletions, cointegrates or insertion, or inversions. These events can take place even in $RecA^-$ hosts. ►Tn*10*, ►cointegrate, ►deletion, ►inversion, ►targeting genes

Position Effect: change in gene expression by a change in the vicinity of the gene.

The new expression may be *stable* or variable (*variegation type position effect*) (see Fig. P117). Stable position effect is observed when promoterless structural genes are introduced by transformation and the transgene is expressed with the assistance of a "trapped" promoter that is regulated differently than the gene's natural (original) promoter. Variegated position effect (PEV) is more difficult to interpret by molecular models. When, however, centromeric heterochromatin was inserted at the *brown* locus of *Drosophila* during larval development the transposed hetero chromatin stochastically associated with the centromeric region and caused PEV (Dernburg AF et al 1996 Cell 85:745). The telomere-linked *ADE2* locus of yeast displayed alternative *ADE* and *ade* phenotypes (Gottschling DE et al 1990 Cell 63:751). It has been assumed that heterochromatin affects the intensity of somatic pairing and variations in somatic association and variations in cross-linking between the homologs by binding proteins bring about the silencing. The *trithorax-like* gene of *Drosophila* encodes a GAGA-homology transcription factor that enhances variegation type position effect (PEV) by decondensation of the chromatin. The mosaicism may also be the result of the spontaneous and random derepression of the promoter in the presence of an activator. The telomeric isochores have been also implicated in position effect (TPE). Position effect may be observed by altering the site or distance of the locus control region. In *Drosophila* over 100 genes were found that affect variegation type position effect (PEV). In *Drosophila* HP1, HP2 proteins of the heterochromatin and histone H3 $lysine^9$ methyltransferase play important role in gene silencing. It seems that RNAi also affects the heterochromatin and several genes encode the RNAi system and their mutations results in loss of silencing (Pal-Bhadra M et al 2004 Science 303:669). It has been hypothesized that these genes control the packaging of the DNA. Many of the cancers develop after translocations or transpositions, indicating the significance of position effect on the regulation of growth. Proteins affecting AT-rich heterochromatin can modify PEV. Transposable elements may also cause position effect (Kashkush K et al 2003 Nature Genet 33:102). Position effects may be exerted even from long distances (2 Mb) and may make difficult to distinguish the position effect causing gene from mutation within the target gene. Such cases may complicate positional cloning. Position effect occurs also in yeasts and other organisms. Some human genetic disorders are due to

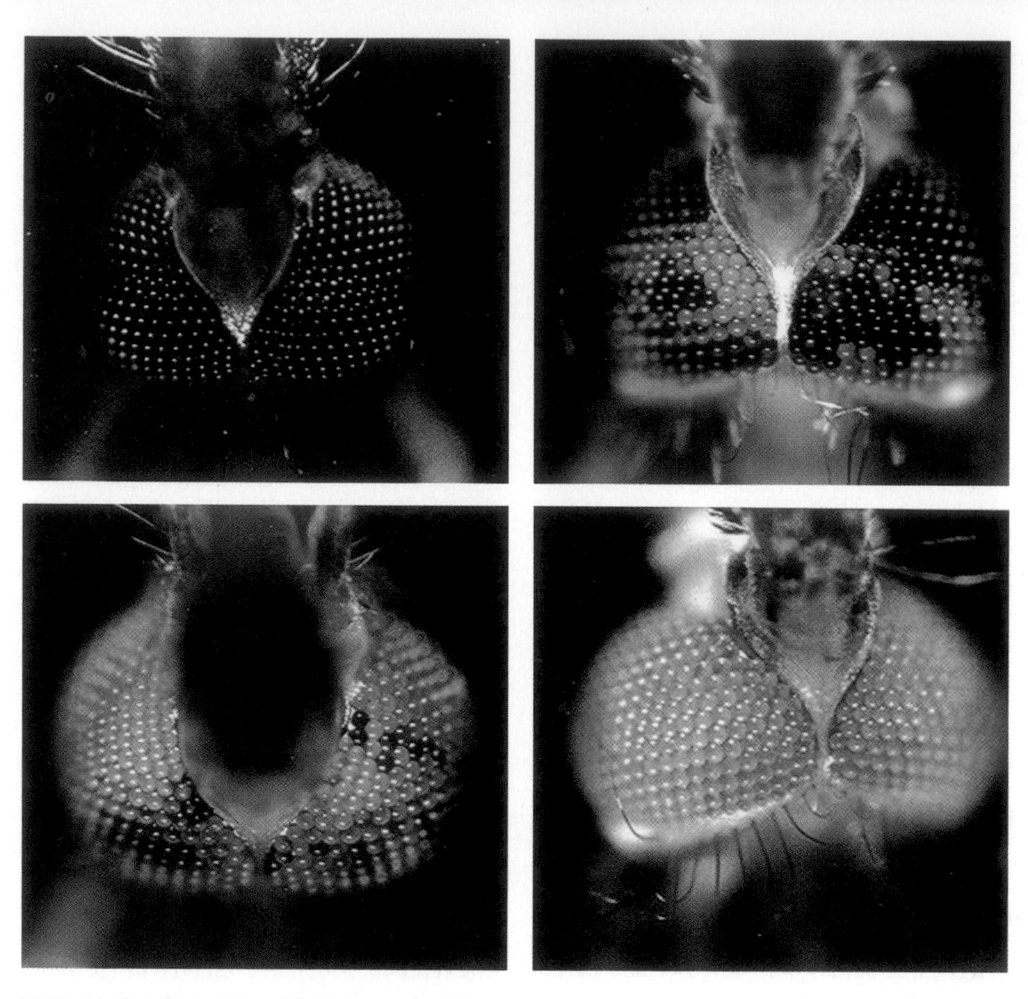

Figure P117. Duplication of the wildtype (*p*+) allele into heterochromatic DNA results in the (*p*) eyes variegated expression of *p*+ in the malaria mosquito *Anopheles gambiae*. (Courtesy of Dr. Mark Benedict, original photograph by James Gathany, CDC)

position effect. ►heterochromatin, ►histone methyltransferases, ►RdMD, ►LCR, ►Offermann hypothesis, ►regulation of gene activity, ►mating type determination in yeast, ►silencer, ►cancer, ►chromosomal rearrangements, ►chromosome breakage, ►locus control region, ►isochores, ►transposable elements, ►epigenesis, ►paramutation, ►positional cloning, ►RPD3, ►developmental-regulator effect variegation, ►RIGS, ►Dubinin effect; Kleinjahn DJ Heyningen V 1998 Hum Mol Genet 7:1611; Baur JA et al 2001 Science 292:2075; Ahmad K, Henikoff S 2001 Cell 104:839; Csink AK et al 2002 Genetics 160:257; suppressors: Ner SS et al 2002 Genetics 162:1763; Monod C et al 2002 EMBO Rep 3:747; Ebert A et al 2006 Chromosome Res 14:377; heterochromatin proteins: Greil F et al 2007 EMBO J 26:741.

Position-Specific Scoring Matrix (PSSM): PSSM represents amino acids at specific positions in a sequence alignment. It can be used for scanning proteins with matches to this tract. ►PWM; Gribskov M et al 1987 Proc Natl Acad Sci USA 84:4355.

Position Weight Matrix: ►PWM

Positional Cloning: ►chromosome walking, ►chromosome landing, ►map-based cloning

Positional Information: Positional information is provided to some cells by signal transducers in a multicellular organism and has an important influence on differentiation and development. ►morphogenesis, ►differentiation

Positional Sensing: Positional sensing provides information for specific differentiation functions. ►morphogenesis

Positive Control: In positive control, gene expression is enhanced by the presence of a regulatory protein (in contrast to negative control, where its action is reduced). The arabinose operon of *E. coli* is a classic example. The regulator gene *araC* produces a repressor (P_1) in the absence of the substrate arabinose. If arabinose is available, P_1 is converted to P_2 (by a conformational change), which is an activator of transcription in the presence of cyclic adenosine monophosphate (cAMP). While the negative control (P_1) is correlated with a low demand for expression, the activator (P_2) appears in response to the demand for high level of expression. In general cases, the addition of an activator protein to the DNA makes possible normal transcription but adding a special ligand to the system removes the activator and the gene is turned off. ►*arabinose* operon, ►negative control, ►*lac* operon, ►autoregulation, ►catabolite activator protein, ►regulation of gene activity

Positive Cooperativity: Binding of a ligand to one of the subunits of a protein facilitates the binding of the same to other subunits.

Positive Interference: ►interference, ►coincidence

Positive/Negative Selection: Selection may be used to isolate cloned constructs containing the desired integrated sequence (positive selection). Negative selection is expected to eliminate integration sites containing the entire vector inserted at non-targeted sites and vector components that have no relevance to cloning. Negative selection is usually less efficient—if it takes place at all—than positive selection. For positive selection in case of hypoxanthine/guanine phosphoribosyl transferase marker, one may use hypoxanthine, aminopterin, and thymidine (HAT) chemicals, whereas in the same experiment for negative selection 6-thioguanine or 5-bromodeoxyuridine may be used.

Positive Selection: In general, it indicates the selection of a desirable type in a population rather than the elimination of the undesirable phenotype/genotypes. ►selection entries

Positive Selection of Lymphocytes: A process of maturation of lymphocytes into functional members of the immune system. In contrast, negative selection eliminates, by apoptosis, early lymphocytes with autoreactive receptors. ►immune system, ►lymphocytes

Positive Selection of Nucleic Acids: Positive selection of nucleic acids isolates and enriches desired types of nucleic acid sequences. The desired (tracer) sequences are digested by restriction endonucleases that generate cohesive ends. The rest of the nucleic acids (driver) are exposed to sonication (or the ends may be even dephosphorylated) and so much sticky ends are not expected. Thus, mainly the tracer-tracer sequences are annealed when the mixture is treated with a ligase enzyme. ►subtractive cloning, ►genomic subtraction, ►RFLP subtraction, ►ligase DNA, ►cohesive ends, ►sonicator

Positive-Strand Virus: The genome of a positive-strand virus is also a mRNA. Upon transcription, the virus may directly produce an infectious nucleic acid. This is a very large class of RNA viruses including the Brome Mosaic Virus, the Hepatitis C Virus, West Nile Virus, Corona Viruses, etc. Their replication is affected by at least 100 genes (Kushner DB et al 2003 Proc Natl Acad Sci USA 100:15764). ►replicase, ►plus strand, ►mRNA, ►negative strand virus

Positive Supercoiling: The overwinding follows the direction of the original coiling, i.e., it takes place rightward. ►supercoiling, ►negative supercoil

Positron Emission Tomography (PET): ►tomography

Post-Adaptive Mutation: Post-adaptive mutation is supposed to arise de novo in response to the conditions of selection. Actually, post-adaptive mutation may not be found if the data are well scrutinized. ►directed mutation, ►pre-adaptive mutation

Post Coitum (p.c.): During embryonal development, the days that follow mating.

Posterior: Pertaining to the hind part of the body or behind a structure toward the tail end.

Posterior Distribution: A summary of random variables collected after new empirical data became available. It is the product of likelihood and prior distribution. ►prior distribution

Posterior Probability: ►Bayes theorem

Post-Genome Analysis: The post-genome analysis studies the experimental results and the informatics of the sequential function (metabolic pathways) and interactions of genes and their products. ►annotation of the genome, ►genetic networks; Lin J et al 2002 Nucleic Acids Res 30:4574, http://www.genome.ad.jp; http://www.genome.ad.jp/kegg/comp/GFIT.html.

Postmeiotic Segregation: Postmeiotic segregation takes place when the DNA was a heteroduplex at the end of meiosis. Among the octad spores of ascomycetes, this may result in 5:3 and 3:5 or other types of aberrant ratios instead of the normal 1:1. Postmeiotic segregation may be an indication of failures in mismatch or excision repair. ►DNA repair, ►tetrad analysis, ►gene conversion

Postnatal: Postnatal refers to that which occurs after birth; generally one to 12 months after birth.

Postprandial: After consuming a meal, a process, e.g., protein anabolism modifies protein synthesis due the change in the amino acid pool or change in insulin supply after eating (postprandially).

Postreduction: As per postreduction, the segregation of the alleles takes place at the second meiotic division. ►tetrad analysis, ►meiosis, ►prereduction

Postreplicational Repair: ►unscheduled DNA synthesis, ►DNA repair

PostScript: A computer application to handle text and graphics the same time. The PostScript code determines what the graphics look like when printed, although may not be visible on the screen of the monitor.

Post-Segregational Killing: ►plasmid addiction

Post-Transcriptional Gene Silencing (PTGS): As per PTSG, the transcript of a transgene is degraded before translation takes place and thus, its expression is prevented. Also, it may be a defense mechanism against viruses in plants. The viral gene may be integrated into the chromosome and duly transcribed, yet it is not expressed. In addition, since the replication of the virus is mediated through a double-stranded RNA that has been found to be a potent inhibitor, it is conceivable that both the plant defense and the transgene silencing rely on similar mechanism(s). In some plant species, the potyviruses, tobacco etch virus, and cucumber mosaic virus may produce a *helper component protease* (HC-Pro) and may inactivate this plant defense by degradation. The HC-Pro may have another role. When a plant is infected simultaneously by two different viruses, one of them promotes the vigorous replication of the other, and the latter by its production of HC-Pro eventually facilitates the spread of the first type of the virus and thus enhances the symptoms of the viral disease. In some of the silenced plant cells, a 25-nucleotide long antisense RNA has been detected that seems to inactivate the normal transcript or infectious viral RNA. According to other studies, the ~25-nt RNA sequence apparently conveys specificity for a nuclease by homology to the substrate mRNA. Several types of hairpin structures of RNAs involving sense and antisense sequences and introns appeared to silence very effectively viral genes in plants. A calmodulin-related plant protein (rgs-CAM) may also suppress silencing. ►silencing, ►plant viruses, ►RNAi, ►RNA interference, ►co-suppression, ►homology-dependent gene silencing, ►methylation of DNA, ►host–pathogen relations; Bass BL 2000 Cell 101:235; Jones L et al 1999 Plant Cell 11:2291; Waterhouse PM et al 2001 Nature [Lond] 411:834; Mitsuhara I et al 2002 Genetics 160:343.

Post-Transcriptional Processing: The primary RNA transcript of a gene is cut and spliced before translation or before assembling into ribosomal subunits or functional tRNA; it includes removal of introns, modifying (methylating, etc.) bases, adding CCA to tRNA amino arm, polyadenylation of the 3′ tail, etc. ►opiotropin; McCarthy JEG 1998 Microbiol Mol Biol Revs 62:1492; Bentley D 1999 Curr Opin Cell Biol 11:347.

Post-Transcriptional Operons: A hypothesis according to which, functionally related genes may be regulated post-transcriptionally as groups by mRNA-binding proteins that recognize common sequence elements in the untranslated 5′ and 3′ subsets of the transcripts. This conclusion is based on findings that mRNA-binding proteins recognize unique subpopulations of mRNAs, the composition of these subsets may vary depending on conditions of growth and the same mRNA occurs in multiple complexes. These conserved *cis* elements were named USER (untranslated sequence elements for regulation) codes. These systems may permit plasticity during developmental processes or responses to drug treatment. ►operon, ►genetic networks; Keene JD, Tenenbaum SA 2002 Mol Cell 9:1161.

Posttranslational Modification: Enzymatic processing of the newly synthesized polypeptide chain, the product of translation. The modification may include proteolytic cleavage, glycosylation, phosphorylation, farnesylation, conformational changes, assembly into quaternary structure, etc. These modifications may alter function. Mass spectrophotometry is generally used for the identification the alterations. ►protein synthesis, ►protein structure, ►conformation, ►proteomics; Németh-Cawley JF et al 2001 J Mass Spectrom 36:1301; Mann M, Jensen ON 2003 Nature Biotechnol 21:255; http://dbptm.mbc.nctu.edu.tw/; tandem mass spectra interpretation server: http://modi.uos.ac.kr/modi/.

Post-Transplantational Lymphoproliferative Disease (PTDL): In PTDL, after engraftment, the Epstein-Barr virus-infected B cells may continue to proliferate because the immuno-suppressive therapy required to maintain the graft inhibits cytotoxic T lymphocytes. Bone marrow transplantation may alleviate the problems. ►Epstein-Barr virus, ►immuno-suppression, ►CTL

Postzygotic: ►prezygotic

Postzygotic Isolation: Postzygotic isolation arises when in allopatric evolution the taxa diverge from the common ancestor by accumulation of different non-deleterious mutations. Although the divergent forms are well adapted, their hybrids may be inviable or sterile because the negative effects of the alleles in a shared background. ►allopatric speciation; Orr HA, Turelli M 2001 Evolution 55:1085.

Potassium-Argon Dating: Potassium-Argon dating is based on the conversion of K^{40} into Ar^{40}, a stable gas.

It is used for dating rocks over 100,000 years old. ►argon dating, ►radiocarbon dating

Potassium Ion Channel: ►ion channel

Potato (*Solanum tuberosum*): The genus has 170 to 300 related species with basic chromosome number x = 12. In nature, species with diploid, tetraploid, and hexaploid chromosome numbers are found. The cultivated potatoes originated from the *S. brevicaulis* group in the Andes Mountains (Spooner DM et al 2005 Proc Natl Acad Sci USA 102:14694), and secondarily from *Solanum andigena* in Central America where they produce tubers under short-day conditions. The majority of the modern varieties is day-neutral and develops tubers under long-day conditions. The cultivated potatoes are usually cross-pollinating species but many set seeds also by selfing. Generally, the seed progeny is very heterogeneous genetically. Potatoes are rarely propagated by seed, as is a crop. The diploid relatives are usually self-incompatible whereas the polyploids may set seeds by themselves. Among the cultivated groups, the tuber color may vary from white to yellow to deep purple. Also the chemical composition of the tubers shows a wide range, depending on the purpose of the market. Potato, besides being a popular vegetable, is an important source of industrial starch. The related species carry genes of agronomic importance (disease, insect resistance, etc.) that have not yet been fully exploited for breeding improved varieties. The application of the molecular techniques of plant breeding seems promising. ►patatin; Isidore E et al 2003 Genetics 165:2107; http://www.tigr.org/tdb/tgi; https://gabi.rzpd.de/projects/Pomamo/; http://www.sgn.cornell.edu.

Potato Beetle: (*Leptinotarsa decemlineata*, n = 18): One of the most devastating pests of the agricultural production of potatoes (see Fig. P118). Plants transgenic for the δ endotoxin of *Bacillus thüringiensis* are commercially available ►potato, ►*Bacillus thüringiensis*

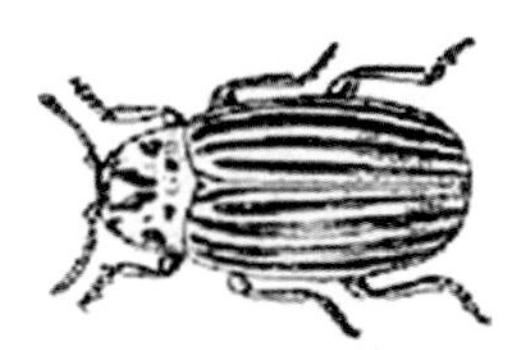

Figure P118. Potato beetle

Potato Leaf Roll Virus: The potato leaf roll virus has double-stranded DNA genetic material.

POTE: A family of genes encoding proteins with an amino-terminal cysteine-rich domain, a central domain with ankyrin repeats, and a carboxyl-terminal domain containing spectrin-like helices. In humans, the POTE gene family is composed of 13 closely related paralogs dispersed among eight chromosomes. These genes are found only in primates, and many paralogs have been identified in various primate genomes. The expression of POTE family is generally restricted to a few normal tissues (prostate, testis, ovary, and placenta but several family members are expressed in breast cancer and many other cancers (Lee Y et al 2006 Proc Natl Acad Sci USA 103:17885).

Potocki-Shaffer Syndrome: ►exostosis

Potocytosis: Moving ions and other molecules into cells by caveola vehicles. ►caveolae

POU: A region with several transcriptional activators of 150–160 amino acids (including a homeo-domain), involved with a large number of proteins controlling development. The acronym stands for a prolactin transcription factor (PIT), an ubiquitous and lymphoid-specific octamer binding protein (OTF), and the *Caenorhabditis* neuronal development factor (Unc-86). A POU domain may directly facilitate the recruitment of TBP and transcriptional activators and may stimulate transcription even when the enhancer is at a distance from the core promoter. POU domain proteins are involved in shuttling between nucleus and cytoplasm (Baranek C et al 2005 Nucleic Acids Res 33:6277). ►homeodomain, ►transcription factor, ►TBP, ►transcriptional activator, ►enhancer, ►core promoter, ►octa, ►unc, ►*Caenorhabditis*, ►deafness; Ryan AK, Rosenfeld MG 1997 Genes Dev 11:1207; Bertolino E, Singh H 2002 Mol Cell 10:397.

pOUT: A strong promoter opposing pIN and directing transcription to the outside end of an insertion element. ►RNA-OUT, ►pIN

Power of a Test: Algebraically, the power of a test is 1 − β, where β = type II error. This test reveals the probability of rejecting a false null hypothesis and accepting a correct alternative. The experimenter needs as large a value of 1 − β as possible, by reducing β to a minimum. To improve the power, the size of the experiment (population) can be increased. In case the size cannot be increased, a more powerful test (statistics) should be chosen. ►error types, ►significance level

Pox Virus: A group of oblong double-stranded DNA viruses of 130–280 kbp (see Fig. P119). Some of these are parasites on insects, others in the family are the chicken pox, cowpox (vaccinia), and smallpox viruses (see Fig. P120).

Figure P119. Pox virus

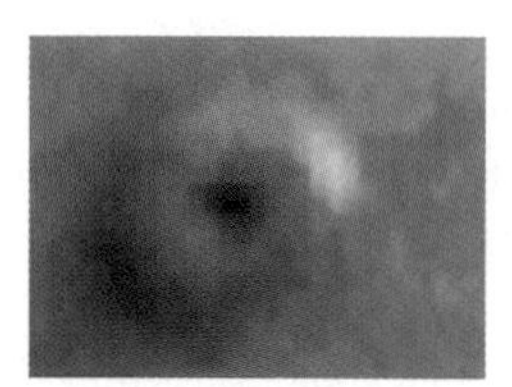

Figure P120. Pox virus lesion

Their transmission takes place through insect vectors or by dust or other particles. Engineered pox virus vectors that are not able to multiply in mammalian cells may have the ability to express passenger genes without the risk of disease. Due to the success of vaccination, smallpox as a disease has now been eradicated and vaccination against is no longer necessary except in case of terrorist attacks (Halloran ME et al 2002 Science 298:1428). The smallpox virus (VARV) linear DNA genome is about 186 kbp with inverted terminal repeats containing 196 to 207 open reading frames. Apparently, there is small variation among the various isolates. The genes of the smallpox virus overlapand its mRNA is not spliced (Esposito JJ et al 2006 Science 313:807).

Poxvirus based vectors are being used orally to protect wild life (red fox) from rabies, for the protection of chickens against the Newcastle virus. Recombinant canarypox virus is employed for the protection of dogs and cats against the distemper, feline leukemia, equine influenza, etc. Highly attenuated derivatives, expressing rabies virus glycoprotein, Japanese encephalitis virus polyprotein, or seven antigens of *Plasmodium falciparum* are used for safe and effective vaccination. Smallpox virus disease has been eradicated and at this time only the Center of Disease Control and Prevention in Atlanta, GA in the USA and the Russian State Research Center of Virology and Biotechnology in Kolsovo, Novosbirsk, Russia, maintain active samples. Limited-scale vaccinations have been performed as protection against terrorism. Large-scale use of the current vaccine may involve side effects such as heart disease in some individuals. A new vaccine developed and used in Japan does not pose serious side effects even at very high doses. It may revert to the wild type progenitor due to mutation of gene *B5R*. Fortunately, this gene can be eliminated without effect on protective immunity (Kidokoro M et al 2005 Proc Natl Acad Sci USA 102:4252). ►malaria, ►*Plasmodium falciparum*, ►variolation; Moss B, Shisler JL 2001 Semin Immunol 13:59; Takemura M 2001 J Mol Evol 52(5):419; Enserink M 2002 Science 296:1592; L1 protein: Su HP et al 2005 Proc Natl Acad Sci USA 102:4240; http://www.poxvirus.org.

POZ: Protein–protein interaction domain of Zinc finger-containing transcriptional regulatory proteins. ►Zinc finger, ►αβ T cells

PP-1, PP-2: Protein serine/threonine phosphatases that are inhibited by okadaic acid. PP-1 may be associated with chromatin through the nuclear inhibitor of PP-1 (NIPP-1). PP enzymes play key roles in many cellular processes. ►okadaic acid, ►DARPP

pp15: A protein factor required for nuclear import. ►membrane transport, ►RNA export

pp125^{FAK}: ►CAM

PP2A: The proline-directed heterotrimeric protein serine-threonine phosphatase dephosphorylates proteins in the MAP pathway of signal transduction and thus balances the effect of kinases. Its deregulation seems to be associated with several types of cancers, Alzheimer disease, and susceptibility to infections by pathogens. The crystal structure of the holoenzyme has been determined (Cho US, Xu W 2007 Nature [Lond] 445:53). PP2A subunit B56 regulates β-catenin signaling and several metabolic processes. PP2A is very sensitive to okadaic acid, a tumor-inducing agent. The non-catalytic α4 subunit of PP2A is a regulator of apoptosis by dephosphorylating c-Jun and p53 transcription factors, which upon phosphorylation promote apoptosis (Kong M et al 2004 Science 306:695). ►Sit, ►MAP, ►signal transduction, ►MAP kinase phosphatase, ►okadaic acid,►catenins, ►cyclin G, ►apoptosis, ►calcineurin, ►TGF

PPAR (peroxisome proliferator-activated receptor, 17q12): A transcription factor in the adipogenic (fat synthetic) pathways. The three types α, γ, and δ show different distribution in human tissues and associate with different ligands. PPARα is the target for the drugs and fibrates (amphipatic carboxylic acids) that reduce triglycerides. Type α also acts as a transcription factor for several genes affecting lipoprotein and fatty acid metabolism. PPARγ is a (3p25) regulator of glucose, lipid, and cholesterol metabolism, may be sensitized by thiazolidinediones (TZD), and offers some hope to be used for the treatment of diabetes mellitus type2 (IDDM). PPARγ2 deficiency dramatically reduces adipogenesis in mouse fibroblasts whereas PPARγ1 affects obesity and diabetes (Zhang J et al 2004 Proc Natl Acad Sci USA 101:10703). The PPARγ 12Ala allele is associated with a small yet significant reduction in the risk for diabetes type II. PPARγ agonists have a controversial—promoting and suppressing—effect on polyposis of the colon and other cancers. In human thyroid carcinoma, PAX8–PPARγ1 has been observed. PPAR-α agonists are also successful for the treatment of some autoimmune diseases. PPARγ deficiency can also lead to hypertension. ►peroxisome, ►ROS, ►diabetes mellitus, ►polyposis, ►farnesoid X receptor, ►leukotrienes, ►leptin, ►obesity, ►Krox20, ►hypertension, ►PAX, ►thiazolidinedione, ►dizygotic twins, ►sirtuin, ►mesenchyma, ►retinoic acid; Lowell BB 1999 Cell 99:239; Kersten S et al 2000 Nature 405:421; Willson TM et al 2001 Annu Rev Biochem 70:341; Michalik L et al 2004 Nature Rev Cancer 4:61; review: Lehrke M, Lazar MA 2005 Cell 123:993.

pPCV: Plasmid plant cloning vector, designation (with additional identification numbers and/or letters) of agrobacterial transformation vectors constructed by Csaba Koncz.

ppGpp: ►discriminator region

PPI (peptidyl prolyl isomerase): An endoplasmic reticulum-bound protein assisting chaperone function. There are 3 PPI families: cyclophilins, FK506, and parvulins. ►chaperone, ►PDI; Dolinski K, Heitman J 1997, p 359 In: Gething MJ (Ed.) Guidebook to Molecular Chaperones and Protein Folding Catalysis, Oxford University Press, Oxford, UK.

ppm: Parts per million.

PP2R1B: PP2R1B at human chromosome 11q22-q24 encodes the β isoform of the PP2A serine/threonine protein phosphatase. The gene displays alterations (LOH) in a variable fraction of lung, colon, breast, cervix, head and neck, ovarian cancers and melanoma, and it is thus a suspected tumor suppressor gene. ►tumor suppressor gene, ►LOH; Mumby MC, Walter G 1993 Physiol Rev 73:673.

PPTs (palmitoyl-protein thioesterases): PPTs hydrolyze long chain fatty acyl CoA and PPT1 may cleave cysteine residues in the lysosomes. Its deficiency may lead to Batten disease. ►Batten disease

Prader-Willi Syndrome (Prader-Labhart-Willi syndrome): A very rare (prevalence 1/25,000) dominant defect involving poor muscle tension, hypogonadism, (hyperphagia [over-eating]) obesity, short stature, small hands and feet, mental retardation, compulsive behavior that sets in by the teens, caused by methylation of the paternal chromosome and by disomy for maternal chromosome 15 (see Fig. P121). The recurrence risk in affected families is about 1/1000. This and cytological evidence indicate that the condition is caused in about 60% of the cases by a chromosomal breakage in the so-called imprinting center (IC) in the long arm of human chromosome 15q11.2-q12. The same deletion (4–5 Mbp or sometimes shorter), when transmitted through the mother, results in the Angelman syndrome. At the breakpoints, the HERC2 gene (encoding a very large protein) may be repeated. The repeats may then recombine and generate the deletions. See two chromosomes shown in Figure P122, with different number of repeats, as detected by FISH. In some cases, there is no deletion but a mutation in an ubiquitin protein ligase gene (UBE3A). Mutations in the proximal part of IC lead to the Angelman syndrome and in the distal part to the Prader-Willi syndrome. Molecular studies indicated in many cases the missing (uniparental disomy) or silencing (imprinting) of a paternal DNA sequences in the patients.

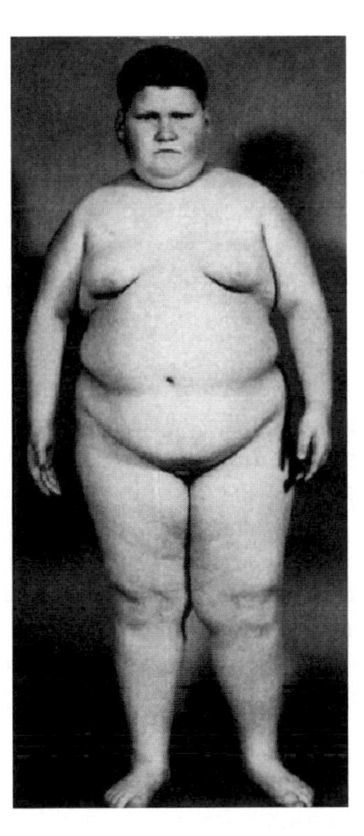

Figure P121. Prader-Willi syndrome at age 15. (From Bergsma, D., ed. 1973 Birth Defects. Atlas and Compendium. By permission of the March of Dimes Foundation)

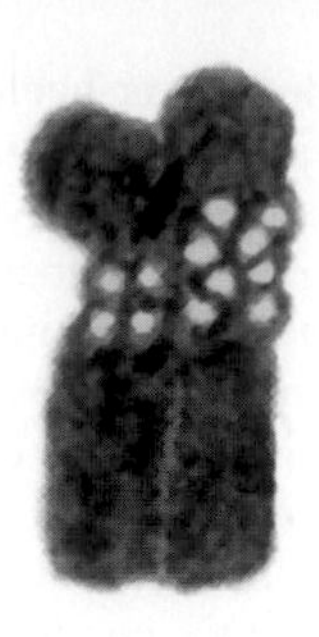

Figure P122. Duplications in the Prader-Willi syndrome. (Redrawn from Amos-Landgraf JM et al 1999 Am J Hum Genet 65:370)

The deletions of this syndrome usually involve the promoter of an snRPN gene, resulting in the silencing (imprinting) of flanking genes (ZNF127 encoding a Zn-finger protein, NDN [necdin], IPW and PAR) on either side. Necdin and Magel2 also interact with Fez (fasciculation) protein and BBS4 (Bardet-Biedl protein) and they affect centrosome function (Lee S et al 2005 Hum Mol Genet 14:627). Lack of expression of snRNP is the most reliable clinical criterion for the syndrome, although snRPN alone does not appear to be the major pathogenic factor in the syndrome. The snoRNA appears to control alternative processing of the serotonin receptor 2C (Kishore S, Stamm S 2006 Science 311:230). Also, exon 1 (1920 bp) includes more than 100 5′-CG-3′ and 5′-GC-3′ dinucleotides liable to methylation. Among the 19 methyl-sensitive restriction enzyme sites within the telomeric region were completely methylated in this syndrome but none of these were methylated in case of the Angelman syndrome. A 2.2 kb spliced and polyadenylated RNA is transcribed 150 kb telomerically to snRPN in human chromosome 15q11.2-q12 and the homologous mouse chromosome 7 region. The transcript is not translated, however. This gene (IPW) is not expressed in individuals with the Prader-Willi syndrome and is therefore said to be imprinted in Prader-Willi syndrome. In the mouse gene *Ipw*, multiple copies of 147 bp repeats are found with retroviral transposons (IAP) insertions. ►obesity, ►imprinting, ►imprinting box, ►epigenesis, ►disomic, ►serotonin, ►alternative splicing, ►Angelman syndrome, ►head/face/brain defects, ►snPRN, ►IAP, ►Bardet-Biedl syndrome; Fulmer-Smentek SB, Francke U 2001 Hum Mol Genet 10:645.

Prairie Dog (ground squirrel, *Sciuridae*): Burrowing mammals with five different species. They are rodents, and not canidae, inhabiting arid areas (see Fig. P123).

Figure P123. *Cynomys ludovicianus Prairie dog*

pRB: Retinoblastoma protein. ►retinoblastoma

PRC1: A spindle midzone-associated kinase. ►midzone

PRD1: An icosahedral, double-stranded-DNA phage (*Tectiviridae*) of Gram-negative bacteria. It lacks the common phage tail and it acquires an injection device from the host membrane during phage assembly. The mature virion is 66 Mda, containing 20 protein species. It is evolutionarily related to adenovirus. ►phage, ►adenovirus; Abrescia NGA et al 2004 Nature [Lond] 432:68.

Pre-Adaptive: A pre-adaptive trait or mutation is that which occurs before selection would favor it but it becomes important when the conditions become favorable for this genotype. ►adaptation, ►post-adaptive mutation, ►fluctuation test

Prebiotic: Prebiotic refers to the period before life originated. ►evolution prebiotic

Precambrian: ►Proterozoic, ►Cambrian, ►geological time periods

Precise Excision: In precise excision, the genetic vector or transposon leaves the target site without structural alterations; the initially disrupted gene or sequence can return to the original (wild type) form.

Precursor Ion Scanning: A powerful technique in proteomics in connection with MS/MS and TOF. ►MS/MS, ►TOFMS; Steen H et al 2001 J Mass Spectrom 36:782; Hager JW 2002 Rapid Commun Mass Spectrom 16:512.

Predetermination: In predetermination, the phenotype of the embryo is influenced by the maternal genotypic constitution but the embryo itself does not carry the gene(s) that would be expressed in it at that particular stage. ►delayed inheritance, ►maternal effect genes

Predictive Value: The true estimate of the number of individuals afflicted by a condition on the basis of the tests performed in the population.

Predictivity: The predictivity of an assay system is, e.g., the percentage of carcinogens correctly identified among carcinogens and non-carcinogens, by indirect carcinogenicity tests, based mainly on mutagenicity. ►accuracy, ►specificity, ►sensitivity, ►bioassays for environmental mutagens

Predictome: A database of protein links and networks. ►genetic networks

Predictor Gene: The expression of a predictor gene signals difference(s) among phenotypically similar but functionally different forms of malignancies. ►cancer classification

Predisposition: Susceptibility to disease. It may be based on a large number of alleles and environmental factors may also have a major role. A predispositional testing, based on the genetic constitution, may or may not indicate the probability of a disease.

Preeclampsia: ►ecclampsia

Preferential Repair: Transcriptionally active DNA is repaired preferentially. ►DNA repair

Preferential Segregation: Non-random distribution of homologous chromosomes toward the pole during anaphase I of meiosis. There are four loci in the Abnormal 10 chromosome (carrying a terminal large knob) in maize that affect neocentromere activity, increased recombination, and preferential segregation (Hiatt EN, Dawe RK 2003 Genetics 164:699). If harmful combination of genes (gene blocks) is preferentially included in the gametes, this may constitute a genetic load. ►meiotic drive, ►neocentromere, ►polarized segregation; Rhoades MM, Dempsey E 1966 Genetics 53:989; Buckler ES et al 1999 Genetics 153:415.

Prefoldins (PFDN): Molecular chaperones built as hexamers from the α and β subunits and four β-related subunits in eukaryotes. Prefoldin 1 was assigned to human chromosome 5, and prefoldin 4 to chromosome 7. Prefoldins may be required for gene amplification in tumors. ►chaperone; Siegert R et al 2000 Cell 103:621; prefoldin-like Skp structure: Walton TA, Sousa MC 2004 Mol Cell 15:367.

Preformation: An absurd historical idea supposing that an embryo preexists in the sperm (spermists) or in the egg (ovists) of animals and plants, rather than developing by epigenesis from the fertilized egg. ►epigenesis; Richmond ML 2001 Endeavour 25(2):55.

Pre-Genome RNA: The replication intermediate in retroid viruses. ►retroid virus

Pregnancy, Male: In seahorses (*Hippocampus*, Syngnathidae), the female lays unfertilized eggs in the ventral pouch of the male where he fertilizes them and the fetus develops.

Pregnancy Test: Pregnancy is the formation of a fetus in the womb; there are about 40 known pregnancy tests, based on chemical study of blood and urine or other criteria. The currently used tests rely on estrogen level. ►Aschheim-Zondek test

Pregnancy, Unwanted: The estimated frequency of unwanted pregnancy in the human population of the whole world was estimated between 35 to 53 million per year. ►pregnancy test, ►abortion medical

Pregnenolone: A precursor in the biosynthesis of several steroid hormones: CHOLESTEROL PREGNENOLONE→PROGESTERONE→ANDROSTENEDIONE→TESTOSTERONE→ESTRADIOL These steps are under the control mainly of several cytochrome P450 (CYP) enzymes and their deficiency or misregulation lead to pseudohermaphroditism, hermaphroditism, and various other anomalies of the reproductive system. ►steroid hormones

Preimmunity: ►host-pathogen relation

Preimplantation Genetics: Preimplantation genetics detects genetic anomalies either in the oocyte or in the zygote before implantation takes place. This can be done by molecular and biochemical analyses, and cytogenetic techniques. The status of the egg—in some cases of heterozygosity for a recessive gene—may be determined prior to fertilization by examining the polar bodies. Since the first polar bodies are haploid products of meiosis, if they show the defect, then presumably the egg is free of it. The purpose of this test is to prevent transmission of identifiable familial disorders. The technology permits selection for sex of the embryo but this is ethically controversial. ►gametogenesis, ►in vitro fertilization, ►ART, ►micromanipulation of the oocyte, ►polymerase chain reaction, ►sperm typing, ►PGD; Delhanty JD 2001 Am J Hum Genet 65:331; Wells D, Delhanty JD 2001 Trends Mol Med 7:23; Bickerstaff H et al 2001 Hum Fertil 4:24; Braude P et al 2002 Nature Rev Genet 3:941; ethical and legal considerations: Knoppers BM et al 2006 Annu Rev Genomics Hum Genet 7:201.

Preinitiation Complex: ►PIC, ►open promoter complex

Pre-mRNA (pre-messenger RNA): The primary transcript of the genomic DNA, containing exons and introns and other sequences. ►mRNA, ►RNA processing, ►introns, ►hnRNA, ►post-transcriptional processing, ►RNA editing, ►splicing enhancer exonic

Pre-Mutation: A genetic lesion, which potentially leads to mutation unless the DNA repair system remedies the defect before it is visually manifested. Pre-mutational lesions lead to delayed mutations. UV irradiation or chemical mutagens with indirect effects (that is the mutagen requires either activation or it induces the

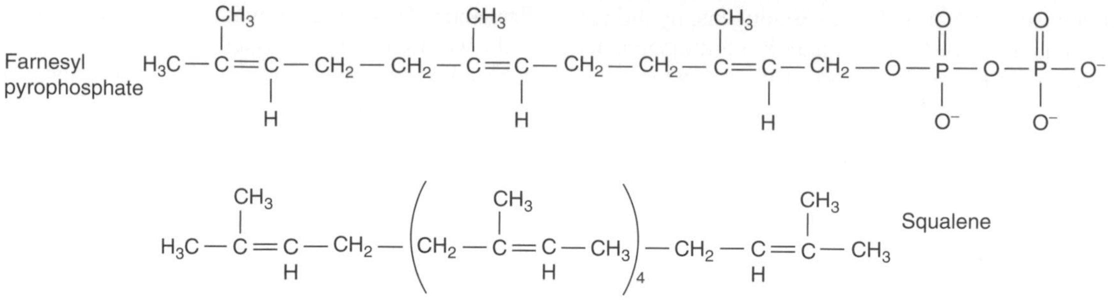

Farnesyl pyrophosphate $H_3C-C(CH_3)=C(H)-CH_2-CH_2-C(CH_3)=C(H)-CH_2-CH_2-C(CH_3)=C(H)-CH_2-O-P(=O)(O^-)-O-P(=O)(O^-)-O^-$

$H_3C-C(CH_3)=C(H)-CH_2-\left(CH_2-C(CH_3)=C(H)-CH_3\right)_4-CH_2-C(CH_3)=C(H)-CH_3$ Squalene

Figure P124. Two molecules of farnesyl pyrophosphate are converted into 30-C squalene in the presence of NADPH

formation of mutagenic radicals, peroxides) frequently cause pre-mutations. Incomplete expansion of trinucleotide repeats may also be considered pre-mutational, e.g., in the fragile X chromosome. ►chromosomal mutation, ►chromosome breakage, ►point-mutation, ►telomutation, ►trinucleotide repeats, ►fragile X, ►Sherman paradox under mental retardation; Auerbach C 1976 Mutation research. Chapman and Hall, London, UK.

Prenatal Diagnosis: Prenatal diagnosis determines the health status or distinguishes among the possible nature of causes of a problem with a fetus before birth. The results of cytological or biochemical analysis permit the parents to prepare psychologically and medically to the expectations. Although chromosomal abnormalities cannot be remedied, for metabolic disorders (e.g., galactosemia) advance preparations can be made. Similarly, fetal erythroblastosis may be prevented. In case of very severe hereditary diseases, abortion may be an option if it is morally acceptable to the parents and does not conflict with the existing laws. Prenatal diagnosis is now available for more than hundred anomalies. Until recently, prenatal diagnosis required mainly amniocentesis or sampling of chorionic villi, now in some instances the maternal blood can be scanned for fetal blood cells and by the use of the polymerase chain reaction, the DNA of the fetus can be examined. ►genetic testing, ►genetic screening, ►genetic counseling, ►amniocentesis, ►polymerase chain reaction, ►RFLP, ►DNA fingerprinting, ►DNA circulating, ►plasma nucleic acid, ►PUBS, ►MSAFP, ►sonography, ►fetoscopy, ►echocardiography, ►hydrocephalus, ►galactosemias, ►chorionic villi, ►pre-implantation genetics, ►ART; Weaver DD, Brandt IK 1999 Catalog of prenatally diagnosed conditions, Johns Hopkins University Press, Baltimore, Maryland; Fetal Evaluation: http://www.cpdx.com/.

Prenylation: The attachment of a farnesyl alcohol, in thioeter linkage, with a cystein residue located near the carboxyl terminus of the polypeptide chain. The donor is frequently farnesyl pyrophosphate. Cytosolic proteins are frequently associated with the lipid bilayer of the membrane by prenyl lipid chains or through other fatty acid chains. Prenyl biogenesis begins by enzymatic isomerization of isopentenyl pyrophosphate (CH_2=C[CH_3]CH_2CH_2OPP) into dimethylallyl pyrophosphate ([CH_3]$_2$C=CHCH_2OPP). These then react to form geranyl pyrophosphate ([CH_3]$_2$C= CHCH_2CH_2C[CH_3]=CHCH_2OPP). Geranyl pyrophosphate is then converted into farnesyl pyrophosphate as shown in Figure P124.

Members of the RAS family proteins, involved in signal transduction, cellular regulation, and differentiation are prenylated at cysteine residues of the COOH-terminus. Prenylation determines the cellular localization of these molecules. Cellular fusions are mediated by prenylated pheromones. The cytoskeletal lamins attaching to the cellular membranes are farnesylated. Prenylation of the C-termini of proteins is generally mediated by farnesyltransferase, a heterodimer of 48 kDa α and 46 kDa β subunit. Protein farnesylation is essential for early embryogenesis and for the maintenance of tumorigenesis (Mijimolle N et al 2005 Cancer Cell 7:313). Squalene is a precursor of cholesterol and other steroids (see Fig. P125). ►lipids, ►abscisic acid, ►lamin, ►RAS, ►cytoskeleton, ►pheromone

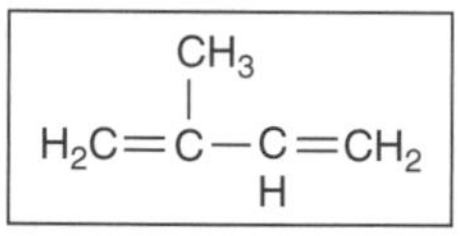

$H_2C=C(CH_3)-C(H)=CH_2$

Figure P125. Isoprene units

Prepatent: The period before an effect (e.g., infection) becomes evident.

Prepattern Formation: The distribution of morphogens precedes the appearance of the visible pattern of particular structures. ►morphogen; Chiang C et al 2001 Dev Biol 236:421.

Prepriming Complex: A number of proteins at the replication fork of DNA involved in the initiation of DNA synthesis. ►DNA replication, ►replication fork

Preprotein: A preprotein is a protein molecule that has not completed yet its differentiation (trimming and processing).

Prereduction: In prereduction, the alleles of a locus separate during the first meiotic anaphase because there was no crossing over between the gene and the centromere. ►tetrad analysis, ►meiosis, ►post-reduction

Pre-rRNA: The unprocessed transcripts of ribosomal RNA genes; they are associated at this stage with ribosomal proteins and are methylated at specific sites. The cleavage of the cluster begins at the 5′ terminus of the 5.8S unit and proceeds to the 18S and 28S units. ►rRNA, ►rrn, ►ribosomal RNA, ►ribosome

Presence-Absence Hypothesis: The presence-absence hypothesis was advocated by William Bateson during the first few decades of the 20th century as an explanation for mutation. The recessive alleles were thought to be losses whereas the dominant alleles were supposed to indicate the presence of genetic determinants. Similar views, in a modified form, were maintained for decades later and were debated in connection with the nature of induced mutations. ►null mutation, ►genomic subtraction; Bateson W et al 1908 Rep Evol Comm R Soc IV, London, UK.

Presenilins (PS): Proteins associated with precocious senility, such the presenilin 1 (S182/AD3) encoded at human chromosome 14q24.3 (442 amino acids) and presenilin 2 (STM2/AD4, 467 amino acids encoded at 1q31-q34.2) proteins of the Alzheimer's disease. Mutations in presenilins account for about 40% of familial cases of the Alzheimer's disease. Presenilins are integral membrane proteases. Presenilin 1 and presenilin 2 increase the production of β-amyloid either directly or most likely by their effect on secretases. They may also promote apoptosis. Presenilins control calcium ion channels of the endoplasmic reticulum and the disruption of these channels can lead to Alzheimer's disease (Tu H et al 2006 Cell 126:981). Mutant PS1 strongly affects both the amplitude of evoked excitatory currents as well as the frequency of spontaneous excitatory synaptic currents by decreasing the number of functional synapses (Priller C et al 2007 J Biol Chem 282:1119). p53 and $p21^{WAF-1}$ promote inhibition of presenilin 1, and that may encourage apoptosis and tumor suppression as well. Presenilin 1 is associated with β-catenin and in the complex β-catenin is stabilized. Mutations in Presenilin 1 may destabilize β-catenin and the latter is usually degraded in Alzheimer's disease. Thus, mutation in Presenilin 1 may predispose to early onset Alzheimer's disease. Presenilin also controls pigmentation of the retinal epithelium and epidermal melanocytes, and mutation may lead to aberrant accumulation of tyrosinase (Wang R et al 2006 Proc Natl Acad Sci USA 103:353). Presenilin 2 contains a domain that is similar to that of ALG3 (apoptosis linked gene) and inhibits apoptosis. Presenilin 1 may also affect various (non-neurodegenerative) cancer-related pathways. The presenilins are involved in the processing of the transmembrane domain of amyloid precursor proteins (APP), and they are essential for normal embryonal development. Protein TMP21 is a component of presenilin complexes and modulates selectively γ-secretase but not ε-secretase (Chen F et al 2006 Nature [Lond] 440:1208). Presenilins also control the transduction of Notch signals. A presenilin locus exists in the third chromosome also (77A-D) of *Drosophila melanogaster*. ►Alzheimer disease, ►prion, ►apoptosis, ►p53, ►p21, ►calsenilin, ►catenins, ►ubiqulin, ►Notch, ►secretase, ►nicastrin; Sisodia SS et al 1999 Am J Hum Genet 65:7; Baki L et al 2001 Proc Natl Acad Sci USA 98:2381; Wolfe MS, Haass C 2001 J Biol Chem 276:5413; Marjaux E et al 2004 Neuron 42:189.

Present: The expressed open reading frames during particular times or conditions when analyzed by microarrays. ►open reading frame, ►microarray hybridization

Presenting: Behavioral signs shown by the female indicating receptivity to mating.

Presequence: A generic name for signal peptides and transit peptides.

Presetting: The penchant of a transposable element to undergo reversible alteration in a new genetic milieu. It may be caused by the methylation of the transposase gene. ►Spm, ►Ac-Ds

Presymptotic Diagnosis: The identification of the genetic constitution before the onset of the symptoms. ►prenatal diagnosis, ►genetic screening

Pre-tRNA: ►tRNA

Prevalence (K, λ): The proportion of a genetic or non-genetic anomaly or disease in a particular human population at a particular time. The percentage of hereditary diseases caused by presumably single nuclear genes in human populations: autosomal dominant 0.75, autosomal recessive 0.20, X-linked 0.05. Besides these, multifactorial abnormalities account for about 6% of the genetic anomalies. In case the general prevalence of the diseases in a population is x, the expected expression among sibs for autosomal dominant is 1/2x, for autosomal recessives it is 1/4x, and for multifactorial control $1/\sqrt{x}$. ►incidence, ►mitochondrial diseases in humans

Prevention of Circularization of Plasmids: ►circularization

Preventive Medicine: Preventive medicine studies the genetic and physiological conditions of individuals and societies in order to take measures to avoid the onset of diseases. ►diseases in humans, ►genetic counseling, ►counseling genetic, ►genetic risk, ►recurrence risk, ►empirical risk, ►heritability

Prey: ►two-hybrid system

Prezygotic: The DNA molecule in the prokaryotic cell before recombination (transduction or transformation); after integration it becomes postzygotic.

Pri: ►cis-acting elements

PriA: Replication priming protein. ►replication fork, ►DNA replication, ►primase

Priapism: An uncommon prolonged engorgement of the penis or erection of the penis or the clitoris. Phosphodiesterase-5A dysregulation is one cause (Champion HC et al 2005 Proc Natl Acad Sci USA 102:1661). It is generally called idiopathic but is some families it occurs repeatedly. It may be evoked by alcoholism or by certain drugs. Chronic hemoglobinopathies and genital cancer may also cause it. ►idiopathic, ►hemoglobinopathies, ►nitric oxide, ►phosphodiesterase

Pribnow Box (TATA box): 5′-TATAATG-3′ (or similar) consensus preceding the prokaryotic transcription initiation sites by 5–7 nucleotides in the promoter region at about −10 position from the translation initiation site. Separated by 17 bp there is another conserved element (called extended promoter) in prokaryotes at −35 (TTGACA). The eukaryotic homolog of the Pribnow box is the Hogness box. ►Hogness box, ►open promoter complex, ►σ; Gold L et al 1981 Annu Rev Microbiol 35:365.

Pride: A living and mating community of animals under the domination of a particular male(s).

Primary Cells: Primary cells are taken directly from an organism rather than from a cell culture.

Primary Constriction: The centromeric region of the eukaryotic chromosome (see Fig. P126).

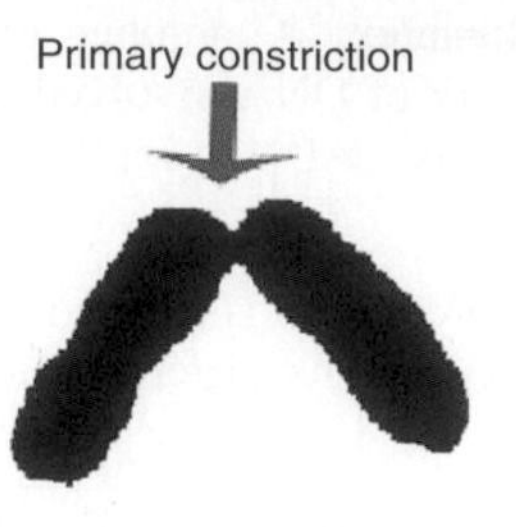

Figure P126. Primary constriction

Primary Nondisjunction: ►nonsdisjunction

Primary Response Genes: The induction of primary response genes occurs without the synthesis of new protein but requires only pre-existing transcriptional modifiers such as hormones. ►sign transduction, ►secondary response genes

Primary Sex Ratio: Ratio of males to females at conception. ►sex ratio

Primary Sexual Characters: The female and male gonad, respectively. ►secondary sexual characters

Primary Structure: The sequence of amino acid or nucleotide residues in a polymer.

Primary Transcript: The RNA transcript of the DNA before processing has been completed. ►processing, ►pre-mRNA, ►pre-rRNA

Primase: Polymerase-α/primase synthesizes an about 30-nucleotide RNA primer for the initiation of replication of the lagging strand of the DNA. In prokaryotes, the primosome protein complex fulfills the function. In bacteriophage M13, an imperfect hairpin is formed at the origin of replication, which is recognized by *E. coli* RNA polymerase σ^{70} holoenzyme and it synthesizes a 18–20 nucleotide primer suitable for synthesis by DNA polymerase III. The RNA polymerase leaves a protruding 3′-end of the RNA but maintains an RNA-DNA hybrid molecule of 8–9 bp. This 3-end of the RNA can then interact with DNA polymerase III. Filamentous phages and bacterial plasmids probably use the priming mechanisms (Zenkin N et al 2006 Nature [Lond] 439:617). The function of the primase is much slower than the processing of the DNA polymerase. Since the Okazaki fragments need priming several times while the leading strand is synthesized by the DNA polymerase, there must be a molecular brake there to assure that the two (leading and lagging) strands are synthesized in concert (Lee J-B et al 2006 Nature [Lond] 439:621). In *E. coli*, gene *DnaG* (66 min) encodes it and it is associated with the replicative helicase. In eukaryotes, the ~60 and ~50 kDa subunits of DNA polymerase α represent the primase.

The latter complex is associated with proteins and forms a mass of ~300 kDa. The primases prime any single-stranded DNA but they are far more effective at specific sequences. DnaG recognizes the 5′-CTG-3′ trinucleotide and synthesizes a 26–29 nucleotide RNA. The mouse primase works at ~17 sites that share either 5′-CCA-3′ or 5′-CCC-3′ at about 10 nucleotides downstream from the priming initiation site at the 3′-end. The active template is usually rich in pyrimidines. In eukaryotes, the primer is directly transferred to the DNA pol α without dissociating from the template. Primase inhibitors (cytosine or adenosine arabinoside, 2′-deoxy-2′-azidocytidine, etc.) have therapeutic potentials. Several binding proteins assist priming. ►replication fork, ►DNA replication, ►PriA, ►DNA polymerases, ►polymerase switching, ►Okazaki fragment, ►primosome, ►replication restart; Keck JL et al 2000 Science 287:2482; Arezi B, Kuchta RD 2000 Trends Biochem 25:572; Frick DN, Richardson CC 2001 Annu Rev Biochem 70:39; Augustin MA et al 2001 Nature Struct Biol 8:57.

Primates: The taxonomic group that includes humans, apes, monkeys, and lemur. To the higher primates, also called anthropoidea or simians, belong the old world monkeys (Cercopithecidae) such as the *Macaca*, *Cercopythecus*, etc., hominoidea (chimpanzee [*Pan*], gorilla [*Gorilla*], orangutan [*Pogo*], and humans, and also the now extinct early evolutionary forms. The anthropoidea includes also the new world monkeys (*Ceboidea*). The lower primates or prosimians mean the genera of the lemur, galago, etc. According to data of D.E. Kohne et al (1972 J Hum Evol. 1:627), on the basis of thermal denaturation of hybridized DNA the numbers in million years of divergence (and the % of nucleotide difference) of various primates from man was estimated to be: chimpanzee 15 (2.4), gibbon, 30 (5.3), green monkey 46 (9.5), capuchin 65 (15.8), galago 80 (42.0). Some of the DNA differences now need revisions (►chimpanzee). Humans have substantially lower variations in the DNA than the great apes, chimpanzees, and orangoutan (see Table P2) (Kaessmann H et al 2001 Nature Genet 27:155).

Table P2. The expressed genes indicate relations among three primate species as follows

	Chimpanzee	Orangutan	Rhesus macaque
Humans	110	120	176
Chimpanzee	-	150	141
Orangutan	-	-	129

(Data from Gilad, Y. *et al.* 2006 Nature [Lond) 440:242)

The taxonomic tree of primates can be outlined as:

PRIMATES: I. Catarrhini. IA1 Cercopithecidae (Old World Monkeys). IA1a Cercopithecinae, IA1b Colobinae. IA1c Cercopithecidae. IB. Hominidae (Gorilla, Homo, Pan, Pongo). IC. Hylobatidae (Gibbons). II. Platyrrhini (New World Monkeys): IIA. Callitrichidae (Marmoset and Tamarins). IIA1. Callimico. IIA2. Callithrix. IIA3. Cebuella. IIA4. Callicebinae. IIA5. Cebinae. IIA6. Pitheciinae. III Strepsirhini (Prosimians) IIIA Cheirogalidae. IIIA1 Cheirogaleus. IIIA2. Microcebus. IIIB. Daubentoniidae (Ayeayes). IIIB1. Daubentonia. IIIC Galagonidae (Galagos). IIIC1. Galago. IIIC2. Otolemur. IIID. Indridae. IIID1 Indri. IIID2. Propithecus (Sifakas). IIIE. Lemuridae (Lemurs). IIIE1. Eulemur. IIIE2. Hapalemur. IIIE3. Lemur. IIIE4. Varecia. IIIF. Loridae (Lorises). IIIF1. Loris. IIIF2. Nycticebus IIIF3. Perodicticus. IIIG Megalapidae. IIIG1. Lepilemur. IV. Tarsii (Tarsiers). IVA Tarsiidae (Tarpsiers). IVA1. Tarsius. ►human races, ►apes, ►prosimii, ►Cebidae, ►Callithricidae, ►Cercopithecideae, ►Colobidae, ►Pongidae, ►*Homo sapiens*, ►Hominidae, ►evolutionary tree; DeRousseau CJ (ed) 1990 Primate Life History and Evolution, Wiley-Liss, New York; Enard W et al 2002 Science 296:340; http://www.primate.wisc.edu/pin; phylogeny database: http://www.hvrbase.org/.

Primatized Antibody: A chimeric antibody constructed using the variable region of monkey antibody linked to the human constant region. ►antibody chimeric

Primer: A short sequence of nucleotides (RNA or DNA) that assists in extending the complementary strand by providing 3′-OH ends for the DNA polymerase to start transcription. In some viruses (hepadna viruses, adenoviruses), the replication of viral DNA, and in some cases viral RNA is primed by proteins. The 3′ OH group of a specific serine is linked to a dCMP and a viral enzyme drives the reaction. Replication may proceed from both ends of the linear molecules without being in the same replication fork. ►nested primers, ►primase, ►PCR, ►Vpg; http://www.genome.wi.mit.edu; primer identification: http://ihg.gsf.de/ihg/ExonPrimer.html; http://web.ncifcrf.gov/rtp/gel/primerdb/; primer design for promoters and exons: http://genepipe.ngc.sinica.edu.tw/primerz/; may better opens by: http://www.citeulike.org/user/sebastien_vigneau/article/1357047; Primer3 primer selection: http://www.bioinformatics.nl/cgi-bin/primer3plus/primer3plus.cgi.

Primer Extension: RNA (or single-strand DNA) is hybridized with a single strand DNA primer (30–40 bases), which is 5′-end-labeled. Generally, the primers are complementary to base sequences within 100 nucleotides from the 5′-end of mRNA to avoid heterogeneous products of the reverse transcriptase which is prone to stop when it encounters tracts of secondary structure. After extension of the primer by reverse transcriptase, the length of the resulting cDNA (measured in denaturing polyacrylamide gel electrophoresis) indicates the length of the RNA from the label to its 5′-end. When DNA (rather than RNA) is used as template DNA-DNA hybridization must be prevented. The purpose of the primer extension analysis is to estimate the length of 5′ ends of RNA transcripts and identify precursors of mRNA and processing intermediates. The cDNA so obtained can be directly sequenced by the Maxam-Gilbert method or also by the chain termination methods of Sanger if dideoxyribonucleoside triphosphates are included in the reaction vessels. Primer extension preamplification (PEP) facilitates the preparation of multiple copies of the genome of a single sperm (Zhang L et al 1992 Proc Natl Acad Sci USA 89:5847). ►DNA sequencing, ►primary transcript, ►post-transcriptional processing, ►chimeric proteins, ►PCR-based mutagenesis, ►amplification; Reddy VB et al 1979 J Virol 30:279; Sambrook J et al 1989 Molecular cloning, Cold Spring Harbor Laboratory Press.

Primer Shift: The primer shift is used for confirmation that a PCR procedure indeed amplified the intended DNA sequence. For this purpose, a primer different from the one initially employed is chosen and attached to the template a couple of hundred bases away from the position of the first. After the completion of the PCR process, the amplified product is supposed to be as much longer as the difference between the position of the first and the second primer if the amplification involved the intended sequence. Such a procedure may be used when a DNA sequence corresponding to a deletion is amplified. ►PCR, ►primer

Primer Walking: A method in DNA sequencing whereby a single piece of DNA is inserted into a large-capacity vector. After a shorter stretch had been sequenced, a new primer is generated from the end of what has been already sequenced and the process is continued until the sequencing of the entire insert is completed. ►DNA sequencing; Zevin-Sonkin D et al 2000 DNA Seq 10(4–5):245; Kaczorowski T, Szybalski W 1998 Gene 223:83.

Primitive Streak: The earliest visible sign of axial development of the vertebrate embryo when a pale line appears caudally at the embryonic disc as a result of migration of mesodermal cells. ►organizer, ►differentiation, ►morphogenesis, ►Hensen's node, ►embryo node; Ciruna B, Rossant J 2001 Dev Cell 1:37.

Primordium: The embryonic cell group that gives rise to a determined structure.

Primosome: The complex of prepriming and priming proteins involved in replication of the Okazaki fragments (see Fig. P127). It moves along with the replication fork in the opposite direction to DNA synthesis. The primosome (containing helicase and primase) unwinds the double-stranded DNA and synthesizes RNA primers. ►DNA replication, ►replication fork, ►Okazaki fragment, ►primase; Marsin S et al 2001 J Biol Chem 276:45818; replication restart primosome PriB component structure: Lopper M et al 2004 Structure 12:1967; Zhang Z et al 2005 Proc Natl Acad Sci USA 102:3254; electron microscopic structure: Norcum MT et al 2005 Proc Natl Acad Sci USA 102:3623.

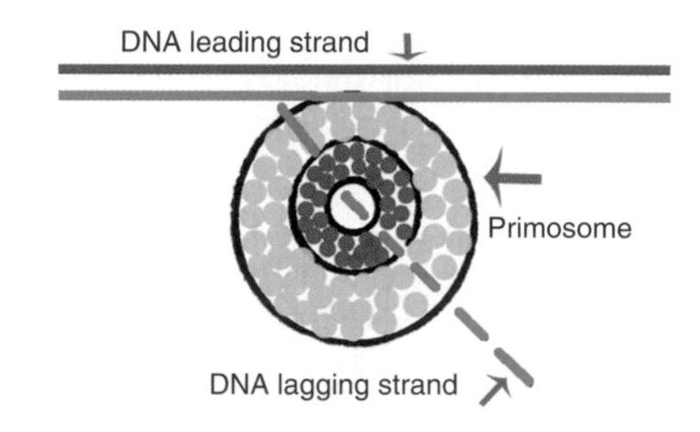

Figure P127. Primosome

Primula (Primrose): An ornamental plant. *P. kewensis* (2n = 36) is an amphidiploid of *P. floribunda* (2n = 18) and *P. verticillata* (2n = 18).

Principal Component Analysis: The aim of the principal competent analysis is to reduce the apparent complexity of the original variables and summarize the information in a simpler manner. The principal components are construed as linear functions of the original variables. ►factorial analysis, ►stratification; Joliffe IT 1986 Principal Component Analysis, Springer, New York.

PRINS (primed in situ synthesis): An in situ hybridization technique bearing some similarities to other methods of probing (e.g., FISH). The PRINS procedure uses small oligonucleotide (18–22 nucleotides) primers from the sequence of concern. After the primer is annealed to denatured DNA (chromosomal or other polynucleotides), a thermostable DNA polymerase is employed to incorporate biotin-dUTP or digoxygenin-dUTP. The procedure is very sensitive to mismatches (because the primer is short) and a mismatch at the 3′-end may prevent chain extension. The concentration of the primer (C) = $Ab_{260}/\varepsilon_{max}$ x L where Ab_{260} = absorbance at 260 nm, ε_{max} = molar extinction coefficient (M^{-1}) and L = the path length of the cuvette of the spectrophotometer. The molar extinction coefficients are determined ε_{max} = (number of A × 15,200) + (number of T × 8400) + (number of G × 12,010) + (number of C × 7050) M^{-1}. (A = adenine, T= thymine, G = guanine, C = cytosine). PRINS are useful for many purposes, including determination of aneuploidy, DNA synthesis, viral infection, etc. ►PCR, ►FISH, ►in situ hybridization, ►LISA, ►biotinylation, ►extinction, ►non-radioactive label; Hindkjaer J et al 2001 Methods Cell Biol 64:55.

PrintAlign: A computer program for graphical interpretation of fragment alignments in physical mapping of DNA. ►physical map

PrintMap: A computer program that produces a restriction map in PostScript code. ►PostScript

PRINTS: A database for the analysis of the hierarchy of protein families on the basis of fingerprints. ►protein families; http://www.bioinf.man.ac.uk/dbbrowser/PRINTS/.

Prion (PrP^C, PrP^{Sc}, PrP*, PRPres): Infective, protease-resistant glycoprotein particles, responsible for the degenerative brain diseases such as scrapie in sheep, chronic wasting disease in deer and elk, BSE in cattle, kuru, Creutzfeldt-Jakob disease, Gerstmann-Sträussler syndrome, fatal familial insomnia of man and, possibly, also Alzheimer's disease. Protease-sensitive prions ($sPrP^{Sc}$) also exist. Prions are transmitted among various animal species although the expression may require a longer lag. Decreased transmission of the prion state between divergent proteins is termed "species barrier" and was thought to occur because of the inability of divergent prion proteins to co-aggregate. Species barrier can be overcome in cross-species infections, e.g., from "mad cows" to humans. The counterparts of yeast prion protein Sup35, originated from three different species of the *Saccharomyces sensu stricto* group exhibit the range of prion domain divergence that overlaps with the range of divergence observed among distant mammalian species. All three proteins were capable of forming a prion in *Saccharomyces cerevisiae*, although prions formed by heterologous proteins were usually less stable than the endogenous *S. cerevisiae* prion. Heterologous Sup35 proteins co-aggregated in the *S. cerevisiae* cells. However, in vivo cross-species prion conversion was decreased and in vitro polymerization was cross-inhibited in at least some heterologous combinations, thus demonstrating the existence of prion species barrier (Chen B et al 2007 Proc Natl Acad Sci USA 104:2791).

Mutations in the gene may result in prion potentiation. The non-familial Creutzfeldt-Jakob disease may be traced to infections by gonadotropins, human growth hormones extracted from cadavers, grafts, improperly-sterilized medical equipment contaminated by prions, or to eating the meat (primarily brain, lymphatic and nerve tissues) of infected animals. In case of chronic inflammation of the kidneys, scapie-infected mice excrete prions by the urine (Seeger H et al 2005 Science 310:324). Normal prion protein, PrPsen (PRP^C protease sensitive), is expressed as a membrane-bound glycophosphatidylinositol (GPI)-anchored protein (see Fig. P128). The GPI anchor may be the requisite for infectious transmission of this protein and the expression of typical scrapie (Chesebro B et al 2006 Science 308:1435). Other amyloidogenic proteins, which are involved in brain degeneration but lack GPI anchor, are not infectious. GPI-anchorless proteins can be secreted from the blood and can form deposits in the amyloid or non-amyloid forms in the brain and heart endothelia. In infected non-transgenic mice, it appears mainly in the non-amyloid form but in transgenic animals, mainly in the amyloid form. The protease-resistant prion causes heart disease (Trifilo MJ et al 2006 Science 313:94).

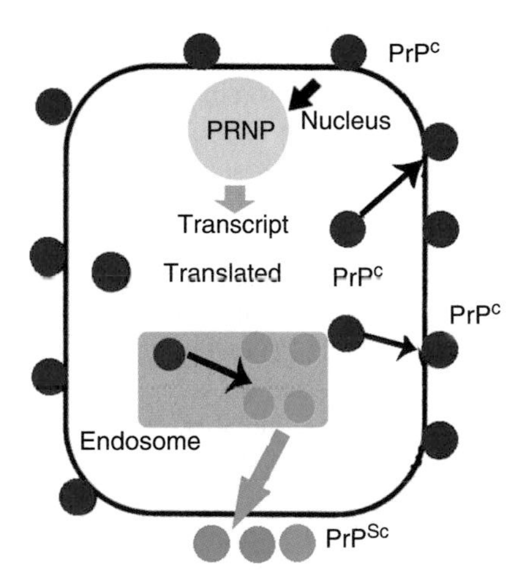

Figure P128. The normal PrP^C protein is encoded in the nucleus by the PRNP gene and after transcription the RNA transcript is translated in the cytoplasm. Some of the PrP^C molecules decorate the surface of the nerve cells and others may be sequestered into the endosomes or lysosomes. Within these compartments, a conformational alteration may take place and the infectious PrP* or PrP^{Sc} protein molecules are released. These altered molecules may then infect other, normal cells and initiate a process of degenerative protein accumulation. The conformational changes may be caused by mutations in PRNP and other genes located in several human chromosomes. (Modified after Weissmann C 1999 J Biol Chem 274:3)

On the basis of the degree and extent of glycosylation, about 400 prions have been distinguished. The N terminus of PrP contains a glycosaminoglycan (GAG)-binding motif. Binding of GAG is important in prion disease. Accordingly, all human mutant recombinant rPrPs bind more GAG, and GAG promotes the aggregation of rPrP more efficiently than wild-type recombinant normal cellular PrP (rPrPC). Furthermore, point mutations in *PRNP* gene also cause conformational changes in the region between residues 109 and 136, resulting in the exposure of a second, normally buried, GAG-binding motif. Importantly, brain-derived PrP from transgenic mice, which express a pathogenic mutant with nine extra octapeptide repeats also binds more strongly to GAG than wild-type PrPC (Yin S et al 2007 Proc Natl Acad Sci USA 104:7546).

Prions appear like virus particles but are free of nucleic acid. The 25 nm virus-like arrays in two cell lines with transmissible spongiform encephalitis (TSE) virions are structurally independent of pathological PrP in the intact cell (Manuelidis L et al 2007 Proc Natl Acad Sci USA 104:1965). It appears that a normal protein is structurally modified; the α helical structure is largely converted into β sheets, leading to the formation of these autonomous disease-causing proteins. Transmission of the disease in the absence of the protease-resistant prion is exceptional (Lasmézas CI 1997 Science 275:402). Although prions are infectious diseases of protein folding, some RNAs of mammals appear adjuvants of the pathogenic alterations in vitro whereas invertebrate RNA has no such effect (Deleault NR et al. 2003 Nature [Lond] 425:717). Experimental protein misfolding cyclic amplification (PMCA) reaction can yield in vitro generated prions that are indistinguishable from prions isolated from scrapie hamster brain in terms of proteinase K resistance, autocatalytic conversion activity, and, most notably, specific biological infectivity (Weber P et al 2006 Proc Natl Acad Sci USA 103:15818).

In order to develop prion disease in mice, the organism must have PrP^C, and if it is absent the animals become resistant to scrapie and show normal neuronal functions (Büeler H et al 1993 Cell 73:1339). PrPC-deficient cattle produced by a sequential gene-targeting system over 20 months of age are clinically, physiologically, histopathologically, immunologically and reproductively normal. Brain tissue homogenates are resistant to prion propagation in vitro (Richt JA et al 2007 Nature Biotechnol 25:132). Also, microglia (cells that surround the nerves and phagocytize the waste material of the nerv tissue) must be present to develop prion disease. Depletion of the endogenous neuronal PrP^C from mice by the Cre recombinase prevents the progression of the disease. The non-neural accumulation of PrP^{Sc} is not pathogenic but leads to an arrest of $PrP^C \rightarrow PrP^{Sc}$ conversion within neurons and prevents neurotoxicity (Malluci G et al 2003 Science 302:871). Depletion of endogenous neuronal prion protein (PrP^C) in mice with early prion infection reversed spongiform change and prevented clinical symptoms and neuronal loss. Thus, early functional impairments precede neuronal loss in prion disease and can be rescued. Further, they occur before extensive PrP^{Sc} deposits accumulate and recover rapidly after PrP^C depletion, supporting the concept that they are caused by a transient neurotoxic species, distinct from aggregated PrP^{Sc} (Malucci GR et al 2007 Neuron 53:325).

If microglia are destroyed by L-leucine-methylester, the neurotoxic PrP fragment, containing amino acids 106–126, does not harm the neurons. The transition from the normal $PrP^C \rightarrow PrP^{Sc}$ (the insoluble scrapie prion) conformation involves changes in amino acid residues 121–231, involved two antiparallel β-sheets and in three α-helices (see Fig. P129).

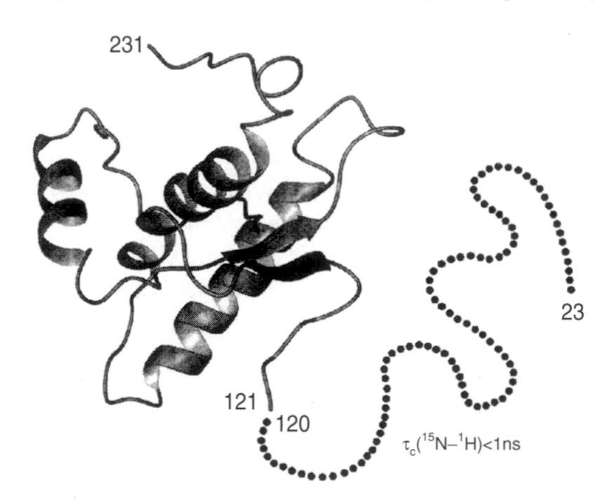

Figure P129. The nuclear magnetic resonance-revealed structure of the PrP^C protein. The amino end displays an about 100 residue flexible sequence that is modified when PRP^{Sc} is formed. (Modified after Riek R et al 1997 FEBS Lett 413:282. By Permission of Elsevier Science and Authors)

Monoclonal antibody 15B3 (Peretz D et al 2001 Nature [Lond] 412:739) and the anti-DNA antibody OCD4 as well the gene 5 protein (Zou W-Q et al 2004 Proc Natl Acad Sci USA 101:1380), both DNA-binding proteins, discriminate between the PrP^C and PrP^{Sc} and may help in the diagnosis of prion diseases or perhaps cure. Early diagnosis of prions (CJD) is possible because two peptides of PrP^C bind 3800 fold more effectively to PrP^{Sc} than to PrP^C (Lau AL et al 2007 Proc Natl Acad Sci USA 104:1151). The Tyr-Tyr-Arg monoclonal antibodies discriminate between PrP^C and PrP^{SC} and hold promise that immunoprophylaxis and/or immunotherapy may eventually become available (Paramithiotis E et al 2003 Nat Med 9:893). Conformation-dependent immunoassay (CDI) can discriminate among different prion strains. This assay quantifies PrP isoforms by simultaneously following antibody binding to the denatured and native prion protein. When the denatured/native PrP is graphed as the function of PrP^{Sc} concentration, each strain occupies a different position indicating a unique conformation (Safar J et al 1998 Nature Med 4:1157). The CDI test is extremely reliable (Safar JG et al 2005 Proc Natl Acad Sci USA 102:3501).

It has been hypothesized—on the basis of experimental observations—that the toxicity of this protein is based on increased oxidative stress. The inactivation of the *PrP* gene in mice does not lead to an immediate deleterious condition, but by the age of 70 weeks, an extensive loss of the Purkinje cells (large neurons in the cerebellar cortex) takes place and the animals have problems with movement coordination (ataxia). In case the normal PrP protein accumulates in the cytosol, a self-perpetuating PrP^{Sc}-like transformation takes place and neurodegeneration results (Ma J et al 2002 Science 298:1781). The disrupted *PrP* genes make them resistant to prions. Susceptibility in mice is affected also by QTLs in chromosomes 4, 6, 8, and 17 (Moreno CR et al 2003 Genetics 165:2085). On the basis of some genetic tests, it was concluded that the period of incubation of the mouse scrapie is controlled by allelic forms of a separate gene (*Sinc/Prni*). Molecular evidence indicates, however, that codons 108 and/or 109 of the *Prp* gene control incubation. In the mouse, there is a second *PrP* locus 16 kb down-stream. This *Prnd* (d for Doppelgänger [alterego in German], downstream prion protein-like) is truncated at the amino end domain and encodes only 179 amino acids.

Although its amino acid sequence shows only 25% homology with PrP, the structure of Prnd is quite similar. Prnd originated probably as an ancient duplication. Mice homozygous knockouts for Prnd and PrC are sterile. Its expression is normally limited to the testes, but if it is expressed in the brain, it causes neurodegeneration.

In case the *PrP* (Prnp) exons are deleted, Doppelgänger exons can be spliced into the PrP mRNAs. In ataxic animals, this intergenic splicing is highly expressed. Apparently, the manifestation of ataxia, the loss of Purkinje cells, and the degeneration of cerebellar granule cells is correlated with the alteration of a ligand-binding site.

A mutant form of PrP, ^{Ctm}PrP, a trans-membrane protein, can also cause prion disease in the absence or presence of PrP^{Sc}. The latter may also modulate the synthesis of the transmembrane form. Actually "the ability of polypeptide chains to form amyloid structures is not restricted to the relatively small number of proteins associated with recognized clinical disorders, and it now seems to be a generic feature of polypeptide chains" (Dobson CM 2003 Nature [Lond] 426:884). PrP^{Sc} molecules could be formed *de novo* from defined components in the absence of preexisting prions. PrP^{Sc} can be formed from a minimal set of components including native PrP^C molecules, co-purified lipid molecules, and a synthetic polyanion. Inoculation of samples containing either prion-seeded or spontaneously generated PrP^{Sc} molecules into hamsters caused scrapie, which was transmissible on second passage (Deleault NR et al 2007 Proc Natl Acad Sci USA 104:9741).

PrP^C may be involved in signal transduction in nerve function. PrP^{Sc} apparently binds plasminogen that selectively imparts neurotoxicity to the prion protein.

In budding yeast, two non-nuclear elements [*URE3*] and [*PSI*] appear (among others) to be the infectious prion forms of the Ure2p protein that is also a regulator of nitrogen catabolism. When Urep was overexpressed in wild type strains, the frequency of occurrence of the [*URE3*] increased 20–200 fold. If the overexpression of Urep was limited only to the amino ends of this protein, the frequency of occurrence of [*URE3*] increased 6000 times. The carboxyl domain of Urep seemed to carry out nitrogen catabolism whereas the amino end induced the prion formation. Both [*URE3*] and [*PSI*] are the prion causing forms of nuclear genes *URE2* and *SUP35*, respectively. The *URE2* gene is involved in the control of utilization of ureidosuccinate as a nitrogen source, while the *SUP35* nuclear gene encodes a subunit (eRF3, eukaryotic release factor) of the yeast translation termination complex. Mutations in both the nuclear genes involve derepression of nitrogen catabolism that is normally repressed by nitrogen. The propagation of [*URE3*] and [*PSI*] depends on *URE2* and *SUP35* nuclear genes, respectively. Guanidine-HCl blocks the propagation of PSI^+. Gdn·HCl-induced loss of the [PSI^+] prion is due to a failure to segregate propagons from daughter cells and not because of degradation of the preexisting propagons (Byrne LJ et al 2007 Proc Natl Acad Sci USA 104:11688). In vitro, the Sup35 protein may show prion-like properties. Normally, translation terminates at a stop codon by an interaction between Sup35 and other proteins such as Sup45. If the Sup35 proteins aggregate, they may assume prion conformation and the translation continues beyond the stop codon and an additional protein sequence is formed (►eRF). The conformation of the SUP35 prion motif varies among different fungal species and is important for its transmissibility. The amyloid fiber morphology and size may vary in different yeast prion strains (Diaz-Avalos R et al 2005 Proc Natl Acad Sci USA 102:10165). A certain conformation of SUP35 of *Saccharomyces cerevisiae* permits transmission to

Candida albicans and such *Candida* can then infect *Saccharomyces*. Thus strain conformation is critical for cross-species transmission (Tanaka M et al 2005 Cell 121:49; Tanaka M et al 2006 Nature [Lond] 442:585). A similar conclusion has been reached in mammals where a single or two residues determine critical requirement for amyloid transmission, yet preformed fibrils may overcome sequence-based structural preferences. For transmission, the amyloid protein conformation is the critical factor (Jones EM, Surewicz WK 2005 Cell 121:63). The cross-β spine structure explains the critical features of stability and self-perpetuation of the amyloid fibers (►cross-β spine). Heat shock protein Hsp104 catalyzes the formation and also the destruction of this yeast prion (Shorter J, Lindquist SW 2004 Science 304:1793). Deletion of Hsp104 eliminates Sup35 and Ure2 prions, whereas overexpression of Hsp104 purges cells of Sup35 prions, but not Ure2 prions. For both Sup35 and Ure2, Hsp104 catalyzes de novo prion nucleation from soluble, native proteins. Hsp104 fragments both prions to generate new prion-assembly surfaces. For Sup35, however, the fragmentation endpoint is an ensemble of noninfectious, amyloid-like aggregates and soluble proteins that cannot replicate conformation (Shorter J, Lindquist S 2006 Mol Cell 23:425). Five glutamine/asparagine-rich oligopeptide repeats at the N-terminus of Sup35 stabilize the aggregated form, and for replication, a chaperone-dependent element is also required (Osherovich LZ et al 2004 PLoS Biol 2(4):E86). In a somewhat different amyloid disease associated with transthyretin, only the highly destabilized molecules are degraded in the endoplasmic reticulum and only in certain tissues, indicating that endoplasmic reticulum-assisted folding depends on energetics, chaperone distribution, and metabolites (Sekijima Y et al 2005 Cell 121:78).

The human PrP repeat (PHGGGWGQ) can substitute for the yeast peptides (Parham SN et al 2001 EMBO J 202111). This feature of prions allows the development of diversity and may have evolutionary significance. Structurally, neither [*URE3*] nor [*PSI*] are similar to the mammalian PrP protein, indicating that there is more than one way for prions to arise. In the fungus *Podospora anserina*, the heterokaryosis incompatibility locus (*Het*) also makes prion-like proteins. The infectious forms of the normal Prp are also called PrP*, and the PrP^{Sc} is designated also PrP^{res} (protease-resistant prion). PrP^{C} and PrP^{Sc} appear to be conformational isomers. PrP* is the misfolded pathological core form. The yeast prion [PSI^{+}] can be reversibly removed, "cured" to [psi^{-}] 100% in seven to eight generations when exposed to guanine hydrochloride or methanol. Guanidin inactivates Hsp104 and Sup35 depolymerizes without a need for cell division (Wu Y-X et al 2005 Proc Natl Acad Sci USA 102:12789). The denaturants induce the expression of chaperones, giving further support to the notion that the prion functions are based on conformational changes. Recent evidence indicates that the $PrP^{C} \rightarrow PrP^{Sc}$ transition may involve the chemical thiol/disulphide exchange between the terminal thiolate of PrP^{Sc} and the disulfide bond of a PrP^{C} monomer and not only a conformational change (Welker E et al 2001 Proc Natl Acad Sci USA 98:4434). The protein chaperones HSP104, and to a lesser extent, HSP70, can affect the expression and transmission of [PSI^{+}] and its conversion to [psi^{-}]. In silico screening for compounds that fitted into a "pocket" created by residues undergoing the conformational rearrangements between the native and the sparsely populated high-energy states (PrP*) and that directly bind to those residues identified 2-pyrrolidin-1-yl-N-[4-[4-(2-pyrrolidin-1-yl-acetylamino)-benzyl]-phenyl]-acetamide (GN8), which efficiently reduced PrP^{Sc} and improved the survival of affected mice (Kuwata K et al 2007 Proc Natl Acad Sci USA 104:11921).

When the *URE2* and *SUP35* genes or the N-terminal domain of their products are deleted, the [*URE3*] and [PSI^{+}] elements permanently disappear. These yeast proteins are different from each other and from the prion proteins of higher eukaryotes, except the N-terminal region where homology exists. The NH_2 domain of SUP35, when fused to the rat glucorticoid receptor protein, can interact with the endogenous Sup35 protein and it undergoes a prion-like change of state. Self-replication requires a conformational conversion of initially unstructured Sup35 protein. Thus, the prion-like behavior is transmissible to another protein (Derkatch IL et al 2001 Cell 106:171). More recently, additional yeast proteins (RNQ1, NEW1) with prion-like properties have been identified (Tuite MF 2000 Cell 100:289; Derkatch IL 2000 EMBO J 19:1942). The *het-s* gene product of *Podospora anserina*, responsible for spore killer properties, also meets the criteria of being a prion (Perkins DD 2003 Proc Natl Acad Sci USA 100:6292). The so-called C hereditary units in *Podospora* have a nature similar to prions; they contain the MAPK cascade and trigger cell degeneration (Kicka S et al 2006 Proc Natl Acad Sci USA 103:13445).

The vCJD (variant of Creutzfeldt-Jakob disease) prions appear to have either single amino acid differences or differences in glycosylation which may also be the cause or consequence of conformational differences. The differences in electrophoretic mobility of the protease-digested prions are expected to shed light upon the problems of tracing the transmission of prions from cattle to man or among different animal species. The PrP gene in humans is in chromosome 20p12, and encodes 253 amino acids by a single exon.

The corresponding mouse gene is in chromosome 2. Other mammalian genes display very substantial homologies, although they may be transcribed by up to three exons. The NH^2-end of the protein displays an 8 amino acid repeat consensus (PHGGGW) in five to six copies, depending on the species. Deletions in these repeats do not involve disease symptoms. Short conserved amino acids downstream of the last repeats are important for $PrP^C \rightarrow PrP^{Sc}$ conversion. Another unique feature of PrP is an alanine-rich tract (AGAAAAGA). The transmission of prions among different species prolongs the incubation period. Mice lacking the gene for PrP^C cannot develop the disease even when inoculated (Büeler H et al 1993 Cell 73:1339). The PrR^C deficient mouse appears normal. Knocking out PrR^C from larger mammals would be an approach to avoid prion formation but because of technical difficulties, the use of shRNA is a viable alternative to block PrP expression and prevent encephalitis. Transgenic goats and cows produced by nuclear transplantation of the cognate shRNA gene resulted in more than 90% reduction in PrP expression (Golding MC et al 2006 Proc Natl Acad Sci USA 103:5285). The most infectious property was attributed to 300–600 kDa particles (14–28 PrP molecules) and less than five molecules, as well as very large aggregates, were much less effective in evoking neurodegenerative disease (Silveira JR et al 2005 Nature [Lond] 437:257). Also, immunodeficient mice, despite the fact that they may accumulate plaques upon scrapie infection, fail do develop the disease. It appears that human individuals who might have been exposed to the same BSE source may not all respond with the development of the disease.

$PrP^C \rightarrow PrP^{Sc}$ conversion by infection with a prion from another species (heterotypic conversion), especially when the inoculum is small or the inoculation occurs rarely, is less likely. The amino acid sequence in the 125–231 sequence displays differences among cow, sheep, dog, cat, pig, mouse, Syrian hamster, and human PrP^Cs and the structure varies as well (Lysek DA et al 2005 Proc Natl Acad Sci USA 102:640). Similarly, structural differences exist in elk compared to cow PrP^C (Gross AD et al 2005 Proc Natl Acad Sci USA 102:646). The chicken, turtle, and *Xenopus* PrP^Cs display only about 30% identity with the mammalian protein amino acid sequence, yet the molecular architectures are similar (Calzolai L et al 2005 Proc Natl Acad Sci USA 102:651).

Wild type mice brain infected with hamster prions did not develop scrapie, although a low level of maintenance of the hamster protein was detectable and reintroduced into hamsters; encephalitis followed. When the human prion is transferred to an animal, the PrP sequences in the new host are determined by the recipient and not by the donor, except when the animal is transgenic for the human PrP. Thus, the prion inoculum acts as a catalyst or as a chaperone. It is known that in Prp-deficient mice the immune system eliminates the PrP^C. It is also conceivable that in mice the hamster PrP^{Sc} is immunologically tolerated. The presence of the PrP gene is a requisite for the development of the PrP^{Sc} protein. PrP^{Sc} exists in multimeric rather than monomeric forms but PrP may become part of the interacting PrP^{Sc} molecular network. There are indications that prion and DNA interaction may modulate the harmful aggregation of the protein (Cordeiro Y et al 2001 J Biol Chem 276:49400). Hamster-adapted prion protein heated up to 600°C for 5 to 15 min (actually ashed) still retained some infectivity and points to the role of an inorganic template in the replication of scrapie. Heating to 1000°C abolished all activity. In case of relatedness between these two proteins, PrP^{Sc} may easily facilitate the conversion to prion. The expression of the PrP^{Sc} may require chaperones. One such protein was named X but its role is unclear. In vitro essay is available for fast and relatively inexpensive assaying of prions using mouse neuroblastoma cell line N2a (Klöhn P-C et al 2003 Proc Natl Acad Sci USA 100:11666).

According to some views, the "protein only" mechanism requires further proof, although all current evidence indicates a "protein only" basis. There is definite proof that the prions of yeast can be caused by the amino-terminal fragment of the Sup-35 protein without the help of any other substance (King C-Y, Diaz-Avalos R 2004 Nature [Lond] 428:319; Tanaka M et al 2004 ibid. 323). Additional recent evidence further supports this protein-only principle. A synthetic peptide (free of nucleic acids), containing mutation at site 102 leucine (see Fig. P130) folded into a β-conformation-rich form, was introduced into mice; and animals developed disease homologous to the Gerstmann-Sträussler syndrome in humans (Tremblay et al 2004 J Virol 78:2088). Furthermore, the peptide-induced disease was serially passaged into healthy mice, which developed symptoms indistinguishable from those appearing spontaneously in PrP^{Sc} leucine mutants. Similarly, recombinant protein consisting of the mouse prion sequence 89–231, rich in β-sheets, was cloned in *Escherichia coli* and then introduced into the brain of animals that over-expressed the normal PrP^C. Mice developed neuropathological symptoms characteristic of specific encephalopathy, and their protease-resistant extract evoked disease symptoms in other animals. There were two essential differences from the normal infection. The responding recipients produced excessive amounts of PrP^C before inoculation, and the period of incubation was substantially extended compared with the normal course of disease development (Legname et al 2004 Science 305:673).

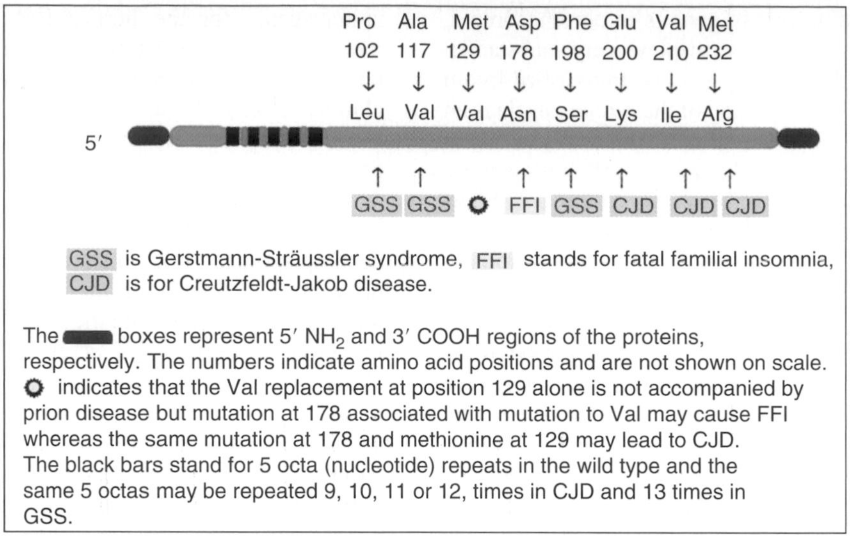

Pro 102 → Leu; Ala 117 → Val; Met 129 → Val; Asp 178 → Asn; Phe 198 → Ser; Glu 200 → Lys; Val 210 → Ile; Met 232 → Arg

5′

GSS GSS ✪ FFI GSS CJD CJD CJD

GSS is Gerstmann-Sträussler syndrome, FFI stands for fatal familial insomnia, CJD is for Creutzfeldt-Jakob disease.

The boxes represent 5′ NH_2 and 3′ COOH regions of the proteins, respectively. The numbers indicate amino acid positions and are not shown on scale. ✪ indicates that the Val replacement at position 129 alone is not accompanied by prion disease but mutation at 178 associated with mutation to Val may cause FFI whereas the same mutation at 178 and methionine at 129 may lead to CJD. The black bars stand for 5 octa (nucleotide) repeats in the wild type and the same 5 octas may be repeated 9, 10, 11 or 12, times in CJD and 13 times in GSS.

Figure P130. Prion mutations (Redrawn after Weissmann, C. 1999. J. Biol. Chem. 274:3.)

Cyclic amplification of protein misfolding (PMCA) in vitro produced protein identical to the protease resistant form in the brain of sick animals. Furthermore, this in vitro produced prion when introduced into healthy hamster caused the same type of disease as infection by PrP^{Sc} (Castilla J et al 2005 Cell 121:195). PMCA amplification technology can be automated and in 140 cycles leads to a 6600-fold increase in sensitivity compared to older techniques. Two successive rounds of PMCA increase the sensitivity of detection 10-million fold and can detect as few as 8000 molecules of PrP^{Sc} at 100% specificity (Castilla J et al 2005 Nature Med 11:982) and permit the detection of prions in the blood before the disease symptoms manifest (Saá P et al 2006 Science 313:92).

The existence of prions seems to be an exception to the "nucleic doctrine." Some evidence seemingly contradicts the infectious nature of the prions and points to accumulation of protein waste. The prion diseases may be familial, with an onset at about 50 years of age in humans. The sporadic forms are attributed to dominant somatic mutations. From cultured scrapie-infected mouse (but not of hamster) neuroblastoma cells, the branched polyamines (polyamidoamide dendrimers, polypropyleneimine, polyethyleneimine) purged PrP^{Sc} prions at non-toxic concentrations. Bis-acridines and a few other compounds appear inhibitory to prion replication in cell cultures (May BCH et al 2003 Proc Natl Acad Sci USA 100:3416). Kastellpaolitine, phenanthridines, 6-aminophenathridine, quinacrine, and chloropromazine appear effective in yeast against mammalian prions (see Fig. P131) (Bach S et al 2003 Nat Biotechnol 21:1075). By 2003, no real cure or preventive measures had emerged for prion diseases. There are some positive cues that proline-rich oligopeptides may restore the conformation of PrP^{Sc} to normal PrP. Preliminary results indicate that the lymphotoxin-β receptor may delay the onset of the symptoms temporarily in mice.

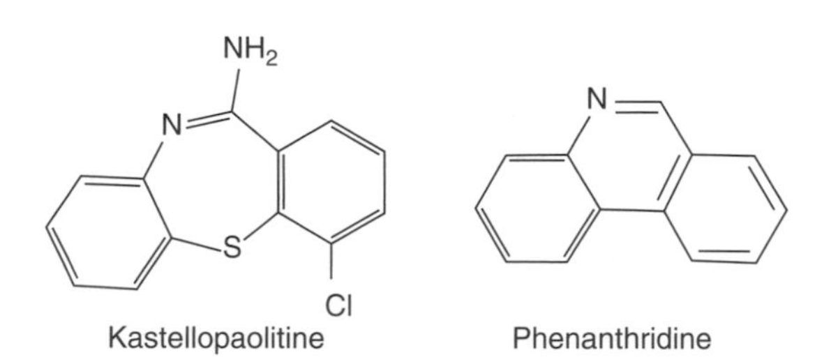

Figure P131. Kastellopaolitine 1 (left) is one of the other similar compounds, which have different substitutions at other positions at the right ring. Phenanthridine (right) basic structure

It seems that the complement component C3 is important for the prions to attach to the follicular dendritic cells, which mediate infection. Antibodies generated against the μ chain of PrP are a promising approach for the prevention of pathogenesis. ►Creutzfeldt-Jakob disease, ►Gerstmann-Sträussler disease, ►presenilin, ►kuru, ►encephalopathies, ►fatal familial insomnia, ►protein structure, ►Protein X, ►tau, ►curing plasmids, ►plasmin, ►chaperones, ►PSI^+, ►PIN^+, ►quinacrine mustard, ►*Cre/loxP*, ►virino hypothesis, ►conformation-dependent immunoassay, ►transthyretin, ►PMCA, ►MAPK,

►polyelectrolyte; Prusiner SB, Scott MR 1997 Annu Rev Genet 31:139; Cohen FE, Prusiner SB 1998 Annu Rev Biochem 67:793; Prusiner SB (ed) 2004 Prion Biology and Diseases, Cold Spring Harbor Laboratory Press, Cold Spring Harbor, New York; Umland T C 2001 Proc Natl Acad Sci USA 98:1459; Heppner FL et al 2001 Science 294:178; Baskakov IV et al 2002 J Biol Chem 277:21140; Kanu N et al 2002 Curr Biol 12:523; Uptain SM, Lindquist S 2002 Annu Rev Microbiol 56:703; Chien P 2004 Annu Rev Biochem 73:617; Curr Mol Med 2004 June issue; valine at site 129 prevents CJD: Wadsworth JDF et al 2004 Science 306:1793; potential therapy: Cashman NR, Caughey B 2004 Nature Rev Drug Discovery 3:874; characterization review: Caughey B, Baron GS 2006 Nature [Lond] 443:803; characterization: Prusiner SB, McCarty M 2006 Annu Rev Genet 40:25.

Prior Distribution: A probability distribution of variables or parameters before empirical information was obtained. Generally, it is part of Bayesian inference. ►Bayes' theorem, ►posterior distribution

Prior Probability: The suspected incidence of a disease before a diagnostic test or change of environmental effects or other extrinsic factors before the onset of a condition are identified.

Prisoner's Dilemma: A game theory, applicable to the interpretation of pairwise competition between two types of organisms using conflicting strategies. The two may cooperate, or either may "defect" for selfish reasons(s) and exploit the other and consequently the fitness may decrease to $1 - s_1$. In case both of them defect (are uncooperative), the population has to pay a cost (c), and the fitness becomes 1 − c. In case the defector gains a fitness advantage $(1 + s_2)$, it may invade the cooperators territory. If *c* is high $([1 - c] < [1 - s_1]$, a stable polymorphism may result. This theory is applicable also to studies on economic activities and to other fields. ►snowdrift game, ►cooperation, ►tragedy of the common; Page KM, Nowak MA 2001 J Theor Biol 209:173; Neill 2001 J Theor Biol 211(2):159; Doebeli M et al 2004 Science 306:859; Imhof LA et al 2005 Proc Natl Acad Sci USA 102:10797.

Pristionchus pacificus: The *Pristionchus pacificus* nematode is somewhat similar to *Caenorhabditis elegans* http://www.pristionchus.org. ►Caenorhabditis

Privacy Rule: An individual's privacy is legally protected against unwanted disclosures, yet medical research has a legitimate need to use, access, and disclose protected health information with certain limitations. ►GWA; http://privacyrulesandresearch.nih.gov; http://privacyruleandresearch.nih.gov/pr_02.asp; European guidelines for human data: http://ec.europa.eu/justice_home/fsj/privacy/docs/wpdocs/2007/wp136_en.pdf; recommended confidentiality certificate of the US National Institutes of Health: http://grants.nih.gov/grants/policy/coc; human molecular genetic data: Lowrance WW, Collins FS 2007 Science 317:600.

Private Blood Groups: A collective name of various blood groups with low frequencies compared to *public blood* group systems that occur frequently.

Private Mutation: Private mutation occurs very rarely in a very limited number of families.

Privilege: ►immune privilege

PRL: ►prolactin

PRL-3 (PTP4A3, 8q24.3): A 22 kDa tyrosine protein phosphatase situated at the cytoplasmic membrane; its elevated expression is associated with metastasis of colorectal cancer. ►colorectal cancer, ►metastasis; Saha S et al 2001 Science 294:1343.

PRM (pattern recognition proteins): PRMs are involved in the regulation of transcription.

PRMT: Protein-arginine methyltransferase. (See Boisvert FM et al 2003 Mol Cell Proteomics 2:1319).

P_{RNP}: The human gene encoding the normal isoform of the prion protein; the same in mouse is P_{rnp}.

Proaccelerin: A labile blood factor (V); its deficiency may lead to parahemophilia and excessive bleeding during menstruation or after surgery or bruising. ►antihemophilic factors

P

Probabilistic Graphical Models of Cellular Networks: In biological modeling we may be interested in different attributes, e.g., in the random expression of the genes observed and the hidden attributes of the model, such as the cluster assignment of a gene. The model includes the joint probability distribution of all relevant random attributes. The probabilistic graphical model represents multivariate joint probability distributions by a product of terms, each involving only a few variables. In Bayesian Networks, the joint distribution is represented as a product of conditional probabilities of the genotype of each individual, given the genotypes of its two parents. In pedigree analysis, the joint distribution of genotypes is the product of conditional probabilities. In phylogenetic models, the probability, over all evolutionary sequences, is the product of the conditional probability of each sequence, given its latest ancestral sequence in the phylogeny. Another classes of models are Markov Networks representing joint distribution as product of potentials. Cellular networks are based upon gene expression data observed on thousands of genes over a large number of microarrays. Then $GeneCluster_g$ denotes the cluster assignment of gene g and

$ArrayCluster_a$ denotes the cluster assignment of array a. Co-expression of genes is assumed to be due to co-regulation. The regulation is mediated by the transcription factors attached to specific sequences of the promoter during transcription, and to interaction of protein products of the gene clusters. The interaction of the proteins may be dependent on modifications. The regulatory networks can be partitioned to expression modules rather than to the study of individual genes. Because of the large number of factors involved, appropriate statistics are necessary to evaluate the reality of the observations and the evaluations. Validation of the models against all biochemical information available is highly desirable. New experimental procedures, high-throughput systems, and bioinformatics are under continuous development. This field is impossible to summarize adequately in the frame of this work. A good overview of the status of the field in 2004 is provided by Nir Friedman in Science 303:799. ►joint probability, ►multivariate analysis, ►Bayes' theorem, ►Markov chain statistics, ►microarray hybridization, ►networks, ►genetic networks, ►model, ►small-world networks

Probability: The statistical measure of chance on a scale between 0 to 1, inclusive. 0 means the lack of chance for an event to occur, 1 indicates a certainty that it will occur, and any value expressed as a decimal or fraction indicate the intermediate chances. The probability function indicates the value of a frequency predicted from the observations related to the parameter. The *simple probability* reveals the chance of a single event; the *compound probability* is the chance of multiple events. When two events are independent, their *joint probability* is the product of their independent probabilities. *Alternate probability* exists in case of sex in dioecious species, when an individual is either female or male; no intermediates are considered. One must keep in mind that probability does not absolutely prove or disprove a point; it simply indicates the chance of its occurrence. ►binomial probability, ►conditional probability, ►likelihood, ►maximum likelihood

Proband: Person(s) through which a family study of the inheritance of a human trait is initiated (also called propositus if male, or proposita if female). Determining the pattern of inheritance on the basis of families chosen by probands may display an excess of affected individuals relative to Mendelian expectations because of the bias in sampling of the population. ►ascertainment test, ►pedigree analysis

Probasin: A secreted and nuclear protein abundant in the prostate epithelium. Its expression is regulated by androgens (two receptors at 5′ of the 17.5 kb gene) and zinc. Its promoter is extensively used with various modifications for the study of prostate function/cancer. ►androgen, ►prostate cancer; Logg CR et al 2002 J Virol 76:12783.

Probe: A labeled nucleic acid fragment used for identifying or locating another segment by hybridization. Similarly, immunoprobes using primarily monoclonal antibodies or enzyme probes or enzymes linked to antibodies can also be employed (see Fig. P132). The probe binds a reporter protein and another binding protein (binding p). For enzymatic detection of a probe, most commonly alkaline phosphatase or horseradish peroxidase are used. The tissue is incubated with the appropriate substrate of the enzyme and the colored precipitate formed through its action identifies its location. ►synthetic DNA probes, ►heterologous probe, ►recombinational probe, ►immunoprobe, ►labeling, ►nick translation, ►padlock probe, ►histochemistry

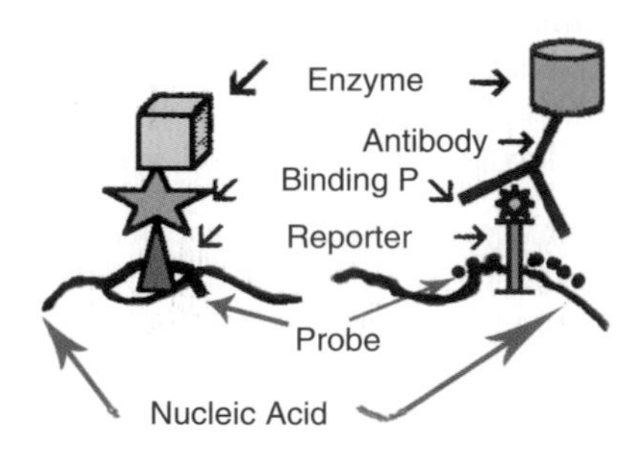

Figure P132. Protein-mediated detection of probe

Probe (primer oligo base extension, PO-BE): A diagnostic procedure for the identification of localized variation in DNA. It is a primer extension procedure using a polymerase, three different deoxyribonucleotide triphosphates, and a dideoxynucleotide triphosphate (ddNTP). The primer is extended until the variable SNP site is reached where the ddNTP is incorporated. The synthetic product is then analyzed by matrix-assisted laser desorption time of flight mass spectrometry. ►primer extension, ►MALDI-TOF, ►SNIP, ►dideoxyribonucleotide; Braun A et al 1997 Clin Chem 43:1151.

Probe Arrays: Oligonucleotides immobilized on silicon wafers in order to study simultaneously the functions of many genes. ►DNA chips, ►microarray

ProbeMaker: A computer program that converts DNA sequence files in FASTA format to digital restriction maps used for MapSearch Probes.

Probiotics: Beneficial bacteria in the body.

Probit: A cumulative normal frequency distribution is represented by an S curve. A cumulative curve can become a straight line by *probit transformation*. We may represent the probability scale in units of standard deviations. Thus the 50% point is the 0 standard deviation, the 84.13 unit becomes +1, and the 2.27 point the −2 standard deviations. The cumulative percentages are also called *normal equivalent deviates* (NED). If the ordinates are in NED units and we plot the cumulative normal curve, a straight line results. Probits are thus the NEDs with 5.0 added and thus we do not get negative values for the majority of deviates. The probit value of 5.0 indicates a cumulative frequency of 50% and a probit value of 6.0 means a cumulative frequency of 84.13% whereas probit 3.0 indicates a cumulative frequency of 2.27%. Probit value tables are available (Fisher RA, Yates F 1963 Statistical Tables. Hafner, New York). Probit transformations are frequently used for dosage mortality responses to chemicals indicating the regression of cumulative mortalities on dosage. The graphs can be plotted on probit papers with abscissa on logarithmic scale. ►normal distribution, ►logit

Proboscis: Tubular snout (nose-like emergence) on the head such as the feeding apparatus of *Drosophila*, elephant trunk, snout of tapirs, shrews, etc. ►morphogenesis in *Drosophila*

Procaine Anesthetics: Benzoic acid derivatives with local numbing of nerves or nerve receptors.

Procambium: The primary meristem that gives rise to the cambium and the primary vascular tissue of plants. ►cambium, ►meristem, ►root

Procapsid: The empty capsid precursor of phage into which the DNA can be packaged. ►development, ►phage

Procarcinogen: A procarcinogen requires chemical modification to become carcinogenic. ►carcinogen, ►phorbol esters, ►activation of mutagens

Procaryote: ►prokaryote

Procentriole: An immature centriole that upon maturing becomes the anchoring site of the spindle fibers, cilia, and flagella. ►centriole, ►centromere, ►spindle fibers

Process, Genetic: Gene product(s) mediated changes to reach a certain goal in the cell.

Processed Genes: Processed genes are obtained by reverse transcriptase from mRNA and therefore are free of all elements (e.g., introns) removed during processing of the primary transcript. They are widely expressed, highly conserved, and short and low in GC. ►cDNA, ►intron, ►primary transcript

Processed Pseudogene (retropseudogene): A processed pseudogene is similar to mRNA, lacks introns, and may have a polyA tail, yet it is non-functional. The faulty reverse transcription of mRNA may have produced processed pseudogenes. Pseudogenes are widely expressed, generally short, highly conserved, and low in GCs. Many of the processed pseudogenes lack promoters and cannot be transcribed. About half of the human pseudogenes are the processed type. Some of processed pseudogenes are actually retroelements. Processed pseudogenes can be mapped in the genome and provide information on ancestral transcripts (Shemesh R et al 2006 Proc Natl Acad Sci USA 103:1364). ►reverse transcriptases, ►cDNA, ►processed genes, ►pseudogene, ►LINE; Gonçalves I et al 2000 Genome Res 10:672; http://pbil.univ-lyon1.fr/.

Processing: The trimming and modifying of the primary transcripts of the DNA into functional RNAs or cutting and modifying polypeptide chains prior to becoming enzymes or structural proteins. ►primary transcripts, ►protein synthesis, ►posttranslational modification

Processing Body: ►P body

Processivity: Processivity defines the number of nucleotides added to the nascent DNA chain before the polymerase is dissociated from the template. The processivity for *E. coli* DNA polymerase I, II, and III is 3–200, >10,000, >500,000, respectively. ►error in aminoacylation, ►clamp-loader, ►DNA polymerases, ►polymerase switching, ►replication fork

Processor: Data-processing hardware or a computer program (software) that compiles, assembles, and translates information in a specific programming language.

Prochiral Molecule: An enzyme substrate that after attaching to the active site undergoes a structural modification and becomes chiral. ►chirality, ►active site

Prochloron: ►evolution of organelles

Pro-Chromatin: A state of the chromatin that is conducive to transcription.

Prochromosome: Heterochromatic blocks detected during interphase. In this interphase nucleus of *Arabidopsis*, the centromeric heterochromatin was stained by fluorescent isothiocyanate and displayed yellow-green color (see Fig. P133). Courtesy of Drs. Maluszinszka J, Heslop-Harrison JH. ►heterochromatin, ►Barr body, ►mitosis

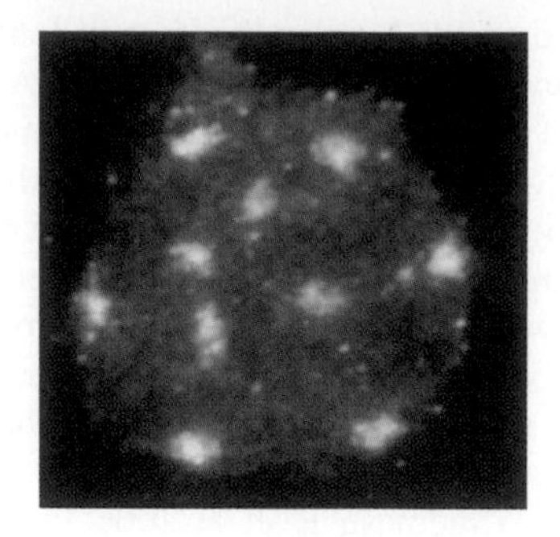

Figure P133. Prochromosomes

Procollagen: Precursor of collagen. ►collagen

Proconsul: A fossil ape that lived about 23–17 million years ago.

Proconvertin: Antihemophilic factor VII, and the deficiency of which may lead to excessive bleeding and hypoproconvertinemia. ►hypoproconvertinemia, ►antihemophilic factors

Proctodeum: An invagination of the embryonal ectoderm where the anus is formed later.

Procyclic: The stage at which *Trypanosoma* is in the gut of the intermediate host (tse-tse fly) and at is not infectious to higher animals. ►metacyclic *Trypanosoma*, ►*Trypanosoma*

Prodroma (prodrome): Ominous sign(s) of a looming disease before the actual onset.

Prodrug: A prodrug is processable to a biologically active compound. ►suicide vector, ►activation of mutagens, ►ADEPT

Producer Cell: An infected cell continuously produces recombinant retrovirus.

Product-Limit Estimator: The product-limit estimator is based on a number of conditional probabilities, e.g., the probability of survival after surviving for one day, then for the next day, and so on. Where $\hat{S}(t)$ the survival function at subsequent times, r_j = the number of individuals at risk at time $t_{(j)}$,, d_j = the number of individuals involved in the event at risk time $t_{(j)}$.

$$\hat{S}(t) = \prod_{j|t_{(j)} \leq t} \left(1 - \frac{d_j}{r_j}\right)$$

Product-Moment Correlation (Pearson's product-moment correlation coefficient): The correlation coefficient of a sample that is used as an estimator for the correlation coefficient of the population:

$$r_{XY} = \frac{\sum_i (X_i - M_X)(Y_i - M_Y)}{N s_X s_Y}, \quad s_X = \sqrt{\frac{\sum_i (X_i - M_X)^2}{N}},$$

$$s_Y = \sqrt{\frac{\sum_i (Y_i - M_Y)^2}{N}}$$

Where the N pairs of values (X_i, Y_i) represent the size of the sample and s_X and s_Y are the standard deviation of the respective variables as shown above. ►correlation

Product Ratio Method: ►F_2 linkage estimation

Product Rule: ►joint probability

Productive Infection: In productive infection, the virus is not inserted into the eukaryotic chromosome and can propagate independently from the host DNA and can destroy the cell while releasing progeny particles. ►lysis

Proembryo: The minimally differentiated fertilized egg.

Profile: A nucleotide or amino acid sequence probability motif. ►motif

Profilin: Profilin mediates actin polymerization. ►actin, ►Bni1, ►cytoskeleton, ►formin; Carlsson L et al 1977 J Mol Biol 115:465.

Proflavin: An acridine dye, capable of inducing frameshift mutations. ►acridine dye, ►frameshift mutation; Brenner S et al 1958 Nature [Lond] 182:933.

Progenitor: An ancestor or an ancestral cell of a lineage. Progenitor cells—unlike stem cells—may lose their ability of self-renewal yet they retain their mitotic ability and may generate one or different types of differentiated cells. ►stem cells, ►cancer stem cell; Reya T et al 2001 Nature [Lond] 414:105; Weissman IL et al 2001 Annu Rev Cell Dev Biol 17:387.

Progenote: The evolutionarily common, primitive ancestor of the eukaryotic cytoplasm and the bacterial cell (Woese CR, Fox GE 1977 J Mol Evol 10:1).

Progeny Test: A procedure for determining the pattern of inheritance. ►Mendelian segregation, ►Mendelian laws

Progeria (premature aging): ►aging, ►Hutchinson–Gilford syndrome

Progeroid Syndromes: ►aging

Progesterone: ►animal hormones, ►steroid hormones, ►testosterone, ►estradiol, ►progestin (see Fig. P134), formula

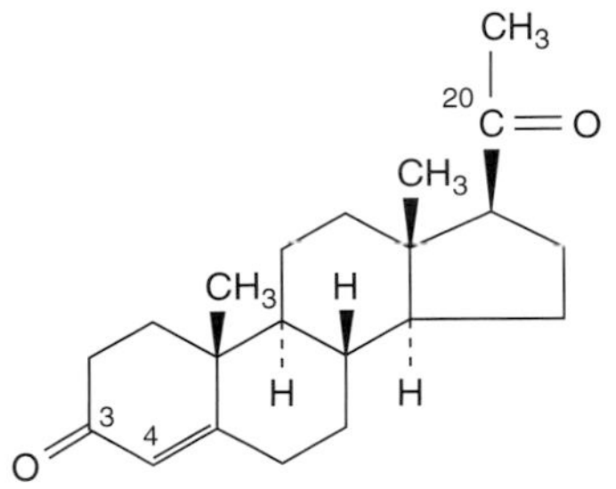

Figure P134. Progesterone (progestin)

Progesterone Receptors (PR): PRs are assembled with the cooperation of at least eight chaperones, including Hsp40, Hsp70, Hsp90, Hip, p60, p23, FKBPs, and cyclophilins. PRs are transcriptional regulators of progesterone-responsive genes. named proteins under separate entries; Hernández P et al 2002 J Biol Chem 277:11873.

Progestin: A steroid hormone, used as medication for the prevention of repeated spontaneous abortion. When added to estrogen, it reduces the risk of endometrial cancer. ►progesterone

Prognosis: The prediction of the course or outcome of a process, e.g., disease, cancer, etc.

Program: A set(s) of instructions in computer language (software) that permits the user to carry out specified tasks. In biology, program refers to the development proceeds according to a genetically determined pattern, realized by environmental effects.

Programmed Cell Death: ►apoptosis

Programming, Dyanamic: In dynamic programming, large groups of data are broken down to subsets to facilitate programming of the complex. ►genetic networks

Progression: A process involved in oncogenic transformation; after the initial mutation of a proto-oncogene progression changes it into an active oncogene. ►cancer, ►phorbol esters

Prohibitin: A 30 kDa tumor-suppressor protein localized mainly in the mitochondria, although it is encoded at human chromosome 17q21. The well-conserved protein is present in other mammals, *Drosophila*, the plant *Arabidopsis*, and several microbes (*Pneumocystis carinii*, the cyanobacterium *Synechocystis*).

Projectin: A myosin-activated protein kinase. ►myosin

Projection Formula: Modeling configurations of groups around chiral centers of molecules. ►chirality

Prokaryon: Same as prokaryote.

Prokaryote: An organism without membrane-enveloped (cell nucleus) genetic material (such as the case in bacteria). The majority of prokaryotic bacteria have circular double-stranded DNA chromosomes. However, *Borrelia*, *Streptomyces*, and *Agrobacterium tumefaciens* have a linear chromosome. The GC content varies between ~72% to ~27%. The genome size of prokaryotes varies 20-fold. The majority of bacteria carry their genes in the leading strand of the DNA. The pathogenic strain may have reduced genome size and/or increased number of pseudogenes. (http://www.cbs.dtu.dk/databases/DOGS/. ►cell comparisons, ►GC skew, ►*Borrelia*, ►*Streptomyces*, ►*Agrobacterium tumefaciens*; Bentley SD, Parkhill J 2004 Annu Rev Genet 38:771; regulation: http://regtransbase.lbl.gov.

Prolactin (PRL): A 23 kDa mitogen, stimulating lactation and the development of the mammary glands. Prolactin receptors are present on human lymphocytes and prolactin may form complexes with IgG subclasses. A prolactin releasing peptide was identified in the hypothalamus. ►lymphocytes, ►immunoglobulins, ►mitogen, ►brain, ►cathepsins; Mann PE, Bridges RS 2001 Progr Brain Res 133:251.

Prolamellar Body: The crystalline-like, lipid-rich structure in the immature plastids that upon illumination develops into the internal lamellae of the proplastids and into the thylakoids of the chloroplasts (see Fig. P135). ►chloroplast

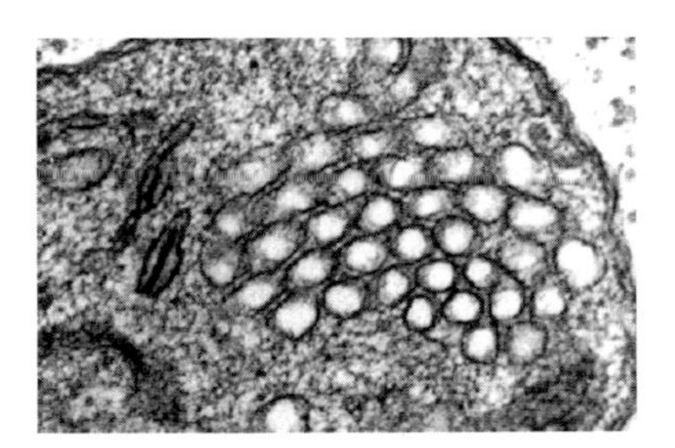

Figure P135. Prolamellar body

Prolamine: ►zein, ►high lysine corn

Proliferating Cell Nuclear Antigen: ►PCNA

Proliferation: The multiplication of cells or organisms. In cells, cytotoxic agents that may induce first cell death may cause proliferation and then regenerative growth, or it may be the result of the action of mitogens. ►mitogen, ►cancer

Proline Biosynthesis: Proline biosynthesis proceeds from glutamate through enzymatic steps involving glutamate kinase, glutamate dehydrogenase, and finally Δ'-pyrroline-5-carboxylate, which is converted to proline by pyrroline carboxylate reductase (see

Fig. P136). In some proteins, e.g., collagen, prolyl-4-hydroxylase generates 4-hydroxyproline from proline. The latter enzyme is coded in human chromosomes 10q21.3-q23.1 (α-subunit) and 17q15 (β-subunit). ►amino acid metabolism, ►hyperprolinemia

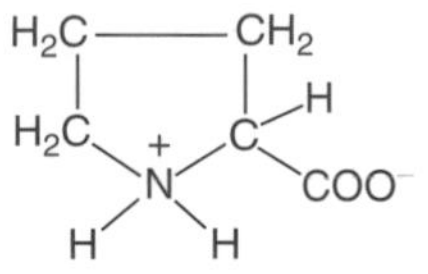

H_2C—CH_2, H_2C, N^+, C—H, COO^-, H, H

Figure P136. Proline

Prolog: A database management and query system in physical mapping of DNA. (See http://portal.acm.org/citation.cfm?id=711875&dl=ACM&coll=&CFID=15151515&CFTOKEN=6184618).

Prolyl Isomerase: ►PPI, ►immunophilin

Promastigote: ►*Trypanosoma*

Prometaphase: Early metaphase. ►mitosis

Prominin: ►CD133

Promiscuous DNA: Homologous nucleotide sequences occurring in the various cell organelles (nucleus, mitochondrion, plastid). They are assumed to owe their origin to ancestral insertions during evolution. ►insertion elements; Ayliffe MA et al 1998 Mol Biol Evol 15:738; Lin Y, Waldman AS 2001 Nucleic Acids Res 29:3975.

Promiscuous Plasmids: ►plasmids promiscuous

Promiscuous Protein: A promiscuous protein has affinity to more than one substrate. ►conformational diversity; Copley SD 2003, Curr Opin Chem Biol 7:265.

Promitochondria: Organelles in anaerobically grown (yeast) cells that can differentiate into mitochondria in the presence of oxygen. ►mitochondria

Promoter: The site of binding of the transcriptase enzyme (RNA polymerase), transcription factor complexes, and regulatory elements, including also the ribosome-binding untranslated sequences (see Fig. P137). Usually, the basal promoter is situated in front of the genes although pol III may rely on both upstream and downstream promoters.

The promoters of the 5S and tRNA genes are internal. Also, some *E. coli* genes have some weak promoters within open reading frames (Kawano M et al 2005 Nucleic Acids Res 33:6268). The arrangement of the promoter used by pol II is outlined.

The promoters used by RNA polymerase II may encompass several hundred nucleotides in yeast but in higher eukaryotes it may extend to several thousand bases. In yeast, UAS (upstream activating sequences) and URS (upstream repressing sequences) are regular binding sites. The transcription start site is usually within a stretch of 30 to 120 nucleotides downstream of the TATA box (see Fig. P138). The pol II enzyme frequently uses in mammals multiple promoters, within which there are multiple start sites, and alternative promoter usage generates diversity and complexity in the mammalian transcriptome and proteome (Sandelin A et al 2007 Nature Rev Genet 8:424). Some of the mammalian promoters are localized more than 100 kb upstream or located downstream, may be multiple, may overlap several genes, and are shared by other genes (Denoeud F et al 2007 Genome Res 17:746).

At the ends of the genes, insulators (boundary elements) separate the genes or the used promoters from the others. The DNase hypersensitive site(s) (also called locus control region) may permit the attachment of sequence-specific transcriptional activators, making the gene competent for transcription. The competence may involve histone acetylation.

Among 1031 human protein-coding genes, the Pol II-like enzymes commonly (~32%) use a TATA box both in prokaryotes and eukaryotes. The TATA box ca. 25 bp upstream from the initiation point of transcription is usually surrounded by GC-rich tracts (97%). Near the transcription initiation site (−3 to +5), there may be an initiator (Inr, 85%) with an ~average type of sequence: $(Pyrimidine)_2CA(Pyrimidine)_5$. Many eukaryotic genes do not have Inr but the TATA box directs the initiation. CAAT box is also a frequent (64%) element in the promoter. Some large eukaryotic genes utilize more than one promoter and the transcripts may vary. Some housekeeping and RAS genes do not use the TATA box. DNA-dependent RNA polymerase I synthesizes ribosomal RNAs; it has a core sequence adjacent to the transcription initiation site and upstream regulator binding sites (UCE). Pol III promoters facilitating the transcription of tRNA usually have split

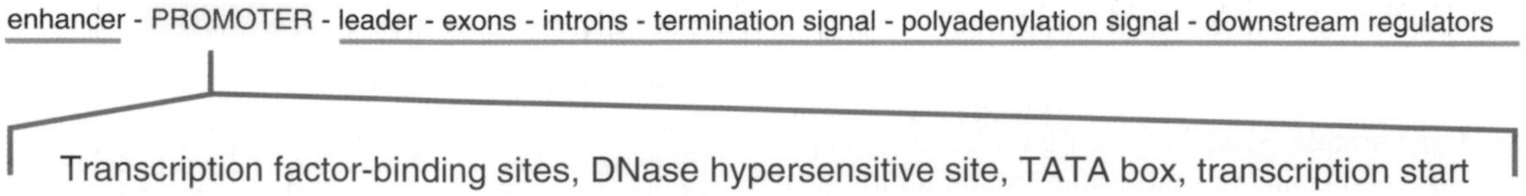

enhancer - PROMOTER - leader - exons - introns - termination signal - polyadenylation signal - downstream regulators

Transcription factor-binding sites, DNase hypersensitive site, TATA box, transcription start

Figure P137. Organization of the promoter and other genic elements

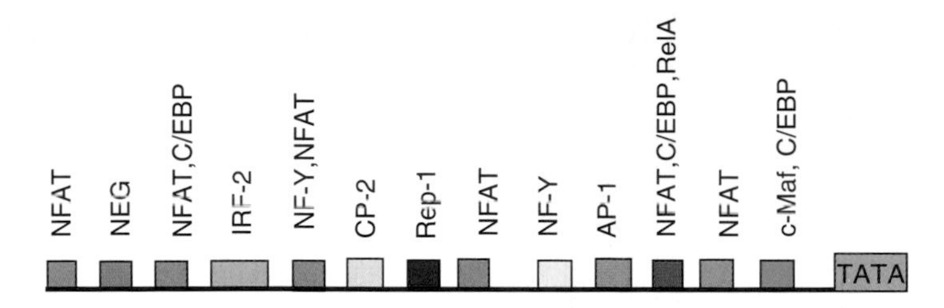

Figure P138. The promoter usually includes several regulatory boxes to which protein factors are recruited. (Modified after Guo, J. *et al.* 2001 J. Biol. Chem. 276:48871)

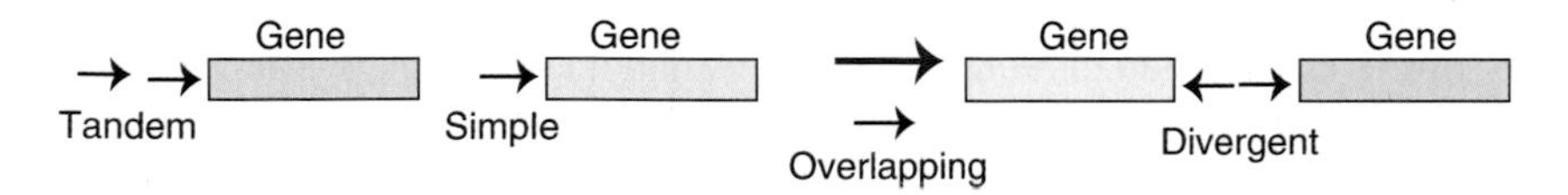

Figure P139. Arrows symbolize promoters; boxes represent structural genes

promoters with an A-box and a B-box about 40 bases apart, situated inside the transcription unit 20 and 60-base downstream from the transcription initiation site. The pol III promoter of some U RNAs has, however, a TATA box 30–60 bases upstream from the transcription initiation site and further upstream a proximal sequence element (PSE) near the TATA box. Synthetic promoters can be constructed with increased activity. The Promoter Scan II program identifies pol II promoters in genomic sequences and is available through Internet: http://www.cbs.umn.edu/software/software.html.

Promoters (→) may be of different types and some genes may rely on multiple promoters (see Fig. P139): TFD, TRANSFAC, or IMD databases can use the Signal Scan to find transcription factor binding sites. A high-resolution analysis revealed 10,567 promoters corresponding to 6763 genes in the human genome. Almost half of all mammalian genes have evolutionarily conserved alternative promoters (Baek D et al 2007 Genome Res 17:145). About 11% of the human promoters are bidirectional/divergent, are more active in transcription than other promoters, and are involved with RNA polymerase II and the modified histones H3K4me2, H3K4me3, and H3ac (Lin JM et al 2007 Genome Res 17:818). This information resulted by mapping the preinitiation complexes labeled by the attached TATA box associated protein and analysis by microarray hybridization of immunoprecipitated complexes (Kim TH et al 2005 Nature [Lond] 436:876). Libraries of engineered promoters can provide a fruitful approach for quantitative study of gene expression (Alper H et al 2005 Proc Natl Acad Sci USA 102:12678). The promoter of the *lac* operon of *E. coli* is controlled by cis elements that integrate signals coming from the cAMP receptor and the Lac repressor (Mayo AE et al 2006 PLoS Biol 4:e45).

Active promoters are marked by trimethylation of Lys4 of histone H3 (H3K4), whereas enhancers are marked by monomethylation, but not trimethylation, of H3K4. Computational algorithms, using these distinct chromatin signatures to identify new regulatory elements, predicted over 200 promoters and 400 enhancers within the 30 Mb region of the vertebrate genome (Heintzman N et al 2007 Nature Genet 39:311).

In vivo spatiotemporal analysis for approximately 900 predicted *C. elegans* promoters (~5% of the predicted protein-coding genes), each driving the expression of green fluorescent protein (GFP) using a flow-cytometer adapted for nematode profiling, generated "chronograms," two-dimensional representations of fluorescence intensity along the body axis and throughout development from early larvae to adults. Automated comparison and clustering of the obtained in vivo expression patterns show that genes coexpressed in space and time tend to belong to common functional categories (Dupuy D et al 2007 Nature Biotechnol 25:663).

►basal promoter, ►core promoter, ►DPE, ►minimal promoter, ►complex promoter, ►UAS, ►URS, ►portable promoter, ►cryptic promoter, ►divergent dual promoter, ►divergent transcription, ►transcription complex, ►transcription factors, ►open promoter complex, ►closed promoter complex *Lac* operon, ►*Tryptophan* operon, ►*Arabinose* operon, ►pol I, ►pol II, ►pol III, ►regulation of gene activity, ►promoter clearance, ►promoter trapping, ►TATA box, ►TBP, ►TAF, ►insulator, ►enhancer, ►LCR, ►chromatin remodeling, ►histone acetyl-transferase, ►promoter inducible, ►promoter tissue-specific, ►antisense DNA, ►microarray hybridization, ►preinitiation complex; analysis of ~900 putative human promoters: Cooper SJ et al 2006 Genome Res 16:1; Chalkley GE, Verrijzer CP 1999 EMBO J 18:4835; Suzuki Y et al 2001 Genome Res 11:677; Pilpel Y et al 2001 Nature Genet 29:153; Schuettengruber B et al 2003 J Biol Chem 278:1784; Ohler U et al 2002 Genome Biol 3:research0087.1;eukaryotic promoters: http://www.epd.isb-sib.ch; eukaryotic promoters: http://cmgm.stanford.edu/help/manual/databases/epd.html; eukaryotic promoters: http://doop.abc.hu/; transcriptional start sites: http://dbtss.hgc.jp/; human promoter binding sites: http://genome.imim.es/datasets/abs2005/index.html; transcription factor binding sites: http://www.isrec.isb-sib.ch/htpselex/; mammalian promoters/transcription factors/regulation: http://bioinformatics.med.ohio-state.edu/MPromDb/; mammalian regulatory promoters:

http://bioinformatics.wustl.edu/webTools/portalModule/PromoterSearch.do; tissue specific promoters: http://tiprod.cbi.pku.edu.cn:8080/index.html; knowledge-based promoter search: http://bips.u-strasbg.fr/PromAn/; promoter motif search: http://melina2.hgc.jp/public/index.html.

Promoter Bubble: ►promoter clearance

Promoter Clearance (promoter escape): In promoter clearance, the RNA polymerase complex (promoter bubble) starts moving forward from the promoter as the first ribonucleotides are transcribed. The RNA polymerase can synthesize a few bases without leaving the promoter site, but after that tension develops, which discontinues the contact between the DNA and the RNA polymerase. Clearance is regulated by both positive and negative elongation factors. Negative elongation requires four polypeptides and two polypeptides sensitivity inducing factor (DSIF). Inhibition takes place when about 18 nucleotides were added to the growing transcript. The transcript length is regulated also by the inhibition of the transcript cleavage factor TFIIS; the latter can be active along the entire length of the transcript (Palangat M et al 2005 Proc Natl Acad Sci USA 102:15036).

The movement may be represented by an inchworm model or a moving domain model (the translocation involves the entire transcription box with minimal stretching) or the tilting model without a flexible polymerase, which is tilted along the axis of the DNA. ►bubble, ►replication bubble, ►inchworm model; Pal M et al 2001 Mol Cell Biol 21:5815; Liu C, Martin CT 2002 J Biol Chem 277:2725.

Promoter Conversion (Pro-Con): Promoter conversion changes the promoter to a heterologous one.

Promoter Escape: ►promoter clearance

Promoter, Extended: ►Pribnow box

Promoter, Inducible: Inducible promoters turn on genes in response to biological, chemical, or physical signals. ►metallothionein, ►*Lac*

Promoter Interference: Promoter interference may occur when within a single viral vector two genes are placed under separate controls, e.g., the strong promoter within the long terminal repeat (LTR) may suppress the function of an internal promoter irrespective of its orientation. This problem may be overcome by utilizing an IRES for the second gene in the common transcript:

LTR — 1st Gene — IRES — 2nd Gene —. ►IRES

Promoter Melting: In promoter melting, the double-stranded DNA unwinds (forms a promoter bubble) to allow access to the template strand for the RNA polymerase enzyme. In *E. coli*, the N-terminal 1–314 amino acids of the β′ subunit and the 94–507 amino acids of the σ subunit cooperate in the melting. ►RNA polymerase, ►promoter; Young BA et al 2004 Science 303:1382.

Promoter Occlusion: Promoter occlusion occurs when, in retroviral elements with direct LTR repeats, the promoters at the 3′-end are inactivated and prevented from binding enhancers or transcription factors because they cannot facilitate transcription due to their wrong orientation. ►LTR, ►enhancer, ►transcription factors

Promoter Swapping: An exchange of promoter by, e.g., reciprocal chromosome translocation. ►translocation, ►pleiomorphic adenoma

Promoter, Tissue-Specific: A tissue-specific promoter permits the transcription of genes only or mainly in specific tissue(s) (see Fig. P140). ►promoter, ►tissue specificity

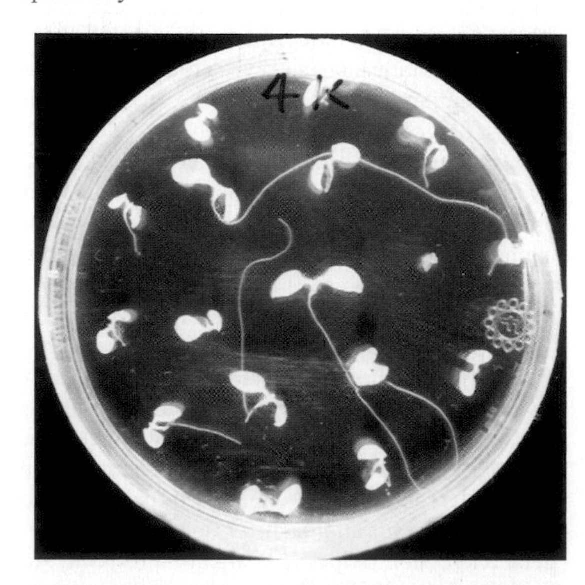

Figure P140. Tobacco seedlings segregating for kanamycin resistance on a root-tissue-specific promoter. Transformation was made by a promoter-less vector construct. The non-transgenic plants cannot grow roots on kanamycin medium. (From Y. Yao & G.P. Rédei)

Promoter Trapping: ►trapping promoters, ►transcriptional gene fusion vectors, ►translational gene fusion vectors, ►gene fusion, ►promoter; Medico E et al 2001 Nature Biotechnol 19:579.

Promoters of Tumorigenesis: Environmental substances or gene products that guide a group of precancerous cells toward malignant growth. The promoters themselves do not initiate cancer. ►carcinogenesis, ►phorbol esters, ►cancer, ►conversion

Promutagen: A promutagen requires chemical modification (activation) to become a mutagen. ►mutagen, ►activation of mutagens

Promyelocytic Body (PML): A nuclear product of the promyelocytic leukemia gene; it mediates the degradation of ubiquitinated proteins and is regulated by the nucleolus. PMLs contain also TRF1 and TRF2 telomeric proteins required for the maintenance of telomeres. PMLs may be targeted by viruses, may act as suppressors of growth and tumors, and mediate apoptosis. PML body may also occur in the cytoplasm and modulates TGF-β signaling. The PML protein also regulates centrosome duplication by the suppression of the Aurora protein. ►leukemia, ►ubiquitin, ►telomerase, ►apoptosis, ►transforming growth factor β, ►centrosome

Pronase: A powerful general (non-specific) proteolytic enzyme isolated from *Streptomyces*.

Pronucleus: The male and female gametic nucleus to be involved in the sexual union.

Proof-of-Concept (proof of principle): Experimental evidence that an idea works in practical application. ►validation

Proofreading: Bacterial DNA polymerase I (and analogous eukaryotic enzymes) can recognize replicational errors and remove the inappropriate bases by its editing 3′ - 5′ exo-nuclease function. In case the editing function is diminished by mutation, mutator activity is gained. In case of gain in editing, function antimutator attributes are observed. In bacteria, proofreading is performed also by the *dnaQ* gene encoding the ε subunit (an exonuclease) of the DNA polymerase III holoenzyme. The product of gene dnaE carries out the base selection. The enzymes MutH, MutL, and MutS and the corresponding homologs in higher organisms repair mismatches. The fidelity of replication due to the combined action of the sequentially acting bacterial genes was estimated to be in the range of 10^{-10} per base per replication. During the process of translation, the EF-Tu•GTP → EF-Tu•GDP change releases a molecule of inorganic phosphate (P_i) and allows a time window to dissociate the wrong tRNA from the ribosome. A similar correction is made also by the aminoacyl synthetase enzyme, by virtue of its active site specialized for this function. DNA polymerase η lacks exonuclease function required for proofreading, but correction is still accomplished by recruiting an extrinsic exonuclease to the error site. ►DNA polymerase I, ►DNA polymerase III, ►DNA polymerases, ►exonuclease, ►proofreading paradox, ►DNA repair, ►error in replication, ►error in aminoacylation, ►ambiguity in translation, ►protein synthesis, ►DNA repair; Friedberg EC et al 2000 Proc Natl Acad Sci USA 97:5681; Livneh Z 2001 J Biol Chem 276:25639; Shevelev IV, Hübscher U 2002 Nat Rev Mol Cell Biol 3:364.

Propagule: A part of an organism that can be used for propagation of an individual by asexual means.

Propeller Twist in DNA: The surface angle formed between individual base-planes viewed along the $C^6–C^8$ line of a base pair.

Properdin (Factor P): A serum protein of three to four subunits (each ca. 56 kDa, encoded in human chromosome 6p21.3). It is an activator of the complement of the natural immunity system that works by stabilizing the convertase. ►convertase, ►complement, ►complement, ►immune system; Perdikoulis MV et al 2001 Biochim Biophys Acta 1548:265.

Prophage: The proviral phage is in an integrated state in the host cellular DNA and it is replicated in synchrony with the host chromosomal DNA until it is induced and thus, becomes a vegetative virus (see Fig. P141). ►prophage induction, ►temperate phage, ►lysogeny, ►lambda phage

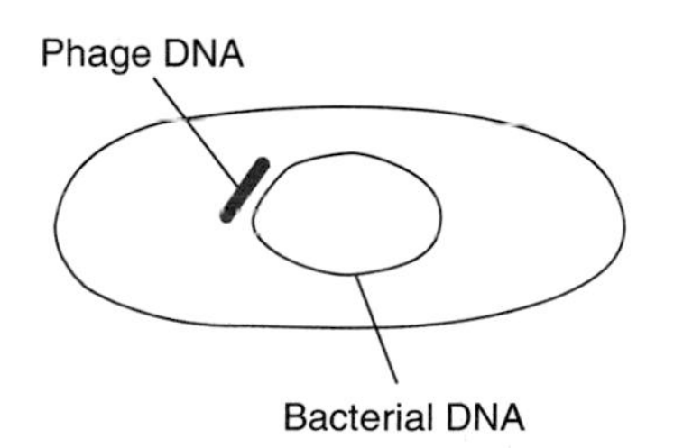

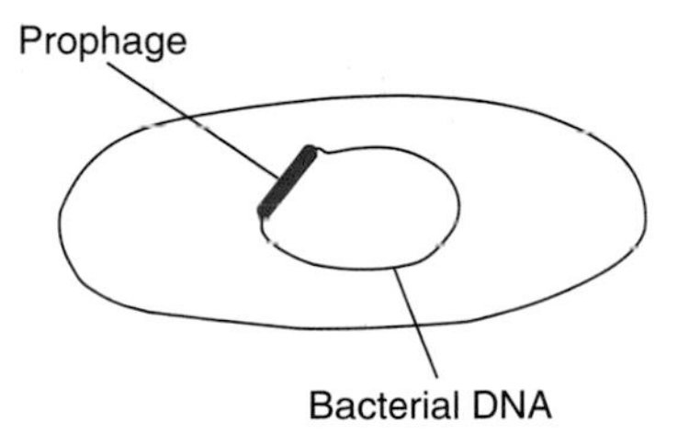

Figure P141. Phage DNA incorporated into bacterial DNA becomes prophage

Prophage Induction: Treating the bacterial cells by physical or chemical agents that cause the moving of the phage into a vegetative lifestyle resulting in asynchronous, independent replication from the host and eventually the lysis and liberation of the phage. ►prophage, ►lysogeny, ►zygotic induction

Prophage-Mediated Conversion: In prophage-mediated conversion, the integrated prophage causes genetic changes in the host bacterium, and it is expressed as an altered antigenic property, etc.

Prophase: ►meiosis, ►mitosis

Prophylaxis: Disease prevention.

Propionicacidemia: ►glycinemia ketotic, ►methylmalonicaciduria, ►isoleucine-valine biosynthetic pathway, ►tiglicacidemia; Chloupkova M et al 2002 Hum Mut 19:629.

Propionyl-CoA-Carboxylase Deficiency: ►glycinemia ketotic

Proplastid: The young colorless plastid without fully differentiated internal membrane structures; it may differentiate into a chloroplast (see Fig. P142). ►etioplast, ►chloroplasts

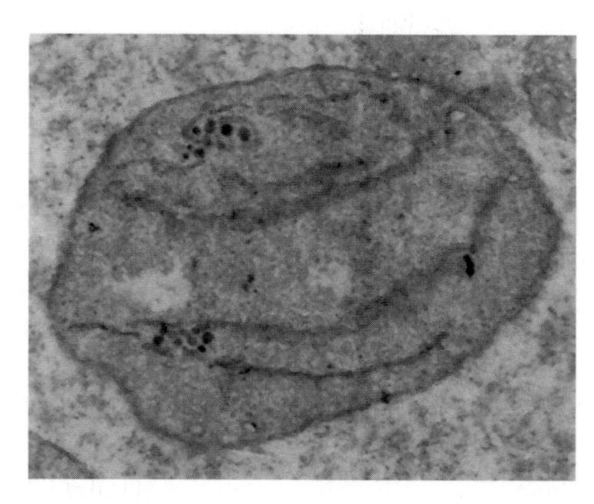

Figure P142. Proplastid

Proportional Counters: Proportional counters are used for measuring radiation-induced ionizations within a chamber. The voltage changes within are proportional to the energy released. It may be used for measuring neutron and α radiations with an efficiency of 35–50%. The equipment must be calibrated to the radiation source. ►radiation measurement, ►radiation hazard assessment

Propositus (Proposita): ►proband, ►pedigree analysis

Propyne (CH_3-C≡CH): An alkyne used as a modifier at the C-5 position of pyrimidines in antisense oligonucleotides, frequently in combination with other modifications such as phosphorothioate. ►antisense technologies

Prosimii (prosimians): A suborder of lower primates, including Galago, Lemur, Tarsius, and Tupaia. Lorisidae. ►primates, ►Lemur, ►Tupaia, ►Lorisidae

Prosite: A protein sequence database, searchable by PROSCAN (Bairoch A et al 1977 Nucleic Acid Res 25:217); http://pbil.univ-lyon1.fr/pbil.html; PROSITE for uncharacterized proteins: http://www.expasy.org/prosite/; improved PROSITE: http://www.expasy.org/tools/scanprosite/.

Prosody: The inability of sensing or expressing variations of the normal rhythm of speech. It seems to be independent of processing musical pitch. ►amusia, ►musical talent, ►pitch

Prosome: Small ribonucleoprotein body. It is identical with the ~20S multifunctional protease complex of the proteasome of eukaryotes and prokaryotes. ►proteasome

Prospective Study: Prospective study involves the epidemiological surveillance of a population after the occurrence of a disease or other harmful exposure. The exposed or involved individuals are compared with a concurrent control cohort. Prospective cohort studies improve on case control information. ►case control, ►concurrent control, ►cohort

PROST (pronuclear state embryo transfer): Basically, very similar to intrafallopian transfer of zygotes but the zygote here is at a very early stage. ►intrafallopian transfer, ►ART

Prostacyclins: Prostacyclins may be derived from arachidonic acid or prostaglandins, regulate blood platelets, cause vasodilation, and are antithrombotic. ►thrombosis, ►prostaglandins, ►COX

Prostaglandins: Long-chain fatty acids in different mammalian tissues with hormone-like muscle-regulating, inflammation-regulating, and reproductive functions; they exist in several forms. They occur in the majority of the cells and act as autocrine and paracrine mediators. Fever development is controlled by prostaglandin E_2 and EP_3 receptors. Prostaglandin synthesis is regulated by cyclooxygenases. Prostaglandin E (cyclopentenone prostaglandins) appears to be an inhibitor of IκB kinase. ►animal hormones, ►autocrine, ►paracrine, ►cyclooxygenases, ►IκB, ►leukotrienes, ►eicosanoids, ►pain-sensitivity, ►implantation, ►misoprosto, ►colorectal cancer; Rudnick DA et al 2001 Proc Natl Acad Sci USA 98:8885.

Prostanoids: Bioactive lipids such as the prostaglandins, prostacyclin, and thromboxane. Aspirin-like drugs may inhibit prostanoid biosynthesis, reduce fever and inflammation, and interfere with female fertility. ►prostaglandins, ►prostacyclins

Prostate Cancer (HPC): About 9–10% of USA males eventually develop prostrate cancer. The autosomal dominant gene has a high penetrance: about 88% of the carriers become afflicted by the age of 85. Several other genes involved in prostate function may mutate and cause cancer. Recurrent fusion of the androgen-responsive promoter element of TMPRSS2 (a transmembrane protease serine 2, 21q22.3) with members of the ETS oncogene family ETV1 at 7p21.2) leads to prostate cancer (Tomlins SA et al 2005 Science 310:6744). Relative to low-grade prostate cancer (Gleason pattern 3) and high-grade cancer (Gleason pattern 4) shows an attenuated androgen signaling signature, similar to metastatic prostate cancer, which may reflect dedifferentiation and explain the clinical association of grade with prognosis (Tomlins SA et al 2007 Nature Genet 39:41). Androgen receptor and PTEN–AKT signaling may initiate and maintain prostate cancer. For therapeutic intervention, androgen receptor and AKT and/or growth factor receptor tyrosine kinases that activate AKT can be targeted (Xin L et al 2006 Proc Natl Acad Sci USA 103:7789). The high level of testosterone may increase the chances for this cancer. Reduced level of testosterone may not slow down advanced prostate cancer growth and metastasis. There is a possible therapeutic approach to prostate cancer in overexpressing an androgen receptor by ligand-independent activation of the N-terminal domain peptide to create decoy molecules that competitively bind the interacting proteins required for activation of the endogenous full-length receptor.

A genetic variant in the 8q24 region, identified by GWA, in conjunction with another variant, accounts for about 11–13% of prostate cancer cases in individuals of European descent and 31% of cases in African Americans (Gudmundsson J et al 2007 Nature Genet 39:631). Seven risk variants, five of them previously unreported, spanning 430 kb and each independently predicting risk for prostate cancer ($P = 7.9 \times 10^{-19}$ for the strongest association, and $P < 1.5 \times 10^{-4}$ for five of the variants, after controlling for each of the others). The variants define common genotypes that span a more than fivefold range of susceptibility to cancer in some populations. None of the prostate cancer risk variants aligns to a known gene or alters the coding sequence of an encoded protein (Haiman CA et al 2007 Nature Genet 39:638).

There is evidence that in vivo expression of the receptor decoys decreased tumor incidence and inhibited the growth of prostate cancer tumors (Quayle SN et al 2007 Proc Natl Acad Sci USA 104:1331). Growth hormone-releasing hormone (GHRH) antagonists increased the intracellular Ca^{2+} and activated tumoral GHRH receptors and induced apoptosis (Rékási Z et al. 2005 Proc Natl Acad Sci USA 102:3435). Metastatic prostate cancer cells may show high levels of caveolin-1 and reduced amount of testosterone. Caveolin-1 antisense RNA promoted apoptosis and increased testosterone. A metastasis suppressor gene, KAI1, in human chromosome 11p11.2, has been identified. The Kal1 protein appears to contain 267 amino acids with four transmembrane hydrophobic and one large hydrophilic domains. This glycoprotein is expressed in several human tissues and also in rats. A negative regulator of the MYC oncogene, MXI1 (encoded in human chromosome 10q24-q25), is frequently lost in prostate cancer. Kal1 is involved in chromatin remodeling by suppressing Tip60, β-catenin, and reptin, and in metastasis of prostate cancer cells (Kim JH et al 2005 Nature [Lond] 434:921). Cytokine-activated IKKα controls metastasis by repressing Maspin (Luo J-L et al 2007 Nature [Lond] 446:690). In the chromosome 10pter-q11 region, a prostate cancer suppressor gene, causing apoptosis of carcinoma, has been detected from loss of heterozygosity mutations (LOH). A major susceptibility locus was identified in human chromosome 1q24-q25 and at Xq27-q28. Candidate genes are expected in human chromosomes 3p, 4q, 5q, 7q32, 8p22-p23 and 8q, 9q, 10p15 (KLF6), 13q, 16q, 17p11, 18q, 19q12, 20q13, and 22q12.3. At 16q22 the transcription factor ATBF1 is transcribed, which negatively regulates AFP and MYB but transactivates CDKN1A and it may be reduced in about a third of the prostate cancer cells (Sun X et al 2005 Nature Genet 37:407). Insulin-like growth factor (IGF-1) levels may be predictors of prostate cancer risks before cancerous growth is observed but some other data are at variance with the claim. In prostate tumors, the prostate-specific cell-surface antigen (STEAP), in human chromosome 7p22.3, is highly expressed in different organs and tissues, except the bladder. A predisposing gene in 17p has been cloned. The various types of prostate cancers can be classified on the basis of DNA microarray and the result may assist treatment and prognostication (Lapointe J et al 2004 Proc Natl Acad Sci USA 101:811). In Northern Europe, about 42% of the cases were found to be hereditary and 58% were sporadic (Lichtenstein P et al 2000 N Engl J Med 343:78). The first-degree relative risk is 1.7–3.7 or more. American blacks have higher and Asians lower risks than Caucasians (Stanford JL, Ostrander EA 2001 Epidemiol Rev 23:19).

Prostate stem cell antigen (PSCA) may be the target for immunological therapy. About 11–12% of all prostate cancer patients harbored mitochondrial mutations in cytochrome oxidase I subunit; mutations

P

inhibiting oxidative phosphorylation can increase ROS and tumorigenicity (Petros JA et al 2005 Proc Natl Acad Sci USA 102: 719). Prostate cancer is the second most frequent cause of cancer mortality in the US but it is very rare in other animals, except dogs. ►PSA, ►cancer, ►tumor suppressor gene, ►IKK, ►Maspin, ►MYC, ►MYB, ►fetoprotein-α, ►CDKN1A, ►insulin-like growth factor, ►caveolin, ►antisense technology, ►automaton, ►testosterone, ►apoptosis, ►microarray hybridization, ►gene fusion, ►probasin, ►ROS, ►mitochondrial genetics, ►mitochondrial diseases in humans, ►Gleason score, ►automaton, ►chromatin remodeling, ►*erbB*, ►androgen, ►PTEN, ►AKT; Ostrander EA, Stanford JL 2000 Am J Hum Genet 67:1367; Xu J et al 2001 Am J Hum Genet 69:341; Stephan DA et al 2002 Genomics 79:41; Ostrander EA et al 2004 Annu Rev Genomics Hum Genet 5:151; susceptibility markers: http://cgems.cancer.gov/.

Prostates (prostata): Gland in the animal (human) male surrounding the base of the bladder and the urethra; upon ejaculation injects its content (acid phosphatase, citric acid, proteolytic enzymes, etc.) into the seminal fluid. ►PSA, ►prostate cancer; http://www.pedb.org.

Prosthesis: Any type of mechanical replacement of a body part, such as artificial limbs, false teeth, etc.

Prosthetic Group: A non-peptide group (iron or other inorganic or organic group) covalently bound (conjugated) to a protein to assure activity.

Prot: Na^+/Cl^--dependent proline transporter that also transports glycine, GABA, betaine, taurine, creatine, norepinephrine, dopamine, and serotonin in the brain. ►transporters

Protamine: The basic (arginine-rich) protein occurring in the sperm substituting for histones.

The protamine gene cluster is in human chromosome 16p13.2 and it includes genes PRM1, PRM2, and TNP2 (transition protein 2). They are transcribed at the postmeiotic round spermatid stage of spermatogenesis and translated in elongating spermatids (see Fig. P143). These messages are bound as cytoplasmic messenger ribonuclear protein particles until histone replacement is initiated with the transition proteins in late elongating spermatids. At this time the mRNAs are activated and then translated into the peptides that will repackage and compact the male genome in terminally differentiated spermatozoa. In humans and mice, the genes first acquire a DNase I-sensitive conformation in pachytene spermatocytes that is even maintained in human spermatozoa and transcription is facilitated (Martins RP, Krawetz SA 2007 Proc Natl Acad Sci USA 104:8340). Protamine 4 is a minor protein and it is different from PRM2 and PRM3 only by a short extension. The genes are potentiated at late pachytene before the haploid (n, 1C) spermatids are formed. In the mature spermatozoon, the histones are replaced by protamines. Protamine controls both condensation and decondensation of the DNA by anchoring to it at about each 11 bp. After fertilization it is removed. In *Drosophila*, the *Hira* gene is involved in the decondensation. The HIRA gene of mammals is essential for chromatin assembly in the male pronucleus and it uses histone variant H3.3 (Lappin B et al 2005 Nature [Lond] 437:1386). In the somatic cells, protamines constitute less than 5% of the nucleus. The majority of mammals have only a single protamine but mice and men have four. If protamines are deleted missing functional sperm is not produced because of haplo-insufficiency. ►histones, ►transition protein, ►haploinsufficiency, ►spermiogenesis, ►gametogenesis, ►C amount of DNA; Cho C et al 2001 Nature Genet 28:82.

Protandry: in monoecious plants the pollen is shed before the stigma is receptive. ►monoecious, ►stigma, ►protogyny, ►self-sterility

Protanope: ►color blindness

Protease (proteinase): Enzyme, which hydrolyzes proteins at specific peptide bonds; for *protease 3* see antimicrobial peptides. The human genome codes for at least 553 proteases. Proteases may either facilitate or reduce the expression of enzymes. ►proteasome, ►peptidase; Ehrmann M, Clausen T 2004 Annu Rev Genet 38:709; http://cutdb.burnham.org.

Protease Inhibitors: Protease inhibitors, such as leupeptin, antipain, and soybean trypsin inhibitors are credited with anticarcinogenic effects and potential cures for asthma, schistosomiasis. Proteases process the primary proteins into their functional role in viral/microbial or other systems. If this processing

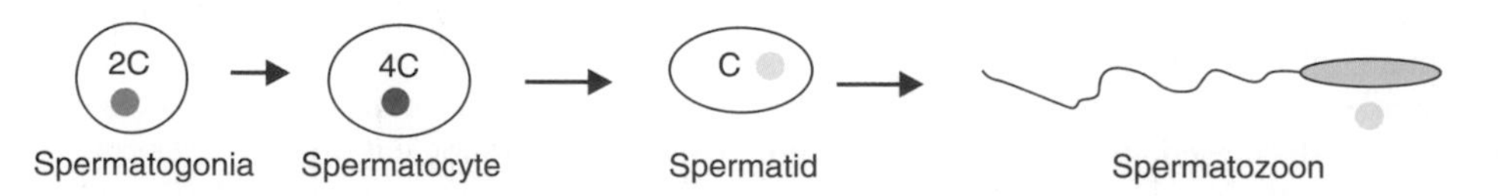

Figure P143. Development of the nuclear DNA in spermatozoa

is prevented, infectious or other agents may not or may have reduced adverse effect to the cells. ►carcinogen, ►cancer, ►AIDS, ►schistosomiasis, ►asthma

Proteasomes: Tools of degradation of intracellular proteins (see Fig. P144). Proteasomes have non-degradatory functions too, such as in transcription, DNA repair, and chromatin remodeling. ATP-dependent ubiquitinated proteins process intracellular antigens into short peptides that are then transported to the endoplasmic reticulum with the aid of TAP, and are responsible for MHC class I-restricted antigen presentation. Proteasomal polymorphism is determined, among others, by LMP2 and LMP7 genes encoded within the MHC class II region in the vicinity of TAPs that are upregulated by interferon γ. The 26S (~2500 kDa) proteasomes (~31 subunits) are hollow cylinders engulfing ubiquitinated proteins and degrade them with proteases. The lid and the base each are 19S (890 kDa). The ATP-dependent dissociation of the 19S subunits from the 26S complex leads to the protein degradation (Babbitt SE et al 2005 Cell 121:553). The ~20S (720 kDa) middle section barrel of the proteasomes contain multiple peptidases. Their active site is at the hydroxyl group of the N-terminal threonine in the β subunit. The PA proteins are proteasome activators. The 26S proteasome is associated with at least 18 ancillary and essential proteins (PSM proteins, including ATPase) and many of these are now genetically mapped to different human chromosomes. Chymostatin, calpain, and leupeptin, etc., are inhibitors. The proteases of the 20S proteasome are activated by the heptameric 11S regulators, which also control the opening of the barrel-shaped structure. The assembly of the 28 subunits of the 20S mammalian proteasomes is mediated by the heteromeric chaperones PAC1 and PAC2 (Hirano Y et al 2005 Nature [Lond] 437:13481). Proteasomes have also ubiquitin-independent function, such as the degradation of the excess amounts of ornithine decarboxylase, a key enzyme in polyamine biosynthesis. The proteasomes have important—although not fully understood—roles in differentiation and development by mediating protein turnover. Proteasomes control also apoptosis and carcinogenesis (Adams J 2004 Nat Rev Cancer 4:349). According to Princiotta MF et al 2003 (Immunity 18:343) each cell of the immune system contains 800,000 proteasomes (immunoproteasomes). The product peptides of the immunoproteasome are different from the regular proteasomes. They degrade 2.5 viral translation product substrates per minute and thus generate one MHC class I peptide complex for each 300 to 5000 viral translation product degraded. The misfolded proteins are removed from the endoplasmic reticulum and in the cytoplasm, after ubiquitination and de-glycosylation, they are degraded by the proteasome. On the average cellular proteins are degraded in about two days and about a third of the new proteins of mammals have less than 10 min half-life. The majority of these have a synthetic defect. In active mammalian cells, about 10 million polypeptides are formed per minute and within seconds these are degraded to amino acids by peptidases (Yewdell JW 2005 Proc Natl Acad Sci USA 102:9089). Membrane proteins US11 and Derlin-1 mediate MHC molecule dislocation from the endoplasmic reticulum (Lilley BN Ploegh HL 2004 Nature [Lond] 429:834; Ye Y et al 2004 *Ibid.* 841). In case the proteasome function is inhibited or lost, inhibitor resistant cells may grow out of the cultures that have a compensating mechanism for proteasome function. The Cop9 signalosome of *Arabidopsis* is functionally homologous to the lid element of the proteasome. It appears that the various elements of the proteasome complex are co-regulated by the RPM4 putative transcription factor. The yeast activators Gcn4, Gal4, and Ino2/4 are actually activated by exposure to the ubiquitin–proteasome system. It appears that after the transcription has started, the removal of the promoter-bound activators is beneficial for the continuation of transcription (Lipford JR et al 2005 Nature [Lond] 438:113). ►ubiquitin, ►LID, ►TAP, ►N-end rule, ►antigen presenting cell, ►MHC, ►antigen processing, ►JAMM, ►DRiP, ►immune system, ►polyamine, ►Skp1, ►tripeptidyl peptidase, ►Clp, ►photomorphogenesis, ►signalosome, ►lysosomes, ►unfolded protein response, ►immunoproteasomes, ►Gcn4, ►Gal4, ►exosome; Voges D et al 1999 Annu Rev Biochem 68:1015; Bochtler M et al 1999 Annu Rev Biophys Biomol Struct 28:295; Kloetzel P-M 2001 Nature Rev Mol Cell Biol 2:179; Ottosen S et al 2002 Science 296:479; Liu C-W et al 2003 Science 299:408; Puente XS et al 2003 Nature Rev Genet 4:544; Goldberg AL 2003 Nature [Lond] 426:895; lid structure: Sharon M et al 2006 PLoS Biol 4(8):e267; minireview: DeMartino GN, Gillette TG 2007 Cell 129:659.

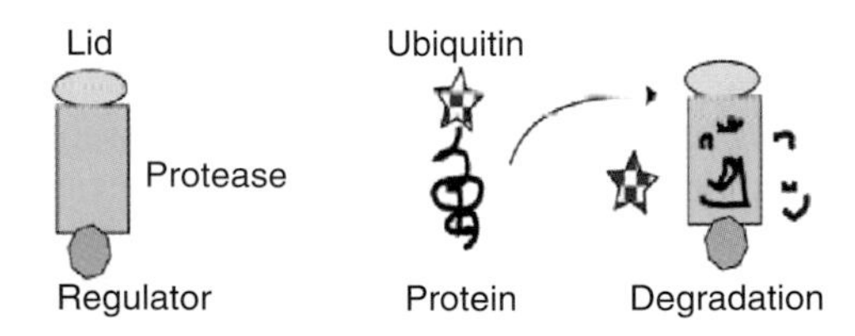

Figure P144. Proteasome and its function

Protectin (CD59): A protein component of the complement encoded at 11p13. ►complement, ►paroxysmal nocturnal hemoglobinuria; Kawano M 2000 Arch Immunol Ther Exp 48(5):367.

Protein: A large molecule (polymer) composed of one or more identical or different peptide chains. The distinction between protein and polypetide is somewhat uncertain; generally a protein has more amino acid residues (50–60) and therefore can fold. In animal cells, there are about 1×10^5 protein species. ►protein synthesis, ►protein structure, ►amino acid sequencing, ►subcellular localization; protein data Bank [PDB]: Westbrook J et al 2002 Nucleic Acids Res 30:245; http://www.rcsb.org/pdb/; http://pir.georgetown.edu/; http://www.ncbi.nlm.nih.gov/gquery/gquery.fcgi.

Protein 14-3-3: A family of 28–33 kDa acidic chaperone proteins named after their electrophoretic mobility. The proteins occur in many forms in different organisms and have roles in signal transduction (RAS-MAP), apoptosis, exocytosis, the regulation of the cell cycle (checkpoint), DNA repair, and oncogenes. They generally bind to phosphoserine/threonine domains. Protein 14-3-3σ isoform binds tumor suppressor p53, regulates translation through eukaryotic translation initiation factor 4B, and reduces the mitotic endogenous ribosomal entry site (IRES)-dependent cyclin Cdk1 (PITSLRE), among performing other functions. In the absence of this isoform, mitotic exit is impaired and causes aneuploidy and tumorigenesis (Wilker EW et al 2007 Nature [Lond] 446:329). ►Chk1, ►cell cycle, ►Cdc25, ►p53, ►CaM-KK, ►checkpoint, ►chaperone, ►longevity, ►PITSLRE, ►IRES, ►cdk, ►eIF-4B; Muslin AJ, Xing H 2000 Cell Signal 12(11–12):703; Masters SC, Fu H 2001 J Biol Chem 276:45193; Tzivion G, Avruch J 2002 J Biol Chem 277:3061; Sehnke PC et al 2002 Plant Cell 14:S339.

Protein A: Protein A is isolated from *Staphylococcus aureus*; it binds the Fc domain of immuno-globulins without interacting with the antigen-binding site. It is used both in soluble and insoluble forms for the purification of antibodies, antigens, and immune complexes. ►antibody, ►immunoglobulins

Protein Abundance: ►genome-wide location analysis

Protein Alignment: http://mozart.bio.neu.edu/topofit/index.php.

Protein Arrays: Protein arrays are used in a manner analogous to microarrays of DNA. On specially treated microscope slide or microtiter plates, samples of a protein or proteins are lined up and exposed to other proteins or to drug molecules or molecular fragments in order to assess their interaction. This new procedure is expected to be useful for analytical purposes and particularly for the development of new drugs. ►microarray hybridization, ►protein chips, ►reverse array; Avseenko NV et al 2001 Anal Chem 73:6047; Brody EN, Gold L 2000 J Biotechnol 74:5.

Protein Assays: ►Bradford method, ►Lowry test, ►Kjeldhal method; for analysis with single cell resolution: Zhang HT et al 2001 Proc Natl Acad Sci USA 98:5497; detection on magnetic nanoparticle-bio-barcode and antibody at 30 attomolar concentration: Nam J-M et al 2003 Science 301:1884.

Protein C (2q13-q14): A vitamin K-dependent serine protease, which selectively degrades antihemophilic factors Va and VIIIa, and it is thus, an anticoagulant. ►protein C deficiency, ►antihemophilic factors, ►thrombin, ►anticoagulation, ►thrombophilia

Protein C Deficiency (thrombotic disease): Protein C deficiency is human chromosome 2q13-q14 dominant and may be a life-threatening cause of thrombosis. ►thrombosis, ►protein C

Protein Chips: A protein mixture (e.g., serum) applied to an about 1 mm^2 surface containing a "bait" that is an antibody, a specific receptor, or other kind of specific molecule, which selectively binds a particular protein (tagged by fluorescent dye) and thus facilitates its isolation even when present only in minute amounts. Alternatively, recombinant proteins are immobilized on the chips and then putative interacting proteins (cell lysates) are applied to it. The unbound material is removed by washing and the bound one(s) are analyzed by mass spectrometry, or phage display or two-hybrid method may be used. These procedures can handle sppedily huge number of samples and bear similarity to DNA chips. ►microarray hybridization, ►ELISA, ►DNA chips, ►mass spectrum, ►MALDI, ►electrospray, ►phage display, ►two-hybrid method, ►gene product interaction, ►proteomics, ►protein microarray; Zhu H et al. 2001 Science 293:2101.

Protein Classification for Machine Learning: ►machine learning; http://hydra.icgeb.trieste.it/benchmark.

Protein Clock: ►evolutionary clock

Protein Complexes: Protein complexes usually play an important role in protein and cellular function. Their study requires enrichment of the complex either by chromatography, co-immunoprecipitation, co-precipitation by affinity-tagged proteins, and SDS-PAGE separation of the components before additional analytical techniques are employed. One study involving 1739 yeast genes, including 1143 human homologous, revealed 589 protein assemblies. Among these, 51% included up to five proteins, 6% more than 40 proteins, 4% 31–40, 6% 21–30, 15% of the complexes 11–20 proteins, and 18% displayed interactions among 6–10 proteins. The technology did not reveal interactions of very short durations.

Obviously, within the cells even more proteins interact. The modules of the interacting systems are better conserved during evolution than are random samples of other proteins. This indicates their importance in specific functions (Wuchty S et al 2003 Nature Genet 35:176). ►immunoprecipitation, ►immunolabeling, ►SDS-PAGE, ►LC-MS, ►mass spectrometry, ►TAP, ►two-hybrid method, ►genetic networks, ►SAGE, ►TAP; Gavin A-C et al 2002 Nature [Lond] 415:141; Ho Y et al 2002 Nature [Lond] 415:180; global surveys of budding yeast cell machineries: Gavin A-C et al Nature [Lond] 440:631; Krogan NJ et al 2006 Nature [Lond] 440:637; http://www.binddb.org/; protein–DNA complexes: http://gibk26.bse.kyutech.ac.jp/jouhou/readout/.

Protein Conducting Channel: Membrane passageways for proteins that interact with the membrane protein and lipid components. ►protein targeting, ►SRP, ►translocon, ►translocase, ►TRAM, ►ABC transporters, ►Sec61 complex; Spahn CM et al 2001 Cell 107:373.

Protein Conformation: ►conformation

Protein Data Bank (PDB): An archive of macromolecular structures. ►protein structure; http://www.pdb.org/; http://www.wwpdb.org/.

Protein Degradation: ►proteasome, ►ubiquitin, ►antizyme, ►lysosomes, ►endoplasmic reticulum, ►endocytosis, ►major histocompatibility complex, ►TAP, ►F-box, ►microRNA, ►RNAi, ►half-life

Protein Degradation within Cells: In protein degradation, endogenous proteins are digested primarily by the proteasomes and exogenous proteins are cleaved mainly by the lysosomal system, although the compartmentalization is not rigid. ►proteasome, ►lysosome, ►N-end rule

Protein Design: Computer programs exist now to design new proteins for physico-chemical potential function and stereochemical arrangements using combinatorial libraries of amino acids. The *designability of a protein* is determined by the amino acids that permit alterations without loss of structure or function. (See Dahiat BI, Mayo SL 1997 Science 278:82).

Protein, Disordered: A disordered protein contains at least one experimentally determined disordered region and lacks fixed structure. Such proteins and regions can carry out important biological functions and may be involved in regulation, signaling, and control. (See http://www.disprot.org).

Protein Domains: Protein domains are generally formed by the folding of 50–350 amino acid sequences for carrying out particular function(s). Small proteins may have only a single domain but larger complexes may have multiple modular units. The alternations of α helices and β sheets constitute a characteristic *motif*. The two β-sheet motifs are shown in Figure P145 in black and red, respectively. The compact motifs are generally covered by polypeptide loops. Domain similarities among proteins from different organisms indicate possible functional relationship (homology) of those proteins. ►protein structure-β sheets, ►α helices, ►helix-turn-helix, ►helix-loop-helix, ►zinc finger, ►binding proteins, ►motif; Ponting CP, Russell RR 2002 Annu Rev Biophys Biomol Struct 31:45; Pearl FM et al 2003 Nucleic Acids Res 31:452; http://smart.embl heidelberg.de/help/smart_about.shtml; http://www.ebi.acuk/interpro; http://smart.embl.de/; domain homology: http://genespeed.uchsc.edu/; conserved domains: http://www.ncbi.nlm.nih.gov/entrez/query.fcgi?db=cdd; conserved domains in new sequences: http://www.ncbi.nlm.nih.gov/Structure/cdd/wrpsb.cgi; three-dimensional structures: http://www.toulouse.inra.fr/prodom.html; domain search on the basis of sequence: http://www.icgeb.trieste.it/sbase; Superfamily domains: http://supfam.org; protein domain prediction: http://www.bioinfotool.org/domac.html.

Figure P145. Protein domains

Protein Engineering: Constructing proteins with amino acid replacements at particular domains and positions (e.g., substrate-binding cleft, catalytic and ligand-binding sites, etc.) or adding a label or another molecule, etc., to explore their effect on function. Incorporation of unnatural amino acids into particular proteins is a common way to accomplish it (Nowak MW et al 1998 Methods Enzymol 293:504). ►directed mutation, ►semisynthesis of proteins, ►DNA shuffling, ►iterative truncation, ►nonsense suppression, ►suppressor tRNA, ►expressed protein ligation, ►proteomics, ►enzyme design; Tao H, Cornish VW 2002 Curr Opin Chem Biol 6:858; Brennigan JA, Wilkinson AJ 2002 Nature Rev Mol Cell Biol 3:964; Wang L et al 2003 Proc Natl Acad Sci USA 100:56.

Protein, Essential: Essential for the viability of the organism in an environment.

Protein Families: Protein families share structural and functional similarities; generally share more than 30% sequence identity. The number of different families in vertebrates is about 750, in invertebrates and plants ~670, in yeast and larger bacteria ~550, and in small parasitic bacteria ~220 (Chotia C et al

P

2003 Science 300:1701). The average family size in higher organisms is about 20 whereas in lower forms it is 8 to 2. *Superfamilies*: (i) catalyze the same chemical reaction or (ii) different overall reactions that share common mechanistic properties (partial reaction, intermediate or transition state) and share 20 to 50% sequence identity. *Suprafamilies*: homologous enzymes but catalyze different reactions. Mutations affecting amino acid sequences in three different evolutionary groups (mammals, chickens, bacteria) are strikingly similar. For categories with the same divergence, common accepted mutations have similar frequencies and rank orders in the three groups. With increasing divergence, mutations increase at different rates in the buried, intermediate, and exposed regions of protein structures in a manner that explains the exponential relationship between the divergence of structure and sequence. This work implies that commonly allowed mutations are selected by a set of general constraints that are well defined and whose nature varies with divergence (Sasidharan R, Chothia C 2007 Proc Natl Acad Sci USA 104:10080). ►gene family, ►PRINTS; Enright AJ et al 2002 Nucleic Acids Res 30:1575; Aravind L et al 2002 Curr Opin Struct Biol 12:392; http://pfam.wustl.edu/; http://www.ebi.ac.uk/interpro; http://www.biochem.ucl.ac.uk/bsm/cath; http://mia.sdsc.edu/mia/html/bioDBs.html; http://systers.molgen.mpg.de; shifts in subfamilies: http://funshift.cgb.ki.se/; families in evolution: http://www.pantherdb.org/.

Protein Folding: The majority of proteins fold to acquire functionality, although some (mainly) surface proteins do not require folding. The pattern of hydrophobic and polar residues of a relatively small number may be required for folding. The native conformation is reached through intermediate stage(s) (see Fig. P146) (Sadqi M et al 2006 Nature [Lond] 442:317). Even at high (88%) amino acid identity, two proteins may have different structures and functions (Alexander PA et al 2007 Proc Natl Acad Sci USA 104:11963).

The native structure is stabilized primarily by hydrogen bonding between amide and carbonyl groups of the main chain. Glycosylation in the endoplasmic reticulum may affect the conformation of proteins.

The folding is determined by the amino acid sequence, however other factors (chaperones) may be needed to facilitate the process. the energetics of backbone hydrogen bonds dominate the folding process, with preorganization in

Besides the primary structure of amino acids, the energetics of backbone hydrogen bonds can dominate the folding process, with pre-organization in the unfolded state. Then, under folding conditions, the resultant fold is selected from a limited repertoire of structural possibilities, each corresponding to a distinct hydrogen-bonded arrangement of α-helices and/or strands of β-sheets (Rose GD et al 2006 Proc Natl Acad Sci USA 103:16623).

The classical diffusion–collision and nucleation–condensation models may represent two extreme manifestations of an underlying common mechanism for the folding of small globular proteins. Characterization of the folding process of the PDZ domain, a protein that recapitulates three canonical steps, is involved in a unifying mechanism, namely: (1) the early formation of a weak nucleus that determines the native-like topology of a large portion of the structure, (2) a global collapse of the entire polypeptide chain, and (3) the consolidation of the remaining partially structured regions to achieve the native state conformation. Classical kinetic analysis identified two activation barriers along the reaction coordinate, corresponding to a more unfolded transition state *TS1* and a more native-like transition state *TS2*. The PDZ2 (PDZ repeat from Protein Tyrosine Phosphatase-Bas Like folding process; Bas for basophil) provides evidence that its folding mechanism is distinct from the pure diffusion–collision as well as from the nucleation–condensation mechanism, but displays characteristic features of both models (Gianni S et al 2007 Proc Natl Acad Sci USA 104:128).

Prokaryotic proteins (which are generally smaller, two to three hundred amino acid residues) fold correctly only after the completion of the entire length of the amino acid chain. Eukaryotic proteins (usually on the average over four to five hundred residues) may

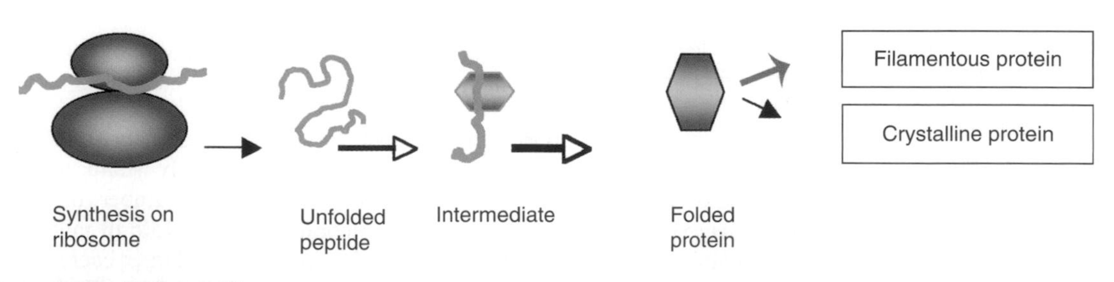

Figure P146. Protein folding

fold the separate domains in a sequential manner during their translation. Both prokaryotic and eukaryotic proteins may start folding before their translation is completed, i.e., co-translationally. Because of this, fusion proteins can also fold and this might have been of an evolutionary advantage. There is evidence that α helices fold faster than β sheets. Local interactions may facilitate speedier folding.

The global pattern of co-evolutionary interactions of amino acids is relatively sparse and a small set of positions in the proteins mutually co-evolves. The co-evolving residues are spatially organized into physically connected networks linking distant functional sites through packing interactions. By the method of statistical coupling analysis (SCA), it was revealed that the amino acid interactions specifying the atomic structure are conserved among the members of protein families. The conservation is not site independent and it occurs due to energetic interactions. The statistical energy functions can be appropriately estimated by the SCA method (Socolich M et al 2005 Nature [Lond] 437:512; Russ WP et al 2005 Nature [Lond] 437:579). Certainly many factors may affect the rate of folding and the rate among different proteins may be nine orders of magnitude. Besides folding, intrinsic plasticity of the enzyme proteins is a characteristic feature of catalysis. The motion is not limited to the active site but a more dynamic network is also involved (Eisenmesser EZ et al 2005 Nature [Lond] 438:117). Diseases may occur due to the misfolding of protein(s) such as in cystic fibrosis, Parkinsonism, prion, Alzheimer's disease, sickle cell anemia, etc. Some of the misfolding problems can be alleviated by inhibitors of the enzyme or by aiding its degradation by small molecules or by stabilizing the conformation (Cohen FE, Kelly JW 2003 Nature [Lond] 426:905). ►chaperones, ►chaperonins, ►conformation, ►calnexin, ►calreticulin, ►protein structure, ►amyloidosis, ►prion, ►encephalopathies, ►GroEL, ►trigger factor, ►endoplasmic reticulum, ►Sec61 complex, ►protein synthesis, ►folding, ►SCA; Bukau B et al 2000 Cell 101:119; Baker D 2000 Nature [Lond] 405:39; Parodi AJ 2000 Annu Rev Biochem 69:69; Klein-Seetharaman J et al 2002 Science 295:1719; Hartl FU, Hayer-Hartl M 2002 Science 295:1852; Myers JK, Oas TG 2002 Annu Rev Biochem 71:783; Gianni S et al 2003 Proc Natl Acad Sci USA 100:13286; Dobson CM 2003 Nature [Lond] 426:884; Selkoe DJ 2003 Nature [Lond] 426:900; evolutionary implications: DePristo MA et al 2005 Nat Rev Genet 6:678; protein misfolding and amyloid disease: Chiti F, Dobson CM 2006 Annu Rev Biochem 75:333; protein misfolding-human disease: Gregersen N et al 2006 Annu Rev Genomics Hum Genet 7:103; http://bioresearch.ac.uk/browse/mesh/D017510.html; protein refolding: http://refold.med.monash.edu.au/; protein folding potential software: http://flexweb.asu.edu/software/; predicting protein folding on the basis of amino acid sequence: http://psfs.cbrc.jp/fold-rate/; folding database: http://www.foldeomics.org/pfd/public_html/index.php.

Protein Function: Protein function is generally determined by biochemical and genetic analyses such as enzyme assays, two-hybrid system, etc. Many proteins are involved in complex functions and interact with several other proteins. These complex functions can be inferred from the known role of proteins in evolutionarily different organisms, from amino acid sequence information, by the rosetta stone sequences, the correlation of mRNA expression, and gene fusion information from sequence data. During evolution, some structural and functional properties of the diverging proteins are retained in the protein families but some groups have acquired new function such as substrate-specificity. These shifts can be analyzed by: http://FunShift.cgb.ki.se. ►rosetta stone sequences, ►microarray hybridization, ►two-hybrid system; http://biozon.org; functional sites, ligands: http://firedb.bioinfo.cnio.es.

Protein G: An immunoglobulin-binding (IgG) streptococcal extracellular cell surface protein.

Protein Grafting: The transfer of a binding epitope in biologically active conformation unto the surface of another protein. Such a procedure may produce an effective antiviral protein or may be used for other biological purposes. ►epitope; Sia SK, Kim PS 2003 Proc Natl Acad Sci USA 100:9756.

Protein H: Streptococcal IgG-binding protein. ►immunoglobulins

Protein Index: Guides to the main databases on proteomes such as Swiss-Prot, RefSeq, Ensembl, etc. ►protein classification; http://kinemage.biochem.duke.edu/~jsr/html/anatax.3a4.html.

Protein Information Resource: ►PIR; http://pir.georgetown.edu; the largest and most comprehensive source is the Swiss-Prot: http://www.expasy.ch/; ►databases

Protein Interactions: Interaction density (PID) is calculated as observed protein interaction/total number of possible pair-wise combinations. Conformational switches detectable by nuclear magnetic resonance relaxation experiments at the microsecond to millisecond time scale may modulate protein interactions (Koglin A et al 2006 Science 312:273). The various proteins may interact in many different ways and the ~6000 proteins of yeast may display about 100,000 relations. In a preliminary attempt to develop a human genome-wide interaction network,

~8100 Gateway-cloned ORFs allowed the detection of 2800 interactions (Rual J-F et al 2005 Nature [Lond] 437:1173). A new method permits prediction of interactions on the basis of protein sequence (Shen J et al 2007 Proc Natl Acad Sci USA 104:4337). Bioinformatic technology is required to separate the genuine from the spurious interactions (Jansen R et al 2003 Science 302:449). In *Drosophila* a draft of 7048 proteins and 20,405 interactions were detected and at a high confidence level 4679 proteins and 4780 interactions were verified by the two-hybrid method and mapped (Giot L et al 2003 Science 302:1727).

Photo-cross-linking permits detection of protein–protein interactions in living cells. Photoreactivable amino acids (e.g., photomethionine, as shown in Figure P147, the critical change is circled by dashed line) are very similar to natural counterparts and can be incorporated into the protein by the translation machinery. Activation by ultraviolet light results in cross-linking of the interacting proteins and it can be detected by western blotting (Suchanek M et al 2005 Nature Methods 2:261).

COOH COOH
NH_2 NH_2
N N S
Photomethionine Methionine

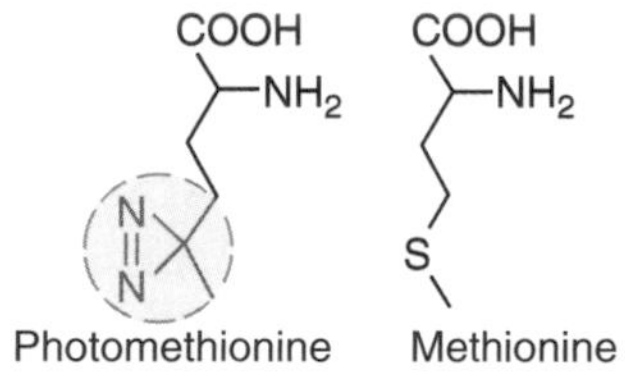

Figure P147. Photomethionine (left); methionine (right). (After Suchanek, M. et al. 2005 Nature Methods 2:261)

The bimolecular fluorescence complementation (BiFC) method permits the visualization of interaction *in situ* within living cells using yellow fluorescent protein variants (Hu C-D, Kerppola TK 2003 Nature Biotechnol 21:539). Protein interaction networks are largely preserved during evolution from prokaryotes to eukaryotes, although some specialization is also evident (Kelley BP et al 2003 Proc Natl Acad Sci USA 100:11394). A recent analysis of more than 70,000 binary interactions in humans, yeast, *Caenorhabditis*, and *Drosophila* showed only 42 were common to human, worm, and fly and only 16 were common to all. An additional 36 were common between fly and worm but not to humans, although by co-immunoprecipitation 9 were present in humans. Proteins known to be involved in similar disorders in humans showed interaction (Gandhi TKB et al 2006 Nat Genet 38:285). ►gene product interaction, ►two-hybrid method, ►protein-DNA interaction, ►affinity tagging, ►protein chips, ►networks, ►genetic networks, ►GRID, ►BIND, ►DIP, ►ORF, ►Gateway cloning, ►interactome; Bock JR, Gough DA 2001 Bioinformatics 17:455; Fernández A, Scheraga HA 2003 Proc Natl Acad Sci USA 100:113; MIPS database: Mewes HW et al 2002 Nucleic Acids Res 30:31; Jansen R et al 2002 Genome Res 12:37; http://bind.ca; http://string.embl.de; human protein reference/interaction database: http://www.hprd.org/; domain–domain interactions: http://3did.embl.de; http://mimi.ncibi.org; interactions from PubMed abstracts: http://cbioc.eas.asu.edu; protein interfaces: http://scoppi.org/; http://pre-s.protein.osaka-u.ac.jp/~prebi; protein domain interaction database: http://mint.bio.uniroma2.it/domino/; http://mint.bio.uniroma2.it/mint/Welcome.do; http://www.hsls.pitt.edu/guides/genetics/tools/protein/interaction/URL1138211431/info; Apid interaction analyzer: http://bioinfow.dep.usal.es/apid/index.htm; protein docking server: http://vakser.bioinformatics.ku.edu/resources/gramm/grammx; interaction software: http://www.ebi.ac.uk/intact/site/index.jsf; molecular ancestry network MANET: http://www.manet.uiuc.edu/.

Protein Intron: ►intein

Protein Isoforms: Closely related polypeptide chain family, encoded by a set of exons, which share structurally identical or almost identical subset of exons. ►family of genes

Protein Kinase: A protein kinase phosphorylates one or more amino acids (frequently threonine, serine, tyrosine) at certain positions in a protein, and thus two negative charges are conveyed to these sites, altering the conformation of the protein. This alteration then involves a change in the ligand-binding properties. The catalytic domain of this large family of enzymes is usually 250 amino acids. The amino acids outside the catalytic domains may vary substantially and specify the recognition abilities of the different kinases and serve in responding to regulatory signals. During the last three to four decades, hundreds of protein kinases have been discovered that can be classified into serine/threonine (TGF-β [transforming growth factor]), tyrosine (EGF [epidermal growth factor receptor], PDGF [platelet-derived growth factor receptor] protein kinases, SRC [Rous sarcoma oncogene product], Raf [product of the Moloney and MYC oncogenes]), MAP kinase, cell cyclin-dependent kinase (Cdk), cell division cycle (Cdc), cyclic-AMP- and cyclic-GMP-dependent kinases, myosin light chain kinase, Ca^{2+}/calmodulin dependent kinases, etc. Protein kinase R (PKR, dsRNA-dependent protein kinase) down-regulates protein synthesis in virus-infected cells. In the N-terminal region, two double-stranded RNA binding domains activate PKR by binding to dsRNA and recruit it to the ribosome where it phosphorylates the eukaryotic elongation factor eIF2α. The consensus sequences for a few protein kinases are shown below:

Protein kinase A

(?)-Arg- (Arg/Lys)-(?)-(**Ser/Thr**)-(?)

Protein kinase G

(?)-{[Arg/Lys] 2x or 3x}-(?)-(**Ser/Thr**)-(?)

Protein kinase C

(?)-([Arg/Lys] 1-3x)-([?] 0-2x)-(**Ser/Thr**)-([?]0-2x)-(Se/[Thr]1-3x)-(?)

Ca^{++}/calmodulin kinase II

(?)-arg-(?)-(?)-(?)-(**Ser-Thr**)-(?)

Insulin receptor kinase

Thr-Arg-Asp-Ile-**Tyr**-Glu-Thr-Asp-**Tyr**-**Tyr**-Arg-Thr

EGF receptor kinase

Thr-Ala-Glu-Asn-Ala-Glu-**Tyr**-Leu-Arg-Val-Arg-Pro

(?) indicates any amino acid, the numbers after the amino acid with an "x" indicate how many times it may occur.

The majority of protein kinases require phosphylation in their activation loop to perform their function. The human genome apparently includes 518 protein kinase genes. Protein kinases play an important role in signal transduction as well in the development of diseases (cancer), behavior and memory. Protein kinase inhibitors have therapeutic potentials. The inhibitors must pass the gatekeeper function of selectivity filters at the site of one or more amino acids (Cohen MS et al 2005 Science 308:1318). ►cAMP-dependent protein kinase, ►epinephrine, ►phosphorylase b kinase, ►signal transduction, ►obesity, ►PKD, ►kinase, ►TGF, ►EGF, ►PDGF, ►RAF, ►MYC, ►MAP, ►tyrosine kinase, ►selectivity filter; Plowman GD et al 1999 Proc Natl Acad Sci USA 96:13603; Ung TL et al 2001 EMBO J 20: 3728; Cohen P 2002 Nature Cell Biol 4:E127; Huse M, Kuryan J 2002 Cell 109:275; Manning G et al 2002 Science 298:1912; Noble MEM et al 2004 Science 303:1800; regulation: Nolen B et al 2004 Mol Cell 15:661; drug targets: Sebolt-Leopolt JS, English JM 2006 Nature [Lond] 441:457; protein kinase locking server: http://abcis.cbs.cnrs.fr/LIGBASE_SERV_ WEB/PHP/kindock.php.

Protein Binding: Protein binding involves binding protein to protein, to RNA, and to DNA. Induced fit, van der Waals interactions, electrostatic interactions, hydrogen bonds, and aromatic stacking (involving mainly tyrosine and phenylalanine) have been implied. Organic chemistry also uses pi–pi (π–π) stacking. terms mentioned; Mignon P et al 2005 Nucleic Acids Res 33:1779; Hunter CA 2004 Angew Chem Int Ed Engl 43:5310; protein ligands: http://www.bindingdb.org.

Protein Knots: Structural sites for ligand binding and enzyme activity.

Protein L: *Peptostreptococcus* bacterial protein binding to the framework of immunoglobulin κ chains. ►immunoglobulins, ►framework amino acids

Protein Length: Protein length shows great differences among individual molecules by the number of amino acids. There is a statistically significant increase along the advancement in the evolutionary rank, e.g., in Archaebacteria 270 ± 9, in bacteria 330 ± 5, and in eukaryotes (budding yeast and Caenorhabditis) 449 ± 25. Some of the mammalian proteins are huge, e.g., dystrophin.

Protein Likelihood Method: The protein likelihood method is used to determine evolutionary distance when the organisms are not closely related and when the non-synonymous base substitutions are higher than the synonymous ones. In such cases, the protein method may provide more reliable information. ►evolutionary distance, ►evolutionary tree, ►least square methods, ►four-cluster analysis, ►un-rooted evolutionary trees, ►transformed distance, ►Fitch-Margoliash test, ►DNA likelihood method; Whelan S Goldman N 2001 Mol Biol Evol 18:691.

Protein Machines: Multimolecular interacting systems such as metabolic circuits, intracellular signal transduction, or cell-to-cell communication. These systems are operated under process control strategies involving integrated feedback control. The input and output of the circuits or modules are coordinated to assure the normal or adaptive function of the cell or organism. ►feedback control, ►microarray hybridization; Baines AJ et al 2001 Cell Mol Biol Lett 6:691; Tobaben S et al 2001 Neuron 31:987.

Protein Mapping: Protein mapping localizes the pattern of expression of genes by identifying the sites of proteins within cells. An automated, multidimensional fluorescence microscopy technology permits mapping and interaction of hundreds of different proteins in a single cell (Schubert W et al 2006 Nat Biotechnol 24:1270). ►gene expression map; Huh W-K et al 2003 Nature [Lond] 425:686; Ghaemmaghami S et al 2003 Nature [Lond] 425:737.

Protein Microarray: Microspots of proteins immobilized on solid support and exposed to samples of binding molecules. In such a system, enzyme–substrate and protein–ligand relations can be visualized by the use of fluorescence, chemiluminescence, mass spectrometry, radioactivity, or electrochemistry. ►protein chips, ►protein profiling, ►antibody microarray, ►chemiluminescence, ►fluorescence,

P

►mass spectrum; Templin MF et al 2002 Trends Biotechnol 20:160.

Protein Network: The protein network detects functional organization of genomes. Two proteins may be related to other in the cell in case the presence of one seems to affect the presence or absence of another. Such a relation exists if both are required to form a structural complex or if they carry out sequential steps in an unbranched pathway. Under natural, biological conditions the presence or absence of multiple proteins exists. Another simple situation is where three proteins are followed. These may display eight different relations, such as C being present only if both A and B are present or A being present if only either B or A are present and so on. The various probabilities or uncertainties of the clusters can then be calculated in a single genome and in phylogenetic relatives to obtain information of the protein network organization (Bowers PM et al 2004 Science 306:2246). ►genetic networks; http://www.cellcircuits.org.

Protein 4.1N: 4.1 N binds to the nuclear mitotic apparatus protein NuMA, a non-histone protein that is associated with the mitotic spindle. It regulates the antimitotic function of the nerve growth factor NGF. ►NGF, ►PIK; Kontragianni-Konstatonopoulos A et al 2001 J Biol Chem 276:20679; Scott C et al 2001 Eur J Biochem 268:1084.

Protein-Nucleic Acid Interaction: ►transcription factors, ►two-hybrid method; thermodynamics of interactions: http://gibk26.bse.kyutech.ac.jp/jouhou/pronit/pronit.html.

Proteins, Number of in a human cell: may exceed that of the number of genes by a factor 5 or more but at this stage it is not known.

Protein Phosphatases: Protein phosphatases remove phosphates from proteins. They include enzymes that reverse the action of protein kinases and have an important role, together with the kinases, in signal transduction. ►protein kinases, ►membrane fusion, ►FK506; Barford D et al 1998 Annu Rev Biophys Biomol Struct 27:133; Terrak M et al 2004 Nature [Lond] 429:780.

Protein pI: isoelectric point of proteins varies between <3 to >12. ►isoelectric point

Protein Profiling: The characterization or identification of proteins on the basis of sequence, structure, mass spectrum, MALDI, MS/MS, high-performance liquid chromatography, protein microarrays, two-dimensional gel electrophoresis, etc. ►proteomics, ►protein chips, ►antibody microarray

Protein Purification: To purify proteins, disrupt cells→separate subcellular organelles by differential centrifugation→wash by buffer the separated bodies→treat the fraction(s) needed by denaturing agents→dialyze to remove the denaturing agent→use reducing agents for protection→concentrate→remove the unneeded or improperly folded protein fractions by ion-exchange chromatography, gel filtration, immunoaffinity, isoelectric focusing, high performance liquid chromatography or other steps→the wanted pure protein. Quantitate the amount or yield of the protein obtained by UV absorption or by the Lowry or Bradford methods. Each of these steps may need detailed operations. ►UV spectrophotometry of proteins, ►Lowry test, ►Bradford method

Protein Quality Control: ►unfolded protein response, ►endoplasmic reticulum-associated degradation

Protein Repair: Protein repair can be managed with assistance of chaperones. If the refolding is not feasible, proteolytic enzymes destroy proteins either directly or by the mediation of ubiquitins. Nascent polypetides, transcribed from truncated mRNAs without a stop codon, acquire a C-terminal oligopeptide (Ala, Ala, Asn, Asp, Glu, Asn, Tyr, Ala, Leu, Ala, Ala or a variant), encoded by an *ssrA* transcript. The ssrA is a 362-nucleotide tRNA-like molecule that can be charged with alanine. The addition of the peptide tag takes place on the ribosome by cotranslational switching from the truncated mRNA to the ssrA RNA. The polypeptide chain so tagged is degraded in the *E. coli* cytoplasm or periplasm by carboxyl-terminal-specific proteases. The Clp chaperone recognizes the peptides by the ssRA tag of AANDENYALAA and targets the proteins to the ClpX and ClpA ATPases. ►amino acids, ►chaperone, ►ubiquitin, ►periplasm, ►protease, ►DNA repair, ►tmRNA, ►Clp, ►ssRA; Wawrzynow A et al 1996 Mol Microbiol 21:895.

Protein S (PROS): The human chromosome 3p11 vitamin K-dependent plasma proteins preventing blood coagulation and a cofactor for Protein C. Their deficiency and dysfibrinogenemia are genetically determined causes of thrombosis. ►protein C, ►antithrombin, ►dysfibrinogenemia, ►thrombosis, ►APC, ►anticoagulation, ►thrombophilia, ►tissue factor

Protein Sequencing: ►amino acid sequencing

Protein Shuttling: The flow of protein within cells or cellular organelles. (See Ando R et al 2004 Science 306:1370).

Protein Similarity Matrix: http://mips.gsf.de/simap/.

Protein Sorting (protein traffic): The mechanism by which the polypeptides synthesized on the ribosomes in the endoplasmic reticulum reach their destination in the cell through secretory pathways by transport, with the aid of endocytotic vesicles. ►endocytosis, ►clathrin, ►Golgi apparatus, ►COP transport vesicle, ►RAFT, ►Sec, ►Fts; Tormakangas K et al 2001 Plant Cell

13:2021; Wickner W, Schekman R 2005 Science 310:1452.

Protein Splicing: ►intein

Protein Structure: The *primary structure* means the sequence of the amino acids. The *secondary structure* is formed by the three-dimensional arrangement of the polypeptide chain. The polypeptides may form α helix or β conformation when hydrogen bonds are formed between pleated sequences of the same chain. The latter type of conformation is frequently found in internal regions of enzyme proteins and in structural protein elements such as silk fibers and collagen. The α helixes are commonly represented by cylinders, whereas the β sheets by ribbons frequently with arrows at their end. The *tertiary structure* involves a folding to either a globular or various types of rope kind of structure of the polypeptide chain that has already secondary structure. Mature proteins may be formed from multiple identical or different polypeptide subunits and this type of association is the *quaternary structure* (PQS) (see Fig. P148): http://pqs.ebi.ac.uk/pqs-doc.shtml. The quaternary structure frequently includes some other molecule(s). For special biological functions, more than one protein may be joined in *supramolecular complexes* such as the myosin and actin in the muscles, and histones and various non-histone proteins in chromatin. The primary structure is genetically determined and all additional structural changes flow from this primary structure although trimming, processing, and association with prosthetic groups may be involved. The information on protein structure is highly relevant for understanding its function (Pal D, Eisenberg D 2005 Structure 13:121). Proteins with greater than 30% sequence homology assume generally the same basic structures. Although the primary structure is relevant for inferences on function and homology, remote homologies are based on folding that indicates their accommodation into particular space required for performance even when the sequence has relatively low homology (see Fig. P149) (Hou J et al 2005 Proc Natl Acad Sci USA 102:3651). Protein structure information can be obtained through several Web sites: http://astral.berkeley.edu/; http://www.pdbj.org; or http://www.ncbi.nlm.nih.gov/entrez/query.

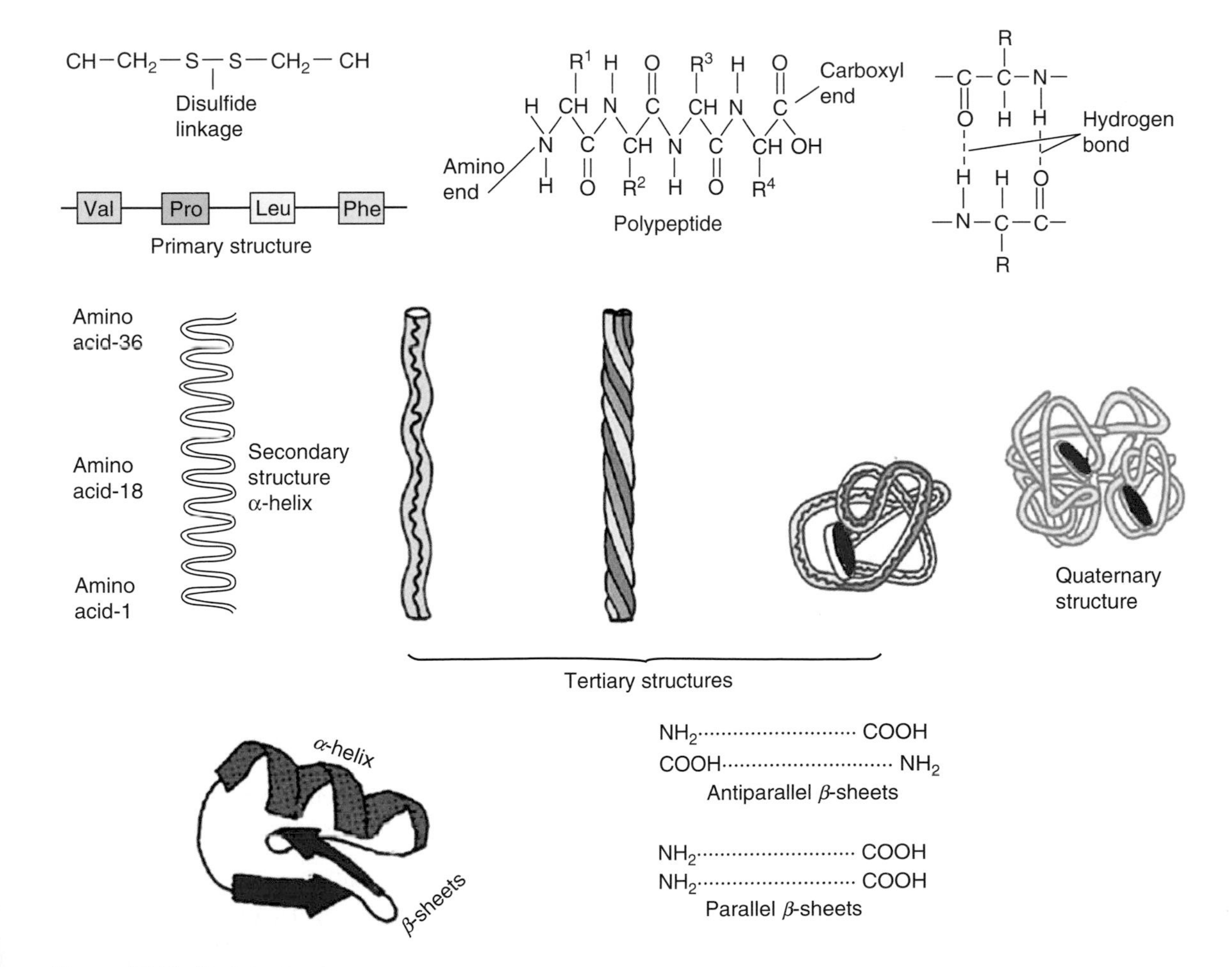

Figure P148. Protein structure

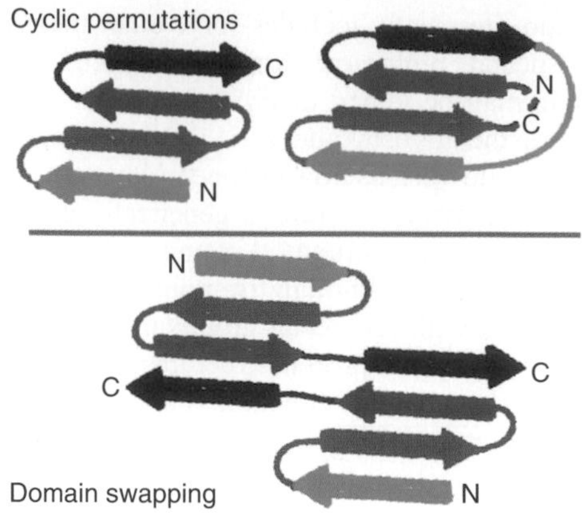

Figure P149. Cyclic permutations of the secondary structure or domain swapping of the α and β strands is tolerated in many proteins. The essential feature of a protein fold is the complementary packing of the secondary structural elements and not the precise manner of connection of the elements. Some of these changes retain stability of the protein and binding ability and can be used in protein engineering. (Diagram is modified after Tabtiang RK et al 2005 Proc Natl Acad Sci USA 102:2305)

fcgi?db=Structure; http://astral.stanford.edu/; http://www.imb-jena.de/IMAGE.html; http://scop.mrc-lmb.cam.ac.uk/scop/; ►protein synthesis, ►protein domains, ►molecular modeling, ►conformation, ►databases, ►CATH, ►SCOP, ►FEMME, ►electron density map of proteins, ►x-ray diffraction analysis, ►MANET, ►MOLSCRIPT, ►block; Goodsell DS, Olsen AJ 2000 Annu Rev Biophys Biomol Struct 29:105; Martí-Renom MA et al 2000 Annu Rev Biophys Biomol Struct 29:291; Koonin EV et al 2002 Nature [Lond] 420:218; Ouzounis CA et al 2003 Nature Rev Genet 4:508; review on structure predictions and biological significance: Petrey D, Honig B 2005 Mol Cell 20:811; tertiary structure matching of proteins: http://proteindbs.rnet.missouri.edu/index.php; molecular structure database tool: http://bip.weiz mann.ac.il/oca-bin/ocamain; structural neighbors: http://fatcat.burnham.org/fatcat-cgi/cgi/struct_neibor/fatcatStructNeibor.pl; structural and functional annotation of protein families: http://cathwww.biochem.ucl.ac.uk:8080/Gene3D/; functional site prediction: http://sage.csb.yale.edu/sitefinder3d/; functional sites from sequence alignment: http://zeus.cs.vu.nl/ programs/seqharmwww/; interacting protein motifs: http://caps.ncbs.res.in/imotdb/; comparative structure models: http://modbase.compbio.ucsf.edu/modbase-cgi-new/index.cgi; protein modeling: http://a.caspur.it/PMDB/; 3D structures: http://molprobity.biochem.duke.edu/; 3D conserved residues: http://3dlogo.uniroma2.it/; annotated three-dimensional structures: http://swissmodel.expasy.org/repository/; automated prediction: http://pcons.net/; tertiary structure: http://prokware.mbc.nctu.edu.tw/; protein short sequence motif search:

http://past.in.tum.de/; computing physicochemical properties on the basis of amino acid sequence: http://jing.cz3.nus.edu.sg/cgi-bin/prof/prof.cgi; stability of mutant proteins: http://cupsat.uni-koeln.de/; http://www.ces.clemson.edu/compbio/protcom; interactive structures: http://www.compbio.dundee.ac.uk/SNAPPI/downloads.jsp; solvability and interfacing: http://pipe.scs.fsu.edu/; protein structure modeling: http://manaslu.aecom.yu.edu/M4T/; unstable (disorder) regions: http://prdos.hgc.jp/cgi-bin/top.cgi; http://biominer.cse.yzu.edu.tw/ipda/; structure animation (movie): http://bioserv.rpbs.jussieu.fr/~autin/help/PMGtuto.html.

Protein Synthesis: Has many basic requisites and a large number of essential regulatory elements. It intertwines with all cellular functions. The blueprint for protein synthesis in the vast majority of organisms (DNA viruses, prokaryotes and eukaryotes) is in the nucleotide sequences of the DNA code. In RNA viruses the genetic code is in RNA. However, the viruses do not have their own machinery for the actual synthesis of protein, rather they exploit the host cell for this task. The genetic code specifies individual amino acids by nucleotide triplets, using one or several synonyms for each of the 20 natural amino acids. The triplet codons are in a linear sequence of the nucleic acid genes. In the organisms with DNA as the genetic material, the process of transcription produces a complementary RNA sequence from one or both strands of the anti-parallel strands of the DNA. The double-strands unwind and the RNA polymerase(s) synthesize(s) a complementary RNA copy of the sequence in the DNA. In the single stranded DNA and RNA viruses, the DNA or RNA may serve both purposes of being the genetic material and the transcript for protein synthesis. In cellular organisms three main classes of RNAs are made, messenger RNA (mRNA), transfer RNA (tRNA) and ribosomal RNA (rRNA), and all three are indispensable for protein synthesis. In addition to these RNAs, a large number of proteins are required for the transcription process (transcription factors), for the organization of the ribosomes (50–80 ribosomal proteins), for the termination of transcription, for the activation of the tRNAs, etc. A broad overview (without details) is shown in Figure P150. Some of the details of the transcriptional process are different in prokaryotes from that in eukaryotes. In the latter group, one DNA-dependent RNA polymerase is responsible for the synthesis of all RNAs. In eukaryotes, pol I synthesizes rRNAs with the exception of the 5S and 7S rRNA, pol II transcribes mRNA and the small nuclear RNAs (snRNA) and pol III synthesizes tRNAs and 5S and 7S rRNA.

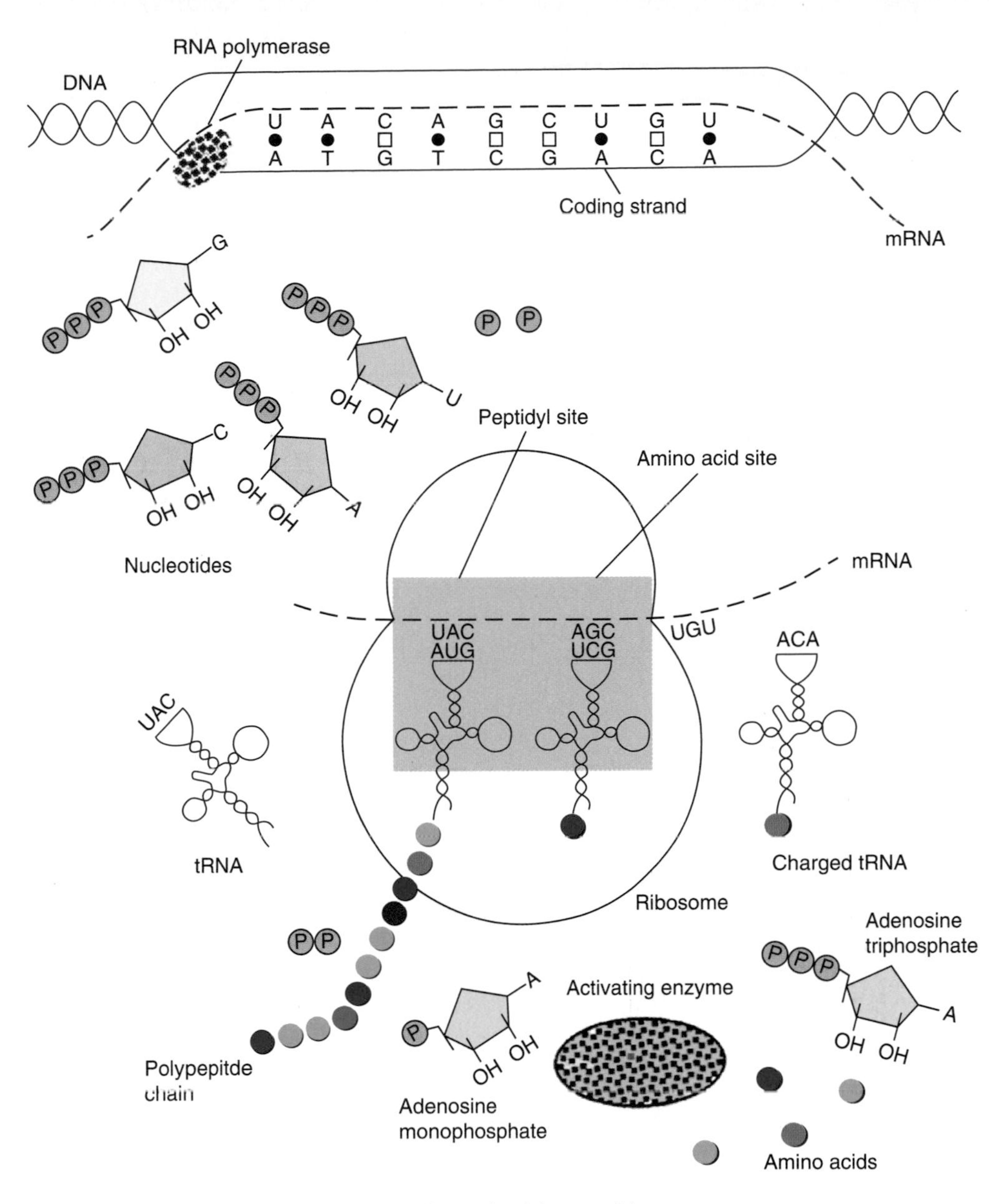

Figure P150. An over-simplified view of the protein synthesizing machinery

In prokaryotes, the process of transcription and translation are *coupled*, i.e., as soon as the chain of mRNA unwinds from the DNA it is associated with the ribosomes and protein synthesis begins. The primary RNA transcripts must be processed to functional size molecules in all categories that may require splicing and other post-transcriptional modifications (capping, formylation, etc.).

In eukaryotes when the mRNA is released from its DNA template it moves into the cytosol where protein synthesis takes place. A small fraction of polypeptides may be synthesized also in the nucleus of eukaryotes (Iborra FJ et al 2001 Science 293:1058). There is evidence for the association of ribosomal components into ribonucleoprotein complexes at the transcription sites of salivary gland chromosomes (Brogna S et al 2002 Mol Cell 10:93). The fate of the mRNA can be monitored by electronmicroscopy in both groups and these pictures show the elongation of RNA and protein strands (see Fig. P151). The first products of both display long strands and the short ones indicate the stage and place where they were started. The ribosomes are captured by the mRNA and form an association of multiple units in the form called *polysomes* (see Fig. P152).

The prokaryotic mRNA is directed to the proper position in the 30S ribosomal subunit by the Shine-Dalgarno nucleotide sequence within 8 to 13-base

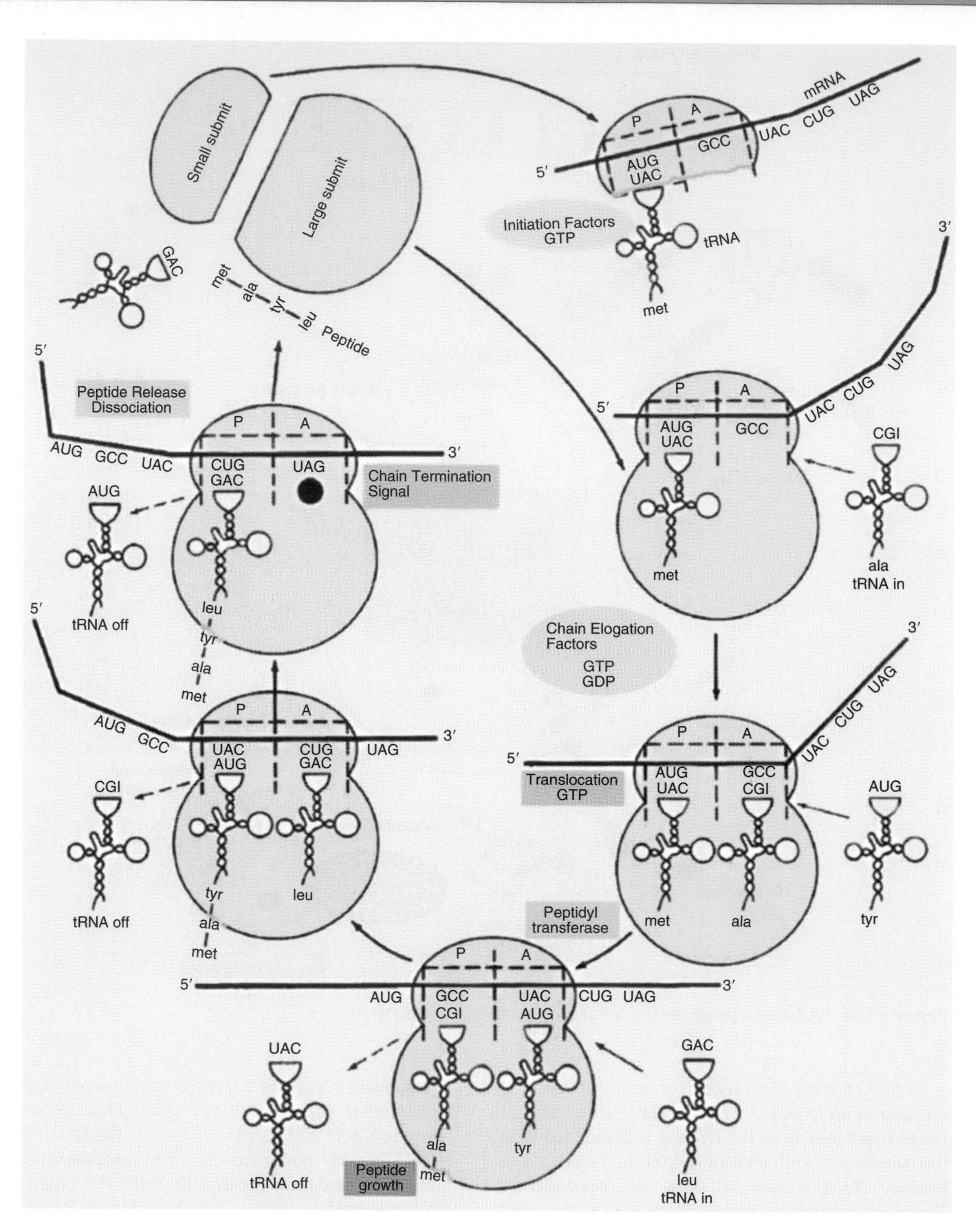

Figure P151. Classical model of translation on ribosomes

area upstream from the initiation codon. In eukaryotes, such a sequence does not exist and the mRNA is simply scanned by the ribosome until the first methionine codon is found.

The ribosomal units then slide from the 5′-end of the mRNA toward the 3′-end and thus, the amino end of the polypeptide chain corresponds to the 5′-end of the mRNA. The ribosomes in both prokaryotes and

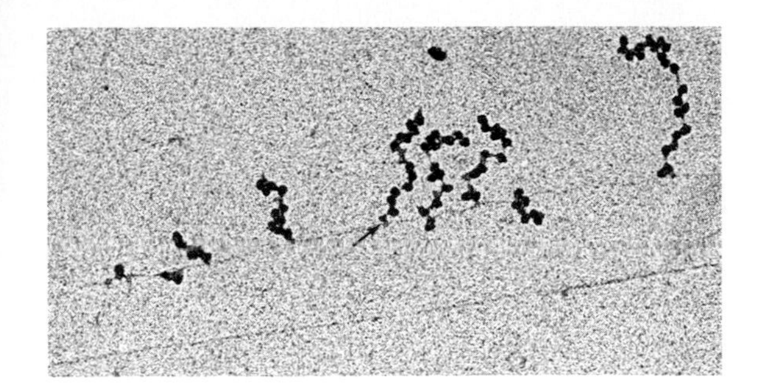

Figure P152. Transcription and translation coupled in *E. coli*. The thin thread is the DNA, the dark round structures are polysomes. The transcriptase attachment → is indicated. (From Hamkalo BA et al 1974 Stadler Symp 6:91)

eukaryotes are composed of a small and a large subunit. The size of these units is somewhat different in the two major taxonomic categories. The small and large subunits of the ribosomes jointly form two compartments, the so-called P (peptidyl-tRNA binding site) and the A (aminoacyl-tRNA binding site). A newer *hybrid-states model* of the translational process is described under the entry "ribosomes." The ribosomes actually do not look like as shown in these diagrams, because they are three-dimensional and have a more elaborate structure. Before protein synthesis (translation) begins and the primary structure of the mRNA is translated from the nucleotide triplet codon words into the singular amino acid word language of the protein, the tRNA molecules must be charged with amino acids. This process is also called activation of tRNA. (►aminoacyl-tRNA synthetase).

The amino-acid-charged methionine-tRNA ($tRNA^{Met}$) in eukaryotes and the formylated $tRNA^{fMet}$ in prokaryotes seek out the cognate codon in the mRNA at the P site of the ribosome through the complementary anticodon. This event requires the presence of protein initiation factor(s) and GTP as energy source. The GTP is cleaved to GDP + inorganic phosphate (Pi) and thus liberates some of the needed energy. The elongation factor proteins and GTP and GDP complexes also police the system to prevent the wrong charged tRNA to go to an A site (proofreading function). Actually a similar correction mechanism is carried out earlier in the process by one of the active sites of the aminoacyl synthetase (activating) enzyme that usually dissociates the amino acid—tRNA link in case of a misalliance. With the double checks available, misincorporation of amino acids is approximately in the 10^{-4} range. Protein synthesis in the mitochondria and chloroplasts is essentially patterned after the prokaryotic systems.

The 5′-base of the anticodon triplet may not be the exact and conventional base, yet it may function normally (►wobble). The two subunits of the ribosomes are combined and the second charged tRNA can now land at the A ribosomal site. The carboxyl end of the methionine forms a peptide bond with the amino terminus of the next incoming amino acid at the A site. This process is mediated by the enzyme peptidyl transferase. For this transferase function a 23S rRNA in the large subunit (a ribozyme) is responsible and not a protein. Again energy donors and elongation protein factors are cooperating in the process of peptide chain growth (►initiation and elongation factors IF, ►eIF, ►EF, ►EF-T, ►EF-Tu). When each peptide bond is completed the tRNA is released and recycled for another tour of duty. The *open reading frame* of the gene is terminated by a nonsense or chain-termination codon. When the ribosome slides to this point the mRNA is released from the ribosomes with the assistance of release factors (►transcription termination in eukaryotes, ►transcription termination in prokaryotes). Protein synthesis proceeds at a rather rapid rate; it has been estimated that in *E. coli* 50–200 amino acids may be incorporated into peptides in 5–10 s. The process is slower in eukaryotes (3–8 s) (see Fig. P153). According to Princiotta MF et al 2003 (Immunity 18:343), the cells of the immune system produce 40 million proteins/min on the 6 million ribosomes.

The ribosomes have an important role in the regulation of protein synthesis. It appears that the availability of active ribosomes is controlled at the level of the transcription of the rRNA genes. In most of the cases, the number of ribosomes is not a limiting factor of translation. Some of the bacterial ribosome proteins have dual roles and participate in transcription and translation (Squires CL, Zaporojets D 2000 Annu Rev Microbiol 54:775). When the supply of ATP and GTP is adequate, rRNA genes are activated for transcription. In case the level of these nucleotide triphosphates is low, rRNA transcription is reduced or halted. Abundance of free ribosomal proteins may feedback-inhibit ribosomal production. The ribosome-associated Rel-A protein may mediate the formation of ppGpp from GTP (and possibly from other nucleotides). Then ppGpp may shut off rRNA and tRNA synthesis by binding to the promoter of RNA polymerase or to its antitermination signal.

Some of the nascent peptides are segregated into the endoplasmic reticulum through the Sec61 conductance opening of the large subunit of the ribosomes. Within the endoplasmic reticulum, the translation continues and the protein is folded by the appropriate chaperones. In prokaryotes, only the completed polypeptide chains are folded whereas in eukaryotes, the separate domains of the large polypeptides are folded as the chain grows.

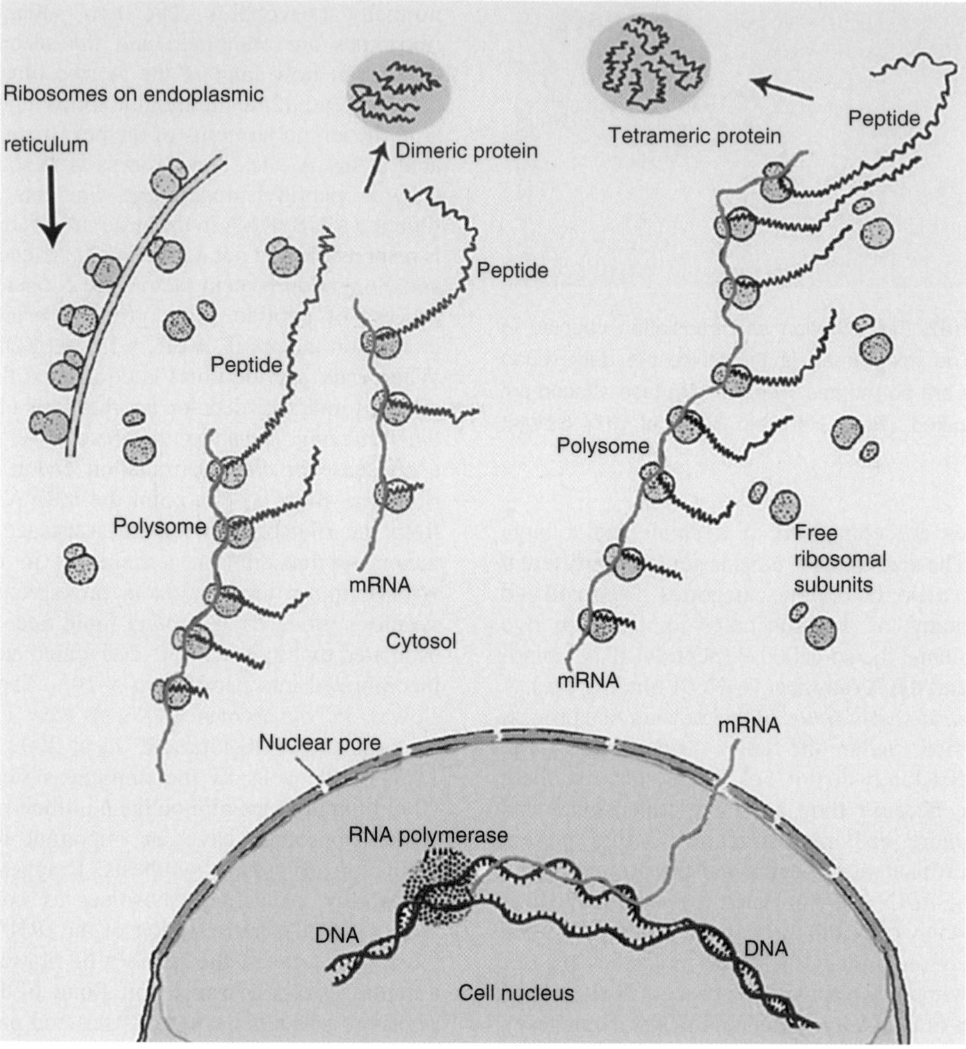

Figure P153. An overview of eukaryotic translation

The dimeric NAC (nascent-polypetide associated complex) interacts with the emerging polypeptide chains before 30 or fewer residue long chain is formed, and protects the nascent chain from becoming associated with other cytosolic proteins until the signal peptide fully emerges and then the signal recognition particle (SRP) crosslinks to the polypeptide. The purpose of the NAC is to assure that the polypeptide would be oriented to the proper SRP and the endoplasmic reticulum. Alternatively, if the protein does not carry a signal peptide, the nascent chain may be folded by chaperones such as heatshock proteins Hsp40 Hsp70 and TRiC. The completed amino acid sequences, the polypeptides, must be then converted to biologically active forms. This post-translational process may involve trimming (removal of some amino acids), proteolytic cleavage, folding to a tertiary structure, aggregation of different polypeptide chains to form the quaternary structure, addition of prosthetic groups (such as heme, lipids, metals), and other non-amino-acid residues such as acyl, phosphate, methyl, isoprenyl and sugar groups.

Some proteins are expressed at very low level and by the classical methods of biochemistry or molecular biology the synthesis may not be detectable. A microfluidic device can, however, detect protein expression at the level of a single molecule (Cai L et al 2006 Nature [Lond] 440:358).

►code genetic, ►mRNA, ►tRNA, ►rRNA, ►ribosomes, ►aminoacyl-tRNA synthetase, ►wobble, ►cap, ►Shine-Dalgarno sequence, ►ribosome recycling, ►RNA polymerases, ►transcription factor, ►transcription initiation, ►elongation initiation factors, ►eIF, ►transcription termination, ►rho factor, ►transcription complex, ►signal sequences, ►transit peptide, ►signal peptides, ►regulation of gene activity, ►antibiotics, ►toxins, ►ambiguity in translation, ►chaperone, ►SRP, ►signaling to translation, ►translation initiation, ►initiation complex, ►polysome, ►introns, ►prenylation, ►TRiC, ►heatshock, ►E site, ►EF-TU•GTP, ►discriminator region, ►protein folding, ►Sec61 complex, ►nonribosomal peptides, ►tmRNA, ►translation in vitro, ►translation nuclear, ►subcellular localization, ►microfluidics; Sonenberg N et al (Eds.) 2000 Translational Control of Gene Expression, Cold Spring Harbor Laboratory Press, Cold Spring Harbor, New York; Fredrick K, Noller HK 2002 Mol Cell 9:1125.

Protein Synthesis, Chemical: Building proteins from peptide domains by non-biological means. The peptides may be synthesized by the methods or organic chemistry and then ligated, spliced and folded in order to assure some specific function. Changing individual amino acids in the sequence may lead to new proteins. This procedure may become cumbersome because aggregation may create problems for proper folding. In contrast, solid phase synthesis is practical and any type of non-natural protein can now be produced using synthetic amino acids and peptides. Chemical synthesis of peptides requires stepwise addition of amino acids on solid support. Unfortunately, with the current technology (2005), routinely the synthetic proteins can be made only of about 40 residues although, e.g., 238-residue precursor of the green fluorescent protein has been made years ago (Nishiuchi Y et al 1998 Proc Natl Acad Sci USA 95:13549. [Bradley L et al. 2005 reviewed the principles of the various available peptide ligation techniques in Annu Rev Biophys Biomol Struct 34:91]. See Dawson PE, Kent SBH 2000 Annu Rev Biochem 69:923; Wei Y et al 2003 Proc Natl Acad Sci USA 100:13270).

Protein Synthesis Inhibitors: ►antibiotics, ►toxins, ►interferons

Protein Targeting: Can be co-translational, i.e., newly synthesized proteins are delivered to specific sites (endoplasmic reticulum) in the cell before the chain is completed or post-translational when the transport takes place after the polypeptide is completed. ►signal hypothesis, ►signal sequence recognition particle, ►translocon, ►TRAM, ►protein conducting channel, ►mitosome; Bachert C et al 2001 Mol Biol Cell 12:3152; Zaidi SK et al 2001 J Cell Sci 114:3093; Takayama S, Reed JC 2001 Nature Cell Biol 3.E237.

Protein Trafficking: ►protein sorting

Protein Transduction: The introduction of protein into the blood stream or organs for experimental or therapeutic purposes. This procedure is usually limited to small size (<600 Da) molecules. When, however, the 120 kDa β-galactosidase was fused to an 11-amino acid NH_2 domain of the Tat protein of HIV and introduced into the intraperitonial cavity of the mouse, the protein was detected in a biologically active form in several organs including the brain. ►AIDS, ►BBB, ►galactosidase, ►protein targeting; Embury J et al 2001 Diabetes 50:1706.

Protein Transport: ►protein sorting

Protein Truncation Test: The test may be used to detect the effects of several mutations that do not permit the completion of a polypeptide chain. The gene is transcribed by using polymerase chain reaction and the RNA is translated in vitro and the polypetide is analyzed in SDS minigels. ►PCR, ►SDS-polyacrylamide gel, ►rabbit reticulocyte in vitro translation; Lutz S et al 2001 Nucleic Acids Res 29:E16.

Protein Tyrosine Kinases (PTK): The phosphorylate tyrosine residues in some proteins. This function is frequently coded for by v-oncogenes of retroviruses but cellular oncogenes and other proteins may be involved and are controlling signal transduction and other cellular processes such as cell proliferation and differentiation. Cytosolic tyrosine kinases preferentially phosphorylate their own SH2 domains or related SH2 domains with hydrophobic amino acids at key positions, e.g., Ile or Val at −1 and Glu, Gly or Ala at the +1 position. Receptor tyrosine kinases prefer Glu at −1 position. These preferences specify their signaling role. The RET oncogene's receptor tyrosine kinase product can shift substrate specificity and thereby cause multiple endocrine neoplasia. Quercetin, genistein, lavendustin A, erbstatin and herbimycin are all natural plant products and inhibitors of these enzymes. ►tyrosine kinase, ►receptor tyrosine kinase, ►protein kinases, ►SH2, ►endocrine neoplasia multiple, ►signal transduction; Hubbard SR, Till JH 2000 Annu Rev Biochem 69:373; Blume-Jensen P, Hunter T 2001 Nature [Lond] 411:355.

Protein Tyrosine Phosphatase: ►tyrosine phosphatases

Protein X: A hypothetical chaperone facilitator of the $PrP^C \rightarrow PrP^{Sc}$ conversion in prion diseases. ►prion

Protein Zero: A major part of the nerve cell myelin sheath of vertebrates. Its defect may lead to neurological anomalies.

Proteinase A: An endopeptidase involved in protein folding. ►endopeptidase, ►protein folding

Proteinase K: A proteolytic enzyme, frequently used to remove nucleases during the extraction of DNA and

RNA. With appropriate heat treatment any DNase associated with it can be safely removed. ►protease

Proteinoid: A polymerized mixture of amino acids formed during prebiotic stage of evolution (or simulated conditions in the laboratory). They may resemble primitive cells and display fission like phenomena (see Fig. P154). ►prebiotic

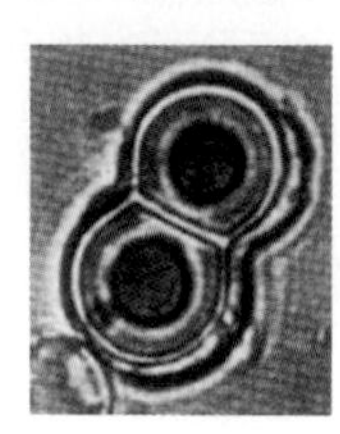

Figure P154. Proteinoid (From S. W. Fox., 1964 BioScience 14(12):13, © Am Inst Biol Sci)

Proteinosis: Anomalous accumulation of protein at particular structures of the body.

Protein-DNA Interaction: Takes place between transcription factors and the DNA template of the RNA. These interactions have been mapped in vivo over the entire mammalian genome (Johnson DS et al 2007 Science 316:1497). ►transcription factors, ►regulation of gene activity

Protein-Protein Interaction: Mediates structural and functional organization of the cells. The knowledge of these processes reveals the essential nature of the biology of organisms. The two-hybrid method may reveal the pair-wise interactions, and by sequential and systematic analysis the interacting systems, the metabolic modules can be identified. ►two-hybrid method, ►microarray hybridization, ►networks, ►gene product interaction, ►networks

Protein-RNA Recognition: Almost all RNA functions involve RNA-protein interactions such as regulation of transcription, translation, processing, turnover, viral transactivation and gene regulatory proteins in general, tRNA aminoacylation, ribosomal proteins, transcription complexes, etc.

Proteobacteria: Gram-negative purple bacteria, putative ancestors of mitochondria. ►NUMTs

Proteoglycan: Heteropolysaccharides with a peptide chain attached through O-glycosidic linkage to a serine or threonine residue. Such molecules are enzymes, animal hormones, structural proteins, basement membranes, cellular lubricants (such as mucin), extracellular matrix proteins and the "antifreeze proteins" of Antarctic fishes. They control plant and animal growth, differentiation, development and signal transduction. The proteoglycan-like xylogen accumulates in the meristem of plants and directs continuous vascular development (Motose H et al 2004 Nature [Lond] 429:873). ►antifreeze proteins, ►amyloids, ►glypican, ►syndecan, ►glycosaminoglycan, ►glycoprotein, ►phloeme, ►xyleme; Selleck SB 2000 Trends Genet 16:206.

Proteolipid Protein: A major part of myelin in the brain. ►myelin

Proteolysis: The hydrolyzing peptide bonds of proteins. The tobacco etch virus (TEV) NIa protease recognizes a seven-residue consensus (Glu-X-X-Tyr-X-Gln/Ser) sequence and does not affect proteins not containing it. The protease attached to the ribosomal exit site is most efficient and permits selective cleavage special target proteins (Heinrichs T et al 2005 Proc Natl Acad Sci USA 102:4246). ►proteasome, ►ubiquitin; Ciechanover A 2005 Nature Rev Mol Cell Biol 6:79.

Proteolytic: Enzymes hydrolyze peptide bonds in proteins. ►proteolysis, ►peptide bond, ►peptidase

Proteome: All the cellular proteins encoded by the cellular DNA; it is the protein complement of the genome. In bacteria 10% of the genes encode 50% of the bulk of the protein in eukaryotes ~90% of the proteome is contributed by 10% of the cellular proteins (Humphery-Smith I 2004, p 5 In: Albala JS, Humphery-Smith I (Eds.) Protein Arrays, Biochips, and Proteomics, Marcel Dekker, New York). The genome is very stable (except rare mutations) and it is the same in practically all cells of an organism. The proteome displays variations according to the developmental stage, organs, metabolic rate and health of the organism, etc. Since the proteins are organized and expressed in interacting systems, their study may be very complicated. While the genome does not reveal the detail of the function of a cell(s), proteomics has exactly this goal. The immediate products of the genome, the RNA is frequently processed in more than one way (alternative splicing and combinatorial assembly) to be translated into more than a single type of polypeptide. The translated product can be further modified by trimming, docking, forming multimeric associations, recruitment of ligand, phosphorylation and/or dephosphorylation, acetylation, glycosylation and various other epigenetic mechanisms. Because of alternatives in transcription (using different promoters and processing of the transcripts) there are in general substantially more proteins than genes in the cells. The proteins have also various regulatory roles at the levels of replication, transcription, translation, etc. The amount and kind of RNAs are correlated with the amount of polypeptides yet this correlation is variable. Proteins may undergo substantial post-translational modifications. Although the genome is essentially constant, the encoded proteins may display great variations during differentiation and development. There are no well-established procedures "fit for all" proteins such as DNA sequencing after cloning, PCR or microarray hybridization. Two-dimensional gel electrophoresis is powerful for the separation of thousands of proteins and monoclonal antibody techniques can be used for the localization of proteins. Although definitive information on the proteome may not come easily it should permit an insight into the function of cells, organisms, evolution and disease that cannot be matched by other means. The size of the human proteome much exceeds that of the number of genes determined by sequencing the genome. The size of the human proteome has been estimated by the formula $N_{CDS} - f_1.f_2.N_{genes}$ where f_1 is the proportion of non-pseudogenic genes and f_2 is the ratio of the total number of protein-coding transcripts to the total number of genes, including those that are spliced alternatively. The estimates so obtained also vary within a wide range (see Harrison PM et al 2002 Nucleic Acids Res 30:1083). ►genome, ►genomics, ►metabolic pathway, ►transcriptome, ►monoclonal antibody, ►two-dimensional gel electrophoresis, ►two-hybrid method, ►protein chips, ►MALDI/TOF/MS, ►electrospray, ►ICAT, ►ACESIMS, ►MS/MS, ►microarray hybridization, ►networks, ►genetic network, ►TOGA, ►core proteome; protein–protein interaction: HUPO: http://www.hupo.org; ►Uniporter; http://www.expasy.ch; https://www.proteome.com/proteome/; http://us.expasy.org; human proteome: http://www.hprd.org/; mass spectrometric characterization of peptide fragments: http://nwsr.bms.umist.ac.uk/cgi-bin/pepseeker/pepseek.pl?Peptide=1; mass spectrum of body proteome: http://www.mapuproteome.com; ►protein, ►genomic sequences, ►exon structure, ►polarity, ►hydrophobicity; Ito T et al 2001 Proc Natl Acad Sci USA 98:4569; Walhaut AJM, Vidal M 2001 Nature Rev Mol Cell Biol 2:55; Harrison PM et al 2002 Nucleic Acids Res 30:1083; Auerbach D et al 2002 Proteomics 2:611; Burley SK, Bonnano JB 2002 Annu Rev Genomics Hum Genet 3:243; Rost B 2002 Curr Opin Struct Biol 12:409.

Proteomic Profiling: Uses chemical labels for the identification of active groups of enzymes in complex mixtures and attempts the identification of the functional role of these groups of proteins. The procedure may reveal the role of protein arrays in the development of disease and may suggest targets for intervention. (See Adam GC et al 2002 Nature Biotechnol 20:805).

Proteomics: The study of the system of the proteome, the modules of metabolism as they carry out cellular functions of the organisms. The new technologies detect the composition/structure of proteins, isoforms, conformational changes, modulatory alterations during development, post-transcriptional and post-translational modifications (phosphorylation, glycosylation), interactions with other proteins or drugs, etc. With low mass tolerance, e.g., 10 ppm single proteins can be identified in a mixture among thousands of molecules. Proteomics has modified the basic approach to investigating biological function. Earlier the experimental design was based on hypotheses. With the aid of the proteomics technologies more direct approaches are possible based on the simultaneous expression patterns of interacting genetic networks. *Expression Proteomics* analyses proteins of the cells by two-dimensional gel electrophoresis (Wagner K et al 2002 Anal Chem 74:809). *Cell-Map Proteomics* is interested in the interaction between/among proteins at various phases of the cell function (Blackstock WP, Weir MP 1999 Trends Biotechnol 17(3):121). *Functional Proteomics* targets specific functions rather than the entire proteome (Graves PR, Haystead TA 2002 Microbiol Mol Biol Rev 66:39). *Structural Proteomics* seeks understanding of protein function on the basis of three-dimensional analysis and modeling (Norin M, Sundstrom M 2002 Trends Biotechnol 20:79; Sali A et al 2003 Nature [Lond] 422:216). *Reverse Proteomics* starts with the genes and proceeds to proteins. Liquid chromatography, two-dimensional polyacrylamide gel electrophoresis and tandem mass spectrometry are important tools of proteomics at large scale. Proteomics is concerned not only with the variability and interactions of proteins but may assist in modifying proteins for new types of interactions. The α-carboxyl group and preceding residues at the C-end of polypetides may offer a useful target for modifications. The PDZ and TPR domains are well qualified for interactions with the C-termini and may facilitate temporal and spatial interactions, degradation, neuronal signaling and other functions (Chung JJ et al 2002 Trends Cell Biol 12:146). The proteome data are expected to be much more complex than that of the genome sequences. The number of proteins and their isoforms far exceeds that of the number of genes. There is a need to develop computer programs that can properly assist in interpreting the "mountain" of information. One of the most complete sources of information on the *E. coli* metabolic system is at: http://ecocyc.org/. The increasing amount of information is fast becoming impossible to integrate for a single human mind and advanced computer models are indispensable. Now proteomic information has important impact of applied biology such as medicine, drug development and agriculture. In painted artwork protein (egg white) has been used since the fourteenth century and before then as binding material. These old paintings now need restoration and for doing the best work, it is necessary to determine in a minimally invasive way the material the artists used. Modern proteomics technology can reveal the nature of the binder used in Renaissance paintings in ~10 μg samples (Tokarski C et al 2006 Anal Chem 78:1494). ►proteome, ►PFAM, ►Atlas human cDNA, ►genomics, ►annotation, ►MALDI, ►HMS-PCI, ►TAP, ►PDZ domain, ►TPR, ►peptide mass fingerprints, ►NMR, ►post-translational modification, ►quadrupole, ►LC-MS, ►FTMS, ►MS/MS, ►ion trap mass analyzer, ►linear ion trap analyzer, ►two-dimensional gel electrophoresis, ►two-hybrid system, ►protein chips, ►protein microarray, ►X-ray crystallography, ►genetic networks, ►networks, ►gene product interaction, ►nucleolomics, ►laser-capture microdissection, ►MCA, ►mass-coded abundance tagging, ►display technologies, ►MudPIT, ►PEDRO, ►protein engineering, ►semisynthesis of proteins, ►bioinformatics, ►International Protein Index; Washburn MP et al 2001 Nature Biotechnol 19:242; Mann M et al 2001 Annu Rev Biochem 70:437; MOWSE 2001 Trends Biotechnol 19(10):Suppl; Fraunfelder H 2002 Proc Natl Acad Sci USA 99(Suppl 1):2479; Altman RB, Klein TE 2002 Annu Rev Pharmacol Toxicol 42:113; Regnier FE et al 2002 J Mass Spectrom 37:133; Laurell T, Mako-Varga G 2002 Proteomics 2:345; Auerbach D et al 2002 Proteomics 2:611; Petricoin EF et al 2002 Nature Rev Drug Discov 1:683; Huber LA 2003 Nature Rev Mol Cell Biol 4:74; Patterson SD, Aebersold RH 2003 Nature Genet 33(Suppl):311; analytical methods: Phizicky E et al 2003 Nature [Lond] 422:208; Zhu H et al 2003 Annu Rev Biochem 72:783; de Hoog CL, Mann M 2004 Annu Rev Genomics Hum Genet 5:267; mass spectrometry methods: Domon B, Aebersold R 2006 Science 312:212; http://www.ebi.ac.uk/interpro; http://dip.doe-mbi.ucla.edu/; Proteomics Identification Database: www.ebi.ac.uk/pride/.

Proterozoic (precambrian): The geological period five billion to 570 million years ago. Aquatic forms of living systems appeared during this era. ►geological time periods

ProtEST: A bioinformatics program tool for protein alignments. ►UniGene; Wasmuth JD, Blaxter ML 2004 BMC Bioinformatics 5:187.

Proteus Syndrome: Involves gigantism of parts of the body probably caused by lipomatosis (abnormally large local fat accumulation). The genetic control is unclear. ►PTEN; Cohen MM Jr 1993 Am J Med Genet 47:645.

Prothallium: The haploid gametophyte generation of ferns.

Prothrombin Deficiency: Caused by autosomal recessive, semidominant defects in the formation of anticoagulation factor VII, Stuart factor, Christmas factor and prothrombin. The human gene for prothrombin was assigned to chromosome sites 11p11-q12. Prothrombin is normally generated in sequential reactions by prothrombinase (Bianchini EP et al 2005 Proc Natl Acad Sci USA 102:10099). These proteins have similar proteolytic properties and the synthesis of all four depends on the presence of vitamin K. The patients have a tendency of bleeding similarly to hemophiliacs. Hereditary deficiency of factor VII itself is rare but it may be fatal if bleeding affects the central nervous system. Stuart factor deficiency has symptoms similar to those in deficiency of factor VII. All of these conditions can be treated by transfusion with blood plasma. ►antihemophilia factors, ►hemophilia, ►vitamin K dependence, ►coumarin-like drug resistance

Protist: A general term for single-cell eukaryotic organisms. The *Monera* including bacteria, blue green algae, viruses are also sometimes called protists although these are prokaryotes.

Protocell: Abiotic ancestor of living cells under prebiotic conditions. ►origin of life

Protochlorophyll: The precursor of chlorophyll ($C_{55}H_{70}O_5N_4Mg$); if the magnesium is removed protophaeophytin results. The NADPH:protochlorophyllide oxidoreductases in the prolamellar body of the etioplast are required for the establishment of the photosynthetic apparatus (deetiolation) and for photoprotection in plants. ►chloroplast, ►etioplast, ►NADP, ►photomorphogenesis, ►photosynthesis; Reinbothe S et al 2003 J Biol Chem 278:800.

Protogyny: In monoecious plants, the stigma is receptive before the pollen is shed. ►protandry, ►monoecious, ►stigma, ►self-incompatibility

Protomer: A polypeptide subunit of an oligomeric protein encoded by a cistron of a gene. ►cistron, ►oligomer

Proton: The positive nucleus of the hydrogen atom. The proton carries a positive charge equal to the negative charge of an electron but its mass is 1837 times larger.

Proton Acceptor: An anion capable of accepting protons. ►anion, ►proton

Proton Donor: An acid

Proton Pump: Mediates transport or exchange of protons across cellular membranes; energy is supplied usually by ATP or light. ►proton, ►ion pumps; Ferreira T et al 2001 J Biol Chem 276:29613.

Protonema: A filamentous stage in the formation of the gametophyte of mosses.

Protonoma: A red-color insensitive color blindness; an X-chromosomal anomaly. ►color blindness

Proto-Oncogenes: These are cellular c-oncogenes, which after genetic alteration(s) may initiate or predispose to cancerous transformation. They generally have their counterparts in oncogenic viruses (v-oncogenes). Also, they may be involved in processes of signal transduction in a variety of organisms in fungi, plants and animals. ►oncogenes, ►signal transduction, ►carcinogenesis, ►tumor suppressors, ►cell cycle

Protoperithecium: ►ascogonia, ►perithecium

Protoplasia: Formation of a new tissue.

Protoplasm: The viscous "live" content of the eukaryotic cell. ►cytoplasm

Protoplast: A cell surrounded by the cell membrane but stripped of the cell wall, generally by a combination of pectin and cellulose digesting enzymes. Protoplasts under appropriate conditions may be regenerated into normal cells and intact plants (see Fig. P155). The bacterial protoplasts are generally called spheroplasts and may have some parts of the cell wall still attached. ►cellulase, ►macerozyme, ►pectinase

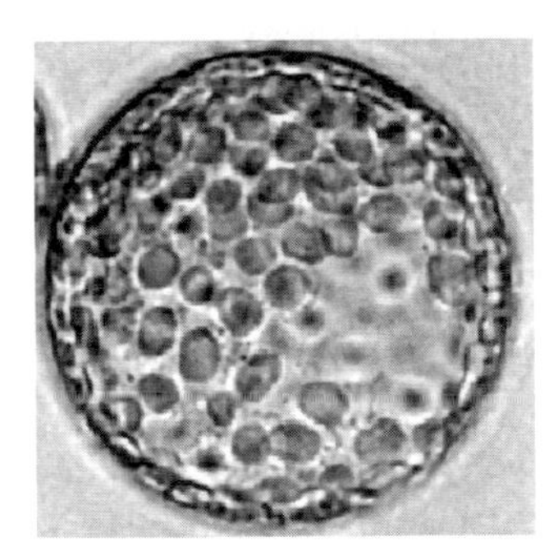

Figure P155. Plant protoplast (Durand J et al 1973 Z. Pflanzenphys. 69:26)

Protoplast Fusion: Protoplasts may fuse in the presence of polyethylene glycol (and some other agents). The fusion may take place within sister cells or with the cells (protoplasts) of any taxonomically distant organisms such as mammalian and plant cells (see Fig. P156). These somatic hybrids, unlike the zygotes derived from the fusion of eggs and sperm, contain all the contents of the two cells, nuclei and cytoplasm, although some cytoplasmic organelles may be lost eventually.

In certain rodent-human cell hybrids even the human chromosomes may be eliminated; similar observations are available for carrot and parsley cell hybrids. When the genetic differences between the fused protoplasts is large, the fused cells may not divide or may not divide continuously. Somatic hybrids between related species may, however, behave like allopolyploids and form fertile or sterile hybrids after regeneration. Fusion of animal cells with bacterial spheroplasts is shown in Figure P157. ►cell fusion, ►polyethylene glycol

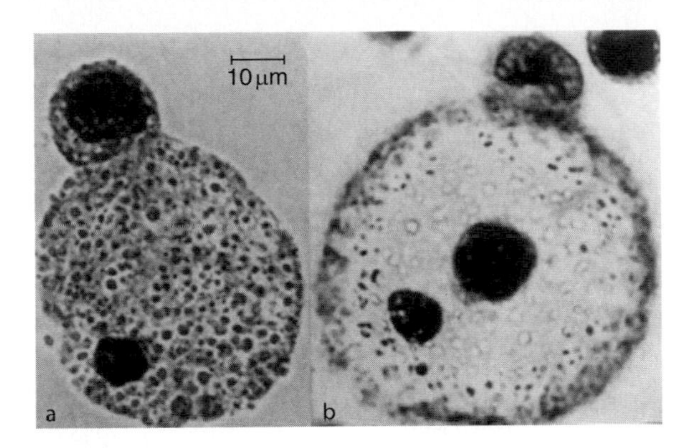

Figure P156. Human HeLa cells attached to tobacco protoplast (a), the HeLa nucleus (larger) inside the tobacco cell (b). (From Jones CW et al Science 193:401)

Protoporphyria, Erythropoietic: An autosomal (human chromosome 18q21.3) dominant (or recessive) disease involving light-sensitive itching, inflammation of the skin. The porphyrin level of the blood may increase by over 16-fold, to 1 g/100 mL. The excess protoporphyrin is deposited in the liver, causing potentially serious damage. The basic defect probably involves a deficiency (10 to 25%) of the mitochondrially located ferrochelatase (FECH). ►light-sensitivity defects, ►mitochondrial disease in humans, ►porphyria; Todd DJ 1994 Brit J Derm 131:751.

Protoporphyrin: The organic part of heme consisting of four pyrroles joined by methylene bridges. ►heme

Protosilencer: On its own, it is incapable of silencing gene(s) or its silencing effect is minimal but it can reinforce and maintain the function of silencers. ►silencers

Protosplice Site: Evolutionarily, the original splice site which is frequently AAG/CAG|GT where | is the insertion site. ►splicing

Protostome: Organisms that develop the mouth from the blastopore such as annelids, molluscs, arthropods. ►blastopore

Prototroph: A genotype that has wild type nutritional requirement. ►autotroph, ►auxotroph

Protozoa: Unicellular animals, mainly free-living (such as the *Paramecia*) some are, however, parasitic

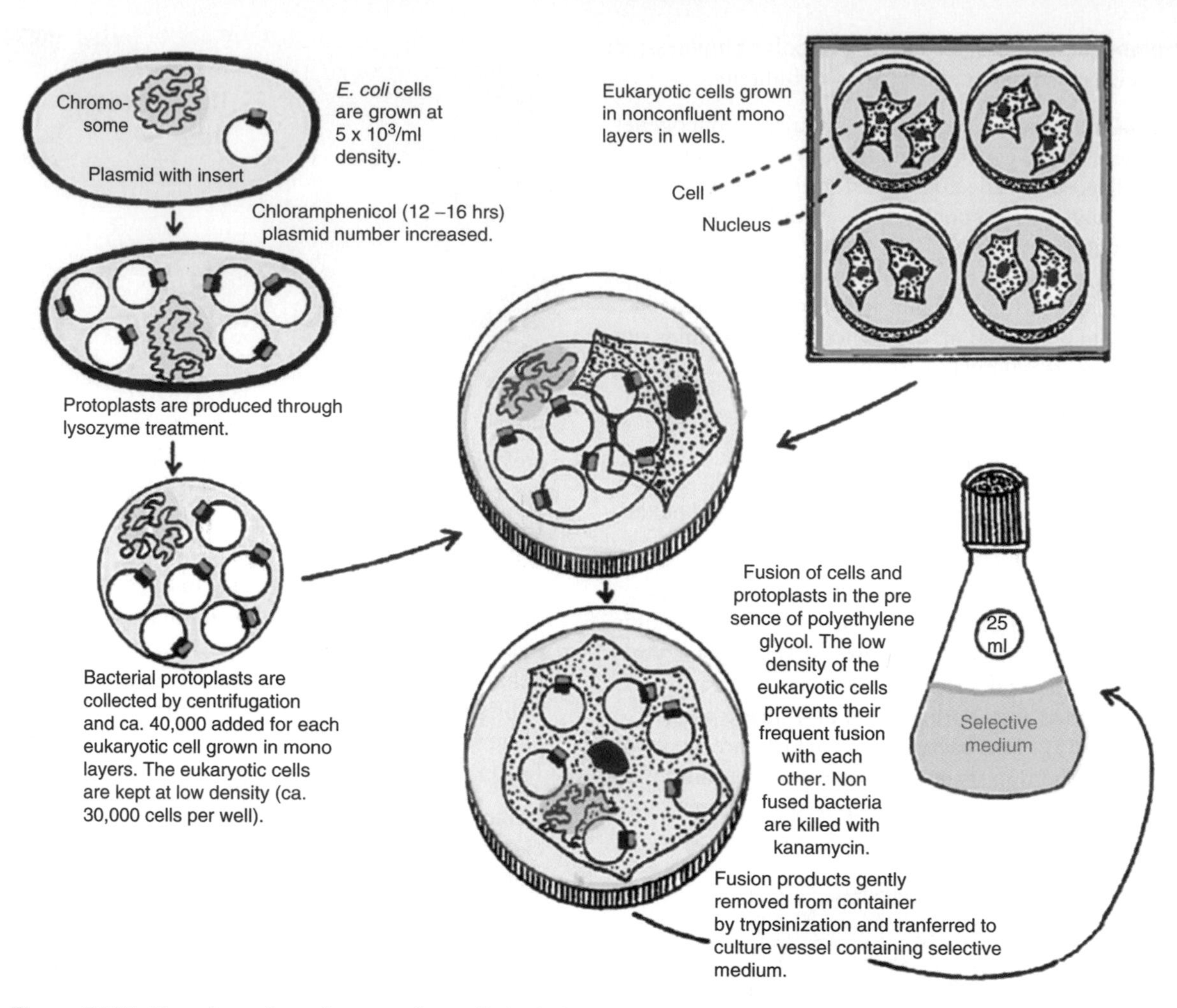

Figure P157. Transformation of mammalian cells by fusion to bacterial spheroplasts. (Modified after Sandri-Goldin, RM et al 1983 Methods Enzymol 101:402)

(such as the *Giardias* which frequently contaminate drinking water sources), the *Trypanosomas* and *Leishmanias* which cause potentially lethal infections in animals and humans. ►*Trypanosoma*, ►*Leischmania*; for the genetic nomenclature of *Tetrahymena* and *Paramecia* see Genetics 149:459; micro- and macronuclear genes: http://oxytricha.princeton.edu/dimorphism/database.htm.

Provenance/Provenience: The origin of a genetic stock. ►accession

Provirus: A DNA sequence in the eukaryotic chromosomal DNA that is a reverse transcriptase product of a retroviral RNA. ►retroviruses, ►reverse transcription, ►prophage

Proximal: Situated in the vicinity of a reference point; e.g., a gene near the centromere is proximal, versus another that is in the direction of the telomere, and thus called distal. In conjugational transfer of bacteria the marker that is transferred before another is the proximal. ►centromere, ►telomere, ►conjugation mapping

Proximal Mutagen: A chemical that has been activated into a mutagenic substance; it may not have reached yet its most reactive state. ►promutagen, ►activation of mutagens, ►ultimate mutagen, ►chemical mutagens, ►activation of mutagens

Proximity Ligation: A protein analysis technique using specific DNA sequences, which bind specific proteins. Sensitivity is much enhanced when polyclonal or monoclonal antibodies are used in connection with oligonucleotide extensions brought in the proximity of the target. ►antibody; Gullberg M et al 2004 Proc Natl Acad Sci USA 101:8420.

PRP: An RNA-splicing factor component of the U snRNP complex. ►splicing

PrP (protease resistant protein): ►prion

Prp73: A mammalian chaperon binding to the first 20 residues (S peptide) of ribonuclease A and stimulates the uptake of polypeptides by lysosomes. ►ribonuclease A, ►Hsp70, ►lysosome

Prp20p: The yeast homolog of RCC1. ►RCC

PrPres (PrP*): A partially protease resistant aggregate of PrP^C and PrP^{SC}. ►prion

PrP-SEN: The general name of the protease-sensitive prion protein. ►prion

PRR: Post-replication repair. ►DNA repair

PRR: Positive regulatory region. ►negative regulation, ►*Arabinose* operon

PRR: Pathogen recognition receptor.

PRTF: Pheromone receptor transcription factors, cooperating with GRM (general regulator mating factor) in the determination of mating type. ►pheromone, ►mating type determination in yeast, ►*Schizosaccharomyces pombe*; Tan S, Richmond TJ 1990 Cell 62:367.

Przewalsky Horse: The Mongolian wild horse but can be found (~1200) only in captivity, although its reintroduction into the wild in Mongolia and China is underway. All existing individuals have descended from the 13 animals captured about a century ago. Its chromosome number is 2n = 66 yet it makes viable hybrids with the domesticated species. ►horse

PSA (prostate-specific antigen): A M_r 33,000 kallikrein type protease glycoprotein (APS) encoded at human chromosome 19q13. High levels of this protein in the serum may be an indication of prostatic carcinoma. The level of PSA varies a great deal and it is high after ejaculation and may provide false positive indication of cancer. It may serve as a target for cancer gene therapy. The six-transmembrane epithelial antigen of the prostate (STEAP, 7p22.3) is also elevated in prostate cancer. Hepsin (transmembrane serine protease) and pim-1 (serine/threonine kinase) levels are strongly correlated with prostate cancer as detected by tissue microarray analysis. The prostate-specific membrane antigen (PSMA) is highly expressed in prostate cancer cells and in other solid tumors. It is a glutamate carboxypeptidase and cuts methotrexate and the neuropeptide *N*-acteyl-L-aspartyl-L-glutamate; its crystal structure may facilitate drug development (Davis MI et al 2005 Proc Natl Acad Sci USA 102:5981). ►prostate cancer, ►cancer gene therapy, ►tissue microarray; Berry MJ 2001 N Engl J Med 344:1373; Dhanasekara S et al 2001 Nature [Lond] 412:822.

PSD-95: A family of membrane associated guanyl kinases; they also anchor K^+ channels by their PDZ domains. ►ion channels, ►GTP

PSE: Proximal sequence element. ►Hogness box

PSE: Pale soft exudative meat is controlled in pigs by the *Halothane* gene.

Pseudoachondroplasia: A dominant human-chromosome 19p12-p13.1 gene mutation controlling the cartilage oligomeric matrix protein (COMP), and it is responsible for short stature (see Fig. P158). ►achondroplasia, ►multiple epiphyseal dysplasia, ►COMP; Hecht JT et al 1995 Nature Genet 10:325; Briggs MD, Chapman KL 2002 Hum Mut 19:465.

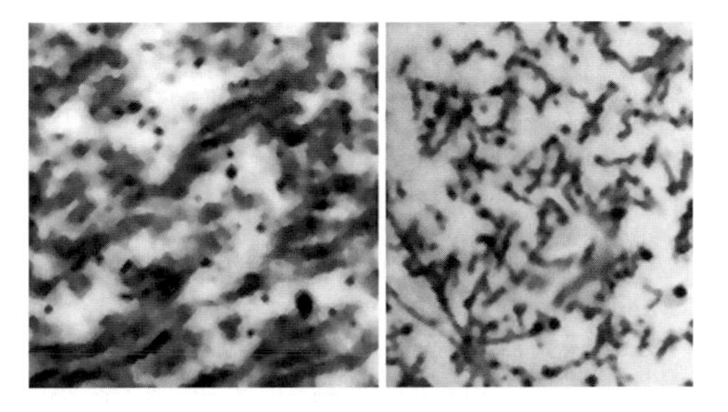

Figure P158. Left: Pseudoachondroplasiac, Right: Normal extracellular cartilage matrix

Pseudoaldostertonism (Liddle syndrome): A human chromosome 4 hypertension associated with hypoaldosteronism, hypokalemia, reduced renin and angiotensin. ►aldosteronism, ►hypokalemia, ►renin, ►angiotensin

Pseudoalleles: A cluster of not fully complementing genes, separable by recombination. Pseudoalleles, e.g., a^1 and a^2 when heterozygous in trans position $a^1\ a^+ // a^+\ a^2$ show mutant phenotype whereas in cis position $a^1\ a^2 // a^+\ a^+$ are complementary (wild type), except when dominant alleles are involved. Since these alleles are closely linked, in order to be able to prove that recombination takes place (rather then mutation), the pseudoalleles must be genetically marked by flanking genes within preferably less than 10 m.u. apart of the locus. ►complex locus, ►step allelomorphism, ►morphogenesis in *Drosophila*, ►cis-trans test, ►SSNC; Carlson EA 1959 Quart Rev Biol 34:33.

Pseudoaneuploid: The chromosome number appears aneuploid but it is not truly the case only, centromere fusion or misdivision of the centromeres have caused the changes in numbers. ►Robertsonian translocation, ►misdivision, ►B chromosomes

P

Pseudoautosomal (PAR): Genes located in both telomeric regions of the X and Y chromosomes (~2.6 Mbp at the short arm [PAR1] and a similar PAR2 site in the long arm in the human genome) where recombination can take place and consequently, despite the sex-chromosomal location, sex-linkage is not obvious. A gene for schizophrenia was suggested to be pseudoautosomal. *SYBL1*, encoding a synaptobrevin-like protein is present in both X and Y chromosomal PAR regions and it displays lyonization in the X-chromosome and inactivation in the Y. The pseudoautosomal boundary is apparently spanned by one or another (depending on the species) 5′- or 3′-truncated gene. The short stature gene (SHOX1/SHOXY), the Leri-Weill dyschondrosteosis and a Hodgkin disease gene are all located in the PAR at Xpter-p22.32. The SHOX2 gene is at 3q25-q26.

All human and chicken homologues of the snake Z-linked genes were located on autosomes, suggesting that the sex chromosomes of snakes, mammals, and birds were all derived from different autosomal pairs of the common ancestor (Matsubara K et al 2006 Proc Natl Acad Sci USA 103:18190). ►autosome, ►sex determinations, ►differential segment, ►holandric genes, ►syntagmin, ►lyonization, ►IL-9, ►Hodgkin disease, ►short syndrome; Ciccodicola A et al 2000 Hum Mol Genet 9:395; Cormier-Daire V et al 1999 Acta Paediatr 88 (Suppl):55.

Pseudobivalent: The chromosomes associated are not homologous. ►synapsis, ►ille-gitimate pairing

Pseudoborder: DNA sequences in certain agrobacterial vectors or within the cloned foreign DNA and may cause deletions and rearrangements within the T-DNA inserts in the transgenic plants. ►T-DNA, ►transformation genetic

Pseudocentromeric: ►supernumerary marker chromosome

Pseudocholinesterase Deficiency (CH1, BCHE): A dominant (human chromosome 3q26.1-q26.2) breathing difficulty (apnea) after treated with the muscle relaxant suxa-methonium (succinylcholine chloride), a drug used for intubation, endoscopy, cesarean section, etc., as an adjuvant to anesthesia. Several allelic forms respond differently to drugs. Individuals with a defective enzyme may be particularly sensitive to cholinesterase inhibitor insecticides (parathion). The frequency of the gene varies a great deal in different populations. In Eskimos, the frequency of the gene controlling the deficiency may be higher than 0.1; in other populations it may be less than 0.0002. The BCHE2 form was assigned to 2q33-35 and the same enzyme was suggested to 16p11-q23.

Pseudodiploidy: Retroviral particles because after infection only a single provirus is detected in the host. Normally retroviruses carry two RNA genomes associated by base pairing at several sites, particularly at the 5′end. It is assumed that the two copies are maintained for the purpose of assured survival and possible repair by recombination. They also contain tRNAs that prime replication. Other RNAs (5S, 7S and cellular mRNA fragments) may also be included. ►retroviruses

Pseudodominance: When a heterozygote loses the dominant allele, the recessive allele is uncovered (expressed) because of the lack of the dominant allele. Treating heterozygotes with mutagens (e.g., ionizing radiation) that cause deletions can readily induce pseudodominance. Before such experiments are conducted, it is advisable to place flanking genetic markers to the chromosome carrying the recessive markers to be able to rule out recombination and reversions. Segregation after somatic recombination may be a common cause of pseudo-dominance. Loss of heterozygosity is a frequent cause of oncogenic transformation. Pseudodominance-like phenomenon occurs in a population when the mating is between some cryptic heterozygotes. ►deletion, ►LOH, ►segregation, ►oncogenic transformation, ►mitotic crossing over

Pseudoextinction: The disappearance of a species by evolution into another form.

Pseudogamy: Apomictic or parthenogenetic reproduction. ►apomixia, ►parthenogenesis

Pseudogene: Has substantial homology with (clustered) functional genes of eukaryotes but it is inactive because of numerous mutations that prevent its full expression and may no longer available for transcription. Some pseudogenes are transcribed but the transcript is degraded by nonsense-mediated mRNA decay (Mitrovich QM, Anderson P 2005 Current Biol 15:963). Of the 201 pseudogenes identified by the 2007 ENCODE project, 20% were found to be transcribed (Zheng D et al 2007 Genome Res 17:839). Although pseudogenes may not have a protein product they may regulate the expression of their normal homolog either by stabilizing the normal transcript by blocking an RNase or by competitively inhibiting a transcriptional repressor (Hirotsune S et al 2003 Nature [Lond] 423:91). The number of pseudogenes is variable in different species. The human genome may contain 20,000 pseudogenes. Organisms with small genomes (e.g., *Drosophila*) have very few and it appears that some organisms eliminated from their genome the DNA sequences that are no longer functional. Pseudogenes may make difficult the estimation of the number of genes on the basis of incomplete sequences and lack of functional information. Pseudogenes originated either from duplication or in case of processed pseudogenes (without intron) by reverse transcription. Their nucleotide sequence is rather well conserved indicating functional significance. Paired-end diTAG (PET) analysis may permit their detection (Ruan Y et al 2007 Genome Res 17:828). ►C-value paradox, ►gene relic, ►processed pseudogene, ►duplications, ►mRNA surveillance, ►paired-end diTAG; Harrison PM et al 2001 Nucleic Acids Res 29:818; Avise JC 2001 Science 294:86; Echols N et al 2002 Nucleic Acids Res 30:2515; Balakirev ES, Ayala FJ 2003 Annu Rev Genet 37:123; human pseudogenes-gene conversion targets: http://genome.uiowa.edu/pseudogenes/.

Pseudohairpin: The overall structure is folded back yet there is not full complementarity along the strands (see Fig. P159).

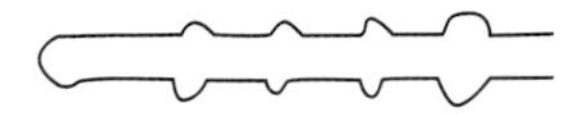

Figure P159. Pseudohairpin

Pseudohemophilia: A bleeding disease, distinct from hemophilia; it is caused by some abnormalities of the platelets. ►hemophilia, ►platelet anomalies, ►hemostasis

Pseudohermaphroditism: ►hermaphrodite

Pseudohermaphroditism, Male: It is determined by a gene in human chromosome 17q12-q21. It is responsible for the deficiency of 17-ketosteroid reductase/17-β-hydroxysteroid dehydrogenase and consequently for feminization in prepubertal males and gynecomastia and virilization after puberty when usually the enzyme is expressed. The affected individuals may be surgically assisted to develop into sterile female phenotype (by removal of the hidden testes) or into male phenotype by reconfiguration of the external male genitalia. Infertility, however, cannot be corrected,. Recessive mutations in the luteinizig hormone receptor gene (LHB, 19q13.32) may also be responsible. The condition may be due to deficiency of steroid 5-α-reductase (SRD5A2, 2p23). The SRDA1 isozyme encoded at 5p15 does not appear to be involved in this disorder. The afflicted XY individuals may have blind vagina and a rudimentary hypospadiac penis but no gynecomastia. They may produce viable sperm although they may sire offspring only by intrauterine insemination because of underdeveloped prostate and seminal vesicles. Several defects in steroid biosynthesis may cause male pseudohermaphroditism. The 17,20 desmolase deficiency is most likely X-chromosome linked. Lipoid adrenal hyperplasia (8p11.2) responsible for complex defects in cortisol or aldosterone may cause even life-threatening conditions. Luteinizing hormone/choriogonadotropin receptor (LHCCGR, 2p21) may cause abnormalities of the Leydig cell differentiation in XY and possibly in XX individuals. Methemoglobinemia and deficiency of cytochrome b5 (18q23) may also cause pseudohermaphroditism. ►gynecomastia, ►polycystic ovarian cancer, ►hermaphroditism, ►infertility, ►testicular feminization, ►luteinization, ►Müllerian ducts, ►anti-Müllerian hormone, ►Reifenstein syndrome, ►hypospadias, ►Wilms tumor, ►adrenal hyperplasia, ►adrenal hypoplasia, ►androgen-insensivity, ►methemoglobin, ►cytochromes

Pseudohitchhiking: Adaptive mutations near neutral loci may simulate genetic drift. ►hitchhiking

Pseudohomothallism: In the fungus (e.g., *Podospora anserina*) binucleate ascospores are formed and each spore contains both mating types and is thus self-fertile. ►homothallism, ►heterothallism

Pseudo-Hurler Syndrome: ►mucolipidosis

Pseudohypha (in *Saccharomyces cerevisiae*): The formation occurs by deficiency of nutrient (N) and may cause polarized growth on the surface of the agar medium favoring delay in mitosis and precocious entry into meiosis. The pseudohyphal growth is symmetric and synchronous in comparison to the regular budding that is asymmetric and asynchronous. Cyclins 1 and 2 promote pseudohyphal growth whereas cyclin 3 is inhibitory in yeast. Alternative controls exist. Protein Ste12, the MAP kinase signal transduction pathways also regulate hyphal growth. Filamentous growth is a requisite for pathogenicity of *Ustilago maydis* and *Candida albicans*. ►cyclin, ►CDK, ►*Ustilago maydis*, ►candidiasis, ►MAP, ►Ste

Pseudohypoaldosteronism (PHA; 1q31-q42, 17p11-q21, 12p13, 16p13-p12): hyperkalemic, hyperchloremic acidosis and hypertension. The genes at chromosomes 17 and 1 encode a threonine/serine kinase, WNK4, localized in the tight junctions. The disease in this protein is due to missense mutations. Mutations in WNK4 in mice cause higher blood pressure, hyperkalemia, hypercalciuria and marked hyperplasia of the distal convoluted tubule (DCT). WNT4 (chromosome 17) regulates the balance between NaCl reabsorption and K^+ secretion (Lalioti MD et al 2006 Nature Genet 38:1124) by altering the mass and function of the DCT through its effect on NCC (Na/Cl co-transporter). In chromosome 12 the cytoplasmic WNK1 is encoded and the defect is due to large intronic deletions that boost the expression of the protein. Both of these proteins are in the distal nephron (a basic morphological and functional unit of the kidney) that is responsible for potassium and pH homeostasis. These two anomalies are dominant. The recessive PHA in chromosome 16 encodes subunits of an epithelial Na^+ ion channel. ►aldosteronism, ►Gordon syndrome, ►hypoaldosteronism, ►hyperkalemic, ►hypertension, ►intron, ►ion channels; Wilson FH et al 2001 Science 293:1107.

Pseudohypoparathyroidism: ►Albright hereditary osteodystrophy

Pseudoknot: Formed when a stem-and-loop RNA structure is bound at the base of the loop by hydrogen bonds or by a ligand resulting in a two-stem two-loop stacking (see Fig. P160). The actual configurations of the pseudoknots may vary. Pseudo-half-knots form only a single loop. Such structures may modulate RNA functions and can be exploited also in designing highly selective drugs. Some insect RNA viruses, which use CAA (glutamine) rather than AUG (methionine) for translation initiation do not require an initiator tRNA but apparently rely on a pseudoknot formed between a 15–43 nucleotide upstream loop and the sequence immediately preceding the CAA codon. Pseudoknot structure is highly conserved in telomerases. Mutations that disrupt the pseudoknot helix abolished telomerase activity whereas intraloop hairpin base-pairing did not reduce telomerase activity (Chen J-L, Greider CW 2005 Proc Natl Acad Sci USA 102:8080). Pseudoknots initiate translational frameshifting in overlapping genes. The Pseudoknot Local Motif Model and Dynamic Partner Sequence Stacking (PLMM_DPSS) algorithm, which predicts all PLM model pseudoknots within an RNA sequence in a neighboring-region-interference-free fashion. The PLM model is derived from the existing Pseudobase (collection of pseudoknots) entries and it is most sensitive. The innovative DPSS approach calculates the optimally lowest stacking energy between two partner sequences (Huang X, Ali H 2007 Nucleic Acids Res 35:656). ►repeat inverted, ►antisense RNA, ►overlapping genes, ►TFO, ►telomerase, ►frameshifting ribosomal; Kim Y-G et al 1999 Proc Natl Acad Sci USA 96:14234; Xayaphoummine A et al 2003 Proc Natl Acad Sci USA 100:15310; pseudoknot folding: http://bibiserv.techfak.uni-bielefeld.de/pknotsrg/.

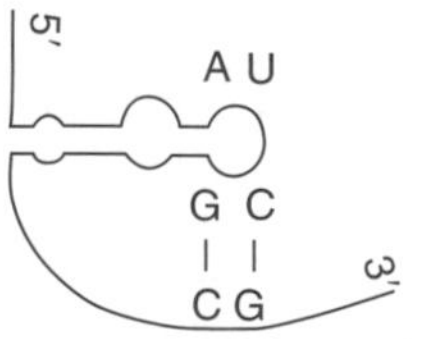

Figure P160. Pseudoknot

Pseudolinkage: The linkage due to translocation between non-homologous chromosomes. ►affinity

Pseudolysogen: Lyses the bacterial cells so slowly as if it would be lysogenic. ►lysogeny

Pseudomonas: *Pseudomonas* bacteria include several species that degrade oil spills, polycyclic hydrocarbons, benzene and other pollutants. ►oil spills, ►biodegradation; Coates JD et al 2001 Nature [Lond] 411:1039.

Pseudomonas aeruginosa: A 6.3 million-bp bacterium and an opportunistic human parasite. It is the most common cause of death in cystic fibrosis but it is involved in some pneumonias and other infections (urinary tract, burn victims, etc.). It grows also on soil and plant and animal tissues. This Gram-negative bacterium is highly resistant to antibiotics and disinfectants. Close to 10% of its genes is regulatory and the large number of its putative pump proteins explains its resistance to drugs. ►cystic fibrosis; Stover CK 2000 Nature [Lond] 407:959; genome, annotations: http://www.pseudomonas.com; http://www.systomonas.de.

Pseudomonas Exotoxin: Kills by irreversible ribosylation of ADP and subsequent inactivation of translation elongation factor, EF-2. Its applied significance is the potential for cancer therapy. ►toxins

Pseudomonas syringae: A plant pathogenic relative of *P. aeruginosa*. The 6.5 megabase sequenced genome includes a circular chromosome plus two plasmids including 5763 open reading frames of which 298 are putative virulence genes. This bacterium may promote secondary infection by the same pathogen rather than display a hypersensitive response in the host by a jasmonic acid structural mimic (coronatine). It can increase susceptibility also to herbivorous insects without relying on coronatine (Cui J et al 2005 Proc Natl Acad Sci USA 102:1791). Related *Pseudomonas* subspecies, distinguished by host-specificity, display differences in genes of antibiotic resistance, DNA repair and ectoin ([4S]-2-methyl-1,4,5,6-tetrahydropyrimidine-4-carboxylacid), a natural protective agent against external harmful effects (Feil H et al 2005 Proc Natl Acad Sci USA 102: 11064). ►ORF, ►host–pathogen relation; Buell CR et al 2003 Proc Nat Acad Sci USA 100:10181.

Pseudomonas tabaci: A bacteria causing "wildfire" disease (necrotic spots) on tobacco leaves (see Fig. P161). The symptoms may be mimicked by methionine sulfoximine, a methionine analog.

Figure P161. Wildfire disease spots (Courtesy of Dr. Peter. Carlson)

Pseudomosaic: May occur in a sample of amniocentesis caused by the conditions of culture rather than the genetic/chromosomal condition of the fetus.

Pseudo-Overdominance: Certain phenotype(s) may appear in excess of expectation in a population because of the close linkage of the responsible gene to advantageous alleles. Also QTL loci may appear overdominant if they are relatively closely linked and display heterosis because the QTL mapping techniques cannot determine the map positions with great accuracy, and the molecular function of the genes involved is not known. ►overdominance, ►fitness, ►QTL, ►interval mapping, ►hitchhiking

Pseudopilus (Ψ-pilus): A bacterial appendage (~55 nm) that may extend beyond the cell surface into the periplasm and may be a conduit for macromolecular transport in bacteria. ►pilus, ►DNA uptake

Pseudoplasmodium: A migrating slug of cellular slime molds. ►*Dictyostelium*

Pseudopodium: ►amoeba

Pseudopregnant: Female (mice) mated with vasectomized males and then implanted with blastocyst stage embryos derived from other matings. ►vasectomy, ►allopheny

Pseudoqueen: In social insects (bees, ants, termites) one worker (XX) may become fertile pseudoqueen after the loss of the queen of the colony. This type of development is promoted by special feeding (royal treatment) of the originally worker caste insects. ►honey bee

Pseudorecombinant: Reassortement of two viral genome components from different viruses, transmitted by the same insect vector.

Pseudoreplication: The samples are not independent replicates and the conclusion based on them may not be statistically reliable.

Pseudoreversion: An apparent back mutation caused by an extra-site suppressor mutation. ►reversion

Pseudorheumatoid Dysplasia: A rare recessive cartilage defect due to mutation in the cysteine-rich secreted protein gene family (see Fig. P162). ►arthritis, ►rheumatic fever

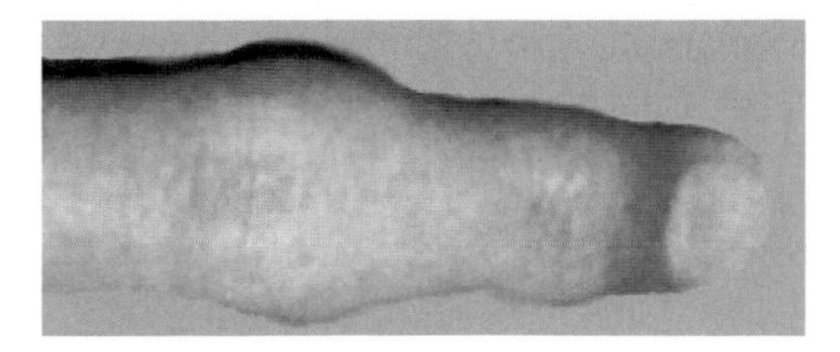

Figure P162. Swollen joints of a finger in pseudorheumatoid dysplasia

Pseudosubstrate: A molecule with similarity to an enzyme substrate but it is actually an inhibitor, and special regulators are required for its removal so the enzyme is permitted to access its true substrate. ►substrate, ►intrasteric regulation

Pseudotemperate Phage: It has a lysogenic cycle yet does not have a stable prophage state, e.g., the PBS1 transducing phage of *Bacillus subtilis*. ►lysogeny, ►prophage

Pseudotransduction: The virus is not integrated into the chromosome and the passenger DNA can be expressed only from the cytoplasm when the appropriate promoter is present in the vector.

Pseudo-Trisomic: It is actually disomic but one of the chromosomes is represented by two telocentric chromosomes, each represent one and the other arm of the same chromosome, thus two telocentrics + one normal chromosome occurs. ►trisomy, ►telocentric chromosome

Pseudotype: The virus carrying foreign protein on his envelope and may expand the normal host range.

Pseudotyping: If two types of viruses invade the same cell, genetic material of one may slip into the capsid of the other and this type of packaging permits the introduction of the viral genome into a host, which otherwise would be incompatible with the virion. This phenomenon may be taken advantage of also during the construction of viral vectors and helper viruses. The ability of a virus to infect a certain type of cell depends on the interaction between the viral glycoprotein and the nature of the cell surface receptors. The vesicular stomatitis virus viral envelope glycoprotein (VSV-G) is highly fusigenic for a wide range of cell types and organisms. Thus, it can be employed for pseudotyped viral vectors to expand their effective host range. The hemagglutinating paromyxovirus of Japan (HVJ) and other viruses can also be used similarly. ►pseudovirus [pseudovirion], ►amphotropic, ►ecotropic, ►packaging cell lines, ►retroviral vectors; Mazarakis ND et al 2001 Hum Mol Genet 10:2109; Peng KV et al 2001 Gene Ther 8:1456.

Pseudouridine (ψ): A pyrimidine nucleoside (5-β-ribofuranosyluracil) occurs in the T arm of tRNA by post-transcriptional modification of a uracil residue. Pseudouridine has been also found in ribosomal RNAs and snRNAs. The modification is mediated by the nucleolar ψ synthase with the assistance of other proteins. A requisite for the process is that a small nucleolar RNA (snoRNA) carrying a single stranded H box (ANANNA) and a ACA-3′ Box would pair with the target RNA at about 12 or less region of complementarity. After the enzyme gained access to the U site, the N1—C1′ bond in a uracil is severed and after a 180° rotation the C5 position becomes available for the formation of a new bond. Thus the N1 and N3 sites may become readily available for hydrogen pairing and pseudouridine can bind easier in inter- or intramolecular reactions. Pseudouridine deficiency is not lethal in yeast yet it adversely affects growth. The crystal structure and function of the H/ACA ribonucleoprotein, a member of pseudouridine synthases, has been determined (Li L, Ye K 2006 Nature [Lond] 443:302). ►ψ for formula, ►tRNA, ►snoRNA; Bortolin M-L et al 1999 EMBO J 18:457; Hoang C, Ferré-D'Amaré AR 2001 Cell 107:929.

Pseudovirion (pseudovirus): Contains non-viral DNA within the viral capsid and can thus be used to unload foreign DNA into a cell if a helper virus is provided. ►virion, ►capsid; Liu Y et al 2001 Appl Microbiol Biotechnol 56:150; Ou WC et al 2001 J Med Virol 64:366.

Pseudowild Type: Displays wild phenotype because a mutation at a site different from the mutant locus that it masks, but most commonly a duplicated segment, compensates for the original and still present recessive mutation. In *Neurospora* it occurs at much higher frequency than expected by back mutation. It may also be due to a suppressor mutation. (See Mitchell MB et al 1952 Proc Natl Acad Sci USA 38:569).

Pseudoxanthoma Elasticum (PXE, 16p13.1): Autosomal recessive or dominant disorders of an ABCC6 (multiple drug resistance) transporter causing by degenerative changes in the skin (peau d'orange = orange rind), veins, eyes, intestines, etc., resulting in heart disease and hypertension. The defect involves dysplasia of elastin fibers and it affects the skin, retina, arteries, teeth, etc. ►coronary heart disease, ►hypertension, ►skin diseases, ►ABC transporters; Le Saux O et al 2001 Am J Hum Genet 69:749; problems of translation and advocacy of research models: Terry SF et al 2007 Nature Rev Genet 8:157.

Pseudo-Zellweger Syndrome: ►peroxisomal 3-oxoacyl-coenzyme A thiolase deficiency, ►Zellweger syndrome

PSI (ψ): Pseudouridine, and also the packaging signal in retrovirions. ►tRNA pseudouridine loop, ►retrovirus, ►retroviral vectors, pseudouridine formula on page at ►ψ

PSI$^+$: A yeast prion, an extrachromosomal protein suppressing nonsense codons. It functions in collaboration with the nuclear genes *SUP*35. Overexpression of this gene induces the formation of PSI$^+$ probably by a conformational change in the protein. Cells deleted in the amino-terminal region of Sup35 are resistant to PSI$^+$. Expansion of imperfect oligopeptide repeats in Sup35 (PQGGYQQYN) and in PrP (PHGGWGQ) seems to be responsible for the abnormality. Overexpression of Hsp104 heat shock protein cures the cells from PSI$^+$. ►prion, ►Hsp; Masison DC et al 2000 Curr Issues Mol Biol 2:51; Jensen MA et al 2001 Genetics 159:527.

Psi Vector: ►E vector

PSI-Blast: ►Blast

PsnDNA: 150–300 bp pachytene DNA sequences flanking 800–3000 bp internal chromosomal segments in eukaryotes, and the two short and the central DNA sequences are called PDNA (pachytene DNA). The PsnDNAs are supposed to be nicked by an endonuclease after homologous small nuclear RNA (snRNA) and a non-histone protein (PsnProtein) have opened the sequences to the action of the enzyme. These molecules appear only during late leptotene to pachytene and are assumed to be mediating recombination. ►crossing over, ►meiosis, ►snRNA, ►ZygDNA; Stern H, Hotta Y 1984 Symp Soc Exp Biol 38:161.

Psoralen Dye: Can combine with the DNA connecting nucleosomal core particles. After irradiation with near-ultraviolet light, cross-link between the two DNA strands occurs. Psoralen-conjugated triple helix forming oligonucleotides have been used to induce site-specific mutations in COS cells at very high frequency (see Fig. P163). Targeting psoralen cross-links with triple helix forming oligonucleotides can induce in base substitution and deletion mutations in mammalian cells. Deficiencies in non-homologous

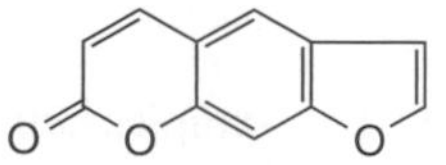

Figure P163. Psoralen

end-joining and mismatch repair did not influence the mutation pattern. In contrast, the frequency of base substitutions depended on ERCC1 and DNA polymerase ζ but it was independent of nucleotide excision repair and transcription-coupled repair genes (Richards S et al 2005 Nucleic Acids Res 33:5382). Some celery stocks may contain higher than normal amounts of psoralen. ►triple helix formation, ►site-specific mutation, ►COS cell, ►DNMA repair, ►DNA polymerases; Cimino GD et al 1985 Annu Rev Biochem 54:1151; Luo Z et al 1997 Proc Natl Acad Sci USA 97:9003; Oh DH et al 2001 Proc Natl Acad Sci USA 98:11271.

Psoriasis (PSOR): A scaly proliferation of keratinocytes, a type of autoimmune skin defect determined either by dominant gene(s) of reduced penetrance or polygenic inheritance involving relatively few genes. Its incidence is common in Caucasian populations (1–3%) but it is much less frequent in Orientals (Eskimos, American Indians and Japanese) and it was almost absent in Africa. Recurrence rate may vary (8–23% among first-degree relatives), depending on the type involved. If both parents are affected the recurrence among children may reach up to 75%. Concordance among monozygotic twins is 35% to 72% and only 12% to 23% in fraternal twins (Duffy DL et al 1993 J Am Acad Dermatol 29:428. The psoriasis haplotype appears to include HLA-BW 17, HLA-C and HLA-A 13 genes. Some observations indicate that bacterial superantigens may trigger psoriasis. Psoriasis-like skin disease and arthritis may be caused by epidermal deletion of Jun proteins in mice (Zenz R et al 2005 Nature [Lond] 437:369). Psoriasis susceptibility genes have been assigned to 16q, 10q, 19p13.3, 3q21, 1q21, 17q25, 4qter, 14q31-q32, 6p21 and 20p. Runx transcription factors may stimulate it. AP1/4/08 and AIRE may cause loss of self-tolerance (Nature Genet 17:399 [1997]). Linkage with other chromosomes is less certain. Psoriasis increases the risk of basal cell carcinoma. Microarray analysis revealed upregulation of transcription of at least 161 genes in psoriasis. Some the transcripts are modulated also in other skin diseases. ►HLA, ►keratosis, ►ichthyosis, ►skin diseases, ►nevoid basal cell carcinoma, ►Hirschsprung disease, ►dermatitis atopic, ►IL-20, ►APEBEC, ►AIRE, ►Runx, ►autoimmune disease, ►Jun; Bhalerao J, Bowcock AM 1998 Hum Mol Gen 7:1537; Bowcock AM et al 2001 Hum Mol Genet 10:1793; Int Psoriasis Genet Consortium 2003 Am J Hum Genet 73:430; review: Schön MP et al 2005 N Engl J Med 352:1899; Bowcock AM 2005 Annu Rev Hum Genet 6:93; review: Lowes MA et al 2007 Nature [Lond] 445:866.

P{Switch}: ►Gene-Switch, ►hybrid dysgenesis

Psychiatric Disorder: ►psychoses

Psychomimetic: Drugs affect the state of mind in a manner similar to psychoses. ►psychoses, ►psychotropic drugs, ►ergot

Psychoses: A group of mental-nervous disorders with variable genetic and environmental components. ►autism, ►manic depression, ►schizophrenia, ►paranoia, ►affective disorders, ►attention deficit hyperactivity, ►Tourette's syndrome, ►IQ, ►dyslexia, ►panic disorder, ►bipolar mood disorder

Psychopathology: The manifestation of a neuronal disease involved in mental and behavioral illness.

Psychotherapy: The treatment/support provided for transient or lasting emotional and behavioral disorders. It may involve verbal support or chemical medication. Genetic counselors need familiarity with the verbal support option. ►counseling genetic

Psychotropic Drugs: These affect the state of mind. They are used as medicine in various types of psychoses and may be very beneficial (e.g., lithium, valium, etc.) if applied under medical monitoring. Possible adverse side effects vary by the chemical nature of the drug and may include heart disease, birth defects, addiction, etc. ►psychoses, ►psychomimetic

Psychrophiles: Organisms that grow under low temperatures. ►antifreeze proteins

PTA Deficiency Disease: Controlled by incompletely dominant (4q35) genes. Plasma thromboplastic antecedent protein deficiency is involved that results in unexpected bleeding after tooth extraction or various surgeries. Nose bleeding (epistaxis) is common but uterine bleeding (menorrhagia) or blood in the urine (hematuria) is rare. The carrier frequency in Ashkenazy Jewish populations is about 8.1%. ►antihemophilia factors, ►pseudohemophilia

PTB (phosphotyrosine-binding domain): PTB is present in proteins involved in signaling. ►SH2, ►SH3, ►WW, ►SCK, ►pleckstrin, ►signal transduction

PTB (polypyrimidine tract binding protein, 58 kDa): Involved in regulation of eukaryotic mRNA metabolism, regulation of splicing, IRES-mediated translation initiation and mRNA stability. ►splicing, ►IRES, ►translation; Oberstrass FC et al 2005 Science 309:2054.

PTC: ►phenylthiocarbamide; also papillary thyroid carcinoma; a variant of the RET oncogene caused neoplasia. ►RET

PtdInsP$_2$: ►phosphoinositides

PTEN (phosphatase and tensin homolog; deleted in chromosome 10 [10ter-q11, 10q24-q26, 10q22-q23], syn. MMAC1 [mutated in multiple advanced cancer]): A tumor suppressor involved in brain, prostate, breast, multiple hamartomas (Lhermitte-Duclos disease/Cowden syndrome), Bannayan-Zonona syndrome and other cancers. It inhibits cell migration and cell adhesion and dephosphorylates FAK, serine, threonine and tyrosine residues in proteins. The primary target of PTEN appears to be phosphatidylinositol-3,4,5 trisphosphate (PIP3) and acts as tumor suppressor by promoting apoptosis. PTEN and p53 mutually promote each other in tumor suppression (Chen Z et al 2005 Nature [Lond] 436:725). In vivo, PTEN may act as a lipid phosphatase and this function may be essential for tumor suppression. PTEN appears to guard centromere-kinetochore integrity and chromosomal stability (Shen WH et al 2007 Cell 128:157). The protein (tyrosine, serine/threonine) phosphatase activity may not be important for tumor suppression. Some cancer cells (glioma, prostate, breast cancer) may be reverted to normalcy by the addition of PTEN. The PTEN-Akt pathway probably governs stem cell activation by helping control nuclear localization of the Wnt pathway effector β-catenin. Akt phosphorylates β-catenin at Ser552, resulting in a nuclear-localized form in intestinal stem cells (ISC). Our observations show that intestinal polyposis is initiated by PTEN-deficient ISCs that undergo excessive proliferation driven by Akt activation and nuclear localization of β-catenin (He XC et al 2007 Nature Genet 39:189).

The catalytic domain identity motif is HCXXGXXRS/T. The two α-helix domains flanking the catalytic domain are encoded in its exon 5, and must be intact for proper function. The tensin-homology domain enables the recognition of the cell adhesion system (actin, integrin, FAK, Src). In mouse the *Pten*$^{+/-}$ heterozygotes are subject to autoimmune disease and FAS-mediated apoptosis. The normal FAS function can be restored by the administration of phosphatidyl inositol 3 kinase. PTEN has an influence on cyclin D1 and signal transduction. Mutations in PTEN may be found in Proteus syndrome or Proteus-like syndrome. Experimental deletion of *Pten* (by using seven doses of polyinosine-polycytidine in *Mx-1 Cre* mice) initiated leukemia cancer stem cell as well as hematopoietic stem cell proliferation. Without *Pten*, the hematopoietic stem cells were depleted by time in a cell autonomous manner. In contrast, the leukemia stem cells became transplantable and progressed to leukemia within 4–6 weeks. Rapamycin, which targets TOR, however, depletes leukemia stem cells and rescues *Pten*-deficient hematopoietic stem cells. Thus the two types of stem cells can be distinguished (Yilmaz ÖH et al 2006 Nature [Lond] 441:475; Zhang J et al 2006 Nature [Lond] 441:518). ►tumor suppressor, ►tensin, ►FAK, ►multiple hamartomas syndrome, ►Bannayan-Zonona syndrome, ►polyposis juvenile, ►AKT, ►catenins, ►*Wingless*, ►phosphat-idylinositol, ►prostate cancer, ►PIP2, ►PIP3, ►PIK, ►TOR, ►rapamycin, ►hematopoiesis, ►stem cell, ►chemotaxis, ►apoptosis, ►wound healing, ►Parkinson disease, ►Proteus syndrome, ►NEDD; Di Cristofano A, Pandolfi PP 2000 Cell 100:387; Wen S et al 2001 Proc Natl Acad Sci USA 98:4622; Maehama T et al 2001 Annu Rev Biochem 70:247; Waite KA, Eng C 2002 Am J Hum Genet 70:829; Zhou X-P et al 2003 Am J Hum Genet 73:404.

Pteridines: Purine derivatives, involved in coloring of insect eyes, wings, amphibian skin, etc. Pteridines may be light receptors. Reduction in tetrahydrobiopterin and related amines may be responsible for nervous disorders (see Fig. P164). ►photoreceptors, ►rhodopsin, ►ommochromes, ►GTP cyclohydrolase deficiency; Blau N et al 1998 J Inherit Metab Dis 21:433.

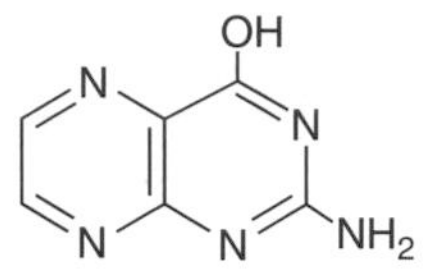

Figure P164. Pterine

Pterygium, Multiple, Syndrome (Escobar syndrome, 2q33-q34): The webbing of the neck and depressed areas (fossae) under the elbow, and other joints and hypogonadism in males, small labia and clitoris in females and other anomalies. ►hypogonadism, ►clitoris, ►Popliteal pterygium

PTG (protein targeting glycogen): Forms complexes of phosphatases, kinases and glycogen synthase with glycogen. ►glycogen, ►kinase

PTGS: Post-transcriptional gene silencing presumably by degradation of the mRNA or inactivation of infectious (viral) RNA. Recent evidence indicates the presence of a 25-nucleotide long antisense RNA in the silenced cells. ►RNAi, ►RIGS, ►methylation of DNA, ►epigenesis, ►post-transcriptional gene silencing, ►RNA surveillance

Ptilinum: An inflatable head of the larva emerging from the puparium that cyclically is inflated/deflated to pry open the puparium by a wedging type of operation.

PTIP (PAX transactivation interacting domain protein): Contains tandem BRCA1 carboxy terminal domains (BRCT) and it is responsible for phosphorylation-dependent protein binding a condition required for DNA repair. The Met^{1775} →Arg mutation in the BRCA1 (breast cancer) gene product fails to bind phosphopeptides and increases the susceptibility to cancer. ►PAX, ►breast cancer; Manke IA et al 2003 Science 3002:636.

PTK: Protein-tyrosine kinase involved in regulation of signal transduction and in growth and differentiation of cells. ►protein kinases

Ptosis: Drooping eyelid(s). ►epicanthus, ►blepharophimosis

PTP: ►tyrosine phosphatase, ►protein-tyrosine phosphatase non-receptor

PTPCR (PicoTiterPlate PCR): A DNA amplification procedure on an extremely small platform in pL quantities. Subsequently the products can be transferred to solid support and transcription, translation or sequencing can be carried out (Leamon JH et al 2004 Electrophoresis 25:1176). ►amplification, ►polymerase chain reaction, ►measurement units

PTPN (protein tyrosine phosphatase non-receptor type, PTPN22, 1p13): Associated with autoimmune diseases (diabetes I, rheumatoid arthritis, lupus, Graves thyroiditis, Addison disease, etc.). In case of a gain-of-function mutation the T cell produce lower amounts of interleukin-2 when stimulated the T cell receptors (TCR) and the phosphatase negatively regulates activation of T lymphocytes. (See Vang T et al 2005 Nature Genet 37:1317).

PTPRC: Protein tyrosine phosphatase receptor type C.

pu (particle units): Used for quantifying the number of potentially infectious virus particles per volume. It is a newer alternative for the *pfu* units but it is supposed to be employed for highly purified preparations. The particle count is determined from the absorbance at 260 nm according to the formula that 1 unit at $A_{260} = 1.25 \times 10^{12}$ particles/mL. Generally, it corresponds to 10–100 times of the *pfu* titer. ►pfu

PU.1 (PU1): A transcription factor in blood-forming cells regulating the differentiation of macrophages, B lymphocytes and monocytes; it belongs to the ETS family of oncogenes. Deletion of an upstream regulatory element (URE) leads to acute myeloid leukemia (Rosenbauer F et al 2006 Nat Genet 38:27). ►ETS, ►monocytes, ►macrophages, ►lymphocytes, ►leukemia, ►transcriptional priming; Dekoter RP, Singh H 2000 Science 288:1439; Lewis RT et al 2001 J Biol Chem 276:9550.

PubChem: Biological activities of small molecules database: http://pubchem.ncbi.nlm.nih.gov/.

Puberty: The time of sexual maturation, accompanied by the appearance of secondary sexual characteristics such as facial hair in males, enlargement of the breast in females, etc. Puberty is initiated by the secretion of gonadotropin releasing hormone by the brain that activates the release of the pituitary hormones required for gonadal functions. It is facilitated by the KiSS-1 peptide and its receptor, GPR54 (Kaiser UB, Kuohung W 2005 Endocrine 26:277). ►gonadotropin, ►pituitary, ►gonads

Puberty Precocious: Autosomal dominant disorders occur in two forms: isosexual, when sexual maturation in both males and females takes place before age 10 and 8.5, respectively but may be even much earlier, especially in females. Another form is male-limited. Testosterone production seems to be independent from gonadotropin releasing hormone production. The disorder is associated with a defect in the luteinizing hormone receptor. ►luteinizing hormone-releasing factor, ►animal hormones, ►hormonal effects on sex expression, ►G-proteins

PubGene: A human gene-to-gene co-citation index involving 13,712 named human genes. (See Jenssen T-K et al 2001 Nature Genet 28:21).

Public Blood Systems: ►private blood groups

Public Opinion: In the underdeveloped world with inadequate educational systems, superstitions greatly affect people's view on all aspects of life and society. In the culturally and technically advanced nations the newspapers, television and Internet resources may influence public opinion to a great degree. Application of scientific principles is commonly decided by plebiscites or by legislative action. In a democratic society the citizens' view must necessarily be considered. The dilemma of how well informed is the general public or the legislative/governmental system regarding the implications of scientific principles is an important problem. In a survey in England the public indicated that automobiles are safer than trains. The actual statistics indicated, however, that the safety of trains is about 100 times better. People generally believe that atomic power plants expose the public to unnecessary health and genetic risks. The hazards burning fossil fuels or using wood fireplaces are much less frequently considered although they generate carcinogenic and mutagenic emissions. Very often even the scientists are unable to predict the future consequences of the scientific achievements they brought about as it was apparent by the consensus reached on recombinant DNA by the historical Asilomar Conference. The problems of using genetically modified organisms, cloning, stem cell applications cannot be resolved by political approaches. The problems created by technology and science can be resolved only by better scientific research. ►gene therapy, ►stem cells, ►GMO, ►recombinant DNA and biohazards, ►atomic radiations, ►informed consent, ►criticism on genetics

Publication Ethics: Subject to the same common sense rules as any other principle of ethics. The detailed guidelines in Human Reproduction 2001, vol 16:1783–1788 contain specific, valid points. Fabricated data in publications have serious effect on scientific research conducted in other laboratories unaware of the misconduct but faked reports are even more critical and dangerous when they influence clinical practice and may endanger life of patients (Unger K, Couzin J 2006 Science 312:38). ►ethics, ►misconduct scientific

PubMed Central: A digital archive of peer-reviewed journals containing >300,000 full text articles. Can be entered through Entrez. ►Entrez

PUBS: Percutaneous (through the skin) umbilical blood sampling, a method of prenatal biopsy for the identification of hereditary blood, cytological and other anomalies. ►amniocentesis, ►prenatal diagnosis

pUC Vectors: Small (*pUC12/13* 1680 bp, *pUC18/19* 2686 bp) plasmids containing the replicational origin (*ori*) and the *Amp*^{r} gene of pBR322, and they carry the *LacZ′* fragment of the bacterial β-galactosidase (see Fig. P165). The *Z′* indicates that within this region there is multiple cloning site (MCS) for recognition by 13 restriction enzymes. The orientation of the MCS is in reverse in pUC18 relative to pUC19. Genes inserted into *Lac* may be expressed under the control of the *Lac* promoter as a fusion protein. Most commonly, the insertion inactivates the *Lac* gene and white colonies are formed in Xgal medium rather than blue when the gene is active. The pUC vectors can be used with JM105 and NM522 *E. coli* strains. ►vectors, ►*Xgal*, ►*Lac*, ►filamentous phages; Messing J 1996 Mol Biotechnol 5:39.

Puccinia graminis: ►stem rust

PUF Proteins: They control mRNA stability by binding to the 3′-untranslated end. (Wickens M et al 2002 Trends Genet 18:150).

Puff: The swollen area of polytenic chromosomes active in transcription. Puffing is induced by expression of transcription factor genes regulated by steroid hormones (ecdysone).

Ecdysone formation comes in sequential pulses and thereby sequential activation of genes involved in metamorphosis of insects can be visualized at the level of the giant chromosomes. The puffs represent active transcription at particular genes, and the pattern of puffing shifts along the salivary gland chromosomes during development (see Fig. P166) and/or activation and the RNA extracted from the puffs reflect the differences in the base sequences of the genic DNA. Puffing has been described also in the rare polytenic chromosomes of some plant species, e.g., *Allium ursinum* or *Aconitum ranunculifolium*. These have been observed in specialized tissues of the chalaza or in the antipodal cells. ►giant chromosomes, ►ecdysone, ►PARP; Beermann W 1961 Verh Dtsch Zool Ges 1961:44; Mok EH et al 2001 Chromosoma 110:186.

Pufferfish, Japanese (*Fugu rubripes, Tetraodontidae*): A small vertebrate with about 365 Mbp DNA, i.e., only somewhat more than 1/10 of that of most mammals, and therefore it is suitable for structural and functional studies at the molecular level (see

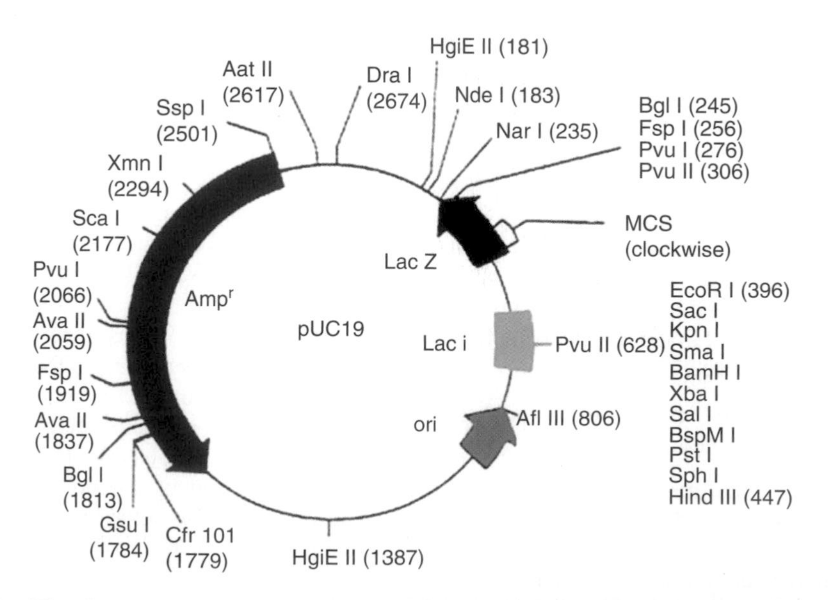

Figure P165. pUC$_{19}$ (The diagram is the courtesy of CLONTECH Laboratories Inc., Palo Alto, CA.)

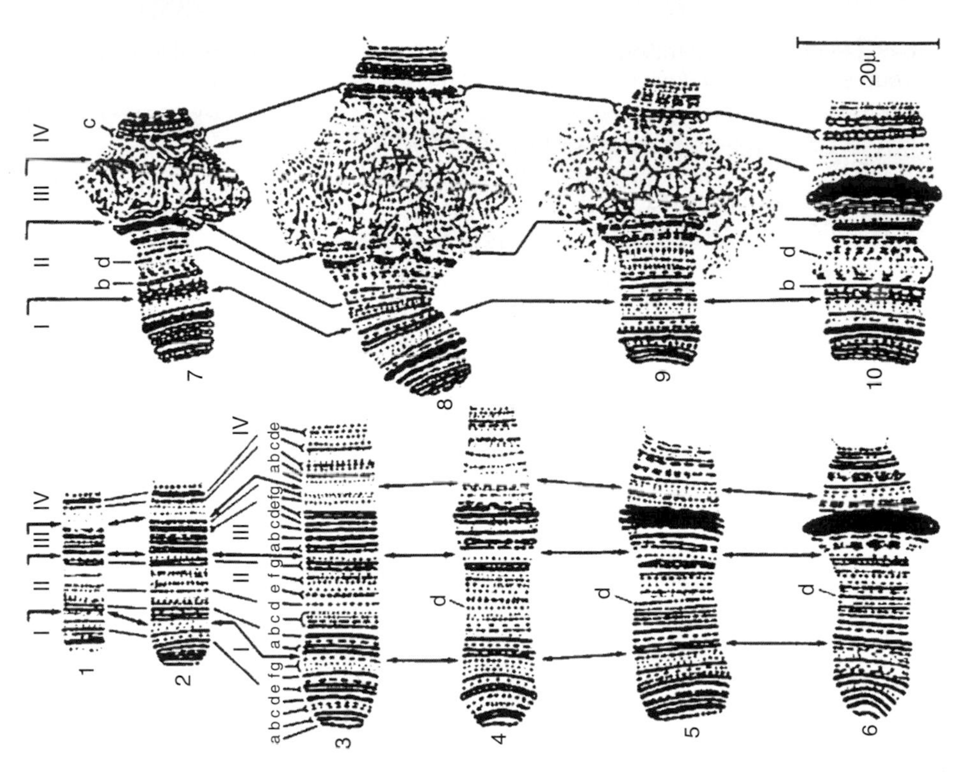

Figure P166. Selective activity of genes during development of the Dipteran fly *Rhynchosciara angelae* is reflected in the puffing pattern of the salivary gland chromosomes. Lower case letters designate bands. Roman numerals indicate regions of the chromosomes. (After Breuer M E, Pavan C 1954. By permission from Kühn A 1971 Lectures on Developmental Physiology, Springer-Verlag, New York)

Figure P167. Takifugu rubripes (courtesy of Wikipedia; author Chris 73)

Fig. P167). More than 95% of the genome has been sequenced by 2002. About one-third of the genome is genic and repetitive sequences occupy less than one-sixth. The species, *Spheroides nephelus* is also used for studies of control of gene expression. *Tetraodon nigroviridis* (n = 21) DNA (27,918 genes) sequences have been used for the determination of human gene number. About 14,500 human ecores are conserved also in this pufferfish species ►ecores; Crollius HR et al 2000 Genome Res 10:939; Aparicio S et al 2002 Science 297:1301; Jaillon O et al 2004 Nature [Lond] 431:946.

Pull-Down Assay: Expected to reveal interacting proteins. One of the proteins is attached to agarose beads and so immobilized. Then the test protein is added and the mixture is incubated to allow time for forming some links. Subsequently the mix is centrifuged. If there is a binding between them both proteins are found in the pellet and interaction is assumed. ►immunoprecipitation, ►genetic networks; Brymora A et al 2001 Anal Biochem 295:119.

Pullulanase: A secreted Klebsiella enzyme (~117 kDa) which cleaves starch into dextrin. It occurs also in the endosperm of cereals and other plant tissues and it is regulated by thioredoxin. ►thioredoxin; Schindler I et al 2001 Biochim Biophys Acta 1548:175.

Pulmonary Adenoma: Lung cancer controlled by QTL. Dense SNP map of mouse identified the pulmonary susceptibility locus (*Pas1*) including within a 0.5 Mb region *Kras2* (Kirsten rat sarcoma oncogene 2) and *Cascl/Las1* susceptibility genes in chromosome 6. The *Glu102* allele of *Casc1* preferentially promotes susceptibility to tumorigenesis by chemicals (Liu P et al 2006 Nat Genet 38:888). Change from in the balance of expression of *Kras2* and *Pas1* entails susceptibility to resistance (To MD et al 2006 Nature

Table P3. Pulse-chase method. Autoradiographic analysis of the replication of the DNA in chromosomes by the pulse-chase procedure. (Drawn after Taylor JH et al 1957 Proc Natl Acad USA 43:122)

Grown without label	Replication in ^{3}H	Labeled chromosomes replicated in ^{3}H-free medium		
		no exchange	sister chromatids exchanged	
				Cytological observation
				Interpretative drawing of the distribution of the radioactive label ■
				Interpretation of the replication of the DNA helices in the two chromatids ^{3}H. . . . .

Genet 38:926). LKB1 hemizygosity or homozygous loss substantially accelerated pulmonary adenoma (Hongbin J et al 2007 Nature [Lond] 448:807). ►QTL, ►RAS, ►small cell lung carcinoma, ►non-small-cell lung carcinoma, ►LKB

Pulmonary Emphysema: The increase in size of the air space of the lung by dilation of the alveoli (small sac-like structures) or by destruction of their walls. Smoking may be a cause.

Pulmonary Hypertension (PPH, FPPH): Characterized by shortness of breath, hypoxemia and arterial hypertension caused by the proliferation of endothelial smooth muscles and vascular remodeling. It is a 2q33 dominant disorder with reduced penetrance. Various drugs (such as the banned anti-obesity drug fen-phen) may trigger it. The basic defect is in gene BMPR2 (bone morphogenetic protein receptor II). Haplo-insufficiency may cause it. The consequence is inappropriate regulation by the serine/threonine kinases of the phosphorylated Smad proteins leading to inadequate maintenance of blood vessel integrity. ►bone morphogenetic protein, ►Smad, ►hypertension, ►haplo-insufficient, ►bone morphogenetic protein, ►Smad; Machado RD et al 2001 Am J Hum Genet 68:92.

Pulmonary Surfactant Proteins: ►respiratory distress

Pulmonary Stenosis: ►stenosis

Pulse-Chase Analysis: Expose cells, for a period of time, to a radioactive compound such as ^{3}H-thymidine (pulse) and examine the labeling of chromosomes in some cells. The culture is then transferred to non-radioactive thymidine and allowed to complete a division (chased to another stage) and study again the distribution of the label and determine its fate in the cells. The experiment permitted the first time the valid conclusion that DNA replication is semi-conservative. ►radioactive tracer, ►radioactive label, see Table P3.

Pulsed Field Gel Electrophoresis (PFGE): A procedure combining static electricity and alternating electric fields with gel electrophoresis for the separation of DNA of entire chromosomes of lower eukaryotes, such as of yeast and *Tetrahymena* or large DNA fragments cloned in YAC vectors of any genome cut by rare-cutting restriction enzymes. ►CHEF, ►FIGE, ►OFAGE, ►YAC, ►PHOGE, ►TAFE, ►RGE; Mulvey MR et al 2001 J Clin Microbiol 39:3481.

Puma Cat (*Felis concolor, Puma concolor*): 2n = 38. (see Fig. P168).

Figure P168. Puma

PUMA (p53 upregulated modulator of apoptosis): ►p53, ►apoptosis, ►BAX, ►SLUG

PUMA2: The evolutionary analysis of metabolism, http://compbio.mcs.anl.gov/puma2/.

Pump: The various transmembrane proteins mediating active transport of ions and molecules through biological membranes. ►sodium pump

Punctuated Equilibrium: ►punctuated evolution

Punctuated Evolution: A theory that evolution would follow alternating periods of rapid changes and relatively stable intervals (punctuations). Natural selection of beneficial mutations appears after some intervals and spread over the population. At the DNA level about 22% of the substitutional changes represent punctuated evolution. Punctuational changes are more common in plants and fungi than in animals (Pagel M et al 2006 Science 314:119). ►speciation, ►beneficial mutation, ►neutral mutation, ►hopeful monster, ►shifting balance theory, ►gradualism; Gould SJ, Eldredge N 1993 Nature [Lond] 366:223; Elena SF et al 1996 Science 272:1797; Voigt C et al 2000 Adv Protein Chem 55:79.

Punctuation Codons (UAA, UGA, UAG): Terminate translation of the mRNA.

Punnett Square: Permits simple prediction of the expected pheno- and genotypic proportions. It is a checkerboard where on top and at the left column the male and female gametic output is represented and in the body of the table the genotypes are found (see Figure P169). If, e.g., the heterozygote has the genetic constitution of *Aa, Bb*, the gametes and genotypes will be AB, Ab aB and ab. In case of linkage and recombination the actual frequency of each type of gamete must be used to obtain the correct genotypic proportions in the body of the checkerboard. ►modified Mendelian ratios, ►Mendelian segregation

Male gametes		AB	Ab	aB	ab
Female gametes	AB	AB AB	AB Ab	AB aB	AB ab
	Ab	Ab AB	Ab Ab	Ab aB	Ab ab
	aB	aB AB	aB Ab	aB aB	aB ab
	ab	ab AB	ab Ab	ab aB	ab ab

Figure P169. Punnett square

Pupa: A stage in insect development between the larval stage and the emergence of the adult (imago) (see Fig. P170). ►*Drosophila*, ►juvenile hormone

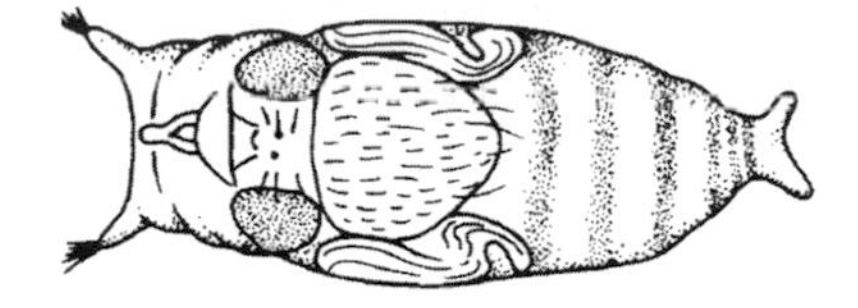

Figure P170. *Drosophila* pupa

Puparium: The case in which the *Drosophila* (and other insect) pupa develops for about four days after hatching of the egg, and in another four days the imago emerges. ►*Drosophila*

Pure Culture: Involves only a single organism. ►axenic culture

Pure Line: Genetically homogeneous (homozygous), and its progeny is expected to be identical with the parental line unless mutation occurs. (See Johannsen W 1909 Elemente der exakten Erblichkeitslehre, Fischer, Jena, Germany).

Pure-Breeding: Homozygous for the genes considered.

Purine: A nitrogenous base composed of a fused pyrimidine and imidazole ring; the principal purines in the cells are adenine, guanine, xanthine, hypoxanthine (but theobromine, caffeine and uric acid are also purines).

5′,8-Purine Cyclodeoxynucleosides: Formed in two diastereoisomers by exposure of DNA to reactive oxygen species. The cyclopurines may cross-link the C-8 adenine or guanine and the 5′ position of 2-deoxyribose (see Fig. P171). These diastereoisomers may block DNA replication and are cytotoxic.

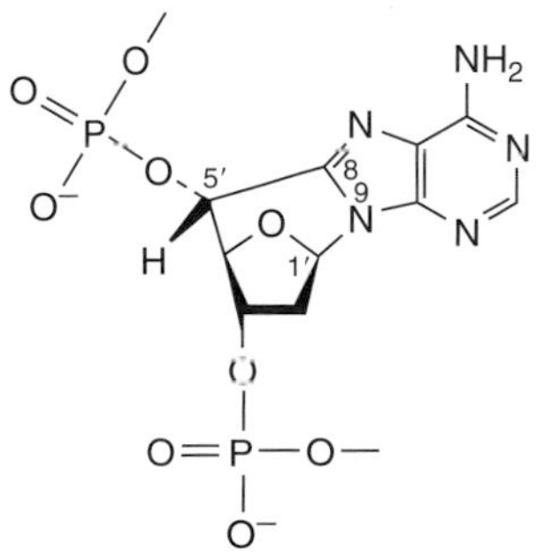

Figure P171. 5′,8-cyclo-2′-deoxyadenosine phosphate

Commonly excision repair may not correct the damage although the xeroderma pigmentosum A protein may cut at both flanks and excises them. These types of damaged nucleosides may accumulate by time and result in progressive neurodegeneration in xeroderma pigmentosum patients. ►cyclobutane, ►excision repair, ►xeroderma pigmentosum; Kuraoka I et al 2000 Proc Natl Acad USA 97:3837.

Purine Repressor (PurR): A member of the *Lac* repressor family of proteins regulating 10 operons involved in the biosynthesis of purine and affecting to some extent 4 genes controling de novo pyrimidine synthesis and salvage. Its ca. 60 amino acids, the NH_2 domain binds to DNA and its ca. 280 residue COOH domain binds effectors and it functions in oligomerization. ►*Lac* repressor, ►salvage pathway; Moraitis MI et al 2001 Biochemistry 40:8109.

Purity of the Gametes: One of the most important discoveries of Mendel. At anaphase I of meiosis of diploids the bivalent chromosomes segregate and at anaphase II, the chromatids separate. Therefore, in the gametes of diploids only a single allelic form of the parents is present with rare exceptions, e.g., nondisjunction and polyploids. ►meiosis, ►nondisjunction, ►gene conversion, ►Mendelian laws

Purkinje Cells: Large pear-shaped cells in the cerebellum, these are connected to multi-branched nerve cells traversing the cerebellar cortex. In the heart they are tightly appositioned cells transmitting impulses. ►cerebellum, ►motor proteins

Puromycin: An antibiotic, it inhibits protein synthesis by binding to the large subunit of ribosomes; its structure resembles the 3′-end of a charged tRNA. Therefore it can attach to the A site of the ribosome and forms a peptide bond but it cannot move to the P site and thus causes premature peptide chain termination. ►antibiotics, ►signaling to translation, (see Fig. P172) formula

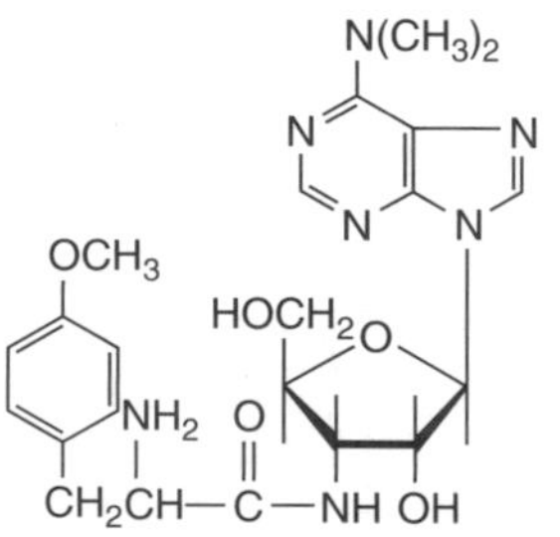

Figure P172. Puromycin

PUS: ►pyogenic

Pushme-Pullyou (pushmi-pullyu) Selection: A positive-negative selection system to isolate engineered chromosomes in somatic cell hybrids, which have retained the segment positively selected for, and lost the regions selected against. (See Higgins AW et al 1999 Chromosoma 108:256; Trimarchi JM, Lees JA 2002 Nature Rev Mol Cell Biol 3:11).

PV16/18E6: A human papilloma virus oncoprotein. (►oncoprotein) Burkitt lymphoma and murine lymphocytomas and may have activating role for MYC that is in the same chromosome. ►MYC, ►oncogenes, ►Burkitt lymphoma

πVX: A microplasmid (902 bp) containing a polylinker and an amber suppressor for tyrosine tRNA. It can be used for cloning eukaryotic genes. ►recombinational probe

PWM (position weight matrix): Used for identification of and search for functional nucleotide sequences, which are highly degenerate, e.g., the TATA boxes in the promoters. The PWM reflects the frequency of the four nucleotides (A, T, G, C) in an aligned set of different sequences sharing common function. After it had been determined in well-characterized core promoter regions, the PWM can be used to scan for TATA boxes in anonymous nucleotide sequences. The similarities between PWM and specific sequences and the matching value (within an accepted range) is determined and called a signal. Bucher (J Mol Biol 212:563) used a PWM for TATA box: GTATAAAGGCGGGG, and when the best fit was designated as 0, the majority of "unknown" TATA boxes scored within 0 to −8.16. Some of these might be, however, false positives. ►TATA box, ►core promoter, ►anonymous DNA segment, ►position-specific scoring matrix; Audic S, Claverie J-M 1998 Trends Genet 14:10.

PX (phox): A 125-amino acid module present in a variety of proteins involved in binding phosphoinositides.

$P2X_1$: A receptor for ATP in the in ligand-gated membrane cation channels. P2X is a component of the contractile mechanism of the vas deferens muscles, which propel the sperm into the ejaculate during copulation. Its defect entails ~90% sterility although without apparent harm to the male or the female mice. It has also several other signaling functions. ►vas deferens, ►ion channels, c►ongenital aplasia of the vas deferens; Khakh BS, North RA 2006 Nature [Lond] 442:527.

PX DNA: A four-stranded molecule where the parallel helices are held together by reciprocal recombination at every site of juxtaposition. Its topoisomer is JX_2 and it also contains adjacent helices but there is no reciprocal exchange at the contact points. (See Yan H et al 2002 Nature [Lond] 415:62).

Pxr: ►SXR

Pycnidium: A hollow spherical or pear-shaped fruiting structure of fungi producing the pycnidiospores, which are released through the top opening, the ostiole. ►stem rust

Pycnodysostosis: A rare autosomal recessive (1q21) human malady characterized by defects in ossification (bone development) resulting in short stature, deformed skull with large fontanelles (soft, incompletely ossified spots of the skull common in fetuses and infants) and general fragility of the bones. The primary defects appears to be in cathepsin K, a major bone protease although interleukin-6 receptor has also been implicated. (see Fig. P173) ►Toulouse-Lautrec, ►cleidocranial dysostosis, ►cathepsins

Figure P173. The famous French artist Henri Toulouse-Lautrec (1864–1901) might have suffered from this malady and his self-portrait reveals some of the characteristics of the malformations. The exact nature of his condition cannot be diagnosed but it is known that his parents were close relatives. (By permission of the St. Martin Press, New York)

Pycnosis (pyknosis): A physiological effect of ionizing radiation on chromosomes expressed as clumping or stickiness. It is dose-dependent and the late prophase stage irradiation is most effective in causing it. Anaphase proceeds but the chromosomes have difficulties in separation, display chromatin bridges and may break up into fragments. ►bridge, ►karyorrhexis, ►heteropycnosis, ►acinus

Pygmy: The Central African human tribe of about 100,000 has an average height of 142 cm. In comparison, the average height of Swiss and Californian is 167–169 and 170–172 cm, respectively. The Pygmies do not respond to exogenous stomatotropin but the concentration of serum somatomedins in the adolescent Pygmies is about a third below that in non-Pygmies of comparable age. Although the shortness of Pygmies appears recessive, intermarriages indicate polygenic determination of height. ►dwarfism, ►stature in humans, ►nanism, ►somatomedin, ►somatotropin

PYK2: Protein tyrosine kinase links Src with G_i and G_q-coupled receptors with Grb2 and Sos proteins in the MAP kinase pathway of signal transduction. Lysophosphatidic acid (LPA) and bradykinin stimulate its phosphorylation by Src. Over-expressing mutants of Pyk or the protein tyrosine kinase Csk reduces the stimulation by LPA, bradykinin or over-expressed Grb2 and Sos. ►CAM, ►MAP, ►Src, ►G_i, ►G_q, ►lysophosphatidic acid, ►kininogen, ►Csk, ►Grb2, ►Sos, ►signal transduction; Felsch JS et al 1998 Proc Natl Acad Sci USA 95:5051; Sorokin A et al 2001 J Biol Chem 276:21521.

Pyknons: Short RNA motifs shared by genic and non-genic transcript regions of the human genome and are presumably mediating post-transcriptional gene silencing and RNA interference. Pyknons are most frequent in the 3′-nontranslated areas of genes (Rigoutsos I et al 2006 Proc Natl Acad Sci USA 103:6605).

Pyknosis: ►pycnosis

Pyloric Stenosis: A smaller than normal opening of the pylorus, the lower gate of the stomach that separates it from the small intestine (duodenum). It docs not appear to have independent genetic control but it is part of some syndromes. It affects males five times as frequently as females; the overall incidence for both sexes is about 3/1000 birth. About 20% of the sons of affected females display this anomaly but only about 4% of the sons if the father has the malady. It may be caused by a deficiency of neuronal nitric acid synthase. ►sex-influenced, ►nitric oxide, ►imprinting

PYO: Personal years of observations; a term used in medical and clinical genetics

Pyocin: A bacteriotoxic protein produced by some strains *Pseudomonas aeruginosa* bacteria. ►bacteriocins

Pyogenic: Producing pus (DNA and protein-rich) excretum upon inflammation containing leukocytes in a yellowish fluid. It is produced abundantly by the body also after *Streptococcus* and *Staphylococcus* and other bacterial infections. ►IRAK, ►leukocyte, ►*Streptococcus*, ►*Staphylococcus*

Pyramidal Cells: Excitatory neurons in cerebral cortex. ►brain, ►neuron

Pyramiding: To build up on a larger base; to accumulate in plant protection introduction of more than one gene and different transgenes into a plant variety conveying resistance to the same agent, in order to slow down or prevent development of resistance in the pathogen or pest.

Pyrene: A fluorochrome, frequently used as a bimolecular excimer (see Fig. P174). Pyrene incorporation into the sugar position of DNA by one carbon linker results in very weak monomer fluorescing because of quenching. Similar was the observation with RNA or RNA–DNA hybridization. Pyrene-modified RNA displayed drastically increased fluorescence however when paired with complementary RNA and is a useful tool for monitoring RNA hybridization (Nakamura M et al 2005 Nucleic Acids Res 33:5887). ►FRET, ►excimer

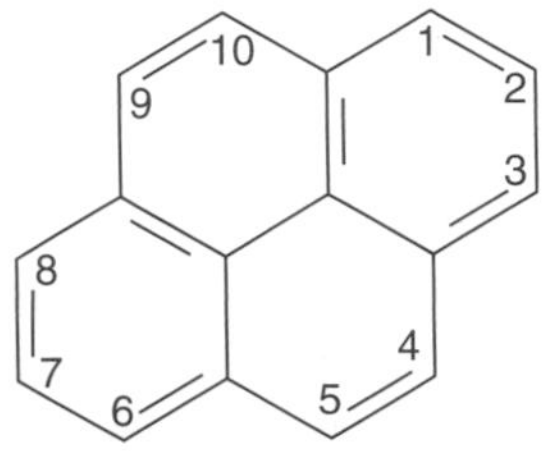

Figure P174. Pyrene

Pyrenoid: A dense, refringent protein structure in the chloroplast of algae and liverworts associated with starch deposition. ►chloroplast

Pyrethrin (pyrethroids, permethrin): Insecticides are natural products of *Pyrethrum* (*Chryanthemum cineraiaefolium*) plants (Compositae). They affect the voltage-gated Na^+ ion channels and humans may have severe allergic reactions to pyrethrins. ►ion channels, ►insecticide resistance

Pyrethrum (*Chrysanthemum* spp): A plant source of the natural insecticide, pyrethrin, with basic chromosome number x = 9 (see Fig. P175). Some species are diploid or tetraploids or hexaploids. ►pyrethrin

Figure P175. Pyrethrum

Pyridine Nucleotide: A coenzyme containing a nicotinamide derivative, NAD, NADP.

Pyridoxine (pyridoxal): Vitamin B_6 is part of the pyridoxal phosphate coenzyme, instrumental in transamination reactions (see Fig. P176). In *E. coli* bacteria vitamin B_6 is synthesized through deoxyxylulose 5-phosphate and phosphohydroxy-L-threonine. In plants (*Arabidopsis*) the synthetic pathway differs in as much as it is biosynthesized from ribose 5-phosphate or ribulose 5-phosphate and from dihydroxyacetone phosphate or glyceraldehyde 3-phosphate in the cytosol rather than in the chloroplasts (Tambasco-Studart M et al 2005 Proc Natl Acad Sci USA 102:13687).

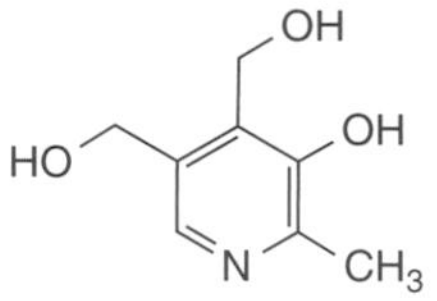

Figure P176. Pyridoxine

An apparently autosomal recessive disorder in humans involving seizures is caused by pyridoxin deficiency because of a deficit in glutamic acid decarboxylase (GAD) activity and consequently insufficiency of GABA, required for normal function of neurotransmitters. Administration of pyridoxin caused cessation of the seizures. The GAD gene was located in the long arm of human chromosome 2. An autosomal dominant regulatory pyridoxine kinase function has also been identified in humans. ►epilepsy

Pyridoxine Dependency: It may manifest as autosomal recessive seizures with perinatal onset (around birth).

Pyrimidine: A heterocyclic nitrogenous base such as cytosine, thymine, uracil in nucleic acids but also the sedative and hypnotic analogs of uracil, barbiturate and derivatives (see Fig. P177). Pyrimidine biosynthesis may follow either a de novo or a salvage pathway. Some of the pyrimidine moieties, e.g., of thiamin are biosynthesized through a route different from that of nucleic acid pyrimidines. ►J base, ►thiouracil, ►pseudouracil, ►*de novo* synthesis, ►salvage pathway, ►formulas; Fox BA, Bzik DJ 2002 Nature [Lond] 415:926.

Pyrimidine Dimer: Cross-linked adjacent pyrimidines (thymidine or cytidine) in DNA causing a distortion in the strand involved and thus interfering with proper functions (see Fig. P178). It is induced by short-wavelength UV irradiation. The thymidine dimers may be split by visible light-inducible enzymatic repair (light repair) or by excision repair (dark repair). ►cyclobutane ring, ►physical mutagens, ►genetic repair, ►DNA repair, ►photolyase, ►CPD, ►glycosylases, ►pyrimidine-pyrimidinone photoproduct,

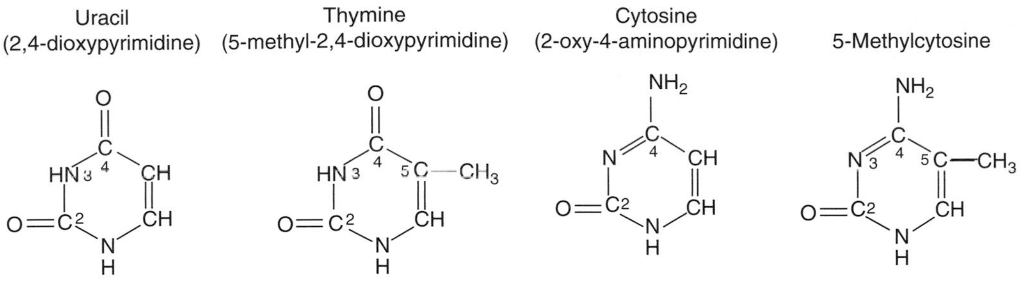

Uracil (2,4-dioxypyrimidine) | Thymine (5-methyl-2,4-dioxypyrimidine) | Cytosine (2-oxy-4-aminopyrimidine) | 5-Methylcytosine

Figure P177. The major cellular pyrimidines

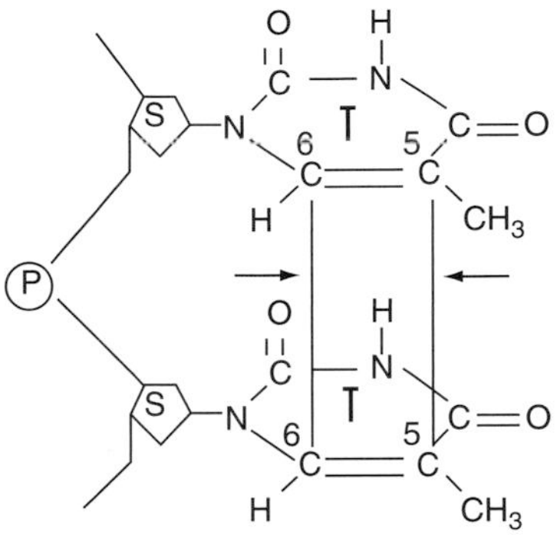

Figure P178. Cross-linked neighboring thymines in the DNA

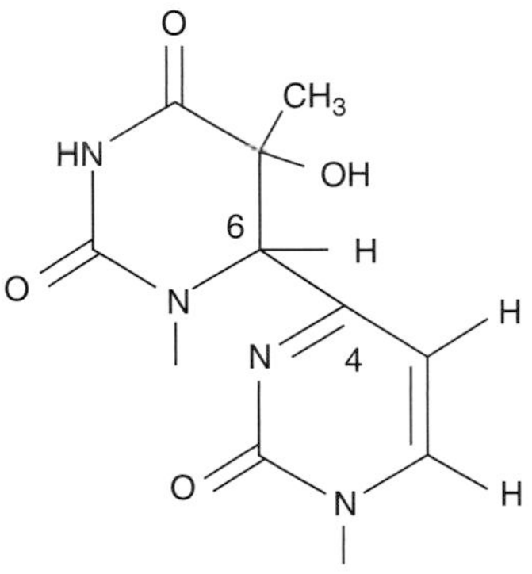

Figure P179. Thymine–Cytosine photoproduct

►photoreactivation; Otoshi E et al 2000 Cancer Res 60:1729.

Pyrimidine Dimer *N*-Glycosylase: A DNA repair enzyme that creates an apyrimidinic site. Then the phosphodiester bond is severed and a 3′-OH group is formed on the terminal deoxyribose. Exonuclease 3′→5′ activity of the DNA polymerase splits off the new 3′-OH end of the apyrimidinic site. After this, the replacement-replication—ligation process repairs the former thymine dimer defect. ►glycosylase, ►DNA repair, ►pyrimidine dimer; Piersen CE et al 1995 J Biol Chem 270:23475.

Pyrimidine 5′-Nucleotidase Deficiency (P5N): May cause hereditary hemolytic anemia as the pyrimidines inhibit the hexose monophosphate shunt in young erythrocytes. There are two isozymes of which P5NI is most commonly the cause of the anemia. ►pentose phosphate pathway, ►anemia; Marinaki AM et al 2001 Blood 97:3327.

Pyrimidine-Pyrimidinone Photoproduct: A pyrimidine dimer involving a 6—4 linkage between thymine and cytosine (see Fig. P179). ►cyclobutane, ►Dewar product, ►cis-syn dimer, ►translesion pathway, ►photolyase; Vreeswijk MP et al 1994 J Biol Chem 269:31858.

Pyrimidone: Hydroxypyrimidine (see Fig. P180).

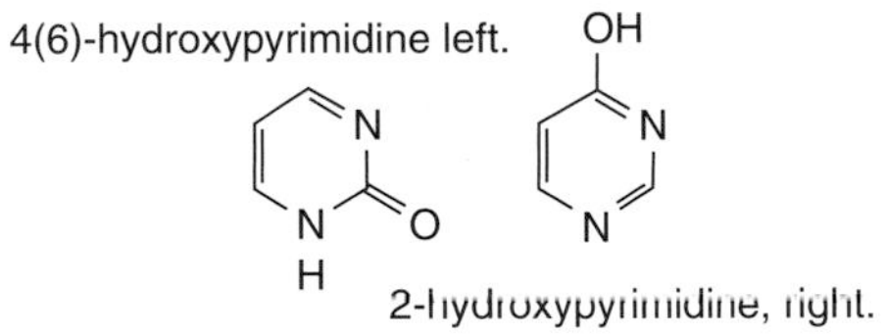

4(6)-hydroxypyrimidine left.

2-hydroxypyrimidine, right.

Figure P180. 4(6)-hydroxyprimidine (left); 2-hydroxypyrimidine (right)

Pyrimidopurinones: A malondialdehyde-DNA adduct derived from deoxyguanosine. ►adduct

Pyronin: A histochemical red stain used for the identification of RNA.

Pyrosequencing: Used for the analysis of the nucleotide sequence of less than 200 base long DNA strands for the detection of mutational alteration(s). Pyrosequencing has been applied to the analysis of the genome of Neanderthals. It uses the enzymes DNA polymerase, sulfurylase, firefly luciferase and apyrase. The incorporation of the nucleotides (which are not labeled) in the growing end is monitored by light flashes in a single tube. Electrophoresis is not used. Nucleotide triphosphates added to the reaction in sequence. Visible light is generated and detected when pyrophosphate is released during incorporation from the nucleotide triphosphates with the cooperative effects of the sulfurylase and the luciferase. This is a very fast procedure and may be automated.

Another fast and convenient method of sequencing by synthesis involves four chemically cleavable fluorescent nucleotide analogues as reversible terminators. Each of the nucleotide analogues contains a 3′-*O*-allyl group and a unique fluorophore with a distinct fluorescence emission at the base through a cleavable allyl linker. These nucleotide analogues are good substrates for DNA polymerase in a -solution-phase DNA extension reaction and that the fluorophore and the 3′-*O*-allyl group can be removed with high efficiency in aqueous solution. By this procedure 20 continuous bases of a homopolymeric DNA template immobilized on a chip were accurately sequenced. Such a methods can be extended eventually to longer sequences (Ju J et al 2006 Proc Natl Acad Sci USA 103:19635).

►DNA sequencing, ►sulfurylase, ►luciferase, ►apyrase, ►Neanderthal; Ronaghi M 2001 Genome Res 11:3; Marziali A, Akeson M 2001 Annu Rev Biomed Engr 3:195; Fakhrai-Rad H et al 2002 Hum Mut 19:479; Goldberg SMD et al 2006 Proc Natl Acad Sci USA 103:11240; Margulies M et al 2005 Nature [Lond] 437:376; note corrigendum Nature [Lond] 441:120.

Pyrrole: A saturated five-membered heterocyclic ring such as found in protoporphyrin. Pyrrole-imidazole polyamides may bind to specific DNA of the transcription factor TFIIIA and regulate the transcription of the 5S RNA. *N*-methylimidazole (Im)—*N*-methylpyrrole (Py) may target G≡C and Py—Im the C≡G base pairs, respectively. The Py—Py combination is specific for T= A and A= T. ►porphyrin, ►porphyria, ►heme

Pyrrolizidine Alkaloids (petasitenine, senkirkine): Occur in several plant species (*Tussilago, Heliotropium*, etc.) and some of which are used as food or medicinal plants but they are mutagenic/carcinogenic. They occur also in some moths and convey protection against predators. ►*Echinacea*; Ober D, Hartmann T 1999 Proc Natl Acad Sci USA 96:14777.

Historical vignette

"...alle essentiellen Merkmale...epigenetisch sind, und da die Determinierung ihrer Specificität durch den Kern erhalten."

"*... all characters are epigenetic and their specificity depends on the cell nucleus.*"

Theodor Boveri 1903 Roux' Arch Entwickl-Mech Org 16:340–363.

Pyrrolysine: The 22nd amino acid encoded in Archaea and Eubacteria by the stop codon UAG. Pyrrolysine is charged to $tRNA^{CUA}$ (encoded by *pylT*) by PylS aminoacyl-tRNA synthetase and thus can be incorporated into *E. coli* proteins (Blight SK et al 2004 Nature (Lond) 431:333). ►amino acids, ►genetic code, ►aminoacyl tRNA synthetase, ►unnatural amino acids, ►selenocysteine; Hao B et al 2002 Science 296:1462; Srinivasan G et al 2002 Science 296:1459; pyrrolysyl-tRNA synthetase crystal structure: Kavran JM et al 2007 Proc Natl Acad Sci USA 104:11268.

Pyruvate Dehydrogenase Complex: Contains three enzymes, pyruvate dehydrogenase, dihydrolipoyl transacetylase and dihydrolipoyl dehydrogenase and the function of the complex requires the coenzymes: thiamin pyrophosphate (TPP), flavine adenine dinucleotide (FAD), coenzyme A (CoA), nicotinamide adenine dinucleotide (NAD), and lipoate. The result of the reactions is oxidative decarboxylation whereby CO_2 and acetyl CoA are formed. ►oxidative decarboxylation; Zhou ZH et al 2001 J Biol Chem 276:21704.

Pyruvate Kinase Deficiency: A recessive (human chromosome 1q21-q22, PK1) hemolytic anemia actually caused by two enzymes that are the products either of differential processing of the same transcript or chromosomal rearrangement. In the presence of some tumor promoters hepatic pyruvate kinase activity decreases. ►anemia, ►hemolytic anemia, ►glycolysis

Pyruvic Acid: A ketoacid (CH3COCOOH) formed from glycogen, starch and glucose under aerobic conditions (under anaerobiosis is reduced to lactate and NAD^+ is formed). Hyperpolarization technology permits imaging of pyruvate metabolic path in real time without invasive procedures (Golman K et al 2006 Proc Natl Acad Sci USA 103:11270). ►Embden-Meyerhof pathway, ►pentose phosphate shunt

PyV: Polyoma virus.

$PYY_{3–36}$: A neuropeptide Y (NPY)-like, but it is a gastrointestinal hormone that inhibits food uptake. ►obesity, ►leptin, ►neuropeptide Y; Batterham RL et al 2002 Nature [Lond] 418:650.

PZD: ►micromanipulation of the oocyte

Q

q: Long arm of chromosomes. ►p

Q Banding: Chromosome staining with quinacrines that reveals cross bands. Due to the availability of newer microtechniques, this procedure is no longer generally used. ►chromosome banding, ►quinacrine mustard; Caspersson TG et al 1971 Hereditas 67:89.

Q2 Domain: ►CREB

Q-β: An RNA bacteriophage of a molecular weight of about 1.5×10^6 Da. The Qβ replicase is an RNA-dependent RNA polymerase that synthesizes the single-stranded RNA genome of the phage without an endogenous primer. The replicase can use both the + and the − strand as a template and therefore it amplifies the genome rapidly. It is a heterotetramer consisting of one viral encoded and three host polypeptides. ►replicase, ►plus strand; Munishkin AV et al 1991 J Mol Biol 221:463.

Q-Q-TOF: Has uses to quadrupoles and time of flight analysis for proteins and proteome.

qORF: An open reading frame with questionable information on its transcription. ►ORF, ►transcription

QSTAR Pulsar: A quadrupole time-of-flight mass spectrometer. ►quadrupole, ►MALDI, ►proteomics; Steen H et al 2001 J Mass Spectrom 36:782.

QTDT: Quantitative transmission disequilibrium test (Abecasis GR et al 2000 Eur J Hum Genet 8:545).

QTL (quantitative trait loci): Control the expression of complex traits such as weight, height, cognitive ability, and many diseases, etc. Their expression is usually not strict and even in the absence of the critical genes a quantitative trait may appear under the influence of extrinsic factors yet their expression is more likely when the appropriate alleles are present. Their physical presence may be traced in restriction fragments separated by electrophoresis because the DNA is independent of extrinsic factors. Their cosegregation is identified, and can be used for improving quantitative traits for plant breeding purposes, and for genetically defining behavioral traits and other polygenic characters. QTLs can be genetically mapped by several procedures, and most commonly using the principles of maximum likelihood for statistical analysis. In a backcross generation the phenotype (ϕ_ι) and genotype (g_i) relations are expressed as: $\phi_\iota = \mu + bg_i + \varepsilon_i$ where g_i corresponds to the homozygous and heterozygous dominants of the QTL and its value may vary between 1 and 0. The mean of ε_i (a random variable) = 0 and its variance is σ^2. The values of μ, b and σ^2 are unknown. The genotypic value of Qq and other contributors to the quantitative trait is μ and b is the effect of a substitution of another allele at the quantitative trait loci. The statistical procedures shown below were adapted from Arús P, Moreno-González J 1993, p 314 In: Hayward MD et al (Eds.) Plant Breeding, Chapman & Hall, London, New York.

The likelihood function $Lg_i(\mu, b, \sigma^2)$

$$= \frac{1}{\sqrt{2\pi\sigma^2}} e^{-\frac{(\phi_i - \mu - bg_i)^2}{2\sigma^2}}$$

and the likelihood that all individuals will be in the flanking parental marker classes (k)$M_1M_1M_2M_2$, $M_1M_1M_2m_2$, $M_1m_1M_2M_2$, and $M_1m_1M_2m_2$ will be $L_k(\mu, b, \sigma^2) = \Pi_i[P_i(1)L_i(1) + P_i(0)L_i(0)]$ where $P_i(1)$, and $P_i(0)$ are the probabilities that QQ and Qq quantitative genes will be in the recombinant classes, respective the flanking markers concerned. The maximum likelihood estimates for incomplete data can also be determined (Dempster AP et al 1977 J R Stat Soc 39:1). The likelihood for all observations is: $L(\mu, b, \sigma^2) = \Pi_k L_k(\mu, b, \sigma^2)$. For the determination of the LOD score, to ascertain that the information obtained is real rather than a false, spurious conclusion would be drawn the following equation has to be resolved:

$$\text{LOD} = \frac{\log L(\mu, b, \sigma^2)}{L(\mu_0, b_0, \sigma_0^2)}$$

One must keep in mind that the estimates are as good as the data collected. Large populations and genes with greater quantitative effects improve the chances to find linkage. For estimating linkage information from multiple marker data, computer assistance is required; various programs are available. For mapping QTL in humans generally sib pairs are used. Statistical analysis indicates that choosing extreme discordant pairs makes the analysis more efficient. Just because of this selection, QTL estimates may be loaded with errors of under and overestimation. Expectation-maximization–likelihood-ratio test may permit assessing the effects of missing data (Niu T et al 2005 Genetics 169:1021). Least squares, Bayesian and non-parametric methods are also available. Allison DB et al (2002 Am J Hum Genet 70:575) are discussing methods useful for eliminating errors. Although most commonly single quantitative traits are analyzed at a time, the simultaneous study of multiple traits may be desirable. Quantitative traits generally are not expressed in isolation and we may have to face epistasis of multi-traits. The abundance of mRNA transcripts can be considered as expression QTL (eQTL). The differences in eQTL may represent different sets of genes and can be mapped (Schadt EE et al 2003 Nature [Lond] 422:297). Microarray analysis can reveal large number of genes that influence phenotypic variation (Morley M et al 2004 Nature [Lond] 430:743). It is highly desirable that QTL would be amenable to isolation and cloning. (see Frary A et al 2000 Science 289:85) to better understand their function. A transition mutation (A→G) in intron 3 of an IGF2 (insulin-like growth factor) locus, despite the fact that it is not translated, may increase muscle growth by about three-fold. It is expressed only in the paternal gene copy and it indicates that regulatory genes may have major phenotypic effect (Van Laere A-S et al 2003 Nature [Lond] 425:832). In yeast, sporulation efficiency was resolved to single-nucleotide changes at the non-coding regulatory region (*RME1*) and to two nonsense mutations (*TAO3* and MKT1). The control of sporulation may be heterogeneous in the different strains (Deutschbauer AM, Davis RW 2005 Nature Genet 37:1333).

In introgression lines of tomato a *Solanum penellii* chromosome segment, containing the flower- and fruit-specific invertase locus, *LIN5* increased sugar yield of the common tomato (*Solanum lycopersicon*). A single amino acid substitution near the catalytic site of the enzyme exerted a critical effect, depending in extent on the genetic milieu of the recombinants (Fridman E et al 2004 Science 305:1786). In yeast crosses, the median variance of heritable QTL appeared 27% and no QTLs were detected for 40% of highly heritable transcripts. Modeling of QTLs indicated that only 3% of the highly heritable traits could be attributed to a single locus, 17 to 18% were apparently determined by one or two loci and half seemed to be under the control of more than five loci. Interaction among the gene products was indicated in 16% of the highly heritable traits (Brem RB, Kruglyak L 2005 Proc Natl Acad Sci USA 102:1572). In most complex traits several gene loci interact to a variable extent. In *two-locus mapping* two-stage procedure was proposed. For each transcript and marker a Wilcoxon test was used for the segregants at a locus. On the basis of the Wilcoxon test the quantitative trait with the most significant rank was named as the primary QTL. Besides the primary locus secondary loci were identified. Then non-parametric empirical Bayes' estimate determined the posterior probability that the primary QTL was a true positive and then the posterior probability of the true positive nature of the secondary QTL was determined. The product of these two probabilities is the joint probabilities for the two estimates. An interaction test for each locus pair used the model

$$t = ax + by + cxy + d$$

where $t = ln$ of the ratio of expression between the strain of interest and the reference sample; a, b, c and d are parameters specific for a given sample of a transcript, x is the inheritance at the first locus and y is inheritance at the second locus. An F-test determined that c + 0 and compared the goodness-of-fit of this model with a pure additive model. The P value of the interaction was evaluated by the Q-VALUE program available free through the Internet http://faculty.washington.edu/~jstorey/qvalue/ (Brem RB et al 2005 Nature [Lond] 436:701). This two-stage test detected many interactions missed by single-locus tests. A robust bootstrap procedure identified 843 QTLs in mouse with an average 95% confidence interval of 2.8 Mb (Valdar W et al 2006 Nature Genet 38:879; http://gscan.well.ox.ac.uk). Multivariate version of the Bayes methodology for joint mapping of QTLs, using the Markov chain–Monte Carlo (MCMC) algorithm is also useful (Liu J et al 2007 Am J Hum Genet 81:304). ►LOD score, ►mapping genetic, ►linkage, ►gene block, ►RFLP, ►liability, ►ASP analysis, ►interval mapping, ►complex inheritance, ►infinitesimal model, ►BLUP, ►bootstrap, ►co-suppression, ►introgression, ►least squares, ►Bayes' theorem, ►non-parametric tests, ►Wilcoxon rank correlation test, ►goodness of fit, ►F-distribution, ►SNIPs, ►Haseman-Elston regression, ►genetic genomics, ►rice; see also Darvasi A 1998 Nature Genet 18:19; Kao C-H 2000 Genetics 156:855; Mackay TFC 2001 Nature Rev Genet 2:11; Flint J, Mott R 2001 Nature Rev Genet 2:437; Mackay TFC 2001 Annu Rev Genet 35:303; Dekkers JCM, Hospital F 2002 Nature Rev Genet 3:22; Korstanje R, Paigen B 2002 Nature Genet 31:235; Feingold E 2002 Am J Hum Genet 71:217; Hoh J, Ott J 2003 Nature Rev Genet 4:701, review on sibling pairs: Cuenko KT et al 2003 Am J Hum Genet 73:863; functional QTL map for developmental traits: Wu R, Lin M 2006 Nature Rev Genet 7:229; QTL mapping, molecular bases in animals: Georges M 2007 Annu Rev Genomics Hum Genet 8 131; QTL comparisons: http://pmrc.med.mssm.edu:9090/QTL/jsp/qtlhome.jsp; miRNA QTL target sites: http://compbio.utmem.edu/miRSNP/; domesticated (agricultural) animal QTL: http://www.animalgenome.org/QTLdb/.

QTN: The nucleotide sequences for quantitative traits.

Quadrant: Consists of four parts, e.g., a tetrad.

Quadratic Check: Used for testing two genes presumed to be required for phytopathogenic infection in the manner shown in Fig. Q1. Resistance in the plants is usually a dominant trait. ►Flor's model

Pathogen / Plant	Low pathogenicity	High pathogenicity
Plant reaction low	No infection	Infection
Plant reaction high	Infection	Infection

Figure Q1. Quadratic check

Quadriplegia: ►tetraplegia

Quadriradial Chromosome: May be produced by cross-linking mutagens (see Fig. Q2).

Figure Q2. Quadriradial

Quadrivalent: Partially or completely identical four chromosomes in a polyploid that display pairing although of the four, at any particular position, only two can be synapsed. During meiosis they show quadrivalent association of four chromosomes. ►synapsis, ►meiosis, ►bivalent

Quadroma (hybrid hybridoma): The fusion product of two hybridomas. ►hybridoma; Lindhofer H et al 1995 J Immunol 155:219.

Quadruplets: Fourfold twins, and in the absence of the use of fertility increasing treatment, their expected frequency is about 1 in $(89)^3$ whereas the expectation for triplets and quintuplets is about $(89)^2$ and $(89)^4$, respectively. ►twinning

Quadruplex: A tetraploid or tetrasomic with four doses of the dominant alleles at a locus. ►autopolyploid, ►G quartet

Quadruplex DNA: Has four parallel and antiparallel strands in vitro, and blocks replication. In some instances G-quadruplex DNA my boost c-Myc gene proliferating activity (Siddiqui-Jain A et al 2002 Proc Natl Acad Sci USA 99:11593). ►Myc, ►tetraplex; Schaffitzel C et al 2001 Proc Natl Acad Sci USA 98:8572; G-quadruplex search tool: http://bioinformatics.ramapo.edu/QGRS/index.php; http://miracle.igib.res.in/quadfinder/.

Quadruplicate Genes: Four genes conveying identical or similar phenotype but segregating independently in F_2 and displaying a dominant recessive proportion of 255:1.

Quadrupole: Used in specific mass spectrometers for the tracking of ion density in proteomic analysis. The quadrupole is made up of four rods, which permit the filtering of the mass that traverses them with the mediation of an oscillating electric field to obtain the mass spectrum. The amplitude of the electric field is scanned and recorded. Triple quadrupole devices usually analyze peptides. The unit mass resolution is excellent with an accuracy of 0.1 to 1 Da. ►mass spectrum, ►proteomics, ►MALDI, ►Q-Q-TOF; Hager JW 2002 Rapid Commun Mass Spectrom 16:512.

Quality of Life: A subjective judgment of the functional ability of a patient.

Quality Protein Maize: ►high-lysine corn

Quantitative Gene Numbers: They are difficult to determine because environmental effects obscure the impact of genes with minor effects. Several statistical procedures have been worked out for approximation. The simplest one is as follows:

$$N = \frac{R^2}{8(s_1^2 - s_2^2)}$$

where N = gene number, R is the difference between the parental means, $s_1{}^2$ = variance of the F1, $s_2{}^2$ = variance of the F2 generations. The most common view is that quantitative traits are determined by large number of genes and each of them contributes only little to the observed phenotype. The association between bristle numbers (a quantitative trait) and the *scabrous* locus of *Drosophila* indicated, however that approximately 32% of the genetic variation in abdominal and 21% of the sternopleural bristle number was associated with DNA sequence polymorphism at this single locus. ►gene number, ►quantitative trait, ►QTL; Mather K, Jinks JL 1977 Introduction to Biometrical Genetics, Cornell University Press, Ithaca, New York; Jones CD 2001 J Hered 92[3]:274.

Quantitative Genetics: Studies genetic mechanisms involved with the expression of quantitative traits and its techniques involve those of population genetics and biometry. ►quantitative trait, ►QTL,

►population genetics, ►biometry, ►selection, ►heritability, ►statistics

Quantitative Trait: Shows continuous variation of expression and can be characterized by measurement or by counting in contrast to qualitative traits, which can be identified satisfactorily by simple description such as black or white. ►gene titration, ►dichotomous trait

Quantitative Trait Loci: ►QTL

Quantized: Using synchronous cultures and time-lapse video tape microscopy showed that the generation time in mammalian cells occurs with intervals of 3–4 h between bursts in division. The length of the cycles is dependent on the prevailing temperature, shorter at higher and longer at lower. This indicates that the cells have an oscillatory clock. ►oscillator; Klevecz RR 1976 Proc Natl Acad Sci USA 73:4012.

Quantum: The unit to quantify energy. ►photon

Quantum Computer: Fundamentally different from the standard binary computers where the information *bits* are either 0 or 1, in the quantum computer the *qubit* can be 0 and 1 and simultaneously both. This concept is different from the existing computers based on physics and it obeys the laws of quantum mechanics. At present there are difficulties in actually building the machines. A new approach to the problem is one-way four-qubit quantum state tomography (Walther P et al 2005 Nature [Lond] 434:169). The quantum computer may not be of general usefulness but may be especially useful to study electronic state of atoms or ions, their structures and reactivity. Quantum computers have the property of *entanglement*, i.e., the qubits can be interdependent. The entangled state may become, unfortunately, quickly unstable (*decoherence*) and one qubit can affect others too. The quantum computers are also much more error prone than the classical ones. There are still both theoretical and engineering problems to solve yet some progress is underway. (See Steane A 1998 Rep Progr Physics; Kim J et al 2004 J Magn Res 166:35; Stix G 2005 Sci Am 292(1):78; Keyes RW 2005 Computer 38(1):65; Ball P 2006 Nature [Lond] 440:398).

Quantum Dot (qdot): Built of semiconductor, luminescent nanometer-size crystals (e.g., of zinc sulfide-capped cadmium selenide). They may bind fluorochromes, organic and macromolecules and thus permit their tracing as stable, very bright, narrow-band emissions can be tuned from ultraviolet to infrared, water-soluble and non-invasive labels. This technology may increase the contrast in MRI, PET, computed tomography, etc. The fluorescence lifetime of the qdots is long and permits the separation from autofluorescence found in cells. The surface molecules can be protected from oxidation or other chemical reactions and this shell conveys to qdots photostability of several orders larger compared to other dyes. The dots are available in a wide range of well-separable colors all of which can be excited by a single wavelength. Confocal microscopy and other devices can track single qdots up to a few hours. The qdots can be tagged by various ligands such as DNA oligonucleotides, aptamers or antibodies. Besides animal and whole cell labeling, special cytoplasmic and nuclear targets, cell lineages, signal transduction pathways, membrane proteins, microtubules, actin, nuclear antigens, chromosomes and pathogens can be monitored. The qdots can be equipped with membrane-crossing and cell internalization or enzymatic functions. Generally qdots are innocuous yet they may adversely effect embryo development at higher concentrations. Quantum dot technology permits relatively fast identification of 10 bacteria/mL samples if amplification is allowed. This procedure facilitates biodefense and identification of other infections. Specific phages are biotinylated in vivo and attached to streptavidin-coated qdots. After infection in 20–45 min 10–1000 phages are released and can be readily detected (Edgar R et al 2006 Proc Natl Acad Sci USA 103:4841). ►nanocrystal semiconductor, ►nanotechnology, ►non-isotopic labeling, ►aptamer, ►semiconductor, ►luminescence, ►confocal microscopy, ►MRI, ►tomography, ►biotin, ►streptavidin, ►pathogen identification, ►bioterrorism; Han M et al 2001 Nature Biotechnol 19:631; Jaiswal JK et al 2003 Nature Biotechnol 21:47; Michalet X et al 2005 Science 307:538.

Quantum Speciation: A rapid formation of a new species by selection and genetic drift. ►selection, ►genetic drift

Quarantine: A state of isolation and observation without any external contact, especially from infection for a period of time in case of a contagious disease.

Quarter-Power Scaling: Biological scaling can be expressed by the formula $Y = Y_0M^b$ where Y is a variable (e.g., life span or metabolic rate), Y_0 is a normalization constant and b = scaling exponent, M = body mass. Y_0 varies with the trait and type of an organism, b is practically constant 1/4 or multiples of it. E.g. blood circulation time and life span are $M^{1/4}$, whole organism metabolic rate is $M^{3/4}$, diameter of tree trunks and aortas $M^{3/8}$, etc. ►synaptic scaling; West GB et al 1999 Science 284:1677.

Quartet: A structure consisting of four elements. ►G quartet

Quartile: One-fourth or 25% of the data or population.

Quasi: In various combinations it indicates almost, resembling or about of the notion that the following word specifies, e.g., quasi-species means that its difference from other form(s) may not qualify it for the status of a separate species with certainty.

Quasidominant: Recessive inheritance is misclassified as dominant because the mating took place between a heterozygote and a homozygous recessive individual.

Quasi Linkage: ►affinity

Quasi-Monoclonal Antibody: Produced by mice heterozygous for the V(D)J IM-imunoglobulin heavy chain (Ig) and the other allele being non-functional. Functional κ chain is also missing. When the heavy chain, specific for the hapten 4-hydroxy-3-nitrophenyl acetyl could join any λ chain, the antibody was monospecific but somatic mutation and secondary rearrangements changed the specificity of 20% of the B cell antigen receptors. Such a system can thus be used to study antibody diversity. ►antibody

Quasi-Species: A small degree of genetic (nucleic acid) variation does not qualify it clearly for separate species status. In RNA viruses due to high mutation rate (low repair) this state seem to exist and play role in their pathogenesis (Vignuzzi M et al 2006 Nature [Lond] 439:344).

Quaternary Structure: The aggregate of multiple polypeptide subunits into a protein or by cross-linking DNA strands into a joint structure. ►cross-linking, ►protein structure; http://www.mericity.com/; http://pqs.ebi.ac.uk.

Queen: The reproductive female in cast insect colonies such as exist in bees, ants. ►pseudoqueen

Queen Victoria: ►hemophilias, ►Romanovs, see Fig. Q3.

Quelling: A gene or chromatin repeat-associated post transcriptional silencing of genes without methylation, but involving RNAi (siRNA). It occurs when into plants or fungi foreign DNA is introduced by transformation. ►co-suppression, ►MSUD, ►transvection, ►sense suppression, ►silencer, ►RNAi; Maine EM 2000 Genome Biol 1(3): Reviews 1018.

Quenched Autoligation Probe (QUAL): QUAL probes use a pair of modified oligonucleotides. An electrophilic probe with internal fluorophore and

Figure Q3. Queen Victoria's family was the most famous to be affected by hemophilia, shown here, at a reunion on April 23, 1894. (1) Kaiser Wilhelm II, grandson, (2) Queen Victoria, (3) daughter Victoria, (4) granddaughter Tsarina Alexandra, (5) granddaughter Irene, (6) granddaughter Alice, (7) son and future king of England Edward VII, (8) daughter Beatrice, (9) son Arthur, (10) granddaughter Marie, (11) granddaughter Elizabeth. For most likely genetic constitutions regarding hemophilia see under hemophilia. (Courtesy of the Humanities Research Center, Gernsheim Collection, University of Texas, Austin, Texas)

a DABOSYL quencher is attached to the 5′ terminus by a sulfonate ester linkage, and a nucleophilic probe containing a 3′-phosphorothioate group is employed. In the presence of a target, the two probes bind side-by-side and nucleophilic displacement of the quencher by the phosphorothioate leads to probe ligation and unquenching of the fluorophore The reaction can be monitored by the gradual increase of fluorescence over minutes or hours. The probe can tell apart 16S RNAs even in very closely related bacteria. ►DABOCYL acid, ►phosphorothioate, ►fluorophore, ►bacteria, ►pathogen identification; Silverman AP, Kool ET 2005 Nucleic Acids Res 33:4978.

Quenching: The suppression of fluorescence, transfer of electrons or suppression of an activator by blocking the binding site of the activator or binding it to another protein which prevents its binding to the activator binding site in the DNA.

Quetelet Index: Essentially the same as body mass index, (weight)/(height)^2. L.A.J Quételet, an astronomer, a pioneer of biometry has already shown in 1835 that human stature follows the normal distribution. Old textbooks of genetics referred to the principle of normal distribution of quantitative traits as Quetelet's Law. According to the Quetelet-Galton Law when a quantitative trait did not follow the normal distribution the role of heredity in the expression of the trait was questioned and the variation was attributed to environmental causes. ►body mass index

Queunine: A rare modified purine. It is a derivative of guanine and it occurs in tRNA. Mammals cannot synthesize it and they obtain it from the intestinal microflora. Cancer cells have less of queunine in the tRNA. ►colicines, ►deazanucleotides

Quick-Stop: The temperature-sensitive DNA replication mutant *dna* of *E. coli* stops DNA replication immediately when the temperature rises to 42°C from the permissive 37°C. ►temperature-sensitive mutation; Rangarajan S et al 1999 Proc Natl Acad Sci USA 96:9224.

Quiescent Zone: A small region at the root tip where no cell division takes place in contrast to the neighboring cells, which are meristematically active (see Fig. Q4). In Arabidopsis roots, a *Retino-Blastoma-Related* gene is the main regulator of the meristematic (stem cell) state (Wildwater M et al 2005 Cell 123:1337). ►root, ►meristem, ►retinoblastoma

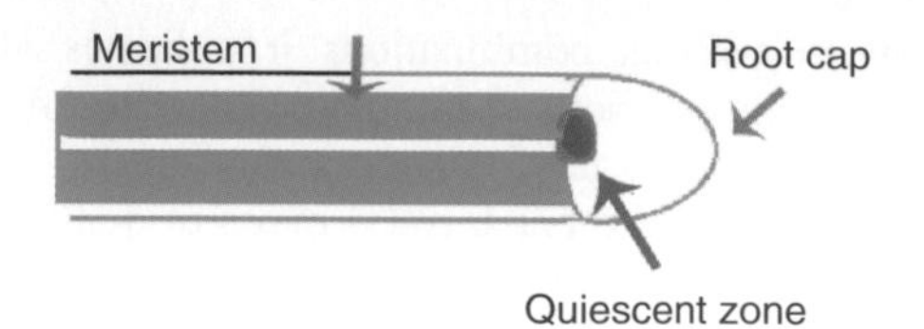

Figure Q4. Quiescent zone

Quinacrine Mustard (ICR 100): A light-sensitive, fluorescent compound used for chromosome staining (see Fig. Q5). Quinacrine (atabrine) staining permitted the first time the visualization of banding in the human chromosomes. It caused particularly bright fluorescence of the long arm of the human Y chromosome and facilitated the recognition of the XYY karyotype (see Fig. Q6). ICR 100 is strongly mutagenic; it is also a highly toxic antihelminthic drug. ►Q banding, ►acridine dyes

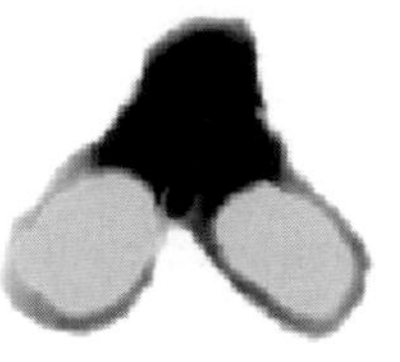

Figure Q5. ICR-stained Y chromosome

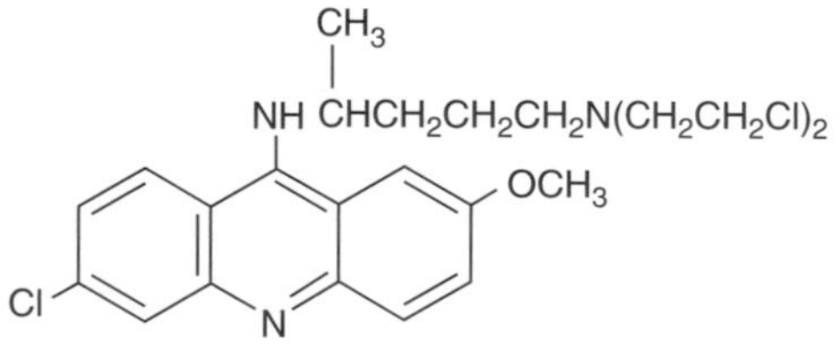

CH_3
NH $CHCH_2CH_2CH_2N(CH_2CH_2Cl)_2$
OCH_3
Cl
N

Figure Q6. Quinacrine mustard

Quintuplex: ►quadruplex

Quormone: Quorum-sensing signaling molecules, acylated homoserine lactones.

Quorum Factors: Signaling molecules in autoinduction. ►autoinduction

Quorum Quenching: *N*-acyl-L-homoserine lactone hydrolase disrupts signals to quorum-sensing (Liu D et al 2005 Proc Natl Acad Sci USA 102:11882).

Quorum-Sensing: A system of cell density-dependent expression of specific gene sets. In the luminescent bacteria *Vibrio fischeri* and *V. harveyi* the quorum-sensing signal is acylated homoserine lactone (AHL). AHL triggers biofilm production in the infectious *Pseudomonas aeruginosa* and that protects the bacteria from antibiotics. AHL production is quite widespread among bacteria, including bacteria living on plant hosts. *P. aeruginosa*—in response to AHL—can produce, e.g., phenazine (mutagen involved also in electron transport), an antibiotic that keeps away other (Gram positive) bacteria and thus may protect its host, e.g., wheat. Besides AHL, other quorum-sensing signals have been detected in various bacteria and also in fungi. The toxic *Enterococcus faecalis* when it detects target cells, releases high level of cytolysin (Coburn PS et al 2004 Science 306:2270). In several species of bacteria quorom-sensing is dependent of an autoinducer (AI-2) product of the LuxS enzyme (4,5-dihydroxy-2,3-pentanedione, Xavier KB, Bassler BL 2005 Nature [Lond] 437:750). AI-2 signal transduction requires the integral membrane receptor LuxPQ, which has a periplasmic component LuxP and a histidine sensor kinase subunit, LuxQ. Light-induced conformation change in LuxPQ seems to regulate quorum sensing (see Fig. Q7) (Neiditch MB et al 2006 Cell 126:1095). ►autoinduction, ►biofilm, ►multicellular, ►cytolysin, ►quorum quenching, ►luciferase; Miller MB Bassler BL 2001 Annu Rev Microbiol 55:165; Fuqua C et al 2001 Annu Rev Genet 35:439; Mok KC et al 2003 EMBO J 22:870; Miller ST et al 2004 Mol Cell 15:677; Neiditch MB et al 2005 Mol Cell 18:507.

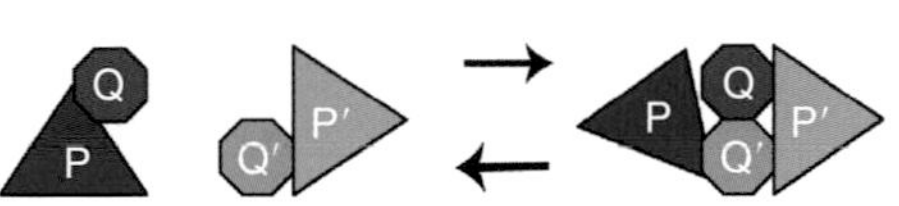

Figure Q7. Cartoon of conformational changes in the four lux subunits leading to quorum sensing

q.v. (quod vide): see it.

Historical vignettes

The majority of geneticists know that Carl Correns was one of the three rediscoverers of the Mendelian principles in 1900. Actually he named them as Mendel's Rules. In the same year in a footnote he also noted that the allelic frequencies comply with the $p^2 + 2pq + q^2$ binomial (nine years before Hardy and Weinberg), and reported linkage in *Matthiola*.. He was also one of the discoverers of cytoplasmic inheritance. In 1902 (Botanische Zeitung 60:64–82) he suggested a mechanism for crossing over nine years before Morgan's paper appeared in J. Exp. Zool. (11:365). "We assume that in the same chromosome the two Anlagen of each pair of traits lie next to each other (A next to a and B next to b, etc.) and that the pairs of Anlagen themselves are behind each other. The picture is shown in Fig. 1.

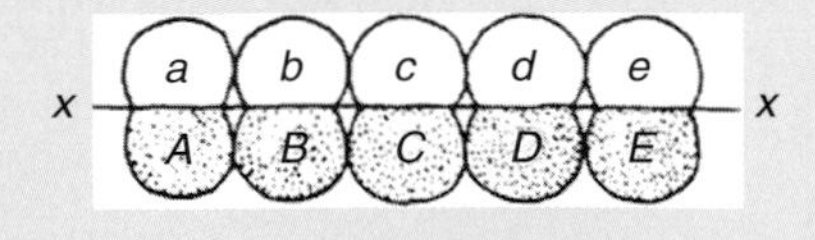

A, B, C, D, E, etc. are the Anlagen of parent I; a, b, c, d, e, etc. are those of parent II.

Through the usual cell- and nuclear divisions the same type of products are obtained as the chromosomes split longitudinally..."

When one pair contains antagonistic Anlagen, while the rest of the pairs are formed of two identical types of Anlagen, or the Anlagen are 'conjugated' as they are in *Matthiola* hybrids, which I have described, then further assumptions are *necessary*... Then AbCdE/aBcDe and aBcDe/AbCdE yield both AbCdE and aBcDe; ABcdE/abCDe and abCDe/ABcdE both ABcdE and abCDe, etc."

Peter Starlinger (discoverer of insertion mutations in bacteria with Heinz Saedler in 1972 [Biochimie 54:177) made the remark below in 2005 in Annu. Rev. Plant Biol. 56:1.

"It was only then that we realized the relation of these element to McClintock's transposable elements, in spite of the fact that I had known McClintock's work since my student days, and in spite of a series of seminars that we had held on this topic in the institute in Cologne. Sometimes we are blind!"

R

R (r, Röntgen, Roentgen): This is a unit of ionizing radiation (1 electrostatic unit of charge in 1 cm^3 dry air at 0 °C and 760 mm pressure; about 93 ergs/living cells). ►Rad, ►Rem, ►rep, ►Gy, ►Sv, ►cR, ►measurement units

R: This is a free software environment for statistical computing and graphics. (See http://www.r-project.org/).

ρ (population recombination parameter): This is inversely proportional to linkage disequilibrium in a population (Hudson RR 2001 Genetics 150:1805). The size of ρ under conditions of selection is generally lower than expected under condition of neutral equilibrium; conversely linkage disequilibrium increases under domestication (Wright SI et al 2005 Science 308:1310). ►linkage disequilibrium

ρ (rho): ►buoyant density, ►petite colony mutants, ►transcription termination in prokaryotes

R1: Refer to methylation sites in the cytoplasmic region of *E. coli* chemotaxis transducer proteins.

R1, R2: These are ubiquitous retroposons in arthropod ribosomal RNA. ►retroposon; Pérez-González CE, Eickbush TH 2002 Genetics 162:799.

R2: ►hybrid dysgenesis I–R

rII: ►rapid lysis mutants of bacteriophages

R Bands: These are heat-denaturation resistant chromosomal bands and half of them have telomeric sites. The bright field R bands usually show the reverse Giemsa pattern of the bands (see Fig. R1). ►isochores, ►C banding, ►G banding, ►chromosome banding; Dutrillaux B, Lejeune J 1974 Adv Hum Genet 5:119.

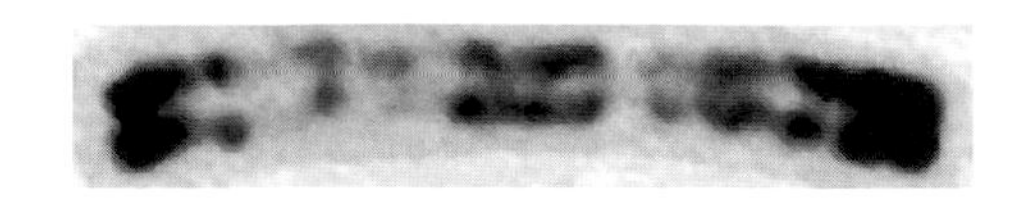

Figure R1. R-banded human chromosome

R Bodies: These refractive bodies are temperate bacteriophages within the κ particles of paramecia. ►symbionts hereditary, ►*Paramecium*

R End: ►packaging of λ DNA

R Factors: Denote resistance factors in bacterial plasmids that may make the host bacteria insensitive to antibiotics and to normally bacteriotoxic drugs. They are common among gram negative bacteria and are readily transmitted to other strains because the plasmids are generally endowed with transfer factors. These plasmids usually do not integrate into the bacterial host genome but can recombine with each other and generate new plasmids with multiple resistance factors. Because of the presence of multiple resistance factors, only simultaneous administration of multiple antibiotics may stop the multiplication of the bacteria. ►plasmid, ►plasmid mobilization, ►plasmid conjugative, ►antibiotics; Watanabe T 1963 Bacteriol Rev 27:87; Falkow S 1975 Infectious multiple drug resistance, London; Patterson JE 2000 Semin Respir Infect 15[4]:299.

R Group (a radix): Abbreviation of an alkyl group or any other chemical substitutions.

R Locus of Maize: Along with the B locus this is involved in the activation of several genes of the anthocyanin biosynthetic pathway. These genes are separable by recombination and give rise to the phenotypes represented in Figure R2. P stands for plant color and S for seed color. In this figure the embryo represents the plant. The R locus includes a large number of different alleles. A detailed structure of the locus is described by May & Dellaporta (1998 Plant J 13:247), and Walker (1998 Genetics 148:1973). Additional references are also provided. ►tissue specificity, ►paramutation

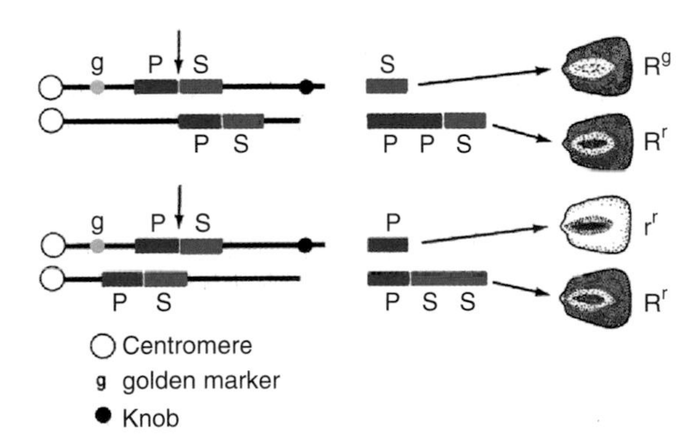

Figure R2. Resolution of the *R* locus of maize

R Loop: Refers to the DNA strand displaced by RNA in a double-stranded DNA-RNA heteroduplex; also the genomic DNA intron forms a R loop when the gene is hybridized with cDNA or mRNA. At the beginning of the replication of the mtDNA a R loop is formed which is synthesized on the light strand of the mtDNA. This R loop is processed into primers for the heavy chain replication. ►DNA replication mitochondrial, ►primer; White RL, Hogness DS 1977 Cell 10:177.

R Plasmids: These carry resistance factors (genes for antibiotic resistance and other agents).

R Point (restriction point): Before the S phase, cells in the G1 phase pause and may or may not continue the cell cycle. Cancer cells bypass the restriction point and continue uncontrolled divisions. Cultured cells may require serum or amino acids to pass from G1 to S. ►cell cycle; Ekholm SV et al 2001 Mol Cell Biol 21:3256.

R Unit: ►r (Röntgen)

RA (rheumatoid arthritis): ►autoimmune disease, ►rheumatoid fever

rAAV: Refers to recombinant adeno-associated virus. ►adeno-associated virus

RAB: RAS oncogene homologs (20–29 kDa guanosine triphosphatase) that regulate transport between intracellular vesicles (Golgi apparatus), and control endosome fusion. It has been traced to human chromosome 19p13.2. In the human genome there are more than 60 Rabs. Rab3A functions in neural synaptic vessels. Rabs have an ever-increasing number of effectors, which control specific functions. ►RAS oncogene, ►Sec, ►Ypt, ►Mss, ►synaptic vessel, ►GTPase, ►endosome, ►SNARE, ►NSF, ►Golgi apparatus, ►RNA export, ►EEA1, ►SNARE, ►Griscelli syndrome, ►Warburg micro syndrome, ►Charcot-Marie-Tooth syndrome; Zerial M, McBride H 2001 Nature Rev Mol Cell Biol 2:107; Rak A et al 2003 Science 302:646; homolog specificities: Eathiraj S et al 2005 Nature [Lond] 436:415; RAB structure: Pan X et al 2006 Nature [Lond] 442:303; review of Rab effectors and functions: Grosshans BL et al 2006 Proc Natl Acad Sci USA 103:11821; Rab family of proteins: http://www.ncbi.nlm.nih.gov/Structure/cdd/cddsrv.cgi?uid=cd00154.

Rabbit: *Oryctolagus cuniculus*, 2n = 44; *Sylvilagous floridanus*, 2n = 42. ►hare; See http://locus.jouy.inra.fr/cgi-bin/lgbc/mapping/common/intro2.pl?BASE=rabbit.

Rabbit Reticulocyte In Vitro Translation: Mammalian mRNA (extracted from cells or transcribed in vitro) can be translated into protein under cell-free conditions using lysates of immature red blood cells of anemic rabbits. Anemia is induced in animals by subcutaneous injections of neutralized 1.2% acetylphenylhydrazine solutions (HEPES buffer) for five days. After the larger white blood cells are removed by centrifugation, the red blood cells are lysed at 0 °C by sterile double-distilled water. Then the endogenous mRNA is destroyed by micrococcal nuclease in the presence of Ca^{2+}. Without calcium the nuclease does not work. The reaction is stopped by EGTA (ethylene glycol tetraacetic acid, which chelates calcium). Hemin [$C_{34}H_{32}ClFeN_4O_4$], dissolved in KOH, is needed for suppressing an inhibitor of eukaryotic translation-initiation factor eIF-2. The translation mixture must contain spermidine or RNasin ribonuclease inhibitors, creatine phosphate (an energy donor), dithiothreitol (a reducing agent to prevent the formation of sulfoxides from the S-labeled amino acids), all normal amino acids (except the one which will carry the radioactive label), buffer, radioactive amino acid (e.g., [^{35}S]methionine), reticulocyte lysate, tRNAs, KCl and magnesium acetate (to enhance translation) and polyadenylate-tailed mRNA (to be translated into protein). It should be ensured that all solutions are made of RNase-free material and the vessels are free of RNase. Incubation is at 30 °C for 30 to 60 min. Before precipitating (by 10% trichloroacetic acid) the synthesized protein, the ^{35}*S*-methionine-tRNA is destroyed either by 0.3 N NaOH or, in case SDS-polyacrylamide gels are used for subsequent analysis, by pancreatic ribonuclease. Immunoprecipitation may also be used for the analysis of the translation product. The amount of protein synthesized can be measured by scintillation counting. Rabbit reticulocyte lysates are also available commercially. Numerous variations of the procedure are available in laboratory manuals. Alternatively, wheat germ extract may be used for in vitro translation. ►wheat germ translation system, ►eIF-2, ►translation repressor proteins, ►polyA mRNA, ►SDS polyacrylamide gels, ►immunoprecipitation, ►scintillation counters; Olliver L, Boyd CD 1998 Methods Mol Biol 86:221; Lorsch JR, Herschlag 1999 EMBO J 18:6705.

Rabies: Encephalomyelitis caused by infection of a non-segmented negative-strand RNA virus. The (−) strand genome is condensed as a nucleoprotein into a nucleocapsid and serves as a template for the RNA-dependent RNA polymerase. Replication produces a full length (+) viral RNA copy, which serves as template for the (−) RNAS genomic copy and is then encapsidated in the virion. Switching from transcription to replication is regulated by the amount of nucleoprotein in the cytoplasm. The 11 oligomers are built from 99-nucleotide RNA segments (see Fig. R3). In some insects the nucleoprotein rings may have 9, 10, 11, 12 or 13 copies (Albertini AAV et al 2006 Science 313:360). The onset of the disease is characterized by inflammation and hyperactivity eventually leading to death. A wide range of wild (raccoons, foxes, mice) and domestic animals (dogs, cats) are susceptible to infection through saliva in their bites. Humans have relatively greater resistance, there is about 15% lethality without treatment. For prevention attenuated virus or genetic immunization (immunoglobulin G) may be used. ►replicase,

►vaccination, ►genetic immunization, ►encephalomyelitis, ►positive strand, ►RNA viruses, ►segmented genomes; new immunotherapy: de Kruif J et al Annu Rev Med 58:359.

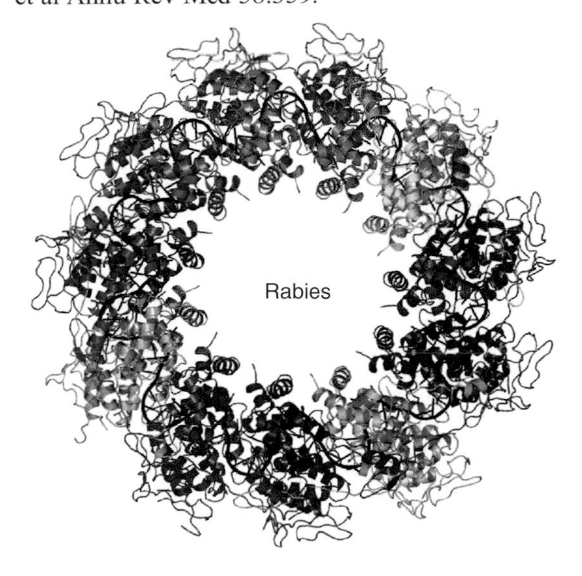

Figure R3. Ribbon diagram of the 11-nucleotide oligomer nucleoprotein-RNA ring viewed from bottom. Nucleoprotein protomers are colored differently; the RNA is represented by thin black coil. (Courtesy of Winfried Weissenhorn and Aurélie A.V. Albertini)

Rabl Orientation: In the late 1800s K. Rabl provided evidence for the continuity of the chromosomes inasmuch as in the very early prophase the chromosomes emerge from the premeiotic interphase in the same configuration as they entered the anaphase and telophase, i.e., the centromeres target the nuclear envelope at special locations and face the centrioles in close proximity. ►co-orientation; Marshall WF et al 1996 Mol Biol Cell 7:825.

Rabson-Mendenhall Syndrome (19p13.2): This is a dominant insulin-receptor defect (degradation) resulting in insulin-resistant diabetes. The condition is characterized by hypertrophy of the pineal gland, dental and skin anomalies (acanthosis nigricans) and early lethality. ►pineal gland, ►brain, ►diabetes, ►insulin-receptor protein

RAC (same as Akt or PKB): A serine/threonine kinase member of the RAS protein family and transmits signals from the cell surface membrane to the cytoskeleton. Human RAC1 (homolog of CED10) mediates the removal of dead cells destroyed by apoptosis. When activated it inhibits transferrin-receptor mediated endocytosis and along with RHO regulates the formation of clathrin-coated vesicles and actin polymerization. It has an important role in RAS-mediated oncogenic transformation. RAC activates NADPH oxidase and thus free radical production as a defense against infections. Rac2 guanosine triphosphatase is selectively expressed in T_H1 lymphocytes (mediating cellular immunity) and in cooperation with NF-κB it induces IFN-γ promoter. Rac mediates the progression of the cell cycle. The prokaryotic RacA protein fastens the chromosomes to the opposite poles for cell division (Ben Yehuda S et al 2005 Mol Cell 17:773). ►serine/threonine kinase, ►RAS, ►RHO, ►PAK, ►Pac, ►p35, ►clathrin, ►signal transduction, ►cell membrane, ►cytoskeleton, ►PKB, ►endocytosis, ►apoptosis, ►transferrin, ►ROS, ►oxidative burst, ►IFN-γ, ►NF-κB, ►T cell, ►dendritic cell; Mettouchi A et al 2001 Mol Cell 8:115; Hakeda-Suzuki S et al 2002 Nature [Lond] 416:438.

RAC: Recombinant DNA Advisory Committee of the National Institute of Health (USA) oversees the application of recombinant DNA technology. (See http://www4.od.nih.gov/oba/rac/aboutrdagt.htm).

Raccoon (*Procyon lotor*): 2n = 38. This is primarily a North American, omnivorous, fury mammal with an average weight of about 10 kg (see Fig. R4).

Figure R4. Raccoon

Race: A group within a species, distinguished by several characteristics, such as allelic frequencies and morphology. For instance, it has been suggested that a combination of isosorbide dinitrate and hydralazine (BiDil) is particularly effective for the treatment of heart failure among American blacks (Taylor A et al 2004 New Engl J Med 351:2049). For many people, talking about race is discriminatory and, therefore, they deny the racial basis of diseases. Certainly many social, environmental and economic factors contribute to the development of a particular disease but if ethnicity also plays a role, it must not be ignored because of "political correctness" as long as genetic factors common in a particular lineage can help in identifying and curing the disease. ►evolutionary distance, ►human races, ►racism, ►ethnicity

RACE: Refers to rapid amplification of cDNA ends by PCR. ►polymerase chain reaction; Schafer BC 1995 Anal Biochem 227:255.

RACEfrags: Denotes rapid amplification of cDNA fragments.

Racemate: This is a Mixture of D and L optical stereoisomers (enantiomorphs) which subsequently becomes optically inactive. All naturally synthesized amino acids are in the L form but degradation generates the D enantiomorph. The degree of racemization of aspartic acid is faster than that of other amino acids. It has been recently used to determine the authenticity of ancient samples of DNA because the degradation of DNA and the racemization of amino acids, particularly Asp, indicate whether the spurious DNA is really ancient or just a contaminant in the archeological sample. In case the D/L Asp ratio exceeds 0.08 the ancient DNA cannot be retrieved. Degradation also depends on a number of factors, most notably the temperature to which the specimen had been historically exposed. The best preservation has been observed in insects enclosed in amber (although there is some controversy about these samples). In specimens where the D/L Asp ratio is about 0.05, up to 340 bp long DNA sequences can be detected using PCR technology. ►enantiomorph, ►radiocarbon dating, ►evolutionary clock

Raceme: Refers to inflorescence with an elongated main stem and flowers on near equal-size pedicels (see Fig. R5).

Figure R5. Raceme

Rachis: Refers to the axis of a spike (grass ear) and fern leaf (frond).

Racial Distance: ►evolutionary distance

Racism: The assumption of superiority of any particular ethnic group or groups and the consequent inferiority of some others. It advocates hatred and social discrimination on the basis of differences. The origin of racism can be traced back to prehistoric times where it served the purpose of exploitation of conquered or minority groups. Some forms of racism may be found in nearly all societies; even the Bible, arguably, is not exempt from racist ideas, and it has been frequently used as a justification by bigots. In the nineteenth century the rise of the eugenics movement gave false scientific encouragement to racism, providing biological and ideological support for colonialism and social exploitation. Racist ideas were used to justify slavery. Racism culminated in the Third Reich of Hitler's Germany resulting not just in discrimination and suppression, but also in mass physical elimination of "Non-Aryan" people in an attempt to establish Rassenhygiene (race hygiene). Racism cannot be justified on the basis of any scientific evidence and it is morally unacceptable to enlightened societies. Human racial differences are based on a limited number of genes and all racial groups share the vast majority of genes. Actually, the world's most successful societies excelled because of their multiracial and multicultural composition. There is ample biological evidence supporting the superiority of hybrids of mammalian and plant species. ►human races, ►evolutionary distance, ►hybrid vigor, ►eugenics, ►miscegenation, ►human intelligence, ►admixture in populations

RACK: Denotes the receptor for activated C kinase. ►C kinase

RAD: A unit of ionizing radiation absorbed dose (100 ergs/wet tissue). ►r, ►rem, ►Gray, ►Sievert

RAD: The genes of yeast are involved in DNA repair and recombination. ►ABC excinucleases

RAD1: Refers to a yeast gene involved in cutting damaged DNA in association with *RAD10* (ERCC1); its human homolog is XPF/ERCC4. ►mismatch repair, ►RAD51, ►xeroderma pigmentosum, ►aging

RAD2: A yeast gene involved in cutting DNA; its human homolog is XPG. ►DNA repair

RAD3: A yeast DNA helicase and a component of transcription factor TFIIH; its human equivalent is XPD. In yeast *RAD3* regulates telomere integrity. A defect in a Rad3 like protein may be responsible for ataxia telangiectasia. ►DNA repair, ►ataxia telangiectasia, ►telomeres

Rad4: ►RAD23

Rad5: This is a member of the SWI/SNF family of ATPases and it also has the characteristic of ubiquitin ligases. ►SWI/SNF, ►ubiquitin

Rad6: A protein with Ubc2 functions involved in both proteolysis and genetic repair in yeast. ►ubiquitin, ►Ubc2, ►DNA repair

RAD10: A yeast homolog of human gene ERCC1, which is involved in nucleotide excision repair. ►nucleotide excision repair, ►mismatch repair

RAD14: This is a yeast gene, its protein product binds to damaged DNA. The human homolog is XPA. ►DNA repair

Rad18: A yeast protein involved in genetic repair in association with Rad6. ►Rad6

RAD21: This controls double-strand break repair caused by ionizing radiation.

RAD23: This is involved in nucleotide exchange repair. It interacts with the 26S proteasome by binding to the RAD4 repair protein. ►DNA repair, ►proteasome, ►xeroderma pigmentosum

Rad24: This is a 14-3-3 protein regulating nuclear export-import. ►Chk1, ►protein 14-3-3

RAD25 (ERCC): A helicase subunit of the general transcription factor TFIIH which is credited with promoter clearance for the onset of transcription following ATP hydrolysis and after the open promoter complex is formed. It is also a DNA repair enzyme encoded by the xeroderma pigmentosum gene F. ►transcription factors, ►open promoter complex, ►regulation of gene activity, ►helicase, ►DNA repair, ►promoter clearance, ►xeroderma pigmentosum, ►progeria

RAD27/FEN1 (Rthp/Fen-1): This 45 kDa 5′→3′ exonuclease/endonuclease removes the RNA primer from Okazaki fragments with the cooperation of other proteins such RNA-DNA junction endonuclease, PCNA and DNA helicase. The FEN-1/DNase IV protein of eukaryotes performs the same functions as carried out by prokaryotic DNA polymerases beyond polymerization. The eukaryotic cells rely for this function on the PCNA-associated FEN-1. FEN-1 can cut also branched DNA molecules. ►Okazaki fragment, ►DNA replication in eukaryotes, ►flap nuclease, ►PCNA; Lieber MR 1997 Bioassays 19:233; Debrauwère H et al 2001 Proc Natl Acad Sci USA 98:8263.

RAD28: This is the yeast homolog of the gene of the Cockayne syndrome. ►Cockayne syndrome

RAD30: This encodes DNA polymerase η and it homologous to *E. coli's* DinB, UmuC *and S. cerevisiae* Rev1.

RAD50, RAD51, RAD52, RAD53, RAD54, RAD55, Rad56, RAD57: These yeast genes are involved in radiation sensitivity, DNA double-strand break, repair and recombination. Rad51 and Rad52 proteins are vital for eukaryotic recombination. Overexpression of RAD51 and RAD52 reduces double-strand break-induced homologous recombination in mammalian cells (Kim PM et al 2001 Nucleic Acids Res 29:4352). Replication protein A interacts with Rad proteins. The human breast cancer gene forms a complex with hRad50-p95-hMre11 proteins. Rad 53 along with chromatin assembly factor Asf1 mediates the deposition of acetylated histones H3 and H4 on to the newly replicated DNA. ►DNA repair, ►replication, ►replication fork, ►replication protein A, ►radiation-sensitivity, ►recombination mechanisms eukaryotes, ►non-homologous end-joining, ►chromatin assembly; Masson J-Y et al 2001 Proc Natl Acad Sci USA 98:8440; Davis AP, Symington LS 2001 Genetics 159:515.

RAD51: A gene of budding yeast regulates double-strand breaks and genetic recombination depending on ATP. RAD51 protein bears resemblance to the human protein (15q151) with similar functions. The human Rad51 protein (hRAD51) forms a helical filament on single-stranded DNA and regulates homologous recombination by controlling ATP activity through its ATPase function stimulated by Ca^{2+}. Ca^{2+} preserves the hRad51-ATP-ssDNA complex by slowing down ATP hydrolysis. Mg^{2+} promotes ATP hydrolysis and converts the complex into a recombination inactive form ((Bugreev DV, Mazin AV 2004 Proc Nature Acad Sci USA 101:9988). Disruption of *RAD51* in mice has embryonic lethal effects. *RAD51* is homologous with the bacterial gene *RecA* and bacteriophage T4 gene Uvsx mediating strand exchange in genetic recombination. In the recombination function RAD52 and its various yeast homologs (RAD55, RAD57 and other proteins) assist RAD1. The RPA (replication protein A, its homologs are the SSB [single strand binding protein] in bacteria and the p32 protein in phage) prepares the broken ends of the DNA to find the proper sequences in the homologous chromosomes that may be suitable for joining. In the plant *Arabidopsis* the homologous locus has almost the same role in recombination as in other eukaryotes but it has no adverse effect on vegetative growth. The *rad51-1* mutation, however, causes male and female sterility (Li W et al 2004 Proc Natl Acad Sci USA 101: 10596). ►DNA repair, ►SRS2, ►RecA1, ►Dmc1, ►recombination mechanisms in eukaryotes, ►RAD54; Fasullo M et al 2001 Genetics 158:959; Yu X et al 2001 Proc Natl Acad Sci USA 98:8419.

RAD53: A yeast kinase gene encoding pRAD53 signal transducer and S phase checkpoint controller; it is also known as SAD1, MEC2 and SPK1. Rad53 is activated by a conserved protein, Mrc1 (mediator of replication checkpoint) in response to DNA damage.

►MEC1, ►DNA replication; Alcasabas AA et al 2001 Nature Cell Biol 3:958.

RAD54: It has been proposed that this has a helicase function but it appears that this protein is DNA-dependent ATPase. It interacts with RAD1 scaffold and promotes homologous DNA pairing at the expense of ATP hydrolysis. Rdh54/Tid1 of yeast–similar to Rad54–also mediates the DNA exchange. ►helicase, ►ATPase, ►RAD1; Solinger JA, Heyer W-D 2001 Proc Natl Acad Sci USA 98: 8447; Ristic D et al 2001 Proc Natl Acad Sci USA 98:8454; Kim PM et al 2002 Nucleic Acids Res 30:2727.

Radiation, Acute: The irradiation is delivered in a single dose at a high rate, in contrast to *chronic radiation* in which the same dose is administered over a prolonged period of time. ►radiation effects, ►physical mutagens

Radiation, Adaptive: ►radiation evolutionary

Radiation, Background: It includes all radioactive (ionizing) radiation in the environment arising from inadequately shielded X-ray machines, cosmic radiation, fallout, laboratory isotope pollution, and radioactive rocks or radon gas that increases the dose delivered by medical treatment or other intended sources.

Radiation, Brain Damage: Actively dividing cells are most likely to suffer from ionizing radiation. The epidemiological data collected from the population of Hiroshima and Nagasaki indicated that the greatest susceptibility was during the first 8–15 weeks of the human fetus.

Radiation Cancer: Many of the different cancers are associated with chromosomal rearrangement(s), and ionizing radiation causes chromosomal breakage and rearrangements. Proximity of the X-ray-induced breakage sites favors rearrangements. Ultraviolet radiation may be responsible for the induction of skin cancer, especially if the body's genetic repair mechanism is weakened. According to estimates, 10 mSv may be responsible for 1 cancer death per 10,000 people. In the USA the permissible legal dose limit for the public is 0.25 mSv/year but it should be reduced to 0.20 mSv/year. Ionizing radiation is used to treat cancer. ►Sievert, ►DNA repair, ►ultraviolet radiation, ►xeroderma pigmentosum, ►excision repair, ►physical mutagens, ►radiation hazard assessment, ►radiation safety hazards

Radiation Chimera: An antigenically different bone marrow transplant is harbored in the body after extensive radiation treatment destroys or substantially reduces the immune reaction of the recipient. Also, mutant sector(s) caused by radiation-induced mutations or deletions (see Fig. R6).

Figure R6. Chimeric Dahlia flower, the consequence of radiation exposure. (Photograph of Dr. Arnold Sparrow. Courtesy of the Brookhaven National Laboratory, Upton, New York)

Radiation Chronic: ►radiation acute

Radiation Density: This is generally measured by LET (linear energy transfer) values, i.e., the average amount of energy released per unit length of the tract. In case the density is low the genetic damage is expected to be discrete. High LET radiation causes extensive damage along a very short path (see Fig. R7). ►physical mutagens, ►radiation effects

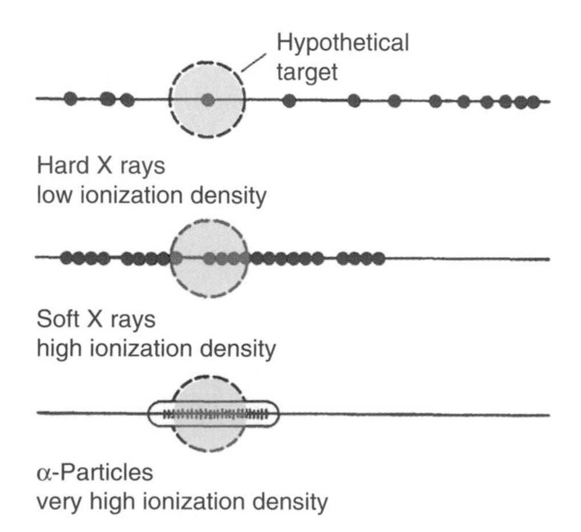

Figure R7. Radiation density. The pattern of ionization density along the track of hard and soft X-rays and α-particles. (After Gray LH from Wagner RP, Mitchell HK 1964 Genetics and Metabolism. Wiley, New York)

Radiation Doubling Dose: ►doubling dose

Radiation Effects: Ionizing radiation may cause gross chromosome breakage (deletions, duplications, inversions, reciprocal translocations, isochromatid breaks, transpositions, change in chromosome numbers if applied to the spindle apparatus) or minute changes including destruction of a single base in the nucleic acids or very short deletions involving only a few base pairs or oxidation of bases. The damage is often clustered. These effects depend on the quality of the ionizing radiation and the status of the biological material involved. The physical effect of radiation is frequently characterized by LET (linear energy transfer in keV/nm path), indicating the amount of energy released per unit tract as ionization and excitation in the biological target (►radiation density). Also, if the dose is delivered at a low rate, most of the damage may be repaired by the metabolic system of the cells. In prophase the chromatid breaks may remain open only for a few minutes but at the interphase they may remain open substantially longer. The frequency of chromosome breakage is considerably increased in the presence of abundant oxygen, whereas anoxia has the opposite effect. Actively metabolizing cells are more susceptible to radiation damage than dormant ones (germinating seeds versus dry ones). Chromosomal aberrations requiring two breaks (e.g., inversions, translocations) occur by second order kinetics whereas the induction rate of point mutations shows first order kinetics. Ultraviolet radiation causes excitation rather than ionization in the genetic molecules. The most prevalent damage is the formation of pyrimidine dimers although chromosome or chromatid breaks may also occur. Radiation-induced malignant growth appears to be mediated by protein tyrosine phosphorylation followed by activation of the RAF oncogene. Suppressing cell proliferation due to interferon regulatory factor (IRF) that arrests the cell cycle may prevent the accumulation of radiation-induced mutations. Independently from IRF, p53 may have a similar effect. Also, ionizing radiation may induce the transcription of p21, a cell cycle inhibitor, by a p53 and IRF dependent mechanism. The adverse effect of radiation therapy may be circumvented by the application of pifithrin (small molecule inhibitor of p53). The potential pathway of events after exposure to radiation is as follows: (i) initial hit, (ii) excitation/ionization, (iii) radical formation and other chemical reactions, (iv) DNA/chromosome damage, (v) DNA repair or mutation, (vi) cell death, (vii) modifying physiological events, and (viii) teratogenesis/carcinogenesis/mutation fixation. For years radiation damage was attributed to damage within the cell exposed, but recent evidence (Azzam EI et al 2001 Proc Natl Acad Sci USA 98:473) has indicated that the damage is also communicated to the neighboring cells through gap junctions with the mediation of connexin 43. As a consequence in the neighboring cells the stress-inducible protein $p21^{Waf1}$ and genetic changes become detectable. The effect of radiation on a wide array of genes (Goss Tusher V et al 2001 Proc Natl Acad Sci USA 98:5116) can be assessed by microarray hybridization. ►ionizing radiation hazards, ►doubling dose, ►hormesis, ►radiation measurement, ►UV, ►ultraviolet light, ►DNA repair, ►RAF, ►signal transduction, ►p21, ►p53, ►interferon, ►radiation hazard assessment, ►X-ray caused chromosome breakage, ►kinetics, ►target theory, ►mental retardation, ►Kudson's two-mutation theory, ►Armitage-Doll model; Hollaender A (Ed.) 1954–55 Radiation Biology, McGraw-Hill, New York, see also General References.

Radiation, Evolutionary: The spread of taxonomic categories as a consequence of adaptation and speciation mediated by forces of selection, mutation, migration and random drift.

Radiation Hazard Assessment: For the exposure of the whole human body to X or γ radiation, the minimal biologically detectable dose in mSv (milliSievert) is: no symptoms 0.01–0.05, chromosomal defects detectable by cytological analysis 50–250, physiological symptoms at acute exposure 500–700, vomiting in 10% of cases 750–1250, disability and hematological changes 1,500–2,000, median human lethal dose 3,000 but at doses above 4,500 the mortality is expected to be over 50%. Prolonged exposure below 1,000 Sv may cause leukemia and death. Single exposure of the spermatogonia to 0.5 Sv may block sperm formation. In mice, the LD_{50}/cGy for primary spermatocytes was 200, meiosis (leptotene-diplotene) 500, diplotene 800, diakinesis 900, secondary spermatocytes 1,000, spermatids 1,500 and spermatozoa 50,000 (mouse data from Alpen, 1998). In mice, 1 Gy may kill the fertilized zygote but the same dose may have no such effect 5–8 days after conception, indicating the potentials of cell replacement. In humans, the most common response of the fetus is mental retardation following intrauterine exposure to 1–5 Gy. Carcinogenesis is well demonstrated by radiation exposure but the course of the initial effect may be greatly influenced by a variety of innate and environmental factors. The risk of very low doses of radiation is very difficult to quantify. Generally, extrapolation is used from higher doses and therefore the risk may be under- or overestimated (Brenner DJ et al 2003 Proc Natl Acad Sci USA 100:13761).

R

A single exposure of women to 3–4 Sv and 10–20 Sv over a longer time (2 weeks) may result in permanent sterility. A fetus in a pregnant woman should not be exposed to any radiation but in the case of an emergency it should not exceed a total of 0.005 Sv and any person under 18 years of age should not receive an accumulated dose over 0.05 Sv/year. A single 0.1 Sv dose may cause cancer in 0.01 fraction of the exposed individuals. For *occupational exposure*, the maximal limit in mSv for the whole body is 50, for lens of the eye 150, and for other specific organs or tissues 500. In the case of cumulative exposure the maximal limit should be below 10 mSv × age in years. For public, educational and training exposure, the recommended maximal limit in mSv/year is 1 for the whole body, and 5.0 for the eye, skin and extremities. In the spring of 1996 the European Union (EU) lowered the permissible dose limits following which members of the public can be exposed to a maximum of 1 mSv every year (the earlier limit was 5 mSv). However, recent analysis has revealed no harmless dose even at low exposure. The limit of exposure for the personnel of radiation industry is now 100 mSv over five consecutive years and an average limit of 20 mSv/year (previous limit was 50 mSv). These guidelines must be implemented within four years by EU member states.

The exposure by routine *medical* X-ray examination is not supposed to be higher than 0.04 to 10 mSv, by fluoroscopy or X-ray movie not more than 25 mSv, and by dental examination involving the entire jaws not above 30 mSv. Nuclear medicine using radioactive tracers or positron emission tomography also involves some exposure (~0.012 Sv). The replacement of ^{131}I by ^{123}I is desirable for thyroid analyses. Smoking tobacco may increase the degree of exposure to ^{210}Pb and ^{210}Po. By comparison, in a normally operated nuclear power plant the exposure may range from 3 to 30 mSv/year. Those living in a granite building may be exposed to 5 mSv/year and a transcontinental flight may involve an exposure of 0.03 mSv. The terrestrial radiation (^{40}K, ^{87}Rb, U, Th series, Rn) may also be a source to reckon with. Radiation from a color television/video display set may be 0.001 mSv/year to the viewer if he/she remains very close to the set, however modern units are safer (2–3 μSv/year). Inspection of luggage at the airports may add 0.002 mrem, and smoke detectors 0.008 mrem to personal exposure. A plutonium-powered cardiac pacemaker may increase the radiation exposure of the wearer by 100 mrem. It should be borne in mind that there may be no threshold below which ionizing radiation would have no effect. It has been assumed until recently that radiation damage is caused by direct hits. There is, however, a significant bystander effect to the cells in the vicinity of the hit cells. Within 1 mm distance α radiation increases the number of micronuclei (broken chromosomes) by a factor of 1.7 and apoptosis increases 2.8-fold. The bystander effect is apparently mediated through gap-junction signals (Belyakov OV et al 2005 Proc Natl Acad Sci USA 102:14203).

The worldwide average exposure from natural sources amounts to ~2.4 mSv, mostly from α particles of radon gas and cosmic rays (muons) and terrestrial γ rays. Moreover, 100 h of air travel adds ~0.5 mSv and medical X-rays contribute ~0.4 mSv annually. The 1945 atmospheric explosion and the following weapon tests constitute a fallout of 0.005 mSv.

The current approximate incidence of mutation in live-born human offspring and the estimated increase per rem per generation in parenthesis are as follows: autosomal dominant 0.0025 – 0.0075 (0.000005 – 0.00002), recessive 0.0025, X-linked 0.0004 (0.000001), translocations 0.0006 (0.0025) and trisomy 0.0008 (0.000001). Approximately 5 Gy (500 rem) is considered the human lethal dose, the bacterium *Deinococcus radiodurans* can recover from doses as high as 30,000 Gy.

The very efficient recombinational repair in this prokaryote can explain this high radiation resistance. The chromosomes are just as well broken into pieces as other DNAs but its genetic material exists in pairs and within 12 to 24 h repair by recombination at the Holliday junctures restores their integrity. Even small doses, such as delivered by therapeutic X-radiation, may increase by about one-third the number of broken chromosomes and radiation by isotope treatment has a similar effect, depending on the dose and the duration of the exposure. Eventually, the broken chromosomes, or at least some of them, may be eliminated from the body. Radiation sensitivity is generally positively correlated with the size of the genetic material although in polyploids the damage may not be readily detectable because of the redundancy of the genes. There is no universal consensus on the hazards involved in exposure to very low levels of ionizing radiation. ►atomic radiation, ►isotopes, ►radiation effects, ►radiation measurement, ►radiation sensitivity, ►radiation protection, ►doubling dose, ►hormesis, ►radiation threshold, ►mental retardation, ►DNA repair, ►Holliday model of recombination, ►X-ray chromosome breakage, ►Sv, ►Gy, ►BERT, ►hemiclonal, ►rem, ►BERT, ►bystander effect; personal exposure: www.umich.edu/~radinfo; Dowd SB, Tilson ER 1999 Practical Radiation Protection and Applied Radiobiology, Saunders, Philadelphia; the book includes more than 100 Internet information addresses; Sankaranarayanan K 1999 Mutation Res

429:45; Mrázek J 2002 Proc Natl Acad Sci USA 99:10943; Health Risks from Exposure to Low Levels of Ionizing Radiation: BEIR VII Phase2 National Acadamic Press, Washington DC, 2005; ►measurement units

Radiation Hybrid (RH): Human chromosomes are broken into several fragments with 8000 rad dose of X-rays. The irradiated cells are quickly fused (with the aid of polyethylene glycol) to somatic cell hybrids with Chinese hamster cells and thus translocations and insertions into the hamster chromosomes are generated. The greater the distance between two human DNA markers, the higher the chances of a breakage. To estimate the frequency of breakage, information is obtained about "recombination" in a manner analogous to classical genetic recombinational mapping. The recombination frequency in radiation hybrids varies between 0 and 1 (no recombination or the markers are always independent, respectively). In meiotic recombination the maximal value is 0.5 for independent segregation. The formula given here and the recombination frequencies (expressed in centiRays [cR]) give an estimate of the frequency of breakage. At 65 Gy the estimated 1 cR ≈ 30 kb, at 90 Gy 1 cR ≈ 55 kb. $\theta = [(A^+B^-) + (A^-B^+)/[T(R_A + R_B - 2R_AR_B)]$ where (A^+B^-) are the hybrid clones retaining A but not B and (A^-B^+) retain B but not A, T = the total number of hybrids and R_A, R_B denote the recombinant fractions. The linkage analysis can be extended to more than two points. The fragments retained can be analyzed by PCR procedures and they are expected to carry with them neighboring sequences and thus provide information on the physical linkage for sequences of about 10 megabase. The recombination process is dose-dependent. In one study 50 Gy permitted the retention of an intact chromosome arm in 10% of the cases whereas 40% had fragments of 3–30 Mb and 50% 2–3 Mb. Using 250 Gy less than 6% of the hybrids involved larger than 3 Mb pieces. If the fragments generated are intended for positional cloning, usually higher doses are used. The retention of fragments varies; centromeric pieces are more likely to be retained. The fragmented DNA can be further analyzed by probing with Southern blots, polymerase chain reaction, using sequence tagged sites (STS), FISH, etc. The fragments are unstable unless they are fused with rodent chromosomes. The radiation hybrids may retain up to a dozen or more fragments. An added special advantage is that genes without allelic variation can also be mapped. The radiation hybrid mapping method has been applied to plants as well. Single maize chromosome additions to oat lines permitted the resolution of 0.5 to 1.0 megabase sequences using 30 to 50 krad γ-rays. The hexaploid oat background assured the survival of the (diploid) maize chromosome fragments. Radiation hybrid transcript maps are available for mouse, rat, human, dog, cat and zebrafish and are useful tools for evolutionary analyses. ►somatic cell hybrid, ►mapping genetic, ►physical mapping, ►framework map, ►Rhalloc, ►Rhdb, ►recombination, ►Gy, ►centiRay, ►STS, ►FISH, ►Southern blot, ►probe, ►positional cloning, ►θ [theta], ►WGRH, ►PRINS, ►IRS-PCR, ►RHKO, ►Rhalloc, ►addition lines; a radiation hybrid map of the mouse genome: http://wwww.ncbi.nlm.nih.gov/genemap; http://www.ebi.ac.uk/RHdb; Van Etten WJ et al 1999 Nature Genet 22:384; Riera-Lizarazu O et al 2000 Genetics 156:327; Olivier M et al 2001 Science 291:1298; Hudson TJ et al 2001 Nature Genet 29:201; Avner P et al 2001 Nature Genet 29:194.

Radiation Hybrid Panel: A set of DNA samples containing radiation hybrid clones derived by the fusion of human and rodent cells. ►radiation hybrid, ►TNG

Radiation Indirect Effects: Radiation generates reactive radicals (e.g., peroxides) in the environment that in turn inflict biological damage. ►radiation effects, ►target theory

Radiation, Ionizing: ►ionizing radiation, ►electromagnetic radiation

Radiation Mapping: ►radiation hybrid; mouse radiation mapping: http://www.jax.org/resources/documents/cmdata/.

Radiation Measurement: ►Geiger counter, ►scintillation counter, ►proportional counter, ►ionization chambers, ►dosimeter pocket, ►dosimeter film, ►thermoluminiscent detectors, ►neutron flux detection, ►radiation hazard assessment, ►autoradiography

Radiation, Natural: ►isotopes, ►cosmic radiation

Radiation vs Nuclear Size: The harmful biological and genetic effects of ionizing radiation depend on the size of the cell nuclei, at the same dose. Larger nuclei present a larger target and suffer more damage than smaller ones. Haploid nuclei are more sensitive because the genes are normally present in a single dose. Polyploids are relatively less sensitive because of the multiple copies of the chromosomes. ►radiation effects, ►ionizing radiation, ►physical mutagens, see Fig. R8.

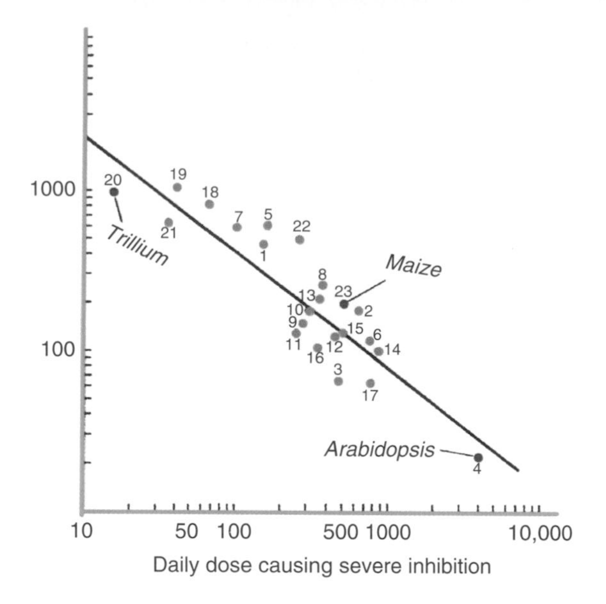

Figure R8. Radiation-sensitivity in plants depending on nuclear volume. 1. *Allium cepa*, 2. *Anethum graveolens*, 3. *Antirrhinum majus*, 4. *Arabidopsis thaliana*, 5. *Brodiaea bridgesi*, 6. *Graptopetalum bartramii*, 7. *Haworthia attenuata*, 8. *Helianthus annuus*, 9. *Impatiens sultanii*, 10. *Luzula purpurea*, 11. *Nicotiana glauca*, 12. *Oxalis stricta*, 13. *Pisum sativum*, 14. *Raphanus sativus*, 15. *Ricinus communis*, 16. *Saintpaulia ionantha*, 17. *Sedum oryzifolium*, 18. *Tradescantia ohiensis*, 19. *Tradescantia paludosum*, 20. *Trillium grandiflorum*, 21. *Tulbaghia violacea*, 22. *Vicia faba*, 23. *Zea mays*. (From Sparrow AH et al Radiation Bot 1: 10)

Radiation Particulate: ►physical mutagens

Radiation Physiological Factors: The effect of radiation may be influenced by the species, age, developmental conditions, type of tissue and cells, metabolic state, genetic back-ground, repair mechanisms, temperature and the chemical environment (presence of oxygen and other enhancing or protective compounds). Actively dividing cells, imbibed seeds are far more sensitive to radiation damage than quiescent tissue or dry material. During pregnancy radiation exposure should be avoided especially during the early stages of gestation when cell division is most rapid. Also, developing children are at a higher radiation risk than adults. Any condition with the possibility of diminished genetic repair increases the chances of chromosomal aberration and gene mutation. ►physical mutagens, ►target theory, ►radiation hazard assessment, ►radiation brain damage, ►radiation vs nuclear size, ►radiation-sensitivity

Radiation Protection: This can be achieved by using isotopes, ventilation should be provided by exhaustion via seamless ducts through the buildings and the particulate material should be trapped in appropriate filters or washed (scrubbed) before environmental discharge.

The source of radiation should be shielded. The transmission of the shielding material depends on the peak voltage of the X-ray or on the energy of the emitting isotope and the thickness of the shield (attenuating material is characterized by half-value [HVL] or tenth-value layers [TVL]) (see Table R1).

Table R1. Data on shielding effectiveness of commonly used radiation insulating material

kV X-ray	Lead in millimeters		Concrete in centimeters	
	HVL	TVL	HVL	TVL
50	0.05	0.16	0.43	0.15
100	0.24	0.8	1.5	5.0
200	0.48	1.6	2.5	8.25
500	3.6	11.9	3.5	11.5
1,000	7.9	26.0	4.38	14.5
4,000	16.5	54.8	9.00	30.00
10,000	16.5	55.0	11.5	38.25
^{60}Cobalt	6.5	21.6	4.75	15.50

Cracks, seams, conduits, filters, ducts, etc. should be regularly checked for possible leaks. Protective clothing (aprons, gloves, etc.) affords very limited protection. Radioactive waste should be disposed of in accordance with the government and local standards, whichever is higher. Caution signs should be used to identify radiation areas. "Radiation Area" is defined by the Occupational Safety and Health Standards of the USA as an area where a major portion of the body could receive in any hour a dose in excess of 5 millirem, or in any 5 consecutive days a dose in excess of 100 millirem. "High Radiation Area" means any area, accessible to personnel, in which radiation is present at such levels that a major portion of the body could receive in any one hour a dose in excess of 100 millirem. For non-ionizing electromagnetic radiation within the range of 10 MHz (megahertz, 10^7 cycles/sec) to 100 Ghz (gigahertz, 10^{11} cycles/sec) the energy density should not exceed 1 mW (milliwatt)/cm^2/0.1 h. [Such radiations are within the realm of radio, microwave and radar range.] Emergency plans must be prepared for spillage, cleanup and fire. Before working with any hazardous material all personnel should be given proper training for safe storage, handling and emergencies involved. ►isotopes, ►atomic radiation, ►ionizing radiation, ►radiation effects, ►radiation hazard assessment, ►radiation measurement, ►electromagnetic radiation, ►sulfhydryl

Radiation Resistant DNA Synthesis: Normally, the ataxia telangiectasia gene/protein (ATM) is a target of the damaging effects of ionizing radiation. ATM activated by radiation results in the activation of the cell cycle (G1→S) checkpoint kinase Chk2. The activation of Chk2 results in the phosphorylation of phosphatase Cdc25A at residue Ser^{125}. Cdc25A prevents dephosphorylation of Cdk2 and this leads to a transient stoppage of DNA synthesis at the S phase. If Chk2 cannot bind or phosphorylate Cdc25A, radiation resistant DNA synthesis proceeds. Thus, Chk2 acts as a tumor suppressor in ataxia telangiectasia. ►Chk2, ►Cdc25, ►cell cycle, ►ataxia telangiectasia; Falck J et al 2001 Nature [Lond] 410:842.

Radiation Response: Deletions (single chromosomal breaks) and mutations occur with 1st order kinetics, whereas the majority of chromosomal rearrangements (inversions, translocations) are proportional to the square of the dose and thus follow 2nd or 3rd order kinetics. The irradiated cell may communicate with the neighboring cells via IL-8 and these neighbors may also suffer oxidative stress as a bystander effect. ►IL8, ►physical mutagens, ►radiation effects, ►kinetics, ►chromosomal aberrations, ►radiation vs nuclear size, ►oxygen effect, ►radiation-sensitivity; Rothkamm K, Löbrich M 2003 Proc Natl Acad Sci USA 100:5057.

Radiation Safety: ►radiation protection, ►radiation hazard assessment, ►atomic radiation, ►cosmic radiation

Radiation Safety Standards: ►radiation hazard assessment, ►radiation protection

Radiation-Sensitivity: This characteristic of the DNA depends on many diverse factors such as the degree of coiling, the status of the nuclear matrix, hydration, copper ions, OH scavengers and thiols. The microbial flora seems to be protective against intestinal radiation effects (Crawford PA, Gordon JI 2005 Proc Natl Acad Sci USA 102:13254). In budding yeast a genome-wide screen of 3670 non-essential genes revealed 107 new loci that influence sensitivity to γ-rays. Further, 50% of these yeast genes display homology to human genes. ►DNA repair, ►chromosome breakage, ►aging, ►microcephaly [Nijmegen breakage syndrome], ►nevoid basal cell carcinoma, ►ataxia telangiectasia, ►xeroderma pigmentosum, ►radiation hazard assessment, ►radiation protection, ►radiation response, ►radiation vs nuclear size, ►*Deinococcus radiodurans*, ►radiation physiological factors; Bennett CB et al 2001 Nature Genet 29:426.

Radiation Sickness: This occurs when the whole human body or parts of it is exposed to ionizing radiation. The symptoms and hazards are dependent on the dose. ►radiation hazard assessment

Radiation Therapy: Ionizing, electromagnetic radiation has anti-mitotic and destructive effects on live tissues and it is used to suppress cancerous growth. The therapeutic effects of radiation in cancer therapy may not have a direct impact on the genetic material of the cancer cells. The effective target may be the surrounding (endothelial) cells that provide angiogenesis to satisfy the cancer cells' increased requirement for blood (Paris F et al 2001 Science 293:293). Radiotherapy has been used to treat lymph nodes to suppress Hodgkin's disease, radioactive isotopes can be injected for localized radiation. Similar effects are expected by the use of magic bullets. Blood withdrawn from the body has been irradiated by UV light and returned to the system. The level of radio-curability varies for tumors of different tissues from 2000–3000 rad for reproductive and nerve tissues, to 5000–6000 rad for lymphatic node and breast cancers, to 8000 rad for melanomas and thyroid cancer. The proper function of the radiation source must be regularly monitored to ensure the safety of patients and operators. It has been assumed that very high doses of radiation may not be as dangerous because they destroy the irradiated cells whereas a lower level of radiation involves higher cancer risks. However, recent observations have indicated that in radiotherapy after the initial killing specific areas in breast and lung cancer are repopulated by new cells, which have a high chance to develop second-cancer growth (Sachs RK, Brenner DJ 2005 Proc Natl Acad Sci USA 102:13040). ►radiation effects, ►magic bullet, ►Hodgkin's disease, ►lymph node, ►radiation hazard assessment; Bharat B et al (Eds.) 1998 Advances in Radiation Therapy, Kluwer, Boston, Massachusetts.

Radiation Threshold: The minimal harmful radiation dose is very difficult to determine because the visible physiological signs may not truly reflect the long-range mutagenic and carcinogenic effects. The maximal permissible doses for medical, diagnostic or occupational radiation exposures reflect only conventional limits that have been revised many times as the sensitivity of physical and biological detection methods as well as the instrumentation of delivery have improved. It is conceivable that no low threshold exists. ►radiation response, ►cosmic radiation, ►mutation frequency-undetected mutations, ►radiation safety standards, ►radiation hazard assessment

Radical: Refers to an atom or a group with an unpaired electron, a free radical.

Radical Amino Acid Substitution: This occurs when nucleotide substitution alters the physicochemical property (e.g., charge, polarity, volume) of the mutant protein. By conservative substitution the physical/chemical characteristics are not altered. The proportions of radical/conservative ratios are positively correlated with the non-synonymous/synonymous replacements. Also, transversions are more likely to cause radical changes. ►synonymous codons; Zhang J 2000 J Mol Evol 50:56; Dagan T et al 2000 Mol Biol Evol 19:1022.

Radical Scavenger: This may combine with free radicals and reduce the potential harm caused by the highly reactive molecules.

Radicle: The seed or primary root of a plant embryo. It also refers to the smallest branches of blood vessels and nerve cells.

Radin Blood Group (Rd): This is encoded in the short arm of human chromosome 1; its frequency is low.

Radioactive Decay: ►isotopes

Radioactive Isotope: ►isotopes, ►radioactive tracer, ►radioactive label

Radioactive Label: Compounds (nucleotides, amino acids) containing radioactive isotopes are incorporated into molecules to detect their synthesis, fate or location (radioactive tracers) in the cells. For detection purposes, scintillation counters or autoradiography is used most frequently. Geiger counters may also qualitatively detect their presence. For cytological analysis, e isotopes that have a short path of radiation and display distinct, sharp marks on the film are used, e.g., tritium (^3H). For molecular biology, most commonly ^{32}P (in DNA, RNA) or ^{14}C and ^{35}S (in proteins) or ^{125}I in immunoglobulins are used. Radioactive labeling for genetic vectors or transgenes is not practical because at each subsequent division the label is diluted to about half. Labeling with integrated genetic markers is more useful because these are replicated along with the genetic material. ►Southern blotting, ►Northern blotting, ►Western blotting, ►nick translation, ►non-radioactive labels, ►isotopes, ►radioactive tracer

Radioactive Tracer: A radioactively labeled compound that permits tracing the biosynthetic transformations of the supplied chemical by determining when the radioactivity appears in certain metabolites after the supply. It also reveals what part of a later metabolite has acquired the label from the supplied substance. The availability of $^{14}CO_2$ permits tracing the path of photosynthesis, and through the use of various isotopes, the metabolism of pharmaceuticals and the role of hormones, etc., can be determined. ►radioactive label, ►radioimmunoassay

Radioactivity: Refers to the emission of radiation (electromagnetic or particulate) by the disintegration of atomic nuclei.

Radioactivity Dating: ►evolutionary clock

Radioactivity Measurement: ►radiation measurements

Radioautography: Same as autoradiography.

Radiocarbon Dating: The age of ancient samples of less than 50,000 years (late Pleistocene to Holocene) is generally estimated by this method in archeology and evolution. Natural radiocarbon is the product of the interaction of cosmic radiation with nitrogen 14 (N^{14}) in the earth's atmosphere. This unstable nitrogen is converted to carbon 14 (C^{14}), which is oxidized to 14 CO_2 and enters plants through photosynthesis and from them to animals. C^{14} is very rare, only $\sim 1 \times 10^{-10}$ fraction of all naturally occurring carbon. Plants absorb this unstable C in proportion to the total C in the environment. When the plant dies, metabolism ceases and in its tissues the proportion of C^{14} decays according to the age of the relic. After every 5568 years the amount of C^{14} isotope is reduced by half. The reduction is measured in terms of the emission of 160 keV β-rays (electrons) by the sample. The wood found in the tomb of the pharaoh in the pyramid Djoser in1949 contained 50% C^{14} indicating that its age was about 5568 years. The validity of this estimate required that the atmosphere contained the same amounts of C^{14} during that period as it did at the time of the analysis. The half-life figure of 5568 was later corrected to 5730 ± 40 (Cambridge half-life). Today improved methods are available such as the accelerator mass spectrometry or liquid scintillation, which provide greater accuracy yet even the most advanced methods struggle with some uncertainties. Nevertheless, the radiocarbon dating principle developed by Willard F. Libby in 1949 earned a much-deserved Nobel Prize in 1960.

Carbon dating is also used for determining human age. The carbon dating method may yield erroneous results if the samples are contaminated by more recent carbon. For example, 1% of such contamination in a 40,000 year old specimen may falsely reduce the age by 7,000 years. Another source of bias is that the C^{14} to C^{12} proportions varied in the earth's atmosphere because the magnetic field varied during the ages and also sunspots affected the amount of cosmic radiation reaching the upper atmosphere. Newer techniques using higher molecular weight of animal and human bone gelatins substantially improved the accuracy level by eliminating contamination by C^{12}. The other source of error was considerably reduced by more accurate determination of C^{14} in deep sea strata during the last 50,000 years. The new calibrations indicate that human dispersal in Europe was thousands of years faster than originally estimated. Also, in Southern France humans appeared around 36,000 BP rather than 31,000 to 32,000 BP. As the methods of calibration improve, changes even in the newer estimates are likely (Mellers P 2006 Nature [Lond] 439:931).

After 1955 (the beginning of the atmospheric nuclear weapons tests) the C^{14} level in the air increased and it was fixed from carbon dioxide by photosynthetic plants and ingested by herbivores through the diet. C^{14} has accumulated in the teeth during the formation of new enamel up to the age of 12. Tooth enamel contains ~0.4% C. In spite of the fact that after the test ban treaties in 1963 the atmospheric C^{14} is decreasing yet the level is sufficient for the determination of its presence in the teeth in proportion to that in the atmosphere at the age of tooth development. Although individual variations in the age of tooth formation and diet affect the incorporated amounts, it is a very sensitive indicator of age (Spalding KL 2005 Nature [Lond] 437:333). ►isotopes, ►half-life, ►evolutionary clock, ►argon dating, ►out-of-Africa hypothesis, ►Eve foremother, ►Lascaux cave; Libby WF 1955 Radiocarbon dating, University Chicago Press; Guilderson TP et al 2005 Science 307:362.

Radioimmunoassay (RIA): The most commonly used isotope for radioimmunoassays, $^{125}I^{+}$ is generated by oxidation of $Na^{135}I$ by chloramine-T (*N*-chlorobenzene sulfonamide). This labels tyrosine and histidine residues of the immunoglobulin without affecting the binding of the epitope and provides an extremely sensitive method for identifying minute quantities (1 pg) of antigen in an experiment. ^{35}S-methionine or ^{14}C can radiolabel target proteins. In the *Competition RIA* an unlabeled target protein competes with a labeled antigen for binding sites on the antibody. The amounts of bound and unbound radioactivity are then quantified. In the *Immobilized Antigen RIA* an unlabeled antigen is attached to a solid support and exposed to a radiolabeled antibody (see Fig. R9). The amount of radioactivity bound then measures the amount of specific antigen present in the sample. In the *Immobilized Antibody RIA* a single antibody is bound to a solid support and exposed to a labeled antigen. Again, the amount of bound radioactivity indicates the amount of antigen present. In the *Double-Antibody RIA* one antibody is bound to a solid support and exposed to an unlabeled antigen. After washing, the target antibody is quantitated by a second radiolabeled antibody (instead of the radiolabel, biotinylation can also be used). This assay is very specific because it involves a step of purification. ►antibody, ►immune reaction, ►immunoglobulins, ►isotopes, ►Protein A, ►IRMA; Eleftherios P et al (Eds.) 1996 Immunoassay. Academic Press, San Diego, California.

Radioisotope: ►isotopes

Radioisotope Dating: ►fossil records

Radiomimetic: These agents, primarily alkylating mutagens, may break single or both chromatids (isochromatids). Although it cannot be ruled out that the two chromatids break at the same place simultaneously, it is believed that isochromatid breaks are due to replicational events involving initially single chromatid breaks. They mimic the effects of ionizing radiation. ►alkylating agents, ►nitrogen mustards, ►sulfur mustards, ►epoxides, ►ionizing radiation; Dustin AP 1947 Nature [Lond] 159:794.

Radiomorphoses: Morphological alterations in plants caused by ionizing irradiation during the life of the irradiated individuals. These effects may not be genetic.

Radionuclide: A nuclide that may disintegrate upon irradiation by corpuscular or electromagnetic radiation. ►nuclides

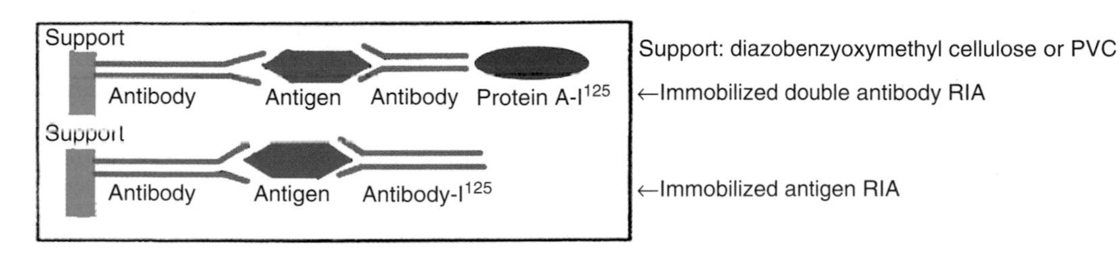

Figure R9. Radioimmuno assay

Radioprotectors: These protect against the harmful effect of ionizing radiation, e.g., sulfhydryl compounds, antioxidants, cysteine, cysteamine, amifostin and melatonin. (See separate entries; Maisin JR 1989 Adv Space Res 9[10]:205).

Radiotherapy: ►radiation therapy

Radioulnar Synostosis with Amegakaryocytic Thrombocytopenia: Refers to dominant (7p15) bone fusion due to mutation in HOXA11 gene associated with abnormal easy bleeding. ►thrombocytopenia; Thompson AA, Nguyen LT 2000 Nature Genet 26:397.

Radish (*Raphanus sativus*): A cruciferous vegetable crop; 2n = 2x = 18. The hybrid of *R. sativus* and *R. raphanistrum* appears to have an unusually superior fitness to both parental species and over a few generations drive to extinction the parental forms within a territory (Hegde SG et al 2006 Evolution 60:1187). ►*Raphanobrassica*, ►*Brassica oleracea*, ►mustards, ►fitness

Radix (root): A multiplier of successive integral powers of a sequence of digits, e.g., if the radix is 4, then 213.5 means 2 times 4 to the second power plus, 1 times 4 to the first power plus 3 times 4 to the zero power plus 5 times 4 to the minus 1 power.

Radon (Rn): ^{219}Rn (An) is a member of the actinium series, ^{220}Rn (Tn) is a member of the thorium emanation group and ^{222}Rn is derivative of uranium, a heavy (generally accumulates in basements), colorless radioactive noble gas which is formed by uranium (radium emanation) contaminated rocks. It may pose health hazards in buildings at some locations and with poor ventilation. In the USA, an estimated 200 mrem of radon is the average annual exposure/person. ►radiation hazards, ►cosmic radiation, ►rem, ►WL; Field RW, Becker K 2001 Radiat Prot Dosimetry 95:75.

RAF1: ►*raf*

raf: v-oncogene (cytoplasmic product is protein-serine/threonine kinase). The cellular homolog RAF1 is closely related to ARAF oncogenes. The v-raf is homologous to the Moloney murine leukemia virus oncogene. The avian MYC oncogene is the equivalent of the murine RAF. RAF1 has been assigned to human chromosome 3 and a pseudogene RAF2 is in chromosome 4. Human renal, stomach and laryngeal carcinoma cells reveal RAF1 sequences. Raf may be recruited to the cytoplasmic membrane by a carboxy-terminal anchor (RafCAAX) and its activation then becomes independent from Raf which is associated with the plasma membrane cytoskeletal elements and not with the lipid bilayer. RAF-1 phosphorylates MEK-1 kinase involved in the signal transduction process of extracellular signal regulated kinases. RAF induces NF-κB through MEKK1. Raf-1 is regulated by RKIP (Raf kinase inhibiting protein). Phosphorylation at the appropriate sites activates/inactivates RAF. The activation of Raf by basic fibroblast growth factor results in phosphorylation of serine 338 and 339 and activation by VGF phosphorylates tyrosine 340 and 341. Both pathways protect against apoptosis (Alavi A et al 2003 Science 301:94). ►ARA, ►MYC, ►Moloney, ►signal transduction, ►RAS, ►BRAF, ►v-oncogene, ►PAK, ►MAP kinase NF-κB, ►MEKK, ►FGF, ►VGF, ►apoptosis; Chong H et al 2001 EMBO J 20:3716.

RAFT: The association of sphingolipids and cholesterol (~50 nm in diameter) and it mediates membrane traffic and cell signaling in mammals. Lipid rafts incorporate glycosylphosphatidyl inositol-anchored proteins, doubly acylated peripheral membrane proteins, cholesterol-linked proteins and transmembrane proteins. Host membrane-derived lipids surround enveloped viruses (e.g., HIV) that are acquired during budding in the replication process. The composition of these lipids is different from that of the host membrane. These lipid-enclosed structures indicate the existence of rafts within living cells (Brügger B et al 2006 Proc Natl Acad Sci USA 103:2641). Each raft does not carry more than 10 to 30 proteins. ►caveolae, ►endocytosis, ►sphingolipids, ►cholesterols, ►TOR, ►protein transport, ►liposome, ►SFK, ►HIV; Langlet C et al 2000 Curr Opin Immunol 12:250; Brown DA, London E 2000 J Biol Chem 275:17221; Simons K, Toomre D 2001 Nature Rev Mol Cell Biol 2:216.

RAFTK: ►CAM

RAG1, RAG2 (recombination activating gene): These are closely linked (11p13) and encode the proteins of lymphocyte-specific recombination of the V(D)J sequences of immunoglobulin genes. The functional part of RAG1 is the core sequence whereas other tracts can be deleted without affecting recombination. Mutation in RAG results in the inability to form functional antigen receptors on B and T cells and antibodies. The recombinational cleavage takes place between the *coding sequence* of the immunoglobulin genes and the so-called *recombinational signal sequence* (RSS) nucleotides. Recombination usually occurs between the original coding sequences or it may take place by the rejoining of one coding end, the signal sequence, which originally belonged to the other coding end (hybrid joint) or the same coding and signal sequence end can be reunited (open-and-shut joint). Besides RAGs, the joining reaction requires the double-strand repair protein XRCC4, a DNA-dependent protein kinase and the Ku protein(s). For recombination, an accessory protein HMG1 or HMG2 is also required. Nucleotides may be added or deleted in each type of joining. RAG1 and RAG2 are actually transposons but after the joining of the V(D)J ends, normally following the formation of the antibody genes, RAGs are inactivated. RAG proteins seem homologous to the Tc*1* transposon of *Caenorhabditis*. The V(D)J recombinase shows a sequence similarity to retroviral integrase superfamily (Zhu L et al 2004 Nature [Lond] 432:995). RAG1 and RAG2 genes are coordinately regulated by cell-type specific elements upstream of RAG2. The 5′ promoter upstream sequences in RAG2 regulate B and T cells differently. For T cells there are four T cell receptors (TCR) and for the B cell to recombine there are three immunoglobulin loci. ►antibody, ►immunoglobulins, ►junctional diversification, ►NHEJ, ►combinatorial diversification, ►RSS, ►V(J)D recombinase, ►XRCC4, ►Ku, ►DNA-PK, ►TCR, ►hybrid dysgenesis, ►reticulosis familial histiocytic, ►integrase; Schultz HY et al 2001 Molecular Cell 7:65; Qiu J-X et al 2001 Molecular Cell 7:77; Raghavan SC et al 2001 J Biol Chem 276:29126; Jones JM, Gellert M 2001 Proc Natl Acad Sci USA 98:12926.

RAGE (receptor for advanced glycation endproduct): A member of the immunoglobulin protein family on the cell surface. It interacts with multiple ligands and mediates homeostasis, development, inflammation, tumor proliferation, development and manifestation of some diseases (diabetes, Alzheimer's disease). RAGE is also a receptor for amphoterin. In cooperation with $p21^{Ras}$, MAP, NF-κB, CDC42, SAP/JNK, p44/p42 may alter cellular programming. By blocking the major factor complex of RAGE—amphoterin may prevent invasiveness of cancer and metastasis. (See individual entries of mentioned terms).

RAGE: Refers to random activation of gene expression.

Ragweed: A largely annual species of the genus *Ambrosia elatior* (Compositae) which is widespread in North America and Central Europe and causes pollen allergy (hay fever) of the nose and eye, skin irritations and even asthma without hay fever. The susceptibility is genetically controlled by a locus, *Ir*, within the HLA complex. The antigen E contained in the ragweed pollen elicits the IgE antibody. ►HLA, ►allergy, ►atopy (See Fig. R10).

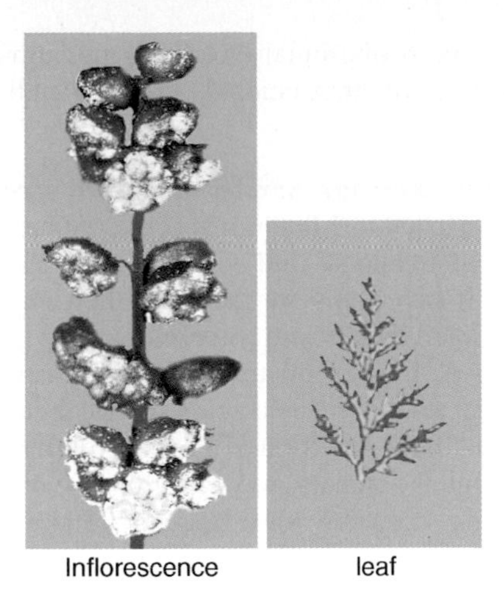

Figure R10. Ragweed *(A. artemisifolia)*

RAIDD: An adaptor protein that joins the ICE/CED-3 apoptosis effector molecules. ►apoptosis, ►ICE

RALA: A RAS-like protein encoded in human chromosome 7. ►RAS oncogene

RALB: A RAS-like protein encoded in human chromosome 13. ►RAS oncogene

Raloxifene: A lipid-like molecule that enters the cell nucleus and can bind to the special raloxifene-response element in the DNA and activate the tumor growth factor-β3 gene (see Fig. R11). 17-epiestriol, an intermediate in the secretion of estrogen, activates Raloxifene. Raloxifene, an anti-estrogenic compound, is used in chemotherapy of breast cancer. Its advantage over the related compound tamoxifen is that it does not pose an appreciable risk for uteral cancer. ►hormone-response elements, ►estrogen receptor, ►tamoxifene, ►breast cancer, ►estradiol; Greenberger LM et al 2001 Clin Cancer Res 7:3166.

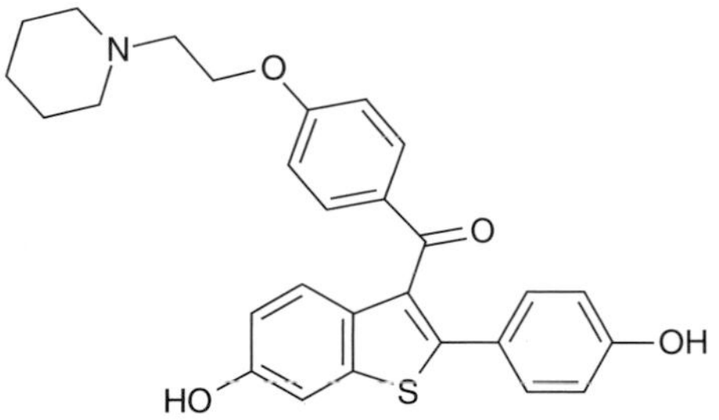

N O O OH HO S

Figure R11. Raloxifene

RAM: ►random-access memory in the computer, also rabbit-antimouse immunoglobulin

ram: Refers to ribosomal ambiguity mutation causing high rate of translational error. ►ambiguity in translation

Raman Spectroscopy: Similar to infrared spectroscopy, it uses 10,000 to 1,000 nm spectral regimes to detect different rotational and vibrational states of molecules. If two atoms are far apart, their interaction is negligible. If they are very close they may show repulsion. It is useful to obtain information on the physical state of nucleic acids. The technique can be applied for the detection of a variety of breast cancers. Although the signals are weaker than in other optical methods, they are very specific (Haka AS et al 2005 Proc Natl Acad Sci USA 102:12371). ►FT-IR; Thomas GJ Jr 1999 Annu Rev Biophys Biomol Struct 28:1.

Ramet: This is a clonal descendant of a plant capable of independent reproduction. ►genet

Ramie (*Boehmeria nivea*): A subtropical-tropical, monoecious fiber plant; 2n = 2x = 14.

Rammer: Refers to the annotation of ribosomal genes (Lagesen K et al 2007 Nucleic Acids Res 35:3100).

RAMP (ribosomally encoded antimicrobial peptide): Refers to the natural defense molecules of animals, plants and fungi. ►antimicrobial peptides, ►defensin

Ramus (plural rami, Latin word): Denotes a branch; it is used in various word combinations.

RAN (RAS-like nuclear G protein, TC4): A guanosine triphosphatase (GTPase) which is required for import and export through nuclear pores and activation of the mitotic spindle. It acts as a switch of GTP→GDP in DNA synthesis and cell cycle progression. The proper balance between RanGTP and RanGDP determines the nucleo-cytoplasmic traffic. Cytoplasmic RanGTP may inhibit transport to the nucleus. The binding protein RanBP enhances the activity of RanGTP-activating protein RanGAP. RanBP1 is a nucleus-localized transporter binding well to RanGTP and weekly to RanGDP. RanBP2 is localized primarily in the cytoplasmic fibers of the nuclear pore complex. RAN also monitors the integrity of tRNAs before they are exported to the cytoplasm. RAN processes the 7S RNA into 5.8 ribosomal RNA of eukaryotes. ►GTPase, ►RAS, ►importin, ►export adaptor, ►signal transduction, ►cell cycle, ►RNA transport, ►RCC, ►GAP, ►nuclear pore, ►nuclear localization signal, ►Pat1, ►meiosis, ►TPX, ►SUMO; Nachury MV et al 2001 Cell 104:95; Carazo-Salas RE et al. 2001 Nature Cell Biol 3:228; Seewald M J et al 2002 Nature [Lond] 415:662.

Rana (genus of frogs): Both the American leopard frog (*Rana pipiens)* and the European frog *R. temporaria* have 2n = 26 chromosomes. Frogs are generally more aquatic than toads. ►frog, ►*Bufo*, ►*Xenopus*, ►toad

Random Access Memory (RAM): This storage of information can be referred to at any order as long the computer is switched on.

Random Amplified Polymorphic DNA: ►RAPD

Random Chromosome Segregation: The gene under study is (absolutely) or very closely linked to the centromere in polyploids. ►maximal equational segregation

Random Fixation: ►random genetic drift, ►founder principle

Random Genetic Drift: A change in gene frequency by chance. Such random changes are most likely to occur when the effective size of the population is reduced to relatively few individuals. Since random drift may occur repeatedly, large changes may eventually result in the genetic constitution of the population. Such changes are likely when a few individuals migrate to a new (isolated) habitat. ►effective population size, ►founder principle; Whitlock MC 2000 Evolution Int J Org Evolution 54:1855.

Random Mating: Each individual in the population has an equal chance to mate with any other of the opposite sex (panmixis). Random mating is assumed in the majority of principles of theoretical population genetics. The rules are based on the Hardy-Weinberg theorem and on that basis the genetic structure of the population can be predicted as long as the allelic frequencies do not change or the change is negligible. Allelic frequencies may be altered primarily by selection, migration and, if the size of the population is very small, by random genetic drift. In the short run mutation does not affect allelic frequencies because of its rarity and the chances of survival of the majority of new mutations are low.

In a population involving two different allelic pairs the frequency of the mating genotypes and the genotypic proportion of their progenies derived from the binomial distribution by expanding $(p + q)^4$ (see Table R2). ►Hardy-Weinberg theorem, ►mating systems, ►Pascal triangle

Random Oligonucleotide Primers Used For Synthesis of Radioactive Probes: Heterogeneous oligonucleotides can anneal to different and many positions along a nucleic acid chain. They can also serve as primers for the initiation of DNA synthesis. If the precursors are one type of radioactive [α-^{32}P]-deoxyribonucleotide (dNTP), and cold dNTPs, highly radioactive

Table R2. Random mating. MATES → (A1A1) × (A1A1), (A1A1) × (A1A2), (A1A2) × (A1A2), (A1A2) × (A2A2), (A2A2) × (A2A2), (A1A1) × (A2A2)

Frequency → p^4	$4p^3q$	$6p^2q^2$	$4pq^3$	q^4
Progeny	A1A1	A1A2	A2A2	
	p^4	$2p^3q$	p^2q^2	
	$2p^3q$	$4p^2q^2$	$2pq$	
	p^2q^2	$2p^3$	q^4	
	4	8	4 → Sum = 16	

probes can be obtained. Single-stranded DNA templates can be copied with the aid of Klenow fragment of DNA polymerase I or in the case of RNA template reverse transcriptase can be used. The primers are usually short (6 to 12 base) and can be generated either by DNase digestion of commercially available DNA (from calf thymus or salmon sperm) or by an automatic DNA synthesizer. ►probe, ►nick translation

Random Sample: This type of sample is withdrawn from a collection without any selection.

Random Variables: These may occur unpredictably in a sample because of unknown factors but their breadth can be described by statistical probability. ►significance level, ►inference statistical

Random Walk: This is a physical theory of material distribution within media such as cell migration within connective tissues, Markov processes in DNA, diffusion in gases, liquids and solids, population changes as a result of birth and death, and evolutionary changes such as the emergence and extinction of species. (Berg HC 1993 Random walks in biology, Princeton University Press, Princeton, New Jersey; Cornette JL, Lieberman BS 2004 Proc Natl Acad Sci USA 101:187).

Range Constraints: The number of repetitive units of a microsatellite has limitations because of viability or adaptability. ►microsatellite

RANK (receptor activator of NF-κB): ►TRANCE

RANTES (regulated on activation normally T-cell expressed and secreted): This is a chemoattractant of cytokines for monocytes and T cells. The chemokine receptors appear to be seven membrane proteins, coupled to G proteins. RANTES is also involved in a transient increase of cytosolic Ca^{2+} as well as Ca^{2+} release. The opening of the calcium channel increases the expression of interleukin-2 receptor, cytokine release and T cell proliferation. In addition to inducing chemotaxis, RANTES can act as an antigen-independent activator of T cells in vitro. RANTES and MIP-1 chemokines along with the receptor CC CKR5 and fusin are believed to suppress replication of HIV. ►MIP-1a, ►fusin, ►chemotaxis, ►cytokine, ►CC CKR5, ►HIV, ►acquired immunodeficiency, ►T cell; Alam R et al 1993 J Immunol 150:3442; Casola A et al 2001 J Biol Chem 276:19715; structure of glycosaminoglycan interaction domain: Shaw JP et al 2004 Structure 12:2081.

RAP: ►RNAIII/rnaiii

RAP: RAP30 and RAP74 are subunits of the general transcription factor TFIIF, which binds RNA polymerase II and recruits TFIID and TFIIB transcription factors to pol II (Conaway JW et al 2000 Trends Biochem Sci 25:375).

RAP1A (1p13.3), **RAP1B** (12q14), **RAP2** (13q34): These are RAS related eukaryotic proteins but unlike RAS they are localized on intracellular membranes. The guanine exchange factor of RAP1, Epac (exchange protein activated by cAMP) is activated by cAMP and it bears homologies to the regulatory subunit of PKA. Rap1A is a suppressor protein of Ras-induced transformation. It has identical amino acid sequences with the effector region of Ras p21. Along with SIR3, RAP1 is a transcriptional repressor of telomeric heterochromatin. RAP74 is a subunit of transcription factor TFIID and RAP74 is involved in binding to the serum response element. RAP1 (repressor/activator protein) of yeast binds to upstream activator sequences (UAS) alone as well as in association with other proteins; it also activates many genes besides silencing the mating type and the telomerase functions. Rap1 activates about 37% of RNA polymerase II initiations in yeast. Rap1 is also a negative regulator of TCR-mediated transcription of interleukin-2 (IL-2) gene. RAP1 may enhance meiotic recombination and opening up of the nucleosomal chromatin. RAP1 may stimulate the formation of boundary elements. The nerve growth factor (NGF) may signal to ERK (environmental regulated kinases) either through RAS or RAP. ►RAS oncogene, ►silencer, ►serum response element, ►transcription factors, ►mating type determination, ►telomerase, ►TRF1, ►TCR, ►IL-2, ►nucleosome, ►PKA, ►cAMP, ►ORC, ►HML and HMR, ►boundary element, ►adherens junction, ►nucleoporin; Rousseau-Merck MF et al 1990 53:2; Morse RH 2000 Trends Genet 16:51; IdrissiF-Z et al 2001 J Biol Chem 276:26090.

Rapamycin (sirolimus): This immunosuppressor may block the cell cycle through the G1 phase by controlling mitogen-activated signal transduction. It regulates the prokaryotic ribosomal protein S6

and the elongation initiation protein eIF-4E (see Fig. R12). ►FK506, ►TOR, ►cell cycle, ►signal transduction, ►S6 kinase, ►eIF-4E, ►immunosuppressant, ►immunophilins, ►non-ribosomal peptide; Cardenas ME et al 1998 Trends Biotechnol 16:427; Rohde, J. et al. 2001 J Biol Chem 276:9583.

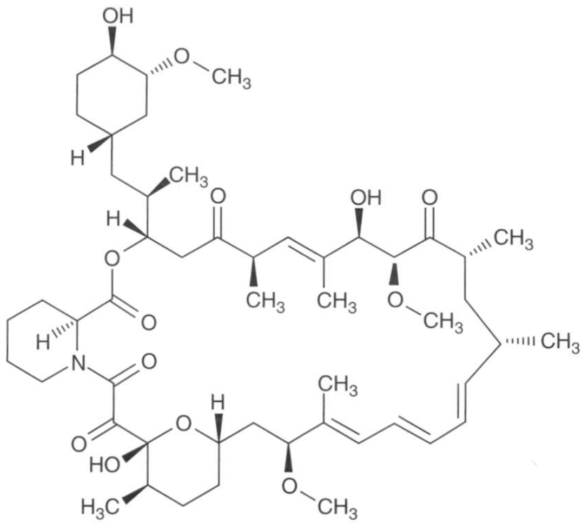

Figure R12. Rapamycin

RAPD (pronounce rapid): These markers are generated by random amplified polymorphic DNA sequences using (on average) 10 base pair primers and the PCR technique for the physical mapping of chromosomes on the basis of DNA polymorphism in the absence of "visible" genes. The map so generated may be integrated into RFLP and classical genetic maps. ►polymerase chain reaction, ►physical mapping, ►sequence-tagged site, ►integrated map; Reiter RS et al 1992 Proc Natl Acad Sci USA 89:1477.

Rape (*Brassica napus*): This is an oil seed crop (2n = 38, AC genomes). The new varieties are low in the toxic erucic acid and glucosinolates. ►canola, ►erucic acid

Raphanobrassica: This is a man-made amphidiploid (2n = 36) of radish (*Raphanus sativus*, n = 9, R genome) and cabbage (*Brassica oleracea*, n = 9, C genome). ►*Brassica oleracea*, ►radish, ►amphidiploid

Raphe: Refers to a ridge on the seeds where the stalk of the ovule was attached; it also refers to the seam of animal tissues.

Raphids: These are needle-like crystals within plant cells (often of oxaloacetic acid).

Raphilin: This is a peripheral membrane protein. It may bind RAB proteins in a GTP-dependent manner and may be phosphorylated by various kinases and may bind Ca^{2+} and phospholipids. ►RAB

Rapid Lysis Mutants: *r* mutants of bacteriophage rapidly lyse the infected bacteria and, therefore, the size of the plaques is much larger than the ones made by wild type phage (see Fig. R13). ►lysis, ►plaque lift, ►lysis inhibition

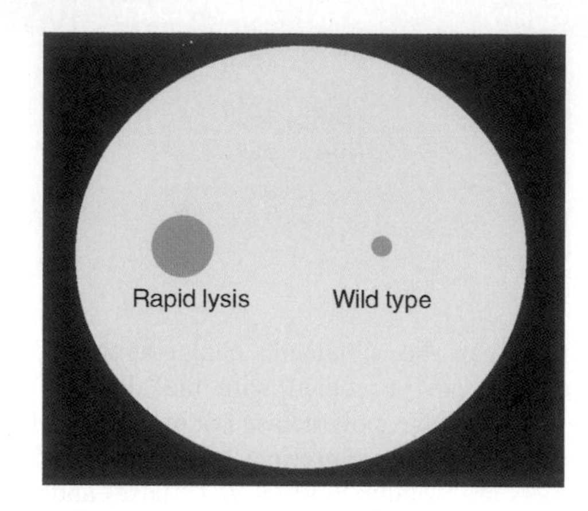

Figure R13 Rapid lysis mutants

Rapp-Hodgkin Syndrome: An anhidrotic ectodermal dysplasia with cleft lip and cleft palate. The latter symptoms do not always co-occur (mixed clefting). ►ectodermal dysplasia, ►anhidrosis, ►cleft palate; Neilson DE et al 2002 Am J Med Genet 108:281.

Rapsyn (43 kDa): A peripheral membrane protein co-localized with the acetylcholine receptors at the neuromuscular synapsis. Mutations in rapsyn may lead to myasthenia. ►acetylcholine, ►myasthenia; Ohno K et al 2002 Am J Hum Genet 70:875.

RAR: Refers to repair and recombination.

RAR, RARE (retinoic acid receptor [element]): RAR and RXR-α, -β, -γ retinoid-X receptors are transducers of ligand-activated morphogenetic and homeostasis signals. RAR and RXR can form homodimers but are usually found as heterodimers. These then bind to the cognate hormone response elements and increase the efficiency of transcription. Docosahexaenoic acid ($CH_3[CH_2CH{=}CH]_6[CH_2]_2CO_2H$) is an activator of RXR. RAR-α ligands can accomplish the binding of the RXR-RAR-α dimers to DNA causing RXR activation and initiating the transcriptional activity of RAR-α. The RXR–RAR complex may also repress transcription. It is encoded in human chromosome 17q12. RAR may play a role in the development of promyelocytic leukemia (PML) when it associates with histone-deacetylase and other co-factors (see Fig. R14). In PML-RARα patients pharmacological doses of retinoic acid lead to cancer remission because of the near normal differentiation of the hematopoietic (red blood-forming) cells but in

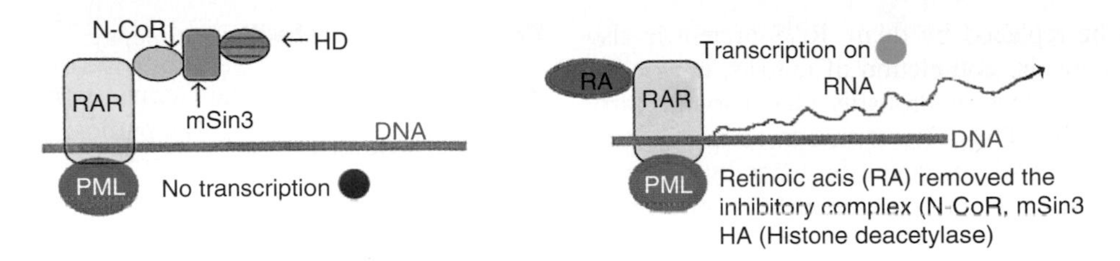

Figure R14. THe action of RAR in promyelocytic anemia (Redrawn after Grignani, F. *et al.* 1998 Nature (Lond) 39:815)

promyelotic leukemia Zinc finger (PLZF) patients this treatment is quite ineffective. Retinoic acid may downregulate telomerase activity. The retinoid system also plays a role in differentiation, motor innervation of the limbs, skeletal development and in the development of the spinal cord. *RAR1* genes in plants play a defense role. ►hormone response elements, ►retinoic acid, ►leukemia, ►PPAR, ►histone deacetylase, ►Zinc finger, ►innervation, ►Sin3; Zhong S et al 1999 Nature Genet 23:287; Pendino F et al 2001 Proc Natl Acad Sci USA 98:6662.

RARE (RecA-assisted restriction endonuclease): At the site of a restriction enzyme recognition in or near a locus a triplex structure, with an oligonucleotide, is generated with the assistance of enzyme RecA. The genome is enzymatically methylated and RecA is removed. Only the site now unprotected will be cut by the restriction endonuclease. This procedure reduces the actual cleavage sites in the DNA. ►recA, ►restriction enzyme, ►DNA methylation Ferrin LJ 2001 Mol Biotechnol 18[3]:233.

Rare-Cutter: This is a restriction enzyme with a longer DNA sequence (>8 nucleotides) recognition site. Therefore it cleaves DNA much less frequently (because of the greater specificity for sites) than enzymes which recognize only 4 bases in a sequence. ►restriction enzyme

Rare-Mating: Cells of non-mating yeast strains are mixed with cells that are expected to mate under favorable conditions. Low frequency of mating may take place and the progeny may be isolated by efficient selection. Rare-mating is presumed to be the result of mitotic recombination or non-disjunction of chromosome III in the non-mating strain. When normal mating fails, protoplast fusion may generate hybrids. The latter procedure, however, produces complex progeny. ►mating type determination, ►protoplast fusion; Spencer JF, Spencer DM 1996 Methods Mol Biol 53:39.

RAS ($p21^{ras}$, 11p15.5): This is a protooncogene originally found in rat sarcoma virus; it codes for a monomeric GTP-binding protein in which point mutation mainly in codons 12, 13, 59 or 61 may lead to oncogenic transformation. Alterations in site 12 in humans and in site 61 in rodents are common in tumors. RAS proteins have an important role in transmembrane signaling. Ras may have an alternate location at the Golgi membrane, depending on de/reacylation (Rocks O et al 2005 Science 307:1746). Protein Spred is an inhibitor of the RAS-MAP signaling pathway. The role of RAS may vary according to the cell type, from stimulation of adenylate cyclase to mating factor signal transduction and from proliferation to differentiation. The RAS protein becomes active only after prenylation by the 15-carbon farnesyl pyrophosphate. The prenylation is a thioether formation with an amino acid, resulting in the association of the protein with a membrane. RAS is not only one of the most important turnstile in signal transduction, but also one of the most common activated oncogen. The activation involves changing of the bound GDP into GTP. GAPs (GTPase activating proteins) inactivate RAS by hydrolysis of GTP. In contrast GNRPs (guanine nucleotide releasing proteins) mediate the replacement of bound GDP by GTP and the activation of RAS. Receptor tyrosine kinases activate RAS either by inactivating GAP or activating GNRP. The guanine-nucleotide exchange reaction is mediated by the SOS (son of sevenless) protein which has a GEF (guanidine exchange factor) domain. RAS is localized to the cell membrane. After ligand binding, SOS attaches to the adaptor protein Grb2 containing an SH3 domain and through the SH2 domain of Grb2 the complex binds to the phosphotyrosine residues of the signal receptor. The complex moving close to RAS makes possible that with the assistance of Cdc25, Sdc25 and GEF/GRF, respectively, the guanidine nucleotide is released. The RAS family is represented in various human chromosomes: NRAS in 1p21, HRAS in 11p15, KRAS in 6p12-p11 and RRAS in 19. Kras2 in mice has a tumor-inhibitory feature (Zhang Z et al 2001 Nature Genet 29:25). On the basis of homology three groups may be classified in mammals: (i) RAS, RAL, RRAS, (ii) RHO, and (iii) RAB. In *Drosophila* there are *Ras1* in 3–49, *Ras2* in 3–15 and *Ras3* in 3–1.4 (the latter has a higher homology to *Rap1* and it is now known as *Rap1*). In yeast, RAS homologs (*RAS1* and *RAS2*) are very closely related to the human protooncogenes

R

and can be replaced by them. RAS protein is also necessary for the completion of mitosis, in association with other factors. Probably, all eukaryotes carry RAS homologs. The p21 protein is also a mobile RAS protein. The RAS oncogene is generally active in tumorigenesis in the presence of the MYC or the E1A "immortalizing" oncogene products. MYC in cooperation with RAS mediates the progression of the cell cycle from G1 to the S phase through induction of the accumulation of active cyclin-dependent kinase and transcription factor E2F. The presence of RAS is necessary for tumor maintenance. The RAS promoter is high in GC and lacks a TATA box. The various RAS genes (human, mouse) may have more than 50% difference at the nucleotide level but the amino acid composition is highly conserved. RAS mutations have been detected in 90% of pancreatic adenocarcinomas, ~40–50% of colon adenocarcinomas and in other cancers. According to a genome-wide survey, RAS affects the expression of more than 250 genes. ►G-proteins, ►RAB, ►raf, ►BRAF, ►RALA, ►RALB, ►RHO, ►RAP, ►RASA, ►oncogenes, ►signal transduction, ►adenylate cyclase, ►farnesyl, ►prenylation, ►p21, ►GTPase, ►MYC, ►cell cycle, ►retinoblastoma, ►GEF, ►Cdc25, ►Sdc25, ►GD, ►SOS, ►SH2, ►SH3, ►Grb2, ►EF-Tu, ►animal models, ►Costello syndrome, ►Seladin-1; Zuber J et al 2000 Nature Genet 24:144; Johnson L et al 2001 Nature [Lond] 410:1111; Stacey D, Kazlauskas A 2002 Current Opin Genet Dev 12:44; Quilliam LA et al 2002 Progr Nucleic Acid Res Mol Biol 71:391; Downward J 2003 Nature Rev Cancer 3:11; Asha H et al 2003 Genetics 163:203.

RASA: This is a guanosine triphosphate activating RAS protein (21 kDa [p21]) encoded by human chromosome 5q13.3 and in mouse chromosome 13. ►RAS oncogene

rasiRNA (repeat-associated siRNA): This silences endogenous (selfish) retroelements and repetitive sequences in the *Drosophila* germ line. rasiRNA (24–29 nucleotides) is produced primarily from the antisense strand of the double-strand precursor unlike siRNA, which is made of both strands. rasiRNA does not require Dicer-1 or Dicer-2 and it functions though the Piwi system rather than through the Argonaute. ►siRNA, ►microRNA piRNA, ►retroelements; Vagin VV et al 2006 Science 313:320; Lalith S et al 2007 Science 315:1587.

Raspberry (*Rubus* spp.): The majority of raspberries are diploid (2n = 14), loganberry is 2n = 42, and blackberries have 2n = 28, 42 and 56 chromosomes. It is likely that some wild blackberries are allopolyploids with 2n = 35 and 2n = 84; the latter is dioecious.

Rasmussen's Encephalitis: ►epilepsy

Rassenhygienie: This is the German term for negative eugenics [often so used]. The purpose was to protect the "purity" of the Aryan (German) race and it was enforced by the laws of the Nazi state. Between 1933 and 1945, it led to 350,000 forced sterilizations, mass murder of millions and ban on marriages between genetically fit and "unfit", and persons whose ancestry included more than 1/4 Jews, Gypsies or other racial groups. ►eugenics, ►racism; Hubbard R 1986 Int J Health Serv 16:227.

Rat (*Rattus norvegicus*, 2n = 42): A genetic linkage map was published by Jacob and others (1995 Nature Genetics 9:63). A radiation hybrid map of 5,255 markers was prepared by Watanabe and associates. (1999 Nature Genet. 22:27). A high quality nucleotide sequence map covers more than 90% of the 2.75 gigabase genome (Nature [Lond] 428:493 [2004]). The chromosome number of other rat species may be different. A strain of albino rats developed by the Sprague-Dawley Animal Company is widely used in experimental work because of the calmness and ease of handling. (For EST map see Scheetz TE et al 2001 Genome Res 11:497). (See knockdowns: Tenenhause DC et al 2006 Proc Natl Acad Sci USA 103:11246; EST map: http://ratEST.uiowa.edu; http://ratmap.gen.gu.se; http://www.tigr.org/tdb/tgi/.; http://rgd.mcw.edu/; genome: http://www.hgsc.bcm.tmc.edu/projects/rat/; physiological studies: http://pga.mcw.edu/; rat resource and research center: www.nrrrc.missouri.edu/; rat genome database, strains: http://rgd.mcw.edu:7778/strains/; http://www.nrrrc.missouri.edu/Straininfo.asp?appn=46).

Rate-Limiting Step: This requires the highest amount of energy in a reaction chain or in a metabolic path, the slowest step.

Rate Matrix: Refers to base pair changes between two genomes. It is used by different algorithms to reveal the most likely evolutionary path of genomes. ►genomics

Ratio Labeling: Using the FISH cytological technology different chromosomes may be labeled by varying proportions of the same fluorochromes to distinguish individual chromosomes in the genome by color. ►FISH, ►combinatorial labeling

Rationale: This is the logical basis of an act, a process or an argument.

Rationalize: Refers to an attempt to make something conform to reason. Sometimes apparent rationalization is used in an effort to explain facts or ideas for which there is no adequate justification.

Raynaud Disease (hereditary cold fingers): This condition is characterized by familial periodic numb and white finger attacks. ►vasculopathy

Raynaud Syndrome: This condition is characterized by scleroderma, cyanosis, cold intolerance, chromosomal aberrations, telangiectasia without a clear pattern of inheritance. ►scleroderma, ►telangiectasia, ►cyanosis

RB: Refers to the right border of T-DNA. ►T-DNA

Rb: ►retinoblastoma

RbAp: These are proteins of the WD family of wide regulators of chromatin, transcription and cell division. ►WD-40; Rossi V et al 2001 Mol Genet Genomics 265:576.

rBAT/4F2hc: These are four membrane-spanning proteins involved in membrane transport or regulation of transport of neutral and positively charged amino acids. ►cystinuria, ►transporters; Malandro MS, Kilberg MS 1996 Annu Rev Biochem 65:305.

rbc: Denotes ribulose bisphosphate carboxylase/oxidase genes. ►chloroplast genetics

RBE (relative biological effectiveness of radiation): This depends on a number of physical (type of radiation, wavelength, dose rate, temperature, presence of oxygen, hydration, etc.), physiological (developmental stage) and biological factors (species, nuclear size and DNA content, level of ploidy, repair system, etc.). The comparison usually relates to ^{60}Co gamma radiation. ►rem, ►radiation effects

RBF (recoverable block of function): This system permits exogenous (conditional) regulation of fertility of plants. For example, when the barnase gene is linked to any gene it inactivates its pollen and therefore cannot be transmitted by sexual means. The expression of barnase in tobacco is under the control of sulfhydryl endopeptidase. Recovery from the barnase effect is managed by barnstar under the control of a heatshock promoter. Such a system may be used to reduce the risk of escape of transgenes from GMO plants into related organisms in the environment. ►barnase, ►barnstar, ►heat-shock proteins, ►terminator technology, ►GMO T-GURT; Kuvshinov VV et al 2001 Plant Sci 160:517.

RBM (RNA-binding-motif, also called RBMY, YRRM): This gene family is found only in the Y chromosome of mammals involved in male fertility. The HNRPG (hnRPG) gene located in human autosome 6p12 shows ~60% homology to RBM. It is assumed that this autosomal locus was retrotransposed in an early ancestor to the Y chromosome. Similarly, in Xq26 sequences virtually identical to exon 12 of hnRPG have been found. It appears that the chromosome 6p12 sequence is a processed pseudogene of the Xq26 sequence and there are similar sequences in chromosomes 1, 4, 9 and 11, all retrotransposed from Xq26. BFLS actually manifests hypogonadism and it is assigned to the same location as RBM. ►DAZ, ►boule, ►PABp, ►chromosome, ►Borjeson-Forssman-Lehmann syndrome [BFLS], ►NRY, ►azoospermia

R2Bm: This is a silkworm retroelement without long terminal repeats. ►retrovirus, ►silkworm; Luan DD et al 1993 Cell 72:595.

RBMY: ►RBM

RBTN (rhombotin): A cystine-rich oncoprotein family (encoded in human chromosome 11) containing a LIM domain. ►LIM domain; Chan SW, Hong W 2001 J Biol Chem 276:28402.

Rbx1: A ring finger protein homolog of APC (anaphase-promoting complex) which is required for SCF and VCB-mediated ubiquitination of Sic1 and probably other proteins. ►ubiquitin, ►APC, ►ring finger, ►SCF, ►Sic1; Carrano AC, Pagano M 2001 J Cell Biol 153:1381.

RCA (regulators of complement activation): ►MCP, ►complement

RCA (replication competent adenovirus): This should be kept as low as possible (<1 RCA/dose) for adenoviral vectors to avoid damage. RCA reduction can be accomplished by deleting the E1 region of the vector and shortening the potentially complementing tract of the host. ►adenovirus

RCAF (replication-coupling assembly factor): Mediates chromatin organization into nucleosomes during replication. It consists of Asf1 protein, H3 and H4 histones. ►nucleosomes, ►CAF, ►ASF1; Tyler JK et al 1999 Nature [Lond] 402:555.

RCC1: This is a chromatin-bound guanine nucleotide release factor that forms complexes with RAN (a G protein). Its deficiency interferes with the cell cycle progression, chromosome decondensation, mating, RNA export and protein import. It is a part of the nuclear pore complex. Its yeast homolog is Prp20p. ►RAN, ►cell cycle, ►nuclear pores, ►RNA transport, ►TPX; Hood J 2001 Trends Cell Biol 11:321; Renault L et al 2001 Cell 105:245.

RCC (renal cell carcinoma, human chromosome 3p14.2): RCC gene product frequently interacts with that of the von Hippel-Lindau (VHL) gene. ►renal cell carcinoma, ►von Hippel-Lindau disease

RCR: Denotes recombination competent retrovirus. ►retrovirus, ►retroviral vectors, ►gene therapy

R

RcsC: Refers to *E. coli* kinase affecting capsule synthesis regulator RcsB. (See Davalos-Garcia M et al 2001 J Bacteriol 183:5870).

RcsG: A 30.6 kDa protein with a C-terminal motif and a N-terminal sequence similar to that of DnaJ. In concert with RcsC/B and DnaK and GrpE, it induces the *cps* capsule polysaccharide operon of *E. coli*. ►DnaJ, ►RcsC

RDA (representational difference analysis): This is a genome scanning procedure for the detection and identification of genetic markers representing disease, other genes and chromosomal aberrations. The cellular DNA is cut by restriction endonuclease(s) and the smaller fragments are amplified by PCR. DNA samples from affected and disease-free samples are denatured and the mixtures of the two samples are allowed to anneal. The sequences which do not match fail to hybridize and are believed to cause the disease. In principle, the process is similar to cascade hybridization. The relatively rapid mass screenings were expected to identify individuals predisposed to particular hereditary differences (diseases). The same procedure may be applicable to non-disease genes of eukaryotes. RDA and GDRDA procedures can be used to generate genetic maps in organisms with a paucity of chromosomal markers. ►cascade hybridization, ►PCR, ►GMS, ►GDRDA, ►genetic screening, ►positional cloning, ►RNA fingerprinting, ►genomic subtraction, ►comparative genomic hybridization; Lisitsyn N et al 1993 Science 259:946; Tyson KL et al 2002 Physiol Genomics 9:121; copy number variations between normal and tumor specimens of mice: Lakshmi B et al 2006 Proc Natl Acad Sci USA 103:11234.

RdDM (RNA-dependent DNA methylation): DNA sequences identical to silenced RNA are methylated at cytosine residues in plants (Cao X et al 2003 Curr Biol 13:2212) but not necessarily in mammalian cells (Park CW et al 2004 Biochem Biophys Res Commun 323:275). ►RNAi, ►methylation of DNA

rDNA: Refers to DNA complementary to ribosomal RNA. The ribosomal RNA genes are in the nucleolar organizer region of the eukaryotic chromosomes and there are multiple tandem repeats of transcriptional units consisting of 18S, 5.8S, 5S and 26S RNAs. The mature rRNAs are cleaved from the large transcripts. In yeast the *FOB1* gene enhances recombination in rDNA. ►ribosome, ►rRNA; Long EO, Dawid IB 1980 Annu Rev Biochem 49:742.

RdRP (RNA directed RNA polymerase, RDR): This may be involved in gene silencing by synthesizing antisense transcripts from aberrant RNA and thereby causing PTGS. RdRP may generate double-stranded RNA of single-strand transcripts and play a role in RNAi production. RdRP may convey virus resistance to plants and various types of post-transcriptional gene silencing (Sugyama T et al 2005 Proc Natl Acad Sci USA 102:152). ►PTGS, ►epigenesis, ►antisense RNA, ►RNAi, ►host-pathogen relation, ►RNAi; Cheng J et al 2001 Virus Res 80:41.

Reaction, Chemical: Denotes a change in the atoms in or between molecules.

Reaction Intermediate: Refers to a short life chemical in a reaction path.

Reaction Norm: This denotes the range of phenotypic potentials of expression of a gene or genotype. Usually, the genes do not absolutely determine the phenotype but they permit a range of expressions, depending on the genetic background, developmental and tissue-specificity conditions and the environment. ►genotype, ►phenotype, ►regulation of gene activity, ►epigenesis, ►plasticity, ►homeostasis, ►fitness, ►adaptation, see Fig. R15; Woltereck R 1909 Verhandl Dtsch Zool Ges p 110.

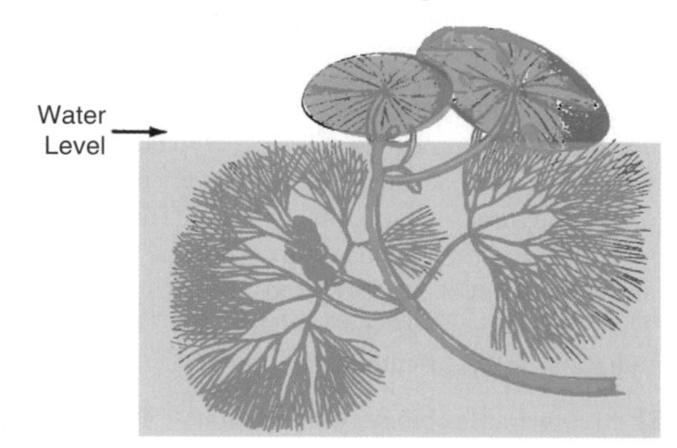

Figure R15. Reaction norm or water lily (farnwort): above water the leaves are different from below water (Goebel, K. 1893 Pflanzenphysiologische Schilderungen Elvert, Marburg, Germany)

Reactivation: ►multiplicity reactivation, ►Weigle reactivation, ►marker rescue

Reactivity in Hybrid Dysgenesis: ►VAMOS

Reactome: Refers to protein interaction pathways database (formerly called the genome knowledge base). ►protein interaction, ►genetic networks; http://www.reactome.org/.

Read: This is a nucleotide sequence determined in a single gel.

Reading Disability: ►dyslexia

Reading Frame: The triplet codons can be read in three different registers, starting with the first, second or third, however, only one may spell the correct protein. ►open reading frame, ►frameshift mutation

Readout: The DNA sequence recognition by proteins may be *direct* (by hydrogen bonding) or *indirect* when the DNA conformation also plays a role. ►binding proteins

Readthrough: The ribosome continues translation downstream of a stop codon. On an average human hereditary disease is due to the inability to readthrough translation-termination nonsense mutation. Some antibiotics (gentamycin) promote readthrough. Unfortunately, it is not useful for clinical purposes because the effective dose produces serious side effects. On the basis of a screening of 800,000 low molecular weight compounds, one report identified PTC124 (3-[5-(fluorophenyl)-[1,2,4]oxadiazol-3yl])-benzoic acid that promoted UGA nonsense suppression at a maximal effective concentration of 3 μM (852 ng mL^{-1}). PTC124 promoted dystrophin production in humans afflicted by muscular dystrophy and in a comparable mutant mice model. Muscle function was restored within 2–8 weeks of exposure to the drug. It was effective to a lesser degree in the case of UAG and UAA nonsense codons. No serious side effects were observed (Welch EM et al 2007 Nature [Lond] 447:87). ►translation termination, ►autogenous suppression, ►nonsense codon, ►gentamycin, ►genetic medicine, ►muscular dystrophy, ►genetic diseases

Readthrough Protein: This is formed when a suppressor tRNA inserts an amino acid at a site where chain termination is normally expected because of a nonsense codon and thus produces a fusion protein from two different "in-frame" cistrons, separated by a nonsense codon. Readthrough may be brought about by mutation in the anticodon of a tRNA or modification of the tRNA, e.g., selenocysteinyl-tRNA inserts selenocysteine into glutathione oxidase by recognizing the UGA (opal) stop codon. ►gene fusion, ►transcriptional gene fusion, ►translational gene fusion, ►trapping promoters

Real Time: This means the actual time during which the physical process takes place.

Realized Heritability: ►gain

Reannealing (reassociation): Double-stranded DNA can be heat denatured (strands sepa-rated) and can be restored to double-stranded form, reannealed, when the temperature falls below 60°C. ►c_0t curve

Rearrangements: These are structural changes of the chromosome(s), e.g., translocation, inversion. ►chromosomal rearrangements

Reasoning: ►inference

Reassociation Kinetics: ►c_0t curve

Reassortant: This is a new virus strain that emerges from a combination of genes of two different strains, e.g., the pandemic influenza strains of 1957 and 1968 contain elements enabling the virus to replicate in humans and the avian segment. The hemagglutinin coding-segment assists in preventing the neutralization of antibodies of humans not previously exposed to the avian flu virus. ►pandemic, ►hemagglutinin, ►influenza virus

Rec8: This meiotic cohesin is cleaved by separin before chiasma are resolved and meiotic anaphase I can proceed. In vertebrates cohesin is removed during prophase but Scc1 remains associated with the centromeres when separin cleaves this protein and thus facilitates the metaphase-anaphase transition. ►cohesin, ►separin; Buonomo SB et al 2000 Cell 103:387; Waizenegger IC et al 2000 Cell 103:399.

rec: Refers to one or another type of recombination-deficient mutation.

RecA Protein: This is a 38.5 kDa polypeptide involved in homologous recombination by promoting pairing. It is a DNA-dependent ATPase which mediates strand exchange. RecA binds to single-stranded DNA and pre-synaptic nucleoprotein molecules mediate the pairing with the duplex DNA target. The paired DNA is inside a 25 Å hole. Inside this cavity projecting toward the axis of the helix are mobile loops L1 and L2 representing the binding sites. RecA appears to play a role in the segregation of the bacterial chromosomes (Ben-Yehuda S et al 2003 Science 299:532). The RecA protein expressed in transgenic plants substantially increases recombinational repair of DNA damage inflicted by mitomycin. The RecA prokaryotic gene when equipped with the nuclear localization signal and transformed into tobacco cells increases sister chromatid exchange by ~two-threefold. Structural and functional homologs of the bacterial RecA are UvsX (phage), Rad51 (eukaryotes) and RadA (archaea). The RecA protein is required for the maintenance of around 40 proteins necessary for the continuation of replication following DNA damage (Courcelle J, Hanawalt PC 2003 Annu Rev Genet 37:611). ►recombination molecular mechanism prokaryotes, ►DNA repair, ►RecA-independent recombination, ►*RecA1*; Kowalczykowski SC, Eggleston AK 1994 Annu Rev Biochem 63:991; Gourves A-S et al 2001 J Biol Chem 276:9613; Bar-Ziv R, Libchaber A 2001 Proc Natl Acad Sci USA 98:9068; Robu ME et al 2001 Proc Natl Acad Sci USA 98:8211; Gasior SL et al 2001 Proc Natl Acad Sci USA 98:8411; Lusetti SL, Cox MM 2002 Annu Rev Biochem 71:71.

recA1: A recombination deficient mutation of *E. coli* (map position 58 min) coding for a DNA-dependent

ATPase, a 3522-amino acid residue enzyme. Plasmids carrying it remain monomeric and do not form multimeric circles. When M13 vectors carry it, the foreign passenger DNA has fewer deletions. The recA protein mediates the association of double-stranded DNAs by synapsis mainly in the major grove but also in the minor grove of the DNA. The RecA-mediated pairing involves a triplex structure, i.e., along parts of the sequences double-stranded DNA associates transiently with a single strand of the other DNA molecule. The pairing of the DNA molecules may be plectonemic (intertwined) and thus may not require stabilization by proteins. (The paranemic coils are only juxtapositioned and require protein to keep them together.) Experimental data have revealed that ATP hydrolysis is not required for the exchange between paired strands rather the removal of RecA requires ATP hydrolysis. In case the homology between the DNAs is not perfect, ATP is needed for the exchange. RecA is also involved in branch migration but with the assistance of RuvAB and RecG proteins. The extension of the DNA heteroduplex (at the rate of 2–10 bp/sec) in the 5′→3′ direction needs ATP hydrolysis. The length of the heteroduplex may increase to 7 kbp. In both prokaryotes and eukaryotes besides RecA (or homologs), a stimulatory exchange protein, binding single strands of the DNA (SSB) is required. The SSB monomers (1/15 base in ssDNA) facilitate synapsis between the heterologous strands. After exchange RecA promotes DNA renaturation. The RecA homologs in yeast, mouse and humans are the RAD51 proteins, and Mei3 in *Neurospora*. (Recombination molecular mechanisms prokaryotes, branch migration, *recB*, other *Rec* genes, DNA repair, RuvABC, RAD).

RecA-Independent Recombination (illegitimate recombination): This may use three pathways (i) simple replication slippage, (ii) sister chromatid-associated replication misalignment, and (iii) single-strand annealing. Single-strand annealing takes place after palindromic sequences within each strand fold into hairpin structures within the strands. When a nuclease (SbcCD) opens the cruciform palindromes resection, followed by annealing of the flanking repeats brings about deletion (see Fig. R16.). RecA-independent recombination occurs at extremely low frequencies and is less responsive to the extent of homology. RecA-independent recombination in *Escherichia coli* is depressed by the redundant action of single-strand exonucleases. In the absence of multiple single-strand exonucleases, the efficiency of RecA-independent recombination events, involving either gene conversion or crossing over, is markedly increased to levels rivaling RecA-dependent events. It seems that RecA-independent recombination is not intrinsically inefficient but is limited by the single-strand DNA substrate availability. Crossing over is inhibited by exonucleases ExoI, ExoVII, ExoX and RecJ, whereas only ExoI and RecJ abort gene-conversion events. In ExoI– RecJ– strains, gene conversion can be accomplished by the transformation of short single-strand DNA oligonucleotides and is more efficient when the oligonucleotide is complementary to the lagging-strand replication template (Dutra BE et al 2007 Proc Natl Acad Sci USA 104:216). ►illegitimate recombination, ►palindrome, ►exonuclease; Bzymek M, Lovett ST 2001 Proc Natl Acad Sci USA 98:8319.

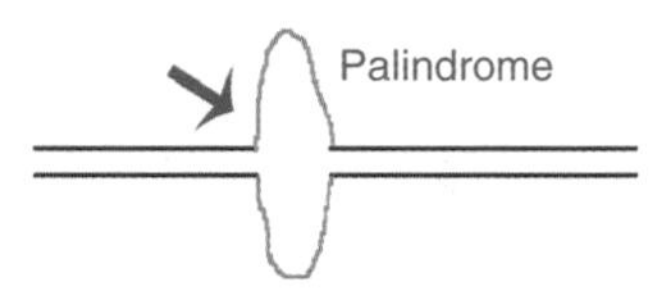

Figure R16. Palindrome

recB: *E. coli* gene (map position 60 min) encoding a subunit of exonuclease, controlling recombination and genetic repair. ►recombination molecular mechanism prokaryotes, ►recombination models, ►DNA repair

RecBCD: An enzyme (ribozyme) complex functioning in recombination of prokaryotes. RecBCD is a helicase and unwinds up to 42,300 bp per molecule, it is a strand specific nuclease. The recognition sequence (also called χ, [chi]) is 5′-GCTGGTGG-3′ promotes recombination in its vicinity by recruiting the RecA protein. The enzyme travels along one of the two strands of the DNA in 3′→5′ direction. ►recombination molecular mechanism prokaryotes, ►recombination models, ►chi, ►DNA repair, ►RecA, ►AddAB; Jockovich ME, Myers RS 2001 Mol Microbiol 41:949; Taylor AF, Smith GR 2003 Nature [Lond] 423:889; Dillingham MS et al 2003 Nature [Lond] 423:893; crystal structure: Singleton MR et al 2004 Nature [Lond] 432:187.

recC: *E. coli* gene (map position 60 min) encoding a subunit of exonuclease V, controlling recombination and genetic repair. ►recombination molecular mechanisms, ►DNA repair; Chen HW et al 1998 J Mol Biol 278:89.

recE: This is the locus of Rac prophage (map position 30 min), encoding exonuclease VIII and promoting homologous binding between single-stranded and double-stranded DNAs. ►RecA; Muyrers JP et al 2000 Genes Dev 14:1971.

RecF: Refers to a single- and double-strand binding recombination protein. (See Nakai H et al 2001 Proc Natl Acad Sci USA 98:8247).

RecG: This unwinds the leading and lagging strands at a damaged replication fork and may contribute to replication restart if it was stalled by the damage. ►replication restart; McGlynn P, Lloyd RG 2001 Proc Natl Acad Sci USA 98:8227.

RecJ: A single-strand (5′→3′) exonuclease used in recombination of *E. coli*. (Hill SA 2000 Mol Gen Genet 264[3]:268).

RecO: A bacterial homolog of the eukaryotic Rad52 protein, which stimulates Rad51 in exchange and mediates single-strand DNA annealing function in recombination. RecO bears functional similarity to bacteriophage T4 protein UvsY. The optimal functioning of RecO necessitates association with the single-strand binding (SSB) proteins in equal proportion of the two. The RecR protein stimulates RecO to facilitate the displacement of SSB by RecA. ►Rad51, ►RecR; Kantake N et al 2002 Proc Natl Acad Sci USA 99:15327.

RecQ: Refers to *E. coli* DNA helicase. Homologs of the prokaryotic enzyme exist in eukaryotes and the enzyme resolves secondary structures of the DNA at stalled or broken replication forks. It is a multidomain enzyme. The catalytic core determines the ATPase and helicase functions. The helicase RNase-D-C-terminal domain shows globular fold binding preferentially to single-stranded DNAs and this latter domain apparently determines the specificities of the enzymes found in different organisms (Bernstein DA, Keck JL 2005 Structure 13:1173). ►helicase; Wu X, Maizels N 2001 Nucleic Acids Res 29:1765; Cobb JA et al 2002 FEBS Lett 529:43; Cui S et al 2003 J Biol Chem 278:1424.

RecR: This mediates DNA renaturation during recombination of *E. coli* (Pelaez AI et al 2001 Mol Genet Genomics 265:663).

RecT: This is encoded by *recE*. It is involved in renaturation of homologous single-stranded DNA and pairing of DNA. ►RecE, ►pairing

RecU: This is a protein in gram positive bacteria. Its absence increases sensitivity to DNA damage, reduces plasmid transformation and affects the segregation of the chromosome in *Bacillus subtilis*. It binds preferentially to single-stranded DNA and cleaves recombination intermediates (Holliday junctures) and anneals single-stranded DNA. ►*Bacillus subtilis*, ►Holliday junctures; Ayora S et al 2004 Proc Natl Acad Sci USA 101:452.

Receptacle: This is the widened end of a flower stalk. It also refers to a container. (See Fig. R17).

Figure R17. Receptacle

Receptin: This is a natural or engineered microbial protein which can bind to a mammalian protein. It is useful for the identification of individual components of a proteome. It is similar to lectins which bind carbohydrates. ►lectins, ►affibody, ►monoclonal antibody, ►MSCRAMM; Kronvall G, Jonsson K 1999 J Mol Recognition 12:38–44.

Receptor: This is also called an operator. The site responding to the controlling element (transposase) as originally called by Barbara McClintock the components of the *Spm* transposable systems in maize. ►transposable elements of maize receptors

Receptor: Refers to proteins that bind to ligands with cellular signaling functions. Receptors may be located within the plasma membrane (transmembrane proteins) or are intracellular and bind ligands, which penetrate cells by diffusion. Some receptors are ligand-gated ion channels. The number of receptors in a cell may run into hundreds. Several agonists and antagonists may affect their function. Cell surface receptors mediate in some cell to cell and virus–cell interactions. Cell surfaces can be tagged by unnatural, exogenous molecules facilitating the therapeutic targeting of surface glycans and imaging disease progression (Prescher J et al. 2004 Nature [Lond] 431:873). ►signal transduction, ►hormone receptors, ►transmembrane proteins, ►serine/threonine kinase, ►receptor tyrosine kinase, ►receptor guanyl cyclase, ►receptor tyrosine phosphatase, ►receptors, ►adaptor proteins, ►T cell, ►nuclear receptor, ►orphan receptor, ►TCR, ►ion channels, ►virus receptor; Xu L et al 1999 Curr Opin Genet Dev 9:h 40; Human plasma membrane: http://receptome.stanford.edu/HPMR/.

Receptor Down Regulation: Epidermal growth factor (EGF) binding receptors concentrate in coated pits after binding with these growth factors. They enter the lysosomes where degradation of the receptor and

EGF occurs. The cell surface has a reduced number of them because of receptor down regulation. ►endocytosis, ►EGF

Receptor Editing: This mechanism modifies the antigen-specificity of antigen receptors in the variable region of the antibody and may lead to immune tolerance or antibody diversification by V(D)J recombination. This mechanism eliminates the autoreactive B cells when confronted with self-antigens. ►antigen receptor, ►immune tolerance, ►V(D)J, ►immunoglobulins, ►clonal selection; Kouskoff V, Nemazee D 2001 Life Sci 69:1105.

Receptor Guanylyl Cyclase: These transmembrane proteins are associated at the cytosolic end with an enzyme that generates cyclic guanosine monophosphate (cGMP). cGMP activates cGMP-dependent protein kinase (G-kinase) that phosphorylates serine/threonine residues in proteins. ►cGMP, ►serine/threonine kinase; Kusakabe T, Suzuki N 2001 Dev Genes Evol 211[3]:145.

Receptor-Mediated Gene Transfer: Cell surface receptors may internalize their ligands by endocytosis. The ligand (peptides, lectins, sugars, antibody, glycoprotein, etc.) may form a conjugate with a polycation, e.g., polylysine. An expression vector plasmid may bind to a ligand—polycation conjugate. A fusogenic peptide or a disabled adenovirus may facilitate the entry of the complex into the cell by endocytosis and transported with the aid of an endosomal vehicle. From the endosome the DNA (gene) may be transferred to the nucleus where it may have a chance for expression. With the assistance of asialoglycoprotein receptor the gene may be targeted to hepatocytes or with a mannose receptor it may be targeted to macrophages. The transferrin receptor facilitates targeting erythrocytes; polymeric immunoglobulin receptors may aim the gene construct at the lung epithelia; other receptor and ligand combinations permit targeting to other cells or tissues. The advantage of this type of transfection is that it is not infectious, the DNA carrying more than a single gene of almost any size can be targeted. Cell division is not a requisite for expression. The transgene functions in the cytoplasm. Unfortunately, the level and duration of expression vary. The system may elicit an undesirable immune reaction. ►endocytosis, ►asialoglycoprotein receptor, ►transferrin, ►macrophage; Varga CM et al 2000 Biotechnol Bioeng 70:593.

Receptor Protein Tyrosine Phosphatase (RPTP): Refers to signaling molecules required for cell development. It dephosphorylates negative regulatory C-terminal tyrosine residues of the Src family kinases. ►signal transduction, ►Src; Carothers AM et al 2001 J Biol Chem 276:39094.

Receptor Serine/Threonine Kinases: These are the receptors of serine/threonine phosphorylating enzymes. They are the major types of plant receptor kinases, unlike in animals where receptor tyrosine kinases dominate. ►serine/threonine kinase; Choudhury GG 2001 J Biol Chem 276:35636.

Receptor Tyrosine Kinases (RTK): These bind protein tyrosine kinase enzymes (59 genes in humans) such as the receptors for the epidermal growth factor (EGF), insulin, insulin-like growth factor-1 (IGF-1), platelet-derived growth factor (PDGF), fibroblast growth factor (FGS), nerve growth factor (NGF), hepatocyte growth factor (HGF), vascular endothelial growth factor (VEGF), macrophage colony stimulating growth factor (M-CSF), RET proteins containing cadherin-like, cysteine-rich extracellular domain. Defects of the RET family of proteins include problems of the glial-cell derived neurotrophic factor, MEN2, Hirschsprung disease, familial medullary thyroid carcinoma, pheochromocytoma, hyperparathyroidism and ganglioneuromatosis. Hepatocyte growth factor receptor defects are responsible for the papillary renal cell carcinoma. Platelet-derived growth factor receptor changes activate the KIT oncogene. The insulin receptor anomalies lead to diabetes type II, leprechaunism and Rabson-Mendenhall disease. Hereditary lymphedema receptor disease is due to the vascular endothelial growth factor. Congenital pain with anhidrosis is caused by defects in the neurotrophin receptors. These receptors are transmembrane proteins and when the receptor is associated with the cognate phosphorylase enzyme both the receptor and the target protein receive γ phosphate groups from ATP at certain tyrosine residues. The phosphorylation results in dimerization or dimerization results in phosphorylation and activation. Activation increases the activity of RAS and subsequently the MAP kinases. This eventually leads to the expression of genes. The various regulatory proteins recognize the different phosphorylated tyrosine residues in the receptor. Upon binding to their specific sites they may also be phosphorylated on their own tyrosine residues and become activated. A cascade of events may follow that activate entire signaling pathways. The different receptors and the associated proteins may control the separate or interacting signaling pathways. Mutations or truncation of RTK that permit dimerization without the proper ligand may lead to carcinogenesis. The specificity of RTK (which are involved in diverse metabolic functions) is determined either by the strength or duration of the signal or it may be qualitative. The various cell types may activate RTK signaling in different ways. In craniosynostosis (Crouzon, Pfeiffer, Apert and Jackson-Weiss syndromes) and some dwarfness (hypochondroplasia, thanatophoric dysplasia) the fibroblast growth factor receptor members of RTK are involved. ►anoikis, ►tyrosine kinase, ►signal transduction, ►SIRP, ►RET oncogene, ►Eph, ►craniosynostosis syndromes, ►biomarkers, ►genetic medicine, ►cancer therapy; Simon MA 2000 Cell 103:13; Robertson SC et al 2000 Trends Genet 16:265; Madhani HD 2001 Cell 106:9; Haj FG et al 2002 Science 295:1708; Baselga J 2006 Science 312:1175.

Receptor Tyrosine Phosphatase: This binds protein tyrosine phosphates and splits off phosphate groups. (See Bateman J et al 2001 Curr Biol 11:1317).

Receptor-Mediated Endocytosis: This is a very efficient delivery system of macromolecules (such as cholesterol), that adheres to coated pits, into cellular organelles. ►coated pits

Receptors: These are proteins that bind other molecules (ligands). They can be *extracellular* receptors that respond to outside signals reaching the cells. They may be located on the *surface* of the cell membrane or *within the membrane* but their ligand-binding domains are exposed to the extracellular space. The *intracellular* receptors respond to ligands that diffuse into the cell. ►ligand, ►signal transduction, ►cargo receptors, ►receptor, ►adaptor proteins, ►T cell

Recessive: Such an expression of a gene means that it is not visible in heterozygotes in the presence of the wild type or other dominant alleles of the locus. Recessivity is not necessarily an absolute lack of expression of the gene (except in null alleles) because extremely low level of transcription/translation may not be observed by a particular type of study but may be detectable by a finer analysis. ►dominance, ►semidominance

Recessive Allele: This does not contribute to the phenotype in heterozygotes in the presence of the dominant allele. ►pseudodominance

Recessive Epistasis: ►epistasis, ►modified checkerboards

Recessive Lethal: Dies when homozygous, and can be maintained only as heterozygote.

Recessive Lethal Tests, *Drosophila*: ►*Basc*, ►*ClB* method, ►sex-linked recessive lethal, ►autosomal recessive lethal assay

Recessive Oncogenes: These are tumor-suppressor genes such as encoding p53. ►tumor suppressors, ►p53

recF: *E. coli gene* (map position 82 min), also called *uvrF*, controls recombination and radiation repair. ►recombination molecular mechanisms, ►DNA repair; Bidnenko V et al 1999 Mol Microbiol 33:846.

recG: *E. coli* gene (map position 82 min) controls recombination. ►recombination molecular mechanism; Qourcelle J, Hanawalt PC 1999 Mutation Res 435:171.

Recipient: Bacterial cells of the F^- state receive genetic material from the donor F^+ strains. This is also the cell to which genetic material is transferred. ►conjugation, ►transformation

Recipient Site: ►donor site

Reciprocal Crosses: For example, A × B and B × A.

In cases when cytoplasmically determined differences exist between the two parents, the F_1 offspring bears greater resemblance to the female parent that usually transmits the cytoplasm. These reciprocal differences may persist indefinitely in the advanced generations. Although reciprocal differences are normally most obvious in plants, inanimal hybrids, e.g., the mule and the hinny they are easily distinguishable. ►mitochondrial genetics, ►chloroplast genetics, see Fig. R18.

Figure R18. Reciprocal hybrids of *Epilobium hirsutum* Essen and *Epilobium parviflorum* Tübingen. Parents are 2n = 36. In the cross at the left an *E. parviflorum* female was crossed by an *E. hirsutum* male. The two plants at the right represent the reciprocal cross when the *E. hirsutum* female provided the cytoplasm. (From Michaelis P Umschau 1965 (4):106)

Reciprocal Interchange: This is the same as reciprocal translocation of chromosomes.

Reciprocal Recombination: This is the most common exchange between homologous chromatids at the 4-strand stage of meiosis in eukaryotes. In the case of single crossing over in an interval, two parental types and two crossover strands are recovered. An exception is gene conversion where the exchange is non-reciprocal. In conjugational transfer in bacteria the reciprocal products of the event are not recovered and their fate is unknown. In sexduction and specialized transduction reciprocal recombination may also take place in bacteria (see Fig. R19). ►crossing over, ►recombination molecular mechanisms prokaryotes, ►conjugation, ►sexduction, ►specialized transduction

Parental	*AB* and *ab*
Reciprocal recombinant	*Ab* and *aB*

Figure R19. Reciprocal recombination

Reciprocal Selection: ►recurrent selection

Reciprocal Translocation: Segments of non-homologous chromosomes are broken off and reattached to each other's place. As a consequence, generally 50% of the gametes of the translocation heterozygotes (formed by adjacent distributions) are defective because they do not have the correct amount of chromatin. ►translocation

Rec-Mutant: This means deficient in recombination and possibly altered in other functions of the DNA. (See individual Rec entries).

Recoding: This mechanism may translate the same DNA sequence in more than one way. It is a common mechanism in viruses with overlapping genes. There are several other ways this can take place. Some genes utilize multiple promoters and depending on the choice of their utilization, the same RNA may code for more than one protein. Frameshifting may take place: e.g., the mRNA may show slippage on the ribosome, a $tRNA^{Leu}$ with an anticodon GAG may recognize CUUUGA in one frame and in a shift it inserts leucine (UUU) for 4 nucleotides: CUUUGA. Similar frameshifting cassettes may be determined by the *E. coli* gene *SF2* and also in other prokaryotes. In the TY3 transposable element of yeast the GCG AGU U instead of the Ala (GCG) and Ser (AGU) it may read GCG **A** GUU Ala (GCG) and Val (GUU). The code words may be interpreted in different ways and stop codons may specify selenocysteine, tryptophan and glutamine. The ribosome may also skip certain sequences, e.g., the T4 phage topoisomerase may bypass 50 contiguous nucleotides and after the long frameshift it continues translation. Variants of phage λ repressor and cytochrome b_{562} when translated from mRNA without a stop codon acquire an unusual COOH end. Co-translation switches the ribosome reading from the defective mRNA to the tRNA-like ssrA transcript that is translated into Tyr-Ala-Leu-Ala-Ala (the normal carboxyl end would have been very similar Trp-Val-Ala-Ala-Ala). Recoding may be of importance in some human diseases, e.g., if in the cystic fibrosis transmembrane conductance regulator a glycine $codon_{542}$ or arginine $codon_{553}$ is replaced by an UGA (opal) stop codon, the disease symptoms are alleviated compared with some missense mutations because this opal codon permits some readthrough leakage. ►overlapping genes, ►frameshift, ►selenocysteine, ►topoisomerase, ►Ty, ►cystic fibrosis, ►set recoding, ►fuzzy logic; Shigemoto K et al 2001 Nucleic Acids Res 29:4079; Harrell L et al 2002 Nucleic Acids Res 30:2011.

Recoding Signal: This is required for translational recoding. ►overlapping genes, ►recoding

Recognition Site of Restriction Enzymes: ►restriction enzyme

Recoil: This means to bounce back; electromagnetic radiation recoils from glass and metal. ►Compton effect

Recombinagenic: This may be involved in genetic recombination at an increased frequency.

Recombinant: Refers to an individual with some of the parental alleles reciprocally exchanged. ►reciprocal recombination

Recombinant Antibody (RAb): This is genetically engineered and usually includes only the variable fragments, which are fused to some other proteins. The appropriate DNA fragments are amplified by PCR and cloned in *E. coli* and a single peptide chain may contain the variable regions of both the light and heavy chains. Animal passage is not required. The antibody gene fragment can be fused to a bacterial signal sequence enabling the direction of the molecule into the periplasmic space where chaperones can properly fold the engineered protein. The procedure may include selection by phage display. Since RAb is produced without an animal and in vitro, sources of contamination by pathogens can be eliminated. RAb is also monoclonal. Recombinant antibodies can be modified with the battery of tools of molecular biology and different properties can be added to them, e.g., the paratope can be specially targeted to tumor cells (bifunctional antibody). It can be obtained by using human gene fragments, thus precluding an immune response against the RAb. ►PCR, ►immunoglobulins, ►antibodies, ►signal sequence, ►monoclonal antibody, ►paratope, ►phage display, ►polyclonal antibody; Kortt AA et al 2001 Biomol Eng 18[3]:95; Karn AE et al 1995 ILAR J 37[3]:132.

Recombinant Congenic: An outcross is followed by several generations of inbreeding in order to minimize the background genetic variations. ►recombinant inbred strain panels

Recombinant DNA: This is a DNA that has been spliced in vitro from at least two sources with the techniques of molecular biology or that results from the replication of such molecules. From the viewpoint of safety regulations, synthetic DNA segments, which yield potentially harmful polynucleotide or polypeptide, if expressed within cells, are subject to the same regulations as any harmful natural product. Transposable elements, unless they include recombinant DNA, are not subject to the US National Institute of Health recombinant DNA regulations. ►vectors, ►cloning vectors, ►transformation genetic, ►genetic engineering, ►restriction enzymes, ►splicing; Fed. Regist. [1999] 64:25361, http://www4.od.nih.gov/oba/rac/guidelines/guidelines.html.

Recombinant DNA and Biohazards: Fears of this were expressed by conscientious scientists way back in the 1970s even before the impact of the new techniques could be fully assessed. Since then evidence has accumulated indicating that some of the fears were not entirely justified, except by the wisdom of caution with a hitherto unknown and unused procedure. To avoid safety risks various levels of containments were made mandatory, depending on the organisms used, to prevent the accidental escape of genetically engineered organisms. Certain types of gene transfers were entirely prohibited to avoid contagions and highly toxic products. Cloning vectors were constructed that would not survive outside the laboratory. Bacterial strain X^{1776} (so designated in honor of the bicentennial anniversary of the national independence of the USA) had an absolute requirement for diaminopimelinic acid, an essential precursor of lysine and absent from the human gut. Cloning bacterial hosts were made deficient for excision repair (*uvrB*), auxotrophic for thymidine (indispensable for DNA synthesis), mutant for recombination (*rec*$^-$) and conjugational transfer of plasmids to other organisms. If reversion frequency of any of, say, 5 defects, each is in the range of 1×10^{-6}, then the joint probability of simultaneous reversion of all 5 would be $(10^{-6})^5 = 10^{-30}$. Since the mass of a single *E. coli* cell is about 10^{-12} g, only in a mass of 10^{11} metric tons of bacteria can one expect to find such a fivefold mutation. Obviously, such a mass of bacteria is not likely to occur because the earth may not support it. To get an idea of what this volume is a comparison can be made: wheat production of the world in 1980 was only 4.5×10^8 tons and the estimated mass of the planet earth 10^{20} tons. To avoid any problem, nevertheless, government authorization is required in all countries where this technology is used, for the release of any genetically engineered species (microbes, plants, animals) for the purpose of economic utilization. Objections to such carefully tested releases are still raised, based not so much on public concern as on personal or political interests and most commonly because of ignorance. During the period spanning more than two decades since recombinant DNA technology has been used, no major accident has been reported and with the guidelines available none is expected. Before recombinant DNA experiments are initiated, the plans are approved by the Institutional Biosafety Committees and Institutional Review Boards. Experiments involving cloning of toxin molecules with LD50 of less than 100 nanograms per kg body weight must be approved by the Office of Biotechnology Activities (National Institute of Health/MSC 7010, 6000 Executive Blvd., Suite 302, Bethesda, MD 20892-7010, Tel. 301-496-9838). Such toxins are botulinum, tetanus, diphtheria *and Shigella dysenteriae* neurotoxin. Specific approval is mandatory for cloning in *Escherichia coli* K12 genes coding for the biosynthesis of toxic substances, which are lethal to vertebrates at 100 ng to 100 mg per kg body weight. Special review and approval by the OBA and the RAC are required for human experimentation or treatment. The OBA sets specific guidelines for different risk categories. Working with plant and animal pathogens requires a permit from the US Department of Agriculture. According to the US National Institute of Health Guidelines, hazardous agents are classified into four groups and the pertinent agents are named according to groups, group 1 is the least hazardous and group 4 is the most dangerous. The principal investigator who is primarily responsible for the observance of the regulations must report all accidents

R

or any potential hazardous events. ►laboratory safety, ►containment, ►biohazards, ►recombinant DNA evolutionary potentials, ►gene therapy, ►cancer gene therapy, ►Institutional Biosafety Committee, ►RAC, ►OBA, ►GMO; http://www4.od.nih.gov/oba/rac/guidelines/guidelines.html.

Recombinant DNA, Evolutionary Potentials: Using molecular biology techniques genes can be transferred among organisms by means not routinely available in nature. However, it cannot be ruled out that during the process of natural evolution fragments of degraded DNA are taken up by direct transformation and exchanged between taxonomically unrelated species.

Recombinant Inbred Strain Panels: These can be used for mapping in mouse. (For information contact Jackson Laboratory Animal Resources, 600 Main Str., Bar Harbor ME 04609, USA. Phone: 1-800-422 MICE or 207-288-3371. Fax: 207-288-3398).

Recombinant Inbreds (RI): These are generated for physical mapping of DNA by selfing F1 hybrid populations and selecting single seed or animal progenies for about 8 generations until only $(0.5)^8$ (≈ 0.0039) fraction remains heterozygous for a particular marker (linkage ignored). The parental lines are chosen on the basis of differences in their DNA sequences, and from the data the map position of these physical markers can be determined genetically by a combination of molecular and progeny tests. In the case of animals, the calculation is as follows: R (the frequency of discordant individuals) is $R = (4r)/(1 + 6r)$ where r is the recombination in any single gamete. Since interference within very short distances is practically complete, the distance in cM is $d = 100r$. The recombination fraction ($\hat{r}$) in function of the size of the sample (N) is $\hat{r} = i/(4N - 6i)$ where i is the number of discordant strains and $\hat{d} = 100 \times \hat{r}$ in cM. In plants, the frequency of recombinant monoploid gametes is calculated by using the same formula $r = R/(2 - 2R)$ where R is the frequency of homozygous recombinant diploid individuals. ►RAPD, ►congenic resistant lines of mice, ►congenic strains; Bailey DW 1971 Transplantation 11:325.

Recombinant Joint: This is the site of connection of two molecules of DNA in a heteroduplex. ►heteroduplex

Recombinant Plasmid: This is generated either from two different DNAs by using the techniques of molecular biology or by spontaneous or induced genetic recombination. ►plasmid

Recombinant Vaccine: This is produced by in vitro modifications of genes/proteins; it does not carry the full complement of the infectious agent. ►vaccine

Recombinase: Refers to enzyme, mediating recombination. ►*FLP/FRT*, ►*Cre/loxP*, ►Rec, ►DMC1, ►Rec51

Recombinase System: ►immunoglobulins

Recombination: It is a process by which the linkage phase (coupling or repulsion) of syntenic genes is altered. In a broad biological sense it means the rearrangement of any molecule. Recombination is most common during meiosis but mitotic recombination also takes place. The mechanism of meiotic and mitotic events is not necessarily identical (see Fig. R20).

Independent segregation and reassortment are outside the realm of this term, according to the original definition given by H.A. Sturtevant, although some textbooks erroneously include these as well (see Fig. R21).

Recombination can be accurately assessed with the aid of sequenced genomes. Data obtained have

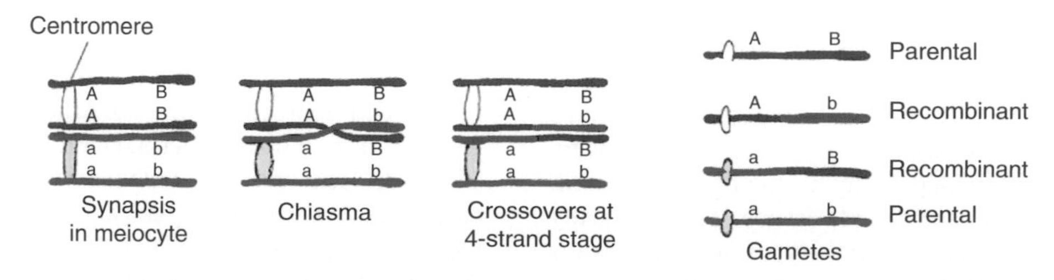

Figure R20. Cytological representation of recombination between homologous chromosomes in coupling phase. Each chiasma leads to crossing over and 50% recombination. The frequency of recombination depends, however, on the distance between the two loci considered. If crossing over takes place in all meiocytes between the bivalents, the frequency of recombination is 50%. If only half of the bivalents undergo crossing over in that particular interval, the recombination frequency will be 25% because, say in 4 meiocytes with (16 chromatids) 4/16 = 0.25. If only two of the meiocytes display crossing over then 2/16 = 0.125 is the frequency of recombination. The maximal recombination frequency by a single crossing over is 50%; the minimal may be extremely rare in case the linkage is tight.

Figure R21. A rare photograph of the three men who independently made basic discoveries in recombination. In the middle is Alfred Henry Sturtevant who as an undergraduate student published the first linkage map of *Drosophila* in 1913 (Science 58:269). On the left is Ernest Gustav Anderson who (with C.B. Bridges in 1925) (Genetics 10:403) demonstrated that crossing over takes place at the four-strand stage. On the right is Curt Stern who demonstrated in 1931 (Biol. Zbl. 51:547) that crossing over involves the physical exchange of chromatids in *Drosophila*. [Harriet B. Creighton and Barbara McClintock also published in 1931 identical proof using maize (Proc Natl Acad Sci USA 17:492)].

revealed considerable (0 to 9 centiMorgan per megabase) variation along each chromosome. The so-called desert" sequences display low and the "jungle" stretches high recombination and this is depicted in the diagram. (See also other recombination entries, ►linkage, ►repulsion, ►coupling, ►recombinational probe, ►flip-flop recombination, ►site-specific recombination, ►hot spot, ►cold spot, ►sex circle model of recombination, ►*Cre/loxP*, ►*FLP/FRT*, ►*rec*, ►ectopic recombination, ►recombination homologous, ►linkage disequilibrium, ►centiMorgan, ►STRP, ►recombination mechanisms of, ►retroviral recombination, ►mitotic crossing-over; Cox MM 2001 Proc Natl Acad Sci USA 98:8173; Sturtevant AH 1913 J Exp Zool 14:43, recombination in populations: Stumpf MPH, McVean GAT 2003 Nature Rev Genet 4:959; http://www.nslij-genetics.org/soft/; yeast meiotic recombination hot spots–cold spots: http://www.bioinf.seu.edu.cn/Recombination/rf dymhc.htm.

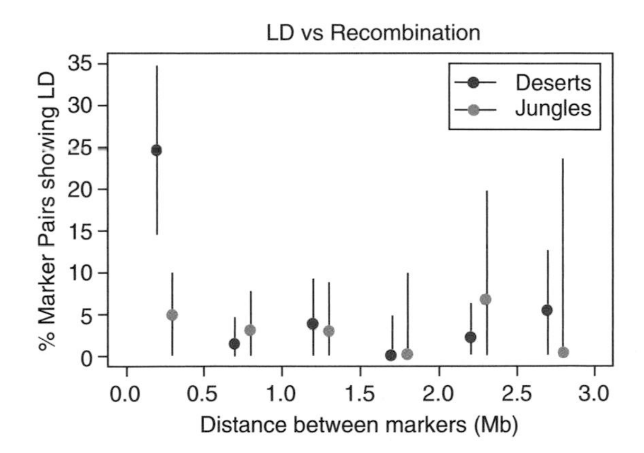

Figure R22. Linkage disequilibrium (LD) among pairs of STRPs (microsatellite repeats) within human autosomal recombination deserts and jungles within 0.25 megabase intervals. (Courtesy of Dr. James L. Weber; see also Yu A et al 2001 Nature [Lond] 409:951)

Recombination by Replication: At the beginning of the twentieth century William Bateson suggested that recombination is basically associated with the process of replication.

At that time neither of these phenomena was sufficiently understood or could even be hypothesized meaningfully. On the basis of cytological evidence (1930s) for marker exchange accompanied by chromosome exchange and later evidence that DNA exchange and phage gene exchange were correlated, the generally accepted view became that recombination does not require replication. However, the discovery of gene conversion remained a puzzling phenomenon although it was observed that about 50% of the gene conversion events involved flanking marker exchange. Holliday and other molecular models of recombination (during the 1960s and 1970s) permitted interpretation of classical crossing over and gene conversion without significant replication. Recently, it has been observed that the mutation or loss of function of the PriA DNA replication protein blocks both replication and recombination in *E. coli*.

The SOS DNA repair activates a replication process that does not require the replicational *oriC* site or the normally functioning DnaA protein but it needs RecA and RecBCD activities. It was assumed and subsequently demonstrated that double-strand breaks may be assimilated into the DNA and result in double D loops in the presence of the nearby chi elements. The *chi* elements block nuclease activity and assist the initiation of replication. It appears that PriA and other proteins of the primosome generate a replication fork at the D loop, and relying on the DnaB helicase and the DnaG primase, replication and recombination can be turned on. First, apparently lagging strand synthesis begins by the replisome and the lagging strand then primes the synthesis of the leading strand. Defective PriA may be compensated for by some elements of the primosome. The processes of double-strand break repair and recombination appear the same with the exception that in repair only the defective region has to be corrected whereas in recombination the entire strand must be replicated in order to recover the recombinants. Some observations have indicated the occurrence of joint processes of replicational repair and recombination in eukaryotes. ►reduplication theory, ►breakage and reunion, ►gene conversion, ►Holliday model, ►recombination molecular models, ►SOS repair, ►DNA repair, ►recA, ►recBCD, ►chi element, ►replication fork, ►D loop, ►replisome, ►lagging strand, ►leading strand, ►gene space; Michel B et al 2001 Proc Natl Acad Sci USA 98:8181.

Recombination by Transcription: Some of the instabilities may be induced by RNA polymerase II, particularly between repeats of the eukaryotic chromosomes. Some yeast mutants may increase the process by over three orders of magnitude. (See Gallardo M, Aguilera A 2001 Genetics 157:79).

Recombination Cloning: This is based on the integration/excision mechanism of the lambda phage and *E. coli* bacterium. Integration involves the λ *attP* site within the bacterial *attB* site. The bacterial *attL* and *attR* sites flank the integrated phage genome, excision reverses the process. Such a procedure can be used for the generation of vectors with DNA-binding and activation domain sites facilitating the study of protein interaction by analogy of the two-hybrid method. ►*att* sites, ►two-hybrid method; Muyrers JP et al 2001 Trends Biochem Sci 26[5]:325.

Recombination Frequency: Linkage is generally noticed in F_2 when independent segregation of the genes does not occur. Two genes in the homologous chromosomes can be at two different arrangements, repulsion (*Ab/aB*) or coupling (*AB/ab*) (see Table R3). Some scholars have described repulsion as trans and coupling as cis arrangement.

Table R3. Recombination detected

	Phenotypic Classes Expected			
	AB	*Ab*	*aB*	*ab*
Independent Segregation →	9/16	3/16	3/16	1/16
Linkage, Repulsion →	less	more	more	less
Linkage, Coupling →	more	less	Less	more

Recombination is commonly calculated as the percentage of recombinants in a test cross population. The maximum frequency is 50% because at this value the frequencies of recombinant and parental chromosomes are equal, i.e., the segregation is independent. Linkage is first observed in F_2 by deviation of the phenotypic proportions from the expectations for independent segregation (see Fig. R23). For example:

Figure R23. Aleurone color (*C*) and shrunken endosperm (*sh*) genes of maize are closely linked in chromosome 9 of maize. On the two ears these markers are in different linkage phase. (From Hutchison CB 1921 J Hered 12:76)

The linkage phase does not affect the frequency of recombination but it affects the frequency of the phenotypic classes (see Table R4). The frequency of recombination is the same in both cases $(5 + 5)/100 = 0.10 = 10\%$ as shown in the table.

Table R4. Phenotypic classes in test crosses in two linkage phases and recombination:

	(A hypothetical case)			
	AB	*Ab*	*aB*	*ab*
Repulsion cross *(Ab/aB) x ab*	5	45	45	5
Coupling cross *(AB/ab) x ab*	45	5	5	45

In F_2 recombination frequencies cannot be calculated by such a simple method because in the heterozygotes the genetic constitution of the individual chromosome strands is concealed but may be revealed in F_3. Nevertheless, recombination frequencies can be calculated (►F_2 linkage estimation). Recombination takes place at the four-strand stage of meiosis (see exception of mitotic recombination). The bivalent pair and in the simplest case two chromatids exchange segments. The maximal frequency of recombination within a chromosomal interval is 50%. Recombination frequencies are converted to map units by multiplication with 100. The realistic conversion of recombination frequencies into map units requires *mapping functions* because some of the recombinational events may not be detectable if the frequency of recombination between markers exceeds 15%. In physical measures 1 map unit has a different meaning in different organisms, depending on the size of the genome in nucleotides (nucleotide pairs) and the genetic length of the genome. Thus, 1 map unit in the plant *Arabidopsis* means about 150 kbp, in maize it is around 2,140 kbp and in humans approximately 1,100 kbp. One study reported that in human male autosomes the mean meiotic recombination frequency was 8.9×10^{-3} per megabase. In human chromosome 3 female recombination frequency was 1.43 cM Mb^{-1} and male was 0.85 cM Mb^{-1} (Muzny DM et al 2006 Nature [Lond] 440:1194). Smaller chromosomes have higher rates of recombination (cM/kb) not only among lower compared to higher eukaryotes, but also within one organism (yeast). The frequency of no recombination is a function (f) of the intensity of linkage and the population size; $f = (1 - r)^n$ where r is the recombination fraction and n is the number of test cross progeny. Data on maize (Fu H et al 2002 Proc Natl Acad Sci USA 99:1082) revealed reduced frequency of recombination in regions containing methylated retrotransposons. Recombination frequency may be affected by sex and may vary in different chromosomal regions; it is also influenced by sex in either plus or minus direction. The total recombinational map length may vary in different studies and by the use of different markers. Using different methods of human recombination, all frequencies corrected by the Kosambi mapping function indicated significantly higher recombination in females than in males (~1.6:1) (Matise TC et al 2003 Am J Hum Genet 73:271). In dogs and pigs, female recombination is higher; in cattle the two are about the same whereas in sheep male recombination is higher. Recombination frequencies may vary according to specific chromosomes. In the centromeric region the frequency of chiasma/recombination is lower than in other regions. In the area near the telomeres recombination increases in human males. In trisomy the recombination frequency is reduced. Many human diseases are associated with chromosomal deletions or duplications and in these cases recombination is reduced. ►mapping, ►mapping function, ►bacterial recombination, ►test cross, ►product ratio method, ►F_2 linkage estimation, ►F_3 linkage estimation, ►maximum likelihood method applied to recombination frequencies, ►recombination modification of, ►recombination variation of, ►sperm typing, ►chiasma, ►hot spot, ►map unit; Lynn A et al 2004 Annu Rev Genomics Hum Genet 5:317; http://www.nslij-genetics.org/soft/.

Recombination Frequencies in Bacteria: ►bacterial recombination

Recombination, Homologous: ►recombination, ►homologous recombination

Recombination Hot Spots: Genetic recombinations do not occur uniformly along the physical length of the DNA (chromosomes). In *Arabidopsis*, 1 cM varied from 30 bp to >550,000 bp. Gene-rich regions display more exchanges than gene-poor sequences. Recombination is usually suppressed around or near the centromere or telomere. In some instances recombination near the telomere is increased. In wheat, 1 cM in gene-rich region is estimated to be 118 kb whereas it is 22,000 kb for gene-poor regions. In humans, 1 cM indicates 1 Mb but there are substantial variations. In humans, 50% of all recombination takes place in less than 10% of the DNA sequence and recombination is preferentially outside the boundary of genes (McVeanan GAT et al 2004 Science 304:581). Human crossing over frequencies are clustered into narrow recombination hot spots (Jeffries AJ et al 2005 Nature Genet 37:601). The human genome-wide hot spot numbers vary between 25,000 and 50,000 and they occur preferentially near genes with 50 kb. Hot spots are determined by a CCTCCCT or larger motif. If in the third base T is changed to C, suppression is observed. L1 elements are underrepresented in hot spots (Myers S et al 2005 Science 310:321). ►coefficient of crossing over,

►hot spot, ►haplotype block, ►L1; JD et al 2000 Genetics 154:823; Arnheim N et al 2003 Am J Hum Genet 73:5; human recombination hot spots: http://www.jncasr.ac.in/humhot.

Recombination, Illegitimate: ►illegitimate recombination

Recombination in Autotetraploids: Measuring linkage and recombination in autotetraploids is far more difficult than in diploids because of the multiplicity of chromatids and alleles and also because the segregation ratios are not simple to predict from genetic data without cytological information. The difficulties are practically insurmountable when F_1 is a duplex or triplex and the genes concerned are far apart. Even in close linkage, large populations are required. In the case of coupling test cross, in simplex individuals the procedure is very similar to that of a test cross in diploids as can be seen in the example (see Table R5). (After deWinton D, Haldane JBS 1931 J Genet 24:121):

Table R5. Autotetraploids. Phenotypes observed in coupling

Parental	Recombinant	Recombinant	Parental	Total
SG	Sg	sG	sg	
336	215	210	353	1114

Recombination frequency

$$(\mathrm{p}) = \frac{\mathrm{Sg} + \mathrm{sG}}{\mathrm{Total}} = \frac{215 + 210}{1114} = \frac{425}{114} \cong 0.38$$

In the case of repulsion, the calculation presupposes knowledge of the possible gametic series that can be derived as shown in the Figure R24 and it is:

$$(1)\frac{\mathrm{SB}}{\mathrm{sb}} : (\mathrm{p})\frac{\mathrm{SB}}{\mathrm{sb}} : (2-\mathrm{p})\frac{\mathrm{SB}}{\mathrm{sb}} : (1+\mathrm{p})\frac{\mathrm{sb}}{\mathrm{sb}}$$

The manipulation of autetraploids with the techniques available for classical genetics is impractical despite the theoretical framework, and molecular analyses are also lagging.

Recombination frequency =

$$\frac{\mathrm{Sg} + \mathrm{sG}}{\mathrm{Total}} = \frac{2(1+\mathrm{p})}{2(1+\mathrm{p}) + 2(2-\mathrm{p})} = \frac{318}{717} \cong 0.44$$

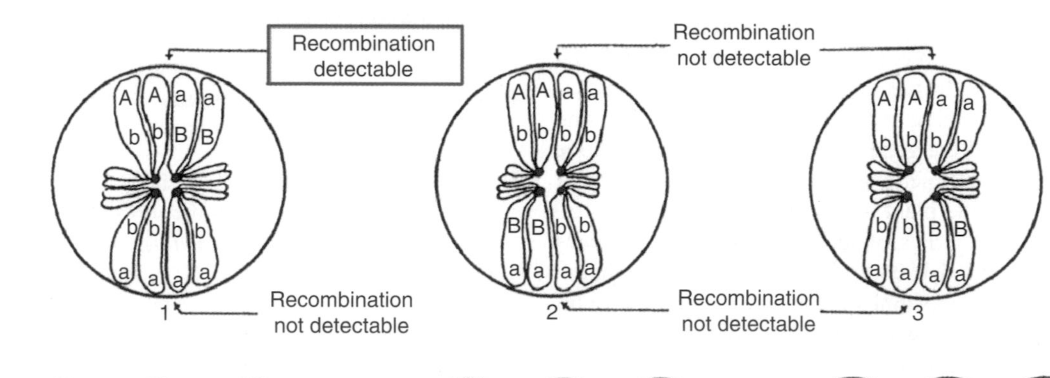

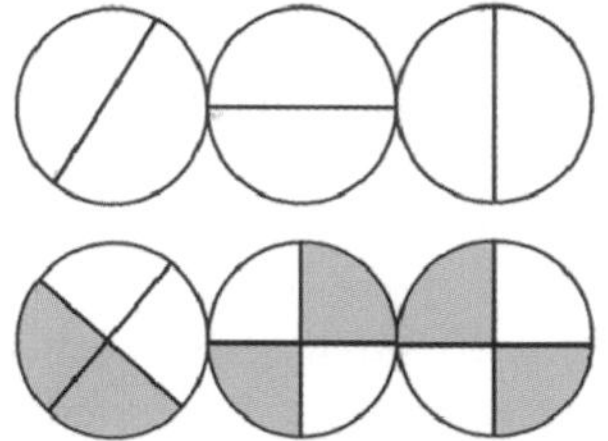

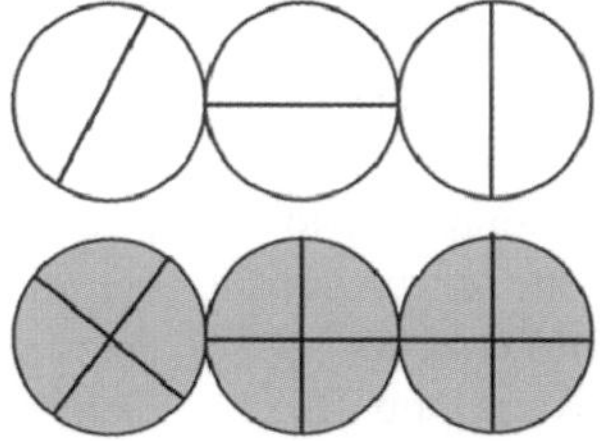

Figure R24. The derivation of the gametic series in a simplex tetrasomic case in repulsion. Recombination is considered only between loci *S* and *B*. In meiosis, three different quadrivalent associations are possible as shown at the top. Recombination is detectable only among the descendants of quadrivalent 1. The second row represents the different types of disjunctions at anaphase I, and the third row of circles shows the types of gametic tetrads formed. Gametes containing recombinant strands are open, gametes with parental strands are shaded

►autopolyploids, ►alpha parameter, see Fig. R24; Luo ZW et al 2001 Genetics 157:1369; Wu SS et al 2001 Genetics 159:1339; Hackett CA et al 2001 Genetics 159:1819.

Recombination In Vitro: ►staggered extension process

Recombination, Intrachromosomal: This occurs when homologous tandem or non-adjacent duplications are present in the chromosome. ►intrachromosomal recombination

Recombination, Intragenic: ►intragenic recombination

Recombination Machine: ►recombination molecular mechanisms

Recombination, Mechanisms, Eukaryotes, Yeast: The Sep 1 (strand exchange protein) 132 kDa fragment of a 175 kDa protein of yeast initiates the transfer of one DNA strand from a duplex to a single-stranded circle with 5′ to 3′ polarity without an ATP requirement. It also has a 5′ to 3′ exonuclease activity and is probably required for the preparation of 3′ end of single- and double-stranded DNA molecules for recombination. Mutation in Sep reduces mitosis, sporulation, meiotic recombination and genetic repair. (The STPβ protein, encoded by gene *DST2/KEM1*, is probably identical to Sep 1). One monomer of Sep 1 binds to nearly 12 nucleotides of single-stranded DNA. This requirement is reduced by the presence of the 34 kDa protein, which at a concentration of 1 molecule per 20 nucleotides reduces the requirement for Sep 1 to about 1/100. The DPA protein (120 kDa) of yeast controls DNA pairing and promotes heteroduplex formation in a non-polar manner independently of ATP. It promotes single-strand transfer from double-strand DNA to single-stranded circular DNA if the former has single-strand tails. Protein STPα (38 kDa) increases 15-fold shortly before yeast cells are committed to recombination during meiosis. If the gene encoding it (*DST1*) mutates, meiotic recombination is greatly reduced without an effect on mitotic recombination. The *RAD50* gene product (130 kDa) has an ATP-binding domain and it binds stoichiometrically to duplex DNA. The *RAD51* gene product is homologous to the RecA protein of *E. coli* (►recombination, ►mechanism, ►prokaryotes) and binds single- and double-stranded DNA. The *DMC1* (*disrupted meiotic cDNA*) gene product appears during meiosis and along with the product of *RAD51* performs functions similar to RecA in prokaryotes. Some organisms, *Drosophila melanogaster*, *Caenorhabditis elegans* and *Neurospora* have Rad51 but lack Dmc1 and in yeast mutation in Dmc1 does not prevent recombination. It is assumed that meiotic recombination in yeast involves double-strand breaks of the DNA. The actual site of exchange may be 25 to 200 kb away from the prominent break (Young JA et al 2002 Mol Cell 9:253). If DNA replication, which normally occurs 1.5 to 2 h before double-strand breaks, is blocked or delayed, recombination does not take place. *Drosophila*: protein Rrp 1 promotes exchanges between single-strand circular and linear duplex DNA. Its C-terminus has homology to *E. coli* exonuclease III and *Streptococcus pneumoniae* exonuclease A. Mammalian Cells: HPP-1 (human pairing protein with 5′ to 3′ exonuclease activity) binds to the DNA and promotes strand exchange in a 5′ to 3′ direction and it does not require ATP. The addition of the hRP-A (human single-strand binding) protein stimulates pairing almost 70-fold and reduces the amount of HPP-1 requirement (cf. SF1 in yeast). The precise mechanism by which the Holliday junction (see Holliday model, steps I to L) is resolved is not clear but endonuclease activity is postulated. Bacteriophage T4 gene *49* encodes endonuclease VII that under natural conditions cuts branched DNA structures. Similarly, bacteriophage T7 gene *3* product encodes endonuclease I which cleaves branched DNAs. In yeast, endonuclease XI (Endo XI, ≈ M_r 200,000 and other Endo proteins) has been found in cells with mutations in the RAD genes and apparently cuts cruciform DNA of the type expected by the Holliday juncture. ►recombination models, ►recombination molecular mechanisms in prokaryotes, ►recombination models, ►RAD51, ►RAG, ►recombination hot spots, ►Sep 1, ►SRS2, ►STPβ, ►synaptonemal complex, ►chiasma, ►sex circle model, ►gene conversion, ►databases; Camerni-Otero RD, Hsieh P 1995 Annu Rev Genet 29:509; Baudat F, Keeney S 2001 Curr Biol 11: R45; Smith GR 2001 Annu Rev Genet 35:243, yeast proteins and models: Krogh BO, Symington LS. 2004 Annu Rev Genet 38:233; exchange of DNA in recombination: Neale MJ, Keeney S 2006 Nature [Lond] 442:153.

Recombination Minimization Map: This is based on a skeletal map. The ordering relies on the smallest number of recombinations for single intervals. ►skeletal map

Recombination Models: ►Holliday model, ►Meselson-Radding model, ►Szostak model

Recombination, Modification of: The frequency of recombination may be altered by any means that affect chromosome pairing such as chromosomal aberrations, by DNA inserts introduced through transformation, temperature (either low or high), physical mutagens, rarely by chemicals, *rec*$^-$ genes, etc. In the heterogametic sex of *Drosophila* and silkworm meiotic recombination is usually absent although mitotic recombination occurs. In animals, recombination may be more frequent in females than in males and it is attributed to imprinting. In plants, in the case of a sex difference in recombination its frequency is usually lower in the megaspore mother cell. (See individual entries, ►coincidence, ►recombination variation of, ►imprinting; Singer A et al 2002 Genetics 160:649; Peciña A et al 2002 Cell 111:173).

Recombination, Molecular Mechanisms of: The RecA protein (M_r 37,842) directs homologous pairing by forming a right-handed helix on the DNA and it catalyzes the formation of DNA heteroduplexes. X-ray crystallography indicates that the DNA rests relaxed in the deep grove of this protein to facilitate scanning for homologous sequences. The RecA protein is also involved in DNA repair function (SOS repair). It digests the LexA bacterial repressor and instrumental indirectly in the derepression of over 20 genes involved in recombination and UV mutagenesis. The mechanism(s) of RecA activities can be studied by in vitro reactions. RecA can interact with 3 or 4 DNA strands by wrapping around the paired molecules. DNA-DNA pairing can take place between linear and circular DNA as well. Strand exchange proceeds at a slow pace (2 to 10 base/sec) in a polar fashion (5′ to 3′). The transfer begins at the 3′ end of the duplex and is transferred to a single strand DNA. Homology is a requisite for the RecA mediated reactions yet it tolerates some mismatches or insertions (up to even 1,000 bases or more) but these slow down the reactions. RecA can mediate pairing between two duplexes as long as there are short single-strand stretches or gaps. Low pH, intercalating chemicals, Z-configuration and other structural changes of the DNA may alleviate the difficulties of binding two duplexes. The RecA protein is a low efficiency ATPase. ATP hydrolysis is not an absolute requirement, for RecA activities in recombination but is more important for the repair reactions. In the presence of ATP the conformation of RecA is altered and in the nucleoprotein complex the DNA is substantially under-wound (the spacing between bases extends from 3.4 Å to 5.1 Å). It is assumed that the paired DNA molecules are not just juxtapositioned, but also one molecule lies in the major groove of the other. The pairing may involve three or four strands.

DNA strand exchange requires that the RecA filament rotates along the longitudinal axis and the DNA molecules are "spooled" inside where they may form the Holliday junction (►Holliday model). ATP stabilizes the RecA-DNA association and when ATP is split into ADT, the heteroduplex is released and RecA is recycled. Besides the RecA protein, recombination requires the presence of a single-strand binding protein (SSB), DNA polymerase I, DNA ligase, DNA gyrase, DNA topoisomerase I and the products of genes *recB*, *recC*, *recD*, *recE*, *recF* (binding protein for single-strand DNA), *recG*, *recJ* (exonuclease acting on single-strand DNA), *recN*, *recO*, *recQ*, *RuvB* (helicases), *recR*, *ruvR*, *ruvB* and *ruvC*. RuvC nicks the DNA at the point of strand exchange. RecBCD is a protein-RNA complex encoded by three genes (mentioned earlier), it performs the activities of (i) ATP-dependent double-strand exonuclease, (ii) ATP-dependent single-strand exonuclease, (iii) unidirectional DNA helicase, and (iv) site-specific endonuclease to nick four to six nucleotides dowstream of *chi*, a recombinational hot spot (5′-GCTGGTGG-3′). It has been suggested that RecBCD generates 3′-tails that are utilized by protein RecA for DNA strand exchange. RecB and RecC mutations can be suppressed by *sbcA* and *sbcB* mutations. Mutations in *sbcA* lead to the activation of the product (exonuclease VIII) of *recE*. Mutation in *scbB* inactivates exonuclease I, an enzyme that digests single-strand DNA, and its inactivation may assist the function of RecA in recombination (►models of recombination). The precise mechanism by which the Holliday junction (►Holliday model, steps I to L) is resolved is not clear but endonuclease (RuvC) activity is postulated.

Bacteriophage T4 gene *49* encodes endonuclease VII that under natural conditions splits branched DNA structures. Similarly, bacteriophage gene *3* encodes endonuclease I and cleaves branched DNAs. Some of the functions of the *ruv* operon of *E. coli* may be involved in the resolution of the Holliday junctions. *E. coli* also has in vivo systems where the molecular mechanism of resolution of recombination intermediates can be studied. Covalently closed plasmid DNA, DNA polymerase I and DNA ligase are transformed into *E. coli recA* mutants. Both monomeric and dimeric plasmid progenies are seen and the available markers permit the conclusion that crossing over occurs in 50% of the progeny. Recombination is not limited to DNA but viral RNA molecules also recombine.

The molecular mechanisms of recombination in eukaryotes have many features in common with those of prokaryotes. It appears that double-strand breaks can stimulate homologous recombination within one kilobase of the site of the break or it may affect recombination at a distance exceeding 30 kb. At the break a *recombination machine* may gain entry and as the machine moves on a heteroduplex of the DNA may be formed. At the broken ends DNA replication may be primed and there is a potential for recombination. The recombination machine is a complex of many enzymes mediating the recombination process. Recombination requires double-strand breaks of the DNA after the S phase and ribonucleotide reductase appears to be the rate-limiting factor for double-strand breaks and it is controlled by checkpoints (Tonami Y et al 2005 Proc Natl Acad Sci USA 102:5797). ►recombination mechanisms eukaryotes, ►recombination models, ►recombinational probe, ►recombination by replication, ►chi elements, ►illegitimate recombination, ►recombination RNA viruses, ►ribonucleotide reductase, ►FK506; Camerini-Otero RD, Hsieh P 1995 Annu Rev Genet 29:509; Barre F-X et al 2001 Proc Natl Acad Sci USA 98: 8189; West SC 1992 Annu Rev Biochem 61:603; Cox MM, Lehman IR 1987 Annu Rev Biochem 56:229; Smith GR 2001 Annu Rev Genet 35:243; Krogh BO, Symington LS 2004 Annu Rev Genet 38:233.

Recombination Nodule: Refers to the suspected site of recombination seen through the electronmicroscope as a 100 nm in diameter densely stained structure adjacent to the synaptonemal complex. There are early nodules seen at the association sites of the paired meiotic chromosomes and the late nodules are visible at pachytene when crossovers are juxtaposed. Non-crossovers do not show nodules after mid-pachytene (see Fig. R25). ►synaptonemal complex, ►chiasma, ►pachytene, ►association point, ►meiosis, ►recombination, ►crossing over, ►recombination RNA viruses, ►synapsis; Zickler D et al 1992 Genetics 132:135; Anderson LK et al 2001 Genetics 159:1259.

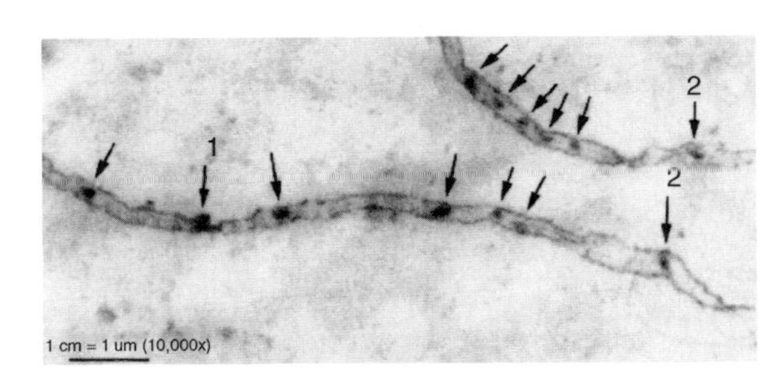

Figure R25. Early recombination nodules on two synaptonemal complexes of *Allium cepa* (onion). At positions marked 2 there is no synapsis yet. (Courtesy of Drs. LK Anderson and SM Stack)

Recombination Proficient: This means that there are no deficiencies involving enzymes mediating recombination. (See *Rec* and *rec* entries).

Recombination Repair: ►DNA repair

Recombination Repeats: These exist in the majority of plant mitochondrial DNAs in 1 to 6 pairs. Recombination can take place between/among these repeats generating various subgenomic DNA molecules. These repeats (264 bp to >5 kbp in maize) may not be indispensable parts of the mtDNA although they may contain genes for rRNA, cytochrome, etc. ►mitochondrial genetics, ►mtDNA, ►rRNA; Chanut FA et al 1993 Curr Genet 23:234.

Recombination, RNA Viruses: After co-infection of a cell by two different viruses recombination may take place by template switching during replication, thus it is more like a copy choice than a breakage and reunion mechanism. Recombination can take place between homologous and non-homologous strands (illegitimate recombination). The latter mechanism may lead to deletions, duplications and insertions. Among picornaviruses the recombination frequency may be as much as 0.9 in the case of high homology. Recombination in RNA viruses helps to eliminate disadvantageous sequences and can generate new variants. The estimated mutation rate per base is 6.3×10^{-4} and per genome is almost 5. The mutation rate is estimated as mutations per replication. Host genes may suppress or their absence may increase viral recombination (Serviene E et al 2005 Proc Natl Acad Sci USA 1032:10545). The double-strand bacteriophages of *Cystoviridae* have their genetic material in three segments. Intra-segment recombination is rare ($\sim 10^{-7}$/segment/generation) but reassortment between segments is high, even higher than in other taxa (Slander OK et al 2005 Proc Natl Acad Sci USA 102:19009). ►copy choice, ►breakage and reunion, ►illegitimate recombination, ►reverse transcription, ►negative interference; Keck JG et al 1987 Virology 156:331; Kirkegaard K, Baltimore D 1986 Cell 47:433; Negroni M Buc H 2001 Annu Rev Genet 35:275.

Recombination, Targeted: ►*Cre/loxP*, ►*FLP/FRT*

Recombination, Variations of: In the heterogametic sex of arthropods (male *Drosophila*, female silkworm) genetic recombination is usually absent or highly reduced. In the latter group of organisms mitotic recombination occurs, and these premeiotic exchanges may account for the observation of recombinants. The most common cause of variation is the presence of *rec*– genes. In Abbott stock 4A × Lindegren's wild type crosses of *Neurospora*, postreduction frequency was found to be 4.6 ± 1.2 whereas in Lindegren's stock it was 13 ± 1.2, and in Emerson's × Lindegren's crosses 27.6 ± 3.7. LJ Stadler, a pioneer of maize genetics, considered recombination as one of the most variable biological phenomena. ►male recombination, ►recombination frequency, ►recombination modification of, ►recombination hot spots, ►tetrad analysis; Browman KW et al 1998 Am J Hum Genet 63:861.

Recombinational Hot Spot: ►hot spot, ►chi site

Recombinational Load: This may emerge from the disruption of favorable, co-adapted gene blocks. ►genetic load, ►fitness-associated recombination

Recombinational Probe: One such short probe is inserted into the 902 bp πVX mini-plasmid containing a polylinker and the *supF* suppressor gene. Lambda phage libraries containing the miniplasmid construct are then propagated. If the phage carries a *supF* suppressible amber mutation, recombination between sequences homologous to the probe can be selectively recovered by forming plaques on an *E. coli* lawn. Recombination may take place even in the absence of perfect homology; less than ca. 8% divergence may be tolerated. Very large populations may reveal recombination within 60 base or longer probes effectively (see Fig. R26). ►*rec*, ►*Rec*, ►miniplasmid, ►*supF*, πVX, ►lawn, diagram on the use of the πVX microplasmid for the selective isolation of eukaryotic genes by recombinational probes; Perry MD, Moran LA 1987 Gene 51:227.

Recombinational Repair: ►DNA repair

Recombinator: Refers to cis-acting chromosomal sites promoting homologous recombination. ►chi

Recombineering: Refers to genetic engineering by homologous recombination. With the aid of a phage vector large DNA molecules can be cloned into bacterial artificial chromosomes. One such system uses the phage lambda genes Gam (inhibits host RecBCD), Exo, which degrades each DNA in 5′→3′ and thus generates single-strand 3′ overhangs, Beta protects the overhangs and anneals with complementary sequences. PCR-generated sequences or single-stranded oligonucleotides can be used as recombination substrates. Double-stranded DNA with 3′ overhangs may not need Exo. For the

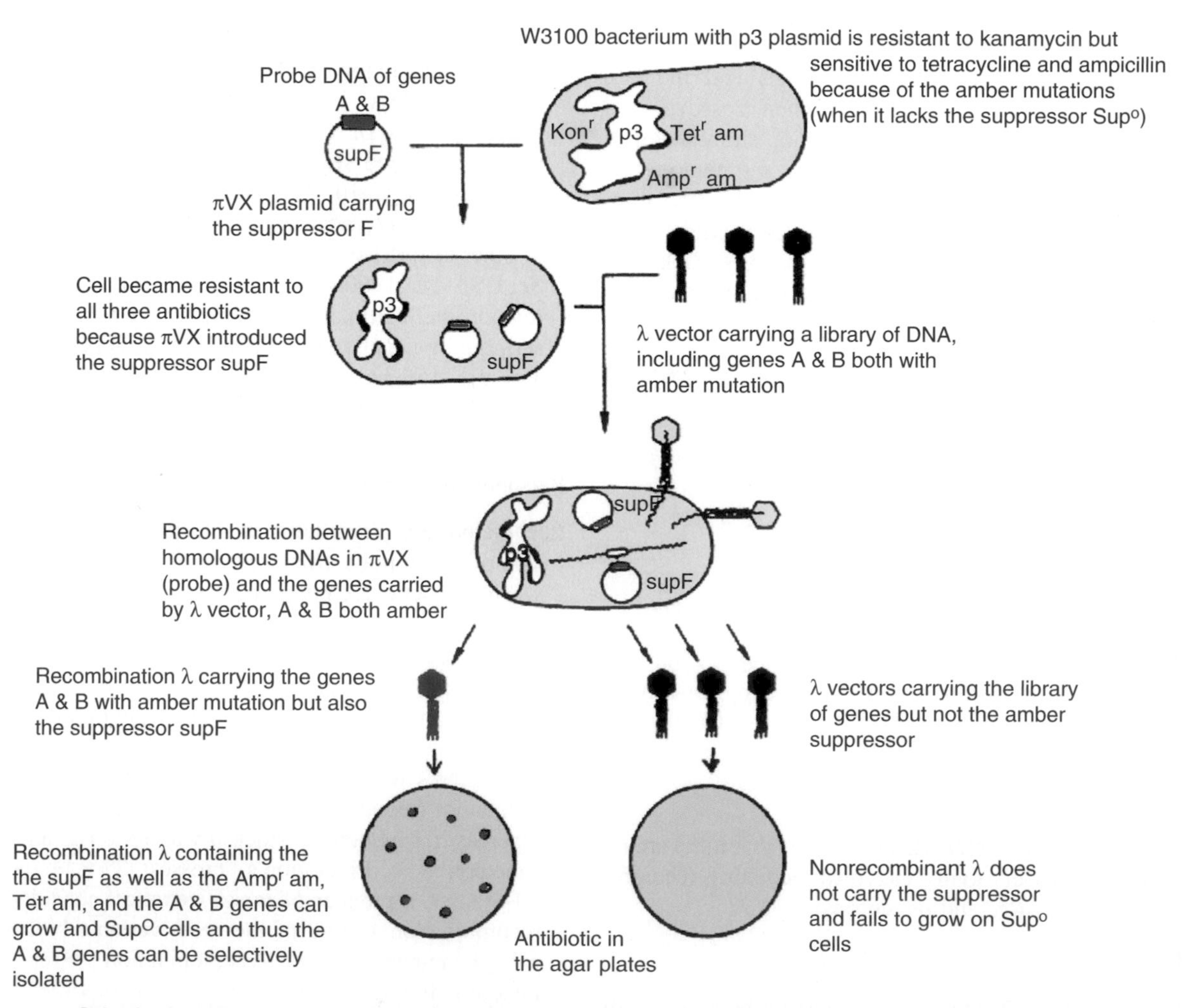

Selective isolation of specific eukaryotic genes with the aid of the πVX microplasmid. Several other plasmids have been constructed for similar purposes.

Figure R26. Recombination probe

procedure PCR-amplified linear or double-stranded DNAs are introduced into targeting cassettes that have either short regions of homology at their end or single-stranded oligonucleotides. The inserted DNA may carry any type of mutation or modification and there is no need for restriction enzyme cuts because the insertion is by homologous recombination. The method is simpler and safer than the somewhat unstable YACs and can be applied to functional genomic studies of higher organisms, e.g., mouse. (See terms under separate entries; Copeland NG et al 2001 Nature Rev Genet 2:769; Court DL et al 2002 Annu Rev Genet 36:351; Yu D et al 2003 Proc Natl Acad Sci USA 100:7207).

Recombinogenic: Refers to any agent (mutagen) which increases recombination (also).

Recombinogenic Engineering: ►recombineering

Recon: This is a historical term for the smallest recombinational unit. Molecular genetics has shown that recombination can take place between two nucleotides within a codon (Benzer S 1957, p 70 In: McElroy WD, Glass B (Eds.) The Chemical Basis of Heredity, Johns Hopkins University Press, Baltimore, Maryland).

Recon: This is a comprehensive literature-based genome-scale metabolic reconstruction that accounts for the functions of ORFs, proteins, metabolites and metabolic and transport reactions. The information on systems biology is a step toward individualized medicine and nutrigenomics, the applications, however, need a context to integrate and analyze data, and models resulting from these reconstructions can play a significant role in fulfilling this need. However, the development of cell-type or context-specific models requires the integration of various types of data, including transcriptomic, proteomic, fluxomic and metabolomic measurements (Duarte NC et al 2007 Proc Natl Acad Sci USA 104:1777). ►annotation, ►transcriptome, ►fluxome, ►metabolome, ►nutrigenetics, ►ORF, ►systems biology

Reconstituted Cell: This is produced by fusing cytoplasts and karyoplasts. ►transplantation of organelles, ►cytoplast, ►karyoplast

Reconstituted Virus: Into an empty viral capsid a complete viral genetic material is introduced, e.g., into the coat of tobacco mosaic virus (TMV) the genome of the related Holmes ribgrass virus (HRV) was introduced and the new particle expressed the characteristic functions of the donor RNA.

This classic experiment proved that the genetic material could also be RNA. Influenza and other viruses can be reconstituted from cloned cDNAs. (See Fig. R27, Fraenkel-Conrat H, Singer B 1957 Biochem Biphys Acta 24:540; Neumann G et al 1999 Proc Natl Acad Sci USA 96:9345).

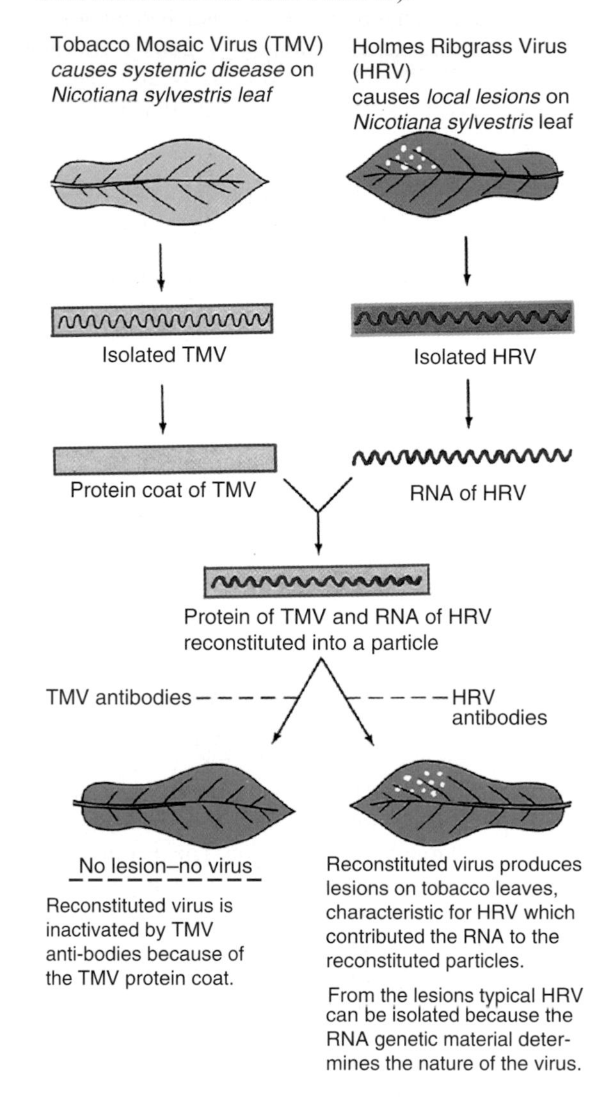

Figure R27. Reconstituted virus

Record: This is a true document of an observation or of a hypothesis which is explicitly stated.

Recoverin: ►rhodopsin

Recovery of DNA Fragments from Agarose Gel: The fragment is driven by electrophoresis on to DEAE cellulose membrane by cutting a slit in front of the band and placing the DEAE sliver in the slit. Alternatively, electroelution can be used or from agar (at low melting temperature) the DNA can be extracted by phenol and precipitated by ammonium acetate in 2-volume ethanol and collected by centrifugation. ►electrophoresis, ►DEAE cellulose

RecQ: Denotes a family of helicase proteins. In the case of mutation in the coding gene, chromosomal instability results in eukaryotes. In prokaryotes the *recQ* is involved in post-replicational repair. ►chromosomal rearrangements, ►helicase, ►*RecA*, ►*RecB* and other *Rec* genes; Enomoto T 2001 J Biochem [Tokyo] 129:501; Wu X, Maizels N 2001 Nucleic Acids Res 29:1765; Cobb JA et al 2002 FEBS Lett 529:43.

RECQL: This is a RecQ-like protein in humans. ►RecQ

Recruitment: For the initiation of transcription some prokaryotic and eukaryotic genes require activators and the transcriptional complex will operate only if these activators are attracted to the transcriptional target. The GAL1 gene of yeast recruits four units of the GAL4 activator at about 250-base upstream to begin transcription. The GAL4 units are blocked, however, unless galactose is available in the culture medium. The activators make contact with some sites of the RNA polymerase subunits. In bacteria, typical activators are the CAP (catabolite activator protein) and also the λ repressor may bind to the σ^{70} subunit of the polymerase. In yeast, the transcription complex includes more than 30 different proteins. ►transcription factors, ►transcription complex, ►activator proteins, ►two-hybrid method; Francastel C et al 2001 Proc Natl Acad Sci USA 98:12120.

Recruitment of Exons: Evolving genes may acquire coding sequences for functional domains by borrowing exons through recombination. The recruited DNA sequences may occur in several protein genes with different function, e.g., the low-density lipoprotein (LDP) receptor (a cholesterol transport protein) has homology in 8 exons with the epidermal growth factor (EGF) peptide hormone gene. ►exon, ►LDP, ►EGF

Recruitment of Genes: Refers to acquiring new genetic information through recombination or transfection (transformation). ►transformation, ►transfection, ►recruitment of exons

Rectification, Inward: Through a voltage gated ion channel the current inward exceeds that of the outward. In the case of outward rectification the opposite holds. ►ion channel

Recurrence Risk: Refers to a couple's chance of having another child with the same defect (see Fig. R28). ►risk, ►empirical risk, ►genetic risk, ►genotypic risk ratio, ►λ_S, ►aggregation familial, ►chart

Recurrent Parent: A plant or an animal is mated with selected line(s) in several cycles for one or more backcrosses. ►recurrent selection

Recurrent Selection (reciprocal recurrent selection): Refers to a variety of methods used for breeding superior hybrids of plants and animals of high productivity. The general procedure is as follows: lines (inbred or not) A and B are crossed in a reciprocal manner, i.e., A males are crossed to B females and B males are mated with A females. The initial lines are expected to be genetically different to assure the sampling of different gene pools. In this manner several lines are mated, not just A and B. The progenies are tested for performance and only the best parents are preserved. The superior parents are mated again with representatives of their own line. On the basis of their progeny, the parents are re-evaluated. The mating cycle is then repeated. Most commonly, each male is mated with several females of the other line in order to assure the availability of a large population of offspring to be able to conduct statistically meaningful tests. The maintenance of the lines requires that females be mated with selected males within their own lines. This procedure results in inbreeding but enhances the chances of further selection. Therefore, the performance of the selected parental lines is expected to decrease but that of the hybrids will increase. An alternative simplified method involves selection for combining ability in only one set of lines. Thus, line A is mated with a previously inbred tester which has an already known combining ability and the selection is restricted to within line A. This latter modification results in faster initial progress but the final gain may be limited. ►combining ability, ►hybrid vigor, ►heterosis, ►QTL, ►heritability, ►diallele analysis; Hull K 1945 Amer Soc Agron 37:134.

Recursive Partitioning: This is a statistical approach used to classify information into alternative classes, e.g., normal or tumorous. It builds classification rules on the basis of feature information. The underlying principle is that observations of *n* units represent a vector feature of measurement. (Vector here means a single class matrix.) Or covariates (e.g., data from a type of condition) and a class label. Unlike linear discriminant analysis, it extracts homogeneous strata and constructs tree-based classification rules. (Strata here means a division of the data into parts.) The information is partitioned into increasingly smaller samples (nodes) to facilitate critical discrimination between the classes. ►matrix algebraic, ►discriminant function, ►cancer classification; Zhang H et al 2001 Proc Natl Acad Sci USA 98:6730.

Recursive PCR: This is a method of DNA amplification. Synthetic oligonucleotide primers (50–90 bases) are used which have only terminal complementarity (17–20 bp). They are annealed at 52 °C to 56 °C. The heating cycle is 95 °C and the cooling is at 56 °C. The Vent polymerase is used at 72 °C. This thermostable polymerase has capability not

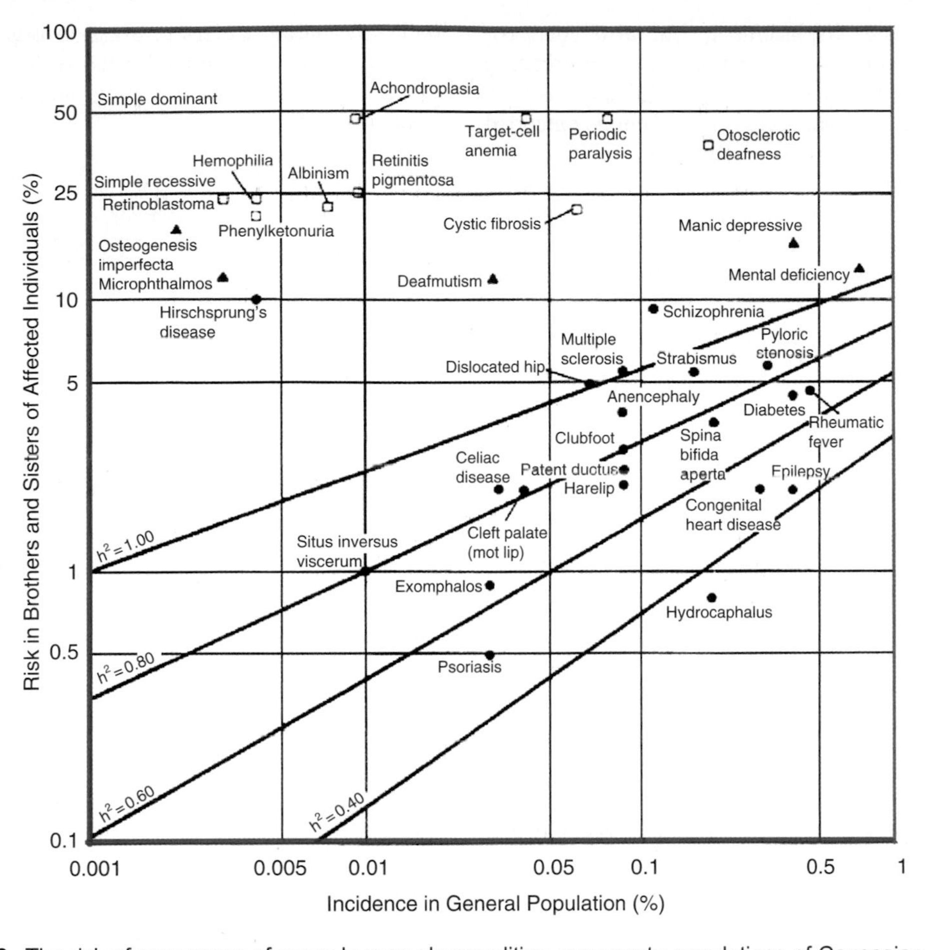

Figure R28. The risk of recurrence of some human abnormalities common to populations of Caucasian descent. The numbers along the horizontal axis refer to the frequency of traits in the population as a whole; the numbers along the vertical axis refer to the risk of occurrence of these defects among brothers and sisters of affected individuals. The heavier lines denote the risk of recurrence according to the nature of the genetic control of these traits. The meaning of dominant and recessive terms is explained in Chapter 4; h^2stands for heritability in the broad sense (see heritability), indicating the expected inheritance of a trait controlled by several genetic factors. Because the realization of many human genetic anomalies is influenced by a number of factors, estimation of the genetic risk requires an experienced genetic counselor. This consultation will secure peace of mind, for it frequently turns out that the risk is less than feared. Open squares stand for high risks, triangles for medium risks, and solid circles for low risks. Albinism: pigment deficiency of hair, skin, and eye; achondroplasia: a type of dwarfism; anencephaly: a deficiency of brain tissue; celiac disease: an intestinal inflammation; cleft palate: an oral fissure; clubfoot: a deformation of the foot; congenital heart disease: a heart anomaly; cystic fibrosis: fibrous tissue overgrowth with cyst formation; deaf-mutism: loss of hearing and speech; diabetes: a defect of carbohydrate metabolism; dislocated hip: a bone displacement; epilepsy: seizures; exomphalos: a hernia; harelip: upper lip fissure; hemophilia: recurrent bleeding due to lack of blood coagulation; Hirschprung's disease: a colon enlargement; hydrocephalus: fluid accumulates in the head; manic depressive: obsessive emotional dejection; mental deficiency: deterioration of the mind; microphthalmos: abnormally small eyes; multiple sclerosis: hardening spots in brain and spinal cord; otosclerotic deafness: spongy bones in the ear; osteogenesis imperfecta: brittleness of the bones; patent ductus: fetal blood vessel defect; periodic paralysis: recurrent impairments in motor functions; phenylketonuria: defect in phenylalanine metabolism; psoriasis: scaling skin plaques; pyloric stenosis: obstruction of the distal end of the stomach; retinitis pigmentosa: atrophy and pigmentation in the eyes, retinoblastoma: an eye tumor; rheumatic fever: inflammation of connective tissues; schizophrenia: a mental defect leading to loss of contact with realities; situs inversus viscerum: visceral transposition; spinal bifida aperta: a spinal defect; strabismus: a deviation in the visual axis; target-cell anemia: abnormally thin erythrocytes. (Adapted from Hartl DL 1977. Our Uncertain Heritage, J. B. Lippincott, Philadelphia; based on data by Newcombe HB in Fishbein M ed., 1964, 2nd Int. Cont. Congenital Malformations, Int. Med. Congr. Ltd.)

only for strand displacement, but also for exonuclease function and it carries out proofreading and therefore the fidelity of the amplification is very good. During the initial steps each 3′ end is extended with the aid of the opposite strand as a template and duplex sections are thus generated. In further cycles one strand of the duplex is displaced by a primer oligonucleotide derived from a neighboring duplex. During the last step high concentration of the terminal oligonucleotides assist in the amplification of the entire duplex. ►polymerase chain reaction, ►Vent; Prodromou C, Pearl LH 1992 Protein Eng 5:827.

red: ►lambda phage, ►Charon vectors

RED613: This is a fluorochrome, a conjugate of R-phycoerythrin and Texas Red. Its excitation maximum is at 488 nm and emission at 613 nm. ►fluorochromes

RED670: This fluorochrome is a conjugate of R-phycoerythrin and a cyanine. Its excitation maximum is at 488 nm from an argon-ion laser, emission is at 670 nm. ►excitation

Red Blood Cell: ►erythrocyte, ►blood, ►sickle cell anemia

Red-Green Color Blindness: ►color blindness

Red King Hypothesis: In mutualistic interaction the slowly evolving species is expected to gain a disproportionate share of the benefits in the population. (See Bergstrom CT, Lachmann M 2003 Proc Natl Acad Sci USA 100:593).

Red Queen Hypothesis: If a population does not continue to adapt at the same rate as its competitors, it will lose ecological niches where it can succeed, and if it stays put long enough it may become extinct. The hypothesis suggests that sex evolved so that the species could respond to changes in the biotic environment. Recent analysis, however, does not support the need for sex evolution in all cases of species interactions (Otto SP, Nuismer SL 2004 Science 304:1018). The name RQ was adapted from Lewis Carroll's (pen name of C. L. Dodgson) 1872 fantasy story about a chess game, *Through the Looking Glass*. ►adaptation, ►extinction, ►beneficial mutation, ►equilibrium in populations, ►genetic homeostasis, ►treadmill evolution, ►co-evolution, ►Kondrashov's deterministic model of evolution of sex, ►Muller's ratchet; Van Valen L 1973 Evol Theory 1:1.

Redifferentiation: Refers to organ or organism formation from dedifferentiated cells, such as from callus (see Fig. R29). ►callus, ►regeneration, ►dedifferentiation

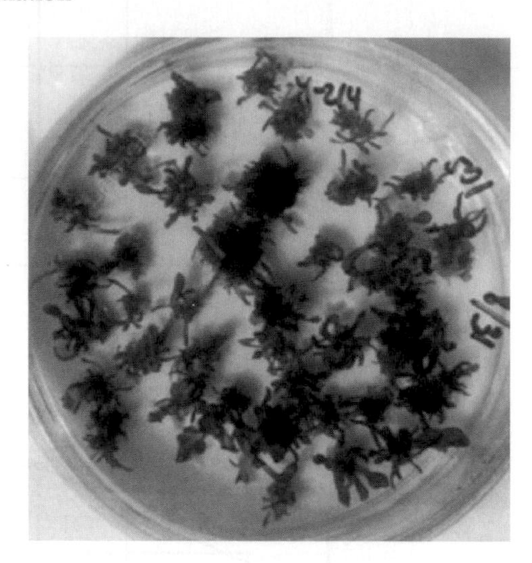

Figure R29. Redifferentiation of leaves from *Arabidopsis* callus (Rédei GP unpublished)

Redox Pair: Refers to an electron donor and the oxidized derivative.

Redox Reaction: ►oxidation-reduction

Reduced Representation Shotgun: ►RRS

Reducing Sugar: Its carbonyl carbon is not involved in glycosidic bond and can thus be oxidized. Glucose and other sugars can reduce ferric or cupric ions, a property that serves for their analytical quantitation (Fehling reaction).

Reductant: An electron donor.

Reduction: Denotes the gain of electrons.

Reductional Division: In meiotic anaphase I half of the chromosomes segregate to each pole and the two daughter cells have *n* number of chromosomes rather than *2n* as in the original meiocyte. In the case of uneven numbers of crossing over between the gene and the centromere the numerical reduction of the chromosomes may not result in the separation of different pairs of alleles, i.e., the reduction does not extend to the alleles. The reductional division at meiosis assures the constant chromosome numbers in the species and serves a basis for Mendelian segregation. ►prereduction, ►postreduction, ►tetrad analysis, ►meiosis

Reductional Separation: In meiotic anaphase I the parental chromosomes separate intact because there is no recombination between the bivalents. ►equational separation

Reductionism: This refers to the practice of reducing ideas to simple forms or making efforts to explain phenomena on the basis of the behavior of elementary units (molecules). This endeavor is frequently criticized because of the complexities of biological systems. It should be borne in mind that without the analytical approach, science (molecular genetics) would not have progressed to its present level. ►model organisms

Reductive Evolution: Some obligate intracellular parasites (*Rickettsia*, *Chlamydia*, *Mycobacterium leprae*) lost during evolution a significant portion (up to 76%) of the coding capacity of their genetic material because the host metabolism provided the essential gene products. These organisms may still have many pseudogenes as well as degraded, non-functional DNA tracts. ►*Rickettsia*, ►leprosy; Pál C et al 2006 Nature [Lond] 440:667.

Redundancy: Refers to repeated occurrence of the same or similar base sequences in the DNA or multiple copies of genes. The repeated gene sequences are considered to have a duplicational origin. About 38–45% of the sampled (ca. 1/3 of all) proteins in *E. coli* are expected to be duplicated and in the much smaller *Haemophilus influenzae* genome, completely sequenced by 1995, 30% appear to have evolved by processes involving duplications. In this small genome some gene families were represented by 10 to over 40 members whereas almost 60% of the genes appeared unique. In yeast almost 60% of the genes are redundant. In the large eukaryotic genomes the repetitious sequences are represented by larger fractions. The number of protein kinases in higher eukaryotic cells may reach 2,000 and that of phosphatases nearly 1,000. Theoretically, true redundancy should succumb to natural selection unless the rate of mutation is extremely low. Besides the shared functions, if the redundant genes have unique roles, they are maintainable. Redundant genes are saved when the *developmental error* rate is high. Redundant genes may affect the fitness of an organism in a very subtle way and may not show a clear independent phenotype. Also, they may serve as an insurance for the loss or inactivation of other members of the gene family. Redundant genes may have the luxury to afford mutation to new functions. Mutation or deletion of redundant genes may not have phenotypic consequences. Some single copy genes can also be removed without any consequences for the phenotype. ►SINE, ►LINE, ►LOR, ►MER, ►MIR, ►tandem repeat, ►inverted repeat, ►polyploid; Goldberg RB 1978 Biochem Genet 1–2:45.

Reduplication Hypothesis: At the dawn of Mendelism, William Bateson postulated that genetic recombination takes place by a differential degree of replication and different associations of genes after they separated at the interphase rather than by breakage and reunion of synapsed chromosomes. ►breakage and reunion, ►copy choice; Bateson W, Punnett RC 1911 J Genet 1:293.

Reelin: This 420 kDa glycoprotein is encoded by the *reeler* gene in mouse and is expressed in the embryonic and postnatal periods. It is similar to extracellular matrix serine proteases involved in cell adhesion. It controls layering and positioning of neurons and mutations in gene impair coordination resulting in tremors and ataxia. In schizophrenia and bipolar psychoses at positions −134 and 139 in the promoter of reelin methylation of cytidine increases as a result of upregulation of methyltransferase (Grayson DR et al 2005 Proc Natl Acad Sci USA 102:9341). ►ataxia, ►CAM, ►schizophrenia, ►bipolar mood disorder, ►chimpanzee; D'Arcangelo G et al 1995 Nature [Lond] 374:719; Keshvara L et al 2001 J Biol Chem 2001 J Biol Chem 276:16008; Quattrocchi CC et al 2002 J Biol Chem 277:303; see retraction 2004 Science 303:1974.

REF: This RNA-binding nuclear protein facilitates the export of the spliced mRNA from the nucleus to the cytoplasm in cooperation with TAP. ►TAP; Le Hir H et al 2001 EMBO J 20:4987.

Reference Library Database (RLDB): Cosmid, YAC, P1 and cDNA libraries for public use on high-density filters. Information: Reference Library Database, Imperial Cancer Research Fund, Room A13, 44 Lincoln's Inn Fields, London WC2A 3PX, UK. Phone: 44-71269-3571. Fax: 44-71-269-3479. INTERNET: genome@icrf.icnet.uk, databases.

Refractile Bodies: *Paramecia* may contain bacterial symbionts and the bacteriophages associated with them may appear as bright (refractile) spots under a phase-contrast microscope. ►symbionts hereditary, ►killer strains, ►*Paramecium*

Refractory Genes: These interfere with the completion of the life cycle of parasites, e.g., *Plasmodium*, within the insect vector, e.g., mosquitos and can therefore be exploited for the control of malaria and other diseases. ►*Plasmodium*, ►malaria; Yan G et al 1997 Evolution 51:441.

Refractory Mutation: This may not be revealed through genetic testing although it may lead to a genetic disease, e.g., mutation in introns, promoters or

3′-downstream regulatory sequences that control transcript levels. ►genetic testing

RefSeq: This is the reference sequence for 2,400 organisms, genomes, transcripts and proteins. (See Pruitt KD et al 2005 Nucleic Acid Res 33:D501; http://www.ncbi.nlm.nih.gov/RefSeq/; http://www.ncbi.nlm.nih.gov/entrez/query.fcgi?db=gene).

Refsum Diseases (10pter-p11.2, 8q21.1, 7q21-q22, 6q23-q24): These autosomal recessive disorders are manifested in adult and early onset forms. Both forms involve phytanic acid accumulation because of the deficiency of an oxidase enzyme in the peroxisomes. The symptoms include polyneuritis (inflammation of the peripheral nerves), cerebellar (hind part of the brain) anomalies and retinitis pigmentosa. The early onset form, in addition, is characterized by facial anomalies, mental retardation, hearing problems, enlargement of the liver, lower levels of cholesterol in the blood and the accumulation of long-chain fatty acids and pipecolate (a lysine derivative). The symptoms of the infantile form overlap with those of the Zellweger syndrome. ►Zellweger syndrome, ►phytanic acid, ►microbodies, ►retinitis pigmentosa; Mukherji M et al 2001 Hum Mol Genet 10:1971; van den Brink DM et al 2003 Am J Hum Genet 72:471.

Refuge: Refers to the area planted with non-transgenic crops (GMO) next to transgenic plants. The purpose is to let non-resistant insects mate with resistant insects and thus dilute out the resistant population. At the moment the best relative size of a refuge is controversial. ►GMO

Regeneration in Animals: This is more limited than in plants where totipotency is preserved in most of the differentiated tissues. Regeneration can actually be classified into two main groups of functions: one is the regular replacement of cells (e.g., epithelia, hairs, nails, feathers, antlers and production of eggs and sperms) in a wide range of animals, and the other is the capacity to regenerate body parts lost by mechanical injuries. The latter type of regeneration may involve the formation of an entire animal from pieces of the body, such as by morphallaxis in sponges, Hydra, flatworms, annelids (preferentially from the posterior segments), echinoderms, etc. A more limited type of regeneration is found in the higher forms. Arthropods may replace lost appendages of the body. Vertebrate fishes can replace lost fins, gills or repair lower jaws. The Zebrafish is capable of heart regeneration after epicardial injury (Lepilina A et al 2006 Cell 127:607). Some mouse strains (MRL) have an exceptional ability to regenerate heart tissues in vivo. Several amphibians (salamanders, newts) readily regenerate lost limbs, tails and some internal organs. Reptile lizards can reproduce lost tails although the regenerated one is not entirely perfect. Regeneration of feathers and repair of beaks may take place in birds. In mammals lost blood cells may be replenished by bone marrow activity, or liver cells may regenerate new ones. More limited regeneration may occur in bone, muscle, skin and nerve cells but unlike in plants, complete organisms cannot be regenerated from any part, except the embryonic stem cells or possibly from other stem cells after special treatments. According to recent evidence, mesoderm, endoderm and ectoderm cell lineages can be reprogrammed, e.g., bone marrow cells can regenerate into nerve cells or muscle-derived cells and cells of the central nervous system can reconstitute other cell types. From embryonic stem cells viable, fertile adult mouse can be differentiated (Eggan K, Jaenisch R 2003 Methods Enzymol 365:25). ►regeneration of plants, ►transdetermination, ►homeotic genes, ►stem cells, ►transplantation of nuclei, ►nuclear transplantation, ►dedifferentiation, ►redifferentiation, ►transdifferentiation, ►*Hydra*, ►zebrafish, ►planarians, ►newt; Leferovich JM et al 2001 Proc Natl Acad Sci USA 98:9830; Alsberg E et al 2002 Proc Natl Acad Sci USA 99:12025; Brockes JP, Kumar A 2005 Science 310:1919; regeneration in planarians: Sanchez AA 2006 Cell 124:241; regeneration of heart review: Laflamme MA, Murry CE 2005 Nature Biotechnol 23:845; comprehensive review of molecular mechanisms involved in regeneration of model organisms: Alvarado AS, Tsonis PA 2006 Nature Rev Genet 7:873.

Regeneration of Plants: Refers to the formation of new organs or entire organisms from dedifferentiated tissues or single cells (see Fig. R30). A higher level of ferredoxin-nitrite reductase quantitative gene locus and nitrate assimilation rice favors regeneration of transgenic plants (Nishimura A et al 2005 Proc Natl Acad Sci USA 102: 11940). ►embryogenesis somatic, ►embryo culture, ►clone, ►vegetative reproduction, ►totipotency, ►dedifferentiation, ►callus, ►root

Figure R30. Regeneration of Arabidopsis from callus

Reglomerate: ►aggregulon

Regression: This is the measure of dependence of one variate on another in actual quantitative terms in contrast to correlation which uses relative terms from 0 to 1. Linear regression involves the independent variate to the first power. Quadratic regression involves the independent variate to the second power and cubic regression to the third power. ►correlation for the calculation of regression coefficient, ►heritability, ►linear regression, ►cluster analysis, ►multiple regression

Regulated Gene: The expression is conditional and affected by genetic and non-genetic factors. Traits associated with dynamic processes may evolve to some extent more readily through regulatory rather than coding mutations. Indeed, the evolution of complex multicellular organisms would have been all but impossible in the absence of *cis*-regulatory systems that allowed context-dependent transcriptional regulation. The expression of many genes is altered through alternative transcription start sites and splicing and post-translational modifications. Quantitative traits are typically regulated by elements in the vicinity of the gene upstream or downstream. Typical examples of cis-regulation are the operons. Certain *duffy* haplotypes segregating in modern human populations offer almost complete resistance to infection with *Plasmodium vivax*. Resistance is due to the lack of Duffy protein expression in erythrocytes, but not in several other cells where it is normally expressed. The causal mutation is a *cis*-regulatory single nucleotide polymorphism (SNP) which disrupts binding of the transcription factor GATA1. Lactose intolerance (LCT) is determined by the presence or absence of an SNP in an intron of minichromosome maintenance deficient 6 homologue (*MCM6*), the next gene 5′ of LCT. Experimental tests demonstrate that this SNP elevates LCT transcription (Wray GA 2007 Nature Rev Genet 8:206). ►housekeeping genes, ►constitutive genes, ►regulation of gene activity, ►operon, ►disaccharide intolerance, ►Duffy blood group, ►SNP, ►GATA, ►MCM1, ►annotation, ►module, ►regulated sequence motifs in humans, ►mouse, ►rat; *Caenorhabditis*: http://www.cisred.org/; annotated regulatory sequences: http://www.swissregulon.unibas.ch; *cis*-regulatory mammalian modules: http://genomequebec.mcgill.ca/PReMod.

Regulation of Enzyme Activity: Enzyme activity is characterized by various measures of enzyme kinetics (►Michaelis-Menten, ►Linweaver-Burk, ►Eadie-Hofstee). The reaction is controlled by the quantity and/or activity of an enzyme. Enzymes may enhance reaction rates by 10^{10} to 10^{23} relative to uncatalayzed transformations in aequeous solutions (Kraut DA et al 2003 Annu Rev Biochem 72:517). The quantity of the enzyme depends on protein synthesis/degradation controlled at the level of transcription, translation, processing of the protein, and its instability (►regulation of gene activity). The substrate of the enzyme may regulate the production of the enzyme protein (►enzyme induction, ►*lac* operon, ►catabolite repression, ►attenuation). *Feedback control* means that the accumulation of the product of an enzyme may shut down the operation of a pathway at any step preceding the final product. Feedback control may be simple or multiple, i.e., more than one enzyme may be affected either simultaneously or sequentially or more than a single product of the pathway may act in a concerted manner (►feedback control). Feedback control may act either at the level of the synthesis (*feedback repression*) or by *inhibition* of the activity of a steady number of enzyme molecules. In general, the inhibitors are either *competitive* (bind to the enzyme and compete with the substrate for the active site) or *non-competitive* (the inhibitors act by attaching to the enzyme at a site other than the active site yet lower enzyme activity [by allosteric effect]). *Uncompetitive inhibitors* operate by binding to the enzyme-substrate complex. *Suicide inhibitors* are converted by the enzyme into an irreversibly binding molecule that permanently damages the enzyme. The inhibitors may simultaneously affect more than one enzyme. *Mechanism-based inhibitors* are highly specific to a single enzyme and as such have great significance for medicinal chemistry. Among these are the *antisense inhibitors* (►antisense RNA). *Allosteric enzymes* may also be stimulated (*modulated*) by allosteric compounds. The modulator may be *homotropic*, i.e., essentially structurally identical to the substrate or *heterotropic* in case it is not identical to the substrate. The activity of an enzyme may require a proteolytic cleavage of the precursor protein, the *zymogen*. Thermostabilization of an enzyme (cytosine deaminase) by site-directed mutagenesis at sites identified on the basis of crystal structure using computations increased half-life at 50 °C 30-fold without lowering catalytic efficiency. The effects of the three mutations were synergistic and indicated that by purposeful design, industrially or biomedically, more useful enzymes can be produced (Korkegian A et al 2005 Science 308:857). ►regulation of gene activity, ►protein synthesis, ►signaling, ►allostery, ►allosteric control, ►feedback; Wall ME et al 2004 Nature Rev Genet. 5:34.

Regulation of Gene Activity: The various types of cells and differentiated tissues of an organism generally contain the same genetic material (►totipotency,

R

►regeneration) yet their differences attest that the genes must function in diverse ways in order to bring about the variety of morphological and functional differences. Genetic regulation accounts for this variety. Many genes are expressed in every cell because they determine the metabolic functions essential for life. Another group of genes is responsible for such generally required structural elements as membranes, microtubules, chromosomal proteins. (►*housekeeping*, ►*constitutive* genes). Other genes are not constitutive, i.e., they are regulated in response to external and internal control signals; in other words, they are expressed only when they are called up for a duty. The latter group of genes is responsible for the differences within an organism. Today highly sensitive computational methods are available (PHYLONET) for the identification of phylogenetically conserved regulatory motifs by analysis of the promoter sequences of several related genomes. By this approach global regulatory networks can be identified (Wang T, Stormo GD 2005 Proc Natl Acad Sci USA 102:17400).

Pretranscriptional Regulation. The expression of genes is regulated by several means, including the structural organization of the eukaryotic chromosome. Although it was earlier believed that the DNA associated with histones was not or not efficiently transcribed. The nucleosomal organization of the DNA may not prevent transcription yet nucleosomal reorganization may be required for the proper expression of genes (►nucleosomes). For efficient transcription of genes chromatin remodeling (histone acetylation) is required. It has been known since the early years of cytogenetics that, e.g., the heterochromatic regions of the chromosomes were not associated with genes that could be mapped by recombinational analysis. It appears that these tightly condensed regions of the chromosome are not suitable for transcription in general. The coiling of the chromosomes is also genetically regulated. Position effect indicates that gene expression is altered or obliterated by transposition into heterochromatin. Similarly, lyonization of the mammalian X chromosome involves heterochromatinization and silencing of genes (►silencer). The insertion of normal genes (by transformation) into the condensed telomeric region (about 10^4 bp in length) interferes with their expression (►heterochromatin, ►position effect, ►lyonization, ►telomeres). Gene expression depends in some way on the presence of nuclease sensitive sites in the chromatin. At these nuclease hypersensitive sites, apparently the DNA is not wrapped around so tightly and is more accessible to transcription initiation (►nuclease sensitive sites). The effects of the chromatin locale on the expression of genes are clear from the large variations in the production of a specific mRNA in various transgenic animals and plants which carry a particular gene inserted at different chromosomal locations (►LCR). Also, in order to make the gene accessible to transcription or replication, in bacteria negative supercoils are formed which must be subsequently relaxed. In eukaryotes, DNA in Z conformation may be preferentially available for initiation of transcription (►supercoiling, ►Z DNA). Some genes are regulated by transposition; this mechanism is common in prokaryotes and eukaryotes for generating defense against the immune system of the host (►phase variation, ►antigenic variation), it is also used for sex determination in yeast (►cassette model). At replication the four basic nucleotides are normally used, several nucleoside analogs (e.g., 5-bromodeoxyuridine) may be incorporated into the DNA with some effects on gene expression. In the T-even (T2, T4, T6) phages in place of cytosine 5-hydroxymethyl cytosine is found as a protection against most of the restriction enzymes. In eukaryotes 5 to 25% of the cytosine residues are 5-methylcytosine. Genes with methylated cytosine are generally not transcribed (►methylation of DNA, ►recruitment, ►SRB, ►nuclear receptors).

Figure R31. Genetic regulation

Regulation of Transcription and Transcripts. The cells have various options for more direct regulation of transcription: (i) control of signal receptor and signal transmission circuits, (ii) construct or take apart assembly lines geared to a particular function, (iii) transcriptional control, (iv) transcript processing and alternative splicing, (v) export of the mRNA to the cytosol in eukaryotes. In prokaryotes and cellular organelles a membrane does not enclose the genetic material and transcription and translation are coupled, and (vi) selective degradation of mRNA or a carboxypeptidase may cleave the transcription factors.

Nucleotide sequences in the DNA (structural gene) specify the primary structure of the transcripts. Upstream cis elements (enhancers, promoters and other protein binding sequences) control the attachment and function of the DNA-dependent RNA polymerases (►pol I, ►pol II and pol III RNA polymerases). Some eukaryotic genes may have more than one promoter and the tissue or cell type and the physiological conditions select the promoter to be used. Transcript length is dependent on the promoter element used and the upstream, non-translated region contains binding sequences for further regulation of gene expression. Various upstream elements of the same gene may respond differently to cytokines, phorbol esters and hormones (►hormone receptors, ►hormone response elements).

The enhancers may be positioned either upstream or downstream. Inducible genes receive cues through membrane receptors and transmitter cascades, generally regulated by kinases and phosphorylases (►signal transduction). Downstream DNA nucleotide sequences control the termination of transcription and in eukaryotes a polyA tail (exceptions are the histone genes) is added enzymatically without the use of a DNA template (►polyadenylation signals, ►transcription termination in eukaryotes and prokaryotes, ►self-cleavage of RNA, ►RNAi, ►microRNA, ►small RNA; Chen K, Rajewsky N 2007 Nature Rev Genet 8:93).

Gene expression begins by the initiation of transcription (►transcription, ►protein synthesis). The DNA displays some specific sequences in the major groves of the double helix that are recognized by DNA-binding proteins. (►*lac* operon, for the *E. coli lac* repressor binding site and the CAP site for binding of the catabolite activator protein). In phage λ the *cI* repressor binding element controls by repression several genes (►lambda phage). The consensus sequence of the budding yeast GAL4 upstream element (►galactose utilization), for the mating type α2 consensus (►mating type determination) and for the transcription factor GCN4 (►GCN4) that regulate specific genes. In plants, the core sequence for a transcriptional activator protein is shown under the G-box element. The binding proteins have a short α helix or a β sheet that fits into the major groove of the DNA at the specific sequence motif (►helix-turn-helix, ►Zinc finger, ►leucine zipper, ►helix-loop-helix). Specific activators may also regulate transcription (Spiegelman BM 2004 Cell1 19:157). The activation may require a positive or negative control process (►arabinose operon, ►*lac* operon, ►CAT). For the initiation of transcription in eukaryotes, the presence of a general transcription factor protein complex is essential (►open transcription complex). Additional specific transcription factors may modulate transcription (►transcription factors inducible). In the DNA there are also a number of *response elements* or regulatory sequences that bind several specific proteins and the proteins in turn may bind additional modules (►response elements, ►hormone response elements). The inducible transcription factors in the eukaryotic nuclei help in the assembly of the transcription complex and activate or repress genes by assembling modules. The interacting elements are responsible for fine-tuning metabolic pathways and regulating morphogenesis. These transcription factors may or may not be syntenic with the genes they act on, and their number may vary depending on the gene concerned. The binding proteins may pile up in a specific way at the promoter after DNA looping brings them to that area. Also, the binding proteins may attract other molecules that act in an activating or silencing manner. In an absolutely abstract form this may be visualized with a few computer symbols.

The bacterial DNA-dependent RNA polymerase (►pol) attaches to the double-stranded DNA, and generates an open promoter complex and proceeds with the transcription (►open promoter complex). The bacterial RNA polymerase may rely on different σ subunits for transcribing various bacterial or viral genes. In some instances bacterial and eukaryotic genes also use activators of transcription to assist the RNA polymerase enzyme to generate the open promoter complex. These proteins may attach to the DNA in an area some distance from the gene (enhancer) and looping may bring the protein to the promoter site (►looping of DNA).

The likelihood of association of two DNA sites by looping reaches an optimum at a distance of about 500 bp and it is considerably reduced when they are very close. Some of the enhancer DNA elements (binding sites for regulatory proteins) may be several thousands of nucleotides apart upstream or downstream of the structural gene (►enhancer). The various binding proteins (symbolized by: ∪, ∩, Ψ,Ω ♠,ζ, •, ∇) may associate with the general transcription factors and with each other in different combinations and numbers either to activate or suppress, or to modulate or silence the gene (see Figs. R32, R33, and R34).

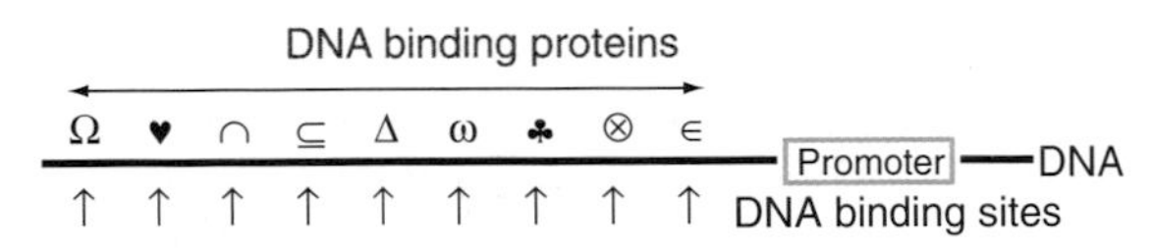

Figure R32. Several different proteins (represented by abstract symbols) can bind upstream of the promoters

The open promoter complex includes the general transcription factors, RNA polymerase II, the TATA box and the transcription initiator (INR). These crude schematic figures do not properly represent the interacting complexes that are required for turning on, turning off and modulating expression as needed

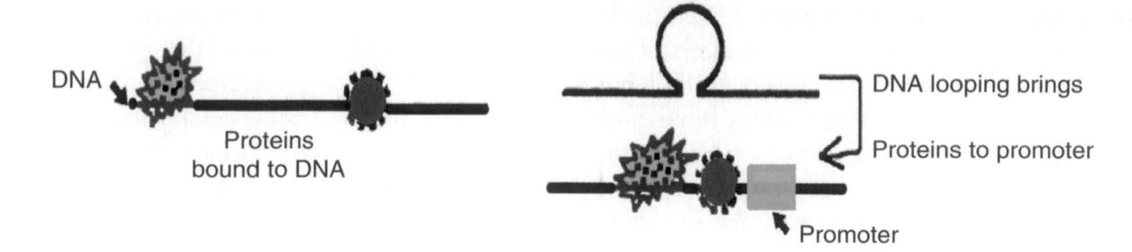

Figure R33. DNA looping

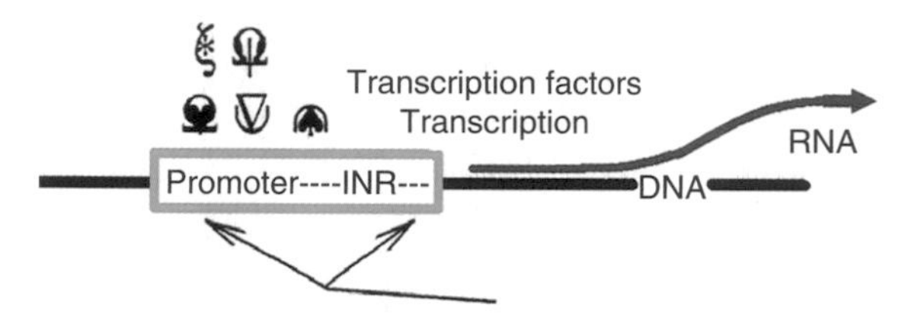

Figure R34. Promoter bound transcription factors regulate transcription

for the orchestration of intricate processes such as the temporal and topological control of morphogenesis (►morphogenesis). The transcription factors regulate these processes but the transcription proteins themselves are subject to regulation by metabolic and environmental cues. These processes include conformational changes, combinatorial assembly of subunits, ligand binding, phosphorylation and dephosphorylation, presence of inhibitors and activators (►signal transduction). In eukaryotes there may be a need for chromatin remodeling to enable the activators and the TATA box binding protein to access the DNA (►nucleosome). For this process a histone acetylase or SWI/SNF complex may have to be recruited in preparation for transcription. In both prokaryotes and eukaryotes special control mechanisms have evolved for the termination of transcription (►transcription termination). The regulation of the transcriptional process and the turnover of the transcripts determine the quantity of the transcripts. Many bacterial genes are organized into coordinated regulatory units employing negative, positive or a combination of these two controls of transcription (►*lac* operon, ►*arabinose* operon). In these operons the genes are either exactly (►*tryptophan* operon) or with some modification (►*histidine* operon) arranged according to the order of the biosynthetic pathway. The amino acid operons use, in addition, *attenuation* for controlling the quantity of the transcripts for maximal economy (►attenuator region, ►*tryptophan* operon). The operons are characterized by coordinated regulation of the transcription of several genes belonging to the same transcriptional unit and transcribe them into a polycistronic mRNA. Eukaryotes usually do not produce polycistronic mRNAs but the rRNA and tRNA transcripts are processed into functional units post-transcriptionally. Elements of a coordinated unit may not all be juxtapositioned (►regulon, ►*arabinose* operon). The small phage (φX174 [can be found under F]) and retroviral genomes may have overlapping genes that specify more than one protein, depending on the register they are transcribed (►overlapping genes, ►recoding, ►retroviruses). The need for the protein products of these overlapping genes transcribed with the aid of the same promoter, may be not the same. Some proteins, e.g., viral coat proteins may be needed in larger quantities than the replicase enzymes. Therefore, mechanisms have evolved to by-pass internal stop signals and produce some fusion proteins that assist in achieving this goal (►overlapping genes, ►recoding). In bacterial, plant and animal viruses, another means of regulation of gene activity at different steps has evolved that involves the use of antisense RNA. This mechanism is being explored so as to develop particular drugs for the highly specific regulation of genes with minimal side effects or for the development of new, selective antimicrobial agents and more desirable crop plants without reshuffling the entire genome (►antisense technologies). Short RNAs use various means to regulate transcription (►repressor, ►RNAi, ►microRNA).

In prokaryotes a special short transcribed stretch of nucleotides, the Shine-Dalgarno box, controls the attachment of the mRNA to the small (30S) ribosomal subunit. For the same task, eukaryotes use "ribosome scanning", i.e., the mRNA tethers a 40S ribosomal subunit and by reeling locates the first initiator codon. Eukaryotic 40S ribosomal subunits can enter circular mRNAs if they contain internal ribosomal entry sites (IRS).

The primary transcripts are generally not suitable for translation into a protein or for a RNA product (rRNA, tRNA). The transcripts are processed to mRNA and/or other RNA units. Introns are excised and the sequences corresponding to exons are spliced and may even be transspliced with the cooperation of spliceosomes (►intron, ►exon, ►spliceosome, ►alternative splicing, ►hnRNA, ►snRNA). The splicing itself may be genetically and organ-specifically regulated. The transposition of the P element of *Drosophila* is relatively rare in the soma but five times more common in the germ line because one intron is not excised from the transposase transcripts in the somatic cells (►hybrid dysgenesis). Tissue-specificity and function-specificity of many proteins is partly controlled by alternative splicing (►immunoglobulins, ►sex determination). Mitochondrial RNA transcripts may be modified by replacing C residues with Us (►RNA editing).

The eukaryotic mRNAs are capped while still in the nucleus. The transcript is cut at the appropriate guanylic residue and it is then modified (►cap, ►capping enzymes). Capping increases the stability of the mRNA, facilitates its transport to the cytosol and assists in the initiation of translation by being recognized by initiation protein factors eIF-4F, eIF-4B, etc. (►cap, ►eIF)

The tail of the eukaryotic mRNAs (with few exceptions, e.g., histones) is equipped with 50–250 adenylic units to increase their stability. Polyadenylation is controlled separately from transcription because a special enzyme adds these nucleotides after processing of the transcript. Generally, the genes carry a short A-rich consensus (►polyadenylation signal) in the DNA that instructs the RNA polymerase to terminate transcription after the enzyme passes through the signal and also indicates the need for polyadenylation. Eventually, the poly-A tail is reduced to about 30 A units. In eukaryotes, the 3′ tail may be substantially regulated by specific, extrinsic genes (Kakoki M et al 2004 Dev Cell 6:597).

Some of the transmembrane proteins have a hydrophobic amino acid sequence in the section that is going to be located within the membrane, whereas the cytosolic end contains a longer hydrophilic carboxyl end. The positioning of the transmembrane proteins shows substantial variations, depending on the intrinsic properties of proteins. The transcript of the same coding sequences is differentially cut in a manner as to assure such a terminus is formed for the membrane-bound proteins whereas a shorter hydrophilic end terminates the otherwise identical circulating immunoglobulin molecules.

After these intricate preparatory processes, the eukaryotic mRNA is transported to the cytosol through the nuclear pores. Prokaryotes do not have membrane-enclosed nuclei but only nucleoids, anchored to the cell membrane, and there the translation proceeds *pari passu* with transcription. (See Carlson M 1997 Annu Rev Cell Dev Biol 13:1; Holstege FC et al 1998 Cell 95:717).

Post-Transcriptional Regulation. The mRNA may be degraded before it is translated into polypeptide chains. About half of the prokaryotic mRNAs may be degraded within 2–3 min after their synthesis. Eukaryotes have long-lived mRNAs, which usually last for at least three times longer but at times in special dormant tissues of plants they may remain intact for years. The degradation is mediated by special endonucleases that recognize mRNAs. Also, A-U sequences in the non-translated downstream regions may remove the poly-(A) tails and thus stability is reduced in both cases.

Translation in eukaryotes begins with the transport of the capped mRNA outside the nucleus, into the cytosol. The mRNA tethers several ribosomes and the polysomal structures are formed. Some mRNAs are equipped with a signal coding sequence, coding for a special tract of 15 to 35 amino acids. That directs it toward the *signal sequence recognition particle* after only a few dozen amino acids are completed on the ribosome. The *signal peptide* then transports the nascent peptide chain into the lumen of the endoplasmic reticulum, Golgi vesicles, lysosomes and mitochondria, plastids, etc. This mechanism facilitates the subcellular localization of the emerging proteins at places where they are most needed and from where they may be diffused in a gradient as required for embryonic differentiation (►signal sequence, ►signal peptide, ►signal sequence recognition particle, ►morphogenesis in *Drosophila*). Various control mechanisms have been involved in the generation of protein products of genes: (i) translational control, (ii) post-translational modification of the polypetides, (iii) control of polypeptide assembly into proteins, (iv) regulation of protein conformation, (v) compartmentalization of proteins, (vi) interaction of protein products and ribozymes, (vii) feedback controls at the level of protein synthesis and function, and recently (viii) the wide scale role of small RNAs was discovered (►induction, ►repression, ►attenuation, ►inhibition, ►silencers, ►small RNAs, ►non-coding RNA, etc.). These may be involved before, during and after the final protein products are made.

The state of phosphorylation of the eukaryotic initiation factor, eIF-2 is critical for the translation process. This protein may form a complex with guanosyl triphosphate (GTP) and can assist in the attachment of the initiator $tRNA^{Met}$ to the P site of the small subunit (40S) of the ribosome and scans the mRNA until it finds a methionine codon (AUG). This occurs after the large ribosomal (60S) subunit joins the small subunit to form the 80S ribosome and at the same time one molecule of inorganic phosphate and the inactivated eIF-2 and GDP are released. Then eIF-2 can acquire another GTP and the initiation process is repeated (►protein synthesis).

Although all polypeptide chains start with a formylmethionine (prokaryotes) or methionine (eukaryotic),

R

the final product is frequently truncated at both the amino and carboxyl termini. Many proteolytic enzymes are translated as large units and become activated only after cleaving off certain parts of the original protein. To become active insulin is initially made as a pre-proinsulin that must be tailored in steps: first pre-, followed by pro-insulin and finally insulin. Several viral proteins, secreted hydrolytic proteins, peptide hormones and neuropeptides are made as polyprotein complexes which have to be broken down into active units in the trans-Golgi network, secretory vesicles or even in the extracellular fluids to become fully functional. The formation of polyproteins appears to be justified as a protective measure against destruction in the cytosol until they can be sequestered and confined into some vesicles. The loaded vesicles then migrate to predetermined sites where upon receiving the cognate signals they release the active protein. The signals can be chemical, physical (electric potentials) or topological. The release of the members of the polyprotein group may be selective regarding the site of release; different proteins can be released at different anatomical sites.

Some proteins are synthesized in separate polypeptide chains but must be folded and/or assume a quaternary structure, e.g., ααββ may even have to acquire a prosthetic group such as heme, a vitamin or other organic or inorganic group(s). The folding in prokaryotes begins after completion of the chain. In eukaryotes the folding may begin before the completion of a polypeptide and thus higher complexity is generated in the large proteins. The mRNA may be degraded before it can be translated into polypeptide chains.

Proteins are commonly acetylated after translation, carbohydrate side chains are added (glycoproteins), prenylated, linked by covalent disulfide bonds, special amino acids (serine, threonine, tyrosine) are phosphorylated by kinase enzymes, lysine residues may be methylated, and extra carboxyl groups may be attached to aspartate and glutamate residues.

Engineered Regulation. Using a genetic vector, it is feasible to introduce into somatic cells a structural gene *A* for a protein of a special need. With the aid of another vector it is possible to introduce gene *B* encoding its special transcription factor. The latter transcription factor gene is equipped with a promoter which responds to a specific drug (or to a specific temperature or to any other conditional factor) regulating its transcription. Thus, supplying the drug at variable dosage, the expression of gene *A* can be modulated by the controlled response of gene *B*. Such a system may permit the controls to be fine-tuned and secure compensation for a genetic defect or improve productivity.

According to some estimates, there are about 2,000 different protein kinases and 1,000 phosphatases in a higher eukaryotic cell. They must be regulated in time, space, and for other specificities. This regulation is an extremely complex task and is expected to be mediated by associations with modular, adaptor, scaffold and anchoring proteins working in sequential cooperation through signal transduction pathways. The availability of complete information on nucleotide sequence of both prokaryotic and eukaryotic genomes as well as microarray hybridization permits the assessment of the simultaneous expression of thousands of genes. Eventually, by using appropriate computer technologies the study of the coordinated regulation of the function of entire genomes will become a reality. ►transcription, ►transcriptional activator, ►co-activator, ►transcriptional modulation, ►mediator complex, ►transcription factories, ►protein synthesis, ►polysome, ►endoplasmic reticulum, ►chromatin, ►chromatin remodeling, ►high mobility group of proteins translation initiation, ►translation, ►regulation of enzyme activity, ►axotomy, ►signal transduction, ►serine/threonine phosphoprotein phosphatases, ►cell cycle, ►LCR, ►RNA polymerase, ►DNA looping, ►insulator, ►transcription complex, ►SL1, ►TBP, ►TAF, ►attenuation, ►open promoter complex, ►RAD25, ►signaling to translation, ►DNA grooves, ►elongation factors, ►DNA chips, ►microarray hybridization, ►genetic network, ►networks, ►regulation of transcription, ►transcription factors, ►RNAi, ►microRNA, ►combinatorial gene control; Tautz D 2000 Curr Opin Genet Dev 10:575; Lemon B, Tjian R 2000 Genes Dev 14:2551; Rao CV, Arkin AP 2001 Annu Rev Biochem Eng 3:391; Emerson BM 2002 Cell 109:267, reviews in Cell 108:439 ff [2002], Wang W et al 2002 Proc Natl Acad Sci USA 99:16893; Pawson T, Nash P 2003 Science 300:445; Alonso CR, Wilkins AS 2005 Nature Rev Genet 6:709; transcriptional regulation in humans: Maston GA et al Annu Rev Genomics Hum Genet 7:29; http://www.gene-regulation.com/; http://regulondb.ccg.unam.mx:80/index.html; Gene Resource Locator: http://www.gene-regulation.com/pub/databases.html#transcompel; regulatory motif detection tool: http://159.149.109.16/modtools/; composite regulatory signatures: http://140.120.213.10:8080/crsd/.

Regulator Gene: This controls the function of other genes through transcription. ►regulation of gene activity, ►enhancer, ►silencer, ►activator, ►co-activator, ►operon

Regulatory Elements: These upstream (enhancer) sequences are located within 100 to 400 bp from the translation initiation nucleotide (+1) and control cell and developmental specificities. Some enhancers may be located at more distant positions and also downstream. The enhancer region provides binding sites for regulatory proteins. ►basal promoter, ►regulation of gene activity, ►regulator gene, ►UAS

Regulatory Enzyme: Allosteric or other modifications alter its catalytic activity rate, thus affecting other enzymes involved in the pathway. The Arg5.6 mitochondrial metabolic enzyme can regulate the expression of genes by association with mitochondrial DNA (Hall DA et al 2004 Science 306:482).

Regulatory Sequence in DNA: This binds transcription factors, RNA polymerase and so regulates transcription. The mammalian genomes contain many short (e.g., 8-mer) sequences, which are binding sites to specific proteins. The TGACCTTG sequence occurs in at least 434 human promoters and has been found 162 times in the mouse, rat and dog genomes at a rate of conservation of 37%. The Err-α (estrogen-related receptor) protein binds this octamer (Xie X et al 2005 Nature [Lond] 343:338) or the TNAAGGTCA element (Sladek R et al 1997 Mol Cell Biol 17:5400). An analysis of the 3′ untranslated region of the four mammalian genomes has revealed 106 short motifs, which probably mediate post-transcriptional regulation. About 20% of the human genes seem to be regulated by microRNAs. ►transcription factors, ►open transcription complex, ►enhancer, ►operon, ►attenuator site, ►UAS, ►nuclear receptors, ►regulation of gene activity, ►microRNA

Regulon: This non-contiguous set of genes is controlled by the same regulator gene. The different sections may communicate through looping of the DNA. Proteins mediate the coordination of mRNA (Keene JD 2007 Nature Rev Genet 8:533). ►looping of DNA, ►arabinose operon, ►regulation of gene activity; Manson McGuire A et al 2000 Genome Res 10:744; Huerta AM et al 1998 Nucleic Acids Res 26:55; conserved microbial regulon targets: http://210.212.212.6/icr/index.html.

Regulome: Refers to the complete set of transcription factors and their co-regulators.

Reifenstein Syndrome (Xq11-q12): In this condition the individual has XY chromosomal constitution but there is an insufficient production of androgen receptor during fetal development. The individual manifests male pseudohermaphroditism with hypospadias, hypogonadism and gynecomastia yet defective germ cells are present and fertility may be possible by early treatment with testosterone. ►androgen-insensitivity, ►hypospadias, ►hypogonadism, ►gynecomastia, ►testosterone, ►pseudohermaphroditism

Reinitiation: The eukaryotic ribosomes can terminate an open reading frame and initiate another downstream (at low efficiency). Reinitiation occurs when the translation of one reading frame is completed and the process moves on to the next cistron. In an unfavorable nucleotide context, translation may be reinitiated not at the first AUG codon but at the next one downstream. Translation factor eIF2 may play an important role in the process. ►backtracking, ►regulation of gene activity, ►transcription, ►eIF2, ►translation, ►cistron; Kozak M 1999 Gene 234:187; Park HS et al 2001 Cell 106:723; Kozak M 2001 Nucleic Acids Res 29:5226.

Reinitiation of Replication: The genome of eukaryotes replicates at many points along the chromosomes. To avoid chaos in the nucleus it is important to prevent restart of replication. Reinitiation is prevented by cyclin-dependent kinases (CDKs) by phosphorylation of origin recognition complex (ORC), down-regulation of Cdc6 and the exclusion of MCM2-7 complex from the nucleus. ►CDK, ►Cdc6, ►ORC, ►MCM; Nguyen VQ et al 2001 Nature [Lond] 411:1068.

Reinitiation of Transcription: For a second cycle of transcription, transcription factors and the RNA polymerase must be re-attracted to the promoter. Reinitiation appears to be a faster process than initiation. TFIID and TFIIA transcription factors do not leave the promoter when the remaining part of the transcription complex is released. The reinitiation intermediate includes TFIID, TFIIA, TFIIH, TFIIE and the Mediator. Subsequently the complete transcription complex, including activators, is reformed depending on ATP and TFIIH. ►backtracking, ►preinitiation complex, ►transcription factors, ►mediator; Hahn S 1998 Cold Spring Harbor Symp Quant Biol 63:181.

Reiter Syndrome: This is a complex anomaly generally accompanied by overproduction of HLA-B27 histocompatibility antigen. It is characterized by arthritis, inflammation of the eyes and the urethra (the canal that carries the urine from the bladder and in males also serves as the genital duct). The inflammations may be related to sexually transmitted and intestinal infections. ►HLA, ►rheumatic fever, ►arthritis, ►connective tissue disorders

Reiterated Genes: These are present in more than one copy, possibly many times.

Rejection: This is an immune reaction against foreign antigens such as may be present in transfused blood or grafted tissue. The rejection of pig organs by humans and Old World monkeys is caused by the presence of α-1,3-galactosyl epitopes on the pig epithelia. During evolution the rejecters lost the appropriate galactosyltransferase gene and as a consequence developed antibodies against the epitope of the foreign tissue transplant. This immune reaction cannot be satisfactorily mitigated through affinity absorption or complement regulators or other means of immunosuppression (drugs) even in transgenic animals. A better solution appears to be the inactivation of the gene and generation of clones by nuclear transfer into enucleated pig oocytes. When fully developed, this procedure may permit xeno-transplantation of pig organs into humans who have serious organ defects. ►immune reaction, ►HLA, ►nuclear transplantation, ►transplantation of organelles, ►xenotransplantation; Lai L et al 2002 Science 295:1089; Prather R et al 2003. Theriogeneology 59:115.

Rejoining: ►breakage and reunion, ►breakage-fusion-bridge cycles

rel (REL) **Oncogene** (2p13-p12, 11q12-q13): This is a turkey lymphatic leukemia oncogene, a transcription factor homologous with NF-κB. c-Rel as a homodimer or as a heterodimer with p50 or p52 is a strong transcriptional activator. In its absence or inactivation, the production of IL-3 and the granulocyte—macrophage colony-stimulating factor is impaired. Rel domains occur in several proteins such as NF-κB, NFAT and another ca. 12. *RelA* encodes the guanosine tetraphosphate synthetase (ppGpp), *RelBE* in *E. coli* encodes the toxin-antitoxin proteins. The toxin severely inhibits bacterial growth as a stringent control whereas the antitoxin is a repressor of the translation of the RelB toxin. ►NF-κB, ►NFKB, ►NFAT, ►IL-3, ►GMCSF, ►oncogenes, ►morphogenesis in *Drosophila*{3}, ►p50, ►stringent response, ►magic spots; Gugasyan R et al 2000 Immunol Rev 176:134; Christensen SK et al 2001 Proc Natl Acad Sci USA 98:14328; Jaque E et al 2005 Proc Natl Acad Sci USA 102:14635.

Relapsing Fever: ►Borrelia

Relaxase: ►relaxosome

Relatedness, Degree of: This term is used in genetic counseling to indicate the probability of sharing genes among family members. In first-degree relatives such as a parent and child half of their genes are in common. In second-degree relatives such as a grandparent and a grandchild 1/4 of their genes are identical. In population genetics, mathematically simpler terms such as the inbreeding coefficient, consanguinity and coefficient of coancestry are preferred. (See these concepts under separate entries, ►relationship coefficient, ►MLS; Weir BS et al 2006 Nature Rev Genet 7:771.

Relational Coiling: ►chromosome coiling

Relationship, Coefficient of: $r = 2F_{IR}/\sqrt{(1+F_I)(1+F_R)}$ where F_I and F_R are the coefficients of inbreeding of I and R. If they are not inbred, F_I and F_R equal 0. ►coefficient of inbreeding, ►relatedness degree

Relative Biological Effectiveness: ►RBE

Relative Fitness: ►selection coefficient

Relative Molecular Mass (M_r): Expresses molecular weight relative to ^{12}C isotope (in $^1/_{12}$ units). It is comparable to molecular weight in daltons but it is not identical to molecular weight (MW) represented by the mass of atoms involved. ►dalton

Relative Mutation Risk: This is equivalent to 1/doubling dose. ►doubling dose, ►genetic risk

Relative Sexuality: The intensity of sexual determination may be expressed in degrees in some organisms. In extreme cases a normal female gamete may behave like a male gamete toward a strong female gamete. ►isogamy, ►pseudohermaphroditism, ►intersex

Relaxed Circular DNA: This is not supercoiled because of one or more nicks. ►nick, ►supercoiled DNA

Relaxed Control Mutants (*relA*): They have lost stringent control and continue RNA synthesis during amino acid starvation of bacteria. ►stringent control, ►fusidic acid

Relaxed Genomes: The organelle DNAs are not replicated in lockstep with the nuclear genome and their replication may be reinitiated during the cell cycle. The distribution of the organelles may not necessarily be equational during cytokinesis. ►cytokinesis, ►stringent genomes

Relaxed Replication Control: The plasmids continue to replicate even when the bacterial divisions stop. ►replication

Relaxin: This water-soluble protein in the corpus luteum mediates the relaxation of the pubic joints and the dilation of the uteral cervix in some mammals. Its two receptors, LGR7 and LGR8, are heterotrimeric G protein binding proteins and are widely distributed among organs indicating their roles in diverse functions. ►corpus luteum; Hsu SY et al 2002 Science 295:671.

Relaxosome: This DNA protein structure mediates the initiation of conjugative transfer of bacterial plasmids. It contains a *nic* site at the origin of transfer (*oriT*). Relaxase catalyzes the nicking and it becomes covalently linked to the 5′ end through a tyrosyl residue. A single strand is then transferred to the recipient by a rolling circle mechanism.

►conjugation, ►rolling circle, ►nick; Xavier Gomis-Rüth F et al 2001 Nature [Lond] 409:637.

Relay Race Model of Translation: A ribosome after passing a chain termination signal of an ORF does not completely disengage from the mRNA and may reinitiate protein synthesis if an AUG codon is within short distance downstream. ►translation, ►regulation of gene activity, ►reinitiation, ►ORF; Ranu RS et al 1996 Gene Expr 5[3]:143.

Release Factor (RF): When translation reaches a termination codon, the release factors allow the polypeptide to be free from the ribosome. In prokaryotes there are two direct release factors RF-1 (specific for UAG/UAA) and RF-2 (specific for UGA/UAA) and a third factor RF-3 stimulates the activity of RF 1 and 2. RF-1 and RF-2 can discriminate between the termination and sense codons by 3 to 6 orders of magnitude effectiveness. In RF-1 a Pro-Ala-Thr and in RF-2 a Ser-Pro-Phe tripeptide, respectively, recognizes the appropriate stop codon. The eukaryotic release factors, eRF and eRF-1 alone can recognize all three stop codons. RF-3 and eRF-3 are GTP-binding proteins. RF-3 is a GTPase on the ribosome in the absence of RF-1 and RF-2; eRF3 requires eRF-1 to act as a GTPase. ERF-1 alone may be sufficient for termination in yeast. It has been suggested that all release factors are homologous to elongation factor G which mimics tRNA in its C-terminal domain and this is the basis of recognition of the RFs of the ribosomal A site. The activity of Class 1 release factors depends on the presence of Gly-Gly-Glu (GGQ) motif in the peptidyl transferase center for the release. In ciliates the stop codons vary and so do the release factors. ►transcription termination, ►protein synthesis, ►regulation of gene activity, ►EF-G, ►eRF; Inagaki Y, Doolittle WF 2001 Nucleic Acids Res 29:921; Zavialov AV et al 2001 Cell 107:115; Ito K et al 2002 Proc Natl Acad Sci USA 99:8494; Klaholz BP et al 2004 Nature [Lond] 427:862.

Releasing Factors: The hormones of the pituitary gland are released under the influence of hypothalmic hormones. ►animal hormones

Relics: These are genes with major lesions (insertions and deletions) in one or more components; they are similar to pseudogenes. ►pseudogenes

REM: An acronym of röentgen equivalent man. It is the product of REB × rad. Generally, 1 rem is considered to be equivalent to 1 rad of 250 kV X-rays; 1 rem is equal to 0.01 Sv (Sievert). ►R unit, ►rad, ►Gray, ►Sievert, ►REB, ►BERT

REM (Ras exchanger motif): This is required in signal transduction for interaction with RAS in GDP exchange for GTP using the GEF motif (Cdc25 homology catalytic unit) of SOS. ►signal transduction, ►RAS, ►SOS, ►GEF, ►Cdc25

REMI (restriction enzyme-mediated integration): An integrating vector is transformed into a cell in the presence of a restriction enzyme that facilitates insertion at the cleavage sites and may bring about insertional mutagenesis. ►insertional mutation, ►restriction enzyme; Thon MR et al 2000 Mol Plant Microbe Interact 13:1356.

Renal Carcinoma, Hereditary Papillary (HPRC): This is frequently caused by triplication (trisomy) and/or mutation of the MET oncogene, a cell surface tyrosine kinase, encoded at human chromosome 7q31.

Renal Cell Carcinoma (RCC): This commonly involves translocation breakage points in human chromosome 3p, each representing a different type. The 3p14.2 region includes the gene for protein tyrosine phosphatase gamma (PTPγ). This region also contains a fragile site, FHIT (fragile histidine triad) and the von Hippel-Lindau syndrome gene. ►hypernephroma, ►papillary renal cell cancer, ►tyrosine phosphatase, ►von Hippel-Lindau syndrome, ►fragile site; Zanesi N et al 2001 Proc Natl Acad Sci USA 98:10250.

Renal-Coloboma Syndrome: Is caused by mutation of the PAX2 (paired-box) gene at 10q24-q25.1 affecting the development of the kidney, eye and ear nerve. ►coloboma; Sanyanusin P et al 1996 Genomics 35:258.

Renal Dysplasia and Limb Defects: This condition is characterized by autosomal recessive underdevelopment of the kidney and the urogenital system, accompanied by defects of the bones and genitalia. ►kidney disease, ►limb defects

Renal Dysplasia and Retinal Aplasia: An autosomal recessive condition kidney developmental anomaly is associated with eye defects. ►kidney disease, ►eye disease

Renal Glucosuria (16p11.2, 6p21.3): Refers to dominant glycosuria which may not be related to diabetes.

Renal-Hepatic-Pancreatic Dysplasia (polycystic infantile kidney disease, ARPKD): Autosomal recessive phenotypes include cystic (sac-like structures) kidneys, liver and pancreas, sometimes associated with other anomalies such as blindness. The polycystic kidney disease of adult type dominant (ADPKD, human chromosome 16) is often associated with internal bleeding or arterial blood sacs (aneurysm). ►kidney disease

Renal Tubular Acidosis: The 17q21-q22 dominant type I defect is primarily in the distal tubules with normal bicarbonate content in the serum. Type II is recessive and the defect is in the proximal tubules and there is a low level of bicarbonate in the urine. Another recessive form involves mutation in the B1 subunit of H^+-ATPase; in human chromosome 2cen-q13 nerve deafness is present. A proximal type is X-linked recessive. Recessive distal tubular acidosis with normal hearing (rdRTA2) has been assigned to gene ATP6N1B at 7q33-q34. The gene encodes an 840 amino acid subunit of a kidney vacuolar proton pump. The excretion of ammonium is reduced and the urine pH is usually above 6.5 in contrast to types I and II where it is around 5.5. Other variations have also been observed. ►kidney diseases

Renal Tubular Dysgenesis: An autosomal recessive defect in the development of kidney tubules. It generally results in perinatal death because of the reduced amount of amniotic fluid (oligohydramnios), absence of urine secretion (anuria) and underdevelopment of the lung (pulmonary hypoplasia) caused by rennin–angiotensin defects. ►rennin, ►angiotensin; Gribouval O et al 2005 Nature Genet 37:964; Allanson JE et al 1992 Am J Med Genet 43:811.

Renaturation: Complementary single DNA strands reform double-strand structure by reannealing through hydrogen bonds. ►c_0t curve, ►denaturation

Renilla GFP (from sea pansy): This is a green fluorescent protein with similarities to aequorin but it has only one absorbance and emission peak and its extinction coefficient is higher. ►aequorin

Renin (chymosin, rennet): Protein hydrolase reacts with casein in cheese making. It is present in the kidneys and splits pro-angiotensin from α-globulin. ►angiotensin, ►pseudoaldosteronism; Kubo T et al 2001 Brain Res Bull 56:23; Krum H, Gilbert RE 2007 J Hypertens 25:25.

Renner Complex: Refers to the chromosomal translocation complex that is transmitted intact. ►translocation, ►translocation complex

Renner Effect: ►megaspore competition; Renner O 1921 Ztschr Bot 13:609.

Reoviruses: Double-stranded RNA viruses cause respiratory and digestive tract diseases and arthritis-like symptoms in poultry and mammals, in humans, however, the infection usually does not involve serious symptoms. The internal capsid particle transcribes (+)-strand copies from the 10 genomic segments. The transcript carries a cap and it is exported to the cytoplasm of the infected cell. ►oncolytic virus, ►PKR, ►rotaviruses, ►cap, ►plus strand; Joklik WK, Roner MR 1996 Progr Nucleic Acid Res Mol Biol 53:249.

REP: Refers to repetitive extragenic consensus of 35 nucleotides, containing inverted sequences, in the bacterial chromosome. There are over 500 copies of it in *E. coli* in intergenic regions at 3′ end of the genes. They are transcribed but not translated and appear to be the bacterial version of "selfish" DNA. ►selfish DNA; Herman L, Heyndrickx M 2000 Res Microbiol 151[4]:255.

Rep: An *E. coli* monomeric or dimeric binding protein and helicase. ►binding protein, ►helicase, ►monomer; Bredeche MF et al 2001 J Bacteriol 183:2165.

rep: The acronym of röntgen equivalent physical, a rarely used unit of X- and γ radiation delivering the equivalent of 1 R hard ionizing radiation energy to water or soft tissues (≈ 93 ergs). ►R

Repair Genetic: ►DNA repair, ►unscheduled DNA synthesis

Repairosome: Refers to the protein complex mediating DNA repair. ►DNA repair

Repbase: Repetitive DNA database http://www.girinst.org/.

Repeat, Direct: This is (tandem) duplication of the same DNA sequence. It may be present at the termini of transposable elements ABC-----ABC. The hexameric CeRep26 repeat of *Caenorhabditis elegans* (TTAGGC) occurs at the telomeres and also at many additional chromosomal regions. The 711 copies CeRep11 are distributed over the autosomes but only one is in the X chromosome. In yeast most of the tandem repeats are in intergenic regions. The majority of them encode cell wall proteins. These repeats recombine frequently with pseudogenes and it is suspected that in pathogenic microbes they contribute to functional diversity of surface antigens, which play an important role in elusion of the host immune defense (Verstrepen KJ et al 2005 Nature Genet 37:986). ►transposable element, ►transposon, ►interspersed repeats, ►tandem repeat, ►homopeptide; interspersed repeat [fossile mobile element] tool: http://repeats.abc.hu/cgi-bin/plotrep.pl; tool for detection of interspersed repeats: http://www.repeatmasker.org.

Repeat-Induced Gene Silencing: ►co-suppression

Repeat, Inverted: The double-stranded DNA carries inverted repeats such as transposable elements. The single strands can fold back and form *stem* (by complementarity) and *loop* (no complementarity) structure (see Fig. R35). In the sequenced *Caenorhabditis elegans* inverted repeats represent 3.6% of the genome and occur on an average 1/4.9 kb, introns contain 45% of them and 55% are in intergenic regions. Inverted repeats may increase inter- and

R

intrachromosomal recombinations by orders of magnitude and are responsible for a large part of the genetic instabilities. Mutation in the *MRE11/RAD50/XRS2* and *SAE2* genes of yeast interferes with the repair of hairpins and contributes to instability of the genome. ►tandem repeats, ►transposable elements, ►LIR; Lobachev KS et al. 1998 Genetics 148:1507; Waldman AS et al 1999 Genetics 153:1873; Lin C-T et al 2001 Nucleic Acids Res 29:3529; Lobachev KS et al 2002 Cell 108:183.

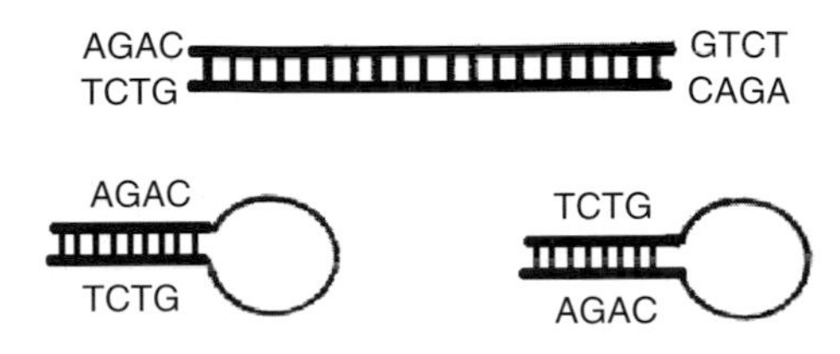

Figure R35. Inverted repeats

Repeat-Associated Short Interfering RNA (rasiRNA): This guides mRNA degradation or chromatin modification and silencing gene expression. ►RNAi

Repeats, Short Tandem (STR): With the aid of polymerase chain reaction, they serve for individual (forensic) discrimination or for the identification of human cell lines, which may or may not be contaminated. ►PCR, ►low-copy repeats; Oldroyd NJ et al 1995 Electrophoresis 16:334; Masters JR et al 2001 Proc Natl Acad Sci USA 98:8012.

Repeats, Trinucleotide: ►fragile sites, ►trinucleotide repeats

Repertoire, Antigenic: This is the complete set of antigenic determinants of lymphocytes.

Repertoire Shift: After a secondary immunization with a hapten, following a primary immunization, the variable heavy/variable light (V_H/V_L) immunoglobulin genes show an altered spectrum of somatic mutations. ►immunoglobulins, ►hapten; Meffre E et al 2001 J Exp Med 194:375.

Repetitive DNA (repetitious DNA): Refers to similar nucleotide sequences occurring many times in eukaryotic DNA. Some of these sequences represent transposable or retrotransposable elements, others such as ribosomal genes are called to duty when there is a special need for high gene activity, e.g., during embryonal development. More than 40% of the human genome appears highly or moderately repetitive and only about 3% may be genetically functional. ►SINE, ►LINE, ►redundancy, ►pseudogenes, ►α-satellite DNA, ►co-suppression, ►microsatellite, ►minisatellite; Britten RJ, Kohne DE 1968 Science 211:667; Toder R et al 2001 Chromosome Res 9[6]:431; Jurka J et al 2007 Annu Rev Genomics Hum Genet 8:241; Repbase: http://www.girinst.org.

Replacement Theory: ►out-of Africa

Replacement Vector: By homology it recognizes and then replaces a particular segment (gene) of the target. It has a pair of restriction enzyme recognition sites within the region of "non-essential" genes. Non-essential means that their removal and replacement do not impair packaging and propagation in *E. coli* by sequences of interest to the experimenter. ►vectors, ►stuffer

Replica Plating: This is designed for efficient selective isolation of haploid microbial mutants. Mutagen-treated cells are spread in a greatly diluted suspension on the surface of complete medium and incubated to allow growth. Because of the dilution, each growing colony represents a single original cell (clone). Then impressions are made of this master plate on minimal medium where only the wild type cells can grow. The absence of growth on the minimal media plates indicates that auxotrophs exist at the spots where no growth was obtained.

The impressions also represent a map of the colonies on the original, complete medium, master plate. Thus, the experimenter can obtain cells from the original colonies and test them for nutritional requirement on differently supplemented media. This procedure thus permits the isolation of mutants and the identification of the nutrient requirement. (See Fig. R36, ►mutant isolation; Lederberg J, Lederberg EM 1952 J Bacteriol 63:399).

Replicase: This is RNA-dependent RNA polymerase enzyme of viruses encoded by the viral RNA and packed to the progeny capsid so that upon entry to a cell replication of the infective *negative-strand RNA* (influenza, Stomatitisvirus [causing inflammation of mucous membranes]) forms the template for replication but does not code for viral proteins. Without the replicase this negative strand would not be able to function. The *positive-strand* RNA viruses (e.g., poliovirus) are directly transcribed into the protein, including the replicase, and in this form it can be infectious. The DNA-dependent DNA polymerase enzymes are also called replicase. ►replication, ►RF, ►positive strand; Tayon R Jr et al 2001 Nucleic Acids Res 29:3576.

Replicating Vector: ►transformation genetic, ►yeast

Replication. ►DNA replication, ►replication fork, ►chromosome replication

Replication Banding: Using sequential 5-bromodeoxiuridine incorporation for the determination of early and late replicating chromosomal regions.

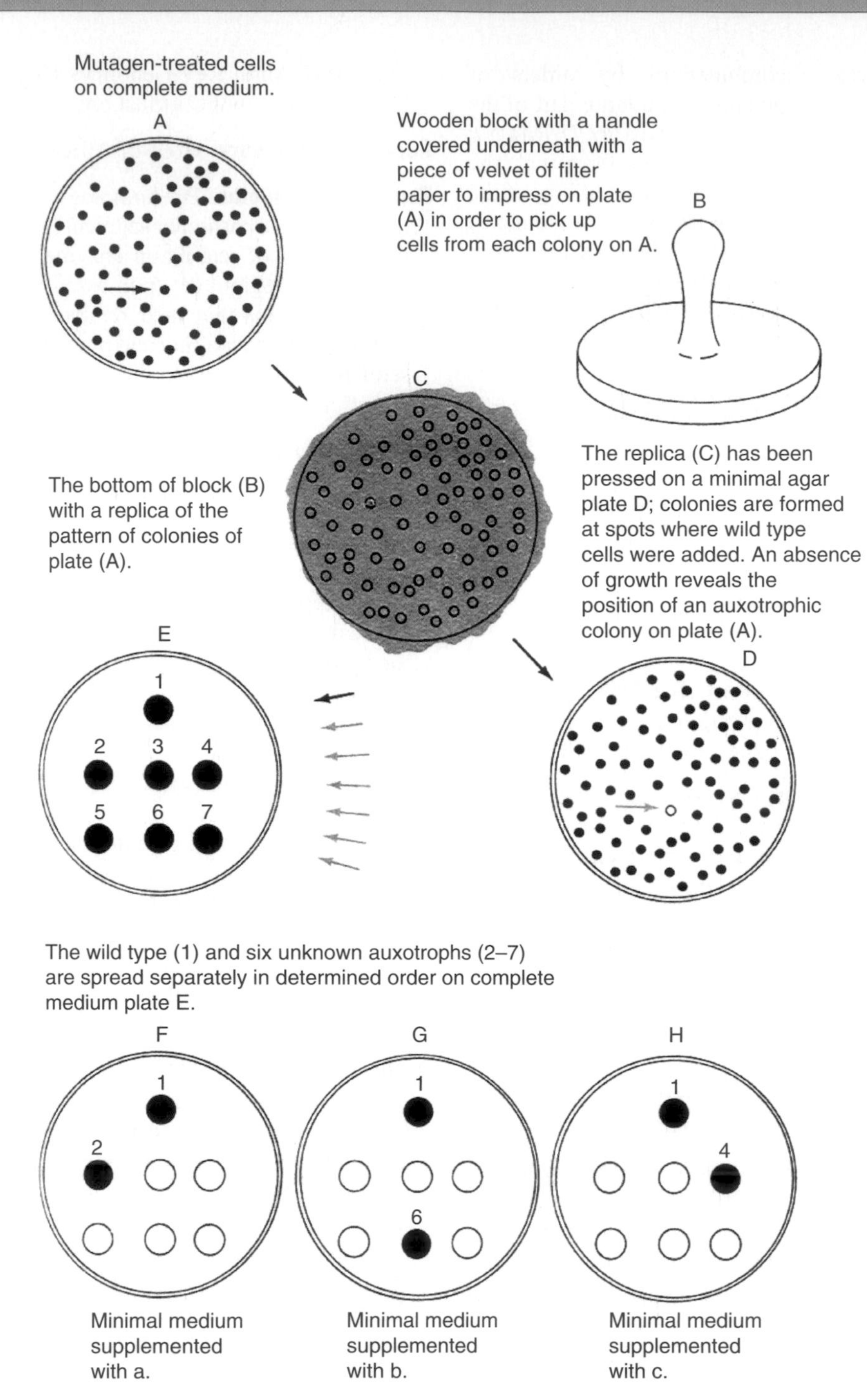
Mutagen-treated cells
on complete medium.
A
Wooden block with a handle
covered underneath with a
piece of velvet of filter
paper to impress on plate
(A) in order to pick up
cells from each colony on A.
B
C
The bottom of block (B)
with a replica of the
pattern of colonies of
plate (A).
The replica (C) has been
pressed on a minimal agar
plate D; colonies are formed
at spots where wild type
cells were added. An absence
of growth reveals the
position of an auxotrophic
colony on plate (A).
E
D
1
2
3
4
5
6
7
The wild type (1) and six unknown auxotrophs (2–7)
are spread separately in determined order on complete
medium plate E.
F
G
H
Minimal medium
supplemented
with a.
Minimal medium
supplemented
with b.
Minimal medium
supplemented
with c.

Figure R36. Replica Plating

Replication, Bidirectional: This is the mode of replication in bacteria as well as in the eukaryotic chromosome. Replication begins at an origin and proceeds in the opposite direction on both the old strands of the DNA double helix. The helicase subunits encoded by the xeroderma pigmentosum genes XPB and XpD of the transcription factor TFIIH unwinds the DNA in both directions. Electron microscope reveals a θ (theta) resembling structure of the circular DNA whereas in the linear eukaryotic DNA bubble-like structures are visible. In prokaryotes this replication is mediated by DNA polymerase III, and in eukaryotes a DNA polymerase α type enzyme. Termination of replication in *E. coli* requires 20 base long Ter elements and the associated protein Tus (termination utilization complex, M_r 36 K) (see Fig. R37). While replicating the template strand T7 RNA polymerase can by-pass up to 24 nucleotide gaps by making a copy of the deleted sequence using the corresponding non-template tract. ►DNA replication eukaryotes, ►DNA replication prokaryotes, ►θ replication, ►replication bubble, ►pol III, ►pol α, ►replication fork, ►xeroderma pigmentosum, ►transcription factors

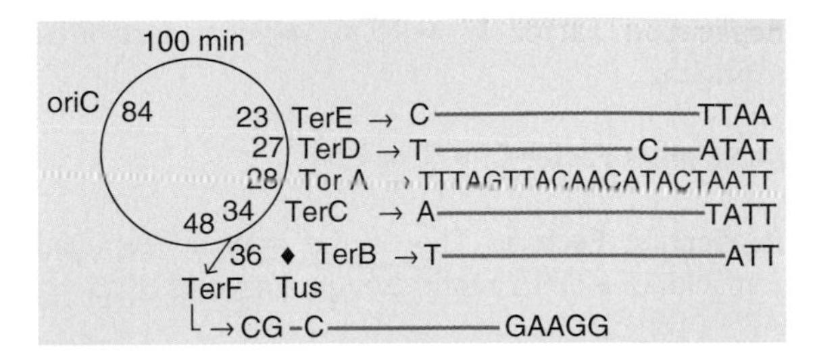

Figure R37. Replication in *E.coli.* (Diagram after Kamada K et al 1996 Nature [Lond] 383:598)

The *TerA, D* and *E* stop the replication in an anticlockwise direction and *TerC, B* and *F* halt replication of the strand elongated clockwise. The Tus-Ter complex probably blocks the replication helicase. Similar mechanisms operate in most bacteria but replication fork arresting sites are also found in eukaryotes, including humans. (See Hiasa H, Marians KJ 1999 J Biol Chem 274:27244; Abdurasidova G et al 2000 Science 287:2023; Gerbi SA, Bielinsky AK 1997 Methods 13:271).

Replication Bubble (replication eye): This is an indication of strand separation in a replicon (see Fig. R38). In an eukaryote nucleus an estimated 10^3 to 10^5 replication initiations occur during each cell cycle without any reinitiation per site, thus ensuring that the gene number is maintained. In yeast the dynamics of chromosome replication can be studied with the aid of DNA microarrays and it appears that the two ends of each chromosome replicate rather synchronously but the replication forks move differently in other regions. ►replication bidirectional, ►DNA replication, ►replication fork, ►replicon, ►promoter bubble, ►ORC, ►replication protein A, ►FFA, ►geminin; Diffley JFX 2001 Curr Biol 11:R367; Raghuraman MK et al 2001 Science 294:115.

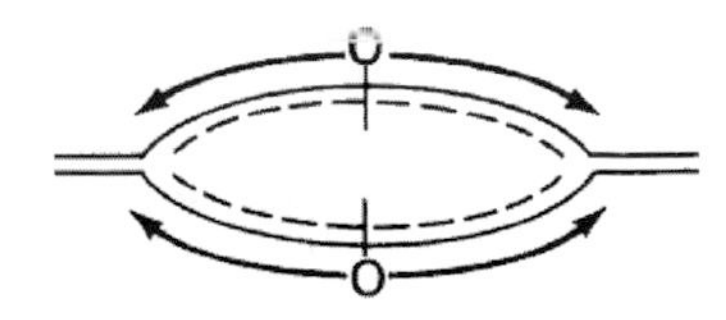

Figure R38. Replication Bubble

Replication, Conservative: A historical model of DNA replication, assuming that the two old (original) DNA strands produce two new copies which then anneal to each other. In other words, double-stranded DNA is not composed of an old and a new strand as revealed by the current and experimentally demonstrated semiconservative replication mechanism. ►semiconservative replication; Delbruck M, Stent GS 1957, p 699 In: McElroy WD Glass B (Eds.) The Chemical Basis of Heredity, Johns Hopkins Press, Baltimore, MD.

Replication Defective Virus: This is a mutant for the replication function or the lost genes required for producing infective particles. ►replicase

Replication, Dispersive: This refers to an unproven old idea that old and new double-strand DNA tracts alternate along the length of the molecule. ►replication, ►replication fork; Delbruck M, Stent GS 1957, p 699 In: McElroy WD, Glass B (Eds.) The Chemical Basis of Heredity, Johns Hopkins Press, Baltimore, MD.

Replication during the Cell Cycle: Eukaryotic DNA replication takes place predominantly during the S phase of the cell cycle although some repair synthesis (unscheduled DNA synthesis) may occur at other stages. In prokaryotes the replication is not limited to a particular stage and DNA synthesis may proceed without cellular fission. Such a phenomenon (endoreduplication) is exceptional in eukaryotes and is commonly limited to certain tissues only, e.g., to the salivary gland chromosomes of insects (*Drosophila, Sciara*) or a rare non-repeating process (endomitosis) that doubles the number of chromosomes. Replication in eukaryotes is an oscillatory process tied to the S phase of the cell cycle. The process of replication shows some variations even among the different eukaryotes and the process described here is modeled after that of *Saccharomyces cerevisiae* (the best known). During the G1 phase, from the pre-origin-of-replication complex (pre-ORC) the ORC (origin of replication complex) is assembled after the cyclosome (APC) proteases degraded the cyclin B-cyclin-dependent kinase (Cyclin B-CDK). The cis-acting *replicator* element and the *initiator* proteins bind at each origin of replication (hundreds or thousands in eukaryotes). The replicator (0.5–1 kb) is a multimeric complex itself and its indispensable component is the A unit but B1, B2 and B3 are also used. The A, B1 and B2 form the core of the replicator and B3 is an enhancer that binds to the *autonomously replicating sequence* (ARS)-binding protein factor 1 (ABF1). The replicator (A + B1) hugs the ORC (origin recognition complex) composed of 6 subunits that form the hub of the replication process and attract other critical regulatory proteins. The site of the initiation in mammals may extend to 50 kb. At the origin, the nucleosomal structure is remodeled and during S, G2 and early M phase DNase hypersensitive sites are detectable that disappear before the anaphase. It appears that protein Cdc7 is needed for the remodeling. CDC6 (or the homologous Cdc18 protein of fission yeast) is required in G1 or S phase for DNA synthesis (the cells may proceed to an abortive mitosis and aneuploidy in its absence). CDC6 seems to be essential for the formation of the pre-ORC complex. Overexpression of this protein leads to polyploidy. The replication also requires a Replication Licensing Protein (RLF) and members of the MCM (minichromosome maintenance) proteins. Cyclin-dependent kinases (CLB5 and CLB6) are also required to establish the pre-initiation complex but after the assembly is completed some of them may be degraded. Some cyclin-dependent kinases block the reinitiation of the complex until the cell passes through mitosis. Cyclin B5—cyclin-dependent kinase (Clb5-CDK) is inhibited by Sic1 (S phase inhibitory complex) that is removed by ubiquitin-mediated proteolysis at the START point before the S phase is fired on. CDC34, CDC53, CDC4 SKP1, CLN1-Cdc28, CLN2-CDC28 and the APC proteins promote ubiquitination. The initiation of DNA replication may also proceed through another pathway mediated by CDC7 and DNA-binding factor 4 (DBF4). ►cell cycle, ►replication fork, ►CLB, ►DNA replication eukaryotes, ►replication protein A, ►DNA replication prokaryotes, ►DNA replication in mitochondria, ►rolling circle replication, ►θ (theta) replication, ►ORC, ►RNA replication, ►reverse transcription, ►bidirectional replication, and other proteins and terms listed under separate entries; Waga S, Stillman B 1998 Annu Rev Biochem 67:721; Kelly TJ et al 2000 Annu Rev Biochem 69:829; Chakalova L et al 2005 Nature Rev Genet 6:669.

Replication Error: This source occurs when a nucleic acid base analog is incorporated into the DNA at the structurally acceptable site but during the following replication, being only an analog, it may cause a replicational error that leads to the replacement of the original base pair by another. E.g., BrU—A base pair in the following replication is converted by error into BrU—G pair which eventually results in the base substitution of C≡G at a site that was formerly T=A. Mispairing—in the absence of base analogs—can also occur, e.g., A=C, and the frequency of such errors is within the range of 10^{-4} to 10^{-6}. ►base substitution, ►incorporation error, ►bromouracil, ►BUdR, ►hydrogen pairing, ►ambiguity in translation, ►error in replication; Ryan FJ 1963 In: Burdette WJ (Ed.) Methodology in basic genetics, Holden-Day, San Francisco, California, p 39; Kunkel TA, Bebenek K 2000 Annu Rev Biochem 69:497.

Replication Eye: ►replication bubble, ►DNA replication eukaryotes, ►replication bidirectional

Replication Factor A: ►RF-A, ►helix destabilizing protein

Replication Factor C: ►RF-C

Replication Factory: This is the same as replication machine. ►DNA replication prokaryotes

Replication Fork: Represents the growing region of the DNA where the strands are temporarily separated. The simplest diagram of the replication fork is reproduced here showing the new (thin line) facing the polymerase with the 3′ end. The leading strand is below and the lagging strand is above (see Fig. R39).

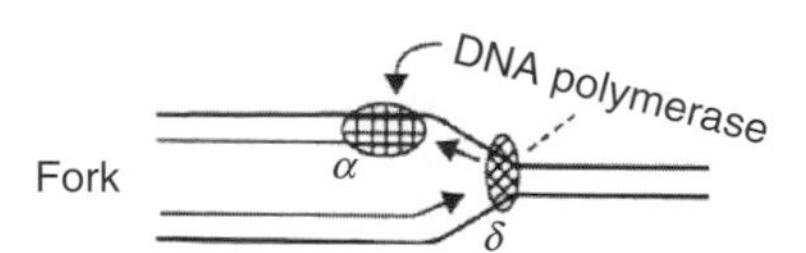

Figure R39. Replication fork

Replication is a very complex process requiring two different polymerases (δ and α for the leading and lagging strand, respectively) and several proteins. ►DNA replication eukaryotes, ►chart, ►DNA replication prokaryotes, ►nucleoid, ►GP32 protein, ►PriA, ►replication bubble, ►replication bidirectional, ►replication licensing, ►primase, ►Okazaki fragment, ►alpha accessory protein, ►processivity, ►replication machine; Waga S, Stillman B 1998 Annu Rev Biochem 67:721.

Replication Intermediate: ►lagging strand, ►Okazaki fragment, ►replication fork

Replication Licensing Factor (RLF): The initiation of replication requires two competency signals: the binding of the RLF and S phase promoting factors. The sequential action of these two signals secures the accurate replication of the chromosomes. RLF has two elements: RLF-M and RLF-B. RLF-M is a complex of MCM/P1. RLF-M protein binds to the chromatin early during the cell cycle but it is displaced after the S phase. ►MCM1, ►MCM3, ►ORC, ►replication, ►cell cycle, ►replication bubble, ►Cdt1, ►geminin; Chong JP, Blow JJ 1996 Progr Cell Cycle Res 2:83; Nishitani H et al 2001 J Biol Chem 276:44905.

Replication Machine: It has been demonstrated that the *replication machine* of bacteria (*B. subtilis* and probably others) occupies a stationary central position in the cell and initially the twin PolC subunits are located in the replicational origin (O) of the bidirectionally replicating double-stranded DNA ring (see Fig. R40). The simultaneously replicating leading and lagging DNA strands are spooled through the twin machines. The machines contain the polymerase and several accessory proteins (also called replication factory). The eukaryotic replication bubbles are operated in a similar manner but there are nearly 100 machines per nucleus and each of them handle about 300 replication forks. ►DNA replication prokaryotes, ►DNA replication eukaryotes, ►replication fork, ►clamp-loader; Ellison V, Stillman B 2001 Cell 106:655; Bruck I, O'Donell M 2000 J Biol Chem 275:28971; Turner J et al 1999 EMBO J 18:771, replicons forming factories: Kitamura E et al 2006 Cell 125:1297.

Replication Origin (o): Refers to the point in the genetic material where replication begins. ►CLB, ►replication bubble; isolation of replication origin in complex genomes: Mesner LD et al. 2006 Mol Cell 21:719.

Replication Protein A (RPA, RFA, HSSB): A complex of three different polypeptides (~70 kDa, human chromosome 17p13.3; ~32 kDa, 1p35; ~14 kDa,

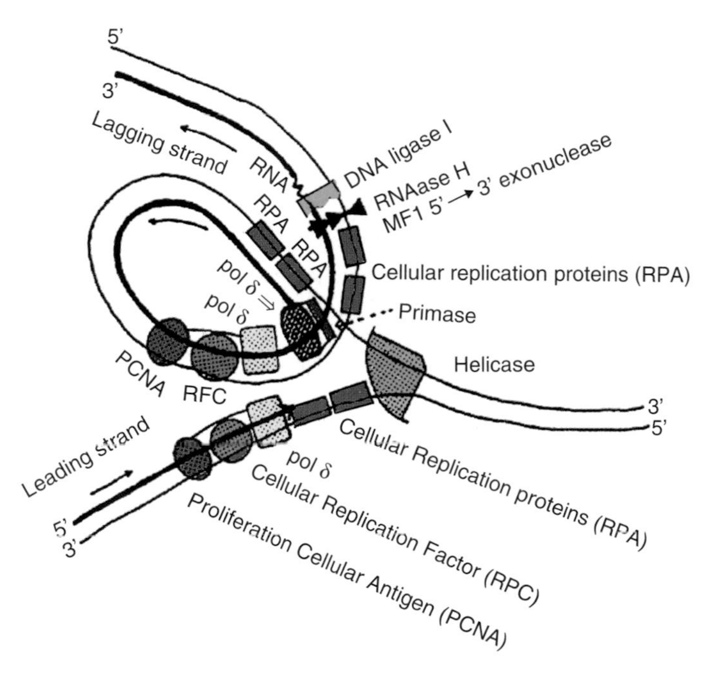

Figure R40. A model of the eukaryotic replication fork based on SV40 studies. Thin lines: old DNA strands; heavy lines: new strands. The replication fork is opened at the replicational origin by helicases. The *cellular replication protein* (RPA) keeps the fork open and brings the pol α DNA polymerase complex to the replication origin. After a short RNA primer is made (not shown on the diagram) at the beginning of the first Okazaki fragment, *cellular replication factor C* (RFC) binds to the DNA and displaces pol α by pol δ. Then RFC, pol δ, and PCNA (proliferating cell nuclear antigen) form a complex on both leading and lagging strands and the two new DNA strands are replicated in concert. The synthesis of the leading strand is straightforward. The lagging strand is made of Okazaki fragments (by "backstitching") because the DNA can be elongated only by adding nucleotides at the 3'-OH ends. After an Okazaki fragment (100–200 bases) is completed, RNase H, MF1 exonuclease remove the RNA primer (jagged line) and DNA ligase I joins the fragment(s) into a continuous new strand. The long arrows indicate the direction of growth of the chain. (Redrawn after Waga S and Stillman B 1994. *Nature* 369:207.)

7p22) binds most commonly to 20–25 nucleotides (with preference for pyrimidines) of single-strand DNA, and may be the first step in DNA replication by participating in DNA unwinding by binding to A-T rich sequences. The size of these subunits varies among different organisms and despite the good homology antigenically not related they cannot be interchanged functionally. The binding to double-stranded DNA is 3–4 orders of magnitude less. RPA binds to other proteins such as the primase subunit of DNA polymerase α, DNA repair proteins. The p53 cancer suppressor gene interferes with its binding to the replication origin. RPA also plays a role in recombination and excision repair. It has affinity to xeroderma pigmentosum damage-recognition protein (XPA) and endonuclease XPG. RPA increases the fidelity of repair polymerases. During the cell cycle RPA is phosphorylated by several kinases such as DNA-PK, CDK proteins, and others. ►RPA, ►DNA replication eukaryotes, ►replication fork, ►xeroderma pigmentosum, ►endonuclease, ►DNA repair, ►replication bubble, ►FFA, ►DNA polymerases, ►DNA-PK, ►CDK, ►somatic hypermutation; Wold MS 1997 Annu Rev Biochem 66:61; Patrick SM, Turchi JJ 2001 J Biol Chem 276:22630; Mass G et al 2001 Nucleic Acids Res 29:3892.

Replication Restart: This is a pathway of the SOS repair system of the DNA bypassing the mismatch and resulting in error-free replication. In bacteria, DNA polymerase II has an important role in the process. After UV irradiation or other damage blocks DNA synthesis, DnaB helicase and in coordination with DnaG primase on both leading and lagging strands provide means to bypass the lesion and a gap is left behind the leading strand *in E. coli* (see Fig. R41). ►DNA repair, ►replication of DNA, ►primase, ►translesion; Rangarajan S et al 1999 Proc Natl Acad Sci USA 96:9224.

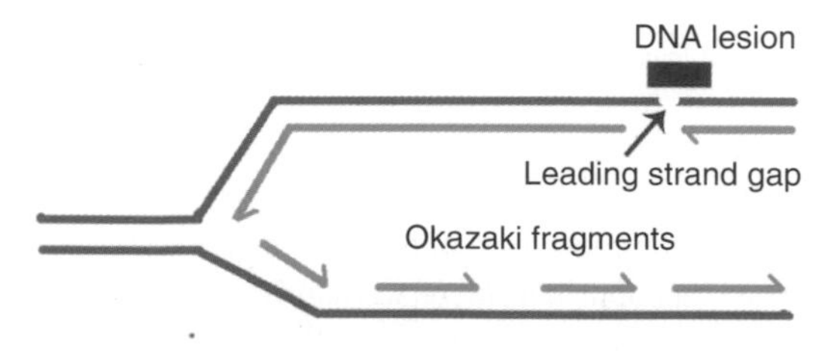

Figure R41. Replication fork is reinitiated after the block leaving a gap behind as primase restarts replication. (Modified after Heller RC and Marians KJ 2006 Nature [Lond] 439:557)

Replication Slippage: One of the several mechanisms generating microsatellite diversity. ►microsatellite; Viguera E et al 2001 J Mol Biol 312:323.

Replication Speed: kbp/min: *E. coli* 45, yeast 3.6, *Drosophila* 2.6, toad 0.5, mouse 2.2.

Replication Timing: Indicates whether chromosomal sequences replicate early or late during the cell cycle. Early tracts are usually more GC-rich and contain a larger number of genes than the AT-rich late replicating zones. Within the early/late transition regions many cancer genes have been found in human chromosomes 11q and 21q. ►heterochromatin; Watanabe Y et al 2002 Hum Mol Genet 11:13; Goren A, Cedar H 2003 Nature Rev Mol Cell Biol 4:25.

Replicational Fidelity: Refers to DNA replication error.

Replicative Aging: In yeast the number of cell divisions during the life of the cell may determine aging. ►chronological aging, ►longevity, ►aging

Replicative Form (RF): Double-stranded form of a single-stranded nucleic acid virus that generates the original complementary type of single-strand (+) nucleic acid. The necessity for the double-stranded replicative form is to generate a minus strand that is the template for the plus strand and is complementary to the "sense" molecule. This assures that all the progeny is identical and of one kind. ►DNA replication, ►RNA replication, ►plus strand

Replicative Intermediate: ►replicative form, ►replication intermediate, ►RNA replication

Replicative Segregation: The newly formed cellular organelles display sorting out in the somatic cell lineages in the case of mutation in organelle DNA. ►sorting out, ►cell lineage

Replicative Transposition: ►transposition, ►transposable elements, ►cointegrate, ►Mu bacteriophage

Replicator: Refers to the origin of replication in a replicon. ►replicon, ►ARS

Replichore: This is the oppositely replicating half of the *E. coli* genome between the origin and the terminus of replication. ►*E. coli*

Replicon: A replicating unit of DNA. The size of the replicational unit varies a great deal. In *E. coli*: ≈ 4.7 Mbp, *Saccharomyces cerevisiae*: (yeast): 40 kb, *Drosophila*: 40 kb, *Xenopus laevis* (toad): 200 kb, mouse: 150 kb, broad bean (*Vicia faba*): 300 kb. ►DNA replication, ►minireplicon; Jacob F et al 1963 Cold Spring Harbor Symp Quant Biol 18:329; Sadoni N et al 2004 J Cell Sci 117(Pt22):5353.

Replicon Fusion: ►cointegration

Replisome: Refers to the enzyme aggregate involved in the replication of DNA of *prokaryotes* (PriA, PrB, PriC, DnaC, DnaB and other proteins). The DNA polymerase III holoenzyme consists of two functional enzyme units, one for the leading and the other for the lagging strand. The *polymerase core* contains one α subunit (for polymerization), the ε subunit (3′→5′ exonuclease for editing repair) and the θ unit. It also includes a ring-like dimer of β *clamp* to hold on leash the DNA strands and a five-subunit *clamp loader* γ complex. There are two subunits that organize the two cores and the clamp loader into a pol III holoenzyme. The asymmetric replication of the leading and lagging strands is determined by the DnaC helicase, unwinding the double helix in front of the replisome. The helicase facilitates the hold of the complex onto the leading strand by the τ unit to make possible the continuous extension of that strand. At the lagging strand, however, the complex goes off and on as the Okazaki fragments are made. ►replitase, ►DNA replication prokaryotes, ►replication fork in prokaryotes, ►GP32 protein, ►DNA polymerases, ►Okazaki fragment; Benkovic SJ et al 2001 Annu Rev Biochem 70:181; Breier AM et al 2005 Proc Natl Acad Sci USA 102:3942; Johnson A, O'Donell M 2005 Annu Rev Biochem 74:283.

Replitase: Refers to the replicational complex at the DNA fork in *eukaryotes*. ►replication fork in eukaryotes, ►replisome, ►DNA polymerases; Reddy GP, Fager RS 1993 Crit Rev Eukaryot Gene Expr 3[4]:255.

Replum: A central membrane-like septum ↑ inside the silique (fruit) of cruciferous plants bearing the seeds (see Fig. R42). The carpels covering it were removed. Recessive pale mutant cotyledons show through the immature seed coats. At maturity the carpels dehisce at the base of the fruit (left end here). ►silique

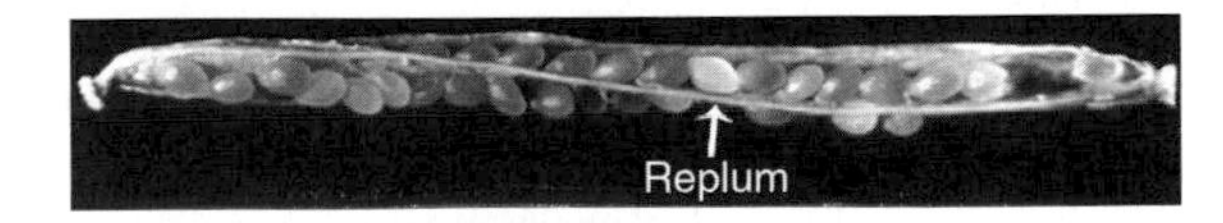

Figure R42. Replum

Reporter Gene: This is a structural gene with easily - monitored expression (e.g., luciferase, β–glucuronidase, antibiotic resistance) that reports the function as differentiation progresses, or any heterologous or modified promoter, polyadenylation or other signals attached to the gene by in vitro or in vivo gene fusion. ►luciferase, ►GUS, ►gene fusion

Reporter Ring: According to the tracking concept of recombination between appropriate *res* (*resolvase*) points, recombination of two DNA molecules would retain during segregation "reporter rings" catenated to one of the two DNA strands of the DNA recombination substrate during synapse and after its resolution in the product. The "reporter rings" were expected to be limited to one of the catenated product molecules but the experimental data did not support this assumption. ►tracking, ►resolvase

Representational Difference Analysis: ►RDA

Repressible: This means subject to potential repression. ►repression

Repressilator: This is a synthetic oscillator system composed of several components that may turn on/off as a natural biological clock. In such a system the product of a gene is a repressor for the promoter of the next one and so on. ►oscillator; Elowitz MB, Leibler S 2000 Nature [Lond] 403:335; Garcia-Ojalvo J et al 2004 Proc Natl Acad Sci USA 101:10955.

Repression: This is the control mechanism interfering with the synthesis (at the level of transcription) of a protein. A general type of repression is attributed to histones, closely associated with the DNA in the nucleosomal structure. When histones are deacylated they reinforce repression by co-repressors (N-CoR, mSin3). Histone acetyl transferases acylate histones with the assistance of pCAF (chromatin assembly factor) which permits the recruitment of transcription factors to the gene. PCIP = p300/CEP co-integrator associated protein (where CEP is a CREB-binding protein), CREB = cAMP-response element, CBP is a CREB-binding protein, p300 = a cellular adaptor and co-activator of some proteins. This scheme of repression is based on some eukaryotic systems. Other mechanisms are also known. Some repression mechanisms interfere with the translation, e.g., threonyl-tRNA synthetase of *E. coli* represses its own synthesis by binding to the operator. $tRNA^{Thr}$ serves as an antirepressor and the balance between the two mechanisms determines the level of translation (see Fig. R43) (Torres-Larios A et al 2002 Nature Struct Biol 9:343). ►feedback control, ►regulation of gene activity, ►signal transduction, ►regulation of enzyme activity, ►MAD, ►co-activator, ►transcription factors, ►*lac* repressor, ►*arabinose* operon, ►repressor, ►*tryptophan* operon, ►suppression; Maldonado E et al 1999 Cell 99:455; Lande-Diner L, Cedar H 2005 Nature Rev Genet 6:648.

Repressor: This is the protein product of the regulator gene that interferes with the transcription of an operon. The majority of the DNA-binding proteins bind the DNA by their α helices but some of the repressors (*met, arc, mnt*) bind by β sheets or a combination of both (*trp*). Repression in eukaryotes

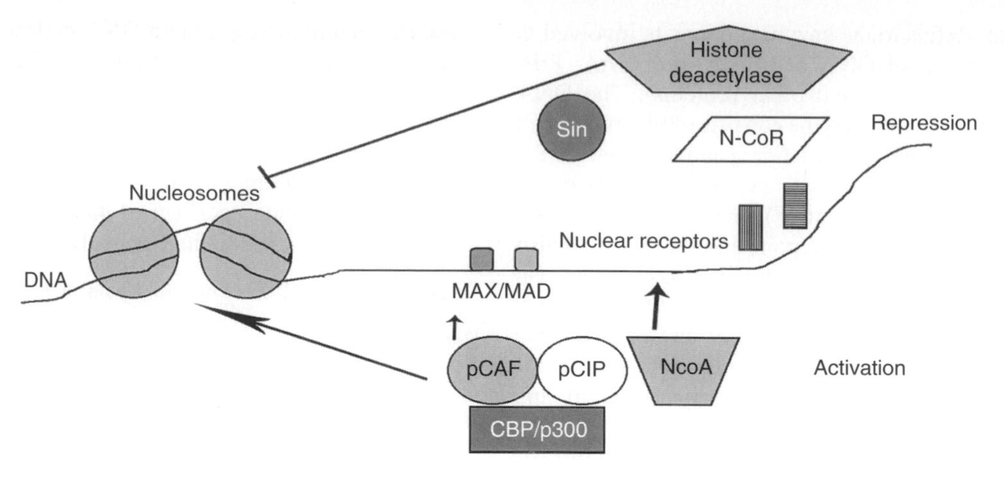

Figure R43. Proteins involved in repression and activation. (Diagram modified after Heinzel T et al 1997 Nature (Lond) 387:43)

has a variety of means for control. The short-range repressors are within 50–150 bp of the transcriptional activators. Since the promoters may be modular (repeating units), the short-range silencers may not affect other than the nearest activators. This organization assures the expression of genes controlling segmentation along the axis of the developing embryo. The long-range repressors may silence the activators from a distance of several kb. The short-range repressor proteins are monomeric whereas the long-range ones are multimeric. The repressors may be either gene-specific or more global. In the latter case in bacteria the repressor binds to the σ subunit of the RNA polymerase. Some of the more recently discovered repressors are not protein molecules but RNAs (ribozymes). In bacteria, glucosamine-6-phosphate (GlcN6P) mRNA encodes within its upstream sequence this ribozyme of ~75 nucleotides, which cleaves itself in the mRNA when GlcN6P reaches a sufficient level resulting in decay of the message. The regulatory response is initiated upon binding of GlcN6P to the mRNA (Winkler WC et al 2004 Nature [Lond] 428:281). ►repression, ►*lac* operon, ►*lac* repressor, ►*arabinose* operon, ►*tryptophan* operon, ►co-repressor, ►morphogenesis in *Drosophila*, ►transcription factors, ►Gene-Switch, ►riboswitch, ►tetracycline, ►suppression, ►suppressor tRNA; Pardee A et al 1959 J Mol Biol 1:165; Hummelke GC, Cooney AJ 2001 Front Biosci 6:D1186; Ryu J-R et al 2001 Proc Natl Acad Sci USA 98:12960.

Reproduction: ►asexual, ►clonal, ►vegetative reproduction, ►dioecious, ►monoecious, ►autogamy, ►apomixia, ►allogamy, ►parthenogenesis, ►hermaphroditism, ►cytoplasmic transfer, ►genetic systems, ►conjugation, ►conjugation *Paramecia*, ►life cycle, ►breeding system, ►fungal life cycle, ►social insects, ►vivipary, ►incompatibility

Reproductive Isolation: Prevents gene exchange between two populations by a hereditary mechanism. Generally reproductive isolation is caused by the inability to mate (pre-mating isolation) but in some instances inviability or sterility of the offspring is the barrier (post-mating isolation). In plants, chromosomal rearrangements normally lead to gametic disadvantage or sterility. In animals, translocated chromosomes are frequently transmitted at fertilization but the duplication-deficiency zygotes are inviable. Mating *Drosophila melanogaster* with *D. simulans* a 3 kb DNA segment is inserted within the *Cyclin E* locus and causes male sterility and inviability but only a low degree of inviability or sterility in females. The *JYAlpha* gene is in chromosome 4 of *Drosophila melanogaster* but during evolution it transposed to the 3rd chromosome of *D. simulans*. In hybrids when this gene is lost hybrid male sterility occurs (Masly JP et al 2006 Science 313:1448). All members of the *Drosophila* species show reproductive isolation from *D. melanogaster*. Reproductive isolation can be associated with the pattern of single nucleotide polymorphism (Payseur BA, Hoekstra HE 2005 Genetics 171:1905). ►incompatibility, ►sexual isolation, ►isolation genetic, ►founder principle, ►drift genetic, ►effective population size, ►translocation, ►inversion, ►infertility, ►transcription, ►hybrid inviability, ►speciation, ►SNP; Harushima Y et al 2001 Genetics 159:883.

Reproductive Rate: (R_0) where l_x = probability that female survives to age x and m_x = expected number of female offspring produced by a female of age x.

▶age-specific birth and death rate, ▶population growth, ▶Malthusian parameter, ▶parity

$$R_0 = \sum_{X=0}^{\infty} 1_X m_X$$

Reproductive Success: ▶fitness, ▶fertility, ▶fecundity

Reproductive Technologies: ▶ART

Reprogramming: During development the genome must be selectively turned on and off by epigenetic modifications such as methylation of nucleotides, acetylation and deacetylation of nucleosomes, and recruiting general and specific transcription factors. Such reprogramming which begins after fertilization of the egg is a natural and indispensable process. Problems arise, however, when diploid nuclei of somatic cells are transplanted into the eggs for the purpose of cloning. The somatic nuclei require dedifferentiation to the totipotent/pluripotent state and after transplantation the cloned embryos must be able to redifferentiate for the intended purpose. The transfer of nuclei from embryonic stem cells poses fewer problems yet in vitro culture conditions may not exactly duplicate the normal programming involved in natural fertilization of the haploid egg with the haploid sperm. Generally cloning involves developmental anomalies such as "large offspring syndrome", various chromosomal anomalies and inviability because of lack of harmony between the donor nucleus and the recipient cytoplasm. Somatic cells may be reprogrammed by transfer into enucleated oocytes or by fusion with pluripotent embryonic stem cells (Tada M et al 2001 Curr Biol 11:1553; Ying QL et al 2002 Nature [Lond] 416:545). In cell fusion the less differentiated cell reprograms the more differentiated ones to its own features. It seems that reprogramming is a nuclear rather than a cytoplasmic process. It is unclear whether reprogramming by the use of cell extracts would be generally feasible. The cell fusions are not very useful for subsequent cloning because the nuclei are tetraploids. No practical procedure is available for reducing them to diploid level. Various culture conditions can also induce reprogramming. Blastocysts can yield pluripotent embryonic stem cells in vitro. The primordial germ cells of the genital ridge can give rise not only to germ cells (oocyte and sperm), but also to embryonic germ cells when used as explants in vitro or they can dedifferentiate into embryonic carcinoma cells. Spermatogonial stem cells can give rise to spermatozoa or in culture they can be reprogrammed into embryonic stem cell-like tissues. Bone marrow-derived adult multipotent progenitor cells can also be reprogrammed into mesenchymal or blood cells.

The *Nanog* gene (4 exons) in mouse chromosome 6 and human chromosome 12 encodes a 305-amino acid protein in humans (Chambers I et al 2004 Cell1 13:643). This protein is essential for the maintenance of pluripotency of embryonic stem cells. The name nanog means the land of the ever young in Celtic mythology. Nanog may increase pluripotency of fused cells up to 100% (Silva J et al 2006 Nature [Lond] 441:997). Several other genetic factors OCTA4, SOX2 and various transcription factors have also been implicated. Mouse primordial germ cells (PGC) undergo erasure of histone 3 lysine 9 dimethylation (H3K9me2) and upregulation of histone H3 lysine 27 trimethylation (H3K27me3) in a progressive, cell-by-cell manner, presumably depending on their developmental maturation. Before or concomitant with the onset of H3K9 demethylation, PGCs enter G2 arrest of the cell cycle, that apparently persists until they acquire high H3K27me3 levels. PGCs repress RNA polymerase II-dependent transcription, which begins after the onset of H3K9me2 reduction in the G2 phase and tapers off after the acquisition of high level H3K27me3. The epigenetic reprogramming and transcriptional quiescence are independent from the function of Nanos3. Before H3K9 demethylation, PGCs exclusively repress an essential histone methyltransferase, GLP, without specifically upregulating histone demethylases (Seki Y et al 2007 Development 134:2627). ▶nuclear transplantation, ▶epigenesis, ▶histone methyltransferase, ▶stem cells; Rideout WM III et al 2001 Science 293:1093; Håkelien A-M et al 2002 Nature Biotechnol 5:460; Hochedlinger K, Jaenisch R 2006 Nature [Lond] 441:1061.

Reprolysin: A metalloproteinase involved in the regulation of morphogenesis. ▶bone morphogenetic protein

Reptation: A theory about the movement of nucleic acid end-to-end in gels.

Reptin: ▶chromatin remodeling

Repulsion: One recessive and one dominant allele are in the same member of a bivalent, such as <u>A b</u> and <u>a B</u>. ▶coupling, ▶linkage; Bateson W et al 1905 Rep Evol Com Roy Soc II:1.

RER: Denotes rough endoplasmic reticulum (endoplasmic reticulum with ribosomes on top of it).

Resection: Nuclease mediated production of single-strand overhangs in double-strand DNA breaks.

Resequencing: is directed to molecularly known gene(s) in order to determine mutations in the sequence (SNP, insertion, deletion) that have potential role in the expression of the gene and the disease it controls.

Capillary electrophoresis, polymerase chain reaction and automated data analysis may be used for the detection of mutations in the tract. For detection of insertions, labeled nucleotide array is hybridized to the target tract and the increase of signal indicates insertions. Loss of hybridization in a similar approach is the sign of deletion. (See Rijk PD, Del-Favero J 2007 Methods Mol Biol 396:331; http://www.resequencing.mpg.de/).

Residue: Refers to the elements of polymeric molecules, such as nucleotides in nucleic acids, amino acids in proteins, sugars in polysaccharides and fatty acids in lipids.

Resilin: This is a member of an elastic protein family, including gluten, gliadin and spider silks. ►glutenin, ►silk fibroin; Elvin CM et al 2005 Nature [Lond] 437:999.

Resistance Transfer Factors (RTF): Plasmids that carry antibiotic or other drug resistance genes in a bacterial host and are capable of conjugational transfer. ►plasmid(s)

Resistin: This is a 12.5 kDa cysteine-rich protein hormone of adipose tissues. The level of resistin is increased in genetically determined and diet-dependent obesity and it may cause insulin resistance and type II diabetes. ►adipocyte, ►adiponectin, ►diabetes mellitus, ►insulin; Way JM et al 2001 J Biol Chem 276:25651.

RESites (related to empty sites): These are genomic sites from where transposable elements moved out but left behind a footprint. Their flanking sequences are similar to insertion targets. (See Le QH et al 2000 Proc Natl Acad Sci USA 97:7376).

Resolution: Refers to the depth of details revealed by the analysis.

Resolution, Optical: Defines the ability of distinguishing between two objects irrespective of magnification. Magnification helps the human eye but does not improve optical resolution. The power of resolution depends on the wavelength of the light and on the aperture of the lens used. (The achromatic lens is free from color distortion, the apochromatic lens is free from color or optical distortions). The naked human eye may discern details larger than 100 μm, the light microscope with a good oil immersion objective lens may resolve 0.2 μm, the lowest limit for an ideal electronmicroscope is 0.1 nm, i.e., 1 Å. Under practical conditions the resolution of the electronmicroscope is about 2 nm. The resolution of the light microscope is generally defined as $\frac{0.61\lambda}{n \sin\theta}$ where λ is the wavelength of the light (for white light it may be 530 nm), *n* stands for the refractive index of the immersion oil or air (when dry lens is used) and θ is half of the angular width of the cone of the light beam focused on the specimen with the condensor of the microscope (sin θ is maximally about 1). Then sin θ is the *numerical aperture* of the lens and using oil immersion its value may increase to 1.4. The immersion oil should be non-fluorescing, slow drying and of right viscosity for vertical or horizontal inverted views. ►fluorescence microscopy, ►phase-contrast microscopy, ►Nomarski, ►confocal microscopy, ►electronmicroscopy, ►oil immersion lens

Resolvases: These are endonucleases mediating site-specific recombination and instrumental in resolving cointegrates or concatenated DNA molecules; transposon Tn3 and γδ encoded proteins promoting site-specific recombination of supercoiled prokaryotic DNA containing replicon fusion, direct end repeats and internal *res* sites (see Fig. R44). ►site-specific recombination, ►recombination site-specific, ►reporter ring, ►TN3, ►concatenate, ►γδ element, ►EMC, ►phase variation, ►recombination molecular mechanism; Kholodii G 2001 Gene 269:121; Croomie GA, Leach DR 2000 Mol Cell 6:815; γδ structure: Li W et al 2005 Science 309:1210; Kamtekar S et al 2006 Proc Natl Acad Sci USA 103:10642.

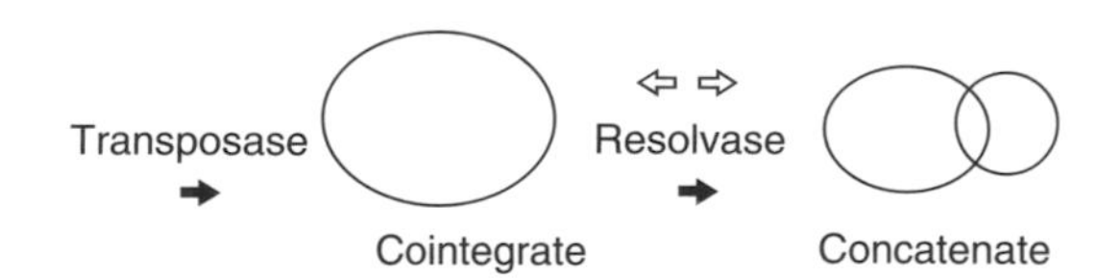

Figure R44. Resolvase functions

Resolving Power: This is the ability to distinguish between two alternatives or detecting an event, the probability of which is expected to be very low, e.g., recombination between two alleles of a gene.

Respiration: Electrons are removed from the nutrients during catabolism and carried to the oxygen through intermediaries in the respiratory chain. Oxygen is taken up and carbon dioxide is produced. ►fermentation, ►Pasteur effect, ►chlororespiration, ►mitochondria

Respiratory Distress Syndrome: This condition is frequently observed in premature birth and is due to deficiency of pulmonary surfactant protein A (SFTPA2, 10q22.2-q22.3) or PFTP3 (2p12-p11.2).

Responder (*Rsp*): A component of the segregation distorter system in *Drosophila*, a repetitive DNA at the heterochromatic site in chromosome 2–62. ►hybrid dysgenesis, ►segregation distorter

Response Elements: ►hormone response elements, ►transcription factors, ►inducible regulation of gene activity

Response Regulator: ►two-component regulatory system

Responsiveness: The establishment of a new steady state readily follows a change in the signal(s). ►signal, ►steady state

REST/NRSF (RE1 silencing transcription factor/neural-restrictive silencing factor): This factor blocks the expression of genes in non-neural tissues. It is a component of the histone deacetylase complex. It is involved in the regulation of chromatin and associated protein complexes. (See Kojima T et al 2001 Brain Res Mol Brain Res 90:174; Ooi L, Wood IC 2007 Nature Rev Genet 8:544).

Restenosis (recurrent stenosis): This is usually caused by therapeutic manipulation of the vascular system as an overcompensation for injury repair, e.g., after angioplasty (surgical/balloon opening of stenosis), or coronary/peripheral bypass surgery. Cell cycle inhibitors, antisense technologies, and modifying the expression of particular genes by using genetic vectors targeting particular genes involved in the cell cycle have been considered for the prevention of this very complex process. ►stenosis, ►cell cycle, ►antisense technologies; Gordon EM et al 2001 Hum Gene Ther 12:1277.

Restitution Chromosomal: Refers to rejoining and healing of broken off chromosomes.

Restitution Nucleus: This is the unreduced product of meiosis.

Restless Leg Syndrome: This age-dependent neurological defect is characterized by compulsive movement of the legs, sleep disorder and increased tendency toward cardiovascular disease. It may affect up to 10% of the population of European descent. A genome-wide association study has revealed that the homeobox gene MEISI1 (chromosome 2p), gene BTBD9 (chromosome 6p), kinase MAP2K5 and transcription factor LBXCOR1 encoded at 15q are involved (Winkelmann J et al 2007 Nature Genet 39:1000).

Restorer Genes: ►cytoplasmic male sterility, ►fertility restorer genes

Restriction Endonuclease: ►restriction enzyme and Table R6.

Table R6. Restriction endonucleases with recognition and cutting sites (Isoschizomers: IS, Cutting site: (‘), N: any base,^{m}: Methylated base)

Aal IS Stu I	Bpu AI GAAGAC$(N)_{2/6}$	Hae III GG’CC	Pma CI IS: Bbr PI
Aat II GACGT’C	Bse AI T’CCGGA	Hgi AI IS: Asp HI	Pml IS Bbr PI
Acc I GT’(A,C)(T,G)AC	Bse PI IS: Bss H II	Hha I IS: Cfo I	Psp 1406 I AA’CGTT
Acc III IS: Mro I	Bsi WI C’GTACG	Hinc II IS: Hind II	Pst I CTGCA’G
Acs I (A,G)’AATT(T,C)	Bsi YI CC$(N)_5$’NNGG	Hind II GT(T,C)’(A,G)AC	Pvu I CGAT’CG
Acy I G(A,D)’CG(C,T)C	Bsm I GAATGCN’N	Hind III A’AGCTT	Pvu II CAG’CTG
Afl I IS: Ava II	CTTAC’GNN	Hinf I G’ANTC	Rca I T’CATGA
Afl II IS: Bfr I	Bsp 12861 IS: Bmy I	Hpa I GTT’AAC	Rsa I GT’AC
Afl III A’C(A,G)(T,C)GT	Bsp 14071 IS: Ssp BI	Hpa II C’CGG	Rsr II CG’G(A,T)CCG
Age IS: Pin AI	Bsp HI IS: Rca I	Ita I GC’NGC	Sac I GAGCT’C
Aha II IS: Acy I	Bsp LU11I A’CATGT	Kpn I GGTAC’C	Sac II IS: Ksp II
Aha III IS: Dra I	Bss HII G’CGCGC	Ksp I CCGC’GG	Sal I G’TCGAC
Alu I AG’CT	Bss GI IS: Bst XI	Ksp 632 I CTCTTC$(N)_{1/4}$	Sau I IS: Aoc I
Alw 44 I G’TGCAC	Bst 1107 I GTA’TAC	Mae II A’CGT	Sau 3A ’GATC
Aoc I CC’TNAGG	Bst BI IS: Sfu I	Mae III ‘GTNAC	Sau 96 I G’GNCC
Aos I IS: Avi II	Bst EII G’GTNACC	Mam I GATNN’NNATC	Sca I AGT’ACT
Apa I GGGCC’C	Bst NI IS: Mva I, Eco RII	Mbo I IS: Nde II	Scr FI CC’NGG

R

Table R6. Restriction endonucleases with recognition and cutting sites (Isoschizomers: IS, Cutting site: (’), N: any base,^{m}: Methylated base) Continued

Apo I IS: Acs I	Bst XI CCA(N)$_5$’NTGG	Mfe I IS: Mun I	Sex AI A’CC(A,T)GGT
Apy I IS: Eco RII, Mva I	Cel II GC’TNAGC	Mlu I A’CGCGT	Sfi I GGCC(N)$_4$’NGGC
Ase I: IS: Asn I	Cfo I GCG’C	Mlu NI TGG’CCA	Sfu I TT’CGAA
Asn I AT’TAAT	Cfr I IS: Eae I	Mro I T’CCGGA	Sgr AI C(A,G)’CCGG(T
Asp I GACN’NNGTC	Cfr 10 I (A,G)’CCGG (T,C)	Msc I IS: Mlu NI	Sma I CCC’GGG
Asp 700 GAANN’NNTTC	Cla I AT’CGAT	Mse I IS: Tru 91	Sna BI TAC’GTA
Asp 718 G’GTACC	Dde I C’TNAG	Msp I C’C^{m}GG	Sno I IS: Alw 44 I
Asp EI GACNNN’NNGTC	Dpn I G^{m}A’TC	Mst I IS: Avi II	Spe I A’CTAGT
Asp HI G(A,T)GC(T,A)’C	Dra I TTT’AAA	Mst II IS: Aoc I	Sph I GCATG’C
Asu II IS: Sfu I	Dra II (A,G)G’GNCC(T,C)	Mun I C’AATTG	Ssp I AA’ATT
Ava I G’(T,C)CG(A,G) (A,G)G	Dra III CACNNN’GTG	Mva I CC’(A,T)GG	Ssp BI T’GTACA
Ava II G’G(A,T)CC	Dsa I C’C(A,G)(C,T)GG	Mvn I CG’CG	Sst I IS: Sac I
Avi II TGC’GCA	Eae I (T,C)’GGCC(A,G)	Nae I GCC’GGC	Sst II IS: Ksp I
Avr II IS: Bln I	Eag I IS: Ecl XI	Nar I GG’CGCC	Stu I AGG’CCT
Bal I IS: Mlu NI	Eam 11051 IS: Asp EI	Nci I CC’(G,C)GG	Sty I C’C(A,T)(A,T)GG
Bam HI G’GATCC	Ecl XI C’GGCCG	Nco I C’CATGG	Taq I T’CGA
Ban I G’G(T,C)(A,G)CC	Eco 47 III AGC’GCT	Nde I CA’TATG	Tha I IS: Mvn I
Ban II G(A,G)GC(T,C)’C	Eco RI G’AATTC	Nde II ‘GATC	Tru 9 I T’TAA
Bbr PI CAC’GTG	Eco RII ‘CC(A,T)GG	Nhe I G’CTAGC	Tth 111 I IS: Asp I
Bbs I IS: Bpu AI	Eco RV GAT’ATC	Not I GC’GGCCGC	Van 91 I CCA(N)$_4$’NTG
Bcl I T’GATCA	Esp I IS: Cel II	Nru I TCG’CGA	Xba I T’CTAGA
Bfr I C’TTAAG	Fnu DII IS: Mvn I	Nsi I ATGCA’T	Xho I C’TCGAG
Bgl I GCC(N)$_4$’NGGC	Fnu 4 HI IS: Ita I	Nsp I (A,G)CATG’(T,C)	Xho II (A,G)’GATC(T,C
Bgl II A’GATCT	Fok I GGATG(N)$_{9/13}$	Nsp II IS: Bmy I	Xma III IS: Ecl XI
Bln I C’CTAGG	Fsp I IS: Avi II	Nsp V IS: Sfu I	Xmn I IS: Asp 700
Bmy G(G,A,T)GC(C,T,A)’C	Hae II (A,G)GCGC’(T,C)	Pin AI AA’CCGGT	Xor II CGAT’CG

Restriction Enzyme: Endonucleases cut the DNA at specific sites (generally) when the bases are not protected (modified, usually by methylation). Bacteria synthesize restriction enzymes as a defense against invading foreign DNAs such as phages and foreign plasmids. They may facilitate recombination and transposition. Three major types have been identified. Type II enzymes are used most widely for genetic engineering. *Type II* enzymes have separate endonuclease and methylase proteins. Their structure is simple, cleave at the recognition site(s); the recognition sites are short (4–8 bp) and frequently palindromic, they require Mg^{2+} for cutting, the methylation donor is SAM. So far more than 3,000 Type II restriction endonucleases have been identified. Type II proteins, unless they have a similar pattern of cleavage, have relatively low sequence homologies.

Type III enzymes carry out restriction and modification with the help of two proteins which share a polypeptide. They have two different subunits.

Cleavage sites are generally 24–26 bp downstream from the recognition site. The recognition sites are asymmetrical 5–7 bp. For restriction, they require ATP and Mg^{2+}. For methylation, SAM, ATP, Mg^{2+}

are needed. *Type I* enzymes are single multifunctional proteins of three subunits, cleavage sites are random and at about 1 kb from specificity sites (crystal structure of the DNA-binding subunit: Kim J-S et al 2005 Proc Natl Acad Sci USA 102:3248). Type I restriction enzymes are encoded in the *hsd* (host-specificity DNA) locus with three components: *hsdS* (host sequence specificity), *hsdM* (methylation) and *hsdR* (endonuclease). Their recognition sites are bipartite and asymmetrical: TGA-N8-TGCT or AAC-N6-GTGC. For restriction and methylation, SAM, ATP and Mg^{2+} are required (see Figs. R45 and R46). Restriction enzymes may create a protruding and a receding end or the two ends may be of equal length, e.g.:

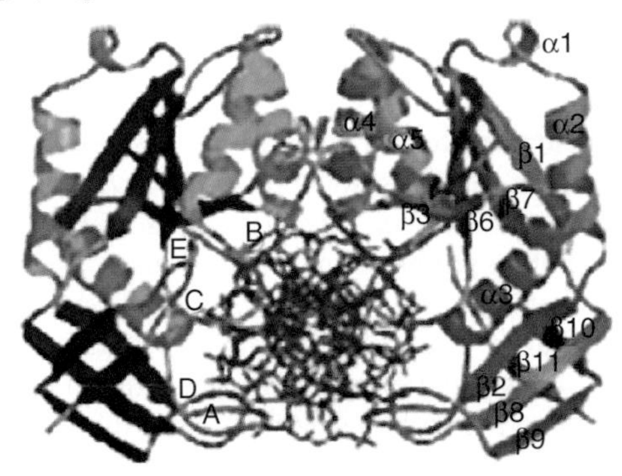

Figure R45. Crystal Structure of *Bgl* II. In the centre is the DNA and α-helices and β-sheets embrace the DNA. Courtesy of Aggarwal AK; From Lukacs CM et al 2000 Nature Struct. Biol. 7:134

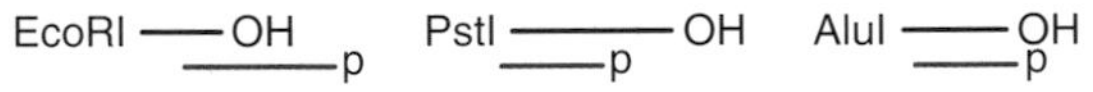

EcoRI ——OH ——p PstI ————OH ——p AluI ——OH ——p

Figure R46. Protruding, receding and blunt ends of restriction fragments

Type II enzymes may be of high specificity, ambiguous, or isoschizomeric, and may be prevented from action on methylated substrates or may be indifferent to methylation. Some restriction enzymes such as McrA, McrBC and Mrr actually cut only methylated DNA. Approximately 3,000 Type II restriction enzymes with nearly 200 different specificities are known. Kilo- and Mega-base DNA substrates can be precisely cleaved by combining a DNA-cleaving moiety, e.g., copper-*o*-phenanthroline with a specific DNA binding protein (e.g., CAP). The complex thus cuts at the 5′-AAATGTGATCTAGATCACATTTT-3′ DNA site of CAP recognition. The great specificity required that the cutting moiety would be attached to an amino acid in such a way that it would bend toward the selected target but not toward unspecific sequences. The IIS restriction endonucleases cleave the DNA at a precise distance outside of their recognition site and produce complementary cohesive ends without disturbing their recognition site. By 2005, over 3,600 restriction endonucleases had been identified representing more than 250 specificities (see Table R6). ►nucleases, ►isoschizomers, ►DNA methylation, ►restriction-modification, ►hsdR, ►CAP, ►antirestriction, ►RNA restriction enzyme; Roberts RJ, Macelis D 2001 Nucleic Acids Res 29:268; Titheradge AJB et al 2001 Nucleic Acids Res 28:4195; Piungoud A, Jeltsch A 2001 Nucleic Acids Res 29:3705; classification: Roberts RJ et al 2005 Nucleic Acids Res 33: D233; historical perspective: Roberts RJ 2005 Proc Natl Acad Sci USA 102:5905; Brownlee C 2005 Proc Natl Acad Sci USA 102:5909; http://www.neb.com/nebecomm/products/category1.asp?; enzymes and genes for restriction and modification: http://rebase.neb.com/rebase/rebase.ftp.html.

Restriction Enzymes, Class-IIs (Enases-IIS): These have 4–7 bp, completely or partially asymmetric recognition sites and the cleavage site is at a distance of 1–20 bp. These enzymes are monomeric. They can be employed for precise trimming of DNAs, retrieval of cloned fragments, assembly of genes, cleavage of single-stranded DNA, detection of point mutations, amplification and localization of methylated bases. ►restriction enzymes, ►indexer; Szybalski W et al 1991 Gene 100:13.

Restriction Factor: This is a general term for various proteins, proteases, hormones and endonucleases involved in limiting or preventing growth. ►APOPBEC3G; retrovirus restriction review: Goff SP 2004 Mol Cell 16:849.

Restriction Fragment: Denotes a piece of DNA released after digestion by a restriction endonuclease. The length of the fragments depends on how many nucleotides are situated between the two cleavage sites. From the same genomic DNA, the same enzyme generates fragments of different lengths because the nucleotide sequence varies along the DNA length. ►restriction enzymes, ►restriction fragment number, ►RFLP

Restriction Fragement Length Polymorphism: ►RFLP

Restriction Fragment Number: This can be predicted on the grounds of the number of bases at the recognition sites. Since the polynucleotide chain has four bases (A, T, G, C), four cutters can have 4^4 = 256 bp average fragment length and six cutters 4^6 = 4096. The average frequency of these fragments is

$0.25^4 = 0.0039$ and $0.25^6 = 0.000244$, respectively. However, these predictions are valid only if the distribution of the bases is random, but that is not the case in the coding sequences, e.g., in the ca. 49.5 kb λ DNA 12 EcoRI fragments would have been predicted but only five have been observed. Four, six cutter indicates that the enzyme cleaves the substrate at a 4 or 6 nucleotide-specified site. ►restriction enzymes, ►restriction fragment

Restriction Landmark Genomic Scanning (RLGS): This method detects cleavable restriction enzyme sites (e.g., NotI) in the genome by direct labeling and high-resolution two-dimensional electrophoresis. NotI is methylation sensitive and does not cleave at methylated sites. Tumor tissues usually display chromosomal aberrations and that may alter the restriction sites and the tissue specific differences in RLGS may be characteristic of the cancer. ►restriction enzyme, ►two-dimensional gel electrophoresis; Hirotsune IH et al 1994 DNA Res 1:239.

Restriction Map: ►RFLP

Restriction Mediated Integration: ►REMI

Restriction-Modification: The bacterial restriction enzymes are endonucleases and the modification enzymes are methyltransferases that recognize the same nucleotide sequence as the endonuclease and transfer a methyl group from *S*-adenosyl methionine either to C-5 of cytidine or to cytidine-N^4 or to adenosine-N^6. For example, HpaII cuts C↓CGG and methylates $C^{m}CGG$, TaqI cuts T↓CGA and methylates $TGC^{m}A$. The biological purpose of this complex is to destroy invading nucleic acids (phages) by cleaving the foreign DNA with the aid of the restriction endonuclease(s), i.e., restrict the growth of the invader and the same time protect the bacterium's own genetic material by methylation. ►restriction enzymes, ►methylation of DNA, ►anti-restriction, ►methyltransferases, ►DNA uptake sequences; Kobayashi I 2001 Nucleic Acids Res 29:3742; http://rebase.neb.com/rebase/rebase.html.

Restriction Point: ►R point, ►checkpoint, ►cell cycle, ►cancer, ►commitment

Restriction Site: Refers to the site where the restriction enzyme cleaves. ►cloning site

Restrictive Conditions: These conditions do not permit the growth or survival of some specific conditional mutants. ►conditional mutation, ►permissive conditions

Restrictive Transduction: ►specialized transduction

Resveratrol (3,5,4′-trihydrostilbene): This is an antioxidant and an anti-inflammatory plant product. The beneficial effect of red wines is attributed to this phytoalexin. Resveratrol is an activator of sirtuin and can rescue polyglutamine-caused symptoms in *Caenorhabditis* as well in the neurons of mice (Parker JA et al 2005 Nature Genet 37:349). Resveratrol improves the health and survival of mice on a high-calorie diet (Baur JA et al 2006 Nature [Lond] 444:337). Resveratrol activates sirtuin and PGC-1α, improves mitochondrial functions and protects against metabolic disease (Lagouge M et al 2006 Cell 127:1109). ►lipoxygenase, ►phenolics, ►phytoalexins, ►sirtuin, ►PGC

RET Oncogene (Rearranged during transfection): This is in human chromosome 10q11.2. In *Drosophila* its homolog is *tor* (see morphogenesis in *Drosophila*). The protein product is a tyrosine kinase, essential for the development of the nervous system; it is a signaling molecule for GDNF. Mutations at the RET locus may be responsible for familial medullary thyroid carcinoma (FMTC), multiple endocrine neoplasia (MEN2A and MEN2B) and Hirschsprung's disease and may involve a dominant negative effect. The RET gene has five important domains: cadherin-binding, cysteine-rich calcium-binding, transmembrane and two tyrosine kinase (TK) domains. The main course of the RET-activated signal pathway is: RET receptor→Grb2→SOS→RAS→RAF→MAPKK→MAPK→NUCLEUS (transcription factors). The RET regulatory function is preserved between zebrafish and humans without any similarity in sequence (Fischer S et al 2006 Science 312:276). ►oncogenes, ►tyrosine kinase, ►TCR, ►GDNF, ►endocrine neoplasia, ►Hirschsprung's disease, ►phaeochromocytoma, ►papillary thyroid carcinoma, ►Grb, ►SOS, ►RAS, ►RAF, ►MAPKK, ►MAPK, ►dominant negative, ►tyrosine receptor kinase, ►multiple endocrine neoplasia; Manie S et al 2001 Trends Genet 17:580.

Retardation: Denotes slower than normal growth and development. ►mental retardation, ►gel retardation assay

Reticulocyte: An immature enucleate red blood cell displays a reticulum (network) when stained with basic dyes. ►rabbit reticulocyte in vitro translation system

Reticulosis: A complex autosomal recessive disease involving anemia, lowered platelet count, nervous disorders, immunodeficiency, etc. The symptoms may overlap different types of leukemias. The prevalence is about 5 × 10^{-5}. Bone marrow transplantation and chemotherapy have proved to be beneficial in some cases. (See separate entries).

Reticulosis, Familial Histiocytic: This disorder involves spleen, liver and lymphnode enlargement; it is caused by mutation in the RAG1 or RAG2 genes at human chromosome 11p13. ►RAG, ►Omenn disease

Retina: This is the inner layer of the eyeball connected to the optic nerve. ►macular degeneration, ►figure of eye at Iris

Retinal: Refers to vitamin A aldehyde. ►vitamin A

Retinal Dystrophy (retinopathies): This is caused by defects in the rod photoreceptor—retinal pigment epithelial complex. The incident light activates rhodopsin which transmits the signal to the G protein transducin and then to a phosphodiesterase (PDE) leading to a decrease in the level of cGMP (required for the activity of the transducin) and the closure of Na^+ ion channels and cellular hyperpolarization. Arrestin and rhodopsin kinase regenerate the photoreceptor rhodopsin. Degeneration of the retinal pigment epithelium and the retinal rod receptors may cause blindness because of failure at any step in this system. Low levels of cytosolic cGMP caused by mutation in the cGMP activating protein may be the basis of amaurosis congenita. A defect in the photoreceptor-specific peripherin/RDS protein (located in the rod and cone photoreceptors' outer membranes) also results in retinal dystrophy, presumably because of the damage to anchoring the structures to the cytoskeleton. Another protein ROM1 (rod outer membrane) homologous to peripherin, associates within a tetrameric form and represents the major outer part of the photoreceptor. Peripherin variants have been implicated in some forms of other eye diseases such as retinitis pigmentosa, macular dystrophy and choroidal dystrophy. Injecting peripherin-2 gene using an adeno-associated viral vector can remedy the complex defect. The Bietti corneoretinal dystrophy due to a 32 kDa fatty acid-binding protein defect eventually causes night blindness and has been assigned to 4q25-4qtel.

Apoptosis may account for some retinal degeneration, which can be arrested by growth substances. Retinal damage may be repaired by transplantation of committed progenitor cells from the peak of the rod genesis. These transplanted cells integrate, differentiate into rod photoreceptors, form synaptic connections and visual function in mice (MacLaren RE et al 2006 Nature [Lond] 444:203). ►arrestin, ►rhodopsin, ►transducin, ►ion channels, ►peripherin, ►phosphodiesterase, ►signal transduction, ►amaurosis congenita, ►Oguchi disease, ►night blindness, ►Stargardt disease, ►eye diseases, ►choroid, ►Usher syndrome, ►apoptosis; Rattner A et al 1999 Annu Rev Genet 33:89; Allikments R 2000 Am J Hum Genet 67:793.

Retinitis Pigmentosa (RP): A group of human autosomal recessives (84%), X-linked recessive (6%, Xp11.3 and Xp21.1) or autosomal dominant (10%, 13q14, 8q11-q13) or mitochondrial diseases entailing visual defects and blindness with an onset during the first two decades of life and a prevalence rate in the range of 10^{-4}. The Xp21 (RP6) gene may form a contiguous gene syndrome in that region. Only a small fraction of cases of pigmentary defect is associated with the rhodopsin receptor, a G-protein receptor (RGR). Mutation in the TULP1 gene, expressed only in the retina, may also involve RP. (Tulp-like proteins occur in vertebrates, invertebrates and plants). Several other diseases, e.g., congenital deafness (Usher syndrome), hypogonadism, mental retardation, other neuropathies and mitochondrial deficiencies may also involve similar defects. Nearly 40 genes are probably involved in RP symptoms. A "digenic retinitis pigmentosa" is due to mutations in the unlinked loci of peripherin-2 (6p21-cen) and the ROM1 (retinal rod outer segment protein-1) in chromosome 11q13 (see Fig. R47) (Loewen C et al 2001 J Biol Chem 276:22388). Mutations in the α subunit of a cGMP phosphodiesterase (human chromosome 5q31.2-q34) gene, PDEA (phosphodiesterase A) may also cause retinitis pigmentosa. In the autosomal dominant disease, when there is single amino acid replacement in rhodopsin (His→Pro) ribozymes may discriminate against the mutant rat mRNA and destroy it when the photoreceptors are transduced by adenovirus-associated ribozyme constructs equipped with a rhodopsin promoter. The rhodopsin-like OPN2 gene has been assigned to 3q21-q24. Recessive mutations at 1q31-q32.1 also cause RP12 due to photoreceptor degeneration. The latter gene is homologous to that of *crumbs* in *Drosophila*. The MERTK receptor tyrosine kinase (2q14.1) may also be responsible for RP. A RP GTPase regulator-interacting gene (RPGIP1) has been mapped to 14q11. Dominant mutations at 17p13.3, 1q21.1 and 19q13.4 encode pre-mRNA splicing factors. Inosine monophosphate dehydrogenase gene type 1 (chromosome 7q) deficiency may also be responsible for dominant RP. ►Lawrence-Moon syndrome, ►Usher syndrome,

►Stargardt disease, ►rhodopsin, ►retinoblastoma, ►choroidoretinal degeneration, ►eye diseases, ►ribozyme, ►contiguous gene syndrome; Bennett J 2000 Curr Opin Mol Ther 2:420; Phelan JK, Bok D 2000 Molecular Vision 6:116; McKie AB et al 2001 Hum Mol Genet 10:1555; Vithana EN et al 2001 Mol Cell 8:375; Chakarova CF et al 2002 Hum Mol Genet 11:87; Kennan A et al 2002 Hum Mol Genet 11:547.

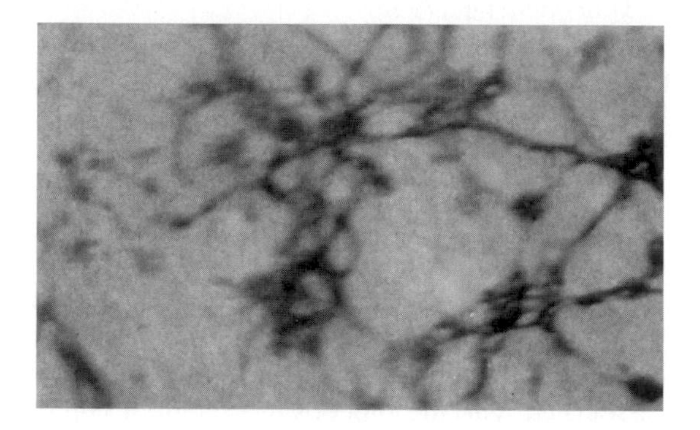

Figure R47. Part of the optic fundus of an eye with dark pigments indicating defects in the photoreceptor cell layer in retinitis pigmentosa. (Courtesy of the March of Dimes Foundation)

Retinoblastoma (RB): This is a tumor arising from the retinal germ cells, a glioma of the retina. The overall incidence is about $1–6 \times 10^{-5}$ per birth, expressed within the first two years. Almost 40% of the cases are genetically determined. The estimated mutation rate is within the range of 10^{-5} to 10^{-6}. The bilateral form is generally familial whereas the unilateral cases may be due to new mutations. A dominant RB gene has been assigned to human chromosome 13q14. The human RB1 allele may express meiotic drive, detectable by sperm typing. The mouse homolog Rb-1 is in mouse chromosome 14. Apparently RB is more frequently traced to the paternal chromosome 13 than to the maternal one. The incidence of sporadic RB is increasing with parental age. It appears that tumor growth factor B is absent in RB cells. The retinoblastoma protein (or homologs) may play a general role in tumorigenesis upon phosphorylation. RB is frequently associated with small cell lung carcinoma (SCLC), osteosarcoma, bladder cancer, breast cancer, leukemia and other types of malignancies and esterase deficiency (in deletions). Retinoblastoma is the first type of cancer with recognized recessive inheritance in humans. About 5–10% of the retinoblastomas are associated with deletions at chromosome 13q14 and rearrangements involving that site. RB binding proteins (RBBP), with homology to the E7 transforming protein of a papilloma virus and to the large T antigen of SV40, have been identified. RBBP2/JUMONJI/JARID1A (human chromosome 12p11) demethylates H3K4, a transcriptional regulator histone (Klose RJ et al 2007 Cell 128:889). The normal allele of the retinoblastoma protein appears to keep in check abnormal proliferation by limiting the activity of pol III and pol I. Besides mutation in the RB gene, tumor suppressor p53 is also inactivated in retinoblastoma (Laurie NA et al 2006 Nature [Lond] 444:61). The retinoblastoma protein stimulates the transcription of several genes primarily by activating the glucocorticoid receptors. RB plays a decisive role at the G1 restriction point decisions (through RAS) in the cell cycle regarding differentiation or continuation of cell divisions. Transcription factor family E2F interacts with RB protein and regulates the cell cycle and cyclins. Retinoblastoma may be unilateral (the majority of the non-hereditary cases) or bilateral (about two-thirds of the hereditary cases). Defect in the RB gene may cause intrauteral death. The normal allele introduced into tumors by adenovirus vectors may slow down cancerous proliferation. Mutations in the retinoblastoma gene homolog of *Arabidopsis* plants cause female and male gametophytic lethality and abnormal nuclear proliferation of the central nucleus of the embryosac (Ebel C et al 2004 Nature [Lond] 429:776). ►eye disease, ►oncogenes, ►oncoprotein, ►pol III, ►deletion, ►CAF, ►papova virus, ►Simian virus 40, ►binding protein, ►sporadic, ►MDM2, ►E2F, ►ARF, ►glucocorticoid, ►restriction point, ►cell cycle, ►tumor suppressor, ►$p110^{Rb}$, ►E2F, ►pocket, ►cyclin, ►p53, ►RAS, ►histone deacetylase, ►papilloma virus, ►adenovirus, ►sperm typing, ►quiescent zone; Nevins JR 2001 Hum Mol Genet 10:699; Chan SW, Hong W 2001 J Biol Chem 276:28402, test for: Richter S et al 2003 Am J Hum Genet 72:253; epigenetic silencers (methyltransferases) and Notch collaborate in developing eye tumors in *Drosophila*: Ferres-Marco D et al 2006 Nature [Lond] 439:430; http://www.es.embnet.org/Services/MolBio/rbgmdb/.

Retinoid: ►RAR

Retinoic Acid (RA): This is a carboxylic acid derivative of vitamin A; the aldehyde form is retinol (see Fig. R48). The 11-cis retinal is the light absorbing chromophore of the visual pigments (carotenoids). Retinoic acids belong to a nuclear receptor family (RARE, RJR) and act as ligand-inducible transcription factors. Retinoids play a role in the anterior-posterior pattern of development of the body axis and limbs of vertebrates (Vermot J et al 2005 Science 308:563). It may also have anti-cancer effects. RA activates RARE and PPARβ/δ, if it acts with RARE it may cause cell growth inhibition and if it acts with PPAR it induces pro-survival genes. A balance between CRABP-II (cellular retinoic acid binding protein) and PPAR (peroxisome proliferator activated receptor) determines these two opposing functions (Schug T et al 2007 Cell 129:723). ►transcription factors, ►RARE, ►opsin, ►PPAR

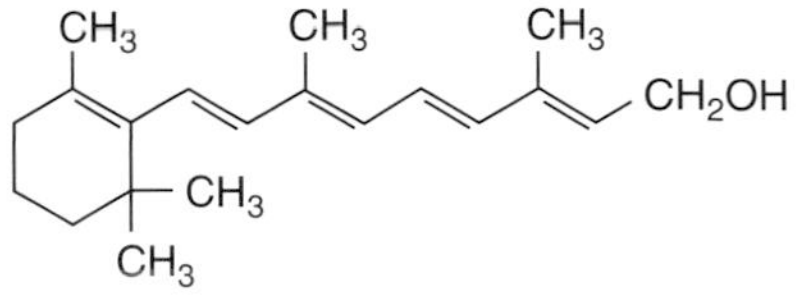

Figure R48. All-trans-retinol (vitamin A)

Retinol: Serum retinol-binding protein 4 contributes to insulin resistance and obesity and type 2 diabetes (Yanjg Q et al 2005 Nature [Lond] 436:356). ►retinoic acid, ►diabetes, formula given here.

Retinopathy: A disease of the retina. ►eye diseases

Retinoschisis: Refers to autosomal dominant, recessive or X-linked (Xp22.3-p22.1) degeneration of the retina involving splitting of that organ. ►eye diseases; Wang T et al 2002 Hum Mol Genet 11:3097.

Retra Genes: These are associated with retroviral sequences and/or transposons in the vertebrate genomes. ►retroviruses, ►transposon; Zdobnov EM et al 2005 Nucleic Acids Res 33:946.

Retroelements: ►retroposon, ►retrotransposon, ►processed pseudogene

Retrogene: This is a pseudogene with a transcriptionally active promoter. Retrogenes may be situated within the retroposon and have no introns. ►pseudogenes, ►SINE

Retrograde: This means backward. ►anterograde

Retrograde Evolution: The deletion of DNA sequences leads to adaptation to new function. Many pathogens lose genes upon becoming pathogenic. Pathogens and symbionts generally have smaller genomes than non-pathogenic or non-symbiotic relatives/ancestors. The reduction in genome size is probably the consequence of lack of need for maintenance of metabolic functions that are available in the host. Pseudogenes represent relics of genes that are no longer required. ►pseudogenes; van Ham RCHJ et al 2003 Proc Natl Acad Sci USA 100:581.

Retrograde Regulation: The expression of nuclear genes is also controlled by mitochondrial and chloroplast factors. Genes in the cellular organelles (mitochondria and plastids) are coordinated in expression. Nuclear genes encode the overwhelming majority of functions in these organelles. In plants, the photosynthetic machinery, other plastid genes, primarily Mg-protoporphyrin IX synthetic complex and the chloroplast-localized pentatricopeptide-repeat protein GUN1 (genomes uncoupled 1) is a node where different chloroplast retrograde signals converge. The GUN complex comprises nearly 450 members and is present across the various plant species. GUN1 is the central mediator of communication with the nucleus. GUN2, 3, 4 and 5 control the MG-protoporphyrin IX system. GUN1 collects the signals and conveys them to the nuclear gene ABI4 and by binding t promoters in the DNA regulates RNA in the nucleus. Defect(s) or stress in the chloroplast leads to repression of some nuclear genes (Koussevitzky S et al 2007 Science 316:715). Similar retrograde regulation is also expressed through the mitochondria. ►chloroplasts, ►chlorophyll, ►pentatrico, ►protoporphyrin, ►photosynthesis; Surpin M et al 2002 Plant Cell 14:S327; retrograde mitochondrial signaling in yeast: Liu Z, Butow RA 2006 Annu Rev Genet 40:159.

Retrohoming: ►intron homing

Retroid Virus: This has a double-stranded DNA genome (e.g., cauliflower mosaic virus, hepadnaviruses) which replicates the DNA with the aid of a RNA intermediate. ►cauliflower mosaic virus, ►hepatitis B virus, ►retroviruses

Retromer: A complex of five proteins: Vps35p, 29p, 26p, 17p and 5p. These proteins are involved in sorting endosomes to the trans-Golgi network and retrieval of the cation-independent mannose-6-phosphate receptor. Nexins are involved in the process of sorting. Retromers are essential for mammalian development. Retromers also function in yeast. ►endosome, ►Golgi; Griffin CT et al 2005 Proc Natl Acad Sci USA 102:15173.

Retron: This is responsible for the synthesis of msDNA. Retrons have several elements: the transcriptase *ret*, the coding regions of msDNA, msdRNA and also requires RNase H. RNase HJ is needed for the maintenance of the proper structure and the termination of the transcription. ►msDNA, ►RNase H, ►reverse transcriptase, ►retronphage; Lampson B et al 2001 Progr Nucleic Acids Res Mol Biol 67:65.

Retronphage: These retrons are parts of different proviruses, e.g., one in *E. coli* inserts within the selenocysteyl gene (*SelC*). ►retron, ►selenocysteine

Retroposon (non-LTR retrotransposon): This transposable element is mobilized through the synthesis of RNA that is again converted to DNA by reverse transcription before integration into the chromosome. Some of the retroposon-derived elements acquired essential functions during evolution. An in silico assay of human genes for transcriptional activity revealed that more than 1,000 retrocopies were transcribed and ~120 developed into bona fide genes (Vickenbosch N et al 2006 Proc Natl Acad Sci USA 103:3220). The *Peg10* (paternally expressed, imprinted gene) of retroposon origin in the mouse is essential for placental development and embryo survival beyond day 10.5 post coitum (see Fig. R49). This gene is present in a wide variety of animals, except chicken and the *Fugu rubripes* fish (Ono R et al 2006 Nature Genet 38:101).

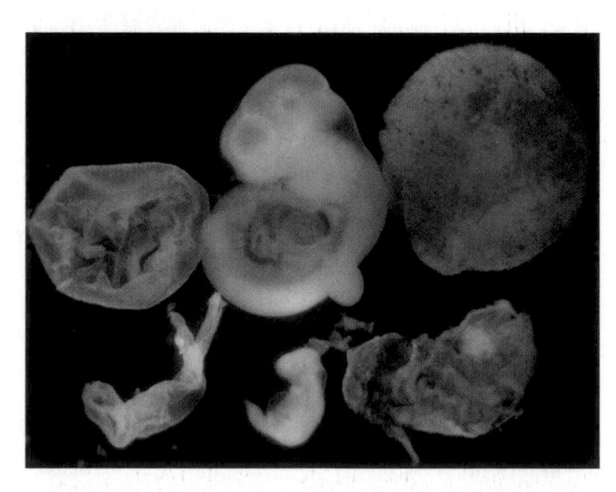

Figure R49. The expression of the *Peg10* gene (10.5. p.c.) wild type (upper row) and the knockout in mouse embryonic development (yolk sac, embryo, placenta). Courtesy of Drs. Fumitoshi Ishino and Ryuichi Ono

The retroposon may be a viral element or it may have originated from an ancient viral element. Retroposons are the long interspersed elements (►LINE) and the short interspersed elements (►SINE), the copia elements in animals and several others of the hybrid dysgenesis factors, and they also occur in several species of plants. The majority of the plant retroposons have lost their ability to move. Retroposons can be distinguished from retrotransposons, the former do not have long terminal repeats (LTR) whereas the latter do. ►retroviruses, ►reverse transcriptase, ►copia, ►hybrid dysgenesis, ►transposable elements, ►LINE, ►SINE, ►processed pseudogene, ►knockout, ►post coitum; Wilhelm M, Wilhelm FX 2001 Cell Mol Life Sci 58:1246.

Retro-Proteins: These have inverse folding. Although folding should theoretically depend only on the amino acid sequence, they may have altered stability. The majority of genetic inversions are detrimental but the deleterious effect may not be the consequence of misfolding of proteins. ►inversions

Retropseudogene: ►processed pseudogene

Retroregulation: RNase III may degrade mRNA from the 3′ end; some mutations in temperate phages may prevent this degradation and thus permit the translation of the mRNA. This control, operating from the end (downstream) forward (upstream), is known as retroregulation. ►retroviruses, ►regulation of gene activity, ►ribonuclease III

Retrosequence: This is transcribed by reverse transcriptase.

Retrospective Study: This study involves the history of diseased individuals or individuals exposed to disease in the past and compares them to concurrent control cohorts. ►prospective study

RetroTet-Art Vector: This has been designed to modulate the expression of the transgene by employing the tetracycline inducible system as well as the p16 growth arrest protein (Activators and repressors expressed together) so that during gene therapy the expression can be varied according to requirement. ►gene therapy, ►tetracycline; Rossi EM et al 1998 Nature Genet 20:389.

Retrotranslocation: Refers to the ejection of misfolded proteins from the endoplasmic reticulum back into the cytoplasm where they are subject to proteasome-mediated degradation. The dislocation is mediated by several Derlin proteins and p97 (Lilley BN, Ploegh HL et al 2005 Proc Natl Acad Sci USA 102:14296).

Retrotransposon: Refers to retrovirus-like transposable elements with long terminal repeats and position within the genome as retroviruses, however, they lack extracellular lifestyle, i.e., they cannot move from cell to cell. In *Arabidopsis* transposon Tag1, a 98 bp 5′ terminal fragment containing a 22 bp inverted repeat and four copies of the AAACCX 5′ subterminal repeat is sufficient for transposition, but a 52 bp 5′ fragment with only one subterminal repeat is not. At the 3′ end, a 109 bp fragment containing four copies of the most terminal 3′ TGACCC repeat, but not a 55 bp fragment, which has no subterminal repeats is sufficient for transposition (Liu D et al 2001 Genetics 157:817).

Usually retrotransposons lack introns because they are propagated through a RNA intermediate and the introns are removed. The *Penelope* retrotransposon of *Drosophila viridis* shows a 75 bp intron with GT/AG donor/acceptor site in the 5′ untranslated region. Similar elements have also been identified in some flatworms, roundworms, crustaceans, fishes and amphibians (Arkhipova IR et al 2003 Nature Genet 33:123).

Retrotransposons are common in the eukaryotic genomes and are apparently not distributed at random in the chromosomes. In mice oocytes retrotransposons, carrying promoters, can developmentally regulate the expression of multiple genes by serving as alternate promoters (Peaston AE et al 2004 Dev Cell 7:597). The VL30 retrotransposons in mice and humans are suppressors of leukemia oncogenes and also play a normal role in steroidogenesis (Song X, Garen A 2005 Proc Natl Acad Sci USA 102:12189).

The five *Ty* elements of *Saccharomyces* congregate in regions about 750 bp upstream of tRNA genes and *Ty5* is found at the telomeres.

Usually the sites of insertion are methylated and not transcribed and that protects the genome from insertional mutations. In a 280 kb region flanking the maize alcoholdehydrogenase gene (*Adh1*, chromosome 1L-128) 10 different retroelements were found crowded with repetition and inserted within each other. The repetitive elements of the maize genome largely represent retrotransposons and constitute at least 50% of the total nuclear DNA. The size of the repeats varies from 10 to 200 kb and they are distributed throughout the genome. The maize retrotransposons with a very high copy number (10,000 to 30,000) usually do not cause insertional mutations. The elements with a small copy number (1–30) preferentially move into genic sequences. The *Arabidopsis* genome, which is about 1/20th of that of maize, contains almost 20 retrotransposons but only with 5–6 copies. In *Vicia faba* plants with a genome size of 13.3 pg (about 1.3×10^{10} nucleotide pairs), 10% of the genome comprises retrotransposable elements but in related species their number is much smaller. In *Allium* and other plants the elements are mainly in the centromeric and telomeric heterochromatin regions whereas in other organisms they may be dispersed (see Table R7). The plant retroposons (with the exception of the Tnt1 of tobacco) fail to move because their transposase is pseudogenic. The major difference between retrotransposons and retroviruses is that the former do not produce envelope (Env) protein. ►retroviruses, ►insertion elements, ►retroposon, see Figs. R50 and R51, ►transposable elements, ►transposons, ►hybrid dysgenesis, ►methylation of DNA, ►Ty, ►copia. It was expected that retroposons and retrotransposons (retroelements) would constitute about 5 to 10% of the human genome but the sequencing data revealed a substantially higher proportion. ►retroposons, ►SINE, ►LINE, ►Ty, ►transposable elements, ►processed pseudogene; Kumar A, Bennetzen J L 1999 Annu Rev Genet 33:479; Wang J et al 2006 Human Mutat 27:323; long terminal repeat (LTR) finder: http://tlife.fudan.edu.cn/ltr_finder/; retrotransposon and L1 signature finder: http://www.riboclub.org/cgi-bin/RTAnalyzer/index.pl.

Table R7. Retrotransposon and retrotransposon-like elements in *Drosophila* (*copia*), *saccharomyces cerevisiae* (*TY 912*), and *Arabidopsis* (*tag-3*) all make 5 bp target site repeats. Their long direct terminal repeats are of different length yet they have highly conserved sequences (aligned in bold). Their internal domains of different length still has a few similarities although *copia* is highly mobile, whereas *tag-3* is no longer moving because its transposase gene underwent too many changes (pseudogenic). (After Voytas & Ausubel 1988 Nature [Lond] 336:242)

Element	5′Long terminal repeat	Internal domain	3′Long terminal repeat
Copia	**TGTTGGA**...TACAA**CA**	G**GTTATGGGCCCAGTC**...TTGA**GGG**GGCG	**TGTTGGA**...TACAA**CA**
	276 bp	4190 bp	276 bp
TY912	**TGTTGGA**...TTTCT**CA**	TG**GTA**GCGCCCTGTGCT...TAT**GCG**TGGTA	**TGTTGGA**...TTTCT**CA**
	334 bp	5250 bp	334 bp
Tag-3	**TGTTGGA**...GGTAA**CA**	AGTG**GTA**TCAGAGCCA....AAGG**TGG**AGAT	**TGTTGGA**...GGTAA**CA**
	514 bp	4190 bp	514 bp

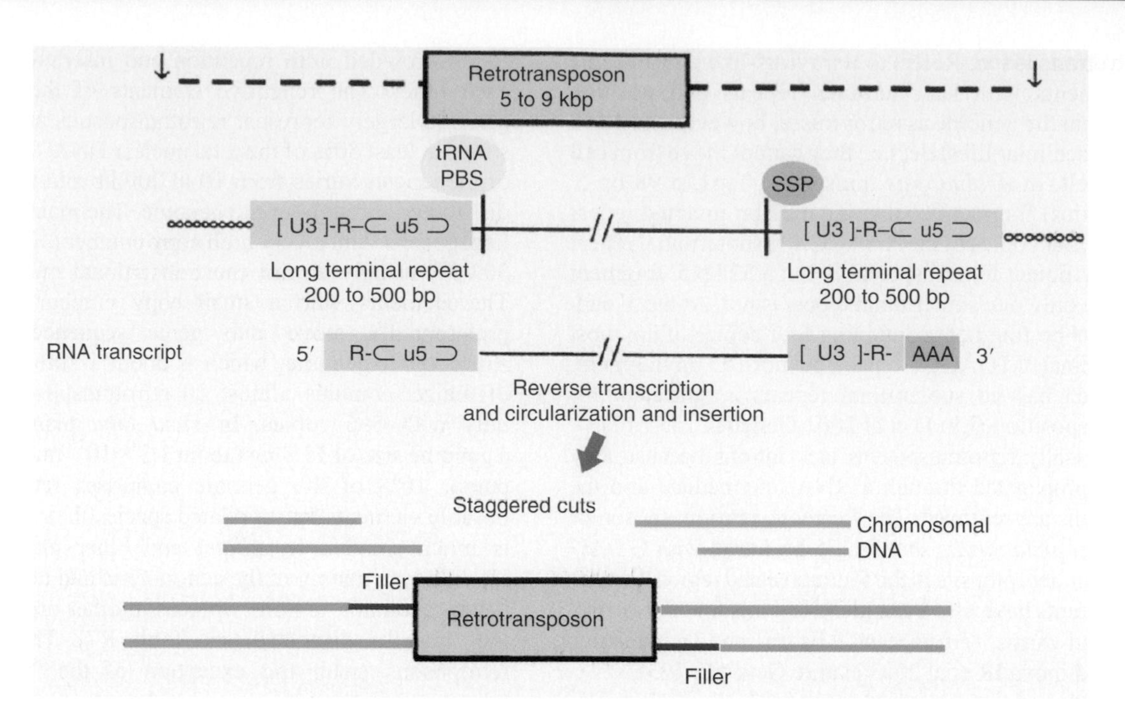

Figure R50. A DNA copy of a retrotransposon in the chromosome is replicated through a RNA transcript. The replication begins at the R segment of the 5′ long terminal repeat (LTR) and proceeds through segment U5 and the non-repetitive internal region of the element (shown within the vertical lines) toward the 3′ boundary of the R segment of the long terminal repeat. The first strand of the DNA is synthesized by the reverse transcriptase, encoded within the retrotransposon and is primed by the CCA-3′-OH end of one or another kind of host tRNA that pairs by complementarity to the upstream LTR. The reverse transcription also employs a protein-tRNA complex that binds to the tRNA-protein site (tRNA-PPS). The second strand is primed at the SSP site (second strand primer). At the target site, duplication occurs

Retroviral Recombination: Retroviruses have dimeric RNA genomes which are transcribed into double-stranded DNA after productive infection. Recombination occurs during the synthesis of the plus and minus strands. Complementarity between the palindromic sequences at the dimerization initiation sites of the hairpins is required for the two RNA chromosomes. The 5′ untranslated sequence forms a kissing-loop structure (Figure R51) that contains essential replication elements. Increased homology and proximity of the homology tracts promote template switching. The diagram depicts a perfect palindrome but the strands may not be perfectly complementary along their entire length. Circles indicate markers. ►retrovirus, ►kissing loop, ►palindrome; Mikkelsen JG et al 2004 Nucleic Acids Res 32:102.

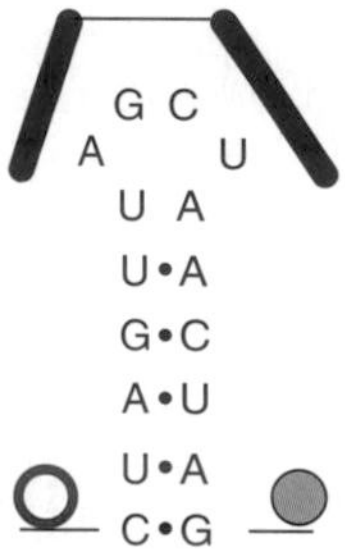

G C
A U
U A
U•A
G•C
A•U
U•A
C•G

Figure R51. Kissing loop

Retroviral Restriction Factors: These mammalian proteins interfere with effective infection or propagation of the retrovirus in the mammalian host cell. The APOBEC3G and ZAP inhibit the replication of the viral nucleic acid. The Friend virus susceptibility factor (Fv1) targets the incoming retroviral leukemia virus capsid. The TRIM5 factor group protects against HIV-1 and murine leukemia virus by acting after the entry of the virus into the host cell. ►retrovirus, ►TRIM, ►leukemia inhibitory factor, ►APOBEC3G, ►ZAP; Hatziioannou T et al 2004 Proc Natl Acad Sci USA 101:10774.

Retroviral Vectors: These are capable of insertion into the chromosomes of a wide range of eukaryotic hosts. The expected important features are: (i) efficiency and selectivity for the target, (ii) safety, (iii) stable maintenance, and (iv) sufficient expression for the purpose employed. Generally, the *gag, pol* and *env* genes are removed but all other elements required for integration and RNA synthesis and the ψ packaging signal are retained. Usually a selectable marker neomycin [*neo*], or dehydrofolate-reductase [*dhfr*] resulting in methotrexate resistance (MTX), or bacterial hypoxanthine-phosphoribosyltransferase [*hprt*], or guanine-hypoxanthine-phosphoribosyl-transferase [*gpt*], or mycophenolic acid (MPA) resistance is inserted in such a way that its initiation codon (ATG) falls into the same place as the ATG of the group-specific antigen protein (*gag*) gene. Transcription will be initiated at the 5′ long terminal repeat (5′LTR) and translation will proceed from the ATG that is at the same place as that of *gag* (see Fig. R52).

The generalized vector shown here is non-infectious and requires superinfection by a helper virus. There are several other types of vector designs. After the DNA has been packaged into a virus particle the transfer of the gene is almost fully efficient as long as the cell has the appropriate receptor. By engineering the knob on the viral surface, the vector may be targeted to special cells. The recombinant DNA integrates into the host as a provirus with aid of the *integrase* protein encoded also by the *pol* gene. The most commonly used murine leukemia virus vector (MuLV) has about 6–8 kb carrying capacity. It is desirable to target the integrase to sites where the danger of insertional mutation is minimized in case the vector is used for therapeutic purposes. This goal may be difficult to achieve. Targeting to the sites of RNA polymerase III attachment with the aid of the yeast *Ty* transposons has been considered. This polymerase (pol III) transcribes ribosomal RNA genes that are present in multiple copies and thus pose a lower risk for deleterious mutations. A better approach may be using targeted homologous recombination. Targeting the transgene may be refined by the use of a cell- specific promoter.

The provirus is generally present in a single copy per cell but this *producer cell* will proceed with the production of recombinant retrovirions. The production of the recombinant virus is enhanced if initially the cell is also co-infected by a wild type proviral plasmid. However, such a system has a disadvantage because of the presence of a helper virus, which competes with a pseudovirion, and it may also be hazardous by being pathogenic. The problems related to the helper virus can be eliminated if it retains all the viral genetic sites except the packaging site (ψ) and is therefore unable to produce infectious particles. There are several means of preventing reconstitution of an infectious viral particle which can cause disease.

Broad host range (amphotropic) viral vectors can be constructed by replacing a narrow range (ecotropic) viral envelope protein with another of an amphotropic virus. The Env protein mediates the virus/vector entry into the cell through the Pit-2 receptor. Env determines the host range. The host range may be extended by pseudotyping, i.e., co-infection of the cells with two different viruses, which results in a mixed envelope glycoprotein. Improved targeting of the retrovirus may be achieved by attaching cell specific ligands or antibodies to Env. Some of the Env proteins may elicit spongiform encephalomyelopathy. Caution must be exercised because the vector or host cell DNA and the helper virus genetic material may recombine and give rise to infectious particles. By genetic engineering additional modifications have been made in the helper virus to prevent the formation of infectious single recombinants. This has been achieved by replacing, e.g., the 3′ LTR with a termination stretch of the SV40 eukaryotic virus DNA.

The *retroviral expression vectors* are more useful because they not only prove that the viral vector is present in the cell, but also propagate desirable genes (e.g., growth hormone genes and globin genes). Since retroviral vectors may carry strong promoters and enhancers within the LTR region, they can over-express some genes. They may also be employed for insertional mutagenesis because they can insert at different chromosomal locations and can be used as tools to study animal differentiation and morphogenesis. The most commonly used retroviral vectors have been derived from the Moloney murine leukemia virus (MoMuLV). In some instances the retroviral vectors within the host are silenced because the viral promoter in the 5′ LTR is methylated by the host enzymes. This *promoter shutoff* may be avoided by using eukaryotic promoters. In some instances the inclusion of a locus control region (LCR) of the host may boost the expression of the transgene. The use of insulators may protect the transgene from host repressors. Also, a heterologous promoter that may drive more efficient expression (SIN vector, double copy vector) may replace the viral promoter. Retroviral vectors may permit the utilization of internal ribosomal entry sites (IRES) to drive the expression

---- pBR322 --- 5′ LTR PBS SD ψ ATG SELECTABLE MARKER 3′ LTR --- pBR322

Figure R52. Retroviral vector. pBR322: bacterial plasmid, 5′ LTR: long terminal repeat, PBS: binding site for primer to initiate first-strand DNA synthesis, SD: splicing donor site, Ψ: packaging signal for the virion, ATG the first translated codon (Met) of the selectable marker (e.g., antibiotic resistance), 3′ LTR: long terminal repeat of more than one gene. By using tissue-specific promoters the expression of the transgene may be limited to a certain cell type, e.g., in tumor cells. The promoter may also be selected for inducibility so that the expression may be regulated by the supply of glucocorticoids or a metal, or tetracycline, etc. Retroviral vectors are useful for gene therapy because they integrate stably into human and animal cells and are normally transmitted through mitoses. ►vectors, ►retroviruses, ►viral vectors, ►double-targeted vector, ►epitope, ►SV40, ►mycophenolic acid, ►MoMuLV, ►lentiviral vectors, ►foamy viral vectors, ►ψ, ►SIN vector, ►E vector, ►double-copy vector, ►IRES, ►packaging cell lines, ►gene therapy, ►gene marking, ►pseudotyping, ►Ty, ►pol III, ►LCR, ►insulator, ►inducible gene expression, ►cancer gene therapy, ►biohazards, ►laboratory safety; Yee J-K 1999, p 21 In: Friedmann T (Ed.) Development of Human Gene Therapy, Cold Spring Harbor Laboratory Press; Smyth TN (Ed.) 2004 Gene and Gene Therapy, Marcel Dekker, New York; Somia N 2004 Methods Mol Biol 246:463.

Retroviruses (Retroviridae): These include onco-, lenti- and spumaviruses. In eukaryotes they contain dimeric single-stranded RNA, which is processed by a stringent switch (Badorrek CS et al 2006 Proc Natl Acad Sci USA 103:13640) as the genetic material that is replicated through a double-stranded DNA intermediate with the aid of a reverse transcriptase enzyme. Retrovirus reverse transcription usually takes place after the host has been infected. In the human foamy virus (spumavirus), the infectious particles already carry a double-stranded DNA, indicating that reverse transcription precedes infection. Besides the polymerase gene (*pol*), they all carry group- specific antigen (*gag*) and envelope (*env*) protein genes. These three viral components are active in trans whereas the other elements require cis position. A provirus introduced into a cell by retroviral infection may become a retrotransposon (e.g., Ty element in yeast, copia in *Drosophila*). They are characterized by long terminal repeats (LTR), measuring a few thousand nucleotides in length. For infection to proceed, the viral capsid protein fiber and knob must find an appropriate receptor on the surface of the target cell.

When the virus enters the host cell its RNA genetic material is converted into double-stranded DNA, and it may be covalently integrated into a host chromosome. Although the targets for integration are spread all over the genome, transcriptionally active regions are preferred. Integration "hot spots" in chicken cells may be used by RSV a million times more often than expected by chance alone, and some sequences in mouse cells (e.g., the HGPRT gene) may be avoided. After a cell is infected by a single virion, in a day, thousands of viral particles may be produced.

Retroviruses are considered to be diploid (dimeric) because they have a pair of genomes, two identical size RNAs of 7 to 9 kb. All retroviruses minimally encode a protease, a polymerase, ribonuclease H activity of the reverse transcriptase and an integrase function. A generalized structure of retroviruses is presented in Figure R53 (individual types may display variations of this scheme):

The three proteins: gag (group-specific antigen) a polyprotein, pol (polymerase, reverse transcriptase), env (envelope protein) are transcribed in different, overlapping reading frames and then for RNA packaging into virion. The genomic subunits are the same as the mRNA, i.e., they are (+) strands. The transcript RNA has a 7-methyl-guanylate group at the 5′ end and a 100 to 200 polyA tract at the 3′ end, similar to eukaryotic mRNAs. The different retroviruses code for various proteins with known or yet to be identified functions. The gag-pol polyprotein complex contains

Figure R53. At the two ends of the viral genome are the *long terminal repeats* (LTR) of 2 to 8 kb. At the left and right termini of the LTRs are *attU3* and *attU5*, respectively for the attachment of the U3 (170 to 1,200 nucleotides) and U5 (80 to 120 nucleotides) direct repeats of the provirus in the host DNA. The *att* sequences at the 3′ end of U5 and at the 5′ end of U3′ contain usually imperfect, inverted repeats where viral DNA joins the host DNA at 2 nucleotides from these ends. The terminal repeats represented by *R* (10 to 230 nucleotides) are used for the transfer of the DNA during reverse transcription. *E*: transcriptional enhancer, *P* promoter, *PA*: signal for RNA cleavage and polyadenylation. *PBS*: binding site for tRNA primer for first strand DNA synthesis (different retroviruses use the 3′-OH end of different host tRNAs for initiation). *PPT:* polypurine sequences, which prime the synthesis of the second strand of DNA. *SD* splice donor (the site where *gag* and *pol* and *env* messages are spliced). *SA*: splice acceptor site (the site where the second splice site joins to the first [donor] site)

information on proteolytic activities that generates a *protease* from the carboxyl end of gag and some other proteins. *Reverse transcriptase* (RT) and *integrase* (IN) and a protease are generated by proteolysis from the translated *pol* gene product. The SU product of the *env* gene recognizes the retroviral receptors which are transmembrane proteins of the host. The nucleocapsid proteins remain attached to the proviral DNA and facilitate the integration of the viral DNA into the host DNA. Some of the murine leukemia viruses can enter the cell nucleus only when the nuclear membrane breaks down during mitosis. Some of the lentiviruses can enter the nucleus with the assistance of the integrase and other proteins. In the human foamy virus the pol protein is translated by splicing mRNA that does not include the gag domain. The reverse transcriptase varies among the different retroviruses. The Rous sarcoma virus RT contains a RNA and DNA-directed polymerase, a RNase H as well as a tRNA-binding protein. It works with either RNA or DNA primers and synthesizes up to 10 kDa molecules from single RNA priming sites. For transcription of the viral proteins, the host RNA polymerase II is utilized. The host machinery translates the viral transcripts. The proteins may be processed in different ways by proteolysis to become functional.

Actinomycin D inhibits the replication on the DNA but not on the RNA template. Azidothymidine (AZT) inhibits polymerization and viral replication. The RNase H activity removes RNA in both 5′ to 3′ and 3′ to 5′ and digests the cap, tRNA and the polyA tail. All retroviruses have an integrase protein derived from the C end of the gag-pol polyprotein complex. Integrase (30 to 46 kDa protein with Zn fingers) inserts the virus into the eukaryotic chromosomes (see Fig. R54).

At the site of integration there is target-duplication as the recessed ends of the target filled in by complementary nucleotides (*in italics*) (see Fig. R55). Inside the chromosomes of the host the retroviral *provirus* still carries the LTRs at both ends. After transcription, viral RNAs are produced. Following proteolytic cleavage, the polyproteins can be converted into viral proteins with the assistance of the cellular machinery. Viral particles (virions) may be assembled from the viral RNA and viral proteins at the surface of the cell and virions can exit from the cell membrane by "budding" and can infect new cells with two single-stranded RNA copies through appropriate cellular receptors.

The retroviral genome has 10^5 times increased mutability compared to cellular genes. Retroviral genomes recombine with a frequency of 10 to 30% during each cycle of replication. Integration may have a profound effect on cellular genes; it may inactivate suppressor genes controlling cellular proliferation or it may activate the transcription of cellular genes with the same effect and thus initiate carcinogenesis.

The major types of retroviruses are: (i) bird's (avian) sarcoma and leukosis viruses such as Rous sarcoma virus (RSV), avian leukosis virus (ALV) and Rous-associated viruses (RAV 1 and 2), (ii) reticuloendotheliosis viruses (hyperplasia of the net-like

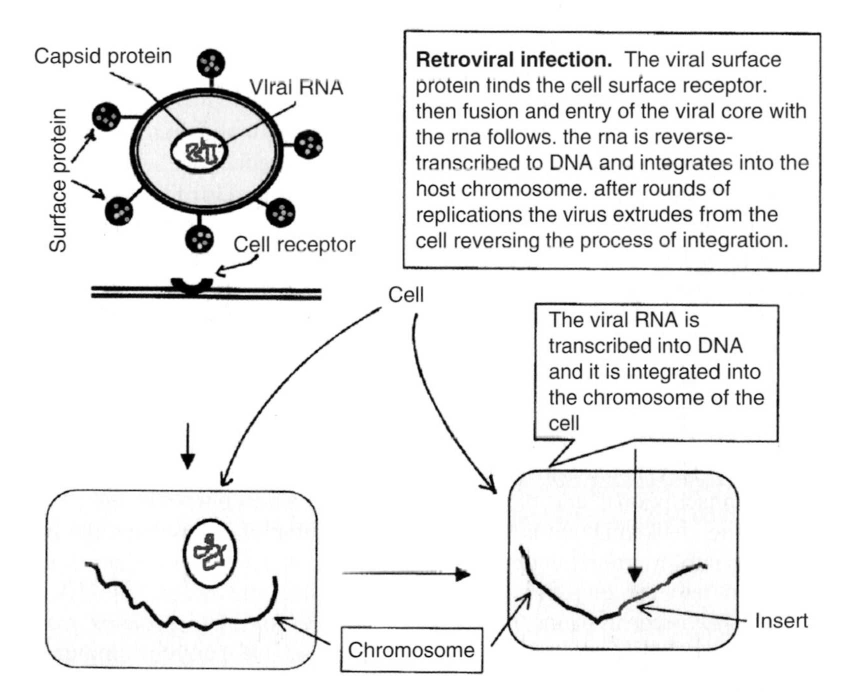

Figure R54. Retroviral integration

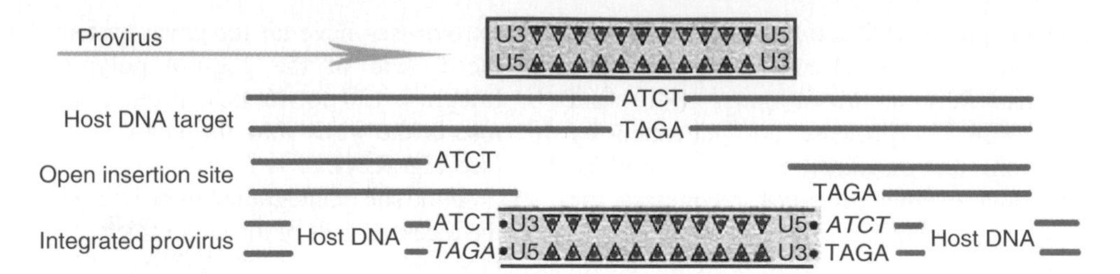

Figure R55. Integration of the provirus into the host DNA

and endothelial [tissues lining organ cavities]), e.g., spleen necrosis virus (SNV), (iii) mammalian leukemia and sarcoma viruses, e.g., Moloney murine sarcoma virus (Mo-MSV), Moloney murine leukemia virus (MoMuLV), Harvey murine sarcoma virus (Ha-MSV), Friend spleen focus-forming virus (FSFFV), feline leukemia virus (FLV), and simian sarcoma-associated virus (SSAV), (iv) mammary tumor viruses (MMTV), (v) primate-type D viruses, e.g., Mason-Pfizer monkey virus (MPMV) and simian retrovirus (SRV-1), (vi) human T-cell leukemia-related viruses (HTLV-1 and HTLV-2), simian T cell leukemia virus (STLV) and bovine leukemia virus (BLV), (vii) immunodeficiency and lentiviruses, e.g., human immunodeficiency viruses (HIV-1 and HIV-2), Visna virus, simian immunodeficency virus (SIV), caprine (goat) arthritis - encephalitis virus (CAEV) and equine (horse) infectious anemia virus (EIAV). (Classification provided by Varmus H, Brown P 1989). About 1% of the human genome includes retroviral sequences (human endogenous retroviruses [HERV]). During evolution these elements presumably inserted themselves and have been extensively modified. These HERV elements gave rise to transposable elements or their pseudogenic forms in the modern genome. The endogenous viruses synthesize the gag and env proteins that not only enables them to become infective, but may also prevent their transposition. The Env protein may also interfere with the viral receptors and thus limit reinfection. The HERV do not seem to have much significance for the genome at present; however, retrotransposition may lead to mutation and loss of gene function, including loss of cancer gene suppression. The resistance alleles block the integration of the viral RNA. The HIV-1 retrovirus usually integrates within the transcriptional unit of a gene. In contrast the murine leukemia virus (MLV) preferentially integrates into or around the promoter of the gene rather than into the translational unit. Typical retroviruses do not occur in plants, however the *Athila* elements of *Arabidopsis* and the *SIRE-1* of soybean come close to retroviruses in as much as they harbor envelop-like genes. ►overlapping genes, ►retroposon, ►retrotransposon, ►retrogene, ►hybrid dysgenesis, ►copia elements, ►LINE, ►reverse transcription, ►retroviral vectors, ►oncogenes, ►tumor viruses, ►cancer, ►HTDV, ►animal viruses, ►plant viruses, ►pararetrovirus, ►nucleocapsid, ►knob; Knipe DM, Howley PM (Eds.) 2001 Fundamental Virology, Lippincott Williams & Wilkins, Philadelphia, Pennsylvania.

Retrovirus Resistance: In general this is controlled by dominant genes in the animal host. In mice a major gene is *Fv1*, which has the $Fv1^n$ allele permitting the replication of the NIH Swiss mice N-tropic viruses but blocking the replication of the B-tropic (Balb/c mouse) viruses. The Fv^b allele allows the replication of the B-tropic virus strains but blocks the N-tropic strains. The *n* and *b* alleles are codominant and their difference is only at two positions of their sequence. The $Fv1^0$ allele does not restrict any virus strains. The resistance genes block the integration of the cDNA of the viral genome into the host nucleus or host chromosome. Tropism may depend on a single amino acid difference of a domain of the group-specific antigen. High infection titer may, however, overcome the restriction. In human and some animal cells REF1 gene is credited with a function similar to that of *Fv1*. TRIM5α family of proteins controls simian resistance to HIV or other lentiviruses. Gene Lv2 provides resistance against HIV-2. APOBEC3G (apolipoprotein B mRNA-editing complex) fights the virus by deaminating the cytosine residues to uracil in the DNA and U-glycosylase generates an abasic site. The deficient provirus fails to code for the required proteins. An active *Vif* gene can, however, deactivate APOBEC3G. The BAF protein (barrier to autointegration factor) may prevent the viral DNA integration into the host chromosome.

Cyclophilin may have the opposite effect and can facilitate retroviral infection. Importin 7 protein may facilitate the access of HIV-1 to the nucleus. A number of other proteins may affect the infection process. ►acquired immunodeficiency, ►tropic, ►provirus, ►cyclophilins, ►APOBEC3, ►RNA editing; Goff SP 2004 Annu Rev Genet 38:61.

Rett Syndrome (dominant, RTT, Xq28, Xp22): This is a neurological disorder with onset after a period of normal early development. Loss of speech, motor skills, constant wringing of the hands, seizures and mental retardation are observed predominantly in girls. The prevalence rate is $\sim$1–2 × 10^{-4} among girls, and in 99.5% cases is caused by new mutations occurring predominantly in the paternal X chromosome. The biochemical basis is missense or nonsense mutation in the meCP2 gene encoding a methyl-CpG-binding protein, which regulates BDNF and thus controls the syndrome (Martinowich K et al 2003 Science 302:890). MeCP2 interacts with transcriptional activators and silencers, chromatin remodeling factors—histone deacetylase and histone methyl transferase. In Rett syndrome homologous pairing in the 15q11-13 (imprinted site of Angelman's syndrome) is impaired (Thatcher KN et al 2005 Hum Mol Genet 14:785). In some atypical cases cyclin-dependent kinase 5 gene is involved (Guy J et al 2001 Nature Genet 27:322). ►autism, ►BDNF, ►meCP2, ►methylation of DNA, ►Angelman's syndrome, ►imprinting; Wan M et al 1999 Am J Hum Genet 65:1520; Meloni I et al 2000 Am J Hum Genet 67:982; Trappe R et al 2001 Am J Hum Genet 68:1093; Chen RZ et al 2001 Nature Genet 27:327; Shahbazian MD et al 2002 Hum Mol Genet 11:115; Bienvenu T, Chelly C 2006 Nature Rev Genet 7:415.

REV1, REV3, REV7: These are subunits of DNA polymerase ζ. REV1 is involved in DNA repair mutagenesis. This polymerase—unlike others—uses aginine^{324} for template after the guanine at its place is ejected and joins it by complementary hydrogen bond to deoxycytidine triphosphate (Nair DT et al 2005 Science 309:2219). Rev1 translesion DNA polymerase is functional primarily during the G_2/M phase of the cell cycle rather than during the S phase (Waters LS, Walker G C 2006 Proc Natl Acad Sci USA 103:8971). ►DNA polymerases, ►translesion, ►SOS repair, ►Y-family DNA polymerases; Lawrence CW, Hinkle DC 1996 Cancer Surv 28:21; Murakumo Y et al 2001 J Biol Chem 276:35644; Masuda Y, Kamiya K 2002 FEBS Lett 520:88.

Rev: This splicing element was originally identified in viruses. It assists exports through the nuclear pore. ►RNA export, ►nuclear pore

Reversal of Dominance: ►dominance reversal

Reverse Array: This is a tissue lysate (rather than cells) blotted on a solid support for protein analysis. ►protein

Reverse Dot Blot: ►colony hybridization

Reverse Endocrinology: Using orphan receptors new hormones and ligands may be searched for and studied. ►orphan receptors, ►endocrinology

Reverse Genetics: ►reversed genetics

Reverse Ligase-Mediated Polymerase Chain Reaction (RL-PCR): When the beginning target mRNA is cleaved at a known location the RL-PCR method generates a product of a predictable length in the presence of an appropriate linker and a nested primer. The linker is a probe for synthesized strand. ►polymerase chain reaction, ►nested primer, ►RACE; Bertrand E et al 1997 Methods Mol Biol 74:311.

Reverse Linkage: ►affinity

Reverse Mosaicism: Secondary mutation(s) restore(s) wild type function to a nucleotide sequence without returning to the wild type nucleotide or amino acid sequence. Such somatic mosaicism may also be produced by intragenic mitotic recombination, gene conversion frameshift mutation or some type of compensatory sequence alterations in the gene as shown in the box. The substituted nucleotides are in lower case, the "mutant" is in bold and the "revertant" alteration in outline letters (see Fig. R56) (Data after Waisfisz Q et al 1999 Nature Genet 22:379).

Wild type DNA...	TTC.CTG.CTC.TGG.GCT				
Amino acids	F	L	L	W	A
Mutant DNA	TTC.CTG.CgC.TGG.GCT				
Amino acids	F	L	R	W	A
Revertant DNA	TTC.CTG.tgC.TGG.GCT				
Amino acids	F	L	C	W	A

Figure R56. Reverse mosaicism

Reverse Mutation (backmutation): This is a change from mutant to wild type allele, $a \rightarrow A$. In therexperiment shown in the photo thiamine prototrophs were selected on soil among thiamine auxotrophs (see Fig. R57). The thiamine mutants died in the absence of thiamine but the revertants grew normally. The material was genetically marked at both flanks 5 and 9 map units, respectively to verify that the apparent revertants were not contaminants. The progeny of the revertants were genetically analyzed. They segregated for auxotrophy and prototrophy in the proportion of 5:3 because at the time of the reversion the diploid germ line consisted of two diploid cells. One of the cells remained homozygous for thiamine requirement and the other became heterozygous and segregated for 3 wild type (2 heterozygotes) and for one homozygous for thiamine auxotrophy. ►Ames test, ►mutant isolation, ►mutation, ►suppressor gene

Figure R57. Reverse mutation of Arabidopsis thiamine mutant

Reverse-Phase Protein Array (RPPA): Samples to be assessed are robotically spotted, and an antibody is used to measure the amount of a particular protein present in the sample. In contrast to 2D-PAGE and antibody arrays, the reverse-phase methodology assesses only one protein per slide, but its advantage is that all the cell or tissue samples can be analyzed side by side in a single array. This is particularly useful in functional studies where protein levels are compared across samples rather than samples compared across protein types. This method—with modifications—has been successfully used for the study of cancer and normal cells and the progression of cancer development as well as prognosis. ►gel electrophoresis, ►microarray hybridization; Nishizuka S et al 2003 Proc Natl Acad Sci USA 100:14229.

Reverse Transcriptases: These enzymes transcribe DNA on a RNA template. An outline of the function of the enzymes within the protein coat using the diploid template is presented here. A similar process is followed in in vitro assays. Reverse transcriptases are commercially available from purified avian myeloblastosis (cancer of the bone marrow) cells or as cloned Moloney murine leukemia virus (MoMLV) gene product. The avian enzymes are dimeric and have strong reverse transcriptase and RNase H activities. The murine polymerase is monomeric and exhibits only weak RNase H activity. Therefore, the murine enzyme is preferred when mRNA is transcribed into cDNA. Also, RNase H can degrade DNA and may reduce the efficiency of cDNA synthesis. The temperature optimum of the avian enzyme is 42° C (pH 8.3) and at this temperature the murine enzyme is already degraded. The pH optimum for the murine enzyme is 7.6. Both enzymes have much lower activity slightly below or above the pH optima. The avian enzyme more efficiently transcribes structurally complex RNAs. Reverse transcriptases are used for generating DNA from mRNA for vector construction or generating labeling probes, for primer extension, and for DNA sequencing by the dideoxy chain termination method. Since reverse transcriptase docs not have an editing (exonuclease) function, it may make errors at the rate of 5×10^{-3} to 1×10^{-6} per nucleotide. This rate is orders of magnitudes higher than the error rate of most eukaryotic replicases. The HIV-1 reverse transcriptase can use as template either RNA or DNA; in the latter case it makes double-stranded DNA. Nucleoside analog-induced inhibition is effective for fighting HIV proliferation. This treatment, however, increases mutation in mtDNA because it also inhibits DNA polymerase γ (Martin AM et al 2003 Am J Hum Genet 72:549). ►retroviruses, ►cDNA, ►central dogma, ►msDNA, ►error in replication. Reverse transcriptase action is given in Figure R58.

Reverse transcription of the retroviruses follows the generalized scheme. A single-stranded viral RNA (vvvv) serves as template for the synthesis of the first strand DNA (→→→). The synthesis is primed by a tRNA attached to the PBS (primer-binding site of the retroviral [-] strand). The host tRNA is base-paired by 18 nucleotides to a sequence next to U5. The first strand DNA (also called strong stop minus DNA) is extended at the rate of about 2 kb per h until the last part of the primer-binding site is copied. When the synthesis is extended, the copying of the second DNA strand (←←←) begins. The process is practically the same in the cell and in vitro conditions. DNA polymerases β and γ also have some reverse transcriptase function in as much as they can copy poly(rA) by using oligo-dT primer. ►retroviruses, ►telomerase, ►transposon, ►DNA polymerases, ►mtDNA, ►HIV; Whitcomb JM, Hughes SH 1992 Annu Rev Cell Biol 8:275; Gao G, Goff SP 1998 J Virol 72:5905; Vastmans K et al 2001 Nucleic Acids Res 29:3154.

Reverse Transfection: Specific cDNAs or RNAs encapsulated in lipid are placed at defined locations on a glass slide and cells are layered on top. The cells can be transfected by the genes or by silencing RNAs. The cells isolated can then express mRNAs or

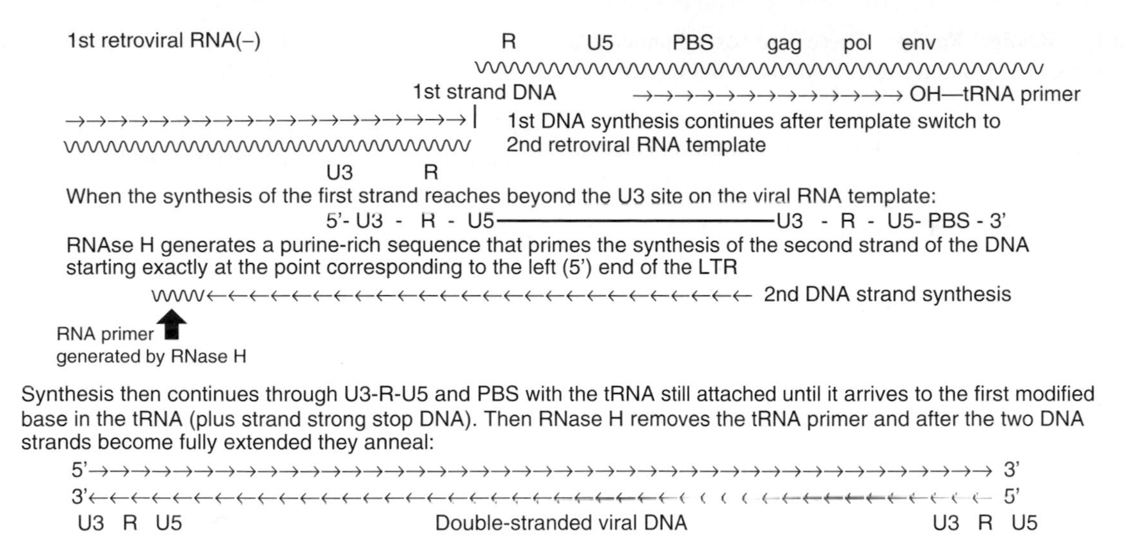

Figure R58. Reverse transcription

the silencing RNAs on microarrays. Nucleic acids when labeled by fluorescent tags can be specifically recognized. This procedure assists in the identification of drug targets or specific gene products controlling metabolism. ►transfection, ►RNAi, ►microarray hybridization; Ziauddin J, Sabatini DM 2001 Nature [Lond] 411:107.

Reverse Translation: If the protein is sequenced and the amino acids of a short segment are coded by non-degenerate or moderately degenerate codons, it is possible to synthesize a few RNAs on the basis of the presumed codon sequences, and one of them may be complementary to the DNA. This short RNA sequence can be hybridized to the DNA that codes for this particular protein. By reversing the translation and generating an appropriate probe, the gene can be isolated. ►probe, ►synthetic probe

Reverse Two-Hybrid System: The system monitors disruptions of protein-protein interactions. ►two-hybrid system

Reversed Genetics (inverse genetics): Nucleic acids and proteins, etc. are first isolated and characterized in vitro by molecular techniques and subsequently their hereditary role is identified. Also, gene expression can be studied by introducing into cells, by transformation, reporter genes with truncated upstream or downstream signals, in vitro generated mutations, site-specific recombination, targeting genes, etc., thereby determining the functional consequences of these alterations. Briefly stated, reversed genetics starts with molecular information and then deals with its biological role. Classical (forward) genetics recognizes genes when mutant forms become available and then studies their transmission, chromosomal location, mechanism of biochemical function, their fate in populations and evolution. Reversed genetics is sometimes called surrogate genetics. ►genetics, ►inheritance, ►heredity; Masters PS 1999 Adv Virus Res 53:245.

Reversion: Refers to backmutation either at the site where the original forward mutation took place or in another (tRNA) gene that may act as a suppressor tRNA or the reversion is caused by correcting the frameshift. The possibility of reversion is frequently considered as evidence that the original forward mutation was not caused by a deletion. Suppressor mutation outside the mutant locus, however, may restore the non-mutant phenotype. Reversion may also take place when the mutation was caused by a duplication and a deletion evicted the duplicated sequence. ►backmutation, ►base substitution, ►suppressor, ►*sup*, ►suppressor tRNA, ►frameshift, ►reverse mutation, ►Ames test

Reversion Assays in *Salmonella* and *E. coli* in Genetic Toxicology: The *Salmonella* assay has been described by the Ames test. The most commonly used *E. coli* test employs strains WP2 and $WP2_{uvrA}$ that are deficient in genetic repair and are auxotrophic for tryptophan. They detect base substitution revertants but the assay does not respond to most frameshift mutagens (unlike some of the *Salmonella* strains TA97, TA98, TA2637 and derivatives). The *E. coli* systems do not offer any advantage over that of the *Salmonella* assay of Ames. ►Ames test, ►bioassays in genetic toxicology, ►mutation detection

Rex Color: Refers to the color of rodent (rabbit) hair which appears in the presence of the recessive fine fur gene, *r,* in certain combinations with black (*B/b*), agouti (*A/a*) and intensifier (*D/d*).

Rex1p, Rex2p, Rex3p: These are exoribonuclease members of ribonuclease D family. ►ribonuclease D

Rexinoids: These are agonists of RXR retinoid X receptor and regulate cholesterol absorption and bile acid metabolism/transport. ►agonist, ►retinoic acid

Reye's syndrome: This disorder is characterized by non-genetic inflammation of the brain in infants and may lead to fever, vomiting, coma and eventually death. The use of aspirin is contraindicated in this condition. ►acetyl-CoA dehydrogenase deficiency

Reynaud Disease: ►Raynaud's disease

Reynolds Number: Characterizes the flow of a liquid-from laminar to turbulent flow or the other way around. Reynolds number is a characteristic of the viscous drag (resistance) on a structure through a fluid medium like the cellular plasma.

RF (release factor): A protein which mediates the release of the peptide chain from the ribosome after it recognizes the stop codons. ►translation termination, ►release factors, ►protein synthesis, ►transcription termination in prokaryotes, ►transcription termination in eukaryotes; Dontsova M et al 2000 FEBS Lett 472:[2–3]:213.

RF: A replicative form of single-stranded nucleic acid viruses (DNA or RNA) where the original single strand makes a complementary copy that serves as a template to synthesize replicas of the first (original) genomic nucleic acid chain. ►replicase, ►plus strand; Buck KW 1999 Philos Trans R Soc Lond B Biol Sci 354:613.

Rf: Refers to fertility restorer genes in cytoplasmic male sterility. ►cytoplasmic male sterility

Rf Value: In paper or thin-layer chromatography, the distance from the baseline of the migrated compound divided by the distance of migration of the solvent (mixture) is the Rf value. This value which is always less than 1 is characteristic of a particular compound within a defined system of chromatography. ►paper chromatography, ►thin-layer chromatography, see Fig. R59.

RF-A: Replication factor A is a human single-stranded DNA binding protein, auxiliary to pol α and pol δ. ►pol, ►helix destabilizing protein, ►replication, ►replication fork eukaryotes

RFA (replication factor A): This is the same as replication protein A (RPA). ►DNA replication eukaryotes

Rfam: This is the database of non-coding RNAs. ►non-coding RNA; http://rfam.janelia.org/.

RF-C: Denotes the DNA replication factor C, a primer/template binding protein with ATPase activity. It plays a primary role in replicating the leading strand DNA in eukaryotes. RF-C loads PCNA on the DNA that tethers the DNA polymerase to the replication fork. RC-F is also called Activator I. ►replication fork, ►PCNA; Mossi R, Hubscher U 1998 Eur J Biochem 254:209.

RFC (also RF-C): This is a cellular replication factor. ►DNA replication eukaryotes; Schmidt SL et al 2001 J Biol Chem 276:34792.

RFLP (restriction fragment length polymorphism): Restriction endonuclease enzymes cut the DNA at specific sites and thus generate fragments of various sizes in their digest, depending on the distances between available recognition sites in the genome. During evolution when base changes occurred at the recognition sites through mutation, the length of fragments (within related strains) may have changed. After electrophoretic separation, a polymorphic pattern may be distinguished. These fragments may constitute co-dominant molecular markers for genetic mapping. Restriction fragment maps can also be generated by strictly physical methods. If a small circular DNA is completely digested by a restriction enzyme yielding fragments, say A, B, C, D, E but incomplete digestion with the same enzyme produces ABD, DB, AD, BC and CE triple or double fragments, respectively but never AB, BE, DC or AC. Thus, the fragment sequence must be ADBCE because the double fragments must be neighbors. Another procedure is to digest by at least two enzymes and determine the overlaps by hybridization in a sequential manner. The overlapping fragments indicate which fragments are next to each other (see Fig. R60).

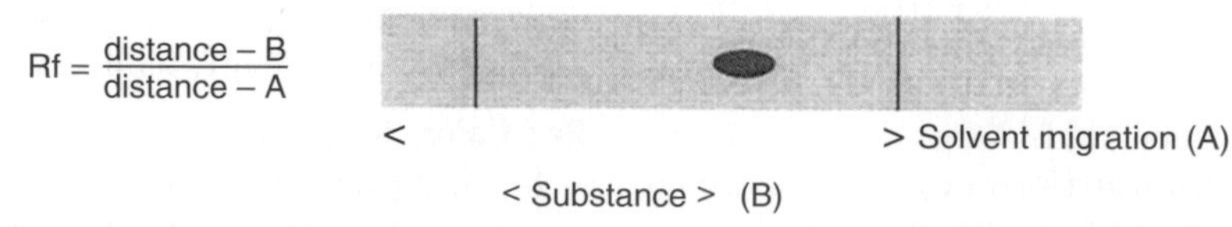

Figure R59. Definition of Rf

Fragments by enzyme 1
Fragments by enzyme 2

Figure R60. Restriction fragment length of the same DNA depending of the enzyme used

Restriction fragments can be used in genetic linkage analysis. They represent "dominant" physical markers because the DNA fragments can be recognized in heterozygotes. RFLP markers are useful for following the inheritance of linked genetic markers, which have variable expressivity and/or penetrance under unfavorable conditions. *Long-range restriction map* (macro-restriction map) represents the cutting sites of restriction enzymes along the chromosome. ►restriction enzymes, ►restriction fragment number, ►restriction fragment length, ►physical map, ►T-RFLP; Sharma RP, Mohapatra T 1996 Genetica 97[3]:313; Wicks SR et al 2001 Nature Genet 28:160.

RFLP Marker: A restriction enzyme-generated DNA fragment which has been or can be mapped genetically to a chromosomal location and can be used for determining linkage to it. They are codominant and always expressed. Their inheritance and recombination can be determined in relatively small populations. ►restriction enzyme, ►RFLP, ►physical map, ►integrated map

RFLP Subtraction: A selective technique for the enrichment of particular polymorphic, eukaryotic genomic unique segments. Small restriction fragments are isolated and purified from one genome containing sequences that are in large fragments in another related genome of mouse. By subtractive hybridization the segments with shared sequences by both genomes are removed. Thus, small fragments unique to one or the other strain are obtained. These sequences then become mappable genetic markers. ►genomic subtraction; Rosenberg M et al 1994 Proc Natl Acad Sci USA 91:6113.

RFLV: This is a RFLP variant. ►RFLP

RFX: A human DNA binding protein that promotes dimerization of MYC and MAX and thus stimulates transcription. A group of transcription factors for the major histocompatibility complex is also designated as RFX. RFX binds cooperatively with NF-Y and X2BP. RFX has four complementation groups (CIITA [16p13], RFX5 [1q21.1-q21.3], RFXAP [13q14], RFXANK [19p12]). Their defects may be responsible for autosomal immunodeficiency syndrome. The RFX factor binds to X-boxes (5′-GTNRCC[0-3N] RGYAAC-3′] where N = any nucleotide, R = purine and Y = pyrimidine. The human RFX1 is a helix-turn-helix protein, which uses a β-hairpin (called also a wing) to recognize DNA. ►MYC, ►MAX, ►MHC, ►immunodeficiency, ►bare lymphocyte syndrome, ►helix-turn-helix, ►MYC, ►MAX, ►NF-X, ►X2BP; Katan-Khaykovich Y, Shaul Y 2001 Eur J Biochem 268:3108.

RGD: An amino acid sequence Arg-Gly-Asp in the extracellular matrix and in fibronectin is recognized by, and bound to, integrin. RGD peptides can activate caspase-3 and initiate the apoptotic pathway. RGD peptides may facilitate cell adhesion, uptake of viral vectors or polycationic synthetic vectors. ►fibronectin, ►integrin, ►apoptosis, ►amino acid symbols in protein

RGE (rotating gel electrophoresis): The gel is rotated 90° at switching the cycle of the electric pulses. ►pulsed field gel electrophoresis

rGH (rat growth hormone): A thyroid hormone. ►animal hormones, ►hormone receptors, ►hormone-response elements

RGR: A yeast gene regulating transcription by RNA polymerase II. ►NAT, ►RNA polymerase

RGR (retinal G protein-coupled receptor): An opsin protein encoded at human chromosome 10q23. (See Chen XN et al 1996 Hum Genet 97:720).

RGS (regulator of G protein signaling): RGS is actually the same as GAP (GTP-ase activating protein). The different RGS proteins have different specificities for the different αβ and γ subunits of the trimeric G proteins. In mammals there are at least 19 members of this family of proteins with a common core, the RGS box. ►G proteins, ►GAP, ►signal transduction, ►conductin; Kehrl JH, Sinnarajah S 2002 Int J Biochem Cell Biol 34:432.

RH: ►radiation hybrid

Rh Blood Group: The name comes from a misinterpretation of the early study, namely that this human antigen would have the same specificity as that of rhesus monkey red cells. It is now known that this was incorrect; the animal antigen is different but the name was not changed. Despite over half a century of research the Rh antigen is not sufficiently characterized. The antigen may be controlled by three closely linked chromosomal sites—*C, D* and *E*—and on this basis eight (2^3) different allelic combinations were conceivable; the triple recessive *cde* being a null combination. The eight combinations are also designated as *R* or *r* with superscripts: *CDe* (R^1 or $R^{1,2,-3,-4,5}$), *cde* (*r*, or $R^{-1,-2,-3,4,5}$), *cDE* (R^2 or $R^{1,-2,3,4,-5}$), *cDe* (R^O or $R^{1,-2,-3,4,5}$), *cdE* (r'' or $R^{-1,-2,3,4,-5}$), *Cde* (r'or $R^{-1,2,-3,-4,5}$), *CDE* (R^Z or $R^{1,2,3,-4,-5}$) and *CdE* (r^y or $R^{-1,2,3,-4,-5}$). The first three of these occur at frequencies about 0.42, 0.39, and 0.14, respectively in England, and the others

R

are quite rare. In some oriental populations, R^1 (0.73%) and R^2 (0.19%) predominate and the recessives have a combined frequency of about 2%. This is in contrast to Western populations where they occur in over 40% of people. Clinically the most important is the D antigen because 80% of the D^- individuals, in response to a large volume of D^+ blood transfusion, make anti-D antibodies. The *d* alleles are amorph. The Rh genes are in human chromosome 1p. In addition, regulatory loci have also been identified in chromosome 3. For phenotypic distinction antisera anti-D, anti-C, anti-E, anti-c, and anti-e are used, and on the basis of the serological reactions 18 phenotypes can be distinguished. Anti-D antibodies are usually immunoglobulins of the G class (IgG). They develop only after immunization by Rh^+ type blood. Anti-C antibodies are generally of IgM type and they occur along with IgG after an Rh^- person is immunized with Rh^+ blood. Anti-E antibodies (IgG) are elicited in E negatives after exposure to E^+ blood. Anti-c antibodies (IgG) occur in CDe/CDe individuals after transfusion with c^+ erythrocytes. Anti-e antibodies are very rare (0.03). The major types of Rh antigens have several different variations. The Rh antigens are probably red blood cell membrane proteins. About 50 different Rh antigens have been identified. An Rh deficiency may also arise by the activity of a special suppressor gene in human chromosome 6p11-p21.1 or by CD47 protein, encoded at human chromosome 3q13.1-q13.2. The RG gene is situated at 1p34.3-p36.1.

It is clinically very significant that about 15% of Western populations are *cde/cde*. In Oriental populations the frequency of this genotype is very low. This type of individuals—called Rh negatives—may respond with erythroblastosis when exposed to Rh positive blood. If an Rh negative female carries a fetus with Rh positive blood type, antibodies against the fetal blood may be produced by the mother. This may then cause severe anemia with a high chance of intrauterine death and abortion. Generally, during the first pregnancy, this hemolytic reaction is absent but the chance in the following pregnancies, by when sufficient immunization has taken place by the fetal blood entering into the maternal bloodstream, the probability of erythroblastosis becomes high. Thus pregnancies of Rh negative females are monitored and appropriate serological treatment provided if antibody production is detected to prevent fetal erythroblastosis. Erythroblastosis may also occur if an Rh negative individual is transfused with Rh positive blood. The rodent antibodies responding to rhesus monkey red cells are now called the LW blood group. The physiological role of the Rh antigens is apparently in the CO_2 gas channel. ►erythroblastosis fetalis, ►blood groups, ►immunoglobulins, ►antibodies, ►schizophrenia, ►eclampsia; Avent ND 2001 J Pediatr Hematol Oncol 23:394; Stockman JA 3rd, de Alarcon PA 2001 J Pediatr Hematol Oncol 23:385, evolution of the Rh proteins: Huang C-H, Peng J 2005 Proc Natl Acad Sci USA 102:15512.

Rhabdomere: A rod-shaped element of the compound eye of insects. There are eight R1 to R8 neuronal photoreceptors in each of the ca. 20 cells of the *Drosophila* eyes, containing about 800 ommatidial clusters. ►compound eye, ►ommatidium, ►*Drosophila*, ►*sevenless*

Rhabdomyosarcoma: A type of cancer involving chromosome breakage in the Pax-3 gene 2q35 and 13q14 (Rhabdomyosarcoma-2) or other translocations involving chromosome 3 and 11 (Rhabdomyosarcoma-1). These break points may also be related to the Beckwith-Wiedemann syndrome or WAGR syndrome. The malignant rhabdoid tumor is associated with deletions in human chromosome 22q11.2 encoding the homologue (hSNF5/INI1) of the yeast chromatin remodeling protein SWI/SNF. Embryonal rhabdomyosarcoma is due to a defect at 11p15.5. Isochromosome 3q has the same effect as mutation at the ATR (ataxia telengiectasia and rad3-related site), i.e., inhibiting MyoD (myogenesis), causing cell cycle abnormalities and predisposition to cancer. ►Pax, ►Beckwith-Wiedemann syndrome, ►WAGR, ►ataxia, ►Wilms tumor, ►chromatin remodeling, ►SWI, ►nucleosome, ►histone, ►nuclease-sensitive sites

Rhabdoviridae: Oblong or rod-shaped (130–380 × 70–85 nm), single-stranded RNA (13–16 kb) viruses with multiple genera and wide host ranges. ►CO_2-sensitivity in *Drosophila*

Rhalloc: The sequences mapped by radiation hybrid methods by various mapping groups. ►radiation hybrids

Rhdb: A database containing the mapping information obtained by radiation hybrids. ►radiation hybrids

Rheology: The study of elasticity, change of shape, viscosity and flow of materials such as blood through the vascular system of veins and heart.

Rhesus Blood Group: ►Rh blood group, ►LW blood group

Rhesus monkey (Macaque, *Macaca mulatta,* 2n = 42, genome size ~2.87 Gb): A representative of mainly South-East Asian and North African species of long-tail monkeys. These small intelligent animals have been used extensively for biological and behavioral studies. Rhesus monkeys separated from the human lineage about 25 million years ago and yet they retained about 93% of their identity to humans. The sequenced genome has revealed 100 different families of DNA transposons and more than half a million recognizable copies of endogenous retroviruses (ERVs). The analysis revealed more than 1,000 rearrangement-induced break points through the HCR (Human-Chimpanzee-Rhesus) lineages, of which 820 occur between rhesus and the reconstructed human-chimpanzee ancestor. As with humans and chimpanzees, the analysis of the macaque assembly revealed an enrichment of segmental duplications near gaps, centromeres, and telomeres. Some segmental duplications contain genes of high biological significance. The statistical approach revealed that 1,358 genes were gained by duplication along the macaque lineage. The average human gene differs from its ortholog in the macaque by 12 nonsynonymous and 22 synonymous substitutions, whereas it differs from its ortholog in the chimpanzee by fewer than three nonsynonymous and five synonymous substitutions. Similarly, 89% of human-macaque orthologs differ at the amino acid level, as compared with only 71% of human-chimpanzee orthologs. Thus, the chimpanzee and human genomes are in many ways too similar for characterizing protein-coding evolution in primates, but the added divergence of the macaque helps substantially in clarifying the signatures of natural selection. Important evolutionary differences from human genes responsible for the diseases, have been detected (Gibbs RA et al 2007 Science 316:222). ►Rh blood group, ►LW blood group, ►*Cercopithecidae*, ►primates, ►chimpanzee

Rheumatic Fever (rheumatoid arthritis, RA, 6p21.3): Rhematic fever consists of ailments affecting mainly the connective tissues and joints, but it may cause also heart and nervous system anomalies. The HLA region accounts for most of the susceptibility but regions associated with other autoimmune diseases (lupus erythematosus, inflammatory bowel disease, multiple sclerosis, ankylosing spondylitis) are also implicated. The disease is complex because environmental and susceptibility factors heavily confound the direct genetic determination. For example, certain streptococcal infections can precipitate rheumatic fever. The familial forms are attributed to dominant genetic factor(s) and susceptibility has been attributed to recessive genes(s). Several antigens have been identified which appeared to be more predominant within affected kindred. One monoclonal antibody, D8/17, was present in 100% of the patients affected with the disease whereas two other monoclonal antibodies showed up between 70% to 90% coincidence and with 17% to 21% presence even among the unaffected people. One susceptibility factor is linked to IL-3. Simultaneously, blocking both B and T cell receptors by the signaling molecule BlyS (B lymphocyte stimulator) and TACI (transmembrane/T-cell-activator and calcium-modulating and cyclophilin ligand interactor) prevented the development of arthritis in mice (Wang H et al 2001 Nature Immunol 2:632; Yan M et al ibid 638). Peptidylarginine deiminases (encoded at 1p36) post-translationally convert arginine into citrulline. Citrullinated epitopes are common targets of arthritis-specific autoantibodies directed against perinuclear factor/keratin and against the Sa system (Suzuki A et al 2003 Nature Genet 34:395). Sa is a hapten-carrier antigen in which vimentin is the carrier and citrulline is the hapten. Various forms of tumor necrosis factor (TNF) inhibitors such as TNF monoclonal antibodies and methotrexate therapy (Keystone EC et al 2004 Arthritis Rheumatism 50:1400) and recombinant TNF receptor (TNFR) and immunoglobulin-G1 fragment crystalline (Fc) fusion protein have been successfully used for clinical treatments (Morteland LW et al 1997 New England J Med 337:141). Injecting an interleukin-1 receptor antagonist (IL-1 Ra) cDNA in a retroviral vector (derived from Moloney leukemia virus) in synovial fibroblasts was beneficial without side effects (Evans CH et al 2005 Proc Natl Acad Sci USA 102:8698). In mice who have been exposed daily to 10% ethanol in drinking water the development of erosive arthritis was almost totally abrogated and they did not display any liver toxicity. In contrast, the antibody-mediated effector phase of collagen-induced arthritis was not influenced by ethanol exposure. Also, the major ethanol metabolite, acetaldehyde, prevented the development of arthritis. This anti-inflammatory and anti-destructive property of ethanol was mediated by (1) down-regulation of leukocyte migration and (2) up-regulation of testosterone secretion, with the latter leading to decreased NF-B activation (Jonsson I-M et al 2004 Proc Natl Acad Sci USA 104:258). ►arthritis, ►rheumatoid, ►pseudorheumatoid dysplsia, ►ankylosing spondylitis, ►HLA, ►Coxsackie virus, ►IL-3, ►IL-32, ►autoimmune disease, ►citrullinemia, ►urea cycle, ►hapten, ►vimentin, ►keratin; Jawaheer D et al 2001 Am J Hum Genet 68:927; Okamoto K et al 2003 Am J Hum Genet 72:303.

Rheumatoid: Resembling a rheumatic condition. ►rheumatic fever

Rhinoceros: *Ceratotherium simum*, 2n = 84; *Rhinoceros unicornis*, 2n = 82; *Diceros bicornis*, 2n = 134.

Rhizobium: ►nitrogen fixation

Rhizofilteration: ►bioremediation

R

Rhizoid: A structure resembling plant roots.

Rhizome: An underground plant stem modified for storage of nutrients and propagation (see Fig. R61).

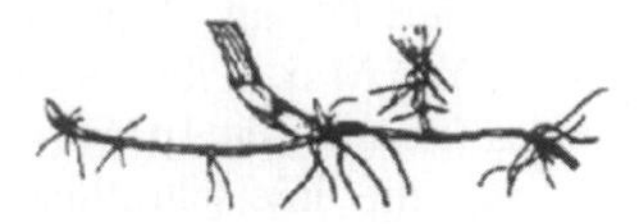

Figure R61. Rhizome

Rhizomorph: ►hypha

Rhizosphere: The environment around the roots of plants. ►microbiome

RHKO (random homozygous knockout): ►knockout

RhlB: A helicase of the DEAD-box family. ►helicase, ►DEAD-box, ►degradosome

RHMAP (radiation hybrid mapping): A multipoint radiation mapping procedure. It analyzes the minimal number of breaks (RHMINBRK) and may provide mapping information and by the use of the maximum likelihood procedure (RHMAXLIK) linkage information. ►radiation hybrid; Am J Hum Genet 49:1174.

RHMAXLIK: ►RHMAP

RHMINBRK: See ►RHMAP

Rho: A GTPase homolog of the RAS oncogene. It relays signals from cell-surface receptors to the actin cytoskeleton. It regulates myosin phosphatase and Rho-associated kinase. In yeast cells, a RHO protein is involved in the stimulation of cell wall β (1→3)-D-glucan synthase and the regulation of protein kinase C, and in mediation of polarized growth, morphogenesis and cell migration (metastasis). Actually, Rho is a subunit of the glucan synthase enzyme complex. Serine-threonine protein kinase and protein kinase N (PKN) are apparently activated by Rho. Rho also mediates endocytosis. Rho, in cooperation with RAF, seems to induce $p21^{Waf1/Cip1}$ protein, which blocks the transition from the G_1 to the S phase of the cell cycle. In human chromosomes, Rhos are designated as ARH6: 3pter-p12, ARH12: 3p21, ARH9: 5q 31-qter. The Rho family includes Rac, CDC42, RhoG, RhoE, RhoL, and TC10 proteins. Increase of RhoC activity accompanies metastasis of melanoma cells. Members of the Rho family of proteins are also involved in the regulation of photoreception and developmental events mediated by light. ►RAS oncogene, ►metastasis, ►melanoma, ►RAF, ►cytoskeleton, ►receptor, ►photoreceptors, ►RAC, ►CDC42, ►endosome, ►CNF, ►p21, ►citron, ►ROCK, ►ROK, ►*Yersinia*; Kaibuchi K et al 1999 Annu Rev Biochem 68:459; Etienne-Mannewill S, Hall A 2002 Nature [Lond] 420:629; Katoh H, Negishi M 2003 Nature [Lond] 424:461.

rho (ρ): A designation of density; high G+C content of DNA increases it while high A+T content decreases it. (ρ = 1.660 + [0.098 × {G+C}] fraction in DNA). The density is determined on the basis of ultracentrifugation in CsCl and refractometry of the bands. ►buoyant density

rho Factor: A protein involved in the termination of transcription in (rho-dependent) prokaryotes. It is about 46 kDa and is a hexamer (~275 kDa). For maximal efficiency it is present in about 10% of the molecular concentration of the RNA polymerase enzyme. It is basically an ATP-dependent RNA-DNA helicase. Rho can stop elongation of the transcript only at specific termination sites in the RNA. In mitochondria, the mtTERM protein can stop transcription on both DNA strands. In yeast, the REB-1 protein terminates transcription and releases RNA from the ribosome. In mouse, the TFF-1 protein terminates the action of RNA polymerase I whereas the La protein controls RNA polymerase III. ►transcription termination in prokaryotes, ►transcription termination in eukaryotes, ►N-TEF, ►antitermination; Yu X et al 2000 J Mol Biol 299:1279; Kim D-E, Patel SS 2001 J Biol Chem 276:13902; crystal structure: Skordalakes E, Berger JM 2006 Cell 127:553.

***rho* Gene**: The *rho* gene is responsible for the suppressive petite (mtDNA) condition in yeast. ►mtDNA

rho⁻ Mutants: The rho mutants of yeast lost from their mitochondrial DNA most of the coding sequences. They are very high in A+T content (the buoyant density of the DNA is low). ►mtDNA

rho-Dependent Transcription Termination: Actually, none of the rho-dependent bacterial strains absolutely require this protein factor for termination. ►rho factor, ►transcription termination, ►rho-independent; Konan KV, Yanofsky C 2000 J Bacteriol 182:3981.

Rhodamine B: A fluorochrome used for fluorescent microscopy; its reactive group forms a covalent bond with proteins (immunoglobulins) and other molecules. It is also a laser dye. Its absorption maxima is 543 (355) nm. Caution: It is carcinogenic. ►fluorochromes

Rhodopseudomonas palustris: A rather ubiquitous purple photosynthetic bacterium containing 5,459,213 bp in the chromosome and one plasmid of 8,427 bp. The estimated gene number is 4,835 (Larimer FW et al 2004 Nature Biotechnol 22:55).

R

Rhodopsin: A light-sensitive protein (opsin, $M_r \approx$ 28,600, human chromosome 3q21-q24) coupled with a chromophore, 11-cis retinal, which isomerizes to all-trans retinal immediately upon the receipt of the first photon.

It functions as the light receptor molecule in the disks of the photoreceptive membrane of the photoreceptor cells of the animal retina of the eye. Rhodopsin has seven short hydrophobic regions that pass through the endoplasmic reticulum (ER) membrane in seven turns. The amino end (with attached sugars) is within the ER lumen and the carboxyl end points out into the cytosol. In the rod shape photoreceptor cells, rhodopsin is responsible for monochromatic light perception at low light intensities. In the cone shape photoreceptor cells color vision is mediated by it in bright light. The photoreceptor cells transmit a chemical signal to the retinal nerves that initiate then the visual reaction series. When the receptor is activated, the level of cyclic guanylic monophosphate (cGMP) drops by the activity of cGMP and bind phosphodiesterase and it is quickly replenished in dark by *guanylyl cyclase*. The activated opsin protein is transducin, an α_t G protein subunit, that activates cGMP phosphodiesterase. When one single photon of light hits rhodopsin, through an amplification cascade, 500,000 molecules of cGMP may be hydrolyzed, 250 Na^+ channels may close, and more than a million Na^+ are turned back from entering the cell through the membrane within the time span of a second. In the dark, the sodium ion channels are kept open by cGMP; in the light the channels are closed. The sodium-calcium channels being shut, in light, the intake of Ca^{2+} is reduced and that leads to the restoration of the cGMP level through the action of the recoverin protein that cannot function well when it is bound to Ca^{2+}. Recoverin is a calcium sensor in the retinal rods. The rhodopsin gene has been assigned to human (see Fig. R62) chromosome 3q21-qter and to mouse chromosome 6. *Drosophila* has three rhodopsin loci (*Rh2* [3-65], *Rh3* [3-70], *Rh4* [3-45]). In flies, too, the *nina* loci are involved in the synthesis of opsins affecting the ommatidia and ocelli. ►phytochrome, ►signal transduction, ►G-proteins, ►retinitis pigmentosa, ►retinoblastoma, ►color blindness, ►color vision, ►ommatidium, ►ocellus, ►opsins, ►circadian rhythm, ►night blindness, ►proton pump; Yokoyama S 1997 Annu Rev Genet 31:315; Palczewski K et al 2000 Science 289:739; Bartl FJ et al 2001 J Biol Chem 276:30161; Sakmar TP 2002 Current Opin Cell Biol 14:189; Garriga P Manyosa J 2002 FEBS Lett 528:17, xanthorodopsin proton pump antenna: Balshov SP et al 2005 Science 309:2061, G protein-coupled receptor: Palczewski K 2006 Annu Rev Biochem 75:743.

Rhoeo discolor: An ornamental plant with large chromosomes (see Fig. R63), 2n = 12; genome x = 14.5×10^9 bp.

Figure R63. Rhoeo haploid set

rho-Independent Transcription Termination: Also it is called intrinsic transcription termination. The original model visualized the involvement of an RNA hairpin followed by a 15-nucleotide T (thymidine-rich)

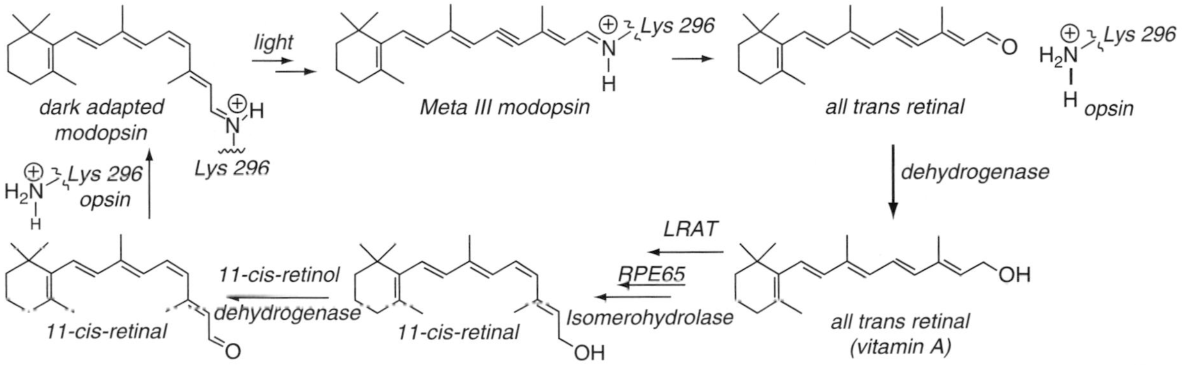

Figure R62. The visual cycle of the rhodopsin pathway in mammals. (From Gollapalli DR, Rando RR 2004 Proc Natl Acad Sci USA 101:10030. Copyright by the National Academy of Sciences USA, 2004)

region. The hairpin may be separated from the T sequences by a 2-nucleotide spacer. Since the *E. coli* genome has been fully sequenced, 135 terminators were identified and 940 putative terminators were found. Some of these are up to 60 nucleotides away from the 3′-end of the transcription units. ►transcription termination; d'Aubenton Carafa Y et al 1990 J Mol Biol 216:835; Lesnik EA et al 2001 Nucleic Acids Res 29:3583.

Rhombocephalon: ►hindbrain

Rhomboid Protease: Phylogenetically widespread membrane/mitochondrial intramembrane serine proteases that activate the epidermal growth factor receptor. In yeast, one rhomboid acts on cytochrome c peroxidase and a dynamin-like GTPase. γ-secretase is also an intramembrane proteolytic enzyme (Urban S, Wolfe MS 2005 Proc Natl Acad Sci USA 102:1883). ►EGF, ►secretase; McQuibban GA et al 2003 Nature [Lond] 423:537; Koonin EV et al 2003 Genome Biol 4:R19; crystal structure of *E. coli* GlpG intramembrane serine protease: Wang Y et al 2006 Nature [Lond] 444:179.

Rhombomeres (neuromeres): Metameric units of eight subdivisional partition of the neuroepithelium of the hindbrain. ►metamerism

Rhoptry (toxoneme): Generally, a club-shaped apical organ of apicomplexan protozoa; it mediates infection by sporozoites. ►malaria, ►*Plasmodium*, ►apicomplex, ►sporozoite

Rhubarb (*Rheum* spp): The plant has about 50 species; 2n = 2x = 44. It is an accessory food plant and some species are used as medicinal herbs (cathartic [laxative]).

Rhynchosciaras: *Rhynchosciaras* are dipteran flies with very clearly banded polytenic chromosomes in the salivary gland nuclei. ►*Sciara*, ►polytenic chromosomes

RI: ►recombinant inbreds

RI Particle: These particles are formed in cold in vitro during the 30S ribosomal subunit reconstitution experiment of rRNA and about 15 proteins. Upon heating, to assume the proper conformation, they become RI* particles. ►ribosome, ►ribosomal RNA, ►ribosomal protein

Ri Plasmid: A root-inciting plasmid of *Agrobacterium rhizogenes*. It can be used for genetic engineering to the same way as the Ti plasmid of *Agrobacterium tumefaciens*. The bacterium is responsible for the hairy root disease of plants. Its T-DNA contains two segments. The right T-DNA (T_R) contains genes for the production of opines, mannopine and agropine, and also for auxin. These auxin genes are highly homologous to the comparable genes in the Ti plasmid of *Agrobacterium tumefaciens*. The left portion of the T-DNA (T_L) includes 11 open reading frames with organization similar to eukaryotic genes; however this segment is different from that of the Ti plasmid. ►Ti plasmid; Moriguchi K et al 2001 J Mol Biol 307:771.

RIA: ►radioimmunoassay

Ribavirin: An antibiotic; its 5′-phosphate inhibits inosine monophosphate (see Fig. R64). ►inosine

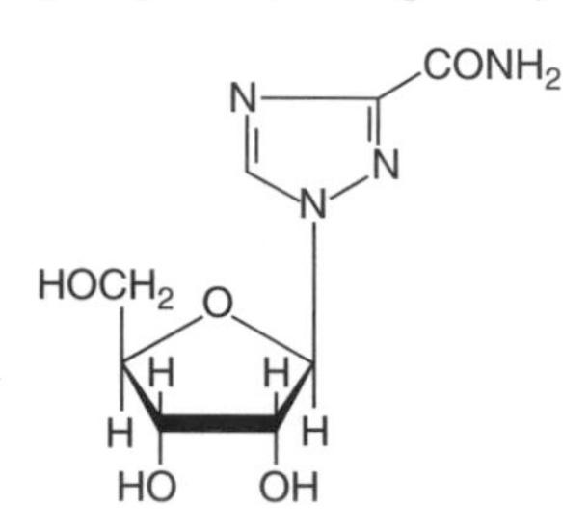

Figure R64. Ribavirin

Ribbon Diagram of Polypeptide Structure: An X-ray structure of a α-helix (at left) and a short β-sheet (at right ending with an arrow) (see Fig. R65). Typically, the protein structure is more complex and contains several of these elements forming multiple domains. ►protein structure

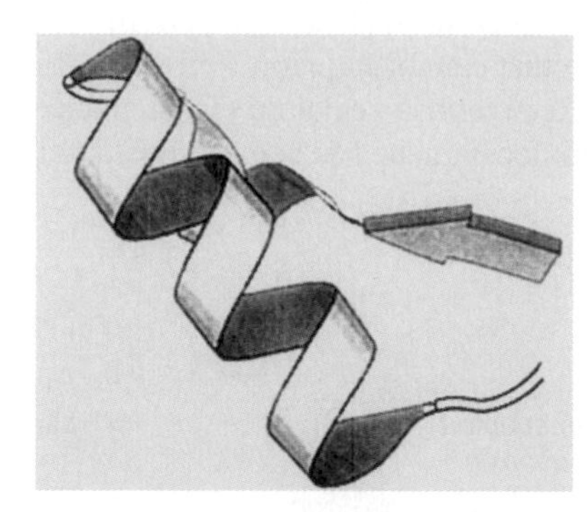

Figure R65. Ribbon diagram of polypeptide structure

Riboflavin (lactoflavin, vitamin B_2): A vitamin precursor of flavin mononucleotide (FMN) and flavin adenine dinucleotide (FAD) oxidation coenzymes. Riboflavin is heat stable but rapidly decomposes in light (see Fig. R66). (See Ritz H et al 2001 J Biol Chem 276:22273).

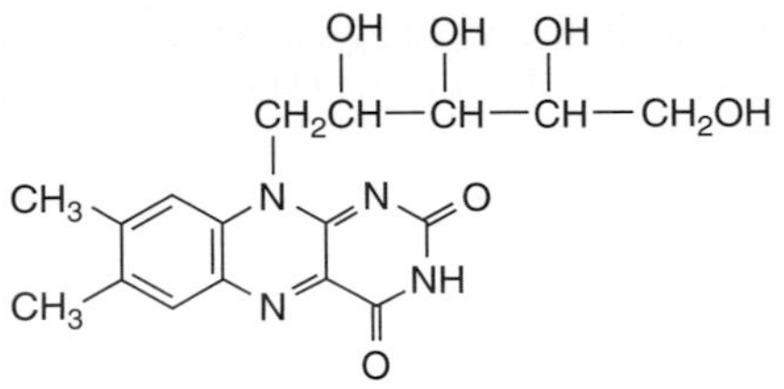

Figure R66. Riboflavin

Ribocyte: An evolutionarily ancestral cell with RNA genetic material. ►RNA world

Ribo-gnome: Small regulatory RNAs, such as micro-RNA, RNAi, etc. (Zamore PD, Haley B 2005 Science 309:1519).

Riboflavin Retention Deficiency: A riboflavin retention deficiency may prevent hatching of eggs in "leaky auxotrophic" chickens. The defect is not in absorption but the vitamin is rapidly excreted by a genetic default, and the *rd/rd* eggs have only about 10 μg of the vitamin rather than the normal level of about 70 μg. If 200 μg is injected into the eggs before incubation, hatching occurs.

Ribonuclease (RNases): Ribonuclease occur in a large number of specificities and they digest various types of ribonucleic acids. The bovine pancreatic ribonuclease is a small (124 amino acids) and very heat-stable enzyme. The pancreatic ribonuclease was the first enzyme chemically synthesized in the laboratory. An autoradiogram (see Fig. R67) permits the distinction of the digestion enzyme control, T_1: G-specific enzyme; U_2: ribonuclease is A-specific; Phy M: *Physarum* enzyme M, specific for U + C; OH: random alkaline digest; and B_c: *Bacillus cereus* enzyme with U + C specificity. On the left side guanosine positions are indicated from the 5′-end, and on the right side the nucleotide sequences are shown as read from the gel. (Courtesy of P-L Biochemicals, Inc.). ►RNases, ►ribonucleases, ►angiogenins; Condon C, Putzer H 2002 Nucleic Acids Res 30:5339.

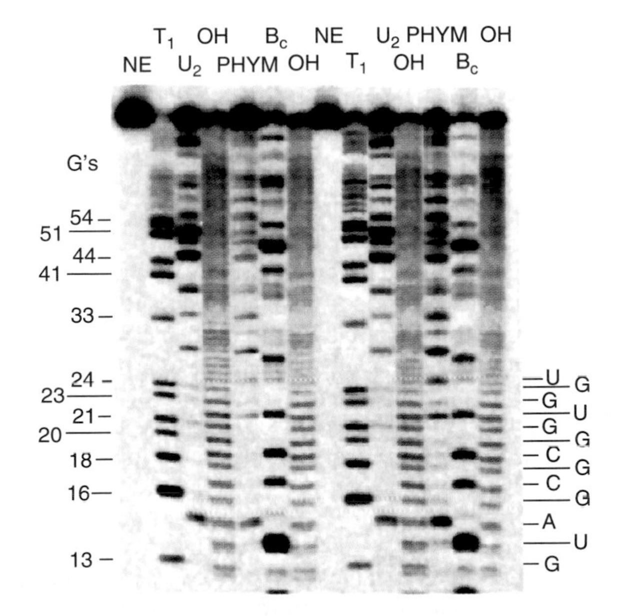

Figure R67. Ribonucleases

Ribonuclease 1: Degrades RNA I. ►RNA I; Cunningham KS, et al 2001 Methods Enzymol 342:28.

Ribonuclease II: Ribonuclease II is similar in action to Ribonuclease D; its role is not just limited to processing; it can also degrade an entire tRNA and mRNA molecule. It is an exonuclease. The enzyme has four domains: two cold-shock domains, the catalytic RNB domain, and one S1 domain. The enzyme contacts RNA in the "anchor" and "catalytic" regions. The catalytic RNB domain includes four conserved sequence motifs. This catalytic pocket is accessible only to single-stranded RNA. The structural features, shown in the Figure R68, are probably characteristic for other members of this family of proteins (Frazão C et al 2006 Nature [Lond] 443:110). ►ribonuclease D; Donovan WP, Kushner SR 1986 Proc Natl Acad Sci USA 83:120.

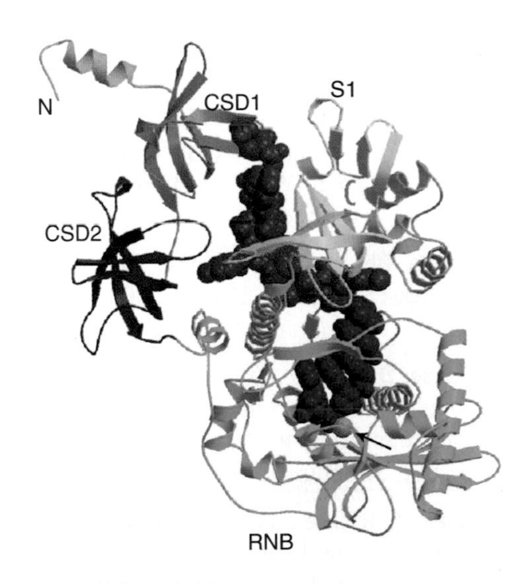

Figure R68. Crystal structure of *E. coli* RNase II. CSD1, CSD2, S1 are oligonucleotide-binding domains. RNB is the catalytic domain. Mg^{2+} is at arrow. The docked RNA is colored from red to blue, according to atomic displacement parameters. (Courtesy of Drs. Carlos Frazão and Maria Arménia Carrondo)

Ribonuclease III (RNase III): A homodimeric phosphodiesterase; an endonuclease cutting double-strand RNA from the 3′ or 5′ end. It cleaves prokaryotic and eukaryotic pre-rRNA at a U3 snoRNP-dependent site. RNase III controls the maturation of cellular and phage RNAs and may determine the translation and half-life of mRNAs. In prokaryotes its cleaving action may be restricted by antideterminants. In yeast, RNA tetraloops (AGNN) are located 13–16 bp from the RNase III recognition sites (see Fig. R69). ►trimming, ►snoRNP, ►antideterminant, ►Dicer; Grunberg-Manago M 1999 Annu Rev Genet 33:193; Conrad C, Rauhut R 2002 Int J Biochem Cell Biol 34:116; crystal structure: Gan J et al 2005 Structure 13:1435; Gan J et al 2006 Cell 124:355.

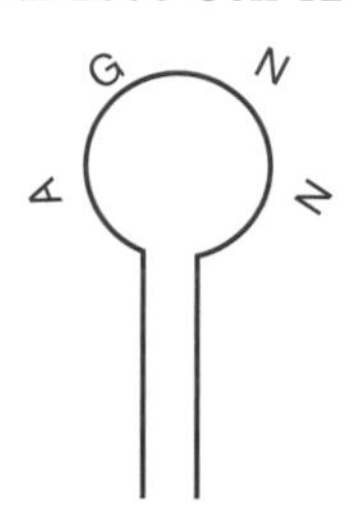

Figure R69. Tetraloops of four nucleotides

Ribonuclease A: A family of RNA digesting enzymes, including pancreatic, brain ribonucleases as well the related eosinophil-derived neurotoxin (EDN), eosinophil cationic protein (ECP) and angiogenin, involved in defense functions. ►eosinophil, ►angiogenesis, ►Prp75; Sheraga HA et al 2001 Methods Enzymol 341:189; Cho S et al 2005 Genomics 85:208.

Ribonuclease B: RNase B cuts at U+C sequences of RNA. (See Zapun A et al 1998 J Biol Chem 273:6009).

Ribonuclease BN: An exonuclease that cuts tRNA. ►exonuclease; Callahan C et al 2000 J Biol Chem 275:1030.

Ribonuclease D: RNase D processes tRNA primary transcripts at the 3′ end into mature tRNA. ►tRNA, ►primary transcript, ►RNAi; crystal structure: Zuo Y et al 2005 Structure 13:973.

Ribonuclease E: RNase E cleaves RNAs with secondary structure within single-stranded regions rich in A and U nucleotides, e.g., RNA I. The N-terminal domain of 1,061 amino acids functions as an endonuclease involved in mRNA and rRNA processing and degradation in *E. coli* and other bacteria. The catalytic activity of RNase E (and RNase G) is enhanced in the multimeric forms by having a monophosphate at the 5′ end despite the fact that mRNA degradation begins endonucleolytically (Jiang X, Belasco JG 2004 Proc Natl Acad Sci USA 101:9211). This enzyme also shortens the polyA and polyU tails of RNA molecules. Its C-terminus may associate with a 3′→5′ exoribonuclease and other proteins in the degradosome complex. ►RNA I, ►*E. coli*, ►endonuclease, ►exonuclease, ►degradosome, ►tmRNA, ►protein repair, ►mRNA, ►tRNA; Grunberg-Manago M 1999 Annu Rev Genet 33:193; Walsh AP et al 2001 Nucleic Acids Res 29:1864, crystal structure: Callaghan AJ et al 2005 Nature [Lond] 437:1187.

Ribonuclease G (CafA): RNase G processes the 5′ end of 16S rRNA with RNase E (Feng Y et al 2002 Proc Natl Acad Sci USA 99:14746).

Ribonuclease H: RNase H digests RNA when paired with DNA but it does not cut single-strand RNA or double-strand RNA or double-strand DNA. This family of proteins includes transposases, retroviral integrase, Holliday structure resolvase and the RISC nuclease Argonaute. RNase H specifically recognizes the A form of RNA and the B form of a DNA strand (Nowotny M et al 2005 Cell 121:1005). RNase HI, as an endonuclease, can remove RNA primers (except 1 nucleotide) from the 5′-end of the Okazaki fragments. RNase H may also cleave "irrelevant sites", i.e., RNA that is imperfectly bound to DNA. RNase H can be used to prevent the translation of mRNA by recruiting it to phosphorothioated DNA. ►DNA replication eukaryotes, ►Rad27/Fen1, ►Okazaki fragment, ►antisense technologies, ►phosphorothioate, ►RNA I, ►microRNA, ►transposase, ►integrase, ►resolvase, ►Aicardi-Goutières syndrome; Wu H et al 2001 J Biol Chem 276:23547, folding of RNase H: Cecconi C et al 2005 Science 309:2057.

Ribonuclease J: An exoribonuclease that can cut 5′ to 3′ both ribosomal and mRNA of *Bacillus subtilis*. RNase J1 plays a role in maturation and 5′ stability of mRNA (Mathy N et al 2007 Cell 129:681).

Ribonuclease L: In dimeric form, RNase L cleaves single-stranded RNA. Its product may reduce viral replication in interferon-exposed cells and suppress prostate cancer. It may be induced by interferon. For activity it depends on 2′,5′-oligoadenylates (2–5A). RNase L may be involved in apoptosis. The RNase L inhibitor plays important role in ribosome biogenesis and HIV capsid assembly (Karcher A et al 2005 Structure 13:649). Compounds bound to the 2–5A-binding domain of RNase L induce RNase L dimerization and activation. Low-molecular-weight activators of RNase L had broad-spectrum antiviral activity against diverse types of RNA viruses, including the human parainfluenza virus type 3, yet these compounds by themselves were not cytotoxic at the effective concentrations (Thakur CS et al 2007 Proc Natl Acad Sci USA 104:9585). ►apoptosis, ►RNAi, ►Ire, ►interferon; Stark GR et al 1998 Annu Rev Biochem 67:227; Carpten J et al 2002 Nature Genet 30:181; Malathi K et al 2005 Proc Natl Acad Sci USA 102:14533.

Ribonuclease MRP (9p21-p12): A mitochondrially-localized enzyme involved in cleavage of pre-mRNA and pre-rRNA. Its defect affects the cell cycle, hair hypoplasia, immunodeficiency, hematological abnormalities and the assembly of the ribosomes and the degree of bone dysplasia, respectively (Thiel CT et al 2007 Amer J Hum Genet 81:519).

Ribonuclease P: RNase processes the 5′ end of transfer RNA transcripts (and cleaves some other RNAs). It may process some pre-tRNAs at the 3′-end. Its catalytic subunit is a ribozyme in bacteria, a 377-nucleotide RNA that can do the processing even without the ~120-amino acid protein. However, the protein may enhance specificity and is required for ribosomal translocation. For catalytic activity the enzyme requires divalent cations (Mg^{2+}). The chloroplast enzyme is not a ribonucleoprotein but is only a protein. The size of the protein subunits in bacteria is about 14 kDa, but in eukaryotes it may exceed 100 kDa. Although at least 10 proteins are associated with RNase P, either protein Rpp29 or C5 are essential for activation of the core RNA and substrate recognition as well as for catalysis (Sharin E et al 2005 Nucleic Acids Res 33:5120). Eukaryotic RNase P RNA is also able to cleave its substrate in the absence of protein(s) although with relatively low efficiency; this suggests that the catalytic activity resides in the RNA subunit of RNase P (Kikowska E et al 2007 Proc Natl Acad Sci USA 104:2062). ►ribozyme, ►external guide sequences, ►KH domain, ►RNase, ►RNA maturases, ►MRP; Kurz JC, Fierke CA 2000 Curr Opin Chem Biol 4:553; Tous C et al 2001 J Biol Chem 276:29059; Gopalan V et al 2002 J Biol Chem 277:6759; Xiao S et al 2002 Annu Rev Biochem 71:165; specificity domain: Krasilnikov AS et al 2003 Nature [Lond] 421:760; structural diversity: Krasilnikov AS et al 2004 Science 306:104; crystal structure of the RNA component: Torres-Larios A et al 2005 Nature [Lond] 437:584; crystal structure of *Bacillus stearothermophilus* P RNase RNA: Kazantsev AV et al 2005 Proc Natl Acad Sci USA 102:13392; structure and function: Marquez SM et al 2006 Mol Cell 24:445.

Ribonuclease R: An RNase II homolog 3′→5′ exoribonuclease of *E. coli*. RNase R and polynucleotide phosphorylase are essential for the proper assembly of ribosomes by eliminating defective rRNA monomers. ►ribonuclease II, ►rRNA, ►polynucleotide phosphorylase; Cheng A-F, Deutscher MP 2002 J Biol Chem 277:21624; Cheng Z-F, Deutscher MP 2003 Proc Natl Acad Sci USA 100:6388.

Ribonuclease S: RNase S enzymes are associated with self-incompatibility of plants (Ma R-C, Oliviera MM 2002 Mol Genet Genomics 267:71).

Ribonuclease T: An exonuclease of tRNA, cutting at the amino acid accepting end (CCA). ►tRNA; Zuo Y, Deutscher MP 1999 Nucleic Acids Res 27:4077; Zuo Y et al 2007 Structure 15:417.

Ribonuclease T_1: RNase T_1 is specific for G (guanine) linkages in RNA. (See Kumar K, Walz FG Jr 2001 Biochemistry 40:3748).

Ribonuclease U1: A guanine-specific RNase. (See Takahashi K, Hashimoto J 1988 J Biochem [Tokyo] 103:313).

Ribonuclease U2: RNase U2 is specific for A+U nucleotides in RNA. (See Taya Y et al 1972 Biochim Biophys Acta 287:465).

Ribonuclease Z (3′ tRNase): A zinc-dependent metallo-hydrolase; an endonuclease, which processes the 3′-end of most prokaryotic, cytoplasmic, and mitochondrial tRNAs (Dubrovsky EB et al 2004 Nucleic Acids Res 32:255; crystal structure: Li de la Sierra-Gallay I et al 2005 Nature [Lond] 433:657).

Ribonucleic Acid: ►RNA

Ribonucleoprotein (RNP): A ribonucleic acid associated with a protein. ►RNP

Ribonucleotide: A ribonucleotide contains one of the four nitrogenous bases (A, U, G, C), ribose and phosphate. It is a building block of RNA. ►deoxyribonucleotide

Ribonucleotide Reductase (RNR): A ribonucleotide reductase converts ribonucleotide di- and triphosphates into deoxyribonucleotide di- and triphosphates. It is required for DNA synthesis, the completion of the cell cycle and malignancy. RNR proteins might have been instrumental in generating DNA in the RNA world. The allotetramer enzyme has a large subunit, R1, which regulates the maintenance of a deoxynucleotidetriphoshate pool; its level is constant throughout the cell cycle. The R2 subunit converts ribonucleotides to deoxyribonucletides and it appears in G1 and vanishes at early S phase. The 351-amino acid R2 subunit is the product of the p53R2 gene, activated by p53. Gene p53R2 apparently has a DNA repair function. ►cell cycle, ►malignant, ►CDC22, ►RNA world, ►p53, ►recombination molecular mechanism of; Jordan A, Reichard P 1998 Annu Rev Biochem 67:71; Tanaka H et al 2000 Nature 404:42; Chimploy K,

R

Mathews CK 2001 J Biol Chem 276:7093; Högbom M et al 2004 Science 305:245; Nordlund P, Reichard P 2006 Annu Rev Biochem 75:681.

Ribose: An aldopentose sugar, present in ribonucleic acids with an OH group at both 2′ and 3′ positions. Its deoxyribose form lacks the O at the 2′ position and it is present in DNA (see Fig. R70). ►aldose, formula given here

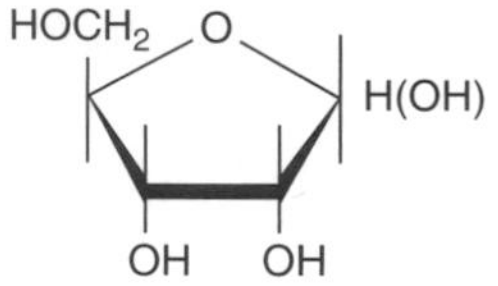

HOCH2 O H(OH) OH OH

Figure R70. D-ribose

Ribose Zipper: When two RNA strands of opposite polarity are situated in close vicinity to each other hydrogen bonds may form between the 2′-OH groups of consecutive riboses in both strands. (See Klostermeier D, Millar DP 2001 Biochemistry 40:37).

Ribosomal DNA: The codes for ribosomal RNAs. ►ribosome

Ribosomal Filter: A hypothesis proposing that cis-regulatory sequences in the mRNA modulate its binding to the 40S ribosomal subunit by complementarity to the 18S or 28S ribosomal subunits or by affinity to specific ribosomal proteins. This binding may filter, i.e., influence the translation in a (+) or (−) manner and may be a factor in differential translation in a tissue- or development-specific manner. ►ribosomes, ►translational control; Mauro VP, Edelman GM 2002 Proc Natl Acad Sci USA 99:12031.

Ribosomal Frame Shifting (translational recoding): ►overlapping genes, ►recoding

Ribosomal Genes: ►*rrn*, ►ribosomes, ►rRNA

Ribosomal Proteins: Ribosomal proteins are generally designated with an S or L indicating whether it is part of the small or large ribosomal subunit. The size of these 55 proteins in *E. coli* range from 6 kDa to 75 kDa. They bind to the RNAs at specific binding sites, either directly or through their association. In *E. coli* the genes for these proteins are scattered among other genes in the chromosome. One of the bacterial ribosomal proteins is present in several copies whereas the other ones occur only once per ribosome. In eukaryotes, about 80 ribosomal proteins exist encoded by a larger number of genes generally occurring in a single or a few copies. Transcription factor Ifh1 of yeast is a key regulator by association with the promoters by the aid of the forkhead-associated factor RAP1 (Wade JT et al 2004 Nature [Lond] 432:1054). The ribosomal proteins assure the proper structural conditions on the ribosomes for translation. About 35% of the bacterial ribosomes are protein. Two-thirds of the chloroplast ribosomal proteins are imported from the cytoplasm and even larger fractions of the mitochondrial proteins are coded by the nucleus. The number of ribosomal proteins in organelles is higher than in prokaryotes. The mitochondrial ribosomes (mitoribosomes) are ~69% protein and ~31% RNA whereas the prokaryotic ribosomes are ~33% protein and ~67% RNA. The number of proteins in the mammalian mitochondrial ribosomes is about 85 and nearly all are imported. The number of ribosomal proteins in the large mitochondria of higher plants is about 65. Proteins bind only single-stranded sequences of RNA. ►nucleolus, ►ribosomes, ►RAP1, ►protein synthesis, ►Diamond-Blackfan anemia; Nomura N et al 1984 Annu Rev Biochem 53:75; Kenmochi N et al 2001 Genomics 77:65; Uechi T et al 2002 Nucleic Acids Res 30:5369; Lecompte O et al 2002 Nucleic Acids Res 30:5382; http://ribosome.miyazaki-med.ac.jp/.

Ribosomal RNA: About 65% of the bacterial ribosomes are RNA. The 16S bacterial rRNA (1.54 kb) has short double-stranded domains and single stranded loops and about ten of the bases near the 3′-end are methylated. The 16S ribosomal RNA undergoes conformational changes (switches) before translation of the mRNA (Lodmell JS, Dahlberg AE 1997 Science 277:1262).

The 16S rRNA is frequently used for taxonomic classification. Actually, both types of base pairings have been found with physiological activity although mutations that favored the Type II conformation favored fidelity of the translation (see Fig. R71). Proteins S5 and S12 facilitate these switches that seem to also play a role in tRNA selection in algae and fungi.

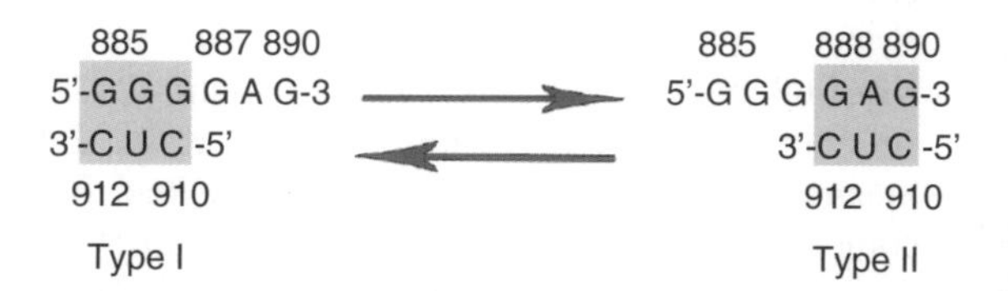

Figure R71. Ribosomal RNA

The 23S rRNA (3.2 kb) carries about 20 methylated bases. The 18S mammalian rRNA (1.9 kb) has more than 40 and the 28S (4.7 kb) has more than 70 methylations. The ribosomal RNAs provide not just a niche for translation, they also interact directly with translation initiation. The 16S rRNA cooperates with the anticodon at both the A and P sites and the 23S rRNA interacts with the CCA end of the tRNA. The ribosomal exit channel (E) plays a regulatory role in peptide elongation (Nakatogawa H, Ito K 2002 Cell 108:629). In the 23S rRNA, two guanine sites are universally conserved (G2252, G2253), and the Cytosine 74 site of the acceptor end of the tRNA (CCA) is required for their functional interaction at the P site of the ribosome for protein synthesis. Methylated sequences in the rRNA mediate, probably, the joining of the small ribosomal subunit to the large subunit after translation is initiated, and holds on to the initiator $tRNA^{fMet}$. Some mutations in the 16S RNA may cause an override through the stop codon and failure of termination of the translation; mutations in the 23S RNA may disturb the A and P ribosomal sites. T 23S ribosomal RNA has six domains (see Fig. R72).

Of these, domain V, in an isolated form, can perform peptide elongation even better than the intact 23S molecule. It seems that ribosomal RNA alone (without protein) is required for peptide elongation (see Fig. R73). Ribosomal ribozymes mediate peptidyl transferase functions. The prokaryotic 23S ribosomal RNA contains the pentapeptide, coding minigene (GUGCGAAUGCUGACAUAAGUA) with a canonical ribosome-binding site and appears to mediate resistance to the antibiotic erythromycin. The 5S bacterial rRNA contains 120 nucleotides and binds three proteins (L25, L18, L5). 5S RNA is also present in eukaryotes where it forms a complex with the L5 ribosomal protein. L25 binds to the E loop of 7 hydrogen-paired nucleotide pairs stabilized by Mg^{2+}. The 3′-end of the small ribosomal subunit is highly conserved from prokaryotes to plants and mammals. For example, in *E. coli*, GAUCACCUCUUA-OH, in yeast, GAUCA—UUA-OH, in maize, GAUCA—UUG-OH, and in rat, GAUCA—UUA-OH, occur (the - signs are inserted for alignment). The function of rRNA synthesis is regulated by homeostasis, feedback, and by many other protein factors. Among them is dksA, a multicopy suppressor of a wide range of apparently unrelated functions. ►nucleolus, ►ribosomes, ►ribosomal genes, ►class I genes, ►class III genes, ►rrn, ►discriminator region, ►introns, ►ribosomal proteins, ►protein synthesis, ►aminoacyl-tRNA synthetase, ►tRNA, ►gene size, ►tmRNA, ►RNase, ►rRNA MRP; Noller HF 1991 Annu Rev Biochem 60:191; Liang W-Q, Fournier MJ 1997 Proc Natl Acad Sci USA 94:2864; Gutell RR et al 2002 Current Opin Struct Biol 12:301; Moore PB, Steitz TA 2002 Nature [Lond] 418:229; Paul BJ et al 2004 Annu Rev Genet 38:749; 5S ribosomal RNA: http://biobases.ibch.poznan.pl/5SData/; diagnostics by 16S rDNA: http://www.ridom.de; post-transcriptionally modified bases in the small subunit ribosomal RNA: http://medlib.med.utah.edu/SSUmods/; rRNA genes: http://rdp.cme.msu.edu; server for 16S ribosomal RNA alignment in prokaryotes: http://greengenes.lbl.gov/cgi-bin/nph-NAST_align.cgi.

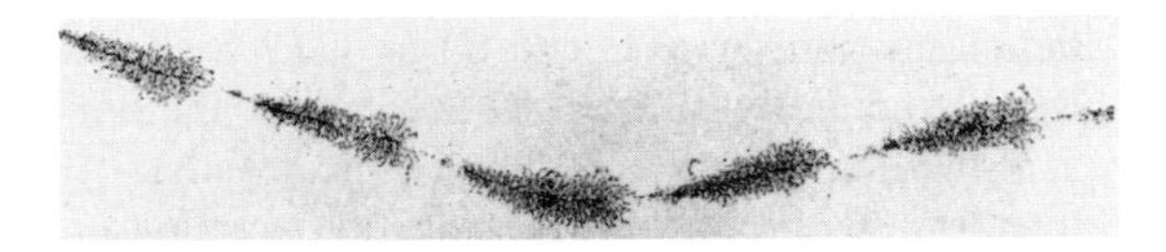

Figure R73. Transcription of ribosomal RNA in the nucleolar organizer region of the chromosomes of *Acetabularia*. The genes are separated by non-transcribed intergenic spacers. One transcription unit is about 1.7 μM. The ribosomal operons in *E. coli* use the proteins S4, L3, L4, and L13 for antitermination with functions similar to the *Nus* genes (Torres M et al 2001 EMBO J 20:3811). (The electronmicrograph is courtesy Spring H et al J Microsc Biol Cell 25:107.)

Ribosome Binding: ►Shine-Dalgarno sequence, ►ribosome, ►mRNA, ►ribosome scanning

Ribosome Binding Assay: The ribosome binding assay was used in the mid 1960s to identify several codons. RNA oligonucleotides that bound to ribosomes attached to those charged tRNA molecules which had the specific anticodons and carried the appropriate amino acids. This way, the relationship between RNA codons and amino acids was revealed. ►genetic code, ►decoding

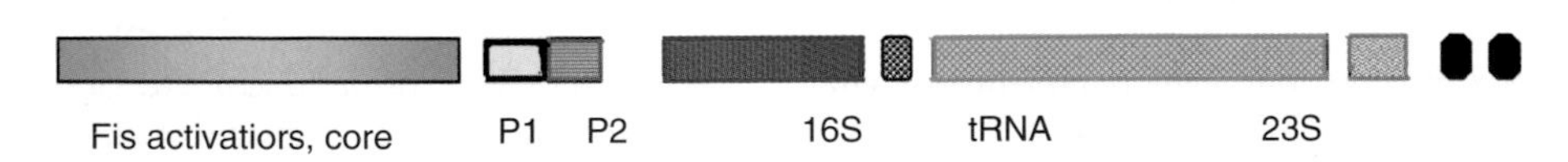

Figure R72. Ribosomal RNA operon in *E. coli*; Fis (factor for inversion stimulation [an old terminology]), core: core promoter, P: promoters, 16S: 16S rRNA genes, tRNA is an intercalataed transfer RNA gene, 23S: 23S ribosomal RNA genes, 5S: 5S rRNA, t1, t2 termination signals. Only 120 bp separates the stronger P1 promoter from P2. The core promoter is the landing site of the RNA polymerase and it includes several hexamers for the recognition of the σ^{70} RNA polymerase holoenzyme. Upstream of the promoters are A-T-rich sequences (UP) that enhance polymerization. The carboxyterminal domain (CTD) of the α subunit also recognizes this area. Promoters in the different systems show variations from the main scheme.

Ribosome Display: The same as RNA-peptide fusions. ►display technologies; Hanes J et al 2000 Nature Biotechnol 18:1287.

Ribosome Hopping: The bypassing of the coding gaps in phage T4 genes with the assistance of a special protein factor. (See Herr AJ et al 2001 J Mol Biol 311:445).

Ribosome Recycling (RRF): After the termination of the translational process, the ribosome is disassembled into the small and large subunits. The post-termination complex in prokaryotes contains release factors RF1, RF2 and RF3, the 70S ribosome with the mRNA still attached to it, a deacylated tRNA at thc P site, and an empty A site. This complex is split up by RRF (ribosome recycling factor, Agrawal RK et al 2004 Proc Natl Acad Sci USA 101:8900) and the elongation factor G (EF-G) by GTP hydrolysis. RRF structurally mimics tRNAs, except the amino acid-binding 3′-terminus. This indicates that RRF interacts with the post-termination complex in a manner similar to that of a tRNA responding to the ribosome. ►ribosomes, ►protein synthesis, ►A site, ►P site, ►EF-G, ►aminoacylation; Kisselev LL, Buckingham RH 2000 Trends Biochem Sci 25:561; Inokuchi Y et al 2000 EMBO J 19:3788; Hirokawa G et al 2002 J Biol Chem 277:35847.

Ribosome Scanning: Eukaryotic mRNAs do not have a Shine-Dalgarno consensus for ribosome binding. They are probably attached by the 5′-m^{7}G(5′)pp. The (5′)mRNA sequence reels on the ribosome until the initiator methionine codon is found. Circular viral or eukaryotic RNAs, if they contain internal ribosome entry sites (IRES), may be translated without a need for a free 5′-end. The RNA helicase eIF4E and the other translation initiation factor eIF4G, as well as the cap-binding protein eIF4E and the tail-binding protein Pab1, are instrumental in a cooperative manner in the initiation of translation. ►Cap, ►Shine-Dalgarno, ►protein synthesis, ►IRES, ►dicistronic translation, ►Kozak rule, ►eIF, ►elongation initiation, ►translation, ►ribosome shunting; Kozak M 1989 J Cell Biol 108:229; Sachs AB 2000 Cell 101:243; Kozak M 2002 Gene 299:1.

Ribosome Shunting: The long leader sequence of viral DNA (adenovirus, cauliflower mosaic virus, etc.) may contain several short open reading frames, which usually interfere with the translation of the downstream ORFs. These impediments may be bypassed (jumped) by the formation some internal structure, e.g., stem-loop structure. It seems that for the proper scanning of the ribosome the ~100 nucleotides at the 5′ and 3′ ends are most essential. ►adenovirus, ►cauliflower mosaic virus, ►ORF, ►stem-loop, ►ribosome scanning; Pooggin MM et al 2001 Proc Natl Acad Sci USA 98:886.

Ribosome Skipping: The same as translational bypassing.

Ribosomes: Ribosomes provide the workshop and some of the tools for protein synthesis in all cellular organisms, including the subcellular organs, mitochondria, and chloroplasts. A yeast cell contains about 200,000–2,000,000 ribosomes. The chloroplastic ribosome genes are situated in the characteristic inverted repeats, except in some *Fabaceae* and conifers. The prokaryotic and organellar ribosomes are similar and their approximate molecular weight is 2.5×10^6 Da with a sedimentation coefficient of 70S. The higher eukaryotic ribosomes, excluding the organellar ones, have a molecular weight of about 4.5×10^6 Da, and they are ≈ 80S. The ribosomes have both a minor and a major subunit built of RNA and protein (see Fig. R74).

The prokaryotic 30S subunit includes 20 proteins, assembled through several steps of conformational transitions as detected by pulse and chase monitored with the aid of quantitative mass spectrometry (Talkington MWT et al 2005 Nature [Lond] 438:628). The crystal structure of the 70S ribosome complexed with mRNA and tRNA has been determined (Selmer M et al 2006 Science 313:1935).

Hrr25 protein kinase-dependent phosphorylation regulates the organization of the pre-40S eukaryotic ribosomal subunit (Schäfer T et al 2006 Nature [Lond] 441:651). The large subunit, 50S is connected to the small subunit, 30S by an RNA-protein bridge. From the middle of the large subunit a tunnel runs toward the small subunit. The formation of the 80S ribosome requires the mediation of eIF5. The mitochondrial ribosomes do not contain 5S subunits but chloroplasts do have them. The number of ribosomes may greatly increase when protein synthesis is very rapid. During early embryogenesis in amphibia, the rRNA gene number may increase three orders of magnitude by a process of amplification and the extra copies of the genes are sequestered into minichromosomes forming micronuclei. Their number in some higher plants regularly runs into thousands. The bacterial ribosomes are about 65% RNA and 35% protein. On the ribosomes, several active centers can be distinguished. The A and P sites receive the tRNAs. This area extends to both the small and the large subunits. The tRNA after unloading of the amino acids leaves the ribosome at the exit site (E) of the large subunit. Some of the peptide chains exit through the tunnel of the large subunit. The translocation factor EF-G seems to occupy a space in between the two subunits. The complex crystal structure of the *E. coli* ribosome is

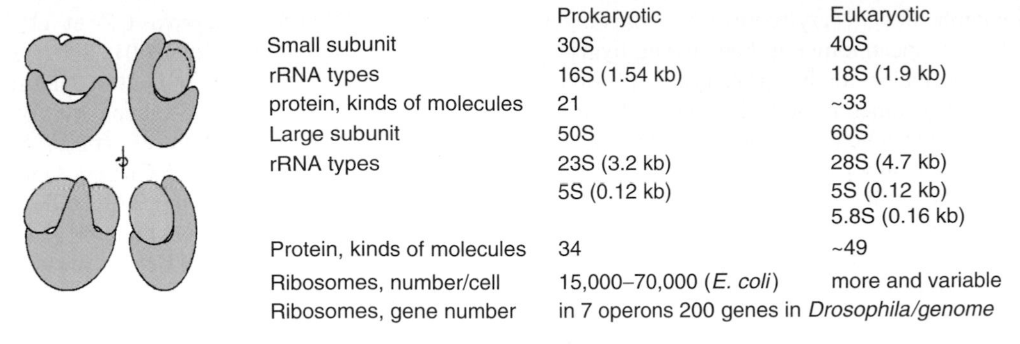

	Prokaryotic	Eukaryotic
Small subunit	30S	40S
rRNA types	16S (1.54 kb)	18S (1.9 kb)
protein, kinds of molecules	21	~33
Large subunit	50S	60S
rRNA types	23S (3.2 kb)	28S (4.7 kb)
	5S (0.12 kb)	5S (0.12 kb)
		5.8S (0.16 kb)
Protein, kinds of molecules	34	~49
Ribosomes, number/cell	15,000–70,000 (*E. coli*)	more and variable
Ribosomes, gene number	in 7 operons	200 genes in *Drosophila/genome*

Figure R74. 70S prokaryotic ribosome viewed from different angles (Courtesy Tischendorf GW et al 1975 Proc Natl Acad Sci USA 72:4870). The ribosome structure is much more complex than shown in the diagram (see Brimacomb R, p 41 In: Eggleston RA et al (Eds.) The Many Facets of RNA. 1998 Acadamic Press, San Diego, CA, USA). By the use of crystallography, nuclear magnetic resonance, neutron diffraction and cryo-electronmicroscopy more detailed three-dimensional structures have been revealed. Also chemical foot printing, mutation and other probes have identified the links between ribosomal components and ribosomal sites of tRNA and mRNA.)

described at 3.5 Å resolution (Schuwirth BS et al 2005 Science 310:827). The *classical* model of translation is discussed under the protein synthesis entry. Binding of the anticodon stem-loop of P-site tRNA to the ribosome is sufficient to lock the head of the small ribosomal subunit in a single conformation, thereby preventing movement of mRNA and tRNA before mRNA decoding (Berk V et al 2006 Proc Natl Acad Sci USA 103:15830).

The newer hybrid-states model is given in the diagram. The elongation factor EF-TU may be located on the small subunit but it communicates with EF-G. The EF-Tu.GTP.tRNA ternary complex is instrumental in the delivery of the aminoacyl-tRNA to the A/T hybrid site of the peptidyl tRNA-ribosome in complex. (A is the amino acid, P the peptidyl, and T is the corresponding large subunit site.) In prokaryotes, then, the anticodon binds to the A site of the 30S ribosomal subunit. Hydrolysis of GTP is followed by the release of the elongation factor EF1A/EF-Tu. After this, the CCA end of the amino acid-charged tRNA moves to the A site of the large (50S) subunit. Peptidyl transferase (located at the end of the tunnel closest to the small subunit) then mediates peptide bond formation between the nascent peptide chain and the next incoming amino acid. After the peptide bond is initiated the peptidyl-tRNA is deacylated and is transferred to the P site on the large subunit. The anticodon stays for a while at the A site of the small subunit while the CCA end is on the P site of the large subunit. Then the anticodon moves to the P site of the large subunit and the CCA site goes to the E site of the large subunit (see Fig. R75) (Márquez V et al 2004 Cell 118:45). Meanwhile, after the recognition of the cognate codon by the anticodon, the EF1A·GTP complex moves to the GTPase center of the ribosome, and EF1A·GDP and the tRNA are released at the E site. After the translocation from the A site to the P site, the elongation factor EF1B mediates the conversion of the inactive EF1A·GDP into the active EF1A·GTP. This and other more recent models indicate that the movement on the ribosomes involves the tRNA, but the peptidyl moiety moves very little. If the A site of a prokaryotic ribosome is unoccupied, water may carry out a nucleophilic attack on the peptidyl-tRNA at the P site. In the absence of an appropriate A-site substrate, the peptidyltransferase center can position the ester link of the peptidyl-tRNA in a conformation that prevents the nucleophilic attack by water. Protein factors may also assist in the protection (Schmeing TM et al 2005 Nature [Lond]:438:520).

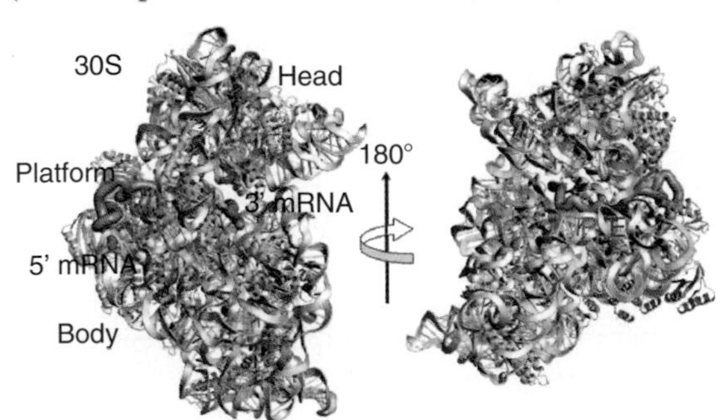

Figure R75. mRNA on the ribosome. Left: Solvent site view of the 30S sunuit with 5′ end (position −36) and 3′-end (position +12) of the mRNA. Right: interface view of the 30S subunit with A,P and E codons of the mRNA. Courtesy of Dr. Marat Yusupov, IGBMC, Illkirch, France; see also Yusupov, G *et al.* 2006 Nature [Lond] 444:391

Chloramphenicol, erythromycin, lincomycin, streptomycin, spectinomycin, kanamycin, hygromycin, etc., normally inhibit prokaryotic ribosomes. Eukaryotic ribosomes are sensitive to cycloheximide, anisomycin, puromycin, tetracyclines, etc. The ribosomes play an important role in the regulation of protein synthesis. It appears that the availability of active ribosomes is controlled at the level of the transcription of the rRNA genes. When the supply of ATP and GTP is adequate, rRNA genes are activated for transcription. In case the level of these nucleotide triphosphates is low, rRNA transcription is reduced or halted. An abundance of free ribosomal proteins may feedback-inhibit ribosomal production. The ribosome-associated Rel-A protein may mediate the formation of ppGpp from GTP (and possibly from other nucleotides). Then, ppGpp (or pppGpp) may shut off rRNA and tRNA synthesis by binding to the promoter of the RNA polymerase or to its anti-termination signal. ►nucleolus, ►protein synthesis, ►ribosomal proteins, ►E site, ►A site, ►P site, ►Sec61 complex, ►ribosomal RNAs, ►EF-G, ►EF-TU-GTP, ►transorientation hypothesis, ►antibiotics, ►discriminator region, ►ribosome recycling, ►ribonuclease R, ►chloroplasts; Green R, Noller HF 1997 Annu Rev Biochem 66:679; Venema J, Tollervey D 1999 Annu Rev Genet 33:261; for crystalline structure of the small and the large subunits of bacterial ribosomes: Clemons WM Jr et al 1999 Nature (Lond.) 400:833; Ban N et al ibid. 841; Cate JH et al 1999 Science 285:2095; Wimberly BT et al 2000 Nature (Lond) 407:327; Ban N et al 2000 Science 289:905; Yusupova GZ et al 2001 Cell 106:233; Ogle JM et al 2001 Science 292:897; LaFontaine DLJ, Tollervey D 2001 Nature Rev Mol Cell Biol 2:514; Moss T, Stefanovsky VY 2002 Cell 109:545; Doudna JA, Rath VL 2002 Cell 109:153; Fatica A, Tollervey D 2002 Current Opin Cell Biol 14:313; Moore PB, Steitz TA 2003 Annu Rev Biochem 72:813; http://rdp.cme.msu.edu; small subunit: http://www.psb.ugent.be/rRNA/ssu/; large subunit: http:/www.psb.ugent.be/rRNA/lsu/.

Riboswitch: In certain mRNAs there are untranslated receptor elements for target metabolites. Highly selective binding of the metabolite permits a conformational switch (somewhat similar to attenuation) that may lead to modulation in the synthesis of a protein or by cutting up mRNA with the aid of a ribozyme (see Fig. R76).

In some instances, tandem riboswitches control complex regulation (Sundarsan N et al 2007 Science 314:300). In some cases, RNase P cleaves the riboswitch (Altman S et al. 2005 Proc Natl Acad Sci USA 102:11284). The crystal structure of *Arabidopsis* thiamine pyrophosphate riboswitch provides information on its basic function (Thore S et al 2006 Science 312:1208). More than 2% of the genes in some species are regulated by riboswitches. Riboswitches have high specificities and can discriminately bind even among guanine (G), adenine (A) or hypoxanthine. A more complex binding pattern has been identified in the thiamine pyrophosphate (TPP) riboswitch (Serganov A et al 2006

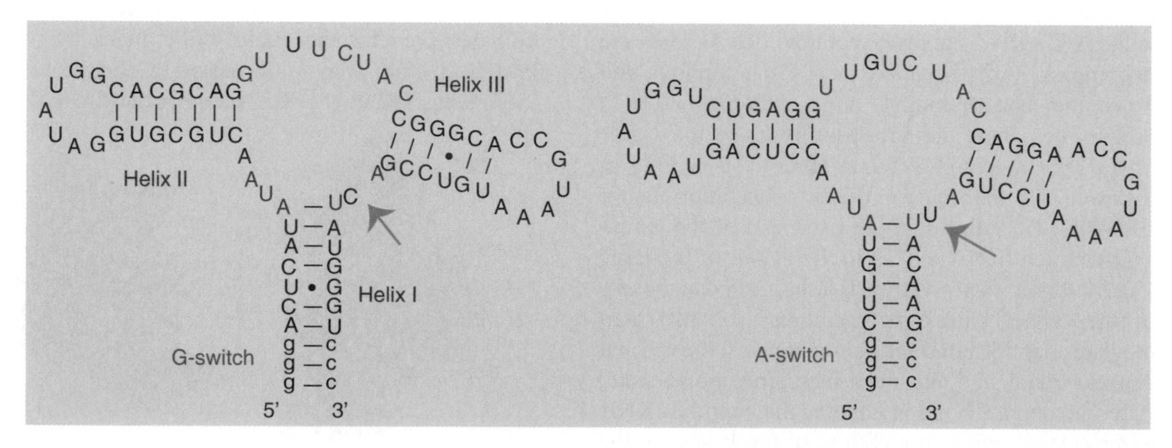

Figure R76. Riboswitch. Secondary structure guanine (G) and adenine (A) switches. The two switches appear highly similar, almost identical, and yet they have high specificity. The red core sequences are well conserved, except the bases represented at purple arrows. The lower-case letters represent residues not found in the original and which are introduced here to facilitate in vitro transcription. (Modified from Noeske J et al 2005 Proc Natl Acad Sci USA 102:1372.)

Nature [Lond] 441:1167) or *S*-adenosylmethionine riboswitch (Montange RK, Batey RT 2006 Nature 441:1172). ►repressor, ►attenuator region, ►aptamer, ►ribonuclease P; Winkler WC et al 2002 Proc Natl Acad Sci USA 99:15908; Mandal M et al 2003 Cell 113:577; Mandal M, Breaker R 2004 Nature Rev Mol Cell Biol 5:451; purine-specific riboswitch structure: Batey RT et al 2004 Nature [Lond] 432:411; riboswitches in genomes: Hammann C, Westhof E 2007 Genome Biol 8:210; alternative splicing: Cheah MT et al 2007 Nature [Lond] 447:497.

Ribothymidine: A thymine in the tRNA which is attached to ribose rather than to deoxyribose as it occurs in the DNA. ►tRNA

Ribotype: The RNA pool (similar to genotype for DNA); the information content of the RNA. It is different from the genotype because by differential processing, splicing, editing, etc., it may convey different meanings. The processed variations—during the course of evolution—may be integrated into the DNA genetic material with the aid of reverse transcriptases and become part of the "hard heredity". (See Herbert A, Rich A 1999 Nature Genet 21:265).

Ribozyme: A catalytic RNA possessing enzymatic activity such as splicing RNA transcripts, cleavage of DNA, amide and peptide bonds, polymerization and limited replication of RNA, etc. Thus, these ribonucleic acids carry out functions similar to those of protein enzymes. Ribozymes are generally metalloenzymes, commonly using Mg^{2+} for catalysis and stabilization.

Some viral ribozymes do not require metal ions to cleave phosphodiester bonds. Most commonly they cleave phosphodiester bonds but they can also synthesize nucleotide chains. The ribozymes are generally large molecules yet the shortest ribozyme is only UUU and it acts on CAAA. A two-base ribozyme may catalyze the formation of 3′,5′ phosphodiester linkages 36,000-fold faster (Reader JS, Joyce GF 2002 Nature [Lond] 420:841). Ribozymes commonly have an internal guide for substrate recognition near their 5′ terminus and a *splice site* (self-cleavage or catalytic site) where they cleave and splice the molecules. Frequently, ribozymes are classified into groups such as the hammerhead ribozymes (see Fig. R78) that are used mainly by plant RNA viruses, the RNase P, the delta, group I, and hairpin ribozymes (see Fig. R77). The hammerhead ribozymes cut at the UCX sequence if the neighboring sequences are complementary and pair. Tertiary contacts distant from the active site prime hammerhead ribozymes for catalysis (Martick M, Scott WG 2006 Cell 126:309). The hairpin ribozymes must have at least ~50 nucleotides in the catalytic domain and ~14 in the substrate domain. The two domains pair in a two-stem form separated with an unpaired loop of the ribozyme and substrate, most commonly containing a 5′-AGUC-3′ sequence. Cleavage is usually between A and G of the substrate.

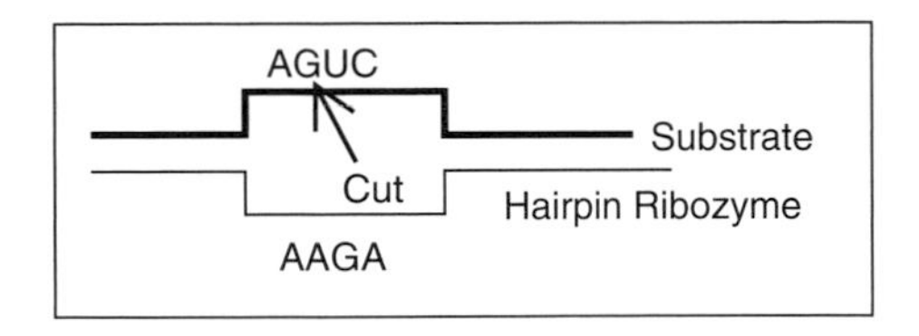

Figure R77. Hairpin ribozyme

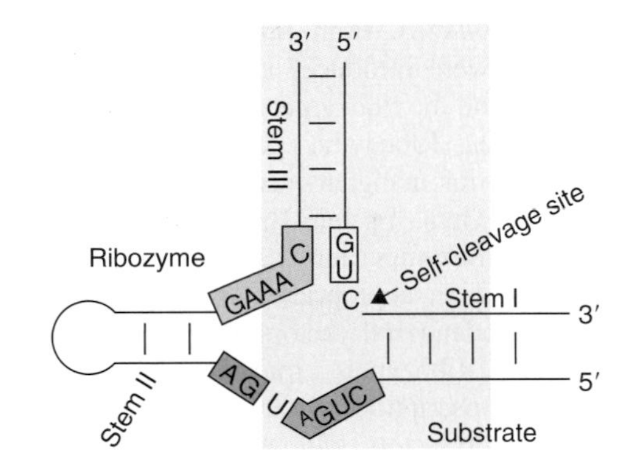

Figure R78. The critical sequences in the hammerhead ribozymes (boxed nucleotides) are conserved

RNA-catalyzed RNA polymerization has also been identified. Various proteins may affect the substrate binding by base-pairing and cleavage product release. Although ribozymes may cleave molecules in trans, their efficiency is usually better for cis substrates. Ribozymes functioning as ligases or polynucleotide kinases in mRNA repair by trans-splicing. Isomerases as well as self-alkylating catalysts have also been isolated from large pools (10^{14}) of diverse RNAs.

From an evolutionary point of view these diverse ribozyme functions lend support to the ideas of the prebiotic RNA world (see Fig. R79). Ribozymes can be engineered to recognize specific mRNAs and by cleaving them prevent the expression of a particular protein. They an have advantage over protein enzymes because it is less likely that they would incite an immune reaction. Because of their small size, their introduction into the cell is facilitated with the aid of transformation vectors. Small RNA transcription units that may accumulate up to 106 copies per cell can propagate them. Transcription units for tRNA, U6 snRNA have been used. Although these units produce high ribozyme titer in the cell, polymerase II transcribed units target the ribozymes more effectively to the desired location. In such Pol II units the ribozyme motif is inserted into the 5′ untranslated sequences. Evolution of ribozymes may be a much faster process than that of protein enzymes. A single polynucleotide sequence may fold into two different conformations and display two different catalytic activities such as such as the Hepatitis Delta Virus self-cleaving ribozyme and a class III self-ligating ribozyme. These ribozymes may not share more than 25% (random) nucleotide identity and yet their folding pattern may satisfy the requirements of the two functions (Schultes EA, Bartel DP 2000 Science 289:448). Both hammerhead and hairpin ribozymes were introduced into human cells infected with HIV and the ribozymes reduced the level of the gag protein. Ribozyme gene therapy has been considered for malignancies caused by the human papilloma virus, Epstein-Barr virus, and hepatitis viruses. Ribozymes thus have various therapeutic potentials if an appropriate targeting system (e.g., retroviral, adenoviral vectors, cationic liposomes) is available. Ribozymes may inactivate tyrosine kinases, transcription factors, cell adhesion molecules, growth factors, telomerase, etc. The ribozyme may recognize complementary RNA targets. Problems may include low efficiency of transfection, poor target recognition, transcriptional silencing, instability, etc. Inadequate target recognition can apparently be remedied by adding a biosensor to the ribozyme, the absence of which the ribozyme is in an inactive conformation. When the sensor recognizes a specific substrate, the ribozyme conformation is changed and it is activated, and it cleaves the specific target (Bergeron LJ, Perrault J-P 2005 Nucleic Acids Res 331240). Ribozymes may be used in the same way as antisense constructs but with the special advantage of catalytic activity. ►introns, ►ribonuclease P, ►peptidyl transferase, ►CBP2, ►deoxyribozyme, ►ligase RNA, ►kinase, ►alkylation, ►RNA world, ►RNA restriction enzyme, ►gene therapy, ►cancer gene therapy, ►HIV, ►SELEX, ►liposomes, ►cytofectin, ►lipids cationic, ►viral vectors, ►transfection, ►antisense technologies, ►transdominant molecules, ►leadzyme, ►DNAzyme; Wadekind JE, Mckey DB 1998 Annu Rev Biophys Biomol Struct 27:475; Doherty EA, Doudna JA 2000 Annu Rev Biochem 69:597; Ferbeyre G et al 2000 Genome Res 10:1011; Takagi Y et al 2001 Nucleic Acids Res 29:1815; Doudna JA, Cech TR 2002 Nature [Lond] 418:222; hairpin ribozyme: Nahas MK et al 2004 Nature Struct Mol Biol 11:1107; structure–function of hammerhead ribozymes: Blount KF, Uhlenbeck OC 2005 Annu Rev Biophys Biomol Struct 34:415.

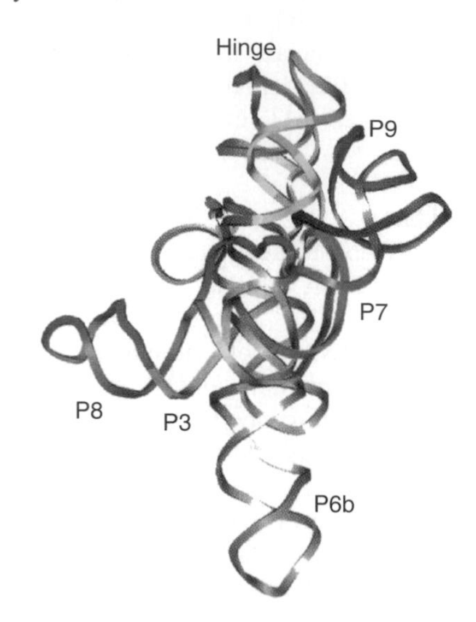

Figure R79. The crystal structure of a 247-nucleotide *Tetrahymena* ribozyme of an rRNA intron. The conserved helical elements are identified by P1 TO P9. (Courtesy of Dr. Tom Cech. From Golden BL et al 1998 Science 282:259)

Ribulose 1,5-Bisphosphate Carboxylase-Oxidase (Rubisco): A chloroplast-located enzyme, generally the largest amount of protein in that organelle. Its large subunit is encoded and translated by the chloroplast system; the small subunit is coded for by a nuclear gene, translated in the cytosol, and imported into the plastids. The abundance of the small subunit affects the translation of mRNA of the large subunit. Rubisco is involved in early steps of photosynthesis and works through an intermediate 3-phosphoglycerol and it is involved in oxidative and reductive carboxylation. In dinoflagellates, Rubisco is encoded by the nucleus. ►chloroplast genetics, ►photosynthesis, ►Rubisco; Suss K et al 1993 Proc Natl Acad Sci USA 90:5514; Taylor TC et al 2001 J Biol Chem 276:48159; Spreitzer RJ, Salvucci ME 2002 Annu Rev Plant Biol 53:449.

Rice (*Oryza*): Gramineae, x = 12, genome size ~430 Mbp; the species is either diploid or tetraploid. The genome size was revised to 389 Mb in 2005 (see Fig. R80).

Figure R80. Oryza

Over 2,000 molecular markers are available in this small genome. The generation time is 90–140 days. Two genomes were sequenced by 2002 (Science 296:79 and 92) and seem to have more open reading frames (32,000–55,000) than other organisms. After sequencing the genome the total nontransposable element-related and protein-coding gene number was revised to 37, 544, and 29% of them are clustered in families. 2,859 of the genes appear unique to rice and other cereals. A transcription map of 35,970 (81.9%) of the genes and the existence of 54,564 transcribed intergenic sequences is available for *indica* rice (Li L et al 2006 Nature Genet 38:124). Homologues to *Arabidopsis* appear to be 71% and 90% of the Arabidopsis genes are homologous to those in rice. Of the 11,487 *Tos17* retroposon sites, 3,243 are within genes. The nuclear genomes showed organellar DNA fragments in 0.38% to 0.43%. There are 80,127 polymorphic sites which distinguish the two cultivated rice subspecies, *japonica* and *indica*. Single nucleotide polymorphism was high, 0.53% to 0.78%, about 20 times higher than in *Arabidopsis* (Matsumoto T et al 2005 Nature [Lond] 436:793).

Two groups of quantitative gene loci (QTL) have been successfully identified with regard to biochemical activity and their significance in breeding more productive crops. A 383-bp deletion of a gibberellin oxidase gene (*Ph1*, plant height) lead to high-yielding short-stem strains (Sasaki A et al 2002 Nature [Lond] 416:701). Gene *Gn1*, encoding cytokinin oxidase/dehydrogenease enzyme at reduced expression and increased cytokinin level in the reproductive organs and increased grain number per panicle by 44% and consequently increased yields. Several *Gn* genes were scattered over five chromosomes (Ashikari M et al 2005 Science 309:741). The completely sequenced chromosomes 11 and 12 contain 289 disease resistance-like and 28 defense-response genes (BMC Biology 2005, 3:20). *O. sativa* cultivars die if completely submerged in water for two weeks. The *Sub1A* allele, near the centromere of chromosome 9, is an ethylene-response-factor-like gene and can convey tolerance to flooding and increases crop security (Xu K et al 2006 Nature [Lond] 442:705). ►cytokinin, ►gibberellins, ►QTL, ►candidate gene, ►*Magnaporthe grisea*; Shimamoto K, Kyozuka J 2002 Annu Rev Plant Biol 53:399; Sasaki T et al 2002 Nature [Lond] 420:312; Feng Q et al 2002 Nature [Lond] 420:316; Kikuchi S et al 2003 Science 301:376; sequenced genome of the pathogen *Xanthomonas oryzae*: Lee B-M et al 2005 Nucleic Acids Res 33:577; Molecular Biological Encyclopedia: http://cdna01.dna.affrc.go.jp/cDNA/; http://rgp.dna.affrc.go.jp/giot/INE.html; http://www.tigr.org/tdb/tgi/; annotation database: http://www.tigr.org/tdb/e2k1/osa1/; reverse genetics: http://orygenesdb.cirad.fr/; annotations: http://rapdb.lab.nig.ac.jp/; mutants: http://rmd.ncpgr.cn/; annotation database: http://rad.dna.affrc.go.jp/; LTR retrotransposons: http://www.retroryza.org/; http://rice.tigr.org.

Richner-Hanhart Syndrome: ►tyrosine aminotransferase

Ricin: An extremely toxic, ribosome-inactivating, dimeric toxin produced by the plant, castor bean (*Ricinus*). The estimated lethal dose is 1–10 μg/kg when delivered as an injection or aerosol to humans. It may be significant in cancer therapy research. The deglycosylated ricin toxin A chain (dgRTA)—when containing three-amino acid mutations (xAspyat, position 97) where x can be Leu, Ile, Gly or Val and y could be Val, Leu, Ser—causes practically no liver damage and shows much reduced vascular leak syndrome. Liver damage and vascular leakage in the lung would be the major obstacles requiring the therapeutic use of this toxin (Smallshaw JE et al 2003 Nature Biotechn 21:387). Limited trials indicate that an effective and low-risk vaccine can be produced against ricin. By using recombinant DNA technology changes can be created at the ribotoxic A chain that can prevent vascular leak effects of ricin. Tests show that 100 μg of the modified antigen provided 100% protection when injected at monthly intervals into five human volunteers (Vitetta ES et al 2006 Proc Natl Acad Sci USA 103:2268). ►magic bullet, ►RIP, ►biological weapons, ►castor bean; Lord JM et al 1991 Semin Cell Biol 2:15; Day PJ et al 2001 J Biol Chem 276:7202.

Ricinosome: A protease precursor vesicle formed from the endoplasmic reticulum in senescing plant tissues. It contains large quantities of a 45 kDa cystin endoprotease and other proteins required for apoptosis. ►apoptosis, ►protease; Schmid M et al 2001 Proc Natl Acad Sci USA 98:5353.

Rickets: Anomalies in bone development caused by defects of calcium and phosphorus absorption and/or vitamin D deficiency. Human autosomal recessive conditions may be caused by defects in the synthesis of calciferol (vitamin D) from sterols; in such cases vitamin D_3 can correct the dependency. In some forms the receptor is defective and vitamin D cannot alleviate the hereditary condition (12q12-q14). Deficiency of pseudovitamin D (25-hydroxycholecalciferol-1-hydroxylase) is also at about the same chromosomal area. Hypophosphatemia (Xp22.2-p22.1) may also cause vitamin D unresponsive rickets. Rickets may then have multiple phenotypic consequences such as alopecia, epilepsy, etc. ►hypophosphatasia, ►hypophosphatemia, ►vitamin D, ►spermine, ►Dent disease

Rickettsia: Small rod-shape or roundish, obligate intracellular, Gram-negative bacteria. They may carry typhus (typhoid fever, accompanied by eruptions, chills, headaches and high mortality) and spotted fever, a tick-borne disease of cerebrospinal meningitis (brain inflammation) from animals to humans by infected arthropods (ticks, lice, fleas). The mammalian receptor for *R. conorii* is Ku70 and the *Ricketsia* protein rOmpB is a ligand for the internalization process. Ubiqiutin ligase c-Cbl is also recruited to the entry foci to block invasion by partial destruction of Ku70 (Martinez JJ et al 2005 Cell 123:1013). The genome of *Rickettsia prowazekii* contains 1,111,523 bp and it shows the closest similarity to mtDNA of eukaryotes (Andersson SG et al 1998 Nature [Lond] 396:109). ►mtDNA, ►mitochondria, ►endosymbiont theory, ►*Wolbachia*, ►Ku70, ►OMP, ►ubiquitin; Nature [Lond] 396:133 for complete physical map; ►reductive evolution, ►*Anaplasma marginale*; Ogata H et al 2001 Science 293:2093.

Rictor: A rapamycin-insensitive mTOR. ►rapamycin, ►TOR

RID (random insertion/deletion): A technique by which certain bases of the DNA can be deleted and/or replaced at various positions, thus generating mutations. ►targeting genes, ►mutagenesis; Murakami H et al 2002 Nature Biotechnol 20:76.

RIDGE (region of increased gene expression): The highly expressed genes appear to be clustered in the chromosome as detected by SAGE. These domains have high G-C content, high SINE and low LINE repeat density and shorter introns. Antiridge domains display opposite characteristics. ►SAGE, ►clustering of genes; Versteeg R et al. 2003 Genome Res 13:1998.

Rieger Syndrome: An autosomal dominant eye, tooth and umbilical hernia syndrome. Its chromosomal location (just as the Nazi discoverer of the disease) was controversial. Human chromosomes 21q22, 4q25, 13q14 and several others have been implicated. The basic cause is also unclear; epidermal growth factor, interleukin-2, alcohol dehydrogenase, fibroblast growth factor deficiencies appear to be involved in chromosome 4. The chromosome 4 gene (RIEG) has now been cloned and it encodes a transcription factor with similarities to the *bicoid* gene of *Drosophila*. Vertebrate homologs *Pitx, Potxlx* and *Apr-1* mediate the left-right development of visceral organs in concert with a number of other genes such as *Sonic hedgehog*, *Nodal*, etc. ►eye diseases, ►tooth agenesis, ►morphogenesis in *Drosophila* {8}, ►left-right asymmetry; Saadi I et al 2001 J Biol Chem 276:23034.

Rifampicin: An antibiotic that inhibits prokaryotic DNA-dependent RNA polymerase (but not the mammalian RNA polymerase) and inhibits replication in *E. coli* and other prokaryotes. Rifamycin has similar effects. The antibiotic effects of the different rifamycins may vary. ►antibiotics, ►maytanisoids; Campbell EA et al 2001 Cell 104:901.

Rifins (repetitive interspersed family): Rifins are *Plasmodium falciparum* proteins in high copy number, involving antigenic variation. They are instrumental in the infection by the parasite. ►*Plasmodium falciparum*, ►antigenic variation

Rigens: A translocation complex in *Oenothera muricata*. If during meiosis this complex goes to the top end of the megaspore tetrad, it may overcome its topological disadvantage and this megaspore may develop into an embryo sac because the other complex, *curvans*, is not functional in the megaspores. ►megaspore competition, ►*Oenothera*, ►zygotic lethal

RIG Oncogene: The RIG oncogene is probably required for all types of cellular growth and it is active in a very wide variety of cancers. The C-terminal domain of RIG-1 binds double-stranded RNA and thus senses infectious viruses. It also stimulates the expression of transcription factors such as NF-κB, IRF3, ATF2, and provides protection against virus infection. The overexpression of MAVS (mitochondrial antiviral signaling) boosts the protection by IFN-β stimulation. ►oncogenes, ►NF-κB, ►IRF3, ►ATF2; Seth RB et al 2005 Cell 122:669; McWhirter SM et al 2005 Cell 122:645; detection of viral RNA by RIG-1:

Hornung V et al 2006 Science 314:994; RIG-1 response to single-strand RNA: Pichlmair A et al 2006 Science 314:997.

RIGS (repeat-induced gene silencing): An apparently epigenetic phenomenon caused by the methylation of cytosine residues. An alternative process is that locally paired regions of homologous sequences are flanked by unpaired heterologous sequences in transgenic plants. Silencing and variegation occurs when transgenes are inserted either in trans or cis position to heterochromatin. The degree of silencing is proportional to the distance between the transgene and the heterochromatin. ►methylation of DNA, ►RIP, ►silencing, ►epigenesis, ►co-suppression, ►heterochromatin, ►position effect; Selker EU 1999 Cell 97:157.

Riken (Genome Exploration Research Group, Genome Science Laboratory): Technologies, cDNA encyclopedias, etc. (http://genome.gsc.riken.jp).

Riley-Day Syndrome (dysautonomia): A human chromosome 9q31-q33 recessive neuropathy involving emotional instability, lack of tearing, feeding difficulties, unusual sweating, cold extremities, etc. The prevalence in Ashkenazic Jews is about 2–3 × 10^{-4} but it is quite rare in other ethnic groups. Defects in the nerve growth factor receptor are suspected. ►neuropathy, ►nerve growth factor, ►IκB, ►pain-insensitivity; Slaugenhaupt SA et al 2001 Am J Hum Genet 68:598.

Ring Bivalent: A ring bivalent has terminalized chiasmata in both arms (see Fig. R81) and thus in early anaphase I the homologous chromosomes appear to be temporarily connected at the telomeric regions of the four chromatids (see Fig. R82). This, however, is not a ring chromosome. ►translocation ring

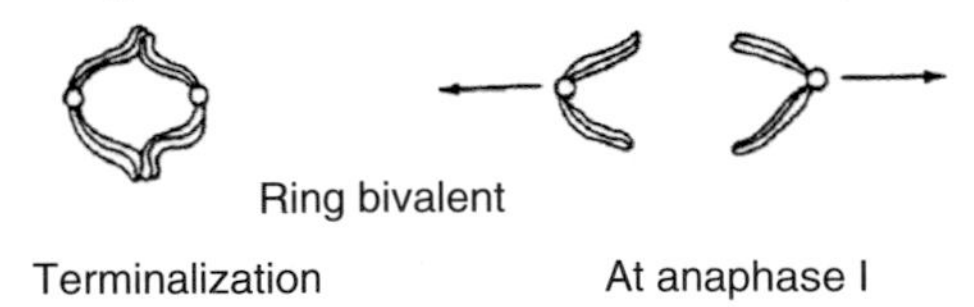

Figure R81. Terminalization of ring bivalents

Figure R82. Ring and rod bivalents

Ring Canal: Ring canals are intercellular bridges during cytoblast differentiation (see Fig. R83). They transport mRNA and proteins from nurse cells to the oocytes. They are composed of actin, the hts (hui-li tai shao), and the kelch proteins. ►cytoblast, ►maternal effect genes, ►RNA localization

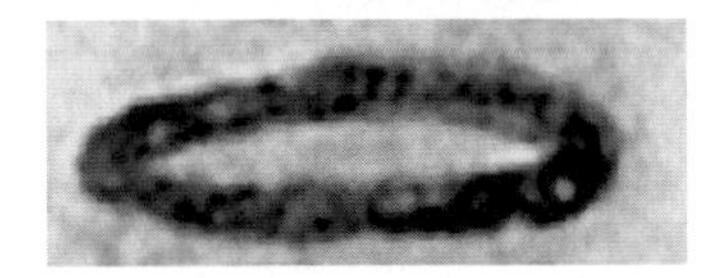

Figure R83. Ring chromosome

Ring Chromosome: A circular chromosome without free ends (o)—such as the bacterial chromosome—as the ring DNAs in mitochondria and plastids. Ring chromosomes may result by different types of chromosome breakage. Simultaneous breaks across the centromere and the chromosome ends (telomeres) may result in the fusion of the two broken termini generating one or two ring chromosomes. Also, crossing over between two ends of the same chromosome may give rise to a centric ring and acentric fragments. Sister chromatid exchange within a ring chromosome may result in a dicentric ring chromosome, which at anaphase separation may break at various points and generate unequal size ring chromosomes and genetic instability (see Fig. R84). ►dicentric ring chromosome, ►ring bivalent, ►translocation ring, ►sister strand exchange, photomicrograph by Dr. D. Gerstel.

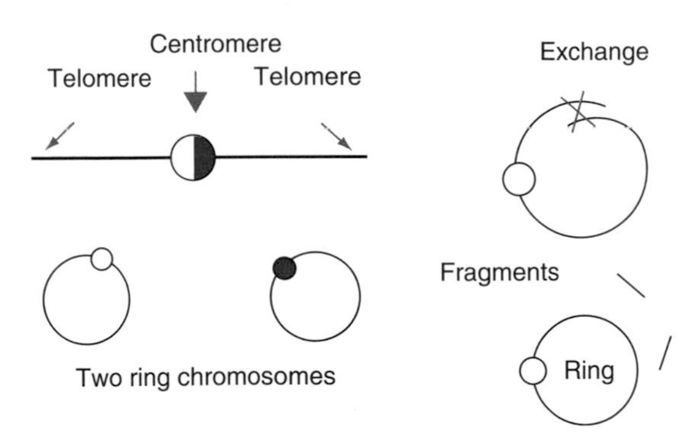

Figure R84. Left: A simultaneous breakage at both the centromere and at the telomeres may result in fusion between the broken ends and the generation of ring chromosomes from both of the arms of the normal chromosome generating two new ring chromosomes. Right: Exchange between the two ends yields a ring chromosome and two acentric fragments. For the sake of simplicity the chromatids of the chromosomes are not represented in the diagrams

RING Finger: A cysteine-rich amino acid motif such as Cys-X2-Cys-X(9–27)-Cys-X(1–3)-His-X2-Cys-X2-Cys-X(4–48)-Cys-X2-Cys. X stands for any amino acids in the numbers shown in parenthesis. These are protein-protein, protein-membrane, protein-DNA interacting elements involved in the regulation of transcription, replication, recombination, restriction, development, cancerous growth, etc. The RING fingers may also bind Zinc and are thus related to Zinc fingers. The name comes from the human gene RING—carrying such a motif—located in the vicinity of HLA. ►Zinc finger, ►DNA-binding protein domains, ►autoimmune disease; Saurin AJ et al 1996 Trends Biochem Sci 21:208.

Ring Species: Ring species are reproductively isolated, yet some connections exist between them (Irwin DE et al 2005 Science 307:414).

Ringer Solution: A ringer solution is prepared in somewhat different concentrations depending on type of tissues it is used for. It is used as a sterilized physiological salt solution, in 100 mL water mg salts: NaCl 860, KCl 30, $CaCl_2$ 33; some formulations also add $NaHCO_3$ 20, NaH_2PO_4 and glucose 200.

RIP: A term for "recombination induced premeiotically" and, alternatively, "repeat induced point mutation". When repeated (generally longer than 400 bp) DNA is introduced into fungi (*Neurospora* and others) by transformation or by other means, and the replicated sequence may be lost premeiotically. Alternatively, the duplications are methylated to reduce recombination or mutations are induced. In *Ascobolus* and *Coprinus*, the repeats may be methylated without mutations to follow. The mutations are attributed to cytidylic acid methylation of 5′-CpG sequences as a defense against duplication introduced by transformation to silence the superfluous genetic material. The methylation results primarily in transition mutations, GC→AT. The distribution of the mutations in the DNA is not random but most commonly occurs 5′ of adenine sites but somewhat less frequently take place 5′ to thymine or guanine but rarely at site 5′ to other cytosines. Generally, within the same chromosome, either C→T or G→A changes occur, but not both. The majority of the RIP mutations are missense or nonsense but occasionally functional alleles also arise. The RIP mutations are frequently unstable and the longer duplications revert at a frequency of about 10^{-4} in the vegetative cells. ►co-suppression, ►ripping, ►RNAi, ►RIGS, ►methylation of DNA, ►MIP, ►position effect, ►suppressor genes, ►integration; Selker EU 1997 Trend Genet 13:296; Hsieh J, Fire A 2000 Annu Rev Genet 34:187; Miao VP et al 2000 J Mol Biol 300:249; Freitag M et al 2002 Proc Natl Acad Sci USA 99:8802.

RIP (ribosome-inactivating protein): Antiviral proteins in plants and animals with glycosylase activity. Thus, they may depurinate RNA of susceptible ribosomes and block protein synthesis. RIPs protect plants against pathogens. ►depurination, ►glycosylases, ►saporins, ►ricin, ►abrin; Nielsen K, Boston RS 2001 Annu Rev Plant Physiol Plant Mol Biol 52:785.

RIP (RNAIII inactivating peptide): ►RNAIII/rnaiii

RIP (regulated intramembrane proteolysis): It may control cell differentiation, lipid metabolism and various proteins in prokaryotes and eukaryotes. The targets included are proteins of the endoplasmic reticulum, sterol regulatory element binding proteins (SREBP), and amyloid precursor protein (APP), Notch). ►SREBO, ►Alzheimer disease; Notch, Brown MS et al 2000 Cell 100:391.

Rip1: ►RNA export

Ripping: The process generated by RIP mutations. RIP may also generate point mutations primarily by G-C → A-T transitions. ►RIP

RISC (RNA-induced silencing complex): RISC is composed of siRNA as well as Argonaute 2, VIG, FXR and the Tudor-SN proteins. ►RNAi, ►named proteins, ►miRNP, ►P body; Caudy AA et al 2003 Nature [Lond] 425:411.

Rise: ►one-step growth

Risk: A combination of the degree of a hazard with its potential frequency of occurrence, e.g., if a recessive gene causes a particular malformation in 30% of the fetuses, the probability of its homozygosity among the progeny of heterozygous parents is 0.25. Thus, the risk of this malformation is $0.3 \times 0.25 \approx 0.075$, i.e., 7.5%. If an individual is known to be heterozygous for a recessive lethal gene and marries a first cousin, the risk of them having a stillborn child may be as high as $0.5 \times 0.25 = 0.125 = 1/8$. If, however, the carrier marries an unrelated person from the general population, where the frequency of this gene is only 0.005, the risk will be $0.5 \times 0.005 = 0.0025 = 1/400$. In some cases, the calculation of the genetic risk is not quite as simple. Let us assume that the penetrance of a dominant gene is 80% but non-transmissible factors (somatic mutation not included into the germline, environmental effects) may also evoke the same symptoms; in this case members of the family do not display the defect. For example, when new dominant germline mutations are responsible for 15% of the cases, in this instance the ancestors are not affected.

The offspring of such a mutant individual, however, has a 0.8 (penetrance) × 0.5 (expected gametic transmission) = 0.4 chance of being affected. The risk of all their offspring not being affected is 1 − 0.15 = 0.85, and if not inherited (0 inheritance), the chance is 0.85 × 0 = 0. The probability of inherited (0.15) × penetrance (0.8) is ≅ 0.12 (12%). Further considerations are necessary for proper genetic counseling if the proband has already normal, not-affected offspring. We designate the hereditary status of the proband, the probability of being a hereditary case: P (H) = 0.15, as specified above. The probability that the first child being normal (not affected), despite the defective parental gene is P(N/H) = 1 − 0.4 = 0.6. The probability that the proband does not have this defective gene in the germline is $P(H^-) = 0.85$. The conditional probability that an offspring would be normal is $P(N/H^-) \cong 1$. From Bayes' Theorem the probability that the first child inherited but did not express the trait is:

$$P(H/N) = \frac{P[H]P[N/H]}{P[H]P[N/H] + P[H^-]P[N/H^-]}$$

$$= \frac{[0.15][0.6]}{[0.15][0.6] + [0.85][1]} = \frac{0.09}{0.94} \cong 0.096$$

With two not-affected children $P(N/H) = 0.6^2 = 0.36$, and for *n* offspring 0.6^n is the probability of the parent being normal in phenotype although having the defective gene.

The second child being normal although carrying the defective gene is:

P(N/H) = 0.096 × 0.4 ≅ 0.0384 and after substitution into the Bayes' formula

$$P(H/N) = \frac{[0.15][0.0384]}{[0.15][0.0384] + [0.85][1]} = 0.0067$$

Attributable risk (AR) reveals the risk of genetically susceptible individuals relative to those who are not susceptible. It is estimated as $AR = \{P_{Aa}(1 - 2q[1 - P_{aa}])\}/(P_{Aa}[1 - 2q])$ where P is the frequency and *A* and *a* are the dominant and recessive alleles, respectively, at a locus. The relative risk can also be estimated by the contingency chi square using an association test. Absolute risk is the excess risk of an agent which causes a difference between exposed and unexposed populations. Background risk is the chance of being afflicted in a population with a certain frequency of disease-causing allele(s). Usually, confounding factors also influence the risk—age, sex, addictions, etc.—and when this is the case more elaborate statistical procedures are required. Life expectancy may be reduced by several factors in a complex manner (smoking a cigarette [10 min], accidents [95 days], obesity [by 20% or 2.7 years], 1 mrem of radiation [1.5 min], medical X-rays [6 days]).

It is of great recent interest to determine the risk of disease by genome-wide association studies (risk engine) (Editorial 2005 Nature Genet. 37:1153). ►Bayes' theorem, ►genetic risk, ►recurrence risk, ►empirical risk, ►genetic hazards, ►genotypic risk ratio, ►λ_S, ►association test, ►mutation in human populations, ►utility index for genetic counseling, ►confidence intervals, ►radiation hazard assessment, ►cosmic radiation, ►linkage

RITS (RNA-induced transcriptional silencing): RITS contains Dicer-generated siRNA and it is required for heterochromatin silencing by attaching to histone-3 methylated at lysine 9 residues. ►RNAi, ►heterochromatin, ►histones; Ekwall K 2004 Mol Cell 13:304; Noma K et al 2004 Nature Genet 36:1174; Bühler M et al 2006 Cell 125:873.

RIZ (retinoblastoma-interacting zinc finger protein, 220 kDa with 8 Zn-finger domains): RIZ has a common loss at human chromosomal site 1p36 and may be responsible for colorectal cancer, breast cancer, and endometrial neoplasias. RIZ also appears as a downstream effector of estrogen action. ►retinoblastoma, ►Zinc finger, ►estradiol, ►colorectal cancer, ►effector; Steele-Perkins G et al 2001 Genes Dev 15:2250.

RK: Rank of utility.

RK2 Plasmids: These plasmids represent a family of broad host-range plasmids (56.4 kb), resistant to tetracycline, kanamycin, and ampicillin. The size and selectability of other members of the family varies. (See Pogliano J et al 2001 Proc Natl Acad Sci USA 98:4486).

RKIP (Raf kinase inhibitor protein): ►RAF

RLDB: ►reference library database

RLGS (restriction landmark genomic scanning): The goal of the procedure is to determine the methylation status of genes that may have undergone epigenetic changes (such as oncogenic transformation) or are imprinted. It also detects tissue-specific methylation patterns in the genome. It is based on digestion of DNA by a rare-cutter restriction endonuclease such as *Not*I, which does not cleave methylated CpG islands. The fragments are then radioactively end-labeled at the *Not*I cut sites and further reduced in size by another restriction enzyme, e.g., *Eco*RV (GAT↓ATC). The fragments are then separated by agarose gel electrophoresis and further cleaved with a third restriction enzyme, e.g., *Hinf*I (G↓ANTC) in the gel. The small fragments are then subjected to electrophoresis in a second dimension in an acrylamide gel and the fragments containing the radioactive label are identified. The procedure thus identifies the difference between methylated and not methylated status at a specific site. The method holds promise for mass scanning potential cancer genes. It has also been used for determining allele-specific methylation of imprinted genes (Plass C et al 1996 Nature Genet 14:106). ►epigenesis, ►imprinting, ►electrophoresis, ►restriction enzymes, ►methylation of DNA; Costello JF et al 2000 Nature Genet 25:132; Costello JF et al 2002 Methods 27:144.

RLK: A Tec family kinase involved in T cell receptor signaling. ►T cell receptor, ►Tec; Yang WC et al 2000 Int Immunol 12:1547.

R-Loop: A quasi three-stranded structure consisting of a double-stranded DNA and a single-stranded RNA, which displaces, at a short section, one of the DNA strands. Such a structure may occur primarily at the replication forks of DNA of prokaryotes and eukaryotes. ►D loop; Nossal NG et al 2001 Mol Cell 7:31; Clayton DA 2000 Hum Reprod 15(Suppl. 2):22; Tracey RB, Lieber MR 2000 EMBO J 19:1055.

RLP (ribosome landing pad): ►IRES

RL-PCR: ►reverse ligase-mediated polymerase chain reaction

RMCE (recombinase mediated cassette exchange): A procedure for large-scale screening of recombinants under selective conditions. Variations exist. The diagram was redrawn after Tsin E et al 2005 Nucleic Acids Res 33(17):e147 (see Fig. R85). ►Cre/LoxP, ►hygromycin, ►gancyclovir, ►tetracycline, ►green fluorescent protein, ►luciferase, ►targeting genes, ►gene replacement

RME1: An inhibitor of yeast meiosis and sporulation. ►mating type determination; Shimizu M et al 1998 Nucleic Acids Res 26:2320.

RMGR (recombinase-mediated genomic replacement): ►gene replacement

RMSA-1 (regulator of mitotic spindle assembly): A protein which is phosphorylated only during mitosis and is a substrate for CDK2 kinase; it is required for the assembly of the spindle. ►spindle, ►CDK; Yeo JP et al 1994 J Cell Sci 107:1845.

RNA: Ribonucleic acid is a polymer of ribonucleotides. There are three main classes of RNAs in the cell: the mRNA, which provides the instructions for protein synthesis, the various ribosomal RNAs, and the tRNAs. Other RNAs are involved in splicing, editing, post-transcriptional modification, ribonucleoproteins that insert proteins into membranes and mediate telomere synthesis, replication priming RNAs, inhibitory RNAs (RNAi), ribozymes, and the noncoding RNAs involved in dosage compensation and imprinting. Maps of nuclear and cytosolic polyadenylated [poly(A)$^+$] RNAs longer than 200 nucleotides (nt) (long RNAs, *l*RNAs) and whole-cell RNAs less than 200 nt (short RNAs, sRNAs) are transcribed over the entire nonrepetitive portion of the human genome. The potential biological function of an appreciable portion of long unannotated transcripts is to serve as precursors for sRNAs. These maps reveal three classes of RNAs that have specific genomic localization at gene boundaries. The biological relevance of these classes of RNAs is supported by a strong correlation with the expression state of the genes they associate with, as well as their syntenic conservation between humans and mouse (Kapranov P et al 2007 Science 316:1484). ►RNA I, ►mRNA, ►tRNA, ►rRNA, ►rrn, ►nucleic acid chain growth, ►non-canonical bases, ►telomerase, ►RNA editing, ►replication, ►RNAi, ►microRNA, ►ribozymes,

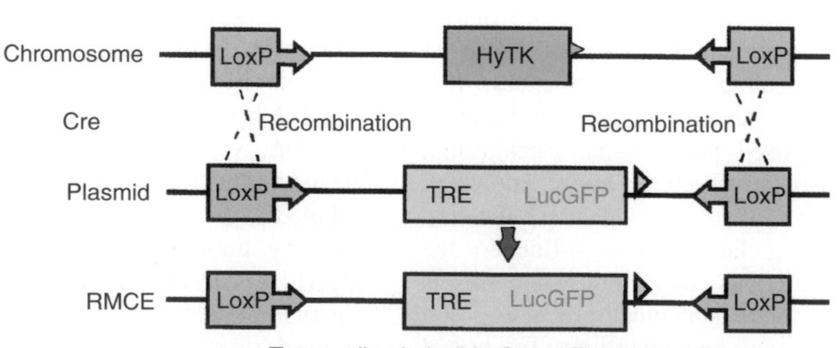

Figure R85. Recombinase-mediated cassette exchange

►dosage compensation, ►imprinting, ►Xist; http://prion.bchs.uh.edu/; structure and motifs: http://uther.otago.ac.nz/v5g.html; RNA structure, function, sequence alignment: http://www.ccrnp.ncifcrf.gov/~bshapiro; RNA secondary structure, folding, neighbors: http://bioinformatics.bc.edu/clotelab/RNAbor/; RNA folding design: http://www.bioinf.uni-freiburg.de/Software/INFO-RNA/start.html; RNA secondary structure predictor: http://compbio.cs.sfu.ca/taverna/.

RNA I: An untranslated bacterial RNA controlling the maturation of RNA II that serves as a primer for plasmid DNA synthesis. RNA I and RNA II are synthesized on opposite DNA strands. RNA I binds to RNA II and thereby prevents its folding into a cloverleaf that is necessary for the formation of a stable DNA:RNA hybrid between RNA II and the plasmid DNA. This binding is promoted by the Rop protein (63 amino acid residues) coded for by 400-base downstream from the origin of replication. A single G → A transition mutation in Rop or upstream from it may contribute to plasmid amplification (see Fig. R86). RNA I, RNA II, and Rop also control plasmid incompatibility. RNase H cuts off the pre-primer section and prepares the primer for the actual DNA synthesis. RNA I may be polyadenylated and then its decay hastened in a way similar to mRNA. RNA I and RNA II may interact initially by base pairing between their 7-nucleotide complementary loops. The Rom/Rop protein may also bind to the transient complex and assures a more stable duplex of the two RNAs causing failure of replication initiation. ►polyadenylation, ►plasmid, ►RNA polymerase, ►ROM; Mruk I et al 2001 Plasmid 46:128.

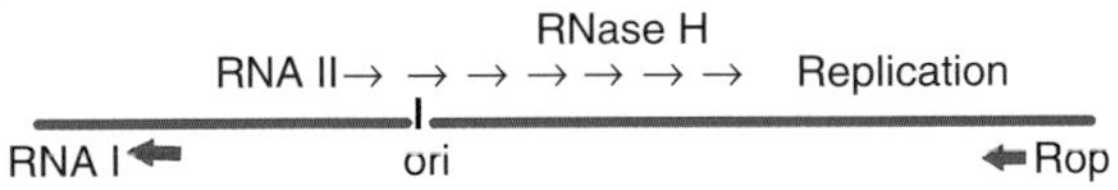

RNase H
RNA II→ → → → → → → Replication
RNA I← ori ←Rop

Figure R86. RNA I and RNA II synthesis

RNA II: ►RNA I

RNA Binding Proteins: These proteins modify the RNA structure locally or globally; they may affect RNA trafficking, mRNA biosynthesis, translation, splicing, polyadenylation, differentiation and diseases. They may be part of the combinatorial tagging of mRNA and may control the localization, translation, and decay of mRNA in the global expression program (Gerber AP et al 2004 PLoS Biol 2:342). The length and composition of the binding domains may vary. A human RNA-binding protein is encoded in chromosome 8p11-p12 (RBP-MS). ►DNA-binding protein domains, ►hnRNP; Burd CG, Dreyfuss G 1994 Science 265:615; Dreyfuss G et al 2002 Nature Rev Mol Cell Biol 3:195; Hall KB 2002 Current Opin Struct Biol 12:283; RNA binding site predictor: http://bindr.gdcb.iastate.edu/RNABindR/.

RNA Cap: ►cap, ►capping enzymes

RNA, Catalytic: ►ribozyme

RNA Chaperone: ►chaperones

RNA Codons: ►genetic code

RNA Computer: See Faulhammer D et al 2000 Proc Natl Acad Sci USA 97:1385; ►DNA computer

RNA Dependent RNA Polymerase (RdRP): This polymerase replicates the RNA genome of viruses. The RNA-directed RNA polymerase, primed by siRNA, amplifies the interference by RNAi. ►RNAi; Ahlquist P 2002 Science 296:1270; Ortin J, Parra F 2006 Annu Rev Microbiol 60:305.

RNA Display: A procedure for the enrichment of specific RNAs. In in vitro the translation system uses a synthetic mRNA, which carries the peptidyl acceptor antibiotic puromycin at the 3′ end. The protein synthesis proceeds until the end of the open reading frame and the translation stalls at the RNA-DNA junction. The puromycin enters the A site of the ribosome and accepts the nascent peptide. Affinity chromatography and immunoprecipitation permit purification of the fused mRNA-peptide from the complex mixture. The RNA serves as a tag for specifying the peptide. ►protein synthesis, ►puromycin, ►affinity chromatography, ►immunoprecipitation, ►phage display; Roberts RW, Szostak JW 1997 Proc Natl Acad Sci USA 94:12297.

RNA, Double-Stranded: A double-stranded RNA is usually a minor fraction of the total cellular RNA (unlike to DNA). It appears that if double-stranded RNA, with sequences homologous to an open reading frame of a gene, is introduced into the cell, the antisense RNA more effectively blocks the function of that gene (in *Caenorhabditis*). It is a surprising fact that even at very low abundance this RNA is a highly effective inhibitor. A double-stranded RNA-binding protein is apparently required for gene activation in the mammalian germ cells. dsRNA is common in virus-infected cells and it is an inducer of interferon. It inhibits the transcription of about one-third of the genes and stimulates the rest to a variable degree.

A substantial fraction of the pairing of RNA molecules is not in compliance with the Watson-Crick model. Hoogsteen and "sugar edge" pairing also occur. The edge indicates the relative orientation of the glycosidic and hydrogen bases. The four RNA bases (A, C, G, U) may associate in 4^2 matrices.

R

These steric associations may be evolutionarily better preserved than the base positions themselves. They facilitate long-range RNA-RNA interactions and create binding sites for proteins and small ligands. ►antisense RNA, ►targeting genes, ►RNAi, ►Watson-Crick model, ►Hoogsteen pairing, ►RNA helicase; Geiss G et al 2001 J Biol Chem 276:30178; Leontis NB et al 2002 Nucleic Acids Res 30:3497.

RNA Driven Reaction: In a DNA-RNA hybridization experiment, the RNA is far in excess compared to single-stranded DNA. This assures that hybridization will take place at all potential annealing sites. ►DNA hybridization, ►nucleic acid hybridization

RNA Editing: A means of posttranscriptional or co-transcriptional altering of the RNA transcript (mRNA, tRNA, rRNA, 7 SLRNA). It is a very common process in the mitochondria of *Trypanosomes* (►*Trypanosoma brucei*). A separately transcribed 40–80-base "guide RNA" with homology to the 5′-end of the RNA to be modified pairs with the target. Then, uracil residues from the 3′-end tract of the guide are transferred into the target sequences. Two enzymes are primarily involved in the editing of exonuclease 1 (REX1) and (REL) RNA ligases (Kang X et al 2005 Proc Natl Acad Sci USA 102:1017). This editing thus changes the content of the message and the amino acids of the translated protein. In the mitochondria of this protozoon, thousands of *U* nucleotides may be inserted into different pre-mRNAs. *U*s may also be removed at a 10-fold lower frequency. It has been hypothesized that the *U* replacements are provided by the 3′-end of the gRNA, but recent experimental evidence indicates that they come from free UTPs. In the mitochondria of plants, in about 10% of the transcripts, U may replace C or C replaces U. The editing of four to 25 RNAs may take place in the chloroplasts. Transacting factors psbE and petB bind to five nucleotides cis upstream from the editing site (Miyamoto T et al 2004 Proc Natl Acad Sci USA 101:48). Although there is no conserved consensus around the edited sequences in plant chloroplasts, in tobacco a 27-nucelotide sequence at the RpoB-2C target plays a critical role (Hageman CE et al 2005 Nucleic Acids Res 33:1454). Editing appears to be rare but not limited to higher animals. One C residue deamination in the apolipoprotein-B results in a U replacement and the creation of a stop codon, and consequently a truncated protein. Another similar deamination in the middle of the transcripts alters the permeability of a Ca^{2+} channel. Thus editing produces two different mRNAs from one. Although C→U is the common change, U→C may also occur exceptionally. Simple deamination, addition or co-transcriptional errors, such as stuttering of the polymerase may also bring about the changes. RNA editing also takes place in different RNA viruses. In HIV-1, G→A editing also occurs besides C→U. Mammalian nuclear RNA editing may involve the deamination of adenosine into inosine in the double-stranded pre-mRNA of the glutamate-receptor subunits. The enzyme responsible for the process is dsRAD (double-stranded RNA adenosine deaminase, also called DRADA or ADAR). The ADAR1 and ADAR2 proteins are associated with spliceosomal components of a 200 S large ribonucleoprotein complex (lnRNP). The deficiency of ADAR in *Drosophila* results in behavioral anomalies during the advanced developmental stages. There are differences in editing among animals. Fruit flies, mosquitoes, and butterflies have similar editing sites within a single exon of the synaptotagmin genes. Honeybees, beetles and roaches, however, do not edit this synaptotagamin site. Comparative genomics of 34 species indicates that complex, multidomain, pre-mRNA structures (e.g., pseudoknots) determine the specificity of adenine editing in synaptotagmin (Reenan RA 2005 Nature [Lond] 434:409).

In the $tRNA^{Asp}$ of marsupials, the GCC anticodon is found that can only recognize the glycine codon. In 50% of these tRNAs the middle base is edited to U, and thus the regular Asp anticodon is generated. The marsupial mitochondria also have the regular $tRNA^{Gly}$ with anticodon GGN that recognizes all four glycine codons, but the edited codon can match up with only two of the glycine codons. Why among the only 22 tRNAs there are two for glycine (normal and edited) is a puzzling observation. RNA editing occurs through the plant kingdom (with few exceptions), in mitochondria as well as in chloroplasts, albeit at lower frequency in the latter. In the plastids there are about 25 editable sites, while in the mitochondria their number may exceed 1,000. RNA editing may generate new initiation and termination codons in plant organelles and thus new reading frames. No U→C editing was observed in the mitochondria or chloroplasts of gymnosperms or in the chloroplasts of angiosperms. The site of editing is apparently selected on the basis of the flanking sequences.

RNA editing also occurs in nuclear genes and contributes to the regulation at an additional level. It appears that the neurofibromas are determined by editing in the neurofibromatosis gene (NF1). Editing may also regulate the mRNA of the serotonin-2C receptor. In the immune system, T cell-independent B cells (B1) also diversify their surface receptors by RNA editing. Defects in RNA editing may potentiate tumorigenesis. ►mtDNA, ►apolipoproteins, ►kinetosome, ►gRNA, ►genetic code, ►stuttering, ►wobble, ►DRADA, ►ADAR, ►Z DNA,

R

►anticodon, ►mooring sequence, ►serotonin, ►editosome, ►B cell, ►RNA ligase, ►synaptotagmin, ►pseudoknot; Gott JM, Emeson RB 2000 Annu Rev Genet 34:499; Raitskin O et al 2001 Proc Natl Acad Sci USA:6571; Aphasizhev R et al 2002 Cell 108:637; Bass BL 2002 Annu Rev Biochem 71:817; Maas B et al 2003 J Biol Chem 278:1391; Blanc V, Davidson O 2003 J Biol Chem 278:1395; http://bioinfo.au.tsinghua.edu.cn/dbRES; http://biologia.unical.it/py_script/search.html.

RNA Enzyme: ►ribozyme

RNA Export: From the nucleus, mRNA requires the presence of the nuclear export factor NES and a cellular cofactor Rip1/Rab. The NES function is part of the Gle1 yeast protein (M_r 62K). Gle1 interacts with Rip1 and nucleoporin (Nup 100) in the nuclear pore. The Rev splicing factor can substitute the Gle1 function. Protein Aly of metazoans is involved in pre-mRNA slicing and mRNA export. The export of unspliced mRNA may be prevented by nuclear retention factors (RF), however some intron-less mRNAs can be exported and efficiently translated in the cytoplasm. The export of mRNA also requires PIP_2 and PIP_3. The abundant hnRNPs (hnRNP C and hnRNP K) have also been implicated in mediating RNA export. It seems, some TAP elements are also involved in the export. tRNA and U snRNA export may require members of the importin β family of proteins. The export of tRNAs may be mediated by the aminoacyl-tRNA synthetases after processing of transcripts. This involves the removal of the not needed 5′ and 3′ sequences, introns, addition to 3′ end of the CCA sequence, and modification of some nucleotides. Some mRNA export protein factors co-ordinate the export of transcriptionally co-regulated functional classes of transcripts. Thus, the yeast export proteins, Yra1 and Mex67, bind 1,000 (20%) and 1,150 (36%of the total) mRNAs, respectively, and usually their specificity does not overlap (Hieronymus H, Silver PA 2003 Nature Genet 33:155). ►RNA transport, ►REV, ►nuclear pore, ►transport elements constitutive, ►export adaptors, ►nuclear export sequences, ►importin, ►TAP, ►aminoacyl-tRNA synthetase, ►Ipk1, ►cell-penetrating peptides; Michael WM 2000 Trends Cell Biol 10:46; Zenklusen D, Stutz F 2001 FEBS Lett 498:150; Sträßer K et al 2002 Nature [Lond] 417:304.

RNA Extraction: An essential requisite that RNase activity be eliminated or prevented during all operations. The glassware can be made RNase free by baking for 8 h or by chloroform washing. A 1% diethyl pyrocarbonate (DEPC [carcinogen!]) washing (2 h, 37 °C) may also be useful. RNase activity in the extraction media can be inhibited by vanadium or by the clay, Macaloid. These are subsequently eliminated by water-saturated phenol extraction. RNases can also be blocked by 4 M guanidium thiocyanate and β-mercaptoethanol. RNA is extracted from the tissues in a buffer containing a detergent (0.5% Nonidet) and a reducing agent (dithiothreitol). Proteins may be removed by proteinase K digestion. RNase-free DNase removes DNA. Finally, chilling in cold ethanol (containing Na-acetate) precipitates RNA. The RNA is taken up in a TE, pH 7.6 buffer. Its quantity can be measured spectrophotometrically at 260 nm. Several variations of these general procedures are being used to isolate RNA. ►DNA extraction, ►RNase-free DNase, ►TE

RNA Factory: The complex associated with RNA polymerase II. It carries out transcription, splicing, and cleavage-polyadenylation of the mRNA precursor. ►mRNA; McCracken S et al 1997 Nature [Lond] 385:357.

RNA Fimger: An RNA element (e.g., CCCH) binding to Zn-finger proteins. It controls various cellular processes. ►DNA binding protein domains

RNA Fingerprinting: The purpose is to identify the differential expression of the total array of genes that constitutes about 15% (or less) of all at a particular time in a mammalian genome. To attain this goal partial cDNAs are amplified by reverse transcription using PCR from a subset of mRNAs. The short sequences are then displayed on a sequencing gel (differential display). Pairs of primers are selected in such a way that each will amplify 50 to 100 mRNAs. One of the primers (5′ TCA) is anchored to the TG upstream of the poly(A) tail of the mRNA. This primer will recognize 1/12 (4!/2!) of the mRNAs with different combination of the last two 3′ bases omitting T as the penultimate base. The primer will then only amplify this subpopulation. As 5′ primers, 6–7 bp arbitrary sequences are used. Such a procedure can be used not just for molecular analysis of development. Eventually the genes producing the transcripts can be cloned. ►PCR, ►reverse transcription, ►DNA finger printing, ►fingerprinting of macromolecules, ►differential display, ►RDA, ►microarray hybridization, ►proteomics, ►SAGE, ►TOGA; Gill KS, Sandhu D 2001 Genome 44:633.

RNA Folding: There is a requirement for Mg^{2+} in maintaining the tRNA tertiary structure but other RNAs also need potassium or magnesium ions for folding. The ionic environment influences the folding of macromolecules and their function depends on the appropriately folded state. RNA folding proceeds through a number of intermediate states. For protein

R

synthesis on the ribosome in the cells the tRNA is aminoacylated in the tertiary fold (see Fig. R87). ►aminoacyl-tRNA synthetase, ►ribozyme, ►RNA structural; Draper DE et al 2005 Annu Rev Biophys Biomol Struct 34:221; single RNA molecule folding: Zhuang X 2005 Annu Rev Biophys Biomol Struct 34:399; Pan T, Sosnick T 2006 Annu Rev Biophys Biomol Struct 35:161; detection of thermodynamically stable and evolutionarily conserved RNA secondary structures in multiple sequence alignments: http://rna.tbi.univie.ac.at/cgi-bin/RNAz.cgi.

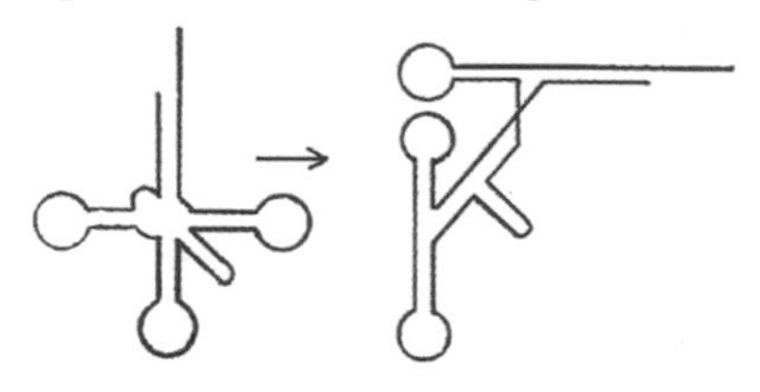

Figure R87. RNA folding

RNA G8: RNA G8 contains about 300 nucleotides and it is associated with the ribosomes in *Tetrahymena thermophila*. It is transcribed by RNA polymerase III and conveys thermal tolerance to the cells. ►thermal tolerance

RNA Helicases: RNA helicases unwind double-stranded RNAs to facilitate transcription. They have two double-stranded RNA domains and have a cis-acting transactivation response (TAR) element binding and a dsRNA activated protein kinase (PKR) domain. TAR enhances the transcription of HIV-1. It appears that the helicase activity resides within the ribosome and that the ribosomal proteins S3, S4 and S5 encircle the incoming RNA on the 30S ribosomal unit and play a role in processivity (Takyar S et al 2005 Cell 120:49). The NPH-II RNA helicase has specificity for the ribose moiety of the loading strand and for a section of the 3′-overhang. Yet, under less stringent conditions it also acts on DNA (Kawaoka J, Pyle AM 2005 Nucleic Acids Res 33:644). ►eIF4A, ►helicase, ►DEAD-box, ►breast cancer, ►acquired immunodeficiency, ►double-stranded RNA; Lüking A et al 1998 Crit Rev Biochem Mol Biol 33:259; Fujii R et al 2001 J Biol Chem 276:5445.

RNA, Heterogenous: ►hnRNA

RNA Insertion Element: An RNA insertion element can be generated by in vitro combination of an endoribonuclease ribozyme with a R3C ribozyme with RNA ligation function. The conjoined bifunctional ribozymes can integrate into a target RNA. Several mutations had to be isolated and selected to generate an efficient modular system. Such structures may be useful for insertional mutagenesis of mRNA. ►insertion element, ►insertional mutagenesis, ►ribozymes; Kumar RM, Joyce GF 2003 Proc Natl Acad Sci USA 199:9738.

RNA Interference: In *Caenorhabditis* nematodes, fed on double-stranded RNA-containing bacteria, gene expression is transiently but specifically suppressed, as long as the dsRNA is available. The interfering dsRNA may be generated by inserting T7 phage RNA polymerase genes at the opposite ends of specific genes in a plasmid. Alternatively, a single copy of the polymerase may be used on inverted duplication of a specific gene. *Note*: some of the papers by Taira, K. had to be corrected (Nature [Lond] 431:211) and another (Nature 423:838) was retracted (Nature 426:100). ►*Caenorhabditis*, ►RNAi, ►inhibition of transcription, ►posttranscriptional gene silencing; Moss EG 2001 Curr Biol 11:R772.

RNA Ligase: An RNA ligase catalyzes the joining of RNA termini, such as generated during the processing of tRNA transcripts, in a phosphodiester bond and may need ATP for the reaction. It is an essential enzyme for RNA editing. ►ligase DNA, ►ligase RNA; Stage-Zimmermann TK, Uhlenbeck OC 2001 Nature Struct Biol 8:863; Ho KC, Shuman S 2002 Proc Natl Acad Sci USA 99:12709; deoxyribozyme RNA ligase: Prior TK et al 2004 Nucleic Acids Res 32:1075.

RNA Localization: Embryonic development is an asymmetric process and the local distribution and enrichment of special RNAs play decisive roles. The RNA transcribed from the bicoid locus of *Drosophila*, encoding a transcription factor, specifies anterior cell fates in the oocytes and embryos. The germcell-less nuclear pore associated factor participates in the definition of posterior development. The growth factor encoded by gurken affects anterior-dorsal differentiation whereas the prospero encoded transcription factor controls apical/basal development in the neuroblasts. In mammals, the transcript of the β-actin genes is involved in the definition of the cytoskeleton and it is localized primarily at the periphery of epithelial cells, fibroblasts, and myoblasts. The different RNAs appear at the predestined locations immediately following transcription. Some of the rather uniformly distributed RNAs are degraded, except at the locations where they exercise morphogenetic function(s). The localization seems to be mediated by signals (usually a few hundred bases or less), generally in the untranslated 3′-end (3′-UTR) of the mRNAs. The mRNAs are not translated until properly localized. Translation without prior localization might cause developmental anomalies. Besides the cis-acting element (e.g., 3′-UTR), trans-acting factors encoded by other genes may also affect localization (e.g., influence the cytoskeletal motor proteins or RNA binding proteins). Localization of the RNA may also be assured by alternative splicing of the transcripts. The pattern of distribution may change as the microtubule-organizing center develops and motor proteins become available. At a later stage, before the nurse cells decay, they dump large amounts of RNA into the oocyte through the ring canals. Later, ooplasmic streaming somewhat mixes up the previously laid down distribution pattern. When the early syncytial blastoderm (~6,000 nuclei) is converted into a cellular blastoderm, the anterior-posterior and the dorsal ventral polarization begins, upon the expression of the genes, specifying the morphogenetic pattern. At this stage the control shifts toward the zygotic effect genes from the maternal effect ones. Another control involved the diffusion pattern of the morphogenetic gene products. ►morphogenesis in *Drosophila*, ►maternal effect genes, ►MTOC, ►ring canal, ►METRO, ►motor proteins; Bashirullah A et al 1998 Annu Rev Biochem 67:335; Palacios IM, St Johnston D 2001 Annu Rev Cell Dev Biol 17:569; Bullock SL, Ish-Horowicz D 2001 Nature [Lond] 414:611; Saxton WM 2001 Cell 107:707.

RNA Maturases: Ribosomal transcripts are processed to size by the U3, U8, U13 independently transcribed small RNAs (snoRNA, small nucleolar RNAs). The U13-U14 snoRNAs and E3 are encoded by the introns of protein-coding genes participating in the process of translation. They are co-transcribed with the pre-mRNA and removed during the processing of the genes. By the intron of the U22 host gene (UHG) seven U RNAs are transcribed. These U RNAs display 12–15 base complementary sequences to rRNAs. ►splicing, ►introns, ►ribonuclease P; Delahodde A et al 1989 Cell 56:431; Claros MG et al 1996 Methods Enzymol 264:389.

RNA Mimicry: ►translation termination

RNA Mimics: 2′-modified oligodeoxynucleotides such as the 2′–O–methyl-modified ones that have enhanced binding affinity to complementary RNA and are resistant to some nucleases. ►antisense technology; Putnam WC et al 2001 Nucleic Acids Res 29:2199.

RNA, Micro (miRNA, mir): A 22-nucleotide inhibitory RNA. ►RNAi; Zeng Y et al 2002 Mol Cell 9:1327.

RNA, Noncoding (ncRNA): There are many mRNA-like-yet-not-translated molecules in the cell. They are polyadenylated and spliced but lack long open reading frames. They have regulatory or signal functions (Szymanski M, Barciszewski J 2003 Int Rev Cytol 231:197). A noncoding RNA transcribed in an intergenic region upstream from the *SER3* gene of yeasts can interfere with the binding of activators to the *SER* promoter and thus regulates its transcription (Martens JA et al 2004 Nature [Lond] 429:571). Noncoding RNAs are the tRNAs, ribosomal RNAs, the snRNAs, microRNAS, and snoRNAs. In whole genome microarrays of *E. coli*, besides 4,052 coding transcripts, 1,102 apparently noncoding transcripts were found in the intergenic regions. In humans, about 98% of the transcripts are noncoding. There are also the tiny noncoding RNAs (tncRNA) that resemble microRNAs but are not transcribed as short hairpin RNA precursors and yet they play regulatory roles in development (Ambros V et al 2003 Current Biol 13:807). The development of the neural stem cells may be modulated by a small double-stranded RNA and protein interaction (Kuwabara T et al 2004 Cell 116:779). On the basis of secondary structure and thermostability, coding and noncoding RNAs can be rapidly identified by a computational procedure (Washietl S et al 2005 Proc Natl Acad Sci USA 102:2454). ►microRNA, ►RNAi, ►antisense RNA, ►RNA regulatory, ►small RNA, ►Xist, ►Rfam; Erdmann VA et al 2001 Cell Mol Life Sci 58:960; Eddy SR 1999 Curr Opin Genet Dev 9:695; Storz G 2002 Science 296:1260; Kiss T 2002 Cell 109:145; Fahey ME et al 2002 Comp Funct Genomics 3:244; Tjaden B et al 2002 Nucleic Acids Res 30:3732; Griffith-Jones S et al 2003 Nucleic Acids Res 31:439; http://biobases.ibch.poznan.pl/ncRNA/; http://www.sanger.ac.uk/Software/Rfam/; http://rfam.wustl.edu/; mammalian non-coding RNA: http://research.imb.uq.edu.au/RNAdb; http://bioinfo.ibp.ac.cn/NPInter/; plants: http://www.prl.msu.edu/PLANTncRNAs/database.html.

RNA Nucleotides, Modified: There are four basic nucleotides in the cell but several others are produced by posttranscriptional modification as needed for tRNAs and various coenzymes. ►modified bases; http://medlib.med.utah.edu/RNAmods.

RNA Plasmids: RNA plasmids exist in some mitochondria that are not homologous to the mtDNA. ►mtDNA

RNA Polymerases (RNAP): DNA-dependent RNA polymerases synthesize RNA on a DNA template. The T7 RNA polymerase is a relatively simple molecule (100 kDa). As a rule, this polymerase directs the incorporation of the first nucleotide in a template-directed manner with a single nucleotide primer. First, it produces several pieces of short RNAs and then these 10–12 nucleotide units polymerize into an elongation complex and exits from the promoter.

Between the DNA template and the 3′-proximal RNA transcript at least a 9-nucleotide long hybrid is formed for efficient processivity during the elongation of the RNA.

The RNA products then separate from the template and the duplex of the DNA is restored. The T7 phage promoter has a binding domain (−17 to −6) and an initiation domain (−6 to +6). After the binding domain of the polymerase recognizes the DNA, melting of the duplex takes place. During the elongation phase, about 200 nucleotides are added per second. The termination is rho independent in T7 and the end forms a stem-loop structure and a stretch of U bases. The termination seems to be a reverse process of the initiation, i.e., a stable isomerization is followed by an unstable sequence. In many respects, the phage enzyme despite its simple structure functions similarly to more complex polymerases. In prokaryotes a single polymerase synthesizes all cellular RNAs. The prokaryotic pol enzyme contains four subunits αα, β, β′, and σ. The large β subunits are evolutionarily highly conserved and they are the main instruments of polymerization with the other subunits.

The prokaryotic RNA polymerases are of about 500 kDA in size whereas the T7 enzyme is only about 40 kDa. The *E. coli*, α subunits (36.5 kDa) recognize the promoter, the β′ subunit (155.2 kDa) binds DNA, and the β (150.6 kDa) is active in RNA polymerization. The σ is essential to start transcription in a specific way by opening the double helix for the action of the RNA polymerization. The eukaryotic organelles contain prokaryotic type RNA polymerases. The eukaryotic RNA polymerase II—transcribing protein-encoding genes—has about 12 subunits and is somewhat larger but highly homologous across wide phylogenetic ranges. The two largest subunits are similar to the β′ and β subunits of the prokaryotic enzyme. The carboxy-terminal domain (CTD) of the largest subunit of the eukaryotic RNAP II carries up to 52 amino acid repeats (YSPTSPS) which are liable to phosphorylation (Meinhart A, Cramer P 2004 Nature [Lond] 430:223). Only the CTD unphosphorylated form participates in a transcription initiation complex but it is phosphorylated when it begins to elongate the RNA transcript. After transcription, the CTD phosphates are removed by phosphatase Fcp1 (Kamenski T et al 2004 Mol Cell 15:399). The CTD repeat is not essential for transcription initiation by RNAP II but for the stability of the initiation complex (Lux C et al 2005 Nucleic Acids Res 33:5139). When transcription is arrested the carboxy-terminal domain of Pol II is ubiquitylated (Somesh BP et al 2007 Cell 129:57). The complex of transcription includes the Mediator and a total of about 60 proteins with a combined mass of 3.5 MDa. In an *E. coli* cell there are about 2,000 core RNA polymerase molecules. The number of protein factors affecting transcription in *E. coli* is about 240–260.

In eukaryotes, there are three different DNA-dependent RNA polymerases (I, II, III) that display substantial homology. Polymerase IV of plants assists in the production of siRNA. It targets de novo methylation of cytosine and contributes to the formation of facultative heterochromatin (Onodera Y et al 2005 Cell 120:613). Besides RNA polymerase II, a single polypeptide nuclear RNA polymerase (spRNAP-IV) alternatively transcribed from the mammalian mitochondrial RNA polymerase gene (*POLRMT*) can also transcribe mRNA. This polymerase lacks terminal 262 amino acids in the nucleus (compared to the mitochondrially located form) and the mitochondrial targeting signal. It is resistant to α-amanitin but remains sensitive to *POLRMT*-specific RNAi. Its promoter is substantially different from that of Pol II and does not respond to transcriptional enhancers. It transcribes several nuclear genes (Kravchenko JE et al 2005 Nature [Lond] 436:735).

TRNA polymerases replicate the genome of the RNA viruses. Mammalian RNA polymerase II may also carry out RNA-dependent RNA synthesis by switching to the RNA genome of the hepatitis delta virus (Chang J, Taylor J 2002 EMBO J 21:157). The viral enzymes—in contrast to the DNA polymerases—lack a proofreading function and, consequently, their error rate may be within the 10^{-3} to 10^{-4} range per nucleotide leading to the extreme diversity of the RNA viruses. The prokaryotic RNA polymerase moves along the DNA at a speed exceeding 10 nucleotides/second. Termination is basically the reverse process of the transcription initiation. In the haploid yeast cell there are 2,000 to 4,000 RNA polymerase II molecules and about ten times more general transcription factor molecules. In 2006, Roger D. Kornberg received the Nobel Prize for his basic study of RNA polymerase structure and function. (See Figs. R88, R89, and R90, ►pol, ►σ, ►sigma factor, ►promoter, ►open promoter complex, ►promoter melting, ►transcription, ►transcription factors, ►transcription termination, ►terminator, ►antitermination, ►error in replication, ►error in transcription, ►pausing transcriptional, ►arrest transcriptional, ►transcript cleavage, ►transcription complex, ►inchworm model, ►nucleic acid chain growth, ►processivity, ►repression, ►mediator complex, ►SRB, ►protein synthesis, ►transcript elongation, ►elongator, ►RNA dependent RNA polymerase, ►RNAi, ►α-amanitin,

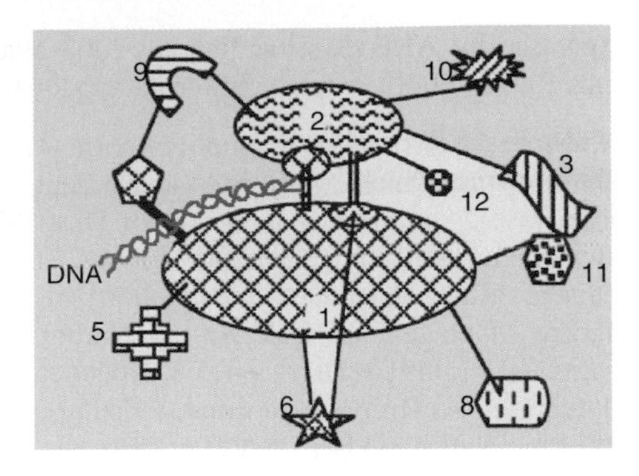

Figure R88. Schematic drawing of the 10/12 subunits of the yeast RNA polymerase II. Subnunits 1, 5 and 9 grip the DNA below the active center of the enzyme. Pores below the active center form the entry of the nucleotides and the exit of the polymerized RNA. (Redrawn after Cramer P et al 2000 Science 288:640)

►heterochromatin, ►transcription factories; Archambault J, Friesen JD 1993 Microbiol Rev 57:703; Barberis A, Gaudreau L 1998 Biol Chem 379:1397; Mooney RA, Landick R 1999 Cell 98:687; Ishihama A 2000 Annu Rev Microbiol 54:499; Cramer P et al 2001 Science 292:1863; Gnatt AL et al 2001 Science 292:1876; Vassylyev DG et al 2002 Nature [Lond] 417:712; Bushnell DA, Kornberg RD 2003 Proc Natl Acad Sci USA 100:6969; single-molecule analysis of transcription: Bai L et al 2006 Annu Rev Biophys Biomol Struct 35:343; structure of human RNA polymerase II: Kostek SA et al 2006 Structure 14:1691; historical steps in the discovery of the yeast RNA polymerase II: Landick R 2006 Cell 127:1087; chromatin and transcription: Li B et al 2007 Cell 128:707; molecular and structural bases of transcription: Kornberg RD 2007 Proc Natl Acad Sci USA 104:12955.

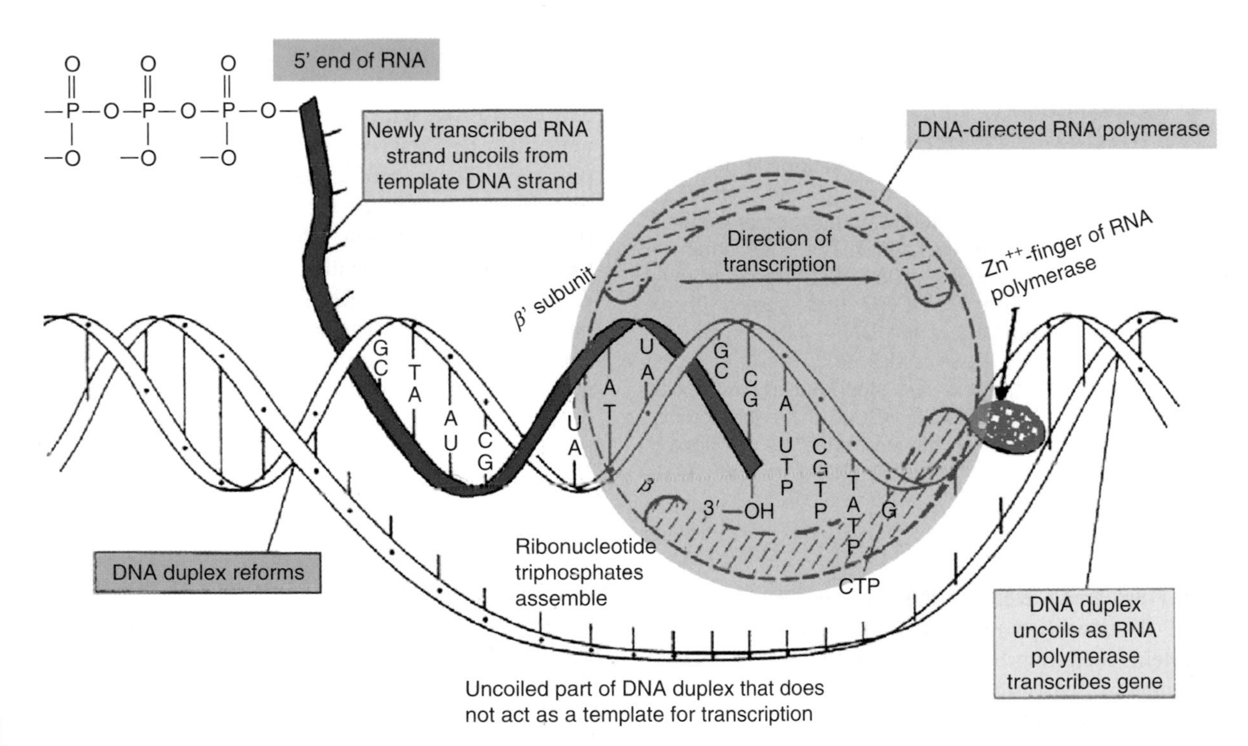

Figure R89. Conceptualization of the general process of transcription of RNA on DNA template from ribonucleoside 5′-triphosphates. The first 5′-nucleotide retains the three phosphates; the following ones split off a pyrophosphate and are then hooked up to the 3′-OH ends. In this process a ternary complex involving DNA, RNA, and protein is involved. The RNA-protein has been located to about 9 bp from the DNA fork into the transcription bubble where the DNA template forms an approximate 8 bp heteroduplex (HBS, heteroduplex-binding site). The RBS (RNA-binding site) together with HBS extends to approximately 14–16 RNA nucleotides from the 3′-OH end. About 7–9 nucleotides further downstream, RBS, HBS, and DBS (DNA-binding site) are situated and stabilize transcription. The DNA entry and the RNA exit sites on the polymerase are in close vicinity to each other. This diagram does not show the general transcription factors and the numerous other proteins regulating the process. (Diagram is modified after Page DS, *Principles of Biological Chemistry.* Willard Grant, Boston MA; see also Nudler E, et al 1998. *Science* 281:424.) For the structure of initiation complex of the T7 enzyme at 2.4 Å resolution, see Cheetham GMT and Steitz TA 1999. *Science* 286:2305.

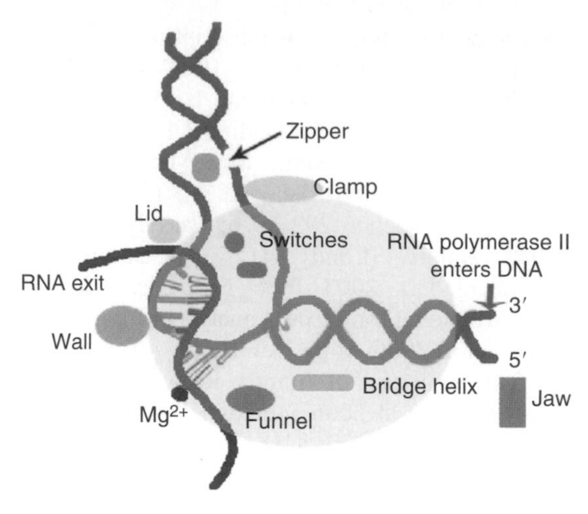

Figure R90. The structures of RNA polymerase II has been determined at 3.3 Å resolution. The schematic of the major functional elements is shown above. The enzyme enters DNA at right and holds onto it by the element designated as *JAW*. The *CLAMP* domain keeps the protein on the DNA. The *Switches* turn on the function. A large protein sununit, the *WALL* causes an approximately 90° turn of the DNA template. This then facilitates the addition of nucleotides to the RNA. The RNA building blocks, the nucleotide triphosphates, enter the complex through the *FUNNEL*, which ends in A *PORE*. Magnesium (or other metals) is required for the polymerization.The RNA chain is symbolized by a dotted line. The DNA and RNA from A 9-base duplex. The *RUDDER* (not shown here) prevents the extension of the DNA—RNA hybrid to go beyond the 9 base pairs and facilitates the separation of the duplex. The synthesized RNA exits under the *LID*. The *BRIDGE HELIX* under the coding strand of the DNA apparently promotes the addition of nucleotides. (Modified after Gnatt AL, et al 2001 *Science* 292:1876.)

RNA Polymerase Holoenzyme: A complex of RNA Pol II, transcription factors and other regulatory proteins including Srb. It is a general transcriptional regulator of eukaryotes. ►transcription factors, ►co-activators, ►TBP, ►TAF, ►RNA polymerase

RNA Polymerase I: It generates ribosomal RNA. In yeast, the UAF includes histones H3 and H4 and four non-histone proteins as essential factors (Tangaonkar P et al 2005 Proc Natl Acad Sci USA 102:10129). ►transcription factors

RNA Polymerase II: ►RNA polymerases

RNA Polymerase III: RNA polymerase III mediates the transcription of tRNAs and other small RNAs. It has efficient proofreading ability and reduces transcriptional errors by 10^3 (Alic N et al 2007 Proc Natl Acad Sci USA 104:10400). This enzyme may be suppressed by ARF (Morton JP et al 2007 Nucleic Acids Res 35:3046). ►transcription factors, ►ARF

RNA Polymerase IV (nuclear RNA polymerase D1A): A plant enzyme which, with RNA-dependent RNA polymerase 2, provides a substrate for Dicer-like 3 endoribonuclease in generating siRNA. It also mediates RNA-dependent methylation of DNA (Matzke M et al 2006 Cold Spring Harbor Symp Quant Biol 71:449). ►Dicer, ►RNA dependent RNA polymerase, ►siRNA, ►epigenesis; Zhang X et al 2007 Proc Natl Acad Sci USA 104:4536.

RNA Polymerizations: ►pol

RNA Primer: ►DNA replication

RNA Processing: Eukaryotic genes are "in pieces", although the DNA is continuous, in between the protein-coding nucleotide sequences (exons), non-translated additional nucleotide sequences, introns occur. The long sequences are transcribed into a long RNA tract but the introns are removed and the exons are spliced to make the mRNA. Similar processing is carried also out with the rRNA and tRNA. The vast majority of prokaryotic genes do not require these processes. These cutoffs, the mRNA, and the primary transcripts constitute the pool of the hnRNA (heterogeneous nuclear RNA). The primary RNA transcripts are coated with different proteins and thus form hnRNP, i.e., hnRNA-protein particles. These particles are instrumental in cutting the transcripts and splicing the exons into mRNA. After the splicing is completed, the methylated guanylic cap and the polyA tail are added and then the mRNA exported into the cytosol. ►introns, ►hnRNA, ►spliceosome, ►post-transcriptional processing, ►exosome; Varani G, Nagai K 1998 Annu Rev Biophys Biomol Struct 27:407.

RNA-Protein Interaction: A common phenomenon in the regulation of transcript elongation, attenuation, and in antitermination and several other functions in the cell. (See Nagai K, Mattaj IW 1994 RNA-Protein interactions, Oxford University Press; ►antitermination; Xia T et al 2005 Proc Natl Acad Sci USA 102:13013).

RNA, Regulatory: See antisense RNA, B2 RNA, Csr, guide RNA, microRNA, noncoding RNA, noncoding sequences, plasmid maintenance, hok-sok-mok, RNAi, 6S RNA, shRNA, siRNA, signal sequence recognition particle, 7SK RNA, snoRNA, SRA RNA, telomerase, U RNA, Xist. (See Storz G et al 2005 Annu Rev Biochem 74:199).

RNA Repair: Among the ~62 naturally occurring RNA modification, 20 involves methylation. Modified RNAs play important roles in the control of gene expression and RNA folding and RNA-Protein interactions, and may affect cellular defenses. The human demethylases AlkB (14q24) and an AlkB-like protein, ABH3 (encoded at 11q11), repair both RNA and DNA. ►DNA repair, ►methylation of DNA, ►methylation of RNA, ►ribozyme; Wei YE et al 1996 Nucleic Acids Res 24:931; Aas PA et al 2003 Nature [Lond] 421:859.

RNA Replication: Some bacteriophages (R17, MS2, fd2, f4, Qβ, M12) and some animal viruses such as the polio virus, the vesicular stomatitis virus, rhinoviruses (influenza viruses) and the vast majority of the plant viruses have RNA genetic material. This RNA is either single- or double-stranded. One of the best known is the Qβ replicase system. The tetramer (210 kDa) is encoded one viral (β subunit about 65 kDa) and three bacterial genes (translation factors). The replicase enzyme does not have an editing exonuclease function. These replicases do not replicate or transcribe host RNAs. The replication of the RNA is similar to the replication of DNA inasmuch as all nucleic acid strands are elongated at the 3′-OH ends. (The retrovirus replication is discussed under reverse transcription.) Most of the bacterial RNA viruses and many animal viruses have the same genetic material as their mRNA (+ strand viruses). They use a replicative intermediate (RI) for the synthesis of the first new (−) strand.

The (−) strand can then generate as many (+) strands as needed. Some viruses are, however (−) strand viruses because their genetic material is not identical to the mRNA but complementary to it. The polio virus (+) strand is the genetic material that is associated with a protein at the 5′-end through a phosphodiester linkage to a tyrosine residue. The (−) strand lacks this feature. Apparently, the OH group of the tyrosine primes the synthesis of the (+) strand but the (−) strand can get by without it. The reoviruses of vertebrates contain 8–10 short double-stranded RNA molecules. When the virus enters the host cell the coat protein is shed and its RNA polymerase is activated and works conservatively and asymmetrically. A virion enzyme copies the (−) strand and a new (+) is released. The original RNA duplex is conserved. On the new (+) strand a (−) strand is made and the double-stranded RNA is reconstituted. ►DNA replication, ►RNA dependent RNA replication, ►replicase, ►replicative intermediate, ►plus strand, ►retroviruses; Tang H et al 1999 Annu Rev Genet 33:133; Knipe DM, Howley PM (Eds.) 2001 Fundamental Virology, Lippincott Williams & Wilkins, Philadelphia, Pennsylvania.

RNA Restriction Enzyme: In ribozymes restriction functions may exist. ►restriction enzyme, ►ribozyme

RNA Rewriting: ►RNA editing

RNA, Ribosomal: ►ribosomes

RNA, 4.5S: A bacterial RNA of 114 nucleotides. It targets signal peptide-equipped proteins to the secretory apparatus where it forms the signal-recognition particle with protein Ffh. It also binds to the translation elongation factor G. It is homologous to the eukaryotic 7S RNA. ►Ffh, ►secretion system, ►RNA 7S, ►elongation factors; Nakamura K et al 2001 J Biol Chem 276:22844.

RNA, 5S; RNA, 5.8S: ►ribosomes; http://biobases.ibch.poznan.pl/5SData/.

RNA 6S: A regulator of the RNA polymerase of *E. coli*. (See Wassarman KM, Storz G 2000 Cell 101:613).

RNA, 7S: A cytoplasmic class of RNA of ~300 nucleotides, present in prokaryotes and eukaryotes. It participates in the function of RNA polymerase III and in the sequence signal recognition particle to direct proteins to the endoplasmic reticulum. The 7S RNA, which is part of the eukaryotic 35S pre-rRNA, is also a precursor of the 5.8S ribosomal RNA and the small GTPase RAN processes it. ►RAN, ►DNA 7S, ►signal sequence recognition particle; Luehrsen KR et al 1985 Curr Microbiol 12:69; Suzuki N et al 2001 Genetics 158:613.

RNA, 10Sa (SsrA/tmRNA): When an mRNA is truncated at the 3′-end and has no stop codon, the ribosome may switch from one RNA to another to terminate the translation. The switching results in the generation of an 11-amino residue at the carboxyl end that destines the protein for degradation. The last 10 amino acids are translated from a stable so-called 10Sa RNA (363 nucleotides). So far, these 10 amino acids were found on murine interleukin-6 translated in *E. coli*, λ phage *cI*, cytochrome b-562, and polyphenylalanine translated on polyU. The 10Sa RNA is similar to a tRNA and it is charged with alanine. This alanine is found as the first amino acid of the tag. This saves the ribosome from the "unfinished" mRNA and permits its quasi-normal operation. The mechanism carries out the same task as ubiquitin does in eukaryotes. ►translation non-contiguous, ►ubiquitin, ►lambda phage, ►tmRNA; Gillet R, Felden B 2001 EMBO J 20:2966; Zwieb C, Wower J 2000 Nucleic Acids Res 28:169; http://www.indiana.edu/~tmrna.

RNA, 20S: A naked, single-stranded RNA phage-like replicon of about 2,500 nucleotides in yeast, encoding a 95 kDa RNA polymerase-like protein. Its termini are 5′-GGGGC and GCCCC-3′ and therefore may circularize. (See Rodriguez-Cusino N et al 1998 J Biol Chem 273:20363).

RNA Sequencing: RNA sequencing can be done by digesting RNAs with different ribonuclease enzymes that cut the phosphodiester linkages at the specific nucleotides and then separate the fragments by gel electrophoresis (ca. 20% polyacrylamide) in one or in two dimensions. The RNA is generally labeled by ^{32}P in order to autoradiograph the dried gels. T_1 ribonuclease (from *Aspergillus oryzae*) generates fragments with 3′ guanosine monophosphate ends. U_2 ribonuclease (from *Ustilago sphaerogena*) cleaves at purine residues (intermediates 2′,3′-cyclicphosphates). RNase CL 3 (from chicken liver) has about 16-fold higher activity for cytidylic than uridylic linkages. RNase B (from *Bacillus cereus*) cuts Up↓N and Ap↓N bonds. RNase Phy M (from *Physarum polycephalum*) has the specificity Ap↓N and Up↓N. When several of these enzymes are employed, from the separate electrophoretic patterns, the sequences from the positions of the fragments with different termini can be deduced. In many ways the sequencing of RNAs with the aid of enzymatic breakage is very similar in principle to the Maxam and Gilbert method of DNA sequencing. Recently, RNA was sequenced by first converting it using reverse transcriptase into cDNA and then using the much easier DNA sequencing methods to determine the RNA nucleotide sequences. ►DNA sequencing, ►ribonucleases; Gilham PT 1970 Annu Rev Biochem 39: 227; Bachellerie J-P, Qu L-H 1993 Methods Enzymol 224:349; Kuchino Y, Nishimura S 1989 Methods Enzymol 180:154; Lane DJ et al 1988 Methods Enzymol 167:138; sequences and motifs: http://uther.otago.ac.nz/Transterm.html.

RNA Silencing: ►RNAi, ►microRNA

RNA 7SL: An approximately 300-base long molecule forming the signal recognition particle (SRP) with six proteins of 10 to 75 kDa size. (See Müller J, Benecke BJ 1999 Biochem Cell Biol 77:431).

RNA, Small: ►snRNA

RNA Splicing: The primary RNA transcript is much larger than required for particular functions and is, therefore, cut up during processing and the transcripts corresponding to introns removed. The remaining pieces are then reattached (spliced together) to form the mature RNA molecule. ►introns, ►RNA processing; Weg-Remers S et al 2001 EMBO J 20:4194.

RNA, Structural: Structural RNA may have a noncoding, regulatory function in the genome (Washietel S et al 2007 Genome Res 17:852). Several computational methods are available for their analysis. ►AlifoldZ, ►EvoFold, ►RNAz, ►RNA folding

RNA Structure (SCOR): See Noller HF 2005 Science 309:1508; ►MS2 phage; structural classification: http://scor.lbl.gov; non-canonical base-base interactions: http://prion.bchs.uh.edu/1/; http://www.rna.icmb.utexas.edu/; RNA tertiary structure server: http://bioinfo3d.cs.tau.ac.il/ARTS/; structural homology search: http://bioinfo.csie.ncu.edu.tw/~rnamst/; regulatory motif prediction: http://regrna.mbc.nctu.edu.tw/.

RNA Surveillance: RNA surveillance monitors for the presence of RNAs that have stop codon mutations, which might lead to the synthesis of truncated proteins. The nonsense mRNA is directed to P bodies for degradation by the NMD (nonsense-mediated decay) pathway (Sheth U, Parker R 2006 Cell 125:1095). In yeast, the NMD process begins by recruiting the Upf1 protein to the ribosomes. Ubf1p can bind ATP. It can also bind nucleic acids without ATP and can hydrolyze ATP in a nucleic-acid-dependent manner. Ubf1p is also an ATP-dependent 5′→3′ RNA/DNA helicase. Upf32p is probably a signal transducer from Ubf1 to Ubf3, and the latter then enlists the nonsense mRNA to the NMD complex. Ubf3p is located mainly in the cytoplasm but it can shuttle between the nucleus and the cytoplasm. In yeast proteins eRF1 and eRF3, termination factors are also required for the abortion of translation. The complex also mediates the decay of the defective polypeptide if any is made. NMD systems operate in a wide range of eukaryotes and may have roles to play in some cancers and hereditary diseases of humans. ►Pab1p, ►Xrn1p, ►exosome, ►degradosome, ►SMG, ►SR motif, ►P body; Culbertson MR 1999 Trends Genet 15:74; Maquat LE, Carmichael GG 2001 Cell 104:173; Ishigaki Y et al 2001 Cell 106:607; Frischmeyer PA et al 2002 Science 295:2258; Lewis BP et al 2003 Proc Natl Acad Sci USA 100:189; Chang Y-F et al 2007 Annu Rev Biochem 76:51.

RNA, Therapeutic: Therapeutic RNAs may be used in medicine as a gene function inhibitory means such as in antisense technologies and RNAi. Ribozymes may cleave pathogenic RNA molecules of infectious agents (e.g., HIV). Trans-splicing of transcripts may correct defective functions (e.g., sickle-cell disease, tumor suppressor transcripts). RNA may be used to prevent the binding of pathogenesis proteins to target (e.g., HIV-1 TAR, REV). Transfection of dendritic, antigen-presenting cells with DNA that generate mRNA for tumor antigens may boost the effectiveness of cytotoxic T lymphocytes. ►decoy RNA, ►RNAi, ►aptamer, ►ribozyme, ►acquired immuno deficiency, ►transsplicing, ►CTL, ►antigen presenting cell, ►tumor antigen, ►tumor suppressor; Sullenger BA, Gilboa E 2002 Nature [Lond]: 418:252.

RNA Transcript: An RNA copy of a segment of DNA.

RNA Transport: mRNA, snRNA, U3 RNA, and ribosomal RNA (but not tRNA) are exported from the nucleus by RCC1 (yeast homolog is PRP20/MTR1). This nuclear protein is a guanine nucleotide exchange factor for the RAS-like guanosine triphosphatase (GTPase). Some plant virus RNAs spread through plasmodesmata to a distance from the site of infection with the aid of protein transporters. Plants can move their own mRNAs through the phloem cells at great distances. ►RAN, ►RCC1, ►nuclear pore, ►plasmodesma, ►phloem, ►transport elements constitutive, ►TAP, ►RNA export, ►export adaptors; Yang J et al 2001 Mol Cell 8:397.

RNA Trap (RNA tagging and recovery of associated proteins): The purpose of the procedure is to test the mechanism of action on gene expression exerted by enhancers, which are distant from a genic site. To this goal, the unprocessed RNA transcript of a specific locus with labeled oligonucleotides is hybridized to isolated cell nuclei. Horseradish peroxidase-conjugated antibodies are localized to the oligonucleotide probe. Peroxidase-activated biotinyl-tyramide covalently labels the electron-rich protein moieties by biotin in this region. From the sonicated cell fragments the biotin-conjugated chromatin is isolated by affinity to a streptavidin column. The specific sequences are then enriched by quantitative PCR. Enrichment of the locus control region—specifically the DNase hypersensitive sites—indicated by immunofluorescence analysis that one of the hypersensitive sites becomes physically associated with the gene transcribed. This technology thus supports the role of DNA looping in gene function. ►biotin, ►enhancer, ►LCR, ►PCR, ►regulation of gene activity, ►streptavidin, ►DNA looping; Carter D et al 2002 Nature Genet 32:623.

RNA Trimming: ►trimming

RNA, Ubiquitous: ►U RNA

RNA Viruses: ►viruses, ►retroviruses, ►animal viruses, ►plant viruses TMV, ►MS2, ►paramyxovirus, ►viral vectors, ►reovirus, ►togavirus, ►ebola virus, ►rabies; virus database: http://www.ncbi.nlm.nih.gov/ICTVdb/ictvdb.htm.

RNA World: RNA world existed in the pre-biotic era when RNA carried out auto- and heterocatalysis without DNA. There is evidence now that ribozymes can even catalyze nucleotide synthesis giving further support to the RNA world concept. In addition, evidence shows that RNA is involved in the catalysis of protein synthesis: (i) by encoding amino acid sequence in protein; (ii) activation of amino acids; (iii) synthesis of aminoacyl-tRNA by a reaction analogous to the aminoacyl-tRNA synthetases; and (iv) formation of peptide bonds. Highly reactive aminoacyl phosphate oligonucleotide adaptors and RNA guides facilitate di- and tripeptide synthesis in a single guide sequence-dependent manner without ribosomes or ribozymes (Tamura K, Schimmel P 2003 Proc Natl Acad Sci USA 100:8666). ►origin of life, ►ribozyme, ►autocatalytic function, ►heterocatalysis, ►peptide nucleic acid, ►protein synthesis; Kumar RK, Yarus M 2001 Biochemistry 40:6998; Joyce GF 2002 Nature [Lond] 418:214; Szabó P et al 2002 Nature [Lond] 420:340; Yarus M 2002 Annu Rev Genet 36:125; Spirin AS 2002 FEBS Lett 539:4; Gilbert W 1986 Nature [Lond] 319:P618.

RNA-Dependent RNA Polymerase: ►RNA-dependent RNA polymerase

RNA-Directed DNA Methylation (RdMD): A sequence-specific silencing of DNA. The methylation may be limited to as short as 30-base sequences but it extends generally to longer stretches (Pélissier T, Wassenegger M 2000 RNA 6:55). Argonaute 4 plays both catalytic and non-catalytic roles in DNA methylation through the RNAi pathway (Qi Y et al 2006 Nature [Lond] 443:1008). Transcriptional silencing via RNA-directed DNA methylation and chromatin modification involves two forms of nuclear RNA polymerase IV (Pol IVa and Pol IVb), RNA-dependent RNA polymerase2 (RDR2), DICER-LIKE3 (DCL3), ARGONAUTE4 (AGO4), the chromatin remodeler, DRD1, and the de novo cytosine methyltransferase, DRM2 (Pikaard CS 2006 Cold Spring Harbor Symp Quant Biol 71:473). ►methylation of DNA, ►epigenetics, ►argonaute, ►RNAi; Wassenegger M et al 1994 Cell 76:567; Mette MF et al 2000 EMBO J 19:5194.

RNAi (RNA-mediated genetic interference): Single or double-stranded RNA, formed or introduced into the cell, may interfere with the translation of the endogenous genes of *Caenorhabditis* or plants, *Drosophila* or mammals. Apparently, *Saccharomyces cerevisiae* lacks miRNA and RNAi capabilities. The miRNAs (~22 nucleotides) are very similar to RNAi but they are produced from foldback DNA. They can cleave hundreds of mRNAs or repress their translation and thereby regulate the function of genes involved in the regulation of plant and animal development (Bartel DP 2004 Cell 116:281). Short synthetic RNAs may also evoke interference. The double-stranded RNA is at least an order of magnitude more potent. In the nematode, the most effective exogenous delivery, to any part of the body, is through the intestines. The effective pre RNAi length is 2,000—1,000 nucleotides. RNAi is ATP-dependent, but it is not linked to mRNA translation. There is now evidence that the effect of RNAi is enzymatic (RNAi nuclease) and involves the degradation of the mRNA rather than some kind of antisense mechanism. The nascent dsRNA, generated by RNA-dependent RNA polymerase, is degraded to eliminate the incorporated mRNA and new cycles of dsRNAs are produced that yield new siRNAs (short interfering RNAs), and secondary siRNAs (Lipardi C et al 2001 Cell 107:297). The secondary siRNAs are the products of the action of RNA-directed RNA polymerase versus the primary siRNAs, which are the products of the action of Dicer nuclease. The secondary siRNAs, in contrast to the primary siRNAs, do not have mismatches to their target (Sijen T et al 2007 Science 315:244). The siRNA (21–22 nucleotides) combines with proteins (RISC [RNA-induced silencing complex]) that degrade the RNAs recognized by siRNA (Martinez J et al 2002 Cell 110:563). siRNA is also involved in chromatin remodeling by the nucleolar RNA processing center (localized in the Cajal bodies) involving RNA-dependent RNA polymerase, DICER-LIKE3, ARGONAUTE4 and the largest subunit RNA polymerase IV b (Pontes O et al 2006 Cell 126:79; Li CF et al 2006 Cell 126:93). RNA silencing by siRNA in both plants and animals may be mediated by the methylation of DNA and histone H3 lysine^9 (Kawasaki H, Taira K 2004 Nature [Lond] 431:211). For high effectiveness of RNAi in the mammalian system the antisense strand should have A/U at the 5′ end and the 5′ end should have a 7-base pair AU-rich sequence. At the 5′ end of the single-strand there should be G/C. There should be no more than a 9 bp long GC stretch present (Ui-Tei K et al 2004 Nucleic Acids Res 32:936). The secondary siRNA may spread 5′ to the original target sequence of the mRNA and this new RNA has been called transitive RNAi. This phenomenon may lead to the silencing of many genes with each siRNA and provide a means for revealing their function (Chi J-T et al 2003 Proc Natl Acad Sci USA 100:6343). Genome-wide analysis of growth and viability of *Drosophila* cells using high throughput RNAi screen with 19,470 double-stranded RNAS identified ~91% of the genes involved in these functions. Many of these genes had no known mutant alleles before (Boutros M et al 2004 Science 303:832).

Theoretically, siRNA appear ideal for therapeutic purposes but their practical application faces several hurdles. Although invertebrate cells readily take up siRNA, the majority of cells of vertebrates do not absorb it efficiently enough to bring about gene silencing (see Fig. R91). Direct injection or delivery

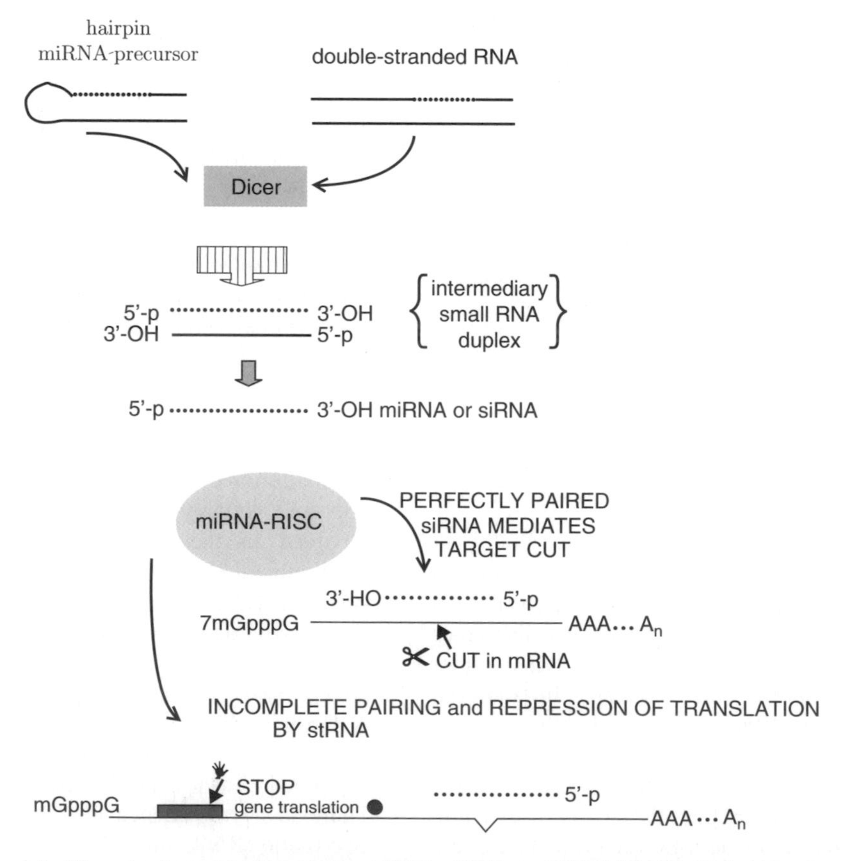

Figure R91. A model of the roles in gene silencing by miRNA, stRNA and RNAi (Modified after Hutvágner G and Zamore PD 2002 *Science* 297:2056)

by cationic lipids into the retina or into the genital tract seems to be more successful and may provide help in age-related macular degeneration or in viral some infections, e.g., corona virus (SARS) in the lung. There are promises for treatment of local inflammatory lesions. New vehicles of delivery to internal organs are under exploration; these include liposomes or nanoparticles. siRNA-binding antibody fragments with fusion proteins may target special cells (tumors, HIV-infected lymphocytes). The free siRNA has a very short half-life in the body but once it is incorporated into the RISC complex it may survive for weeks. Unintended targets may be adversely silenced by siRNA. Chemical modification of the second residue in the active siRNA strand may prevent off-target problems. Other problems may involve triggering inflammatory, immune or interferon reactions by the body to the siRNA resembling viral RNA. The cells may become resistant to siRNA if the viral or cancer targets undergo mutational changes. siRNA may also adversely affect endogenous microRNAs (see, for excellent reviews, Dykxhoorn DM, Lieberman J 2006 Cell 126:231; Kim DH, Ross JJ 2007 Nature Rev Genet 8:173)

An shRNA/siRNA library targeting 9,610 human and 5,563 mouse genes is also available (Paddison PJ et al 2004 Nature [Lond] 428:427; see also Berns K et al 2004 ibid:431). In organisms, which have not been sequenced or annotated there are, in some cases, difficulties with the use of the RNAi technology. To overcome these problems, siRNA libraries were constructed containing all possible permutations. These were obtained by the use of a plasmid vector containing two convergent RNA III polymerase promoters. This procedure facilitated the construction of 5×10^7 siRNA-encoding plasmids in a single vessel. Such a library permitted genome-wide screening on the basis of phenotype (Chen M et al 2005 Proc Natl Acad Sci USA 102:2356). RNA polymerase II can synthesize a short hairpin RNA with a special green fluorescent protein coupled marker and good function (Zhu H et al 2005 Nucleic Acids Res 33(6):262). A pDECAP vector, expressing double-stranded RNA from an RNA polymerase II promoter that lacks the 5′-cap and the 3′-polyA tail transcript is not exported to the cytoplasm where it would encounter interferons, which would destroy it. The induction of an interferon response may cause general rather than specific silencing. The introduction of double-strand RNAs activates protein kinase PKR in the animal cell which phosphorylates and inactivates the eukaryotic translation initiation factor eIF2a. This causes general inactivation of many genes in the same way as an antiviral response. By keeping it within the animal nucleus it can be used to knock down specific genes by the double-stranded RNA transcript of a transcriptional corepressor gene (Shinagawa T, Ishii S 2003 Genes Dev 17:1340). In cases when RNAi inactivates more than the target gene, the rescue of the critical gene is feasible by transforming the cell with the aid of a bacterial artificial chromosome carrying an RNAi-resistant transgene from a related species (Kittler R et al 2005 Proc Natl Acad Sci USA 102:2396). If a gene-specific siRNA sequence is inserted between two opposing promoters such as the mouse U6 and human H1, sense and antisense strands of the same template results in the mammalian cell after transfection. The construction of siRNA expression cassettes can be used in a high-throughput manner to scan targets genome-wide (Zheng L et al 2004 Proc Natl Acad Sci USA 101:135). siRNAs can be generated efficiently by restriction enzymes (Sen G et al 2004 Nature Genet 35:183; Shirane D et al 2004 Nature Genet 36:189). In *Caenorhabditis* genes, *rde–1, –2, –3, –4* control this interference. RDE-1 and RDE-4 proteins seem to practice surveillance for the presence of double-stranded RNA, and transposons and RDE-2 and RDE-3 (and MUT-7) degrade these RNA molecules as a defense mechanism. The mut-7 gene controls RNA interference and transposon silencing in *Caenorhabditis*. The two strands of the targeted double-stranded RNA are cleaved into 21–23- (or 25-) nucleotide long segments (siRNAs [short interfering RNAs] and stRNA [small temporal RNA]). The siRNA degrades its target whereas the regulatory stRNA represses translation of its target mRNA. There appear to be two RNase III motif enzymes involved in their processing from precursors RNAs. One, the initiator enzyme, Dicer, generates a ~22 nucleotide guide that marks the mRNA for further degradation by the RISC (RNA-induced silencing complex), an effector ribonuclease complex. Dicer-1 has specificity for single-stranded RNA and Dicer-2 (R2D2) cuts double-stranded RNA. R2D2 also selects the proper strand for association with the Argonaute component of RISC (Tomari Y et al 2004 Science 306:1377). The silencing is highly specific and yet the interference may affect related genes too; the effect may vary in degree in the various tissues. The dsRNA precursor of RNAi is transported within the body by the transmembrane protein SID-1 (Feinberg EH, Hunter CP 2003 Science 301:1545). The interference can also be transmitted to the progeny although, most probably, posttranslational events are involved. The protein product of the gene is homologous with the 3′-5′ exonuclease domains of RNaseD and the Werner syndrome protein. By establishing a library of DNA clones in a bacterium, which produces double-stranded RNA (RNAi/shRNA) and fed to post-embryonic *Caenorhabditis*, the function of a large number of hitherto unknown

ORF could be identified by the silenced phenotype. Or, in case of genes controlling cell division by the use of in vivo, time-lapse differential interference contrast microscopy. The *eri-1* mutants of *Caenorhabditis* encode an evolutionarily conserved protein domain, homologous to nucleic acid-binding and exonuclease proteins. The *ERI-1* gene product (and its human homolog) degrades siRNA and thus negatively regulates RNAi (Kennedy S et al 2004 Nature [Lond] 427:645). A genome-wide study of *Caenorhabditis* revealed 90 proteins, including Piwi/PAZ, DEAH helicases, DNA binding/processing factors, chromatin-associated factors, DNA recombination proteins, nuclear export/import factors and 11 known components of the RNAi machinery within the interference pathway (Kim JK et al 2005 Science 308:1164). A genome-wide library of *Drosophila melanogaster* RNAi transgenes, extending to 88% of the genome, enables the conditional inactivation of gene function in specific tissues of the intact organism (Dietzl G et al 2007 Nature [Lond] 448:151).

In plants, the same mechanism can account for co-suppression and VIGS. In plants, two types of mechanism account for viral RNA silencing: (i) The helper component-proteinase (HC-Pro) derived from the potyviruses, which is very efficient and can silence a broad range of plant viruses and transgene-induced as well as VIGS silencing; (ii) The potato virus X (PVX) p25, which is much less effective and its action targets systemic silencing. *Arabidopsis* has four Dicer enzymes. The Dicer-like 4 (DCL4) complex confers antiviral immunity but it can be suppressed by a viral factor. Both DCL4 and DCL2 have to be suppressed for a suppressor-deficient virus to achieve systemic infection and cause bleaching (see Fig. R92) (Deleris A et al 2006 Science 313:68).

Figure R92. Photobleaching

These small RNAs may fight (HIV, Hepatitis viruses, cancer) and suppress transposons. The viral suppression of miRNAs may be responsible for some of the disease symptoms (Kasschau KD et al 2003 Dev Cell 4:205). Some endogenous proteins can also mediate gene silencing in plants and in *Caenorhabditis* or *Drosophila*. RNAi technology may also be applied to human pathogenic viruses such as HIV (Jacque J-M et al 2002 Nature [Lond] 418:435). The therapeutic application of RNAi technology appears promising. Systemic delivery of siRNA in liposomes by intravenous injection into nonhuman primates seems very effective. Within 24–48 h after injection it showed precise targeting of apolipoprotein B and the biological effect lasted for 11 days at the highest dose (Zimmermann TS et al 2006 Nature [Lond] 441:111). Interfering RNAs have sequence-specificity for silencing but the specificity is not absolute. The siRNA technology combined with targeting specific dominant disease alleles (Machado-Joseph/spinocerebellar ataxia type 3 or Pick disease/tau) can specifically inhibit a disease symptom in a highly discriminating manner (Miller VM et al 2003 Proc Natl Acad Sci USA 100:7195). RNAi can specifically eliminate new single nucleotide mutations in the SOD1, amyotrophic lateral sclerosis alleles (Ding HG et al 2003 Aging Cell 2:209).

siRNA microbicides targeted to the Herpes simplex virus 2 envelope glycoprotein and DNA-binding protein. siRNA microbicides targeted to Herpes simplex virus 2 envelope glycoprotein and DNA-binding protein and thymidine kinase provided effective protection by vaginal application against the lethal effects of Herpes in mice. The treatment protected the animals even if applied one day after the infection and provided protection for several days. No adverse interferon effects occurred. This approach appears promising in the fight against sexually transmitted diseases (Palliser D et al 2006 Nature [Lond] 439:89). Although targeting of siRNAs to the proper disease sites and their maintenance in the cells is difficult, chemical modifications substantially improve their therapeutic potential (Soutschek J et al 2004 Nature [Lond] 432:173). Recently, reservations emerged concerning the specificity of siRNAs for effective medical treatment because the RNA molecules also hit unintended targets in some studies (Jackson & Linsley 2004 Trends Genet. 20:521). Certain mismatches between siRNA and the target abolish silencing effects. This may be due either to the failure of accessing the RISC complex or the inability of base recognition in the target. A study of the 57 all possible single-nucleotide mutations in the CD46 siRNA (siCD46) revealed that in this case mismatches between A and C bases as well in the G:U wobble pair were well tolerated, but alterations at other sites involved degradation of the target mRNA (Du Q et al 2005 Nucleic Acids Res 33:1671). The RNAi-induced silencing may be transmitted to several generations of *Caenorhabditis* although the penetrance may not be more than 30% and the expressivity may vary (Vastenhouw NL et al 2006 Nature [Lond] 442:882).

The average silencing activity of randomly selected siRNAs is as low as 62%. Applying more than five different siRNAs may lead to saturation of the RNA-induced silencing complex (RISC) and to the degradation of untargeted genes. Therefore, selecting a small number of highly active siRNAs is critical for maximizing knockdown and minimizing off-target effects. To satisfy these needs, a publicly available and transparent machine learning tool is available that ranks all possible siRNAs for each targeted gene. Support vector machines (SVMs) with polynomial kernels and constrained optimization models select and utilize the most predictive effective combinations, thermodynamic, accessibility and self-hairpin features. This tool reaches an accuracy of 92.3% in cross-validation experiments (Ladunga I 2007 Nucleic Acids Res 35:433; http://optirna.unl.edu/).

Curiously, 21-nt dsRNAs targeted to selected promoter regions of human genes of E-cadherin, p21, and VEGF boost the expression of the genes rather than dampen it. The dsRNA mutation at the 5′ end of the antisense strand, or "seed" sequence, is critical for activity. Mechanistically, the dsRNA-induced gene activation requires the Argonaute 2 (Ago2) protein and is associated with a loss of lysine-9 methylation on histone 3 at dsRNA-target sites (Li LC et al 2006 Proc. Natl. Acad. Sci. USA 103:17337). Multiple duplex RNAs complementary to the progesterone receptor (PR) promoter increase the expression of both PR protein and RNA after transfection into cultured human breast cancer cells. Upregulation of PR protein reduced the expression of the downstream gene encoding cyclooxygenase 2 but did not change concentrations of the estrogen receptor. This demonstrates that activating RNAs can predictably manipulate physiologically relevant cellular pathways. Activation decreased over time and was sequence specific. Chromatin immunoprecipitation assays indicated that activation is accompanied by reduced acetylation at histones H3K9 and H3K14 and by increased di- and trimethylation at histone H3K4 (Janowski BA et al 2007 Nature Chem Biol 3:166).

The 2006 Nobel Prize for physiology and medicine was awarded to Andrew Z. Fire and Craig C. Mello for their pioneering research on interfering RNA. ►co-suppression, ►microRNA, ►shRNA, ►siRNA, ►rasiRNA, ►piRNA, ►tncRNA, ►Dicer, ►RNA non-coding, ►antisense RNA, ►small RNA, ►self-cleavage of RNA, ►argonaute, ►PAZ domain, ►heterochronic RNA, ►antisense technologies, ►heterochromatin, ►methylation of DNA, ►deletion mapping, ►posttranscriptional silencing, ►RNA interference, ►DEAH box proteins, ►apolipoproteins, ►liposome, ►interferon, ►quelling, ►inhibition of transcription, ►host-pathogen relation, ►epigenesis, ►Cajal body, ►Werner syndrome, ►Machado-Joseph syndrome, ►Pick's disease, ►ribonuclease D, ►viral encephalitis, ►Nomarski differential phase contrast microscopy, ►RISC, ►VIGS, ►methylation of DNA, ►amyotrophic lateral sclerosis, ►SOD, ►support vector machine; Fire A 1999 Trends Genet 15:358; Tabara H et al 1999 Cell 99:123; Grishok A et al 2000 Science 287:2494; Grishok A et al 2001 Cell 106:23; Elbashir SM et al 2001 Nature [Lond] 411:494; Vance V, Vaucheret H 2001 Science 292:2277; Hutvágner G et al 2001 Science 293:834; Matzke M et al 2001 Science 293:1080; Hammond SM et al 2001 Science 293:1146; Ruvkun G 2001 Science 294:797; Plasterk RHA 2002 Science 296:1263; Zamore PD 2002 Science 296:1265; Mlotshwa S et al 2002 Plant Cell 14:S289; Hutvágner G Zamore PD 2002 Current Opin Genet Dev 12:225; Hannon GJ 2002 Nature [Lond] 418:244; McManus MT, Sharp PA 2002 Nature Rev Genet 3:737; Tijsterman M et al 2002 Annu Rev Genet 36:489; analysis of function of the genome: Kamath RS et al 2003 Nature [Lond] 421:231; Novina CD, Sharp PA 2004 Nature [Lond] 430:161; design and validation RNAi effectors: Huppi K et al 2005 Mol Cell 17:1; Matzke MA, Birchler JA 2005 Nature Rev Genet 6:24; Voinnet O 2005 Nature Rev Genet 6:206; in mammals: Martin SE, Caplen NJ 2007 Annu Rev Genomics Hum Genet 8:81, high-throughput techniques: Echeverri CJ, Perrimon N 2006 Nature Rev Genet 7:373; http://dnaseq.med.harvard.edu/rnai_database.htm; human siRNA: http://itb1.biologie.hu-berlin.de/~nebulus/sirna/; RNAi web site: http://www.rnaiweb.com/RNAi/RNAi_Web_Resources/siRNA_Collections___Databases/; RNAi Codex: http://codex.cshl.org/scripts/newmain.pl; *Drosophila* cell culture RNAi: http://www.flight.licr.org; http://flyrnai.org/cgi-bin/RNAi_screens.pl; RNAi probes: http://rnai.dkfz.de.

RNAIII/rnaiii: A bacterial virulence-controlling molecule. It is induced by the RNAIII activating protein (RAP). The RIP protein produced in non-pathogenic strains competes for activation of rnaIII and the production of *Staphylococcus aureus* toxin. ►RAP; Balaban N et al 2001 J Biol Chem 276:2658.

RNA-IN: The leftward transcript of the bacterial transposase gene in transposable elements, transcribed from the pIN promoter. ►RNA-OUT, ►Pin

RNA Interference: ►RNAi

RNA-Mediated Gene Activation: ►RNAi

R

RNA-Mediated Recombination: RNA-mediated recombination is thought to be involved in the exchange between the reverse transcript and the corresponding cellular allele. The Ty element-mediated recombination is supposed to involve RNA. ►Ty; Derr LK et al 1991 Cell 67:355.

RNA-OUT: A transcript originating from the strong pOUT promoter of bacterial transposable ele-ments. It opposes pIN and it directs transcription toward the outside end of the IS*10* element. ►RNA-IN, ►pIn, ►pOUT, ►Tn*10*

RNAP: RNA polymerases; in prokaryotes there is only one DNA-dependent RNA polymerase whereas in eukaryotes RNA Pol I, Pol II, and Pol III are found. ►RNA polymerases

RNA-PCR: A polymerase chain reaction may amplify rare RNAs after the RNA is reverse-transcribed into DNA. ►polymerase chain reaction, ►reverse transcriptases

RNA-Peptide Fusions: Synthetic mRNAs can be fused to their encoded polypeptides when the mRNA carries puromycin, a peptidyl acceptor antibiotic at the 3′-end. After in vitro enrichment, proteins can be selected in a directed manner. ►directed mutation, ►evolution; Roberts RW, Szostak J 1997 Proc Natl Acad Sci USA 94:12297.

RNA-Protein Interactions: RNA-protein interactions are ubiquitous in all cells in the formation of ribosomes, spliceosomes, in several ribozymes, in posttranscriptional regulation, in translation machinery (in tRNAs, elongation factors), and viral coat proteins. (See Jones S et al 2001 Nucleic Acids Res 29:943).

RNase: ►ribonucleases

RNase-free DNase: Heat, at 100 °C for 15 min, 10 mg RNase A per mL and 0.01 M Na-acetate (pH 5.2). Then cool and adjust pH to 7.4 (1 M Tris-HCl) and store at −20 °C. ►DNase free of RNase

RNase MRP: A ribonuclease that cleaves rRNA transcripts upstream of the 5.8 rRNA. In the mitochondria it cleaves the primers of DNA replication. ►U RNPs, ►cartilage hair dysplasia; Ridanpää M et al 2001 Cell 98:195.

RNASIN: A ribonuclease inhibitor.

RNAz: A computer program which combines comparative sequence analysis and structure prediction for noncoding RNAs. The measure for RNA secondary structure conservation is based on computing a consensus secondary structure and a measure for thermodynamic stability, which, in the spirit of a *z* score, is normalized with respect to both sequence length and base composition but can be calculated without sampling from shuffled sequences (Washietl S et al 2005 Proc Natl Acad Sci USA 102: 2454; http://www.tbi.univie.ac.at/~wash/RNAz). ►AlifoldZ, ►EvoFold, ►Z score

RNKP-1: A homolog of ICE and fragmentin-2. ►apoptosis, ►ICE, ►fragmentin-2

RNP: A ribonucleoprotein; any type of RNA associated with a protein particle as in the hnRNA or in the ribosomes. Proteins bind only single-stranded sequences of RNA. In the recognition of the bases, the shape and charge distribution of the RNA are also important. ►RNA, ►hnRNA, ►ribosomes

RNR: ►ribonucleotide reductase

RNS (reactive nitrogen species): ►nitric oxide, ►ROS

Roadmap: A project of the National Institute of Health. It was initiated in 2002 for the purpose of promoting medical research and its application for the benefit of public health. It provides technical information, support for high-risk research, and other funding (http://nihroadmap.nih.gov).

ROAM Mutation (regulated overproducing alleles under mating signals): Activates yeast Ty elements through the influence of the MAT gene locus. Such a system may operate if the Ty insertion takes place at the promoter of a gene and then the Ty enhancer may be required for the expression of that gene. These genes are expressed only in the *a* or *α* mating type cells but not in the diploid *a*/*α* cells. ►Ty, ►mating type determination in yeast; Rathjen PD et al 1987 Nucleic Acids Res 15:7309.

Roan: A fur color (cattle, horse) with predominantly brown-red hairs interspersed with white ones. It is common sign of heterozygosity for the *R* and *r* alleles of a gene locus. ►co-dominance

Roberts Syndrome: The Roberts syndrome actually overlaps with the *SC phocomelia* (= absence or extreme reduction of the bones of extremities located proximal to the trunk of the body; see Fig. R93) and with the TAR syndrome involving thrombocytopenia (reduction in the number of blood platelets), mental retardation, cleft palate, etc. These three syndromes are all autosomal recessive and seem to be basically the same. They are caused by chromosomal instability. Although linkage to chromosomes 1q, 4q and 8p has been detected, the strongest linkage (lod score 13.4) pointed to 8p12-p21.2. In this tract, gene ESCO2 controls sister chromatid cohesion and the protein supposedly expresses acetyltransferase activity (Vega H et al 2005 Nature Genet 37:468). ►chromosome breakage, ►thrombocytopenia, ►Holt-Oram syndrome, ►Wiskott-Aldrich syndrome, ►mental retardation

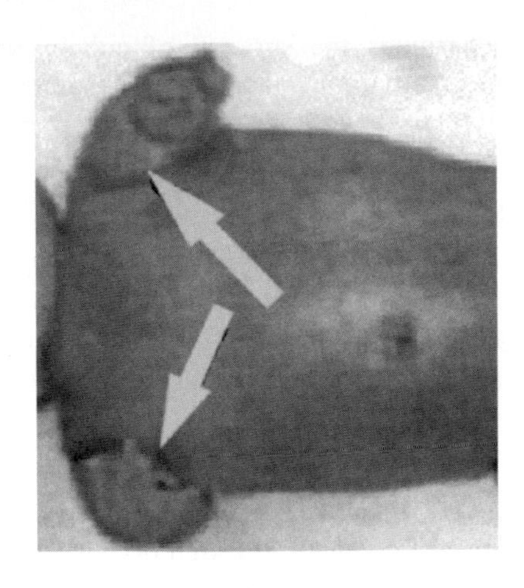

Figure R93. Roberts syndrome

Robertsonian Translocation (Robertsonian change): Two nonhomologous telocentric chromosomes fused at the centromere. Or, more likely, the translocation between two nonhomologous acrocentric chromosomes. The outcome is a replacement of two telo- or acrocentric chromosomes with one clearly bi-armed chromosome. These translocated chromosomes may have preserved the centromeres of both acrocentrics and remain cytologically stable because one of the centromeres is inactivated. Robertsonian translocations are very common in mouse cell cultures but they also occur in wild natural populations, resulting in an apparent change in chromosome morphology and numbers. The telomeric region of different mouse chromosomes include a contiguous linear order of T_2AG_3 repeats that share considerable identity in the centromeric minor satellite DNA, ranging from 1.8 to 11 kb (see Fig. R94). This telomeric domain shows the same polarity and more than 99% identity between nonhomologous chromosomes and may explain both the maintenance of the telomeric state and Robertsonian translocations (see Fig. R95) (Kalitsis P et al 2006 Proc Natl Acad Sci USA 103:8786). If this translocation occurs, generally a minute piece of the acrocentric chromosome is lost, but being genetically inert, it has no consequence for fitness. Robertsonian translocation may affect 1/843 human neonates. ►translocations, ►acrocentric, ►telocentric, ►fitness, ►dicentric chromosome; Pardo-Manuel de Villena F, Sapienza C 2001 Cytogenet Cell Genet 92:342; Bandyopadhyay R et al 2002 Am J Hum Genet 71:1456.

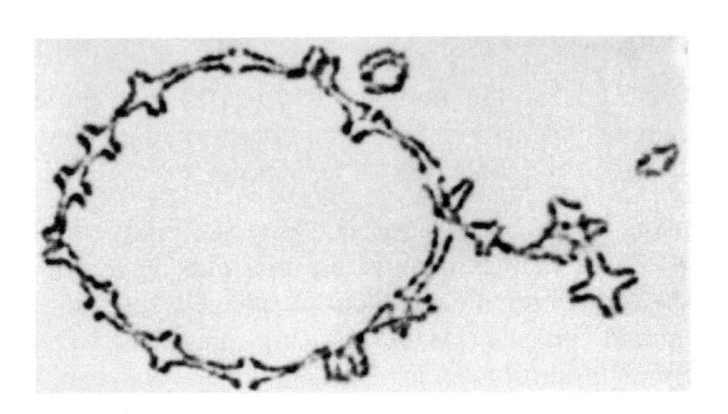

Figure R94. Late diakinesis/early metaphase I in the mouse, heterozygous for different Robertsonian translocations; 15 metacentrics in a superchain, 1 Trivalent, 2 Bivalents. The sex chromosomes are not involved in Robertsonian changes. (Courtesy Capanna E et al 1976 Chromosoma 58:341)

Two Acrocentrics → Robertsonian translocation

Figure R95. Robertsonian translocation

Robinow Syndrome (RRS): A rare autosomal dominant phenotype involving, usually but not always, short stature, normal virilization but micropenis, hypertelorism of the face, etc. In the Robinow-Sorauf syndrome the main characteristic is the flattened and almost doubled big toe. An autosomal recessive Robinow syndrome included the same facial and genital features but, in addition, multiple ribs and abnormal vertebrae were present. The basic defect of the recessive RRS (9q22) is due to premature chain termination in an orphan receptor tyrosine kinase (ROR2). It is allelic to brachydactyly type B. ►limb defeats, ►Chotzen syndrome, ►limb defects, ►stature in humans, ►brachydactyly, ►micropenis

Robo (*roundabout*): A *Drosophila* gene determining the straight movement of axons along the embryonal axis and preventing crossing back over the midline once a cross-over is made. Homologs exist in other eukaryotes. ►axon, ►slit; Simpson JH et al 2000 Cell 103:1019.

Robot Scientist: Implementing the technology of artificial intelligence makes it possible to originate hypothesis, devise experiments for their testing, carry out physical experiments in the laboratory, interpret the results, and falsify hypotheses inconsistent with test. Upon actual trials involving genes, proteins, and metabolites in the aromatic amino acid pathway of yeast, the system appeared competitive with the human performance and proved a cost decrease threefold to 100-fold. ►artificial intelligence; King RD et al 2004 Nature [Lond] 427:247.

Robotic Technologies: Machine operations for the performance of routine tasks in order to carry out analytical methods on a large scale and in a precise manner. For example, the robotic spotting of thousands of nanoliter or picoliter volumes on microscope slides to an area of 1 cm^2. DNA that can be structurally programmed can be inserted within a two-dimensional crystalline DNA array and facilitates the construction of a nanorobotic system. Atomic force microscopy demonstrated that a rotary device was fully functional after insertion (Ding B, Seeman NC 2006 Science 314:1583). ►atomic force microscope

Robustness: A statistical concept that indicates the justification of an assumption concerning the procedures applicable to the data at hand (e.g., normal distribution). A robust statistical method is less liable to violations of the assertion. Usually, but not necessarily, the parametric methods are more robust than the nonparametric ones (because of, e.g., subjective scales). Robustness is also used in physiology to characterize the stability of the steady state under variable conditions; robustness is invariance of the phenotype despite perturbations. Robustness is inherent in all evolvable biological systems (Kitano H 2004 Nature Rev Genet 5:826). Genetic robustness can be associated with redundancy, epistasis, and with the function of genetic networks. The "ordered systems" of the cell are very resistant to perturbations whereas the "chaotic" systems are very susceptible to perturbations (Shumulevich I et al 2005 Proc Natl Acad Sci USA 102:13439). Robustness against mutation can be explained by multiple solutions to specific biological problems. Mutations can be neutral in a "neutral space" where alternative configurations can solve the same biological need. ►parametric methods in statistics, ►non-parametric methods, ►epistasis, ►redundancy, ►genetic networks, ►canalization; Albert R et al 2000 Nature [Lond] 406:378; Stelling J et al 2004 Cell 118:675; Wagner A 2005 Robustness and Evolvability in Living Systems. Princeton University Press, Princeton, New Jersey.

R

Rock: An RHO-associated protein serine/threonine kinase involved with microtubules of nuclear division. It induces the phosphorylation of cofilin by LIM-kinase. Caspase-3 removes an inhibitory domain of Rock 1 and that then phosphorylates the light chain of myosin. This is the main cause of the blebs on cells undergoing apoptosis. ►RHO, ►microtubules, ►ROK, ►LIM, ►citron, ►cofilin, ►myosin, ►apoptosis, ►immunological surveillance for blebs; Sebbagh M et al 2001 Nature Cell Biol 3:346.

Rock-Paper-Scissors Model: An ecological dispersal, movement and interaction model based on a children's game by the same name. A rock crushes the scissors, the scissors cut paper, and the paper covers the rock. There is a complex facultative interaction here that determines the survival of colicin-producing, colicin-sensitive and colicin-resistant strains of *E. coli* bacterium. ►*E. coli*, ►colicin; Kerr B et al 2002 Nature [Lond] 418:171.

Rocks: ►scaffolds in genome sequencing

Rocket Electrophoresis: A type of immunoelectrophoresis where antigens are partitioned against antisera. ►immunoelectrophoresis, ►antigen, ►antiserum, ►electrophoresis; Hansen SA 1988 Electrophoresis 9:101.

Rodents (order *Rodentia*): A large number of species (mouse, rat, hamsters, rabbit) that have been extensively used for genetic research because of the small size of these multiparous mammals and short generation time. Mice and rats reach their sexual maturity in 1 or 2 months, and their gestation period is 19 and 21 days, respectively. They have been exploited as laboratory models for the study of cancer, antibodies, population genetics, behavior genetics, radiation and mutational responses, etc. Rodents are carriers of several human pathogens (bubonic plague [*Pasteurella pestis*], tularemia [*Pasteurella tularensis*], etc.). Several inbred strains of mice have contributed very significantly to the understanding of immunogenetics. ►animal models, ►mouse, ►rat; http://www.niehs.nih.gov/crg/cprc.htm.

Roentgen (röntgen): A unit of ionizing radiation (X-rays). ►R unit, ►Röntgen machine

Rogers Syndrome (TRMA): ►megaloblastic anemia (human chromosome 1q23)

Rogue: An off-type of unknown (genetic) determination.

ROI (reactive oxygen intermediates): ROI are by-products of oxidative metabolism of mitochondria, peroxisomes, and may be formed by ionizing radiation. ►ROS; Ono E et al 2001 Proc Natl Acad Sci USA 98:759.

ROK: One of the multiple RHO-associated protein kinases regulating the microtubules of the spindle. ►RHO, ►spindle, ►ROCK, ►citron

Rolling Circle: Replication is common among circular DNAs (such as conjugative plasmids [e.g., the F plasmid], double-stranded [λ phage] and single-strand phages [M13, φX174], amplified rDNA minichromosomes in the amphibian oocytes). A protein nicks one of the DNA strands and remains attached to the 5′-end. The free 3′-OH terminus serves as the point of extension by DNA polymerase in such a way that the opened old strand is displaced from the circular DNA while the new strand is formed and is immediately hydrogen bonded to the old template strand (see Fig. R96). Thus, the rolling circle remains intact and may generate new single-stranded DNAs that may be doubled later. The displaced single strand may be formed in just a single unit length of the original duplex circle or it may become a single- or double-stranded concatamer. It may also circularize in a single- or double-stranded form with the assistance of a DNA ligase to join the open ends.

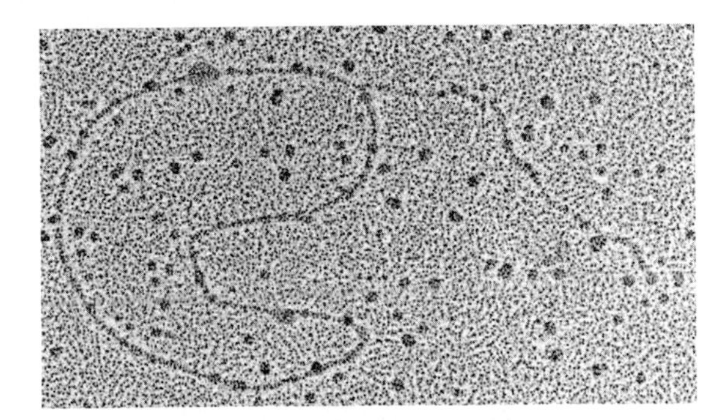

Figure R96. Rolling circle replication. The (−) strand is the template for the (+) strand. The → indicates the direction of growth. The 5′ end of the (+) strand attaches to the membrane (□5′). (Diagram is courtesy Gilbert W, Dressler D 1968 Cold Spring Harbor Symp Quant Biol 33:473. Electronmicrograph is the courtesy of Dr. Nigel Godson)

Rolling circle mechanisms have been identified in the transposition of bacterial insertion elements but they are more common in eukaryotes. The transposons of eukaryotes that utilize rolling circle transposition are called Helitrons. These elements may be quite abundant in eukaryotes but are hard to detect by the common computer programs because they do not generate target site duplications like the majority of transposable elements. Many of them are nonautonomous in transposition. Helitrons do not have inverted terminal repeats. Their 5′ end begins with TC nucleotides and the 3′ end is CTRR [R stands for purine]. Near the 3′ end they have a characteristic 16—20-nucleotide palindrome, which does not have a conserved sequence. Helitrons can pick up several genes and transpose them to other locations and thus gene collinearity is altered within the same species (Lal SK, Hannah C 2005 Proc Natl Acad Sci USA 102:9993). The *Arabidopsis*, rice, and *Caenorhabditis* Helitrons may encode about 1,500 amino acids that embed a 5′ →3′ helicase-like protein and a replication A protein besides some other gene products. For transposition several host proteins are also required. Evolutionarily Helitrons may have originated from geminiviruses. (See Fig. R97) ►conjugation, ►conjugation mapping, ►concatamer, ►padlock probe, ►insertion element, ►palindrome, ►replication protein A, ►geminivirus; del Pilar Garcillan-Barcia M et al 2001 Mol Microbiol 39:494; Kapitonov VV, Jurka J 2001 Proc Natl Acad Sci USA 98:8714; Feschotte C, Wessler SR 2001 Proc Natl Acad Sci USA 98:8923.

Figure R97. Rolling circle (see Figure R96 for explanation)

Rolling Circle Amplification (RCA): A short DNA primer, which is complementary to a segment of a circular DNA, and an enzyme generate many single-stranded, concatameric copies of DNA in the presence of deoxyribonucleotide (see Fig. R98). It can produce sufficient material for microarray analysis from specific locations and can detect single-nucleotide differences. ►microarray hybridization, ►concatamer; Nallur G et al 2001 Nucleic Acids Res 2001 29[23]:e118.

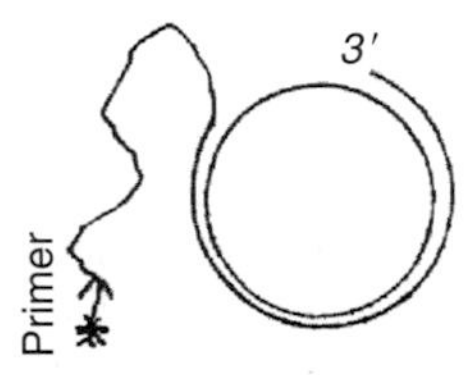

Figure R98. Rolling circle amplification

ROM (RNA-one-Modulator): A protein of 63-amino acid residues which affects the inhibitory activity of an antisense-RNA. A transacting inhibitor of plasmid replication is also named Rop/Rom. ►RNA I, ►antisense technology, ►Fanconi's anemia; Lin-Chao S et al 1992 Mol Microbiol 6:3385.

ROMA: Representational oligonucleotides microarray analysis. ►RDA

ROMA (Romani): ►gypsy

Roman and Arabic Numerals: See Fig. R99.

I = 1, II = 2, III = 3, IV = 4, V = 5, VI = 6, VII = 7, VIII = 8,
IX = 9, X = 10, XX = 20, XXX = 30, XL = 40, L = 50,
LX = 60, LXX = 70, LXXX = 80, XC = 90, C = 100,
CC = 200, CCC = 300, CD = 400, D = 500, DC = 600,
DCC = 700, DCCC = 800, CM = 900, M = 1000

Figure R99. Roman and Arabic numerals

Romanovs: The last Tsar of Russia, Nicholas II, and his wife and three children were executed and buried near Ekaterinburg, Russia, on 16 July 1918 after the

Bolshevik take-over of power. Seventy-five years later the bodies were exhumed and from the bone tissues, the identities of the remains were determined on the basis of mitochondrial and nuclear DNAs. Tsarina Alexandra was granddaughter and Prince Philip (husband of the present Queen of England, Elizabeth) the great grandson of Queen Victoria. Their mitochondrial DNA should be identical, as well as those of the three daughters of Alexandra. Forensic DNA analysis confirmed the expectation. Sex was determined from the bone samples by identifying the X-chromosomal amelogenin gene that is expressed in the enamel of the teeth. It is rich in GC (51%) and codes for a proline-rich protein (24%). The nuclear DNA samples confirmed the identity. The Tsar and his paternity and his mtDNA tied him to his brother (by the exhumed remains of Grand Duke Georgij) on the basis of heteroplasmy. The four other skeletons did not belong to royal family members but were of the physician and other people of the court. The two youngest children, Anastasia and Alexis, were not found in the grave. The purported identity of Anastasia with Anna Anderson was not confirmed by DNA analysis. The results of these studies have been questioned (Stone R 2004 Science 303:753), and the controversy regarding the identity of the skeletal remains continues (Science 306:407 [2004]). ►DNA finger-printing, ►heteroplasmy, ►Tsarevitch Alexis, ►Queen Victoria; Gill P et al 1994 Nature Genet 6:130.

ROMK: A potassium ion channel family member; it is encoded in human chromosome 11 and alternative splicing generates its isoforms. ►ion channels, ►isoform, ►splicing, ►Bartter syndrome

Ron: A cell-membrane tyrosine kinase and receptor of the macrophage stimulating protein. Both are located in human chromosome 3p21. ►macrophage-stimulating protein, ►HGF

Röntgen Machine: An X-ray machine (invented by WK Röntgen), producing ionizing radiation (used for induction of mutation, mainly deletions) and medical examination of the body. The dose delivered is measured by R, Rad, Rem, rep, Sv, Gy. ►R, ►roentgen, ►radiation hazard, ►radiation effects, ►radiation threshold, ►radiation protection, ►radiation measurement

RO•, ROO•: Reactive oxygen, hydroperoxide radical. ►ROS

roo: ►copia

Roof Plate: An organizing center for dorsal neural development in the embryo. ►floor plate, ►organizer

Root: Refer a segment of the transverse cross section in Figure R100. Roots secrete a great variety of chemicals through their roots and transgenic plants may eventually be used to manufacture, in hydroponic cultures, various needed chemicals (see Fig. R101). The advantage of the hydroponic cultures is that the purification of the secreted substances is easier. The Rho GTPase GDP dissociation inhibitor, RhoG-DI, regulates root hair development at special sites (Carol RJ et al. 2005 Nature [Lond] 438:1013). Root regeneration is mediated by auxin distribution and the expression pattern of three transcription factors (Xu J et al 2006 Science 311:385). ►quiescent zone, ►seed germination, ►hydroponic culture, ►transgenic, ►trichostatin, ►rho; Ryan PR, Delhaize E 2001 Annu Rev Plant Physiol Plant Mol Biol 52:527; Xie Q et al 2002 Nature [Lond] 419:1676.

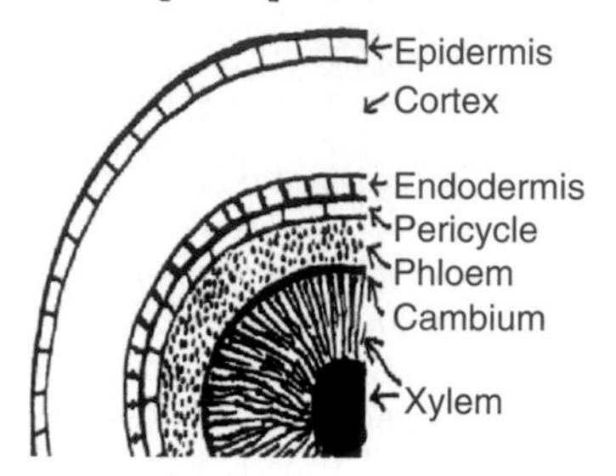

Figure R100. Root cross section segment

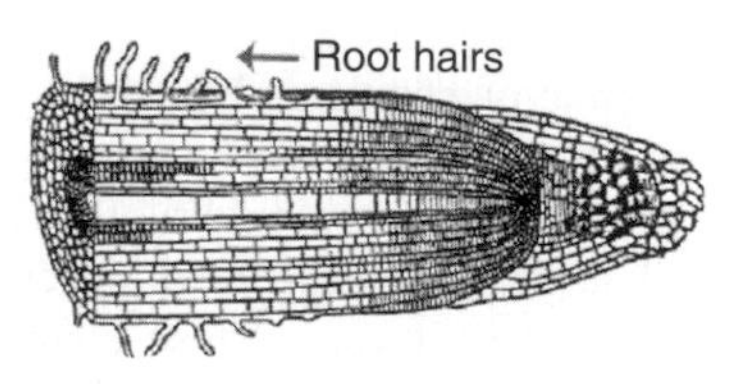

Figure R101. Root cross section segment

Root Cap: A thin membrane-like protective shield at the tip of the roots. Genetic ablation with the aid of transformation by diphteria-toxin transgene, driven by a root cap-specific promoter showed that transgenic plants are viable without several root tip cell layers and develop more lateral roots.

Root Nodule: ►nitrogen fixation

Root Pressure: The guttation of a wounded stem caused by osmosis in the roots of plants.

Rooted Evolutionary Tree: The evolutionary tree indicates the origin of the initial split of divergence. ►evolutionary tree, ►unrooted evolutionary tree

Rop Protein: ►RNA I

ROP (RhoGEF GTP-binding protein): ROP increases nucleotide dissociation (1,000 x) from Rop (RAS-related GTP binding protein) and associates tightly with nucleotide-free Rop. It functions as a molecular switch in plant growth and development (Berken A et al 2005 Nature [Lond] 436:1176). ►RAS, ►GEF

ROR (retinoid-related orphan receptor): RORs are animal hormone receptors regulating Bc*l* and thus the survival of lymphocytes. They are, however, involved in the regulation of several other developmental processes too. The ROR response elements (RORE) have the AGGTCA consensus, preceded by a 5 bp A/T-rich sequence. ►Bc*l*, ►survival factors, ►IL-17; Jetten AM et al 2001 Progr Nucleic Acid Res Mol Biol 69:205.

RoRNP: A ribonucleoprotein involving the so-called Y RNA (transcript of RNA polymerase III), the 60 kDa Ro60 protein and, in some cases, also the La protein (regulator of pol III). RoRNP has been identified in all eukaryotes examined and it appears to be the target of the autoimmune diseases of lupus erythematosus and Sjögren syndrome. ►autoimmune disease; Labbe JC et al 1999 Genetics 151:143.

ROS: The reactive oxygen species that may play a detrimental role in radiation damage, modification of DNA, degenerative human diseases, plant disease, etc. Mitochondrial oxidative phosphorylation, lipid peroxidation, and induced nitric acid synthase (NOS) may produce large amounts of reactive NO (nitric oxide) and thus generate ROS. NO is a known deaminating mutagen. The production of antioxidants (vitamin C, vitamin E, glutathiion, ferritin, β carotene) is the defense against ROS damage. A PPARγ coactivator (PGC-1α) is required for the induction of several ROS-detoxifying enzymes (such as GPx1, SOD2); it protects against neurodegeneration by ROS (St-Pierre J et al 2006 Cell 127:397). The cells may also use SOD, catalases, and peroxidases to degrade ROS. Reactive oxygen may be responsible for half of the human cancer cases. Also, deficiency of ROS (by supplying 2-methoxyestradiol) may trigger the demise of leukemia cells because cancer cells have increased aerobic metabolism. The redox enzyme $p66^{Shc}$ generates mitochondrial hydrogen peroxide and is a signal to apoptosis. Electron transfer between cytochrome c and $p66^{Shc}$ contributes to ROS formation, apoptosis, and aging (Giorgio M et al 2005 Cell 122:221). ►SOD, ►hydrogen peroxide, ►PPAR, ►hydroxyl radical, ►RO$^{\bullet}$, ►ROO$^{\bullet}$, ►oxidative DNA damage, ►nitric oxide, ►Fenton reaction, ►oxidative deamination, ►aging, ►host-pathogen relationship, ►photodynamic effect, ►singlet oxygen, ►ROI, ►apoptosis, ►aging; Møller IM 2001 Annu Rev Plant Physiol Plant Mol Biol 52:561; Lee D-H et al 2002 Nucleic Acids Res 30:3566.

ROS oncogene: The ROS oncogene is in human chromosome 6q22 and it is the c-homolog of the viral v-ros. It appears to be the same as MCF. ►oncogenes

Rosa spp: Ornamentals with 2n = 14, 21, 28. ►*Rosa canina*

Rosa canina (dog rose): A pentaploid species with 35 somatic chromosomes (see Fig. R102). Unlike other pentaploids it is fertile. In meiosis, the plants produce seven bivalents and 21 univalents. The univalents are lost at gametogenesis in the male and so the sperms contain only seven chromosomes, derived from the seven bivalents. During formation of the megaspore all the 21 univalents and the seven chromosomes from the seven bivalents are incorporated into the embryo sac. The addition of the seven male and the 28 female chromosomes to the zygote restores the 35 somatic chromosome number. The female contributes more chromosomes to the offspring; it is therefore matroclinous. Recombination is limited to the seven bivalents. Its breeding system is a unique mixture of generative and apomictic reproduction. ►pentaploids, ►apomixis, ►matroclinous, ►univalent; Gustafson Å 1944 Hereditas 30:405.

Figure R102. *Rosa canina*

Rosacea: A skin disease caused by inflammation due to persistent reddening of the skin (erythema). The antimicrobial peptides cathelicidine and serine protease levels seem to be elevated, which are involved in innate immunity, dilation of the blood vessels, leukocyte migration and wound healing responses. It may be due to candidiasis, Sjögren syndrome and other causes (Bevins, Yamasaki K et al 2007 Nature Med 13:975). ►candidiasis, ►Sjögren syndrome

Rosetta Stone Sequences: Rosetta stone sequences aid in deciphering the function of simple polypeptide sequences. Prokaryotic proteins frequently carry out the same functions as the corresponding eukaryotic ones. In eukaryotes, the homologous proteins are frequently fused with other proteins that are required for their function whereas in prokaryotes they may appear separately as single proteins. Once the function of the fused eukaryotic proteins is known, the function of both the prokaryotic counterparts can also be inferred. For example, in *E. coli*, gyrase A and B are encoded separately but in yeast they have homology to different domains of topoisomerase II. In a sequential manner, then, functional connections can be revealed to other proteins too. ►phylogenetic profile method, ►gene neighbor method; Marcotte EM et al 1999 Science 285: 751.

Rosette: Plant shoots with greatly reduced internodes commonly found in dicots before the stem bolts after induction of flowering; any anatomical structure in animals arranged in a form resembling the petals of a rose (see Fig. R103).

Figure R103. Leaf rosette

ROSI: Also called the round spermatid injection or round spermatid nucleus injection (ROSNI). This and the elongated spermatid injection (ELSI) involve injection of haploid germ cells retrieved from testicular biopsies into recipient oocytes. These procedures help overcome male infertility when the spermatozoon is unable to penetrate the egg (Jurisikova A et al 1999 Mol Hum Reprod 5:323). ►ART

Rossellini-Guienetti Syndrome: ►ectodermal dysplasia

Rossmann Fold: An NAD(P)-binding domain (Gly-XXXGlyXGly) near the N-terminus, encoded by a large number of eukaryote and prokaryote genes. ►$NADP^+$

Rostral: In the direction of the beak, mouth or nose rather than toward the hind position.

r_0t: In a RNA-driven DNA-RNA hybridization reaction, it is the concentration of RNA × the time of the reaction (analogous to c_0t in reassociation kinetic studies with DNA). r_0t sheds information on the RNA complexity of different cells during development. ►c_0t, ►RNA driven reaction

Rotamase: A group of enzymes catalyzing cis-trans isomerization. ►cyclophilin, ►immunophilins

Rotamer: A rotational isomer involving a side chain, i.e., not the backbone of the molecule. The place of atoms is altered by rotation around single bonds.

Rotational Diffusion: A process wherein membrane proteins travel within the membranes by rotation perpendicular to the plane of the lipid bilayer. ►cell membrane

Rotaviruses (*Reoviridae*): Their genomes consist of 10–12 double-stranded RNA and each particle carries a single copy of this genome (see Fig. R104). The terminal sequences control replication and packaging. Through internal deletions the RNAs may become aberrant, called DI RNA (defective interfering), resulting in lower infectious capacity. The rotaviruses may cause gastroenteritis (stomach and intestinal inflammation) and diarrhea in human babies and animals. The introduction of recombinant vaccinia virus T7 RNA polymerase into rotavirus generates artificial viral mRNA and can induce mutation (Komoto S et al 2006 Proc Natl Acad Sci USA 103:4646). ►reovirus, ►vaccinia

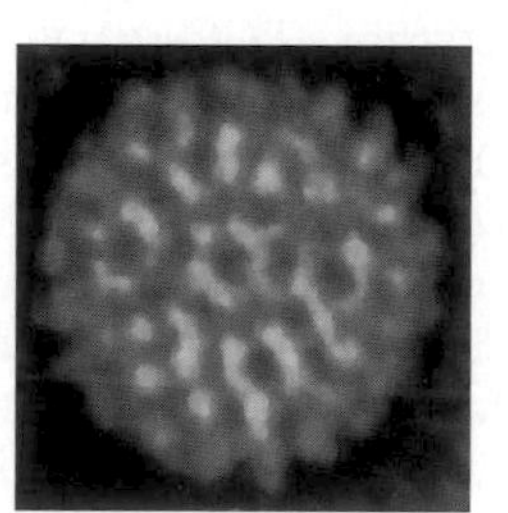

Figure R104. Rotavirus

Rothmund-Thomson Syndrome: An autosomal recessive human disorder involving dermal (skin) lesions (see Fig. R105), dark pigmentation, light-sensitivity, early cataracts, bone and hair problems, and premature aging. It may lead to squamous (scaly) carcinomas. Autosomal dominant genes determine some similar types of diseases. One form is assigned to human chromosome 8q24.3 and another to 17q25. Both genes encode helicases; the homologs of what control recombination in yeast. ►cancer, ►light-sensitivity diseases; helicase: Sangrithi MN et al 2005 Cell 121:887).

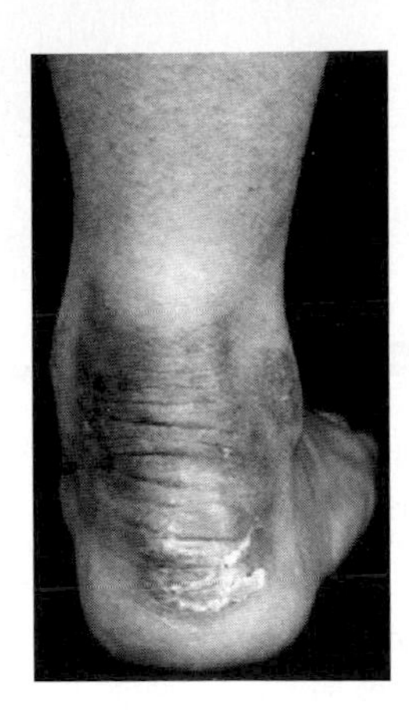

Figure R105. Rothmund syndrome skin lesions. (From Bergsma D (Ed.) 1973 Birth Defects. Atlas and Compendium. By permission of the March of Dimes Foundation)

Rotor Syndrome: ►Dubin-Johnson syndrome

Rough Draft of the Syndrome: A not complete sequence of a genome and yet it displays the sequences of about 90% of the euchromatic parts, containing the coding units. ►human genome, ►genome projects

Rough ER: ►endoplasmic reticulum rough, ►RER

Round of Matings: After bacteriophages have been replicated within the host cell, the newly formed molecules of "vegetative DNAs" may recombine with each other several times. Because of the multiple exchanges (in agreement with the Poisson distribution expectations), it appears that the maximal recombination between markers cannot exceed 30–40%. ►mapping function, ►negative interference, ►coefficient of coincidence; Visconti N, Delbrück M 1953 Genetics 38:5.

Rous Sarcoma: Rous sarcoma was originally detected as a viral RNA cancer in chickens. The protooncogene homolog was detected in rats and other mammals. ►RAS and homologs, ►oncogenes, ►TV/RCAS

Rowley-Rosenberg Syndrome: An autosomal recessive growth retardation, different from dwarfness, and characterized by aminoacidurias. ►dwarfism, ►aminoacidurias

RPA: A single-strand DNA-binding protein (of subunits of 70, 34 and 14 kDa) participating in replication, nucleotide excision repair, repair of double-strand breaks by recombination. ►replication protein A, ►DNA replication eukaryotes, ►NER, ►BER, ►UNG2; Davis AP, Symington LS 2001 Genetics 159:515.

RPB: Subunits (differently numbered, each) of RNA polymerase II. ►RNA polymerase

RPD3/Sin3: A histone deacetylase and a component of the Mad/N-CoR/Sin3/RPD transcriptional protein repressor complex. It may also block the position effect exerted by centromeric and telomeric heterochromatin. ►position effect, ►PEV, ►histone, ►histone deacetylase, ►Mad, ►N-CoR, ►signal transduction, ►repression, ►transcription, ►Sin3; Fazzio TG et al 2001 Mol Cell Biol 2001 21:6450.

R-Phycoerythrin: A phycobiliprotein fluorochrome isolated from algae. Maximal excitation is at 545 and 565 nm but it is also excited at 480 nm. Maximal emission is at 580 nm and hence the red color.

rpo: RNA polymerases such as A, B, C_1, C_2 in (organelles) plastids and mitochondria and resembles bacterial RNA polymerases. ►RNA polymerase, ►σ

RPPA: ►reverse-phase protein array

r-Proteins: Ribosomal proteins. ►ribosome

RPTK (ROS receptor protein tyrosine kinase): A sperm receptor protein tyrosine kinase that may bind to the ZP3 (zona pellucida) protein of the egg matrix. These proteins, in various stages of cooperation, affect several cellular processes. ►sperm, ►egg, ►fertilization; Zeng L et al 2000 Mol Cell Biol 20:9212.

RRAS: ►RAS oncogene

RRE (Rev response element): An RNA export adaptor promoting the export of the HIV-1 transcripts from the nucleus. If an RRE is inserted into an intron (which normally retained within the nucleus), even that could be exported to the cytoplasm. ►export adaptors, ►acquired immunodeficiency; Zhang Q et al 2001 Chem Biol 8:511.

rrn: In *E. coli* there are seven ribosomal transcription units, *rrn-A, -B, -C, -D, F, -G, -H* including the 16S-23S-5S RNAs, spacers and intercalated tRNA genes within the spacers. Maturation involves trimming of the co-transcript (cleavage by RNase III) and processing by other RNases, RNases P and D (See Fig. R106). The number of rRNA genes in eukaryotes is very variable and subject to amplification. At the developmental stages of very active protein synthesis the number of ribosomal genes may be amplified to several thousands, and, e.g., in the amphibian oocytes may be sequestered into mini-nuclei. Eukaryotic rRNA genes are transcribed in a ca. 45S precursor RNA, containing the 18S-5.8S-28S (in this order) and

Figure R106. Segment of a eukaryotic rRNA gene cluster in transcription. (Courtesy Spring H et al 1976 J Microsc Biol Cell 25:107)

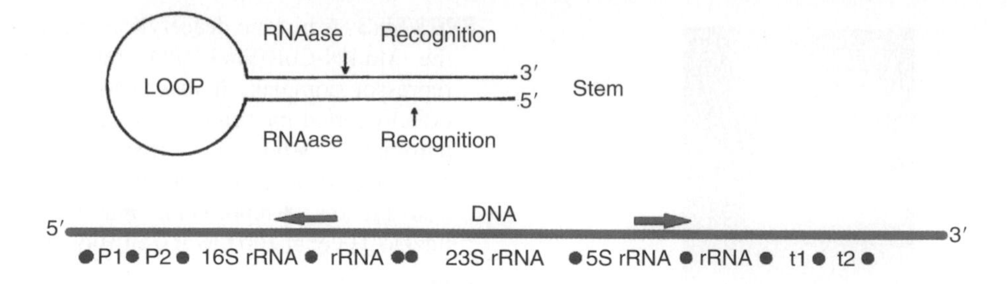

P1 and P2 : promoters, ● : spacers, t1 and t2: termination signals (not on scale)

Figure R107. The DNA region of the rRNA and tRNA gene cluster (*rrn*) in *E. coli* is transcribed into longer than 30S primary transcripts, interrupted by spacers. The ribosomal and transfer RNA genes are clustered and co-transcribed in the order 5′ - 26S-23S-5S-3′ and within the intergenic spacers the tRNA genes are situated. RNase III at the duplex stems trims the individual gene transcripts. The different loops contain ca. 1,600, 2,900 and 120 nucleotides, corresponding to the 16S, 23S and 5S ribosomal RNAs, respectively. (The 1,600 and the 2,900 base sequences are also called p16 and p22, respectively). RNase P further trims each of these rRNA precursors and the tRNA precursors are processed by RNase P at the 5′ and by RNase D at the 3′ end

spacer sequences. The pathway of trimming may vary among the different species. The 18S rRNA is immediately methylated at about 40 sites and a few more methyl groups are added in the cytoplasm after maturation. The 28S rRNA is methylated immediately after transcription at over 70 sites and these methylated sites are saved during the process of maturation. The cleavage takes place at the 5′ side of the genes and between the spacers. The 5.8S rRNA eventually associates with the 28S rRNA by base pairing. (See Fig. R107, ►ribosomal RNA, ►ribosomal proteins, ►ribosomes, ►stringent response, ►stringent control, ►RNA maturases; Hirvonen CA et al 2001 J Bacteriol 183:6305).

rRNA: Ribosomal RNA; a structural component of the ribosomes. rRNAs may be preferentially amplified in the oocytes of amphibians and other organisms. ►ribosome, ►ribosomal RNA

RRS (reduced representation shotgun): A method for mapping single nucleotide polymorphism in a genome. ►SNIPS; Altshuler D et al 2000 Nature [Lond] 407:513.

RS Domain: The RS domain is rich in arginine (R) and serine (S); RS proteins are parts of the spliceosomes. ►spliceosome

RSC (remodel structure of chromatin): ►chromatin remodeling

r-Scan Statistics: Detects anomalous spacing of oligonucleotides and peptides. (Jeff AR et al 2001 Genes, Chromosomes Cancer 32:144).

RSF: ►hybrid dysgenesis I-R system

RSF: A 400–500 kDa protein complex of a 325 kDa protein and the 135 kDa hSNF2h/ISWI. ►chromatin remodeling, ►SNF2; Labourier E et al 1999 Genes Dev 13:740.

RSK (ribosomal protein S6 kinase, RSK-3 at Xp22.2-p22.1, RSK-2 at 6q27, RSK-1 in human chromosome 3, there is also an RSK-4): An epidermal growth factor-regulated protein kinase phosphorylating histone 3; RSK-3 defect is involved in the Coffin-Lowry syndrome. MAPK-phosphorylated RSK prevents parthenogenetic development of unfertilized eggs. RSK integrates MAPK and PDK1 signaling pathways. ►Coffin-Lowry syndrome, ►EGF, ►histones, ►MAPK, ►PDK, ►S6 kinase, ►chromatin remodeling

R-Spondin: A family of proteins binding Wnt ligands to the Frizzled (Fzd) receptor and the low-density lipoprotein-related receptor (LRP) 5 or LRP6 coreceptor. They initiate downstream signaling events leading to gene activation by β-catenin and the T-cell factor (TCF)-lymphoid enhancer factor (LEF) family transcription factor complex (Nam J-S et al 2006 J Biol Chem 281:13247). ►*Wingless*, ►lipoprotein, ►T cell receptor, ►T cells, ►LEF, ►catenins

RSR: The relative survival rate.

RSS (recombinational signal sequence): V(J)D recombinase mediates recombination only in gene segments flanked by tripartite recombination signal sequences consisting of a highly conserved heptamer (7mer), an AT-rich nonamer (9mer), and 12 or 23 base-long intervening nucleotides. ►immunoglobulin, ►V(J)D recombinase, ►RAG

R^{St} (*R-st*): The stippled, paramutable allele of *R* locus of maize in the long arm of chromosome 10. ►paramutation, ►R locus. (See Fig. R108).

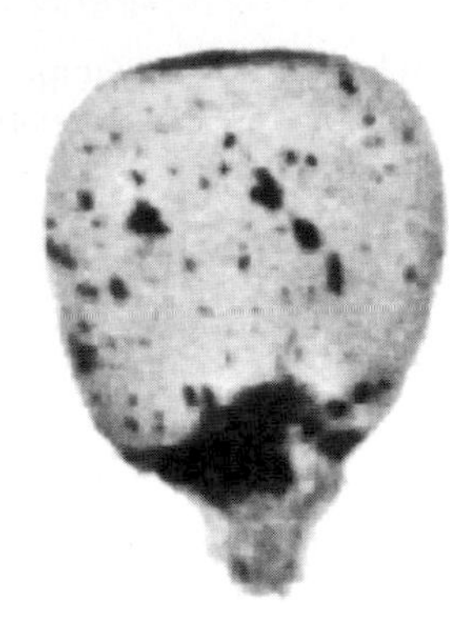

Figure R108. R^{st} Kernel

R_{ST}: The same as F_{ST} but it is based on variation in microsatellites. ►F_{ST}, ►microsatellite

r_0t Value: The measure of RNA-DNA or RNA-RNA hybridization; the product of the concentration of single-stranded RNA and the time elapsed since the beginning of the reaction. ►c_0t value

RTF (resistance transfer factor): Bacterial plasmids carrying various antibiotic and other resistance genes. ►resistance transfer factors, ►conjugation in bacteria, ►plasmid mobilization

Rth: ►Rad 27

RTK: ►receptor tyrosine kinase

RTP (replication termination protein): The RTP is functionally, but not structurally, similar to Tus. ►replication bidirectional, ►Tus gene product; Gautam A et al 2001 J Biol Chem 276:23471.

RT-PCR: A reverse transcription-polymerase chain reaction. The purpose of the procedure is similar to the PCR in general. In this instance it amplifies the small amounts of RNA transcripts as cDNA. The reaction requires reverse transcriptase, mRNA, deoxyribonucleotides and primers that can be random DNA sequences, oligodeoxythymidine or antisense sequences. The method is very sensitive and the RNA of a single cell can be amplified and thus localized gene expression studied. Under well-controlled conditions it can be semi-quantitative. It is now used for clinical diagnostic purposes too. ►RNA fingerprinting, ►PCR, ►in situ PCR, Barlič – Maganja D, Grom J 2001 J Virol Methods 95:101; http://web.ncifcrf.gov/rtp/gel/primerdb/.

rtTA (reverse transactivator tetracycline): Basically a tTA system containing a nuclear localization signal at the 5′-end. It binds efficiently the *tetO* operator—but only in the presence of tetracycline derivatives such as doxycycline or anhydrotetracycline. ►tTA, ►tetracycline; Pacheco TR et al 1999 Gene 229:125.

rtTA-nls: The same as rtTA.

RU486 (mifepristone/mifeprex): A pregnancy prevention drug. It chemically induces abortion. Recently, a small number of women developed excessive bleeding and ectopic pregnancies after its use. Misopristol, the most common vaginally administered drug, in rare cases lead to deadly infection by *Clostridium sordelli* bacteria by weakening the immune system. These drugs are in wide use in Europe. RU486 has been approved in the USA since 2000. ►hormone receptors; Mahajan DK et al 1997 Fertil Steril 68:967; DeHart RM, Morehead MS 2001 Ann Pharmacother 35:707; Schulz M et al 2002 J Biol Chem 277:26238.

RU Maize: RU maize carry plasmid-like elements in their mitochondria and yet they are not male sterile. ►cytoplasmic male sterility, ►mtDNA

Rubella Virus (a toga virus): The Rubella virus causes the disease of German measles. Infection during early pregnancy may cause intrauterine death of the human embryo and/or developmental anomalies in the newborn. ►teratogenesis

Rubinstein Syndrome (Rubinstein-Taybi syndrome, RSTS, 16p13.3): The Rubinstein syndrome is characterized by dominant defects in the heart valve of the pulmonary aorta, collagen scars on skin wounds, an enlarged passageway between the skull and the vertebral column, mental retardation, and broad thumbs (see Fig. R109).

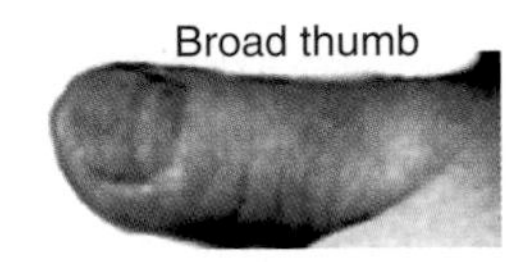

Figure R109. Rubinstein Syndrome

This condition has very low recurrence risk (1%) and about 0.2–0.3% of the inmates of mental asylums are afflicted by it. Haplo-insufficiency for the CBP transcription factor seems to be involved in the abnormal differentiation. Many afflicted individuals have break points at chromosome 16p13.3, and this is the site of the human cyclic AMP response element-binding protein (CBP/CREB). The syndrome may also be elicited by a continued requirement for CREB co-activation and histone acetylation of the CREB-binding protein (Alarcón JM et al 2004 Neuron 42:947). In addition, the p300 protein encoded by the EP300 gene (22q132) is also similar in structure to CBP and can be responsible for RSTS (Roelfsema JH et al 2005 Am J Hum Genet 76:572). ►mental retardation, ►CREB, ►CBP, ►GLI3 oncogene, ►epigenesis; Murata T et al 2001 Hum Mol Genet 10:1071.

Rubisco: Ribulose bisphosphate carboxylase-oxygenase (M_r 550,000), a chloroplast enzyme.

The eight large subunits (each M_r 56,000) are coded for by chloroplast DNA and the small subunits (each M_r 14,000) are under nuclear control (see Fig. R110). The carboxylase function catalyzes the covalent attachment of carbon-dioxide to ribulose-1,5-bisphosphate and then splits into two molecules of 3-phosphoglycerate. The oxygenase function mediates the incorporation of O_2 into ribulose-1,5-bisphosphate and the resulting phosphoglycolate re-enters the Calvin cycle. ►chloroplast, ►chloroplast genetics, ►photosynthesis, ►ribulose bisphosphate carboxylase; Douce R, Neuberger M 1999 Curr Opin Plant Biol 2:214; Spreitzer R, Salvucci ME 2002 Annu Rev Plant Biol 53:449.

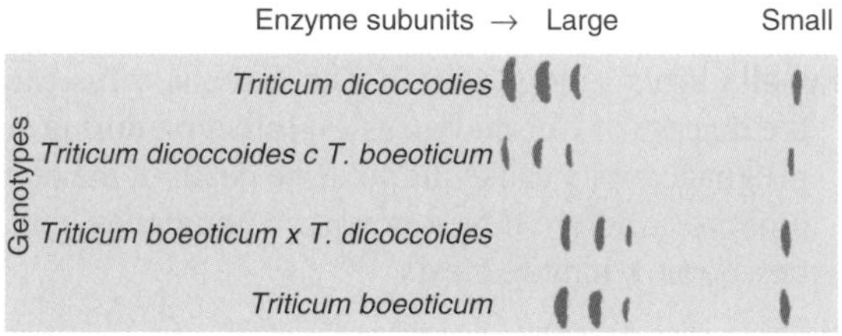

Enzyme subunits → Large Small
Genotypes
Triticum dicoccodies
Triticum dicoccoides c T. boeoticum
Triticum boeoticum x T. dicoccoides
Triticum boeoticum

Figure R110. Genetic and electrophoretic evidence that the large subunits of RUBISCO are maternally inherited, whereas the small subunit is transmitted biparentally. Molecular studies mapped the large subunits in the chloroplast DNA. (After Chen K, Gray JC, and Wildman SG. See also 1975. *Science* 190:1304.)

Rudiment: Either the beginning stage of a developing structure or the remains of a decayed or reduced one.

Rule 12/23: ►immunoglobulins, ►V(D)J, ►RAG

Rumex hastatulus: A North-American herbaceus plant which has variable chromosomal sex determination. The plants found in North Carolina have three pairs of autosomes, one X and two Y chromosomes. The form prevalent in Texas has four pairs of autosomes and one X and one Y chromosome. The middle line connects the positions of the centromeres. ►sex determination, ►chromosomal sex determination, see Fig. R111.

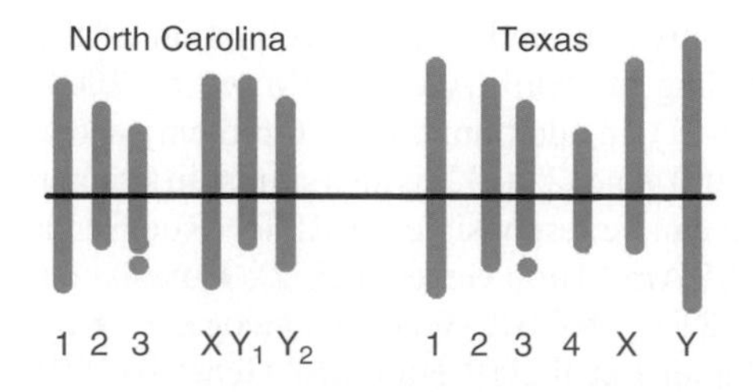

Figure R111. (Redrawn after Smith BW 1964. *Evolution* 18:93.)

Runaway Plasmids: At lower temperature (30 °C) runaway plasmids are present in relatively low copy number and do not interfere with the growth of the host cell. At above 35 °C their copy number rises substantially and so does the DNA they contain. Only after about 2 h (by the time they account for 50% of the DNA of the cell) do they suppress cellular growth. ►vectors, ►copy-up mutation; Uhlin BE et al 1983 Gene 22:255.

Runaway replication: The replication is not restricted to the normal manner. (See Chao YP et al 2001 Biotechnol Progr 17:203).

Run-off Transcription: The inducer of gene activity (e.g., light signal) is withdrawn and the tapering off of transcription is monitored by incorporation of labeled nucleotides. (See Delany AM 2001 Methods Mol Biol 151:321).

Run-on Transcription: Once the transcription is turned on in isolated nuclei it proceeds without further need for enhancers as measured by the incorporation of labeled nucleotides into mRNA. ►transcription, ►regulation of gene activity; Hu ZW, Hoffman BB 2001 Methods Mol Biol 126:169.

Runt (*run*, 1–65): A *Drosophila* pair-rule gene encoding a DNA-binding protein/transcription factor with homologies to, e.g., CBFA1, -2, -3 core-binding factor subunits of human and mouse genes. ►leukemia, ►pair rule genes, ►tandem repeats

Runx: Transcription factors, which may either activate or suppress tumorigenesis. *Runx3* may also aid the proliferation of B lymphocytes and autoimmune disease. The mouse *Shox2* gene upstream of *Runx2* is a key regulator of chondrogenesis and long-bone development (Cobb J et al 2006 Proc Natl Acad Sci USA 103:4511). In hematopoietic stem cells, Smad6, an inhibitor of Bmp4 signaling, binds and inhibits Runx1 activity, whereas Smad1, a positive mediator of Bmp4 signaling, transactivates the *Runx1* promoter. Three key determinants of HSC development are the Scl (small cell lung) transcriptional network, Runx1 activity, and the Bmp4/Smad signaling pathway (Pimanda JE et al 2007 Proc Natl Acad Sci USA 104:840). Runx2 is also involved in the control of RNA polymerase I promoter and associated proteins (Young DW et al 2007 Nature [Lond] 445:442). During mitosis, when transcription is shut down, Runx2 selectively occupies target gene promoters, and Runx2 deficiency alters mitotic histone modifications (Young DW et al 2007 Proc Natl Acad Sci USA 104:3189). ►lymphocytes, ►mesenchyma, ►osteoblast, ►Smad, ►Bmp, ►small cell lung carcinoma; Spender LC et al 2005 Oncogene 24:1873; Sato T et al 2005 Immunity 22:317.

Rupert: The hemophiliac grandson of Leopold of Albany (a hemophiliac himself), son of Queen Victoria. ►hemophilias, ►Queen Victoria

Russell-Silver Syndrome (RSS): Most commonly autosomal recessive (7p11.2-p13, 17q23-24, 15q26.1-qter, 11p15), but X-linked or sporadic cases with low birth-weight dwarfism, frequently with asymmetric body and limbs, deformed fingers, relatively large skull and mental retardation are common. The 7p site is closely associated with the location of the genes encoding growth factor receptor-binding protein 10 and the insulin-like growth factor-binding proteins 1 and 3. Maternal disomy, at least for a 35 Mb sequence, appears to account for the symptoms. ►dwarfism, ►stature in humans, ►imprinting, ►uniparental disomy; Hannula K et al 2001 Amer J Hum Genet 68:247; Nakabayashi K et al 2002 Hum Mol Genet 11:1743.

Rust: Disease of grasses (cereals) caused by *Puccinia graminis. Puccinia* fungi display rust color pustules (see Fig. R112). Resistance in barley is based on a receptor-like serine/threonine kinase protein (Nirmala J et al 2006 Proc Natl Acad Sci USA 103:7518). New aggressive variants of *Puccinia graminis* are threatening existing resistant wheat varieties. ►host-pathogen relations; Staples RC 2000 Annu Rev Phytopath 38:49.

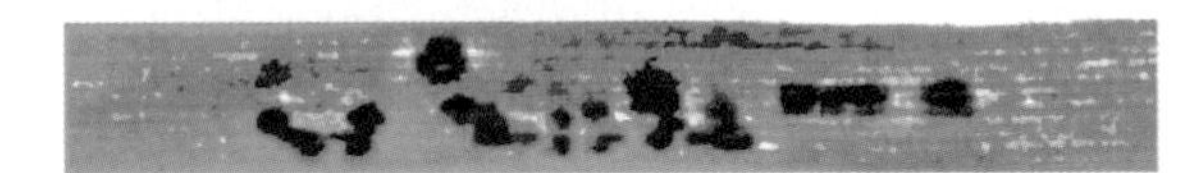

Figure R112. Puccinia triticina (Courtesy of Dr ER Sears)

Rut: ►oestrus

RuvABC: A protein complex operating in the Holliday structure of recombination. The RuvAB helicase/ATPase and motor protein complex mediates branch migration (also through nucleosomes in eukaryotes) and replication. RuvC is an endonuclease. ►branch migration, ►Holliday juncture, ►recombination molecular mechanisms in prokaryotes, ►recA, ►endonuclease, ►AAA proteins; West SC 1997 Annu Rev Genet 31:213; Constantinou A et al 2001 Cell 104:259; Yamada K et al 2001 Proc Natl Acad Sci USA 98:1442.

RVs: ►BIN

RXR: ►RAR

Ryanodine (ryanodol 3-[1H-pyrrole-2-carboxylate]): A toxic extract (insecticide) from the new world tropical shrub *Ryania speciosa*. It regulates calcium ion channels in muscles. The mutation of deletion of the ryanodine receptor gene (RYR1, 19q13.1) may lead to malignant hyperthermia. ►ion channels, ►central core disease, ►hyperthermia; Zalk R et al 2007 Annu Rev Biochem 76:367.

Rye (*Secale cereale*): 2n = 2x = 14 is an outbreeding crop plant used for the production of bread, biscuits, starch, and alcohol.

Its taxonomy is somewhat controversial. Rye can be crossed with a number of other cereals; among them the allopolyploid *Triticales* (wheat × rye hybrids, 2n = 42 and 2n = 56) are most notable. Addition lines and transfer lines carrying rye chromosomes or chromosomal segments have been made. Rye belongs to those exceptional grain crops where autotetraploid varieties have agronomic value. Trisomic lines are known. Some varieties harbor a variable number of B chromosomes (see Fig. R113). Rather unusually, some of the plastids are also transmitted through the pollen. ►chromosome banding, ►*Triticale*, ►alien addition, ►chromosome substitution, ►alien substitution, ►transfer lines, ►holocentric, ►ergot; http://www.tigr.org/tdb/tgi.shtml.

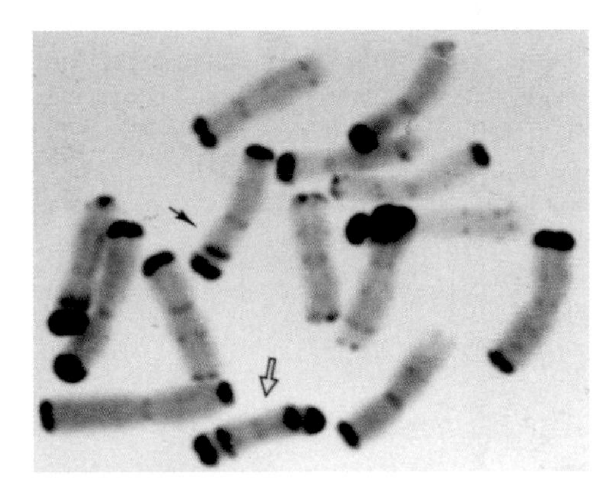

Figure R113. Giemsa-stained rye karyotype displaying an isochromosome (⇨) and the corresponding arm in a normal chromosome (→). (Courtesy Dr. Gordon Kimber)

Ryegrass: *Lolium multiflorum* and *L. perenne*, both 2n = 14 (see Figs. R114 and R115).

Figure R114. Rye ear

Figure R115. Ryegrass ear

Historical vignettes

George W. Beadle recollections (Stadler Genetics Symp. 2:114) about Archibald E. Garrod discoveries of inborn errors of metabolism beginning in 1899 and Richard Goldschmidt's classical book of Physiologische Theorie der Vererbung 1927 (Springer, Berlin).

"I recall giving a lecture at the University of California, Berkeley, in the mid-forties in which I recounted this remarkable story of the neglect of GARROD's work. RICHARD GOLDSCHMIDT, then on the faculty of that university was in the audience. He told me after the lecture that he could not understand how he had omitted mention of GARROD's work in his well-known book *Physiological Genetics* and that he had indeed been well aware of it, but had forgotten about it when he wrote the book. That seems to me a pretty good indication that he had not really appreciated its significance, much as de VRIES had not properly assessed the work of MENDEL when he first read about it."

At the January 6-8, 1909 meeting of the American Breeder's Association the above paper was presented in Columbia, Missouri.

American Breeders' Association. 365

WHAT ARE "FACTORS" IN MENDELIAN EXPLANATIONS?

By Prof. TH Morgan.

Columbia University, New York, N. Y.

In the modern interpretation of Mendelism, facts are being transformed into factors at a rapid rate. If one factor will not explain the facts, then two are invoked; if two prove insufficient, three will sometimes work out. The superior jugglery sometimes necessary to account for the results may blind us, if taken too naively, to the common-place that the results are often so excellently "explained" because the explanation was invented to explain them. We work backwards from the facts to the factors, and then, presto! explain the facts by the very factors that we invented to account for them. I am not unappreciative of the distinct advantages that this method has in handling the facts. I realize how valuable it has been to us to be able to marshal our results under a few simple assumptions, yet I cannot but fear that we are rapidly developing a sort of Mendelian ritual by which to explain the extraordinary facts of alternative inheritance. So long as we do not lose sight of the purely arbitrary and formal nature of our formulae, little harm will be done; and it is only fair to state that those who are doing the actual work of progress along Mendelian lines are aware of the hypothetical nature of the factor-assumption. But those who know the results at second hand and hear the explanations given almost invariably in terms of factors, are likely to exaggerate the importance of the interpretations and to minimize the importance of the facts.

S

S (such as in 5S RNA): ►sedimentation coefficient

s: ►selection coefficient

s: Standard deviation of a set of experimental observations. ►σ parametric

σ: The measure of superhelical density of DNA.

σ: A subunit of prokaryotic RNA polymerase enzyme, essential to start transcription in a specific way. This factor opens the double helix for the action of the RNA polymerase. Also, the σ^{70} is involved in pausing of transcription. The σ^{70} recognition sites consist of two hexamers located at −10 and −35 positions from the transcription start point. The sigma factor can melt promoter independently from the RNAP (polymerase) protein (Hsu H-H et al 2006 Cell 2006 127:317). The σ^{70}s are in excess of the polymerase enzyme and there is a competition for the enzyme molecules among the σ^{70} units (Grigorova I et al 2006 Proc Natl Acad Sci USA 103:5332). The σ^{38} subunits are used at the stationary phase of growth. σ^{28} is a minor subunit transcribing only less than two dozen genes. σ^{54}, another minor subunit binds to the promoter even in the absence of the core polymerase. In the synthesis of stress proteins σ^{32} and σ^{24} are used. Usually the σ^{70} is released from the polymerase (RNAP) at the beginning of the elongation of the transcript or shortly afterwards. However, some of them stay on throughout elongation and regulate gene expression depending on the cellular circumstances. The σ elements are complexed with anti-σ proteins when not in use. In some algae the protein present in the chloroplast is encoded by the nucleus. In several plant species, the same polypeptide is coded for by the chloroplast DNA. ►RNA polymerase, ►open promoter complex, ►sigma factor, ►chloroplast, ►chloroplast genetics, ►rpo, ►Pribnow box, ►transcription factors, ►UP elements, ►DnaJ; Dartigalongue C et al 2001 J Biol Chem 276:20866; Marr MT et al 2001 Proc Natl Acad Sci USA 98:8972; Bar-Nahum G, Nudler E 2001 Cell 106:443; Kuznedelov K et al 2002 Science 295:855; Mekler V et al 2002 Cell 108:599; Nickels BE et al 2005 Proc Natl Acad Sci USA 102:4488; review: Mooney RA et al 2005 Mol Cell 20:335; anatomy of *E. coli* σ^{70}: Shultzaberger RK et al 2007 Nucleic Acids Res 35:771.

σ: The parametric designation of standard deviation. ►standard deviation, ►standard error

σ: Yeast transposable element. ►Ty

σ: A viral infectious hereditary agent of *Drosophila*. ►CO_2 sensitivity, ►infectious heredity

σ: 387 bp region intercalated between the two S elements in opposite orientation of the complex *R* locus of maize. ►paramutation, ►tissue specificity, ►R locus of maize

S8: A ribosomal protein with binding site at the 597–599/640–643 at the hairpin of the 16S rRNA and it is required for the assembly of the 30S small ribosomal subunit (see Fig. S1). ►ribosomes

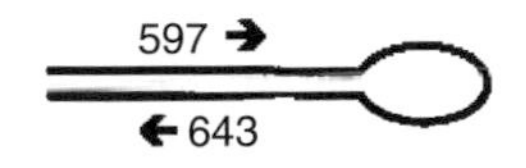

Figure S1. Hairpin

S-9: ►microsomes, ►Ames test

S^{35} or ^{35}S: Sulfur isotope. ►isotopes

S49: A mouse lymphoma cell line.

S Alleles: Control self-sterility in plants. ►self-incompatibility, ►incompatibility alleles

S Cytoplasm: Present in some cytoplasmically male sterile lines. ►cms

S Factor: A mitochondrial plasmid-like element in male sterile plants. ►cms

S6 Kinase (RSK, S6K): The collective name for the cytosolic $p70^{s6k}$ and the nuclear $p85^{s6k}$ kinases that phosphorylate the S6 ribosomal protein before the initiation of translation. The supply of amino acids affects the process. Mutation in the genes results in reduced cell and body size. Mice deficient for S6K are glucose intolerant and hypoinsulinemic. S6K is an effector of mammalian TOR. The carboxyl end of S6 and the phosphorylation sites within are highly conserved from *Drosophila* to humans. ►translation initiation, ►platelet-derived growth factor, ►phosphatidylinositol, ►S6 ribosomal protein, ►5′-TOP, ►cell size, ►insulin, ►TOR; Duffer A, Thomas G 1999 Exp Cell Res 253:100; Um SH et al 2004 Nature [Lond] 431:200.

S Locus: ►selfsterility alleles

S1 Mapping: When genomic DNA is hybridized with the corresponding cDNA or mRNA the non-homologous sequences cannot find partners to anneal with, and the single-stranded loops can be digested with S1 nuclease (see Fig. S2). The remaining DNAs that formed double-stranded structure can then be isolated by gel electrophoresis or their position and length can be determined by autoradiography if appropriately labeled material was used. Thus, intron positions are revealed. ►introns, ►S1 nuclease, ►DNA hybridization, ►genomic DNA; Favaloro J et al 1980 Methods Enzymol 65:718; Dziembowski A, Stepien PP 2001 Anal Biochem 294:87.

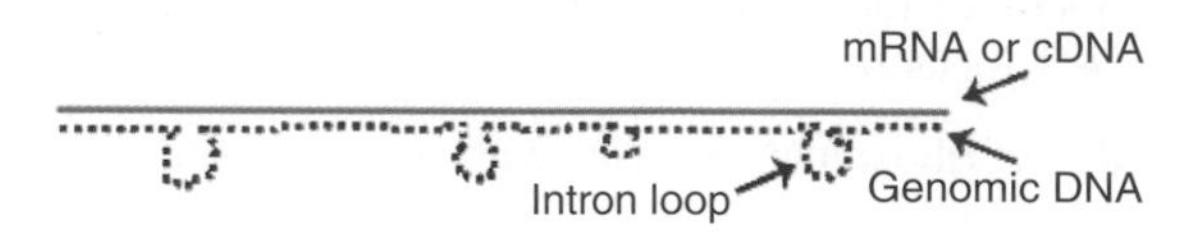

Figure S2. S1 mapping

S_1 Nuclease: S_1 Nuclease from *Aspergillus oryzae* cleaves single-stranded DNA (preferentially), and single-stranded RNA. Double-stranded molecules and DNA–RNA hybrids are quite resistant to it unless used in very large excess. The enzyme produces 5′-phosphoryl mono- and oligonucleotides. S_1 has many applications in molecular biology: mapping of transcripts, removal of single-stranded overhangs from double-stranded molecules, analysis of the pairing of DNA–RNA hybrids, opening up "hairpin" structures. Its pH optimum is 4.5 and this may cause unwanted depurination. ►nucleases, ►S1 mapping, see figure of hairpin structure; Kormanec J 2001 Methods Mol Biol 160:481.

S Phase: S phase of the cell cycle when regular DNA synthesis takes place. ►cell cycle

S Protein: ►vitronectin

S1 Ribosomal Protein: Binds to U-rich sequences upstream of the Shine-Dalgarno sequence and may promote translation. ►Shine-Dalgarno sequence, ►translation; Boni IV et al 2001 EMBO J 20:4222.

S6 Ribosomal Protein: Phosphorylated at about 5 serine residues near its C terminus, and the phosphorylated state is correlated with the activation of protein synthesis on the ribosomes; the phosphorylation is stimulated by mitogens and growth factors. ►ribosome; Recht MI, Williamson JR 2001 J Mol Biol 313:35.

5S RNA: ►ribosomal RNA, ►ribosomes; Artavanis-Tsakonas S et al 1977 Cell 12:1057.

6S RNA: Binds σ^{70} RNA polymerase in response to limited nutrient supply of bacteria and represses its transcriptional activity (Wassarman KM, Storz G 2000 Cell 101:613) and assures cell survival (Trotochaud AE, Wassarman KM 2004 J Bacteriol 186:4978). ►B2 RNA, ►7SK RNA, ►SRA RNA, ►RNA regulatory

7S RNA: ►RNA 7S

SAA: Serum amyloid A. ►amyloidosis

SABE (serial analysis of binding elements): A method for identification of DNA-binding transcription factors. ►SAGE; Chen J, Sadowski I 2005 Proc Natl Acad Sci USA 102:4813.

***S*-Adenosylmethionine** (SAM): A methyl donor for restriction-modification methylase enzymes, and general methylation of DNA; synonymous with Adomet. Folic acid and other compound contribute indirectly to SAM synthesis and methylation. ►methylation of DNA

***S*-Adenosylmethionine Decarboxylase** (AdoMetDC): An enzyme involved in the biosynthesis of spermidine and spermine. For its activity, it is essential to contain a covalently bound pyruvoyle end group to the α-subunit of the dimeric enzyme (see Fig. S3). This enzyme is found in both prokaryotes and eukaryotes. ►spermidine, ►spermine; Li Y-F et al 2001 Proc Natl Acad Sci USA 98:10578.

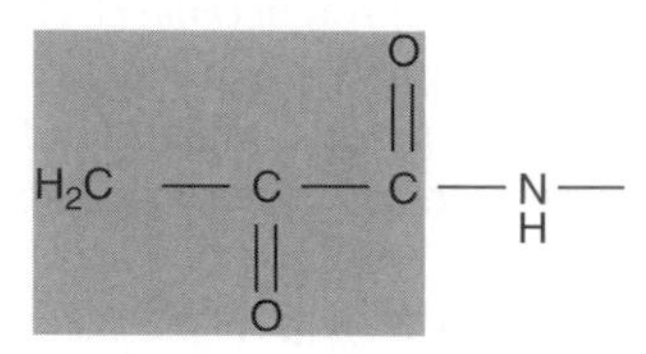

Figure S3. Pyruvoyle

SAC: A regulator of nuclear export and the cell cycle. ►nuclear pore, ►nuclear export factors

Saccades: Abrupt changes in the fixation of the eyes during scanning objects directed by the reflex center of the brain (superior colliculus).

Saccharin: A non-caloric sweetener; hundreds of times sweeter than sucrose. Oral LDLo for humans is 5 g/kg; it has been a suspected carcinogen and mutagen but later studies did not confirm its classification as a carcinogen. ►LDLo, ►aspartame, ►fructose

Saccharomyces cerevisiae: The eukaryotic budding yeast has chromosome number n = 16, its genome sizes is about 1.2×10^7 bp, approximately three times that of the prokaryotic *E. coli*. Recent sequencing and knockout information contradicted earlier estimates that less than 10% of its genome would be repetitive. A newer study estimates gene duplication (concerted evolution) and gene conversion rate to be about 25 million years and about 28 times of the mutation rate (Gao L-Z, Innan H 2004 Science 306:1367). In chromosome III of 55 open reading frames only 3 appeared indispensable for growth on a rich nutrient medium. Of 42 other genes, only 21 displayed a phenotype. This information points to redundancy even in this small genome. This conclusion may be misleading, however, because some genes are called to duty only under specific circumstances. Its entire genome was sequenced by 1996. The 5885 open reading frames are encoded by 12,068 kb. About 140 genes code for rRNA, 40 for snRNA and for 270 tRNA. About 11% of the total protein produced by the yeast cells (proteome) has metabolic function, 3% is involved in DNA replication and energy production, respectively; 7% is dedicated to transcription, 6% to translation and 3% (ca. 200) are different transcription factors. About 7% is concerned with transporting molecules. About 4% are structural proteins. The original gene number estimate has been since reduced to 6128 or fewer, and some studies consider the number only 5726 (Kellis M et al 2003 Nature [Lond] 423:241). Promoters, terminators, regulatory sequences and intergenic sequences with unknown functions occupy about 22% of the genome. The majority of yeast genes are not absolutely essential for survival or function (Giaever G et al 2002 Nature [Lond] 418:387). Many proteins are involved with membranes. In rich nutrient media its doubling time is about one and half-hour. The organism has regular meiosis and mitosis. The vegetative multiplication is by budding (budding yeast), i.e., the new (daughter) cell is formed as a small protrusion (bud) on the surface of the mother cell. Haploid cells may fuse to generate diploidy and the diploid cells may undergo meiosis (sporulation), and the four haploid products are retained in an ascus as an unordered tetrad. The haploid cells may have *a* or *α* mating type. Although budding yeast is eukaryotic it can be cultured much like prokaryotes, and thus it combines many of the advantages of both groups of organisms.

The yeast cell can be spherical (oblong) as represented in Figure S4 or they may become psdeudohyphal (filament-like) in appearance. MAPK, PKA and AMPK proteins mediate this switch of growth type. Approximately 25–30% of the human genes have significant homology with a yeast gene. The present genome might have evolved through extensive duplications. Baker's yeast is not a pathogen but may be harmful for immune-compromised individuals (Wheeler RT et al 2003 Proc Natl Acad Sci USA 100:2766) and it is considered to be an opportunistic pathogen. ►mating type determination in yeast, ►fungal life cycles, ►mtDNA, ►tetrad analysis, ►YAC, ►duplication, ►yeast vectors, ►yeast transformation, ►yeast transposable elements, ►gene replacement, ►synthetic genetic array, ►synthetic lethal, ►transcript mapping, ►*Schizosaccharomyces*, ►databases, ►MAPK, ►PKA, ►AMPK; sequencing: ftp://ftp.ebi.ac.uk/pub/databases/yeast; protein coding sequences: ftp://ftp.ebi.ac.uk/pub/databases/lista; http://genome-www.stanford.edu/Saccharomyces; http://www.yeastgenome.org/; http://mips.gsf.de/genre/proj/yeast/; yeast introns: http://www.cse.ucsc.edu/research/compbio/yeast_introns.html; Triples: http://bioinfo.mbb.yale.edu/e-print/genome-transposon-nature/text.htm; phenome: http://prophecy.lundberg.gu.se; transcriptional regulation: http://www.yeastract.com; protein–protein interaction: http://mips.gsf.de/genre/proj/mpact; proteomics, fluorescence microscopy: http://www.yeastrc.org/pdr/; Dwight SS et al 2001 Nucleic Acids Res 30:69; Barnett JA, Robinow CF 2002 Yeast 19:151 and 745; Mackiewicz P et al 2002 Yeast 19:619; 4000 interactions among 1000 genes: Tong AHY et al 2004 Science 303:808.

SACO (serial analysis of chromatin occupancy): Detects the binding sites of proteins along the chromosome. (See Impey S et al 2004 Cell 119:1041).

Sacral Agenesis (Currarino triad): A malformation of the caudal (tail) end of the notochord. Genes at human chromosomes 7q36 and 1q41-q42 may affect its rare dominant expression. ►Currarino triad, ►notochord

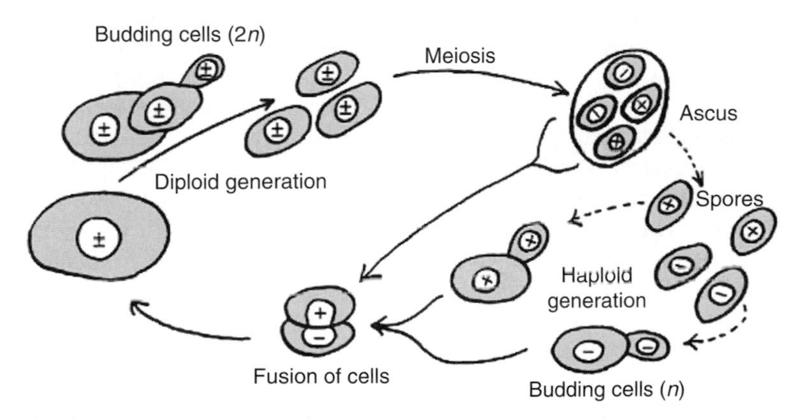

Figure S4. Life cycle of budding yeast

SAD Mouse: An animal model for human sickle cell anemia. It produces polymerized hemoglobin due to mutations in β globin, Hb^{SAD}. ►sickle cell anemia; Martinez-Ruiz R et al 2001 Anesthesiology 94:1113.

SADDAN (severe achondroplasia with delayed development and acanthosis nigricans): ►achondroplasia, ►acanthosis nigricans

SADR (serious adverse drug reaction): Considered the fourth leading cause of human death, it claims 100,000 fatalities and affects about 2,000,000 persons each year in the USA due to marketed medicine. Individual reaction to specific drugs varies and certain ethnic groups include more vulnerable genotypes than others. Besides beneficial effect, many drugs have unknown side effects. It is highly desirable to develop test that would predict the genetic bases of SADR. With the progress of genomic information there are potentials for clinical application of such tests. The European pharmacogenomics project EUDRAGENE, the Canadian Genotypic Adjustment of Therapy in Childhood (GATC) and various programs of the US National Institute of Health (Pharmacogenetics Research Network) are seeking solutions for the problems. ►drug development, ►metabonomics, ►genetic medicine, ►pharmacogenetics, ►pharmacogenomics; Giacomini KM et al 2007 Nature [Lond] 446:975.

Saethre-Chotzen Syndrome: ►Chotzen syndrome

SAF: SKP-associated factors, F-box proteins. ►SKP, ►F-box

Safety: ►laboratory safety, ►chemicals hazardous, ►recombinant DNA and biohazards, ►radiation hazard assessment, ►cosmic radiation, ►gloves, ►environmental mutagens. (See Fleming DO, Hunt DL (Eds.) 2000 Biological Safety. Principles and Practices, ASM Press, Washington DC).

S

Safflower (*Carthamus tinctorius*): An oil crop of warmer climates, 2n = 24; other related species have 2n = 20 or 2n = 44 chromosomes.

SAGA: A histone acetyltransferase complex of about 20 different proteins that interact with TBP (TFIID) and with gene-specific transcriptional activators. In yeast SAGA, SLIK and TFIID complexes appear to have some redundant functions. The Chd1 (chromo-ATPase/helicase-DNA binding domain) links histone (H3, H2B) methylation with SAGA and SLIK-dependent acetylation (Pray-Grant MG et al 2005 Nature [Lond] 433:434). ►histone acetyltransferase, ►histones, ►TBP, ►bromo-domain, ►transcription, ►transcription factors, ►chromatin remodeling, ►SNF/SWI; Sterner DE et al 1999 Mol Cell Biol 19:86.

SAGE (Serial Analysis of Gene Expression): A procedure that permits the analysis of the function of many genes by a sweeping procedure. The first step is to isolate all the mRNAs that are produced in a single organ, at a particular developmental stage. By reverse transcription they are converted into cDNA. The 3′ ends are then tagged by biotin. The cDNAs are digested by a restriction enzyme and the end fragments are trapped on streptavidin beads. After that, a second restriction enzyme is applied which cuts at least 9 bp from the fragments. In a following step, each short (9 bp or longer) tag is amplified by PCR and the tagged pieces are linked into a single DNA molecule. Then each tag is sequenced by an automatic sequencer and the tags are counted. In this sweeping manner 20,000 genes can be monitored in a month. The same effort would require years if it would be conducted on separate genes. This procedure reveals not just the number of expressed genes in that organ but reveals also the level of their activity. Some may be expressed in a single copy, others may be very active and are represented by multiple copies. Long SAGE is a variant of the SAGE procedure (Saha H et al 2002 Nature Biotechnol 20:508). ►genome project, ►genomics, ►biotin, ►streptavidin, ►DNA sequencing automated, ►electrospray MS, ►laser desorption MS, ►DNA chips, ►expressed-sequence tag, ►microarray hybridization, ►microarray analysis, ►SABE, ►RIDGE; Lash AE et al 2000 Genome Res 10:1051; Polyak K, Riggins GJ 2001 J Clin Oncol 19:2948; review: Wang CM 2007 Trends Genet 23:42; http://www.sagenet.org; 5′ end SAGE rags: http://5sage.gi.k.u-tokyo.ac.jp; DEGSAGE, accurate mapping of SAGE tags: http://dna.bio.puc.cl/SAGExplore.html.

SAGE Genie: An innovative new bioinformatics tool based on SAGE but providing a single platform for acquiring, annotating and interpreting large sets of gene expression data. The new technology measures gene expression by the frequency of 3′ signature SAGE tags of 10 bases specifique and unique to each transcript. The method permits the analysis of gene expression that are up- or down-regulated and allows automatic matching of SAGE tags to known transcripts. The incorrectly linked or those occurring only once or obtained by sequencing errors are filtered out from millions of tags. Then *confident SAGE tags* (CST) are obtained. The SAGE Genie permits *horizontal comparisons* (e.g., normal versus cancerous expression) and in addition *vertical comparisons* (e.g., expression profiles in different tissues or organs) under any desired conditions. This technology appears simpler and has far greater specificity than microarrays. The SAGE Genie provides an automatic link between gene names and SAGE transcript levels accessible by the Internet: http://cgap.nci.nih.gov/SAGE. ►SAGE, ►microarray hybridization; Boon K et al 2002 Proc Natl Acad Sci USA 99:11287.

Sagittal (adjective): In the anterior-posterior body plan.

Saguenay—Lac-Saint-Jean Syndrome (2p16): A rare, recessive, morbid cytochrome oxidase deficiency with symptoms similar to Leigh disease. ►Leigh's encephalopathy

SAHA (suberoylanilide hydroxamic acid): An inhibitor of histone deacetylase.

Saimiri (squirrel monkey): ►*Cebidae*

Sainfoin (*Onobrychis vicifolia*): A leguminous forage plant; 2n = 14 or 28 (see Fig. S5).

Figure S5. Onobrychist

Sal I: Restriction endonuclease with recognition site G↓TCGAC.

Salamander: *Salamandra salamandra*, 2n = 24, *Ambystoma mexicanum*, *A. tigrinum tigrinum* (n = 14). The estimated genome size of the ambystomas is 7291 map units, the largest known. (See Voss SR et al 2001 Genetics 158:735).

Salicylic Acid (*O*-hydroxybenzoic acid): A painkiller, keratolytic and fungicidal agent; it also mediates the expression of disease defense-related responses in plants (see Fig. S6). Its methylsalicylate derivative, a volatile compound, may carry out airborn signaling after infection of plants by pathogens. The PAD genes encode lipase-like molecules involved in salicylic signaling. Salicylic acid activity may be modulated by desaturation. ►host-pathogen relation, ►hypersensitivity reaction, ►desaturase, ►aspirin; Dempsey DM et al 1999 Crit Rev Plant Sci 18:547; Wildermuth MC et al 2001 Nature [Lond] 414:562.

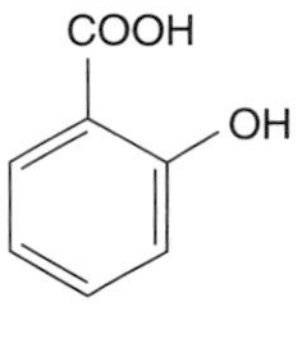

COOH
OH

Figure S6. Salicylic acid

Saline: The water solution of NaCl; the "physiological saline" is 0.9% salt solution for humans.

Salivary Gland Chromosomes: These are polytenic and because of their large size and clear landmarks, have been used extensively for cytogenetic analyses of *Drosophila* and other flies. The cultures to be used for this type of studies should be less crowded and moist. Third instar larvae can be used as they crawl out of the medium before the cuticle hardens. The larvae are placed in aceto-orcein or into 7% aqueous NaCl on microscope slides. A needle is used to hold the larva in place, and with a second needle placed behind the mouth parts the larva is decapitated and the salivary gland is pulled out. In aceto-orcein on a clean slide, the nuclei are stained in 5 to 10 min. The chromosomes may be spread by gentle pressure on the cover slide and after sealing the edge with wax, they can be examined under a light microscope. ►polytenic chromosomes, ►*Drosophila*, see Fig. S7; Lifschytz E 1983 J Mol Biol 164:17; Cold Spring Harb Symp Quant Biol 38 1974.

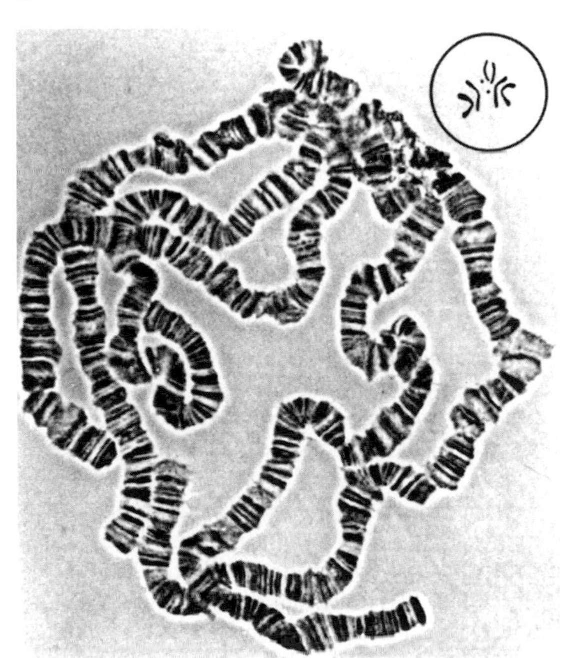

Figure S7. Salivary gland chromosomes of *Drosophila* (Courtesy of Dr. HK Mitchell). Upper right (circled) a regular mitotic set of chromosomes at about the same scale. (Redrawn after Painter TS 1934 J Hered 25:465)

S

Salivary Gland Chromosomes and Mapping: Since the salivary chromosomes display clear topological markers (bands), they can be used to locate deletions, duplications, inversions, translocations and can be used to associate mutant phenotypes with physical alterations. The genetic and cytological maps are collinear yet they are not exactly proportional by distance. ►coefficient of crossing over, ►*Drosophila*; Painter TS 1943 J Hered 25:465; Bridges CB 1935 J Hered 26:60.

Salla Disease: ►sialic acid, ►sialuria

Salmon: *Salmo gardneri*, 2n = 58–65. A transgenic Atlantic salmon strain carrying the growth hormone of the Pacific chinook salmon (*Oncorhynchus tshawytscha*) and the promoter of the eel-like ocean pout (*Macrozoarces americanus*), for an antifreeze protein grows twice as fast as its normal counterpart and actually thrives on less food. Some other transgenic strains do not perform as well (see Fig. S8). ►animal hormones, ►antifreeze protein; http://pewagbiotech.org/research/fish.

Figure S8. Coho Salmon (*Oncorhyncus kisutch*)

Salmonella: A member of the enteric Gram-negative bacteria and as such related to *E. coli*, and its handling ease is similar to it. It is a human pathogen and even in the laboratory it requires some caution when manipulated. Several of the related species contaminate food and feed supplies and create health hazards. F $^+$, F′ and Hfr strains are available. Its best known transducing phage is P22. The 4,809,037 bp genome of *Salmonella enterica* CT18 and two plasmids, pHCM1 (multiple-drug-resistance incH 218,159 bp) and the cryptic plasmid pHCM2 (106,516 bp) have been sequenced. It includes >200 pseudogenes and hundreds of insertions and deletions. *Salmonella enterica* serovar Typhimurium LT2 contains 4857 kb chromosome and a 94 kb virulence plasmid. *Salmonella cholerae suis* genome of 4.7 Mb has been sequenced. This bacterium primarily infects swine but is also pathogenic to humans (Chiu C-H et al 2005 Nucleic Acids Res 33:1690). ►histidine operon, ►Ames test, ►phase variation; Parkhill J et al 2001 Nature [Lond] 413:848; McClelland M et al 2001 Nature [Lond] 413:852; Edwards RA et al 2002 Trends Microbiol 10:94; http://www.salmonella.org.

Salpiglossis variabilis (Solanaceae): A plant species with the rather unusual characteristics where the four pollen grains, product of a single meiosis, stick together and, therefore can be used for tetrad analysis and gene conversion in higher plants. ►tetrad analysis, ►gene conversion

Salpingectomy: The sterilization of mammals by removal of the Fallopian tube (salpinx) leading to the uterus. ►sterilization humans, ►vasectomy, ►tubal ligation, ►birth control, ►uterus

Salt Bridges: Non-covalent ionic bonds in multimeric proteins.

Saltation: The unproven evolutionary proposition that species (and even higher taxonomic categories) arise by non-Darwinian sudden, major alterations. ►hopeful monster

Saltatory Replication: The sudden amplification of DNA segments during evolution.

Salt-Tolerance: Salt-tolerance of plants is regulated by the HKT1 (high affinity potassium [K^+] transporter). This protein is actually a Na^+ and K^+ cotransporter and at high K^+ level results in low level of Na^+ uptake thus conveying some salt tolerance. Ca^{2+} is beneficial for salt tolerance supposedly under the control of a protein sensor displaying about 50% similarity to calcineurin and neuronal calcium sensors. The Na^+/H^+ antiport protein (NH1X) mediates another salt control. Low proline content in the cells causes high salt-sensitivity whereas an increase in the level of sugar alcohols favors salt tolerance. Another potential mechanism for salt tolerance would be the improved elimination of the salt from the cells. *Arabidopsis* plants carrying a mutant NHX1 gene displayed higher tolerance for sodium and, in addition, because of the accumulation of NaCl in the vacuoles acquired better exploitation of the soil moisture through osmosis. Having salt tolerant crops may extend the usage of alkaline soils and may make it possible to use seawater for irrigation. Increased glyoxylase activity improves salt tolerance (Singla-Pareek SL et al 2003 Proc Natl Acad Sci USA 100:14672). Tobacco plant transgenic for the pea DNA helicase 45 (a homolog of the translation initiation factor eIF-4A) increased salt tolerance without reducing yield (Sanan-Misra N et al 2005 Proc Natl Acad Sci USA 102:509). The overlapping gene pair of *Arabidopsis* encoding Δ^1-pyrroline-5-carboxylate dehydrogenase (*P5CDH*) and *SRO5* of unknown function can generate a 21-nucleotide siRNA that can increase salt tolerance (Borsani O et al 2005 Cell 123:1279). ►ion channels, ►calcium signaling, ►calcineurin, ►glyoxylate cycle, ►eIF-4a,

►*Priformospora indica*, ►overlapping genes, ►RNAi; Zhang H-X, Blumwald E 2001 Nature Biotechnol 19:765; Zhu J-K 2002 Annu Rev Plant Biol 53:247; Shi H et al 2003 Nature Biotechnol 21:81.

Salvage Pathway: A recycling pathway, in contrast to the de novo pathway, e.g., nucleotide synthesis from nucleosides after removal of the pentose followed by phosphorybosylation. ►phosphoribosyl transferase, ►HGPRT, ►HAT medium

SAM (significance analysis of microarrays): Permits statistical distinction between microarray signals. ►microarray hybridization; Tusher VG et al 2001 Proc Natl Acad Sci USA 98:5116.

SAM: ►*S*-adenosyl-L-methionine, ►substrate adhesion molecules

SAM68 (SRC-associated in mitosis): A phosphoprotein and a target of SRC, FYN and ITK kinases during mitosis. It also binds to the SH domains of the Grb adaptor protein of the cytoplasm and NCK in the nucleus. When SAM68 is tyrosine dephosphorylated, it binds RNA. A homolog of SAM68 is ÉTOILE. ►STAR, ►signal transduction, ►Src

SAM (system for assembling markers): http://www.sanger.ac.uk/Software/sam/.

Sampling Distribution: The probability distribution of a statistic estimated from a random sample and a certain size. The sampling distribution is somewhat different from the normal distribution that is characterized by the mean (μ) and the standard deviation (σ) inasmuch as mean (M) has a standard deviation of σ/√n.

Sampling Error: Can occur when the population is too small or when few individuals are sampled. The proper sample size *n* can be statistically estimated as shown in the formula. $n = \frac{2(z_\alpha - z_\beta)^2 \sigma^2}{(\mu_1 - \mu_2)^2}$ where z_α is the normal deviate for a level of significance α, and z_β is the normal deviate for β and 1 – β is the power required, σ^2 is the variance of each population with means μ_1 and μ_2 chosen or whenever the selection is not random. ►drift genetic, ►genetic drift, ►effective population size, ►founder principle, ►normal deviate, ►normal deviation, ►Z distribution

Sancho: A non-viral retrotransposable element. ►transposable elements

Sandhoff's Disease: Characterized either by the absence of both β-hexosaminidase α and β activity or by β-hexosaminidase β subunit only. This autosomal recessive hereditary defect has very similar symptoms to those of the Tay-Sachs disease that is caused by β-hexosaminidase α subunit deficiency. β-Hexosaminidase β subunit defect blocks the degradation of β-galactosyl-*N*-acetyl-(1→3)-galactose-galactose-glucose-ceramide to β-(1→4)-galactosyl-galactose-glucose-ceramide and hexosaminidase α normally converts GM_2 ganglioside (neuraminic-*N*-acetyl-galactose-4-galactosyl-*N*-acetyl-glucose-ceramide) into GM_3 ganglioside (neuraminic-*N*-acetyl-galactose-glucose) ceramide. The genes for β-hexosaminidase α and β are in human chromosomes 15q23-q24 and 5q13, respectively. Their structural similarity indicates evolution by duplication. ►sphingolipidoses, ►sphingolipids, ►gangliosides, ►Tay-Sachs disease

Sandwich-Like Protein: Two β-sheets of 3–5 β-strands packed face-to-face.

Sanfilippo Syndrome: ►mucopolysaccharidosis

Sanger Method of DNA Sequencing: ►DNA sequencing

SANT (SWI3, ADA-2, N-COR, TFIIB): The chromatin remodeling complex sharing among them an ATPase and transcription and chromatin remodeling domains. ►chromatin remodeling, ►SWI, ►ADA, ►N-COR, ►TFIIB

Santa Gertrudis Cattle: Originated by a cross between the exotic-looking Brahman x shorthorn breeds (*Bos taurus*). The Brahman cattle descended from *Bos indicus*. ►cattle

SAP-1 (stress activated protein): SAP-1 is involved in the activation of MAP kinases in binding to their serum response factor (SRF). SAP is required for the activation of SLAM. ►MADS box, ►MAP, ►SRF, ►Epstein-Barr virus, ►SLAM, ►Elk, ►lymphoproliferative diseases X-linked, ►XPL; Latour S et al 2001 Nature Immunol 2:681; Cannons JL et al 2004 Immunity 21:693.

SAPK: A stress-activated protein kinase of the ERK family. SAPK can be activated by SEK a protein kinase, related to MAP kinase kinases. The signaling cascade involved is targeted to JUN. ►signal transduction, ►ERK, ►MAPKK, ►MEKK, ►JUN, ►JNK, ►Ire

Saponification: Alkaline hydrolysis of triaglycerols to yield fatty acids. ►triaglyceride, ►fatty acids

Saponins: Glycosylated triterpenoid or steroidal alkaloids in plants where they may be protective against fungal pathogens. Some fungal pathogens detoxify this defense molecule in a two-step process. ►host-pathogen relation; Bourab K et al 2002 Nature [Lond]: 418:889.

S

Saporins: Plant glycosidases, which remove adenine residues from RNA and DNA but not from ATP or dATP. ►RIP

Saposins (10q22.1): Glycoproteins involved in the activation of galactosylceramidase. Pro-saponins are precursors of the endosomal lipid transfer proteins, which in association with CD1 present lipid antigens to the natural killer lymphocytes (NKT). ►leukodystrophy, ►shingolipids, ►killer cell, ►CD1; Qi X, Grabowski GA 2001 J Biol Chem 276:27010.

Saprophytic: Lives on non-living organic material. ►biotrophic

SAR (structure-activity relationship): A field of study in carcinogen, mutagen and drug research. Structural modifications may or may not affect activity and it would be important to know what are the decisive factors and how to design more effective drugs. The biological assays of inhibition not only permit classification of the compounds (potential drugs) but also provide information on the structure of the proteins they interact with (Fliri AF et al 2005 Proc Natl Acad Sci USA 102:261). ►TD_{50}, ►CASE, ►MULTICASE, ►biophore

SAR (systemic acquired resistance): SAR may be induced in plants by pathogens, salicylic acid, etephon, a compound releasing ethylene. The SAR-related genes may be members of regulons under common promoters binding specific transcription factors. Inactivation of MAP kinase 4 by the maize *Ds* transposon increased SAR and salicylic acid level and boosted the expression of pathogenesis-related proteins in *Arabidopsis*, although it reduced the size of the plants. In *Arabidopsis* for the expression of SAR a protein secretory pathway, including many genes, requires activation (Wang D et al 2005 Science 3008:1036). Salicylic acid binding protein (SABP2) can convert methylsalicylic acid to salicylic acid and activates SAR (Forouhar F et al 2005 Proc Natl Acad Sci USA 102:1773). ►host-pathogen relationship, ►ethylene, ►salicylic acid, ►MAPK, ►desaturase; Maleck K et al 2000 Nature Genet 26:403; Petersen M et al 2000 Cell 103:1111; Maleck K et al 2000 Nature Genet 26:403.

SAR: ►scaffold

SAR1: A low activity GTPase, related to Arf. It regulates the traffic between the endoplasmic reticulum and the Golgi apparatus. ►endoplasmic reticulum, ►Golgi, ►GTPase, ►Arf, ►GTPase; Takai Y et al 2001 Physiol Rev 81:153.

SAR by NMR (structure-activity relationship by nuclear magnetic resonance-based methods): An essential and sophisticated method to synthetic drug design. Natural or synthetic molecules are screened and optimized analogs are synthesized to identify high-affinity ligands to develop effectively targetable drugs. ►SAR, ►NMR

SARA (Smad anchor for receptor activation): Recruits SMADs to the TGF-β receptor and thus regulates TGF signaling. ►SMAD, ►TGF

SarA: A pleiotropic staphylococcal accessory regulator of virulence. It binds to multiple AT-rich sequences. (See Schumacher MA et al 2001 Nature 409:215).

Saran Wrap: A thin sheet of plastic that clings well to most any surface and suitable for covering laboratory dishes, gels, etc.

Sarcoglycans (SGCA, 17q12-q21.33; SBCB, 4q12): A part of the dystrophin—glyco-protein complex in the sarcolemma. The complex protects the muscles and connects the cytoskeleton and the extracellular matrix. Sarcoglycan mutations may cause dystonia and myoclonous. ►dystroglycan, ►muscular dystrophy, ►dystonia, ►myoclonous; Zimprich A et al 2001 Nature Genet 29:66.

Sarcoidosis (susceptibility locus at 6p21.3, HLA-DRB1, BTNL2 [butyrophilin-like 2], Löfgren syndrome): A polygenic disorder of the immune system affecting lung, lymph nodes and eyes. Prevalence is $\sim 1.2 \times 10^{-4}$ among the majority of people of European extraction, except Swedes whose risk is about 5 times larger. The relative risk of first-degree relative is 2.8 to 18. ►HLA; Valentonyte R et al 2005 Nature Genet 37:357.

Sarcolemma: Membranes covering the striated muscle fibers. ►caveolin, ►dystrophin

Sarcoma: A solid tumor tissue with tightly packed cells embedded in a fibrous or homogeneous substance; sarcomas are frequently malignant. Sticker sarcoma is canine transmissible venereal tumor. ►Rous sarcoma, ►RAS, ►oncogenes; Bonnicelli JL, Barr FG 2002 Curr Opin Oncol 14:412.

Sarcomere: Muscle units of thick myosin, thin actin filaments between two plate-like Z discs. These units are repetitive. ►titin, ►nebulin, ►myosin

Sarcoplasmic Reticulum: The membrane network in the cytoplasm of muscle cells containing high concentration of calcium, which is released when the muscle is excited.

Sarcosinemia: Sarcosine (methylglycine, CH_3NHCH_2-COOH) is normally converted to glycine (NH_2CH_2-COOH) by the enzyme sarcosine dehydrogenase. A defect at this step increases the level of sarcosine in the blood and in the urine (hypersarcosinemia), and may result in neurological anomalies but it may have almost no effect at all. Glutaric aciduria and defects in folic acid metabolism may also cause hypersarcosinemia. ►glycine, ►folic acid

Sarin: A nerve gas, which inhibits acetylcholinesterase. ►organophosphates

Sarkosyl (*N*-lauroylsarcosine): A detergent for solubilizing membranes and is used for extraction of tissues.

SARS (severe acute respiratory syndrome): The disease symptoms resembling flu develop by exposure to a corona virus of about 30,000-nucleotide, positive single-stranded RNA genome (see Fig. S9) (Marra MA et al 2003 Science 300:1399). The genome displays regional variations at several sites. The different strains show essential variations in the proteins decorating the capsid that determine virulence. During the first week high fever occurs without much additional symptoms. During the following four days, pneumonia develops in the lung that subsequently may become destructive to this organ. Protection is prevention, hygiene, avoiding breathing the airborne particles but no effective drug therapy exists. It is a zoonosis. Apparently the Himalayan palm civet (*Paguma larvata*) and the raccoon dog (*Nyctereutes procyonoides*) transmitted the virus to humans. In these animals the genomic sequence of the virus is 99.8% identical to that found in human (Song H-D et al 2005 Proc Natl Acad Sci USA 102:2430). Mortality rate is 5 to 15%. In mice DNA vaccine is effective (Yang Z-y et al 2004 Nature [Lond] 428:561). Angiotensin-2 converting enzyme (ACE2) is the only known receptor for this virus. L-SIGN is similar to SIGN/DC-SIGNR (dendritic cell specific ICAM-3 grabbing non-integrin) is expressed in the lymph nodes and placenta is encoded by *CLEC4M* gene in mouse. L-SIGN and DC-SIGN bind to high-mannose oligosaccharides of viruses including the SARS virus. Human individuals heterozygous for the *CLEC4M* tandem repeats are less susceptible to SARS. L-SIGN homozygotes bind even better the virus and degrade it and show lower capacity for trans infection (Chan VSF et al 2006 Nature Genet 38:38; the conclusions of this paper were questioned by Teng NL-S et al 2007 Nature Genet 39:691; Zhi L et al 2007 Nature Genet 39:692; Chang KVK et al 2007 Nature Genet 39:694 did not concede). The coronavirus variants SARS-CoV and other coronaviruses encode two RNA-dependent replicases, a specific spike protein, a small envelope protein, a membrane protein, a nucleocapsid protein and nine unidentified other proteins. Diagnostic approaches involve RT-PCR, ELISA and an IIFT kit for the detection of IgG antibody response and all of them display some advantage and shortcomings for diagnosis. A microarray-based technology appears rapid, sensitive and accurate and it is adaptable to other viral infections (Zhu H et al 2006 Proc Natl Acad Sci USA 103:4011). ►zoonosis, ►immunization genetic, ►plant vaccine, ►angiotensin, ►ICAM, ►dendritic cell, ►RT-PCR, ►ELISA, ►immunoglobulins, ►Newcastle disease; Navas-Martin S, Weiss SR 2003 Viral Imunol 16(4):451; Webby RJ, Webster RG 2003 Science 302:1519; antiviral drugs: Wu C-Y et al 2004 Proc Natl Acad Sci USA 101:10012.

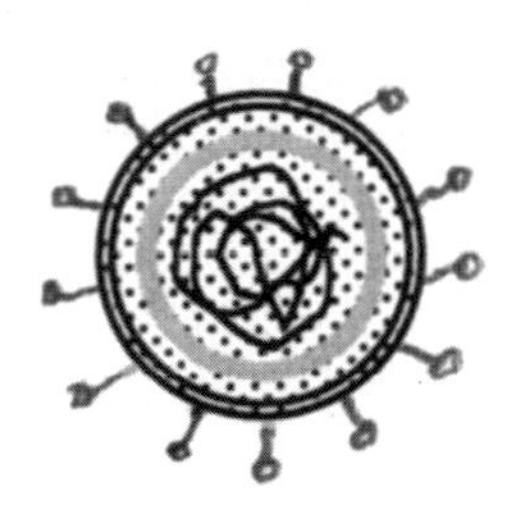

Figure S9. Corona virus

SAS (statistical analysis system): A computer software for the many types of data analyses.

SAS (switch-activating site): ►*Schizosaccharomyces pombe*

SAT: ►satellited chromosome

SATB1 (special AT-rich sequence binding protein 1): Regulates chromatin folding and anchores special DNA sequences (ATC) into its network. It facilitates histone acetylation at many specific loci and it controls tissue-specific gene expression. ►tissue-specificity, ►chromatin remodeling; Cai S et al 2003 Nature Genet 34:42; controling cytokine gene expression: Cai S et al 2006 Nature Genet 38:1288.

SAT-DAC (satellite-DNA-based artificial chromosome): SAT-DAC of mammals is expected to replicate independently from the genome and to express its gene content as a huge vector. It may serve also for transferring any gene into mammals (humans) without the risk of disruption resident genes. SAT-DACs composed mainly of AT sequences may be readily isolated from the rest of the genome. Such a structure was stably inherited in mice. (See Hadlaczky G 2001 Curr Opin Mol Ther 3:125).

Satellite Cells: Surround each myofiber beneath the basal lamina and are precursors of muscle growth and repair. As few as seven satellite cells, associated with myofibers, are sufficient sources of muscle regeneration (Collins CA et al 2005 Cell 122:289; Montarras D et al 2005 Science 309:2064). ►basement membrane, ►myofibril, ►stem cells

Satellite DNA: DNA fraction with higher or lower density (during ultracentrifugal preparations) than the bulk DNA (see Fig. S10). Generally, it contains substantial repetitive DNA (in up to 5 Mb tracts). These satellite sequences generally pose problems in genome sequencing because of their instability in the cloning vectors. Very little information is available regarding the function of SAT DNA. Dimeric oligopyrroles-imidazoles target adenine-thymine-rich satellite sequences and the scaffold-associated region (SAR) of *Drosophila*. Unexpectedly these polyamides induce gain- or loss-of-function phenotypes. It is assumed that these chemicals facilitate accessibility to chromatin. ►repetitious DNA, ►ultracentrifugation, ►heterochromatin, ►α-satellite, ►polyamides, ►satellite, ►pyrrole, ►imidazole; Janssen S et al 2000 Mol Cell 6:999; Janssen S et al 2000 Mol Cell 6:1013.

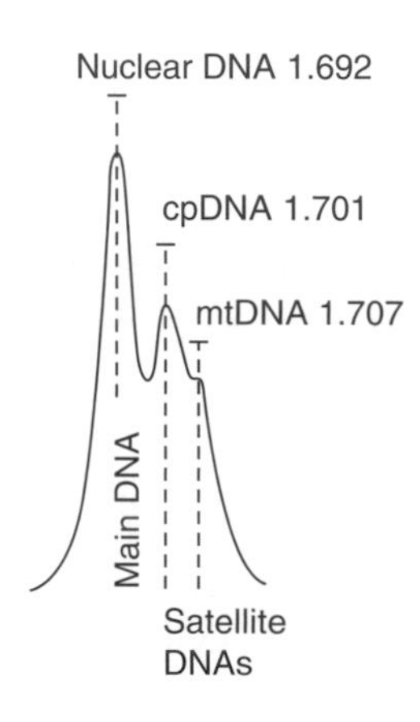

Figure S10. Satellite DNAs are identified by ultracentifugation on the basis of their densities in CsCl

Satellite RNA: The transcript of satellite DNA: ►satellite DNA

Satellite Virus: A defective virus co-existing with another (helper) virus to correct for its insufficiency. The unrelated helper is required for infection.

Satellited Chromosome: Carries an appendage to one arm by a constriction (see Fig. S11). The bridge between the main body of the chromosome and appendage is the site of the nucleolar organizer. It was named originally SAT as an abbreviation for *sine acido thymonucleinico,* i.e., a place where there was, then, no detectable DNA (= thymonucleic acid, as it was called in the 1930s) in that part of the chromosome. The appendages were also called trabants.

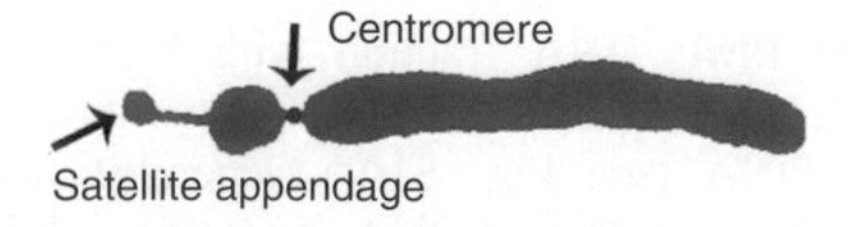

Figure S11. Satellited chromosome

Satsuma Mandarin: A pollen sterile citrus variety that, if no foreign pollen can reach the flowers, produces seedless oranges. This is the characteristic also for navel oranges although some navel oranges have many seeds because in those the abnormal carpel development may interfere with pollination but they are not basically self-sterile. ►seedless fruits

Saturated Fatty Acids: All their chemical affinities satisfied, have higher energy content than the unsaturated fatty acids that contain one or more double bonds. ►fatty acids, ►cholesterols, ►omega-3 fatty acids

Saturation Density: of a mammalian cell culture before contact inhibition takes place. ►cancer, ►malignant growth, ►contact inhibition

Saturation Hybridization: One component in the nucleic acid annealing reaction mixture has excessive concentration to allow finding all possible sites of homology and hybridization. ►nucleic acid hybridization

Saturation Mutagenesis: Induce mutations at all available sites to reveal the relative importance of these sites. ►mutagenesis, ►localized mutagenesis, ►linker scanning, ►GAMBIT

Saturation of Molecules: The carbon–carbon attachments of single covalent bonds (no double bonds when entirely saturated). ►saturated fatty acids

Sau 3A: Restriction endonuclease with recognition sequence ↓GATC; *Sau* 96 I: G↓GNCC.

SAUR (small auxin-up RNA): RNAs encoded by several genes that are induced by auxins. ►auxins; Guilfoyle TJ 1995 ASGSB Bull 8:39.

SBD: Single-strand DNA binding domains of ~120 amino acids in the large subunits (and possibly in the small) of the RPA protein of eukaryotes. ►DNA replication in eukaryotes, ►RPA

SBF (Saccharomyces binding factor): Composed of Swi4 and Swi6 along with MBF (composed of Swi4 and Mbp1) initiate the transcription of genes cyclin 1 (*CLN1*) and cyclin 2 (*CLN2*) required for activation of Cdc28 enabling the progress of the cell cycle from G1 to S phase. ►cell cycle, ►cyclin, ►Cdc28, ►Swi, ►MBF

Sbf1 (SET-binding factor 1): A myotubularin-like protein (but without phosphatase activity) binding to SET domains. ►SET, ►myopathy

SBMA: ►Kennedy disease

sc: Indicating *Saccharomyces cerevisiae* (budding yeast) DNA, RNA or protein as a prefix.

SC1: An immunoglobulin superfamily cell adhesion molecule, it is transiently expressed during avian embryogenesis by a variety of cell types.

SC Phocomelia: ►Roberts syndrome

SC35: An SR protein. ►SR motif

SCA: ►spinocerebellar ataxia, ►ataxin, ►ataxia

SCA (statistical coupling analysis): Maps energetic interactions in proteins and measures the statistical interactions between amino acid positions in order to shed light on folding patterns (Lockless SW, Ranganathan R 1999 Science 286:295). ►protein folding

Scaffold: The cytoskeleton of the cell or residual protein fibers left in the chromosome after the removal of histones. The bulk of the scaffold is of two proteins, Sc1 (a topoisomerase II) and Sc2. The scaffold is attached to SAR (scaffold attaching regions) of the chromatin. Cytoskeletal scaffolds facilitate the transport of various molecules along its network and secure them at the proper cellular positions. Scaffold proteins—in connection with other molecules—can also fine tune quantitatively input-output properties (Bhattacharyya RP et al 2006 Science 311:822). Artificial polymer scaffolds have been constructed and used as an alternative to viral vectors to deliver therapeutic proteins or gene constructs to cells with defective or deleted genes and release their cargo in their target environment. In animal models, implanted platelet-derived growth factor greatly enhanced vascularization. ►cytoskeleton, ►topoisomerase, ►loop domains model, ►genetic engineering, ►MAR, ►WGS; Dietzel S, Belmont AS 2001 Nature Cell Biol 3:767.

Scaffold Protein: A platform that accepts, assembles and delivers cofactors to biological targets.

Scaffolds in Genome Sequencing: A set of ordered, oriented contigs, assembled relative to each other by mate pairs in adjacent contigs. For assembling the over 99% accurate sequences of the *Drosophila* genome, Celera group used 8 Compaq Alpha ES40 computers with 32 gigabyte memory. In the assembly, the repetitive sequences pose real challenges because the overlaps can be *true overlaps* that belong together or the overlaps are parts of repeated sequences that may occur multiple times and being scattered in the genomes and therefore, do not belong together. A collection of fragments whose arrangement is uncontested by overlaps from other fragments are *unitigs*. The unitigs that represent unique (non-repetitive) sequences—although some extending into repeats—are called *U-unitigs*. In Figure S12 X′ and X″ represent repeated sequences and in the box, they are "overcollapsed" in a unitig because they consistently subassemble as the interior of the repeats. In the repeat boundary box, A and B or B and C do not overlap and they are computationally resolved and help in extending of the assembly of the U-unitigs into scaffolds (see Fig. S13).

If mate pairs and overlaps consistently appear in bundles, the ordering of the non-repetitive euchromatic segments is facilitated. Usually gaps appear between the U-unitigs and scaffolds. Smaller fragments called *rocks*, *stones* and *pebbles* fill in these gaps. The shorter unitigs still have two or one mate links. The correct *tiling path* (a minimally overlapping DNA fragment map spanning that length of the genome) is also verified by statistics. The *reads* (base sequences) are verified also by multiple *base-callings* (identifying the correct nucleotide in a sequence) and the sets are evaluated by Bayesian statistics.

The *Drosophila* genome was sequenced and assembled in 838 *firm scaffolds* (containing at least one U-unitig). As of 2000, the WGS still has 1887 gaps with a total length of 2322 Mbp varying in size up to 150 Mbp and zero. The number of U-unitigs was 7164 (8.007 Mpb), rocks 1787 (0.927 Mbp), stones 132 (0.118 Mbp) and pebbles 25,101. STS (sequence-tagged site) mapping and BAC/P1/cosmid clone tiling

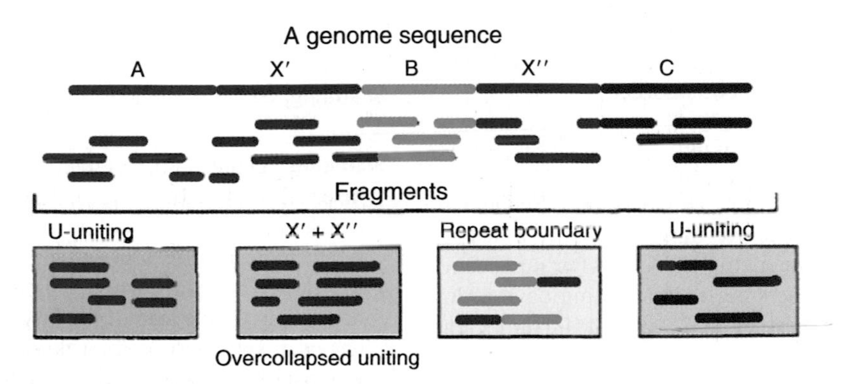

Figure S12. Scaffold. (Modified after Myers EW, et al. 2000 Science 287:2196)

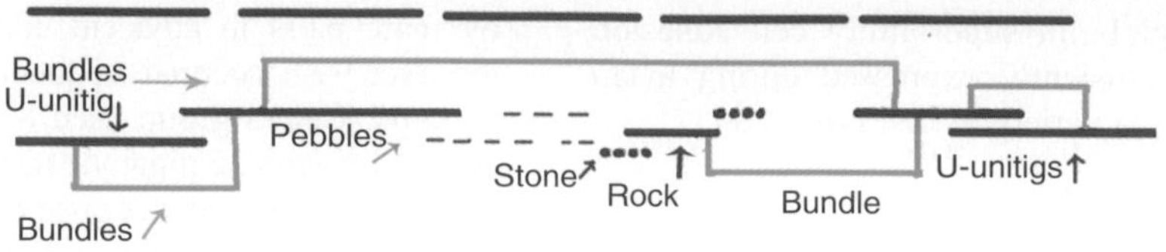

Bundles
U-unitig
Pebbles
Stone
Rock
Bundle
U-unitigs
Bundles

Figure S13. Scaffold of a collection of ordered contigs built of unitigs. (Modified after Myers EG et al. 2000 Science 287:2196)

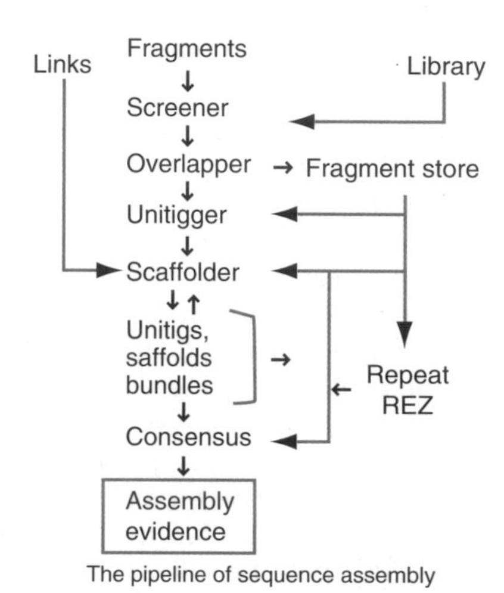

Figure S14. The overall strategy of the Celera operations. (Modified after Myers EW et al. 2000 Science 287:2196)

path (CTP) has validated the WGS map. Yet some refinement will continue, especially in the heterochromatic regions (around the centromere, etc.). ►WGS, ►genome projects, ►contig, ►mate pair, ►Bayes' theorem, ►STS, ►BAC, ►P1, ►cosmid, ►tiling, ►DNA sequencing, see Fig. S14; Myers EW et al 2000 Science 287:2196.

Scaffold-Mediated Activation: The assembly of a molecular platform in response to death stimuli and the recruitment of pro-caspases. It is initiated through cell surface receptors such as the Fas receptors, TNFα receptors and their bound ligands leading to the formation of caspases and ending up in apoptosis. ►caspase, ►TNF, ►apoptosis; Earnshaw WC et al 1999 Annu Rev Biochem 68:383.

SCAIP (single condition amplification/internal primer sequencing): Amplifies large number of exons at a single set of PCR temperatures. Sequencing specificity is gained by uniform use of a second, internal set of sequencing primers. After purification of the DNA sequencing requires three days at low cost (Flanigan KM et al 2003 Am J Hum Genet 72:931).

Scale-Free Networks: networks.

Scaling: ►quarter-power scaling, ►synaptic scaling

SCAM (substituted-cysteine accessibility method): The method used for studying cysteine substitution and covalent modification on the structure-function relationship of proteins.

SCAMP (secretory carrier membrane proteins): Integral membrane proteins of secretory and transport vesicles, like the synaptic vessels, etc. (See Wu TT, Castle JD 1998 Mol Biol Cell 9:1661).

Scanning Electronmicroscopy (SEM): In contrast to transmission electron microscopy, the electron beam is reflected from the surface of the specimen, coated with a heavy metal vaporized in a vacuum, through a process called *shadowing*. As the electron beam scans the specimen, secondary electrons are reflected according to the varying angles of the surface of the object and generate a *three-dimensional image* corresponding to the grade of reflections. The maximal resolution is 50–100 times less than with the transmission electronmicroscopy but the image obtained can be highly magnified. It is an important technique for developmental studies. ►electronmicroscopy, ►stereomicroscopy, ►scanning tunneling microscopy, ►SPM, for picture see ►petals

Scanning Force Spectroscopy (SFS): Measures association and dissociation constants of biological molecules. ►optical tweezer; Bonin M et al 2002 Nucleic Acids Res 30(16):e81.

Scanning, Genetic: Genetic scanning.

Scanning Linker Mutagenesis: The transposon is inserted within the gene in-frame at an appropriate restriction enzyme recognition site (see Fig. S15). After transcription and translation of the construct the target protein carries a new peptide of a size depending on the size of the insertion. The procedure has been applied to both prokaryotes and eukaryotes. ►transposon, ►restriction enzyme, ►insertional mutation; Hayes F 2003 Annu Rev Genet 37:3.

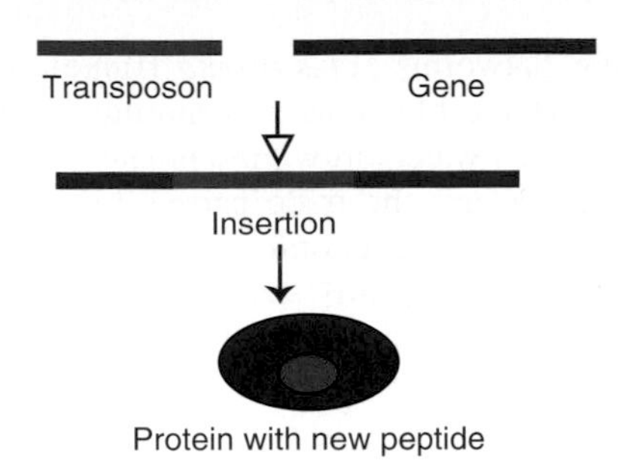

Figure S15. Scanning linker mutagenesis

Scanning Mutagenesis: ►homolog-scanning mutagenesis, ►linker scanning

Scanning of mRNA: The eukaryotic mRNA does not have a special consensus (such as the Shine-Dalgarno box in prokaryotes) for attachment to the ribosomes, so the mRNA leader adhers by its methylated guanylic cap to the ribosomes and the ribosome runs on it until it finds an initiation codon (generally AUG) to start translation. A preferred sequence, however, may be around the AUG codon AG——CC**AUG**G. The 3′ terminus of the 18S rRNA of mammals bears some similarity to the prokaryotic 16S terminus:

A^{Me2} A^{Me2} CCUGCGG**A**AGGAUGA——UUA-3′-0H.

The presence of the m^7G-cap facilitates the initiation but is not absolutely indispensable. The average length of the 5′-untranslated (50–70 nucleotides) sequence and its structure may favor translation but only large differences from it (very short tracts) decrease substantially the initiation. A ~12-nucleotide hairpin secondary structure between the cap and AUG improves the efficiency of translation. Similarly, the polyA tail is advantageous for translation possibly by facilitating the recycling of the ribosomes on the same mRNA. *Leaky scanning* indicates that more than one initiator codons or only the second or a later AUG is used for translation and the preceding one(s) are skipped (Chen A et al 2005 Nucleic Acids Res 33:1169). In this case different polypeptides may result. ►Shine-Dalgarno sequence, ►eIF, ►initiation codon; Kozak M 1989 J Cell Biol 108:229; Samuel CE 1989 Progr Nucleic Acid Res Mol Biol 37:127.

Scanning Tunneling Microscope (STM): STM can resolve biological molecules at the atomic level; its use has been proposed for DNA sequencing. With STM, vibrational spectroscopy is possible permitting the analysis of molecules adsorbed on a surface. The vibrational energies may reveal adsorption sites, orientation and adsorption changes. ►electrospray MS, ►laser desorption, ►nanotechnology, ►SPM

SCAP (SREBP cleavage-activating protein): Regulates cholesterol metabolism by promoting the cleavage of transcription factors SREBP-1 and -2 (sterol regulatory element binding proteins). In low-sterol cells, discrete proteolysis cuts of the amino-terminal of SREBPs. As a consequence, these proteins enter the nucleus and activate the LDL receptor and cholesterol and fatty acid biosynthetic enzymes. The system can be studied by mutant CHO cells that either cannot synthesize cholesterols or LDL receptors in response to sterol depletion, or are sterol resistant and cannot terminate the synthesis of sterols or their LDL receptor. ►sterol, ►LDL, ►CHO, ►Niemann-Pick disease, ►lipodystrophy; Shimano H 2001 Progr Lipid Res 40(6):439.

SCAR (sequence-characterized amplified region): Physical markers obtained by polymerase chain reaction-amplified RAPD bands. ►amplification, ►RAPD, ►PCR; Iturra P et al 2001 Heredity 84:412.

SCAR: The suppressor of cAMP receptor. ►cAMP

SCARMD: ►muscular dystrophy

Scatter Diagram: A two-dimensional graphic representation of the characteristics of two variables, two traits, which may be or may not be correlated, e.g., people with a certain eye and hair color.

Scatter Factor (hepatocyte growth factor): Cellular responses of scatter factor are mediated by the Met tyrosine kinase receptor. It has multiple cell targets and it is probably involved in mesenchym-alepithelial interactions and liver, kidney development, organ regeneration, metastasis, etc. ►hepatocyte growth factor, ►metastasis, ►macrophage-stimulating protein; Tacchini L et al 2001 Carcinogenesis 22:1363.

Scattering: Deflection of electrons by collision(s). ►Compton effect

Scavenger Molecules: They clean up the cells of substances that are no longer needed.

SCC: Yeast integral membrane protein with partial structural homology to cyclophilins. ►cyclophilins, ►double-strand breaks

Scc1: ►separin, ►Rec8

SCD25: A suppressor of gene *cdc25* mutations in yeast; increases the dissociation of Ras•GDP but does not affect Ras•GTP. ►cell cycle, ►*cdc25*

SCE: ►sister chromatid exchange

Scent: ►fragrances

SCEUS (smallest conserved evolutionary unit sequences): Reveal evolutionary similarities of the DNA in the genomes across taxonomic boundaries. ►unified genetic map

SCF (Skp1—CDC53—F-box protein): A complex of ubiquitination function. The SCF protein complexes have regulatory role in the cell cycle and development. ►Skp, ►CKS1, ►CDC53, ►F-box, ►ubiquitin, ►jasmonic acid, ►E1, ►E2, ►E3, ►cell cycle, ►von Hippel-Lindau syndrome, ►glucose induction; see also Patton EE et al 1998 Trends Genet 14:237; Schwab M, Tyers M 2001 Nature [Lond] 413:268; Zheng N et al 2002 Nature [Lond] 416:703.

SCF: ►stem cell factor

ScFv (single chain fragment variable): A variable portion of the antibody molecule that may still be expressed, e.g., in plantibodies or bacteria or other cells. ►antibody, ►plantibody, ►single-chain Fv antibody; Norton EJ et al 2001 Hum Reprod 16:1854.

Scheie Syndrome: ►Hurler syndrome

Schiff Base: α-amino groups of amino acids may react reversibly with aldehydes and form a Schiff base; these are labile intermediates in amino acid reactions. ►Schiff reagent

Schiff's Reagent: Retains a blue color in the presence of aldehydes. Aldehydes are exposed to a fuchsin solution (0.25 g/L H_2O) and decolorized by SO_2. ►aldehyde, ►fuchsin

Schimke Immuno-Osseus Dysplasia (SIOD): An apparently recessive disease displaying spondyloepiphyseal dysplasia, lentigines and progressive immune and other defects. The basic anomaly is due to mutation of chromosomal matrix-associated protein (SMARCAL1, 2q34-q36) controlling chromatin remodeling. The gene contains 17 exons and encodes a 954 amino acid protein. Variant forms exist. ►spondyloepiphyseal dysplasia, ►lentigines, ►chromatin remodeling; Boerkoel CF et al 2002 Nature Genet 30:215.

Schindler Disease: An α-*N*-acetylgalactosaminidase deficiency (human chromosome 22), a lysosomal storage abnormality(?) leading to a neurological disease with onset before age one and progressive deterioration of motor and talking skills. ►Kanzaki disease

Schinzel Syndrome (ulnar-mammary syndrome, 12q24.1): An autosomal dominant ulnar-mammary syndrome shows complex symptoms including malformation of the hand and shoulder, mammary glands, delayed puberty, obesity, etc. The T box gene TBX3 causes it. ►ulna

Schinzel-Geidion Syndrome: An autosomal recessive malformation of the face and head, heart, growth retardation, telangiectasia, supernormal hair development. ►hypertrichosis, ►telangiectasis

Schistosomiasis: The state of disease of animals and humans seized by one or another species of the parasitic flatworms *Schistosoma* (fluke). The parasites infect the blood vessels through contact with contaminated waters in warm climates of the world. In these species, the male carries the female in a ventral sac (gynecophoral canal) (see Fig. S16). Intermediate hosts are snails and molluscs. The *S. haematobium* is primarily a human parasite.

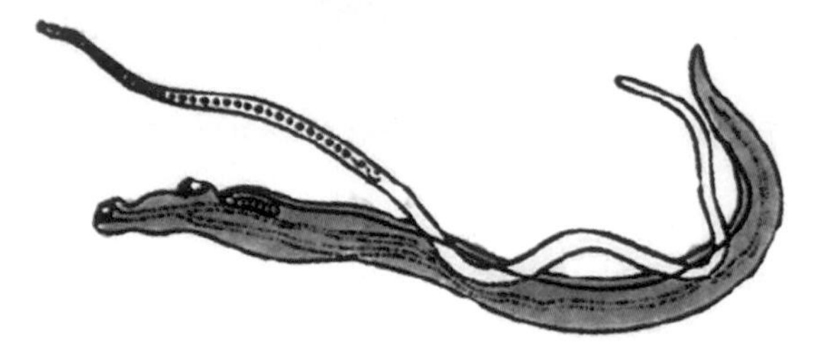

Figure S16. *Schistosoma haematobium* male carrying the female

The *S. japonicum* infects several animals as well. The symptoms of the disease may vary according to the various species and may include systemic irritations, cough, fever, eruptions, tenderness of the liver, diarrhea, and bladder carcinoma, etc. Without medication, the infected intestines, liver, kidneys, brain, etc., may be seriously damaged. The therapeutic antimony derivatives are highly toxic to humans. The parasite is present in millions of people of the tropics including the ancient Egyptian mummies. A codominant locus in human chromosome 5q31-q33 provides some protection against *Schistosoma mansoni*. The interferon-γ receptor (IFN-γR1) locus (6q22-q23) also controls *S. mansoni* infection. Vaccination with attenuated larvae is feasible. ►IFN; http://www.tigr.org/tdb/tgi/; Verjovski-Almeida S et al 2003 Nature Genet 35:148; Hu W et al 2003 Nature Genet 35:139.

Schizencephaly (10q26.1): A very rare brain disorder. The hemisphere is partly missing and cerebrospinal fluid fill the void. Type one is mild but type II is severe involving mental retardation, seizure, lack of speech and blindness. EMX2 linkage and EMX2 independent case are known (Tietjen I et al 2005 Am J Med Genet 135:166.

Schizocarp: A fruit where the carpels split apart in order to free the seeds.

Schizophrenia (dementia praecox): A behavioral disorder. The afflicted individuals cannot distinguish well reality from dreams and imagination. Hallucinations and paranoid behavior, delusions, inappropriate emotional responses, lack of logical thought and concentration are common symptoms. The precise mechanism of inheritance is unknown yet in about 13% of the children of afflicted parents it reoccurs, and ~45% of the identical twins are concordant in this neurological disease but only ~15% of dizygotic twins.

The general incidence in the population varies from 1 to 4%. Both autosomal, pseudoautosomal single and multiple recessive and dominant loci were implicated. There were some indications that schizophrenia genes are in chromosomes 1q32.2 (?), 4q34.3, 5q33.2, 6q21.31, 6q25.2, 6pter-p22, 7p15.2, 7q11, 8p21-p22, 9q21, 10p, 10q22.3, 12q24, 13q34, 14q13, 15q11.2, 17p11-q25.1, 20, 21q21.2, 22q12.1-q11.23 (proline dehydrogenase), 22q21, 4q31 and in the long and short arms of chromosome 5. The 22q11 site encodes catechol-*O*-methyltransferase (COMT) and its deletion involves velocardiofacial syndrome and high incidence (20–30%) of psychiatric diseases, including schizophrenia. COMT controls the metabolism of catecholamine neurotransmitters (Shifman S et al 2002 Am J Hum Genet 71:1296). Some evidence indicates chromosomal sites 13q34 and 8p21-p22 as the most important for the determination of susceptibility. A schizophrenia locus was assigned to human chromosome 1q21-q22 with a lod score of 6.5 (Brzustowicz LM et al 2000 Science 288:678). This gene encodes the selenium-binding protein (SELENBP1) of unknown function and its upregulation is the most consistent biomarker of the disease (Glatt SJ et al 2005 Proc Natl Acad Sci USA 102:15533). DISC1 (disrupted in schizophrenia) is a major factor in developing this affective disorder. DISC is disrupted either by balanced translocation t (1;11)(q42;q14), or t(1;16)(p31.2;q21). The 1p31.2 breakpoint disrupts the B1 isoform of phosphodiesterase 4B gene (PDE4B). The N-terminal domain (amino acids 219-283) of DISC binds PDE4B. PDE4B apparently inactivates adenosine 3′,5′ cyclic monophosphate (cAMP), which has important function in learning, memory and mood. DISC1 interacts with the UCR2 domain of PDE4B and a raise in cAMP disrupts the association of PDE4B and DISC1; concomitantly this increases PDE4B activity (Millar JK et al 2005 Science 310:1187). *Disc1* missense mutations in mice give rise to phenotypes related to depression and schizophrenia, thus supporting the role of *DISC1* in major mental illness and can serve as an animal model for the disease (Clapcote SJ et al 2007 Neuron 54:387).

The various manifestations of schizophrenia have strong environmental components. Maternal malnutrition during the first trimester, maternal influenza during the second trimester, perinatal complications, intrauterine fetal hypoxia, or maternal pre-eclampsia may substantially aggravate the risk (Tsuang T 2000 Biol Psychiatry 47:210). Maternal-fetal incompatibility due to the presence of the Rh D protein in the pregnancy increases the chance for the psychiatric condition by a factor of ~2.6 (Palmer CGS et al 2002 Am J Hum Genet 71:1312). Retroviral sequences have been detected in the cerebrospinal fluid in ~28% of the patients with recent onset and in ~5% of patients with chronic affliction (Karlsson H et al 2001 Proc Natl Acad Sci USA 98:4634). *Oligodendrocyte lineage transcription factor 2* (OLG2) abnormality increases susceptibility to schizophrenia by itself and by affecting other genes namely CNP, NRG1 and ERBB4 (Georgieva L et al 2006 Proc Natl Acad Sci USA 103:12469).

MAO (monoamine oxidase) level is reduced in the afflicted individuals. MAO enzyme removes amino groups from neurotransmitters. The overproduction of dopamine (3,4-dihydroxyphenyl-ethylamine), a precursor of neurotransmitters may be suspected in causing it. Glutamate decarboxylase 67 (GAD67), a marker for this system has been found by many studies to show decreased expression in schizophrenia and bipolar disorders. In the latter, suppression of transcription factors involved in cell differentiation may contribute to GABA dysfunction (Benes FM et al 2007 Proc Natl Acad Sci USA 104:10164). Chloropromazine (a peripheral vasodilator, antiemetic drug) and reserpine (alkaloid) may alleviate the psychological symptoms by inhibiting dopamine receptors. According to the glutamate-dysfunction hypothesis the disease is caused by an imbalance between dopamine and glutamate or the glutamate receptor NMDA. In schizophrenia and other affective disorders frequently an expansion of trinucleotide repeats is detectable. Schizophrenia, other mental illnesses and drug addiction predispose to violent crime 2–6 times in excess of normal individuals (Friedman RA 2006 N Engl J Med 355:2064). ►paranoia, ►psychoses, ►affective disorders, ►neurological disorders, ►pseudoautosome, ►concordant, ►twinning, ►trinucleotide repeats, ►obsessive-compulsive disorder, ►catatonia periodic, ►eclampsia, ►Rh blood factor, ►hypoxia, ►MAO, ►NMDA; Baron M 2001 Am J Hum Genet 68:299; Gurling HMD et al 2001 Am J Hum Genet 68: 661; neuregulin and susceptibility: Stefansson H et al 2002 Am J Hum Genet 71:877; Chumakov I et al 2002 Proc Natl Acad Sci USA 99:13675; Levinson DF et al 2003 Am J Hum Genet 73:17; Lewis CM et al 2003 Am J Hum Genet 73:34; review: O'Donovan MC et al 2003 Hum Mol Genet 12:R125; research forum: http://www.schizophreniaforum.org.

Schizosaccharomyces pombe (fission yeast): The 3 chromosomes are of 5.7, 4.7 and 3.5 Mb size, respectively (see Fig. S17). It has the lowest number of genes among eukaryotes, 4944. The cells are 7 × 3 μm. This eukaryote, an ascomycete, has both asexual (by fission) and sexual life cycles (each meiosis producing 8 ascospores). Under good growing conditions it reproduces by mitotic divisions.

At starvation (for any factors of growth) the plus (P) and minus (M) type cells fuse and meiosis follows. The mitotic cell cycle (2.5 h) has the typical G_1, S, G_2 and M phases. The G_2 phase takes 70% of the total time whereas the other phases equally share the rest. Under severe nutritional limitations instead of sexual development, the cells are blocked in either of the G phases and this dormant state is called "GO".

Figure S17. *S. pombe*

The mating type is determined by which of the *P* or *M* alleles is switched (transposed) from their silent position to the *mat1* locus where they are expressed. Actually both *P* and *M* genes have two alleles with different number of amino acids in their polypeptide products: *Pc* (118), *Pi* (159) and *Mc* (181) and *Mi* (42). The *c* alleles (required for meiosis and conjugation) are transcribed rightward from the centromere when nitrogen is available, and the *i* alleles (required only for meiosis) are transcribed in the opposite direction in N starvation. The product of the *Pi* allele has a protein-binding domain, whereas that of *Mc* shows some homology to the *Drosophila Tdf* (testis-determining factor) and the mouse *Tdy.* Homothallic strains can switch between mating types but the heterothallic ones are either *P* or *M*. The *MAT* site is comparable to the disk drive of a computer (or the slot of a tape player) where either the *P* or the *M* disk (or tape cassette) is plugged in and that determines whether the mating type in the heterothallic strain will be *P* or *M.* The *P* (1113 bp) and *M* (1127 bp) sites are actually the storage sites for the *P* and *M* mating type information, respectively. The recombination-promoting complex contains proteins Swi2 and Swi5 that are located near the silent mating type region and exercise long-range effects (Jia S et al 2004 Cell 119:469).

The mating type region in the right arm of chromosome II can be represented as shown in Figure S18.

The *DSB* (double-strand break) near the *MAT* site—in about 25% of the cells—is probably required so that the chromosome would permit the insertion of one or the other cassette. This breakage is probably transient and quickly restored so that the continuity of the chromosome would not be compromised. According to new data the break at the *mat1* site is actually an artifact arising during purification of alkali-labile DNA due to a genetic imprint occurring while the lagging strand is synthesized. The imprints are one or two RNA nucleotides in the DNA (Vengrova S, Dalgaard JZ 2004 Genes Dev 18:794). This imprint may then reverse the *mat1* locus or introduce an origin of replication. The ≈15 kb *L* and *K* sequences are spacers where meiotic recombination is not observed. The H_1 (59 bp) and H_2 (135 bp) homology boxes are flanking the disk drive (*MAT*) and both floppy disks while the H_3 (57 bp) only occurs at left of the *P* and *M* sequences. It has been supposed that the reason why the *P* and *M* elements are silent at the storage sites is because of the H_3 presence there but not at the *MAT* site where they are expressed. The switching (transposition) is controlled by *SAS1* and *SAS2* (switch-activating sites) right of *DSB* (within 200 kb). In addition, at least 11 other transacting (*swi*) loci regulate switching (transposition). Mating type determination in the budding yeast (*Saccharomyces cerevisiae*) is also controlled by transposition, albeit in a different way, and the homology of the DNA sequences in the elements is low. Fission yeast has contributed to learning many aspects of the cell cycle control. Its genome sequence (Wood V et al 2002 Nature [Lond] 415:871) revealed 4824 protein-coding genes, so far the smallest number in eukaryotes. Its promoter sequences are longer than those in budding yeast indicating extended control of functions. About 43% of the genes contain introns. About 50 genes seem homologous to some extent to human genes controlling disease and half of these are cancer-related. ►*Saccharomyces cerevisiae*, ►mating type determination, ►SWI, ►cell cycle, ►imprinting; Vengrova S et al 2002 Int J Biochem Cell Biol 34:1031; http://www.sanger.ac.uk/Projects/S_pombe/; http://pingu.salk.edu/~forsburg/pombeweb.hml.

Schneckenbecken Chondrodysplasia: Autosomal recessive, lethal defect of the cartilage. It bears similarity to thanatophoric dysplasia.

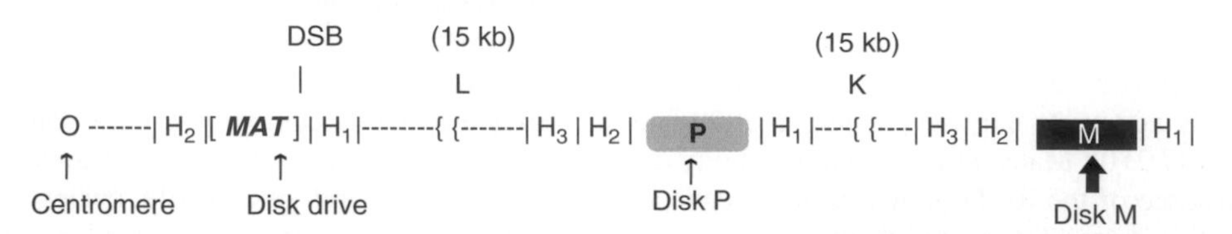

Figure S18. Mating type controls in *Schizosaccharomyces*

Schnipsel Database: Facilitates the identification of protein relationships by using Schnipsels (fragments) using SMART. ►SMART

Schwachman-Diamond Syndrome (human chromosome 7q11): Involves recessive pancreatic lipomatosis, bone marrow dysfunction and skeletal anomalies. (See Goobie S et al 2001 Am J Hum Genet 68:1048; Boocock GRB et al 2003 Nature Genet 33:97).

Schwann Cell: A glial cell (forms myelin sheath for the peripheral nerves). Its differentiation requires the Oct-6 POU factor, Ca^{2+}/calmodulin kinase, MAPK, cAMP response element-binding protein, expression of *c-fos* and *Krox24* genes. ATP arrests the differentiation of the Schwann cells before myelination. ►myelin, ►Oct-1, ►POU, ►calmodulin, ►MAPK, ►cAMP, ►fos, ►Krox-24

Schwannoma: ►neurofibromatosis

Schwartz-Jampel Syndrome (chondrodystrophic myotonia, 1p34-p36.1): A rare recessive failure to relax muscles, reduced height, skeletal dysplasia (abnormal development). The affection is due to an altered proteoglycan (perlecan) of the basement membranes. ►basement membrane; Arikawa-Hirasawa E et al 2002 Am J Hum Genet 70:1368.

Scianna Blood Group (Sc): Represented by antigenic groups Sc-1 and Sc-2, located in human chromosome 1. ►blood groups

Sciara: Dipteran flies with polytenic, giant chromosomes in their salivary glands. The basic chromosome number in *Sciara coprophylla* is 3 autosomes and 1 X chromosome but there are also the heteropycnotic so called *limited chromosome*(s), which are present only in the germ-line. They are eliminated from the nuclei during the early cleavage divisions. The egg pronucleus contains three autosomes, an X chromosome and one or more limited chromosomes. The sperm contributes three autosomes, two X chromosomes and some limited chromosomes that are all of maternal origin. The first division of the spermatocyte is monocentric and separates the maternal chromosome set from the paternal one. The maternal chromosomes move to a single pole whereas the paternal set is positioned away from the pole and are never transmitted to the progeny. The single secondary spermatocyte displays an unusual unequal type division. The X-chromosome divides longitudinally and both copies are included into the same cell, and only this cell survives. From the cleavage nuclei first the limited chromosomes are eliminated and subsequently, from the cells that become male one of the X-chromosomes is also evicted thus the males become XO and the females are XX. ►*Rhynchosciara*, ►polytenic chromosomes, ►salivary gland chromosomes, ►sex determination, ►chromosomal sex determination

SCID (severe combined immunodeficiency): A heterogeneous group's genetically determined diseases. It involves defective V(J)D recombination of immunoglobulin genes. Several frameshifts, point mutations or deletions in the IL-2Rγ (γc) chain (human chromosome Xq13) were found to be associated with SCID-X1. Actually, this part of the γ-chain is shared by the receptors of IL-4, IL-7, IL-9 and IL-15. Because of the deficiency, the T cells and the natural killer cells also become abnormal. The anomaly may involve also the Jak/STAT and other signaling pathways. Jak3 defects cause SCID symptoms. Introduction of genetically engineered macrophages expressing IFN-γ into the lung of mice provided marked protection against infection. IFN-γ upregulates the expression of major histocompatibility class I and class II molecules that lead to the activation of killer lymphocytes. Apparently after successful gene therapy, using IL2RG interleukin-2 receptor component in genetically engineered vector, caused lymphoma in several cases. Ora1 protein (encoded at 12q24) controls CRAC channel function, i.e., Ca^{2+}-release-activated Ca^{2+} influx. Sustained calcium supply is essential for the adaptive immune response. Some SCID patients have defect in CRAC and consequently loss of immune protection (Feske S et al 2006 Nature [Lond] 441:179). In a mouse model with ablated *Arf* tumor suppressor and IL-2Rγ high frequency of insertional mutations occurred near or within several protooncogenes; thus this strain could be used as a model for testing potential risks of human gene therapy for XSCID constructs (Shou Y et al 2006 Proc Natl Acad Sci USA 103:11730). ►adenosine deaminase deficiency, ►severe combined immunodeficiency, ►gene therapy, ►signal transduction, ►immunoglobulins, ►RAG, ►DNA-PK, ►T cell, ►killer cell, ►MHC, ►Jak/STAT, ►IL-2, ►IL-4, ►IL-7, ►IL-9, ►IL-15, ►gene therapy, ►calcium signaling; Wu M et al 2001 Proc Natl Acad Sci USA 98:14589; French Gene Therapy Group 2003 J Gene Med 5:82.

Science: A systematic study of natural phenomena with the explicit purpose to prove or negate a working hypothesis or hypotheses by experimental means. According to K.R. Popper (1962 Conjectures and Refutations, Basic Books, New York, p 218), for science the "criterion of potential satisfactoriness is thus testability, or improbability: only a highly testable or improbable theory is worth testing and is actually (and not merely potentially) satisfactory if it withstands severe tests—especially those tests to which we could point as crucial for the theory before they were ever undertaken. I refuse to accept the view that there are statements in science which we have, resignedly, accept as true merely because it does not seem possible…to test them". Science seeks to store, classify and evaluate certainties. Research is exploring uncertainties to provide facts for science. Applied science seeks to find economic use of the principles discovered by basic or pure science. Gathering information is one of the tools of science. The information becomes science when it can be integrated into a proven theoretical framework. The framework may need modification as information accrues. Science should never be subjected to ideology, politics and unproved or improvable ideas. Such influences are disastrous for science as the famous Galileo trial had shown in 1663 when the great scientist was forced to recant his valid thesis that the Earth is not the center of the Universe: "I abjure, curse, and detest the aforesaid errors and heresies, and generally every other error". Using big words and overcomplicated sentences are against the aim of sciences, which calls for lucidity. Coining new terms unless they facilitate communication must be avoided. It is important to explore the exact contents and meanings of terms before application in communication to be sure they are not preempted. ►experiment, ►genetics, ►misconduct scientific, ►lysenkoism, ►creationism; theory of scientific discovery: Koshland DE Jr 2007 Science 317:761.

Scientific Freedom: The right to pursue researches and publish the outcome. This right is limited, however, by the need to regulate activities that are directed against the well-being of humans, human economic activities and specifically the freedom to create biological and chemical terror weapons and publishing papers that may jeopardize public safety. Serious concerns arise about how to implement the regulation without fettering scientific inquiry and safeguarding society. ►bioterrorism, ►informed consent, ►bioethics; Fraser CM, Dando MR 2001 Nature Genet 29:253.

S

Scientific Misconduct: According to the NSF (National Science Foundation, USA) "fabrication, falsification, plagiarism, or other serious deviation from accepted practices." ►publication ethics, ►ethics, ►misconduct scientific; Federal Register 18, March, 2002.

Scintigraphy: The photographic location of radionuclides within the body after introduction of radioactive tracers. ►radioactive label, ►radioactive tracer

Scintillation Counters: Can be liquid scintillation counters or crystal scintillation-counters for solids. The counter is an electronic appliance where the sample is placed in a solution of organic compounds (cocktail). The radiation coming from the isotopes (even from the weak β–emitters) causes flashes in the fluorescing cocktail that are directed to a photoelectric cell. The cell then releases electrons that are amplified and registered (counted). Each flash corresponds to a disintegration of an atom of the isotope so the equipment displays (or prints out) the disintegrations per minute (dpm) or counts per minute (cpm), generally with background radiation subtracted. This information provides measures of the quantity of the label (or labeled compound). In the crystal scintillation counter, the radiation (usually energetic γ-rays, X-rays or β-rays) emitted by the isotope hits a crystal of sodium iodide containing traces of thallium iodide, and again the disintegrations are registered similarly to the liquid scintillation counter. ►radioisotopic tracers, ►radiolabeling, ►isotopes, ►dpm, ►radiation hazard assessment

Scission: Cuts in both strands of a DNA molecule at the same place. ►nick

Scissors Grip: ►Max

Scissors, Molecular: ►Cre/loxP, ►excision vector, ►targeting genes

Scj1: A 40 kDa budding yeast chaperone in the endoplasmic reticulum, inducible by tunicamycin antibiotic but not by heat. ►chaperones; Nishikawa S, Endo T 1997 J Biol Chem 272:12889.

SCK: A protein with phosphotyrosine binding domain but different from the SH2 domain of SRC. ►SRC, ►SH2, ►PTB; Kojima T et al 2001 Biochem Biophys Res Commun 284:1039.

SCLC: ►small cell lung carcinoma

Sclerenchyma: Plant tissues with tough, hard cell walls.

Scleroderma: A probably autosomal dominant disease involving scaly hardening of the skin and increased frequency of chromosome breakage. ►skin diseases; Whitfield ML et al 2003 Proc Natl Acad Sci USA 100:12319.

Scleroids: Cells with unusually hardened walls.

Sclerosis: The hardening caused by inflammation or hyperplasia of the connective tissue.

Sclerosteosis (SOST/BEER, 17q12-q21): A rare recessive dysplasia of the bones displaying progressive sometimes-massive overgrowth distinct from osteopetrosis. It may distort the face and may be accompanied by syndactyly. Its highly conserved glycoprotein product, sclerostin appear to be an antagonist of the bone morphogenetic protein and other members of the TGFβ family. It contains cystine-rich "knots." ►BMP, ►syndactyly, ►TGF, ►van Buchem disease; Brunkow ME 2001 Am J Hum Genet 68:577; Balemans W et al 2001 Hum Mol Genet10:537.

Sclerotia: ►hypha, ►ergot

Sclerotome: Mesenchymal embryonic precursor of the vertebral column and ribs.

SCMRE: A cis-acting element in the c-fos protooncogene and it is responsible for induction by some mitogens. ►FOS, ►protooncogene, ►cis, ►mitogen

SCN: Sodium ion channel.

scnDNA: A single-copy nuclear DNA.

Scoliosis: In this condition, the spine is not straight but laterally curved (see Fig. S19); it may be due to polygenic causes or it can be part of skeletal syndromes. Its incidence is about 1–6% of the adult human populations. Kyphoscoliosis is a similar condition in mice. ►horizontal gaze palsy, ►lordosis, ►kyphosis; Blanco G et al 2001 Hum Mol Genet 10:9.

Figure S19. Scoliotic pig

S-Cone Syndrome (ESCS, encoded by PNR/NR2E3 gene in human chromosome15q23): A night blindness and blue light hypersensitivity because of defect in the retinal cones in the eye. ►eye diseases

SCOP: The structural classification of proteins. (See Murzin AG et al 1995 J Mol Biol 247:536; Przytycka T et al 1999 Nature Struct Biol 6:672; http://scop.mrc-lmb.cam.ac.uk/scop.

Scopes Trial: In 1925, John Scopes was convicted for breaking Tennessee state law (repealed in 1967) by teaching evolution in a public school. The verdict was later overturned on a technicality. ►intelligent design, ►creationism

Scopolamine: ►*Datura*, ►alkaloids

Score Test (Wald-Wolfowitz test): A non-parametric test to compare unmatched samples and all information is paired with a numerical score. It is assumed that the scores represent a continuous distribution. The null hypothesis is that the runs represent identical populations. (See Hays WL, Winkler RL 1971 Statistics: Probability, Inference and Decision. Holt. Rinehart and Winston, New York).

Scotomorphogenesis: The differentiation of plants in the absence of light, e.g., etiolated growth. ►brassinosteroids

SCP$_2$: A sterol carrier protein 2; probably the same as nsl-TP. ►sterol

Scrapie: ►encephalopathies, ►Creutzfeldt-Jakob disease, ►prion, ►kuru

Scratchy: An in vitro method of combinatorial protein engineering independent of sequence identity. It involves incremental truncation of protein coding genes and then reshuffling and DNA fusion. The purpose is to improve protein function. ►iterative truncation; Lutz S et al 2001 Proc Natl Acad Sci USA 98:11248.

Screening: Selective classification of cell cultures for mutation or for special genes, conveying auxotrophy, antibiotic or other resistance, selecting antibodies by cognate antigens, plant populations for disease- or chemical-resistance, animal progenies for blood groups, etc. Narrow sense screening may not involve selective isolation but special criteria are used to distinguish in the growing population those individuals who have a special trait. Genetic screening in animals is feasible by the use of RNAi. The RNA is introduced into pronuclear cells by injection and it is transmitted for several generations. The RNAi can knock down important genes and thus their function is revealed (Peng S et al 2006 Proc Natl Acad Sci USA 103:2252). ►genetic screening, ►genetic testing, ►knock-down, ►RNAi; Forsburg SL 2001 Nature Rev Genet 2:659; Jorgensen EM, Mango SE 2002 Nature Rev Genet 3:356.

Screwworm: ►*Cochliomya hominivorax*, ►genetic sterilization, ►myasis

Scribble: ►PDZ

Scripton (transcripton): A unit of lambda phage transcription. ►lambda phage

scRNA: Small cytoplasmic RNAs. ►snRNA

scRNAP: Small cytoplasmic ribonucleoprotein.

Scrotum: The pouch containing the testes and accessory sex organs of mammals.

Scrunching (according to the Oxford Dictionary squeezing into a compact shape): At abortive or productive initiation of transcription RNA polymerase (RNAP) "scrunches" the DNA at a stationary stage to unwind DNA and pulls it into itself. The process requires RNA synthesis. Promoter escape also involves scrunching. (See Revyakin A et al 2006 Science 314:1139; Kapanidis AN et al 2006 Science 314:1144).

Scutellum: The single cotyledon of the grass embryo (see Fig. S20).

Figure S20. S = scutellum, E = embryo

SD: ►*Segregation distorter*, ►standard deviation

Sdc25: A guanine-nucleotide release factor, similar to Cdc25. ►RAS, ►Cdc25, ►EF-Tu

SDF-1 (stromal cell-derived factor): A chemokine and natural ligand of fusin. With its receptor CXCR-4 it mobilizes and promotes the proliferation of $CD34^+$ cells. ►CD34, ►fusin [LESTR], ►chemokines, ►CXCR; Weber KS et al 2001 Mol Biol Cell 12:3074.

S-DNA: A mechanically stretched DNA that might have been extended 1.7 times of its normal contour length. ►slipped-structure DNA, ►trinucleotide repeats; Cluzel P et al 1996 Science 271:792.

SDP: ►short-day plants

SDR (short dispersed repeats): Organellar DNA sequences of 50–1000 bp that may occur in direct or inverted forms and may represent more than 20% of the chloroplast genomes of *Chlamydomonas reinhardtii alga* and various land plants. Similar or shorter redundant sequences occur in the mitochondria of plants, animals and fungi. In *Saccharomyces cerevisiae* eight 200–300 bp *ori* and *rep* sequences and 200 G + C sequences of 20–50 bp may be clustered into several mtDNA gene families. Similar mobile G + C elements may occur in other fungi. SDRs have been exploited for forensic population studies. ►chloroplasts, ►mtDNA, ►mobile genetic elements, ►organelle sequence transfers; Seidl C et al 1999 Int J Legal Med 112(6):355.

SDR: ►strain distribution pattern

SDS (sodium dodecyl sulfate): A detergent, used for electrophoretic separation of protein and lipids.

Sds: A leucine-rich protein and regulates protein phosphatase 1C during mitosis of yeast. ►protein phosphatases; Peggie MW et al 2002 J Cell Sci 115:195.

SDS-PAGE: ►SDS-polyacrylamide gel

SDS-Polyacrylamide Gels: Electrophoretic gel containing sodium dodecyl sulfate (SDS, also called sodium lauryl sulfate [SLS], detergents) and polyacrylamide. This medium dissociates proteins into subunits and reduces aggregation. Generally, the proteins are denatured with heat and a reducing agent before loading on the gel. The polypeptides become negatively charged by binding to SDS and are separated in the gel according to size (rather than by charge). On the basis of the mobility, the molecular weight of the subunits can be estimated with the aid of appropriate molecular size markers (ladder) but caution is required because glycosylated proteins may not reflect the molecular mass of the protein. The concentration of the polyacrylamide determines the size of the polypeptides that can be separated. Polyacrylamides (bisacrylamide: acrylamide, 1:29) separates [kDa proteins] as follows: 15% [12–43], 10% [16–68], 7.5% [36–94], 5% [57–212]. ►gel electrophoresis, ►electrophoresis

SDT Test: A non-parametric sign test used in pedigrees to compare the average number of candidate alleles between affected and non-affected siblings. ►TDT, ►sib TDT; Rieger RH et al 2001 Genet Epidemiol 20(2):175; http://www.sdtcorp.com/companyprofile.html.

SE: ►standard error

Sea Lion (*Eumetopias jubatus*, *Zalophus californianus*): Other species occur also in Australia and New Zealand and other parts of the coastal oceans of the world. The chromosome numbers are 2n = 32 to 36 in the different species. Some of the species form hybrids (see Fig. S21).

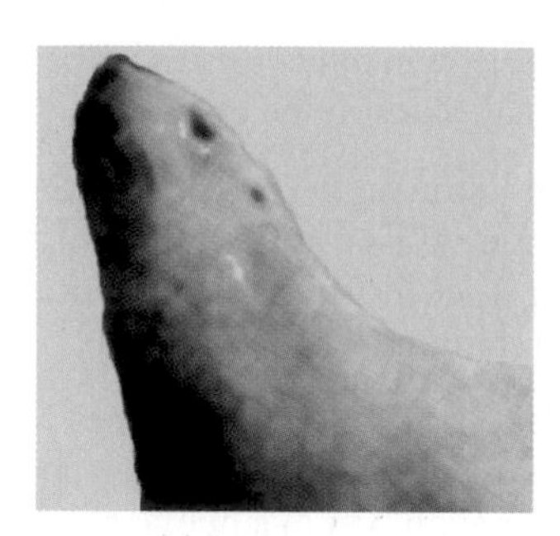

Figure S21. Sea lion

Sea Oncogene: An avian erythroblastosis virus oncogene. The human homolog is at human chromosome 11q13 in very close vicinity to INT2 and BCL1. ►oncogenes

Sea Urchins: *Strongylocentrotus purpuratus* (see Fig. S22) and *Toxopneustus lividus*, both 2n = 36, and other echinodermata have been favorable objects of cell cycle studies, fertilization, embryogenesis, development and evolution. They have very long life (>100 years) and produce annually millions of gametes. Their large-size eggs can be easily collected and handled in the laboratory. Their non-adaptive immune system includes a highly complex system of receptors. The 814 megabase includes ~23,300 genes. The average transcript length is 8.9 kb and the average gene length is 7.7 kb. The average exon were about 100–115 nucleotides whereas the introns ~750 nucleotides. Most of the genes are in regions of 35 to 39% GC regions. Cytogenetic maps are not available. This outbreeding species is highly heterozygous and individuals display ~5% single nucleotide polymorphism (SNP), about ten times higher than that in humans. The sequenced *S. purpuratus* genome has orthologs of vertebrate genes for vision, hearing, chemosensation and several human diseases. genome: 2006 Science 314:941; immune system: Rast JP et al 2006 Science 314:952; transcriptome: Samanta MP et al 2006 Science 314:960; for genome: http://www.ncbi.nlm.nih.gov/genome/guide/sea_urchin/; for embryology: http://www.stanford.edu/group/Urchin/; proteins: http://www.expasy.org/cgi-bin/get-entries?OC=Strongylocentrotus.

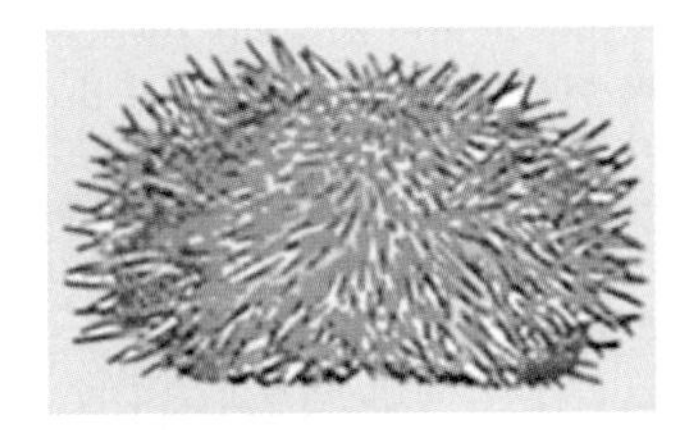

Figure S22. *Strongylocentrotus*

Seal. *Callorhinus ursinus*, 2n = 36; *Zalophus californianus*, 2n = 36; *Crystophora crystata*, 2n = 34; *Erignatus barbatus* 2n = 34; *Helichoereus grypus* 2n = 32.

Search Engine: Uses the Internet as a source for finding keywords and documents. Frequently used search engines are Google, Yahoo, AltaVista and others.

Sebastian Syndrome: ►May-Hegglin anomaly

Sec6/8: A multiprotein complex that mediates cell-to-cell contacts and transport vesicle delivery. ►Golgi apparatus; Matern HT et al 2001 Proc Natl Acad Sci USA 98:9648.

Sec63 (NPL1/PTL1): A 663-amino acid yeast trans-membrane chaperone protein with partial homology at the near-N-end of DnaJ. It interacts with Kar2, Ces61, Sec71, and Sec72 proteins. ►chaperones, ►DnaK, ►Kar2, ►Sec61 complex, ►Mtj1; Young BP et al 2001 EMBO J 20:262.

Sec61 Complex (Sec Y complex): Built of the three Sec subunits (α, β, γ) and other proteins form a protein-conducting channel across the endoplasmic reticulum (ER) membrane. It associates with the large subunit of the ribosome of prokaryotes and eukaryotes and transports some of the nascent proteins into the lumen of the ER. A small molecule, cotransin (see Fig. S23), inhibits protein translocation into ER in a discriminatory manner of signal sequence (Garrison JL et al 2005 Nature [Lond] 436:285). ►Sec proteins, ►protein-conducting channel, ►ribosomes, ►endoplasmic reticulum, ►protein synthesis, ►Sec63, ►endoplasmic reticulum-associated degradation, ►unfolded protein response, ►CAM, ►chloroplast import; Mori H, Ito K 2001 Proc Natl Acad Sci USA 98:5128; Beckmann R et al 2001 Cell 107:361; van den Berg B et al 2004 Nature [Lond] 427:36.

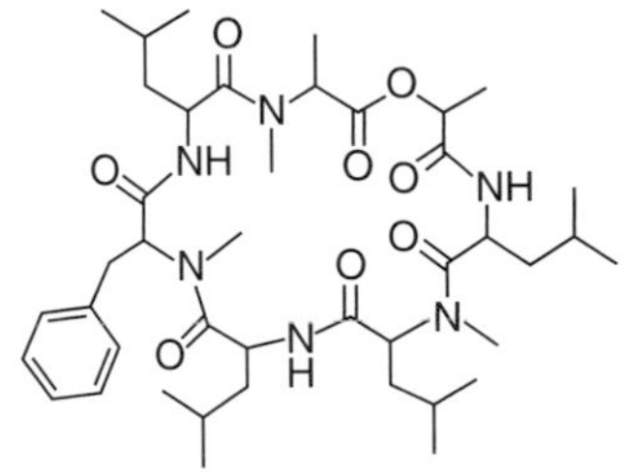

Figure S23. Cotransin

SecA (α) **Protein**: (seven-component complex): A peripheral membrane domain of the translocase enzyme and it is the primary receptor for the SecB/pre-protein complex by recognizing the leader domain of the pre-protein. Hydrolyzes ATP, GTP, promotes cycles of translocations and pre-protein release. ►membranes, ►translocase, ►Mss, ►Ypt, ►Rab, ►translocase, ►translocon, ►protein targeting, ►SRP, ►ARF, ►protein synthesis, ►endoplasmic reticulum, ►exocytosis; Hsu SC et al 1999 Trends Cell Biol 9:150; Jilaveanu LB 2005 Proc Natl Acad Sci USA 102:7511.

SecB (β) Protein: A 17-subunit chaperone involved in translocation of pre-proteins by com-plexing and keeping them in the right conformation and binding to membrane surface of the endoplasmic reticulum. Recognizes both the leader and mature protein domains. ►chaperones, ►membranes, ►endoplasmic reticulum, ►SRP, ►translocon, ►translocase, ►protein targeting; Driessen AJ 2001 Trends Microbiol 9:193.

Secis: The selenocysteine regulatory element in coding and non-coding regions of selenocystein genes in prokaryotes and eukaryotes (Mix H et al 2007 Nucleic Acids Res 35:414).

Seckel's Dwarfism (bird-headed dwarfism, 3q22.1-q24): A microcephalic autosomal recessive condition with reduced intelligence. The basic defect is in a phosphatidylinositol 3-kinase-like kinase. ►dwarfism, ►microcephaly, ►phosphatidylinositol; O'Driscoll M et al 2003 Nature Genet 33:497.

Second Cycle Mutation: Caused by the excision or movement of a transposable element and leave behind a footprint which still causes some type of alteration in the expression of the gene. This type of alterations may be connected with the defective nuclear localization of a transcription factor. ►transposon footprint, ►transposable elements, ►nuclear localization sequences

Second Division Segregation: ►tetrad analysis, ►postreduction

Second-Harmonic Imaging Microscopy: Visualizes biomolecular arrays in cells, tissues and organisms. ►microscopy; Campagnola PJ, Loew LM 2003 Nature Biotechnol 21:1356.

Second-Male Sperm Preference: ►last-male sperm preference

Second Messenger: Molecules with key roles in signal transduction pathways such as cyclic-AMP, cyclic-GMP, and others. Animal physiologists call hormones first messengers. In animal and plant cells Ca^{2+} is also considered to play the role of second messenger. Inositol triphosphate is also a second messenger. ►cAMP, ►cGMP, ►mRNA, ►PIK, ►PIP, ►signal transduction, ►G proteins

Second Site Non-Complementation: ►non-allelic non-complementation

Second Site Reversion: A suppressor mutation at a site different from that of the original lesion but is capable of restoring the normal reading of the mRNA. ►suppressor tRNA, ►suppressor gene, ►reversion, ►compensatory mutation

Second Strand Synthesis: ►reverse transcriptase

Secondary Constriction: ►nucleolar organizer, ►satellited chromosome, ►SAT

Secondary Immune Response: An immune reaction conditioned by the memory cells when antigenic exposure occurs repeatedly. ►immunological memory

Secondary Lymphoid Tissues: The lymph nodes and spleen (in contrast to primary lymphoid tissues, the bone marrow and thymus). ►immune system

Secondary Metabolism: Produces molecules that are not basic essentials for the cells and their products occur only in specialized tissues, e.g., anthocyanin, hair pigments. Many of the secondary plant metabolites, e.g., phytoalexins, phenolics, flavone, pterocarpan, chlorogenic acid, sesquiterpenes, diterpene, saponins, furanoacetylene, alkaloids, indole-derivatives, etc., are defense molecules against microbial pathogens. Jasmonate-mediated genetic reprogramming of the transcriptome may reveal some basic aspects of the complex processes (Goossens A et al 2003 Proc Natl Acad Sci USA 100:8595). (See Dixon RA 2001 Nature [Lond] 411:843).

Secondary Nondisjunction: ►nondisjunction

Secondary Response Genes: Their transcription is preceded by protein synthesis; probably primary response genes are involved in their induction. They are stimulated by mitogens alone without cycloheximide. ►mitogens, ►primary response genes, ►signal transduction

Secondary Sex Ratio: The proportion of males to females at birth. ►age of parents, ►sex ratio, ►primary sex ratio

Secondary Sexual Character: Usually accompany the primary sexual characters but they are not integral part of the sexual mechanisms, e.g., facial hair in human males, red plumage of the male cardinal birds, increased size bosoms in females and higher pitch voice, etc. ►primary sexual characters, ►accessory sexual characters

Secondary Structure: The steric relations of residues that are next to each other in a linear sequence within a polymer such as α−helix, a pleated β-sheet of amino acids. ►protein structure; http://cmgm.stanford.edu/WWW/www_predict.html.

Secondary Trisomic (isotrisomic): The third chromosome has two identical arms, originated by misdivision of the centromere or by the fusion of identical telochromosomes. ►trisomics, ►misdivision, ►telosome

Secondary-Ion Mass Spectrometry: A focused ion beam removes neutral and ionized atoms and molecules from a solid surface. These secondary ions are then accelerated and separated according to mass-to-charge ratio in a mass spectrometer. ►mass spectrum

Secretagogue: A compound or factor that stimulates secretion. ►ghrelin; Pombo M et al 2001 Horm Res 55 (Suppl. 1):11.

Secretases: α-secretase enzyme cleavage of APP (amyloid precursor protein) interferes with the production of α amyloids whereas cleavage by β- and γ-secretase contributes to the formation of amyloid plaques. γ-Secretase (transmembrane aspartyl protease) activity requires the presence of the protein nicastrin, which binds presenilin near the active site, and there is an interaction among these and the fragment generated by β-secretase. The γ-secretase complex has four components: presenilin, nicastrin, aph-1 and pen-2 (Kimberly WT, Wolfe MS 2003 J Neurosci Res 74:353). The cytoplasmic tail of APP, released intracellularly by secretase γ, teams up with histone acetyl transferase and other proteins and may promote gene expression. Cleavage of APP by β-secretase at the N-end releases APPsβ (~100 kDa, N-terminal fragment) and a membrane-bound 12 kDa C-end fragment (C99). Cleavage by α-secretase generates the N-end APPsα and a membrane-bound 10 kDa piece (C83). C99 and C83 can be further split by secretases and yield 4 kDa Aβ (Alzheimer plaque material) and the harmless 3 kDa p3 peptides, respectively. γ-Secretase generates the 40-residue and in smaller proportion, 42-residue Aβ fragments. The latter forms the tangled brain fibers and the 40-residues accumulate as brain plaques. β-Secretase cuts at Asp1, Val3 and Glu11 and most commonly, Aβ begins with Asp. Met→Leu replacements favor the generation of amyloid plaques common in the early onset Alzheimer disease. Pin1 prolyl isomerase binds the phosphorylated Thr 668-Pro motif in APP and increases its polymerization by three orders of magnitude. Overexpression of Pin1 reduces Aβ secretion but knockout of Pin1 increases its secretion. Pin1 knockout increases amyloidogenic APP processing and selectively elevates the insoluble and toxic Aβ42 (Pastorino L et al 2006 Nature [Lond] 440:528). The BACE transmembrane aspartic protease is a β-secretase. Secretase γ has a role also in Notch-controled developmental processes. ►Alzheimer disease, ►β-amyloid, ►BACE, ►TACE, ►ADAM, ►presenilins, ►memapsin, ►nicastrin, ►rhomboid protease, ►CD147, ►Notch; Kopan R, Goate A 2000 Genes Dev 14:2799; Cao X, Südhof TC 2001 Science 293:115; Esler WP, Wolfe MS 2001 Science 293:1449; Fortini ME 2002 Nature Rev Mol Cell Biol 3:673; Takasugi N et al 2003 Nature [Lond] 422:438.

Secretion Machine (injectisome): The bacterial pathogens of both animals and plants secrete about 20 different proteins mediating infection of the host. Of these, nine are conserved across phylogenetic boundaries and appear to have a universal mRNA targeting signal. ►host-pathogen relation; Lee VT, Schneewind O 1999 Immunol Rev 168:241; injectisome structure: Merlovits TC et al 2006 Nature [Lond] 441:637.

Secretion Systems: Type III (~20 proteins) is a contact-dependent mechanism of transfer of bacterial pathogenicity island genes and toxins to other organisms. Type III secretion system (T3SS) has a complex needle structure (Galán JE, Wolf-Watz H 2006 Nature [Lond] 444:567). Bacterial plasmids may encode the pathogenicity island. Type IV/V: the autotransporters are the bacterial conjugation and the agrobacterial T-DNA transfer systems. Type II (12–14 proteins) is the general secretion system. The Type I system requires three proteins and it is encoded in pathogenicity islands or plasmids. ►pathogenicity island, ►T-DNA, ►conjugation bacterial; Galán JE, Collmer A 1999 Science 284:1322; Burns DL 1999 Curr Opin Microbiol 2:25; Cornelis GR, Van Gijsegem F 2000 Annu Rev Microbiol 54:735; Fadouloglou VE et al 2004 Proc Natl Acad Sci USA 101:70; secretion type III system assembly: Yip CK et al 2005 Nature [Lond] 435:702; secreted protein database: http://spd.cbi.pku.edu.cn.

Secretion Trap Vector: Inserts a reporter gene, which is expressed in a transmembrane region. The constructs may carry a region of the CD4 gene fused in-frame to the 5′-end of the reporter, e.g., *LacZ*, which can be than identified by Xgal at the membrane. ►CD4, ►Xgal; Shirozu M et al 1996 Genomics 37(3):273.

Secretion Vector: Besides an expressed structural gene, it carries a secretion signal to direct the gene product to the appropriate site. (See Bolognani F et al 2001 Eur J Endocrinol 145:497).

Secretome: Type III secretion system of bacteria serve as effectors for causing disease. ►secretion systems, ►effector; Guttman DS et al 2002 Science 2002 295:1722.

Secretor: Secretes the antigens of the *ABH* blood group into the saliva. Fucosyltransferase, FUT is the same gene locus as SEC/SE but there is a difference in tissue-specific expression. ►ABH antigen, ►fucosyltransferase; D'Adamo PJ, Kelly GS 2001 Altern Med Rev 6(4):390.

Secretory Immunoglobulin A: An IgA dimer with a secretory component. ►immunoglobulins

Secretory Proteins: Mainly glycoproteins that are released by the cell after synthesis, such as hormones, antibodies and some enzymes. Secretory proteins (~1400) are located in the endoplasmic reticulum (~64%) and in the Golgi-derived COP transport vesicles (~14%); about 29% occur in both (Gilchrist A et al 2006 Cell 127:12165).

Secretory Vesicle (secretory granule): Releases stored molecules such as hormones within the cell. Chromogranin A appears to control the biogenesis of the granules. ►Golgi apparatus, ►endocytosis; Kim T et al 2001 Cell 106:499.

Sectioning: A generally required procedure for the preparation of biological specimens for histological examination. The material may need embedding before they are cut either free hand or by microtomes. Some microtomes cut tissues frozen by CO_2. The 1–20 μm thin sections are subsequently placed on microscope slides and subjected to a series of manipulations (paraffin wax removal, dehydration, staining) before examination. ►microtome, ►embedding

Sectored-Spore Colonies: The colonies arise when the haploid spores carried heteroduplex DNA with different alleles in the heteroduplex region (see Fig. S24). ►heteroduplex

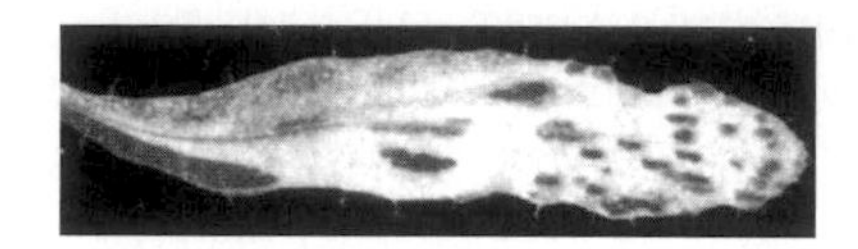

Figure S24. Sectorial plant leaf

Sectorial: Displays sectors, e.g., mitotic recombinant, sorting-out of organelles, somatic mutation, etc. ►mosaic, ►chimeric

Securins: They control the onset of the separation of sister chromatids during mitosis. The anaphase-promoting complex (APC) destroys securin before the sister chromatids can separate. Deletion of securin leads to chromosomal instability as it is common in cancer cells. ►sister chromatid cohesion, ►APC, ►cohesin, ►separin, ►meiosis I, ►CDC2, ►spindle; Sjögren C, Nasmyth K 2001 Curr Biol 11:991; Jallepalli PV et al 2001 Cell 105:445; Zhou Y et al 2003 J Biol Chem 278:462.

SecY Protein: An integral membrane component of bacteria involved in chaperoning the assembly of membrane and some soluble proteins. ►membrane; Veenendaal AK et al 2001 J Biol Chem 276:32559; Osborne AP et al 2007 Cell 129:97.

SecY/E Protein: The membrane embedded domain of translocase enzyme consisting of SecY and SecE polypeptides. Stabilizes and activates SecA and facilitates membrane binding. ►membrane, ►SecY

SED: Spondyloepiphyseal dysplasia, bone diseases.

Sedimentation Analysis: ►satellite DNA

Sedimentation Coefficient: The rate by what a molecule sediments in a solvent. It is characterized by the Svedberg unit (S) that is a constant of 1×10^{-13}. S is derived from the equation $s = (dx/dt)/\omega^2x$, where x = the distance from the axis of rotation in the centrifuge, ω is the angular velocity in seconds ($\omega = \theta/t$, where θ is the angle of rotation and t is time). At a constant temperature (20 °C) in a solvent *s* depends on the weight, shape and hydration of a molecule. The S value is used for the characterization of macromolecules, e.g., RNA such as 16S ($= 16 \times 10^{-13}$).

Seed: ►microRNA

Seed Alignment: Uses only one of each pair of homologs represented by a CLUSTAL W-obtained phylogenetic tree linked by a branch length <0.2. ►CLUSTAL W, ►phylogenetic tree, ►MOST

Seed Coat: ►integument

Seed Development: ►embryogenesis in plants, ►endosperm; seed biology: http://www.seedbiology.de/index.html.

Seed in Genome Sequencing: ►genome projects

Seed Germination: Requires the activation of >2000 genes. The process begins by activation of the root apical meristem, followed by activation of the cotyledons, the shoot apical meristem and secondary meristems. Six D-type and two A-type cyclins have important roles. ►meristem, ►root, ►cyclins, ►abscisic acid; Masubelele NH et al 2005 Proc Natl Acad Sci USA 1002:15694.

Seed Size: In *Arabidopsis*, seed size is controlled by at least 3 genes acting in the same pathway regulated by a kinase (Luo M et al 2005 Proc Natl Acad Sci USA 102:17531).

Seed Storage: Generally viability (germination and survival) of seed can be maintained by storage below −20 °C and low moisture content (<7–8%) well beyond the conditions of normal ambient temperature and humidity. ►freeze drying, ►artificial seed; Buitink J et al 2000 Proc Natl Acad Sci USA 97:2385.

Seeding: To use a set of DNA fragments (in e.g., BACs) as the beginning points for chromosome walking. ►BAC, ►genome projects, ►chromosome walking, ►parking

Seedless Fruits: May be the result of different genetic mechanisms. Aneuploids, triploids, self-incompatibility or gametic sterility genes may be most commonly responsible for this condition. ►seedless watermelon, ►bananas, ►pineapple, ►naval orange, ►stenospermocarpy, ►seedless grapes, ►parthenocarpy

Seedless Grapes: The result of a gene that causes early embryo abortion although fertilization occurs normally (stenospermocarpy). ►seedless fruits

Seedless Watermelons: Triploids (produced by crossing tetraploids with diploids). They are more convenient to eat since seeds do not have to be removed. Their flavor and sweetness may make them superior to conventional varieties. ►seedless fruits, ►watermelon

Segment Polarity of the Body Genes: Segment polarity of the body genes in *Drosophila* mutations are involved in the alteration of the characteristic body pattern and are often accompanied by inverted repetition of the remaining structures. By cell-cell communication, they maintain the pattern imposed on them through subsequent developmental processes. ►morphogenesis in *Drosophila*

Segmental Aneuploid: Contains an extra chromosomal fragment(s) in addition to the normal chromosome complement; it is a partial hyperploid. ►hyperploid, ►aneuploid

Segmental Interchange: ►translocation chromosomal

Segmentation Clock: An ensemble of numerous cellular oscillators located in the unsegmented pre-somitic mesoderm, controls segmentation as it travels anteriorly in the embryo (Horikawa K et al 2006 Nature [Lond] 441:719).

Segmentation Genes: Control the polarity of body segments in animals. ►morphogenesis in *Drosophila*, ►metamerism; Zákány J et al 2001 Cell 106:207; Dubrulle J et al 2001 Cell 106:219; Tautz D 2004 Dev Cell 7:301.

Segmentation of Microarray Spots: The separation of the not entirely circular signal spots from the background.

Segmented Genomes: For example, the DNA of T5 phage is in four linkage groups, the RNA genetic material of alfalfa mosaic virus is in four segments.

Segregation: The separation of the homologous chromosomes and chromatids at random to the opposite pole during meiosis and carrying in them the alleles to the gametes. Independent segregation of non-syntenic genes (or ones, which are more than 50 map units apart within a chromosome) is one of the basic Mendelian rules. In the haploid products of meiosis of diploids, the 1:1 segregation of the alleles may be identified. Sometimes "x-segregation" is distinguished when after mitotic crossing over at the four-strand stage, each of the daughter cells carries one recombinant and one parental strand, and this is most common. At "z-segregation" the distribution of the chromosomes into the daughter cells is biased inasmuch as one of them carries two parental and the other two recombinant strands. Mitotic nondisjunction has been called "y-segregation." ►Mendelian laws, ►autopolyploid, ►tetrad analysis, ►segregation distorter, ►preferential segregation, ►epistasis, ►meiosis, ►mitotic crossing over, ►nondisjunction, ►chromosome segregation, ►partitioning, ►condensin, ►cohesin, ►separin; Ghosh SK et al 2006 Annu Rev Biochem 75:211.

Segregation Analysis: The analysis reveals the pattern of inheritance whether it is autosomal recessive, autosomal dominant, X-linked recessive or dominant, multifactorial, penetrance, expressivity of a gene(s). This the first, sometimes laborious step, especially in human genetics where controlled matings are not available, before gene frequencies, genetic risk, recombination, etc. can be meaningfully estimated. (See mentioned terms at separate entries; http://hasstedt.genetics.utah.edu/).

Segregation, Asymmetric: The differentiation and morphogenesis require that the progeny cells would differ from the mother cell after cell division. This difference is brought about by the unequal distribution of cellular proteins. ►morphogenesis in *Drosophila*

Segregation Distorter: A dominant mutation (*Sd*, map position 2-54 in *Drosophila*). When it is present, the homologous chromosomes (and the genes within) are not recovered in an equal proportion after meiosis. The second chromosomes that carry it are called SD and may be involved in chromosomal rearrangement and other (lethal) mutations. At the base of the left arm they may have *E(SD)* [*enhancer of Sd*] or *Rsp* (*Responder of SD*, at the base of the right arm), and more distally *St(SD)*, *Stabilizer of SD* and other components of the system. *SD/+* males transmit the *SD* chromosome to about 99% of the sperm. When *Rsp* is in the homologous chromosome, *Sd* is preferentially recovered. *Sd* is actually a tandem duplication of a 6.5 kb segment, and transformation by a 11.5 kb stretch of its DNA confers full *Sd* activity to the recipient flies. The sequence carries two nested genes, (*dHS2ST*), a homolog of the mammalian heparan-sulfate 2-sulfotransferase and *dRanGAP,* a guanine triphosphatase activator of the Ras-related Ran protein. It appears that the truncation of RanGAP is responsible for the poor transmission of the spermatids. Some of the other elements of the system act in a modifying manner and may cause recombination in the male. In many species of insects, the infection of males by the bacterium *Wolbachia* kills the offspring of the infected males x uninfected females but the viability of the infected eggs is normal. In mice, the transmission ratio distorter system (TRD) that impairs sperm flagellar motility and the *Tcr* (*t complex responder*) in cooperation with other genes located within chromosome 17 (t haplotype) affects chromosome segregation. The chromosome carrying the *Tcr* gene (even as a transgene at another location) enjoys high transmission in the presence of *Tcd*. The *Tcd* genes can act in both cis and trans position whereas *Tcr* (80 kb protein kinase) acts in cis position. *Tcr* represents a fusion between a part of the ribosome S6 protein kinase (*Rsk3*) and another gene of the microtubule affinity-regulating (MARK) Ser/Thr protein kinase family. The *Tcr* gene appears to have descended from a rearrangement between a member of the *Smok* (sperm motility kinase) family and a *Rsk* allele. Eventually, a practical method can be devised to increase one or the other sex by moving *Tcr* either to the X or to the Y chromosome of farm animals without relying on sperm sorting and artificial insemination. ►lethal factors, ►meiotic drive, ►transmission disequilibrium, ►certation, ►preferential segregation, ►polarized segregation, ►megaspore competition, ►tetrad analysis, ►gene conversion, ►epistasis, ►pollen-killer, ►infectious heredity, ►RAS, ►RAN, ►RSK, ►heparan sulfate, ►sex-ratio, ►spirochetes, ►dosage compensation; Powers PA, Ganetzky B 1991 Genetics 129:133; Schimenti J 2000 Trends Genet 16:240; Pardo-Manuel De Villena F, Sapienza C 2001 Mamm Genome 12:331.

Segregation Index: The gene number in quantitative traits, also called effective number of loci. ►gene number in quantitative traits

Segregation Lag: The mutation or transformation is expressed only by third division of the bacteria until all chromosomes are sorted out. ►phenotypic lag; Angerer WP 2001 Mutation Res 479:207.

Segregation Ratio: The phenotypic (genotypic) proportions in the progeny of a heterozygous mating. ►segregation, ►Mendelian segregation, ►modified Mendelian ratios, ►ascertainment test

Segregational Petite: ►petite colony mutants, ►mtDNA

Segregational Sterility: A heterozygote produces unbalanced gametes. ►inversion, ►translocation, ►gametophyte factor, ►hybrid dysgenesis, ►segregation distorter

Seip Disease: ►lipodystrophy familial

Seitelberger Disease: A degenerative encephalopathy (infantile neuroaxonal dystrophy). ►encephalopathies

Seizure: A sudden attack precipitated by a defect in the function of the nervous system. *Audiogenic seizures* are due to multifactorial mutations in the mouse, upon exposure to loud high-frequency sound. ►epilepsy, ►double cortex, ►periventricular heterotopia

Sek1 (MKK): A tyrosine and threonine dual-specificity kinase involved in the activation of SAPK/JNK families of kinases and it protects T cells from Fas and CD3-mediated apoptosis. ►SAPK, ►JNK, ►Fas, ►CD3, ►apoptosis, ►T cell; Yoshida BA et al 1999 Cancer Res 59:5483.

Seladin-1: A mediator of Ras-induced senescence. ►RAS, ►senescence; Wu et al 2004 Nature [Lond] 432:640.

Selaginella (club moss): Primitive green plant with small genome size (~132 Mb) (see Fig. S25). It is a suitable model plant for evolutionary and developmental studies. It has both female and male gametophytes but rather than producing seed, the fertilized megaspore (embryo) remains embedded in the vegetative tissue of the female gametophyte and starts its development at that place. (See http://www.genome.arizona.edu/BAC_special_projects/).

Figure S25. *Selaginella*

SelB: A prokaryotic translation factor homologous to eIF-2A and eIF-2γ. ►eIF-2A

SELDI-TOF (surface enhanced laser desorption/ionization–time of flight spectrophotometry): A special form of MALDI–TOF procedure for the study of protein–protein interactions. ►MALDI–TOF; Bane TK et al 2000 Nucleic Acids Res 30(14):e69.

Selectable Marker: Permits the separation of individuals (cells) that carry it from all other individuals, e.g., in an ampicillin or hygromycin medium (of critical concentration) only those cells (individuals) can survive that carry the respective resistance genes (selectable markers). The aequorin gene equipped with an appropriate promoter may light up the tissue of expression in insects. Many different genes may serve as selectable markers (mutant dihydrofolate reductase [DHFR)], methylguanine methyltransferase [MGMT)] multidrugresistance [MDR1]). Selectable markers such as antibiotic (used in medical practice) resistance are undesirable in producing transgenic crops. ►aequorin

Selectins: These are cell surface carbohydrate-binding, cytokine-inducible, transmembrane proteins. They also bind to endothelial cells in the small blood vessels along with integrins and enable white blood cells and neutrophils to ooze out at the sites of small lacerations to combat infection. L-selectin facilitates the entry of lymphocytes into the lymph nodules by binding to CD34 cell adhesion molecules. The ESL-1 selectin ligand is a receptor of the fibroblast growth factor. The selectins contain an amino-terminal lectin domain, an element resembling epidermal growth factor and a variable number of complementary regulatory repeats and a cytoplasmic carboxyl end. The P (1q23-q25) and E (1q23-q25) selectins recruit T helper-1 but not T helper-2 cells to the site of inflammation. L-selectin may promote metastasis of cancer cells. ►integrins, ►cell adhesion, ►cell migration, ►FGF, ►EGF, ►lectins, ►T cell, ►TACE, ►leukocyte adhesion deficiency, ►metastasis, ►integrins; Dimitroff CJ et al 2001 J Biol Chem 276:47623.

Selection: The unequal rate of reproduction of different genotypes in a population. Bacteria are particularly well suited for the study of the consequences of long-term selection because on simple culture media the evolving population after 20,000 generations can be compared to the ancestral ones. In an experiment of similar nature, two genes affecting DNA supercoiling (*topA*, *fis*) proved competitive. DNA supercoiling affects cell viability, replication, repair, transcription, recombination and transposition (Crozat E et al 2005 Genetics 169:523). In the constant environment of a chemostat a bacterial population substantially diverged in several directions within less than a month (Maharjan R et al 2006 Science 313:514). ►selection coefficient, ►allelic fixation, ►genetic load, ►fitness of hybrids, ►selection and population size, ►mutation pressure and selection, ►genetic drift, ►selection types, ►non-Darwinian evolution, ►selection conditions, ►selection purifying, ►screening, ►natural selection, ►chemostat; Brookfield JFY 2001 Curr Biol 11:388.

Selection and Population Size: In very small population, chance (random drift) may be more important than the forces of selection in determining allelic frequencies. When the selection pressure becomes very large, even in small populations there is a good chance for the favored allele to become fixed. ►allelic fixation, ►random drift, ►genetic drift; in human populations: Kreitman M 2000 Annu Rev Genomics Hum Genet 1:539.

Selection, Balanced: ►balanced polymorphism

Selection Coefficient: The measure of fitness of individuals of a particular genetic constitution relative to the wild type or heterozygotes in a defined environment. If the fitness is zero the selection coefficient is 1. In other words, if an individual does not leave offspring (fitness is zero), the selection against it in a genetic sense is 1 (100%). The selection coefficient is generally denoted as (s) or (t), where the former indicates the selection coefficient of the recessive class and (t) indicates the selection coefficient of the homozygous dominants. The meaning and relation of fitness and selection coefficients in a population complying with the Hardy-Weinberg theorem is best illustrated in Table S1.

In case we take the fitness of one genotype as unity (in this case we choose the heterozygotes, e.g., $w_2 = 1$), then the *standardized fitness* becomes:

$$\frac{w_1}{w_2} = 1 - s \text{ and } s = 1 - \frac{w_1}{w_2}$$

$$\text{similarly } \frac{w_3}{w_2} = 1 - t \text{ and } t = 1 - \frac{w_3}{w_2}$$

►fitness, ►allelic frequencies, ►Hardy-Weinberg theorem, ►advantageous mutation

Selection Conditions: Selection may operate at different levels beginning in meiosis (distorters, gametic factors) or at any stage during the life of the individual beginning with the zygote through the entire reproductive period. The intensity of the selection depends on the genes, the overall genetic constitution of the individual and the environment, including the behavioral pattern of the population (e.g., protecting the young and infirm). The formulas representing the various means of selection were derived from the Hardy-Weinberg theorem $p^2 + 2pq + q^2 = 1$ (see Table S2) and their use is exemplified.

The genotypic contributions to the *AA*, *Aa* and *aa* phenotypes are p^2, $2pq$ and $q^2(1 - s)$, respectively, where *s* is the selection coefficient (►selection coefficient for derivation of *s*).

The total contribution is now reduced from 1 to $1 - sq^2$ because sq^2 individuals are eliminated by selection. Thus, the new frequency of the recessive alleles becomes q_1 and the

Table S1. Selection and fitness in a Hardy–Weinberg population

Genotypes	AA	Aa	aa	Total
Zygotic frequencies	p^2	2pq	q^2	1
Fitness	w_1	w_2	w_3	
Gametes produced	$(w_1) \times (p^2)$	$(w_2) \times (2pq)$	$(w_3) \times (q^2)$	1

Table S2. Formulas to calculate the change in allelic frequencies per generation at various conditions of selection. In case *s* or *q* is very small, omission of the *sq* product from the denominator may be of very little consequence for the outcome

Type of selection	Increase (+) or decrease (-) in the rate of change of an allele (Δq) per generation
1. Selection against gametes	$-\frac{sq(1-q)}{1-sq}$
2. Differential selection in males and females (without sex linkage)	$q^2[1-\frac{1}{2}(s_{\mathrm{male}}+s_{\mathrm{female}})]$
3. Selection at X-chromosome linkage	gametic in the heterogametic and zygotic in the homogametic sex
4. Selection against recessive lethals	$-\frac{q^2}{1+q}$
5. Selection against the allele in absence of dominance	$-\frac{\frac{1}{2}sq(1-q)}{1-sq}$
6. Selection against dominant lethals	$+(1-q)$
7. Partial selection against homozygous recessives in case of complete dominance	$-\frac{sq^2(1-q)}{1-sq^2}$
8. Partial selection against completely dominant alleles	$+\frac{sq^2(1-q)}{1-s(1-q)^2}$
9. Selection against recessives in autotetraploids	$-spq^4$
10. Selection against intermediate heterozygotes	$-\frac{sq(1-q)}{1-2sq}$
11. Selection against heterozygotes	$+2spq(q-\frac{1}{2})$
12. Selection against both homozygotes (heterozygote advantage)	$+\frac{pq(s_1p-s_2q)}{1-s_1p^2-s_2q^2}$

$$q_1 = \frac{q^2[1-s]+pq}{1-sq^2}$$

change in the frequency of q, is $\Delta_q = \frac{q^2[1-s]+pq}{1-sq^2} - q$ which reduces to $\frac{sq^2(-q)}{1-sq^2}$ The effectiveness of the selection also depends on the frequency of the allele involved. Selection against rare recessive alleles may be very ineffective because only the homozygotes are affected. Selection most frequently works at the level of the phenotype and the recessives may not influence the fitness of the heterozygotes. At low values of q the majority of the recessive alleles is in the heterozygotes and thus sheltered from the forces of selection. Dominant, semidominant and codominant alleles may be, however, very vulnerable if they lower the fitness of the individuals. As an example let us assume that the frequency of a recessive allele is q = 0.04 and by using formula 4 shown in Table S2 the change in allelic frequency per generation is $\Delta_q = -\frac{[0.04]^2}{1+0.04} \cong -0.00154$ and after $n = 25$ generations, the initial frequency of the gene, $q_0 = 0.04$ changes to: $q_n = \frac{q_0}{1+nq_0} = \frac{0.04}{1+[25\times0.04]} = 0.02$ meaning that complete elimination of all homozygotes (s = 1) for 25 generations reduces only to half the frequency of that recessives allele. The initial zygotic frequency of $(0.04)^2 = 0.0016$ (1/625) will thus change to $(0.02)^2 = 0.0004$ (1/2500).

If the same deleterious allele would be semidominant and it conveys a fitness of 0.5 relative to the homozygotes for the other allele, according to formula 5 in Table S2, the change of the frequency of this semidominant allele in a generation becomes: $\Delta q = -\frac{[1/2][0.5][0.04][0.96]}{1-\{[0.5]\times[0.04]\}} \cong -0.0098$. Thus, a semidominant allele with 0.5 selection coefficient will be selected against more than six times as effectively as a recessive lethal factor because $0.0098/0.00154 \cong 6.4$ in these examples.

The number of generations required to bring about a certain change in gene frequencies can be calculated in the simple case when the homozygous recessives are lethal, i.e., the selection coefficient is, s = 1: $T_{\mathrm{generations}} = \frac{q_0-q_T}{q_0q_T} = \frac{1}{q_T} - \frac{1}{q_0}$ where q_0 is the initial frequency of the allele and q_T is its frequency after T generations. If it is assumed that the genotypic frequency is $(q_0)^2 = 0.0001$ and $q_0 = 0.01$ then the number of generations required to reduce the initial frequency to $q_T = 0.005$ is $T = \frac{1}{0.005} - \frac{1}{0.01} = 100$. After 100 generations then the frequency of the recessive lethal allele becomes $(q_T)^2 = (0.005)^2 = 0.000025 = 1/40{,}000$ compared to the initial frequency of 1/10,000. The effectiveness of selection is much influenced by the heritability of the allele concerned. In complex cases more elaborate computations are required that cannot be illustrated here. ►selection coefficient, ►balanced polymorphism, ►mutation pressure opposed by selection, ►selection and population size, ►allelic fixation, ►genetic load, ►fitness of hybrids, ►gametophyte, ►QTL, ►heritability, ►gain, ►mutation neutral

Selection, Cyclic: Different phenotypes are selected depending on seasonal variations in the environment.

Selection Differential: ►gain

Selection Index in Breeding: ►gain

Selection Inferred from DNA Sequences: Different types of natural selection (negative selection, recurrent positive selection, balancing selection) affect the pattern of variation in DNA sequences although recent population growth, bottlenecks and population subdivisions confound the pattern of genetic variation and may mimic the effects of natural selection. If synonymous mutations are neutral then natural selection at non-synonymous sites can be effectively tested. A non-parametric test can compare allele frequency spectrum of segregating sites among regions of functional classes. Because of being non-parametric the comparisons encounter difficulties in biological interpretation with other methods. A maximum likelihood method can infer, however, both selection and demographic changes. Assuming that non-coding single nucleotide polymorphism (SNP) is selectively neutral and correction can be made to determine selection among non-synonymous changes. Experimental data indicate negative selection on non-synonymous alterations (as expected) and the strength of selection is most explicit when the amino acid substitutions involve radical changes (Williamson SH et al 2005 Proc Natl Acad Sci USA 102:7882). ►non-parametric test, ►non-synonymous codon, ►maximum likelihood, ►selection types, ►SNIPs, ►McDonald-Kreitman hypothesis

Selection Intensity: ►gain

Selection—Medical Care: The progress of effective medical care saves increasing number of human lives. Some of the saved individuals will have a chance to transmit deleterious genes to their offspring so eventually some increase in detrimental alleles is expected. Non-synonymous mutations reduce fitness by an average of 4.3% and selection acting against non-synonymous polymorphism is ~9 x10^{-5}. This also indicates that medical care may not affect the extent of fitness too much but it is very difficult to identify the gene involved in complex diseases. (Eyre-Walker A et al 2006 Genetics 173:891). ►selective abortion, ►genetic risk, see Fig. S26.

Selection, Long Term: Plant and animal breeders carry on selection for many generations and the effectiveness is reduced less than expected based on selection for single desirable genes/traits. The cause of the effectiveness is that multiple interacting genes determine the majority of traits of agricultural interest (Carlborg Ö et al 2006 Nature Genet 38:418). ►polygenic inheritance, ►QTL, ►gain

Selection, Negative: Maintains disadvantageous alleles. The term is also used in the sense of elimination disadvantageous alleles. ►drift genetic, ►hitchhiking, ►selection purifying

Selection, Natural: ►natural selection

Selection Positive: Maintains the advantageous alleles. (See selection motif tool: http://oxytricha.princeton.edu/SWAKK/).

Selection Pressure: The intensity of selection affecting the frequency of genes in a population. ►volatility

Selection, Purifying: Acts against the heterozygotes of a new allele with lower fitness. Asexual reproduction favors the accumulation of deleterious mutations. Mitochondrial mutations in asexual lineages of microcrustacean *Daphnia pulex* accumulated deleterious mutation four times the rate compared to sexual lineages (Paland S, Lynch M 2006 Science 311:990). ►hitchhiking, ►genetic drift, ►Muller's ratchet

Selection Response: (heritability) × (selection differential). ►heritability, ►gain

Selection Types: (i) *Stabilizing* favors the intermediate forms that have ability to survive under the most common but opposite conditions (such as cold and heat, draught and excessive precipitation). (ii) *Directional selection* shifts the mean values of a population either toward higher or lower values than the current mean. (iii) *Disruptive selection* breaks up the population into two or more subpopulations that each has adaptive advantage in particular niches of a larger habitat. (iv) *Frequency-dependent selection* favors an allele when it is relatively rare and may turn against it when it becomes abundant. Common examples are found in host-parasite, predator-prey relationships or in resource utilization (see Fig. S27) (Carius HJ et al 2001 Evolution 55:1136).

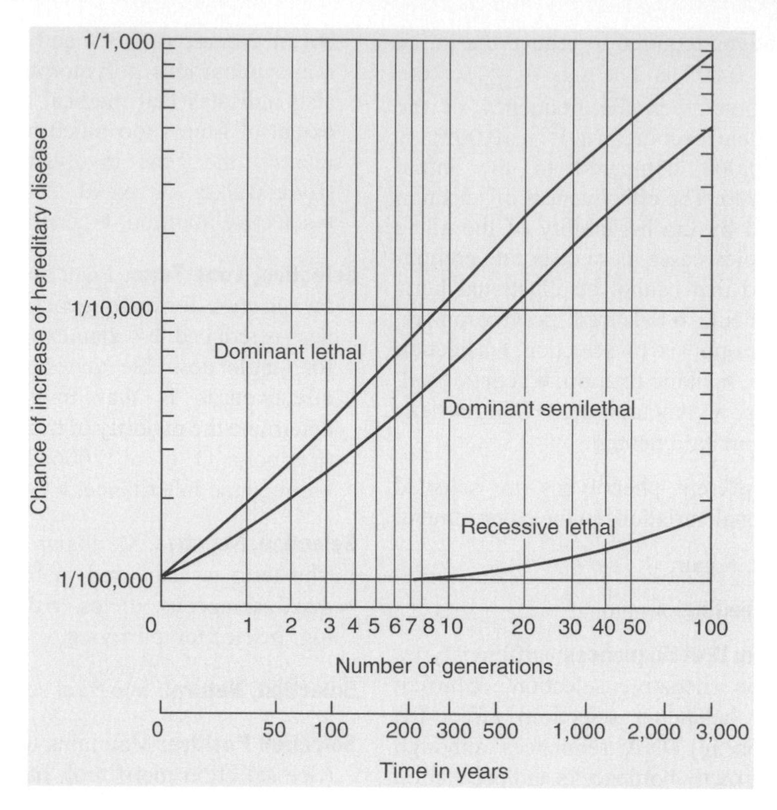

Figure S26. The effect of medical care on the incidence of human diseases. Thus the improvement of medical services may lead to deterioration of the gene pool. If the initial frequency of a detrimental recessive allele is 0.001 or less, the consequences of the selection may not be evident for about 300 years. Even after thousands of years of selection the increase of incidence is relatively modest. The frequency of dominant lethal or dominant semi-lethal alleles may increase much faster. (Redrawn after Bodmer WF, Cavalli-Sforza LL 1976 Genetics, Evolution, and Man. Freeman, San Francisco, California)

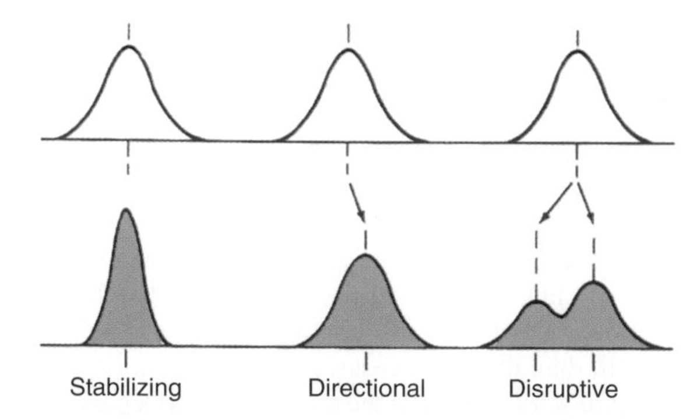

Figure S27. Top: Original frequency distribution of populations. Below: The shifts in distribution after selection. (After Mother K 1953 Symp Soc Exp Biol 7:66)

When the number of predators increase beyond a point there will not be enough prey to maintain the predators and their number will decrease. Similarly, when animals overgraze in the natural habitat, ultimately their population decreases. Similarly, highly virulent viruses may outcompete the less aggressive types even when they may kill the host faster. ►fitness, ►selection apostatic selection, ►selection cyclic, ►competition, ►artificial selection

Selection Value: ►selection index, ►gain

Selective Abortion: The termination of a pregnancy by precocious removal of the fetus from the womb if the condition of the mother or of the fetus medically justifies it and the legal system permits it. The genetic constitution or condition of the fetus may be tested with the aid of amniocentesis or sonography. From the viewpoint of genetics, selective abortion may pose biological problems. If all families would compensate for the abortions elected based on genetic defects, the frequency of these defective genes may actually rise in the population. This may happen because it assists heterozygotes for genetic abnormalities to leave offspring that—although may not display the morbid trait—can again transmit the undesirable genes to future generations. If all carriers would refrain from reproduction, the frequency of the deleterious genes may sink to the level of new mutations. Selective abortion may involve also ethical, moral and political problems but these are beyond the scope of genetics. ►abortion spontaneous, ►abortion medical, ►pregnancy unwanted, ►amniocentesis, ►counseling genetic, ►selection and medical care, ►sterilization humans, ►ethics, ►bioethics, ►genetic screening

Selective Advantage: In population genetics, selective advantage is expressed by the relative fitness of bearers of (two) genotypes. Generally the wild type has greater fitness (W_N) than a mutant type (W_M) and their relative fitness is $(W_N)/(W_M)$. *W* (fitness) is the reproductive success. Usually, $(W_N)/(W_M) = 1 - s$, where *s* is the selection coefficient indicating the disadvantage of the mutant type. In case the fitness of a genotype exceeds 1 it has an advantage in survival. ►selection coefficient, ►fitness, ►beneficial mutation, ►codon usage

Selective Fertilization: Some sperms, because of their gene content, may be at a disadvantage in competition with other sperms for penetrating the egg, or where multiple eggs or megaspore cells are formed, their success depends on their genetic constitution. Because of this, the genetic segregation may deviate from the standard Mendelian expectation. ►gamete competition, ►certation, ►megaspore competition, ►meiotic drive, ►sperm

Selective Medium: Only permits the propagation of individuals or cells that carry a selectable marker, such as high or low temperature, antibiotic, drug resistance, etc. (see Fig. S28) ►selectable marker

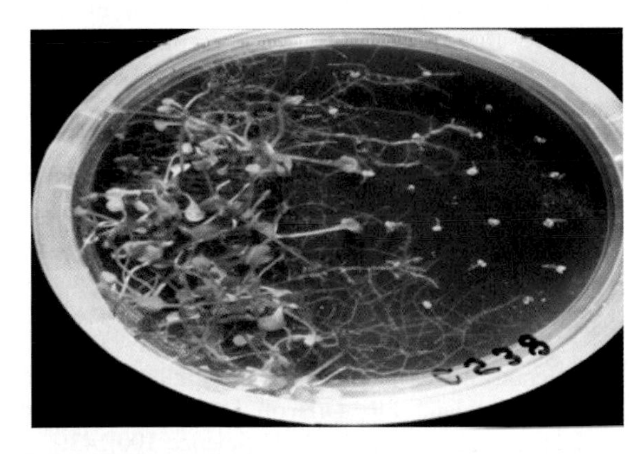

Figure S28. Hygromycin medium. Left: *Arabidopsis* plants transgenic for resistance; right: Wild type fails to grow (Rédei, unpublished)

Selective Neutrality: Assumes that random drift is responsible for the allelic frequencies in a particular population, i.e., all alleles at a locus have the same selective value. ►random genetic drift, ►allelic frequencies

Selective Peak: Determined by the genetic homeostasis of a population, i.e., the gene frequencies are maintained at this optimum as long as catastrophic changes in the environment do not occur. ►homeostasis, ►adaptive landscape

Selective Screening: ►screening, ►selective medium, ►mutation detection

Selective Sieve, Extent of: The rate of substitution at a gene site/mutation rate for the gene.

Selective Sweep: The rapid establishment of advantageous allele(s) in a population. Such alleles may carry along other linked genes too by a mechanism of hitchhiking. Less advantageous mutations may also be swept along with linked advantageous ones. The selective sweep may reduce the genetic variation in the population. ►hitchhiking, ►selection types, ►selective advantage, ►mutation beneficial; Nurminsky DI et al 1998 Nature [Lond] 396:572.

Selective Value: ►fitness, ►selection coefficient

Selectivity Factor (SF): A human general transcription factor homologous to TFIIB of other eukaryotes. ►transcription factors

Selectivity Filter: Functions as a gatekeeper to allow or prevent the access of another (small) molecule into s structural pocket of a protein. Aromatic amino acids (being relatively bulky) permit the access of larger structures whereas smaller amino acids are more restrictive. Selective filters have significance for the development and effectivity of special drugs.

Selector Genes: These are supposed to specify segmental differences during morphogenesis, such as anterior/posterior or dorsal/ventral. The selector genes after receiving morphogenetic signals may team up with the relevant other transcription factor genes and recruit a number of specific activators to turn on the transcription of specific morphogenetic genes (see Fig. S29). The process of differentiation of an organ may take place by two major steps. First, the body position is specified and then the pattern of differentiation is determined. Selector genes are used also for genes that facilitate selection of transformed cells, e.g., neomycin, kanamycin, hygromycin resistance and others. ►morphogenesis, ►morphogenesis in *Drosophila*, ►signal transduction, ►differentiation, ►imaginal disc; Affolter M, Mann R 2001 Science 292:1080; Guss KA et al 2001 Science 292:1164; Ao W et al 2004 Science 305:1743.

Selenium-Binding Protein Deficiency (SP56, 1q21-q22): Causes neurological disorder and sterility due to the deficiency of a 56 kDa protein and other selenoproteins. Selenium may have anticarcinogenic property.

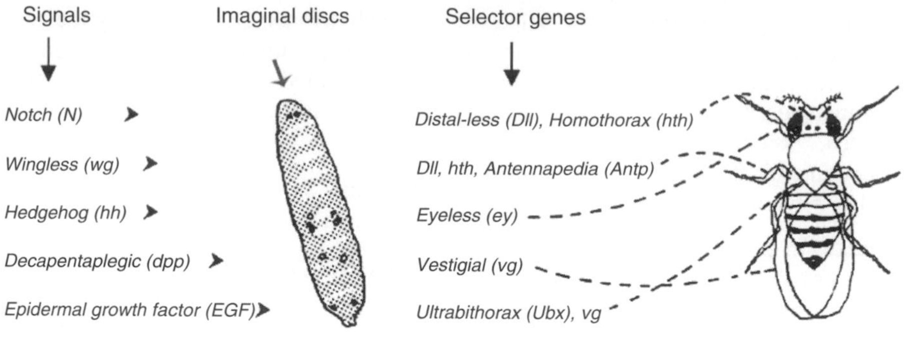

Figure S29. Selector gene responses and expression

►selenocysteine; Martin-Romero FJ et al 2001 J Biol Chem 276:29798.

Selenocysteine (SC, $C_6H_{12}N_2O_4Se_2$): A reactive, very toxic, oxygen-labile amino acid. Selenophosphate is the donor of selenium for the synthesis of selenocysteyl-tRNA, which has the UCA anticodon and SC incorporation is directed by the UGA (usually a stop) codon. In bacteria, a special mRNA fold immediately following UGA mediates the incorporation of SC, carried by its own tRNA. The loop binds the SelB protein, which acts like a somewhat unusual elongation factor. In mammals, the loop (called SECIS) is locked at the end of the mRNA. Two proteins carry out the function of SelB. SBP2 binds the SECIS element whereas eEFSec binds the selenocysteine tRNA. Incorporation of selenocystein into proteins requires a selenocystein-specific translation elongation factor (eEFSec) and a SECIS binding protein (SBP2). SBP2 associates with the 28S RNA of the ribosome (Kinzy SA et al 2005 Nucleic Acids Res 33:5172). Thus, SC is the 21st "natural" amino acid. A putative selenocysteine synthase had been identified as a pyridoxal phosphate-containing protein called soluble liver antigen. The activity of this synthase was characterized using selenophosphate and a tRNA aminoacylated with phosphoserine as substrates to generate selenocysteine. Identification of selenocysteine synthase allowed the delineation of the entire pathway of biosynthesis in mammals. Selenocysteine synthase is present only in those archaea and eukaryotes that make selenoproteins (Xu X-M et al 2007 PloS Biol 5(1):e4).

Selenocysteine-containing proteins (glutathione peroxidase, 5′-deiodinases, formate and other dehydrogenases, glycine reductase) may have substantially higher or lower catalytic activity and may have theoretical and applied interest. Selenoproteins may be scavengers of heavy metals and may favor survival of cultured neurons. Removal of the selenocysteine $tRNA^{Sec}$ causes embryonic lethality in mice. Deiodination enzymes catalyze deiodination of thyroid hormones and modulate the level of these hormones. Selenoprotein-binding protein binds to deiodinase 2 and causes thyroid disease (Dumitrescu AM et al 2005 Nature Genet 37:1247). In a marine gutless worm, *Olavius algarvensis* 99 selenoprotein genes that clustered into 30 families were found. In addition, several pyrrolysine-containing proteins were identified in this dataset. Most selenoproteins and pyrrolysine-containing proteins were present in a single deltaproteobacterium, δ1 symbiont (Zhang Y, Gladyshev VN 2007 Nucleic Acids Res 35:4952). ►code genetic, ►antideterminant, ►protein synthesis, ►SelB, ►SECIS, ►unnatural amino acids, ►pyrrolysine; Gladyshev VN, Kryukov GV 2001 Biofactors 14:87.

Selenoproteins: They appear to have roles in antioxidant defenses (protection against UV, peroxides), against tumor formation, viral resistance, thyroid and reproductive functions. The selenoprotein phospholipid, hydroperoxide glutathione peroxidase (PHHGPx) is an active, soluble enzyme during sperm maturation but in mature spermatozoa, it is enzymatically inactive and insoluble. There its role is structural, securing the stability of midpiece where mitochondria are abundant. The human genome may encode 25 selenoproteins. Deficiency of selenoproteins may accelerate the development of cancer (Diwadkar-Navsariwala V et al 2006 Proc Natl Acad Sci USA 103:8179). ►selenocysteine, ►UV, ►sperm, ►glutathione peroxidase; Copeland PR, Driscoll DM 2001 Biofactors 14:11; Kryukov GV et al 2003 Science 300:1439.

SELEX (systematic evolution of ligands by exponential enrichment): A method of aptamer selection from random oligonucleotide sequences (10^{15} to 10^{18}) resulting in the isolation of aptamers and ligands. The oligonucleotide library is synthesized by commercial DNA synthesizers and used in 20–30 or 200 base long sequences ($4^{20} \sim 10^{12}$ whereas $4^{75} \sim 1.4 \times 10^{45}$ combinations). The phage R17 coat protein provides an excellent RNA recognition site. It binds to a loop

S

(AUUA) and a four-base stem structure with a bulged A residue. The synthetic nucleotide sequence may use modified ribose residues such as 2′ F (fluoride) or $2'NH_2$ in the triphosphate nucleotides. This makes the oligonucleotides stable for several hours whereas unmodified RNA may be degraded by nucleases immediately. By "partition" (gel electrophoresis, nitrocellulose filtration) the few good sequences must be separated from the large pool. The aptamers may be truncated to limit to the minimal effective binding sites and may be further modified by extra nucleotide analogs for increased stability. The technology may be applied to the search for new drugs, diagnostic compounds and ribozymes of diverse functions. ►aptamer, ►nitrocellulose filter, ►gel electrophoresis, ►ribozyme, ►DNA binding proteins; Wilson DS, Szostak JW 1999 Annu Rev Biochem 68:611; http://wwwmgs.bionet.nsc.ru/mgs/systems/selex/.

Self-Antigen: Antigens synthesized within the system that may stimulate autoimmune reaction by the T cells although normally there is no immune response to the self-antigens. ►T cells, ►autoantigen, ►MHC, ►self-tolerance, ►immune tolerance

Self-Assembly: A process of reconstituting a structure from components when there is enough information within the pieces to proceed without outside manipulation, e.g., the self-assembly of ribosomes from RNA and protein subunits.

Self-Cleavage of RNA: Some ribozymes may cleave more than 99% of the mRNAs and thereby preventing gene expression. By genetic engineering, effective ribozymes can be introduced into vectors and thus prevent transcription of functional mRNA or its translation. ►ribozyme, ►RNAi; Yen L et al 2004 Nature [Lond] 431:471.

Self-Compatibility: Self-fertilization can take place and the offspring is normal. ►self-incompatibility, ►incompatibility alleles

Self-Cross: A misnomer for self-fertilization. ►cross

Self-Deleting Vector: A lentiviral vector may integrate into the cell and deliver a gene (Cre), which is expressed in the target and subsequently deleted in a Creb-dependent manner. The undesirable effect of the vector is thus avoided. ►viral vectors, ►lentivirus, ►Cre/LoxP; CREB Pfeifer A et al 2001 Proc Natl Acad Sci USA 98:11450.

Self-Destructive Behavior: ►Lesch-Nyhan syndrome, ►Smith-Maganis syndrome

Self-Fertilization: Can take place between the gametes in hermaphroditic individuals; autogamy is the strictest means of inbreeding. Self-fertilization occurs in ~20% of plant species and ~33% are intermediates between selfing and outcrossing (Vogler DW, Kalisz S 2001 Evolution 55:202). The sperm of XO males preferentially fertilizes *Caenorhabditis elegans* hermaphrodites. ►autogamy, ►inbreeding, ►*Caenorhabditis*; Herlihy CR, Eckert CG 2002 Nature [Lond] 416:320.

Self-Immunity: Mediated by protein factors that bind to, for e.g., Moloney virus and thus prevent the integration into the sequences by another viral element. ►cis-immunity, ►phage immunity

Self-Inactivating/Self-Activating Vector: ►E vector

Self-Incompatibility: The failure of fusion of male and female gametes produced by the same individual or lethality of the embryo formed by such fusion. Self-incompatibility may be controlled either by the sporophyte or by the gametophyte. The incompatibility for effective pollination is determined by secreted glycoproteins (SLG, on the surface of the stigma) and membrane-bound receptor protein kinases (SRK, attached to the stigma cells). The stigma cells preferentially synthesize these proteins and only low level of expression is found in the anthers of cruciferous plants. In several solanaceous plants (tomato, *Nicotiana alata* and *Petunia inflata*), an S (selfcompatibility) RNase may attack in the style the RNA in self-pollen. The S-RNase is normally compartmentalized in vacuoles of the pollen tube. Primarily protein HT-B keeps S-RNase sequestered. When HT-B is degraded RNase is released and incompatibility is established (McLure B 2004 Plant Cell 16:2840; Goldraij A et al 2006 Nature [Lond] 439:805).

In poppy plants, a receptor on the pollen recognizes a stylar protein. The binding between this ligand and receptor leads to release of calcium ions that in turn inhibit pollen-tube growth. Two soluble inorganic pyrophosphatases regulate the incompatibility reaction (de Graaf BHJ et al 2006 Nature [Lond] 444:490). In *Brassica* plants self-incompatibility is overcome if the aquaporin gene was disrupted. This indicates that functional water channels are one of the requirements for the manifestation of self-incompatibility in *Brassica*. Recent information indicates that in the stigma cell membranes of crucifers the single *S* locus encodes an S locus glycoprotein (SLG) and a single-pass transmembrane serine/threonine S receptor kinase (SRK). The latest evidence indicates that SRK and not SLG is responsible for the haplotype specificity and SLG reinforces the selfincompatibility. Transfer of the SRK and SCR genes from *Arabidopsis lyrata* to *A. thaliana* confers self-incompatibility to the latter species, which is normally autogamous (Nasrallah ME et al 2002 Science 297:247). The secretion of a cysteine-rich protein (SCR, encoded also at the

S locus) by the self-pollen catalyzes the autophosphorylation of the SLG—SRK complex. The SRK protein then interacts with ARC1 (autorejection component) stigma protein and as a consequence, through a series of events the self-pollen is incapacitated on the stigma by ubiquitination (Stone SL et al 2003 Plant Cell 15:885). It has been shown that an S-RNase with another component is responsible for the pollen tube breakdown in the incompatible combinations. The factor PiSNF encoding 389 amino acids ~161 kb downstream of the S_2 allele may protect the pollen from the decay (Sijacic P et al 2004 Nature [Lond] 429:302). In poppy, a caspase-3 type protein prevents self-fertilization in self-incompatibility (Thomas SG, Franklin-Tong VE 2004 Nature [Lond] 429:305). Accordingly, self-incompatibility in the different species may have different controls.

Self-incompatibility is usually determined by a large array of alleles in the populations (~100 haplotypes in *Brassicas*) and there are substantial differences among the species regarding the number of such alleles maintained depending on the effective population size. Heteromorphic self-incompatibility is based on mechanical barriers to self-fertilization. The stigma is located to a higher position in the flowers than the anthers, and the pollen usually does not reach the stigmatic surface. In such cases, artificial pollination is successful. Sporophytic incompatibility means that the incompatibility factors operate not at the gametophyte (pollen - egg) level but the pistil or pollensac tissues and control the growth of the pollen tube. In *Brassica rapa* sporophytic selfincompatibility is controlled methytlation of the promoter of the recessive allele in the tapetal tissues (Shiba H et al 2006 Nature Genet 38:297). Selfincompatibility may break down by epigenetic mechanisms; in *Arabidopsis* species hybrids the S-locus receptor kinase transcript may be aberrantly processed and in *Capsella* hybrids the S-locus cysteine-rich protein may be suppressed (Nasrallah JB et al 2007 Genetics 175:1965). ►incompatibility alleles, ►*S* alleles, ►HLA, ►unilateral incongruity, ►population effective size, ►gametogenesis, ►Ribonuclease-S, ►RNA I, ►ligand, ►gametophyte, ►pistil, ►pollen, ►tapetum, ►aquaporin; Nasrallah JB 2005 Trends Immunol 26:412; McClure B 2006 Curr Opin Plant Biol 9:639; Dickinson HG 2000 Trends Genet 16:373; Kachroo A et al 2001 Science 293:1824; Takayama S et al 2001 Nature [Lond] 413:534; Tong N 2002 Trends Genet 18:113; Kachroo A et al 2002 Plant Cell 14:S227; Stone JL 2002 Q Rev Biol 77:17; Takayama S, Isogai A 2005 Annu Rev Plant Biol 56:467; self/nonself discrimination in pre- and post-zygotic systems: Boehm T 2006 Cell 125:845.

Self-Regulation: The process of regeneration of embryos from bisected (cut into halves) blastulas. This can take place in lower and higher animals including human, when identical twins are formed by spontaneous events. The dorsal–ventral morphogenetic gradients in the two halves are mediated by the bone morphogenetic proteins (BMP) and extracellular proteins such as Chordin, Sizzled and Bambi and others, which are situated at opposite poles and are under opposite transcriptional regulation (Reversade B, De Robertis EM 2005 Cell 123:1147). ►Spemann's organizer, ►organizer

Self-Renewal: An ability of cells to also produce stem cells that are not just able to divide mitotically and generate progenitor cells. ►stem cells, ►progenitor; Smith AG 2001 Annu Rev Cell Dev Biol 17:435.

Self-Reproduction: The general property of living systems. It means that the system can produce detached self-reproducing copies of itself. Self-reproducing machines—if they can function—may have special human interest for working in environments hazardous to humans such as outer space, mutagenic conditions, etc. Important requirement would be not just self-assembly, which would make only a copy of itself from parts but would continue—if parts provided—making additional copies continuously. The self-reproducing feature has now been achieved in principle by the use of a four-module system. It is still very far from generating a system comparable to a biological unit, which can generate 10^{20} amino acid combinations not to consider the many other biological elements of a living organism (Zykov V et al 2005 Nature [Lond] 435:163).

Selfing: Self-fertilization; it is symbolized by ⊗.

Selfish DNA: An assumption for certain DNA sequences (introns, repetitive non-coding sequences, transposable elements) that they have no selective (adaptive, evolutionary) value for the carrier, therefore, the presence of such sequences is of no advantage to the cells concerned, and are propagated only for selfish (parasitic) purposes. Some of the originally "selfish DNAs" (1979–80) turned out to have some functions, e.g., as maturases, and others represent transposable elements and continuously reshape the genome and are thus significant for mutation and evolution. Alu sequences appear to be more common within the gene-rich GC regions hinting some regulatory functions. The minisatellite DNAs and the trinucleotide repeats are implicated in an increasing number of hereditary diseases. The majority of the Y-chromosomal sequences of *Drosophila* do not seem to have any identifiable function yet male fertility may be impaired if they are deleted. ►junk DNA, ►introns, ►ignorant DNA, ►copia, ►trinucleotide repeats, ►REP, ►plasmid addiction, ►transposable elements, ►Alu;

S

van der Gaag M et al 2000 Genetics 156:775; Hurst GDD, Werren JH 2001 Nature Rev Genet 2:597; Hammerstein P, Hagen EH 2006 Genes in Conflict. The Biology of Selfish Genetic Elements. Harvard University Press, Cambridge, Massachusetts.

Selfish Genes: These have selective advantage over comparable ones and ensure their own propagation. The selfishness contrasts altruism when genes of the non-reproductive casts of social insects promote the welfare of the colony at the expense of their own work although they themselves are sterile. ►altruism

Selfish Replicon: The small circular plasmids in eukaryotic nuclei (maize) without any apparent function beyond perpetuating themselves.

Self-Organizing Map: ►cluster analysis

Self Protein: ►self antigen, ►molecular mimics, ►bystander activation, ►immune system

Self-Primed Synthesis: The synthesis from single-strand DNA obtained through reverse transcription, one primer may be used at the 5′ end to produce the second strand by extension at the 3′-end in a hairpin like structure. The synthesis by self-priming is slow (see Fig. S30).

Figure S30. Self-priming

Self-Replicating DNA: A replication mechanism for short double-stranded molecules that do not require assistance by proteins. Similar mechanisms may have operated during prebiotic evolution, and it can be reproduced in the laboratory. ►replication, ►self-assembly

Self-Replicating Peptide: An autocatalytic molecule capable of assembling amino acids into oligopeptides. The yeast transcription factor GCN4 leucine-zipper domain can promote its own synthesis of 15–17 amino acid residues. ►GCN4, ►leucine zipper

Selfrestriction: Self restriction lymphocyte recognizes foreign antigen bound to self MHC molecule. ►lymphocyte, ►MHC

Self-Splicing Introns: Group I and group II introns can fold into catalytic structures capable of removing their own sequences from the RNA transcripts of genes. ►intron

Self-Tolerance: The unresponsiveness of the immune system to self-antigens. Apoptosis is a means for the maintenance of self-tolerance; lymphocytes and dendritic cells play an important role (Chen M et al 2006 Science 311:1160). *Central self-tolerance* may be caused by death of the lymphocytes when encountering autoantigens. Peripheral self-tolerance takes place among the mature lymphocytes in the peripheral lymphatic organs. *Clonal ignorance* fails to recognize the autoantigens because of their sequestration or the failure to stimulate the indispensable secondary signals, such as cytokines, etc. ►immune tolerance, ►self-antigen, ►autoimmune diseases, ►immune system, ►at least one hypothesis; Rubin RL, Kretz-Rommel A 2001 Crit Rev Immunol 21:29.

Selvin: A rarely used unit of absorbed radiation dose. ►Sievert

SEM (scanning electronmicroscopy): ►electronmicroscopy

SEM-5: A homolog of *Grb2* in *Caenorhabditis* nematodes. ►*Grb2*

Semantics: The differentiating meaning of words and sentences. Also used by information-extraction programs for interpreting the correct meaning.

Semaphorin: A family of membrane-associated, secreted protein factors, required for axonal pathfinding in neural development. Semaphorin 5A can be bifunctional, promontory and inhibitory, to axon guidance depending on heparan and chondroitin sulfate proteoglycans (Kantor DB et al 2004 Neuron 44:961). Transduction by semaphorin III is mediated by the neuropilin-1 receptor. Semaphorins also regulate the development of the right ventricle and the right atrium of the heart as well as various cartilagineous and other tissues. Human semaphorins IV and V genes reside at the 3p21.3 chromosomal site; it is deleted in small cell lung carcinoma. Semaphorin 4A primes T cells and regulates Th1/Th2 (Kumanogoh A et al 2005 Immunity 22:305). Semaphorin 7A controls both axon guidance and T cell reaction (Czopik AK et al 2006 Immunity 24:591) and initiates T cell-mediated inflammator7 through $\alpha 1\beta 1$ integrin (Suzuki K et al 2007 Nature [Lond] 446:680). Semaphorin 4D links axon guidance and tumor-induced angiogenesis (Basile JR et al 2006 Proc Natl Acad Sci USA 103:9017). ►axon, ►collapsin, ►neuropilin, ►netrin, ►neurogenesis, ►small cell lung carcinoma, ►plexin, ►fasciclin, ►axon guidance, ►integrin; Tessier-Lavigne M, Goodman CS 1996 Science 274:1123; Pasterkamp RJ, Verhaagen J 2001 Brain Res Rev 35:36; Serini G et al 2003 Nature [Lond] 424:391; Pasaterkamp RJ et al 2003 Nature [Lond] 424:398.

SEMD: ►PAPS

Semelparity: The organism reproduces only once during its lifetime, e.g., *Palingea longicauda*

(Ephemeroptera) or some marsupials, which die after one mating season.

Semen: The viscous fluid in the male ejaculate composed of the spermatozoa and secreted fluids of the prostate and other glands. The seminal fluid of *Drosophila* reduces the propensity of the females to mate with another male. Its higher quantity lowers the viability of the females. Thus, the semen per se may have a role in fitness. ►testis, ►prostate

Semenogelin: A protein in the seminal fluid that promotes the viscosity of the ejaculate. It tends to prevent successful, additional insemination in promiscuous females (Dorus S et al 2004 Nature Genet 36:1326).

Semiconductor: The materials of germanium, silicon and others are characterized by increased electric conductivity as temperature increases to room temperature. These materials are called semiconductors because their conductivity is much lower than that of metals. In metals, the increase in temperature lowers conductivity. In the semiconductor material, the electronic motion is turned on through the crystal lattice structure. (The crystal lattice is a complex of atoms and molecules held together by electrons and atomic nuclei into an extremely large molecule-like structure.) The energy states are in so-called bands. When all the sites in an energy band are completely occupied by electrons, there is no flowing electric current because none of the electrons can accept increased energy even if exposed to an electric field of ordinary magnitude. This is thus a non-conductor state. When the energy gap between two bands is small, the electrons can be thermally excited into a conduction band and the electrons under the influence of an external electric source can initiate an electric current. This state of the crystal is an *intrinsic semiconductor*. The carriers of the current are called positive holes. Transistors (electronic amplifying devices utilizing single-crystal semiconductivity) operate by the principles of conduction electrodes and mobile positive holes. Industrially used semiconductors are *extrinsic semiconductors*, which mean that when small amounts of other material is introduced into them, it results in enhanced conductive properties. These devices are essential components of electronic laboratory equipments and communication systems, computers and television sets.

Semi-Conservative Replication: The regular mode of DNA replication where one old strand serves as template for the synthesis of a complementary new strand and this then with an old strand becomes the daughter double helix. ►DNA replication, ►replication, see Table P3 at ►pulse-chase entry and Fig. W4 in ►Watson and Crick model

Semi-Dominant: The dominance is incomplete, and therefore such genes may be useful because the heterozygotes can be phenotypically recognized. ►incomplete dominance, ►codominance

Semigamy: Occurs when the egg and sperm do not fuse, rather they contribute separately to the formation of the embryo that may become thus a paternal-maternal chimera. ►apomixis, ►androgenesis, ►parthenogenesis

Semi-Lethal: Genes that reduce the viability of the individual, and may cause premature death. ►lethal equivalent, ►lethal factors, ►LD50, ►LD*lo*

Seminal Fluid: ►semen, ►sperm

Seminal Root: The root of the embryo in plant seeds. ►root

Seminoma: Same as spermatocytoma.

Semio: In association with additional terms indicates signals or symptoms, e.g., semiochemicals such as pheromones or semiology, i.e., symptomatology.

Semi-Ortholog: The duplicated copy of a single copy ortholog. ►orthologous loci

Semiotics: The study of signs and symbols. It is used in bioinformatics and communication. Organisms interpret their environment by the signs encountered. Medicine makes diagnoses on the bases of signs.

Semisynthesis of Proteins: Non-natural amino acids, e.g., norvaline and homoserine can be inserted at specific protein sites to explore the conformational and functional consequence of the alteration. Similarly, photoactivatable cross-linkers, fluorophores, alteration of the active site of enzymes by specific amino acid replacement, cassette mutagenesis, introduction of stable isotopes have been accomplished. Selective chemical ligation of unprotected peptides may be useful to preserve solubility. Proteins can be changed posttranslationally by removing inteins and religation the flanks through different chemical reactions. Protein transsplicing can use inteins that were split into N- and C-terminal sections, which are separately inactive but activity is gained by ligation. By chemical synthesis, the D and L enantiomorphs of the HIV-1 proteases have been prepared displaying reciprocal chiral specificity as a proteases and similar specificity for enzyme inhibitors (Milton RC et al 1992 Science 256:1445). Ligation via an intein permits the production of new types of proteins from different elements and insertion of amino acid analogs may result in a cytotoxic molecule (Evans TC et al 1998 Protein Sci 7:2256). Sequential steps can link more than two elements. These engineered proteins have great theoretical and applied utility. ►homoserine, ►norvaline, ►active site, ►fluorophore, ►intein, ►cassette mutagenesis, ►expressed protein ligation, ►HIV-1, ►enantiomorph, ►chirality; Wallace CJ, Clark-Lewis I 1997 Biochemistry 34:14733; Muir TW 2003 Annu Rev Biochem 72:249.

Semisterility: Indicates that in an individual some gametes or gametic combinations are not viable when others are normal. Semisterility is common after deletion and duplication in the offspring of inversion and translocation heterozygotes but it may be caused by self-incompatibility, incompatible non-allelic combinations, cytoplasmic factors, fungal or viral infections, adverse environmental conditions, etc. ►chromosomal aberrations, ►mtDNA

Semisynthetic Compounds: Natural products but chemically modified.

Semliki Forest Virus: A member of the alphavirus group. ►alphavirus

Sendai Virus: A parainfluenza virus. In an ultraviolet light-inactivated form, it has been used to promote fusion of cultured mammalian cells or uptake of liposomes by its modifying effect on the lipids of the plasma membrane. ►cell genetics, ►cell fusion, ►cell membranes, ►polyethylene glycol, ►fusigenic liposome, ►alpha viruses

senDNA: Mitochondrial DNA (ca. 2.5 kb), excised from the first intron of the *cox1* gene (cytochrome oxidase) and amplified in *Podospora anserina.* This and similar structures appear to be responsible for aging in vegetative cultures. ►aging, ►killer plasmids

Senescence: The process of aging of organisms. At the cellular level, it has a somewhat different meaning. Cell senescence indicates how many cell divisions are expected on an average from isolated mammalian cells. Generally, cell senescence is correlated with the age of the individual and organism from where it was explanted. Human fibroblast cells under normal conditions cease to proliferate after about ±50 divisions although individual lineages may vary. Normal human mammary epithelial cells are different as they fail to senesce as fibroblasts do but eventually they develop telomerase problems and chromosomal anomalies. It has been suggested that the activity of the telomerase enzyme slows down and causes this phenomenon. The irreversible arrest of proliferation is *replicative senescence.* The tumor-suppressor protein p53, the retinoblastoma protein (Rb1), cyclin-dependent kinase (Cdk) inhibitors such as $p21^{CIP1/WAF1}$ and $p16^{INK}$ are also involved. The disruption of $p21^{CIP1/WAF1}$ leads to an escape of senescence by human fibroblasts. Some rodent cell lines (glia oligodendrocyte precursor cells) may not senesce. In addition, cancer cells and human fibroblast cells fused to cancer cells may divide indefinitely. Hyper-replication of the DNA induced by oncogenes may also invoke senescence (Micco RD et al 2006 Nature [Lond] 444:638). Senescence involves the upregulation of genes that are clustered in the chromosomes (Zhang H et al 2003 Proc Natl Acad Sci USA 100:3251). Plant cells when provided with an appropriate regime of phytohormones may be maintained continuously and can even be regenerated into differentiated organisms. Senescence of plant cells is regulated by cytokinins, and the increase in cytokinin level inhibits the process of senescence. ►aging, ►agonescence, ►monoclonal antibody, ►apoptosis, ►tissue culture, ►embryogenesis somatic, ►cell cycle, ►hybridoma, ►senDNA, ►killer plasmids, ►Hayflick's limit, ►telomerase, ►telomeres, ►p53, ►p21, ►$p16^{INK4a}$, ►Ets oncogene, ►Id proteins, ►Cdk, ►retinoblastoma, ►mole, ►plant hormones; Romanov SR et al 2001 Nature [Lond] 409:633; Karlseder J et al 2002 Science 295:2446; He Y, Gan S 2002 Plant Cell 14:805.

Senescence, Replicative: After a certain number of replications due to dysfunction of the telomeres, proliferation ceases. When this process goes out of control, it may involve tumorigenesis.

Senior-Loken Syndrome (NPHP1, 2q13; NPHP3, 3q22; NPHP4, 1p36; IQCB1, 3q13.31-q21.2): A heterogeneous syndrome involved in chronic kidney disease of children. NPHP1 encodes nephrocystin protein. NPHP3, responsible for nephronophtisis recessive kidney disease, encodes a 1330 amino acid protein interacting with nephrocystin. NPHP4 encodes the 1426 amino acid nephroretinin protein, which interacts with nephrocystin. IQCB1/NPHP5 also encodes nephrocystin and the patients display simultaneously retinitis pigmentosa. (See Otto EA et al 2005 Nature Genet 37:282.

Sense-Antisense Genes: Overlapping genes. ►overlapping genes

Sense Codon: Specifies an amino acid. ►genetic code

Sense Strand: The DNA strand that carries the same nucleotide sequences as the mRNA, tRNA and rRNA (of course in the RNAs U stand in place of T). It does not carry an absolute meaning because of some cases both strands are transcribed although in context of a particular RNA it is correct. ►template strand, ►coding strand

Sense Suppression: ►co-suppression, ►quelling, ►RNAi

Sensillum: A cuticular sensory element; *sensilla campaniformia* are small circular structures along longitudinal veins of *Drosophila* wings (see Fig. S31). ►*Drosophila*

Figure S31. Sensillum

Sensitivity: The percentage of correct identification of (carcinogens) on the basis of the (mutagenicity or other rapid) assay system. In DNA sequencing: the correctly predicted bases divided by total length of the cDNA. ►accuracy, ►specificity of mutagen assays, ►predictability, ►bioassays for environmental mutagens

Sensor Gene: Responsible for perceiving a signal. ►signal transduction

Sensoryneural: Affecting the nerve mechanism of sensing.

Sensory Neuropathy 1 (HSN1): A dominant (human chromosome 9q22.1-q22.3) de-generative disorder of the sensory neurons, ulcerations and bone defects. A serine palmitoyletransferase subunit maps within the HSN1 gene and it is expressed in the dorsal root ganglia. ►neuropathy, ►pain-insensitivity, ►hypomyelination, ►ganglion; Bejaoui K et al 2001 Nature Genet 27:261.

Sensory Transduction: Mediates touch, heat and pain responses in the skin. Sensory neurons of Aβ, Aδ or differently myelinated C fibers respond to the stimuli. Thermosensation is mediated through Transient Receptor Potential (TRP) families among them is the TRPV1 vanilloid transducer protein of capsaicin (present also in pungent peppers). Neuronal fibers evoke heat and pain sensation (nociception). TRPV2 is activated at temperatures above 52 °C. Other TRPVs respond to chemical stimuli. Cooling sensation (below 32 °C) is transduced by other TRPVs, TRPA (ankyrin family) and TRPNs. Mechanotransduction (touch responses) are sensed by Aβ or Aδ nociceptors, depending on intensity of the stimulus and other factors. Different channels exit for osmolarity, stretch, different chemicals, auditory (stereocilia) and other functions. Keratinocytes, Merkel (tactile) cells respond also to shape and texture. Some of the mechanisms have polymodality and are integrated into circuits. Many of these functions can be now be studied by mutations (Lumpkin EA, Caterina MJ 2007 Nature [Lond] 445:858). ►nociceptor, ►capsaicin, ►ion channel, ►osmolarity, ►deafness

Sentinel Phenotypes: These are used in human genetics to detect newly occurring mutations. These traits are supposed to be relatively easily detectable by direct appearance or clinical laboratory data can be obtained through routine examinations. Their frequencies are statistically evaluated for epidemiological information regarding possible increase in mutagenicity/carcinogenicity in an environment. ►mutation in human populations, ►epidemiology; Czeizel A 1989 Mutat Res 212:3.

Sentrin: A ubiquitin-carrier protein (known also under other names). ►ubiquitin, ►UBL, ►SUMO, ►PIC; Kahyo T et al 2001 Mol Cell 8:713.

Sep 1: A pleiotropic eukaryotic (yeast) strand exchange protein. ►recombination in eukaryotes, ►STPβ

Sepal: The whorl of (usually green) leaves below the petals in a flower (see Fig. S32). ►flower differentiation

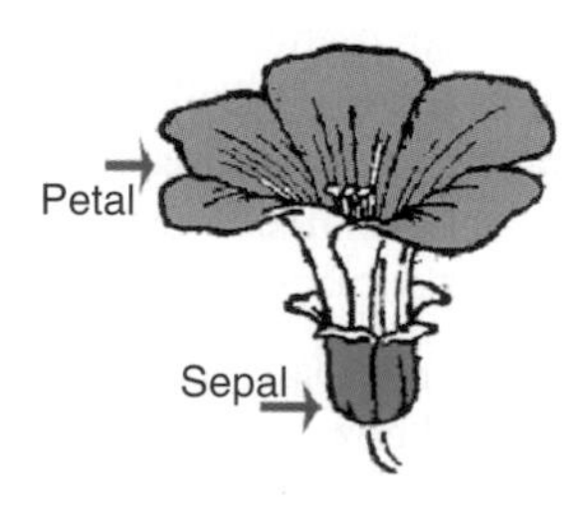

Figure S32. Sepal

Separins (Scc1, separase): Ubiquitous Esp1/Cut1-like proteins (∼180–200 kDa), which are removed from their inhibitory association with securin by APC in order to separate the sister chromatids at anaphase. Separins may have endopeptidase-(cysteine proteases) like function. The Rec8 subunit of cohesin is cleaved by separin and consequently the meiotic chiasmata are resolved. The disjunction of the homologous chromosomes takes place during meiosis I. Separase also cleaves the kinetochore-associated protein Slk19 at the onset of anaphase. Slk stabilizes the anaphase spindle and assures an orderly exit from anaphase. The inner centromere-like protein–Aurora complex is regulated by the CDC14 phosphatase (Pereira G, Schiebel E 2003 Science 302:2120). ►sister chromatid cohesion, ►co-orientation, ►Rec8, ►Scc1, ►mitosis, ►chiasma, ►cohesin, ►securin, ►shugoshin, ►APC, ►anaphase, ►Aurora, ►CDC14, ►kinetochore, ►spindle, ►FEAR; Hauf S et al 2001 Science 293:1320; Sullivan M et al 2001 Nature Cell Biol 3:771.

Sephadex: An ion exchanger or gel-filtration medium on cross-linked dextran matrix. ►dextran, ►gel filtration

Sephardic: Jews who moved to Spain after the Roman occupation of Israel and then in the Middle Ages to Western European countries. ►Ashkenazim, ►Jews and genetic diseases

Sepharose: The anion exchanger of agarose matrix such as DEAE (diethylaminoethyl) sepharose.

Sepsis: Bacteria or bacterial toxins in the blood stream, resulting in potentially fatal condition. Sepsis induces apoptosis of the lymphocytes by the action of caspases. Caspase inhibitors or introduction or stimulation of Bcl-2 may improve survival. Caspase 12 exists in both short and long forms by genetic determination. The long form may be less effective in controlling the innate reaction of inflammation and thus in sepsis. Inflammation is the first step in the immune reaction and it is controlled by cytokines. Bacterial clearance and sepsis resistance is found in mice deficient in caspase-12 (Saleh M et al 2006 Nature [Lond] 440:1064). ►apoptosis, ►Bcl, ►lymphocyte, ►caspase, ►sepsis, ►caspase; Hotchkiss RS et al 2000 Nature Immunol 1:496; Saleh M et al 2004 Nature [Lond] 429:75.

Septal: The adjective for septum (dividing structure, a wall).

Septate: Separated by cross walls (septa) (see Fig. S33).

Figure S33. Septate fungal mycelium

Septation: ►tubulins

Septic Shock: A bacterial lipopolysaccharide endotoxin-induced hypotension leading to inadequate blood supply to several organs; it is potentially fatal. Neutralizing MIF may alleviate it. The inflammatory responses are amplified by the triggering receptors (TREM) on neutrophils and monocytes. ►MIF; Patel BM et al 2002 Anesthesiology 96:576.

Septins: Rather ubiquitous polarizing GTPase (38–52 kDa GTP-binding) proteins, found in organisms from fungi to humans. In yeast, septins form hourglass-shaped pure proteins in between mother and daughter cells and mechanically mediate cytokinesis and growth (see Fig. S34) (Vrabioiu AM, Mitchison TJ 2006 Nature [Lond] 443:466). A human septin gene is at 17q25.3. Septin deficient spermatozoa are defective in movement because the kinesin-mediated intraflagellar transport of sperm stalls in this case, although fertility is rescued if the spermatozoa are injected into the oocytes (Ihara M et al 2005 Dev Cell 8:343). ►congression, ►cytokinesis; McIlhatton MA et al 2001 Oncogene 20:5930.

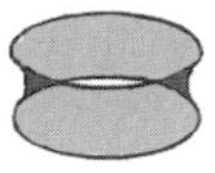

Figure S34. Hourglass-shaped septin

Septooptic Dysplasia (SOD, De Morsier syndrome, 3p21.2-p21.1): A relatively rare disorder involving optic nerve and pituitary gland hypoplasia and absence of the septum pellucidum (the double membrane separating the anterior horn and the lateral ventricles of the brain in the median plane bounded by the corpus callosum). A phenotype is quite variable. The molecular basis is deficiency in pituitary hormone production due to mutation in the HESX1 gene. ►pituitary; Carvaljho LR et al 2003 J Clin Invest 112:1192.

Sequatron: An automated high performance DNA sequencing apparatus (Hawkins TL et al 1997 Science 276:1887). ►sequenator, ►DNA chips, ►SAGE, ►DNA sequencing

Sequenase: A genetically engineered DNA polymerase. It combines the 85 kDa protein of phage T7 gene 5 and the 12 kDa *E. coli* thioredoxin protein (the latter keeps it associated with the template). The $3' \rightarrow 5'$ exonuclease activity is suppressed. It synthesizes about 300 nucleotides per second, and it is used for DNA sequencing and oligolabeling. ►DNA sequencing, ►oligo-labeling probes

Sequenator: An automated equipment that breaks up a protein sequentially, starting at the NH_2 terminus, into amino acids, identifies them by chromatography, and thus determines their sequence. ►amino acid sequencing

Sequence Alignment: ►CLUSTAL, ►Sequin; multiple sequence alignments algorithm: http://msa.cgb.ki.se/cgi-bin/msa.cg; multiple alignment tools: http://www.igs.cnrs-mrs.fr/Tcoffee/; multiple alignment including codons, mismatches, pseudogenes: http://coot.embl.de/pal2nal/; multiple alignment tool with shuffled and repeated sequences: http://aba.nbcr.net/.

Sequence Analysis Toolkit, for genes: http://www.migenas.org/home/index.jsp.

Sequence Logo: A sequence logo graphically represents DNA and amino acid sequence patterns from a set of aligned sequences (see Fig. S35). A column of stacked symbols represents each position of the alignment with its total height, reflecting the information content in this position by the Delila Software package. It displays aligned base sequences in bits, at each position by the height of the symbols http://biodev.hgen.pitt.edu/enologos/. (Diagram redrawn after Shaner M et al from web site: http://www-lecb.ncifcrf.gov/~toms/logoprograms.html).

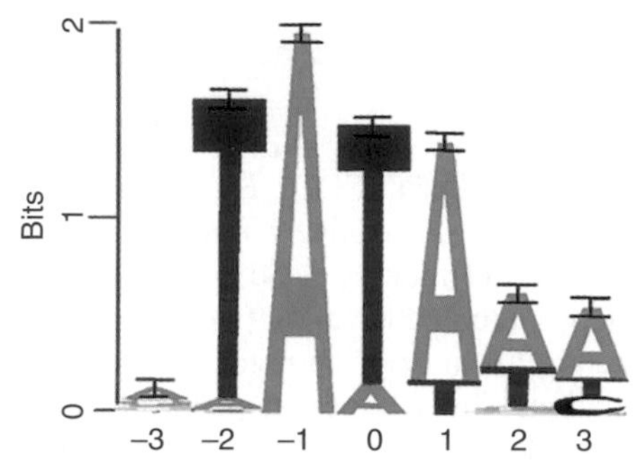

Figure S35. Sequence logo of 40 yeast TATA sites

Sequence Saturation Mutagenesis (SeSaM): To carry out SeSaM: 1. Generate DNA fragments of random length, 2. With the aid of terminal transferase add to the tails universal bases, 3. Elongate the fragments in a PCR to full length genes using a single-strand template and replace the universal base by standard one. Random mutations occur because the universal bases are promiscuous in pairing. ►terminal deoxynucleotidyl transferase, ►universal bases, ►polymerase chain reaction, ►mutagenesis; Wong TS et al 2004 Nucleic Acids Res 32(3):e26.

Sequence Skimming: In sequence skimming, a long DNA fragment is probed with some known genes in order to test whether the probes have homology and thus, a site within this long fragment chosen at random. ►probe; Elgar G et al 1999 Genome Res 9:960.

Sequence Space: The number of possible sequences of a particular length.

Sequence-Based Taxonomy: http://www.ncbi.nlm.nih.gov/Taxonomy/taxonomyhome.html.

Sequence-Tagged Connector (STC): ►genome project; Mahairas GG et al 1999 Proc Natl Acad Sci USA 96:9739.

Sequence-Tagged Site: ►sequenced tagged sites

Sequenced Tagged Sites: Single-copy DNA regions (100–500 bp), for which polymerase chain reaction (PCR) primer pairs are available and can be used for DNA mapping. ►PCR, ►primer, ►expressed-sequence tag; Venichanon A et al 2000 Genome 43:47.

Sequencing: ►DNA sequencing, ►survey sequencing, ►protein sequencing, ►RNA sequencing, ►genome projects, ►deep sequencing

Sequest: A software package for the analysis of mass spectral data of proteins/peptides (http://fields.scripps.edu/sequest/).

Sequester: To lay away or separate (into a compartment).

Sequin: A software tool for submitting nucleic acid sequence information to GenBank, EMBL, or DDBJ. It can be reached at http://www.ncbi.nlm.nih.gov/Sequin/index.html. ►GenBank, ►EMBL, ►DBJ

SER (smooth endoplasmic reticulum): An internal flat vesicle system in the cytoplasm, involved in lipid synthesis. ►RER

Serca: Sarcoplasmic reticulum Ca^{2+} ATPase. ►Brody disease, ►Darier-White disease

SEREX (serological analysis of tumor antigens by recombinant expression cloning): SEREX screens cancer patients' own sera for (autologous) tumor cells in order to determine antigens (cDNAs), which may be used for antibody-mediated immunotherapy. ►cancer gene therapy, ►immunotherapy; Okada H et al 2001 Cancer Res 61:2625.

Serial Analysis of Gene Expression: In the serial analysis of gene expression, short DNA sequence tags are prepared from many cDNA clones and after forming concatamers the entire cDNA is sequenced.

Serine (Ser, S): An amino acid (β-oxy-α-aminopropionic acid, MW 105.09); it is soluble in water. RNA codons: UCU, UCC, UCA, UCG, AGU, AGC. Serine is derived from the glycolytic pathway (see Fig. S36).

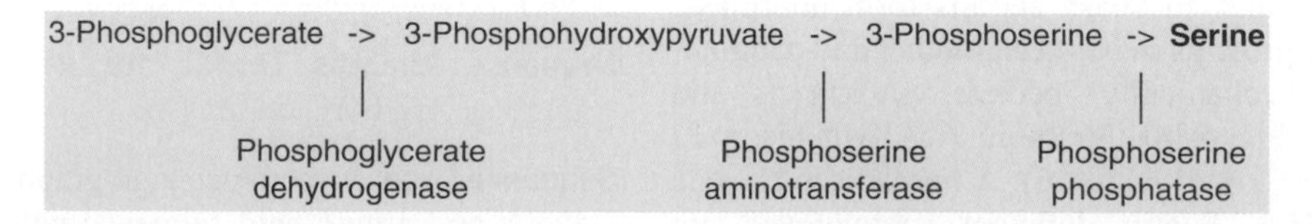

Figure S36. Serine biosynthetic path

Serine dehydratase enzyme, with pyridoxalphosphate prosthetic group, degrades serine into pyruvate and NH_4^+. ►amino acids, ►amino acid metabolism, ►oxalosis, ►3-phosphoglycerate dehydrogenase

Serine Kinase: ►MCF2 oncogene for serine phosphoprotein

Serine Protease: Serine protease degrades proteins in extracellular matrix and the family includes about 98 members and regulates many physiological reactions. ►matrix, ►kallikrein; Stoop AA, Craik CS 2003 Nature Biotechnol 21:1063.

Serine/Threonine Kinase: Serine kinase phosphorylates serine and tyrosine residues in proteins. Their receptors are transmembrane proteins and the kinases attach to the cytosolic carboxyl end of the receptors. ►transforming growth factor β, ►PIM oncogene, ►PKS oncogene, ►activin, ►bone morphogenetic protein, ►membrane proteins, ►protein kinases, ►receptor guanylyl cyclase, ►signal transduction, ►SMAD

Serine/Threonine Phosphoprotein Phosphatases: Serine phosphoprotein phosphatases remove phosphate from serine and threonine residues of proteins. *Protein phosphatase-I* is inhibited by cAMP by promoting the phosphorylation of a *phosphatase inhibitor protein* through protein kinase A. *Protein phosphatase IIA* is the enzyme most widely involved in dephosphorylation of the products of serine/threonine kinases. *Protein phosphatase-IIB* (calcineurin) is most common in the brain where Ca^{2+} activates it. *Protein phosphatase-IIC* plays only a minor role in the cells. The catalytic subunit of the first three is homologous but they also contain special regulatory subunits. *Phospholipase C* (PLC) may be coupled to G-proteins and upon its activation the level of Ca^{2+} increases. This cation mediates numerous cellular reactions. ►serine/threonine kinases, ►phosphorylases, ►signal transduction, ►regulation of gene activity

Serine/Threonine Protein Kinase: ►serine/threonine kinase

Seripauperines (PAU): A large group of proteins in eukaryotes, conspicuously low in serine and having amino-terminal signal sequence. Their function is still unknown. They are encoded at subtelomeric sites in all yeast chromosomes. ►signal sequence; Coissac E et al 1996 Yeast 12:1555.

Seroconversion: As per seroconversion, new antibody production against an antigen alters the serological state.

Serodeme: A particular type of antigen produced by a clone. ►antigen

Serology: Serology deals with antibody levels and with the reactions of antigens. ►serum; serological classification: http://fred.bioinf.uni-sb.de/sepacs.html.

Seronegative: A seronegative fails to display antibodies against the antigen in question.

Seropositive: In a serpositive, reactive antibody to an antigen is present in the serum.

Seroswitch Vector: In a seroswitch vector, the epitope of the viral coat protein can be changed to evade the adverse serological reaction against it, in cases when the same passenger DNA must be used repeatedly, e.g., in cases of adenoviral vectors. ►adenovirus, ►epitope, ►serotype

Serotonin (5-hydroxytryptamine): A tryptophan-derived neuro-transmitter modulates sensory, motor, and behavioral processes (including also feeding behavior) controlled by the nervous system (see Fig. S37). Hydroxytryptamine 2B receptor also regulates the cell cycle by interacting with the tyrosine kinase pathway through the phosphorylation of the retinoblastoma protein and the activation of cyclin D1/Cdk4 and cyclin E/Cdk2. Cyclin D1, in concert with other proteins, induces also the MAPK pathway. The 5-hydroxytryptamine receptor 5-HT_{1B} binds protein p11 and in a mouse model seems to alleviate depression. Triptan drugs used for treatment of migraine seem to act as a 5-HT_{1B} agonist (Sharp T 2006 Science 311:45; Svenningsson P et al 2006 Science 311:77). Serotonin transporter (SERT) is a polytopic membrane transporter with 12 transmembrane domains. The 5-HT_{3A} receptor is a large-conductance neuronal serotonin channel. Low activity of 5-HT serotonin transporter may cause a propensity to psychiatric disorders (Ansorge MS et al 2004 Science 306:879). ►neurotransmitters, ►glucocorticoid, ►obesity, ►substance abuse, ►alcoholism, ►cocaine, ►cyclins, ►Cdk, ►MAPK, ►retinoblastoma, ►migraine

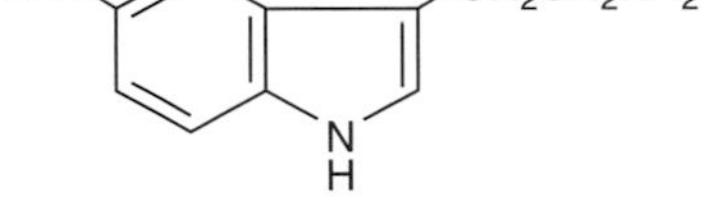

HO — $CH_2CH_2NH_2$ — N H

Figure S37. Serotonin

Serotype: A serotype is distinguished from other cells by its special antigenic properties. ►antigenic variation, ►capsule

Serovar: Same as serotype.

Serpentine Receptors: Seven-membrane spanning receptors. ►seven membrane proteins

Serpines (serine-protease inhibitors, 14q32.1): Serpines, when mutated, may be responsible for emphysema, thrombosis, and angioedema. Viral infection may counteract them. The accumulation of neuroserpin caused by mutation leads to a familial encephalopathy with neuroserpin inclusion bodies (FENIB), a dementia. There are about 500 serpins of 350–500 amino acid residues. They occur in all organisms from mammals, to plants, to viruses. Serpin, SRPN6 seems to control innate resistance in *Anopheles* against *Plasmodium* (Abraham EG et al 2005 Proc Natl Acad Sci USA 102:16327). ►maspin, ►C1 inhibitor, ►Hsp [Hsp47], ►L-DNase II, ►emphysema, ►thrombosis, ►angioedema, ►encephalopathy, ►antitrypsin, ►malaria, ►*Plasmodium*; Atchley WR et al 2001 Mol Biol Evol 18:1502; Silverman GA et al 2001 J Biol Chem 276:33293; Crowther DC 2002 Hum Mut 20:1; Lomas DA, Carrell RW 2002 Nature Rev Genet 3:759.

Serprocidin: ►antimicrobial peptides

Sertoli Cells: ►Wolffian ducts

Serum: The clear part of the blood, from which the cells and the fibrinogen have been removed; the clear liquid that remains of the blood after clotting. The immune serum contains antibodies against specific infections. It differs from plasma, which is the non-particulate portion of cells. ►plasma, ►antibody production, ►serology

Serum Dependence: As per serum dependence, animal cells may grow or differentiate only or preferentially in cultures containing serum.

Serum Response Element (SRE): A DNA tract that assures transcriptional activation in response to growth factors in the serum. ►SRE, ►CArG box

Server (web server): The server delivers information available on the internet with the aid of computer programs.

Sesame (*Sesamum indicum*): An oil seed crop with about 37 related species; the cultivated form is 2n = 2x = 26 but related species may have x = 8 and different levels of ploidy.

Sesquidiploid: A sesquidiploid contains a diploid set of chromosomes derived from one parent and a higher-number set from the other. ►allopolyploid, ►allopolyploid segmental

Sessile: A sessile plant is attached directly to a base without a stalk.

SET Motifs (Su[var3-9]-enhancer-of-zeste-trithorax): originally named after the three *Drosophila* regulatory proteins where they occur and modulate chromatin structure and thus gene expression. Set2 is a methyltransferases enzyme. So far the products of at least 20 genes display such a domain with an apparent role in epigenesis/development. The *polycomb* gene of *Drosophila*, yeast telomeric silencing, heterochromatin-mediated gene silencing, variegation position effects (PEV), as well as, the Mll factors involve SET domains. ►chromodomain, ►integration, ►transcription, ►Mll, ►*w* locus, ►Polycomb, ►Sbf1, ►epigenesis, ►methyltransferase; Baumbusch LO et al 2001 Nucleic Acids Res 29:4319.

Set Point, Viral: The level of circulating virus in the plasma during the nonsymptomatic phase preceding the progression to AIDS (Fellay J et al 2007 Science 317:944). ►acquired immunodeficiency

Set Recoding: ►fuzzy inheritance

Seta: Bristles, stiff hairs of animals or plants.

SETGAP (selectable expression of transient growth-arrest phenotype): A system suitable for the isolation of genes that interfere with the expression of certain other genes or genetic pathways. ►negative regulator; Pestov DG, Lau LF 1994 Proc Natl Acad Sci USA 91:12549.

Sevenless (sev): X-chromosomal gene (1-33.38) of *Drosophila* controlling the R7 rhabdomeres and thus altering the photoreceptivity of the eye. The carboxy terminal of the protein product of the wild type allele shows homology to the tyrosine kinase receptor of *c-ras*, *v-src*, and EGF. ►photoreceptor, ►rhodopsin, ►signal transduction, ►ommatidium, ►compound eye, ►daughter of sevenless, ►RAS, ►EGF, ►BOSS, ►rhabdomere

Seven-Membrane Proteins (7tm): Integral parts of the plasma membrane that span the membrane by seven helices; they are important in signal receptor binding and in association with G-proteins. More than 800 genes encode these most versatile receptors of chemokines, hormones, neurotransmitters, odorants, and taste and light signals. Heteromeric G proteins, β-arrestin, and GRK family of proteins mediate the signal transmission to 7tm protein receptors. ►signal transduction, ►G-protein, ►transmembrane receptors, ►arrestin, ►GRK; Pierce KL et al 2002 Nature Rev Mol Cell Biol 3:639.

Seven-Pass Transmembrane Proteins: Same as seven membrane protein.

Severe Combined Immunodeficiency (SCID): A less frequently occurring (0.00001-0.00005) autosomal disease than the X-linked agammaglobulinemia, but it is generally lethal before age two. The thymus is abnormally small and therefore there is a severe deficiency of the T- and sometimes also the B-lymphocytes. The afflicted infant cannot overcome infections. SCID-X1 may be effectively treated by gene therapy but cancerous growth may ensue when the retroviral vector is inserted in the vicinity of the promoter of the LMO2 proto-oncogene (Hacein-Bey Abina S et al 2003 Science 302:415). In some cases, viral infection may severely damage the thymus, and this non-hereditary disease may closely mimic the symptoms of SCID. The DNA-dependent kinase (p350), encoded in human chromosome 8q11, is most likely responsible for SCID-1. A gene in chromosome 10p of humans interferes with the V(D)J recombination system and thus prevents normal function of both B- and T-lymphocytes (Moshous D et al 2001 Cell 105:177). Several other gene loci may also be involved in the development of the disease. SCID mouse devoid of T and usually also B-lymphocytes can accept human grafts and can be used to create a partial human immune system in the mouse. The most common form of it is X-chromosome-linked (Xq13). In about 40% of the cases, there is adenosine deaminase deficiency. T cell deficiency may be treated by the transplantation of hematopoietic cells. ADA may be corrected by gene therapy (Fischer A et al 2001 Immunity 15:1). In some cases, transplantation of thymus tissues may lead to improvement. ►agammaglobulinemia, ►hypogammaglobulinemia, ►immunodeficiency, ►SCID, ►DNA-PK, ►adenosine deaminase deficiency, ►gene therapy

Sewall Wright Effect: Same as drift genetic.

Sex: In eukaryotes, sex makes possible the production of two kinds of gametes and it is the requisite of syngamy. By recombination and by promoting linkage equilibrium, sex facilitates selection of adaptive variation and provides a means for elimination of deleterious genes. In prokaryotes and viruses, "sex" is recombination. The majority of species reproduce sexually and the relatively few asexual species seem to represent dead ends in evolution. The tiny *Bdelloid* freshwater rotifers are exceptional because they survived and evolved for 35–40 million years without sex. The *genetic sex* in the female is determined by the X chromosome(s) whereas the Y chromosome carries the male determining genes. The *gonadal sex* is represented by the ovarian differentiation in the female and the testicular differentiation in the male. *Somatic sex* is gonadally controlled. Female differentiation takes place—irrespective of the chromosomal constitution—if the gonads are removed during early fetal development. The anti-Müllerian hormone synthesized by the Sertoli cells and the fetal androgens (testosterone, androstenedione) synthesized by the Leydig cells normally suppress female differentiation. The fetal female gonads have no affect on the female somatic sex development. In the gonadless genital tract of both sexes, the Müllerian ducts are maintained but the Wolffian ducts degenerate. Estrogen synthesized by the female may adversely affect the male type differentiation. The anti-Müllerian hormone apparently blocks, however, an enzyme (aromatase) required for feminization and in this case the somatic sex shifts towards the direction of masculinization. ►gender, ►sex cell, ►syngamy, ►copulation, ►sex determination, ►recombination, ►linkage, ►meiosis, ►Wolffian ducts, ►Müllerian ducts, ►gonads, ►sex hormones

Sex Allocation: The variation in sex ratio in favor of males or females due to non-chromosomal sex-determining mechanisms such as exist in social insects, and are caused by colony size, mating behavior, and available resources. ►sex determination, ►sex ratio

Sex Bias in Disease Phenotype: As per the sex bias, one or the other sex is more likely to express the disease. ►Rett syndrome, ►imprinting

Sex Bias in Mutation: ►mutation rate

Sex Bivalent: In the sex bivalent, the X and Y chromosomes have homology only in the short common segment where they can pair and recombine. ►pseudoautosomal

Sex Body (XY body): A structure associated with both the X and the Y chromosomes. For its formation, the H2AX (a histone 2A variant) is required apparently for chromatin remodeling. The sex body inactivates both the X and Y chromosome during the pachytene stage of sperm formation and is involved in gene silencing. The sex body presumably protects against illegitimate chromosome association and thus, aneuploidy (McKee BD, Handel MA 1993 Chromosoma 102:71). ATR localizes, under BRCA1, to the XY chromatin and after phosphorylating histone H2AX, the sex body is formed (Turner JMA et al 2004 Curr Biol 14:2135). In H2AX deficiency the males but not the females are infertile. ►ATR, ►breast cancer, ►Meisetz, ►histone variants, ►pachytene; Fernandez-Capetillo O et al 2003 Dev Cell 4:497.

Sex Cell: A gamete that can fuse with another sex cell of the opposite mating type (sperm, egg) to form a zygote. ►zygote, ►mating type, ►gamete, ►isogamy

Sex Chromatin: ►Barr body

Sex Chromosomal Anomalies in Humans: Sex chromosomal anamolies are of various types and they may occur at a frequency of 0.002 to 0.003 of all births. *Females*: X0, XXX, XXXX, XXXXX, X0/XX, X0/XXX, X0/XXX/XX, XX/XXX, X0/XYY, XXX/XXXX, XXX/XXXX/XX and *males*: XX, XYY, XXY, XXYY, XXXY, XXXYY, XXY/XY, XYY/XYYY, X0/XXY/XY, XXYYY/XY/XX, XXXY/XXXXY. Other, even more complicated types have been reported. The most common mechanism by which these anomalies occur is nondisjunction in meiosis and mitosis. The more complex type mosaics (indicated by /) are the result of repeated nondisjunctional events. The X0 condition is called *Turner syndrome*, the XXX is *triplo-X,* while XXY and other male conditions with multiple X and Y(s) are generally referred to as Klinefelter syndrome along with the XX males, which have a Y-chromosome translocation to another chromosome. Similar sex-chromosomal anomalies have been identified in various other mammals. The X0 condition results in an abnormal female in humans and mice but in a normal male in grasshopper or *Caenorhabditis*, and in an abnormal male in *Drosophila*. ►trisomy, ►chromosomal sex determination, ►Turner syndrome, ►Klinefelter syndrome, ►triplo-X, ►XX males, ►gynandromorph, ►testicular feminization

Sex Chromosome: The sex chromosome is unique in number and/or function to the sexes (such as X, Y or W, Z); see ►chromosomal sex determination. In the heterogametic sex, the X and Y chromosomes pair and may recombine in a relatively short terminal region although in the heterogametic sex in insects recombination is practically absent, except when transposable elements function. ►PAR, ►sex determination

Sex Circle Model of Recombination: The basic tenets of the sex circle model of recombination in fungi, according to F.W. Stahl (1979 Genetic Recombination. Freeman, San Francisco, California), are: *1.* Any marker can recombine either by reciprocal exchange or by gene conversion. *2.* Close markers are more likely to recombine non-reciprocally. *3.* Gene conversion observes the principle of parity. *4.* Gene conversion is polar. *5.* In half of the cases, gene conversion is accompanied by classical exchange of outside markers. *6.* Reciprocal recombination is always accompanied by exchange of outside markers. *7.* Conversion that does not involve outside exchange shows no interference of flanking genes. *8.* Gene conversion accompanied by outside marker exchange may also involve interference. *9.* Conversion asci (5:3, 6:2) obey the principles listed under 1 to 8. *10.* All markers (except deletions, and a small fraction of conversion alleles) can segregate post-meiotically. *11.* The very rare aberrant 4:4 conversion asci may be the results of two events. ►recombination, ►gene conversion

Sex Comb: Special structures on the metatarsal region of the foreleg of *Drosophila* male (see Fig. S38). ►*Drosophila*

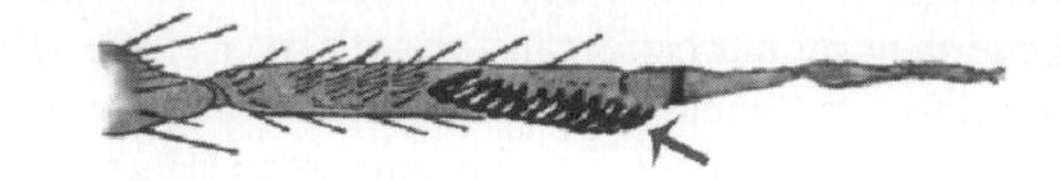

Figure S38. Sex comb

Sex Controlled (sex influenced): The degree of expression of a gene is determined by the sex (e.g., baldness is more common in human males than females). ►Hirschsprung's disease, ►Huntington's disease, ►imprinting

Sex Determination: In dioecious animals and plants, sex is usually determined by the presence of two X (female) and XY (male) chromosomal constitution, respectively. In other words, the females are homogametic (i.e., the eggs all carry an X chromosome) and the males are heterogametic (i.e., they can produce sperm with either X or Y chromosomes). Exceptionally XX individuals may be males if the sex-determining section of the Y chromosome is translocated to an X. An XY individual can be a female if from the Y chromosome the sex-determining part of her Y chromosome was lost. In some species, e.g., birds and moths, the females are heterogametic (WZ) and the males are homogametic (ZZ). In the nematode *Caenorhabditis,* some grasshoppers, and some fishes, the females are XX and the males are of XO (single X) constitution. In *Drosophila*, the proportion of the X chromosome(s) and autosomes (A sets) determines sex. Normally, if the ratio is 1 X:2 sets of autosomes, the individual is male; if there are 2 Xs:2 sets of autosomes, the fly is female. All individuals with a sex ratio above 1 are also females and those with a ratio between 0.5 and 1 are intersexes. XO human and mouse individuals are females, however, and irrespective of the number of X chromosomes, as long as there is at least 1 Y chromosome, they appear male. In hermaphroditic plants the development of the gynoecia and androecea are determined by one or more gene loci. Actually, in *Drosophila* three major and some minor genes are known to control sex. *Sexlethal* (*Sxl*, 1-19.2) can mutate to recessive *loss-of-function* alleles that are deleterious to females but inconsequential to males. The dominant *gain-of-function* mutations do not affect appreciably the females but are deleterious to males. The *Sxl* locus may produce ten different transcripts. Three transcripts (4.0, 3.1 and 1.7 kb) are expressed at the blastoderm stage. Adult females have four transcripts (4.2, 3.3, 3.3, and 1.9 kb); the latter two are missing or reduced if the germline is defective. Adult males display three transcripts (4.4, 3.6 and 2.0 kb). The *Sxl* transcripts are alternatively spliced and functional in the female and are non-functional in the male. The *Sxl* gene product is apparently required for the maintenance of sexual determination and the processing of the downstream *tra* (*transformer*, 3-45) gene product. The Sxl protein (354 amino acids) controls alternative splicing of the *tra* pre-messenger RNA by binding to a polypyrimidine tract (UGUUUUUUU) of a non-sex-specific 3′ splice site of one *tra* intron. This binding then prevents the binding of the U2AF general splicing factor binding to the site, and U2AF is forced then to a female-specific 3′ splice site. The Sxl protein binds also to its own pre-mRNA and promotes its female-specific splicing. The *Sxl* locus is regulated by other known genes: *da* (*daughterless*, 2-41.5) is a positive activator of *Sxl*, and it is suppressed by the gain-of-function mutations of the latter gene. The expression of da^+ is necessary for the proper development of the gonads of the female in order to form viable eggs. In both sexes, the product of da^+ is required also for the development of the peripheral and central nervous systems and the formation of the cells that determine the adult cuticle. Thus, the da^+ gene has both maternal and embryonic influence. Females heterozygous for the da^1 mutations produce sterile or intersex males and masculinize the exceptional daughters, which are homozygous for *male* (*maleless*, 2-55.2); *male* is lethal to single X males but has no affect on XX females. The DA gene product is a helix-loop-helix protein with extensive homology to the human kE2 enhancer (human chromosome 19p13.3-p13.2) of the κ-chain family of immunoglobulins. Chromosomally, female (XX) flies homozygous for the third-chromosome recessive *tra* mutations become sterile males. XXY *tra/tra* individuals are also sterile males but XY *tra/tra* males are normal males. A 0.9 kb transcript of the locus is female-specific and is required in the female, and another 1.1 kb RNA is present in both sexes but no functions are known and is probably not essential. The splicing of the *tra* transcripts is controlled by *Sxl* gene products. When the 0.9 transcript is expressed in a XY fly, the body resembles that of a female. Another *tra* locus (*transformer 2*, 2-70) regulates spermiogenesis and mating in normal males. Null mutations of *tra2*, when homozygous, transform XX females into sterile males. Actually, the *tra2* gene products seem to mediate the splicing of the *dsx* (*double sex*, 3-48.1) transcripts. Dominant mutations at the *dsx* locus when heterozygous with the wild type allele change XX individuals into sterile males but they have no effect on XY males. Null alleles of *dsx*, when homozygous, transform XX flies into intersexes. The recessive allele *dsx11* transforms XY flies into intersexes, and the null alleles change both XX and XY flies into intersexes. Germline sexual differentiation is not affected by the normal allele of this gene but it is controlled by the X: autosome ratio. A 3.5 kb female-specific transcript is present in the larvae and adults. In the larvae, 3.8 and a 2.8 kb male-specific transcripts are detectable and by adult stage, in addition, a 0.7 kb RNA also appears. The *ix* (*intersex*, 2-60.5) mutations when homozygous also change the XX flies into intersexes. Homozygous *ix* XY males appear normal morphologically but their courtship and mating behavior is altered. Thus, sex determination in *Drosophila* appears to follow the cascade and *fru* regulates mating behavior and sexual orientation through the *tra* and *tra2* genes:

In summary: the X:autosome ratio is the trigger mechanism for the alternate sex developmental pathways. In the males, the *Sxl* and *tra* genes are expressed but their transcript is not spliced to functional forms (see Fig. S39). The critical male sex-determining function is attributed to locus *dsx*, which in the wild type produces a protein blocking the genes required for female development. In the females, with a chromosomal constitution of 2X:2A sets, a functional *Sxl* product is manufatured that mediates the female-specific splicing of its own transcripts. The *Sxl* protein mediates then the splicing of the *tra* transcripts, leading to the synthesis of a Tra protein, which along with the Tra2 protein directs the female-specific splicing of the *dsx* transcript. The synthesized DSX protein blocks then all the genes with functions that would be conducive to male development. Sex determination in *Caenorhabditis* is different from that in *Drosophila*, probably because the XX individuals are hermaphrodites and the nondisjunctional gametes lead to the development of the rare XO males. The level of expression of the known sex-determination genes is shown in Figure S40.

$$\overline{\hspace{4em}\textit{fru}\hspace{4em}}$$
$$\rightarrow \textit{Sxl} \rightarrow \textit{tra} \rightarrow \textit{tra2} \rightarrow \textit{dsx} \rightarrow \textit{ix}$$

Figure S39. Major sex genes in *Drosophila*

Early in the pathway, *fox* (female X) acts as a numerator of the X-chromosomes, and 5 X-linked *dpy* (*dumpy*) alleles regulate dosage compensation. The hermaphroditic XX females of *Caenorhabditis* originally may produce sperm, oocyte production is then switched on. The *fem-3* gene turns on sperm production in the XX animals. The 3′-untranslated region of the mRNA of the *fem-3* gene mediates the switch after the cytoplasmic binding factor FBF protein binds to this region.

Six *mog* genes are important regulators of female ⇌ male switching. In XX *Caenorhabditis*, the sex determination complex protein (SDC-2) blocks the expression of the male-determining gene *her-1* and in that state hermaphrodites are formed. The SDC-2

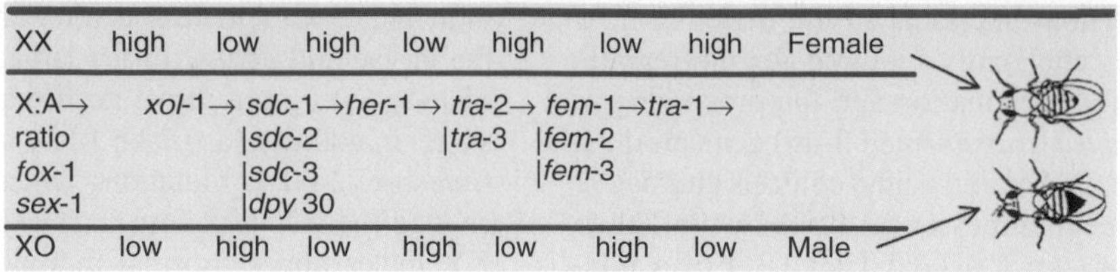

XX high low high low high low high Female

X:A→ *xol*-1→ *sdc*-1→*her*-1→ *tra*-2→ *fem*-1→*tra*-1→
ratio |*sdc*-2 |*tra*-3 |*fem*-2
fox-1 |*sdc*-3 |*fem*-3
sex-1 |*dpy* 30

XO low high low high low high low Male

Figure S40. Sex chromosome: autosome ratio. (Diagram modified after Kuwabara, PE & Kimble, J 1992. *Trends Genet* 8:164.)

recruits to the X chromosome SDC-3, DUMPY (dpy), MIX-1 (mitosis and X), and other proteins, resulting in the reduced expression of the X chromosomal genes; thus, dosage compensation is realized. In the XO males, *her-1* is transcribed and the SOC complex does not attach to the single X chromosome and all its genes are expressed normally. Sex determination in mammals is much more complex and the pathway is not entirely clear. For a number of years, the H-Y antigen was thought to have a major role but this turned out to be an incorrect notion. A major critical difference was found in an 11 amino acid segment of SMCX (structural maintenance chromosome X) and SMCY proteins encoded within the X and Y homology region. The genes DMRT1 and 2 in the short arm of human chromosome 9 have homologs in other mammals. Also, the Z chromosomes in chicken, alligators, *Drosophila*, and *Caenorhabditis* display higher expression in the male gonads than in the female gonads and appear to be basic regulators of sexual dimorphism. Sex chromosomal anomalies in humans generally also lead to mental retardation. In several reptiles, sex is determined by the temperature the eggs are exposed to during incubation in the sand they are laid in. Thus climate change and vegetation change can affect population structure in these animals (Kamel SJ, Mrosovsky N 2006 Ecol Appl 16:923). The actual manifestation of sex may be deeply affected by endocrine hormones directly or indirectly through environmental pollutants. In *Plasmodium* (causing malaria), induction of blood formation favors an increased production of the male parasite. In both plants and some hermaphroditic animals, stress conditions promotes maleness (Hughes RN et al 2003 Proc Natl Acad Sci USA 100:10326). ►mating type determination in yeast, ►freemartins, ►hormones in sex determination, ►*fru*, ►gynandromorphs, ►chimera, ►intersex, ►fat body, ►hermaphrodite, ►testicular feminization, ►sex hormones, ►pheromones, ►accessory sexual characters, ►sex phenotypic, ►mental retardation, ►chromosomal sex determination, ►X-chromosome counting, ►mealy bug, ►dosage compensation, ►*Msl*, ►*Mle*, ►transsexual, ►H-Y antigen, ►SRY, ►sex-selection, ►mating type determination in yeast, ►*Schizosaccharomyces pombe*, ►F plasmid, ►sex plasmid, ►Hfr, ►sex-chromosomal anomalies in humans, ►social insects, ►*Sciara*, ►*schisotomiasis*, ►sex determination in plants, ►*Rumex*, ►sex-reversal, ►pseudoautosomal, ►amelogenin test, ►arrhenotoky, ►complementary sex determination, ►numerator, ►*Plasmodium*, ►mealy bug, ►haploid; for sex determination in *Caenorhabditis* see Meyer BJ 2000 Trends Genet 16:247; Mittwoch U 2001 J Exp Zool 290:484; Koopman P 2001 Cell 105:843; Vilain E 2000 Annu Rev Sex Res 11:1; Goodwin EB, Ellis RE 2002 Curr Biol 12:R111; Hodgkin J 2002 Genetics 162:767; detailed review of the molecular mechanism of sex determination in *Drosophila*: Black DL 2003 Annu Rev Biochem 72:291; sex chromosome evolution in mammals: Marshall Graves JA 2006 Cell 124:901.

Sex Determination in Plants: In dioecious plants, sex determination is very similar to that in animals (►sex determination). In monoecious and hermaphroditic plants, sex is controlled without the presence of special chromosomes and a number of genes (nuclear, mitochondrial, and plastidic) involved in morphogenesis, phytohormone synthesis, and environmental responses determine the differentiation of the flowers and the oogenesis (female) and microsporogenesis (male), and therefore, sexuality. Genes are known that are similar to those of sex reversal in animals and feminize or masculinize, respectively, the monoecious or hermaphroditic flowers (e.g., *tassel seed*, *silkless* [in maize], *superman*, *gametophyte female* [in *Arabidopsis*], etc.). *Tasselseed 2* encodes a short-chain alcohol dehydrogenase that is involved in stage-specific floral organ abortion. Gibberellic acids, brassinosteroids, ethylene, chromosome-breaking agents, and mutagens may also influence the expression of sexual development as well as temperature regimes and other environmental factors. ►gametophyte, ►gametophyte factors, ►vernalization,

►photoperiodism, ►self-incompatibility, ►flower differentiation, ►phytohormones; Juarez C, Banks JA 1998 Curr Opin Plant Biol 1:68.

Sex Differences: Sex differences vary among phylogenetic groups in both morphology and function. In humans, the chromosomal constitution (XX, XY) is different. The more than single X chromosome generally undergoes lyonization. The ribosomal protein RPS4Y encoded by the Y chromosome is different from the X chromosome RPS4X. The relative level of hormones depends on sex; the ovaries produce more estrogen and progesterone (correlated with incidence of breast cancer). Reduction of the natural supply of estrogen may involve reduced memory and increase autoimmune disorders in females. Other, phenotypic differences are obvious. A survey of the expression of 13,977 mouse genes in males and females indicated differences besides the reproductive tissues; differences were also found in the kidney and liver genes involved in drug and steroid metabolism and osmotic regulation (Rinn JL et al 2004 Dev Cell 6:791). Sex hormones quantitatively affect the expression of many genes involved in different metabolic functions and disease (Weiss LA et al 2006 Nature Genet 28:218). ►sex, ►sex determination, ►lyonization, ►estrogen, ►androgen, ►autoimmune diseases, ►gender; human sex differences between men and women: Federman DD 2006 N Engl J Med 354:1507.

Sex Differentiation: ►sex, ►sex determination

Sex, Evolutionary Significance: It has generally been assumed that sex facilitates the purging of linked deleterious parental genes by recombination. But Keightley, P.D. and Eyre-Walker, A. (Science 290:331) arrived to the conclusion that sex is not maintained by its ability to purge deleterious mutations. Sex may be disadvantageous for evolution. Experimental evidence in yeast supports the advantage of sex in selective environment (Goddard MR et al 2005 Nature [Lond] 434:636). Sex selects for robustness and evolutionary advantage (Azevedo RBR et al 2006 Nature [Lond] 440:87). Acquisition of sex led to the evolution of diploidy, which is protective against the consequences of deleterious recessive mutations. ►sex, ►Kondrashov's deterministic model of evolution of sex; Rice WR, Chippindale AK 2001 Science 294:555; Kondrashov FA, Kondrashov AS 2001 Proc Natl Acad Sci USA 98:12089; Bachtrog D 2003 Nature Genet 34:215.

Sex Factor: A transmissible plasmid in bacteria that carries the fertility factor(s) F. ►F^+, ►Hfr, ►F plasmid

Sex Hormones: Sex hormones have either estrogenic (female) or androgenic (male) influence. These are steroids of the ovaries and placenta (estradiol, progesterone), the testes (testosterone), or of the adrenal cortex (cortisol and aldosterol). Gene expression patterns of the uterine luminal epithelial cells might be regulated by estradiol-17β and inhibited by progesterone. Progesterone rapidly downregulated about 20 genes associated with DNA replication. Among the down-regulated group were all six minichromosome maintenance proteins (MCM), suggesting that replication licensing is a key in sex steroid hormone regulation of cell proliferation in the uterus (Pan H et al 2006 Proc Natl Acad Sci USA 103:14021). Testosterone is also required in females, although in smaller amounts. Androgens control the reproductive organs but also affect hair growth (beard) and the early death of the hair follicles causing preferential male baldness. Androgens promote bone and increased muscle growth as well. Some of the synthetic "anabolic hormones," without androgenic effects, are used (illegally) by athletes to boost performance. Testosterones are also precursors of estrogens. Estrogens are formed in the female-specific organs and their targets include the mammary glands, bones, and fat tissues. Estrogen synthesis is regulated by the follicle-stimulating hormone (FSH) of the anterior pituitary. The pituitary luteinizing hormone mediates the release of the egg, and progesterone is required for the maintenance of pregnancy. The administration of exogenous estrogens and progestins inhibit ovulation and can be used as contraceptives. Other compounds act by prevention of the fusion of the sperm with the egg or implantation of the egg in the uterus, e.g., the drug RU486. Before the development of the contraceptive pills, in the ancient world herbal medicine had been used, prepared from plants (berries of *Juniperus sabina*, gentiana) (see Fig. S41) that contained estrogens. Sperm production may be stopped by injection of progestin and androgen combinations, while inhibiting epididymal functions can prevent sperm maturation or may interfere with the release of enzymes required for breaking through the protective coat of the egg. Antiprogestins, antiestrogens, or other inhibitors of steroid biosynthesis and non-peptide antigonadotropin-releasing hormone antagonists may serve as female contraceptives. The steroid hormone-controlled sexual behavior is also mediated by neuronal activity. Prolonged use of steroid contraceptives or androgenic or anabolic steroids may increase the risk of liver, ovarian, or uterine carcinomas. There are several diseases or conditions (heart diseases, thromboses, embolism, diabetes, skin irritations, *Chlamydia* infection, etc.) where contraceptive drugs are not or conditionally permissive. In some conditions (ovarian and endometrial cancer, uterine myoma, rheumatoid arthritis, etc.), oral contraceptives may be beneficial. Ethinylestridiol may occur in human females taking oral contraceptives and if they become pregnant accidentally (3%), the fetus is exposed to this compound. Bisphenol leached out from resin-coated lining of food and beverage containers, dental sealers, and polycarbonate plastic is also an estrogen. Male mouse fetus, when exposed to these substances, develops prostate duct abnormalities (Timms BG et al 2005 Proc Natl Acad Sci USA 102:7014). ►animal hormones, ►hormone receptors, ►RU486, ►bisphenol, ►estradiol, ►progesterone, ►MCM, ►hyperlipoproteinemia, ►epididymis, ►transsexual, ►sex, ►fertilization, ►fertility, ►infertility, ►ART, ►contraceptives

Figure S41. *Juniper*

Sex Influenced: ►sex-influenced, ►sex controlled, ►imprinting

Sex Linkage: In sex linkage, various genes in the sex chromosomes are inherited with the transmission of that chromosome. Sex linkage in females is generally partial because the two X-chromosomes may recombine. The recombination between the X and Y chromosomes is limited only to the homologous (pseudoautosomal) regions. In some insects (*Drosophila*, silkworm), recombination even between autosomes is usually absent in the heterogametic sex. ►recombination frequency, ►recombination mechanisms, ►crossing over, ►hemophilia, ►autosexing, ►genetic equilibrium, ►pseudoautosomal, ►criss-cross inheritance; Morgan TH 1910 Science 32:120; Morgan TH 1912 Science 36:719.

Sex Mosaic: The sex-chromosomal constitution in the body cells may vary in a sectorial manner. Typical examples are the gynandromorphs in insects, which have body sectors with both XX and XO constitution. Sex mosaicism also occurs in humans with variable numbers of X and Y chromosomal sectors. The mosaicism is generally the result of non-disjunction or chromosome elimination. ►sex determination, ►non-disjunction, ►gynandromorphs

Sex, Phenotypic: The phenotypic manifestations of the influence of the steroid sex hormones, such as facial hair, increased phallic size in males, horns or special plumage in animals, and enlarged breast and mammary glands development in females. ►sex determination, ►animal hormones

Sex Pilus: ►pilus

Sex Plasmid: The bacterial F plasmid. ►F element, ►F plasmid

Sex Proportion: The proportion of male individuals in a population. ►sex ratio

Sex Ratio: The *primary sex ratio* is the number of male conceptuses relative to that of females. The *secondary sex ratio* indicates the number of females:males at birth. The *tertiary sex ratio* states the ratio among adult males and females. Since XX females are mated with XY males, the proportion—just as in a testcross—should be 1:1. In the United States, at birth the proportion is about 105–106 males:100 females. In the West Indies the proportion is about 1:1 or the number of females is slightly more. In China and Korea, the sex ratio in newborns is about 115 males to 100 females. Generally the female:male ratio shifts in favor of females by progressing age. By about 21–22, in the USA, the female:male proportion becomes about 1:1, and because of mortality differential of the sexes, by age 65 there are about 145 females for 100 males. In dioecious plants, the sex ratio may vary a great deal because of modifier genes and physiological factors (hormone supply). Since the 1930s, population geneticists have repeatedly considered the problem whether infanticide alters the sex ratio by selection. Infanticide generally biases the childhood ratio against females. This might have the consequence that the genes of families producing males would be favored, would have greater fitness, and the secondary sex ratio would tend to be biased in favor of males. The problem is more complicated, however, in human societies because systems of mating and the socio-economical conditions have a substantial influence. Statistical data seems to indicate that under stressful living conditions, women preferentially abort male fetuses and this reduces the secondary sex ratio (Catalano R, Bruckner T 2006 Proc Natl Acad Sci USA 103:1639).

In some *Drosophila* stocks, infection by filiform bacteria results in female offspring. The *sex-ratio* genes in the *Drosophila* X chromosome cause an excess of females in the progeny of the males carrying this gene(s). Usually, drive suppressors in the autosomes and in the Y chromosome balance the sex-ratio gene expression. Some gynandromorphs with very small XO sectors may also survive. Triploid intersexes live, as do females sex-transformed by *tra*, *ix*, and *dsx* genes (i.e., phenotypically males, although XX). The sex ratio in the Seychelles warbler's (*Acrocephalus sechellensis*) eggs may vary according to the availability of food supply; in low-food territories they may have 77% male offspring, whereas with high food supply the proportion of sons may be only 13%. This difference is not due to the different viability of the eggs. In some species of the *Cyrtodiopsis* flies in Malaysia, the sex ratio may be biased either toward the females or toward the males. In the cases of female bias, the males carry a *driver-X* chromosome that by some means eliminates from fertilization most of the Y-bearing sperms resulting in predominantly female (XX) offspring. In the male biased stocks, the Y chromosome carries a suppressor for the driver-X and actually somewhat increases the chances of function for the Y-bearing sperms, resulting in more than 50% male progeny. Irrespective of the actual mechanism of sex determination, natural selection tends to promote the 1:1 proportion of the two sexes because as long as sexual reproduction is maintained, both males and females contribute to the offspring. ►hermaphroditism, ►sex determination, ►sex-reversal, ►spirochete, ►infectious heredity, ►sex proportion, ►gynandromorph, ►age of parents and secondary sex ratio, ►male-stuffing, ►sex selection, ►segregation distorter, ►meiotic drive, ►dosage compensation, ►*Wolbachia*; Jaenike J 2001 Annu Rev Ecol Syst 32:25; Hardy ICW (Ed.) 2002 Sex Ratios. Concepts and Research Methods. Cambridge University Press, Cambridge, UK.

Sex Realizer: A substance that determines whether male or female gonads would develop.

Sex Reversal (sex change): Sex reversal involves sex phenotypes that do not match the expectation based on chromosomal sex determination. Gene *fru* splices differently in males and females of *Drosophila*. If the splicing in males is according to the female pattern or in the females in the male pattern, the sexual behavior (courtship) is reversed and males and females court the same sex (Demir E, Dickson BJ 2005 Cell 221:785) because of altered response to pheromones through the olfactory system (Stockinger P et al 2005 Cell 1231:795). It has been suggested that a certain number of trinucleotide repeats (glutamine) in the *Sry* gene would be responsible for this reversal but other studies could not confirm the mechanism in mice and it appears that alteration in the function of autosomal regulator genes may be involved. There are apparently sex-determining autosomal (17q24.3-q25.1 and 9p24) factors, which may cause sex reversal in 46XY individuals. Other autosomal and X-chromosomal genes (Xp21.3-p21.2) may also be responsible for sex reversal. In young mice with knockout for the estrogen receptors α and β (αβERKO), the development of the sexual organs is near normal. By adult stage in the ovaries of the females, seminiferous tubule-like structures develop, Müllerian inhibitory substance is formed at an elevated level, and Sox9 protein has been found indicating that the estrogen receptors (ER) are essential for the maintenance of the normal ovarian phenotype. The αβERKO males displayed some spermatogenesis but also became sterile. In both males and females, the αER is most essential for normal sexuality. Male-to-female sex change in mice seems to be controlled by fibroblast growth factor 9. Defect in the nuclear localization of SRY may lead to gonadal dysgenesis in humans. Mutation in the R-spondins family of human growth gene results in a recessive syndrome in the absence of the testis-determining SRY gene, characterized by complete XX sex reversal, palmoplantar hyperkeratosis, and predisposition to squamous cell carcinoma of the skin (Parma P et al 2006 Nature Genet 38:1304). R-spondins are ligands interacting with Fzd/LRP (Frizzled/Lipoprotein related) receptor complexes and inducing beta-catenin-T cell factor (TCF) gene activation in different species both in vitro and in vivo. ►SRY, ►*tra*, ►sex determination, ►campomelic dysplasia, ►hermaphroditism, ►pseudohermaphroditism, ►homosexual, ►testicular feminization, ►SF-1, ►estradiol, ►SOX, ►knockout, ►gonads, ►Müllerian ducts, ►*Wingless*, ►adrenal hypoplasia congenital, ►FGF, ►gonadal dysgenesis, ►thrombospondin, ►R-spondin; Ostrer H 2000 Semin Reprod Med 18:41; Colvin JS et al 2001 Cell 104:875; Li B et al 2001 J Biol Chem 276:46480; sex reversal–sex determination review: Camerino G et al 2006 Curr Opin Genet Dev 16:289.

Sex Selection: Sex selection is possible with the use of cell sorters, followed by artificial insemination. Sex determination is feasible also by FISH or PCR before implantation. The mammalian X-chromosome-bearing sperm has 2.8 to 7.5% more DNA than the Y bearing sperms. The sperm can be classified and selected with high efficiency and at high speed (18 million sperm/h). It is used in animal husbandry for artificial insemination and its effectiveness is 85–95%. It would be technically feasible also in human artificial insemination and ethically less objectionable than the preimplantation selection of fertilized eggs. Sex selection involves not just ethical problems but it may also affect the sex ratio in the population. It has been estimated that in China the males exceed females by ~30% because of selective abortion. This also leads to a social imbalance because many males may not find mates. ►cell sorter, ►segregation distorter, ►sex determination, ►ART, ►PGD, ►social selection, ►sex ratio; Garner DL 2001 J Androl 22:519; Johnson LA 2000 Anim Repr Sci 60–61:93; van Munster EB 2002 Cytometry 47:192; Welch GR, Johnson LA 1999

Theriogenology 52:1343; international survey of frequency: Sermon K et al 2005 Hum Reprod 20:19.

Sex Vesicle (XY body): The meiotically paired mammalian sex chromosomes in the males may be heterochromatinized and form the sex vesicle, a special, visible structure. ►sex chromosomes, ►heterochromatin

Sexduction (F-duction): F-duction takes place when genes carried in the bacterial sex element (F′ plasmid) recombine with the bacterial chromosome. ►F′ plasmid, ►Hfr, ►conjugational mapping, ►transduction; Jacob F et al 1960 Symp Soc Gen Microbiol 10:67; Lederberg EM 1960 Symp Soc Gen Microbiol 10:115; see Fig. S42.

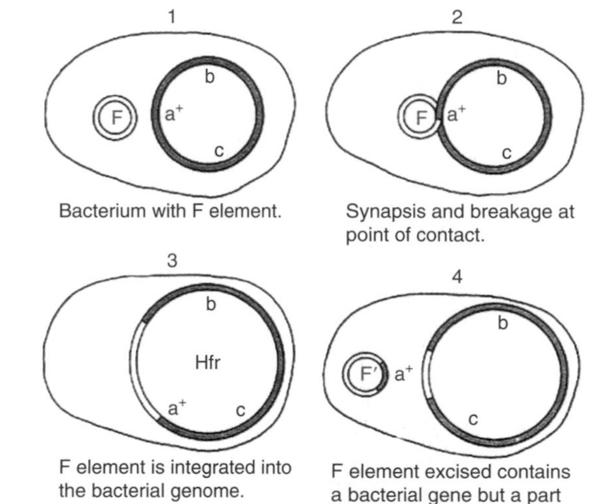
1
F
a+
b
c
Bacterium with F element.
2
F
a+
b
c
Synapsis and breakage at point of contact.
3
b
Hfr
a+
c
F element is integrated into the bacterial genome.
4
F′
a+
b
c
F element excised contains a bacterial gene but a part of its own DNA is left behind within the main genophore of the cell.

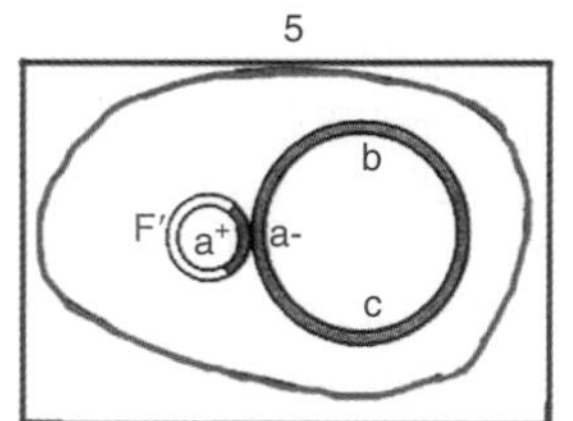
5
F′
a+
a-
b
c
The F′ element containing a segment homologous to the main genophore of the cell may participate in reciprocal recombination.
Another cell acquires thus the sexduced a* gene

Figure S42. Sexduction

Sex-Influenced: A sex-influenced trait is that, which has a different degree of expression in male and female individuals, e.g., facial hairs in humans, color of plumage in birds, horns in deer, etc. ►hare lip, ►pyloric stenosis, ►Hirschsprung's disease, ►lupus erythematosus, ►imprinting

Sexing: Distinguishing female from male forms of animals. This procedure may be difficult in young birds because the genitalia may appear ambiguous to those who do not have special expertise (►autosexing). The avian males are homogametic (WW) and the females are heterogametic (ZW). Sexing may be carried out also by the karyotyping of the tissues or Barr body detection. Molecular sexing may make possible the identification of sex on the basis of DNA markers from any tissue sample. On the Z chromosomes, there are both the CHD-W and the CHD-NW genes whereas the W chromosome carry only the CHD-NW gene. The base sequences of these two genes are very similar, except in a short tract. When a restriction enzyme cuts within this segment of CHD-W, the females display three electrophoretic bands but the males show only one.

A non-invasive sexing may be carried out on preimplantation embryos by inserting a green fluorescent transgene into the X chromosome. The male offspring of green fluorescent mammalian males will not display fluorescence. Such transgenic animals are apparently normal. ►sex determination, ►autosexing, ►genetic sexing lines, ►aequorin

Sex-Limited: The expression of a sex-limited trait is limited to one sex (e.g., lactation in females, Wildervanck syndrome).

Sex-Linked Lethal Mutations: Sex-linked lethal mutations in *Drosophila* served as the first laboratory test to quantitate mutation frequency and assess the mutagenic properties of physical and chemical agents. The old procedure was called the *ClB* test (C stands for crossover exclusion brought about by the presence of usually three inversions, *l* is a recessive lethal gene, and *B* indicates the dominant *Bar eye* mutation.) The principle of the techniques is diagrammed at *ClB*. If any new recessive mutation takes place in the X chromosome of a male, then in the F_2 only females may occur because the original recessive *l* gene present in the inverted *B l* chromosome will kill the hemizygous male progeny. If a new lethal mutation occurs, it may kill (or much reduce the proportion) of the other type of males in F_2. Nowadays instead of the *ClB* method, generally the improved *Basc* chromosome is used (*B: Bar, a: apricot* eye color [*w* locus] and *sc*: *scute* inversions). Any recessive lethal mutation in the X-chromosome of the grandfather's sperm results in the death of one of the grandsons. Rarely, some exceptional females are also found, which are the result of unequal sister chromatid exchange in the inversion heterozygote mother. Somewhat similar manner autosomal recessive lethals can also be detected in *Cy L/Pm* stocks. ►bioassays in genetic toxicology, ►autosomal recessive lethal assay, ►*Basc*, ►*Clb* and diagrams there.

Sex-Reversal: As per sex-reversal, the sex by karyotype does not always correspond to sex phenotype. By translocation, a testis-determining factor (TDF/SRY) may move from the Y chromosome to an X chromosome, and thus 46 autosome XX males can occur. Also, it has been claimed that an autosomal testis-determining factor (TDFA) may be responsible for some of the intersexes and sex reversion cases. Terminal deletions of human chromosome 9p and 10q may result in sex-reversal. ►intersexes, ►H-Y antigen, ►SRY, ►hermaphroditism, ►pseudohermaphroditism, ►testicular feminization, ►Swyer syndrome, ►gonadal dysgenesis, ►sex determination, ►campomelic dysplasia, ►DSS, ►*Polycomb*; Osterer H 2001 J Appl Physiol 91:2384; Li B et al 2001 J Biol Chem 276:46480.

Sex-Selection: As per sex selection, X-bearing sperms can be separated with reasonably good efficiency from their Y-bearing counterparts with the aid of cell sorters and fluorescent labeling, on the basis of the increased DNA content of X-bearing sperms. This process then can be used in artificial insemination in animal breeding. For humans, sex-selection is primarily an ethical issue. ►sex determination, ►sex ratio; Johnson LA 2000 Anim Reprod Sci 60–61:93; Welch GR, Johnson LA 1999 Theriogenology 52:1343.

Sexual Conflict: In sexual conflict, the enhanced reproductive success of one of the sexes reduces the fitness of the other, e.g., polyspermic fertilization. The seminal fluid of the male (mating success) may be toxic to the female and reduces the lifespan of the female *Drosophila*. ►polyspermic fertilization, ►sexual selection; Rice WR 1996 Nature [Lond] 381:232.

Sexual Differentiation: The realization of sex-determination (gonads) and development of secondary sexual characters, such as facial hair in men, differential plumage in birds, etc. In *Drosophila*, for the male gonad differentiation and normal sexual development the expression of the JAK-STAT signal transduction pathway is required in the somatic cells. For the female gonads, JAK-STAT activity is not necessary (Waversik M et al 2005 Nature [Lond] 436:563). ►gonad, ►signal transduction; Wedell A et al 2000 Lakartidningen 97:449; Burtis KC 2002 Science 297:1135; Reiner WG, Gearhart JP 2004 N Engl J Med 350:333; gonadal and functional abnormalities: MacLaughlin DT, Donahoe PK 2004 N Engl J Med 350:367.

Sexual Dimorphism: As per sexual dimorphism, the two sexes are morphologically distinguishable. In some species, the differences appear only during later development or by the time of sexual maturity. The differences are not limited to morphology but also various functions may differ. In the males, the language centers are localized in the left inferior frontal gyrus region of the brain, while in females both left and right regions are active. Conversion of testosterone to estradiol by neuronal tissue critically affects sexual differentiation of the brain. GABA and calcium-binding proteins are more abundant in the newborn male rats relative to females. This difference appears to be a switch from excitatory action in the males to inhibitory signals in the females. GABA would increase phosphorylation of CREB at Ser^{133} by protein kinase A, calcium-activated calmodulin kinase, ribosomal S6 kinase 2, and mitogen-activated kinase-activated protein kinase 2 in the brain of the male. This appears to be the initial signal to sexual dimorphism. Wnt-7a regulates sexual dimorphism of mouse by controlling the Müllerian inhibitory substance. ►human intelligence, ►autosexing, ►sex determination, ►GABA, ►protein kinases, ►calmodulin, ►CREB, ►*wingless*, ►Müllerian duct, ►gonads; Auger AP et al 2001 Proc Natl Acad Sci USA 98:8059; Vincent S et al 2001 Cell 106:399.

Sexual Dysfunction: ►erectile dysfunction, ►priapism, ►Peyronie disease, ►vaginismus

Sexual Incompatibility: ►incompatibility

Sexual Isolation: Sexual isolation has significance in speciation by, either preventing mating (isolation by life cycle, behavior or generative organs) between certain genotypes, or because of gametic or zygotic death or inviability of their offspring. A very common form of sexual isolation is the inviability of the recombinants of chromosomal inversions. The cross-breeding at the incipient speciation may be prevented by pheromone gene(s) responsible for the production of differences in the female cuticular hydrocarbons in *Drosophila*. ►incompatibility, ►inversions, ►speciation, ►fertility, ►pheromone, Majestic J 2001 FEMS Lett 199:161; Fang S et al 2002 Genetics 162:781.

Sexual Maturity: The developmental stage by which reproductive ability is attained. It varies in different organisms (m: month, y: year): cat 6–12 m, cattle 6–12 m, chimpanzee 8–10 y, dog 9 m, elephant 8–16 y, horse 12 m, humans 12–14 y, mouse 1 m, rabbit 3–4 m, rat 2 m, sheep 6 m, swine 5–6 m. Sexual maturity is affected also by environmental factors. ►gestation

Sexual Orientation: ►homosexual

Sexual Reproduction: The production of offspring by mating of gametes of opposites sex or mating type. Sexual reproduction seems to have evolved because it facilitates segregation of alleles and recombination among linked loci, resulting in increased fitness in different environments. The fungus *Cryptococcus neoformans* has both *a* and α mating types, yet the majority of the populations are represented by α mating type only. In these cases, the same mating type spores may fuse, undergo meiosis and produce four α spores (see Fig. S43) (Lin X et al 2005 Nature [Lond] 434:1017). ►meiosis, ►gametogenesis; http://www.germonline.org/index.html.

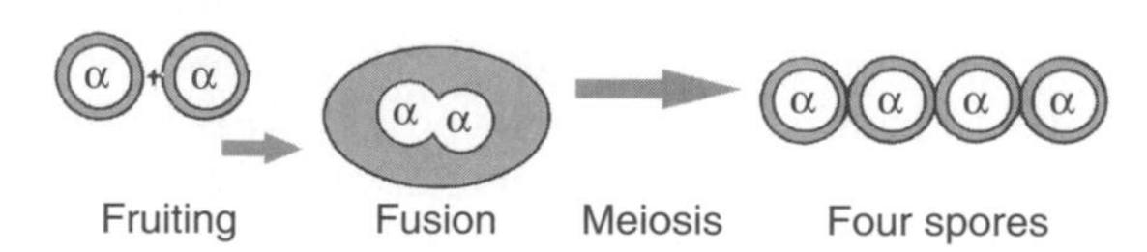

Figure S43. Sexual reproduction between same mating type

Sexual Selection: Competition among mates of the same sex or gametes or preferential choice of one or another type of mates. The general purpose of sexual selection is to find mates with selective advantage for the offspring. Some body ornaments of the male birds may increase their sexual attractiveness to females and provides a selective advantage in mating and producing offspring. Some body features may be associated with high viability genes and are evolutionarily advantageous to the bearer as a means of higher chances of leaving offspring by mate choice. A quantitative analysis of the correlation between body-ornament (white head-spot) of the male flycatcher bird (*Ficedula albicollis*) indicated that the ornament did not evolve by hitchhiking with genes of otherwise higher fitness but by its own merit (see Fig. S44) (Qvarnström A et al 2006 Nature [Lond] 411:84). In some instances, the actual value of the selected trait, e.g., the fancy tail of the peacock may not easily be rationalized (see Fig. S45). The gynogen Sailfin Molly fish females reproduce clonally, yet they rely on sperm of heterospecific males to initiate embryogenesis; thus, it appears that the males do not contribute to the progeny. Yet the sexual forms of the females prefer those males, which mate with the gynogens. Consequently, the males exploited by the gynogens still benefit from the unusual sexual selection. Although sexual selection most frequently involves the selection of the males, in the sand lizard (*Lacerta agilis*), the females achieve selection. The

Figure S44. *Ficedula albicollis*

Figure S45. Peacock, Chinese handiwork

females may copulate with several closely or distantly related males but it appears that the share of offspring sired by the remotely or unrelated males is higher in the same clutch. Thus, the females selecte apparently the sperm of the more distantly related males. Females of feral fowl may eject the sperm acquired through coerced mating by inferior males. Mate preference in male and female rats whose progenitors had been treated with the antiandrogenic fungicide vinclozolin (LD_{50} 10 mg/kg, rat) has been observed. This effect is sex-specific, and females three generations removed from the exposure discriminate (transgenerational effect) and prefer males who do not have a history of exposure, whereas similarly epigenetically imprinted males do not exhibit such a preference (Crews D et al 2007 Proc Natl Acad Sci USA 104:5942). Polymorphism exists in binding a protein that facilitates the attachment of the sperm to the egg and that may affect male selection by the egg. In some instances, the sexual selection is based on meiotic drive. In *Drosophila* species, mate recognition/preference is influenced by cuticular hydrocarbon composition. In guppy fish, sexual attractiveness of the males is positively correlated with ornamentation (encoded in the Y chromosome) but negatively correlated with survival. Asexual populations may have higher fitness than the sexual ones because every individual has a chance to reproduce, especially when the deleterious mutation rate is higher in males than in females, as is most commonly the case. ►meiotic drive, ►sexual dimorphism, ►certation, ►megaspore selection, ►assortative mating, ►disassortative mating, ►gynogenesis, ►heterospecific, ►conflict evolutionary, ►sexual conflict, ►transgenerational effect; mate choice: Chenoweth SF, Blows MW 2006 Nature Rev Genet 7:681; Swanson WJ et al 2001 Proc Natl Acad Sci USA 98:2509; Snook RR 2001 Curr Biol 11:R337; Knight J 2002 Nature [Lond] 415:254; Pizzari T, Birkhead TR 2002 Biol Rev 77:183.

Sexual Swelling: A large, conspicuous reddish structure in between the vulva and the anus of the females of many Old World primates at the time of ovulation. It is a mating attractant for the males and has a value in sexual selection. ►vulva, ►anus, ►sexual selection

Sexuparous: The sexual production of offspring in species where parthenogenetic reproduction co-exists with sexual reproduction. ►parthenogenesis

SF: ►hybrid dysgenesis I-R system. Also ►splicing factor protein. ►introns

SF1 (stimulatory factor): A yeast protein (33 kDa) reducing the binding requirement (at the concentration of 1 SF/20 nucleotides) of protein Sep 1 to DNA during recombination, by two orders of magnitude. It probably has a role in DNA pairing. ►synapsis

SF1 (splicing-transcription factor): SF1 represses transcription. (See Goldstrohm AC et al 2001 Mol Cell Biol 21:7617).

SF-1 (Ftz-F1, fushi tarazu factor homolog, nuclear receptor subfamily 5 group A member 1[NR5A1]): The steroidogenic factor (adrenal 4 binding protein [Ad4bp]) at human chromosome 9q33 (30 kb genomic DNA) regulates cytochrome 450 steroid 21-hydroxylase, aldosterone synthase, anti-Müllerian hormone (MIS), XY sex reversal, and adrenal failure and plays key role in steroid biosynthesis. SF-1 function is modulated by phospholipids through mediating ligand binding (Li Y et al 2005 Mol Cell 17:491). ►steroid hormones, ►fushi tarazu, ►Müllerian ducts; Whitworth DJ et al 2001 Gene 277:209; crystal structure: Wang W 2005 Proc Natl Acad Sci USA 102:7505.

SF2/ASF: ►SR protein, ►SR motif

SFF: A cell cycle regulating yeast protein. ►cell cycle

SFK: A group of tyrosine protein kinases, SFKs, reside on the interior part of the cell membrane and respond to external signals. They are members of the Src family. When phosphorylated by Csk or other kinases, the proteins assume an inactive conformation and their activation requires phosphatases. Cbp attracts Csk to the membrane. The active SFK then phosphorylates other proteins. Rafts localized on the outer surface of the membrane modify SFKs. ►CBP, ►Csk, ►RAFT, ►Src; Vara JA et al 2001 Mol Biol Cell 12:2171.

SFM: Serum-free medium.

Sfpi1: A protooncogene; probably the same transcription factor as PU.1 and Spi*1*. ►*Spi1*, ►PU.1

SGLT (sodium/glucose co-transporter): A defect in SGLT results in Glucose-Galactose Malabsorption (GGM), a potentially fatal neonatal (human chromosome 22) recessive disorder, unless the diet is made sugar-free. (See Xie Z et al 2000 J Biol Chem 275:25959).

SGP-1: The syntenic gene predictor in homologous sequences of genomes. ►gene predictor; Wiehe T et al 2001 Genome Res 11:1574; http://soft.ice.mg.de/spg-1.

SGT1: A proteolysis factor controlled by ubiquitin. ►ubiquitin; Tor M et al 2002 Plant Cell 14:993.

SH_2 (src homology domain): An about 100 amino acid long binding site for tyrosine phosphoproteins. These phosphoproteins such as the SRC and ABL cellular oncogenes, phosphotyrosine phosphatases, GTPase activating protein, phospholipase C, and the Grb/Sem 5 adaptor protein have important role in signal transduction. In the human genome, there are about 115 SH2 domains. ►SH3, ►SRC, ►WW, ►PTB, ►pleckstrin, ►signal transduction

SH_3: (*src* homology domain): The binding site for the proline-rich motif (Arg-X-Leu-Pro-Pro-Z-Pro [the latter is Leu for the Src oncoprotein and Z is Arg for phosphoinositide kinase] or it can be X-Pro-Pro-Leu-Pro-X-Arg) in an adaptor or mediator protein in the signal transduction pathway through RAS. By binding, conformational and functional changes take place. The activity of the cellular SRC protein increases during normal and neoplastic mitoses. Protein p68, closely related to the GAP associated p62, is bound to the SH3 domain of SRC. The SH3 binding is specific in vivo, yet of low affinity. The human genome encodes about 250 SH3 domains (see Fig. S46). ►SH2, ►SRC, ►RAS, ►signal transduction, ►GAP; SH3 site detection: http://cbm.bio.uniroma2.it/SH3-Hunter/.

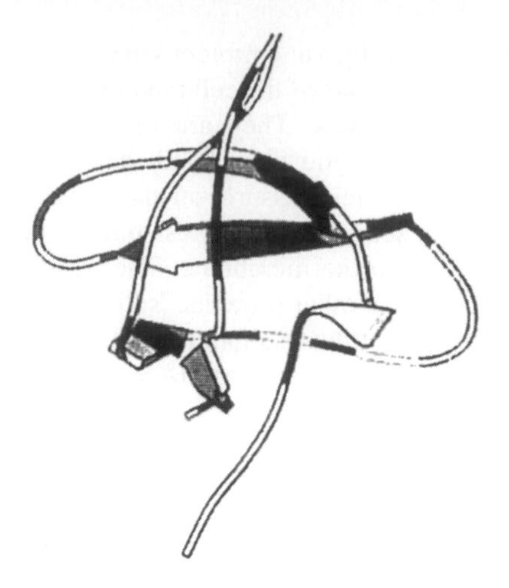

Figure S46. Structure of the SH3 domain. From Alm E and Baker D 1999 Proc. Natl. Acad. Sci. USA 96:11305

Shade-Avoidance Syndrome: The shade-avoidance syndrome of plants results in elongated stem, reduced leaf expansion, and accelerated flowering under reduced light. Several phytochrome genes mediate it. ►phytochrome, ►floral evocation; Cerdán PD, Chory J 2003 Nature [Lond] 423:881.

Shadow Bands: Same as stutter band.

Shadowing: An electronmicroscopic preparatory procedure by which the surface of the specimen is coated with a vaporized metal, such as platinum. The shadowed objects display a three-dimensional effect in scanning, but in some cases, even in transmission electronmicroscopy. ►scanning electronmicroscopy

Shadowing, Phylogenetic: A cross-species base sequence comparison in order to find important functional, structural, and regulatory elements in different evolutionary categories. (See Boffelli D et al 2003 Science 299:1391).

Shah-Waardenburg Syndrome (SOX10/WS4, 22q13; EDN3/ET3 20q13.2-q13.3): The Shah-Waardenburg Syndrome involves the endothelin-3 signaling pathway; it shows the combined symptoms of the Hirschsprung disease and the Waardenburg syndrome. The embryonic neural crest appears to have recessive (EDN3) or dominant (SOX10) defects. ►Hirschsprung disease, ►Waardenburg syndrome, ►endothelin, ►RET oncogene, ►SOX; Touraine RL et al 2000 Am J Hum Genet 66:1496.

Shaking (shak): Alleles at several chromosomal locations in *Drosophila* cause shaking of the legs under anesthesia to a variable degree, depending on the locus and allele involved. Some mutants may display hyperactive behavior in a temperature-dependent manner. Some may be viable, while others are homozygous lethals. The *shakB* mutants may cause a defect in the synapse between the giant fiber neuron, postsynaptic interneuron, the dorsal longitudinal muscle, and the nerves operating the tergotrochanter (back-neck) muscle (see illustrations at ►*Drosophila*). The shaking may be caused by a defect in a protein of a potassium ion channel. ►ion channel

Shannon-Weaver Index: ►diversity

Shark Cartilage: Sharks, mainly sea-inhabiting cartilaginous fishes, are presumably rarely afflicted by tumors. The prevailing assumption was that their cartilage protects from cancer. Experimental evidence—contrary to some early studies—indicates, however, that shark cartilage products are ineffective against advanced cancer (Loprinzi CL et al 2005 Cancer 104:176).

SHARP (SMRT and histone deacetylase-associated repressor protein): A regulator of PPAR. ►PPAR, ►SMRT; Shi Y et al 2002 Proc Natl Acad Sci USA 99:2613.

Shasta Daisy (*Chrysanthemum maximum*): An ornamental plant; 2n ≈ 90 (see Fig. S47).

Figure S47. Shasta daisy

SHC: An adaptor protein involved in RAS-dependent MAP kinase activation after stimulation by insulin, epidermal (EGF), nerve (NGF), platelet derived growth factor (PDGF), interleukins (IL-2,-3,-5), erythropoietin and granulocyte/macrophage colony-stimulating growth factor (CSF), and lymphocyte antigen receptors. It binds to tyrosine-phosphorylated receptors and when phosphorylated at tyrosine it interacts with the SH2 domain of Grb2, which interacts with SOS in the RAS signal transduction pathway. The phosphotyrosine-binding (PTB) domain can also recognize the tyrosine-phosphorylated protein. The latter is similar to the pleckstrin homology domain and most likely binds the acidic phospholipids of the cell membrane. SHC is also an oncogene, involved in the development of phaeochromocytoma neoplasias. ►signal transduction, ►phaeochromocytoma,

►adenomatosis endocrine multiple, ►neoplasia, ►insulin, ►EGF, ►NGF, ►PDGF, ►SOS, ►interleukins, ►CSF, ►pleckstrin domain; Ravichandran KS 2001 Oncogene 20:6322.

Shearing: Cutting DNA to fragments by mechanical means, e.g., by rapid stirring or brusque pipetting.

Sheats: Any tube-like structure surrounding another, also the part of a leaf that wraps the stem.

Sheep, Domesticated (*Ovis aries*): 2n = 54; some wild sheep have higher number of chromosomes. A medium-density linkage map of 1062 unique loci became available by 2001 and has helped in mapping the cattle and goat genomes (Maddox JF et al 2001 Genome Res 11:1275; Cockett NE et al 2001 Physiol Genomics 7:69; http://www.wool.com.au/LivePage.aspx?pageId=116#mainBody; http://www.angis.org.au/Databases/BIRX/mis/; http://www.ncbi.nlm.nih.gov/genome/guide/sheep).

Sheep Hybrids: Domesticated sheep (*Ovis aries*, 2n = 54) forms fertile hybrids with muflons but the goat x sheep hybrid embryos rarely develop normally. ►transplantation nuclear; Ruffing NA 1993 Biol Reprod 49:1260.

Sheldon-Hall Syndrome (distal arthgrogryposis): ►Freeman-Sheldon syndrome

Shelterin: ►telosome

Sherman Paradox: The various recurrence risks among relatives caused by expansion of fragile X sites transmitted by non-symptomatic males. ►mental retardation, ►recurrence risk, ►fragile X, ►trinucleotide repeats; Sherman SL et al 1985 Hum Genet 69:289; Fu YH et al 1991 Cell 67:1047.

Shift: An internal chromosomal segment, generated by two break points, translocated within the same or into another chromosome, within a gap opened by a single break. It is a rare phenomenon. Shifting of the relative proportion of mitochondrial recombination products commonly occurs in plants. ►transposition, ►reciprocal interchange, ►translocation

Shifting-Balance Theory of Evolution: According to the Shiftin-balance theory of evolution, polymorphism in a population is determined by a dynamic interplay of the forces of pleiotropy, epistasis, genotypic values, fitness, and population structure. Alternative ideas would be the discredited neo-Lamarckian internal drive or the well-documented neutral mutation concepts. ►polymorphism, ►pleiotropy, ►genotype, ►fitness, ►neo-Lamarckism, ►neutral mutation, ►population structure; Wade MJ, Goodnight CJ 1991 Science 253:1015.

Shigella: A group of gram-negative enterobacteria causing dysentery (intestinal inflammation and diarrhea) in humans and higher monkeys. Butyrate treatment effectively mitigates shigellosis by up-regulation of a natural antibiotic (LL-37/cathelicidin) in the intestinal epithelium (Raquib R et al 2006 Proc Natl Acad Sci USA 103:9178). ►primates; database: http://www.mgc.ac.cn/ShiBASE/.

Shikimic Acid: An intermediate in aromatic amino acid-, flavonoid-, fragrance-, alkaloid- and plant pigment biosynthesis (see Fig. S48).

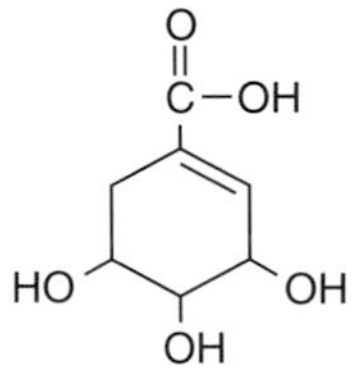

Figure S48. Shikimic acid

Shine-Dalgarno Sequence: Nucleotide consensus (AGGAGG) in the non-translated 5′ region of the prokaryotic mRNA (close to the translation initiation codon), complementary to the binding sites of the ribosomes. The terminus of the 16S ribosomal RNA is generally:

A^{Me2} A^{Me2} CCUGCGG**U**UGGAUGA**CCUCC**-UUA-3′-0H. The eukaryotic mRNAs do not have this sequence and the mRNA attaches to the ribosomes by means of ribosomal scanning. In the polycistronic messages, each cistron generally has a Shine-Dalgarno sequence. Mitochondrial and ribosome mRNAs, generally but not always, have this or a modified Shine-Dalgarno. ►anti-Shine-Dalgarno, ►ribosome scanning, ►IF, ►initiation codon, ►polycistronic mRNA; Shultzaberger RK et al 2001 J Mol Biol 313:215.

Shingles: The herpes (varicella-zooster) virus responsible for chicken pox may emerge from a latent stage, later in the life of an individual when immunity has waned. It then causes shingles (sore eruptions) in parts of the body innervated by ganglions harboring the earlier latent virus (see Fig. S49). About 20% of adults become affected. Chickenpox is highly contagious but shingles are transmitted only from the open eruptions and not by sneezing or casual contact and the risk of infection is very low. Hand washing is helpful to prevent infection from afflicted individuals. Guanosine analogs (acyclovir) are commonly used for treatment, which interfere with replication by viral DNA or RNA polymerases (see Fig. S50). ►herpes, ►chickenpox, ►Varicella zooster virus, ►ganglion; Dwyer DE, Cunningham AL 2002 Med J Austral 177(5):267; vaccine: Vázquez M et al 2005 N Engl J Med 352:439.

S

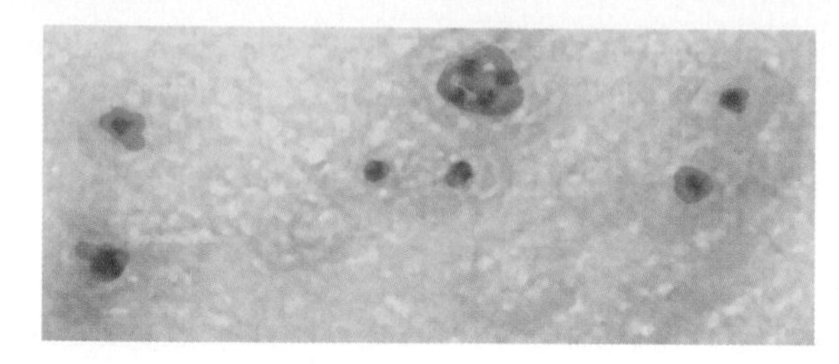

Figure S49. Shingles

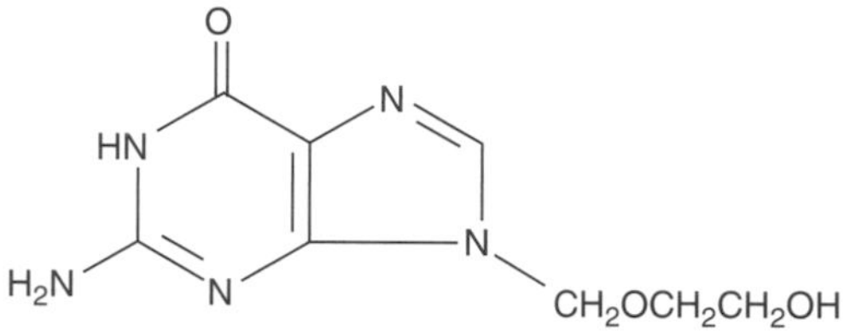

O

HN N

N

H_2N N $CH_2OCH_2CH_2OH$

Figure S50. Acydovir

SHIP: Inositol phosphatase with SH_2 domain. SHIP proteins in tyrosine phosphorylated form signal to hematopoiesis, cytokines, PTEN, etc. ►SH_2, ►inositol, ►cytokines, ►hematopoiesis, ►PTEN, ►ITIM; Rohrschneider LR et al 2000 Genes Dev 14:505.

SHIRPA: A protocol for phenotypic characterization of mutation in mice (Rogers DC et al 1997 Mammalian Genome 8:711).

SHIV: A simian immunodeficiency virus (SIV) engineered to carry an HIV coat protein and thus capable of infection of, and symptoms of AIDS in macaque monkeys. ►HIV, ►SIV, ►acquired immunodeficiency syndrome, ►primates

SHMOOs: Mating projections in yeast.

SHOM (sequencing by hybridization to oligonucleotide microchips): One of the automated (robotic) procedures developed for nucleotide sequence diagnostics. (See Yershov G et al 1996 Proc Natl Acad Sci USA 93:4913; ►DNA chips, ►microarray hybridization).

Shoot: The plant part(s) above ground or a branch of a stem.

Shope Papilloma: A viral disease of rabbits causing nodules under the tongue. The double-stranded DNA virus of about 8 kbp has 49 mole percent G + C content. ►bovine papilloma

Short-Day Plants: Short-day plants require generally less than 12–15 h daily illumination for flowering; at longer light periods they usually remain vegetative. ►photoperiodism, ►long-day plants

Short Dispersed Repeats: ►SDR

Short Patch Repair: An excision repair removing, and then replacing, about 20 nucleotides. ►DNA repair, ►excision repair, ►mismatch repair; Mansour CA et al 2001 Mutat Res 485:331.

Short RNA (sRNA): ►microRNA, ►RNAi

SHORT Syndrome: An autosomal recessive phenotype characterized by the initials of the SHORT acronym: short **S**tature, **H**yperextensibility of joints and hernia, **O**cular depression, **R**ieger anomaly (partial absence of teeth, anal stenosis [narrow anus], hypertelorism [increased distance between organs or parts], mental and bone deficiencies, and **T**eething delay). ►stature in humans, ►dwarfism, ►pseudoautosomal

Shortstop Promoter: The shortstop promoter mediates the transcription of only some of the exons although some repeats within the exons may be expanded.

Shotgun Cloning: In shotgun cloning, the DNA of an entire genome is cloned without aiming at particular sequences. From the cloned array of DNA fragments (library), the sequences of interest may be identified by appropriate genetic probes. ►cloning, ►DNA probe, ►DNA library; Matsumoto S et al 1998 Microbiol Immunol 42:15.

Shotgun Sequencing: In shotgun sequencing, random samples of cloned DNA, e.g., the segments of a cosmid are sequenced at random. *Whole-genome pairwise shotgun* procedure sequences paired ends of cloned DNAs of varying sizes and fragmented them into a larger number of contigs ordered with the aid of high-power computers. If there are still gaps between the contigs those are filled in by "finishing." The *hierarchical shotgun sequencing* procedure is based on mapped clones generated by BACs. The *Double-barrel shotgun* sequences the DNA from both ends. The short sequences are arranged into longer tracts by computers. The *Full shotgun sequence* indicates that the cloned inserts have been covered about 8–10 times. *Half shotgun coverage* is only 4–5-fold random sequence. ►DNA sequencing, ►first-draft sequence, ►contig, ►completion, ►WGS, ►scaffolds in genome sequencing, ►human genome, ►genome projects; Bankier AT 2001 Methods Mol Biol 167:89; whole genome shotgun (WGS) products: http://www.ncbi.nlm.nih.gov/projects/WGS/WGSprojectlist.cgi.

Shoufflons: Clustered (generally 6 to 7) recombination/inversion sites and a shoufflon-specific recombinase in bacteria. These elements determine the nature of the bacterial pili, thick rigid or thin pilus. A typical shoufflon is plasmid R64, a 120.8 kb conjugative plasmid encoding streptomycin and tetracycline resistance and at least 49 genes in the 54 kb transfer region. Various types of shoufflons occur in different bacteria. ►site-specific recombination, ►pilus; Komano T 1999 Annu Rev Genet 33:171.

SHP-1 (synonymous with SH-PTP1, PTP1C, HCP): Tyrosine phosphatase; it contains the SRC homology domain SH2. Upon activation of T cells, it binds to the kinase ZAP-70 resulting in increased phosphatase activity but in a decrease in ZAP-70 kinase activity. It is a negative regulator of the T cell antigen receptor. It is activated by radiation stress. ►T cell, ►ZAP-70, ►ITIM; Kosugi A et al 2001 Immunity 14:669.

Shprintzen-Goldberg Syndrome: ►Marfanoid syndromes

SHREC: A multienzyme effector complex, which regulates nucleosome positioning to assemble higher-order chromatin structures critical for heterochromatin functions and gene silencing (Sugiyama T et al 2007 Cell 128:491).

Shrew: The smallest insectivorous mammals; *Blarina brevicauda*, 2n = 50; *Cryptotis parva*, 2n = 52; *Neomys fodiens*, 2n = 52; *Notiosorex crawfordi*, 2n = 68; *Sorex caecutiens*, 2n = 42; *Suncus murinus*, 2n = 40; *Tupaia belangeri* tree shrew, 2n = 62 (see Fig. S51).

Figure S51. Tree shrew

Shrinkage: Shrinkage occurs when multiple regression data are applied to new information and the regression decreases. ►correlation, ►multiple regression

shRNA (short heterochromatic RNA): shRNA is instrumental in the formation of heterochromatin and the epigenetic remodeling of chromatin. ►heterochromatin, ►epigenesis, ►RNAi; Jenuwein T 2002 Science 297:2215.

shRNA (short hairpin RNA): shRNA can be used for global identification of genes involved in certain functional pathways by blocking their expression (Moffat J et al 2006 Cell 124:1283). By intravenous infusion of 49 Adeno-associated virus vectors carrying different shRNAs, 36 caused dose-dependent liver injuries and 23 caused death in adult mice. Liver-derived microRNA pathway was down-regulated by nuclear karyopherin/exportin-5. (Grimm D et al 2006 Nature [Lond] 441:537). Reversible gene inactivation was attained by expression of an insulin receptor (*Insr*)-specific shRNA. Upon induction by doxycycline, mice developed severe hyperglycemia within seven days. The onset and progression of the disease correlates with the concentration of doxycycline, and the phenotype returns to the baseline, shortly after the withdrawal of the inductor doxycycline (Seibler J et al 2007 Nucleic Acids Res 35(7):e54). ►RNAi, ►microRNA, ►karyopherin; Paddison PJ 2002 Genes Dev 8:948.

sHsp (small heat-shock proteins): Diverse ubiquitous proteins (15–80 kDa) formed in response to heat or other stress. Their transcriptional activation requires three inverted repeats of NNGAAN motif (HSE) where the heat shock transcription factor (HSF) binds. Their regulation may require other factors (e.g., estrogen, ecdysterone, etc.). The homology among the different types resides in the COOH-terminal half (α-crystalline domain). They may form large oligomers. Their role is heat and chemical protection of cells. Plant sHPS chaperones respond to various stresses to resist irreversible protein denaturation. ►heat-shock proteins, ►ibp; Haley DA et al 2000 J Mol Biol 298:261.

Shuffling: ►DNA shuffling

Shufflon: A region in the chromosome with structural rearrangements.

Shugoshin: The shugoshin cohesin keeps sister chromatids together during anaphase I of meiosis. At meiosis I, Rec8, which replaces SCC1/RAD21 in meiosis, mediates the separation of the chromatids in the chromosome arms but keeps cohesion intact around the centromere until meiosis II. In yeast and probably in other eukaryotes as well, shugoshin protects the separation of chromatids from separase cleavage in meiosis I. Shugoshin is associated with serine/threonine protein phosphatase 2A (PP2A) protein in humans. Shugoshin and PP2A protect the separation of chromatids at the centromere. Shugosin probably dephosphprylates cohesin and protects separation at the centromere (see Fig. S52). ►meiosis I, ►meiosis II, ►cohesin, ►separin; Kitajima TS et al 2006 Nature [Lond] 441:46; Riedel CG et al 2006 Nature [Lond] 441:53.

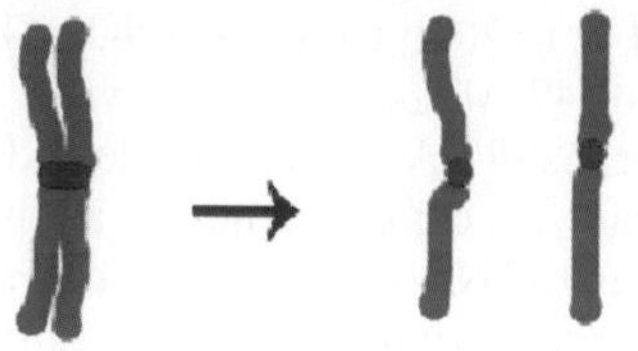

Figure S52. Sister chromatids separated at the centromere

Shunting: In shunting, during scanning the 40S ribosomal subunit jumps from the region of capture of the mRNA at the 5′ m7 GpppN cap or at another sequence until it locates the translation initiation codon. Thus, the ribosome bypasses the 5′ leader. For efficient translation of the mRNA with an internal ribosomal entry site (IRES), the homeodomain protein Gtx translated from IRES and a 9 to 7 nucleotide long sequence complementary to a site in the 18S eukaryotic ribosomal subunit was necessary (Chappell SA et al 2004 Proc Natl Acad Sci USA 101:9590). The presence of the Gtx translation element and the complementary 8 nucleotide sequence facilitates shunting even across an upstream uAUG codon in the leader and across a hairpin structure in the leader (Chappell SA et al 2006 Proc Natl Acad Sci USA 103:9488). ►scanning, ►translational hopping, ►IRES, ►leader sequence, ►Cap

Shuttle Vector: A "promiscuous" plasmid that can carry genes to more than one organism and can propagate the genes in different cells, e.g., in *Agrobacterium, E. coli*, and plant cells. ►vectors, ►cloning vectors, ►transformation genetic, ►promiscuous DNA; Perez-Arellano I et al 2001 Plasmid 46:106.

Shwachman-Diamond Syndrome (Shwachman-Bodian-Diamond syndrome, 7q11): Pancreatic insufficiency, blood and bone abnormalities, and increased hematological cancer risk; probably recessive. The yeast ortholog, Sdo1 is critical for the release and recycling of the nucleolar shuttling factor Tif6 (ribosome biogenesis factor) from pre-60S ribosomes, a key step in 60S maturation and translational activation of ribosomes and its defect predisposes to bone marrow failure and leukemia (Menne TF et al 2007 Nature Genet 39:486).

SI: The unit of absorbed dose (1 Joule/kg) of electromagnetic radiation; it is generally expressed in Gray (Gy) or Sievert (Sv = 1 rem) units. Earlier rad (= 0.01 Gy) was used. ►r, ►rem, ►Gray, ►Sievert, ►Curie, ►Becquerel

Sialic Acid: An acidic sugar such as *N*-acetylneuraminate or *N*-glycolylneuroaminate; it is present in the gangliosides (see Fig. S53). Polysialic acid is involved in cell and tissue type differentiation, learning, memory, and tumor biology. The synthesis is mediated by polysialyl transferase under the regulation of neural cell adhesion molecules. The recessive sialic acid storage diseases (e.g., Salla disease, 6q14-q15) involve hypotonia and cerebellar ataxia, while mental retardation is caused by a family of anion/cation symporters.

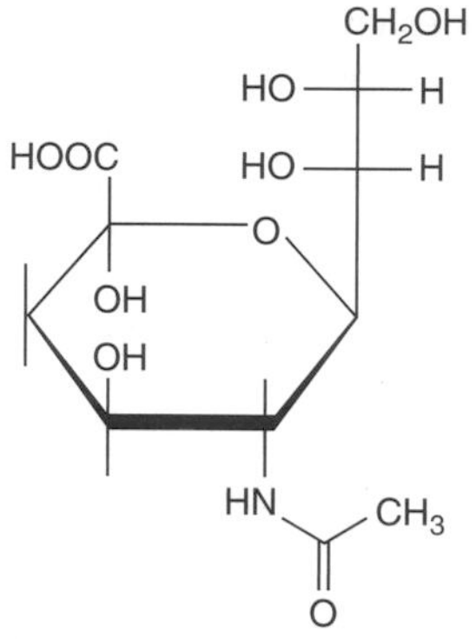

CH_2OH HO H HOOC HO H O OH OH HN CH_3 O

Figure S53. Sialic acid

The enzyme CMP-sialic acid hydroxylase changes *N*-acetylneuraminic acid into *N*-glycolylneuraminic acid. This enzyme is active in all mammals, except humans, where the 6p22.3-p22.2 locus suffered a 92-base deletion in the 5′ region after the separation of humans from the primate lineage of evolution. Also, the human lineage—in contrast to that of chimpanzee—has only pseudogenes for the receptor (Siglec-L1) of *N*-glycolneuraminate. ►gangliosides, ►gangliosidoses, ►CAM, ►neuraminidase deficiency, ►fusogenic liposome, ►lysosomal storage disease, ►symport, ►sphingolipids, ►cytidylic acid, ►sialuria; Chou H-H et al 1998 Proc Natl Acad Sci USA 95:11751; Angata T et al 2002 J Biol Chem 277:24466; Aula N et al 2000 Am J Hum Genet 2000 67:832.

Sialidase Deficiency: ►neuraminidase deficiency

Sialidosis: ►neuraminidase deficiency

Sialolipidosis: ►mucolipidosis IV

Sialuria (9p12-p11): Sialuria is caused by a semidominant/recessive gene defective in the feedback-sensitivity of uridine diphosphate *N*-acetylglucosamine 2-epimerase enzyme by cytidine monophoshphate-neuroaminic acid. The afflicted have defects in bone (dysostosis and psychomotor [movement and psychic activity]) development and infantile death may occur. Salla disease (6q14-q15) is a sialic acid storage disease with mental and psychomotor retardation. ►neuroaminidase deficiency

Siamese Cat: A Siamese cat displays darker fur color at its extremities because during slow blood circulation, more pigment develops at specific locations of the body. This is due to the presence of a temperature-sensitive gene, similar to Himalayan rabbits and other animals. The pattern distribution is due to a transition mutation from G→A in the 2nd exon of a tyrosinase gene, resulting in replacement of glycine by arginine (Lyons LA et al 2005 Anim Genet 36(2):119). The mutant allele produces pigment at the lower temperature of the extremities; it is thus a modification of albinism when pigment formation is absent in the entire body. Normal function of tyrosinase is required for the production of melanin, a pigment of fur color. ►pigmentation of animals, ►Himalayan rabbit, ►temperature-sensitive mutation, see Fig. S54.

Figure S54. Siamese cat

Sib: Same as sibling. ►full sib, ►half sib

Sib Pair Method: ►affected-sib-pair method

Sib TDT (s-TDT): A transmission disequilibrium test to detect genetic linkage/association on the basis of analysis of close genetic markers and disease among sibs. ►SDT; Spielman RS, Ewens WJ 1998 Am J Hum Genet 62:450.

Sibling: Natural children of the same parents. ►λ_S, ►risk, ►genetic risk, ►genotypic risk ratio

Sibling Species: Sibling species are morphologically very similar and frequently share habitat but they are reproductively isolated. ►species, ►speciation, ►fertility

Sibpal: A linkage analysis computer program for sib pairs. ►sibling; Fann CS et al 1999 Genet Epidemiol 17(Suppl. 1):S151.

Sibship: Natural brothers and sisters. ►kindred, ►sibpal

SIC1: Cell cycle S-phase cyclin-dependent kinase (CDK, Cdc28-Clb) inhibitor. ►CDC34, ►CDC6, ►cell cycle [START], ►mitotic exit, ►APC; Verma R et al 2001 Mol Cell 8:439.

Sickle-Cell Anemia: A human hereditary disease caused by homozygosity of a recessive mutation(s) or deletions in the hemoglobin β chain gene. Heterozygosity causes the sickle cell trait. Under low oxygen supply, the red blood cells lose their plump appearance and partially collapse into sickle or odd shapes because of the aggregation of the abnormal hemoglobin molecules (see Fig. S55).

The disease is not absolutely fatal but crises may occur when the blood vessels are clogged. Complications may arise due to poor blood circulation. In the classical form of the sickle cell disease, in the hemoglobin S a valine residue replaced a glutamine residue of the normal beta chain (hemoglobin A). In the hemoglobin C at the same position, a lysine replacement occurs, and this condition causes less severe clinical symptoms. Other forms, hemoglobin D and E are less common. This disease provided the first molecular evidence that mutation leads to amino acid replacement. Sickle cell anemia affects more than 2 million persons worldwide. About 10% the population of African descent in the USA are carriers (heterozygous) for this mutation and about 1/400 is afflicted with homozygosity at birth. In populations of European (except southern Europeans) descent, the frequency of this mutant gene is about 1/20 of that in Mediterraneans and Africans. The high frequency of the genetic condition in areas of the world with high infestation by malaria is correlated. The individuals without sickle cell anemia gene have about 2–3 times higher chance to be infected by *Plasmodium falciparum*. The mutation is selectively advantageous by protecting heterozygotes against malaria. The globin gene cluster has been located to human chromosome 11p15. The order is 5′—γG—γA—δ—β—3′. Correction of the defective allele is possible by transduction of the defective cells with retroviral or adenoviral vectors that can deliver the normal gene to the hematopoietic stem cells. However, these viral vectors may have deleterious consequences for the body. There is a problem of the low expression of the transgene. In a mouse model, embryonic stem cells containing the normal β^A allele, when introduced into β^S embryonic mouse blastocysts produced both types of globins indicating a promise that this approach may help in curing human β-thalassemia (Chang JC et al 2006 Proc Natl Acad Sci USA 103:1046).

Another method is to introduce into the lymphoblastoid cells chimeric DNA-RNA oligonucleotides with a correction for the β^S allele mutation, brought about through gene conversion in the target cells. This procedure may eventually be clinically applicable.

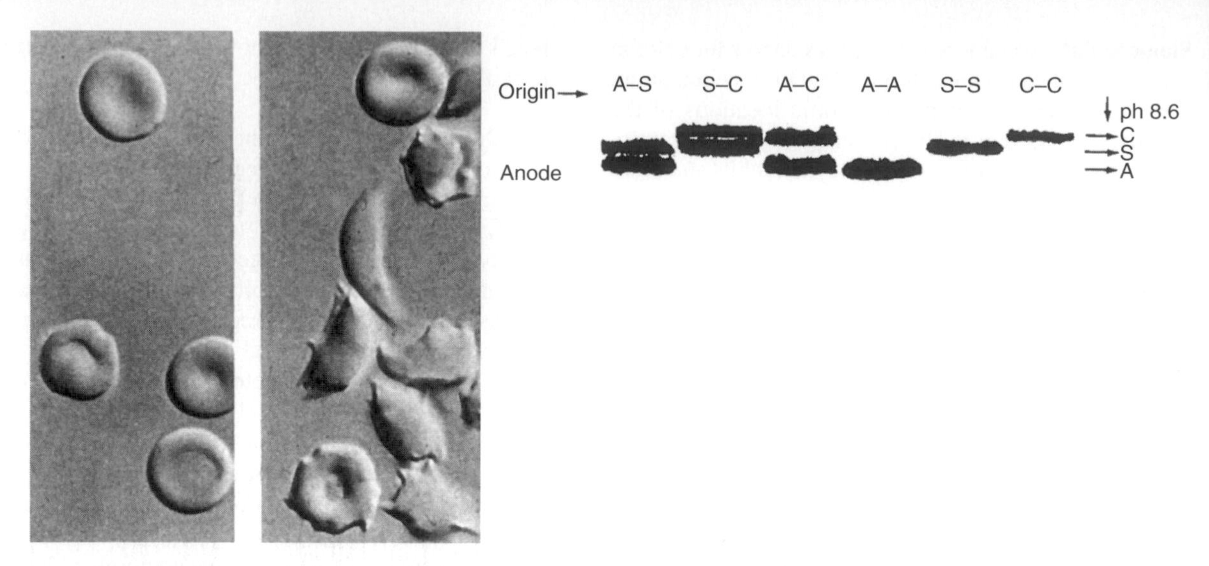

Figure S55. Human red blood cells from a sickle cell anemia patient. Left: In the presence of normal oxygen supply, Right: At low oxygen supply. Normal red blood cells look like biconcave discs very similar to that at left here. The sickling cells are unable to hold oxygen and that condition is responsible for the disease. (Photographs are the courtesy of Cerami A, Manning JM 1971 Proc Natl Acad Sci USA 68:1180). At far right: Carriers or sufferers can be unambigously identified by electrophoretic separation of the blood protehomozygotes for the normal blood protein (A-A), sickle cell anemia (S-S), hemoglobin C (C-C). In the heterozygotes (A-S, S-C and A-C)—because of codominance—both types of parental proteins are detected. (From Edington GM, Lehman H 1954 Trans R Soc Trop Med Hyg 48:332)

Another possible therapeutic approach involves correction of the defective β-globin mRNA by the similar (fetal) normal (anti-sickling) protein transcript using a transsplicing ribozyme and the generation of a normally functional transcript. When the hemoglobin α and β genes were knocked out of mouse and mated with animals transgenic for the human sickle cell gene, an animal model was generated for experimentation with this human disease. Attempts are being made to activate the silent fetal hemoglobin genes—by urea compounds—at later developmental stages in order to compensate for the defective adult hemoglobin. ►malaria, ►hemoglobin, ►thalassemia, ►sickle cell trait, ►genetic screening, ►viral vectors, ►gene therapy, ►gene conversion, ►ribozymes, ►transsplicing, ►introns, ►SAD mouse; Pawliuk R et al 2001 Science 294:2368; Vichinsky E 2002 Lancet 360:629.

Sickle Cell Trait: The sickle cell trait is due to heterozygosity of the recessive mutation in the gene controlling the β-chain of hemoglobin. Normally, these heterozygotes do not suffer from this condition but under low oxygen supply, e.g., at high elevations, adverse consequences may arise. ►sickle cell anemia, ►hemoglobin

Side-Arm Bridge: An attachment of chromatids resembling chiasma but actually only an anomaly in mitosis or meiosis, usually arising when chemicals or radiation disturb division (see Fig. S56). ►bridge, photo is the courtesy of Dr. BR Brinkley & WN Hittelman.

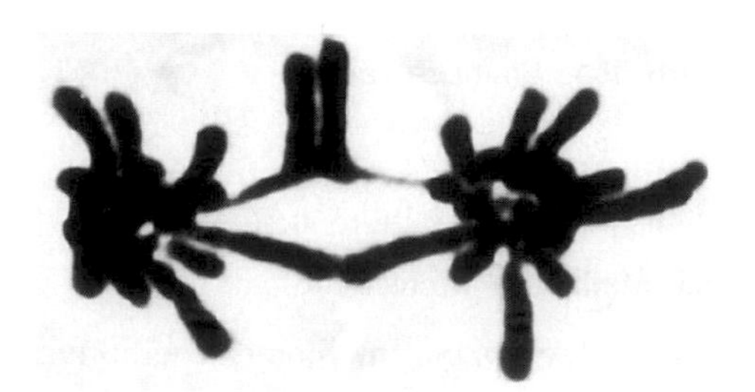

Figure S56. Side-arm bridge

Siderocyte Anemia (sideroblastic anemia): An anemia with erythrocytes containing non-hemoglobin iron; it may be controlled by either autosomal recessive or dominant or X- linked genes and based on defects in erythroid β-aminolevulinate synthase. ►anemia

Siderophore: Iron transporter. Transformation into rice plants highly efficient siderophore system genes (for nicotinamine aminotransferase) from barley facilitates iron uptake on alkalic soils and substantially improves growth (Takahashi M et al 2001 Nature Biotechnol 19:466). Mammals sequester excess iron into bacterial siderophores through the mediation of the host protein lipocalin 2. When the Toll receptor senses infection and the system fights bacteria. This constitutes a novel means of innate immunity (Flo TH et al 2004 Nature [Lond] 432:917). ►immunity innate, ►Toll

Siderosis: Iron overload in the bloodstream.

SIE: STAT-inducible element. ►Jak-Stat pathway

Siemens (SI): A unit of conductance; 1 Ampere/Volt in a tissue of 1 Ohm resistance. ►Ampere, ►Volt, ►Ohm

SIEVE TUBE: A plant food transporting tube-shape, tapered, long cells; they may be connected by sieve plates that hold them together.

SIEVERT (Sv): The name for Sv (unit of absorbed dose equivalent [J/kg]) = 100 rem = 1 Sv. It is commonly used for measuring the occupational radiation hazards. ►rem, ►Gray, ►rad, ►R

SIGLECs (sialic acid recognizing Ig-superfamily lectins, 19q13.3): Immune inhibitory receptors with sialic acid ligands. During evolution, T cell Siglec expression was largely lost from humans while it was retained in apes. This fact explains why human are more at risk by T-cell-mediated diseases, such as AIDS and hepatitis B and C, compared to apes (Nguyen DH et al 2006 Proc Natl Acad Sci USA 103:7765). ►sialic acid, ►AIDS, ►hepatitis

Sigma Factor: A subunit of DNA-dependent bacterial RNA polymerase, required for the initiation of transcription and for promoter selection. It has been reported that in *Bacillus subtilis*, the σ factor alone is sufficient for melting the DNA around site −10 of the promoter (Hsu HH et al 2006 Cell 127:317). The documentation of this finding has been, however, questioned (Xin H 2006 Science 314:1669) and the paper was retracted (2007 Cell 128:211). ►open promoter complex, ►RNA polymerase, ►σ

Sigma Replication: ►rolling circle

Sigma Virus of *Drosophila*: ►CO_2 sensitivity

Sign Mutations: Frameshift mutations because an equal number base addition(s) (+) and deletion(s) (−) at the gene locus may restore the reading frame, although may not always restore normal function. ►frameshift

Signal: A molecular determinant, emitted by an extracellular source, or by an intracellular organizer and is directed either to an adjacent tissue (vertical signal) or to adjacent cells of the same tissue (planar signal). The response to a signal frequently involves change in transcription. ►signal transduction; Camili A, Bassaler BL 2006 Science 311:1113.

Signal End: ►immunoglobulins

Signal Joint: ►signal end

Signal Hypothesis: The signal hypothesis postulated that the signal peptide of the nascent polypeptide chain guides it to the endoplasmic reticulum (and to other) membranes where the signal peptidase cleaves it off, and subsequently the peptide chain is completed within the lumen of the membranes. It has now been validated for transport through bacterial cell membranes, mitochondria, plastids, peroxisomes, etc. ►signal peptides, ►transit peptide, ►protein targeting; Loureiro J et al 2006 Nature [Lond] 441:894.

Signal-Noise Ratio: The signal-noise ratio determines the sensitivity of an instrument or procedure to detect the presence or the function of a molecule in a reliable manner.

Signal Peptidase: ►signal hypothesis

Signal Peptides (signal sequence): 15 to 35-amino acid long sequences generally occurring at the NH_2 terminus of the nascent polypeptide chains of proteins that have a destination for an intra-organellar or transmembrane location. They are made in eukaryotes and prokaryotes but not all the secreted proteins possess signal peptides. At the beginning of the sequence, generally there are one or more positively charged amino acids, followed by a tract of hydrophobic amino acid residues that occupy about three-fourth of the length of the chain. This hydrophobic region may be required to pass into the lipoprotein membrane. The amino acid sequences among the various signal peptides are not conserved, indicating that the secondary structure is critical for recognition by the signal peptide recognition particles and for the function within the membrane. The eukaryotic signal sequences are recognized by the prokaryotic transport systems and the prokaryotic signal peptides can function in eukaryotes. After the passage of the nascent peptide has started the signal, peptides are split off by peptidases on the carboxyl end of (generally) glycine, alanine, and serine. Consequently, the majority of the proteins in the membrane or transported through the membranes, have the nearest downstream neighbor in one of these three amino acids at the amino end. One major characteristic of the signal peptide is the co-transcriptional targeting, whereas the transit peptides are targeted posttranslationally. The signal sequences may show polymorphism that might result in incorrect targeting. ►signal sequence recognition particle, ►transit peptide, ►leader peptide, ►endoplasmic reticulum; von Heijne G et al 1989 Eur J Biochem 180:535; Watanabe N et al 2001 J Biol Chem 276:20474; transmembrane and signal peptide server: http://phobius.cgb.ki.se/.

Signal Recognition Particle: ►signal sequence recognition particle, ►SRP

Signal Sequence: The amino terminal of some proteins signals the cellular destination of these proteins, such as the signal peptides. ►signal peptide

Signal Sequence Recognition Particle (SRP): In mammals, a complex of six proteins and an RNA (7SL RNA) that recognizes the *SRP receptor protein* on the surface of the endoplasmic reticulum and the *signal peptides* of the nascent proteins translated on the ribosomes associated with the endoplasmic reticulum (rough endoplasmic reticulum) and facilitate the transport of these polypeptides into the lumen of the Golgi apparatus and lysosomes (co-translational transport) (see Fig. S57). Signal recognition particle proteins have been located to human chromosomes, 5q21, 15q22, 17q25 and 18.

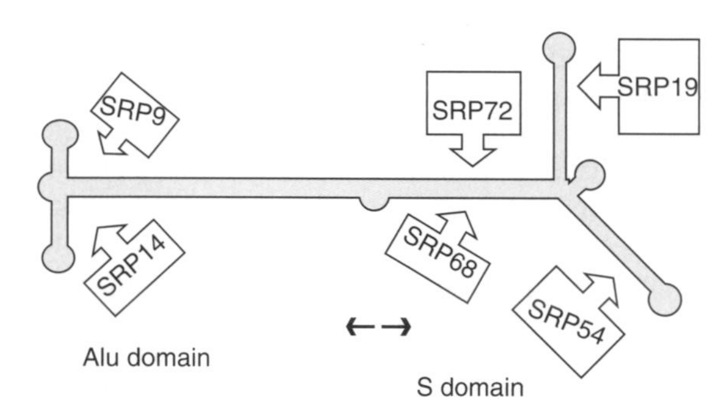

Figure S57. Schematic illustration of a signal recognition particle. The boxes represent the six proteins around the RNA. SRP9—SRP14 and SRP72—SRP68, respectively form dimeric structures. (Modified after Weichenrieder O et al 2000 Nature [Lond] 408:167)

The SRP binds to the signal peptide after a about 70-amino acid chain is completed at the beginning of translation. The polypeptide chain elongation is somewhat relaxed until the SRP attaches to the SRP receptor. Then, the SRP comes off the amino acid chain, elongation resumes its normal rate, and the entry of the chain through the membrane proceeds. In the meantime, a peptidase inside the endoplasmic reticulum cuts off the 15–30 amino acid long signal peptide sequences. ►lysosomes, ►Golgi, ►signal peptides, ►protein synthesis, ►signal sequence particle, ►RNA 7S, ►protein targeting, ►translocon, ►translocase, ►TRAM, ►protein conducting channel; Keenan RJ et al 2001 Annu Rev Biochem 70:755; Fulga TA et al 2001 EMBO J 20:2338; Wild K et al 2001 Science 294:598; Hainzl T et al 2002 Nature [Lond] 417:767; Pool MR et al 2002 Science 297:1345; structure: Egea PF et al 2004 Nature [Lond] 427:215; structure of *E. coli* signal recognition particle bound to ribosome: Schaffitzel C et al 2006 Nature [Lond] 444:503; http://psyche.uthct.edu/dbs/SRPDB/SRPDB.html.

Signal Transduction: A system of proteins transforming various stimuli into cellular responses. The process requires four major categories of elements: signals, receptors, adaptors, and effectors.

Signals. The extracellular signals interact with cell membrane receptors (EGFR, FGFR) and subsequently make contacts with intracellular target molecules to stimulate a cascade of events leading to the formation of effector molecules that switch genes on and off and control cellular differentiation in structure and time by regulating transcription. The *signals* are proteins, peptides, nucleotides, steroids, retinoids, fatty acids, hormones, gases (ethylene, nitric oxide, carbon monoxide), inorganic compounds, light, etc. The target cells accept the signals by special sensors, *receptors*. The receptors are generally specific proteins with high binding specificities and positioned on the cell surface or within the plasma membranes and thus readily accept the signal *ligands*. The receptors may be also inside the cells, and the ligands may have to pass the cell membranes to reach them. As a consequence of the binding, a cascade of events is triggered and eventually the instruction reaches the cell nucleus and the relevant genes. The *paracrine signals* are restricted in movement to the proper, a generally nearby, target. The nerve cells are communicating by *synaptic signals*. The *endocrine hormone* signals may affect also distant targets in the entire body. The neurotransmitters are activated through long circuits of the nervous system by electric impulses emitted by neurons in response to the environment. The travel of the electric signals through the neurons is very fast, may pass through meters per second. The neurotransmitters have only a few nanometers to pass and the process takes only a few milliseconds. The local concentration of the endocrine hormones is extremely low. In contrast, the neurotransmitters may be quite concentrated at very small target area. The neurotransmitters also may very rapidly be removed either by re-absorption or by enzymatic hydrolysis. Generally, the hydrophobic signals persist longer in the cells than the hydrophilic ones. The membrane anchored growth factors and cell adhesion molecules are signaled through the *juxtacrine* mediators.

Receptors. The target cells respond by *receptor proteins*. These receptors are endowed with specificities regarding the signal they respond to. Also, the same signal may have different receptors in differently specialized cells. In addition, the interpretation and use of the signal within similar cells may vary. The signals may act also in a combinatorial manner: several signals together may be involved in the cellular decisions and influence the length and quality of the effect of a signal received. The various receptors, despite substantial chemical differences of the signals (e.g., cortisol, estrogen, progesteron, thyroid hormones, retinoic acid, vitamin D), may bind ligands that control through the signal transduction path closely related and interchangeable upstream DNA consensus elements, involved in the regulation of transcription of different genes. Steroid hormone receptors, after bound to cognate hormones, may activate the transcription of the so-called *primary response genes*. These proteins then repress the further transcription of the primary response genes and turn on the transcription of *secondary response genes* (►regulation of gene activity).

Receptors can be (i) *ion-channel* or (ii) *G-protein* or (iii) *enzyme-linked* types. Group (i), also called transmitter-regulated ion channels, are involved in transmitting neuronal signals (►ion channels). Group (ii) receptors are transmembrane proteins of the so-called seven-membrane type (►seven-membrane proteins) associated with guanosine phosphate-binding G proteins (►G-proteins).

When GTP is bound to the G-protein, a cascade of enzymes or other proteins may be activated or an ion channel may become more permeable. G-protein linked receptors represent a large family of proteins, more than 100 of which, have been already identified in a variety of eukaryotes. The receptors, generally monomeric and evolutionarily related proteins, respond to a variety of signals, such as hormones, mitogens, light, pheromones, etc. Activation of group (iii) receptors may directly or indirectly lead to the activation of enzymes. These three different types of signal transduction may not be entirely distinct because the function of the ion channels may interact with the pathways mediated through G-proteins and various kinases. Some of the receptors are *protein tyrosine phosphatases* (e.g., CD45 protein) and *serine/threonine phosphatases* residing within the membrane or in the cytosolic domain of transmembrane proteins or in the cytosol.

Pathways. Pathways may show a great variation depending on the signals and receptors involved. Ca^{2+} is a general regulator. It may enter nerve cell terminals through voltage-gated Ca^{2+} channels in the cell membranes and stimulates the secretion of neurotransmitters (►ion channels, ►voltage-gated ion channel). Alternatively, Ca^{2+} may play a more general role by binding to G-protein linked receptors in the metabolism of inositol phospholipids, PIP (phospho-inositol phosphate), and PIP_2 (phosphoinositol bisphosphate). The specific trimeric G-protein, Gq is involved in the activation of *phospholipase C*-β, that is specific for phosphoinositides, and splits PIP_2 into inositol triphosphates and diaglycerol. Hydrolysis of PIP_2 yields IP_3 (inositol 1,4,5-triphosphate). The latter sets free calcium from the endoplasmic reticulum through IP_3-gated channels, ryanodine receptors (►ryanodine). Upon further phosphorylation, IP_3 may give rise to IP_4 (inositol 1,3,4,5-tetrakisphosphate) that slowly yet steadily replenishes cytosolic calcium. The calcium level in the cytosol rises and subsides in very short bursts according to how phosphatidyl-inositols regulate it (calcium oscillations). These oscillations still may assure increased secretion of second messengers and spare the cell from a constant level of the toxic Ca^{2+} in the cytosol. Besides pumping out Ca^{2+} shielded in the endoplasmic reticulum, diaglycerol and eicosanoids (arachidonic acid) may be produced.

Diaglycerol and the latter lipid derivatives may activate the Ca^{2+}-dependent enzymes, serine/threonine kinases, which have key role in activating proteins that mediate signal transduction (►protein kinases). *Protein kinase C* may activate the cytosolic *MAPK* (mitogen activated protein kinase) by phosphorylation and may phosphorylate a cytoplasmic inhibitor complex such as IκB + NF-κB. Thus, MAPK may phosphorylate DNA-binding proteins such as SRF (serum response factor) and Elk (member of the ETS oncoprotein family), sitting already at the upstream regulatory regions of a gene(s), such as the serum-response element (SRE). The phosphorylation then initiates transcription (►regulation of gene activity). The released protein factor NF-κB may migrate to the nucleus and, by binding to its cognate DNA site, sets into motion transcription (if other factors are also present). The response of the genes to the transducing signals depends on the number of regulatory proteins responding to the signal. Monomolecular reactions display relatively slow response to the concentration of the signal molecules, whereas if the number of effectors is multiple, the reaction to them may follow 3rd or multiple order kinetics. Similarly, prompt response is expected if the signal activates one reaction (e.g., phosphorylation) and at the same time deactivates an inhibitor or suppressor (e.g., by phosphatase action).

The enzyme-linked signal receptors do not need G-proteins. The transmembrane receptor binds the ligand at the cell surface and the cytosolic domain functions as an enzyme or it associates with an enzyme and the transfer of the signal to the cell nucleus is more direct. An example is the cytokine-activated cell membrane receptor Jak *tyrosine kinases*, which when dimerize, can combine with cytoplasmic STATs (signal transducers and activators of transcription) and chromosomal responsive elements. In response of interferon or other cytokine signals, the JAK phosphorylates tyrosine of the SH2 domains in a variety of STAT proteins. This may be followed by dimerization and the transfer of these proteins to the nucleus where they may turn on transcription of particular genes (see Fig. S58). It has been estimated that about 1% of the human genes code for protein kinases.

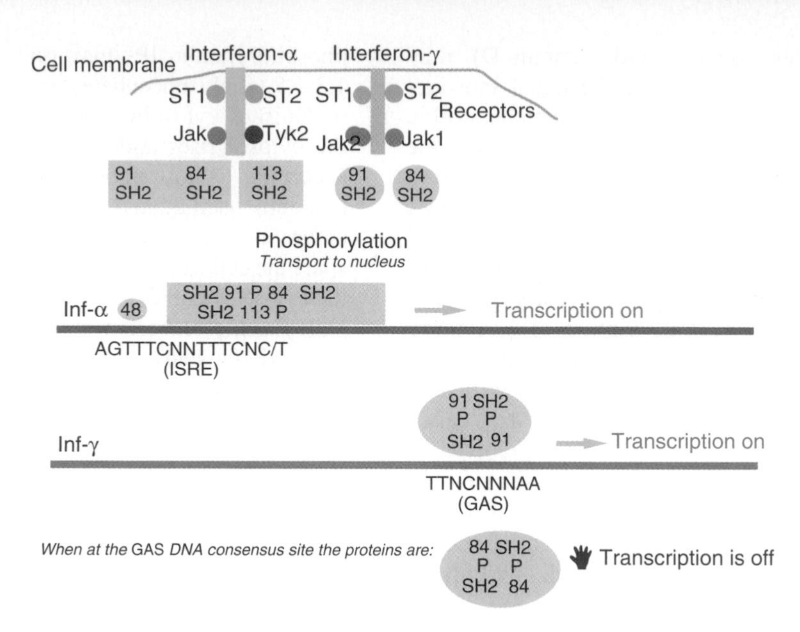

Figure S58. The JAK-STAT signal transduction pathway, based on interferon (IFN)-mediated receptors. *JAK* is a family of the Janus tyrosine protein kinase proteins, and *STAT* stands for signal transducer and activator of transcription. When the ligand binds to the signal-transducing receptors (ST), the receptor-attached *JAK kinases* modify the *STAT proteins*. After dimerization (using SH2 domains), they are transported directly to *ISRE* (interferon-α-response element) or to *GAS* (γ-interferon activation site) In the chromosomal DNA.These two elements vary, yet consensus sequences exist (as shown in the diagram). The 84, 91, 113, and 48 are proteins (in kDa), but additional ones may also be involved, depending on the nature of the signals received. The diagram does not display IFNGR1 and IIFNGR2 integral membrane proteins, which are also essential parts of the γ-receptor. (After Darnell JE Jr., et al. 1994. *Science* 264:14125 and Heim MH et al. 1995. *Science* 267:1347.)

In case of autophosphorylation or other reactions involving enzymes, which may bind their own products, the activity of the enzyme may be increased in the course of time with the increase of the number of product molecules, through a positive feedback.

These kinases may be mainly either serine/threonine or tyrosine kinases. Some proteins may phosphorylate all three of these amino acids. The reliance on protein tyrosine kinases for signal transduction is rather general. Epidermal growth factor (EGF), nerve growth factor (NGF), fibroblast growth factor (FGF), hepatocyte growth factor (HGF), insulin, insulin-like growth factor (IGF), vascular endothelial growth factor (VEGF), platelet-derived growth factor (PDGF), macrophage colony stimulating factor (M-CSF), etc., function with the assistance of transmembrane *receptor tyrosine kinases.* Upon the arrival of the ligand (signal), the receptor is dimerized either by cross-linking two receptors by the dimeric ligand or by inducing autophosphorylation and linkage of two cytosolic domains of the receptors. The different phosphorylated sites may bind different cytoplasmic proteins. The insulin itself is a tetramer (ααββ) and thus does not need dimerization. After autophosphorylation, it phosphorylates an insulin receptor (IRS-1) at tyrosine sites and that may bind to other proteins, which also may become phosphorylated and may form different complexes, thus, generating a variety of transcription factors. Alternatively, the *tyrosine kinase-associated receptors* themselves are not tyrosine kinases but associate with proteins of this capability. Some of the enzyme-linked receptors are *serine/threonine kinases* with specificities for these two amino acids. The phosphorylated tyrosine residues are binding sites for proteins with SH2 domains. The *receptor tyrosine phosphatases* may activate or inhibit the signal pathways by the removal of phosphate from tyrosine residues. The receptor guanylate (guanylyl) cyclases operate in the cytosolic domain of the receptors and function by serine/threonine phosphorylation in association with trimeric G-proteins. Within the last decade, discoveries about signal transduction have changed the views about cellular functions and added a new dimension to biology by integrating reversed and classical genetics. The signal transduction mechanisms have a large variety of means to regulate diverse functions of metabolism, differentiation, and development (see Fig. S59). The different signaling molecules are organized into separate pathways. The different protein components appear to be regulated and coordinated into signaling complexes by SH, pleckstrin-homology, phosphotyrosine, and PDZ (post-synaptic density, disclarge, zo-1) protein domains through protein-protein interactions.

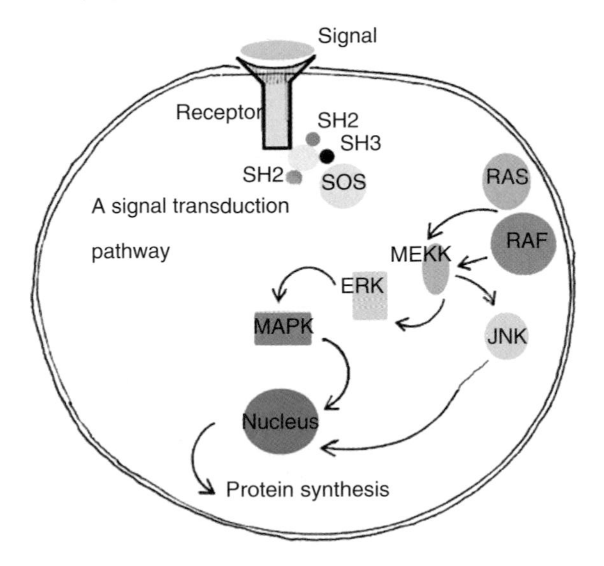

Figure S59. A signal transduction pathway leading through the G-protein RAS. This pathway controls cell division, differentiation, development of cancer, mating type, cell wall biosynthesis, and a variety of other processes. The signal can be a variety of molecules such as epidermal growth factor (EGF), nerve growth factor (NGF) or their homologs in various other animals such as *LET-23* in *Caenorhabditis* or *DER* in *Drosophila* the RAS-mediated pathway operates also in fungi and plants. The process begins with the (mitogen) signal arriving to the double membrane of the cell (see at 12 O'clock). The signal is recognized by a signal receptor protein, which may vary from signal to signal. This is a transmembrane protein with a hydrophobic tract forming generally seven turns within the cell membrane. When the signal arrives, the receptor is activated. The protein tyrosine kinase receptors (e.g., SEV) recruit then the down-stream receptor kinases or adaptor proteins such as GRB (or homologs e.g., SEM-5 in *Caenorhabditis*, DRK in *Drosophila*). The GRB proteins have an SH2 domain, which is a binding site for tyrosine kinase proteins and the SH3 domains are binding sites for proline-rich motif proteins. The SH domains were originally identified in the SRC protein (product of the Rous sarcoma oncoprotein) and both are characteristic for mediators of the signal transduction path. The GRB protein then binds another mediator or adaptor protein, SOS (named after the product of a *Drosophila* gene, called *son of sevenless*). The sevenless gene encodes Rhabdomore seven light receptors in the eyes of the flies. In place of SOS, there may also be SHC (an oncoprotein of the phaeochromocytoma tumor of the adrenal medulla or in paraganglia and thus causing increased secretion of the hormones epinephrine and norepinephrine). Upon the influence of the GRB-SOS complex, the membrane-bound RAS G-protein (see at 3 O'clock) becomes activated. The membrane association of RAS is mediated by a carboxy-terminal CAAX box, a signal for farnesylation, proteolysis and carboxymethylation and by the neighboring six lysine residues. RAS (named after rat sarcoma) is also a GTPase. The RAS serves as a turnstile for a series of processes. It is shut down when the situation is RAS*GDP (RAS guanosine diphosphate) and it opens when it becomes RAS*GTP (RAS-guanosine triphosphate). RAS activation may be prevented by another gene, encoding a GTP-ase activating protein (GAP). The RAF protein (its homologs in budding yeast is STE11, and BYR 2 in fission yeast) is also membrane bound by a CAAX box. RAF is a Ser-Thr kinase and it phosphorylates, independently of RAS, also MEK, a protein of the extra-cellular signal regulated kinase (ERK) family. The extracellular regulators can be growth factors (e.g., EGF, NGF), receptor kinases and TPA (12-*O*-tetradecanoyl-phorbol-13-acetate). MEK kinase (MEKK) is phosphorylated on Thr and Ser residues by RAF. Protein MAPK (mitogen-activated protein kinase) may be capable of autophosphorylation and it may be phosphorylated by the ERK family of kinases at Tyr and Thr residues. MAPK homologs are KSS, HOG-1, FUS3, SLT-2, SPK-1, SAPK (a stress activated kinase), FRS (FOS regulating kinase), etc. MAPK may then activate the FOS and JUN oncogene complex that is also known as the AP-1 heterodimeric transcription factor of mitogen-inducible genes. The RAS to MAPK route may branch downstream through several effector proteins and may control the transcription of several different genes. The specificity of activation depends on combinatorial arrangements of the effectors. (The size, shape or shading of the symbols is not intended to represent their structure.)

Plant Signals. Plant signals are somewhat different from the signals in animals. The plant hormones, similarly to animal hormones, are signaling molecules but most of them—except the brassinosteroids—are very different molecules. In plants, light and temperature signals (photoperiodism, vernalization, phytochrome) are very important for growth and differentiation. Salicylicacid is a signaling molecule for defense genes, etc. In plants, signaling depends a great deal on positional cues.

Several signal transduction pathways may be simultaneously operational within an organism and they may interact at various levels. In addition, the pathways and components may operate differently in different cellular compartments (e.g., plasma membrane, cytosol, nucleus, organelles). The cytoskeleton network may serve in various capacities to direct the flow of reactions through the signaling pathways. The current view is that signaling proteins translocate in the cytoplasm and bind in a reversible manner and dynamic fashion (soft-wired signaling concept). The

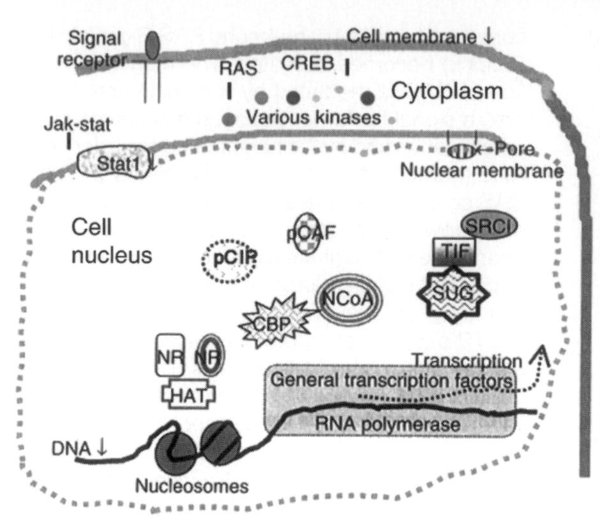

JAK-STAT: see more above
RAS: see more above
CREB: cAMP-response element-binding protein
pCIP: p300/CBP/cointegrator-associated protein
CBP: CREB-binding protein
pCAF: chromatin assembly factor
NcoA: nuclear receptor coactivator
SRCI, TIF, SUG: nuclear coactivator proteins
NR: nuclear receptors
HAT: histone acetyltransferase
Besides the molecules shown and named, several other proteins may be involved. Some of the proteins are activated only after binding with their ligands. The size or shape of the structures shown does not reflect the actual nature of these molecules.

Figure S60. Transactions inside the nucleus after the signals arrived

earlier idea was that receptors and other signalling proteins occupy fixed positions in the cell and second messengers mediate the connections with aid diffusion (hard-wired signaling). As opposed to electric circuits, a model signal of transduction is that of a gradient of quantitative information propagated outwards throughout the dense protein network following an input through an individual receptor. A central maximum of information transfer can include the old canonical pathways but the propagation throughout the signaling network may be in part stochastic and influenced by the local network composition. Although, large-scale, global static interaction maps are useful initially, these rarely recapitulate the often-transient connections of known signaling modules. Understanding signal transduction from a network perspective allows an appreciation of how variants of proteins (by mutation or concentration) within the network can lead to quantitatively different outputs. Thus, for diseases linked to disruption of particular signaling networks, comparison of QTLs with functional genomic RNAi screens of those networks synergistically provides a mechanistic insight into disease etiology. Sequencing of a large number of genes from breast and colorectal cancers (Sjöblom T et al 2006 Science 314:268), projected an average of 90 mutated genes per tumor, 11 of which occurred at significant frequency, suggesting that many mutations are needed to attack the robust cell signaling network. It is logical that signal transduction and transcriptional machinery components are extremely common functional categories mutated in these cancers (Friedman A, Perrimon N 2007 Cell 128:225).

The introduction of GFP labeling now facilitates the tracing of the signaling traffic. The interacting signals are usually expressed in a quantitatively variable manner. Understanding how different cells work under the control of the genetic potentials and the environment will be the most challenging task of the research on growth, differentiation, development, the nervous system, behavior, productivity, pathogenesis, evolution, etc.

►G-proteins, ►diffusion, ►JAK-STAT pathway, ►interferons, ►histidine kinase, ►regulation of gene activity, ►hormones, ►morphogenesis, ►AKAP79, ►T cells, ►integrin, ►ciliary neutrotrophic factor, ►adaptor proteins, ►morphogen, ►photomorphogenesis, ►vernalization, ►photoperiodism, ►phytohormones, ►salicylic acid, ►host-pathogen relationship, ►signaling to translation, ►nuclear pore, ►SUG, ►TIF, ►TRIP, ►cell membranes, ►PP2A, ►MPK, ►MPK phosphatase, ►MLK, ►REM, ►GEF, ►CBP, ►GFP, ►cross-talk, ►arrestin, ►desensitization, ►SMAD, ►NF-κB, ►genetic networks, and the other terms mentioned, ►feedback, ►selector genes; http://www.signaling-gateway.org; http://www.mshri.on.ca/pawson/research.html; Milligan G (ed) 1999 Signal transduction. A practical approach. Oxford University Press, New York; Morris AJ, Malbon CC 1999 Physiol Revs 79:1373; Hunter T 2000 Cell 100:113; Dohlman HG, Thorner JW 2001 Annu Rev Biochem 70:703; Heldin C-H 2001 Stem Cells 19:295; review articles in 2001 Nature [Lond] 413:186–230; Brivanlou AH, Darnell JE Jr 2002 Science 295:813; for newer reviews: 2002 Science 296:1632 ff; Dorn GWII, Mochly D 2002 Annu Rev Physiol 64:407; Ernstrom GG, Chalfie M 2002 Annu Rev Genet 36:411; Pires-daSilva A, Sommer RJ 2003 Nature Rev Genet 4:39; Pawson T, Nash P 2003 Science 300:445; signal transduction networks: Bhattacharyya RP et al 2006 Annu Rev

Biochem 75:655; signal transduction proteins: http://www-wit.mcs.anl.gov/sentra; signaling database: http://www.signaling-gateway.org/molecule/; signals: http://www.grt.kyushu-u.ac.jp/spad/; http://www.gene-regulation.com/pub/databases.html; http://bibiserv.techfak.uni-bielefeld.de/stcdb/; prokaryotic signal transduction proteins: http://compbio.mcs.anl.gov/sentra; microbial signal transduction database: http://genomics.ornl.gov/mist.

Signal Transfer Particle: PDGF associated with phospholipase C-γ, phosphatidyl inositol 3-kinase, and the RAF protooncogene product regulates signaling. ►platelet derived growth factor, ►phosphatidyl inositol kinase, ►RAF

Signaling: The phosphorylation of enzymes mediated by second messengers leads to their activation through outfolding of the pseudosubstrate domains of the enzymes and thus opens the active sites for the true substrate. ►signal transduction

Signaling Molecule: A signaling molecule alerts cells to the behavior of other cells and environmental factors. ►autoinduction

Signaling to Translation: Signaling transduction is directed toward the 5′-untranslated region (UTR) and involves the translation of ribosomal proteins and elongation factors (eEF1A, eEF2). The target of the signals appears to be the 5′-terminal oligopyrimidine sequences (5′-TOP) and the UTR polypyrimidine tracts. Secondary structure formation with long UTRs may also be regulatory. Growth factors may phosphorylate the eukaryotic initiation factor eIF4E-binding protein, 4E-BP-1, and cause its dissociation from eIF4E. In order that the translation would proceed, the initiation factors eIF4A, -4B, -4G, -4E attach to the methylguanine cap and the secondary structure of the RNA is untwisted by the helicase action of eIF4A. Phosphorylation by the $p70^{S6k}$ kinase activates the S6 protein of the 40S ribosomal subunit, a process subject to enhancement by mitogens. This process is not a general requirement for translation, indicating that it affects only special genes with 5′-TOP and polypyrimidine tracts in their UTRs. Cycloheximide and puromycin are also involved in the phosphorylation of S6. Both of these phosphorylations are inhibited by rapamycin. ►F506, ►$P70^{S6k}$, ►secondary structure, ►cycloheximide, ►puromycin, ►cap, ►regulation of gene activity, ►regulation of enzyme activity, ►protein synthesis, ►signal transduction; Wilson KF, Cerione RA 2000 Biol Chem 381(5–6):357.

Signalosome: A molecular complex transmitting various cues. ►signal transduction; Lyapina S et al 2001 Science 292:1382; Zhang SQ et al 2000 Immunity 12:301.

Signature, Evolutionary: Within the various genomes (mammalian, mitochondrial, plants, prokaryotes, etc.), there appears to be a characteristic distribution of dinucleotide sequences, which is different from that of the other species. This constitutes an evolutionary signature. (See Campbell A et al 1999 Proc Natl Acad Sci USA 96:9184).

Signature of a Molecule: The characteristic feature(s) convenient for identification. Discriminating sequences of DNA, RNA, or the proteins of organisms may serve as signatures. Molecular signatures can predict the course of diseases (O'Shaughnessy JA 2006 N Engl J Med 355:615).

Signature-Tagged Mutagenesis (STM): The induction of mutation by the insertion of plasmids, transposable elements, or passengers of specially constructed vectors into the genetic material. An insertional mutagenesis system that uses transposons carrying unique DNA sequence tags—flanked by PCR-amplifiable short tracts—was developed for the isolation of bacterial virulence genes. Originally in a murine model of typhoid fever caused by *Salmonella typhimurium*, mutants with attenuated virulence were revealed by use of tags that were present in the inoculum but not in bacteria recovered from infected mice (Hensel M et al 1995 Science 269:400). ►targeting genes, ►vectors, ►insertional mutation; Nelson RT et al 2001 Genetics 157:935; Shea JE et al 2000 Curr Opin Microbiol 3:451; Mazurkiewicz P et al 2006 Nature Rev Genet 7:929.

Significance Level: The significance level indicates the probability of error by rejecting a null hypothesis that is valid (Type I error, α) or accepting one that is not correct (Type II error, β). By convention, 5% (*, significant), 1% (**, highly significant), and 0.1% (***, very highly significant) levels are used most commonly. These are not sacrosanct limits. In field experiments with crops, the 5% level may be a satisfactory measure for comparative yields but even the 0.1% may not be acceptable for pharmaceutical tests because the chance of harming 1/1000 persons is unacceptable. In general experimental practice, levels above 5% and below 0.1% are not considered meaningful although they may have relevance for pharmacology. ►goodness of fit, ►*t* test, ►probability, ►power of the test, ►inference

Sign-R1: A lectin, which captures microbial polysaccharides in the spleen and interacts with complement component C1q. ►lectins, ►complement

SIL (short insert library): The SIL is generated by the restriction cleavage of gap-bridging clones (used in the final stages of physical mapping) into 0.5 kb or smaller fragments to break up secondary structures of the DNA that complicate the determination of continuity in sequencing. ►chromosome walking, ►physical mapping, ►restriction enzymes

SILAC (stable isotope labeling by amino acid in cell culture): A proteomics technique used for separate identification of similar proteins in high performance liquid chromatography–mass spectrometry (Foster LJ et al 2003 Proc Natl Acad Sci USA 100:5813).

Silencer: A negative regulatory element reducing transcription of the region involving the target genes. Their action bears similarity to the heterochromatic chromosomal regions, which reduce transcription of genes transposed to their vicinity. The Sir proteins may interact with the amino-terminal of histones 3 and 4. Sir2 deacetylates histone H4 lysine^{16} tail and mediates the synthesis of 0-acetyl-ADP-ribose. As a consequence, Sir2, Sir3, and Sir4 bind to the histone tail and mediate chromatin silencing (Liou G-G et al 2005 Cell 121:515). Sir2 homotrimer is required for histone deacetylation and rDNA suppression, whereas telomere silencing requires the heterotrimeric complex of Sir2, Sir3, and Sir 4 (Cubizolles F et al 2006 Mol Cell 21:825). Silencing requires combination of a protein(s) and the site where silencing takes place. There is evidence that Sir-generated heterochromatinization interferes with the assembly of the components of the pre-initiation complex (Sekinger EA, Gross DS 2001 Cell 105:403). Some evidence in yeast indicates that TFIIB, RNA polymerase II, and TFIIE occupancy is reduced by a silencer at downstream of the gene activator protein (Chen L, Widom J 2005 Cell 120:37). For example, the *MATa* and *MATα* genes of yeast encode regulatory proteins that permit the expression of *a* and *α* mating types, respectively, of the haploid cells and the non-mating phenotype of the sporulation-deficient *a/α* diploid cells. When these genes are at the *HMLa* and *HMRα* sites, they are silenced until they are transposed to the *MAT* locus. The mating type switch is catalyzed by a cut mediated through HO endonuclease when the mating type alleles are at the MAT locus, but not at the *HMLa* and *HMRα* locations. This indicates that the silencing is under the dual control of the repressed domains, which appear to extend to 0.8 kb proximal to the centromere from *HML-E* and a silencer protein and a specific site. Inactivation of *SIR2*, *SIR3*, and *SIR4* derepresses *HML* and *HMR*. These genes affect the telomeric position effect of other genes as well. SIR2, SIR3, and SIR4 are also involved in DNA repair and recombination in cooperation with the *HDF1* locus of yeasts (a *Ku* homolog). Mutations at the amino terminus of the *HISTONE 4* gene also have a similar effect. Over-expression of *SIR2* causes its hypoacetylation, while *SIR3* mutations may alter the conformation of this histone bound to *HMR*. Loci *HML* and *HMR* both are flanked by *HML-E* and *HML-I* silencer elements. These silencers are similar to the autonomously replicating elements of yeast that are involved in DNA synthesis and apparently also in silencing. *HML-E* is capable of repression only in the presence of *HML-I* and is 0.4 kb distal from *HML-I* (based on Loo, Rine 1994 Science 264:1768). *HMR-E* is a very potent silencer endowed with binding sites for ORC (origin recognition complex), Rap1 (a suppressor of RAS-induced replication), and Abfl (another silencer) suppressors of the S phase of the cell cycle. Recent evidence indicates that transcriptional silencing does not require DNA replication, although it seems to require some cell cycle events. Silencer elements are also present in animal and plant systems. In plants, when multiple copies of a gene are introduced into the genome by transformation, all or most copies of the gene are inactivated (trans-inactivation). The mechanism of this phenomenon is unclear. It has been suggested that when the level of a particular RNA is increased, a degradative process is initiated. This has been attributed to a defense mechanism, since the majority of plant viruses are RNA viruses. Some of the silencing appears, however, post-transcriptional. In fungi, silencing has been attributed to premeiotic methylation when multiple copies are present in cis position. This view is supported by the long-standing knowledge that the repetitive sequences of the heterochromatin are not expressed. Position effect has also been known as a type of silencing. The reversible type of paramutation can also be considered a transinactivation mechanism. Silencing of genes may be accomplished by moving the region of the intact chromosome toward the centromere (position effect). Transposition to the vicinity of heterochromatin may also result in silencing. Sir1 protein is most common in the telomeric region of yeast chromosomes but it may be present also in the nucleolus. Sir3 and Sir4 normally are absent from the nucleolus, but are present in case Sir2 is mutant. The SIR2 homolog in mammalian cell is SIRT2, which controls caloric intake and thereby extends life span. Absence of SIRT6 in mammals results in genomic instability, autoimmune disease (lymphopenia), metabolic defects, and premature death as a consequence of deficiency in excision repair of DNA (Mostoslavsky R et al 2006 Cell 124:315).

Although silencers have some similarities to insulators, the latter are different because they must be situated in between the enhancer and the target promoter. Unpaired DNA may cause meiotic silencing in *Neurospora*. Hypermethylation of CpXpG nucleotides by chromomethylase 3 in *Arabidopsis* silences the expression of some genes and its mutation restores wild phenotype to the epigenetically silenced genes and reactivates retrotransposons. ►targeting genes, ►enhancer, ►position effect, ►mating type determination in yeast, ►*Schizosaccharomyces pombe*, ►antisense technologies neuron-restrictive silencer factor, ►CREB, ►co-suppression, ►quelling, ►epigene conversion, ►paramutation, ►heterochromatin, ►dominant-negative mutation, ►methylation of DNA, ►epigenesis, ►histone deacetylase, ►Ku, ►Hst1p, ►Abf, ►ORC, ►RAP1, ►post-transcriptional gene silencing, ►transcriptional gene silencing, ►RNAi, ►microRNA, ►HML and HMR, ►nucleosome, ►chromatin remodeling, ►MSUC, ►telomeric silencing, ►looping of DNA, ►insulator, ►methylation of DNA, ►pre-initiation complex, ►sirtuin, ►PIC, ►ascus-dominant, ►DNA repair; Guareente L 1999 Nature Genet 23:281; Kirchmaier AL, Rine J 2001 Science 291:646; Lindroth AM et al 2001 Science 292:2077; Sijen T et al 2001 Curr Biol 11:436; Ogbourne S, Antalis TM 1998 Biochem J 331:1; Moazed D 2001 Mol Cell 8:489; Mlotshwa S et al 2002 Plant Cell 14: S289; Béclin C et al 2002 Curr Biol 12:684; Shiu PKT, Metzenberg RL 2002 Genetics 161:1483; Meister G, Tuschl T 2004 Nature [Lond] 431:343; Talbert PB, Henikoff S 2006 Nature Rev Genet 7:793.

Silene: ►*Melandrium*

Silent Information Regulators: Silent information regulators are involved in the assembly of the silent chromatin domains. ►silencer; Moazed D et al 1997 Proc Natl Acad Sci USA 94:2186.

Silent Mutation: Base-pair substitution in DNA that does not involve amino acid replacement in protein and entails no change in function. ►mutation

Silent Sites: Silent sites are where mutations in the DNA base sequence have no consequence for function. ►synonymous codons

Silicon (Si): The second most abundant element (27.6%) in nature. Si occurs in quartz, sand, and sandstone, or as silicate, e.g., in kaolinite. Plants take up 0.1 to 10% of their dry weight as silicic acid (H_2SiO_3) by using genetically determined silicon transporter proteins. It increases firmness of plant tissues and disease resistance (Ma JF et al 2006 Nature [Lond] 440:688). Silicons, organosilicon oxides polymers, are used in the laboratory as lubricants and seals. Silica gels are lustrous granules used for absorption and column chromatography.

Siliconization: In siliconization, glassware used in genetic laboratories is treated in a vacuum by dichlorodimethylsilane in order to prevent DNA molecules sticking to the vessel, and resulting in loss of recovery. ►DNA extraction

Silique: A typical fruit of cruciferous plants; two carpels, dehiscing at the base at maturity, enclose the placentae which sit in one row on each of the opposite sides of the replum. In Figure S61, the carpels were removed before maturity.

Figure S61. Silique

Silk Fibroin: A protein rich in glycine and alanine residues arranged largely in β sheets (β keratin). It is synthesized within the silk gland to protect the pupa. The pupa is called also chrysalis or cocoon. A fibroin gene is transcribed into about 10,000 long-life molecules of mRNA within a few days and they are translated several times into about a billion protein molecules. Each gland manufactures about 10^{15} fibroin molecules (300 µg) in four days. Actually, the gland is a single cell but it contains polytenic chromosomes and thus the fibroin locus is amplified about a million fold ($10^9 \times 10^6 = 10^{15}$ fibroins). In spiders, there is great diversity and also conservation in the silk fibroin genes (Garb JE et al 2006 Science 312:1762). Some of the spiders' silks are tougher than that of silkworm and rival the best man-made fibers. The spider dragline silk, besides its outstanding tensile strength, has unrivalled torsional quality that stop the spider from twisting and swinging and thus makes the animal less conspicuous to predators (Emile O et al 2006 Nature [Lond] 440:621). Spider dragline silk can be synthesized in transgenic tobacco, potato plants, and also in transgenic mammalian cells. ►polytenic chromosomes, ►silk worm, ►resilin; Vollrath F, Knight DP 2001 Nature [Lond] 410:541; Scheller J et al 2001 Nature Biotechnol 19:573; Lazaris A et al 2002 Science 295:472; Jin H-J, Kaplan DL 2003 Nature [Lond] 424:1057.

Silk: The botanical term for the pistils of the maize female inflorescence (see Fig. S62).

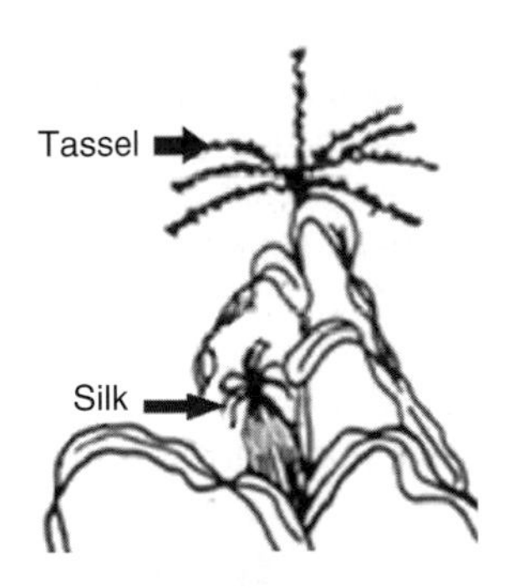

Figure S62. Maize female inflorescence

Silkworm (Bombyx mori, 2n = 56): One of the best-studied insects in genetics (see Fig. S63). There are about 1000 markers in the genetic map spacing at ~2 cM. Its RAPD map (~2000 cM), includes ~1018 markers scattered over all chromosomes.

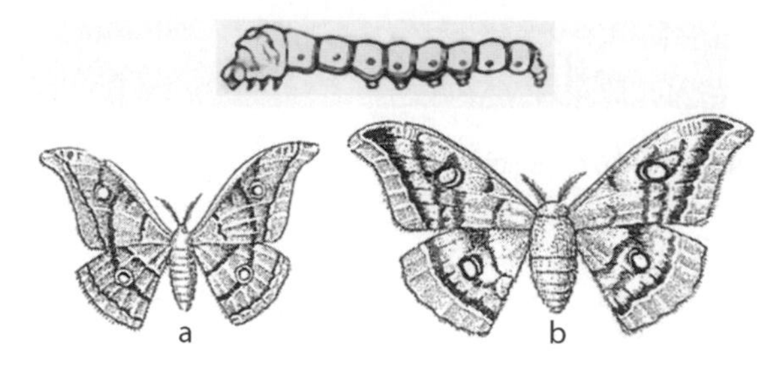

Figure S63. Silkworm larva; at top (a) male (ZZ), at right ZW female (b); wild type Asian imagoes. The pattern varies in the different wild insects and also in the domesticated varieties

About 91% of the genome of the domesticated silkworm has been sequenced and it contains ~18,510 genes (Xia Q et al 2004 Science 306:1397). Its genome contains a special type of transposable element, R2Bm, that is present also in some other insects. R2Bm has no long terminal repeats. It is inserted in the 28S rRNA genes only and encodes an integrase and a reverse transcriptase function within one protein molecule. The R2 protein nicks one of the DNA strands and uses it also as a primer to transcribe its RNA genome, which is then integrated as DNA-RNA heteroduplex. Subsequently, a host polymerase synthesizes the second DNA strand. ►transposon, ►complete linkage, ►autosexing, ►tetraploidy, ►polyhedrosis virus, ►silk fibroin, ►RAPD, ►pheromone; EST database: Mita K et al 2003 Proc Natl Acad Sci USA 100:14121; silkworm microsatellite database: http://210.212.212.7:9999/PHP/SILKSAT/index.php; http://www.cdfd.org.in/silksatdb; http://www.ab.a.u-tokyo.ac.jp/silkbase/; http://silkworm.genomics.org.cn.

Silver Syndrome (Silver spastic paraplegia, SPG17, 11q12-q14): A neurodegenerative disease involving amyotrophy (muscle weakness) in the hands. Hereditary dominant spastic paraplegia is a highly variable disease and it is encoded in several other chromosomes and locations. ►paraplegia, ►spastic paraplegia, ►Berardinelli-Seip congenital lipodystrophy; Patel H et al 2001 Am J Hum Genet 69:209.

Silver-Russel Syndrome: ►Russel-Silver syndrome

Silverman-Handmaker Syndrome: ►dyssegmental dwarfism

Silyl-Phosphite Chemistry: Silyl-phosphite chemistry is used in oligoribonucleotide synthesis. (See Agarwal S (Ed.) 1995 Methods in Molecular Biology. Humana, Totowa, New Jersey, p 81).

Simian: Ape or monkey type. ►primates, ►hominidae

Simian Crease: See ►Down's syndrome for illustration. It can be rarely observed (1–4%) in normal infants but it is characteristic for human trisomy 21, De Lange, Aarskoog, and other syndromes. ►Down syndrome, ►Aarskoog syndrome, ►De Lange syndrome

Simian Sarcoma Virus (SSAV): A gibbon/ape leukemia retrovirus with a homologous element in human chromosome 18q21. The long terminal repeat (535 bp) appears to contain transcriptional control and signal sequences. The human chronic lymphatic type leukemia seems to be associated with a break point of chromosome 18. ►leukemia

Simian Virus 40: A eukaryotic virus of a molecular weight of 3.5×10^6 with double-stranded, supercoiled DNA genetic material of 5243 bp (see Fig. S64). The DNA is organized into a nucleosomal structure that does not have H1 histone. The DNA around the nucleosome cores is 187 ± 11 bp and the cores are separated by 42 ± 39 bp linkers. The viral particles are skewed icosahedral capsids and have 72 protein units.

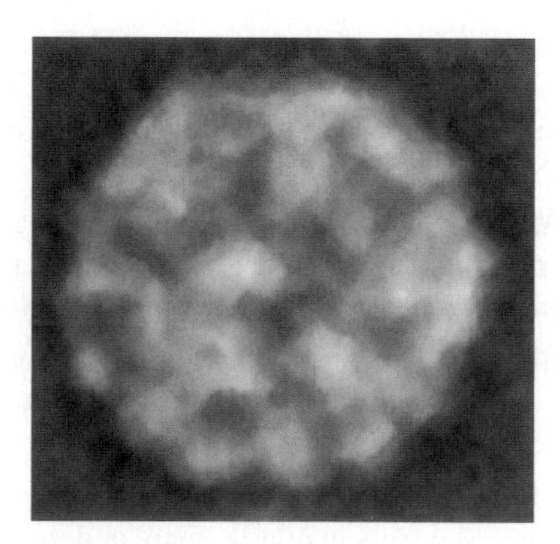

Figure S64. SV40

In primates, the virus generally follows a lytic lifestyle and the virions multiply in the cytoplasm, i.e., primates are *permissive hosts* for replication. SV40 encodes microRNAs that protect to some extent against the host cytotoxic T cells (Sullivan CS et al 2005 Nature [Lond] 435:682).

Occasionally, in humans the viral DNA integrates into the chromosomes. Such an event may lead to cancerous transformation. Rodent cells are *non-permissive* hosts for viral replication and the viral DNA integrates into the chromosomes leading to cancerous tumor formation. The correlation between

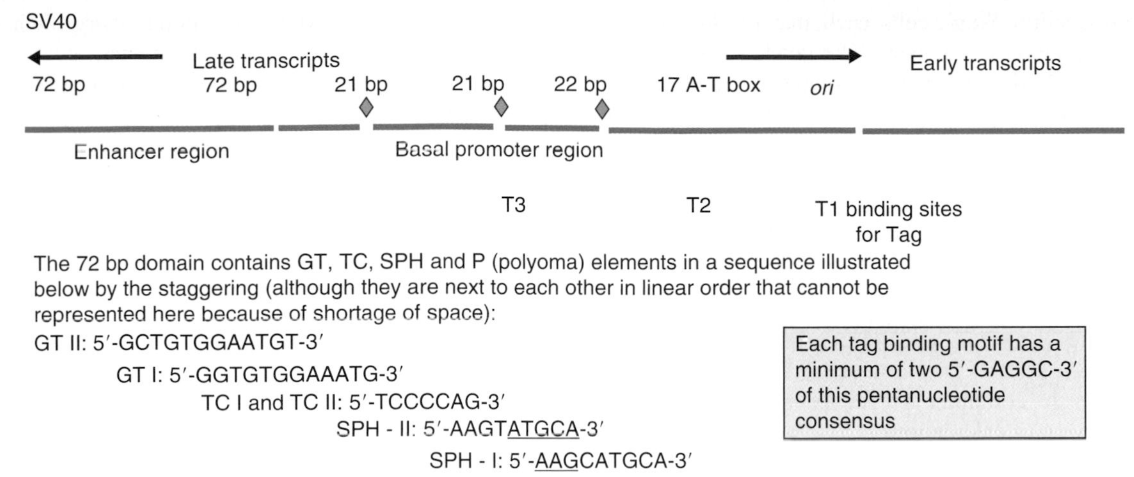

Figure S65. Organization of SV40

SV40 infection and cancer has been questioned (Paulin DL, DeCaprio JA 2006 J Clin Oncol 24:4356). The virus codes for early (t and T antigens) and late (VP1, 2, 3) viral proteins (see Fig. S65).

The viral replication and transcription are bidirectional (see Fig. S66). In the non-permissive host, only the early genes are expressed that are needed for replication of the genetic material before integration, but there is no need for the coat proteins. The integration can take place at different sites, therefore it uses a mechanism of illegitimate recombination. The few integrated copies may be rearranged and may cause continued chromosomal rearrangement in the host. The infectious cycle spans about 70 h. The joint replication and transcriptional origin (*ori*) area extends to about 300 bp and includes a rather sophisticated control system. The replication of the SV40 DNA begins at the 27 bp palindrome of the *ori* site that is adjacent to a region consisting of 17 A-T base pairs. Next to it, on the side of the late genes, there are three other units of 22, 21, and 21 GC-rich repeats that also promote replication, although are not absolutely essential to the process. The SPH elements have an overlapping *octamer* that is present in other eukaryotic genes as well. Several other sequence motifs are similar to those in other promoters. The 72 elements include the 47 bp B and the shorter 29 bp A domains that are parts of the essential enhancer region. The A-T box is essentially a TATA box: 5′-TATTTAT-3′. For the start of replication, the large T has to bind to the Tag binding sites Δ. At the initiation, when low amounts of Tag are available, binding begins at the T1 site located at the pre-mRNA region (right of *ori*). As more Tag will become available, Tag binds to T2 and becomes an ATP-dependent helicase and with the cooperation of

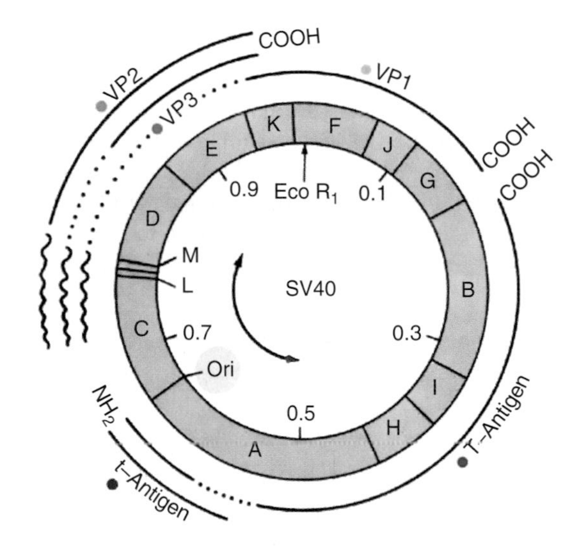

Figure S66. Gene and polypeptide map of SV40

cellular proteins (DNA primase, polymerase, etc.), DNA replication proceeds.

Transcription. For the function of *ori* in transcription the 17 bp A-T sequence is needed, although this TATA box does not affect the rate of transcription. The 21◇21◇22◇ sequences promote transcription and the two hexamers 5′-GGGCGG-3′ within these elements are essential for transcription. Their orientation and inversion does not interfere with transcription. The 72 bp element of SV40 is a capable enhancer also for mammalian, amphibian, plant, and fission yeast genes. The natural host of SV40 is the rhesus monkey (*Macaca mulatta*). In the laboratory, the kidney cell cultures of the African green monkey (*Cercopithecus aethiops*) are used primarily for its propagation. Single cells, each, may produce 100,000 viral genomes after lysis. The general assumption is that SV40 does not cause human cancer, yet in many tumors its presence was detected by some laboratories, but not by others. ►SV40 vectors, ►COS cell, ►CTL, ►microRNA; Butel JS, Ledniczky JA 1999 J Natl Cancer Inst 91(2):119; see reviews in Semin Cancer Biol 2001 11:5–85; replication initiation complex: Simmons DT et al 2004 Nucleic Acids Res 32:1103; review of virus entry into mammalian cell: Damm EM, Pelkmans L 2006 Cell Microbiol 8:1219.

SimIBD: A computer procedure to assess affected-relative pair calculations.

Similarity Index: ►character index

SIMLINK: A simulation based computer program for estimating linkage information. ►SLINK, ►simulation

Simple Protein: Upon the hydrolysis of a simple protein, only amino acids are produced.

Simple Sequence Length Polymorphism (SSLP): In SSLP, variations in microsatellite sequences can be used for DNA mapping. PCR primers are designed for the unique flanking sequences and their length can thus be determined in the PCR products. ►microsatellite, ►PCR; Cai W-W et al 2001 Nature Genet 29:133.

Simple Sequence Repeats (SSR): SSRs are common in genomes and can be used for mapping genes and for taxonomic studies. (See SSR primer tool: http://bioinformatics.pbcbasc.latrobe.edu.au/ssrdiscovery.html).

Simplesiomorphy: The primitive features retained during evolution. It is not very useful to trace evolutionary development. ►synapomorphy, ►homology

Simplex: A polyploid having only a single dominant allele at a particular gene locus; the other alleles are recessive at the locus. ►duplex, ►triplex, ►quadruplex

Simplexvirus: A member of the herpes family of viruses infecting humans and other mammals. ►herpes

Simpson-Golabi-Behmel Syndrome (Simpson Dysmorphia, SGBS): A human Xq26 (500 kilobase stretch) syndrome, encoding the GPC3 gene responsible for the synthesis of glypican (cell surface molecules of heparan sulphate proteoglycans), associated with the insulin-like growth factor (IGF2). Individuals afflicted by this condition are generally very tall, usually have facial anomalies, heart and kidney defects, cryptorchidism, hypospadia, hernias, and bone anomalies and show susceptibility to cancer. Many of the symptoms involving overgrowth are shared with the Beckwith-Wiedeman syndrome. ►Beckwith-Wiedeman syndrome, ►IGF, see terms at corresponding entries.

SIMS (secondary ions MS): A method for the production of intact molecular ions for mass spectrometry. ►mass spectrometer

Simulation: The representation of a biological system by a mathematical model generated frequently by a computer program. ►modeling; http://www.nrcam.uchc.edu; ►Monte Carlo method

Simulon: ►origon

Sin3/RPD: A repressor protein complex probably involved in chromatin remodeling by being recruited to histone deacetylase. It may be associated with SAP18, SAP30, and the retinoblastoma binding protein. ►Mad, ►chromatin remodeling, ►histone deacetylase, ►NuRD; Brubaker K et al 2000 Cell 103:655.

SIN Vector (self-inactivating vector): The SIN vector has a deletion in the U3 element of the 3′-LTR of the retroviral construct, and after replication results in a deletion also in the 5′-LTR promoter and enhancer and prevents the transcription from the cell-specific internal promoter, which may otherwise activate silent cellular oncogenes. This happens because the viral polymerase enzyme uses the 3′-U3 as template for the replication of both 3′- and 5′-U3 sequences. A disadvantage of this construct is the generally slow replication. A high-efficiency heterologous promoter to enhance the expression of the transgene in the retroviral vector may also replace the deleted viral promoter. ►retroviral vector, ►double-copy vector, ►E vector, ►gene therapy; Gatlin J et al 2001 Hum Gene Ther 12:1079.

Sindbis Virus: A single-stranded RNA virus.

sine: The ratio of the side opposite (a) to an acute angle (A) of a right triangle and the hypotenuse (c): a/c, sine of angle A. ►arcsine, ►angular transformation

SINE: Short (≈0.5 kbp) interspersed repetitive DNA sequences that may occur over 100,000 times in the mammalian genomes. The B1 SINE of mice is 130 to 150 bp in length and constitutes nearly 1% of the genome; it is homologous to the human Alu sequences. The B2 SINE (∼190 bp) constitutes about 0.7% of the mouse genome but it apparently has no human homolog or its abundance is very low. In the dog genome, 3 to 5% of the genes have SINE insertions, may differ in the different breeds, and may be a major source of allelic differences (Wang W, Kirkness EF 2005 Genome Res 15:1798). RNA polymerase III transcribes the B2 SINE elements into short sequences that are not translated. The mouse gene carries an active RNA polymerase II promoter and can support transcription by pol II (Ferrigno O et al 2001 Nature Genet 28:77). The SINE elements are retroposons, but lack reverse transcriptase function. SINE type elements occur in all eukaryotes, including birds, fungi, insects, and higher plants. They may be pseudogenes of small RNA genes. The SINE sequences can also be used for fingerprinting and evolutionary studies. These are remnants of ancient retroviral insertions but once they were inserted—because of the loss of the LTR [transposase] function they remained at the position of insertion. ►retroposons, ►transposable elements, ►LINE, ►Alu family, ►DNA fingerprinting, ►reverse transcriptase; Cantrell MA et al 2001 Genetics 158:769; Weiner AM 2002 Curr Opin Cell Biol 14:343.

Singing Ability: The ability to sing has genetic determination and some of the "song genes" of birds have been mapped for expression in different parts of the brain. (See Marler P, Doupe AJ 2000 Proc Natl Acad Sci USA 97:2965).

Single Burst Experiment: Virus-infected bacterial population diluted and distributed into vessels in such a way that each vessel would contain a single infected bacterial cell. (See Ellis EL, Delbrück M 1939 J Gen Physiol 22:365).

Single Cell Analytical Methods (chemical/physical): ►MALDI/TOF/MS, ►FISH, ►immunocytochemistry, ►immuno-electronmicroscopy, ►immunoelectrophoresis, ►immuno-fluorescence, ►SMART; Cannon DM Jr et al 2000 Annu Rev Biophys Biomol Struct 29:239; Slepchenko BM et al 2002 Annu Rev Biophys Biomol Struct 31:423; Subkhankulova T, Livesey FJ 2006 Genome Biol 7:R18; Xie XS et al 2006 Science 312:228.

Single-Chain Fv Fragment: A monoclonal single heavy plus light chain immunoglobulin that can be encoded by a single transgene. Because of its single structure, it is a monovalent antibody in contrast to the common antibodies, which are divalent. It lacks effector function. ►antibody monovalent, ►antibody effector function, ►monoclonal antibody, ►ScFv, ►immunostimulatory DNA

Single Copy Plasmids: ►plasmids

Single Copy Sequence: DNA sequences containing non-redundant, genic portions.

Single Cross: ►double cross

Single-Feature Polymorphism: ►allele; Borevitz JO et al 2007 Proc Natl Acad Sci USA 104:12057.

Single Gene Trait: A single gene trait is controlled by one gene locus, and shows monogenic inheritance.

Single Nucleotide Polymorphism: ►SNIPS

Single Strand Assimilation: A single strand displaces another homologous strand and then takes its place during a recombinational event. ►recombination molecular models

Single Strand Binding Protein: A single strand binding protein binds to both separated single stands of DNA and thus stabilizes the open region to facilitate replication, repair, and recombination. ►recombination molecular mechanisms, ►binding proteins; Witte G et al 2005 Nucleic Acids Res 33:1662.

Single Strand Conformation Polymorphism (SSCP): In SSCP, when small deletions or even single base substitutions take place in one of the DNA strands of a gene locus, the alteration may be detectable by the electrophoretic mobility of the DNA in denaturing polyacrylamide gels. The two strands, the normal and the affected, may differ. If the individual is heterozygous for the amplified segment of the locus concerned, the electrophoretic analysis may indicate three or more band differences. In some cases, even the homozygotes may show multiple bands. With this method, nearly all of the alterations are detected in fragments of 200–300 bp. ►gel electrophoresis, ►polymerase chain reaction, ►gene isolation, ►DGGE, ►mutation detection, ►dideoxy fingerprinting, ►MASDA; Orita M et al 1989 Proc Natl Acad Sci USA 86:2766.

Single-End Invasion (SEI): A meiotic recombination intermediate during the transition from double-strand breaks to double-Holliday junction. SEIs are formed by strand exchange between one and then the other double strand. The appearance of SEI coincides with that of the synaptonemal complex. SEI is preceded by a nascent double-strand partner intermediate that differentiates into a crossover and a non-crossover type after the synaptonemal complex has formed. Strand exchange occurs relatively late after synapsis and recombination may be avoided between homeologous and structurally rearranged partners. ►Holliday model, ►synaptonemal complex, ►homeologous; Hunter N, Kleckner N 2001 Cell 106:59.

Single-Feature Polymorphism (SPF): SPF reveals detailed information about the genomic variations between/among species or different accessions of a species. The genomic DNA is hybridized to an RNA expression platform (gene chip) and single-base differences prevent the hybridization of 25 mer probes. By such a procedure, 4000 SPFs were found between the Columbia wild type and the Ler genotype *Arabidopsis* plants. On the basis of SPFs, gene map locations, including QTLs, can be readily identified. It can also reveal organizational differences along the chromosomes, e.g., in centromeric or telomeric tracts versus the rest of the yeast chromosomes or various functional regions. (See Borevitz JO et al 2003 Genome Res 13:513; Winzeler EA et al Genetics 163:79).

Single-Molecule Chemistry: With the currently available optical facilities, it is now possible to observe the dynamic behavior of single biomolecules and to study their kinetics. Within a group of molecules (e.g., in an enzyme), static and dynamic heterogeneity exists among the different molecules; there is also an inherent and ubiquitous fluctuation in the structure and function of these molecules. Classical chemistry could detect only the behavioral average of these molecules. Recent approaches open new vistas in biological chemistry. Conformational change could be detected by fluorescence of an added fluorophore within the T7 DNA polymerase ternary complex, upon binding of a dNTP substrate. This fluorescence change is believed to reflect the closing of the T7 pol fingers domain, which is crucial for polymerase function (Luo G et al 2007 Proc Natl Acad Sci USA 104:12610). The single-molecule spectroscopy method directly probes kinetic reversibility and the chaperone role of the nucleocapsid of the HIV-1 immunodeficiency virus strand transfer to cell at various stages along the reaction sequence, giving access to previously inaccessible kinetic processes and rate constants (Zeng Y et al 2007 Proc Natl Acad Sci USA 104:12651).

Single-Positive T Cell: The single-positive cell expresses either the CD4 or the CD8 surface proteins. ►CD4, ►CD8, ►T cell

S

Single-Strand Annealing Repair: ►SSA

Singlet Oxygen ($^1O_2$): A highly reactive O_2 molecule produced during inflammation, by photosensitization in UV light, chemiexcitation in dark, decomposition of $NDPO_2$, etc. $^1O_2$ may be toxic to molecules in the cell, oxidizes DNA, and produces mutagenic 7-hydro-8-oxodeoxyguanosine. It may affect gene expression and carcinogenesis. ►ROS, ►8-oxodeoxyguanosine, ►photodynamic effect

Singleton: Singly occurring whole-body mutations; the spontaneous frequency in mice for seven standard loci is 6.6×10^{-6} per locus. ►mutation rate

Singleton-Merten Syndrome: A rare disease involving aortic calcification but defects in bone development.

Singlets: Genes that occur only once in the genome.

Singular Value Decomposition (SVD) of RNA: SVD uncovers in the mRNA data matrix of genes, x arrays, i.e., electrophoretic migration length (Alter O, Golub GH 2006 Proc Natl Acad Sci USA 103:11828).

Sink: The storage of metabolites from where they can be mobilized on need.

Sink Habitat: A habitat in which some individuals contribute less to the future generations than the average individual. ►source habitat, ►habitat

Sinndakiss: A receptor internalization signal.

Sinorhizobium: ►nitrogen fixation

Siphonogamy: In siphonogamy, the immotile microgametes of higher plants are delivered to the archegonia through the elongating pollen tube. ►pollen tube, ►embryosac, ►zoidogamy

Sipple Syndrome: ►phaeochromocytoma

SIR: ►silencer

Sire: The male mammal; the term used primarily in animal breeding and applied animal genetics. ►dam

Sirenomelia: A developmental malformation showing fused legs and usually lack of feet.

SIRM: The sterile insect release method. ►genetic sterilization

siRNA (silencing RNA): A 29-amino-acid peptide specifically binds to the acetylcholine receptor expressed by neuronal cells. To enable siRNA binding, a chimeric peptide was synthesized by adding nine arginine residues at the carboxy terminus of rabies virus glycoprotein (RVG.) This RVG-9R peptide was able to bind and transduce siRNA to neuronal cells in vitro, resulting in efficient gene silencing (Kumar P et al 2007 Nature [Lond] 448:39). Low-copy promoter-associated siRNAs transcribed through RNAPII are recognized by the antisense strand of the siRNA and function as a recognition motif to direct epigenetic silencing complexes to the corresponding targeted promoters, in order to mediate transcriptional silencing in human cells (Han J et al 2007 Proc Natl Acad Sci USA 104:12422). ►RNAi, ►rasiRNA, ►microRNA, ►RNA polymerase IV, ►BBB; Pikaard CS 2006 Cold Spring Harb Symp Quant Biol 71:473; human siRNA database: http://siRNA.cgb.ki.se; http://itb.biologie.hu-berlin.de/~nebulus/sirna/v2/.

Sirolimus: ►rapamycin

SIRPs (signal regulatory proteins, 20p13): Members of the SIRP family inhibit signaling through tyrosine kinase receptors and represent immune inhibitory receptors expressed on macrophages or other blood cells. ►tyrosine kinase receptor, ►macrophage; Latour S et al 2001 J Immunol 167:2547.

Sirtuin (Sirt): NAD-dependent histone deacetylase and ADP ribosylase proteins of the Sir2 family. Sirtinol also activates many auxin-inducible plant genes. More than 65% of the 138 sirtinol-induced genes are auxin-inducible. Both auxin- and sirtinol-induction are apparently mediated by ubiquitin-activated protein degradation (Zhao Y et al 2003 Science 301:1107). Resveratrol may activate sirtuins and prolong life (Howitz K et al 2003 Nature [Lond] 425:191). Sirtuin mediates the mobilization of fat by repressing genes controlling the peroxisome proliferator-activator receptor-γ. Overproduction of Sirt reduces adipogenesis, and interference with Sirt RNA enhances fat production. SIRT4 functions in the mitochondria, represses glutamate dehydrogenase by AD2064-ribosylation, downregulates insulin secretion, and opposes the effect of caloric restriction in pancreatic β cells (Haigis MC et al 2006 Cell 126:941). SIRT1 and SIRT3 activate acetyl-CoE synthetase by deacetylation of the cytoplasmic or the mitochondrial enzyme, respectively, by targeting a lysine residue (Hallows WC et al 2006 Proc Natl Acad Sci USA 103:10230). Sirt1 docks with nuclear receptor co-repressor (NcoR) and silencing mediator of retinoid and thyroid hormone receptors (SMRT) leading to PPAR-γ repression. Reduction of fat by lipolysis enhances lifespan (Picard F et al 2004 Nature [Lond] 429:771). Sirtuins regulate aging and age-related diseases such as cancer, diabetes, and neurodegeneration (Longo VD, Kennedy BK 2006 Cell 126:257). In cell-based models of mouse for Alzheimer disease, amyotrophic lateral sclerosis and other tauopathies resveratrol, a SIRT1-activating molecule, promotes neuronal survival (Kim D et al 2007 EMBO J 26:3169). ►adipocyte, ►PPAR, ►longevity, ►aging, ►obesity, ►ARF, ►neurodegenerative diseases, ►silencer, ►resveratrol, ►PGC1, ►acetyl-CoA; Grozinger CM et al 2001 J Biol Chem 276:38837; Pandey R et al 2002 Nucleic Acids Res 30:5036; signaling and regulation of protein deacetylation: Sauve AA et al 2006 Annu Rev Biochem 75:435.

SIS: The simian sarcoma virus oncogene is located in human chromosome 22q12.3-q13.1 and mouse chromosome 15. The SIS protein has high homology to the β chain of the platelet derived growth factor (PDGF), KIT oncogene, FOS oncogene, and the colony stimulating factor. ►oncogenes, ►PDGF, ►colony stimulating factor; Liu J et al 2001 Nucleic Acids Res 29:783.

Sis1: The DnaJ structural homolog of budding yeast indispensable protein with multiple chaperone functions, including initiation of translation. ►chaperones, ►DnaJ, ►DnaK

Sister Chromatid Cohesion: The juxtaposition of the sister chromatids until the end of metaphase in mitosis and until the end of metaphase II in meiosis. The inner centromere proteins (INCENP) and the centromere-linking proteins (CLiP) provide the physical basis of the cohesion. The multiprotein cohesion complex, which binds most tightly to the centromere, was named cohesin. A separation protein (separin, a cysteine protease) mediates sister chromatid cohesion, and the dissociation is achieved when the Scc1/Mcd1/Rad21 subunit of cohesin dissociates from the chromatids upon proteolytic cleavage. Rec8 is also a component of the meiotic cohesin complex. The Esp1 (separin) is tightly bound to the chromosomes by the anaphase inhibitor Pds1 (mammalian homolog Securin). Pds1 is ubiquitinated by the triggering effect of the anaphase-promoting complex (APC) and Cdc20. The sister chromatids are closely juxtapositioned until anaphase, indicating the presence of inter-sister connector structures. Sister chromatid cohesion affects proper disjunction of the mitotic chromatids but it appears important also for meiotic recombination. In mitosis, the separation of the sister chromatids and the splitting of the centromere take place during the single anaphase. In meiosis, at anaphase I, the sister chromatids separate but the centromere does not until anaphase II. This timing is apparently under the control of a specific protein(s). For the orderly segregation together of the sister chromatids during meiosis I, the protein monopolin is required in yeast.

Dominant and recessive mutations have been identified in plants, animals, and yeast that are defective in chromatid cohesion. In yeast, centromeric element III (CDEIII) is essential for sister chromatid cohesion and for kinetochore function. ►mitosis, ►meiosis, ►synapsis, ►asynapsis, ►desynapsis, ►sister chromatids, ►sister chromatid exchange, ►DNA polymerases, ►chiasma, ►centromere, ►cell cycle, ►cohesin, ►ORC, ►condensin, ►adherin, ►checkpoint; Nasmyth K et al 2000 Science 288:1379; Tóth A et al 2000 Cell 103:1155; Carson DR, Christman MF 2001 Proc Natl Acad Sci USA 98:8270; Lee JY et al 2001 Annu Rev Cell Dev Biol 17:753.

Sister Chromatid Exchange (SCE): Sister chromatid exchanges are detectable in eukaryotic cells provided with 5-bromo-deoxyuridine for (generally) one cycle of DNA replication (see Fig. S67). Subsequently, at metaphase the chromosomes are stained with either the fluorescent compound Hoechst 33258 (harlequin staining) or according to a special Giemsa procedure.

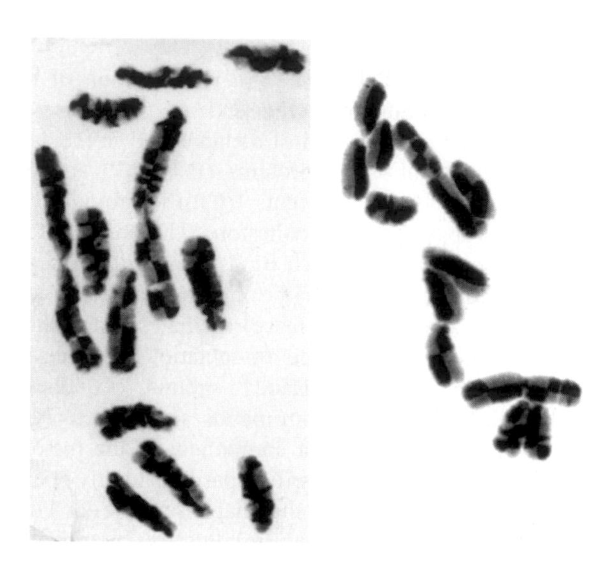

Figure S67. Sister chromatid exchange. Right: Untreated control. Left: Exposed to the alkylating compound thiotepa during DNA synthesis. (Courtesy of Professor BA Kihlman)

If sister chromatids are reciprocally exchanged, sharp bands appear in mirror image-like fashion. The frequency of sister chromatid exchange is boosted by about a third by potential carcinogens and mutagens. This method has been successfully used in various animal and plant cells for identifying genotoxic agents. The data must be evaluated with care in comparison with the concurrent control because BrdU itself may break chromosomes under UV-B light.

In *Saccharomyces cerevisiae*, molecular and genetic evidence are available for meiotic sister chromatid exchange. When one of the bivalents had different number of ribosomal RNA repeats with an embedded LEU2 gene, duplication and deficiency of LEU2 and the repeats were detected. Similar observations were made with other chromosomes and markers. After the DNA double-strand breaks, histone H2AX is phosphorylated at serine 139 and facilitates the homologous recombination of chromosomal double-strand breaks by using the sister chromatids as template (Xie A et al 2004 Mol Cell 16:1017). ►harlequin staining, ►Giemsa staining, ►bioassays in genetic toxicology, ►ring chromosomes, ►BrdU, ►ultraviolet light, ►genotoxic, ►sister chromatids, ►chiasma, ►crossing over, ►cohesin, ►double-strand breaks, ►DNA repair, ►histone variants; Shaham J et al 2001 Mutat Res 491:71.

Sister Chromatids: Sister chromatids are attached to the same side of the same centromere but they seem to be coiled in opposite directions (see Fig. S68). Their separation in mitosis requires the activation of a proteolytic enzyme encoded by the *Cut2* gene in *Schizosaccharomyces pombe*. ►chromatids; Nasmyth K 2001 Annu Rev Genet 35:673.

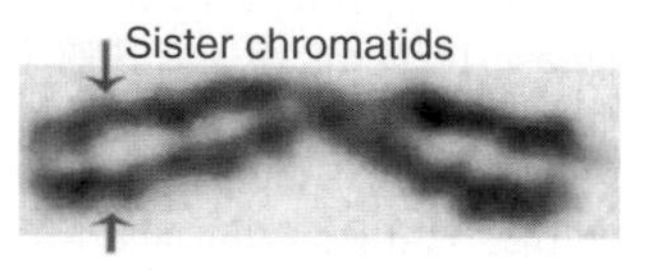

Figure S68. Sister chromatids

Sister-Strand Exchange: Same as sister chromatid exchange.

SIT: A family of protein phosphatases regulating diverse metabolic pathways. ►PP2A

SIT (sterile insect technique): ►genetic sterilization, ►GSM

Site-Directed Immunization: ►immunization genetic

Site-Directed Mutagenesis: ►directed mutation, ►localized mutagenesis, ►targeting; Storici F et al 2001 Nature Biotechnol 19:773.

Site-Specific Cleavage: The site specific cleavage of nucleic acids is accomplished by restriction endonucleases, some special RNases, and oligonucleotide-phenanthroline conjugates, which may cut both strands of the DNA in the presence of Cu^{2+} and a reducing agent. EDTA-Fe^{2+} may do the same if tethered to triplex molecules, albeit with low efficiency. In the presence of light, ellipticine attached to homopyrimidines may cleave a double helix within a triplex (see Fig. S69). ►restriction endonucleases, ►triplex, ►tethering; Gimble FS 2001 Nucleic Acids Res 29:4215.

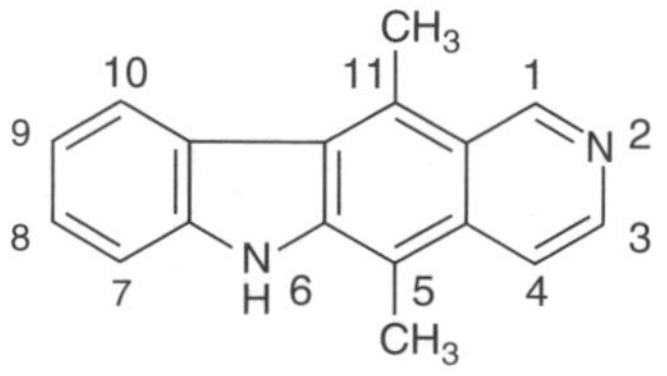

Figure S69. Ellipticine

Site-Specific Mutations: Site-specific mutations occur at particular nucleotides in the DNA and RNA, respectively. ►base substitution, ►localized mutagenesis, ►gene replacement, ►site-specific recombination, ►PCR-based mutagenesis, ►cassette mutagenesis, ►homolog-scanning mutagenesis, ►alanine-scanning mutagenesis, ►TAB mutagenesis, ►cysteine-scanning mutagenesis, ►Kunkel mutagenesis, ►targeting vector, ►oligonucleotide-directed mutagenesis, ►degenerate oligonucleotide directed mutagenesis, ►look-through mutagenesis

Site-Specific Recombinases: Site-specific recombinases are resolvases that attach at the two-base staggered cut sites. The enzyme is then covalently linked to the 5′ ends and the PO_4 of the DNA is covalently linked to the OH group of the recombinase. Subsequently, the broken DNA strand releases the deoxyribose hydroxyl group. The PO_4 is joined to another deoxyribose OH group and the DNA backbone is reconstituted. The members of the integrase group of enzymes attach at sites 6–8-bases apart. The first breakage results in a Holliday juncture, that may lead to branch-migration and after a second strand exchange and rotation isomerization (►Holliday model steps H-J) the strands may be resolved either with an outside marker exchange (classical recombination) or in gene conversion (the constellation of the outside markers retained). It is conceivable that the broken ends are reconstituted without any change, or deletions may also take place, or the position of the broken ends are inverted by 180° resulting in what classical cytology called inversion. Resolvase and integrase reactions can be very specific for the sites and the reaction is secured by the assistance of additional proteins that bring into contact only the appropriate DNA stretches. These two enzymes act only on supercoiled DNA. The integrase family of recombinase enzymes is more liberal in choice, yet affected by various conditions. The Mu phage or the HIV integration does not require covalent association between the DNA and a protein. The phosphodiester bond of the donor DNA is hydrolyzed to generate an OH group. This group and a phosphodiester group of the receiving DNA then join, and thus the strand is integrated. Site-specific recombinase enzymes with new specificities have been engineered using bacterial resolvase domains combined with substrate recognition domains borrowed from a mouse transcription factor (Akopian A et al 2003 Proc Natl Acad Sci USA 100:8688). ►site-specific recombination, ►Holliday juncture, ►resolvase, ►*Cre/loxP*, ►*FLP/FRT*, ►integrase, ►phosphodiester linkage, ►transesterification, ►homing endonucleases; Woods KC et al 2001 J Mol Biol 313:49; Dhar G et al 2004 Cell 119:33; structural and biochemical mechanisms of tyrosine and serine types of recombinases: Grindley NDF et al 2006 Annu Rev Biochem 75:567.

Site-Specific Recombination: Site-specific recombination occurs when the recombination is limited to a specific few nucleotide sequences. Homology may be present at the exchange region in both recombining molecules like at the integration–excision site of the temperate phage. Alternatively, the specificity is limited to only one of the partners like at the 25 bp termini of the T-DNA or the direct and indirect repeats of the transposable elements. In the latter cases, the recombinational target sites may have no or only minimal similarity. The IS*30* insertion element can serve as site-specific recombinases when replacing the integration/excision genes of phage λ (Kiss J et al 2003 Proc Natl Acad Sci USA 100:15000). Peptide nucleic acids enhance site-specific recombination and DNA repair. ►lambda phage, ►gene replacement, ►*Cre/Lox*, ►*FLP/FRT*, ►switching, ►site-specific recombinase, ►peptide nucleic acids, ►knockout, ►targeting genes, ►chromosomal rearrangement, ►ligand-activated site-specific recombination, ►T-DNA, ►recombination, ►shoufflons, ►integrases, ►DD(35)E; Sauer B, Henderson N 1988 Proc Natl Acad Sci USA 85:5166; Pena CE et al 2000 Proc Natl Acad Sci USA 97:7760; Christ N et al 2002 J Mol Biol 319:305.

Sitosterolemia (phytosterolemia, STSL, 2p21): A rare, recessive hypercholesterolemia resulting in more than 30-fold increase of the level of this plant cholesterol in the plasma. Intestinal absorption of sterols is increased and the excretion of sterols into the bile is impaired. Initially, it causes xanthomatosis and later premature coronary artery disease. Actually, two genes encoding sterolin 1 and sterolin 2 are involved in opposite orientation separated by a short interval. Sterolins apparently regulate sterol transport. ►cholesterol, ►low-density lipoprotein, ►VLDL, ►xanthomatosis, ►familial hypercholesterolemia, ►coronary heart disease; Lee M-H et al 2001 Nature Genet 27:79; Lu K et al 2001 Am J Hum Genet 69:278.

Situs Inversus Ambiguus: Some of the organs situated at the common side of body axis, others are at a misplaced site regarding the axis. ►left-right asymmetry

Situs Inversus Totalis: Complete inversion of left-right body axis. ►left-right asymmetry

Situs Inversus Viscerum (7p21): A malformation of mammals, including humans, where the internal organs such as the heart are shifted to the right side of the chest (thorax). It is frequently accompanied by chronic dilation of the lung passages (bronchi) and inflammation of the sinus; the latter disorder is also called Kartagener syndrome, which is characterized also by the immotility of sperm and cilia. The anomaly may by either autosomal or X-linked recessive. Its incidence in the general population may be about 1/10,000. In the mouse, the genes *iv* (chromosome 12) and the *inv* (chromosome 4) disturb left-right axis formation and cause 50 and 100% manifestation of situs inversus, respectively. In the chicken, the fibroblast growth factor (FGF8) mediates the determination of the right side and in Sonic hedgehog (SHH), of the left side. In the mouse, FGF8 is instrumental in the left side and SHH in the right side specification. ►dynein, ►heterotaxy, ►isomerism, ►Kartagener syndrome, ►asymmetry of cell division, ►axis of asymmetry, ►FGF, ►sonic hedgehog, ►left-right asymmetry, ►ciliary dyskinesia; Bartoloni L et al 2002 Proc Natl Acad Sci USA 99:10282; Bisgrove BW et al 2003 Annu Rev Genomics Hum Genet 4:1.

SIV (Simian immunodeficiency virus): a relative of HIV. ►acquired immunodeficiency, ►HIV

Size: Size depends primarily on cell number and cell size and it is developmentally and genetically determined. (See Conton I, Raff M 1999 Cell 96:235). According to Kleiber's rule (1932 Hilgardia 6:315), the size of an organism (body mass) follows the ~¾ power of the metabolic/respiratory rate. A morphogen gradient may determine the size of an organ in animals. It is not entirely resolved which way the gradient is controlled. One set of data of *Drosophila* wing size determination indicates that imaginal disk size is determined relative to the fixed morphogen distribution by a certain threshold level of morphogen required for growth. When disk boundary reaches the threshold, the arrest of cell proliferation throughout the disk is induced by mechanical stress in the tissue. Mechanical stress is expected to arise from the non-uniformity of morphogen distribution that drives growth. This stress, through a negative feedback on growth, can compensate for the non-uniformity of morphogen, achieving uniform growth with the rate that vanishes when the disk boundary reaches the threshold (Hufnagel L et al 2007 Proc Natl Acad Sci USA 104:3835).

It seems that plants differ in metabolism from animals extensive data (500 observations on 43 plant species) indicates that respiration (relative to nitrogen content) at a scaling exponent of ~1 better represents mass (Reich PB et al 2006 Nature [Lond] 439:457). ►body mass, ►body size, ►morphogen, ►imaginal disk

Size-Exclusion Chromatography: In size-exclusion chromatography, molecules are separated by size; large molecules may not enter the surface of the matrix but small molecules may penetrate the core. The penetration depends on size and shape of the analyte and the nature of the matrix.

SJL Mouse: Non-inbred strain.

Sjögren-Larsson Syndrome: ►ichthyosis

Sjögren (sicca = dry) **Syndrome** (SS) Autosomal recessive autoimmune disease leading to the destruction of the salivary and lacrimal glands by the production of autoantibody against the SS-A (RoRNA) and SS-B (La Sn RNA) particles. The affected individuals have dry mouth and dry eyes (no tears). The autoantigens have been identified and purified. The Ro autoantigen appears to be encoded in human chromosome 19pter-p13.2. The La autoantigen may be involved with RNA polymerase III. The 120 kDa α-fodrin appears to be the critical autoantigen that elicits the disease. Aberrant T cells with impaired class IA phosphoinositide 3-kinase signaling can lead to organ-specific autoimmunity in mice and resemble human SS (Oak JS et al 2006 Proc Natl Acad Sci USA 103:16882). ►autoimmune disease, ►fodrin, ►RoRNP, ►PIK, ►rosacea

SK: Calcium-activated potassium ion channels. ►ion channels

SK Oncogene: The SK oncogene probably regulates tumor progression; it was assigned to human chromosome 1q22-q24.

7SK RNA: A small (330 base) nuclear RNA (snRNA) of ubiquitous presence and involvement in the control of transcription by interfering with the RNA elongation factor P-TEFb. ►snRNA, ►RNA regulatory; Yang Z et al 2001 Nature [Lond] 414:317; Michels AA et al 2004 EMBO J 23:2608.

Skeletal Map: The skeletal map uses only microsatellite marker data. ►framework map, ►recombination minimization map, ►integrated map, ►genetic map, ►physical map, ►radiation mapping

Skewed Distribution: In skewed ditribution, the data are not symmetrical around the mean; either one or the other extreme flank is predominant (see Fig. S70). ►normal distribution, ►kurtosis

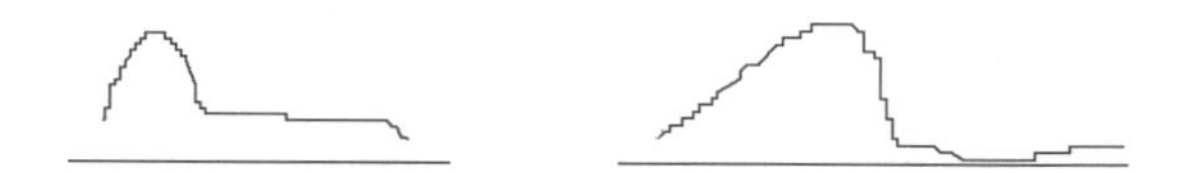

Figure S70. Skewed distribution

Skewness: Asymmetry in the distribution frequency of the data. ►kurtosis, ►normal distribution, ►moments

Ski (Sloan Kettering Institute): A protein discovered at the Sloan Kettering Institute, as a viral factor in tumorigenesis. Ski occurs in vertebrates and insects and, along with Sno (Si-related novel gene), regulates the effect of Smad4 and Smad3 proteins that in response to the phosphorylation signals coming from TGF-β may negatively control gene expression. It interacts with Skip, a transcriptional activator. ►Smad, ►TGF; Prathapam T et al 2001 Nucleic Acids Res 29:3469.

Skin Cancer: Skin cancer constitutes about 40% of all newly diagnosed cancers. The incidence of melanoma is 4%, basal cell carcinoma 80%, and squamous-cell carcinomas 16% of the skin cancers. Sunscreen provides limited protection; conversely, it may even increase the risk because the assumed protection allows more exposure to the sun. The sensitive initial target of carcinogenesis by UV light is mutation in gene p53 and various protooncogenes. The UV-light damage starts with *initiation* in single cells that may be followed by the effects of tumor promoting agents (*expansion*), eventually leading to the *progression* of the tumor cells, resulting in cancer. UVB can enhance these three steps, each, by acting on the signaling molecules, epidermal growth factor (EGF), mitogen-activated protein kinases (MAPK), and phosphatidylinositol 3-kinase (PI3K). EGF activation leads to the production of reactive oxygen species (ROS). ROS effects can be mitigated by antioxidants. UV induces several transcription factors such as AP-1, JUN, FOS, etc., and cyclooxygenase-2. AP-1 activation may be inhibited by salicylate (aspirin) or perillyl alcohol (monoterpene). Several other promising chemical protective agents are under study. (See terms under separate entries; Bowden GT 2004 Nature Rev Cancer 4:23).

Skin Color: ►pigmentation of animals; Sturm RA et al 1998 Bioessays 20(9):712; Barsh G 2003 PLoS Biol 1(1):e7; Lin JY, Fisher DE 2007 Nature [Lond] 445:843.

Skin Diseases: ►acne, ►epidermolysis, ►keratosis, ►ichthyosis, ►psoriasis, ►blisters, ►porphyria, ►pemphigus, ►acrodermatitis, ►familial hypercholesterolemia, ►Fabry disease, ►pseudoxanthoma elasticum, ►nevus, ►vitiligo, ►ectodermal dysplasia, ►focal dermal hypoplasia, ►scleroderma, ►lupus erythematosus, ►dermatitis, ►eczema, ►Gardner syndrome, ►Kindler syndrome, ►cutis laxa, ►pigmentation defects, ►light-sensitivity, ►glomerulonephrotis, ►Rothmund-Thompson syndrome, ►Werner syndrome, ►epithelioma, ►dyskeratosis, ►erythrokeratoderma variabilis, ►skin cancer, ►xeroderma pigmentosum, ►connexin

Skotomorphogenesis: Morphogenesis without dependence on light. ►photo-morphogenesis, ►de-etiolation

SKP (cyclin A-CDK2 associated protein): An intrinsic kinetochore protein (22.3 kDa) widely conserved among species. It coordinates centromere, centrosomes, and other cell cycle factors. The Skp1p is a proteasome-targeting factor. Mice Skp2$^{-/-}$ is viable, yet has reduced growth rate, has polyploid cells, and accumulates cyclin E and p27^{Kip1} proteins that it cannot efficiently eliminate during the S and G2 phases of the cell cycle (see Fig. S71). For the degradation, SCF^{Skp2} is required. Skp2 is up-regulated in some types of epithelial carcinogenesis. ►kinetochore, ►cell cycle, ►CDC4, ►proteasome, ►F-box, ►cyclin A, ►CDK, ►SCF, ►von Hippel-Lindau disease; Nakayama K et al 2000 EMBO J 19:2069; Latres E et al 2001 Proc Natl Acad Sci USA 68:2515.

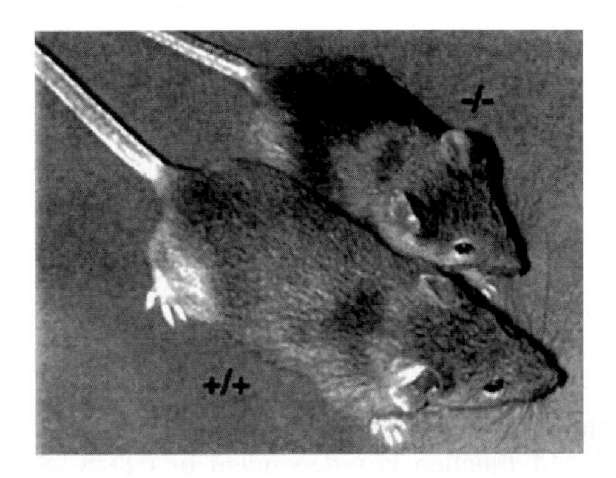

Figure S71. Skp2 disrupted mice is deficient in the F box protein and SCF ubiquitin ligase has enlarged nuclei, multiple centrosomes and reduced growth. They accumulate cyclin E and p27Kip. (From Nakayama K et al 2000 EMBO J 19:2069)

Skunk (*Mephitis mephitis*): 2n = 50; (*Spilogele putorius*), 2n = 64.

Sky: Same as spectral karyotyping. ►spectral karyotyping

Sky: A cellular tyrosine kinase. It regulates B cell development. ►B lymphocyte; Kishi YA et al 2002 Gene 288:29.

SL1, SL2 (spliced leader): SL1 and SL2 are involved in the transsplicing in *Caenorhabditis*. The 100-nucleotide leader donates its 5′ end 22 nucleotides to a splice acceptor site on the primary transcript. Trans-splicing is very common (70%) among the nematode's genes. This mechanism is used for the coordinately regulated gene clusters, transcribed in polycistronic RNA. The nematode operons use SL2 whereas other genes use SL1. ►transsplicing, ►coordinate regulation, ►operon

SL1: A transcription factor complex of RNA polymerase I. It is a complex of the TATA-box-binding protein (TBP) and the three TATA box associated factors (TAF). The TBP protein binds exclusively, either SL1 (RNA pol I) or TFIID (RNA pol II). In the case of RNA pol III, TFIIIB is required for the recruitment of the polymerase to the promoter complex. ►pol I, ►pol II, ►TBP, ►TAF, ►transcription factors

Slicer: A structural homolog of ribonuclease H; it is also a domain of Piw. ►Ribonuclease H, ►piRNA

7SL RNA: An RNA component in the signal recognition protein (SRP) complex. ►signal sequence recognition particle, ►Alu

Slalom Library: The slalom library is based on a combination of the principles of linking and jumping libraries. ►jumping library, ►linking library; Zabarovska VI et al 2002 Nucleic Acids Res 30(2):e6.

SLAM (signaling lymphocyte activation molecule, CDw150): A T cell receptor protein (M_r 70K) of the immunoglobulin family, constitutively and rapidly expressed on activated peripheral blood memory T cells, immature thymocytes, and on some B cells. It is a receptor also for the measles virus. T cells carrying the $CD4^+$ antigens produce increased amounts of interferon γ without an increase of interleukins 4 or 5. SLAM function is independent of CD28. ►T cell, ►interferon, ►interleukin, ►CD28, ►Epstein-Barr virus, ►SAP; Bleharski JR et al 2001 J Immunol 167:3174.

SLAM: The gene predictor program for the detection of homologous sequences in different species. ►gene prediction; Alexandersson M et al 2003 Genome Res 13:496.

SLAP (Fyb/Slap): One of the adaptor proteins that regulate TCR-mediated signal transduction. Cb*l* has an inhibitory effect. SLAP interacts with Sky, ZAP-70 and LAT. ►Fyb, ►TCR, ►signal transduction, ►CBL; Peterson EJ et al 2001 Science 293:2263.

SLD: A yeast chromosomal replication protein acting after phosphorylation during the S phase. ►GINS; Masumoto H et al 2002 Nature [Lond] 415:651.

Sleep: A circadian organization of rest after activity, controlled by several neural genes. Apparently, mutation of the human homolog of *Per2* (2q) may be responsible for the familial advanced sleep phase syndrome. In *Drosophila*, some *Shaker* (1.57.6) null mutants (Sh^{102}, *minisleep*) involved in a voltage-dependent K^+ channel controlling membrane polarization and transmitter release have reduced requirement for sleep and display apparently normal functions but shorter life span (Cirelli C et al 2005 Nature [Lond] 434:1087). The point mutation (Ser→Gly) is within the casein kinase Iε and alters the circadian clock. Sleep may have a weak role in the consolidation of memory and frequently inspires insight. Insight is a mental restructuring leading to sudden gain of explicit knowledge (Wagner U et al 2004 Nature [Lond] 427:352). Sleep-deprivation seriously affects job-performance and cognitive abilities. The Ampakine (α-amino-3-hydroxy-5-methyl-4-isoxazolepropropionic acid, AMPA) drugs appear promising in non-human primates for alleviating the problems of sleeplessness (Porrino LJ et al 2005 PLoS Biol 3(9):e299). In aging organisms (humans or *Drosophila*) the sleep cycles are generally fragmented and it has been attributed of age-related oxidative damage (Koh K et al 2006 Proc Natl Acad Sci USA 103:13843). ►apnea, ►narcolepsy, ►circadian rhythm, ►memory; Siegel JM 2001 Science 294:1058; Shaw PJ et al 2002 Nature [Lond] 417:287; Pace-Schott EF, Hobson JA 2002 Nature Rev Neurosci 3:591; 2005 Nature [Lond] 437:1253–1289.

Sleeping Beauty (SB): An artificially constructed human mariner transposable element, equipped with a salmon transposase function enabling the otherwise non-mobile element to move in HeLa or other somatic cells by a cut-and-paste mechanism. Another type of transposon vector (pTnori) is outlined here in Figure S72. Another transposon (T2/Onc2), containing a larger fragment of splice acceptor sequence, is flanked by optimized transposons binding sites. It also contains a murine stem cell virus (MSCV) long terminal repeats and a splice donor site to promote gene expression when integrated upstream or within the gene. Low methylation in MSCV promoter and high copy numbers are also an advantage. A more active transposase was constructed to increase transposition frequency and to be expressed in all tissues. This new system generated transposon mutagenesis in many cancer genes and appears promising to shed light on cancer etiology of mammalian cells (Dupuy AJ et al 2005 Nature [Lond] 436:221; Collier LS et al 2005 Nature [Lond] 436:272). Transposition of Sleeping Beauty requires the presence of the Miz-1 transcription factor and the slow-down of the G1 phase of the cell cycle. The slow-down decreases D1/cdk4-specific phosphorylation of the retinoblastoma protein (Walisko O et al 2006 Proc Natl Acad Sci USA

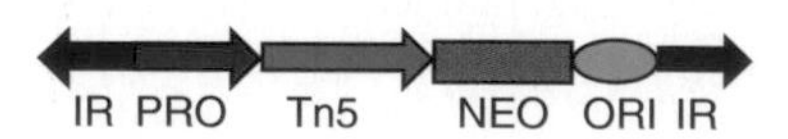

Figure S72. IR: inverted repeats, PRO: promoter, Tn5: bacterial transposon, NEO: selectable marke, ORI: origin of replication (p Tnori)

103:4062). ►mariner, ►transposase, ►piggyBac, ►cut-and-paste, ►HeLa, ►cyclin D, ►CDK, ►retinoblastoma; Horie K et al 2001 Proc Natl Acad Sci USA 98:9191; Izsvák Zs et al 2002 J Biol Chem 277:34581.

Sleeping Sickness: A potentially fatal disease caused by *Trypanosomas*. The tse-tse fly spreads the disease. ►*Trypanosomas*

SLG: Self-compatibility locus-secreted glycoprotein. ►self-incompatibility

Sliding Clamp: A gp45 gene of phage T4 controls high speed of replication (processivity) by gp43 (DNA polymerase gene); gp44 and gp62 are the clamp loaders for gp45. The clamp hugs the DNA and slides along the duplex DNA. The replication complex is then attached to the clamp. The cellular homologs for p45 are the β units of the DNA polymerase III of prokaryotes and the PCNA of eukaryotes. The transcription may be coupled to replication and regulated by protein-protein and protein site-specific DNA interactions. In eukaryotes, the sliding clamp is PCNA. The τ subunit of DNA polymerase complex switches off the polymerase from the DNA as one Okazaki fragment is finished and then switches on again the β subunit to start a new Okazaki fragment (López de Saro FJ et al 2003 Proc Natl Acad Sci USA 100:14689). ►DNA polymerases, ►PCNA, ►replication factor, ►clamp loader, ►Okazaki fragment; Fishel R 1998 Genes Dev 12:2096; Trakselis MA et al 2001 Proc Natl Acad Sci USA 98:8368; crystal structure: Jeruzalemi D et al 2001 Cell 106:417; Johnson A, O'Donnell M 2005 Annu Rev Biochem 74:283.

SLIK: ►SAGA

Slime Molds: Slime molds are either of plasmodial (*Myxomycetes*) or cellular type (*Acrasiomycetes*) eukaryotes, of which *Dictyostelium discoideum* is perhaps the most important object for research. ►plasmodium, ►*Dictyostelium*, ►*Physarum*

SLINK: A computer program for estimating linkage information by a simulation approach. ►SIMLINK

Slippage: Usually, when homopolymeric sequences are embedded in the template DNA strand, the RNA polymerase may synthesize RNA strands that are much longer (by a few to thousands nucleotides) than the template; this is known as slippage. The slippage can be inhibited, however, when nucleotides, representing the next one to the homopolymeric stretch, are added. Frame shift mutations may be interpreted as the result of slippage. Slippage may occur during decoding at translation on the ribosome in case the codon–anticodon interactions are weak at the ribosomal P site (Hansen TM et al 2003 EMBO Rep 4:499). Slippage during DNA replication may generate trinucleotide repeats, which can lead to neurodegenerative disease. Slippage may be the origin of base mismatches, which generate instabilities (Chi LM, Lam SL 2005 Nucleic Acids Res 33:1604). ►slipping, ►attenuator region, ►overlapping genes, ►unequal crossing over, ►microsatellite, ►replication slippage, ►decoding, ►mismatch repair, ►transcript elongation, ►trinucleotide repeat; Viguera E et al 2001 EMBO J 20:2587.

Slipped-Structure DNA (S-DNA): Incomplete pairing within intrastrand folds (hairpins) in not exactly opposite position to each other. Such a structure may form when there are variable number trinucleotide repeats in the DNA (see Fig. S73). ►trinucleotide repeats, ►hairpin

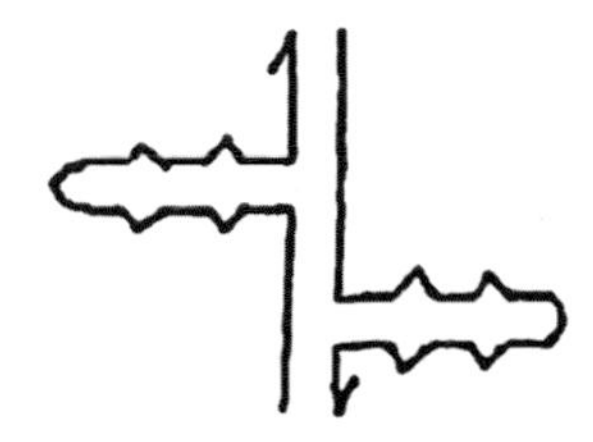

Figure S73. S-DNA

Slipping: A shifting of the translational reading frame. ►recoding, ►hopping, ►slippage

Slip-Strand Mispairing: Slip-strand mispairing may cause replicational error if not corrected by repair; it is frequently the cause of micro- and minisatellite instability. The slippage may cause either additions to or deletions from the repeat sequences. This process is independent from the mechanism(s) of recombination (it is not unequal crossing over), and flanking markers are not exchanged here. In yeast, its frequency is about the same as at mitosis and meiosis. Defects in the genes controlling replication and repair of the DNA may increase the instability and cause more mutations in microsatellites than in other types of DNA sequences. The packaging of the DNA, the temperature, methylation state, base composition, cell cycle stage, etc., may affects its frequency. Its rate may vary among different species because of differences in the replicase and mismatch repair enzymes. CTG repeats in the lagging strands are less stable than in the leading strand. ►microsatellite, ►minisatellite, ►MVR, ►unequal crossing over,

►slippage; Lewis LA et al 1999 Mol Microbiol 32:977.

Slit: An axon-repellent molecule with Robo as its receptor. ►axon, ►axon guidance, ►Robo

Slithering: A creeping-like motion of the recombination sites towards each other. A recombinase enzyme within a supercoiled DNA molecule may mediate slithering in case of site-specific recombination. ►site-specific recombination; Huang J et al 2001 Proc Natl Acad Sci USA 98:968.

Slot Blot: Binding cDNA or RNA onto slots→ ● on membrane filters for the analysis of transcripts by hybridization to specific sequences.

Slow Component: During a reassociation reaction of single-stranded DNA, the unique sequences anneal slowly. ►c_0t value, ►annealing

Slow Stop: In a slow stop, bacterial mutant *dna* may complete slowly the replication underway but cannot start a new cycle at 42 °C. ►replication, ►strong-stop DNA

SLP-76 (a 76 kDa specific leukocyte protein adaptor): SLP-76 binds to TCR, is phosphorylated at tyrosines near the N-terminus, and provides SH2 binding sites for the VAV protein. SLP also binds the SH3 domain of Grb2. At the C-end, it associates also with SLAP (SLP-76 associated phosphoprotein), resulting in activation of NF-AT. SLP-76 is essential for TCR activity and T cell development in signaling pathways through the activity of phosphotyrosine kinases (PTK), but it is not required for macrophage and natural killer cells. ►signal transduction, ►TCR genes, ►T cell receptor, ►SH2, ►VAV Grb2, ►NF-AT, ►macrophage, ►killer cell, ►BASH, ►CD3, ►GADS; Pivniouk VI et al 1999 J Clin Invest 103:1737.

SLS: Sodium lauryl sulfate. ►SDS

SLT: ►specific locus mutations assay

SLT-2: A protein kinase of the MAPK family. ►signal transduction, ►protein kinase, ►MAPK

Slug: Slug, in general usage, means a land mollusk but see also ►*Dictyostelium*

Slug: A zinc-finger transcription factor activated by p53. It represses Puma and antagonizes apoptosis (Wu W-S et al 2005 Cell 123:641). ►apoptosis, ►p53, ►zinc fingers, ►PUMA

Sly Disease: ►mucopolysaccharidosis type VII

Sma I: A restriction endonuclease; recognition site CCC↓GGG. ►restriction enzymes

SMAC (supramolecular activation cluster): CD4 T cells with antigen presenting cells, receptors, and intracellular proteins form SMAC. (See terms at separate entries; Potter TA et al 2001 Proc Natl Acad Sci USA 98:12624).

Smac (second mitochondria-derived activator of caspases): Smac inhibits IAPs and facilitates apoptosis by caspase-3. A small, synthetic Smac mimic potentiates the activation of TRAIL and TNFα, inhibits IAP activity, and promotes apoptosis (Li L et al 2004 Science 305:1471). Smac is homologous with Diablo. ►IAP, ►apoptosis, ►DIABLO; Zhang XD et al 2001 Cancer Res 61:7339.

Smad: Signal transducing proteins, which when stimulated by TGF-β can enhance gene transcription and tumor formation (TGF-β signaling). The Smad binding element is GTCTAGAC. The facilitating effects of the bone morphogenetic protein (BMP, acting *via* the serine/threonine kinase receptor) determine the SMAD protein function and opposing epidermal growth factor (EGF, acting via the receptor tyrosine kinases). Smad2 is essential for the formation of the early embryonic mesoderm of the mouse. Smad3 is normally phosphorylated by the TGF receptor TβRI, and Evi-1 represses its transcriptional activator function. Smad4 controls the mesoderm and visceral endoderm. Other SMADs associated with various ligands of the TGF family control the expression of genes involved in embryonal tissue differentiation. Smad3 and Smad4 cooperate with Jun/Fos (A1) and bind to the TPA-responsive gene promoter elements. SARA (Smad anchor for receptor activation) retains Smad2 and 3 in the cytoplasm. Pancreatic, colon, and other cancers are frequently associated mutation(s) in SMAD2 and SMAD4/DPC4. Smad4 is a tumor suppressor in the gastrointestinal tract of mice (Kim B-G et al 2006 Nature [Lond] 441:1015). SMADs 6 and 7 are modulators/inhibitors of signaling by some SMADs and their mutation may cause hyperplasia of the cardiac valves and other structural anomalies of the heart as well as ossification of the aorta and high blood pressure in mice. The SMAD acronym was derived from the human SPA (spinal muscular atrophy genes, 5q12.2-q13.3) and the *Drosophila* gene *Mad* (*mothers against decapentaplegic*). Smad proteins are classified as R-Smads (receptor regulated), Co-Smads (common Smads), and I-Smads (inhibitory). ►TGF, ►EGF, ►bone morphogenetic protein, ►serine/threonine kinase, ►receptor tyrosine kinase, ►activin, ►Evi oncogenes, ►TPA, ►AP1, ►SARA, ►DPC4, ►*Mad*, ►spinal muscular atrophy, ►Gli, ►Ski, ►osteopontin; Wrana JL 2000 Cell 100:189; Zauberman A et al 2001 J Biol Chem 276:24719;

López-Rovira T et al 2002 J Biol Chem 277:3176; Derynck R, ZhangYE 2003 Nature [Lond] 425:577.

Small Cell Lung Carcinoma (SCLC): SCLC is associated with a deletion of the human chromosomal region 3p14.2; the susceptibility is dominant. It accounts for about 1/3 of all lung cancers. Lung cancer genes were located also to 3p21, 3p25, and to several other chromosomes Smoking may be the major cause of the development of this condition. Surgical remedies are usually not applicable because of the rapid metastasis but it generally responds to radiation and chemotherapy. Deregulation of the MYC oncogene is the suspected cause. Recently, a gene for fragile histidine triad (FHIT) was found to be associated with SCLC and with some of the non-small cell lung carcinomas (NSCLCs). The product of FHIT splits Ap_4A substrates asymmetrically into ATP and AMP. Its metastasis may be inhibited by CC3/TIP30. The heterozygotes (T/C) for the checkpoint kinase gene (CHECK 2/CDS1, human chromosome 22q12.1) was associated with a highly significantly *lower* incidence of lung cancer than the common T/T genotype [relative risk (RR), T/C versus T/T, 0.44, with 95% confidence interval (CI) 0.31–0.63, $P < 0.00001$] and with a significantly *lower* incidence of upper aero-digestive cancer (RR 0.44, CI 0.26–0.73, $P = 0.001$; $P = 0.000001$ for lung or upper aero-digestive cancer). The results of this study, involving 4015 smoking patients and 3050 non-smoking individuals in several East-European countries, were surprising because earlier, mutation in the same gene showed an increase in the incidence of the Li-Fraumeni syndrome (Brennan P et al 2007 Hum Mol Genet 16:1794). ►oncogenes, ►cancer, ►MYC, ►p53, ►ATP, ►AMP, ►semaphorin, ►metastasis, ►non-small lung cell carcinoma suppressor, ►neuroendocrine cancer, ►Li-Fraumeni syndrome, ►smoking, ►checkpoint; Zöchbauer-Müller S et al 2002 Annu Rev Physiol 64:681; Tonon G et al 2005 Proc Natl Acad Sci USA 102:9625.

Small Molecule Microarray: In small molecule microarray, polystyrene beads covered by presumed biologically active ligands are arrayed in micro-well plates. The molecules are released from the beads in a solution and are spread over glass plates and tested for biological function(s) by high-throughput technology. (Uttamchandani M et al 2005 Curr Opin Chem Biol 9:4; MacBeath G et al 1999 J Am Chem Soc 121:7967; Clemons PA et al 2001 Chem Biol 8:1183).

Small Nuclear RNA: ►snRNA

Smallpox: ►pox virus, ►variola

Small RNA (sRNA): sRNA includes microRNA, RNAi, transcripts of small genes, pseudogenes, intergenic regions, and transposons. By the use of massively parallel signature sequencing in *Arabidopsis* plants, a library of a total of 104,800 distinct signatures of these RNAs have been identified. Of these, 77,434 could be matched with the genome (Lu C et al 2005 Science 309:1567). The enormous number of these sequences regulates several ways the expression of the genomes. In *Arabidopsis*, ~2% of the genes may be under the control of microRNAs. Many of these sequences match intergenic regions, indicating that so far, unannotated protein-coding genes, pseudogenes, and transposons are located there. Single miRNAs may regulate on the average five–six times as many genes (Vaughan MW, Martienssen R 2005 Science 309:1525). The small RNAs usually regulate genes by interference (RNAi). Transfection of some dsRNAs into human cell lines was found to cause long-lasting and sequence-specific induction of targeted genes. dsRNA mutation studies reveal that the 5′ end of the antisense strand, or "seed" sequence, is critical for activity (Li L-C et al 206 Proc Natl Acad Sci USA 103:17337). ►microRNA, ►RNAi, ►piRNA, ►U-RNA, ►21U-RNA, ►pseudogene, ►intergenic region, ►transposons, ►RNA non-coding, ►massively parallel signature sequencing, ►transitivity; sRNA for cereals: http://sundarlab.ucdavis.edu/smrnas/.

Small t Antigen: ►SV40

Small-Pool PCR: Small-pool PCR amplifies 20–100 molecules of minisatellite DNA from single individuals within a population, and thus reveals a mutation rate that is $>10^{-3}$ in the sperm germline at a number of loci. ►minisatellite, ►PCR, ►MVR; Crawford DC et al 2000 Hum Mol Genet 9:2909.

Small-World Networks: Small-world networks can represent models of many types of self-organizing biological systems, including interaction of gene products. Although the networks can be completely regular or completely random, biological interaction networks are usually connected neither in a completely regular nor in a completely random manner, but in a fashion in between these extremes and thus represent "small-world networks." Small-world networks display enhanced speed of propagation of signals and coordinated regulation and can be subjected to computational analysis (see Fig. S74). ►networks, ►genetic networks, ►probabilistic graphical models of cellular networks, ►synthetic genetic array, ►model; diagram modified after Watts DJ, Strogatz SH 1998 Nature [Lond] 393:440.

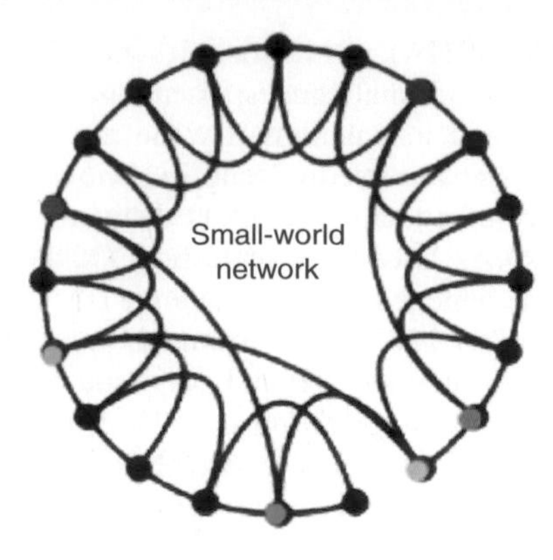

Figure S74. Small-world network

SMART (Simple Modular Architecture Research Tool): SMART facilitates annotation of protein domains. ►domain, ►annotation; http://smart.embl-heidelberg.de/; http://smart.embl.de/.

Smart Ammunition: *Drosophila* P transposable element vectors with selectable markers to produce selectable (e.g., neomycin resistance) insertions. ►hybrid dysgenesis, ►insertional mutation; Engels WR 1989, p 437. Mobile DNA. In: Berg DE, Howe MM (Eds.) Am Soc Microbiol, Washington DC.

Smart Cells: A generalized concept that cells (genes) have the ability to sense internal and external cues and respond to them in a purposeful manner, such as shown in signal transduction.

Smart Linkers: Synthetic oligonucleotides with multiple recognition sites for restriction enzymes; they can be ligated to DNA ends to generate the desired types of cohesive ends (see Fig. S75). ►cloning vectors, ►cohesive ends, ►blunt end, ►blunt end ligation

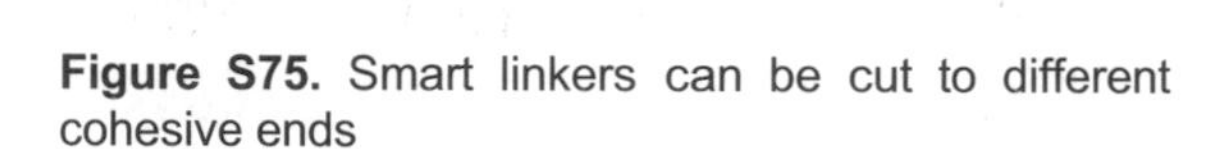

Figure S75. Smart linkers can be cut to different cohesive ends

Smart-PCR: See Villalva C et al 2001 Biotechniques 31:81.

SMC: Proteins involved in the structural maintenance of the chromosomes, including condensation (increasing coiling), cohesion of sister chromatids, and segregation. The top of Figure S76 represents a SMC monomer. The amino and the carboxyl ends are globular and the two long α-helixes and a short hinge domain may facilitate a foldback of the monomer. Two folded monomers may form heterodimers as shown. The DNA single strand may wrap around the termini of the folded dimers and condensation is facilitated by SMC or the two chromatids may be hold together in cooperation with cohesin and the kleisin proteins. SMC proteins are ubiquitous among prokaryotes and eukaryotes. The three Muk gene products of *E. coli* are functionally homologous. ►sex determination, ►chromosome coiling, ►condensin, ►cohesin, ►kleisin, ►achiasmate; Ball AR, Yokomori K 2001 Chromosome Res 9(2):85; Kitajima TS et al 2003 Science 300:1152; Milutinovich M, Koshland DE 2003 Science 300:1101; Nasmyth K, Haering CH 2005 Annu Rev Biochem 74:595.

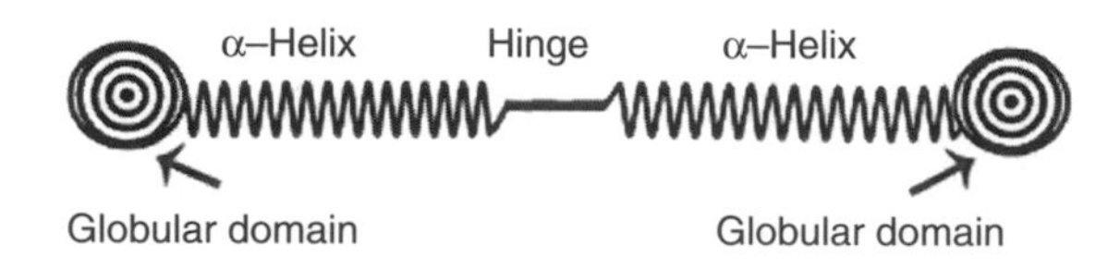

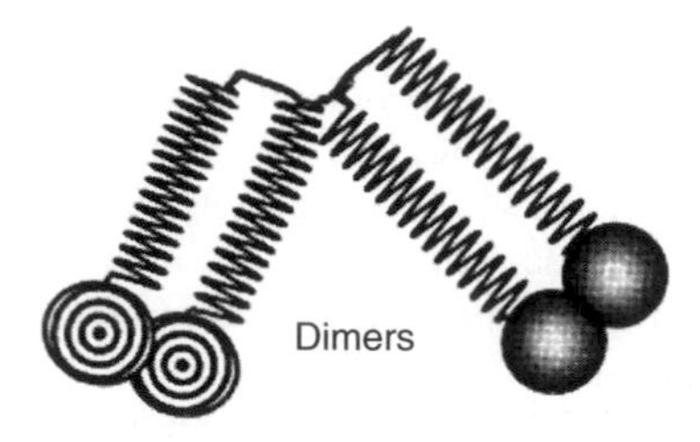

Figure S76. Chromosome structural maintenance proteins

Smear: Preparing a soft specimen for microscopic examination by gentle spreading directly on the microscope slide. ►squash, ►sectioning, ►microscopy

Smg p21: A protein similar to Rap 1. ►Rap

SMGT (sperm-mediated gene transfer): In SMGT, the sperm internalizes DNA and by artificial insemination, transgenic pigs have been produced that carry the human decay accelerating (hDAF) gene. The efficiency of transformation is high (64%) and the expression is very good (83%). The presence of dDAF is expected to help in overcoming hyperacute rejection of xenotransplanted organs. ►decay accelerating factor, ►xenotransplantation, ►transformation genetic; Lavitrano M et al 2002 Proc Natl Acad Sci USA 99:14230.

Smith-Lemli-Opitz Syndrome (SLOS/RSH): A high prevalence (2×10^{-4}) autosomal recessive anomaly involving microcephalus, mental retardation, abnormal male genitalia, polydactily, etc., encoded in human chromosomes 11q12-q13 (Type I) and 7q32.1 (Type II). Prevalence in Northern European populations is 1×10^{-4} to 2×10^{-5}. It is caused by Δ^7-reductase deficiency in the cholesterol pathway and by the accumulation of 7-dehydrocholesterol. Deficiency of the Δ^{24}-reductase results in similar symptoms. The afflicted individuals show mevalonic aciduria. This condition may also involve holoprosencephaly. Some of the symptoms may overlap with those of the Pallister-Hall syndrome. Prenatal diagnosis in the amniotic fluids after 15 weeks of gestation is feasible. A non-invasive urine analysis may also be practical. ►mental retardation, ►dwarfness, ►head/face/brain defect, ►cholesterol, ►sonic hedgehog, ►holoprosencephaly, ►chondrodysplasia; Wassif CA et al 2001 Hum Mol Genet 10:555.

Smith-Magenis Syndrome (SMS): SMS involves head malformation, short brain, growth retardation, hearing loss, and self-destructive behavior, such as pulling off nails, inserting foreign objects into the ear, etc. The incidence is >4×10^{-5}. In 80–90% of the patients, an approximately 4 Mb interstitial deletion heterozygosity is found at the chromosome 17p11.2 region. The locus contains or is flanked by low-copy number repeats, which favor the occurrence of localized recombinations that generate deletions and duplications (Bi W et al 2003 Am J Hum Genet 73:1302). In individuals without cytologically detectable deletions, a 29 nucleotide loss was observed in exon 3 of the RAI1 (retinoic acid induced, 17p11.2) gene. ►mental retardation, ►self-destructive behavior; Slager RE et al 2003 Nature Genet 33:466.

Smith-McCort Dysplasia: ►Dyggve-Melchior-Clausen dysplasia

Smith–Waterman Algorithm: Computer analysis for nucleic acid sequences (Waterman MS 1988 Methods Enzymol 164:765).

SMM: ►stepwise mutation model, ►IAM, ►two-phase model

Smoking: Smoking is responsible for a wide variety of ailments such as heart disease, respiratory problems, cancer, etc., but it may decrease the risk of Parkinson's disease. The inhalation of tobacco smoke by the mother may initiate cancer also in the fetus. Although cancer may be induced by a variety of genotoxic agents in the environment, the smoking induced alteration spectrum in the genetic material is different and thus can be distinguished from the effects of other agents. Tobacco smoke adducts induce a higher proportion of transversion mutations of the p53 gene in the lung and also increases loss of heterozygosity by deleting introns particularly at the fragile site 3 (FRA3B) region including FHIT (fragile histidine triad) in human chromosome 3p14.2. According to one report, 19/31 newborns of smoking mothers had the carcinogen 4-methylnitrosamino-1(3-pyridyl)-1-butanone in their urine.

The smoking habit is particularly prevalent in affective disorders. In the brain of smokers, the level of monoamine oxidase B (MAOB) is 40% lower relative to that in non-smokers. MAOB degrades the neurotransmitter dopamine. Subcortical regions, such as the amygdala, the nucleus accumbens, and the mesotelencephalic dopamine system, have been shown in animal models to promote the self-administration of drugs of abuse. Functional imaging studies have shown that exposure to drug-associated cues activates cortical regions such as the anterior cingulate cortex, the orbitofrontal cortex, and the insula. In nicotine addiction, the insula, a narrow island within the brain, seems to play a critical role; persons with disrupted insula are more likely to be able to quit the addition without relapse (Naqvi NH et al 2007 Science 315:531). The nicotinic acetylcholine receptors play, however, very important roles in the cognitive processes of the brain. Tolerance to smoking seems to be influenced by diet and ethnic background. The carcinogenic effect of smoking tobacco is primarily due to specific N-nitrosamines. Second hand smoking may also stimulate angiogenesis and thus tumor growth (Zhu B et al 2003 Cancer Cell 4:191). ►Parkinson's disease, ►dopamine, ►affective disorders, ►nicotinic acetylcholine receptors, ►nicotine, ►transversion, ►intron, ►fragile site, ►p53, ►chemical mutagens, ►tobacco, ►MAO, ►infertility, ►mortality, ►small cell lung carcinoma; Hecht SS 1999 Mutat Res 424:127; Schuller HM 2002 Nature Rev Cancer 2:455; environmental exposure: Besaterinia A et al 2002 Carcinogenesis 23:1171; Allan M Brandt AM 2007 The Cigarette Century. The Rise, Fall, and Deadly Persistence of the Product that Defined America. Basic Books, New York; effect on respiratory tract genes: http://pulm.bumc.bu.edu/siegeDB.

Smooth Endoplasmic Reticulum: The smooth endoplasmic reticulum has no ribosomes on its surface. ►SER, ►endoplasmic reticulum

Smooth Muscle: Smooth muscles lacks sarcomeres; they are associated with arteries, intestines, and other internal organs, except the heart. ►sarcomeres, ►striated muscles

SMRT: A silencing-mediator of retinoid and thyroid hormone receptors. It is also corepressor of PPARδ. ►retinoic acid, ►animal hormones, ►nuclear receptors, ►PPAR; Becker N et al 2001 Endocrinology 142:5321.

Smut: Infection of grasses by *basidiomycete* fungi, causing black carbon-like transformation of the inflorescence (by *Ustilago*, loose smut) or seed tissues (by *Tilletia*, covered smut).

Snail: *Helix pomatia univalens,* 2n = 24.

S

Snail: A family of zinc-finger transcription factors and a negative regulator of E-cadherin. ►cadherin; Betlle E et al 2000 Nature Cell Biol 2:84; Nieto MA 2002 Nature Rev Mol Cell Biol 3:155.

Snakes: Reptiles represented by a large number of cosmopolitan species of diverse sizes up to 30 ft in length. Generally, they do not have legs, except some vestigial remnants in a few species. They use their protruding tongues to smell the environment. Their heat-sensory organs are located between the eyes and the nostrils, are highly sensitive, and are used to detect potential prey. Their teeth (fangs) with groves conduct their venom to the body of the prey. The venom sacs (in the venomous species) are modified from the salivary gland. Snakes are carnivorous. Their sexual organs are located at the end of the cloaca (alimentary channel). Fertilization is internal but the eggs are laid in the environment, although in some species the eggs hatch within the female body. Snake venom is used primarily for killing the prey. The prey is swallowed when dead, without chewing. Snake protein venoms are very diverse and have evolved from acetylcholinesterase, ADAM, AVIT, complement C3, crotasin/β-defensin, cystatin, endothelin, factor V, factor X, kallikrein, Kunitz-type proteinase inhibitor, LYNX/SLUR, L-amino oxidase, lectin, natriuretic peptide, β-nerve growth factor, phospholipase A_2, SP1a/Ryanodine, vascular endothelial growth factor, and whey acidic protein/secretory leukoproteinase inhibitor (Fry BG 2005 Genome Res 15:403). Knowledge of the physiological/molecular nature of these proteins has evolutionary interest and is important because of its therapeutic relevance.

Snake Venom Phosphodiesterase: Snake venom phosphodiesterase releases 5′-nucleotides from the 3′ end of nucleic acids. ►phosphodiester bond, ►phosphodiesterases

Snap: ►NSF, ►membrane fusion, ►SNAREs

Snap-Back: Inverted repeat sequence in nucleic acids. ►repeat inverted, ►lollipot structure

Snapdragon (*Antirrhinum majus*): 2n = 16, a dicotyledonous plant (*Scrophulariaceae*) much employed for the study of mutation (transposable elements) and flower pigments. It is also a popular ornamental. Snapdragon also has many beautiful flower morphology mutants as shown. ►TAM, ►*Antirrhinum*, ►peloric, flower morphology mutants at ►mutation spectrum; Schwarz-Sommer Zs et al 2003 Nature Rev Genet 4:655; http://www.antirrhinum.net/.

SNAREs (soluble *N*-ethylmaleimide-sensitive factor attachment protein receptor): Binding protein attaching vesicles (v-SNAREs) to target membranes (t-SNAREs) (see Fig. S77).

They mediate, among others, transport through the Golgi compartments. SNAREs also mediate membrane fusions, alongwith plant cell wall penetration resistance to fungal pathogens on normally non-host species. (Collins NC et al 2003 Nature [Lond] 425:973). The name has been attributed to surgical wire tools, by which polyps and projections are removed. It could also be linked with bird traps (snares). SNARE seems to be activated by Ypt1p. ►NSF, ►RAB, ►EEA1, ►Ypt, ►snare, ►synaptobrevin, ►VAMP, ►syntaxin, ►synaptogamin, ►NSF, ►Golgi apparatus, ►membrane fusion, ►exocytosis, ►Munc1, ►Ipk1; Bock JB, Scheller RH 1999 Proc Natl Acad Sci USA 96:12227; Peters C et al 2001 Nature [Lond] 409:581.

SNF: SNF yeast genes are helicases involved in chromatin remodeling (*SNF2, SNF5, SNF6, SNF11*) and *SNF1* is an AMP-activated kinase. SNF1 senses depletion of ATP and increase of AMP in the cell. This is a ubiquitous enzyme family involved in carbohydrate and lipid metabolism, phosphorylation of transcription factors, regulating stress responses in plants, etc. An SNF-6 protein is an acetylcholine transporter in *Caenorhabditis* (►muscular dystrophy). ►chromatin remodeling, ►*SWI*, ►*SUC2*, ►bromodomain; Eisen JA et al 1995 Nucleic Acids Res 23:2723; Lo W-S et al 2001 Science 293:1142.

SNIPs (single nucleotide polymorphism, SNP): SNPs refer to the difference in a single nucleotide at a particular DNA site; these are used as genomic

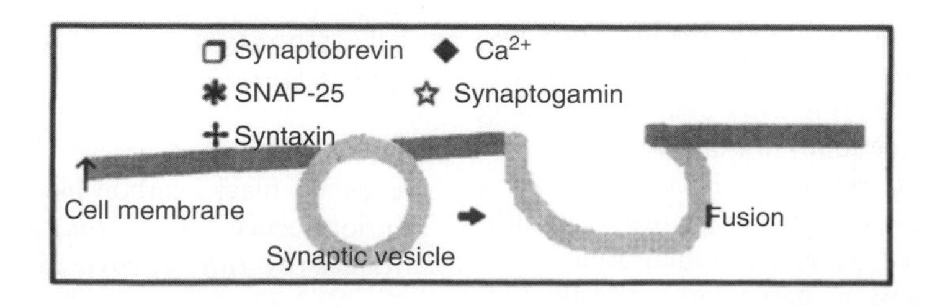

Figure S77. SNARE

markers for human (or other) populations. The most common variation involves C↔T transition in CpG sequences. The analysis uses DNA chips or gel-based sequencing and biotin-labeled probes (VDA, variant detector array). Several different methods exist for the detection of SNPs.

The presence of a single mismatch decreases the electrochemical potential of the DNA. On this basis, transitions and transversions can be detected by the electrochemical response (Inouye M et al 2005 Proc Natl Acad Sci USA 102:11606). A survey of 2748 SNPs indicated high degree of polymorphism (4.58×10^{-4}) and mutation rate $\mu \sim 10^{-8}$ to the range of about 10% of the confirmed SNPs. Other estimates for average nucleotide diversity (π) were higher: 9.01×10^{-4} for African-Americans and 6.97×10^{-4} for European-Americans (Crawford DC et al 2005 Annu Rev Genomics Hum Genet 6:287). Other studies indicated one SNIP/600 base pairs in the human genome (Kruglyak L, Nickerson DA 2001 Nature Genet 27:235). The annotated human chromosome 6 (166,880,988 bp) contains 2761 SNPs in the protein-coding genes (Mungall AJ et al 2003 Nature [Lond] 425:805).

SNPs can be mapped to chromosomal location by radiation hybrid cell lines. If the SNP is not within the gene, recombination may lead to false positive identification. SNPs can be generated for the identification of the critical base substitutions responsible for human disease. The majority of SNPs occur in non-coding regions of the genome and are non-informative regarding human disease (Sachidanandam R et al 2001 Nature [Lond] 409:928). Some SNPs in non-coding regions may, however, regulate gene expression.

Mapping of SNPs can be carried out by the reduced representation shotgun sequencing (RRS) and by locus-specific polymerase chain reaction amplification (LSA).

Frequently within a gene or within a haplotype, several SNPs exist and several of them may contribute to the disease phenotype. In order to determine their significance for a disease, their location within the haplotype requires mapping (see Fig. S78) (Carlson CS et al 2004 Am J Hum Genet 74:106; Livingston RJ et al 2004 Genome Res 14:1821). If polony amplification is used from a buccal smear, a single microscope slide permits appropriate genotyping (Mitra RD et al 2003 Proc Natl Acad Sci USA 100:5926).

However, when many special cases of the same disease are analyzed, the significance of the base substitutions may be statistically or even causally determined (see Fig. S79). Generally, the number of SNPs is much higher in introns than in exons. For population genetics and linkage studies, the SNPs are frequently classified into types I (involving

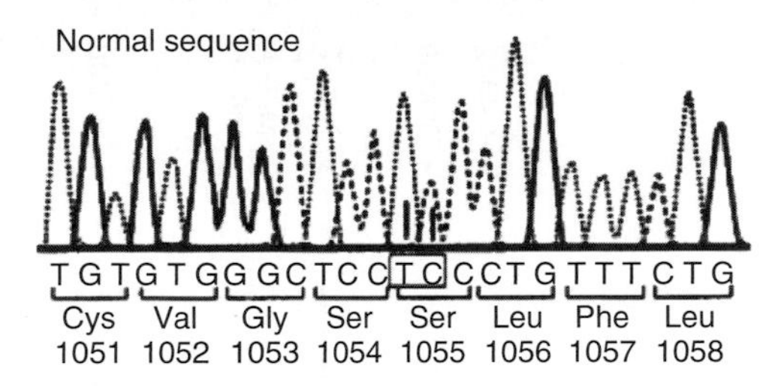

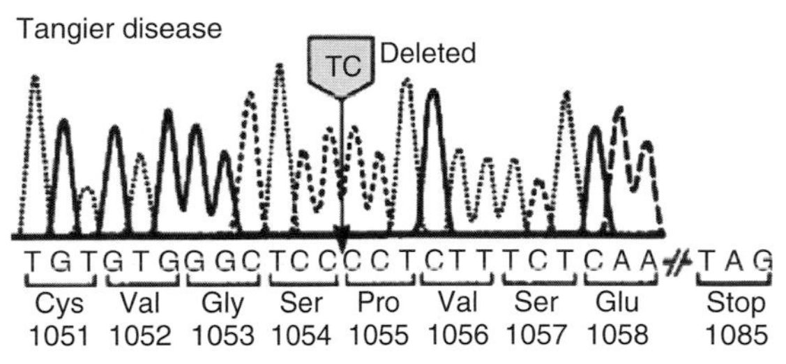

Figure S78. Two-nucleotide deletion in the ABC1 transporter gene results in Tangier disease. Note the frameshift. (Courtesy of H. Bryan Brewer, Jr. Modified after Remaley AT et al 1999 Proc Natl Acad Sci USA 96:12685)

non-synonymous alterations regarding its coding property and being a non-conservative change), II (within coding region and non-synonymous yet conservative), III (in coding sequence but synonymous), IV (within the non-coding 5′ sequence), V (within the non-coding 3′ sequence), or VI (in other non-coding regions). The type I SNPs are most useful for genetic analyses because they have phenotypic (functional) characteristics. Preliminary information indicates that the majority of the SNIP haplotyes (~80%) occurs in all ethnic groups and only 8% are population-specific (Patil N et al 2001 Science 294:1719). MALDI analysis applied to SNPs may facilitate the analysis of QTLs (Mohlke KL et al 2002 Proc Natl Acad Sci USA 99:16928). By March 2001, 2.84 million SNPs had been deposited in the public databases and they represented 1.64 million non-redundant mutations (Marth G et al 2001 Nature Genet 27:371). The human SNP map includes an ever-increasing number (~8 million by 2005; actually 6 million are validated by 2007) of variants of the world population (http://www.hapmap.org). Regulatory SNPs were detected in the germline of several breast cancer genes controlling the somatic function, the reactive oxygen species (ROS) pathway (Kristensen VN et al 2006 Proc Natl Acad Sci USA 103:7735). ►allele, ►DNA chips, ►MALDI/TOF/MS, ►genotyping, ►MRD, ►radiation hybrid, ►STS, ►QTL, ►RRS, ►LOS, ►biotinylation, ►DASH, ►DNA repair, ►padlock probe, ►linkage disequilibrium, ►association test, ►polony, ►haplotype, ►giSNP, ►haplotype block, ►Tangier disease, ►breast cancer, ►ABC transporters, ►allele-specific

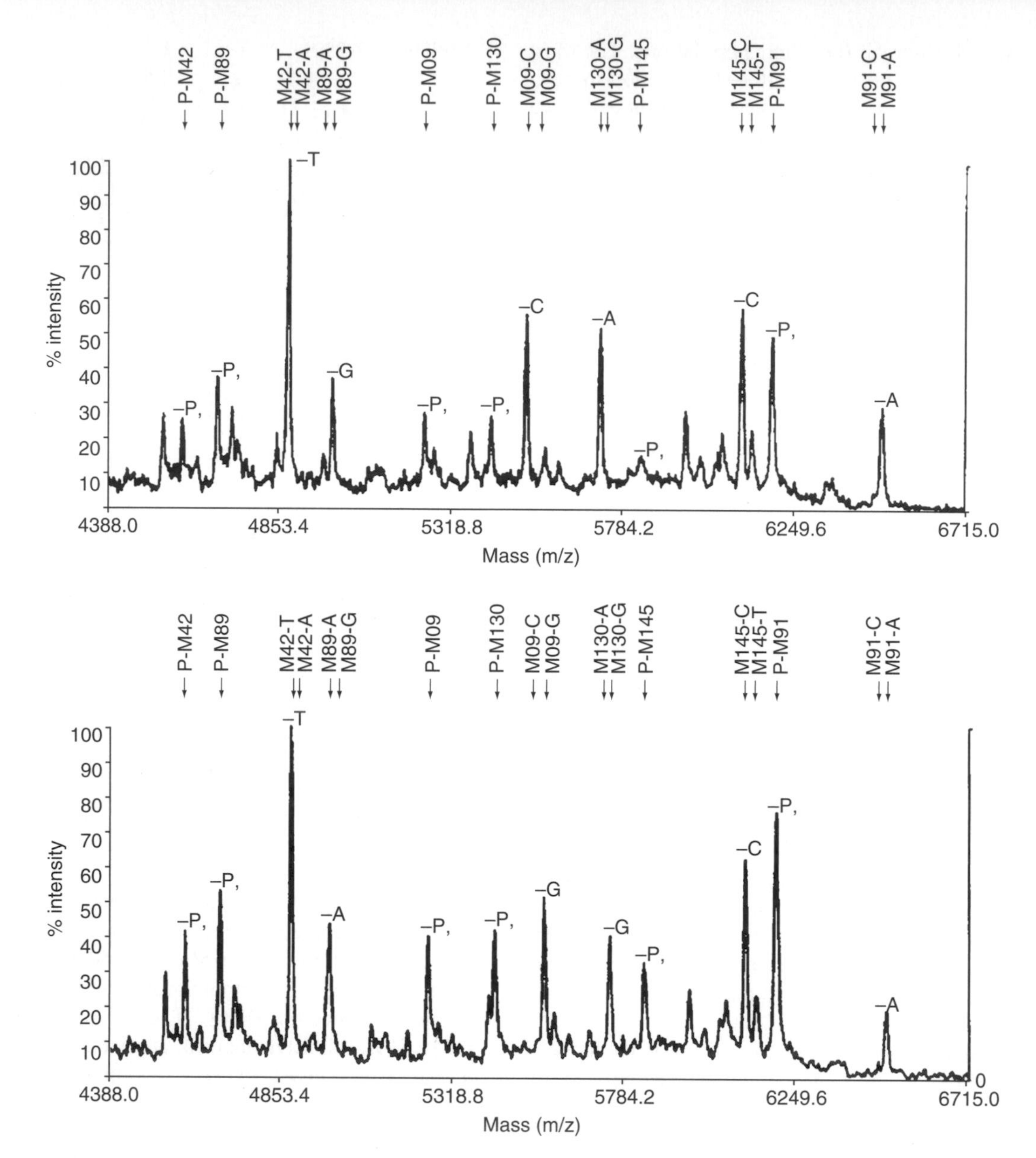

Figure S79. Multiplex primer extension products of the human Y chromosome of two individuals analyzed by MALDI-TOF mass spectrometry displaying allelic differences at the sites indicated at the top. -P indicate the primers and -A, -C, -G, -T stand for the nucleotides. (Courtesy of Silvia Paracchini, Barbara Arredi, Rod Chalk, and Chris Tyler-Smith, 2002.)

probe, ►dynamic allele-specific hybridization, ►genotyping, ►ROS, ►PolyPhred, ►cSNP; Wang DG et al 1998 Science 280:1077; Sunyaev S et al 2000 Trends Genet 16:198; Mullikin JC et al 2000 Nature [Lond] 407:516; Buetow KH et al 2001 Proc Natl Acad Sci USA 98:581; Grupe A et al 2001 Science 292:1915; Roger A et al 2001 Genome Res 11:1100; Miller RD, Kwok P-Y 2001 Hum Mol Genet 10:2195; Gut IG 2001 Hum Mutat 17:475; Werner M et al 2002 Hum Mutat 20:57; Kirk BW et al 2002 Nucleic Acids Res 30:3295; Paracchini S et al 2002 Nucleic Acids Res 30(6):e27; Coronini R et al 2003 Nature Biotechnol 21:21; genotyping SNPs in complex DNA: Kennedy GC et al 2003 Nature Biotechnol 21:1233; human genetic variation: http://www.ncbi.nlm.nih.gov/SNP; http://hgbase.interactiva.de; http://www.genomic.unimelb.edu.au/mdi/dblist/ccent.html; tools: http://bio.chip.org/biotools; medical SNP information: http://snp.cshl.org/; SNP: http://www.ncbi.nlm.nih.gov/entrez/query.fcgi?db=Snp; SNP effect: http://snpeffect.vib.be/; browser tools: http://genewindow.nci.nih.gov; SNP with phenotypic effects: http://pupasuite.bioinfo.cipf.es/; medically important SNP search: http://fastsnp.ibms.sinica.edu.tw; SNP associations with disease phenotypes: http://gmed.bu.edu/; SNP in protein domains: http://snpnavigator.net/; coding SNP tool: http://www.pantherdb.org/tools/; detection of human nonsynonymous SNPs: http://coot.embl.de/PolyPhen/; sorting intolerant from tolerant amino acid substitutions in proteins: http://blocks.fhcrc.org/sift/SIFT.html; http://www.ncbi.nlm.nih.gov/projects/SNP/; SNP-short tandem repeat software for human, dog, mouse and chicken: http://www.imperial.ac.uk/theoreticalgenomics/data-software; non-synonymous SNPs with relevance to disease: http://polydoms.cchmc.org/; SNAP annotation platforms: http://snap.humgen.au.dk/; http://snap.genomics.org.cn/; ethnicity –SNP – disease databases: http://variome.net; http://bioportal.net/; http://bioportal.kobic.re.kr/SNPatETHNIC/; SNP effect on protein structure: http://glinka.bio.neu.edu/StSNP/; markers for genotyping: http://bioinformoodics.jhmi.edu/quickSNP.pl.

SNO Oncogenes: Two SKI-related oncogenes. ►SKI, ►oncogene

Snorbozyme: A ribozyme within the nucleolus, processing or degrading nucleolar RNA. (See Samarsky DA et al 1999 Proc Natl Acad Sci USA 96:6609).

snoRNA (small nucleolar RNA): snoRNA is 60–300 nucleotide in length and assists in maturation of ribosomal RNAs in the nucleolus, the folding of RNA, RNA cleavage, base methylation, assembly of pre-ribosomal subunits, export of RNP, etc. A family of snoRNAs of 10–21 nucleotides, complementary to the methylation sites of rRNA, guides the methylation within the nucleolus. Some of the snoRNA genes are situated in introns. The majority of the snoRNAs are either box C/D or H/ACA snoRNA family members. All CD boxes contain fibrillarin and (spliceosomal) Snu13. They are apparently involved in splicing and alternative splicing (Kishore S, Stamm S 2006 Science 311:230). ►RNA maturase, ►pseudouridine, ►introns, ►fibrillarin, ►spliceosome, ►non-coding RNA; Hirose T, Steitz JA 2001 Proc Natl Acad Sci USA 98:12914; Song X, Nazar RN 2002 FEBS Lett 523:182; snoRNA database: http://www-snorna.biotoul.fr; snoRNA and Cajal body-specific RNA: http://gene.fudan.sh.cn/snoRNAbase.nsf.

Snowdrift Game: One of the theories of evolutionary population dynamics and natural selection. An oversimplified analogy for the principle is that of drivers trapped on either side of a snowdrift. The drivers can either cooperate and both start clearing the road from opposite sides or one of them fails to work (defects). In case of cooperation, they benefit (b) by sharing the labor (c). The result is R = b – c/2. If both defect, the probability to get through is P = 0. If only one works, they can both get through, but the defector avoids the cost of labor and gets all the benefits (b – c). ►prisoner' dilemma; Hauert C, Doebeli M 2004 Nature [Lond] 428:643; Nowak MA et al ibid. p 646.

SNP: ►SNIPS

snRNA (small nuclear RNA): Low molecular weight RNA in the eukaryotic nucleus, rich in uridylic residues. When associated with protein, sRNA mediates the splicing of primary RNA transcripts and frees them from introns through assistance of the lariat. The spliceosomal snRNP can be exported to the cytoplasm if appropriately capped, i.e., if it possesses the nuclear cap-binding complex, the export receptor CRM1/Xpo1. RanGTP and the phosphorylated adaptor of RNA export (PHAX) are also required for this process. ►hnRNA, ►U1-RNA, ►RNP, ►Ran, ►CRM1/SXPO1, ►export adaptor, ►introns, ►spliceosome, ►lariat, ►Ohno's law, ►7SK RNA; M et al 2000 Cell 101:187; Kiss T 2001 EMBO J 20:3617; 2002 Gene Expr 10(1/2).

snRNP: Small nuclear ribonucleoprotein, (also pronounced as "snurp"). It is involved in the processing of RNA and the assembly of spliceosomes. ►KH domain, ►imprinting, ►spliceosome; Nagengast AA, Salz HK 2001 Nucleic Acids Res 29:3841.

Snurportin: An α importin-like transport protein handling snRPN import to the cell nucleus. ►importin, ►nuclear pore, ►snRPN; Paraskeva E et al 1999 J Cell Biol 145:255.

Snurposomes: A complex of the five snRNP particles ("snurps") that process RNA transcripts in the Cajal bodies. ►U RNAs, ►Cajal body, ►coiled body; Gall JG et al 1999 Mol Biol Cell 10:4385.

SOAP (Simple Object Access Protocol): http://www.w3.org/TR/soap/.

SOB Bacterial Medium: The SOB bacterial medium consists of: H_2O 950 mL, bacto tryptone 20 g, bacto yeast extract 5 g, NaCl 0.5 g plus 10 mL of 250 mM KCl; its pH is adjusted to 7 with 5 N NaOH in a total volume of 1 L. Just before use add 5 mL of 2 M $MgCl_2$.

SOC (store-operated channel, synonym: TRP transient receptor potential): a group of plasma membrane ion channels controling the release of ions stored in the lumen of the endoplasmic reticulum. ►ion channels; Ma R et al 2001 J Biol Chem 276:25759.

SOC Bacterial Medium: The same as SOB but contains glucose (20 mM). ►SOB

Social Darwinism: The application of Darwinian views (survival of the fittest) to social order. Many anthropologists, sociologists, and ethicists rejected social darwinism and portend that it is an attempt to justify inequalities, harsh competition without adherence to ethics, aggression, imperialism, racism, and unbridled capitalism as necessities for the survival of the fittest. Social darwinism of the 19th century is no longer accepted in the developed world. ►Darwinism, ►IQ, ►social engineering; Rogers JA 1972 J Hist Ideas 33:265.

Social Engineering: The Utopian idea that the genetic determination of an individual is unimportant in defining the abilities and their realization but education, welfare, medical service, etc., may determine how a person will function in a society. Therefore, the state and its institutions must actively control human life from cradle to death. However, even though the importance of compassion, education, caring, and multiple social safety nets are indispensable in a modern society, the significance of individuality cannot be ignored. ►social Darwinism, ►IQ; Graebner W 1980 J Am Hist 67:612.

Social Insects: Insects like bees, wasps, ants, and termites that live in a colony and generally divide various tasks among different castes, such as workers (soldiers), queen, and drones. The queen (gyne) and the workers and soldiers are diploid. In some species, the workers may produce males. The drones are different because they hatch from unfertilized eggs. The major differences between the queen and the workers is that the queens have predominantly 9-hydroxy-(*E*)2-decanoic acid and 9-keto-(*E*)2-decanoic acid in their mandibular glands. The workers have predominantly 10-hydroxy-(*E*)2-decanoic acid. These pheromones then determine their respective functional roles in the colony. The workers and soldiers have ovaries and under certain circumstances (especially the soldiers) may produce haploid eggs. Cuticular hydrocarbons may regulate sex expression and convert workers into egg-laying females (gamergate), although usually smaller in size. Such a change in some groups may take place by physical contact of individuals transmitting the hydrocarbon molecules. Recently, the status of soldiers has been questioned as being genetically equivalent with the workers. The soldiers' (not found in all species of social insects) role is protection of the colony. In the fire ant *Wasmannia auropunctata*, a widely distributed species, sex determination is unique. The queens produce other queens clonally (without sexual reproduction). The diploid sterile workers are the products of sexual reproduction. The males are derived from unfertilized eggs containing only the paternal genome (the maternal genome is eliminated from the diploid eggs). Therefore, the male and female genomes are completely separated (Fournier D et al 2005 Nature [Lond] 435:1230). ►sex determination, ►male-stuffing, ►ant, ►honey bee, ►wasp; Bourke AF 2001 Biologist [Lond] 48(5):205; Parker JD, Hedrick PW 2000 Heredity 85(pt 6):530; Thorne BL, Traniello JFA 2003 Annu Rev Entomol 48:283; development of caste differences: Hoffman EA et al 2007 BMC Biol 5:23.

Social Selection: Social selection is not based on Darwinian fitness but on the social status of the individual carrying the trait. A particular Y-chromosome is found in a large region from the Pacific to the Caspian Sea where the Mongols of Genghis Khan ruled. It originated about the same time—about 1000 years ago—as the Mongolian Empire. Thus, the rapid spread of this chromosome may be due to the social privileges of the carriers. Preimplantation selection of the sex of human offspring is also called social selection. Its purpose can be the avoidance of defective children caused by chromosomal anomalies and hereditary disease. Family balancing, i.e., having children in a certain proportion of boys and girls is generally an ethically unacceptable goal. In the USA, many institutions object to it and refuse the medical procedure required. ►Y chromosome, ►sex selection; Zerjal T et al 2003 Am J Hum Genet 72:717.

Sociobiology: The study of the biology and behavior of social insects and other animal communities. ►social insects; Wilson EO 2000 Sociobiology: The new synthesis. Harvard University Press, Cambridge, Massachusetts.

Sociogenomics: Sociogenics studies the genetic and molecular biology of the functions involved in social life. ►kin selection, ►altruism, ►mutualism; Robinson GE et al 2005 Nature Rev Genet 6:257.

SOCS-Box (suppressor of cytokine signaling): The SOCS-box contains a protein docking site(s) for protein-protein interaction and seems to be involved in transduction signal attenuation and ubiquitination. SOCS-2$^{-/-}$ mice display greatly increased body weight caused apparently by inappropriate regulation of the Jak/Stat signal transduction and the IGF-I pathways. ►cytokine, ►signal transduction, ►CIS, ►JAB, ►SSI-1, ►CSAID, ►ubiquitin, ►insulin-like growth factors; Zhang JG et al 2001 Proc Natl Acad Sci USA 98:13261.

SOD (superoxide dismutase): ►superoxide dismutase, ►amyotrophic lateral sclerosis

SODD: ►TNFR

S

Sodium Azide (N_3Na): An inhibitor of respiration that blocks the electron flow between cytochromes and O_2. It is also a potent mutagen for organisms that can activate it.

Sodium Channel: ►ion channels

Sodium Dodecyl Sulfate: A synonym for sodium lauryl sulfate; ►dodecyl sulfate sodium salt

Sodium Pump: A plasma membrane protein that moves Na^+ out and K^+ into the cells with the energy obtained by hydrolyzing ATP; it is also called Na^+—K^+ ATPase. ►ion channels

Soft Inheritance: Soft inheritance has been a rather controversial notion; its theory claims that environmental influences mold heredity and acquired characters are inherited. ►acquired characters, ►hard heredity of inheritance of acquired characters, Lamarckism, Neo-Lamarckism, Lysenkoism, and other ideas claiming that environmental factors modify the genetic material at a non-random manner (not including mutation or transformation) but by direct manner shaping adaptive traits. Epigenetics has shown however that some traits are transmitted to the progeny by their acquired status of methylation. (See terms in alphabetical order, ►hard heredity).

Software: A computer program that tells the hardware (the computer) the applications, i.e., which instructions to carry out and how. ►databases; http://www.sanger.ac.uk/Software/Pfam/.

Sog: ►bone morphogenetic protein

Soil Remediation: Soil remediation involves the removal of environmental pollutants from soil and water. It requires sensitive identification of small amounts of toxic and carcinogenic and mutagenic substances, such as arsenic, heavy metals, polycyclic hydrocarbons, etc. ►arsenic, ►phytoremediation, ►DNA-zyme

Solenoid Structure: A coiling electric conductor used for the generation of a magnetic field; by analogy, the coiled nucleosomal DNA fiber is frequently described as a solenoid, although it has no relation to electricity or magnetism. It merely resembles those coils. The DNA solenoids are about 30 nm in diameter, contain about six nucleosomes per turn, and are packed with a large number of structural and catalytic proteins. Actually two different models have been proposed for the 30 nm solenoid structure of the chromosome sub-fibers although neither of them has been fully and universally confirmed. One has been based on the crystal structure of a four-nucleosome core array lacking the linker histone and the other, far more compact structure, has been derived from the electron microscopic analysis of long nucleosome arrays containing the linker histone. The first model is of the two-start helix type, the second a one-start helix with interdigitated nucleosomes. Both of these models assume that the fiber is regulated by histones (Robinson PJ, Rhodes D 2006 Curr Opin Struct Biol 16:336; Tremethick DJ 2007 Cell 128:651). ►nucleosome, ►nuclear matrix, ►zig-zag model of chromosome fibers

Solid Phase Synthesis: In a solid-phase synthesis, the synthesized molecules are continuously added to a solid support.

Solid State Control: Solid state control is exercised by electric or magnetic means in solids, e.g., in a transistor. ►semiconductors

Solitary LTR: Long terminal repeats that have lost their internal (transposase) sequences by re-combination and excision. ►long terminal repeats, ►retrovirus, ►retroposon, ►retrotransposon; Domansky AN et al 2000 FEBS Lett 472(2–3):191.

Solo Elements: Terminal sequences (LTR) of retroposons that can exist in multiple copies without the coding sequences between the two direct repeats. ►Ty

Soluble RNA: A somewhat outdated term for transfer RNA. ►tRNA

Solute: Any substance dissolved in a solvent.

Solution Hybridization: Molecular hybridization in a liquid medium. ►nucleic acid hybridization

SOM (self-organizing map): A type of mathematical cluster analysis that is particularly well suited for recognizing and classifying features in complex, multidimensional data of microarray hybridization. The method has been implemented in a publicly available computer package, GENECLUSTER, that performs the analytical calculations and provides easy data visualization. ►microarray hybridization, ►cluster analysis; Sinha A, Smith AD 1999 Microgravity Sci Technol 12(2):78; Haese K, Goodhill GJ 2001 Neural Comput 13:595; Unneberg P et al 2001 Proteins 42:460.

Soma: Body cells distinguished from those of sexual reproduction (germinal cells). ►disposable soma

Somaclonal Variation: Genetic variation occurring at a frequency higher than spontaneous mutation in cultured plant cells (see Fig. S80). (See Kaeppler SM et al 2000 Plant Mol Biol 43(2–3):179).

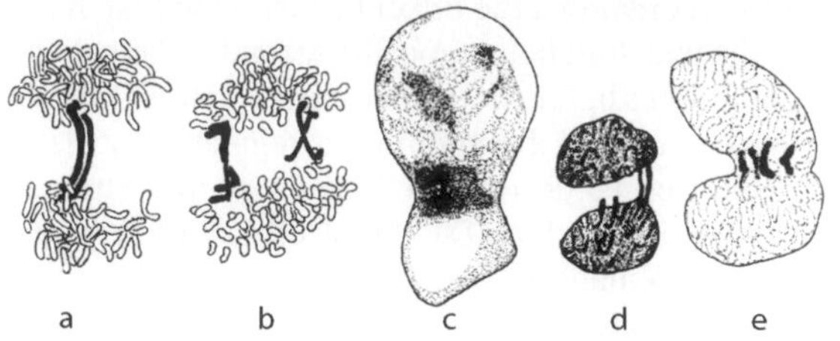

Figure S80. Mitotic anomalies in a clone of tobacco maintained in cell culture. (a), (b) anaphase bridges, (c) telophase bridge, (d) the same as (c) but enlarged nuclei, (e) early interphase with conjoined nuclei. Polyploidy is also of common occurrence. (From Cooper LS et al 1964 Am J Bot 51:284)

The causative mechanism is poorly understood. It is conceivable that the asynchrony between nuclear and cell divisions are accompanied by chromosomal damage. Also, there is evidence that movement of endogenous transposable elements is involved. The mobility is attributed to the "stress" imposed by the culture (see Fig. S81).

Somatic Cell: The majority of the body cells (reproduced by mitosis), including those of the germline but not the products of meiosis, the sex cells (gametes). ►germline, ►cell lineages, ►mitosis, ►cell genetics, ►parasexual mechanisms

Somatic Cell Hybrids: Somatic cell hybrids are formed through fusion of different somatic cells of the same or different species. Somatic cell hybrids contain the nucleus of both cells and in addition all cytoplasmic organelles from both parents, in contrast to the generative hybrids where generally mitochondria and plastids are not transmitted through the male.

For the fusion of cultured somatic cells, the use of various special techniques is necessary. Most commonly, protoplasts are used and the fusion medium is:

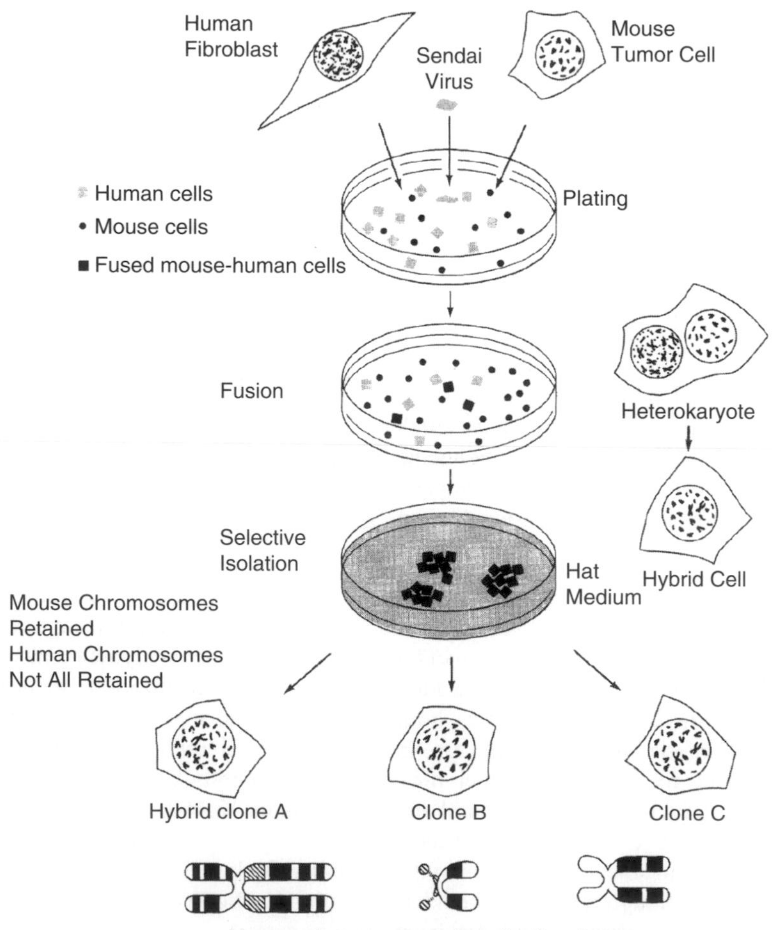

Figure S81. Selective isolation of mouse + human cell hybrids. The fused cell usually retains all mouse chromosomes but the human fibroblast chromosomes may be partially eliminated and in some clones only one or another human chromosome is maintained. (Modified after Ruddle FH, Kucherlapati RS 1974 Sci Am 231(1):36)

polyethylene glycol [MW 1300–1600] 25 g, $CaCl_2.2H_2O$ 10 mM, KH_2PO_4 0.7 mM, glucose 0.2 M in 100 mL H_2O, pH 5.5 for plants. The best media may vary according to species. The isolation of somatic animal cell hybrids was greatly facilitated by using selective media (see Fig. S82). (►HAT medium).

The fusion of animal cells is promoted by polyethylene glycol, by attenuated Sendai virus, or by calcium salts at higher pH. Immediately after fusion, the somatic cells may become heterokaryotic but eventually the nuclei may also fuse. The availability of fused cells along with loss of one set of chromosomes or partial deletion of chromosomes make possible the localization of many human genes (see Fig. S84). In human genetics, mouse + human cells have been used most commonly. In such cultures, the human chromosomes are gradually eliminated. however, if one of the mouse chromosomes carries a defective gene, the human chromosome carrying a functional wild type

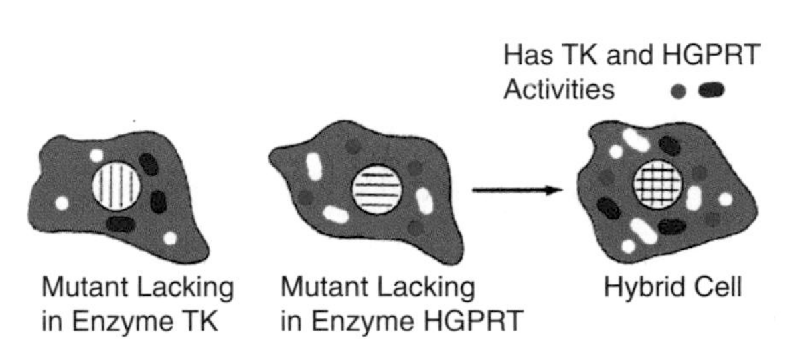

Figure S82. Selective isolation of animal somatic cell hybrids deficient in thymidine kinase and hypoxanthine-guanine phosphoribosyltransferase. On "HAT" medium only the complementary heterokaryons survive. (Modified after Ephrussi B, Weiss MC 1969 Sci Am 220 (4):26)

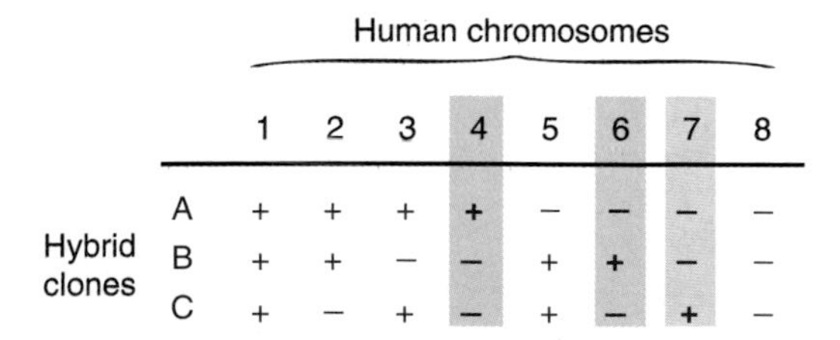

Hybrid clones	Human chromosomes 1	2	3	4	5	6	7	8
A	+	+	+	+	–	–	–	–
B	+	+	–	–	+	+	–	–
C	+	–	+	–	+	–	+	–

Figure S83. Chromosome assignment of genes on the basis of incomplete clones of mouse + human cell hybrids. A panel of 3 clones, each containing a set of 4 of 8 human chromosomes. If a gene is expressed uniquely in clone (C) but not in (A) or (B), the locus must be in chromosome 7. Because only C carries chromosome 7. Additional panels are needed to test genes in the entire human genome. (After Ruddle FH, Kucherlapati RS 1974 Sci Am 231(1):36)

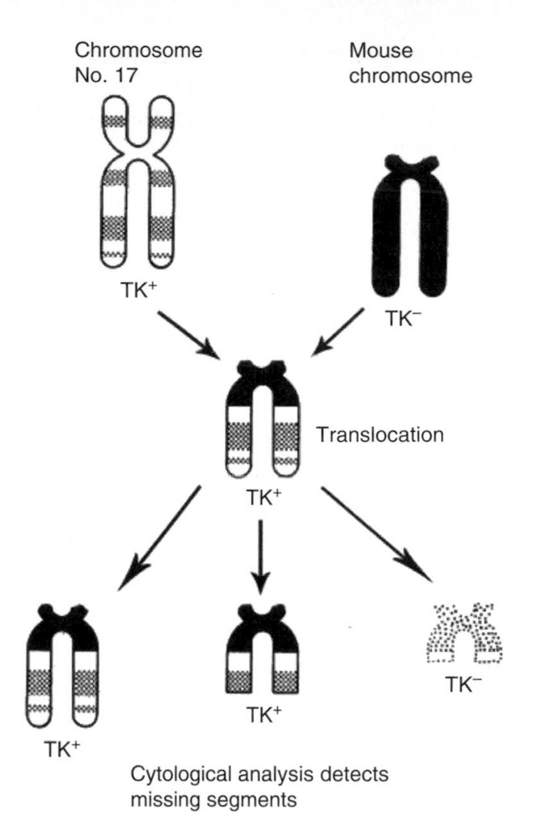

Figure S84. Regional mapping of the thymidine kinase (TK) gene to a segment of human chromosome 17. A translocation was obtained between human chromosome 17 and a mouse chromosome that was deficient for TK. The translocation chromosome was exposed to a chromosome-breaking agent that cleaved off segments of various lengths from the end of the translocation. When the segment containing the gene (TK) was removed, the cells carrying this broken chromosome no longer displayed TK activity and the gene's location was revealed at 17q23.2-q25.3. (►human chromosome map; Modified after F.H. Ruddle and R.S. Kucherlapati)

allele must be retained in order that the culture be viable in the absence of a particular supplement, required for normal function (see Fig. S83). The genetic constitution of the retained human chromosome can be also verified by enzyme assays, electrophoretic analysis of the proteins, or immunological tests. For genetic analysis of higher mammals, and particularly in humans, where controlled mating is not feasible, the availability of somatic hybrids has opened a new and very productive approach. In the somatic hybrid cells, allelism and also synteny or linkage can be shown. If two or more human genes are consistently expressed in a particular chromosome retained, it is a safe conclusion that they are located in the same chromosome. The diagram (see Fig. S83) explains the principle of gene assignment. Genes can be mapped to particular chromosome bands by deletions. By somatic cell fusion, hybrids can be obtained between taxonomically very distant species.

All kinds of animal cells may be fused with each other, and in addition plant and animal cells can also be fused. Some of these exotic hybrids may have, however, difficulties in continuing cell divisions.

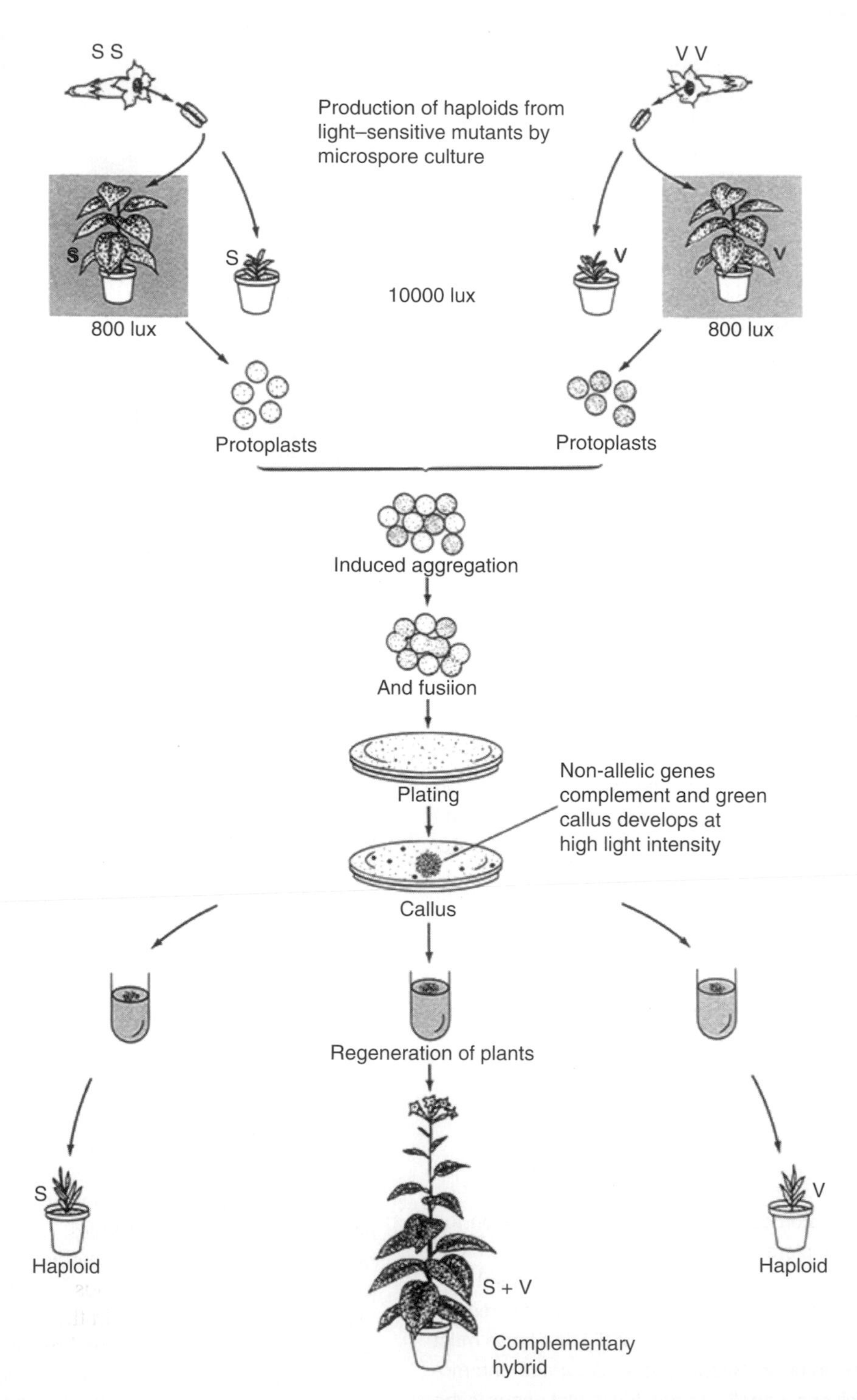

Figure S85. Somatic-cell hybrid production between non-allelic tobacco mutants. (Redrawn by permission after Melchers G & Labib G 1974 Mol. Gen. Genet. 135:277)

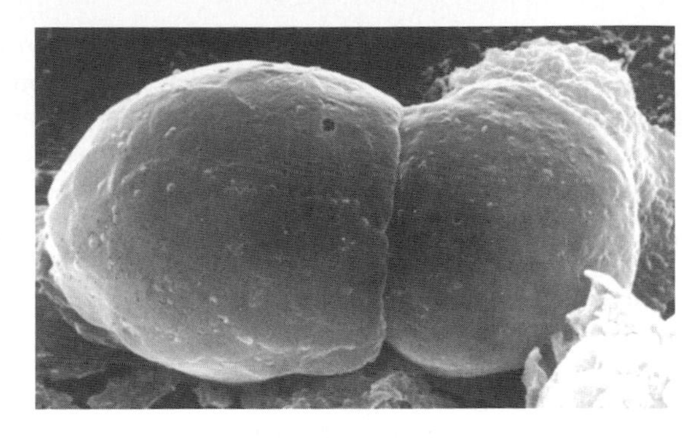

Figure S86. The process of fusion of plant protoplasts. (From Fowke LC et al 1977 Planta 135:257)

Somatic cell fusion may also make possible the study of recombination between various mitochondria or chloroplast DNAs in cells. However, in generative hybrids, because of the uniparental (female) transmission of the organelles, such analyses are not feasible. Since plant cells generally retain totipotency in culture, the hybrid cells may be regenerated into intact organisms and can be further studied in favorable cases by the methods of classical progeny analysis.

►HAT medium, ►cell fusion, ►fusion of somatic cells, ►monoclonal antibody, ►human chromosome maps, ►in situ hybridization, ►mapping genetic, ►somatic embryogenesis, ►microfusion, ►radiation hybrids, ►mitochondrial mutation, ►IFGT; Vasil IK (Ed.) 1984 Cell Culture and Somatic Cell Genetics of Plants. Academic Press, San Diego, California; Rédei GP 1987 McGraw-Hill Enc Sci Technol 6th ed 16:628.

Somatic Cell Nuclear Transfer: After the enucleation of oocytes, the nucleus of somatic cells from the same or different animals is transferred into the oocytes and this way the donor cell or individual is cloned. This procedure was successful initially for sheep; successful procedures for mice, cows, goats, pigs, rabbits, mule, horse, rats, and dogs followed. Eventually, the oocytes with a new donated diploid nucleus is transferred into the oviduct of a female, which carries the embryo to term. Applying such techniques to human cells is restricted or prohibited in many countries. It may be permitted in the UK, Belgium, China, Singapore, South Korea and Australia under regulated conditions. In the USA, such research cannot be supported by federal funds although there are less and variable limitations for private of state-funded activities. ►nuclear transplantation, ►cloning; Lee BC et al 2005 Nature [Lond.] 436:641.

Somatic Crossing Over: ►mitotic crossing over

Somatic Embryogenesis: The formation of embryos either directly, or by first passing through a callus stage, from cultured adult plant cells or protoplasts. LEAFY COTYLEDON2 gene controls proteins involved in somatic embryogenesis of *Arabidopsis* (Braybrook SA et al 2006 Proc Natl Acad Sci USA 103:3468). In about nine animal phyla, the germline is not separated from the somatic line, therefore gametes develop from the somatic cell lineage. The situation in protists and fungi is essentially similar. A strict germline does not exist in plants either. ►embryogenesis somatic, ►embryo culture, ►apomixis, ►germline; Sata SJ et al 2000 Methods Cell Sci 22:299.

Somatic Hybrids: ►somatic cell fusion, ►somatic cell hybrids, ►graft hybrids, ►microfusion

Somatic Hypermutations: Somatic hypermutations generally occur in a region of one to two kilobases around the rearranged V-J regions of the immunoglobulin genes and very rarely extend into the C (constant) sections. These mutations, usually transitions, and predominantly involving guanine, are most common in the complementarity-determining region (CDR) and the events usually take place in the germinal centers. The preferred hot spots are purine-G-pyrimidine-(A/T) sequences, although not all these sequences are hot spots for mutation. The serine codons AGC and AGT can represent a hot spot but the TCA, TCC, TCG and TCT are not. The codon usage in the CDR region appears to be evolutionarily determined to secure the maximum complementarity to antigens. It is not entirely clear what determines the targeting of the hypermutations to the special area but transcriptional enhancers appear to be involved. Hypermutations are limited to the coding strand and occur in a downstream polarity. Deletions and insertions are rare. For somatic hypermutation V(D)J rearrangement is a prerequisite. Activation-induced cytidine deaminase (AID) is apparently targeted by replication protein A to the sites of V(J)D recombination and is one factor of the hypermutations of the immunoglobulin genes (Chaudhuri J et al 2004 Nature [Lond] 430:992). C is deaminated to U by activation-induced deamination (AID) and either mutagenic repair of the mismatch takes place or the broken strand may lead to somatic hypermutation. Error-prone mismatch repair machinery preferentially targets non-template uracils in a way that promotes somatic hypermutation during the antibody response (Unniraman S, Schatz DG 2007 Science 317:1227). Somatic hypermutation, along with recombination, is the major source of antibody variation and increases the defense repertory in mammals. The Rad6/Rad18 protein may recruit error-prone translesion polymerases to the DNA

replication of immunoglobulin G (Ig) genes (Bachl J 2006 Proc Natl Acad Sci USA 103:12081). ►immunoglobulins, ►CDR, ►germinal center, ►lymphocytes, ►transition, ►hot spot, ►enhancer, ►hypermutation, ►mosaic, ►cytosine deaminase, ►UNG, ►replication protein A, ►translesion pathway; Harris RS et al 1999 Mutat Res 436:157; Meffre E et al 2001 J Exp Med 194:375; Papavasiliou FN, Schatz DG 2002 Cell 109:S35; Di Nola J, Neuberger MS 2002 Nature [Lond] 419:43; Honjo T et al 2005 Nature Immunol 6:655; Di Noia JM, Neuberger MS 2007 Annu Rev Biochem 76:1.

Somatic Mutation: Somatic mutation occurs in the body cells. It is undetectable if recessive, unless the individual is heterozygous. The mutation is manifested by sector formation. This procedure for testing mutability is particularly effective if multiple-marker heterozygotes are treated. It is not inherited to generative progeny unless it occurs or expands also to the germline. It has been successfully used for studies of mutation in the stamen hairs of the plant *Tradescantia.* In fungal cultures mutation in mitotic cells may appear as sectorial colonies. Somatic mutation can be studied in in vitro cell cultures. Recessive mutations are detectable for X-linked genes in hemizygous cells such as those of the male. The procedure is effective if it is directed toward loci with selectable products. If *loxP* is introduced into mammalian cells in a near centromeric region and if *neo* is inserted distal to the selectable marker, after random mutagenesis the *Cre* expressing cells—through recombination—produce homozygotes for some the mutations induced in recessive genes (Koike H et al 2002 EMBO Rep 3:433). Molecular methods (PCR) may permit the identification of the mechanisms involved such point mutations, chromosomal deletions and rearrangement. The frequency of mutation may be affected by age, and various environmental factors and these can be analyzed. ►pseudodominance, ►LOH, ►bioassays in genetic toxicology, ►*Tradescantia* stamen hairs, ►somatic hypermutations, ►paramutation, ►transposable elements, ►mosaic, ►PCR, ►cancer, ►SNIP, ►nucellus, ►Knudson's two-mutation theory, ►*Cre/LoxP*, ►*neo*; Orive ME 2001 Theor Popul Biol 59(3):235; p53 mutation database: http://www-p53.iarc.fr/; somatic mutation in cancer: http://www.sanger.ac.uk/genetics/CGP/cosmic/.

Somatic Pairing: Generally, only the meiotic chromosomes pair during prophase (possibly late interphase), but in some tissues, such as the salivary glands, the chromosomes are always tightly associated. Also, mitotic association of the chromosomes is a requisite for mitotic (somatic) crossing over in a few organisms where this phenomenon has been analyzed. (*Drosophila*, some fungi, *Arabidopsis*, etc.). ►mitotic recombination, ►parasexual mechanisms, ►pairing, ►intimate pairing, ►polytenic chromosomes; Burgess SM et al 1999 Genes Dev 13:1627.

Somatic Recombination: ►mitotic crossing over

Somatic Reduction: The reduction of chromosome numbers during mitosis in polyploids.

Somatic Segregation: The unequal distribution of genetic elements during mitoses. ►mitotic crossing over, ►somatic mutation, ►sorting out, ►chloroplast genetics, ►mosaic

Somatogamy: The fusion of sexually undifferentiated fungal hypha tips. ►fungal life cycles

Somatomedin: A second messenger-type polypeptide. In association with other binding proteins, it is involved in the stimulation of several cellular functions. ►insulin-like growth factor, ►pituitary dwarfness, ►second messenger; Deng G et al 2001 J Cell Physiol 189:23.

Somatoplastic Sterility: In certain plant hybrids, the nucellus may show excessive growth, and chokes the embryo to death. The embryo can be rescued if excised early and transferred to in vitro culture media. ►embryo culture

Somatostatin: A 14–20-amino acid long hypothalmic neuropeptide inhibits the release of several hormones (somatotropin, thyrotropin, corticotropin, glucagon, insulin, gastrin) in contrast to the growth hormone-releasing factor that stimulates the production of the growth hormone. The somatostatin gene has been mapped to human chromosome 3q28 and to mouse chromosome 16. Two somatostatin receptors SSTR1 and 2 (391 and 369 amino acids, respectively) with very substantial homology have been identified. They are sevenpass-transmembrane proteins frequently bound to G-proteins and distributed all over the body, at particularly high levels in the stomach, brain, and kidney. The somatostatin and dopamine receptors may co-oligomerize. ►animal hormones, ►G-proteins, ►seven membrane proteins, ►signal transduction, ►GHRH, ►dopamine, ►Langerhans islets, ►pancreas, ►ovarian cancer; Patel YC 1999 Front Neuroendocrinol 20:157.

Somatotroph: The growth hormone-secreting cell in the adenohypophysis (in the anterior lobe of the pituitary of the brain).

Somatotropin: The mammalian growth hormone (GH, $M_r \approx 21500$), it can correct some dwarfisms when its level is increased; also it stimulates milk production.

Actually, there are three human growth hormones, all coded at chromosome 17q23-q24. Their mRNAs display about 90% homology and their amino acid sequences are shared as well. The placental lactogen protein is an even more effective growth hormone, also located nearby, and it is highly homologous. The human growth hormone gene is transcribed only in the pituitary, whereas the homologs are expressed in the placental tissues. Human and other somatotropin genes have been cloned. Transformation of mice with rat growth hormone genes increased body size substantially. ►dwarfism, ►pituitary dwarfism, ►stature in humans; Yin D et al 2001 J Anim Sci 79:2336.

Somites: Somites are paired mesoderm blocks along the longitudinal axis (notochord) of an embryo that give rise to the vertebral column and other segmented structures. After migration, they may form the skeletal muscles. The development of the somites is controlled mainly by the Notch Wnt family and associated proteins. ►Notch, ►*wingless*; Pourquié O 2001 Annu Rev Cell Dev Biol 17:311; Pourquié O 2003 Science 301:328.

Somitogenesis: The differentiation of the somites. ►somites

Son of Sevenless: ►*SOS*

Sonic Hedgehog (Shh, human chromosome 7q36): A vertebrate gene that provides information in the head–tail direction for development; it is a rather general signaling protein of animal differentiation. *Shh* contributes to the specification of the notochord, brain, lung, and foregut. It is homologous to the *Drosophila hedgehog* (*hh*) gene; its receptors are *patched* (*ptc*) and *smoothened* (*smo*), signaling factor genes. A freely diffusible *Shh* is modified by cholesterol and a balance of Patched and the Hedgehog-interacting proteins (Hip) regulates it. Defects in *patched* may cause various carcinomas, medulloblastoma, and rhabdomyosarcoma. These genes have regulatory functions in oncogenic development. Patched activates apoptosis in neuroepithelia if its Sonic Hedgehog ligand is missing (Thibert C et al 2003 Science 301:843). ►*hedgehog*, ►notochord, ►holoprosencephaly, ►GLI, ►medulloblastoma, ►rhabdomyosarcoma; Villavicencio EH et al 2000 Am J Hum Genet 67:1047; Cohen MM Jr 2003 Am J Med Genet 123A:5; Ingham PW, Placzek M 2006 Nature Rev Genet 7:841.

Sonicator: An ultrasonic (≈20 kHz) equipment used for disrupting cells to extract contents.

Sonography: A method of ultrasonic prenatal analysis of possible structural and other defects in heart, kidney, bone, sex organs, umbilical chord, body movements, and for verifying pregnancy, etc. Presumably, it entails no appreciable risk. ►ultrasonic, ►prenatal diagnosis, ►fetoscopy

SopE: A guanyl-nucleotide-exchange factor for Rho and Rac GTPase proteins. It participates in the reorganization of the cytoskeleton and mediates bacterial entry into mammalian cells. After entry of the bacteria, SptP (GTPase-activating protein) restores the cytoskeleton. ►Rho, ►Rac, ►GTPase, ►cytoskeleton

Sordaria fimicola (n = 7): An ascomycete with linear spore octads. It has been extensively used for genetic recombination. A large number of mutants is available. In the 1940s, it was suspected that sexuality is relative in this fungus, but now it's been found that the poor maters are just weak mutants.

SORF: The short ORF generally includes not many more than 15 codons. ►codon, ►ORF

Sorghums: Arid, warm region crops. *S. bicolor* (and kaoliang), 2n = 2x = 20. *S. halepense* (Johnson grass) is tetraploid. Their nutritional value compared to other grain crops is lower because of the tannin content and low lysine level of the grain. The folding of its proteins (kafirin) lowers digestibility. Mutant varieties exist, however, that correct these deficiencies (see Fig. S87). (See Oria MP et al 2000 Proc Natl Acad Sci USA 97:5065; Klein PE et al 2000 Genome Res 10:789; molecular cytogenetics: Kim J-S et al 2005 Genetics 179:1963; http://www.tigr.org/tdb/tgi/).

Figure S87. Sorghum

Sori: ►sorus

Sorrel: A light chestnut fur color of horses, determined by homozygosity for *d* gene; similar brownish color in other mammals. The sorrel plants *Rumex acetosella*, *R. scutatus* are used as tart vegetables. The sorrel tree is *Oxydendron.*►*Rumex*

Sorsby Syndrome: ►night blindness, ►Stargardt disease

Sortase: A bacterial enzyme that anchors surface proteins to the bacterial cell wall. These surface proteins promote interaction between the pathogen and the animal cell. Also, these proteins mediate escape from the immune defense system of the animal cell. (See Muzmanian SK et al 2001 Mol Microbiol 40:1049).

Sortins: Proteins that inhibit the vacuolar sorting of small molecules. (See Zouuhar J et al 2004 Proc Natl Acad Sci USA 101:9497).

Sorting: The mechanism that ensures that molecules (isozymes) are directed into the appropriate cellular compartments (cytosol, nucleus, mitochondria, chloroplasts). The sorted proteins are generally equipped with special NH_2-end signals, generated by differential transcription, and/or translation. This process is assisted by trans-acting transmembrane proteins and glycosylphosphatidylinositol-linked proteins. ►transit peptide, ►chloroplasts, ►mitochondria

Sorting Out: In sorting-out, genetically different organelles (plastids, mitochondria) segregate into homogeneous groups of cells or during embryogenesis cells of common origin reaggregate in order to form certain cell types and/or structures (see Fig. S88).

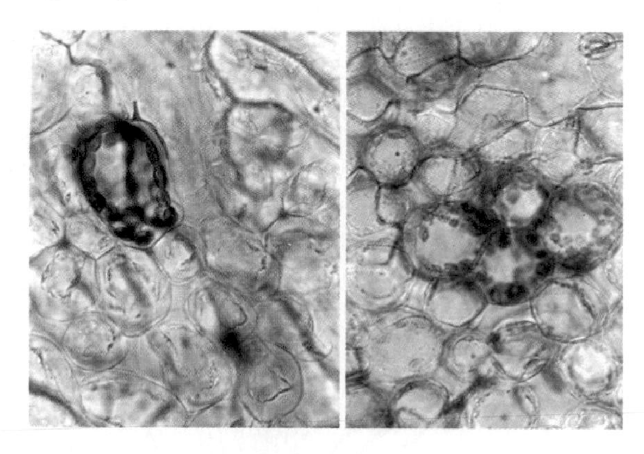

Figure S88. When variegation is caused by a mutation in a nuclear gene all the plastids within a cell are either green or colorless (left). Mutation in the chloroplast genetic material display sorting out at the boundary of sectors, i.e., cells with both green and colorless plastids occur (mixed cells) and in a non-stochastic process cells with all colorless and all green plastids appear in later sectors (right).

The segregation of two different types, A and B, of mitochondria in a heteroplasmic cell line may be characterized by the formula of Solignac, M. et al (Mol Gen Genet 197:183): $V_n = \rho_0 (1 - \rho_0)(1 - [1 - 1/N]^n)$ where V_n is the variance of ρ (the fraction of A within a cell) at the nth cell generation and ρ_0 = the fraction of A in the original cell line, N is the number of sorting out units. In *Schizosaccharomyces*, the distribution of the mitochondria is mediated by microtubules.

In humans, the frequency of mutation in mtDNA is 10 times higher than that in the nuclear DNA, yet heteroplasmy is very rare except in some diseases. Despite the fact that the number of mitochondria in mammalian cell runs to the thousands per cell, usually the replication switches to one type. The number of the founder mtDNAs has been estimated within wide ranges of 1–6 and 20–200 in cattle, and in *Drosophila* 370–740. These founders then undergo a restriction/amplification type of replication, i.e., they pass through a bottleneck and therefore heteroplasmy is very limited. ►ctDNA, ►plasmone mutation, ►plastid number, ►mtDNA, ►mitochondrial genetics, ►heteroplasmy, ►Romanovs; for mitochondrial sorting, see also Kowald A, Kirkwood TBL 1993 Mutat Res 295:93.

Sorus (plural sori): A group of sporangia, such as found on lower surface of fern leaves.

SOS (son of sevenless, human homolog at 2p22-p21): A *Drosophila* gene that functions downstream from *sevenless*, encoding a receptor tyrosine kinase (RTK) in the light signal transduction pathway. SOS is a guanine nucleotide releasing protein (GNRP/GNRF), it is also called guanine-nucleotide exchange factor, GEF, and it interacts with RTK through the protein Drk receptor kinase, a homolog of the vertebrate Grb2, and SEM-5 in *Caenorhabditis*. These proteins function in a variety of signal transduction pathways involving EGF. SOS is frequently also called a mediator protein. ►BOSS, ►DRK, ►Grb2, ►receptor tyrosine kinase [RTK], ►rhodopsin, ►signal transduction, ►GNRP, ►daughter of sevenless, ►Noonan syndrome; Hall BE et al 2001 J Biol Chem 276:27629; Sondermann H et al 2004 Cell 119:393; crystal structure–function: Freedman TS et al 2006 Proc Natl Acad Sci USA 103:16692.

SOS Recruitment System (SRS): A tool in proteomics. A temperature-sensitive cdc25–2 allele of yeast permits grows at 25 °C, but not at 36°. Normally this protein, when localized to the plasma membrane, facilitates Ras guanyl nucleotide exchange and via signal transduction events promotes cell growth. The human homolog (hSOS) is complementary for the mutant and secures growth at the otherwise non-permissive regime. The hSOS function requires protein–protein interaction that is secured by fusing a bait to the C-end of the truncated protein. The co-expressed bait and prey are targeted to the membrane. The prey is either an integral membrane protein or a soluble protein, which after myristoylation signaling attaches to the membrane, and then growth is restored. ►two-hybrid system, ►CDC25, ►proteomics; Auerbach D et al 2002 Proteomics 2:611.

SOS Repair: An error-prone repair. ►DNA repair

Sotos Syndrome: ►cerebral gigantism

Source Habitat: In a source habitat, some individuals contribute more to the future generations than do the average individuals. ►sink habitat, ►habitat

Southern Blotting: In Southern blotting, DNA fragments cut by restriction endonucleases are separated on agarose gel by electrophoresis, then transferred to membrane filters by blotting, to hybridize the pieces with radioactively (or fluorescent) labeled DNA or RNA and then identify the physical sites of restriction fragments and genes. The transfer to membranes may be achieved by capillary action of wicks and DNA is sucked through layers of filterpapers on top or by vacuum-driven devices (see Fig. S89). ►restriction enzyme, ►RFLP, ►autoradiography, ►biotinylation, ►fluorochromes, ►nucleic acid hybridization, ►membrane filters

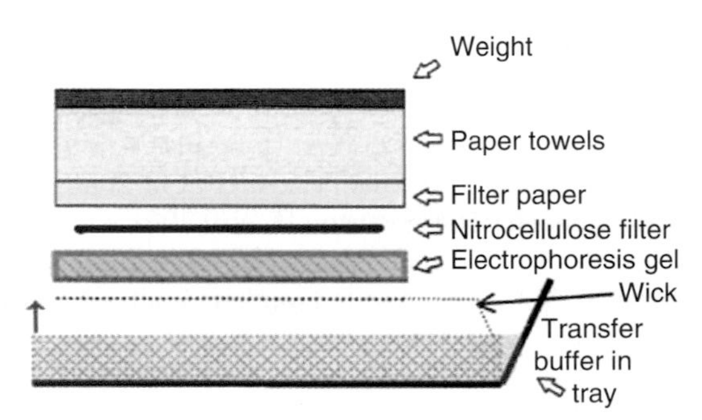

Autoradiography reveals the probed spots on the filter on the photographic film

Figure S89. Southern blot

Southern Hybridization: ►Southern blotting

South-Western Method: The simultaneous labeling of cDNA and binding proteins (transcription factors). The screening is carried out by hybridizing labeled probes of DNA to bind to polypeptides immobilized on nitrocellulose filters. ►Southern blot, ►Western blot

Soviet Genetics: The term Soviet Genetics is not meant to define a special genetics because science does not have ideological, political, or ethnic attributes and it transcends all boundaries. A sad exception is "soviet genetics," a misnomer because it did not involve genetics at all and it collapsed before the implosion of the political system that nurtured and enforced it. Genetics in the Soviet Union had a very remarkable and successful beginning. In 1944, L.C. Dunn, Professor of Zoology at Columbia University noted: "There are today literally hundreds of trained genetical investigators in the U.S.S.R., certainly more than in any other country outside the U.S.A." (Science 99:2563). This outstanding research and teaching establishment was destroyed, however, in 1948 and geneticists suffered humiliation, persecution, and almost total physical annihilation for a period of over 20 years by lysenkoism. ►lysenkoism, ►Mitchurin

SOX (SRY type HMG box): Mammalian genes (~30) encoding proteins with over 60% similarity to the HMG box of SRY. The SOX9 gene is critical for the differentiation of the Sertoli cells and chondrocytes. Sox genes apparently encode transcription factors, are capable of transactivation of genes involved in gonadal differentiation, are involved in cartilage formation, and induce testis development in chromosomally XX mice (Vidal VPI et al 2001 Nature Genet 28:216). The bacterial *Sox* genes are involved in superoxide responses. The mammalian SOXs bind to an 5′-(A/T)(A/T)CAA(A/T)G site in the DNA and regulate the expression of the LINE-1 retrotransposons in human cells. The different SOX genes use different short DNA sequences for binding specific transcription factors. SOX3 transcription factor defects may involve X-linked mental retardation and growth hormone deficiency (Laumonnier F et al 2002 Am J Hum Genet 71:1450). Several human syndromes have mutation in Sox genes. ►SRY, ►HMG, ►campomelic dysplasia, ►Wolffian duct, ►Shah-Waardenburg syndrome, ►Hirschsprung disease, ►LINE, ►anti-Müllerian hormone, ►Sertoli cells, ►lymphedema, ►stem cells; Kamachi Y et al 2000 Trends Genet 16:182; Takash W et al 2001 Nucleic Acids Res 29:4274; Wilson M, Koopman P 2002 Curr Op Genet Dev 12:441.

Soybean: *Glycine max* (2n = 40, ~1100 Mb); the basic chromosome number may be x = 10. Some related species are tetraploid. This crop is of great economic significance because of the high oil (20–23%) and high protein content (39–40%) of the seed. ►legumes; polyploidy in the genome: Walling JG et al 2006 Genetics 172:1893; for gene index: http://www.soybase.org; http://soybeangenome.siu.edu.

sp: sp as a prefix indicates *Schizosaccharomyces pombe* (fission yeast) DNA, RNA or protein.

Sp1: A mammalian protein, a general transcription factor for many genes recognizing the DNA sequence: $\frac{GGGCGG}{CCCGCC}$. Sp1 elements also protect CpG islands of housekeeping genes from methylation. Sp1 requires the general transcription factor TFIID and the complex CRSP. Sp3 and Sp4 are very similar; the latter is expressed in the brain. ►transcription factors inducible, ►transcription factors, ►CRSP, ►DNA methylation, ►LCR, ►MDM2; Nicolás M et al 2001 J Biol Chem 276:22126.

Space Flight: The genetic effects of space flight are not entirely clear. The microgravity in cooperation with space radiation may have adverse stress consequences. (See White RJ, Averner M 2001 Nature [Lond] 409:1115).

Space Search: An experimental study of large-scale protein matrix for detection of interactions. ►ORFEeome

Spaced Dyads: Many regulatory sites include a pair of conserved trinucleotides, which are spaced by a non-conserved tract of fixed length. Genes with function(s) coordinated by a common regulatory element are expected to share upstream binding sequences. These elements contain fixed sites for a linker domain and the dimerization domain of transcription factors. Their analysis facilitates the identification of co-regulated genes. (See van Helden J et al 2000 Nucleic Acids Res 28:1808).

Spacer DNA: Non-transcribed nucleotide sequences between genes (IGS) in a cell. In the rDNA, there are spacers within (internal transcribed spacers, ITS) and between gene clusters (external transcribed spacers, ETS). These were thought of earlier as not-transcribed tracts but actually they represented very short transcripts. Animal mtDNA genes generally have very short (few bp) spacers. In fungi, mtDNA spacers are common and variable in length because of recombination and slippage during replication; they may also be mobile. Plastid genes of higher plants are organized into operons without interruptions. Insertions, deletions, and unequal recombinations determine the length of the spacers. ►ribosomal RNA, ►rrn; ITS2 database: http://its2.bioapps.biozentrum.uni-wuerzburg.de/.

S

Spanandric Males: Spanandric males occur rarely within hermaphroditic species and are sterile.

Spasmoneme: A rod-like thin bundle of filaments (2 nm each), forming a cytoplasmic organelle of 2–3 mm in extended form. When exposed to calcium, it contracts and serves as an engine for the movement of different structures within the cell of celiated protozoa. ►centrin; Maciejewski JJ et al 1999 J Eukaryot Microbiol 45(2):163.

Spasticity: Muscle tone; its increase causes lack of coordination in movement as it occurs in various diseases affecting the brain, such as stroke, spinal cord injuries, etc.

Spastic Paraplegia (Strumpell disease, SPG): A collection of paralytic diseases encoded by at least 14 loci under a variety of genetic control, recessive, dominant autosomal, and Xq28-linked. Genes responsible for the disease have been mapped to 14q (SPG3), 2p21-p22 (SPG4), 15q (SPG6), 12q13, and 8q (SPG8). Currently, 20 genes have been mapped and eight more have been identified. The common denominator of these various genes appears to be aberrant cellular transport (Crosbyl AH, Proukakis C 2002 Am J Hum Genet 71:1009). The most prevalent form (40–50%) is SPG4, which encodes an AAA protein, spastin, in the microtubules. Atlastin, a Golgi-localized GTPase protein interacts with spastin in axon maintenance (Evans K et al 2006 Proc Natl Acad Sci USA 103:10666). A mitochondrial ATP-dependent protease (m-AAA) defect interferes with ribosome assembly in the mitochondria and with the maturation of the L32 mitochondrial protein (Nolden M et al 2005 Cell 123:277). SPG13 is formed due to a mutation in mitochondrial Hsp60. ►heatshock proteins, ►Pelizaeus-Merzbacher disease, ►Silver syndrome, ►AAA proteins, ►ribosomal proteins, ►mitochondrial diseases in humans; Vazza G et al 2000 Am J Hum Genet 67:504; Svenson IK et al 2001 Am J Hum Genet 68:1077; mutations in the receptor-enhancing protein (2p22) in the mitochondria: Zuchner S et al 2006 Am J Hum Genet 79:365.

Spaying: The neutering of a female. ►castration

SPB: ►spindle pole body

Spearman Rank-Correlation Test (SPC): SPC determines the relations between two variables without elaborate calculations (Table S3).

The SR correlation coefficient, $r = 1 - \frac{6\,Sum\,d^2}{n^3 - n}$, for example, $1 - \frac{6 \times 14}{7^3 - 7} = 0.75$

In case of ties, correction can be used, $T = \frac{m^3 - m}{12}$ where m stands for the number of measurements or classifications of identical values. If, for example, there are three measurements of 5, two measurements of 7, and two amounting to 8, the correction for ties among the X trait variables becomes: $\text{Sum Tx} = \frac{3^3 - 3}{12} + \frac{2^3 - 2}{12} + \frac{2^3 - 2}{12}$ and in a similar way, the correction for ties can be determined in the Y series. Now we can obtain the terms: $\text{Sum X}^2 = \frac{n^3 - n}{12} - \text{SumTx}$, and $\text{Sum Y}^2 = \frac{n^3 - n}{12} - \text{SumTy}$, and $r = \frac{\text{SumX}^2 + \text{SumY}^2 - \text{Sumd}^2}{2\sqrt{\text{SumX}^2\,\text{SumY}^2}}$

It is a non-parametric test and can also be used for the comparison of traits that cannot be measured but can be classified subjectively (on the basis of their appearance). If the traits are quantified by measures, they are ranked according to their relative magnitude,

Table S3. Example of hypothetical data for the calculation of the Spearman Rank Correlation test

Pairs	Trait X	Rank	Trait Y	Rank	Difference of Ranks (d)	d^2
1	8	3	14	2	1	1
2	10	5	20	7	−2	4
3	12	7	17	5	2	4
4	9	4	15	3	1	1
5	6	1	13	1	0	0
6	11	6	18	6	0	0
7	7	2	16	4	−2	4
					Sum d = 0	Sum d^2 = 14

i.e., by assigning the highest rank to the largest. If an exact measurement is impractical (e.g., degree of susceptibility), they are simply ranked. In case of ties, an equal rank is assigned to both. The differences of the rank scores are squared and summed. Tables constructed for various degrees of freedom can be used to determine the probabilities. ►correlation, ►non-parametric tests, ►statistics

Specialized Transduction: In specialized transduction, the transducer temperate phage picks up a piece of the host DNA at the immediate vicinity of its established prophage site (and generally leaves behind a comparable length of its own). When this modified phage infects another bacterium, it may integrate into the host genome the gene that it picked up from the previous host. Lambda phage (λdgal; meaning a defective lambda that carries *gal*) is a typical specialized transducer with preferred site near 17 min on the map next to the *gal* locus, and it is a specialized transducer of this gene. Another well-known specialized transducer phage is ϕ80trp that can carry the tryptophan operon (at about 27 and 1/2 min). The specialized transducer phage must be a temperate phages, which transduces (only) the special gene located at its integration site. Gene transfer in eukaryotes by transposable elements, the genetic vectors, is a process similar to prokaryotic transduction. ►transposable elements, ►transformation genetic, flowchart of specialized transduction on Figure S90, ►high-frequency lysate, ►double lysogenic, ►helper virus, ►bacterial recombination; Morse ML 1954 Genetics 39:984.

Speciation: The process by which a new species diverges from an ancestral species. Evolutionists may distinguish conventional methods of speciation when geographic isolation and accumulated mutations cause eventually reproductive isolation. According to the *quantum model* of speciation, the divergence begins with spatial isolation followed by the survival of a few new types of individuals that give rise to reproductively isolated new forms as mutations accumulate. *Saltational* speciation comes about by sudden major mutations. *Parapatric* (or stasipatric) speciation occurs without geographic separation and it is initiated from a relatively small number of individuals that produce divergence under continued natural selection. *Sympatric* speciation occurs within the original area of dispersal, due to the emergence of a genetic isolation mechanism or sexual selection. In *homoploid hybrid speciation*, two ancestral taxa give rise to a third by hybridization without a change of chromosome number, and it may occur in the parasitic species of animals (Schwarz D et al 2005 Nature [Lond] 436:546). Reproductive isolation required for speciation may involve hybrid sterility, hybrid inviability, or special mate preference (Ortíz-Barrientos D, Noor MAF 2005 Science 310:1467). Strong assortative mating among the hybrids, favored by the wing pattern of *Heliconius* butterflies, favors hybrid speciation (Mavárez J et al 2006 Nature [Lond] 441:868). The mechanism of speciation in plants and animals differs because in animals, in contrast to plants, behavioral traits may also lead to speciation. In plants, sterility may not necessarily hinder propagation because some species reproduce primarily by vegetative means and they can practice both autogamy and cross-pollination. Chromosomal rearrangements and the action of transposable elements may lead to hybrid sterility, sexual isolation, and eventually speciation. ►evolution, ►sibling species, ►phylogeny, ►co-speciation, ►theory of evolution, ►saltation, ►neo-Darwinism, ►neo-Lamarckism, ►species, ►fertility, ►reproductive isolation, ►Müllerian mimicry, ►hybrid zone,

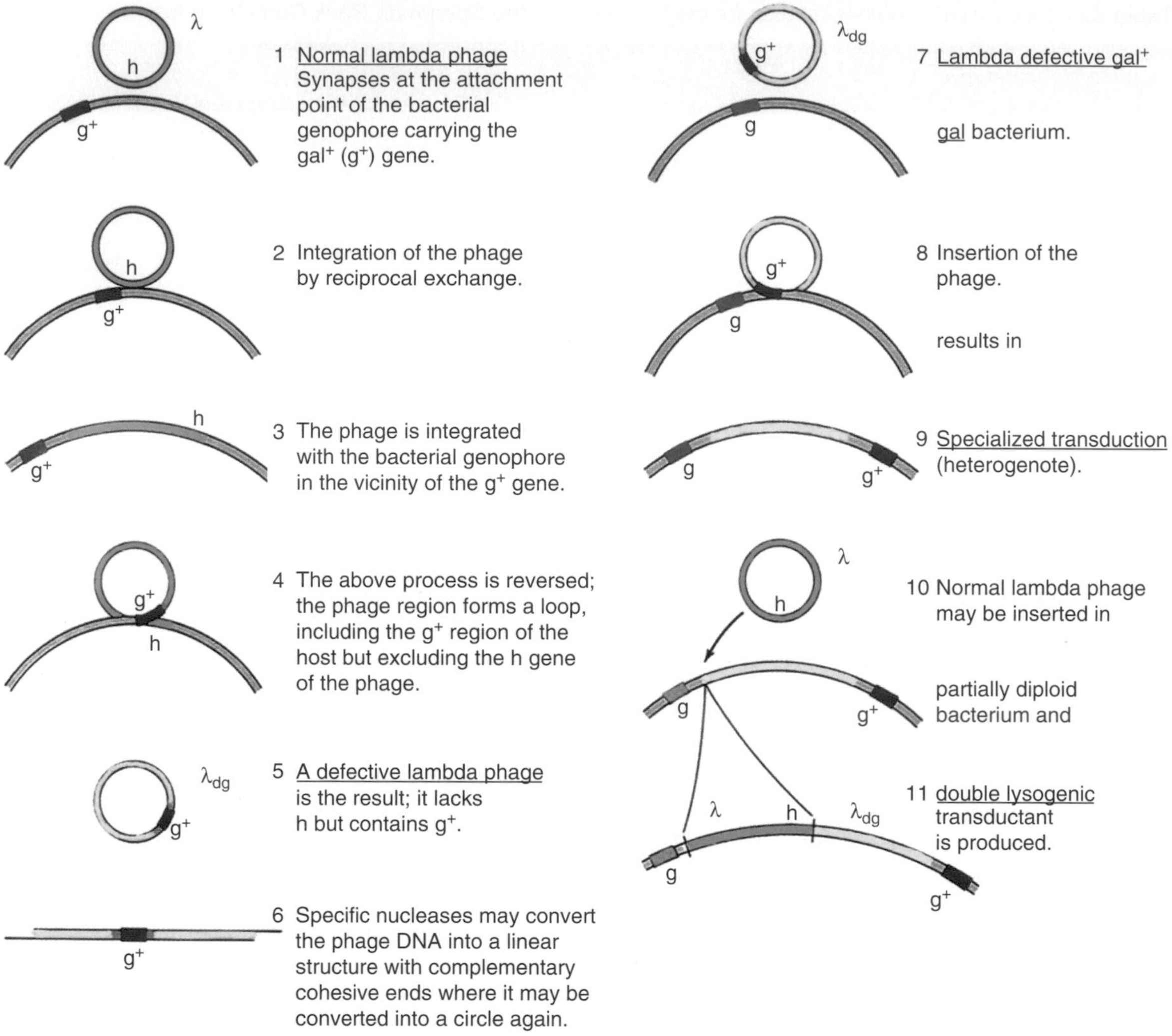

1 Normal lambda phage Synapses at the attachment point of the bacterial genophore carrying the gal+ (g+) gene.

2 Integration of the phage by reciprocal exchange.

3 The phage is integrated with the bacterial genophore in the vicinity of the g+ gene.

4 The above process is reversed; the phage region forms a loop, including the g+ region of the host but excluding the h gene of the phage.

5 A defective lambda phage is the result; it lacks h but contains g+.

6 Specific nucleases may convert the phage DNA into a linear structure with complementary cohesive ends where it may be converted into a circle again.

7 Lambda defective gal+ gal bacterium.

8 Insertion of the phage. results in

9 Specialized transduction (heterogenote).

10 Normal lambda phage may be inserted in partially diploid bacterium and

11 double lysogenic transductant is produced.

Figure S90. Specialized transduction (Modified after Stent GS 1971 Molecular Genetics. WH Freeman, San Francisco)

►introgressive hybridization; Noor MAF, Feder JL 2006 Nature Rev Genet 7:851; plant speciation review: Rieseberg LH, Willis JH 2007 Science 317:910; comprehensive compendium of species: http://species.wikipedia.org.

Species: A potentially interbreeding population, which shows reproductive isolation from other species (see discussion of the definition on p 371 and ff in Dobzhansky T 1949 Genetics and the Origin of Species. Columbia University Press, New York). Horizonmtal transfer of genes among bacteria mediated primarily by phages or direct DNA uptake in natural environment challenge the classical concept of species (Goldenfield N, Woese C 2007 Nature [Lond] 445:369). Also, recombination in mixed populations may be a force of speciation (Fraser C et al 2007 Science 315:476). The transgenic technology somewhat confuses the classical definition because of the possibility of exchanging genes among totally different taxonomic categories. The number of eukaryotic species is not known but about 10^6 have been described. Expectations are that between 3 to 10 fold more existing have not been discovered yet. By shotgun genome sequencing of mass collection of seawater microbes in the Sargasso Sea, 148 new bacterial phylotypes were discovered representing >1.2 million so far unknown genes (Venter JJ et al 2004 Science 304:66). ►speciation, ►sibling species, ►phylotype, ►shotgun sequencing, ►fertility, ►sexual isolation, ►species extant, ►transgene, ►OTU, ►endangered species act, ►horizontal transmission; identification by genome profile: Watanabe T et al 2002 Genome Biol 3(2):res.0010; description of ~270,000 species of all types of organisms: http://www.discoverlife.org/.

Species Extant: With the increase of agricultural, industrial, and changing land use the currently alive (A), % extinct (E), and % threatened (T) species indicate an alarming reduction of biodiversity: molluscs (A) 10^5, (E) 0.2, (T) 0.4; crustaceans: (A) $4–10^3$, (E) 0.01, (T) 3; insects(A) 10^6, (E) 0.005, (T) 0.07; vertebrates total (A) 4.7×10^4, (E) 0.5, (T) 5; mammals (A) 4.5×10^3, (E)1, (T) 11; gymnosperms (A) 758, (E) 0.3, (T) 32; dicots (A) 1.9×10^5; (E) 0.2, (T) 9; monocots (A) 5.2×10^4, (E) 0.2, (T) 9. (Data from Smith FDM et al 1993 Nature [Lond] 364:494). Within these larger categories, some species are much more endangered that the figures indicate. These numbers are not exactly valid because new species are being discovered. Between 1980–90 more than 100 new mammals have been identified. The extinction percentages date from the year 1600 A. D. According to Dobson, A.P. et al 1997 (Science 275:550) in the United States of America, the number of endangered species is 503 plants, 84 molluscs, 57 arthropods, 107 fish, 43 herptiles, 72 birds, and 58 mammals. Important factors affecting the survival of the feral species is the annual increase of the human populations (1.7%), urbanization, industrial and livestock production, pollution, and natural disasters.

According to the World Conservation Union 1996 biennial report (http://www.iucn.org/themes/ssc/index.html), currently 5205 species are endangered and risk extinction. Actually, the number of species is still unknown and the latest estimates vary from 3.5 to 10.5 million, about a third lower than some earlier estimates. The number of species in an environment has been extrapolated on the basis of the diversity of the 16S RNA of microorganisms. A new complex type of statistical approach is expected to provide better estimates than the most commonly used extrapolation from the lognormal distributions (Hong S-H et al 2006 Proc Natl Acad Sci USA 103:117). Around the world, 595 sites are in danger of imminent extinction. Of these, 203 are within partially protected and 87 in declared protected areas; 27 sites are known to be unprotected, and 48 sites have unknown protection status. In 2005, 794 species were in imminent danger, three times as many as have gone extinct since 1500 (Ricketts TH et al 2005 Proc Natl Acad Sci USA 102:18497). ►species, ►lognormal distribution, ►conservation genetics; 2000 Nature [Lond] 405:207 ff; 2001 Proc Natl Acad Sci USA 98:5389 ff; Alroy J 2002 Proc Natl Acad Sci USA 99:3706; Pitman NCA, Jørgensen PM 2002 Science 298:989; Convention on International Trade of Endangered Species (CITES) 1973 Public law 93–205.

Species, Synthetic: Synthetic species are produced in the laboratory by crossing putative ancestors of amphidiploid forms, which are partially isolated genetically. Doubling the chromosome number of the hybrids prevents the sterility of the offspring. The synthetic species permit a verification of the putative evolutionary path of existing polyploid species by studying the pairing behavior of chromosomes, the frequency of chiasmata, the number of univalents and multivalents formed, etc. These classical cytogenetic methods may be supplemented with nucleic acid hybridization, in situ hybridization, study of proteins, nucleic acid sequences, etc. Some of the hybrid species if they ever occurred in nature, were not maintained by natural selection (see Fig. S91). ►allopolyploids, ►*Triticale*, ►*Raphanobrassica*, ►*Mycoplasma*, ►synthetic virus, ►speciation, ►evolution

Figure S91. Middle: *Triticale* (2n = 56), Left: Wheat (2n = 42), Right: Rye (2n = 14). (Courtesy of Dr. Árpád Kiss)

Specific Activity: The number of molecules of a substrate (μmol) acted on by enzymes (mg protein) per time (minutes) at standard temperature (25 °C). Also, it means the relative amount of radioactive molecules in a chemical preparation.

Specific Combining Ability: ►combining ability

Specific Heat: Joules or calories required to raise the temperature of 1 g substance by 1 °C.

Specific Locus Mutations Assays (SLT): SLTs have been used to detect X-chromosomal mutations in hemizygous males (mouse), in autosomal heterozygotes for special fur color genes such as the Oak Ridge stock heterozygous for *a, b, p,* c^{ch}, *se, d, s* or the Harwell tester stock (HT) heterozygous for six loci (*a, bp, fz, ln, pa, pe*), or heterozygotes for thymidine kinase or hypoxanthine-guanine phosphoribosyltransferase genes in different mammalian cell cultures (mouse, Chinese hamster ovary cells). Mutation of the wild type allele is then immediately revealed in the first generation. Similar procedures are applicable also in plants and any other diploid systems. The animal assays are not as simple as the microbial assays, e.g., the Ames test, but they are considered to be more relevant to human studies. ►bioassays in genetic toxicology, ►mutations in human populations, ►doubling dose; Cattanach BM 1971, p 535. In: Chemical Mutagens. Hollaender A (Ed.) Plenum, New York.

Specific Rotation (α): The degrees of the plane of polarized light rotated by an optically active compound of specific concentration at 25°C: $(\alpha) = \frac{rv}{nl}$, where the r = rotation in degrees, v = volume in cubic centimeter of the solution, l = length in centimeters of the light path.

Specificity: A measure of discrimination among compounds. In DNA sequencing, it is the correctly predicted bases divided by the sum of the correctly and incorrectly predicted bases.

Specificity of Mutagen Assays: The percentage of correct identifications by presumably "non-carcinogenic" ("non-mutagenic") agents. ►sensitivity, ►accuracy, ►predictivity, ►bioassays in genetic toxicology

Speckles, Intranuclear (IGCs): IGCs are storage areas for primary RNA transcript processing factors; the speckles may not be the compartments of the eukaryotic splicing, which takes place rather in association with transcription. ►splicing, ►introns, ►paraspeckles; Eilbracht J, Schmidt-Zachmann MS 2001 Proc Natl Acad Sci USA 98:3849.

Spect: ►tomography

Spectinomycin: Spectinomycin inhibits the translocation of charged tRNA from the A site to the P site of the prokaryotic ribosome by interfering with elongation factor G (see Fig. S92).

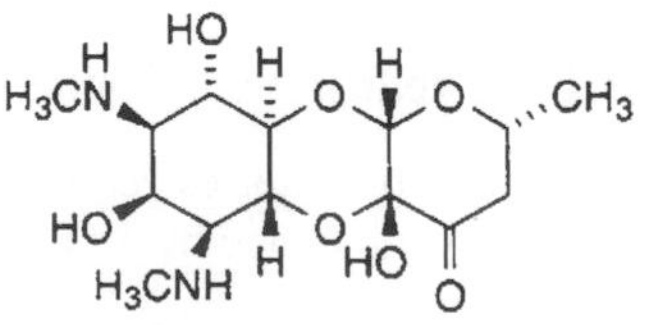

Figure S92. Spectinomycin

Spectral Genotyping: Spectral genotyping uses molecular beacons labeled with different fluorophores. When the probes for the wild type (green) and for the mutant with one base substitution (red) are fluorophore labeled, the homozygous wild type (green) can be distinguished from the heterozygote's DNA (displaying both green and red) and the homozygous mutant showing only the other color (red). The potential color of the fluorophores can be chosen arbitrarily. This procedure can distinguish genotypes in clinical diagnostic analyses without DNA sequencing. Electrochemical examination of conformational alterations is also detectable at very low concentrations (10 pM). ►beacons molecular, ►fluorophore; Kostrikis LG et al 1998 Science 279:1228; Fan C et al 2003 Proc Natl Acad Sci USA 1009134.

Spectral Karyotyping: In spectral karyotyping, the chromosomes or chromosomal segments are hybridized with probes of fluorescent dyes in a combinatorial manner. Five dyes (Cy2, Spectrum Green, Cy3, Texas Red, and Cy5) provide enough combinatorial possibilities to "paint" each chromosome with a different color or shade. Although the naked human eye cannot distinguish these different hues, by the use of optical filters, Sagnac interferometer, a CCD camera, and Fourier transformation, it was possible to discriminate the special differences in standard classification colors using a computer program. The approach permitted classification of translocations that were not identifiable by other staining techniques. The lowest limit of differentiation by this technology is 500 to 1500 kbp. The procedure is applicable to clinical laboratory testing and evolutionary analyses. (See also Schröck E et al 1996 Science 273:494; ►chromosome painting, ►FISH, ►GISH; http://www-ermm.cbcu.cam.ac.uk/0000199Xh.htm).

Spectrins: Filamentous tetrameric proteins (220–240 kDa) present in the red blood cells; they may constitute 30% of the membranes. Spectrins may mediate muscle and neuron organization by facilitating the binding of other proteins. They are also involved in the formation of a network on the cytoplasmic surface in cooperation with actin, dynein, ankyrin, and *band III* protein. The βII-spectrin is the same as fodrin. Several human chromosomes encode spectrins. Spectrin defects may cause auditory and motor neuropathies. ►ankryn, ►cytoskeleton, ►elliptocytosis, ►spherocytosis, ►poikilocytosis, ►dynein, ►fodrin, ►glycophorin, ►spinocerebellar ataxia; Parkinson NJ et al 2001 Nature Genet 29:61.

Spectrophotometry: Spectrophotometry estimates the quality and quantity of a substance in solution on the basis of absorption of monochromatic light passing through it.

Spectrosome: A cytoplasmic organelle anchoring the mitotic spindle, and thus defining the spatial direction of cell division. It contains spectrins, cyclin A and other regulatory proteins.

Speech and Grammar Disorder (SPCH1): A human FOXP2 ([forkhead box P2]7q31) dominant gene (locus is more than 600 kb) is involved in the lack of coordination of face and mouth muscles (verbal dyspraxia), some cognitive impairment (low IQ), and articulation and expressive language. Heterozygosity for nonsense mutation leads to the truncated protein product affecting neurodevelopment (MacDermot KD et al 2005 Am J Hum Genet 76:1074). ►language impairment specific, ►stuttering, ►human intelligence, ►dyspraxia, ►dyslexia, ►autism, ►aphasia, ►MASA syndrome; Lai CSL et al 2000 Am J Hum Genet 67:357; Enard W et al 2002 Nature [Lond] 418:869.

Spemann's Organizer: The embryonic tissue site of signals that mediate the organization of the body. Its signaling center is at the blastopore of the gastrula and releases a variety of polypeptides controlling neural and dorsal or ventral mesodermal differentiation. ►morphogenesis, ►organizer, ►gastrula, ►blastopore; Niehrs C 2001 EMBO J 20:631; De Robertis EM et al 2000 Nature Rev Genet 1:171.

Sperm: Sperm refers to the animal seminal fluid; geneticists generally understand it as the spermatozoon or plant male generative cells. In the first meaning, the word has no plural, in the other sense both singular and plural are justified. The *Drosophila* spermatozoon far exceeds in length that of any other animal's (up to 58 mm in *D. bifurcata*, although that of *D. melanogaster* is about 1.91 mm; the human spermatozoon is about 0.0045 mm (see Fig. S93), Karr TL, Pitnick S 1996 Nature [Lond] 379:405). In a single normal human ejaculate, the number of spermatozoa is within the range of 20–40 million in the seminal fluid volume of >2 mL. The human spermatozoon count shows a decreasing tendency during the last decades of life. The cause of this appears to be the increase of environmental pollutants with hormone-like effects, such as PCB, (polychlorinated biphenyl) an industrial carcinogen and pesticide, dioxin (a solvent), phthalates (may be present in cosmetics), and bisphenols (used in manufacturing resins and fungicides).

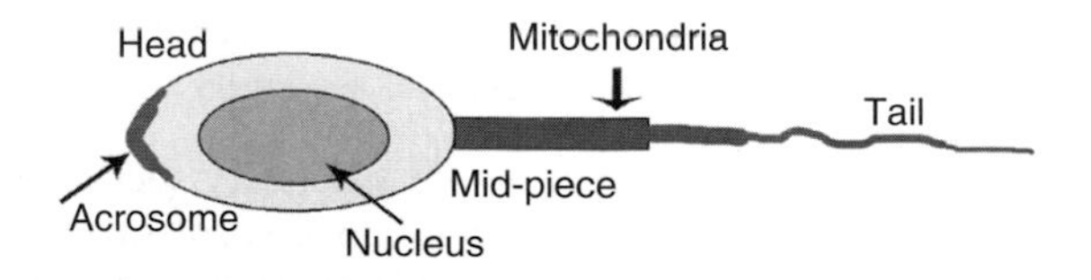

Figure S93. Outline of a human spermatozoon

During fertilization, the animal spermatozoon is attracted to the egg by chemotactic peptides, causing changes in the voltage of the membrane and altering the concentration of cAMP, cGMP, Ca^{2+}, and the activity of K^+ ion channels. In the tail of the spermatozoa, tetrameric proteins (CatSper1) form six transmembrane calcium ion channels, which are hyperactivated during ejaculation to assure appropriate motility and fertilization (Kirichok Y et al 2006 Nature [Lond] 439:73). A sperm chemoattractant protein, a member of the cysteine-rich secretory protein (CRISP) family, allurin has been isolated (Olson JH et al 2001 Proc Natl Acad Sci USA 98:11205). The sperm contains many different RNAs and proteins (Miller D et al 2005 Trends Mol Med 11(4):156), and they have roles in fertilized oocytes. Contrary to earlier views, mammalian spermatozoa can translate proteins but use the mitochondrial ribosomes for the process. These proteins have roles, while the sperm travels through the female reproductive canal and are essential for motility and fertilization (Gur Y, Breitbart M 2006 Genes Dev 20:411). Cytoskeletal proteins of the fertilizing sperm trigger oocyte maturation and ovulation of animals. The number of pollen grains (containing two sperms) released by a plant may vary between a few hundreds to tens of millions. The male poultry reduces the sperm when mating with promiscuous females but allocate more of it in matings with new and attractive females (Pizzari T et al 2003 Nature [Lond] 426:70). ►oocyte primary, ►environmental mutagens, ►fertilization, ►polyspermic fertilization, ►gametogenesis, ►RPTK, ►DDT, ►semen, ►seleno-protein, ►infertility, ►acrosomal process, ►cryopreservation; Wassarman PM et al 2001 Nature Cell Biol 3: E59; Ren D et al 2001 Nature [Lond] 413:603; Swallow JG, Wilkinson GS 2002 Biol Rev 77:153; http://www.med.unipi.it/agp/siti/siti.htm.

Sperm Activation: The conversion of the spermatids into moving spermatozoa. In *Caenorhabditis,* four gene loci (*spe-8*, *-12*, *-27*, and *spe-29*) control the process.

Sperm Bank: A (human) sperm depository for the purpose of artificial insemination in case of sterility of the husband or other conditions, which may warrant their use. Thousands of sperm banks (gene banks) exist in the world, most of the deposited samples are anonymous and the genetic constitution of the donors is not known completely. Since the 1930s, HJ Muller advocated the use of the sperm banks for positive eugenics purposes. Accordingly, the donors should be selected on the basis of superior talents, mental ability, and physical constitution. This idea has not been widely accepted, however, because of moral objections and biological shortcomings. Most of the

S

"superior phenotypes" cannot be evaluated by generally accepted criteria, and the phenotype may not fully represent the heritability of particular traits. Since artificial insemination of domestic animals became routine, sperm banks are exploited for animal breeding programs in order to produce maximal number of progeny of high-performance males. Even for animals, this technology should be used with thorough consideration of population genetics principles in order to avoid narrowing the gene pool and inbreeding. ►in vitro fertilization, ►ART, ►bioethics; Critser JK 1998 Hum Reprod (Suppl. 2):55; Deech R 1998 Hum Reprod 13(Suppl. 2):80.

Sperm Competition: In animals where the females may mate repeatedly during the period of receptivity, the spermatozoa of different genotypes may have selective advantage or disadvantage in fertilization. The success of a particular type of spermatozoa is influenced by a large number of seminal proteins. The accessory gland proteins appear to be more polymorphic than other proteins, yet this may not distinguish between the interpretations of rapid evolution or limited selection. ►multipaternate litter, ►sperm displacement, ►certation, ►accessory gland; Fu P et al 2001 Proc R Soc Lond B Biol Sci 268:1105; Simmons LW 2001 Sperm Competition and its Evolutionary Consequences in the Insects. Princeton University Press.

Sperm Displacement: In animals where the females practice sperm storage, depending on the genetic constitution of the sperm, one or another may interfere with the fertilization of a competing sperm already present in the spermatheca of the female. In this process, the genetic constitution of the female has also a selective role. ►sperm storage, ►sperm competition, ►sperm precedence, ►spermatheca; Gilchrist AS, Partridge L 2000 Evolution Int J Org Evolution 54:534.

S

Sperm Morphology Assays in Genetic Toxicology: Sperm morphology assays are based on the expectation that mutagens and carcinogens may interfere with normal spermiogenesis, resulting in abnormal head shape, motility, and viability of the treated or exposed sperm. This expectation is met with some agents but not with all. Thus, sperm alterations may indicate mutagenic and/or carcinogenic properties but not all mutagens/ carcinogens seem to affect these sperm parameters within non-lethal doses. ►bioassays in genetic toxicology; Baccetti B et al 2001 Hum Reprod 16:1365.

Sperm Precedence: In multiple copulation of a female with different males within a period of receptivity, one type of sperm (usually the last mating male's) may have selective advantage in producing offspring. ►sperm competition, ►certation, ►last-male sperm precedence; Price CS et al 2000 Evolution Int J Org Evolution 54:2028.

Sperm Receptor: ►fertilization

Sperm Selection: ►sex selection

Sperm Storage: Insects commonly store sperm in the spermatheca for two weeks, and fertilization may follow any time within this period. This is why geneticists tend to use virgin females for controlled matings. Sperm can be stored at the temperature of liquid nitrogen (−195.8 °C) and retain its ability of fertilization. The efficiency of storage depends on seminal fluid proteins that the male transmits along with the sperm. Mouse spematozoa can be stored by freeze-drying and the function is maintained. ►sperm bank, ►sperm competition, ►sperm displacement, ►sperm precedence; Yin HZ, Seibel MM 1999 J Reprod Med 44(2):87; Mortimer D 2000 J Androl 21:357; Corley-Smith GE, Brandhorst BP 1999 Mol Reprod Dev 53:363; Kusakaba H et al 2001 Proc Natl Acad Sci USA 98:13501.

Sperm Typing: Sperm typing permits analysis of recombination in diploids, using the recombinant gametes. The method of analysis requires the analysis of PCR-amplified gamete DNA sequences of known paternal types. This type of analysis can resolve recombination even between single base pairs. The results must be scrupulously studied because the PCR method may have inherent errors. Single sperms can be separated with the aid of fluorescence activated cell sorters. If the sperms are subjected to "primer extension preamplification" (PEP) before analysis, enough material can be obtained to carry out multipoint tests. This procedure is not practical with egg cells. ►recombination frequency, ►crossing over, ►maximum likelihood applied to recombination, ►polymerase chain reaction, ►cell sorter, ►prenatal diagnosis, ►genetic screening; Hubert R et al 1994 Nature Genet 7:420; Shi Q et al 2001 Am J Med Genet 99:34.

Spermatheca: A storage facility of insect females, for sperm to be used at a later fertilization following an initial mating. ►sperm precedence, ►sperm competition, ►sperm storage, ►sperm displacement

Spermatia: Male sperms produced at the tip of hyphae within the spermatogonia of rust fungi; they are comparable to the microconidia. ►conidia

Spermatid: A cell formed by the secondary spermatocyte and which differentiates into a spermatozoon. ►spermatozoon, ►sperm, ►gametogenesis, ►spermiogenesis

Spermatocyte (primary): A diploid cell that by the first division of meiosis gives rise to the haploid secondary spermatocytes. By cell division, these cells form the spermatids, which differentiate into spermatozoa in animals. In plants, the spermatocytes function in a similar manner and thus produce the microspores, which develop into pollen grains and within them the two sperms are formed either before or after the pollen tubes begins to elongate. ►gonads, ►gametogenesis

Spermatocytoma (spermocytoma): The malignancy of the testis or of undifferentiated male gonads. ►seminoma

Spermatogenesis: ►gametogenesis, ►spermiogenesis

Spermatogonia: The primordials of the sperm cells; the secondary spermatogonia, produce the primary spermatocytes. Spermatogonial stem cells can be transformed/transfected by genetic vectors in culture. If introduced into seminiferous tubules of infertile mouse, they undergo spermatogenesis and could produce mutant offspring (Kanatsu-Shinohara M et al 2006 Proc Natl Acad Sci USA 103:8018). ►gametogenesis, ►spermatocyte; in vitro culture: Feng L-Xin et al 2002 Science 297:392; Zhao G-Q, Garbers DL 2002 Dev Cell 2:537.

Spermatophyte: Seed-bearing plant. ►cryptogamic plants

Spermatozoon: The fully developed (differentiated) male germ cell, a sperm. ►sperm, ►spermatid, ►oocyte primary, ►fertilization, see Fig. S94 of a human spermatozoon.

Figure S94. Spermatozoon

Spermidine ($C_7H_{19}N_3$, *N*-[3-aminopropyl]-1,4-butanediamine): A polyamine regulating (+ or −) the binding of proteins to DNA, condensation of DNA, controlling gene expression, etc. Spermidine synthase is encoded by human chromosome 1p36-p22 and 3p14-q21 (see Fig. S95). ►spermine

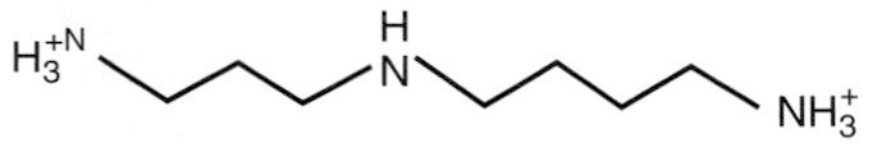

Figure S95. Spermidine

Spermine ($C_{10}H_{26}N_4$, *N,N'*-bis[3-aminopropyl]tetramethylenediamine, a polyamine): The oxidative cleavage product of spermine is spermidine. Spermine synthase was assigned to human chromosome Xp22.1 (see Fig. S96). Its mouse homolog *Gyro* [*Gy*]) is suspected in hypophosphatemia, resulting in rickets, hearing disorders, etc. ►hypophosphatemia, ►hypophosphatasia, ►rickets, ►*S*-adenosylmethionine decarboxylase, ►polyamine

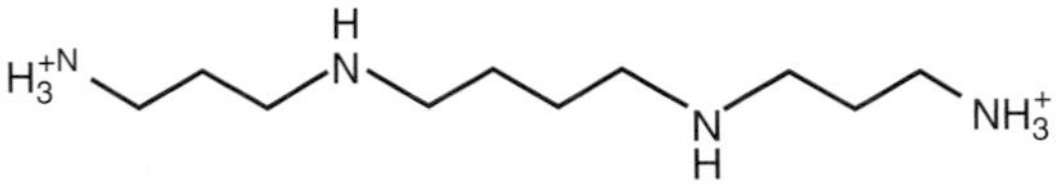

Figure S96. Spermine

Spermiogenesis: A postmeiotic process of differentiation of mature spermatozoa. It is supposed that CREM is involved in the control of genes required for the process because CREM-deficient mouse (obtained by recombination) cannot complete the first step of spermiogenesis and late spermatids are not observed. The defective spermatids are apparently eliminated by apoptosis. Histone 1 (H1) variant H1T2 localizes to the apical chromatin domain at the apical pole and it is required for the reconstruction of chromatin and the replacement of histones by protamines during spermiogenesis (Martianov I et al 2005 Proc Natl Acad Sci USA 102:2808). More than 2000 genes seem to be expressed postmeiotically in the mouse sperm (Schultz N et al 2003 Proc Natl Acad Sci USA 100:12201). ►sperm, ►spermatozoon, ►CREM, ►apoptosis, ►gametogenesis, ►transition protein, ►protamines, ►histones, ►adenovirus; Macho B et al 2002 Science 298:2388.

Spermist: ►preformation

SPF: Cell cycle S phase promoting factor, a kinase. ►cell cycle

SPF Condition: The specific pathogen-free condition of organisms maintained in a quarantine. ►quarantine; Yanabe M et al 2001 Exp Anim 50(4):293.

Spfi-1: An ETS family transcription factor. ►ETS oncogenes

SPH: Protein-binding DNA elements. ►Simian virus 40

Spherocytosis, Hereditary (HS): The most common hemolytic anemia in Northern Europe. The basic defect involves both recessive and dominant mutations affecting ankyrin-1 (8p11.2) and spectrin. The β-spectrin gene was assigned to human chromosome 14q22-q23. The spectrin α-chain is responsible for elliptocytosis II (1q21). An autosomal dominant form was assigned to human chromosome 15q15. There is also an autosomal recessive type. ►ankyrin, ►spectrin, ►elliptocytosis, ►anemia, ►poikilocytosis

Spheroplast: Spherical bacterial cell after (partial) removal of the cell wall. ►protoplast

Sphingolipid Activator Protein (SAP, Saposin): A cofactor for the physiological degradation of sphingolipids. The GM2 activators are encoded in different chromosomes and their deficiencies cause metachromatic leukodystrophy-like and Gaucher disease-like symptoms. ►Gaucher disease, ►metachromatic leukodystrophy; Matsuda J et al 2001 Hum Mol Genet 10:1191.

Sphingolipidoses: Hereditary diseases involving the metabolism of sphingolipids with the enzyme defects indicated in parenthesis. (See Table S4, diseases under separate entries, ►sphingolipids).

Sphingolipids: Sphingosine-containing lipids with the following general structure and depending on the particular substitutions at X.

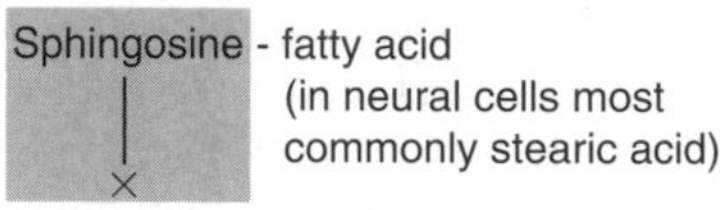

Sphingosine - fatty acid (in neural cells most commonly stearic acid)
|
X

hydrogen (ceramide)
glucose (glucosylcerebroside), [neutral glycolipid]
glucose and galactose (lactosylceramide), [neutral glycolipid]
complex of sialic acid, glucose, galactose, galactose amine (ganglioside G_{M2})
phosphocholine (sphingomyelin).

If sialic acid (acetyl neuraminic acid (see Fig. S97), glycolylneuraminic acid) is removed, asiogangliosides result. Sphingolipids mediate signal transduction, calcium homeostasis and signaling, traffic of secretory vesicles, cell cycle, etc. They are the lipid moiety of glycosylphosphatidylinositol anchoring proteins. ►sphingosine, ►cerebrosides, ►gangliosides, ►sphingolipidoses, ►Sandhoff disease, ►ceramide; Dickson RC, Lester RL 1999 Biochim Biophys Acta 1438:305; Leipelt M et al 2001 J Biol Chem 276:33621; sphingolipid metabolic map in yeast: Alvarez-Vasquez F et al 2005 Nature [Lond] 433:425.

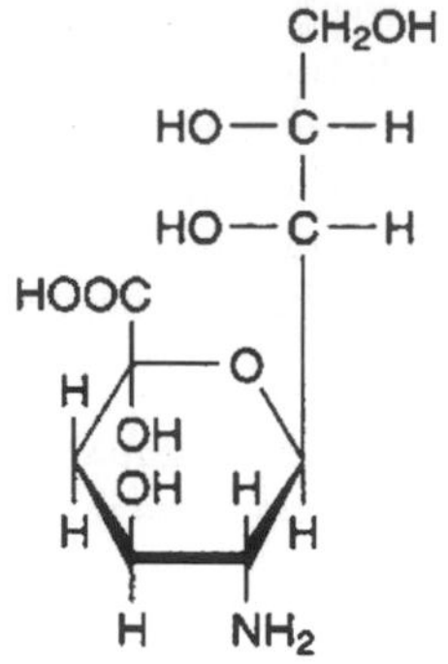

Figure S97. Neuraminic acid

Sphingomyelin: A phospholipid with the sphingosine amino group linked to fatty acids and the terminal OH group of sphingosine esterified to phosphorylcholine. ►fatty acids, ►sphingosine, ►Niemann-Pick disease, ►osteogenesis imperfecta

Sphingosine: A solid fatty acid-like component of membranes. The principal naturally occurring sphingosine is D(+) erythro-1,3-dihydroxy-2-amino-4-transoctadecene, $CH_3(CH_2)_{12}CH{=}CH(OH)$-$CH(NH_2)$-$CH_2OH$. In addition to the C_{17} sphingosines shown, the molecules may have 14, 16, 18, 19, and 20 carbons. The molecules may also be branched or may contain an additional OH group. Sphingosine-1 phosphate is a rather universal signaling molecule for cell proliferation, chemotaxis, differentiation, senescence, and apoptosis. The action of sphingosine kinase is reversed by sphingosine phosphatase. Sphingosine 1-phosphate receptor mediates the egression of lymphocyte from thymus and lymphoid organs. Its activity is increased by two orders of magnitude upon blocking S1P lyase and causes

Table S4. Sphingolipidoses

Farber's disease (ceramidase)	Lactosyl ceramidosis (β-galactosyl hydrolase)
Fabry's disease (α-galactosidase)	Metachromatic leukodystrophy (sulfatase)
Gaucher's disease (β-glucosidase)	Niemann-Pick disease (sphingomyelinase)
Generalized gangliosidosis (β-galactosidase)	Sandhoff's disease (hexosaminidases A and B)
Krabbe's leukodystrophy (galactocerebrosidase)	Tay-Sachs disease (hexosaminidase A)

See These diseases under separate entries.

lymphopenia in mouse (Schwab SR et al 2005 Science 309:L1735). ►sphingomyelin, ►lymphopenia; Brownlee C 2001 Curr Biol 11:R535; Merrill AH 2002 J Biol Chem 277:25843; Spiegel S, Milstien S 2003 Nature Rev Mol Cell Biol 4:397.

Spi⁺: The wild type λ phage is sensitive to phage P2 inhibition. Phage λ lacking the functions of *red* and gam can grow in P2 carrying-lysogens if it has chi (recombination sites for the RecBC system). ►lambda phage

Spi1 Oncogene: The spleen focus forming retrovirus homolog; it is located in human chromosome 11p11.22. Spi1 is an ETS family transcription factor. ►oncogenes, ►ETS, ►PU.1

Spiders (*Aranae*): An extremely large group of ancient genera of animals. ►silk; http://research.amnh.org/entomology/spiders/catalog/index.html.

Spidey: A mRNA-to-genomic DNA alignment program that can be used for draft sequences, finished sequences, and inter-species comparisons. ►mRNA; http://www.ncbi.nlm.nih.gov/spidey/.

Spielmeyer-Sjögren-Vogt Disease: ►ceroid lipofuscinosis

Spike: An inflorescence, which is alternatively called head or ear; a typical example is the spike of wheat or some other grasses. The spikelets or flowers are sessile on opposite sides of the axis that is called rachis. Also, spike refers to short duration electrical variations along the nerve axon, a peak in electric potential.

Spike: The claw-like structures on the base plate of bacteriophages. ►development

Spikelet: A group of florets sitting on a common base in a spike (see Fig. S98). ►spike

Figure S98. Wheat spikelet

Spiked Oligos: Phosphoramidates are incorporated into deoxyribonucleotides and when these are used as primers for PCR-based mutagenesis random mutations may be selected that slow down the activity of the mutant gene products (proteins) and opens a chance to screen for extragenic suppressors that restore better full function. ►phosphoramidate, ►PCR-based mutagenesis, ►suppressor gene; Hermes JD et al 1989 Gene 84:143.

Spiking: Information transfer by neurons.

Spin Infection: In spin infection, adherent cells are exposed to transformation vector mix by centrifuging.

Spina Bifida: A developmental disability in various mammals, including humans, determined by autosomal dominant inheritance and reduced penetrance. The spinal column is incompletely closed and in some instances this involves no serious problem and is detectable only by X-ray examination (spina bifida occulta). The more serious case is the spina bifida aperta when the spinal cord, the membranes (meninges), spinal cord, and nerve ends are protruding (*myelomeningocoele*). Hydrocephalus, incontinence, etc., frequently accompany this condition. The *meningocele* form involves only membrane extrusion and consequently it is a less severe defect. Defects in cadherins may be responsible for the anomalous differentiation. The overall frequency of these anomalies may be 0.2–0.3% in the general population. The anomaly may be the consequence of folic acid deficiency (Lucock M 2004 Br Med J 328:211). ►neural tube defects, ►prenatal diagnosis, ►genetic screening, ►MSAPF, ►mental retardation, ►hydrocephalus, ►cadherins

Spinach (*Spinacia oleracia*): A dioecious plant, 2n = 12 (XX or XY, included). Many of the trisomics can be identified without cytological examinations. All six pairs of the chromosomes have distinct morphology.

Spinal and Bulbar Muscular Atrophy (SBMA): ►Kennedy disease

Spinal Muscular Atrophy (SMA): A degeneration of the spinal muscles; it occurs in different forms and under different genetic controls. The adult type, proximal, is autosomal dominant and may be associated with different chromosomes. The juvenile (Kugel-berg-Welander syndrome) affects primarily the proximal limb muscles and frequently involves twitching. This form was assigned to human chromosome 5q11.2-q13.3. The prevalence of this type is about 1/6000 in newborns. The literature also distinguishes the Werdnig-Hoffmann disease type, however, the various juvenile forms map to the same chromosomal segment, although the expressions may differ. Deletions of this area are frequently the basis of the disease. The combined, estimated gene frequency is about 0.014. The spinal muscular dystrophy with microcephaly and mental retardation, the spinal muscular dystrophy distal (11q12-q14), SMA proximal adult type, and other variations of it appear autosomal recessive. SMA expression is positively associated with the activity of the *survival motor neurons* controlled by the centromeric SMN2 gene (human chromosome 5q12.2-q13.3.) SMN proteins (38 kDa) are part of spliceosomal small ribonuclear simplex. The highly homologous SMN1 gene has telomeric location. The transcript of SMN2 is frequently deficient in exon 7 due to a single nucleotide substitution; otherwise it is identical with SMN1. The two proteins function in association and the truncated protein can be compensated for by higher amounts and can ameliorate the phenotype (Le TT et al 2005 Hum Mol Genet 14:865). Mutation in SMA may affect the U snRNA nuclear import system. Spinal muscular atrophy with respiratory distress (SMARD) is caused by mutation, if an immunoglobulin (IgG) binding protein (IGHMBP2) is encoded at 11q13.2-q13.4. ►muscular atrophy, ►Kennedy disease, ►dystrophy, ►atrophy, ►lipodystrophy familial, ►muscular dystrophy, ►neuromuscular diseases, ►dynein, ►Kugelberg-Welander syndrome, ►Silver syndrome, ►Werdnig-Hoffmann disease, ►U RNA, ►SMAD, ►gemini of coiled bodies, ►NAIP; Jablonka S et al 2001 Hum Mol Genet 10:497; Grohmann K et al 2001 Nature Genet 29:75; Feldkötter M et al 2002 Am J Hum Genet 70:358; Frugier T et al 2002 Curr Opin Genet Dev 12:294; Paushkin S et al 2002 Curr Opin Cell Biol 14:305; review of SMN1: Monani UR 2005 Neuron 48:885.

Spindle: A system of microtubules (≈20–30 nm) emanating from the poles (centrioles of the centrosomes in animals) during mitosis and meiosis, attaching to the kinetochores within the centromeres, and pulling the chromosomes toward the poles. In plants and also in some animal oocytes, there are no centrosomes and the chromosomes assume some of the centrosome functions. In yeast, the spindle forms within the nucleus. Some of the microtubules reach from one pole to the other without attaching to the kinetochore. In *Drosophila* meiosis, the spindle originates from each of the chromosomes and as the prophase progresses, a bipolar spindle emerges. This stage uses a kinesin-like protein (NCD). The arrangement of the microtubules also requires the motor protein dynein. The meiotic pole is different from the mitotic centrosomes (DMAP60, DNAP190 and γ-tubulin are apparently absent). According to a model, for the capture of microtubules and development of the spindle apparatus, the chromosomes distribute a gradient of RAN guanosine triphosphate and importin-β toward the cytoplasm, in the direction of the microtubules emanating from the centrosomes. This eventually leads to self-organization of the bipolar spindle (Caudron M et al 2005 Science 309:1373). About 200 genes determine the spindle assembly in *Drosophila* (Goshoma G et al 2007 Science 316:417).

The orientation of the spindle may be controlled in yeast by myosin V. In case of univalents, only monopolar spindle is formed. In some species, the spindle origination from the chromosome is apparently suppressed by the centrosome. In some other species, the chromosomes and the centrosomes cooperate in developing the meiotic spindle. The mitotic spindle may also need both chromosomes and the centrosomes (echinoderms) (see Fig. S99). Microtubule assembly is not an exclusive property of the kinetochores as holocentric chromosomes indicate. Anaphase and cytokinesis, however, can take place in cells after the chromosomes have been removed. On the oocyte chromosomes of several species, surface proteins (NOD, Xklp1) are found that stabilize premetaphase chromosomes, and in achiasmate meiosis, substitute for the chiasmata. According to recent views, the information for meiotic disjunction resides within the chromosomes and not in spindle apparatus. The kinetochore determines the transition from metaphase to anaphase. The spindle checkpoint ensures the fidelity of chromosome segregation by preventing cell-cycle progression until all the chromosomes make proper bipolar attachments to the mitotic spindle and are subjected to tension. Some experiments indicate that the checkpoint recognizes the lack of microtubule attachment to the kinetochore, others indicate that the checkpoint senses the absence of tension generated on the kinetochore by microtubules. Both of these alternative explanations may be true (Pinsky BA, Biggins S 2005 Trends Cell Biol 19:486). Tension of the kinetochore generates a checkpoint signal (Cdc20, fizzy, Cdc55), and it is supposed that a phosphorylated kinetochore protein attracts proteins. At least six known genes seem to be involved in kinetochore functions. The X-chromosomes in XO cases of sex determination do not involve such a pause. The cytoskeleton is also involved in the correct organization of the spindle. The mRNA export protein Rae1 is a microtubule-associated protein that binds directly to importin β

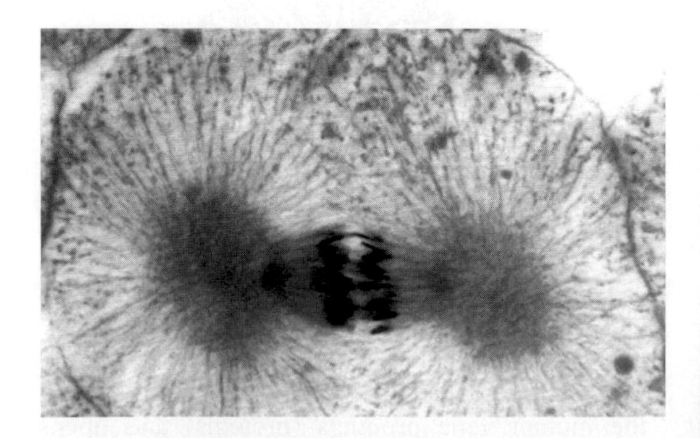

Figure S99. Spindle with chromosomes in the middle

and in egg extract it stabilizers microtubules by RanGTP/importin. Rae 1 is located in a large RNA-containing complex and thus indicates that RNA also has a role in spindle formation (Blower MD et al 2005 Cell 121:223). In some large cells, such as oocytes, the spindle fibers are not long enough to reach and capture the centromeres. In such cases, a contractile nuclear actin network is formed that facilitates the attachment of the centromeres to the spindle fibers (Lénárt P et al 2005 Nature [Lond] 436:812). Spindle assembly requires the presence of lamin B, which is activated by RanGTP (Tsai M-Y et al 2006 Science 311:1887). ►nucleus, ►centromere, ►centromere proteins, ►chromokinesins, ►spindle fibers, ►cytoskeleton, ►tubulins, ►myosin, ►kinetochore, ►chromatid, ►mitosis, ►meiosis, ►BTB, ►securin, ►NuMA, ►dynactin, ►kinesin, ►Bim, ►univalent, ►holocentric chromosome, ►achiasmate, ►Swi6p, ►microtubule, ►PARP, ►RMSA-1, ►CENP, ►Stathmin, ►APC, ►multipolar spindle, ►acenaphthene, ►Cdc2, ►Cdc20, ►Cdc55, ►Mad2, ►RAN, ►ASAP; Sharp DJ et al 2000 Nature [Lond] 407:41; Shah JV et al 2000 Cell 103:997; Compton DA 2000 Annu Rev Biochem 69:95; Karsenti E, Vernos I 2001 Science 294:543; Musachio A, Hardwick KG 2002 Nature Rev Mol Cell Biol 3:731; Lew DJ, Burke DJ 2003 Annu Rev Genet 37:251; Kline-Smith SL, Walczak CE 2004 Mol Cell 15:317; Gadde S, Herald R 2004 Curr Biol 14:R797; orientation of the mitotic spindle: Théry M et al 2007 Nature [Lond] 447:493; Illustration by courtesy of Dr. P.C. Koller.

Spindle Fibers: Microtubules clearly visible (by appropriate techniques) during mitotic and meiotic nuclear divisions. The microtubules originate at the spindle poles, the asters in animals. Three classes of fibers emanate from the poles: the astral microtubules that radiate from the centrioles, the polar microtubules that meet at the divisional plane and appear to stabilize the spindle, and the kinetochore microtubules that are anchored at the centromere of the chromosomes and at anaphase pull them toward the opposite poles. The eukaryotic spindle fibers are made largely of tubulin. The analogous prokaryotic system is built mainly of actin filaments. ►spindle, ►mitosis, ►meiosis, ►aster, ►centrioles, ►centromere, ►kinetochore, ►centrosome, ►tubulins, ►Pac-Man model

Spindle Poison: Spindle poisons block the formation of spindle fibers and as a consequence polyploidy may result. ►polyploidy, ►colchicine, ►acenaphthene, ►Stathmin

Spindle Pole Body (SPB): The fungal equivalent of the centrosome. Cyclins and cyclin-dependent kinases promote and regulate their duplication. SPBs assists in the assembly of membrane proteins in the meiotic pro-spores of yeast. Their defect results in genetic instability, aneuploidy. SPBs affect also mRNA metabolism. ►centrosome, ►mitochondrial genetics; Haase SB et al 2001 Nature Cell Biol 3:38; Bajgier BK et al 2001 Mol Biol Cell 12:1611; Lang BD et al 2001 Nucleic Acids Res 29:2567.

Spinobulbar Muscular Atrophy: ►Kennedy disease

Spinocerebellar Ataxia (SCA): Autosomal dominant defects involving CAG repeats (polyglutamine) in the coding region of the ataxin genes (SCA1, SCA2, SCA3, SCA7) and causing nerve degeneration and loss of Purkinje cell and neurons in the brain. As a consequence, motor functions deteriorate. SCA5 is due to mutation in spectrin B, which destabilizes glutamate transporter EAAT4 on the surface of the plasma membrane (Ikeda Y et al 2006 Nature Genet 38:184). SCA6 affects a calcium ion channel. In a normal state, the number of repeats is 6–44, while in the diseases state it is 40–93. The ataxin gene (SCA1) encoded at 6p23 is transcribed into 792–869 amino acids, depending on the number of CAG repeats. The mouse homolog is *Mathl*. Ataxin-1 binds to RNA and controls the expansion of the polyglutamine sequences. The insect transcription factors Sens and the mammalian homolog Gfi-1 interact with the AXH domain of ataxin and mediate neurodegeneration (Tsuda H et al 2005 Cell 122:633). The SCA8 involves CTG repeats. SCA10 (22q13-qter) displays ATTCT repeats in variable numbers in intron 9. SCA12 (5q31-q33) is caused by CAG repeats in the regulatory subunit of protein phosphatase PP2A. SCA13 childhood ataxia with mental retardation and delayed motor functions is encoded at 19q13.3-q13.4. The SCA13 area contains the KCNC3 voltage-gated ion channel, and its mutation leads to neurodegeneration (Waters MF et al 2006 Nature Genet 38:447). SCAN11 (spinocerebellar ataxia with axonal neuropathy, 14q31-q32) is caused by a defect in tyrosyl-DNA phosphodiesterase (TDP1), involved in normally transient single-strand DNA break and repair generated by topoisomerase function (El-Khamisy SF et al 2005 Nature [Lond] 434:108). ►trinucleotide repeats, ►topoisomerases, ►spectrin, ►GLAST, ►GLT, ►Huntington chorea, ►Kennedy disease, ►Machado-Joseph disease, ►myotonic dystrophy, ►ataxia, ►migraine, ►CACNA1A, ►RNAi; Stevanin G et al 2000 Eur J Hum Genet 8:4; Yue S et al 2001 Hum Mol Genet 10:25; Libby RT et al 2003 Hum Mol Genet 12:41.

Spiracle: The breathing hole on the insect body.

Spiralization: A pattern of winding of molecules or chromosomes. ►coiling

Spirochetes (Spirochaeta): Filiform bacteria (5–6 μm) that cause sex ratio distortion in *Drosophila* by the fatal effect of their toxin on the male flies. For the killing of the male flies by *Spiroplasma poulsoni*, the normal function of the *Msl* genes of *Drosophila* (involved in dosage compensation) is required (Veneti Z et al 2005 Science 307:1461). Leptospirosis, a flu-like disease with frequent hepatic hemorrhage and jaundice, is caused *Leptospira interrogans* (see Fig. S100). Its sequenced genome consists of two chromosomes of 4,322,241 and 358,943 bp and encoding 4360 and 367 proteins, respectively. Its physiological characteristics are quite different from *Treponema* or *Borrelia*. Some lower animals (e.g., *Schistosoma*) are also called spirochetes. ►*Schistosoma*, ►*Treponema*, ►*Borrelia*, ►dosage compensation, ►killer genes, ►segregation distorter, ►endosymbiont, ►symbionts hereditary; Ren S-X et al 2003 Nature [Lond] 422:888.

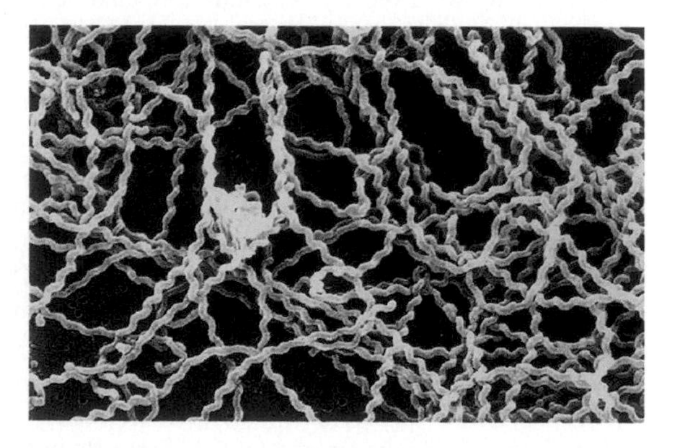

Figure S100. *Leptospires* bacteria. (Courtesy of CDC Rob Weyant and Janice Carr)

Spironucleoside (hydantocydin): A herbicidal growth regulator of *Streptomyces hygroscopicus*. ►herbicide

Spiroplasma: ►symbionts hereditary, ►sex ratio, ►spirochetes

Spite: An inimical action toward other individuals without benefit to the actor.

spitz (*Drosophila* chromosome 2–54): The 2–54 locus encodes the Spitz protein, a ligand of DER. ►DER

SPK-1: A protein kinase of the MAPK family. ►signal transduction

Spleen: An upper left abdominal, oblong (ca. 125 mm) ductless gland. At the embryonal stage, it participates in erythrocyte formation, in adults it also manufactures lymphocytes. By decomposing the erythrocytes, it provides to the liver, the hemoglobin to form bile. The red pulp contains red blood cells and macrophages, the white pulp carries the lymphocytes. It has a role in the defense mechanism of the body. It is generally very enlarged in some lysosomal storage diseases, e.g., Gaucher's disease. ►asplenia, ►lysosomal storage diseases, ►hemophilias, ►Gaucher's disease

Splice: The resolution of a recombination intermediate (Holliday junction), resulting in an exchange of the flanking markers. ►Holliday model, ►introns, ►spliceosome

Spliceosomal Intron: Spliceosomal introns utilize spliceosomes in eukaryotes for the removal of introns in contrast to the prokaryotic group I and II introns, which do so themselves. ►introns, ►spliceosome, ►splicing; origins and evolution: Rodríguez-Trelles F et al 2006 Annu Rev Genet 40:47.

Spliceosome: A protein-U snRNA (there are five main uridine-rich oligonucleotides) complex required for the folding of pre-mRNA into the proper conformation for removal of introns and splicing the transcripts of exons. The spliceosome contains about 300 proteins (Rappsilber J et al 2002 Genome Res 12:1231). The *majority* of eukaryotic primary transcripts contain introns with 5′ GU and AG 3′ splice sites, and the mammalian consensus is: (A/C) **AG↓GU**A/GAGU. In yeast, the consensus is: **(A/G)↓GU**AUGU. Less than 1% of the mammalian splice junctures are GC–AG. In the excision, the spliceosome complex U1, U2, U4-U6, and U5 snRNAs and non-snRNAs work together. Initially, the U1 and U2 snRNA attaches by base pairing to the splice and to the branch sites, respectively. The U4-U6•U5 tri-snRNP complex joins in pre-splicing complexes. This is followed by snRNP–snRNP and mRNA–snRNP interactions. U6 base-pairs with U4. During spliceosome assembly, U1 and U4 are displaced and U6 pairs with the 5′ splice site and with U2 snRNA. Through the coordination of a divalent metal ion (Mg^{2+}), the U6 snRNA contributes to the splicing of the RNA transcript. The 5′ splice site and the branch nucleotide then move toward each other and the 2′-OH group of the latter serves as an electron donor for the first step of splicing. The excised 5′-exon and the lariat intron-3′-exon are the reaction intermediates. This first step is followed by a reaction between the electron donor nucleophile and the electron-deficient electrophile at the 3′ splice junction by the 3′-OH of the freed 5′-exon, resulting in the ligation of the exons, and the removal of the intron.

A *minority* (0.1%) of the introns, the AT-AC introns, occurs in some animal genes such as encoding PCNA. The 5′ splice site of such introns has the consensus **AT↓AC**CTT and their branch site is TCCTTAAC. Their splicing complex includes U11 and U12 snRNP and one or more U5 snRNP variants (Russell AG et al 2006 Nature [Lond] 443:863). The

PCNA branch site pairs with U12 and a loop of U5 aligns the exons of PCNA for ligation. U4 and U6 snRNAs are not used, but the highly divergent U4atac and U6atac take over their role. The reaction at the 5′ splice site is mediated by a metalloenzyme but not at the 3′site. The spliceosome is a huge and variable complex of proteins. For the recognition of the 3′ AG splice site, $U2AF^{35}$ (the 35K subunit of the heterodimeric [M_r 65K] $U2AF^{65}$ protein) is required in vivo, it is needed for viability and it is present in different organisms. U2AF was considered to be only an auxiliary factor (because in vitro it was not indispensable) to recognize the 5′-polypyrimidine sequences. For splicing, the U2AF protein must be in the close vicinity of the polypyrimidine tract that is recognized by its 35K subunits. The *Drosophila* protein Sex-lethal (SXL) controls dosage compensation by inhibiting splicing of the *male-specific-lethal-2* transcripts. When the large subunit of U2AF is displaced from the polypyrimidine tract—3′AG interval by SXL, the 35K subunit can mediate the removal of the intron. In the neurons the Nova proteins regulate alternative splicing. Nova binding the YCAY nucleotide sequences (Y stands for either pyrimidine) in exons blocked U1 snRNP binding and thereby exon inclusion, whereas intronic YCAY clusters promoted spliceosome assembly and exon inclusion. Thus, mapping Nova binding sites predicts mRNA regulation in neurons (Ule J et al 2006 Nature [Lond] 444:580). ►introns, ►exons, ►splicing, ►PCNA, ►snRNA, ►dosage compensation; Hastings ML, Krainer AR 2001 Curr Opin Cell Biol 13:302; Nagai K et al 2001 Biochem Soc Trans 29:15; Valadkhan S, Manley JL 2001 Nature [Lond] 413:701; Sträßer K, Hurt E 2001 Nature [Lond] 413:648; Nilsen TW 1998, p 279. In: RNA Structure and Function. Simons RW, Grunberg-Manago (eds). Cold Spring Harbor Laboratory Press; Villa T et al 2002 Cell 109:149; Zhu Z et al 2002 Nature [Lond] 419:182; Brow DA 2002 Annu Rev Genet 36:333; Patel AA, Steiz JA 2003 Nature Rev Mol Cell Biol 4:960; Shukla GC, Padgett RA 2004 Proc Natl Acad Sci USA 101:93; structure: Azubel M et al 2004 Mol Cell 15:833; core crystal structure: Schellenberg M et al 2006 Proc Natl Acad Sci USA 103:1266; Xue S et al 2006 Science 312:906; spliceosome components: Stark H, Lührmann R 2006 Annu Rev Biophys Biomol Struct 35:435; supraspliceosomes: ChenY-IG et al 2007 Nucleic Acids Res 35:3928.

Splicing: Joining of RNA with RNA or DNA with DNA at the sites of previous cuts. Constitutive splicing indicates that the exons are spliced in the same order as they occur in the primary RNA transcript, in contrast to alternative splicing when the exons may be joined in alternative manners, thus providing mRNAs for different proteins transcribed from the same gene. The general scheme of pre-mRNA splicing is shown in Figure S101. The *Dscam* axon guidance gene of *Drosophila* has more than 38,000 alternative splice variants and dozens of different forms may be present within single cells (Neves G et al 2004 Nature Genet 36:240). About 74% of the human genes are alternatively spliced and about 15% of the human hereditary diseases are caused by mutation in the splicing mechanism (Johnson JM et al 2003 Science 302:2141).

The splicing factor family SR (named so because they are rich in serine [S] and arginine [R]), contain an RNA recognition motif (RRM/RNP) and Ψ-RRM domain Ser-Trp-Gln-Asp-Leu-Lys-Asp, separated by a Gly rich tract. The Ser-Arg domains are well phosphorylated by the serine kinase SRPK1. The 3′ splice site is recognized by the $U2AF^{65}/U2AF^{35}$ heterodimer. The former binds to the polypyrimidine sequence in the RNA, whereas the latter associates with the RS domain of other RS-containing factors or the U4/U6.U5 small ribonuclear complexes at the 5′ splice site. These proteins select the splice sites and are active parts of the spliceosome complex (►spliceosome). The ASF/SF2 family recruits the U1 snRNP to the RNA transcript. SR proteins communicate also between introns across exons, and affect alternative splicing. The SR proteins are subject to regulation. Dephosphorylation of the splicing factor SRp38 represses splicing (Shinj C et al 2004 Nature [Lond] 427:553). Between the branch site and AG, there are located polypyrimidine sequences, which

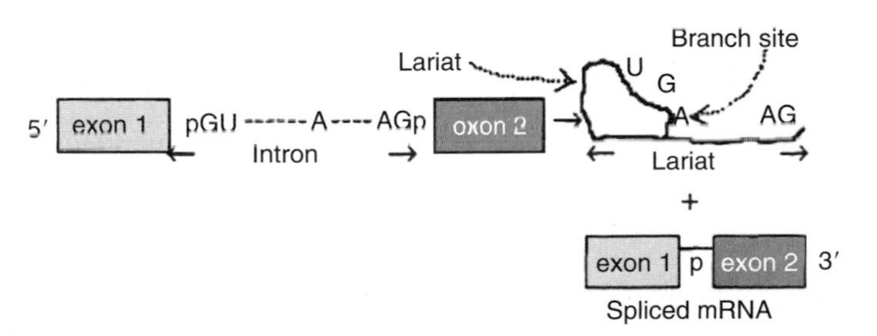

Figure S101. Pre-mRNA splicing

assist with the aid of various proteins to define the 3′ splice site. Inactivation of ASF/SF2 results in genetic instability due to the formation of an R loop between RNA and the non-template strand of DNA (Li X, Manley JL 2005 Cell 1222:365). Some of the many relevant proteins are PTB (polypyrimidine tract-binding protein) and PSF (PTB-associated splicing factor). The splicing reaction is driven by various nucleotidetriphosphatases. Genome-wide analysis revealed 285 human (ESR) exonic splicing regulatory sequences (Goren A et al 2006 Mol Cell 22:769). Plants also have splicing factors but these are more variable than those of animals and this is the cause why animal introns are not spliced out in plants. Splicing precedes the export of the RNA to the cytoplasm and requires the association of the mRNA with the splicing factor Aly. ►restriction enzyme, ►exon junction complex, ►vectors, ►introns, ►speckles intranuclear, ►lariat, ►spliceosome, ►alternative splicing, ►U1 RNA, ►snRNA, ►alternative splicing; Madhani HD, Guthrie C 1994 Annu Rev Genet 28:1; Kramer A 1996 Annu Rev Biochem 65:367; Kim N et al 2001 EMBO J 20:2062; Thanaraj TA, Clark F 2001 Nucleic Acids Res 29:2581; Luo M-J et al 2001 Nature [Lond] 413:644; Clark TA et al 2002 Science 296:907; splicing factors: Sanford JR et al 2005 Proc Natl Acad Sci USA 102:15043; splice site prediction: http://spliceport.cs.umd.edu:2000/SplicingAnalyser.html.

Splicing Enhancer, Exonic (ESE): The repetitive GAA sequences associated with proteins; they facilitate the joining of exons. Pre-messenger RNA, containing introns, is retained within the nucleus. Intronless mRNAs containing ESEs were found to be poorly exported from the nucleus; spliced mRNAs produced from ESE-containing pre-mRNAs were found to be efficiently exported to the cytoplasm (Taniguchi I et al 2007 Proc Natl Acad Sci USA 104:13684). ►pre-mRNA; Fairbrother WG et al 2002 Science 297:1007.

Splicing Inhibition: Splicing inhibition may be brought about by (phosphorothioate) 2′-*O*-methyl-oligoribonucleotides or morpholino oligonucleotides that are resistant RNase H (see Fig. S102).

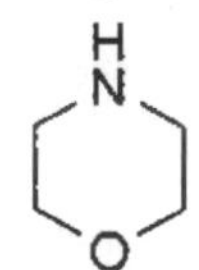

Figure S102. Morpholine

However, sometimes these analogs activate cryptic splice sites within exons. (See Wang Z et al 2004 Cell 119:831).

Splicing Juncture (splice junction): Sequences at the exon-intron boundaries. They are needed for the selective export of mature mRNAs from the nucleus. In addition, they assist in RNA surveillance that leads to degradation of mRNAs with premature transcription termination. Mutations at splice junctions may cause the removal or the ordering of some exons. ►introns, ►exons, ►spliceosome, ►RNA surveillance, ►nonsense-mediated decay; Lykke-Andersen J et al 2001 Science 293:1836.

Splicing of Proteins: Protein splicing involves post-translational excision by an endopeptidase of polypeptides and ligation the resulting carboxy- and amino-terminal sequences. Such a mechanism may occur in prokaryotes, plants, and animals. The shorter peptide can be then sliced up by a proteasome, further processed, and short pieces may be delivered with the aid of a transporter of MHC class I molecules to the cell surface where cytotoxic T lymphocytes (CTL) may destroy them. Such a mechanism can get rid of the FGF-5 fibroblast growth factor fragment responsible for renal cancer. ►splicing, ►inteins, ►CTL, ►proteasome, ►MHC; Hanada K-i et al 2004 Nature [Lond] 427:252.

Splign: A tool for computing cDNA-to genomic or spliced alignements. (See http://eu-transcoder.usablenet.com/tt/www.ncbi.nlm.nih.gov/sutils/splign/; http://www.ncbi.nlm.nih.gov/sutils/splign/splign.cgi?textpage=downloads).

Splinkerette: A modified vectorette for PCR walking. In a PCR reaction, the free 3′ end of the bottom strand flips back on itself forming a hairpin and begins elongation further along the bottom strand. The resulting double-stranded structure is stable and it is functionally removed from further reaction (see Fig. S103). (Diagram redrawn after Devon RS et al 1995 Nucleic Acids Res 23:1644). The system is well-suited fort large-scale identification of mouse gene trap events (Horn C et al 2007 Nature Genet 39:933). ►vectorette, ►trapping promoters

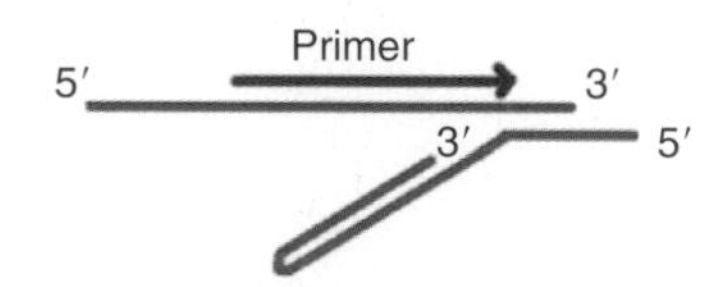

Figure S103. Splinkerette

Split Gene: The split gene is discontinuous because introns are intercalated between exons; the majority of the eukaryotic genes contains introns. ►introns, ►exons, ►splicing, ►spliceosome

Split-Hybrid System: The split-hybrid system provides means for positive selection for molecules that disrupt protein-protein interactions. The genetically engineered construct is shown in Figure S104.

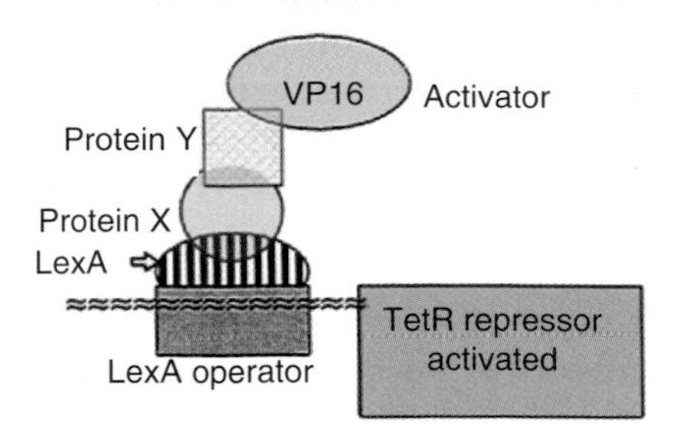

Figure S104. Split-hybrid method. If binding of X and Y is prevented the VP16 transactivator cannot turn on the TetR repressor and thus permitting the expression of *HIS3* gene and can be seen by the growth of the yeast cells in histidine-free medium

►two-hybrid method, ►VP16, ►one-hybrid binding assays, ►split-hybrid system, ►three-hybrid system, ►tTA, ►rtTA; Goldman PS et al 2001 Methods Mol Biol 177:261.

SPM (scanning probe microscope): SPM images biological macromolecules under a thin layer of aqueous solution. ►scanning electronmicroscopy

Spm (suppressor mutator): A transposable element system of maize of an autonomous *Spm* and a non-autonomous (originally unnamed) element. The non-autonomous element cannot insert or excise by its own power because it is defective in the transposase enzyme. This system is the same as the *En* (*Enhancer*)-*I* (*Inhibitor*) system. The non-autonomous component has also been called *dSpm* (*defective Spm*). The original name represents the fact that insertional mutations caused by the non-autonomous *dSpm* (*I*) element revert at high frequency only when *Spm* (*En*) is introduced into the genome because the latter has a functional transposase gene. The insertion does not always eliminate the function of the target gene and both a recessive mutation and the *Spm* element may be expressed. Such a case at the *A* (anthocyan) locus of maize (chromosome 3L-149) was designated as *a-m2* (*a* mutable) because of its frequent reversion to the dominant allele and displaying sectors in the presence of *Spm* (see Fig. S105). The *Spm*-dependent alleles harbor a non-autonomous (*dSpm* or *I*) insertion element that may jump out (and cause reversion), only when the functional *Spm* is introduced. The *Spm suppressible* alleles indicate the presence of a non-autonomous element that may or may not permit the expression of the gene, but the presence of the active *Spm* allows the insert's removal. The terms *Spm-w* and *Spm-s* indicate

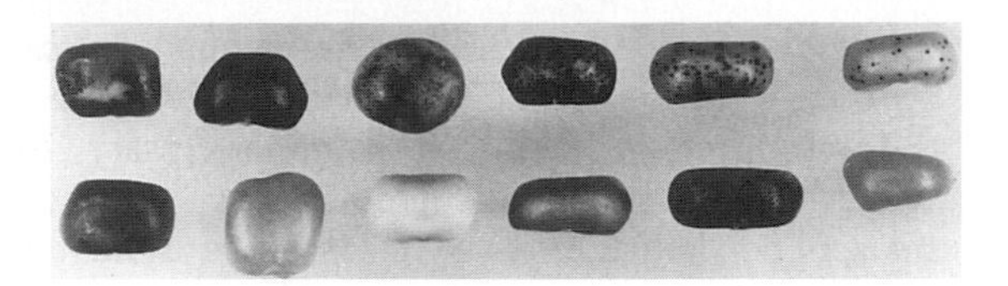

Figure S105. The dominant alleles of maize normally form dark color but the a^{m-1} allele possessing the Spm element may display alternative states. In the absence of an inactive Spm any shade of solid color may appear (bottom row) but in the presence of an active Spm a variety of sectors are displayed (top row). (Courtesy of Barbara McClintock)

weak and strong *Spm* transposase elements, respectively. *Inactivated Spm* has been called *Spm-i,* and elements that have alternating active and inactive phases were called *cycling Spm* (*Spm-c*), whereas very stable inactive forms were named *cryptic Spm* (*Spm-cr*). In addition, a *Modifier* factor has been named that enhances the activity of *Spm.* The various forms of the *Spm* alleles were found to alter their activity with time, according to developmental stage and extraneous factors, which cause chromosome breakage (radiation, tissue culture, etc.). The term *presetting* was applied to the phenomenon, in which *Spm* determined the expression of a gene even after its removal from the genome; but this effect later may fade away. Presetting is supposed to occur during meiosis and it is attributed to methylation. Indeed, the first exon of *Spm* is rich in cytidylic residues, the most commonly methylated base in DNA. Methylation may be instrumental in the variations and the inactivation of the various *Spm* elements named. A *Regulator* element is credited with the control of the extent of methylation. Although, the various *Spm* alleles display some heritable qualities, they appear to represent labile "changes in states" of the element, except the *dSpm* or other deletional forms. Transposition of the *Spm* elements is not tied to DNA replication and the transposition favors new insertion sites within the same chromosome, although it may move to any other part of the genome. Integration of *Spm* results in a 3 bp duplication at the target site. The frequency of insertion and excision is influenced also by the base sequences flanking the target site. The *Spm* appears to be modified by the process of excision and involves various lengths of terminal deletions, but the subterminal repetitious sequences may also be involved. Deletion of the terminal repeats abolishes transposition of the element. The abilities of *Spm* to transpose and to affect gene expression are inseparable.

The *Spm* element is transcribed into a 2.4 kb RNA with 11 exons; it includes two close open reading frames. The transcript may be alternatively spliced into four different open reading frames, called *tnpA*, *tnpB*, *tnpC* and *tnpD*, the latter being the longest.

Transcripts *tnpA* and *tnpD* are necessary for transposition. The sequences downstream from the transcription initiation site are rich in GC base pairs and are susceptible to methylation. The active *Spm* elements, located at about 0.6 kb around the transcription start site, are not methylated. The methylated elements are, however, inactive. The entire element (8.4 kb) is flanked by 13 bp inverted repeats (CACTACAAGAAAA and TTTTCTTGTAGTG). In a region 180 bp from the 5′-end and 299 bp from its 3′-end, several copies of a receptive consensus CCGACATCTTA occur. The defective elements, *dSpm* have various lengths of internal deletions covering the entire length or part of the two open reading frames. Some partially deleted elements may still function as a weak *Spm-w* element. The function of an Spm element is regulated by sequences within the element and the location of the transposable element within the target genes. The function of the target genes depends on when, where, and how the *Spm* element and the target gene's transcripts are spliced. ►transposable elements, ►hybrid dysgenesis, ►insertional mutation, ►*Ac-Ds*; Fedoroff N 1989, p 375. In: Mobile DNA. Berg DE, Howe MM (Eds.) Am Soc Microbiol, Washington DC.

SPN (single nucleotide polymorphism): SPN indicates single nucleotide difference between two nucleic acids. ►SNIP

SPO: A group of sporulation mediating proteins in yeast. (See Kee K, Keeney S 2002 Genetics 160:111).

Spondylocostal Dysostosis (SD): Non-syndromal short stature and vertebral and rib defects encoded at 19q13.1-q13.3 (see Fig. S106). The defect in the homolog of *Drosophila* gene *Delta* (*Dll3*) is involved. ►Alagille syndrome, ►Simpson-Golabi-Behmel syndrome

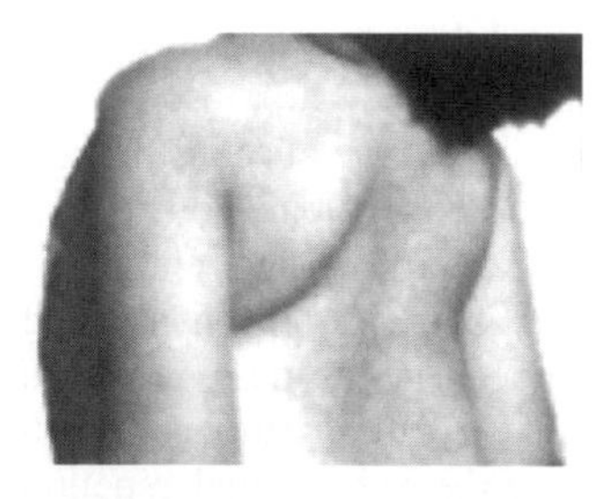

Figure S106. Spondylocostal dysostosis may involve increased size abdomen and deformed vertebral column. (Modified after Turnpenny PD, Kusumi K 2004)

Spondyloepimetaphyseal Dysplasia (SEMD): The heterogeneous group of hereditary skeletal bone and cartilage diseases involving defects in protein sulfation. A recessive type was assigned to chromosome 10q23-q24.

Spondyloepiphyseal Dsyplasia (SED): The *autosomal dominant* phenotype of SED includes flattened vertebrae, short limbs and trunk, barrel-shaped chest, cleft palate, myopia (near-sightedness), muscle weakness, hernia, and mental retardation. Collagen defects are incriminated in many cases. *Autosomal recessive* forms mimic arthritis-like symptoms (arthropathy), besides the short stature. The autosomal recessive forms may not involve flat vertebrae and the defect was attributed to a deficiency of phosphoadenosine-5′-phosphosulfate and thus to undersulfated chondroitin. An *X-linked* SED (Xp22.2-p22.1) was also described. SED type diseases were located to human chromosomes 5q13-q14.1, 12q13.11–13.2, and 19p13.1. The prevalence of the X-linked form is about 2×10^{-6}. ►achondroplasia, ►dwarfness, ►arthropathy-campylodactyly, ►chondroitin sulfate, ►collagen, ►Schimke immuno-osseus dysplasia; Gedeon AK et al 2001 Am J Hum Genet 68:1386.

Spontaneous Generation, Current: Until Louis Pasteur (1859–61), it was assumed by many scientists that microorganisms were formed from abiotic material even during the present geological period. However, Pasteur demonstrated that the organisms found in broth and other rich nutrients grew only when the solutions were not heated to sufficiently high temperature, were maintained for a certain duration of time, and exposed to unfiltered air. His discovery has been fundamental to modern microbiology and medicine and proved that spontaneous generation is not responsible for the current variations in microbial cultures. ►spontaneous generation unique or recurrent, ►lysenkoism, ►pleomorphism, ►biogenesis; Farley J 1972 J Hist Biol 5:285, ibid. 95.

Spontaneous Generation, Unique or Repeated: An explanation for the abiotic origin of life. It is assumed that after the earth was formed, simple molecules, such as water, carbon dioxide, ammonia, and methane were first formed. Subsequently, in a reducing atmosphere or at deep oceanic vents containing reducing minerals (iron and nickel sulfides) and high temperature (300–800°C), molecular nitrogen could have been reduced to ammonia. When energy sources became available at the surface (ultraviolet light), simple organic acids (acetic acid, formic acid) arose. In the presence of ammonia, methane, hydrogen, hydrogen cyanide, and lightning energy amino acids could be formed. In following steps, nucleotides could arise. Actually, under simulated early earth conditions chemists could synthesize amino acids, polypeptides, carbohydrates, and nucleic acids. These simple organic molecules might have aggregated into some sorts of micellae (bubbles), and after self-replicating mechanisms came about the possibility for the generation of an ancestral cell with a primitive RNA as the genetic material. It is not entirely clear when, where, how, and how many times these events took place. Life is estimated to have begun 3–4 billion years ago. It is not known whether on other planets, under similar conditions to that of the earth, living cells evolved. ►biopoesis, ►exobiology, ►spontaneous generation current, ►evolution prebiotic, ►origin of life, ►abiogenesis; Lennox J 1981 J Hist Philos 19:19; Harris H 2002 Things Come to Life: Spontaneous Generation Revisited. Oxford University Press, New York.

Spontaneous Mutation: Spontaneous mutation occurs at a relatively low frequency when no known mutagenic agent is or was present in the environment of the cell or organism; the cause of the mutation is thus unknown. The spontaneous frequency of mutation in mice for seven standard loci is 6.6×10^{-6} per locus. The frequency in man is in the range of 10^{-5} to 10^{-6}, in *Drosophila* 10^{-4} to 10^{-5}, in yeast *Neurospora* 10^{-5} to 10^{-9}, in bacteria 10^{-4} to 10^{-9}, and in bacteriophages, 10^{-4} to lower to 10^{-11} range has been reported. In maize, the frequency is comparable to that in other eukaryotes. At some other loci and in other organisms, it may be substantially higher or lower. In higher eukaryotes, the frequencies appear lower than in microorganisms but this does not seem to be due to intrinsic biological differences; rather it reflects the limitations of the size of the populations, which were amenable to screening. The extent of DNA repair may have profound influence on the mutations recovered. ►mutation rate, ►mutation spontaneous, ►diversity; Wloch DM et al 2001 Genetics 159:441.

Spooling: Spooling assumes that the triple- or quadruple-stranded naked DNA molecules are wound into the RecA protein filament for pairing and exchange. After the formation of heteroduplexes, the DNAs are released. ►recombination molecular mechanism

Sporadic: Sporadic refers to the rare occurrence of an off type, which does not show a clear familial pattern, and the etiology (cause or origin) is unknown. ►epidemiology

Sporangiophore: A sporangium-bearing branch. ►sporangium

Sporangium (plural sporangia): The spore-producing and -containing structure in lower organisms (fungi, protozoa) (see Fig. S107).

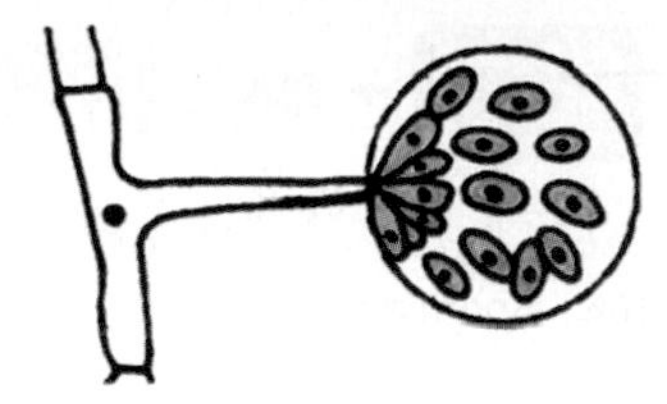

Figure S107. Sporangiophore and sporangium

Spore: A reproductive cell; generally the product of the eukaryotic *meiosis*, or it may arise *mitotically* as fungal conidiospores (conidia). The *bacterial spores* are metabolically dormant cells, surrounded by a heavy wall for protection under very unfavorable conditions.

Spore Mother Cell: ►sporocyte

Sporidium: The "sexual" spore of basidiomycetes fungi. ►basidium

Sporocyte: A diploid cell that produces haploid spores as a result of meiosis. ►gametogenesis

Sporogenesis: The mechanism or process of spore formation. ►meiosis, ►conidia

Sporophore: The fruiting body capable of producing spores in fungi. ►spore

Sporophyte: The generation of the plant life cycle that produces by meiosis the (1n) gametophytes. The common form of plants (displaying leaves and flowers, etc.) is the (2n) sporophytic generation. ►gametophyte

Sporopollenin: Sporopollenins are mainly polymerized carotenoids forming the exine of the pollen grains, facilitating its adhesion to the stigma of the female reproductive structure of plants.

Sporozoite: The infective stage of protozoan life cycle; in malaria, they are formed within the mosquito. ►malaria, ►*Plasmodium*, ►*Anopheles*

Sporulation: In bacteria, the process of formation of morphologically altered cells that can survive adverse conditions and assure the survival of the sporulating bacteria. *Bacillus subtilis* is a typical spore-forming bacterium. Within its cell, a new cell is pinched off to create the spore that can eventually develop into a new regular cell (see Fig. S108). Also, ascospore formation through meiosis in fungi is sporulation. In budding yeast, 334 sporulation-essential genes have been identified. ►meiosis; Enyenihi AH, Saunders WS 2003 Genetics 163:47; Fujita M, Losick R 2003 Genes Dev 17:1166; http://cmgm.stanford.edu/pbrown/sporulation/.

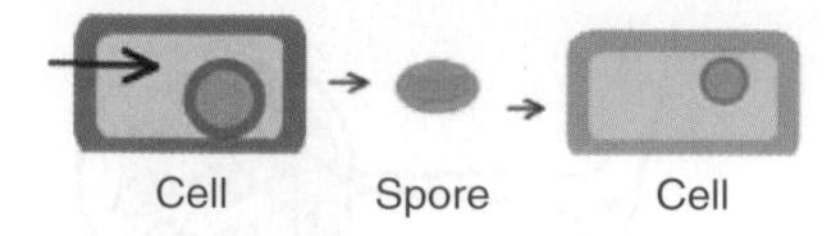

Figure S108. Sporulation

Spot Test: A variation of the mutagen/carcinogen bioassays when the compound to be tested by bacterial reversions is added to the surface as a crystal or a drop after the Petri plates have been seeded by the bacterial suspension and the S9 microsomal fraction added. If the substance to be tested is mutagenic, a ring of revertant cells should appear around the spot where it was added (see Fig. S109). This type of mutagenic assay is no longer much in use. ►Ames test, ►plate incorporation test, ►reversion assays of *Salmonella*; Ames BN 1971, p 267. In: Chemical Mutagens I. Hollaender A (Ed.) Plenum, New York.

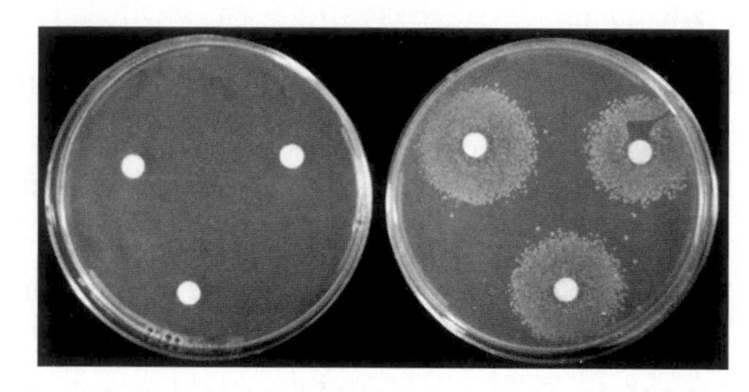

Figure S109. Reversion assay of *Salmonella* bacterium by the spot test. (Courtesy of Dr. G. Ficsor)

Spotting: Spotting at specific body parts is frequently found in various species (see Fig. S110). The basic colors (melanin or melanin precursors) are generally uniformly distributed at various intensities. Mutations in regulatory cis elements create novel opportunities for transcription factors for association with the mutant sequences and determine the intensity of the local deposition of the pigment. Genes in trans locations in the genome encode the transcription factors. (Gompel N et al 2005 Nature [Lond] 433:481). ►cis-acting element, ►transacting element, ►piebaldism, ►transcription factors, ►KIT oncogene, ►variegation

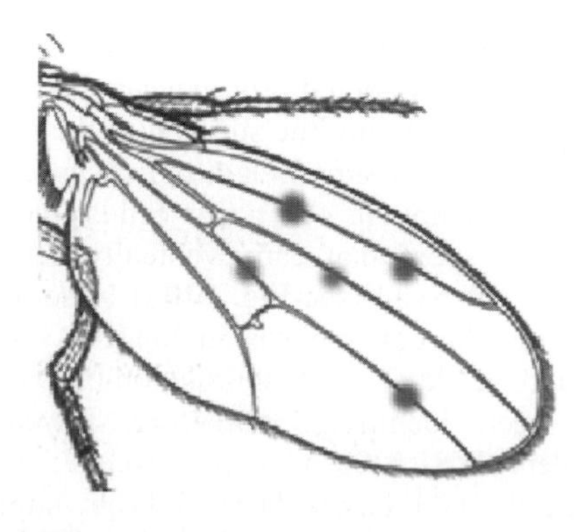

Figure S110. *Drosophila* wing spotting

spp: The abbreviated plural of the word species.

Spreader: A simple instrument for distributing microorganisms on the surface of agar plates (see Fig. S111).

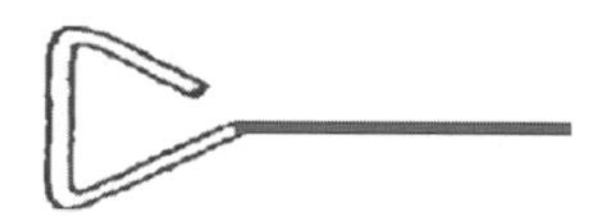

Figure S111. Spreader

Spreading: Moving silencing proteins along the chromosome. ►silencer

Sprekelia formosissima (Amaryllidecea): Subtropical plant, n = ca. 60; genome size 1.8×10^{11} bp.

spretus: *Mus spretus*, a species of mouse, commonly used for genetic analyses.

Springer: ►*copia*

Spruce (*Picea* spp): Timber tree species with 2n = 2x = 24. ►pine, ►douglas fir

Sprue: A tropical intestinal disease, apparently due to infection(s).

SP-Score: ►MP-score

SPT: A suppressor of transposition (e.g., of a Ty element) and a regulator in SAGA function in yeast. The *Drosophila* homologs are *Dspt4* and *Dspt6*. This family of genes encodes histones H2A and H2B and TATA-binding proteins. There is evidence for their regulation of transcription, replication, recombination, and some developmental processes. ►SAGA, ►histones, ►transcription, ►TBP; Yamaguchi Y et al 2001 J Biochem 129:185.

Spt Proteins: Spt proteins are involved in the elongation of RNA transcripts on DNA. ►RNA polymerase; Winston F 2001 Genome Biol 2(reviews):1006.

SptP: ►SopE

Squalene: ►prenylation

Squamous: Squamous denotes scaly, e.g., as in squamous cell carcinoma, a form of epithelial cancer. Sunscreen provides some protection against it. C57BL/6 strain of mice is resistant to skin squamous carcinomas (SCCs) induced by an activated Ras oncogene, whereas FVB/N mice are highly susceptible. Susceptibility is under the control of a carboxy-terminal polymorphism in the mouse *Ptch* gene. F1 hybrids between C57BL/6 and FVB/N strains are resistant to Ras-induced SCCs, but resistance can be overcome either by elimination of the C57BL/6 *Ptch* allele (*PtchB6*) or by overexpression of the FVB/N *Ptch* allele (*PtchFVB*) in the epidermis of *K5Hras*-transgenic F1 hybrid mice. The human Patched (PTCH1, 9p22.3) gene is a classical tumor suppressor gene for basal cell carcinomas and medulloblastomas. Apparently, *Ptch* is critical for determining both basal or squamous cell lineage, and both tumor types can arise from the same target cell depending on carcinogen exposure and host genetic background (Wakabayashi Y et al 2007 Nature [Lond] 445:761). ►epidermolysis, ►nevoid basal cell carcinoma, ►medulloblastoma; Lin JY, Fisher DE 2007 Nature [Lond] 445:843.

Squash: *Cucurbita maxima* (winter squash) 2n = 24, 40; *C. pepo* (summer squash) 2n = 40; *C. moschata* (pumkin), 2n = 24, 40, 48.

Squash Preparation: For microscopic analysis of chromosomes in soft (softened) tissues, there may be no need for sectioning but the fixed and stained material can be examined after smearing it directly on the microscope slide. In some cases, the fixation is done by gentle heating on the slide followed by adding a small drop of aceto-carmine or aceto orceine stain or even just placing the specimen into a drop of acetic stain and heating. These rapid procedures may permit an estimation of the stage of meiosis or mitosis. ►microscopy, ►stains; Belling J 1921 Am Nature 55:573.

Squelching: The suppression or silencing of adverse effects.

Squirrel: *Tamiasciurus hudsonicus streatori*, 2n = 46; *Callospermophilus lateralis*, 2n = 42; *Ammospermophilus*, 2n = 32; *Citellus citellus*, 2n = 40.

SR1: A cytoplasmic mutant of tobacco resistant to streptomycin. ►*Nicotiana*, ►streptomycin

SR Motif: Serine/arginine-rich domains in RNA-binding proteins, involved in splicing pre-mRNA transcripts. They are required in the early steps of spliceosome assembly. One SR protein (SC35) alone is sufficient to form a committed complex with human β−globin pre-mRNA. Different single SR proteins commit different pre-mRNAs to splicing and different sets of SR proteins may determine the alternative and tissue-specific splicing within an organism. The SR proteins are regulated by phosphorylation/dephosphorylation. The ASF/SF2 SR-binding protein is required for genomic stability (Li X, Manley JL 2005 Cell 122:365). ►splicing, ►spliceosome, ►processing, ►primary transcript, ►introns, ►tissue specificity, ►DEAD-box proteins, ►DEAH box proteins, ►ESE; Tian H, Kole R 2001 J Biol Chem 276:33833; Shopland LS et al 2003 J Cell Biol 162:981.

SR Proteins: ►SR motif

SRA (steroid-receptor RNA activator): An RNA coactivator of steroid hormone receptors. ►nuclear receptors, ►RNA regulatory; Lanz RB et al 2002 Proc Natl Acad Sci USA 99:16081.

SRB: Protein stabilizing RNA polymerase II binding to general transcription factors and it phosphorylates pol II RNA polymerase. SRBs occur in eukaryotes from yeast to humans and have the forms, SRB 2, 4, 5, 6, 7, 10, and 11. The CDK-like *SRB* genes involve mutation control and promotion/suppression of transcription by phosphorylating the carboxyl-terminal of RNA pol II. ►transcription factors, ►transcription complex, ►kinase, ►CDK, ►RNA polymerase holoenzyme, ►open promoter complex, ►regulation of gene activity, ►TUP, ►mediator complex; Hampsey M, Reinberg D 1999 Curr Opin Genet Dev 9:132; Carlson M 1997 Annu Rev Cell Dev Biol 13:1.

SR-BI (scavenger receptor class BNI): A high-density lipoprotein receptor mediating cholesterol uptake and secretion into the bile. PDZK1 protein regulates its expression and controls HDL level. ►high-density lipoprotein, ►cholesterol; Nakamura T et al 2005 Proc Natl Acad Sci USA 102:13404.

SRBC: Sheep red blood cell.

SRC: Rous sarcoma virus oncogene of chicken. Its product is a protein-tyrosine kinase, a cellular signal transducer. Its homology domains SH2 and SH3 are present in several cytoplasmic mediator and adaptor proteins in the signal transduction pathways in different organisms. These domains bind phosphotyrosine or proline-rich residues. Phosphorylation, dephosphorylation and proteolysis regulate SRC. In humans, SRC is in chromosome 20q12-q13. SRC may be involved in both RAS-dependent and RAS-independent signaling pathways and may lead through either FOS or MYC to transcription factors. This family of non-receptor kinases includes Src, Yes, Fgr, Fyn, Lck, Hck, Blk, Zap70, Tec. Csk$^-$ cells display increased Src, Fyn and Lyn activity. The Src proteins may also have autoinhibitory function. The Cbl oncogene acting downstream of Src is responsible for bone resorption in osteoporosis. c-SRC has been implicated in various types of cancers. Src enzyme with mutation Arg-388 to alanine with about 5% enzyme activity can be rescued to about half of normal activity by the use of the tautomerism-inducing small molecule of imidazole under in vivo conditions (Qiao Y et al 2006 Science 311:1293). ►oncogenes, ►SH2, ►SH3, ►signal transduction, ►Tec, ►Zap-70, ►Yes, ►Fgr, ►Fyn, ►Lck, ►Hck, ►Blk, ►Csk, ►Cbl, ►osteoporosis, ►imidazole, ►tautomeric shift, ►TCR; Schlessinger J 2000 Cell 100:293.

SRC-1 (steroid receptor coactivator-1): Enhances the stability of the transcription complex controled by the progesteron receptor. It is actually a co-activator of histone acetyltransferase and mediates the access of the transcription complex within the nucleosome. ►transcription, ►progesteron, ►histone acetyltransferase, ►N-CoR, ►TGF; Liu Z et al 2001 Proc Natl Acad Sci USA 98:12426; Auboef D et al 2002 Science 298:616.

SRE: A cis-acting enhancer element responding to serum induction: CC(AT)$_6$GG (CarG box) is present in all serum response factor regulated genes. ►MADS box, ►TCF, ►serum response element, ►serum response factor

SREBP (sterol regulatory element binding proteins): Hairpin shaped it is found in the membrane of the endoplasmic reticulum (ER). The N-terminal domain is in the cytosol and acts as a basic helix-loop-helix transcription factor. The C-terminus is also in the cytosol and it is complexed with the cleavage activating protein SREBP-SCAP, which has 8 membrane-spanning regions. When the ER is low on sterols the complex is transferred to the cleavage compartment. In case of sterol overload the complex is sequestered into the ER and there is no cleavage. SREBP controls cholesterol and lipid homeostasis (Yang F et al 2006 Nature [Lond] 442:700). Cholesterol and oxysterol bind to a hexapeptide (MELADL) of the SREBP-escort protein Scap, and causes Scap to bind to Insig anchor proteins. Oxysterols bind to Insigs, causing Insigs to bind to Scap. Mutational analysis of the six transmembrane helices of Insigs reveals that the third and fourth are important for binding Insigs to oxysterols and to Scap. Thus Insigs are oxysterol-binding proteins, explaining the long-known ability of oxysterols to inhibit cholesterol synthesis in animal cells (Radhakishnan A et al 2007 Proc Natl Acad Sci USA 104:6511). S1P and S2P process SREBP. SREBP may suppress the insulin regulator IRS-2 and has a potential therapeutic role as a drug target in some diabetes (Ide T et al 2004 Nature Cell Biol 6:351; Sun L-P et al 2007 Proc Natl Acad Sci USA 104:6519). ►sterols, ►cholesterol, ►oxysterol, ►statins, ►Insig, ►Rip, ►SCAP, ►lipodystrophy, ►diabetes, ►ATF2; Shimano H 2001 Progr Lipid Res 40:539; Dobrosotskaya IY et al 2002 Science 296:879.

SRF (serum response factor): A transacting regulatory protein binding to SRE and regulating serum-induced gene expression. ►trans-acting element, ►serum response element; Kim SW et al 2001 Oncogene 20:6638.

SRK: Self-compatibility protein receptor kinase. An Srk1 kinase associated with Cdc25 phosphatase controls mitotic checkpoints (López-Avilés S et al 2005 Mol Cell 17:49). ►self-incompatibility, ►Cdc25, ►checkpoint; Takayama S et al Nature [Lond] 413:534.

sRNA: ►suppressor RNA

sRNA: ►RNAi

S-RNase: A ribonuclease responsible for pollen rejection (with factor HT) in self-incompatible plants, ►self-incompatibility; Luu DT et al 2001 Genetics 159:329.

SRP (signal recognition particle): An element of polypeptide transport systems through the membranes of the endoplasmic reticulum in eukaryotes and to the plasma membranes in prokaryotes. Some of the polypeptides are inserted in the endoplasmic reticulum and others are destined toward the cell membrane for secretion. The subunit (Ffh), which recognizes the signal sequence as well as the α subunit (FtsY) of its receptor (SR) are GTPases. The SRP has a variable length RNA component (4.5 S in prokaryotes and 7SL in eukaryotes). The Ffh subunit has an N domain where the signal peptide binds and the G domain of the GTPases. Adjacent to the signal-binding pocket the methionine-rich M domain forms a small globular structure, which folds into helix-turn-helix type motif for binding the RNA component. ►7SL RNA, ►endoplasmic reticulum, ►signal sequence recognition particle, ►signal peptide, ►Fts; Batey RT et al 2000 Science 287:1232; Oubridge C et al 2002 Mol Cell 9:1251.

SRS2: A helicase protein; displays single-strand Dna-dependent ATPase activity. When mutant, it increases sensitivity to genotoxic agents and may cause chromosome loss. Normally, it suppresses RAD51-dependent recombination by dislodging RAD51 protein but when it is defective, it increases recombination. ►RAD51, ►PCNA; Krejci L et al 2003 Nature [Lond] 423:305; Veaute X et al 2003 Nature [Lond] 423:309.

SRY (sex-determining region Y, called earlier TDF [testis determining factor]): A mammalian gene in the short arm of the Y chromosome (Yp11.3) responsible for testis determination and for the development of pro-B lymphocytes. The protein (223 amino acids) is a member of the high-mobility group proteins. In mice this HMG protein includes a large CAG trinucleotide repeat tract, which functions as a transcriptional trans-activator and it is required for male sex expression. The expression of SRY initiates the formation of the Müllerian inhibiting substance (MIS) and the synthesis of testosterone. For normal function of SRY three insulin or insulin-like receptor tyrosine kinases are required in mouse (Nef S et al 2003 Nature [Lond] 426:291). SRY contains the DNA minor groove-binding domain, the HMG box that is conserved among mammals. Mutations affecting human sex-reversal are generally within the HMG box. Sex reversal may be the result of defect in the nuclear localization signals of SRY and the N-terminal signal is not recognized by the nuclear receptor protein β-importin (Harley VR et al 2003 Proc Natl Acad Sci USA 100:7045). A human chromosome 17q24.3-q225.1-located gene SRA-1 (sex reversal autosomal) may also be controlled by SRY. Another sex-reversal gene was identified in human chromosome 9p24. In rodents, several *tda* (testis-determining autosomal) alleles exist. In the short arm of the human X-chromosome (Xp21) there is the DSS (dosage-sensitive sex reversal) locus that may be involved in sex reversal (male→female) in case of its duplication or in case the SRY alleles are weak. Deletion in the 160 kb DSS region (DAX) does not affect male development but may cause adrenal hypoplasia. Apparently the DAX gene encodes nuclear hormone receptors. DAX1 (Nr0b1) expression ceases early in testis development but persists through the development of the ovaries. There is/are the SRYIF inhibitory factor(s) involved in gonadal differentiation. The voles (rodents) *Ellobius lutescens* 2n = 17, XO constitution in males, and females and *E. tancrei* 2n = 32–54, XX in both males and females, as a normal condition do not have SRY, whereas SRY is present in other rodents as well as in other eutherian and marsupial species. The *Sry* gene is specifically expressed in the substantia nigra of the brain of rodent males in tyrosine hydroxylase-expressing neurons and it controls specific motor behavior. This condition is not the consequence of gonadal hormone action yet it contributes to the sexual differentiation of the brain (Dewing P et al 2006 Curr Biol 16:415). ►high mobility group proteins, ►SOX, ►TDF, ►Müllerian ducts, ►ZFY, ►animal hormones, ►Wolffian ducts, ►campomelic dysplasia, ►eutherian, ►adrenal hypoplasia, ►Swyer syndrome, ►trinucleotide repeat, ►LINE; Murphy EC et al 2001 J Mol Biol 312:481; Yuan X et al 2001 J Biol Chem 276:46647.

σ^S (RpoS): Required for the expression of many growth phase and osmotically regulated prokaryotic genes. ►σ subunit of RNA polymerase.

SS Blood Group: ►MN blood group

SSA (single-strand annealing): A repair mechanism apparently employed by prokaryotes and eukaryotes when one or more units of tandem repeats are eliminated by bacterial RecBCD-mediated degradation or by other nucleases (see Fig. S112). ►DNA repair; Van Dyck E et al 2001 EMBO Rep 2:905; Paques F, Haber JE 1999 Microbiol Mol Biol Rev 63:349, Diagram redrawn after Sugawara N et al 2000 Mol Cell Biol 20:5300.

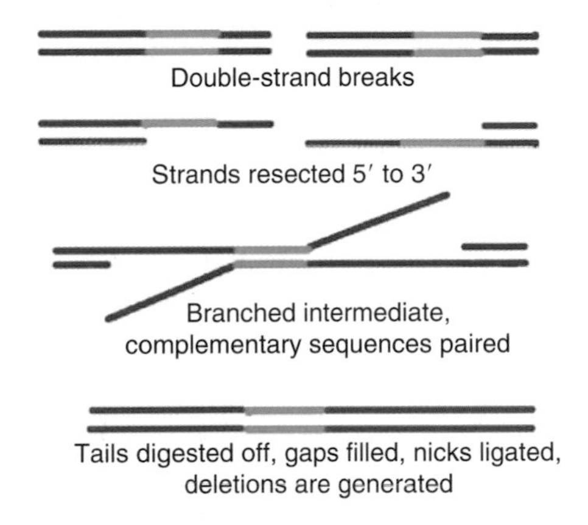

Figure S112. SSA repair

SSA: ►Hsp70

Ssb: ►Hsp70

SSB: A single-strand (DNA) binding protein; in yeast it is encoded by gene *RPA1*. ►recombination molecular mechanism, ►replication protein A, ►DNA repair (SOS repair), Reddy MS et al 2001 J Biol Chem 276:45959.

SSC: Somatic stem cell. ►ESC, ►stem cell

SSC: 1 × SSC is a solution of 0.15 M NaCl + 0.015 M sodium citrate, frequently used as a solvent for nucleic acids. ►DNA extraction

Ssc: ►Hsp70

SSCA (single-strand conformation analysis): ►single-strand conformation polymorphism

SSCP: ►single-strand conformation polymorphism

ssDNA: A single-strand DNA.

SSE: same as HSP, ►HSP

SSEA (stage-specific embryonic antigen): Can be used as surface markers for undifferentiated embryonic stem cells.

Ssh: ►Hsp70

Ssi: ►Hsp70

SSI-1 (STAT-induced STAT inhibitor): ►STAT

SSLP: ►simple sequence length polymorphism

SSM (slipped-strand mispairing): ►unequal crossing over

SSN6: The yeast factor abolishing glucose repression of SUC2 invertase and regulator of nucleosome positioning in the chromatin. Other *SSN* genes are components of the RNA polymerase II complex and are negative regulators of transcription. ►*SUC2*, ►SNF, ►catabolite repression; Li B, Reese JJ 2001 J Biol Chem 276:33788.

SSNC (second-site non-complementation): Two heterozygous recessive mutations at different chromosomal sites exhibit mutant phenotype by some sort of interaction in contrast to expectation that non-allelic recessive mutations would be complementary. ►cis-trans test, ►allelic complementation; Halsell SR, Kiehart DP 1998 Genetics 148:1845.

SSPA (significant segment pair alignment): Each sequence is used to align with a standard other sequences. It accommodates indels and other gaps. The information obtained can be used for searches for similarities in databases. ►indel, ►BLOSUM, ►ITERALIGN; Brocchieri L, Karlin S 1998 J Mol Biol 276:249.

SSR (small segment repeat or simple sequence repeat, microsatellite): These clusters of single to multiple nucleotides within the genome occurring per ~6 to ~30 to ~80 kb in plants and per ~6 kb in mammals have been used as chromosomal markers in various types of genetic studies. They may show high degree of mutability. In bacteria, only very short repeats occur (contingency loci) and they facilitate bacterial adaptation. ►microsatellite, ►phase variation; Qi X et al 2001 Biotechniques 31:358; Bacon AL et al 2001 Nucleic Acids Res 29:4405; Moxon R et al 2006 Annu Rev Genet 40:307.

S

SsrA (tmRNA, 10Sa RNA): ►protein repair

SSRP1 (structure-specific recognition protein): A high-mobility group protein probably targets FACT to the nucleosomes and facilitates gene transcription by RNA polymerase II. ►FACT, ►high-mobility proteins, ►transcription factors, ►nucleosome remodeling; Bruhn SL et al 1993 Nucleic Acids Res 21:1643.

SSV: A simple sequence variation in DNA.

ST-1: A single-stranded DNA phage, related to ϕX174 and G4. ►map, ►ϕX174 and, ►G4, Figure S113 is by courtesy of D. N. Godson.

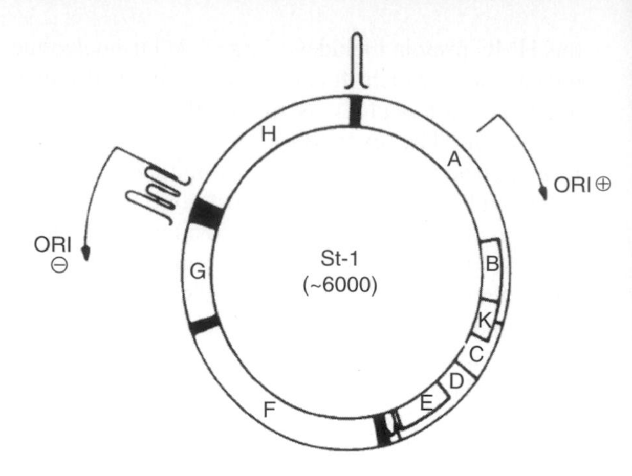

Figure S113. St-1 single-stranded DNA phage

Stab Culture: The microbial inoculum is introduced into agar medium by a stabbing motion of the inoculation needle or loop for the purpose of propagation.

Stabilizing Selection: ►selection types

Stable RNA: Ribosomal and tRNA that persists long in the cell in comparison to mRNA that may be degraded in minutes. ►rRNA, ►tRNA, ►mRNA

Stack: Sequence Tag Alignment and Consensus Knowledgebase: http://ziggy.sanbi.ac.za/stack/stacksearch.htm. ►gene indexing

Stacking Gel: A porous gel on top of the SDS polyacrylamide electrophoresis running gel. It concentrates large volumes into a thin, sharp band at the beginning of the run and thus permits sharper separation of proteins. ►electrophoresis, ►SDS

STAGE (sequence tag analysis of genomic enrichment): A method similar to ChIP-chip for detecting protein factor binding regions but use extensive short sequence determination rather than genomic tiling arrays. ►ChIP, ►ChIP-chip, ►tiling, ►STAT; mapping chromosomal STAT: Bhing AA et al 2007 Genome Res 17:910.

Staggered Cuts: After a double-stranded DNA is cut, the length of the two polynucleotides is unequal.

Staggered cuts

Staggered Extension Process (StEP): A method for in vitro mutagenesis and recombination of polynucleotides. By using it, mutant proteins can be generated in vitro. The template sequences are primed, then repeated cycles of denaturation and very short annealing and extension follows. In each cycle, the extended fragments anneal to different templates depending on complementarity, and the extension continues until full length is formed. The template switching generates recombined sequences from different parental sequences. ►molecular evolution, ►directed mutation, ►RNA-peptide fusions; Zhao H et al 1998 Nature Biotechnol 16:258; Xia G et al 2002 Proc Natl Acad Sci USA 99:6597.

Stains: For light microscopic examination of chromosomal specimens aceto-carmine, aceto-orcein or Feulgen stains are commonly used. Preparation: 0.5–1 g dry *carmine* powder is boiled for about half an hour under reflux in 100 mL 45% acetic or propionic acid. Orcein 1.1 g is dissolved in 45 mL glacial acetic acid or propionic acid and filled up to 100 mL by H_2O. Filter and store stoppered at about 5°C. Feulgen: 1 g leuco-basic fuchsin is dissolved by pouring over 200 mL boiling H_2O, shake, cool to 50°C, filter, add 30 mL 1/N HCl, then 3 g $K_2S_2O_5$, allow to bleach in dark for 24 h stoppered. Decolorize by 0.5 g carbon, shake 1 min then filter and store stoppered in refrigerator. For carmine or orcein staining, fix specimens in Carnoy and stain. For Feulgen, fix in Farmer's solution for a day, rinse with water, hydrolyze at 60 °C for 4–10 min. (The duration of hydrolyzation is critical and may need adjustment for each species). Rinse with stain for 1–3 h. Tease out tissue in 45% acetic acid, remove debris, flatten by coverslip and examine. May need overstaining with carmine if Feulgen staining is poor. For histological staining a variety of other stains may be used such as haematoxylin, methylene blue, ruthenium red, malachite green, sudan black, coomassie blue, fluorochromes, etc. may be employed. ►fixatives, ►sectioning, ►light microscopy, ►C-banding, ►G-banding, ►Q-banding, ►chromosome painting, ►harlequin stain, ►FISH, ►fluorochromes, ►aequorin

Stamen: The male reproductive organs of plants composed of the anther, which contains the pollen and the filament (see Fig. S114). ►anther, ►pollen

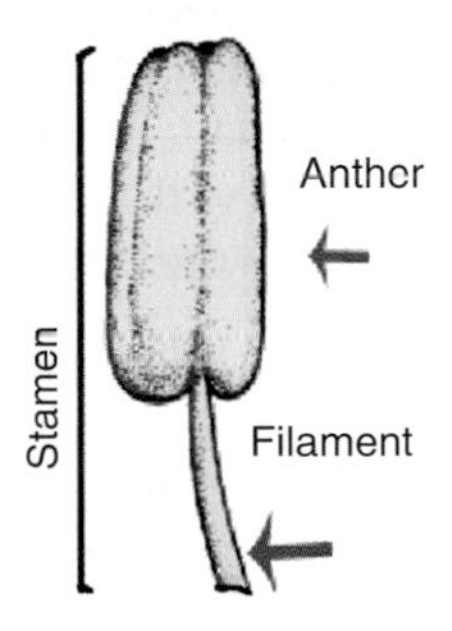

Figure S114. Stamen

Stamen Hair Assay: ►*Tradescantia*

Stamina: The capacity to endure, vigor.

Staminode(s): Infertile stamen(s).

Staminate Flower: The male flower of monoecius or dioecious plants.

Stammering: A most serious form of speech defects, *stuttering*. The sufferers usually cannot talk fluently and after involuntary stops repeat syllables or entire words. The frequency of stammering is very frequent among Japanese and very unusual among American Indians. Autosomal dominant inheritance seems to be involved. Apparent linkage to chromosomes 1, 5 and 7 was detected and to chromosome 12 the linkage displayed lod scores 4.61 to 3.51 (Riaz N et al 2005 Am J Hum Genet 76:647). ►lod score, ►stuttering

Standard: An accepted point of reference, e.g., standard wild type. ►control

Standard Deviation (parametric symbol σ): A measure of the variability of members of the populations $s = \sqrt{\text{variance}}$. ►variance, ►standard error, ►normal deviate

Standard Error: Measures the variation of the means of various samples of a population $s\sqrt{m}$ (by some authors standard error and standard deviation are used as synonyms); standard error of proportions or fractions or frequencies is shown by the boxed formula where p = proportion or frequency and *n* is the population size (or number of measurements).

$$\sqrt{\frac{p(1-p)}{n}}$$

Standard Type: Generally means the wild type used as a genetic reference. ►wild type

Standardized Fitness: ►selection coefficient

Stanford–Binet Test: A modified Binet intelligence test. ►Binet test, ►human intelligence

Stanniocalcin: The protein hormone inhibiting Ca^{2+} uptake and stimulating phosphate adsorption in fishes and mammals. (See Varghese R et al 1998 Endocrinology 139:4714).

Staphylococcus aureus: A bacterium responsible for toxic shock, scarlet fever and hospital-acquired infections and various inflammations. In some strains a conjugative plasmid has been detected that harbors trimethoprim, β-lactam, aminoglycoside as well the Tn*1546* transposon carrying also vancomycin resistance. The genome sequence and partial annotations of 27 bacteria phages of *S. aureus* are available (Kwan T et al 2005 Proc Natl Acad Sci USA 102:5174). ►toxic shock syndrome, ►aminoglycosides, ►β-lactamase, ►trimethoprim, ►vancomycin,

►inflammasome; Kuroda M et al 2001 Lancet 357:1225; Weigel LM et al 2003 Science 302:1569.

STAR (signal transduction and activation of RNA): Generally, a 200 amino acid protein domain associated with the cell and RNA splicing. The tripartite domain has a single RNA-binding site, a KH module of different length (2–14) and flanked by QUA1 (80) and QUA2 (30) amino acid sequences. These complexes have been detected in a wide range of eukaryotic organisms and they appear to be regulators of translation. STAR family proteins may repress *tra-2* and cause masculinization in females. ►Sam 68, ►sex determination; Stoss O et al 2001 J Biol Chem 276:8665.

STAR (signature-tagged allele replacement): May facilitate genetic analysis in cases where mutational studies are impractical due to the essentiality of the gene for viability. ►signature of a molecule, ►targeting genes; Yu Y et al 2001 Microbiology 147:431.

STAR (subtelomeric antisilencing region): An insulator type sequence that protects a gene from the action of a silencer if it lies in between the silencer and the gene. ►silencers, ►insulator

StAR (steroidogenic acute regulatory protein, 8p11.2): Enhances mitochondrial conversion of cholesterol into pregnenolone, an intermediate in steroid biosynthesis. Mutation in the coding gene results in deficiency of adrenal and gonadal steroidogenesis, and leads to congenital lipoid adrenal hyperplasia, an autosomal recessive disorder. ►cholesterol, ►steroid, ►pregnenolone, ►hormones, ►adrenal hyperplasia; Petrescu AD et al 2001 J Biol Chem 276:36970.

Starch: ►amylopectin

Starfish: *Asterias forbesi*, 2n = 36 (see Fig. S115).

Figure S115. Starfish

Stargardt Disease: A complex recessive degenerative disease of the retina evoked by environmental factors such as smoking and high cholesterol. In both, the early onset form (macular dystrophy with flecks) and the age-related macular degeneration (AMD), mutations in the ATP-binding cassette transporter of the retina (ABCR) gene are involved. ABCR contains 51 exons in chromosome 1p13-p21. Physically the product is located in the outer segment of the retinal rods (called also rim protein). The autosomal dominant form is caused by mutation in the ELOVL4 gene, involved in fatty acid chain elongation. In mice, so affected undigested phagosomes and lipofuscin and other fluorophores accumulate and that is followed degeneration of the retinal pigment epithelium (Karan G et al 2005 Proc Natl Acad Sci USA 102:4164). In many cases, the disease is responsible for blindness. The Sorsby fundus dystrophy, an autosomal dominant disease, appears somewhat similar but it is caused by a malfunction of the tissue inhibitor metalloproteinase-3 gene (TIMP3). ►ABC transporters, ►macular degeneration, ►eye diseases, ►macular dystrophy, ►retinal dystrophy, ►retinitis pigmentosa, ►lipofuscin, ►phagosome, ►fluorophore

Start: ►cell cycle

Start Codon: AUG in RNA that specifies either formylmethionine (in prokaryotes) or methionine (in eukaryotic cells). Note that the Met codon in mitochondria varies in different organisms. In some organisms other triplets may also initiate translation. ►genetic code

Start Point: The position in the transcribed DNA where the first RNA nucleotide is incorporated. ►transcription

Stasis: An equilibrium state without change.

STATs (signal transducers and activators of transcription): Cytoplasmic proteins, which become activated by SRC (SH2, SH3) mediated phosphorylation of tyrosine at around residue 700 and serine residues at the C-terminus through the action of Jak kinases, which are receptors for cytokine signals and enzymes at the cytosolic termini. PDGF, EGF and CSF catalyze phosphorylation also. The latter process requires the action of the 42 kDa MAPK or ERK2. Stat 1 mediates reactions to microbial and viral infections. Stat4 protein is essential for interleukin-12 mediated functions such as the induction of interferon-γ, mitogenesis and T lymphocyte killing and helper T lymphocyte differentiation. Disruption of STAT 2 & 3 cause embryonic lethality. Synthetic Sta3-inhibitory peptide (corresponding of the Tyr^{705} phosphorylation site of immunoglobulin G) reduced embryo transplantation by 70%. Normally phosphorylation of the luminal epithelium of the uterus by LIF (leukemia inhibitory factor) is required for implantation (Catalano RD et al 2005 Proc Natl Acad Sci USA 102:8585). STAT 5A & B are required for breast development in mice, and for the stimulation of T cell proliferation. Stat5 tetramerization is associated with leukemogenesis (Moriggi R et al 2005 Cell Metab 1:87). Stats 1, 3, 4 recognize and activate different genes by binding to the TTCC(C/G) GGAA (TTN5AA) sequences. Rac1 GTPase may regulate STAT 3 activation through phosphorylation. Stat6 prefers TTN6AA. Cooperative binding of Stats makes possible the recognition of variations at the different binding sites. The sequence-selective recognition resides in their amino terminal domains. In humans, there are at least seven STAT genes encoding proteins of 750 to 850 amino acids. The 130-amino-terminal residues bind to multiple sites in the DNA. Residues 600–700 are homologous to SH2 domains and mediate dimerization. Disruption of Stat activity leads to the loss of interferon-controlled immunity to pathogens. STAT genes can be found in mouse (m) and human (h) chromosomes: STAT1, STAT4: m 1, h 2q12-q33, STAT 3, STAT 5A & B m 11, h 12q13-q14–1, STAT 2 & 6: m 10, h 17q11.1-q22. Stat genes occur also in *Drosophila* and *Dictyostelium.*

The PIAS family of proteins includes negative regulators of STATs. Stat3 is active in many tumors, including breast cancer. It probably activates Bc*l*2 protein and is thus anti-apoptotic. Sta-21 protein inhibits breast cancer cells by reducing Stat3 binding to DNA (see Fig. S116) (Song H et al 2005 Proc Natl Acad Sci USA 102:4700). ►signal transduction, ►Jak-STAT pathway, ►signal transduction, ►SRC, ►lymphocytes, ►leptin, ►PDGF, ►EGF, ►CSF, ►SH2, ►SH3, ►SSI-1, ►interferon, ►RAC, ►Bc*l*2, ►apoptosis, ►STAGE; Davey HW et al 1999 Am J Hum Genet 65:959; Naka T et al 1999 Trends Biol Sci 24:394; Levy DE, Darnell JE 2002 Nature Rev Mol Cell Biol 3:651; Dupuis S et al 2003 Nature Genet 33:388.

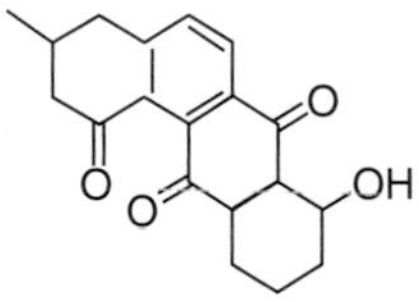

Figure S116. STA-21

Stathmin (op18): Regulates microtubule polymerization by affecting tubulins. The activity of Stathmin is controlled through phosphorylation by chromatin. Stathmin expressed in the amygdala of the brain modulates the psychological responses of innate and learned fear in mice (Shumyatsky GP et al 2005 Cell 123:697). ►spindle; Gavet O et al 1998 J Cell Sci 111:3333; Charbaut E et al 2001 J Biol Chem 27 6:16146; Niethammer P et al 2004 Science 303:1862.

Stathmokinesis: Mitotic arrest. ►spindle poison

Statins: Inhibitors of 3-hydroxy-3-methylglutaryl-coenzyme A reductase, an important enzyme of cholesterol biosynthesis. Cholesterol-lowering drugs reduce the incidence of Alzheimer disease too. Statins are useful also for the treatment of hypertension and autoimmune diseases by inhibiting MHC class II molecules and by shifting from T_h1-type pro-inflammatory cytokines to T_h2-type cytokines. ►cholesterol, ►lovastatin, ►SREBP, ►Alzheimer disease, ►autoimmune disease, ►MHC, ►T_h, ►cytokine; Fassbender K et al 2001 Proc Natl Acad Sci USA 98:5856.

Stationary Phase: The population size is maintained without increase or decrease. At this stage the genetic material may not replicate and the mutations occurring in bacteria are expected to be due to recombination. Mutation (chromosome breakage) in the stationary phase cells of higher eukaryotes may be one of the causes of cancer. ►growth curve; Bull HJ et al 2001 Proc Natl Acad Sci USA 98:8334.

Stationary Renewal Process: An assumption—based on a paper of R.A. Fisher (1947 Phil Trans Roy Soc B 233:55)—that crossing over is formed as a regular sequence starting from the centromere and the length between two adjacent crossovers always following the same distribution. The idea that crossing overs begin at the centromere and proceed toward the telomere turned out to be incorrect yet the stationary renewal process gained entry into several later models of recombination, interference and mapping functions. ►mapping function; Zhao H, Speed TP 1996 Genetics 142:1369.

Statistic: An estimate of a property of an observed set of data (e.g., mean) that bears the same relation to the data as the parameter does to the population. ►statistics

Statistics: A mathematical discipline that assists in collecting and analyzing data; guides to make conclusions and predictions based on the analysis and reveals their trustworthiness by determining probability or likelihood. Authors with inadequate training in this discipline might have incorrectly used statistical procedures in some published papers (Giles J 2006 Nature [Lond] 443:379; ►fluctuation test). Classical genetic analyses usually require statistical methods. Statistics is used for collections of facts on demography, industrial productivity, trade, etc., such as in statistical yearbooks. Statistics cannot replace the need for sound biological data (Spence MA et al 2003 Am J Hum Genet 72:1084). *Sufficient statistics* reduces the information to that needed for the parameter of interest and avoids nuisance parameters. ►probability, ►Bayes' theorem, ►likelihood, ►maximum likelihood, ►non-parametric tests, ►inference statistical, ►nuisance parameter; Robbins LG 2000 Genetics

S

154:13; assistance for the use of the most frequently needed statistical procedures: http://faculty.vassar.edu/~lowry/VassarStats.html; http://www.ruf.rice.edu/~lane/rvls.html.

Statistical Mechanics: Extrapolates from the microscopic structure to macroscopic features.

Statolith: Granules that are believed to sense gravity in cells. ►gravitropism

Stature in Humans: Influenced by environmental causes such as nutrition, disease, and injuries, by the use of various medications and by simple or complex genetic factors. Heritability of human height commonly exceeds 80%. Prenatal anomalies of bone length may be determined by prenatal diagnosis. QTL analysis (Hirschorn JN et al 2001 Am J Hum Genet 69:106) of Scandinavian and Canadian populations involving 2327 individuals revealed linkage of height to 6q24-q25 (lod score 3.85), 7q31.3-q36 (lod score 3.40), 12p11.2-q14 (lod score 3.35), 13q32-q33 (lod score 3.56) and 3.26 (lod score 3.17, Wiltshire S et al 2002 Am J Hum Genet 70:543). The most common types of genetically determined human dwarfism and other defects involving reduced stature are: achondroplasia, hypochondroplasia, achondrogenesis, osteochondromatosis, dyschondrosteosis, Russel-Silver syndrome, Smith-Lemli-Opitz syndrome, Opitz-Kaveggia syndrome, dwarfism, SHORT, Aarskog syndrome, Noonan syndrome, Hirschsprung disease, Turner syndrome, Trisomy, Mulibrey nanism, hairy elbows, Pygmy. ►growth hormone, ►growth retardation, ►limb defects, ►exostosis, ►regression, ►lod score QTL

STC (sequence-tagged connector): ►genome project

STCH: ►Hsp70

STE (sterile): Proteins are pheromone receptors and scaffolding proteins, and coordinate and organize the signal transduction paths in budding yeast. Ste3 and Ste2 are receptors of the α and a mating type factors, respectively. Ste7 is MAPKK, Ste4 is the β subunit, Ste18 is the γ subunit of the G trimeric proteins, Ste5 is a member of the MAPK cascade, Ste11 is MAPKKK, Ste 20 is a PAK/MEKKK protein and it is required for the MAPK activation of Gβγ. A Ste20-like protein kinase Mst regulates chromatin structure and apoptosis. Ste12 is a transcription factor. In fission yeast, the homologs *Byr* are extra-cellular signal regulated kinase homologs of MEK and ERK. (The *Ste* [*Stellate*, 1–45.7] locus of *Drosophila* encodes protein crystals in the primary oocytes). ►signal transduction, ►G proteins, ►pheromones, ►MAPK, ►MEK, ►MEKK, ►MP1, ►FUS3, ►KSS1, ►PAK, ►mating type determination in yeast; Graves JD et al 2001 J Biol Chem 276:14909; Ura S et al 2001 Proc Natl Acad Sci USA 98:10148; Ge B et al 2002 Science 295:1291.

Steady State: In a reaction enzyme-substrate concentration and other intermediates appear constant over time but input and output are in a flow.

Steel Factor (stem cell factor): A 40–50 kDa dimeric protein produced by the bone marrow and other cells and migrates to the hematopoietic stem cells. Its receptor is a transmembrane protein tyrosine kinase. ►cell migration, ►hematopoiesis, ►tyrosine kinase, ►microphthalmos, ►Kit oncogene, ►stem cell factor

Steer: An emasculated bovine male animal. ►emasculation

Steinernema carpocapsae: A nematode. ►*Caenorhabditis*, ►*Xenorhabdus*

Steinert Disease (dystrophia myotonica 1, 19q13.2-q13.3): Dominantly inherited diseases affecting different tissues and it is caused by duplication of the CTG triplet in the DNA. ►myotonic dystrophy

Stele: The core cylinder of vascular tissues in plant stems and roots. ►root

Stem Cell Factor (SCF/M-CSF, 5q33.2-q33.3): Required for the normal development of B lymphocytes. The Kit oncogene product is its receptor. ►lymphocytes, ►B cells, ►Kit oncogene, ►M-SCF; Broudy VC 1997 Blood 90:1345; Smith MA et al 2001 Acta Haematol 105:143.

Stem Cells: Stem cells of animals are not terminally differentiated, can divide without limit and when they divide the daughter cells, can remain stem cells or terminally differentiate in one or more ways (Hock H, Orkin SH 2005 Nature 435:573), or they may have a restricted potential for differentiation (*transit amplifying cell*) and with time only produce differentiated cells. When human embryonic stem cells (hESC) were encapsulated in (3D HA) hyaluronic hydrogels (but not within other hydrogels or in monolayer cultures on HA), hESCs maintained their undifferentiated state, preserved their normal karyotype, and maintained their full differentiation capacity (Gerecht S et al 2007 Proc Natl Acad Sci USA 104:11298). Among the 300–500 stem cell lines (existing in 2007), 59 human embryonic stem cell lines from 17 laboratories worldwide were characterized. The lines were not identical, however, and differences in expression of several lineage markers were evident, and several imprinted genes showed similar allele-specific expression patterns, but some gene-dependent variation was observed. Some female lines expressed readily detectable levels of XIST, whereas others did not. No significant contamination of the lines with mycoplasma, bacteria or cytopathic viruses was detected. Despite diverse genotypes and different techniques used for derivation and maintenance, all lines exhibited similar expression patterns for several markers of human embryonic stem cells (International Stem Cell Initiative 2007 Nature Biotechnol 25:803). Differentiation could be induced within the same hydrogel by simply altering soluble factors. Mammalian embryonic stem cells, derived from the inner cell mass of the blastocyst, are considered *pluripotent*, i.e., endowed with the potential to develop any type of cells under appropriate conditions. Pluriptency of embryonic stem cells is mediated by transcription factors Oct4, Sox2 and NANOG, which co-occupy substantial portion of their target genes, encoding mainly other transcription factors and by transcriptional regulatory circuits maintain the cell's self-renewal (Boyer LA et al 2005 Cell 122:947; Wang J et al 2006 Nature [Lond] 444:364). Nanog expression in mouse is upregulated in embryonic stem cells by binding brachyury (T) and STAST3 to its enhancer element. Nanog then blocks bone morphogenetic protein-induced differentiation of mesoderm by interacting with Smad1 and interfering with the recruitment of co-activators to the Smad transcriptional complexes (Suzuki A et al 2006 Proc Natl Acad Sci USA 103:10294). Under conditions of culture suitable for embryonic stem cells, adult mouse fibroblasts required Oct3/4, Sox2, c-Myc and Kfl4 for the formation of pluripotent stem cells but Nanog was not necessary. These induced pluripotent stem cells (iPS) caused tumors in nude mice containing tissues from all three germlayers. When iPS was injected into blastocysts, mouse embryonic development ensued (Takahashi K, Yamanaka S 2006 Cell 126:663). Laser-assisted injection of ES cells into eight-cell–stage embryos efficiently generates viable and healthy mice that contain no more than 0.1% host cell contamination. These mice, derived from either heterozygous or homozygous mutant ES cells, can be used directly in phenotypic analyses. The mutant phenotypes in these mice are indistinguishable from those observed in mice derived by conventional breeding (Poueymirou WT et al 2007 Nature Biotechnol 25:91).

An important development in stem cell research is the simultaneous and independent discovery of new methods for the production of pluripotent stem cells from adult somatic cells. These methods apparently obviate the need for using human embryos to which moral, ethical and legal objections exist. Furthermore the technology required for harvesting and processing human eggs has several impediments. The new methods reprogram somatic cell nuclei to an undifferentiated state by the introduction of four transcription factors Octa4, Sox2, Nanog and Lin28 (Yu J et al 2007 Science 318:1917) or Oct3/4, Sox2, Klf4 and c-Myc (Takahashi K et al 2007 Cell 131:861). In both sets of experiments the results were comparable to that obtainable by embryonic stem cells in cell morphology, proliferation, surface antigens gene expression, epigenetic status of pluripotent cell-specific genes and telomerase activity. The reprogrammed cells have the ability to differentiate into all three primary germ layers or form teratomas and appear suitable for all the uses embryonic stem cells. One cautionary note: the somatic cells (skin fibroblasts) were transformed by a retroviral vector containing a mouse viral receptor (Sic7a1). The vector integrated into more 20 sites of the human chromosomes and increased the risk of tumorigenesis by more than 20% in mice. The c-Myc reactivation might have been the major cause of the problem. c-Myc may not be absolutely necessary and other constructs can be used; the report in Science actually does not use it. The technical problems will certainly be ironed out in the future.

Identification of stem cells requires appropriate cell autonomous markers that distinguish them from other cells in a tissue. Only a few of such markers have been found and they seem to suppress translation as a means of preventing their differentiation (Siddall NA et al 2006 Proc Natl Acad Sci USA 103:8402).

It appears that in *Drosophila,* the double-stranded RNA processing enzyme (Dicer) controls stem cell development in as much as microRNAs are required for the germline stem cells to bypass the G1/S checkpoints and maintain ability for continuous cell division (Hatfield SD et al 2005 Nature [Lond] 435:974). They occur in various tissues. *Drosophila* ovarian germ cells may dedifferentiate into stem cells (Kai T, Spradling A 2004 Nature [Lond] 428:564). In *Drosophila* ovaries ISWI chromatin remodeling protein controls germline stem cell status maintenance whereas in the ovarian somatic stem cells the DOM chromatin remodeling factor is critical (Xi R, Xie T 2005 Science 310:1487). Adult mouse ovaries can rapidly generate hundreds of oocytes despite the small number of germ cells. Spermatogonial stem cells of adult mouse testis also display pluripotency in 27% and acquire embryonic stem cell properties (Guan K et al 2006 Nature [Lond] 440:1199). Bone marrow transplantation restores oocytes production in chemosterilized or ataxia telangiectasia mutated gene-deficient sterile animals. Also, oocytes carrying donor-derived genetic markers were detected after peripheral blood transplantation (Johnson J et al 2005 Cell 122:303) The conclusions of the Johnson et al 2005 paper have been questioned (Powell K 2006 Nature [Lond] 441:795, ►oocyte primary). Although some studies indicate karyotypic stability in stem cell cultures (Amit M et al 2000 Dev Biol 227:271), more recently aneuploidy (gain of human chromosomes 17q and 12) has been observed (Draper JS et al 2004

Nature Biotechnol 22:63). Spontaneous mutation rate/nucleotides in embryonic stem cell cultures after repeated passage is $\sim 10^{-9}$ and the alterations may involve also genomic copy number (45%), mtDNA sequence (22%) and promoter methylation (90%). These mutations may or may not have serious consequences depending on their selective value (Maitra A et al 2005 Nature Genet 37:1103).

Stem cells, by definition, are expected to multiply in an undifferentiated state, besides giving rise to specialized cells, in practice, after a period of time they undergo aging and decline (see Fig. S117). In mouse embryonic stem cells both X chromosomes are demethylated or have reduced methylation and this indicates that both X chromosomes are active although one is frequently lost (Zvetkova I et al 2005 Nature Genet 37:1274). A chromosome 2 (odd ratio 4.4) factor of mouse controls aging of hematopoietic stem cells. Apparently, inadequate DNA repair is responsible for the process (Geiger H et al 2005 Proc Natl Acad Sci USA 102:5102).

The gene expression profile in embryonic, neural and hematopoietic stem cells displays some overlaps yet distinct specificities are also evident (Ramalho-Santos M et al 2002 Science 298:597; Ivanova NB et al 2002 Science 298:601). More recent data indicate that embryonic stem cells obtained by fertilization or nuclear transfer are functionally equivalent (Brambrink T et al 2006 Proc Natl Acad Sci USA 103:933). The favorable environment (niche) of hematopoietic stem cells was located to the surface of the bone marrow at the surface of the cancellous (spongy) material of the trabeculae ossis (anastomosing spicules of the bones). At this location, the number of spindle-shaped, N-cadherin^{+}–CD45^{-} osteoblasts positively correlates with the number of hematopoietic stem cells (Zhang J et al 2003 Nature [Lond] 425:836). Notch activation is required for the increase of osteoblast number (Calvi LM et al 2003 Nature [Lond] 425:841). The frequency of hematopoietic and epidermal stem cells within these tissues is $\sim 1 \times 10^{-4}$ (Schneider TE et al 2003 Proc Natl Acad Sci USA 100:11412). Cryopreserved hematopoietic cord blood stem cells may remain functional after 15 years of storage (Broxmeyer HE et al 2003 Proc Natl Acad Sci USA 100:645). Umbilical cord blood is rich in hematopoietic and other stem cells are gaining increasing attention especially for HLA-mismatched recipients (Newcomb JD et al 2007 Cell Transplant 16:150). Human umbilical cord blood hematopoietic stem/progenitor cells transplanted in utero into mouse may differentiate into human hepatocyte-like cells with evidence of the expression of human hepatocyte-specific proteins as well as partially repair or protect liver damage induced by CCl_4 (Qian H et al 2006 Int J Mol Med 18:633). Pre-immune fetus develops a no injury human-rat xenograft in which the in utero transplantation of low-density mononuclear cells (MNCs) from human umbilical cord blood (hUCB) into fetal rats at 9–11 days of gestation led to the formation of human hepatocyte-like cells (hHLCs) with different cellular phenotypes (Sun Y et al 2007 Biochem Biophys Res Commun 357:1160).

Several investigators reported that the bromodeoxyuridine or other nucleoside (analog)-labeled bone marrow cells transplanted into mouse brains transformed into new neurons and supporting glial cells (e.g., Kopen GC et al 1999 Proc Natl Acad Sci USA 96:10711). These findings may be flawed (Burns TC et al 2006 Stem Cells 24:1121), however, because the original bone marrow cells so labeled may decay and other cells can pick up the label or even cell markers and the observations may be misleading (Kuan C-Y et al 2004 J Neurosci 24:10763).

Embryonic mouse stem cells differentiate into T lymphocytes in vitro if co-cultured with OP9 cells with the Notch receptors engaged by Delta-like 1 ligand. The T cells are functional in immunodeficient mice (Schmitt TM et al 2004 Nature Immunol 5:410). Also, human embryonic stem cells, transferred into human thymic tissue growing in immunodeficient mouse, resulted in the differentiation of T lymphoid cells that appeared functional by being able to express appropriate markers (Galic Z et al 2006 Proc Natl Acad Sci USA 103:11742). Adult neural stem cells may differentiate into cells of diverse germ layers with broad developmental potentials. The balance between self-renewal of stem cells and differentiation seems to be determined by the signal receptors or their phosphorylation in the neighboring cells. In the ovaries of mammals, the initial stem cells are arrested after a finite number of divisions and then enter into meiotic prophase. Functionally they resemble the meristem in plants. The *embryonic germ cells* (EG)

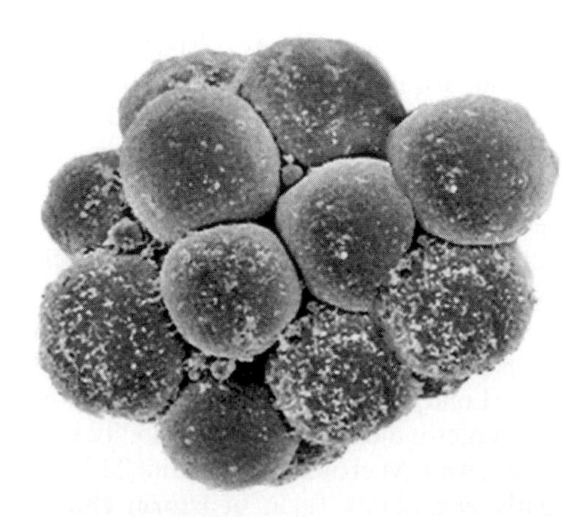

Figure S117. Blastocyst

have been used for transfer into mouse blastocysts and are capable to differentiate into most types of fetal tissues, including the germline.

Oocyte differentiation takes place in vitro in mouse embryonic stem cell medium in the presence of Oct4 transcription factor. Follicular structures are formed and blastocyst-like structures were observed. After 16 days, the oocytes were ready for meiosis but did not proceed beyond prophase (Hübner K et al 2003 Science 300:1251). Mouse embryonic stem cells, however, differentiated into primordial germ cells, erased methylation, characteristic for *Igf-2* and *H19* gene methylation in imprinting, and developed into haploid male gametes and were capable of—apparently normal—fertilization of eggs (Geijsen N et al 2004 Nature [Lond] 427:148). From embryonic stem cells the regeneration of viable, fertile adults, using tetraploid embryo complementation, has been reported (Eggan K, Jaenisch R 2003 Methods Enzymol 365:25).

The synthetic heterocyclic compound, SC1, permits the propagation of murine embryonic stem cells in an undifferentiated, pluripotent state under chemically defined conditions in the absence of feeder cells, serum, and leukemia inhibitory factor. Long-term SC1-expanded murine ES cells can be differentiated into cells of the three primary germ layers in vitro and can generate chimeric mice and contribute to the germ line in vivo. Biochemical and cellular experiments suggest that SC1 works through dual inhibition of RasGAP and ERK1 (Chen S et al 2006 Proc Natl Acad Sci USA 103:17266).

Sterile, hermaphrodite male mouse could produce fertile offspring when nuclear transfer cells were injected into normal, diploid blastocysts. Many of the offspring were chimeric but one was found, which transmitted genes from the sterile hermaphrodite to fertile daughters and thus overcame the problem of the "father," which originally lacked germ cells (Wakayama S et al 2005 Proc Natl Acad Sci USA 102:29). The feasibility of culturing pluripotent (endoderm, mesoderm, ectoderm) human embryonic stem cells (ES) opens new potential for the generation of tissues (Smith AG 2001 Annu Rev Cell Dev Biol 17:435). They may be used for the purpose of transplantation and study and treat neurodegenerative diseases (Jakel RJ et al 2004 Nature Rev Genet 5:136), spinal cord injuries, liver diseases, diabetes, immunological diseases, cancer and find means for the repopulating of hematopoietic cells, etc. (Krause DS et al 2001 Cell 105:369). Human central nervous system cells grown as neurospheres survive, migrate and express differentiation markers for neurons and oligodendrocytes after long-term engraftment in mice with spinal cord injuries. Locomotor activity was also restored (Cummings BJ et al 2005 Proc Natl Acad Sci USA 102:14069).

Stems cells derived from another individual may lead to adverse immune reaction. Neural stem cells repeatedly provided protection against inflammation of the central nervous system of mouse with an immune-like mechanism (Pluchino S et al 2005 Nature 436:266). Allogeneic, hematopoietic bone marrow cell transplantation into host cells treated by ionizing radiation or other immunosuppressive techniques, however, may become permanently tolerant of the foreign stem cells. The intolerance to a third-party donor, however, is not entirely solved (Sykes M, Nikolic B 2005 Nature [Lond] 435:620).

The immunological intolerance can be prevented when blastocysts formed from donated human oocytes (preferably by less than 30 years old donors) are used into which the nucleus of potential female or male recipient's own skin cells is transferred. The blastocysts were generated by this nuclear transfer by an average of ~24% success. The immunological identity of the stem cells to that of the donor was confirmed. In addition, this newer procedure solved another problem, i.e., rather than using mouse feeder cells it employed human feeders (Hwang WS et al 2005 Science 308:1777). These developments may assure progress toward medical use of embryonic stem cells. Some of the results, unfortunately, have been questioned and there is evidence on faking and fabricating some or much of the data. Authors have withdrawn the paper after an international uproar (Kennedy D 2006 Science 311:36) and on 20 January 2006, the Editor withdrew the invalid papers of this research group (Kennedy D 2006 Science 311:335).

The production of embryonic stem cells by somatic nuclear transfer into oocytes remained unsuccessful for primates. The failure has been attributed to an effect of the nuclear spindle in the oocyte due to maturation factor deficiency that prevented proper reprogramming of the introduced nucleus. Complete removal of the meiotic spindle from the karyoplasts was a key factor for reprogramming. Then the donor fibroblast nuclei were introduced into cytoplasts by electrofusion, incubated for 2 h to allow nuclear remodeling to occur, and subsequently activated and cultured to the blastocyst stage and 16% (35 out of 213) blastocyst formation took place. This new protocol prevented premature cytoplast activation and maturation promoting. The rate of successful development of stem cells was relatively low but their quality appeared identical to that of embryonic stem cells. Factor decline resulted in robust nuclear envelope breakdown and premature chromosome condensation and in significantly increased blastocyst development in *Macaca mulatta* monkeys.

The earlier failures were attributed to the detrimental effect of fluorochrome bisbenzimide (Hoechst

33342) and ultraviolet on the relatively transparent primate oocyte. The dye and the UV light were used for the removal of the spindle. These factors apparently reduced cytoplast mitochondrial DNA function. The rate of successful development of stem cells was relatively low, 0.2–3.4% per oocyte and 4–10% per blastocyst but their quality appeared identical to that of embryonic stem cells (Byrne JA et al 2007 Nature [Lond] 450:497).

In mouse pluripotent embryonic stem cells were generated from blastocysts, which were deficient in the Cdx2 transcription factor. Such blastocysts—generated by nuclear transfer—were unable to form functional trophectoderm, necessary for the development of the implanted embryo but formed normal internal cell mass, a suitable source for pluripotent embryonic stem cells. This procedure was named ANT (for altered nuclear transfer). The mouse model thus excluded the development of embryos from the blastocysts yet provided good source of pluripotent stem cells overcoming the ethical objections—in case of humans—to destroying and manipulating embryos (Meissner A, Jaenisch R 2006 Nature [Lond] 439:212). Withdrawing single cells from mouse blastomeres or using trophoblast cells permits the normal development of the embryo and the isolated cells form embryonic stem cell lines with pluripotent capabilities (Chung Y et al 2006 Nature [Lond] 439:216). Arrested human embryos can express pluripotency marker genes such OCT4, NANOG and REX1 and others, and can differentiate under in vitro and in vivo conditions into three germ layers. All the new lines derived from late arrested embryo have normal karyotype. Such stem cells may obviate the need for using normal embryonic stem cell lines (Zhang X et al 2006 Stem Cells 24:2669). Undifferentiated amniotic fluid stem cells (AFS) cells expand extensively without feeders, double in 36 h and are not tumorigenic. Lines maintained for over 250 population doublings retained long telomeres and a normal karyotype. AFS cells are broadly multipotent. Clonal human lines verified by retroviral marking were induced to differentiate into cell types representing each embryonic germ layer, including cells of adipogenic, osteogenic, myogenic, endothelial, neuronal and hepatic lineages (De Coppi P et al 2007 Nature Biotechnol 25:100).

Cells derived from single human blastomeres display the characteristics of embryonic stem cells and the biopsies show normal differentiation (Klimanskaya I et al 2006 Nature [Lond] 444:481). Adult bone marrow co-purifying with mesenchymal stem cells when injected into early blastocysts can develop into most types of cells (Jiang Y et al 2002 Nature [Lond] 418:41). (The results of this paper by Catherine Verfaille laboratory was difficult to replicate by other laboratories because of the complicated technique used [Nature 442:344] and there are some errors in details but the main conclusions seem valid) (Check E 2007 Nature [Lond] 447:763). Bone-marrow-derived stem cells regenerate into liver cells primarily after fusion with hepatocytes (Vassilopoulos G et al 2003 Nature [Lond] 422:901). There is evidence that bone marrow-derived stem cells develop into neurons, cardiomyocytes (heart muscle cell) and liver cells by fusion rather than by a transdetermination type mechanism (Alvarez-Dolado M et al 2003 Nature 425:968). Recent studies could not confirm the transdifferentiation of hematopoietic stem cell into cardiac myocytes (Murry CE et al 2004 Nature [Lond] 428:664; Balsam L et al ibid. p 668). Bone marrow-drived mesenchymal stem cells—despite their allogeneic origin—successfully repaired damage inflicted by myocardial infarction when administered through a catheter intra-myocardially in pigs, closely following the time (three days after) of the infarction. The procedure provided good healing without rejection (Amado LC et al 2005 Proc Natl Acad Sci USA 102:11474). Steel factor positive bone marrow-derived stem cells can repair the heart muscles by neovascularization and myogenesis after myocardial infarction in mice (Ayach BB et al 2006 Proc Natl Acad Sci USA 103:2304). Pluripotent murine stem cells can be converted to adipocyte lineage by bone morphogenetic protein (BMP-4) if methylation is prevented (Bowers RR et al 2006 Proc Natl Acad Sci USA 103:13022). A complex differentiation process converts human embryonic stem cells to endocrine cells capable of synthesizing the pancreatic hormones insulin, glucagon, somatostatin, pancreatic polypeptide and ghrelin. This process mimics, in vivo, pancreatic organogenesis by directing cells through stages resembling definitive endoderm, gut-tube endoderm, pancreatic endoderm and endocrine precursors to cells that express endocrine hormones (D'Amour KA et al 2006 Nature Biotechnol 24:1392).

Normal embryonic stem cells, however, can rescue mouse embryos with cardiac defects because of being deficient in the ld proteins, which are dominant antagonists of basic helix-loop-helix transcription factors (Fraidenraich D et al 2004 Science 306:247). Some postnatal rodent cardioblasts can develop into fully differentiated cardiomyocyte lines (Laugwitz K-L et al 2005 Nature [Lond] 433:647). Cardiac stem cells ameliorated myocardial infarction when delivered by intravascular injection in rats (Dawn B et al 2005 Proc Natl Acad Sci USA 102:3766) and similar observations were made in dogs (Linke A et al 2005 Proc Natl Acad Sci USA 102:8966). Using adenoviral virus-derived vectors and homologous recombination, defects in hypoxanthine phosphoribosyl transferase mutations could be corrected at high efficiency in mouse embryonic stem cells by random integration within and between genes (Obhayashi F et al 2005 Proc Natl Acad Sci USA 102:13628).

On polymer scaffolds the development of three-dimensional structures may be facilitated (Levenberg S et al 2003 Proc Natl Acad Sci USA 100:12741). It is conceivable that co-transplantation of hematopoietic stem cells and stem cell for other tissues using the same donor may become feasible. In order to maintain stem cells in culture in the pluripotent state, transcription factor Oct4 and the leukemia inhibitory factor (Lif), both must be expressed. When Lif is withdrawn various types of differentiation may begin. TCF and LEF may also be important. Actually this may be regarded as a teratocarcinoma type of growth. Hypoxia (ca. 5% oxygen)—rather than normal atmospheric conditions (21% O_2)—favors the maintenance of the embryonic, pluripotent stem cell state (Ezashi T et al 2005 Proc Natl Acad Sci USA 102:4783).

In order to force the cells into a specific type of differentiation appropriate and special culture conditions must be established. In bone marrow-derived stem cells the presence of mesenchymal stem cells, sonic hedgehog and retinoic acid signals synergistically promote the differentiation glutamatergic sensory neuron markers (Kondo T et al 2005 Proc Natl Acad Sci USA 102:4789). Retinoic acid, insulin (triiodothyronine) passage may lead to the differentiation of adipocytes. Employing c-Kit (a transmembrane tyrosine kinase) + Erythropoietin may lead to the development of erythrocytes. Macrophage colony stimulating factor + IL-3, IL-1 lead to the differentiation of macrophages. Fibroblast growth factor and epidermal growth factor combinations coax the ES culture to form astrocytes and oligodendrocytes. There will be a potential for genetically engineering embryonic stem cells (ES) for special medical purposes. Undifferentiated mouse ES cells may become tumorigenic, developing into teratomas or teratocarcinomas when introduced into an animal. A rich source of embryonic stem cells is the umbilical cord at birth. Epithelial cells, hair follicles, intestinal epithelium, multipotent brain cells, hematopoietic cell may be employed as potential ES. The full clinical exploitation of stem cell technology requires technical improvements (Humpherys D et al 2001 Science 293:95).

On 19 December 2000, the British Parliament approved greater freedom in embryonic stem-cell research (Ramsay S 2000 Lancet 356:2162). On 15 June 2006, the parliament of the European Union voted for lifting the ban on funding for human embryonic stem cells (Vogel G 2006 Science 312:1732). In the U.S., human embryonic stem cell research can be funded by government agencies only on the existing ca. 60 cell lines (9 August 2001). Many research workers are dissatisfied with this restriction. Although embryonic stem cells are supposed to have the potential to develop into any kind of differentiated cells—in fact, there are differences among embryonic stem cells because practically no two human beings are genetically identical—therefore, there is a need for the development of additional embryonic cell lines. (Monozygotic twins may not remain entirely identical genetically because of epigenetic changes during development). Stem cell research, thus, face great promises and great challenges; some of the problems are biological, others are ethical and regulatory. It is unfortunate that ethicists and public policy makers are not fully aware of the biology and some of the research workers are inclined to take political stands. The objection against the use of human embryonic stem cells (and destroying human embryos) could be eliminated if appropriate and effective means would be available for reprogramming somatic cells to embryonic state. One recent investigation fused human fibroblasts ($2n = 46$) with pluripotent embryo cells ($2n = 46$) into 92-chromosome cells, successfully securing their proliferation as reprogrammed somatic hybrid cells. Both component cells were appropriately marked and their fusion was verified. Similarly the embryonic state of the fusion product was established. This procedure of reprogramming would be of great significance if the chromosome number could be reduced to the normal level ($2n = 46$). Unfortunately, this is still a formidable obstacle to overcome (Cowan CA et al 2005 Science 309:1369).

There are many technical problems with human embryonic stem cell cultures because they—unlike mouse embryonic stem cells—require mouse fibroblast feeder cells and a lipid-rich bovine serum (ALBUMAX) for maintenance of stem cell condition. It appears that human cells take up from the animal products *N*-glycosyl neuraminic acid, which may cause rejection in humans if transplanted. In addition, the animal medium may transmit viruses of potential health hazard.

The mouse cells also require LIF (leukemia inhibitory factor). LIF binds a LIF receptor and glycoprotein 130, which activate the Jak/Stat3 signaling pathway sustaining the cells in undifferentiated state. These factors, however, are not sufficient for human embryonic stem cells. Human stem cells need FGF2 (fibroblast growth factor) and noggin (an inhibitor of bone morphogenetic protein, BMP) in addition to serum, to keep the human stem cells in undifferentiated state. This new medium is, however, a substantial progress in developing useful human stem cell cultures (Xu R-H et al 2005 Nature Methods 2:185).

No human embryos are supposed to be generated for the purpose of extracting stem cells or no fertilized and unused eggs generated by the process of in vitro artificial fertilization should be used for stem cell research. The objection to stem cell research originates from the belief that life begins at fertilization and extracting embryonic stem cells from 4–5 days old blastomeres amounts to violation of the sanctity of life (►ensoulment). The ethical problems may be dispelled if somatic stem cells occurring in placental, umbilical or fat tissues are used. Another, albeit tenuous, possibility is to generate blastomeres by inserting diploid somatic nuclei into enucleated human eggs and cloning them. Transplantation of nuclei harboring human disease genes into oocytes may make it possible to generate large quantities of embryonic tissues for laboratory studies of the mechanism and potentially the treatment of the disease. Not all cells originate from stem cells as some terminally differentiated animal cells may retain significant proliferative capacity (Dor Y et al 2004 Nature [Lond] 429:41). Stem cell antigen (Sca-1) is expressed in a small fraction of various cells such as of hematopoietic tissue, cardiac tissue, mammary gland, skin, muscle, testis, murine prostatic duct and the anti-apoptotic protein Bc*l*-2 may protect them to survive. The expression of Sca-1 and Bc*l*-2 may help in isolation and enriching the stem cell populations for therapeutic purposes (Burger PE et al 2005 Proc Natl Acad Sci USA 102:7180). Post-natal muscle-derived stem cells maintained proliferating ability for 300 doublings and displayed regenerating ability after transplantation into a mouse model of Duchenne muscular dystrophy even after 200 doublings. After that lower muscle regeneration occurred and loss of CD34 expression and loss of myogenic activity were observed. Nevertheless, this appeared a remarkable long-term self-renewal for non-embryonic stem cells (Deasy BM et al 2005 Mol Biol Cell 16:3323).

The exact mechanisms of the differentiation from stem cells into specific somatic cells are not entirely clear, yet compelling evidence is available for some cases of effectiveness in clinical applications. Therefore, continued research is indispensable (Quesenberry PJ et al 2005 Science 308:1121). In 2007, three laboratories succeeded to some extent in producing pluripotent stem cells without the use of eggs or embryos by applying specific transcription factors in culture introduced into cells by viral vectors (Okita K et al 2007 Nature [Lond] 448:313; Wernig M et al 2007 Nature [Lond] 448:318; Maherali M et al 2007 Cell Stem Cell 1:55). Unlike interphase zygotes, mouse zygotes temporarily arrested in mitosis can support somatic cell reprogramming, the production of embryonic stem cell lines and the full-term development of cloned animals. Thus, human zygotes and perhaps human embryonic blastomeres may become useful for stem cell research (Egli D et al 2007 Nature [Lond] 447:679). Earlier researchers concluded (McGrawth J, Solter D 1984 Science 226:1317) that nuclei transferred to enucleated zygote cannot support development in vitro.

There are various ethical and moral problems in transplantation of human neural stem cells into animals, particularly into non-human primates. These brain cells may alter the cognitive development of animals, particularly if the implantation takes place at early embryonal or post-natal periods (Greene M et al 2005 Science 309:385). To overcome the objections against the use of human embryonic stem cells and avoid the adverse immunological reactions to foreign cells, in 2007, pluripotent human embryonic stem cell (hESC) lines were obtained from blastocysts of parthenogenetic origin. Parthenogenesis was chemically induced. The parthenogenetic human embryonic stem cells (phESC) demonstrate typical hESC morphology, express appropriate markers, and possess high levels of alkaline phosphatase and telomerase activity. The phESC lines had a normal 46, XX karyotype and had been cultured from between 21 to 35 passages. The phESC lines form embryoid bodies in suspension culture and teratomas after injection to immunodeficient animals and give differentiated derivatives of all three embryonic germ layers (Revazova ES et al 2007 Cloning Stem Cells 9:432). ►meristem, ►microRNA, ►regeneration in animals, ►activin, ►nuclear transplantation, ►cryopreservation, ►niche, ►embryo research, ►oocyte, ►transplantation, ►grafting in medicine, ►graft rejection, ►transplantation of nuclei, ►pluripotent, ►Polycomb, ►therapeutic cloning, ►GSK3β, ►teratoma, ►hematopoiesis, ►hematopoietic stem cells, ►diabetes, ►Krabbe's leukodystrophy, ►mesenchyma, ►reprogramming, ►transcriptional priming, ►hepatocyte, ►bone marrow, ►genetic engineering, ►tissue engineering, ►leukemia inhibitory factor, ►Oct, ►Sox, ►NANOG, ►Smad, ►Stat, ►brachyury, ►bone morphogenetic protein, ►KIT, ►LIF, ►TCF, ►LEF, ►erythropoietin, ►PTEN, ►retinoic acid, ►adipocyte, ►insulin, ►macrophage colony stimulating factor, ►IL-1, ►IL-3, ►fibroblast growth factor, ►epidermal growth factor, ►metaplasia, ►public opinion, ►transdetermination, ►MAPCs, ►plasticity, ►cancer stem cell, ►EC, ►Parkinson disease, ►neurogenesis, ►helix-loop-helix, ►transcription factors, ►adoptive cellular therapy, ►satellite cells, ►trophoblast, ►transformation genetic, ►sickle cell anemia; patents: Loring JF, Campbell C 2006 Science 311:1716; Fuchs E, Segre JA 2000 Cell 100:143; Weissman IL 2000 Science 287:1442; Mezey É et al 2000 Science 290:1779; Edwards

BEBA et al 2000 Fertil Steril 74:1; Lennard AL, Jackson GH 2001 West J Med 175:42; Wakayama T et al 2001 Science 292:740; Odorico JS et al 2001 Stem Cells 19:193; Blau HM et al 2001 Cell 105:829; Nichols J 2001 Curr Biol 11:R503; Toma JG et al 2001 Nature Cell Biol 3:778; Edwards RG 2001 Nature [Lond] 413:349; Nature [Lond] 2001 414:87–131; Hochedlinger K, Jaenisch R 2002 Nature [Lond] 415:1035; 2002 J Cell Biochem 85: S38; Board of Life Sciences, National Research Council and Board on Neuroscience and Behavioral Health, Institute of Medicine 2002 Stem Cells and the Future of Regenerative Medicine. National Academic Press, Washington DC; 2003 Proc Natl Acad Sci USA 100(Suppl. 1); germline stem cell regulation: Wong MD et al 2005 Annu Rev Genet 39:173; historical background of stem cell research: Solter D 2006 Nature Rev Genet 7:319; stem cell niches: Moore KA, Lemischka IR 2006 Science 311:1880; review and evaluation of major methods: Yamanaka S 2007 Cell Stem Cell 1:38; International Society for Stem Cell Research: http://www.isscr.org/; intestinal stem cells: Crosnier C et al 2006 Nature Rev Genet 7:349; stem cells for neurological disorders: Lindvall O, Kokaia Z 2006 Nature [Lond] 441:1094; stem cell therapy for heart: Srivastava D, Ivey KN 2006 Nature [Lond] 441:1097; USA stem cell registry: http://stemcells.nih.gov/research/registry; Guidelines for Human Embryonic Stem Cell Research 2005 (National Academies Press; free online): http://newton.nap.edu/books/0309096537/html/; database of embryonic stem cell lines: http://www.stemcellcommunity.org; mutant mouse stem cell lines: http://www.genetrap.org/.

Stem-Loop Structure: Any DNA or RNA that may have non-paired single strand loops associated with a double-strand stem similar to a lollipop (see Fig. S118). In the stem, the Watson-Crick pairing may not be perfect along its length. ►palindrome, ►repeat inverted

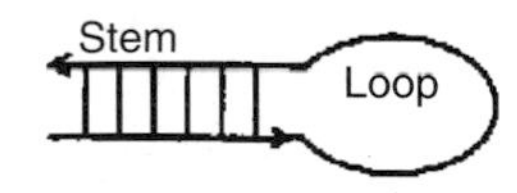

Figure S118. Stem-loop structure

Stem Rust: Caused by infection of the basidiomycete fungus *Puccinia graminis* on cereal plants. The haploid spores produced on the wheat plant germinate on the leaves on barberry shrubs and form pycnia (pycnidium). The pycniospores of different mating types undergo plasmogamy and form dikaryotic aecia (aecidia) on the lower surface of the barberry leaves. The aeciosopores infect the wheat leaves and form the dark brown rust postules, called uredia. The dikaryotic uredospores reproduce then asexually and spread the disease. At the end of the growing season karyogamy takes place and the diploid teliospores are formed. The teliospores overwinter and eventually undergo meiosis and liberate the haploid basidiospores that germinate on barberry and restart the cycle. Stem rust may cause very substantial crop loss in wheat and other Gramineae. The newer varieties are genetically more or less resistant to the fungus. *P.* spp have chromosome numbers 3–6. ►host-pathogen relationship, ►rust

Stenosis: The narrowing of a body canal or valve such as in the aorta, heart valve, pulmonary artery, vertebral canal, etc. ►restenosis

Stenospermocarpy: Genetically determined abortion of the embryo soon after fertilization resulting in seedless normal size berries, a desirable trait of table grapes. ►seedless fruits

Stenting: Medical use of some device that keeps a graft or a structure in place.

Step: Single-target expression profile.

Step Allelomorphism: A historically important concept that paved the way to allelic complementation and to the study of gene structure. In the late 1920s, Russian geneticists discovered that partial complementation among allelic genes may occur in a pattern and that was inconsistent with the then prevailing idea that the gene locus is the ultimate unit of function, mutation and recombination and that alleles are stereochemical modifications of an indivisible molecule. ►allelic complementation, ►Offermann hypothesis; Carlson EA 1966 The gene: A Critical History. W.B. Saunders, Philadelphia, Pennsylvania.

Step Gradient Centrifugation: In the centrifuge tube usually three different concentrations of CsCl or sugar are layered without allowing mixing. The highest concentration is at the bottom of the tube. The different components of the mix, layered at the top, accumulate at the boundaries, which have higher density than the separated component.

Stepwise Mutation Model (SMM): In this model, microsatellites may change by repeated gain or loss of small number of nucleotide repeats. Electrophoretic variations of enzymes may be also fitted to such a model. ►microsatellite, ►infinite allele mutation model, ►trinucleotide repeats; Moran PAP 1975 Theor Popul Biol 8:318; Ohta T, Kimura M 1973 Genet Res 22:201.

Stereocilia: Protoplasmic thin filaments like the ones in the inner ear. Whirlin, a PDZ domain protein of the stereocilia interacts with a membrane-associated protein kinase (Ca^{2+}-calmodulin serine kinase) and erythrocyte protein p55 (4.1 R) and plays similar roles (actin cytoskeletal assembly) in erythrocytes as well as in stereocilia (Mboru P et al 2006 Proc Natl Acad Sci USA 103:10973). ►deafness, ►PDZ domain; Frolenkov GI et al 2004 Nature Rev Genet 5:489.

Stereoisomers: Molecules of identical composition but with different spatial arrangement.

Stereomicroscopy (dissecting microscopy): Used for visual analysis under relatively low magnification of natural specimens without sectioning. It has special advantage for dissecting structural elements with binocular viewing and top or side illumination without fixation and/or staining. ►confocal microscopy, ►scanning electronmicroscopy, ►microscopy

Stereotactic: An action precisely positioned in space such as irradiation of a small spot in the body, surgical introduction of cells, a genetic vector at a defined location of the brain, etc.

Steric-Exclusion Model: This model of DNA replication states that the Watson-Crick hydrogen pairing is not an absolute necessity for the faithful replication of DNA but it is essential that the building block (not necessarily a purine or pyrimidine) would fit into the frame of the DNA double helix. A pyrene nucleoside triphosphate, with the size close to a nucleotide pair, has sufficient steric complementarity to fit into an abasic site and permits DNA replication (see Fig. S119). ►Watson and Crick model, ►DNA replication, ►hydrogen pairing, ►abasic sites; Matray TJ, Kool ET 1999 Nature (Lond) 399:704.

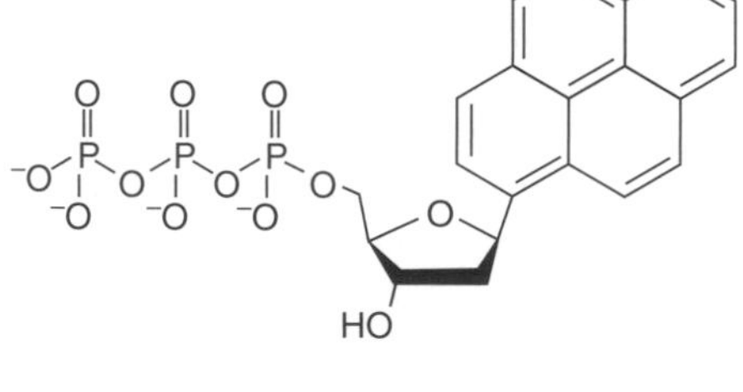

Figure S119. Pyrene nudeoside triphosphate

Sterigma: A small stalk at the tip of a fungal basidium where spores come off. ►basidium

Sterile Insect Technology (SIT): ►genetic sterilization

Sterile RNA: ►germline transcript

Sterility: Either the male, the female, or both types of gametes (haplontic), or the zygotes (diplontic) have reduced or no viability caused by lethal or semilethal genes, chromosomal defects, differences in chromosome numbers or incompatible cytoplasmic organelles. ►infertility, ►semisterility, ►somatoplastic sterility, ►cytoplasmic male sterility, ►incompatibility, ►self-incompatibility, ►hybrid sterility, ►azoospermia, ►infertility, ►deletion, ►inversion, ►translocation

Sterilization: ►autoclaving, ►filter sterilization, ►pasteurization, ►aseptic, ►axenic, ►radiation effects, ►ethanol, ►hypochlorate, ►ethylene oxide, ►genetic sterilization, ►*Cochliomya hominivorax*, ►sterilization humans, ►birth control

Sterilization, Genetic: ►genetic sterilization

Sterilization, Humans: Practiced by various societies for different reasons. The eunuchs of the Chinese imperial courts and of the Osmanic harems served as guardians of the privileges of tyrannical social structures. The castration of male Italian opera artistes were performed for singing in female roles, in an era when women were banned from the performing arts. In the 1880s, by the publications of Sir Francis Galton sought scientific justifications for negative eugenics in order “to produce a highly gifted race of men by judicious marriages during consecutive generations.” 1890s initiated sporadic sterilization of institutionalized, mentally retarded persons. Starting in 1907, in about 14 states (USA), laws were enacted for systematic sterilization of mentally retarded, blind, deaf, crippled or afflicted by tuberculosis, leprosy, syphilis, and chronic alcoholism. This “practical, merciful and inevitable solution” eventually degenerated into legal suggestions to eliminate criminal behavior, disease, insanity, weaklings and other defectives and “ultimately to worthless race types.” By the time strong moral objections gained noticeable ground in the year 1956, nearly 60,000 human individuals were legally sterilized. Interestingly, the Oklahoma law exempted from mandatory sterilization offenses against prohibition, tax evasion, embezzlement and political crimes. Several state laws advocated mandatory sterilization also as a guard against illegitimacy, particularly by unwed recipients of the Aid to Families with Dependent Children (AFDC). Until 1965, judicial approval could be obtained for a “good cause” for forced sterilization of mentally retarded individuals whose family had hardship in supporting the offspring of promiscuous children. Although not all states rescinded yet the old laws, sterilization of humans is now practiced only voluntarily by ligation of the vas deferens (vasectomy), tubal constriction, ovariectomy or by the use of various types of mechanical and hormonal contraceptives. Mandatory sterilization was practiced for eugenic and social reasons in several enlightened countries (e.g., Sweden) until the 1970s. Compulsory sterilization is objectionable on moral ground because reproduction is a basic human right although the society cannot support irresponsible reproductive behavior in cases of certain genetic defects. Reproductive rights must be balanced with the right of born and unborn children with potential severe genetic load (►wrongful birth)

Sterilization is particularly reprehensible when advocated as a selective measure against certain human races. The Third Reich annihilated millions and sterilized thousands for eugenic and other evil reasons. From genetic perspective it is controversial since 83% of the mentally retarded children are born to non-retarded parents. In addition, selection against the majority of human defects is quite inefficient, since the vast majority of the defective genes are in heterozygotes and many of the conditions are under polygenic control or are non-hereditary. Furthermore, there are no objective scientific or practical measures for the evaluation of most of the human traits. ►selection, ►selective abortion, ►eugenics, ►polygenic, ►salpingectomy, ►vasectomy, ►ovariectomy; Reilly P 1977 Genetics, Law, Social Policy. Harvard University Press, Cambridge, Massachusetts.

Sternites: The ventral epidermal structures of the abdomen. ►*Drosophila*

Steroid 5-Beta Reductase (SRD5B1, 7q32-q33): Catalyzes reduction of bile acid intermediates and steroid hormones.

Steroid Dehydrogenase-Like Protein (NSDHL, XDq28): The mutations affect cholesterol biosynthesis and may cause male lethality. Hydroxysteroid dehydrogenase (HSD3B1, 1p13.1) deficiency may involve adrenal hyperplasia, hypospadias and gynecomastia. ►cholesterol, ►hypospadias

Steroid Doping: Used by athletes to boost performance. Amphetamines increase alertness and may reduce onset of fatigue. Side effects are insomnia, exhaustion, violence and potential heart disease. The health hazards are increased if anabolic steroids, insulin, insulin-like growth hormone, etc., are used simultaneously. Even non-steroid anti-inflammatory drugs may be risky because they mask pain and may aggravate injuries. ►anabolic steroids

Steroid Hormones: Derived by the pathway shown in Figure S120. [3] is the principal hormone of the endocrine gland, corpus luteum, in the ovarian follicle after the release of the ovum.

They regulate the expression of the secondary sexual characters of females. [4] is the main male sex hormone that is produced in 6–10 mg quantities daily in men and ca. 0.4 mg in women. It is responsible for the production of facial hair and baldness and the regulation of growth [5] is formed by oxidative removal of C-129 from its precursor; primarily a female hormone occurring in the ovaries and placenta and it is responsible for regulating, among other functions, bone growth, increased fat content and smoother skin of females compared to men. This hormone is present also in the testes. In cooperation with progesterone it regulates also the menstrual cycles. [6] and [7] are synthesized in the kidney cortex and regulate, among others, mineral (Na^+Cl^-, HCO_3^-) reabsorption and are frequently called as mineralcorticoid hormones. [8] is a glucocorticoid affecting protein, carbohydrate metabolism regulates the immune system, allergic reactions, inflammations, etc. [9] is also an anti-inflammatory glucocorticoid with a role in activating the glucocorticoid receptors. The number of steroid hormones is about 50 and they are present in practically every cell of the body, besides those mentioned, and they, along with thyroid hormones, have important roles in activation of genes. Up to 1966, the general assumption was that plants do not use steroid hormones. It has been demonstrated that brassinolids (related to cholesterol, ecdysone) mediate several developmental processes in plants, such as elongation, light responses, etc. The steroid receptor superfamily includes receptors for estrogen, progesterone, glucocorticoid, mineralcorticoid, androgen, thyroid hormone, vitamin D, retinoic acid, 9-cis retinoic acid and ecdyson. The steroid hormone receptors stimulate the formation and then stabilize the pre-initiation complex of transcription. Most commonly the condition of their binding to the hormone response element is the binding to their appropriate ligands.

Some, such as the thyroid hormone receptor, can bind to DNA in the absence of a ligand. In the

| → [4] Testosterone → [5] Estradiole
[1] Cholesterol → [2] Prognenolone → [3] Progesterone → [6] Corticosterone → [7] Aldosterone
↓ |→ [8] Cortisol
[9] Dexamethasone

Figure S120. Steroid hormone biosynthetic pathway

absence of the ligand, they function as silencers through interaction with the TFIIB transcription factor. ►hormone-response elements, ►silencer, ►PIC, ►transcription factors, ►transcriptional activator, ►coactivator, ►regulation of gene activity, ►animal hormones, ►estradiol, ►aromatase, ►plant hormones, ►brassinosteroids, ►anabolic steroids, ►prenylation, ►steroids, ►SRC-1; Lösel R, Wehling M 2003 Nature Rev Mol Cell Biol 4:46.

Steroid Receptor: ►hormone receptors

Steroid Sulfatase Deficiency: ►ichthyosis

Steroidogenic Factor-1: ►SF-1

Steroids: Contain a four-ring nucleus consisting of three six-membered rings and one five-membered ring. (See structural formula in Figure S121, ►steroid hormones, ►brassinosteroids).

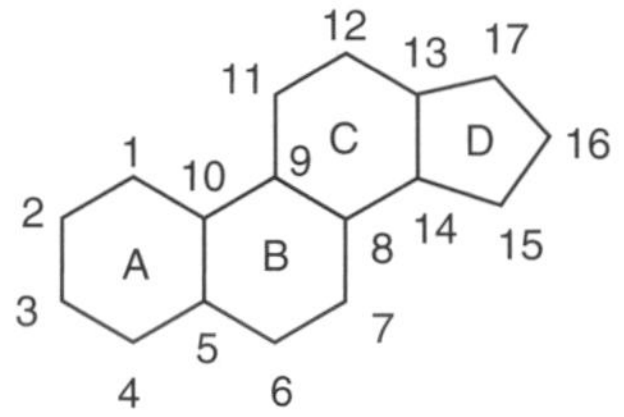

Figure S121. General structural formula of steroids

Sterols: Lipids with a steroid nucleus. The concentration of free sterols determines the fluidity of the eukaryotic cell membranes. Esterification of sterols prevents their participation in membrane assembly. The process is mediated the ACAT complex (acyl-CoA:cholesterol acyltransferase). Increase of ACAT activity may lead to hyperlipidemia and atherosclerosis. Sterol esterification may modify the LDL receptors and potentiates atherogenic processes. The ACAT inhibitor, CP-113-818 reduces amyloid plaques in mice (Hutter-Paier B et al 2004 Neuron 44:227). It may limit intestinal sterol absorption. ►hyperlipidemia, ►amyloids, ►atherosclerosis, ►LDL, ►membranes, ►cholesterol, ►SREBP, ►phytoestrogen, ►oxysterol; Kelley RI, Herman GE 2001 Annu Rev Genomics Hum Genet 2:299; Xu F et al 2005 Proc Natl Acad Sci USA 102:14551.

Stevens-Johnson Syndrome (toxic epidermal necrolysis susceptibility, 6p21.3): The drugs carbamazepine, phenobarbital and allopurionol may evoke cutaneous blistering, epidermal detachment, fever and potentially death. Han Chinese people are more likely to show this reaction than Caucasians. The major histocompatibility complex Class I B5801 allele may be responsible for it. ►allopurinol, ►gout, ►HLA; Hung S-I et al 2005 Proc Natl Acad Sci USA 102:4134.

stg (string, map position 3-99): *Drosophila* gene locus controlling the first ten embryonic divisions (similarly to gene *Cdc28* in *Schizosaccharomyces pombe*); it is a cyclin gene. ►cell cycle

Stick-and-Ball Model: A representation of chemical structure (see Fig. S122).

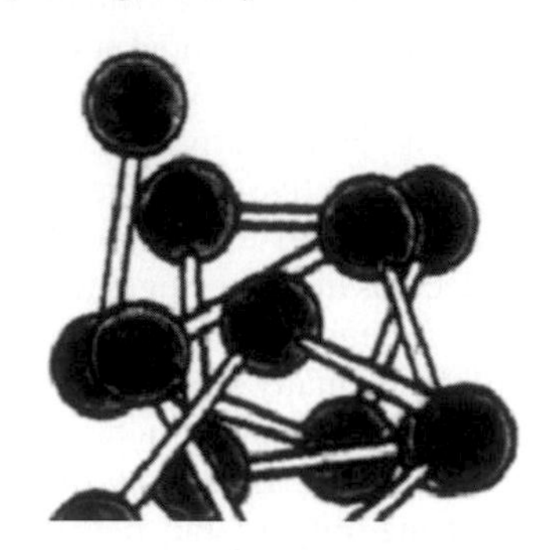

Figure S122. Stick- and ball model

Stickiness of Chromosomes: Observed as some sort of adhesion between any chromosomes within a cell. ►side-arm bridge

Stickleback (*Gasterosteus aculeatus*, 2n = 42 visible chromosomes, XXIII linkage groups): Small, common fresh and seawater fish (see Fig. S123). (See http://cegs.stanford.edu/index.jsp).

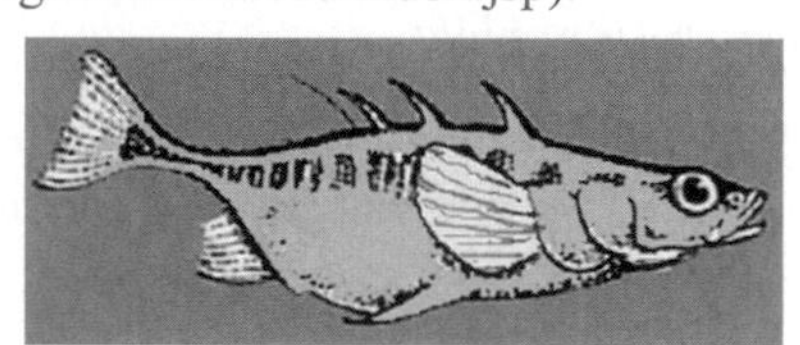

Figure S123. Stickleback

Stickler Syndrome (arthroophthalmopathy, AOM): An early and strong progressive myopia (nearsightedness) and hearing deficit. Retinal detachment may result in blindness, caused probably by a dominant mutation in the collagen (COL11A1) gene (human chromosome 1p21). Overlapping mutations are responsible for the Marshall syndrome at 1p21. Stickler syndrome 3 is located at 6p21.3. The collagen type II (COL2A1, 12q13.11-q13.2) defects of the Stickler syndrome involve achondroplasia, skeletal dysplasia, eye and hearing defects. ►collagen, ►eye disease, ►connective tissue disorders, ►skin diseases; Annunen S et al 1999 Am J Hum Genet 65:974; Richards AJ et al 2000 Am J Hum Genet 67:1083.

Sticky Ends: Double-stranded DNA with a single-stranded overhang to what complementary sequences are available and so they can stick by base pairing (see Fig. S124).

Figure S124. Sticky ends

Stigma: The tip of the style that is normally receptive to the pollen of plants. In zoology, it means spot, such as a hemorrhagic small area on the body. ►gametophyte female, ►gametophyte male, ►protogyny, ►protandry

Stigmasterol: A plant lipid derivative formed by methylation of ergosterol. For guinea pigs, it is a vitamin necessary to avoid stiffness of the joints. ►ergosterol, ►cholesterol

Stilbene: ►resveratrol

Still-Birth: The birth of a dead offspring. It is caused by chromosomal defects in ca. 7% of the stillborn or by other pathological conditions. ►chromosomal breakage

Stipule: A leaf like bract at the base of a leaf (see Fig. S125).

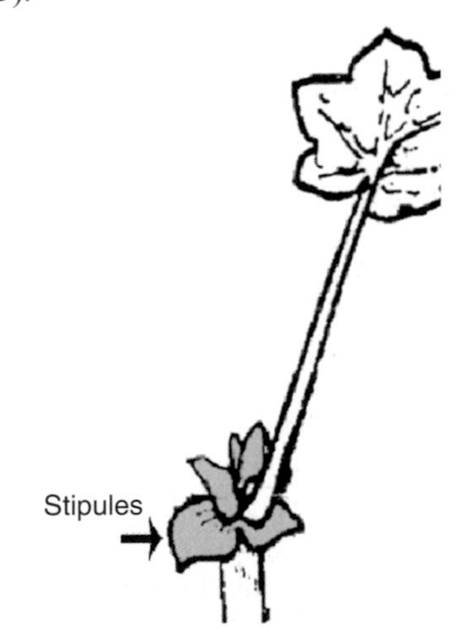

Figure S125. Stipules

Stk: A macrophage-stimulating factor receptor. ►macrophage

STM: ►scanning tunneling microscope

Stn1: A telomere length determining protein factor of yeast working in concert with Cdc13. ►Cdc13; Grandin N et al 1997 Genes Dev 11:512.

Stochastic: Corresponds to a random process; a process of joint distribution of random variables. In a population—in contrast to a deterministic model—random drift and other chance events may determine the gene frequencies. Mutations are assumed to occur at random, and selective forces acting upon these random alterations shape evolution. Generally, deterministic and stochastic processes run parallel and simultaneously. ►deterministic model

Stochastic Detriment of Radiation: The combined risk of cancer, genetic damage and life shortening due to radiation exposure. The figures may vary according to tissues: for gonads it may be 1.33, for bone marrow 1.04, for breast 0.24, for liver 0.16 (in 10^{-2} Sv^{-1}), etc. ►radiation hazards

Stock: A genetically defined strain of organisms; also a root stock on what a scion is grafted.

Stock, Garden (*Matthiola incana*): ►*Matthiola*

Stoke: A unit of kinematic viscosity (the ratio of viscosity to density). ►viscosity

Stolon: A horizontal underground stem such as the tuber-bearing structures of potatoes.

Stoma (plural stomata): A small pore on the leaf surface surrounded by two guard cells, which control opening and closing. Basic helix-loop-helix proteins under the control of three genes (Plitteri LJ et al 2007 Nature [Lond] 445:501) control stoma differentiation, although several other factors also have regulatory role. The stoma permits gas exchange (CO_2 uptake), and release of water vapors (transpiration). The opening of the stomata requires an increase in the turgor of the guard cells (see Fig. S126). It had been suggested that the process could be promoted by opening of K^+ and Cl channels and the subsequent influx of K^+ and Cl^-. A light controlled proton pump activates the opening of the K^+ channel. The closure of the stomata is controlled by the hormone ABA and the influx of Ca^{2+} and the efflux of K^+ and Cl^-.

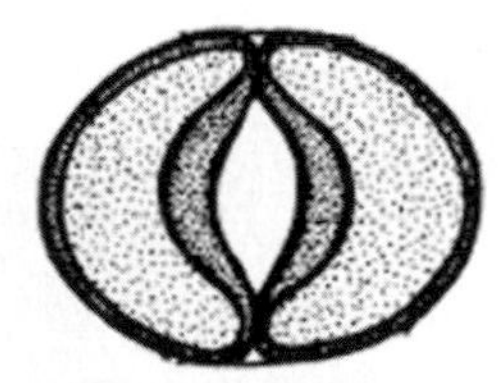

Figure S126. Open stoma

The calcium level is sensed by a cyclin-dependent protein kinase (CDPK). Blue light photoreceptors *CRY1* and *CRY2* and phototropin genes *PHOT1* and *PHOT2* receptors regulate blue light response but the quadruple mutants *cry1*, *cry2*, *phot1*, *phot2* mutants barely responded. Mutation in COP1 (constitutive photomorphogenesis) permitted opening of the stomata in darkness and in blue light the triple recessive mutants displayed open stomata (Mao J et al 2005 Proc Natl Acad Sci USA 102:12270). Pattern of stoma development is controlled by a MAPKK kinase (Bergmann D et al 2004 Science 304:1494).

Erecta mutants (*er*, *erl1*, *erl2*) defective in the regulation in leucine-rich repeat receptor kinases cause clustering of the guard cells of the leaves (see Fig. S127) (Shpak ED et al 2005 Science 309:290). Besides stoma density in the *erecta* mutants, epidermal cell expansion, mesophyll cell proliferation and cell-cell contacts regulate transpiration and photosynthesis in plants and determine the efficiency of carbon fixation (Masle J et al 2005 Nature [Lond] 436:866). (For the phenotype of *er/er* see ►*Arabidopsis thaliana* entry.)

In the regulation of stomata, Ca^{2+}-dependent ATP-ases and GTP-ases have major role. The processes involve changes in the electric potentials (depolarization). In the control of the ABA response, syntaxin-like proteins play a role. Sphingosine-1-phosphate level signals to calcium mobilization. Phospholipase Dα1-produced phosphatidic acid signals to abscisic acid-promoted stomatal closure. Phospholipase Dα1 and phosphatidic acid interact with the Gα subunit of the heterotrimeric G protein to mediate abscisic acid inhibition of stoma opening (Mishra G et al 2006 Science 312:264). Open stomata provide passive entry for bacterial infection of plants. Stomatal guard cells—as an innate immune reaction—perceive bacterial surface molecules by the FLS2 receptor, production of nitric oxide and guard cell-specific OST1 kinase and respond with stomatal closure. Some plant pathogenic bacteria have developed, however, a virulence mechanism for reopening the entry port (Melotto M et al 2006 Cell 126:969). ►ion channels, ►cell cycle, ►aequorin, ►calmodulin, ►cyclin, ►ATPase, ►MAP kinase, ►GTPase, ►proton pump, ►ABA, ►abscisic acid, ►phospholipase, ►syntaxin, ►sphingolipids, ►G protein, ►abscisic acid, ►phosphatidate, ►host-pathogen relation; Blatt MR 2000 Annu Rev Cell Dev Biol 16:221; Schroeder JI et al 2001 Nature [Lond] 410:327; Wang X-Q et al 2001 Science 292:2070; Schroeder JI et al 2001 Annu Rev Plant Physiol Plant Mol Biol 52:627; Hetherington AM 2001 Cell 107:711; Nadeau JA, Sack FD 2002 Science 296:1697; Hosy E et al 2003 Proc Natl Acad Sci USA 100:5549; Hetherington AM, Woodward FJ 2003 Nature [Lond] 424:901.

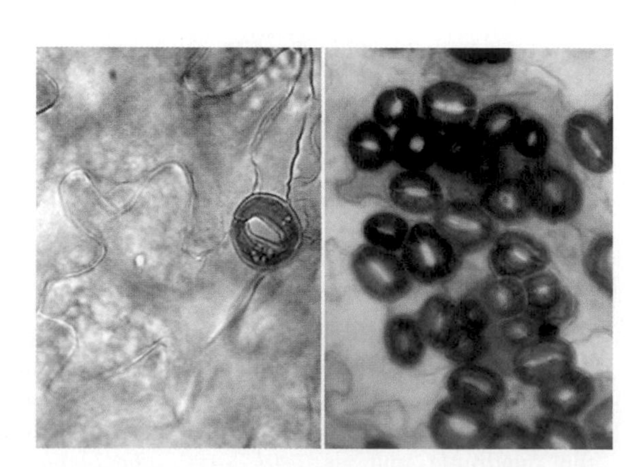

Figure S127. Rosette leaf epidermis wild type (left) and erecta triple mutant (right). Stomata express GUS activity. (Courtesy of Drs. Jessica McAbee and Keiko Tori)

Stomatin: A cation conductance protein in the cell membrane. ►anesthetics

Stone: ►scaffolds in genome sequencing

Stop Codon: ►nonsense codon, ►genetic code

Stop Signal: ►transcription termination in eukaryotes, ►transcription termination in prokaryotes, ►stop codon, ►release factor [RF]

Stoppers: Mitochondrial mutations in *Neurospora* displaying stop-start growth. ►poky

STP: ►signal transfer particle

STPβ (second strand transfer protein): ►recombination mechanisms eukaryotes, ►Sep 1

STR: Short/single tandem repeats, such as found in micro- and minisatellites. Using the profiles of only 13 STRs it is possible to provide a rapid test for crime scenes; STR are used for forensic analysis and for population studies. ►microsatellite, ►minisatellite, ►forensic genetics, ►DNA fingerprinting

Strabismus: An anomaly of the eyes; they may be either divergent or convergent or one directed up, the other down because of the lack of coordination of the muscles concerned. Some persons display this anomaly only periodically. The pattern of inheritance is not entirely clear; most likely dominant factor(s) are involved. The recurrence among the offspring of convergent probands is higher than that among children of the divergent type. Its incidence in the general population is ~0.002. ►eye diseases, ►Duane retraction syndrome, see Fig. S128.

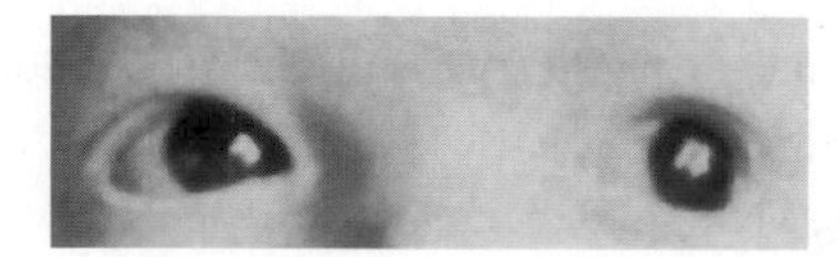

Figure S128. Strabismus

Strain: An isolate of an organism with some identifiable difference from other similar groups. This term does not imply any stringent other criteria.

Strain Distribution Pattern (SDR): The distribution of two alleles of a diploid among the progeny where linkage is studied either by a backcross or by recombinant inbred procedure. ►backcross, ►linkage, ►recombinant inbred

Strand Assimilation: The *exo* gene of lambda phage codes for a 5′-exonuclease (M_r 24,000) that can convert a branched DNA structure to an unbranched nicked duplex by the process called strand assimilation during recombination. A progressive incorporation of one DNA strand into another during recombination (••••>) (see Fig. S129). ►recombination, ►lambda phage

Strand Bias: ►gene distribution

Strand Displacement: A type of viral replication involving the removal of the old strand before the new strand is completed. Similar mechanism is used by mtDNA, ►D loop

Stratification: Layering; in statistical analysis studying the population by, for e.g., age groups, ethnicity or other suitable attributes besides some other criteria of comparison, such as onset of a disease. Stratification may lead to false positive association; comparison of gene frequencies among population mandates appropriate case controls. Principal component analysis and stratification is used by the method of Eigenstrat for minimizing spurious associations in disease studies (Price AL et al 2006 Nature Genet 38:904). Lactase persistence and tall stature was strongly associated in Americans of European descent (Campbell CD et al 2005 Nature Genet 37:868). ►lactose intolerance, ►case control, ►descent; Hoggart CJ et al 2003 Am J Hum Genet 72:1492; Reich DE, Goldstein DB 2001 Genet Epidemiol 20:4.

Stratification Artefact: The disease and control alleles dealt with are from different (ethnic) populations in case-control studies. Sib, parent or other family comparisons may correct the problems. Lower gene frequencies favor reliable result. ►case—control method

Stratified Random Sample: Represents the entire population (including subpopulations) in a reliable manner.

Stratocladistics: The study of evolution on the basis of fossil records. It minimizes the significance of homoplasy and lack of preservation of lineages that would preserve other lineages under examination. ►cladistic

Strauss Family: Viennese composers and conductors of three generations. Johann Strauss the Elder (1804–1849) became celebrated for his light waltzes and other dance music. His son Johann Strauss the Younger (1825–1899) is the author of the Blue Danube and many other waltzes was the most celebrated composer (see Fig. S130). His brothers Josef Strauss (1827–1870) and Eduard Strauss (1835–1916) were also famous conductors and composers. Son of Eduard, Johann (1866–1939) was also a renowned conductor. ►musical talent

Figure S130. Johann Strauss Jr.

Strawberry (*Fragaria ananassa*): About 46 *Fragaria* species with x = 7; the wild European *F. vesca* is diploid (2n = 14), *F. moschata* (2n = 42), some east Asian species are tetraploid, the American strawberries as well as the garden strawberries are 2n = 56. (See http://bioinformatics.pbcbasc.latrobe.edu.au/index.htm).

Streak (primitive streak): A sign on the early embryonal disc indicating the movement of cells and the beginning of the formation of the mesoderm and an embryonal axis. ►organizer

Streaking: Spreading microbial cells on the surface on a nutrient agar medium to observe growth or lack of it. (See Fig. S131).

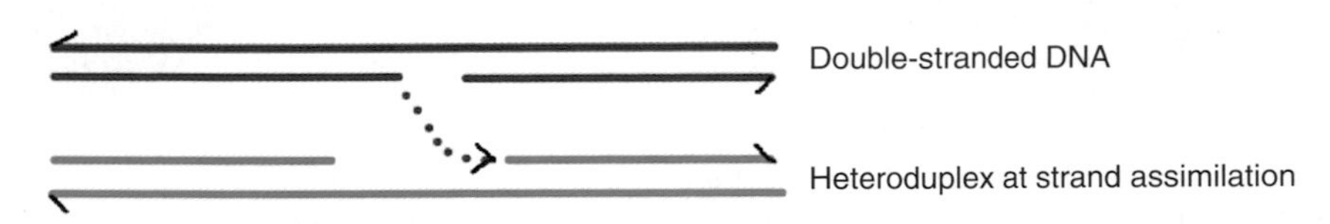

Figure S129. DNA strand assimilation

Figure S131. Streaking of wild type (left top) and mutant strains of yeast

Streptavidin: Conjugated with rhodamine, it specifically binds to biotin (biotinylated nucleic acids, immunoglobulins) and permits their detection by fluorescence. The binding constant for biotin is $k_a = 10^{15}\ M^{-1}$. ►avidin

Streptavidin-Peroxidase: Identifies biotinylated antibodies in ELISA, in immunochemistry in general and in protein blots. ►genomic subtraction, ►ELISA, ►biotinylation

Streptococcus pneumoniae A (*Diplococcus pneumoniae*): The common pathogenic bacterium causing pharyngitis (sore throat) (see Fig. S132). About 5–10% of the infections may involve necrotic lesion of various severities. In rare extreme cases, it may cause death. Some strains secrete substantial amount of a pyrogenic (fever-producing) exotoxin A, which stimulates the immune system as a superantigen. The excessive stimulation results in the overproduction of cytokines that may damage the lining of the blood vessels and thus cause fluid leakage, reduced blood flow and necrosis of the tissues because of the lack of oxygen. As a further consequence, fasciitis (inflammation of the fibrous tissues) and myositis (inflammation of the voluntary muscles) may follow. Immunity to *S. pneumoniae* may be independent of the capsular antigens and protection requires the presence of $CD4^+$ T cells at the time of infection (Malley R et al 2005 Proc Natl Acad Sci USA 102:4848).

The destruction of the tissues may result in death within a very short period after infection by the extremely virulent strain of "flesh-eating bacteria." *Streptococcus* (B) *agalactiae* is a serious threat to diabetic, cancerous or elderly people; it may be responsible for neonatal sepsis occurring during vaginal delivery. *Streptococcus pneumoniae* (*Pneumococcus*) provided the first information on genetic transformation in 1928. Its sequenced genome (in 2001) of 2,160,837 bp contains 2236 ORFs (see Fig. S133). Approximately 5% of its genome is insertion sequences. The completely sequenced genome of *S. pyogenes* M1 has 1,852,442 bp and encodes ~1752 proteins. *Streptococcus mutans* UA159, the major cause of tooth decay contains 2,030,936 bp and 1963 ORF; its genome has several insertion elements and transposons. Group B *Streptococcus* pathogens display multiple serotypes and they are life threatening to newborns in the first week after birth. The infection comes generally from the healthy mothers who (in 25–40%) harbor the bacteria in anogenital area. This bacterium as many other pathogens are resistant to several antibiotics. Vaccine production has problems because of the serotype variation. However, four proteins provide rather general type of protective antibodies against the various types of *Streptococcus* B bacteria. The protective antibody (immunoglobulin G, IgG) is transmitted to the baby though the placenta (Maione D et al 2005 Science 309:148). ►transformation genetic, ►necrosis, ►superantigen, ►toxic shock syndrome, ►streptolysin, ►fratricide, ►host-resistance genes; Tettelin H et al 2001 Science 293:498; Ferretti JJ et al 2001 Proc Natl Acad Sci USA 98:4658; Hoskins J et al 2001 J Bacteriol 183:5709; Tettelin H et al 2002 Proc Natl Acad Sci USA 99:12391; Ajdic D et al 2002 Proc Natl Acad Sci USA 99:14434.

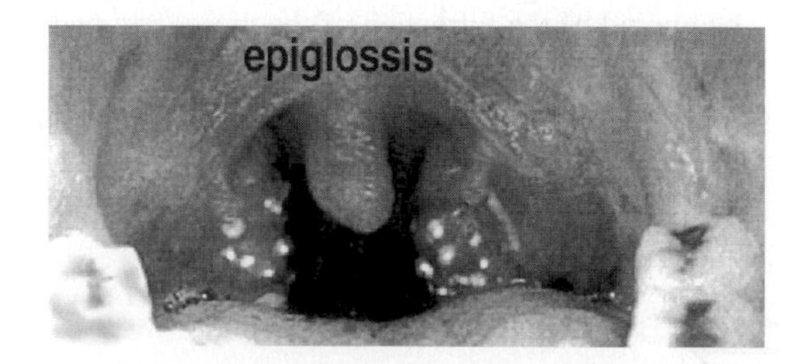

Figure S132. *Streptococcus* exudates on tonsils appear as white spots in severe sore throat. (Modified after Nimishikavi S & Stead L 2005 New England J. Med. 352: p e10)

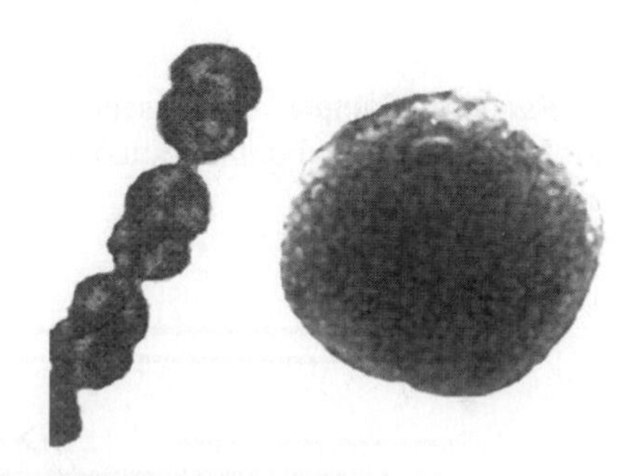

Figure S133. *Streptococcus* colonies; Single cell at high magnification

Streptokinase: An activator of plasminogen. ►plasminogen activator, ►plasmin

Streptolygidin: An antibiotic that blocks the action of prokaryotic RNA polymerase.

Streptolysin: A cholesterol-binding bacterial exotoxin, which forms large holes through the mammalian plasma membrane. At low concentration, it is suitable for introducing proteins through living cell membranes without irreversible damage to the cell. Streptolysin O prevented *Streptococcus* internalization into lysosomes, killing the extracellular pathogen (Håkansson A et al 2005 Proc Natl Acad Sci USA 102:5192). ►cytolysin; Walev I et al 2001 Proc Natl Acad Sci USA 98:3185.

Streptomyces: A group of Gram-negative bacteria of the actinomycete group, characterized by mycelia-like septate colonies. On these mycelial colonies spore-bearing organs develop. These bacteria somewhat simulate a multicellular type of development. Their genetic material, unlike the majority of prokaryotes, is a linear DNA. The sequenced genome of *S. coelicolor* A3(2) is 8,667,507 bp, containing an estimated 7825 genes (Bentley SD et al 2002 Nature [Lond] 417:141). *S. avermitilis* has linear chromosome built of 9,025,608 bp encoding at least 7574 open reading frames (Ikedo H et al 2003 Nature Biotechnol 21:526; chromosome: Hopwood D 2006 Annu Rev Genet 40:1).

Streptomycin: An antibiotic compound, precipitates nucleic acids, inhibits protein synthesis and interferes with proofreading, and thus causes translational errors (see Fig. S134). Mutation in its S12 ribosomal protein binding sites leads improved translational precision. Some mutations may lead to streptomycin-dependence. Streptomycin resistant mutations in the ctDNA are maternally inherited; such mitochondrial DNA mutations may lead to hearing loss in humans. ►antibiotics, ►mitochondrial diseases in humans, ►mtDNA

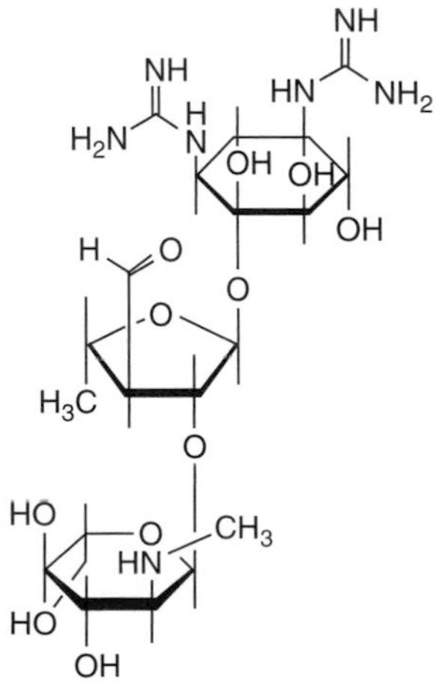

Figure S134. Streptomycin

Streptozotocin (a nitrosamide, 2-deoxy-2-[3-methyl-3-nitrosoureide]-D-glucopyranose): A methylating, carcinogenic, antibiotic agent (effective even against fungi). Induces diabetes and poisons B lymphocytes.

Stress: A condition when living beings must cope with difficult mental or physiological conditions. A major gene duplication for panic and phobic disorders appeared to be at human chromosome 15q24-q26 but another study failed to confirm this finding (Tabiner M et al 2003 Am J Hum Genet 72:535). Stress or anxiety activates the corticotropin releasing factor (CRF/CRH) synthesis in the hypothalmus. CRF then stimulates the CRF receptors (CRHR) in the pituitary and this turns on the adrenocorticotropin hormone (ACTH) in the kidneys leading to the production of glucocorticoids, which hinder by feedback to the brain the stress reaction. In case of a failure to respond successfully with some type of a homeostatic mechanism, death or substantial harm may result. Phosphoinositide 3-kinase (PIK) related kinases mediate a variety of cellular stress responses (Bakkenist CJ, Kastan MB 2004 Cell 118:9). Stress activates sphingomyelinase to generate ceramide and the latter initiates apoptosis. Disruption of the glucocorticoid receptor gene may lead to reduced anxiety. Stress also activates heatshock proteins and glucose-regulated proteins (GRP). The GRP proteins are highly active during tumor progression. Their suppression may lead to apoptosis and rejection of the tumor cells. The stress signals may mediate the activation of genetic repair systems or, in animals, may proceed through three main pathways, the c-Abl or the JNK or the p53 routes. The first two are specific to different types of genotoxic agents, the p53 protein responds rather generally to various chemical stresses. The signal transducers eventually reach the DNA by the activation of transcription factors. Ionizing or excitatory (UV) radiations may directly cause chromosome breakage resulting in either repair or apoptosis. In plants, stress stimulates the formation of elicitors and pathogenesis-related proteins. *Arabidopsis* plants exposed to ultraviolet radiation (UV-C), to stress caused by infection by pathogens, or to flagellin, displayed increased level of homologous somatic recombination; the genomic instability persisted in following generations and the condition was transmitted as a dominant trait by both female and male gametes to progeny (Molinier J et al 2006 Nature [Lond] 442:1046). ►homeostasis, ►p38, ►GADD153, ►CAP, ►ceramides, ►sphingolipids, ►sphingolipidoses, ►SAP, ►SAPK, ►apoptosis,

►JNK, ►cAbl, ►p53, ►pathogenesis-related proteins, ►heat-shock proteins, ►adrenocorticotropic hormone, ►human brain, ►glucocorticoid, ►drought resistance, ►salt tolerance, ►host-pathogen relations, ►flagellin, ►ultraviolet light, ►instability genetic, ►Pik; Smith MA et al 1995 Proc Natl Acad Sci USA 92:8788; Gratacòs M et al 2001 Cell 106:L367; Dolan RJ 2002 Science 298:1191.

Stress Granules: May accumulate in mammalian and other cells under physical or chemical stress and regulate metabolism (Andreson P, Kedersha N 2006 J Cell Biol 172:803).

Stress Proteins: Most commonly mean heat shock proteins. ►heat-shock protein

Stretching Chromosomes: For more precise localization of FISH labels the chromosome can be extended 5–20 times their highly coiled length using hypotonically treated, unfixed metaphase chromosomes and centrifugation. ►FISH; Bennink ML et al 2001 Nature Struct Biol 8:606.

Striated Muscles: The heart and skeletal muscles are made of sarcomeres and thus striated transversely. ►smooth muscles

Striatum: A layered tissue region, e.g., in the subcortical area of the brain where cognitive and movement regulation functions are mediated. ►brain human

STRING (search tool for the retrieval of interacting genes/proteins): ►genetic network; http://string.embl.de/.

String Edit Distance: Determined upon addition, deletion or replacement one base symbol in order to transform a DNA sequence (a string) into another. ►tree edit distance

Stringent: Rigidly controlled.

Stringent Control: In amino-acid-starved bacteria (auxotrophs), the product of the *relA*$^+$ gene shuts off ribosomal RNA synthesis as an economical device. In the presence of *relA* amino acid synthesis is promoted (relaxed control) because ppGpp regulates the discriminator regions of the promoters. ►fusidic acid, ►relaxed control, ►discriminator region, ►ribosomes; Chatterji D et al 1998 Genes Cells 3:279; Chatterji D, Ojha AK 2001 Curr Opin Microbiol 4:160; Barker MM et al 2001 J Mol Biol 305:673.

Stringent Genomes: These are the nuclear ones because each chromosome is normally replicated once during the cell cycle and normally during mitosis, each member of a diploid (or other euploid) chromosome set is partitioned equally between the daughter cells. ►relaxed genomes

Stringent Plasmid: The low copy number of it is genetically controlled.

Stringent Replication: Limited replication of the low copy number plasmid DNA.

Stringent Response: Under poor growth condition prokaryotic cells may shut down protein synthesis by limiting tRNA and ribosome formation. The synthesis of the *rrn* genes is mediated by binding ppGpp or pppGpp sequences to the *rrn* promoters. Proteins DksA or GreA and Mdf are also regulators (Trautinger BW et al 2005 Mol Cell 19:247). ►stringent control, ►magic spot, ►*rrn*, ►ribosomes, ►transfer RNA, ►guanosine tetraphosphate, ►Rel oncogene; Chatterji D Ojha AK 2001 Curr Opin Microbiol 4(2):160; van Delden C et al 2001 J Bacteriol 183:5376.

STRL: Co-receptor of HIV and SIV. ►acquired immunodeficiency syndrome

stRNA: ►RNAi, ►microRNA

Strobe: see http://www.strobe-statement.org/; ►HuGENet, ►risk

Stroke: Causes 150,000 death/year and afflicts three times as many in the USA. Stroke results from occlusion of a cerebral artery. A mouse model indicates that newborn neurons can migrate to the ischemic regions and may affect recovery in some cases (Jin K et al 2006 Proc Natl Acad Sci USA 103:13198). The major genetic factors involved are telangiectasia, the Osler-Weber-Rendu syndrome, CADASIL, Ehlers-Danlos syndrome, polycystic kidney disease, Marfan syndrome, cardiovascular diseases, hypertension, MELAS syndrome. Gretarsdottir, S. et al (2002 Am J Hum Genet 70:593) reported a susceptibility locus at 5q12 encoding phosphodiesterase 4D acting probably though atherosclerosis in ischemic stroke (Gretarsdottir S et al 2003 Nature Genet 35:135). The genetic association of stroke with the *PDE4* marker or SNP45 could not be generally replicated not due to entirely clear bases, but because of the complexity of the major types of stroke (ischemic [blockage of vessels by blood clots], hemorrhagic [rupture of blood vessels] or other subtypes [cardiogenic or large vessels]) may account for the discrepancies (Rosand J et al 2006 Nature Genet 38:1091; Gulcher JR et al 2006 Nature Genet 38:1092). Protein kinase PRKCH appears to be liable for the pathogenesis of cerebral infarction in Japanese populations (Kubo M et al 2007 Nature Genet 39:212). (See terms mentioned under separate entries).

Stroma: The aqueous solutes within an organelle. A pseudoparenchymatous association of fungal mycelia is also so named. The supportive tissue of an organ;

stroma cells in the bone marrow may produce collagen and extracellular matrix. ►parenchyma

Stroma Cell-Derived Factor-1α: A chemoattractant for hematopoietic stem cells and has role in cell migration within or to the bone marrow. It may affect cell proliferation and cell migration in cancer metastasis.

Stromelysin: ►metalloproteinases, ►transin

Strong-Stop DNA: When reverse transcriptase initiates transcription of the first strand DNA—from the RNA—then a second strand is made on the first strand DNA template. The transcriptase pauses after the transcription of the R U5 segments (first strand). The U5 R, U3 (second strand) "strong stop" DNA species accumulate (in the latter case including small portions of the RNA primer) before transcription continues to completion of the first, and then of the second strand, respectively. ►retroviruses, ►reverse transcriptase; Driscoll MD et al 2001 J Virol 75:672.

Strontium: An earth metal with several isotopic forms, the ^{90}Sr, a β-emitter radioactive component of nuclear fallout is readily substituted for calcium and, thus, may be concentrated in the milk if the cows grazed on contaminated pastures. Its half-life is 28 years. After the bomb testings in the 1950s, it became especially threatening to children whose bones accumulated 2.6 μμCi in contrast to adults (0.4 μμCi). ►radiation hazards, ►Ci, ►isotopes

STRP: Short tandem repeat polymorphism. ►microsatellite

Structural Classification of Proteins: ►SCOP, ►ASTRAL

Structural Gene: A primarily non-regulatory DNA sequence that codes for the amino acid sequence in a protein or for rRNA and tRNA.

Structural Genomics: Develops high-resolution models of protein structure in order to understand catalytic and other functional mechanisms, ligands, domains, reveal critical targets for site-directed mutagenesis, and develop means for therapeutic interventions. The main tools are x-ray crystallography and nuclear magnetic resonance analysis. (See Baker D, Sali A 2001 Science 294:93; http://www.nysgrc.org/).

Structural Heterozygosity: Involves normal and rearranged homologous chromosomes within cells. ►inversion, ►translocation, ►aberration chromosomal

Structural Variants of Chromosomes: involve deletions, duplications, inversions, transpositions, translocation, complex rearrangements and copy number variations. The Database of Structural Variants of the human genome in 2007 includes a total of 29,289 entries. Among those 11,784 were copy number variants of 4357 loci, 182 inversions, and 17,323 indels (http://projects.tcag.ca/variation/). These variants are more numerous than initially expected and seriously affect the function of many loci. Paired-end mapping (PEM) is an effective means for their identification. PEM involves the preparation and isolation of paired ends of 3 kb fragments and their massive sequencing with 454 technologies and computationally. In two individuals 1300 structural variants (SVs) were identified suggesting that humans may differ to a greater extent in SVs than in single nucleotide polymorphism (Korbel JO et al 2007 Science 318:420). ►paired-end sequence method; 454 sequencing: http://www.454.com/products-and-reagents/genome-sequencer.asp.

Structure-Directed Combinatorial Mutagenesis: PCR-based mutagenesis.

Strumpell Disease: ►spastic paraplegia

STS: ►sequenced tagged sites

STS-Content Mapping: In physical mapping, the large sequences used (YAC clones) contain STS tracts and thus their position can be mapped. ►sequenced tagged sites, ►YAC; Chen YZ et al 2001 Genomics 74:55.

Stuart Factor Deficiency: ►prothrombine deficiency, ►antihemophilic factors

Student's t Distribution: A statistical test for assessing a hypothesis about the means of two populations. This distribution enables the statisticians to compute confidence limits for μ (the true mean of a population) when σ (the standard deviation of the true mean) is not known, and only the standard deviation of the sample *s* is available.

The quantity of *t* is determined by the equation:

$$t = \frac{\overline{X} - \mu}{s/\sqrt{n}}$$

where $\overline{X}$ is the experimental mean and *n* is the population size. The critical *t* values are generally read from statistical tables after it had been quantified by the calculated *t* value,

$$t = \frac{d}{\sqrt{V}}$$

where *d* s the difference between means and V = variance. Under practical conditions the significance of the difference between two means is calculated by the formula $t = (\bar{x}_1 - \bar{x}_2)/\sqrt{[s_1]^2 + [s_2]^2}$ where the *x* values stand for the two means and s^2 values are the variances of the two populations. ►arithmetic mean, ►standard deviation, ►variance, ►paired t test, see Table S5.

S

Table S5. The calculated value at the determined degrees of freedom (*df*) must be identical or greater than the closest value found on the pertinent df line in order to qualify for the probability shown at the top of the columns. E.g., for *df* = 10, and *t* = 3.169 P = 0.01 but if the *t* would be only 3.168 *P* would be only 0.05 according to table and statistical conventions. The use of t charts or linear interpolation using the logarithms of the two-tailed probability values (Simaika 1942 Biometrika 32:263) can obtain more precise P values. Remember that the *t test* indicates the probability of the null hypothesis that the two means would be identical

df	P → 0.900	0.500	0.400	0.300	0.200	0.100	0.050	0.010	0.001
1	0158	1.000	1.376	1.963	3.078	6.314	12.706	63.654	636.620
2	0.142	0.816	1.061	1.386	1.886	2.920	4.303	9.925	31.599
3	0.137	0.765	0.978	1.250	1.638	2.353	3.182	5.841	12.924
4	0.134	0.741	0.941	1.190	1.533	2.132	2.776	4.604	8.610
5	0.132	0.727	0.920	1.156	1.476	2.015	2.571	4.032	6.869
6	0.131	0.718	0.906	1.134	1.440	1.943	2.447	3.707	5.959
7	0.130	0.711	0.896	1.119	1.415	1.895	2.365	3.500	5.408
8	0.130	0.706	0.889	1.108	1.397	1.860	2.306	3.355	5.041
9	0.129	0.703	0.883	1.100	1.383	1.833	2.262	3.250	4.781
10	0.129	0.700	0.879	1.093	1.372	1.812	2.228	→ 3.169	4.587
11	0.129	0.697	0.876	1.088	1.363	1.796	2.201	3.106	4.437
12	0.128	0.696	0.873	1.083	1.356	1.782	2.179	3.054	4.318
13	0.128	0.694	0.870	1.080	1.350	1.771	2.160	3.012	4.221
14	0.128	0.690	0.868	1.076	1.345	1.761	2.145	2.977	4.140
15	0.128	0.691	0.866	1.074	1.341	1.753	2.131	2.947	4.073
16	0.128	0.690	0.865	1.071	1.337	1.746	2.120	2.921	4.015
17	0.128	0.689	0.863	1.069	1.333	1.740	2.110	2.898	3.965
18	0.127	0.688	0.862	1.067	1.330	1.734	2.101	2.878	3.922
19	0.127	0.688	0.861	1.066	1.328	1.729	2.093	2.861	3.883
20	0.127	0.687	0.860	1.064	1.325	1.725	2.086	2.845	3.850
21	0.127	0.688	0.859	1.063	1.323	1.721	2.080	2.831	3.819
22	0.127	0.686	0.858	1.061	1.321	1.717	2.074	2.819	3.792
23	0.127	0.685	0.858	1.060	1.320	1.714	2.069	2.807	3.768
24	0.127	0.685	0.857	1.059	1.318	1.711	2.064	2.797	3.745
25	0.127	0.684	0.856	1.058	1.316	1.708	2.060	2.787	3.725
26	0.127	0.684	0.856	1.058	1.315	1.706	2.056	2.779	3.707
27	0.127	0.684	0.855	1.057	1.314	1.703	2.052	2.771	3.690
28	0.127	0.683	0.855	1.056	1.312	1.701	2.048	2.763	3.674
29	0.127	0.683	0.854	1.055	1.311	1.699	2.045	2.756	3.659
30	0.127	0.683	0.854	1.055	1.310	1.697	2.042	2.750	3.646
40	0.126	0.681	0.851	1.050	1.303	1.684	2.021	2.704	3.551
60	0.126	0.679	0.848	1.046	1.296	1.671	2.000	2.660	3.460
120	0.126	0.678	0.845	1.041	1.289	1.658	1.980	2.617	3.373
∞	0.126	0.674	0.842	1.036	1.282	1.645	1.960	2.576	3.290

Study Bias: A difference in being studied because of a certain property of the entity attracted more interest for research workers.

Stuffer DNA: Part of the phage λ genome is not entirely essential for normal functions of the phage. Sequences between gene *J* and *att* representing about one-fourth of the genome can be removed and replaced (stuffed in) by genetic engineering without destroying viability of the phage. ►lambda phage, ►vectors; Parks RJ et al 2001 J Virol 73:8027.

Sturge-Weber Syndrome (phakomas): Ektodermal hamartomas, angiomas of the meninges. ►hamartoma, ►angioma, ►meninges, ►von Hippel-Lindau disease, ►tuberous sclerosis

sturt: The unit of fate mapping; named after Alfred Sturtevant who first used fate mapping. ►fate maps

Stutter Bands: DNA slippage may occur during PCR replication, (especially of long dinucleotide repeats) and may create shorter sequences than expected, making the identification of the heterozygotes for microsatellite sequences difficult. ►microsatellite, ►PCR; Miller MJ, Yuan BZ 1997 Anal Biochem 251:50; Walsh PS et al 1996 Nucleic Acids Res 24:2807.

Stuttering: A transcription termination phenomenon; poly U may easily break U-A associations. ►stammering

Stylopodium: The bones of the humerus and femur.

Stylus (style): The slender structure leading from the stigma to the ovary of plants through which the pollen tube grows to the embryosac. ►gametophyte female, ►gametophyte male

su1 = supD, su2 = supE, su3 = supF, su4 = supC, su5 = supG* and *su7 = supU: ►su^-

su^-: The wild type allele of a suppressor mutation; the suppressor allele is su^+.

su^+: The suppressor allele at a locus in contrast to the wild type that is designated su^-.

Su Blood Type: Occurs in pigs and resembles the Rh blood type in humans. ►Rh blood type, ►erythroblastosis fetalis

Subcellular: An organelle or other structure or site within a cell. (See subcellular localization of mouse proteins: http://locate.imb.uq.edu.au/; proteins of several organisms: http://bioinformatics.albany.edu/~ptarget; http://gpcr.biocomp.unibo.it/esldb/; Arabidopsis subcellular proteins: http://www.suba.bcs.uwa.edu.au).

Subcellular Localization: The expression levels of many genes are correlated with their subcellular site(s). Gene expression level is generally high in the cytoplasm, low in the nuclear membrane, and intermediate level in the secretory pathways (endoplasmic reticulum and Golgi). In each group fluctuations occur. It has been estimated that in yeast, 47% of the proteins are located in the cytoplasm, 13% in the mitochondria, 27% in the nucleus and nucleolus, and 13% in the endoplasmic reticulum and secretory vesicles (Kumar A et al 2002 Genes Dev 16:707). ►mitochondria, ►mitochondrial diseases in humans, ►chloroplast genetics, ►FL-REX, ►gene ontology, ►organelle, ►tissue-specificity; Drawid A et al 2000 Trends Genet 16:426; Feng Z-P, Zhang C-T 2002 Int J Biochem Cell Biol 34:298; protein targeting signals: Schneider G, Fechner U 2004 Proteomics 4:1571; Gene Ontology Molecular Function and Subcellular Localization: http://www.cs.ualberta.ca/~bioinfo/PA/GOSUB/; subcellular localization of mammalian proteins: http://locate.imb.uq.edu.au/.

Subcloning: Recloning a piece of DNA. ►cloning molecular, ►cloning vectors

Subculturing: Transferring of a culture into a fresh medium.

Subcutaneous: Beneath/under the skin.

Suberin: A corky, complex polymeric material (of fatty acids but no glycerol associated with it) on the surface and within plant cells. In many plants, there is a subepidermal layer of suberin in air-filled cells and in various scar tissues. Suberin is frequently associated with cellulose, tannic acid, dark pigments (phlobaphenes) and inorganics. The commercial cork produced by the oak, *Quercus suber*, is suberin. ►host-pathogen relation

Subfunctionalization: Gene duplication involving somewhat different function of the duplicated copy, e.g., expression at a different location of the body. In *Arabidopsis* and *Antirrinum*, the *AGAMOUS* and the *PLENA* genes, respectively, descended in a non-orthologous manner from the *SHATTERPROOF* and *FARINELLI* genes, respectively (Causier B et al 2005 Curr Biol 15:1508). Subfunctionalization model assumes that the paralogous genes are selectively neutral and thus permitted to mutate and explore new adaptive functionalities without extinction. Eventually, one of the copies may be converted to the status of pseudogenes or is entirely lost. The duplicated genes may have lost one of the function, therefore, both copies became necessary and favored by evolution or they might be preserved by drift.

►duplication, ►paralogous, ►pseudogene, ►neo-functionalization; Force A et al 1999 Genetics 151:1531; Tochini-Valentini GD et al 2005 Proc Natl Acad Sci USA 102:8933.

Sublethal: Only about 50% of the affected my live until sexual maturity.

Sublimon: Sub-stoichiometric molecules of mtDNA that are supposed to be the products of recombination within short repeated sequences in this organelle. ►mtDNA; Kajander OA et al 2000 Hum Mol Genet 9:2821.

Subline: New colony of rodents set up in a new laboratory. ►substrain, ►inbred

Submetacentric: Chromosome with two arms clearly unequal in length (see Fig. S135). ►chromosome morphology

Figure S135. Submetacentric

Submission: ►data submission standards

Submission Signal: In the majority of vertebrates, aggressive behavior generally ends when the weaker partner in the conflict displays the submission signal, e.g., dogs lie on their back. The human race does not employ such definite signals and thus, the conflicts frequently end in violence. ►aggression, ►behavior genetics, ►human behavior, ►ethology

Subspecies: A group of organisms within a species distinguishable by gene frequencies, chromosomal morphology and/or rearrangement(s), and may show some signs of reproductive isolation from the rest of the species. ►species

Substance Abuse: The proclivity is genetically controlled. Morphine preference has at least three known QTLs; alcoholism has also several QTLs. Some of these conditions may be associated with variations in the serotonin transporter. A genetic factor for the latter is linked to the human ALPC2 locus, controlling vulnerability to alcohol. The conditions leading to depression may generally affect substance abuse, some of the quantitative trait loci mentioned, however, do not appear to act globally. ►alcoholism, ►serotonin, ►QTL, ►steroid doping; Uhl GR et al 2001 Am J Hum Genet 69:1290.

Substantia Nigra: A site in the middle part of the brain (mesencephalon) with dark pigment deposits. The pigments are the products of the dopaminergic neurons that control movement and coordination; dopamines are essential neurotransmitters. In Parkinson disease, the cells degenerate and cannot make sufficient amounts of the dopamine pigment. Somatic mutations (deletions) in the mitochondrial DNA are associated with respiratory deficiency and selective loss of neurons that are observed in aging and Parkinson disease (Bender A et al 2006 Nature Genet 38:515; Kraytsberg Y et al 2006 Nature Genet 38:518). ►Parkinson disease, ►neurotransmitter, ►aging, see Fig. S136.

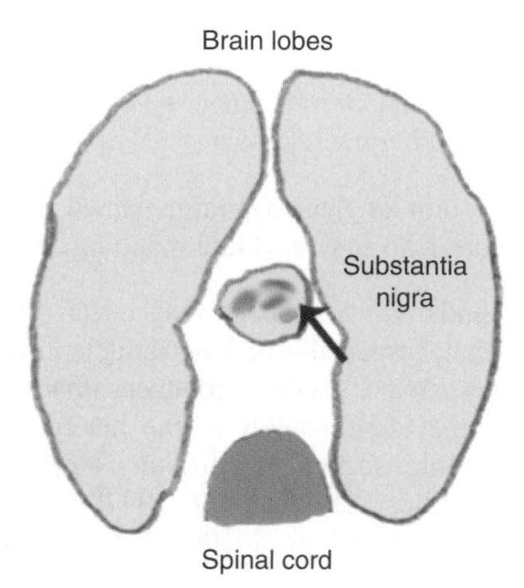

Figure S136. Substantia nigra location in the brain

Substitution, Disomic: Two homologous chromosomes replaced by two others. ►alien substitution, ►intervarietal substitution

Substitution Line: One of its chromosomes (or pair) is derived from a donor variety or species. ►alien substitution, ►intervarietal substitution

Substitution, Monosomic: One entire chromosome is substituted for another. ►substitution line

Substitution Mutation: When a base pair is replaced by another. The evolutionary mutation rate in the human X chromosome is lower than in the Y chromosome and the ratio is about 1.7 at the 95% confidence level according to one study based on molecular analysis (see Fig. S137). ►transition and transversion, ►base substitution, ►point mutation, ►Li-Fraumeni syndrome, ►SNIP, ►base sequences

Substoichiometric Shift: The mtDNA of some species undergoes rounds of unequal recombinations and, thus, generates mutations in this organelle and possibly also in the plastids. The organelle mutations are then maternally inherited. ►mutator genes, ►mitochondrial genetics

S

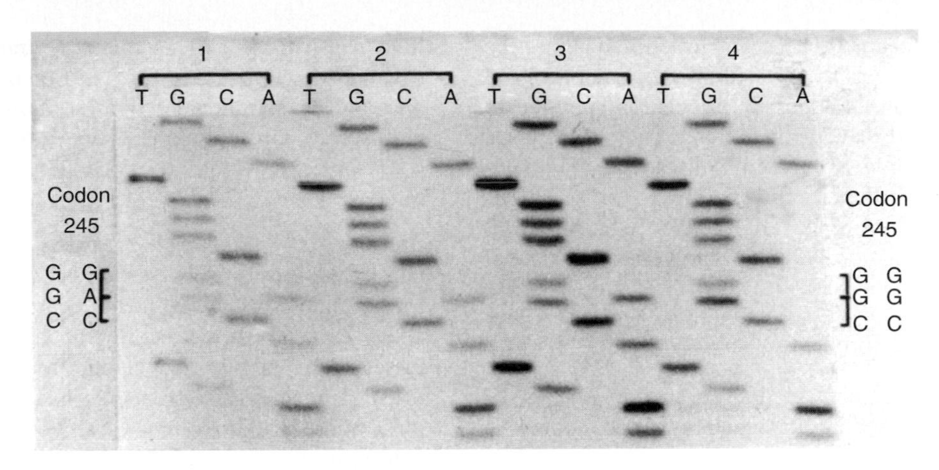

Figure S137. Base substitution mutations in the p53 gene in the noncancerous fibroblast cells in a family with the Li-Fraumeni syndrome. (1) Proband, (2) his brother, (3) their father and (4) a normal control. In codon 245 G**G**C→G**A**C mutations are evident in the two generations investigated. (Courtesy of Professor Esther H. Chang; see also Nature [Lond] 348: 747)

Substrate: The compound an enzyme can act on, or the culture medium for an organism, or a particular surface. ►pseudosubstrate

Substrate Adhesion Molecules (SAM): Bind as extracellular molecules to independent receptors on adhering cells.

Substrate Cycle: ►futile cycle

Substrate Induction: Enzyme synthesis is stimulated by the presence of the substrate of the enzyme. ►*lac* operon, ►substrate

Substrate Ordering: The order of degradation—by multiubiquitination—of the cell cycle regulators by the anaphase-promoting complex (APC). (See Rapc M et al 2006 Cell 124:89; ►cell cycle, ►APC, ►ubiquitin).

Subtelomeric Duplications: ►telomeres

Subtiligase: An enzyme capable of ligating esterified peptides in aqueous solutions.

Subtilisin: A protease enzyme that cuts at serine in the context Gly-Thr-**Ser**-Met-Ala-Ser; chymotrypsin also cuts at serine but within a different sequence. Subtilisin is translated as a pre-pro-polypeptide containing the IMC (intramolecular chaperon) sequence between the signal peptide and the mature enzyme. IMC is responsible for the folding of the final enzyme without being present in the functional subtilisin, which is a general scavenger molecule. ►chymotrypsin, ►scavenger molecules, ►signal peptide

Subtraction Genomic: ►genomic subtraction

Subtractive Cloning (driver excess hybridization): Provides information of genes selectively expressed in different tissues. A single-stranded RNA (or DNA, the so-called tracer) is hybridized to another nucleic acid (the driver), which is present in the reaction mixture at least 10-fold in excess. Usually, from the tester RNA a cDNA is generated by the use of poly (dT) primers. Then from the single strand cDNA, double-strand DNA is generated with the aid of poly (dT) primer. The tester cDNA is mixed with an excess of driver cDNA and the cDNAs are allowed to reanneal after denaturation. The double-strand DNA is removed by adsorption to a hydroxyapatite column. The hybrids and the unhybridized driver are selectively removed (subtraction) and the remaining single-stranded, unhybridized tracer is further enriched by hydroxyapatite, biotinylation and selection by streptavidin, chemical cross-linking, RNase H and PCR. The process may be repeated until pure tracer-specific nucleic acid becomes available. This can then be used to screen a library for tracer-specificity. The procedure bears much similarity to c_0t or r_0t analysis. Hyperchromicity or digestibility by S1 nuclease can monitor the single-stranded molecules. The technique may provide information on the number or kind of genes expressed in specific tissues. ►hydroxyapatite, ►streptavidin, ►genomic subtraction, ►RFLP subtraction, ►positive selection, ►c_0t, ►r_0t analysis, ►S1 nuclease, ►hyperchromicity, ►tissue-specificity, ►nucleic acid hybridization, ►cascade hybridization, ►subtractive hybridization, ►normalization; Sagerström CG et al 1997 Annu Rev Biochem 66:751.

S

Subtractive Hybridization: ►subtractive cloning

Subtractive Suppression Hybridization (SSH): First subtracted cDNA libraries are generated. Then selective and/or suppressive cycles of PCR is combined with normalization and subtraction in a single procedure. The normalization equalizes the abundance of cDNAs within the samples and the subtraction excludes the sequences common to the target and driver populations. The procedure is well suited for identification of disease, developmental or other differentially expressed genes. ►PCR, ►subtractive cloning; Diatchenko L et al 1999 Methods Enzymol 303:349; den Hollander AI et al 1999 Genomics 58:240; Nishizuka S et al 2001 Cancer Res 61:4536; selective transcriptome extraction: Li L et al 2005 Nucleic Acids Res 33:(16):e136.

Substrain: A line separated from another after 8–19 cycle of inbreeding or when a single colony is different from the rest of the strain of rodents. ►subline, ►inbred

Subunit Vaccine: Contains only a fragment of the antigenic protein, sufficient to stimulate an immune response. ►vaccines, ►antigen

Subunits of Enzymes (protomers): The polypeptides that make up the oligomeric proteins.

Subvital: Has reduced viability, yet has 50% chance to survive up to the reproductive period of the species. ►sublethal

SUC2: Yeast invertase gene; it mediates glucose effect and may be suppressed by *ssn* (suppressor of *snf*). ►*SSN6*, ►*SNF*, ►glucose effect

Succinate Dehydrogenase: A non-heme iron, inner mitochondrial dehydrogenase enzyme (subunits encoded at human chromosomes 11q23 and 1q21). It converts succinate to fumarate while flavin adenine dinucleotide serves as a H_2 acceptor. ►phaeochromocytoma, ►mitochondrial diseases in humans

Sucrose Gradient Centrifugation: ►density gradient centrifugation

Sucrose Intolerance: ►disaccharide intolerance

Sucrose Transporters: Deliver the photosynthetically produced sucrose within the plant body. ►photosynthesis; Truernit E 2001 Curr Biol 11:R169.

Sudan Black: In microscopic use, it stains fatty tissues, wax, resins, cutins, etc., red.

Sudden Infant Death Syndrome (SID): Unexpected death of healthy, normal infants within the first year of life during sleep. It appears to be associated with a deficiency in the binding of the muscarinic cholinergic receptors of the brain resulting in accumulation of carbon dioxide or lack of oxygen in the blood. ►muscarinic acetylcholine receptors

Sufficiency, Statistical: Indicates that all necessary information about a parameter has been obtained.

SUG1: An ATPase and activator of transcription; it can substitute in yeast for Trip1 and can interact with the transcriptional activation domain of GAL4 and herpes virus protein VP16. ►ATPase, ►transcriptional activator, ►Trip1, ►GAL4, ►signal transduction

Sugar Beet (*Beta vulgaris*): 2n = 18 (see Fig. S138). One of the greatest success stories of plant breeding. In the middle of the eighteenth century, the average sugar content of the plant was about 2%. This increased by the twentieth century to about 20%, and the sugar yield per hectare increased to about 4 metric tons. Some of the modern varieties have numerous agronomically important features (disease resistance, monogermy), and this is a rare plant where triploid varieties (besides bananas) are grown commercially. ►banana, ►monogerm seed

Figure S138. Sugar beet

Sugar Code: Hypothesized to involve proteoglycan modifications on the basis that heparan sulfate and chondroitin sulfate proteoglycans would control axon guidance. ►heparan sulfate, ►chondroitin sulfate, ►axon guidance; Holt CE, Dickson BJ 2005 Neuron 46:169.

Sugarcane: (*Saccharum,* x = 10); its diploid forms are unknown and the cultivated varieties have high and variable number of chromosomes (*S. spontaneum* 2n = 36-128, *S. robustum* 2n = 60-170, *S. officinarum* 2n 70-140), including polyploids and aneuploids. The modern cultivated forms are natural hybrids of *S. spontaneum* × *S. officinarum* (this hybrid is also called *S. barberi* in India and *S. sinense* in China). Genetic mapping requires DNA markers, which—unlike the gene markers—can be studied despite the complex chromosomal situations. It is the most important source of saccharose or common sugar. ►sugar beet

Sugars: Mono- and oligosaccharides, components of polysaccharides and have important role in nutrition, and in complexes of critical significance in immunology, cancer, other diseases, several pigments, etc. Database for chemistry, classification, structure, physiological role: http://www.glycosciences.de/sweetdb/index.php.

Suicidal Antibody: The antibody is equipped with a targeting signal that can direct it to a degradative cellular compartment (e.g., lysosome, proteasome) where the bound antigen is destroyed. ►lysosome, ►proteasome; Larbig D et al 1979 Pharmacology 18:1; Hsu KF et al 2001 Gene Ther 8(5):376.

Suicidal Behavior: Appears to have a familial component. The suspected association between tryptophan hydroxylase and suicidal behavior is not supported by recent analysis (Lalovic A, Turecki G 2002 Am J Med Genet 114:533).

Suicide Inhibitor: A molecule that inhibits enzyme action after the enzyme acted upon it. In the original form, it is only a weak inhibitor but after reacting with the enzyme it binds to it irreversibly and becomes a very potent inhibitor. Allopurinol, fluorouracil are such molecules. ►allopurinol, ►regulation of enzyme activity

Suicide Mutagen: Uses a ^{32}P-labeled or other radioactive nucleotide that is incorporated into the genetic material and causes mutation by localized radiation. ►magic bullet

Suicide Vector: Delivers a transposon into the host cells in which the vector itself cannot replicate but the transposon can be maintained and used for transposon mutagenesis. Additional use involves the delivery of Herpes simplex virus thymidine kinase gene (HSV-TK) and administering ganciclovir or acyclovir. The HSV-TK is about three orders of magnitude more effective than the cellular TK in phosphorylating these DNA base analogs and thus blocking DNA synthesis. The phosphorylated ganciclovir is also impaired in moving through cell membranes and has longer-lasting local effects. Eventually some ganciclovir resistance may develop. The delivery of the gene for cytosine deaminase (CD) and the compound 5-fluorocytosine may produce within the target 5-fluorouracil, a DNA inhibitor. Xanthine-guanine phosphoribosyltransferase (XGPRT) may make the tumor cells more sensitive to 6-thioxanthene. P450–2B1 gene encoding a cytochrome, converts cyclophosphamide into the toxic phosphoramide mustard. The *deoD* bacterial gene (purine-nucleoside phosphorylase) converts 6-methylpurine deoxyribonucleoside into the deleterious 6-methylpurine. The bacterial nitroreductases convert the relatively non-toxic monofunctional N-chloroethylamine-derivative, C1954, into 10^4 times more active bifunctional alkylating agent. Employing HSV-TK, CD and XGPRT to the same tumor, three different transgenic cell populations, a mosaic may result. Theoretically, this could be used in a preventive program in putative cancer-prone cases. If the HSV-TK cells would become cancerous, ganciclovir would eliminate that cell subpopulation and the healthy tissues carrying the CD and XGPRT can take over as a prophylactic measure. Unfortunately, this mosaic approach has serious limitations at this time. Targeting a foreign antigen (e.g., HLA-B7 protein) to the cancer cells may stimulate the immune system of the host and kill the tumor cells. The immune system may be stimulated by transformation with cytokine genes. ►transposon mutagenesis, ►ganciclovir, ►gene therapy, ►cancer gene therapy, ►immunotherapy adoptive, ►cytokines, ►HLA, ►bystander effect, ►cytosine deaminase, ►thymidine kinase; Takamatsu D et al 2001 Plasmid 46:140.

Sulfatase Deficiency: ►mucosulfatidosis

Sulfhydryl Group: —SH, and when two are joined the linkage is a disulfide bond. Sulfhydryl compounds, such as cysteine and cysteamine are protectors against ionizing radiation. ►thiol, ►cysteine, ►cysteamine

Sulfocysteinuria: A deficiency of sulfite oxidase. Restricted intake of sulfur amino acids in the diet greatly reduces sulfocysteine and thiosulfate in the urine and improves conditions.

Sulfolobus (Archaea): genomes database: http://www.sulfolobus.org.

Sulfonylurea: A group of compounds that stimulate insulin production by regulating insulin secretion and lower blood sugar level; it is used to treat patients with non-insulin dependent diabetes. These drugs interact with the sulfonylurea receptor of pancreatic β cells and inhibit the conductance of ATP-dependent K^+ ion channels. Reduction of potassium exit activates the inward rectifying Ca^{2+} channels and promotes exocytosis. Sulfonylureas are inhibitors of the acetolactate synthase enzyme and there are also sulfonylurea herbicides. ►ion channels, ►diabetes, ►herbicides

Sulfur Mustard: See Fig. S139 where Al is an alkyl group. If the two alkyl groups are chlorinated (Cl) the compound is called bifunctional. If only one alkyl group is chlorinated, the compound is monofunctional. ►nitrogen mustard, ►alkylating agents; Michaelson S 2000 Chem Biol Interact 125:1.

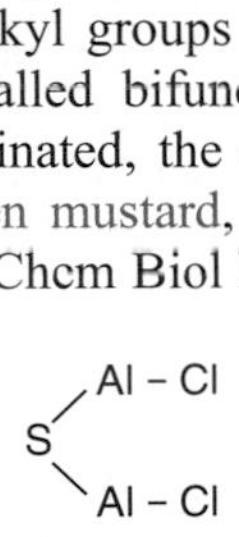

Al - Cl
S
Al - Cl

Figure S139. Sulfur mustard

S

Sulfurylase: Catalyzes the reaction $ATP^{3-} + SO_4^{-2} \rightarrow$ adenylyl sulfate (adenosine-5′-phosphosulfate, $C_{10}H_{14}N_5O_{10}PS$).

Sulston Score: Used for comparative microsynteny analysis. It displays matching status of fingerprints of a chosen genomic clone and its matching sequences in others by a given probability of coincidence and a given tolerance. The analysis permits physical mapping of genomes. The scores may vary from 04 to 38 and the lower is more stringent. (See Sulston J et al 1988 Comput Appl Biosci 4:125).

Summary Statistic: Defines characteristics of a diverse population by the mean and variances. ►mean, ►variance

Sum of Squares: The sum of the squared deviations from their mean of the observations; it is used in statistical procedures to estimate differences. ►analysis of variance, ►intraclass correlation

SUMO (small ubiquitin-like modifier, 2q32.2-q33): Ubiquitin-like carrier proteins (11 kDa). Sumos are conjugated to a variety of proteins (RanGAP1-Ubc9-Nup358) and seem to be involved in multiple functions such as cell cycle progression, DNA repair, spontaneous mutation, heterochromatin stabilization and counteract recombinogenic events at damaged replication forks, etc. Histone sumoylation may repress transcription. ►UBL, ►monoubiquitin, ►sentrin, ►PIC, ►IκB, ►IKK, ►RAN, ►Ubc9, ►nuclear pores, ►Huntington's chorea, ►NF-κB; Melchior F 2000 Annu Rev Cell Dev Biol 16:591; Müller S et al 2001 Nature Rev Mol Cell Biol 2:202; Shiio Y, Eisenman RN 2003 Proc Natl Acad Sci USA 100:13225; Reverter D, Lima CD 2005 Nature [Lond] 435:687; crystal structure of thymine DNA glycosylase–SUMO: Baba D et al 2005 Nature [Lond] 435:979; heterochromatin: Shin JA et al 2005 Mol Cell 19:817; sumoylation site server: http://bioinformatics.lcd-ustc.org/sumosp/.

Sunburn: Sunburn (UV) causes keratoses in about 60% of the cases. The p53 gene may suffer mutation(s), mainly C→T transitions. These p53 mutant cells, after clonal propagation, may develop into squamous cell skin cancer (SCC). In over 90% of the SCC cells, the p53 gene is mutant. The increased melanin production involved with ultraviolet light exposure may be correlated with the DNA repair system. ►p53, ►keratoses, ►DNA repair, ►ultraviolet light, ►pigmentation in animals, ►melanoma

Sunds: ►Brugada syndrome

Sunflower (*Helianthus*): An oil crop with about 70 xenogamous species, 2n = 2x = 34. (See Burke JM et al 2002 Genetics 161:1257).

Sun-Red Maize: Develops anthocyanin pigment when the tissues are exposed to sunshine (e.g., through a stencil) (see Fig. S140).

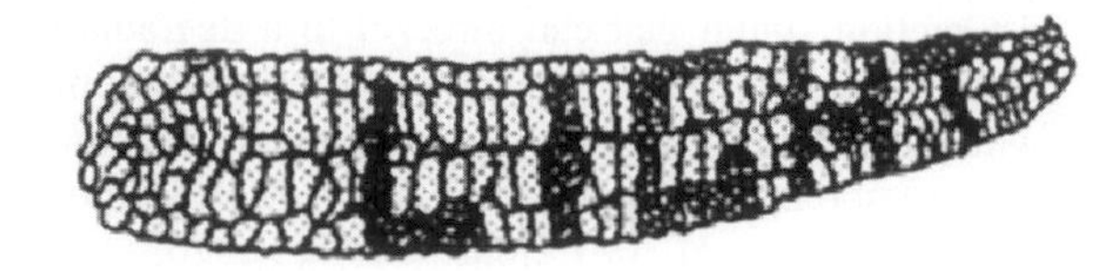

Figure S140. Sun-red maize ear

SUP35: ►prion

supB: Ochre/amber suppressor; it inserts glutamine.

supC: A suppressor mutation for amber (5′-UAG-3′) and ochre (5′-U3′) chain-terminator codons. The mutation causes a base substitution in the anticodon of tyrosine tRNA (5′-GUA-3′) and it changes to 5′-UUA-3′ that can recognize both the ochre and amber codons as if they were tyrosine codons. Note: pairing is 5′-3′ and 3′-5′. For the recognition of the former, wobbling is required. ►suppressor, ►wobbling, ►anticodon, ►ochre, ►amber

supD: An amber suppressor mutation that reads the amber (5′-UAG-3′) chain termination codon as if it would be a serine (5′-UCG-3′) codon because the normal serine tRNA anticodon (5′-CGA-3′) mutates to (5′-CUA-3′), therefore, instead of terminating translation, a serine is inserted into the amino acid chain. (Remember, the pairing is antiparallel). ►suppressor, ►anticodon

supE: An amber-suppressor mutation that reads the chain-termination codon 5′-UAG-3′ (amber codon) in the mRNA as if it would be a glutamine codon (5′-CAG-3′) because the glutamine tRNA anticodon mutates to 5′-CUA-3′ from the 5′-CUG-3′. As a consequence, the translation proceeds and a glutamine is inserted at the site where in the presence of an amber codon in the mRNA the translation would have terminated. (Note: the pairing is antiparallel).

supF: Amber suppressor; it inserts tyrosine.

Superantigen: Native bacterial and viral proteins that can bind directly (without breaking up into smaller peptides) to MHC class II molecules on antigen-presenting cells. The variable regions of T cell receptor β chains thus activate more T cells against, e.g., enterotoxins of *Staphylococci*—causing toxic shock syndrome or food poisoning—than normal antigens. Cellular superantigens are likely to be responsible for such diseases as diabetes mellitus. ►MHC, ►TCR, ►enterotoxin, ►toxic shock syndrome, ►diabetes mellitus, ►antigen, ►endogenous virus; Muller-Alouf H et al 2001 Toxicon 39:1691; Alam SM, Gascoigne NR 2003 Methods Mol Biol 214:65.

Superchiasmatic Nucleus: A region near the optical center of the hypothalmus in the brain controlling circadian signals. ►brain, ►circadian rhythm

Supercoiled DNA: May assume the *positive supercoiled* structure by twists in the same direction as the original, generally (right handed) coiling of the double helix, or it may be twisted in the opposite direction, *negative supercoiling*. Negative supercoiling (Z DNA) may be required for replication and transcription (see Fig. S141).

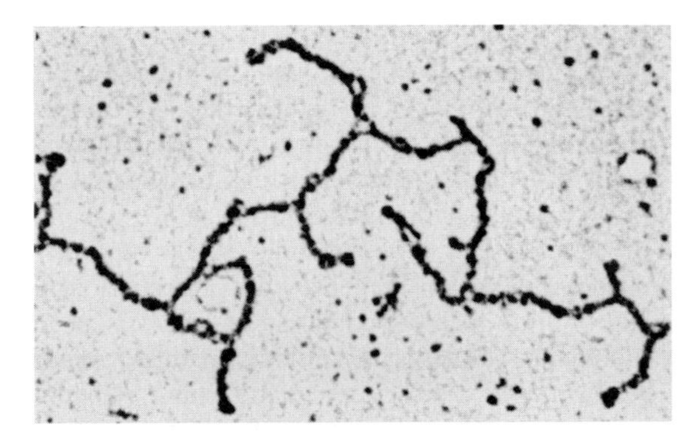

Figure S141. Supercoiled DNA plasmid of *Streptococcus lactis*. (Courtesy of Dr. Claude F. Garon)

The superhelical density expresses the superhelical turns per 10 bp and it is about 0.06 in cells as well as in virions. Loss of positive supercoiling may be lethal. Localized negative super coiling may be lethal. Localized negative supercoiling is essential for gene expression, replication and many other functions of the DNA. ►DNA replication prokaryotes, ►transcription, ►Z DNA, ►packing ratio, ►linking number; Holmes VF, Cozarelli NR 2000 Proc Natl Acad Sci 97:1322.

Superdominance: Same as overdominance or monogenic heterosis. ►hybrid vigor, ►overdominance

Superfamily of Genes: A group of genes that are structurally related and may have descended from common ancestors although their present function may be different. (See http://supfam.mrc-lmb.cam.ac.uk/SUPERFAMILY/index.html).

Superfamily of Proteins: Evolutionarily related proteins. (See http://cathwww.biochem.ucl.ac.uk/latest/index.html).

Superfemale (metafemale): *Drosophila*, trisomic for the X-chromosome (XXX) but disomic for the autosomes; she is sterile. ►aneuploidy, ►supermale *Drosophila*

Superfetation: Due to an apparently rare autosomal dominant gene ovulation may continue after implantation of the fertilized egg and an unusual type of twinning results. In animals, when a female may mate repeatedly with several males, the same litter may become multipaternal. Human dizygotic twins may also be of different paternity when during the receptive period the female had intercourse with two different males. ►twinning, ►multipaternate litter; Fontana J, Monif GR 1970 Obstet Gynecol 35:585.

Supergenes: Linked clusters of genes that are usually inherited as a block because inversion(s) prevent(s) the survival of the recombinants for the clusters and thus have evolutionary and applied significance in plant and animal breeding. Supergenes may be adaptively linked polymorphic loci.

Superinfection: A bacterium is infected by another phage. This is generally not possible in a lysogenic bacterium because of immunity, i.e., the superinfecting phage cannot enter a vegetative cycle within the lysogen. Infection by two T even phages may fail in such an attempt by "superinfection breakdown." Some higher organisms may be infected by parasites of different genotypes. The number of clones of the parasites may not increase with the progression of the disease indicating competition among the superinfecting strains (e.g., *Plasmodium*) ►immunity, ►phage, ►*Plasmodium*; Hayes W 1965 The Genetics of Bacteria and their Viruses. Wiley, New York; Vogt B et al 2001 Hum Gene Ther 12(4):359.

Supermale *Asparagus*: Can be obtained by regenerating and diploidizing plants obtained from Y chromosomal microspores or pollen by the techniques of cell culture. Thus, their chromosomal constitution is 18 autosomes + YY. These plants are commercially advantageous because of the higher yield of the edible spears. ►YY *Asparagus*

Supermale *Drosophila*: Has 1 X and 3 sets of autosomes, i.e., the fly is monosomic for X but he is trisomic for all the autosomes; he is sterile. ►superfemale, ►sex determination

Superman: A homeotic mutation in *Arabidopsis* resulting in excessive development of the androecium at the expense of other flower parts. The DNA base sequences are the same as in the wild type but the mutation evokes methylation. ►flower differentiation, ►androecium; Koshimoto N et al 2001 Plant Mol Biol

46:171; Cao X, Jacobsen SE 2002 Curr Biol 12:1138; Dathan N et al 2002 Nucleic Acids Res 30:4945.

Super-Mendelian Inheritance: The transmission of one of the alleles in a diploid is preferential because of the advantage of a gene at meiosis or post-meiotic steps of gametogenesis. This advantage may not be due an intrinsic superiority of the allele. Homing endonuclease genes (HEG) may be preferentially transmitted horizontally, but may or may not be propagated at a selective advantage. ►homing endonuclease, ►segregation distorter, ►meiotic drive, ►preferential segregation; Goddard MR, Burt A 1999 Proc Natl Acad Sci USA 96:13880.

Supermutagen: An efficient mutagen that causes primarily point mutations without inducing frequent chromosomal defects. ►point mutation, ►ethylmethane sulfonate, ►nitrosoguanidine

Supermutation: ►hypermutation, ►somatic hypermutation, ►antibody gene switching

Supernatant: The non-sedimented fraction after centrifugation of a suspension or in general, any floating fraction derived of a mixture.

Supernatural: ►paranormal

Supernumerary Chromosome: ►B chromosome, ►cat eye syndrome

Supernumerary Marker Chromosome: Rare ($\sim 3 \times 10^{-4}$) abnormal chromosomes with internal duplications and deletions. The majority (~60%) involves two inverted copies of the short arm, centromere and the proximal segment of the long arm of human chromosome 15 (SMC[15]). The majority of the latter are dicentric but one of the centromeres is inactive (pseudocentromeric 15). ►chromosomal rearrangements; Roberts SE et al 2003 Am J Hum Genet 73:1061.

Superoperon: A regulatory complex tied together, e.g., photoreceptor pigment synthesis and photosynthesis. ►operon, ►überoperon, ►supraoperon

Superovulation: By injection of gonadotropic hormones (into mice), the number of eggs produced may increase several fold. Fertilization follows after about 13 h and the eggs can be surgically collected to study, in vitro, the preimplantation development. Hormonal treatment may cause superovulation also in usually monoparous animals too. ►twinning

Superoxide ($O_2{-}\bullet$): A highly reactive species of oxygen. In the hypersensitivity reaction defense mechanism, it appears to play an important role along with nitric oxide. ►nitric oxide, ►hypersensitivity reaction, ►SOD, ►hydrogen peroxide, ►ROS, ►Fenton reaction, ►peroxynitrite, ►peroxides

Superoxide Dismutase (SOD): Catalyzes the reaction $2O_2 + 2H^+ \rightarrow O_2 + H_2O_2$ and thus participates in the detoxification of the highly reactive (mutagenic) superoxide radical O_2^-. These enzymes are the main detoxificants of the free radicals. SOD overexpression confers resistance to ionizing radiation under aerobic conditions. There are three isozymes of SOD: the cytosolic SOD-1 (human chromosome 21q22.1, the mitochondrial SOD-2 [6q25.3] and the extracellular SOD-3 [4pter-q21]). In SOD deficiency, cardiomyopathy, brain damage, mitochondrial defects, Lou Gehrig's disease and precocious aging may result. A non-peptidyl manganese complex with bis (cyclohexylpyridine)-substituted macrocyclic ligand may mimic SOD activity. Inhibition of SOD enzymes may trigger suicide of leukemia cells, which have a higher rate of oxidative metabolism. Cell suicide may also occur when the level of hydrogen peroxide is high or when the latter is converted to hydroxyl radicals in a Fe2+ dependent process. Using the cytosolic SOD gene in an adenoviral vector may provide protection against ROS injury. ►amyotrophic lateral sclerosis, ►granulomatous disease, ►cardiomyopathies, ►aging, ►ROS, ►host-pathogen relationship, ►oxidative stress, ►AMPA; Fridovich I 1975 Annu Rev Biochem 44:147; Danel C et al 1998 Hum Gene Ther 9:1487.

Superparamagnetic Scale Particles: ►magnetic relaxation switches, ►magnetic targeting

Super-Repressed: Bacterial operon cannot respond to inducer. ►repression, ►inducer

Supershift: ►gel retardation assays

Supersuppressor: A dominant suppressor acting on more than one allele or even on different gene loci. ►suppressor; Gerlach WL 1976 Mol Gen Genet 144:213.

Supervillin: ►androgen receptor

Supervised Learning: Finding shared patterns or motifs common to all "positive" sequences and absent from all "negative" ones in a curated way. ►motif, ►unsupervised learning, ►curated; Moler EJ et al 2000 Physiol Genomics 2(2):109.

Supervital: Fitness exceeds that of the standard (wild) type.

supF: An amber-suppressor mutation in $tRNA^{Tyr}$ that recognizes the chain-termination codon (5′-UAG-3) as if it would be a tyrosine codon (5′-UAC-3′ or

5′-UAU-3′) because a mutation at the anticodon sequence in the tyrosine tRNA changes the 5′-GUA-3′ into a 5′-CUA-3′ and thus the tyrosine tRNA inserts a tyrosine into the growing peptide chain where translation would have been terminated. ►suppressor tRNA, πVX.

supG: Suppressor mutation for both amber (5′-UAG-3′) and ochre (5′-UAA-3′) chain termination codons, by a base substitution mutation in the anticodon of lysine tRNA from (5′-UUU-3′) to 5′-UUA-3′ that can recognize the amber and ochre codons in the mRNA, as if they would be lysine (5′-AAA-3′ or 5′-AAG-3′) codons and thus inserting in the peptide chain a lysine rather than discontinuing translation. (Note: the pairing is antiparallel).

Support: A statistical concept it is synonymous with log likelihood. ►likelihood

Support Vector Machine (SVM): A computer program based on the principle of supervised learning techniques. It uses a training set to determine which data should be clustered a priori. The operations start with a set of genes that are already known to have common function, e.g., genes that encode ribosomal proteins. It picks another set of genes that are not members of the functional group specified. These two sets form the training examples and the genes are marked positive in case they fall in the specified functional class or negative if they are not. The SVM thus learns from the two groups how to discriminate and classify any "unknown" on the basis of function. This procedure can classify genes based on microarray hybridization data and identify functional clusters genome-wide. ►microarray hybridization, ►machine learning, ►supervised learning, ►cluster analysis; Brown MPS et al 2000 Proc Natl Acad Sci USA 97:262; Hua S, Sun Z 2001 Bioinformatics 17(8):721; Cristianini N, Shawe-Taylor J 2000 An Introduction to Support Vector Machines and Other Kernel-Based Learning Methods. Cambridge University Press, New York.

Supportive Counseling: To alleviate the psychological problems involved with the discovery of hereditary disorders and birth defects in families. ►counseling genetic

Suprachiasmatic Nucleus: A small site in the hypothalamus above the optic region controlling rhythmic function of the body and memory of all vertebrates. ►brain human

Supramolecular/Supermolecular: Literally means beyond molecular, but it is also used for structures assembled from molecules held together by bonds weaker than those within molecules are. The number of molecules, how they are connected, and the way they are oriented relative to each other, characterizes the system structurally. Supermolecular structure is important for understanding how cells function. (See Turro NJ 2005 Proc Natl Acad Sci USA 102:10765).

Supraoperon: A region of the prokaryotic genome where operons of similar function are located, e.g., in the aromatic (*aro*) amino acid pathway including tryptophan (*trp*) and histidine (*his*): ►operon, ►überoperon, ►superoperon; Berka RM et al 2003 Proc Natl Acad Sci USA 100:5682.

Supravalvar Aortic Stenosis (SVAS): A human chromosome 7q11 dominant mutation in an elastin gene with a prevalence of about 5×10^{-5} causing obstruction of the aortic blood vessels. The basic defect is due to a deficiency of elastin. It is pleiotropic and part of the Williams syndrome. ►coarctation of the aorta, ►cardiovascular diseases, ►William's syndrome, ►elastin

Suppressor, Bacterial: Protein product of the bacterial regulator gene (e.g., *i* in the *Lac* operon) that when associated with the operator prevents transcription. ►*Lac* operon, ►ara operon

Suppressor, Extragenic: The suppressor mutation is outside the boundary of the suppressed gene. ►suppressor tRNA, ►second-site suppressor, ►second site reversion

Suppressor Gene: Restores function lost by a mutation without causing a mutation at the site suppressed. The suppressor can be intragenic (within the cistron but also at another site) or at any other locus, e.g., in the signal transduction pathway. The suppressor may act by reducing and also by overexpression of a gene product. ►suppressor extragenic, ►suppressor intragenic, ►suppressor informational, ►frameshift, ►mutation in organelle DNA

Suppressor, Haplo: Dominant suppressor that can act in a single dose.

Suppressor, Informational: Interferes with the expression of another gene at the level of translation by affecting either tRNAs or ribosomes or peptide elongation factors.

Suppressor, Intragenic: The suppressor site is within the gene where suppression takes place. ►frameshift suppressor

Suppressor Mutation: ►suppressor gene, ►*sup*, ►sup pressor tRNA, ►suppressor RNA, ►frameshift suppressor, ►mitochondrial suppressor

Suppressor RNA (sRNA): Prevents translation of a mRNA by partial base pairing with a specific sequence of the target. Two 22 and 40 nucleotide long tracts of the *lin-4* transcripts control the translation of gene *lin-14*. The latter is an early expressed nuclear protein in *Caenorhabditis* but its expression is blocked during later stages by the *lin-4* RNA that itself is not operating through a protein. Several other genes are subject to RNA suppressors in other organisms. The mechanism of this suppression seems to be different from that of the suppressor tRNA. ►suppressor tRNA, ►suppressor gene, ►antisense RNA, ►antisense DNA, ►RNAi, ►repressor; Olsthoorn RC et al 2004 RNA 10:1702.

Suppressor, Second Site: A suppressor outside the boundary of the locus but usually within the same chromosome. ►suppressor gene, ►second site reversion

Suppressor Selection Gene Fusion Vector: Carries nonsense codon(s) in the structural gene and can be expressed only if the target genome carries nonsense suppressors. ►vectors gene fusion, ►transcriptional gene fusion vector

Suppressor T Cell: Can suppress antigen-specific and allospecific T cell proliferation by competing for the surface of antigen-presenting cells. ►T cell, ►allospecific; Maloy KJ, Powrie F 2001 Nature Immunol 2:816.

Suppressor tRNA: Makes the translation of nonsense or of missense codons in the original, normal sense possible because a mutation in the anticodon of the tRNA recognizes the complementary sequence in the codon but its specificity resides in the tRNA molecule (see Fig. S142). Exceptions are possible, however. In the anticodon 5′-CCA-3′ of the tryptophan codon (5′-UGG-3′) is not mutant yet it may deliver to the opal position a tryptophan if at position 24 of the D loop of the tryptophan tRNA a guanine is replaced by an adenine. Similarly, if in the GGG codon of glycine another G base is inserted by a frameshift mutagen but subsequently, in its anticodon (CCC) an extra C is inserted then the tRNA may read the four bases normally as a glycine codon rather than nonsense. Beyond the codon–anticodon binding, mutation elsewhere in $tRNA^{Trp}$ can effect the correct decoding (Cochella L, Green R 2005 Science 308:1178). ►tRNA, ►Hirsh suppressor, ►code genetic, ►mutation in organelle DNA, ►phenotypic reversion, ►translation in vitro; Smith JD et al 1966 Cold Spring Harb Symp Quant Biol 31:479; Murgola EJ 1985 Annu Rev Genet 19:57; Beier H, Grimm M 2001 Nucleic Acids Res 29:4767.

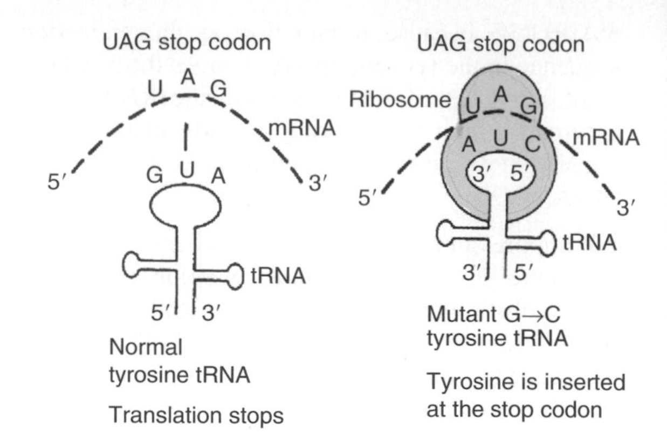

Figure S142. Suppressor tRNA

supU: A suppressor of the opal (5′-UGA-3′) chain terminator codon with an anticodon sequence of 5′-UCA-3′. The mutation changes the anticodon of the tryptophan tRNA from 5′-UCA-3′ to 5′-CCA-3′ and that permits the insertion of a tryptophan residue into the polypeptide chain where it would have been terminated without the suppressor mutation (codon-anticodon recognition is antiparallel). This suppressor mutation is unusual because it recognizes both the tryptophan codon (5′-UGG-3′) and the opal suppressor codon, i.e., its action is ambivalent.

SurA: Periplasmic parvulin type peptidyl-prolyl isomerase involved in chaperoning outer membrane proteins. ►peptidyl-prolyl isomerase, ►parvulin, ►periplasma

Surface Antigen: Generally, glycoprotein molecules on the cell surface that determine the identity of the cells for immunological recognition. The display of surface antigens is regulated at the level of transcription. ►VSA, ►antigen, ►*Trypanosomas*, ►*Borrelia*

Surface Plasmon Resonance (SPR): Incoming light, in a certain angle, hits a hydrophilic dextran layer covered gold surface. The interacting molecule is immobilized on this surface. As the molecule injected binds to this surface the refractive index of the medium is increased and the change in the angle at which the intensity change occurs is measured in resonance units. The size of the macromolecule is positively correlated with the response. This analysis has numerous biological applications for the determination of protein-protein interactions such as antibody affinities, epitope, growth factor, signal transducing molecule, receptor, ligand binding, etc. ►biosensors, ►microcalorimetry, ►immunoprecipitation; Heaton RJ et al 2001 Proc Natl Acad Sci USA 98:3701.

Surfectant: Common surfectants are soaps and detergents. In gene therapy, different surfectants such as perfluorochemical liquids, phospholipids may facilitate gene delivery to pulmonary tissues.

Surfection: A procedure to transfer genes to cells layered of the surface of polyethylenimine/collagen-coated wells. It can deliver multiple plasmids into cells at large-scale arrays. The procedure can assay RNAi-mediated silencing, live-cell imaging and cell-based drug screening. ►RNAi; Chang F-H et al 2004 Nucleic Acids Res 32(3):e33.

Surrogate Chains (ΨL, ΨH): Of immunoglobulins are found in the progenitor (pro-) B lymphocytes. In the human and mouse fetal liver and adult bone marrow an 18 kDa protein with (~45%) homology to the variable regions of the κ, λ and heavy chains and a 22 kDa protein with ~70% homology to the constant region of the λ chain have been found. The synthesis of these proteins (ΨL) ceases after the IgM chains appear on the lymphocyte surface. They participate in the formation of the B cell receptor with other immunoglobulins and a short μ chain containing only the N-terminal D(J)C sequences. The ΨL light chains are associated with a complex of 130 kDa/35–65 kDa glycoproteins (ΨH) and together make the surrogate receptors. ►immunoglobulins, ►B lymphocytes, ►immune system; Meffre E et al 2001 J Immunol 167:2151.

Surrogate Genetics: ►reversed genetics

Surrogate Mother: A female who carries to term a baby for another couple. She may actually contribute the egg or may just be a gestational carrier of a fertilized egg and has no genetic share in the offspring. In either case, moral, ethical, psychological and legal problems must be pondered before this method of child bearing is chosen. Civil law generally keeps the maternal right of the gestational mother despite any contractual agreement contrary to natural parenthood. In case, however, the gestational mother is not the donor of the ovum, the maternal right pertains to the biological ovum donor. Obviously, there are serious ethical problems here beyond the principles of genetics. The society must protect the best interest of the child. ►oocyte donation, ►ART, ►paternity testing

Surveillance, mRNA: A mechanism of quality control for the elimination of defective proteins. (See Hilleren P et al 1999 Annu Rev Genet 33.229, ►immunological surveillance).

Survey Sequencing: Covers only 1x-2x of the genome, providing information on 60–80% of the sequences in a fragmented form, compared with complete sequencing that employs 6x-8x coverage. It is a temporary compromise mandated by monetary, physical and manpower limitations. It still provides substantial information because several other genomes with homologies supplement and complement the missing information. (See Hitte C et al 2005 Nature Rev Genet 6:643).

Survival: The total survival probability per genetic damage is expressed by the equation:

$$S(\rho,\partial) = \sum_{h=0}^{h=\partial} P(\rho, h, \partial!)$$

where ρ = hit probability, ∂ = VD, V = total cell volume, D = density of the active events [dose], P = likelihood of survival. (See more in Alpen EL 1998 Radiation Biophysics. Academic Press; ►target theory, ►neutral mutation, ►beneficial mutation, ►cost of evolution, ►genetic load, ►fitness)

Survival Factors: Interfere with apoptosis. The balance between survival and apoptosis is very complex and a large number of proteins and ligands are involved through different pathways. (See Fig. S143, ►apoptosis, ►BAD, ►survivin)

Survival Estimator: ►Kaplan-Meier estimator

Survival of the Fittest: ►fitness, ►neo-Darwinism, ►social engineering, ►social Darwinism

Survival Phenotype: Under poor nutritional condition the fetus develops a lean body, undersized visceral organs, fewer muscle cells, undersized nephrons and liver. Because of the maternal nutritional deficiency, the promoters of many genes remain unmethylated in the early embryo, resulting in altered phenotype persisting into adulthood. After birth, they may gain more weight when food becomes available and may become diabetic. ►nephron, ►diabetes; Gluckman P, Hanson M 2005 The fetal matrix: Evolution, development and disease. Cambridge University Press.

S

Survivin: An inhibitory protein (16.5 K) of caspases that are activated by cytochrome c. It is antiapoptotic. Survivin is expressed as a passenger protein at the G2—mitosis phases of the cell cycle and it is associated with the microtubules of the mitotic spindle. Apoptosis may be caused by the disruption of this association. Overexpression of survivin may lead to cancer. Disruption of survivin action may reduce melanoma growth. Survivin is essential for the development of erythroid cells; but overexpression of survivin interfered with megakaryocyte development but did not affected erythroid cells (Gurbuxani S et al 2005 Proc Natl Acad Sci USA 102:11480). The *survivin* gene promoter contains several Sp1 canonical, Sp1-like, and p53-binding elements, suggesting participation of the Sp1

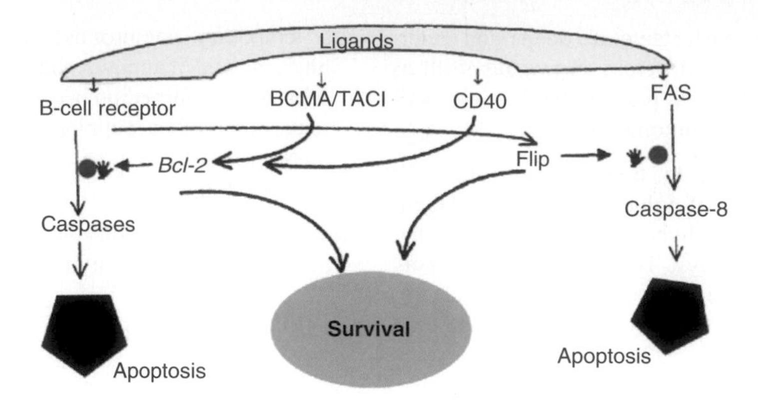

Figure S143. Survival factors

transcription factor and/or p53 in gene regulation. Furthermore, *survivin* transcription is down-regulated by the DNA-damaging agent doxorubicin, which mediates p53 induction in acute lymphoblastic leukemia (Estève P-O et al 2007 J Biol Chem 282:2615). ►apoptosis, ►cell cycle, ►spindle, ►survival factor, ►passenger protein, ►Stat, ►megakaryocyte, ►erythrocyte, ►p53, ►doxorubicin, ►leukemia; Wheatley SP et al 2001 Curr Biol 11:886; Hoffman WH et al 2002 J Biol Chem 277:3247; Altieri D 2003 Nature Rev Cancer 3:46.

Suspension Culture: The cells are grown in liquid nutrient medium.

Suspensor: A line of cells through which the plant embryo is nourished by the maternal tissues. In general, anatomy ligaments may be called suspensors.

Sutent: ►Gleevec

Suture: A junction of various solid animal and plant tissues.

Su(var): Suppressor of variegation. ►heterochromatin; Rudolph T et al 2007 Mol Cell 26:103; Haynes KA et al 2007 Genetics 175:1539.

SUV39H1 (Clr): ►histone methyltransferases

Sv: ►Sievert

SV2s: Glycosylated transmembrane proteins, homologous to prokaryotic and eukaryotic transporters. ►transporters

SV40 Tag: Simian virus 40 large T antigen. ►Simian virus 40

SV40 Vectors: SV40 plasmids (vectors) can be packaged only if their DNA is within the range of 3900 to 5300 bp. Since these small genomes do not have much dispensable DNA, it is almost impossible to construct a functional vector with any added genes to it. Fortunately, functions provided by helper DNA molecules might help to overcome these problems. Simian Virus 40 cannot replicate autonomously if the replicational origin (*ori*) is defective, yet it can integrate into chromosomal locations of green monkey cells and can then be replicated along the chromosomal DNA (such a cell is COS [cell origin simian virus]). Also, since the early genic region is normal (►Simian Virus 40 for structure), it may produce the T antigen within the cell. If such a cell is transformed by another SV40 vector in what the viral early gene region was replaced by a foreign piece of DNA, the COS cell may act as a helper and replicate multiple copies of the second, the engineered SV40 DNA. Since the late gene region of this plasmid is normal, the viral coat proteins can be synthesized within the cell. The availability of the coat proteins permits the packaging of the engineered SV40 DNA into capsids.

The virions so obtained can be used to infect other mammalian cells where the passenger DNA can be transcribed and translated and the foreign protein can be processed. The transformed cell thus can acquire a new function. Also, an SV40 plasmid can be constructed with insertion into the late gene cluster, a foreign gene with a desired function. This plasmid can then use inactivated (deleted) early genes but with a good *ori* site. Upon coinfection of a mammalian cell with these two plasmids, the SV40 plasmids can replicate to multiple copies and the inserted foreign gene can be expressed. Also, it is feasible to insert into a prokaryotic pBR322 plasmid the *ori* region of SV40 and another piece of DNA including all the necessary parts of a foreign gene. When this plasmid is transfected into a COS cell, the passenger gene can be transcribed, translated and processed thanks to the multiple copies replicated within this mammalian cell. With the assistance of SV40 based constructs, mammalian and other genes can be shuttled between mammalian and bacterial cells.

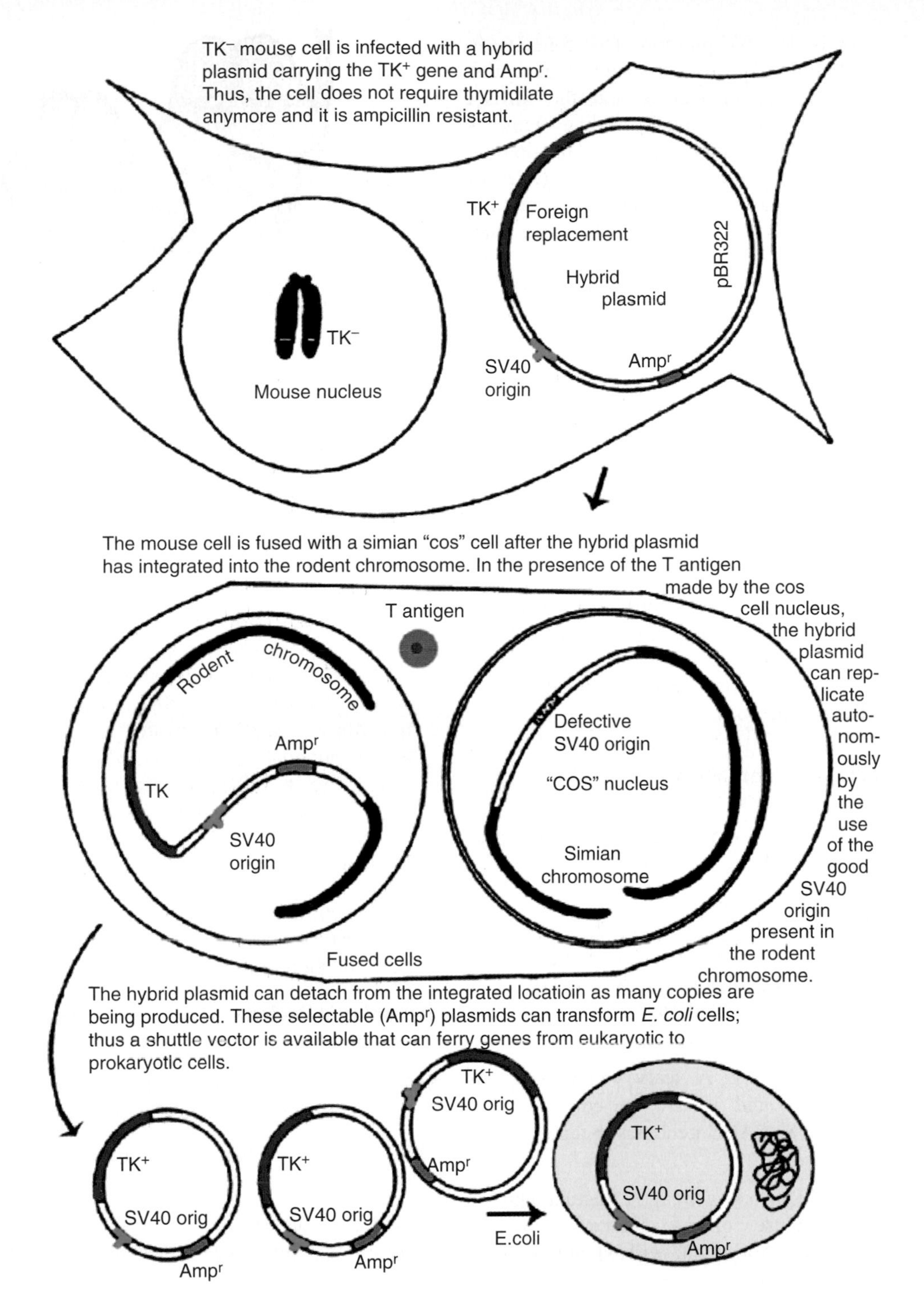

TK⁻ mouse cell is infected with a hybrid plasmid carrying the TK⁺ gene and Ampʳ. Thus, the cell does not require thymidilate anymore and it is ampicillin resistant.
TK⁺
Foreign replacement
Hybrid plasmid
pBR322
TK⁻
Mouse nucleus
SV40 origin
Ampʳ
The mouse cell is fused with a simian "cos" cell after the hybrid plasmid has integrated into the rodent chromosome. In the presence of the T antigen made by the cos cell nucleus, the hybrid plasmid can replicate autonomously by the use of the good SV40 origin present in the rodent chromosome.
T antigen
Rodent chromosome
Ampʳ
TK
SV40 origin
Defective SV40 origin
"COS" nucleus
Simian chromosome
Fused cells
The hybrid plasmid can detach from the integrated locatioin as many copies are being produced. These selectable (Ampʳ) plasmids can transform E. coli cells; thus a shuttle vector is available that can ferry genes from eukaryotic to prokaryotic cells.
TK⁺
SV40 orig
Ampʳ
TK⁺
SV40 orig
Ampʳ
TK⁺
SV40 orig
Ampʳ
E.coli
TK⁺
SV40 orig
Ampʳ

Figure S144. SV40 shuttle vectors can ferry genes from mammals to prokaryotes

The procedure: a thymidine kinase-deficient (*tk*⁻) rodent cell is transformed by pBR322 bacterial plasmid carrying the *ori* of SV40 and the TK^+ (functional thymidine kinase), and an ampicillin resistance gene (amp^R) for bacterial selectability. Within the rodent cell the bacterial plasmid is integrated into a chromosome of the rodent cell and this cell is then fused with a COS cell. The integrated pBR322 plasmid is replicated into many copies thanks to the presence of the COS nucleus. The hybrid plasmid carrying a bacterial replicon, an SV40 *ori*, the TK^+ and amp^R can infect *E. coli* cells and can be selectively propagated there. Thus, the shuttle function is achieved. Another SV40 and pBR322 based vector is the pSV plasmid. This contains, in a Pvu II and HinD III restriction enzymes generated fragment, the promoter signals and the mRNA initiation site. When any open reading frame is attached to it, transcription can proceed. In addition, an intron of the early region provides splicing sites for other genes. The region also contains the transcription termination and polyadenylation signals. Some other pBR322 parts may be equipped with additional specific, selectable markers. ►Simian virus 40, ►vectors, ►viral vectors, ►shuttle vector; Jayan GC et al 2001 Gene Ther 8:1033; see Fig. S144.

SVA: Non-autonomous transposable elements in the human genome that are mobilized by L1. ►transposable element, ►LINE; Ostertag EM et al 2003 Am J Hum Genet 73:1444.

SVC: ►carbon dioxide (CO_2) sensitivity

SVD (singular value decomposition): A comparative mathematical procedure for genome-scale study of expression of two data sets, containing "genelets," shared by both sets. (Alter O et al 2003 Proc Natl Acad Sci USA 100:3351).

Svedberg Units: ►sedimentation

SVF-2: A member of the tumor necrosis factor receptor family. ►Fas, ►TNF

Swapping Genes: exchanging genes by horizontal transfer via plasmids. ►transmission, ►site-specific recombination, ►targeting genes; Nebert DW et al 2000 Ann N Y Acad Sci 919:148.

Swede (*Brassica napus*): A leafy fodder crop in Northern climates and it also an edible human vegetable; 2n = 38, AABB genomes. ►rape

Sweet Clover (*Melilotus officinalis*): A fragrant leguminous plant because of its coumarin content. The coumarin-free forms (*M. albus*) are used as hay, 2n = 16. ►coumarin

Sweet Pea (*Lathyrus odoratus*, 2n = 14): An ornamental that had been exploited for studies on the genetic determination of flower pigments (see Fig. S145). Another peculiarity is that the "long"/"disc" pollen shape (*L/l*) is determined by the genotype of the sporophytic (anther) tissue rather than by the gametophyte. Therefore, delayed inheritance is observed. ►delayed inheritance

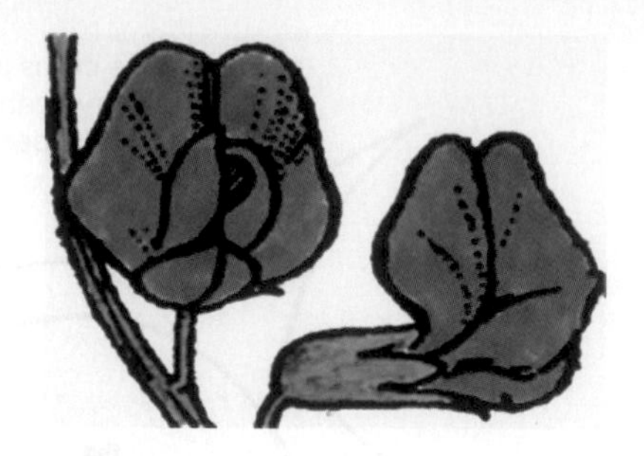

Figure S145. Sweet pea

Sweet Potato (*Ipomoea batatas*): Primarily a warm climate vegetable with about 25 species with basic chromosome number x = 15 and a single genome; the most common cultivated form is hexaploid, although related species may be diploid or tetraploid.

Swept Radius: ►linkage

SWI*, *SWI2/SNF2*, *SWI3: Yeast genes encoding transcriptional activators by chromatin remodeling. The homologous protein in *Drosophila* is BRAHMA. SWI2/SNF2 is a DNA-dependent ATPase. A heterodimer of Swi4 and Swi5 is SBF. A heterodimer of Swi6 and Mbp1 is MBF. ►activator genes, ►co-activator, ►transcriptional activator, ►*Polycomb*, ►chromatin remodeling, ►NURF, ►CHRAC, ►ACF, ►nucleosome, ►nuclease-sensitive sites, ►Mbp1, ►SBF; Muchardt C, Yaniv M 1999 J Mol Biol 293:187; Sengupta SM et al 2001 J Biol Chem 276:12636.

Swi6: A centromeric repressor chromodomain protein in *Schizosaccharomyces pombe*, essential for centromere function. ►centromere

Swimming in Bacteria: Movement by counterclockwise rotation of the flagella.

Swine: Most commonly used name for the female animals of the *Sus* species (hogs). ►pig

SWI/SNF: ►nucleosome, ►chromatin remodeling, ►SWI

SwissProt Database: protein information: http://www.expasy.org/; http://www.ebi.ac.uk/swissprot/.

Switch, Genetic: An individual cell may initially express immunoglobin gene Cμ but in its clonal progeny it may change to the expression of Cα as a result of somatic DNA rearrangement. Similar DNA switch may occur in the variable region of the light-chain genes. Switching occurs during the mating type determination of yeast and phase variation in prokaryotes. The operation of genetic switch was first thoroughly studied in phage lambda. ►immunoglobulins, ►antibody gene switching, ►epigenesis,

►*Trypanosomas*, ►*Borrelia*, ►lambda phage, ►site-specific recombination, ►mating type determination in *Saccharomyces*, ►phase variation, ►Gene-Switch; Ptashne PM 2004 Genetic Switch, Cold Spring Harbor Laboratory Press, Cold Spring Harbor, New York; ►riboswitch

Switching, Phenotypic: May occur without any change in the genetic material and involves only altered regulation of transcription resulting in different phenotypes.

Swivelase (topoisomerase type I): After a single nick in a supercoiled DNA, it permits the cut strand to make a turn around the intact one to relieve tension. ►DNA replication, ►prokaryotes, ►topoisomerase; Zhu Q et al 2001 Proc Natl Acad Sci USA 98:9766.

SWR: ►chromatin remodeling

SWR Mouse: Prone to cancer and autoimmunity.

Swyer Syndrome (gonadal dysgenesis XY type, Xp22.1-p21.2): Apparently mutation or loss in the SRY gene is responsible for the anomaly. The afflicted individuals appear normal females until puberty but display only deficient ("streak") gonads and fail to menstruate. ►gonadal dysgenesis, ►testicular feminization, ►SRY, ►sex determination, ►Denys-Drash syndrome

Sxl (sex lethal): (chromosomal location 1-19.2) Controls sexual dimorphism in *Drosophila*; it is required for female development. ►sex determination

SXR (steroid xenobiotic receptor in humans): Its mouse homolog is Pxr. This protein is an inducer of the cytochrome P450 detoxifying enzyme CYP3A4 (7p22.1). Mice transgenic for SXR displayed enhanced protection against environmental and drug carcinogens. This system also has important role in the oxidative activation of procarcinogens such as aflatoxin B and many different drugs. Apparently, the herb St. John's wort, which may lower the effectiveness of the antiseizure phenobarbital, breast cancer drug tamoxifen, the contraceptive ethinyl estradiol, etc., activates SXR/Pxr. ►cytochromes; Synold TW et al 2001 Nature Med 7:536.

Sybase: A computer program that links various databases for macromolecules such as GeneBank, EMBL. ►data management system, ►databases; https://login.sybase.com/login/userLogin.do.

Sycamore (*Platanus* spp): Large attractive monoecious trees, 2n = 42.

SYK (M_r 72 K): A signal protein for the interleukin-2, granulocyte colony-stimulating factor and for other agonists. It is a protein tyrosine kinase of the SRC family and indispensable for B lymphocyte development. Syk is downregulated by the Cbl protein. Syk seems to modulate epithelial cell growth and may suppress human breast carcinomas. ►B cell, ►ZAP-70, ►agonist, ►B lymphocyte receptor, ►Cbl, ►BTK; Sada K et al 2001 J Biochem 130:177.

Syllogism: A form of deductive reasoning using a *major premise*, e.g., mice show graft rejection, the rejection of transplants is based on the presence of the MHC system (*minor premise*), and therefore mice must have a major histocompatibility system (*conclusion*). ►logic

Symbionts: Mutually interdependent cohabiting organisms, such as the *Rhizobium* bacteria within the root nodules of leguminous plants, or algae within green hydra animals. The boundary between pathogenesis and mutualistically advantageous situation is not always clear. Many symbionts are essential for nutrition, digestion, development, and defense against other pathogens whereas others are definitely harmful for the health or reproduction of the host.

The leaf-cutting ants rear and feed only a single species of fungus through millions of years and develop mutualistic relations with it (Poulsen M, Boomsma JJ 2005 Science 307:741). Some wasps (*Philanthus triangulum*) cultivate *Streptomyces* bacteria in their antennal glands and apply them to the blood and this protects the cocoon from fungal infestation (Kaltenpoth M et al 2005 Curr Biol 15:475). Symbiosis may lead to loss or inactivation of genes that are no longer necessary in a cohabiting situation. The marine worm *Olavius algarvensis* lack mouth, gut and nephridia (excretory organs) and the nutritional functions are supplied by the cohabiting sulphur-oxidizing and sulphate reducing bacteria (Woyke T et al 2006 Nature [Lond] 443:950). On the other hand, new functions may be acquired or pre-existing gene functions may be enhanced that are mutually beneficial. Symbiosis may be interpreted either as mere cohabitation or a mutually meaningful association (advantageous or parasitic). Bacterial nitrogen fixation and mycorrhizal symbiosis is controlled by a ligand-gated cation (Ca^{2+}, calmodulin) channel (Lévy J et al 2004 Science 303:1361). Calcium spiking is one of the initial steps in the symbiotic process. The proteins for establishing the symbiotic relationship for both fungi and bacteria reside in the plastids of the plant roots (Imaizumi-Anraku H et al 2005 Nature [Lond] 433:527). The plant pathogenic fungus *Rhizopus* harbors intracellular *Burkholderia* bacteria, which are responsible for the production of a polyketide, rhizoxin toxin. The toxin poisons β-tubulin and has strong antimitotic effect and, thus, can have medical use as an anticancer agent (Partida-Martinez LP, Hertweck C 2005 Nature [Lond] 437:884). ►mycorrhiza, ►nitrogen fixation, ►mutualism, ►*Piriformospora indica*; Bermudes D, Margulis L 1987 Symbiosis 4:185; Currie CR 2001 Annu Rev Microbiol 55:357; mutualistic symbiosis among insects, bacteria and viruses: Moran NA et al 2005 Proc Natl Acad Sci USA 102:16919.

Symbionts, Hereditary: Occur in a wide range of eukaryotic organisms and their maternal transmission simulates extranuclear inheritance. They may be more wide spread than recognized. The temperate viruses of prokaryotes and the retroviruses may also be classified along these groups. In some strains of the unicellular protozoa, *Paramecium aurelia*, carrying the *K* gene, bacteria (e.g., *Caedobacter taenospiralis*) live in the cytoplasm and is transmitted to the progeny. The first such infectious particles were named kappa (κ) particles before their bacterial nature was recognized, and were supposed to be normal extranuclear, hereditary elements. Many of these kappa particles contain R (refractive) bodies that are bacteriophages. The κ particles with R bodies appear "bright" under the phase-contrast microscope. The non-bright cells may give rise to bright, indicating that the phages are in the free and infectious stage and the brights are in the integrated, proviral stage. The virus directs the synthesis of toxic protein ribbons that are responsible for killing kappa-free paramecia (κκ). The strains carrying the dominant *K* gene are immune to the toxin. Other strains have been discovered carrying different infectious particles lambda, sigma, mu that also make toxins. The mu particles do not liberate free toxin and kill only the cells with which they mate (mate killer). Other symbionts delta, nu and alpha are not killer symbionts.

In *Drosophila* strains Rhabdovirus σ may be in the cytoplasm and responsible for CO_2-sensitivity. Normal flies can be anesthetized with the gas for shorter periods without any harm. Those, which carry the virus may be paralyzed and killed by the same gas treatment. This virus is similar to the vesicular stomatitis virus of horses that cause fever, eruptions and inflammation in the mouth of horses, and to the fish rhabdovirus. In the non-stabilized strains, only the females transmit the virus to part of the progeny (depending on whether a particular egg does or does not contain the virus). Some of the non-stabilized may become stabilized. Stabilized strains transmit it through nearly 100% of the eggs and even some of the males transmit σ with some of the sperm yet the offspring of the male will not become stabilized. The *ref* mutants in chromosomes 1, 2 and 3 are refractory to infection. In some strains, there are mutants of the virus that are either temperature-sensitive or constitutively unable to cause CO2 sensitivity although they are transmissible. Different ribosomal picornaviruses can be harbored in *Drosophila* that may reduce the life and fertility of the infected females. Females with the sex ratio (SR) condition produce no viable sons and the transmission is only maternal. In their hemolymph (internal nutrient fluids) the females carry spiroplasmas, bacteria without cell wall. If the infection is limited to the XX sector of gynanders, they may survive but not if the infection is in the XO sector. Triploid intersexes or females, phenotypically sterile males because of the genes *tra* (*transformer*, chromosome 3-45), *ix* (*intersex* gene located at 2-60.5) or *dsx* (*double-sex*, intersexes, chromosome 3-48.1) are not killed. Their special viruses may destroy the spiroplasmas. In plants (petunia, sugar beet), cytoplasmically inherited male sterility can also be transmitted by grafting. Some of the variegated tulips (broken tulips) are infected by viruses and had special ornamental value. During the seventeenth and eighteenth century "tulip mania," some rich Europeans paid the mainly Turkish and Persian merchants for the bulbs of the most attractive varieties, in equal weights in gold. ►lysogeny, ►cytoplasmic male sterility, ►extranuclear inheritance, ►*Drosophila*, ►aphids, ►meiotic drive, ►segregation distorter, ►broken tulips, ►sex determination, ►*Wolbachia*, ►*Spirochetes*, ►*Paramecia*, ►endosymbiont theory, ►pathogenicity island; Preer JR 1975 Symp Soc Exp Biol 29:125; Ehrman L, Daniels S 1975 Aust J Biol Sci 28:133; review of bacterial symbionts: Dale C, Moran NA 2006 Cell 126:453.

Symbiosis: ►symbiont

Symbiosome: In legume root nodules 2 to 5 μm structures enclosing (by peribacteroid membrane) 2 to 10 bacteroids. The fixed nitrogen is released through this membrane to the plant and reduced carbon is received by the bacteroids from the plant. ►nitrogen fixation

Symbols: ►gene symbols, ►*Drosophila*, ►pedigree analysis

Symmetric Heteroduplex DNA: ►Meselson-Radding model of recombination

Symmetry: ►axis of asymmetry, ►asymmetric cell division, ►bilateral symmetry, ►left-right asymmetry, ►retinoic acid

Sympathetic Nervous System: Communicates with the central nervous system (CNS, brain) through the thoracic and lumbar parts of the spinal cord, controls the blood vessels in the various organs, and the involuntary movements (reflexes) of the body.

Sympatric: Populations have overlapping habitats; it may be a beginning of speciation. ►speciation

Sympatric Speciation: Species that live in the same, shared area, for some reason become sexually isolated. This is a relatively rare phenomenon because most commonly geographical isolation is the major factor in speciation. ►allopatric, ►parapatric; palm trees: Savolainen V et al 2006 Nature [Lond] 441:210; cichlid fish: Barluenga M et al 2006 Nature [Lond] 439:719.

Sympetaly: Fused petals in a flower, see e.g., ►snapdragon

Symphalangism, Proximal: Fusion of the carpal and tarsal bones (SYM1, 17q22) caused by a defect in the noggin protein. ►noggin, ►bone diseases

Symplast: Multinucleate giant cell.

Symplastic Domain: A regulatory (supracellular) unit of several cells within the body.

Symplekin: A CPEB and CPSF binding protein scaffold, essential for cytoplasmic polyadenylation of mRNA in connection with xGLD-2 poly(A) polymerase. ►polyadenylation signal; Barnard DC et al 2004 Cell 119:641.

Symplesiomorphic: Two or more species sharing a primitive evolutionary trait. ►plesiomorphic, ►apomorphic, ►synapomorphic

Symport: Co-transportation of different molecules through membranes in the same direction.

SYN: A prefix indicating union of tissues named after.

SYN: ►ANTI

Synaesthesia: Involuntary physical experience of a cross-modal linkage, e.g., hearing a tone (an inductive stimulus) and seeing a color or perception of smell or taste (Beeli G et al 2005 Nature [Lond] 434:38).

Synapomorphic: Species sharing an apomorphic trait. ►apomorphic, ►plesiomorphic, ►symplesiomorphic

Synapomorphy: Shared derived characters that can be used to advance phylogenetic hypotheses. ►simplesiomorphy; Venkatesh B et al 2001 Proc Natl Acad Sci USA 98:11382.

Synaps (synapse): The site of connection between neural termini at which either a chemical or an electric signal is transmitted from one neuron to another (or to another type of cell) (see Fig. S146). Neurotransmitter release occurs at the presynaptic active zones and at postsynaptic densities. There is now evidence that extrasynaptic (ectopic) transmission also occurs (Coggan JS et al 2005 Science 309:446).

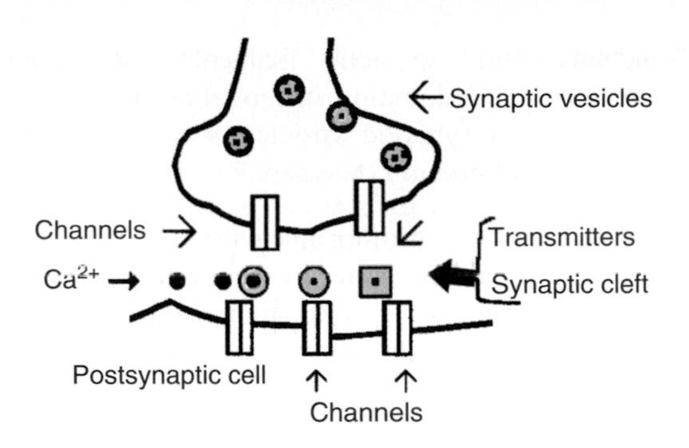

Figure S146. Synaps

The neurotransmitter may diffuse across the synaps or the electric signal may be relayed from one cytoplasm to the other through a gap junction. The neurons must interpret the postsynaptic potentials (PSP) and integrate the excitatory (EPSP) and inhibitory (IPSP) paired pulse potentials. The leukocyte common antigen related (LAR) protein, liprin (encoded by gene *syd-2* in *Caenorhabditis*), regulates the differentiation of the presynaptic vessels. This protein has tyrosine phosphatase activity. Mitochondria normally accumulate close to the site of synaps and their movement is controlled by Miro mitochondrial GTPase (Verstreken P et al 2005 Neuron 47:365; Guo X et al 2005 Neuron 47:379). ►neurotransmitters, ►tyrosine phosphatase, ►gap junction, ►memory, ►ionotropic receptor, ►metabotropic receptor, ►neuregulin, ►heregulin, ►synaptic scaling, ►*Dscam*, ►NMDA; Ziv NE 2001 Neuroscientist 7(5):365; 2002 Science 298:770–791; Yamada S, Nelson WJ 2007 Annu Rev Biochem 76:267; http://syndb.cbi.pku.edu.cn.

Synapse, Immunological: The adhesion (gap) between the T lymphocyte receptor and the antigen presenting cell carrying the MHC—antigen complex or the killer cell inhibitory molecules and the special T cell receptor. Cytotoxic killer cells destroy virus-infected and tumorigenic cells by delivering secretory lysosome granules to the synaptic targets. The delivery is mediated by transient contact of the centrosome and plasma membrane and it is driven by the reorganization of the actin cytoskeleton (Stinchcombe JC et al 2006 Nature [Lond] 443:462). ►T cell, ►MHC, ►killer cell; Khan AA et al 2001 Science 292:1681; Bromley SF et al 2001 Annu Rev Immunol 19:375.

Synapse, Informational: A specialized cell-cell junction mediating chemical or electric communication and displays a supramolecular structure to mediate information transfer between cells. (See Dustin ML et al 2001 Annu Rev Cell Dev Biol 17:133).

S

Synapsins: Bind to actin filaments, microtubules, annexing, SH3 domains, calmodulins and important regulators of synaptic vesicles. ►actin, ►annexin, ►SH3, ►calmodulin, ►synaptic vesicles, ►synaps

Synapsis: Intimate chromosome pairing during meiosis between homologous chromosomes that may lead to crossing over and recombination (see Fig. S147). In some instances, non-homologous chromosomes or chromosomal regions may also associate. Protein-mediated synapsis involves recombination facilitated by integrases. Unpaired DNA causes meiotic silencing in *Neurospora* (Shiu PK et al 2001 Cell 107:905). Similarly in mouse in unsynapsed chromosomes the tumor suppressor BRCA1 and the kinase ATR co-localize to meiotic nodules (recombination nodules) and apparently because of the lack of phosphorylation of H2AX histone associated with the X and Y chromosomes asynapsis, sterility and apoptosis results (Turner JMA et al 2005 Nature Genet 37: 41). ►pairing, ►illegitimate pairing, ►meiosis, ►crossing over, ►synaptonemal complex, ►*res*, ►topological filter, ►tracking, ►integrase, ►asynapsis, ►histone variants, ►breast cancer, ►recombination nodule, ►ATR; McClintock B 1930 Proc Natl Acad Sci USA 16:791; Romanienko PJ, Camerini-Otero RD 2000 Mol Cell 6:975; Baudat F et al 2000 Mol Cell 6:989; minireview: McKim KS 2005 Cell 123:989.

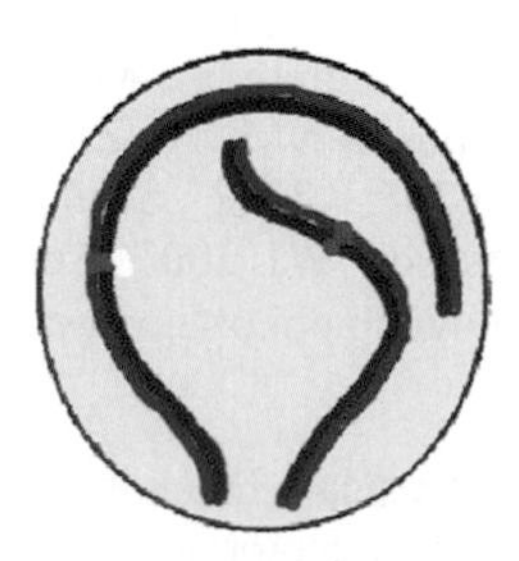

Figure S147. Synapsed homologous chromosomes (2 pairs)

Synapsis, Bimolecular: Occurs between complementary ends of different/separate transposable elements. ►transposable elements

Synaptic Adjustment: The degree of synapsis may change during meiosis in certain chromosomal regions. ►synapsis

Synaptic Cleft: Electric neuronal signals are transmitted from the presynaptic cell to the postsynaptic cell by the gap called the synaptic cleft. ►neuron, ►synaptosome

Synaptic Scaling: A bidirectional phenomenon in which excitatory synapses scale up in response to activity reduction but scale down in response to increases in activity of neurons. Synaptic scaling is mediated by the pro-inflammatory cytokine, tumor-necrosis factor-α (TNF-α) of the glia. ►neuron, ►glia, ►synaps, ►TNF; Stellwagen D, Malenka RC 2006 Nature [Lond] 440:1054.

Synaptic Vesicles: Originate from endosomes, and store and release the neurotransmitters and other molecules required for signal transmission between nerve and target cells. Ca^{2+} regulates their function. ►neurotransmitters, ►syntaxin, ►RAB, ►synaptotagmins, ►synaptophysin, ►NSF, ►neuromodulin, ►neurogenesis; Cousin MA, Robinson PJ 2001 Trends Neurosci 24:659.

Synaptinemal Complex: ►synaptonemal complex

Synaptobrevins (VAMP): ►syntaxin, ►SNARE

Synaptogamins: Integral membrane proteins, binding calcium and interacting with other membrane proteins in the synaptic vessels of the nerves. Synaptogamins may interact with the different types of PtdIns and Ca^{2+}. ►neurexin, ►botulin, ►synaptic vesicles, ►SNARE, ►phosphoinositides, ►clathrin, ►Ipk1; Li C et al 1995 Nature [Lond] 375:594; Südhof TC 2002 J Biol Chem 277:7629.

Synaptogyrin: Membrane proteins; may be phosphorylated on tyrosine. (See Zhao H, Nonet ML 2001 Mol Biol Cell 12:2275).

Synaptojanin: A neuron-specific phosphatase (M_r 145,000) working on phosphatidylinositol and inositols and its putative role is in the recycling of synaptic vesicles. Its defect is responsible for Lowe's oculocerebrorenal syndrome. It binds to the SH3 domain of Grb2. ►synaptotagmin, ►syntaxin, ►Lowe's oculocerebrorenal syndrome, ►Grb2, ►dynamin; Khvotchev M, Südhof TC 1998 J Biol Chem 273:2306; Ha SA et al 2001 Mol Biol Cell 12:3175.

Synapton: same as synaptonemal complex.

Synaptonemal Complex: Proteinacious element between paired chromosomes in meiosis. In yeast—analyzed by green fluorescent protein labeling—the complex is initiated at zygotene and by pachytene it forms a ribbon-like structure. At diplotene the complex falls apart yet the association of the homologs continues because of the mature chiasmata (White E et al 2004 Genetics 167:51). They consist of two lateral and a central element (see Fig. S148). Denser spots within it are called recombination nodules and were supposed to have a role in genetic recombination. It is supposed that the complex holds in place the recombination intermediates rather than actively promoting the process. Actually, in some fungi (*Saccharomyces*, *Aspergillus*) no synaptonemal complex is observed and concomitantly chromosome interference is absent. These observations led to the assumption that the complex is responsible for interference rather than recombination.

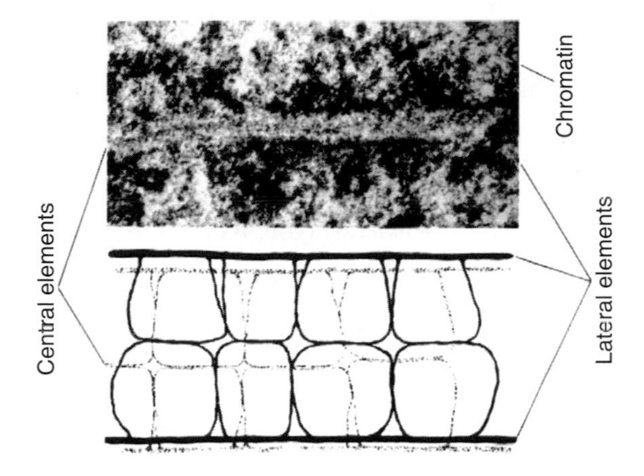

Figure S148. The tripartite structure may be visible from interphase through diplotene. (Electronmicrogph by the courtesy of Dr. H.A. McQuade. Interpretative drawing after Comings DE, Okada TA 1970 Nature [Lond] 227:451)

Actually, recombination may precede the formation of the synaptonemal complex, thus, the role of the complex in interference may be questionable. In *Drosophila,* synaptonemal complex may be formed in the absence of meiotic recombination (crossing over or gene conversion), however, in the male flies both the synaptonemal complex and meiotic recombination are absent yet mitotic recombination may be observed.

►synapsis, ►meiosis, ►interference, ►recombination mechanism eukaryotes, ►association site, ►recombination nodule, ►single-end invasion, ►crossing over, ►mitotic crossing over, ►male recombination; Westergaard M, von Wettstein D 1972 Annu Rev Genet 6:71; Schmekel K et al 1993 Chromosoma 102:669; Solari AJ 1998 Methods Cell Biol 53:235; Heng HH et al 2001 Genome 44:293; Page SL, Hawley RS 2004 Annu Rev Dev Biol 20:525; Lynn A et al 2002 Science 296:2222.

Synaptophysin: Membrane-spanning proteins in the synaptic vessel involved in neurotransmitter release. ►synaptic vessel, ►neurotransmitter, ►ceroid lipofuscinosis

Synaptosome: The protein complex mediating interactions among neurotransmitters and receptors across the synaptic cleft. ►neurotransmitter; Husi H, Grant SG 2001 Trends Neurosci 24:259.

Synaptotagmins: Proteins in the synaptic vesicles and have a role in Ca^{2+}-involved release of neurotransmitters, and in general in exo- and endocytosis. Exocytosis of synaptic vesicles is controlled by complexin and synaptotagmin (Tang J et al 2006 Cell 126:1175). ►synaptojanin, ►syntaxin, ►complexin, ►synapse, ►SNARE; Fernandez-Chacon R et al 2001 Nature [Lond] 410:41; Hui E et al 2005 Proc Natl Acad Sci USA 102:5210.

Synchronous Divisions: The cells are at the same stage of the cell cycle.

Synchrotron: Radiation emitted by high-energy, high-speed electrons accelerated in magnetic fields. The range varies from infrared to hard X-rays. Their high energy permits to study their effects also in monochromatic forms. ►ionizing radiation; http://www.srs.ac.uk/srs/.

Synclinal: ►anticlinal

Syncytial Blastoderm: An early stage of embryogenesis when the single layer of nucleated cytoplasmic aggregates do not yet have a cell membrane. ►blastoderm

Syncytium: A collection of nuclei surrounded by cytoplasm without the formation of separate membranes around each, such as in early embryogenesis or among the progeny of a single spermatogonium or abnormal multinucleate cells or the plasmodia of slime molds. ►imaginal discs, ►blastoderm, ►*Dictyostelium*

Syndactyly: Webbing or fusion between fingers and toes (see Fig. S149). In polysyndactyly, mutation in the polyalanine extension of the amino terminal of human homeotic gene (HOX13) is the responsible factor. Syndactyly 1 was located to 2q34-q36. ►Poland syndrome, ►limb defects, ►Rubinstein-Taybi syndrome, ►Greig's cephalopolysyndactyly, ►GLI3 oncogene, ►homeotic genes, ►polysyndactyly, ►Pallister-Hall syndrome, ►Guttmacher syndrome; Bosse K et al 2000 Am J Hum Genet 67:492.

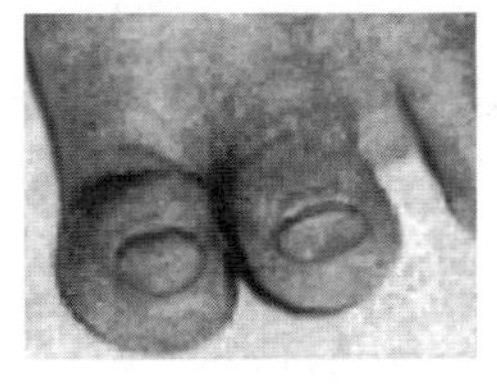

Figure S149. Syndactyly

Syndecans: Heparan sulfate proteoglycans, membrane-spanning cell adhesion molecules and co-factor receptors bearing heparan sulphate proteoglycans distal from the plasma membrane. Syndecan promotes over-eating and obesity in mice. Syndecan-1 opposes the effect of the melanocyte-stimulating hormone. Syndecan-1, -2, -3, -4 are encoded in human chromosome 2p23-p24, 8q23, 1p32-p36 and 20q12-q13, respectively. ►CAM, ►selectin, ►heparan sulphate, ►glypican, ►obesity, ►melanocyte-stimulating hormone; Reizes O et al 2001 Cell 106:105; Couchman JR 2003 Nature Rev Mol Cell Biol 4:926.

Syndrome: A collection of symptoms, traits caused by a particular genetic constitution. In humans, there are about 2000 syndromes determined mainly by single genes. The individual symptoms of different syndromes, however, may overlap among a large number of genetic and non-genetic disorders. Therefore, the precise identification is often an extremely difficult task. More accurate identification will probably be possible when the genome projects will provide structural and topological evidence for all the loci concerned. The *unknown genesis syndromes* usually occur sporadically yet they may have genetic bases. ►genome projects, ►association, ►non-syndromic, ►physical mapping, ►microarray, ►epistasis; Brunner HG et al 2004 Nature Rev Genet 5:545; human medical databases: http://www.lmdatabases.com/.

Synergids: Two haploid cells in the embryosac of plants flanking the egg. The pollen tube first penetrates one of the synergides, after the rupture of that synergid and rupture of the pollen tube the tube elongation is arrested, and the vegetative and the sperm nuclei fertilize the polar nuclei and the egg nucleus, respectively. The *FER* gene of *Arabidopsis thaliana* encodes a synergid-expressed, plasma membrane–localized, receptor-like kinase. The FER protein accumulates asymmetrically in the synergid membrane at the filiform apparatus. Interspecific crosses using pollen from *Arabidopsis lyrata* and *Cardamine flexuosa* on *A. thaliana* stigmas resulted in a *fer*-like phenotype that correlates with sequence divergence in the extracellular domain of FER. Our findings show that the female control of pollen tube reception is based on a *FER*-dependent signaling pathway, which may play a role in reproductive isolation barriers (Escobar-Restrepo J-M et al 2007 Science 317:656). ►gametogenesis, ►gametophyte, ►pollen tube; Higashiyama T et al 2001 Science 293:1480.

Synergistic Action: The participating elements enhance the reaction above the sum of the separate strength of their separate actions. ►interaction variance, ►epistasis

Synexpression Groups: Genes that are expressed together either spatially, temporally or developmentally. Their identification is facilitated by microarray hybridization. These groups do not need to be genetically linked and therefore are different from gene clusters or operons of prokaryotes that are linked. ►macroarray analysis, ►microarray hybridization, ►operon, ►clustering of genes; Nirehrs C, Pollet N 1999 Nature [Lond] 402:483.

Syngamy: The union of two gametes in fertilization leading potentially to the fusion of the two nuclei in the cell. ►plasmogamy, ►karyogamy, ►synkaryon, ►semigamy

Syngen: A reproductively isolated group of ciliates.

Syngeneic: Antigenically similar type cells (in a chimera). ►antigen, ►immunoglobulins

Synkaryon: A cell (zygote, fused conidia or spores) with a nucleus originated by the union of two nuclei. ►karyogamy, ►heterokaryon, ►dikaryon

Synonymous Codons: These have different bases in the triplet yet they specify the same amino acids. The 61 sense codons stand for 20 common amino acids. Some amino acids have up to 6 codons. Thus, mutation may not have any genetic consequence, except when exon splice site is involved. A synonymous single nucleotide polymorphism (SNP) in the *Multidrug Resistance* 1 (*MDR*1) gene, leads to altered function of the *MDR*1 gene product P-glycoprotein (P-gp) and results in altered drug and inhibitor interactions. Although the mRNA and protein levels are similar, conformation was altered. Supposedly, the presence of a rare codon affects the timing of cotranslational folding and insertion of P-gp into the membrane, thereby altering the structure of substrate and inhibitor interaction sites (Kimchy-Sarfaty C et al 207 Science 315:525). Some bases may affect the intensity of translation. In functional areas of the genome, synonymous substitutions are more common than non-synonymous ones. Purifying selection usually eliminates the deleterious non-synonymous mutations. Slightly deleterious non-synonymous substitutions can be maintained in bacterial populations with high effective population size (Hughes AL 2005 Genetics 169:533). ►genetic code, ►splicing, ►exon, ►intron, ►non-synonymous codon, ►radical amino acid substitution, ►effective population size, ►Grantham's rule

Synostosis: Bone fusion. ►noggin

Synovial Sarcoma: ►SYT

Synpolydactyly: ►polydactyly, ►polysyndactyly, ►syndactyly

Syntax: Sentence structure; organization of groups of words, phrases and clauses in the correct manner. The term is also used by computerized information retrieval.

Syntaxin: A synaptic membrane protein forming part of the nerve synaptic core complex along with the synaptosome associated protein and synaptobrevin (VAMP), a vesicle associated membrane protein. Omega-3 and omega-6 fatty acids act on syntaxin and stimulate membrane expansion (Darios F, Davletov B

2006 Nature [Lond] 440:813). Syntaxins are involved in vesicular transport between the endoplasmic reticulum and the Golgi apparatus as target membrane receptors. Syntaxin-5 is an integral part of the endoplasmic reticulum-derived transport vesicles. A synaptobrevin-like gene (SYBL1) was located to the pseudoautosomal region of the human X chromosome. It recombines with Y-chromosomal homolog, displays lyonization in the X chromosome, and is inactivated in the Y chromosome. A score of syntaxins have been identified in humans encoded at chromosomes 7q11.2, 17p12, 16p11.2, etc. ►synaptotagmin, ►SNARE, ►synaps, ►endoplasmic reticulum, ►Golgi apparatus, ►pseudoautosomal region, ►lyonization, ►Munc, ►cystic fibrosis, ►stoma, ►omega-3 fatty acids; Bennett MK et al 1992 Science 257:255; Mullock BM et al 2000 Mol Biol Cell 11:3137.

Syntelic Distribution: The kinetochores of the two sister chromatids are attached to spindle fibers that pull them to the same pole during mitotic anaphase.

Syntenic Genes: Within the same chromosome; they may, however, freely recombine if they are 50 or more map units apart. Gene blocks may display synteny among related taxonomic entities even when their chromosome number varies. Syntenic gene sets may provide information on phylogeny of the species. ►linkage, ►crossing over

Syntenin: 32 kDa adaptor protein with two PDZ domains involved in cytoskeleton-membrane organization. ►PDZ, ►IL-5; Zimmermann P et al 2001 Mol Biol Cell 12:339; Cierpicki T et al 2005 Structure 13:319.

Synthases: Mediate condensation reactions of molecules without ATP. ►synthetases

Synthetases: Mediate condensation reactions that require nucleoside triphosphates as energy source. ►synthases

Synthetic Biology: A new area of science aiming either to introduce new (synthetic) molecules into existing organisms, or interchange existing molecules between organisms or chemically synthesizes new systems from known compounds, e.g., create synthetic viruses on the basis of the genetic code. This field includes from de novo organic synthesis, to genetic engineering, stem cell research, etc. Networks have been designed from natural transcription factors and binding sites. Its aims include creation of toggle switches, cellular oscillators, new forms of cell-to-cell communications and cell pattern formation. A beautiful paper reports the duplication and parallel modifications of the 16S prokaryotic ribosomal RNA (the anti-Shine-Dalgarno sequence, CCUCC) and the Shine-Dalgarno sequence (GGAGG) of the mRNA. The new ribosome may or may not function anymore as the original one; the modified ribosome however can translate the modified mRNA. Through mutations in both the ribosomes and in the mRNA, various combinations of enormous variety of potential functions can be established and tried (Rackham O, Chin JW 2005 Nature Chem Biol 1:159).

Although this type of research can generate altered or new organisms that may pose unknown hazards, the potential advantages for agriculture and medicine outweigh the risks. ►genetic engineering, ►stem cells, ►synthetic genes, ►gene circuits, ►synthetic virus, ►influenza virus, ►protein engineering, ►oscillators; Benner SA, Sismour AM 2005 Nature Rev Genet 6:533; http://syntheticbiology.org.

Synthetic DNA Probes: If the amino acid sequence in the protein is known but the gene was not yet isolated, a family of synthetic probes may be generated to tag the desired gene. This probe is generally no longer than 20 base because of the difficulties involved in their synthesis. The genetic code dictionary reveals which triplets spell the amino acids. An amino acid sequence that uses few synonymous codons is selected. A computer match generally chooses the possible combinations. E.g., a probe for the His-Thr-Met peptide sequence would require the following 8 polynucleotide sequences to consider all possible sequences for a probe (see Fig. S150). The inclusion of **Methionine** (having a single codon) is simplifying the task. **Histidine** is relatively advantageous because it has only two synonymous codons. **Threonine** with four codons make the work more difficult; leucine, serine and arginine containing parts of the proteins should be avoided (because they have six codons) but tryptophan, also with a single codon would be highly desired. Insertion of ambiguous deoxyinosine nucleotides at some positions may facilitate the design of probes. ►probe, ►functional cloning, ►gene isolation; Ohtsuka E et al 1985 J Biol Chem 260:2605; Lichtenstein AV et al 2001 Nucleic Acids Res 29(17):E90.

His Thr Met
5′ CACACUAUG 3′
CACACCAUG
CACACAAUG
CACACGAUG
CAUACUAUG
CAUACCAUG
CAUACAAUG
CAUACGAUG

Figure S150. Synthetic probe

Synthetic Enhancement: Basically an epistatic process by increasing or reducing interaction between gene products by using crossing, knockouts, transformation, etc.

Synthetic Genes: Produced in vitro by the methods of organic chemistry, by systematic ligation of synthetic oligonucleotides into functional units, including upstream and downstream essential elements. The first successful synthesis involved the relatively short tyrosine suppressor tRNA gene. By 2007, more than three-dozen commercial enterprises can provide synthetic genetic sequences worldwide. These synthetic genetic elements can be used to modify genomes and create new biological functions for the benefit and for the hazards to human societies (http://www.etcgroup.org/upload/publication/602/01/synbioreportweb.pdf). ►suppressor tRNA, ►poliovirus, see Fig. S151, ►synthetic virus, ►*Mycoplasma genitalium*, ►synthetic biology; synthetic genes for optimal protein expression: http://www.evolvingcode.net/codon/sgdb/index.php; sequence comparison tool: http://www.evolvingcode.net/codon/sgdb/aligner.php; synthetic gene browser: http://www.evolvingcode.net/codon/sgdb/browse.php; new data submission: http://www.evolvingcode.net/codon/sgdb/submit.php.

Synthetic Genetic Array: Determines the functional relation between two genetic sites.

A particular gene is crossed to a large number of different deletions. If the double mutants are inviable, the functional relationship of the two is revealed. On this basis, network maps can be constructed (see Fig. S152). ►genetic network, ►small-world network; Tong AHY et al 2001 Science 294:2364; Tong AHY et al 2004 Science 303:808.

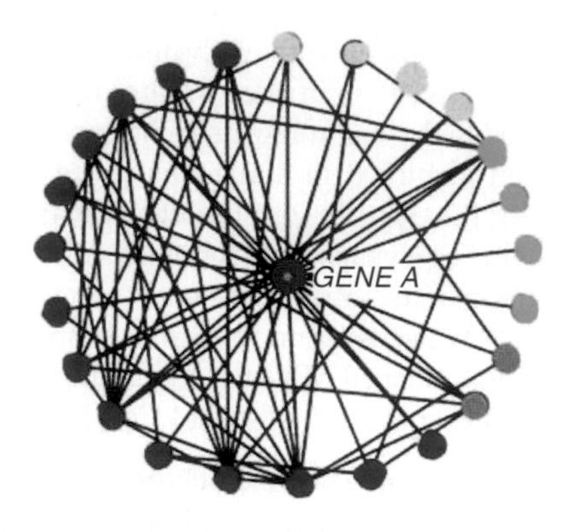

Figure S152. Synthetic genetic array. The hypothetical query gene (gene A) in the center (red) is connected to nodes representing color-coded functions according to their fitness in synthetic combination. The chart also shows some interactions among pairs of synthetic lethals. (Modified after Tong AHY et al 2004 Science 303:808)

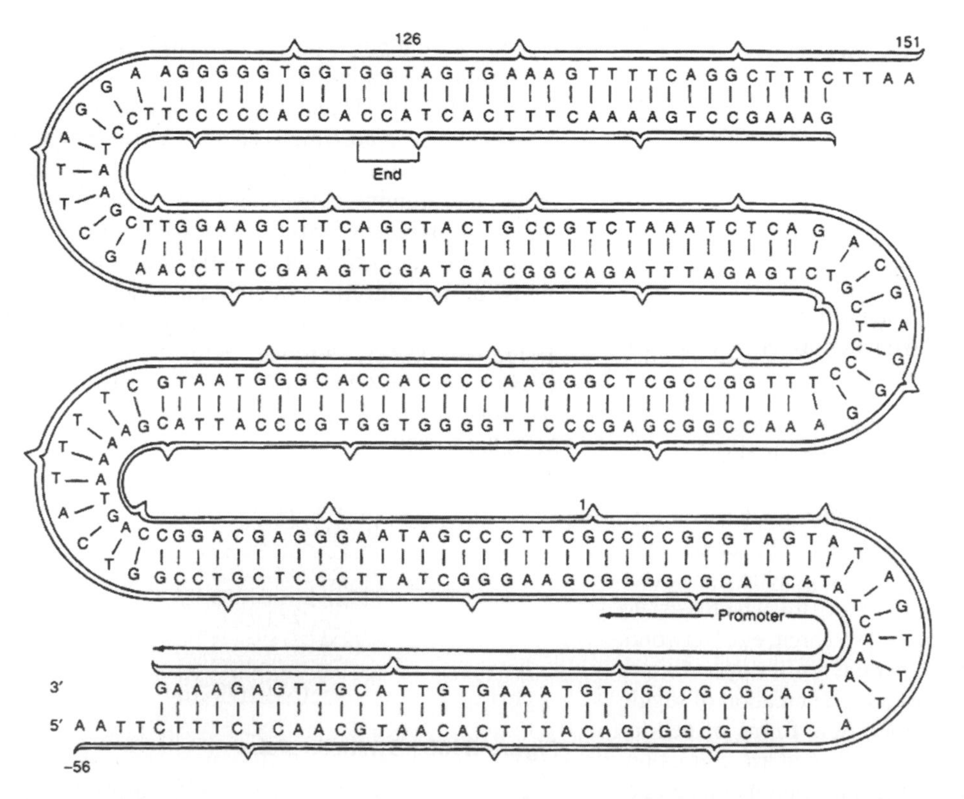

Figure S151. The complete (including promoter and terminator) synthetic structural gene of the tyrosine suppressor tRNA of *E. coli*. The projections along the sequence bracket the size of the fragments ligated together to form the complete gene. The gene when transformed into bacteria actually worked. (Redrawn after Khorana HG 1974. *Proc. Int. Symp. Macromol*. p. 371, Mano EB., ed. Elsevier, Amsterdam: NL and Macaya G, ed. 1976. *Recherche (Paris)* 7: 1080.)

Synhthetic Genetic Networks: Constructs to simplify the understanding of regulated gene complexes. It is based on tools of nonlinear dynamics, statistical physics and molecular biology. ►genetic networks; McMillen D et al 2002 Proc Natl Acad Sci USA 99:679.

Synthetic Genomes: ►synthetic virus, ►*Mycoplasma genitalium*, ►potentials of functional; synthetic microbial genomes: Holt RA et al 2007 BioAssays 29:580.

Synthetic Lethal (synthetic enhancement): Gene is inviable only in certain genetic constitutions. Thus, two single mutations have no or insignificant phenotypic consequence separately but the double mutant may be lethal. In yeast, only ~1000 of the genes are essential and 5000 are viable even when deleted. The synthetic lethals cells include more than one defective gene. Partners in synthetic lethal systems can be essential as well as non-essential genes. In gene networks, at least one of the genes must be essential although most of the non-essential genes can compensate for each other. Synthetic lethals may be involved in inbreeding depression and may be the cause of some hybrid inviabilities and sterility. A synthetic lethal test may be used in anticancer drug design by combining two different mutations that in combination, seriously impair cells to gain information on how to kill cancer cells. ►inbreeding, ►synthetic genetic array, ►synthetic genetic networks, ►*Saccharomyces cerevisiae*; Jacobson MD et al 2001 Genetics 159:17; Davierwala AP 2005 Nature Genet 37:1147.

Synthetic Organisms: ►synthetic biology

Synthetic Polynucleotides: Nucleic acid oligomers or polymers generated in the laboratory by enzymatic or other synthetic methods. Nucleic acid synthesizer machines produce some of them. (See Benner SA et al 1998 Pure Appl Chem 70:263).

Synthetic Seed: Somatic embryos encapsulated into a protective capsule (e.g., calcium alginate) and used for propagation in cases when regular seed is not available or homozygotes are difficult to obtain. ►artificial seed

Synthetic Species: Amphidiploids of presumed progenitors of existing species obtained by crossing and diploidization. Some synthetic species have never existed in nature before as the *Raphanobrassica*, 2n = 36, an amphidiploid of radish (*Raphanus sativus*, n = 9) and cabbage (*Brassica oleracia*, n = 9) (see Fig. S153). *Triticales* are similarly new amphidiploids either 2n = 48 or 2n = 56, obtained by crossing tetraploid (2n = 28) or hexaploid (2n = 42) wheat (*Triticum*) with diploid rye (*Secale cereale*, 2n = 14). Some synthetic species are only reconstructions of the evolutionary form, e.g., *Nicotiana tabacum*, *Hylandra suecica*, *Primula kewensis*, etc. The entire *Mycoplasma* genome of one species has been transferred intact to another and entirely replaced that of the recipient, creating a new species. ►alien substitution, ►amphiploid, ►*Mycoplasma*, ►*Triticale*; *Raphanobrassica*: Karpetchenko GD 1928 Z Indukt Abstammungs Vererbungsl 48:1.

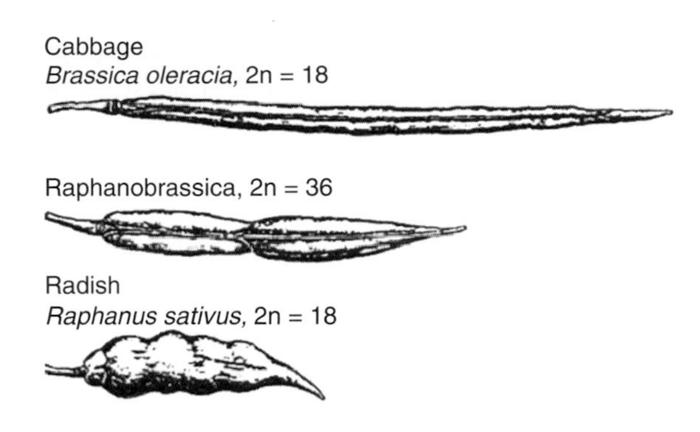

Figure S153. Synthetic *Raphanobrassica*

Synthetic Variety: Composed of several selected lines, which may reproduce by out-crossing within the group. ►polycross

Synthetic Vector: ►liposome

Synthetic Virus: The functional RNA poliovirus and the single-stranded DNA virus ϕX174 have been assembled from synthetic oligonucleotides and the technology is already available to generate cellular genomes within the laboratory. ►polio virus, ►ϕX174, ►influenza virus; Smith HO et al 2003 Proc Natl Acad Sci USA 100:15440.

Synthon: A synthetically produced molecule, in the laboratory.

Syntrophic: Can be maintained (only) by cross-feeding. Based on cross-feeding metabolic pathways of microorganisms could be identified on culture media. In a metabolic pathway A → B → C→ D mutation blocked before D may cross-feed mutants blocked before C and B. Mutation blocked before C may facilitate the growth of mutants inactive in step B, and so on. A syntrophic hypothesis is that the mitochondria of eukaryotes evolved by the fusion of an achaeon, a δ-proteobacterium and an α-proteobacterium. ►cross-feeding, ►channeling

Synuclein (α-synuclein, 4q21): A 140-amino acid protein in the presynaptic neurons, a major constituent of the Lewy bodies. All synucleins display imperfect KTKEGV amino acid motifs and a variable C-terminus. Synucleins are phosphorylated at Ser and

Tyr tail residues and this reduces aggregation. Hydrophobic residues 71-82 promote aggregation. α-synuclein promotes the fibrillization of tau in neurodegenerative diseases (Giasson BI et al 2003 Science 300:636). A truncated α-synuclein accumulates in the neuronal cells and promotes Parkinson disease (Li W et al 2005 Proc Natl Acad Sci USA 102:2162). It interacts with synphilin-1 for normal function. In the oxidized or nitrated forms synuclein aggregates and may cause the synucleinopthies such as Parkinson disease, Alzheimer disease, amyotrophic lateral sclerosis and Huntington disease. In the majority of cases of sporadic Parkinson disease (PD) several mutant genes have been identified. The principal genes that cause (PD) are alpha-synuclein, parkin, leucine-rich repeat kinase 2 and PTEN-induced putative kinase 1 (Wood-Kaczmar A et al 2006 Trends Mol Med 12:521). Beta synuclein gene, also expressed in the brain in Alzheimer disease, was assigned to 5q35, gamma synuclein/persyn (in breast cancer) to 10q23.2-q23.3. ►Lewy body, ►neurodegenerative diseases, ►Parkinson disease, ►alcoholism, ►PTEN; Touchman JW et al 2001 Genome Res 11:78; Shimura H et al 2001 Science 293:263; Cuervo AM et al 2004 Science 305:1292.

SYP: A tyrosine phosphatase.

Syphilis: ►*Treponema pallidum*

Sypro Ruby: A sensitive protein stain used in 2-dimensional gel electrophoresis.

Syringa: ►lilac

Syringe Filter: A syringe equipped with a commercially available sterilizing filter block (0.45 or 0.20 μm pores) and removes microbial contaminations instantly without heating (see Fig. S154).

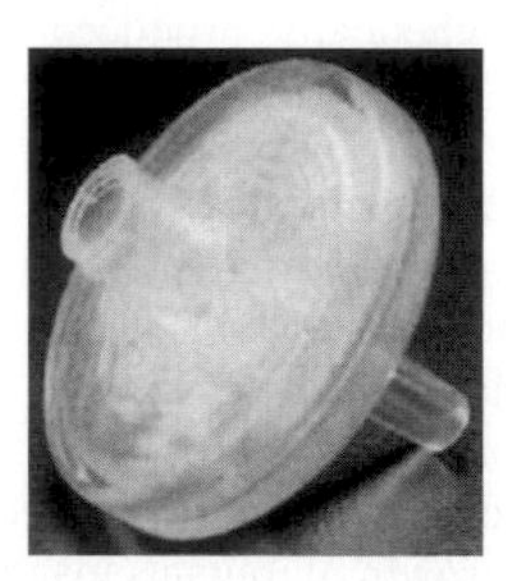

Figure S154. Syringe filter

Syringomelia: A rare autosomal dominant or autosomal recessive cavitations (formation of cavities) in the spinal cord. It may be also due non-hereditary causes.

System Biology: Biological phenomena can be fully understood only by learning their complex interacting networks. (Kitano H 2002 Curr Genet 41:1).

Systematic Error: The same bias affects all measurements or observations.

Systematics: A method of classification such as taxonomy. Carl Linné (Linnaeus) Swedish botanist (1707–1778) initiated the classification and binomial nomenclature of organisms.

Systemic: Affects the entire cell or the entire body of an organism.

Systemic Acquired Resistance: ►SAR

Systemic Amyloidosis, Inherited: An extracellular deposition of fibrous proteins in the connective tissues under autosomal dominant control. ►amyloidosis

Systemic Genes: Cell autonomous versus genes, regulated by intercellular communication.

Systemin: An 18-amino acid signaling peptide for plant defense mechanisms. ►plant defense, ►host-pathogen relation

Systemoid: Similar to a system, alternatively it is used to denote tumors that include different types of tissues. ►teratoma

Systems Biology: Studies biological mechanisms by monitoring gene, genome, protein, proteome and informational pathways systematically using all possible, suitable means. The genes control metabolic functions and are associated with regulatory elements. The proteins may function in complexes and networks and through systems of interacting networks. The simplest controls of the genes are exercised through cis elements such as general and specific transcription factors. The genes are also subject to external signals (signal transduction). The different elements of these systems may change dynamically and thus physiological and developmental modifications arise such as in the healthy or pathologically affected conditions (Hood L et al 2004 Science 306:640). Systems biology separates the effects of noise, and permits the development of rational models and offers new approaches for controlling complex system of the cells, tissues and organisms using bioinformatics, biological and molecular methods. It permits prediction regarding emergent complex phenomena involving, and may enable preventive, predictive and personalized medicine. The components of the system may respond dynamically to different stimuli in time and space and follow different trajectories before steady state is reached (Barbano PE et al 2005 Proc Natl Acad Sci USA 102:6245). A new method has been developed for integration of multiple datasets by using a free software (POINTILLIST) package (Hwang D et al 2005 Proc Natl Acad Sci USA 102:17296). The method was successfully applied to 18 datasets of galactose utilization of yeast involving mRNS, protein abundance, genome-wide protein interaction data, etc. (Hwang D et al 2005 Proc Natl Acad Sci USA 02:17302). Because of the rapid progress in experimental research and the increasing number of databases containing important relevant information, new bioinformatic tools are indispensable for integration of the knowledge. The databases must be found and the data must be converted into forms suitable for integration by electronic means. Modern database management system must be expanded and appropriate interfaces must be used. The Trace Ensemble Server only on Unix system (http://trace.ensembl.org/) already contains more than one billion information about 759 species. ►genetic networks, ►networks, ►drug development, ►ONDEX, ►databases; Recon H-InvDB Hood L, Galas D 2003 Nature [Lond] 421:444; Ideker T et al 2001 Annu Rev Genomics Hum Genet 2:343; Shogren-Knaak MA et al 2001 Annu Rev Cell Dev Biol 17405; Philippi S Köhler 2006 Nature Rev Genet 7:482; software standardization: Swertz MA, Jansen RC 2007 Nature Rev Genet 8:235; http://bind.ca/; http://www.cytoscape.org; network tool: http://biological networks.net/; system biology tool: http://babelomics.bioinfo.cipf.es/.

Systems of Breeding: Sexual reproduction may be allogamous, autogamous, inbreeding, assortative mating, hermaphroditic, monoecious and dioecious but reproduction may also be asexual. ►breeding system

Systole: Contraction of the heart and the forcing of the blood into the arteries. ►hypertension

SYT (synovial sarcoma): Oncogene in human chromosome 18q11.2. Translocations t(X;18) (p11;q11) are common. Gains of 8q and 12q as well as losses of 13q and 3p are frequent.

SZI: ►micromanipulation of the oocyte

Szostak Model of Recombination (Szostak JW et al 1983 Cell 33:25): A double break and repair model. It is applicable to transformational insertion of DNA molecules as well, as it can account for gene conversion and/or conventional recombination by outside marker exchange. The version of the model shown in Figure S155 does not explain why in yeast 5:3 gene conversion does not occur. (*Saccharomyces cerevisiae* has 4 spores per ascus but *Schizosaccharomyces pombe* has 8 unordered ascospores.)

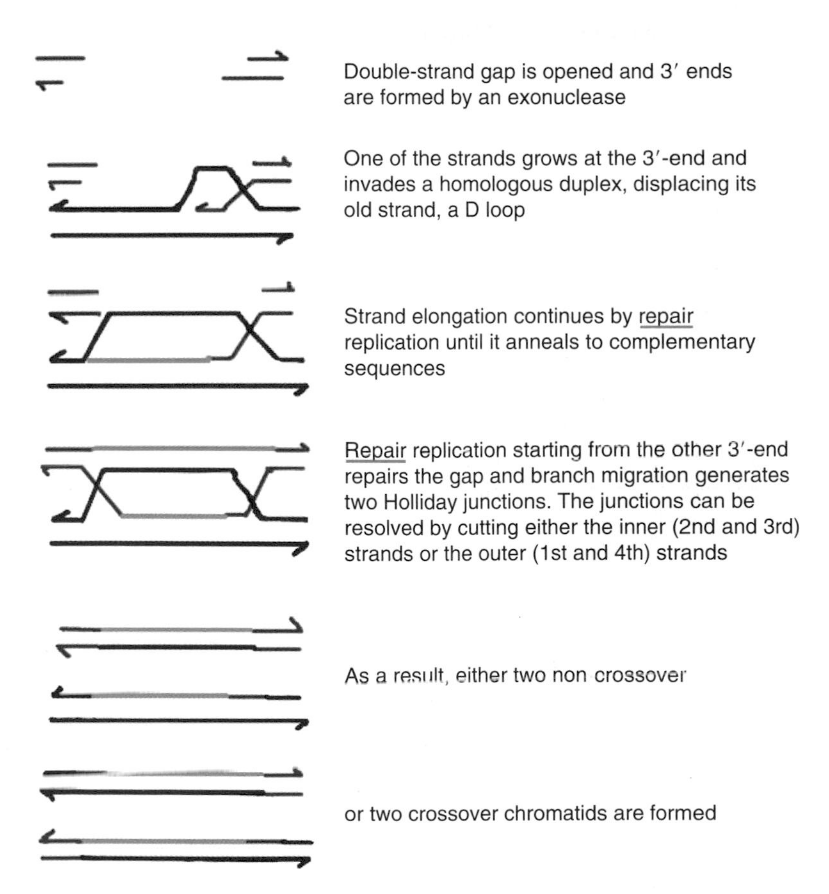

Double-strand gap is opened and 3′ ends are formed by an exonuclease
One of the strands grows at the 3′-end and invades a homologous duplex, displacing its old strand, a D loop
Strand elongation continues by repair replication until it anneals to complementary sequences
Repair replication starting from the other 3′-end repairs the gap and branch migration generates two Holliday junctions. The junctions can be resolved by cutting either the inner (2nd and 3rd) strands or the outer (1st and 4th) strands
As a result, either two non crossover
or two crossover chromatids are formed

Figure S155. The Szostak et al model of recombination is based on double-strand breaks in contrast to the Holliday or the Meselson—Radding models that suggest single-strand breaks in the DNA

Other modified models of Szostak can account for these types of experimental data. Genetic recombination in eukaryotes generally occurs by double-strand break. Radiation and various chemicals can increase the frequency of double-strand breaks. Double-strand breaks can also be increased in plant cells by transformation of a restriction endonuclease into the cell. ►recombination molecular models; Szostak JW et al 1983 Cell 33:25; Smith GR 2000 Annu Rev Genet 34:243, see Fig. S155.

Historical vignettes

Wilson EB 1896 The Cell in Development and Heredity. Macmillan, New York.

"Now, chromatin is known to be closely similar to, if not identical with, a substance known as nuclein—which analysis shows to be a tolerably definite compound composed of nucleic acid (a complex organic acid rich in phophorous) and albumin [protein]. And thus we reach the remarkable conclusion that the inheritance may, perhaps, be effected by the physical transmission of a particular chemical compound from parent to offspring."

Linus Pauling's letter to S Leonard Wadler on August 15 1966. (Cited after the Oregon State University Manuscript Collection). Pauling received Nobel Prize in 1954 for the nature of the chemical bond and again in 1962 for peace.

"I have suggested that the time may come in the future when information about heterozygosity in such serious genes as the sickle cell anemia gene would be tattooed on the forehead of the carriers, so that young men and women would at once be warned not fall in love with each other."

Kingman JFC 2000 in Genetics 156, p 1463
"Those who analyze stochastic models always lift their eyes from their equations to ask what they actually mean."

T

T: ►thymine (see Fig. T1).

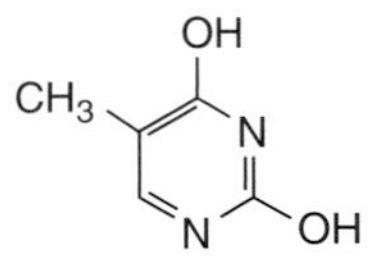

Figure T1. Thymine

t: Time.

τ: Yeast retroelement. ►Ty

θ (population mutation parameter): θ is characterized by the number of segregating alleles in the population (Watterson GA 1975 Theor Popul Biol 7:256).

θ: The symbol of recombination (and some other) fractions. ►w

T2: A virulent bacteriophage. ►bacteriophages, ►T4

T4: A virulent (lytic) bacteriophage of *E. coli*. It has double-stranded DNA genetic material of 1.08 × 10^6 Da (about 166 kbp) with a total length of about 55 μm. Its cytosine exists in hydroxymethylated and glycosylated forms. The linear DNA is terminally redundant and cyclically permuted. The redundancy occupies more than 1% of the total DNA. It was an unexpected discovery that its thymidylate synthetase gene contains an intron. Introns are common in eukaryotes but exceptional in prokaryotes. The phage has over 80 genes involved in metabolism but only about a fourth of them are indispensable. The metabolic genes control replication, transcription, and lysis. Other metabolic genes have functions overlapping those of the host. After infection, the phage turns off or modifies bacterial genes, degrades host macromolecules, dictates the transcription of its own genes, and utilizes the host machinery for its own benefit. For the synthesis of its own DNA, it relies on the nucleotides coming from the degradation of the host DNA. The bacterial RNA polymerase transcribes the phage genes. The viral protein Alc interacts with the β subunit of the host and terminates transcription on templates with cytosine residues. The viral DNA contains hydroxymethyl cytosine, which can serve as a template. The cytidylic acid residues of the host are prevented from being incorporated into the phage DNA by phosphatases. A deaminase converts them to thymidylic acid or a methylase enzyme converts them to hydroxymethyl cytosine through a few steps. The hydroxymethylcytidylate is then glucosylated by a glucosyl transferase enzyme. The molar proportion of thymidylate is higher in T4 than in *E. coli*, presumably because of the conversion of cytidylate into thymidylate. A series of more than 50 genes are involved in morphogenesis. At least 40% of the genome is required for the synthesis and assembly of the viral particle. 24 genes mediate head assembly, while the baseplate and tail require at least 31 genes. ►development, ►one-step growth, ►bacteriophages, ►sliding clamp; Calendar R (Ed.) 1988 The Bacteriophages, Plenum, New York; Knipe DM, Howley PM (Eds.) 2001 Fundamental Virology, Lippincott Williams & Wilkins, Philadelphia, Pennsylvania.

T7: A virulent bacteriophage with DNA genome size 39,937 bp, which encodes 59 proteins. ►bacteriophages

T18: A transcriptional activator protein.

2,4,5-T: ►agent orange

***T* Alleles**: ►*Brachyury*

T Antigen: The T antigen of SV40 virus is a 708-amino acid multifunctional protein and an effector of DNA polymerase α function. It assists in separating the DNA strands for replication and generates the replication bubble as the polymerase moves on. It has a special role in SV-induced tumors after tumor–suppressor proteins are eliminated or weakened. The T antigen and sequentially similar proteins are also factors in nuclear localization. ►Simian virus 40, ►SV40, ►nuclear localization sequences; Li D et al 2003 Nature [Lond] 423:512; Sullivan CS, Pipas JM 2002 Microbiol Mol Biol Rev 66:179.

t Antigen: The t antigen shares the same N-terminal sequences with the T antigen but the carboxyl end is different. The t antigen has homology to the G_t protein, an α subunit of the trimeric G-proteins involved in the activation of cGMP phosphodiesterase involved in photoreception and other processes. ►Simian virus 40, ►G-proteins

T Band: ►T-band

T Box (Tbx): A conserved ~14-nucleotide DNA domain upstream of the transcription terminator of Gram-positive bacteria in about 250 genes encoding aminoacyl-tRNA synthetases, and amino acid biosynthetic and transport enzymes. Uncharged tRNAs appear to interact with the leader sequence of these genes by their amino acid-accepting termini at the middle of the T box (UGGN') to stabilize antitermination at the expense of the terminator. T box genes (although named differently) occur in all species. In humans sequencing, the genome identified at least 18. These are generally transcription factors and are specific to diverse downstream genes. T box proteins interact with each other (through heterodimerization) and with many other proteins. They are involved in Wtn (winged), TGF-β (transforming growth factor), hedgehog, FGF (fibroblast growth factor), Notch, and receptor guanylate phosphate signaling. Several developmental disorders and human diseases are due to T box genes, which show both positive and negative effects on development. ►antitermination, ►MAR, ►Brachyury, ►eomesodermin, ►Holt-Oram syndrome, ►DeGeorge syndrome, ►Scheinzel syndrome, ►T-bet; Papaioannou VE, Silver LM 1998 Bioessays 20:9; Smith J 1999 Trends Genet 15:154; Putzer H et al 2002 Nucleic Acids Res. 30:3026; Showell C et al 2004 Dev Dyn 229:201; Naiche LA et al 2005 Annu Rev Genet 39:219.

T Cell Receptor (TCR): T cell surface glycoproteins recognize antibodies. The differentiation of the T cells begins with the differentiation of their receptors. At the beginning, the TCR is double negative, i.e., it is $CD4^-$ $CD8^-$. The disulphide-linked heterodimers have α and β or γ and δ chains, containing variable and constant regions and are homologous to the corresponding antibodies. First, the TCRβ is rearranged, followed by the rearrangement of the α chain. At this stage, the β chain may signal allelic exclusion and this means the end of rearrangements. After this, the double positive (DP) stage follows, i.e., $CD4^+$ $CD8^+$ TCR appears. A selection process ensues, resulting in an array of self MHC-restricted and self-tolerant TCRs. The TCR α chain is a transmembrane protein and the cytoplasmic (carboxyl) end has two potential phosphorylation sites and a Src homology 3 (SH3) domain (see Fig. T2). The TCRβ chain regulates the development of the T cells in the absence of the α chain. The αβ TCR generally recognizes antigens bound to the major histocompatibility (MHC) molecules. The TCR complex also includes the CD3 protein, required for signal transduction. It is made of the γ, δ, ε, and ζ chains. The ζ chain plays a role in thymocyte development but it is not indispensable for signal transduction. The chromosomal site of the human α (14q11), β (6q35), γ (7p15-p14), δ (14q11.2), and ζ (1p22.1-q21.1) chains are shown in parenthesis. Marsupials also have TCRμ that does not have a known homolog in eutherian mammals but has features analogous to a TCR isoform in sharks (Parra ZE et al 2007 Proc Natl Acad Sci USA 104:9774).

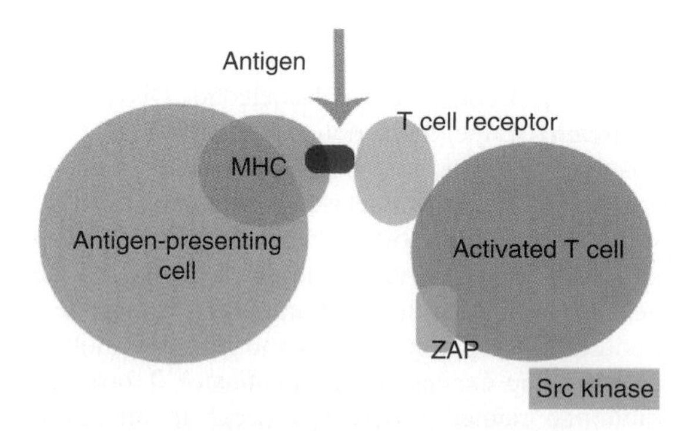

Figure T2. T cell receptor in T cell function

Activation is triggered when the T cell receptor binds the MHC associated ligand on the antigen presenting cell (Choudhuri K et al 2005 Nature [Lond] 436:578). Co-stimulating signals may help. The CD3 component of the TCR may be altered, and protein tyrosine kinases (PTK) are activated. The PTKs may turn on the calcium-calcineurin, the RAS-MAP, and the protein kinase signaling pathways (see Fig. T3).

These pathways then may activate the transcription factors NFAT, NF-κB, JUN, FOS (AP1), and ETS. They, in turn, activate new genes, some of which are specific transcription factors that facilitate the release of cytokines that further activate the clonal expansion of T cells. This is followed by the production of antibodies by B cell or T cell cytotoxicity. Immune memory, immune tolerance, anergy, and apoptosis are alternative functions that follow. Regulation of the development of TCR requires the cooperation of the protein tyrosine kinases (Src family), phospholipase C (PLCγ), CD5, CD28, CD48, CD80, VCP, ezrin, VAV, SHC, PtdIns, PIP2, PIP3. TCR may also regulate apoptosis of T cells. For the TCRs, there are 42 variable (V) and 61 joining (J) segments at the α chain immunoglobulin locus, and 47 V, 2 diversity (D), and 13 J segments for the β chain genes. During rearrangements, V_α—J_α and V_β—D_β—J_β deletions and additions of nucleotides and dimerization may increase the variations. In blood, there appears to be ~10^6 different β chains that may combine on the average with ~25 different α chains. In the memory subsets, the diversity appears about 1/3 less. The number of estimated distinct TCR receptors may be 10^{12}. ►TCR genes, ►ICAM, ►LFA, ►T cells, ►lymphocytes, ►immunoglobulins, ►RAG, ►Src, ►Yes, ►Fgr, ►Fyn, ►Lck, ►Hck, ►Blk, ►Zap70, ►Tec, ►antibody, ►MHC, ►CD3, ►LCK, ►ITAM, ►B lymphocytes, ►phosphoinositides, ►PIP, ►γδ T cells, ►immune system, ►immunological synaps, ►integrin, ►signal transduction, ►thymus, and the other factors under separate entries, ►SLP-76, ►α-CPM, ►caveolae, ►ICOS, ►CTLA-4; Hennecke J, Wiley DC 2001 Cell 104:1; Germain RN 2001 J Biol Chem 276:35223; Isakov N, Altman A 2002 Annu Rev Immunol 20:761; Natarajan K et al 2002

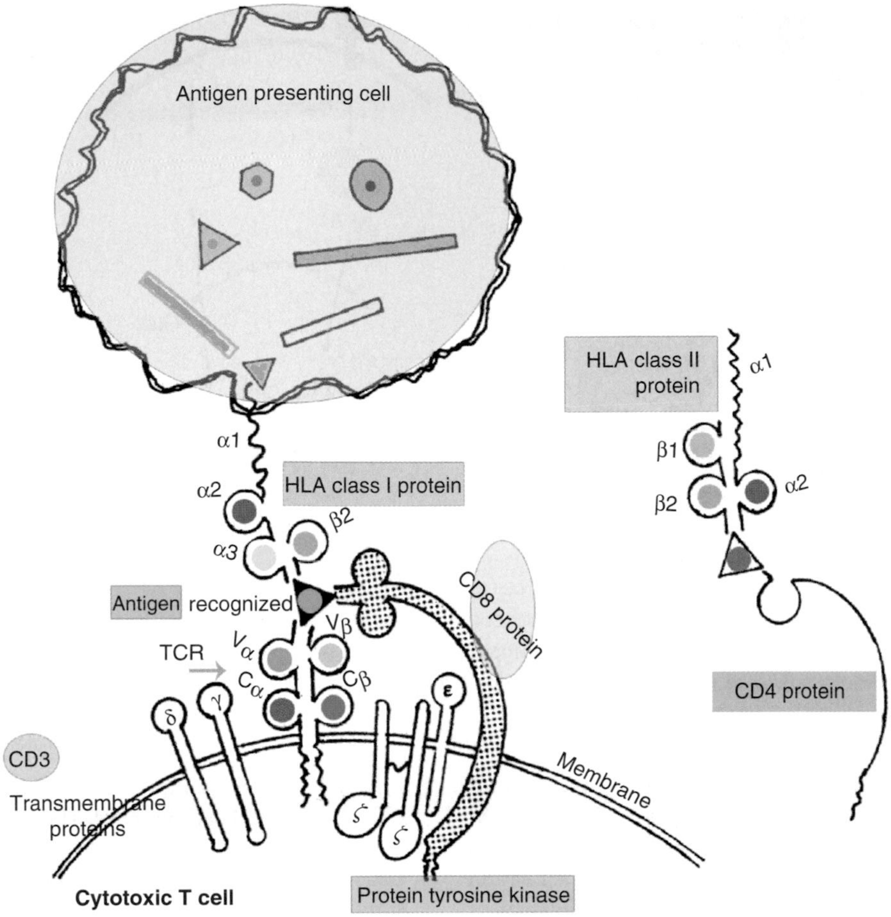

Figure T3. A general outline of the functions of the T cell receptor (TCR) complex. Although diagrams always generalize beyond reality, they may be helpful to obtain a broad understanding. The foreign antigens are presented to the T lymphocytes either by the antigen-presenting cells or by macrophages. The macrophages are capable of partially degrading the large molecules or the invading cells. These cells associate then with either class II or class I MHC proteins, which are encoded by the HLA genes. The MHC molecules recognize the foreign antigens and bring them to the TCR and to the CD protein complex associated with the T cell surface. The class I gene products use the CD4 transmembrane proteins, whereas the class II molecules rely on CD8. The COOH ends of the TCR chains are inside the T cell's double membrane and the NH_2 end is involved in the recognition of the antigen in association with the MHC and CD elements. The TCR complex includes also the CD3 transmembrane proteins. The ζ subunit serves also as an effector in signal transduction. Protein tyrosine kinase is an important element in several signal transduction pathways. ICAM-1 on the antigen-presenting cell and LFA-1 on the $CD8^+$ also cooperate in mediating the adhesion of the MHC complex. The right side of the incomplete diagram shows the association of the class II and CD4 proteins with an antigen. The other elements of this system are very similar to that shown at the left main part of the outline

Annu Rev Immunol 20:853; Werlen G et al 2003 Science 299:1859; Call ME et al 2006 Cell 127:355; http://imgt.cines.fr/.

T Cell, Regulatory (Treg/Tr). Regulatory T cells may disarm non-tolerant (self-reactive) naive lymphocytes (*dominant tolerance*) and bring about *infectious tolerance* of tissue grafts. These T cells occur in $CD4^+$ $CD25^+$ and $CD4^+$ $CD25^-$ cells. FOX3 controls Treg by cooperation with NFAT. Ca^{2+} regulate NFAT transcription factors. In activated T cells, NFAT forms co-operative complexes with AP-1. Structural mutation in FOXP3 progressively disrupts its interaction with NFAT, FOXP3 has the reduced ability to repress T cell growth factor IL-2; Treg markers CTLA4 and CD25 are upregulated. Thus, murine autoimmune diabetes is suppressed (Wu Y et al 2006 Cell 126:375). FOXP3 (Forkhead box P3; Xp11.23-q13.3) mutations involve aggressive autoimmuninity mediated by regulatory T cells. Interaction of FOXP3

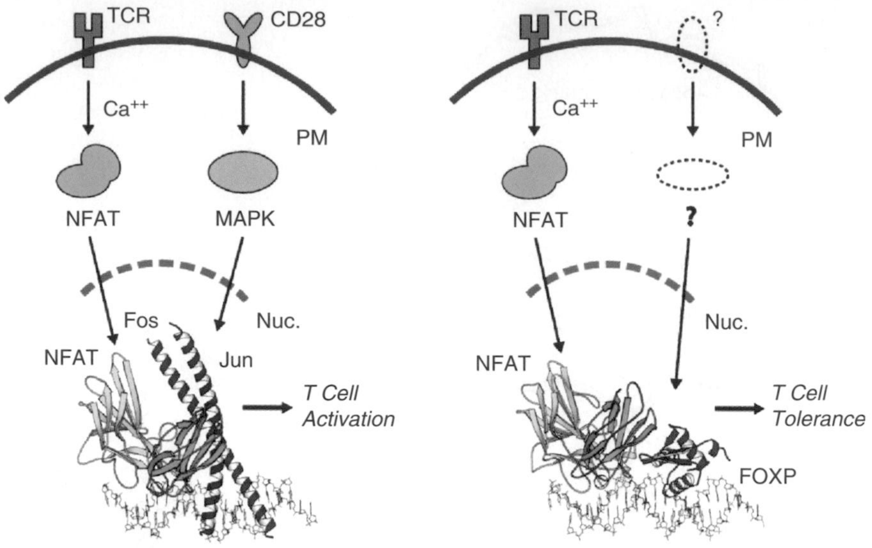

Figure T4. Left: Schematic representation of the T cell activation program mediated in part by the assembly of the NFAT/Fos-Jun/DNA complex on specific promoters. Right: schematic representation of the T cell tolerance program mediated in part by the assembly of the NFAT/FOXP3/DNAcomplex on specific promoters. Signals involved in the regulation of FOXP3 are currently unknown and indicated by ? mark. (Courtesy of Lin Chen and James stroud)

with AML1/Runx1 (acute myeloid leukemia 1/Runt-related transcription factor 1) controls Treg (Ono M et al Nature [Lond] 446:685). Mast cells are recruited by IL-9 to immune-tolerant tissues through activation by Treg and are essential for immune tolerance of allografts (Lu L-F et al 2006 Nature [Lond] 442:997). Autoimmune disease susceptibility and resistance alleles on mouse chromosome 3 (*Idd3*) correlate with differential expression of the key immunoregulatory cytokine interleukin-2 (IL-2). Reduced IL-2 production correlates with reduced function of $CD4^+$ $CD25^+$ regulatory T cells (see Fig. T4) (Yamanouchi J et al 2007 Nature Genet 39:329). ►T cells, ►autoimmune disease, ►immune tolerance, ►Mast cell, ►IL-2, ►IL-9, ►forkhead, ►NFAT, ►MAP kinase [MPK], ►AP1, ►*fos*, ►CD28, ►T cell receptor [TCR], ►T cells, ►Cd25, ►CTLA-4, ►IL-2, ►IL-9, ►leukemia, ►Runx; Graca L et al 2002 J Exp Med 195:1641; Rudensky AY et al 2006 Cell 126:253; Wan YY, Flavell R 2007 Nature [Lond] 445:766; Gavin MA et al 2007 Nature [Lond] 445:771.

T Cell Replacing Factor: A lymphokine. ►lymphokines

T Cell Vaccination: Immunization with irradiated autologous T cells. This type of vaccination is expected to deplete circulating autoreactive T cells and has therapeutic significance for autoimmune diseases. ►autologous, ►autoimmune disease; Hong J et al 2006 Proc Natl Acad Sci USA 103:5024.

T Cells: Thymic lymphocytes control cell-mediated immune response (the foreign antigens are attached to them). The T lymphocytes originate in the bone marrow, differentiate in the thymus, and later migrate to the peripheral lymph nodes. While the early T cells are located in the thymus, a *negative selection* eliminates those T cells that react with self-antigens. At about the same stage of T cell development, a positive selection takes place under the influence of the MHC complex, securing the survival of those T cells that can interact with antigens associated with either of the two types of MHC molecules presented to these cells. These encounters lead to the differentiation and activation of T cells. Cytotoxic T cells (CTL) are the major elements of the immune system; they cooperate with the natural killer cells (NK) and degrade foreign antigens with immunoproteasomes. The heterodimeric PA28a and PA28b (proteasome activating protein) activate the immunoproteasomes.

The helper T_H ($CD4^+$ T cells) and the T suppressor (T_S) cells mediate the humoral (secreted) immune responses. On the surface of the T cells are the T cell surface receptors (TCR). The T_H cells stimulate the proliferation of the B (bursa) cells when they recognize their cognate antigens. The joint action of T_H cell surface receptors (TCR) and the B cell's antigens bring about in the B cells the formation, growth, and differentiation of the proteins named lymphokines, which stimulate the propagation of B cells and the secretion of the humoral antibody. The T cell surface receptors recognize foreign antigens only if they become associated with major histocompatibility (MHC) molecules carried to the TCR by the antigen presenting cells (APC) or macrophages. The TCR links to the various intracellular signaling pathways. The activation of the T cells requires phosphorylation by SRC tyrosine kinase of the CD3 immunoglobulin chains. The activation involves co-stimulatory molecules CD28, ICAM-1, and LFA-1. For full activation, ZAP-70 protein-tyrosine kinase is also needed. The T cells are presented with a variety of antigens, including self-antigens that are carried by the APCs without discri-mination. T cells distinguish self/foreign antigens. This ability begins to develop while the T cells are still in the thymus. Discrimination is a difficult task to achieve and complex phosphorylations are required for surveying the very large array of ligands of varying degree of specificity (affinity). Alternatively, it is conceivable that one peptide–MHC complex interacts first with one TCR. Then, in a contact cap this TCR detaches from the ligand, thus making possible for the ligand to bind another TCR. The process is repeated in a serial manner and assembles sufficient number of TCRs in the contact cap for productive signaling.

Chemokines enhance immunity by guiding naïve $CD8^+$ T cells to sites of $CD4^+$ T cell and antigen presenting dendritic cell interaction (Castellino F et al 2006 Nature [Lond] 440:890). The activation of the T cell may be only partial in case there is a subtle change (e.g., one amino acid replacement) in the peptide–MHC. The subtle variants may also inhibit the $CD4^+$ helper T cells from responding to the real antigen. Altered ligands or lack of co-stimulatory signals may cause anergy of the T cell, and cannot be stimulated by the TCR but may or may not proliferate under the influence of interleukin 2 (IL2). The ligands thus can be *agonists* that fully or partially activate the T cells, or altered ligands may be weakened agonists or even *antagonists* and reduce activation. The *null ligands* provoke no response. The weak agonists may not activate ZAP-70 and may have a different pattern of phosphorylation of the CD3 ζ chain. For the fully active immune reaction, all the elements of the complex T cell activation must be in place. Some viral infections (HIV-1, hepatitis B, etc.,) may lead to the production of antagonist ligands and then the cytotoxic T cells (CTL) cannot protect the body against the invader. Some aspects of the T cell activation and regulation are outlined.

On the cell membrane is the *T cell receptor* associated with CD3 and CD4 immunoglobulins. This receptor TCR mediates phosphorylation with the aid of a *protein tyrosine kinase* (PTK), *phospholipase C-γ1*. The so activated PLC-γ1 cleaves phosphatidylinositol-4,5 bis-phosphate (PIP2) and generates *inositol trisphosphate* (IP3) and *diaglycerol* (DAG), which are second messengers. These second messengers activate *protein kinase C* and make available cytoplasmic Ca^{2+} for *calcineurin* (also called protein phosphatase IIB). As a consequence, the transcription factor NFAT (*nuclear factor activated T cell*) promotes the transcription of the *interleukin-2 gene*. Calcineurin also contributes to the activation of *ERK* (member of the mitogen-activated protein kinase family, MAPK). Co-stimulation is provided through the CD28 immunoglobulin system, which activates protein tyrosine kinase C that, in collaboration with the RAS G protein, stimulates the *Jun NH_2-terminal kinase* (JNK). It is also known that the *FOS* and *JUN* oncogenes contribute to the formation of the *AP* group of transcription factors, probably by acting on protein kinase C (PKC). *Phorbol myristyl acetate* (PMA) is an adjuvant for the stimulation of the *IL*-2 gene. These processes of development of effector T cells may be down-regulated when the 2*E*1 *antibody* attaches to the *effector cell protease receptor*-1. Immunosuppressive agents such as FK506, cyclosporine, and OKT3 may cause further down-regulation.

Another important player in the T cell response is CTLA (cytotoxic T lymphocyte antigen), a molecule with about 75% homology to CD28. While CD28 is a co-stimulator of T cell activation, CTLA-4 is a negative regulator and protects against rampant lymphoproliferative disorders. The recognition of the role of CTLA offers an opportunity to neutralize its effect by specific antibodies and thereby accelerate the action of the antitumor interleukin production without serious side effects. T cell differentiation is accompanied by the appearance of surface markers in an order: CD4, CD25, CD44, CD8, and CD3. The commitment of T cell differentiation requires tumor necrosis factor-α (TNF-α) and interleukin-1α [IL-1α]). About 5–10% of the T cells are regulatory (T_R) and may express the CD25 activation marker. The $CD4^+$ T_R cells may prevent autoimmune disease, transplant rejection, and inflammatory bowel disease (Hori S et al 2003 Science 299:1057). The understanding of these circuits may lead to designing better drugs against infections and the suppression of the rejection of tissue grafts and cancer. (See Fig. T5, ►TCR, ►T cell regulatory, ►SLAM TIL, ►HLA, ►MHC, ►killer cells, ►proteasomes, ►TAP, ►vaccines, ►antibody, ►ZAP-70, ►CD3, ►CD4, ►CD8,

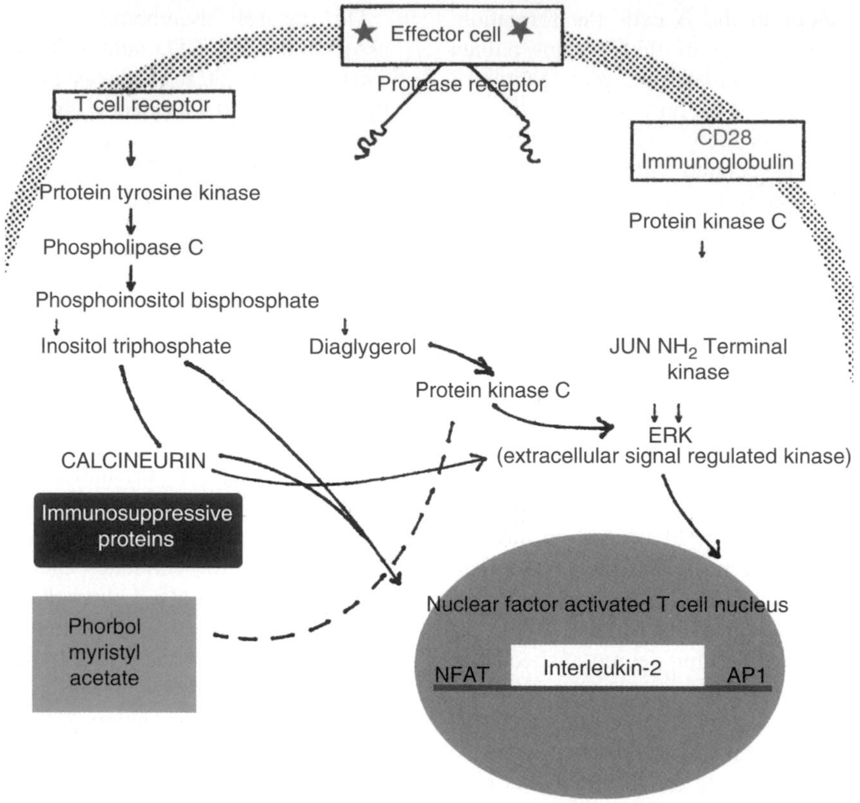

Figure T5. T cell signaling pathways. (Modified after Trucco M, Stassi G 1996 Nature (Lond.) 380:284)

►CD28, ►CTLA-4, ►ICOS, ►GATA, ►Cd117, ►SRC, ►HIV, ►immune system, ►immunoglobulins, ►protein tyrosine kinase, ►phospholipase C, ►PIP, ►IP_3, ►EPR, ►calcineurin, ►NFAT, ►interleukin, ►ERK, ►MAPK, ►RAS, ►AP, ►phorbol esters, ►FK506, ►cyclosporin, ►OKT, ►LCK, ►RANTES, ►immunophilins, ►signal transduction, ►ICAM, ►AKAP79, ►TAP, ►RAD25, ►NF-AT, ►NF-κB, ►Fas, ►autoantigen, ►αβ T cells, ►γδ T cells, ►T_H, ►thymus, ►ICAM, ►LFA; Dustin ML, Chan AC 2000 Cell 103:283; Staal FJT et al 2001 Stem Cells 19:165; Barry M, Bleackley RC 2002 Nature Rev Immunol 2:401; Sadelain M et al 2003 Nature Rev Cancer 3:35).

T Complex: The products of the virD2 and virE2 agrobacterial virulence genes associated with the 5′-end of the transferred strand of agrobacteria. ►agrobacterial virulence genes, ►T-DNA

t Complex: ►*brachyury*

T Cytoplasm: Texas male sterile cytoplasm of maize; almost 100% of the pollen is incapable of fertilization (sporophytic control of male sterility). ►cytoplasmic male sterility

t Distribution: ►Student's t distribution, ►t value

t Haplotype: ►brachyury

T Helper Cell: ►T_H

t Loop: ►telomeres

T Lymphocytes: ►T cells

T1 Phage: A double-stranded DNA (48.5 kbp) virulent phage, a general transducer. Only 0.2% of its cytosines and 1.7% of the adenines are methylated. The terminal redundancy is about 2.8 kb. It infects *E. coli* and *Shigella* strains. ►bacteriophages

T2, T4, and T6 Phages: T-even virulent phages that are closely related. The 166 kbp linear genome of T4 contains glycosylated and hydroxymethylated cytosine and 1–5% of its DNA is terminal redundancy; it encodes about 130 genes with known function, and about 100 additional open reading frames have been revealed. ►bacteriophages, ►development

T5 Phage (relatives BF23, PB, BG3, 29-α): The genetic material of T5 phages is linear double-stranded DNA (≈121.3 kbp) with terminal repeats (≈10.1 kbp) but without methylated bases. Three internal tracts can be deleted without loss of viability. They are virulent phages with long tails. ►bacteriophages, ►development

T7 Phage (T3 is related): A virulent phage; it does not tolerate superinfection (would be required for recombination). Its 39.9 kbp genetic material is enclosed in an icosahedral head with a very short tail. It codes for about 55 genes. The T7 promoter is used in genetic vectors for in vitro transcription. Its replication requires DNA and RNA polymerase, helicase-primase complex, single-strand-binding protein, and endo- and exonuclease activities. ►bacteriophages

θ Type Replication: θ type replication generally occurs in small circular DNA molecules starting at a single origin and following a bidirectional course (see Fig. T6). ►replication, ►bidirectional replication

Figure T6. Bidirectional (theta) replication

t Value (t test): The ratio of the observed deviation to its estimate of standard error: $t = \frac{d}{\sqrt{V}}$. It is used as a statistical device to estimate the probabilty of difference between two means. In its most commonly used form, the t is calculated as,

$$t = \bar{x}_1 - \bar{x}_2 / \sqrt{[s\bar{x}_1]^2 + [s\bar{x}_2]^2}$$

where $\bar{x}$ indicates the mean of the two sets of data and s stands for the standard deviation. ►standard deviation, ►Sudent's t distribution, ►paired t test

Ta: A copia-like element (5.2 kbp) in *Arabidopsis*, flanked by 5 bp repeats but without transposase function. ►copia, ►retroposon, ►Tat1

Tα: A non-viral retrotransposable element. ►retroposon, ►retrotransposon

TAB1: A human TAK1 kinase-binding protein. Overproduction of TAB1 enhances the activity of the promoter of the inhibitor of the plasminogen activator gene, which regulates TGF-β and increases the activity of TAK1 human kinase. ►TAK1, ►plasmin [plasminogen], ►TGF-β

TAB Mutagenesis: (two amino acid Barany): In TAB mutagenesis, a DNA segment containing two sense codons is introduced at a certain position into a gene in vitro; thus, probing the effect of the two amino acid modification regarding the function of the protein product. A brief outline of the essence of the procedure is shown in Figure T7. ►directed mutagenesis, ►cassette mutagenesis, ►localized, ►site-specific mutagenesis; Barany F 1985 Proc Natl Acad Sci USA 82:4202.

Tabatznik Syndrome: Heart and hand disease II. ►Holt-Oram syndrome

TAC: Transcriptionally active complex. ►open promoter complex, ►transcription complex, ►transcription factors, ►transcription factors inducible

TAC: Transformation-competent bacterial artificial chromosome. ►BAC

TACC: A centrosomal protein that in combination with other factors regulates spindle formation. ►centrosome, ►spindle; Gergely F et al 2000 Proc Natl Acad Sci USA 97:14352.

TACE (tumor necrosis factor α converting enzyme): A member of the ADAM metalloproteinase family of proteins involved in inflammatory responses. It removes ectodomains from cell surface TNF-α receptors and ligands, selectins, etc., and cleaves amyloid precursor proteins. It proteolytically makes available soluble growth factors synthesized on the cell surface. ►metalloproteinases, ►ADAM, ►TNF, ►selectins, ►secretase, ►β-amyloid, ►FAS, ►TACI; Skovronsky DM et al 2001 J Neurobiol 49:40; Black RA 2002 Int J Biochem & Cell Biol 34:1.

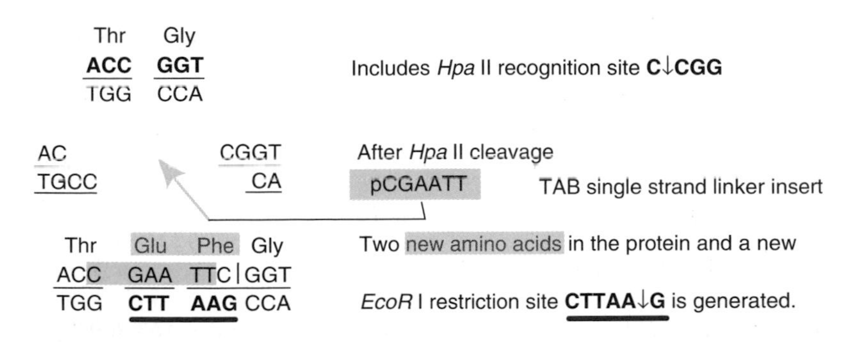

Figure T7. TAB mutagenesis

Tachykinins: Peptides mediating secretion, muscle contraction, and dilation of veins. ►angiotensin; Labrou NE et al 2001 J Biol Chem Oct 12 Online.

Tachytelic Evolution: ►bradytelic evolution

TACI (transmembrane activator and calcium modulator and cyclophilin ligand [CAML]): One of the TNF receptors, which regulate the expression of transcription factors NF-AT, NF-κB, and AP-1. ►TNF, ►TNFR, ►TACE; Wang H et al 2001 Nature Imunol 2:632; Pan-Hamarström Q et al 2007 Nature Genet 39:429.

TACTAAC Box: A highly conserved consensus in mRNA introns of *Saccharomyces* yeast.

Tadpole: Amphibian (frog/toad) larva at an early developmental stage. As the tadpole matures, it loses its gills, a more pronounced tail appears, and legs develop. Subsequently, the tail is lost before the frog/toad emerges from the free-swimming larva (see Fig. T8).

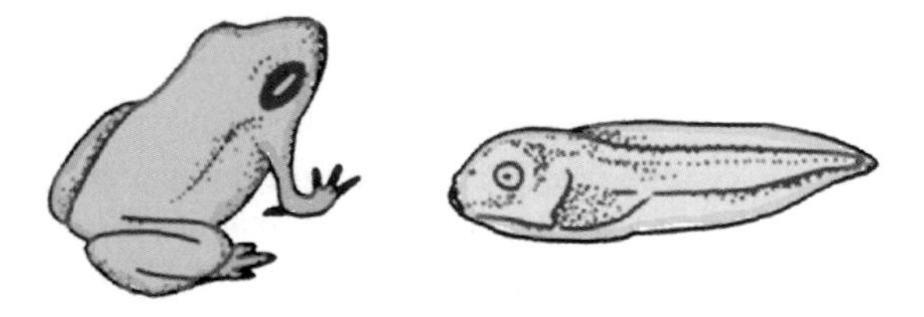

Figure T8. Left: frog; right: tadpole

TAE (Tris-acetate-EDTA): ►electrophoresis buffers

TAF: TATA box-associated factors TAF250, TAF150, TAF110, TAF60, TAF40, TAF30α, and TAF30β. In yeast there are promoters that require TATA box binding protein (TBP) and TAFs (i.e., they are TAF-dependent). TAFs may require TIC for efficient function. Alternatively, other promoters are independent of TAFs and require only TBP. ►TBP, ►TIC, ►core promoter, ►fermentation; Wassarman DA, Sauer F 2001 J Cell Sci 114:2895; Green MR 2000 Trends Biochem Sci 25:59.

TAF1: TAF1 subunit, TFIID carboxy-terminal kinase, activates transcription by phosphorylation of Ser^{33} residue in histone H2B (Maile T et al 2004 Science 304:1010). ►transcriptioin factors

TAF_{II} (transcription activating factors): TAF_{II} serve as coactivators of enhancer-binding proteins. TAF_{II} appear to have sufficient homology to transcription factor TFIID that in the absence of TAF_{II} transcription can still proceed in yeast. TFIID is actually a complex of TBP and TAFs. TAF causes a conformational change in transcription factor TFIIB. The $TAF_{II}250$ subunit apparently modifies the H1 histone protein to facilitate the access of the RNA polymerase to the DNA in the chromatin. The acidic activator disrupts the amino and carboxy-terminal interactions within this molecule and this results in an exposure of the binding sites for the general transcription factors to enter into a preinitiation complex with TFIIB. TFIIB initiates the formation of an open promoter complex. ►regulation of gene expression, ►TF, ►open promoter complex, ►transactivator, ►co-activator, ►transcription factors, ►histones, ►SAGA; Frontini M et al 2002 J Biol Chem 277:5841.

TAF_{II} 230/250: TAF_{II} 230/250 has histone H3, H4 acetyltransferase and protein kinase domains. The two bromodomains of the protein apparently accommodate acetyl lysines and support the activity of TFIID. The TFIID transcription complex regulates gene expression and several critical developmental processes. ►histone acetyltransferase, ►bromodomain

TAFE (transverse alternating field electrophoresis): TAFE is used for pulsed field gel electrophoresis, when a current is pulsed across the thickness of the gel. ►pulsed field gel electrophoresis

Taffazzin: A fibroelastin group of proteins in the muscles. ►Barth syndrome

Tag: Large T antigen, an early-transcribed gene of SV40. It functions as an ATP-dependent helicase in the replication of the DNA. The two Tag binding sites, each, include two consensus sequences 5′-GAGCC-3′, separated by six or seven A = T base pairs. In order to start replication, Tag must bind to the replicational origin, *ori*, and to neighboring sequences of the virus. ►Simian virus 40

tag: The small t antigen of SV40, an early-transcribed gene product.

Tag1: An autonomous transposable element (3.3 kb with 22 bp inverted repeats) of (Le-) *Arabidopsis*. It is a member of the Ac (maize), Tam3 (*Antirrhinum*), and hobo (*Drosophila*) family of elements (see Fig. T9). ►*Arabidopsis*, ►retrotransposons, ►Ac-Ds, ►Tam, ►hybrid dysgenesis, ►GUS, ►gametophyte; Galli M et al 2003 Genetics 165:2093.

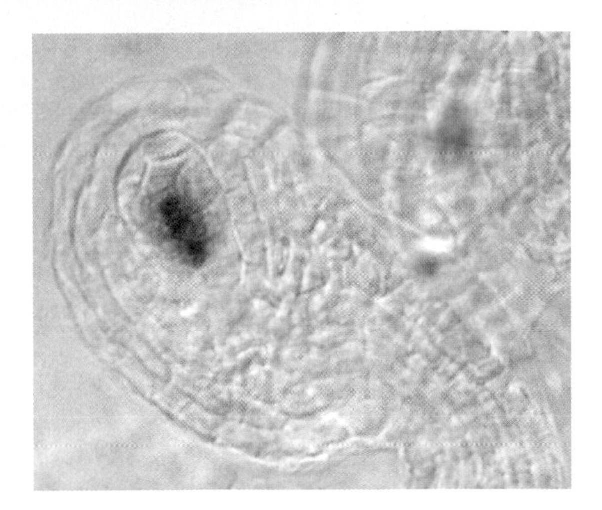

Figure T9. The *Tag1* element displays late excision during vegetative and reproductive development. When its transposase is fused to the GUS gene and transformed by such a construct, the expression (detected by the blue color) is limited to the male and female gametophytes. The illustration shows its expression in the megagametophyte. Similar is its behavior in the pollen. If however an enhancer is in its vicinity, its expression in the vegetative parts of the plants is restored. (Courtesy of Mary Galli and Nigel M. Crawford)

TAG SNP (tagSNP): A single-nucleotide(s) polymorphism characteristic for a haplotype. ►SNIPs, ►haplotype; Mueller JC et al 2005 Am J Hum Genet 76:387.

Tagging: Identifying a gene by the insertion of a transposon, an insertion element, a transformation vector, or by annealing with a DNA probe. These tags have known DNA sequences and can be detected on the basis of homology. When they are inserted within a structural gene or a promoter or other regulatory element of a gene, the expression of the gene may be modified or abolished. Therefore, their location may be detected by alteration in a specific function and may also assist in the isolation and cloning of the target DNA sequence. ►transposon tagging, ►transformation genetic, ►insertional mutation, ►probe, ►chromosome painting, ►FISH

Tagmosis: The segmental organization of the insect body. ►homeotic genes; Angelini DR, Kaufman TC 2005 Annu Rev Genet 39:95.

Tail Bud: The tail bud gives rise to the tail of the animal from epithelial cells of the mesenchymal tailbud. Its differentiation is usually completed after the basic organization pattern of the embryo has been realized.

Tailing Homo-A: The eukaryotic mRNA generally contains a post-transcriptionally added polyA tail. Poly-A or other homopolynucleotide sequences may be added to DNAs by terminal transferases. ►mRNA, ►terminal transferase

Tail-Less: ►*Brachyury*, ►*Manx*

Tail-PCR (thermal asymmetric interlaced-PCR): A polymerase chain reaction resembling inverse PCR. A non-specific primer is paired with specific primers to obviate the need for circular genomic fragments. ►polymerase chain reaction, ►inverse PCR; Liu YG, Whittier RF 1995 Genomics 25:674.

Tailpiece, Secretory: Immunoglobulins IgM and IgA possess an 18-amino acid heavy chain C-terminal extension, and 11 of these amino residues are identical between them. This tailpiece interacts with the J chain. ►immunoglobulins, ►membrane segment; Olafsen T et al 1998 Immunotechnology 4(2):141.

Tajima's Method: Tajima's method statistically tests the neutral mutation theory on the basis of DNA polymorphism. D is the normalized difference of the two estimates shown.

$D_t = \frac{\pi - \theta_s}{\sqrt{Var(\pi - \theta_s)}}$, where θ_s is the expected number of polymorphic sites and π is the average number nucleotide differences. ►mutation neutral; Tajima F 1989 Genetics 123:585.

TAK1: A human homolog of the kinase MAPKKK, an activator of the TGF-β signal. ►TGF-β, ►TAB, ►MAP; Wang C et al 2001 Nature [Lond] 412:285.

Talin: A cytoskeletal protein binding integrin, vinculin, and phospholipids. ►adhesion, ►integrin; Xing B et al 2001 J Biol Chem 276:44373; Wegener KL et al 2007 Cell 128:171.

Talipes: ►clubfoot

TAM (transcription associated mutation): In TAM, the rapid rate of transcription may involve increase of mutation, apparently because translesion, nucleotide excision, or recombination may not repair the DNA damage. ►translesion, ►DNA repair, ►Cockayne syndrome

TAM (transposable element *Antirrhinum majus*): *TAM* are responsible for the high mutability of genes controlling the synthesis of flower pigments, known in this plant since pre-Mendelian times. There are several *TAM* elements. The termini of the *TAM*1 and *TAM*2 are homologous and their insertion results in a 3 bp target site duplication. Their termini are almost identical to those of the *Spm/En* transposons of maize and somewhat homologous to the termini of the *Tgm1* transposon of soybean. The *TAM3 transposon* is different from *TAM1* and *TAM2* and 7/11 bps of its terminal repeats are homologous to the *Ac* element of maize. Both *TAM3* and *Ac* may generate 8 bp target site duplications, although *TAM3* may be flanked also by 5 bp repeats. *TAM* elements, similarly to the maize transposons, seem to move by excision and relocation. The excision is usually imprecise. The insertion within genes results in mutation and the resulting mutant phenotype depends on the site of the insertion. The excision results in more or less faithful restoration of the non-mutant phenotype depending on the extent of alteration left behind at the insertion site (see Fig. T10). ►transposable elements plants, ►transposons; Coen ES et al 1989 Mobile DNA. In: Berg DE, Howe MM (Eds.) Amer Soc Microbiol Washington, DC, pp 413 photograph is by courtesy of BJ Harrison and Rosemary Carpenter.

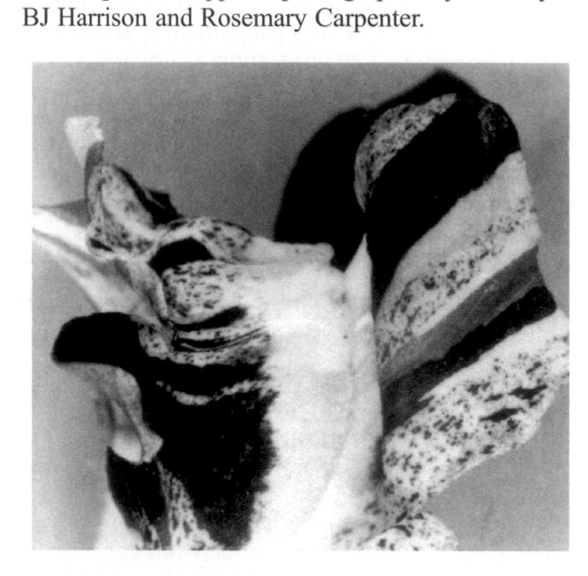

Figure T10. TAM

Tamarins: New world monkeys. ►*Callithricidae*

TAMERE (trans-allelic meiotic recombination): TAMERE is mediated by a targeted mechanism involving *loxP* sites in between two markers situated in trans, i.e., in homologues pairs of chromosomes. The recombinations may result in deletions and duplications. ►*Cre/LoxP*; Herault Y et al 1998 Nature Genet 20:381.

Tamoxifen (2-[4-{1,2-diphenyl-1-butenyl}phenoxy]-*N, N* dimethyl-ethanamine): A selective estrogen modulator drug used for the treatment of breast cancer; it may cause endometrial cancer although the benefits may outweigh the risk (Michalides R et al 2004 Cancer Cell 5:597). Tamoxifen binds to AP1 and to other estrogen response elements in the uterus. Tamoxifen and estrogen both activate PAX2 by hypomethylation of the PAX2 promoter (see Fig. T11). This may be the cause of endometrial cancer (Wu H et al 2005 Nature [Lond] 438:981). A multigene assay has predictive value for cancer recurrence in tamoxifen-treated node-negative breast cancer (Paik S et al 2004 New Engl J Med 351:2817]). ►breast cancer, ►raloxifene, ►AP1, ►estrogen receptor, ►estrogen response element, ►antiestrogens, ►estradiol; Brewster A, Helzlsouer K 2001 Curr Opin Oncol 13:420.

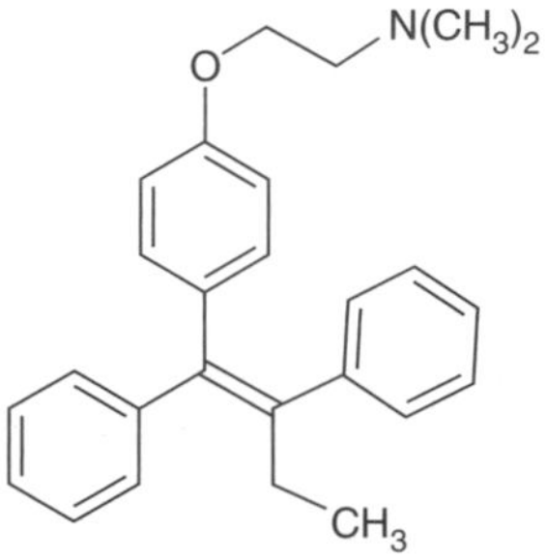

$N(CH_3)_2$, O, CH_3

Figure T11. Tamoxifen

TAN (translocation-7–9-associated *Notch* homolog): TAN is located in human chromosome 9q34. It is involved in T cell acute leukemia. ►morphogenesis in *Drosophila* [*Notch*], ►*Notch*, ►leukemia, ►T cell; Suzuki T et al 2000 Int J Oncol 17:1131.

Tandem Duplications: ►tandem repeats

Tandem Fusion: In tandem fusion, elements are associated head-to-tail, following each other in the same direction.

Tandem Mass Spectrometry: The tandem mass spectrometry procedure selects one kind of peptide in a mixture by collision with argon or nitrogen gas. The fragments are then processed in the tandem mass spectrometer, thus obtaining the MS/MS spectrum. This analytical method has been very useful for the detection of inborn metabolic errors in small complex blood samples. ►electrospray, ►MALDI, ►MS/MS, ►proteomics; Kinter M, Sherman NE 2000 Protein Sequencing and Identification Using Tandem Mass Spectrometry, Wiley-Interscience, New York; Chace DH et al 2002 Annu Rev Genomics Hum Genet 3:17.

Tandem Repeat: Adjacent direct repeats (such as ATG ATG ATG) of any size and number. In *Caenorhabditis elegans*, 2.7% of the genome involves tandem repeats and these occur on the average once per 3.6 kb. Comparative genomic analysis of breeds of dogs and other species indicate that increase or decrease of tandem repeats affect the apparently rapid evolution of morphological traits, particularly in the *Alx*-4 (aristaless-like 4) and *Runx*-2 (runt-related transcription factor) involved in skull and limb morphology, respectively (Fondon JW Garner HR 2004 Proc Natl Acad Sci USA 101:18058). ►repeat inverted, ►*Drosophila*, ►runt; https://tandem.bu.edu/cgi-bin/trdb/trdb.exe.

Tangier Disease (HDL deficiency): Tangier disease has a human chromosome 9q22-q31 recessive phenotype caused by a deficiency of the α-I component of apolipoproteins. The afflicted individuals have enlarged orange tonsils, liver, spleen, and lymph nodes and deficiency of the beneficial high-density lipoproteins. They accumulate cholesterol in their cells because of a defect in the ABC transporter (cholesterol-efflux regulatory protein [CERP]) that normally pumps out excessive amounts of cholesterol into the low-density lipoprotein fraction. Therefore, these patients are prone to develop coronary heart disease. ►apolipoprotein, ►high-density lipoprotein, ►cardiovascular disease, ►ABC transporters, ►HDL, ►cholesterol, ►SNIPS; Brooks-Wilson A et al 1999 Nature Genet 22:336; McManus DC et al 2001 J Biol Chem 276:21292.

Tankyrase (TRF1-interacting ankyrin-related ADP-ribose polymerase, 8q13): A telomere associated protein binding a negative regulator of telomere length (TRF1) through the 24 ankyrin repeats of 33 amino acids involved in binding to TRF1; at its C terminal domain it exhibits homology to poly (ADP-ribose) polymerase. TRF2 is located at 10q23.2. Tankyrase1 inhibition in human cancer cells enhances telomere shortening by a telomerase inhibitor, hastens cell death, and offers a therapeutic potential (Seimiya H et al 2005 Cell Metabolism 1:25). Upon DNA damage in humans, TRF2 is transiently phosphorylated by ataxia telangiectasia mutated locus (ATM) and in the phosphorylated state it does not bind the telomere (Tanaka H et al 2005 Proc Natl Acad Sci USA 102:15539). ►telomere, ►cohesin, ►ankyrin, ►shelterin, ►TRF, ►PARP, ►ATM; Lyons RJ et al 2001 J Biol Chem 276:17172.

Tanning: The ability to produce darker skin color depends largely on the activity of the melanocyte stimulating hormone (MSH) and its receptor (MC1R). Red-haired individuals—low in these activities—do not tan easily and are susceptible to UV damage (skin cancer). ►pigmentation in animals, ►albinism, ►UV, ►melanoma, ►melanin, ►hair color melanocortin

TAP: (transporter associated with antigen processing, encoded by loci TAP1 [6p21.3] and TAP2 [6p21.3] within the HLA complex): TAP delivers major histocompatibility class I molecule-bound peptides with the cooperation of β_2 microglobulin to the endoplasmic reticulum. When these peptides exit from the endoplasmic reticulum, they are carried to the T cell receptors (TCR). Mutation in TAP may prevent antigen presentation. One member of the TAP proteins is the retroviral constitutive transport element (CTE), which may mediate RNA (hnRNP, tRNA, mRNA) export from the nucleus. Efficient export, however, requires that RNA transcript would be processed within the nucleus, i.e., introns would be removed. Some retroviruses, using CTE can transport RNA. The TAP gene contains a functional CTE element in alternatively spliced intron 10 and if this is present, mammalian mRNAs can also be exported to the polyribosomes (Li Y et al 2006 Nature [Lond] 443:234). ►HLA, ►major histo-compatibility antigen, ►microglobulin, ►immune system, ►TCR, ►proteasomes, ►endoplasmic reticulum, ►RNA export, ►hnRNP, ►DriP, ►REF, ►introns; Karttunen JT et al 2001 Proc Natl Acad Sci USA 98:7431.

TAP (tandem affinity purification): A method for rapid purification of protein complexes under native conditions. A tag is affixed to either the N or C end of one of the target proteins to facilitate the process. Proteins interacting with the tags can be purified as complexes. ►proteomics, ►affinity purification; Puig O et al 2001 Methods 24:218; Ghammaghami S et al 2003 Nature [Lond] 425:737.

Tap: A bacterial transducer protein responding to dipeptides.

TAPA-1: A membrane-associated protein that in association with CD19 protein and the complement receptor 2 (CR2) mediates early immune reaction by the B-lymphocytes. ►CD19, ►CD81, ►complement, ►B lymphocyte, ►immunity; Dijkstra S et al 2000 J Comp Neurol 428:266.

Tapasin: A transmembrane glycoprotein, encoded by an HLA-linked gene. Tapasin is involved with the endoplasmic reticulum chaperone, calreticulin, in processing the MHC class I restricted antigens, and in oncogenesis. ►MHC, ►HLA, ►major histocompatibility complex, ►antigen processing and presentation, ►TAP, ►chaperone, ►calnexin, ►calreticulin; Turnquist HR et al 2001 J Immunol 167:4443; Vertegaal ACO et al 2003 J Biol Chem 278:139.

Tapetum: The lining, nutritive tissue of the anther, sporangia, or other plant or animal organs (see Fig. T12). During meiosis, they may become bi- or even multinucleate cells. The presence of B chromosomes may increase these irregularities.

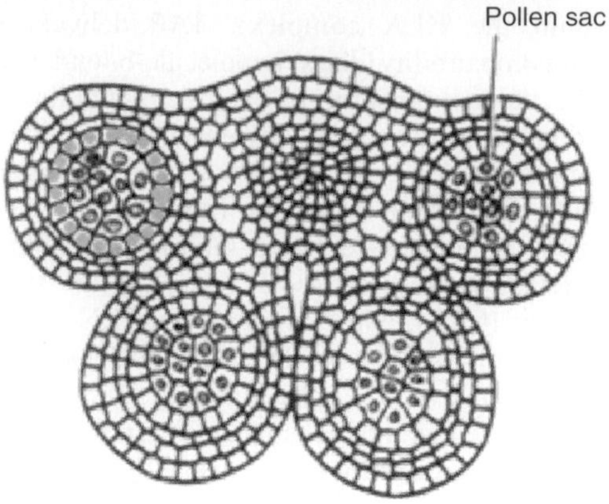

Figure T12. Anther cross section with tapetum colored green at left

Taphonomy: Taphonomy considers the processes of organisms becoming part of the fossil record. ►fossil record

Tapir: *Tapirus terrestris*, 2n = 80.

Taq DNA Polymerase: A single polypeptide chain, 94 kDa enzyme that extends DNA strands 5′→3′; it has also shows a 5′→3′ exonuclease activity. The enzyme is obtained from the bacterium *Thermus aquaticus*. The commercially available, genetically engineered enzyme AmpliTaq has temperature optima of 75 to 80°C. The enzyme is used for DNA sequencing by the Sanger method, for cloning, and for PCR procedures of DNA amplification. For the latter applications, it is particularly useful because during the heat denaturation cycles it is not inactivated and it is not necessary to add a new enzyme after each cycle. Phosphate buffers and EDTA are inhibitory to polymerization. ►DNA sequencing, ►polymerase chain reaction; Kainz P 2000 Biochim Biophys Acta 1494:23.

TaqMan (RT-PCR, real time PCR): A fluorogenic 5′-nuclease assay using a FRET probe, generally consisting of a green fluorescent dye at the 5′-end and an orange quencher dye at the 3′-end of a DNA. In a PCR process when the probe anneals to the complementary strand, the Taq polymerase cleaves the probe and the dye molecules are separated. Thus, the quencher can no longer suppress the reporter (e.g., the green dye) and a fluorescence detector can quantitate the green emission and the green fluorescence directly correlates with the yield of the PCR product. In seven minutes, sufficient quantities of DNA can be produced for the identification of pathogenic microbes. This technique may be used also for the molecular definition of deletions and SNIPs. ►Taq DNA polymerase, ►FRET, ►PCR, ►quenching, ►SNIPs; Medhurst AD et al 2001 Brain Res Mol Brain Res 90(2):125; Ranade K et al 2001 Genome Res 11:1262.

TAR (trans-activation responsive element): ►transcription factors, ►hormone response elements, ►regulation of gene activity, ►DNA binding proteins, ►acquired immunodeficiency

TAR: ►transformation-activated recombination

Tar: Bacterial chemotaxis transducer protein with aspartate and maltose being attractants and cobalt and nickel, repellents.

TAR Syndrome: ►Robert's syndrome

TARDP (TDP43, Transcription of RNA activating protein/TAR DNA binding protein, human chromosome 1p36.2): TARDP activates the long terminal repeat of the HIV-1 virus and regulates transcription of some other viruses. ►amyotrophic lateral sclerosis

Target: Anything that is the site for an action, e.g., target cell, target organ, target for DNA insertion. The target of an X-ray machine is the surface hit by the electrons, following which electromagnetic radiation is emitted in the cathode tube. ►insertion element, ►transposons, ►probe, ►x-rays

Target Immunity (transposition immunity): As per target immunity, transposable elements usually do not insert within themselves or not even within close vicinity of an existing element. ►insertional mutation

Target Site Duplications: The insertion, transposable, and other mobile genetic elements generally make staggered cuts in the DNA where they move; on completion of the process the gaps are filled by complementary nucleotides, creating duplications at the flank (see Fig. T13).

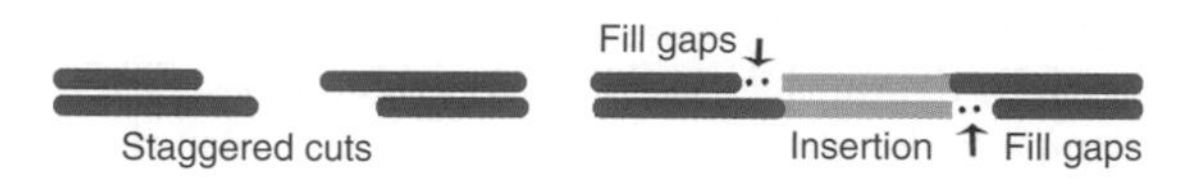

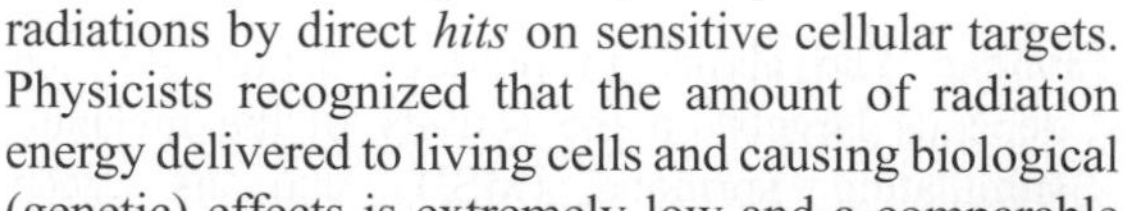

Figure T13. Target site duplication after insertion (grey) and filling gaps

Target Theory: The target theory interprets the effect of radiations by direct *hits* on sensitive cellular targets. Physicists recognized that the amount of radiation energy delivered to living cells and causing biological (genetic) effects is extremely low and a comparable dose of heat energy would have no effect at all. Therefore, there must be ceratin special sensitive targets in the cells that respond highly to ionizing radiations. Studies with irradiated sperm and cytoplasm of *Drosophila* indicated that the targets are the chromosomes and the genes. These experiments in the period between 1920s and 1940s paved the way to physical inquiries into the nature of genetic material.

At this early period, it was hoped that the different radiation sensitivities among genes would permit the estimation of the size of these genes. It turned out, however, that radiation-sensitivity of the same genes varied according to the physiological stage of the tissues (higher in imbibed seeds than in dry, dormant ones) and it was higher in spermatozoa than in spermatogonia. Furthermore, temperature, genetic background, and irradiation of only the culture media of microorganisms affected radiation-sensitivity, indicating that radiation sensitivity is a more complex phenomenon and it does not precisely reveal the molecular nature of the gene. The direct action of radiation is proportional to the molecular weight of the target molecule = $(7.28 \times 10^{11})/D_{37}$, where D_{37} is the dose required to reduce the number of undamaged molecules to 37% of the initial total at a radiation dose of Gy^{-1}. DNA ($\times$ 10^8 mol s^{-1}) interacts with radiolytic products of water: OH• 3, H•: 8 $\times$ 10^7, e^-_{aq} (hydrated electrons): 1.4 $\times$ 10^8. ►radiation effects, ►physical mutagens, ►radiation indirect effects, ►survival; Dessauer F 1954 Quantenbiologie, Springer, Heidelberg, Germany; Timofeéff-Ressovsky NW et al 1935 Nachr Ges Wiss Göttingen Math Phys Kl Biol 1:189; Lea DDA, Catcheside DG 1945 J Genet 47:41.

Targeted Gene Transfer: Targeted gene transfer is used for "knockouts" ►targeting genes, ►knockout

Targeted Gene Trap: As per targeted gene transfer, the targeted gene flanks in the vector assures precise local insertion by homologous recombination into the gene (intron). The target promoter should be avoided. The targeting cassette contains selectable marker(s). The selectable/selected cells contain inactivated copies of the target and display the selected marker. Apparently, the efficiency of the procedure is very high, ~50% (Friedel RH et al 2005 Proc Natl Acad Sci USA 102:13188). ►trapping promoters, ►translational gene fusion

Targeted Mutation Recovery: In targeted mutation recovery, mutations are induced by chemical or physical mutagens. The genetic alterations are then determined by polymerase chain reaction combined with denaturing high performance liquid chromatography to distinguish between homo- and heteroduplex DNA sequences. If homozygotes were mutagenized, the presence of heteroduplexes indicates mutation at the selected locus. ►PCR, ►HPLC; Bentley A et al 2000 Genetics 156:1169.

Targeted Nucleotide Exchange (TNE): In TNE, a oligonucleotide sequence homologous to a gene and carrying a nucleotide that is critically different from a single nucleotide within the gene is aligned by homology with the gene after transformation. At a low frequency, recombination may take place and the (mutant) nucleotide within the gene is replaced by the one delivered by the oligonucleotide sequence. ►targeting genes; Liu L et al 2002 Nucleic Acids Res 30:2742.

Targeted Recombination: ►*Cre/loxP*, ►*FLP/FRT*, ►targeting genes

Targeting: Aiming at or transporting to a site of some molecules. The homing of free cancer cells recognizes endothelial surfaces by their peptide markers and permits organ selectivity. ►lymphocytes, ►metastasis, ►transit peptide, ►transit signal, ►site-specific mutagenesis, ►mRNA targeting

Targeting Frequency: The number of insertions formed at homologous or quasi homologous site in a genome by a transforming vector.

Targeting Genes: Gene targeting can be accomplished either by insertional mutagenesis or gene replacement. *Inducible gene targeting* can be carried out by first introducing into an embryonic stem cell of a mouse by homologous recombination the gene *loxP*, to a flanking position of the desired target gene; *lox* facilitates the recognition of the sites for the *Cre* recombinase of phage P1 (see Fig. T14). Then the

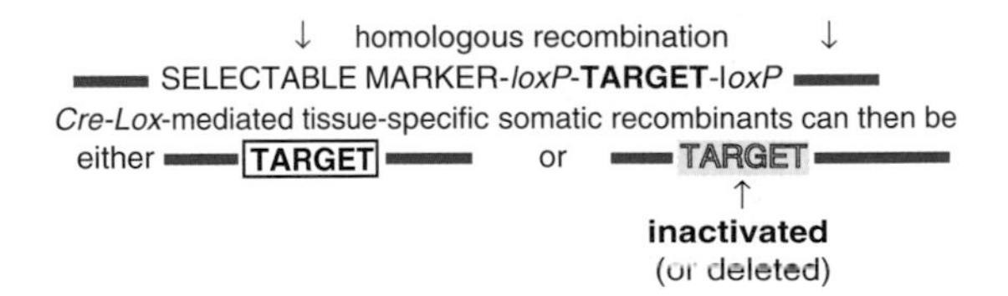

Figure T14. Targeting genes

mouse is crossed with a transgenic line expressing the *Cre* recombinase under the control of an interferon-responsive cell type-specific promoter. The tissue-specific recombinase (fused to a tissue-specific promoter), thus can remove from a particular type cell the targeted gene ("floxing"). The same procedure is applicable to other eukaryotic organisms using either the phage *Cre/loxP* or the yeast *FLP/FRT* system. Since the introduction of this site-specific alteration procedure, thousands of genes have been targeted, and in mouse alone several thousand targeted stocks have been generated. Lately gene targeting became one of the most powerful tools in genetic analysis of eukaryotes. The general principles of the procedures are illustrated. By gene targeting through double crossover within the flanking chromosomal region, different copies of the gene can be inserted (replaced) or the gene can be placed under

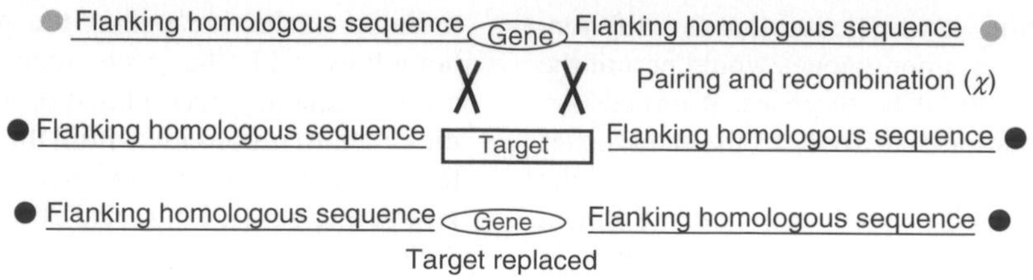

Flanking homologous sequence (Gene) Flanking homologous sequence

Pairing and recombination (χ)

Flanking homologous sequence [Target] Flanking homologous sequence

Flanking homologous sequence (Gene) Flanking homologous sequence

Target replaced

Figure T15. Target replacement by homologous recombination

the control of a specific endogenous or foreign promoter (see Fig. T15).

Targeting mammalian genes is feasible but the efficiency is fairly low (10^{-2} to 10^{-5}) compared to embryonic chicken stem cells (ES). In the transfection of avian leukosis virus (ALV)-induced chicken pre-B cells, the efficiency of recombination between the exogenous DNA and target locus may be as high as 10 to 100%. When a single mammalian chromosome is transferred to chicken cells by microcell fusion, in the somatic hybrid cell the recombination proficiency of the mammalian chromosome at the selected locus may increase up to 10–15%. The recombined chromosome can then be shuttled back to mammalian cells for analysis.

Another targeting procedure takes advantage of the bacterial tetracycline repressor gene that attaches to the promoter of some genes and keeps them silent unless tetracycline is applied that binds to the repressor; by inactivating it the genes are turned on. When this prokaryotic tetracycline repressor gene is inserted into a murine activator gene by transformation, the activator is incapacitated and the gene silenced. Alternatively, by inserting the tetracycline suppressor into a viral activator gene, all the genes of the transgenic mice that recognize the tetracycline suppressor—activator construct are turned on in the absence of tetracycline. On adding tetracycline to such a system, the antibiotic combines with the repressor—activator in the hybrid construct and the genes are now shut off because the suppressor—activator construct is removed from the activation position. Such a targeting construct can thus be used for on/off switching of particular genes. The success of targeting may be increased if the vector is an RNA-DNA hybrid molecule that pairs more efficiently with the target. By the PCR targeting procedure, a 20 bp DNA sequence tag may be generated using the photolithography procedure and DNA chips. The tag sequences are as different as possible, yet possess hybridization properties to be identified simultaneously on high-density oligonucleotide arrays. Genomic DNA is isolated from a pool of deletions tagged and used as templates for amplification. For selectability, a resistance gene (aminoglycoside phosphotransferase) may be used. The targeting sequence is amplified by PCR that have primers at the 3′-end homologous to the marker and at the 5′-end homologous to the target. This system is introduced into the cells by transformation. After homologous recombination at two flanks of the targeted open reading frame, the target is replaced by a construct including the 20-base tag, the selectable marker, and the deletion mutation sequence (see Fig. T16).

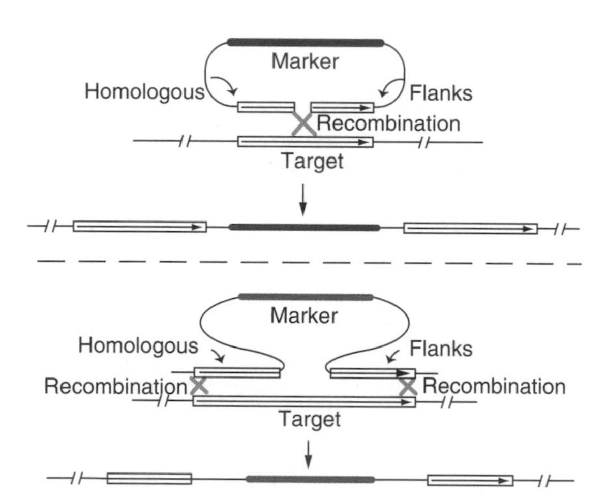

Figure T16. In *Drosophila* gene targeting can be achieved by "ends-in" (upper part of the diagram) and by "ends-out" (lower part of the diagram) procedures. The lower diagram indicates that parts of the 3′ and 5′ sequences of the target are deleted by the recombination. The requisites are (i) expressing a site-specific recombinase transgene, (ii) a transgene expressing a site-specific endonuclease, and (iii) a transgenic donor construct carrying recognition sites for both enzymes and the DNA of the locus targeted. This gene targeting mutates genes, which are not known by function but only by sequence. (Modified after Rong YS, Golic KG 2000 Science 288:2013)

The large number of tagged deletion strains can then be pooled and tested under a variety of conditions to test how deletion affects the function of the gene. The molecular tags are amplified and hybridized to a high-density array of known oligonucleotides, complementary to the tags. The relative intensity of hybridization reveals the relative proportion of the individual deletion strains in the pool and their fitness. Phenotypic methodological and other relevant information is contained in the TBASE http://www.jax.org/. Chimeric, site-specific nucleases when used with appropriate nuclear localization signals, greatly enhance targeting by the generation of double-strand breaks in the human somatic cells (Porteus MH, Baltimore D 2003 Science 300:763; Bibikov M et al 2003 Science 300:764). A precisely placed double-strand break induced by engineered zinc finger nucleases (ZFNs) can stimulate integration of long DNA stretches into a predetermined genomic location, resulting in high-efficiency site-specific gene addition. Using an extrachromosomal DNA donor carrying a 12 bp tag, a 900 bp ORF, or a 1.5 kb promoter-transcription unit flanked by locus-specific homology arms, we find targeted integration frequencies of 15%, 6%, and 5%, respectively, within 72 h of treatment, and with no selection for the desired event. Importantly, we find that the integration event occurs in a homology-directed manner and leads to the accurate reconstruction of the donor-specified genotype at the endogenous chromosomal locus, and hence presumably results from synthesis-dependent strand annealing repair of the break using the donor DNA as a template. This site-specific gene addition occurs with no measurable increase in the rate of random integration. Remarkably, we also find that ZFNs can drive the addition of an 8 kb sequence carrying three distinct promoter-transcription units into an endogenous locus at a frequency of 6%, also in the absence of any selection (Moehle EA et al 2007 Proc Natl Acad Sci USA 104:3055). Gene targeting should be extended to any animal in which embryonic stem cells can be successfully managed. Unfortunately, this has not been realized with the majority of mammals, with the exception of mice and sheep. An alternative approach by nuclear transplantation appears more practical (Kubota C et al 2000 Proc Natl Acad Sci USA 97:990). Initially, the success of gene targeting in plants was generally in the 10^{-4}–10^{-3} range but this has been greatly improved recently. Locus-specific deletions (52 to 1 bp), which were meiotically transmissible, could be obtained in *Arabidopsis* plants by generating double-strand breaks with the aid of zinc-finger nucleases (Lloyd A et al 2005 Proc Natl Acad Sci USA 102:2232). When the promoterless GFP (green fluorescent protein) gene was inserted into the *CRUCIFERIN* seed protein gene of *Arabidopsis* and it acquired the functional promoter, the expression of the green fluorescence could be easily detected in the large seed-output of the plants. In plants transgenic for the yeast chromatin-remodeling and recombination-promoting protein gene, RAD54, the success of targeting increased from 10^{-2} to 10^{-1} and could be readily detected by capturing the promoter in the Cruciferin-GFP fusion protein (Shaked H et al 2005 Proc Natl Acad Sci USA 102:12265).

Actually, the INO80 protein encoded in *Arabidopsis* also displays an effect similar to Rad54 (see Fig. T17) (Fritsch O et al 2004 Mol Cell 16:479). ►insertional mutation, ►GAMBIT, ►local mutagenesis, ►gene replacement, ►RMCE, ►knock-out, ►homologous recombination, ►site-specific recombination, ►ends-in ends-out recombination, ►adeno-associated virus, ►gene therapy, ►chromosomal rearrangement, ►homing endonucleases, ►chromosome uptake, ►IRES, ►photolithography, ►DNA chips, ►RNA double-stranded, ►RNAi, ►*Cre/loxP*, ►*Flp/FRT*, ►GMO, ►nuclear transplantation, ►conditional targeting, ►RID, ►TFO, ►targeted nucleotide exchange, ►MICER, ►gene targeting, ►trapping promoters; Thomas KR et al 1986 Cell 44:419; Sauer B 1998 Methods 14:381; Vasquez KM et al 2001 Proc Natl Acad Sci USA 98:8403; Rong YS et al 2002 Genes Dev 16:1568; a historical account of targeting complex disease genes: Smithies O 2005 Nature Rev Genet 6:419.

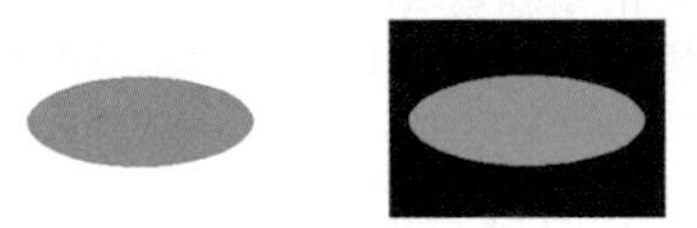

Figure T17. GFP-labeled cruciferin in *Arabidopsis* seed

Targeting Physiological: Physiological targeting locates regulatory factors to the appropriate intracellular site. Protein domains may control these functions and may be hampered by chromosomal translocation and gene fusion.

Targeting Proteins: ►two-hybrid system, ►one-hybrid binding assay, ►three-hybrid system, ►microcalorimetry, ►surface plasmon resonance, ►gel retardation, ►gel filtration, ►immunoprobe, ►immunolabeling, ►prenylation, ►myristic acid; Fischer W et al 2001 Infect Immun 69:6769.

Targeting Signal: ►signal peptide, ►signal sequence, ►signal sequence recognition particle

Targeting, Transcriptional: Transcriptional targeting is generally directed to the enhancer/promoter area of a specific gene. The goal is to avoid non-specific genes that may have deleterious consequences. (See Sadeghi H, Hitt MM 2005 Curr Gene Ther 5:411; Izumo T et al 2007 Int J Oncol 31:379).

T

Targeting Vector: In a viral vector, a section of the envelope protein gene is replaced by the coding sequences of, e.g., 150-amino acids of erythropoietin (EPO) and thus, the replacement improves its ability to recognize the EPO receptor. Other approaches involve pseudotyping or attaching special ligands to the envelope (Müller OJ et al 2003 Nature Biotechnol 21:1040). Liposomal vehicles may be conjugated with special antibodies for target recognition. Bifunctional antibodies that recognize both viral epitopes and target cell antigens have been constructed. Another approach was found to covalently link biotin to recombinant adenovirus. The Kit receptor and the stem cell factor (SCF) were then linked through an avidin bridge to the target to assure the proper tropism. It is highly desirable that ligand–receptor pairs be limited to specific functions rather than cross over to members of regulatory networks. Such a goal can be reached by stepwise, individual, site-specific saturation mutagenesis followed by phenotypic screening based on the yeast two-hybrid system. The nuclear hormone estrogen receptor so modified can favor a synthetic 4,4′-dihydroxybenzil by more than a million-fold over the natural ligand 17 β-estradiol. Such a technique can specifically target human endometrial cancer (Chockalingam K et al 2005 Proc Natl Acad Sci USA 102:5691). ►liposome, ►vectors, ►pseudotyping, ►magic bullet, ►gene therapy, ►KIT oncogene, ►stem cell factor, ►biotin, ►avidin, ►epitope, ►dihydoxybenzil, ►nuclear receptor, ►estradiol, ►estrogen receptor, ►targeting transcriptional; Peng KW et al 2001 Gene Ther 8:1456; Yu D et al 2001 Cancer Gene Ther 8:628; therapeutic applications: Waehler R et al 2007 Nature Rev Genet 8:573.

Target-Primed Reverse Transcription: Retrotransposons without long terminal repeat sequences are inserted into eukaryotic genomes by a process in which cleaved DNA targets are used to prime reverse transcription of the RNA transcript of the element. ►LINE, ►reverse transcription

TART: A telomere-specific retroposon of *Drosophila* with an about 5.1-kb 3′-non-coding tract, which is homologous to HeT-A; it also encodes a reverse transcriptase. ►telomere, ►Het-A, ►LINE, ►retroposon, ►reverse transcription; Haoudi A, Mason JM 2000 Genome 43:949.

Tarui Disease: ►glycogen storage diseases (Type VII)

TAS (termination associated sequences): Signals for ending transcription. Also, telomere-associated sequences involved in the silencing of genetic functions. ►transcription, ►telomere, ►silencing

tasiRNA: Transacting siRNA produced by Dicer4. ►Dicer, ►RNAi, ►siRNA

Tasmanian Devil (*Sarcophilus harrisii*, 2n = 12 + XY): Australian, large, black, white-spotted, carnivorous marsupial frequently with large tumors on its face (⇨) that generally become fatal because the animals cannot feed (see Fig. T18). Tissue fragments acquired during fighting with infected individuals probably transmit the tumors. In the tumor cells in each of the 11 animals examined, the sex chromosomes, a pair of chromosome 2, one chromosome 6, and one arm of chromosome 1 were missing, but four chromosomes of unidentified origin were added so the tumor cells had 13, rather than the normal 14 chromosomes (Pearese A-M, Swift K 2006 Nature [Lond] 439:549; Owen D, Pemberton D 2006 Tasmanian Devil: A Unique and Threatened Animal. Natural History Museum, Allen & Unwin. London, UK; McCallum H, Jones M 2006 PLoS Biol 4(10): e342). ►canine transmissible venereal tumor

Figure T18. Tasmanian devil

Tassel-seed: Mutations (*ts*) in maize result in kernels on the normally male inflorescence (tassel) as a result of the effemination. ►sex determination

Taste: Taste is controlled by a signal transducing G protein, gustducin. Both bitter sweet and unami-tasting is mediated by gustducin. Ionic stimuli of salts and acids interact directly with ion channels and depolarize taste-receptors. Sugars, amino acids, and most bitter stuff bind to specific receptors outside the cell membrane and these are then connected to G proteins. It is assumed that gustducin is involved with a phosphodiesterase. Phospholipase C also appears to play a role in the taste circuits. Gustducin receptors are present in the tongue and also in the stomach and the intestines. Taste buds are only a few thousand in number, with 3 to 5 × 10^4 taste receptors. The TRPM cation channel family greatly enhances sweet perception at temperatures between 15 °C and 35 °C (Talavera K et al 2005 Nature [Lond] 438:1022). A locus in human chromosome 5p15 was associated with sensing the bitter taste of 6-n-propyl-2-thiouracil. The phenythiocarbamide taste locus (PTC/PROP) is a receptor, TAS2R, located in human chromosome 7q. There is great variation among the TASR genes. Some individuals, however, fail to identify its bitter taste. Seven transmembrane domain taste receptor sequences were attributed to clusters also in human chromosomes 7q31-q32 and 12p13. The T1R-1/TAS1R taste receptor gene (1p36) family is apparently specific for sweetness, whereas T1R2 for bitter taste. The latter may include 50–80 genes. Mammals can taste sour, salty, sweet, bitter, and unami (monosodium glutamate). Genes for different taste receptors have been isolated. A distinct and separate sour taste receptor is PKD2L1 (polycystic kidney disease-like ion channel; Huang AL et al 2006 Nature [Lond] 442:934). ►ion channels, ►degenerin, ►signal transduction, ►olfactogenetics, ►fragrances, ►phenylthiocarbamide; Dulac C 2000 Cell 100:607; Nelson G et al 2001 Cell 106:381; Lindemann B 2001 Nature [Lond] 413:219; Margolskee RF 2002 J Biol Chem 277:1; Bufe B et al 2002 Nature Genet 32:397; brain circuits of bitter and sweet: Sugita M, Shiba Y 2005 Science 309:781; Drayna D 2005 Annu Rev Genomics Hum Genet 6:217; mammalian taste receptors: Chandrashekar J et al 2006 Nature [Lond] 444:288.

TAT (twin-arginine translocase): A ~600 kDa protein complex that moves proteins through thylakoid and prokaryotic membranes. ►thylakoid; Robinson C, Bolhuis A 2001 Nature Rev Mol Cell Biol 2:350.

Tat: 14 kDa primary regulator of the HIV virus. ►acquired immunodeficiency

Tat1: A transposon-like element (431 bp) in *Arabidopsis*, flanked by 13 bp inverted repeats and 5 bp target-site duplications, but without any open reading frame and thus incapable of movement by its own power. There is also a Tat1 human sulfate transporter and TAT 1 and TAT2 yeast amino acid permeases. ►*Arabidopsis*, ►transposons, ►open reading frame; Peleman J et al 1991 Proc Natl Acad Sci USA 88:3618; Toure A et al 2001 J Biol Chem 276:20309; Schmidt A et al 1994 Mol Cell Biol 14:697.

TAT-GARAT: The TAATGARAT enhancer motif of Herpes simplex virus. ►cigar

TATA Box: Thymine (T) and adenine (A) containing binding sites for transcription factors and the RNA polymerase complex. The bases, their numbers, and the exact base sequences in the TATA boxes vary. In yeast, the TATA box is 40–120 bp, in the majority of other eukaryotes 25–30 bp ahead of the transcription initiation site (Struhl K 1989 Annu Rev Biochem 58:1051). Many housekeeping genes and the RAS oncogene do not have this sequence (Suzuki Y et al 2001 Genome Res 11:677). In case of the absence of TATA, some other A and T nucleotides may associate with the TATA-box-binding proteins (TBP). A downstream promoter element (DPE) has also been identified. DPE may bind TFIID. In *Drosophila*, some core promoters have both TATA box and DPE or only one of the two (Kutach AK, Kadonaga JT 2000 Mol Cell Biol 20:4754). ►Pribnow box, ►Hogness (Goldberg) box, ►transcription factors, ►transcription complex, ►open promoter complex, ►asparagine synthetase, ►core promoter, ►promoter, ►TBP, ►PWM, ►DPE; Smale ST, Kadonaga JT 2003 Annu Rev Biochem 72:449.

TATA Box Binding Protein: ►TBP

TATA Factor (TF): ►transcription factors, ►TATA box

TATA Inr: Core promoters may or may not contain these pyrimidine-rich transcription initiator elements. ►TATA box, ►open promoter complex, ►transcription factors, ►core promoter

Tatsumi Factor: A blood-clotting factor required for the activation of Christmas factor by activated PTA. It is controlled by an autosomal locus. ►antihemophilic factors, ►blood clotting pathways, ►PTA deficiency disease, ►hemostasis

TAU (MAPT, 17q21.1): The size of the tau gene can vary from ~352 to 441 amino acids in the isoforms of the microtubule-associated proteins by alternative splicing of the mRNA (Margittai M, Langen R 2004 Proc Natl Acad Sci USA 101:10278).

By reducing motor reattachment rates, tau affects cargo travel distance, motive force, and cargo dispersal. Different isoforms of tau, at concentrations similar to those in cells, have dramatically different potency.

These defined mechanism show how altered tau isoform levels could impair transport and thereby lead to neurodegeneration without the need of any other pathway (Vershinin M et al 2007 Proc Natl Acad Sci USA 104:87). It seems to form tangles by virtue of the ^{306}Val-Gln-Ile-Val-Tyr-Lys^{311} motif in the 6 tau monomers in several types of nerve degenerative diseases (e.g., Pick disease, Alzheimer disease, progressive supranuclear palsy, corticobasal degeneration).

A pseudogene exists at 6q21. Base substitution and splice site mutations may lead to Pick disease, parkinsonism and Alzheimer's disease. In the tangle of the paired helical filaments, tau is hyperphosphorylated causing defects in microtubule assembly and mitotic arrest. The use of a kinase inhibitor (at lysine 252) can prevent hyperphosphorylation of tau and the aggregation of tau without reducing the tangles; such a treatment reduces severe motor function impairment in transgenic mice indicating that aggregation rather than tangling is the cause of the development of tau pathology (Le Corre S et al 2006 Proc Natl Acad Sci USA 103:9673).

The adverse effect of hyperphosphorylation may be prevented by trimethylamine N-oxide (TMAO) because this natural compound lowers the concentration of tubulin needed for assembly. The hyperphosphorylated tau may self-assemble and the causes the fibrillary tangle observed in the degenerated brain of Alzheimer's patients. Cyclic-AMP-dependent protein kinase (PKA), glycogen synthase kinase 3β (GSK-3β), or Cdk5 may carry out the phosphorylation.

Phosphorylation of a Ser or a Thr amino acid preceding a Pro creates a binding site for the prolyl-isomerase Pin1. When Pin1 binds to this site in tau, it may deplete it in the brain leading to some of the problems related with Alzheimer's disease. Amyloid-β immunotherapy can clear the fibrillar tangle with the aid of proteasomes, if applied before the hyperphosphorylation of tau (Oddo S et al 2004 Neuron 43:321). In human Parkinsonism, mutation can affect aberrant splicing of exon 10. Using a spliceosome-mediated trans-splicing, the mRNA can be reprogrammed. Thus, creating a new exon 9–exon 10 junction indicates the feasibility of therapeutic trans-splicing (Rogriguez-Martin T et al 2005 Proc Natl Acad Sci USA 102:15659). Neurofibrillary degeneration is increased when both Aβ and tau are expressed the same time. In a mouse model of Alzheimer's disease, the chronic supply of nicotine exacerbates tau tangles (Oddo S et al 2005 Proc Natl Acad Sci USA 102:3046). In mouse transgenic for a suppressible tau, memory was recovered and neuron numbers were stabilized after suppression of tau, yet neurofibrillary tangles continued to grow, indicating that the tangles are not sufficient to account for the degenerative phenomena (see Fig. T19) (SantaCruz K et al 2005 Science 309:476). Fragments of the repeat domain of tau, containing mutations of FTDP17 (frontotemporal dementia with Parkinsonism linked to chromosome 17 also called Pick disease), produced by endogenous proteases, can induce the aggregation of full-length tau. Fragments are generated by successive cleavages, first N-terminally between lysine257 and serine258, then C-terminally around residues 353–364; conversely, when the N-terminal cleavage is inhibited, no fragmentation and aggregation takes place. The C-terminal truncation and the

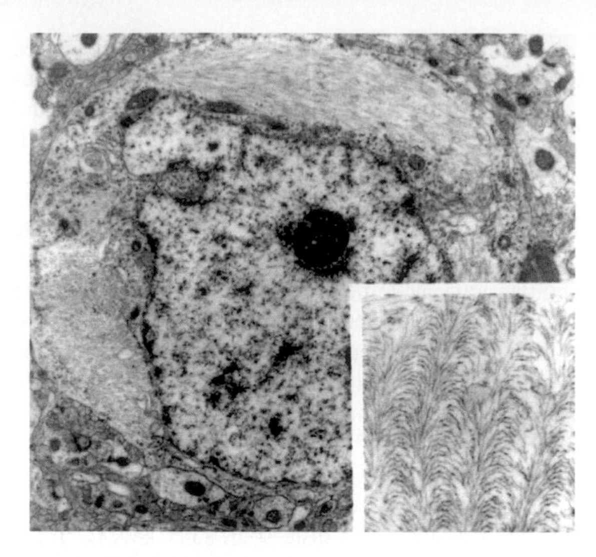

Figure T19. Transgenic mouse brain expressing neurofibrillary tangle (enlarged in insert) and Aβ plaques similar that occurs in humans afflicted by Alzheimer disease. (Courtesy of Drs. Dennis W. Dickson and Wen-lang Lin, Mayo Clinic, Jacksonville, Florida; I am indebted also to Mike Hampton, Maryland)

coaggregation of fragments with full-length tau depend on the propensity for β-structure. The aggregation is modulated by phosphorylation but does not depend on it. Aggregation but not fragmentation is toxic to cells; conversely, inhibiting either aggregation or proteolysis can prevent toxicity (Wang YP et al 2007 Proc Natl Acad Sci USA 104:10252). In a *Drosophila* model of tauopathy, neurodegenerative symptoms appeared without the fibrillary tangle. ►FTDP-17, ►prion, ►amyloids, ►secretase, ►sirtuin, ►Pick disease, ►Alzheimer disease, ►p35, ►parkinsonism, ►microtubule, ►CDK, ►palsy, ►corticobasal degeneration, ►synuclein, ►RNAi, ►trans-splicing; von Bergen M et al 2000 Proc Natl Acad Sci USA 97:5129; Wittmann CW et al 2001 Science 293:711; Lewis J et al 2001 Science 293:1487; Liou Y-C et al 2003 Nature [Lond] 424:556.

Taurine (H_2N-CH_2-CH_2-SO_3H): An amino acid absent from muscle proteins but present in several body fluids (it was first isolated from the bile of cattle [*Bos taurus*]; hence the name). It is important for membrane transport and as an antioxidant and is thus an essential nutrient. In its absence from the tRNA, Leu^{UUR} (5-taurinomethyluridine) may cause mitochondrially-determined disease. ►mitochondrial disease

Tautomeric Shift: The reversible change in the position of a proton in a molecule, affecting its chemical properties; it may trigger base substitution and thus mutation in DNA. ►hydrogen pairing, ►enol form, ►substitution mutation; Watson JD, Crick FHC 1953 Cold Spring Harbor Symp Quant Biol 19:123.

TAX1: A human T-cell leukemia virus (40 kDa) protein gene (1q32.1) with three 21 bp CRE-like sites; it increases DNA binding of transcription factors containing a basic leucine zipper domain. ►leucine zipper, ►leukemia, ►CRE, ►CREB, ►HTLV; Soda Y et al 2000 Leukemia 14:1467.

Taxol (Paclitaxel): A spindle fiber blocking natural substance isolated from the yew *Taxus brevifolia*; it is a carcinostatic and radio-sensitizing drug. It induces apoptosis. ►spindle, ►carcinostasis, ►epothilone, ►microtubule, ►apoptosis, ►Herceptin, ►Abraxane; Okano J-i, Rustgti AK 2001 J Biol Chem 276:19555; Jennewein S et al 2001 Proc Natl Acad Sci USA 98:13595; Ganesh T et al 2004 Proc Natl Acad Sci USA 101:10006; 19 proteins involved in taxol synthesis: Jennewein S et al 2004 Proc Natl Acad Sci USA 101:914.

Taxon (plural taxa): The collective name of taxonomic categories.

Taxonomy: (biological) classification with a number of different systems. The bases of this classification are morphology, anatomy, genetics, biochemistry, physiology, cytology and macromolecular structure (DNA, RNA and proteins). Generally, five broad categories are recognized: prokaryotes and viruses, protists, fungi, plants, and animals. The taxonomic categories of eukaryotes include Phylum, Class, Order, Family, Genus, Species, and Subspecies (such as varieties, cultivars, breeds). In the past, the classification was more rigid because the species was considered the mark of genetic isolation. Today, with somatic cell hybridization and transformation (transfection) there is no limit to genetic exchange between various categories. There are over 300,000 plant and over 1,000,000 animal species named and classified by rules of nomenclature. For naming, binomial nomenclature is used. The capitalized first letter of the first name identifies the genus and the species is designated by the second name in lower case letters. This is sometimes followed by the name of the first taxonomists who classified the organisms, e.g., *Arabidopsis thaliana* (*L.*) *Heynh.*, indicating *Arabidopsis* as the genus, *thaliana* as the species, L. stands for Linnaeus and Heynh. is the abbreviation for Heynhold who suggested the current name. Plant taxonomy: http://www.itis.gov/; virus taxonomy: http://www.ncbi.nlm.nih.gov/ICTVdb; NCBI taxonomy database: http://0-www.ncbi.nlm.nih.gov.library.vu.edu.au/Taxonomy/. For sequence-based taxonomy: http://www.ncbi.nlm.nih.gov/Taxonomy/taxonomyhome.html; sequence database: http://www.ebi.ac.uk/seqdb/Projects.html.

Tay-Sachs Disease: One of the most thoroughly studied biochemical diseases in human populations; it is controlled by an autosomal recessive gene (see Fig. T20). This defect occurs in all ethnic groups, but it is particularly common among Ashkenazi (eastern European) Jews where the

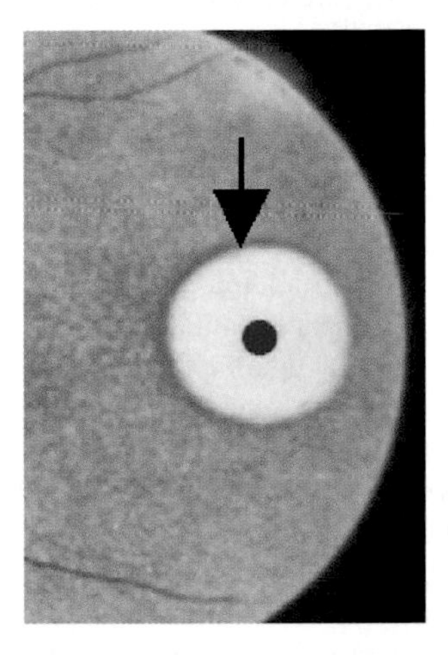

Figure T20. Cherry-red spot on the macula in Tay-Sachs disease

frequency of the gene is approximately 0.02 and the frequency of heterozygotes may be over 3%. The prevalence is about 1/2,500 to 1/5,000 per birth. Among the Sephardim Jews and other ethnic groups, the frequency is about 1/1,000,000. Since the afflicted individuals generally die by age three to four, the high frequency indicates that some heterozygote advantage must have existed for this gene. The effective genetic screening for heterozygotes in the USA has greatly reduced the prevalence of this disease (Kaback MM 2001 Adv Genet 44:253). The onset is at six months when general weakness, extension of the arms in response to sounds, a scared look, and muscular stiffness and retardation appear in an earlier apparently normal child. Then in rapid succession, paralysis, reduction of mental abilities, and vision problems leading to blindness become evident. One characteristic symptom is a cherry-red spot (↓) on the macula (gray opaque part of the cornea [eye]), caused by cell lesions. All these symptoms are the results of a deficiency of β-hexosaminidase enzyme α subunit that controls the conversion of ganglioside G_{M2} into G_{M3}. As a consequence, G_{M2} ganglioside accumulates leading to degeneration of the myelin of the nervous system. Hexosaminidase A is composed of the α subunits (human chromosome 15q23-q24) and hexosaminidase B is a multimer of the β subunits (human chromosome 5q13). Sandhoff's disease involves both hexosaminidase A and B deficiencies or only hexosaminidase B deficiency, and has somewhat similar symptoms with more rapid progression. Another (Type 3), milder form of G_{M2} gangliodisosis (5q31.3-q33.1) with some hexosaminidase A activity may permit survival up to age ~15. In a mouse model, *N*-butyldeoxynojirimycin prevents the accumulation of G_{M2}. In Sandhoff's disease model mouse, stereotactic (precisely positioned) brain injection by adeno-associated virus vector containing the α and β subunits of the human β-hexosaminidase gene, including also the tat sequence of the HIV virus (for enhancing protein expression and distribution), resulted in the transgene being expressed and consequently GM_2 ganglioside storage and inflammation being reduced. The survival of the animals expanded to more then twice as long, to over a year, and motor function was restored, indicating that this type of gene therapy may be eventually applicable for this incurable human disease (Cachón-Gonzalez MB et al 2006 Proc Natl Acad Sci USA 103:10373).

The older common name of these gangliosidoses was amaurotic familial idiocy. ►Sandhoff's disease, ►hexosaminidase, ►gangliosidoses, ►gangliosides, ►sphingolipids, ►lysosomal storage disease, ►genetic screening, ►Jews and genetic disease, ►gene therapy; Mahuran DJ 1999 Biochim Biophys Acta 1455:105; Myerowitz R et al 2002 Hum Mol Genet 11:1343.

Tay Syndrome: ►trichothiodystrophy

T-Bam: Same as CD40 ligand.

T-Band: The telomeric regions of chromosomes with the highest concentrations of genes and G + C in the genome. ►band, ►chromosome banding, ►isochores

T-bet (T-box expressed in T cells): A 530-amino acid transcription factor that promotes the differentiation and activity of T_H1 cells and represses the formation of T_H2 lymphocytes. The repression of T_H2 is brought about by tyrosine kinase-mediated interference of GATA-3 binding to its target DNA (Hwang ES et al 2005 Science 307:430). T-bet deficiency reduces atherosclerosis in mice (Buono C et al 2005 Proc Natl Acad Sci USA 102:1596). T-bet is aided by CpG DNA oligonucleotides in innate immunity (Lugo-Villariono G et al 2005 Proc Natl Acad Sci USA 102:13248). ►T_H, ►T box, ►GATA, ►eomesodermin, ►innate immunity, ►memory immunological; Mullen AC et al 2001 Science 292:1907; Lovett-Racke AE et al 2004 Immunity 21:719.

TBL1: transducin-beta-like protein. ►transducin; Tomita A et al 2004 Mol Cell Biol 24:3337.

TBP (TFIIτ): TATA box binding protein is a subunit of the general transcription factors, TFIID, SL1, and TFIIIB proteins that bind to DNA like a saddle in a two-fold symmetry (see Fig. T21) (Burley SK, Roeder RG 1996 Annu Rev Biochem 65:769). The TFIIIB complex shares amino-terminal homology with TFIIB. TFIIIB has anti-parallel β sheets at the concave area where it forms a reaction with the special A and T rich sequence of the DNA. TFIIB-related factor (Brefl) at the carboxy-terminal mediates the binding of TBP to the DNA. The convex surface provides opportunities for interaction with other proteins. The TBP is highly conserved among different organisms from yeast to the plant *Arabidopsis* to mammals. TBP may form a complex with several TAF proteins. It apparently nucleates the preinitiation complex of all three DNA-dependent RNA polymerases—pol I, II, and III—but in a somewhat different manner (Fan X et al 2005 Nucleic Acids Res 33:838). The TBF-like factors (TLFs) variously denoted as

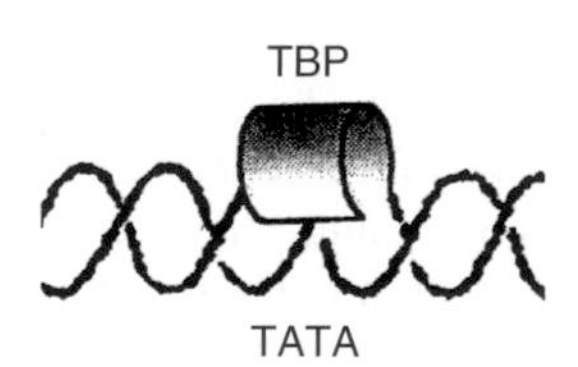

Figure T21. TBP—TATA box schematic representation. TBP of *Arabidopsis* recognizes the minor grove of TATAAAAG

TRF2 or TRP are orthologs of TFL and appear functionally different from TBP. TLF appears to be required for differentiation and is needed for the transcription of some special sets of genes. The TBP may also bind SL1 transcription factor of pol I but this binding is exclusively either with TFIID or SL1. The pol III TBP complex is called TFIIIB. The Dr1 repressor binds to TBP and selectively inhibits pol II and pol III action but pol I is not affected, however, because when pol II and III are repressed the relative output of pol I appears higher. The TFIID complex activity is restricted by the nucleosomal organization of the chromatin. The C-terminus of TBP is highly conserved, whereas the N region displays great variations. In $TBP^{-/-}$ mouse cells, DNA-dependent RNA polymerase II seems functional but pol I and pol III are arrested. ►transcription factors, ►pol, ►TAF; Berk AJ 2000 Cell 103:5; Magill CP et al 2001 J Biol Chem 276:46693; Zhao X, Herr W 2002 Cell 108:615; Martianov I et al 2002 Science 298:1036.

TB-parse: A program for the identification of protein-coding sequences. (See Nucleic Acids Res 22:4768).

TBR: Transforming growth factor receptors. ►TGF

Tc: A family of transposons in *Caenorhabditis*. *Tc*1 is silenced in the germline by a natural RNAi (Sijen T, Plasterk RHA 2003 Nature [Lond] 426:310). ►transposon, ►*Caenorhabditis elegans*

TC4: A small nuclear G protein with CD28 being its primary ligand. It is involved in the regulation of T lymphocytes. ►G proteins, ►RAN; Nieland JD 1998 Cancer Gene Ther 5:259.

TC Motif: ►AP1, ►AP2, ►transcription factors

TCC (terminal complement complex): TCC includes C5b-9, C5b-8, and C5b-7 complement components involved in the complement-mediated killing of foreign cells. ►complement, ►immune system

TCC (transitional cell carcinoma): See Dal Cin P et al 1999 Cancer Genet Cytogenet 114(2):117.

TCCR (T cell cytokine receptor): TCCR mediates adaptive immune response of T_H1 lymphocytes. (See Chen Q et al 2000 Nature [Lond] 407:916).

TCF (ternary complex factors): TCF are co-activators of transcription such as Elk, Sap-1a, Sap-1b, ERP-1, and other members of the ETS family of transcription factors and oncoproteins. Some are members of the high mobility group proteins. TCF usually binds to the AACAAAG sequences of the promoter. In case of defect in the tumor suppressor APC gene (adenomatous polyposis) and/or in the β-catenin gene, Tcf-4 (responsible for crypt stem cells) is activated and malignancy may result. TCF4 when interacting with phosphorylated Jun regulates intestinal cancer (Nateri AS et al 2005 Nature [Lond] 437281). CBP seems to be a repressor of TCF. ►high mobility group proteins, ►ETS, ►CBP, ►Gardner syndrome, ►melanoma, ►LEF, ►catenins, ►mtTAF, ►CBP, ►*Wingless*; Morin S et al 2001 Mol Cell Biol 21:1036.

TCGF: ►interleukin 2, ►IL-2

$TCID_{50}$: Tissue culture infective dose causing (viral) infection to 50% of the cells.

Tcl: ►hybrid dysgenesis

TCL1: An oncogene (14q32.1). It is normally expressed in fetal thymocytes, in pre-B, and immature B cells and weakly in $CD19^+$ peripheral blood lymphocytes. Chromosomal translocations or inversions near the enhancer element of TCR may cause leukemia or lymphoma. It enhances the Akt kinase activity and promotes nuclear transport. ►B lymphocyte, ►CD19, ►TCR, ►leukemia, ►lymphoma, ►Akt; Pekarsky Y et al 2001 Oncogene 20:5638.

TCLo: Toxic concentration low; the lowest concentration of a substance in air that produces toxic, neoplastic, and carcinogenic effects in mammals.

t-Complex: The t-complex of mouse consists of six complementation groups in chromosome 17, affecting tail development and viability. Homozygous mutants, of the same complementation group, are generally lethal. ►brachyury

TCP-1 (CCT): The cytoplasmic α subunit chaperonin coded for by the t-complex in mouse chromo-some 17. Its homolog occurs also in the pea leaf cytosol. A TCP protein has been suggested to mediate the curvature of the leaf surface of crinkly mutations (Nath U et al 2003 Science 299:1404). ►chaperonins, ►chaperone, ►brachyury

Tcp20: Synonymous with CCTζ and Cct6. ►chaperonins; Li WZ et al 1994 J Biol Chem 269:18616.

TCR Genes: The T cell receptors are glycoproteins and are similar to the antibody molecules. The TCR α chain has variable (V), diversity (D), junction (J), and constant (C) regions in the polypeptides, which are generated with rearrangements of the gene clusters in human chromosome 14, in the proximity of the immunoglobulin heavy chain genes (IgH). These recombinations take place in the switching regions and the TCR genes also have the same hepta- and nanomeric sequences as the Ig genes. In the mouse, TCRA (α) genes are in mouse chromosome 14 whereas the Ig heavy chains of the mouse are in chromosome 12. The TCR δ chain locus is situated within the human α gene between the V_α and J_α regions. When the δ gene is excised, an excision circle is generated which includes also the excision signal joints (TREC). The human β chain of TCR is encoded in chromosome 7q22–7qter. (In mouse, its homolog is in chromosome 6). The expression of the β genes also requires rearrangement of the VDJ genes next to the C genes. The γ1 and γ2 chain genes are located in the human chromosome 7p15-p14 area, whereas their homologs are in mouse chromosome 6. The early lymphocytes carry TCR built of γ and δ polypeptides, whereas in the later, about 95% of the TCRs are built of α and β chains. The size of the αβ TCR is about 80-kDA, built of four subunits. After the αβ TCRs are formed, the γδ TCRs are eliminated. The αβ TCR is part of the protein complex CD3 γ, δ, ε, and ζ. These chains also contain 1 ITAM motif except ζ which has 3. Phosphorylation of the ITAM motif facilitates signal transduction from TCR. Subunit ζ determines largely the specificity by specific series of phosphorylations. In contrast to the αβ antigen receptors, which recognize only peptides (fragments) bound to MHC molecules, the γδ TCRs recognize the polypeptides without the

MHC molecules and also the MHC molecules without bound peptides. Also the γδ TCRs may use non-peptide ligands such as phosphate-containing molecules. The TCR molecules are of a great variety, and are determined by rearrangements, just like in immunoglobulins. It appears that the TCR does not mutate somatically, in contrast to the Ig genes. The association of TCR either with Class I and Class II HLA gene products (MHC) augments the specificity of the TCR. The Class I HLA antigens are associated with the cytotoxic (killer) cells, whereas the helper lymphocytes attach to the Class II antigens. In the function of the TCRs, an important role is played by the CD4 (in Class I) and CD8 (in Class II associations). Both types of TCRs also require the CD3 transmembrane proteins complex involved in signal transduction. The CD peptides also activate some protein tyrosine kinases, important elements of several signal transduction pathways. Many forms of cancer are associated with chromosome breakage in the regions where the TCR proteins are coded. The human TCRβ locus, consisting of 685 kb, has been sequenced. This large family includes, besides the TCR elements, other genes too, such as a dopamine-hydroxylase-like gene and eight trypsinogen genes. The large locus involves, besides the 46 functional genes, 19 pseudogenes and 22 relics (genes with major lesions in one or more components). A portion of the locus is translocated from chromosome 7q22–7qter to 9. The $V_β$ segments include promoters, the first exon as a signal peptide, with RNA splicing signals, the second exon is the V element and DNA rearrangment signal sequence. In some $V_β$ families a conserved decamer interacts with binding proteins. ►T cell receptor, ►immunoglobulins, ►immune response, ►antibody, ►HLA, ►lymphocytes, ►CD4, ►CD8, ►SLAM, ►antigen-presenting cell, ►MHC, ►ITAM; Mak TW et al 1987 J Infect Dis 155:418; Weiss A 1990 J Clin Invest 86:1015; Moffatt MF et al 2000 Hum Mol Genet 9:1011; Willemsen RA et al 2003 Hum Imunol 64:56.

TCV: ►Turnip crinkle virus

TD_{50} (toxic dose): A dose causing toxic (carcinogen) effects in 50% of the experimental organisms. Frequently identified as mmol/kg/day. ►CASE

TDF (testis determining factor): ►sex determination, ►SRY, ►H-Y antigen

TDM: Tissue-specific differentially methylated regions in the genome (Song F et al 2005 Proc Natl Acad Sci USA 102:3336). ►methylation of DNA

T-DNA (transferred DNA): The T-DNA of the Ti plasmid is bordered by 24–25 bp incomplete direct repeats: TGGCAGGATATATT$^{C}_{G}$X$^{G}_{A}$ TTGTAAA for the left and TGGCAGGATATATT$^{C}_{G}$X$^{G}_{A}$ TTGTAAA for the right in the octopine plasmids of *Agrobacteria*. The border sequences of the nopaline plasmids are somewhat different. The left part of the sequences within the borders, T_L (14 kb) and right, T_R (7 kb) are distinguished. The left segment carries, among others, genes for plant oncogenicity (coding for the plant hormones indole acetic acid and the cytokinin, isopentenyl adenine) and either octopine or nopaline, the right segment contains genes for other opines and others with unknown functions. The integration of the T-DNA into plant chromosomes is mediated by the virulence genes (VirA/G) of the Ti plasmid and some chromosomal loci rather than by the T-DNA sequences. In the plant tissues, phenolic compounds such as acetosyringone, cell wall sugars, and pH5.5 favor the transfer of the T-DNA. The transferred DNA of about 20 kb encodes the enzymes for the synthesis of indoleacetic acid and for cytokinin, required for tumorous growth in the plant. In addition, opine genes, which provide carbon and nitrogen nutrition for the growing tumor are transferred. In monocots (maize), 2,4-dihydroxy-7-methoxy-2H-1,4-benzoxazin-3(4H)-one (DIMBOAS) inhibits *vir* genes as well as growth. The related compound 2-hydroxyl-4,7-dimethoxy-benzoxazin-3-one inhibits *vir* genes but not the growth of the tumor. These compounds may be the reason why it is so difficult to transform maize by *Agrobacteria*. After transformation is completed in dicots, the no longer needed *vir* genes are shut down by indoleacetic acid (Piu P, Nester EW 2006 Proc Natl Acad Sci USA 103:4658).

The T-DNA can integrate into the chromosome of plants by a process of illegitimate recombination at practically random locations. The integration involves most likely only one of the strands, called T-strand. The Ku80 protein of the plants appears to be involved in the recombination between an intermediate double-stranded DNA and the plant DNA (Li J et al 2005 Proc Natl Acad Sci USA 102:19231). Because of the transfer feature, the T-DNA can be utilized as the most efficient plant transformation vector. The oncogenes or other sequences can be deleted and replaced by any desired DNA sequences (genes) and they are still inserted into the plant chromosome as long as the border sequences are retained. In Agrobacteria, besides the T-DNA, multisubunit protein complexes are also transmitted to the eukaryotic cell. ►*Agrobacterium*, ►Ti plasmid, ►transformation plants, ►virulence genes of *Agrobacterium*, ►binary vectors, ►cointegrate vectors, ►overdrive, ►opines, ►Ku; Koncz C et al 1992 Methods in Arabidopsis Research. In: Koncz C et al (Eds.) World Scientific, Singapore, pp 224; Szabados L et al 2002 Plant J 32:233; T-DNA *Arabidopsis*: http://www.GABI-Kat.de.

TDT: ►transmission disequilibrium test

TE: *Drosophila* transposable elements cytologically localized in chromosomes 1, 2, and 3.

TE Buffer: TE buffer contains Tris-EDTA, the pH range is 7.2–9.1. ►Tris-HCl buffer, ►EDTA

TEC: A family of non-receptor tyrosine kinases required for signaling through the T cell receptor. ►TCR genes, ►T cell receptor, ►Zap-70, ►Src, ►MAPK; Mao J et al 1998 EMBO J 17:5638.

Technology Transfer: Converting basic or laboratory research results into industrial, agricultural, medical, pharmaceutical, or other applications.

Tectorin: A protein encoded at human chromosome 11q and mouse chromosome 9, and defective in a non-syndromic deafness. ►deafness

Tectum: The dorsal part of the midbrain, controlling visual reflexes and hearing stimuli. ►brain

TEFb: A positive transcript elongation protein factor (P-TEFb). The *Drosophila* dimer has one ~43 and one ~120 kDA subunits. The small subunit is homologous to PITALRE and the large subunit is called also cyclin T. Protein TEFb phosphorylates the carboxyl-terminal domain of the largest subunit of DNA-dependent RNA polymerase II and thereby assures that the transcipt elongation proceeds with few or no pauses. ►TFIIS, ►DRB, ►DSIF, ►NELF, ►NusG,►transcript elongation, ►PITALRE, ►cyclins; Lee DK et al 2001 J Biol Chem 276:9978.

Tegument: A protein layer between the viral capsid and envelope (see Fig. T22).

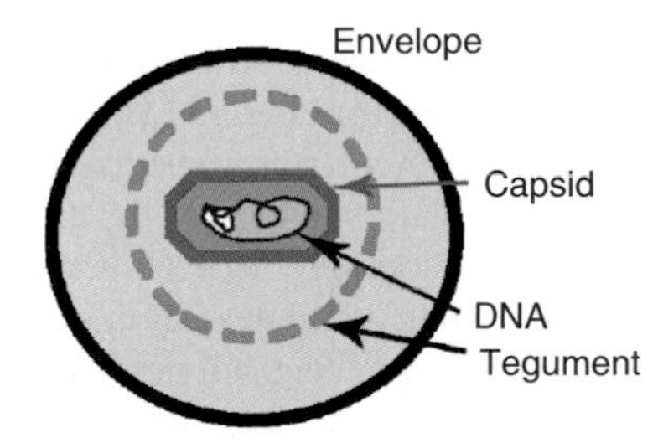

Figure T22. Tegument

Teichoic Acid: A constituent of the cell membrane and cell of some bacteria. The membrane techoic acid contains polyglycerol phosphate, linking glycerol units through phosphodiester, and there are glycosyl substitutions and alanine residues at some positions. The cell teichoic acids are more variable polymers of 6 to 20 units and include polyribitol phosphate chains. ►gram-negative/positive

TEL: The phosphatidylinositol kinase of yeast. ►PIK

Telangiectasia, Hereditary Hemorrhagic (Osler-Rendu-Weber syndrome): A generally non-lethal bleeding disease (except when cerebral or pulmonary complications arise) caused by lesions of the capillaries due to weakness of the connective tissues. The gene ORW1 in human chromosome 9q13 encodes a receptor (endoglin) for the transforming growth factor β, expressed on vascular endothelium. ORW2 was mapped to chromosome 12, and it encodes an activin receptor-like kinase 1, a member of an endothelial serine/threonine kinase family. The prevalence of ORW syndrome in the USA is about 2×10^{-5}. ►hemostasis, ►Rothmund-Thompson syndrome, ►transforming growth factor β, ►endoglin, ►activin, ►ataxia telangiectasia, ►lymphedema

Telangiectasis (telangiectasia): Defective veins causing red spots of various sizes. ►poikiloderma telangiectasia, ►glomerulonephrotis

Telemicroscopy: In telemicroscopy, a microscope linked to a computer transmits JPEG compressed images through the Internet network to distant viewers.

Teleology: A dogma attributing a special vital force and ultimate purpose to natural processes beyond the material scientific evidence. (See Lennox J 1981 J Hist Philos 19:219).

Telethonin: A Z-disc protein in the sarcomeres. It mediates the assembly of titin molecules. ►sarcomere; Faulkner G et al 2000 J Biol Chem 275:41234.

Teliospores: Fungal spores protected by a thick wall; these are either dikaryotic or diploid (see Fig. T23). ►stem rust, ►telium

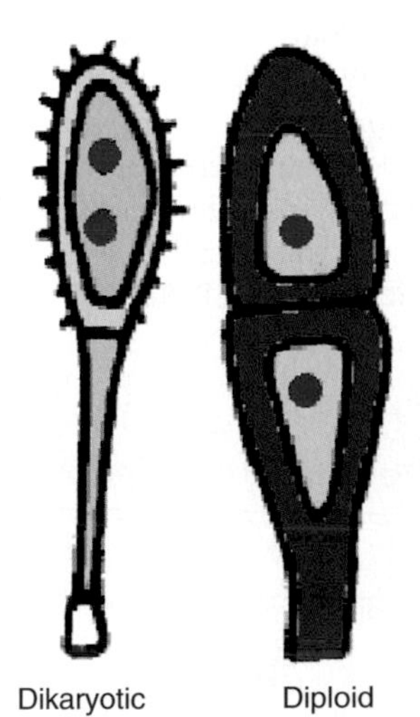

Figure T23. Teliospores

Telium: A fruiting structure (sorus) of fungi producing dikaryotic teliospores.

Telocentric Chromosome: A telocentric chromosome has terminal centromere (has one arm) (see Fig. T24). Telochromosomes can be used to determine which genes are located in that particular single chromosome arm if a telotrisomic female is used according to the scheme represented in Figure T25.

Figure T24. Telochromosome

The segregation of *B:b* in both the 2n and 2n + telo progeny is expected to be 1:1 (testcross). Among the 2n offspring, none or very few are expected to be *A* by phenotype because the telocentric egg cannot remain functional because of dosage effect of the essential genes in the missing chromosome arm. The *A* phenotype is usually due to recombination between the telo and the biarmed chromosomes. The 2n + telo offspring should be all *A* by *a* because of the dominant phenotype and none *a* because the dominant *A* allele is in the trisomic arm (see Fig. T25).

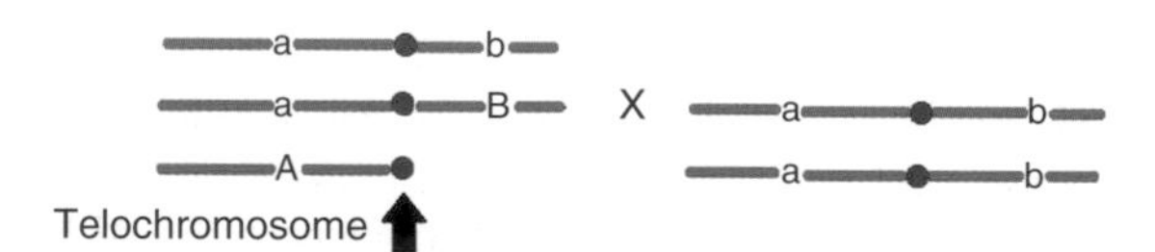

Figure T25. Gene localization with the aid of telochromosome

In allopolyploids, telochromosomes can be used to assign genes to the telochromosome and to determine recombination frequency between genes and the centromere. ►centromere mapping in higher eukaryotes, ►misdivision of the centromere [see Fig. T26], ►Robertsonian translocation, ►tetrad analysis, photo is the courtesy of Dr. ER Sears.

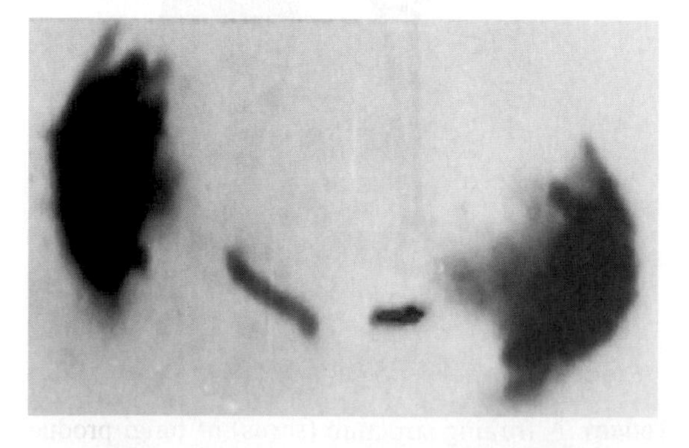

Figure T26. Misdivision generates telocentrics

Telochromosome: Telocentric chromosome.

Teloisodisomic: In wheat, 20" + ti", 2n = 42, [" = disomic, t = telosomic, i = isosomic].

Teloisotrisomic: In wheat, 20"+ (ti)1"', 2n = 43, [" = disomic, "' = trisomic, t = telosomic, i = isosomic].

Telomerase (TERT [telomere reverse transcriptase]): The enzyme synthesizing telomeric DNA. This enzyme is different from other replicases inasmuch as a RNA template (TERC) that is a part of the telomerase (ribozyme) specifies the telomeric DNA. An RNA polymerase II enzyme transcribes this RNA and it has a 5′-2,2,7-trimethyl guanosine cap. It has a binding site for Sm proteins that are characteristics for snRNPs. TERT (telomere reverse transcriptase) is the enzyme synthesizing telomeric DNA. The transcription of the human TERT (*hTERT*) is activated by the joint action of Sp1 and cMyc. It seems that the occupancy of the E box by either Myc or Mad1 determines the acetylation/deacetylation of the chromatin in HeLa cells and thus the regulation of human hTERT. This enzyme is different from other replicases inasmuch as a RNA template (TERC) that is a part of the telomerase (ribozyme), which specifies the telomeric DNA. The RNA has A and C repeats and therefore T and G repeats characterize the shortening of the telomeres that leads to p53-dependent senescence of the cells in vivo and limits tumor progression (Feldser DM, Greider GW 2007 Cancer Cell 11:461). The human telomerase template has 11 nucleotides: 5′-CUAACCCUAAC and the telomere has $(5'\text{-TTAGGG-}3')_n$ repeats. In *Oxytricha*, the telomere consists of 36 nucleotides of which 16 form a single strand of $3'\text{-}G_4T_4G_4T_4$ overhang protruding from the double-stranded remainder. The telomere-end-binding protein (TEBP) binds to the $3'\text{-}G_4T_4$ tract and protects it from degradation. During replication this TEBP is displaced for the replication to proceed. By rejoining the end it displaces the telomerase and thus regulates telomere length. In murine embryonic stem cells but not in other types of cells, sister chromatid exchange at the telomeres may contribute to shortening of telomeres (Wang Y et al 2005 Proc Natl Acad Sci USA 102:10256). In budding yeast, the *TLC*1 gene is responsible for telomerase activity and gene *EST*1 is also needed for the maintenance of the telomeres. The *TLC*1 (telomerase component) gene is also required for the preservation of the RNA template and normal telomerase function. The telomeres are made mainly of double-stranded DNA repeats, however the far end has only single strand G repeats. When the telomerase stays at the end of the chromosomes (capped state) the cells—even when the telomeres are short—remain viable. When the telomeres become uncapped the cells exit from cycling and senesce. An automated highthroughput quantitative telomere FISH platform allows the quantification of telomere length as well as percentage of short telomeres in large human sample sets. This technique provides the accuracy and sensitivity to uncover associations between telomere length and human disease (Canela A et al 2007 Proc Natl Acad Sci USA 104:5300). Stronger and positive correlation and association seems to occur between telomere length (TL) in the offspring and paternal TL ($r = 0.46$,) than offspring and maternal TL ($r = 0.18$) (Njajou T et al 2007 Proc Natl Acad Sci USA 104:12135).

The replication of the telomere takes place near the end of the cell cycle. Telomerase elongates only the G-rich strand and the C-rich strand is filled in later. In budding yeast, p23 molecular chaperone Sba1p controls telomere length maintenance and that Sba1p can modulate telomerase DNA binding and extension activities in vitro (Toogun OE et al 2007 Proc Natl Acad Sci USA 104:5765). Protein RCF binds also to the telomeric DNA and it is required for the replication of the leading strand by pol δ. The Pif1 helicase inhibits yeast telomerase by removing it from the DNA and may suppress healing of double-strand breaks by telomerase (Boulé J-B et al 2005 Nature [Lond] 438:57). Yeast cells recognize telomere replication as double-strand breaks and recruit to the telomeres the MRX complex of Mre11, Rad 50, and Xrs proteins. At late S phase, Mec1 is recruited to the telomeres where it assembles the Cdc13 and Est telomerase regulators (Takata H et al 2005 Mol Cell 17:573). Telomere-binding proteins (TEBP) bind either to the single-strand terminal repeats (*Oxytricha* proteins) or to the double strand sequences (Rap1). The telomeres of the ciliate *Oxytricha fallax* terminate in duplex DNA loops. The TEBP heterodimer α subunit (M_r 56K) binds to the 3′-T_4G_4 single-stranded end whereas the β subunit (M_r 41 K) attaches to other proteins. The heterogeneous nuclear ribonucleoprotein A1 (hnRNP A1) is also required for the maintenance of the normal length of the telomeres. The E6 protein of the Human Papilloma Virus-16 (HPV16) activates telomerase but does not immortalize the cells, although it expands their life span. The newly replicated telomeres are processed to size by proteins. Telomerase activity in tumor tissues is high but it is low in somatic cells or in benign neoplasias (Artandi SE, De Pinho RA 2000 Curr Opin Genet Dev 10:39). Intact telomere length and telomerase activity are required for the maintenance of function of epidermal stem cells (Flores I et al 2005 Science 309:1253). Targeting the telomerase by antisense technologies may be an approach to anticancer therapy. The telomerase RNAs of *Tetrahymena*, *Euplotes*, and *Oxytricha* seem to be transcribed by polymerase III and the human telomerase RNA may be the product of pol II because it is sensitive to α-amanitin. The telomere end of *Tetrahymena* where it was originally discovered is somewhat different. In the RNA template, 5′-CAACCC-3′, and in the telomere 5′-GGGTG-3′ repeats occur. The Cdc13 protein of yeast mediates the access of the telomerase to the telomere. In the catalytic subunit of a telomerase of yeast (*EST2*, *ever-shorter telomeres*), reverse transcriptase-like motifs were identified. The telomeres are usually shortened as nuclear division proceeds and aging has been attributed to reduced telomerase activity. Lack of telomerase does not prevent tumorigenesis, however telomerase-deficient mice are resistant to skin tumorigenesis. Maintenance of normal telomere function is critical for the prevention of carcinogenesis. Usually in normal cells telomerase activity is barely detectable but it is well-expressed in tumor cells. The catalytic subunits bear structural similarities to reverse transcriptases. In mammals reduction in telomerase activity is harmful to the reproductive and blood-generating systems. On introduction into mammalian cells, an active telomerase may substantially extend the life of the cell. Telomerase defects increase the susceptibility to damage by ionizing radiation. Nevertheless, sustained telomerase activity does not result in malignancy. The ubiquitous c-MYC protein normally activates telomerase. The promoter of the TERT catalytic subunit has several c-MYC binding sites. Sheep produced by nuclear transplantation displayed apparently shorter telomeres than normally generated animals yet in the small-scale experiments no premature aging was observed. Mutations in the telomerase genes may result in senescence and late-onset sterility. Further, when the telomeres are substantially lost the chromosomes may fuse end-to-end. One class of *Caenorhabditis* telomerase mutants (*mrt, mortal germline*) is pleiotropic because in addition to the symptoms displayed by the other mutants in yeast and mice, it involves chromosomal loss and sensitivity to DNA-damaging agents. Mutation *mrt*-2 was identified to have a defect in checkpoint function that is apparently repaired when the telomerase is normal. Loss of the telomerase function leads to swift progressive deterioration in animals but *Arabidopsis* plants may survive up to ten generations without telomerase function. However, eventually due to chromosomal aberrations both vegetative and reproductive fatal damage ensues. The human TERT expression is necessary for repairing chromosome breaks (Masutomi K et al 2005 Proc Natl Acad Sci USA 102:8222). In the presence of manganese, both human and yeast TERT can switch to RNA- and template-independent mode of DNA synthesis and act as a terminal transferase (Lue NF et al 2005 Proc Natl Acad Sci

USA 102:9778). TERT induction can also cause the proliferation of hair follicles—without a need for TERC, the RNA component of telomerase—and as a result abnormal heavy hair growth is observed in mice (Sarin KY et al 2005 Nature [Lond] 436:1048). Damaged telomeres can be repaired even in the absence of telomerase by relying on a break-induced recombination process (BIR). The pathways and mechanisms of BIR have been reviewed (McEachern MJ, Haber JE 2006 Annu Rev Biochem 75:111). ►telomeres, ►ribozyme, ►RAP, ►tankyrase, ►pol II, ►pol III, ►α-amanitin, ►*Tetrahymena*, ►aging, ►p16^{INK}, ►ribozyme, ►tumorigenesis, ►antisense technologies, ►hTERT, ►malignant growth, ►MYC, ►Mad2/Mad1, ►E box, ►RNP, ►cap, ►SP1, ►Myc, ►nuclear transplantation, ►MRX, ►Mre11, ►RAD50, ►Xrs, ►Cdc13, ►Est, ►pseudoknot, ►p23; Collins K 1999 Annu Rev Biochem 68:187; Herbert B-S et al 1999 Proc Natl Acad Sci USA 96:14276; Tzfati Y et al 2000 Science 288:863; Cooper JP 2000 Curr Opin Genet Dev 10:169; Betts DH et al 2001 Proc Natl Acad Sci USA 98:1077; Wenz C et al 2001 EMBO J 20:3526; Antal M et al 2002 Nucleic Acids Res 30:912; Arai K et al 2002 J Biol Chem 277:8538; Neidle S, Parkinson G 2002 Nature Rev Drug Discovery 1:383; structure and function: Autexier C, Lue NF 2006 Annu Rev Biochem 75:493; history of discovery of telomerase: de Lange T 2006 Cell 126:1017; telomere replication review: Verdun RE, Karsleder J 2007 Nature [Lond] 447:924.

Telomere Crisis: The demise of the telomerase reverse transcriptase and the loss of telomere replication. The chromosome becomes unstable. ►telomerase; Chin K et al 2004 Nature Genet 36:984.

Telomere Mapping: The eukaryotic chromosome has characteristic telomeric repeats and can be used for mapping RFLPs relative to the telomeres. ►telomeres, ►RFLP

Telomere Position Effect: ►telomeric silencing

Telomere Switching: In *Trypanosoma*, the variable surface antigens (VSG) are encoded at the telomeres and frequent recombination of the chromosome ends can generate new variations and may affect the expression of the VSG glycoproteins. ►*telomeres*, ►*Trypanosoma*; Rudenko G et al 1998 Trends Microbiol 6:113.

Telomere Terminal Transferase: The ribozyme involved in replication telomeric DNA. ►telomerase

Telomeres: Special terminal structural elements of eukaryotic chromosomes rich in T and G bases.

Electronmicroscopic observations indicate that in mice and humans the end of the telomeres bend backward and form a D loop/t-loop as the telomeric single-strands pair with the double-stranded telomeric DNA (Figure T28). T loops normally prevent (NHEJ) non-homologous end joining recombination and the loss of the T loops may cause loss of telomere integrity (Wang RC et al 2004 Cell 119:355). In human chromosomes and all vertebrates, *Trypanosomas*, and fungi the many times repeated telomeric box is CCCTAA/TTAGGG (ca. 300 bp). The telomeric region is highly conserved among diverse eukaryotes; in some species, however variations exist. In *Drosophilas* (and other insects), the telomeres are unusual inasmuch as rather than having the common repeats, they have transposable elements, e.g., *HeT*-A and *TART* in *Drosophila* (Casacuberta E, Pardue M-L 2003 Proc Natl Acad Sci USA 100:3363). This transposable element may assist the replication of the telomeric DNA. In most organisms proximal to the telomere, middle-repetitive DNA is found (telomere-associated DNA, TA), which may display some similarity to the *Drosophila* transposable sequences (Levis RW et al 1993 Cell 75:1083). The subtelomeric regions are repetitive and yet highly variable. Ectopic recombination in this tract may lead to restoring defective ends but it may be the cause also of undesirable rearrangements (Heather C et al 2002 Nature Rev Genet 3:91; Linardopoulou EV et al 2005 Nature [Lond] 437:94). The length of the telomeric sequences may vary among the organisms but variations exist during the life of the cells, according to developmental stage. By aging the telomeres usually become shorter in cultured cells although in mice telomere length may not be correlated with lifespan (Hemann MT, Greider CW 2000 Nucleic Acids Res 28:4474). The *Rtel* helicase-like gene (chromosome 2q) controls telomere length in mice and it essential for survival beyond days 10–11.5 of gestation (Ding H et al 2004 Cell 117:873). In maize, telomere-mediated truncation of chromosome ends appeared (Yu W et al 2006 Proc Natl Acad Sci USA 103:17331).

Mutants of shorter telomeres may function reasonably well, although a transient pause may occur in the cell division. The length of the human telomeres may be different between homologous chromosomes (Londoño-Vallejo JA et al 2001 Nucleic Acids Res 29:3164). Telomere length in humans is apparently linked to the X chromosome and the transmission occurs from father to son or father to daughter but no correlation was found between mother and son or mother and daughter (Nordfjäll K et al 2005 Proc Natl Acad Sci USA 102:16374). Various techniques are available for measuring the length of the telomeres. (Baird DM et al 2003 Nature Genet 33:203).

The non-nucleosomal special chromatin, including the repeats of yeast, is called telosome, which contains six core proteins: TRF1, TRF2, RAP1, TIN2, POT1, and TPP1 (O'Connor MS et al 2006 Proc Natl Acad Sci USA 103:11874). TPP1 (telomere binding human homolog of TEBP) and POT1 (protection of telomere) form a complex with telomeric DNA that increases the activity and processivity of the human telomerase core enzyme. POT1–TPP1 seems to switch from inhibiting telomerase access to the telomere, as a component of *shelterin*, to serving as a processivity factor for telomerase during telomere extension (Wang F et al 2007 Nature [Lond] 445:506). However, in mammals the large telomeric DNA is nucleosomal. The major structural protein associated with the yeast telomeres is Rap1. TPP1 and TIN2 (TRF-interacting factor) are critical for the assembly of the telomeric complex. The same protein may also be present at other locations of the chromosomes and acts either as a repressor or activator of transcription. Telomeres are required for the proper replication of the linear eukaryotic chromosomes. This is probably the reason why among the diverse types of chromosomal aberrations cytologists failed to find attachments of internal chromosomal pieces to the telomeres or telomere to telomere fusions although chromosomes with broken ends may fuse into dicentric chromosomes after replication. Shortened telomeres may fuse and involve elevated frequencies of different types of chromosomal aberrations (Mieczkowski PA et al 2003 Proc Natl Acad Sci USA 100:10854). Normally, the telomeric repeat binding factor, TRF2, or DNA-dependent protein kinase (DNA-PK) prevents telomeric fusions. In ciliates, a cap is built at the telomere by binding TEBPα and β to 16-nucleotide single-strand DNA overhang. In fission yeast, the Taz1 protein protects the telomeric ends of the chromosomes from fusion, which otherwise might be mediated by the Ku proteins recognizing double-strand breaks (ends) of the DNA. In *taz*⁻ cells *rad*22 may also promote telomere fusion (Godinho Ferrira M, Promisel Cooper J 2001 Molecular Cell 7:55). In the budding yeast, telomere capping is mediated by Cdc13 protein that recruits other proteins (Ten1/Stn1) to build a protective cap. DNA-dependent protein kinase may play a critical role in capping (Gilley D et al 2001 Proc Natl Acad Sci USA 98:15084). In *Drosophila*, the HP1 protein interacts with histone methyltransferases and histone H3 methylated lysine 9 and participates in telomere capping, elongation, and silencing (Perrini B et al 2004 Mol Cell 15:467). In fission yeast, Pli1p is a SUMO E3 ligase and mutants are impaired for global sumoylation, yet they are viable, but exhibit deregulated homologous recombination, marked defects in chromosome segregation, and centromeric silencing, as well as a consistent increase in telomere length. Telomere lengthening induced by lack of sumoylation is not due to unscheduled telomere–telomere recombination. Instead, sumoylation increases telomerase activity, suggesting that this modification controls the activity of a positive or negative regulator of telomerase (Xhemalce B et al 2007 Proc Natl Acad Sci USA 104:893).

In mammals, the TRF2 protein has an end-protective role. It may also remodel the chromosome ends by forming the so-called *t loop*. The t loop of mammals is double-stranded DNA. In both fission yeast and mammals, the *POT1* (protection of telomere) wild type allele encodes a telomere-cappingprotein, which binds to the G-rich telomeric single-strand tail and also to the *t* loop. POT1 in humans has some role in elongating the telomeres and may disrupt the G-quadruplexes to allow proper elongation (Zaug AJ et al 2005 Proc Natl Acad Sci USA 102:10864). Mice have two POT proteins (POT1 and POT2) whereas humans have only one (Hockermeyer D et al 2006 Cell 126:63). In budding yeast, the evolutionarily conserved KEOPS complex, including Cdc13, protein kinase and peptidase, is a telomere regulator (Downey M et al 2006 Cell 124:1155). In *Drosophila*, a zinc-finger protein, a putative transcription factor, encoded by gene *Woc*, prevents telomeric fusion (Raffa GD et al 2005 Mol Cell 20:821). The mammalian single-strand DNA with 3′-end may encompass 300 nucleotides and is located where the loop folds onto the main double-strand DNA (Baumann P, Cech TR 2001 Science 292:1171).

The structure of the telomere may bear some similarity to the centromere and in some instances (e.g., rye) it may function as a neocentromere. It may be somewhat of a puzzle how the normally fragmented somatic chromosomes of *Ascaris* are still able to produce their telomeres and how the germline chromosomes control the apparently multiple intercalary telomeres. In the polyploid ciliate macronucleus, millions of centromeres may exist in some species. HIV-infected individuals appear to have shorter telomeres than healthy individuals of comparable age. Telomere loss or inactivation may play a role in senescence and telomerase activation may be a mechanism of cellular immortalization. The replication of the telomeric DNA is carried out by the ribozyme, telomerase. A specific telomere-binding human protein hTRF (60 kDa), recognizing the TTAGGG (and mammalian) sequences, has been isolated. TRF1 (8q13) is regulated by TIN2 (TRF1-interacting nuclear protein). The corresponding repeat in *Tetrahymena* is TTGGGG and in yeast, TG_{1-3}. One of its domains is similar to the Myb oncogene product. Telomeric sequences may be shortened in several types of cancer cells, although in about 90% of the malignant cell types the activity of the telomerase enzyme is higher than in normal cells. The enzyme poly(ADP)-ribose polymerase (PARP) that recognizes DNA interruptions, adds ADP-ribose units to the DNA ends. Defects in the encoding gene (ADPRP) and the protein PARP results in chromosomal instability and shortening of the telomeres. In *Caenorhabditis*, all chromosomes are capped by the same 4–9 kb tandem repeats of TTAGGC, but the sequences next to it differ among the chromosomes. Double-strand break repair proteins mediate the capping of the chromosomes. Intact telomeres cannot fuse but a repair pathway may generate non-homologous end-joining (NHEJ). When, however NHEJ fuses two telomeres, unstable dicentric chromosomes are produced. Chromosomes lacking telomeres can perform most functions but lack stability and undergo fusion, degradation, and loss at a high rate and the chromatids may not be able to separate normally. If yeasts lose the *TRT*-1 telomerase subunit gene, the cells may survive either by circularizing the chromosome or by restoring the function through recombination controlled by *RAD*50 or *RAD*51 genes. In the majority of tumors, the telomeres are not shortened during consecutive divisions, whereas normal somatic human cells may lose 100 bp by each cell division and that may lead to discontinuation of cell divisions and cellular aging. In ciliates, telomeres may arise de novo but in yeast this rarely occurs. In humans, telomerase may rarely replace lost telomeres. Genes that would be normally transcribed by either pol I, pol II, or pol III are frequently repressed at the telomeric location (see Figs. T27 and T28). ►telomerase, ►breakage-fusion-bridge cycle, ►neocentromere, ►chromosome diminution, ►immortalization, ►senescence, ►Myb, ►RAP, ►telomeric probe, ►HeT-A, ►TART, ►plasmid telomere, ►pol, ►Ku, ►non-homologous end-joining [NHEJ], ►tankyrase, ►dyskeratosis, ►aging, ►Werner syndrome, ►TRF, ►UbcD1, ►ALT, ►G-quadruplex, ►Cdc13; Griffith JD et al 1999 Cell 97:503; Pardue M-L, DeBaryshe PG 1999 Chromosoma 108:73; Varley H et al 2000 Am J Hum Genet 67:610; Knight SJL et al 2000 Am J Hum Genet 67:320; McEachern MJ et al 2000 Annu Rev Genet 34:331; Hodes R 2001 Proc Natl Acad Sci USA 98:7649; Shay JW et al 2001 Hum Mol Genet 10:677; Blackburn EH 2001 Cell 106:661; McEachern MJ et al 2002 Genetics 160:63; Ren J et al 2002 Nucleic Acids Res 30:2307; Phan AT, Mergny J-L 2002 Nucleic Acids Res 30:4618; Pardue M-L, DeBaryshe PG 2003 Annu Rev Genet 37:485; telomeres in disease, cancer and aging: Blasco MA 2005 Nature Rev Genet 6:611; human disease: Artandi SE 2006 New England J Med 355:1195.

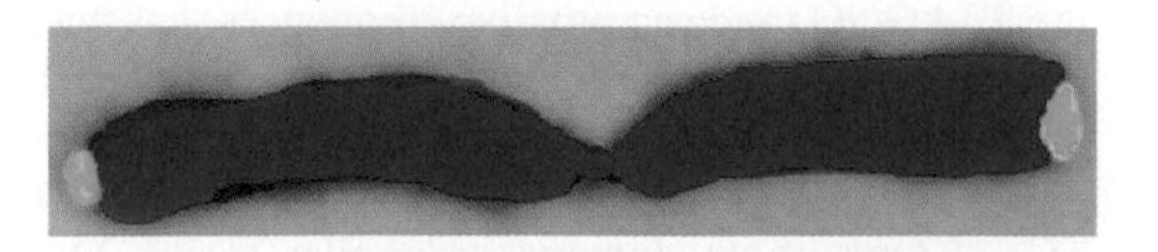

Figure T27. Glowing telomeres probed by a fluorescent sequence

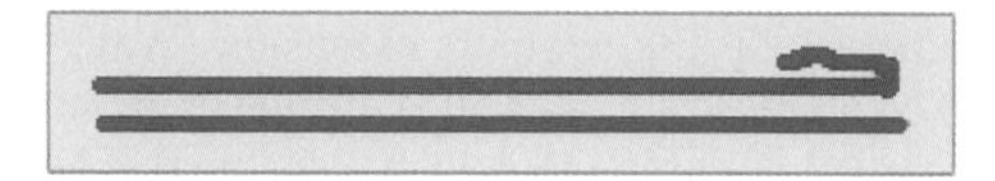

Figure T28. D- or t-loop formation of mammalian telomeres

Telomeric Fusion: As per telomeric fusion, end-to-end fusion of chromosomes does not take place between intact, telomere-capped chromosomes. However, it may take place in double-strand repair deficient cells. ►non-homologous end-joining

Telomeric Probe: Telomeric probes are fluorochrome-labeled DNA sequences, complementary to the telomeric repeats TTAGGG; they can be used for the cytological identification of short (cryptic) translocations. ►chromosome painting, ►telomere

Telomeric Silencing (telomeric position effect): As per telomeric silencing, telomeres frequently reduce transcription of associated genes. This effect is similar to the silencing or position effect exercised by heterochromatin, although it may occur in this region even in the absence of heterochromatin. It may be due to higher order chromatin arrangement. Rap1 (repressor and activator protein) recruits the Sir3 and Sir4 (silent information regulators) and Rif1 (Rap1-interacting factor). Histones 3 and 4 and the Ku proteins are among the best known regulators. Mec1 (mitotic entry checkpoint) protein controls S-phase arrest of the cell cycle in reaction to DNA damage and regulates telomere length; some forms may also affect telomeric silencing. These genes/proteins are highly conserved but may have different names in the various organisms. In yeast, *DOT*1 (disruptor of telomeric silencing) encodes a methyltransferase specific for histone 3 in the nucleosome globular domain and trimethylates lysine 79. The Rmt1 methyltransferase dimethylates arginine at position 3 in histone-4 termini. ►heterochromatin, ►position effect, ►telomeres, ►RAP1A, ►Ku, ►silencers, ►checkpoint, ►MEC1; de Bruin D et al 2001 Nature 409:109; Lacoste N et al 2002 J Biol Chem 277:30421.

Telomutation: A (dominant) premutation occurring in and transmitted through both sexes but expressed only in the offspring of the heterozygous female. ►premutation, ►delayed inheritance; Aleck KA, Hadro TA 1989 Am J Med Genet 33:155.

Telophase: The final major step of the nuclear divisions. The chromosomes have been pulled by the spindle fibers all the way to the poles. The microtubules attached to the kinetochores fade from view and the nuclear envelope reappears. The chromosomes relax and the nucleoli become visible again. ►mitosis, ►meiosis

Telosome: A telocentric chromosome; the non-nucleosomal, six-protein chromatin complex at the end of the yeast, ciliate, and mammalian chromosomes is also called telosome or *shelterin*. ►telomeres, ►telochromosome; Liu D et al 2004 J Biol Chem 279:51338; de Lange T 2005 Genes Dev 19:2100.

Telotrisomic: A trisomic having two bi-armed and one telochromosome. ►trisomy

Telson: The most posterior part (opposite to the head) of the arthropod body. ►*Drosophila*

TEM (transmission electronmicroscopy): ►electronmicroscopy

TEM (triethylenemelamine): An alkylating clastogen. ►alkylating agent, ►alkylation, ►clastogen

TEM1: A GTPase. ►GTPase

TEMED: ►acrylamide

Temperate Phage: A temperate phage has both lysogenic and lytic life-styles. The temperate phages may be of different types: (i) those that insert their DNA into one or few preferred sites like λ phage, (ii) those that insert their DNA into the host bacterial chromosome with the aid of a transposase at different sites like Mu-1, (iii) those, e.g., P1, that do not insert their DNA into the host chromosome but are maintained as a plasmid and (iv) those, e.g., P4, which can be either plasmids or prophages. ►bacteriophage, ►lysogen, ►lysogeny, ►prophage, ►plasmid, ►lambda phage, ►specialized transduction

Temperature Conversion: 0.556°F − 17.8 → °C, 1.8°C + 32 → °F, K − 273.15 → °C

Temperature-Sensitive Mutation (ts): The ts mutation causes such an alteration in the primary structure of the polypetide chain that its conformation varies according to temperature, and it is functional at either high or low temperature but is non-functional at the other (non-permissive temperature). Temperature-sensitive conditional lethal mutants are very useful for various analyses because the biochemical/molecular basis of the genetic defect can be analyzed at the permissive temperature range, within which, the cells or organisms grow (normally). ts condition is correlated with the buried hydrophobic residues in the protein. The majority of organisms (mesophiles) normally thrive best at temperatures <40 °C; hyperthermophiles may exist at temperatures around 100 °C. Temperature variations are sensed by the transient receptor potential activation channels (Voets T et al 2004 Nature [Lond] 430:748). Inflammation and elevated temperature caused pain is controlled by the calcitonin gene-related neuropeptide and it is correlated with the differences in heat/pain response variations in mouse strains (Mogil JS et al 2005 Proc Natl Acad Sci USA 102:12938). ►hyperthermia, ►cold hypersensitivity, ►pain-sensitivity, ►trichothiodystrophy

Template: A template determines the shape or structure of a molecule because it serves as a "mold" for it (in some way similar to the mold to cast iron). Old DNA strands serve as templates for the new, or one of the DNA strands may be a template for the mRNA and the other for another sense or nonsense RNA. Molecular biologists frequently call template (or antisense) the strand of the DNA that serves for the synthesis of mRNA by complementary base pairing. T7 RNA polymerase can bypass up to 24-nucleotide gaps in the template strand by copying a faithful sequence of the deletion using the non-template strand. ►semiconservative replication, ►replication fork, ►sense strand, ►coding strand

Template Switch: As per the template switch, during replication the polymerase enzyme may jump to another DNA sequence and copy elements, which were not present in the original DNA tract. Such a switch may occur when plasmid DNA (T-DNA) is inserted into a chromosomal target site. As a consequence, deletions and rearrangements may follow. During primer extension, the strand displaced may reanneal onto the template and the extended strand may be partially dissociated. A single-stranded sequence, attached to the 5′ end of the displaced strand and complementary to the dissociated segment of the extending strand, can thus serve as an alternative template ►switch, ►copy choice, ►T-DNA, ►primer extension; Negroni M, Buc H 2000 Proc Natl Acad Sci USA 97:6385.

Tenascin: A large glycoprotein complex with disulfide-linked peptide chains. It either promotes or interferes with cell adhesion, depending upon the type of cell and the different protein domains. It also controls cell migration, axon guidance embryogenesis, and oncogenic pathways. ►cell migration, ►cell adhesion; Hicke BJ et al 2001 J Biol Chem 276:48644; Daniels DA et al 2003 Proc Natl Acad Sci USA 100:15416.

Tendril: A plant organ that coils around objects and provides support (see Fig. T29). ►circumnutation

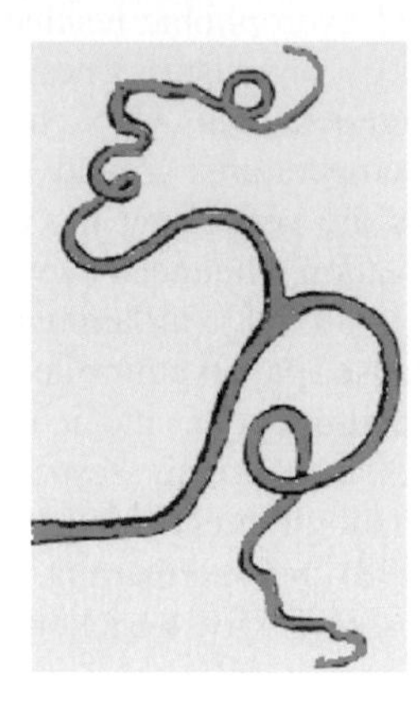

Figure T29. Tendril

Tensin: A cytoskeletal protein binding vinculin and actin; it contains an SH2 domain. ►actin, ►vinculin, ►multiple hamartomas, ►PTEN; Chen H et al 2002 Proc Natl Acad Sci USA 99:733.

Tension: The force(s) generated by pulling the mitotic/ meiotic chromosomes to the opposite poles, opposed by the attachment between the homologs. ►spindle, ►kinetochore, ►pole

Teosinte: ►maize

TEP1 (TGFβ-regulated and epithelial cell-enriched phosphatase): A tumor suppressor function of PTEN. TEP1 is also an RNA-binding protein, which interacts with the mammalian and other telomerases. ►PTEN, ►vaults; Sharrard RM, Maitland NJ 2000 Biochim Biophys Acta 1494:282.

***ter* Sites**: ►DNA replication prokaryotes

tera-: 10^{12}

Terahertz Radiation (T-rays): Electromagnetic radiation between microwaves and infrared. Semiconductor crystals of alternating layers of gallium arsenide and aluminum gallium arsenide emit light when electrons are boosted to higher level of energy by exposure to laser. Terahertz devices are expected to detect hidden objects of industrial and biological importance, such as anthrax pores within an envelope beyond what is detectable by other technologies. Apparently the T-rays are not harmful to living material. ►electromagnetic radiation

Teratocarcinoma: Malignant tumor containing cells of embryonal nature; common in testes. The teratocarcinoma cells may differentiate into various types of tissues in vitro and have been used for studies of differentiation. ►cancer, ►teratoma, ►stem cells; Silver LM et al (Eds.) 1983 Teratocarcinoma Stem Cells, Cold Spring Harbor Laboratory Press, Cold Spring Harbor, New York.

Teratogen: Teratogens are agents causing malformation during differentiation and development.

Teratogenesis: Malformation during differentiation and development. The inducing agents may be genetic defects, physical factors and injuries, infections, drugs and chemicals. In frogs, limb developmental anomalies have been traced to trematode (*Planorbella campanulata* and *Ribeiroira* sp.) infection stimulated by agricultural pesticides. (See Kiesecker JM 2002 Proc Natl Acad Sci USA 99:9900).

Teratoma: A mixed tissue group with cells of different potentials for development. It may be formed in various early or late animal tissues and may eventually become a malignant tumor. Teratomas are most common in the germinal tissues. In plants amorph, undifferentiated tumor tissue may be interspersed with differentiated elements giving rise to either shoots or roots. Teratomas may be formed in the tumors induced by *Agrobacterium rhizogenes*. ►*Agrobacterium*

Tergite: The dorsal epidermis developed from histoblasts not from imaginal disks. ►histoblast

Terminal Deoxynucleotidyl Transferase (TdT): An enzyme that can elongate DNA strands—at the 3-OH end—with any base that is present in the reaction mixture. It is used in genetic vector construction to generate homopolymeric cohesive tails for the purpose of splicing a passenger DNA to a plasmid. The passenger and the plasmid are equipped with complementary bases, such as polyA and polyT, respectively, and can readily anneal and be ligated. Terminal nucleotidyl transferase is also expressed in the adult bone marrow and generates immunoglobulin diversity. The enzyme occurs primarily as a 58 kDa protein but much smaller molecules were also found. Its N-terminus is homologous to the C-terminus of the breast cancer gene, BRCA1. The N-terminus of TdT interacts with the heterodimeric Ku autoantigen, which is the DNA-binding component of a 460 kDa DNA-dependent protein kinase complex. This complex apparently plays a role (besides RAGs) in the generation of immunoglobulin diversity and DNA double-strand break repair. ►vectors, ►cloning vectors, ►immunoglobulins, ►T cell, ►breast cancer, ►Ku, ►RAG, ►immunoglobulins, ►double-strand break; Delarue M et al 2002 EMBO J 21:427.

Terminal Differentiation: Terminal differentiation is usually irreversible; in plants, however, under culture of high levels of phytohormones, dedifferentiation may be possible. ►dedifferentiation, ►differentiation

Terminal Nucleotidyl Transferase: The enzyme that can add homopolymeric ends to DNA. It is useful in generating cohesive termini (in genetic plasmid vectors), if one of the reaction mixtures contains adenylic acids and the other (e.g., passenger DNA), thymidylic acid (or G and C, respectively). ►terminal deoxynucleotidyl transferase, ►cloning vectors

Terminal Protein (TP): Serine, threonine, or tyrosine residues of the terminal proteins provide OH groups for the initiation of replication, instead of 3′ OH group of a nucleotide in a linear double-stranded viral DNA.

Terminal Redundancy: Repeated DNA sequences at the ends of phage DNA. (See permuted and non-permuted terminal redundancy.)

Terminalization of Chromosomes: During meiosis, the bivalents may display one or more chiasma(ta), which eventually (during anaphase) move toward the telomeric region in the process called terminalization (see Fig. T30). ►meiosis

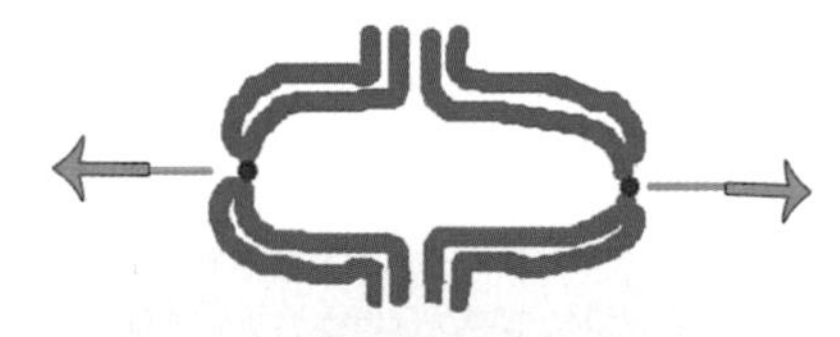

Figure T30. Terminalization of chromosomes

Terminase: The terminase phage enzyme binds to specific nucleotide sequences and cuts in the vicinity of the binding at cohesive sites (*pac* or *cos*). ►lambda phage; de Beer T et al 2002 Mol Cell 9:981.

Termination Codons: ►nonsense mutation, ►genetic code, ►terminator codons

Termination Factors: The proteins that release the polypeptide chains from the ribosomes.

Termination of Replication in Bacteria: The termination of replication in bacteria is mediated by the homodimeric replication terminator protein (RTP). Two dimers bind to the two *Ter* inverted repeat sites in the DNA. The binding site has a strong core and an auxiliary site. First one dimer binds to the core and the binding of the second dimer follows this to the auxiliary site. When the replication fork encounters the RTP-*Ter* site, it is blocked, but not when it encounters it from the auxiliary site. The terminator prevents the unwinding of the DNA by the helicase. There are several *Ter* sites. (See Lemon K et al 2001 Proc Natl Acad Sci USA 98:212).

Terminator: The sequence at the end of a gene that signals for the termination of replication or transcription. The RNA polymerase recognizes this signal directly or indirectly by various proteins. The intrinsic terminator of *E. coli* (rho) uses two sequence motifs for the release of RNA from the DNA template: (i) a stem-loop hairpin and (ii) a tract of 8–10 nucleotides immediately downstream at the end of the released RNA. Before termination, there is generally a pause, mediated by U-rich transcribing DNA segment. ►transcription termination, ►antitermination, ►protein synthesis, ►release factor, ►rho factor, ►pausing transcriptional, ►RNA polymerase

Terminator Codons: UAA, UAG, and UGA in RNA (same as nonsense codons).

Terminator Technology: Terminator technology uses three different transgenes in plants to control their ability to bear seed, to protect the interest of seed companies by preventing unauthorized use of their genetic stocks. One of the genes is a repressor of a recombinase, the other is a recombinase that is capable of deleting internal spacers, and the third is a toxin gene. The seed company can grow seed-bearing plants because the toxin gene is inactivated by a spacer sequence between the promoter and the structural toxin gene as long as the recombinase is suppressed. Treating the commercially sold seed by the antibiotic tetracycline, the synthesis of the suppressor is prevented, the recombinase then deletes the spacer, and the toxin gene becomes activated to cause seed failure. This trick of biotechnology may become a safeguard for the property of the company but may hurt the poor potential consumer who cannot afford buying the seeds annually. Similar, although simpler, technology has been practiced by the seed industry for decades in the commercially available double-flowered garden stocks. Although social activists condemn the terminator technology, its principles may offer promises for shutting off undesirable genes or activating useful ones, and serve beneficial roles in improving agricultural productivity ►T-Gurt, ►suppressor, ►recombinase, ►tetracycline, ►promoter, ►structural gene, ►*Matthiola*; Kuvshinov VV et al 2001 Plant Sci 160:517.

Termisome: Nucleic acid terminal protein complexed with other proteins and DNA. ►terminator, ►protein synthesis

Ternary: Ternary has two meanings; either that it is made up of three elements, or it is third in order.

Ternary Complex Factors: ►TCF

Terpenes: Hydrocarbons or derivatives (>30,000) with isoprene repeats; occur as animal pheromones and diverse types of plant fragrances. ►pheromones, ►fragrances, ►prenylation, ►*Gossypium*; Trapp SC, Croteau RB 2001 Genetics 158:811.

T

Terrestrial Radiation: Terrestrial radiation is emitted from the unstable isotopes in the soil, such as uranium-containing rocks. ►cosmic radiation, ►radiation hazard assessment, ►radiation effects, ►ionizing radiation, ►isotopes

Terrific Broth (TB): Bacterial nutrient medium containing bacto-tryptone 12 g, bacto-yeast extract 24 g, glycerol 4 mL, H_2O 900 mL, and buffered by 100 mL phosphate buffer.

Tertiary Structure: The three-dimensional arrangement of the secondary structure of the polypeptide chain into layers, fibers, or globular shapea. It is also a third order of complexity, folding or coiling the secondary structure once more. ►protein structure

Test Statistics: The specific procedure used to test a parameter or a null hypothesis.

Testa (seed coat): Maternal tissue in hybrids, therefore seed coat characters display delayed expression of recessive markers by one generation (e.g., to F_3) rather than in F_2.

Testcross: A cross between a heterozygote and a homozygote for the recessive genes concerned, e.g., (*AB/ab*) × (*ab/ab*); segregation is 1:1. ►recombination frequency; *Arabidopsis* testcross photo (see Fig. T31); Rédei unpublished)

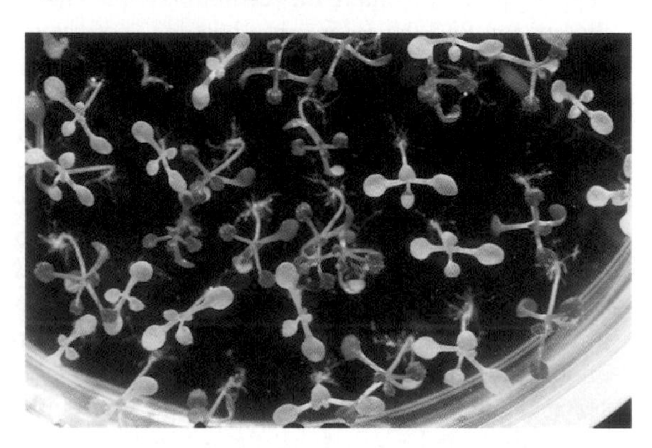

Figure T31. Segregation of a test cross progeny

Tester: A tester in a genetic cross is intended to reveal either the qualitative or quantitative gene content of the individual(s) tested. ►testcross, ►hybrid vigor, ►combining ability

Testes (sing. testis): The male gonads of animals, producing the male gametes. ►gonad, ►gamete, ►sex determination; Raymond CS et al 2000 Genes Dev 14:2587.

Testicles: Same as testes.

Testicular Feminization: Is a developmental anomaly, which occurs in humans and other mammals. The chromosomally XY individuals display female phenotype including the formation of a blind vagina (no uterus), female breasts, and generally the absence of pubic hairs (see Fig. T32). Usually, the individuals affected by this recessive disorder (human chromosome Xq11.1-q12) appear very feminine but sterile. Generally, they develop abdominal or somewhat herniated small testes. About 1.5×10^{-5} of the chromosomally males have this disorder. The condition is the result of a complete or partial deficiency or instability of the androsterone receptor protein (917 amino acids). The function of this androgen receptor protein may be either totally missing or only partial (Reifenstein syndrome). This receptor appears to be highly conserved among mammals. The gene extends to about 90 kb DNA. The protein binds to DNA by two domains encoded by exons 2 and 3, five exons code for androgen binding, while exon 1 has a regulatory function. ►chromosomal sex determination, ►hermaphrodite, ►hormone receptors, ►pseudoherma-phroditism, ►dihydrotestosterone, ►Swyer syndrome, ►sex reversal, ►SF-1; Boehmer AL et al 2001 J Clin Endocrinol Metab 86:4151.

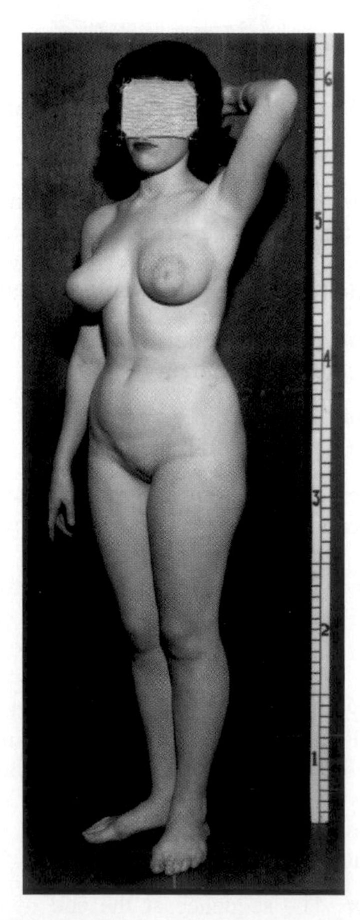

Figure T32. Testicular feminization. (Courtesy of Dr. McL. Morris)

Testicular Germ-Cell Tumor: The TGCT gene at Xq27 conveys susceptibility to this cancer affecting about 2×10^{-3} of men of Western European descent between ages 15–40.

Testin (TES, human chromosome 7q31-q31.2): A tumor suppressor in humans and mice. The protein contains three LIM motifs and is localized along actin fibers of the cytoskeleton. ►LIM domain; Drusco A et al 2005 Proc Natl Acad Sci USA 102:10947.

Testing, Genetic: ►genetic screening

Testosterone: The most important androsterone. Testosterone increases the risk of coronary heart disease and atherosclerosis. If testosterone is converted to estrogen by aromatase, the risk is reduced (see Fig. T33). ►animal hormones, ►hormonal effects on sex expression, ►progesterone

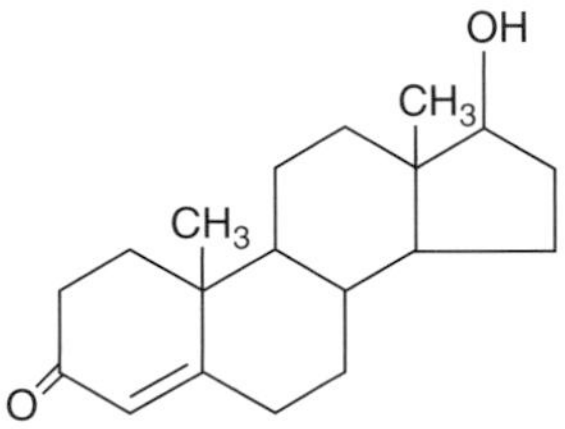

Figure T33. Testosterone

Testtube Baby: ►in vitro fertilization, ►intrauterine insemination, ►intra-fallopian transfer, ►ART

Testweight: The mass of 1,000 seeds or kernels randomly withdrawn from a sample. ►absolute weight

TET: Tetracycline antibiotic, inhibits protein synthesis, genetic (selectable) marker in the pBR plasmids. ►pBR322, ►tetracycline

TET (tubal embryo transfer): Essentially, the same as intrafallopian transfer of zygotes. ►intra-fallopian transfer, ►ART

Tet A: The tetracycline resistance protein that controls the efflux of the antibiotic from the cells. ►tetracycline, ►TetR

Tetanus: Tetanus toxin, ►tetany

Tetany: A highly stimulated condition of the nervous and muscular system caused by low levels of calcium due to various diseases and to infection by *Clostridium tetani*. Vaccines are available. The anaerobic, spore-forming soil bacterium with a genome size of 2,799,250 bp includes 2,368 open reading frames; it also has a low G + C content plasmid of 74,082 bp. ►biological weapons, ►*Clostridium perfringens*; Brüggemann H et al 2003 Proc Natl Acad Sci USA 100:1316.

Tethering: Bringing together two distantly located nucleic acid sequences either by DNA looping, or catenanes or RNA lariats. ►introns, ►lariat RNA

Tet-OFF: ►tetracycline

Tet-ON: ►tetracycline, ►targeting genes

TetR: The tetracycline repressor (homodimer) regulates its own expression as well as that of the antiporter (TetA, which exports the drug from the cells) at the level of transcription and it is activated by $[Mg - Tc]^+$ complex. The tetracycline repressor binds to the dual operators of the TetR (repressor) and the TetA genes and blocks their transcription. Tetracycline alters the conformation of the repressor protein, which then is not capable of blocking transcription by the RNA polymerase at the operators even in some eukaryotes. This system is thus suitable for exogenous regulation—by tetracycline—of a gene promoter fused by genetic engineering to the TetR system. The failure of the system in eukaryotes is determined by the toxicity of tetracycline to a particular organism. ►tetracycline, ►antiport; Scholz O et al 2001 J Mol Biol 310:979.

Tetracycline (Tet): An antibiotic that prevents the binding of amino acid-charged tRNAs to the A site of the ribosomes (see Fig. T34). The tetracycline repressor gene (TetR) interferes with the expression of tetracycline resistance. Tetracycline in the medium prevents the binding of TetR to the DNA and relieves repression. Tetracycline is widely used as a tool in turning genes ON and OFF, respectively. When the TetR gene is fused to the activation domain of the Herpes simplex virus protein VP16, it becomes a very effective Tet-responsive transactivator (tTA) of other genes. In the absence of Tet or Tet analogs, the tTA binds the Tet operator (TetO) and initiates transcription. When Tet is present in the culture medium, the transactivator (TA) binds the antibiotic, the DNA binding is disrupted, and transcription is prevented. A mutant form of TetR protein binds to the TetO only in the presence of the antibiotic doxycycline, a Tet

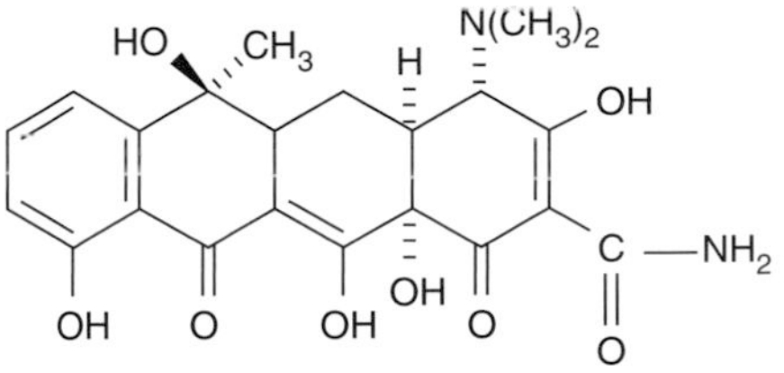

Figure T34. Tetracycline

analog, and when fused to VP16 (rtTA) transcription is activated. Modification f the D ring yielded highly effective new enantiomorphs against some antibiotic resistant pathogens (Charest MG et al 2005 Science 308:395). ►doxycycline, ►pBR322, ►ribosomes, ►protein synthesis, ►targeting genes, ►tTA, ►rtTA, ►transactivator, ►Gene-Switch, ►antibiotic resistance; Chopra I, Roberts M 2001 Microbiol Mol Biol Revs 65:232; Stebbins MJ et al 2001 Proc Natl Acad Sci USA 98:10775.

Tetrad, Aberrant: The allelic proportions deviate from 2:2 because of polysomy, gene conversion, non-disjunction, suppression, etc. ►polysomy, ►gene conversion, ►non-disjunction, ►suppression

Tetrad Analysis: The meiotic products of ascomycetes (occasionally some other organisms) stay together as the four products of single meiosis, as a *tetrad*. In some organisms, tetrad formation is followed by a post-meiotic mitosis within the ascus, resulting in spore *octads*. If the four spores are situated in the same linear order as produced by the two divisions of meiosis it is an *ordered tetrad.*

In the ordered tetrad, considering two genes *A* and *B*, three arrangements of the spores (parental ditype [PD], tetratype [TT], non-parental ditype [NPD]) can be distinguished as seen in the Figure T35. The parental ditype (PD) indicates no crossing over; tetratype (TT) reveals one recombination between the two genes and the second division segregation of the B/b alleles reveals recombination between the B/b gene and the centromere (see Fig. T36).

The nonparental ditype (NPD) is an indication of double crossing over between the two gene loci. The PD, TT, and NPD may appear even if the genes are in separate chromosomes. An excess of PD over NPD is an indication of linkage. If the deviation from the 1:1 ratio between PD and NPD is small, a *chi square test* may be used to test the probability of linkage by the formula: $\chi^2 = (PD - NPD)^2/(PD + NPD)$.

By counting the number of tetrads of the above three types, *recombination frequency between the two loci* can be calculated as $\frac{[1/2]TT + all\ NPD}{all\ tetrads}$, and recombination frequency between the B/b *gene and the centromere can be calculated* as $\frac{TT[1/2]}{all\ tetrads}$. (Dr. Fred Sherman recommended to me to use for the gene-gene map distance the formula of Dr. David Perkins [Genetics 34:607], i.e. 100{[0.5(TT) + 6NPD]/[PD + TT + NPD]}.)

The recombination frequencies (if they are under 0.15) multiplied by 100 provide the map distances in centiMorgans. If the recombination frequencies are larger, mapping functions should be used. From the genetic constitution of the tetrads, a great deal of information can be revealed about recombination. When the four meiotic products are not in the order brought about by meiosis, the tetrad is *unordered.* For the estimation of gene–centromere distances from unordered tetrad data, one must rely on three markers, from which no more than two are linked, and algebraic solutions are required (e.g., Whitehouse 1950 Nature 165:893, see unordered tetrads). Tetrad analysis is most commonly used in ascomycetes (*Neurospora, Aspergillus, Ascobolus, Saccharomyces,* etc.) (see Fig.T37) yet it can be applied to higher plants where the four products of male meiosis stick together (*Elodea, Salpiglossis*, orchids, *Arabidopsis* mutants). Using transgene constructs encoding pollen-expressed fluorescent proteins of three different colors in the *qrt1* mutant, which retains pollen in the tetrad stage, segregation of the fluorescent alleles in 92,489 pollen tetrads could be

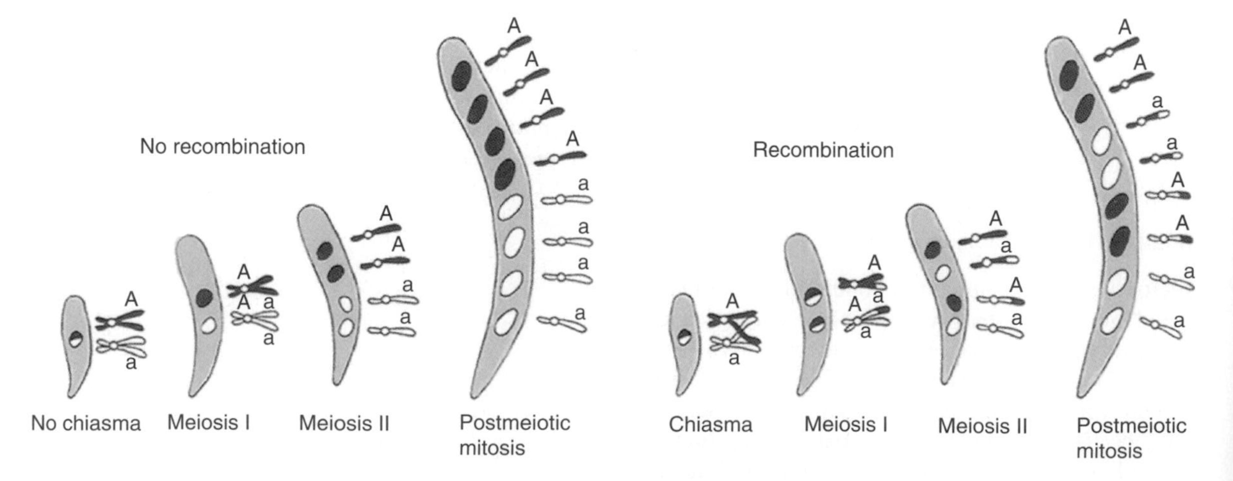

Figure T35. Spore tetrads and octads without and with recombination

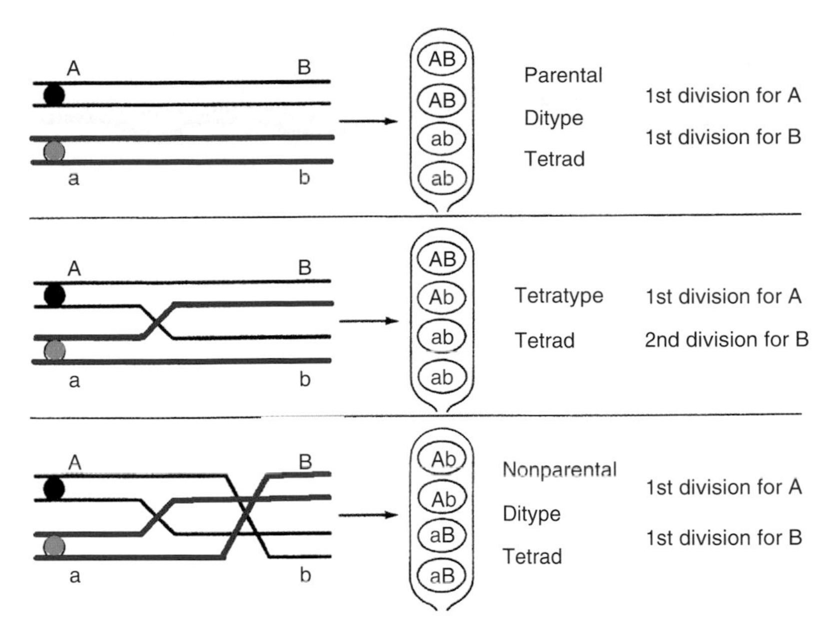

Figure T36. Transactions and results examplified by two gene loci in an ordered tetrad. Gene *A* is so close to the centromere that practically no recombination occurs between them. (Diagram after Barratt RW et al 1954 Adv Genet 6:1)

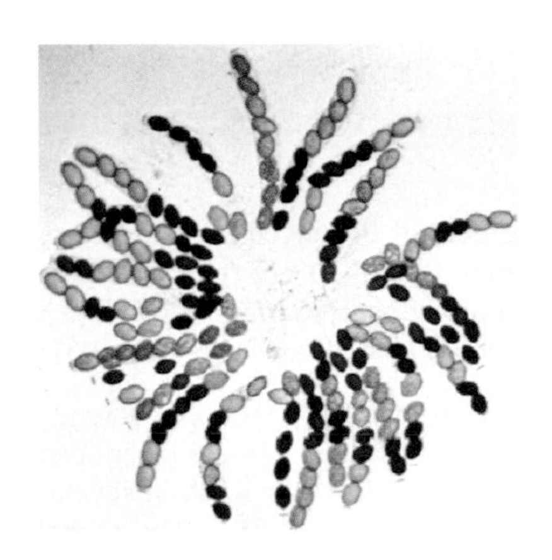

Figure T37. *Neurospora* octads. (Courtesy of Dr. David Stadler)

observed (see Fig. T38). Correlation between developmental position and crossover frequency, temperature dependence for crossingover frequency, meiotic gene conversion, as well as interference were detectable (Francis KE et al 2007 Proc Natl Acad Sci USA 104:3913). In *Drosophila* with attached X-chromosomes half-tetrad analysis is feasible. Since several genomes of higher eukaryotes have been sequenced, molecular markers are available for tetrad analysis for the cases when the products of individual meioses can be identified. ►unordered tetrads, ►half-tetrad analysis, ►meiosis, ►mapping, ►linkage, ►mapping function, ►four-point analysis of tetrads

Tetrahydrofolate (THF): The active (reduced) form of the vitamin folate, a carrier of one-carbon units in oxidation reactions, and a pteridine derivative. ►hyper-homocysteinemia

Tetrahymena pyriformis: (2n = 10) is a ciliated protozoon with linear mtDNA. Its rRNA transcripts are self-spliced. ►splicing, ►mtDNA, ►telomerase, ►micronucleus, ►macronucleus, ►chromosome breakage programmed; Turkewitz AP et al 2002 Trends Genet 18:35; http://www.lifesci.ucsb.edu/~genome/Tetrahymena/; http://www.ciliate.org/.

Tetralogs: Paralogous gene groups originated by duplications (polyploidy), e.g., the families of multiple kinases or other genes. Some of these genes can be deleted without serious phenotypic consequence, yet their point mutations (dominant negative) may be quite deleterious, especially if they have multiple interactive domains and function in protein complexes. Some of the evolutionarily redundant genes can further evolve by selectable mutations. ►polyploidy, ►paralogous loci, ►point mutation, ►dominant negative, ►duplication, ►co-ortholog, ►semi-ortholog; Spring J 1957 FEBS Lett 200:2.

Tetralogy of Fallot: ►Fallot's tetralogy

Tetraloop: In structured RNAs, duplex runs are connected by loops of 5′GNRA (or UNCG or CUYG) tetranucleotides (N = any nucleotide, R = purine, Y = pyrimidine). They are involved in long-range molecular interactions, such as in hammerhead ribozymes and introns. ►ribozymes, ►introns,

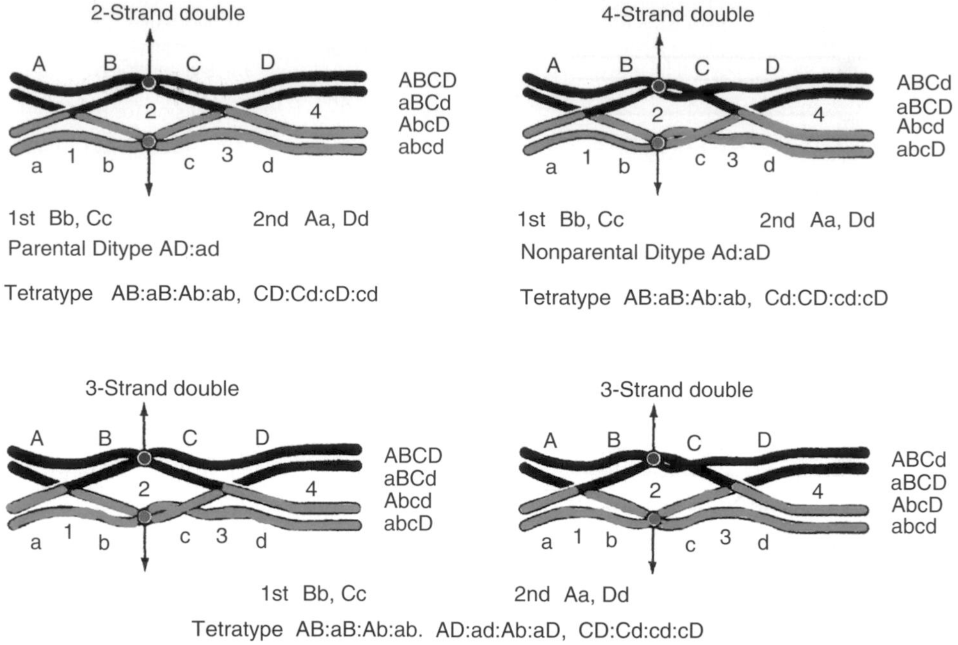

2-Strand double

A B C D ABCD aBCd AbcD abcd

a 1 b c 3 d 2 4

1st Bb, Cc 2nd Aa, Dd

Parental Ditype AD:ad

Tetratype AB:aB:Ab:ab, CD:Cd:cD:cd

4-Strand double

A B C D ABCd aBCD Abcd abcD

a 1 b c 3 d 2 4

1st Bb, Cc 2nd Aa, Dd

Nonparental Ditype Ad:aD

Tetratype AB:aB:Ab:ab, Cd:CD:cd:cD

3-Strand double

A B C D ABCD aBCd Abcd abcD

a 1 b c 3 d 2 4

3-Strand double

A B C D ABCd aBCD AbcD abcd

a 1 b c 3 d 2 4

1st Bb, Cc 2nd Aa, Dd

Tetratype AB:aB:Ab:ab. AD:ad:Ab:aD, CD:Cd:cd:cD

Figure T38. Four-point cross with genes in both arms of the chromosomes. It is a five-point cross if we consider the centromere as a genetic marker. From the spore order we can determine even if the chromatids rotated 180° after the exchange. (After Emerson S 1963 p 167. In: Methodology in Basic Genetics. Burdette WJ (Ed.) Holden-Day, San Francisco)

►stem-loop, ►ribose zipper, for illustration of a tetraloop see ►ribonuclease III; Baumrook V et al 2001 Nucleic Acids Res 29:4089; Koplin J et al 2005 Structure 13:1255.

Tetranucleotide Hypothesis: A historical assumption that nucleic acids are made of repeated units of equal numbers of the four bases (A, T/U, G and C), and because of this monotonous structure could not qualify for being the genetic material. It was probably first proposed around the turn of the twentieth century by the great German chemist, Albrecht Kossel, but became a rather widely accepted view during 1909 to 1940 through the brilliant yet erroneous work of Phoebus Levine. (See Levine PA 1909 Biochem Zeit 17:120; Levine PA 1917 J Biol Chem 31:591).

Tetraodon: ►*pufferfish*

Tetraparental Offspring: Tetraparental offspring results if in vivo fused blastulas of different matings are implanted together into the uterus. ►allophenic, ►multiparental hybrids, ►chimera

Tetraplegia (quadriplegia): Paralysis of the four limbs, caused by brain injury. Various mechanical devices have been designed to alleviate the incapacitation. Recent experiments inserting small neuromotor prostheses into the brain seem promising in restoring control of movement (Hochberg LR et al 2006 Nature [Lond] 442:164). Brain-computer interfaces can provide means to operate computer cursors (Santhanam G et al 2006 Nature 442:195).

Tetraplex (quadruplex): Consecutive guanine sequences may take four-stranded parallel or antiparallel conformations in the DNA or RNA in several different configurations (see Fig. T39). The tetraplex structure may have biological significance for the telomeres, specific recombination of the immunoglobulin genes, dimerization of the HIV genome, etc. ►telomere, ►G-quadruplex, ►immunoglobulins, ►HIV; Simonsson T 2001 Biol Chem 382:621; Weisman-Shomer P et al 2002 Nucleic Acids Res 30:3672; Saccà B et al 2005 Nucleic Acids Res 33:1182; ►tetrasomic

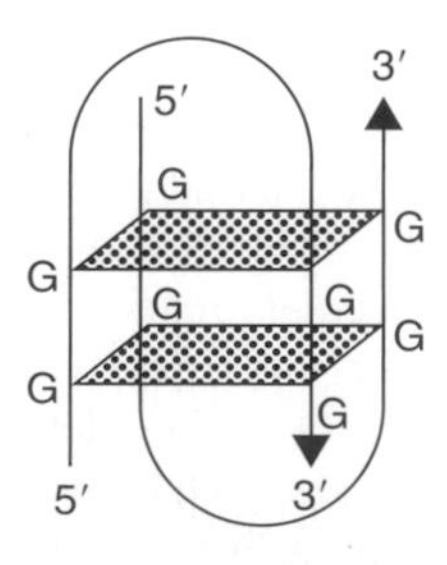

Figure T39. Guanine (G) tetraplex

Tetraploid: Tetraploids have four sets of genomes (4x) per nucleus. Tetraploid plants have broader leaves and petals, and larger seeds (see Fig. T40). ►auto-polyploidy, ►tetrasomic

Figure T40. *Cardaminopsis petraea* tetraploid (4x) and diploid (2x); Rédei, unpublished

Tetraploid Embryo ComplEmentation: Embryonic stem cells have a unique ability to complete embryonic development after nuclear transplantation or after injection into tetraploid host blastocysts. Tetraploid mouse blastocysts cannot autonomously develop into embryos normally, but they can do so when complemented by diploid embryonic stem cells. The success is substantially better if the nuclei are not derived from inbred cells. Tetraploid embryo complementation assay has shown that mouse embryonic stem cells (ES) cells alone are capable of supporting embryonic development and adult life of mice. Newly established F1 hybrid ES cells allow the production of ES cell-derived animals at a high enough efficiency to directly make ES cell-based genetics feasible (George SHL et al 2007 Proc Natl Acad Sci USA 104:4455). ►stem cells, ►nuclear transplantation, ►embryo culture; Ewggan K et al 2001 Proc Natl Acad Sci USA 98:6209.

Tetrapod: The collective evolutionary designation of four-limbed vertebrate animals in contrast to lobe-finned fishes (sarcopterygians). ►missing link, Long JA et al 2006 Nature [Lond] 444:199.

Tetrasomic: In tetrasomics, one chromosome is present in four doses in the nuclei. Tetrasomy exists in tetraploids for all chromosomes and may not involve any serious anomaly although generally the fertility is reduced. Tetrasomy for individual chromosomes may have more serious consequences because of genic imbalance. Tetrasomy is usually not tolerated by animals. Tetrasomic mosaicism for human chromosome 12p leads to developmental anomalies, mental retardation, defects in the central nervous system and speech, etc. ►polyploidy, ►autopolyploid, ►trisomy, ►sex-chromosomal anomalies in humans, ►cat-eye syndrome

Tetrasomic-Nullisomic Compensation: In allotetraploids, the homoeologous chromosomes may compensate for each other, and thus if two chromosomes are missing (nullisomy), and one type or their homoeologues is present in four copies (tetrasomy), the individuals may function rather well, depending on the particular chromosomes involved. When, however, in the presence of nullisomy a non-homoeologous chromosome is substituted, the condition is worsened. ►chromosome substitution, ►nullisomic compensation; Morris R, Sears ER 1967 p 19. In: Wheat and Wheat Improvement. Quisenberry KS, Reitz LP (Eds.) Am Soc Agron Madison, Wisconsin.

Tetraspanins: Cell membrane molecules functioning in cell adhesion, motility, proliferation, differentiation signal transduction, and fertilization. The TM4SF2 protein encoded at Xp11.4 and involved in determining a non-syndromic mental retardation also belongs to the tetraspanin family of proteins. ►CD9, ►CD63, ►CD82, ►mental retardation; Cannon KS, Cresswell P 2001 EMBO J 20:2443.

Tetratrico Sequences (TPR): Amphipathic α-helical amino acid tracts punctuated by proline-induced turns. CDC16, CDC23, and CDC27 of yeast and their homologs in other eukaryotes all contain tandem repeated ~34-residues. Such motifs exist in more than 100 proteins and may control mitosis and RNA synthesis in the cells. ►APC, ►CDCs; Wendt KS et al 2001 Nature Struct Biol 8:784; Cortajarena AL, Regan L 2006 Protein Sci 15:1193.

Tetratype: The meiotic products concerned with two genes show four types of combinations (e.g., *AB Ab aB* and *ab*). ►tetrad analysis

Tetrazolium Blue: Tetrazolium blue detects oxidation-reduction enzyme activity and thus identifies living cells and cancerous metabolism.

Tetrodotoxin: ►toxins

T-Even Phages: The designation has even numbers such as T2, T4, etc. ►bacteriophages

Texas Red: A fluorochrome with excitation at 580 nm and emission peak at 615 nm. ►fluorochromes

Text-Mining: The search for specific (overlooked) connections in the literature for the purpose of revealing specific trends/rules in multiple papers. ►literature mining

Textpresso: The information retrieval and extraction system for *Caenorhabditis*. (See http://www.textpresso.org/).

TF: Transcription factors such as TF I, TF II, or TF III, involved in the control of transcription by pol I, pol II, or pol III, respectively; TFs assist transcription by cooperation with other binding proteins. ►transcription factors, ►TFIIS, ►open transcription complex

TF: ►transferrin

Tfam: ►mtFAM

TFD (transcription factor decoy): A double-stranded nucleotide sequence (e.g., GGG ACTTT CC / CCC TGAAAGG) that may sequester a transcription factor, e.g., NF-κB, within the cytoplasm by virtue of the attachment of the TF to a sequence homologous to its recognition site in the upstream region of the gene(s) in the nucleus. As a consequence, the nuclear gene may be (partially) silenced. Such an approach may be exploited for the control of oncogenes. ►NF-κB, ►antisense technologies; Mann MJ, Dzau VJ 2000 J Clin Invest 106:1071.

TFE: Helix-loop-helix leucine zipper transcription factor. ►DNA binding protein domains

TFG: ►transforming growth factor

TFII: ►transcription factors, ►TBP, ►PIC

TFIIS: The eukaryotic transcript elongation stimulatory factor. It makes yeasts hypersensitive to 6-azauracil, which reduces the intracellular UTP and GTP pool. In *Drosophila*, the transcripts of two protein factors, N-TEF (negative transcript elongation) and P-TEF, regulate elongation. The positive effect is mediated by ATP-dependent phosphorylation of the RNA polymerase. ►backtracking, ►elongation factors, ►transcription factors, ►azauracil, ►DSIF, ►NELF, ►TEF, ►DRB, ►PITSLRE, ►transcript elongation; Lindstrom DL, Hertzog GA 2001 Genetics 159:487; Ubukata T et al 2003 J Biol Chem 278:8580.

TFII: Synonymous with TBP.

TFM: ►testicular feminization

TFO (triple-helix-forming oligonucleotide): TFO may be used to interfere with the activity of genes.

The effective TFO needs to recognize 10 to 17 bases in order to be gene specific. Some problems arise in vivo by the fact that genes in the chromosomes are organized as chromatin. TFO seems to mediate mutation, nucleotide excision repair, and transcription-coupled repair and consequently enhances mutation rate induced by UV-A in the presence of psoralen dye sensitization at 365 nm. Psoralen is a bifunctional photoreagent. TFO may also stimulate targeted gene conversion using nucleotide excision repair and may be antiviral. The triplex-directed, site-specific mutagenesis is much improved when cationic phosphoramidate linkages replace the phosphodiester backbone, particularly by employing *N,N*-diethyl-ethylenediamine to target short polypurine tracts (see Fig. T41). ►antisense RNA, ►antisense DNA, ►triple helix formation, ►triplex, ►inhibition of transcription, ►peptide nucleic acid, ►targeting genes, ►directed mutation, ►DNA repair, ►ultraviolet light, ►gene conversion, ►psoralen, ►pseudoknot; Vasquez KM et al 2000 Science 290:530; Vasquez KM et al 2001 J Biol Chem 276:38536; Besch R et al 2002 J Biol Chem 277:32473.

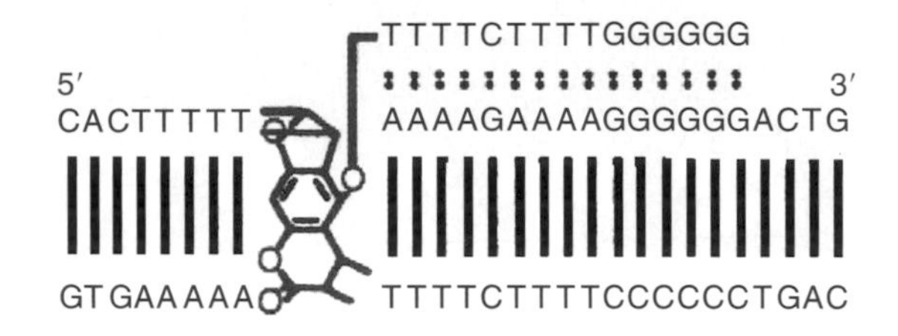

Figure T41. TFO. Modified after Barre FX et al 2000 Proc Natl Acad Sci USA 97:3084

TFR: Transferrin receptor. ►transferrin

TFTC (TBP-free TAF_{II}-containing complex): ►transcription factors, ►TAF; Walker AK et al 2001 EMBO J 20:5269.

TGF: The transforming growth factors are cell proliferation inhibitors and their loss may be involved in cancerous growth. In humans, the TGFA gene is in chromosome 2p13, TGFB1 in 19q13.1-q13.3, TGFB2 in 1q41, and TGFB3 in 14q24. A member of the family, Vg1, a factor localized to the vegetal pole of the animal embryo, may be involved in the induction of mesoderm formation in the embryo. The TGF-β family cytokines activate the receptors of the heterodimeric serine-threonine kinases. TGF-β binds directly to TGF receptor II, a kinase. The bound TGF-β then, binds by the TGF receptor I and becomes phosphorylated (P) by it. Type I receptor contains a repeated GS (Gly.Ser) domain and a binding site for FKBP12. TGFR I then transmits the cytokine signal along the signaling cascade and the target genes may be turned on. In humans, about 30 members of the TGFβ protein family exist and various homologs—although in smaller numbers—occur in other species. TGFβ acts on cytotoxic T lymphocytes (CTL) and specifically inhibits perforin, granzyme A and B, Fas ligand, and interferon (Thomas DA, Massagué J 2005 Cancer Cell 8:369).

Ligand induced phosphorylation of receptor-activated Smads (R-Smads) is catalyzed by TGFβ type I receptor kinase. The PPM1A/PP2Cα phosphatase dephosphorylates and promotes nuclear export of Smad2/3. Ectopic expression of PPM1A abolishes TGFβ-induced antiproliferative and transcriptional responses but depletion of this phosphatase boosts

TGFβ signaling in mammalian cells (Lin X et al 2006 Cell 125:915). The SMAD3 protein affecting the pathway outlined in the diagram is also a co-activator of the vitamin D receptor by forming a complex with the steroid receptor co-activator-1 protein (SRC-1) in the cell nucleus (see Fig. T42).

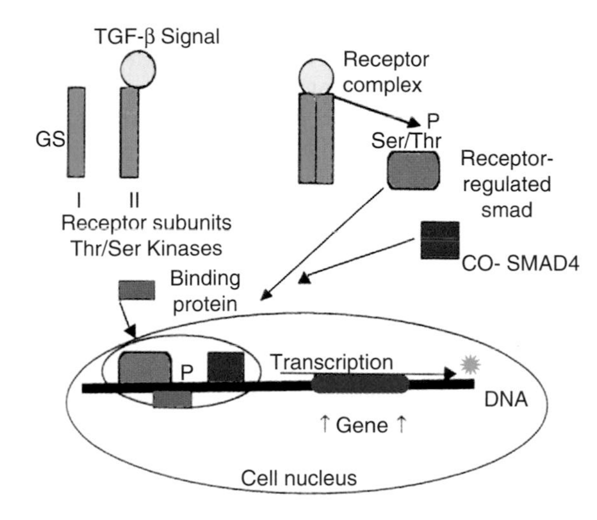

Figure T42. TGF signaling

TGF-α is an epidermal growth factor-like molecule. Mutations leading to chondrodysplasia involve members of the TGF family (CDMP1). Beta-glycan and endoglin cell surface proteins are also involved somehow with TGF. Endoglin and ALK1 (a type I receptor) mutations are found in hereditary hemorrhagic telangiectasia. TGFβ receptor I mutations were found in different types of cancers and TGFβ mutations may be responsible for heart disease, hypertension, osteoporosis, and fibrosis. TGFβ mutations upregulate the transforming growth factor in the arterial wall in Loeys-Dietz syndrome and in aortic aneurism cause arterial tortuosity. The human genome includes 42 TGFβ genes compared with nine in *Drosophila* and six in *Caenorhabditis*. ►activin, ►bone morphogenetic protein, ►leukemia inhibitory factor, ►ectoderm, ►serine/threonine kinase, ►signal transduction, ►SMAD, ►immunophilins, ►SRC-1, ►cytokines, ►EGF, ►MDM2, ►FK506, ►PP2A, ►telangiectasia hereditary hemorrhagic, ►Müllerian ducts, ►Loey-Dietz syndrome; Massagué J 1998 Annu Rev Biochem 67:753; Massagué J, Chen Y-G 2000 Genes & Development 14:627; Massagué J et al 2000 Cell 103:295; Zwaagstra JC et al 2001 J Biol Chem 276:27237; Wakefield LM, Roberts AB 2002 Current Opin Genet Dev 12:22; Waite KA, Eng C 2003 Nature Rev Genet 4:763.

TGN: ►trans-Golgi network

T-GURT (trait-specific genetic use restriction technology): T-GURT is applied to products of plant breeding with the aid of biotechnology and makes the plants express special traits such as salt or drought tolerance on condition that a special (chemical) treatment is applied. The plants will grow without the special treatment but do not express the agronomically advantageous special trait to the same extent. Such seed stocks protect the proprietary interests of the seed companies, which developed them. ►terminator technology, ►GMO, ►RBF

T_H (or Th): T helper cell. They can be T_H1 type activating macrophages and T_H2 type, which activate primarily B cells. T-bet transcription factor regulates the activity of T_H1 cells. T_H1 cells produce IL-2, interferon-γ (IFN-γ), and tumor necrosis factor-β (TNF-β) and offer defense against intracellular pathogens by aiding cellular immunity. T_H2 cells make interleukin-4, -5, -6, -10, and -13 (IL-4, IL-5, IL-13) against extracellular pathogens and allergens, promote antibody formation, and have direct cytolytic activity. Interleukin-12 (IL-12) favors the development and maintenance of T_H1, whereas IL-4 promotes the differentiation of T_H2. IL-18 may both promote and hinder T_H2 activity by its effect on IL-13. T_H3 cells may express transforming growth factors. A new type of T helper cells T_H17 provides protection against extracellular bacteria and is involved in autoimmune diseases. T_H17 cells produce IL-17. Its development is linked to IL-23 and IL-12 and transforming growth factor TGFβ, which is antagonized by interferon γ and IL-4 (Mangan PR et al 2006 Nature [Lond] 441:231). The differentiation factor for T_H17 is not IL-23 but IL-6 and TGF-β together. There is a different path for the development of the regulatory T_H17 cells that inhibit autoimmune tissues and T_H17 cells, which induce autoimmunity (Betteli E et al 2006 Nature [Lond] 441:235). ►T cells, ►B cell, ►T-bet, ►dendritic cell, ►macrophage, ►IL-4, ►IL-5, ►IL-6, ►IL-10, ►IL-12, ►IL-13, ►IFN-γ, ►TNF, ►TCCR; O'Garra A, Arai N 2000 Trends Cell Biol 10:542; Mullen AC et al 2001 Science 292:1907; Helmby H et al 2001 J Exp Med 194:355; Ho I-C, Glimcher LH 2002 Cell 109: S109; minireview: Reiner SL 2007 Cell 129:33.

Thalamus: A double-egg-shaped area deep within the basal part of the brain, involved in transmission of sensory impulses. ►fatal familial insomnia, ►brain human

Thalassemia: Hereditary defects of the regulation (or deletions) of hemoglobin genes, causing anemia. In the thalassemias, generally the relative amounts of the various globins is affected because of deletions of the hemoglobin genes. *Thalassemia major* is the most severe form of the disease in patients homozygous for a defect in the two β-chains and who have an excessive amount of the F hemoglobin (HPFH: *hereditary persistence of fetal hemoglobin*). *Cooley's anemia* is also caused by β chain defects. *Thalassemia minor* is a relatively milder form with some hemoglobin A_2 present; usually, slight elevation of the F hemoglobin is characteristic for the heterozygotes. In the *β–thalassemias*, different sections of the β globin gene family from human chromosome 11p15.5 (5′- ε G_γ ψβ δ β − 3′) are missing. The deletions may involve only 600 bp from the 3′ end as in β^0 or about 50 kbp, eliminating most of the family beginning from the 5′ end and retaining only the β gene at the 3′ end in human chromosome 11p. In the β chain gene, about 100 point mutations and various deletions have been analyzed. In severe cases, the symptoms of the β thalassemias are anemia, susceptibility to infections, bone deformations, enlargement of the liver and spleen, iron deposits, delayed sexual development, etc. These may appear within a few months after birth. In the *α-thalassemias* (in human chromosome 16p13), various members of the α chain gene cluster (5′- ζ ψζ ψα ψα α2 α1 ∂- 3′) or even all four α chains are defective; the latter case results in the lethal hereditary disease, hydrops fetalis (Bart's hydrops), entailing accumulation of fluids in the body of the fetus and severe anemia, resulting in prenatal death. In α-thalassemia-1, both the α1 and the α2 genes are deleted, in thalassemia α-*thal*-2 only either the left (α *thal* 2*L*) or right end (α *thal* 2*R*) of the α-gene(s) is lost. The *Hb Lepore hemoglobin* is the consequence of unequal crossing over within the α gene cluster, resulting in N-terminal-δβ-C protein fusion. The N-βδ-C reciprocal protein fusion is called *Hb anti-Lepore* or *F thalassemia*. One of these types of hemoglobins was first reported in 1958 in the Lepore Italian family. Since then, a large number a Lepore type hemoglobins were discovered in various parts of the world and named Greece, Washington, Hollandia, etc., hemoglobin. The Xq13.1-q21.1 locus (ATRX) is a regulator of α-thalassemia and it usually involves mental retardation and a variety of developmental abnormalities. Thalassemias have much higher incidence worldwide in areas where malaria is a common disease. Homozygotes for thalassemias rarely contribute to the gene pool, thus the heterozygotes appear to have selective advantage in the maintenance of this condition. Some recent studies, however, indicate that α thalassemia homozygotes have a higher incidence of uncomplicated childhood thalassemia and splenomegaly (enlargement of the spleen an indication of infection by *Plasmodium*). This increase of susceptibility to the relatively benign *P. vivax* may provide some degree of immunization in the later stages of life against the more severe disease caused by *P. falciparum* infection. This relatively high incidence of thalassemias in tropical and subtropical areas and the well-known molecular genetics mechanism makes this disease a candidate for somatic gene therapy by either bone marrow transplantation or transformation. The prevalence of thalassemias varies from about 10 to 5 × 10^{-4}. However, in some geographically isolated areas in the Mediterranean region, the frequency of the thalassemias may be much higher. Thalassemias are frequently associated with sickle cell anemias that are β globin defects and co-morbidity aggravates the conditions (see Fig. T43). Prenatal diagnosis is possible by the use of protein or DNA technologies. Fetal mutation may be identified after PCR amplification of chorionic samples (9–12 weeks) or in directly withdrawn cells from the amniotic fluid after the third month of pregnancy. Hydrops can be identified by ultrasonic techniques during the second trimester. Transformation by a functional β-globin gene may ameliorate the condition in a mouse model. Erythroid specific expression of the transduced human α-globin gene and relatively high levels of expression of the human α-globin gene were observed in mice receiving the lentiviral vector by yolk sac vessel injection at midgestation. However, the expression decreased to low levels on long-term follow up (Han X-D et al 2007 Proc Natl Acad Sci USA 104:9007). ►hemoglobin, ►methemoglobin, ►sickle cell anemia, ►plasma proteins, ►hemoglobin evolution, ►unequal crossing over, ►genetic screening, ►PCR, ►trimester, ►*Plasmodium*, ►antisense RNA, ►Juberg-Marsidi syndrome, ►ATRX, ►hydrops fetalis; Weatherall DJ 2001 Nature Rev Genet 2:245; β-thalassemia review: Rund D, Rachmilewitz E 2005 New England J Med 353:1135.

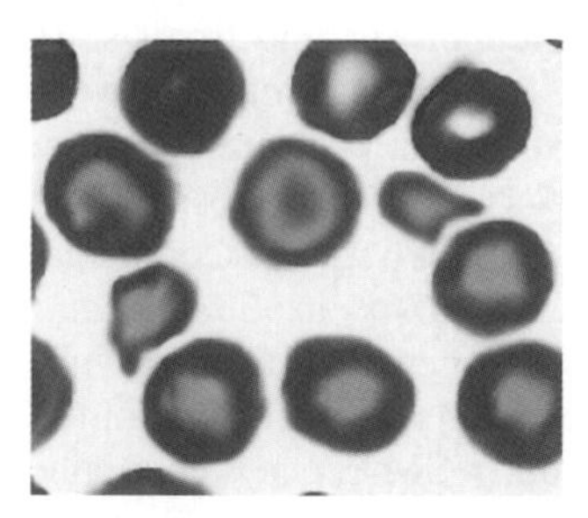

Figure T43. Thalassemia blood smear. Irregular cells are characteristic

Thalidomide (*N*-phtaloyl glutamimide): A sedative and hypnotic agent of several trade names. It had been used experimentally also as an immunosuppressive and antiinflammatory drug. Its medical use as a tranquilizer during pregnancy caused one of the most tragic disasters in medical history, resulting in severe malformations of primate embryos and newborns. It does not appear to be mutagenic and it is not teratogenic for other mammals. It is effective against multiple myeloma and possibly some other cancers. ►teratogen, ►phocomelia, ►amelia; Ashby J et al 1996 Mutation Res 396:45; Meierhofer C et al 2001 BioDrugs 15(10):681; Richardson P et al 2002 Annu Rev Med 53:629.

Thallophyte: A plant, fungus, or algal body, a thallus. ►thallus

Thallus: A relatively undifferentiated colony of plant cells without true roots, stem or leaves.

Thanatophoric Dysplasia: A dominant lethal human (4p16.3) dwarfism caused by deficiency of the fibroblast growth factor receptor 3, FGFR3. FGFR3 has constitutive tyrosine kinase activity, which activates the STAT1 transcription and nuclear translocation. This system also controls the $p21^{WAF1/CIP1}$ protein involved in cell cycle suppression. Its prevalence is about 2×10^{-4}. ►dwarfism, ►fibroblast growth factor, ►STAT, ►p21, ►receptor tyrosine kinase, ►achondroplasia, ►Schneckenbecken dysplasia

Thea (*Camellia sinensis*): 82 species, 2n = 2x = 30. The epigallocatechin-3 component of green tea appears to be an angiogenesis inhibitor and thus can have anticancer effect. ►angiogenesis, ►angiostatin, ►endostatin, ►polyphenols; Kuroda Y, Hara Y 1999 Mutation Res 436:69.

Theileriosis: An African or East Cost Fever parasitic *Theileria* infection of the cattle's immune system, causing disease and death to the infected animals. In the T lymphocytes casein kinase II, a serine/threonine protein kinase is markedly higher. This protein is not identical with the common casein kinase found in lactating animals. *Theileria annulata* and *T. parva* are tick-borne protozoa (n = 4, genome size 8,351,610 and 8,308,027 bp, respectively). (See Pipano E, Shkap V 2000 Ann NY Acad Sci 916:484; Pain A et al 2005 Science 309:131; Gardner MJ et al 2005 Science 309:134).

Thelytoky: Parthenogenesis from eggs resulting in maternal females. ►ant, ►arrhenotoky, ►deuterotoky, ►sex determination, ►chromosomal sex determination; Normark BB 2003 Annu Rev Entomol 48:397.

Theobromine: The principal alkaloid in cacao bean, containing 1.5 to 3% of the base. It is also present in tea and cola nuts. The TDLo orally for humans is 125 mg/kg. It is a diuretic, smooth muscle relaxant, cardiac stimulant, and a vasodilator (expands vein passages). ►TDLo, ►caffeine

Therapeutic Cloning: The generation of (human) stem cells that induces them to differentiate into a certain type of tissue which can be used to replace damaged tissues without the risk of rejection. The transfer stem cells are prepared by dislodging the nucleus of an oocyte and replacing it by the nucleus of an adult cell of the individual who is supposed to be treated by the stem cells. Such a modified oocyte (any type, possibly even from a different species) is then grown in culture until the blastocyst stage. The inner cell mass is then excised and cultivated in vitro until differentiated cells (either muscle, or neural, or hematopoietic cells) develop under the direction of the experimental conditions applied. Subsequently, the cells can be transplanted into the body of an individual. Nuclear transfer is possible between different species but the few experiments carried out so far could not maintain steady normal development. An alternative to the use of oocytes is to fuse an embryonic cytoplast with an appropriate karyoplast. The ES cells do not develop into an embryo even it is transplanted into a uterus because the extra-embryonic membranes required would not be formed. If such a stem cell is introduced into a blastocyst, chimeric embryos (host + ES) may form in the mouse. These cells may form different types of tissues (pluripotent) but are apparently not totipotent. The therapeutic cloning is thus similar in some respects to reproductive cloning but current laws prohibit the latter in the majority of countries. Some of the cloned embryos suffered from (lethal) developmental anomalies. Although the majority of the implanted stem cells develop rather normally, some may become teratogenic and carcinogenic when transferred into mice. ►stem cell, ►nuclear transplantation, ►graft rejection, ►grafting in medicine, ►cytoplast, ►karyoplast, ►blastocyst, ►totipotency, ►cloning animals and humans; Mitalipov SM 2000 Ann Med 32:462; Illmensee K 2002 Differentiation 69:167; Rideout WM III et al 2002 Cell 109:17.

Therapeutic Index: The therapeutic index reveals the dose of a drug with optimal, threshold, maximal tolerated or lethal, etc., effect in a particular organism or tissue. At high therapeutic index, a low dose gives the desirable effect without the possible toxic effects possible at higher doses.

Therapeutic Radiation: ►radiation hazard assessment

Therapeutic Vaccine: vaccination may force down the level of an infective agent (e.g., HIV) and then eventually the natural immune system may overpower the infection. ►vaccines

Thermal Asymmetric Interlaced-PCR: ►tail-PCR

Thermal Cycler: An automatic, programmable incubator use for the polymerase chain reaction. ►PCR

Thermal Neutrons: ►physical mutagens

Thermal Tolerance: Thermal tolerance is genetically determined and in plants may be affected by fatty acid biosynthesis and abscisic acid deficiency, etc. Heatshock proteins have also been credited with it. In *Tetrahymena thermophila*, a small cytoplasmic RNA (G8 RNA) is responsible specifically for thermotolerance independently from a heat shock response. ►temperature-sensitive mutation, ►heat-shock proteins, ►RNA G8, ►antifreeze protein

Thermatoga maritima: A thermophil (80° C optimum) bacterium; it has a completely sequenced circular DNA genome of 1,860,725 bp with 46% G + C content; 24% of its genes appear homologous to archaea bacteria. (Nelson KE et al 1999 Nature 399:323).

Thermodynamic Values: The thermodynamic values of chemicals are measured in calories; 1 calorie = 4.1840 absolute joule. ►joule

Thermodynamics, laws of: 1). When a mechanical work is transformed into heat, or the reverse, the amount of work is equivalent to the quantity of heat. 2). It is impossible to continuously transfer heat from a colder to a hotter body.

Thermoluminiscent Detectors: Thermoluminiscent detectors are used for personnel monitoring. In a small crystalline detector (lithium fluoride, lithium borate, calcium fluoride or calcium sulfate, and metal ion traces as activators), radiation is absorbed in the crystalline body and upon heating, is released in the form of light. The useful range is 0.003–10,000 rem. Radiophotoluminiscent (RPL) and thermally stimulated exoelectron emission (TSEE) detectors may be used instead of film badges in radiation areas. ►radiation measurements

Thermoplasma acidophilum: An archaea bacterium with a completely sequenced genome of ~1.5×10^6 bp. Its temperature optimum is 59 °C, pH opt. 2. Its cells are without cell wall. ►Archaea; Ruepp A et al 2000 Nature [Lond] 407:508.

Thermosome: ►chaperonins

Thermotolerance: Some mutants of plants display increased growth relatively to the wild type at higher temperature. Some kinds of temperature-sensitivity are associated with differences in fatty acid biosynthesis. Thermosensation neurons in mammals are located in the keratinocytes of the skin (Moqrich A et al 2005 Science 307:1468). The tolerance of higher temperature in plants is correlated with reduced level in trienoic fatty acids. Trienoic acid synthesis is controlled by omega-3 fatty acid desaturase in the chloroplast membrane. If this desaturase is silenced, photosynthesis is improved at higher temperature. ►cold-regulated genes, ►temperature-sensitive mutation, ►antifreeze protein, ►fatty acids; Mirkes PE 1997 Mutation Res 396:163.

THF: ►tetrahydrofolate

Theta Replication: θ is a stage of bidirectional form of replication in a circular DNA molecule (see Fig. T44). When the two replication forks reach about half way toward the termination sites, the molecule resembles the Greek letter θ. ►DNA replication, ►θ replication, ►bidirectional replication, ►replication, ►prokaryotes

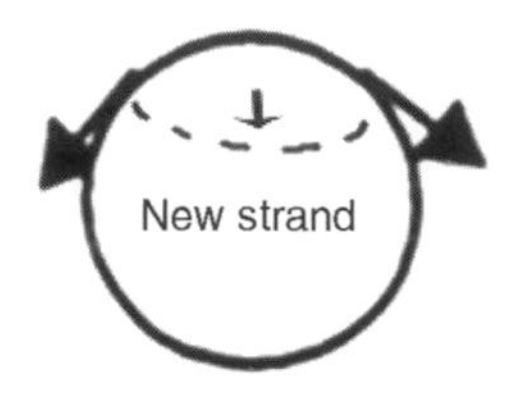

Figure T44. Theta replication

Thiamin (vitamin B_1): The absence of thiamin from the diet causes the disease beri-beri, alcoholic neuritis, and the Wernicke-Korsakoff syndrome. Its deficiency was the first auxotrophic mutation in *Neurospora,* and it is the most readily inducible auxotrophic mutation at several loci in lower and higher green plants, such as algae and *Arabidopsis* (see Fig. T45). The vitamin is made of two moieties: 2-methyl-4-amino-5-aminomethyl pyrimidine and 4-methyl-5-β-hydroxyethyl thiazole. The pyrimidine requirement of the thiamin mutants cannot be met by the precursors of nucleic acids. In *Arabidopsis*, more than 200 absolute auxotrophic mutations for thiamin have been isolated at five loci. This is remarkable because absolute auxotrophs in other metabolic pathways still need to be found in higher plants. ►thiamin pyrophosphate, ►Wernicke-Korsakoff syndrome, ►megaloblastic anemia, ►vitamins; Rédei GP, Koncz C 1992 Methods in Arabidopsis

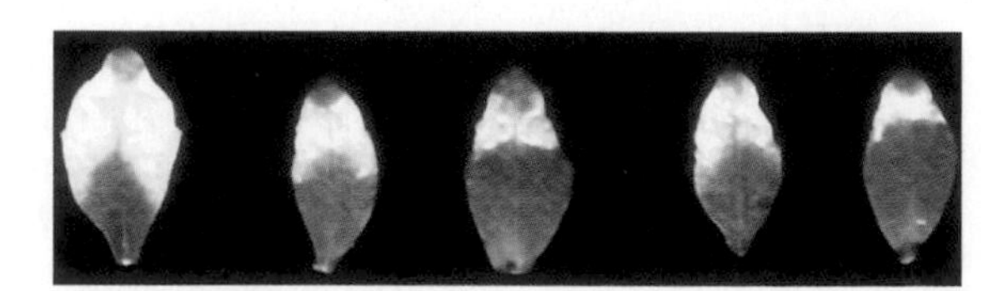

Figure T45. Thiamin deficiency symptoms on *Arabidopsis* leaves, appearing after the vitamin supplementation ceased. A bleached band appears at about the upper third-fourth part of the leaf blade where the metabolism is most active (From Rédei, G.P., unpublished)

research, pp 16–82. In: Koncz C et al (Eds.) World Scientific; Miranda-Ríos J et al 2001 Proc Natl Acad Sci USA 98:9736.

Thiamin Pyrophosphate: The coenzyme of vitamin B_1. It is an essential cofactor for pyruvate decarboxylase (alcoholic fermentation), pyruvate dehydrogenase (synthesis of acetyl Co-A), α–ketoglutarate dehydrogenase (citric acid cycle), transketolase (photosynthetic carbon fixation), and acatolactate synthetase (branched chain amino acid biosynthesis). ►thiamin; Frank RAW et al 2004 Science 306:872.

Thiazolidinediones: Antidiabetic (Type 2) drugs enhancing sensitivity to insulin. They are agonists and ligands to PPAR-γ receptors of adipocytes. They may have modest beneficial effects on the level of the desirable high-density lipoprotein (see Fig. T46). ►diabetes, ►PPAR, ►high-density lipoprotein, ►adipocytes, ►insulin; Yki-Järvinen H 2004 New England J Med 351:1106.

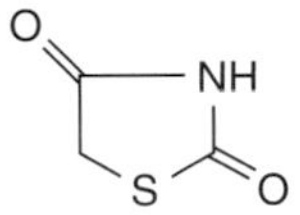

O NH S O

Figure T46. Thiazolidinedione

Thin-Layer Chromatography (TLC): The separation of (organic) mixtures within a very thin layer of cellulose or silica gel layer applied uniformly onto the surface of a glass plate or firm plastic sheet and used in a manner similar to paperchromatography. The material is applied at about 2 cm from the bottom and the plate is then dipped into an appropriate solvent (mixture). Generally, the substances are separated rapidly and with excellent resolution. Identification is made generally on the basis of natural color or with the aid of special color reagents. ►Rf

Thioester: Acyl groups covalently linked to reactive thiol (see Fig. T47); they are high-energy acyl carriers in coenzyme A; also thioester may assist the formation of oligopeptide without any cooperation by ribosomes). There are suggestions that thioethers are the relics of prebiotic conditions when sulfurous (volcanic) environment existed on the Earth and might have played a role in the origin of living cells by being energy carriers. ►sulfhydryl, ►disulfide

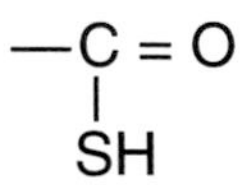

—C = O
|
SH

Figure T47. Thioester

Thioguanine (2-amino-1,7-dihydro-6H-purine-6-thione): An antineoplastic/neoplastic agent. Its delayed cytotoxicity is attributed to postreplicative mismatch repair. After incorporation into the DNA, it is methylated and pairs with either thymine or cytosine. The immuno-suppressant, azathioprine (6-[1-methyl-4-nitroimidazol-5-yl]-thiopurine), used in organ transplantation, can be converted to thioguanine and may be the cause of the increased incidence of cancer after transplantation. ►DNA repair, ►mismatch, ►base analogs, ►hydrogen pairing; Nelson JA et al 1975 Cancer Res 35:2872.

Thiol (SH^-): Thiol compounds may reduce DNA breakage by scavenging radiation induced hydroxyl radicals and may repair the radicals chemically on the DNA by hydrogen atom transfer. Thiols have various regulatory roles in oxidative metabolism and nitrosative stress. ►sulfhydryl; Paget MSB, Buttner MJ 2003 Annu Rev Genet 37:91.

Thiopurine-S-Methyltransferase (TPMT): An enzyme, which can inactivate mercaptopurine and azathiopurine drugs, and is used for controlling cancerous growth. If the gene encoding TPMT is deficient for two of the same alleles (homozygous), excessive amounts of the drugs may accumulate and lead to potentially lethal hematopoietic toxicity unless much reduced (5–10% of the usual) doses are administered to the affected individuals. ►azathiopurine, ►mercaptopurine; Evans WE 2002 Pharmacogenetics 12:421.

Thioredoxins (TRX): ~12 kDa dithiol proteins that mediate the reduction of disulfide bonds in proteins. Also, light regulates photosynthesis through reduced thioredoxin, linked to the electron transfer chain by ferredoxin. Light also modulates translation in the chloroplast by redox potential. In insects it may substitute for glutathione reductase. Histone deacetylase inhibitors (suberoylanilide hydroxamic acid, and a benzamide) cause an accumulation of reactive oxygen species (ROS) and activation of caspases in cancer cells but not in normal cells. In normal cells, thioredoxin, a reducing agent, accumulates due to histone deacetylase inhibition but this does not happen in transformed cells. Thus, a selective apoptosis results, indicating therapeutic potentials (Ungerstedt JS et al 2005 Proc Natl Acad Sci USA 102:673). The thioredoxin domain of *E. coli* processivity factor represents a molecular switch, which regulates interaction of T7 DNA polymerase with other proteins of the replisome (Hamdan SM et al 2005 Proc Natl Acad Sci USA 102:5096). ►photosynthesis, ►photosystems, ►glutaredoxin, ►lysosomes, ►pullulanase, ►oxidative stress,

►processivity, ►replisome; Yano H et al 2001 Proc Natl Acad Sci USA 98:4794.

Thiostrepton ($C_{72}H_{85}N_{19}O_{18}S_5$): An antibiotic, binding to *E. coli* rRNA and 10 folds less effectively to yeast ribosomes. (See Porse DT et al 1998 J Mol Biol 276:391).

Thiotepa: An alkylating agent, formerly also used as an antineoplastic drug. It induces a very high frequency of sister-chromatid exchange. ►alkylating agent, ►sister chromatid exchange

Thiouracil: RNA base analog, carcinogen (?). Thiouracil—just as uracil—can base-pair with both adenine and guanine, while the analog-containing nucleic acids show enhanced resistance to nucleases and display increased thermostability. 2-Thiouridine regularly occurs at the first anticodon position (s^2-U34) in place of uridine in the glutamate, lysine, and glutamine tRNAs. Thiouridine assures appropriate wobble for precise translation on the ribosome. The modified s^2-U is also required for the proper assembly of the HIV-1 viral genome and for the viral reverse transcriptase primer. 4-Thiouridine (s^4-U8) can be found at the junction of the acceptor stem and D stem of the tRNA (see Fig. T48).

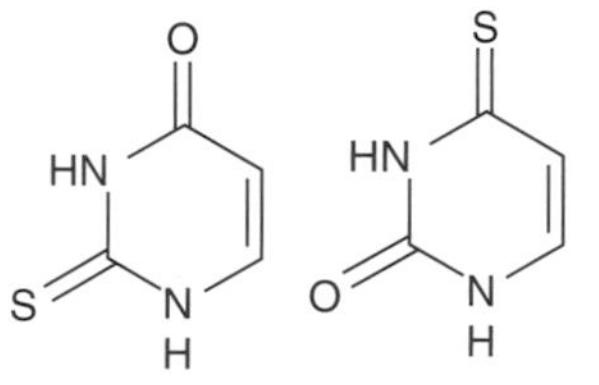

2-Thiouracyl 4-Thiouracyl

Figure T48. Left: 2-thiouracyl; right: 4-thiouracyl

In *E. coli*, the MnmAS thiouridylase synthesizes s^2-U with the aid of IscS desulphurase, which eliminates sulphur from L-cysteine. S is then transferred to MnmA via the Tus relay and it is incorporated in a precise manner into the correct tRNA site. The structural bases of the complex process was revealed by Numata T et al 2006 (Nature [Lond] 442:419). ►antisense technology, ►anticodon, ►tRNA, ►wobble, ►Tus, ►acquired immunodeficiency syndrome; Diop-Frimpong B et al 2005 Nucleic Acids Res 33:5297.

Third Messengers: Third messengers propagate signals of the second messenger and thus activate or deactivate a series of genes. ►second messenger

Thomsen Disease: ►myotonia

Thorax: The chest of mammals, the segment behind (posterior) the head in *Drosophila*.

Threading: A combinatorial alignment of amino acid sequences and DNA base sequences. On this basis a quantitative measure of protein-DNA binding can be derived statistically using, e.g., the Z score. The process permits a conclusion with regard to macromolecular structure. ►Z, ►binding protein; Papaleo E et al 2005 J Mol Model 21:1.

Three-Hybrid System: A construct that detects the role of RNA-protein interactions (see Fig. T49). ►two-hybrid system, ►one-hybrid binding assay, ►split-hybrid system; Koloteva-Levine N et al 2002 FEBS Lett 523:73.

Three M Syndrome (6p21.1): An intrauterine and postnatal growth retardation anomaly caused by impaired endovascular trophoblast invasion and reduced placental perfusion. It is caused by mutations in gene CUL7, which assembles the E3 ubiquitin

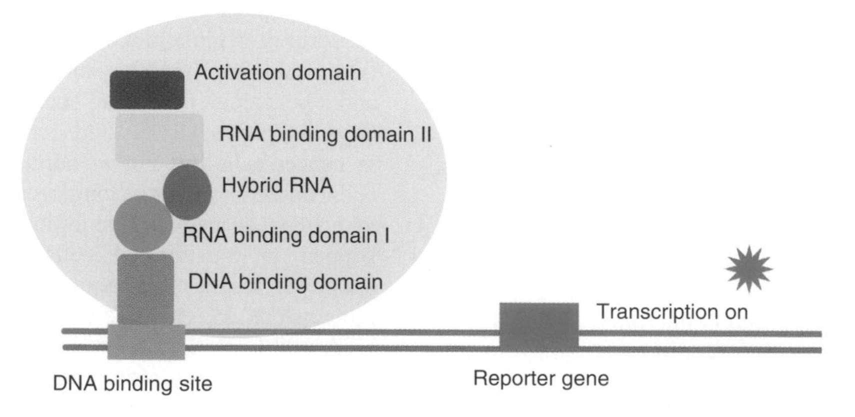

Figure T49. Three-hybrid system. The DNA-binding domain (e.g., LexA bacterial protein), the RNA-binding domain (e.g., MS2 phage coat protein), the top-binding proteins contain an RNA binding domain (e.g., IRP1[iron-response element]), and an activation domain (e.g., Gal4). The hybrid RNA (two MS2 phage RNAs) links the proteins together and in case of function the reporter gene (e.g., *LacZ*) is turned on. (Modified after Gupta DJS et al 1996 Proc Natl Acad Sci USA 93:8496)

ligase complex (Skp1, Fbx29 and ROC). ►ubiquitin; Huber C et al 2005 Nature Genet 37:1119.

Three-Point Testcross: The three-point testcross uses three genetic markers and thus permits the mapping of these genes in a linear order. ►mapping genetic, ►testcross

Three-Way Cross: A three-way cross is one in which a single-cross (the F_1 hybrid of two inbred lines) is crossed with another inbred. The purpose is to test the performance of the single-cross. ►combining ability

Threonine: This amino acid is derived from oxaloacetate via aspartate and immediately from:

$(HO)_2(PO)OCH_2CH(NH_2)COOH$
O-phosphorylhomoserine
→ $CH_3CH(OH)CH(NH_2)COOH$
threonine synthase threonine

Threonine dehydratase degrades threonine into α-ketobutyrate and NH_4^+; and then, after another five steps, isoleucine is produced from α-ketobutyrate. The first of these steps is mediated by threonine deaminase. ►isoleucine-valine biosynthetic steps, ►threoninemia

Threonyl tRNA Synthetase (TARS): TARS charges $tRNA^{Thr}$ by threonine. It is encoded in human chromosome 5 in the vicinity of leucyl tRNA synthetase gene. ►aminoacyl-tRNA synthetase

Threose: A D-aldose sugar synthesized from D-glyceraldehyde (see Fig. T50).

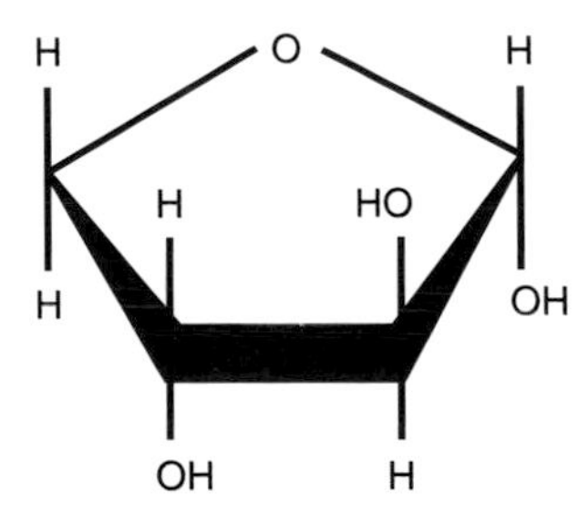

Figure T50. Threose

Threose DNA: A synthetic molecule containing threose in place of deoxyribose (see Fig. T51).

Threose DNA and RNA are capable of base pairing despite the presence of the sugar analog.

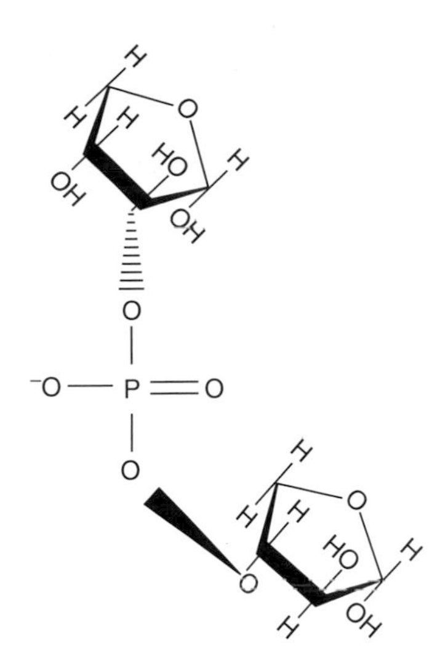

Figure T51. Threose DNA

Threshold Traits: Threshold traits are expressed conditionally when the liability reaches a certain level. These characters are frequently under polygenic control and yet they may fall into two recognizable classes—irrespective of whether they exhibit it or not—although with special techniques more classes may be revealed. Many pathological syndromes may fall into this category thereby making it very difficult to determine the genetic control mechanisms. ►syndrome, ►liability

Thrombasthenia: Autosomal dominant and autosomal recessive forms involving a platelet defect causing anemia and bleeding after injuries. The recessive form was assigned to human chromosome 17q21.32. ►platelet, ►anemia, ►thrombophilia

Thrombin (fibrinogenase): A serine protease enzyme that converts fibrinogen to fibrin and causes hemostasis (blood clotting). Thrombin also plays another role of an anticoagulant by the proteolytic activation of protein C in the presence of Ca^{2+}. These seemingly conflicting functions are modulated by allosteric alteration brought about by Na^+ and the thrombomodulin protein, changing the substrate-specificity of thrombin. The level of the activated protein C is further increased by a compound named LY254603. The protease-activated G protein-coupled receptor of thrombin (PAR1) is vitally important for embryonic development of mice. ►fibrin, ►fibrin-stabilizing factor deficiency, ►hemostasis, ►antihemophilic factors, ►plasmin, ►protein C, ►PAR, ►anticoagulation; Coughlin SR 2000 Nature [Lond] 407:258; Griffin CT et al 2001 Science 293:1666; carcinogenesis regulation by thrombin: Nierodzik ML, Karpatkin S 2006 Cancer Cell 10:355.

Thrombocytes: The same as platelets.

Thrombocytopenia (TAR syndrome): An autosomal dominant bleeding disease (10p11.2-p12) caused by low platelet counts. Autosomal recessive (11q23.3-qter deletion and 21q22.1-q22.2) forms are associated with heart and kidney anomalies and absence of the radius (the thumb side arm bone), phocomelia, etc. Another thrombocytopenia is linked in human chromosome Xp11. It is allelic to the Wiskott-Aldrich gene, WAS. Mutation in GATA1 may also cause the disease. The May-Hegglin anomaly, Fechtner syndrome, and the Sebastian syndrome involve macrothrombocytopenia (giant platelets) and leukocyte inclusions besides some specific symptoms. All three of these diseases show mutations in non-muscle myosin heavy chain 9. In some instances, the Epstein syndrome and the Alport syndrome display macrothrombocytopenia. ►Holt-Oram syndrome, ►Roberts syndrome, ►phocomelia, ►Wiskott-Aldrich syndrome, ►thrombopathic purpura, ►May-Hegglin anomaly, ►Alport disease, ►Kassabach-Merritt syndrome, ►GATA, ►dyserythropoietic anemia, ►platelets, ►giant platelet syndrome, ►radioulnar synostosis; Heath KE et al 2001 Am J Hum Genet 69:1033.

Thrombomodulin: ►thrombin, ►anticoagulation

Thrombopathia (essential athrombia): A collection of blood clotting anomalies caused by problems in aggregation of the platelets. ►platelet abnormalities, ►hemophilias, ►hemostasis

Thrombopathic/Thrombocytopenic Purpura (TTP, called also von Willebrand-Jürgen's syndrome, 9q34): TTP is caused by an autosomal recessive condition with symptoms resembling the bleeding and platelet abnormalities present in the Glanzmann's disease. The primary defect may involve the platelet membrane. Other symptoms may vary from case to case. Blood transfusion may shorten the time of bleeding. The basic defect is in the ADAMS13 zinc metalloproteinase genes. ►platelet abnormalities, ►hemophilias, ►hemostasis, ►thrombocytopenia, ►Glanzmann's disease; Levy GG et al 2001 Nature [Lond] 413:488.

Thrombophilia: A complex blood clotting disease that may occur as a consequence of mutation in any of the genes of antithrombin III, protein C, protein S, antihemophilic factor V, plasminogen, plasminogen activator inhibitor, fibrinogens, heparin cofactor, thrombomodulin, etc. Mutation at the 3′-untranslated region of the prothrombin gene may result in increased mRNA levels and protein synthesis due to this gain-of-function and to the disease. See separate entries, ►giant platelet syndrome, ►prothrombin deficiency; Gehring NH et al 2001 Nature Genet 28:389.

Thromboplastin: ►antihemophilic factors

Thrombopoietin: Thrombopoietin regulates blood platelet formation and megakaryocytopoiesis. It is member of a cytokine receptor superfamily and it is similar to erythropoietin and granulocyte colony-stimulating factor receptors. In the brain, it is pro-apoptotic (Ehrenreich H et al 2005 Proc Natl Acad Sci USA 102:862). ►megakaryocytes, ►platelet, ►erythropoietin, ►thrombocytopenia, ►G-CSF, ►apoptosis; Kato T et al 1998 Stem Cells 16:322; structure of receptor-binding domain: Feese MD et al 2004 Proc Natl Acad Sci USA 101:1816.

Thrombosis: A type of obstruction in blood flow caused by the aggregation of platelets, fibrin, and blood cells. Thrombosis may be caused by mutation in serpin. Antihemophilic factors V, VII, VIII, IX, XI, XII, the von Willebrand disease, tissue plasminogen activator, homocysteine and the activated protein C ratio display a genetic correlation with the incidence of thrombosis. ►antihemophilic factors, ►serpin, ►protein C, ►protein C deficiency, ►thrombocyclins, ►protein C, ►protein S, ►APC, ►cyclooxygenase, ►serpin, ►plasminogen activator; Souto JC et al 2000 Am J Hum Genet 67:1452.

Thrombospondin (THBS): Thrombospondin are glycoprotein inhibitors of angiogenesis. THBS1 (15q15) is a 180-K MW platelet membrane protein occurring on the endothelium, fibroblasts, and smooth muscles. THBS2 (6q27) binds thrombin, fibrinogen, heparin, plasminogen, etc. THBS2 functions as a tumor inhibitor by curtailing angiogenesis. Immature astrocytes express THBS and thereby promote synaptogenesis in the central nervous system (Christopherson KS et al 2005 Cell 120:421). ►angiogenesis, ►apoptosis, ►COMP; Hawighorst T et al 2001 EMBO J 20:2631; Rodríguez-Manzaneque CC et al 2001 Proc Natl Acad Sci USA 98:12485; Adams JC 2001 Annu Rev Cell Dev Biol 17:25.

Thrombotic Disease: ►protein C deficiency

Thromboxanes: Thromboxanes induce the aggregation of platelets and act as vasoconstrictors. They are antagonists of prostacyclin G_2 and may mediate intrauterine growth retardation. ►cyclooxigenase, ►prostaglandins

Thumb: ►DNA polymerase

Thy-1: A 19–25 kDa, single-chain glycoprotein expressed on mouse thymocytes but not on mature T cells of humans or rats.

Thylakoid: Thylakoids are flat, sac-like internal chloroplast membranes; when stacked they look like grana. Diacylglycerol galactolipids are common in these membranes but are absent from others. An estimated 80 proteins constitute the intra-thylakoid lumen. ►grana, ►chloroplast, ►girdle bands, (see Fig. T52); Dalbey RE, Kuhn A 2000 Annu Rev Cell Dev Biol 16:51; Schubert M et al 2002 J Biol Chem 277:8354; Kóta Z et al 2002 Proc Natl Acad Sci USA 99:12149.

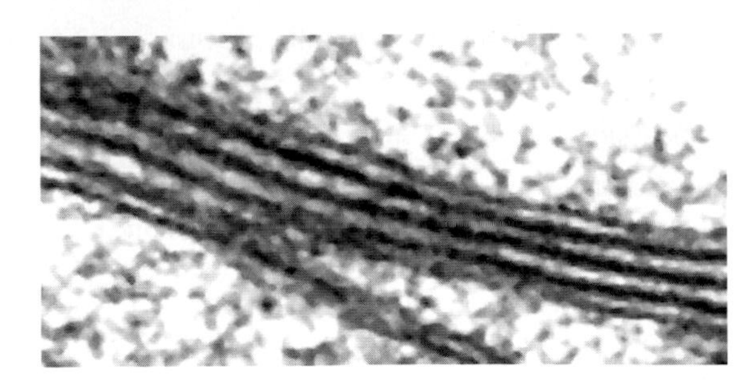

Figure T52. Thylakoids

Thymidine: A nucleoside of thymine; thymine plus pentose (deoxyribose or ribose). (See formula of thymine at T).

Thymidine Kinase (thymidine phosphorylase): ►TK

Thymidylate Synthetase: Thymidylate synthetase mediates the synthesis of thymidylic acid (dTMP) from deoxyuridine monophosphate (dUMP). Its gene is assigned to human chromosome 18p11.32. Its inhibition may lead to apoptosis. A number of nonsymbiotic archaea and bacteria lack this enzyme and use another protein, dependent on reduced flavin nucleotides rather than tetrahydrofolate as the source for reduction, as is the case in the majority of organisms. This is not a part of the salvage pathway. Thymidylate synthase may play an indirect role in carcinogenesis. ►apoptosis, ►salvage pathway, ►synthase; Myllykallio H et al 2002 Science 297:105.

Thymidylic Acid (nucleotide of thymine): Thymine + pentose + phosphate.

Thymine: A pyrimidine base occurring almost exclusively in DNA. An exception is the T-arm of tRNA. ►tRNA, ►pyrimidines, formula at T.

Thymine Dimer: UV light-induced damage in DNA, covalently cross-linking adjacent thymine residues through their 5 and 6 C atoms. This cyclobutane structure interferes with the replication and other functions of the DNA. Cytidine and uridine can also form similar dimers. The dimers can be eliminated by enzymatic excision and replacement replication (excision or dark repair). Alternatively, the dimers can be split by DNA photolyase, an enzyme that is activated by visible light maximally at 380-nm wavelength (light repair). ►DNA repair, ►cyclobutane dimer, ►pyrimidine-pyrimidinone photoproduct, ►cys-syn dimer, ►DNA polymerases; Medvedev D, Stuchebrukhov AA 2001 J Theor Biol 210:237.

Thymocytes: Thymocytes are precursors of T lymphocytes. Bone marrow stem cells migrate into the thymus and, in response to antigens, develop from naïve $CD4^-CD8^-$ cells with the cooperation of the T cell receptors into double positive $CD4^+CD8^+$ cells that undergo clonal selection with the assistance of MHC molcules and eventually become specific T cells. ►T cell, ►T cell receptor, ►MHC, ►antigen presenting cells, ►immune system, ►memory immunological, ►thymus

Thymoma: A cancer of the thymus. ►AKT oncogene

Thymosins: A group of about 40 different proteins with an apparent immune-boosting ability. Thymosin α1 consists of 28 amino acids and thymosin β4 contains 43. They may have therapeutic values. β4 binds monomeric actin and prevents its polymerization into filaments. Its expression increases when cancer cells metastase. ►actin, ►metastasis; Hall NR, O'Grady MP 1989 Bioessays 11:141; Goldstein AL, Garaci E (Eds.) 2007 Thymosins in Health and Disease Ann NY Acad Sci 1112.

Thymus: A bilobal organ with three functionally important compartments—the subcapsular zone, the cortex and the medulla. The immature lymphoid cells coming from the bone marrow invade it. The subcapsular space contains most of the immature $CD4^-$ and $CD8^-$ lymphoid stem cells. When they enter the cortex they begin to express the CD molecules and they rearrange the T cell receptors and form the different TCR αβ heterodimers. Defects in the genes of *Foxn1* and *Aire* of mouse cause immunodeficiency and autoimmunity, respectively. The epithelial cells in the cortex express MHC I and MHC II molecules. The fate of the T cell will be determined here in response to the MHC molecules carried by the antigen-presenting cells (thymic selection). The $CD4^+$ $CD8^-$ cells with TCR recognizing the MHC II complex, or the $CD4^-$ $CD8^+$ with TCR specific for MHC I will leave the thymus and populate the secondary lymphoid tissue. Those cells that do not acquire self-peptide-MHC specificity are eliminated. Those with strong avidity for self-peptide MHC die by apoptosis because they would be autoreactive (non-autoimmune). Aging (involution) reduces the thymic tissues and it was thought that this lead to a loss in their activity too. In the thymus, developing, autoreactive T cells undergo negative selection. Vascularized thymic lobe grafts from juvenile donors were capable of inducing tolerance in thymectomized juvenile hosts. Also, the aged, involuted thymus, transplanted as a vascularized graft into juvenile recipients, leads to rejuvenation of both thymic structure and function, suggesting that factors extrinsic to the thymus are capable of restoring juvenile thymic function to aged recipients. A rejuvenated aged thymus has the ability to induce transplant tolerance across class I MHC barriers indicating that it may be possible to manipulate thymic function in adults to induce transplantation tolerance after the age of thymic involution (Nabori S et al 2006 Proc Natl Acad Sci USA 103:19081).

Newer evidence indicates the thymus can generate new peripheral T cells after antiviral treatment of HIV patients. In mice, a second lymphoid thymic structure may be present in the neck (cervical thymus) and its function is very similar to the thoracic thymus although its number of thymocytes is $\sim 1.6 \times 10^5$ whereas in the thoracic organ it is $\sim 10^8$ (Terszowski G et al 2006 Science 312:284). In the C57BL/6 strain, cervical thymus occurs in about 50% whereas in BALB/c mice its occurrence is in over 90%. ►T cells, ►T cell receptor, ►TCR genes, ►MHC, ►autoimmune, ►spleen, ►HIV, ►nude mice

Thyroglobulin: An iodine containing protein in the thyroid gland and has a hormone-like action upon the influence of the pituitary hormone. ►animal hormones

Thyroid Carcinoma: The transforming sequence was localized to human chromosome 10q11-q12, and a tumor suppressor gene in human chromosome 3p may be involved. Susceptibility loci to nonmedullary thyroid carcinoma have been revealed at 2q21, 19p13.2 and 1q21 (McKay JD et al 2001 Am J Hum Genet 69:440).

Thyroid Hormone Resistance: Dominant hyperthyroxinemia mutations (ERBA2) due to defects in the thyroid hormone receptor (3p24.3). (See Tsai MJ, O'Malley BW 1994 Annu Rev Biochem 63:451).

Thyroid Hormone Responsive Element: ►TRE, ►hormones, ►hormone response elements, ►regulation of gene activity, ►goiter

Thyroid hormone unresponsiveness: Autosomal recessive in humans. ►hyperthyroidism

Thyroid Peroxidase Deficiency (TPO): A group of recessive human chromosome 2p13 defects involving the incorporation of iodine into organic molecules.

Thyroid Stimulating Hormone (TSHB, 1p13): The β-chain and its deficiency leads to hypothyroidism, goiter, and cretenism. It also regulates both bone formation and bone resorption. ►goiter, ►osteoporosis

Thyroid Transcription Factor: The TTF-2 defect is responsible for thyroid agenesis and cleft palate in mice and the human homologue, FKHL1 (9q22), also controls the same functions and also cochanal atresia (closure of the nasal passageways).

Thyronine (3p-[p(p-hydroxyphenoxy)-phenyl]-L-alanine): A component of the thyroglobulin of the thyroid hormone. It generally occurs as 3,5,3′-triiodothyrosine. ►VDR

Thyrotropic: Affecting (targeting) the thyroid gland.

Thyrotropin: A thyrotropin deficiency is apparently due to a defect in the thyroid stimulating hormone β-chain defect at 1p13. Deficiency of the thyrotropin-releasing hormone (TRH, 3q13.3-q21) results in hypothyroidism and various malformations including defects in the development of the central nervous system. Hyperthyroidism may involve fast pulsation of the heart and goiter and adenoma. Hyperthyroidism may be transient during pregnancy. ►goiter, ►adenoma

Thyroxin: ►goiter (see Fig. T53)

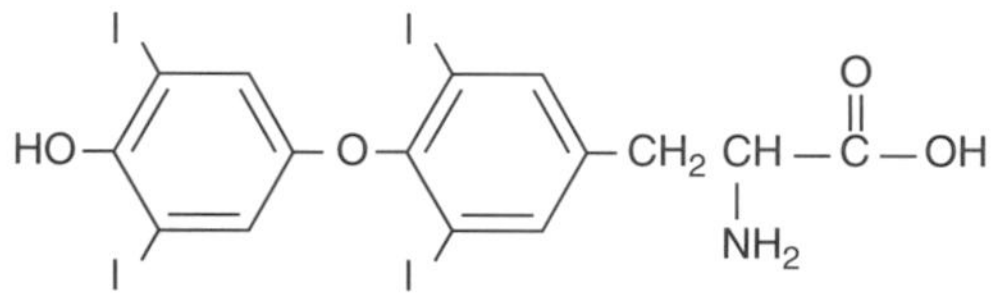

Figure T53. Thyroxin

Thyroxin-Binding Globulin (TBG, Xq22.2): A deficiency of the thyroxin-binding globulin generally does not lead to disease although in Graves disease it may be more common. ►goiter, ►Graves disease

TI Antigens: TI antigens stimulate antibody production independently of T cells and in the absence of MHC II molecules. The TI type 2 (TI-2) has polysaccharide antigens, which are usually large molecules with repeating epitopes that activate the complement and are rather stable. The TI-1 type is mitogenic for mature and neonatal B cells. ►antigen, ►antibody, ►T cell, ►MHC; Vinuesa CG et al 2001 Eur J Immunol 31:1340.

Ti Plasmid: A large (about 200 kbp) tumor-inducing plasmid of *Agrobacterium tumefaciens.*

It is responsible for the crown gall disease of dicotyledonous plants. There are a few particularly important regions in this plasmid. The T-DNA of the octopine plasmid shown is divided into the left (T_L) and right (T_R) segments and is flanked by the two border sequences (B_L and B_R); in-between are several genes. The nopaline-type Ti plasmid (pTiC58) has only a single 20 kb T-DNA. The virulence gene cascade, and its function, is described under virulence genes of *Agrobacterium*. The origin of vegetative replication (*oriV*) is functioning during proliferation of the cells; the nearby Inc (incompatibility) site determines host-specificity. The oriT is the origin of replication operated on during conjugation. The latter is often called bom (base of mobilization) or CON

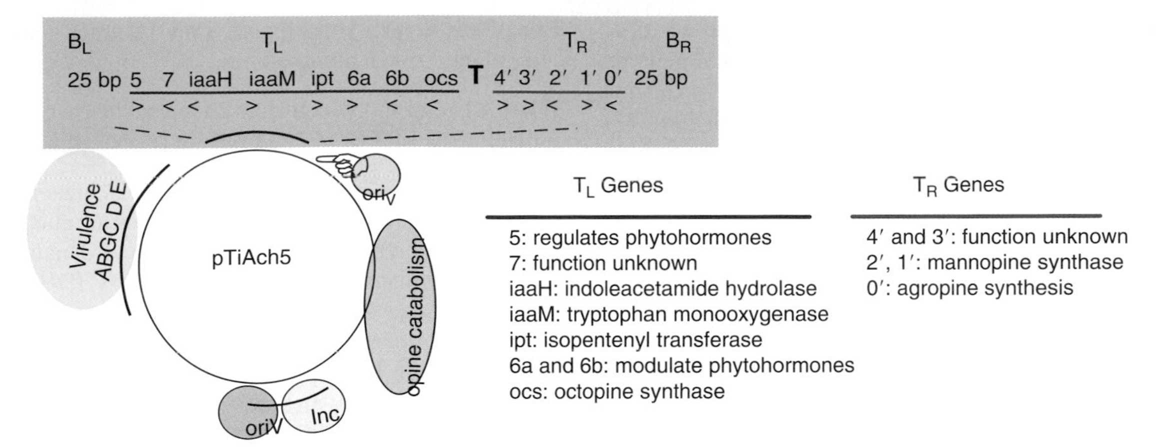

Figure T54. Major landmarks of the Ti octopine plasmid, pTiACH5. Arrowheads indicate the direction of transcription. **T** marks the T_L-T_R boundary

(conjugation) because the synthesis of the transferred DNA begins there. The actual transfer starts when the mob site encoded protein (Mob, synonym Tra) attaches to a specific nick site, a single-strand cut, and one of the strands (the T-strand) is transported to the recipient bacterial cell through the cooperation of some pore proteins. The oriT and Mob complex includes proteins TraI, TraJ, and TraH. TraJ attaches to a 19 bp sequence in the vicinity of the nick and binds also TraI, which is a topoisomerase. TraH promotes these bindings. The process is a rolling circle type replication and transfer. The integrity of virulence genes *virD* and *virB* are absolutely required for productive conjugation (see Fig. T54). ►*Agrobacterium*, ►T-DNA, ►virulence genes of *Agrobacterium*, ►transformation, ►rolling circle, ►conjugation; Hellens R et al 2000 Trends Plant Sci 5:446; Suzuki K et al 2000 Gene 242:331.

TIBO (O-TIBO and Cl-TIBO): The non-nucleoside (benzothiadiazepin-derivatives) inhibitors of HIV-1 reverse transcriptase. ►Nevirapine, ►acquired immunodeficiency syndrome, ►AZT

TIC (translation initiator-dependent cofactor): ►TAF

TID50: A 56 kDa mitochondrial protein of *Drosophila*, encoded at 2–104 in the nucleus by the *l*(2)*Tid* tumor suppressor gene. It is homologous to the DnaJ chaperone. ►chaperones, ►DnaK, ►tumor suppressor factors

Tie1, Tie2: Receptor tyrosine kinases, expressed during endothelial cell growth and differentiation of the blood vessels. ►vascular endothelial growth factor, ►Flk-1, ►Flt-1, ►tyrosine kinase, ►transmembrane proteins, ►signal transduction; Lin TN et al 2001 J Cereb Blood Flow Metab 21:690.

Tier: An ordered arrangement by increasing stringency of a series of tests, e.g., the first tier provides an overview of potential mutagens but subsequent tiers (using different techniques) may be necessary for clearance even when the first test might have been negative. Each tier may provide a different weight of evidence.

TIF1-α: Interacts with the steroid hormone receptor. ►steroid hormones, ►hormone receptors

TIF-1β: A transcription initiation factor of mouse binding to the core promoter of rRNA genes and controlling RNA polymerase I function. TIF1β is a component of the histone deacetylase complex. ►transcription factors

TIF1-γ (TRIM33/RFG7/PTC7/Ectodermin): A Transcriptional Intermediary Factor, a protein that selectively binds receptor-activated Smad 2 and Smad 3. It mediates erythroid differentiation in response to TGFβ (He W et al 2006 Cell 125:929). ►Smad, ►TGF, ►erythrocyte

Tight Junction (zonula occludens): A tight junction forms a seal between adjacent plasma membranes and provides a barrier to paracellular leakage of membrane lipids and proteins and thus guards cellular polarity (see Fig. T55). Tight junctions are apical domains of polarized epithelial and endothelial cells. The family of PDZ genes includes CLAUDIN 1, encoding a senescence-associated membrane protein; CLAUDIN 3, a *Clostridium perfringens* receptor [7q11]; CLAUDIN 4, a *C. p.* enterotoxin receptor; CLAUDIN 5, a velocardiofacial syndrome protein [22q11.2]; CLAUDIN 11, which mediates sperm and nerve functions [3q26.2-q26.3]; CLAUDIN 14, a deficiency responsible for deafness [21q22.3]; and CLAUDIN 16, which encodes paracellin, a paracellular conductance protein associated with hypo-magnesia [3q27]; and others that are distributed over the genome. ZO-1 and ZO-2 proteins determine where the claudins are polymerized (Umeda K et al 2006 Cell 126:741). ►deafness, ►velocardio facial syndrome, ►infertility, ►PDZ; Tsukamoto T, Nigam SK 1999 Am J Physiol 276:F737; Tsukita A et al 2001 Nature Rev Mol Cell Biol 2:285; Gonzalez-Mariscal L et al 2003 Progr Biophys Mol Biol 81:1.

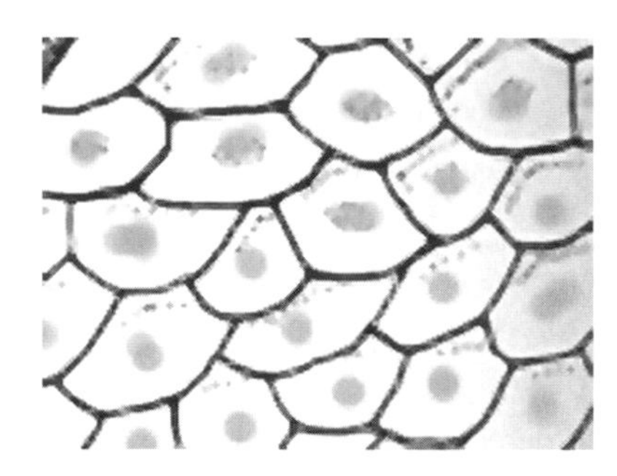

Figure T55. Tight juncture of epithelial cells

Tiger (*Panthera tigris*): 2n = 38. The tiger cat (*Leopardus tigrina*): 2n = 36.

Tiglic Acid: A methyl-2-butenoic acid. It is present with geranyl in the ornamental geranium and as an ester in the oil of chamomile (*Matricaria chamomilla*), a herbal medicine used as a tea or disinfectant.

Tiglicacidemia: A defect in the degradation of isoleucine to propionic acid leading to large amounts of tiglic acid accumulating in the urine.

TIGR: The Institute for Genomic Research in Rockville, MD, USA. ►databases

TIL: ►tumor infiltrating lymphocytes

Tiling: Generating a longer (minimally overlapping) DNA fragment map spanning the overall length of the genome or chromosome. By tiling, the transcribed sequences of the nucleus, chloroplast and mitochondria have been identified in *Arabidopsis* (Stolc V et al 2005 Proc Natl Acad Sci USA 102:4453). ►physical map, ►contig, ►DNA crystals, ►sequence-tagged connectors, ►genome project, ►scaffold in genome sequencing, ►TUF, ►noncoding RNA; Siegel AF et al 1999 Genome Res 9:297; Bertone P et al 2004 Science 306:2242; transcript maps: Cheng J et al 2005 Science 308:1149; Emanuelsson O et al 2007 Genome Res 17:886.

Tiling Microarrays: In microarray hybridization overlapping nucleotide sequences are used. ►microarray hybridization

Tiling Path Array: A tiling path array includes a set of minimally overlapping clones.

Tiller: A lateral shoot of grasses arising at the base of the plant. (See Li X et al 2003 Nature [Lond] 422:618).

Tilling (targeting induced local lesions in genomes): Attempts to induce mutations by chemical mutagens (e.g., ethylmethane sulfonate) and use of denaturing high-performance liquid chromatography (DHPLC) to detect base alterations by heteroduplex analysis. By such a procedure 246 *waxy* alleles were identified in wheat. This holds promise of useful modifications without transgenic methods (Slade AJ et al 2004 Nature Biotechnol 23:75). ►ethylmethanesulfonate, ►HPLC, ►*wx* gene, ►heteroduplex; McCallum CM et al 2000 Nature Biotechnol 18:455; Till BJ et al 2003 Genome Res 13:524; http://tilling.fhcrc.org:9366/.

TIM (transfer inner membrane): A protein complex that regulates the import of proteins into mitochondria. ►mitochondria, ►TOM; Meinecke M et al 2006 Science 312:1523.

Time of Crossing Over: The time of crossing over appears to coincide with the meiotic prophase (late leptotene and early diplotene, probably at zygotene). Some experimental data indicate that treatments at S phase have an effect on the outcome. It is difficult to assess, however, whether these effects are direct or indirect. Meiosis is under the control of a long series of genes acting sequentially and cooperatively and any of these may affect crossing over. ►crossing over, ►recombination, ►recombination mechanisms; Allers T, Lichten M 2001 Cell 106:45.

Time-Lapse Photography: Time-lapse photography records events continuously as they take place in time.

Timeless (Tim): A protein of *Drosophila* involved in the circadian clock. ►circadian clock

Timothy Syndrome (long QT syndrome, 12p13.3): A calcium ion channel defect leading to Ca^{2+} accumulation in the cells. It causes a variety of abnormalities such as heart arrhythmia, webbed fingers and toes, immune deficiency, intermittent hypoglycemia, cognitive abnormalities, and autism (Splawski I et al 2004 Cell 119:19; Splawski I et al 2005 Proc Natl Acad Sci USA 102:8089). ►QTL

TIMP (tissue inhibitor metalloproteinase): ►metalloproteinases, ►night blindness [Sorsby], Brew K et al 2000 Biochim Biophys Acta 477:267.

Tinkering in Evolution: The idea that evolution does not precede on the basis of purposeful plans; rather, it uses the means at hand at a particular stage. ►evolution; Jacob F 2001 Ann NY Acad Sci 929:71.

TIP: Also known as tumor-inducing principle. ►*Agrobacterium*

TIP (tail-interacting protein): TIP47 (47 kDa) recognizes the cytoplasmic domains of mannose-6-phosphate and binds to Rab9. It facilitates the endosome to Golgi transport and recruits effector proteins for appropriate membrane targeting. ►RAB, ►mannose-6-phosphate receptor, ►Golgi, ►endosome; Caroll KS et al 2001 Science 292:1373.

TIP (T cell immunomodulatory protein): The T cell immunomodulatory protein causes secretion of IFN-γ, TNF-α and IL-10 in vitro; in vivo it protects against the host-versus-graft disease. ►IFN, ►TNF, ►IL-10, ►host-versus-graft disease; Fiscella M et al 2003 Nature Biotechn 21:302.

TIP60 (HIV-1 Tat-interacting protein, 11q13): The 60 kDa regulator of HIV gene expression is a product of TIP60 (Kamine J et al 1996 Virology 216:357). The *Drosophila* TIP60 phosphorylates histone variant H2AX in case of DNA damage and remodels the chromatin by acetylation and replacement of H2A by an H2A variant (Kusch T et al 2004 Science 306:2084). ►acquired immunodeficiency, ►prostate cancer

Tipping Point: An epidemiological concept when treatment reaches a critical level resulting in a qualitative change, a turning of the tide of infections or inflammation (immune reaction) or the spread of the disease. It is a critical physiological stage in between health and disease.

TIR: Terminal inverted repeats as occuring in transposons ABCD——DCBA. It appears that hobo, Ac, TAM transposable elements have similar amino acid sequences although they occur in fungi, animals, and plants (Calvi BR et al 1991 Cell 66:465).

TIRAP (Mal): A PIP2 domain-containing adaptor controlling Toll-like receptor (TLR) signaling. It recruits MyD88 to TLR (Kagan JC, Medzhitov R 2006 Cell 125:943). Heterozygosity for leucine substitution at Ser180 in TIRAP attenuates Toll-like receptor signaling and increases resistance to infectious disease conveyed by *Plasmodium*, pneumococcal disease, bacteremia, and tuberculosis in diverse populations (Khor CC et al 2007 Nature Genet 39:523). ►Toll, ►Toll-like, ►*Plasmodium*, ►*Diplococcus/Pneumococcus*, ►mycobacteria, ►PIP2, ►MyD88

TIS1: ►nur77 and ►NGFI-B

TIS-8: A mitogen-induced transcription factor. ►NGFI-A, ►egr-1

TIS11: A transcription factor, inducible by various hormones (similar to Nup475). ►Nup475

TIS11b: A murine homolog of cMG1. ►cMG1

TIS11d: A transcription factor with 94% identity to 367 amino acids in TIS11b. ►TIS11b

Tiselius Apparatus: An early model of electrophoretic separation equipment.

Tissue Culture: The in vitro culture of isolated cells of animals and plants. ►cell culture, ►cell genetics, ►organ culture, ►cell fusion, ►somatic cell fusion, ►embryogenesis somatic, ►axenic, ►aseptic

Tissue Engineering: The purposeful culture of stem cells for producing tissues and organs. For example, long-lasting blood vessels can be formed in mice by co-implantation of vascular endothelial cells and mesenchymal precursor cells. Submillimeter size collagen rods, seeded with endothelial cells, could be assembled into a man-made vascular system that permits percolation of blood through it. Such a construct is expected to be suitable for supplying blood into tissue without thrombosis, would delayed clotting time and inhibit loss of platelets (McGuigan AP, Sefton MV 2006 Proc Natl Acad Sci USA 103:11466).

Another approach is to deliver signaling molecules and cells on a three-dimensional scaffold that supports cell infiltration and tissue organization. Artery bypass experiments have shown that nanofibrous scaffolds allowed efficient infiltration of vascular cells and matrix remodeling. Acellular grafts without mesenchymal stem cells (MSCs) resulted in significant intimal thickening, whereas cellular grafts (with MSCs) had excellent long-term patency and exhibited well-organized layers of endothelial cells (ECs) and smooth muscle cells (SMCs), as in native arteries. Short-term experiments showed that nanofibrous scaffolds alone induced platelet adhesion and thrombus formation, which was suppressed by MSC seeding. Cell-seeded scaffolds provide proper environment for native tissue-like environment and can be well aerated at a small scale (Hashi CK et al 2007 Proc Natl Acad Sci USA 104:11915). Soft lithography permits the control of surface topography and spatial distribution of molecules on scaffold surfaces.

Replica molding of biocompatible polymers from patterned silicon wafers constructed from poly (dimethylsiloxane) (PDMS) or poly(DL-lactide coglycolide) (PLGA) or poly(glycerol sebacate) (PGS) are used. Hydrogels with gradients of signaling or adhesive molecules or various cross-linking densities can provide means for migration, adhesion, and differentiation. Microfluidic technology can be employed on PLGA surfaces for the generation of two-dimensional patterns. A microtextured and nanotextured substrate may facilitate the generation of topographical features favoring cell adhesion, gene expression, and migration. For engineering bone tissues, porous ceramic or demineralized bone matrix support with bone marrow-derived mesenchymal cells and/or bone morphogenetic protein have been used with appropriate growth factors. Alternatively, in a space (called bioreactor) between the surface of a long bone and the membrane-rich periosteum (a connective tissue around the bone with the potentials of forming bone tissue) there is a niche for the injection of biocompatible calcium-alginate gel cross-linked in situ and it facilitates reconstitution of the functional living bone. The new microbioreactors provide a suitable niche, proper nutrients, and aeration for tissue fabrication. The engineered bone or liver tissue can then be transplanted to an area where bone tissue replacement is required. By inhibiting angiogenesis and promoting hypoxic conditions cartilage formation occurs (Stevens MM et al 2005 Proc Natl Acad Sci USA 102:11450). The availability of various types of stem cells and regulated culture conditions opens new approaches for the fabrication of special tissues. Experiments are in progress for the use of high throughput technology for tissue engineering. The goal is to replace defective human tissues with man-made tissues of high quality and without the danger of immune rejection. One product, Carticel, has been licensed for therapeutic use. Since 1997 more than 10,000 patients with injured knee cartilage have been treated with this autologous chondrocyte procedure (Parson A 2006 Cell 125:9). This is a very rapidly expanding field and by early 2006 more than 6,600 papers were published worldwide. By 2007, this number exceeded 8,500 although clinical applications are in infancy. ►stem cells, ►progenitor, ►anchorage dependence, ►microfluidics, ►lithography, ►biotechnology, ►wound healing; Shin H et al 2003 Biomaterials 24:4353; Koike N et al 2004 Nature [Lond] 428:138; Nature Mater 3:249; Jakab K et al 2004 Proc Natl Acad Sci USA 101:2864; review: Khademhosseini A et al 2006 Proc Natl Acad Sci USA 103:2480; osteoblast formation: Datta N et al 2006 Proc Natl Acad Sci USA 103:2488; muscle degeneration after in vivo delivery of myoblasts on a scaffold: Hill E et al 2006 Proc Natl Acad Sci USA 103:2494; smooth muscle regeneration was greatly improved by ectopic expression of human telomerase reverse transcriptase transfection: Klinger RY et al 2006 Proc Natl Acad Sci USA 103:2500; review: Atala A 2006 Current Opin Pediatr 18(2):167; review: MacNeil S 2007 Nature [Lond] 445:874.

Tissue Factor (TF): A cell surface glycoprotein mediating blood clotting after injuries by its interaction with the clotting factor VIIa. It is also important for hemostasis, inflammation, angiogenesis, atherosclerosis, and cancer. Protein S has an inhibitory effect. ►antihemophilic factors, ►blood clotting pathways, ►Protein S; Hackeng TM et al 2006 Proc Natl Acad Sci USA 103:3106.

Tissue Microarray (TMA): An adaptation of the microarray hybridization to tissue samples in order to reveal the cellular location of gene activity at the DNA, RNA or protein level. Cylindrical core specimens (biopsies) are acquired from formalin-fixed paraffin-embedded tissues and arrayed into high-density TMA blocks. Archival specimens are also suited for this analysis. Then, up to 300 5 μm sections are prepared for probing with DNA, RNA or protein using in situ hybridization (or FISH) or immunostaining. A single TMA allows the simultaneous study of the targets in thousands of specimens on microscopic slides By TMA on a single side the activity of one gene can be analyzed in 1000 tissue specimens. By microarray hybridization thousands of genes can be probed from a single tissue. TMA can be used to study cancerous tissue but it is not suitable for clinical diagnostics. This technique permitted the identification of some common transcriptional mechanisms in various types of cancer (Rhodes DR et al 2004 Proc Natl Acad Sci USA 101:9309) ►microarray hybridization, ►FISH, ►immunostaining; Kallioniemi O-P et al 2001 Hum Mol Genet 10:657; ONCOMINE: http://www.oncomine.org/meta.

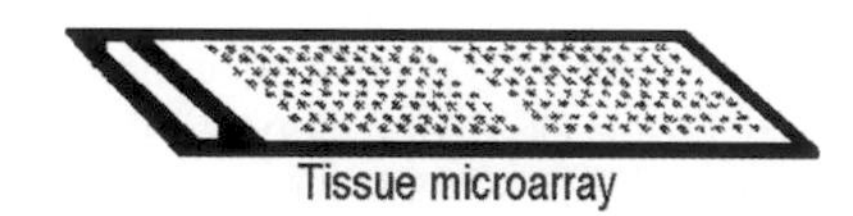

Figure T56. Tissue microarray

Tissue Plasminogen Activator (tissue plas): A tissue plasminogen activator cleaves plasminogen to plasmin and enhances fibrinolysis; it controls blood coagulation. ►blood clotting pathways, ►PARs

Tissue Remodeling: A coordinated series of events regulated by a balance between tissue proliferation and anti-proliferation factors such as cyclins, cyclin inhibitors and other metabolic regulators. Morphogenesis and various diseases involve alterations/anomalies in these processes. (See Nabel EG 2002 Nature Rev Drug Discovery 1:587).

Tissue-Specific Promoter: A tissue-specific promoter facilitates gene expression limited to certain tissues or organs. ►promoter

Tissue-Specificity: The expression of a gene is limited to certain tissue(s). Tissue-specificity may be controled at many levels (transcription initiation, differential recruitment of of repressors or activators to the gene targets, promoter clearance, attenuator, transcription termination, etc.). The macrophage colony stimulating factor receptor gene is constitutively expressed in many cell types but only in the macrophages is the elongation of the transcripts permitted (see Fig. T57).

The evolutionarily determined size of a gene may have something to do with tissue-specificity. In the extremely fast dividing embryonic tissues the very large genes may not have enough time to be fully transcribed. The proportion of tissue-specific genes has been estimated by transformation of *Arabidopsis* with transcriptional and translational gene fusion vectors. Of the 200 transgenic plants about 10% displayed some degree of tissue-specific expression of the reporter gene (aph[3′]). A large number of genes are expressed only in the central nervous system but repressed in other tissues. The REST/NRSF (repressor element silencing transcription factor/neuronal restricted silencing factor) along with the CoREST corepressor recruit histone deacetylase to chromosomal regions in the non-neuronal tissues to accomplish the task (Lunyak VV et al 2002 Science 298:1774). A computational model for the identification of cell-specific genes from the EST database is available (Nelander S et al 2003 Genome Res 13:1838). The identification of cis-regulatory elements and cognate transcription factors, expressed or inhibited in 45/56 different human and mouse tissues, permits a statistically significant predictions for tissue-specificity (Smith AD et al 2006 Proc Natl Acad Sci USA 103:6275). A global survey of specificity of 3,274 mouse proteins in six organs (brain, heart, kidney, liver, lung, and placenta) provides a searchable store for proteomics (see Fig. T58) (Kislinger T et al 2006 Cell 125:173; http://genome.dkfz-heidelberg.de/menu/tissue_db/faq.html). ►transcription illegitimate, ►housekeeping genes, ►macrophage colony stimulating factor, ►constitutive mutation, ►aph, ►subtractive cloning, ►cascade hybridization, ►chromatin remodeling, ►SATB1, ►microarray hybridization, ►isoenzymes, ►EST, ►organelle, Appendix II-10; Su AI et al 2002 Proc NatlAcad Sci USA 99:4465.

Tissue Typing: Determining the genetic constitution of a potential graft before transplantation. Blood typing is used. Even better, it can be done by DNA analysis (RFLP) or by polymerase chain reaction. ►blood typing, ►DNA fingerprinting, ►polymerase chain reaction

Titer: The amount of a reagent in titration required for a certain reaction. The number of phage particles per volume determines phage titer by counting the pfu number on a bacterial lawn after a series of dilution. The vector titer is assessed by the expression of a reporter gene (e.g., GFP, galactosidase, luciferase) in the target cells on a plate. Alternatively, special RNA or DNA can be quantitated by PCR. ►pfu, ►reporter gene, ►GFP, ►galactosidase, ►luciferase, ►PCR

Titin (connectin): One of the largest protein molecules (3×10^6 M_r) along with nebulin; it forms a network of fibers around actin and myosin filaments in the skeletal muscles, and may also be involved in the condensation of the chromosomes. The human titin gene contains 178/234 exons. Titin keeps myosin within the sarcomeres by being anchored to the Z discs (membrane bands in the striated muscles) and assures that the stretched muscles spring back. Titin is made up mainly of repeated modules but at its C-end it also has threonine/serine kinase domain specifically phosphorylating myosin. Calmodulin is required for its activation. The titin kinase domain itself is activated by phosphorylation at a tyrosine residue. Mutation in titin at 2q31 may cause dilated cardiomyopathy and tibial muscular dystrophy. The titin N2B region is dispensable for cardiac development and systolic properties but is important to integrate trophic and elastic functions of the heart (Radke MH et al 2007 Proc Natl Acad Sci USA 104:3444). ►nebulin, ►glutenin, ►sarcomere, ►dystrophin, ►calmodulin, ►cardiomyopathy, ►CaMK, ►exon; Labeit S, Kolmerer B 1995

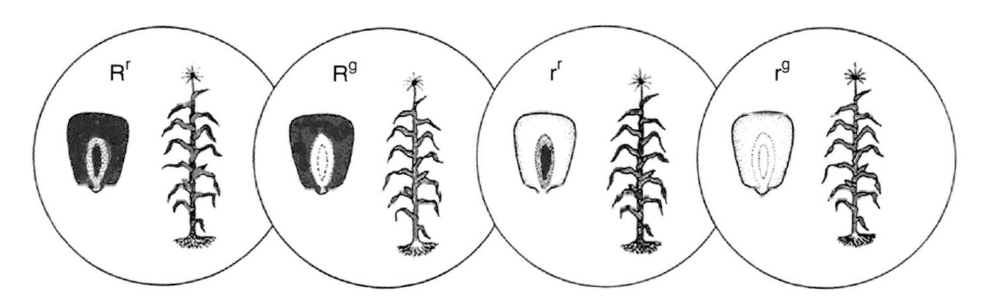

Figure T57. Tissue-specificity of expression (anthocyanin formation) of four different *R* alleles of maize in the aleurone, embryo, stalk, tassel and leaf tips

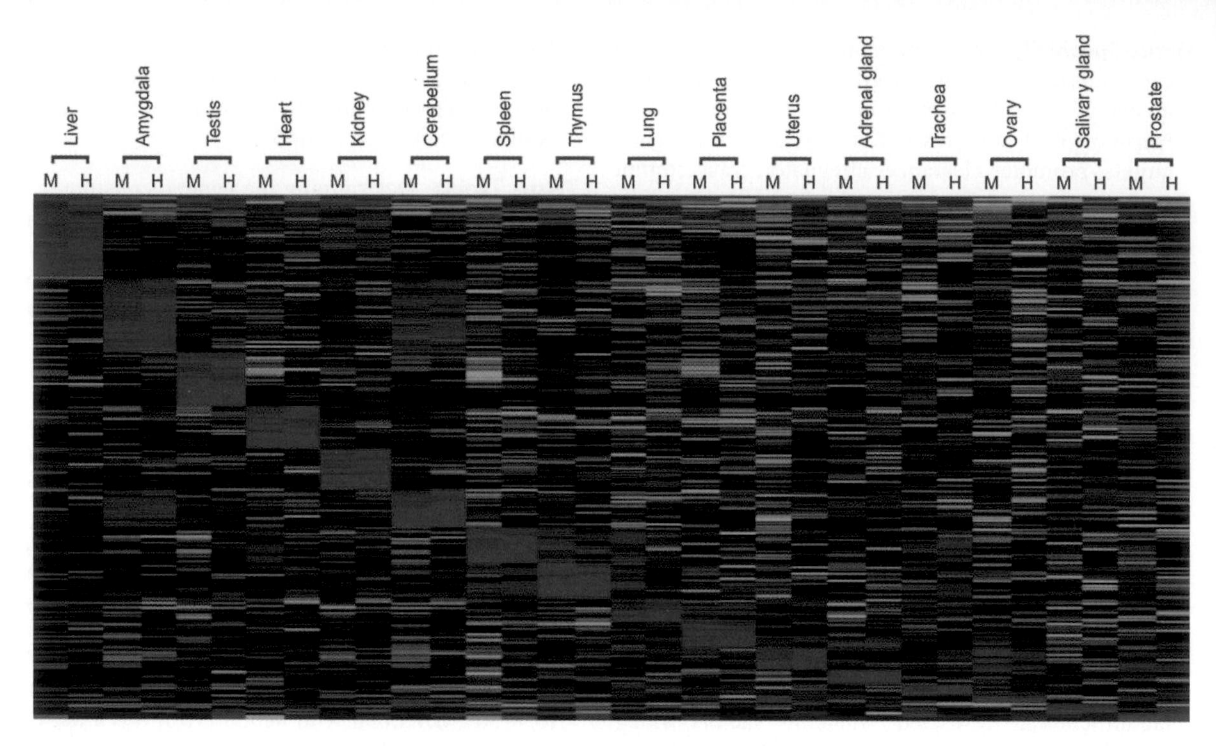

Figure T58. Transcriptome of 427 human and mouse gene pairs. Gene expression profile in 16 different types of tissues detected by Affymetrix GENECHIPs. Individual tissues express about 30 to 40% of the genes. Statistical analysis indicated that 78 to 82% of the genes were expressed differently in the mouse (M) and humans (H). The expression of the genes may be strictly tissue-specific. In this particular study, 85 human genes were expressed in the testis alone. (From Su AI et al 2002 Proc Nat Acad Sci USA 99:4465–4470; Copyright 2002 National Academy of Sciences, USA Courtesy of Dr. JB Hogenesch The Genomics Institute of the Novartis Research Foundation, San Diego, California)

Science 270:293; Machado C, Andrew DJ 2000 J Cell Biol 151:639; Gerull B et al 2002 Nature Genet 30:201; Hackman P et al 2002 Am J Hum Genet 71:492; Lange S et al 2005 Science 308:1599.

Titration: The addition of a measured amount of a solution of known concentration to a sample of another solution for the purpose of determining the concentration of the target solution on the basis of the appearance of a color or agglutination, etc. Also, by using this method, the number of cells or phage particles in a series of dilutions can be determined. ►titer, ►gene titration

TK: Thymidine kinase, which phosphorylates thymidine. Its gene in humans encodes the cytosolic enzyme that contains 7 exons in the short arm of human chromosome 17q25-q25.3; but the mitochondrial TK gene is in chromosome 22q13.32-qter. The latter enzyme defect may be responsible for myoneurogastrointestinal encephalopathy. The TK sequences are apparently highly conserved in different species. The herpes virus TK gene activates acyclovir and ganciclovir drugs and is used in gene therapy. ►HAT medium, ►gene therapy, ►ganciclovir, ►myoneurogastrointestinal encephalopathy

T_L: The left border of the T-DNA. ►T-DNA, ►Ti plasmid

TLC: ►thin layer chromatography, ►telomerase

TLE: ►groucho, ►Tup

TLF: ►TBP

TLR: Toll-like receptor. ►Toll

TLV (threshold limit value): The upper limit or time-weighted average concentration (TWA) of a substance that people can be exposed to without adverse consequences.

T_m: ►melting temperature

TM1, TM2: Transmembrane amino acid domains of *E. coli* transducers, spanning the membrane layer inner space. TM1 is well conserved among the various transducers; TM2 is variable. ►transducer proteins

TMF: TATA box modulatory factor. ►TATA box, ►promoter, ►transcription

TMHMM2: A program for the detection of transmembrane topology helices.

TMP: Thymidine monophosphate (see Fig. T59). (See formula at right)

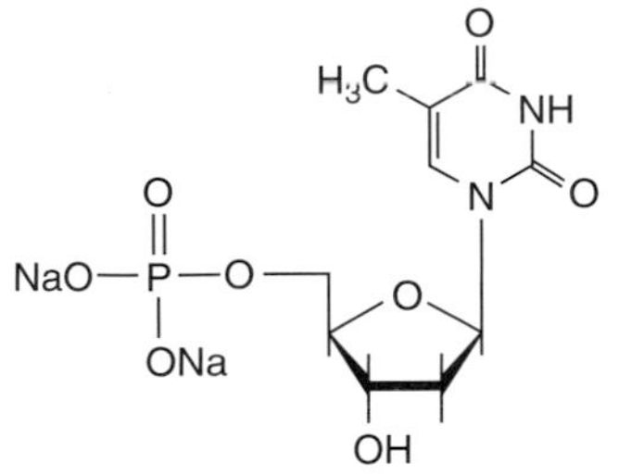

Figure T59. Thymidine monophosphate

TMRCA (time back to the most recent common ancestor): The concept for the study of the evolutionary descent of species or other taxonomic entities. It is estimated on the basis of the frequency of microsatellite variations associated with a particular mutation. If the mutation is recent the microsatellite variation linked to it is none and as time goes by the variation in microsatellites increases. The estimate may be biased by several factors, e.g., some variations may be lost by genetic drift, the generation time may be variable, etc.

tmRNA (10Sa RNA, SsrA): tmRNA basically is a tRNA and an mRNA hybrid molecule. It is used in bacteria to protect against the detrimental effects of mRNAs that lack stop codons and then stall translation. The SmpB (small protein B) protein binds to tmRNA and promotes its stable association with the 70S ribosome. The tmRNA is first charged with alanine and then it associates with EF-Tu. It binds to the A site of the ribosome and the alanine is incorporated into the growing peptide chain. Simultaneously, the mRNA-like domain of tmRNA substitutes for the defective mRNA on the ribosome. The tmRNA facilitates the inclusion of an additional 10 amino acids into the growing chain and then when the translation is stopped, the polypeptide is released. Proteins so formed are recognized by proteases because of the tmRNA tag and destroyed. This thus protects against defective proteins in the cell. This RNA is generated by RNase E cleavage. ►aminoacylation, ►regulation of gene activity, ►eEF, ►EF, ►EF-Tu, ►GTP, ►proteasomes, ►ribonuclease E, ►protein repair, ►ARAGORN, ►ribosome recycling; Zwieb C et al 1999 Nucleic Acids Res 27:2063; Lee S et al 2001 RNA 7:999; Moore SD et al 2003 Science 300:72; structure and ribosome binding: Gutmann S et al 2003 Nature [Lond] 424:699; Asano K et al 2005 Nucleic Acids Res 33:5544; Moore SD, Sauer RT 2007 Annu Rev Biochem 76:101; http://www.indiana.edu/~tmrna; http://www.ag.auburn.edu/mirror/tmRDB/; http://psyche.uthct.edu/dbs/tmRDB/tmRDB.html.

TMS (tandem mass spectrometry): A method used to detect defects in fatty acid metabolism, organic acidemias and other human anomalies. ►mass spectrum, ►MALDI-TOF

TMV (tobacco mosaic virus): A single-stranded RNA virus of about 63,900 bases. Its cylindrical envelope contains 2,130 molecules of a 158-amino acid protein. Its rod-shape particles are about 3,000 Å long and 180 Å in diameter (see Fig. T60). The historical reconstitution experiments from coat protein and RNA demonstrated that RNA could be genetic material. Its mutagenesis by nitrous acid contributed substantially to the genetic confirmation of RNA codons. Its genome encodes four open reading frames. Foreign genes attached to the regulatory tract of the coat protein or to the coat protein itself can be expressed in tobacco plants. Thus, tobacco plants may be eventually used to manufacture proteins, e.g., α-glycosidase that is deficient in Fabry disease patients or the antigen of Hodgkin's lymphoma. ►genetic code, ►Fabry disease, ►Hodgkin's disease, ►plant vaccines; Goelet P et al 1982 Proc Natl Acad Sci USA 79:5818; Culver JN 2002 Annu Rev Phytopathol 40:287.

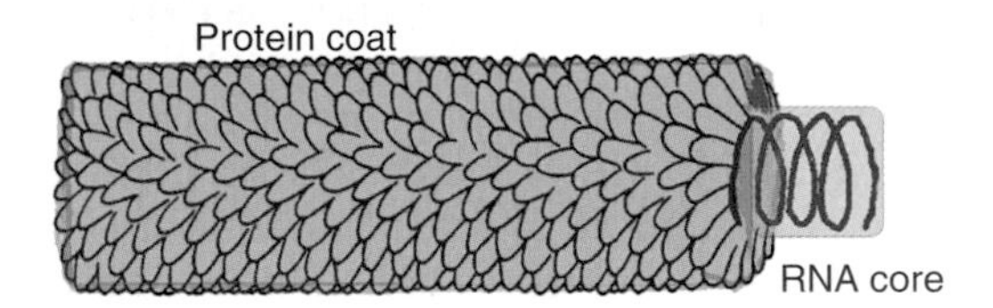

Figure T60. TMV

Tn3 Family of Transposons: Genetic elements that can move within a DNA molecule and from one DNA molecule to another and carry genes besides those required for transposition. The best-known representative is the Tn*3* element carrying ampicillin resistance Ap^r. Tn*3* is 4,957 bp long with 38 bp inverted terminal repeats and leaving behind—after moving at the transposition target site—5 bp direct duplication. Although the terminal repeats of the various Tn elements vary, some sequences are well conserved. The GGGG sequence is generally present outside the repeats and ACGPyTAAG is common inside of one or both terminal repeats. In the Tn3 group an internal ACGAAAA is common. Normally transposition requires the presence of both terminal repeats; the presence of only one of them may still allow a lower frequency transposition. The sites of integration may vary and yet AT-rich sequences are preferred and some homology between the terminal repeats and the insertional target may be needed. Some proteins of the host cell (IHF = integration host factor and FIS = factor for inversion stimulation) may facilitate the expression of the transposase. Other members of the family are Tn*1*, Tn*2*, Tn*401*, Tn*801*, Tn*802*, Tn*901*, Tn*902*, Tn*1701*, Tn*2601*, Tn*2602*, and Tn*2660*. All are about 5 kb length with 39 bp terminal repeats and are found in various Gram-negative bacteria. Related to these is the γδ (Tn*1000*) element in the F plasmid of 5.8 kb and with 36/37 bp terminal repeats and IS*101* (insertion element *101*), a cryptic element. Tn*501* (8.2 kb) was the source of mercury resistance (Hg^r), Tn*1721*, Tn*1771* (11.4 kb) for tetracycline resistance (Tc^r), Tn*2603* (22 kb) for resistance to oxacillin (Ox^r), and hygromycin (Hg^r), streptomycin (Sm^r), and sulfonamide (Su^r). Tn*21* (19.6 kb) carried resistance to Su^r, Hg^r, Ap^r, Sm^r, and Su^r. Tn*4* (23.5 kb) was endowed with genes for Ap^r, Sm^r, and Su^r. Tn*2501* (6.3 kb) was cryptic (i.e., expressed no genes besides the transposase). Tn*551* and Tn*917* (both 5.3 kb) from *Staphylococcus aureus* and *Streptococcus fecalis*, respectively, carried the erythromycin resistance (Ery^r) gene. The cryptic Tn*4430* (4.1-kb) was isolated from *Bacillus thüringiensis*, R46 from enterobacteria, and pIP404 from *Clostridium perfringens* plasmids coded for the resolvase protein. The terminal repeats are generally within the range of 35–48 bp. The transposition is usually replicative and its frequency is about 10^{-5} to 10^{-7} per generation. The integration of the element requires the presence of a specific target site, called *res* site or IRS (internal resolution site), and genes *tpnA* (a transposase of about 110 MDa) and *tpnR*, encoding a resolvase protein (ca. 185 amino acids). The *res* site (about 120 bp) is where the resolvase binds and mediates site-specific recombination and protects the DNA against DNase I. Within the *res* site are located the promoters of *tnpR* and *tpnA* genes, functioning either in the same or in opposite direction, depending on the nature of the *Tn* element. The recombination between two DNA molecules requires the presence of at least two *res* sites in a negatively supercoiled DNA. The resolvase apparently has type I DNA topoisomerase function too. After synapses, mediated by resolvase and multiple *res* sites, strand exchange and integration may result. The recipient molecule thus acquires the donor transposon. The transposition event requires replication of the transposon DNA and then a fusion of the donor and the recipient replicons. This must be followed by a resolution of the cointegrate into a transposition product. ►transposable elements, ►gram-positive bacteria, ►transposon, ►cointegration, ►topoisomerase, ►antibiotics, ►resolvases; Sherratt D 1989, p 163. In: Mobile DNA. Berg DE, Howe MM (Eds.) Amer Soc Microbiol, Washington, DC.

Tn5: A bacterial transposon of 5.8 kb of the structure shown on Fig. T61.

The inverted termini represent the IS*50* insertion element that includes the *tnp* (58 kDa protein) and *inh* genes (product 54 kDa). The left (L) and right (R) IS*50* elements are almost identical except that the L sequences contain an ochre stop codon in the *tnp* gene, rendering it non-functional, save when the bacteria carry an ochre suppressor (see Fig. T61). At the I (inside ends) site binds the IHF (integration host factor) protein. *O* is the outside end of the IS sequence. Within the repeat beginning at nucleotide 8 is the bacterial DnaA protein-binding site TTAT$\genfrac{}{}{0pt}{}{C}{A}CA\genfrac{}{}{0pt}{}{C}{A}$A. The DnaA product controls the initiation of DNA synthesis. Within the 2,750 bp central region are the genes for resistance to kanamycin (*kan*) and G418, bleomycin [phleomycin] (*ble*), and streptomycin (*str*) antibiotics; they are transcribed from the *p* promoter located within the Is *L* element at about 100 bp from the *I* end. These antibiotic resistance genes may not convey resistance in some cells, e.g., *str* may be cryptic in *E. coli*. The activation of this antibiotic resistance operon is contingent on the ochre mutation in the *tnp* gene. The inhibitor protein (product of *inh*) apparently interacts with the terminal repeats rather than with the transcription or translation of the *tnp* gene.

Tn*5* can insert one copy at many potential target sites within a genome. The IS terminal repeats are capable of transposition themselves without the internal 2.8 kb element. In *direct transposition* the complete Tn*5* will occur in the same sequence as shown in the diagram. In *inverse transposition*, mediated by the *I* ends, the 2.8 kb central element is left behind, away from the termini. Inverse transposition occurs 2–3 orders of magnitude less frequently than the direct one. In*5* can form cointegrates with plasmids or bacteriophages and in these the IS elements and the entire Tn*5* may occur and the orientation of the termini may be either direct or indirect as shown in the diagram. The transposon may be present as a monomer or as a dimer. Transposition may not require special homologous target sequences yet some targets represent "hot spots" (displaying

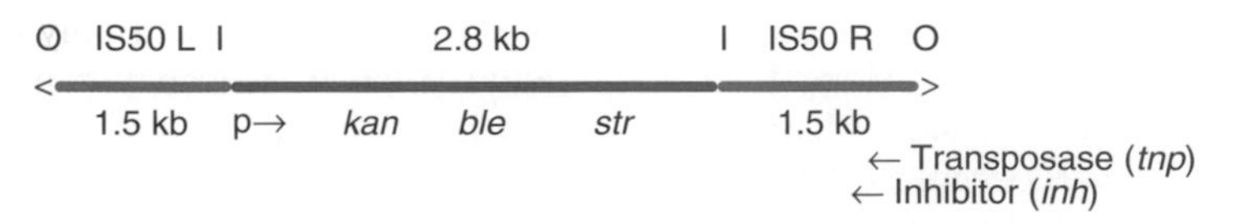

Figure T61. Structure of transposon Tn*5*

G•C or or C•G pairs next to the 9 bp target duplications) because they are preferred for insertion. The less frequently used targets generally show G•C and A•T pairs at the ends. Tn*5* insertions, in general, are almost random and yet it appears that transcriptionally active promoters may present favorable targets. Insertion of the transposon into active genes usually results in inactivation because of the interruption of the coding sequences. Excision of the transposon may revert the gene to the original active state. This may occur at frequencies of 10^{-8} to 10^{-4} per cell divisions. The transposase gene alone does not mediate excision. It is independent from the bacterial *recA* gene but it depends on the structure of the inverted terminal repeats. Replicational errors involving slippage of pairing between the new and template DNA strands may occur. The presence of some sequences in the target may also promote Tn excision. The excision may also involve flanking DNA sequences and in this case the wild type function of the target gene is not restored. Several mutations in *E. coli* (*recB, recC, dam, mutH, mutS, mutD, ssb*) may promote excision, and mutation to *drp* reduces excision. Tn*5*, unlike Tn*3*, does not have a resolvase function. DNA gyrase, DNA polymerase I, DnaA protein, IHF and Lon (a protease cleaving effectively the SulA protein, a cell division inhibitor), may affect transposition. Transposition of Tn*5* is substantially increased in *dam* mutant strains that are deficient in methylating GATC sequences. The *I* ends contain GATC sequence and can thus be affected by methylation. The *O* ends do not have a methylation substrate yet they are also affected by methylation in the *I* sequences (19 bp) (see Fig. T62). ►transposon, ►Tn*3*, ►Tn*7*, ►Tn*10*, ►transposable elements bacteria, ►transposable elements, ►cut-and-paste, ►resolvase; Berg DE 1989 Mobile DNA. In: Berg DE Howe MM (Eds.) Amer Soc Microbiol, Washington, DC, pp 185; Naumann TA, Reznikoff WS 2002 J Biol Chem 277:17623; Peterson G, Reznikoff W 2003 J Biol Chem 278:1904.

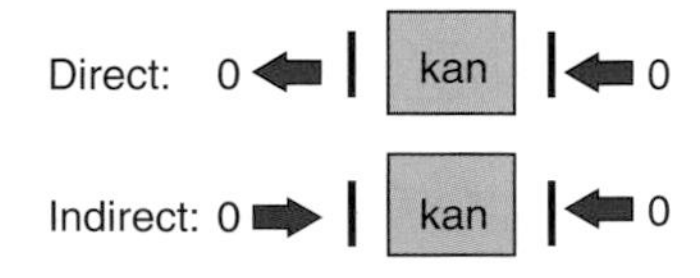

Figure T62. Tn*5* termini can be reverse oriented

Tn*7*: A 14 kb bacterial transposon of the following general structure. The Tn*7* element has a high capacity to insert into the specific *att* Tn*7* site of *E. coli* (25 kb counterclockwise from the origin of replication at map position 83). The target DNA required for Tn7 transposition is within the C-terminus of the glucosamine synthetase (*glmS*) gene in bacteria and it has similar specificity for the human homolog of *glmS*, GFPT-1 and GFPT-2, at chromosomal location 2p13 (Kuduvalli P et al 2005 Nucleic Acids Res 33:857). When this site is not available it may transpose—at about two orders of magnitude lower frequency—to a *pseudo-att* Tn*7* or to some other unrelated sites (see Fig. T63). At the *att* Tn*7* the right end is situated proximal to the bacterial *o* gene. Genes *tnsABC* mediate all transpositions but through different pathways; for transposition to *att* Tn*7* and *pseudo-att* Tn*7*, the function of gene *D* is also required, whereas for transposition to all other sites the expression of the *ABC* + *E* genes are needed. (The name *tns* abbreviates transposon **s**even.) The insertion at *att* Tn*7* is also in a consistent orientation and it is within an intergenic region and thus does not harm the host. Insertion of the Tn*7* element protects the cell from an additional Tn*7* insertion (immunity). Yet under some conditions the immunity may not work. Integration results in 5 bp target site duplication; these duplications are different at *att* Tn*7* and other sites. Several bacterial species besides *E. coli* have specific *att* Tn*7* sites in their genomes. For transposition both *L* and *R* terminal repeats are required. Elements with two *L* repeats do not move whereas two complete *R* termini can assure insertion to *att* Tn*7* sites. Tn*7*, because of the site specificity, is not very useful for insertional mutagenesis, but it has an advantage of inserting genes at the standard map position. Tn*7* inserts preferentially in the vicinity of triple-helical sites. The Tn*7* family of transposons includes Tn*73*, Tn*1824*, and Tn*1527* with substantial similarity or even identity. Tn*1825* is a clearly distinct member. ►transposons bacterial, ►transposable elements, ►triplex; Craig N 1989, p 211. In: Mobile DNA. Berg DE, Howe MM (Eds.) Amer Soc Microbiol, Washington, DC.

Tn*10*: A bacterial (*E. coli, Klebsiella, Proteus, Salmonella, Shigella, Pseudomonas*, etc.) transposable element of 9.3 kb. It may move within a bacterial

Figure T63. Tn*7*. *dhfr* = dehydrofolate reductase gene with much reduced sensitivity to trimethoprim, an inhibitor of the enzyme involved in the biosynthesis of both purines and pyrimidines and thus nucleic acids. *aada* = adenylyl transferase gene, encoding an enzyme, which inactivates aminoglycoside antibiotics, streptomycin and spectinomycin, and thus conveys resistance to the cells. The *tns* genes are responsible for transposition

genome, from the bacterial chromosome to a temperate phage or plasmid and among different bacterial species. Its overall structure can be represented as in Figure T64.

The boxed left insertion element IS*10* (L) is a defective transposase. The boxed right IS*10* element (R) is a functional transposase (see Fig. T64). Thus Tn*10* is a composite transposable element. The *tetR, A, C and D* genes are involved in resistance against tetracycline. The *tetR* is a negative regulator and *tetA* encodes a membrane protein. The arrows indicate orientation and direction of transcription. The promoter of the Is*10* element may serve as promoter for adjacent outside genes (pOUT).

Insertion of Tn*10* within a gene, operon or upstream regulatory element may abolish the activity of the gene(s) or may modify their transcription. In the so-called polar insertions transcription is initiated within the Tn element but it may be terminated when rho signals are encountered. In the nonpolar insertions there are no rho sequences downstream to halt transcription. This type of *read-through* transcription may only take place in some Tn*10* derivatives but not in the wild type element. The rate of transposition for IS*10* is 10^{-4} and for Tn*10*, 10^{-7} per cell cycle.

Both the IS*10* and the Tn*10* elements can cause chromosomal rearrangements (see Fig. T65). Insertion and transposon sequence may show the portable region of homology and can undergo homologous recombination (see Fig. T66). These recombinations may generate deletions between, or inversions in, the regions in-between them at rates two orders of magnitude less frequent than transposition or lead to the formation of cointegrates.

All DNA segments flanked by IS*10* can become transposable and thus may represent new composite transposons. IS*10* and Tn*10* may assist the fusion of different replicons and transfer information between bacterial chromosomes, plasmids, and phages. The transpositions may also generate new units of regulated gene clusters by the movement of structural genes under the control of other regulatory sequences.

The excision of the transposon may be "precise" if the nucleotide sequence is restored to its pre-insertion condition. Precise excision (average frequency 10^{-9}) may remove one copy of the 9-bp target site inverted duplications.

The "near-precise excision" events involving removal of most of the internal sequences of Tn*10* occur at a frequency of about 10^{-6} and may later be followed by precise excision of the remaining sequences. These non-transposase mediated events are also independent from the host RecA recombination functions. Precise excision is mediated by RecBC and RecF pathways. Tn*10* and IS*10* may transpose either by a non-replicative or a replicative mechanism. In the former case, the whole double-stranded element is lifted from its original position and transferred to another site. In the second case, only one of the old strands of the transposon is integrated into the new position and the other strand represents the newly replicated one.

At the target site apparently two staggered cuts are made, 9 bases apart. This causes then the 9 bp target duplications when the gap is filled and the protruding ends are used as templates Figure T67.

The transposon (ca. 46 kDa) is coded within the IS*10* right terminus by about 1,313 nucleotides. The frequency of transposition may increase up to five orders of magnitude by increasing the expression of the transposase. The transposase action prefers being

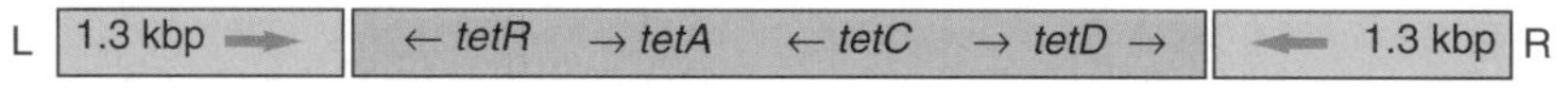

L | 1.3 kbp → | ← *tetR* → *tetA* ← *tetC* → *tetD* → | ← 1.3 kbp | R

Figure T64. Structure of Tn*10*

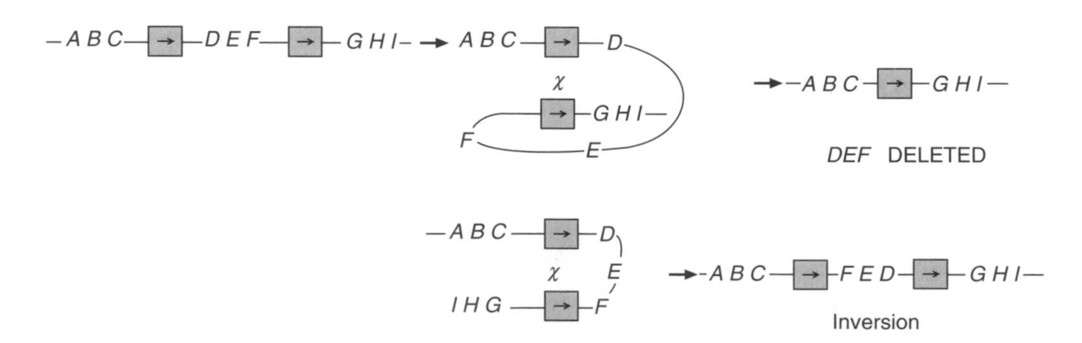

Figure T65. Generation of deletions and inversions by a portable region of homology represented by transposon Tn*10* (→ OR ←). Genes or sites are shown by capital letters in italic; χ indicates recombination

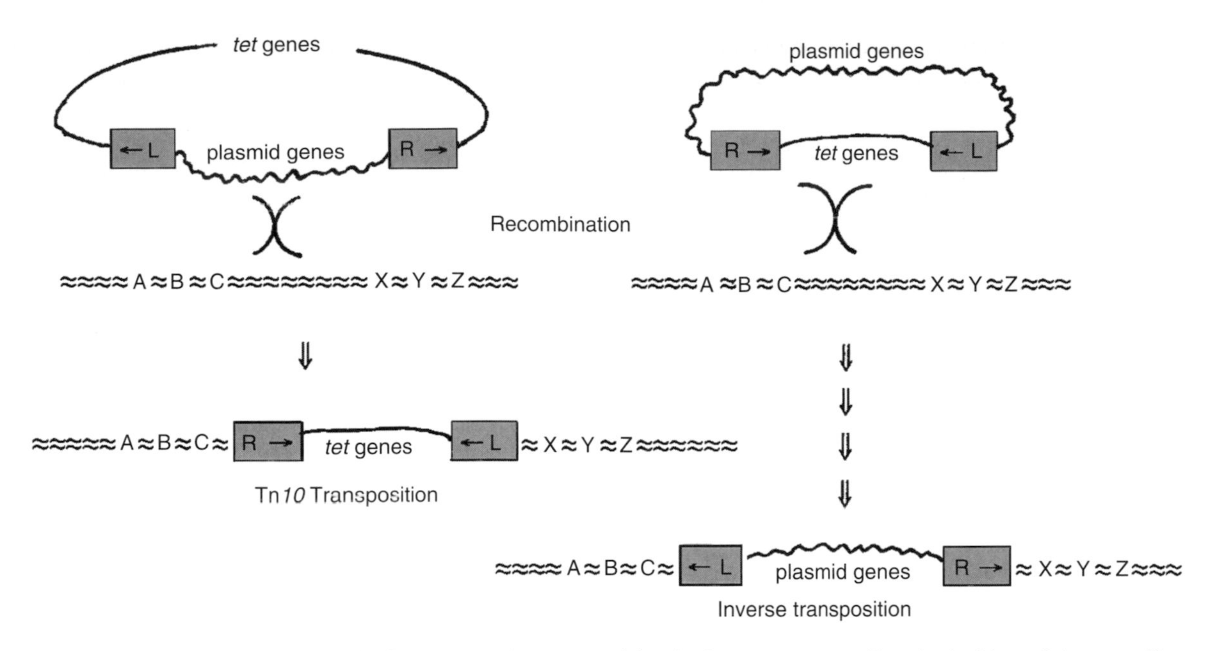

Figure T66. Tn*10* transpositions. Left: the normal event, at right: the inverse transposition (or inside-out) transposition. Some other events may also lead to deletions or deletions and inversions. The L and R boxes represent Tn*10* termini. The thin line stands for the transposon sequences whereas the single jagged line indicates the sequences flanking the transposon in the original location of the plasmid. The ≈≈≈≈ symbolizes the DNA sequences of the target

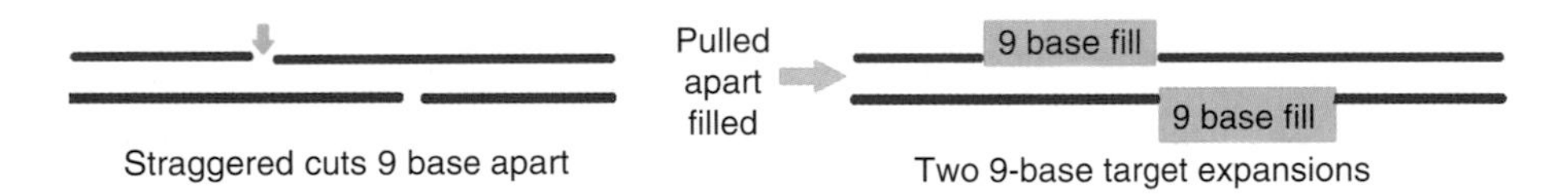

Figure T67. Target expansion

in the vicinity of the transposed sequences and its action is reduced by distance. Also, longer transposons are moved less efficiently than shorter ones. Each kb increase in length involves about 40% reduction in movement within transposon sizes 3 kb to 9.5 kb. Proteins IHF, HU and DNA gyrase also regulate transposition. The transposition frequency may also be modified (within three orders of magnitude) by the chromosomal context, i.e., cis-acting sequences. The integration hot spots appears to have some consensus sequences within the 9 bp target site in three bases (see Table T1).

Transposition seems to avoid being actively transcribed and thus the most essential genes of the

Table T1. Sites of integration hot spots

1st	2nd	3rd position			
G	C				
A	T	T	C	A	G
90%	98%	63%	23%	12%	2%, respectively

recipient. The transposase gene is transcribed from a very low efficiency pIN promoter origin-ating near the *O* end of the right *IS* element. The divergent pOUT promoter opposes its transcription also. Apparently, on average, each cell generation may not produce one molecule of transposase. Two *dam* (adenine) methylation sites (GATC) are located within pIN and the activity of this promoter is facilitated in the absence of methylation by the host *dam* methylase (located in *E. coli* at 74 min). Actually, the transposase promoter is usually hemi-methylated. An increase in the number of Tn*10* copies per cell reduces transposase activity because the pOUT promoter generates an antisense RNA transcript that may pair to a 35-base complementary region of the pIN promoter region. The transposase activity is also regulated by fold-back inhibition (FBI), hindering the attachment of the Shine-Dalgarno sequence of the transcript to the ribosome. The Tn*10* system is protected from transposase activation by inhibition of read-through.

Transposons inserted into a particular gene cause mutations because of the disruption in the continuity of the coding sequences. Transposon mutagenesis has considerable advantage over chemical or radiation mutagenesis because it induces mutation only at the site of insertion while chemicals may affect several genes simultaneously. In addition, mutagenesis with Tn*10* labels the gene by tetracycline resistance, an easily selectable marker. Transposons may also label foreign genes cloned in *E. coli* or *Salmonella*. The tetracycline-resistance gene has been extensively used for monitoring insertions that result in the loss of tetracycline resistance but retain other (selectable) markers in the cloning vector. ►IHF, ►HU, ►gyrase, ►antisense RNA, ►FBI site, ►readthrough, ►transposable elements bacteria; Craig N 1989 Mobile DNA. In: Berg DE, Howe MM (Eds.) Amer Soc Microbiol, Washington, DC, pp 227.

Tn Syndrome: A rare autoimmune disease involving exposure on the surface of erythrocytes, platelets, granulocytes, and lymphocytes of the normally hidden Tn antigen. In patients, both Tn+ and Tn– hematopoietic cells show up and leukemia, thrombocytopenia and nephropathy appear. The Tn membrane glycoprotein is incompletely glycosylated. The altered Tn loses a terminal galactose due to a defect in the T-synthase enzyme, encoded at 7p14-p13. A somatic mutation in the Xq23 gene, COSMC, alters a molecular chaperone required for the proper folding of the enzyme. (See Vainchenker W et al 1985 J Clin Invest 75:541; Ju T, Cummings RD 2005 Nature [Lond] 437:1252).

TNA: (L)-α-Threo-furanosyl-('-2')-oligonucleotides. These molecules contain 4-carbon (tetrose) sugars rather than pentose and are capable of pairing with RNA and DNA.

TncRNA (tiny noncoding RNA): ►RNA noncoding

TNF (tumor necrosis factor): Proteins (M_r ≈17 K) that are selectively cytotoxic or cytostatic to the cancer cells of mammals. They are relatively harmless to normal cells by being lymphokines. They are defense molecules against intracellular pathogens. The TNF family includes diverse transmembrane proteins with high homology in the receptor-binding regions. TNF was originally discovered in rodent cells infected by bovine *Mycobacterium* and then with endotoxin. The serum of these animals produced hemorrhagic necrosis and occasionally complete regression of transplanted tumors. The mature human TNFα consists of 157 amino acids after trimming off 73 amino acids from the pre-TNF. It has receptor sites on the surface of tumors and its action is synergistic with γ interferon. TNFα and TNBβ have similar functions and display ≈30% homology. TNFα plays a promoting role in obesity and inflammatory metabolic diseases (diabetes, atherosclerosis, hypertension, asthma, osteoarthritis, cancer, etc.). TNFα inhibitors are used for antiinflammation (rheumatid arthritis) therapy. The relatively small molecule, trifluoromethyphenyl indole, causes a very high (600-fold) dissociation of the trimeric structure of TNFα. This observation may be exploited for the development of new drugs (He MM et al 2005 Science 310:1022).

The two TNF genes each have 3 introns in their 3 kb sequence. In mice, it is localized within the *H-2* (histocompatibility cluster) and in humans in the homologous *MHC* region in chromosome 6p23, either between HLA-DR and HLA-A or proximal to the centromere. The genes were cloned and sequenced in the mid-1980s. TNFα and IL-1 are the major inflammatory cytokines whereas IL-10 and TGFβ, IL-R and TNF-R are anti-inflammatory. TNF is produced mainly by macrophages. A TNFβ deficiency may increase chromosomal instability and may lead to cancerous transformation. The tumor necrosis family of proteins includes FASL, CD40L, LT and β, CD30L, CD27L, 4–1BBL, OX40L, TRAIL, OPGL, LIGHT, APRIL, and TALL. Estrogen deficiency and ovariectomy stimulate osteoclastogenesis (bone loss) by T cell-produced tumor necrosis factor (TNFα). This process is mediated by enhanced production of interleukin-7 (Ryan MR et al 2005 Proc Natl Acad Sci USA 102:16735). ►lymphokines, ►cytokines, ►endotoxin, ►hemorrhage, ►necrosis, ►interferons, ►histocompatibility, ►HLA, ►TNF-R, ►TRAF, ►Crohn's disease, ►arthritis, ►macrophage, ►TACE, ►TACI, ►TNFR, ►TRAIL, ►NGF, ►lymphotoxin, ►LTα, ►HVEM, ►transposase, ►IL-7, ►IL-32, ►asbestos; Kassiotis G, Kollias G 2001 J Clin Invest 107:1507; Liu Y et al 2003 Nature [Lond] 423:49; review of inflammation and diseases: Hotamisligil GS 2006 Nature [Lond] 444:860.

TNFR (tumor necrosis factor receptor, p55, p75): Similar receptors in both animals and plants may be involved in processes of differentiation. TNFR-1 and -2 are distinguished; TNFR-1 mediates different effector functions through separate pathways. The intracellular portion of TNFR-1 contains a 70-amino acid death domain, which mediates signals for apoptosis and the activation of NF-κB. TNF binds to the extracellular domain of TNFR-1 resulting in its trimerization. The TNF1 attracts the adaptor TRADD (tumor necrosis factor receptor associated death domain) and TRAF2 and TRAF1. Subsequently, the TNFR1 recruits FADD (Fas associated death domain). TRAF2 (tumor necrosis factor associated protein 2) is contacted by TRAF1 and RIP (receptor interacting protein). This complex then signals for apoptosis, JNK/SAPK and NF-κB activation. Protein SODD (457-amino acid silencer of death domains) is expressed in all human tissues and interacts with the intracellular domain of TNFR-1 but not with TNFR-2, TRADD, FADD or RIP, and interferes with all TNF signaling through TNFR-1. JNK does not participate in the apoptotic pathway and the activated NF-κB works against apoptosis though inflammation may result. ►TNF, ►TRAF, ►NF-κB, ►nitrogen fixation [ENOD], ►apoptosis, ►Jun, ►Fas, ►NGF, ►Paget's disease, ►APRIL, ►BAFF, ►Blys; Locksley RM et al 2001 Cell 104:487.

TNG: A panel of 90 independent radiation hybrid clones constructed at Stanford University by irradiation of 50 Krad x-rays. This panel is used for the construction of a high-resolution STS map. ►radiation hybrid panel, ►STS; Robic A et al 2001 Mamm Genome 12:380; Olivier M et al 2001 Science 291:1298.

Tnp: A transposase enzyme. ►Tn*5*

Tnt: A nonviral, retrotransposable element. ►transposable elements

TNT: A solution containing 10 mM Tris.HCl buffer (pH 8.0), 150 mM NaCl, 0.05% Tween 20.

Toad: *Bufo vulgaris,* 2n = 36*; Xenopus laevis,* 2n = 36 (see Fig. T68). ►frogs

Figure T68. Toad

TOASTS: Traced orthologous amplified sequence tags.

Tobacco (*Nicotiana* spp): The smoking tobacco is an allotetraploid (*N. sylvestris x N. tomentosiformis*) 2n = 48. The basic chromosome number generally is x = 12; however, diploid forms with 2n = 20 (*N. plumbaginifolia*) and 2n = 18 (*N. langsdorfii*), and in the *Suavolens* group, 2n = 36 (*N. benthamina*), 2n = 46 (*N. caviola*), 2n = 32 (*N. maritima*), 2n = 36 (*N. amplexicaulis*), 2n = 40 (*N. simulans*), 2n = 44 (*N. rotundifolia*) are also found. For genetic studies the diploid species are most useful (*N. plumbaginifolia*). For transformation, *N. tabacum* is used most commonly because of the easy regeneration of plants from single cells. Antibiotic resistant chloroplast mutations are available. The S strain is a streptomycin resistant mutant of the variety Petite Havana. Mutation, recombination and transformation techniques are available for its plastid genome, using primarily antibiotic resistant ctDNA mutations. ►*Nicotiana*, ►smoking

Tobacco Mosaic Virus: ►TMV

Tobacco Necrosis Virus (TNV): An icosahedral single-stranded RNA virus of ≈4 kb. It is a root plant pathogen. ►icosahedral

Tobacco Satellite Necrosis Virus (TSNV): A 17 nm diameter, single-strand RNA (≈1.2 kb) virus that depends on TNV for its replication. ►tobacco necrosis virus

Tocopherol α (vitamin E): Tocopherol,α deficiency results in nutritional muscular dystrophy and sterility, although, normally, humans with regular diet do not show any need for it. Vitamin E may be beneficial as it is an antioxidant and regulates nerve functions and atherosclerosis. Vitamin E is a lipid-soluble molecule and plant oils are a major source. ►vitamin E, ►atherosclerosis

T-Odd Virus: Bacteriophages with odd number designations such as T3, T5, T7, etc. ►bacteriophages

Toeprinting: A procedure for mapping the translation initiation ternary complex (EI1A•GTP•tRNA) to the ribosome. In one approach, a ^{32}P-labeled oligonucleotide is annealed to the mRNA, downstream (3′) from the presumed site of initiation, and reverse transcriptase is used for the extension of the radioactive primer up to the position of the bound ribosome where the chain growth stops. Various types of purification may be used for the isolation and identification of the initiation complex. ►elongation factors, ►protein synthesis, ►ribosomes, ►mRNA surveillance; Ringquist S, Gold L 1998 Methods Mol Biol 77:283; Dmitriev SE et al 2003 FEBS Lett 533/C:99.

TOFMS (time-of-flight mass spectrophotometry): A powerful technique used for the analysis of the primary structure of proteins. ►mass spectrum, ►matrix-assisted laser desorption time of flight mass spectrometry, ►MALDI; Verentchikov AN 1994 Anal Chem 66:126; She Y-M et al 2001 J Biol Chem 276:20039.

TOGA (total gene expression analysis): A completely automated technology for the simultaneous analysis of the expression of nearly all genes. Basically, it selects a four-base recognition endonuclease site and an adjacent four nucleotide parsing sequence (a syntactical determinant, e.g., for *Msp*I CCGGN_1N_2-N_3N_4) and their distance from the 3′-end of an mRNA (from the polyA tail). These generate a specific, single identity label for each mRNA. The parsing sequences also serve as parts of the PCR primer-binding sites in 256 PCR-based assays, which determine the presence and concentration of that mRNA in a tissue. ►microarray hybridization, ►SAGE, ►RNA fingerprinting; Sutcliffe JG et al 2000 Proc Natl Acad Sci USA 97:1976; expanded TOGA: http://www.tigr.org/tdb/tgi/ego.

TOGA (Tiger Orthologous Gene Alignment): A database generated by pair-wise comparison between tentative consensus sequences in different organisms: http://www.tigr.org/tdb/tgi.shtml.

Togavirus: RNA plus strand viruses of about 12 kb. The capsid proteins are synthesized only after completion of replication. This viral family includes rubella, yellow fever, and encephalitis viruses. ►RNA viruses, ►plus strand

Toggle Switch, Genetic: ►Gene-Switch cassette, ►synthetic biology

Toilet: The medical meaning is cleansing, clearing.

Tolerance of antibiotics: The infectious organism does not die but stops reproducing. ►antibiotic resistance

Tolerance, Immunological: Non-reactivity to an antigen that under other conditions would evoke an immune response. Antigens provided to fetuses or neonates with immature immune systems can induce tolerance. In adults, a very high or very low dose of the antigen may cause tolerance. The tolerance is the result of either clonal elimination or inactivation of lymphocytes in the thymus. Liver transplantation may induce systemic immune tolerance for certain (kidney, heart) allografts. ►lymphocytes, ►immune response, ►antigen, ►allograft, ►immunosuppressants; Gaunt G, Ramin K 2001 Am J Perinatol 18(6):299; Salih HR, Nussler V 2001 Eur J Med Res 6(8):323; Chang CC et al 2002 Nature Imunol 3:237.

Tolerization: $CD4^+$ $CD25^+$ T cells secure tolerance to organ-specific self-antigens. The Foxp3 transcription factor mediates the process and its defect cause immune reaction in humans and mice. Elimination of these factors is tolerization of both autoreactive and alloreactive lymphocytes. This may be a precondition to successful stem cell therapy. ►T cell, ►B cells, ►immune reaction, ►autoimmune diseases, ►FKH, ►stem cells

Toll (*Drosophila*, 3–91, human homolog TRAF6): Toll encodes a signaling protein operating through the NF-κB receptor in both the fly and humans, and the MyD88 adaptor protein mediates it. For the activation of NF-κB, IRAK (interleukin associated receptor kinase) and the TRAF6 (tumor necrosis factor associated receptor) protein, activated by IL-1, are required. This pathway is essential for the immune system. The human Toll-like receptors (hTLR) are transmembrane glycoproteins that respond to bacterial lipoproteins (TLR2), lipopolysaccharides (TLR4), unmethylated CpG-DNAs (TLR9), etc., with activation of apoptosis, bacterial killing, tissue injuries and the induction of the innate immune response (Beutler B 2004 Nature [Lond] 430:257). The toll-like receptors TLR7 and TLR8 mediate the recognition of guanine and uridine-rich single-stranded RNA oligonucleotides, which stimulate dendritic cells and macrophages to secrete interferon-α and cytokines (Hell F et al 2004 Science 303:1526). The Toll-like receptor includes recognition sites for the innate immune system. TLRs signal through TRAF3 and TRAF6 effectors (Häcker H et al 2006 Nature [Lond] 439:204). TLR receptor signaling may not be required for adjuvant-enhanced antibody response as thought earlier (Gavin AL et al 2006 Science 314:1936). The *Toll/spaetzle/cactus* gene cassette is involved in the control of the antifungal peptide, drosomycin production. ►morphogenesis {52} in *Drosophila*, ►IRAK, ►TRAF, ►IL-1, ►IL-12, ►MyD88, ►NF-κB, ►ATF3, ►apoptosis, ►CD14, ►innate immunity, ►antimicrobial peptides, ►pattern recognition receptors, ►inflammation; Kobayashi K et al 2002 Cell 110:191; Pasare C, Medzhitov R 2003 Science 299:1033; Takeda K et al 2003 Annu Rev Immunol 21:335; crystal structure of TLR3: Choe JH et al 2005 Science 309:581; toll receptors: Gay NJ, Gangloff M 2007 Annu Rev Biochem 76:141.

Tolloid: ►bone morphogenetic protein

TOM (transfer outer membrane): A protein complex that regulates transport through the outer layer of the mitochondrial membrane. ►mitochondria, ►TIM

Tomato (*Solanum lycopersicum*, 2n = 24): The tomato has about 8–10 related species. It is one of the cytologically and genetically best-known autogamous plants and is suitable for practically all modern genetic manipulations. Its genome size is bp/n $\cong 6.6 \times 10^8$. ►coffee; for gene index: http://www.tigr.org/tdb/tgi.shtml; http://zamir.sgn.cornell.edu/mutants/; expression database: http://ted.bti.cornell.edu/; EST database: http://biosrv.cab.unina.it/tomatestdb.

Tomato Bushy Stunt Virus: A single-strand RNA virus of about 4,000 bases enveloped by an icosahedral shell consisting of 180 copies of a 40-kDa polypeptide. ►icosahedral

Tomography (body section radiography): Tomography is conducted by a tomograph in which a source of X-radiation moves in the direction opposite to that of a film, recording the image clearly only in one plane and blurring the remaining images. In computerized axial tomography (CAT scan), the scintillations produced by the radiation are recorded on a computer disk and the cross section of the body is analyzed electronically. Positron emission tomography (PET) involves the use of positron-labeled metabolites (e.g., γ-ray emitting glucose) (see Fig. T69). Along the path of the radiation, positrons and electrons collide and the local concentration of the isotopes is recorded electronically. Electron tomography permits the three-dimensional reconstruction of cells, organelles, and supramolecular assemblies (Lució V et al 2005 Annu Rev Biochem 74:833). Single-photon emission computed tomography (SPECT) takes γ-ray photographs around the body and a computer reconstructs three-dimensional images resulting in great resolution even of overlapping organs. Optical coherence tomography (OCT) resolves details in tissues at 1–15 μm in situ and it is thus one to two orders of magnitude finer than the ultrasound (Fujimoto JG 2003 Nature Biotechnol 21:1361). Ultrasonic tomography uses ultrasound scanning. The radiation may not be without risk (Löbrich M et al 2005 Proc Natl Acad Sci USA 102:8984). Gene expression tomography uses sliced brain sections in a cryostat. It obtains three-dimensional images of the expression pattern of particular genes using axial rotation (Brown VM et al 2002 Physiol Genomics 8:159). The RNA transcripts may be amplified by quantitative reverse PCR. ►X-rays, ►ultrasonic, ►sonography, ►nuclear magnetic resonance spectroscopy, ►imaging, ►RT-PCR; Czernin J, Phelps ME 2002 Annu Rev Med 53:89; Cristofallini M et al 2002 Nature Rev Drug Discovery 1:415.

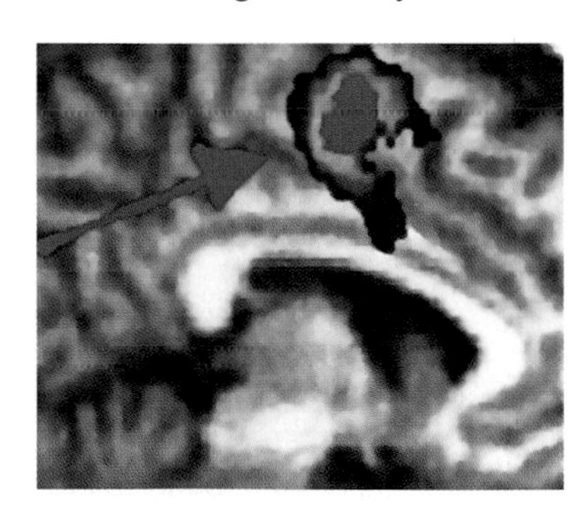

Figure T69. Pet scan of the brain; note different activity area at the arrow

Tongue Rolling: A human hereditary trait. Although it is frequently considered to be dominant, its inheritance is somewhat complex. Sturtevant AH (1940 Proc Natl Acad Sci USA 26:100) discovered it; he was the same person who first constructed a map of genes on a chromosome. Identical twins may be discordant for the trait (Matlock P 1952 J Hered 43:24). This trait seems to also be associated with moving the ears.

Tonoplast: An elastic membrane (≈8 nm) that surrounds the vacuoles. Several proteins control the traffic through this membrane. ►vacuoles; Maeshima M 2001 Annu Rev Plant Physiol Plant Mol Biol 52:469.

Tooth (pl. teeth): See the Figure T70. The crown (at top) is covered by enamel and it is remodeled after eruption. The cusp pattern varies in different species and it is regulated by an ectodin protein, secreted as an inhibitor of bone morphogenetic protein (see Fig. T70) (Kassai Y et al 2005 Science 309:2067). Teeth and bone DNA, present in small quantities, can be amplified by PCR and used for the analysis of forensic and ancient samples (Alonso A et al 2001 Croatian Med J 42:260; Kemp BM, Smith DG 2005 Forensic Sci Int 154:53). High-resolution micro-computed tomography permits detection of small differences in adult tooth morphology that are determined easily in embryonic development. Such a technology detected that the molars of Neanderthals grew very much the same way as in modern humans (Macchiarelli R et al 2006 Nature [Lond] 444:748). ►bone morphogenetic protein, ►tomography; gene expression in tooth: http://bite-it.helsinki.fi/.

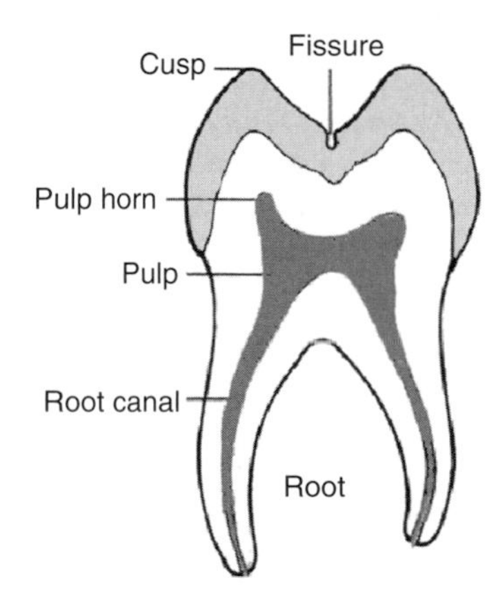

Figure T70. Posterior tooth of vertebrate

Tooth Agenesis: Tooth agenesis is caused by a dominant (chromosome 4p) mutation in the homeodomain of transcription factor MSX1. It involves failure to form the second premolars and the third molars. Tooth development is under the control of many genes. Among them fibroblast growth factor (FGF) and bone morphogenetic (BMP) play pivotal and multiple roles. FGF and BMP regulate MSX1 and PAX9 transcription factors. ►homeodomain, ►Rieger syndrome, ►amelogenesis imperfecta, ►FGF, ►BMP, ►PAX, ►MSX1, ►oligodontia, ►hypodontia

Tooth Development: ►anodontia; Tucker A, Sharpe P 2004 Nature Rev Genet 5:499.

Tooth Malposition: Apparently, an autosomal dominant gene controls the various misplacements or underdevelopment of the incisors and the canine teeth. ►Hallermann-Streiff syndrome, ►Jackson-Lawler syndrome, ►amelogenesis imperfecta, ►cherubism

Tooth Size: The tooth size appears to be influenced by the human Y chromosome as indicated by observations on various sex-chromosomal dosages.

Tooth-and-Nail Dysplasia (Witkop syndrome, TNS): In tooth-and-nail dysplasia, most commonly, the incisors, canine teeth and some of the molars are not or poorly developed. In some cases the condition is accompanied with abnormal toenails in children. Mutations at an MSX1 (4p161) subunit—a transcriptional repressor—seem to account for the phenotype. ►hypodontia, ►Hallermann-Streiff syndrome, ►dental no-eruption, ►dentin dysplasia, ►denticle, ►MSX1; Jumlogras D et al 2001 Am J Hum Genet 69:67.

5′-TOP (5′-terminal oligopyrimidine tract): mRNAs are part of the protein synthetic machinery and they are translated under the control of S6 kinases that are targets of insulin signaling in mammals. Cell growth (not cell division) is controlled by this process. ►S6, ►insulin-like growth factors; Crosio C et al 2000 Nucleic Acids Res 28:2927.

Top Agar: An agar solution of ~0.6% generally containing 0.5% NaCl and some organic supplements. About 2 mL of the solution, with suspended bacterial cells, is spread over the agar medium (30 mL/10 cm Petri plates) to initiate selective bacterial growth, e.g., in Ames tests. ►Ames test

Top-Down Analysis: Starting with a mutant phenotype, the physiological or molecular mechanism responsible for the alteration is investigated. In contrast, the bottom-up analysis first studies the molecules and then the analysis is extended to their relationships to the phenotype. Currently, the major endeavor is to go beyond the role of individual molecules and major interest is being focused on the critical domains of the molecules. ►reversed genetics

Top-Down Mapping: Top-down mapping uses either traditional genetic recombinational analysis or radiation hybrid maps. ►mapping genetic, ►radiation hybrids, ►bottom-up map

Topical Reversion: The back mutation is the result of alterations within the gene rather than due to extragenic suppression. ►reversion, ►suppressor extragenic, ►suppressor intragenic

Topoisomerase (TOP): Topoisomerases are enzymes which alter the tertiary structure of DNA without a change in the secondary or primary structure. The monomeric topoisomerase I (10q12-q13.1) nicks and closes single strands of DNA and changes the linking number in one strand. The dimeric topoisomerase II can cut and reattach both strands of the DNA and affect linking number in both strands.

Topoisomerase II (17q21-q22) disentangles DNA strands and plays an important role in DNA replication, transcription and recombination, suppression of mitotic recombination, stabilization of the genome (chromosome breakage), regulation of supercoiling, eukaryotic chromosome condensation, control of segregation of the chromosomes, regulation of the cell cycle, and nuclear localization of imported molecules. The mitochondrially located Top1mt is encoded at 8q24.3. Topoisomerases are important objects in cancer therapy research. Charvin G et al 2005 (cited in the caption in Figure T71), outlines mechanisms of action of the different topisomerases. The prokaryotic topoisomerases I—in contrast to the eukaryotic ones—require Mg^{2+} and single-stranded DNA segments and relax only the negatively supercoiled molecules. TOP I activity is required for the proper segregation of prokaryotic chromosomes (Zhu Q et al 2001 Proc Natl Acad Sci USA 98:9766) and has been located to human chromosome.

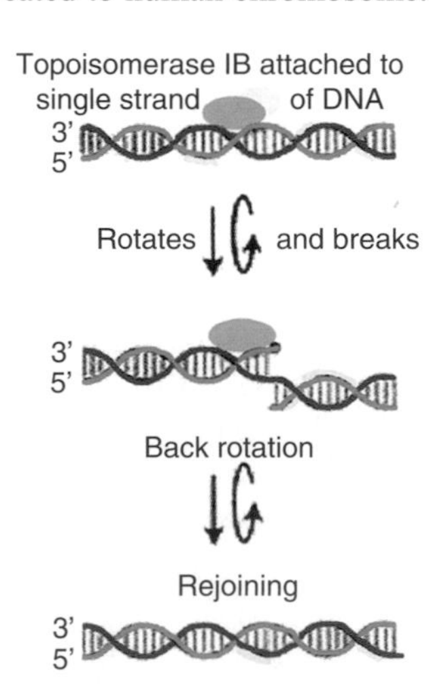

Figure T71. Redrawn after Charvin G et al 2005 Annu. Rev. Biophys. Biomol. Struct. 34:201

The human topoisomerase I consists of 765 amino acids. Prokaryotic DNA topoisomerase III supports the movement of the DNA replication fork and can also function as an RNA topoisomerase and can interconvert RNA circles and knots. The eukaryotic topoisomerase III (17p12-p11.2) is homologous with prokaryotic topoisomerase. Mice with knocked-out DNA topoisomerase IIIβ are viable but senesce earlier and have about a 40% shorter life span. Chromosomal abnormalities in *top3β*$^{-/-}$ mice might lead to a persistent increase in apoptotic cells, which might in turn lead to the progression of autoimmunity (Kwan KY et al 2007 Proc Natl Acad Sci USA 104:9242). ►DNA replication, ►gyrase, ►linking number, ►linking number paradox, ►mtDNA, ►camptothecin, ►p53; Wang JC 1996 Annu Rev Biochem 65:635; Changela A et al 2001 Nature [Lond] 411:1077; Champoux JJ 2001 Annu Rev Biochem 70:369; Wang JC 2002 Nature Rev Mol Cell Biol 3:430; evolution of topoisomerases: Forterre P et al 2007 Biochemie 89:427; topoisomerases as therapeutic targets of infection: Tse-Dinh YC 2007 Infect Disord Drug Targets 7:3; see Fig. T72.

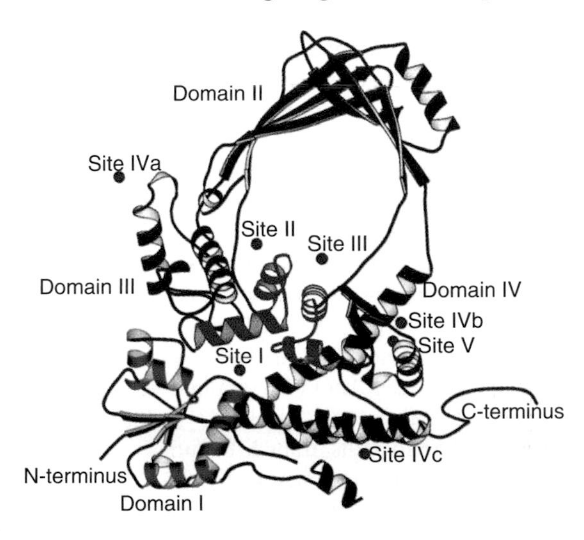

Figure T72. The 67-kDa fragment of DNA topoisomerase I of *E. coli* displaying major domains and nucleotide-binding sites. (From Feinberg H, Changela A, Modrigón A 1999 Nature Struct. Biol. 6:961)

Topological Filter: The topological filter is synonymous with two-step synapsis. The assumption is that synapsis requires an interaction between DNA *res* sites and the three subunits of a resolvase enzyme (step 1). After this initial step, the II and III subsites of the resolvase-*res* dimers pair in a parallel manner and the subunits interwrap. Then, the two I subsites of the resolvase dimer bind resulting in a productive synaptic complex (Step 2), capable of initiating DNA change. ►synapsis, ►*res*, ►resolvase; Watson MA et al 1996 J Mol Biol 257:317.

Topological Isomers of DNA: ►linking number

Topomap: A topomap displays genes that show correlated expression across a large set of microarray experiments.

Toponome: A single-cell-localized organization of proteins, carbohydrates, lipids, and nucleic acids. ►MELK, ►genetic networks

TORC (transducer of regulated CREB): A coactivator of CREB, a CRE (cyclic AMP-response element) binding protein. The glucose level is regulated in animal cells by insulin and glucagons, and TORC plays a central role in the homeostasis of glucose levels as outlined in the oversimplified chart (see Fig. T73). ►gluconeogenesis, ►CRE, ►glucagons, ►diabetes, ►cAMP, ►PKA, ►CREB, ►AMPK; Koo S-H et al Nature [Lond] 438:1109.

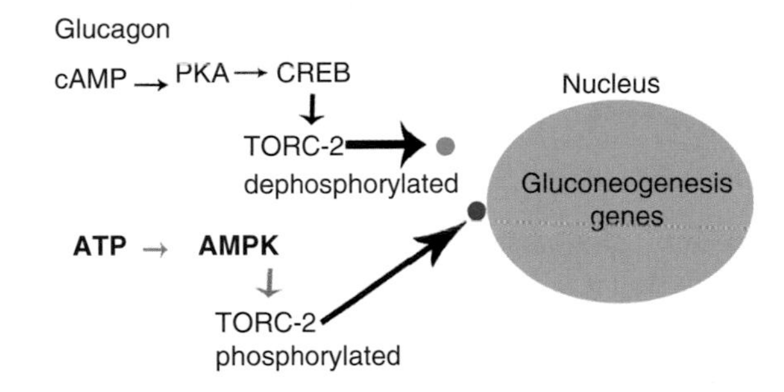

Figure T73. When TORC-2 is dephosphorylated gluconeogenesis genes are turned on in the nucleus. When TORC-2 is phosphorylated gluconeogenesis is attenuated

TORF: An open reading frame identified with the aid of a transposon. ►ORF, ►transposon

TORs (target of rapamycin): TORs are phosphatidylinositol kinases of yeast (TOR1, TOR2). They are also called RAFT1 and FRAP. TORs convey resistance to FKBP. The C-terminal domains of the TOR family are homologous to the catalytic domain of PI(3)K. TORs block the turnover of nutrient transporters, autophagy, and the interphase of the cell cycle but promote protein kinase C signaling, actin and cytoskeleton organization, and tRNA and ribosome formation, transcription and translation. TOR may be involved in the regulation of life span of *Drosophila* through affecting nutrition (Kapahl P et al 2004 Current Biol 14:885). The movement of TOR from the cytoplasm to the nucleus is controlled by the nutritional status in yeast (Li H et al 2006 Nature [Lond] 442:1058). The TORC1 complex is sensitive to rapamycin but the TORC2 is not. TORC1 positively regulates ribosome biogenesis, translation and nutrient import and blocks stress-responsive transcription. TORC2 mediates cell growth through actin organization (Wullschleger S et al 2006 Cell 124:471). TORs play a central role in the regulation of several human diseases such as hamartomas and other hypertrophic diseases (Inoki K et al 2005 Nature Genet 37:19). ►rapamycin, ►PIK, ►FK506, ►PI[3]K, ►S6K, ►autophagy, ►cell cycle, ►aging, ►cytoskeleton, ►protein kinases, ►cisplatin, ►Akt, ►leukemia; Schmelzle T, Hall MN 2000 Cell

103:253; Dennis PB et al 2001 Science 294:1102; Schalm SS, Blenis J 2002 Current Biol 12:632.

Tormogen: A component cell of the bristle that secretes the bristle socket. The other bristle cells are the trichogens that secrete the bristle shaft, also a neuron, which contacts the shaft and through its axon connects with the nervous system.

Toroid/Toroidal: A body or surface generated by the rotation of a plane curve or circle. It may assume the shape of a ring/doughnut or may resemble a barrel. ►torus; Hud NV, Vilfan UD 2005 Annu Rev Biophys Biomol Struct 34:295.

Torpedo Model: ►transcription termination in eukaryotes

Torsion Dystonia: ►idiopathic torsion dystonia, ►dystonia

Tortoiseshell Fur Color: Develops in female (cats) heterozygous for the X-chromosome-linked genes black and yellow colors because of selective, alternate inactivation of the two mammalian X-chromosomes. The tortoiseshell animals have mixed patches of black and yellow fur (Fig. T74). These fur patterns occur in the XX (female) or exceptionally in the XXY Klinefelter male cats. (See Lyon hypothesis, lyonization, calico cat).

Figure T74. Back of a tortoiseshell cat

Torus: A ring-shaped emergence, swelling, or a bordered pit. ►toroid

Totipotency: A characteristic of zygotic cells that permits differentiation into any type of cell or structure, including the whole organism. Initially, the paternal genome of animals is highly condensed and it is bound by protamines, which are replaced by histones before the S phase. Histone H3.1 is, however, absent from the paternal genome before DNA replication. The paternal genome displays H3K4me1, H3K9me1, and H3K27me1 (single-methylated lysines) but their dimethylation is largely postponed after DNA replication. The paternal genome is demethylated globally but not the maternal genome. For totipotency, the maternal transcription factors Oct3/4, Sox2, and the Polycomb group proteins are essential. By the two-cell stage the differences diminish yet the blastomeres maintain totipotency to the eight-cell stage. Between the eight and 16-cell stages polarization is initiated and this signals a switch to transition and to pluripotency. From pluripotency, totipotency is reestablished by the differentiation of the germline from the soma. These processes require numerous genetic and epigenetic factors (Surani MA et al 2007 Cell 128:247).

In the germline of *Caenorhabditis* the expression of translational regulators is required for the maintenance of totipotency. In *mex-3* and *gld-2* double mutants, germ cells transdifferentiate into somatic cells of muscles and neurons (Ciosk R et al 2006 Science 311:851). Plant cells maintain their totipotency in diverse adult tissues and after dedifferentiation may initiate other types of differentiations and of somatic cells entire organisms may be regenerated in cell cultures. Totipotency of animal cells is much more limited although lower animals such as hydra and earthworms may regenerate from differentiated tissues. The embryonic stem cells (ES) of mouse come close to totipotency/pluripotency inasmuch as they can be transferred to mouse embryos and can contribute to the formation of various cell types, including the germline. After special treatments, differentiated animal cells may also revert to stem cell status. ►morphogenesis, ►somatic embryogenesis, ►redifferentiation, ►multipotent, ►pluripotency, ►ES, ►nuclear transplantation, ►stem cells

Touch-and-Go Pairing: The end-to-end synapsis of the sex chromosomes in the heterogametic sex of some insects (see Fig. T75).

Figure T75. Touch-and-go-pairing. After Schrader F 1940 Proc Natl Acad Sci USA 26:634

Touch-Down PCR: ►hot-start PCR

Touch-Sensitivity: In mammals, the brain sodium channel BNC1 mediates touch sensitivity. ►degenerin; Welsh MJ et al 2002 J Biol Chem 277:2369.

Toulouse-Lautrec, Henri (1864–1901): One of the most remarkable painters of the end of nineteenth-century Paris life. He was a colleague of van Gogh and an influential precursor of modern art and was thought to suffer from pycnodysostosis. His parents were first cousins. This diagnosis of his genetic malady has been questioned [Nature Genetics 11:363], and it appears that deficiency of cathepsin K might have been involved. ►pycnodysostosis, ►cathepsins

Tourette's Syndrome (Gilles de la Tourette disease, GTS): A human behavioral anomaly causing motor and vocal incoordination (tic, twitching), stuttering (echolalia), the use of foul language (coprolalia), obsessions, and hoarding. The onset is between 7 to 14 years of age and three-fourths of those affected are male. The genetic determination is apparently dominant with incomplete penetrance and expressivity. Male:female ratio of 4.3:1 has been reported. Genes in several chromosomes (4q, 5q, 8p, 18q, 7q, 9, 3, 11q23, 17q) have been implicated. 238 nuclear families yielding 304 "independent" sibling pairs and 18 separate multigenerational families, for a total of 2,040 individuals using whole-genome screen for 390 microsatellite markers, indicated strong linkage for a region on chromosome 2pA (Tourette Syndrome Assoc. Internat. Consort. Genet 2007 Am J Hum Genet 80:265). A paracentric inversion involving chromosome 18q22 causes delayed replication in a sequence across at least 500 kb (State MW et al 2003 Proc Natl Acad Sci USA 100:4684). A sequence variant in Slit and Trk-like1 (SLITRK1)—encoding a single-pass transmembrane protein—has been identified at 13q31.1 (Abelson JF et al 2005 Science 310:317).

The frequency of the defective gene(s) was estimated to be 0.4 to 0.9%, and a prevalence of 1 to 0.02% has been observed in different populations. Some of the mild cases are suspected to be responsible for male alcoholism and female obesity. The expression of the condition is frequently much influenced by the environment. ►affective disorders; Zhang H et al 2002 Am J Hum Genet 70:896.

Townes-Brocks Syndrome (SALL1, 16q12.1): A highly variable dominant malformation with good penetrance. The symptoms include imperforate anus, polydactyly, syndactyly, abnormal earlobes, hearing-deficit, kidney and heart anomalies, and mental retardation. The majority of the individuals display mutation in the SALL gene encoding a 1,325 amino acid protein with double Zn-finger domains. The normal protein is a transcriptional repressor and it is associated with heterochromatin. ►penetrance, ►Zinc finger, ►heterochromatin; Netzer C et al 2001 Hum Mol Genet 10:3017.

Toxic Shock Syndrome: TSS is caused by infection of *Staphylococcus aureus/Streptococcus pyogenes* Gram-positive bacteria. It affects, primarily, menstruating women. It begins with sudden high fever, vomiting, diarrhea, and muscle pains (myalgia). Later, rash, hypotension, and potentially death may follow. Although septic shock has been attributed mainly to the cell wall lipopolysaccharides (present in Gram-negative bacteria) and peptidoglycan and lipoteichoic acid (present in Gram-positive bacteria), actually the unmethylated 5′-CpG-3′ prokaryotic DNA may be responsible for the immune reaction. These bacterial superantigens (SAG) bind directly to the class II histocompatibility complex on the antigen-presenting cells and to specific variable regions of the β-chain of the T cell receptor and evoke very severe response. Plasmids or phages carry the bacterial enterotoxin genes. ►Gram-negative/positive, ►peptidoglycan, ►teichoic acid, ►*Streptococcus*, ►*Staphylococcus*; McCormick JK et al 2001 Annu Rev Microbiol 55:77.

Toxicogenomics: Toxicogenomics applies microarray hybridization techniques for the detection of genes that are turned on or off upon exposure of animals to toxic substances. ►microarray hybridization, ►genetox; Simmons PT, Portier CJ 2002 Carcinogenesis 23:903.

Toxin Targeting: ►toxins, ►magic bullet, ►immunotoxin; Goodsell DS 2001 Stem Cells 19:161.

Toxins: Toxins are organic poisons. A plasmid gene of *E. coli* produces colicins and *Shigella* bacteria may affect sensitive bacteria in several ways. The cholera toxin is produced by the bacterium *Vibrio cholerae* and it interferes with the active transport through membranes thus causing the loss of excessive amounts of fluids and electrolytes in the gastrointestinal system. The *Bordatella pertussis* toxin, pertussin, contributes to high levels of adenylate cyclase and thereby to the symptoms of whooping cough. Bungarotoxin (from *Bungarus multicinctus*) and cobrotoxin (from Formosan cobra) in snake venoms block the acetylcholine receptors (McCann CM et al 2006 Proc Natl Acad Sci USA 103:5149) or interfere with ion channel functions. Bungarotoxin has an LD50 value in mice of 0.15–0.21 μg/g. Other bacterial toxins include the tetanus toxin, diphtheria toxin, botulin, etc. The diphtheria toxin (of *Corynebacterium diphtheriae* if it carries a temperate phage with the *tox* gene) is one of the most dreaded human poisons (sensitivity is encoded in human chromosome 5q23) with a minimal lethal dose of 160 μg/kg in guinea pigs. Mice and rats are, however, insensitive to this toxin. It inactivates eukaryotic initiation factor eIF-2 in translation on the ribosomes. The fungus *Amanita phalloides* toxin, amanitin (amatoxin), is a poison of RNA polymerase II. About two dozen cytochalasins are synthesized by different fungi and are composed of substituted hydrogenated isoindole rings fused with a macrocyclic ring (large organic compound). Colchicine is a plant alkaloid poison of the spindle fibers. The seed toxin of the plants *Strophantus* and *Acocanthera* is a blocking agent for membrane transport and is used as a selective agent in mammalian cell cultures. Abrin (from a legume) and ricin (from castor bean seeds) are inhibitors of the attachment of aminoacylated t-RNAs to the ribosomes. The piscine toxin, tetrodotoxin (from *Spheroides rubripes*; LD_{50} in mice 10 μg/kg), and the dinoflagellate, *Gonyalux* species, saxitoxin (LD50 in mice 3.4–10 μg/kg), locks the sodium ion channels and blocks neurotransmission. The curare toxins were obtained originally as an arrow poison from the bark of the trees *Strychnos* and *Chondodendron*. The sources of other curare toxins, bamboo curare, pot curare, gourd curare, etc., are members of the *Menispermaceae* family and are highly poisonous muscle relaxants. They block the acetylcholine receptors and some ion channels. Some of the curare toxins were used medically for treatment of tetanus shock and in surgery to alleviate muscle rigidity. ►colicins, ►amatoxin, ►diphtheria toxin, ►cytochalasins, ►aflatoxins, ►anthrax, ►colchicine, ►abrin, ►ricin, ►antibiotics, ►ion channel, ►acetylcholine receptors, ►neurotransmitters, ►LD_{50}, ►laboratory safety; Schiavo G, van der Goot G 2001 Nature Rev Mol Cell Biol 2:530; Nesioy D et al 2004 Nature [Lond] 429:429; http://www.epa.gov/iris.

Toxmap: A geographic information system which provides facts of toxic substances released to the environment by the major manufacturing industries based on government mandated reports: http://toxmap.nlm.nih.gov/toxmap/main/index.jsp.

Toxoid: An inactivated bacterial toxin, which may still incite the formation of antitoxins and retain antigenicity.

Toxoplasmosis: Opportunistic infection of about 25% of the human populations by *Toxoplasma* protozoa reducing nerve connections. It is frequently lethal in AIDS and other immune-compromised people. Meiotic recombination among different strains may greatly enhance their virulence. The high virulence of the parasite *Toxoplasma gondii* is due to a secreted serine-threonine kinases (ROP) encoded in chromosome VIIa (Taylor S et al 2006 Science 314:1776; Saeij JPJ 2006 Science 314:1780). ►AIDS, ►apicoplast; Grigg ME et al 2001 Science 294:161; Su C et al 2003 Science 299:414.

TP53 (tumor protein): TP53 is now called p53, a tumor suppressor. ►p53

TPA: A phorbol ester (12-o-tetradecanoyl phorbol 13-acetate) that promotes neoplastic growth after induction has taken place. ►carcinogens, ►cancer, ►phorbol esters

TPA: An inducer protein; a mitogen activating protein kinase C. ►protein kinases

TPA (Third Party Annotation Sequence Database): An annotation for a sequence that the submitter derived from GenBank primary data: http://www.ncbi.nlm.nih.gov/Genbank/TP.html; http://www.ebi.ac.uk/embl/Submission/align_top.html.

TPEA (3′ end amplification) PCR: TPEA PCR permits the detection of mRNA at the single cell level (Dixon AK et al 1998 Nucleic Acids Res 29:4426; Subkhankulova T, Livesey FJ 2006 Genome Biol 7: R18). ►PCR, ►mRNA

TPK1, TPK2, TPK3: Catalytic subunits of A-kinase. ►protein kinases

TPM: ►two-phase mutation model

TPN: TPN is a triphosphopyridine nucleotide; TPNH is the reduced form. They are synonymous with the analogs α-NADP and α-NADPH, respectively. Many enzymatic reactions require β-NADP and β-NADPH (nicotinamide adenine dinucleotide phosphates). ►NAD, ►$NADP^+$

TPO: TPO modulates megakaryocyte differentiation along with EPO and various cytokines. ►EPO

TPR: ►tetratrico sequences

TPX2: A microtubule-associated protein. It is required for spindle assembly. Binding to importin α inactivates it but RAN•GTP reverses the binding. ►microtubule, ►importin, ►RAN, ►spindle; Wittmann T et al 2000 J Cell Biol 149:1405; Gruss OJ et al 2001 Cell 104:83.

T_R: A regulatory T cell. ►T cells

TR3: ►nur77

tra (*transformer*): A gene (chromosomal location 3–45) of *Drosophila* that controls sterile male development in XX flies; XY *tra/tra* males are, however, normal males. *tra* acts in cooperation with *tra2* (2–70). Tra and Tra2 proteins mediate sex-specific processing pre-mRNAs of *dsx* (*doublesex*, 3–48.1) and *fru* (*fruitless*, 3–62.0). The transcripts occur in both sexes but are processed differently. In the male germline, Tra2 is required for normal spermatogenesis. Tra2 also mediates the sex-specific processing of *exu* (*exuperentia*, 2–93) and *att* (*alternative-testes-transcript*, chromosome 3 at 92E3–92E4). The prokaryotic *tra* genes in conjugative plasmids control the conjugal transfer of DNA. ►sex determination, ►*tra* genes

***tra* Genes**: (more than 17) mediate the conjugal transfer of the F and other conjugative plasmids. ►*tra*, ►conjugation, ►ori_T, ►relaxosome, ►plasmids

Trabant: A terminal chromosomal appendage (see Fig. T76). ►satellited chromosome

Figure T76. Trabant

Trace Elements: Trace elements are only required in minute amounts.

Tracer: A (radioactively) labeled molecule that permits the identification of the fate of the molecules into which it has been incorporated. ►isotopes

Trachea: The duct leading from the throat (larynx) to the lungs of animals or the duct system of insects through which air is distributed into the tissues. ►FGF

Tracheid: A long, lignified xylem cell specialized for transport and support in plants.

Trachophyte: A vascular plant endowed with xylem, phloem, and (pro)cambium in-between. ►xylem, ►phloem, ►cambium

Tracking: A mechanism to ensure that transposition occurs between two appropriate *res* (recombination sites). Experimental proofs for successful tracking (reporter rings) are not unequivocal. ►reporter ring

Tracking Dyes: In electrophoresis, loading buffers permit the visualization of the front migration toward the anode. Bromophenol blue in 0.5 × TBE buffer moves at the same rate as double-stranded linear DNA of 300 bp length. Xylene cyanol FF moves along with 4 kb linear double-strand DNA. ►electrophoresis, ►electrophoresis buffers

***Tradescantia* Species**: The *Tradescantia* species occur in the polyploidy range of 2x to 12x. The plants develop ca. 100 stamen hairs in their flowers (see Fig. T77). Each hair represents a single cell line (see Fig. T78). When plants heterozygous for anthocyanin markers are exposed to mutagen, somatic mutations can be assessed as differently colored cell lines in the large number of flowers that single plants may form. Some of the species have been favorites for cytologists. ►bioassays in genetic toxicology, ►somatic mutation; pictures are the courtesy of A Sparrow, and GP Rédei unpublished.

Figure T77. *Tradescantia* stamens

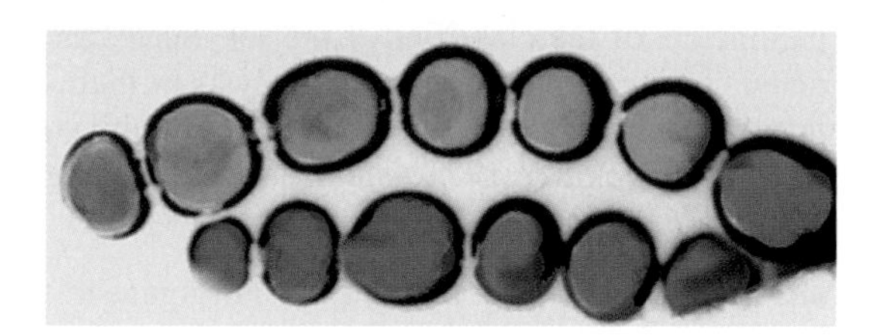

Figure T78. *Tradescantia* stamen hairs

TRADD (tumor necrosis factor receptor associated death domain): ►tumor necrosis factor, ►death domain, ►apoptosis, ►TRAF

TRAF: A tumor necrosis factor-associated receptor and a signal transducer for some interleukins. TRAF5 is involved in CD40 and CD27-mediated signaling to lymphocytes. TRAF6 activates IκB kinase through a polyubiquitin chain. ►TNF, ►TNFR, ►CRAF, ►interleukins, ►TRADD, ►IRAK, ►NF-κB, ►IκB, ►ASK1, ►Toll, ►MATH, ►Ire, ►CD27, ►CD40, ►ubiquitin; Tada K et al 2001 J Biol Chem 276:36530; Li X et al 2002 Nature [Lond] 416:345.

Tragedy of the Commons: The title of an article published in 1968 by Garrett Hardin (Science 162:1243) where he called attention to the limitations of the "common"—the earth—and the unlimited use of its resources and "freedom to breed" (overpopulation) which will inevitably lead to tragedy because there are no other solutions than the use of temperance to avoid the consequences of the inevitable necessity. ►human population growth, ►cooperation, ►prisoner's dilemma

TRAIL (Apo-2L, 3q26.1-q26.2): A tumor necrosis factor related apoptosis-inducing ligand which attaches to the death receptor DR4 and mediates apoptosis. DR4 does not respond to FADD like the Fas, TNFR-1, and DR3 system. It has five receptors. TRAIL also activates NF-κB. The nontoxic TRAIL causes preferential killing of neoplastic cells in animal models, especially in combination with radiation therapy of cancer. It appears, however, that it does not discriminate sufficiently between normal human liver cells and liver cancer cells. ►apoptosis, ►CD4, ►death domain, ►death receptor, ►FAS, ►FADD, ►TNF, ►TNFR, ►TWEAK, ►NF-κB, ►imaging; Kontny HU et al 2001 Cell Death Differ 8:506; collection of papers in Vitam Horm 2004 vol 67:1–453.

Trailer Sequence: A trailer sequence follows the termination codon at the 3′-end of the mRNA. It is not translated and yet it may have a regulatory function. ►polyA mRNA

Training Set: A set of observations used to fit the parameters of the classifier of the set. Such sets are generally used for discriminant analysis or multivariate analysis. ►discriminant function, ►multivariate analysis; Tsypkin Y 1971 Adaptation and learning in automatic systems, Academic Press, New York.

Trait: A distinguishable character of an organism that may or may not be inherited.

TRAM (translocating-chain associating membrane proteins): TRAMs are membrane-spanning glycoproteins associated with the nascent peptide chain while the SRP mediates its transfer to the endoplasmic reticulum. ►SRP, ►protein targeting, ►translocons, ►translocase, ►signal hypothesis

TRAM: Transverse rectus abdominis muscle.

Tramp: An Apo-3 type TNF/NGF receptor. ►Apo-3, ►TNF, ►NGF

Tramp: A polyadenylation cofactor complex of Trf4p (poly A polymerase)-Mtr4p (RNA helicase)-Air2p (zinc knuckle protein) involved with the 3′–5′ exonuclease participating in RNA maturation and quality control (LaCava J et al 2005 Cell 121:713; Wyers F et al 2005 Cell 121:725).

Trance (RANK/ODF/OPGL): A tumor-necrosis factor-related activation-induced cytokine ligand (encoded at 13q14) that regulates T cell-dependent immune reactions and bone differentiation, bone mass, and Ca^{2+} metabolism. High levels of the RANK ligand (RNKL) may lead to bone breakdown (osteoporosis), and estrogen treatment reduces bone loss. The RANKL cytokine triggers migration of human epithelial cancer cells and melanoma cells and leads to metastasis of cancer where RANK is expressed. Cancer metastasis to bones can be blocked using osteoprotegerin but not to other organs (Jones DH et al 2006 Nature [Lond] 440:692). Osteoprotegerin is a TGF-family osteoblast-secreted decoy receptor binding to osteoclasts; it inhibits their maturation. ►TNF, ►osteoclast, ►osteoblast, ►osteoporosis; Pearse RN et al 2001 Proc Natl Acad Sci USA 98:11581; Theill LE et al 2002 Annu Rev Immunol 20:795.

Trans: A trans position indicates that two genetic markers are not on the same molecule or not on syntenic parts of the chromatids. ►cis arrangement, ►synteny, ►chromatid

Trans Arrangement of Alleles: The trans arrangement of alleles indicates that they are not in the same chromosome (DNA) strand (they are in repulsion). This is in contrast with the cis arrangement (coupling) where the two alleles are within the same strand. ►coupling, ►repulsion, ►cis

Transacetylase: A protein which transfers an acetyl group from an acetyl coenzyme A (Acetyl-CoA) to another molecule. The third structural gene of the lactose operon (*lacA*) encodes a 275-amino acid polypeptide that forms a dimer of 60 kDa that is a transacetylase. ►*lac* operon

Trans-Acting Elements: Proteins that are synthesized anywhere in the genome but which regulate transcription by attachment to specific sites of a gene. ►cis-acting element

Transactivation Responsive Element (TAR): In the HIV transcript the Tat protein binds to the TAR sequence near the 5′-end. This binding then mediates an increased expression of the viral genes and the synthesis of more mRNA. The Rev protein binds to a specific RNA site and to the rev-responsive element (RRE), and facilitates the export of the unspliced transcript to the cytoplasm where viral structural proteins and enzymes are made. ►acquired immunodeficiency, ►VP16, ►transactivator, ►VDR

Transactivator: A protein domain attached to a specific inhibitory protein that may prevent blocking of transcription and may increase the transcription of the target gene(s) by several orders of magnitude. ►transactivation responsive element, ►Switch-Gene, ►VP16, ►p53, ►tetracycline, ►STAT, ►two-hybrid method; Devaux F et al 2001 EMBO Rep 2:493; Lottmann H et al 2001 J Mol Med 79:321.

Transactive Catastrophy: The similar base sequences in different organisms or even within the same genome that may have different functional meanings.

Transaldolase Deficiency (TALDO1, 11p15.5-p15.4, a pseudogene is at 1p34.1-p33): Transaldolase deficiency causes cirrhosis of the liver and infantile hepatosplenomagaly (enlargement of the spleen and liver) and accumulation of metabolites of the pentose phosphate pathway (ribitol, D-arabitol and erythrol) in the urine and blood plasm. Male mice lacking transaldolase (TAL$^{-/-}$) are sterile because of defective forward motility. TAL$^{-/-}$ spermatozoa show loss of mitochondrial transmembrane potential and mitochondrial membrane integrity because of diminished NADPH, NADH, and GSH (Perl A et al 2006 Proc Natl Acad Sci USA 103:14813). ►cirrhosis of the liver, ►pentose phosphate pathway, ►GSH; Verhoeven NM et al 2001 Am J Hum Genet 68:1086.

Transaminases: ►aminotransferases

Transcapsidation (heteroencapsidation): The viral coat protein and the enclosed genetic material are of different origin (wolf in a sheep's skin). ►pseudovirion; Quasba PK, Aposhian HV 1971 Proc Natl Acad Sci USA 68:2345.

Transchromosomic (transchromosomal, TC): A cell which contains foreign chromosomal segments, e.g., a mouse cell with human chromosomal fragment(s). Human chromosome 21 or its fragments in mouse cells provide means of experimentation with Down's syndrome in mice (O'Doherty A et al 2005 Science 309:2033). ►chromosome uptake, ►somatic cell hybrid, ►alien addition, ►alien transfer, ►alien substitution, ►microcell, ►Down's syndrome; Tomizuka K et al 2000 Proc Natl Acad Sci USA 97:722.

***trans*-Cinnamic Acid**: *trans*-cinnamic acid is formed from phenylalanine by phenylalanine ammonia-lyase and it is converted to a large number of compounds in plants (see Fig. T79). ►chalcones

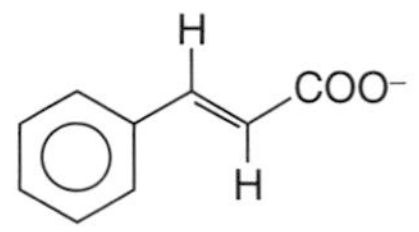

H
COO−
H

Figure T79. *trans*-Cinnamic acid

Transcobalamin (TCN1): A ligand and transporter of the B12 vitamin; it is encoded in human chromosome 11q11-q12.

Transcobalamin (TC2): A recessive condition which causes megaloblastic anemia, located at human chromosome 22q11 qter. ►megaloblastic anemia, ►anemia

Transconjugant: The bacterial genetic material was (partly) derived by recombination during conjugation. ►conjugation

Transcortin Deficiency (CBG): A dominant, human chromosome 14q31-q32 reduction in a corticosteroid-binding globulin. ►corticosteroid

Transcribed spacer: A DNA element between genes that is transcribed but eliminated during the processing of the primary transcript. ►primary transcript

Transcript: An RNA copied on DNA and complementary to the template. The average number of transcripts per genes is variable and it has been estimated to range from 0.3 to more than 9,400 copies in single human cells. The ENCODE project found 5.4 transcripts per human gene loci formed by alternative splicing. The low copy transcripts are very difficult to recognize. In the relatively small yeast genome, 606 genes showed no mRNA transcripts and 3,796 (~69%) displayed one or fewer transcripts per cell and only ~0.04% had more than 50 transcripts (Akashi H 2003 Genetics 164:1291). Highly important genes have biased usage of the synonymous codons and an abundance of isoacceptor tRNAs. Natural selection for efficient synthesis appears to favor shorter proteins in yeast but not necessarily in other eukaryotes. ►transcription, ►isoacceptor tRNAs, ►codon usage, ►intergenic transcript, ►transfrags, ►antisense transcript, ►ENCODE, ►alternative splicing; Velculescu VE et al 1999 Nature Genet 23:387; ►human transcript [database]

Transcript Cleavage Factor: The transcript cleavage factor induces cleavage and then the release of the 3′-end of a 1- to 17-nucleotide RNA transcript while the 5′-end is still associated with the polymerase. The process may not terminate transcription and transcription may be reinitiated without a deletion in the transcript. In prokaryotes, *GreA* and *GreB*, in eukaryotes transcription factor TFIIS carry out such functions. The role of the process may be to remove wrong sequences, facilitating the transition from initiation to the elongation phase and the escape of the polymerase from the promoter complex. ►RNA polymerases, ►promoter, ►transcription complex, ►transcription factors; Conaway JW et al 1998 Cold Spring Harbor Symp Quant Biol 63:357.

Transcript Elongation: Transcription has four basic phases: 1) promoter binding and activation of the RNA polymerase; 2) RNA chain initiation, followed by escape of polymerase from the TATA site; 3) transcript elongation; 4) termination of transcription and transcript release. The elongation phase in *E. coli* is marked by the release of the σ subunit of the polymerase, the exit of the polymerase from the promoter, and the tight association between the polymerase, DNA template and the nascent transcript ternary complex. Transcript elongation in eukaryotes is quite similar. The rate of elongation varies because of sites of pause, arrest, and termination. In eukaryotes, in vivo the rate of transcription is ~1,200–2,000 nucleotides/minute but in vitro it is only 100–300/minute. Some of the DNA-binding proteins (such as repressors, CCAAT-binding proteins, etc.), gaps in the DNA, or drugs may interfere with elongation although few generalizations are possible at this time. In some instances, the RNA elongation and DNA replication may conflict. During development a variety of controls regulate transcript elongation. The rapid early embryonal development involves relatively shorter transcripts. The crystal structure and function of the RNA elongation complex of *Thermus thermophilus* bacterium has been elucidated (Vassylyev DG et al 2007 Nature [Lond] 448:157). ►promoter, ►transcription factors, ►σ, ►walking, ►transcription rate, ►mediator complex, ►elongator, ►elongin, ►ELL, ►RNA polymerase, ►error in transcription, ►tryptophan operon, ►transcription termination, ►attenuation, ►chromatin remodeling, ►TRAP, ►operon, ►inchworm model, ►DRB, ►NELF, ►TFIIS, ►TEFb, ►leukemias [MLL], ►Cockayne syndrome, ►von-Hippel-Lindau syndrome; Weliky Conaway J, Conaway RC 1999 Annu Rev Biochem 68:301; Wind M, Reines D 2000 Bioassays 22:327; Toulokhonov I et al 2001 Science 292:730; Pal M, Luse DS 2003 Proc Natl Acad Sci USA 100:5700; Shilatifard A et al 2003 Annu Rev Biochem 72:693.

Transcript Mapping: RNA transcripts are hybridized with specific DNA probes and subsequently the (not annealed) single strands are digested by S1 nuclease. Thus, the resistant (annealed) fragments represent the homologous tracts and demarcate the transcript. For the process of mapping an entire genome, sequence-tagged sites (STS) of cDNAs are used. By the PCR method, the STS sequences can be amplified and their position can be mapped either by YAC clones or by radiation hybrid panels. Using a high throughput system in yeast, 6.5 million probes of both strands of the DNA were interrogated. On rich media, 85% of the yeast genome is expressed. In addition to the expected transcripts, operon-like transcripts were also found. The intergenic regions did not separate some transcripts of adjacent genes, and different parts of the same gene expressed at different levels. Transcript mapping revealed an unexpectedly complex organization of this relatively simple eukaryotic genome (David L et al 2006 Proc Natl Acad Sci USA 103:5320). ►nucleic acid hybridization, ►S1 nuclease, ►STS, ►radiation hybrid, ►YAC, ►PCR, ►differential hybridization mapping, ►transfrag; Barth C et al 2001 Curr Genet 39(5–6):355; Rinn JL et al 2003 Genes Dev 17:529; global transcript tiling: Bertone P et al 2004 Science 306:2242.

Transcriptase: DNA-dependent RNA polymerase or RNA-dependent DNA polymerase or RNA-dependent RNA polymerase enzyme. ►RNA polymerase, ►reverse transcription

Transcription: The synthesis of RNA complementary to a strand of a DNA molecule. In prokaryotes the majority of the transcriptionally active genes are located in the leading strand of replication and are transcribed in the same direction as the DNA synthesis. In the absence of a functional DNA helicase, genes involved in the replication of the lagging strand are hampered by the transcription complex fork and stalled for many minutes. If, however, the DNA helicase is present the replication fork on the lagging strand can quickly pass the RNA polymerase complex. In prokaryotes, transcription and translation are coupled unlike in eukaryotes where the mRNA must be released through the nuclear pore complex into the cytoplasm. Transcription is most commonly regulated by a variety of proteins (transcription factors). In the red clover necrotic mosaic virus (RCNMV), a subgenomic portion (sgRNA) of one of the two RNA genomes (RNA-1), a 34-base portion of the RNA-2, is required for the transactivation of transcription of sgRNA. In eukaryotes, transcription taker place in higher order transcriptional domains (16 in the mouse cell), which are usually independent from the replicational domains. In yeast, transcription and replication appear independent. Based on 678 loci, the Spearman rank correlation coefficient was found to be 0.45 (Washburn MP et al 2003 Proc Natl Acad Sci USA 100:3107). In *Drosophila*, transcriptional activity is correlated with DNA replication at the early S phase.

Transcription in eukaryotes is regulated by the intragenic pattern of the nucleosomal structure. In order to access the gene by RNA polymerase the nucleosomal arrangement must be loosened at the 5′-end of the gene but nucleosomal assembly is maintained downstream to prevent illegitimate transcription initiation. As transcription proceeds the nucleosomal structure is modified in front of the polymerase and reformed as it passes through a particular region. Histone acetylation at the promoter and other critical regions assures the beginning and continuation (elongation) of transcription. Histone deacetylation is mediated by methylation of histone H3 at lysine 36. H3 methylation at lysine 36 is brought about by protein complex Set2, which interacts with RNA polymerase 2 during elongation but not during transcription initiation. The Rpd3 smaller unit is attracted to H3Lysine36

T

(H3K36) and is involved in deacetylation of the 3′-coding sequences. The larger unit mainly carries out the deacetylation and suppression of the 5′-end promoter region. The Rpd3S and Set2 have functions similar to Spt in chromatin remodeling and function. The histone variant H2A.Z is localized to the nucleosome-free region ca. 200 bp upstream of the first codon of the promoter of many genes. It assists in chromatin remodeling by SWR1 (Lieb JD, Clarke ND 2005 Cell 123:1187). DNA topoisomerase IIβ (Topo IIβ), which breaks double-stranded DNA, is required in a signal-dependent manner for the initiation of transcription. Other essential factors include DNA-binding transcription factors, activating protein 1, DNA-dependent protein kinase 1, Ku86, Ku70, CBP coactivator and Pol II that were also recruited. The PARP-1 co-repressor complex (nucleolin, nucleophosmin, heat-shock protein 70) was rapidly eliminated after induction (Ju B-G et al 2006 Science 312:1798).

►pol, ►class II and class III genes of eukaryotes, ►transcription complex, ►transcription factors, ►transcription termination, ►regulation of gene activity, ►open transcription complex, ►chromatin remodeling, ►nucleosomes, ►histones, ►histone variants, ►signal transduction, ►mitochondria, ►mitochondrial genetics, ►chloroplasts, ►chloroplast genetics, ►replication fork, ►transcription complex, ►regulation of gene activity, ►Set motifs, ►Rpd3/Sin3, ►Spt, ►inchworm model, ►antitermination, ►RNA polymerase, ►transcription rate, ►pause transcriptional, ►Spearman rank-correlation test, ►antisense transcription, ►topisomerases, ►DNA-PK, ►Ku86, ►Ku70, ►CBP, ►PARP, ►nucleolin, ►nucleophosmin, ►heat-shock proteins; Cold Spring Harbor Symp Quant Biol vol 63, 1999; Lee TI, Young RA 2000 Annu Rev Genet 34:77; Johnson KM et al 2001 Curr Biol 11:R510; Nature Rev Mol Cell Biol 2002 3:11; Kapranov P et al 2002 Science 296:916; Schübeler D et al 2002 Nature Genet 32:438; transcription/translation tool: http://www.cbs.dtu.dk/services/VirtualRibosome/.

Transcription Coactivator: A transcription coactivator activates RNA polymerase II but does not bind to DNA. (See Oswald F et al 2001 Mol Cell Biol 21:7761).

Transcription Cofactor: A transcription cofactor links a transcription factor to the transcription complex without binding directly to the DNA.

Transcription Complex: The TATA box-associated complex has the components TFIIA, TFIIB, TFIID, TFIIE, TFIIF, TFIIH, TFIIJ, TFIIK, and RNA polymerase II. After the Pol II enzyme moves downstream and away from the preinitiation complex and is phosphorylated by a CTDK and TFIIH kinase action, it can continue transcription in the absence of other regulatory factors although generally specific transcription factors (transactivators) may boost and regulate its activity (see Fig. T80).

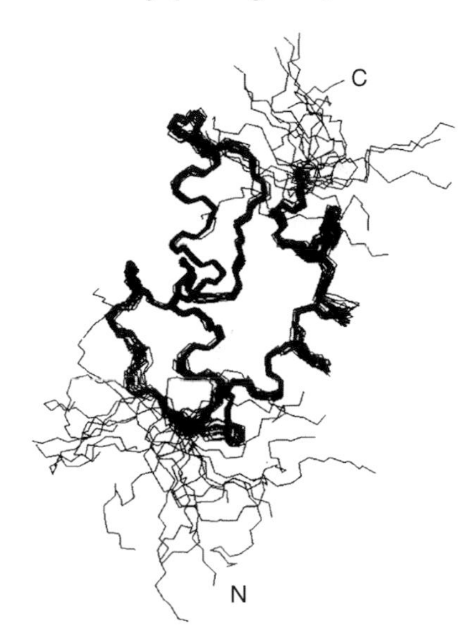

Figure T80. Human TFIIEβc structure as determined by nuclear magnetic resonance analysis. This transcription factor binds to the DNA where the promoter opens up when RNA polymerase II initiates transcription. (From Okuda M et al 2000 EMBO J 19:1346. Courtesy Professor Y. Nishimura.)

Usually, a low level of phosphorylation of the heptapeptide repeats of the C terminus (CTD) of the largest subunit of the RNA polymerase is conducive to attraction to the promoter sequences and to the initiation of transcription. The yeast protein FCP is phosphatase, associated with the RNA polymerase II complex, and it facilitates the association of the polymerase to the preinitiation complex. Some transcription factors also play a role in translation. ►RNA polymerase, ►transcription factors, ►regulation of gene activity, ►SRB, ►TBP, ►snRNA, ►transcription factors, ►open promoter complex, ►Hogness (-Goldberg) box, ►Pribnow box, ►transcription shortening, ►CTD, ►elongation factors, ►TCF, ►nucleic acid chain growth; Wolfberger C 1999 Annu Rev Biophys Biomol Struct 28:29; transcriptional regulatory networks: Lee TI et al 2002 Science 298:799.

Transcription Corepressor: A transcription corepressor represses RNA polymerase II without binding to DNA.

Transcription, Ectopic: ►transcription, ►illegitimate

Transcription Factories: A transcription factory is formed when active genes move to shared nuclear subcompartments for transcription and move apart as transcription ceases (Osborne CS et al 2004 Nature Genet 36:1065). Regions of different chromosomes may make contact within the nucleus and influence gene regulation (Dillon N 2006 Chromosome Res 14:117). Although the chromosomes have some special territories within the nucleus, the intermingling of territories can be visualized at finer resolution (Branco MR, Pombo A 2006 PLoS Biol 4(5):e138). Although some research has indicated that the nuclear periphery is a repressive compartment, newer work indicates that transcription of the mammalian β major-globin gene (responsible for thalassemia) begins at the periphery and that its transcription increases as it moves toward the interior. The locus control region assists in moving the β-globin gene from the nuclear periphery toward the nuclear interior during the maturation of erythroid cells (see Fig. T81). The DNA-dependent RNA polymerase II becomes hyperphosphorylated as it moves to the interior transcription factory (Ragoczy T et al 2006 Genes Dev 20:1447). In yeast, the *GAL* genes move to the nuclear periphery—near the nuclear pore complex—upon becoming transcriptionally activated by members of the SAGA histone acetyltransferase complex (Sus1, Ada2) (Cabal GG et al 2006 Nature [Lond] 441:770). Similarly, the activated subtelomeric hexokinase isoenzyme 1 (*HXK*1) gene moves to the nuclear periphery. This move, controlled by the 3′-untranslated region, is consistent with efficient processing and export of mRNA through the nuclear pore. If the gene was activated by VP16, it moves away from the periphery and induction by galactose abrogated. Accordingly, the nuclear position plays an active role in optimal gene expression (Taddei A et al 2006 Nature [Lond] 441:774). ►chromosome territories, ►nucleolus, ►LCR, ►RNA polymerase, ►pol II, ►SAGA, ►VP16, ►hexokinase; Lanctôt C et al 2007 Nature Rev Genet 8:104.

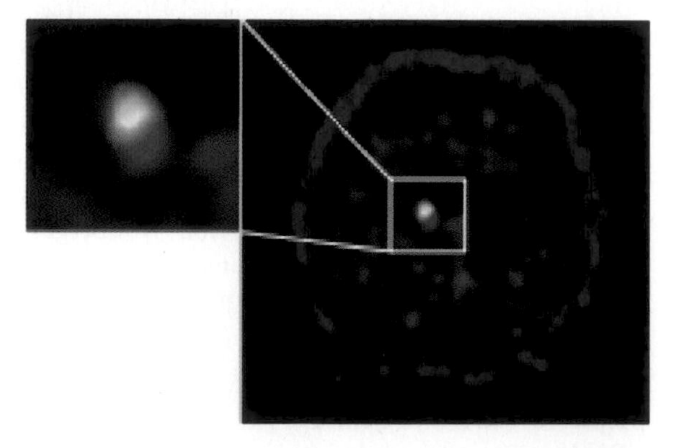

Figure T81. The β-globin locus (green) in the interior of the nucleus (blue) is associated with hyperphosphorylated Pol II (red). Courtesy of Tobias Ragoczy and Mark Groudine, Fred Hutchinson Cancer Research Center

Transcription Factors: A large number of different proteins that bind to either short upstream elements, terminator sequences of the gene, or to the RNA-polymerases and modulate transcription. These specific transcription factors occur in very large numbers and are essential for the function of genes in a particular manner. The ~6,300 genes of budding yeast are regulated by about 200 transcription factors. Zinc-finger proteins equipped with specific DNA-binding abilities are suitable for engineering functional roles and altered phenotypes such as drug resistance, temperature tolerance, and various developmental pathways (Park K-S et al 2003 Nature Biotechnol 21:1208). The general transcription factors have highly conserved sequences and are interchangeable even among such diverse organisms as mammals, *Drosophila*, yeast, and plants. The TAF proteins, by virtue of their different domains, can modulate the transcription of different genes. Also, the large TAF_{II} protein associated with the TFIID complex, by acetyl-transferase activity, can remodel the chromatin structure and using its amino-terminal kinase activity can transphosphorylate TFIIF and thus modulate transcription of protein genes in two different ways. TFIIB is also capable of self-acetylation as a means of regulating transcription (Choi CH et al 2003 Nature [Lond] 424:965). The specific transcription factors may form a large-combinatorial network with various promoter elements.

The transcription factors for RNA polymerase I, transcribing ribosomal RNA genes, show a relatively simple organization (see Fig. T82).

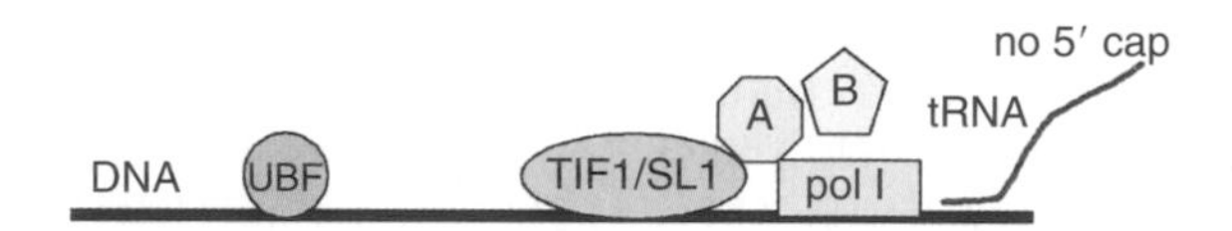

Figure T82. Some transcription factors of RNA polymerase I

The UBF (upstream binding factor) binds upstream in the promoter and regulates transcription by acting as an assembly factor for the transcription complex. TIF-1 or its vertebrate homolog SL1 is required for the attachment of Pol I to the promoter and the accessory proteins A and B assist in the transcription.

The tRNAs, 5S rRNAs and some other small RNAs are transcribed by RNA polymerase III (Pol III). This enzyme has a requirement for protein factor TFIIIB and in case of transcribing the 5S rRNA it also requires TFIIIA and TFIIIC for the assembly of the transcription complex. Proteins TFIIIA and C, however, are detached after TFIIIB binds and only the latter stays on the DNA when Pol III lands and the transcription begins. TFIIIA is not required for the transcription of the tRNAs. The Pol III transcription units also have internal control regions A (box 5′-TG GCNNAGTGG-3′) and B (box 5′-GGTCGANNC-3′) or similar sequences. The U6 snRNA gene of yeast has an elaborate promoter (TATA box, downstream T_7 tract, upstream segment) in the non-transcribed strand. It provides a scaffold for the recruitment for the RNA polymerase III but it is deficient at a subsequent step of promoter opening by TFIIIB. It is thus a rare type of single-strand promoter (Schröder O et al 2003 Proc Natl Acad Sci USA 100:934).

The transcription factors required for RNA polymerase II (Pol II), which transcribe protein-coding genes form the most elaborate complex.

The general transcription complex is initiated by binding the TBP (TATA-box-binding protein) subunit of general transcription factor TFIID to the TATA box of eukaryotes (Hogness box) (see Fig. T83). The TFIID complex exists in different forms (α and β) depending on its association with the TAF_{II}-30 protein (in β) or its absence (in α). The TATA box is present in upstream region genes, coding for protein and transcribed by RNA polymerase II. TFIID also attracts TFIIB. TFIID also brings the cleavage-polyadenylation specificity factor (CPSF) to the preinitiation complex and this assists in the formation of the 3′-end of the mRNA. After this, TFIIF, TFIIE, and TFIIH proteins attach to RNA polymerase II (Pol II) to the TATA box. TFIIH expresses a DNA helicase function (encoded by xeroderma pigmentosum genes XPB and XPD) and a cyclin-dependent protein kinase activity (encoded by CDK7). TFIIF also stimulates transcript elongation by stimulating the phosphorylation of polymerase II and it plays a role in Pol I transcription. In addition, TFIIF stimulates a phosphatase, specific for the largest subunit of the RNA polymerase. The TFIIS protein factor in eukaryotes and *GreA* and *GreB* stimulates transcript cleavage and readthrough. It is a coactivator and regulator of transcription and it binds 3′ to the TATA box. The Pol II is inactive at this stage until TFIIH phosphorylates the bound Pol II using ATP as a phosphate donor. The targets of phosphorylation are several sites near the COOH-end of the largest subunit of pol II. TFIIH also has a helicase, ATPase, and nucleotide excision repair activities. It also mediates promoter melting and promoter clearance. Some genes may be transcribed without the kinase activity of TFIIH yet they need its helicase function. Eukaryotic RNA polymerase II generally contains nine or 12 subunits. The largest subunit is usually about 200 kDa. There are 26 (yeast) to 52 (mammals) repeats (Tyr-Ser-Pro-Thr-Ser-Pro-Ser) close to the carboxyl end. The phosphorylated Pol II then moves out of the complex and can now initiate transcription. The specific transcription factors are operative in special genes and at special tissues and time frames. The transcription factors may also bind a variety of other proteins before or during transcription and thus provide a great variety of fine-tuning of gene expression (Hager GL et al 1992 Curr Opin Genet Dev 12:137). The not absolutely essential TFIID TATA box-binding protein associated factors $dTAF_{II}$ 42 and $dTAF_{II}$ 62 form a heterotetramer resembling the heterotetrameric core of the histone octamer in the nucleosome. The TBP protein subunit of TFIID may also be dispensable in case $TAF_{II}30$ is present (named TFTC [TBP-free TAF_{II}-containing complex]). The general transcription factors are the basic instruments of transcription initiation but the modulation and regulation of transcription requires a large number of specific factors, activators, coactivators, suppressors and their interactions. Transcription factors with altered specificities can be generated in the laboratory using structure-based design and molecular technology. Besides the general transcription factors that are involved with almost all RNA polymerase II transcribed genes, specific and inducible transcription factors, activators and coactivators as well as chromatin reorganization factors also play an important role. Transcription factors participate in regulatory networks and display altered interactions to varying degrees. A few transcription factors serve as permanent hubs but most act transiently only during certain conditions (Luscombe NM et al 2004 Nature [Lond] 431:308). Quantifying the affinities of molecular interactions is

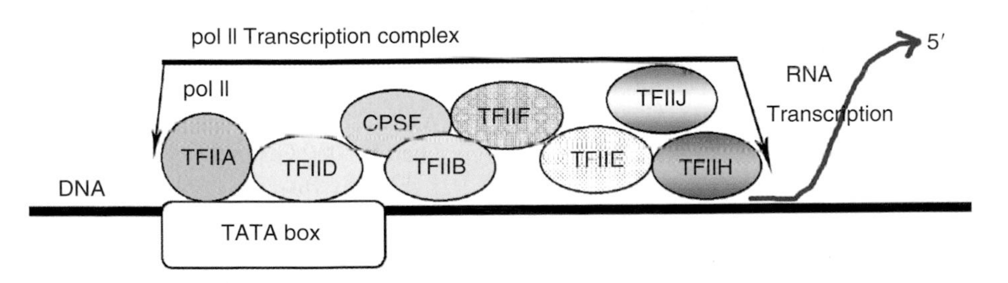

Figure T83. General transcription complex

difficult because of the large number of variables involved. Also, the interactions are transient and exhibit nanomolar to micromolar affinities, leading to rapid loss of bound material or little bound material. Protein-protein and protein-DNA binding microarrays are especially susceptible to loss because of their stringent wash requirements and may lose weakly bound material. A high-throughput microfluidic platform capable of detecting low-affinity transient binding events on the basis of the mechanically induced trapping of molecular interactions eliminates the off-rate problems facing current array platforms and allows for absolute affinity measurements. Basic helix-loop-helix (bHLH) transcription factors generally bind to a consensus sequence of 5′-CANNTG-3′ (N is any nucleotide) called enhancer box or E box, which is the second most conserved motif in higher eukaryotes. Members of the bHLH family show mid- to low nanomolar DNA binding affinities and have off rates above 10^{-2} s^{-1} for their consensus sequences with orders of magnitude higher off rates for non-consensus sequences. The transcription factor binding energy topographies were measured with highly integrated microfluidic devices. The biological function of two yeast transcription factors could be successfully predicted by combining purely in vitro biophysical measurements with informatics knowledge of the genome (Maerkl SJ, Quake SR 2007 Science 315:233).

In the roots of *Arabidopsis*, 80% of the transcription factors seem to be regulated by their upstream noncoding sequences and about one-fourth of the transcription factors are regulated posttranscriptionally by microRNAs or intercellular protein movement (Le J-Y et al 2006 Proc Natl Acad Sci USA 103:6055) (see Fig. T84).

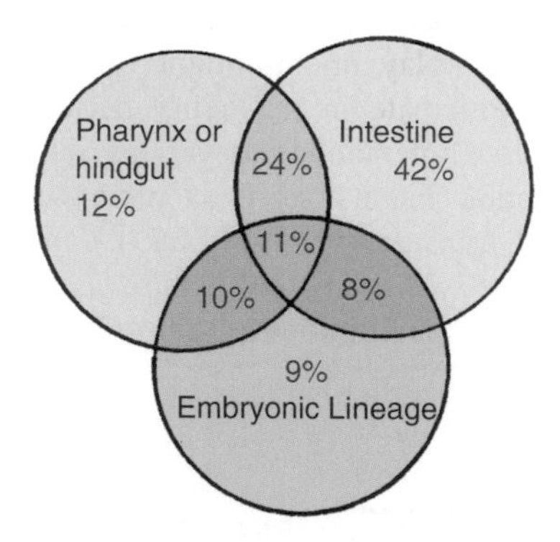

Figure T84. Transcription factor (117) interactions with 72 genes in *Caenorhabditis*. (After Deplancke B et al 2006 Cell 125:1193)

The human genome has ~1,850 transcriptional regulators; that of *Arabidopsis* has 1,533 (~5.9%) whereas *Drosophila, Caenorhabditis* and budding yeast have 635 (4.3%), 669 (3.5%) and 209 (3.5%), respectively. Tissue-specific transcription factors (FOXA2, HNF1A, HNF4A and HNF6) bound to 4,000 orthologous gene pairs in hepatocytes have been purified from human and mouse livers. Despite the conserved function of these factors, 41 to 89% of their binding events seem to be species specific (Odom DT et al 2007 Nature Genet 39:730). These molecules may be further explored for basic research and gene therapy. Viral repressors may prevent the association of RNA polymerase II with the transcriptional preinitiation complex. Using multiphoton microscopy special transcription factor functions can be visualized in vivo (Yao J et al 2006 Nature [Lond] 442:1050). Procedures have been developed for probabilistic predictions for the identities of binding sites and their physical binding energies between transcription factors and DNA (Kinney JB et al 2007 Proc Natl Acad Sci USA 104:501). The nucleosomal structure in eukaryotes must accommodate transcription factors so they can reach the DNA. It seems that spontaneous unwrapping of nucleosomes rather than histone dissociation or chromatin remodeling provides the means for access to the DNA (Bucceri A et al 2006 EMBO J 25:3123).

►TF, ►TBP, ►mtTF, ►transcription complex, ►DNA binding, ►SRB, ►CTD, ►CDK, ►SWI, ►class II and class genes, ►Pol II [RNA polymerase II], ►RNA polymerase, ►regulation of gene activity, ►genetic networks, ►genome-wide location analysis, ►protein synthesis, ►gene therapy, ►ABC excinuclease, ►transcriptional activators, ►co-activators, ►nuclear receptors, ►transcriptional modulation, ►transcript elongation, ►STAGE, ►TFIIS, ►mtTFA, ►CAAT box, ►enhancer, ►regulation of gene activity, ►chromatin, ►chromatin remodeling, ►nucleosome, ►FOX, ►HNF, ►FACT, ►coactivator, ►transactivator, ►SAGA, ►HMG, ►promoter, ►open promoter complex, ►GAGA, ►gene number, ►transcription rate, ►inchworm model, ►pause transcriptional, ►NAT, ►mediator, ►elongator, ►FACT, ►PITSLRE, ►reinitiation of transcription, ►JASPAR, ►xeroderma pigmentosum, ►multiphoton microscopy, ►Hogness (-Goldberg) box, ►Pribnow box, ►transcription complex; Myer VE, Young RA 1998 J Biol Chem 273:27757; Riehmann JL et al 2000 Science 290:2105; Conaway RC, Conaway JW 1997 Progr Nucleic Acid Res Mol Biol 56:327; Hampsey M 1998 Microbiol Mol Biol Rev 62:465; Pauli MR, White RJ 2000 Nucleic Acids Res 28:1283; Lemon B, Tjian R 2000 Genes & Development 14:2551; Misteli T 2001 Science 291:843; Svejstrup JQ 2002 Current Opin Genet Dev 12:156; Burley SK, Kamada K 2002 Current Opin Struct Biol 12:225; Warren AJ 2002 Current Opin Struct Biol 12:107; Roulet E et al 2002 Nature Biotechnol 20:831; Ponting CP 2002 Nucleic

Acids Res 30:3643; Gerland U et al 2002 Proc Natl Acad Sci USA 99:12015; Cosma MP 2002 Mol Cell 10:227; Lee TI et al 2002 Science 298:799; Bushnell DA et al 2004 Science 303:983; transcriptional regulatory code: Harbison CT et al 2004 Nature [Lond] 431:99; statistical detection of cis-regulatory motifs in eukaryotes: Gupta M, Liu JS 2005 Proc Natl Acad Sci USA 102:7079; transcription factor binding site detection: Li X et al 2005 Proc Natl Acad Sci USA 102:16945; transcription factor binding site identification by immunoprecipitation: Euskirchen GM et al 2007 Genome Res 17:898; http://www.ifti.org; http://compel.bionet.nsc.ru/FunSite.html; http://www.gene-regulation.com/pub/databases.html; transcriptional start site database: http://dbtss.hgc.jp; transcription factor binding sites: http://bio.chip.org/mapper; binding sites: http://genome.imim.es/datasets/abs2005/index.html; transcription factor binding site tools in humans and mouse: http://hscl.cimr.cam.ac.uk/TFBScluster_genome_portal.html; comparative genomics-based resource for initial characterization of gene models and the identification of putative cis-regulatory regions of RefSeq Gene Orthologs: http://genometrafac.cchmc.org; mammalian binding site prediction: http://pdw-24.ipk-gatersleben.de:8080/VOMBAT/faces/pages/choose.jsp; transcription binding site prediction: http://www.bioinfo.biocenter.helsinki.fi/poxo; promoter models and binding sites: http://www.gene-regulation.com/pub/programs/cma/CMA.html; transcription factor binding sites in yeast: http://cg1.iis.sinica.edu.tw/~mybs/; over-represented binding sites: http://www.cisreg.ca/oPOSSUM/.

Transcription Factors, Designed: These include DNA-binding Zinc-finger domains fused to either gene activation or suppression domains. Such constructs may up- or down-regulate the expression of selected targets by affecting chromatin remodeling. ►zinc finger, ►polyamide, ►fermentation; Urnov FD et al 2002 EMBO Rep 3:610; Rebar EJ et al 2002 Nature Med 8:1427; Blancafort P et al 2003 Nature Biotechn 21:269.

Transcription Factors, Inducible: Inducible transcription factors are proteins that are synthesized within the cell in response to certain agents or metabolites. They bind to short upstream or dowstream DNA sequences and affect transcription, frequently by looping the bound DNA back to the promoter area and forming an association with the other protein factors and the general transcription factors. Such transcription factors may be hormones, heat-shock proteins (DNA binding site consensus: CNNGAANNTCCNNG), phorbol esters (TGACTCA), serum response elements (CCATATAGG), etc. These protein factors exert their specificity in gene regulation not only by discriminative ability for individual genes (since their numbers must be lower than that of the genes), but by their modular assembly. ►transcription factors, ►hormone response elements, ►heat-shock proteins, ►HSTF, ►regulation of gene activity; Mathew A et al 2001 Mol Cell Biol 21:7163, regulated combinations: Setty Y et al 2003 Proc Natl Acad Sci USA 100:7702.

Transcription Factors, Intermediary: Intermediary transcription factors do not associate directly with the promoter but either affect the conformation of the DNA or "adapt" other proteins to the transcription complex. ►open promoter complex, ►regulation of gene activity; Steinmetz AC et al 2001 Annu Rev Biophys Biomol Struct 30:329.

Transcription Factor Map: The transcription factor map identifies the binding sites of transcription factors in the genome. The binding sites are important in the regulation of development, differentiation, and carcinogenesis. The positions can be revealed by enhancer-binding assays (Hallikas O et al 2006 Cell 124:47) or alternatively a paired-end ditag procedure can be used. The latter procedure found 542 binding sites for p53 (Wee C-L et al 2006 Cell 124:207). ►enhancer, ►transcriptome, ►p53, ►DNA binding protein domains, ►paired-end diTAG

Transcription Illegitimate (ectopic transcription): Illegitimate transcription takes place when very low-level transcripts are detected in organs, tissues or developmental stages where these special transcripts are not expected to occur. Usually, nested primers are employed for the amplification. ►housekeeping genes, ►tissue-specificity, ►nested primer; Salbe C et al 2000 Int J Biol Markers 15:41.

Transcription Initiation (transcription start): The DNA forms a transcription bubble when transcription begins. In about 60% of the human genes CpG sequences are located 5′ in the early-replicating, highly acetylated gene-rich regions (Cross SH et al 2000 Mamm Genome 11:373). The phage T7 RNA polymerase undergoes a major conformational change at the amino-terminal 300 residues. This then entails the loss of the promoter-binding site and facilitates promoter clearance when the initiation is followed by transcript elongation. The RNA transcript peels off of a seven-base pair heteroduplex and an exit tunnel is created for the enhanced processivity of the elongation complex. ►replication bubble, ►promoter clearance, ►processivity, ►TSS; Yin YW, Steitz TA 2002 Science 298:1387; Young BA et al 2002 Cell 109:417; Pokholok DK et al 2002 Mol Cell 9:799; http://elmo.ims.u-tokyo.ac.jp/dbtss.

Transcription Rate: The transcription rate may vary from gene to gene and site. Based on fluorochrome labeling of the β-actin RNA and serum induction indicated 1.1 to 1.4 kb per minute. Yeast RNA polymerase II can synthesize 1.2 kb long sequences/min. The single-subunit bacteriophage RNA polymerase can incorporate in vitro 12–24 kb nucleotides/minute. The bacterial enzyme can build in 3–6 kb/min in vivo and 0.6–2 kb/min in vitro. ►RNA polymerase, ►transcription factors, ►transcript elongation, ►error in transcription; Brem RB et al 2002 Science 296:752.

Transcription Shortening: RNA polymerase II (RNAP II) hydrolyzes the 3′ end of the transcript as part of the process of reading through pause signals and also secures fidelity of the transcription. It requires a TFIIS protein.

Transcription Termination in Eukaryotes: The ribosomal gene cluster is generally terminated much beyond the 28S rRNA gene and at about 200 bp upstream from the core promoter of the following pre-rRNA cluster. In mouse, the Pol I termination signal contains a Sal I box (5′-AGGTCGACCAG [T/A][A/T]NTCCG-3′) preceded by T-rich clusters. The actual termination is within the T-rich area and it is assisted by the Sal I box and the T-rich sequences around it. In humans, the conserved repeats (5′-GACTTGACCA-3′) terminate pre-tRNA transcription. In *Xenopus* (5′-GACTTGC-3′), repeats and T-rich sequences in the spacer region bring about termination. Probably some proteins bind to the Sal I box. Recently, polypeptide chain release factors (RF) have also been identified in eukaryotes. In yeast, the carboxy-terminal domain (CTD) of the DNA-dependent RNA polymerase II (Pol II) contains a consensus sequence YSPTSPS in multiple copies. Different phosphorylation patterns recruit RNA processing factors after phosphorylation of CTD serine 2. These factors are required for both transcription and termination of transcription. Although Pol II proceeds with transcription beyond the polyadenylation signal, the polyadenylation system cleaves the transcript and the RNA is degraded by the Rtt103 complex and the Rat1/Rai 5′→3′ exonucleases joined to the 3′-end of the protein genes, ending transcription (Kim M et al 2004 Nature [Lond] 432:517). The termination mechanism in humans is very similar to that in yeast and it occurs co-transcriptionally (CoTC) by an exonuclease called Xrn2 (West S et al 2004 Nature [Lond] 432:522). The termination in this manner is called torpedo model because the transcription is "torpedoed" at the end. The *Drosophila* N-TEF protein releases the Pol II transcript in an ATP-dependent manner. The Reb-1 yeast protein stops the polymerase and mediates the transcript release. The mouse TTF-1 protein, similarly to Reb-1, binds to the DNA and brings about termination. Thus the termination mechanisms in different species vary substantially.

In case of DNA, template damage during transcription ubiquitylation at the carboxy-terminal repeat domain (CTD) takes place. If the serine 5 residue (involved in chain elongation) and CTD is phosphorylated, ubiquitylation is inhibited (Somesh BP et al 2005 Cell 121:913).

In the mtDNA a 34 kDa protein (mTERM) is bound to a 13-residue sequence embedded in the $tRNA^{Leu(UUR)}$ and the complex is required for termination of transcription. The MAZ protein may regulate the transcription from different promoters in closely spaced genes. The poly(A) signal is required for the termination of transcription but before the actual termination pretermination cleavage of the RNA transcript takes place. Therefore, before Pol II is released from the DNA it must transcribe the pretermination cleavage site and also the poly(A) signal. ►polyadenylation signal, ►transcription termination in prokaryotes, ►mRNA, ►rho factor; Langst G et al 1998 EMBO J 17:3135; Dye MJ, Proudfoot NJ 2001 Cell 105:669.

Transcription Termination in Prokaryotes: Transcription termination in prokaryotes can be rho-independent (intrinsic terminators exist in the RNA polymerase) and rho-dependent, i.e., the RNA polymerase requires the cofactor rho for termination of transcription. The terminator regions in various systems have similar structures. They consist of palindromic sequences that can fold back into a hairpin. In the rho-independent terminator there are one or more G≡C rich sequences in the stem; at the base of the stem there are about six consecutive U residues. This structure mediates a pause in the movement of the RNA polymerase thus causing dissociation from the DNA template because the ribosyl-U of the transcript can make only weak hydrogen bonds with the deoxyribosyl-A in the DNA. In the rho-dependent termination, rho recognizes 50 to 90 bases before the hairpin facilitates termination. The *E. coli* protein NusA promotes folding of the hairpin and termination. The λN protein promotes antitermination. Polypeptide release factors (RF) may also be used in both prokaryotes and eukaryotes. On lambda phage templates, the N-terminal of the 109-amino acid Nun protein of phage HK022 blocks transcription by binding to BOXB on the nascent RNA transcript of the pL and pR operons; and the C-terminal domain interacts with the RNA polymerase. If the RNA polymerase ternary complex (at 3′-OH end) cleaves the transcript, transcription may be reinitiated as long the upstream sequences remain firmly aligned with the DNA. The prokaryotic proteins *GreA* and *GreB* and the eukaryotic TFIIS may favor transcript cleavage.

(See Fig. T85, ►antitermination, ►lambda phage; Washio T et al 1998 Nucleic Acids Res 26:5456; Gusarov I, Nudler E 2001 Cell 107:437; Unniraman S et al 2000 Nucleic Acids Res 30:675; Kashlev M, Komissarova N 2002 J Biol Chem 277:14501).

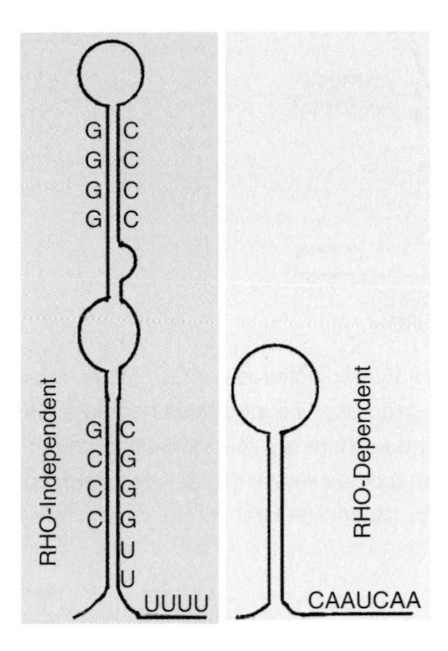

Figure T85. Transcription terminator sequences in prokaryotes

Transcription Unit: The DNA sequences between the initiation and termination of transcription of a single or multiple (co-transcribed, multi-cistronic) gene(s).

Transcriptional Activators: Transcriptional activators are proteins that facilitate the activity of DNA-dependent RNA polymerase(s) in prokaryotes and eukaryotes. Transcription activators may have two independent domains: DNA-binding domain and transcriptional activating domain.

These domains may not have to be covalently associated and dimeric association may carry out their function. In the abstract sketch (see Fig. T86), the dimeric activators stimulate transcription by recruiting the components of the transcription complex to the TATA box, even when transcription may be hindered in their absence or in the presence of the monomers. Employing several of the activators in a non-covalently bound bundle may enhance the potency of transcriptional activation. The addition of six or eight extraamino acids to the yeast Gal4 activator has generated artificial activators. The new activators may have higher or more specific activity. Many of the transcriptional activators (Myc, Jun, etc.) are unstable in the cell and are destroyed by calpains, lysosomal proteases and, most commonly, by ubiquitin-mediated proteolysis. Actually, the activation and the ubiquitination domains functionally overlap. ►positive control, ►negative control, ►transcription factors, ►POU, ►DNA binding proteins, ►catabolite activator, ►regulation of gene activity, ►FK506, ►FKBP12, ►transcription complex, ►GCN5, ►transcription termination, ►transcriptional modulation, ►SW, ►VDRI, ►suppression, ►degron, ►N-degron, ►N-end rule, ►ubiquitin, ►calpain, ►destruction box, ►lysosomes, ►chromatin remodeling, ►co-activators; Ptashne M, Gann A 1997 Nature (Lond) 386:569; Hermann S et al 2001 J Biol Chem 276:40127; Lu Z et al 2002 Proc Natl Acad Sci USA 99:8591.

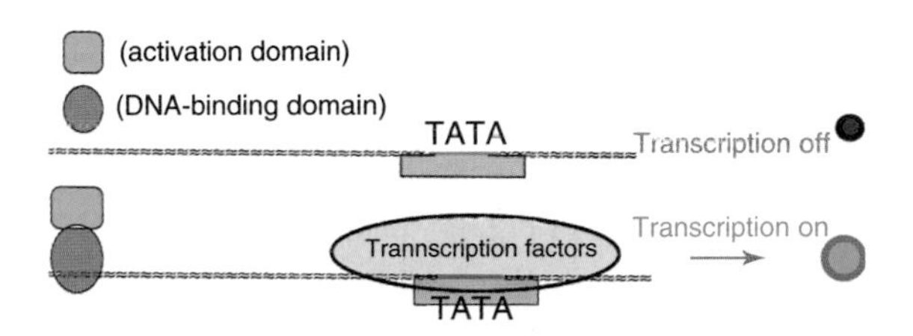

Figure T86. Transcriptional activation

Transciptional Adaptor: A histone acetyltransferase, acylating histone in the chromatin before activation of transcription. The yeast gene *GCN5* affects, specifically, histones 3 and 4. ►histone, ►nucleosome, ►transcription, ►transcription initiation

Transcriptional Coactivator: ►TAF_{II}, ►co-activator, ►GCN, ►coactivators

Transcriptional Control: The regulation of protein synthesis at the level of transcription. ►regulation of gene activity, ►signal transduction, ►transcription factors, ►operon, ►regulon

Transcriptional Error: When cytosine is deaminated to uracil in the template DNA strand and the error is not repaired, adenine is incorporated in the RNA transcript in place of guanine (to be inserted by the RNA polymerase). This may be translated into a defective or aberrant protein. Such a mutation may take place in nondividing cells. ►error in translation

Transcriptional Gene Fusion Vector: The transcriptional gene fusion vectors carry transcription-termination codons (stop codons) in front of the promoterless structural gene, so when the structural gene fuses to a host promoter and is thereby expressed (transcribed with the assistance of a host promoter), it will contain only the amino acid sequences specified by the inserted DNA. ►gene fusion, ►read-through proteins, ►translational gene fusion, ►trapping promoters, see Figs. T87 and T88.

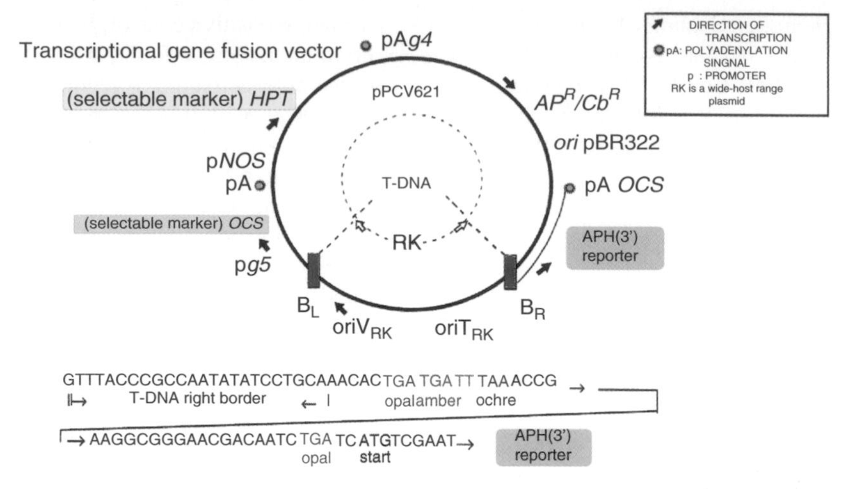

Figure T87. In the transcriptional gene fusion vector the reporter (APH[3']II, or luciferase or GUS) has no promoter and it is fused to the right border of the T-DNA. The reporter gene is expressed only if it integrates behind a plant promoter that can provide the promoter function. In front of the structural gene here, there are four nonsense codons to prevent the fusion of the protein with any plant peptides. Transcriptional fusion vectors are similar in other groups of organisms. For other symbols and abbreviation see translational gene fusion vectors. (Based on oral communications by Dr. Csaba Koncz)

Figure T88. Transcriptional *in vivo* gene fusion of bacterial luciferase genes to a flower-bud-specific promoter of *Arabidopsis*. The highest level of expression is indicated in red (false color). The background is photographically cleared. (From Rédei GP, Koncz C, Langridge WHR, unpublished)

Transcriptional Gene Silencing (TGS): Transcriptional gene silencing is based on mechanisms of methylation of DNA and chromatin remodeling by proteins or RNA. ►methylation of DNA, ►chromatin remodeling, ►silencer, ►dosage compensation, ►imprinting

Transcriptional Landscape: The pattern of transcriptional control signals and the transcripts generated.

Transcriptional Modulation: In many transcription factors proline- and glutamine-rich activation domains exist and modulating effects are attributed to them. ►transcriptional activators, ►transcription factor, ►transcriptional suppressor, ►WD-40; Rushlow C et al 2001 Genes Dev 15:340.

Transcriptional Noise: The variation in the rate of transcription due to the assembly of the preinitiation complex, TATA box sequence, TATA box-binding proteins, activators, coactivators, regulatory elements, reinitiation of transcription and thus translation. This noise determines the heterogeneity of gene expression in different cells of the organism and contributes to differentiation and phenotypic differences. ►noise; Blake et al 2003 Nature [Lond] 422:633.

Transcriptional Pause: Pause transcriptional.

Transcriptional Priming: Transcriptional priming occurs when multipotential progenitor cells can be induced to develop into different cell types. GATA-2 and PU.1 transcription factors may have antagonistic or cooperative effects in the determination of cell fate. GATA-2 is required for the generation of mast cells but is not required for differentiation of macrophages. The Ets family member PU.1 is necessary for myeloid and lymphoid cells but not for erythroid or megakaryocyte lineages. In addition, PU.1 directs the differentiation of hematopoietic cells into macrophages. ►GATA, ►PU.1, ►stem cells; Walsh JC et al 2002 Immunity 17:665.

Transcriptional Profiling: ►microarray hybridization

Transcriptional Regulation: ►transcription factors, ►regulation of gene activity, ►quelling, ►methylation of DNA, ►co-suppression, ►posttranscriptional gene silencing, ►RNA noncoding; Carlson M 1997 Annu Rev Cell Dev Biol 13:1; Vaucheret H, Fagard M 2001 Trends Genet 17:29; Myers LC, Kornberg RD 2000 Annu Rev Biochem 69:729; Arnosti DN 2003 Annu Rev Entomol 48:579; http://rulai.cshl.edu/TRED.

Transcriptional Slippage: Transcriptional slippage occurs when the RNA polymerase transcribes longer or shorter RNA sequences than the actual template, either during initiation or the elongation process. Slippage generally occurs when the template has homopolymeric sequences. (See Larsen B et al 2000 Proc Natl Acad Sci USA 97:1683).

Transcriptional Start Site: ►TSS, ►transcription initiation

Transcriptional Suppressor: A transcriptional suppressor can tightly bind to the operator or to other upstream elements of the DNA and thus prevent the initiation of transcription by elements (activators) of the transcription complex. Mot1 is an ATP-dependent inhibitor of the TATA box-binding protein and the members of the NOT complex inhibit the transcription machinery by various ways (TBP, TAF, etc.). Heterochromatin protein (HP1) and histone methyl transferase may also repress genes. ►transcriptional activator, ►transcriptional coactivator, ►transcriptional modulation, ►nucleosome, ►mating type determination in yeast, ►silencer, ►MADS box, ►inhibition of transcription; Ma Y et al 2001 J Biol Chem Online Sept 27; Hwang K-K et al 2001 Proc Natl Acad Sci USA 98:11423.

Transcriptional Synergy: Multiple transcription factors exert greater effect together than the sum of their individual contributions to gene regulation. ►transcription factors; Green MR 2005 Mol Cell 28:399.

Transcriptional Targeting: Transcriptional targeting intends to limit the expression of therapeutically used transgenes to specific cells, e.g., tumor cells. This goal is achievable by the use of tissue-specific promoters, e.g., the carcinoembryonic antigen is expressed mainly in colon and liver cells. The α-fetoprotein promoter is specific for hepatocellular carcinoma. The erbB2 promoter is expressed predominantly in breast tumors and the prostate-specific antigen is active in prostate cancer. Inducible promoters may be protected from a viral enhancer by insulator elements. ►insulator; Brand K 2004, p 531. In: Gene and Cell Therapy. Smyth Templeton N (Ed.) Marcel Dekker, New York.

Transcriptional-Coupled Repair: ►DNA repair, ►excision repair, ►colorectal cancer

Transcriptome: The collection of RNAs transcribed from the genome; transcriptomics is the generation and study of the mRNA profile of the cell. In a single human cell about half of all the genes may be expressed and a total of about 25,000 to 30,000 unique genes are expressed in a human body. The total number of transcripts in the different tissue cells of these unique genes was found to be 134,135. Some genes were expressed only at 0.3 copies per cell while others had up to 9,417 transcripts. About 1,000 transcripts were expressed in five copies in all cells. In some cancer cells some transcripts were present in about 10 copies but absent in normal cells. 40 genes were expressed in all types of cancer cells in about three copies, and this was twice the number for normal cells. The length of the transcripts can be determined by the use of GIS (gene identification signature). The 5′ and 3′ ends (18 nucleotides each) of full-length cDNAs are extracted and concatenated into paired-end ditags (PETs). These are then mapped to genome sequences to tag the ends of every gene and then these signatures indicate the complete transcription units between the end signatures (Ng P et al 2005 Nature Methods 2:105).

The majority of yeast genes have two or more *transcription start sites*. The analysis revealed 667 transcription units in the intergenic regions and transcripts derived from antisense strands of 367 known features. It turned out that 348 open reading frames carry start sites in their 3′-halves to generate sense transcripts starting from inside the reading frame. Thus, the budding yeast transcriptome is considerably more complex than previously thought, and it shares many recently revealed characteristics with the transcriptomes of mammals and other higher eukaryotes (Miura F et al 2006 Proc Natl Acad Sci USA 103:17846).

►expression profile, ►SAGE, ►MPSS, ►RIDGE, ►GIS, ►Atlas human cDNA, ►non-ribosomal peptides; Velculescu VE et al 1999 Nature Genet 23:387; Caron H et al 2001 Science 291:1289; Camargo AA et al 2001 Proc Nat Acad Sci USA 98:12103; Appendix II-10; Su AI et al 2002 Proc Natl Acad Sci USA 99:4465; Bono H, Okazaki Y 2002 Current Opin Struct Biol 12:355; Wu LF et al 2002 Nature Genet 31:255; autoannotation tool: http://jbirc.jbic.or.jp/tact/; plant transcriptome: http://plantta.tigr.org.

Transcriptome Map: 2–30 genes occupy a co-regulated domain in the genome and represent 20 to 2,000 kb. The co-regulated expression domains may control metabolic, developmental and organ-specific pathways. They also occupy nuclear sub-compartments, such as the nucleolus where ribosomal genes and proteins congregate. Neighboring genes mediate histone and nucleosome modifications involved in epigenetic changes. The nucleolus contains the ribosomal proteins and RNA. In cancer, neighboring loci can also be activated and this permits the construction of transcriptome correlation maps (Reyal F et al 2005 Cancer Res 65:1376). ►chromatin remodeling, ►chromosome territories; Kosak ST, Groudine M 2004 Science 306:644; Stolc V et al 2004 Science 306:655.

Transcripton: A unit of genetic transcription.

Transcriptosome: A complex of RNA processing proteins (capping enzymes, splicing factors, etc.) within the nucleus, associated with the COOH-terminal domain of the large subunit of RNA polymerase II. ►RNA polymerase, ►transcription factors, ►post-transcriptional processing, ►capping enzymes, ►splicing; Halle JP 7 Meisterernst M 1996 Trends Genet 12:161.

Transcytosis: Transcytosis is a process by which immunoglobulin (or other molecules) is/are transported within a vesicle from a secreting cell across the epithelial layer to another domain of the plasma membrane and also receptor-mediated processes of the capillary veins, by a type of endocytosis. ►endosome; McIntosh DP et al 2002 Proc Natl Acad Sci USA 99:1996.

Transdetermination: A particular pathway of differentiation is overruled by genetic regulation thus e.g., at the *Antp* (*Antennapedia* locus, 3–47.5) "gain of function mutants" in *Drosophila* the antenna is transformed into a mesothoracic leg or a wing in the place of an eye, etc. These changes in the developmental pattern may be associated with chromosomal rearrangements. The breakpoints may be within the promoters.

They may be altered as a cause of transcript heterogeneity and alternate splicing of the transcripts.

Transdetermination occurs in plants (snapdragon, *Arabidopsis* and others) by transforming anthers and pistil into petals and by producing sterile full flowers (see Fig. T89). In animals, transdetermination occurs by changing *Drosophila* antennae into legs, etc. Suppression of the Polycomb group of proteins by JNK signaling induces the changes in the imaginal disks of *Drosophila* (Lee N et al 2005 Nature [Lond] 438:234). ►morphogenesis, ►homeotic genes, ►imaginal disk, ►polycomb, ►JNK, ►stem cells, ►determination; Hadorn E 1978 Sci Am 219:110; Maves L, Schubiger G 1998 Development 125:115; Wei G et al 2000 Stem Cells 18:409; Glotzer M et al 2001 Annu Rev Cell Dev Biol 17:351.

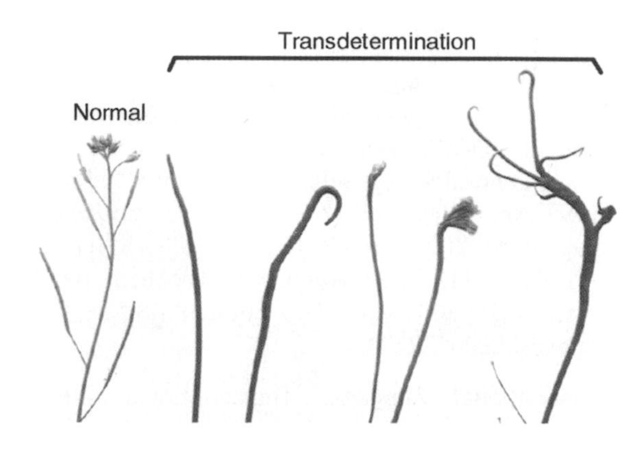

Figure T89. The apical meristem developed into horn- or antler-shape structures rather than into a normal inflorescence (at left) in *Arabidopsis* mutants (Rédei GP, unpublished)

Transdifferentiation: Transdifferentiation is a rare biological phenomenon when one type of differentiated cells is converted into another discrete type. ►transdetermination, ►regeneration

Transdominant Molecules: Transdominant molecules are used to selectively inhibit gene expression, e.g., antisense RNA, decoy RNA, ribozymes, suppressor proteins, single-chain antibodies. ►antisense technologies, ►suppressor RNA, ►RNAi, ►suppressor gene, ►ribozymes, ►tumor infiltrating lymphocytes, ►tumor suppressor factors, ►TNF; Kamb A, Teng DH 2000 Curr Opin Mol Ther 2:662.

Transducer Proteins: Transducer proteins respond to effectors to relay information to cytoplasmic components of the excitation path to switch molecules. After the effector is diluted out, the adaptation pathway leads the restoration to the base condition. (See Wang C et al 2001 Nature [Lond] 4121:285).

Transducianism: ►creationism

Transducin: A G-protein, G_t, involved in transduction of light signals (RAS-related proteins) regulating cyclic GMP phosphodiesterase. It is activated by cholera toxin and inhibited by the pertussis toxin. ►G_t protein, ►cholera toxin, ►pertussis toxin, ►transduction, ►retinal dystrophy; Norton AW et al 2000 J Biol Chem 275:38611.

Transducing Phage: ►transduction

Transductant: A transduced cell. ►transduction

3′ Transduction: Through retrotransposition, the 3′ flanking sequences of L1 transposons may move to a position downstream to the poly(A) signal of the parental L1 sequence and it is retained there. This may occur in 10 to 20% of the new retrotranspositions. Much less frequently, 5′ transduction can also occur in a similar manner. Such events may rearrange exons and regulatory sequences and may lead to the evolution of new genes. ►LINE, ►poly(A) signal, ►exon shuffling; Moran JV et al 1999 Science 283:1465.

Transduction: The transfer of DNA from one cell to another. In human molecular genetics, the transduced gene, or the transgene, is expected to be expressed in the new location. ►transformation, ►transfection

Transduction, Abortive: Transduced DNA is not integrated into the bacterial chromosome and it therefore fails to replicate among the bacterial progenies and is diluted out during the subsequent cell divisions. The non-replicating abortively transduced genetic material is transmitted in a unilinear fashion. See Fig. T90, ►transduction generalized, ►transduction specialized; Stocker BAD 1956 J Gen Microbiol 15:575.

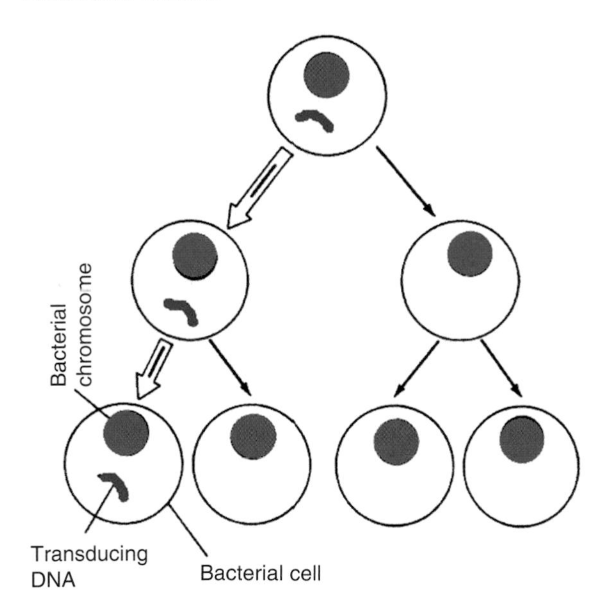

Figure T90. Abortive transduction

Transduction, Generalized: A phage-mediated transfer of *unspecified* genes among bacteria (lysogenic or non-lysogenic). The virulent phage breaks down the host cell DNA into various size fragments by the process called lysis. The transducing phage coat may then scoop up fragments of the DNA that fit into the capsule (head) that may not contain any phage genetic material. The DNA fragments enclosed in the phage head are picked up at random from the proper size group of the bacterial DNA without regard to the genes located in the fragments. For productive transduction the transferred fragments must be stably integrated into the chromosome of the recipient bacteria. ►specialized transduction, ►transduction abortive, ►transduction mapping, ►pac site, ►marker effect, see Fig. G23 in generalized transduction; Lederberg J et al 1952 Cold Spring Harbor Symp Quant Biol 16:413.

Transduction Mapping: Determines the map position of very closely linked bacterial genes on the basis of co-transduction frequencies; e.g., if the donor DNA is $a^+ b^+$ and the recipient bacterium is $a^- b^-$ then the recombination frequency is $[(a^+ b^-) + (a^- b^+)]/[(a^+ b^-) + (a^- b^+) + (a^+ b^+)]$. ►transduction generalized [see Fig. G23].

Transduction, Specialized (or restricted): A temperate-phage-mediated transfer of *special* genes between bacteria. Transduction may be mediated in higher eukaryotes by horizontal transmission, transposable and retrotransposable elements. ►specialized transduction [see Fig. G23], ►transposable elements; Morse ML 1954 Genetics 39:984.

Transesterification: An esterase enzyme catalyzes a replacement reaction. A nucleophile displaces an alcohol during the hydrolysis of an ester. A similar reaction occurs when, in nucleic acid processing, phosphodiester exchanges take place at the splice junctions of exons and introns. ►splicing, ►introns

TRANSFAC: A database for transcription factors and their binding sites. ►transcription factors; http://www.gene-regulation.com/; http://www.gene-regulation.com/pub/databases.html.

Transfection: Originally, the term was coined for introduction of viral RNA or DNA into bacterial cells and the subsequent recovery of virus particles. Today, it is used for introduction of foreign DNA into animal cells where the genes may be expressed. The delivery system may be through microinjection, bombardment (gene gun), electroporation, DEAE dextran, calcium phosphate precipitation of the target cell membranes, liposomes, and polyamidoamine dendrimers (highly branched cationic polymers). After delivery, the DNA needs protection from elimination and degradation by nucleases, opsonins, and endocytosis. Nuclear targeting may be facilitated by polyethylene glycol. Nuclear localization signals, combined with peptide nucleic acid, may facilitate homing in on the nucleus. There may be other hurdles to overcome such as cytotoxicity, condensation of the DNA, tissue targeting, etc. ►transduction, ►transformation genetic, ►microinjection, ►reverse transfection, ►DEAE dextran, ►biolistic transformation,

►electroporation, ►liposomes, ►receptor-mediated gene transfer, ►opsonin, ►endocytosis, ►nuclear localization sequences, ►peptide nucleic acid, ►polyethylene glycol; Földes J, Trautner TA 1964 Z Vereb-Lehre 95:57; Lug D, Saltzman WM 2000 Nature Biotechn 18:33.

Transfectoma: A hybridoma cell producing a specific mouse/human chimeric antibody. ►hybridoma, ►antibody; Sun LK et al 1991 J Immunol 146:199.

Transfer Clockwise/Counterclockwise: The bacterial F plasmid may be integrated in different orientations and at different locations in the bacterial chromosome and thus Hfr strains may be formed that transfer the chromosome either clockwise or counterclockwise during conjugation. ►Hfr, ►conjugation, ►clockwise, ►counterclockwise

Transfer Factors: Bacterial plasmids capable of transferring information from one bacterial cell to another through conjugational mobilization. Some of the factors (e.g., ColE1) may not have genes for transfer yet they may be transferred to other cells by helper function of conjugative plasmids. These transfer factors may contain genes for resistance (transposable elements) and have great medical significance because of the transfer of antibiotic resistance and thus make the defense against pathogenic infection difficult. ►antibiotics, ►transposable elements bacterial, ►colicins, ►resistance transfer factors

Transfer Horizontal: Same as transfer lateral.

Transfer, Lateral: In lateral transfer, genetic information is transmitted by "infection" (horizontally) or by plasmids rather than by sexual means (vertical transfer). This transfer accounts for a great deal of the variation in prokaryotes. Many species of bacteria take up and maintain extracellular DNA, especially under conditions of starvation. In 22 prokaryotic genomes, the average horizontally acquired sequences constitute ~6% of the genome (Dufraigne C et al 2005 Nucleic Acids Res 33(1):e6). *Salmonella* synthesizes a histone-like nucleoid structuring protein (H-NS) that selectively silences horizontally acquired genes, which contain lower GC content than the recipient genome and thereby protects its fitness against potentially less favorable genes (Wiley Navarre W et al 2006 Science 313:236). Mitochondrial and plastid genes can also transmit horizontally during evolution of higher organisms (Rice DW, Palmer JD 2006 BMC Biol 4:31). The fraction of horizontally transferred genes is not equally frequent among and between phylogenetically distant organisms. In Archaea the Pfam family of protein domains may be laterally acquired in >50%, and in three taxonomic ranges of bacteria, the horizontal transfer was found to be 30–50% when examined at three taxonomic ranges. In Eukarya, it was <10%. In certain gene families the horizontal transfer was more prevalent than in others (Choi IG, Kim S-H 2007 Proc Natl Acad Sci USA 104:4489). ►incongruence, ►lateral transmission; Pfam, Finkel SE, Kolter R 2001 J Bacteriol 183:6288; Koonin EV et al 2001 Annu Rev Microbiol 55:709.

Transfer Line: A polyploid species carrying a relatively short foreign chromosomal segment in its genome. The transfer is generally made either by crossing over between homoeologous chromosomes, in the absence of a gene or chromosome (chromosome 5B in wheat) that would normally prevent homoeologous pairing. It can be obtained also by (X-ray) induced translocation. Construction of such lines may have agronomic importance for introducing disease resistance or any other gene(s) that are not available in the cultivated varieties or their close relatives. ►alien addition, ►chromosome substitution, ►alien substitution, ►homoeologous chromosome; Sears ER 1972 Stadler Symp 4:23.

Transfer RNA (tRNA): Genes coding for tRNAs are clustered in both prokaryotes and eukaryotes. Some of the tRNA genes are located within the spacer regions of the ribosomal gene clusters.

The majority of tRNA genes is clustered as a group in the DNA, and frequently occurs in –two to three copies. In *Drosophila*, 284 tRNA genes have been identified. In humans there are 497 (plus 324 pseudogenes), in *Caenorhabditis* 584 tRNA genes. Some *E. coli* tRNA (86) gene clusters include genes for proteins. The tRNA genes within the cluster are separated by intergenic sequences and are transcribed as long pre-tRNA sequences. The primary transcript is processed at the 5′-end by RNase P and at the 3′-end by RNase D, BN, T, PH, RNase II, and polynucleotide phosphorylase. The pre-tRNA processing also involves joining the fragments after introns are removed by tRNA ligase multifunctional enzymes (Englert M, Beier H 2005 Nucleic Acids Res 33:388). In the *Nanoarchaeum equitans* bacterium, separate sequences code for the 5′ and 3′ halves of some tRNAs (Randau L et al 2005 Nature [Lond] 433:537). Before the tRNAs are released to the cytoplasm their integrity is ascertained and only the mature and structurally correct molecules are exported in a Ran-guanosine triphosphate dependent manner. Newer information indicates that in yeast tRNA can shuttle between nucleus and cytoplasm (Takano A et al 2005 Science 309:140).

Aminoacylation takes place before export from the nucleus. The tRNAs are small molecules (70–90 nucleotides). They assume a "clover leaf" secondary structure formed by single-stranded loops and

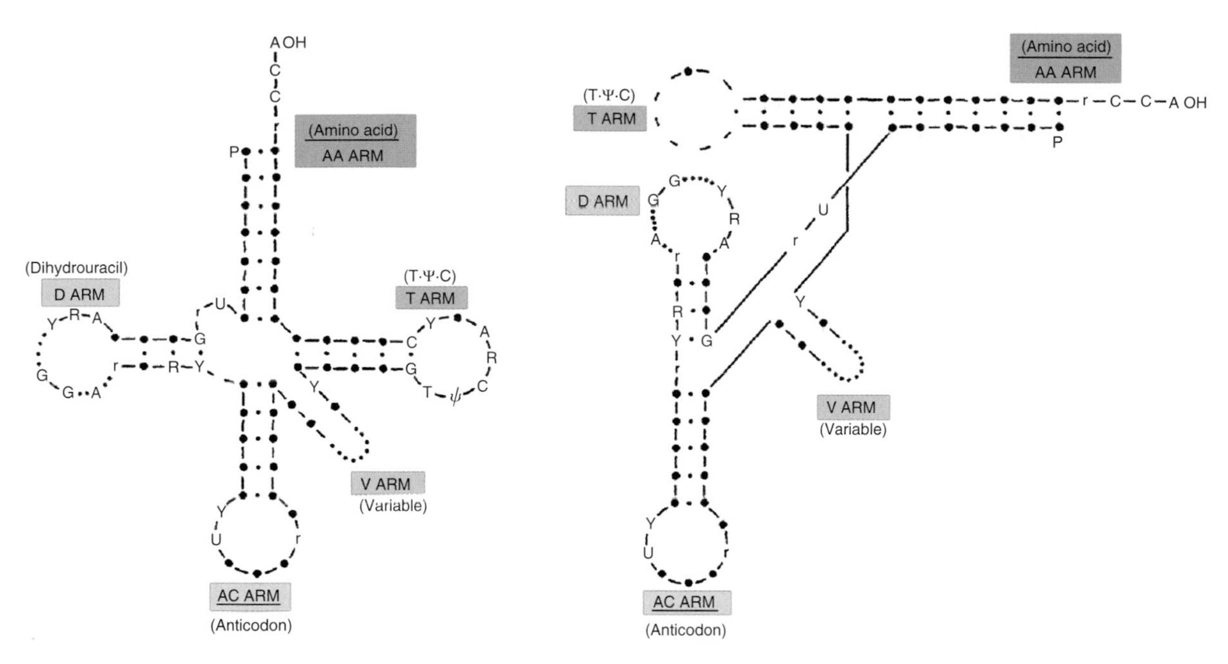

Figure T91. The general structural features of tRNAs. Left: The cloverleaf; Right: The L-shaped tertiary conformation. R stands for purine anf Y for pyrimidine bases in all tRNAs; R and Y indicate the occurence of these bases in many tRNAs; ψ is pseudouridine. (After Kim SH 1976. *Progr. Nucleic Acid Res. Mol. Biol.* 17:181.)

double-stranded sequences. *Caenorhabditis* and some other nematodes have two unusual, yet functional, types of tRNAs: one lacks the T arm and the other lacks the D arm but has a short T arm (Watanabe Y et al 1994 J Biol Chem 269:22902).

The functioning tRNAs assume an L-shape configuration (see Fig. T91). After being charged with amino acids (aminoacylation), they haul the amino acids to the ribosomes for translation of the genetic code (protein synthesis). The amino acids are attached to the protruding C-C-A-(OH) amino acid arm and one of the C residues interact directly at the P site with the G2252 and G2253 of the 23S ribosomal subunit of prokaryotes. The 3′-CAA is generally added enzymatically to the tRNA after transcription (Tomita K et al 2004 Nature [Lond] 430:700) but some bacterial tRNA genes also encode this sequence (Xiong Y, Steitz TA 2004 Nature [Lond] 430:640). The crystal structure and mechanism of function of the CCA-adding RNA polymerase has been elucidated (Tomita K et al 2006 Nature [Lond] 443:956). The anticodon loop contains a triplet complementary to the amino acid code word. This anticodon recognizes the code in the mRNA on the surface of the ribosome. The D-arm (dihydrouracil loop) is the recognition site for the aminoacyl-tRNA synthetase enzyme, whereas the T-arm (a thymine-pseudouracil [ψ]-C consensus loop) recognizes the ribosomes. There is also a small variable loop (V arm). Besides the anticodon site, the tRNAs have differences in bases at other positions too. The $tRNA^{Glu}$ and $tRNA^{Gln}$ can be quite similar yet different as shown in the Figure T92 of the *Staphylococcus aureus* molecule. In eukaryotes, glutaminyl-tRNA synthetase directly acylates $tRNA^{Gln}$. Bacteria and archaea uses a tRNA-dependent transamidation process. The first step for Gln-tRNA synthesis is the formation of misacylated Glu-$tRNA^{Gln}$ by a glutamyl-tRNA synthetase, which is used for the formation of Glu-$tRNA^{Glu}$. This tRNA is converted through transamidation by Glu-$tRNA^{Gln}$, an amidotransferase. The crystal structure of this bacterial enzyme has been determined (Oshikane H et al 2006 Science 312:1950; Nakamura A et al 2006 Science 312:1954).

The presence of modified nucleotides is characteristic for tRNAs and they modulate the anticodon domain structure for many tRNA species to accurately translate the genetic code (Yarian C et al 2002 J Biol Chem 277:16391). These modifications take place right after transcription or during processing. A small fraction of the base pairs in tRNA are non-conventional, i.e., G–U or A–C and these mispairings are widespread among species and have functional significance, i.e., they enhance aminoacylation and translation (McClain WH 2006 Proc Natl Acad Sci USA 103:4570).

The number of tRNAs in prokaryotes is higher then the number of genetic code words (64); in prokaryotes the number of different tRNA molecules may run into hundreds in the different species. In the

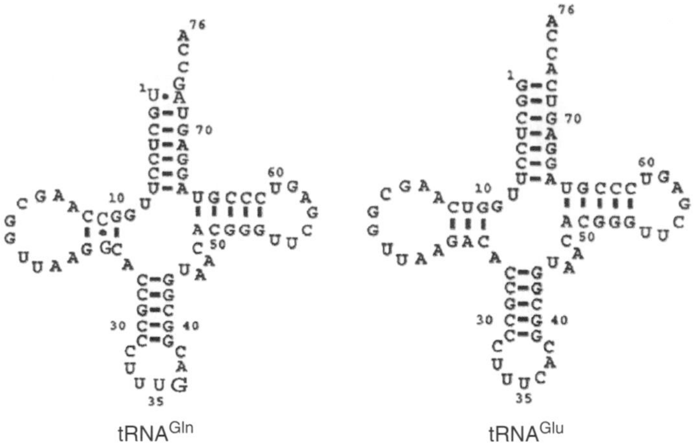

tRNA^{Gln} tRNA^{Glu}

Figure T92. Differences in glutamine and glutamic acid tRNAs in *Staphylococcus*. (Modified from references cited)

nematode *Caenorhabditis elegans*, sequencing the entire genome identified 659 tRNA genes and 29 pseudogenes. Since there are only 20 amino acids, of the high number of tRNAs several deliver the same amino acid to the site of translation (isoaccepting tRNA). The amino acid accepting *identity* of transfer RNAs resides within the base sequences and structural features of the tRNAs that are then recognized by aminoacyl tRNA synthetase enzymes. The tRNAs possess dual functions: synthetase and editing, i.e., the removal of the wrong amino acid if attached. The mRNA site recognition is the property of the anticodon. The vast majority of animal and fungal mitochondria synthesize their own tRNAs (about 22). Their anticodon reads the codons by the first two bases and thus do not need isoaccepting tRNAs. Their structure usually lacks the pseudouridine loop, and there are other minor variations. The mitochondria of land plants, algae, *Paramecium*, *Tetrahymena* and *Trypanosomes* partially rely on tRNA import. Organelle tRNAs similarly to those of plants, add post-transcriptionally the 3′-CAA amino acid accepting terminus, in contrast to some cases of *E. coli*. The mitochondrial tRNAs in plants are generally quite variable and show similarities also to the chloroplast tRNAs. The chloroplast tRNAs of higher plants (about 30) may frequently be coded by more than one tract. The universal genetic code requires a minimum of 32 different tRNAs. However, in actuality 26–24 tRNAs (with special wobbles) may suffice for protein synthesis. The mammalian and some fungal mitochondria do not code for all the tRNAs required and import these nuclearly-coded molecules from the cytosol. Many fungi code for 25–27 tRNA genes, however. In *Caenorhabditis*, the tRNA gene number and the number of tRNAs are highly correlated, and the estimated number of tRNA genes is 579 plus 207 tRNA like pseudogenes. Prokaryotes and chloroplasts have a special tRNA$^{F\text{-}Met}$ for the initiation of translation and the same anticodon, CAU may recognize not only the Met codon (AUG) but three other initiator codons as well. Nematodes initiate translation with UUG, which is a leucine codon. tRNAs are used as primers for reverse transcription. During amino acid starvation in bacteria, the guanosine tetraphosphate (ppGpp) is manufactured on the ribosomes with the aid of cognate, uncharged tRNA. Attenuation requires the cooperation of tRNA. In eukaryotes, a unique glutamyl-tRNA reductase mediates the formation 5-amino-levulinic acid, which contributes to the porphyrin ring. ►tRNA, ►fMet, ►*rrn* genes for tRNAs within ribosomal gene clusters, ►aminoacyl-tRNA synthase, ►isoaccepting tRNAs, ►code genetic, ►wobble, ►isoacceptor tRNA, ►ribosome, ►mitochondrial genetics, ►mtDNA, ►chloroplasts, ►chloroplast genetics, ►ribonuclease P, ►Ran, ►stringent response, ►attenuation, ►porphyrin, ►modified bases, ►codon usage, ►ARAGORN; Giege R et al 1993 Progr Nucleic Acid Res Mol Biol 45:129; Morl M, Marchfelder A 2001 EMBO Rep 2:17; Grosshans H et al 2000 J Struct Biol 129:288; Beunning PJ, Musier-Forsyth K 1999 Biopolymers 52:1; Intine RV et al 2002 Mol Cell 9:113; tRNA genes: http://www.tRNA.uni-bayreuth.de; tRNA functional [initiator, suppressor] classification: http://tfam.lcb.uu.se/.

Transfer, Vertical: As per vertical transfer, genetic information is transmitted by sexual means rather than by horizontal, infection type mechanisms.

Transferase Enzymes: Transferase enzymes move a chemical group(s) or molecule(s) from a donor to an acceptor.

Transferrin (TF): A β–globulin (M_r ~75,000–76,000) that transports iron. The encoding gene is located in human chromosome 3q21; the transferrin receptor

(TfR) gene was located nearby. Adenosine ribosylation factors affect the cellular redistribution of transferrin and endocytosis. Transferrin receptor 1 facilitates the infection by New World hemorrhagic viruses (Radoshitzky SR et al 2007 Nature [Lond] 446:92). ►endocytosis, ►BLYM, ►RAC, ►receptor-mediated gene transfer, ►hemochromatosis, ►aceruloplasminemia, ►atransferrinemia, ►ferroportin, ►blood-brain barrier; Aisen P et al 2001 Int J Biochem Cell Biol 33:940.

Transformant: A cell or organism that has been genetically transformed by the integration of exogenous DNA into its genetic material. ►transformation

Transformation Associated Recombination (TAR): Yeast cells are transformed simultaneously by a YAC vector (TAR) with terminal human genomic repeats, such as Alu and a long piece of human genomic DNA containing interspersed repeats (Alu). Recombination between the YAC and the human genomic DNA within the homologous repeats yields large, stable circular YACs that can be used for mapping or cloning. If the TAR vector contains an *E. coli* F-factor cassette, the vector can be propagated also in bacterial cells. ►YAC, ►Alu, ►F factor, ►vector cassette; González-Barrera S et al 2002 Genetics 162:603; Kouprina N, Larionov V 2006 Nature Rev Genet 7:805.

Transformation by Protoplast Fusion: ►protoplast fusion

Transformation, Genetic: Information transfer by naked DNA fragments or plasmid, obviating traditional sexual or asexual processes in prokaryotes and in eukaryotes. Transformation procedures may be transient when the introduced DNA is not being integrated into the genome of the cell. It may also be permanent when the exogenous DNA becomes an integral part of the recipient's genetic material.

Bacterial Transformation. Genetic transformation was discovered in bacteria in the late 1920s. It became a widely used genetic method only in the 1950. Originally, only naked bacterial DNA was used in fragments of 1/200 to 1/500 of the genome. This was provided to *competent* bacterial cells at a concentration of 5–10 μg/mL culture medium. The exogenous DNA can synapse with the bacterial genome and generally only one strand of the transforming DNA is integrated into the recipient, although some bacteria (e.g., *Haemophilus influenzae*) preferentially take up double-stranded DNA from their own species but integrate only one of the strands. Recognition of the homospecific DNA is mediated by uptake signal sequences (USS).

5′-AAGTGCGGT in the plus and 5′-AC CGCACTT in the minus strand. In the completely sequenced genome of 1,830,137 bp, 1,465 such USS were recognized. *Neisseria gonorrhoeae* also has USS elements (5′-GCCGTCTGAA).

Bacterial transformation generally may not involve an addition, rather a replacement of a part of the DNA of the recipient cell, except when plasmids are used. The non-integrated parts of the donor DNA are degraded and the rest replicates along the genes as a permanent integral part of the bacterial chromosome. The frequency of bacterial transformation may be in the range of 1% or as low as 10^{-3} to 10^{-5}, however, using bacterial protoplasts (spheroplasts) up to 80% transformation is attainable. For bacterial transformation most commonly various genetic vectors are used. Although different empirical procedures are employed in different laboratories, some general features of the methods are obvious. For transformation, either high molecular weight (DNase-free) DNA or plasmids (phage), dissolved in 1 × SSC, is used. Competence in recipient bacteria is induced by $CaCl_2$, $MnCl_2$, reducing agents and hexamine cobalt chloride or competent cells are purchased in a frozen state from commercial sources. Highly competent cells may yield ca. 10^7 to 10^9 colonies per 1 μg plasmid DNA. The success of transformation is improved by highly nutritious culture media and good aeration. The recognition of transformant cells is greatly facilitated by selectable markers. Transformation of bacteria by electroporation may be extremely efficient (10^{10} transformants/μg DNA). Cultures in mid-log phase are chilled and washed by centrifugation in low-salt buffer. The cells (3×10^{10}/mL) are suspended then in 10% glycerol and can be stored on dry ice or –70°C for up to half year. Thawed aliquots of the cells are mixed with properly prepared donor DNA and exposed to high voltage electric field in small volumes (20–40 μL). Gram-positive bacteria such as *Bacillus subtilis*, is more difficult to transform genetically than the Gram-negative bacteria, e.g., *E. coli*. *B. subtilis* attracted interest for cloning because it is not pathogenic for humans. In the presence of the *B. subtilis recE* transformation is facilitated if the cell contains already a plasmid homologous to the vector. Also, spheroplasts in the presence of polyethylene glycol, take up exogenous DNA much easier. Vectors derived from *Staphylococcus aureus* containing tetracycline- (pT127) or chloramphenicol- (pC194) resistance has been successfully used for the development of vectors. *Staphylococcus aureus* is a serious pathogen. Shuttle vectors containing *E. coli* pBR322 and *S. aureus* plasmid elements were also used. Some of the antibiotic resistance genes, e.g., β-lactamase, have very different expression in different species of bacteria. *Streptomyces* were of substantial interest for transformation because of their efficient production of antibiotics. They can be transformed by a sex plasmid, liposomes and phage vectors. ►competence, ►SSC, ►DNA extraction, ►electroporation, ►vectors, ►cloning vectors, ►liposome, ►β-lactamase, ►antibiotics, ►sex plasmid

Fungal Transformation. Is not entirely different from that in prokaryotes. Transformation of *Neurospora* started already during the early 1970 and caused genetic instabilities in the gen-ome (see RIP). Transformation of budding yeast begun in the late 1970s and became very use-ful for various types of studies (cloning, YACs, gene replacement, etc.). Yeast cells are grown to about 10^7 density/mL and then suspended in a stabilizing buffer containing 1 M sorbitol. Subsequently the cell wall is removed by digestion with β–glucanase (an enzyme hydrolyzing glucan, the polysaccharide of the cell wall [yeast cellulose]). To the washed spheroplasts in sorbitol, in the presence of $CaCl_2$ and polyethylene glycol (PEG4000), the donor DNA is ad-ded. After about 10 min incubation of the mixture, the cells are gently embedded in 3% agar and layered over a selective medium in a Petri plate. The frequency of transformation depends a great deal on the type of vector used (between 1 to 10^6 colonies per μg DNA). Alternatively — although with lower yield — intact yeast cell have been treated with lithium salts before the DNA and polyethylene glycol is applied. This is followed by selection after spreading the cells onto the surface of selective media. This procedure does not require the production of spheroplasts and agar embedding. Both of these procedures may cause mutations. Similar methods of transformation have been used also in other fungi (*Neurospora, Aspergillus, Podospora*) and also in green algae. Shuttle vectors were advantageous for the transfer of genes between various fungi and between fungi and bacteria. ►YAC, ►gene replacement, ►episomal vector, ►integrating vector, ►replicating vector, ►centromeric vector, ►shuttle vector

Transformation of Animal Cells. The most commonly used procedures involve precip-itation of the donor DNA with calcium phosphate or DEAE-dextran. The precipitated granules may enter animal cells by phagocytosis and up to about 20% of the cells may integrate the donor DNA into the chromosomes. By precipitation, physically unlinked DNA molecules can also be transformed (cotransfected) into the cultured animal cells. The polycation Polybrene (Abbott Laboratories trade name for hexadimethrine bromide) is also used to facilitate the transformation by relatively low molecular weight DNA (plasmid vectors) when some other procedures are not working. Electroporation has also been successfully applied for stable or transient introduction of DNA into the cells. Bacterial (or even plant) protoplasts can also be used to bring about fusion of cell membrane (in the presence of polyethylene glycol) and this may be followed by transfer of plasmid DNA into animal cell nuclei. This procedure is less efficient than endocytosis mediated by calcium phosphate and the plasmids are frequently integrated in tandem into the chromosome(s) of vertebrates. The exogenous DNA may be introduced also by direct *microinjection* into (pro)nuclei or into embryonic stem cells (ES) and thus generate chimeras (see gene transfer by microinjection). In the latter case the transformed cells can be screened for insert copy number or the insert can be targeted to a specific site by homologous recombination. *Infection* of stem cells, bone marrow, zygotes, early embryos by vectors or by isolated chromosomes is also feasible (►gene transfer by microinjection). Transformation of cultured sperm cells followed by fertilization is another alternative (Kurita K et al 2004 Proc Natl Acad Sci USA 101:1263). Transformation of spermatogonial stem cells of rats with DNA containing geneticin (G418) constructs are capable of self-renewal and retain the ability to colonize recipient testes, remain euploid and thus provide a means for gene targeting and obviate the need for embryonic stem cells (Hamra FK et al 2005 Proc Natl Acad Sci USA 102:17430).

Gene replacement by homologous recombination (►targeting) is also an option. In the latter case sufficient information is required about the needs for critical cis-acting elements. The success of transformation of animal cells varies a great deal according to cell types used. Transformation of vertebrate cells became a very important tool of molecular biology and reversed genetics but unfortunately the transformed cells cannot be regenerated into complete individuals, except when germline or ES cells are transformed. In *Drosophila*, into the cloned *P* element an isolated gene can be inserted with the aid of genetic engineering and the element may be then microinjected into a young embryo where the DNA can integrate into the chromosome resulting in a stably transformed individual fly. The procedure is particularly effective if the P element vector is equipped also with a selectable marker. In mosquitos microinjection into the egg cells is feasible but was not very effective. A more successful approach was using viral vectors with the vesicular stomatitis virus glycoprotein envelope, which binds to the cell membrane and delivers the foreign DNA. Transgenic rhesus monkeys were produced by injecting pseudotyped replication-defective retroviral vector into the perivitelline space and later fertilized by intracytoplasmic injection of sperm (ICSI) into mature oocytes. ►hybrid dysgenesis, ►ammunition, ►smart ammunition, ►SV40 vectors, ►adenoma, ►Bovine Papilloma Virus vectors, ►retroviral vectors, ►DEAE-dextran, ►Polybrene, ►electroporation, ►polyethylene glycol,

►liposome, ►gene replacement, ►targeting genes, ►surfection, ►transgenic, ►gene therapy, ►ES, ►vesicular stomatitis virus, ►*Anopheles*, ►pseudotyping, ►ICSI, ►SMGT; Chan AW et al 2001 Science 291:309; for protocols: Ravid K, Freshney RI (Eds.) 1998 DNA Transfer to Cultured Cells. Wiley-Liss, New York.

Transformation of Plants. Can be carried out by a variety of procedures. Most extensive-ly used were the techniques of infecting leaf or root explants or protoplasts or seeds by agrobacteria, carrying genetically engineered plasmids.

Practically all dicots can be readily transformed by agrobacteria but some monocots (*Dioscorea, Narcissus and Asparagus*) could also be transformed. The difficulty with monocot transformation by agrobacteria is apparently caused by the lack of secretion of substances needed for the activation of the virulence gene cascade or monocot cells fail to develop competence in response to infection by *Agrobacterium*. The DNA can be introduced, however by the biolistic methods.

The vector plasmids are either cointegrate or binary. Cointegrate plasmids contain in cis also the virulence genes of the Ti plasmids whereas in the binary vectors the virulence genes are carried by a separate small helper plasmid. Common features of all the vectors are that the genes to be integrated into the plant chromosome are in-between the two 25 bp inverted repeats of the T-DNA. The virulence genes and all other DNA sequences are not integrated into the host genetic material. The left and right border sequences are important for successful transformation but only a few bp of the left border and either none or 1 to 3 bp or the right border are retained in the host (►*Agrobacterium*, ►Ti plasmid). The insertional target is not strictly specified in plants yet a few base similarities are frequently found. It appears that the border repeats scout for appropriate target sites and they are appositioned there. The plasmid virulence genes direct this process also and the bacterial chromosome has also some genes that assist the transformation. The target suffers initial staggered nicks, followed by degradation, and the DNA within the 25 bp borders of the T-DNA is integrated into the chromosome (see Fig. T93). In *Arabidopsis* the histone-2A gene appears to be required for the integration of the T-DNA.

In order to be expressed, the genes within the T-DNA generally carry appropriate (plant compatible) eukaryotic promoters and polyadenylation signals to be expressed in the plant cells. Some vectors may lack promoters and can be expressed only when fused (upon integration) in vivo with plant promoters. These may be translational or transcriptional fusion vectors. The translational fusion vectors lack the translation initiation methionine codon in order to facilitate the fusion of the structural gene in the T-DNA with some amino acid sequences of the plant host. The purpose of these types of transformation is to study the strength and tissue-specificity of different plant promoters and study the function of fusion proteins. The transcriptional fusion vectors carry one or more translational stop codons (nonsense codons) in the nucleotide tract preceding the ATG (translation initiation codon). Because of this, the structural transgene will be expressed if a genuine plant promoter will drive it and no fusion protein is obtained. These vectors are also shuttle vectors, they can be propagated in agrobacteria and *E. coli*. Being shuttle vectors greatly facilitates various manipulations. The vectors can be replicated in *E. coli* because they have the origin of replication of the pBR322 plasmid and outside the boundaries of the two T-DNA sequences carry genes *oriV* (required for replication) and *oriT* (required for transfer) in *Agrobacterium*. The latter genes were derived from the promiscuous (wide host range) RK plasmid. Aseptically (axenically) grown plant tissues may start the transformation procedure. The vector cassettes generally carry selectable markers, most commonly for resistance against hygromycin B or kanamycin. The gene fusion reporter genes may be (bacterial or firefly) luciferase or GUS (β-glucuronidase) because of the easy monitoring, and the time and space of expression of the fused reporter gene. ►transcriptional gene fusion vector, and ►translational gene fusion vector

Scalpels generally wound the plant explants as they are harvested. Then the tissues dipped into a fresh bacterial suspension (grown to a density of about 10^6, washed and diluted to about half or less in plant nutrient solution). After the bacteria are blotted off the plant material is incubated for 2 days on the surface of an agar medium in Petri plates (see embryogenesis somatic). Following incubation the bacteria are stopped either by claforan (syn. cefotaxime) or carbenicillin and the plant cells are grown further in media containing also hygromycin or kanamycin (G418) or other selective agent depending of the vector constructs. The regenerated plants may then be grown axenically to maturity in test tubes (in case of *Arabidopsis*) or in soil (in case of larger plants). The transgenes usually segregate as dominant alleles if the plant does not have a corresponding native locus that might mask their expression.

Alternatively, the agrobacterial infection can be applied to presoaked seeds of plants and even-tually a small fraction of the embryos developing after meiosis will carry the transgene. Also, seedlings or plants can be infiltrated by agrobacterial suspensions and selection carried out at large scale in soil cultures treated with an appropriate herbicide (Basta) against

T

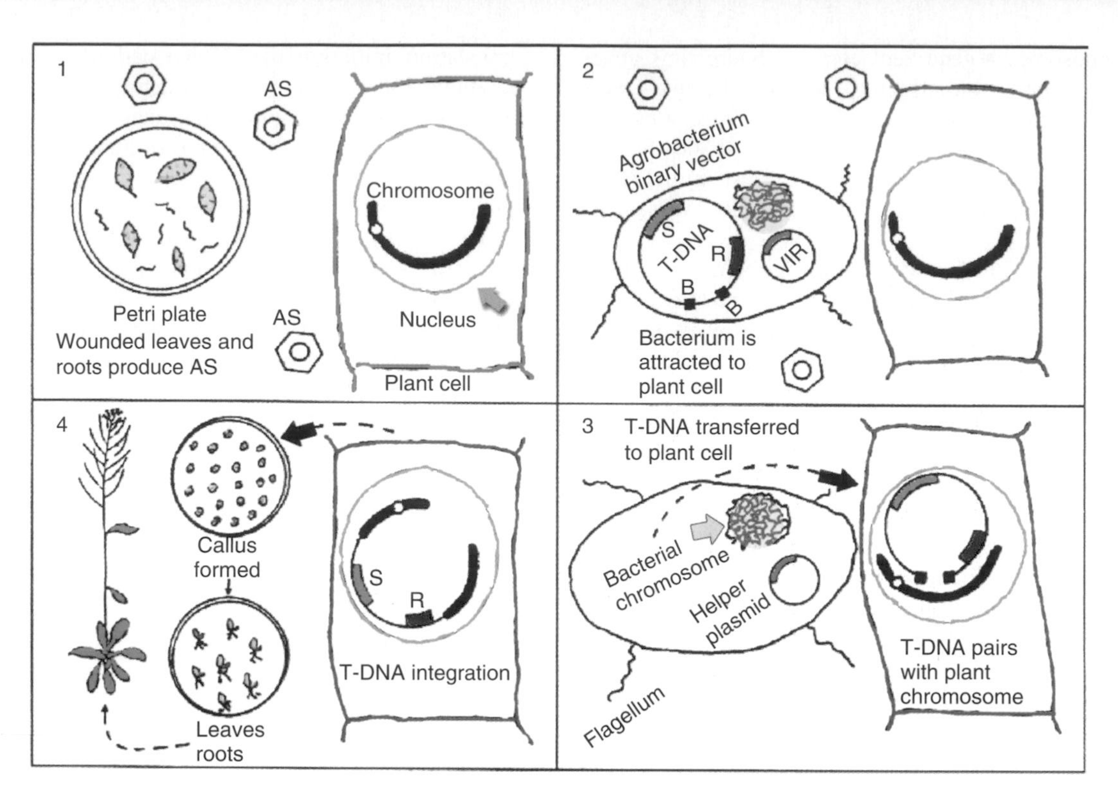

Figure T93. The major steps of transformation of plants by the use of agrobacterial binary vectors. Axenically grown leaves, roots, or stem segments are wounded. The wounding stimulates the production of phenolics such as acetosyringone (AS) and attracts the bacteria carrying the engineered plasmids with selectable markers (S) and a reporter gene (R) placed between the two border sequences (B) of the T-DNA. A small helper plasmid carries the *vir* genes required for transfer and integration of the T-DNA. Two days after infection, the bacterial growth is stopped by antibiotics. The isolated plant organs develop callus, roots, and shoots, and eventually complete plants on the selective media only if the transformation was successful. The transformation is confirmed by the expression of the reporter gene. The transformants produce seed that develops into heterozygous progeny. The antibiotic markers are dominant in the plants

what the vector carries resistance (*in planta* transformation). Plants can be transformed also by electroporation and by biolistic methods. Also cauliflower mosaic virus (CaMV) and geminivirus vectors have been developed but these are not widely used. Generally the vectors carry one desirable gene, for plant breeding purposes it may be desirable the simultaneous transfer of several genes. Assembling in a single transformation vector up to ten genes is feasible. To make such a vector construct the acceptor vector is sequentially altered by the use of two donor vectors and advantage is taken by the Cre/loxP recombination system as well by homing endonucleases (Lin L et al 2003 Proc Natl Acad Sci USA 100:5962). ►*Agrobacterium*, ►floral dip, ►electroporation, ►biolistic transformation, ►microinjection, ►gene transfer by microinjection, ►transformation of organelles, ►genetics of chloroplasts, ►genetics of mitochondria, ►gene trap vectors, ►Cre/loxP, ►homing endonucleases; Tzfira T et al 2004 Trends Genet 20:375.

Transformation Mapping: A procedure in prokaryotes for determining gene order within the genetic region of integration of the homologous transforming DNA. It can be used as a three-point cross but the additivity of recombination is generally imperfect in transformation. The general principle is outlined at the “bacterial recombination frequency” entry. ►bacterial recombination frequency

Transformation of Organelles: Mitochondria and plastids are also amenable to transformation by exposing protoplasts to appropriate vectors in the presence of polyethylene glycol (PEG) or even more effectively by employing the biolistic procedures. The efficiency of transformation is generally somewhat low. Transformation of organelles is still struggling with methodological difficulties yet it will have great potentials upon further improvements. The number of mitochondria (and plastids) within single cells may run into around hundred or even thousands and that would make possible the amplification of economically important proteins. Isolated mitochondia can be transformed by conjugation with bacteria. Into a DNA construct the origin of DNA transfer (oriT) is inserted and *E. coli* is transformed first and then though conjugation the construct is mobilized into mitochondria (Yoon YG, Koob MD et al 2005 Nucleic Acids Res 33:(16):2139). Transformation of organelles has the additional safety of containment of the transgenic organisms because mitochondria and plastids are usually not transmitted through the male. The rate of transfer of genes among mitochondria, plastids and nucleus is relatively rare, however it takes place. It may provide a possibility of transferring genes from, e.g., chloroplast to nucleus and the spread the gene in the population by pollination at not negligible rate (10^{-5}). ►chloroplast genetic, ►mitochondrial genetics, ►metabolite engineering, ►genetic engineering, ►protein engineering, ►biolistic transformation, ►electroporation, ►transformation, ►human gene transfer, ►gene therapy, ►organelle sequence transfer, ►Bellophage; Bogorad L 2000 Trends Biotechnol 18:257; Ruf S et al 2001 Nature Biotechnol 19:870.

Transformation, Oncogenic: change to malignant (cancerous) cell growth. Transformed animal cells are anchorage independent and typically free of contact inhibition and may initiate tumors when implanted into immune-compromised animals. Oncogenic transformation results when tumor suppressor genes are inactivated and oncogenes are activated. Spontaneous oncogenic transformation may occur on the same genetic background during repeated passage and proliferation of rodent cells. When these cells are injected into animals they become tumorigenic in the absence of any carcinogen, presumably due to large cell density and selection (Rubin H 2005 Proc Natl Acad Sci USA 102:9276). ►cancer, ►carcinogen, ►gatekeeper, ►phorbol esters, ►tumor suppressor, ►oncogenes; Hunter T 1997 Cell 88:333.

Transformation Rescue: introducing a viable allele or DNA sequences with the aid of transformation compensates for recessive lethal phenotype. By narrowing the transforming DNA to the minimal length that restores a viable phenotype (using restriction enzymes), the site of the damage can be estimated, cloned and sequenced. ►transformation genetic; de Vries J, Wackernagel W 1998 Mol Gen Genet 257:606.

Transformation, Stable: produces cells which carry the transforming DNA in an integrated form and thus the acquired information is consistently transmitted to the progeny. ►transformation transient, ►transformation genetic

Transformation, Transient: The introduced DNA may be expressed only for a limited time (for 1 to 3 days) in the recipient cell because it is not integrated into the host genetic material. Electroporation is most commonly results in this kind of transformation. ►transformation stable, ►electroporation

Transformation Vectors: ►cloning vectors, ►vectors, ►transformation genetic

Transformation-Competent Artificial Chromosome Vector (TAC): can carry large (40–80 kb) inserts and can be maintained in *E. coli*, *Agrobacterium tumefaciens* and expressed in plant cells. (See Liu Y-G et al 1999 Proc Natl Acad Sci USA 96:6535).

Transformed Distance (TD): is an UPGMA procedure for species *i* and *j*:
$d_{ij'} = (d_{ij} - d_{ir} - d_{jr})/2 + c$ where *r* is a reference species within or outside the group and *c* is a constant to make $d_{ij'}$ positive. This TD formula is unsuitable for the estimation of branch length in evolutionary trees. ►evolutionary tree, ►evolutionary distance, ►UPGMA, ►Fitch—Margoliash method for TD

Transforming Growth Factor: ►TGF

Transforming Growth Factor β: A superfamily of proteins that induces the change of undifferentiated tissues into specific types of tissues. TGF-β1 peptide factor causes reversible arrest in G1 phase of the cell cycle and thus it is considered as a tumor suppressor. TGF-β1 had been detected in cis position to several genes. These proteins are serine/threonine kinases. ►activin, ►bone morphogenetic protein, ►tumor suppressor, ►TGF

Transforming Principle: A historical term used in the early bacterial transformation reports when it was not yet proven that DNA is the agent of transformation. (See Alloway JL 1931 J Exptl Med 55:91).

Transformylase: Enzyme that adds a formic acid residue to the methionine-charged fMet tRNA ($tRNA^{fMet}$) in prokaryotes. ►protein synthesis

TRANSFRAG: Transcribed fragment. On the average number of transfrags in ten human chromosomes were 16,864 with average length of 115 to 78 nucleotides. The average transcripts size varied substantially from 173 to 4650 nucleotides (average 680). On the average of ten chromosomes 31% of the transfrags come from intergenic regions, 26% are intronic and 5% are mRNA. 60.8% of the transcripts are from both genomic strands. Of the 178 cloned transcripts 64% are spliced of 3.2 exons and the average exon length was 238 nucleotides; 26 (14.6%) was spliced from antisense transcripts. Exclusively non-polyadenylated transcript were twice as abundant as the exclusively polyadenylated ones. Many transcripts (36.9%) were bimorphic, i.e., they had both polyA+ and polyA- forms. The polyA- transcripts were mainly intronic. The function of many transcripts still remains unknown in 2005. ►transcription, ►polyadenylation signal, ►splicing, ►exon, ►intron, ►ENCODE; Cheng J et al 2005 Science 308:1149.

Transgene: A gene transferred to a cell or organism by isolated DNA in a vector rather than by sexual means. ►transformation genetic

Transgene Mutation Assay: In a transgene mutation assay, a mouse transgenic for a prokaryotic reporter gene is exposed to mutagenic conditions (spontaneous or treated with an agent). The genomic DNA is isolated and rescued in phage lambda vector or used in a plasmid rescue system. The cloning bacteria are then plated and the number of mutant reporter genes compared with all the reporter genes analyzed to provide mutation frequency. Under experimental conditions, spontaneous mutation rates (*lacZ*) were observed within the range about 6 to 80 $\times$ 10^{-6}, depending on the tissues from where the DNA was extracted. Some transgenic lines carry p53 tumor suppressor or the RAS oncogene to test their effects on mutagenesis. Inserted genes of the P450 cytochromes may be helpful for the studies of the metabolism of promutagens and procarcinogens. ►host-mediated assays, ►plasmid rescue, ►*lac* operon, ►β galactosidase, ►vectors, ►bioassays in genetic toxicology, ►mutation detection, ►promutagen, ►procarcinogen, ►P450, ►p53, ►RAS; Chroust K et al 2001 Mutation Res 498:169; McDiarmid HM et al 2001 Mutation Res 497:39.

Transgenerational Effect: Epimutations transmitted to the progeny. The agouti viable yellow allele A^{vy} of mice may be affected in a mosaic pattern by a retrotransposon in the female germline and the alteration is not cleared during subsequent meiosis. Gestating rodent females exposed to antiandrogenic compound (vinclozolin) or an estrogen compound (methoxychlor) (see Fig. T94) reduced spermatogenic cell number and increased infertility in the male offspring and this epigenetic trait reappeared in the four following generations studied (Anway MD et al 2005 Science 308:1466). ►epimutation, ►directed mutation, ►sexual selection

Transgenesis: Introducing a gene by genetic transformation. *Conditional transgenesis* introduces the desired gene by *Cre*-mediated recombinase under the control of developmentally-regulated promoter. Germline stem cell transplantation may also lead to transformation. ►transformation genetic, ►*Cre*; Moon AM, Capecchi MR 2000 Nature Genet 26:455; Brinster RL 2002 Science 296:2174.

Transgenic: A transgenic carries gene(s) introduced into a cell or organism by transformation. Transgenic animals can potentially produce therapeutically needed proteins such as human tissue plasminogen activator (tPA) or α_1-antitrypsin (ATT), human monoclonal antibodies, etc. Transgenic plants may have direct use in agriculture by virtue of their resistance to herbicides, pathogens or even by the production of biodegradable plastics, nutritionally safer fats and carbohydrates, various antigens that may be substituted for the standard type vaccines by eating them. The availability of transgenic crops and farm animals raised concerns by consumers and environmentalists about transfer of herbicide resistance to weeds, to introduce antibiotic resistance genes into the food chains and their potential hazards for fighting microbial infections, affecting the immune system of animals and humans, etc. Although the long-term consequences of the new technologies cannot be precisely assessed yet, there appear more advantages than risks of these new technologies. The wide-scale application of antibiotics in medicine eventually was followed by the appearance of resistance to many antibiotics. It must not be forgotten, however, that antibiotics saved and saving millions of life since the introduction of penicillin after World War II. Also, pharmaceutical research has produced and producing an ever-increasing variety of new antibiotics in order to compete with the evolutionary changes in the microbial world. While the danger of emergence of antibiotic resistant pathogens must not be ignored, the reasonable medical use of these drugs remains

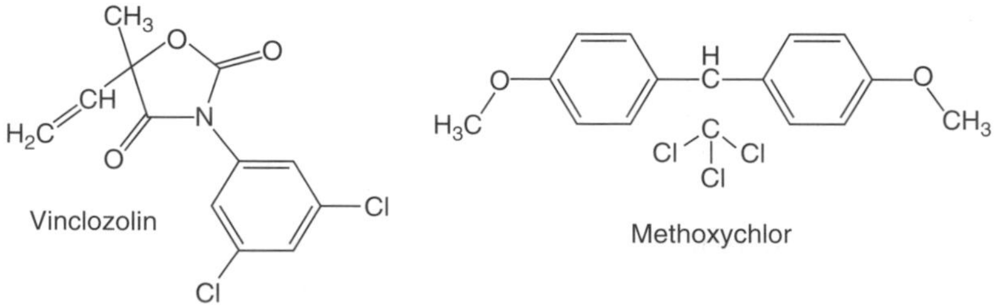

Vinclozolin

Methoxychlor

Figure T94. Left: vinclozolin; right: methoxychlor

a necessity. Viral vectors that integrate into the nucleus and can cause unavoidable deleterious modification at the site of insertion most commonly produce transgenic animals. An episomal vector (pEPI) can be delivered with high efficiency to the pig embryo by sperm-mediated gene transfer (SMGT), and the transgene is expressed in all tissues of positive fetuses that were tested, and that is retained as an episome in the course of embryogenesis and fetal development. This procedure successfully produced genetically modified animals (Manzini S et al 2006 Proc Natl Acad Sci USA 103:17672). ►transformation genetic, ►nuclear transplantation, ►antibiotic resistance, ►ANDi, ►paratransgenic, ►bitransgenic regulation, ►genetic engineering, ►gene therapy, ►GMO; http://www.transgenic-animals.com/.

Transgenome: Transformed eukaryotic cells by isolated whole or parts of chromosome(s) containing transgenes in their nuclei. ►transformation genetic [animal cells]; Porteous DJ 1994 Methods Mol Biol 29:353.

Transgenosis: An alternative term for (a controversial means) of transfer of genes by non-sexual means. (See Doy CH et al 1973 Nature New Biol 244:90).

Trans-Golgi Network (TGN): The connection of the Golgi complex exit face to transport vesicles so the molecules coming from the endoplasmic reticulum will be transported to their proper destination. ►cis-Golgi, ►Golgi

Transgression: Some segregants exceed both parents and the F1 hybrids. (See Burke JM, Arnold ML 2001 Annu Rev Genet 35:31).

Transheterozygous Non-complementation: ►non-allelic non-complementation

Transient Amplifying Cells: In transient amplifying cells, the progeny of stem cells replicate but do not revert to stem cell status; rather they generate differentiated cells. ►stem cells

Transient Expression DNA: ►transformation transient

Transilience, Genetic: Rapid changes in fitness by a multilocus complex in response to changes in the genetic environment. (Templeton AR 1980 Genetics 94:1011).

Transin: Cell-secreted metalloproteinase, a homolog of stromelysin. (See Luo D et al 2002 J Biol Chem 277:25527).

Transinactivation: ►co-suppression

Transistor: ►semiconductor

Transit Amplifying Cell: ►stem cells

Transit Peptide: Dozen to five dozen amino acid residue leader sequences directing the import of proteins synthesized in the cytosol into mitochondria and chloroplasts. These peptides are generally rich in basic and almost free of acidic amino acids. Serine and threonine are usually very common. The transit peptide recognizes special membrane proteins, but itself is not transferred into the target organelle and it is cut off by a peptidase. The different transit-peptides do not appear to have conserved sequences. The transit peptide is targeted posttranslationally. Some mitochondrial and plastid proteins do not have these cleavable N-terminal sequences. The routing within the target seems to be influenced by the carboxy terminus or inner sequences of the proteins. Mitochondria can import several plastid proteins but the plastids do not import mitochondrial proteins. This mitochondrial import is not physiological because this organelle does not have essential specificity factors (Cleary SP et al 2002 J Biol Chem 277:5562). The transit peptide engages several proteins localized in the organelle membranes. Subsequently the protein inside the cell folds with the assistance of chaperonin 60. Some proteins targeted to the endoplasmic reticulum, Golgi membrane, peroxisome, etc., may carry the transit peptide at the C-end. ►signal peptide, ►chaperones; Jean-Benoît P et al 2000 Plant Cell 12:319.

Transition Matrix: The conditions of probabilities for a certain type of amino acid substitution during an evolutionary period.

Transition Mismatch: In transition mismatch, purine mispairs with a wrong pyrimidine. ►transversion mismatch, ►mismatch, ►transition mutation

Transition Mutation: In transition mutation, either a pyrimidine is replaced by another pyrimidine, or a purine by another purine in the genetic material leading to mutation. ►base substitutions, ►transversion; Freese E 1963, p 207. In: Molecular Genetics Taylor JH (Ed.) Acad Press, New York.

T

Transition Proteins: Basic proteins that replace (temporarily) histones during spermiogenesis with protamines. ►protamine, ►spermiogenesis; Finney LA, O'Halloran TV 2003 Science 300:931.

Transition State: An unstable (life time $\sim 10^{-13}$ second) intermediate between reactants and products of an enzymatic reaction: REACTANT → TRANSITION STATE → PRODUCT(S). Factors, which stabilize the transition state relative to the reactant are expected to lower the activation energy. ►Φ value; Komatsuzaki T, Berry RS 2001 Proc Natl Acad Sci USA 98:7666.

Transitivity: A bioinformatics concept for sequence analysis of macromolecules. It aids in the identification of distant repeat homologues, for which no alignment has been found, provides confidence about consistently well-aligned regions, and recognizes and reduce the contribution of non-homologous repeats (Szklarczyk R, Heringa J 2004 Bioinformatics 20 (Suppl. 1):1311).

Translation: Converting the information contained in a mRNA nucleotide sequence into amino acid sequences of polypeptides on the ribosomes. mRNA is threaded through a channel wrapping around the 30S subunit of the prokaryotic ribosome and translation initiation, polypeptide chain elongation, and other functions follow (Yosupova GZ et al 2001 Cell 106:233). The translational apparatus of eukaryotes consists of more than 200 macromolecules of varying importance. Some proteins, e.g., bacterial ribosomal proteins S10 and L4, particiwhich pate both in transcription and translation. ►protein synthesis, ►ribosomal proteins, ►rabbit reticulocyte in vitro translation, ►wheat germ in vitro translation, ►translation nuclear, ►decapping; Sonnenberg N et al (Eds.) 2000 Translational Control of Gene Expression, Cold Spring Harbor Laboratory Press, Cold Spring Harbor, New York.

Translation: The term "translation" is also used for the transfer of basic scientific information into clinical (bench-to-bedside) or other practical applications. ►attrition, ►roadmap

Translation Error: ►error in aminoacylation

Translation Factors: ►translation initiation

Translation Initiation: Translation initiation is usually triggered by growth factors through signaling to the RAS/RAF G proteins and MEK/MAPK proteins. Upon phosphorylation, 4E-BP1 releases the cap-binding protein eIF-4F and the mRNA cap associates with the 40S subunit of the eukaryotic ribosome. The phosphorylation of ribosomal protein S6 by protein kinase p70^{s6k} is also required. Eukaryotic initiation factor eIF-2B is active when it is bound to GTP and ensures the supply of tRNA^{Met}. Insulin and other growth factors keep eIF2B attached to GTP, whereas glycogen-synthase kinase (GSK) inactivates it because GSK inactivates insulin. In prokaryotes, the initiation begins when the ribosomal binding site of the mRNA (including the Shine-Dalgarno sequence and AUG^{fMet} codon) binds to anti-Shine-Dalgarno sequence in the 16S RNA of the 30S ribosomal subunit. The AUG codon thus is directly placed into the P pocket of the ribosome and can interact with the formyl-methionine-charged fMet-tRNA. Before the 30S subunit combines with the 50S subunit, a number of other interactions also take place. In eukaryotes, the first step of the initiation is the (*1*) dissociation of the 60S + 40S ribosomal subunit mediated by eIF6 and the attachment of eIF6 to the 60S subunit. Then (*2*) eIF3 attaches to the 40S subunit, (*3*) followed by the attachment of eIF1A. Next, (*4*) eIF2+ GTP combines with tRNA^{Met} and the complex joins the 40S subunit with eIF1A and eIF3 already in place, forming the **43**S subunit. (*5*) The capped mRNA plus elongation initiation factors eIF-4F, eIF-4A, eIF-4B energized by ATP→ADP mediate the attachment of the capped mRNA to the small ribosomal subunit (now **48**S pre-initiation complex). (*6*) While eIF5 and GTPase activating protein (GAP) mediate the release of eIF2•GDP•eIF3 (*7*) the small subunits scans the mRNA until the AUG^{Met} is located. (*8*) Capturing the free 60S ribosomal subunit and restoration of the 80S ribosome and beginning of translation follows this event. ►eIF, ►protein synthesis, ►DEAD box, ►cap, ►S6 kinase, ►ribosome, ►Shine-Dalgarno sequence, ►IRES, ►initiator tRNA, ►ribosome scanning, ►PABp, ►mRNA circularization; Gingras A-C et al 1999 Annu Rev Biochem 68:913; Kimball SR 2001 Progr Mol Subcell Biol 26:155; Pestova TV et al 2001 Proc Natl Acad Sci USA 98:7029; Walker M et al 2002 Nucleic Acids Res 30:3181; prokaryotic translation initiation sites detection: http://tico.gobics.de/.

Translation In Vitro: Translation in vitro used to be accomplished by employing isolated mRNA and other factors required for translation (see wheat germ, rabbit reticulocyte). When the isolated gene is included in an appropriate expression vector, transcription and translation may be obtained in a single step from the plasmid construct. A far more efficiently defined system of in vitro translation employs highly purified and tagged protein factors that permit an efficient purification of the products by affinity chromatography. It may yield 160 μg protein per mL/hr. It can produce modified proteins by incorporating amino acid analogs with the aid of suppressor tRNA. Radioactive tags or non-radioactive fluorescent dyes may label the translation products. ►translation, ►rabbit reticulocyte in vitro translation system, ►wheat germ in vitro translation system, ►suppressor tRNA; Shimizu Y et al 2001 Nature Biotechnol 19:751; Traverso G et al 2003 Nature Biotechnol 21:1093.

Translation, Non-contiguous: In non-contiguous translation, generally the mRNA is translated in a collinear manner into amino acid sequences without skipping any parts of the former. There are few exceptions to this continuity; 50 nucleotides in bacteriophage T4 *gene 60* are skipped during translation.

Translation, Nuclear: Translation usually takes place on the ribosomes. In prokaryotes that do not have membrane-enclosed nucleus, transcription and translation are coupled. In eukaryotes intact ribosomes are limited to the cytoplasm and according to the traditional view, translation is limited to the cytoplasmic compartment. Recently, evidence has been accumulating in favor of the notion that even in mammals some translation may take place within the nucleus. The evidence for nuclear translation is based on the observation that in isolated, purified nuclei, fluorescence-labeled proteins were not present outside the nuclei. Electronmicroscopy indicated the co-localization of the nuclear translation sites with the eIF4E polypeptide elongation factor, the ribosomal subunit L7, and a β-subunit of proteasome. The presence of proteasomal activity was surprising inasmuch as that it degrades proteins. It is conceivable that most of the nuclear-translated proteins are degraded normally. Increasing the concentration of ribonucleotides lead to increased protein synthesis in the nucleus, indicating the possibility of coupled transcription and translation in the nucleus. Despite the critical evidence provided, the role and significance of nuclear translation is not entirely clear. ►protein synthesis, ►eIF4, ►proteasome; Iborra FJ et al 2001 Science 293:1139.

Translation Reinitiation: ►reinitiation

Translation Repressor Proteins: Translation repression proteins may be attached to a site near the 5′ end of the mRNA and prevent the function of the peptide chain initiation factors. ►protein synthesis, ►eIF-2, ►rabbit reticulocyte, ►aconitase, ►trinucleotide repeats

Translation Termination: Translation termination takes place in the decoding A pocket of the ribosome, where the polypeptide release factors, RF1 recognizing prokaryotic stop codons UAG and UAA, and RF2 specific for UGA and UAA or RF3 without selectivity (and may cause misreading of all three stop signs) release the polypeptide chains. In eukaryotes, the eRF1 termination factor recognizes all three stop codons and eRF1 and eRF3 are interactive. Several other proteins modulate the function of the RFs. RF3 has homology to elongation factors EF-G and EF-Tu. This fact seems to indicate that termination and chain elongation processes bear similarities; in one case the stop codon is read, in the other the sense codons. The Rfs may have additional homology domains, e.g., with the acceptor stem, the anticodon helix, and T stem of tRNAs, called "tRNA mimicry". There is a conserved GGQ (Gly-Gly-Gln) group in eRF1s corresponding to the amino acyl group attached to the CCA-3′ end of the tRNA. These homologies may assist their function. The yeast eRF3 is a prion-like element, psi$^+$. Interestingly, the heatshock protein 104, a molecular chaperone can cure the cell from it. After protein synthesis is terminated, the termination complex and the ribosome are recycled. In the 16S rRNA of *E. coli*, mutation at nucleotide position C1054 causes translational suppression. Similarly at the corresponding site in the 18S eukaryotic rRNA, substitutions of A or G resulted in dominant nonsense suppression while the T substitution was a recessive antisuppressor. The deletion of the site was found to have a lethal effect. Although translation termination is mediated at the ribosomes, premature termination may result not just from nonsense codons but also by decay of the mis-spliced transcripts. Bacterial mRNA truncated at the 3′-OH and without a termination codon may stall on the ribosomes. In such cases, the tRNA-like 10Sa RNA transcribed from gene *ssrA* gene of *E. coli* indirectly causes the degradation of the nascent peptide chain. The 10Sa binds to the ribosome and the ANDENYALAA amino acid sequence is added to the C-end of the peptide making it a target for carboxyl-end-specific proteases. Translation termination may regulate gene expression with the aid of some weak internal termination codons, which can be facultatively transpassed. Many human diseases are caused by improper translation termination and modulating the process may have therapeutic potentials. ►translation initiation, ►release factor, ►stop codon, ►sense codon, ►autogenous suppression, ►readthrough, ►recoding, ►EF-G, ►EF-Tu-GTP, ►chaperone, ►prion, ►protein synthesis, ►ribosome, ►phenotypic reversion, ►PABp; Song H et al 2000 Cell 100:311.

Translational Bypassing: Translational bypassing processes two separate open reading frames into one protein. Its mechanism follows. (i) The charged peptidyl-tRNA and mRNA complex arrives to the P site of the ribosome and after dissociation the mRNA slides through the ribosome (take-off). (ii) The peptidyl-tRNA searches the mRNA through the decoding center of the ribosome (scanning). (iii) The peptidyl-tRNA pairs with the appropriate codon after skipping some others (landing). (See Herr AJ et al 2000 Annu Rev Biochem 69:343; Gallant J et al 2003 Proc Natl Acad Sci USA 100:13430).

Translational Control: In translational control, protein synthesis is regulated during the process of translation on the ribosome; e.g., attenuation. The mRNA of some vitamins (B_1, B_{12}), besides the Shine-Dalgarno sequence, is equipped with a vitamin-binding site. When the vitamin is sensed by the RNA, the Shine-Dalgarno sequence is occluded by a conformational change and thus translation of the vitamin mRNA is suspended in the final outcome reminding to attenuation (Szostak JW 2002 Nature (Lond) 419:890). ►attenuation, ►termination factors, ►translation termination, ►terminator codons, ►regulation of protein synthesis, ►translational termination, ►suppressor RNA, ►closed-loop model of translation, ►masked RNA, ►ribosomal filter; Gale M Jr et al 2000 Microbiol Mol Biol Revs 64:1092; Johnstone O, Lasko P 2001 Annu Rev Genet 35:365; http://uther.otago.ac.nz/Transterm.html.

Translational Coupling: As per translational coupling, when the secondary structure of the mRNA is such that the AUG site or the Shine-Dalgarno sequence is not readily amenable for translation at the first cistron, translation started at the initiator codon of another cistron may open up for translation. Such a situation may occur in phages but rarely also in eukaryotes. (See Herr AJ et al 2000 Annu Rev Biochem 69:343).

Translational Error: ►ambiguity in translation, ►error in aminoacylation

Translational Gene Fusion Vectors: Translation gene fusion vectors carry promoterless, 5′-truncated structural Genes. When the trapped host promoter drives these, the vectors direct the synthesis of fusion proteins containing amino acid residues coded for by both host and vector DNA sequences. See Fig. T95, ►gene fusion, ►transcriptional gene fusion, ►read-through proteins, ►trapping promoters, ►transformation genetic

Translational Hopping: Translational hopping occurs when a peptidyl-tRNA dissociates from its first codon and then reassociates with another dowstream. ►aminoacyl-tRNA synthetase, ►protein synthesis, ►overlapping genes, ►translational frameshift, ►shunting, ►hopping; Herr AJ 2001 J Mol Biol 309:1029.

Translational Recoding (same as ribosomal frame shift): ►overlapping genes

Translational Research: Translational research applies basic research to a patient and determines the outcome of the treatment (Birmingham K 2002 Nature Med 8:647).

Translational Restart: ►reinitiation

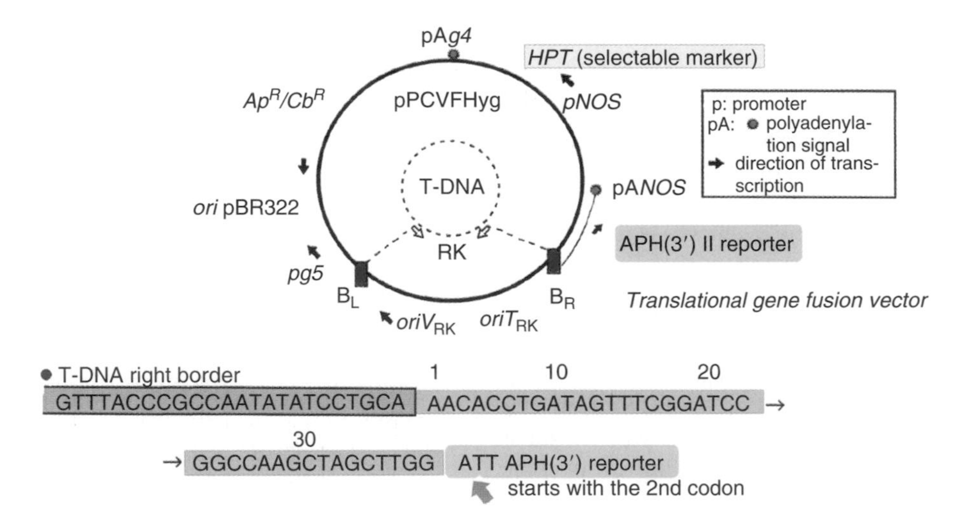

Figure T95. Agrobacterial translational in vivo gene fusion vector for plants. The critical feature is that the reporter gene is fused to the right border sequence of the T-DNA in such a manner that the translation initiator codon (AUG^{Met}) is deleted and the structural gene of the reporter begins with its second codon. It does not have a promoter either. The reporter gene can be expressed only when it is inserted and fused in the correct register into an active plant promoter. Since the AUG codon is missing, there is a good chance that the reporter protein will be fused with some plant (poly-) peptides. The use of such a vector permits an analysis of the expression of various fusion proteins on the reporter. Besides the structure, shown in detail at the lower part of the diagram, the transformation cassette contains selectable markers (e.g., HPT, permitting selective isolation of transformant on hygromycin media), ampicillin (Ap^R) and carbencillin (Cb^R) resistance for selectability in bacteria and also the replicational origin of the *E. coli* plasmid pBR322. Outside the boundaries of the T-DNA, there are genes for both vegetative (oriV) and conjugational transfer (oriT) derived from the multiple-host range RK plasmid. Only the genes between two border sequences (B_L and B_R) are inserted into the plant genome. The basic principle of this diagram has been exploited in designing vectors for other organisms. (After oral communication by Dr. Csaba Koncz.)

Translational Selection: Optimized codon usage. ►codon usage; selection of best codons in DNA sequence: http://genomes.urv.es/OPTIMIZER/.

Translational Termination: Translational termination may take place by encountering stop codons, endonucleolytic cleavage, shortening the poly(A) tail, and premature decapping of the mRNA. ►translational control

Translesion Pathway: An SOS repair system of DNA; replication may lead to targeted mutation at the site of mismatches, such as at a thymine dimer, and at bases chemically modified or deleted by mutagens. The repair of the cis-syn cyclobutane dimers is mutational in about 6% of the cases, whereas the pyrimidine 6–4 pyrimidinone adducts are repaired in a mutagenic manner in almost 100%. The Rev proteins of yeast and the UmuC (DNA pol V) protein of *E. coli* are involved with translesion. The Rev polypeptides are subunits of DNA polymerase ζ involved in damage-induced mutagenesis. In humans, DNA polymerase η may protect cells against damage that may lead to skin cancer. In bacteria, DNA polymerases II, IV, and V are involved in translesion (Napolitano R et al 2000 EMBO J 19:6259). In bacteria, LexA represses translesion and RecA induces it. Which polymerase is selected depends on the damaging agent. The repair frequently involves mutation. In yeast, the Rad6 and Rad18 proteins are elevated upon UV irradiation and the *REV3*-encoded polymerase ζ unit and a polymerase η carry out the repair. The products of genes *REV7* and *REV1* are also required. In humans, pol ι and pol κ are additional repair polymerases. Werner syndrome protein (WRN) and the translesion polymerases, Polμ, Polκ, and Polι interact. In vitro, WRN stimulates the extension activity of these polymerases on lesion-free and lesion-containing DNA templates, and alleviates pausing at stalling lesions but increases mutation (Kamath-Loeb AS et al 2007 Proc Natl Acad Sci USA 104:10394). This replicative bypass requires multiple switching to various repair polymerases but after the bypass is completed the high-fidelity polymerases are restored to the replication primer terminus (Friedberg EC et al 2005 Mol Cell 18:499). ►DNA repair, ►DNA polymerases, ►Y-family DNA polymerase, ►REV, ►ultraviolet photoproducts, ►UMU, ►cis-syn dimer, ►pyrimidine-pyrimidinone photoproduct, ►somatic hypermutation, ►Werner syndrome; Livneh Z 2001 J Biol Chem 276:25639; Pham P et al 2001 Proc Natl Acad Sci USA 98:9350; Friedberg EC 2001 Cell 107:9; Rattray AJ, Strathern JN 2003 Annu Rev Genet 37:31; Prakash S et al 2005 Annu Rev Biochem 74:317.

Translin: A protein binding to GCAGA[A/T]C and CCCA[C/G]GAC sequences at the translocation breakpoint junctions in lymphoid malignancies; it supposedly has a role in the rearrangement of immunoglobulin—T cell receptor. ►immunoglobulins, ►T cell, ►T cell receptor, ►TCR, ►lymphoma; VanLock MS et al 2001 J Struct Biol 135:58.

Transloading: The modification of cancer vaccines by including non-self peptides so they would boost immunogenicity. ►cancer gene therapy; Buschle M et al 1997 Proc Natl Acad Sci USA 94:3256.

Translocase: A protein complex mediating transport of proteins through cell membranes. ►SecA, ►SecB, ►SecY/E, ►translocon, ►ABC transporters, ►protein targeting, ►signal hypothesis, ►SRP, ►ARF, ►Mori, ►DNA translocase; H, Ito K 2001 Trends Microbiol 9:494.

Translocation: Translocation transfers codons of the mRNA on the ribosomes as the peptide chain elongates. In general, it also refers to any type of transfer of molecules from one location to another. ►protein synthesis

Translocation, Chromosomal: Segment interchange between two nonhomologous chromosomes (see Fig. T96).

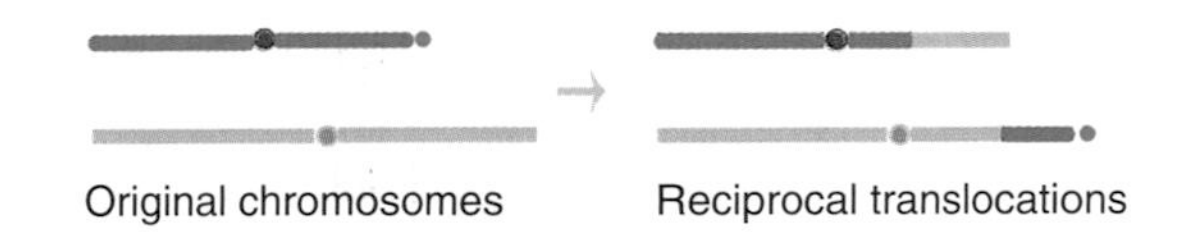

Figure T96. Chromosomal interchange

Broken chromosomes do not stick however to telomeres; the interchange must involve internal regions. Fragments may be inserted in-between two ends of an internal breakpoint, and such an aberration is called *shift*. Translocations are detectable by the light microscope, if the length of a chromosome arm is substantially altered. Translocations are usually reciprocal but during subsequent nuclear divisions one of the participant chromosomes, which carries no essential genes may be lost. Heterozygotes for reciprocal translocations display cross-shaped configuration in meiotic prophase (see Fig. T97).

Figure T97. Pairing of reciprocal translocations. Courtesy of Brinkley BR and Hittelman WN, 1975

Translocations repeatedly occur at the same locations of two non-homologous chromosomes and this may be due to the vicinity of these chromosomal sites in the nuclear architecture rather than to the special nature of the DNA sequences (Roix JJ et al 2003 Nature Genet 34:287). Translocation between human chromosomes 11q23 and 22q11 occurs repeatedly at frequencies in the 10^{-4} to 10^{-7} range and the breakpoints are most common within ~450 bp palindromic sequence of AT-rich repeats of these two regions (Kato T et al 2006 Science 311:971).

Translocation heterozygotes generally display 50% pollen sterility in plants because alternate and adjacent-1 distributions occur at about equal frequency in the absence of crossing over; adjacent-2 distributions, being non-disjunctional, are very rare. Note that inversion heterozygotes may also produce 50% male sterility but this occurs only if recombination within the inverted segment occurs freely (see inversion). In animals, the gametes of translocation heterozygotes may succeed in fertilization but the zygotes or early embryos resulting from such a mating are generally aborted. Translocation homozygosity may not have any phenotypic consequence; however, it has been shown that many types of cancerous growth are associated with translocation breakpoints. Apparently, the DNA rearrangements in the vicinity of the genes interfere with the normal regulation of their activity as a kind of position effect. Translocation breakpoints reduce the frequency of crossing over between the breakpoint and the centromere (*interstitial segment*). Recombination in translocation homozygotes may be normal. Because the reciprocal interchange physically alters the synteny of genes, linkage groups may be reshuffled as a consequence of the exchange. Because translocations partially join two linkage groups, they can be exploited for assigning genes to chromosomes. The number of crosses to localize a gene to a chromosome may thus require fewer crosses. Also, the reduction of recombinations around the breakpoints may call attention to linkage over a somewhat larger chromosomal tract rather than a single marker. Furthermore, the association of certain genes with sterility may also be used as a chromosomal marker for the breakpoints. The sterility marker may not always be very useful because it can be recognized only later during development (after sexual maturity). (See Tennyson RB et al 2002 Genetics 160:1363). Continued.

Translocation Complex: The interchanged chromosomes of eukaryotes; members of the group are inherited as a complex that alone contributes viable gametes to the progeny.

Translocation Heterozygote: At least two of the chromosomes of the genome are reciprocally exchanged (mutually translocated) whereas the corresponding homologous chromosomes are not involved in translocation within the same cell nucleus. ►translocation chromosomal, ►genome, ►homologous chromosomes, see Fig. T98.

According to some estimates, there is an about 0.004 chance that a human baby will carry a translocation. Many types of tumors carry translocations and the pattern involved is not a haphazard one. Potential oncogenes (MYC, RAS, SRC) are frequently translocated into the 14q11 region, the location of the T cell receptors (TCR) α and δ. MYC translocated to immunoglobulin genes is common in B cell neoplasia and Burkitt's lymphoma. The chain formation question is intriguing as in how these special translocations are controlled. One interpretation may be that the gene fusions involved may lead to the creation of highly selective combinations to stimulate proliferation. ►adjacent disjunction, ►position effect, ►synteny, ►multiple translocations, ►cancer, ►telomere, ►B chromosomes, ►trisomic tertiary, ►chromosomal rearrangement, ►oncogenes, ►Burkitt lymphoma, ►promoter swapping, ►unbalanced chromosomal constitution; Generoso WM 1984, p 369. In: Mutation, Cancer and Malformation. Chu EHY, Generoso WM (Eds.) Plenum, New York; translocation break-point mapping by in wheat by microarrays: Bhat PR et al 2007 Nucleic Acids Res 35:2936.

Translocation Ring: Multiple reciprocally translocated chromosomes, after terminalization, are attached end-to-end forming a ring of several chromosomes. (See Fig. T99, ►ring bivalent, ►complex heterozygotes, ►terminalization).

Translocation Test, Heritable: ►heritable translocation tests under bioassays in genetic toxicology. In some organisms, translocation testers have been developed to expedite linkage analysis. A clear and early marker is translocated to several (tester) chromosomes. Then, the gene to be identified regarding its chromosomal position is crossed with all the translocation testers available. If according to previously obtained information from crosses involving non-translocated chromosomes, the marker in the tester showed independent segregation but was crossed with the translocation testers it is linked to a single specific chromosome, its chromosomal position is revealed. The frequency of translocation may vary a great deal according to the species. In the *Oenothera* plants translocations are widespread and in *Oenothera lamarckiana* all the chromosomes are involved in translocations.

Translocon: The multiprotein complex (SecYp, SecGp, SecE in bacteria, Sec61p, Sbh1p, Ssh1p in yeast,

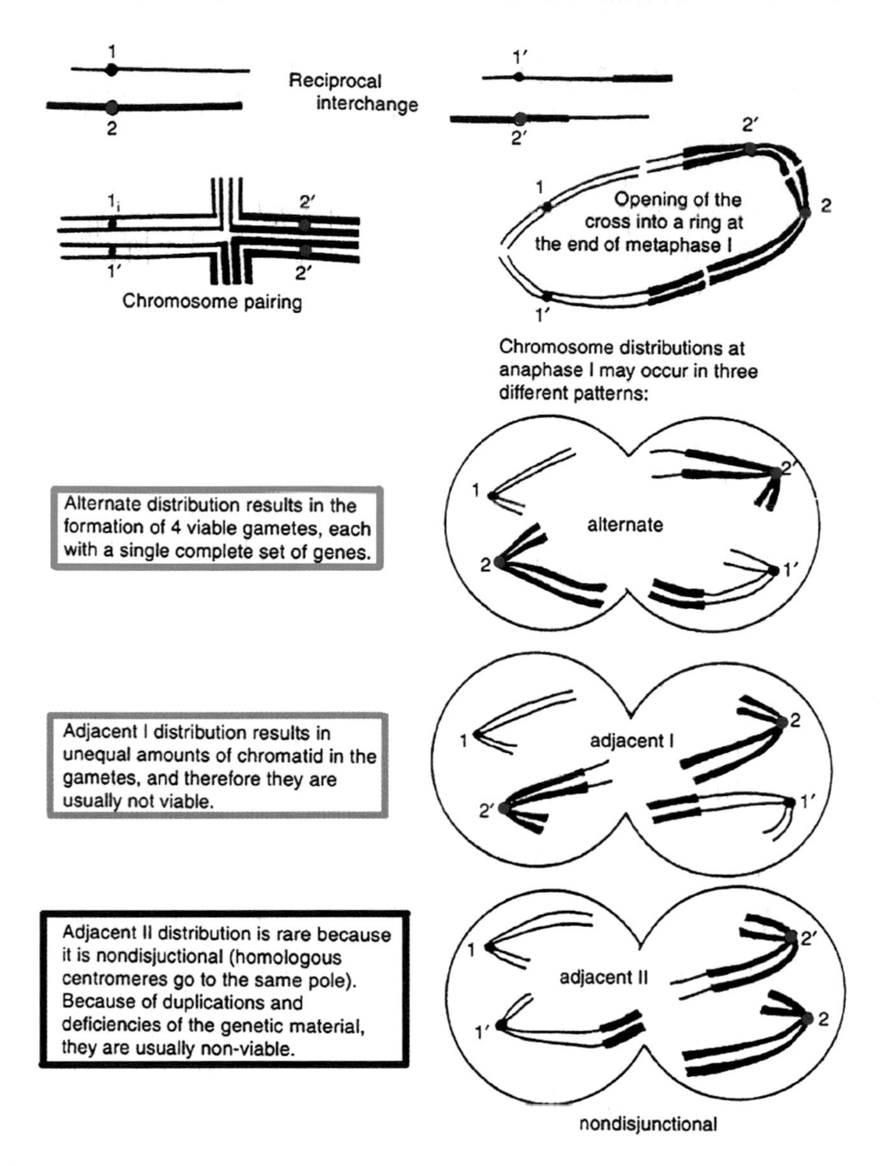
1
2
Reciprocal
interchange
1′
2′
2′
1
Opening of the
cross into a ring at
the end of metaphase I
2
1′
1ᵢ
2′
1′
2′
Chromosome pairing
Chromosome distributions at
anaphase I may occur in three
different patterns:
Alternate distribution results in the
formation of 4 viable gametes, each
with a single complete set of genes.
1
2′
alternate
2
1′
Adjacent I distribution results in
unequal amounts of chromatid in the
gametes, and therefore they are
usually not viable.
1
2
adjacent I
2′
1′
Adjacent II distribution is rare because
it is nondisjuctional (homologous
centromeres go to the same pole).
Because of duplications and
deficiencies of the genetic material,
they are usually non-viable.
1
2′
adjacent II
1′
2
nondisjunctional

Figure T98. Consequences of recombination in translocation heterozygotes

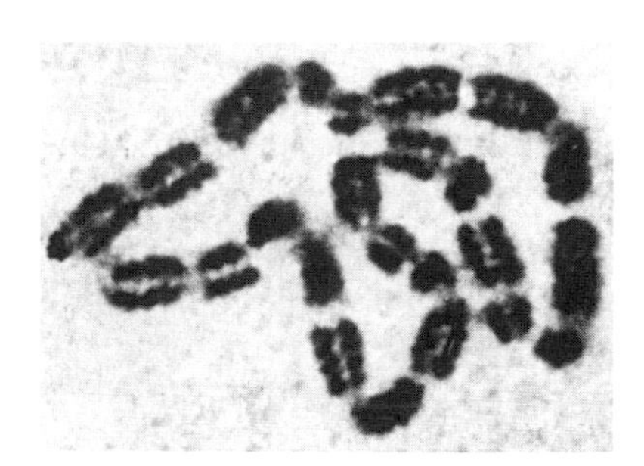

Figure T99. Six reciprocal translocations resulting in a ring of 12 in *Rheo discolor* (Sax KJ 1935 Arnold Arboretum 16:216)

sec61α, β, γ in mammals) involved in the transport of proteins through 40–60 Å diameter aqueous pores in membranes, the endoplasmic reticulum. In bacteria, the DnaK chaperones keep the nascent polypeptide chains in shape until their synthesis is completed. ►translocase, ►protein targeting, ►SRP, ►TRAM, ►ABC transporter, ►ARF, ►DnaK; Hamman BD et al 1997 Cell 89:535; Heritage D, Wonderlin WF 2001 J Biol Chem 276:22655; cryolectronmicroscopic revelation of structure of *E. coli* protein-conducting channel bound to translating ribosomes: Mitra K et al 2005 Nature [Lond] 438:318.

Transmembrane Proteins: Transmembrane proteins generally have three main domains; the amino terminus reaches into the cytoplasm where it usually associates with other cytosolic proteins, the hydrophobic domain generally makes seven turns within the cell membrane and the carboxylic end serves as a receptor for extracellular signals (see Fig. T100).

►cell membrane, ►membrane proteins, ►signal transduction, ►INT3 oncogene, ►KIT oncogene, ►MAS1 oncogene, ►seven membrane proteins, ►receptors; Baldwin SA (Ed.) 2000 Membrane Transport. Oxford University Press, New York; Hessa T et al 2005 Nature [Lond] 433:377; http://pdbtm.enzim.hu; transmembrane helix localization: http://localodom.kobic.re.kr/LocaloDom/index.htm.

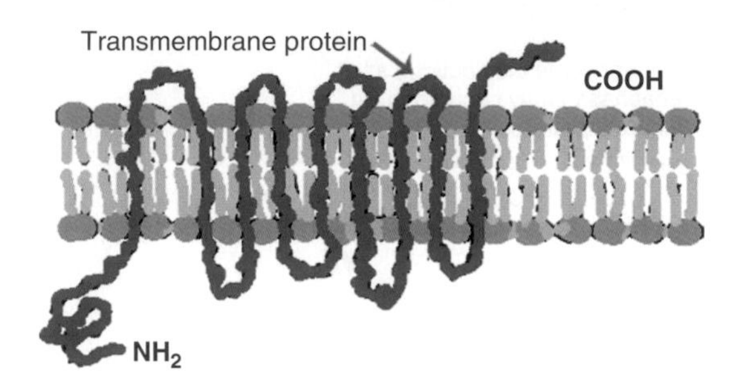

Figure T100. Transmembrane protein

Transmethylation: ►methylation of DNA

Transmission: Transmission indicates whether a particular gene or chromosome survives meiotic or post-meiotic selection and is recovered in the zygotes, embryos, or adults. Apoptotic mechanisms eliminate defective differentiating spermatogonia and elongating spermatids. During oogenesis, apoptosis is less active. This may be the cause of the long-standing knowledge that many (chromosomal) defects are preferentially or to a large extent transmitted through the egg. Transmission is generally reduced if the chromosomes have deficiencies, duplications or structural rearrangements, meiotic drive, maternal-fetal incompatibilities, etc. Monosomes and trisomes also have impaired transmission as well as defective genes. Segregation distorter genes and gametophyte factors may cause reduced transmission. Genetic factors located within the cellular organelles and infectious heredity usually display uniparental (maternal) transmission. *Vertical transmission* indicates transfer by the gametes and *horizontal transmission* means spread of a condition through infectious agents without the involvement of the host genetic system. Horizontal transmission is indicated by sequenced genomes, e.g., in the archaebacterium *Holobacterium halobium* the dihydrolipoamide dehydrogenase gene displayed 50% homology to that of Gram-positive eubacteria but only 25% to other archaebacteria. In the *Bacillus subtilis* genome, 10 nucleotide sequences were detected that are of infectious prophage origin.

Transformation by foreign DNA presents the best-documented case for horizontal transfer. Apparent homologies among genes carried by taxonomically different organisms may be interpreted as convergent evolution. Although horizontal transmission of genes is a fact, during the evolution of higher organisms its role might have been only marginal compared to the Darwinian process (Kurland CG et al 2003 Proc Natl Acad Sci USA 100:9658). The findings of transposable element (SINE) homologies in organisms widely separated taxonomically may be assumed to have origin in retroviral infections. Some of the homologous retrotransposons still carry the characteristic terminal repeats in, e.g., *Vipera ammodytes* and the bovine genome. The transfers might have also been mediated by parasites, such as ticks (*Ixodes*), common to a very wide variety of vertebrates (from reptiles to humans). ►megaspore competition, ►meiotic drive, ►preferential segregation, ►gene conversion, ►certation, ►self-incompatibility, ►infectious heredity, ►transformation, ►transposable elements, ►transposon, ►retroposon, ►retrotransposon, ►intein, ►apoptosis, ►lateral transmission

Transmission Disequilibrium Test (TDT): The TDT is used to ascertain whether a tentative association between two traits is or is not transmitted from the heterozygous parents. This test is not applicable meaningfully to a population where one of the alleles has very high frequency because in such a case the association will appear high, although no causal relationship may be present. The TDT test is not a genetic linkage test. The TDT test can be used to estimate quantitatively the distribution of *k* offspring carriers of a mutation among *r* affected progeny by truncated binomials.

$$\text{if } k = r \quad t^k(1 - t^{s-r}) / \{1 - [t^s + (1 - t)^s]\}$$

$$\text{if } 0 < k < r \quad \binom{r}{k} t^k (1 - t)^{r-k} / \{1 - [t^s + (1 - t)^s]\}$$

$$\text{if } k = 0 \quad (1 - t)^r [1 - (1 - t)^{s-r}] / \{1 - [t^s + (1 - t)^s]\}$$

where t = the segregation parameter, s = the size of the sibship.

Segregation distortion can also be determined on the basis of m carriers of the mutation among s genotyped progeny:

if parent is <u>typed</u>: $\binom{s}{m} t^m (1 - t)^{s-m}$, if parent is <u>inferred</u>: $\binom{s}{m} t^m (1 - t)^{s-m} / \{1 - [t^s + (1 - t)^s]\}$

(Formulas adopted from Hager J et al 1995 Nature Genet 9: 299) ►association test, ►association mapping, ►segregation distortion, ►binomial distribution, ►binomial probability, ►sib TDT, ►SDT, ►genomic control; McGinnis R 2000 Am J Hum Genet 67:1340.

Transmission Genetics: Transmission genetics is actually a misnomer because all genetics deals with inherited (transmitted) properties of organisms and in case there is no transmission there is no genetics. The term has been used to identify those aspects of genetics that deal only with the transmission of genes and chromosomes from parents to offspring involving also the study of segregation, recombination, mutation and other genetic phenomena without the use of biochemical and molecular analyses. It is used in the same sense as classical or Mendelian genetics. ►molecular genetics, ►reversed genetics

Transmitochondrial: Cells containing mitochondrial DNA introduced exogenously. The procedure is suitable to produce animal models for human mitochondrial diseases. ►mtDNA, ►heteroplasmy, ►mitochondrial diseases in humans; Hirano M 2001 Proc Natl Acad Sci USA 98:401.

Transmitter-Gated Ion Channel: The transmitter-gated ion channel converts chemical signals, received through neural synaptic gates, to electric signals. The channels in the postsynaptic cells receive the neurotransmitter. The process results in a temporary permeability change and a change in membrane potential, depending on the amount of the neurotransmitter. Subsequently, if the membrane potential is sufficient, voltage-gated cation channels may be opened. ►ion channels

Transmogrification: A complete metamorphosis of living creatures, such as in the case of mythological chimeras, satyrs, mermaids, etc. Genetic engineering and organ transplantation in medicine now bring into reality—in some way—the formerly imaginary beings; animals and plants expressing bacterial genes or vice versa. ►chimera, ►homeotic genes, ►gene fusion, ►allografts

Trans-Morphism: Variations in the low-copy repeats are situated in trans position (repulsion), i.e., one variant in one, the other is in the other homologous chromosome. ►cis-morhism

Transmutation: Changing one species into another (an unproven idea). Also, changing one isotope into another by radioactive decay or changing the atomic number by nuclear bombardment.

Transomic: ►transsomic

Transorientation Hypothesis: The transorientation hypothesis suggests a fourth site (D [decoding]) on the ribosome (besides A, P, and E) and assumes that the EF-G-GTP and tRNA ternary complex rotation (transorientation) moves the tRNA from the D site to the A site during protein synthesis. ►ribosomes; Simonson AB, Lake JA 2002 Nature [Lond] 416:281.

Transpeptidation: The transfer of an amino acid from the ribosomal A site to the P site. ►protein synthesis, ►aminoacyl-tRNA synthetase

Transpiration: Releasing water by evaporation through the stoma in plants, and through exhalation, through the skin, etc., in animals. ►stoma

Transplacement: Gene replacement with the aid of plasmid vectors. ►gene replacement vector, ►localized mutagenesis, ►targeting genes

Transplantation: The transfer of tissues or subcellular organelles or other cellular components from one site to another within an organism or between related or unrelated organims. Transplantation among genetically non-identical organisms may have serious immunological barriers. ►grafting, ►grafting in medicine, ►graft rejection, ►graft-versus-host disease, ►stem cells

Transplantation Antigens: Proteins on the cell surface, encoded by the major histocompatibility (MHC) genes; they play a major role in graft (allograft) rejection in mammals. The rejection may depend on the perception of the foreign antigens (tissues) by the lymphoid organs. Transplantation of lost or severely damaged body parts (limbs, face) is technically feasible but the reconstruction surgery—unless from the patient's own body—mandates life-long treatment with immunosuppressors. The French facial tissue transplantation also used twice hematopoietic stem cells from the facial donor tissues in the hope that immune tolerance may develop. Despite these efforts, acute rejection was experienced three weeks after the surgery. Even with the use of all immunosuppressive methods available, rejection may occur within one or two years after the transplantation. These plastic surgeries involve also psychological problems along with the medical ones (Okie S 2006 New Eng J Med 354:889). ►HLA, ►mixed lymphocyte reaction, ►microcytotoxicity assay, ►transplantation

Transplantation In Utero: Postnatal transplantation of foreign tissue generally results in adverse immunological reaction and rejection of the engraftment. This is very serious problem in blood transfusion, grafting, and stem cell therapy. Immature fetuses do not have yet fully developed immune systems. Therefore, attempts have been made to introduce foreign tissue (human hemopoietic stem cells) into developing fetuses at about the first third or even shorter period of pregnancy in the goat (Zeng F et al 2005 DNA Cell Biol 24:403), sheep (Zanjani ED et al 1992 J Clin Invest 89:1178), monkeys (Harrison MR et al 1989 Lancet 2(8677):1425) and other animals. Such transplants were highly successful and various types of human tissues differentiated and were identified in the animals after two years without rejection. Such a procedure obviates the use of immunosuppressive treatment. By the use of green fluorescent protein markers the right topography of the differentiated tissue could be followed (Zeng F et al 2006 Proc Natl Acad Sci USA 103:7801). ►stem cells, ►hemopoiesis, ►immunosuppression, ►immune tolerance

Transplantation of Organelles: As per transplantation of organelles, nuclei, isolated chromosomes, mitochondria, and plastids can be transferred into other cells by cellular (protoplast) fusion and by microinjection. In case of nuclear transplantation, the resident nucleus is either destroyed (by radiation) or evicted by the use of the fungal toxins, cytochalasins. The enucleated cell is called *cytoplast* and the nucleus surrounded by small amount of cytoplasm is a *karyoplast*. After introduction into the cytoplast of another nucleus by fusion, a *reconstituted cell* is obtained. The individual components are labeled either genetically or by radioactivity or by staining or even mechanically (by 0.5 μm latex beads). The transferred organelles may express their genetic information and can be isolated efficiently and identified if selectable markers (e.g., antibiotic resistance) are used. Defective livers may be repopulated with normal liver cells expressing transgenic BCL-2 because of its protection against apoptosis mediated by FAS. Without BCL-2, the transplanted liver cells do not survive. Such a procedure may eventually become an alternative to liver transplantation. ►cell fusion, ►transformation genetic, ►nuclear transplantation, ►apoptosis, ►paternal leakage, ►rejection; Kagawa Y et al 2001 Adv Drug Deliv Rev 49:107; Kuhholzer B, Prather RS 2000 Proc Soc Exp Biol Med 224:240.

Transplastome: The plastid genome, containing DNA introduced by transformation. ►plastome, ►chloroplasts, ►chloroplast genetics

Transponder (microtransponder): A few hundred micrometer wide silicon chip-based device for memory storage. When prompted by laser light it emits a radio signal that transmits its identification number. In a manner similar to DNA chips, it may assist identification of DNA sequences that are recognized by a probe. ►DNA chips, ►probe

Transport Elements, Constitutive (CTE): CTE permit the transport of spliced and non-spliced RNAs (such a viral RNAs, U snRNAs, tRNAs) although unspliced mRNAs are not exported from the nucleus. ►splicing, ►nuclear pore, ►RNA export, ►RNA transport

Transportan: ►cell-penetrating peptide

Transporters: Permease proteins that assist the transport of various molecules and ions through membranes. ►membranes, ►receptors, ►G-proteins, ►ABC transporters, ►CAT transporters, ►GLAST, ►PROT, ►GLYT, ►rbat/4F2hc, ►ASCT1, ►TAP, ►DNA transport; Sprong H et al 2001 Nature Rev Mol Cell Biol 2:504; Alper SL 2002 Annu Rev Physiol 64:899; http://plantst.sdsc.edu; transporter classification: http://www.tcdb.org/.

Transportin: A 90 kDa protein, distantly related to importin, which mediates nuclear transport with the M9, 38-amino acid, transport signal by a mechanism different from that of the importin complex. ►importin, ►karyopherin, ►nuclear localization sequences, ►nuclear pore, ►RNA export, ►export adaptor; Lai M-C et al 2001 Proc Natl Acad Sci USA 98:10154.

Transposable Elements: Transposable elements occur in the majority of organisms. Their major characteristic is that they are capable of changing their position within a genome or may move from one genome to another. Transposable elements are classified into two major groups. Class I elements transpose with an RNA intermediate. Class II elements rely on a cut-and-paste mechanism. Elements that lack terminal repeats are unable to transfer horizontally. The estimated rate of transposition is about 10^{-5} to 10^{-4} per element in *Drosophila*. The frequency of transposition may be regulated also by the host genome in lower and higher organisms as well as by methylation of the transposase. Transposable elements can remain silent in the host genomes as of cryptic elements. In such cases, epigenetic mechanisms suppress their activity (Slotkin RK, Martienssen R 2007 Nature Rev Genet 8:272). Also, the various types of elements may have intrinsic differences in mobility. The eukaryotic elements can be either retrotransposons (retrovirus-like) and have long direct terminal repeats (Class I.1) or do not have long terminal repeats (Class I.2), also called retroposons. Both types of Class I elements also possess active or inactive reverse transcriptase. The Class II elements have inverted terminal repeats that code for transposase. Transposition may take place through an RNA intermediate or directly by DNA. Transpositions may take place by homologous recombination between elements located at different map positions in the genome. Close to 50% of the human genome consists of transposons and about 70% of the maize genome is transposable. Transposable elements may have an evolutionary role in the remodeling of the genomes. Transposable elements are transmitted from generation to generation but selection may act against them because insertions may damage the genes; and an equilibrium may generally be reached. The I element of hybrid dysgenesis in *Drosophila*

carries an internal sequence that regulates copy number. After about 10 generation of the inclusion of the first I element, transposition is 'tamed', i.e., this internal sequence slows down the movement of the transposon. Elements with regulated transposition rate are successful in invading a population and after a burst their activity is limited and their presence is secured (Le Rouzic A, Capy P 2005 Genetics 169:1033). The strength of selection for host alleles controlling transposition may be estimated according to Charlesworth and Longley (Genetics 112:359):
$s \approx -\delta u = \left[\frac{\bar{n}(u-v)}{2H} + \frac{\bar{n}\pi}{2(1-2\pi)}\right]$ where δu = change in the rate of transposition, n = copy number, u = rate of transposition, v = rate of excision per element, H = harmonic mean of the rate of transposition, π = sterility or lethality caused by the transposition. The spliceosomes, telomerases, and the ability of immunoglobulin genes to transpose may have originated from transposons. The whole-genome sequencings revealed the existence of transposable elements that were not detectable by the methods of classical genetics. About 3% of the *Drosophila* genome is transposable. A survey of 13,799 human genes revealed that 533 (~4%) included some type of a transposable element (Nekrutenko A, Li W-H 2001 Trends Genet 17:619). A survey of 25,193 human proteins revealed that 4,653 human genes carry similarities to sequences in putative transposable elements. Since during evolution many rearrangements have taken place, the exact number of contribution of the transposable elements is not easy to ascertain (Britten R 2006 Proc Natl Acad Sci USA 103:1798). Transposable elements frequently generate chromosomal aberrations such as deletions, duplications, inversions, translocations, etc. *Doc1420* LINE element of *Drosophila* can generate an adaptive pesticide resistance through truncation of a gene (Aminetzach YT et al 2005 Science 309:764). ►transposable elements bacterial, ►retroposon, ►transposable elements fungal, ►transposable elements animal, ►transposable elements plants, ►transposable elements viral, ►transposons, ►transposase, ►Polintons, ►cut-and-paste, ►isochores, ►transposon footprint, ►second cycle mutation, ►hybrid dysgenesis, ►transposon conjugative, ►spliceosome, ►telomerase, ►immunoglobulins, ►selfish DNA, ►transposon — recombination, ►genome-defence model; Helitron, Berg DE, Howe MM (Eds.) 1989 Mobile DNA. Amer Soc Microbiol Washington, DC; Kidwell MG, Lisch DR 2000 Trends Ecol Evol 15:95; Lönning W-E, Saedler H 2002 Annu Rev Genet 36:389; ecological/evolutionary significance: Brookfield JFY 2005 Nature Rev Genet 6:128, classification: Wessler SR 2006 Proc Natl Acad Sci USA 103:17600; transposon insertion site profiling chip (TIP-chip): Whelan SJ et al 2006 Proc Natl Acad Sci USA 103:17632.

Transposable Elements Animal: ►copia, ►P element, ►LINE, ►SINE, ►hybrid dysgenesis, ►R2Bm, ►immunoglobulins

Transposable Elements, Bacterial: Bacterial transposable elements may be classified according to the Gram-negative host (Tn*3*, Tn*5*, Tn*7*, Tn*10*) or gram-positive host (Tn*554*, Tn*916*, Tn*1545*, Tn*55 1* and Tn*917*, Tn*4556*, Tn*4001*). The large transposon such as Tn*916* and Tn*1545* are capable of conjugative-like transfer to other cells. ►insertion elements, ►non-plasmid conjugation

Transposable Elements Fungal: ►Ty, ►transposable elements yeast

Transposable Elements, Plants: ►Ac-Ds, ►Spm (En), ►Dt, ►Mu, ►Tam, ►TAG, ►controlling elements, ►somaclonal variation, ►retroposons, ►retroposon, ►transposons, ►Helitron, ►pack-MULEs; Wessler SR 2001 Plant Physiol 125:149; Jurka J, Kapitonov VV 2001 Proc Natl Acad Sci USA 98:12315; Feschotte C et al 2002 Nature Rev Genet 3:329.

Transposable Elements, Yeast: ►Ty [including δ, ►σ, ►τ], ►Ω, ►mating type determination, ►*Schizosaccharomyces pombe*

Transposant: An individual/line generated with the aid of a gene trap vector. ►gene trap vector

Transposase: An enzyme mediating the transfer of transposable genetic elements within the genome. The transposase function may be a part of the transposable element or it may be provided from trans position for elements that are defective in the enzyme. The transposase of phage MU first cuts away the transposons from the flanking DNA. Then it inserts the cleaved ends into the new DNA target site. The conserved DD35E motif is essential for both of these reactions and a single active site is responsible for the cleavage as well as for the transfer (D: aspartic acid, 35: amino acids, E: glutamic acid). Before transposition the multimeric transpososome is assembled, including the transposon ends and proteins (such as the tetrameric MuA protein of bacteriophage Mu). ►transposon, ►transposable element, ►Tn*10*, ►cut-and-paste; Goldhaber-Gordon I et al 2003 Proc Natl Acad Sci USA 100:7509.

Transposition: The transfer of a chromosomal segment to another position. The transposition may be *conservative* when the segment (transposon) is simply transferred to another location or it may be *replicative* when a newly synthesized copy is moved to another place while the original copy is still retained where it was (see Fig. T101).

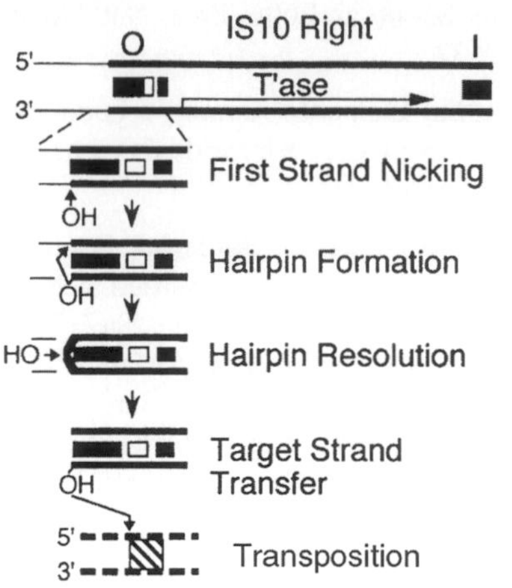

Figure T101. A model of transposition by IS10 of transposable element Tn*10*. (Courtesy of Mizuuchi K. From Kennedy AK et al 2000 Cell 101:295)

Transposition usually requires that both terminal repeats of the transposon would be intact. *One-sided transposition*—when one terminal repeat is lost—may still be feasible by replicative transposition. In *Drosophila*, ~80% of the mutations were attributed to transpositions. *Non-linear transposition* takes place when the transposon ends are located in different molecules. The latter type of event may generate diverse chromosomal rearrangements. ►insertion element, ►transposons, ►transposable elements, ►cut-and-paste, ►Tn, ►hybrid dysgenesis, ►immunoglobulins, ►mating type determination in yeast, ►*Schizosaccharomyces pombe*, ►transposon—recombination

Transposition Immunity (target immunity): In target immunity, the transposable element does not move into a replicon, which already carries another transposon or the inverted terminal repeats of a transposon. Transposition immunity is overcome by high expression of the transposase or defects in the terminal repeats of the resident transposon. Transposition of Tn*7* is inhibited by the presence of Tn*7* sequences within the same replicon. ►transposition, ►transposable elements, ►Tn*3*, ►insertional mutation; Manna D, Higgins NP 1999 Mol Microbiol 32:595.

Transposition Induction: In transposition induction, the normal rate of transposition of the Ty1 yeast retrotransposon is about $10^{-5} - 10^{-7}$ per element per cell division, its rate can be substantially increased if in a multicopy plasmid pGTy1. The Ty element is under the control of an inducible *GAL1* promoter. Transposition of Ty is controlled also by *RAD25* and *RAD3*. ►Ty, ►galactose utilization, ►*RAD3*, ►*RAD25*, ►transposase; Staleva L, Venkov P 2001 Mutation Res 474:93; Eichenbaum Z, Livneh Z 1998 Genetics 149:1173.

Transposition Site: The target of insertion is generally not random. The bacterial Tn10 prefers 5′-NGCTNAGCN-3′. The mariner transposon selects CAYA-**TA**-TRTG environment. Tn5 prefers a palindrome-like sequence flanked by A and T: **A**-GNTYWRANC-**T**. The insertion element IS231A likes a site within an S-shaped DNA. Retroviral elements show predilection for DNA around nucleosomes and cruciform DNA. Bacteriophage Mu is inserted nearly at random yet some preferences for a pentamer within a 23–24-bp tract and avoidance of the *lac*Z control region has been noted. ►transposons, ►cruciform DNA, ►nucleosomes, ►Mu bacteriophage; Haapa-Paananen S et al 2002 J Biol Chem 277:2843.

Transposome: A complex of a transposase, the transposon, and other proteins mediating the insertion of the transposon into a target DNA. ►transposon, ►transposase; Hoffman LM et al 2000 Genetica 108:19.

Transposon: ►Tn, ►transposable elements, ►retroposon, ►retrotransposon, ►TRIPLES

Transposon, Conjugative: A diverse group of broad host-range transposons varying in size from 18 to over 150 kb double-stranded DNA. They occur in different Bacteroides species. After excision they form a circular intermediate molecule that can integrate into the DNA of another cell by a conjugation-like process. When a transposon excises it carries along a 6 bp adjacent sequence of the host. The excision may be followed by restitution—without duplication—at the original host site or it may leave a footprint at the original location. Transposons may trigger the movement of other transposable elements. ►transposable elements, ►plasmid, ►conjugation; Hinerfeld D, Churchward G 2001 Mol Microbiol 41:1459.

Transposon Footprint: Short insertions left behind in the original target after the transposon exited from the sequences. These nucleotides may be the consequence of the genetic repair after excision, e.g., the sequence before the insertion CTGGTGGC after excision may become: CTGGTGGC-TGGTGGGC or CTGGTGG**gc**TGGTGGC. ►transposable elements, ►second cycle mutation; Plasterk RH 1991 EMBO J 10:1919; Hare RS et al 2001 J Bacteriol 183:1694.

Transposon Mutagenesis: As per transposon mutagenesis, transposable and insertion elements can move in the genome (mobile genetic elements) and may insert within the boundary of genes. Such an insertion, by virtue of interrupting the normal reading frame, may eliminate, reduce, or alter the expression of the gene and the event is recognized as a mutation. Recently, it has been shown that many of the insertions do not lead to observable changes in the expression of the genes, or their effect is minimal and only sequencing of the target loci reveals their presence. Mutations so generated have great advantage for genetic analysis because the insertion serves as a tag on the gene permitting its isolation and molecular study. Many of the insertions are retrotransposons and in plants commonly located within introns. In animals the comparable elements are frequently within intergenic regions. In *Caenorhabditis*, transposons—but not retrotransposons—tend to be located within sequences of high recombination. In *Drosophila*, such differences were not confirmed. Coupling a site-specific DNA-binding domain (DBD) of a polydactyl zinc-finger protein E2C to the *Sleeping Beauty* transposase produced a transposable element specific to human chromosome 17 (Yant SR et al 2007 Nucleic Acids Res 35(7):e50). ►gene tagging, ►insertional mutation, ►scanning transposon mutagenesis, ►transformation, ►TraSH, ►labeling, ►gene isolation, ►plasmid rescue, ►suicide vector, ►retrotransposon, ►retroposon, ►sleeping beauty; Mills DA 2001 Curr Opin Biotechnol 12:503; Dupuy AJ et al 2001 Genesis 30:82.

Transposon—Recombination: In transposon recombination, transposons may induce various types of chromosomal rearrangements and deletions in both prokaryotes and eukaryotes. The bacterial transposons (Tn), the Drosophila P elements, the budding yeast Ty elements, and under some conditions (but not under others) the plant transposons too may enhance a somewhat or even dramatically homologous recombination at the sites of their insertion. The recombination in plants may precede meiosis. ►transposon, ►Tn, ►hybrid dysgenesis, ►Ty; Xiao Y-L et al 2000 Genetics 156:2007.

Transposon Tagging: Tagging a gene by the insertion of a transposon. The insertion disrupts the continuity of the gene, causing a mutation and thereby the success of the tagging is identified by the phenotype. Subsequently, using the labeled transposon as a probe can aid the isolation of the gene. ►transposon mutagenesis, ►probe, ►gene isolation; Long D, Coupland G 1998 Methods Mol Biol 82:325; Pereira A, Aarts MG 1998 Methods Mol Biol 82:329; Kumar A et al 2000 Methods Enzymol 328:550.

Transposon Vector: A transposon vector can be used for introducing genes into the somatic cells of animals by microinjection into embryos of vertebrates or into invertebrates. These vectors must include transposase function, selectable marker(s) and a chosen gene. The *Drosophila* Mariner-like, Tc1-like or Sleeping Beauty vectors appeared more successful and safer than viral vectors. ►vectors, ►mariner, ►hybrid dysgenesis, ►piggyBAC, ►Sleeping Beauty, ►P-element vector; Izsvak Z et al 2000 J Mol Biol 302:93; Grossman GL et al 2000 Insect Biochem Mol Biol 30:909.

Transposon-based Sequencing: The transposon-based sequencing is used primarily for sequencing cDNA. Various transposons (Mu, Tn5) or repetitive DNAs are introduced at random into the cells and isolated on the basis of the selective markers (antibiotic resistance) within the transposon. Sequencing employs primers, which are specific for the ends of the transposon. ►EST, ►DNA sequencing; Yaron SN et al 2002 Nucleic Acids Res 30:2460; Shevchenko Y et al 2002 Nucleic Acids Res 30:2469.

Transposons, Animal: ►transposable elements animal

Transposons, Bacterial: DNA segments which can insert into several sites of the genome and contain gene(s) besides those required for insertion; they are generally longer than 2 kilobases. It has been suggested that the introns of eukaryotic cells might have been introduced into the genes by broad host-range phages or transposons. ►Tn, ►insertion elements, ►accessory proteins

Transposons, Fungal: ►Ty

Transposons-Controlling Elements, Plant: The major transposable elements in maize are *Ac-Ds*, *Spm*, *Dt*, *and Mu*. Besides these, there are much less well defined controlling elements: *Bg* (*Bergamo*), *Fcu* (*Factor Cuna*), *Mr* (*Mutator of R*), *Mrh* (Mutator of *a1-m-rh*), *Mst* (*Modifier of allele R-st*), *Mut* (controlling element of *bz1-m-rh*), *Cy* (regulatory element of *bz1-rcy*). ►controlling elements, ►Ac-Ds, ►Spm, ►Dt, ►Mu, ►Tam, ►Ta

Transresponder: The ABL oncogene is activated (transresponds) by translocation to BCR in the Philadelphia chromosome and causes chronic myelogenous leukemia in more than 90% of the cases. ►BCR, ►ABL, ►Philadelphia chromosome, ►leukemia; Gardner DP et al 1996 Transgenic Res 5:37.

Trans-Sensing: Trans-sensing is an interaction between somatically "paired" homologous chromosomes affecting gene expression in diploids. ►transvection; Tartof KD, Henikoff S 1991 Cell 65:201.

Transsexual: A transsexual has an innate desire to change her/his anatomical sex to the other form. The volume of the central subdivision of the bed nucleus of the strial terminals of the brain is larger in males than in females. In male-to-female transsexuals, this particular area of the brain is female sized. Thus this anatomical condition may be a determining factor for transsexualism and sex hormone production. Estrogen family drug treatment may cosmetically help to improve the size of the breast. ►sex determination

Transsomic Line: The transsomic line carries microinjected chromosomal fragments in the cell nucleus.

Trans-Splicing: Trans-splicing is the splicing together of exons that are not adjacent within the boundary of the gene but are remotely positioned and may be even in different chromosomes. ►introns, ►regulation of gene activity, ►tau, ►SL1, ►SL2; Vandenberghe AE et al 2001 Genes Dev 15:294; Denker JA et al 2002 Nature [Lond] 417:667.

Transthyretin: A homotetrameric 55 kDa protein in the brain facilitating thyroxine transport and mediating association of retinol to retinol-binding protein. It also binds the β fragment of amyloid proteins and is instrumental in Alzheimer disease and other amyloid diseases. Genistein is an inhibitor of the dissociation of the transthyretin tetramers and also inhibits amyloidogenesis (Green NS et al 2005 Proc Natl Acad Sci USA 102:14545). Thus consumption of soybean products may be preventive or therapeutic. ►goiter, ►retinol, ►Alzheimer disease, ►prion, ►amyloidosis; Li MD et al 2000 J Neurosci 20:1318.

Trans-Translation: Trans-translation may occur if the stop codon and the preceding end of the mRNA are lost and using another template RNA completes the translation. ►recoding, ►tmRNA; Lee S et al 2001 RNA 7:999; Bessho Y et al 2007 Proc Natl Acad Sci USA 104:8293.

Trans-Vection: ►cis-vection, ►transacting element, ►cis-trans effect

Transvection: A synapsis-dependent modification of activity in "pseudoalleles." In paired chromosomes genes in trans position may affect the expression of an allele. It has also been called trans-sensing. It has been interpreted as the result of interaction between DNA binding proteins attached to the two synapsed promoters. ►co-suppression, ►RIP, ►trans-sensing, ►pseudoalleles; Lewis EB 1951 Cold Spring Harbor Symp Quant Biol 16:159; Matzke M et al 2001 Genetics 158:451; Duncan IW 2002 Annu Rev Genet 36:521.

Transversion Mismatch: A mispairing involving either two purines or two pyrimidines. ►mismatch, ►transition mismatch

Transversion Mutation: The substitution of a purine for a pyrimidine, or a pyrimidine for a purine in the genetic material. ►base substitutions, ►base substitution mutations; Freese E 1959 Brookhaven Symp Biol 12:63.

Trap: The same as CD40 ligand.

TRAP (testis-specific cytoplasmic poly[A] polymerase): TRAP controls germ cell morphogenesis. ►polyadenylation signal; Kashiwabara S-I et al 2002 Science 298:1999.

TRAP (tryptophan RNA-binding attenuation protein): In *Bacillus subtilis* when activated by L-tryptophan, this protein binds to the mRNA leader causing a termination of transcription. This is in contrast to the situation in *E. coli* where the attenuation is brought about by an altered secondary structure of the nascent RNA transcript. Some sort of attenuation takes place also in eukaryotes but the mechanism of that is not entirely clear yet.

The *mtrB* gene in *B. subtilis* encodes the TRAP protein containing 11 identical subunits and it binds single-stranded RNA. The β-sheet subunits form a wheel-like structure with a hole in the center and tryptophan is attached to the clefts between the β-sheets resulting in circularization of the RNA target in which eleven U/GAG repeats are bound to the surface of this ondecamer (11 subunit) protein modified by tryptophan. TRAP may regulate both transcription and translation. Similar mechanisms occur also in some other bacterial species. ►tryptophan operon, ►attenuator region; structure in: Antson AA et al 1999 Nature (Lond) 401:235; Yakhnin AV, Babitzke P 2002 Proc Nat Acad Sci USA 99:11067; Chen G, Yanofsky C 2003 Science 301:211; Gollnick P et al 2005 Annu Rev Genet 39:47.

TRAP/DRIP/ARC: A part of a multiprotein complex of transcriptional regulators. (See Crawford SE et al 2002 J Biol Chem 277:3585).

Trapoxin: An inhibitor of histone deacetylase. ►histone deacetylase

TRAPP: A Golgi-associated protein for docking vesicles. ►Golgi, ►vesicles

Trapping Poly-A Tails: In the process of trapping poly-A tails, the UPAS Trap suppresses nonsense-mediated RNA surveillance and permits targeting transcriptionally silent genes in embryonic stem cells. ►nonsense-mediated RNA surveillance, ►stem cells; Shigeoka T et al 2005 Nucleic Acids Res 33(2):e20.

Trapping Promoters: When a promoterless structural gene is inserted into a host genome with the assistance of a transformation vector, the inserted sequences may become "in-frame" located within the host chromosome and a host promoter may drive the transcription of the foreign gene that in the vector had no promoter. Since the promoter and upstream regulatory elements control the transcription, directly or in association with transcription factors, the expression pattern (timing, tissue site) may be altered and the intensity of expression may be increased or decreased according to the nature of the promoter (see Fig. T102). ►gene fusion, ►transcriptional gene fusion vectors, ►translational gene fusion vectors, ►read-through proteins, ►gene trapping, ►expression trapping, ►targeted trapping, see photo; Medico E et al 2001 Nature Biotechnol 19:579; mouse gene trap consortium (IGTC): www.genetrap.org; splinkerette; mouse embryonic stem cell trap library: http://baygenomics.ucsf.edu.

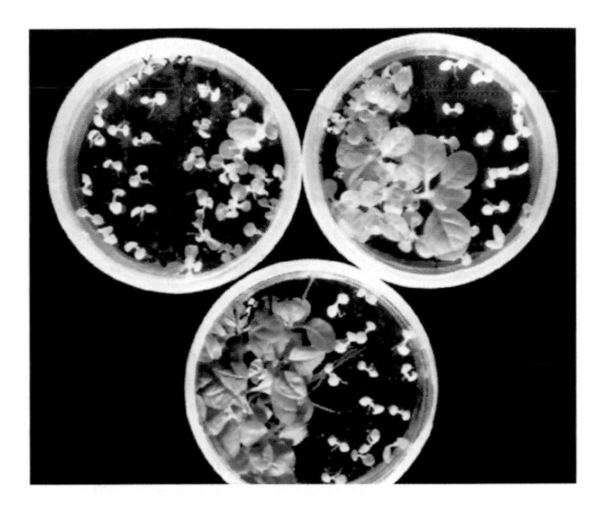

Figure T102. Transgenic tobacco seedlings segregate for kanamycin-sensitivity and resistance. A promoterless vector introduced the aminoglycoside gene into the cells. The structural gene was expressed only when it trapped a tobacco promoter. The strength of the promoters varied and consequently the degree of resistance too. Each Petri plate was divided into two sections and in one section all the small, bleached, sensitive seedlings died. (From Yao Y, Rédei GP, unpublished)

TraSH (transposons site hybridization): Genes are disrupted by transposons and changes in functions are compared—by microarray hybridization—to that in the wild type genome. ►transposons mutagenesis, ►microarray hybridization

Trastuzumab/Herceptin: is a monoclonal antibody used for specific cancer therapy. A new generation of HER2 breast cancer drug is Lapatinib. ►biomarker, ►Herceptin

Traveler's Diarrhea: is caused generally by bacterial (*E. coli*, *Salmonella*) infection.

TRCF (transcription repair coupling factor): A eukaryotic repair helicase corresponding to *UvrA* in *E. coli*; it is encoded by yeast gene *MFD* (mutation frequency decline, Li BH et al 1999 J Mol Biol 294:35).

TRD (transmission ratio distortion): ►meiotic drive, ►segregation distorter

TRE: Thyroid hormone responsive element in the rat growth hormone gene with a consensus of AGGTCA....TGACCT. ►ERBA, ►hormone response elements, ►regulation of gene activity; Oofusa K et al 2001 Mol Cell Endocrinol 181:97.

Treacher Collins Syndrome (TCOF1): A dominant (human chromosome 5q32-q33) complex defect of the face with an incidence of $\sim 2 \times 10^{-5}$. Mutation in the treacle, a 26-exon phosphoprotein is involved in neural crest cell formation. Treacle affects mammalian ribosomal DNA transcription by interaction with an upstream binding factor, which controls RNA polymerase I (Valdez BC et al 2004 Proc Natl Acad Sci USA 101:10709; Dixon J et al 2006 Proc Natl Acad Sci USA 103:13403).

Treadmill Evolution: ►Red Queen hypothesis

Treadmilling: Is the addition of microtubule subunits to the growing plus end and loss of subunits at the minus end. ►microtubules, ►dynamic instability; Shaw SL et al 2003 Science 300:1715.

TREC (TCR excision unit including the recombination signals): ►TCR genes

Tree of Life: ►evolutionary tree

Trees: ►forest trees

Tree Edit Distance: The minimal weighted number of changes required to change one tree of descent into another. ►string edit distance, ►evolutionary tree

Trehalose (α-D-glucopyranosyl-α-D-glucopyranoside): A non-reducing disaccharide, which accumulates in the yeast cell wall under conditions of stress. The use of trehalose for platelet and other cell preservation by freeze-drying has applied significance. Trehalose 6-phosphate regulates starch synthesis in plants by post-translationally activating ADP-glucosephosphorylase (see Fig. T103) (Kolbe A et al 2005 Proc Natl Acad Sci USA 102:11118). ►inflorescence; Darg AK et al 2002 Proc Natl Acad Sci USA 99:15898.

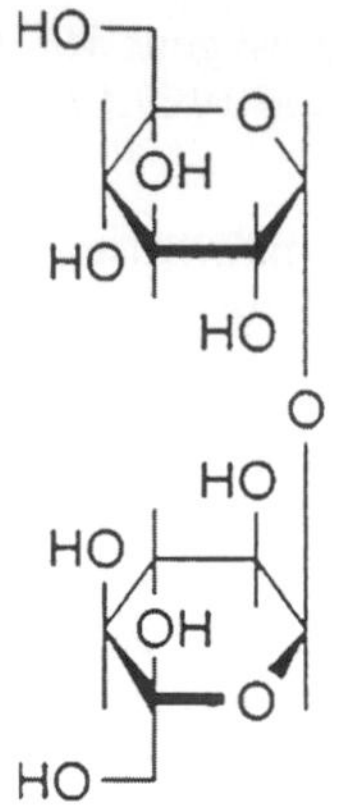

Figure T103. Trehalose

TREMBL: A computer annotated protein sequence extension of the SWISS-PROT. ►SWISS-PROT in databases

Treponema pallidum: Spirochete bacterium with completely sequenced (1998) genome of 1,138,006 bp including 1041 open reading frames. It is responsible for the potentially deadly disease of syphilis. Due to protective immunity, waves of infections occur with 8–11 year periods in the populations (see Fig. T104) (Grassly NC et al 2005 Nature [Lond] 43 3:417).

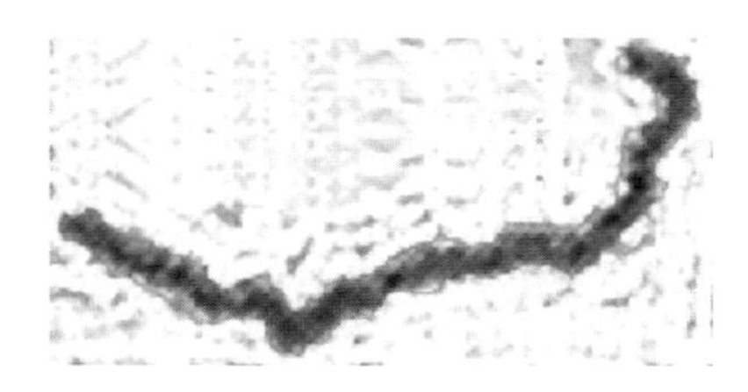

Figure T104. Treponema

Many of the nineteenth to twentieth century artists (Maupassant, Daudet), the famous philosopher Friedrich Nietzsche suffered from mental illness caused by the infection. *Treponema denticola* causes periodontal (bleeding gum) disease. ►leptospirosis, ►spirochetes; Fraser CM et al 1998 Science 281:375.

Trexon (transposed exon): Short duplicated modular units in the DNA with inverted terminal repeats. ►transposon, ►exon

TRF: ►T cell replacing factor, ►a lymphokine

TRF (thyrotropic release factor): ►corticotropin

TRF1 (TATA-box-binding protein related factor): is a tissue- and gene-specific binding protein with preference for one of the two *Tudor* locus (2-[97]) promoters of *Drosophila*. The protein product of *tud* has maternal effect and it is expressed mainly in the embryos and pupae. (See Takada S et al 2000 Cell 101:459; ►histones).

TRF1: 60 kDA telomeric TTAGGG repeat binding protein negatively regulates telomere extension and facilitates its interaction with the telomerase enzyme. ►telomeres, ►telomerase, ►RAP, ►tankyrase; Nakamura M et al 2001 Curr Biol 11:1512.

TRF2: (telomeric repeat-binding factor-2): One of the proteins, which protect telomeric ends of the chromosomes. TRF2 inhibition may lead to apoptosis, mediated by p53 and mutated ataxia telangiectasia genes. TRF2 is associated with RAD50, MRE11 and BS1 proteins. Oxidative damage (e.g., 8-oxogunanmine lesion) greatly reduces the affinity of both TRF1 and TRF2 (Opresko PL et al 2005 Nucleic Acids Res 33:1230). TRF2 mutations predispose to premature aging, cancer and ultraviolet sensitivity (Muñoz P et al 2005 Nature Genet 37:1063). ►TRF1, ►telomeres, ►apoptosis, ►ataxia telangiectasia, ►p53, ►RAD50, ►Mre, ►oxidative DNA damages, ►Nijmegen breakage syndrome; Fairall L et al 2001 Mol Cell 8:351.

T-RFLP: The procedure has been successfully exploited for the estimation of genetic diversity within bacterial populations. It involves PCR amplification of a gene of interest (often 16S rRNA genes) with fluorescent dye-labeled primers, followed by multiple single restriction digests done in parallel. The resulting fragments are then separated by capillary electrophoresis with an internal size standard to determine the lengths of the terminal (fluorescently labeled) fragments. Each distinct terminal restriction fragment is considered an operational taxonomic unit (OTU), thus the choice of restriction enzymes can impact the number of OTUs observed in each sample and the calculation of diversity statistics (Collins RE, Rocap G 2007 Nucleic Acids Res 35:W58). ►RFLP; restriction enzyme choice: http://rocaplab.ocean.washington.edu/tools/repk.

Trg: Bacterial transducer protein with attraction to ribose and galactose. (See Beel BD, Hazelbauer GL 2001 Mol Microbiol 40:824).

Triabody: Trimeric antibody built of three single chain pairs of the variable heavy and light chain regions of antibody. ►antibody chimeric, ►diabody, ►recombinant antibody; Le Gall F et al 1999 FEBS Lett 453 (1–2):164.

Triacylglycerols (synonym triglycerides): are uncharged esters of glycerol and thus called also neutral fats. Triglycerides are energy storage compounds and contain four times as much energy in the human body than all the proteins combined. By lipase, they are hydrolized into glycerol and fatty acids. Lipolysis is controled by cAMP in the adipose (fat) cells. Insulin inhibits lipolysis. Impaired long-chain fatty acid oxidation, triglyceride breakdown defects, triglyceride transfer (MTP, 4q22-q24) in abetalipoproteinemia, hypertriglyceridemia (15q11.2-q13.1) a dominant hyperlipidemia and it is a risk for heart disease, etc. are diseases involved in triglyceride metabolism. ►epinephrine, ►norepinephrine, ►glucagon, ►adrenocorticotropic hormone, ►fatty acids, ►triglycerol; Zimmermann R et al 2004 Science 306:1383.

Triad Test: is an association mapping procedure where the two parents and the proband are tested, and it may involve unrelated control(s). ►association mapping; Epstein MP et al 2005 Am J Hum Genet 76:592.

Triage: Assignment of priorities in medicine or in the regulation of cellular metabolism.

Triallelic Inheritance: The manifestation of the recessive disease e.g., the Bardet-Biedl syndrome may require the expression of three mutant alleles. ►digenic disease, ►Bardet-Biedel syndrome, ►epistasis; Katsanis N et al 2001 Science 293:2256.

Triangulation Number: Represents the number of protein subunits (facets) in an icosahedral viral capsid. ►icosahedral, ►capsid; Paredes AM et al 1993 Proc Natl Acad Sci USA 90:9095.

Tribe: Descendants of a female progenitor or a taxonomic group below a suborder or a group of primitive people with a common origin, culture and social system.

Tribolium castaneum (n = 10, 200 Mb): red flour beetle, object of cytological and population genetics studies. The body segmentation gene (*mlpt*) produces polycistronic mRNA (Savard J et al 2006 Cell 126:559). For a genetic map see Beeman RW, Brown SJ 1999 Genetics 153:333; Klingler M 2004 Current Biol 14:R639; http://www.ksu.edu/tribolium/; ftp://ftp.ncbi.nlm.nih.gov/genomes; http://www.bioinformatics.ksu.edu/BeetleBase/; http://www.intlgenome.org/viewOrganisms.cfm?organismID=1000185.

TRiC: is a ring complex of eukaryotic chaperonin. TRiC-P5 is synonymous with CCTγ, Bin2p, Cct3p. ►chaperonins; Dunn AY et al 2001 J Struct Biol 135:176.

Tricarboxylic Acid Cycle: ►Krebs-Szentgyörgyi cycle

Trichocyst: An organ of protozoa that may extrude fibrous shafts and may serve as an anchor or defensive or offensive tool.

Trichogen Cell: ►tormogen

Trichogyne: Hypha emanating from the protoperithecium, to what the conidia are attached prior to fertilization in some ascomycetes. ►hypha, ►conidia

Trichome: Hair or filament in plants, algae and animals; some plant hairs may be single filaments or they may have tripartite termini (see Fig. T105). (See Szymanski DB et al 2000 Trends Plant Sci 5:214, photo above).

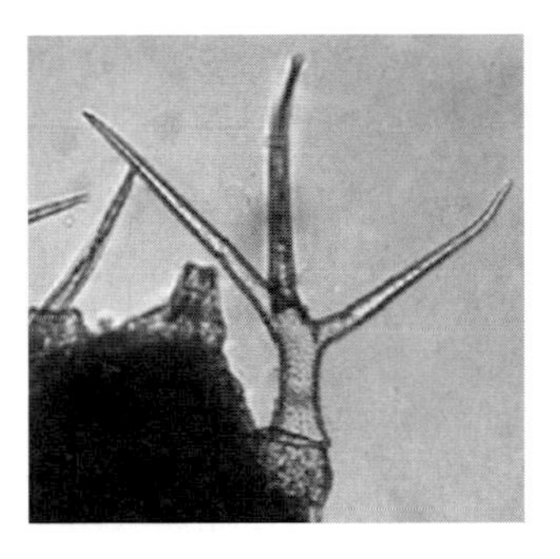

Figure T105. Trichome

Trichomonas vaginalis: Is an infectious, non-mitochondrial protozoan parasite of the human female and male sexual organs and urinary tract and it is transmitted by sexual intercourse (see Fig. T106). It may cause severe discomfort. Other species may infect domestic animals and also birds and invertebrates. *Trichomonas* and other protists do not have the regular type mitochondria but only mitosomes. The sequenced genome of *T. vaginalis* is about 176 megabase with a predicted protein-coding gene number of 59,681, 479 tRNA and about 25O rRNA genes (Carlton JM et al 2007 Science 315:207). ►mitosome, ►*Entamoeba*, ►*Giardia*, ►*Protozoa*

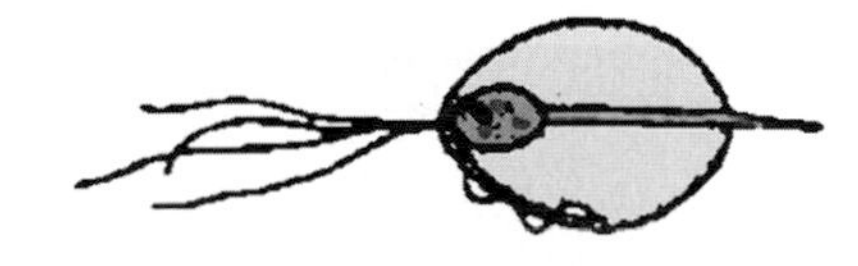

Figure T106. Trichomonas

Trichorhinophalangeal Syndrome (TRPS1): Dominant or recessive human chromosome 8q24 defect involving multiple exostoses (bone projections), mental retardation, protruding ears, sparse hair on the scalp, bulbous nose and short stature. Mutation in a zinc-finger protein gene is responsible for TRPS1. TRPSIII is most severe ►Langer-Giedion syndrome

Trichostatin A: An antifungal antibiotic, an inhibitor of histone deacetylase of yeast. Trichostatin may reverse the effect of methylation and activate methylated genes. In some instances low doses of

5-aza-2′-deoxycytidine along with trichostatin are required for substantial expression of originally methylated and silent cancer genes. Trichostatin increases the number of root hairs and their pattern of *Arabidopsis* (Xu C-R et al 2005 Proc Natl Acad Sci USA 102:14469). ►histone deacetylase, ►fragile X, ►root; Marks PA et al 2001 Curr Opin Oncol 13:477.

Trichothiodystrophy (TTD, 19q13.2-q13.3): is a collective name for autosomal reces-sive human diseases involving low-sulfur abnormalities of the hair. The *Tay syndrome* involves also ichthyosiform erythroderma (scaly red skin), mental and growth retardation, etc. The *Pollitt syndrome* (trichorrhexis nodosa or trichothiodystrophy neurocutaneus) displays low cystin content of the hair, and the nails, and the head and the nervous system are also defective. Xeroderma pigmentosum IV includes trichothiodystrophy and sun- and UV-sensitivity. Also called PIBIDS. This type of mutation lacks helicase and excision repair activity because of the defect in the interaction between one of the xeroderma pigmentosum and the p44 protein subunit of the transcription factor TFIIH. Some TDD mutations are temperature-sensitive. ►hair-brain syndrome, ►xeroderma pigmentosum, ►Cockayne syndrome, ►excision repair, ►transcription factors, ►temperature-sensitive mutation, ►ichthyosis; Vermeulen W et al 2001 Nature Genet 27:299; de Boer J et al 2002 Science 296:1276.

Triclosan (trichlorinated diphenyl ether): Antibacterial and antifungal agent (blocking lipid biosynthesis) used in antiseptics, soaps, and other cosmetics.

Tricotyledony: is a relatively rare and generally not inherited developmental anomaly in plants (see Fig. T107). In some families of plants there is a higher than average tendency for the condition.

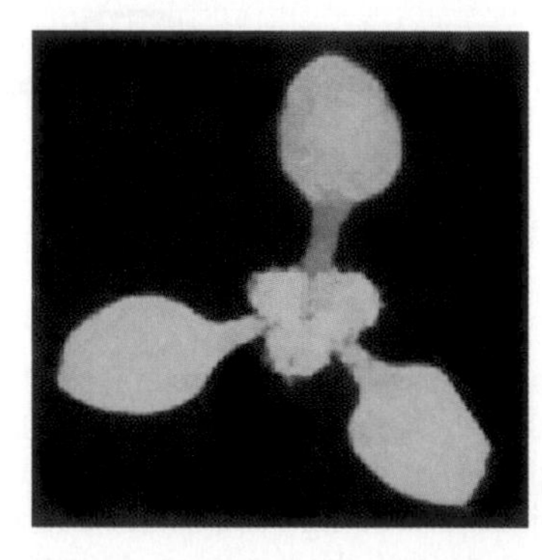

Figure T107. Tricotyledonous *Arabidopsis* seedling

Tricyclo-DNA: Tricyclo-DNA and -RNA can be used in antisense technologies to block selectively the expression of genes (see Fig. T108). ►antisense technologies; Renneberg D et al 2002 Nucleic Acids Res 30:2751.

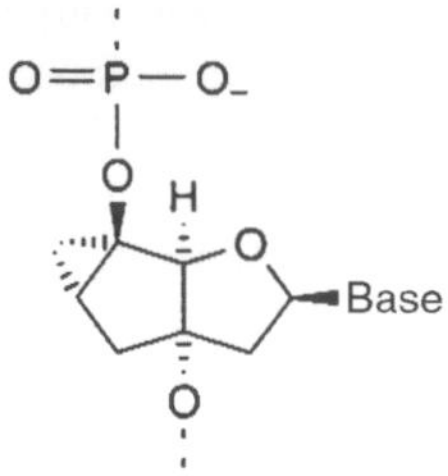

Figure T108. Tricyclo-DNA

TRID: is a TRAIL decoy receptor. ►TRAIL

Tricuspid Atresia: is an agenesis of the tricuspid valve, which connects the right atrium to the right ventricle of the heart. Some other heart defects may be associated with it. The condition is generally sporadic but some cases are familial and involve defects of the Zfpm2/Fog2 Zinc-finger protein.

Triethylene Melamine (TEM): Alkylating agent (see Fig. T109). ►TEM

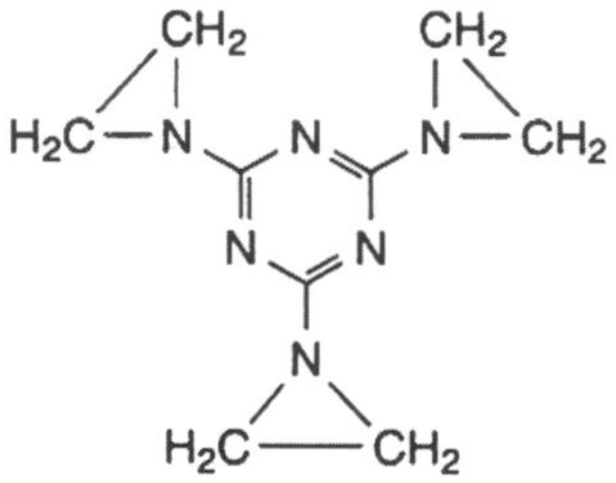

Figure T109. Triethylene melamine

Trigger Factor (TF, ~48 kDa): is a prolyl isomerase enzyme (PPI) associated with the 50S ribosome unit of bacteria or with the GroL chaperone. It may aid translocation of molecules through the cytoplasmic outer membrane. In bacteria TF and DnaK may cooperate in protein folding but they are not indispensable at intermediate temperatures. Cyclosporin or FK506 does not inhibit TF and it is only moderately related to cyclophilins of FKBPs. ►PPI, ►GroL, ►cyslosporin, ►FK506, ►cyclophilins, ►chaperone, ►GroEL, ►DnaK, ►Clp, ►protein sorting; Liu Z et al 2001 FEBS Lett 506:108; Patzelt H et al 2001 Proc Natl Acad Sci USA 98:14244; Ferbitz L et al 2004 Nature [Lond] 431:590.

Trigger Loop: is a part of RNA polymerase II complex just beneath the A site of ribosome; it triggers the formation of a phosphodiester bond during transcription of RNA and assures fidelity (Wang D et al 2006 Cell 127:941).

Triglyceride: Same as triaglycerol.

Triglyceridemia: ▶familial hypertriglyceridemia

Trihybrid Cross: The parental forms are homozygous altogether for 3 allelic pairs at (unlinked) loci, e.g., AABBdd × aabbDD and therefore in the F_2 8 phenotypic classes may be distinguished. ▶gametic array, ▶Mendelian segregation

Tri-isosomic: in wheat, 20"+ i"', 2n = 43, [" = disomic, "' = trisomic, i = isosomic]

Trillium: Species have one of the largest normal chromosomes (2n = 10) in plants, about 50 times larger than *Arabidopsis*, a plant with one of the smallest chromosomes (2n = 10) (see Fig. T110).

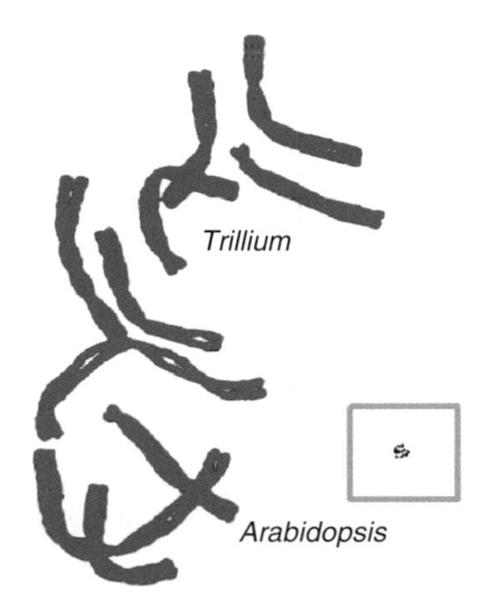

Figure T110. *Trillium* is from Sparrow AH, Evans HJ 1961 Brookhaaven Symp Biol 14:76; *Arabidopsis* karyotype is the courtesy of Lotti Sears

TRIM (tripartite interaction motif): Retrovirus/lentivirus can be restricted after entry into the cells by the TRIM5α protein variants (Sayah DM et al 2004 Nature [Lond] 430:569). The TRIM proteins form a family and different members may occur in different mammals. ▶retrovirus restriction factors; Song B et al 2005 J Virol 79:6111.

Trimester: Period of three months; the human pregnancy of 9 months includes 3 trimesters.

Trimethoprim (3,4-diamino-5-[3,4,5-trimethoxyben zylpyri-midine]): An antibiotic.

Trimethylaminuria (1q23-q25): Defect in flavin monooxygenase and failure of detoxification of drugs and endogenous amines. Afflicted persons exhale fish odors.

Trimming: The processing of the primary RNA transcripts to functional mRNA or ribosomal and tRNA. The cleavage of pre-rRNA transcripts by RNase III into 16S, 23S and 5S rRNA as well as into the tRNAs contained within the spacer sequences of the co-transcripts. The cleavage takes place at the duplex sequences forming the stem of the rRNA loops. ▶post-transcriptional processing, *rrn* genes for diagram, ▶introns

Trinomial Distribution: $(1 + 2 + 1)^n$ can be expanded to predict the segregation of the genotypic classes (note that the quotients within parentheses must not be added!)

$$1(1+2)^n + \frac{n!}{1(n-1)!}(1+2)^{n-1} + \ldots + \frac{n!}{(n-1)!1}(1+2)^{n-(n-1)} + 1(1+2)^{n-n}$$

An example for three pairs of alleles:

$$1(1+2)^3 + \ldots + \frac{3!}{1(2!)}(1+2)^2 + 1$$

$1 + (3 \times 2) + (3 \times 4) + 8 + 3 \times (1 + 2 + 2 + 4) + 3 \times (1 + 2) + 1$

When rewritten in a symmetrical distribution: 1:2:1:2:4:2:1:2:1:2:4:2:4:8:4:2:4:2:1:2:1:2:4:2:1:2:1 the 27 terms indicate that triple heterozygotes are 8, double heterozygotes 4, and single heterozygotes are 2 in a distribution in compliance with Mendel's law. ▶binomial, ▶multinomials

Trinucleotide Repeats: are microsatellite sequences displaying some clustering in yeast and other genes. In more than ten human neurodegenerative diseases CAG (glutamine codons) are repeated many times. The resulting polyglutamine (polyQ) oligomers seem to be the cause of the diseases or the disease accumulates the polyglutamine tracts. Diminished folding capacity of proteins seems to be involved in the trinucleotide repeat diseases (Gidalevitz T et al 2006 Science 311:1471). Autophagy more readily eliminates the polyglutamine tracts from the cytoplasm than from the nucleus (Iwata A et al 2005 Proc Natl Acad Sci USA 102:13135). The long polyglutamine sequences (most frequently beyond 35–40) interfere with CREB-dependent transcription by interacting with transcription factor $TAF_{II}130$. The polyglutamine proteins are more resistant to decay. Transglutaminase inhibitors (cystamine [←decarboxylated cysteine], monodansyl cadaverin [←decarboxylated lysine] may alleviate apoptosis of the cells. RNAi may alleviate the problems normally associated with polyglutamine tracts (Caplen NJ et al 2002 Hum Mol Genet 11:175). Several small molecules are potent inhibitors of polyglutamine aggregation and suppress neurodegeneration in vivo (Zhang X et al 2005 Proc Natl Acad Sci USA 102:892). Unusual, commons feature of these diseases that in successive generations the symptoms appear earlier and with greater severity (anticipation) as gain-of-function mutations. The repeats form a hairpin structure and interfere with DNA replication. Also, the CpG sequences are likely to be methylated. Cis elements may contribute to instability (Cleary JD, Pearson CE 2003 Cytogenet Genome Res 100:25). The nature of the repeats may vary and may involve CGG, GCC, CAG, CTG sequences in different autosomes and the X chromosome, respectively. These repeats may expand (from a few [5–50] in the normal to hundreds of copies) in an unstable manner in the 5′-untranslated region of the FMR1 gene and cause translational suppression by stalling on the 40S ribosomal RNA. Defects in DNA repair may contribute to instability. Flanking sequences of the polyglutamine tracts affect the degree of misfolding and toxicity (Duennwald ML et al 2006 Proc Natl Acad Sci USA 103:11045). More than 40 human anomalies are attributed to changes in the repeats.

In *E. coli*, the larger expansions occur predominantly when the CTG trinucleotides are in the leading strands and deletions are mainly on the opposite lagging strands. The toxic effect of polyglutamine tracts can be genetically suppressed in *Drosophila* by proteins homologous to heat shock protein 40 and a tetratricopeptide, both containing chaperone-like a domain. The instability caused by the repeats is more common in meiosis than in mitosis. In yeast the CAG/CTG repeat meiotic instability is based on double-strand DNA break repair (see Fig. T111).

It has been suggested that the DNA polymerase stalls within the CTG-CAG repeat sequences and cause nicks and double-strand breaks that promote homologous recombination in lower and higher eukaryotes at 10% or higher frequency. Polyglutamine expansions can affect transcription and can be one of the causes of the disease (Helmlinger D et al 2006 Trends Genet 22:562). ►anticipation, ►resveratrol, ►Huntington's chorea, ►Machado-Joseph disease, ►ataxia, ►Kennedy disease, ►polysyndactyly, ►spinocerebellar ataxias, ►dentatorubral-pallidoluysian atrophy, ►Jacobsen syndrome, ►fragile sites, ►fragile X, ►FRAXA, ►FRAXE, ►Friedreich's ataxia, ►myotonic dystrophy, ►epilepsy, ►myoclonic epilepsy, ►muscular dystrophy, ►methylation of DNA, ►translation repressor proteins, ►SRY, ►human intelligence, ►FMR1 mutation, ►schizophrenia, ►neurodegenerative diseases, ►pre-mutation, ►ERDA1, ►dinucleotide repeats, ►microsatellites, ►tetratrico sequences, ►chaperone, ►heat-shock proteins, ►TAF_{II}, ►CREB, ►slipped-structure DNA, ►RNAi, ►homopolymeric amino acids, ►autophagy, ►oxoguanine; Claude T et al 1995 Annu Rev Genet 29:703; Orr HT, Zoghbi HY 2000 Cell 101:1; Cummings CJ, Zoghbi HY 2000 Annu Rev Genomics Hum Genet 1:281; Cleary JD et al 2002 Nature Genet 31:37; Hum Mol Genet 2002 11:1909–1985; Pluciennik A et al 2002 J Biol Chem 277:34074; Napierala M et al 2002 J Biol Chem 277:34087; Pearson CE et al 2005 Nature Rev Genet 6:729; Gatchel JR, Zoghbi HY 2005 Nature Rev Genet 6:743; therapeutic approaches: Di Prospero NA, Fischbeck KH 2005 Nature Rev Genet 6:756; Orr HT, Zoghbi HY 2007 Annu Rev Neurosci 30:575; repeats and human disease review: Mirkin SM 2007 Nature [Lond] 447:932.

Trinucleotide-Directed Mutagenesis (TRIM): Introduction into the coding sequences of a gene(s) trinucleotide analogs such as 9-fluorenylmethoxycarbonyl (Fmoc) trinucleotide phosphoramidites. The synthetic analogs convey resistance to nucleases and are effective in induction of specific mutations. ►trinucleotide repeats, ►phosphoramidates; Sondek J, Shortle D 1992 Proc Natl Acad Sci USA 89:3581.

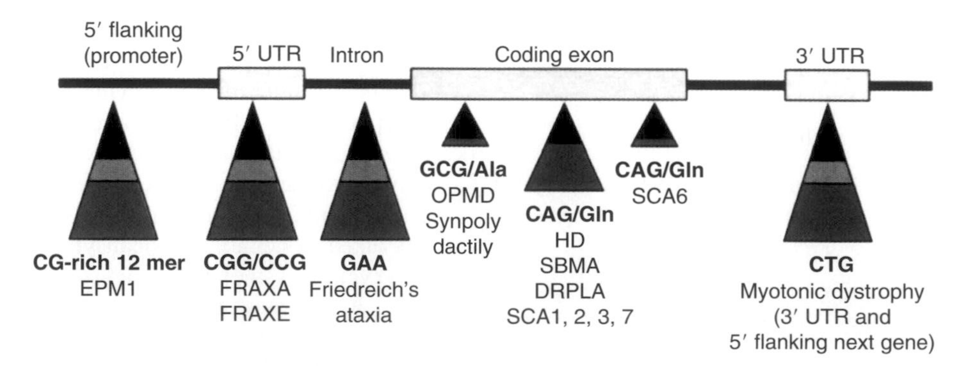

Figure T111. Location of expanded trinucleotide repeats in diseases. EPM: myoclonous epilepsy; FRAXA: fragile X syndrome; OPMD: Muscular atrophy; HD: Huntington chorea; SBMA: Kennedy disease; DRPLA: dentatorubralpallidoluysian atrophy; SCA: spinocerebellar ataxia. The size of the triangles indicates the size of the repeat expansions. Top triangle: normal, gray in the middle: unstable premutational, bottom of the triangle: pathological condition. (Courtesy of Lunkes A et al 1998, p 149. In: Molecular Biology of the Brain. Higgins SJ (Ed.) Biochemical Society, London, UK)

TRIO (in human genetics): mother + father + child are available for study.

Trioma: ►heterohybridoma

Triose: is a sugar with 3 carbon backbone.

Triosephosphate Isomerase Deficiency (TPI1): is encoded in human chromosome 12p13 but pseudogenes seem to be present at other locations. The level of activity of the enzyme varies. Null mutations are not expected to be viable since this is a key enzyme in the glycolytic pathway. The symptoms may be quite general weakness, neurological impairment, anemia, recurrent infections, etc. ►glycolysis

Trip1 (thyroid hormone receptor): ►Sug1

Triparental Human Embryo: Can be produced if a fertilized human egg nucleus is transferred into the egg of another female and implanted into a foster mother. The purpose of such an experiment would be to get rid of defective mitochondria of the first female and secure a healthy offspring. Such a procedure is not allowed in the USA but it has been approved in the United Kingdom in 2005. ►multiparental offspring, ►ART, ►mitochondrial diseases in humans; Check E 2005 Nature [Lond] 437:305.

Tripeptide Discriminator (tripeptide anticodon): is part of the prokaryotic and eukaryotic translational release factors that recognize nonsense codons in mRNA and terminate protein synthesis. ►release factor; Nakjamura Y, Ito K 2002 FEBS Lett 514:30.

Tripeptidyl Peptidase (TPPII): is a large protein complex outside the lysosomes with activity resembling the proteasome. ►proteasomes

Triplasmy: is heteroplasmy for three different types of mtDNA. ►heteroplasmy, ►mtDNA

Triple A Syndrome: ►achalasia-addisonianism-alacrima

Triple Helix Forming Oligonucleotides (TFO): May bind to polypurine-polypyrimidine tracts in the major groove of the DNA helix by Hoogsteen or reverse Hoogsteen bonding and prevent the access of transcription factors. This may block transcription and cleave the DNA but may enhance repair DNA synthesis. In the triplex sequences, mutation rate in SV40 increased more than an order of magnitude in the suppressor gene, supFG1, employed as reporter, with 30 nucleotide long AG sequences (AG30). Shorter sequences or oligonucleotides of all four bases were either not or were much less effective. The triplex structure in xeroderma pigmentosum or in the Cockayne syndrome cells was not effective for mutation enhancement, indicating the requirement of excision repair for the events. TFOs may be used for targeting specific genes and prevent their transcription or to induce mutation. ►TPO, ►LNO, ►DNA kinking, ►triplex, ►DNA repair, ►SV40, ►supF, ►xeroderma pigmentosum, ►Cockayne syndrome, ►Hoogsteen pairing, ►antisense technologies, ►psoralen dyes, ►locked nucleic acids, ►pseudoknot, ►targeting genes, ►inhibition of transcription, ►helix; Grimm GN et al 2001 Nucleosides Nucleotides Nucleic Acids 20:909.

Triple Test: used for the identification of Down syndrome by assaying chorionic gonadotropin, unconjugated estriol and α-fetoprotein levels. ►Down's syndrome, ►gonadotropin, ►estriol, ►fetoprotein

Triple-A Syndrome: ►ALADIN

Triple-Stage Quadrupole/Ion-Trap Mass Spectrometry: is proteome analytical procedure. ►mass spectrometer, ►electrospray, ►ESI, ►CID

TRIPLES (transposon-insertion phenotypes): is a database with information on phenotype, protein localization and expression on the results of transposon mutagenesis in yeast. ►insertional mutation, ►transposon; http://ygac.med.yale.edu.

Triplet Binding Assay: A historically important method to determine the meaning of genetic triplet codons. A single type of radioactively labeled amino acid, charged to the cognate tRNA was allowed to recognize and bind to ribosomes with mRNA attached. Each type of charged tRNA then recognized only the corresponding code and the ribosomes were then trapped on the surface of a filter. Synthetic polynucleotides (mRNA) of known base composition retained only the cognate aminoacylated tRNA and thus provided the base composition and sequence of the true coding triplets. (See Fig. T112, ►genetic code; Nirenberg MW, Leder P 1964 Science 145:1399).

Triplet Code: ►genetic code

Triplet Expansion: ►trinucleotide repeats

Triplex: Three-stranded nucleic acid structure, e.g., an RNA oligo-nucleotide may bind within the strand or to double-stranded DNA and result in antisense effects. Triplex strands occur transiently in genetic recombination. Triplex forming oligonucleotides may increase somatic mutation and recombination by the mechanism of nucleotide exchange repair (see Fig. T113). Some DNA polymerases (T7, Klenow fragment) can elongate a DNA strand from primers forming triple helices of 9–14 deoxyguanosine-rich residues. Triplex-forming oligonucleotides (TFO) may be used for sequence-specific control of gene expression. ►antisense RNA, ►Hoogsteen pairing, ►helix, ►peptide nucleic acid, ►TFO, ►H-DNA,

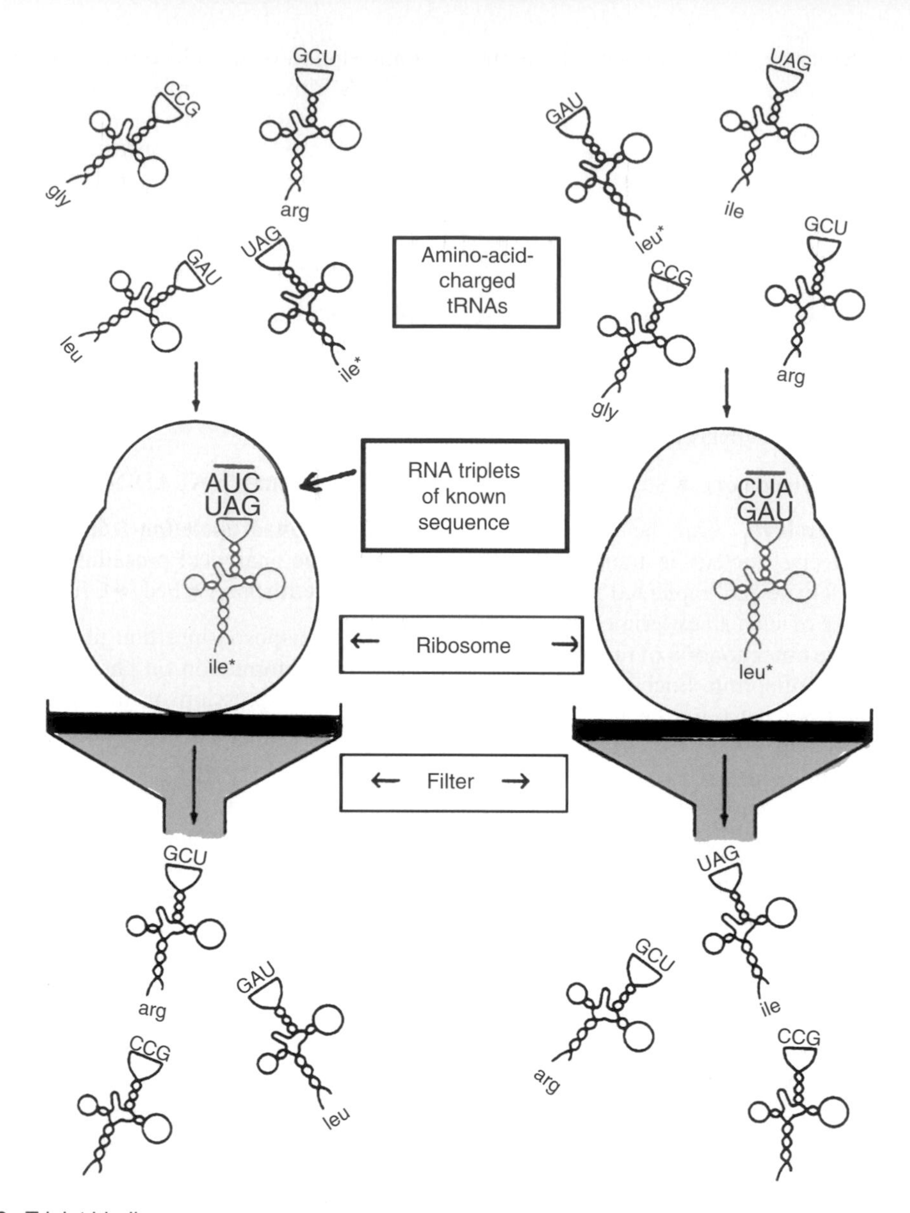

Figure T112. Triplet binding assay

Figure T113. Triplex. **H**: Hoogsteen pairing, **RH**: reverse Hoogsteen R: purine, Y: pyrimidine. Other configurations are also possible. (Hoyne PR et al 2000 Nucleic Acids Res 28:770)

►nodule-DNA, ►excision repair, ►Klenow fragment, ►isomerization of strands; Vasquez KM, Wilson JH 1998 Trends Biochem Sci 23:4; Luo Z et al 2000 Proc Natl Acad Sci USA 97:9003; Datta HJ et al 2001 J Biol Chem 276:18018; Rocher C et al 2001 Nucleic Acids Res 29:3320; Knauert MP, Glazer PM 2001 Hum Mol Genet 10:2243; Coman D, Russu IM 2004 Nucleic Acids Res 32:878.

Triplex: A polyploid with three dominant alleles at a gene locus. ►autopolyploid, ►trisomy

Triplicate Genes: Triplicate genes convey identical or very similar phenotype and in a diploid, when segregate independently, display an F_2 phenotypic ratio of 63 dominant and 1 recessive.

Triploblasts: Animals with ecto-, meso- and endodermal germlayers. ►diploblasts

Triploid: A cell or organism with three identical genomes. Triploids (3x) are obtained when a (4x) is crossed with a diploid (2x). The majority of edible bananas, several cherry and apple varieties, and many sterile ornamentals (chrysanthemums, hyacinths) are triploid. The seedless watermelons, produced by crossing tetraploids with diploids are triploids and have commercial value. Triploid sugar beets are produced in a large-scale agriculturally because they offer of ≈10% or higher sugar yield per acre than the parental diploid varieties. Triploidy in humans may occur by fertilizing a diploid egg (digyny) by a monosomic (normal) sperm or by the union of a normal haploid egg and two spermatozoa (diandry). The green toad (*Bufo viridis*) is triploid but surprisingly reproduces bisexually. ►trisomic; Zaragoza MV et al 2000 Am J Hum Genet 66:1807; Stöck M et al 2002 Nature Genet 30:325.

Triplo-X: Triplo-X females (XXX) occur in about 0.0008 of human births. Their phenotype is close to normal and they can conceive. Yet they are somewhat below average in physical and mental abilities although they tend to be somewhat tall. With an increasing number of X-chromosomes beyond 3, the adverse effects are further aggravated. ►sex determination, ►sex chromosomal anomalies in humans, ►trisomy, ►Turner syndrome; Barr ML et al 1969 Can Med Assoc J 101:247.

Tripsacum: ►maize

Triradial Chromosome: A triradial chromosome may be formed by fusion of broken translocated chromatids in a three-armed way (see Fig. T114). (See Jenkins EC et al 1986 Am J Med Genet 23:531).

Figure T114. Triradial

Tris-HCl Buffer: The tris-Hcl buffer contains tris-[hydroxymethyl]aminomethane and hydrochloric acid; it is used in various dilutions within the pH range 7.2–9.1.

Triskelion: Three-legged proteins on the surface of vesicles built of three clathrin and three smaller proteins (see Fig. T115). ►clathrin, ►endocytosis; Ungewickell E 1983 EMBO J 2:1401; Fotin A et al 2004 Nature [Lond] 432:573.

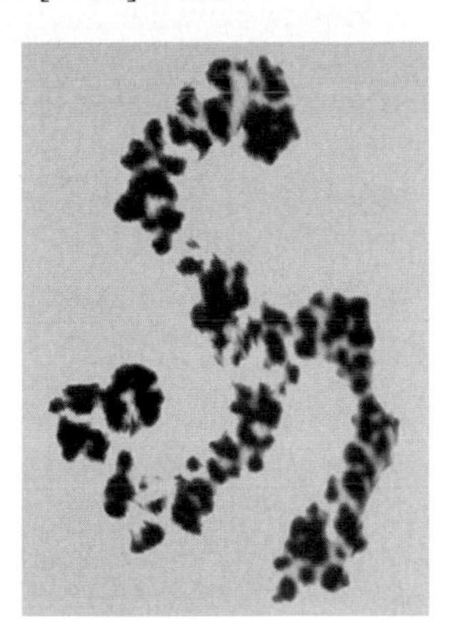

Figure T115. Triskelion (Redrawn after Ungewickell E, Branton D 1981 Nature [Lond] 289:420)

Trismus-Pseudocamptodactyly Syndrome (17p13.1): Camptodactyly, arthrogryposis, foot deformity, and failure to open the mouse completely due to defect in the myosin heavy chain gene (MYH8). ►camptodactyly, ►arthrogryposis, ►myosin; Veugelers M et al 2004 New England J Med 351:460.

Trisomic: A cell or organism with one or more chromosomes (but not all) represented three times. ►trisomy, ►trisomic analysis, ►triploid

Trisomic Analysis: Trisomics have one or more chromosomes in triplicate (*AAA*) and thus produce both disomic (*AA*) and monosomic (*A*) gametes for gene loci in those chromosomes (see Table T2).

Male transmission of the disomic gametes is usually poor, it rarely exceeds 1%.

The transmission of disomic gametes through females varies according to the chromosome (genes) involved and depends also upon environmental conditions; most commonly this transmission is 1/4 or 1/3 of the normal (monosomic) gametes. The few viable human sex-chromosome trisomics (XXX, XXY, XYY) are either sterile or have very poor fertility because of the failure of normal development of the ovaries and testes (gonadal dysgenesis). From a single alternative allelic pair (*A/a*), disomics produce either *A* or *a* gametes. In contrast, trisomics potentially produce a maximum of five kinds of gametes, three disomic and two monosomic. The frequency of these five types of gametes depends also on the distance of the gene from the centromere, i.e., whether *chromosome segregation* (no crossing over between gene and centromere) or *maximal* (see Table T3).

Table T2. Gametic output of trisomics

Genotype ⬇	Chromosome segregation					Maximal equational segregation				
	AA	*Aa*	*aa*	*A*	*a*	*AA*	*Aa*	*aa*	*A*	*a*
AAa (duplex)	1	2	0	2	1	5	6	1	8	4
Aaa (simplex)	0	2	1	1	2	1	6	5	4	8

Table T3. Phenotypic proportions in the F_2 of trisomics (disomic and trisomic pooled)

Transmission[‡] & genotype ⬇	Chromosome segregation			Maximal equational segregation		
	Dom.	Rec.	*aaa* (%)	Dom.	Rec.	*aaa* (%)
DUPLEX (*AAa*)						
Male and female	35	1	0	22.04	1	1.39*
Female only	17	1	0	13.40	1	1.39
None	8	1	0	8	1	0.00
Simplex (*Aaa*)						
Male and female	3	1	11[§]	2.41	1	13.9*
Female only	2	1	11	1.77	1	13.9
None	5	4	0	1.25	1	0.

[‡]refers to transmission of the disomic gametes
*1/576 tetrasomics (*aaaa*) are not included
[§]1/36 tetrasomic (*aaaa*) is not included
*25/576 tetrasomics (*aaaa*) are not included

In order to obtain two identical recessive alleles from a duplex, the chromosomes must form trivalents and recombination must take place between the gene and the centromere, following which the anaphase I distribution must move the two exchanged chromosomes toward the functional megaspore (equivalent to secondary oocyte in animals). At the most favorable coincidence of these events, in only half of the cases can we expect the two identical alleles to move into the same megaspore (equivalent to egg in animals).

Thus, the maximal chance of having a tetrad with a double recessive gamete (*aa*) will be $1/3 \times 1/2 = 1/6$. These fractions are based on one reductional and two equational disjunctions at meiosis I (1/3), while two alternative disjunctions at anaphase II (1/2) are the determining factors of the outcome.

The phenotypic segregation can be derived from the gametic output by random combinations. One factor, the transmission difference between the female and male gametes, seriously alters the theoretically expected proportions (see Table. T4).

Table T4. Phenotypic ratios in test-cross progenies (trisomics + disomics)

Genotypes of cross ⬇	Only female transmission of disomic gametes Dom.: Rec.	No transmission of disomic gametes Dom.: Rec.
AAa x aa	5:1	2:1
Aaa x aa	1:1	1:2

The genetic behavior of all chromosomes not present in triplicate in the trisomics is consistent with disomy. When the trisomic individuals are phenotypically distinguishable per se (do not require cytological identification of the chromosomal constitution), trisomy may be used very effectively to assign genes to chromosomes, irrespective of their linkage relationship. The segregation in a duplex will not be 3:1 but it will vary (most commonly) between 17:1 to 8:1, as predicted by the table. In this case, the gene *a* is located in the chromosomethat is triplicated. Genes in the disomic set of chromosomes are expected to display 3:1 segregation. In case telotrisomics are crossed with disomics (___o___*a*__, ___o___*a*___, o___*A*___ x normal disomic *a/a*), the dominant allele will be very rare among the diploid offspring but may be expressed in every individual of 2n + telochromosome. Mapping can be done also by properly constructed isotrisomics, tertiary trisomics, and compensating trisomics. Genetic mapping with the aid of multipoint trisomic data is feasible also in humans (see Fig. T116). ►disomic, ►monosomic, ►gonadal dysgenesis, ►chromosome segregation, ►maximal equational segregation, ►mapping by dosage effect, ►trisomy, ►Down's syndrome, ►Pätau syndrome, ►Edwards syndrome; Rédei GP 1982 Genetics. Macmillan, New York; Li J et al 2001 Am J Hum Genet 69:1255.

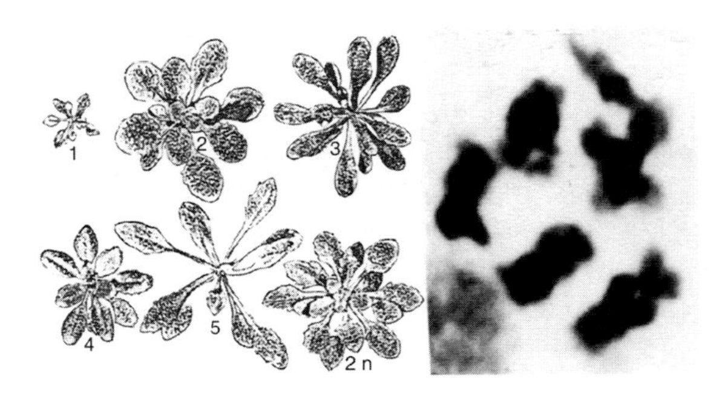

Figure T116. Left: The four primary trisomics (1–4) and a telotrisomic (5) of *Arabidopsis*. At the lower right corner is a normal disomic individual (2n). All are in columbia wild type background, grown under short daily light periods. At right: The chromosome complement of a primary trisomic. Note the trivalent association at the upper right area. (The photomicrograph is the courtesy of Dr. Lotti Steinitz-Sears)

Trisomic, Complementing: Trisomic complementing can be of three types; one is an apparent trisomic only because it involves a normal biarmed chromosome, while the two other chromosomes are actually telosomes, representing the left and right arms of the intact, normal chromosome. The second complementing type also involves one biarmed normal chromosome, but the two telosomes represent a pair of identical arms; thus, in essence this case is very similar to that of a secondary trisomic, having three identical arms and one single different arm. In the third type, in a normal biarmed chromosome, the second chromosome has an arm identical to the first but its second arm is a translocated segment from a non-homologous chromosome, and the third chromosome has a different translocated arm linked to the centromere and to an arm of a normal chromosome. ►trisomic analysis; scheme from GP Rédei's Lecture Notes 1980.

Trisomic, Primary: In a primary trisomic, the chromosome number is 2n + 1, and the triplicate homologous chromosomes are structurally normal. ►trisomic analysis

Trisomic, Secondary: In a secondary trisomic, the extra chromosome is an isochromosome, i.e., its two arms are identical; such a chromosome is an isochromosome. ►trisomic analysis, ►isochromosome

Trisomic, Tertiary: In a tertiary trisomic, the extra chromosome is involved in a reciprocal translocation and is partly homologous with two standard chromosomes. ►trisomic analysis

Trisomy: Trisomy involves one or more but not all chromosomes of a genome in triplicate; trisomics are aneuploids. They may arise by selfing triploids when some of the extra chromosomes are lost. Nondisjunction during meiosis may also lead to trisomic progeny. Trisomy may exist in various forms. The phenotype of the trisomics varies, depending upon what genes are in the trisome. The phenotype may be very close, almost indistinguishable from that of normal disomics, and in other cases it may be lethal. Only a few of the autosomal human trisomies permit growth and development beyond infancy. Trisomy 9 allows near normal life expectancy although it also involves developmental and mental retardation. Individuals with trisomy 21 (Down's syndrome) also reach adulthood but remain mentally subnormal. Trisomy 22 may exist in mosaic form and the afflicted individuals are retarded in growth and mental abilities. About 3/4 of the trisomy 8 cases are mosaics, mentally retarded, and affected by head abnormalities to a variable degree, according to the extent of mosaicism. Trisomics for chromosome 11 have a variable extra long arm in chromosome 11 and accordingly exhibit brain and other internal organ damage of variable extent. Trisomics for the long arm of chromosome 3 have a very short life and variable abnormalities in internal organs and body development and severe mental retardation. Trisomy for the short arm of chromosome 4 is characterized by serious malformations of the brain, head, extremities, genitalia, and such individuals usually die early. About 1/3 of the abortuses, due to autosomal trisomy, involve chromosome 16; these are never carried to term. Trisomy 13 (Pätau's syndrome) and trisomy 18 (Edwards syndrome) are live-born but die very early. Trisomy 16 is responsible for about 7.5% of all spontaneous abortions. XYY trisomies are apparently never found in spontaneously aborted or stillborn fetuses. All other autosomal human trisomies lead to abortion at various stages after conception. Sex-chromosomal trisomy generally does not compromise viability of humans yet such individuals are usually sterile and display gonadal dysgenesis and a variety of physical and most commonly mental retardation as well. The vast majority of trisomics are the result of disomic female gametes. Paternal nondisjunction accounts for only a very small frequency of autosomal trisomy. However, 80% of 45 + X monosomics involve the loss of a paternal sex chromosome. Analytical procedures of about 65% efficiency exist for the detection of trisomy during the first trimester (Wapner R et al 2003 New England J Med 349:1405).

Trisomy is useful in plants for assigning genes to chromosomes. Microcell-mediated chromosome transfer permits the inclusion into mouse cell of only fragments of the third chromosome. The fragments are produced by chromosome breakage with the aid of ionizing radiation. The size and content of the fragments so generated is beyond precise experimental control; nevertheless, information can be obtained about the effects of segments smaller than an entire extra chromosome. Homologous recombination by gene targeting would appear to be a more desirable tool for generating segments of designed contents. Unfortunately, the frequency of this type of recombination is low. A chicken pre-B cell line (DT40) is known, however with 3–4 orders of magnitude higher homologous recombination. Technology is available for the transfer of human chromosomes to these chicken cell lines. The truncated human chromosomes from the chicken cell lines can then be reintroduced into human cells in culture. Another approach is the construction of human artificial chromosomes and minichromosomes (see Fig. T117).

The presence of an extra chromosome in yeast involves a number of phenotypes that are independent of the identity of the individual extra chromosome; all of these include defects in cell cycle progression, increased glucose uptake, and increased sensitivity to conditions interfering with protein synthesis and protein folding. These phenotypes reflect the consequences of additional protein production, imbalances in cellular protein composition and proliferative disadvantage (Chan LY et al 2007 Science 317:916). ►trisomics, ►trisomic analysis, ►Down syndrome, ►Edwards, ►Pätau's, ►Klinefelter syndromes, ►sex determination, ►sex-chromosomal anomalies in humans, ►cat-eye syndrome, ►aneuploidy, ►targeting genes, ►chromosome uptake, ►human artificial chromosome, ►minichromosome, ►microcell; Hassold TJ, Jacobs PA 1984 Ann Rev Genet 18:69; Robinson WP et al 2001 Am J Hum. Genet 69:1245.

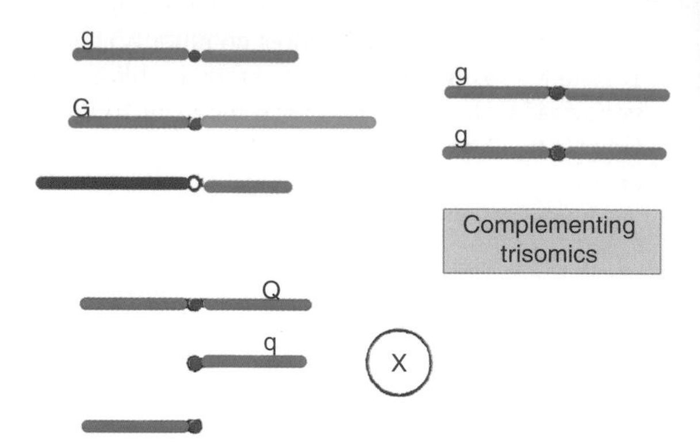

Figure T117. In the complementing trisomics there is only one normal representative of a particular chromosome. For the homologous arms it may be compensated by translocations containing jointly the entire chromosome (1) or by two complementary isochromosomes (2). The complementing trisomics (2). The complementing trisomics can also be used to assign genes to chromosome arms. In case (1) the predominant fraction in the 2n progeny is expected to be recessive because only the non-translocation strand is transmitted by the monosomic gamete. The wild-type allele (G) can be transmitted only after recombination. In contrast, the 2n+1 progeny should be almost all wild type (except when recombination occurs between gene and centromere). In selfing of genotype (2) the 2n and most of the 2n+1 progeny are expected to be of dominant phenotype; however, when recombination occurs between (q) and centromere, the recessive allele is included in the normal chromosome. Such a setup as (2) can be exploited only if the two telo chromosomes are not distributed together in the majority of anaphases I's of meiosis

Trisomy, Segmental: ►segmental aneuploidy

Tristetraprolin (TTP): A zinc-finger protein, inhibitory to TNF-α by destabilizing its mRNA, although TNF and agents that stimulate TNF stimulate its production too. TTP binds to A-U-rich (AUR) 3′ untranslated sequences and responds to cytokines and other regulatory molecules. The Dicer and Argonaute proteins of the RNAi and microRNA pathways are also required for the AUR decay (Jing Q et al 2005 C ell 120:623). TTP and several homologous proteins control also apoptosis. ►TNF, ►zinc fingers, ►DNA-binding protein domain, ►RNAi, ►antiviral protein; Johnson BA, Blackwell TK2002 Oncogene 21:4237.

Tritanopia: ►color blindness

Tritelosomic: Tritelosomic occurs in wheat, 20" + t"', 2n = 43, [" = disomic, "' = trisomic, t = telosomic].

Table T5. The major species of *Triticum*

Diploids (2n = 14)	genomic formula	former designation* and species, respectively
T. monococcum	A	*T. boeoticum**, *T. aegilopoides, T. thoudar. T. urartu*
T. speltoides	S (= G?)	*Aegilops speltoides**, *Aegilops ligustica**
T. bicorne	S^b	*Aegilops bicornis**
T. searsii	S^s	*Aegilops searsii**
T. longissimum	S^l	*Aegilops longissima** (*Aegilops sharonensis ?*)
T. tauschii	D	*Aegilops squarrosa**, *T. aegilops*
Tetraploids (2n = 28)		
T. turgidum	AB	*T. dicoccoides**, *T. dicoccon, T. durum**, *T. polonicum**, *T. carthlicum**, *T. persicum**
T. timopheevi	AG	*T. araraticum, T. dicoccoides var. nudiglumis, T. armeniacum*
Hexaploids		
T. aestivum	ABD	*T. vulgare**, *T. spelta**, *T. macha**, *T. sphaerococcum**, *T. vavilovii**

Trithorax: *Drosophila* gene (3–54.2) regulating embryonic development (head, thorax, abdomen). Some of the alleles cause homeotic developments. The heterozygotes for the trx^D mutation display patchy variegation. ►developmental-regulator effect variegation, ►homeotic genes, ►mll, ►*bithorax*, ►*polycomb*, ►SET motif, ►epigenesis, Petruk S et al 2001 Science 294:1331.

Triticale: A synthetic species produced by crossing wheat (*Triticum*) and rye (*Secale*) and doubling the chromosome number. The wheat parent may be tetraploid (2n = 28) or hexaploid (2n = 42) and the rye is generally diploid (2n = 14). Therefore, the amphidiploids may have the chromosome number of either 2n = 42 (hexaploid) or 2n = 56 (octaploid). As a crop variety, the former is used. On some soils suitable primarily for rye, the *Triticale* may provide grains with a milling quality approaching that of wheat. The new hybrids generally have shrunken kernels because some rye genes cause poor development of the endosperm. Breeding efforts have largely solved these problems and the commercial varieties have rather plump grains. ►*Triticum*, ►*Secale*, ►amphidiploid, ►chromosome doubling, ►species synthetic; Gustafson JP 1987 Plant Breed Rev 5:41.

Triticolsecale: ►*Triticale*

Triticum (common wheat and relatives): *Triticum* form a series of allopolyploids (see Table T5) (Huang S et al 2002 Proc Natl Acad Sci USA 99:8133). The common bread wheat, *T. aestivum* (both winter and spring forms), and the *T. durum* (*T. turgidum*) are used mainly for various pastas (macaroni, spaghetti, etc). The wheat kernel contains ca. 60–80% carbohydrates (starch), 8–15% protein that is usually low in lysine, tryptophan, and methionine (see Fig. T118). Some wild relatives of the cultivated wheats produce kernels with higher amounts of protein than the commercial varieties, and chromosomal engineering can transfer these favorable traits. Wheats are the staple food for about 1/3 of the human population. ►allopolyploids, ►polyploidy, ►monosomic analysis, ►nullisomic, ►glutenin, ►celiac disease; Sears ER 1974 Handbook of Genetics. In: King RC (Ed.) Vol 2. pp 59–91, Plenum; Peng J et al 2003 Proc Nat Acad Sci USA 100:2489; extensive genomic information: Genetics 168 (2) 2004; http://compbio.dfci.harvard.edu/tgi/; http://www.ksu.edu/wgrc/; hexaploid heat transcript estimation: http://www4.rothamsted.bbsrc.ac.uk/whets.

Figure T118. Wheat species: From right to left: *Triticum monococcum, T. turgidum (durum), T. timopheevi, T. aestivum* (Chinese Spring [most widely used for cytogenetic studies]), *T. compactum. T. spelta*. (Courtesy of Dr. Carlos Alonso-Arnedo)

Tritium (H^3): ►isotopes

Triton X-100: A non-ionic detergent (MW 537). It used for the extraction of proteins and the solubilization of biological materials. The reduced form is preferred when spectrophotometric measurement is required. Molar absorption is 1.46×10^3 at pH 8, 20°C in 1% sodium dodecylsulfate (SDS) at 275 nm. Ammonium cobaltthiocyanate reacts with it as a blue precipitate.

Tritordeum: An allohexaploid hybrid of tetraploid wheat (A, B genomes) and barley (H) of allooctaploid (ABDH) constitution. ►*Triticum*, ►*Hordeum*, ►allopolyploid; Hernandez P et al 2002 Genome 45:198.

Tritryp: ►*trypanosomatids*

Trituration: The disintegration into cells or homogenization of tissues or mixing different components of various materials.

Trivalent: In a trisomic or polyploid individual, three chromosomes may associate during meiotic prophase (see Fig. T119). In the trivalent association at any particular place, only two homologous chromosomes can pair, however, at a specific time and site.

Figure T119. Trivalent

***trk* (TRK) Oncogene**: The product of the TRK gene is a receptor in membranes (tyrosine kinase activity) for NGF (nerve growth factor). It is localized to human chromosome 1q23-q24. TRK affects axonal and dendritic growth, synapsis, cytoskeleton assembly, membrane traffic, signal transduction, etc. ►nerve growth factor, ►ovary; Nakagawara A 2001 Cancer Lett 169:107; Huang EJ, Reichardt LF 2003 Annu Rev Biochem 72:609.

tRNA (transfer RNA): The shortest RNA molecule in the cell (ca. 3.8S), consisting of about 76 to 86 nucleotides. tRNAs carry amino acids to the ribosomes during protein synthesis. The majority of cells have 40 to 60 types of tRNAs because most of the 61 sense codons have their own tRNA in the eukaryotic cytosol. The tRNAs, which accept the same amino acid are known as *isoaccepting tRNAs*. In the human mitochondria, there are only 22 different tRNAs and in plant chloroplasts, about 30. tRNA is frequently called an adaptor molecule because it adapts the genetic code for the formation of the primary structure of protein. Rarely (ca. 1/3000), a tRNA is charged with the wrong amino acid, and in these cases the complex is usually disrupted and the tRNA, recycled. The tRNA genes may occur in between ribosomal RNA genes (promoter - 16S - - tRNA - 23S - 5S - tRNA - tRNA...) or they frequently form independent clusters of different tRNA genes (promoter – $tRNA^1$ – $tRNA^2$ – $tRNA^3$ – •). The different cognate tRNA groups are generally identified by a superscript of the appropriate amino acid, e.g., $tRNA_{Met}$. Sometimes, individual tRNA genes may be present in multiple copies. The tRNA gene clusters are transcribed into large primary molecules that require successive cleavage and trimming to form the mature tRNA. The endonuclease, RNase P (a ribozyme) recognizes the primary transcript, whether it is a single tRNA sequence or a cluster of rRNA or tRNA genes, and cleaves at the 5′ terminus where the tRNA begins. For the tRNA to become functional, RNase D cuts at the 3′-end, and stops at a CCA sequence if any, or the 3′-end receives through post-transcriptional synthesis (by tRNA nucleotidyl synthetase) one, two, or three bases; thus, tRNA always ends with a 3′-CCA_{-OH} sequence. This 3′-OH end becomes the amino acid attachment site of the tRNA (►transfer RNA). During the process of maturation, through modification of the original bases, thiouridine (S4U), pseudouridine (ψ), ribothymidine, dihydrouridine (DHU), inosine (I), 1-methylguanine (m^1G), 1-dimethyl-guanine (m^1dG), and N^6-isopentenyl adenosine (i^6A) may be formed within the tRNA sequence. ►transfer RNA, ►amino acylation, ►aminoacyl-tRNA synthetase, ►protein synthesis, ►trimming, ►ribozyme, ►intron, ►pseudouridine, ►ribothymidine, ►dihydrouridine, ►isopentenyladenosine, ►hypoxanthine [for inosine], ►mitochondrial diseases in humans, ►tRNA nucleotidyl transferase; Söll D, Rajbahandary UL (Eds.) 1994 tRNA, Structure, Biosynthesis and Function. AMS Press, Washington, DC; http://medlib.med.utah.edu/RNAmods/.

tRNA Cleavage: In tRNA cleavage, T even phages may cripple a bacterial host tRNA and the viral-coded isoaccepting tRNA is then substituted for that of the host. (See Amitsur M et al 1987 EMBO J 6:2499; Kaufmann G 2000 Trends Biochem Sci 25(2):70; Meidler R et al 1999 J Mol Biol 287:499).

tRNA Deacylase: tRNA deacylase cuts off the tRNA D-amino acids from the polypeptide chain after the complex reaches the P site on the ribosome. ►protein synthesis, ►aminoacyl-tRNA synthetase; Ferri-Fioni ML et al 2001 J Biol Chem 276:47285.

tRNA Mimicry: In tRNA mimicry, certain set of translation factor proteins resemble tRNA in shape and may even mimic tRNA in deciphering the genetic code. (See Nakamura Y 2001 J Mol Evol 53(4–5):282).

tRNA Nucleotidyl Transferase: Attaches after transcription CCA-3′-OH to the 3′-end of the amino acid accepting arm of tRNA without relying on a template. Before building this 3′-end of the tRNAs a nuclease must remove the tail of the primary transcript at a 'discriminator" position, which may be the 73th base. Some prokaryotic tRNA transcripts already contain CCA ends and thus the tRNA nucleotidyl transferase is not indispensable yet advantageous because it may repair the amino acid acceptor. The CCA transfer enzyme is not choosy; it recognizes all tRNAs irrespective their amino acid specificity. Most of the U2 snRNAs in humans also carries CCA ends. ►tRNA, ►transfer RNA; Vasil'eva IA et al 2000 Biochemistry [Moscow] 65:1157; Cho HD et al 2002 J Biol Chem 277:3447.

tRNA-SE: is a fast computer program capable of identifying 99–100% of tRNA genes in DNA sequences with extremely low false positives. It may be applied also to the detection of unusual tRNA homologs such as selenocysteine tRNA, tRNA pseudogenes, etc.

Trophectoderm: An extra-embryonic tissue at the blastocyst stage of mammalian development. It is also the founder cell population of the chorionic cells of the placenta. Its function is nutrient exchange and protection of the embryo against the maternal immune system. It also signals to gastrulation. ►blastocyst, ►chorion, ►gastrula

Trophoblast: The surface cell layers of the blastocyst embryo connecting to the uterus.

Trophozoite: Growing and actively metabolizing cells of unicellular organisms vs. the cysts.

Trophophoresy: Propagating of species and disseminating them into new habitats by tranferring them as food. ►co-evolution

Tropic: Indicates after a word that something is aimed at it. e.g., T-tropic means that a virus targets T cells or directed at a site in some way.

Tropic Hormone: Stimulates the secretion of another hormone at another location.

Tropism: Growth of plants in the direction of some external factors.

Tropomodulin: Maintains actin filament growth by capping the pointed ends of the actin filaments. (See Littlefield R et al 2001 Nature Cell Biology 3:544).

Tropomyosin (TPM): is a skeletal muscle fiber protein with two chain α-helical coiled coil is essential for muscle filament stabilization and regulation of contraction. It bonds to actin. TPM3 is encoded at 1q22-q23. Endostatin binds tropomyosin. ►troponin, ►endostatin, ►actin; MacDonald NJ et al 2001 J Biol Chem 276:25190; Brow JH et al 2005 Proc Natl Acad Sci USA 102:18878.

Troponin: Ca^{2+}-binding, regulatory polypeptides in the muscle tissue. Troponin C binds four molecules of calcium. Troponin I has an inhibitory effect on myosin and actin, and troponin I binds tropomyosin, an accessory protein. In the relaxed muscles, troponin I binds to actin and moves tropomyosin to the position where actin and myosin would interact at muscle contraction. When the level of Ca^{2+} is high enough troponin I action is blocked so myosin can bind actin again allowing the muscle to contract. Troponin C is related in function to calmodulin. In the *nemaline* (thread-like) *myopathy* at 19q13.4 the sarcomeric thin-filament protein (TNNT1) is truncated. This recessive/dominant, infant-lethal disease has an incidence of ~0.002 in the Amish populations. ►calmodulin, ►signal transduction, ►myotonic dystrophy, ►receptor tyrosine kinase, ►myopathy, ►arthrogryposis; Johnston JJ et al 2000 Am J Hum Genet 67:814; Hinkle A, Tobacman LS 2003 J Biol Chem 278:506, structure: Takeda S et al 2003 Nature [Lond] 424:35.

Trospa: ►Lyme disease

TRP (transient receptor potential): plasma membrane ion channel components in control of active and passive Ca^{2+} stores and are activated by 1,4,5-trisphosphate receptors. ►phosphoinositides, ►SOC, ►ion channels; Montell C et al 2002 Cell 108:595; Clapham DE 2003 Nature [Lond] 426:517; Vanketechalam K, Montell C 2007 Annu Rev Biochem 76:387.

Trp: ►tryptophan operon, ►tryptophan

TRPC (transient receptor potential channel): Protein sensors for temperature, osmolarity, mechanical stress, taste and axon guidance (Wang GX, Poo M-m 2005 Nature [Lond] 434:898).

TRRP (transactivation/transformation-domain associated protein): is a member of the ATM protein superfamily and it is a co-factor of cMYC-mediated transformation. The yeast homolog (Tra1) is a component of the histone acetyltransferase (HAT), SAGA, PCAF and NuA4. Trrap is essential for normal development. (See terms mentioned; Herceg Z et al 2001 Nature Genet 29:206).

True Breeding: absence of segregation among the offspring.

Truncation: is a cut-off point; e.g., in artificial selection individuals before or beyond an ar-bitrarily determined point are discarded or maintained, respectively. Also a cloned eukaryotic-gene without a polyadenylation signal or an incomplete upstream control element may be called truncated. ►selection; Crow JF, Kimura M 1979 Proc Natl Acad Sci USA 76:306.

Truncation Selection: Individuals with more mutations are more likely to be removed from the population and individuals with fewer mutations may survive and maintained.

Trypanosomatids: are three major species of protozoa, *Trypanosoma brucei* (~genes 9,068), *T. cruzi* (~12,000) and *Leishmania major* (8,311) protozoa spread by the tse-tse fly (*Glossina*) and other bug bites and cause sleeping sickness, Chagas disease and other serious diseases of animals and humans. As the Venn diagram below shows that about two third of their genes are very closely related (see Fig. T120) (El-Sayed NM et al 2005 Science 309:404). The three species (Tritryp) have variable numbers of transposons; most of them are not intact, however. Various developmental stages are distinguished on the basis of the relative position of the flagella (basal flagella: trypomastigote, median: epimastigote, apical: promastigote, no flagella: amastigote). Flagellar motility of *Trypanosomes* in the blood stream is essential for survival and flagellar proteins may be targeted as a control measure of the parasites (see Fig. T121) (Broadhaed R et al 2006 Nature [Lond] 440:224). The *Trypanosomas* may reach a level of 10^9–10^{10} individuals per mL blood of mammals. The chromosomes are small and variable in number because in addition to the stable chromosomes mini-chromosomes are also found. *T. brucei* has only 11 pair larger chromosomes whereas *T. cruzi* and *L. major* have ~28 and 36 pairs, respectively. In addition they have various, undetermined numbers of minichromosomes. The ca. 100 minichromosomes (50 to 150 kb) contain open reading frames for the variable surface glycoproteins (VSG). These sequences are transcribed only when transposed to expression sites in the 0.2 to 6 megabase long 20 maxichromosomes. Since some of the genes may be present in more than single copy in different chromosomes, these organisms may resemble allodiploids. The VSG genes are pseudogenes, which generate mosaics by ectopic recombination. The genes do not have introns and some of the tandem arrays of genes are transcribed as long polycistronic pre-mRNA. At the 5′-end each primary transcript is capped by a 39-nucleotide, spliced leader RNA (SLRNA). This cap itself is transcribed as a 139 base sequence but the 100 base sequence is not used in this trans-splicing reaction. The *Trypanosomas* have homologs of the mammalian U2, U4 and U5 small nuclear RNAs in the form of ribonucleoprotein particles (RNP). The mRNAs are polyadenylated. In *T. brucei* there is a family of about 1,000 genes involved in the recurring production of a great repertory of *variant antigen type* (VAT). At each flare-up of division of the parasite, it switches on the production of a different type of antigen (serodeme). The more than million molecules of antigens on the surface of its cells are the phosphatidylinositol-anchored (about 60 kDa) *variable surface glycoproteins* (VSG). These proteins are linked to the membrane by glycosylphosphatidyl inositol (GPI) anchors. Analogs of these proteins may be used for therapeutic purposes (Smith TK et al 2004 EMBO J 23:4701). All the different VSGs have at least one N-linked oligosaccharide and several cysteine residues near the N-end, and some similarities within the 50–100 amino acids at the C-terminus. Because of the rapid switches in production to new antigens, the vertebrate cell's immunological surveillance system cannot adapt rapidly enough to contain the infection. In chronic infections 50–100 different antigens may be produced. Although a particular *Trypanosoma* has only 1/million or less chance per cell division to switch to the production of a different antigen yet because of their immense number, the parasite has a good chance to escape the immune system of the mammalian host. At any one time only one of the VSG genes is expressed in the protozoon. The switching is not a direct response to the host antibody rather that acts

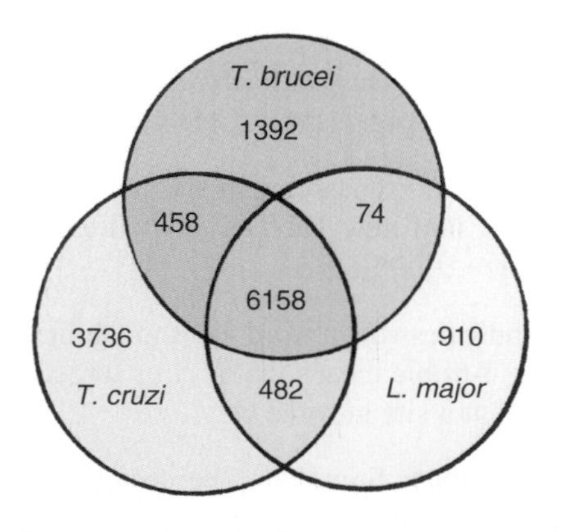

Figure T120. Clusters of orthologous genes

Figure T121. Trypanosoma

only as a selecting mechanism. The switching of transcription may require a shut-off of a gene or a gene conversion type process takes place. The expressed surface antigen gene, the so-called *expression-linked copy* (ELC), is always located in the telomeric regions of the chromosomes. Its promoter is located, however, 50 kbp upstream. In addition, non-VSG genes are also located in the expression region; these are called *expression site associated genes* (ESAG). The expression of these basic, silent genes requires a transposition into the activation region, in a manner similar to the mating type switching in yeast. The switching apparently depends on expression-linked copies (ELC) of the mini exon dependent transcription into 140-base long eukaryotic type mRNA (called also medRNA). Because of this effective switching, there are serious problems in developing vaccines against the *Trypanosomas*. *Trypanosomas* may also display genetic recombination, duplications, deletions and uniparental inheritance mediated by the kinetoplast (Gaunt MW et al 2003 Nature [Lond] 421:936). Trypanosomas practice RNA editing by a high molecular mass RNP complex, which contains a 42 kDa protein that recognizes both dsDNA, dsRNA but not the DNA/RNA hybrid molecules. This protein is an endo- and exoribonuclease and functions in also in the mitochondria (Brecht M et al 2005 Mol Cell 17:621). Mutants can be produced that turn on simultaneously more than one type surface glycoproteins. *Trypanosomas* take up transferrin (a β-globulin) from the host cells through about 20 homo- or heterodimeric transferrin receptors, which are activated alternatively just as the VSG genes. The different transferrin receptors make possible for the parasite to adapt to different hosts. The TbAT1 nucleoside transporter confers melaminophenyl arsenical susceptibility whereas defects in TbAT1 render the cells resistant to the trypanocide. ►kinetosome, ►*Leishmania*, ►mating type determination in yeast, ►flagellar antigen, ►intron, ►transporters, ►RNA editing, ►Chagas disease, ►Apolipoproteins; Coppens I, Courtoy PJ 2000 Annu Rev Microbiol 54:129; Navarro M, Gull K 2001 Nature [Lond] 414:759; Spadiliero B et al 2002 J Cellular Biochem 85:798; Beverly SM 2003 Nature Rev Genet 4:11; *T. cruzi* sequenced genome: El-Sayed NM et al 2005 Science 309:409; *T. brucei* genome sequence: Berriman M et al 2005 Science 309:416, *T. cruzi* proteome: Atwood JA III et al 2005 Science 309:473; http://www.tigr.org/tdb/tgi.shtml; *T. cruzi* database: http://tcruzidb.org/.

Tryphine (pollenkitt): Lipids and proteins filling the depressions of the pollen surface.

Trypomastigote: ►*Trypanosoma*

Trypsin: A proteolytic enzyme synthesized as an enzymatically inactive zymogen, trypsinogen that is activated by proteolytic cleavage. The 6,000 Mr pancreatic trypsin inhibitor inactivates it. Trypsin specifically cleaves polypeptides at the carbonyl sides of Lys and Arg. Ser 195 and His 57 are at its active site.

Trypsinogen Deficiency (TRY1): is a 7q22-ter recessive hypoproteinemia and insufficient amino acid level. It is very similar to enterokinase deficiency. ►enterokinase deficiency

Tryptases: in the form of heparin-stabilized tetramers function similarly to trypsin-like serine proteases mainly in the mast cells. ►mast cell, ►trypsin, ►heparin

Tryptic Peptides: The products of digestion of a protein by trypsin. ►trypsin

Tryptophan: Is an essential aromatic amino acid (MW 204.22) (see Fig. T122), soluble in dilute alkali, insoluble in acids and it is degraded when heated in acids. Its biosynthetic pathway (*with enzymes involved in parenthesis*): Chorismate- > (*anthranilate synthase*) - > Anthranilate - > (*anthranilate-phosphoribosyltransferase*) - > *N*-(5′-Phosphoribosyl)-anthranilate - > (*N-(5′-phosphoribosyl)-anthranilate isomerase*) - > Enol-1-*o*-carboxyphenylamino-1-deoxyribulose phosphate - > (*indole-3-glycerol phosphate synthase*) - > Indole-3-glycerol phosphate - > Indole - > (*tryptophan synthase*) - > Tryptophan. In *E. coli* the first two enzymes shown constitute a single anthranilate synthase complex. Tryptophan is converted to formylkynurenine by *tryptophan dioxygenase* (tryptophan pyrrolase). Through the action of an *aminotransferase* tryptophan gives rise to indole-3-pyruvate which after *decarboxylation* forms indole-3-acetic acid, one of the most important plant hormone (auxin). Tryptophan and phenylalanine contribute to the formation of lignins, tannins, alkaloids (morphine), cinnamon oil, cloves, vanilla, nutmeg, etc., flavors. ►chorismate, ►tyrosine, ►phenylalanine, ►melanin, ►plant hormones, ►pigments in animals, ►Hartnup disease, ►attenuator, ►*tryptophan* operon, ►fragrances

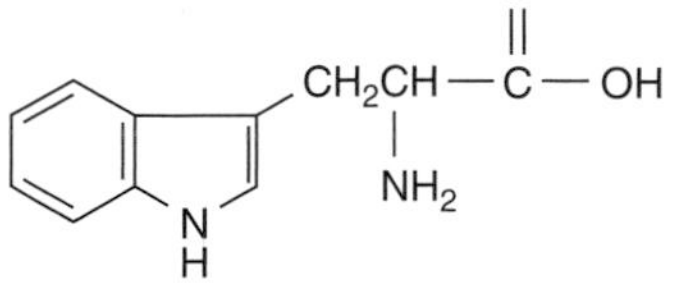

Figure T122. Tryptophan

***Tryptophan* Operon**: Contains five structural genes, in *E. coli* they have been mapped (at 27 min) in exactly

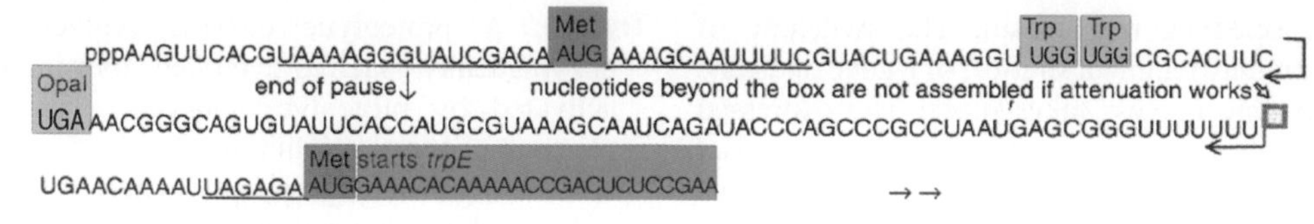

Figure T123. Part of the bacterial trplophan operon

the same order as their sequence of action in the biosynthetic path (►tryptophan). This operon has a principal promoter and a secondary low-efficiency one. Between the operator and the proximal gene there is a 162 bp leader sequence (*trpL*) including an attenuator site *a*. (►temperature sensitive controling sequences of transcription). When there is sufficient amount of tryptophan in the cell and all the $tRNA^{Trp}$ are charged with the amino acids, transcription of the leader sequence stops at base 140. Thus, synthesis of the specific mRNA temporarily ceases when there is no immediate need for tryptophan (see Fig. T123).

The site of attenuation (*a*) is within the *trpL* (tryptophan leader) sequences. Before attenuation becomes effective, the RNA polymerase pauses at the *tp* site in the *trpL* (►attenuation). Two main mechanisms, repression and attenuation regulate expression of this operon. Repression prevents the initiation of transcription. The repressor is transcribed from the *trpR* gene (located at 100 min) and the $tRNA^{Trp}$ is coded by gene *trpT* (84 min) whereas the aminoacylation of this tRNA is determined by gene *trpS* (74 min). The product of *trpR* is an aporepressor, i.e., it becomes active only when combined with tryptophan its corepressor (see tryptophan repressor).

The primary sequence of the leader sequence is as shown below (the underlined sequences indicate ribosome binding, tryptophan codons and critical codons in outline); see above. The tryptophan repressor alone may reduce transcription by a factor of 70 and attenuation may decrease it 8 to 10 fold but by the combination of these two controls transcription may be reduced by 8 × 70 (=560) or 10 × 70 (=700) fold. In *Rhodobacterium sphaeroides* the tryptophan operon is shared between the two chromosomes of this bacterium.

Bacillus subtilis regulates seven genes of tryptophan biosynthesis; six of them (*TrpEDCFBA*) are clustered within a 12-gene aromatic supraoperon. The seventh gene (*TrpG*) is unlinked and it is in the folate operon. TRAP regulates by attenuation as it binds to a specific site in leader sequence of the *trp* operon and facilitates a terminator formation of the RNA transcript. In addition, TRAP binds to the ribosome-binding sequence in the *trpG* mRNA and thus inhibits translation. An anti-TRAP (AT) protein signals to the $tRNA^{Trp}$ and then it is not charged with tryptophan. The uncharged tRNA induces the synthesis of AT. Linking AT to TRAP prevents TRAP binding to the RNA. As a consequence tryptophan biosynthesis is promoted (Valbuzzi A, Yanofsky C 2001 Science 293:2057; Gutierrez-Preciado A et al 2005 Trends Genet 21:432).

In *Neurospora* there are also 5 distinct genetic loci controling tryptophan biosynthesis. These genes are derepressed coordinately with histidine, arginine and lysine loci and this phenomenon is called *cross-pathway regulation*. Humans do not have tryptophan synthetic genes and depend on it (essential amino acid) in the food. ►tryptophan, ►helix-turn-helix, ►*lac* operon, ►repressor, ►tryptophan repressor, ►essential amino acids, ►TRAP, ►antitermination, ►attenuation; Yanofsky C 1981 Nature [Lond] 289:271; *Bacillus subtilis*: Gollnick P et al 2005 Annu Rev Genet 39:47.

Tryptophan Repressor: Consists of a repressor protein (using a helix-turn-helix motif) for binding to the operator site of the tryptophan operon and becomes active only when it is associated also with tryptophan. Actually the binding of tryptophan to the repressor protein facilitates the more intense binding of the complex due to a conformational change of the repressor protein. In prokaryotes besides the repressor attenuation is another regulatory mechanism. *E. coli* and *Bacillus subtilis* use somewhat different mechanisms of control. In *E. coli* tryptophan activates a repressor, which binds to the promoter–operator region and inhibits the initiation of transcription. In *B. subtilis* tryptophan activates the RNA-binding protein TRAP. TRA bound to the leader sequences terminates transcription. In *E. coli* uncharged $tRNA^{Trp}$ accumulation stalls when tries to translate the two Trp codons in the leader-peptide coding region. This causes transcription termination. In contrast *in B. subtilis* uncharged $tRNA^{Trp}$ actually activates transcription and translation of this operon. The antitermination protein AT then inhibits TRAP, activated by tryptophan. ►negative control, ►*tryptophan* operon, ►attenuator region, ►*Lac* operon, ►positive control, ►*arabinose* operon, ►helix-turn-helix motif, ►conformation; Khodursky AB et al 2000 Proc Natl Acad Sci USA 97:12170; Yanofsky C 2004 Trends Genet 20:367.

T

Tryptophan Zipper: is a structural motif that can stabilize the hairpin structure of short peptide chains of 12 to 16 amino acids with the aid of cross-strand pairs of indole rings. It does not require metal or disulphide links.

Tryptophanase (*tna*): is a bacterial operon including major genes A, B and a permease with a 319 bp leader encoded (by tnaC) preceding gene tnaA. It degrades L-tryptophan to indole, pyruvate and ammonia. Tryptophan regulates it by induction and antitermination. ►induction, ►antitermination; Gong F et al 2001 Proc Natl Acad Sci USA 98:8997.

Tryptophanyl tRNA Synthase (WARS): charges $tRNA^{Trp}$ by the amino acid tryptophan. The gene WARS is in human chromosome 14. ►aminoacyl-tRNA synthetase

ts: indicates temperature-sensitivity of an allele. ►temperature-sensitive mutation

T_S: ►T cell

Tsarevitch Alexis: great-grandson of Queen Victoria who inherited her new mutation causing classic hemophilia. ►hemophilias, ►Queen Victoria

tsBN2: is a temperature-sensitive baby hamster kidney (BHK) cell line. ►hamster

TSC: is a reduced representation shotgun sequencing data set. ►shotgun sequencing

TSE (transmissible spongiform encephalopathy): ►encephalopathy, ►prion

Tse-Tse Fly: African fly of the genus *Glossina*, host of the parasitic *Trypanosomas*, causing sleeping sickness, a disease characterized by relapsing fever, enlargement of the lymph glands, anemia, severe emaciation and eventually death in humans and domestic animals. ►*Trypanosomas*; Akman L et al 2002 Nature Genet 32:402.

TSG (tumor susceptibility gene): Acts as transcriptional cofactor and also as a nuclear hormone receptor-mediated transactivation. TSG101 protein regulates cell growth and division, MDM2 protein and affects ubiquitination. It controls also epidermal growth factor receptor trafficking by interacting with the endosomal transport. There is ~94% homology between the mouse and the human proteins. ►MDM2, ►ubiquitin, ►EGFR, ►endocytosis; Teh BT et al 1999 Anticancer Res 19(6A):4715; Lu Q et al 2003 Proc Natl Acad Sci USA 100:7626.

TSHB: thyroid-stimulating hormone, thyrotropin. ►animal hormones

Tsix: ►Xist

Tsr: Bacterial transducer protein recognizing serine as an attractant and leucine as a repellent. ►transducer proteins

TSS: Transcription start site. (See http://dbtss.hgc.jp/).

T-Strand: A single-stranded intermediate of the T-DNA that is transferred from the Ti plasmid of *Agrobacterium* to the plant nucleus through the nuclear pores under the guidance of a virulence gene-encoded protein which is covalently attached to its 5′-end. ►transformation, ►T-DNA, ►Ti plasmid

tTA (tetracycline transactivator protein): is a fusion protein of the *tet* (tetracycline) repressor of *E. coli* and the transcriptional activation domain VP16 of herpes simplex virus. This system is generally driven by the *tetP* promoter which is actually a minimal immediate early cytomegalovirus (CMVIE) promoter, preceded by 7 copies of *tetO*, the tetracycline resistance operator of transposon 10 (Tn*10*). In the presence of tetracycline, this system is expressed at very low level and by removal of tetracycline the gene(s) under its control (e.g., luciferase, β-galactosidase) may be expressed at three orders of magnitude higher level. The system can be used also under the control of other promoters that suit best for regulating the expression pattern of the gene of interest. ►rtTA, ►tetracycline, ►split-hybrid system; Gossen M, Bujard H 1995 Science 268:1766.

t-Test: is used for the estimation of the statistical significance of the difference(s) between means. The *t* value is the ratio of the observed difference to the corresponding standard error: $t = (\bar{x} - m)/(s/\sqrt{n})$ where $\bar{x}$ and *m* are the two means, *n* = population size, *s* = standard deviation. More commonly the significance of the difference between two means is calculated as $t = \frac{m_1 - m_2}{\sqrt{[e_1]^2 - [e_2]^2}}$ where m_1 and m_2 are the two means ($\bar{x}$) and e_1 e_2 are the standard errors of the two means determined as $e = \frac{s}{\sqrt{n}}$ and $s = \sqrt{V}$ where $V = \text{variance} = \frac{\sum[(x-\bar{x})^2]}{n-1}$. When the *t* value is available the probability of the difference is determined with the aid of a "t table". ►Student's t distribution

TTKs: are tubulin-associated kinases. ►tubulin

TTP (tris-tetrapolin): cDNA shares 102 amino acid sequences in its product with TIS11 (insulin and serum-responsive transcription factor). ►TIS11, ►insulin, ►serum response element, ►transcription factors

T-Tropic: The T lymphocyte is targeted (e.g., by a virus).

TU: Protein elongation factor (EF-TU) in prokaryotes binds aminoacyl-tRNA to the ribosomal A (acceptor) site. It is a guanine nucleotide-binding RAS-like protein. EF-TU.GTP, ►aminoacyl-tRNA, ►ribosome, ►protein synthesis, ►RAS; Zvereva MI et al 2001 J Biol Chem 276:47702.

Tubal Ligation: Surgical fertility control by constricting of the Fallopian tube, usually by placing a plastic ring on it. ►sterilization humans, ►salpingectomy

Tube Nucleus: It is in the vegetative cell at the tip of the growing pollen tube. It has only physiological and no genetic role because it does not enter the embryosac. ►gametophyte

Tuber: An underground enlarged stem, specialized for food storage, e.g., in potato.

Tubercidine (7-deazadenosine): ►deazanucleotides

Tuberculosis: ►mycobacteria

Tuberous Sclerosis: ►epiloia

Tuboplasty: Surgical repair of a defect on an internal tube such as the Fallopian tube.

Tubulins: Globular polypeptides of α and β subunits (50 kDa, each with about 40% homology and very similar in structure [β sheets surrounded by α helices]) are G-proteins concerned with signal transduction of nerve cells and are components of microtubules such as the spindle fibers and the cytoskeleton. Mutations in α-tubulin cause abnormal migration of neurons in mice and lissencephaly in humans (Keays DA et al 2007 Cell 128:45). Both subunits can bind one guanine nucleotide, which is an exchange cable on the β but not on the α binding site. The folding of the tubulins requires cytosolic chaperonins, cofactors A, C, D and E, ATP and GTP. The bacterial plastid septation protein FtsZ filamenting temperature-sensitive septal peptidoglycan (Z ring) displays structures similar to tubulins and participates in the septation of the cell (see Fig. T124). FtsZ may have a GGGTGTG motif and has GTPase activity. FtsZ3 with the GGGAGTG motif has no and FtsZ84 with the AGGTGTG sequence have reduced GTPase activity. FtsZ recruits then addition cell division proteins. In the spore-forming bacteria (*B. subtilis*) initially two Z rings are formed, one at both poles but only one is activated as the genetic material passes to the spore. In the chloroplasts a nuclear encoded ftsZ protein and dynamin function in fission (Yoshida Y et al 2006 Science 313:1435). The mitochondria apparently do not need ftsZ. ►spindle, ►cytoskeleton, ►chaperonins, ►peptidoglycan, ►septum, ►GTPase, ►chloroplasts, ►mitochondria, ►FtsZ, ►lissencephaly; Oakley BR 2000 Trends Cell Biol 10:537; Thanbichler M, Shapiro L 2006 Cell 126:147; Z ring regulation: Lutkenhaus J 2007 Annu Rev Biochem 76:539.

Figure T124. Septation by bacterial Z ring

Tudor (*tud*, 2–97): *Drosophila* gene locus involved in embryonic development. ►TRF1

Tudor-SN: Protein containing the Tudor protein of *Drosophila* and five staphylococcal nuclease domains. It is a component of the RISC silencer complex. ►RISC; Maurer-Stroh S et al 2003 Trends Biochem Sci 28:69.

TUF (transcripts of unknown function): The number of RNA transcripts (from intergenic regions, introns and antisense DNA strands) far exceeds the number of protein-coding mRNAs (Johnson JM et al 2005 Trends Genet 21:903; Willingham AT, Gingeras TR 2006 Cell 125:1215). They are parts of the operational system of the genome. ►non-coding RNA, ►antisense RNA, ►RNAi, ►microRNA

Tukey's Test: is an analysis of variance test for determining whether there is a significant difference between three or more group means, two comparisons made at a time. This is a non-additivity test.

$$S_{AB} = \frac{\left[\sum_{i=1}^{r}\sum_{j=1}^{c}\{y_{ij}(\bar{y}_i - \bar{y}_{..})(\bar{y}_j - \bar{y}_{..})\}\right]^2}{S_A S_B}$$

Where the sum of squares is S_{AB}, r is the number of rows, c = number of columns, y_{ij} = observations in the ij^{th} cell, $\bar{y}_i$ = mean of the ith row, $\bar{y}_{.j}$ = mean of the jth column and $\bar{y}_{..}$ = mean of all observations. (See Tukey JW 1949 Biometrics 5:232; ►analysis of variance).

Tularemia: is an extremely dangerous infection causing sudden fever, weakness, bodyache in humans and animals. It is caused by as few as 10 cells of the bacterium *Francisella tularensis* transmitted by arthropod vectors. It may also spread by airborne dust. Without antibiotic treatment the mortality is 5–30%. Currently no vaccine is available. The sequenced genome of the bacterium is 1,892,819 bp. ►biological warfare, ►inflammasome; Larsson P et al 2005 Nature Genet 37:153.

Tulip Mania: ►symbionts hereditary, ►tulips broken

Tulips, Broken: Variegation caused by infection with the tulip-breaking virus. Floriculturists value the plants displaying this sectoring. ►tulip mania, ►broken tulips

Tulips-PCR (touch up and loop incorporated primers PCR): See Ailenberg M, Silverman M 1999 Biotechniques 29:1018.

Tumbling: results when the bacterial flagella (singular: flagellum) rotate clockwise.

TU/ml: Transforming unit per mL, i.e., the number of cells expressing a transgene. ►transgene, ►transformation genetic

Tumor: An abnormal clump of cells originated by benign or malignant growth. Malignant tumors may be invasive and show metastasis, common in many types of cancer. Mutant genes, chromosome breakage and viral infections altering the normal regulation of cell proliferation may cause it (see Fig. T125). Tumorigenesis is generally a multi-phase process involving activation of cell cycle promoting genes and inactivation of tumor suppressors. Active MAPKK may stimulate tumorigenesis if transfected into mouse cells. The type of the tumor is different in different cancers and even within a single cancer may vary a great deal because of the frequent chromosomal breakage and other mutations as a consequence of cancerous growth. The origin and maintenance of the tumorous condition requires the emergence of special tumor stem cells brought about by asymmetric division of normal somatic cells (Gonzalez C 2007 Nature Rev Genet 8:462). Their characterization, using molecular markers may be facilitated by microarray analysis of the biopsies. Plant tumorous growth is often called callus that is never metastatic although the crown gall tumors of plants may be spread by new foci of infection of the inciting bacterium, *Agrobacterium tumefaciens*. Plant viruses and higher concentrations of phytohormones may also cause tumors. ►oncogenes, ►cancer, ►SV40, ►adenoviruses, ►*Agrobacterium*, ►genetic tumors, ►tumor in situ, ►cancer, ►MAPKK, ►habituation, ►CATR1, ►microarray hybridization, ►cancer gene therapy, ►RAS, ►PIK, ►TOR; Shaw RJ, Cantley LC 2006 Nature [Lond] 441:425.

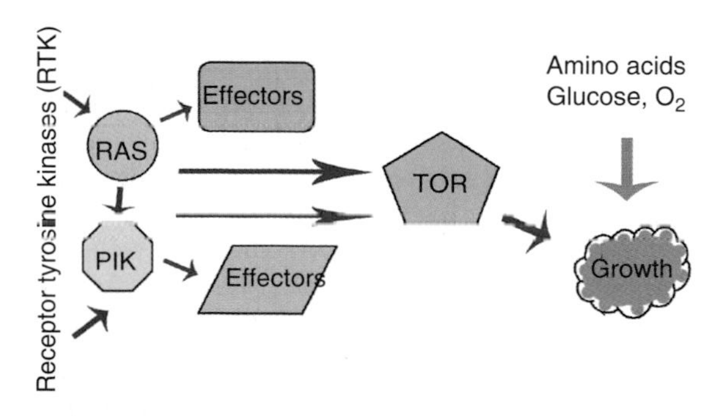

Figure T125. General growth regulatory pathways

Tumor Antigens: are MHC-associated peptides recognized by CTLs as tumor antigens. The MHC class I tumor peptides are also called CTL epitopes. Mutant chaperone can convert a glycosyltransferase into a tumor-specific glycopeptidic neo-epitope (Schietinger A et al 2006 Science 314:304). ►MHC, ►CTL, ►epitope, ►glycosyltransferases, ►dendritic cell vaccine, ►tumor vaccination, ►cancer

Tumor, In Situ: Dormant, tiny clumps of tumor cells, usually revealed by autopsies of breast, prostate and thyroid glands. The microscopic cells may never develop into cancer because the lack of appropriate blood supply (angiogenesis). The in situ tumors may originate by genetic instability but unless they succeed in recruiting new blood vessels, they are unable to proliferate into neoplasias. There is a balance between the angiogenic growth factors such as FGF, VEGF, IL-8, and PDGF and the angiogenic inhibitors of the body (thrombospondin, canstatin [24 kDA human basement membrane-derived antiangiogenesis protein], tumstatin [cleavage fragment of the α-3-chain of collagen type IV that prevents blood circulation], endostatin, angiostatin, interferon α/β. Genetical and environmental factors control the balance among the pro- and anti-angiogenic factors. (See terms at corresponding entries, ►tumor).

Tumor Infiltrating Lymphocytes (TIL): seek up tumors. TIL are isolated from solid tumors and cultured in single-cell suspension in a medium containing interleukin 2 (IL-2). Through genetic engineering they may be equipped with the gene of tumor necrosis factor through transformation by retroviral or other vectors. The transformed cells may selectively kill then cancerous tumor cells. ►cancer gene therapy, ►tumor necrosis factor, ►retroviral vectors, ►lymphocytes; Smyrk TC et al 2001 Cancer 91:2417.

Tumor Necrosis Factor: ►TNF

Tumor Necrosis Factor Receptor: ►TNFR

Tumor Progression: ►evolutionary clock

Tumor Promoter: ►phorbol ester

Tumor Suppressor Gene: Its loss, inactivation or mutation (even haploinsufficiency) permits neoplastic growth by deregulation. They most commonly have a role in the cell cycle or in the regulation of RNA polymerase II or III. Also, inhibition of peptide chain elongation may be a mechanism of tumor suppression. (Genes with cytostatic or cytotoxic effects are excluded from this category of tumor suppressors). Actually, the majority of cancer cells display deletions that may indicate the loss of a tumor suppressor gene; for the direct proof further evidence is required for the loss of a tumor suppressor (see Fig. T126). Animal models are available for many human tumor suppressors (Hakem R, Mak TW 2001 Annu Rev Genet 35:149). Tumor suppressor genes

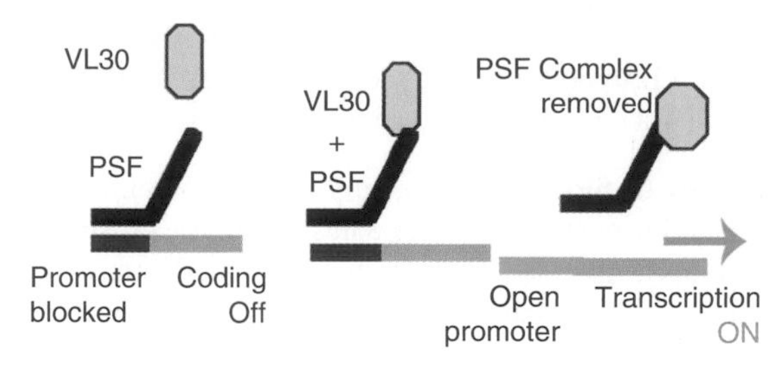

Figure T126. A tumor suppressor path

are widely scattered in the human genome: p53/TP53 (17p13.1), retinoblastoma (RB1, 13q14.1-q14.2), adenomatous polyposis of colon (APC, 5q21-q22), deleted in colorectal cancer (DCC, 18q21.3), neurofibromatosis (NF1, 17q11.2), von Hippel-Lindau syndrome (VLH, 3p26-p25), Wilms tumor (WT1, 11p13), breast cancer (BRCA1, 17q21), Cyclin-dependent kinase inhibitor CDKN2A, (9p21), patched homolog PTCH (9q22.3), tuberous sclerosis (TSC2, 16p13.3; TSC1, 9q34), etc. Mutation in the polycomb group protein enhancer of zeste homolog (EZH2) may increase its expression and represses the function of many genes that are apparent suppressors of tumorigenesis and thus promotes lethal metastasis of prostate cancer (Varambally S et al 2002 Nature [Lond] 419:624). The methyltransferase Suv39h1 normally methylates histone H3 lysine 9 and may silence growth-promoting genes. By the use of RNAi, tumor suppressor genes can be inactivated and the procedure permits their identification (Westbook TF et al 2005 Cell 121:837; Kolfschoten IG et al 2005 Cell 121:849). Ras transgenic mice carrying intact, X-linked, functional Suv39h1 developed lymphomas only at a median of 305 days. Suv39h1 deficient homozygotes (female) or hemizygotes (male) succumbed to death at a median time of 66 days. The methyltransferase acted as a tumor suppressor by promoting senescence (Braig M et al 2005 Nature [Lond] 436:660). The protein spliceosome factor (PSF) has an RNA and DNA binding domain and can bind to the P450ssc gene, which initiates also the steroidogenic pathway. The retrotransposon VL30 – a multiple element in the mouse genome and also in the human genome–regulates the function of PSF (Song X, Garen A 2005 Proc Natl Acad Sci USA 102:12189) by binding its RNA transcripts to the RNA binding domain (RBD) of PSF. Subsequently PSF is removed from the promoter and the oncogene is transcribed (See diagram redrawn and modified after Desisseroth A 2005 Proc Natl Acad Sci USA 102:12292). ►cancer, ►malignant growth, ►transformation oncogenic, ►p53, ►p73, ►p16, ►p16^{INK4} p21, ►retinoblastoma [Rb], ►ELL, ►elongin, ►DPC4, ►PTEN, ►PPP2R1B, ►pol III, ►tumor suppressor factors, ►RNAi, ►polycystic kidney disease, ►breast cancer, ►prostate cancer, ►oncogenes, ►cell cycle, ►apoptosis, ►colorectal cancer, ►tumor suppressor factors, ►LOH, ►oncogenes, ►immunosuppression, ►histone deacetylase, ►gatekeeper genes, ►caretaker genes, ►hypermethylation; Robertson GP et al 1999 Mol Cell Biol Res Comm 2:1.

Tumor Suppressor Factors: The development of tumors follows multiple routes yet genes involved in the general control of differentiation, as revealed by studies of *Drosophila* morphogenesis, seem to be involved. Mammalian homologs of *Drosophila hedgehog* (*hh*), sonic hedgehog (SHH), Indian hedgehog (IHH) and desert hedgehog (DHH) seem to be entailed in holoprosencephaly, a developmental anomaly connected to cancer. The receptor of *hedgehog* of *Drosophila* is *smoothened*, and its suppressor is *patched* which may be concerned with the development of basal cell carcinoma and defects in the central nervous and skeletal systems. The *cubitus interruptus* (*ci*) *Drosophila* gene appears to function as an effector of *hh* and similarly to the GLI genes of humans are responsible for a type of brain tumor and for the cephalopolysyndactyly syndrome. The decapentaplegic (*dpp*) *Drosophila* protein bears si-milarity to the mammalian tumor growth factor (TGF) and to the bone morphogenetic protein 4 (BMP4), causing defects in limb and gut formation. The effector of the mammalian TGF-β receptor (DPC4) involved in pancreatic tumors has its homolog in the *Drosophila mad* gene. The *Drosophila wingless* locus corresponds to the mammalian WNT loci controling mammary tumors. The *Drosophila zeste-white-3* gene codes for a signal molecule similar to a mammalian glycogen synthase kinase, and the *Drosophila armadillo* gene, encoding β-catenin, is also an oncoprotein, controling intestinal tumors. Naïve $CD4^+$ T cells respond to MHC-II tumor antigenic signal carried by macrophages and along with interferon-γ inhibit tumor cell growth (Corthay A et al 2005 Immunity 22:371). Fe chelators have potent and broad antitumor activity and can overcome resistance to established chemotherapeutics because of their unique mechanism of action (Whitnall M et al 2006 Proc Natl Acad Sci USA 103:14901). (Based to some extent on Dean M 1996 Nature Genetics 14:245, ►immunological surveillance, ►Tid50, see terms mentioned in the alphabetical list).

Tumor Susceptibility: May be determined by mutation of some major tumor suppressor (RAS, CyclinD1,

RB, pt3, p16, etc.) genes, favorable combination(s) of minor genes may be responsible for the "sporadic" cases of cancer. ►tumor suppressor gene, ►oncogenes; for a list of mouse susceptibility loci see Balmain A, Nagase H 1998 Trends Genet 14:139.

Tumor Vaccination: Immunization by increasing the efficiency of tumor-specific antigen presentation or enhancing the activity of tumor-infiltrating T cells. The purpose is to generate antitumor immune reaction. Cancer cells may be ex vivo genetically modified and reintroduced into the body. Introducing into the cancer cells a range of cytokines e.g., IL-12 may enhance T cell response. The T cells of the tumor can be genetically modified to secrete effector molecules to enhance the immune response against the tumor or increase their potentials for binding of tumor antigens. Introduction of the wild type tumor-suppressor genes may also be an option. Transformation by GM-CSF may recruit monocytes/macrophages and APCs by a process called *cross priming*. Some tumors may mask the immunogenic potentials within the cells. In such cases, antisense technologies (against IGF or TGF or IL-10), intrabodies, triple helix formation, and inactivation by the use of ribozymes may help in silencing the endogenous inhibitors. It is also possible to introduce into the tumor activation enzymes that can convert pro-drugs into highly cytotoxic anticancer compounds or transfect the cells with toxin genes.

Because of pleiotropy and protein interactions, the genetic modifications may not always be favorable for goals. Within the same individual, molecular heterogeneity may exist among the cancer cells. Therefore, only multivalent cancer vaccines may achieve relevant goals. Favorable results were obtained in animal models with the combinations of IL-4 + IL-12, co-stimulatory molecules + IL-12, interferon-γ + IL-2, etc. Often, tumor cells have no or very weak antigen-presenting mechanism(s), and do not express the co-stimulatory proteins. Melanoma cells over-produce IL-10, a cytokine that down-regulates the path of CTL formation. The Fas-ligand expressing lymphocytes may be the targets of the up-regulated Fas, produced by tumor cells, especially when exposed to certain types of chemotherapy. As a consequence, the apoptotic process may weaken immunotherapy against cancer. Cancer may involve tumor-specific immunodeficiency but low doses of IL-2, administered over long periods may strengthen the immune system. It is essential that in therapy against cancer the various treatments be effective against cancer cells without attacking the normal cells and balance the antitumor and autoimmune responses. ►cancer gene therapy, ►tumor infiltrating lymphocytes, ►CTL, ►antigen presentation, ►immune system, ►interferon, ►vaccines, ►dendritic cell vaccine, ►cross-priming, ►GM-CSF, ►co-stimulator, ►antisense RNA, ►intrabody, ►triple helix formation, ►ribozyme, ►IGF, ►TGF, ►IL-10, ►lipids cationic, ►co-stimulator, ►LAK, ►Fas, ►apoptosis, ►autoimmune disease, ►effector cell; Gunzer M, Grabbe S 2001 Crit Rev Immunol 21 (1–3):133; Tada Y et al 2003 Cancer Gene Ther 10:134.

Tumor Viruses: Tumor viruses induce or participate in the formation of tumors (cancer). Their gentic material can be DNA (such as, SV40, adenoviruses, papilloma virus, hepatitis-B virus, Epstein-Barr virus, adenoma virus, herpes virus, pox virus) or RNA (such as the retroviruses causing leukemia, lymphoma, AIDS, Kaposi's sarcoma, avian leukosis virus, mammary tumor viruses, etc). The DNA viruses can integrate into the mammalian genetic material and activate cell replication by overwhelming the function of the tumor suppressor genes. The genetic material of the DNA tumor virus has no counterpart in the host genetic material. The genetic material of the RNA tumor virusis replicated by reverse transcription and produces a double-stranded DNA counterpart of the genome. The viral RNA is then transcribed from the cellular DNA template. Most of the oncogenes in the cells (c-oncogenes) correspond to the v-oncogenes in the virus. The class I RNA tumor viruses themselves do not induce tumors unless they pick up growth-regulating genes from cell. The class II type RNA viruses do not contain oncogenes and induce cancer only when the proviral DNA integrates in the vicinity of cellular oncogenes (c-oncogenes). ►SV40, ►adenovirus, ►Epstein-Barr virus, ►papilloma virus, ►polyoma, ►Rous sarcoma, ►retroviruses, ►oncogenes; Barbanti-Brodano G et al (Eds.) 1995 DNA Tumor Viruses, Plenum, New York; McCance DJ (Ed.) 1998 Human Tumor Viruses. AMS Press, Washington, DC.

Tumor-Associated Antigen: Over-expressed normal self-proteins of cancer cells. They may incite immune reaction by breaking the self-tolerance limit. Examples: oncofetal differentiation, nuclear proteins, such as carcinoembryonic, melanoma-associated proteins, etc. ►MAGE, ►HER2

Tumorigenesis: The formation of tumors. It is usually based either on the activation of proto-oncogenes or the inactivation of tumor suppressor genes. The former mechanism generally involves dominant-acting genes, the latter is usually based on recessive loss of function. Before genetic instability, characteristic for cancer or cancer-prone cells appears, an ATR/ATM-regulated DNA damage control system may be activated. Mutations in ATM-Chk2, p53 pathway might allow, however cell proliferation, cell survival, increased chromosomal instability, and eventually progression to cancer (Bartkova J et al 2005 Nature [Lond] 434:864). Fourier-transform infrared spectra (FT-IR) reveal structural modifications at many points in the DNA and marked differences between the primary and metastatic states. ►cancer, ►neoplasia, ►FT-IR, ►telomerase, ►tumor suppressor, ►proto-oncogenes, ►chromosome breakage, ►apoptosis, ►DNA repair, ►radiation effects, ►environmental mutagens, ►carcinogenesis, ►ATM, ►ATR, ►CHK2, ►p53

Tumorous Hybrids: ►genetic tumors

Tumor-Specific Antigen: Endogenous tumor cell-surface antigens that can be presented by major histocompatibility molecules to T cells. These antigens are absent from normal cells and are modified in response of viral transformation or genetic or somatic mutations in oncogenes. ►MHC, ►T cell, ►antigen presenting cell

TUNEL ASSAY (terminal deoxytransferase-mediated deoxyuridine nick end-labeling): Tunnel essay uses usually biotin-labeled uridine and a streptavidin or avidin-labeled enzyme, and at the reaction site color develops. ►biotinylation, ►terminal nucleotidyl transferase; Maciorowski Z et al 2001 Cytometry 46(3):159; Yamamoto-Fukud T et al 2000 Histochemistry J 32(11):697.

Tunica: The cell layer in the plant apical meristem wrapping the inner corpus. (See Fig. T127).

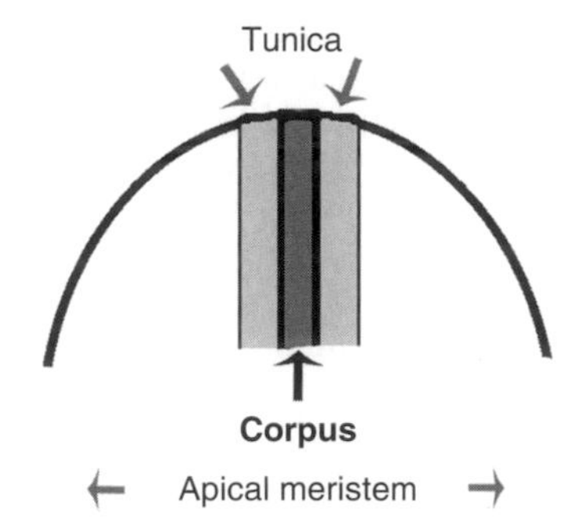

Figure T127. Tunica

Tunneling: connecting through a structured path. ►channeling

TUP1: A general suppressor of sugar (and other) metabolism of yeast; its product is a trimeric G-protein. It is similar to AAR1 and AAR2. TUP forms a repressor complex with yeast protein Cyc8, RNA polymerase II, and Srb10. ►proteins, ►glucose effect; Groucho, Tel, SRB, Wu J et al 2001 Mol Cell 7:117.

Tupaia: A family of prosimii. *Tupaia*, 2n = 62; *Tupaia glis*, 2n = 60; *Tupaia montan*, 2n = 68. ►prosimii

Turbid Plaque: ►plaque

Turbidity: Turbidity is commonly used for measuring total cell numbers in liquid cultures as long as the cell density is not excessive. It is measured as optical density (OD) = x*l*, where x = cell density and *l* = the light path in the cuvette of the spectrophotometer. Turbidity is expressed also as log (I_0/I) where I_0 = incident light, I = transmitted light (at e.g., 550 nm wavelength). The number of viable cells is determined by plating. ►plating efficiency, ►cell growth

Turcot Syndrome (9p22, 54q21-q22, 3p21.3): An autosomal recessive malignant tumor of the central nervous system (glioma), associated with polyposis. Defects in the mismatch repair system predispose for this tumor. ►polyposis adenomatous, ►Gardner syndrome, ►PMS1, ►mismatch repair

Turgid: Expanded because of water uptake.

Turgor: Intracellular pressure caused by water absorption.

Turing, Alan (1912–1954): A mathematician, philosopher, and pioneer in computing and bioinformatics who intended to develop machines and software that work like the human brain.

Turkey: *Meleagris gallopavo*, 2n ≈ 80.

Turner's Rule: Turner's rule deals with the thermodynamics of helix formation of ribonucleotides, which also involves mismatches (Freier SM et al 1986 Biochemistry 25:3209).

Turner Syndrome: Based on an X0 chromosomal constitution in some female mammals. Its incidence in humans is ≈ 0.0003 but the frequency in abortuses may be 0.01 to 0.02. The missing short arm is critical for the Turner symptoms. Turner females usually have short stature, webbed skin on a broad neck, under-developed genitalia and sterility, heart problems, proclivity to kidney disease, diabetes, and hypertension but are generally of normal or near normal intelligence (see Fig. T128). Mathematical abilities are generally reduced (dyscalculia). The retention of the paternal X usually results in better cognitive abilities (imprinting).

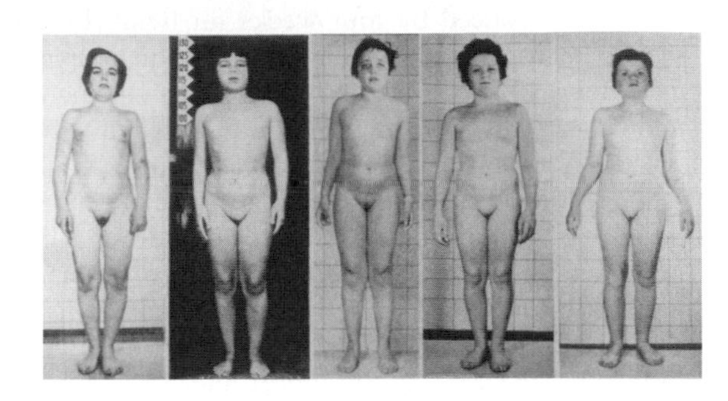

Figure T128. Turner syndrome females of different ages. With three doses of the long arm but only one short arm of the X chromosome. Despite the variations in appearance the similarities of the unrelated individuals is obvious. (Courtesy of Lindsten J et al 1963 Ann Hum Genet 26:383)

In some instances they are also fertile; most of these cases are probably X0/XX mosaics. In a study among 410 Danish women with Turner syndrome, only the 45, X/46,XX or 46,XX with structural abnormality of one of the X-chromosomes were found to give birth after spontaneous pregnancy. After egg donation, some Turner patients can deliver children although there may be a higher chance for chromosomal anomalies among the offspring (Birkabaek N et al 2002 Clinical Genet 61:35). The symptoms are similar even when only the one of the short arms of one of the X-chromosomes is missing. The single X chromosome is maternal in 75% of the cases. The underlying mechanism is puzzling because males normally have a single X, while even in females one of the X-chromosomes is inactive, except during oogenesis and the first few weeks of embryogenesis. The majority of studied female animals with single X (pigs, cattle, horse) are abnormal but the XO female mice appears rather normal and fertile. In *Drosophila*, the XO individuals are sterile males. In Turner syndrome, the most critical is the loss of the distal (pseudoautosomal) part of the short arm of the X chromosome (see Fig. T129). Some of the human developmental difficulties may be normalized by estrogen treatment. Artificial insemination or intrauterine implantation may permit reproduction. About 15% of the X chromosomes in Turner patients are dicentric. ►trisomy, ►sex determination, ►sex chromosomal anomalies in humans, ►human chromosome map, ►prenatal diagnosis, ►MSAFP, ►Noonan syndrome, ►imprinting, ►pseudoautosomal, ►ART, ►dicentric chromosome; Zinn AR et al 1993 Trends Genet 9:90; Sybert VP, McCauley E 2004 New England J Med 351:1227.

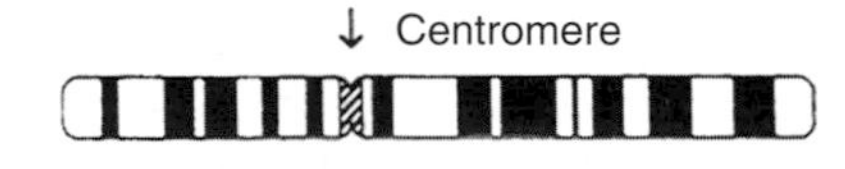

Figure T129. Normal human X-chromosome

Turnip (*Brassica campestris*): A polymorphic species with chromosome number 2n = 2x = 20, *Brassica napus* is 2n = 4x = 38, AACC genomes. ►*Brassica oleracea*

Turnip Crinkle Virus (TCV): A plant RNA virus related to the Tomato Bushy Stunt Virus.

Turnover: The depletion-repletion cycle of molecules or cells. The age of cells in the human body is not easy to determine. Nuclear weapon testing after World War II increased the ^{14}C level in the atmosphere but after 1963 an exponential decrease followed international agreements of testing. The level of ^{14}C in the DNA paralleled this trend and thus it became possible to infer the age of cells, i.e., the time when the DNA and the cell replicated. Occipital neurons in the adult human brain were as old as the individual, indicating that postnatal neurogenesis did not take place in this region. In contrast, non-neural cells were replaced by cell division. This type of procedure makes possible the study of cell turnover that may be important for normal physiology and pathology (Spalding KL et al 2005 Cell 122:133).

Turnover Number: The number of times an enzyme acts on a molecule in a unit of time at saturation.

Turnover Rate: The pace of decay and replacement of a molecule.

***Tus* Gene Products**: Proteins that sense the *Ter* (transcription termination) sequence signals in DNA replication in *E. coli*. ►DNA replication, ►replication bidirectional, ►prokaryotes, ►RTP; Henderson TA et al 2001 Mol Genet Genomics 265:941.

TUSC (Trait Utility System for Corn): TUSC generates insertional mutations in maize with the aid of transposable elements like *Mutator*. ►insertional mutation, ►*Mu*

Tv/Rcas Vector: The Rous sarcoma avian virus proviral vector (maximum carrying capacity 2.5 kb) first mediates in vitro the production the avian leukosis virus (AVL) coat protein. The protein is required for the recognition of the avian retroviral receptor (tv-a), transfected into mouse cells. The replication competent virus thus can produce high titer in the mouse cells infected and can be injected into mice. ►Rous sarcoma, ►retroviruses; Fisher G et al 1999 Oncogene 18:5253.

Tweak (Apo3-L, 17p13.3, TNF-related weak inducer of apoptosis): A TNF ligand with similarity to CD120 and CD95. The related motif of its mouse homolog displays greater similarity. It is a factor in the Klippel-Trenauney syndrome. ►TNF, ►CD95, ►CD120, ►Klippel-Trenauney syndrome; Kaplan MJ et al 2000 J Immunol 164:2997; Schneider P et al 1999 Eur J Immunol 29:1785.

Tween 20 (polyoxyethylene sorbitan, monolaurate): An anionic biological detergent with about 50% lauric acid; the rest comprises of myristic, palmitic, and stearic acids.

Twigdam: Technical Working Group on DNA Analysis Methods. It is concerned with forensic application of DNA analytical and statistical techniques ►DNA fingerprinting

Twin Hybrids: ►complex heterozygotes

Twin Meiosis: Twin meiosis may occur in diploid *Schizosaccharomyces* when after copulation the two nuclei undergo separate meioses.

Twin Spots: Twin spots are visible if (mitotic) somatic crossing over takes place between appropriately marked chromosomes, or may be caused by nondisjunction (see Fig. T130). ►mitotic crossing over; Rédei GP, Sandhu SS 1988 Mutation Res 201:337.

Figure T130. Wheat leaf displaying white and dark green twin sectors on pale green background in a heterozygote for the hemizygous ineffective *Neatby's Virescent* gene. (See ►hemizygous ineffective)

twine: *Drosophila* homolog of *Cdc25* meiosis-specific gene. ►cdc25, ►azoospermia

Twinning: The phenomenon of developing two (or multiple) zygotes from a single impregnation in usually uniparous mammals. The frequency of twins is different in various ethnic groups. Among Nigerians it may be as high as 4.5% and in some South American and Far-East populations it may be as low as 0.8%. In the USA, its frequency among whites was found to be about 0.89 to 0.94% and among blacks, 1.37%, before the widespread use of fertility promoting drugs.

Following in vitro fertilization about half of the births are multiple. Between 1980 and 1997, twinning increased by 52% and the frequency of triplets and higher order gestations has quadrupled in the USA. Multiple births involve considerable medical risk to the babies and mother. The gestation time is frequently reduced by four weeks for twins, by six weeks for triplets and by 10 weeks for quadruplets. The mortality rate (~16%) and developmental anomalies are higher. The lower birth weight involves the risk for physical and mental handicaps. "Multifetal reduction" during the first 9–12 weeks of gestation by injecting potassium chloride into one or more of the fetuses may alleviate medical problems but may cause the loss of the pregnancy entirely. The procedure also involves serious ethical dilemmas, religious conflicts, psychological and other trauma (Elster NJD 2000 FertilSteril 74:617.) Twins are either *monozygotic* (identical) or *dizygotic* (fraternal). The former are derived from a single fertilized egg and the latter develop from two separate eggs fertilized by different sperms. In the overwhelming majority of cases, monozygotic twins are genetically identical whereas the dizygotic twins are comparable to any other siblings. It is possible that some dizygotic twins display higher similarity if one unfertilized egg gave rise to two blastomeres and two separate sperms fertilized each. Another possibility is if one of the polar bodies (identical to the egg) becomes an egg due to a developmental mishap. The identity of monozygotic twins may be somewhat reduced due to epigenetic changes (epigenetic shift).

In an unusual case of twins (Souter VL et al 2007 Human Genet 121:179), both had a 46, XX/46,XY chromosome complement in peripheral lymphocytes, skin fibroblasts, and gonadal biopsies. The proportion of XX to XY cells varied between the twins and their tissues. The cells were chimeric, and shared 100% of maternal alleles and approximately 50% of paternal alleles in the DNA analysis of skin fibroblasts. Possibly, earlier the egg divided and each was fertilized by a different sperm or a single egg was fertilized by two spermatozoa resulting in a triploid zygote, which later split and lost one or another set of chromosomes and became cytologically euploid (see Fig. T131).

Although after birth monozygotic twins appear entirely identical, during development, due to differences of methylation and histone acetylation, phenotypic differences may arise (see Fig. T132) (Fraga MF et al 2005 Proc Natl Acad Sci USD 102:10604). In a unique incident, a monozygotic twin suffering from ovarian failure was able to conceive and deliver apparently healthy offspring, after transfer of ovarian cortical tissue from her healthy sister. (Sherman J et al 2005 New England J Med 353:58).

The frequency of twinning has a genetically determined component as the studies of various ethnic groups indicate. There are indications that the genetic component plays a greater role in dizygotic twin birth than in monozygotic births.

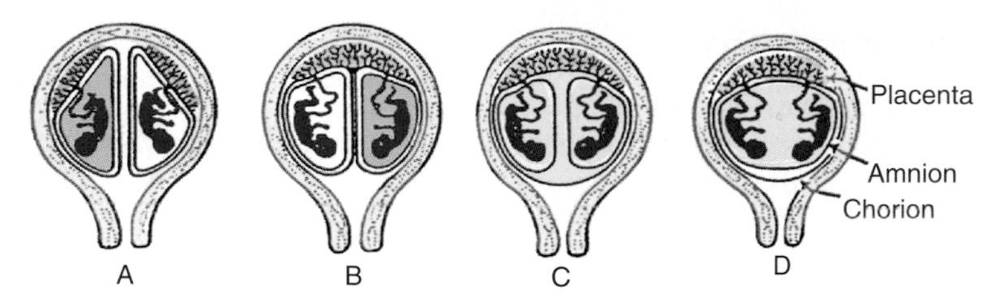

Figure T131. Human twins can be either mono- or dizygotic and the two conditions can be distinguished also on anatomical basis of the developing fetus. Monozygotic twins are surrounded by a common chorion (c and d) or even by a single amnion (D). (A) and (B) are most likely dizygotic. (Modified after Stern, C. Principles of Human Genetics, Freeman. San Francisco)

Figure T132. Monozygotic (1st and 3rd) and dizygotic twins (2nd and 4th)

Monozygotic twins are expected to be of the same sex. Exceptionally, one of the XY male twin embryos may lose the Y chromosome and develop into an XO (possibly mosaic) female. The XO human cells develop into Turner syndrome females, unlike in *Drosophila* where XO individuals are sterile males. Non-concordance for sex in identical twins may be the result of mutation in autosomal or X-chromosomal sex-reversal genes. Sometimes it is difficult to distinguish between identical and non-identical twins on the basis of phenotypic similarities. DNA fingerprinting may resolve the problem. Dermatoglyphics may not always be conclusive due to developmental differences in digital, palmar, or plantar (sole) ridge counts in monozygotic twins. Mono- and dizygotic twins provide useful tools for the study of inheritance of polygenically determined human traits. Among women of families with dizygotic twins, the rate of twinning is about double that of the general population and may be attributed to hereditary hormone levels. The inheritance of twinning through the male parent is lower than through the female. Monozygotic twinning may not or may have very low genetic component. There are statistical methods for the discrimination between identical (MZ) and non-identical (DZ) twins based on concordance of alleles (DNA or any other type).

Maynard Smith and Penrose (Ann. Human Genet. 19:273) worked out the following formula for the probability of concordance for DZ twins: $P = \left(1 + 2\sum_{i=1}^{n} p_i^2 + 2\left[\sum_{i=1}^{n} p_i^2\right]^2 - \sum_{i=1}^{n} p_i^4\right)/4$, where i stands for the phenotype of the markers, n = number of alleles, p_i = allelic frequencies calculated on the basis of the binomial distribution for the various types of matings. ►DNA fingerprinting, ►fingerprints, ►forensic genetics, ►heritability estimation in humans, ►freemartin, ►superfetation, ►multipaternate litter, ►multiparous, ►zygosis, ►concordance, ►discordance, ►co-twin, ►quadruplex, ►ovulation ovary, ►SRY, ►sex reversal; Jones H, Schnorr JA 2001 Fertil Steril 75:11; Boomsma D et al 2002 Nature Rev Genet 3:972; epigenetics of twins: Petronis A 2006 Trends Genet 22:347.

Twins: ►twinning

Twinscan: A gene-finder program (similar to GENSCAN) suitable to analyze homologous sequences and genes of two related genomes. ►GENSCAN, ►SGP-1; Korf I et al 2001 Bioinformatics 17, Suppl. 1:S40; Wei C et al 2005 Genome Res 15:577.

Twintron: ►intron group III

Twist (*Drosophila* gene *twi*, 2–100): The lethal embryo is twisted in the egg case as the germ cell layers are defective. Mutation in the human homolog is responsible for the Saethre-Chotzen syndrome. Twist mediates also metastasis by reducing e-cadherin-controlled cell adhesion (Yang J et al 2004 Cell 117:927). ►cadherins, ►metastasis, ►Chotzen syndrome

Twisting Number: The twisting number characterizes DNA supercoiling by indicating the number of contortions (writhing) and the number of twists, i.e., the number of nucleotides divided by the number of nucleotides per pitch. ►supercoiling, ►DNA

Twitchin: A myosin-activated protein kinase.

Two-Component Regulatory Systems: In bacteria, pairs of proteins transduce environmental signals. In one of the proteins, ca. 250 amino acids at the C-terminus and ca. 120 amino acids at the N-terminus are conserved. They control chemotaxis, virulence, nitrogen assimilation, dicarboxylic acid transport, sporulation, etc. They function by autophosphorylation of a histidine residue by the γ phosphate of ATP (see Fig. T133). The phosphate is then transferred to an aspartate, in the *response regulator*, that modifies regulatory activity of the C-terminal output domain. The system is also called a phosphorelay. A specific phosphatase may reset the system. Frequently, more than two components are involved. ►signal transduction; Itou H, Tanaka I 2001 J Biochem 129:343.

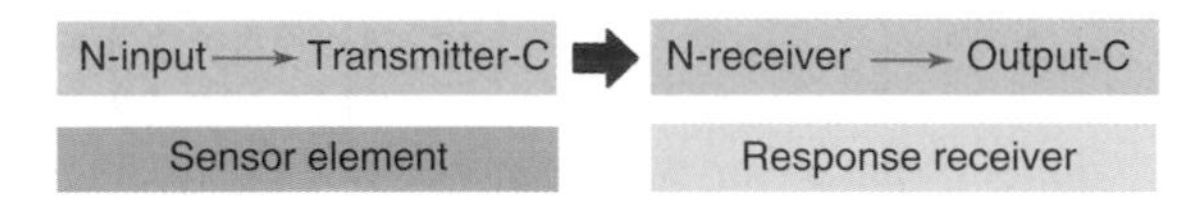

Figure T133. Two-component regulatory system

Two-Dimensional Gel Electrophoresis: In two-dimensional gel electrophoresis, the (protein) mixture is first separated by isoelectric focusing, and then separated by size using a slab of SDS polyacrylamide gel. Thus all proteins, except those rare molecules that have identical charge and molecular size, are distinguished. For the detection of small quantities of the molecules, they are either labeled radioactively or by non-radioactive means. A single two-dimensional gel slab permits the separation of hundreds or thousands of proteins at a time (see Fig. T134).

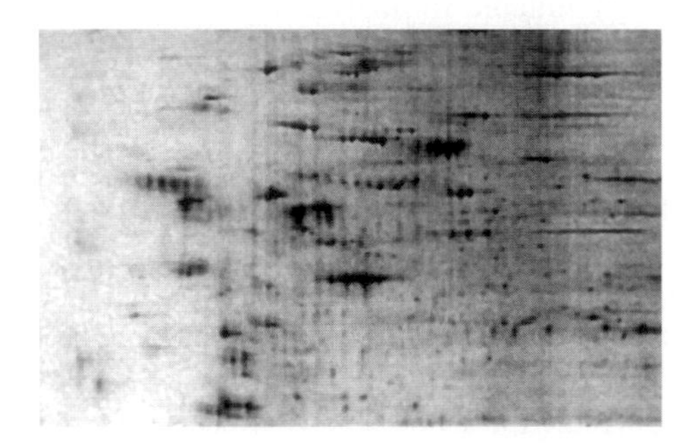

Figure T134. Two-dimensional gel electrophoresis of proteins. (Courtesy of ESA Inc. Chelmsford MA 01824–4771)

For the identification of proteins of very low abundance affinity purification may be required. The technique is used extensively in proteomic analyses. The definitive identification of proteins is hindered often by the fact that the concentration of individual proteins may vary 5–7 orders of magnitude. The two-dimensional electrophoretic analyses of the proteome may have important potential for the understanding of disease development (Hanash S 2003 Nature [Lond] 422:226). In the immobilized pH gradient gel electrophoresis (IPG) the pH gradient is fixed in the acrylamide gel. In the differential in-gel electrophoresis (DIGE) uses two pools of proteins labeled with different fluorescent dyes and then separated in two dimensions. ►electrophoresis, ►isoelectric focusing, ►microchannel plate, ►FTICR, ►proteome, ►normalome; Hoving S et al 2002 Proteomics 2:127; Ros A et al 2002 Proteomics 2:151; Gromov PS et al 2002 Progr Biophys Mol Biol 80:3; http://www.expasy.ch/ch2d/; http://proteomics.cancer.dk/; http://us.expasy.org/; http://www.wzw.tum.de/proteomik/lactis.

Two-Hit Hypothesis: The two-hit hypothesis assumed that in order to develop cancer, two genetic alterations must take place in succession. Some chromosomal aberrations are also known as two-hit causes because two breaks are necessary to bring about an inversion or translocation. ►kinetics, ►Knudson two-mutation theory

Two-Hybrid Method: Genetic constructs of yeast facilitate the study of protein-protein interactions. The GAL4 protein is both an enzyme and an inducer (see Fig. T135). The native GAL4 protein contains an N-terminal UAS (upstream activator sequence) DNA-binding region and a carboxyl-terminal transcription-activating region. These regions—in close vicinity—are required for the activation. Thus fusing the N-terminal of a protein (bait) and the C terminal of another protein (prey) can help study interaction between two proteins. If the two proteins interact, they reconstitute the link between the binding, and the activating domains and transcription (expression) of the reporter gene (represented by the turned on light bulb) may proceed. Thus to see expression, the DNA binding domain (DBD) must bind to the UAS element and contact established with the other protein element, which is often called *prey* and is attached to a transcriptional activator. The DBD + bait (hybrid I) and the prey + activator (hybrid II) are separately inactive. The expression of the downstream reporter gene requires interaction between the *two-hybrid* proteins. The most commonly used binding component is derived from the Gal4 or LexA proteins and the bacterial LacZ or luciferase is employed as a reporter. The receptor domain, usually called bait, may occur on a plasmid, which may also carry sequences to promote dimerization and thus the required protein interaction. The Gal bait (in contrast to the majority of LexAs) usually also contains nuclear localization sequences.

Efficient prey vectors may contain the VP16 activation domain or the Gal4 region II. The B42 bacterial activator is weaker than the other two but has affinity for a wider range of proteins and suppresses (squelches) the toxic effects that Gal4 and strong transcriptional activators may have on yeast. The two-hybrid method in yeast is very simple to use. Strains with the bait can be mixed and mated with strains with the prey and then plated on selective media where the interactions are readily detectable. The advantage of the two-hybrid method is that it can be used to test protein interactions, determine the amino acid sequences critical for interactions, and screen gene libraries for binding proteins or activators.

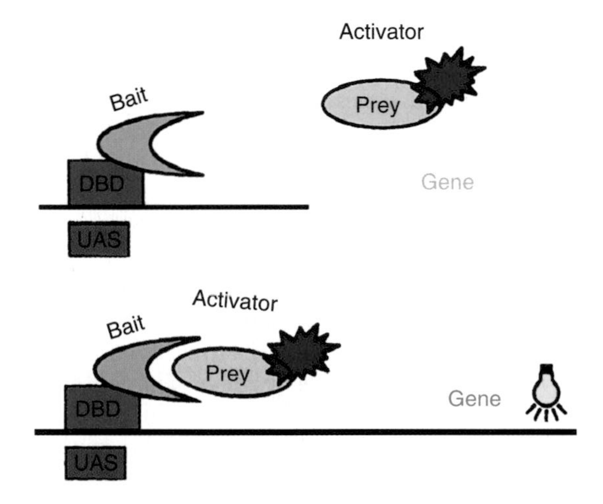

Figure T135. Two-hybrid method

The system is also suitable for testing any molecule (including aptamers) that may affect interactions, e.g., in the development of specific drugs. The system can be used in studies of the cell cycle and transcription factors, tumorigenesis, tumor suppression, etc. In some instances, positive (i.e., non-relevant for the purpose of the study) activation of the reporter gene may also occur. This interference can be reduced or eliminated by selective systems. False negative interactions are observed when the protein-protein interactions are low, there are problems with the intracellular folding of the proteins, or when other domains of larger proteins hinder proper interactions. The cases of false negatives may be low, yet may be very important in pharmacological studies. Some proteins may require a third (protein or non-protein) element for stabilization, bridging, or modification and then only a ternary complex is active. Systems have been constructed that prevent interactions between proteins. The URA3 reporter gene product converts fluoroorotic acid—with assistance of other genes—into fluorodeoxyuridine monophosphate, an analog of thymidylate synthase, which inhibits DNA synthesis and can be used for contraselection. The two-hybrid method applied sequentially or simultaneously to large gene pools may detect interacting systems of genes. It may also place functionally unknown open reading frames into a biological context of a metabolic role. *Two-bait* systems have been developed for the possible detection of (allelic) variants of the same protein. Mammalian two hybrid procedures may facilitate the detection of interacting proteins in mammalian systems that depend on post-translational modifications not available in yeast cells. Protein-RNA interactions are studied by the *three-hybrid* systems. The *reverse two-hybrid* system detects mutations that are unable to bring the activation element to the DNA binding domain and thus cannot convert a potentially toxic compound to a toxic compound or into a suicide inhibitor. ►galactose utilization, ►split-hybrid system, ►three-hybrid system, ►reverse two-hybrid system, ►four-hybrid system, ►suicide inhibitor, ►recruitment, ►epistasis, ►pleiotropy, ►suppressor second site, ►VP16, ►one-hybrid binding assay, ►aptamer, ►microarray hybridization, ►interaction trap, ►genetic networks, ►networks, ►proteomics; Brent R et al 1997 Annu Rev Genet 31:663; Fields S, Song O 1989 Nature [Lond] 340:245; Shioda T et al 2000 Proc Natl Acad Sci USA 97:5220; Hirst M et al 2001 Proc Natl Acad Sci USA 98:8726; Fang Y et al 2002 Mol Genet Genomics 267:142; Uetz P 2002 Current Opin Chem Biol 6:57; Stelzl U et al 2005 Cell 122:957; http://dip.doe-mbi.ucla.edu/.

Two-Phase Mutation Model (TPM): In the TPM, microsatellite instability may be due to the gain or loss of a number of repeats. ►IAM, ►SMM, ►microsatellite

Two-Photon Microscopy: In two-photon microscopy, a fluorochrome is excited almost simultaneously by two photons of lower energy. This allows reduced light scattering and less photo damage of the sample.

Two-Point Cross: The two-point cross involves differences at two gene loci, e.g., *AB* × *ab* or *Ab* × *aB*.

Two-Step Synapsis: ►topological filter

Two-Tailed Test: The two-tailed test estimates deviations in both directions from the mean.

Ty: Non-viral retroelements (retrotransposons) in the nucleus of budding yeast (*Saccharomyces cerevisiae*). Ty elements occur in several related forms and are designated Ty1 (25–30 copies), Ty2 (10 copies), Ty3 (2–4 copies), etc. The elements are flanked by long terminal direct repeats (251 – 371-bp) that are designated as δfor Ty1 and Ty2, and σfor Ty3. The open reading frame between the repeats (LTR) in Ty1 and Ty2 is identified as ε. Elements resembling these terminal repeats, δ (ca. 100 copies), σ (20–30 copies), and τ (15–25 copies) are also found. The terminal repeats contain sequences identified as U3 (unique for 3′), R (repeated), and U5 (unique for 5′ end) similarly to the designations in retroviruses. These LTRs contain the upstream gene activation sequence, TATA box, and polyadenylation, transcription, and termination signals. They also include the TG...CA bases involved in the integration of the reverse transcripts into the chromosomes. All the retroelements are generally flanked in the host by 5-bp target duplications. The open reading-frames, distinguished as TYA and TYB between the ends, are also similar and resemble retroposons of other organisms. The TYA protein may be processed through proteolysis into several smaller proteins involved in the formation of the shell of the VLP (virus-like particle). TYB contains genes for protease (*pro*), integrase (*int*), reverse transcriptase (*rt*) and RNase H (*rnh*) but no *env* gene is present as would be expected for retroviruses. Although most of the Ty elements and the independently standing termini are intact, some contain deletions up to a few kb, and the LTR sequences may be truncated. There may be apparent insertions and duplications within the Ty elements or inversion(s) involving parts of a LTR. The heterogeneity of the coding regions is due to base substitutions.

Ty retrotransposons transpose by synthesizing an RNA that is reverse-transcribed into DNA for integration. The Ty1 transcripts represent about 0.8% of the total RNA in the cell but the Ty cDNA is present in less than one copy per haploid cell before integration because of the inefficient processing of the transcript. The rate of Ty1 transposition is about 10^{-5} to 10^{-7} per element per cell division. The Fus3 protein kinase, the transcription factor TFIIH and nucleotide exchange repair modulate the transposition of Ty. The reverse-transcribed DNAs can integrate into Ty elements with the aid of recombination factors of the host and form tandem elements. RNA polymerase II transcribes them and the transcripts are polyadenylated. Transcription may be prevented by mutations symbolized as *spt* (suppressor of Ty). The transcription of Ty elements may be induced by sex pheromones synthesized by the *MATa* or *MATα* genes but not in the *MATa/Mat α* diploids. *MAT* homozygotes do not affect the transcription of Ty RNA. The Ty RNA can be packaged into virus-like particles (VLP). Proteolysis is involved in the processing of the TYA and TYB products, required for the completion of VLPs. The reverse transcription appears to be primed by $tRNA^{Met}$. Degradation of the template RNA is due to RNase H. During reverse transcription, recombination, sequence modifications, and deletions may occur at high frequency. Transcription of Ty is regulated by several *PST* (suppressor of transposition), *ROC* (reducer of overproduction of transcripts), and *TYE* (Ty enhancer) genes. Presumably, about 100 host genes affect the Ty1 retrotransposition cycle. About half of them are involved in the production of the cDNA and the rest affect steps following Ty replication (Griffith JL 2003 et al 2003 Genetics 164:867). Transcription may be increased by exposure to UV light, ionizing radiation, chemical mutagens, and the culture media and it is elevated 20 fold at low temperature. The RAD52 group of recombinational repair genes inhibits the transposition of Ty. Insertion into and activation of these particular genes measure the frequency of transposition by genetically tagged Ty elements.

The VLP is made in the cytosol but must enter the nucleus for transposition to take place. Insertion of Ty into structural genes eliminates their function and reversion by excision is very rare because the target site is modified by the event. Insertion within non-coding sequences may activate or silence previously inserted elements. Ty1 and Ty2 elements appear to be distributed at random throughout the genome but the family of Ty3 elements is far more restricted. The Ty3 element inserts at the transcription initiation sites of tRNA genes by pol III. The transcription factors TFIIIB and TFIIIC are required for this type of insertions. The Ty1 element has preference for integration into the 5′-sequences of pol II-transcribed genes. Ty5 integrates preferentially into heterochromatin. Ty, δ, σ, and τ are commonly targeted to the vicinity of tRNA genes (hot spots). The insertions are usually directed into the leader sequences (promoter sites) or near the 5′-end of the coding region. Among the abundant Ty elements, recombination may occur and cause chromosomal rearrangements and deletions. Recombination is most common between LTR (δ) repeats. Gene conversion may also occur between these elements. Ty elements may incite deletions, inversions, and translocations with all the phenotypic consequences. The mutation rate within Ty elements was estimated to be 0.15 per Ty per replication cycle, i.e., about 1/15 of the retrotranspositions result in some types of mutation. Multiple copies of Ty1 may involve co-suppression, i.e., silencing of the elements (Garfinkel DJ et al 2003 Genetics 165:83). This rate of mutation is approximately 4 to 6 orders of magnitude higher than that in the rest of the nuclear genome but is comparable to that in RNA viruses. The high mutation rate is attributable to the lack of proofreading ability of the reverse transcriptase. The mutations in Ty (about 30% of all) occur within the seven bases of the primer-binding site (PBS) where minus strand replication is primed by a tRNA. Imprecise cutting by RNase H causes some mutations. Another error-prone event is the addition of nucleotides immediately adjacent to the tRNA primer-binding site. After the replication system reaches the 5′-end, nucleotide additions take place at the 3′-end. After the second-strand transfer, partial DNA•DNA duplexes are formed and the recessed 3′-ends prime the completion of the double strand synthesis. As a consequence of the addition of nucleotides, the 3′-end of the minus strand cannot anneal precisely with the plus-strand DNA template. In order to reconstitute the 5′-LTR of the Ty, the mispaired 3′-primer ends are extended by reverse transcription. This process incorporates then the mispairs into the Ty element that will integrate into the yeast chromosome. The mismatches must then be fixed by DNA repair. Additional mutational mechanisms may also occur. These mutational processes are not unique features of the Ty elements but are also used by other retroelements. Retroviruses are somewhat different, however, because they show frameshift mutations and complex rearrangements. The Ty elements have been utilized as vectors (pGTy), as insertional mutagenic agents (transposon tagging), and for fusion of TyA proteins with certain proteins of interest, to facilitate the purification of epitopes. Similar transposable elements occur also in other fungi (*Candida*, *Pichia*, *Hansenula*). ►retroposon, ►retrotransposon, ►transposition induction, ►reverse transcription, ►strong-stop DNA, ►mating type determination in yeast, ►protease, ►integrase, ►epitope, ►*Saccharomyces cerevisiae*, ►pol II, ►pol I, ►DNA repair, ►fus3, ►TFIIH, ►VLP; Boeke JD et al 1988 Mol Cell Biol 8:1432; Jordan IK, McDonald JF 1999 Genetics 151:1341; Wickner RB 2001, p 473. In: Fundamental Virology. Knipe DM, Howley PM (Eds.) Lippincott Williams & Wilkins, Philadelphia, PA, USA; Umezu K et al 2002 Genetics 160:97; Zhu Y et al 2003 Proc Natl Acad Sci USA 100:5891.

Ty δ: A terminal repeat of the Ty insertion element. ►Ty

Tyk2: A non-receptor tyrosine kinase. ►Janus kinases

Tylosis: The formation of animal callus. ►keratosis

Type III Secretion System: ►secretion systems

Typing: The determination of the blood group antigens and HLA. It is a general classification by type. Also, DNA typing by fingerprinting using restriction enzyme-generated fragment pattern. ►blood groups, ►HLA, ►DNA fingerprinting

Tyrosinase: ►albinism

Tyrosine: A non-essential aromatic amino acid (MW 181.19), soluble in dilute alkali (see Fig. T136). Its biosynthetic pathway is *with enzymes involved in parenthesis*): Chorismate - > (*chorismate mutase*) - > Prephenate - > (*prephenate dehydrogenase*) - > 4-Hydroxyphenyl pyruvate - > (*amino trans-ferase* with glutamate NH_3 donor) - > Tyrosine. Tyrosine can be derived also from phenylalanine by dehydroxylation. Tyrosine is a precursor for norepinephrine, epinephrine, 3,4-dihydroxy-phenylalanine (dopa), dopamine, and catechol that form the catecholamine family of animal hormones. ►phenylalanine, ►chorismate, ►pigmentation of animals, ►alkaptonuria, ►tyrosinemia, ►tyrosine aminotransferase, ►tyrosine kinase, ►animal hormones, ►goiter, formula above.

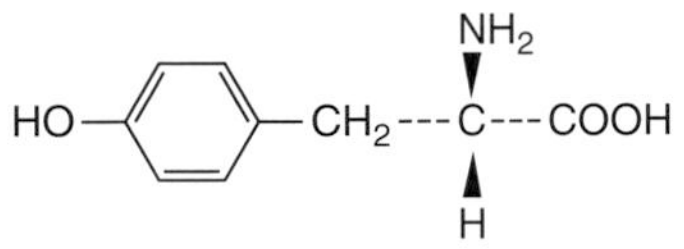

Figure T136. Tyrosine

Tyrosine Aminotransferase (TAT, Richner-Hanhart syndrome): TAT converts tyrosine into p-hydroxyphenylpyruvate. Glucocorticoid hormones induce it. TAT deficiency occurs rarely in humans and is controlled by a recessive gene (ca. 11 kb, 12 exons) in human chromosome 16q22.1-q22.3 (and mouse in chromosome 8). The condition involves elevated level of tyrosine and in some cases an increased urinary excretion of p-hydroxyphenylpyruvate and hydroxyphenylacetate. The disease generally involves corneal ulcer, palm keratosis (callous skin), and mental and physical retardation. The mitochondrial enzyme is under the control of another gene. A TAT regulator gene may be located in the human X-chromosome and glutamic oxaloacetic transaminase (16q21) also regulates its activity. ►tyrosine, ►tyrosinemia

Tyrosine Hydroxylase (TYH): TYH is encoded in human chromosome 11p15 (mouse chromosome 7), which controls the synthesis of dopamine from phenylalanine. Dopamine is a hormone involved with adrenergic neurons, the sympathetic nerve fibers that liberate norepinephrine when an impulse passes the nerve synapse. This enzyme may play a key role in fetal development and in manic depression. The microsatellite sequence in the first intron serves normally as a transcriptional enhancer of the gene. ►tyrosine, ►manic depression, ►DOPA; Albanèse V et al 2001 Hum Mol Genet 10:1785.

Tyrosine Kinases (protein tyrosine kinase): Tyrosine kinase activity is essential for many processes of the signal transduction, tissue differentiation involving oncogenesis, and signaling to immunoreceptors. The most frequently activated proteins are phospholipase C-γ, phosphatidylinositol 3-kinase, and GTPase activating kinase. ►oncogenes ABL, ►ARG, ►Blk, ►Btk, ►FES, ►FGR, ►FLT, ►FHS, ►ERBB1, ►Fyn, ►RAF, ►SRC, ►KIT, ►LCA, ►Lyn,

►MET, ►RET, ►Syk, ►YES, ►ZAP-70, ►morphogenesis in *Drosophila* [*tor*], ►*sevenless*, ►*SOS*, ►insulin [receptor β–chain], ►TCR [T cell receptor], ►EGFR, ►PDGF, ►MCSF, ►VEGF, ►Steel factor, ►hepatocyte growth factor, ►trk, ►neutrotrophin, ►FGF, ►receptor tyrosine kinase, ►signal transduction, ►protein kinases, ►kinome; Latour S, Veillette A 2001 Curr Opin Immunol 13:299; Wang Z et al 2004 Science 304:1164; tyrosine kinase inhibitors: Levitzki A, Mishani E 2006 Annu Rev Biochem 75:93.

Tyrosine Phosphatases (protein tyrosine phosphatase, PTP): The PTP family of enzymes has at least 40 known members. Some have properties of transmembrane receptor proteins (RPTP) and the cytosolic forms have a characteristic ca. 240-amino acid catalytic domain. Each PTP has also a phosphate-binding, 11-residue motif containing catalytically active Cys and Arg. The proteins have critical roles in signal transduction relevant to growth, immune response, proliferation, and differentiation. Dimerization negatively regulates the activity of receptor tyrosine phosphatase-α. ►signal transduction, ►PTK, ►synaps; Stetak A et al 2001 Biochem Biophys Res Commun 288:564; substrates: Tiganis T, Bennett AM 2007 Biochem J 402:1; Liu Y et al 2007 Nature Rev Immunol 7:202.

Tyrosine Protein Kinase: ►tyrosine kinase, ►phosphorylase

Tyrosine Transaminase: ►tyrosine aminotransferase

Tyrosinemia (FAH): The recessive gene responsible for the anomaly is in human chromosome 15q23-q25. Its prevalence is about 1/2,000 live births. FAH involves increased levels of tyrosine in the blood and a lack of p-hydroxyphenylpyruvate oxidase, but the primary enzyme defect appears to be fumarylacetoacetase deficiency leading to accumulation of succinylacetone and succinylacetoacetate. Defects in porphyrinogenesis appear secondary. Because of the deterioration of the liver, as secondary symptoms, accumulation of methionine and other amino acids in the blood and urine is frequent. The particular odor of the body fluids may be due to α-keto-γ-methiolbutyric acid. Prenatal diagnosis of FAH relies on amniotic fluid analysis for succinylacetolactone or measurement of fumarylacetoacetase in cultured amniotic cells. Low tyrosine diet alleviates the symptoms but to avoid liver cancer, liver replacement before age two is advisable. This type of tyrosinemia involves various kinds of chromosomal anomalies. Tyrosinemia II (16q22.1-q22.3) is a deficiency of tyrosine aminotransferase. Tyrosinemia Type III (12q24-qter) is due to a deficiency of 4-hydroxyphenyl-pyruvate dioxigenase. This disease does not usually involve dysfunction of the liver. The condition can be detected prenatally but carriers cannot be identified. Mice mutants homozygous for FAH but heterozygous for alkaptonuria were partially normalized in their hepatocytes. The tyrosine degradation pathway: tyrosine →(*tyrosinemia type II*)→ 4-OH-phenylpyruvate→(*tyrosinemia type III*) →homogentisic acid→ (*alkaptonuria*)→ maleylacetolactate→ fumarylacetolactate→ (*tyrosinemia type I*)⇒fumarate and acetolactate. ►tyrosine aminotransferase, ►genetic screening, ►methionine adenosyltransferase deficiency, ►methionine biosynthesis, ►amino acid metabolism, ►liver cancer, ►mosaic, ►gene therapy, ►alkaptonuria; Jorquera R, Tanguay RM 2001 Hum Mol Genet 10:1741.

Tyrosinosis: ►tyrosine aminotransferase

Historical vignettes

Aristotle reviews the ancient theories of sex determination, but finds them unsatisfactory:

"Some suppose that the difference [between sexes] exists in the germs from the beginning; for example, Anaxagoras and other naturalists say that the sperm comes from the male and that the female provides the place [for the embryo], and that the male comes from the right, the female from the left, since in the uterus the males are at the right and the females at the left. According to others, like Empedocles, the differentiation takes place in the mother, because, according to them, the germs penerating a warm uterus become male, and a cold uterus female…" (Generation of Animals, Book IV, Part I, Para. 2).

Peter Starlinger (discoverer of insertion mutations in bacteria with Heinz Saedler in 1972 [Biochimie 54:177) noted in 2005 in Annu Rev Plant Biol 56:1

"It was only then that we realized the relation of these element to McClintock's transposable elements, in spite of the fact that I had known McClintock's work since my student days, and in spite of a series of seminars that we had held on this topic in the institute in Cologne. Sometimes we are blind!"

U

U: Denotes uracil. ►pyrimidines

U: Symbol for uranium.

U-937: Refers to human monocyte line. ►monocytes

U Protein (unwinding protein): Unwinds the DNA strands at a distance from a ►nick. (See Basak S, Nagaraja V 2001 J Biol Chem 276:46941).

U RNA (ubiquitous RNA): Denotes a snRNA. Its transcript has a 2,2,7-trimethyl guanosine cap and may have modified U residues; it does not have a poly-A tail. ►snRNA, ►snurposome, ►cap, ►poly-A tail

U1 RNA: This has complementary sequences to the 5′ consensus sequences of the splice sites and probably plays a role in mRNA processing from the primary transcript. Mutated U1 snRNA may be a strong inhibitor of gene expression (Fortes P et al 2003 Proc Natl Acad Sci USA 100:8264). ►hnRNA, ►RNA; McNamara-Schroeder KJ et al 2001 J Biol Chem 276:31786.

U2 RNA (snRNA): Apparently, this recognizes the 3′ end of introns at the lariat and is involved in splicing. ►lariat, ►splicing, ►introns

U3 snRNP: Most abundant of the U RNA (≈10^6 molecules/cell) processes near the 5′ end of the ribosomal RNA transcripts. It stays on and generates a 5′ knob characteristic of rRNA only. Binding U3 of the snRNP initiates processing, especially the transcripts of 18S rRNA. ►rRNA, ►snRNP, ►ribosome; Venema J et al 2000 RNA 6:1660.

U7 snRNA: Involved in the processing of the 3′ end of the histone pre-mRNAs. In *Drosophila*, it contains 71 nucleotides. In vertebrates it is located in the Cajal bodies. ►Cajal body; Dominski Z et al 2003 Proc Natl Acad Sci USA 100:9422.

U8 snRNP: Required for the upstream cleavage of 5.8S and for cutting off 28S RNA ca. 500 nucleotides at the 3′ end. ►rRNA, ►snRNP, ►ribosome; Peculis BA, Steitz JA 1994 Genes Dev 8:2241.

U12 snRNP: A spliceosomal component along with other U snRNAs. ►spliceosome; Otake LR et al 2002 Mol Cell 9:439.

U14 snRNP: A maturase for 18S rRNA. ►ribosome; Newman DR et al 2000 RNA 6:861.

U22 snRNP: Essential for processing both ends of 18S rRNA, separated by about 2,000 nucleotides.

UAA: Refers to the ochre codon of translation termination. ►code genetic, ►translation termination

U2AF: A protein assisting U2 snRNA recognition. U2AF family proteins regulate both steroid hormone receptor-mediated transcription and alternative splicing (Dowhan DH et al 2005 Mol Cell 17:429). Phosphorylated DEK protein assists U2AF in intron removal (Mendel Soares LM et al 2006 Science 312:1961). ►U2 RNA, ►alternative splicing; Guth S et al 2001 Mol Cell Biol 21:7673.

UAG: Refers to the amber codon of translation termination. ►code genetic, ►amber suppressor

UAS: Denotes upstream activating sequences (which regulate gene transcription). They behave in the same manner as enhancers. UAS encode DNA-binding proteins, e.g., the GAL4 UAS codes for a 100 kDa protein and protects its 17 bp palindromic sequence against DNase I digestion or methylation. These proteins bind to special DNA sequences and to other proteins in the transcriptional complex. Transcriptional activation and binding may rely on more than one tract of amino acids. ►promoter, ►galactose utilization, ►two-hybrid method; Blackwood EM, Kadanoga JT 1998 Science 281:61.

uAUG: In 3 to 10% of the RNA transcripts translation initiation codons occur in the 5′-untranslated sequences (5′-UTR) of viruses, fungi, plants and mammals. The open reading frame due to uAUG may have an untranslated intercistronic region relative to the downstream ORF or they may overlap or it may be in-frame. uAUGs may have regulatory functions. (See Rose JK, Iverson L 1979 J Virol 32:404).

UBC2 (E2): Refers to ubiquitin-conjugating enzymes. From UBC ubiquitin is transferred to a lysine residue of the target protein. Ubc2/Rad6 - Rad18 proteins have functions in both proteolysis and DNA repair. Ubc2 proteins occur in many isoforms within the same organism. ►ubiquitin, ►E2, ►Rad 18, ►isoform; Ptak C et al 2001 Mol Cell Biol 21:6537.

UBC3/Cdc34: A ubiqitination enzyme required for G1→S phase transition in the cell cycle. ►ubiquitin, ►Cdc34

Ubc9: A ubiquitin interacting protein. It is involved in the control of the cell cycle from G2→M, and in the degradation of diverse proteins in a variety of organisms. ►ubiquitin, ►UBL; Kaul S et al 2002 J Biol Chem 277:12541.

UbcD1: This ubiquitin protein of *Drosophila* degrades some telomere associated proteins and thus controls the proper detachment and attachment of the telomeres during mitosis and meiosis. ►telomere,

U

►ubiquitin; Bocca SN et al 2001 Biochem Biophys Res Commun 286:357.

UBE3A: A ubiquitin ligase gene plays a role in Angelman syndrome. ►ubiquitin, ►Angelman syndrome; Kishino T et al 1997 Nature Genet 15:70.

Überoperon: Includes gene neighborhoods where individual operons may show rearrangement in different species, yet remain in the functional and regulatory context (Lathe WC III et al 2000 Trends Biochem Sci 25:474). Rogozin and associates (Rogozin IB et al 2002 Nucleic Acids Res 30:2212) named this phenomenon genome hitchhiking. The largest such neighborhood in prokaryotes includes 79 genes. Most, albeit not all these genes, share a known functional role. ►operon, ►supraoperon

UBF (upstream binding factor): In association with protein SL-1 UBF, controls the transcription of rRNA genes by RNA polymerase I. These protein factors differ even among closely related species; they are members of the high mobility group proteins. ►high mobility group proteins, ►transcription factors, ►SL1; Santoro R, Grummt I 2001 Mol Cell 8:719; Chen D, Huang S 2001 J Cell Biol 153:169; Stefanovsky VY et al 2001 Mol Cell 8:1063.

Ubiquilin: A protein that stimulates the biosynthesis of neurofibrillary tangles in Alzheimer's disease and Lewy bodies in Parkinson's disease. ►Alzheimer's disease, ►Parkinson's disease, ►Lewy body, ►tau; Mah AL et al 2000 J Cell Biol 151:847.

Ubiquinones (coenzyme Q): Lipid-soluble benzoquinones mediate electron transport. The reduced form is a potent lipid-soluble antioxidant capable of inhibiting lipid peroxidation. ►cytochromes, ►lipids, ►peroxides; Elias M et al 2001 J Biol Chem 276:48356.

Ubiquitin: This acidic polypeptide (~76 amino acids) is ubiquitously present in prokaryotic and eukaryotic cells. It may be associated with H2A histone and the conjugate is called UH2A. It binds to other proteins and proteolytic enzymes and then degrades the ubiquitinylated proteins. In the majority of cases the degradation takes place in the cavity of the (20S–26S) *proteasomes*, cylindrical complexes of proteases, or the ligand-bound molecule can be internalized into vacuoles and degraded by lysosomal enzymes. Ubiquitination is involved in many aspects of cellular regulation, DNA repair, stress response, cell cycle progression, formation of the synaptonemal complex, signal transduction and apoptosis. Ubiquitin stress alters proteasome composition by enhanced loading of deubiquitinating Ubp6 to the proteasome (Hanna J et al 2007 Cell 129:747). A cascade of conjugating, activating and carrier enzymes recognizes ubiquitin. The E1 enzyme forms a thiol ester with ubiquitin using ATP energy. The E1-ubiquitin complex then transfers UB to the carrier E2 which with the aid of E3 ligases mediates the association of E2 with the target proteins. For catalysis, the E3 ligases use either a HECT or a RING finger domain. The ubiquitin E3 ligase complexes may be SCF (Skp1-Cdc53/CUL1-F-box proteins), APC (anaphase promoting complex) and the VCB-like family (VHL-ElonginC/ElonginB) proteins. SCF-βTrCP is also a ubiquitin ligase. Several other molecules are also involved. The ubiquitin complex is degraded by deubiquitinating cysteine protease enzymes (DUBP). The signals for degradation are PEST sequence at the C-ends, N-end rule domains and phosphorylation. The human ubiquitin genes are encoded at chromosome 17p11.1-p12 and polyubiquitin at 12q24.3. Polyubiquitin is a chain of several ubiquitin molecules linked by lysine (K-48 or K-63). Some protein molecules may be degraded co-translationally in case they are defective or if their folding is slow. Ubiquitin may have non-destructive functions as well such as processing signaling molecules in the cell and facilitating the expression of genes (Finley D 2001 Nature [Lond] 412:283; Mukhopadhyay D, Riezman H 2007 Science 315:201). ►monoubiquitin, ►histones, ►proteolytic, ►lactacystin, ►lysosomes, ►proteasomes, ►PEST, ►Ubc, ►UBP, ►CDC34, ►Ubl, ►PIC, ►sentrin, other entries listed separately, ►SCF, ►Skp1, ►Cullins, ►D box, ►K-box, ►F-box, ►APC, ►VHL, ►E1, ►E2, ►E3, ►Socs-box, ►Rbx1, ►PEST, ►N-end rule, ►destruction box, ►multivesicular body, ►IκB, ►antigen processing and presentation, ►lid, ►SUMO, ►Huntington's chorea; Hershko A, Ciechanover A 1998 Annu Rev Biochem 67:425; Pickart CM 2001 Annu Rev Biochem 70:503; Weismann AM 2001 Nature Rev Mol Cell Biol 2:169; Conaway RC et al 2002 Science 296:1254; Walden H et al 2003 Nature [Lond] 422:330; Liu YC 2004 Annu Rev Immunol 22:81; regulation: Gao M, Karin M 2005 Mol Cell 19:581; unfolding: Irbäck A et al 2005 Proc Natl Acad Sci USA 102:13427; ubiquitin chain assembly: Hochstrasser M 2006 Cell 124:27; ubiquitin binding domains: Harper JW, Schulman BA 2006 Cell 124:1133.

Ubiquitous RNA: ►URNA

Ubistatins: These are block binding ubiquitinated substrates to the proteasome. ►ubuiquitin; Verma R et al 2004 Science 306:117.

UBL: Refers to ubiquitin (Ubc9) interacting protein operating on many different proteins under various names such as sentrin and SUMO-1. ►ubiquitin, ►Ubc, ►E2, ►SUMO

UBP: A group of (16 in budding yeast) ubiquitin degrading enzymes or de-ubiquitinating enzymes or isopeptidases. UBPs may regulate gene silencing, differentiation and cell division (cyclins, Cdks, MPF, etc.). ►ubiquitin, ►cell cycle, ►cyclins, ►CDK, ►silencer, ►MPF, ►morphogenesis; Lin H et al 2000 Mol Cell Biol 20:6568.

UBR: An enzyme that mediates the formation of thioesther with ubiquitin and UBC. ►UBC; Kwon YT et al 1998 Proc Natl Acad Sci USA 95:7898.

UCE (upstream control elements): These regulate transcription. ►regulation of gene activity

UDG (uracil-DNA-glycosylase): This repair enzyme is capable of removing accidentally incorporated U or deaminated C, which is followed by the removal of the apyrimidinic site with the aid of AP endonuclease. There are two nuclear-coded UDG enzymes in the mammalian cells; one of them is in the mitochondria. ►DNA repair, ►excision repair, ►AP endonuclease; Dinner AR et al 2001 Nature [Lond] 413:752.

UDP: Uridine diphosphate is formed from uridine monophosphate and gives rise to uridine triphosphate by using ATP as the phosphate donor (see Fig. U1).

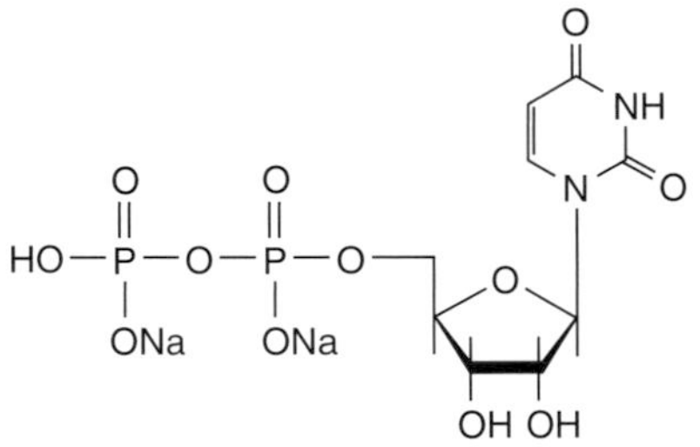

Figure U1. UDP-5′diphosphate Na

UEP: ►unit evolutionary period

UEP (unique event polymorphism): Includes SNIPS, INDELs and microsatellite sites that facilitate characterization of haplotypes.

UGA: Refers to opal stop codon. ►code genetic

UGT/UGPGT (uridinediphosphate-glucuronosyltransferase): Detoxifies various lipid-soluble toxins by conjugation with glucuronic acid, makes them water-soluble, excretable and may prevent mutation and cancer. Its phosphorylation controls substrate-specificity (Basu NK et al 2005 Proc Natl Acad Sci USA 102:6285). ►Crigler-Najjar syndrome

UH2A: ►ubiquitin

UhpB: *E. coli* kinase involved in the regulation of sugar phosphate transport.

uidA: The gene for β-glucuronidase. ►β-glucuronidase

Ulcer: A condition characterized by corruption of the surface or deeper cell layers of the body, e.g., gastric ulcer, an ulceration inside the stomach or diabetic ulcers of the legs, or genital ulcers in the case of sexually transmitted disease. ►Crohn's disease

Ulcerative Colitis: This ia an inflammatory bowel disease (IBD) like Crohn's disease. It is apparently under the control of several chromosomes and genes. ►inflammatory bowel disease

Ullrich Congenital Muscular Dystrophy (Ullrich disease): ►muscular dystrophy

Ullrich-Turner Syndrome: This is the the same as Turner's syndrome.

Ulna: The larger bone of the forearm opposite the thumb.

Ulnar-Mammary Syndrome: ►Schinzel syndrome

Ultimate Mutagen: Some chemical substances (promutagens) become mutagenic only after activation. In this process, first proximal mutagens are formed that may be subsequently converted to a form (ultimate mutagen) that is genetically most reactive with the DNA. ►activation of mutagens

Ultrabar: ►*Bar* mutation

Ultracentrifuge: Laboratory equipment suitable for the separation of cellular organelles and macromolecules by high (over 20,000 revolution/minute) centrifugal force in an evacuated and refrigerated chamber. At maximal speed it may exceed the gravitational force several hundred thousand times. The *analytical ultracentrifuge* monitors continuously or intermittently the boundary of movement of macromolecules in a solute (e.g., Cs_2SO_4). The more commonly used *preparative ultracentrifuge* fractionates organelles or macromolecules in sucrose or CsCl. The suspended material forms a band corresponding to its density and the density of the solute. Chloroplasts in sucrose gradients can be identified by color; DNA bands in CsCl can be identified by staining with the dye ethidium bromide. The bands, containing homogeneous material, can be removed by careful suction from the centrifuge tubes or by drop-wise collection through a hole punctured at the bottom with a syringe needle. (See Laue TM, Stafford WF III 1999 Annu Rev Biophys Biomol Struct 28:75; ►Lamm equation).

Ultraconserved DNA Elements (UCE): 200-base pair or shorter or longer and entirely or almost entirely identical sequences in the human, rat and mouse genomes and are well conserved in chicken, dog and fish as well. They occur frequently in overlapping exons in genes involved in RNA processing or in introns or nearby genes regulating transcription and development. Sequences longer than 100 nucleotide are common in both RNA and DNA binding and in transcriptional regulation. Less than 50 nucleotide ultraconserved elements are common in microRNA, other non-coding RNA and transcription factor binding sites. The ultraconserved sequence elements in whole human and mouse genomes follow the power law distribution and apparently have evolutionary significance (Salerno W et al 2006 Proc Natl Acad Sci USA 103:13121). Their distribution along the chromosomes is not random but they are absent in human chromosome 21 and their numbers are significantly depleted in the case of segmental duplications and copy number variants. Their conservation across phylogenetic entities indicates positive selection rather than neutrality in evolution. UCEs indicate enhancer-like activity of these elements (Derti A et al 2006 Nature Genet 38:1216). ►mutation neutral, ►networks; Bejarano G et al 2004 Science 304:1321.

Ultradian: Occurring periodically but less frequently than once a day. ►circadian

Ultrasonic (ultrasound): Refers to radiation in excess of 2×10^4 hertz/second, generally 5×10^5 hertz/sec. It is used for breaking down cells, treatment of arthritis and for tomographic examination or sonography, etc. Prenatal ultrasonic scanning of fetuses is widespread in the developed world and there is no convincing evidence of adverse effects. There has been much speculation about increase of left-handedness, lower cognitive development and cancer but experimental data have not confirmed these conjectures in large human populations. Ultrasound increases the delivery efficiency of exogenous nucleic acid to the intended target. The ideal system enhances gene expression in the target without having any effect on non-target tissues. Ultrasound may be able to provide this localization in somatic gene therapy as well. Low doses of about 100 mW cm^{-2} do not appear to have lasting and harmful effects. ►arthritis, ►tomography, ►sonography; see reviews in Prog Biophys Mol Biol 93:2007.

Ultrastructure: A fine structure beyond the resolution of the light microscope.

Ultrathin-Layer Gel Electrophoresis: Along with capillary microelectrophoresis it is well suited for rapid automated separation of biopolymers, e.g., nucleic acids as needed for the genome sequencing projects. (See Guttman A, Ronai Z 2000 Electrophoresis 18:3952).

Ultraviolet Light (UV): Refers to emission below the wavelength of violet (400–424 nm). UV-A has the most effective emission between 315 and 400 nm, UV-B between 280 and 315 nm and UV-C between 200 and 290 nm wavelengths. Nucleic acids have maximal absorption at about 260 nm. The absorption maximum may depend on base composition and pH. The maximal genetic effects of UV light coincide with the absorption maximum of nucleic acids. The major genetic effect of UV-B light is the generation of pyrimidine dimers and pyrimidinones. UV-A causes the production mainly of oxoguanine and reactive peroxides (Besaratinia A et al 2005 Proc Natl Acad Sci USA 102:10058) and this part of the UV spectrum is particularly carcinogenic if the DNA repair system is inefficient (Kozmin S et al 2005 Proc Natl Acad Sci USA 102:13538). Another study has indicated that UV-A-induced cyclobutane pyrimidine dimers are the major cause of DNA lesions in the human skin (Mouret S et al 2006 Proc Natl Acad Sci USA 103:13765). These compounds may damage the DNA and interfere with replication and transcription and can cause mutation and induce cancer. UV-B has immunosuppressive potential whereas UV-A is inert in this respect but it may suppress the UV-B effects. This protective effect is due to the induction of skin heme oxygenase. Oxidation of the membrane proteins may affect the signal transduction pathways and activate genes of the cellular defense systems. In prokaryotes, the LexA repressor is inactivated leading to the upregulation of genes mediating mutation, recombination and DNA repair. In yeast, cell cycle genes and various kinases are activated. In animal cells, AP and NF-κB transcription factors may be activated through the RAS signal transduction pathways. In plant cells, UV exposure may increase the synthesis of UV-absorbing flavonoids and phenylpropanoids and activates the octadecanoid defense pathway of fatty acids and generates less than 0.2 cyclobutane dimers per gene. In plants, transposons like *Mu* may enhance the mutagenic effects of UV-B radiation. Ultraviolet light in general has low penetration into biological material and a few cell layers may completely trap it. Nearly 100 genes of *Arabidopsis* are activated by UV-B and 7 are repressed (Ulm R et al 2004 Proc Natl Acad Sci USA 101:1397). The DNA must have a protection mechanism to convert dangerous electronic excitations by UV into less harmful vibrational energy and assure photostability (Satzger H et al 2006 Proc Natl Acad Sci USA 103:10196). ►DNA repair, ►genetic repair, ►physical mutagens, ►light response elements, ►light-sensitivity diseases, ►signal transduction, ►RAS, ►cyclobutane, ►pyrimidine dimer, ►8-oxodeoxyguanine, ►flavonoids, ►phenylpropanoids, ►fatty acids, ►electromagnetic radiation, ►*Mu*, ►JUN, ►excimer, ►melanocortin

Ultraviolet Photoproducts: ►cyclobutane ring, ►pyrimidine-pyrimidinone, ►Dewar product, ►cis-syn photoproduct, ►translesion pathway, ►DNA repair

Ultraviolet-Sensitivity Syndrome (UV^S S): RNA synthesis is inhibited in the cells by ultraviolet light. It appears to be different from xeroderma pigmentosum but is similar to the Cockayne syndrome. Ultraviolet-sensitivity is caused by a defect in the RAD genes required for genetic repair. ►xeroderma pigmentosum, ►Cockayne syndrome

Ultraviolet Spectroscopy: Used to measure the absorption of (≈ monochromatic) UV light and thus qualitatively or quantitatively identifies molecules such as DNA and RNA.

Umbilical Cord: Cord, which attaches the fetus to the mother during pregnancy. The placental blood in the umbilical cord contains stem cells suitable for repopulating the bone marrow. ►stem cells

UME: Denotes a unique mutational event which is a marker of evolutionary significance.

UML: Provides guidelines for data macromolecular structure and function presentation (Booch G et al 1997 The Unified Modelling Language User Guide, Addison-Wesley, Boston, Massachusetts).

UMP: This denotes uridine monophosphate. ►UDP

UMU (UV mutagenesis): Genes *UmuC* and *UmuD* are involved in repairing the DNA damaged by ultraviolet light. UmuD' is a post-translationally processed active form of the UmuD protein. ►DNA repair [SOS repair], ►translesion pathway

Unami: Describes the taste of monosodium glutamate, a flavor enhancer in some oriental foods. ►taste

Unbalanced Chromosomal Constitution: Parents heterozygous for inversions or translocations are functionally normal but may transmit to their offspring (unbalanced) duplication-deficiency gametes resulting in various physical and mental disabilities depending on the chromosomal region(s) involved (see Fig. U2). The frequency of unbalanced gametes (duplication/deficiency and duplication and deficiency) is more than 2% in the conceptuses.

This is, however, the lowest estimate because very frequently pregnancy is terminated before the current methods of analysis can detect fertilization. Even greater is the uncertainty about the frequency of chromosomal anomalies in the human sperm and egg that are extremely difficult to analyze cytologically. An unbalanced karyotype may be due to trisomy (in ~20–30% of spontaneous abortions). Monosomy has been observed in about 5 to 10%, triploidy in 6 to 7%, and tetraploidy in 2 to 4% of the spontaneously aborted fetuses. Although spontaneous abortion is extremely undesirable, suffering due to chromosomal anomalies experienced by live offspring is far more. There is no effective cure for unbalanced chromosomes. ►translocation, ►inversion; Levy B et al 1998 Genet Med 1:4.

UNC-6: A protein of the netrin family in *Caenorhabditis elegans* which guides ventral migration of the axon growth cone. The 407-amino acid TGF-β-like protein

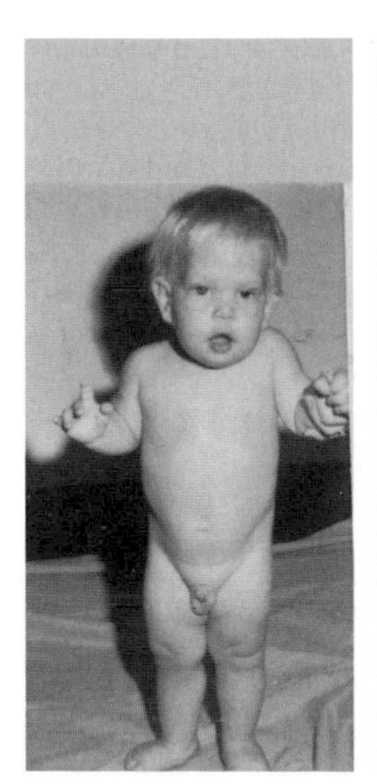
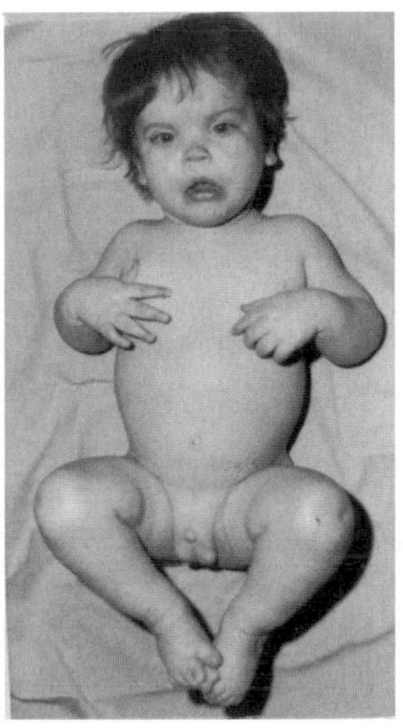
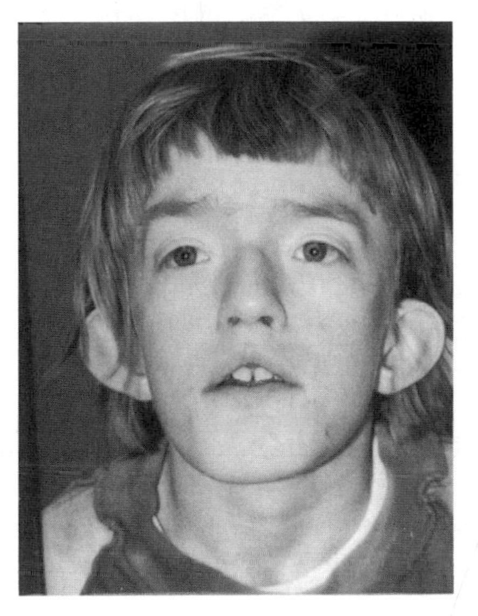

Figure U2. Unbalanced Chromosomal constitution inherited from phenotypically normal carrier parents. Left: Deficiency for the short arm of Chromosome 18. Middle: Duplication of the tip of the short arm of Chromosome 3. (Courtesy of Dr. Judith Miles). Right: Duplication of part of the long arm of Chromosome 4, resulting from a translocation heterozygosity involving Chromosomes 20 and 4. (Courtesy of Dr. D.L. Rimoin)

encoded by the UNC-129 gene has a similar function. ►netrins, ►semaphorins, ►TGF; Merz DC et al 2001 Genetics 158:1071.

UNC-33: A *Caenorhabditis* protein regulating axon extension. ►CRMP, ►axon; Ricard D et al 2001 J Neurosci 21:7203.

UNC-43: Encodes a calcium/calmodulin-dependent serine/threonine kinase type II and its mutation is responsible for multiple behavioral defects in *Caenorhabditis*. (See Sagasti A et al 2001 Cell 105:221).

unc-86 (uncoordinated): The *Caenorhabditis* gene has a pivotal role in determining neural identities. (See Burglin TR, Ruvkun G 2001 Development 128:779).

Uncertainty: ►Heisenberg's uncertainty

Uncharged tRNA: This has no amino acid attached to it. ►tRNA

Uncoating: During the initiation of infection the viral genome dissociates from the other constituents of the virus particle and is injected into the bacterium (see Fig. U3).

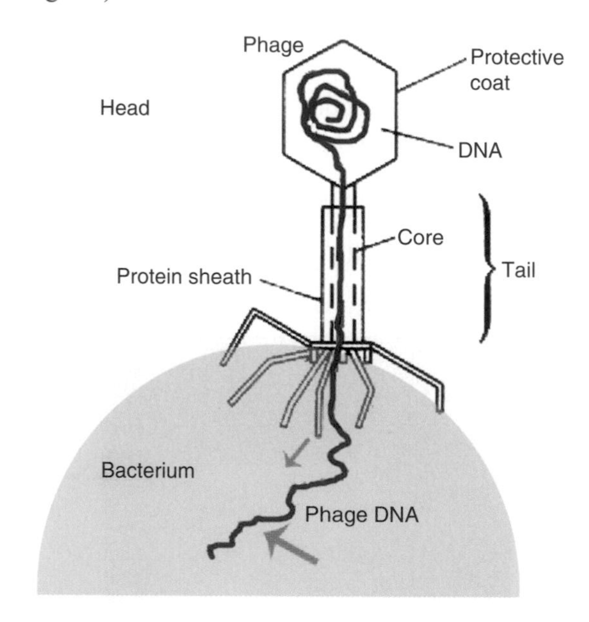

Figure U3. Uncoating of phage DNA

Uncompetitive Inhibitor: ►regulation of enzyme activity

Uncoupling Agent: This uncouples electron transfer from phosphorylation of ADP, e.g., dinitrophenol. ►cold-shock proteins

Uncoupling Protein (mitochondrial UCP): Uncouples oxidative phosphorylation from ATP synthesis and generates heat to safeguard against cold. Noradrenaline and adrenalin regulate it. UCP-3 is a member of the mitochondrial transporter family in the skeletal muscles and controls metabolic rate and glucose homeostasis. UCP-1 is present in the brown adipose tissue and is involved in fat combustion and heat generation. UCPs play an apparent role in obesity, macrophage-mediated immunity and in the control of ROS. ►ROS; Jarmuszkiewicz W 2001 Acta Biochim Pol 48:145.

Underdominance: Refers to inferiority in performance or fertility or proportion of hybrids in feral populations. It may be caused by Robertsonian fusion, transposable element, meiotic drive, post-implantation selection, etc. ►overdominance, ►Robertsonian translocation, ►meiotic drive; Davis S et al 2001 J Theor Biol 212:83.

Underwinding: Characterizes negative supercoiling. ►supercoiling

Unequal Crossing Over: Occurs between repeated sequences in the gene that may pair obliquely and therefore in the recombinant strands may have either more or less copies of the sequences than in the parental ones (see Fig. U4).

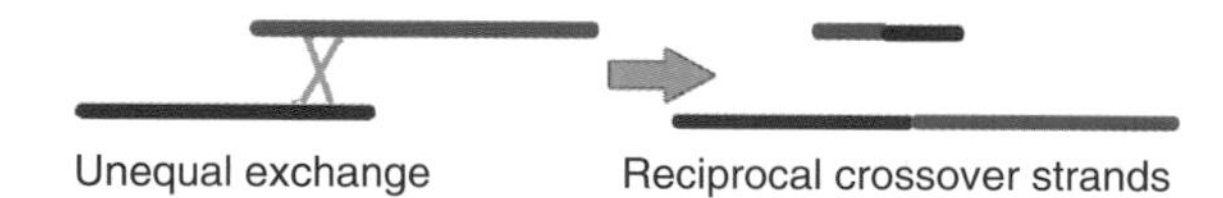

Figure U4. Unequal crossing over

The extra DNA material may not be needed for the species and thus can be used for evolutionary experimentation and may contribute to the evolution of new function(s). The frequency of such events may vary, e.g., in *Arabidopsis* ~3×10^{-6} has been observed. This estimate is within the range of mutation frequency and lower than that found at the *Bar* locus of *Drosophila* where unequal crossing over was first identified at a variable rate of 0.03 to 0.11. ►duplication, ►*Bar* locus, ►intragenic recombination; Bridges CB 1936 Science 83:210.

UniBLAST: Provides information on UniGene clusters, which are similar to the one interested in. ►UniGene, ►BLAST

Unidentified Reading Frame: ►URF

Unidirectional Replication: The replication fork moves only in one direction, left or right from the origin. ►replication fork, ►bidirectional replication; Maisnier-Patin S et al 2001 J Bacteriol 183:6065.

Unified Genetic Map: Such a map represents similarities in the distribution of nucleotide sequences across phylogenetic groups and is expected to not only provide tools for evolutionary studies, but also assist in transferring economically advantageous genes to crops. The ancestral chromosomal pattern is shared between many species. Unique sequences characterize species that are more closely related. ►SCEUS, ►evolution, ►comparative maps, ►CATS, ►mapping genetic, ►physical map, ►integrated map; O'Brien SJ et al 1993 Nature Genet 3:103.

Unfolded Protein Response (UPR): The accumulation of unfolded proteins in the endoplasmic reticulum (ER) activates the transcription of molecular chaperones such as BiP, GRP, calreticulin and protein disulfide isomerase in the nucleus so that protein folding can proceed as needed. Misfolded proteins may, however, lead to cell death if UPR cannot alleviate the problems and reactive oxygen species accumulate (Haynes CM et al 2004 Mol Cell 15:767). UPR signaling is mediated by endoplasmic reticulum transmembrane protein, Ire1 (inositol-requiring kinase 1), which has both kinase and nuclease domains, and by PERK (protein kinase-like endoplasmic reticulum kinase). Protein $P58^{IPK}$ is a cytoplasmic co-chaperone and is associated with the endoplasmic reticulum protein translocation channel Sec61 complex. $P58^{IPK}$ recruits heatshock protein 70 (HSP70) chaperone to the cytosolic face of Sec61. Proteins that enter slow ER can be exposed to proteasome in a co-chaperone dependent manner and thus the unfolded protein overload is reduced (Oyadomari S et al 2006 Cell 126:727). BAX and Bak proapoptotic proteins activate IRE1α signaling and provide a physical link between the core apoptotic pathway and UPR (Hetz C et al 2006 Science 312:572). Various adverse physiological conditions, toxins, inhibitors of the calcium pump, genetic defects altering protein structure, etc. evoke UPR. Proteasome inhibitors disrupt UPR in myeloma cells (Lee A-H et al 2003 Proc Natl Acad Sci USA 100:9946). Sialoglycolipid ganglioside can induce UPR and neuronal apoptosis (Tessitore A et al 2004 Mol Cell 15:753). Mutations in the editing domain of alanyl-tRNA synthetase disrupt fidelity of translation leading to misfolded proteins, cell death and neurodegeneration (Lee JW et al 2006 Nature [Lond] 443:50). ►endoplasmic reticulum, ►BiP, ►GRP, ►calreticulin, ►PDI, ►UPR, ►proteasome, ►gangliosides, ►apoptosis, ►BAK, ►BAX, ►endoplasmic reticulum-associated degradation, ►Sec61, ►heat-shock proteins; Rüegsegger U et al 2001 Cell 107:103; Ma Y, Hendershot LM 2001 Cell 7:827; Calfon M et al 2002 Nature [Lond] 415:92; Kaufman RJ et al 2002 Nature Rev Mol Cell Biol 3:401; Harding HP et al 2002 Annu Rev Cell Dev Biol 18:575; Ma Y, Hendershot LM 2004 Nature Rev Cancer 4:966; Schröder M, Kaufman RJ 2005 Annu Rev Biochem 74:739; protein quality control: Bakau B et al 2006 Cell 125:443.

UNG (uracil nucleotide DNA glycosylase): This DNA repair enzyme removes mis-incorporated uracil from the DNA. Deaminase may convert cytosine to uracil and then UG mismatch is formed that is either repaired by UND or the DNA strand is broken but either way it can result in mutagenic alteration. The removal of uracil from viral RNA by host UNG is detrimental to the pathogen. ►DNA repair, ►glycosylases, ►immunoglobulins; Krokan HE et al 2001 Progr Nucleic Acid Res Mol Biol 68:365.

Uniformity Principle: ►Mendelian laws

UniGene: A database of human and other organisms' ESTs and sequences of known genes to assist in identifying those which belong to the same cluster and have similar functions. This collection is suitable for microarrays and large-scale study of gene expression. ►EST, ►locuslink/ENTREZ GENE; Schuler GD et al 1997 J Mol Med 75:694; Zhuo D et al 2001 Genome Res 11:904; http://www.ncbi.nlm.nih.gov/entrez/query.fcgi?db=unigene; sets of transcript sequences: http://www.ncbi.nlm.nih.gov/UniGene/build2.html; transcriptome of various eukaryotic species: http://www.ncbi.nlm.nih.gov/UniGene/.

Unikont: ►bikont

Unilateral Incongruity: A special type of incompatibility when only the females of one of the related species are compatible with the males of the other species. In other words, it is a unidirectional compatibility. ►self-incompatibility; Bernacchi D, Tanksley SD 1997 Genetics 147:861.

Unimodal Distribution: Has only one major peak (see Fig. U5).

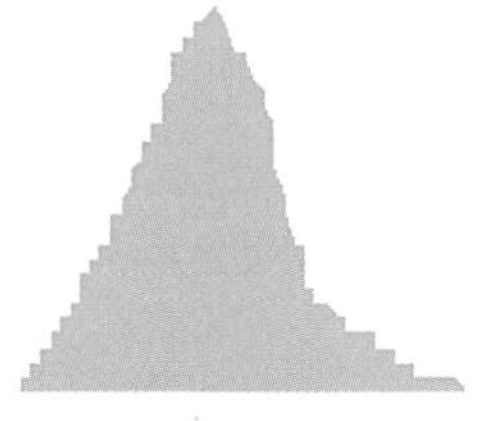

Figure U5. Unimodal distribution

Unineme: Refers to a single strand (e.g., of DNA or post-mitotic chromosome before the S phase).

Uninformative Mating: Does not shed light on the genetic constitution of the parents because it lacks clarity regarding linkage of the markers that may have poor penetrance or expressivity or both or the traits concerned are under polygenic control.

Uniparental Disomy (UPD): Both homologous chromosomes are inherited from only one of the parents in a diploid. UPD involves an epigenetic change like imprinting. When the two chromosomes are identical the case is called *isodisomy*, if they are different it is known as *heterodisomy*. For e.g., in 10% of the Russel-Silver syndrome cases a 35 Mb segment at 7q31-ter is maternally inherited as an isodisome. Uniparental disomy may be caused by the loss of one chromosome in a trisomics or duplication of a monosome. In a female cystic fibrosis patient markers on chromosome 7 indicate non-paternal disomy (Spence JE et al 1988 Amer J Hum Genet 42:217). In human UPD acrocentric isochromosomes and Robertsonian translocations between non-homologous chromosomes predominate. ►imprinting, ►epigenesis, ►nondisjunction, ►trisomy, ►monosomic, ►isochromosome, ►Robertsonian translocation, ►hydatidiform mole; Kotzot D 2001 J Med Genet 38:497.

Uniparental Inheritance: In the absence of male transmission of plastids and mitochondria, the genetic material of these organelles is transmitted to the progeny only through the egg. Also, telochromosomes and large deletions can generally be transmitted only through the female, if transmitted at all. In some species of *Mytilus* (mussel) one set of mtDNA is preferentially transmitted through the female whereas the other set is inherited paternally (Kenchington E et al 2002 Genetics 161:1579). The latter phenomenon is *doubly uniparental inheritance*. ►mtDNA, ►ctDNA, ►doubly uniparental inheritance, ►paternal leakage, ►mitochondrial genetics, ►chloroplast genetics, ►imprinting

Unipolar Depression: Refers to periods of depression with generally debilitating consequences but usually last for a shorter duration. A single nucleotide replacement (Arg441→His) in the tryptophan hydroxylase-2, a rate-limiting step in serotonin biosynthesis, causes 80% loss in serotonin production and it may be one of the causes of unipolar depression (Zhang X et al 2005 Neuron 45:11). ►bipolar mood changes, ►affective disorders

Uniporter: A transporter, which transports a type of molecule without coupling it to any other.

UniProt (universal protein resource): The most comprehensive information source on proteins from several resources. (See http://www.uniprot.org).

Unique DNA: DNA present in a single copy per genome. ►singlet

UniSTS: A marker information database, which is a unified non-redundant view of sequence-tagged sites (STS); it can be used for e-PCR. ►STS, ►e-PCR; http://www.ncbi.nlm.nih.gov/entrez/query.fcgi?db=unists.

Unit Evolutionary Period (UEP): Refers to time in million years (MY) required for the fixation of 1% divergence in two initially identical nucleotide sequences.

Unitig: ►scaffolds in genome sequencing, ►U-unitig

Univalent: Denotes an eukaryotic chromosome without a pair. In case the chromosomes in a hybrid are not sufficiently homologous, they may form univalents and their distribution to the poles in meiosis may be disorderly. The frequency of univalents may permit conclusions regarding the lack of relatedness of the parental forms (see Fig. U6). ►triticale

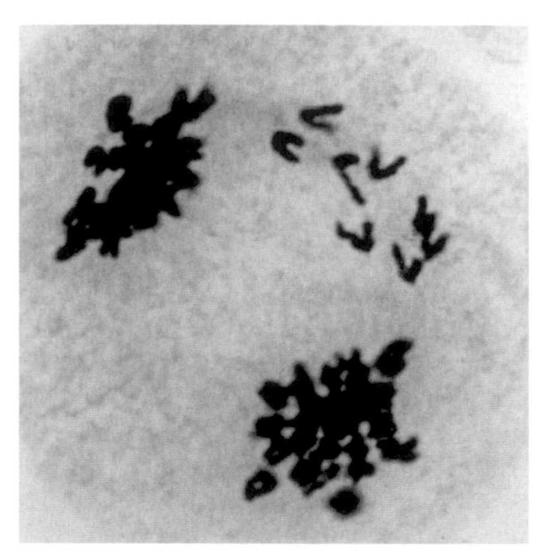

Figure U6. Hybrid of an octaploid (2n = 56) and hypoploid (2n = 41) hexaploid triticale. The female gamete contributed 28 and the male only 20 chromosomes. Therefore, there are 20 bivalents and 8 univalents in meiosis. Seven out of 8 univalents represent the *D* genome of wheat. Generally, the lagging univalents are not incorporated into the functional gametes because they fail to reach the poles. Usually, they divide belatedly. (Courtesy of Kiss Á 1966 Z Pflanzenzücht 55:309).

Universal Bases: The natural nucleoside of hypoxanthine or the synthetic 6-hydroxy- and 6-amino-5-azacytosine nucleosides or 1-(2′deoxy-β-D-ribofuranosyl)-3-nitropyrrole may serve for modified nucleotides in recognition of G, T and U with equal efficiency for Watson-Crick pairing (see Fig. U7).

The latter maximizes stacking while minimizing hydrogen pairing without sterically disrupting the double helix. Some of the analogs lower, whereas others increase the melting temperature of the polynucleotides that contain them. These may be used to synthesize oligonucleotide probes and primers when the exact sequence needed cannot be inferred because of the redundancy of the genetic code. Oligonucleotides containing the 5-nitroindole base may be easily detected by an antibody. In SNIPs the mismatched base

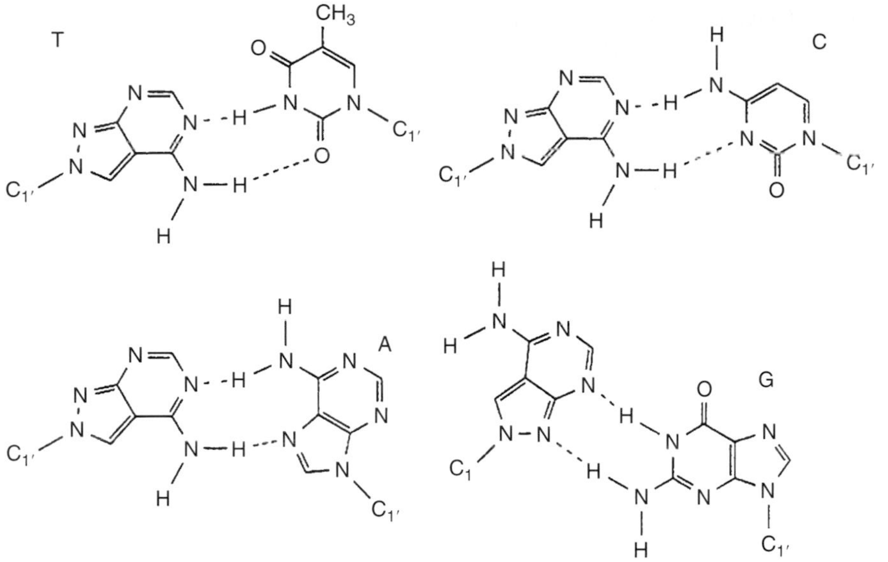

Figure U7. Pairing of the bases T (thymine), C (cytosine), A (adenine) and G (guanine) with N^8-(2′-deoxyribofuranoside) of 8-AZA-7-deazaadenine, a universal base.

can be identified easily. The addition of analogs to the 5′ end may enhance the stability of the analog oligonucleotide on the chips. The presence of these universal bases may alter nucleic acid-protein interactions. When in an A:C mismatch 3-nitro-pyrrole replaced C the fidelity of ligation increased substantially. The incorporation of nitroazole analogs may increase the stability of triplex sequences. Klenow fragment polymerase inefficiently incorporated nitropyrrole derivatives and the analog generally terminated further chain extension. ►hydrogen pairing, ►probe, ►primer, ►DNA chips, ►azacytidine, ►PCR, ►SNIPs, ►sequence saturation mutagenesis; Loakes D 2001 Nucleic Acids Res 29:2437; Seela P, Debelak H 2000 Nucleic Acids Res 28:3224.

Universal Code: The majority of DNAs across the entire phylogenetic range use DNA codons in the same sense. Notable exceptions exist in the mitochondria and in a few species. ►genetic code; O'Sullivan JM et al 2001 Trends Genet 17:20.

Universal Donor: ►ABO blood group

Universal Recipient: ►ABO blood group

Universal Trees: Display the evolution of orthologous proteins across taxonomic boundaries. (See Brown JR et al 2001 Nature Genet 28:281).

Univoltine: Having one generation annually.

Unix: A multi-user, multi-tasking computer system for servers, desktops and laptops. (See http://www.ee.surrey.ac.uk/Teaching/Unix/).

Unnatural Amino Acids: These are not coded for by the natural genetic system. The amber suppressor tyrosyl-tRNA synthetase of *E. coli* can translate the suppressor codon as tyrosine in the bacterium. When the enzyme is expressed in *Saccharomyces cerevisiae* it can aminoacylate the $tRNA_{CUA}$ but it does not aminoacylate any of the regular cytoplasmic tRNAs. The bacterial $tRNA_{CUA}$ is transported into the yeast cytoplasm and participates efficiently in translation. Mutations can be selected, which under specific conditions can translate certain unnatural phenylalanine and tyrosine amino acid derivatives in prokaryotic and eukaryotic systems. Thus, by such a technology the classic 20-amino acid code can be expanded and the consequences of these unnatural amino acids on the proteins can be determined. The incorporation of unnatural amino acids into proteins requires a unique codon suppressor tRNA and the corresponding aminoacyl-tRNA synthetase. The specificity of the aminoacyl tRNA can be modified by mutations at the active site and selection of the enzyme, which can specifically acylate the tRNA with the unnatural amino acid but not the conventional amino acid. At present more than 30 unnatural amino acids can be added to proteins using the procedure (Turner JM et al 2006 Proc Natl Acad Sci USA 103:6483). Actually, >190 unnatural amino acids are available commercially and >90 unnatural backbone and side-chain analogs can be enzymatically charged to tRNA (Hartman MCT et al 2006 Proc Natl Acad Sci USA 103:4356). Using an appropriately designed frameshift suppressor, tRNAs with four-base anticodons can deliver simultaneously two or three unnatural amino acids in response to the quadruplet codons CGGG and GGGU into a neuroreceptor of *Xenopus* oocytes. The frameshift suppressor effectively competes with an amber suppressor if delivered in excess (Rodriguez EA et al 2006 Proc Natl Acad Sci USA 103:8650). A genetically engineered ribosome, ribo-X, is more efficient at readthrough of amber codons with an attached unnatural amino acid, because of reduced affinity of the 30S ribosome for release factor RF-1 that is normally supposed to terminate the translation (Wang K et al 2007 Nature Biotechnol 25:770). ►amino acids, ►aminoacylation, ►genetic code, ►suppressor tRNA, ►aminoacyl-tRNA synthetase, ►selenocystein, ►pyrrolysine, ►orthogonal mRNAs; Chin JW et al 2003 Science 301:964; structural requirement in tRNA synthetase: Kobayashi T et al 2005 Proc Natl Acad Sci USA 102:1366; expanded genetic code: Wang W et al 2006 Annu Rev Biophys Biomol Struct 35:225; Liao J 2007 Biotechnol Progr 23:28.

Unnatural Bases: ►modified bases, ►base analogs

Unordered Tetrads: These tetrads do not contain the spores in a linear order as generated in the first and second meiotic divisions (see Fig. U8). In contrast to ordered tetrads, unordered tetrads require the presence of three genetic markers and two of them must be in

Figure U8. Unordered tetrad

different chromosomes to be able to calculate gene—centromere distances. Again—as in ordered tetrads—it is important to determine the frequencies of tetratype tetrads for at least three markers considered and for this p is designated as the tetratype frequency of say *a and b,* and q is the tetratype frequency of b and c, and similarly r is the tetratype frequency for a and c. The exchange frequency between a and its centromere = x, between b and its centromere = y, and between c and its centromere = z. Furthermore, the following three equations are needed:

$$\mathrm{p} = \mathrm{x} + \mathrm{y} - 3/2\mathrm{xy} \quad \mathrm{q} = \mathrm{y} + \mathrm{z} - 3/2\mathrm{yz}$$

$$r = x + z - 3/xz$$

The values, p, q and r being known, the unknown quantities, the recombination frequencies (x, y and z) between the three genes and their centromeres can be determined by solving the three equations:

$$\mathrm{x} = 2/3\left(1 \pm \sqrt{\frac{4 - 6p - 6r + 9pr}{4 - 6q}}\right),$$

$$\mathrm{y} = 2/3\left(1 \pm \sqrt{\frac{4 - 6p - 6q + 9pq}{4 - 6r}}\right) \text{ and}$$

$$z = 2/3\left(1 \pm \sqrt{\frac{4 - 6q - 6r + 9qr}{4 - 6p}}\right).$$

Once the recombination frequency between a gene and its centromere becomes available, the exchange frequencies between additional genes and their centromeres can also be calculated by the formula: $\mathrm{s} = \frac{2(\mathrm{v}-\mathrm{t})}{2-3\mathrm{t}}$ where s = the unknown recombination frequency between marker d and its centromere, t = the known recombination frequency between gene e and its centromere, and v = the tetratype frequency for the unmapped gene d and the mapped gene e. ►tetrad analysis; Emerson S 1963, p. 167 Methodology in Basic Genetics, Burdette, WJ, [Ed.] Holden-Day, San Francisco.

Unphased Diploid Population: The linkage phase of the heterozygotes is unknown.

Unrooted Evolutionary Trees: These do not indicate the initial split of the branching. ►evolutionary tree; Steel M, McKenzie A 2001 Math Biosci 170:91.

Unsaturated Fatty Acid: This contains one or more double bonds. ►Omega-3-fatty acids

UNSCARE: United Nations Committee on Effects of Atomic Radiation.

Unscheduled DNA Synthesis: The replication of the DNA is outside the normal S phase and this indicates a repair replication. The tests are generally carried out on cultured hepatocytes or fibroblasts exposed to certain treatment(s) and the incorporation of radioactive thymidine is monitored by either autoradiography or scintillation counting. The data are compared with concurrent controls that have not been exposed to any mutagen. Although the procedure appears attractive it is not very effective and practical for the identification of mutagens or carcinogens. ►bioassay in genetic toxicology, ►DNA polymerases; Zbinden G 1980 Arch Toxicol 46:139; Hoege C et al 2002 Nature [Lond] 419:135.

Unstable Genes: These have higher than average mutation rate. Usually instability is caused by the movement of insertion or transposable elements (see Fig. U9). Higher mutation rate may also be due to deficiency of genetic repair or defects in DNA replication. ►DNA repair, ►insertional mutation, ►transposable elements, ►error in replication, ►fractional mutation, ►variegation, ►genomere

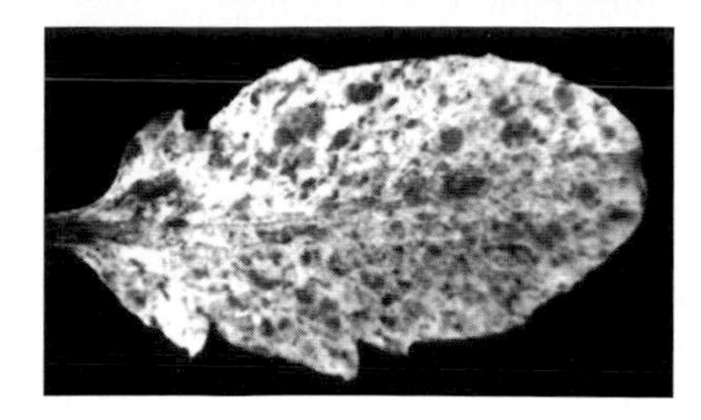

Figure U9. Unstable genes

Unstructured Proteins: These lack three-dimensional (globular) structure ion free state but may become structures upon binding to other macromolecules. The adaptability of their structure is an advantage in regulatory functions. ►phosphorylation; Wright PE, Dyson HJ 1999 J Mol Biol 293:321; Iakoucheva LM et al 2004 Nucleic Acids Res 32:1037.

Unsupervised Learning: Identifies new, so far undetected, shared pattern(s) of sequences in macromolecules and determines the positive and negative representatives of the pattern(s). The information permits correlations between structure and function in languages as well as in proteins without direct human intervention in the details. ►supervised learning, ►support vector machine; Wallis G, Baddeley R 1997 Neural Comput 9:883; Hatzivassiloglou V et al 2001 Bioinformatics 17 Suppl 1:S97; Solan Z et al 2005 Proc Natl Acad Sci USA 102:11629.

Untoward Pregnancy: Pregnancy that terminates with congenital malformation or stillbirth or infant death within 14 days after delivery. ►congenital

Untranslated Regions (UTR): These are the leader sequences upstream from the first methionine codon and the downstream sequences beyond the stop codon of the mRNA. The upstream and downstream regions include various control elements and at the 3′ end in eukaryotes the polyadenylation signal is situated. The untranslated regions of the mRNA do not code for any amino acid sequence. The 3′-UTRs in RNAs control translation, the fate of germ line cells, life cycle, anterior-posterior axis of development, meiotic cycles by binding proteins similarly to the DNA binding proteins. The ENCODE project observed that most of the longer protein-coding transcripts differ only in their UTRs but the coding regions are shared. ►upstream, ►downstream, ►stop codon, ►polyadenylation, ►ENCODE; Kuersten S, Goodwin EB 2003 Nature Rev Genet 4:626; eukaryotic regions: http://www.ba.itb.cnr.it/UTR/.

Unusual Bases: Refer to modified forms of the normal DNA or RNA bases; they may be common in tRNA. Their incorporation into the DNA may lead to base substitution mutations. Methylation of C and A nucleotides may lead to imprinting, transient genetic variations in expression and alteration of RFLP. DNA containing an unnatural base pair can be amplified and function as a template for the site-specific incorporation of base analog substrates into RNA via transcription. Base-pairing in such a case takes place by specific hydrophobic shape complementation between the bases, but lacks hydrogen-bonding interactions. In replication, this unnatural base pair exhibits high selectivity in combination with the usual triphosphates and modified triphosphates, gamma-amidotriphosphates, as substrates of 3′ to 5′ exonuclease-proficient DNA polymerases, allowing PCR amplification. In transcription, the unnatural base pair complementarity mediates the incorporation of these base substrates and their analogs, such as a biotinylated substrate, into RNA by T7 RNA polymerase (Hirao I et al 2006 Nature Methods 3:729). With this system, functional components can be site-specifically incorporated into a large RNA molecule. ►tRNA, ►RFLP, ►imprinting, ►methylation of DNA, ►Bar, ►modified bases; Barciszewski J et al 1999 Mol Biol Rep 26:231.

Unverricht-Lundborg Disease (EPM1): ►myoclonic epilepsy

Unwindase: Refers to double-stranded RNA helicases that unwind in a single step the entire length of a molecule in an ATP-dependent process. ►DexH, ►NPH; Nishikura K 1992 Ann NY Acad Sci 660:240.

Unwinding Protein: Facilitates unwinding of a DNA double helix and stabilizing single strands. The nucleotide triphosphate (NTP)-dependent RNA helicases are required for unwinding of double-stranded RNA and replication of many pathogenic viruses. ►DNA replication, ►helicase; Dillingham MS et al 2001 Proc Natl Acad USA 2001 98:8381.

Unzipping of DNA: Refers to the separation of the double-stranded structure. Depending on the base composition, the threshold force required is about 12 pN (piconewton). One joule is the work done when a force of 1 N acts through a distance of 1 meter. ►joule; Cocco S et al 2001 Proc Natl Acad Sci USA 98:8608.

uORF: Denotes upstream open reading frame which may play a regulatory role in protein synthesis.

UP Elements (upstream elements): In the DNA, (−40 to −150 sites from the initiation of transcription) are recognition sites for the α subunit of the prokaryotic RNA polymerase which boost (30 to 300-fold) the frequency of transcription initiation. The −46 to −38 element, 5′-AAAAAARNR-3′

stimulates transcription up to 170-fold, the −57 to −47 element, 5′-AWWWWWTTTTT-3′ stimulates transcription up to 16-fold (Estrem ST et al 1999 Genes & Develop 13:2134). [W = A or T, R = A or G, N = no single base pair is present in 70% of the population and no 2 bp make up 95% of the population]. The α subunit C-terminal domain and the σ subunit bind initially to the −35 sequence and in the final isomerization the downstream double helix is embedded in the β/β′ jaws leading to the transcriptionally active complex (Sclavi B et al 2005 Proc Natl Acad Sci USA 102:4706). ►transcription factors, ►promoter, ►σ, ►FIS, ►CRP, ►RNA polymerase, ►isomerization of strands, ►gyrase, ►topoisomerase, ►triplex

Up-and-Down: The structural arrangement of helical protein bundles comparable to the meander of β-sheets. ►meander, ►protein structure

UPE: Refers to the upstream promoter element. ►promoter

UPGMA (unweighted pair group method with arithmetic means): These are formulas for determining evolutionary distances. ►transformed distance, ►evolutionary distance; Kim KI et al 2002 Animal Genet 33:19.

UPR (unfolded protein response): Regulates gene expression when the endoplasmic reticulum does not function properly. It may control chaperones, phospholipid biosynthesis, secretory pathways and degradation of proteins associated with the endoplasmic reticulum under stress or even without stress. ►endoplasmic reticulum, ►chaperones, ►unfolded protein response; Bertolotti A, Ron D 2001 J cell Sci 114:3207.

Upregulation: Refers to increasing activity by regulation. ►downregulation, ►regulation of gene activity

Upstream: This is in the direction of the 5′ end of polynucleotides (DNA). ►downstream

Upstream Activation Sequence: ►UAS

Upstream Regulatory Sequence (USR): This is the regulatory element in the promoter region.

Uptag: ►bar code genetic

Uptake: Eukaryotic cells may incorporate nuclei, plastids and mitochondria, pseudovirions, plasmids, liposomes and various other macromolecules besides smaller organic and inorganic molecules. Viruses move from cell to cell through the plasmodesmata and through the vascular system of plants. Plant viruses generally encode a movement protein that modifies the plasmodesmata, binds to single-stranded nucleic acids and associates with the cytoskeleton and the endoplasmic reticulum. (See Tzfira T et al 2000 Annu Rev Microbiol 54:187).

Uptake, Selective: Inorganic or organic molecules are not taken up by a stochastic process but the uptake is regulated by special mechanisms. ►ion pumps, ►transformation, ►conjugation

Uracil: A pyrimidine base in RNA; 2,4-dioxypyrimidine, MW 112.09, soluble in warm water but insoluble in ethanol. ►pyrimidines

URE (ureidosuccinate utilization): Cytoplasmic proteins involved in nitrogen metabolism in yeast are responsible for the production of a protein analogous to prion. ►prion; Baxa U et al 2002 Proc Natl Acad Sci USA 99:5253.

Urea Cycle: Describes the formation of urea ($[NH_2]_2CO$) from amino acids and CO2; ornithine is converted to citrulline which in turn is converted to arginine. The hydrolytic cleavage of arginine produces urea and regenerates ornithine and thus the cycle is complete. The urea cycle in the mitochondria secures homeostasis for ammonium with some independence of the nitrogen intake. The mutation in the ornithine transporter (encoded at 13q14) involves the symptoms of hyperornithinemia-hyperammonemia-homocitrullinuria syndrome. ►arginine, ►ornithine, ►citrulline carbamoylphosphate synthetase deficiency, ►amino transferase

Uredium: A uredospore producing sorus, a type of sporangium of fungi and protozoa (see Fig. U10). ►stem rust, ►sorus

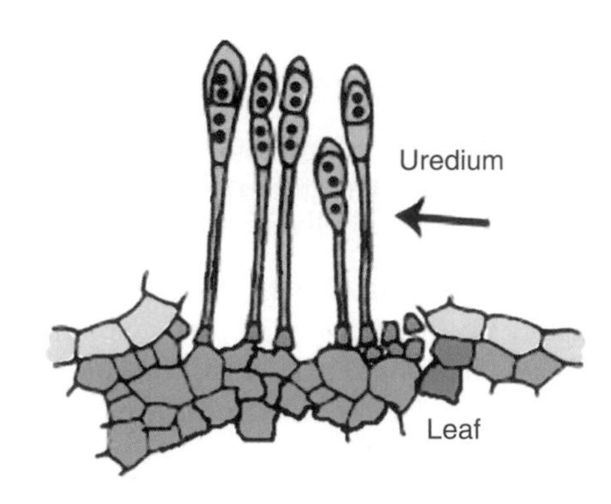

Figure U10. Uredium

Uremia: Refers to urine in the blood caused by a variety of factors.

Ureotelic: Excreting urea. ►Urea (NH_2CONH_2), ►Urea cycle

Urethan ($CH_3H_7NO_2$): This is a toxic (lethal dose 2g/kg in rabbits) liquid (at 48–50 C°). It causes chromosomal breakage and mutation; it is also antineoplastic.

URF (unidentified reading frame): Capable of transcription (open) but the gene product has not been identified. ►reading frame

Uric Acid: A degradation product of xanthine and it is excreted in the urine (see Fig. U11). It is secreted in particularly high amounts in hyperuricemic individuals under dominant and polygenic control, as well as in glycogen storage diseases, HPRT deficiency and gout. Birds and reptiles normally excrete high amounts. ►glycogen storage diseases, ►gout, ►HPRT, ►Lesch-Nyhan syndrome, ►xanthinuria

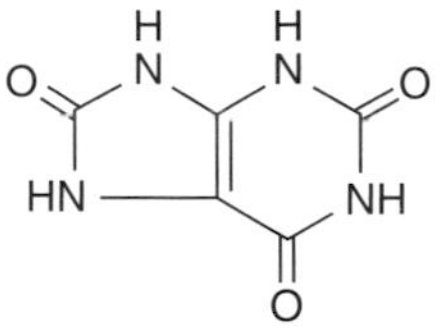

Figure U11. Uric acid

Uricotelic: Excretes uric acid. ►uric acid

Uridine: Uracil + ribose, an RNA nucleoside. ►pyrimidines

Uridine Diphosphate Glucuronosyl Transferase: Encoded in human chromosome 2; this group of enzymes is glucuronate steroid hormones. ►Crigler-Najjar syndrome

Uridine Monophosphate Synthetase Deficiency (UMPS): ►oroticaciduria

Uridylate: Refers to the nucleotide of uracil. It contains uracil + ribose + phosphate.

URL (uniform resource locator): The generic name for an Internet resource, such as WWW page; Gopher menu (http://www.sc.edu/bck2skol/fall/lesson12.html), a file transfer protocol server, etc. ►bookmarks

URNA: This is a uridine-rich nuclear, non-coding RNA involved in transcript processing within the cell nucleus. (See Dye MJ et al 2007 Cold Spring Harbor Symp Quant Biol 71:275; Matera AG et al 2007 Nature Rev Mol Cell Biol 8:209).

21U-RNA: 21U-RNAs are precisely 21 nucleotides long, begin with a uridine 5′-monophosphate but are diverse in their remaining 20 nucleotides, and appear modified at their 3′ terminal ribose. 21U-RNAs originate from more than 5,700 genomic loci dispersed in two broad regions of chromosome IV of *Caenorhabditis*—primarily between protein-coding genes or within their introns. These loci share a large upstream motif that enables accurate prediction of additional 21U-RNAs. The motif is conserved in other nematodes, presumably because of its importance for producing these diverse, autonomously expressed, small RNAs (Ruby JG et al 2006 Cell 127:1193). ►small RNA

U-RNP: U-RNA associated with proteins involved in processing the primary transcripts of genes within the nucleus.

Urocortin: A neuropeptide similar to urotensin and corticotropin releasing factor. It elicits the synthesis adrenocorticotropic hormone and is therefore involved in stress-related endocrine, autonomic and behavioral responses. ►adrenocorticotropin, ►urotensin, ►corticotropin releasing factor; Parkes DG, May CN 2000 News Physiol Sci 15:264.

Urogenital: Denotes both the system of urine secretion and the reproductive organs.

Urogenital Adysplasia: Autosomal dominant failure to develop one or both kidneys, frequently coupled with wide set eyes, low set ears and other anomalies. The incidence among newborns may be as high as 4.5% and among adults 0.3%. ►kidney disease

Urokinase (uPA): A plasminogen activator. The urokinase receptor is a glycosyl-phosphatidyl-inositol-linked cell surface protein that regulates cell adhesion. It has been suggested that urokinase is a requirement for some cancerous growths. Some catechins (present in green teas) appear inhibitory to uPA. ►plasminogen, ►CAM, ►intravasation, ►anthrax; Andreasen PA et al 1997 Int J Cancer 72:1; Plesner T et al 1997 Stem Cells 15:398; crystal structure: Huai Q et al 2006 Science 311:656.

Uropathy: Refers to diseases of the urogenital system.

Urotensins: These are short bioreactive peptides of 41 (urotensin I) and 12 (urotensin II) amino acids, respectively. ►urocortin; Lewis K et al 2001 Proc Natl Acad Sci USA 98:7570.

URS: These are upstream regulatory/repressing sequences which are the binding sites for various transcription factors. ►promoter, ►transcription factors, ►regulation of gene activity, ►UAS; Hanna-Rose W, Hansen U 1996 Trends Genet 12:229.

Urticaria, Familial Cold (Muckle-Wells syndrome, 1q44): Autosomal dominant disease is characterized by reoccurring, transient burning papules and macules on the skin, cold-sensitivity, and is frequently associated with progressive deafness and renal amyloidosis (fibrillary protein deposits in the kidneys). ►cold hypersensitivity

Use and Disuse: ►Lamarckism

Usher Syndrome (USH): The condition is hereditary yet the pattern of inheritance is quite variable although in most cases it is probably autosomal recessive. It is characterized by deafmutism, retinitis pigmentosa, mental disabilities and ataxia. Three types of the syndrome are usually distinguished on the basis of the severity and onset of the disease. The prevalence is around 4 – to 5 × 10^{-5}. USH1 has a pre-pubertal onset; seven loci are associated with it. About 75% of the afflicted individuals have a severe defect in the USH1B gene in human chromosome 11q13.5, encoding myosin VIIA. USH1A is in chromosome 14q32 (Heilig R et al 2003 Nature [Lond] 421:601). Other types of the disease all involving hearing deficit and some other anomalies have been reported: USH1D and DFNB12 (10q21-q22), USH1F (chr. 10q21-q22, protocadherin), USH1E (21q21) USH2A (1q41). USH1G encodes a scaffolding protein. USH1D encodes a type of cadherin (CDH23) that causes stereocilia disorganization in waltzing mouse with hearing deficit (Bolz H et al 2001 Nature Genet 27:108). A milder form has been assigned to 1q32. Myosin VIIA may be involved in the transport between the outer and inner layers of the photoreceptors of the eye. In type IIa form a 171.5 kDa protein is involved that has a laminin epidermal growth factor and fibronectin type III motifs pointing to the involvement of defects in cell adhesion. USH1C involves a defect in the PDZ domain (involving PSD-95, DLG and ZO-1 proteins) of harmonin. The PDZ modules interact with all other proteins concerned with signaling and the cytoskeleton. Deletions of the USH1C gene may lead to infantile hyperinsulinism, enteropathy (intestinal disease) and deafness. USH3 (3q21-q25) encodes a 120-amino acid protein concerned with recessive, progressive hearing loss and severe retinal degeneration. ►retinitis pigmentosa, ►ataxia, ►deafmutism, ►deafness, ►waltzing mouse, ►myosin, ►cadherin, ►hypoglycemia, ►EGF, ►laminin, ►fibronectin, ►CAM; Verpy E et al 2000 Nature Genet 26:51; Bolz H et al 2001 Nature Genetics 27:108; Bork JM et al 2001 Am J Hum Genet 68:26; Ahmed ZM et al 2001 Am J Hum Genet 69:25; Joensuu T et al 2001 Am J Hum Genet 69:673; Adato A et al 2005 Hum Mol Genet 14:347.

USM (ubiquitous somatic mutations): These are attributed to defects in DNA repair (mismatch repair) and in DNA replication. ►DNA repair, ►unstable genes

U-snRNP: A splicing factor of RNA transcript. ►splicing, ►spliceosome; Xue D et al 2000 EMBO J 19:1650.

USP (chromosome-specific unique sequence probes): Employ locus-specific fluorescent DNA sequences which are suitable for the identification of small deletions and duplications. ►FISH, ►chromosome painting, ►WCPP, ►telomeric probes

Ustilago maydis (n = 2): Basidiomycete that has been extensively used for meiotic and mitotic analysis of recombination and for the isolation of biochemical mutations, etc. The 20.5-million base genome contains 6,902 protein-coding genes (see Fig. U12). Its pathogenicity is controlled by 12 clusters of genes encoding secreted proteins (Kämper J et al 2006 Nature [Lond] 444:97). This fungus causes the ear smut of maize. The haploid forms are non-pathogenic. Several other *Ustilago* species are pathogenic to other Gramineae. ►fungal life cycles; Banuett F 1995 Annu Rev Genet 29:179.

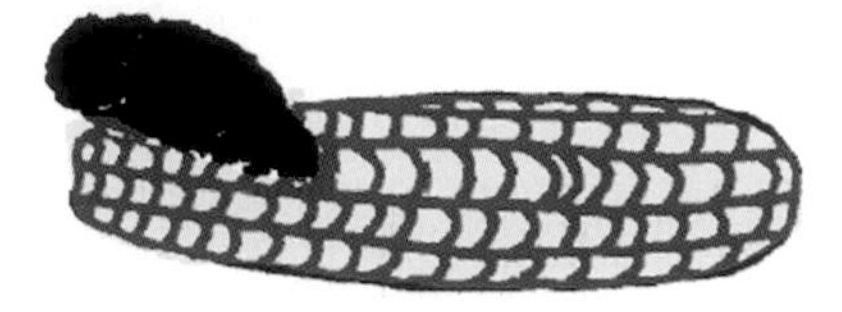

Figure U12. Ustilago on maize ear

UTase (uridylyl transferase): Catalyzes the transfer of uridyl group to the P_{II} regulatory subunit of adenylyl transferase (ATase), an enzyme which transfers an adenylyl group from ATP to a tyrosine-hydroxyl in glutamine synthetase. The complex ATase • P_{II}-uridylyl catalyzes phosphorolytic deadenylylation of glutamine synthetase. Glutamine synthetase is an enzyme involved in many functions. ►glutamine synthetase, ►glutamate dehydrogenase

UTE (untranslated exon): May play a role in alternative splicing of transcripts. ►alternative splicing; Chen C et al 2002 Proc Natl Acad Sci USA 99:2930.

Uterine Cancer: Cancer of the uterus may be treated by anti-estrogens similarly to breast cancer. The response to tamoxifen is very limited, however. Cervical cancer usually does not respond to such treatment and surgery and radiation are used. ►breast cancer

Uterus: The hollow female abdominal organ where the fertilized egg is embedded for the development of the embryo. The pear-shaped uterus (~5 × 7.5 cm in humans) is connected to the vagina through the narrow neck-like passage, the cervix, permitting the entry of the spermatozoa, to retain the conceptus and open to release the embryo at birth. The ovaries are connected to the uterus by the oviducts (see Fig. U13). The uterus is lined by endometrium that feeds the early embryo. Its outer layer is shed during menstruation and regenerated thereafter. When the blastocysts reach a state ready for implantation, the uterus becomes receptive under the influence of the ovarian steroid hormones—progesterone and oestrogen. Luteinizing hormone, secreted by the ovaries, is essential for ovulation and for the programming of progesterone and oestrogen. The duration of this oestrus cycle is about 4 days in mice and almost a week in humans. ►gonads, ►ovary, ►blastocyst, ►oestrus, ►luteinizing hormone, ►fallopian tube, ►morsus diaboli, ►fertilization, ►embryogenesis in animals, ►menstruation, ►vagina, ►endometrium, ►tamoxifen, ►cervical cancer; Wang H, Dey SK 2006 Nature Rev Genet 7:185.

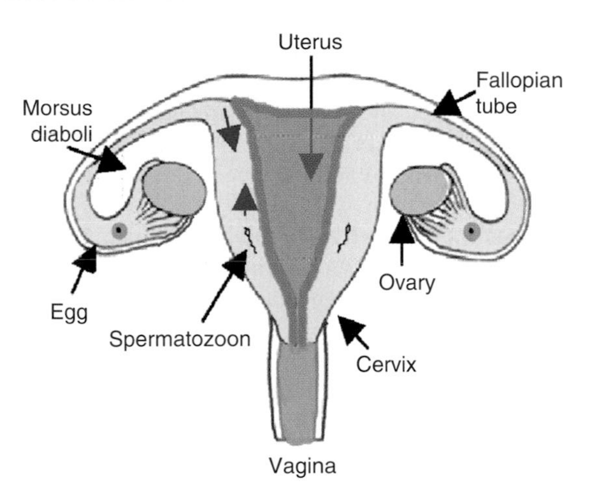

Figure U13. The abstract outline of the female reproductive tract; the red arrow indicates the path of the egg and the blue arrow indicates the direction of movement of the spermatozoa

Utility Index for Genetic Counseling: If the mother is heterozygous for a recessive disease allele d/D and other linked alleles (M_1/M_2), i.e., she is DM_1/dM_2 and the frequency of recombination between the two loci = r, half of her sons will be afflicted by the disease and 1 − r frequency of M_2 sons is expected to express the disease (as long as penetrance and expressivity are high). A small *r* helps predictability. If her husband's genotype is DM_1, among her M_1M_2 daughters 1 − r will be a carrier of the recessive disease allele *d* but one must know for sure which of the codominant *M* alleles is syntenic with *d*. In the case of *X linkage*, for prediction on the basis of the *M* markers, the genetics counselor should know the genotype of the affected grandfather (dM_1 or dM_2). If the frequencies of the M_1 and M_2 markers is x_1 and x_2, respectively and the grandfather is dM_2, then the probability that the mother being informative for genetic counseling is = $2x_1x_2$. Roychoudhury and Nei have described this as the *utility index of a polymorphic locus* for genetic counseling. If the grandfather has the recessive allele *m* and the grandmother is either MM or Mm, their heterozygous (Mm) daughter must have inherited the *M* allele maternally and the expected frequency of informative mothers is (1 − x)x. In case both *M* and *m* grandfathers are considered, the utility index of the mother becomes: $(1 - x^2)x$.

Genetic information about the mother can also be obtained from her children. In case the X-linked disease gene deemed to be in coupling with another marker, the Bayesian probability for coupling is $(1-r)^2/[(1-r)^2+r^2]$ and the probability for repulsion is $r2/[(1-r)^2+r^2]$ if the mother has already two afflicted sons. If the mother has both normal and afflicted sons: n_1 (DM_1), n_2 (DM_2), n_3 (dM_1) and n_4 (dM_2), the probability for coupling that she has 4 (= n) sons with the genotypes above is: $r^{n2+n3}(1-r)^{n1+n4}/2^n$; in case of repulsion the probability is: $r^{n1+n4}(1-r)^{n2+n3}/2n$. The posterior probability that she is in coupling is $1/(1+\rho^{\alpha})$ where $\rho = r/(1-r)$ and $\alpha = n_1 + n_4 - (n_2 + n_3)$, and in case $\alpha = 0$, the linked markers will not help in the prognosis. For repulsion, the probability is $1/(1+\rho^{-\alpha})$. The probability of the genotype of the next offspring depends on the tightness of linkage.

In the case of *autosomal dominant* disease in the presence of *D*, the disease is expected and both parents are informative. If the mother is M_1M_2 and the father is M_1M_1 and the recombination frequency between the *M* and the *D* loci is r, the probability that the offspring of M_1M_1 genotype will have *D* is 1 − r. *DD* homozygotes are very rare because their occurrence depends on the product of the frequency of the *D* gene, which is usually in the 10^{-5} range. In case there are multiple alleles at a locus, the total frequency of informative parents is

$$1 - \sum X_i^2 - (\sum X_i^2)^2 + \sum X_i^4.$$

In a mating of (*Dm/dM*) × (*dM/dm*), the offspring homozygous for *m* has a probability of 1 − r to carry the dominant disease gene, *D* 1 − r but the one with the dominant *M* phenotype is expected to be *D* at a frequency of (1 + r)/3. If the affected parent *D* is heterozygous and the other parent is *dd*, the frequency of informative families is $2(1 - x)x^4$. In case the mating is (DdM_1M_2) × (ddM_1M_1), the children are $n_1(DdM_1M_1)$, $n_2(DdM_1M_2)$, $n_3(ddM_1M_1)$ and n_4-(ddM_1M_2), respectively. The linkage phase can be estimated as discussed earlier for X linkage [coupling: $1/(1 + \rho^{\alpha})$, repulsion $1/(1 + \rho^{-\alpha})$]. All children of parent DdM_1M_2 will be informative, except when the spouse is ddM_1M_2 and all the progeny will be heterozygous for the *M* locus but the probability that all children would be of such genetic constitution is $(0.5)^n$. The proportion of informative families when n > 1 is: $2x_1x_2(1 - 0.5^{n-1}\,x_1x_2)$. If there are multiple markers, the proportion of informative families (with x_i frequency of the ith allele) is:

$$2\sum_{i<j} X_iX_j(1 - 0.5^{n-1}X_iX_j).$$

In the case of autosomal recessive diseases the calculation of the mathematical probabilities of informative families is more complex and biochemical or molecular (DNA) analyses are preferred.

►DNA fingerprinting, ►genetic counseling, ►counseling genetic, ►risk, ►paternity testing; Roychoudhury AK, Nei M 1988 Human Polymorphic Genes, Oxford University Press, New York.

UTP: Denotes uridine triphosphate (see Fig. U14). ►UDP

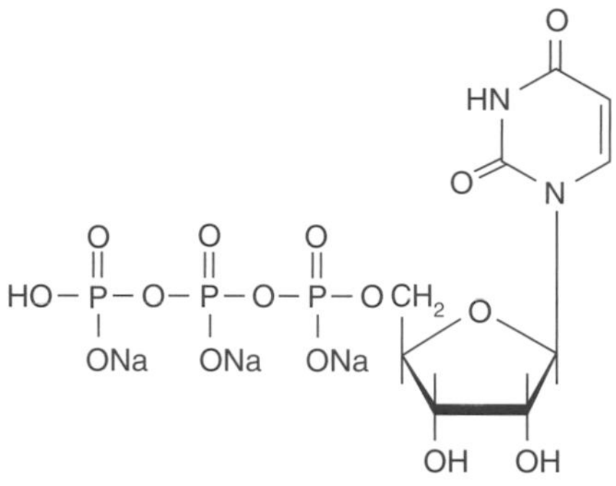

Figure U14. Uridine triphosphate

UTR: ►untranslated region

Utrophin: ►dystrophin (DRP2)

U-Unitig: ►scaffolds in genome sequencing, ►unitig

Historical vignettes

"The attainment…of fundamental knowledge is usually of the utmost immeasurable practical importance in the end".

HJ Muller 1916, quoted by Elof Carlson

(30 years later [in 1946] Muller received the Nobel Prize)

Priority has been of important concern to authors from the early days of genetics. In his book "A History of Genetics" (Harper & Row, New York, 1965) AH Sturtevant writes on p. 27:

In a letter received and published by Roberts, de Vries later stated that he had worked out the Mendelian scheme for himself, and was then led to Mendel's paper by reading Bailey's copy of Focke's reference. In 1954, nineteen years after the death of de Vries, his student and successor Stomps reported the de Vries had told him that he learned of Mendel's work through receiving a reprint of the 1866 paper from Beijerinck, with a letter saying that he might be interested in it. This reprint is still in the Amsterdam laboratory, as has been stated.

UV: ►ultraviolet light, ►cyclobutane dimers

Uvomorulin (UM): A transmembrane glycoprotein, also known as E-cadherin. ►cadherins

UvrABC: An endonuclease complex of uvrA, uvrB and uvrC and these enzymes are involved in the excision of ultraviolet light-induced pyrimidine dimers. After the dimer is recognized, cuts are made on both sides of the dimer and thus the damaged area of about 12 nucleotides is excised. ►DNA repair, ►ABC excinucleases, ►excision repair

UV Spectrophotometry of Proteins: Prepare a series of dilutions (20 to 3,000 μg/mL) from a pure 3 mg/mL standard bovine serum albumin (BSA). Make a blank and a series of dilutions of the sample to be tested. Determine UV absorption at 280 nm. BSA, 3 mg/mL is expected to have an absorption of 1.98. Calibrate the sample relative to the standard. In case the absorptivity of the protein to be tested is known use the formula for the calculations (a_{280} = the absorption in the units of mg/mL per centimeter path b).

$$\frac{O.D_{280}}{a_{280} x b}$$

V

V Gene: Codes for the variable region of the antibody molecule.

V or V_{max}: The maximal velocity of the reaction when the enzyme is saturated with substrate.

v-mos: ►Moloney mouse sarcoma

v-Oncogene: ►oncogenes

V-point: The progression of the cell cycle beyond this point (≈6 h before the S phase) requires no insulin but only IGF-1. ►cell cycle, ►insulin, ►insulin-like growth factor

Vaccines: A suspension of killed or attenuated pathogens or recombinant protein or DNA or capsular polysaccharides conjugated to a carrier protein for generating an immune defense system. The most successful vaccines (measles, mumps, rubella) are composed of antigens generated against disease-causing microorganisms and injected into the blood-stream to stimulate the development of circulating or serum antibodies (immunoglobulin G). More recently efforts have been made to develop vaccines that activate mucosal immunity. Membranes of the body, covering the gastrointestinal tract, the air-intake organs and the reproductive system are covered with mucosa. The aim of this procedure is to trap infectious agents at the port of entry. The mucosa can develop sufficient quantities of immunoglobulin A. These new vaccines may be orally delivered and do not have to be injected. Dendritic cells, which are most active in antigen presentation, can be targeted by microparticles conjugated with monoclonal antibodies. These can then reach the lymph nodes and stimulate both cytotoxic T cells as well naïve T cells and the enhanced immune response is proven by increased interferon-γ secretion (Kwon YJ et al 2005 Proc Natl Acad Sci USA 102:18264).

Some vaccines use live microorganisms that have been genetically engineered by removal of part of their genome so that they do not cause the disease yet promote the production of IgA and some also IgG (see Fig. V1). Other approaches include the introduction of the genes of cytotoxic lymphocyte epitopes into the cells. The presence of CpG oligonucleotide motifs in the vector plasmid appears to enhance the immunogenicity of DNA vaccination in humans. Genetically engineered vaccines may be produced in transgenic plants as well and they may eventually be edible. After infection by the lymphocytic choriomeningitis virus, mice may clear the LCMV and gain lifelong immunity to reinfection because the DNA containing LCMV sequences is transcribed in the animals represents a new type of DNA vaccination. Prophylactic vaccination against retroviral diseases is hampered by the apparent poor immune reaction against these agents. Bovine leukemia virus (BLV), which produces a relative paucity of variants that confound the cytotoxic T cells (CTL) can, however, be vaccinated successfully by using a viral glycoprotein, gp51 epitope. Several poxviruses, modified vaccinia virus, replication-defective adenovirus, etc., appear to boost the DNA vaccines' effectiveness for primed CD^+ T cells. Some adjuvants (alum, monophosphoryl lipidA, oil/water emulsion, etc.) have also revealed beneficial effects in some cases. A new approach to vaccination is the use of antibodies against cytokines that accumulate in inflamed or tumor tissues (Zagury D et al 2001 Proc Natl Acad Sci USA 98:8024). Synthetic polysaccharide antigen conjugated with protein (tetanus toxoid) yields an effective vaccine at an economically low cost (Verez-Bencomo V et al 2004 Science 305:522). The availability of completely sequenced genomes of viral and bacterial pathogens permits vaccine design by bioinformatics seeking out unique, potentially antigenic sequences without growing the target organisms. This approach is known as *reverse vaccinology* (Mora M et al 2003 Drug Discovery Today 10:459). The post-exposure prophylactic vaccines are administered after acute infections by tetanus, diphteria and rabies. The latter type of vaccination follows a somewhat different logic than the prophylactic type (and it is somewhat controversial) yet its efficacy is corroborated by medical practice. The effectiveness of the vaccines may be boosted by the increase in CD4 and CD8 T lymphocyte activity and helps to reduce the proliferation of the pathogen (Autran B et al 2004 Science 305:205). When administered with various interleukins and other chemokines, their effectiveness may be enhanced. After successful vaccination, the T cell may clonally expand in 7–10 days and antigen-specific $CD8^+$ T cells may increase 100,000-fold and differentiate into effector helper T (TH) cells and cytotoxic killer cells (NK). After the removal of the antigen the number of the antigen-specific cells contract (in 2–4 weeks) and a fraction of the T cells differentiate into memory cells for long-term protection. In germinal centers and plasma foci short-lived antibody cells and longer lasting memory B cells are formed. Once an antigen presenting dendritic cell locates an antigen-specific T cell by immunological synapse, various proteins (TCR-MHC, CD2, CD8 LFA-1, PKC, Lck, Fyn, CD4, CD45) are recruited within the central SMAC region and this further strengthens the immunological response. The number of T cells may be stimulated to increase 100 to

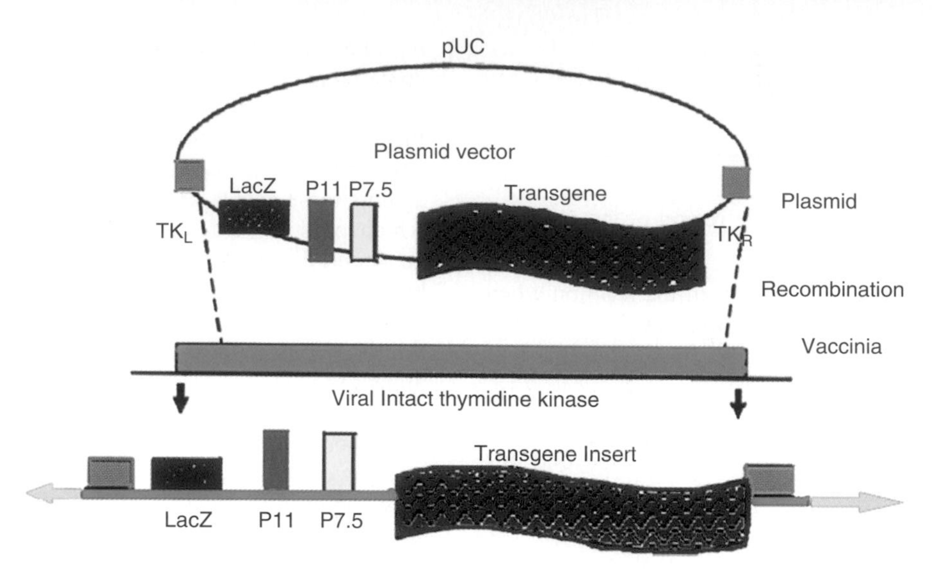

Figure V1. Construction of a recombined vaccinia genome (LacZ bacterial reporter, P11 and P7.5 are promoters, TK: thymidine kinase fragments, pUC plasmid provides origin of replication and ampicillin resistance [β-lactamase])

1,000-fold within a few days. The innate immune system helps to direct the activated T cells to the sites of pathogen invasion. B1 B cells produce small amounts of immunoglobulin M (IgM) and immunoglobulin A (IgA) even in the absence of a specific antigen. High affinity B cells are produced by affinity maturation. These B cells may migrate to the plasma foci, induce strong germinal center responses and affinity maturation and probably generate long-life memory B cells. Antigen presenting cells may take up attenuated bacterial pathogens (*Listeria monocytogenes*), which do not replicate and are rapidly destroyed. After destroying them, the phagolysomes release the antigens for endogenous processing and presentation and stimulate CD8$^+$ effector T cells. The intracellular replication deficient vaccine strain is rapidly cleared from normal and imunocompromised animals but antigen-specific T cells are stimulated and the animals acquire resistance to the wild type bacteria in a directed manner (Bouwer HGA et al 2006 Proc Natl Acad Sci USA 103:5102).

Some vaccines are highly effective and long-lasting, but vaccination against many other infections demands substantial improvements (Pulendran B, Ahmed R 2006 Cell 124:849). ►immune system, ►immunization, ►lymphocytes, ►mucosal immunity, ►CTL, ►epitope, ►immunization genetic, ►peptide vaccine, ►cancer prevention, ►therapeutic vaccine, ►plant vaccines, ►subunit vaccine, ►memory immunological, ►TCR, ►TLR, ►PAMP, ►PRR, ►CD2, ►CD4, ►CD8, ►CD45, ►LFA, ►PKC, ►Lck, ►Fyn, ►immunoglobulins, ►adjuvant immunological, ►acquired immunity, ►innate immunity, ►tumor vaccination, ►cytokines, ►variolation, ►germinal center, ►SMAC, ►immunological synapse, ►Listeria; Nature Medicine 4(5), 1998 May supplement; Burton DR 2002 Nature Rev Immunol 2:706.

Vaccinia Virus: Closely related to small pox (variola) and cowpox viruses. It does not occur in nature and is found in the laboratory only where these two viruses are handled. Thus, it appears to have been somehow derived from variola and cowpox. Vaccinia vectors have been used to express antigens of unrelated pathogens (AIDS virus, hepatitis B) and employ them for immunization. The virus contains a DNA genome of about 190 kbp with nearly 260 potential open reading frames and around 200 bp telomeric sequences (see Fig. V2). These viruses direct replication and transcription in the cell cytoplasm by viral-encoded enzymes. Therefore, insertion of the viral genome by recombination is not a major threat. Vaccinia virus appears to be relatively safe yet periodically (every 10 years) laboratory workers should be vaccinated against it and the handling should be under containment level 2. Vaccinia virus with a deletion in the serpin gene and expressing IFN-γ replicated to high titer in vivo were avirulent in both immunocompetent and imunocompromised

Figure V2. Vaccinia virus of ~360 × 270 × 250 nm (Redrawn after Cyrklaff, M. et al. 2005 Proc. Natl. Acad. Sci. USA 102:2772)

animals, indicating its potential for the production of efficacious and safe vectors against smallpox and other diseases (Legrand FA et al 2005 Proc Natl Acad Sci USA 102:2940). Vaccinia vectors appear to be effective against some tumors and metastasis when injected directly into the neoplasia. Vaccinia virus infects productively the majority of mammalian and avian cells but the Chinese hamster ovary cells are not infected and the replication of the virus may not be completed in primary lymphocytes or macrophages. For infection, the virus requires the presence of the E3L vaccinia protein or a homolog of it. These proteins facilitate viral binding to Z-DNA. Vaccinia protein F11L interacts with RhoA and blocks downstream effectors and RhoA signaling is required for both vaccinia morphogenesis and virus-induced cell motility (Valderrama F et al 2006 Science 311:377).

Mouse antibody constructed from the Chimpanzee Fab domain completed with the human γ1 heavy chain constant region provided an effective response to vaccinia virus B5 envelope protein. It protected mice from intranasal challenge against the virus even 2 days after exposure. The chimpanzee/human mAb provides effective prophylaxis and immunotherapy of smallpox (Chen Z et al 2006 Proc Natl Acad Sci USA 103:1882). ►immunization, ►immunotherapy, ►Fab, ►antibody, ►Xgal, ►β-galactosidase, ►biohazards, ►serpins, ►interferon, ►Rho; Kim Y-G et al 2003 Proc Natl Acad Sci USA 100:6974, diagram of the genome redrawn after Chakrabarti S et al 1985 Mol Cell Biol 5:3403.

Vacuoles: These vesicles within plant and fungal cells are filled with various substances (nutrients, products of secondary metabolism, enzymes, crystals, solutes). Vacuoles may occupy minimal space in meristematic cells of plants whereas in older cells they occupy up to 90% of the cell inner volume. The elastic tonoplast membrane surrounds the vacuoles and permits a change in their size. They may regulate the osmotic pressure of the cytosol by releasing smaller molecules or polymerizing them as needed to maintain a constant value in the cytoplasm. Vacuoles also regulate pH by a similar balancing action. They supply the cells with storage nutrients, hydrolytic enzymes, anthocyanin pigments, and in some cases various toxic substances such as tannins, phenolics and alkaloids. Vacuoles are inherited by the partition of the cytoplasm but if a particular cell becomes vacuole free its progeny may generate them de novo. ►lysosomes, ►cells, ►membrane fusion; Klionsky DJ, Ohsumi Y 1999 Annu Rev Cell Dev Biol 15:1; Rojo E et al 2001 Developmental Cell 1:303; Weisman LS 2003 Annu Rev Genet 37:435.

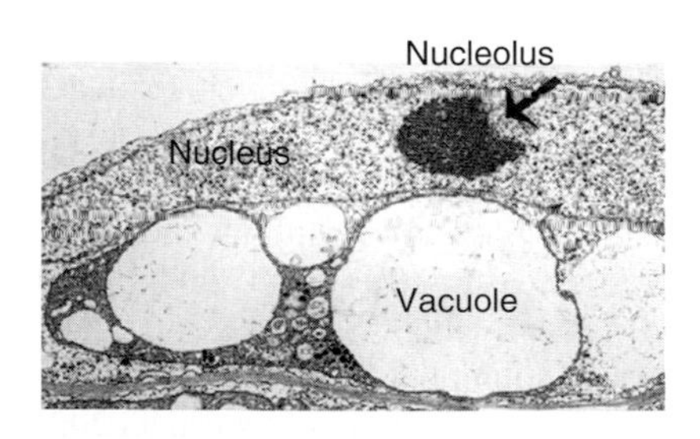

Figure V3. Large vacuoles in a plant cell (bottom) and an oblong nucleus with nucleolus (top)

Vagility: Refers to the ability of organisms to disperse in a natural habitat, it is thus a factor of speciation and survival.

Vagina: The female organ of copulation beginning at the vulva and extending to the cervix of the uterus. *Vaginismus* is an involuntary painful contraction of the vaginal muscles and may cause severe pain especially in intercourse. An autosomal recessive condition of vaginal atresia (absence of vagina) is known. In general, the term denotes an anatomical sheath. The vaginal epithelium may harbor several species of bacteria even under completely healthy conditions and without any symptoms; the most common of such bacteria are various lactobacilli (Hyman RW et al 2005 Proc Natl Acad Sci USA 102:7952). ►clitoris, ►fallopian tube, ►ovary, ►egg, ►migraine, ►uterus, ►vulvovaginitis, ►microbiome

Vaginal Plug: After successful copulation in mice, part of the male ejaculate forms a vaginal plug that closes the vagina for 16–24 hours and in some strains even for a few days. The presence of the plug reveals to the breeder the success of the mating. ►mating plug

Valence (valency): The valence of an antibody indicates the number of antigen-binding sites. In chemistry, it indicates the number of covalent bonds an atom can form and it is also called oxidation number, e.g., O^{2-} or Na^+ or Ca^{2+}.

Validation: Refers to the confirmation of experimental results or working hypotheses by repeated tests (proof of principle). It is the process of determining the degree to which a model is an accurate representation of real life facts. Validation is important as one depends increasingly on predictions by computer models. Validation of a target may be based on the expression of genes in the pathway, the use of microRNA or RNAi blocks or mutation or knockout in the system. ►proof-of-concept; Benson JD et al 2006 Nature [Lond] 441:451.

Valine Biosynthesis: ►isoleucine-valine biosynthetic pathway

Valinemia: ►hypervalinemia

Valyl tRNA Synthetase (VARS): This is encoded in human chromosome 9. VARS charge $tRNA^{Val}$ by the amino acid valine. ►aminoacyl-tRNA synthetase

VAMOS (variability modulation system): Denotes the expression of *reactivity* of females, measured on the basis of the percentage of sterility in daughters in hybrid dysgenesis. Reactivity is modulated by various factors such as temperature, age of the females over generations and inhibitors of DNA synthesis and ionizing radiation. ►hybrid dysgenesis

VAMP (vesicle associated membrane protein, synaptobrevin): This is a synaptic protein. ►syntaxin, ►synaptobrevin, ►SNARE

Van Buchem Disease (17q11.2): A form of osteosclerosis. ►sclerosteosis; Loots GG et al 2005 Genome Res 15:928.

Van der Waals Force: Refers to weak, short-range attraction between non-polar (hydrophobic) molecules.

Van der Woude Syndrome: A dominant, human chromosome 1q32-q41-located cleft lip and palate syndrome. ►cleft palate, ►harelip, ►epithelial cell, ►popliteal pterygium; Schutte BC et al 2000 Genome Res 10:81.

Van Gogh, Vincent (1853–1890): The famous Dutch painter may have suffered from intermittent porphyria and periods of deep depression (see Fig. V4). The beautiful Starry Night painting was attributed to a hallucinatory spell experienced during a bout of bipolar disorder and heavy drinking of absinth. The *Drosophila* gene named *vangogh* modulates the expression of *wingless*. The mammalian homolog (*Vangl2*) of this *Drosophila* gene encodes a four-transmembrane PDZ protein, which along with Scrb1 (scribble) regulates a signaling pathway involved in planar orientation of the epithelium and the stereocilia. ►depression, ►bipolar mood disorder, ►PDZ, ►stereocilia, ►porphyria, ►*wingless*; Montcouquiol M et al 2003 Nature [Lond] 423:173; http://www.med.wayne.edu/elab/vangogh/MainIndex.htm.

Figure V4. Sorrowful Old Man (Van Gogh, 1890). Reproduced with permission of the Kröller-Müller Museum, Otterlo, The Netherlands

Vancomycin ($C_{66}H_{75}Cl_2N_9O_{24}$): An extremely potent glycopeptide antibiotic effective against gram-positive bacteria by blocking the cross-linking of adjacent peptidoglycan strands by peptide bonds in bacterial cell wall synthesis (see Fig. V5). It also inhibits transglycosylation, which connects the existing glycan strands. In recent years *Streptococcus pneumoniae*, responsible for pneumonia, bacterial meningitis and ear infection, as well as several *Enterococcus* species have acquired tolerance or resistance to this antibiotic. Vancomycin tolerance is due to mutation in the Vancomycin signal transduction sensor kinase (VncS). Upon autophosphorylation in the presence of ATP, VncS-P is generated. As a consequence the Vancomycin sensor regulator (VncR) is phosphorylated to VncR-P and the defense genes (resistance to the infection) of the cells are turned off. In Vancomycin (and some other antibiotic) sensitive strains VnsS dephosphorylates VncR. The efficacy of Vancomycin as an antibiotic is due to its binding to the D-Ala—D-Ala moiety of the bacterial peptidoglycan precursors and thereby hindering growth of the cell wall. In the resistant strains, the dipeptide is replaced by the depsipeptide D-Ala—D-Lac reducing sensitivity to the antibiotic by three orders of magnitude. (Depsipeptide is a member of the bicyclic peptide class of histone deacetylase [HDAC] inhibitors that was first isolated as a fermentation product from *Chromobacterium violaceum*). The level of resistance depends on the number of peptidoglycan precursor molecules that carry this replacement. Vancomycin resistance may be overcome by chlorobiphenyl vancomycin derivatives, which interact with penicillin-binding protein-1 (PBP1b) although they do not bind the enzyme (Chen L et al 2003 Proc Natl Acad Sci USA 100:5658). If D-Ala—D-Lac is selectively disrupted by small molecules (prolinol derivatives) (see Fig. V6) sensitivity to Vancomycin is restored. Reengineering Vancomycin for dual D-Ala-D-Ala and D-Ala-D-Lac binding can effectively counter resistance of the bacteria (Crowley BM, Boger DL 2006 J Am Chem Soc 128:2885). A neoglycosylatiom procedure increases efficacy 40-fold (Griffithm BR et al 2007 J Am Chem Soc 1298150). ►antibiotics, ►tolerance of antibiotics, ►antibiotic resistance, ►non-ribosomal peptide; Chiosis G, Boneca IG 2001 Science 293:1484; Eggert US et al 2001 Science 294:361; Reynolds AM et al 1996 J Infect 32:11.

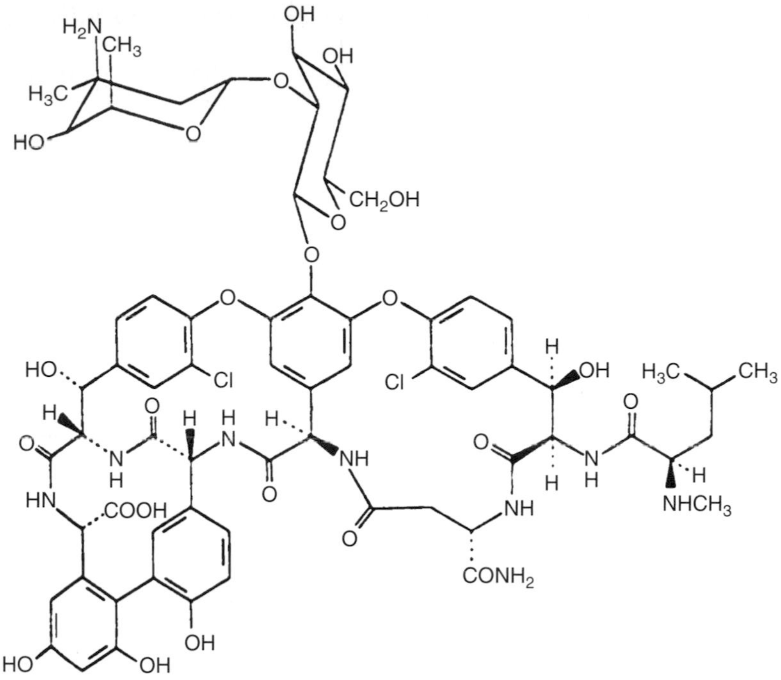

Figure V5. Vancomycin

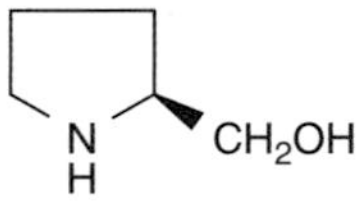

Figure V6. L-Prolinol [(s)-(+)2-pyrrolidinemethanol]

Vanilla (*Vanilla planifolia*): This is a tropical spice tree; 2n = 2x = 32.

Vanillin ($C_8H_8O_3$): A flavoring agent and an analytical reagent. It is an antimutagen, anticlastogen, anticarcinogen and an inhibitor of non-homologous end-joining and DNA-PK. ►clastogen, ►NHEJ, ►DNA-PK; Durant S, Karran P 2003 Nucleic Acids Res 31:5501.

Váradi-Papp Syndrome: ►orofacial-digital syndrome V

Variability: Denotes the condition of being able or apt to vary.

Variable: A condition or treatment that may be replaced by the experimenter or that may change during the study. The pre-analytical variables are known or unknown factors that may be present before or arise during the analysis. In clinical studies the pre-analytical variables may contribute to serious errors in the evaluation. In experimental research the variables may be under more stringent control than in observational studies.

Variable Number Tandem Repeats: ►VNTR

Variable Regions: These regions of the antibody are situated at the amino end of both light and heavy chains, and this region determines antibody specificity and antigen binding. ►immunoglobulins, ►antibody

Variable Surface Glycoprotein: ►*Trypanosoma*

Variance: The mean of the squared deviations of the variates from the mean of the variates: $V = \sum[(x - \bar{x})^2]/n - 1$ where x are the variates and $\bar{x}$ is the mean of the variates and *n* is the number of variates (individuals). ►variate, ►invariance, ►standard deviation, ►standard error, ►analysis of variance, ►intraclass correlation, ►genetic variance

Variance Analysis: ►analysis of variance

Variant: Refers to a cell or individual different from the standard type.

Variant Detector Array (VDA): Uses oligonucleotide labeled probes to locate/identify particular genes in microarrays. ►microarray hybridization, ►oligo-labeling probes

Variate: A variable quantity measured in a sample of a population.

Variation: Can be *continuous* and individual measurements do not fall into discrete classes, e.g., traits determined by polygenic systems. In the case of *discontinuous* variation the measurements can be classified into distinct classes such as the qualitative traits (black and white [no gray]) in a segregating population (see Fig. V7). ►variance, ►genetic variance, ►continuous variation, ►discontinuous variation, ►SNIP; Crawford DC et al 2005 Annu Rev Genomics Hum Genet 6:287.

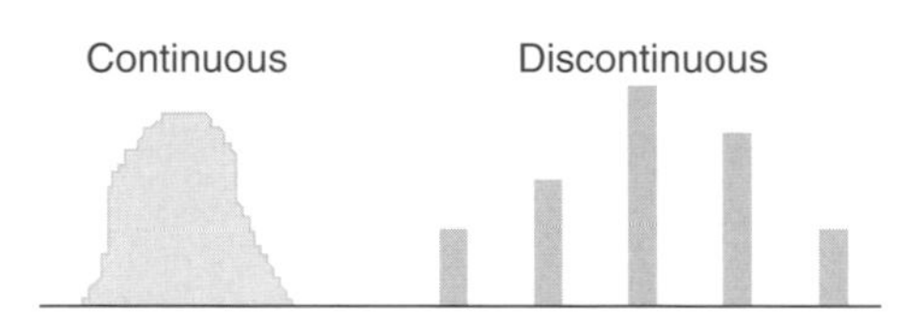

Figure V7. Variation

Variation, Mathematical: This is illustrated by an example. How can a cell express 3 different receptors of a pool of 10 receptors? The solution is $\frac{10\times9\times8}{1\times2\times3}=120$. ►combination

Varicella-Zooster Virus: This member of the Herpes virus family causes chickenpox and shingles. Vaccination effectively reduces mortality due to chickenpox but its effect on shingles is still unknown. The Oka vaccine strain is a live attenuated virus that is routinely administered to about 40 million children in the United States and Europe to prevent chickenpox. It is effective and safe but occasionally produces a rash. The vaccine virus has accumulated mutations at more than 30 loci during its attenuation and after administration, but the rashes are not explained by their reversion, unlike complications reported in the case of other viral vaccines. The host genotype and immune status also affect the extent of response to the vaccine (Quinlivan ML et al 2007 Proc Natl Acad Sci USA 104:208). Insulin degrading enzyme (IDE) is a cellular receptor of the virus and mediates cell-to-cell spread (Li Q et al 2006 Cell 127:305). ►herpes, ►shingles, ►chickenpox

Variegation: Refers to sector formation or mosaicism of the somatic cells due a number of different mechanisms such as nondisjunction, somatic mutation, segregation of organelles (chloroplasts) deletion and disease (see Figs. V8 and V9). Heterochromatin may cause variegation of genes in its close vicinity by transient modulation of the exposure of the DNA to transcription factors. In *Drosophila* more than 15 genes act as suppressors of variegation (*Su*[*va*]). ►uniparental inheritance, ►lyonization, ►broken tulip, ►transposable elements, ►position effect, ►piebaldism, ►spotting, ►mitotic recombination, ►nondisjunction, ►developmental-regulator effect variegation, ►heterochromatin; Ahmad K, Henikoff S 2001 Cell 104:839; Sjiv IS et al 2007 Nature [Lond] 447:399.

Figure V8. Mosaicism in cattle

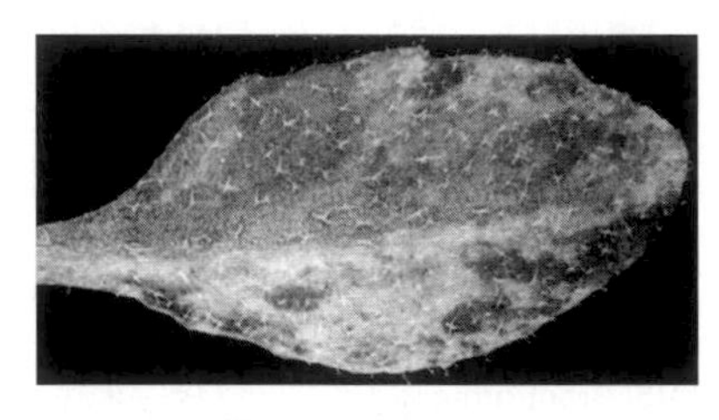

Figure V9. Variegation in a plant leaf

Variety: Refers to an organism of a distinct form or function. ►cultivar, ►cultigen

Variogram: A plot of genetic distance relative to geographic distance. ►genetic distance

Variola: This is the technical term for smallpox virus, pox virus, vaccinia and biological weapons.

Variolation: The earliest form of vaccination known at least by the sixth century. Infective material from the smallpox lesions was transferred to healthy individuals to provide resistance against the disease with an expectation of not more than ~4% fatality versus 20–30% from natural infection. ►vaccine, ►pox virus

Varkud Plasmid: ►*Neurospora* mitochondrial plasmids

Vas Deferens: An excretory channel of the testis connected to the ejaculatory duct of the sperm. ►CBAVD, ►P2X

Vascular Cell Adhesion Molecule (VCAM): ►metastasis

Vascular Diseases: ►cardiovascular diseases

Vascular Endothelial Growth Factors (VEGF): These are required for vasculogenesis and are particularly active in some tumor tissues to provide the necessary blood supply for proliferation. VEGFs depend on their appropriate (VEGFRs) integrin cell adhesion receptors (De S 2005 Proc Natl Acad Sci USA 102:7589). The receptor (VEGFRT2) is a tyrosine kinase. In human primary lymphedema (Milroy disease) VEGFR-3 tyrosine kinase activity is deficient. Mouse mutants with a defect in receptor-3 of VEGF die of cardiovascular failure (9.5 d) before birth. VEGF has been considered as a potential benefit in case heart disease damages the blood vessels. The peptide QK binding to the interface of VEGF receptor can induce endothelial cell proliferation and blood capillary formation. Therefore, QK has biomedical (therapeutic and diagnostic) potential (D'Andrea LD et al 2005 Proc Natl Acad Sci USA 102:14215). VEGF can also cause vascular permeability and the accumulation of fibrin barrier around tumors and wounds. The leakage can reduce metastasis of tumors and can induce obstruction and necrosis of blood vessels (infarction). VEGF not only promotes tumor growth and reduces metastasis, but it may also facilitate atherosclerosis (Weis SM & Cheresh DA 2005 Nature [Lond] 437:497). The endocrine-gland-derived endothelial growth factor (EG-VEGF) is selectively expressed only in the ovary and other steroid-producing tissues (testis, adrenal and placental tissues). ►angiogenesis, ►Flk-1, ►Flt-1, ►neuropilin, ►tyrosine kinase, ►VEGF, ►polyposis hamartomatous, ►KDR, ►*Peg*-3, ►angiopoietin, ►atherosclerosis, ►hypoxia, ►wound healing, ►lymphedema, ►phospholipase, ►macular degeneration; Ferrara N 1999 J Mol Med 77:527; Bellamy WT et al 1999 Cancer Res 59:728; LeCouter J et al 2001 Nature [Lond] 412:877; Niethammer AG et al 2002 Nature Med 8:1369; http://www.researchvegf.com.

Vascular Targeting: For the development and maintenance of tumors, an ample supply of blood is a requisite. The monoclonal antibody to bFGF and VEGF may block the required angiogenesis factors. Similarly, integrins (CD51/CD61) are also required for angiogenesis and they can be interfered with by their cognate monoclonal antibodies. Anti-endoglin antibody, especially with conjugated ricin, may lead to anti-tumor effects. ►angiogenesis, ►angiostatin, ►endostatin, ►ADEPT, ►VEGF, ►FGF, ►integrin, ►endoglin

Vascular Tissue: In plants, this includes the xylem, the phloem, the (pro)cambium and the surrounding fibrous parenchyma. In animals the blood vessels are the primary vascular tissue. ►phloem, ►xylem, ►cambium, ►parenchyma, ►proteoglycan

Vascularization: Refers to the development of veins and other vessels. ►vasculogenesis

Vasculogenesis: Refers to the differentiation of mesodermal cells into hemangioblasts. ►hemangioblast, ►blood formation, ►angiogenesis, ►CXCR

Vasculopathy: Vascular retinopathy, cerebroretinal vasculopathy, endotheliopathy with retinopathy, nephropathy and stroke all map to 3p21.1-p21.3. The abnormalities of the vascular system cause Raynaud's disease, migraine, retinal vein impairment, visual disease, renal disease, neurological problems and possibly premature death. ►Raynaud's disease; Ophoff RA et al 2001 Am J Hum Genet 69:447.

Vasectomy: The surgical removal of the vas deferens (ductus deferens), the excretory channel of the semen. It is a method of fertility control for males. ►birth control drugs; Sandlow J et al 2001 Fertil Steril 75:544; Weiske WH 2001 Andrologia 33:125.

Vasodilator: Causes expansion of (blood) vessels.

Vasopressin: ►antidiuretic hormone, ►oxytocin (see Fig. V10)

```
┌────S──────S───┐
Cys-Tyr-Phe-Gln-Asn-Cys-Pro-Arg-GlyNH2
```

Figure V10. Human vasopressin

VASP: A profilin-binding protein. ►profilin

VAT: Denotes variant antigen type. ►*Trypanosoma*

V-ATPases: These are vascular ATPases in control of ionic homeostasis. ►APPase, ►homeostasis

Vaults: These are large (42 × 75 nm, 12.9 MDa) predominantly cytoplasmic ribonuclein particles present across phylogenetic ranges. The mammalian vaults include the vault poly(A)DP-ribose polymerase, telomerase associated protein 1 and one or more non-translated RNAs (Poderycki MJ et al 2005 Nucleic Acids Res 33:893). Vaults are involved in detoxification and many cellular and developmental processes (Gopinath SCB et al 2005 Nucleic Acids Res 33:4874). (See Stephen AG et al 2001 J Biol Chem 276:23217).

Vav Oncogene: Found in human chromosome 19p13.2-p12, and appears to be GDP→GTP exchange factor requiring tyrosine phosphorylation. It regulates lymphocyte development and activation. ►oncogenes, ►B lymphocytes, ►T cell receptor

vBNS (very high-speed Backbone Network Service): This is a computer network linking five supercomputer centers to facilitate fast scientific communication and remote control by the use of special equipment.

vCJD: A variant of the Creutzfeldt-Jakob disease, a possible contagious form of the mad cow disease; it infects humans as well, especially those homozygous for codon 129 methionine of the prion. ►Creutzfeldt-Jakob disease, ►encephalopathies, ►prion

VCP (vasoline containing protein): Involved in lipid metabolism. ►T cell receptor

V(D)J (variable[diversity]juncture): Sequences in immunoglobulins where antibody diversity is generated by recombination at the RSS (recombinational signal sequence) sites. Several factors are important for V(D)J recombination: DNA-dependent protein kinase, Artemis, CRCC4 and ligase IV. The recombination may show tissue/organ specificity. ►immunoglobulins, ►T cell receptor, ►RSS, ►RAG, ►Artemis, ►DNA-PK, ►CRCC, ►ligase DNA, ►non-homologous end-joining; Gellert M 1992 Annu Rev Genet 26:425; Gellert M 1997 Adv Imunol 64:39; Bassing CH et al 2002 Cell 109:S45; Dai Y et al 2003 Proc Natl Acad Sci USA 100:2462; Xiong N et al 2004 Proc Natl Acad Sci USA 101:260.

VDR (vitamin D3 receptors): Receptors that control homeostasis, growth and differentiation. They preferentially bind to response elements of direct repeats, palindromes and inverted palindromes of hexameric core-binding domains, particularly well when they are spaced by three nucleotides. They can dimerize with 3,5,3′-triiodothyronine, a thyroid hormone receptor that can direct sensitivity of ligands for transactivation. ►vitamin D, ►transactivator, ►hormone response elements

Vector: Generally an insect or other organism transmitting parasites and/or pathogens. It also refers to any gene carrier derived from plasmids, viruses or produced synthetically such as liposomes or polyethylene particle driven by motor proteins along the microtubules. ►vectors, ►vectorette, ►vector cassette, ►liposome; Suh J et al 2003 Proc Natl Acad Sci USA 100:3878.

Vector, Algebraic: ►matrix algebra

Vector Cassette: A transformation construct carrying all essential elements (including reporter genes, selectable marker, replicator, etc.), and it can be used for the insertion of different DNA sequences. ►vectors, ►reporter gene, ►transformation, ►knockout, ►targeting genes

Vectorette: A short DNA sequence serving as a specific linker-primer for PCR amplification. It generally contains an inner, non-complementary sequence (bubble), flanked by two short pieces of duplex DNA (see Fig. V11). The 5′ end may be either blunt or complementary to a restriction site, depending on the restriction enzyme used to digest the DNA. An overhang may prevent ligation of the 3′ end. ►polymerase chain reaction, ►amplification, ►linker, ►primer, ►ligase, ►blunt end, ►overhang, ►restriction enzyme, ►splinkerette; Eggert H et al 1998 Genetics 149:1427; Devon RS et al 1995 Nucleic Acids Res 23:1644.

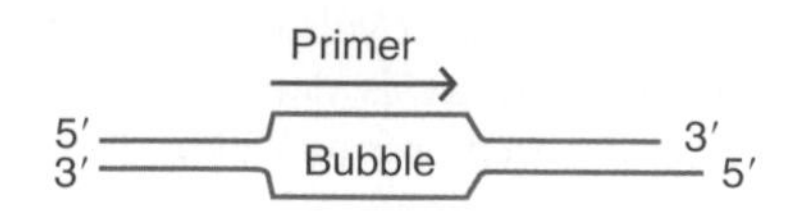

Figure V11. Vectorette

Vectors: These molecular genetic constructs are generally circular plasmids that can introduce exogenous genetic material into prokaryotic and eukaryotic cells. They may be *cloning vectors* that only replicate the DNA according to the plasmid replicon. The *expression vectors* carry genes complete with all the elements required for expression (promoter, structural gene, termination signals, etc.). *Gene fusion vectors* do not have promoters and the expression of the transferred gene is contingent on fusion with a host cell promoter. *Shuttle vectors* can carry the DNA among different hosts (such as *E. coli* and COS cells, *Agrobacterium* and plants). All vectors must have at least a replicator site, selectable marker(s) and mechanisms for the introduction of parts of their sequences into the host genetic material. ►cloning vectors, ►cosmids, ►phagemids, ►fosmids, ►ColE1, ►plasmovirus, ►yeast vectors, ►viral vectors, ►excision vectors, ►plasmids, ►transposable elements, ►NOMAD, ►BAC, ►BIBAC, ►YAC, ►HAC, ►PAC, ►transcriptional gene fusion vector, ►translational gene fusion vector, ►targeting vector, ►polyethyleneimine, ►pBR322, ►pUC vectors, ►transposon vector, ►liposomes, ►nanoparticles, ►cancer gene therapy, ►RetroTet-Art vector, ►polyplexes; vector database: http://seq.yeastgenome.org/vectordb/.

Vectors for Pathogens: See human pathogen vectors: http://www.vectorbase.org/.

VEGA (vertebrate genome annotation): A central depository for manual annotation of different vertebrate (human, mouse zebrafish) finished genome sequences. ►annotation of the genome; http://vega.sanger.ac.uk/index.html.

Vegetal Pole: Refers to the lower end of the animal egg where the yolk is concentrated. The opposite end of the egg is known as the animal pole. After fertilization the yolk moves to the central position and becomes the starting site of the differentiation of axes (anterior-posterior, dorsal-ventral, median-lateral) of the embryo. ►morphogenesis *Drosophila*, ►pole cell

Vegetative Cell: Involved in metabolism but not in sexual reproduction. ►gametogenesis in plants

Vegetative Hybrids: ►graft hybrids

Vegetative Incompatibility: ►fungal incompatibility

Vegetative Nucleus (macronucleus): ►*Paramecium*, ►tube nucleus

Vegetative Petite: ►petite colony mutants

Vegetative Reproduction: Commonly observed in many species of plants (grafting, rooting) and lower organisms that use fission for propagation. The advantage of this type of reproduction is that the progeny forms a genetically homogeneous clone unless or until mutation takes place. ►somatic embryogenesis, ►grafting, ►regeneration, ►clone, ►tissue culture

Vegetative State: Denotes asexual, unconscious, non-replicating, non-infectious, etc., depending on the context.

VEGF: Smooth muscle cells synthesize vascular endothelial growth factor, which is somewhat related to PDGF. The VEGF genes are divided among eight exons and by alternative splicing three different proteins are produced. It may enhance the growth of new blood vessels. FLT-1 is one of the VEGF receptors. ►signal transduction, ►PDGF, ►vascular endothelial growth factor, ►neuropilin, ►angiopoietin, ►vascular targeting

Vehicle: ►vectors

Velans: ►*gaudens*

Velocardiofacial Syndrome (VCFS): A heart and face, kidney, parathyroid and thymus defect caused by deletion in human chromosome 22q11. The gene most commonly affected is UFD1-l (ubiquitin fusion degradation; 22q11.2). ►DiGeorge syndrome, ►face/heart defects, ►claudin, ►tight junction, ►schizophrenia, ►deletion 22q11.2

VENA (plural venae, adj. venous): A vein that carries blood towards the heart.

Veneering of Antibody: Refers to the generation of a humanized antibody where some of the surface regions of the mouse antibody framework are replaced by human sequences in order to reduce immunogenicity. ►antibody, ►humanized antibody

Venn Diagrams: Data are represented by circles or ovals according to their common features, overlapping functions or exclusion (see Fig. V12). The three circles generate seven ($2^3 - 1$) areas that may be shaded or otherwise marked. Such diagrams may represent interactions of proteins or interactions among environmental effects.

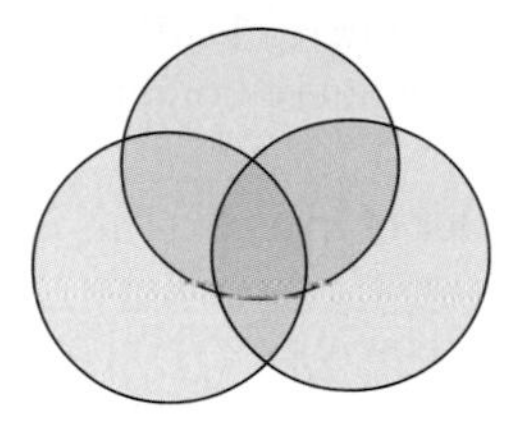

Figure V12. Venn diagram

Venom: A (highly) toxic secretion of some serpents and insects. Some types of venoms have medical applications, e.g., for hemostasis. ►hemostasis, ►mast cells; toxin database: http://ntrc.tamuk.edu/cgi-bin/serpentarium/snake.query.

Venous Malformation (VM): This is caused by mutation at loci 1p21-p22 and 9p21. VM can occur in any tissue but is most common in the skin and muscle; it causes pain and bleeding, and even death at times.

Vent: A DNA polymerase extracted from *Thermococcus litoralis*. ►recursive PCR

Ventral (from the Latin venter meaning abdomen): This is related to the position on the side opposite to the back.

Ventricular: An adjective that means belonging to a ventriculus (a cavity such as in the heart).

Venture: Variable number repeating elements (40 to >150) of 14–15 nucleotides, rich in guanine. The shorter VENTR alleles are associated with susceptibility to insulin-dependent diabetes mellitus (IDDM) but one of the long repeats ($14 \times 50 = 700$ nucleotides) appears to be protective against IDDM. ►diabetes mellitus, ►imprinting, ►VNTR

Venules: These are small vessels that collect blood from the capillary veins.

Venus: A yellow derivative (YFP) of the green fluorescent protein (GFP). ►aequorin

Venus Mirror: ►pedigree, ►female (see Fig. V13)

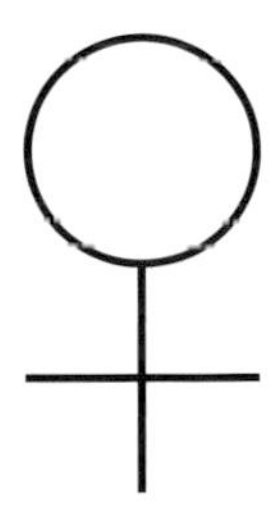

Figure V13. Venus mirror

Vermiculite: A commercial silicate medium to grow plants under greenhouse conditions.

Vermilion Eye Color of Insects: This is controlled by the *v* locus of *Drosophila* encoding tryptophan pyrrolase, an enzyme that converts tryptophan to formylkynurenin. The recessive vermilion eye color is actually bright scarlet because the brown ommochrome is not formed. The *v* eye discs when transplanted into normal tissues develop wild-type eye color. ►animal pigments, ►ommochromes, ►tryptophan

Vernalization: Some biannual or winter annual species of plants have a low temperature requirement for the induction and completion of the bolting and flowering stage of development. This need can be satisfied in spring planting by exposing the germinating seeds to near freezing temperatures for a genetically determined and variable period. It is also known as yarowization (in Russian yarowie kchleba means spring cereal). In *Arabidopsis*, the *FLC* is one of the loci to repress vernalization response. The other vernalization gene *FRIGIDA* (*FRI*) has been cloned. VRN2 is a nuclear zinc finger protein that mediates the level of FLC. The VRN/VIN proteins deacylate methylated histones H3K27me and H3K9me during vernalization-induced epigenetic silencing of the FLC (Mylne JS et al 2006 Proc Natl Acad Sci USA 103:5012). The LHP1 (like heterochromatin protein) also assists in the maintenance of epigenetic silencing of the FLC. The VRN1 gene of the A genome of wheat has also been cloned (Yan L et al 2003 Proc Natl Acad Sci USA 100:6263). Vernalization in cereals is controlled by MADS box transcription factors (Trevaskis B et al 2003 Proc Natl Acad Sci USA 100:13099). Apparently, low temperature and other gene products modulate the level of its transcript supposedly by demethylation of specific DNA sites. ►photoperiodism, ►photomorphogenesis, ►floral induction, ►histones; Reeves PH, Coupland G 2001 Plant Physiol 126:1085; Gendall AR et al 2001 Cell 107:525; Macknight R et al 2002 Plant Cell 14:877; Sung S, Amasino RM 2004 Nature [Lond] 426:159; Bastow R et al 2004 Nature [Lond] 426:164; Henderson IR et al 2003 Annu Rev Genet 37:371; molecular mechanisms: Sung S, Amasino RM 2005 Annu Rev Plant Biol 56:491.

Versenes (EDTA): These widely used laboratory chelating agents may cause chromosome breakage at higher concentrations (see Fig. V14). ►EDTA

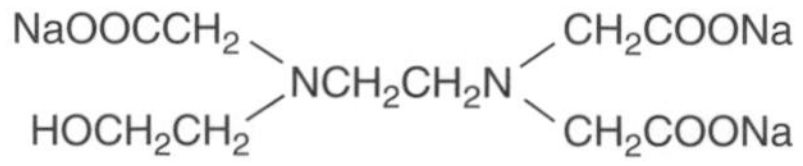

$NaOOCCH_2$ \ / CH_2COONa
NCH_2CH_2N
$HOCH_2CH_2$ / \ CH_2COONa

Figure V14. EDTA

Verterbrates: These are animals with a spinal column (chordates) such as fishes, amphibians, reptiles, birds and mammals. (See VEGA for genomes vertebrate browser: http://genome.ucsc.edu/; http://vega.sanger.ac.uk/index.html).

Vertex (plural: vertices; node): ►networks

Vertical Resistance: The host plant is resistant to a specific race of the pathogenic microorganism.

Vertical Transmission: ►transmission

Very Low Density Lipoprotein: ►VLDL

Vesicles: These are membrane-surrounded sacs in the cell, generally performing storage and transport functions.

Vesicular Stomatitis Virus (VSV): A negative-strand RNA virus of 11,161 nucleotides, enclosed by a nucleocapsid (N). N is a 35-turn helix within the membrane-surrounded oval particles. There is a transmembrane G protein on the surface of the virion for binding cell surface receptors, required for infection. VSV viruses can be engineered into useful genetic vectors. An oncolytic variant is involved in antitumor activity and at the same time establishes an antiviral state that protects against toxicity in healthy cells (Bell JC et al 2002 Curr Gene Ther 2:243). ►CO_2 sensitivity, ►viral vectors, ►lentiviruses

Veterinary Medicine: ►databases; http://netvet.wustl.edu/vetmed/health.htm.

Veto Cell: This recognizes T cells and inactivates them. ►immune suppression; Reich-Zeliger S et al 2000 Immunity 13:507.

Vg-1: A protein that sends signals in animals to develop head and other nearby organs.

VG5Q: 4,049 bp DNA tract (chromosome 5) with a 2,145 ORF encoding 714 amino acids responsible for susceptibility to the Kippel-Trenauney syndrome. ►ORF, ►Kippel-Trenauney syndrome

VHL: ►von Hippel-Lindau syndrome

Viability: Refers to the ability to survive; this is a property of organisms depending on genetic, developmental and environmental factors. A normal human fetus may become viable outside the womb after its weight reaches about 500 g and around 20 weeks after gestation. The viability of a mutant is often expressed as the survival rate relative to the wild type.

Vibrio cholerae: Cause of cholera (see Fig. V15). The toxin is encoded in a filamentous phage (CTX φ) and the same extracellular protein secretion pathway that facilitates the horizontal spread of the phage, releases it. The integration of this virus requires host recombinase XerC and XerD. It takes place at a 28 bp

Figure V15. Diagram of *Vibrio cholerae*

site (*dif1*) in bacterium chromosome 1. The single-stranded phage genome forms a hairpin structure and thus creates a recombination site for XerCD. XerC mediates a single pair of strand exchanges and integration at *dif1* (Val M-E et al 2005 Mol Cell 19:559). *Vibrio* contains two chromosomes (~2.9×10^6 and ~1.1×10^6 bp) encoding 2770 and 1115 ORF, respectively (see Fig. V16). The major virulence factor, CT and the toxin-coregulated, TCP are in the longer chromosome. The bacterium also contains a pathogenicity island (VPI) which apparently does not have a viral origin. The integron island involved in gene integration and dissemination is in the smaller chromosome. Chitin, present in the natural aquatic environment of the pathogen due to the shedding of crustacean exoskeleton, induces natural competence for transformation (Meibom KL et al 2005 Science 310:1824). Virstatin inhibits the expression of cholera toxin and the toxin coregulated pilus of the bacterium and protects mice from intestinal colonization (see Fig. V17) (Hung DT et al 2005 Science 310:670). Several members of the *Vibrios* are pathogenic and their genomes have been sequenced (Ruby EB et al 2005 Proc Natl Acad Sci USA 102:3004). ►cholera toxin; Heidelberg J et al 2000 Nature [Lond] 406:477; transcriptome: Xu Q et al 2003 Proc Natl Acad Sci USA 100:1286.

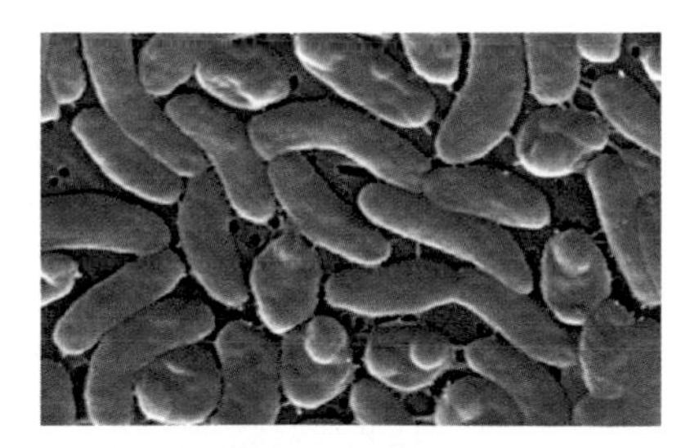

Figure V16. *Vibrio vulnificus*, saltwater parasite (Courtesy of CDC Public Health Image Library)

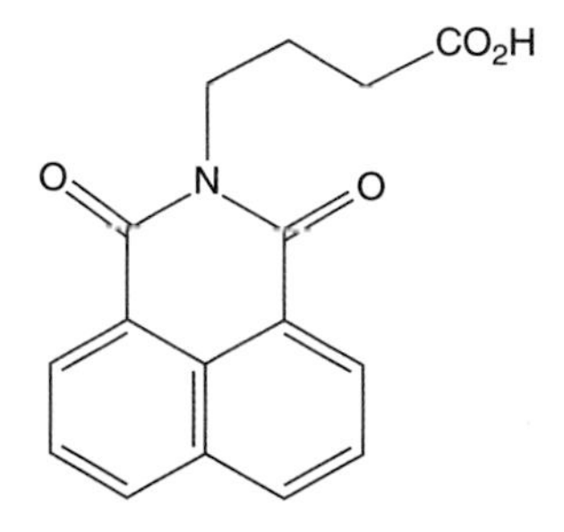

Figure V17. Virstatin

Vicariance: The occurrence of a species in a habitat other than expected or a function not expected by an organ.

Vicia faba: ►broad bean, 2x = 2n = 12; its large chromosomes are well suited for cytological study. With the exception of the two largest chromosomes, the rest are acrocentric (see Fig. V18) (Courtesy of Dr. A. Sparrow). ►favism, ►vicin

Figure V18. *Vicia faba*

Vicine: An alkaloid produced by *Vicia sativa* (vetch) may be the source of this cyanoalanine toxin which is especially hazardous for people and animals on a low sulphur diet. ►favism

VIGS (virus induced gene silencing): Infecting viruses may silence the host plant genes as well as transgenes in plants may silence viral genes by the mechanism of RNAi. ►RNAi; Fagard M, Vaucheret H 2000 Plant Mol Biol 43:285; Voinnet O 2005 Nature Rev Genet 6:206.

Villus: Refers to vesicular projections on a membrane (see Fig. V19). The amniotic villi, near the end of the umbilical cord, are sampled for genetic examination during prenatal amniocentesis. The intestinal villi facilitate food absorption and their structure may be affected by intestinal cancer. ►amniocentesis

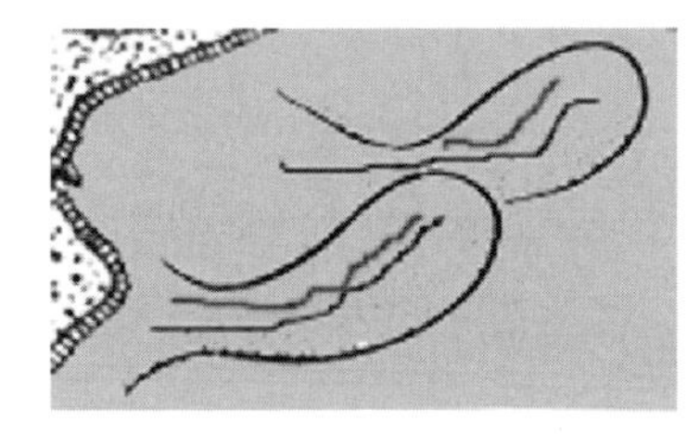

Figure V19. Intestinal villi with blood vessels inside

Vimentin: A constituent of the filament network extending through the cytoplasm of eukaryotic cells. ►intermediate filaments, ►aggresome; Perez-Martinez C et al 2001 J Comp Pathol 124:70; Mor-Vaknin N et al 2003 Nature Cell Biol 5:59.

Vinblastine and Vincristine: These are antineoplastic alkaloids (interfere with microtubules and the cytoskeleton) from the shrub *Vinca rosea.* (See Fig. V20, ►microtubule, ►cytoskeleton; Gigant B et al 2005 Nature [Lond] 435:519).

Figure V20. *Vinca rosea*

Vinculin: A protein binding α-actinin, talin, paxillin, tensin, actin filaments and phospholipids; it also mediates the assembly of the cytoskeleton. (See mentioned items under separate entries, ►adhesion; Kálmán M, Szabó A 2001 Exp Brain Res 139:426; crystal structure of human vinculin: Borgon RA et al 2004 Structure 12:1189.

Violent Behavior: In humans, impulsive aggression has been attributed to reduced levels of 5-hydroxyindole-3-acetic acid in the cerebrospinal fluid and a nonsense mutation in MAOA enzyme results in aggressive behavior in a kindred. ►MAOA, ►behavior in humans

VIP16: A herpes simplex virus-encoded protein; it is a potent activator that controls the transcription of immediate early viral genes through interaction with the host cell factors. It is used for studies of molecular interaction with promoters and transcription factors.

VIR Genes: ►virulence genes *Agrobacterium*

VirA: Denotes agrobacterial kinase phosphorylating the product of virulence gene *VirG.*

Viral Budding: ►budding

Viral Cancer: ►cancer, ►oncogene

Viral Capsid: ►capsid; http://viperdb.scripps.edu/.

Viral Encephalitis: Caused by neurotropic flaviruses such as the Japanese encephalitis virus and the West Nile virus. Targeting a common domain of the envelope protein by short hairpin or RNAi delivered to the brain of mice by lentiviral vector secured complete protection against the disease (Kumar P et al 2006 PLoS Med 3(4):e96). ►RNAi, ►West Nile virus, ►lentiviruses

Viral Envelope: A protein-lipid coat of viruses. ►virus

Viral Ghost: Refers to empty viral capsids, without their own genetic material but they can be filled with DNA and become a genetic vector. ►transformation genetic, ►generalized transduction

Viral Oncogene: ►v-oncogene

Viral Vectors: These are *in vitro* genetically modified viral DNA (e.g., adenovirus, SV40, bovine papilloma, Epstein-Barr, BK viruses and Baculovirus [Polyhedrosis] virus), containing non-viral genes to be introduced into eukaryotic cells. A minimal viral vector contains only the replicational origin of the virus and the packaging signals but the other viral genes involving replication, virion structure and pathogenicity sites are removed. The removed viral DNA or RNA is replaced by a cassette of transgenes required for experimental purposes. *Autonomous stable viral vectors* have also been constructed that replicate in the cytoplasm. In order to prevent destruction of the cells, their copy number is limited by introducing copy number regulators. From the *bovine papilloma virus* (BPV) autonomous (episomal) and shuttle vectors have been constructed that maintain low (10–30) copies in the cytoplasm. The *shuttle vectors* can be rescued from the mammalian cells and can propagate various protein genes in another cell depending on a number of intrinsic and extrinsic factors. The *Epstein—Barr virus* (EBV) vector can be propagated in the cytoplasm of various types of mammalian cells at low copy number (2–4) and is suitable for the study of gene expression, regulatory proteins, etc. It can also be maintained in the nucleus of, e.g., B lymphocytes. The vector has up to 35 kb carrying capacity and can be rescued. The *BK* (baby kidney) *virus* has been advantageously used for human cells. Truncated retroviral HIV vectors (may not be able to recombine and restitute pathogenic forms) may be useful because they can be introduced into non-dividing cells. Lentivirus vectors can also pass into non-dividing cells such as hepatocytes, hematopoietic stem cells and neurons, in contrast to the most widely used mouse leukemia retrovirus (MuLV) vectors, which require DNA replication for integration. For human applications, *replication-deficient retroviral* vectors can accommodate 9 kb exogenous DNA and they are generally used in *ex vivo* studies. *Adenovirus vector* can carry 7.5 kb DNA and can be taken up by the cell, by a specific virus receptor and the $\alpha_V\beta_3$ or $\alpha_V\beta_5$ surface integrins. The adenoviral, herpes simplex, vaccinia and the autonomous parvovirus vectors are not integrated into the human genome and therefore do not lead to permanent genetic change and the treatment has to be reapplied periodically (in weeks or months). The *adeno-associated virus* can integrate into the chromosomes in dividing cells but it is episomal in stationary stage cells. Most of the current vectors may cause inflammation because of antivector cellular immunity. The *Baculovirus* (*Polyhedrosis*) *virus* is used as a vector for insect cells. The targeting of viral vectors to specific tissues can be increased by genetically engineering into the envelope protein a special receptor for a target ligand, e.g., inserting erythropoietin into the Moloney murine leukemia retrovirus or adding an integrin sequence to the avian retrovirus envelope. The vesicular stomatitis virus (VSV) and the gibbon ape leukemia retrovirus (GALV) offer some target specificity but they form only 10^7 to 10^9 cell forming units (CFU)/mL and therefore all the target cells of the body cannot be reached. In order to target the vector to particular cell type the envelope protein of the virus must be so modified that it would recognize the target cell membrane receptor. In order to pass the membrane and move into the cell, the envelope protein—receptor complex must undergo a conformational change. The modified envelope protein, however, may not form an effective complex resulting in very low level of passing of the vector into the target cell. New technology (ligands specific for fibronectin and collagen of the cell matrix) may facilitate the enrichment of the vector in the extracellular matrix of the host cell facilitating a more effective uptake. *Non-viral vector*, organically modified silica (ORMOSIL) after appropriate treatment can bind the DNA and effectively deliver it to neural stem cells of the brain (Bharali DJ et al 2005 Proc Natl Acad Sci USA 102:11539). ►virus, ►vectors genetic, ►parvoviruses, ►vaccinia virus, ►retroviral vectors, ►Bellophage, ►shuttle vector, ►HIV, ►lentiviruses, ►episomal vector, ►adenovirus, ►adeno-associated virus, ►herpes, ►plasmid rescue, ►gene therapy, ►erythropoietin, ►ex vivo, ►microinjection, ►biolistic transformation, ►liposomes, ►transfection, ►transformation genetic, ►packaging cell line, ►biohazards, ►laboratory safety, ►self-deleting vector, ►packaging cell lines; Pfeifer A, Verma IM 2001, pp 353. In: Knipe DM, Howley PM (Eds.) Fundamental Virology. Lippincott Williams & Wilkins, Philadelphia, Pennsylvania; Porcellini A Blasi AD 2004 Methods Mol Biol 259:155; http://www.brc.riken.jp/lab/dna/rvd/.

Viremia: Refers to viruses in the blood.

Virgin: Refers to an individual who has not been mated or has not had prior sexual intercourse. In *Drosophila* the females can store the sperm received during prior mating and, therefore, the paternal identity can be determined only if virgin females are used.

Virgin T Cell: ►immune response

Virile: An adjective for having the characteristics of an (adult) male, masculine.

Virino Hypothesis: This suggests that prions are caused by small nucleic acid-containing agents. ►prion; Dickinson AG, Outram GW 1988 Ciba Found Symp, London 135:63.

Virion: A complete virus particle (coat and genetic material).

Virion RNA: Refers to cytomegalovirus/herpes virus encoded transcripts which are packaged within the virion and delivered to the host cell to assure their immediate expression after infection.

Virioplasm (virus factory): An intercellular compartment associated with the replication of some viruses.

Viroceptors: These are parts of the viral attack mechanisms directed against the host immune system. They mimic the cell receptors and tie up cytokines and chemokines that are destined to stimulate the immune defense and weaken the antibody production. ►antibody, ►immune system, ►interferons; Upton C et al 1991 Virology 184:370.

Viroid: A non-encapsidated RNA (ca. 1.2×10^5 daltons) which is capable of autonomous replication and (plant) pathogenesis, such as the potato spindle tuber viroid. Viroids are single-stranded, circular or linear RNA molecules with extensive intramolecular complementarity. These agents are localized in the nuclei of plants and do not seem to occur in animals although it was earlier assumed that prions were viroids. They are probably the smallest nucleic acid agents causing infectious diseases. (See Elena SF et al 2001 J Mol Evol 53(2):155; Rezaian MA 1999 Curr Issues Mol Biol 1:13; viruses and virions: http://www.ncbi.nlm.nih.gov/genomes/VIRUSES/viruses.html; http://www.dpvweb.net/).

Virosomes: These are liposomes with associated viral proteins and are expected to be used as vehicles for gene therapy. ►liposome; Yomemitsu Y et al 1997 Gene Ther 4:631.

Virtual: This has the same properties as the real but it is not real; it is an imaginary concept. ►digital gene, ►genetics digital

Virtual Reality Training: Medical students learn surgery on three-dimensional models rather than on patients.

Virulence: Determines or indicates the infectivity or pathogenicity of an organism. It has recently been discovered that several bacterial species of different structure and functions acquired a shared mechanism for virulence. About 15–20 protein genes with relatively low G-C contents (below 40%) are assembled in a "pathogenicity island" of either the bacterial chromosome or in a plasmid. These genes encode the molecular machinery (Type III virulence) to produce and transmit the bacterial toxins to their target. Inflammatory responses of the host fight *Shigella* bacteria. The O antigen on the surface of the lipopolysaccharide coat protects the bacteria. Glycosylation of this antigen enhances type III secretion, aids the injection of the toxin and contributes to serotype divergence (West NP et al 2005 Science 307:1313). In *Salmonella* species a 65–100 kb plasmid carries the genes required for systemic infection. A *Yersinia* effector protein (Yopj), a cysteine protease, specifically blocks the host signal transduction system at MAPKK. Yopj-related proteins occur in other bacterial pathogens of animals and plants. ►neurovirulence, ►signal transduction, ►avirulence, ►SAR, ►host-pathogen relations, ►pathogenicity island, ►secretion systems, ►*Shigella*, ►*Yersinia*; Mahan MJ et al 2000 Annu Rev Genet 34:139; Cotter PA, DiRita VJ 2000 Annu Rev Microbiol 54:519; bacterial virulence factor database: http://www.mgc.ac.cn/VFs/.

Virulence Genes of *Agrobacterium*: The Ti plasmid carries—in an about 35 kb DNA—major (*A*, *B*, *C*, *D*, *E*) and minor (*F*, *H*) virulence genes that mediate the process of infection and T-DNA transfer. The *VirA* gene codes for a single protein that is a transmembrane receptor. Its N-terminal periplasmic region responds to sugars and pH whereas the periplasmic loop between the two membrane layers responds to phenolic compounds (e.g., acetosyringone, secreted by the wounded plant tissues). This substance plays an important role in the induction of the cascade of all *Vir* genes although *VirA* itself is constitutive yet it is modulated by several factors. The C-terminus of VirA protein is autophosphorylated. *VirG* also codes for a single protein, a transcriptional regulator. For expression, it requires phosphorylation by VirA. *VirG* regulates by feedback the phosphate metabolism, mediated by *VirA*, and these two genes together are involved in conjugational transfer. The *VirB* operon encoding 11 proteins is also a conjugational mediator. The VirB complex is assembled at the cell pole and has an essential role in targeting the plant cell. VirB7 to VirB10 are minimally required for the formation of the pole VirB complex (Judd PK et al 2005 Proc Natl Acad Sci USA 102:11498). VirB1 is a lysozyme-like protein. VirB4 serves as a docking protein, along with other proteins, for strand transfer during infection (Middleton R et al 2005 Proc Natl Acad Sci USA 102:1685). *VirC* determines the host range and the C1 protein binds to the *overdrive* repeats near the right border of some octopine plasmids. The *VirD* genes are responsible for four polypeptides. VirD1 is a topoisomerase, VirD2 is an endonuclease; in addition, this locus codes for a binding protein and a pilot protein guiding the T-DNA to the plant chromosome. VirD2 interacts with the TATA box-binding protein and a nuclear protein kinase of plants (Bakó L et al 2003 Proc Natl Acad Sci USA 100:10108). VirD4, VirB4 and VirB11 also perform an ATPase function. *VirE*2 codes for a binding protein, which coats the T-strand, mediates the transfer of single strand DNA into the plant nucleus (Zupan JR et al 1996 Proc Natl Acad Sci USA 93:2392). For stable transformation, viral protein 1 (VIP1) interacts with host cell H2A histone (Li J et al 2005 Proc Natl Acad Sci USA 102:57330). *VirF* probably encodes an extracellular protein that regulates *Vir* functions. *VirH* product may metabolize plant phenolics. Nopaline plasmids contain the gene *tzs* (trans-zeatin secretion) and *pin* (plant inducible) *F* loci. The chromosomal virulence loci (*ChvA*, *ChvB*, *Chv*, *psc*) are involved in the production of bacterial surface polysaccharides. Chromosomal genes *cbg*, *pgl*, when present, may enhance virulence. A single-stranded T-DNA complexed with proteins is transferred to the plant cells. Plant proteins are also involved in the transformation process (Tzfira T, Citovsky V 2002 Trends Cell Biol 12:121). ►*Agrobacterium*, ►Ti plasmid, ►T-DNA, ►overdrive, ►transformation genetic, ►crown gall, ►CRAFT, ►MPF; Winans SC et al 1999, p. 289. In: Kaper JB, Hacker J (Eds.) Pathogenicity Islands and Other Mobile Virulence Elements. Amer Soc Microbiol Washington, DC; Dumas F et al 2001 Proc Natl Acad Sci USA 98:485; Christie PJ et al 2005 Annu Rev Microbiol 59:451.

Virulent: In general, this is the poisonous form of prokaryotes. The virulent bacteriophages do not have the prophage life style and after reproduction they destroy the host bacteria by lysis. ►prophage, ►lysis

Virus: This is a small particle containing either double- or single-stranded DNA or RNA as genetic material. A generalized structure is diagrammed here; the architecture of viruses shows great variations from icosahedral to filamentous forms (see Fig. V21).

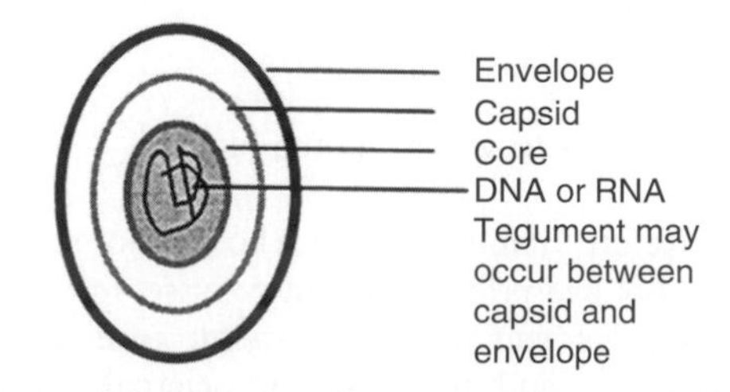

Figure V21. Generalized diagram of a viral structure

Viruses are the ultimate parasites because they lack any element of the metabolic machinery and are totally dependent on the host for assistance to express their genes. They are generally so small in size (15–200 nm) that light microscopy cannot reveal them, except the pox viruses which may be up to 450 nm in length. There are about 1,500 virus species. Bacterial, animal and plant viruses exist. The majority of plant viruses have single-stranded RNA genetic material but a few have double-stranded DNA, e.g., CaMO. Viruses have attracted new interest in nanotechnology as biological platforms, as vehicles for delivery into cells, for encapsulation of organic and inorganic molecules, etc. (Douglas T, Young M 2006 Science 312:873). ►animal viruses, ►herpes, ►acquired immunodeficiency, ►plant viruses, ►bacteriophages, ►oncogenic viruses, ►retroviruses, ►cauliflower mosaic virus, ►mimivirus, ►TMV, ►nanotechnology; Whittaker GR et al 2000 Annu Rev Cell Dev Biol 16: 627; for a chronology of virology: see Oldstone MBA, Levine AJ 2000 Cell 100:139; Sharp PM 2002 Cell 108:305; Virus Particle Explorer: http://viperdb.scripps.edu/; virus genome and protein sequence bioinformatics: http://athena.bioc.uvic.ca/; viral genomes: http://gibv.genes.nig.ac.jp/; http://www.ncbi.nlm.nih.gov/genomes/VIRUSES/viruses.html; http://www.biochem.ucl.ac.uk/bsm/virus_database/VIDA.html; plant viruses and viroids: http://www.dpvweb.net/.

Virus, Computer: A deliberately generated destructive program that can be spread through borrowed computer disks as well as network services. The damage to the files can be usually prevented by the use of continuously updated virus monitoring and eliminating programs.

Virus Hybrid: When the *2b* gene of cucumber mosaic RNA virus is replaced by its homolog of tomato, virulence of the interspecific hybrid virus increases. ►vaccinia virus

Virus Morphology: ►development, ►T4, ►lambda phage, ►retroviruses

Virus Receptors: These are cell surface proteins that mediate viral entry into the host cell. The proteins can be cell adhesion molecules (CXCR4, CD4, dystroglycan, integrins, ICAM, major histocompatibility antigens, etc.), extracellular matrix proteins (heparan sulfate glycoaminoglycan, sialic acid derivatives) and complement control proteins (CD46, CD55), aminopeptidase-N, lipoprotein receptors, coxackie virus and adenoviral receptors (CAR1), etc. Some receptors may serve several different viruses and some viruses can take advantage of more than one type of receptor. Single amino acid replacements in the receptor either abolish or facilitate the uptake of the virus. (See Baranowski E et al 2001 Science 292:1102; Bomsel M, Alfsen A 2003 Nature Rev Mol Cell Biol 4:57).

Virus Reconstitution: ►reconstituted virus

Virus Resistance: The Fv and Rfv genes of mouse provide resistance against mouse leukemia (Friend virus) by encoding a retroviral envelope protein or the viral gag (group specific antigen). In the latter case the Fv gene product blocks the entry of the virus into the nucleus or these proteins may exercise a dominant negative effect on the virus that happens to be there. Introducing into the plant genome tobacco mosaic virus coat protein genes restricts the virulence of the virus in normally susceptible plants. ►CRISPR

Virus Transport to the Cell Nucleus: This is mediated by GAG matrix associated protein (MA) and the VPR gene product in the case of HIV. MA includes a nuclear localization sequence (NSL) and a signal targeting the cell membrane. When the NLS is phosphorylated, the MA becomes part of the pre-integration complex. Further phosphorylation detaches the MA from the membrane. ►acquired immunodeficiency, ►nuclear localization sequence; Bell P et al 2001 J Virol 75:7683.

Virus-Free Plants: Plants regenerated under axenic conditions from apical meristems are generally free of virus disease until they are reinfected. Plants so obtained are commercially useful for the production of virus-free seed stocks. Antisense RNA constructs may render plants virus resistant. (See Hammond J, Kamo KK 1995 Mol Plant Microbe Interact 8:674).

Virus-like Particles (VLP): These are artificial constructs of a viral protein capsid enclosing a core particle passenger. Such protein cage structures have shown increasing promise as therapeutic and diagnostic vectors, imaging agents, and as templates and microreactors for advanced nanomaterials synthesis. This type of biomimetic self-organization can combine the natural characteristics of virus capsids with the exquisite physical properties of nanoparticles. The principle challenges for nanoparticle delivery include limited lifetime in body fluids, nanoparticle transduction across the cellular membrane, avoidance of the exocytotic pathways and target specificity. A gold core functionalized with a coating of carboxylated polyethylene glycol (PEG) can allow efficient assembly of VLPs (Sun J et al 2007 Proc Natl Acad Sci USA 104:1354).

Virusoid: 300–400 nucleotide long RNAs, pathogenic to plants and accompany other plant viruses. ►viroid

Viscosity: The internal friction of fluids expressed as dyne-seconds/cm^2 called poise unit. ►stoke, ►dyne

Visfatin: A cytokine secreted by adipocytes that mimics the effects of insulin. ►adipocyte, ►insulin; Fukuhara A et al 2005 Science 307:426; this paper has been retracted. (Science 318: 565 [2007]).

Visible Mutation: Can be identified by the phenotype seen.

Vision: ►rhodopsin

Vista: A visualization tool for alignment; also a computer program.

Vital Genes: ►lethal mutation

Vital Stain: Colors living cells without serious damage to viability.

Vital Statistics: Pertains to birth, marriage and death registrations. Such information may assist in constructing human pedigrees and provide important facts about family histories, congenital and hereditary disease, longevity, etc.

Vitalism: A theory of the nineteenth century and earlier, which postulates that living beings are controlled not only by physical and chemical mechanisms, but life is also associated with a transcendental vital force. However, there still remain phenomena like the development of an embryo or regeneration from single cells that cannot be fully explained in physico-chemical terms although vitalism is no longer a viable idea since Friedrich Wöhler synthesized urea in 1828 from inorganic ingredients. (See Hein H 1972 J Hist Biol 5:159).

Vitamin A: Also known as retinol, it is synthesized from carotenoids. Its deficiency results in visual and skin anomalies (see Fig. V22). However, excessive amounts may be harmful. With the aid of three transformation vectors the β-carotene pathway has been introduced into the carotenoid-free rice endosperm enabling the production of provitamin A and correcting the deficiency in the diet. ►retinol, ►retinal

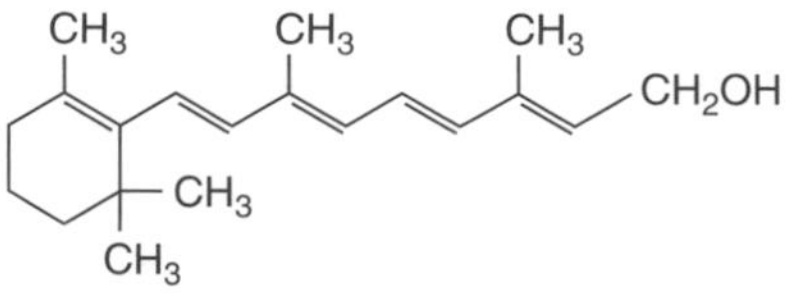

Figure V22. Vitamin A (all trans retinol)

Vitamin B_1: ►thiamin (see Fig. V23)

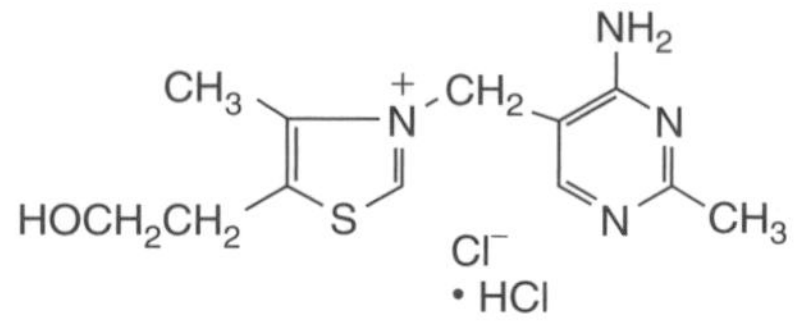

Figure V23. Vitamin B_1

Vitamin B_2: ►riboflavin, ►riboflavin retention deficiency

Vitamin B_6: ►pyridoxine

Vitamin B Complex: Includes thiamin, riboflavin, nicotinic acid (amide), panthotenic acid, pyridoxin and vitamin B_{12}.

Vitamin B_{12} Defects: Vitamin B_{12} or its coenzyme form has a MW of about 1,355. It is composed of a core ring with Cobalt (Co^{3+}) at its center and to it, through isopropanol, a dimethyl benzimidazole ribonucleotide is joined; it also contains a 5′-deoxyadenosine. It is usually isolated as a cyanocobalamin because during the process of purification a cyano group may be attached to the cobalt at the place where the 5′-deoxyadenosyl group is positioned in the coenzyme. B_{12} is not synthesized in plants and animals

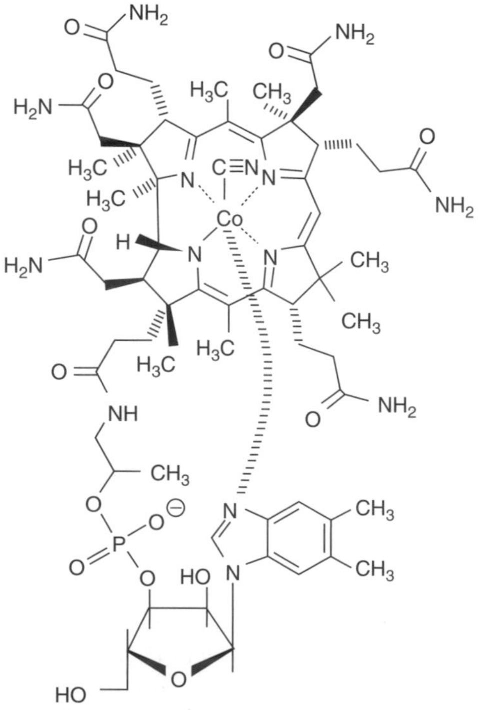

Figure V24. Vitamin B_{12}

and is not usually present in the diet. Intestinal microorganisms produce it from the meat consumed and it is absorbed when the so-called "intrinsic factor," a glycoprotein, is available in satisfactory amounts. If the intake is less than 3 μg/day, pernicious anemia develops in humans (see Fig. V24). Autosomal recessive mutation may prevent the release of the lysosomally stored vitamin and B_{12} deficiency may result. Cysteinurias is often called cobalamin F (cbl F) disease. In methylmalonicacidemia combined with homocystinuria, methylmalonyl-CoA mutase and homocysteine methyl-tetrahydrofolate methyltransferase (cbl C) deficiencies are involved. ►methylmalonicaciduria, ►cysteinuria, ►amino acid metabolism, ►cobalamin; Banerjee R, Ragsdale SW 2003 Annu Rev Biochem 72:209.

Vitamin C: ►ascorbic acid, formula given here (see Fig. V25)

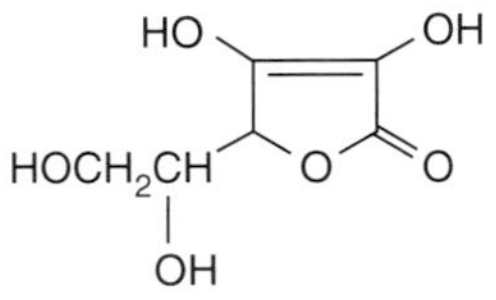

Figure V25. Vitamin C

Vitamin D: An antirachitic, fat soluble vitamin, whose deficiency leads to rickettsia and defects in bone development and maintenance. Vitamin D^2 (ergocalciferol) (see Fig. V26) is formed upon irradiation of ergosterol and vitamin D^3 (cholecalciferol) from 7-dehydro-cholesterol. Children need about 20 μg/day in their diet. Autosomal recessive human defects in vitamin D receptors (12q12-q14) do not respond favorably to vitamin D fortified diet. Several cancer cells are inhibited by vitamin D^3 and it may induce apoptosis. The use of a sunscreen may reduce the amount of vitamin D below a desirable level. ►Williams syndrome, ►VDR, ►hormone-response elements; McGuire TF et al 2001 J Biol Chem 276:26365; formula.

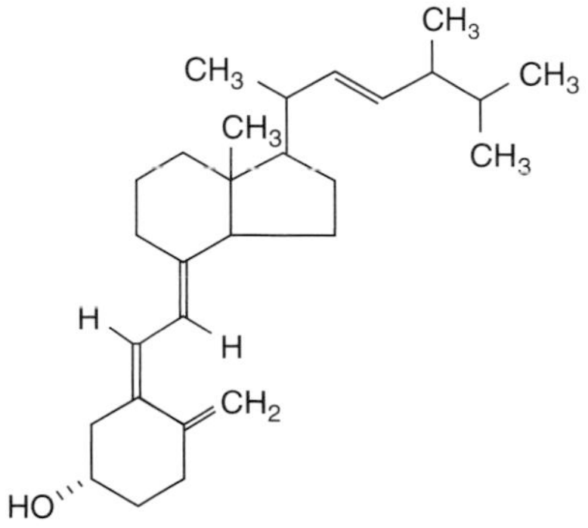

Figure V26. Ergocalciferol

Vitamin E (tocopherols): Vitamin E is an antioxidant (in several different forms) normally present in satisfactory amounts in a balanced diet (see Fig. V27). Transgenic plants expressing homogentisic acid geranylgeranyl transferase display increased amounts of vitamin E and antioxidant ability (Cahoon EB et al 2003 Nature Biotechnol 21:1082). In an autosomal recessive condition vitamin E malabsorption has been observed, causing intestinal and nervous anomalies and accumulation of cholesterol. The severe symptoms could be alleviated by 400–1,200 international units (IU) of vitamin E. It appears that the affected individuals lack an α-tocopherol binding protein, required to build it into very low-density lipoproteins. In a transgenic mouse model vitamin E reduced chromosomal damage and hepatic tumors. The familial vitamin E deficiency is under autosomal recessive control in humans. Doses of 400 IU may be very harmful. ►tocopherols, ►VLDL, ►atherosclerosis; Azzi A et al 2002 FEBS Lett 519:8.

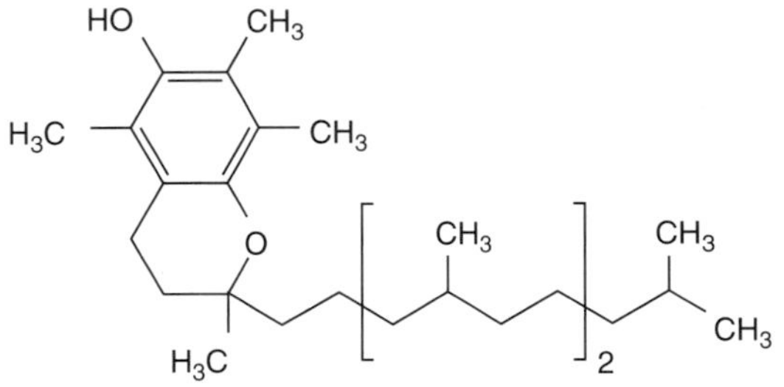

Figure V27. Tocopherol

Vitamin K (phylloquinone): A plant lipid cofactor of blood coagulation (vitamin K_1) and a related substance, menaquinone (vitamin K_2) is synthesized by intestinal bacteria of animals; the synthetic menadione (vitamin K_3) also has some vitamin K activity. The name K was derived from the Danish word koagulation. ►vitamin K-dependent blood clotting factors

Vitamin K-Dependent Blood Clotting Factors: Some autosomal recessive bleeding diseases respond favorably to the administration of vitamin K (see Fig. V28). It is required for post-translational modification of at least six proteins involving the conversion of the NH_2-end of glutamic acid into γ-carboxyglutamic acid. Deficiency of this process may occur as a consequence of treatment by coumarin drugs—such as warfarin—used as anticoagulants. A vitamin K-dependent blood coagulation factor is encoded at 2p12 and its mutation is caused by vitamin K-dependent carboxylase (GGCX) in the microsomes of the liver and includes bleeding disease. Similarly, vitamin K 2,3-epoxide reductase (VKOR) mutations (16p12-q21) in a transmembrane protein of the endoplasmic reticulum lead to familial bleeding. ►prothrombin deficiency, ►resistance to coumarin-like drugs, ►antihemophilic factors, ►warfarin, ►γ-glutamyl carboxylase, vitamin K-dependent blood clotting factors; Li T et al 2004 Nature [Lond] 427:541.

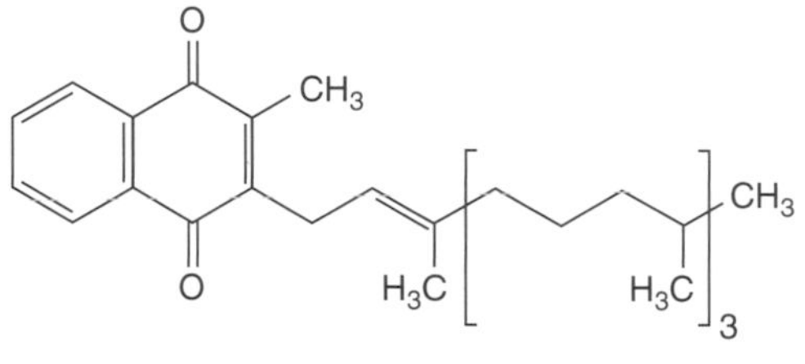

Figure V28. Vitamin K

Vitamins: These are dietary supplements, without measurable caloric value. They serve the role of coenzymes. Widespread dietary supplements can cause selective accumulation of vitamin deficient genotypes in human populations (Lucock M, Yates Z 2005 Nature Rev Genet 6:235).

Vitelline Layer: A yolk (heavier) layer around the eggs. In mammals it is (a thinner) zona pellucida. ►egg

Vitellogenin: Refers to yolk protein. In honeybees the amount of yolk protein is proportional to longevity because it protects from oxidative stress. ►longevity, ►oxidative stress; Seehuus A-C et al 2006 Proc Natl Acad Sci USA 103:962.

Vitiligo: Dominant (1p31.3-p32.2) halo skin spots (may be identical on opposite sides of the body) present after birth that may spread or regress. Apparently, loci in several autosomes are also involved in susceptibility. It is an autoimmune disease with an incidence of ~1%. ►piebaldism, ►nevus, ►skin diseases; Alkhateeb A et al 2002 Hum Mol Genet 11:661; Fain PR et al 2003 Am J Hum Genet 72:1560.

Vitrification: A procedure of protecting sensitive biological material (enzymes, seeds) from deterioration by coating with sugar mixtures (e.g., sucrose and raffinose) which, however, readily dissolve when needed. Vitrification literally means the formation of a glass-like structure. Human oocytes can be preserved in an appropriate vitrification medium and frozen in liquid nitrogen. ►cryopreservation; Yokota Y et al 2001 Steril Fertil 75:1027.

Vitronectin (S protein): A fluid phase protein in the cell binding the C5b-9 complement component which prevents its attachment to the membrane and thus interferes with lytic activity. ►complement; Tomasini BR, Mosher DF 1991 Progr Hemost Thromb 10:269; Seger D et al 2001 J Biol Chem 276:16998.

Vivipary: Refers to giving birth to live offspring. In plants the seed germinates before shedding from the fruits. This phenomenon is genetically controlled through abscisic acid metabolism. ►ABA; Paek NC et al 1998 Mol Cell 8(3):336; Jones HD et al 2000 Plant J 21:133.

V(J)D Recombinase: Assembles immunoglobulins and T cell receptors. The activity of this enzyme is facilitated by histone acetylation. ►immunoglobulins, ►T cell receptors, ►antibody gene switching, ►RSS, ►accessibility, ►SCID, ►ligase, ►RAG, ►histone acetyltransferase; Fugmann SD et al 2000 Annu Rev Immunol 18:945; Gellert M 2002 Annu Rev Biochem 71:101; Jung D, Alt FW 2004 Cell 116:299.

VLDL (very low density lipoprotein): This is the 55 nm precursor of LDL and triglycerides. Its core contains cholesteryl ester. In muscle capillaries and adipose tissues the VLDL-triglycerides are removed and exchanges take place with other lipoproteins resulting in the loss of proteins except apolipoprotein B-100 and a smaller (22 nm) particle size. It is now called LDL (low-density lipoprotein). The VLDLR receptor gene (9p24) is very similar to the LDLR but it contains an additional exon. ►low-density lipoproteins, ►familial hypercholesterolemia, ►sitosterolemia, ►sterol, ►Alzheimer's disease, ►hypertension, ►apolipoproteins; Merkel M et al 2001 Proc Natl Acad Sci USA 98:13294.

VLP: A virus-like particle, such as some transposable elements (retroposons). ►Ty, ►retroposons

V_{max}: The maximal level of enzyme activity (maximal velocity reaction).

v-mil: ►MIL, ►MYC

vMIP: ►herpes virus, ►chemokines

VNC: Refers to viable but non-cultivable microbial cells. They may be transgenic (genetically modified) and they may be protected by nucleic acids.

VNO: ►vomeronasal organ, ►olfactogenetics

VNTR (variable number tandem repeats): Loci are used in forensic DNA fingerprinting. These repeats may display hundreds of alleles per single locus and are extremely polymorphic in restriction fragments. They are useful as physical markers for mapping. Also, because matching patterns occur by a chance of 10^{-7} to 10^{-8} only, they are well suited for criminal, personal and legal identification on the basis of minute amounts of DNA extracted from drops of body fluids, blood or semen. VNTR can also be used for taxonomic and evolutionary studies of animals and plants. ►DNA fingerprinting, ►diabetes mellitus, ►MVR, ►MLVA, ►trinucleotide repeats, ►SSM; Le Stunff C et al 2001 Nature Genet 29:96.

Vogt-Spielmeyer Disease: This is also known as Batten disease. ►ceroid lipofuscinosis

Volatility: Refers to the proportion of point mutations that encodes a different amino acid. The arginine codon CGA has eight potential ancestor codons, i.e., non-stop codons that differ from CGA by one point mutation. Four of the potential ancestor codons of CGA encode another amino acid. Thus, the volatility is 4/8. If the selection pressure for a particular protein is high, the demand for amino acid replacement increases. ►selection pressure; Plotkin JB et al 2004 Nature [Lond] 428:942.

Vole: *Microtus agrestis*, 2n = 50; *Microtus arvalis*, 2n = 46; *Microtus montanus*, 2n = 24.

Volicitin: ►biological control

Volt (V): The unit of electric potential. In case the resistance is 1 Ω (Ohm = 1 V/A = 1 m^2 kg sec^{-3} A^{-2}) and the electric current is 1 ampere, the voltage is 1. ►ampere, ►watt

Voltage-Gated Ion Channels: These channels are opened/closed for the transport of ions in response to a change in voltage across cell membranes. ►ion channels, ►signal transduction

Volunteer: This plant sprouts from the seed shed on the field and appears after the regular harvest season. The farmer does not plant it.

Vomeronasal Organ (VNO): This is a pair of chemosensory organs at the base of the nasal cavity in the majority of higher animals, except birds and some monkeys and apes (humans). VNO is stimulated by pheromones. MHC class I peptides act as chemosensory signals to the sensory neurons (Leinders-Zufall T et al 2004 Science 306:1033). The signaling through GTP-binding proteins requires 1,4,5-trisphosphate rather than cAMP. In humans, pheromones play either no role or only a subordinate role and the VNOs are apparently non-functional. Apparently, the vision-based signaling-sensory mechanism has replaced the chemical-based system in the social/reproductive activities of hominoids and Old World monkeys (Zhang J, Webb DM 2003 Proc Natl Acad Sci USA 100:8337). In mouse 187 and in rat 102 V1R genes have been identified but in dogs only 8, cows 32 genes and in the marsupial opossum 49 function (Grus WE et al 2005 Proc Natl Acad Sci USA 102:5767). ►olfactogenetics, ►phosphoinositides, ►cAMP, ►pheromones, ►major histocompatibility complex, ►marsupial; Lane RP et al 2002 Proc Natl Acad Sci USA 99:291.

v-Oncogene: Viral oncogene, homologous to a c-oncogene, but carried by oncogenic viruses, capable of causing cancerous growth. ►oncogenes, ►cancer, ►c-oncogene, ►retrovirus

Von Gierke Disease: ►glycogen storage disease

Von Hippel-Lindau Syndrome (VHL): A dominant phenotype (human chromosome 3p26-p25) involving tumorous growth primarily of the blood vessels of the eye (hemangioma) and the brain (hemangioblastoma). The central nervous system, kidneys (phaeochromocytoma) and the pancreas may also become tumorous. The incidence estimates vary between 0.00002 and 0.00003, and the mutation rate appears to be about 2–4 × 10^{-6}. The primary cause of the tumor is the inactivation of the VHL tumor suppressor gene. The VHL protein forms ternary complexes with Elongin C and Elongin B and CUL2 (see Fig. V29). This complex marks HIF (hypoxia-inducible factor) for degradation by the proteasome. RBX helps to recruit the ubiquitinating proteins. With the assistance of the normal pVHL, the α subunit of HIF is degraded and the level of oxygen increases. This process results in the activation of the CXCR4 chemokine receptor. HIF, with the cooperation of other proteins, e.g., VEGF, promotes the formation of blood vessels that are required for cancerous growth. When pVHL protein formation is reduced or prevented, HIF remains active and creates conditions for angiogenesis and cancerous growth. It is believed that VHL is a negative regulator of VEGF by association with complexes (SCF) which target proteins for degradation. The VHL gene falls in the same area as the RCC gene. The VHL inactivation seems to affect TGF-α and that may stimulate the renal carcinogenic path. ►ELONGIN, ►VEGF, ►CXCR,

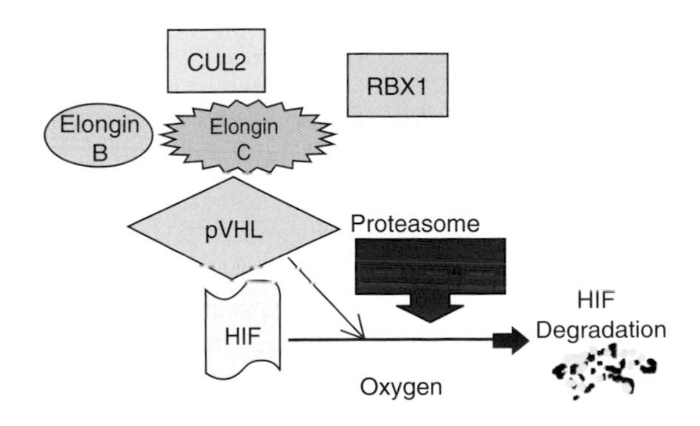

Figure V29. HIF pathway in VHL

►eye diseases, ►kidney diseases, ►phaeochromocytoma, ►polycythemia, ►mutation rate, ►renal cell carcinoma, ►hypernephroma, ►RCC, ►SCF, ►SKP, ►cullin, ►angiogenesis, ►wound healing, ►proteasome, ►elongin, ►Rbx1, ►ubiquitin, ►HIF, ►hypoxia; Iwai K et al 1999 Proc Natl Acad Sci USA 96:12436; de Paulsen N et al 2001 Proc Natl Acad Sci USA 98:1387; Ivan M et al 2001 Science 292:464; Friedrich CA 2001 Hum Mol Genet 10:763.

Von Neumann–Morgenstern Gamble: Seriously or terminally ill patients may chose an experimental drug or treatment that may improve the quality of life but may further aggravate the illness and even shorten life.

Von Recklinghousen Disease (NF1): A dominant, human chromosome 17q11.2 neurofibromatosis, a skin tumor with characteristic café-au-lait spots. Some other features include are pseudoarthritis, glioma, mental retardation, hypertension, hypoglycemia, etc. The mutation rate is ~10^{-4}. The basic defect appears to be in a cytoplasmic GAP protein. ►neurofibromatosis, ►GAP, ►elephant man, ►café-au-lait spot

Von Willebrand's Disease (VWD): A complex, hereditary bleeding condition due to a deficiency of a large antihemophilic cysteine-rich glycoprotein (2813-amino acids) in the blood plasma, platelets and subendothelial connective tissue. It is different from hemophilias inasmuch as the bleeding from the gastrointestinal, urinary system and the uterus is prolonged. Several forms of this disease have been identified on the basis of which component of the large gene is affected. The type III form is characterized by very severe symptoms and is the least common type of VWD. The most common type VWD protein has a binding domain to antihemophilia factor VIII and its defect appears as a relatively rare recessive. There is also an X-linked form. The bleeding can be readily stopped upon supplying normal blood to the patients. The most common dominant gene which leads to a reduction of this glycoprotein has been assigned to human chromosome 12pter-p12.3. In its ~180 kb it contains 52 exons. Exon 28 is the largest encoding domains A1 and A2. Domains D1-D2-D'-D3-A1-A2-A3-D4-B1-B2-B3-C-1-C2-CK are distinguished. Its highly homologous pseudogene is located in 22q11.2. At least 22 human genes carry homologies to the A domain of chromosome-12 gene. The frequency of heterozygotes has been estimated to vary from 1.4 to 5%, and thus VWD appears quite frequent although some other estimates are much lower. Recurrence risk for the dominant form, if one of the parents is affected, is about 50%. Both heterozygotes and homozygotes may express VWD. Percutaneous umbilical blood sampling (PUBS) may permit prenatal diagnosis. ►hemophilias, ►antihemophilic factors, ►Glanzmann's disease, ►hemostasis, ►prenatal diagnosis, ►epiphyseal dysplasia; Sadler JE 2002 Science 297:1128; Manucci PM 2004 New England J Med 351:683.

Von Willebrand-Jürgen's Syndrome: ►thrombopathic purpura

VP: Refers to viral proteins such as the VP1, VP2 and VP3 of Simian virus 40. ►SV40

VP16 (α-transinducing factor, α-TIF): Refers to a herpes simplex virus transcription activation domain that can boost the expression of other genes by a factor of 10^5. The normally minimal activation is due a peptide module of VP16 called ATF14. A small synthetic molecule, less than 1/5 of ATF14, is almost as efficient and more resistant to proteolytic degradation (Minter RA et al 2004 J Am Chem Soc 126:10504). ►transactivator; Hirst M et al 2001 Proc Natl Acad Sci USA 98:8726.

Vpg: A genome-linked (polio) viral protein (22 amino acids) attaches to the 5′ end of RNA viruses and acts as a primer in the replication of the nucleic acid. ►primer

VRE (ventral response element): This is involved in the regulation of ventral development by preventing switching on an activator that operates dorsal development.

vRNA: ►vault

Vrolik Disease: ►osteogenesis imperfecta Type II

VSG: Refers to variable surface glycoprotein. ►*Trypanosoma*, ►*Borrelia*, ►telomere switching

VSP (very short patch repair): A prokaryotic repair involving T-G mismatches by restoring the original base pairs. (See Lieb M et al 2001 J Bacteriol 183:6487).

VSV: ►CO_2 sensitivity

V-type ATPases: These are responsible for the acidification of cellular organelles (vacuoles, lysosomes, Golgi complex) by the maintenance of the vacuolar-type ATPase—proton pump in plant and animal cells. ►Golgi complex, ►lysosomes, ►ATPase; Gruber G et al 2001 J Exp Biol 204:2597.

V-type Position Effect: A variegated expression of genes transposed into the vicinity of heterochromatin. This phenomenon is common in *Drosophila* but only a few cases have been observed in plants. One of the most common causes of variegation is the movement of insertion or transposable elements or sorting out of plastid-DNA encoded mutations. ►position effect,

►heterochromatin, ►variegation, ►chloroplast genetics, ►transposable elements

Vulva: This is the outer region of the external female genital organ, the vaginal orifice and organs associated with it.

Historical vignettes

RC Punnett (Heredity 4:9) gives the following account: "I was asked why it was that, if brown eye were dominant to blue, the population was not becoming increasingly brown eyed: yet there was no reason for supposing such to be the case. I could only answer that the heterozygous browns also contributed their quota of blues and that somehow this lead to equilibrium. On my return to Cambridge I at once sought out GH Hardy with whom I was then very friendly. For we had acted as joint secretaries to the Committee for the retention of Greek in the Previous Examination and we used to play cricket together. Knowing that Hardy had not the slightest interest in genetics, I put my problem to him as a mathematical one. He replied that it was quite simple and soon handed to me the now well-known formula $pr = - q^2$ (where p, $2q$ and r the proportions of AA, Aa, and aa individuals in the population varying for the A-a difference). Naturally pleased at getting so neat and prompt an answer I promised him that it should be known as 'Hardy's Law'— a promise fulfilled in the next edition of my Mendelism. Certain it is…that 'Hardy's Law' owed its genesis to a mutual interest in cricket."

Vannevar Bush, Director of the US Office of Scientific Research and Development, 1938 on science policy: "Scientific progress on a broad front results from the free play of free intellects, working on subjects of their own choice, in the manner dictated by their curiosity for exploration of the unknown."

Vulvovaginitis: An autosomal dominant allergy to semen resulting in vaginal inflammation lasting from a couple to several hours after coitus. The most common cause is, however, infection or candidiasis. ►candidiasis; Eschenbach DA 2004 New England J Med 351:851.

V

W

w: A symbol of map distance. ►θ, ►map distance, ►cM, ►recombination, ►mapping function

W Chromosome: This corresponds to the Y chromosome in heterogametic females, WZ in birds and butterflies. ►chromosomal sex determination

***w* Locus**: The first mutation discovered in *Drosophila* by Morgan is involved in the control, production and distribution of brown (ommochrome) and red (pteridine) pigments of the eyes and ocelli and some other anatomical structures. The gene (at 1–1.5) apparently encodes an ATP-binding membrane transport protein for the precursors of the pigments. More than 200 alleles have been identified within a 0.03 centiMorgan region, which has been mapped by intragenic recombination into 7 domains. The wild type allele is incompletely dominant over many mutant alleles. The alleles do not show partial complementation with the exception of the w^{sp} (white spotted) allele that displays allelic complementation with the majority of other alleles in the presence of the z^{a} (*zeste*). The latter is a regulatory gene at 1–1.0 location and *zeste* encodes a specific protein binding to the promoters of *w*, *Ultrabithorax* (*Ubx*) and *decapentaplegic* (*dpp*). ►map unit, ►recombination, ►morphogenesis in *Drosophila*, ►eye color, ►Tangier disease

W Mutagenesis: A tendency of increased mutation after Weigle reactivation. ►Weigle; Yatagai F et al 1983 Adv Space Res 3(8):65.

W Point: A stage just before the S phase when animal cells still have serum growth factor requirement to enter the S phase. ►cell cycle

W Reactivation: This is the same as Weigle reactivation.

Waardenburg Syndrome: Autosomal dominant forms may be distinguished on the basis of displacement (type I) and without displacement (type II) of the eyelids. Variegation in the color of the iris, white forelock, eyebrows and eyelashes, syndactyly, heart problems, hearing defects may occur as autosomal recessive anomaly. Dominant mutations in the SOX10 gene may affect the neural crest-derived cell lineages. The Waardenburg syndrome type 2 gene (MITF [microphthalmia-associated transcription factor]) converts fibroblasts into melanocyte-like cells by transactivation of a tyrosinase gene. If MITF is inactive hypopigmentation occurs. Mutation in the PAX3 transcription factor is responsible for hearing and pigmentation defects of type 1 form caused by failure of transactivation of MITF. Human chromosomal locations are type I 2q35, type IIA 3p14.1-p12.3 and type IIB 1p21-p13.3. The type III form also called Klein-Waardenburg syndrome is at 2q35 location. The syndrome may be haplo-insufficient. The merle phenotypes observed in several breeds of dogs are characterized by hypopigmentation and by hearing and eye defects is caused by insertion of SINE elements into the *SILV* (silver) gene, responsible for pigmentation. The merle phenotype is similar to the Waardenburg type II anomaly (Clark LA et al 2006 Proc Natl Acad Sci USA 103:1376). ►PAX, ►DiGeorge syndrome, ►eye defects, ►eye color, ►polydactily, ►microphthalmos, ►haplo-insufficient, ►Shah-Waardenburg syndrome, ►dog; Sánchez-Martín M et al 2002 Hum Mol Genet 11:3231.

Waardenburg-Shah Syndrome: ►Shah-Waardenburg syndrome

WAF: ►p21

WAGR: ►Wilms tumor

Wahlund's Principle: When two populations, each with different allelic frequencies and both in Hardy-Weinberg equilibrium, are mixed by migration, there is an overall decrease in heterozygotes: $\overline{H} = 2\overline{p}\,\overline{q}[1 - (\sigma^2/\overline{p}\,\overline{q})]$. The decrease in overall heterozygosity indicates the degree of heterogeneity between the two populations and $(\sigma^2/\overline{p}\,\overline{q}\ )$ is the Wahlund's variance of gene frequencies. ►allelic frequencies, ►migration; Yasuda N 1968 Am J Hum Genet 20:1.

Wald-Wolfowitz Test: Used for comparison of two unmatched, supposedly continuous distributions and the null hypothesis is that the two samples are distributed identically. ►null hypothesis, ►logistic regression; Hays WL, Winkler RL 1970 Statistics: Probabilitity, Inference, and Decision. Holt, Reinhart and Winston, New York.

Waldemar of Prussia: A hemophiliac great-grandson of Queen Victoria. ►hemophilia A, ►Queen Victoria

Waldenström Syndrome: ►macroglobulinemia

Walker Boxes (P-loops): These are nucleotide triphosphate-binding amino acid sequences in several proteins. Box A promotes branch migration in Holliday junctions during recombination mediated by the Ruv B protein. Walker A: GlyXXGlyXGly**Lys**Thr, Walker B: AspGluXAsp. The Lys residue binds the γ-phosphate of nucleotides directly. ►branch migration, ►Holliday junction, ►RuvABC; Walker JE et al 1982 EMBO J 1:945; Hishida T et al 2001 J Biol Chem 274:25335.

Walker-Wagner Syndrome: An autosomal recessive hydrocephalus (accumulation of fluid in the enlarged head) generally associated with retinal detachment, congenital muscular dystrophy and lissencephaly. ►Miller-Dieker syndrome, ►hydrocephalus, ►prenatal diagnosis, ►head/face/brain defects, ►lissencephaly

Walker-Warburg Syndrome (HARD, 9q31): Refers to autosomal recessive hydrocephalus agyria and retinal dysplasia that was originally described as lissencephaly. ►Miller-Dieker syndrome; Beltrán-Valero de Barnabé D et al 2002 Am J Hum Genet 71:1033.

Walking: ►chromosome walking

Walking of Transcriptase: Describes transient halting of the movement of the RNA polymerase ternary complex (polymerase, DNA, transcript) by sequentially providing subsets of the four ribonucleotides. ►transcript elongation

Wallaby: *Wallabia bicolor*, 2n = 11 in males and 10 in females; *Wallabia eugenii* 2n = 16.

Wallaby: A non-viral retrotransposable element named after the jumping small Australian kangaroo. ►retroposon

Walnut (*Juglans* spp): Occurs in both wild and cultivated forms; 2n = 2x = 32.

Walnut Comb: In poultry, it is determined by the genetic constitution *RrPp*, and as a result of epistasis, occurs in 9/16 frequencies in F2 after brother-sister mating of the same double heterozygotes (see Fig. W1). The other phenotypes in the segregating F_2 are rose *Rr/RR, pp* (3), pea *rr, PP/Pp* (3), and single *pp, rr* (1).

Walrus: *Odobenus rosmarus*, 2n = 32.

Waltzing Mouse: A chromosomal deletion causing involuntary movements.

Wanda: ►fish orthologous genes

Wandering Spots: An older method of sequencing short oligonucleotides. (Le Gall O et al 1988 J Gen Virol 69:423).

Warburg Effect: ►glycolysis

Warburg Micro Syndrome (2q21.3): A condition characterized by autosomal recessive microcephaly, micropupil, congenital cataract, mental and physical retardation, underdeveloped genitalia, facial hairiness, hypoplasia of the corpus callosum and other anomalies. The molecular defect appears to be in the RAB3 GTPase activating protein. ►corpus callosum, ►RAB, ►Griscelli syndrome; Aligianis IA et al 20005 Nature Genet 37:221.

Ward-Romano Syndrome (WRS, 11p15.5): An autosomal dominant or recessive LQT disease involving anomalous heart muscle fibrillations, fainting (syncope) and possibly sudden death. It is the same as long QT syndromes. ►LQT, ►electrocardiography, ►ion channels, ►HERG, ►Jarvell and Lange-Nielson syndrome, ►long QT syndrome, ►Andersen syndrome

Warfarin ($C_{19}H_{16}O_4$, named after Wisconsin Alumni Research Foundation): This is a slightly bitter, water and alcohol soluble compound. It depresses the formation of prothrombin, necessary for blood clotting, and may cause fragility of the capillary veins leading to hemorrhages. It is used in certain surgeries and treatment of diseases that block arteries by blood clots. It is also a rodent poison. A single ingestion may not necessarily be very hazardous to humans but rats or mice consuming it repeatedly in baits suffer internal bleeding and die. The antidote of warfarin is vitamin K. Warfarin inhibits vitamin K epoxide reductase. Mutations in rodents may make them resistant to warfarin. ►anticoagulation factors, ►prothrombin deficiency, ►coumarin-like drug resistance, ►vitamin K-dependent clotting factors, ►chondrodysplasia; Van Aken H et al 2001 Clin Appl Thromb Hemost 7:195; Loebstein R et al 2001 Clin Pharmacol Ther 70(2):159.

Wasp: Wiskott-Aldrich syndrome proteins that regulate the assembly of actin monomers into filaments and regulate the cytoskeletal organization and motility of cells. WASF1 encoded at 6q21-q22 and WASF3 at 13q13 are effectors for the signal transmission from tyrosine kinase receptors to the cytoskeleton. The function of WASF2 (1p36.11-p34.3) is also similar. The latter has a pseudogene at Xp11.22. ►Wiskott-Aldrich syndrome, ►actin, ►cytoskeleton; Ward ME et al 2004 Proc Natl Acad Sci USA 101:970.

Wasp: *Habrobracon* spp. 2n = 20 for female, 2n = 10 for male. The males hatch from unfertilized eggs and are haploid. The females come from fertilized eggs and are diploid. They are heterozygous for any of the

Figure W1. From left to right: Walnut, Rose, Pea, Single comb

nine sex factors. As a result of inbreeding, sex factor homozygotes occasionally arise that are sterile "biparental males" (see Fig. W2). Gynandromorphs occur in wasps but these are different from those in *Drosophila* because the haploid sectors are not necessarily male sectors as expected from the loss of one set of chromosomes. Although in insects circulating sex hormones do not seem to exist, some type of diffusible substance affects the chromosomally male sectors. Exceptional gynandromorphs may arise by fertilization of binucleate eggs. Gynandromorphic tendency is genetically determined. ►social insects, ►honey bee; Page RE et al 2002 Genetics 160:375; Cowan DP, Stahlhut JK 2004 Proc Natl Acad Sci USA 101:10374.

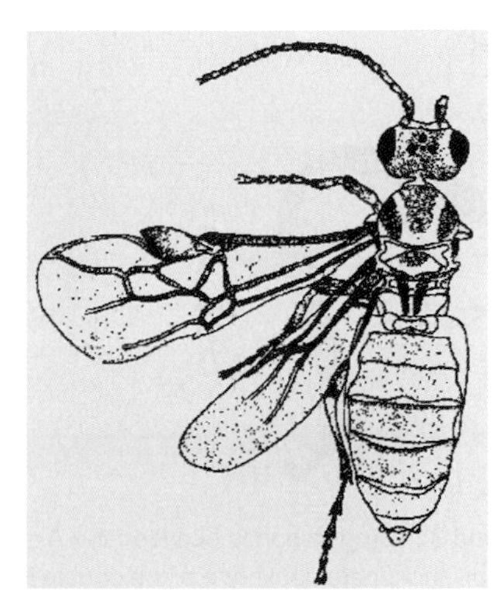

Figure W2. Wasp (Right-side wings and legs removed)

Watercress (*Rorippa nasturtium-aquaticum*): This is a northern European vegetable with x = 16 but the species may be diploid, sterile triploid or tetraploid.

Watermelon (*Citrullus vulgaris*): An annual fruit, 2n = 22. Triploids are grown commercially and crossing tetraploid plants with diploids produces the seeds. The fruits of the triploid plants are practically seedless and are therefore easier to eat (see Fig. W3). According to some reports, triploids have a higher sugar content than either of the parental forms.

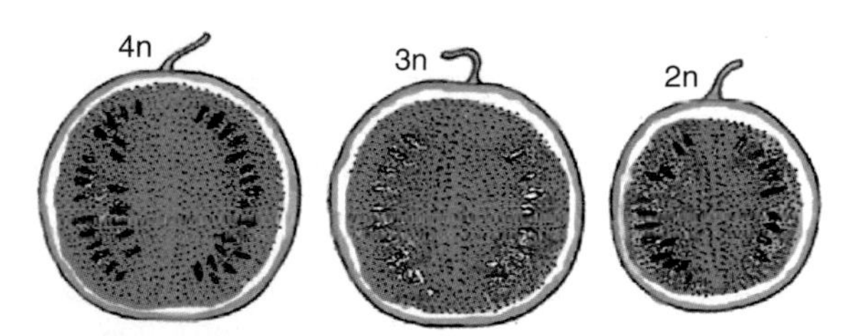

Figure W3. Watermelon. (Diagram of watermelons modified after Eigsti OJ, Dustin P Jr 1955 Colchicine, Iowa State College Press, Ames, Iowa)

Watson and Crick Model: This has been described by Watson JD, Crick FHC in 1953 (Nature [Lond] 171:964) and it became the world's most famous biological model ever conceived. (See Fig. W4 of the model, and Fig. W6 for historical documents: http://osulibrary.orst.edu/specialcollections/coll/pauling/dna/).

It is interesting to note in the first DNA models of Watson and Crick (1953) only two hydrogen bonds have been shown between G and C. The third hydrogen bond between G and C was first mentioned in a paper by Pauling and Corey in 1956 (Waine-Hobson S 2006 Nature [Lond] 439:539). The genome of the first human being that was entirely sequenced in 2007 belongs to James Watson (Project Jim). He agreed that his DNA sequence should be added to public databases but requested that his *ApoE* gene status – which is indicative of a risk for Alzheimer's disease – be blanked out. It is likely that other disease susceptibility gene sequences will not be made public either (Marshall E 2007 Science 315:1780).

Watt (W): The product of volts and amperes in the case of direct current. 1 W = 1 joule/sec = 0.293 calories/sec = 1/735 HP (horse power). In other words, 1 W power is generated by the electric potential between two points of 1 volt and 1 ampere current. ►volt, ►ampere

Wax Coat: ►eceriferum

WCPP (whole chromosome painting probe): Contains a combination of many probes, specific for a single chromosome and thus may label with color its entire length. The multicolor labeling probes may permit the differentiation of all chromosomes in a single karyotype. ►chromosome painting, ►USP, ►FISH, ►GISH

WD-40: Refers to a repeat (N) motif of tryptophan (W) - aspartic acid (D) in several eukaryotic regulatory proteins (absent in prokaryotes) (see Figs. W5): WD repeats are involved in signal transduction, RNA processing, developmental regulation, cell cycle, vesicular traffic, etc. ►signal transduction; Neer EJ et al 1994 Nature [Lond] 371:2987; Smith TF et al 1999 Trends Biochem Sci 24:181.

W-DNA: Refers to a left-handed zig-zag duplex with the same directions as B-DNA but other characteristics match those of the Z-DNA. ►DNA types

Weasel: *Mustela erminea*, 2n = 44; *Mustela frenata*, 2n = 44.

WEB Service: This is a standard system of communication of machines in a network.

WEBB (WB): A very rare blood group involving an altered glycosylation of glycophorin. ►glycophorin,

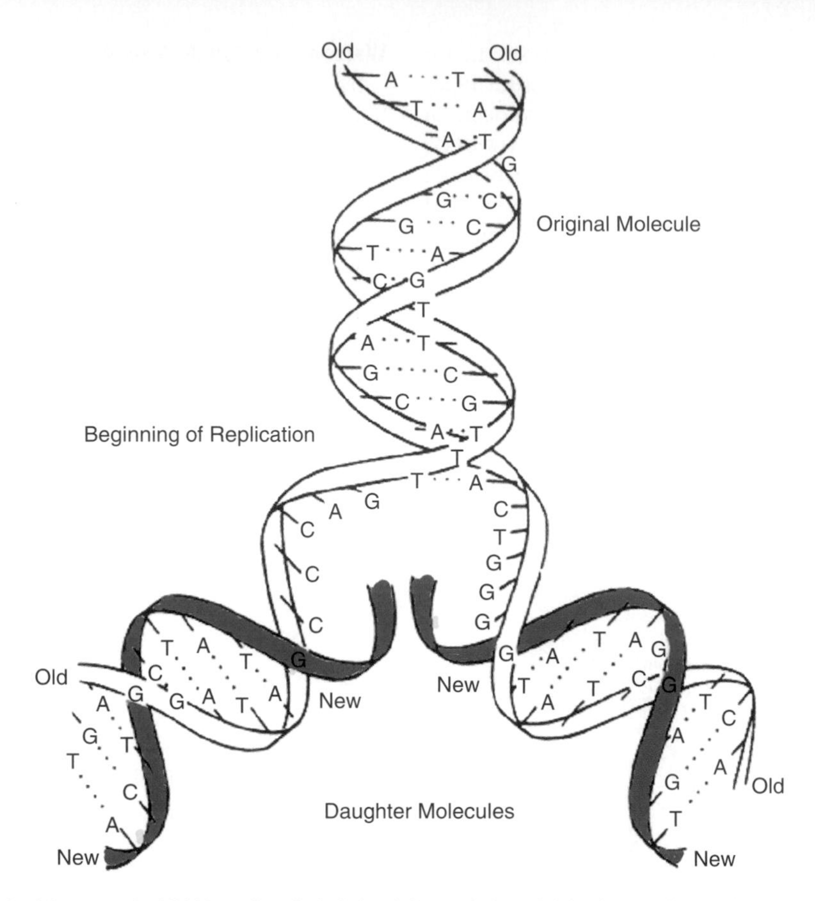

Figure W4. The double-stranded DNA molecule is joined through 2 and 3 hydrogen bonds between the A = T and G≡C nucleotides, respectively. The staircase-like ribbons represent the sugar-phosphate backbone of the double helix. During replication the old plectonemic strands unwind and each old separated strand serves as a template for the formation of the new molecules that are composed from one old and one new single strand (►DNA replication, ►replication fork). This is, therefore, known as the semi-conservative mode of replication. The model is consistent with most genetic phenomena (mutation, recombination, gene expression, etc.). Since the model was originally proposed, the details of the mechanisms of the DNA transactions have been worked out in greater detail but basically none of the essential features had to be revised. This model served as a basis for the *central dogma of genetics* indicating that the flow of information is from the DNA to the RNA and to protein. During the 1960s it was discovered that through reverse transcriptase information can be directed by reverse transcription from the RNA to the DNA but not from protein to the RNA and the DNA. The discovery of prions makes the role of proteins in heredity somewhat ambiguous. All other proposals concerning hereditary molecules (besides the DNA and the RNA) have now faded into oblivion

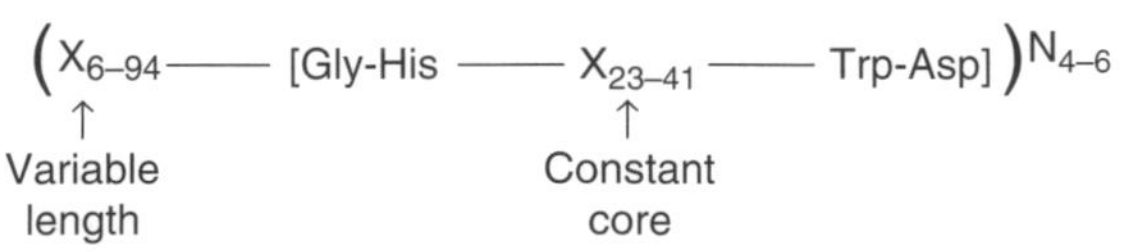

$$\left(X_{6\text{–}94} \text{ —— [Gly-His —— } X_{23\text{–}41} \text{ —— Trp-Asp]}\right)_{N_{4\text{–}6}}$$

↑ Variable length ↑ Constant core

Figure W5. WD-40

►Gerbich, ►En, ►MN, ►blood groups; Reid ME et al 1985 Biochem J 232:289.

WEBIN: A nucleotide sequence information form to the EMBL database. It is offered by the European Bioinformatics Institute: http://www.ebi.ac.uk/embl/Submission/webin.html; Webin qualifiers: http://www.ebi.ac.uk/embl/WebFeat/index.html.

WEE1: A protein kinase which inactivates the *CDK1/cdc2* gene product through phosphorylation of the tyrosine-15 residue. Wee1 is subject to proteolysis in a Cdc34-dependent way before the S phase can be completed. ►kinase, ►Mik1, ►cell cycle, ►*cdc*, ►checkpoint; Tzivion G et al 2001 Oncogene 20:6331; Bartholomew CR et al 2001 Mol Cell Biol 21:4949; crystal structure: Squire CJ et al 2005 Structure 13:541.

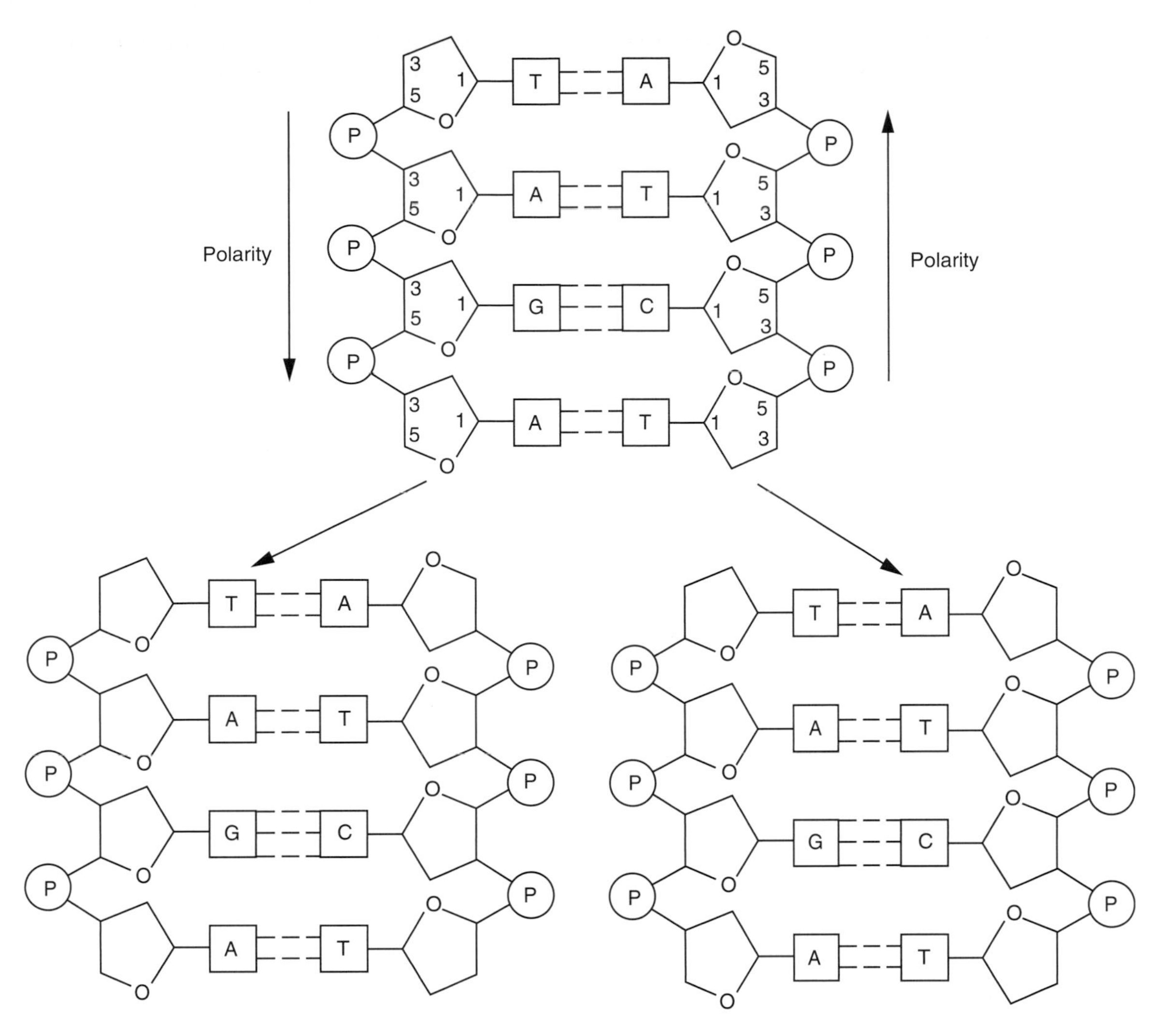

Figure W6. The Watson & Crick model as represented by Josse JJ, Kaiser AD, and Kornberg A, 1961 J Biol Chem 236:864

Wegener Granulomatosis: An autosomal dominant condition characterized by the presence of a ~29 kDa serine protease antigen, identical to myeloblastin. It is a very serious ailment primarily affecting the upper and lower airways and kidneys. ►myeloblast, ►Goodpasture syndrome

Weibel-Palade Bodies: These are 0.1–0.2 μm × 4 μm membrane-enclosed bodies found around the platelets or in the endothelium of all animals. They store in their 150–200 Å tubules the von Willebrand glycoprotein. ►von Willebrand disease; Denis CV et al 2002 Proc Natl Acad Sci USA 98:4072.

Weighted Mean: Refers to the calculated mean multiplied by the pertinent frequency, e.g., in a population the mean value of the homo- and heterozygous dominants (*AA and Aa*) is 250 and that of the homozygous recessives (*aa*) is 200, and the respective frequencies are 0.8 and 0.2, then the weighted mean of the population is (250 × 0.8) + (200 × 0.2) = 240. ►mean

Weigle Mutagenesis: Denotes an increase in mutation of phage by mutagenic treatment of the host. (See Weigle JJ 1953 Proc Natl Acad Sci USA 39:628; Bhattacharyya SC et al 1991 Can J Microbiol 37:265).

Weigle Reactivation: An increase in phage survival when mixed with host cells exposed to low doses of UV light. ►DNA repair, ►marker rescue, ►multiplicity reactivation; Calsou P, Salles B 1991 Mol Gen Genet 226:113.

Weismannism: Inheritance takes place by the transmission of the genetic determinants through the germ line (Keimbahn) and environmentally-induced phenotypic variations are not inherited. ►germ line; Weismann A 1985 Die Continuität des Keimplasmas als Grundlage einer Theorie der Vererbung. Fischer, Jena, Germany.

Weissenbacher-Zweymuller Syndrome (WZS, 6p21.3): A condition characterized by neonatal small jaws (micrognathia), hip and/or shoulder bone defects (rhizomelic chondrodysplasia), general underdevelopment of bones that may improve in later years optic nerve defects, myopia and deafness. Some of the symptoms resemble those of the Stickler syndrome. The basic defect is in the collagen gene COL11A2. ►collagen, ►Stickler syndrome

Welander Distal Myopathy (WDM, 2p13): A dominant late-onset muscular dystrophy. ►muscular dystrophy, ►Miyoshi myopathy

Werdnig-Hoffmann Disease: A recessive (5q12.2-q13) infantile muscular dystrophy primarily affecting the spinal cord muscles. The prevalence is approximately 1×10^{-4}. The gene frequency has been estimated to be around 0.014 and the frequency of heterozygotes about 0.02. It encodes a protein that shows homology to dystrophin. The survival rate varies according to the severity of symptoms. A subunit of transcription factor TFIIH usually suffers deletions. ►dystrophin, ►muscular dystrophy, ►neuromuscular diseases, ►spinal muscular atrophy, ►transcription factors

Werner Syndrome (WRN): Involves premature aging, hardening of the skin, cataracts, atherosclerosis, diabetes mellitus, etc. The gene expression profile is similar to that seen in normal aging (Kyng KJ et al 2003 Proc Natl Acad Sci USA 100:12259). Cultured cells have higher chromosome breakage and mutability. Neoplasias develop frequently. Ulceration around the ankles and soft tissue calcification are symptoms of this condition, unrelated to aging. It was earlier held that the recessive defect was controlled by human chromosome Xp12-p11.2 but recent linkage information, including complete sequencing on the basis of positional cloning, confirms its location at 8p12. WRN contains four structurally folded domains comprising an exonuclease, a helicase, a winged-helix and a helicase-and-ribonuclease D/C-terminal domain. The protein is required for focus forming activity in DNA replication (FFA). It contains 1,432 amino acids and resembles RecQ type ATP-dependent helicases (homologous to budding yeast *SGS1* and *SRS2* genes). Defects in the helicase (frameshift, nonsense mutation) explain most of the symptoms in terms of a flaw in DNA metabolism. WRN also interacts with DNA polymerase δ. However, the amino end domain of WRN shows 3′→5′ exonuclease activity and it may be involved in genetic recombination and/or repair. Forced telomerase activity substantially extends the life span of cultured WRN cells. The WRN gene may be inactivated by CPG island hypermethylation at its promoter. Since WRN involves DNA repair, inactivation may lead to loss of exonuclease-mediated DNA repair and increased chromosomal instability as well as to apoptosis, induced by a camptothecin analog (irinothecan), a topoisomerase inhibitor. Thus, cancer cells may become more amenable to chemotherapy (Agrelo R et al 2006 Proc Natl Acad Sci USA 103:8822). Normal fibroblasts and U20S osteosarcoma cells rendered deficient in WRN exhibit reduced phosphorylation of p53 and histone H2AX in response to T-oligo (telomere single-strand overhang) treatment indicating its role in the processing of telomeric DNA and subsequent activation of DNA damage responses (Eller MS et al 2006 Proc Natl Acad Sci USA 103:15073). There is evidence that the genome instability in WS cells depends directly on telomere dysfunction, linking chromosome end maintenance to chromosomal aberrations in this disease (Crabbe L et al 2007 Proc Natl Acad Sci USA 104:2205). ►aging, ►progeria, ►helicase, ►exonuclease, ►FFA, ►Bloom syndrome, ►co-suppression, ►RNAi, ►atherosclerosis, ►telomeres, ►translesion; Kamath-Loeb AS et al 2000 Proc Natl Acad Sci USA 97:4603; Shen J-C, Loeb LA 2000 Trends Genet 16:213; Kawabe Y et al 2001 J Biol Chem 276:20364; Kipling D et al 2004 Science 305:1426.

Wernicke-Korsakoff Syndrome: A chromosome 3p14.3 recessive disorder involving transketolase deficiency caused by the lowered ability of the enzyme to bind thiamin pyrophosphate. Another transketolase deficiency has been traced to Xq28. This dysfunction of energy metabolism leads to neuropsychiatric anomalies, especially in alcoholics and when the level of thiamin is low in the diet. ►thiamine pyrophosphate

West Nile Virus (WNV, Flaviviridae): A single-stranded RNA virus. E-glycoprotein is its most important protein, both structurally and for infection (see Fig. W7). Of the two strains, lineage 1 has caused most of the health problems in the US. Outbreaks of the epidemic have been described in Africa, the Mid East and southern Europe. About 29 species of mosquitos transmit the virus to birds and from birds to humans. In North America the human- and bird-biter forms of the *Culex* mosquitos form hybrids and may facilitate the severity of transfer between these species (Fonseca DM et al 2004 Science 303:1535). If infected mosquitos feed in close vicinity of non-infected individuals, the secreted saliva is sufficient for transmission of the virus in an appreciable frequency (5.8%) even when the blood of vertebrates does not contain the virus (non-viremic). This contributes to the rapid spread of the disease (Higgs S et al 2005 Proc Natl Acad Sci USA 102:8871). The symptoms include

W

fever, headache, gastrointestinal and mental changes (memory loss, depression); rash, cough and other symptoms may also be manifested. The virus appears genetically quite stable. The fatality ranges from 4 to 14% but in older or immune-compromised persons the risks may be more than double. The most effective diagnosis reveals immunoglobulin M (IgM) in the cerebrospinal fluids. RNAi holds promise for effective medication against viral encephalitis at least in mice. However, mosquito control is the key to prevention.

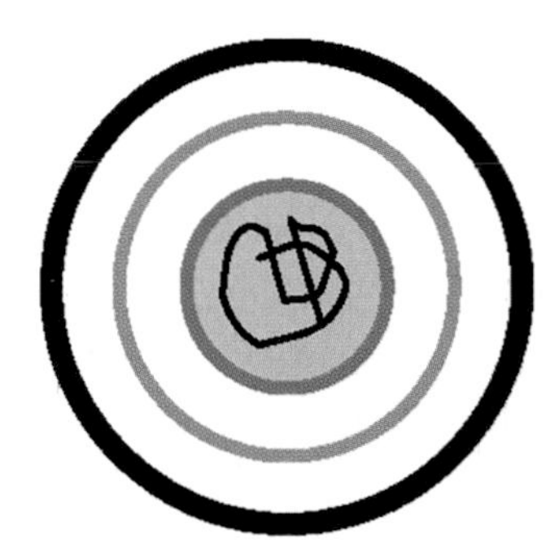

Figure W7. West Nile virus with heavy capsule and two concentric lipid layers surrounding the core with RNA

DNA vaccine coding for the RNA transcript of the related Kunjin flavovirus has protective effects against the West Nile Virus (Hall RA 2003 Proc Natl Acad Sci USA 100:10460). ►immunization genetic, ►viral encephalitis; Petersen LR, Marfin AA 2002 Annals Int Med 137:E173; Brinton MA 2002 Annu Rev Microbiol 56:371; http://westnilemaps.usgs.gov; http://www.cdc.gov/ncidod/dvbid/westnile/.

West Syndrome (ARX, Xp22.13; ISSX, Xp21.3-p22.1): This involves an X-linked central nervous defect, hypsarrhythmia (electroencephalographic anomaly) resulting in infantile spasms and mental retardation. Its incidence is 2–5 × 10^{-4} at birth. The basic defect may be in the serine-threonine kinase 9 region. ►epilepsy, ►mental retardation X-linked; Kalscheuer VM et al 2003 Am J Hum Genet 72:1401.

Western Blot (Western hybridization): Refers to the identification of polypeptides separated electrophoretically on SDS polyacrylamide gels, then transferred to nitrocellulose filter and labeled by immunoprobes such as radioactive or biotinylated antibodies. ►immunoblotting, ►gel electrophoresis, ►far-western hybridization; Towbin H et al 1979 Proc Natl Acad Sci USA 76:4350; Burnette WN 1981 Anal Biochem 112:95.

Weyers Acrofacial/Acrodental Dysostosis: An allelic variation in the gene locus at 4p16 encoding the Ellis-van Creveld syndrome. ►Ellis-van Creveld syndrome

WGRH (whole genome radiation hybrid): Formed when the donor is a diploid cell line for the generation of radiation hybrids. ►radiation hybrid; Vignaux F et al 1999 Mamm Genome 10:888.

WGA (whole genome association, whole genome assembly, whole genome array). Whole genome association study attempts to identify all the genetic factors involved in the development of a particular disease. The PLINK software tool can handle a very large set of data involving five main domains of function: data management, summary statistics, population stratification, association analysis, and identity-by-descent estimation (Purcell S et al 2007 Am J Hum Genet 81:559). ►microarray hybridization, ►genome annotation

WGS (whole-genome shotgun sequencing): Developed by Craig Venter and associates at Celera Genomics, Rockville, MD, USA. The genome is sheared into a few thousand base pair long pieces and cloned into sequencing vectors. Each fragment end (~500 bp) is covered by the sequencing several times and then assembled into overlapping segments by end-sequences. In this way oriented, contiguous sequences (contigs or scaffolds) are generated. Between the scaffolds a physical gap may remain until the physical map is "finished" and the complete genome is reconstructed with the aid of high-power computers. The scaffolds are also mapped to eukaryotic chromosomes and chromosome arms with the aid of sequence-tagged sites. The heterochromatic regions (mainly around the centromeres and near the telomeres) are unstable when cloned and thus resist sequencing. These regions contain transposons and redundant ribosomal RNA genes with a few interspersed ORFs. For the repeated sequences the WGS method provides less resolution (She X et al 2004 Nature [Lond] 431:927). The function of the sequences is indicated by annotation using Genscan or Genie computer software. ►shotgun sequencing, ►mate pair, ►sequence-tagged sites, ►transposon, ►ribosomal RNA, ►ORF, ►annotation, ►gene ontology, ►scaffolds in genome sequencing, ►genome projects; Adams MD et al 2000 Science 287:2185; http://www.ncbi.nlm.nih.gov/projects/WGS/WGSprojectlist.cgi.

Whale: *Baleonoptera* species are 2n = 44; *Kogia breviceps*, 2n = 42.

Wheat: ►*Triticum*

Wheat Germ In Vitro Translation: Cell-free wheat germ extracts can be used for the translation of viral, prokaryotic and eukaryotic mRNAs into protein. The supernatant of the extract must be chromatographically purified from inhibitory endogenous amino acids and pigments before translation. Tritin, one of the endosperm proteins, inactivates the ribosomes and thus reduces the efficiency of the system unless carefully removed. The extract contains tRNA, rRNA and other factors required for protein synthesis. Phosphocreatin and phosphocreatine kinase additions are needed for supplying energy. Spermidine is added to stimulate the translation efficiency and prevent premature termination of the polypeptide chain. Magnesium acetate and potassium acetate, mRNA (to be translated) and amino acids (including one in a radioactively labeled form) are also necessary. Incubation is at 25°C for 1 to 2 hours. In general, the procedure is very similar to the rabbit reticulocyte system. ►rabbit reticulocyte in vitro translation system; Erickson AH, Blobel G 1983 Methods Enzymol 96:38.

Whim Syndrome (2q21): Refers to hypogammaglobulinemia, neutropenia, myelokathexis (retention of bone marrow neutrophils) and papilloma virus susceptibility due to mutations in the chemokine receptor CXCR4. ►gammaglobulin, ►neutropenia, ►neutrophil, ►CXCR, ►papilloma virus; Hernandez PA et al 2003 Nature Genet 34:70.

Whipple's Disease: An infectious disease caused by *Tropheryma whipplei* bacteria.

White Blood Cell: ►leukocyte

White Forelock: ►forelock, ►white

White Leghorn (chicken): Has genes controlling the color of the plumage *CC*, *OO*, II; the color is white (see Fig. W8). *C* and *O* both are needed for pigmentation but *I* is an inhibitor of color. ►White Silkie, ►White Wyandotte

Figure W8. White leghorn rooster

White Matter: ►gray matter

White Silkie (chickens): They are of the constitution *cc*, *OO*, *ii and* have white feathers because only one of the two dominant genes, *O*, required for pigmentation is present. ►White Leghorn, ►White Wyandotte

White Wyandotte (chickens): They have the genes *CC*, *oo*, *ii* and are white because only one, *C*, of the two dominant genes is present but not the other, *O*. ►White Leghorn, ►White Silkie

Whooping Cough: A respiratory disease in humans caused by the pertussis toxin of the bacterium *Bordatella pertussis* (3,816 genes). Species such as *B. parapertussis* (4,404 genes) and *B. bronchiseptica* (~5,007 genes) are related pathogenic species for various mammals (Parkhill J et al 2003 Nature Genet 35:32). ►pertussis toxin, ►microfluidics

Whorl: Refers to a circular or spiral arrangement of structures, such as various parts of flowers or the dermal ridges in a human fingerprint. ►fingerprinting, ►flower differentiation

Wide Cross: Refers to hybridization between plants of different species or genera.

Widow's Peak: An autosomal dominant pointed hairline in humans (see Fig. W9).

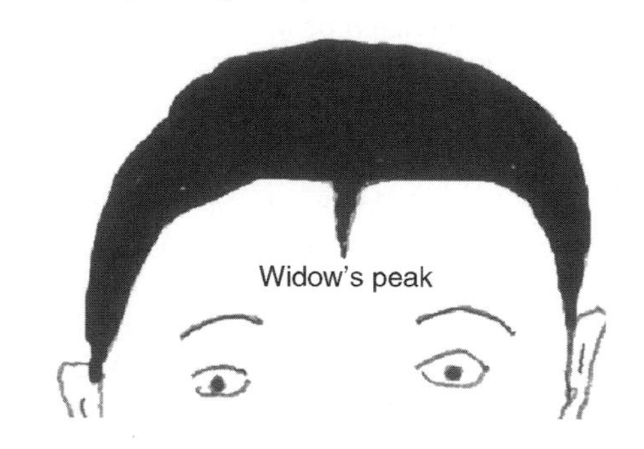

Figure W9. Widow's peak

Wilcoxon's Signed-Rank Test: A non-parametric substitute for the t-test for paired samples (see Table W1). The desirable minimal number of paired samples is 10 and it is expected that the population would have a median, be continuous and symmetrical. The differences between the variates are tabulated and ranked; the largest receives the highest rank. In the case of ties, each should be assigned to a shared rank. The smaller group of signed-rank values is then summed as the T value. This T value is compared with figures in a statistical table. If the value obtained is smaller than that in the body of the table under probability and on the line corresponding to the number of pairs tested, then the null hypothesis is rejected and the conclusion is justified that the two samples are different.

Example: Since T being 8.0 according to the first line of Table T1, the difference between the two sets of data is significant at the level of 0.05 probability but not at the 0.01 level.

Table W1. Wilcoxon's Signed-Rank Test

Pairs	Difference	Signed	Ranks	Probability		
		+	–	*n*	0.05	0.01
1	+6	7		10	10	5
2	+5	6		11	13	7
3	+10	10		12	17	10
4	−3		4	14	25	16
5	+4	5		16	35	23
6	+7	8		18	47	33
7	−2		3	20	60	43
8	−1		0.5	22	75	55
9	+9	9		24	91	69
10	−1		0.5	26	110	84
			T = 8.0			

A more general procedure for determining the probabilities depends on determining the Z value either for threshold probabilities or more precisely by using a table of the cumulative normal variates (such as the *Biometrika Tables for Statisticians*, Vol. 1, Pearson ES, Hartley HO, (Eds.), Cambridge University Press, Cambridge). Z values larger than 1.960, 2.326 and 3.291 correspond to P 0.05, 0.01 and 0.001, respectively. These probabilities rule out the null hypothesis.

$$Z = \frac{\mu - T - 0.5}{\sigma} \text{ and } \mu = \frac{n(n+1)}{4} \text{ and } \sigma = \sqrt{\frac{[2n+1]\mu}{6}}$$

where n = the number of paired data. ►non-parametric statistics, ►Mann-Whitney test, ►Student's *t* distribution, ►QTL

Wild Type: The standard genotype (that is most common in wild [feral] populations). ►isoalleles

Wildervanck Syndrome: This appears to be a X-linked dominant deafness, frequently associated with other disorders. Approximately 1% of deaf females is affected by it. It does not occur in males, presumably because it is lethal when homo- or hemizygous. It has been suggested that it is polygenically determined but the mechanism of male exclusion is as yet unknown. ►deafness, ►sex-limited, ►imprinting

Wildfire Disease of Plants: Caused by the toxin (methionine analog) of the bacterium *Pseudomonas tabaci*, and it leads to necrotic spots on the leaves. ►*Pseudomonas*

Williams Factor (Flaujeac factor deficiency): An autosomal recessive mutation in human chromosome 3q26-qter causing deficiency of a high molecular weight kininogen, a precursor of a blood clotting factor. ►kininogen, ►blood clotting pathways, ►antihemophilic factors

Williams Syndrome (Williams-Beuren syndrome, WMS): This autosomal dominant condition involves stenosis (narrowing) of the aorta, arteries and lungs, elfin face (elfins are diminutive mythological creatures), malformation of teeth and stature, mental deficiency and excessive amounts of calcium in the blood (hypercalcemia) and in some tissues. Cognitive abilities are impaired in an unusual fashion. During infancy patients are poor in language skills but their performance is relatively better in numerical intelligence tests. By adulthood the trend is reversed indicating that these abilities are under the control of two different developmental modules. Various types of deletions in different chromosomes (15, 4, 6) have been suspected. Chromosome-specific probes have revealed a ~1.5 Mb deletion of 7q11.23. This region harbors several genes including the GTF21RD1 basic transcription factor and the Williams syndrome transcription factor (WSTF/WCRF/ACF), which bears structural similarity to a 180 kDa chromatin-remodeling factor. The differences in the phenotypes of affected individuals may be due to the dosage effect of the chromosomal segments of this tract. Lower calcium in the diet may alleviate some of the symptoms. Vitamin D2 anomaly is suspected. The function of the Williams factor is unrelated. Recent findings have indicated the involvement of LIMK protein kinases (carrying two LIM domains) that are serine/threonine/tyrosine kinases and regulate actin in the cytoskeleton. Under normal conditions RAC-GTP activates the LIM kinases that in turn phosphorylate cofilin and inactivate it. Dephosphorylation permits the formation of active cofilin that is associated with actin. The actin—cofilin association is apparently mediated by phosphoinositides ($PtdInsP_2$). Defects (heterozygosity for a mutation) in LIMK lead to abnormal neuronal connections. Elastin defects characterize the symptoms of the Williams syndrome. ►supravalvular aortic stenosis, ►cutis laxa, ►cardiovascular diseases, ►dwarfism, ►vitamin D, ►Williams factor, ►LIM domain, ►face/heart defects, ►cofilin, ►actin, ►phosphoinositides, ►human intelligence, ►module, ►Marfan syndrome, ►BTK; Peoples R et al 2000 Am J Hum Genet 66:47; MorrisCA, Mervis CB 2000 Annu Rev Genomics Hum Genet 1:461; Sumi T et al 2001 J Biol Chem 276:23092; Urbán Zs et al 2002 Am J Hum Genet 71:30; Bayés M et al 2003 Am J Hum Genet 73:131; Tassabehji M et al 2005 Science 310:1184.

Wilms Tumor (WT): This is usually associated with a deletion in the short arm of human chromosome 11 extending from 11p13 to 11p15.5 (WT1) spanning several genes and is frequently referred to as the WAGR syndrome (Wilms tumor-aniridia-genitourinary anomalies and RAS oncogene-like function). The condition is characterized by symptoms of all or parts of the functions implied by WAGR. Wilms tumor is caused by a mutation in a cancer-suppressor transcription factor and splicing factor with 4 Zn finger domains. Wilms tumor is extremely complex and additional WT genes in chromosomes 16q, and 1p, 4p, 8p, 14p, 17p and q, 18q have also been implicated. The transcript displays alternative splicing. Wilms tumor suppressor gene is expressed only in the maternally transmitted allele. Prevalence is about 10^{-4} during the first five years of age. Recent evidence has revealed a WT gene in the 17q12-q21 region. The WT1 gene also regulates muscle differentiation. The WT1 gene affects the SRY locus in the Y chromosome and that explains the symptoms associated with genitourinary development when it is altered. The different functions are based on isoforms of the protein. ►aniridia, ►RAS, ►deletion, ►hypertension, ►kidney diseases, ►Zinc finger, ►breast cancer, ►imprinting, ►isoform, ►Rhabdosarcoma, ►Denys-Drash syndrome, ►Frasier syndrome, ►acatalasemia; Hossain A, Saunders GF 2001 J Biol Chem 276:16817; Hammes A et al 2001 Cell 106:319.

Wilson Disease (WD): Recessive disease encoded in human chromosome 13q14.2-q14.3. It affects primarily persons of age 30 and above although juvenile forms have also been described. The major symptoms are cirrhosis of the liver and psychological ailments caused by a deficiency of ceruloplasmin resulting in copper accumulation. In this as well as other diseases involving cirrhosis of the liver, at the upper and lower margin of the cornea (see Fig. W10) a greenish narrow ring (↑) occurs. The basic defect is in a copper transporting ATPase of the mitochondria. Inactivation of the enzyme (ATP7B or ATP7A) causes neurogeneration and overexpression can lead to resistance to therapeutic drugs. Mutations in the N domain of the enzyme prevent tight ATP binding and can lead to >30 forms of Wilson disease (Dmitriev O et al 2006 Proc Natl Acad Sci USA 103:5302).

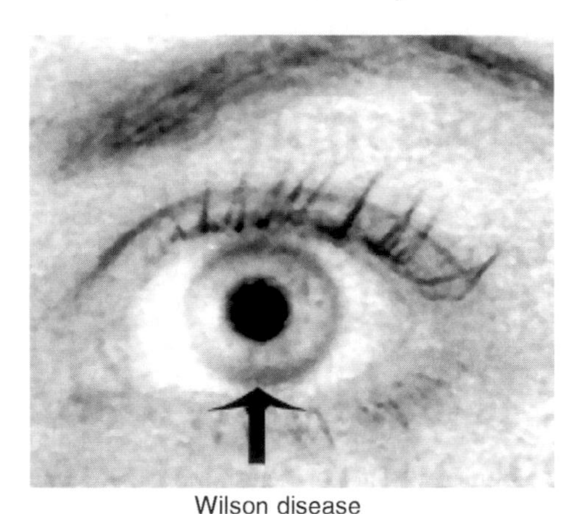

Figure W10. Wilson disease

XIAP (X-linked inhibitor of apoptosis) inhibits caspases and copper metabolism in WD and other copper toxicosis diseases. Intracellular copper alters the conformation of XIAP and lowers its ability to inhibit caspase. This results in sensitivity to apoptosis (Mufti AR et al 2006 Mol Cell 21:775). The prevalence of WD in the USA is about 3×10^{-5}. A similar anomaly has been observed in Long-Evans Cinnamon (LEC) rats that may serve as an animal model for the study of the disease. Prenatal diagnosis of offspring of carrier parents may be possible by using linkage with DNA markers. ►Menkes disease, ►acrodermatitis, ►hemochromatosis, ►mitochondrial disease in humans, ►neurodegenerative diseases

Winged Helix Protein: This class of the helix-turn-helix proteins uses a β-hairpin (a wing) to bind the DNA. ►DNA-binding protein domains; Gajiwala KS, Burley SK 2000 Curr Opin Struct Biol 10:110.

Wingless (*wg/Dint-1*; 2–30): *Drosophila* gene is involved in morphogenetic signaling and its homologs are found in all vertebrates and invertebrates. The mouse homolog is *Wnt*. Components of the *Wnt* cascade are altered in breast and colon cancers in mice and in human melanomas. The path of *Wg* function is depicted (see Fig. W11).

Fz (*frizzled trichomes*) and Dally (*division abnormally delayed*, controls heparan sulfate) are

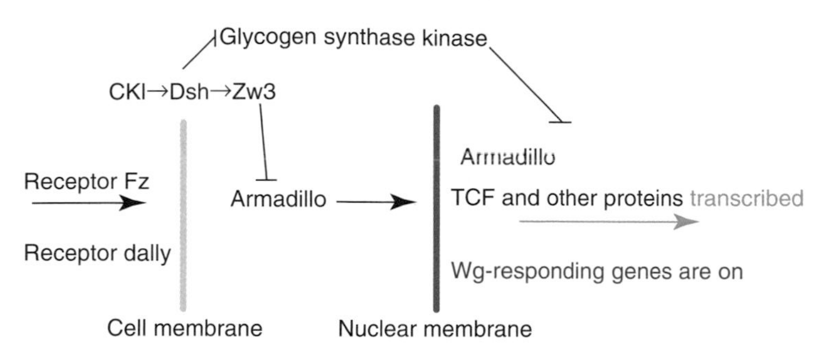

Figure W11. Path of Wg function

G protein-like receptors. *Dsh* (*dishevelled*) mitigates the block by *Zw*3 (*zeste white*), Armadillo is a β-catenin-like protein. *TCF* (*ternary complex factors*) co-activate transcription. Dsh protein blocks glycogen synthase kinase (GSK3) that would be a negative regulator of Armadillo. CK1 (a serine kinase, called casein kinase) phosphorylates Dsh~Zw. Actually, Wg is involved in more complex functions. Wg also signals to the epidermal growth factor (EGF) receptor. WNT-4 duplication in humans and mice masculinizes XX individuals. In the Sertoli and Leydig cells DAX1, an antagonist of SRY, is upregulated. The *Wnt* pathway is involved in signaling to somatic stem cells and the maintenance of cancer (Reya T, Clevers H 2005 Nature [Lond] 434:843). One study has revealed 238 potential regulators of the Wnt pathway and half of them have human orthologs and several of them are involved in human disease (DasGupta R et al 2005 Science 308:826). The secretion of Wnt ligand requires a transmembrane protein (Wentless [Wls]/ Eveness interrupted [Evi]). *Wls* is in chromosome 3 of *Drosophila* within 68A/B band; homologs are present in the *Caenorhabditis* and human genomes (Bänziger C et al 2006 Cell 125:509; Bartscherer K et al 2006 Cell 125:523). ►organizer, ►EGF, ►*Armadillo*, ►adrenal hypoplasia congenital, ►sex reversal, ►catenins, ►GSK, ►TCF, ►heparin sulfate, ►morphogenesis in *Drosophila*, ►zeste, ►SRY, ►Wolffian duct, ►Gardner syndrome; Kühl M et al 2000 Trends Genet 16:279; Peifer M, Polakis P 2000 Science 287:1606; Wilkie GS, Davis I 2001 Cell 105:209; van de Wetering M et al 2002 Cell 109:S13; Peifer M, McEwen DG 2002 Cell 109:271; Mosimann C et al 2006 Cell 125:327; review: Clevers H 2006 Cell 127: 469; http://www.stanford.edu/~rnusse/wntwindow.html.

Winner's Curse: A finance/business concept. In competitive bidding, the winner may be the loser. This may be the case when the highest bidder overestimates the value of an item/stock or any other object in a bidding competition (e.g., at an auction) and it turns out that, even if some preferential values are added, the value of the purchase is lower than estimated/expected. A similar situation may be observed in population genetics when screening is conducted for a low expression, rare allele. Before the tests are carried out the frequency of carriers or homozygotes for the disease can be only guessed. If the size of the population studied is too small, the information gathered may not be reliable and useful. If a very large population is tested—at great expense—and the results are still ambiguous, the investment in labor and other means is not worthwhile. Thus, the winner (the thorough investigator) suffers the curse of being a loser. ►game theory, ►critical population size, ►mutation rate; examples for winner's curse: http://www.techcentralstation.com.

Wiskott-Aldrich Syndrome (WAS, Xp11.23-p11.22): This X chromosomal immunodeficiency disease causes eczema, reduced platelet size, bloody diarrhea, greater susceptibility to infections, lymphocyte malignancies and usually death before the age of 10. The prevalence is about 4×10^{-6}. The affected individuals are deficient in a 115 kDa lymphocyte membrane protein and the platelets are abnormally low in a glycoprotein (sialophorin). Carriers may be identified by linkage, lymphocyte analysis and nonrandom inactivation of the X chromosomes. WAS is allelic to the human Xp11.23 thrombocytopenia gene. This gene includes 12 exons in 9 kb genomic DNA and encodes 502 amino acids. The involvement of CDC42 signaling defect is likely. Cdc42-GTP is a far better agonist for the protein WASP than Cdc4 2-GDP (Leung DW, Rosen MK 2005 Proc Natl Acad Sci USA 102:5685). The WAS protein (WASP) mediates cytoskeletal rearrangement and transcriptional activation of T cells. ►immunodeficiency, ►T cell, ►cytoskeleton, ►thrombocytopenia, ►thrombopathic purpura, ►cancer, ►CDC42, ►agonist, ►podosome, ►WASP, ►platelet; Silvin C et al 2001 J Biol Chem 276:21450; Devriendt K et al 2001 Nature Genet 27:313; Caron E 2002 Current Opin Cell Biol 14:82.

Witkop Syndrome: ►tooth-and-nail dysplasia

WL (working level): Used for the characterization of short-lived radon decay products in 1 liter of air resulting in the ultimate emission of 1.3×10^{-5} MeV of alpha radiation energy. WL is also defined as 2.08×10^{-5} joule h m^{-3}. WLM (working level/month) = exposure of 170 h at 1 WL. ►radon

WNT1 (cysteine-rich glucoprotein ligand): The ~18 vertebrate genes have numerous and diverse roles in signaling to development, particularly along the anterior-posterior body axis. ►INT1 oncogene in mouse, ►*wingless* gene product in *Drosophila*, ►morphogenesis in *Drosophila* {63}, ►pattern formation, ►organizer, ►Gardner syndrome, ►gonads; Skromme E, Stern CD 2001 Development 128:2915.

Wobble: The 5′-base of the anticodon can recognize more than one kind of base at the 3′ position of the codon, e.g., both U or C in the mRNA may pair with G, and both G and A may pair with U, or A or U, or C may recognize I (inosinic acid at the 5′-position in the anticodon). There is no AAA anticodon for Phe but the GAA anticodon recognizes both UUU and UUC codons in the mRNA. The GUU and GUC codons of Val are decoded by an anticodon AAC. Inosinic acid occurs in the anticodon of 8 tRNAs of higher eukaryotes, in 7 of yeast and in the $tRNA^{Arg}{}_2$ of prokaryotes and plant chloroplasts. Inosine may be formed from adenosine by adenosine deaminases. In tRNAs, however, a different dimeric deaminase, encoded in yeast by *Tad*2 and *Tad*3 genes, is active. This deaminase is related to the cytidine deaminase (CDA). According to the classical or universal genetic code of 61 sense and 3 missense codons, a minimum of 32 tRNAs would be required to recognize all the amino acids. Further simplifications permit protein synthesis, however, by 22–24 tRNAs. In the human genome tRNA genes have been found all over the genome yet 140 tRNA genes are crowded in a 4 Mb region of chromosome 6. The altered mitochondrial code requires modifications in the anticodon wobble. ►genetic code, ►tRNA, ►anticodon, ►isoacceptor tRNA, ►mtDNA, ►hypoxanthine, ►decoding; Crick FHC 1966 J Mol Biol 19:548; Agris PF 1991 Biochemie 73:1345; Lim VI 1994 J Mol Biol 240:8; Sibler AP et al 1986 FEBS Lett 194:131.

Wolbachia: Refers to an endocellularly infectious group of bacteria of arthropods, nematodes and crustaceans that are transmitted maternally and may cause feminization, cytoplasmic incompatibility and thelytoky. Wolbachia is localized to the stem cell niche of *Drosophila* germarium (Frydman HM et al 2006 Nature [Lond] 441:509). Infected males cannot produce viable offspring with uninfected females because of cytoplasmic incompatibility. They are compatible, however, with infected females and produce offspring. In some *Culex* mosquitos variations have been found in cytoplasmic incompatibility controlled by two prophage genes of *Wolbachia* involving ankyrin repeats (Sinkins SP et al 2005 Nature [Lond] 436:257). Antibiotics can cure incompatibility in many instances. The removal of *Wolbachia* by antibiotics from parasitic wasps stops oogenesis. *Wolbachia*s as endosymbionts of filarial nematodes are responsible for the *river blindness* disease (filariasis). *Wolbachia* infection of flies with some oogenesis defect causing *Drosophila Sex-lethal* alleles surprisingly restores fertility (Starr DJ, Cline TW 2002 Nature [Lond] 418:76). *Wolbachia* may serve as an agent to block the spread of malaria by causing sterility in the insect host (Ito J et al 2002 Nature [Lond] 417:452). There are several related species. The *W. pipientis* genome of 1,267,782 bp has been sequenced and it appears to contain an unusually high number of mobile elements as well as essential difference in metabolism compared to related species e.g., *Ricketsia* (Wu M et al 2004 PloS Biol 2:327). ►cytoplasmic incompatibility, ►thelytoky, ►symbionts hereditary, ►segregation distorter, ►pronucleus, ►Ricketsia, ►ankyrin; Stouthammer R et al 1999 Annu Rev Microbiol 53:71; Zimmer C 2001 Science 292:1093; Dedeine F et al 2001 Proc Natl Acad Sci USA 98:6247; Saint André v A et al 2002 Science 295:1892; http://troi.cc.rochester.edu/~wolb/FIBR/database.html.

Wolcott-Rallison Syndrome (WRS, 2p12): Involves mutation in the translation initiation factor EIF2AK3. The recessive disorder affects neonatal or infantile insulin-dependent diabetes. Later, bone defects (epiphyseal dysplasia, osteoporosis), retarded growth, liver and kidney malfunction, mental retardation and heart disease may complicate the condition. ►EIF2

Wolf: (*Canis lupus*, 2n = 78), a carnivorous mammal, which can form fertile hybrids with domesticated dogs (*Canis familiaris*, 2n = 78) as well as with the coyote (*C. latrans*, 2n = 78) but not with foxes (*Vulpes vulpes*, 2n = 36). Domesticated dogs appear to be much closer evolutionarily to wolves (on the basis of mtDNA) than to coyotes. ►fox

Wolf-Hirschhorn Syndrome: A condition, which involves deletion (unequal crossing over, insertion) in one of the short arms of human chromosome 4p16.1 (usually the paternal) resulting in severe growth, mental, face and genitalia defects, etc. The deletion generally eliminates HOX7 (homeobox 7), responsible for normal development in humans and mice. Hemizygosity for the gene is sufficient for the disease. The severity of the disease is affected by alteration at adjacent sites (Bergmann D et al 2005 Trends Genet 21:188). ►deletion, ►hemizygous, ►homeobox, ►homeotic genes, ►Huntington's chorea; Näf D et al 2001 Am J Hum Genet 10:91; Zollino M et al 2003 Am J Hum Genet 72:590.

Wolffian Ducts: Develop as a precursor of the male gonads of vertebrates. This development is enhanced under the influence of testosterone hormone. The male gonad (testes) is formed by the Sertoli cells that eventually surround the spermatogonia. The Leydig cells (Yao H-C et al 2002 Genes Dev 16:1433) secrete the steroid testosterone and the Sertoli cells produce the anti-Müllerian hormone (human chromosomal location 12q13), which causes regression of the uterus and the Fallopian tubes. ►gonads, ►Müllerian ducts, ►DSS

Wolf-Parkinson-White Syndrome: A heart disease with a short P and a long QRS phase of electrocardiography. It is also called preexcitation syndrome because the heart ventricles are excited prematurely. It may cause increased palpitations and sudden death. The causes are complex; dominant defects at chromosome 7q3 have been implicated and a mitochondrial component may also be involved. ►electrocardiography, ►mitochondrial diseases in humans [Leber hereditary optic dystrophy].

Wolfram Syndrome (DIDMOAD): Mutation in human chromosome 4p16.1 gene encoding a ~100 kDa transmembrane protein involves diabetes insipidus, diabetes mellitus, optic atrophy and deafness. It has been proposed that a large mitochondrial deletion extending over several coding sequences (7.6 kb) is responsible for these diseases but it has not been confirmed. ►diabetes, ►optic atrophy, ►deafness, ►mitochondrial diseases in humans; Strom TM et al 1998 Hum Mol Genet 7:2021.

Wolman Disease (lysosomal acid lipase deficiency): A condition due to autosomal recessive genes in the long arm of human chromosome 10q24-q25 and in mouse chromosome 19. The early onset forms are due to a deficiency of this enzyme (cholesteryl ester hydrolase) and are characterized by liver and spleen enlargement, failure to feed normally and death by the age of 2 to 4 months. The accumulation of cholesterol esters is caused by mutant alleles of the same locus. In some forms survival is till the teens. ►cholesterol, ►lysosomes, ►lipase; Du H et al 1998 Hum. Mol Genet 7:1347.

Woodchuck: ►*Marmota monax*

Woodrats: There are many species mainly with 2n = 52.

Woods' Light: Refers to an ultraviolet light source with nickel oxide filter and with a maximal transmission at about 365 nm while most other spectral regions are blocked. ►ultraviolet light

Woolly Hair: This may be black (and autosomal dominant) or blond (and autosomal recessive). In both cases, the hair is short and tightly curled.

Wordmatch: Finds exact matches of a given size between two protein sequences. (http://ocgc.ca/programs/emboss/wordmatch.html).

Working Hypothesis: This is an experimentally testable assumption regarding a problem.

Worm Genetics: An informal reference to *Caenorhabditis elegans*. ►*Caenorhabditis*

Woronin Bodies: These occur around fungal pores supposedly to protect against excessive leakage if the cells are damaged. (See Tenney K et al 2000 Fungal Genet Biol 31:205).

Wortmannin: A protein from *Penicillium fumiculosum/Talaromyces wortmanni* which is a stimulator of neutrophils and inhibitor of PIK, DNA-PK and thus inhibits the repair of double-breaks in the DNA (see Fig. W12). Wortmannin modulates the phosphatidylinositol metabolic pathway. ►neutrophil, ►PIK, ►DNA-PK, ►phosphatidylinositol; Wang H et al 2001 Nucleic Acid Res 29:1653.

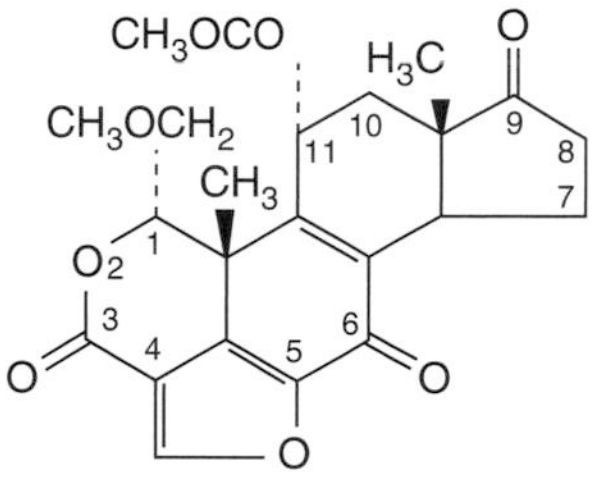

Figure W12. Wortmannin

Woude Syndrome: ►van der Woude syndrome

Wound Healing: Requires angiogenesis, which can be provided by administering angiogenic proteins such as vascular endothelial growth factor (VEGF) or placental growth factor. To ensure a stabilizing effect, the delivery of multiple growth factors may be helpful. Hypoxia-inducible factor (HIF) can stimulate angiogenesis. HIF-1 heterodimer has two subunits α and β. The former has an oxygen-sensitive degradation domain (ODD). This domain is prolyl hydroxylated in an oxygen-dependent manner and can bind the von Hippel-Lindau protein and consequently the HIF-1α subunit is degraded by the proteasome under normal oxygen tension. Under hypoxia, HIF-1α can heterodimerize with HIF-β and upon entering the nucleus it induces gene expression. The prevention of HIF-α degradation or its overexpression results in nuclear translocation and VEGF expression under normal oxygen level. If the ODD domain is removed, VEGF and other genes are upregulated. In a specially engineered vector complex on a fibrin surgical matrix, entrapped in a DNA nanoparticle delivered to the dermal wound of mice resulted in potent therapeutic effects (Trentin D et al 2006 Proc Natl Acad Sci USA 103:2506). The disruption of the epithelial cell layer generates endogenous electric field and cell migration. Phosphatidylinositol-3-OH kinase-γ and PTEN genes are essential for electrical signal-induced cell migration and wound healing (Zhao M et al 2006 Nature [Lond] 442:457). ►tissue engineering, ►phosphoinositides, ►PTEN, ►von Hipple-Lindau syndrome, ►VEGF, ►angiogenesis

Wound-Healing Assay: Used to analyze cell migration. A confluent mammalian cell monolayer is scratched by a blunt pipette tip and the "wound" so generated heals by cell migration, which can be monitored by automatic devices (Todaro GJ et al 1965 J Cell Physiol 66:325). It also detects the metastatic ability of neoplasias (see Fig. W13). ►metastasis, ►neoplasia

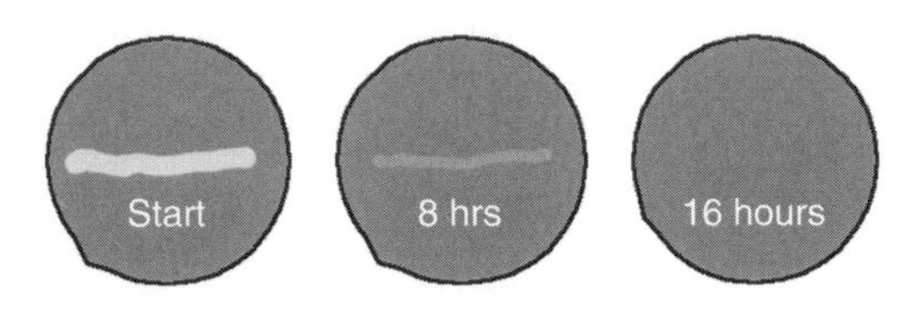

Figure W13. Wound-healing time course

Wound Response: Upon wounding one leaf physiological changes take place in other parts of the plant body and the expression of proteinase inhibitors (Pin) is triggered by glycan, jasmonate and peptide signals. The 18-amino acid peptide signal, systemin, in tomato plants, wounded by herbivorous insects, stimulates proteinase expression. Salicylic acid may block the wound response. Wounding of plants may also induce a large number of genes of different function. The plants may produce peroxides against the microbes infecting the wounds. Phenolics may accumulate, photosynthesis may be reduced and ethylene biosynthesis may be induced. ►glycan, ►jasmonic acid, ►insect resistance in plants, ►host-pathogen relationship, ►phenolics, ►ethylene; Zhou L, Thornburg R 1999, p 127. In: Reynolds PHS (Ed.) Inducible Gene Expression In Plants. CAB, New York.

Wrapping Choice: A generalized transducing phage "chooses" to scoop up the host rather than the viral DNA and "wraps" it into the phage capsid. ►transduction

Wright Blood Group: A very rare blood group; the frequency of the Wr(a) antigen is about 3×10^{-4} in Europe. ►blood groups

Wright-Fisher Model: ►genetic drift, ►drift genetic

Wrinkled/Smooth: This is a gene locus of pea, immortalized by Mendel's discovery of monogenic inheritance. The recessive "wrinkled allele" turned out to be an insertional mutation. ►pea for photograph of wrinkled and smooth seeds.

Writhing Number: This indicates the contortion of a DNA double helix in a supercoiled state. It measures the helix axis in space. ►linking number; Kobayashi S et al 2001 Chem Pharm Bull Tokyo 49:1053.

Wrod Score (wrong lod score): Obtained when a linkage is estimated as lod scores on the assumption of an incorrect genetic model (ϕ). The genetic model in complex traits may easily be misspecified. ►lod score, ►mod score, ►model genetic; Hodge SE, Elston RC 1994 Genet Epidemiol 11(4):329.

Wrongful Birth: Potential responsibilities of a physician or a genetic counselor for negligence in informing or prenatal care of prospective parent(s) about risks involving childbirth. (See Randall KC 1979 Hofstra Law Rev 8:257).

Wrongful Life: A potential responsibility of parents, physicians and genetic counselors for not preventing the birth of a child with a serious hereditary disease or in the case of illegitimacy, which may carry a social stigma. The affected offspring may sue. ►counseling genetic, ►genetic privacy, ►confidentiality, ►paternity test; Foutz TK 1980 Tulane Law Rev 54:480.

WW Domain: A two-tyrosine motif (38–40 amino acids) of signaling proteins and a binding site for proline-rich peptides. The binding of a WW domain by Pin1 and Nedd4 proteins apparently does not require prolines for binding, rather the WW domains are binding sites for phosphoserine and phosphothreonine. Among the many diverse proteins, their ligands include Cdc25C phosphatase, microtubule associated tau, carboxy-terminal of RNA polymerase II, etc. ►SH2, ►SH3, ►pleckstrin, ►PTB, ►signal transduction, ►Pin1; Nedd Sudol M, Hunter T 2000 Cell 103:1001.

www (world wide web): The system is available through the Internet with the aid of a browser program, which makes possible the search through a computer linked to the system. ►Internet, ►HTML

***wx* Gene** (*waxy*): Occurs in various cereal plants and is used as a chromosome marker (in the short arm of chromosome 9–56 of maize). In the presence of the dominant allele starch is formed (stained blue by iodine stain). If it is replaced by recessive allele, amylopectin is formed (stained red-brown) because of a defect in NDP-starch glucosyltransferase. ►iodine stain, ►amylopectin, ►TILLING

Historical vignettes

Edith Kramer 2002 American J Art Therapy 40(4):218

"How can we characterize [the opposite of kitsch] good art? Three elements seem essential; evocative power, inner consistency, and an economy of means so that the quality of the work would be diminished if anything would be added or omitted."

Apparent mutations were recorded by the ancient literature. Aristotle says, "whoever does not resemble the parent is, in some respects, a monster, because in this case nature has deviated, to a certain degree, from the hereditary type" (*Generation of Animals*, Book IV, Part 3, Para. 1).

In 1844 WH Prescott retells the *History of the Conquest of Mexico* (Dutton, New York): "I must not omit to notice a strange collection of human monsters, dwarfs, and other unfortunate persons, in whose organization Nature had capriciously deviated from her regular laws. Such hideous anomalies were regarded by the Aztecs as a suitable appendage of state. It is even said, they were in some cases the result of artificial means, employed by unnatural parents desirous to secure a provision for their offspring by thus qualifying them for a place in the royal museum."

Thus even induced mutation had been anticipated much before it had been experimentally demonstrated by HJ Muller and LJ Stadler in the mid 1920s.

X

x: Refers to basic chromosome number. ►polyploids, ►n

x̄: Denotes the arithmetic mean of the sample.

χ: A symbol for crossing over or chiasma.

χ^2: ►chi square

X^{1776} (bicentennial): A bacterial strain with an absolute requirement for diaminopimelic acid (a lysine precursor) and needed for the growth of viable bacteria, and therefore it cannot survive outside the laboratory. It was so named in the US bicentennial year at the Asilomar Conference in 1976. ►Asilomar Conference

χ Element: ►chi elements

Xa: A blood coagulation factor. It is also a plasmid factor that specifically cleaves protein after Arg of the tetrapeptide Ile-Glu-Gly-Arg that connects the 31 amino terminal of phage λ cII protein. ►clotting, ►lambda phage; Verner E et al 2001 J Med Chem 44:2753.

Xanthine: A purine derived from either adenine through hypoxanthine by xanthine oxidase or from guanine by deamination (see Fig. X1). By xanthine oxidase it is converted to uric acid. ►uric acid

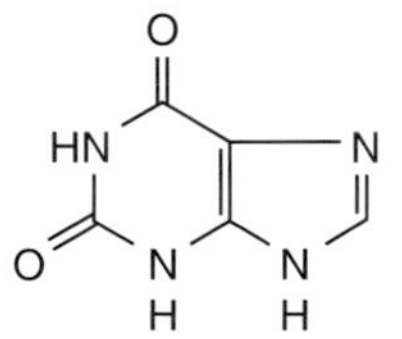

Figure X1. Xanthine

Xanthinuria (XDH, 2p23-p22): Refers to a recessive, xanthine dehydrogenase/oxidase deficiency resulting in the excretion of excessive amounts of xanthine and xanthine stones in the kidneys. The uric acid content of the urine and serum is reduced. The 36-exon gene spans ~60 kb DNA. ►kidney diseases, ►uric acid

Xanthoma Cell (foam cell): Highly vacuolated because of excessive amounts of lipids.

Xanthomatosis: ►cerebral cholesterinosis

Xanthophyll: Refers to yellow carotenoid pigments that play an accessory role in light absorption. ►photosystems; Ruban AV et al 2001 J Biol Chem 276:24862.

X-Box: The sequence–with some variations–of the X-boxes widely present in eukaryotes: GTTTCCAT-GGAAAC. Using stringent criteria for X-boxes within 250 bp upstream of the start codon, 293 genes have been identified in *Caenorhabditis* (Chen N 2006 Genome Biol 7:R1216). X-box binding protein (XBP-1) is a transcription factor essential for hepatocyte growth, differentiation of cilias as well as for plasma cell differentiation. Overexpression of XBP-1 has been reported in breast cancer and other carcinomas (Fujimoto T et al 2007 Anticancer Res 27(1A):127). XBP1 has evolved a mechanism for "on-demand" switching of translation between two overlapping reading frames but it can provide functionality to both reading frames (Nekrutenko A, He J 2006 Trends Genet 22:645).

XBP: Encodes the 89 kDa subunit of human transcription factor TFIIH, corresponding to yeast *SSL2*, encoding a 105 kDa polypeptide. XBP-1 is required for the differentiation of plasma cells into B lymphocytes. ►transcription factors, ►B lymphocytes

X2BP: MHC-II promoter binding protein, along with NF-Y and RFX. ►RFX, ►MHC; Reimold AM et al 2001 Nature [Lond]:412:300.

XCAP-C and XCAP-E: Proteins involved in the condensation of chromosomes. ►condensin complex; Neuwald AF, Hirano T 2000 Genome Res 10:1445.

X Chromosomal Inactivation: ►lyonization, ►Barr body, ►dosage compensation, ►Xic

X Chromosome: One of the sex chromosomes generally present in two doses in females and in one in males. The mammalian X chromosome developed from autosomes ~300 million years ago. In some species females are XY whereas in males either XX or XO or other chromosomal doses may be found. In species where the female is heterogametic, her sex chromosomal constitution is often designated as WZ and that of the male as ZZ. The sex chromosomes of the semi-aquatic, duck-billed platypus (*Ornitorynchus anatinus,* 2n = 56), an egg-laying Australian mammal, share genes with the bird's Z chromosomes and the mammalian X (Grützner F et al 2004 Nature [Lond] 432:913). Sex chromosomes apparently evolved from autosomes. The PAR (pseudoautosomal) region is a relic of autosomal origin. It is assumed that the X chromosome underwent substantial evolution in humans and sequences fairly well conserved in other mammals but display only small homology with the bird's Z chromosome. Some of the genes were acquired from autosomes, other genes were retained or transposed to autosomes because even in males (with only one X) two doses were required for certain developmental processes. Initially, in humans 1,098 genes (7.1/Mb) and 700 pseudogenes (4.6/Mb) have been revealed in the X chromosome.

More recent analysis has revealed 696 protein-coding genes and 652 X-linked and 432 X-derived pseudogenes. About 5% of the pseudogenes displayed some transcripts, indicating their role in genetic regulation. On an average, the pseudogenes differed from the parental gene by 70 changes/1,000 nucleotides. New exons were detected in 22% of the genes and 35% displayed exon skipping. Of the earlier reported 142 putative non-coding RNAs, 64 (44%) were untranslated segments of known genes (Harsha HC et al 2005 Nature Genet 37:331). Slightly different conclusions were reached by another contemporaneous publication (Ross MT et al 2005 Nature [Lond] 434:325).

The gene density and gene size is generally smaller than in the autosomes although the largest human gene, the muscular dystrophy gene is X-linked. The frequency of CpG islands is about half in the X compared to the genome average. Approximately 10% of the so-called cancer-testis antigens are in the human X chromosome. The cancer-testis antigens are specifically expressed in many different types of cancer cells (see e.g., MAGE). Apparently 153,146 single nucleotide polymorphisms occur in the X chromosome. The euchromatic part of the X chromosome contains 57% interspersed repeats versus 45% of the genome average. The L1 family constitutes 29% of the X versus only 17% of the genome average. Several human diseases are associated with deletions or inversion in X chromosomal duplications. There is a relatively longer 2.7 Mb pseudoautosomal (PAR1) and a shorter 330 kb PAR2 region shared with the Y chromosome where recombination can occur, but in some other homologous regions (XAR) there is no recombination between X and Y. The latter types can be mapped only by radiation hybrids. Only 54 of the annotated X chromosomal genes display homologies with Y. In the PAR1 region there are 24 and in PAR2 5 genes that may recombine. The latter ones are located in three groups on both sex chromosomes but they do not recombine. About 15 of the protein-coding Y chromosomal genes lack homologies in the X chromosome. The human X chromosome has lower variability (~60%) than the autosomes. Both human and mouse X chromosomes show very close synteny. One or more of the double (or multiple) X chromosome(s) of mammals are inactivated early in female development. Yet more than 15% of the genes escape inactivation (Ross MT et al 2005 Nature [Lond] 434:325). ►Y chromosome, ►chromosomal sex determination, ►dosage compensation, ►Barr body, ►lyonization, ►Ohno's law, ►monotrene, ►pseudoautosomal, ►sex chromosome, ►autosome, ►Xist, ►Tsix, ►Xic, ►radiation hybrid, ►synteny, ►L1, ►Y chromosome, ►heterochromatin; Vallender EJ et al 2005 Nature Genet 37:343; Chow JC et al 2005 Annu Rev Genomics Hum Genet 6:69.

X Chromosome Counting: This mechanism is used for sex determination in the case of XX versus XO. The *xol*-1 gene of *Caenorhabditis* constitutes a switch mechanism that specifies male developmental course if it is inactive. In XX nematodes *xol*-1 is repressed post-transcriptionally or the RNA-binding protein encoded by the fox-1 gene reduces its level of expression. The transcription of gene *xol*-1 is suppressed primarily by SEX-1 hormone receptor protein, which binds to the promoter in XX nematodes. ►sex determination, ►chromosomal sex determination, ►dosage compensation; Maxfield Boumil R, Lee JT 2001 Hum Mol Genet 10:2225.

X Chromosome Inactivation: ►lyonization, ►*Xic*

XCID: X chromosome-linked severe combined immunodeficiency. ►SCID, ►immunodeficiency

xDNA (expanded DNA): Differs from the natural DNA double helix in that some of the natural bases are expanded by a benzene ring (B, red) and, therefore, the molecule is broader (A: adenine [purple], T: thymine [blue]) (see Fig. X2). Besides the bases shown, the others (C and G) can also be expanded; the sugar–phosphate backbone is omitted for the sake of simplicity. The designed duplex is right-handed and antiparallel, and hydrogen-bonded in a way analogous to that of Watson-Crick DNA. The sugar-phosphate backbone adopts a regular conformation similar to that of B-form DNA. The xDNA can encode more information than the natural one because it can use 8 instead of 4 nucleotides. It is more heat resistant than the natural DNA and it fluoresces. This property may make it useful for the detection of genetic defects. ►base analogs, ►hydrogen pairing, ►Watson and Crick model, ►DNA types; Kool ET 2002 Acc Chem Res 2002 35(11):936; Liu H et al 2003 Science 302:868; Lynch SR et al 2006 J Am Chem Soc

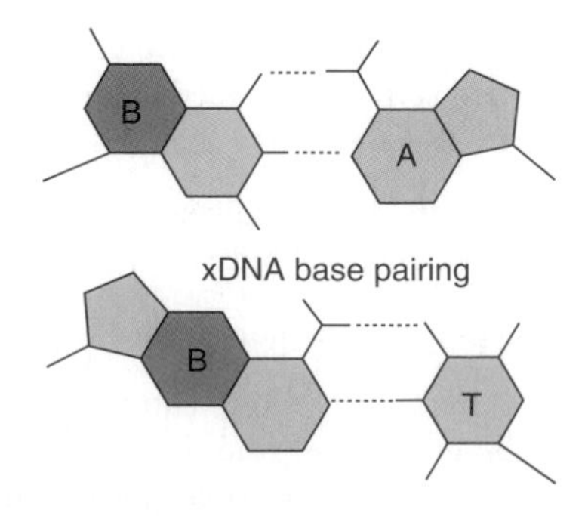

Figure X2. xDNA base pairing

128:14704; Leconte AM, Rosenberg FE 2006 Nature [Lond] 444:553.

Xenia: This is the expression of the gene(s) of the male in the endosperm (e.g., purple) following fertilization; the expression in the embryo may not yet be visible (see Fig. X3). ►metaxenia

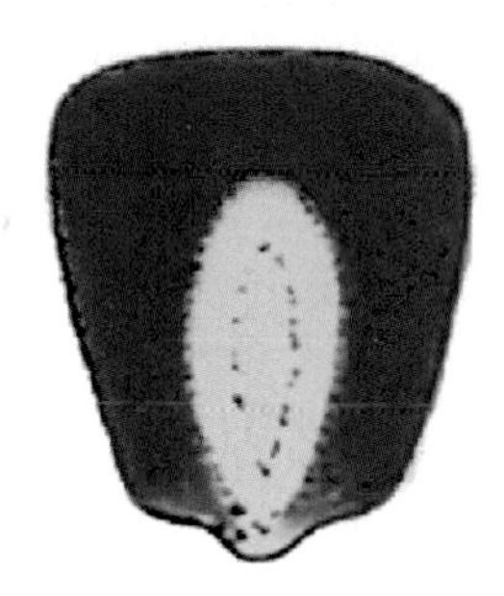

Figure X3. Xenia

Xenobiotics: Refers to compounds that do not naturally occur in living cells. They are often toxic or harmful substances.

Xenogamy: Meaning fertilized by different neighboring plants.

Xenogeneic: Refers to transplantation from another species (xenotransplantation). ►allogeneic

Xenogenetics: The study of the effect of environmental factors on conditions under polygenic control. ►polygenic

Xenograft: Refers to the transplantation of tissue from another species (e.g., animal→human). It poses problems of rejection and the likelihood of viral infection (e.g., porcine endogenous proviruses [PERV]) as well as the emergence of new diseases. The complement cascade of the immune system mediates the rejection. If swine, transgenic for the human complement, is used as a heart donor to baboons, the function of the heart is prolonged by several hours. The limited clinical evidence, however, does not indicate substantial risk in pig→human grafts. ►complement, ►immune system, ►transgenic, ►epitope, ►xenotransplantation, ►gene therapy, ►αGT, ►PERV; Matsunami K et al 2001 Clin Exp Immunol 126:165.

Xenology: A study of the combination of original and foreign genetic sequences within an organism or a group of organisms caused by horizontal transfer and transformation. ►homology, ►transmission, ►transformation, ►infectious heredity, ►orthologous loci, ►paralogous loci; Gogarten JP 1994 J Mol Evol 39:541.

Xenopus (*X. laevis*, South African clawed toad, 2n = 4x = 36, DNA 3.1 x 10^9 bp): This is a species of frogs. Being tetraploid and having a long generation time of several years does not make it a favorite object of genetic manipulations although it is an excellent object of embryological studies. The oocyte may reach 1.5 mm in size and it is thus visible by the naked eye and lends itself to various manipulations. Pioneering research was conducted with nuclear transplantation. Transformation of embryonic tissues was not as successful as expected because too few cells expressed the transgenes. Now techniques have been developed for transformation of sperms, and this will permit the study of dominant transgenes on various developmental and physiological processes. *Xenopus* (*Silurana*) *tropicalis* is the only diploid species (1.7 × 10^9 bp, generation time 4–6 months) but it is somewhat distantly related. *X. tropicalis* is very prolific (1,000–3,000 eggs per ovulation) and transformation is very efficient. ►toad, ►frog, ►transformation, ►transgene, ►*Xenopus* oocyte culture; genetic regulatory networks in embryonic development: Koide T et al 2005 Proc Natl Acad Sci USA 102:4943; http://www.dkfz-heidelberg.de/abt0135/axeldb.htm; http://www.nih.gov/science/models/; http://www.tigr.org/tdb/tgi/.

***Xenopus* Oocyte Cultures**: *Xenopus* frog oocytes are about 1 μL in volume. They contain large amounts of DNA (12 pg in the nucleus, 25 pg in the nucleoli, 4 ng in the mitochondria). They can synthesize daily 20 ng of RNA and 400 ng of protein. About 10^6 to 10^7 bp DNA can be directly injected into the *germinal vesicle* (the nucleus). The injected DNA is packaged into the nucleosomal structure, and it is replicated and translated by the cellular machinery and the products are ready for analysis within a few hours. Recombinant DNA can be introduced and studied the same way. Exogenous DNA is replicated according to the cell cycle of the oocytes and no re-initiation of the foreign DNA replication occurs. All foreign DNAs are replicated according to the oocyte cell cycles and all the foreign replicational signals are overridden. Replication in the oocytes also is extremely rapid because transcription is halted during replication. *Xenopus* oocytes have been very successfully used for the structural, morphological analysis of vertebrate embryogenesis. They have proved to be extremely useful for the molecular analysis of transcription, regulation, cell-to-cell communication and early morphogenesis. ►oocyte, ►protein synthesis, ►morphogenesis; Romero MF et al 1998 Methods Enzymol 296:17.

Xenorhabdus: *X. nematophila* is the best understood of the five identified species of *Xenorhabdus*. The bacterium colonizes the intestine of a non-feeding stage of *Steinernema carpocapsae* nematodes. The nematode is the vector that shelters the bacteria

from the competitive soil environment, and shuttles *X. nematophila* into insect hosts. The bacterium functions as a potent pathogen that infects and kills diverse insect species, which serve as the nutrient source for the development and reproduction of the nematode. Some *Xenorhabdus* species produce effective antibiotics against plant and animal pathogens. ►*Caenorhabditis*; Goodrich-Blair H, Clarke DJ 2007 Mol Microbiol 64:260.

Xenotransplantation: Refers to the transplantation of organs/tissues among different species and may involve the potential danger of transferring new viruses. On the other hand, xenotransplantation may have potentials in curing disease, e.g., the use of resistant baboon livers in the case of hepatitis B infections and baboon bone marrow for AIDS victims. The transplanted foreign cell or tissue may not function in the recipient or stimulate an adverse immune reaction. The transplanted tissue may incite the complement within minutes in hyperacute rejection. In some instances the reaction may be delayed for weeks but an acute vascular rejection may follow by interaction of the xenoreactive antibodies with the donor blood vessels. Galactose α-1,3-transferase knockout lines may overcome the immune reaction of the recipient against the graft but does not entirely solve the problem because the antibodies may attack other antigens. Genetically modified heme oxidase (HO-1) may provide some protection but it may generate toxic side effects (oxidant, bilirubin). The modification of the recipient antibodies against the transplant by genetic engineering may come into consideration. Also, e.g., the pig donor may be modified by human gene knock-ins. The porcine endogenous retrovirus (PERV) and some as yet unknown viruses may pose great risks not only to the recipient, but more importantly to the human populations also. Deleting the virus or inactivating them may become a means of protection. More research is required in this field. ►xenogeneic, ►xenograft, ►immune tolerance, ►immunosuppression, ►grafting in medicine, ►rejection, ►baboon, ►hyperacute reaction, ►αGT, ►knock-in, ►PERV, ►OBA, ►microchimerism, ►1,3-galactosyltransferase α; Sim KH et al 1999 Can J Gastroenterol 13:11; Lai L et al 2002 Science 295:1089; Cooper DKC et al 2002 Annu Rev Med 53:133; http://www.fda.gov/cber/gdlns/clinxeno.htm.

Xenotropic Retroviruses: These are replicated only in the cells of species other than the species from which the virus originated. ►ecotropic and polytropic virus

Xeroderma Pigmentosum (XP): The primary symptoms may develop during the first year of life as extremely pronounced freckles induced by sunshine proliferate and appear as skin cancer (see Fig. X4). In some cases the central and peripheral nervous systems are also affected. Usually A (9q22.2-q31), B (2q21), C (3p25), D (19q13.2-q13.3), E (11p12-p11, F (16p13.2-p13.1) and G (13q32-q33) types are distinguished, the other complementation groups (H and I) are less clear. Complementation group A encodes DNA damage binding protein B, and D codes for helicases, C initiates global nucleotide excision repair and selectively repairs cyclobutane pyrimidine dimers rather than 6–4 photoproducts. The XP-D gene product is a subunit of the TFIIH transcription factor and it is regulated by vitamin D receptor-responsive genes (Drané P et al 2004 Mol Cell 16:187). XPC is homologous to yeast genes RAD23A and B and it is sometimes referred to as XPC-HR23A and B. Types F and G are defective in a DNA repair endonuclease (homologous to yeast RAD1).

Figure X4. A mild form of xeroderma pigmentosum. The initial signs are usually very heavy freckles that gradually develop into different types of skin cancer. The progress of the disease is enhanced by sunshine. A variety of ailments may accompany the main symptoms and the afflicted persons rarely reach adulthood. (Courtesy of the March of Dimes—Birth Defects Foundation)

One type of XP, XP-V patients, representing about 25% of the clinical cases of the disease, do not have a defect in nucleotide exchange repair and are not sensitive to UV radiation. In such cases the defect is caused by failure of replication of the leading strand through pyrimidine dimers. The replication (by polymerase ζ) of the lagging strand associated with the photoproduct can slowly proceed and results in an asymmetrical replication fork with an extended single strand leading strand of the parental DNA molecule. Unfortunately, pol ζ is error prone and the error-free pol η (encoded by *RAD*30) is defective in XP-V individuals. Humans have two homologs of RAD30.

As a consequence the patients'cells become susceptible to mutation and carcinogenesis.

Another form involving milder symptoms is also known which has an apparently dominant inheritance. In some forms the defect is not in excision repair but a post-replicational anomaly causes light sensitivity. Similar gene(s) occur(s) in mouse and five *RAD* genes of yeast also have defects in excision repair. ►DNA repair, ►excision repair, ►Bloom syndrome, ►ataxia telangiectasia, ►Fanconi syndrome, ►Cockayne syndrome, ►*RAD*, ►light-sensitivity diseases, ►trichothiodystrophy, ►helicase, ►complementation groups, ►DDB, ►DNA polymerases, ►ultraviolet-sensitivity syndrome, ►cyclobutane dimer, ►pyrimidine-pyrimidinone photoproduct, ►5′,8-purine cyclodeoxynucleosides, ►transcription factors, ►vitamin D; Bootsma D et al 1998, p 245. In: Vogelstein B, Kinzler KW (Eds.) The Genetic Basis of Human Cancer. McGraw-Hill, New York.

Xerophyte: A draught-tolerant plant which thrives in low precipitation regions.

X-Family of DNA Polymerases: polβ-like nucleotidyl-transferases. Polλ is similar to polβ. TdT is a non-templated polymerase of immature lymphocytes in the bone marrow. Polμ is homologous to TdT. TdT promotes diversity in non-homologous end joining (HEJ). Polμ and Pol λ are also involved in NHEJ (Nick McElhinny SA et al 2005 Mol Cell 19:157). Polσ is a templated or non-templated enzyme establishing sister chromatid cohesion. ►DNA polymerases, ►nucleotidyl-transferase, ►NHEJ; Aravind L, Koonin EV 1999 Nucleic Acids Res 27:1609; Rattray AJ, Strathern JN 2003 Annu Rev Genet 37:31.

XG (Xg[a]): A blood group antigen determining dominant factor in the short arm of the human X chromosome. It appears that this end of the X chromosome can recombine with the Y chromosome, and it does not undergo lyonization like the majority of the X chromosomal genes. ►lyonization, ►blood groups, ►ichthyosis; Fouchet C et al 2000 Immunogenetics 51(8–9):688.

Xgal (5-bromo-4-chloro-3-indolyl-β-D-galactopyranoside): When β-galactosidase enzyme hydrolyzes this chromogenic substrate blue color is formed in a bacterial culture plate. ►β−galactosidase

XIAP: A caspase inhibitor which is a member of the human IAP family of proteins. ►IAP, ►caspase; Riedl SJ et al 2001 Cell 104:791.

XGLD: Denotes poly(A) polymerase. ►symplekin

Xic (X chromosome inactivation center, Xq13.2): Its effect is somewhat similar to but unlike *Xist,* which has a fairly localized cis effect, *Xic* extends globally on the X chromosome. *Xic* delays the replication of the X chromosome and mediates the hypoacylation of histone 4 and contributes to heterochromatinization of that chromosome. *Xic* controls chromosome counting in dosage compensation and selects one or more of the X chromosomes for inactivation. In case there is a short transient pairing between two normal X chromosomes, which both carry the *Xic* locus, the inactivation of the X chromosomes is random. In case *Xic* and *Tsix* are deleted, X chromosome inactivation may become erratic, i.e., either it does not take place at all or both are inactivated or one is inactivated and the other is not. In case an additional copy of *Xic* is introduced by transformation and the two X chromosome cannot pair at the *Xic* site then neither of the two X chromosomes is inactivated (Carrel L 2006 Science 311:1107; Xu N et al 2006 Science 311:1149). ►lyonization, ►*Xist*, ►methylation of DNA; Prissette M et al 2001 Hum Mol Genet 10:31.

XID: Refers to sex chromosome-linked immunodeficiency. ►scid, ►immunodeficiency

Xist (X chromosome inactive sequence transcript at Xq13.2; antisense of *XIST* is *TSIX* at Xq13.2): This gene has no protein product but its 15-kb RNA covers the genes of the inactive mammalian X chromosome beginning from the *Xic* site (Xq13.2). The *Xist* RNA gene evolved only in placental mammals by pseudogenization (functional relaxation) of a protein-coding gene (Duret L et al 2006 Science 312:1653). Deletion analysis demonstrates that the first exon of human *XIST* is sufficient for both transcript localization and the induction of silencing and that, unlike in mice, the conserved repeat region is essential for both functions. This indicates that there is a difference in inactivation between humans and mice (Chow JC et al 2007 Proc Natl Acad Sci USA 104:100104).

Underacetylated histones, methylated CpG islands and late replication of the DNA cause inactivation of the X chromosome. *Xist*-triggered inactivation is preceded by methylation of histone-3 at lysine 9. *Xist* acts in cis and may also be expressed in autosomes or in transgenes inserted in the autosomes and affect the neighboring genes. Since the X chromosomal inactivation is restricted to the copies of the X chromosome present more than once, it has been suggested that *Xist* counts (senses) X chromosomal dosage. Deletions downstream of *Xist* make it constitutive, i.e., it is expressed and causes inactivation of the X chromosome concerned. From 15 kb downstream location a 40 kb apparently antisense RNA *Tsix* is transcribed across the *Xist* locus in both X chromosomes of diploids until the determination of inactivation. Prior to the onset of the inactivation *Tsix* is transcribed in an antisense orientation only on the

X chromosome destined for inactivation and it is shut down again after the inactivation begins. Apparently, *Tsix* determines the selection of the X chromosome for inactivation by the *Xce* controlling element without affecting the silencing itself. Knockout of the paternal *Tsix* does not affect embryonal development. If the *Tsix* knockout is transmitted through the female, embryos of both sexes die because both X chromosomes are silenced in the female offspring and the single X in males. The role of *Tsix* in humans and mice is debated. The site *Xite* promotes *Tsix* persistence on the active X chromosome (Ogawa Y, Lee JT 2003 Mol Cell 11:731). In X chromosome: autosome translocations the inactivation is inefficient because the spreading of the inactivating signal is not favored by the chromatin structure (Popova BC et al 2006 Proc Natl Acad Sci USA 103:7706). It has been reported that the breast cancer gene BRCA1 alters the Xist function but this finding has not been confirmed (Xiao C et al 2007 Cell 128:977), it was, however, substantiated later (Silver DP et al 2007 Cell 128:991). ►lyonization, ►*Xic*, ►dosage compensation, ►histones, ►imprinting; Lee JT 2000 Cell 103:17; Matsui J et al 2001 Hum Mol Genet 10:1393; Migeon BR et al 2001 Am J Hum Genet 69:951; Heard E et al 2001 Cell 107:727; Wutz A et al 2002 Nature Genet 30:167; Plath K et al 2002 Annu Rev Genet 36:233; Sibata S, Lee JT 2003 Hum Mol Genet 12:125; Migeon BR 2003 Nature Genet 33:337; Lee JT 2003 Nature Genet 33:337; mechanism of *Xist* effect on its partner: Carrel L 2006 Science 311:1107.

Xklp1: ►NOD

xl: The prefix for *Xenopus laevis* toad protein or DNA or RNA, e.g., xlRNA.

XLA: X-linked agammaglobulinemia. ►agammaglobulinemia

XLF/Cernunnos: A protein similar to XRCC4 with a role in NHEJ. ►NHEJ, ►XRCC4, ►X-ray repair; Ahnesorg P et al 2006 Cell 124:301.

X-Linked: The gene is within the X chromosome (see Fig. X5). ►Z linkage

XLNO 38: A protein with homologies to nucleoplasmin. Although it is not part of the mature ribosomes, it is associated with both small and large subunits of ribosomes. Seemingly its role is chaperoning the assembly of basic proteins on ribosomal RNA precursors. ►chaperone

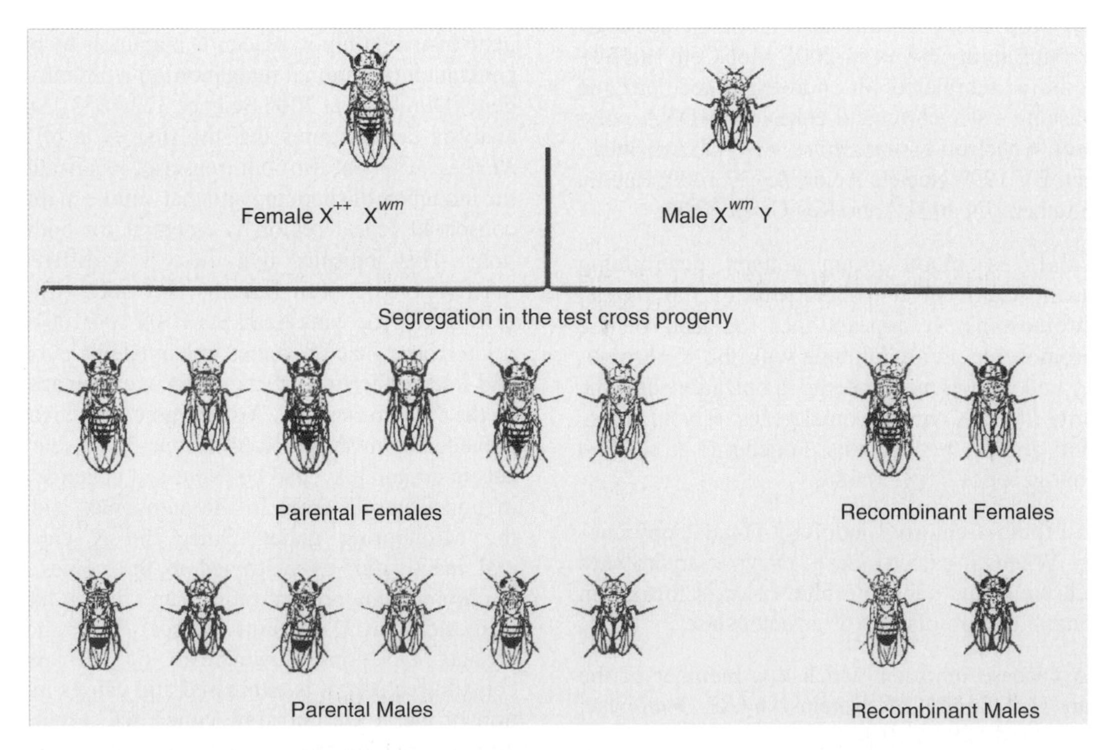

Figure X5. The female fly is heterozygous for the recessive white eye and the miniature body-size genes, the male X chromosome carries the two recessive alleles. Since in the female fly recombination is normal (unlike in the *Drosophila* male), free recombination yields 16/4 = 0.25 recombinants in both sexes

XLP: The lymphoproliferative syndrome encoded at Xq25 causes extreme sensitivity to infectious mononucleosis and Epstein-Barr virus infection. ►Epstein-Barr virus, ►mononucleosis

XML (extensible markup language): A flexible communication system that is capable of sharing format and data through the World Wide Web. ►www, ►XSL; http://www.w3.org/XML/.

XML for INSD sequence submission: http://www.insdc.org/documents.html; ►INSD

X-Numerator: The number of X chromosomes relative to autosomes (X:A) in the determination of sex. ►chromosomal sex determination

XO: Having a single X chromosome in a diploid cell. ►sex determination, ►Turner syndrome

Xolloid: ►bone morphogenetic protein

XPA, XPD, XPF: ►ABC excinucleases

XPB: Denotes the DNA helicase. It is the equivalent of Ss12. ►DNA repair, ►excision repair

XPC-HR23A or B: ►Xeroderma pigmentosum

XPG: Refers to xeroderma pigmentosum endonuclease. ►xeroderma pigmentosum, ►endonuclease, ►DNA repair

X-Ray Caused Chromosome Breakage: The major effects of ionizing radiations on living cells are chromosome breakage (see Fig. X6). Chromosome breakage has been extensively utilized for the genetical analyses of knocking out genes, the production of radiation hybrids, deletion mapping, genomic subtraction, etc. The destructive effects of the radiation depend on the nature of radiation. Soft X-rays having higher density of linear energy transfer, break the chromosome more effectively than the shorter wavelength and high-energy hard X-rays. The destructive effect also depends on the species exposed, the type of tissues irradiated, the physiological conditions during the delivery of the radiation, the repair system, etc. In *Drosophila* and locusts usually 1000 – 5000 R doses are employed for chromosome breakage in adults and embryos. In the case of gonadal radiation, 100 to 500 R may be effective. In the testis cells of *Macaca mulatta* monkey an increase of radiation dose from 25 R to 400 R resulted in close to exponential increase of chromosome breakage reaching about 1.5% of the chromosomes at the highest dose. In the large nuclei and chromosomes of *Trillium* and *Vicia faba* generally 100 to 500 R break the root tip chromosomes and lower doses may be sufficient if applied to pollen mother cells. For the smaller nuclei of maize 800 to 1500 R may be chosen. The very small somatic nuclei of *Arabidopsis* may tolerate 5–10 fold higher doses of irradiation than maize. ►radiation sensitivity, ►radiation hazard assessment, ►radiation effects, ►LET, ►radiation hybrid, ►cathode rays; Brinkley BR, Hittelman WN 1975 Int Rev Cytol 42:49.

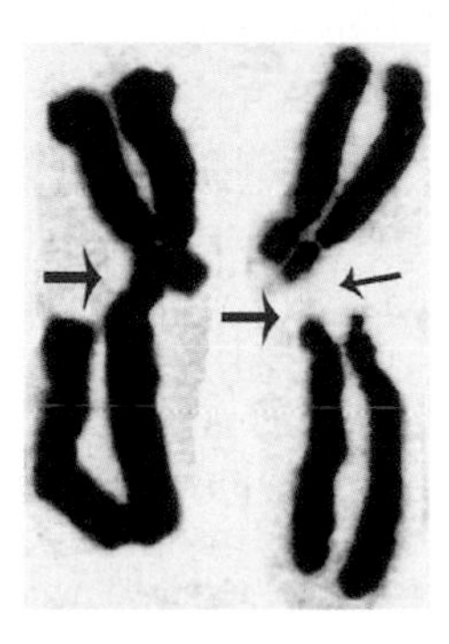

Figure X6. Single- and double-strand breaks. Courtesy of Brinkley BR, and Hittelman WN

X-Ray Crystallography: ►X-ray diffraction analysis

X-Ray Diffraction Analysis: Used for the analysis of the structure of molecules by determining the angles of electron scattering upon exposure to X-rays. When a large number of a particular type of molecules in an array is irradiated they scatter the incident electrons. Where the scattered beams cancel each other, no bright image is formed. Where however the scattered electrons reinforce each other because they diffracted by a certain common arrangement of crystals or molecules, a bright image is formed on the screen. Thus, the intensity of the spots on the screen provides a basis for calculating and determining the internal, three-dimensional structure of the object. ►nuclear magnetic resonance spectrography, ►X-rays; http://www.rcsb.org/pdb/.

X-Ray Hazard: ►X-ray, ►radiation hazard assessment, ►radiation protection, ►radiation effects, ►X-ray caused chromosome breakage

X-Ray Repair: Involves excision repair mechanisms. A human X-ray repair gene locus has been isolated through complementation of excision repair-deficient Chinese hamster ovary cells by human DNA. XRCC2 repairs DNA double-strand breaks by homologous recombination. The locus XRCC1, encoding a ligase III polypeptide, has been assigned to human chromosome 19q13.2-q13.3. XRCC4 locus (human chromosome 5) is involved in the determination of X-ray sensitivity in the G1 phase of the cell cycle but its response in the S phase appears normal, indicating that the defective mutants (in Chinese hamster ovary cells) are deficient in the repair of DNA double strands. The XRCC9 allele is in the FANC-G complementation group of Fanconi anemia. ►radiation sensitivity, ►excision repair, ►DNA repair, ►physical mutagens, ►X-rays, ►Fanconi anemia, ►X-ray caused chromosome breakage, ►XRCC, ►NHEJ, ►XLF; Tebbs RS et al 1999 Dev Biol 208:513.

X-Ray Sensitivity: ►X-ray repair, ►X-ray caused chromosome breakage, ►nuclear size

X-Ray Therapy: ►radiation therapy

X-Rays: Ionizing electromagnetic radiation emitted by the cathode tubes of Röntgen machines within the range of 10^{-11} and 10^{-8} m wavelength (see Table X1). The shorter ones (hard rays) have greater penetration and lower ionization density while the longer wavelengths (soft rays) have reduced penetration and denser dissipation of the energy. Hard rays cause more discrete lesions to the genetic material, soft rays are expected to produce greater chromosomal breakage. The spectrum of the radiation may be controlled by filters. The effectiveness of the filters depends on the attenuation of the radiation by the nature of the filter. The fluence of the radiation can be defined as $I = (I_0)e - \mu x$ where I = fluence at a certain depth *x;* I_0 = the fluence rate at the surface and μ = the specific attenuation coefficient and μ/p is mass attenuation coefficient (p = density of the absorbing material). At photon energy MeV = 0.1, the mass attenuation coefficients (cm^2/g) are for aluminum (0.171), iron (0.370), lead (5.400), water (0.171) and concrete (0.179). ►cathode rays, ►ionizing radiation, ►X-ray repair, ►Compton effect, ►radiation hazard assessment, ►radiation protection, ►radiation effects, ►Ku

XRCC: The genes XRCC2 (7q36.1), XRCC3 (14q32.3), XRCC4 (5q13-q14) and XRCC5 (Ku70, 2q35) in humans along with RAD51 (15q15.1, 17q11-q12) and DMC control recombination and repair. XRCC1 stimulates human polynucleotide kinase and promotes the repair of single-strand DNA breaks. XRCC3 protein, in concert with RAD51C, facilitates homologous pairing and recombinational repair. ►X-ray repair, ►RAG, ►RAD, ►Ku, ►DNA repair, ►DMC, ►XLF; Whitehouse CJ et al 2001 Cell 104:107; Masson J-Y et al 2001 Proc Natl Acad Sci USA 98:8440; Brenneman MA et al 2002 Mol Cell 10:387.

Table X1. Thickness of the protective shield needed in mm using lead, depending on voltage and amperage of the X-radiation

	milliAmpere		
Kilo Volt	<5	5–10	30
50	0.5	0.6	0.7
125	1.5	3.0	3.5
250	6.0	7.0	8.0
400	16.0	18.0	21.0

Xrn1p: An exonuclease degrading RNA beginning at the 5′ end.

XREF: Cross-referencing model organism genes with human disease and other mammalian phenotypes. ►databases

XRS: Mutation conveys radiation sensitivity, DNA double-strand break repair and V(D)J recombination. ►DNA repair, ►V(DJ); Matheos D et al 2003 J Cell Sci 116:111.

X-SCID (X-linked severe combined immunodeficiency): ►SCID

XSL (XSLT): A software for transforming XML documents. ►XML; http://www.org/TR/xslt.

X-Stain: A fluorochrome which stains the X chromosome differently from other chromosomes. ►chromosome painting, ►FISH

XX Males: A normal condition in birds and some other species, but rare 5×10^{-5} condition of it is a male mammalian (human) births. The recurrence risk is, however, about 25%. Most of them (90%) have a X chromosome—Y chromosome short arm translocation. Their phenotype and infertility resemble those of Klinefelter syndrome cases although the afflicted persons tend to be shorter in stature. Their distinction from hermaphrodites requires the use of an appropriate, fluorochrome-labeled cytogenetic probe for the critical Y chromosomal segment. ►sex determination, ►sex chromosomal anomalies in humans, ►Klinefelter syndrome, ►hermaphrodite; Vidal VP et al 2001 Nature Genet 28:216.

XXX: ►triplo-X, ►metafemale, ►sex chromosomal anomalies in humans

XXY: ►Klinefelter syndrome, ►sex chromosomal anomalies in humans

Xylan: This is poly(β-D-xylopyranose[1→4]) polysaccharide of plant cell walls. The enzyme xylanase facilitates maize pollen tube penetration into silk by xylan hydrolysis (Suen DF, Huang HC 2007 J Biol Chem 282:625). ►silk

Xylella fastidiosa:: A bacterium that causes citrus variegated chlorosis which affects a broad spectrum of crop species. It has a 2,679,305 bp sequenced genome and two plasmids (51158 bp and 1285 bp). (See Nature 406:151).

Xylem: The transporting tracheid vessels of plants carrying nutrients and water from the roots toward the leaves. ►vascular tissue, ►phloem, ►proteoglycan

Xylene ($C_6H_4[CH_3]_2$): Synonymous with xylol, this highly flammable and irritant, narcotic liquid is used as a solvent in microtechnique. The permissible threshold of vapors in the air is 100 ppm.

Xylene Cyanole FF: ►tracking dyes

Xylose (wood sugar, $C_5H_{10}O_5$): An epimer of aldopentoses. It is used in the tanning industry, as a diabetic carbohydrate food, and in clinical tests of intestinal absorption. ►epimer

Xylulose ($C_5H_{10}O_5$): An intermediate in the pentose phosphate pathway which accumulates in the urine of pentosuriac patients (see Fig. X7). ►pentose phosphate pathway, ►pentosuria

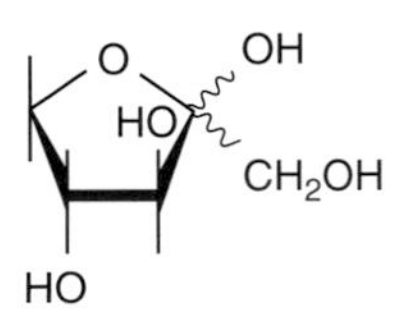

Figure X7. D-Xylulose

XY Body: ►sex vesicle

XYY: ►sex chromosomal anomalies in humans

Historical vignettes

AH Sturtevant commented on human genetics in 1954 (Science 120:405)

"Man is one of the most unsatisfactory of all organisms for genetic study. The time interval between successive generations is long, at best individual families are too small to establish ratios within them, and the test-matings that a geneticist might want cannot be made. Obviously no geneticist would study such a refractory object, were it is not for the importance that a knowledge of the subject has in other fields."

Holger Breithaupt remarked in 2002 (EMBO Reports 3:391)

"Now imagine if Bush, Watson, Venter or Steffánsson had had to have their proposals passed by a democratically elected committee. Most likely, their obviously great ideas would have died an early and lonely death."

Calvin B Bridges (1889–1938) was one of the most talented geneticists of all time. Most everybody can identify him as the founder of cytogenetics through his studies on nondisjunction, or with his work on salivary maps. Fewer people remember that he initiated the currently used gene symbols in Drosophila. Bridges also discovered the basic types of chromosomal aberrations:

"The general term 'deficiency' is used to designate the loss or inactivation of an entire, definite, and measurable section of genes and framework of a chromosome. A case of deficiency in the X chromosome of *Drosophila ampelophila* [currently *melanogaster*] occurred in September 1914, and has given rise to a whole series of correlated phenomena. The first indication of this deficiency was the occurrence of a female which had failed to inherit from her father his sex-linked dominant mutant 'bar,' though she inherited in a normal manner his sex-linked recessive mutant 'white.' This female, when bred, gave only about half as many sons as daughters, the missing sons, as shown by the linkage relations, being those which had received that X which was deficient for bar" (*Genetics* 2:445, 1917).

Two years later Bridges reported duplications and translocations in the Abstracts of the Zoological Society (published in *Anatomical Record*):

"... a section of the X-chromosome, including the loci for vermilion and sable, became detached from its normal location in the middle of the X-chromosome and became joined on to the 'zero' end (spindle fiber) of its mate. For certain loci this latter chromosome carries two sets of genes—those present in the normal location and also the duplicating set. If a male carries the recessive genes for vermilion and for sable in the normal loci and the wild-type allelomorphs in the duplicating loci, he is wild-type in appearance precisely as though he were an XX female heterozygous for vermilion and sable. ...

"A third case is the transposition of a piece of the second chromosome to the middle (spindle fiber) of the third chromosome. The genes of this duplication piece show linkage to both the second and the third chromosome at the same time" (*Anat. Record* 15:357–358, 1919).

The basis of a crossover reducer as an inversion of the third chromosome of *Drosophila* was identified by Alfred H. Sturtevant (1891–1970) in 1926 (*Biol. Zbl.* 46:697).

Y

Y: The symbol of pyrimidines in nucleic acid sequences.

Y Box Proteins: Members of a family of transcription factors that bind to an inverted CCAAT box (Y box) that activates genes involved in cell proliferation and growth. Y box proteins interact with other proteins and modulate transcription. The absence of mouse MSY2 protein causes infertility due to developmental defects in gametogenesis in otherwise healthy-appearing male and female animals (Yang J et al 2005 Proc Natl Acad Sci USA 102:5755). (See Ladomery M, Somerville J 1995 Bioessays 17:9).

Y Chromosome: One of the sex chromosomes present in males (XY) generally. However, in some species (birds, insects, fishes) females are XY and males may be XX or XO. In organisms with homogametic males their chromosomal constitution is commonly identified as WW. In the majority of the species the Y chromosome has only genes for sex differentiation or sex determination. These genes in the non-recombinant (NRY, ~35 Mb) part are involved in sex and fertility determination whereas the other Y chromosomal genes have homologies with a common segment of the X chromosome (Xq21, 3.4 Mb with two genes) as well as sequences scattered in autosomes. Some X-transposed sequences are degenerate and contain 27 pseudogenic homologs of the X chromosomal genes. It was earlier held that the major part (~95%) of the Y chromosome lacked homology to the X chromosome and was thus unable to recombine with the X chromosome. Recent research, however, has revealed that this male-specific region (MSY, 8 Mb in the short and 14.5 Mb in the long arm) is flanked on both sides by pseudoautosomal sequences where recombination is actually a normal and frequent occurrence (Skaletsky H et al 2003 Nature [Lond] 423:825). The MSY also includes the so-called ampliconic sequences (high density of genes and repeats) with highly conserved tracts maintained by gene conversion. Of the eight large palindromes at least six contain testes determining genes. The MSY includes 156 transcription units in its euchromatic sequences. Further, 78 of the protein-coding units correspond to nine different MSY-specific gene families, encoding at least 27 distinct proteins or protein families. The TTY2 gene family includes at least 26 members arranged in tandem repeats. They are transcribed in the testis and in the adult kidney but some of them are not translated. Evolutionists have suggested that the general low gene number is a consequence of the absence of recombination. R.A. Fisher assumed that the absence of recombination and genes is due to the fact that recombination would mess up the system of sex determination and would lead to intersexes. The diminution of the gene content of the Y chromosome has been attributed to Muller's ratchet. Molecular markers for the Y chromosome (>250) are increasing and it is now possible to trace paternal lines of evolutionary descent on the basis of variations in these chromosomes. This analysis is analogous to the use of mtDNA for the development of evidence for (mitochondrial) Eve's origin (Su B et al 1999 Am J Hum Genet 65:1718). Data on the Y chromosome should be carefully evaluated in pedigree analysis because the paternity may be equally likely for grandfather, brothers, cousins, etc., in the same family or even illegitimate male relatives. On the bases of the Y chromosomal constitution, 10 lineages appear to account for more than 95% of the current European human populations. The distribution of the Y haplotype shows more geographic than linguistic diversity because language influences were acquired more frequently than genes. The diversity of the human Y chromosome is lower than that of any other chromosome. The diversity between the human Y chromosome and that of the chimpanzee is larger (1.78%) than between the whole genomes where the difference is 1.23% (Kuroki Y et al 2006 Nature Genet 38:158). The Y chromosome of *Drosophila pseudoobscura* has no homology to that of *D. melanogaster* (Carvalho AB, Clark AG 2005 Science 307:108). ►sex determination, ►SRY, ►recombination variations of, ►Muller's ratchet, ►Eve foremother of molecular mtDNA, ►human evolution, ►F_{ST}, ►mutation rate, ►holandric genes, ►UEP, ►azoospermia, ►chimpanzee; for Thomas Jefferson's paternity analysis see Nature [Lond] 396:27; Nature [Lond] 397:32 and the rare K2 type Y chromosome (King TA et al 2007 Amer J Phys Anthropol 132:584; Owens K, King M-C 1999 Science 286:451; Kayser M et al 2001 Am J Hum Genet 68:173; Stumpf MPH, Goldstein DB 2001 Science 291:1738; Underhill PA et al 2001 Ann Hum Genet 65:43; Tilford CA et al 2001 Nature [Lond] 409:943; Hurles ME et al 2002 Genetics 160:289; Bachtrog D, Charlesworth B 2002 Nature [Lond] 416:323; Jobling MA, Tyler-Smith C 2003 Nature Rev Genet 4:598.

YAC: Refers to yeast artificial chromosome vectors equipped with a yeast centromere and some (*Tetrahymena*) telomeres in a linear plasmid containing selectable markers, ARS (autonomously replicating sequence) for the maintenance and propagation of eukaryotic DNA inserts in cloning of larger than 200 kb size sequences. YACs play an important role in identifying contigs in physical mapping of larger genomes, as well as in in situ hybridization, map based gene isolation, etc. The rate of instability of YAC has been estimated to be about 2%. In mitotic yeast cells

YACs behave like other chromosomes. Meiosis can be analyzed by tetrads although recombination appears to be reduced. YACs may be maintained through some cell divisions in the mouse cytoplasm and behave like *double minute* chromosomes or they may be integrated into the mouse chromosomes. ►anchoring, ►contigs, ►YAC library, ►in situ hybridization, ►chromosome walking, ►pulsed field gel electrophoresis, ►DM; Peterson KR 1999 Methods Enzymol 306:186; Brown WR et al 2000 Trends Biotechnol 18:218; Adam G et al 1997 Plant J 11:1349; Ragoussiz J, Monaco AP 1996 Methods Mol Biol 54:157, see Fig. Y1.

YAC Library: Contains large restriction fragments of genomic DNA cloned in YAC vectors and separated by pulsed field gel electrophoresis. A YAC library generally comprises 100 to 250 kb (or larger) size DNA fragments in multiple (at least five) copies if possible, covering the entire genome of an organism. Selecting the appropriate YAC clone for the purpose of finding the region of interest, the YAC library in yeast colony filter hybridization experiments is probed by RFLPs, PCR, inverse polymerase chain reaction probes, plasmid rescue or by other means. ►YAC, ►contigs, ►pulsed field gel electrophoresis, ►colony hybridization, ►RFLP, ►PCR, ►inverse polymerase chain reaction, ►plasmid rescue, ►probe, ►genome project; Larin Z et al 1997 Mol Biotechnol 8:147.

YAK (*Bos grunniens*): This is a "wild ox" and also an Asian draft animal, 2n = 60.

YAM (*Dioscorea* spp): A tropical food crop. The Asian and African species are x = 10 but the American are x = 9. The actual chromosome numbers vary from diploid to decaploid.

Yama: A Ced-3-like protease, also known as CPP32β. ►apoptosis

Yang Cycle: The pathway of biosynthesis of ethylene from 2-keto-4-methylthiobutyrate (KMB)→ S-adenosyl-L-methionine (AdoMet)→5′-methylthioadenosine (MTA)→5′-methylthioribose (MTR)→5′-methylthioribose-1-phosphate (MTR-1-P)→KMB. AdoMet is then converted to 1-amino-cyclopropane-1-carboxylic acid (ACC) by ACC synthase and ACC oxidase generates ethylene. ►ethylene,

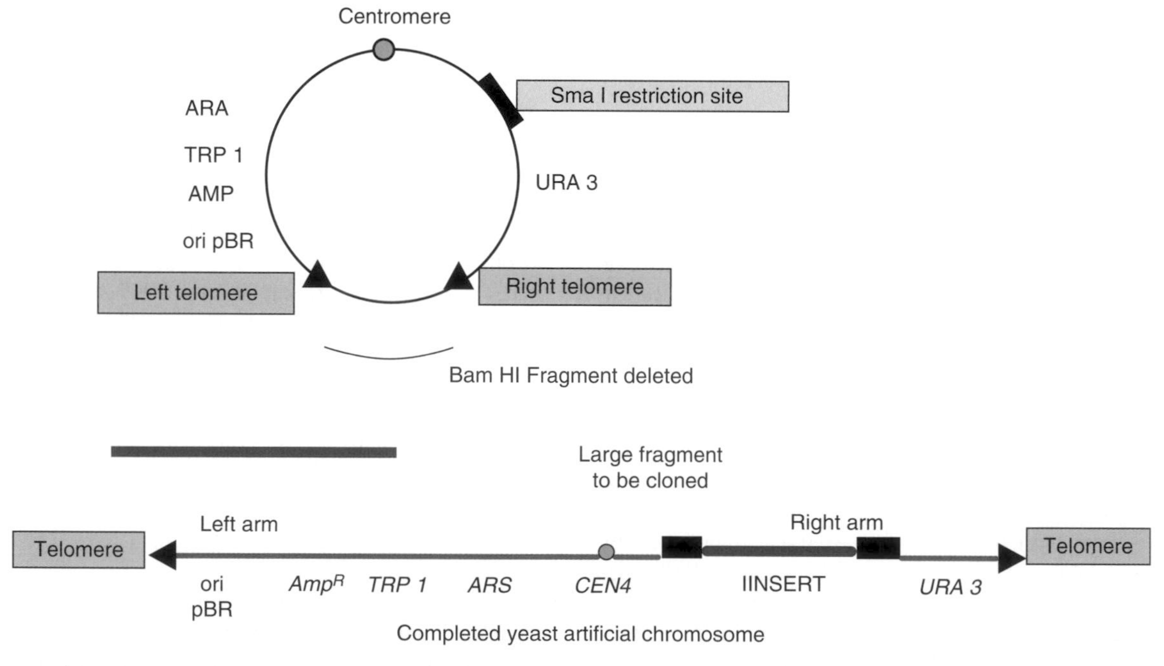

Figure Y1. Top: A circular yeast plasmid with *Tetrahymena* telomeres. A segment of the pBR322 prokaryotic plasmid contains the replicational origin and the Amp^R gene (selectable in *E. coli* for ampicillin resistance). The *ARS* (autonomous replication sequence), cloned centromere of yeast chromosome 4 (CEN 4), restriction enzyme recognition sites where the plasmid can be opened for insertion and later ligation are shown. The *TRP1* (tryptophan) AND *URA 3* (uracil) genes of yeast serve to ascertain that both arms are present and the transformant can synthesize tryptophan and uracil. A piece of the plasmid between the two Bam sites and also the yeast *HIS* (histidine) gene are deleted. Because of the pBR322 sequences, this vector replicates in *E. coli* as well as in eukaryotic cells. All details are not shown and the diagram is not to scale. Other YAC vectors are similar but not identical with this model. Bottom: Completed linear YAC. (Modified after Burke DT et al 1987 Science 236:806)

►plant hormones; AdoMet, Pardee AB 1987 J Cell Physiol Suppl 5:107.

YAP1: Refers to yeast transcription factor which regulates oxidative stress. ►oxidative stress; Wood MJ et al 2004 Nature [Lond] 430:917.

Yarowization: ►vernalization

YBP: Years before present.

Y Chromosomal Linkage: ►holandric genes

Ycp: ►yeast centromeric vectors

Ydj1: Refers to the yeast homolog of DnaJ. ►DnaJ, ►chaperones

Yeast (budding yeast): ►*Saccharomyces cerevisiae*

Yeast (fission yeast): ►*Schizosaccharomyces pombe*

Yeast Artificial Chromosomes: ►YAC

Yeast Centromeric Vectors (Ycp): These carry centromeres, telomeres and ARS, and they can be linear or circular. Their stability improves with increasing length (see Fig. Y2). ►YAC, ►vectors

Yeast Cell Density Test: Cell suspension at 0.1 O.D. (optical density) at 600 nm corresponds to ~3 × 10^6 cells per mL.

Yeast Episomal Vector: Carries the origin of replication of the 2 μm yeast plasmid and can be propagated independently in the cytoplasm or integrated into a chromosome. ►*Saccharomyces cerevisiae*, ►yeast plasmid 2 μm, ►vectors

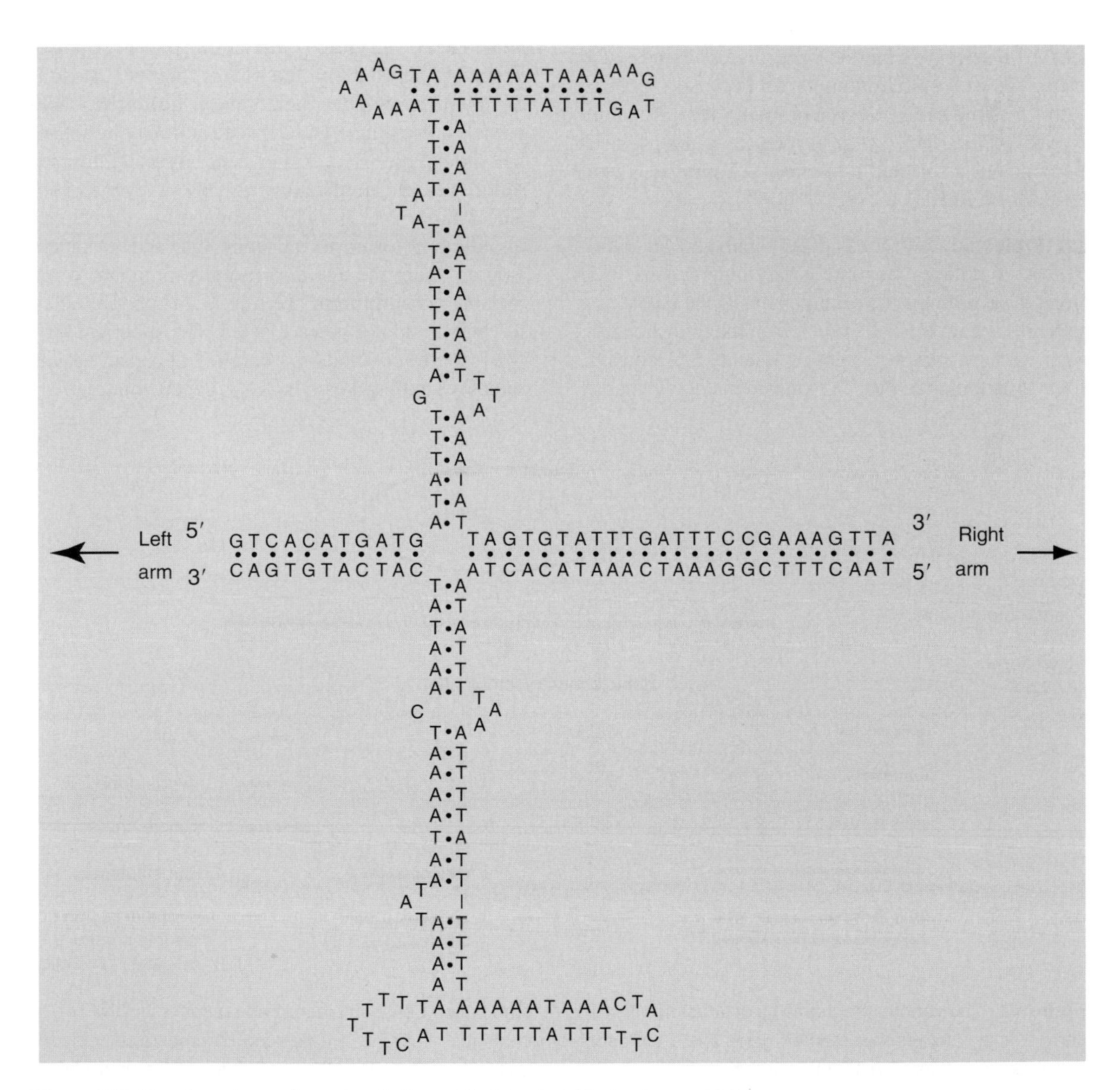

Figure Y2. Yeast centromeric region 3. From Clark L et al Stadler Symp 13:9

Yeast Hybrids: ►two-hybrid method, ►split-hybrid system, ►three-hybrid system, ►reverse two-hybrid system, ►four-hybrid system, ►one-hybrid binding assay

Yeast Integrating Vector (YI): Replicated only by the host, it behaves like a gene in the chromosome, can show homologous recombination, duplications and substitutions like bacterial episomes. Although it can generate chromosomal rearrangements its stability is extremely high but its ability to transform is extremely low. ►vectors

Yeast Plasmid 2 μm: This circular duplex DNA plasmid of 6,318 bp contains genes for replication and can thus maintain a copy number of about 50/cell but it lacks any selectable marker in native form. By attaching it to the pBR322 *E. coli* plasmid shuttle vectors (ca. 8.5 kbp) have been generated that carry the bacterial histidine operon (as selectable marker), and it is also expressed in bacteria as well as in budding yeast because a segment of this plasmid can serve as a promoter for the operon. ►yeast vectors, ►*Saccharomyces cerevisiae*, ►histidine operon, ►selectable marker; Velmurugan S et al 1998 Mol Cell Biol 18:7466.

Yeast Proteome: YPD has been analyzed as multiprotein complexes beyond binary interactions. The proteome represents the arrangement of the interacting networks within the cell and include the mitochondria. ►genetic networks, ►proteome; Wuchty S, Alamaas E 2005 Proteomics 5:444; Wiwatwattana N, Kumar A 2005 Nucleic Acids Res 33:D598; http://www.proteome.com/.

Yeast Replicating Vectors (Yrp): These carry autonomously replicating sequences (ARS), have moderate stability and relatively low copy number, and they may be integrated into chromosomes. ►vectors

Yeast Transformation: Can involve a number of different changes in the yeast chromosome at the site of transformation. (See Fig. Y3, ►transformation genetic fungal transformation).

Yeast Transposable Elements: ►Ty, ►Ω

Yellow Crescent: Cytoplasmic motions in the fertilized ascidian embryo may generate a yellowish area that is broken up and shared by the cleavage cells thus revealing the origin of the differentiating cell lineage.

Yellow Fever: A viral (*Flaviviridae*) tropical disease primarily caused by the *Aedes aegypti* mosquito. Symptoms include high fever, headache, aching muscles, vomiting and backache and may be followed by shock, bleeding, kidney and liver failure. Liver failure causes jaundice (giving yellow fever its name). The fatality rate is ~50%. Preventive measures are avoiding exposure to mosquito bite and vaccination. Infants under the age of 6 months, pregnant women, immunocompromised persons and people allergic to egg should not be vaccinated. The spherical virions (40–50 nm) covered by a polyhedral capsid are decorated by small spikes. The RNA genetic material of the

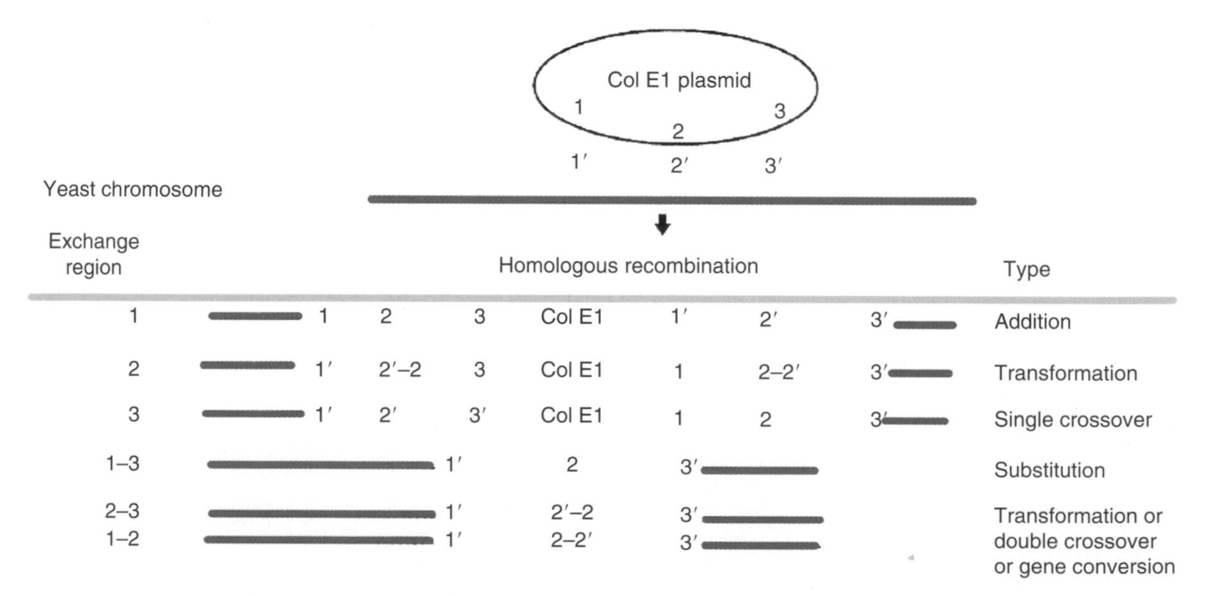

Figure Y3. The various mechanisms of transformation and recombination between plasmid and chrosomal DNA in yeast. Gene 2′ is an auxotrophic marker; gene 2 is a prototrophic homologue; 1′ and 3′ are flanking chromosomal sequences homologous to 1 and 2 in the plasmid with a Col E1 or other replicator. (Modified after Hinnen, Botstein & Davis in *Molecular Biology of the Yeast Saccharomyces*, Strathern, et al., eds. 1981 Cold Spring Harbor Lab. Press, Cold Spring Harbor, NY.)

virus consists of a single long open reading frame of 10,233 nucleotides, which can encode a polypeptide of 3,411 amino acids. The structural proteins are located within the amino-terminal 780 residues of this polyprotein; the remainder of the open reading frame consists of non-structural viral polypeptides. This genome organization implies that mature viral proteins are produced by post-translational cleavage of a polyprotein precursor (Rice CM et al 1985 Science 229:726).

Yep: ►yeast episomal vector

YEPD: Refers to yeast-peptone-dextrose culture medium.

Yersinia: A group of gram-negative bacteria responsible for bubonic plague (characterized by enlarged lymph nodes [bubo]), gastroenteritis, etc. in rodents and humans acquired through fleabites. Generally, but not always, infectiveness of the flea carriers may be blocked during the early stages of the incubation period (Eisen RJ et al 2006 Proc Natl Acad Sci USA 103:15380). The sequenced genome of *Y. pestis* includes a 4.65 Mb chromosome and three plasmids of 96.2, 70.3 and 9.6 kb. It contains 150 pseudogenes. The genome shows frequent insertion sequences and intragenic recombinations. The bacteria secrete the Yop protein that targets primarily dendritic cells, macrophages and neutrophils of the immune system but rarely the B and T lymphocytes (Marketon MM et al 2005 Science 309:1739). YopJ acetylates and inhibits MAPK kinase activation and inhibits NF-κB signaling, thus disarming the innate immune system. YopJ acts as an acetyltransferase, using acetyl CoA, and modifies critical serine and threonine residues in the activation loop of MAPKK6 and prevents phosphorylation (Mukherjee S et al 2006 Science 312:1211). Yersinia protein kinase A (YpkA) disrupts the eukaryotic actin cytoskeleton. Its Rac1 binding domain mimics the host guanidine nucleotide dissociation inhibitors of Rho GTPases (Prehna G et al 2006 Cell 126:869). Pneumonic plague development requires the presence of plasminogen-activating protease (Lathem WW et al 2007 Science 315:509). ►virulence, ►biological weapons, ►plague, ►plant vaccines, ►herpes, ►cytoskeleton, ►Rac, ►RHO; Parkhill J et al 2001 Nature [Lond] 413:323; Hinnebusch BJ et al 2002 Science 296:733; Cornelis GR 2002 Nature Rev Mol Cell Biol 3:742; genome: http://bbrp.llnl.gov/bbrp/html/microbe.html.

YES: Refers to the yeast – *E. coli* shuttle vector. ►shuttle vector; Elledge SJ et al 1991 Proc Natl Acad Sci USA 88:1731.

YES1 Oncogene: This is in human chromosome 18q21. Its protein product is homologous to that of Roux' sarcoma virus (SRC) and it is also a protein tyrosine kinase. In association with other proteins (YAP), it acts as a transcriptional co-activator. ►tyrosine kinases, ►oncogenes, ►palmitoylation

Y-Family DNA Polymerases: These are error-prone enzymes such as the prokaryotic UmuD'$_2$C (pol v), Din B (pol IV) and the yeast proteins Rev1 (cooperating with polζ) and Rad30. They copy damaged DNA efficiently, make frequent incorporation error ($\sim 10^{-1}$ to 10^{-3}), rely on mispairing, use mismatches, misaligned templates, etc. Polη is defective in xeroderma pigmentosum-V. Polι incorporates nucleotides opposite purines much more efficiently and faithfully than against pyrimidines. Human Polι incorporates dCTP opposite template G through the G.C + Hoogsteen base pair (Nair DT et al 2005 Structure 13:1569). It has the ability to bypass deaminated cytosine but it is error prone when the DNA is undamaged and it inserts G opposite to T or U templates; it appears to be responsible for somatic hypermutation. Human Polι replicates the DNA by Hoogsteen base pairing (Nair DT et al 2004 Nature [Lond] 430:377). Polκ is activated by an arylhydrocarbon receptor (benzo[a]pyrene) and may be a repair enzyme by bypassing DNA lesions but it may insert A across abasic sites. The Rev1 polymerase uses a protein template. ►DNA repair, ►error-prone repair, ►DNA polymerases, ►error in replication, ►X family of DNA polymerases, ►Rev1, ►Hoogsteen pairing; Goodman MF 2002 Annu Rev Biochem 71:17; Rattray AJ, Strathern JN 2003 Annu Rev Genet 37:31.

YFP (yellow fluorescent protein): Used as a vital stain and is studied by FRET. ►FRET; Galperin E, Sorkin A 2003 J Cell Sci 116:4799.

YI: ►yeast integrating vector

yIF-2: Refers to the yeast eIF-5B; it is homologous to the prokaryotic IF2. ►eIF-5B

Yin Yang (YY1, NF-E1): A multifunctional zinc finger transcription factor of growth factors, hormones and cytokines. YY1 is also a potential negative regulator of the tumor suppressor p53 (Sui G et al 2004 Cell 117:859). ►DNA-binding protein domains, ►animal hormones, ►cytokines; Santiago FS et al 2001 J Biol Chem 276:41143.

Yin Yang Haplotypes: These two haplotypes display differences in each SNIPs. ►haplotype, ►SNIPs; Zhang J et al 2003 Am J Hum Genet 73:1073.

Yin-Yang Crosses: A QTL is mapped by crossing to strains A and B. New strains are expected to segregate in crosses with one (A) or the other (B) and thus QTL fine mapping becomes feasible. ►QTL; Shifman S, Darvasi A 2005 Genetics 169:849.

Y-Linked: ►holandric gene

Yohimbine: An alkaloid produced by African plants of the *Rubiaceae* and *Apocinaceae* families; their extracts have been used as an adrenergic blocking medicine for arteriosclerosis and hypertension. It is supposedly an aphrodisiac (see Fig. Y4). ►ferritin, ►aphrodisiac; Morales A 2001 World J Urol 19(4):251.

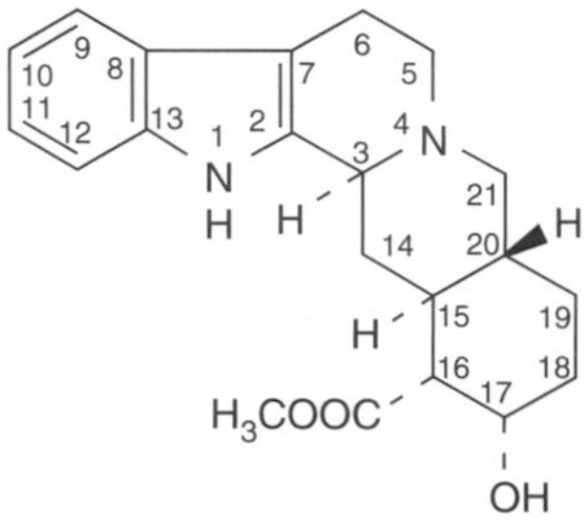

Figure Y4. Yohimbine

Yolk: The complex nutrients embedding the animal egg. ►egg, ►vitellogenin

YPD: ►yeast proteome

YPD: A yeast nutrient medium containing g/L yeast extract 10, glucose 20, Bacto-Peptone 10 or 20.

YPGE: A nutrient medium containing yeast extract (10 g), Bacto-Peptone (10 g), glycerol (20 g) and ethanol (10 g) per liter of H_2O.

YPL.db (Yeast Protein Localization database): http://ypl.tugraz.at.

YPT: A homolog of Sec and RAB. ►Sec, ►RAB

Y RNAs: These have a phylogenetically conserved secondary structure consisting of at least three stems and small internal loops. In humans, Y1, Y3, Y4 and Y5 have been identified but their function is unclear. It is suspected that they regulate ribosomal protein synthesis.

Yrp: ►yeast replicating vectors

YT Blood Group: Coded in human chromosome 7q.

YT Bacterial Medium: H_2O 900 mL, bacto-tryptone 16 g, bacto-yeast extract 10 g, NaCl 5 g, pH 7.0 (adjusted by 5N NaOH), filled up to 1 L and diluted to half before use.

Yttrium (Y): An extremely rare metal; the ^{90}Y has a half-life of 64 hours and has been used for internally exposing cancerous tissues to β-radiation. ►ionizing radiation, ►isotopes, ►magic bullet

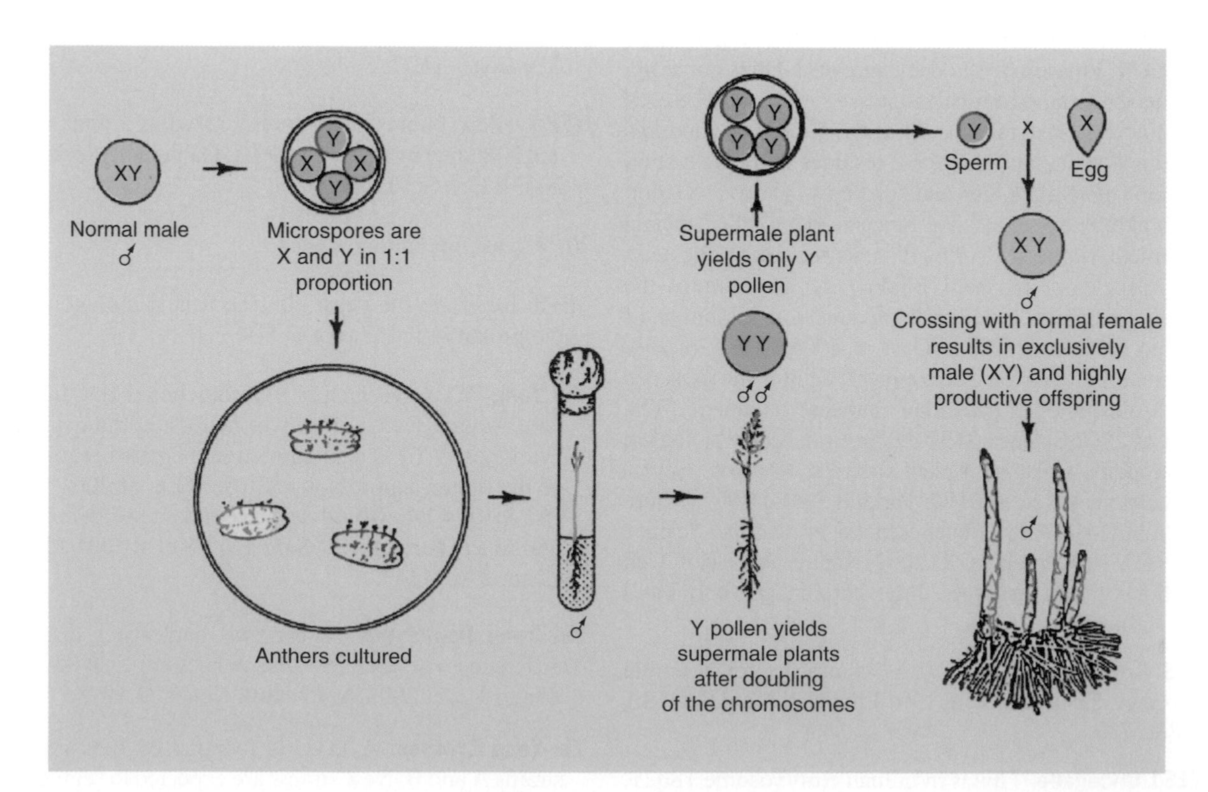

Figure Y5. Production of (YY) all-male asparagus

Yuasa Oncogene: Its human homolog is in chromosome 7.

YY Asparagus: By regeneration of plants from microspores and chromosome doubling, plants can be obtained with this chromosomal constitution. The practical advantage of this vegetatively reproducible plant is the approximately 30% higher yield of edible spears. (See Fig. Y5, for embryogenesis, somatic).

Historical vignettes

Oscar Hertwig on genetics: "The hypothetical idioblasts… are according to their different composition, the bearers of different properties, and produce by direct action, or by various methods of cooperation, the countless morphological and physiological phenomena, which we perceive in the organic world. Metaphorically they can be compared to the letters of the alphabet, which, though small in number, when combined form words, which in their turn, combine to form sentences or to sounds, which produce endless harmonies by their periodic sequence and simultaneous combination." (*The Cell, Macmillan, New York,* 1895, *page* 340)

Although Friedrich Miescher a contemporary of Mendel described nucleic acids in 1871, there remained an almost century-long disconnect between his findings and Mendel's. It is interesting to note that the nature of the mechanisms of fertilization remained an enigma even to Miescher. "So werden wir von allen seiten genöthigt, es mit Bestimmheit auszusprechen: *Es giebt keine spezifischen Befruchtungsstoffe*. Die chemischen Thatsachen haben secundäre Bedeutung; sie sind einem höheren Gesichtspunkt untergeordnet." (Miescher. 1874). [There is no specific fertilization-substance. The chemical realities have secondary meaning; they are subject to a higher order of strategy.] (Miescher F 1874 Die Spermatozoen einiger Wirbelthiere. Ein Beitrag zur Histochemie. Verhandl Naturforsch Ges Basel 6, (issue I):138–208.)

Y

Z

Z (or z or Z score): The standard normal probability density function calculated by the formula:

$$Z = \frac{1}{\sigma\sqrt{2\pi}} e - (Y - \mu)^2/2\sigma^2$$

where Y is the normal variate, $\pi \cong 3.14159$ and $e \cong 2.71828$, μ is the mean and σ is the standard deviation. The z values are generally read from statistical tables. Z indicates the height of the ordinate of the curve and thereby the density of the items. ►standard deviation, ►normal distribution, ►confidence intervals, ►F distribution

Z Buffer: $Na_2HPO_4.7H_2O$ 0.06 M, NaH_2PO_4 0.04 M, KCl 0.01 M, $MgSO_4.7H_2O$ 0.001 M, β-mercaptoethanol 0.05 M, pH 7. Do not autoclave it.

Z Chromosome: Refers to the sex chromosome present in both sexes of a species with heterogametic females (thus comparable to the X chromosome). Males are ZZ and females are WZ. ►W chromosome, ►sex determination

Z Disc (Z-disk): ►sarcomere

Z Distribution: A statistical device for testing the significance of the differences between correlation coefficients in case the null hypothesis is not r = 0. The relation of z to r has been elaborated by R.A. Fisher as $z = (1/2)[ln(1 + r) - ln(1 - r)]$ and its standard error as $\sigma_z = \frac{1}{\sqrt{(n-3)}}$. For routine calculations, tables are available in statistical textbooks. ►covariance for correlation coefficient

Z DNA (zig-zag DNA): A relatively rare left-handed double helix that may be formed in the short (8–62 bp) regions of alternating purines and pyrimidines (ATGTGTGT, GCATGCAT). The polyGC.polyGC sequences are most favorable for B→Z transition. In some species CA/TG repeats are most conducive for Z DNA formation. Alternating purine-pyrimidine tracts may also modulate the transcription in a plus or minus direction. Base sequences near the transcription starting point favor the formation of Z DNA and it may stimulate transcription. Nucleosomes are not formed in the Z DNA tract, thus facilitating the access of transcription factors. In some genes, however, such as in the constitutive nucleolin gene, Z DNA in the promoter downregulates transcription. Vaccinia virus is no longer lethal in mice if the Z DNA binding site is lost. Four families of proteins bind to the crystal structure of Z DNA: ADAR1, an editing enzyme, LM1, an interferon inducible protein, E3L, a pox virus virulence factor and an orthologue of PKR, interferon-induced protein kinase. Z DNA occurs in a dynamic state; it is formed and then it can revert to the B form of the DNA. There is a transition point between the common B and Z DNA forms (see Fig. Z1). The handedness changes abruptly from left to right by a sharp turn of the phosphate backbone. At the transition point an adenine (A0) and thymine (T0), respectively, are extruded as shown by the crystal structure (Ha SC et al 2005 Nature [Lond] 437:1183). B-Z DNA conformational polymorphism is optically detectable on single-walled carbon nanotubes in whole blood, tissue and within living mammalian cells (Heller DA et al 2006 Science 311:508). The figure shows space-filling models (modified after Dickerson RE et al 1982 Science 216:475). In mammals, Z DNA forming sequences induce a high level of chromosomal instability by generating double-strand breaks and large deletions. These deletions at the Z DNA may raise the incidence of certain cancers by increased transcription (Wang G et al 2006 Proc Natl Acad Sci USA 103:2677). In bacteria, the Z DNA related deletions are short.

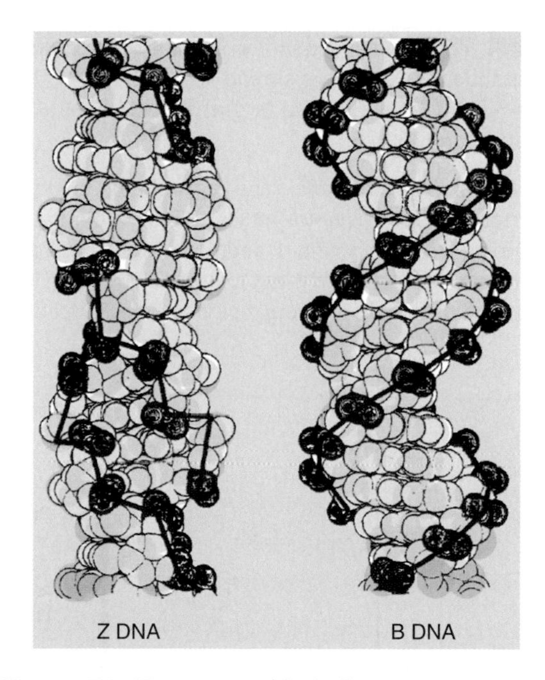

Figure Z1. The heavy black line represents the sugar-phosphate backbone

An alternating deoxycytidine-deoxyguanosine dinucleotide repetitive sequence [d(CG)*n*] can potentiate transcription in yeast when placed approximately three helical turns (28 bp) upstream of the cytochrome c 1 (CYC1) TATA box. Transcriptional activation by the d(CG)9 repeat sequence depends on the Z conformation. The activation is core promoter-specific and most effective when situated 28 bp or below upstream of the TATA box. Changing the distance between the d(CG)9 repeat sequence and the TATA box modulates the extent of activation. Linker-DNA region is formed at the Z DNA structure with two flanking nucleosomes. Z DNA creates an open chromatin state at the promoter by displacing nucleosomes from its environs and establishing the boundaries of its neighboring nucleosomes (Wong B et al 2007 Proc Natl Acad Sci USA 104:2229). ►DNA types, ►ADAR, ►RNA editing, ►nanotechnology; Liu R et al 2001 Cell 106:309; Rothenburg S et al 2001 Proc Natl Acad Sci USA 98:8985; Rich A, Zhang S 2003 Nature Rev Genet 4:566.

Z Inactivation: The Z chromosome of birds has the same role in sex determination as the X chromosome in mammals yet its genes are not inactivated in homogametic individuals. ►dosage compensation; Kuroda Y et al 2001 Chromosome Res 9:457.

Z Linkage: Refers to genes syntenic in the Z chromosome. ►X-linked

Z Ring: Same as Fts Z ring. ►tubulin

Z RNA: This left-handed molecule may occur in double-stranded RNA and is similar to Z DNA. ►Z DNA, ►ADAR; structure and origin: Placido D et al 2007 Structure 15:395.

Z Scheme: An additional (zig-zag) scheme to generate enough ATP by photosynthesis. Through a two-step process (photosystem I and II) an electron passes from water but there is not enough energy in a single quantum of light to energize the electron directly and efficiently all the way from PS II to the top of PS I and make $NADP^+$. The leftover energy makes pumping H^+ possible across the membranes to capture some light energy for synthesizing ATP. The redox state of plastoquinone regulates the transcription of genes encoding the reaction center proteins of both photosystem I and II. ►photosystems; Prince RC 1996 Trends Biochem Sci 21:121; Allen JF 2003 Science 299:1530; Allen JF, Martin W 2007 Nature [Lond] 445:610, see Fig. Z2.

Z Score: ►Z

z Value: The natural logarithm of the ratio of two estimated standard deviations. ►variance

ZAG (Zn-α_2-glycoprotein): Though this resembles MHC class I heavy chain molecules, it is different because it cannot bind β_2-microglobulin. It occurs in the majority of body fluids and is apparently involved in inducing fat loss in adipocytes. It accumulates in breast cancer cells and other cells in serious distress (cachexia). ►MHC, ►HLA, ►obesity; Kennedy MW et al 2001 J Biol Chem 276:35008.

Zanier-Roubicek Syndrome: ►ectodermal dysplasia, ►hypohidrotic

ZAK (zipper sterile-α motif kinase): A mixed lineage kinase (MLK) in signal transduction.

ZAP-70 (zeta-associated protein 70): A cytosolic protein tyrosine kinase expressed only in T cells and natural killer cells. By binding to phosphorylated ζ-chains of the CD3 T cell antigen-receptor complex, it assists in the activation of T cells. Its defect may lead to severe combined immunodeficiency. ZAP-70 tyrosine kinase signals to the CXCR4 chemokine receptor

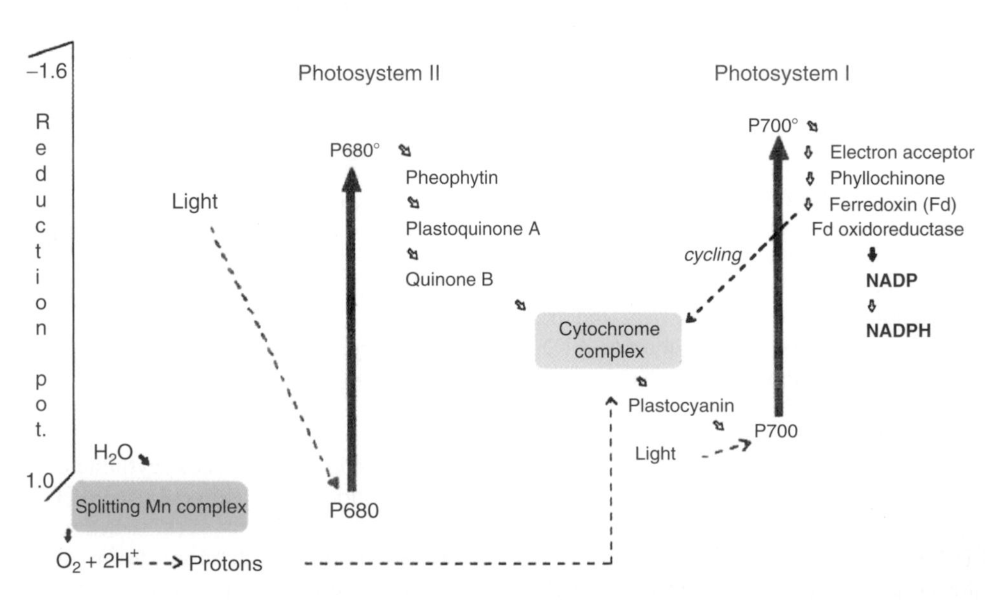

Figure Z2. Z Scheme of photosystems

Z

and thus regulates the migration of T lymphocytes. It seems to be involved in arthritis in mice. ►SHP-1, ►CD3 T cell, ►CXCR, ►killer cell, ►immunodeficiency, ►tyrosine kinases, ►retroviral restriction factors; Ottoson NC et al 2001 J Immunol 167:1857; Sakaguchi N et al 2003 Nature [Lond] 426:454, crystal structure: Deindl S et al 2007 Cell 129:735.

ZBP1 (Zipcode binding protein): ►mRNA migration

Zea mays (*L.*): ►maize

Zeatin (6[4-hydroxy-3-methyl-cis-2-butenylamino] purine): Refers to a cytokinin plant hormone. In *Agrobacterium* T-DNA-located *ipt* gene encodes an isopentenyl transferase that mediates the synthesis of transzeatin and isopentenyl adenosine. In plant tissue culture media it is a commonly used alternative to kinetin, benzylamino purine or isopentenyl adenosine. ►plant hormones

Zebra: *Equus quagga* 2n = 44; it can form hybrids with both horses (Equus caballus, 2n = 64) and donkeys (2n = 62). *Equus grevyi*, 2n = 46 (see Fig. Z3); *Equus zebra hartmanniae*, 2n = 32. The Zebra duiker (*Cephalophus zebra*, 2n = 58) is not a member of the Equidae family but is the male member of a bovine species.

Figure Z3. Equus grevyi

ZEBRA (Zta): A non-acidic activator protein of the lytic cycle of the Epstein-Barr virus; it promotes the assembly of the DA complex (TFIID-TFIIA) of transcription factors. ►transcription factors, ►open transcription complex, ►DA

Zebrafish (*Brachydanio rerio*, 2n = 50): A 3–4 cm tropical freshwater fish; the genome size is about 2×10^9 bp (see Fig. Z4). It is easy to breed and becomes sexually mature in 2–3 months (4 generations/year). The embryogenetic pattern is laid down in 12 hours, and it is well suited for the analysis of developmental pathways and cell lineages in this small vertebrate. Hundreds of eggs are laid externally. The embryos are transparent and permit observation of gastrulation, development of the brain and heart, etc. Haploids survive for several hours and mutants are available. About 600 genes involved in its development have been identified, and in 1998 a 3350 cM map of 3.3 cM resolution became available. The zebrafish genome has extensive homology to that of humans. ►EP, ►medaka, ►GloFish; Fishman MC 1997 Methods Cell Biol 52:67; Detrich HW III et al 1999 (Eds.) The ebrafish: Biology. Academic Press; radiation hybrid map: Geisler R et al 1999 Nature Genet 23: 86; Hukriede NA et al 1999 Proc Natl Acad Sci USA 96:9745; Woods IG et al 2000 Genome Res 10:1903; Shin JT, Fishman MC 2002 Annu Rev Genomics Hum Genet 3:311; hemetopoiesis: de Jong JLO, Zon LI 2005 Annu Rev Genet 39:481; human disease model: Lieschke GJ, Currie PD 2007 Nature Rev Genet 8:353; http://zebra.sc.edu; http://zfish.uoregon.edu/ZFIN/; http://zfin.org; http://depts.washington.edu/fishscop/; http://www.tigr.org/tdb/tgi.shtml; http://zfin.org; http://zf-espresso.tuebingen.mpg.de.

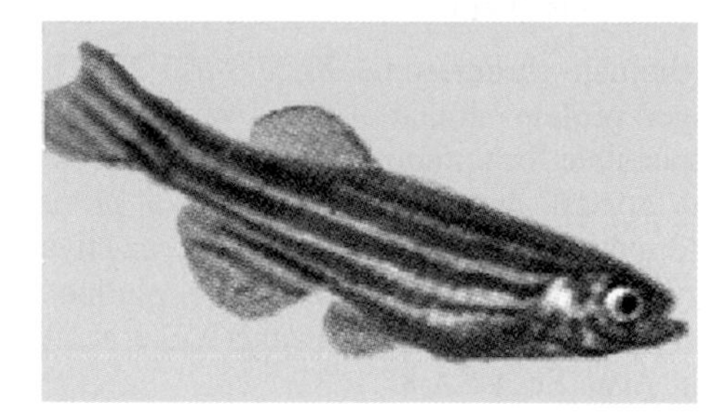

Figure Z4. Zebrafish

Zebu (*Bos indicus*): 2n = 60. ►Santa Gertrudis cattle

Zein: A prolamine protein in maize (homologous to gliadin in wheat) that may account for up to 50% of the grain proteins. It does not have high nutritional value (and is therefore undesirable) because of the low lysine and tryptophan content. It is deposited in zein bodies at the place of synthesis. In the high-lysine maize varieties (*opaque*, *floury*) prolamines are very low and non-prolamine proteins increase. ►high lysine corn, ►glutenin

Zeitgeber: A rhythmic external signal for the circadian change. ►circadian rhythm

Zeitnehmer: An internal signal for rhythmicity in response to the zeitgeber or even in its absence. ►circadian rhythm, ►zeitgeber

Zellweger Syndrome (ZS): A brain-liver-kidney (cerebrohepatorenal) recessive disease involving human chromosome 7q11.12-q11.13; but over a dozen other loci may have similar effects. The basic defect is due to peroxisome anomalies. ►microbodies, ►neuromuscular diseases, ►chondrodysplasia punctata, ►cataract, ►pseudo-Zellweger syndrome

Zero, Absolute: The minimum lowest temperature; Kelvin 0° = Celsius − 273.15°.

Zero Time Binding: Refers to the status of reassociation of two single-strand palindromic DNAs at the beginning of an annealing kinetics experiment. These are the fastest reassociating fractions because they are repeats and are close to each other. ►c_0t curve

zeste (*z*, chromosome 1.10 of *Drosophila*): The protein kinase product apparently alters the chromatin structure and affects the expression of *w*, *Ubx*, *dpp* by attaching to their promoters. Its action bears similarity to *Polycomb*. The zeste-white3 complex regulates the spindle attachment to the cortical actin. ►*w* locus, ►*Polycomb*; McCartney BM et al 2001 Nature Cell Biol 3:933.

Zeugopodium: Corresponds to the radius, the ulna, the tibia and the fibula.

ZFX (human chromosome Xp22.3-p21.2): This Zinc finger protein similar to Zfy was believed to be responsible for feminization of some XY individuals and gondal dysgenesis. It is not a primary sex determining protein. ZFX escapes inactivation in humans but not in mice. It controls pluripotency of stem cells. ►pluripotency; Palmer MS et al 1990 Proc Natl Acad Sci USA 87:1681.

ZFY (zinc finger Y, Yp11.3): A sequence in the Y chromosome assumed to be involved in the maturation of testes or sperm. A 729 bp intron located immediately upstream of the Zinc finger exon shows no or ery little sequence variation in worldwide human samples. ►Zinc finger, ►ZFX

Zidovudin: ►AZT

zif: ►NGFI-A

Zig-Zag Inheritance: ►criss-cross inheritance

Zig-Zag Model of Chromosome Fiber: Radiation-induced DNA breaks revealed fragment sizes of 78 bases, which corresponds to one turn of the DNA around the nucleosome. Additional peaks between 175 and 450 bases reflect the relative position of the nearest neighbor nucleosomes. Calculations and other considerations seem to support a zig-zag model of the chromatin fiber rather than a simple helical model (Rydberg B et al 1998 J Mol Biol 284:71). ►solenoid

Zinc Finger Nucleases: These are limited-specificity nucleases that can recognize certain DNA sequences and cause small deletions at selected targets. The Zn fingers prefer guanine-rich tracts, particularly 5′-GNN-3′ triplets. Each nuclease unit may have three fingers and each grabs one triplet, i.e., a total of nine bases. The dimeric structure of the nuclease cleaves both strands in a tract lying between the G-rich sequences (see Fig. Z5). Non-homologous end-joining may follow the double-strand breaks and thus somatic and germ line mutations may occur. Zinc finger nucleases are associations of the Zinc finger with, e.g., a FokI restriction endonuclease (target GGATG(N)$_{9/13}$) and can thus be targeted to specific mammalian genes (Porteus MH 2006 Mol Ther 13:438) or can be used to repair specific defective genes in plants by targeted homologous recombination (Wright DA 2005 Plant J 44:693). ►non-homologous end-joining, ►gene therapy; Bibikova M et al 2002 Genetics 161:1169; Miller JC et al 2007 Nature Biotechnol 25:778.

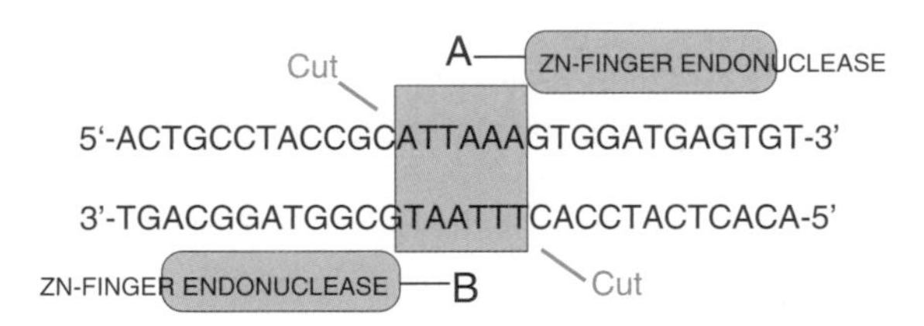

Figure Z5. Diagram of a dimeric (A and B) zinc-finger endonuclease and its mode of cleavage (at the green lines) of DNA. The subunits recognize of the DNA sequence by the green shaded square domain. The zinc-fingers bind to the G-rich repeats. (Modified after Bibikova M et al 2002 Genetics 161:1169)

Zinc Fingers: These are binding mechanisms of transcription factors and other regulatory proteins containing tandemly repeated cysteine and histidine molecules and they fold in a "finger-like" fashion cross-linked to Zn. They contain other highly conserved amino acids—phenylalanine (F), leucine (L) and tyrosine (Y). Some Zinc finger proteins bind the RNA to the DNA. About 0.7% of the proteins in budding yeast and the nematode *Caenorhabditis elegans* contain Zn finger motifs; in yeast most commonly 2/molecule and in *Caenorhabditis* up to 14/molecule. Using DNA microarray designs, putative Zn finger transcription factor binding sites can be identified. By designing new Zinc finger motifs, specific regulation of gene expression is possible (Reynolds L et al 2002 Proc Natl Acad Sci USA 100:1615). Zinc is an essential element for life and it is a cofactor for several enzymes. ►hormone receptor, ►DNA-binding protein domains, ►microarray hybridization, ►acrodermatitis enteropathica, ►ZFY, ►RING finger, ►binuclear Zinc cluster, ►Cys_4 receptor, ►GATA, ►LIM; Rubin GM et al 2000 Science 287:2204; Bulyk ML et al 2001 Proc Natl Acad Sci USA 98:7158; Pabo CO et al 2001 Annu Rev Biochem 70:313; binding tools: http://www.scripps.edu/mb/barbas/zfdesign/zfdesignhome.php; Zn finger protein design: http://bindr.gdcb.iastate.edu/ZiFiT/.

Zinc Knuckle (Zf-CCHC): A Zn-binding motif commonly found in the nucleocapsids of retroviruses (e.g., HIV) and in regulatory proteins of diverse eukaryotes. Its sequence is CX2CX4HX4C where X can be different amino acids. These protein motifs are involved in binding RNA or single-strand DNA. (See Le Thuy et al 2001 Mol Cell Biol 21:8346).

Zinc Ribbon Fold (CXXC(H)-15/17-CXXC): This is present in diverse binding proteins. ►binding proteins, ►transcription factors

Zinc Ring Finger: ►RING finger

ZIP1, ZIP2: Protein factors that mediate phosphorylation of potassium ion channels by protein kinase C. ►protein kinase, ►ion channel

Zipcode: ►mRNA migration

Zipper Domain: Refers to part of dimeric DNA-binding proteins where the two subunits are held together by repeating amino acid residues; in the other parts the subunits are separated from each other. ►leucine zipper

Zoidogamy: In this fertilization takes place by motile antherozoids. ►siphonogamy

Zollinger-Ellison Syndrome (Wermer syndrome MEN 1): A multiple endocrine adenomatosis (neoplasia) encoded in human chromosome 11q13. ►adenomatosis multiple endocrine

Zona Pellucida: A yolky layer around the mammalian egg. Zona pellucida domains are present in dauer larvae of nematodes and mechanotransducers in *Drosophila* and are involved in other functions. The domain ~260 amino acids have eight conserved cysteines near the carboxyl end and are often glycosylated. ►fertilization [animals], ►vitelline layer, ►egg, ►dauer larva, ►mechanosensory genes; Jovine L et al 2002 Nature Cell Biol 4:457; Jovine L et al 2005 Annu Rev Biochem 74:83.

Zonoskeleton: This includes the scapula, the clavicle and the hip bones.

Zonula Occludens: A tight cell junction mediated by ZO proteins associated with a 210–255 kDa membrane-tied guanylate kinase. ►MAGUK, ►ELL; Meyer TN et al 2002 J Biol Chem 277:24855.

Zoo Blot: Southern hybridization experiments with probes derived from a variety of different species to test potential homologies. (See Rijkers T, Ruther U 1996 Biochim Biphys Acta 1307:294).

Zoo FISH: Fluorescent in situ hybridization maps for several species to study the evolutionary relations of their chromosomes. ►FISH

Zoonosis (zoonotic infection): Animal pathogens transferred to humans (by e.g., grafts, food) may lead to the development of new human diseases under natural conditions. ►grafting in medicine, ►acquired immunodeficiency, ►xenotransplantation, ►Dengue fever, ►Ebola virus, ►Pox Virus, ►West Nile Virus, ►sleeping sickness, ►influenza, ►SARS, ►encephalopathies, ►mycobacteria, ►plague, ►laboratory safety; Weiss RA 1998 Nature Med 4:391.

Zoospore: This is a motile (swimming) spore.

Zootype Hypothesis: According to this, HOX-type homeobox genes are present in all metazoa. ►homeotic genes

ZP1, ZP2, ZP3: Refer to zona pellucida glycoproteins. ►fertilization, ►zona pellucida

ZPA (zone of polarizing activity): Determines the anterior/posterior differentiation of the limbs and is located behind AER in the limb bud. ►AER, ►limb bud

ZPK: Denotes a leucine zipper protein kinase. ►DNA-binding protein domains

ZPR: Refers to zinc finger protein binding to the epidermal growth factor receptor. ►zinc fingers, ►EGFR

Zta: ►ZEBRA

Zuotin: Z DNA and tRNA binding protein of yeast. It has structural homology with DnaJ and is functionally related to the mammalian chaperone MIDA1, required for cellular growth. ►chaperones, ►DnaK, ►MIDA1; Braun EL, Grotewold E 2001 Mol Biol Evol 18:1401.

Zwischenferment: ►G6PD

Zwitterion: A dipolar ion with separated positive and negative poles.

ZygDNA: It has been reported that 0.1 to 0.2% of the eukaryotic DNA may not be replicated until late leptotene–zygotene. This delayed replication involves dispersed 4–10 kb stretches and it has been assumed that these segments (ZygDNA) code for genes with products aiding chromosome pairing. It is assumed that a leptotene (L) lipoprotein is involved in the delayed replication. (L) ►meiosis; Hotta Y et al 1985 Cell 40:785.

Zygomeres: Refer to hypothetical initiators of chromosome pairing in the DNA.

Zygomorphic: A structure of bilateral symmetry, like a snapdragon flower (see Fig. Z6).

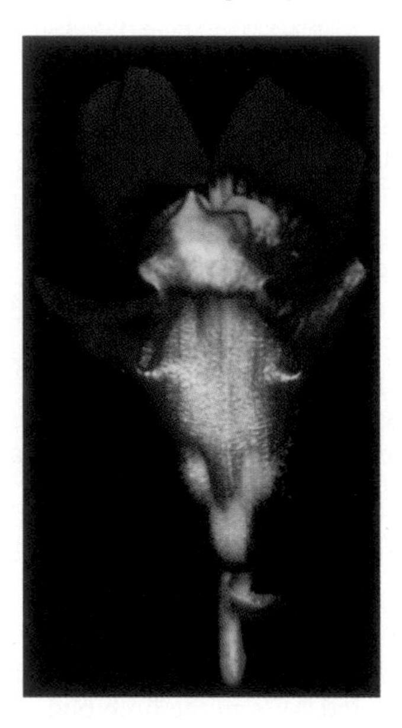

Figure Z6. Zygomorphic. Courtesy of Dr. Z. Schwarz-Sommer

Zygonema: Denotes the chromosome at the zygotene stage. ►meiosis, ►zygotene stage

Zygosis: Twins can be identical (monozygotic, MZ) or non-identical (dizygotic, DZ). The distinction is not always simple because dizygotic twins (like any siblings) may show several identical features, depending on the genetic constitution (consanguinity) of the parents. If the probability of monozygosis between twins with identity in a genetic marker is designated as $P(A_1/B)$ where A_1 and B are different markers and the probability of the twins being either dizygotic or erroneously assumed to be monozygotic is: $1 - P(A_1/B)$. The calculation may be based on the formula:

$$P(A_1/B) = \frac{1}{1 + [Q \times L]}$$

where Q is the dizygotic:monozygotic proportion in the population (DZ/MZ), and L = likelihood ratio of the conditional probabilities for DZ and MZ twins would be identical for a particular genetic condition. The conditional probabilities that if one of the twins is of a particular dominant type, the second would also be of the same type at dizygosis is 0.5 – 1.0, and at monozygosis 1.0. In the case of recessive markers, these probabilities are 0.25 and 1.0, respectively. The probabilities also depend on the genetic constitution of the parents. Recessive markers may be expressed only if both parents carry that particular allele. If either of the parents is homozygous for a dominant marker, then both twins must carry that marker, irrespective of zygosity. L can be computed as $L = \frac{P[DZ]}{P[MZ]} L_1 \times L_2 \ldots L_n$ where $\frac{P[DZ]}{P[MZ]}$ is the empirical probability of the DZ:MZ proportions in the general population, and $L_1 \times L_2 \ldots L_n$ are the conditional probabilities for the genetic markers 1 to *n* used, either 0.25 or 1.0. Further complications may arise if either the penetrance or expressivity of the markers varies. The DZ:MZ proportions may vary generally from 0.65:0.35 to 0.70:0.30 but are somewhat different in various ethnic groups, depending on the age of the mother, the use of fertility drugs and artificial insemination, etc. DNA markers, either by RFLP or the PCR method of typing, can better resolve the problem than using blood types or other genetic analyses. ►conditional probability, ►likelihood, ►RFLP, ►PCR, ►DNA fingerprinting, ►twinning, ►monozygotic, ►dizygotic, ►concordance, ►discordance

Zygosity: ►zygosis

Zygospore: This is formed by the fusion of two spores or two multinucleate gametangia. ►heterothallism, ►homothallism, ►gametangia

Zygote: A cell resulting from the union of two gametes of opposite sexes. ►gamete

Zygotene Stage: This stage of meiosis occurs when the homologous chromosomes (supposedly) begin to synapse. The pairing usually begins at the termini and proceeds toward the centromeric region. At this stage the synaptonemal complex is detectable by electronmicroscopy. The intimate bivalent pairing appears to be a requisite for chiasma formation as well as for genetic recombination (see Fig. Z7). ►meiosis, ►chiasma, ►synaptonemal complex, ►association point, illustration after K. Bela ř.

Figure Z7. Zygotene

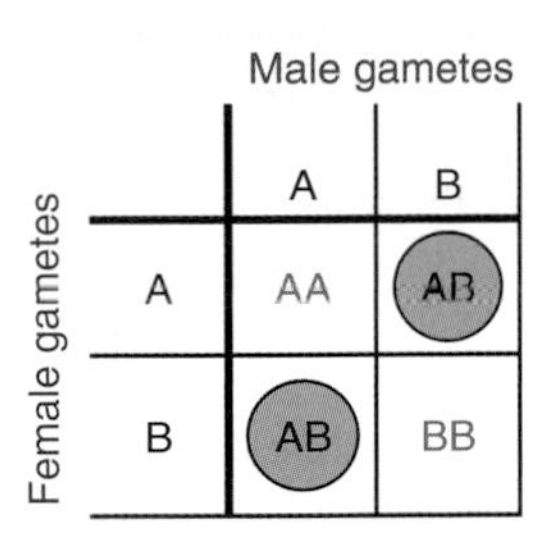

Figure Z8. In zygotic lethality only the heterozygotes are viable and both types of homozygotes are lethal

Female gametes \ Male gametes	A	B
A	AA	AB
B	AB	BB

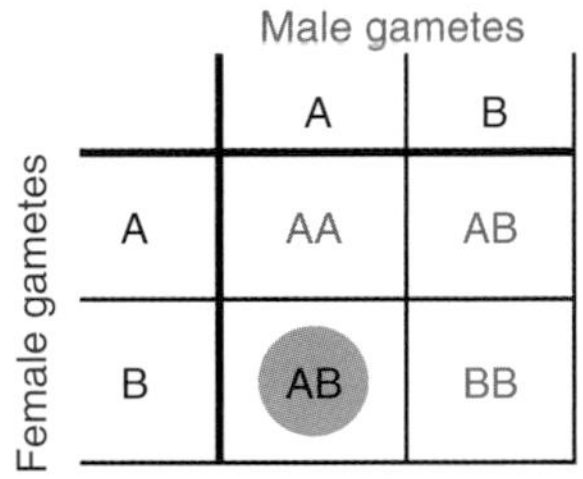

Figure Z9. Gametic lethal factors A and B are unable to form selfed zygotes and only the heterozygous embryos are viable. Only the female can contribute viable B gamete and the male produces viable A gametes

Zygotic Combinations: ►Punnett square, ►allelic combinations

Zygotic Gene: During embryo development this gene is involved in the early control of differentiation in contrast to the maternal effect genes, which are transcribed from the maternal genome and their product is transfused to the embryo. ►maternal effect genes, ►morphogenesis

Zygotic Gene Activation: After fertilization there is a transition from maternal to zygotic control of development. This process in mouse begins after G_2 of the one cell embryo (Schultz RM 1993 Bioessays 15:531). In humans it seems to be somewhat delayed.

Zygotic Induction: Usually the integrated λ prophage is inherited as an integral gene of the bacterial chromosome. However, when a lysogenic Hfr cell carrying λ is crossed to a *non-lysogenic* F^- recipient, the prophage leaves the chromosomal position (induction) and becomes an infectious vegetative phage after replicating to about 100–200 particles. After zygotic induction, only the markers transmitted before the position of the prophage has changed are recovered in the recombinants. Zygotic induction does not occur if the *F^- cells are lysogenic*, irrespective of whether or not the Hfr is lysogenic. The F^- recipient is immune to superinfection by free λ phage and it cannot support the vegetative development of the chromosomal λ phage transmitted by the Hfr donor. This indicates that the F^- cytoplasm carries an immunity substance or a repressor. ►lysogeny, ►Hfr, ►F^-, ►lysogenic repressor, ►lambda phage; Hayes W 1965 The genetics of bacteria and their viruses. Wiley, New York.

Zygotic Lethal: A genetic factor that permits the function of the gametes but kills the zygote. The two systems presented here occur in the plant *Oenothera*, which carry translocation complexes. Letters printed in gray indicate lack of zygote formation (see Figures Z8 and Z9) ►translocation, ►complex heterozygote, ►translocation [in animals], ►*Oenothera*

Zymogen: An inactive enzyme precursor. ►regulation of enzyme activity

Zymogram: Electrophoretically separated isozymes are identified in the electrophoretic medium (starch, agarose) by supplying a chromogenic substrate in situ and this reveals the position of the functionally active enzyme bands. ►electrophoresis, ►isozyme

Zymolase: This hydrolyzes 1→3 glucose linkages such as those existing in yeast cell walls; it is not an exactly defined mixture of proteins extracted from *Athrobacter luteus.*

Zymosan: A cell wall extract with a variety of components interfering with the C3 complement. ►complement; Ohki K et al 2001 Immunol Cell Biol 79:462.

Zymotype: This is an electrophoretically determined pattern of enzymes (proteins) which is characteristic of individuals or group of individuals. ►isozymes, ►electrophoresis, ►zymogram

Historical vignettes

"I have a rather strange feeling about our DNA structure. If it is correct, we should obviously follow it up at a rapid rate. On the other hand it will at the same time be difficult to avoid the desire to forget completely about nucleic acid and to concentrate on other aspects of life." (*JD Watson's letter to Max Delbrück on March* 22, 1953. *Quoted after HF Judson 1979 The Eighth Day of Creation. Simon and Schuster, New York, p.* 229)

Evelyn Witkin (2002) in reminiscing about the pre-Watson and Crick and Hershey and Chase era at the Cold Spring Harbor Laboratory writes (Annu. Rev. Microbiol. 56:1):

"Although Avery et al (1) had demonstrated in 1944 that the genetic material is DNA, the prevailing attitude at Cold Spring Harbor had been respectful skepticism. Some suggested that the transforming DNA in their experiments had activated genetic information already present or somehow caused a directed mutation. Others believed that the minuscule trace of protein still contaminating the DNA was the active agent. Delbruck declared DNA to be a "stupid" molecule, incapable of carrying genetic information."

In 1962 Francis Harry Compton Crick, James Dewey Watson and Maurice Hugh Frederick Wilkins were jointly awarded the Nobel Prize in Physiology or Medicine "for their discoveries concerning the molecular structure of nucleic acids and its significance for information transfer in living material".

In 1969 the Nobel Prize in Physiology or Medicine was shared by Max Delbrück, Alfred D Hershey and Salvador Luria.

General References

Given below are a list of books, published during the last few years, and selected on the basis of relevance to the topics covered in this volume. The grouping is somewhat arbitrary because of the overlaps in scope and content. The selection does not necessarily coincide with my views regarding the technical accuracy of the works listed. All books have positive and negative aspects and the evaluation may be partly subjective.

CONTENTS

Aging ►*molecular biology*, ►*cancer*, ►*diseases*

Arking R (1998) Biology of Aging: Observations and Principles. Sinauer, Sunderland, MA

Barnett YA and Barnett CR (eds) (2000) Aging Methods and Protocols. Humana Press, Totowa, NJ

Bellamy D (1995) Aging: A Biomedical Perspective. Wiley, New York

Birren J (2006) Encyclopedia of Gerontology. Elsevier, Philadelphia, PA

Clark WR (1999) A Means to the End: The Biological Basis of Aging and Death. Oxford University Press, New York

Crews DF (2003) Human Senescence. Cambridge University Press, New York

De Grey ADNJ (1999) Molecular Biology Intelligence Unit: The Mitochondrial Free Radical Theory of Aging. Landes, Austin, TX

De Gray A and Rae M (2007) Ending Aging: The Rejuvenation Breakthroughs that Could Reverse Human Aging in Our Lifetime. St. Martin Press, New York

Fossel MB (2004) Cells, Aging, and Human Disease. Oxford University Press, New York

Goate A and Ashall F (1995) Pathobiology of Alzheimer's Disease. Academic Press, San Diego, CA

Guarente L (2003) Ageless Quest. One Scientist's Search for Genes that Prolong Youth. Cold Spring Harbor Laboratory Press, Cold Spring Harbor, NY

Guarente L (ed) (2008) Molecular Biology of Aging. Cold Spring Harbor Laboratory Press, Cold Spring Harbor, NY

Harris J (2007) Enhancing Evolution. The Ethical Case for Making Better People. Princeton University Press, Princeton, NJ

Hekimi S (2000) The Molecular Genetics of Aging. Springer, New York

Holbrok NJ, et al (1995) Cellular Aging and Cell Death. Wiley, New York

Holliday R (1995) Understanding Aging. Cambridge University Press, New York

Holliday R (1997) Aging: The paradox of Life. Springer, Berlin, D
Kanungo MS (1994) Genes and Aging. Cambridge University Press, New York
Lockshin RA and Zakeri Z (eds) (2004) When Cells Die II: A Comprehensive Evaluation of Apoptosis and Programmed Cell Death. Wiley, Hoboken, NJ
Medina JJ (1996) The Clock of Ages: Why We Age—How We Age—Winding Back the Clock. Cambridge University Press, New York
Rattan SIS and Toussaint O (eds) (1996) Molecular Gerontology. Plenum Press, New York
Rose MR and Finch CE (1994) Genetics and Evolution of Aging. Kluwer, Norwell, MA
Smith DWE 1993 Human Longevity. Oxford University Press, New York
Tomel DL and Cope FO (eds) (1994) Apoptosis II: Molecular Basis of Apoptosis in Disease. Cold Spring Harbor Laboratory Press, Cold Spring Harbor, NY
Vijg J (2007) Aging of the Genome: The Dual Role of DNA in Life and Death. Oxford University Press, New York

Behavior ►*neurobiology*

Adkins-Regan E (2005) Hormones and Animal Behavior. Princeton University Press, Princeton, NJ
Alcock J (2001) The Triumph of Sociobiology. Oxford University Press, New York
Alcock J (2005) Animal Behavior: An Evolutionary Approach. Sinauer, Sunderland, MA
Avital E and Jablonka E (2001) Animal Traditions: Behavioral Inheritance in Evolution. Cambridge University Press, New York
Barondes SH (1998) Mood Genes: Hunting for the Origins of Mania and Depression. Freeman, New York
Begleiter H and Kissin B (1995) The Genetics of Alcoholism. Oxford University Press, New York
Bekoff B (2005) Animal Passions and Beastly Virtues: Reflections on Redecorating Nature. Temple University Press, Philadelphia, PA
Boake CB (ed) (1994) Quantitative Genetic Studies of Behavioral Evolution. University of Chicago Press, Chicago, IL
Bock GR and Goode JA (eds) (1996) Genetics of Criminal and Antisocial Behavior. Wiley, New York
Bouchard TJ and Propping P (1993) Twins as a Tool of Behavioral Genetics. Wiley, New York
Bouton ME (2006) Learning and Behavior: A Contemporary Synthesis. Sinauer, Sunderland, MA
Burkhardt RW Jr (2005) Patterns of Behavior: Konrad Lorenz, Niko Tinbergen, and the Founding of Ethology. University of Chicago Press, Chicago, IL
Carew TJ (2000) Behavioral Neurobiology: The Cellular Organization of Natural Behavior. Sinauer, Sunderland, MA
Carson RA and Rothstein MA (eds) (1999) Behavioral Genetics: The Clash of Culture and Biology. Johns Hopkins University Press, Baltimore, MD
Cassidy SB and Allanson JE (eds) (2005) Management of Genetic Syndromes. Wiley, Hoboken, NJ
Clark WR and Grunstein M (2000) Are We Hardwired? The Role of Genes in Human Behavior. Oxford University Press, New York
Cloninger CR and Begleiter H (eds) (1990) The Genetics and Biology of Alcoholism. Cold Spring Harbor Laboratory Press, Cold Spring Harbor, NY
Crawley JN (2000) What is Wrong with My Mouse? Behavioral Phenotyping of Transgenic and Knockout Mice. Wiley-Liss, New York
Daston L and Mitman G (eds) (2005) Thinking with Animals: New Perspectives on Anthropomorphism. Columbia University Press, New York
de Waal F (1997) Good Natured: The Origins of Right and Wrong in Humans and Other Animals. Harvard University Press, Boston, MA
Dixson AF (1999) Primate Sexuality: Comparative Studies of the Prosimians, Monkeys, Apes, and Human Beings. Oxford University Press, New York
Dowling JE (2001) Neuron Networks: An Introduction to Behavioral Neuroscience. Harvard University Press, Cambridge, MA
Everitt B and Howell D (eds) (2005) Encyclopedia of Statistics in Behavioral Science. Wiley, Hoboken, NJ
Grandin T (ed) (1997) Genetics and the Behavior of Domestic Animals. Academic Press, San Diego, CA
Hurst JL et al (eds) (2008) Chemical Signals in Vertebrates 11. Springer, Berlin, D
Jensen AR (1998) The g Factor. The Science of Mental Ability. Praeger, Westport, CN
Jones BC and Mormède P (eds) (2007) Neurobehavioral Genetics. Methods and Applications. CRC Press, Boca Raton, FL
Levy F and Hay D (eds) (2001) Attention, Genes and ADHD. Brunner and Routledge, Hove, East Sussex, UK
LeVay S and Valente SM (2005) Human Sexuality. Sinauer, Sunderland, MA
Mackintosh NJ (1998) IQ and Human Intelligence. Oxford University Press, New York
Matsuzawa T (ed) (2001) Primate Origins of Human Cognition and Behavior. Springer, Tokyo
Novartis Foundation Symposium (2005) Molecular Mechanisms Influencing Aggressive Behavior. Wiley, Hoboken, NJ
Parens E et al (eds) (2006) Wrestling with Behavior Genetics: Science, Ethics, and Public Conversation. Johns Hopkins University Press, Baltimore, MD
Plomin R et al (1997) Behavioral Genetics. Freeman, New York
Quiatt D and Reynolds V (1995) Primate Behavior. Cambridge University Press, New York
Rodgers JL et al (eds) (2000) Genetic Influences on Human Fertility and Sexuality. Kluwer, Dordrecht, the Netherlands
Rutter M (ed) (1995) Genetics of Criminal and Antisocial Behavior. Wiley, New York

Rutter M (2006) Genes and Behavior: Nature–Nurture Interplay Explained. Blackwell, Cambridge, MA
Scott JP and Fuller JL (1998) Genetics and the Social Behavior of the Dog. University of Chicago Press, Chicago, IL
Simmons P and Young D (1999) Nerve Cells and Animal Behavior. Cambridge University Press, New York
Sternberg RJ and Ruzgis P (eds) (1994) Personality and Intelligence. Cambridge University Press, New York
Tollefsbol TO (ed) (2007) Biological Aging. Methods and Protocols. Humana, Totowa, NJ
Turner JR et al (eds) (1995) Behavior Genetic Approaches in Behavioral Medicine. Plenum, New York
Waring GH (2002) Horse Behavior. Noyes, Norwich, New York
Wasserman D and Wachbroit R (eds) (2001) Genetics and Criminal Behavior. Cambridge University Press, Cambridge
Wilson EO (1975) Sociobiology: The New Synthesis. Harvard University Press, Cambridge, MA

Biochemistry and Physiology ►*molecular biology*, ►*laboratory technology*, ►*biotechnology*

Aidley DJ and Stanfield PR (1996) Ion Channels: Molecules in Action. Cambridge University Press, New York
Alberts B et al (2003) Essential Cell Biology. Garland, New York
Alberty RA (2003) Thermodynamics of Biochemical Reactions. Wiley, Hoboken, NJ
Annual Review of Biochemistry, Palo Alto, CA
Annual Review of Physiology, Palo Alto, CA
Banaszek LJ (2000) Foundations of Structural Biology. Academic Press, San Diego, CA
Barrett AJ et al (eds) (1998) Handbook of Proteolytic Enzymes. Academic Press, San Diego, CA
Bauerlein E (ed) (2004) Biomineralization. Wiley, Hoboken, NJ
Beavo J et al (2006) Cyclic Nucleotide Phosphodiesterases in Health and Disease. CRC Press, Boca Raton, FL
Belfort M et al (eds) (2005) Homing Endonucleases and Inteins. Springer, New York
Bender DA (2003) Nutritional Biochemistry of the Vitamins. Cambridge University Press, New York
Benz R (ed) (2004) Bacterial and Eukaryotic Proteins. Wiley, Hoboken, NJ
Berg JM et al (2006) Biochemistry. Freeman, New York
Berg JM et al (2007) Stryer Biochemie. Springer, Berlin, D
Berger A and Roberts MA (eds) (2005) Unraveling Lipid Metabolism with Microarrays. CRC Press, Boca Raton, FL
Berger S and Braun S (2004) 200 and More NMR Experiments. Wiley, Hoboken, NJ
Beynon R and Bond JS (eds) (2001) Proteolytic Enzymes. Oxford University Press, New York
Bisswanger H (2004) Practical Enzymology. Wiley-VCH, Hoboken, NJ
Blackburn GM and Gait MJ (eds) (1996) Nucleic Acids in Chemistry and Biology. Oxford University Press, New York
Bodansky M (1993) Principles of Peptide Synthesis. Springer, New York
Bolander FK (2004) Molecular Endocrinology. Elsevier Academic Press, San Diego, CA
Bowles DJ (ed) (1997) Essays in Biochemistry. Cell Signalling. Portland Press, London
Boullata J and Armenti VT (eds) (2004) Handbook of Drug–Nutrient Interactions. Humana Press, Totowa, NJ
Branden C and Tooze J (1999) Introduction to Protein Structure. Garland, New York
Breslow R (2005) Artificial Enzymes. Wiley, Hoboken, NJ
Brigelius-Flohé R and Joost H-G (eds) (2006) Nutritional Genomics. Impact on Health and Disease. Wiley, Hoboken, NJ
Brito-Arias M (2007) Synthesis and Characterization of Glycosides. Springer, Berlin, D
Buchner J and Kiefhaber T (eds) (2005) Protein Folding Handbook. Wiley-VCH, Hoboken, NJ
Burrell MM (ed) (1993) Enzymes of Molecular Biology. Humana Press, Totowa, NJ
Buslig BS and Manthey JA (eds) (2002) Flavonoids in Cell Function. Kluwer/Plenum, New York
Campbell JL (ed) (1995) DNA Replication. Methods in Enzymology, vol 262. Academic Press, San Diego, CA
Cao L and Schmid RD (2005) Carrier-Bound Immobilized Enzymes. Wiley, Hoboken, NJ
Carafoli E and Klee C (eds) (1999) Calcium as a Cellular Regulator. Oxford University Press, New York
Chapman BJ (ed) (2002) Advances in DNA Sequence Specific Agents. Elsevier Science, St. Louis, MO
Chawnshang C (ed) (2002) Androgens and Androgen Receptors. Mechanisms, Functions, and Clinical Applications. Kluwer, Boston, MA
Cohen NC (ed) (1966) Guidebook on Molecular Modeling in Drug Design. Academic Press, San Diego, CA
Conly EC (1995) Ion Channel Factsbook. Extracellular Ligand-Gated Ion Channels. Academic Press, San Diego, CA
Copeland RA (ed) (2005) Evaluation of Enzyme Inhibitors in Drug Discovery. Wiley, Hoboken, NJ
Copeland RA (ed) (2006) Enzymes. A Practical Introduction to Structure, Mechanism, and Data Analysis. Wiley-VCH, Hoboken, NJ
Couvreur P and Malvy C (eds) (2000) Pharmaceutical Aspects of Oligonucleotides. Taylor & Francis, London
Cox GS (ed) (2005) Drug Discovery Handbook. Wiley, Hoboken, NJ
Crich D (ed) (2005) Reagents for Glycoside, Nucleotide and Peptide Synthesis. Wiley, Hoboken, NJ
Csermely P and Vigh L (eds) (2007) Molecular Aspects of the Stress Response. Chaperones, Membranes and Networks. Springer, Berlin, D
Dalbey R et al (eds) (2002) Protein Targeting, Transport, and Translocation. Academic Press, San Diego, CA
D'Alessio G and Riordan JF (eds) (1997) Ribonucleases. Structures and Functions. Academic Press, San Diego, CA

De Maeyer E and De Maeyer-Guignard J (1988) Interferons and Other Regulatory Cytokines. Wiley, New York
Devi LA (ed) (2005) The G Protein-Coupled Receptors Handbook. Humana Press, Totowa, NJ
Devlin TM (ed) (2006) Textbook of Biochemistry with Clinical Correlations. Wiley, Hoboken, NJ
Dickinson JR and Schweizer M (eds) (2004) The Metabolism and Molecular Physiology of *Saccharomyces cerevisiae*. CRC Press, Boca Raton, FL
Diederich M (ed) (2003) Apoptosis. From Signaling Pathways to Therapeutic Tools. Annals New York Academic Science, vol 1010
Dingermann T et al (eds) (2004) Molecular Biology in Medicinal Chemistry. Wiley, Hoboken, NJ
Doonan S (2003) Peptides and Proteins. Wiley, Hoboken, NJ
Doublie S (ed) (2007) Preparation and Crystallization of Macromolecules. In: Macromolecular Crystallography Protocols, vol 1. Humana Press, Totowa, NJ
Doublie S (ed) (2007) Structure Determination. In: Macromolecular Crystallography Protocols, vol 2. Human Press, Totowa, NJ
Duttaroy AK and Spener F (eds) (2003) Cellular Proteins and Their Fatty Acids in Health and Disease. Wiley-VCH, Hoboken, NJ
Ebashi S and Ohtsuki I (eds) (2007) Regulatory Mechanisms of Striated Muscle Contraction. Springer, Berlin, D
Elliott WH and Elliott DC (1997) Biochemistry and Molecular Biology. Oxford University Press, New York
Ellis RJ (ed) (1996) The Chaperonins. Academic Press, San Diego, CA
ElSohly MA (ed) (2006) Marijuana and the Cannabinoids. Humana Press, Totowa, NJ
Endo M et al (eds) (2006) Endoglycosidases-Biochemistry, Biotechnology, Application. Springer/Kodansha Scientific, Berlin/Japan
Enna SJ (ed). (eds) (2007) The GABA Receptors. Humana Press, Totowa, NJ
Fenniri H (ed) (2000) Combinatorial Chemistry. A Practical Approach. Oxford University Press, New York
Findeis MA (ed) (2001) Nonviral Vectors for Gene Therapy. Methods and Protocols. Humana Press, Totowa, NJ
Fink AL and Golo Y (1998) Molecular Chaperones in the Life Cycle of Proteins. Marcel Dekker, New York
Finkelstein AV and Ptitsyn OB (2002) Protein Physics. Academic Press, San Diego, CA
Flockerzi VM and Nilius B (eds) (2007) Transient Receptor Potential (TRP) Channels. Springer, Berlin, D
Fruton JS (1999) Proteins, Enzymes, Genes: The Interplay of Chemistry and Biology. Yale University Press, New Haven, CT
Futai M et al (eds) (2004) Handbook of ATPases. Wiley-VCH, Hoboken, NJ
Gellissen G (ed) (2005) Production of Recombinant Proteins. Wiley, Hoboken, NJ
George SR and O'Dowd BF (eds) (2005) G Protein-Coupled Receptor-Protein Interactions. Wiley, Hoboken, NJ
Goldstein DJ (ed) (2005) The Management of Eating Disorders and Obesity. Humana Press, Totowa, NJ
Griffiths W (ed) (2008) Metabolomics, Metabonomics and Metabolite Profiling. Springer, Berlin, D
Grisshammer R and Buchanan SK (eds) (2006) Structural Biology of Membrane Proteins. Springer, New York
Hagemeyer A et al (eds) (2004) High-Throughput Screening in Chemical Catalysis. Technologies, Strategies and Applications. Wiley, Hoboken, NJ
Hall A (ed) (2000) GTPases. Oxford University Press, New York
Hamdan MH (2005) Proteomics Today. Protein Assessment and Biomarkers Using Mass Spectrometry, 2D Electrophoresis, and Microarray Technology. Wiley, Hoboken, NJ
Hammes GH (2000) Thermodynamics and Kinetics for Biological Sciences. Wiley, New York
Hammes GH (2005) Spectroscopy for the Biological Sciences. Wiley, Hoboken, NJ
Hardie GD and Hanks S (1995) The Protein Kinase Factsbooks. Academic Press, San Diego, CA
Harford JB and Morris DR (eds) (1997) mRNA Metabolism and Post-Transcriptional Gene Regulation. Wiley, New York
Harrison JK and Lukacs NW (eds) (2007) The Chemokine Receptors. Humana Press, Totowa, NJ
Hartmann RK et al (eds) (2005) Handbook of RNA Biochemistry. Wiley-VCH, Hoboken, NJ
Hecht SM ed (1996) Bioorganic Chemistry. Nucleic Acids, Oxford University Press, New York
Hellwinkel D (2001) Systematic Nomenclature of Organic Chemistry. Springer, New York
Helmreich EJM (2001) The Biochemistry of Cell Signalling. Oxford University Press, New York
Hesse M (2002) Alkaloids. Wiley, Hoboken, NJ
Hill RW et al (2004) Animal Physiology. Sinauer, Sunderland, MA
Holmes RM (2007) Cell Biology in the Post-Genomics Era. Wiley, Hoboken, NJ
Höltje H-D et al (2003) Molecular Modeling. Basic Principles and Applications. Wiley, Hoboken, NJ
Hunter GK (2000) Vital Forces: The Discovery of the Molecular Basis of Life. Academic Press, San Diego, CA
Hurtley S (ed) (1996) Protein Targeting. IRL Press, New York
James P (ed) (2001) Proteome Research: Mass Spectrometry. Springer, New York
Jeon K (2006) International Review of Cytology. A Survey of Cell Biology. Elsevier, Philadelphia, PA
Juo PS (ed) (1996) Concise Dictionary of Biomedicine and Molecular Biology. CRC Press, Boca Raton, FL
Kaput J and Rodriguez RL (eds) (2006) Nutritional Genomics. Discovering the Path to Personalized Nutrition. Wiley, Hoboken, NJ
Kent JA (ed) (2007) Kent and Riegel's Handbook of Industrial Chemistry and Biotechnology. Springer, Berlin, D
Kirk-Othmer Encyclopdia of Chemical Technology. Wiley, Hoboken, NJ
Klein DR (2006) Organic Chemistry as a Second Language. Wiley, Hoboken, NJ
Klussmann S (ed) (2006) The Aptamer Handbook. Functional Oligonucleotides and Their Applications. Wiley-VCH, Hoboken, NJ
Kobilinsky L et al (2005) DNA. Forensic and Legal Applications. Wiley, Hoboken, NJ
Kornberg A and Baker TA (1992) DNA Replication. Freeman, New York
Kovacheva SD and Velev P (2006) Elsevier's Dictionary of Chemistry. Elsevier, Philadelphia, PA
Kricka LJ (ed) (1995) Nonisotopic Probing, Blotting and Sequencing. Academic Press, San Diego, CA

Kubinyi H and Müller G (eds) (2004) Chemogenomics in Drug Discovery. Wiley, Hoboken, NJ
Kyte J (1995) Structure in Protein Chemistry. Garland, New York
Landecker H (2007) Culturing Life. How Cells Became Technologies. Harvard University Press, Cambridge, MA
Lappalainen P (ed) (2007) Actin Monomer Binding Proteins. Springer, Berlin, D
Lennarz WJ (ed) (2004) Encyclopedia of Biological Chemistry. Elsevier, St. Louis, MO
Lundblad RL (2006) The Evolution from Protein Chemistry to Proteomics. CRC Press, Boca Raton, FL
Lundstrom KH (ed) (2006) Structural Genomics on Membrane Proteins. CRC Press, Boca Raton, FL
Marangoni AJ (2003) Enzyme Kinetics. A Modern Approach. Wiley, Hoboken, NJ
Matsumoto RR et al (eds) (2007) Sigma Receptors. Chemistry, Cell Biology and Clinical Implications. Springer, Berlin
Mayer RJ et al (eds) (2005) Protein Degradation. Ubiquitin and the Chemistry of Life. Wiley-VCH, Hoboken, NJ
McIntosh TJ (2007) Lipid Rafts. Humana Press, Totowa, NJ
Melmed S and Conn PM (eds) (2005) Endocrinology. Basic and Clinical Principles. Humana Press, Totowa, NJ
Merck Index (2006) An Encyclopedia of Chemicals, Drugs and Biologicals. 14th edn. Wiley Indianapolis, IN
Messerschmidt A (ed) (2004) Handbook of Metalloproteins. Wiley, Hoboken, NJ
Metzler DE (2003) Biochemistry. The Chemical Reactions of Living Cells. Academic Press, San Diego, CA
Meyers RA (ed) (2007) Proteins. From Analytics to Structural Genomics. Wiley-VCH, Hoboken, NJ
Michal G (ed) (1999) Biochemical Pathways. An Atlas of Biochemistry and Molecular Biology. Wiley, Hoboken, NJ
Mozayani A and Noziglia C (eds) (2006) The Forensic Laboratory Handbook. Humana Press, Totowa, NJ
Neidle S (ed) (1999) Oxford Handbook of Nucleic Acid Structure. Oxford University Press, New York
Neidle S and Balasubramanian S (eds) (2007) Quadruplex Nucleic Acids. Springer, Berlin, D
Nelson DL and Cox MM (2004) Lehninger Principles of Biochemistry. Freeman, New York
Niederhaus KH and Wilson DN (2004) Protein Synthesis and Ribosome Structure. Translating the Genome. Wiley, Hoboken, NJ
Nielsen J and Jewett MC (eds) (2007) Metabolomics. A Powerful Tool in Systems Biology. Springer, Berlin, D
Nielsen PE (ed) (2004) Pseudo-Peptides in Drug Discovery. Wiley, Hoboken, NJ
Niemeyer CM and Mirkin CA (eds) (2004) Nanobiotechnology. Wiley, Hoboken, NJ
Oettel M and Schillinger E (eds) (1999) Estrogens and Antiestrogens. Springer, New York
Oprea TI (ed) (2005) Chemoinformatics in Drug Discovery. Wiley, Hoboken, NJ
Orkin M et al (eds) (2005) The Vocabulary and Concepts of Organic Chemistry. Wiley, Hoboken, NJ
Pain RH (ed) (2001) Mechanisms of Protein Folding. Oxford University Press, New York
Papa S and Tager JM (eds) (1995) Biochemistry of Cell Membranes. A Compendium of Selected Topics. Birkhäuser, Cambridge, MA
Phoenix DA (ed) (1999) Protein Targeting and Translocation. Princeton University Press, Princeton, NJ
Pratt CW and Cornely K (2004) Essential Biochemistry. Wiley, Hoboken, NJ
Purich D and Allison R (eds) (2002) The Enzyme Reference: A Comprehensive Guidebook to Enzymes Nomenclature, Reactions, and Methods. Academic Press, San Diego, CA
Rabilloud TH (ed) (2000) Proteome Research: Two-Dimensional Gel Electrophoresis and Identification Methods. Springer, New York
Regazzi R (ed) (2006) Molecular Mechanisms of Exocytosis. Springer, Berlin, D
Remedios CD and Chhabra D (eds) (2008) Actin-Binding Proteins and Disease. Springer, Berlin, D
Rensselaer Polytechnic Institute (eds) (2007) Protein Structure Prediction. Humana Press, Totowa, NJ
Ridley A et al (eds) (2004) Cell Motility. From Molecules to Organisms. Wiley, Hoboken, NJ
Rimbach GH et al (eds) (2005) Nutrigenomics. CRC Press, Boca Raton, FL
Rognan D et al (eds) (2006) Ligand Design for G Protein-Coupled Receptors. Wiley-VCH, Hoboken, NJ
Schomburg D and Schomburg I (2001) Handbook of Enzymes. Springer, Heidelberg, Germany
Schomburg D et al (eds) (2007) Class 2 Transferases VI,2.4.2.1-2.5.1.30. Springer, Berlin, D
Schreiber SL et al (2007) Chemical Biology. From Small Molecules to Systems Biology and Drug Design. Wiley-VCH, Hoboken, NJ
Sen CK et al (eds) (1999) Antioxidant and Redox Regulation of Genes. Academic Press, San Diego, CA
Sigel A et al (eds) (2007) The Ubiquitous Roles of Cytochrome p450 Proteins. Metal Ions in Life Sciences. Wiley, Hoboken, NJ
Simons RW and Grunberg-Manago M (1998) RNA Structure and Function. Cold Spring Harbor Laboratory Press, Cold Spring Harbor, NY
Sipe D (ed) (2005) Amyloid Proteins. The Beta Sheet Conformation and Disease. Wiley, Hoboken, NJ
Smith RM (2004) Understanding Mass Spectra. Wiley, Hoboken, NJ
Soyfer VN and Potaman VN (1996) Triple-Helical Nucleic Acids. Springer, New York
Sperelakis N (ed) (1998) Cell Physiology Source Book. Academic Press, San Diego, CA
Springer Handbook of Enzymes. Springer, New York
Stafstrom CE and Rho JM (2004) Epilepsy and the Ketogenic Diet. Humana Press, Totowa, NJ
Storey KB (2004) Functional Metabolism. Wiley, Hoboken, NJ
Stryer L (2003) Biochemistry. Freeman, New York
Sundrstrom M et al (eds) (2006) Structural Genomics and High Throughput Structural Biology. CRC Press, Boca Raton, FL
Tamanoi F and Clarke S (2006) The Enzymes. Protein Methyltransferases. Elsevier, Philadelphia, PA
Temple NJ et al (eds) (2005) Nutritional Health. Strategies for Disease Prevention. Humana Press, Totowa, NJ
Toone EJ (2007) Advances in Enzymology and Related Areas of Molecular Biology. Protein Evolution. Wiley, Hoboken, NJ
Torrence PF (2005) Antiviral Drug Discovery for Emerging Diseases and Bioterrorism Threats. Wiley, Hoboken, NJ

Tramontano A (2006) Protein Structure Prediction. Concepts and Applications. Wiley-VBCH, Hoboken, NJ
Tsai CS (2002) An Introduction to Computational Biochemistry. Wiley, New York
Tsai CS (2006) Biomacromolecules. Wiley, Hoboken, NJ
Van Eldik L and Watterson DM (eds) (1998) Calmodulin and Signal Transduction. Academic Press, San Diego, CA
Vandenbroeck K (ed) (2006) Cytokine Gene Polymorphisms in Multifactorial Conditions. CRC Press, Boca Raton, FL
Voet D and Voet JG (2005) Biochemistry. Wiley, Hoboken, NJ
Wagenknecht H-A (ed) (2005) Charge Transfer in DNA. Wiley, Hoboken, NJ
Wallsgrove RM (ed) (1995) Amino Acids and Their Derivatives in Higher Plants. Cambridge University Press, New York
Walsh C (2003) Antibiotics: Action, Origins, Resistance. American Society for Microbiology Press, Washington, DC
Walz W (ed) (2005) Integrative Physiology. In the Age of Genomics and Proteomics. Humana Press, Totowa, NJ
Waring MJ (ed) (2007) Sequence-Specific DNA Binding Agents. Springer, Berlin, D
Weaver CM and Heaney RP (eds) (2004) Calcium in Human Health. Humana Press, Totowa, NJ
Webb EC (1992) Enzyme Nomenclature. Academic Press, San Diego, CA
Weckström M and Tavi P (eds) (2007) Cardiac Mechanotransduction. Springer, Berlin, D
Wendisch VF (ed) (2007) Amino Acid Biosynthesis—Pathways, Regulation and Metabolic Engineering. Springer, Berlin, D
Whiford D (2005) Proteins. Structure and Function. Wiley, Hoboken, NJ
White D (2000) The Physiology and Biochemistry of Prokaryotes. Oxford University Press, New York
White JS and White DC (2002) Proteins, Peptides and Amino Acids Sourcebook. Humana Press, Totowa, NJ
Whitford D (2005) Proteins. Structure and Function. Wiley, Hoboken, NJ
Wikstrom M (ed) (2006) Biophysical and Structural Aspects of Bioenergetics. Springer, New York
Wilson K and Walker J (eds) (2005) Principles and Techniques of Biochemistry and Molecular Biology. Cambridge University Press, New York
Wong C-H (ed) (2003) Carbohydrate-Based Drug Discovery. Wiley, Hoboken, NJ
Woodgett JR (ed) (1995) Protein Kinases. IRL/Oxford University Press, New York
Yehuda S and Mostofsky DI (eds) (2005) Nutrients, Stress and Medical Disorders. Humana Press, Totowa, NJ
Zoref-Shani E and Sperling O (eds) (2000) Purine and Pyrimidine Metabolism in Man X. Kluwer/Plenum, New York
Zubay GL et al (1995) Principles of Biochemistry. Brown, Dubuque, IA

Biology and Cell Biology ►*molecular biology*, ►*microbiology*

Agutter PS and Wheatley DN (2007) About Life. Concepts in Modern Biology. Springer, Berlin, D
Alberts B et al (2002) Molecular Biology of the Cell. Garland, New York
Alon U (2006) Introduction to Systems Biology: Design Principles and Biological Circuits. CRC Press, Boca Raton, FL
Annual Review of Cell Biology. Palo Alto, CA
Asai DJ and Forney JD (eds) (2000) *Tetrahymena thermophila*. Academic Press, San Diego, CA
Bertrand E and Faupel M (eds) (2007) Subcellular Proteomics. From Cell Deconstruction to System Reconstruction. Springer, Berlin, D
Blatch GL (ed) (2007) The Networking of Chaperones By Co-Chaperones. Springer, Berlin, D
Boal D (ed) (2002) Mechanics of the Cell. Cambridge University Press, New York
Bolsover SR et al (2004) Cell Biology. A Short Course. Wiley, Hoboken, NJ
Bonucci E (2007) Biological Calcification Normal and Pathological Processes in the Early Stages. Springer, Berlin, D
Bourtzis K and Miller T (eds) (2003) Insect Symbiosis. CRC Press, Boca Raton, FL
Bringmann P et al (eds) (2007) Systems Biology. Applications and Perspectives. Springer, Berlin, D
Broach JR et al (eds) (1991) Molecular and Cellular Biology of the Yeast *Saccharomyces*. Cold Spring Harbor Laboratory Press, Cold Spring Harbor, NY
Calderwood SK (ed) (2007) Cell Stress Proteins. Springer, Berlin, D
Calisher CH and Fauqquet CN (eds) (1992) Stedman's/ICTV Virusworlds. Williams & Williams, Baltimore, MD
Cassman M et al (eds) (2006) Systems Biology. International Research and Development. Springer, Berlin, D
Cheresh DA and Mecham RP (eds) (1994) Integrins. Molecular and Biological Responses to the Extracellular Matrix. Academic Press, San Diego, CA
Choi S (ed) (2007) Introduction to Systems Biology. Humana Press, Totowa, NJ
Clemente CD (ed) (1985) Gray's Anatomy. Lea & Febiger, Philadelphia, PA
Clifford HT and Bostock PD (2007) Etymological Dictionary of Grasses. Springer, Berlin, D
Coluccio LM (ed) (2008) Myosins. A Superfamily of Molecular Motors. Springer, Berlin, D
Colwell RK (2004) Biota 2: The Biodiversity Database Manager. Sinauer, Sunderland, MA
Cooper GM and Hausman RE (2006) The Cell: A Molecular Approach. Sinauer/ASM, Sunderland, MA
Cowin SC and Doty SB (2007) Tissue Mechanics. Springer, Berlin, D
Cross PC and Mercer KL (1993) Cell and Tissue Ultrastructure. A Functional Perspective. Freeman, New York
Deb S and Deb SP (eds) (2003) Protocols. Humana Press, Totowa, NJ, p 53
Derynck R and Myazono K (eds) (2007) The TGF-β Family. Cold Spring Harbor Laboratory Press, Woodbury, NY
Desjardins C and Ewing LL (eds) (1993) Cell and Molecular Biology of the Testis. Oxford University Press, New York

Devor EJ (ed) (1993) Molecular Application in Biological Anthropology. Cambridge University Press, New York

Dulbecco R (ed) (1997) Encyclopedia of Human Biology. Academic Press, San Diego, CA

Epstein HF and Shakes DC (eds) (1995) *Caenorhabditis elegans*. Modern Biological Analysis of an Organism. Academic Press, San Diego, CA

Ernst EJ and Rogers PD (eds) (2005) Antifungal Agents. Methods and Protocols. Humana Press, Totowa, NJ

Fabbro D and McCormick F (eds) (2005) Protein Tyrosine Kinases. From Inhibitors to Useful Drugs. Humana Press, Totowa, NJ

Ferraris JD and Palumbi SR (eds) (1966) Molecular Zoology. Advances, Strategies and Protocols. Wiley-Liss, New York

Ferretti P and Geraudie J (1998) Cellular and Molecular Basis of Regeneration. Wiley, New York

Fielding CJ (ed) (2006) Lipid Rafts and Caveolae. From Membrane Biophysics to Cell Biology. Wiley, Hoboken, NJ

Gilbert LI et al (eds) (2004) Comprehensive Molecular Insect Science. Elsevier, San Diego, CA

Goeldner M and Givens R (eds) (2005) Dynamic Studies in Biology. Wiley-VCH, Hoboken, NJ

Graham PH et al (1994) Symbiotic Nitrogen Fixation. Kluwer, Norwell, MA

Grimaldi D and Engel MS (2005) Evolution of the Insects. Cambridge University Press, New York

Groombridge B and Jenkins MD (2002) World Atlas of Biodiversity: Earth's Living Resources in the 21st Century. University of California Press, Berkeley, CA

Grundzinskas JG and Yovich JL (eds) (1994) Gametes. The Oocyte. Cambridge University Press, New York

Grundzinskas JG and Yovich JL (eds) (1995) Gametes. The Spermatozoon. Cambridge University Press, New York

Guan J-L (ed) (2005) Cell Migration. Developmental Methods and Protocols. Humana Press, Totowa, NJ

Gübitz G and Schmid MG (eds) (2004) Chiral Separations. Humana Press, Totowa, NJ

Hall BK (ed) (1994) Homology. The Hierarchial Basis of Comparative Biology. Academic Press, San Diego, CA

Harris JR et al (2004) Cell Biology Protocols. Wiley, Hoboken, NJ

Harvati K and Harrison T (eds) (2006) Neanderthals Revisited. New Approaches and Perspectives. Springer, Berlin, D

Hedrich H and Bullock G (eds) (2004) The Laboratory Mouse. Elsevier, San Diego, CA

Ho AD and Zanjani ED (eds) (2006) Stem Cell Transplantation. Biology, Process, Therapy. Wiley, Hoboken, NJ

Holt JG et al (eds) (1993) Bergey's Manual of Determinative Bacteriology. Williams & Williams, Baltimore, MD

Humber JM and Almeder RF (eds) (2004) Stem Cell Research. Humana Press, Totowa, NJ

Hutchison C and Glover DM (eds) (1995) Cell Cycle Control. IRL, New York

Jacobson MD and McCarthy N (eds) (2002) Apoptosis: The Molecular Biology of Programmed Cell Death. Oxford University Press, Oxford

Jong S-C et al (eds) (1993) Stedman's ATCC Fungus Names. Williams & Williams, Baltimore, MD

Kaneko K (2006) Life: An Introduction to Complex Systems Biology. Springer, Berlin, D

Karp G (2007) Cell and Molecular Biology. Concepts and Experiments. Wiley, Hoboken, NJ

Kessin RH (2001) *Dictyostelium*. Evolution, Cell Biology, and the Development of Multicellularity. Cambridge University Press, New York

Kitano H (ed) (2001) Foundations of Systems Biology. MIT Press, Cambridge, MA

Kohen E et al (1995) Photobiology. Academic Press, San Diego, CA

Krauss G (2003) Biochemistry of Signal Transduction and Regulation. Wiley-VCH, Hoboken, NJ

Krishnashwamy G and Chi DS (2005) Mast Cells. Humana Press, Totowa, NJ

Krstíc RV (1994) Human microscopic anatomy. An Atlas for Students of Medicine and Biology. Springer, New York

Lackie JM et al (eds) (1999) Cell Behaviour: Control and Mechanism of Motility. Princeton University Press, Princeton NJ

Lanza R et al (2006) Essentials of Stem Cell Biology. Elsevier/Academic Press, San Diego, CA

Larsen WJ (1993) Human Embryology. Churchill Livingstone, New York

Lenz P (ed) (2008) Cell Motility. Springer, Berlin, D

León R et al (eds) (2008) Transgenic Microalgae as Green Cell Factories. Springer, Berlin, D

Lieberman HB (ed) (2004) Cell Cycle Checkpoint Protocols. Humana Press, Totowa, NJ

Lodish H et al (2007) Molecular Cell Biology. Freeman, New York

Mabberly DJ (1997) The Plant-Book. A Portable Dictionary of the Vascular Plants. Cambridge University Press, New York

Morgan DO (2006) The Cell Cycle: Principles of Control. Sinauer, Sunderland, MA

Markow T and O'Grady P (2005) Drosophila. A Guide to Species Identification and Use. Elsevier, San Diego, CA

Mayr E (1997) This is Biology: The Science of the Living World. Harvard University Press, Cambridge, MA

McLaren A (2007) Impotence: A Cultural History. University of Chicago Press, Chicago, IL

Merz KM and Roux B (eds) (1966) Biological Membranes. A Molecular Perspective from Computation and Experiment. Birkhäuser, Boston, MA

Moore D (1998) Fungal Morphogenesis. Cambridge University Press, New York

Netter F and Drake R (2006) Atlas of Human Anatomy 4E and Gray's Anatomy. Saunders, Philadelphia, PA

Newton AC (2003) Protein Kinase C Protocols. Humana Press, Totowa, NJ

Ostlund-Nisson S et al (eds) (2007) Biology of the Three-Spined Stickleback. CRC Press, Boca Raton, FL

Palsson BO (2006) Systems Biology. Properties of Reconstructed Networks. Cambridge University Press, New York

Patel TB (ed) (2006) Epidermal Growth Factor. Methods and Protocols. Humana Press, Totowa, NJ

Peracchia C (ed) (1994) Handbook of Membrane Channels. Academic Press, San Diego, CA

Phillips IR (ed) (2005) Cytochrome P450 Protocols. Humana Press, Totowa, NJ

Pollard TD and Earnshaw WC (2002) Cell Biology. Saunders, Philadelphia, PA

Potten C and Wilson J (2004) Apoptosis. The Life and Death of Cells. Cambridge University Press, Cambridge
Purves WK et al (2004) Life: The Science of Biology. Sinauer/Freeman, New York
Quick MM (2002) Transmembrane Transporters. Wiley, Hoboken, NJ
Raghavan V (1997) Molecular Evolution of Flowering Plants. Cambridge University Press, New York
Raven PH and Johnson GB (1992) Biology. Mosby, St. Louis, MO
Roberts K (ed) (2008) Handbook of Plant Science. Wiley, Hoboken, NJ
Sadava D et al (2007) Life: The Science of Biology. Sinauer/Freeman, Sunderland, MA
Schliwa M (2003) Molecular Motors. VCH-Wiley, Weinhein, Germany
Scott TA (Translator) (1966) Concise Encyclopedia. Biology. Walter de Gruyter, New York
Smrcka AV (ed) (2004) G Protein Signaling. Humana Press, Totowa, NJ
Stein GS et al (eds) (1999) The Molecular Basis of Cell Cycle and Growth Control. Wiley, New York
Stein GS and Pardee AB (eds) (2004) Cell Cycle and Growth Control. Biomolecular Regulation and Cancer. Wiley, Hoboken, NJ
Storey KB (ed) (2004) Functional Metabolism. Regulation and Adaptation. Wiley, Hoboken, NJ
Tamm LK (ed) (2005) Protein-Lipid Interactions. From Membrane Domains to Cellular Networks. Wiley-VCH, Hoboken, NJ
Thorpe TA (ed) (1995) In Vitro Embryogenesis in Plants. Kluwer, Norwell, MA
Wheeler MJ (2005) Hormone Assays in Biological Fluids. Humana Press, Totowa, NJ
Wilson J and Hunt T (2002) Molecular Biology of the Cell. A Problems Approach. Garland Science, New York
Wilson RA (2005) Genes and the Agents of Life. The Individual in the Fragile Science Biology. Cambridge University Press, Cambridge
Wood JW and de Gruyter A (1954) Dynamics of Human Reproduction. Biology, Biometry, Demography. Hawthorne, New York
Yu H-S (1994) Human Reproductive Biology. CRC Press, Boca Raton, FL

Biometry, Statistics and Bioinformatics

Aitken CGG (1995) Statistics and the Evaluation of Evidence for Forensic Scientists. Wiley, New York
Amaratunga D (2004) Exploration and Analysis of DNA Microarray and Protein Array Data. Wiley, Hoboken, NJ
Ambrosius WT (ed) (2007) Topics in Biostatistics. Human Press, Totowa, NJ
Amos M (2006) Genesis Machines: The New Science of Biocomputing. Atlantic Books, New York
Armitage P and Colton T (eds) (2005) Encyclopedia of Biostatistics. Wiley, Hoboken, NJ
Azuaje F and Dopazo J (2005) Data Analysis and Visualization in Genomics and Proteomics. Wiley, Hoboken, NJ
Bailey NTJ (1995) Statistical Methods for Biologists. Cambridge University Press, New York
Balding DJ et al (eds) (2007) Handbook of Statistical Genetics. Wiley, Hoboken, NJ
Barnes M (ed) (2007) Bioinformatics for Geneticists. A Bioinformatics Primer for the Analysis of Genetic Data. Wiley, Hoboken, NJ
Baxevanis AD and Quellette BFF (eds) (2005) Bioinformatics. A Practical Guide to the Analysis of Genes and Proteins. Wiley, Hoboken, NJ
Belle van G et al (2004) Biostatistics. A Methodology for the Health Sciences. Wiley, Hoboken, NJ
Berger V (2005) Selection Bias and Covariate Imbalances in Randomized Clinical Trials. Wiley, Hoboken, NJ
Bishop M (ed) (1998) Guide to Human Genome Computing. Academic Press, San Diego, CA
Bolstad WM (2004) Introduction to Bayesian Statistics. Wiley, Hoboken, NJ
Bower JM and Bolouri H (eds) (2001) Computational Modeling of Genetic and Biochemical Networks. MIT Press, Cambridge, MA
Buehler LK and Rashidi HH (eds) (2005) Bioinformatics Basics. Applications in Biological Science and Medicine. CRC Press, Boca Raton, FL
Bürger R (2000) The Mathematical Theory of Selection, Recombination, and Mutation. Wiley, New York
Chow S-C and Liu J-P (2004) Design and Analysis of Clinical Trials. Concepts and Methodologies. Wiley, Hoboken, NJ
Cogdon P (2005) Bayesian Models for Categorical Data. Wiley, Hoboken, NJ
Daniel WW (2005) Biostatistics. Wiley, Hoboken, NJ
Dardel F and Képès F (2006) Bioinformatics. Genomics and Post-Genomics. Wiley, Hoboken, NJ
Dawson B and Trapp RG (2004) Basic & Clinical Biostatistics. Lange Medical Books/McGraw-Hill, New York
Deutsch A et al (eds) (2008) Mathematical Modeling of Biological Systems. Epidemiology, Evolution and Ecology, Immunology, Neural Systems and the Brain, and Innovative Mathematical Methods. Springer, Berlin, D
Dorogovtsev SN and Mendes JFF (2003) Evolution of Networks: From Biological Nets to the Internet and WWW. Oxford University Press, New York
Dowdy S et al (2004) Statistics for Research. Wiley, Hoboken, NJ
Draghici S (2003) Data Analysis Tools for DNA Microarrays. CRC Press, Boca Raton, FL
Dubitzky W et al (eds) (2007) Fundamentals of Data Mining in Genomics and Proteomics. Springer, D
Dunn OJ and Clark VA (2000) Basic Statistics: A Primer for Biomedical Sciences. Wiley, New York
Durbin R et al (1998) Biological Sequence Analysis: Probabilistic Models of Proteins and Nucleic Acids. Cambridge University Press, New York
Elands-Johnson RC (1971) Probability Models and Statistical Methods in Genetics. Wiley, New York

Elston RC (ed) (2002) Biostatistical Genetics and Genetic Epidemiology. Wiley, New York

Emery AEH (1986) Methodology in Medical Genetics. An Introduction to Statistical Methods. Churchill Livingstone, Edinburgh

Everitt BS (2006) The Cambridge Dictionary of Statistics. Cambridge University Press, New York

Evett IW and Weir BS (1998) Interpreting DNA Evidence: Statistical Genetics for Forensic Scientists. Sinauer, Sunderland, MA

Fall CP et al (2002) Computational Cell Biology. Springer, New York

Fielding AH (2006) Cluster and Classification Techniques for the Biosciences. Cambridge University Press, New York

Forbes N (2004) Imitation of Life: How Biology is Inspiring Computing. MIT Press, Cambridge, MA

Forthofer RN and Lee ES (1995) Introduction to Biostatistics. A Guide to Design, Analysis, and Discovery. Academic Press, San Diego, CA

Friedman A (ed) (2006) Tutorials in Mathematical Biosciences III. Springer, Berlin, D

Gigerenzer G (2002) Calculated Risk. How to Know When Numbers Deceive You. Simon & Schuster, New York

Good PI (2005) Introduction to Statistics Through Resampling Methods and R/S-plus. Wiley, Hoboken, NJ

Good PI and Hardin JW (2006) Common Errors in Statistics (and How to Avoid Them). Wiley, Hoboken, NJ

Govindarajulu Z (2001) Statistical Techniques in Bioassay. Karger, Farmington, CN

Hancock JM and Zvelebil MJ (eds) (2004) Dictionary of Bioinformatics and Computational Biology. Wiley-Liss, Hoboken, NJ

Hays WL and Winkler RL (1971) Statistics: Probability, Inference and Decision. Holt, Rinehart and Winston, New York

Hedeker D and Gibbons RD (2006) Longitudinal Data Analysis. Wiley, Hoboken, NJ

Howson C and Urbach P (1993) Scientific Reasoning. The Bayesian Approach. Open Court, La Salle, IL

Jones NC and Pevzner PA (2004) An Introduction to Bioinformatics Algorithms. MIT Press, Cambridge, MA

Kanehisa M (2000) Post-Genome Informatics. Oxford University Press, New York

Keedwell E and Narayanan A (2005) Intelligent Bioinformatics. Application of Artificial Intelligence Techniques to Bioinformatics Problems. Wiley, Hoboken, NJ

Kempthorne O (1957) An Introdution to Genetic Statistics. Wiley, New York

Klipp E et al (2005) Systems Biology in Practice. Concepts, Implementation and Application. Wiley-VCH, Hoboken, NJ

Konopka K and Crabbe MJ (eds) (2005) Compact Handbook of Computational Biology. CRC Press, Boca Raton, FL

Kotz S et al (eds) (2005) Encyclopedia of Statistical Sciences. Wiley, Hoboken, NJ

Lange K (2002) Mathematical and Statistical Methods for Genetic Analysis. Springer, New York

Larose DT (2006) Data Mining Methods and Models. Wiley, Hoboken, NJ

Larson RS (ed) (2005) Bioinformatics and Drug Discovery. Humana Press, Totowa, NJ

Lawson AB and Kleinman K (2005) Spatial and Syndromic Surveillance for Public Health. Wiley, Hoboken, NJ

Lengauer T (ed) (2002) Bioinformatics—From Genomes to Drugs. Wiley, New York

Lengauer T (ed) (2007) Bioinformatics—From Genomes to Therapies. Wiley-VCH, Hoboken, NJ

Lesk M (2002) An Introduction to Bioinformatics. Oxford University Press, New York

Letovsky S (ed) (1999) Bioinformatics. Databases and Systems. Kluwer, Boston, MA

Lindley DV (1965) Introduction to Probability and Statistics from a Bayesian Viewpoint. Cambridge University Press, New York

Liu BH (1998) Statistical Genomics. Linkage, Mapping, and QTL Analysis. CRC Press, Boca Raton, FL

Lucy D (2006) Introduction to Statistics for Forensic Scientists Wiley, Hoboken, NJ

Machin D et al (2004) Textbook of Clinical Trials. Wiley, Hoboken, NJ

Malécot G (1969) The Mathematics of Heredity. Freeman, San Francisco, CA

Mangel M (2006) The Theoretical Biologist's Toolbox. Quantitative Methods for Ecology and Evolutionary Biology. Cambridge University Press, New York

Marubini E and Valsecchi MG (2004) Analysing of Survival Data from Clinical Trials and Observational Studies. Wiley, Hoboken, NJ

Mather K (1995) Statistical Analysis in Biology. Methuen, London

Matthews DE and Farewell VT (2007) Using and Understanding Medical Statistics. Karger, Farmington, CT

McLachlan GJ and Ambroise C (2004) Analyzing Microarray Gene Expression Data. Wiley, Hoboken, NJ

Mikkelsen SR and Cortón E (2004) Bioanalytical Chemistry. Wiley, Hoboken, NJ

Miller GK (2006) Probability. Modeling and Applications to Random Processes. Wiley, Hoboken, NJ

Mitchell M (1996) An Introduction to Genetic Algorithms. MIT Press, Cambridge, MA

Mount D (2001) Bioinformatics: Sequence and Genome Analysis. Cold Spring Harbor Laboratory Press, Plainview, NY

Naidoo JJ et al (eds) (2007) Modelling Molecular Structure and Reactivity in Biological Systems. Springer, Berlin, D

Norton NE et al (1983) Methods in Genetic Epidemiology. Karger, New York

Paterson AH (ed) (1997) Molecular Dissection of Complex Traits. CRC Press, Boca Raton, FL

Peltz G (ed) (2005) Computational Genetics and Genomics. Tools for Understanding Disease. Humana Press, Totowa, NJ

Pevsner J (2003) Bioinformatics and Functional Genomics. Wiley, Hoboken, NJ

Pevzner PA (2000) Computational Molecular Biology: An Algorithmic Approach. MIT Press, Cambridge, MA

Pevzner PA (2006) Bio-Informatique Moléculaire. Une Approche Agorithmique. Springer, Berlin, D

Piantadosi S (2005) Clinical Trials. A Methodologic Perspective. Wiley, Hoboken, NJ

Porter TM (2004) Karl Pearson. The Scientific Life in a Statistical Age. Princeton University Press, Princeton, NY

Rao DC and Province MA (eds) (2002) Genetic Dissection of Complex Traits. Academic Press, San Diego, CA
Rashidi HH and Buehler LK (2005) Bioinformatics Basics. Application in Biological Science and Medicine. CRC Press, Boca Raton, FL
Ricard J (2004) Emergent Collective Properties: Networks and Information in Biology. Elsevier, San Diego, CA
Riffenburgh RH (1999) Statistics in Medicine. Academic Press, San Diego, CA
Salzberg S et al (eds) (1999) Computational Methods in Molecular Biology. Elsevier Science, New York
Seref O et al (eds) (2008) Data Mining, Systems Analysis and Optimization in Biomedicine. Springer, Berlin, D
Sham P (1997) Statistics in Human Genetics. Oxford University Press, New York
Shipley B (2001) Cause and Correlation in Biology: A User's Guide to Path Analysis, Structural Equations and Causal Inference. Cambridge University Press, New York
Shortliffe EH et al (eds) (2000) Medical Informatics. Springer, New York
Shoukri MM and Edge VL (1995) Statistical Methods for Health Sciences. CRC Press, Boca Raton, FL
Snedecor GW and Cochran WG (1967) Statistical Methods. Iowa State University Press, Ames, IA
Sokal RR and Rohlf FJ (1969) Biometry. Freeman, San Francisco, CA
Speed T (ed) (2003) Statistical Analysis of Gene Expression Microarray Data. CRC Press, Boca Raton, FL
Spiegelhalter DJ et al (2004) Bayesian Approaches to Clinical Trials and Health-Care Evaluation. Wiley, Hoboken, NJ
Srivastava S (ed) (2006) Informatics in Proteomics. CRC Press, Boca Raton, FL
Stekel D (2003) Microarray Bioinformatics. Cambridge University Press, New York
Tanizaki H (2004) Computational Methods in Statistics and Econometrics. Marcel Dekker, New York
Thomas DC (2004) Statistical Methods in Genetic Epidemiology. Oxford University Press, New York
Townend J (2002) Practical Statistics for Environmental and Biological Scientists. Wiley, New York
Ulin PR et al (2005) Quantitative Methods in Public Health. A Field Guide for Applied Research. Wiley, Hoboken, NJ
Waller LA and Gotway C (2004) Applied Spatial Statistics for Public Health Data. Wiley, Hoboken, NJ
Wang JTL et al (eds) (1999) Pattern Discovery in Biomolecular Data. Tools, Techniques, and Applications. Oxford University Press, New York
Wang JTL et al (eds) (2005) Data Mining in Bioinformatics. Springer, London
Wastney ME et al (1998) Investigating Biological Systems Using Modeling. Strategies and Software. Academic Press, Orlando, FL
Waterman MS (1995) Introduction to Computational Biology: Maps, Sequences and Geneomes. Chapman & Hall, New York
Wit E and McClure J (2004) Statistics for Microarrays. Design, Analysis and Inference. Wiley, Hoboken, NJ
Woodsworth GG (2004) Biostatistics. A Bayesian Introduction. Wiley, Hoboken, NJ
Woolson RF and Clarke WR (2002) Statistical Methods for the Analysis of Biomedical Data. Wiley, New York
Wu R et al (2007) Statistical Genetics of Quantitative Traits Linkage, Maps and QTL. Springer, Berlin, D
Xiong J (2006) Essential Bioinformatics. Cambridge University Press, New York
Yan W and Kang MS (2003) GGE Biplot Analysis. A Graphical Tool for Breeders, Geneticists, and Agronomists. CRC Press, Boca Raton, FL
Yang MC (2000) Introduction to Statistical Methods in Modern Genetics. Taylor & Francis, New York
Young ID (1991) Introduction to Risk Calculation in Genetic Counselling. Oxford University Press, New York
Zelterman D (2004) Discrete Distributions. Application in the Health Sciences. Wiley, Hoboken, NJ
Zhang W and Shmulevich I (2006) Computational and Statistical Approaches to Genomics. Springer, Berlin, D
Zhang W et al (2004) Microarray Quality Control. Wiley, Hoboken, NJ
Ziegler A and Koenig IR (2006) A Statistical Approach to Genetic Epidemiology. Concepts and Applications. Wiley, Hoboken, NJ
Zimmermann K-H (2003) An Introduction to Protein Informatics. Kluwer, New York
Zomaya AY (ed) (2006) Parallel Computing for Bioinformatics and Computational Biology. Wiley, Hoboken, NJ
Zwillinger D (1996) CRC Standard Mathematical Tables and Formulae. CRC Press, Boca Raton, FL

Biotechnology ► *laboratory technology*, ► *molecular biology*, ► *biochemistry*

Allahabadia GN et al (eds) (2003) The Art and Science of Assisted Reproductive Techniques. Taylor & Francis, London, New York
Appasani K (ed) (2005) RNA Interference Technology. From Basic Science to Drug Development. Cambridge University Press, New York
Atherton KT (ed) (2002) Genetically Modified Crops. Assessing Safety. Taylor & Francis, New York
Bajaj YPS (ed) (1999) Transgenic Medicinal Plants. Biotechnology in Agriculture and Forestry, vol 45. Springer, New York
Bajaj YPS (ed) (1999) Transgenic Crops I. Springer, New York
Bajaj YPS (ed) (1999) Transgenic Trees. Springer, New York
Bajaj YPS (ed) (2001) Transgenic Crops II. Springer, New York
Balbás P and Lorence A (eds) (2004) Recombinant Gene Expression. Reviews and Protocols. Humana Press, Totowa, NJ
Baneyx F (ed) (2004) Protein Expression Technologies. Current Status and Future Trends. Horizon Biosience. Wymondham, Norfolk, UK
Baxevanis AD and Quellette BFF (2001) Bioinformatics. A Practical Guide to the Analysis of Genes and Proteins. Wiley, New York
Borrebaeck CAK (ed) (1995) Antibody Engineering. Oxford University Press, New York

Budisa N (2005) Engineering the Genetic Code. Expanding the Amino Acid Repertoire for the Design of Novel Proteins. Wiley-VCH, Hoboken, NJ

Burden DW and Whitney DB (1995) Biotechnology. Proteins to PCR. A Course in Strategies and Lab Techniques. Birkhäuser Boston, MA

Channarayappa PD (2006) Molecular Biotechnology: Principles and Practices. CRC Press, Boca Raton, FL

Craig A et al (eds) (1999) Automation. Academic Press, Orlando, FL

Crooke ST and Lebleu B (eds) (1993) Antisense Research and Applications. CRC Press, Boca Raton, FL

Darvasi F et al (2004) Chemical Genomics. CRC Press, Boca Raton, FL

Demidov VV and Broude NE (eds) (2004) DNA Amplification. Current Technologies and Applications. Horizon biosience. Wymondham, Norfolk

Eder K and Dale B (2000) In Vitro Fertilization. Cambridge University Press, New York

Elçin YM (ed) (2003) Tissue Engineering, Stem Cells, and Gene Therapies. Kluwer, New York

Endreß R (1994) Plant Cell Biotechnology. Springer, New York

Factor PH (ed) (2001) Gene Therapy for Acute and Acquired Diseases. Kluwer, Boston, MA

Fiechter A and Sautter C (eds) (2007) Green Gene Technology. Springer, Berlin, D

Friedmann T and Rossi J (eds) (2006) Gene Transfer. Delivery and Expression of DNA and RNA. Cold Spring Harbor Laboratory Press, Woodbury, NY

Gelissen G (ed) (2005) Production of Recombinant Proteins. Novel Microbial and Eukaryotic Expression Systems. Wiley-VCH, Hoboken, NJ

Glazer AN and Nikaido H (1995) Microbial Biotechnology. Fundamentals of Applied Microbiology. Freeman, New York

Gresshoff PM (ed) (1997) Technology Transfer of Plant Biotechnology. CRC Press, Boca Raton, FL

Grotewold E (ed) (2003) Plant Functional Genomics. Humana Press, Totowa, NJ

Halford N (ed) (2006) Plant Biotechnology. Current and Future Applications of Genetically Modified Crops. Wiley, Hoboken, NJ

Handler AM and James AA (2000) Insect Transgenesis. Methods and Applications. CRC Press, Boca Raton, FL

Hauser H and Fussenegger MM (eds) (2007) Tissue Engineering. Humana Press, Totowa, NJ

Heikki MT and Lynch JM (eds) (1995) Biological Control. Cambridge University Press, New York

Heiser WC (ed) (2004) Gene Delivery to Mammalian Cells vol 1 & 2. Humana Press, Totowa, NJ

Ho AD et al (2006) Stem Cell Transplantation. Wiley, Hoboken, NJ

Hollander AP and Hatton PV (eds) (2004) Biopolymer Methods in Tissue Engineering. Humana Press, Totowa, NJ

Howe C (1995) Gene Cloning and Manipulation. Cambridge University Press, New York

Ikada Y (2006) Tissue Engineering. Fundamentals and Applications. Elsevier, Philadelphia, PA

Kabanov A et al (eds) (1998) Self-Assembling Complexes for Gene Delivery: From Laboratory to Clinical Trial. Wiley, New York

Kingsman SM and Kingsman AJ (1988) Genetic Engineering. An Introduction to Gene Analysis and Exploitation in Eukaryotes. Blackwell Scientific, Oxford

Kreuzer H and Massey A (2001) Recombinant DNA and Biotechnology. A Guide for Teachers. ASM Press, Washington, DC

Kreuzer H and Massey A (2005) Biology and Biotechnology. Science, Applications, and Issues. ASM Press, Washington, DC

Krishnarao A (ed) (2007) Bioarrays. From Basics to Diagnostics. Human, Totowa, NJ

Lanza R (ed) (2004) Stem Cells. 2 Volumes. Elsevier, San Diego, CA

Lanza R et al (eds) (2007) Principles of Tissue Engineering. Academic Press, San Diego, CA

Lattime EC and Gerson SL (eds) (2002) Gene Therapy of Cancer. Academic Press, San Diego, CA

Lauritzen P (ed) (2001) Cloning and the Future of Human Embryo Research. Oxford University Press, New York

Lee K and Kaplan DL (eds) (2007) Tissue Engineering II. Basics of Tissue Engineering and Tissue Applications. Springer, Berlin

Letourneau DK and Burrows EB (eds) (2002) Genetically Engineered Organisms. Assessing Environmental and Human Health Effects. CRC Press, Boca Raton, FL

Machida CA (ed) (2002) Viral Vectors for Gene Therapy. Methods and Protocols. Humana Press, Totowa, NJ

Maclean N (ed) (1995) Animals with Novel Genes. Cambridge University Press, New York

Makrides SC (ed) (2003) Gene Transfer and Expression in Mammalian Cells. Elsevier, Philadelphia, PA

Meager A (ed) (1999) Gene Therapy Technologies. Applications and Regulations. From Laboratory to Clinic. Wiley, New York

Meager A (ed) (2006) The Interferons. Wiley, Hoboken, NJ

Meyers RA (ed) (1995) Molecular Biology and Biotechnology. A Comprehensive Desk Reference. VCH, New York

Minuth WW et al (2005) Tissue Engineering. From Cell Biology to Artificial Organs. Wiley-VCH, Hoboken, NJ

Monsatersky GM and Robl JM (eds) (1995) Strategies in Transgenic Animal Science, ASM Press, Washington, DC

Murray TH and Mehlman MJ (2000) Encyclopedia of Ethical and Policy Issues in Biotechnology. Wiley, New York

Newell-McGloughlin M and Re E (2007) The Evolution of Biotechnology: From Natufians to Nanotechnology. Springer, Berlin, D

Nielse J (ed) (2001) Metabolic Engineering. Springer, New York

Novartis Foundation Symposium (2005) Stem Cells. Nuclear Reprogramming and Therapeutic Applications. Wiley, Hoboken, NJ

Oksman-Caldentey K-M and Barz WH (eds) (2002) Plant Biotechnology and Transgenic Plants. Marcel-Dekker, New York

Old RW and Primrose SB (1994) Principles of Gene Manipulation. An Introduction to Genetic Engineering, Blackwell Scientific, Cambridge, MA

Parekh SR (ed) (2004) The GMO Handbook. Genetically Modified Animals, Microbes and Plants in Biotechnology. Humana Press, Totowa, NJ

Pells S (ed) (2005) Nuclear Reprogramming. Methods and Protocols. Humana Press, Totowa, NJ
Percus J (2002) Mathematics of Genome Analysis. Cambridge University Press, New York
Pinkert CA (ed) (2002) Transgenic Animal Technology. Academic Press, San Diego, CA
Pisano GP (2006) Science Business: The Promise, the Reality and the Future of Biotech. Harvard Business School Press, Cambridge, MA
Pörtner R (ed) (2007) Animal Cell Biotechnology. Humana Press, Totowa, NJ
Potrykus I and Spangenberg G (eds) (1995) Gene Transfer to Plants, Springer, New York
Quesenberry PJ et al (eds) (1998) Stem Cell Biology and Gene Therapy. Wiley, New York
Rajasekaran K et al (eds) (2002) Crop Biotechnology. American Chemical Society, Washington, DC
Ratledge C and Kristiansen B (eds) (2006) Basic Biotechnology. Cambridge University Press, New York
Rawlings DE and Johnson DB (eds) (2007) Biomining. Springer, New York
Scanlon KJ (ed) (1998) Therapeutic Application of Ribozymes. Humana Press, Totowa, NJ
Sgaramella V and Eridani S (eds) (2004) Mammalian Artificial Chromosomes. Humana Press, Totowa, NJ
Shuler ML et al (eds) (1994) Baculovirus Expression System and Biopesticides. Wiley-Liss, New York
Simpson RJ (2003) Purifying Proteins for Proteomics: A Laboratory Manual. Cold Spring Harbor Laboratory Press, Cold Spring Harbor, New York
Smith JE (1996) Biotechnology. Cambridge University Press, New York
Springer CJ (ed) (2004) Suicide Gene Therapy. Methods and Reviews. Humana Press, Totowa, NJ
Stein CA and Krieg AM (eds) (1998) Applied Antisense Oligonucleotide Technology. Wiley, New York
Strauss M and Barranger JA (eds) (1997) Concepts in Gene Therapy. Walter de Gruyter, New York
Suhai S (ed) (1997) Theoretical and Computational Methods in Genome Research. Plenum, New York
Suhai S (ed) (2000) Genomics and Proteomics. Kluwer, Dordrecht, the Netherlands
Sullivan NF (1995) Technology Transfer: Making Most of Your Intellectual Property. Cambridge University Press, New York
Swindell SR et al (eds) (1996) Internet for the Molecular Biologist. Horizon Scientific, Portland, OR
Thomas JA and Fuchs RL (eds) (2002) Biotechnology and Safety Assessment. Academic Press, San Diego, CA
Tzotzos GT (ed) (1995) Genetically Modified Organisms. A Guide to Biosafety. CAB International, Oxford
Vettel E (2006) Biotech: The Countercultural Origins of an Industry. University of Pennsylvania Press, Philadelphia, PA
Vining LC and Stuttrad C (eds) (1994) Genetics and Biochemistry of Antibiotic Production. Butterworth-Heinemann, Stoneham, MA
Vunjak-Novakovic G and Freshney RI (2006) Culture of Cells for Tissue Engineering. Wiley, Hoboken, NJ
Wang K et al (eds) (1995) Transformation of Plants and Soil Microorganisms. Cambridge University Press, New York
Wu J et al (eds) (2007) Complex Medical Engineering. Springer, Berlin, D
Zhao S and Stodolsky M (eds) (2004) Bacterial Artificial Chromosomes. Humana Press, Totowa, NJ

Cancer ▶*diseases*, ▶*aging*, ▶*viruses*, ▶*molecular biology*, ▶*mutation*, ▶*cytology and cytogenetics*

Adams J (ed) (2004) Proteasome Inhibitors in Cancer Therapy. Humana Press, Totowa, NJ
Adams VR and Burke TG (2005) Camptothecins in Cancer Therapy. Humana Press, Totowa, NJ
Advances in Cancer Research (Serial). Elsevier, Philadelphia, PA
Albitar M (ed) (2007) Monoclonal Antibodies. Methods and Protocols. Humana Press, Totowa, NJ
Ali-Osman F (ed) (2005) Brain Tumors. Humana Press, Totowa, NJ
Berenson JR (ed) (2004) Biology and Management of Multiple Myeloma. Humana Press, Totowa, NJ
Bertino JR (ed) (2002) Encyclopedia of Cancer, Academic Press, San Diego, CA
Bonadonna G et al (eds) (2006) Textbook of Breast Cancer: A Clinical Guide to Therapy. Taylor & Francis, London
Boultwood J and Fidler C (eds) (2001) Molecular Analysis of Cancer. Humana Press, Totowa, NJ
Bradbury R (ed) (2007) Cancer. Springer, Berlin, D
Bronchud MH et al (eds) (2007) Principles of Molecular Oncology. Humana Press, Totowa, NJ
Brooks SA and Schumacher U (eds) (2001) Metastasis Research Protocols. Humana Press, Totowa, NJ
Brooks SA and Harris A (eds) (2005) Breast Cancer Research Protocols. Humana Press, Totowa, NJ
Brown DM (2004) Drug Delivery Systems in Cancer Therapy. Humana Press, Totowa, NJ
Caligiuri MA and Lotze MT (2007) Cytokines in the Genesis and Treatment of Cancer. Humana Press, Totowa, NJ
Canellos GP et al (2006) The Lymphomas. Saunders, Philadelphia, PA
Carr BI (ed) (2005) Hepatocellular Cancer. Diagnosis and Treatment. Humana Press, Totowa, NJ
Chang AE et al (eds) (2006) Oncology. An Evidence-Based Approach. Springer, Berlin, D
Choi S-W (ed) (2006) Nutrient-Gene Interactions in Cancer. CRC Press, Boca Raton, FL
Chung LWK et al (eds) (2007) Prostate Cancer Biology, Genetics, and the New Therapeutics. Humana Press, Totowa, NJ
Clayson DB (2001) Toxicological Carcinogens. Lewis Publishers, Boca Raton, FL
Coleman WB and Tsongalis GJ (eds) (2001) The Molecular Basis of Human Cancer. Humana Press, Totowa, NJ
Colotta F and Mantovani A (eds) (2008) Targeted Therapies in Cancer. Myth or Reality? Springer, Berlin, D
Cress AE and Nagle RB (eds) (2006) Cell Adhesion and Cytoskeletal Molecules in Metastasis. Springer, New York
Curiel DT and Douglas JT (eds) (2004) Cancer Gene Therapy. Humana Press, Totowa, NJ

Daoud SS (ed) (2008) Cancer Proteomics. From Bench to Bedside. Humana Press, Totowa, NJ
Degos L et al (eds) (2005) Textbook of Malignant Hematology. Taylor & Francis, London
Dietel M (ed) (2007) Targeted Therapies in Cancer. Springer, Berlin, D
Disis ML (ed) (2006) Immunotherapy of Cancer. Humana Press, Totowa, NJ
Edler L and Kitsos C (eds) (2005) Recent Advances in Quantitative Methods in Cancer and Human Health Risk Assessment. Wiley, Hoboken, NJ
Eeles R et al (1996) Genetic Predisposition to Cancer. Chapman and Hall, New York
Ehrlich M (ed) (2000) DNA Alterations in Cancer: Genetic and Epigenetic Changes. Eaton Publishing Co., Natick, MA
El-Deiry WS (ed) (2003) Tumor Suppressor Genes. Humana Press, Totowa, NJ
El-Deiry WS (ed) (2005) Death Receptors in Cancer Therapy. Humana Press, Totowa, NJ
Ellis CN (ed) (2004) Inherited Cancer Syndromes: Current Clinical Management. Springer, New York
Figg WD (ed) (2004) Handbook of Anticancer Pharmacokinetics and Pharmacodynamics. Humana Press, Totowa, NJ
Fisher DE (ed) (2001) Tumor Suppressor Genes in Human Cancer. Humana Press, Totowa, NJ
Fisher PB (ed) (2007) Cancer Genomics and Proteomics. Methods and Protocols. Humana Press, Totowa, NJ
Fojo AT (ed) (2007) Microtubule Targets in Cancer Therapy. Humana Press, Totowa, NJ
Frank SA (2007) Dynamics of Cancer: Incidence, Inheritance, and Evolution. Princeton University Press, Princeton, NJ
Gabriel JA (ed) (2007) The Biology of Cancer. Wiley, Hoboken, NJ
Garcea R and DiMaio D (eds) (2007) The Papillomaviruses. Springer, Berlin, D
Gasparini G and Hayes DF (eds) (2005) Biomarkers in Breast Cancer. Molecular Diagnostics for Monitoring Therapeutic Effect. Humana Press, Totowa, NJ
Giordano A et al (eds) (2007) Molecular Pathology of Gynecologic Cancer. Humana Press, Totowa, NJ
Giordano A and Romano G (2004) Cell Cycle Control and Dysregulation Protocols. Cyclins, Cyclin-Dependent Kinases, and Other Factors. Humana Press, Totowa, NJ
Giordano A and Soprano KJ (eds) (2003) Cell Cycle Inhibitors in Cancer Therapy. Humana Press, Totowa, NJ
Greaves M (2000) Cancer: The Evolutionary Legacy. Oxford University Press, New York
Greene JN (ed) (2004) Infections in Cancer Patients. Marcel Dekker, New York
Greene FL et al (eds) (2007) AJCC Cancer Staging Atlas. Springer, Berlin, D
Groner B (ed) (2007) Targeted Interference with Signal Transduction Events. Springer, Berlin, D
Habib NA (ed) (2000) Cancer Gene Therapy. Kluwer, Dordrecht, the Netherlands
Hayat H (2006) Immunohistochemistry and In Situ Hybridization of Human Carcinomas. Molecular Genetics, Gastrointestinal Carcinoma, and Ovarian Carcinoma. Elsevier, Philadelphia, PA
Hearing VJ and Leong SPL (eds) (2005) From Melanocytes to Malignant Melanoma. Humana Press, Totowa, NJ
Henderson BE et al (eds) (2003) Hormones, Genes, and Cancer. Oxford University Press, New York
Hesketh RT (1997) The Oncogene and Tumor Suppressor Gene Factsbook. Academic Press, San Diego, CA
Hodgson S and Maher E (1999) A Practical Guide to Human Cancer Genetics. Cambridge University Press, New York
Holland EC (ed) (2004) Mouse Models of Human Cancer. Wiley, Hoboken, NJ
Hunt KK et al (eds) (2007) Gene Therapy for Cancer. Humana Press, Totowa, NJ
Hunt KK et al (eds) (2008) Breast Cancer. Springer, Berlin, D
ICRP (1999) Genetic Susceptibility to Cancer. Pergamon/Elsevier Science, New York
Iland H (2006) Myeloid Leukemia. Methods and Protocols. Humana Press, Totowa, NJ
Illidge T and Johnson PWM (eds) (2005) Lymphoma. Methods and Protocols. Humana Press, Totowa, NJ
Karp J (ed) (2007) Acute Myelogenous Leukemia. Humana Press, Totowa, NJ
Kastan MB (ed) (1997) Checkpoint Controls and Cancer. Cold Spring Harbor Laboratory Press, Cold Spring Harbor, NY
Kelloff GJ et al (eds) (2004/2005) Cancer Chemoprevention. Vol I. Promising Cancer Chemopreventiver Agents. Vol II. Strategies for Cancer Chemoprevention. Humana Press, Totowa, NJ
Kelly K et al (eds) (2007) Gene Therapy for Cancer. Humana Press, Totowa, NJ
Kitchin KT (ed) (1999) Carcinogenicity. Testing, Predicting, and Interpreting Chemical Effects. Marcel Dekker, New York
Knudsen S (2006) Cancer Diagnostics with DNA Microarray. Wiley, Hoboken, NJ
Langdon SP (ed) (2004) Cancer Cell Culture. Methods and Protocols. Humana Press, Totowa, NJ
LaRochelle WJ and Shimkets RA (eds) (2005) The Oncogenomics Handbook. Humana Press, Totowa, NJ
La Thangue NB and Bandara LR (2002) Targets for Cancer Chemotherapy. Transcription Factors and Other Nuclear Proteins. Humana Press, Totowa, NJ
Lattime EC and Gerson SL (1998) Gene Therapy of Cancer. Translational Approaches from Preclinical to Clinical Implementation. Academic Press, San Diego, CA
Liebler DC (ed) (2005) Proteomics in Cancer Research. Wiley, Hoboken, NJ
Lindahl T (ed) (1996) Genetic Instability in Cancer. Cold Spring Harbor Laboratory Press, Cold Spring Harbor, NY
Mansel RE et al (eds) (2007) Metastasis of Breast Cancer. Springer, Berlin
Marcus AI (1994) Cancer from Beef. DES, Federal Regulation, and Consumer Confidence. Johns Hopkins University Press, Baltimore, MD
McKinnell RG et al (eds) (2006) The Biological Basis of Cancer. Cambridge University Press, New York
Mendelsohn J et al (eds) (1994) The Molecular Basis of Cancer. Saunders, Philadephia, PA
Meyers C (ed) (2007) AIDS-Associated Viral Oncogenesis. Springer, Berlin, D
Meyers RA (2007) Cancer. From Mechanisms to Therapeutic Approaches. Wiley, Hoboken, NJ

Minson AC et al (eds) (1954) Viruses and Cancer. Cambridge University Press, New York

Mitelman F (1997) Catalog of Chromosomal Aberrations in Cancer. Wiley, New York

Mocellin S (ed) (2007) Microarray Technology and Cancer Gene Profiling. Springer, Berlin, D

Mor G and Alvero A (eds) (2007) Apoptosis and Cancer. Methods and Protocols. Humana Press, Totowa, NJ

Morrison PJ et al (eds) (2002) Familial Breast and Ovarian Cancer: Genetics, Screening and Management. Cambridge University Press, Cambridge

Morse MA et al (2004) Handbook of Cancer Vaccines. Humana Press, Totowa, NJ

Nagl S (2005) Cancer Bioinformatics. From Therapy Design to Treatment. Wiley, Hoboken, NJ

Nathan DG (2007) The Cancer Treatment Revolution: How Smart Drugs and Other New Therapies Are Renewing Our Hope and Changing the Face of Medicine. Wiley, Hoboken, NJ

Nakamura RM et al (2004) Cancer Diagnostics, Current and Future Trends. Humana Press, Totowa, NJ

Newton R et al (1999) Infections and Human Cancer. Cold Spring Harbor Laboratory Press, Cold Spring Harbor, NY

Nickoloff BJ (ed) (2001) Melanoma Techniques and Protocols. Molecular Diagnosis, Treatment and Monitoring. Humana Press, Totowa, NJ

Offit K (1998) Clinical Cancer Genetics. Wiley, New York

Panasci LC and Alaoui-Jamali MA (eds) (2004) DNA Repair in Cancer Therapy. Humana Press, Totowa, NJ

Pandolfi PP and Vogt PK (eds) (2007) Acute Promyelitic Leukemia. Molecular Genetics, Mouse Models and Targeted Therapy. Springer, Berlin, D

Parsonnet J (ed) (1999) Microbes and Malignancy. Infection as a Cause of Human Cancer. Oxford University Press, New York

Peters G and Vousden KH (1997) Oncogenes and Tumour Suppressor. IRL/Oxford University Press, New York

Pettit GR et al (1994) Anticancer Drugs from Animals, Plants and Microorganisms. Wiley, New York

Pfragner R and Freshney RI (eds) (2004) Culture of Human Tumor Cells. Wiley, Hoboken, NJ

Pinedo HM (2006) Drugs Affecting Growth of Tumours. Birkhäuser, Basel, Switzerland

Ponder BAJ et al (eds) (1996) Genetics and Cancer. A Second Look. Cold Spring Harbor Laboratory Press, Cold Spring Harbor, NY

Prendergast GV (ed) (2004) Molecular Cancer Therapeutics: Stratagies for Drug Discovery and Development. Wiley, Hoboken, NJ

Priestman T (2008) Cancer Chemotherapy in Clinical Practice. Springer, Berlin, D

Pui C-H (ed) (1999) Childhood Leukemias. Cambridge University Press, New York

Retting RA et al (2007) False Hope: Bone Marrow Transplantation for Breast Cancer. Oxford University Press, New York

Rollins BJ (ed) (1999) Chemokines and Cancer. Humana Press, Totowa, NJ

Roulston JE and Bartlett MS (eds) (2004) Molecular Diagnosis of Cancer. Methods and Protocols. Humana Press, Totowa, NJ

Russo J and Russo IH (2004) Molecular Basis of Breast Cancer: Prevention and Treatment. Springer, Berlin, Germany

Saltz LB (2006) Colorectal Cancer. Humana Press, Totowa, NJ

Schneider K (2001) Counseling About Cancer. Strategies for Genetic Counseling. Wiley, New York

Schönthal AH (ed) (2004) Checkpoint Controls and Cancer. Vol I. Reviews and Model Systems. Vol II. Activation and Regulation Protocols. Humana Press, Totowa, NJ

Schwab M (ed) (2001) Encyclopedic Reference to Cancer. Springer, New York

Schwartz GK (ed) (2005) Combination Cancer Therapy. Modulators and Potentiators. Humana Press, Totowa, NJ

Senn H-J and Morant R (eds) (2003) Tumor Prevention and Genetics. Springer, New York

Sherbet G (2003) Genetic Recombination in Cancer. Elsevier, Philadelphia, PA

Sherbet GV and Lakshmi MS (1997) Genetics of Cancer. Genes Associated with Cancer Invasion, Metastasis, and Cell Proliferation. Academic Press, San Diego, CA

Shields AF (ed) (2007) In Vivo Imaging of Cancer Therapy. Humana Press, Totowa, NJ

Silva O and Zurrida S (eds) (2006) Breast Cancer. Elsevier, Philadelphia, PA

Singh G and Rabbani SA (eds) (2005) Bone Metastasis. Experimental and Clinical Aspects. Humana Press, Totowa, NJ

Soiffer RJ (ed) (2004) Stem Cell Transplantation for Hematological Malignancies. Humana Press, Totowa, NJ

Steinberg K (ed) (2001) The Genetic Basis of Cancer. Oxford University Press, New York

Stillman B and Stewart D (eds) (2006) Molecular Approaches to Controlling Cancer. Cold Spring Harbor Symposium on Quantitative Biology. vol 70. Cold Spring Harbor Laboratory Press, Cold Spring Harbor, NY

Stuhler G and Walden P (2002) Cancer Immune Therapy: Current and Future Strategies. Wiley, Hoboken, NJ

Su GH (ed) (2005) Pancreatic Cancer. Methods and Protocols. Humana Press, Totowa, NJ

Swansbury J (ed) (2003) Cancer Cytogenetics. Methods and Protocols. Humana Press, Totowa, NJ

Teicher BA (ed) (2002) Tumor Models in Cancer Research. Humana Press, Totowa, NJ

Teicher BA and Ellis LM (2007) Antiangiogenic Agents in Cancer Therapy. Humana Press, Totowa, NJ

Thompson JF et al (eds) (2003) Textbook of Melanoma. Martin Dunitz, New York

Tomatis L (1990) Air Pollution and Human Cancer. Springer, Berlin

Tooze J et al (eds) (1999) Infections and Human Cancer. Cold Spring Harbor Laboratory Press, Cold Spring Harbor, NY

Utsunomiya J et al (eds) (1999) Familial Cancer Prevention. Molecular Epidemiology: A New Strategy Toward Cancer Control. Wiley, New York

Vogelzang NJ et al (eds) (2006) Comprehensive Textbook of Genitourinary Oncology. Lippincott, Williams & Wilkins, Philadelphia, PA

Vogt PK and Verma IM (eds) (1995) Oncogene Techniques. Academic Press, San Diego, CA

Walther W and Stein U (eds) (2000) Gene Therapy of Cancer. Humana Press, Totowa, NJ

Warshawsky D and Landolph JR Jr (2002) Molecular Carcinogenesis. CRC Press, Boca Raton, FL
Wartofsky L and Nostrand D (eds) (2006) Thyroid Cancer. A Comprehensive Guide to Clinical Management. Humana Press, Totowa, NJ
Weber GF (2007) Molecular Mechanisms of Cancer. Springer, Berlin, D
Weinberg RA (1996) Racing to the Beginning of the Road. The Search for the Origin of Cancer. Crown, New York
Weinberg RA (2006) The Biology of Cancer. Garland Science, New York
Welschof M and Krauss J (eds) (2002) Recombinant Antibodies for Cancer Therapy. Methods and Protocols. Humana Press, Totowa, NJ
Wickstrom E (ed) (1998) Clinical Trials of Genetic Therapy with Antisense DNA and DNA Vectors. Marcel Dekker, New York
Wolman SR and Sell S (eds) (1997) Human Cytogenetic Cancer Markers. Humana Press, Totowa, NJ.
Zhou J (ed) (2007) Microtubule Protocols. Humana Press, Totowa, NJ
zur Hausen H (2006) Infections Causing Cancer. Wiley, Hoboken, NJ

Conservation ►*evolution*, ►*population genetics*

Avise JC and Hamrick JL (eds) (1996) Conservation Genetics: Case Histories from Nature. Chapman and Hall, New York
Beattie A and Ehrlich PR (2001) Wild Solutions: How Biodiversity is Money in the Bank. Yale University Press, New Haven, CN
Caughly G and Gunn A (1995) Conservation Biology in Theory and Practice. Blackwell Science, Cambridge, MA
Clark TW (1997) Averting Extinction. Reconstructing Endangered Species Recovery. Yale University Press, New Haven, CN
Conner JK and Hartl DL (2004) A Primer of Ecological Genetics. Sinauer, Sunderland, MA
Costa LG and Eaton DL (2006) Gene–Environment Interactions. Wiley, Hoboken, NJ
Ehrenfeld D (ed) (1995) Plant Conservation. Blackwell Science, Cambridge, MA
Ehrlich PR (2000) Human Natures: Genes, Cultures and the Human Prospect. Island Press, Washington, DC
Frankel OH et al (1995) The Conservation of Plant Biodiversity. Cambridge University Press, New York
Frankham R et al (2004) A Primer of Conservation Genetics. Cambridge University Press, New York
Freeland JR (2005) Molecular Ecology. Wiley, Hoboken, NJ
Gotelli NJ and Ellison AM (2004) A Primer of Ecological Statistics. Sinauer, Sunderland, MA
Hawkes JG et al (2000) The *Ex Situ* Conservation of Plant Genetic Resources. Kluwer, Dordrecht, the Netherlands
Hunter ML Jr (1995) Fundamentals of Conservation Biology. Blackwell Science, Cambridge MA
Huston MA (1994) Biological Diversity. The Coexistence of Species in Changing Landscapes. Cambridge University Press, New York.
Juo ASR and Freed RD (eds) (1993) Agriculture and Environment. Bridging Food Production and Environmental Protection in Developing Countries. American Society of Agronomy, Madison, WI
Kammen DM and Hassenzahl DO (1999) Should We Risk It? Exploring Environmental, Health, and Technological Problem Solving. Princeton University Press, Princeton, NJ
Kjellsson G and Amman K (1997–1999) Methods for Risk Assessment of Transgenic Plants. Birkhäuser, Basel, Switzerland
Krupnick GA and Kress J (eds) (2005) Plant Conservation. A Natural History Approach. University Chicago Press, Chicago, IL
Landweber LF and Dobson AP (eds) (2000) Genetics and the Extinction of Species. Princeton University Press, Princeton, NJ
Lawton JH and May RM (eds) (1995) Extinction Rates. Oxford University Press, New York
Levin DA (2000) The Origin, Expansion, and Demise of Plant Species. Oxford University Press, New York
Levin SA (ed) (2000) Encyclopedia of Biodiversity. Academic Press, San Diego, CA
Loreau M et al (eds) (2002) Biodiversity and Ecosystem Functioning. Synthesis and Perspectives. Oxford University Press, Oxford.
Meffe GK and Carrol CR (1995) Principles of Conservation Biology. Sinauer, Sunderland, MA
O'Riordan T and Stoll-Kleeman S (2002) Biodiversity, Sustainability and Human Communities: Protecting Beyond the Protected. Cambridge University Press, New York
Paehlke R (ed) (1995) Conservation and Environmentalism. An Encyclopedia. Garland, New York
Primack RB (2006) Essentials of Conservation Biology. Sinauer, Sunderland, MA
Primack RB (2004) A Primer of Conservation Biology. Sinauer, Sunderland, MA
Purvis A et al (eds) (2005) Phylogeny and Conservation. Cambridge University Press, New York
Reaka-Kudle ML et al (eds) (1996) Biodiversity II. Understanding and Protecting Our Biological Resources. Joseph Henry Press (National Academy of Science, USA), Washington, DC
Samways MJ (1994) Insect Conservation Biology. Chapman and Hall, New York
Samways MJ (2005) Insect Diversity Conservation. Cambridge University Press, New York
Sax DF et al (eds) (2005) Species Invasions. Insight into Ecology, Evolution, and Biogeography. Sinauer, Sunderland, MA
Shaw IC and Chadwick J (1996) Principles of Environmental Toxicology. Taylor & Francis, London
Singh SN and Tripathi RD (eds) (2007) Environmental Bioremediation Technologies. Springer, Berlin, D
Watson PF and Holt WV (eds) (2000) Cryobanking the Genetic Resource? Wildlife Conservation for the Future? Taylor & Francis, New York
Wilson EO (2002) The Future of Life. Knopf, New York
Wilson EO (2006) The Creation. An Appeal to Save Life on Earth. Norton, New York

Winston JE (2000) Describing Species: Practical Taxonomic Procedure for Biologists. Columbia University Press, New York

Woodroffe R et al (eds) (2005) People and Wildlife. Conflict or Co-existence? Cambridge University Press, New York

Cytology and Cytogenetics ►*molecular biology*, ►*evolution*, ►*cancer*

Allan V (2000) Protein Localization by Fluorescence Microscopy. A Practical Approach. Oxford University Press, New York

Andrews L et al (eds) (2007) Telomerase Inhibition. Strategies and Protocols. Humana Press, Totowa, NJ

Appels R et al (1998) Chromosome Biology. Kluwer Academic, Norwell, MA

Berezney R and Riordan JF (eds) (1997) Nuclear Matrix. Structural and Functional Organization. Academic Press, San Diego, CA

Berrios M (ed) (1997) Nuclear Structure and Function. Methods in Cell Biology, vol 53. Academic Press, San Diego, CA

Bickmore WA (ed) (1999) Chromosome Structural Analysis —A Practical Approach. Oxford University Press, Oxford

Blackburn EH, Greider CW, Borgaonkar, D (eds) (1997) Chromosomal Variation in Man. A Catalog of Chromosomal Variants and Anomalies. Wiley-Liss, New York

Braga PC and Ricci D (eds) (2004) Atomic Force Microscopy. Humana Press, Totowa, NJ

Chen B (2001) Grauzone and Completion of Meiosis During *Drosophila* oogenesis. Kluwer, Dordrecht, the Netherlands

Choo KH (1997) The Centromere. Oxford University Press, New York

Cook P (2001) Principles of Nuclear Structure and Function. Wiley, New York

Darlington CD and Wylie AP (1955) Chromosome Atlas of Flowering Plants. Allen & Unwin, London

Davis KE and Warren ST (eds) (1993) Genome Rearrangement and Stability. Cold Spring Harbor Laboratory Press, Cold Spring Harbor, NY

DeGrouchy J and Turleau C (1984) Clinical Atlas of Human Chromosomes. Wiley, New York

De Lange T et al (eds) (2005) Telomeres. Cold Spring Harbor Laboratory Press, Cold Spring Harbor, NY

Double JA and Thompson MJ (2002) Telomeres and Telomerase. Methods and Protocols. Humana Press, Totowa, NJ

Drlica K and Riley M, (eds) (1990) The Bacterial Chromosome. American Society for Microbiology, Washington, DC

Egel R and Lankenau D-H (eds) (2008) Recombination and Meiosis. Crossing-Over and Disjunction. Springer, Berlin, D

Elgin SCR and Workman JL (eds) (2001) Chromatin Structure and Gene Expression, Oxford University Press, New York

Epstein CJ (1986) Consequences of Chromosome Imbalance: Principles, Mechanisms and Models. Cambridge University Press, New York

Exbrayat J-M (2001) Genome Visualization by Classic Methods in Light Microscopy. CRC Press, Boca Raton, FL

Fan Y-S (ed) (2003) Molecular Cytogenetics. Protocols and Applications. Humana Press, Totowa, NJ

Fantes P and Beggs J (eds) (2001) The Yeast Nucleus. Oxford University Press, New York

Fukui K and Nakyama S (eds) (1996) Plant Chromosomes. Laboratory Methods. CRC Press, Boca Raton, FL

Fukui K and Ushiki T (eds) (2007) Chromosome Nanoscience and Technology. CRC Press, Boca Raton, FL

Gersen SL and Keagle MB (eds) (2005) The Principles of Clinical Cytogenetics. Humana Press, Totowa, NJ

Gécz J and Sutherland GR (eds) (2003) Nucleotide and Protein Expansions and Human Disease. Karger, Farmington, CT

Glauert AN and Lewis PR (1999) Biological Specimen Preparation for Transmission Electron-Microscopy. Princeton University Press, Princeton, NJ

Goodhew P et al (1997) Introduction to Scanning Transmission Electron Microscopy. Springer, New York

Gosden JR (ed) (1994) Chromosome Analysis Protocols. Humana Press, Totowa, NJ

Gustafson JP and Flavell RB (eds) (1996) Genomes of Plants and Animals. 21st Stadler Genetics Symposium. Plenum, New York

Hagerman R and Hagerman P (eds) (2002) Fragile X Syndrome. Johns Hopkins University Press, Baltimore, MD

Harman OS (2004) The Man Who Invented the Chromosome: A Life of Cyrill Darlington. Harvard University Press, Cambridge, MA

Harris H (1999) The Cell: The Development of an Idea. Yale University Press, New Haven, CT

Heim S and Mitelman F (1995) Cancer Cytogenetics. Wiley-Liss, New York

Henderson DS (ed) (2004) *Drosophila* Cytogenetics Protocols. Humana Press, Totowa, NJ

Henriquez-Gil N et al (eds) (1997) Chromosomes Today. Chapman and Hall, New York

Herzberg AJ et al (1999) Color Atlas of Normal Cytology. Harcourt Brace & Co., New York

Hsu TC and Benirschke K (eds) (1967–1977) An Atlas of Mammalian Chromosomes. Springer, New York

Humphrey T and Brooks G (eds) (2005) Cell Cycle Control. Mechanisms and Protocols. Humana Press, Totowa, NJ

Hutchison C and Glover DM (eds) (1995) Cell Cycle Control. Oxford University Press, New York

Hyams JS and Lloyd CW (1994) Microtubules. Wiley-Liss, New York

Jauhar PP (ed) (1996) Methods of Genome Analysis of Plants. CRC Press, Boca Raton, FL

Javois LC (ed) (1995) Immunocytochemical Methods and Protocols. Humana Press, Totowa, NJ

Jena BP (2002) Atomic Force Microscopy in Cell Biology. Academic Press, San Diego, CA.

Jena BP and Hoerber JK (eds) (2006) Force Microscopy. Wiley, Hoboken, NJ

John B and Lewis KR (1968) The Chromosome Complement. Springer, New York

Kiernan JA and Mason I (eds) (2002) Microscopy and Histology for Molecular Biologists. A User Guide. Portland, London
Kipling D (1995) The Telomere. Oxford University Press, New York
Kundu TK and Dasgupta D (eds) (2007) Chromatin and Disease. Springer, Berlin, D
Kuo J (ed) (2007) Electron Microscopy. Methods and Protocols. Humana Press, Totowa, NJ
Lacey AJ (ed) (1999) Light Microscopy in Biology. A Practical Approach. Oxford University Press, New York
Lance V (ed) (2003) Vertebrate Sex Determnination. Karger, Farmington, CT
Macgregor HC (1993) An Introduction to Animal Cytogenetics. Chapman and Hall, New York
Mark HFL (ed) (2000) Medical Cytogenetics. Marcel Dekker, New York
Maunsbach AB and Afzelius BA (1998) Biomedical Electron Microscopy. Academic Press, San Diego, CA
Migeon BR (2007) Females Are Mosaics: X Inactivation and Sex Differences in Disease. Oxford University Press, Oxford, UK
Miller OJ and Thurman E (2000) Human Chromosomes. Springer, New York
Moore RJ (1973) Index to Plant Chromosome Numbers 1967–1971. Oosthoek, Utrecht, the Netherlands
Morel G and Raccurt M (2003) PCR/RT-PCR In-Situ Light and Electron Microscopy. CRC Press, Boca Raton, FL
Nigg EA (ed) (2004) Centrosomes in Development and Disease. Wiley, Hoboken, NJ
Obe G and Natarayan AT (eds) (1994) Chromosomal Aberrations. Origin and Significance. Springer, New York
Obe G and Vijayalaxmi (eds) (2007) Chromosomal Aberrations. Methods, Results and Importance in Human Health. Springer, Berlin, D
O'Brien SJ et al (2006) Atlas of Mammalian Chromosomes. Wiley, Hoboken, NJ
Olmo E and Redi CA (eds) (2000) Chromosomes Today. Birkhäuser, Basel, Switzerland
Palasso RE and Scahatten GP (eds) (2000) The Centrosome in Cell Replication and Early Development. Academic Press, San Diego, CA
Peltz G (ed) (2005) Computational Genetics and Genomics. Tools for Understanding Disease. Humana Press, Totowa, NJ
Polak J and Van Noorden S (1997) Introduction to Immunocytochemistry. Springer, New York
Popescu P et al (eds) (2000) Techniques in Animal Cytogenetics. Springer, New York
Rautenstrauß BW and Liehr T (2002) FISH Technology. Springer, Berlin
Rieder CL and Matsudaira P (eds) (1999) Mitosis and Meiosis. Academic Press, San Diego, CA
Rooney DE (ed) (2001) Human Cytogenetics. Oxford University Press, New York
Rooney DE and Czepulkowski BH (1997) Human Chromosome Preparations: Essential Techniques. Wiley, New York
Rooney DE and Czepulkowski BH (eds) (2001) Human Cytogenetics. Oxford University Press, New York
Rudolph KL (ed) (2008) Telomeres and Telomerase in Aging, Disease, and Cancer. Molecular Mechanisms of Adult Stem Cell Ageing. Springer, Berlin, D
Ruffolo RR Jr et al (eds) (1997) Cell Cycle Regulation. Harwood Academic Langhorne, PA
Ruzin SE (1999) Plant Microtechnique and Microscopy. Oxford University Press, New York
Shaffer LG and Tommerup N (eds) (2005) An International System for Human Cytogenetic Nomenclature (2005) Karger, Farmington, CT
Shapiro HM (1994) Practical Flow Cytometry. Wiley-Liss, New York
Sharma AK and Sharma A (1994) Chromosome Techniques. Harwood, Langhorne, PA
Sharma AK and Sharma A (eds) (1999) Plant Chromosomes: Analysis, Manipulation and Engineering. Harwood, Amsterdam
Sheppard C and Shotton D (1997) Confocal Laser Scanning Microscopy. Springer, New York
Singh RJ (2003) Plant Cytogenetics. CRC Press, Boca Raton, FL
Speel E et al (eds) (2003) Chromosome Analysis Protocols. Humana Press, Totowa, NJ
Stein G (ed) (1998) The Molecular Basis of Cell Cycle and Growth Control. Wiley, New York
Taatjes DJ and Mossman BT (eds) (2005) Cell Imaging Techniques. Methods and Protocols. Humana Press, Totowa, NJ
Takagi N (ed) (2003) Vertebrate Sex Chromosomes. Karger, Farmington, CT
Takeyasu K and Nagata K (eds) (2007) Nuclear Dynamics. Molecular Biology and Visualization of the Nucleus. Springer, Berlin, D
Therman E (1993) Human Chromosomes: Structure, Behavior, Effects. Springer, New York
Turner BM (2001) Chromatin and Gene Regulation. Mechanisms in Epigenetics. Blackwell Science, Cambridge, MA
Van Driel R and Otte AP (1997) Nuclear Organization, Chromatin Structure and Gene Expression. Oxford University Press, New York
Wagner RP et al (1993) Chromosomes. A synthesis. Wiley-Riss, New York
Wang XF and Herman B (eds) (1996) Fluorescence Imaging Spectroscopy and Microscopy. Wiley, New York
Wegener R-D (ed) (1999) Diagnostic Cytogenetics. Springer, New York
Wolffe A (1998) Chromatin. Structure and Function. Academic Press, San Diego, CA
Zlatanova J and Leuba SH (eds) (2004) Chromatin Structure and Dynamics. State-of-the-Art. Elsevier, San Diego, CA

Development ▸*molecular biology*, ▸*neurology*

Allis CD et al (eds) (2007) Epigenetics. Cold Spring Harbor Laboratory Press, Cold Spring Harbor, NY
Al-Rubeai M and Fussenegger M (eds) (2007) Systems Biology. Springer, Berlin, D

Annual Review of Cell and Developmental Biology. Palo Alto, CA

Arias AM and Stewart A (2002) Molecular Principles of Animal Development. Oxford University Press, New York

Arthur W (2000) The Origin of Animal Body Plans. Cambridge University Press, New York

Bab I et al (2007) Micro-Tomographic Atlas of the Mouse Skeleton. Humana Press, Totowa, NJ

Bainbridge D (2001) Making Babies: The Science of Pregnancy. Harvard University Press, Cambridge, MA

Bainbridge D (2003) The XC in Sex: How the Chromosome Controls Our Lives. Harvard University Press, Cambridge, MA

Barry JM (2002) Molecular Embryology: How Molecules Give Birth to Animals. Taylor & Francis. Hamden, CT

Bate M and Martinez Arias A (eds) (1994) The Development of *Drosophila melanogaster*. Cold Spring Harbor Laboratory Press, Cold Spring Harbor, NY

Batygina TB (ed) (2002) Embryology of Flowering Plants. Terminology and Concepts. Science Publishers, Inc., Enfield, NH

Beck S and Olek A (eds) (2003) The Epigenome. Molecular Hide and Seek. Wiley, Hoboken, NJ

Birkhead TR and Moller AP (1998) Sperm Competition and Sexual Selection. Academic Press, Orlando, Fl

Bowman J (ed) (1994) Arabidopsis. An Atlas of Morphology and Development. Springer, New York

Brookes M and Zietman A (1998) Clinical Embryology. A Color Atals and Text. CRC Press, Boca Raton, FL

Butler MG and Meaney FJ (eds) (2005) Genetics of Developmental Disabilities. Taylor & Francis, Boca Raton. FL

Campos-Ortega JA and Hartenstein V (1997) The Embryonic Development of *Drosophila melanogaster*. Springer, New York

Cardew G and Goode JA (2001) The Molecular Basis of Skeletogenesis. Novartis Foundation Symposium 232. Wiley, New York

Carrell DT (2006) Genetics of Male Infertility. Humana Press, Totowa, NJ

Carroll SB et al (2001) From DNA to Diversity. Molecular Genetics and the Evolution of Animal Design. Blackwell Science, Malden, MA

Chadwick DJ and Cardew G (eds) (1998) Epigenetics. Wiley, New York

Csillag A (2005) Atlas of Sensory Organs. Functionl and Clinical Anatomy. Humana Press, Totowa, NJ

Coen E (1999) The Art of Genes: How Organisms Make Themselves. Oxford UniversityPress, New York

Cohen M et al (2002) Overgrowth Syndromes. Oxford University Press, New York

Cronk QCB et al (eds) (2002) Developmental Genetics and Plant Development. Taylor & Francis. London

Davidson EH (2001) Genomic Regulatory Systems: Development and Evolution. Academic Press, San Diego, CA

Davidson EH (2006) The Regulatory Genome. Gene Regulatory Networks in Development and Evolution. Academic Press/Elsevier, San Diego, CA

Davies JA (2005) Mechanisms of Morphogenesis: The Creation of Biological Form. Academic Press/Elsevier, San Diego. CA

De Jonge C and Christopher BC (eds) (2006) The Sperm Cell: Production, Maturation, Fertilization, Regeneration. Cambridge University Press, New York

DeWitt TJ and Schreiner SM (eds) (2004) Phenotypic Plasticity. Oxford University Press, New York

Dickson RB and Salomon DS (eds) (1998) Hormones and Growth Factors in Development and Neoplasia. Wiley-Liss, New York

Dreger DA (2004) One of Us: Conjoined Twins and the Future of Normal. Harvard University Press, Cambridge, MA

Duboule D (ed) (1994) Guidebook to the Homeobox Genes. Oxford University Press, New York

Engel E and Antonarakis SE (2001) Genomic Imprinting and Uniparental Disomy in Medicine. Clinical and Molecular Aspects. Wiley, New York

Epstein CJ et al (eds) (2004) Inborn Errors of Development. The Molecular Basis of Clinical Disorders of Morphogenesis. Oxford University Press, New York

Fernandes J et al (eds) (2006) Inborn Metabolic Diseases. Diagnosis and Treatment. Springer, New York

Ferretti P et al (eds) (2006) Embryos, Genes and Birth Defects. Wiley, Hoboken, NJ

Finch CE and Kirkwood TBL (2000) Chance, Development, and Aging. Oxford University Press, Oxford

Findley JK (ed) (1994) Molecular Biology of the Female Reproductive System. Academic Press, San Diego, CA

Fisher JP (ed) (2007) Tissue Engineering. Springer, New York

Fleming T and Muse S (eds) (2002) Cell-Cell Interactions. Oxford University Press, New York

Forgacs G and Newman SA (2005) Biological Physics of the Developing Embryo. Cambridge University Press, New York

Fosket DE (1994) Plant Growth and Development: A Molecular Approach. Academic Press, San Diego, CA

Foster RG and Kreitzman L (2004) Rhythms of Life. The Biological Clocks that Control the Daily Lives of Every Living Thing. Profile, London

Freshney RI et al (2007) Culture of Human Stem Cells. Wiley, Hoboken, NJ

Gehring WJ (1998) Master Control Genes in Development and Evolution. The Homeobox Story. Yale University Press, New Haven, CT

Gilbert LI et al (eds) (1996) Metamorphosis. Postembryonic Reprogramming of Gene Expression in Amphibian and Insect Cells. Academic Press, San Diego, CA

Gilbert SF (2006) Developmental Biology. Sinauer, Sunderland, MA

Gilbert SF and Raunio AM (eds) (1997) Developmental Diversity. Sinauer, Sunderland, MA

Glover TD and Barratt CLR (eds) (1999) Male Fertility and Infertility. Cambridge University Press, New York

Gu J (1997) Analytical Morphology. Theory, Applications and Protocols. Springer, New York

Gundelfinger ED et al (eds) (2006) Cell Communication in Nervous and Immune System. Springer, New York

Hall BK (1999) The Neural Crest in Development and Evolution. Springer, New York

Hall DH and Altun ZF (2006) *C. elegans* ATLAS. Cold Spring Harbor Laboratory Press, Cold Spring Harbor, NY

Hall JC (ed) (2003) Genetics and Molecular Biology of Rhythms in *Drosophila* and Other Insects. Adv Genet 48:244
Hall MN et al (eds) (2004) Cell Growth: Control of Cell Size. Cold Spring Harbor Laboratory Press, Woodbury, NY
Handel MA (ed) (1997) Meiosis and Gametogenesis. Academic Press, San Diego, CA
Harding R and Bocking AD (2001) Fetal Growth and Development. Cambridge University Press
Harrison LG (1993) Kinetic Theory of Living Patterns. Cambridge University Press, New York
Heath JK (2001) Principles of Cell Proliferation. Blackwell, Oxford, UK
Held LI Jr (2002) Imaginal Discs. The Genetic and Cellular Logic of Pattern Formation. Cambridge University Press, New York
Howell SH (1998) Molecular Genetics of Plant Development. Cambridge University Press, New York
Hunter RHF (1995) Sex Determination, Differentiation and Intersexuality in Placental Mammals. Cambridge University Press, New York
Inui A (ed) (2006) Epigenetic Risks of Cloning. CRC Press, Boca Raton, FL
Jirasek JE (2001) An Atlas of the Human Embryo and Fetus. A Photographic Review of Human Prenatal Development. Pathenon, New York
Johné Liu X (ed) (2005) Xenopus Protocols. Cell Biology and Signal Transduction. Humana Press, Totowa, NJ
Jones S (2003) The Descent of Men. Houghton Mifflin, Boston, MA
Jowett T (1997) Tissue *In Situ* Hybridization: Methods in Animal Development. Wiley, New York
Kalter H (ed) (2007) Mortality and Maldevelopment. Congenital Cardiovascular Malformations. Springer, Berlin, D
Kalthoff K (1995) Analysis of Biological Development. McGraw-Hill, New York
Kaufman MH (1992) The Atlas of Mouse Development. Academic Press, San Diego, CA
Kaufman MH and Bard JBL (1999) The Anatomical Basis of Mouse Development. Academic Press, San Diego, CA
Kiessling AA and Anderson SC (2003) Human Embryonic Stem Cells: An Introduction to the Science and Therapeutic Potential. Jones & Bartlett, Boston, MA
Kolata G (1997) Clone: The Road to Dolly and the Path Ahead. Morrow, Allen Lane, New York
Lash J (1999) Interactive Embryology: The Human Embryo Program. Sinauer, Sunderland, MA
Lawrence PA (1992) The Making of a Fly: The Genetics of Animal Design. Blackwell Scientific, Cambridge, MA
Le Douarin NM and Kalcheim C (1999) The Neural Crest. Cambridge University Press, New York
Leroi AM (2003) Mutants. On Genetic Variety and the Human Body. Viking, New York
Lester LB (ed) (2005) Stem Cells in Endocrinology. Humana Press, Totowa, NJ
Lieberman JR (ed) (2005) Bone Regeneration and Repair. Biology and Clinical Applications. Humana Press, Totowa, NJ
López-Gracia ML and Ros MA (2007) Left-Right Asymmetry in Vertebrate Development. Springer, Berlin, D
Lubichler M and Maienschein J (eds) (2007) From Embryology to Evo-Devo: A History of Developmental Evolution. MIT Press, Cambridge, MA
Lucas PW (2004) Dental Functional Morphology: How Teeth Work? Cambridge University Press, New York
Macieira-Coelho A (ed) (2007) Asymmetric Cell Division. Springer, Berlin, D
Madhani HD (2007) From a to α: Yeast as a Model for Cellular Differentiation. Cold Spring Harbor Laboratory Press, Cold Spring Harbor, NY
Martini Neri ME et al (eds) (1996) Gene Regulation and Fetal Development. Wiley-Liss, New York
McElreavey K (ed) (2000) The Genetic Basis of Male Infertility. Springer, New York
Minelli A (2003) The Development of Animal Form. Ontogeny, Morphology, and Evolution. Cambridge University Press, Cambridge, UK
Moody SA (ed) (1998) Cell Lineage and Fate Determination. Academic Press, San Diego, CA
Moody SA (ed) (2007) Principles of Developmental Genetics. Academic Press, San Diego, CA
Müller WA (1997) Developmental Biology. Springer, New York
Nijhout HF (1991) The Development and Evolution of Butterfly Wing Patterns, Smithonian Institution, Washington and London
Novartis Foundation Symposium (2005) Nuclear Organization in Development and Disease. Wiley, Hoboken, NJ
Nüsslein-Volhard C (2006) Coming to Life: How Genes Drive Development. Kales Press, Carlsbad, CA
Nüsslein-Volhard C and Dahm R, (eds) (2002) Zebrafish. A Practical Approach. Oxford University Press, New York
Odorico J et al (eds) (2005) Human Embryonic Stem Cells. Garland Science, New York
Ohlsson R et al (eds) (1995) Genomic Imprinting—Causes and Consequences. Cambridge University Press, New York
Ohlsson R (ed) (1999) Genomic Imprinting: An Interdisciplinary Approach. Springer, New York
Osiewacz HD (2002) Molecular Biology of Fungal Development. Marcel Dekker, New York
Pablo de F et al (1997) Cellular and Molecular Procedures in Developmental Biology. Academic Press, Orlando, FL
Penn MS (ed) (2006) Stem Cells and Myocardial Regeneration. Humana Press, Totowa, NJ
Pigliucci M (2001) Phenotypic Plasticity: Beyond Nature and Nurture. John Hopkins University Press, Baltimore, MD
Piontellin A (2002) Twins—From Fetus to Child. Taylor & Francis, New York
Potten C (ed) (1996) Stem Cells. Academic Press, San Diego, CA
Reik W and Surani A (1997) Genomic Imprinting. Frontiers in Molecular Biology. IRL/Oxford University Press, New York
Riddle JM (1992) Contraception and Abortion from the Ancient World to the Renaissaince. Harvard University Press, Cambridge, MA
Rossant J and Pedersen RA (eds) (1988) Experimental Approaches to Mammalian Embryonic Development. Cambridge University Press, New York
Rossant J and Tam PPL (eds) (2002) Mouse Development: Patterning, Morphogenesis and Organogenesis. Academic Press, San Diego, CA

Russo VEA et al (1992) Development. The Molecular Genetic Approach. Springer, New York
Russo VEA et al (eds) (1997) Epigenetic Mechanisms of Gene Regulation. Cold Spring Harbor Laboratory Press, Cold Spring Harbor, NY
Schatten G (2006) Current Topics in Developmental Biology. Elsevier, Philadelphia, PA
Scherer G and Schmid M (eds) (2001) Genes and Mechanisms in Vertebrate Sex Determination. Birkhäuser, Basel, CH
Schulkin J (ed) (2005) Allostasis, Homeostasis, and the Costs of Physiological Adaptation. Cambridge University Press, New York
Sell S (ed) (2004) Stem Cells Handbook. Humana Press, Totowa, NJ
Sernagor E et al (eds) (2006) Retinal Development. Cambridge University Press, New York
Sharpe PT and Mason I (eds) (1999) Molecular Embryology: Methods and Protocols. Humana Press, Totowa, NJ
Shatten H (ed) (2004) Germ Cell Protocols. Humana Press, Totowa, NJ
Slack J (2001) Essential Developmental Biology. Blackwell, Boston, MA
Sive HL et al (1999) Early Development of *Xenopus laevis*: A Laboratory Manual. Cold Spring Harbor Laboratory Press, Plainview, NY
Soares MJ and Hunt JS (eds) (2005) Placenta and Trophoblast. Methods and Protocols. Humana Press, Totowa, NJ
Solari AJ (1994) Sex Chromosomes and Sex Determination in Vertebrates. CRC Press, Boca Raton, FL
Stern CD (2004) Gastrulation: From Cells to Embryo. Cold Spring Harbor Laboratory Press, Wodbury, NY
Sullivan S et al (2007) Human Embryonic Stem Cells. The Practical Handbook. Wiley, Hoboken, NJ
Sutovsky P (ed) (2006) Somatic Cell Nuclear Transfer. Springer, Berlin, D
Tollefsboll TO (2004) Epigenetics Protocolls. Humana Press, Totowa, NJ
Tomanek RJ and Runyan RB (eds) (2001) Formation of the Heart and Its Regulation. Birkhäuser, Boston, MA
Toriello HV et al (eds) (2004) Hereditary Hearing Loss and Its Syndromes. Oxford University Press, New York
Tuan RS and Lo CW (eds) (1999–2000) Developmental Biology Protocols. Humana Press, Totowa, NJ
Turksen K (ed) (2004) Adult Stem Cells. Humana Press, Totowa, NJ
Turksen K (ed) (2006) Embryonic Stem Cell Protocols. Vol I. Isolation and Characterization. Vol II. Differentiation Models. Humana Press, Totowa, NJ
Unsicker K and Kriegelstein K (eds) (2005) Cell Signaling and Growth Factors in Development. Wiley-VCH, Hoboken, NJ
Veeck LL (1999) An Atlas of Human Gametes and Conceptuses, Parthenon, New York
Wachtel SS (1993) Molecular Genetics of Sex Determination. Academic Press, San Diego, CA
Walsh WR (ed) (2005) Repair and Regeneration of Ligament, Tendons and Joint Capsule. Humana Press, Totowa, NJ
Wassarman PM (ed) (1999) Advances in Developmental Biochemistry, vol 5. JAI Press, Stamford, CT
Wedlich D (ed) (2005) Cell Migration in Development and Disease. Wiley, Hoboken, NJ
Wilkins AS (1993) Genetic Analysis of Animal Development, Wiley-Liss, New York
Wilkins AS (2002) The Evolution of Developmental Pathways. Sinauer, Sunderland, MA
Williams EG et al (eds) (1994) Genetic Control of Self-Incompatibility and Reproductive Development in Flowering Plants. Kluwer, Norwell, MA
Winters SJ (ed) (2004) Male Hypogonadism. Basic Clinical and Therapeutic Principles. Humana Press, Totowa, NJ
Wizemann T and Pardue M-L (eds) (2001) Exploring the Biological Contributions to Human health—Does Sex Matter? National Academic Press, Washington, DC
Wolffe A (ed) (1998) Epigenetics. Wiley, New York
Wolpert L et al (2007) Principles of Development. Oxford University Press, New York
Wyszynski DF (ed) (2002) Cleft Lip and Palate: From Origin to Treatment. Oxford University Press, New York
Zhang J and Rokosh G (eds) (2007) Cardiac Gene Expression. Springer, Berlin, D
Zhiyuan G and Korzh V (eds) (2004) Fish Development and Genetics. The Zebrafish and Medaka Models. World Scientific, Singapore
Zon LI (ed) (2001) Hematopoiesis: A Developmental Approach. Oxford University Press, New York
Zucker-Franklin D and Grossi CE (eds) (2004) Atlas of Blood Cells: Function and Pathology. Ermes, Milan, Italy

Diseases ►*cancer*, ►*aging*, ►*molecular biology*, ►*immunology*

Abraham T (2005) Twenty-First Century Plague: The Story of SARS. Johns Hopkins University Press, Baltimore, MD
Advances in Clinical Chemistry (Serial). Elsevier, Philadelphia, PA
Alfaro DV III et al (eds) (2006) Age-Related Macular Degeneration: A Comprehensive Textbook. Lippincott Williams & Wilkins, Philadelphia, PA
Anaissie EJ et al (eds) (2003) Clinical Mycology. Churchill Livingstone, New York
Ashcroft FM (2000) Ion Channels and Disease. Academic Press, San Diego, CA
Bagasra O (1999) HIV and Molecular Immunity: Prospects for the AIDS Vaccine. Eaton, Natick, MA
Baker HF (ed) (2001) Molecular Pathology of Prions. Humana Press, Totowa, NJ
Baker HF and Ridley RM (eds) (1996) Prion Diseases. Humana Press, Totowa, NJ
Bakris GL (ed) (2004) The Kidney and Hypertension. Martin Dunitz, London
Banja JD (2005) Medical Errors and Medical Narcissism. Jones and Bartlett, Sudbury, MA
Barker J and McGrath J (eds) (2001) Cell Adhesion and Migration in Skin Diseases. Gordon & Breach, Amsterdam, the Netherlands
Barrow CJ and Small DH (eds) (2007) Abeta Peptide and Alzheimer's Disease. Celebrating a Century of Research. Springer, D

Barry JM (2004) The Great Influenza: The Epic Story of the Deadliest Plague in History. Penguin, New York

Barter P and Rye K-A (eds) (1999) Plasma Lipids and Their Role in Disease. Harwood Academic, Cooper Station, NY

Battler A and Leor J (eds) (2007) Stem Cell and Gene-Based Therapy. Factors in Regenerative Medicine. Springer, Berlin, D

Bazin H (2000) The Eradication of Small Pox: Edward Jenner and the First and Only Eradication of a Human Infectious Disease. Academic Press, San Diego, CA

Becker R (1996) Alzheimer Disease. From Molecular Biology to Therapy. Springer, New York

Beer MH et al (2006) The Merck Manual of Diagnosis and Therapy. Wiley, Hoboken, NJ

Bellamy R (ed) (2004) Susceptibility to Infectious Diseases. The Importance of Host Genetics. Cambridge University Press, New York

Bewley CA (2006) Protein-Carbohydrate Interactions in Infectious Diseases. Springer, New York

Bignold LP (ed) (2006) Cancer: Cell Structures, Carcinogens and Genomic Instability. Birkhäuseer, Basel, Switzerland

Blum HE et al (eds) (2007) Genetics in Liver Disease. Springer, Berlin, D

Boshoff HI et al (eds) (2007) Systems Biological Approaches in Infectious Diseases. Springer, Berlin, D

Boué A (1995) Fetal Medicine. Prenatal Diagnosis and Management. Oxford University Press, New York

Bowcock AM (ed) (1999) Breast Cancer: Molecular Genetics, Pathogenesis, and Therapeutics. Humana Press, Totowa, NJ

Brioni JD and Decker MW (eds) (1997) Pharmacological Treatment of Alzheimer's Disease. Molecular and Neurobiological Foundations. Wiley, New York

Brodell RT and Johnson SM (eds) (2003) Warts: Diagnosis and Management—An Evidence-Based Approach. Martin Dunitz, New York

Brosens I (ed) (2006) Uterine Leiomyomata: Pathogenesis and Management. Taylor & Francis, Abingdon

Bross P and Gegersen N (eds) (2003) Protein Misfolding and Disease. Humana Press, Totowa, NJ

Brown RD and Ho PJ (eds) (2005) Multiple Myeloma, Methods and Protocols. Humana Press, Totowa, NJ

Bynum WF and Bynum H (eds) (2007) Dictionary of Medical Biography. Greenwood Press, Westport, CN

Carson C and Resnick M (2005) Erectile Dysfunction. An Issue of Urologic Clinics. Elsevier, Philadelphia, PA

Chadwick R et al (eds) (1999) The Ethics of Genetic Screening. Kluwer, Dordrecht, Holland

Clarke JTR (1996) A Clinical Guide to Inherited Metabolic Diseases. Cambridge University Press, Cambridge

Cohen WI et al (eds) (2002) Down Syndrome: Vision for the 21st Century. Wiley, New York

Colemen WB and Tsongalis GJ (eds) (2005) Molecular Diagnostics. For the Clinical Laboratorian. Humana Press, Totowa, NJ

Collinge J and Palmer MS (1997) Prion Disease. Oxford University Press, New York

Cope AP (ed) (2007) Arthritis Research. Methods and Protocols. Humana Press, Totowa, NJ

Cossart P et al (eds) (2000) Cellullar Microbiology. American Society for Microbiology, Washington, DC

Crocker J and Burnett D (eds) (2005) The Science of Laboratory Diagnosis. Wiley, Hoboken, NJ

Crosby AW (2002) The Deadly Truth: A History of Diseases in America. Harvard University Press, Cambridge, MA

Dean RT and Kelly DT (eds) (2001) Atherosclerosis: Gene Expression, Cell Interactions and Oxidation. Oxford University Press, New York

Dec GW et al (eds) (2005) Heart Failure: A Comprehensive Guide to Diagnosis and Treatment. Marcel Dekker, New York

DeCeuninck F et al (eds) (2004) Cartilage and Osteoarthritis. Vol I. Cellular and Molecular Tools. Vol II. Structure and In Vivo Analysis. Humana Press, Totowa, NJ

Decker J and Reischl U (2004) Molecular Diagnosis of Infectious Diseases. Humana Press, Totowa, NJ

DeFronzo RA et al (eds) (2004) International Textbook of Diabetes Mellitus. Wiley, Chichester UK

Dismukes WE et al (eds) (2003) Clinical Mycology. Oxford University Press, New York

Doolan DL (ed) (2002) Malaria Methods and Protocols. Humana Press, Totowa, NJ

Dragani TA (ed) (1998) Human Polygenic Disease. Animal Models. Harwood Academic, Marston Book Services, Abingdon, Oxon

Drislane FW (ed) (2005) Status Epilepticus. A Clinical Perspective. Humana Press, Totowa, NJ

Dropulic B and Carter B (2008) Concepts in Genetic Medicine. Wiley, Hoboken, NJ

Ebadi M and Pfeiffer RF (eds) (2005) Parkinson's Disease. CRC Press, Boca Raton, FL

Edwards RG (ed) (1993) Preconception and Preimplantation Diagnosis of Human Genetic Diseases. Cambridge University Press, New York

Elles R and Mountford R (eds) (2004) Molecular Diagnosis of Genetic Diseases. Humana Press, Totowa, NJ

Emery AEH (ed) (1998) Neuromuscular Disorders. Clinical and Molecular Genetics. Wiley, New York

Emery AEH (ed) (2001) The Muscular Dystrophies. Oxford University Press, New York

Emilien G et al (eds) (2004) Alzheimer Disease: Neuropsychology and Pharmacology. Birkhäuser, Basel, Switzerland

Evans DAP (1994) Genetic Factors in Drug Therapy. Clinical and Molecular Pharmacogenetics. Cambridge University Press, New York

Ewald PW (1994) Evolution of Infectious Diseases. Oxford University Press, New York

Factor PH (ed) (2001) Gene Therapy for Acute and Acquired Diseases. Kluwer, Dordrecht, the Netherlands

Fennell JP and Baker AH (eds) (2005) Hypertension. Methods and Protocols. Humana Press, Totowa, NJ

Finkel T and Gutkind JS (eds) (2003) Signal Transduction and Human Disease. Wiley, Hoboken, NJ

Fong IW and Alibek Ken (eds) (2007) New and Evolving Infections of the 21st Century. Springer, Berlin, D

Fry M and Usdin K (eds) (2006) Human Nucleotide Expansion Disorders. Springer, New York

Futerman AH, Zimran A (eds) (2007) Gaucher Disease. CRC Press Boca Raton, FL

Gehlbach SH (2005) American Plagues: Lessons from Our Battles with Disease. McGRaw-Hill, New York

Gilbert-Barness E and Barness LA (2000) Metabolic Diseases. Foundations of Clinical Management, Genetics, and Pathology. Eaton, Natick, MA

Gillespie SH et al (eds) (2004) Microbe-Vector Interactions in Vector-Borne Diseases. Cambridge University Press, New York

Giraldo G et al (eds) (1996) Development and Applications of Vaccines and Gene Therapy in AIDS. Karger, Basel, Switzerland

Glynn I and Glynn J (2004) The Life and Death of Smallpox. Cambridge University Press, New York

Goodman JL et al (eds) (2005) Tick-Borne Diseases of Humans. American Society for Microbiology Press, Washington, DC

Goozner MK (2004) The $800 Million Poll: The Truth Behind the Cost of New Drugs. University California Press, Berkeley, CA

Gorbach SL et al (eds) (2004) Infectious Diseases. Lippincott Williams & Wilkins, Philadelphia, PA

Gorlin RJ et al (eds) (1995) Hereditary Hearing Loss and Its Syndromes. Oxford University Press, New York

Greenwell P and McCulley M (2007) Molecular Therapeutics. 21st Century Medicine. Wiley, Hoboken, NL

Grob G (2002) The Deadly Truth: A History of Diseases in America. Harvard University Press, Cambridge, MA

Gupta S (ed) (1996) Immunology of HIV Infection. Plenum, New York

Haas C (ed) (1999) Molecular Biology of Alzheimer Disease. Genes and Mechanisms Involved in Amyloid Generation. Gordon & Breach, Newark, NJ

Haines JL and Pericak-Vance MA (2006) Genetic Analysis of Complex Disease. Wiley, Hoboken, NJ

Hall LL (ed) (1996) Genetics and Mental Illness: Evolving Issues for Research and Society. Plenum, New York

Harper JC et al (2001) Preimplantation Genetic Diagnosis. Wiley, New York

Harper PS and Clarke AJ (1997) Genetics, Society and Clinical Practice. Bios Scientific, Cambridge, UK

Harper PS and Perutz M (2001) Glutamine Repeats and Neurogenerative Diseases: Molecular Aspects. Oxford University Press, New York

Hassold TJ and Patterson D (eds) (1998) Down Syndrome. A Promising Future, Together. Wiley, New York

Hebert CA (1999) Chemokines in Disease: Biology and Clinical Research. Humana Press, Totowa, NJ

Hertl M (ed) (2005) Autoimmune Diseases of the Skin: Pathogenesis, Diagnosis, Management. Springer, Vienna, Austria

Hooper NM (ed) (2000) Alzheimer's Disease. Humana Press, Totowa, NJ

Hunt KK et al (eds) (2001) Breast Cancer. Springer, New York

Iannaccone PM and Scarpelli DG (eds) (1997) Biological Aspects of Disease: Contributions from Animal Models. Taylor & Francis, New York

Iliffe J (2006) The African AIDS Epidemic: History. Ohio University Press, Athens, OH

James DG and Zumla A (eds) (1999) The Granulomatous Disorders. Cambridge University Press

Jeffery S et al (1997) Nucleic Acid-Based Diagnosis. Springer, New York

Johnstone MT and Veves A (eds) (2005) Diabetes and Cardiovascular Diseases. Humana Press, Totowa, NJ

Joos TO (ed) (2005) Microarrays in Clinical Diagnostics. Humana Press, Totowa, NJ

Kalichman SC (ed) (2005) Positive Prevention: Reducing HIV Transmission Among People with HIV/AIDS. Kluwer, New York

Kalow W et al (eds) (2001) Pharmacogenomics. Marcel Dekker, New York

Kaper JB and Hacker J (eds) (1999) Pathogenicity Islands and Other Mobile Virulence Elements. American Society for Microbiology Press, Washington, DC

Karpati G et al (eds) (2001) Disorders of the Voluntary Muscle. Cambridge University Press, Cambridge, UK

Kase H et al (eds) (1999) Adenosine Receptors and Parkinson's Disease. Academic Press, San Diego, CA

Kashima H et al (eds) (2002) Comprehensive Treatment of Schizophrenia. Linking Neuro-Behavioral Findings to Psychosocial Approaches. Springer, New York

Khachigian LM (ed) (2005) High-Risk Atherosclerotic Plaques: Mechanisms, Imaging, Models, and Therapy. CRC Press, Boca Raton, FL

Khan MG (2006) Encyclopedia of Heart Diseases. Elsevier, Philadelphia, PA

Khoury MJ et al (eds) (2000) Genetics and Public Health in the 21st Century. Oxford University Press, New York

Killeen AA (ed) (2004) Principles of Molecular Pathology. Humana Press, Totowa, NJ

King FD (2006) Progress in Medicinal Chemistry. Elsevier, Philadelphia, PA

King RA et al (eds) (2002) The Genetic Basis of Common Diseases. Oxford University Press, New York

Kipshidze NN and Serruys PW (eds) (2004) Handbook of Cardiovascular Transplantation. Martin Dunitz, London

Kleinbaum DG et al (eds) (2007) A Pocket Guide to Epidemiology, Springer, Berlin, D

Koch TR (ed) (2003) Colonic Diseases. Humana Press, Totowa, NJ

Kohwi Y (ed) (2004) Trinucleotide Repeat Protocols. Humana Press, Totowa, NJ

Kuschner HI (1999) A Cursing Brain? The Histories of Tourette Syndrome. Harvard University Press, Cambridge, MA

Lee CA et al (eds) (2005) Textbook of Hemophilia. Blackwell, Malde, MA

Lehmann S and Grassi J (2004) Techniques in Prion Research. Birkhäuser, Heidelberg, Germany

Lewis WH (2003) Medical Botany. Plants Affecting Human Health. Wiley, Hoboken, NJ

Licinio J and Wong ML (eds) (2002) Pharmacogenomics. The Search for Individualized Therapeutics. Wiley, New York

Lieberman P (ed) (2005) Allergic Diseases. Diagnosis and Treatment. Humana Press, Totowa, NJ

Lindee S (2005) Moments of Truth in Genetic Medicine. Johns Hopkins University Press, Baltimore, MD

Lörincz A (ed) (2006) Nucleic Acid Testing for Human Disease. CRC Press, Boca Raton, FL

Lowe WL Jr (ed) (eds) (2001) Genetics and Diabetes Mellitus. Kluwer, Dordrecht, the Netherlands

Luo ZD (ed) Pain research. Methods and protocols. Humana Press, Totowa, NJ

Lupski JR and Stankiewicz PT (eds) (2006) Genomic Disorders. The Genomic Basis of Disease. Humana Press, Totowa, NJ

Mantzoros CS (ed) (2006) Obesity and Diabetes. Human Press, Totowa, NJ
Marcus FI et al (eds) (2007) Arrythmogenic Right Ventricular Cradiomyopathy/Dysplasia. Recent Advances. Springer, Berlin, D
Mastroeni P and Maskell D (eds) (2006) Salmonella Infections. Clinical, Immunological and Molecular Aspects. Cambridge University Press, New York
McCance KL and Huether SE (1994) Pathophysiology: The Biologic Basis of Disease in Adults and Children. Mosby, St. Louis, MO
McCrae MA et al (eds) (1997) Molecular Aspects of Host-Pathogen Interactions. Cambridge University Press, New York
Meniru G (2001) Cambridge Guide to Infertility Management and Assisted Reproduction. Cambridge University Press, New York
Miller MS and Cronin M (eds) (2000) Genetic Polymorphisms and Susceptibility to Disease. Taylor & Francis, New York
Moldin SO and Rubenstein JLR (eds) (2006) Understanding Autism. From Basic Neuroscience to Treatment. CRC Press, Boca Raton, FL
Mooney MP and Siegel MI (eds) (2002) Understanding Cranifacial Anomalies. Etiopathogenesis of Craniosynostoses and Facial Clefting. Wiley, New York
Morabia A (ed) (2004) A History of Epidemiologic Methods and Concepts. Birkhäuser, Basel, Switzerland
Morrow DA (ed) (2006) Cardiovascular Biomarkers. Pathophysiology and Disease Management. Humana Press, Totowa, NJ
Mozayani A (ed) (2004) Handbook of Drug Interactions. A Clinical and Forensic Guide. Humana Press, Totowa, NJ
Neote K et al (eds) (2007) Chemokine Biology—Basic Research and Clinical Application. Pathophysiology of Chemokines. Birkhäuser, Basel, Switzerland
Okazaki I (ed) (2003) Extracellular Matrix and the Liver. Academic Press, San Diego, CA
Oxford Medical Databases. Oxford University Press, New York
Panayiotoulos CP (2005) The Epilepsies: Seizures, Syndromes and Management. Bladon Medical Publishing, Chipping Norton, UK
Pasternak JJ (2005) An Introduction to Human Molecular Genetics. Mechanisms of Inherited Disease. Wiley, Hoboken, NJ
Pawlowtzki I-H et al (1997) Genetic Mapping of Disease Genes. Academic Press, Orlando, FL
Phillips MI (ed) (2005) Antisense Therapeutics. Humana Press, Totowa, NJ
Pinsky L et al (1999) Genetic Disorders of Human Sexual Development. Oxford University Press, New York
Piscitelli SC and Redvold KA (eds) (2005) Drug Interactions in Infectious Diseases. Humana Press, Totowa, NJ
Pottier R et al (eds) (2007) Photodynamic Therapy with ALA. A Clinical Handbook. Springer, Berlin, D
Prusiner SB (1966) Prions, Prions, Prions. Springer, New York
Prusiner SB (ed) (2004) Prion Biology and Diseases. Cold Spring Harbor Laboratory Press, Cold Spring Harbor, NY
Pulst S-M (ed) (2002) Genetics of Movement Disorders. Academic Press, San Diego, CA
Quinn S (2001) Human Trials: Scientists, Investors and Patients in the Quest for a Cure. Perseus, Cambridge, MA
Raizada MK et al (eds) (2005) Cardiovascular Genomics. Humana Press, Totowa, NJ
Rapley R and Harbron S (eds) (2004) Molecular Analysis and Genome Discovery. Wiley, Hoboken, NJ
Rattan S and Kassem M (eds) (2006) Prevention and Treatment of Age-Related Diseases. Springer, Berlin, D
Read SJ and Virley D (eds) (2005) Stroke Genomics. Humana Press, Totowa, NJ
Reilly PR (2004) Is it in Your Genes? The Influence of Genes on Common Disorders and Diseases that Affect You and Your Family. Cold Spring Harbor Laboratory Press, Woodbury, NY
Rho JM et al (eds) (2004) Epilepsy. Marcel Dekker, New York
Robinson A and Linden MG (1993) Clinical Genetics Handbook. Blackwell Scientific, Oxford
Roland PE (1997) Brain Activation. Wiley, New York
Ross DW (1997) Introduction to Molecular Medicine. Springer, New York
Royce PM and Steinmann B (eds) (2002) Connective Tissue and its Heritable Disorders. Molecular, Genetic and Medical Aspects. Wiley, New York
Runge MS and Patterson C (2005) Principles of Molecular Cardiology. Humana Press, Totowa, NJ
Runge MS and Patterson C (eds) (2006) Principles of Molecular Medicine. Humana Press, Totowa, NJ
Scriver CR et al (eds) (2001) The Metabolic and Molecular Bases of Inherited Disease. McGraw-Hill, New York
Semenza GL (1998) Transcription Factors and Human Disease. Oxford University Press, New York
Serjeant GR and Serjeant BE (2001) Sickle Cell Disease. Oxford University Press, New York
Sharpe NF and Carter RF (2006) Genetic Testing. Care, Consent and Liability. Wiley, Hoboken, NJ
Shaw KJ (ed) (2002) Pathogen Genomics: Impact on Human Health. Humana Press, Totowa, NJ.
Smyth TN (ed) (2004) Gene and Cell Therapy. Therapeutic Mechanisms and Strategies. CRC Press, Boca Raton, FL
Srivastava R (ed) (2007) Apoptosis, Cell Signaling, and Human Diseases. Human Press, Totowa, NJ
Stamatoyannopoulos G et al (1994) The Molecular Basis of Blood Diseases. Saunders, Philadelphia, PA
Steinberg MH et al (eds) (2000) Disorders of Hemoglobin: Genetics, Pathophysiology, and Clinical Management. Cambridge University Press, New York
Stevenson RE et al (eds) (1993) Human Malformations and Related Anomalies. Oxford University Press, New York
Stine GJ (1994) Acquired Immunodeficiency Syndrome. Biological, Medical, Social, and Legal Issues. Prentice Hall, Englewood Cliff, NJ
Stoelting R and Miller R (2007) Basics of Anesthesia. Elsevier, Philadelphia, PA
Stone WS et al (eds) (2004) Early Clinical Intervention and Prevention in Schizophrenia. Humana Press, Totowa, NJ
Tanzi RE and Parson AB (2000) Decoding Darkness: The Search for the Genetic Causes of Alzheimer's Disease. Perseus, New York
Teebi AS and Farag TI (eds) (1996) Genetic Disorders Among Arab Populations. Oxford University Press, New York
Terry RD et al (eds) (1994) Alzheimer Disease. Raven, New York
Thakker RV (ed) (1997) Molecular Genetics of Endocrine Disorders. Arnold/Oxford University Press, New York

Theofilopoulos AN and Bona CA (eds) (2002) The Molecular Pathology of Autoimmune Diseases. Taylor & Francis, New York
Thoene JG (ed) (1996) Physician's Guide to Rare Diseases. Dowden, Montvale, NJ
Tibayrenc M (ed) (2007) Encyclopedia of Infectious Diseases. Modern Methodologies. Wiley, Hoboken, NJ
Traboulsi EI (ed) (1999) Genetic Diseases of the Eye. Oxford University Press, New York
Uversky VN and Fink AL (eds) (2006) Protein Misfolding, Aggregation and Conformational Diseases. Springer, Berrlin, D
Van Eyk JE and Dunn MJ (eds) (2007) Clinical Proteomics. From Diagnosis to Therapy. Wiley, Hoboken, NJ
Varga J and Brenner DA (eds) (2005) Fibrosis Research. Methods and Protocols. Humana Press, Totowa, NJ
Vile RG (1997) Understanding Gene Therapy. Springer, New York
Vivanco F (ed) (2007) Cardiovascular Proteomics. Methods and Protocols. Humana Press, Totowa, NJ
Vos J-MH (1995) DNA Repair Mechanisms. Impact on Human Diseases and Cancer. Springer, New York
Wahlgren W and Perlmann P (eds) (1999) Malaria. Molecular and Clinical Aspects. Gordon and Breach-Harwood Academic, Cooper Station, NY
Weinberg JM and Buchholz R (eds) (2006) TNF-Alpha Inhibitors. Birkhäuser, Basel, Switzerland
Weiss KM (1999) Genetic Variation and Human Diseases. Principles and evolutionary Approches. Cambridge University Press, New York
Wilding JPH (ed) (2008) Pharmacotherapy of Obesity. Birhäuser, Basel, Switzerland
Winter R (1997) Oxford Medical Databases. Dysmorphology Photo Library 2.0. CD-ROM, Windows. Oxford University Press, New York
Wood PA (2006) How Fat Works. Harvard University Press, Cambridge MA
Wootton D (2006) Bad Medicine: Doctors Doing Harm Since Hippocrates. Oxford University Press, New York
Wormser GP (ed) (2004) AIDS and Other Manifestations of HIV Infections. Elsevier-Academic Press, San Diego, CA
Yam P (2003) The Pathological Protein: Mad Cow, Chronic Wasting, and Other Deadly Prion Diseases. Copernicus Books/Springer, New York
Young DB (1998) Genetics and Tuberculosis. Wiley, New York
Zaven SK and Mesulam M-M (eds) (2000) Alzheimer's Disease. A Compendium of Current Theories. Annals of the New York Academy of Science #924, New York Academy of Science, New York
Zitelli BJ and Davis HW (eds) (2002) Atlas of Pediatric Physical Diagnosis. Mosby, St. Louis, MO

Evolution ►*population genetics*, ►*cytology and cytogenetics*

Alters BJ and Alters SM (2001) Defendng Evolution: A Guide to the Creation/Evolution Controversy. Jones and Bartlett, Boston, MA
Anderson E (1949) Introgressive Hybridization. Wiley, New York
Andersson M (1994) Sexual Selection. Princeton University Press, Princeton, NJ
Arnold FH and Georgiou G (eds) (2003) Directed Evolution Library Creation. Humana Press, Totowa, NJ
Arnold FH and Georgiou G (eds) (2003) Directed Enzyme Evolution. Humana Press, Totowa, NJ
Arnold ML (1997) Natural Hybridization and Evolution. Oxford University Press, New York
Arnqvist G and Rowe L (2005) Sexual Conflict. Princeton University Press, Princeton, NJ
Avise J (2000) Phylogeography: The History and Formation of Species. Harvard University Press, Cambridge, MA
Avise JC (2004) Molecular Markers, Natural History and Evolution. Sinauer, Sunderland, MA
Avise JC (2006) Evolutionary Pathways in Nature: A Phylogenetic Approach. Cambridge University Press, New York
Barbieri M (ed) (2008) The Codes of Life. The Rules of Macroevolution. Springer, Berlin, D
Barton NH et al (eds) (2007) Evolution. Cold Spring Harbor Laboratory Press, Woodbury, NY
Bell G (1997) Selection: The Mechanism of Evolution. Chapman and Hall, New York
Birkhead T (2000) Promiscuity: An Evolutionary History of Sperm Competition and Sexual Conflict. Faber & Faber, Winchester, MA
Bökönyi S (1974) History of Domestic Mammals in Central and Eastern Europe. Akadémiai Kiadó, Budapest, Hungary
Brack A (ed) (1999) The Molecular Origins of Life: Assembling Pieces of the Puzzle. Cambridge University Press, New York
Briggs D and Walters SM (1996) Plant Variation and Evolution. Cambridge University Press, New York
Brockman J (ed) (2006) Intelligent Thought: Science Versus the Intelligent Design Movement. Vintage Books, Victoria, BC, Canada
Browne J (2002) Charles Darwin: The Power of Place. A.A. Knopf, New York
Cachel S (2006) Primate and Human Evolution. Cambridge University Press, New York
Carroll RL (1997) Patterns and Processes of Vertebrate Evolution. Cambridge University Press, New York
Carroll SB (2005) Endless Forms Most Beautiful. The New Science of Evo Devo and Making the Animal Kingdom. Norton, New York
Cavalier-Smith T (1985) The Evolution of Genome Size. Wiley, New York
Cavalli-Sforza LL (2000) Genes, People, and Languages. Farrar, Straus & Giroux, New York
Cavalli-Sforza LL and Cavalli-Sforza F (1995) The Great Human Diasporas: The History of Diversity and Evolution. Addison-Wesley, Boston, MA
Cavalli-Sforza LL et al (1994) The History and Geography of Human Genes. Princeton University Press, Princeton, NJ

Cavalli-Sforza LL et al (eds) (1999) Human Evolution. Cold Spring Harbor Laboratory Press, Cold Spring Harbor, NY

Cavalli-Sforza LL and Moroni A, Zei G (2004) Consanguinity, Inbreeding and Genetic Drift in Italy. Princeton University Press, Princeton, NJ

Charlesworth B and Charlesworth D (2003) Evolution: A Very Short Introduction. Oxford University Press, Oxford

Clark M (ed) (2000) Comparative Genomics. Kluwer, Dordrecht, the Netherlands

Cockburn A et al (1998) Mummies, Disease and Ancient Cultures. Cambridge University Press, New York

Coyne JA and Orr HA (2004) Speciation. Sinauer, Sunderland, MA

Cracraft J and Donoghue MJ (eds) (2004) Assembling the Tree of Life. Oxford University Press, New York

Crandall KA (ed) (1999) The Evolution of HIV. Johns Hopkins University Press, Baltimore, MD

Crow TJ (ed) (2001) The Speciation of Modern Homo Sapiens. Oxford University Press, New York

Crozier RH and Pamilo P (1996) Evolution of Social Insect Colonies. Oxford University Press, New York

Dayhoff MO (ed) (1972) Atlas of Protein Sequences and Structure, vol 5. National Biomedical Research Fund, Washington, DC

Delsemme AH (2001) Our Cosmic Origins. Cambridge University Press, New York

DeSalole R et al (eds) (2001) Techniques in Molecular Systematics and Evolution. Birkhäuser, Basel, Switzerland

Dobzhansky TH et al (1977) Evolution. Freeman, San Francisco, CA

Dronamraju KR (ed) (2004) Infectious Disease and Host-Pathogen Evolution, Cambridge University Press, New York

Dugatkin LA (2006) The Altruism Equation: Seven Scientists Search for the Origin of Goodness. Princeton University Press, Princeton, NJ

Dyson FJ (1999) Origins of Life. Cambridge University Press, Cambridge, UK

Eldredge N (2000) The Triumph of Evolution and the Failure of Creationism. Macmillan, New York

Felsenstein J (2004) Inferring Phylogenies. Sinauer, Sunderland, MA

Fichman M (2004) An Elusive Victorian. The Evolution of Alfred Russel Wallace. The Univesity of Chicago Press, Chicago, IL

Fix AG (1999) Migration and Colonization in Human Microevolution. Cambridge University Press, Cambridge

Fleagle JG and Kay RF (eds) (1994) Anthropoid Origins. Plenum, New York

Ford EB (1971) Ecological Genetics. Chapman and Hall, London

Forsdyke DR (2006) Evolutionary Bioinformatics. Springer, Berlin, D

Futuyama DJ (2005) Evolution. Sinauer, Sunderland, MA

Gerhart J and Kirschner M (1997) Cells, Embryos and Evolution: Developmental Understanding of Phenotypic Variation and Evolutionary Adaptability. Blackwell Science, Cambridge, MA

Gesteland RF et al (eds) (2006) The RNA World. Cold Spring Harbor Laboratory Press, Cold Spring Harbor, NY

Gibbs AJ et al (eds) (1995) Molecular Basis of Virus Evolution. Cambridge University Press, New York

Gillespie JH (1992) The Causes of Molecular Evolution. Oxford University Press, New York

Godfrey A (2000) Life Without Genes. HarperCollins, New York

Gould SJ (2002) The Structure of Evolutionary Theory. Harvard University Press, Cambridge, MA

Grafen A and Ridley M (eds) (2006) Richard Dawkins: How a Scientist Changed the Way We Think. Oxford University Press, New York

Graur D and Li W (2000) Fundamentals of Molecular Evolution. Sinauer, Sunderland, MA

Gregory T (ed) (2005) The Evolution of the Genome. Elsevier, Philadelphia, PA

Hall BK (ed) (1994) Homology. The Hierarchial Basis of Comparative Biology. Academic Press, San Diego, CA

Hall BK (1999) Evolutionary Developmental Biology. Kluwer, Dordrecht, the Netherlands

Hall BG (2004) Phylogenetic Trees Made Easy: A How Manual for Molecular Biologists. Sinauer, Sunderland, MA

Hammerstein P (ed) (2004) Genetic and Cultural Evolution of Cooperation. MIT Press, Cambridge, MA

Hanski I and Gaggiotti O (eds) (2004) Ecology, Genetics, and Evolution of Metapopulations. Academic Press, San Diego, CA

Harris H (2002) Things Come to Life. Spontaneous Generation Revisited. Oxford University Press, New York

Hazen RM (2005) Genesis: The Scientific Quest for Life's Origin. Joseph Henry Press, Washington, DC

Henke W and Tattersall I (eds) (2007) Handbook of Paleoanthropology. Principles, Methods and Approaches. Primate Evolution and Human Origins. Phylogeny of Hominids. Springer. Berlin, D

Herrmann B and Hummel S (eds) (1994) Ancient DNA. Recovery and Analysis of Genetic Material from Paleontological, Archeological, Museum, Medical and Forensic Specimens. Springer, New York

Hey J (2001) Genes, Categories, and Species. The Evolutionary and Cognitive Causes of the Species Problem. Oxford University Press, New York

Hillis DM et al (eds) (1996) Molecular Systematics. Sinauer, Sunderland, MA

Hochachka PW and Somero GN (2002) Biochemical Adaptation: Mechanism and Process in Physiological Evolution. Oxford University Press, Oxford

Howard DJ and Berlocher SH (eds) (1998) Endless Forms—Species and Speciation. Oxford University Press, New York

Hughes AL (2000) Adaptive Evolution of Genes and Genomes. Oxford University Press, New York

Hummel S (2002) Ancient DNA Typing. Springer, New York

Hurles M et al (eds) (2003) Human Evolutionary Genetics: Origins, Peoples and Disease. Garland, New York

Jablonka E and Lamb MJ (1999) Epigenetic Inheritance and Evolution: The Lamarckian Dimension. Oxford University Press, New York

Jones M (2002) The Molecule Hunt: Archeology and the Search for Ancient DNA. Arcade Publishing Co., New York

Jones S et al (eds) (1994) The Cambridge Encyclopedia of Human Evolution. Cambridge University Press, New York

Kirschner MW and Gerhart JC (2005) The Plausibility of Life. Resolving Darwin's Dilemma. Yale University Press New Haven, CT
Klein J and Takahata N (2001) Where We Do Come From? The Molecular Evidence of Human Descent. Springer, New York
Koonin EV and Galperin MY (2003) Sequence-Evolution-Function. Kluwer Academic, Boston, MA
Lahav N (1999) Biogenesis. Theories of Life's Origin. Oxford University Press, New York
Laubichler MD and Maienschein J (eds) (2007) From Embryology to Evo-Devo. MIT Press, Cambridge MA
Lawton JH and May RM (eds) (1995) Extinction Rates. Oxford University Press, New York
Lerner IM and Libby WJ (1976) Heredity, Evolution and Society. Freeman, San Francisco, CA
Levinton JS (2001) Genetics, Paleontology, and Macroevolution. Cambridge University Press, New York
Lewin R (1997) Patterns in Evolution: The New Molecular View. Freeman, New York
Lewin R (1998) The Origin of Modern Humans. Freeman, New York
Li W-H (1997) Molecular Evolution. Sinauer, Sunderland, MA
Luisi PL (2006) The Emergence of Life. From Chemical Origins to Synthetic Biology. Cambridge University Press, New York
Margulis L (1993) Symbiosis in Cell Evolution. Freeman, New York
Margulis L and Sagan D (2002) Acquiring Genomes: A Theory of the Origin of Species. Basic Books, New York
Maynard SJ (1978) The Evolution of Sex. Cambridge University Press, Cambridge
Maynard SJ (1982) Evolution and the Theory of Games. Cambridge University Press, Cambridge
Maynard SJ (1998) Evolutionary Genetics. Oxford University Press, New York
Maynard SJ and Szathmáry E (1995) The Major Transitions in Evolution. Freeman, New York
Maynard SJ and Szathmáry, E. (1999) The Origins of Life: From the Birth of Life to the Origin of Language. Oxford University Press, New York
Mayr E (1963) Animal Species and Evolution. Harvard University Press, Cambridge, MA
Mayr E (2001) What Evolution Is. Basic Books, New York
McElreath R and Boyd R (2007) Mathematical Models of Social Evolution. A Guide for the Perplexed. University of Chicago Press, Chicago, IL
McGhee GR (2006) The Geometry of Evolution. Adaptive Landscapes and Theoretical Morphospaces. Cambridge University Press, New York
Mindell DP (2006) The Evolving World: Evolution in Everyday Life. Harvard University Press, Cambridge, MA
Mousseau TA et al (eds) (2000) Adaptive Genetic Variation in the Wild. Oxford University Press, New York
Myamoto ME and Cracraft J (eds) (1991) Phylogenetic Analysis of DNA Sequences. Oxford University Press, New York
Nei M (1987) Molecular Evolutionary Genetics. Columbia University Press, New York
Nei M and Kumar S (2000) Molecular Evolution and Phylogenetics. Oxford University Press, New York
Niklas KJ (1997) The Evolutionary Biology of Plants. University of Chicago Press, Chicago, IL
Ohno S (1970) Evolution by Gene Duplication. Springer, New York
Okasha S (2006) Evolution and the Levels of Selection. Clarendon Press, Oxford
Osawa S (1995) Evolution of the Genetic Code. Oxford University Press, New York
Page RDM and Holmes EC (1998) Molecular Evolution: A Phylogenetic Approach. Backwell Science, Oxford
Pagel M (ed) (2002) Encyclopedia of Evolution. Oxford University Press, New York
Pagel M and Pomiankowski A (eds) (2007) Evolutionary Genomics and Proteomics. Sinauer, Sunderland, MA
Palumbi SR (2001) The Evolution Explosion: How Humans Cause Rapid Evolutionary Change. Norton, New York
Patthy L (1999) Protein Evolution. Blackwell, Oxford
Piatigorsky J (2007) Gene Sharing and Evolution: The Diversity of Protein Function. Harvard University Press, Cambridge, MA
Powell JR (1997) Progress and Prospects in Evolutionary Biology: The *Drosophila* Model. Oxford University Press, New York
Raff RA (1996) The Shape of Life. Genes, Development, and the Evolution of Animal Form. University Chicago Press, Chicago, IL
Relethford JH (2001) Genetics and the Search for Modern Human Origins. Wiley, New York
Relethford JH (2003) Reflections of Our Past: How Human History Is Revealed in Our genes. Westview Press, Boulder, CO
Renfrew C and Boyle K (eds) (2000) Archaeogenetics: DNA and the Population Prehistory of Europe. McDonald Institute, Archaelog. Res., Cambridge, UK
Rice SH (2004) Evolutionary Theory: Mathematical and Conceptual Foundation. Sinauer, Sunderland, MA
Richerson PJ and Boyd R (2004) Not by Genes Alone: How Culture Transformed Human Evolution. Chicago University Press, Chicago, IL
Ridley M (2000) The Cooperative Gene. How Mendel's Demon Explains Evolution of Complex Beings. Free Press, New York
Robert D et al (eds) (1996) Evolution of Microbial Life. Cambridge University Press, New York
Roff DA (1997) Evolutionary Quantitative Genetics. Chapman and Hall, New York
Roff DA (2001) Life History Evolution. Sinauer, Sunderland, MA
Rose KD and Archibald JD (eds) (2005) The Rise of Placental Mammals. Origins and Relatioships of the Major Extant Clades. Johns Hopkins University Press, Baltimore, MD
Rose MR and Lauder GV (eds) (1996) Adaptation. Academic Press, San Diego, CA
Rosenberg A (2006) Darwinian Reductionism: Or, How to Stop Worrying and Love Molecular Biology. University of Chicago Press, Chicago, IL
Roughgarden J (2004) Evolution's Rainbow. Diversity, Gender, and Sexuality in Nature and People. University of California Press, Berkley, CA
Ruse M (1997) Monad to Man: The Concept of Progress in Evolutionary Biology. Harvard University Press, Cambridge, MA

Salemi M and Vandamme A-M (eds) (2003) The Phylogenetic Handbook. A Practical Approach to DNA and Protein Phylogeny. Cambridge University Press, New York

Sankoff D (ed) (2000) Comparative Genomics. Kluwer, Dordrecht, the Netherlands

Schlichting CD, Pigliucci M (1998) Adaptive Phenotypic Evolution. Freeman, New York

Scott EC (2005) Evolution vs. Creationism: An Introduction. University of California Press, Berkeley, CA

Seckbach J (ed) (2006) Life As We Know It: Cellular Origin, Life in Extreme Habitats and Astrobiology. Springer, New York

Shermer M (2002) In Darwin's Shadow. The Life and Science of Alfred Russel Wallace. Oxford University Press, New York

Simmonds NW (ed) (1976) Evolution of Crop Plants. Longman, London

Simmons LW (2001) Sperm Competition and Its Evolutionary Consequences in the Insects. Princeton University Press, Princeton, NJ

Slotten R (2004) The Heretic in Darwin's Court: The Life of Alfred Russel Wallace. Columbia University Press, New York

Smith JM (1998) Evolutionary Genetics. Oxford University Press, New York

Stebbins LG (1950) Variation and Evolution in Plants. Columbia University Press, New York

Steele AJ (1981) Somatic Selection and Adaptive Evolution: On the Inheritance of Acquired Characters. University of Chicago Press, Chicago, IL

Strick E (2000) Sparks of Life: Darwinism and the Victorian Debates Over Spontaneous Generation. Harvard University Press, Cambridge, MA

Strickberger MW (1995) Evolution. Jones and Bartlett, Boston, MA

Syvanen M and Kado CL (eds) (2002) Horizontal Gene Transfer. Academic Press, San Diego, CA

Thomson K (2005) Before Darwin. Reconciling God and Nature. Yale University Press, New Haven, CN

Valentine JW (2004) On the Origin of Phyla. University of Chicago Press, Chicago, IL

Wagner A (2005) Robustness and Evolvability in Living Systems. Princeton University Press, Princeton, NJ

Weiss KM and Buchanan AV (2004) Genetics and the Logic of Evolution. Wiley, Hoboken, NJ

Williams RJP and Fraústo da Silva JJR (2006) The Chemistry of Evolution. Elsevier, Philadelphia, PA

Wills C and Bada J (2000) The Spark of Life: Darwin and the Primeval Soup. Perseus, Cambridge, MA

Wolf JB et al (eds) (2000) Epistasis and the Evolutionary Process. Oxford University Press, New York

Yeats DK and Wiegmann BM (eds) (2005) The Evolutionary Biology of Flies. Columbia University Press, New York

Yockey HP (2006) Information Theory, Evolution, and the Origin of Life. Cambridge University Press, New York

Zeuner FE (1963) A History of Domesticated Animals. Harper & Row, New York

Genetics, Genomics ►*genetics textbooks*, ►*molecular biology*, ►*cytology*, ►*transposable elements*, ►*mapping*, ►*recombination*, ►*mutation*, ►*radiation*, ►*organelles*, ►*evolution*

Advances in Genetics (Serial). Elsevier, Philadelphia, PA

Annual Review of Genetics. Palo Alto, CA

Annual Review of Genomics and Human Genetics, Palo Alto, CA

Allis CD et al (eds) (2006) Epigenetics. Cold Spring Harbor Laboratory Press, Cold Spring Harbor, NY

Arora DK and Berka R (2005) Genes and Genomics. Elsevier, Philadelphia, PA

Baker DL et al (eds) (1998) A Guide to Genetic Counseling. Wiley, New York

Baraitser M (1997) The Genetics of Neurological Disorders. Oxford University Press, New York

Bartel PL and Fields S (1997) The Yeast Two-Hybrid System. Oxford University Press, New York

Bennett RL (2008) The Practical Guide to the Genetic Family History. Wiley, Hoboken, NJ

Bergman NH (ed) (2007) Comparative Genomics. Humana Press, Totowa, NJ

Bishop J (1999) Transgenic Mammals. Longman, Harlow, UK

Bishop MJ (ed) (1999) Genetics Databases. Academic Press, San Diego, CA

Bowling AT and Ruvinsky A (eds) (2000) The Genetics of the Horse. Oxford University Press, New York

Bradshaw JE (ed) (1994) Potato Genetics. Oxford University Press, New York

Brenner S and Miller JH (eds) (2002) Encyclopedia of Genetics. Academic Press, San Diego, CA

Bridge PJ (1994) The Calculation of Genetic Risks. Worked Examples of DNA Diagnostics. Johns Hopkins University Press, Baltimore, MD

Brown AJP and Tuite MF (eds) (1998) Yeast Gene Analysis. Academic Press, San Diego, CA

Brown SM et al (2002) Essentials of Medical Genomics. Wiley, Hoboken, NJ

Bulmer MG (1980) The Mathematical Theory of Quantitative Genetics. Oxford University Press, New York

Camp NJ and Cox A (eds) (2002) Quantitative Trait Loci. Methods and Protocols. Humana Press, Totowa, NJ

Charlebois RL (ed) (1999) Organization of the Prokaryotic Genome. American Society for Microbiology, Washington, DC

Chaudharty BR and Agarwal SB (eds) (1996) Cytology, Genetics, and Molecular Biology of Algae. SPB Academic, Amsterdam, the Netherlands

Chen H (2005) Atlas of Genetic Diagnosis and Counseling. Humana Press, Totowa, NJ

Cooper DN (1999) Human Gene Evolution. Academic Press, San Diego, CA

Cooper D (ed) (2003) Nature Encyclopedia of the Human Genome. Nature Publishing Group, London

Cullis CA (2004) Plant Genomics and Proteomics. Wiley-Liss, Hoboken, NJ

Dale JW and von Schantz M (2008) From Genes to Genomes. Concepts and Applications. Wiley, Hoboken, NJ

Danielli GA (2002) Genetics and Genomics for the Cardiologist. Kluwer, Boston, MA
Dawkins R (1999) The Extended Phenotype: The Long Reach of the Gene. Oxford University Press, New York
Detrich HW et al (eds) (1998) The Zebrafish: Genetics and Genomics. Academic Press, San Diego, CA
DeSalle R and Yudell M (2004) Welcome to the Genome. Wiley, Hoboken, NJ
de Winde JH (ed) (2003) Functional Genetics of Industrial Yeasts. Springer, New York
Dunn MJ et al (eds) (2006) Encyclopedia of Genetics, Genomics, Proteomics and Bioinformatics. Wiley, Hoboken, NJ
Edwards AWF (2000) Foundations of Mathematical Genetics. Cambridge University Press, New York
Epstein HF and Shakes DC (eds) (1995) *Caenorhabditis elegans*: Modern Biological Analysis of an Organism. Academic Press, San Diego, CA
Esser K and Kuenen R (1967) Genetics of Fungi. Springer, New York
Falconer DS and Mackay TEC (1996) Introduction to Quantitative Genetics. Longman/Addison Wesley, White Plains, NY
Fernandez J and Hoeffler JP (eds) (1998) Gene Expression Systems. Using Nature for the Art of Expression. Academic Press, San Diego, CA
Freeling M and Walbot V (eds) (1994) The Maize Handbook. Springer, New York
Freies R and Ruvinsky I (ed) (1999) The Genetics of Cattle. Oxford University Press, New York
Goldstein DB and Schlötterer C (eds) (1999) Microsatellites. Evolution and Applications. Oxford University Press, Oxford, UK
Goldstein LSB and Fryberg EA (eds) (1994) *Drosophila melanogaster*. Practical Uses in Cell and Molecular Biology. AP Professional, Cambridge, MA
Goodwin W et al (2007) An Introduction to Forensic Genetics. Wiley, Hoboken, NJ
Guthrie C et al (eds) (2004) Guide to Yeast Genetics and Molecular and Cell Biology. Methods in Enzymology 350. Elsevier Science, St. Louis, MO
Haines JL and Pericak-Vance MA (eds) (2006) Genetic Analysis of Complex Disease. Wiley, Hoboken, NJ
Haydon J (ed) (2007) Genetics in Practice. A Clinical Approach for Healthcare Practitioners. Wiley, Hoboken, NJ
Harper PS (1998) Practical Genetic Counseling. Butterworth-Heinemann/Oxford University Press, New York
Hartwell L (2006) Genetics: From Genes to Genomes. McGraw-Hill, New York
Hawley RS and Mori CA (1998) The Human Genome. A User Guide. Academic Press, San Diego, CA
Hayes JD and Wolf CR (1997) Molecular Genetics of Drug Resistance. Harwood Academic, Cooper Station, NY
Heß D (1968) Biochemische Genetik. Eine Einführung Unter Besonderer Berücksichtigung Höherer Pflanzen. Springer, Berlin
Hrabé de Angelis MC and Brown SP (eds) (2006) Standards of Mouse Phenotyping. Wiley, Hoboken, NJ
Jackson JE and Linskens HF (eds) (2002) Testing for Genetic Manipulations in Plants. Springer, New York
John B and Miklós GLG (1988) The Eukaryotic Genome in Development and Evolution. Allen & Unwin, London, UK
Kahl G (2005) The Dictionary of Gene Technology. Genomics, Transcriptomics, Proteomics. Wiley, Hoboken, NJ
Kang M (ed) (2002) Quantitative Genetics, Genomics and Plant Breeding. CABI, Wallingford, Oxon, UK
Kearsey MJ and Pooni HS (1996) The Genetical Analysis of Quantitative Traits. Chapman and Hall, New York
King RC (ed) (1974) Handbook of Genetics. Plenum, New York
Klotzko AJ (2001) The Cloning Sourcebook. Oxford University Press, New York
Kohane IS et al (2003) Microarrays for an Integrative Genomics. MIT Press, Cambridge, MA
Koncz C et al (eds) (1993) Methods in *Arabidopsis* Research. World Scientific, Singapore
Lankenau D-H (ed) (2006) Genome Integrity: Facets and Perspectives. Springer, NewYork
Liebler DC (2001) Introduction to Proteomics. Tools for the New Biology. Humana Press, Totowa, NJ
Lindsley DL and Zimm GG (1992) The Genome of *Drosophila melanogaster*. Academic Press, San Diego, CA
Long S (2006) Veterinary Genetics and Reproductive Physiology—A Textbook for Veterinary Nurses and Technicians. Mosby, St. Louis, MO
Lynch M (2007) The Origins of Genome Architecture. Sinauer, Sunderland, MA
Lynch M and Walsh B (1998) Genetics and Analysis of Quantitative Traits. Sinauer, Sunderland, MA
Lyon MF et al (1996) Genetic Variants and Strains of the Laboratory Mouse. Oxford University Press, New York
Malcolm S and Goodship J (eds) (2001) Genotype to Phenotype. Academic Press, San Diego, CA
Martinelli SD and Kinghorn JR (eds) (1994) Aspergillus: 50 years on. Elsevier, Amsterdam, the Netherlands
Mather K and Jinks JL (1977) Introduction to Biometrical Genetics. Cornell University Press, Ithaca, NY
McKusick VA (ed) (1978) Medical Genetic Studies of the Amish. Johns Hopkins University Press, Baltimore, MD
McKusick VA (1994) Mendelian Inheritance in Man: A Catalog of Human Genes and Genetic Disorders. Johns Hopkins University Press, Baltimore, MD
Meyerowitz EM and Somerville CR (eds) (1994) Arabidopsis. Cold Spring Harbor Laboratory Press, Cold Spring Harbor, NY
Meyers RA (ed) (2007) Genomics and Genetics. From Molecular Details to Analysis and Techniques. Wiley, Hoboken, NJ
Mielke JH et al (2006) Human Biological Variation. Oxford University Press, New York
Miesfeld RL (1999) Applied Molecular Genetics. Wiley, New York
Morling N (ed) (2006) Handbook of Forensic Genetics. Humana Press, Totowa, NJ
Neale MC and Cardon LR (1992) Methodology for Genetic Studies of Twins and Families. Kluwer, Dordrecht, the Netherlands
Nicholas FW (1987) Veterinary Genetics. Clarendon Press, Oxford, UK
Nicholas FW (1996) Introduction to Veterinary Genetics. Oxford University Press, New York
Nicholl DST (2002) An Introduction to Genetic Engineering, Cambridge University Press, New York

Nicholoff JA and Hoekstra MF (eds) (1998) DNA Damage and Repair: DNA Repair in Prokaryotes and Lower Eukaryotes. Humana Press, Totowa, NJ

Ohlsson R et al (eds) (1995) Genomic Imprinting. Causes and Consequences. Cambridge University Press, New York

Ostrander EA et al (eds) (2005) The Dog and Its Genome. Cold Spring Harbor Laboratory Press, Cold Spring Harbor, NY

Ostrer H (1998) Non-Mendelian Genetics in Humans. Oxford University Press, New York

Palzkill T (2002) Proteomics. Kluwer, Boston, MA

Papageorgiou S (ed) (2007) Hox Gene Expression. Springer, Berlin, D

Papaioannou VE and Behringer RR (2005) Mouse Phenotypes: A Handbook of Mutation Analysis. Cold Spring Harbor Laboratory Press, Woodbury, NY

Passarge E (2006) Color Atlas of Genetics. Thieme, Stuttgart, D

Peña L (ed) (2004) Transgenic Plants. Methods and Protocols. Humana Press, Totowa, NJ

Phillips MI (ed) (2002) Gene Therapy Methods. Elsevier Science, St. Louis, MO

Piper L and Ruvinsky A (eds) (1997) The Genetics of the Sheep. CAB/Oxford University Press, New York

Powell KA et al (eds) (1994) The Genus *Aspergillus*: From Taxonomy and Genetics to Industrial Applications. Plenum, New York

Pringle JR et al (eds) (1997) Cell Cycle and Cell Biology. Cold Spring Harbor Laboratory Press, Cold Spring Harbor, NY

Rader K (2004) Making Mice: Standardizing Animals for American Biomedical Research. Princeton University Press, Princeton, NJ

Progress in Forensic Genetics. Elsevier Science, New York

Reece RJ (2004) Analysis of Genes and Genomes. Wiley, Hoboken, NJ

Reeve ECR (ed) (2001) Encyclopedia of Genetics. Fitzroy Dearborn, London, UK

Riddle DE et al (eds) (1997) *C. elegans* II. Cold Spring Harbor Laboratory Press, Cold Spring Harbor, NY

Rimoin DL et al (eds) (2002) Emery and Rimoin's Principles and Practice of Medical Genetics. Churchill Livingstone, New York

Rolland A (ed) (1999) Advanced Gene Delivery. From Concepts to Pharmaceutical Products. Harwood Academic, Villiston, VT

Rotschild MF and Ruvinsky A (eds) (1998) The Genetics of the Pig. Oxford University Press, New York

Rothstein MA (ed) (2003) Phramacogenomics: Social, Ethical and Clinical Dimensions. Wiley, Hoboken, NJ

Salinas J and Sanchez-Serrano JJ (eds) (2005) Arabidopsis Protocols. Humana Press, Totowa, NJ

Schindhelm K and Nordon R (eds) (1999) *Ex Vivo* Cell Therapy. Academic Press, London

Sensen CW (ed) (2005) Handbook of Genome Research. Genomics, Proteomics, Metabolomics, Bioinformatics, Ethical and Legal Issues. Wiley, Hoboken, NJ

Silver LM (1995) Mouse Genetics. Concepts and Applications. Oxford University Press, New York

Simpson J and Elias S (2003) Genetics in Obstetrics and Gynecology. Elsevier, Philadelphia, PA

Singer M and Berg P (eds) (1997) Exploring Genetic Mechanisms. University Science Books, Sausalito, CA

Somerville CR and Meyerowitz EM (eds) (2003) The *Arabidopsis* Book. American Society of Plant Biology, Rockville, MD

Spector TD et al (eds) (2000) Advances in Twin and Sib-Pair Analysis. Greenwich Medical Media, London, UK

Stevens L (1991) Genetics and Evolution of the Domestic Fowl. Cambridge University Press, New York

Stillman B and Stewart D (eds) (2005) Epigenetics. Cold Spring Harbor Symposium on Quantitative Biology, vol 69. Cold Spring Harbor Laboratory Press, Cold Spring Harbor, NY

Strachan T and Read AP (2003) Human Molecular Genetics. Garland Science, New York

Stubbe H (1966) Genetik und Zytologie von *Antirrhinum* L. Sect. *Antirrhinum*. Fischer, Jena, Germany

Sussman HE and Smit MA (eds) (2006) Genomes. Cold Spring Harbor Laboratory Press, Cold Spring Harbor, NY

Sykes B (ed) (1999) The Human Inheritance: Genes, Language, and Evolution. Oxford University Press, Oxford

The genome of the human species. (CSHSQB LXVIII) (2004) Cold Spring Harbor Laboratory Press, Woodbury, NY

Tsonis PA (2002) Anatomy of Gene Regulation. A Three-Dimensional Structural Analysis. Cambridge University Press, New York

Volff J-N (ed) (2006) Vertebrate Genomes. Karger, Basel, Switzerland

Volff J-N (ed) (2006) Genome and Disease. Karger, Basel, Switzerland

Volff J-N (ed) (2007) Gene and Protein Evolution. Karger, Basel, Switzerland

Weber WW (1997) Pharmacogenetics. Oxford University Press, New York

Weil J (2000) Psychosocial Genetic Counseling. Oxford University Press, New York

Weir BS (1996) Genetic Data Analysis II. Sinauer, Sunderland, MA

Wheals AE et al (eds) (1995) The Yeasts, vol 6. Yeast Genetics. Academic Press, San Diego, CA

Young ID (1999) Introduction to Risk Calculation in Genetic Counseling. Oxford University Press, New York

Genetics Textbooks ►*genetics genomics*, ►*history of genetics*, ►*molecular biology*, ►*cytology*

Birge EA (2000) Bacterial and Bacteriophage Genetics. Springer, New York

Brown SM et al (2003) Essentials of Medical Genomics. Wiley-Liss, Hoboken, NJ

Brown TA (2006) Genomes. Garland Science, New York

Brown TA (2007) Genome und Gene. Springer, Berlin, D

Brown TA (2007) Gentechnologie für Einsteiger. Springer, Berlin, D

Caldwell G et al (2006) Integrated Genomics. A Discovery-Based Laboratory Course. Wiley, Hoboken, NJ

Campbell AM and Heyer LJ (2003) Discovering Genomics, Proteomics, and Bioinformatics. Cold Spring Harbor Laboratory Press, Cold Spring Harbor, NY
Dales JW and Park SF (2004) Molecular Genetics of Bacteria. Wiley, Chichester
Datta A and Dougherty ER (2006) Introduction to Genomic Signal Processing with Control. CRC Press, Boca Raton, FL
Emery AEH and Mueller RF (1992) Elements of Medical Genetics. Churchill Livingstone, New York
Emery AEH and Malcolm S (1995) An Introduction to Recombinant DNA in Medicine. Wiley, New York
Fincham JRS et al (1979) Fungal Genetics. Blackwell, Oxford
Gibson G and Muse SV (2004) A Primer of Genome Science. Sinauer, Sunderland, MA
Griffiths AJF et al (2004) An Introduction to Genetic Analysis. Freeman, New York
Griffith AJF et al (2002) Modern Genetic Analysis. Freeman, New York
Hartl DL and Jones EW (2002) Essential Genetics: A Genomic Perspective. Jones and Bartlett, Boston, MA
Hartwell L et al (2004) Genetics: From Genes to Genomes. McGraw-Hill, Dubuque, IA
Hill WE (2002) Genetic Engineering. A Primer. Taylor & Francis, New York
Johanssen ME and Mange AP (1998) Basic Human Genetics. Sinauer, Sunderland, MA
Jorde LB et al (2003) Medical Genetics. Mosby, St. Louis, MO
Korf BR (2000) Human Genetics: A Problem-Based Approach. Blackwell, Malen, MA
Kowles R (2001) Solving Problems in Genetics. Springer, New York
Kresina TF (ed) (2000) An Introduction to Molecular Medicine and Gene Therapy. Wiley, New York
Lewin B (2004) Genes VIII. Prentice Hall, New York
Nussbaum RL et al (2004) Thompson & Thompson Genetics in Medicine. W.B. Saunders, Philadelphia, PA
Old HW and Primrose SB (1995) Principles of Gene Manipulation. Blackwell Science, Cambridge, MA
Pasternak JJ (2005) An Introduction to Human Molecular Genetics: Mechanisms of Inherited Diseases. Wiley, Hoboken, NJ
Pierce B (2003) Genetics. A Conceptual Approach. Freeman, New York
Primrose SB et al (2001) An Introduction to Genetic Engineering. Blackwell Science, Cambridge, MA
Rédei GP (1982) Genetics. Macmillan, New York
Ringo J (2004) Fundamental Genetics. Cambridge University Press, New York
Rose PW and Lucassen A (1999) Practical Genetics for Primary Care. Oxford University Press, New York
Russel PJ (1999) Fundamentals of Genetics. Addison-Wesley, Reading, MA
Schleif R (1993) Genetics and Molecular Biology. Johns Hopkins University Press, Baltimore, MD
Serra JA (1965) Modern Genetics. Academic Press, New York
Singer M and Berg P (1991) Genes & Genomes. A Changing Perspective. University Science Books, Mill Valley, CA
Strickberger MW (1985) Genetics. Macmillan, New York
Thompson JN et al (1997) Primer of Genetic Analysis. Cambridge University Press, New York
Turnpenny P (2005) Emery's Elements of Medical Genetics. Elsevier, Philadelphia, PA
Vogel F and Motulsky AG (1996) Human Genetics. Springer, New York
Whitehouse HLK (1972) Towards an Understanding of the Mechanism of Heredity. St. Martin's Press, New York
Wilson GN (2000) Clinical Genetics: A Short Course. Wiley, New York
Winter PC et al (2002) Genetics. Bios Scientific, Oxford, UK

History of Genetics ►*genetics genomics*

Allen GE (1978) Thomas Hunt Morgan. The Man and His Science. Princeton University Press, Princeton, NJ
Ashburner M (2006) Won for All. How the *Drosophila* Genome Was Sequenced. Cold Spring Harbor Laboratory Press, Cold Spring Harbor, NY
Bearn AG (1996) Archibald Garrod and the Individuality of Man. Oxford University Press, New York
Berg P and Singer M (2003) An Uncommon Farmer: George Beadle and the Emergence of Genetics in the 20th century. Cold Spring Harbor Laboratory Press, Cold Spring Harbor, NY
Brenner S (2001) My Life in Science. BioMed Central, London
Brock TD (1990) The Emergence of Bacterial Genetics. Cold Spring Harbor Laboratory Press, Cold Spring Harbor, NY
Brown A (2003) In the Beginning was the Worm. Finding the Secrets of Life in a Tiny Hermaphrodite. Columbia University Press, New York
Bulmer M (2003) Francis Galton: Pioneer of Heredity and Biometry. Johns Hopkins University Press, Baltimore, MD
Cairns J et al (1966) Phage and the Origins of Molecular Biology. Cold Spring Harbor Laboratory Press, Cold Spring Harbor, NY
Cambrosio A and Keating P (1996) Exquisite Specificity: The Monoclonal Antibody Revolution. Oxford University Press, New York
Carlson EA (1966) The Gene: A Critical History. Saunders, Philadelphia, PA
Carlson EO (1981) Genes, Radiation, and Society. The Life and Work of H.J. Muller. Cornell University Press, Ithaca, NY
Carlson EO (2001) The Unfit: A History of a Bad Idea. Cold Spring Harbor Laboratory Press, Cold Spring Harbor, NY
Carlson EO (2004) Mendel's Legacy: The Origin of Classical Genetics. Cold Spring Harbor Laboratory Press, Cold Spring Harbor, NY
Childs B (1999) Genetic Medicine: A Logic of Disease. Johns Hopkins University Press, Baltimore, MD

Comfort NC (2001) The Tangled Field. Barbara McClintock's Search for the Patterns of Genetic Control. Harvard University Press, Cambridge, MA

Crow JF and Dove WF (eds) (2000) Perspectives on Genetics. Anecdotal, Historical, and Critical Commentaries. University Wisconsin Press, Madison, WI

Davies K (2001) Cracking the Genome. Inside the Race to Unlock Human DNA. Free Press, New York

Davis RH (2000) Neurospora: Contributions of a Model Organism. Oxford University Press, New York

de Chaderevian S (2002) Design for Life. Molecular Biology After World War II. Cambridge University Press, New York

Dunn LC (1965) A Short History of Genetics. The Development of Some of the Main Lines of Thought: 1864–1939. McGraw-Hill, New York

Echols H and Gross CA (2001) Operators and Promoters. The Story of Molecular Biology and Its Creators. University California Press, Berkeley, CA

Emery AEH and Emery MLH (1995) The History of a Genetic Disease: Duchenne Muscular Dystrophy or Meryon Disease. Royal Society of Medicine Press, London, UK

Fedoroff N and Botstein D (eds) (1992) Barbara McClintock's Ideas in the Century of Genetics. Cold Spring Harbor Laboratory Press, Cold Spring Harbor, NY

Fisher EP and Lipson C (1988) Thinking About Science. Max Delbrück and the Origins of Molecular Biology. WW Norton, New York

Focke WO (1881) Die Pflanzenmischlinge. Borntraeger, Berlin, Germany

Fredrickson DS (2001) The Recombinant DNA Controversy: A Memoir, Science, Politics and the Public Interest. American Society for Microbiology Press, Washington, DC

Friedberg EC (1997) Correcting the Blueprint of Life. An Historical Account of the Discovery of DNA Repair Mechanisms. Cold Spring Harbor Laboratory Press, Cold Spring Harbor, NY

Friedberg E (2003) The Writing Life of JamesWatson: Professor, Promoter, Provocateur. Cold Spring Harbor Laboratory Press, Cold Spring Harbor, NY

Friedmann T (ed) (1999) The Development of Human Gene Therapy. Cold Spring Harbor Laboratory Press, Cold Spring Harbor, NY

Gall JG and McIntosh JR (eds) (2001) Landmark Papers in Cell Biology. American Society for Cell Biology Press, Bethesda, MD

Gillham NW (2001) Sir Francis Galton: From African Exploration to the Birth of Eugenics. Oxford University Press, New York

Goldschmidt RB (1956) Portraits from Memory. Recollections of a Zoologist. University of Washington Press, Seattle, WA

Hall MN and Linder P (eds) (1993) The Early Days of Yeast Genetics. Cold Spring Harbor Laboratory Press, Cold Spring Harbor, NY

Harris H (1997) The Cells of the Body: A History of Somatic Cell Genetics. Cold Spring Harbor Laboratory Press, Cold Spring Harbor, NY

Harwood J (1993) Styles of Scientific Thought—The German Genetics Community 1900–1933. University Chicago Press, Chicago, IL

Hayes W (1964) The Genetics of Bacteria and Their Viruses. Wiley, New York

Holmes FL (2002) Meselson, Stahl, and the Replication of DNA: A History of the "Most Beautiful Experiment in Biology." Yale University Press, New Haven, CT

Holmes FL (2006) Reconceiving the Gene: Seymour Benzer's Adventures in Phage Genetics. Yale University Press, New Haven, CT

Inglis JR et al (eds) (2003) Inspiring Science. Jim Watson and the Age of DNA. Cold Spring Harbor Laboratory Press, Cold Spring Harbor, NY

Jacob F (1995) The Statue Within: An Autobiography. Cold Spring Harbor Laboratory Press, Cold Spring Harbor, NY

Jacob F (1997) La Souris, la Mouche et l'Homme. Odile Jacob, Paris, France

Johannsen W (1909) Elemente der Exakten Erblichkeitslehre. Fisher, Jena, Germany

Judson HF (1996) The Eighth Day of Creation. Makers of the Revolution in Biology. Cold Spring Harbor Laboratory Press, Cold Spring Harbor, NY

Kay LE (2000) Who Wrote the Book of Life? A History of the Genetic Code. Stanford University Press, Stanford, CA

Knudson AG (1965) Genetics and Disease. McGraw Hill, New York

Lerner IM (1950) Genetics in the U.S.S.R. An Obituary. University of British Columbia Press, Vancouver, Canada

Lewis RA (2001) Discovery—Windows on the Life Sciences. Blackwell Science, Malden, MA

Maas W (2001) Gene Action. A Historical Account. Oxford University Press, New York

Maddox B (2002) Rosalind Franklin. Dark Lady of DNA. Harper Collins, New York

Magner LN (1993) A History of the Life Sciences. Dekker, New York

McCarty M (1985) The Transforming Principle: Discovering that Genes are Made of DNA. Norton, New York

McElheny VK (2003) Watson and DNA: Making a Scientific Revolution. Perseus, Cambridge, MA

Medvedev ZS (1969) The Fall and Rise of T.D. Lysenko. Columbia University Press, New York

Morange M (1998) A History of Molecular Biology. Harvard University Press, Cambridge, MA

Morgan TH (1919) The Physical Basis of Heredity. Lippincott, Philadelphia, PA

Morgan TH et al (1915) The Mechanism of Mendelian Heredity. Holt, New York

Müller-Hill B (1996) The *lac* Operon. A Short History of a Genetic Paradigm. De Gruyter, New York

Müller-Hill B (1998) Murderous Science. Elimination by Scientific Selection of Jews, Gypsies and Others in Germany 1933–1945. Cold Spring Harbor Laboratory Press, Plainview, NY

Neel JV (1994) Physician to the Gene Pool: Genetic Lessons and Other Stories. Wiley, New York

Olby RC (1966) Origins of Mendelism. Schocken Books, New York

Olby R (1974) The Path to the Double Helix. University of Washington Press, Seattle, WA

Orel V (1996) Gregor Mendel the First Geneticist. Oxford University Press, New York

Perutz MF (1998) I Wish I'd Made You Angry Earlier: Essays on Science, Scientists, and Humanity. Cold Spring Harbor Laboratory Press, Cold Spring Harbor, NY
Portugal FH and Cohen JS (1977) A Century of DNA. MIT Press, Cambridge, MA
Potts DM and Potts WTW (1995) Queen Victoria's Gene: Hemophilia and the Royal Family. Alan Sutton, Dover, NH
Provine WB (2001) The Origins of Theoretical Population Genetics. University of Chicago Press, Chicago, IL
Rabinow P (1996) Making PCR. A Story of Biotechnology. University of Chicago Press, Chicago, IL
Ridley M (1999) Genome: The Autobiography of a Species in 23 Chapters. Harper-Collins, New York
Ridley M (2006) Francis Crick, Discoverer of the Genetic Code. Atlas Books/HarperCollins. New York
Roberts HF (1965) Plant Hybridization Before Mendel. Hafner, New York
Rushton AR (1994) Genetics and Medicine in the United States, 1800–1922. Johns Hopkins University Press, Baltimore, MD
Sarkar S (ed) (1992) The Founders of Evolutionary Genetics: A Centennary Reappraisal. Kluwer Academic, Dordrecht, NL/Boston, MA
Schrödinger E (1944) What is Life? Cambridge University Press, Cambridge
Shnoll SE (1997) Heroes, Martyrs and Villains in Russian Life Sciences. Springer, New York
Sinsheimer RL (1994) The Strands of Life: The Science of DNA and the Art of Education. University of California Press, Berkeley, CA
Sirks MJ and Zirkle C (1964) The Evolution of Biology. Ronald Press, New York
Stahl FW (ed) (2000) We Can Sleep Later: Alfred D. Hershey and the origins of Molecular Biology. Cold Spring Harbor Laboratory Press, Cold Spring Harbor, NY
Stent GS (ed) (1981) The Double Helix: Text, Commentary, Reviews, Original Papers. Norton, New York
Stubbe H (1972) History of Genetics from Prehistoric Times to the Rediscovery of Mendel's Laws. MIT Press, Cambridge, MA
Sturtevant AH (1965) A History of Genetics. Harper and Row, New York
Sulston J and Ferry G (2002) The Common Thread: A Story of Science, Politics, Ethics and the Human Genome. Joseph Henry Press, Washington, DC
Summers WC (1999) Félix d'Hérelle and the Origins of Molecular Biology. Yale University Press, New Haven, CN
Watson JD (1997) The Double Helix. Weidenfeld & Nicholson, New York
Watson JD (2000) A Passion for DNA: Genes, Genomes and Society. Oxford University Press, New York
Watson JD (2001) Genes, Girls and Gamow. Oxford University Press, New York
Watson JD and Berry A (2003) DNA: The Secret of Life. Alfred A. Knopf. New York
Watson JD and Tooze J (1981) The DNA Story: A Documentary History of Gene Cloning. Freeman, San Francisco, CA
Watson JD et al (eds) (2005) Rereading Heredity in Relation to Eugenics: Contemporary Reflections on the Promise of Human Genetics. Cold Spring Harbor Laboratory Press, Cold Spring Harbor, NY
Weiner J (1999) Time, Love, Memory: A Great Biologist [Benzer] and His Quest for the Origin of Behavior. Random House, New York
Weir RF et al (eds) (1994) Genes and Human Self-Knowledge. Historical and Philosophical Reflections on Modern Genetics. University Iowa Press, Iowa City, IA
Wilkins M (2003) The Third Man of the Double Helix: The Autobiography of Maurice Wilkins. Oxford University Press, New York
Wilmut I et al (2000) The Second Creation: Dolly and the Age of Biological Control. Farrar, Straus & Giroux, New York
Wilson EB (1925) The Cell in Development and Heredity. Macmillan, New York
Witkowski J (ed) (2000) Illuminating Life. Cold Spring Harbor Laboratory Press, Cold Spring Harbor, NY
Witkowski J (ed) (2005) The Inside Story: DNA to RNA to Protein. Cold Spring Harbor Laboratory Press, Cold Spring Harbor, NY
Wood R and Orel V (2001) Genetic Prehistory in Selective Breeding. A Prelude to Mendel. Oxford University Press, New York
Zirkle C (1935) The Beginnings of Plant Hybridization. University of Pennsylvania Press, Philadephia, PA

Immunology ►*molecular biology*, ►*biology*

Abbas A and Lichtman A (2006) Basic Immunology. Elsevier, Philadelphia, PA
Advances in Immunology (Serial). Elsevier, Philadelphia, PA
Albitar M (ed) (2007) Monoclonal Antibodies. Methods and Protocols. Humana Press, Totowa, NJ
Annual Review of Immunology. Palo Alto, CA
Asea AAA and De Maio A (eds) (2007) Heatshock Proteins: Potent Mediators of Inflammation and Immunity. Springer, Berlin, D
Austyn JM and Wood KJ (1993) Principles of Cellular and Molecular Immunology. Oxford University Press, New York
Badolato R and Sozzani S (eds) (2006) Lymphocyte Trafficking in Health and Disease. Birkhäuser, Basel, Switzerland
Beck G et al (eds) (2001) Phylogenetic Perspectives on the Vertebrate Immune System. Kluwer, Dordrecht, the Netherlands
Bell JI et al (eds) (1995) T Cell Receptors. Oxford University Press, New York
Benjamini E et al (1996) Immunology. A Short Course. Wiley, New York
Birch JR and Lennox ES (eds) (1995) Monoclonal Antibodies. Wiley-Liss, New York
Bloom BR and Lambert PH (eds) (2002) The Vaccine Book. Academic Press, San Diego, CA
Bona A and Bonilla FA (1996) Textbook of Immunology. Gordon and Break-Harwood Academic, Toronto, Canada
Bona C (ed) (2004) Neonatal Immunity. Humana Press, Totowa, NJ

Bona C (2005) Contemporary Immunology. Humana Press, Totowa, NJ
Bot A and Bona AA (2000) Genetic Immunization. Kluwer, Dordrecht, the Netherlands
Breitling F and Dübel S (1999) Recombinant Antibodies. Wiley, New York
Burns R (ed) (2005) Immunochemical Protocols. Humana Press, Totowa, NJ
Carr DJJ (ed) (2005) Interferon Methods and Protocols. Humana Press, Totowa, NJ
Cattaneo A and Biocca S (eds) (1997) Intracellular Antibodies. Springer, New York
Chen L (ed) (2006) CD137 Pathway: Immunology and Diseases. Springer, Berlin, D
Cochet O et al (eds) (1998) Immunological Techniques Made Easy. Wiley, New York
Coffman RL and Romagnani S (eds) (1999) Redirection of Th1 and Th2 Responses. Springer, New York
Coico R et al (2003) Immunology. A Short Course. Wiley, Hoboken, NJ
D'Ambrosio D and Sinigaglia F (eds) (2003) Cell Migration in Inflammation and Immunity. Humana Press, Totowa, NJ
Delves PJ and Roitt I (eds) (1998) The Encyclopedia of Immunology. Academic Press, San Diego, CA
Durum SK and Muegge K (eds) (1998) Cytokine Knockouts. Humana Press, Totowa, NJ
Ewbank J and Vivier E (eds) (2007) Innate Immunity. Humana Press, Totowa, NJ
Fairchild PJ (ed) (2007) Immmunological Tolerance. Methods and Protocols. Humana Press, Totowa, NJ
Falus A (2006) Immunogenomics and Human Disease. Wiley, Hoboken, NJ
Fernandez N and Butcher G (1997) MHC1. A Practical Approach. Oxford University Press, New York
Flint SJ et al (2000) Principles of Virology. Molecular Biology, Pathogenesis, and Control. American Society for Microbiology Press, Washington, DC
Flower DR and Timmis J (eds) (2007) In silico Immunology. Springer, Berlin, D
Flower DR (ed) (2007) Immunoinformatics. Predicting Immunogenicity in Silico. Humana Press, Totowa, NJ
Frank SA (2002) Immunology and Evolution of Infectious Disease. Princeton University Press, Princeton, NJ
Frelinger JA (ed) (2006) Immunodominance. Wiley, Hoboken, NJ
Gershwin ME et al (eds) (2007) Liver Immunology. Principles and Practices. Humana Press, Totowa, NJ
Gessani S and Belardelli F (eds) (2007) The Biology of Dendritic Cells and HIV Infection. Springer, Berlin, D
Graca L (ed) (2008) The Immune Synapse As a Novel Target for Therapy. Birkhäuser, Basel, Switzerland
Gregory CD (ed) (1995) Apoptosis and the Immune Response. Wiley-Liss, New York
Gu H, Rajewsky K (eds) (2004) B Cell Protocols. Humana Press, Totowa, NJ
Gupta S et al (eds) (2007) Mechanisms of Lymphocyte Activation and Immune Regulation XI. B cell biology. Springer, Berlin, D
Hamann A and Engelhardt B (eds) (2006) Leukocyte Trafficking. Molecular Mechanisms, Therapeutic Targets, and Methods. Wiley, Hoboken, NJ
Harlow E and Lane D (1999) Using Antibodies: A Laboratory Manual. Cold Spring Harbor Laboratory Press, Plainview, NY
Herbert WJ et al (eds) (1995) The Dictionary of Immunology. Academic Press, San Diego, CA
Holgate ST et al (2006) Allergy. Mosby, St. Louis, MO
Honjo T and Alt FW (eds) (1995) Immunoglobulin Genes. Academic Press, San Diego, CA
House RV and Descotes J (eds) (2007) Cytokines in Human Health. Immunotoxicology, Pathology, and Therapeutic Applications. Humana Press, Torowa, NJ
Hughes DA et al (eds) (2004) Diet and Human Immune Reaction. Humana Press, Totowa, NJ
Janeway CA Jr et al (2004) Immonobiology. The Immune System in Health and Disease. Garland, New York
Katsikis PD et al (eds) (2007) Crossroads Between Innate and Adaptive Immunity. Springer, Berlin, D
Kaufmann SHE et al (eds) (2004) The Innate Immune Response to Infection. ASM Press, Washington, DC
Kaufmann SHE et al (2002) Immunology of Infectious Diseases. ASM Press, Washington, DC
Kauser K and Zeiher AM (eds) (2007) Bone Marrow-Derived Progenitors. Springer, Berlin, D
King DJ (1998) Application and Engineering of Monoclonal Antibodies. Taylor & Francis. Philadelphia, PA
Kontermann R and Dübel S (eds) (2001) Antibody Engineering. Springer, New York
Krakauer T (ed) (2002) Superantigen Protocols. Humana Press, Totowa, NJ
Kropshofer H and Vogt AB (eds) (2006) Antigen Presenting Cells. From Mechanisms to Drug Development. Wiley-VCH, Hoboken, NJ
Kuby J (1997) Immunology. Freeman, New York
Lanzetta M and Dubernard J-M (eds) (2006) Hand Transplantation. Springer, Berlin, D
Law SKA and Reid KIM (1995) Complement. IRL/Oxford University Press, New York
Lambris JD (ed) (2007) Current Topics in Innate Immunity. Springer, Berlin/D
Leffell MS et al (1997) Handbook of Human Immunology. CRC Press, Boca Raton, FL
Lefranc M-P and Lefranc G (2001) The T Cell Receptor Factsbook. Academic Press, San Diego, CA
Lefranc M-P and Lefranc G (2001) The Immunoglobin Factsbook. Academic Press, San Diego, CA
Liu MA et al (eds) (1995) DNA Vaccines. New York Academy of Science, p 772
Lo BKC (ed) (2004) Antibody Engineering. Methods and Protocols. Humana Press, Totowa, NJ
Lowrie DB and Whalen R (eds) (1999) DNA Vaccines. Humana Press, Totowa, NJ
Ludewig B (ed) (2005) Adoptive Immunotherapy. Humana Press, Totowa, NJ
Luttmann W (2006) Immunology. Elsevier, Philadelphia, PA
Mak TW and Saunders ME (2006) The Immune Response. Basic and Clinical Principles. Elsevier/Academic Press, San Diego, CA
Male D et al (1996) Advanced Immunology. Mosby, St. Louis, MO
March SGE et al (2000) The HLA Facts Book. Academic Press, San Diego, CA

McCafferty J et al (1996) Antibody Engineering. A Practical Approach. IRL/Oxford University Press, New York
Meyers RA (2007) Immunology. From Cell Biology to Disease. Wiley, Hoboken, NJ
Morgan BP and Harris CL (1999) Complement Regulatory Proteins. Academic Press, San Diego, CA
Morley BJ and Walport MJ (eds) (2000) The Complement Factsbook. Academic Press, San Diego, CA
Nezlin R (1998) The Immunoglobulins. Structure and Function. Academic Press, San Diego, CA
Nijkamp FP and Parnham MJ (eds) (2005) Principles of Immunopharmacology. Birkhäuser, Basel, Switzerland
Ochs HD et al (eds) (1999) Primary Immunodeficiency Diseases: A Molecular and Genetic Approach. Oxford University Press, New York
Parham P (2004) The Immune System. Garland, New York
Parmiani G and Lotze MT (eds) (2002) Tumor Immunology: Molecularly Defined Antigens and Applications. Taylor & Francis, London
Paul WE (1993) Fundamental Immunology. Raven, New York
Perrl A (2004) Autoimmunity. Methods and Protocols, Humana Press, Totowa, NJ
Pillai S (2000) Lymphocyte Development: Cell Selection Events and Signals During Immune Ontogeny. Birkhäuser, Boston, MA
Pitha PM (ed) (2007) Interfereon: The 50th Anniversary. Springer, Berlin
Pollards KM (ed) (2006) Autoantibodies and Autoimmunity. Molecular Mechanisms in Health and Disease. Wiley, Hoboken, NJ
Powis SH and Vaughan RW (eds) (2002) MHC Protocols. Humana Press, Totowa, NJ
Pulendran B and Ahmed R (eds) (2006) From Innate Immunity to Immunological Memory. Springer, New York
Radbruch A et al (eds) (2007) Immunotherapy in 2020. Visions and Trends for Targeting Inflammatory Disease. Springer, Berlin, D
Raz E (ed) (1998) Gene Vaccination: Theory and and Practice. Springer, New York
Reid ME and Lomas-Francis C (1997) The Blood Group Antigen. Academic Press, San Diego, CA
Ritter MA and Ladyman HM (eds) (1995) Monoclonal Antibodies. Production, Engineering and Clinical Application. Cambridge University Press. New York
Roitt I (1991) Essential Immunology. Blackwell Science, Oxford
Rose NR et al (eds) (2002) Manual of Clinical Laboratory Immunology. AMS Press, Washington, DC
Rose N and Mackay I (eds) (2006) The Autoimmune Diseases. Elsevier, Philadelphia, PA
Rowland-Jones SL and McMichael AJ (eds) (2000) Lymphocytes: A Practical Approach. Oxford University Press, New York
Rother K et al (eds) (1998) The Complement System. Springer, New York
Salomon DR and Wilson C (eds) (2003) Xenotransplantation. Springer, New York
Shepherd PS and Dean C (eds) (2000) Monoclonal Antibodies. Oxford University Press, New York
Shurin MR and Smolkin YS (eds) (2007) Immune Mediated Diseases. From Theory to Therapy. Springer, Berlin, D
Sell S (2001) Immunology, Immunopathology, and Immunity. ASM Press, Washington, DC
Solheim JC (ed) (2000) Antigen Processing and Presentation Protocols. Humana Press, Totowa, NJ
Stevenson CS et al (eds) (2006) In Vivo Models of Inflammation. Springer, New York
Stites DP, Terr AI, Parslow TG (eds) (1994) Basic & Clinical Immunology. Appleton & Lange, Norwalk, CT
Theofilopoulos AN and Bona CD (eds) (2002) The Molecular Pathology of Autoimmune Diseases. Taylor & Francis, London, UK
Thèze J (ed) (1999) The Cytokine Network and Immune Functions. Oxford University Press, New York
Tilney NL et al (eds) (1996) Transplantation Biology. Cellular and Molecular Aspects. Lippincott-Raven, Philadelphia, PA
Weir D (ed) (1996) The Handbook of Experimental Immunology. Blackwell Science, Oxford, UK
Westwood O and Hay F (eds) (2001) Epitope Mapping. Oxford University Press, New York
Wingard JR and Anaissie EJ (eds) (2005) Fungal Infections of the Immunocompromised Patient. Taylor & Francis, Boca Raton, FL
Winkler JD (ed) (1999) Apoptosis and Inflammation. Birkhäuser, Basel, Switzerland
Wodarz D (2007) Killer Cell Dynamics. Mathematical and Computational Approaches to Immunology. Springer, Berlin, D
Yefenof E (ed) (2008) Innate and Adaptive Immunity in the Tumor Microenvironment. Springer, Berlin, D
Zanetti M and Capra JD (eds) (1995) The Antibodies. Harwood, Langhorne, PA
Zhang J (ed) (2007) Immune Regulation and Immunotherapy in Autoimmune Disease. Springer, Berlin, D

Laboratory Technology ►*molecular biology*, ►*biotechnology*, ►*cytology*

Adams DS (2003) Lab Math: A Handbook of Measurements, Calculations, and Other Quantitative Skills for Use at the Bench. Cold Spring Harbor Laboratory Press, Cold Spring Harbor, NY
Aguilar M-I (ed) (2004) HPLC of Peptides and Proteins. Humana Press, Totowa, NJ
Akay M (ed) (2007) Genomics and Proteomics. Engineering in Medicine and Biology. Wiley, Hoboken, NJ
Alfa C et al (eds) (1993) Experiments with Fission Yeast. A Laboratory Course Manual. Cold Spring Harbor Laboratory Press, Cold Spring Harbor, NY
Amberg DC et al (2005) Methods in Yeast Genetics. Cold Spring Harbor Laboratory Press, Cold Spring Harbor, NY
Andreeff M and Pinkel D (eds) (2000) Introduction to Fluorescence In Situ Hybridization. Principles and Clinical Applications. Wiley, New York

Andrews L and Tollefsbol TO (eds) (2007) Telomerase Inhibition. Strategies and Protocols. Humana Press, Totowa, NJ

Anson D (ed) (2007) Reporter Genes. A Practical Guide. Humana Press, Totowa, NJ

Ansorge W et al (eds) (1996) DNA Sequencing Strategies. Automated and Advanced Approaches. Wiley, New York

Aquino de Muro M and Rapley R (eds) (2001) Gene Probes. Principles and Protocols. Humana Press, Totowa, NJ

Armour M-A (1996) Hazardous Laboratory Chemicals Disposal Guide. CRC Press, Boca Raton, FL

Ausubel FM et al (eds) (1987) Current Protocols in Molecular Biology. Wiley, New York

Ausubel FM et al (2002) Short Protocols in Molecular Biology. Wiley, Hoboken, NJ

Azuaje F and Dopazo J (eds) (2005) Data Analysis and Visualization in Genomics and Proteomics. Wiley, Hoboken, NJ

Bagasra O and Hansen J (1997) In Situ PCR Techniques. Wiley-Liss, New York

Baldi P and Hatfield GW (2002) DNA Microarrays and Gene Expression. Cambridge University Press, New York

Baldwin SA (ed) (2000) Membrane Transport. A Practical Approach. Oxford University Press, New York

Barbas CF et al (2001) Phage Display: Laboratory Manual. Cold Spring Harbor Laboratory Press, Cold Spring Harbor, NY

Barker K (2002) At the Helm. A Laboratory Navigator. Cold Spring Harbor Laboratory Press, Cold Spring Harbor, NY

Baron MH (ed) (2005) Developmental Hematopoiesis. Methods and Protocols. Humana Press, Totowa, NJ

Baxevanis AD et al (eds) (2003) Current Protocols in Bioinformatics. Wiley, New York

Bernstam VA (1992) Handbook of Gene Level Diagnostics in Clinical Practice. CRC Press, Boca Raton, FL

Bhojwani SS and Razdan MK (1996) Plant Tissue Culture: Theory and Practice. Elsevier Science, New York

Birren B and Lai E (1993) Pulsed Field Gel Electrophoresis: A Practical Guide. Academic Press, San Diego, CA

Birren B and Lai E (eds) (1996) Nonmammalian Genomic Analysis. A Practical Guide. Academic Press, San Diego, CA

Birren B et al (eds) (1997) Genome Analysis: A Laboratory Manual. Cold Spring Harbor Laboratory Press, Cold Spring Harbor, NY

Birren B et al (1998) Detecting Genes: A Laboratory Manual. Cold Spring Harbor Laboratory Press, Cold Spring Harbor, NY

Birren B et al (eds) (1999) Cloning Systems: A Laboratory Manual. Cold Spring Harbor Laboratory Press, Cold Spring Harbor, NY

Birren B et al (1999) Mapping Genomes: A Laboratory Manual. Cold Spring Harbor Laboratory Press, Cold Spring Harbor, NY

Bishop M and Rawlings C (1997) DNA and Protein Sequence Analysis. A Practical Approach. Oxford University Press, New York

Bjornsti M-A and Osheroff N (eds) (1999) DNA Topoisomerase Protocols. Humana Press, Totowa, NJ

Blumenthal RD (ed) (2005) Chemosensitivity. Vol I. In Vitro Assays. Vol II In Vivo Models, Imaging and Molecular Regulators. Humana Press, Totowa, NJ

Bonifacino JS et al (eds) (1999) Current Protocols in Cell Biology. Wiley, New York

Bowtell D and Sambrook J (eds) (2003) DNA Microarrays: A Molecular Cloning Manual. Cold Spring Harbor Laboratory Press, Woodbury, NY

Box GEP et al (2005) Statistics for Experimenters. Wiley, Hoboken, NJ

Bozzola JJ and Russel LD (1992) Electron Microscopy. Principles and Techniques for Biologists. Jones & Bartlett, Boston, MA

Brady HJM (ed) (2004) Apoptosis Methods and Protocols. Humana Press, Totowa, NJ

Bretherick L (1990) Bretherick's Handbook of Reactive Chemical Hazards. Butterworth, London

Brown AJP and Tuite MF (eds) (1998) Yeast Gene Analysis. Academic Press, San Diego, CA

Brown T (1998) Molecular Biology Labfax. Recombinant DNA. Gene Analysis. Academic Press, San Diego, CA

Brown T (ed) (2001) Essential Molecular Biology. Oxford University Press, New York

Buckleton J et al (eds) (2005) Forensic DNA Evidence Interpretation. CRC Press, Boca Raton, FL

Bulte JW and Modo M (eds) (2008) Nanoparticles in Biomedical Imaging. Springer, Berlin, D

Burger S and Braun S (2004) 200 and More NMR Experiments. A Practical Course. Wiley, Hoboken, NJ

Burke D et al (2000) Methods in Yeast Genetics: A CSHL Course Manual. Cold Spring Harbor Laboratory Press, Cold Spring Harbor, NY

Butler M (2004) Animal Cell Culture and Technology. Garland Science/BIOS, New York

Buxton RB (2002) An Introduction to Functional Resonance Imaging: Principles and Techniques. Cambridge University Press, New York

Cambell J and Modrich P (2006) DNA Repair. Elsevier. Philadelphia, PA

Carmichael G (ed) (2005) RNA Silencing. Methods and Protocols. Humana Press, Totowa, NJ

Carraway KI and Carothers Carraway CA (eds) (2000) Cytoskeleton: Signalling and Cell Regulation. A Practical Approach. Oxford University Press, New York

Celis J (ed) (2005) Cell Biology: A Laboratory Handbook. Elsevier, San Diego, CA

Chalfie M and Kain S (eds) (2006) Green Fluorescent Protein: Properties, Applications and Protocols. Wiley, Hoboken, NJ

Chan WCW (ed) (2008) Bio-Applications of Nanoparticles. Springer, Berlin, D

Chen B-Y and Janes HW (2002) PCR Cloning Protocols. Humana Press, Totowa, NJ

Chrispeels MJ and Sadava DE (1994) Plants, Genes and Agriculture. Jones and Bartlett, Boston, MA

Clark M (1997) Plant Molecular Biology. A Laboratory Manual. Springer, New York

Clegg RA (ed) (1998) Protein Targeting Protocols. Humana Press, Totowa, NJ

Clynes M (ed) (1998) Animal Cell Culture Techniques. Springer, New York

Coligan JE et al (eds) (1991) Current Protocols in Immunology. Wiley, New York

Conn PM (ed) (2003) Handbook of Proteomic Methods. Humana Press, Totowa, NJ

Corti A and Ghezzi P (eds) (2004) Tumor Necrosis Factor. Methods and Protocols. Humana Press, Totowa, NJ
Cotterill S (ed) (1999) Eukaryotic DNA Replication: A Practical Approach. Oxford University Press, New York
Coutts AS (ed) (2007) Adhesion Protein Protocols. Humana Press, Totowa, NJ
Craig AG and Hoheisel JD (1999) Automation: Genome and Functional Analyses. Academic Press, San Diego, CA
Crowther JR (1995) ELISA. Humana Press, Totowa, NJ
Current Protocols (editors vary, frequently updated) Bioinformatics, Cell Biology, Cytometry, Human Genetics, Immunology, Magnetic Resonance Imaging, Molecular Biology, Neuroscience, Nucleic Acid Chemistry, Pharmacology, Protein Science, Toxicology. John Wiley & Sons, Somerset, NJ
Cutler P (ed) (2004) Protein Purification Protocols. Humana Press, Totowa, NJ
D'Ambrosio D and Sinigaglia F (eds) (2004) Cell Migration in Inflammation and Immunity. Methods and Protocols. Humana Press, Totowa, NJ
Dangler CA (ed) (1996) Nucleic Acid Analysis. Principles and Bioapplications. Wiley, New York
Darby IA and Hewitson TD (eds) (2006) In Situ Hybridization Protocols. Humana Press, Totowa, NJ
Dass C (2000) Principles and Practice of Biological Mass Spectrometry. Wiley, New York
Davenport AP (eds) (2005) Receptor Binding Techniques. Humana Press, Totowa, NJ
Davis L (ed) (2002) Basic Cell Culture. Oxford University Press, New York
Davis LG et al (1994) Basic Methods in Molecular Biology. Appleton and Lange. Norwalk, CN
Day JG and Stacey G (eds) (2007) Cryopreservetion and Freeze-Drying Protocols. Humana Press, Totowa, NJ
deBoer AG and Sutanto W (eds) (1997) Drug Transport Across the Blood-Brain Barrier. In Vitro and In Vivo Techniques. Gordon & Breach, Amsterdam, the Netherlands
Decker J and Reischl U (eds) (2004) Molecular Diagnosis of Infectious Diseases. Humana Press, Totowa, NJ
De Ley M (ed) (2004) Cytokine Protocols. Humana Press, Totowa, NJ
Dickson RC and Mendenhall MD (eds) (2004) Signal Transduction Protocols. Humana Press, Totowa, NJ
Didenko VV (ed) (2002) In Situ Detection of DNA Damage. Humana Press, Totowa, NJ
Dieffenbach CW and Dveksler GS (eds) (2003) PCR Primer. A Laboratory Manual. Cold Spring Harbor Laboratory Press, Cold Spring Harbor, NY
Dodds JH and Roberts LW (1995) Experiments in Plant Tissue Culture. Cambridge University Press, New York
Dolezel J et al (2007) Flow Cytometry with Plant Cells. Wiley, Hoboken, NJ
Dopico A (ed) (2007) Methods in Membrane Lipids. Humana Press, Totowa, NJ
Downing AK (ed) (2004) Protein NMR Techniques. Humana Press, Totowa, NJ
Doyle A and Griffith JB (1998) Cell and Tissue Culture. Laboratory Procedures in Biotechnology. Wiley, New York
Dracopoli NC et al (eds) (2005) Short Protocols in Human Genetics. Wiley, Hoboken, NJ
Draper J et al (eds) (1988) Plant Genetic Transformation and Gene Expression. A Laboratory Manual. Blackwell, Oxford
Echalier G (1997) Drosophila Cells in Culture. Academic Press, San Diego, CA
Edwards D (ed) (2007) Plant Bioinformatics. Methods and Protocols. Humana Press, Totowa, NJ
Edwards R (ed) (2000) Immunodiagnostics. A Practical Approach. Oxford University Press, New York
Edwards S and Collin HA (1997) Plant Cell Culture. Springer, New York
Eidhammer I et al (2004) Protein Bioinformatics. Wiley, Hoboken, NJ
Elles R and Mountford R (eds) (2004) Molecular Diagnosis of Genetic Diseases. Humana Press, Totowa, NJ
Evans IH (ed) (1966) Yeast Protocols. Methods in Cell and Molecular Biology. Humana Press, Totowa, NJ
Farrell RE Jr (1998) RNA Methodologies. A Laboratory Guide for Isolation and Characterization. Academic Press, San Diego, CA
Fennell JP (ed) (2005) Hypertension. Methods and Protocols. Humana Press, Totowa, NJ
Ferré F (1997) Gene Quantification. Springer, New York
Fields GB (ed) (2007) Peptide Characterization and Application Protocols. Humana Press, Totowa, NJ
Fleming DO and Hunt DL (eds) (2000) Biological Safety. Principles and Practices. AMS Press, Washington, DC
Floriano PN (ed) (2008) Microchip-Based Assay Systems. Methods and Applications. Humana Press, Totowa, NJ
Frank J (ed) (2007) Electron Tomography. Methods for Three-Dimensional Visualization of Structures in the Cell. Springer, Berlin, D
Freshney RI (2005) Animal Cell Culture. Manual of Basic Technique. Wiley, Hoboken, NJ
Friedmann T and Rossi J (eds) (2006) Gene Transfer: Delivery and Expression of DNA and RNA. A Laboratory Manual. Cold Spring Harbor Laboratory Press, Cold Spring Harbor, NY
Fu H (ed) (2004) Protein–Protein Interactions. Humana Press, Totowa, NJ
Fung E (ed) (2004) Protein Arrays. Methods and Protocols. Humana Press, Totowa, NJ
Furr AK (ed) (1989) CRC Handbook of Laboratory Safety. CRC Press, Boca Raton, FL
Gallagher S and Wiley EA (2007) Current Protocols. Essential Laboratory Techniques. Wiley, Hoboken, NJ
Gardner DK et al (2004) A Laboratory Guide to the Mammalian Embryo. Oxford University Press, New York
Garman A (1997) Non-Radioactive Labelling. Academic Press, San Diego, CA
Gibbins JM and Mahaut-Smith MP (2004) Platelets and Megakaryocytes. Vol I. Functional Assays. Vol II. Perspectives and Techniques. Humana Press, Totowa, NJ
Gillespie SH (ed) (2000) Antibiotic Resistance Methods and Protocols. Humana Press, Totowa, NJ
Giordano A and Romano G (eds) (2004) Cell Cycle Control and Dysregulation Protocols. Cyclins, Cyclin-Dependent Kinases, and Other Factors. Humana Press, Totowa, NJ
Givan AL (2001) Flow Cytometry: First Principles. Wiley, New York

Glasel JA and Deutscher MP (eds) (1995) Introduction to Biophysical Methods for Protein and Nucleic Acid Research. Academic Press, San Diego, CA
Glass DJ (2006) Experimental Design for Biologists. Cold Spring Harbor Laboratory Press, Cold Spring Harbor, NY
Glezen M, van der (ed) (eds) (2007) Protein Targeting Protocols. Humana Press, Totowa, NJ
Glover DM and Hames BD (eds) (1996) DNA Cloning: A Practical Approach. IRL/Oxford University Press, New York
Gold LS and Zeigler E (eds) (1997) Handbook of Carcinogenic Potency and Genotoxicity Databases. CRC Press, Boca Raton, FL
Goldman RD and Spector DI (eds) (2005) Live Cell Imaging: A Laboratory Manual. Cold Spring Harbor Laboratory Press, Woodbury, NY
Goldstein LSB and Fryberg EA (eds) (1994) Drosophila melanogaster. Practical Uses in Cell and Molecular Biology. Academic Press, San Diego, CA
Golemis EA (ed) (2005) Protein-Protein Interactions: A Molecular Cloning Manual. Cold Spring Harbor Laboratory Press, Woodbury, NY
Goligosky MS (ed) (2003) Renal Disease. Techniques and Protocols. Humana Press, Totowa, NJ
Goodrich JA and Kugel JF (2006) Binding and Kinetics for Molecular Biologists. Cold Spring Harbor Laboratory Press, Cold Spring Harbor, NY
Gosden JR (ed) (1996) Prins and In Situ PCR Protocols. Humana Press, Totowa, NJ
Gosling JP (ed) (2000) Immunoassays. A Practical Approach. Oxford University Press, New York
Gott JM (ed) (2004) RNA Interference, Editing, and Modification. Humana Press, Totowa, NJ
Gottschall WC (2001) Laboratory Health and Safety Dictionary. Wiley, New York
Goulden NJ and Steward CG (eds) (2004) Pediatric Hematology. Methods and Protocols. Humana Press, Totowa, NJ
Graham CA and Alison JM (2001) DNA Sequencing Protocols. Humana Press, Totowa, NJ
Grandi G (ed) (2007) In Vitro Transcription and Translation Protocols. Human Press, Totowa, NJ
Greco RS et al (eds) (2005) Nanoscale Technology in Biological Systems. CRC Press, Boca Raton, FL
Greenspan R (2004) Fly Pushing: The Theory and Practice of *Drosophila* Genetics. Cold Spring Harbor Laboratory Press, Woodbury, NY
Grout B (1995) Genetic Preservation of Plant Cells In Vitro. Springer, New York
Hackett CJ and Harn DA (2005) Vaccine Adjuvants. Immunological and Clinical Principles. Humana Press, Totowa, NJ
Hamatake RK and Lau JYN (eds) (2004) Hepatitis B and D Protocols. Humana Press, Totowa, NJ
Hames BD and Higgins SJ (eds) (1996) Gene Probes 2. Oxford University Press, New York
Hames BD (ed) (1998) Gel Electrophoresis of Proteins: A Practical Approach. Oxford University Press, New York
Han J and Kamber M (2006) Data Mining. Elsevier, Philadelphia, PA
Hardie DG (ed) (2000) Protein Phosphorylation. A Practical Approach. Oxford University Press, New York
Hardin C et al (2001) Cloning, Gene Expression and Protein Purification. Experimental Procedures and Process Rationale. Oxford University Press, New York
Harrison MA and Rae IF (1997) General Techniques of Cell Culture. Cambridge University Press, New York
Harwood AJ (ed) (1994) Protocols for Gene Analysis. Humana Press, Totowa, NJ
Hau J and Van Hoosier GL Jr (eds) (eds) (2003) Handbook of Laboratory Animal Science. CRC Press, Boca Raton, FL
Hawley TS and Hawley RG (eds) (2004) Flow Cytometry. Humana Press, Totowa, NJ
Haynes LW (ed) (eds) (1999) RNA—Protein Interactive Protocols. Humana Press, Totowa, NJ
Helfrich MH and Ralston RH (eds) (2003) Bone Research Protocols. Humana Press, Totowa, NJ
Helgason CD and Miller CL (2005) Basic Cell Culture Protocols. Humana Press, Totowa, NJ
Henderson DS (2005) DNA Repair Protocols. Mammalian Systems. Humana Press, Totowa, NJ
Herdwijn P (ed) (2005) Oligonucleotide Synthesis. Methods and Applications. Humana Press, Totowa, NJ
Higgins D and Taylor W (2000) Bioinformatics: Sequence, Structure and Databanks. Oxford University Press, New York
Higgins SJ and Hames BD (1999) Protein Expression. A Practical Approach. Oxford University Press, New York
Hilario E and Mackay JF (eds) (2007) Protocols for Nucleic Acid Analysis by Non-Radioactive Probes. Humana Press, Totowa, NJ
Hofker MH and Deursen van J (eds) (2003) Transgenic Mouse. Methods and Protocols. Humana Press, Totowa, NJ
Hogan B et al (1994) Manipulating the Mouse Embryo. A Laboratory Manual. Cold Spring Harbor Laboratory Press, Cold Spring Harbor, NY
Hope IA (2000) C. elegans—A Practical Approach. Oxford University Press, New York
Horabin JI (ed) (2007) Hedgehog Signaling Protocols. Humana Press, Totowa, NJ
Hotz CZ and Bruchez M (eds) (2007) Quantum Dots. Applications in biology, Humana Press, Totowa, NJ
Houdebine L-M (2003) Animal Transgenesis and Cloning. Wiley, Hoboken, NJ
Howard GC and Kaser MR (2006) Making and Using Antibodies: A Practical Handbook. CRC Press, Boca Raton, FL
Howl J (ed) (2005) Peptide Synthesis and Applications. Zanders ED (ed). Chemical Genomics. Reviews and Protocols. Humana Press, Totowa, NJ
Hu X and Pan Y (2007) Kowledge Discovery in Bioinformatics. Wiley. Hoboken, NJ
Huang Z (2007) Drug Discovery Reseaerch. New Frontiers in the Post Genomic Era. Wiley, Hoboken, NJ
Hunt S and Livesey F (eds) (2000) Functional Genomics. A Practical Approach. Oxford University Press, New York
Innis M et al (eds) (1999) PCR Applications: Protocols for Functional Genomics. Academic Press, Orlando, FL
Innocenti F (ed) (2005) Pharnacogenomics. Methods and Applications. Humana Press, Totowa, NJ

Isaac PG (ed) (1994) Protocols for Nucleic Acid Analysis by Nonradioactive Probes. Humana Press, Totowa, NJ

Jackson IJ and Abbott CM (eds) (2000) Mouse Genetics and Transgenics: A Practical Apoproach. Oxford University Press, New York

Jenkins N (ed) (1999) Animal Cell Biotechnology: Methods and Protocols. Humana Press, Totowa, NJ

Jezzard P et al (eds) (2001) Functional Magnetic Resonance Imaging: An Introduction to Methods. Oxford University Press, New York

Johnston JR (1994) Molecular Genetics of Yeast. A Practical Approach. IRL Press, New York

Jones GE (ed) (1996) Human Cell Culture Protocols. Humana Press, Totowa, NJ

Jones P (ed) (1997) Cloning Applications. Wiley, New York

Jordan BR (ed) (2001) DNA Microarrays: Gene Expression Applications. Springer, New York

Joyner AL (ed) (2000) Gene Targeting: A Practical Approach. Oxford University Press, New York

Kamberova G and Shah S (eds) (2002) DNA Array Image Analysis: Nuts & Bolts. DNA Press, Eagleville, PA

Kannicht C (2002) Posttranslational Modification of Proteins. Tools for Functional Proteomics. Humana Press, Totowa, NJ

Kavanagh K (ed) (2007) Medical Mycology. Cellular and Molecular Techniques. Wiley, Hoboken, NJ

Keedwell E and Narayanan A (2005) Intelligent Bioinformnatics. Wiley, Hoboken, NJ

Kendall DA and Hill SJ (eds) (1995) Signal Transduction Protocols. Humana Press, Totowa, NJ

Kinter M and Sherman NE (2000) Protein Sequencing and Identification Using Tandem Mass Spectrometry. Wiley, New York

Kmiec EB (ed) (2000) Gene Targeting Protocols. Humana Press, Totowa, NJ

Kneale GG (ed) (1994) DNA—Protein Interactions. Principles and Protocols. Humana Press, Totowa, NJ

Knudsen S (2004) A Guide to Analysis of DNA Microarray Data. Wiley, Hoboken, NJ

Kobayashi T et al (eds) (1994) Animal Cell Technology. Basic and Applied. Kluwer, Norwell, MA

Kohler JM and Mejewaia T (1999) Microsystem Technology: A Powerful Tool for Biomolecular Studies. Birkhäuser, Basel, Switzerland

Kohwi Y (ed) (2004) Trinucleotide Repeat Protocols. Humana Press, Totowa, NJ

Korenberg MJ (ed) (2007) Microarray Data Analysis. Methods and Applications. Humana Press, Totowa. NJ

Körholz D and Kiess W (eds) (2002) Cytokines and Colony Stimulating Factors. Methods and Protocols. Humana Press, Totowa, NJ

Krieg PA (ed) (1996) A Laboratory Guide to RNA. Isolation, Analysis, and Synthesis. Wiley-Liss, New York

Khudyakov YE and Fields HA (eds) (2003) Artificial DNA. Methods and Applications. CRC Press, Boca Raton, FL

Kwok P-Y (ed) (2003) Single Nucleotide Polymorphisms. Methods and Protocols. Humana Press, Totowa, NJ

Lambert DG (ed) (2005) Calcium Signaling Protocols. Humana Press, Totowa, NJ

Landegren U (1996) Laboratory Protocols for Mutation Detection. Oxford University Press, New York

Langdon SP (ed) (2004) Cancer Cell Culture. Humana Press, Totowa, NJ

Larijani B et al (eds) (2006) Chemical Biology. Techniques and Applications. Wiley, Hoboken, NJ

Larrick JW and Siebert PD (eds) (1995) Reverse Transcriptase PCR. Ellis Horwood, London

Lash LH (ed) (2005) Drug Metabolism and Transport. Molecular Methods and Mechanisms. Humana Press, Totowa, NJ

Lasick DD (1997) Liposomes in Gene Delivery. CRC Press, Boca Raton, FL

Laudet V and Gronemeyer H (2001) The Nuclear Receptor Factsbook. Academic Press, San Diego, CA

Lehmann NK (2007) Serial Analysis of Gene Expression (SAGE). Methods and Protocols. Humana Press, Totowa, NJ

Levy ER and Herrington CS (1995) Non-Isotopic Methods in Molecular Biology—A Practical Appoach. Oxford University Press, New York

Liang P et al (eds) (2005) Differential Display Method and Protocols. Humana Press, Totowa, NJ

Lincoln PJ and Thomson J (eds) (1998) Forensic DNA Profiling Protocols. Humana Press, Totowa, NJ

Liu Q and Weiner MP (eds) (2001) Cloning and Expression Vectors for Gene Function Analysis. Eaton, Westborough, MA

Lo YMD (1998) Clinical Applications of PCR. Humana Press, Totowa, NJ

Lowestein PR and Enquist LW (eds) (1996) Protocols for Gene Transfer in Neuroscience: Towards Gene Therapy of Neurological Disorders. Wiley, New York

Ludewig B and Hoffmann MW (eds) (2005) Adoptive Immunotherapy. Humana Press, Totowa, NJ

Lugnier C (ed) (2005) Phosphodiesterase Methods and Protocols. Humana Press, Totowa, NJ

Luo ZD (ed) (2004) Pain Research. Methods and Protocols. Humana Press, Totowa, NJ

Macey MG (ed) (2007) Flow Cytometry. Principles and Applications. Humana press, Totowa, NJ

MacKenzie A (ed) (2006) YAC Protocols. Humana Press, Totowa, NJ

Mak TW (ed) (1998) The Gene Knockout Factsbook. Academic Press, San Diego, CA

Maliga P et al (eds) (1995) Methods in Plant Molecular Biology. A Laboratory Course Manual. Cold Spring Harbor Laboratory Press, Cold Spring Harbor, NY

Malik VS and Lillehoj EP (eds) (1994) Antibody Techniques. Academic Press, San Diego, CA

Markie D (ed) (1995) YAC Protocols. Humana Press, Totowa, NJ

Marsh S (ed) (2007) Pyrosequencing Protocols. Humana Press, Totowa, NJ

Martin SR (2004) Understanding Mass Spectra. A Basic Approach. Wiley, Hoboken, NJ

Mason WT (1999) Fluorescent and Luminescent Probes for Biological Activity. Academic Press, San Diego, CA

Masters JRW (ed) (2000) Animal Cell Cultures. Oxford University Press, New York

Mattson MP (ed) (2005) Membrane Microdomain Signaling. Lipid Rafts in Biology and Medicine. Humana Press, Totowa, NJ

Matson RS (2005) Applying Genomic and Proteomic Microarray Technology in Drug Discovery. CRC Press, Boca Raton, FL

Matzuk MM et al (eds) (2001) Transgenics in Endocrinology. Humana Press, Totowa, NJ
Maunsbach AB and Afzelius BA (1998) Biomedical Electronmicroscopy: Illustrated Methods and Interpretations. Academic Press, San Diego, CA
McCarthy DA and Macey MG (eds) (2001) Cytometric Analysis of Cell Phenotype and Function. Cambridge University Press, New York
McCulley M and Greenwell P (2007) Molecular Therapeutics. 21st Century Medicine. Wiley, Hoboken, NJ
McPherson A (2003) Introduction to Macromolecular Crystallography. Wiley, Hoboken, NJ
Meier T and Fahrenholz F (1997) A Laboratory Guide to Biotin-Labeling in Biomolecule Analysis. Springer, New York
Mellick AS and Rodgers L (2006) A Handbook of Recipes, Reagents, and Other Reference Tools for Use at the Bench. Cold Spring Harbor Laboratory Press, Cold Spring Harbor, NY
Melville SE (ed) (2004) Parasite Genomics Protocols. Humana Press, Totowa, NJ
Methods in Cell Biology. Elsevier Science, St. Louis, MO
Methods in Enzymology: Continuous Series Including Wide Areas of Biology. Academic Press, San Diego, CA
Methods in Molecular Biology: Continuous Volumes. Humana Press, Totowa, NJ
Methods in Neurosciences: Continuous Volumes. Academic Press, San Diego, CA
Michels CA (2001) Genetic Techniques for Biological Research. Wiley, New York
Micklos D and Freyer G (2002) DNA Science: A First Course in Recombinant DNA. Cold Spring Harbor Laboratory Press, Woodbury, NY
Miller CH (ed) (2005) Forensic Botany. Principles and Applications to Criminal Caswork. CRC Press, Boca Raton, FL
Miller CH (ed) (2007) Nonhuman DNA Typing. Theory and Casework Applications. CRC Press, Boca Raton, FL
Milligan G (ed) (1999) Signal Transduction: A Practical Approach. Oxford University Press, New York
Mills KI and Ramsahoye BH (2002) DNA Methylation Protocols. Humana Press, Totowa, NJ
Minor L (ed) (2006) Handbook of Assay Development in Drug Discovery. CRC Press, Boca Raton, FL
Minteer SD (ed) (2005) Microfluidic Techniques. Reviews and Protocols. Humana Press, Totowa, NJ
Monaco AP (ed) (1995) Pulsed Field Gel Electrophoresis. Oxford University Press, New York
Montesano R et al (eds) (1982) Handling Chemical Carcinogens in the Laboratory— Problems of Safety. Biol Zbl 101:653–670
Morgan JR (ed) (2002) Gene Therapy Protocols. Humana Press, Totowa, NJ
Mousa SA (ed) (2004) Anticoagulants, Antiplatelets, and Thrombolytics. Humana Press, Totowa, NJ
Murray GI and Curran S (eds) (2005) Laser Capture Microdissection. Methods and Protocols. Humana Press, Totowa, NJ
Murray JC (ed) (2001) Angiogenesis Protocols. Humana Press, Totowa, NJ
Mülhardt C (1999) Der Experimentator: Molekularbiologie. G. Fischer Verlag, Stuttgart, Germany
Nag S (ed) (2003) Blood–Brain Barrier. Humana Press, Totowa, NJ
Nagy A et al (2003) Manipulating the Mouse Embryo. A Laboratory Manual. Cold Spring Harbor Laboratory Press, Woodbury, NY
Nguyen DT et al (2003) Flow Cytometry in Hematopathology. A Visual Approach to Data. Analysis and Interpretation. Humana Press, Totowa, NJ
Nielsen PE and Egholm M (eds) (1999) Peptide Nucleic Acids: Protocols and Applications. Horizon, Wymondham, UK
New RRC (ed) (1990) Liposomes: A Practical Approach. Oxford University Press, New York
Nguyen DT et al (2007) Flow Cytometry in Hematopathology. A Visual Approach to Data Analysis and Interpretation. Humana Press, Totowa, NJ
Nunnally BK (ed) (2005) Analytical Techniques in DNA Sequencing. CRC Press, Boca Raton, FL
Nuovo GJ (1994) PCR In Situ Hybridization. Raven, New York
O'Brien PM and Aitken R (eds) (2002) Antibody Phage Display. Humana Press Totowa, NJ
Ochs M (ed) (2007) Gene Function Analysis. Humana Press, Totowa, NJ
O'Neill LA and Bowie A (eds) (2001) Interleukin Protocols. Humana Press, Totowa, NJ
Ormerod MG (ed) (2000) Flow Cytometry: A Practical Approach. Oxford University Press, Oxford
Papaioannou VE and Behringer RR (2005) Mouse Phenotypes: A Handbook of Mutation Analysis. Cold Spring Harbor Laboratory Press, Woodbury, NY
Patterson C and Cyr DM (2005) Ubiquitin-Proteasome Protocols. Humana Press, Totowa, NJ
Perl A (ed) (2004) Autoimmunity. Methods and Protocols. Humana Press, Totowa, NJ
Phillips MI (ed) (2005) Antisense Therapeutics. Humana Press, Totowa, NJ
Picard D (ed) (1999) Nuclear Receptors. A Practical Approach. Oxford University Press, Oxford, UK
Picot J (ed) (2005) Human Cell Culture Protocols. Humana Press, Totowa, NJ
Poirier J (ed) (1997) Apoptosis Techniques and Protocols. Humana Press, Totowa, NJ
Pollard JW and Walker JM (eds) (1997) Basic Cell Culture Protocols. Humana Press, Totowa, NJ
Prasad PV (ed) (2005) Magnetic Resonance Imaging. Methods and Biolgical Applications. Humana Press, Totowa, NJ
Rampal JB (ed) (2001) DNA Arrays. Methods and Protocols. Humana Press, Totowa, NJ
Rampal JB (ed) (2007) Microarrays. Applications and Data Analysis. Humana Press, Totowa, NJ
Rapley R (ed) (1996) PCR Sequencing Protocols. Humana Press, Totowa, NJ
Ravid K and Freshney RI (eds) (1998) DNA Transfer to Cultured Cells. Wiley-Liss, New York
Razdan MK (2003) Introduction to Plant Tissue Culture. Science Publishers, Enfield, NH
Read SJ and Virley D (eds) (2005) Stroke Geneomics. Methods and Reviews. Humana Press, Totowa, NJ
Ream W and Field KG (1999) Molecular Biology Techniques. An Intensive Laboratory Course. Academic Press, San Diego, CA

Reinert J and Bajaj YPS (1977) Plant Cell, Tissue, and Organ Culture. Springer, New York

Reymond J-L (ed) (2006) Enzyme Assays. High-Throughput Screening, Genetic Selection and Fingerprinting. Wiley-VHC, Hoboken, NJ

Rickwood D and Hames BD (1990) Gel Electrophoresis of Nucleic Acids: A Practical Approach. IRL Press, New York

Ridley DD (1996) Online Searching. A Scientist's Perspective. A Guide for the Chemical and Life Sciences. Wiley, New York

Roberts DB (ed) (1998) Drosophila—A Practical Approach. Oxford University Press, New York

Robinson A et al (eds) (2003) Vaccine Protocols. Humana Press, Totowa, NJ

Roe S (ed) (2001) Protein Purification Techniques. A Practical Approach. Oxford University Press, New York

Rosato E (ed) (2007) Circadian Rhythms: Methods and Protocols. Humana Press, Totowa, NJ

Rosenthal SJ and Wright DW (eds) (2005) Nanobiotechnology Protocols. Humana Press, Totowa, NJ

Ross J (ed) (1997) Nucleic Acid Hybridization. Wiley, New York

Roulston JE and Bartlett JMS (eds) (2004) Molecular Diagnosis of Cancer. Humana Press, Totowa, NJ

Rudin N and Inman K (2001) Introduction to Forensic DNA Analysis. CRC Press, Boca Raton, FL

Russell PJ et al (2003) Prostate Cancer. Methods and Protocols. Humana Press, Totowa, NJ

Saccone C and Pesole G (2003) Handbook of Comparative Genomics: Modern Methodology. Wiley-Liss, Hoboken, NJ

Safety Sense: A Laboratory Guide. (2007) Cold Spring Harbor Laboratory Press, Woodbury, NY

Sambrook J and MacCallum P (2006) Molecular Cloning: A Laboratory Manual. Cold Spring Harbor Laboratory Press, Cold Spring Harbor, NY

Sanchez J-C et al (eds) (2004) Biomedical Applications of Proteomics. Wiley, Hoboken, NJ

Sansone EB and Tewari YB (1978) The permeability of laboratory gloves to selected Nitroamines. In: Walker EA, Griciute L, Castenegro M, Lyle RE, Environmental Aspects of N-nitroso Compounds, International Agency for Research on Cancer, Sci. Publ. 19, Lyon, France, pp 517–543

Schaefer BC (1997) Gene Clonig and Analysis: Current Innovations. Horizon, Wymondham, UK

Schena M (2003) Microarray Analysis. Wiley-Liss, Hoboken, NJ

Schena M, Knudsen S (2004) Guide to Analysis of DNA Microarray Data. Wiley, Hoboken, NJ

Schepers U (2005) RNA Interference in Practice. Principles, Basics, and Methods for Gene Silencing in *C. elegans*, *Drosophila*, and Mammals. Wiley, Hoboken, NJ

Schmidtke J and Krawczak M (1998) DNA Fingerprinting. Springer, New York

Schoenberg DR (ed) (2004) mRNA Processing and Metabolism. Humana Press, Totowa, NJ

Schumann W et al (eds) (2001) Functional Analysis of Bacterial Genes: A Practical Approach. Wiley, New York

Sebatini M et al (eds) (2004) Cartilage and Osteoarthritis. Humana Press, Totowa, NJ

Sechi S (ed) (2007) Quantitative Proteomics by Mass Spectrometry. Humana Press, Totowa, NJ

Seeley HW et al (1991) Microbes in Action: A Laboratory Manual of Microbiology. Freeman, New York

Seger R (ed) (2004) MAP Kinase Signaling Protocols. Humana Press, Totowa, NJ

Seitz H (ed) (2007) Analytics of Protein-DNA Interactions. Springer, Berlin, D

Serdyuk IN et al (2007) Methods in Molecular Biophysics. Structure, Dynamics, Function. Cambridge University Press, New York

Serre J-L, (ed) (2006) Diagnostic Techniques in Genetics. Wiley, Hoboken, NJ

Settle F (ed) (1997) Handbook of Instrumental Techniques for Analytical Chemistry. Prentice Hall, Upper Saddle River, NJ

Shapiro HM (2003) Practical Flow Cytometry. Wiley, Hoboken, NJ

Sharpe NF and Carter RF (2006) Genetic Testing. Care, Consent, and Liability. Wiley, Hoboken, NJ

Shepherd PS and Dean C (eds) (2000) Monoclonal Antibodies. A Practical Approach. Oxford University Press, New York

Shimkets RA (ed) (2004) Gene Expression Profiling. Humana Press, Totowa, NJ

Shorte SL and Frischknecht F (ed) (2007) Imaging Cellular and Molecular Biological Functions. Springer, Berlin, D

Shoseyov O and Levy I (eds) (2007) Nanobiotechnology. Bioinspired Devices and Materials of the Future. Humana Press, Totowa, NJ

Sigurdsson EM (ed) (2005) Amyloid Proteins. Methods and Protocols. Zanders ED (ed) (2005) Chemical Genomics. Reviews and Protocols. Humana Press, Totowa, NJ

Simpson R (2003) Proteins and Proteomics: A Laboratory Manual. Cold Spring Harbor Laboratory Press, Woodbury, NY

Sioud M (ed) (2004) Ribozymes and siRNA Protocols. Humana Press, Totowa, NJ

Slater R (ed) (2002) Radioisotopes in Biology. Oxford University Press, New York

Smales CM (ed) (2005) Therapeutic Proteins. Methods and Protocols. Humana Press, Totowa, NJ

Smith BJ (ed) (2002) Protein Sequencing Protocols. Humana Press, Totowa, NJ

Smith M and Sockett E (eds) (1999) Genetic Methods for Diverse Prokaryotes. Academic Press, San Diego, CA

Spangler BD (2002) Methods in Molecular and Protein Biochemistry: Cloning and Characterization of an Enterotoxin Subunit. Wiley, Hoboken, NJ

Spector DL et al (eds) (1998) Cells: A Laboratory Manual. Cold Spring Harbor Laboratory Press, Woodbury, NY

Spector D and Goldman R (eds) (2003) Essentials from Cells: A Laboratory Manual. Cold Spring Harbor Laboratory Press, Woodbury, NY

Speed TP (ed) (2003) Statistical Analysis of Gene Expression Microarray Data. Chapman & Hall, New York

Sperry AO (ed) (2007) Molecular Motors. Methods and Protocols. Humana Press, Totowa, NJ

Springer CJ (ed) (2005) Suicide Gene Therapy. Methods and Reviews. Humana Press, Totowa, NJ

Strange K (ed) (2006) *C. elegans*: Methods and Applications. Humana Press, Totowa, NJ

Starkey MP and Elswarapu R (eds) (2001) Genomics Protocols. Humana Press, Totowa, NJ

Strege MAS and Lagu AL (eds) (2004) Capillary Electrophoresis of Proteins and Peptides. Humana Press, Totowa, NJ

Studzinski GP (1999) Apoptosis: A Practical Approach. Oxford University Press, New York

Su GH (ed) (2005) Pancreatic Cancer. Methods and Protocols. Humana Press, Totowa, NJ

Sullivan W et al (eds) (2000) *Drosophila* Protocols. Cold Spring Harbor Laboratory Press, Cold Spring Harbor, NY

Sundstrom M et al (eds) (2006) Structural Genomics and High Throughput Structural Biology. CRC Press, Boca Raton, FL

Thompson T and Black S (2006) Forensic Human Identification: An Introduction. CRC Press, Boca Raton, FL

Thongboonkerd V (ed) (2007) Proteomics of Human Body Fluids. Principles, Methods and Applications. Humana Press, Totowa, NJ

Thornhill A (ed) (2007) Single Cell Diagnostics. Methods and Protocols. Humana Press, Totowa, NJ

Tollefsbol TO (ed) (2004) Epigenetics Protocols. Humana Press, Totowa, NJ

Trower MK (1996) In Vitro Mutagenesis Protocols. Humana Press, Totowa, NJ

Tymms MJ (ed) (1995) In Vitro Transcription and Translation Protocols. Humana Press, Totowa, NJ

Tymms MJ (ed) (2000) Transcription Factor Protocols. Humana Press, Totowa, NJ

Tymms MJ and Kola I (eds) (2001) Gene Knockout Protocols. Humana Press, Totowa, NJ

Vasil IK (ed) (1984) Cell Culture and Somatic Cell Genetics of Plants. Academic Press, San Diego, CA

van den Berg A and Andersson H (2007) Lab-on-Chips for Cellomics. Micro and Nano-Technologies for Life Science. Humana Press, Totowa, NJ

Vemuri MC (ed) (2007) Stem Cell Assays. Humana Press, Totowa, NJ

Vo-Dinh T (ed) (2005) Protein Nanotechnology. Protocols, Instrumentation and Applications. Humana Press, Totowa, NJ

Walker JM (ed) (2002) The Protein Protocols Handbook. Humana Press, Totowa, NJ

Walker JM (ed) (2005) The Proteomics Protocols Handbook. Humana Press, Totowa, NJ

Walker JM (2005) PCR Protocols on CD. Humana Press, Totowa, NJ

Walker JM and Rapley R (eds) (2005) Medical Biomethods Handbook. Humana Press, Totowa, NJ

Walz W et al (2002) Patch-Clamp Analysis. Advanced Techniques. Humana Press, Totowa, NJ

Ward A (2001) Genomic Imprinting. Methods and Protocols. Humana Press, Totowa, NJ

Wasserman PA and DePamphilis ML (eds) (1993) Guide to Techniques in Mouse Development. Methods in Enzymology 225. Academic Press, San Diego, CA

Weigel D and Glazebrook J (2002) *Arabidopsis*: A Laboratory Manual. Cold Spring Harbor Laboratory Press, Woodbury, NY

Weiner MP et al (eds) (2007) Genetic Variation. A Laboratory Manual. Cold Spring Harbor Laboratory Press, Cold Spring Harbor. NY

Weissenssteiner T et al (eds) (2004) PCR Technology. Current Innovations. CRC Press, Boca Raton, FL

Weising K et al (1995) DNA Fingerprinting in Plants and Fungi. CRC Press, Boca Raton, FL

Westermeier R and Naven T (2002) Proteomics in Practice: A Laboratory Manual of Proteome Analysis. Wiley-VCH, New York

White BA (ed) (1996) PCR Cloning Protocols. Humana Press, Totowa, NJ

Wilkinson DG (ed) (1998) In Situ Hybridization: A Practical Guide. Oxford University Press, New York

Willars GB and Challiss RAJ (eds) (2004) Receptor Signaling Transduction Protocols. Humana Press, Totowa, NJ

Wilson ZA (ed) (2000) Arabidopsis. A Practical Approach. Oxford University Press, New York

Wolf K et al (eds) (2003) Non-Conventional Yeasts in Genetics, Biochemistry and Biotechnology. Practical protocols. Springer, New York

Wong DWS (2006) The ABCs of Gene Cloning. Springer, Berlin, D

Wu W et al (2004) Gene Biotechnology. CRC Press, Boca Raton, FL

Wu W et al (2005) Gene Biotechnology. CRC Press, Boca Raton, FL

Xiao W (ed) (2005) Yeast Protocols. Humana Press, Totowa, NJ

Yang N-S and Christou P (1994) Particle Bombardment Technology for Gene Transfer. Oxford University Press, New York

Yuryev A (ed) (2007) PCR Primer Design. Humana Press, Totowa, NJ

Zachariou M (ed) (2007) Affinity Chromatography. Methods and Protocols. Humana Press, Totowa, NJ

Zanders ED (ed) (2005) Chemical Genomics. Reviews and Protocols. Humana Press, Totowa, NJ

Mapping ►*recombination*

Boultwood J (ed) (1996) Gene Isolation and Mapping Protocols. Humana Press, Totowa, NJ

Cantor CR and Smith CL (1999) Genomics: The Science and Technology Behind the Human Genome Project. Wiley, New York

Collins AR (ed) (2007) Linkage Diseqilibrium and Association Mapping. Analysis and Applications. Humana Press, Totowa, NJ

Dear PH (1997) Genome Mapping. A Practical Approach. Oxford University Press, New York

Haines JL and Pericak-Vance MA (eds) (1998) Approaches to Gene Mapping in Complex Human Diseases. Wiley, New York

Hunter W and Chittaranjan K (eds) (2008) Genome Mapping and Genomics in Arthropods. Springer, Berlin, D

Liu BH (1997) Statistical Genomics. Linkage, Mapping, and QTL Analysis. CRC Press, Boca Raton, FL

Mather K (1957) The Measurement of Linkage in Heredity. Methuen, London

Meksem K and Kahl G (eds) (2005) The Handbook of Plant Genome Mapping. Genetic and Physical Mapping. Wiley, Hoboken, NJ

O'Brien SJ (ed) (1993) Genetic Maps. Locus Maps of Complex Genomes. Cold Spring Harbor Laboratory Press, Cold Spring Harbor, NY

Oraguzie NC et al (eds) (2007) Association Mapping in Plants. Springer, Berelin, D

Ott J (1999) Analysis of Human Genetic Linkage. Johns Hopkins, University Press, Baltimore, MD

Paterson AH (ed) (1996) Genome Mapping in Plants. Academic Press, San Diego, CA

Pawlowitzki I-H et al (eds) (1997) Genetic Mapping of Disease Genes. Academic Press, San Diego, CA

Primrose SB (1995) Principles of Genome Analysis. A Guide to Mapping and Sequencing DNA from Different Organisms. Blackwell Science, Cambridge, MA

Schook LB et al (eds) (1991) Gene-Mapping Techniques and Applications. Marcel Dekker, New York

Siegmund D and Yakir B (2007) The Statistics of Gene Mapping. Springer, Berlin, D

Speed T and Waterman MS (1996) Genetic Mapping and DNA Sequencing. Springer, New York

Microbiology ►*molecular biology*, ►*biology*, ►*viruses*, ►*laboratory technology*

Adolph KW (ed) (1995) Microbial Gene Techniques. Academic Press, San Diego, CA

Advances in Applied Microbiology (Serial). Elsevier, Philadelphia, PA

Advances in Microbial Physiology (Serial). Elsevier, Philadelphia, PA

Aktoris K and Just I (2000) Bacterial Protein Toxins. Springer, Berlin, Germany

Alouf J and Popoff M (eds) (2006) The Comprehensive Sourcebook of Bacterial Protein Toxins. Elsevier, Philadelphia, PA

Amyes SGB (2001) Magic Bullets, Lost Horizons: The Rise and Fall of Antibiotics. Taylor & Francis, New York

Annual Review of Microbiology. Palo Alto, CA

Antosia R and Cahill J (eds) (2006) Handbook of Bioterrorism and Disaster Medicine. Springer, New York

Atlas RM (1993) Handbook of Microbiological Media. CRC Press, Boca Raton, FL

Atlas RM and Philp J (eds) (2005) Bioremediation: Applied Microbial Solutions for Real-World Environmental Cleanup. ASM, Washington, DC

Baltz RH et al (eds) (1993) Industrial Microorganisms. Basic and Applied Molecular Genetics. American Society for Microbiology Press, Washington, DC

Barnett JA et al (2000) Yeasts. Characteristics and Identification. Cambridge University Press, New York

Baron EJ et al (eds) (1994) Medical Microbiology. Wiley, New York

Barredo JL (ed) (2005) Microbial Enzymes and Biotransformations. Humana Press, Totowa, NJ

Barredo JL (ed) (2005) Microbial Processes and Products. Humana Press, Totowa, NJ

Baumberg S (1999) Prokaryotic Gene Expression. Oxford University Press, Oxford

Baumberg S et al (eds) (1995) Population Genetics of Bacteria. Cambridge University Press, New York

Berg HC (2003) *E. coli* in Motion. Springer, New York

Black JG (2004) Microbiology. Principles and Explorations. Wiley, Hoboken, NJ

Boyd RF (1995) Basic Medical Microbiology. Little Brown, New York

Brown GD and Netea MG (eds) (2007) Immunology of Fungal Infections. Springer, D

Bruijn de FJ et al (1998) Bacterial Genomes: Physical Structure and Analysis. Chapman and Hall, New York

Brun YV and Shimkets LJ (eds) (2000) Prokaryotic Development. American Society for Microbiology Press, Washington, DC

Bryant DA (1994) The Molecular Biology of Cyanobacteria. Kluwer, Norwell, MA

Casali N and Preston A (eds) (2003) *E. coli* Plasmid Vectors. Humana Press, Totowa, NJ

Clayton CL and Mobley HLT (1997) *Helicobacterium pylory* Protocols. Humana Press, Totowa, NJ

Clewell DB (ed) (1993) Bacterial Conjugation. Plenum, New York

Collier L (ed) (1998) Topley and Wilson's Microbiology and Microbial Infections. Arnold/Oxford University Press, New York

Cooper GM et al (eds) (1995) The DNA Provirus. American Society for Microbiology Press, Washington, DC

Cossart P et al (eds) (2000) Cellular Microbiology. American Society for Microbiology, Press, Washington, DC

Dale JW and Park SF (2004) Molecular Genetics of Bacteria. Wiley, Hoboken, NJ

Dangl JL (ed) (1994) Bacterial Pathogenesis of Plants and Animals. Molecular and Cellular Mechanisms. Springer, New York

Day INM (2002) Molecular Genetic Epidemiology—A Laboratory Perspective. Springer, Berlin, Germany

Donnenberg MS (2002) *Escherichia coli*: Virulence Mechanisms of a Versatile Pathogen, Academic Press, San Diego, CA

Dorman CJ (1994) Genetics of Bacterial Virulence. Blackwell Scientific, Cambridge, MA

Doyle MP and Beuchat LR (eds) (2007) Food Microbiology: Fundamentals and Frontiers. ASM Press, Washington, DC

Dunn BM (ed) (1999) Proteases of Infectious Agents. Academic Press, San Diego, CA

Dworkin M et al (eds) (2006) The Prokaryotes. A Handbook on the Biology of Bacteria. Springer, New York

Elliott CG (1993) Reproduction of Fungi: Genetical and Physiological Aspects. Chapman and Hall, New York

El-Sharoud W (ed) (2008) Bacterial Physiology. A Molecular Approach. Springer. Berlin, D

England R et al (eds) (1999) Microbial Signalling and Communication. Cambridge University Press, New York
Falkow S et al (eds) (2006) The Prokaryotes. Springer, New York
Fraser CM (ed) (2004) Microbial Genomes. Humana Press, Totowa, NJ
Frosch M and Maiden MCJ (eds) (2006) Handbook of Meningococcal Disease. Wiley, Hoboken, NJ
Fugelsang KC and Edwards CG (2007) Wine Microbiology. Practical Applications and Procedures. Springer, Berlin, D
Funell BE (1996) The Role of Bacterial Membrane in Chromosome Replication and Partitition. Chapman and Hall, New York
Gartland KMA and Davey M (1995) Agrobacterium Protocols. Humana Press, Totowa, NJ
Ghannoum M and O'Toole G (eds) (2004) Microbial Biofilms. ASM Press, Washington, DC
Gilmore MS (2002) The *Enterococci:* Pathogenesis, Molecular Biology, and Antimicrobial Resistance. ASM Press, Washington, DC
Gnanamanickam SS (ed) (2007) Plant-Associated Bacteria. Springer, Berlin, D
Goset F and Guespin-Michel J (1994) Prokaryotic Genetics: Genome Organization, Transfer and Plasticity. Blackwell Science, Cambridge, MA
Gould IM and van der Meer JWM (eds) (2007) Antibiotic Policies: Fighting Resistance. Springer, Berlin, D
Gow NAR and Gadd GM (eds) (1994) The Growing Fungus. Chapman & Hall, New York
Griffin DH (1996) Fungal Physiology. Wiley, New York
Guillemin J (2005) Biological Weapons: From Invention of State-Sponsored Program to Contemporary Bioterrorism. Columbia University Press, New York
Hacker J and Heesemann J (eds) (2002) Molecular Infection Biology. Interactions Between Microorganisms and Cells. Wiley-Liss, Hoboken, NJ
Hacker J and Dobrindt U (2006) Pathogenomics. Genome Analysis of Pathogenic Microbes. Wiley, Hoboken, NJ
Hawkey PM and Lewis DA (eds) (1989) Medical Bacteriology: A Practical Approach. IRL Press, New York
Herbert MA et al (eds) (2002) Hemophilus Influenzae Protocols. Humana Press, Totowa, NJ
Heritage J et al (1999) Microbiology in Action. Cambridge University Press, New York
Higgins NP (ed) (2005) The Bacterial Chromosome. ASM Press, Washington DC
Holt JG (ed) (1993) Bergey's Manual of Determinative Bacteriology. William & Wilkins, Baltimore, MD
Howard RJ and Gow NAR (eds) (2007) Biology of the Fungal Cell. Springer, Berlin, D
Jungblut PR and Hecker M (eds) (2007) Proteomics of Microbial Pathogens. Wiley, Hoboken, NJ
Jurkevitch E (ed) (2007) Predatory Prokaryotes. Biology, Ecology and Evolution. Springer, New York
Keyes DC et al (eds) (2005) Medical Response to Terrorism: Preparedness and Clinical Practice. Lippincott Williams & Wilkins, Philadelphia, PA
Khardori N (ed) (2006) Bioterrorism Preparedness. Wiley, Hoboken, NJ
Koch AL (2007) The Bacteria: Their Origin, Structure, Function and Antibiosis. Springer, Berlin, D
Koehler TM (ed) (2002) Anthrax. Springer, New York
Kwon-Chung KJ and Bennett JE (1992) Medical Mycology. Lea & Febiger. Philadelphia, PA
Larone D (2002) Medically Important Fungi. ASM Press, Washington, DC
Lax AJ (ed) (2005) Bacterial Protein Toxins. Role in the Interference with Cell Growth Regulation. Cambridge University Press, New York
Leadbetter J (2005) Environmantal Microbiology. Elsevier, Philadelphia, PA
Lederberg J (ed) (2000) Encyclopedia of Microbiology. Academic Press, San Diego, CA
Levine J et al (eds) (1994) Bacterial Endotoxins. Wiley-Liss, New York
Lin ECC and Lynch AS (eds) (1996) Regulation of Gene Expression in *Escherichia coli.* Chapman and Hall, New York
Lindler LE et al (eds) (2005) Biological Weapons Defense: Infectious Diseases and Counterbioterrorism. Human Press, Totowa, NJ
Maloy SR and Taylor RK (1996) Genetic Analysis of Pathogenic Bacteria: A Laboratory Manual. Cold Spring Harbor Laboratory Press, Cold Spring Harbor, NY
McCrae MA et al (eds) (1997) Molecular Aspects of Host-Pathogen Interactions. Cambridge University Press, New York
McKane L and Kandel J (1995) Microbiology. Essentials and Applications. McGraw-Hill, New York
Mendez-Villas A (ed) (2006) Modern Multidisciplinary applied microbiology. Wiley, Hoboken, NJ
Miller VL et al (eds) (1994) Molecular Genetics of Bacterial Pathogenesis. American Society for Microbiology Press, Washington, DC
Moat AG et al (2002) Microbial Physiology. Wiley, New York
Mobley HTL et al (eds) (2001) *Helicobacter pylori*: Physiology and Genetics. ASM Press, Washington, DC
Murhammer DW (ed) (2007) Baculovirus and Insect Cell Expression Protocols. Humana Press, Totowa, NJ
Murray PR et al (eds) (1995) Manual of Clinical Microbiology. ASM Press, New York
Nakano MM and Zuber P (eds) (2005) Strict and Facultative Anaerobes. CRC Press, Boca Raton, FL
Neidhardt FC et al (eds) (1966) *Escherichia coli* and *Salmonella.* American Society for Microbiology, Washington, DC
Nester EW et al (2004) Microbiology: A Human Perspective. McGraw-Hill, Boston, MA
Nickoloff JA (ed) (1995) Electroporation Protocols for Microorganisms. Humana Press, Totowa, NJ
Nies DH and Silver S (eds) (2007) Molecular Microbiology of Heavy Metals. Springer, Berlin, D
Oliver R and Schweizer M (eds) (1999) Molecular Fungal Biology. Cambridge University Press, New York
Ollivier B and Magot M (eds) (2005) Petroleum Microbiology. ASM Press, Washington, DC
Owens RC Jr et al (eds) (2005) Antibiotic Opitimization: Concepts and Strategies in Clinical Practice. Marcel Dekker, New York

Parish T and Stoker NG (eds) (2001) Mycobacterium Tuberculosis Protocols. Humana Press, Totowa, NJ

Perry RD and Fetherston JD (eds) (2007) The Genus Yersinia: From Genomics to Function. Springer, Berlin, D

Pilch RF and Zilinskas RA (eds) (2005) Encyclopedia of Bioterrorism Defense. Wiley, Hoboken, NJ

Riley MA and Chavan MA (eds) (2007) Bacteriocins. Ecology and Evolution. Springer, Berlin, D

Rosa C and Péter G (eds) (2006) Biodiversity and Ecophysiology of Yeasts. Springer, New York

Sachse K and Frey J (eds) (2002) PCR Detection of Microbial Pathogens. Humana Press, Totowa NJ

Salyers AA and Whitt DD (2002) Bacterial Pathogenesis: A Molecular Approach. AMS Press, Washington, DC

Schaechter M (ed) (2003) Desk Encyclopedia of Microbiology. Elsevier, Philadelphia, PA

Schatten H and Eisenstar KA (eds) (2007) Salmonella. Methods and Protocols. Humana Press, Totowa, NJ

Schroten H and Wirth S (eds) (2007) Pediatric Infectious Diseases Revisited. Birkhäuser, Basel, Switzerland

Schumann W (2006) Dynamics of the Bacterial Chromosome. Structure and Function. Wiley, Hoboken, NJ

Schüler D (ed) (2007) Magnetoreception and Magnetosomes in Bacteria. Springer, New York

Simjee S (2007) Foodborne Diseases. Humana Press, Totowa, NJ

Singleton P (2004) Bacteria in Biology, Biotechnology and Medicine. Wiley, Chichester

Singleton P and Sainsbury D (2006) Dictionary of Microbiology and Molecular Biology. Wiley, New York

Smith I et al (eds) (1989) Regulation of Prokaryotic Development. Structural and Functional Analysis of Bacterial Sporulation and Germination. American Society for Microbiology Press, Washington, DC

Smith MCM and Sockett RE (eds) (1999) Genetic Methods for Diverse Prokaryotes. Academic Press, London

Snyder L and Champness W (2002) Molecular Genetics of Bacteria. ASM Press, Washington, DC

Somasegaran P and Hoben HJ (1994) Handbook of Rhizobia. Methods in Legume-Rhizobia Technology. Springer, New York

Sonenshein AL et al (eds) (2001) *Bacillus subtilis* and Its Closest Relatives: From Genes to Cells. American Society for Microbiology, Washington, DC

Spencer JFT and Ragout de Spencer AL (eds) (2004) Public Health Microbiology. Methods and Protocols. Humana Press, Totowa, NJ

Spencer JFT and Ragout de Spencer AL (eds) (2004) Environmental Microbiology. Methods and Protocols. Humana Press, Totowa, NJ

Staley JT et al (2007) Microbial Life. Sinauer, Sunderland, MA

Storz G and Hengge-Aronis R (eds) (2000) Bacterial Stress Responses. American Society for Microbiology, Washington, DC

Streips UN and Yasbin RE (eds) (2002) Modern Microbial Genetics. Wiley, New York

Summers DK (1996) The Biology of Plasmids. Blackwell Science, Cambridge, MA

Thomas CM (ed) (2000) The Horizontal Gene Pool. Bacterial Plasmids and Gene Spread. Taylor & Francis, New York

Tortora GJ et al (1998) Microbiology. An Introduction. Addison Wesley Longman, Menlo Park, CA

Truant AL (ed) (2002) Manual of Commercial Methods in Clinical Microbiology. ASM, Washington, DC

Vaillancourt PE (ed) (2002) *E. coli* Gene Expression Protocols. Humana Press, Totowa, NJ

Varma A et al (eds) (2008) Plant Surface Microbiology. Springer, Berlin, D

Wackett LP and Hershberger CD (2001) Biocatalysis and Biodegradation. ASM Press, Washington, DC

Waldor MK et al (eds) (2005) Phages: Their Role in Bacterial Pathogenesis and Biotechnology. ASM Press, Washington, DC

Wang K (ed) (2007) Agrobacterium Protocols. Humana Press, Totowa, NJ

Wheelis M et al (eds) (2006) Deadly Cultures: Biological Weapons Since 1945. Harvard University Press, Cambridge, MA

White D (1999) The Physiology and Biochemistry of Prokaryotes. Oxford University Press, New York

Wilkins CL et al (2005) Identification of Microorganisms by Mass Spectrometry. Wiley, Hoboken, NJ

Wilson M (ed) (2002) Bacterial Adhesion to Host Tissues: Mechanisms and Consequences. Cambridge University Press, Cambridge

Wilson M (2005) Microbial Inhabitants of Humans. Their Ecology and Role in Health and Disease. Cambridge University Press, New York

Wilson M et al (2002) Bacterial Disease Mechanisms: An Introduction to Cellular Microbiology. Cambridge University Press, Cambridge, UK

Woodford N and Johnson A (1998) Molecular Bacteriology. Protocols and Clinical Applications. Humana Press, Totowa, NJ

Woodford N and Johnson A (eds) (2004) Genomics, Proteomics, and Clinical Bacteriology. Humana Press, Totowa, NJ

Yang CS and Hensohn P (2007) Sampling and Analysis of Indoor Microorganisms. Wiley, Hoboken, NJ

Yinduo J (ed) (2007) Methicillin-Resistant *Staphylococcus aureus* (MRSA) Protocols. Humana Press, Totowa, NJ

Zhao S and Stodolsky M (eds) (2004) Bacterial Artificial Chromosomes. Vol I. Library Construction, Physical Mapping, and Sequencing. Vol II. Functional Studies. Humana Press, Totowa, NJ

Molecular Biology ►*laboratory technology*, ►*biotechnology*, ►*biochemistry*

Abelson JN (ed) (1996) Combinatorial Chemistry. Methods in Enzymology 267. Academic Press, San Diego, CA

Agoulnik AI (ed) (2008) Relaxin and Related Peptides. Springer, Berlin, D

Agrawal S (1996) Antisense Therapeutics. Humana Press, Totowa, NJ
Albala JS et al (eds) (2004) Protein Arrays, Biochips, and Proteomics. The Next Phase of Genomic Discovery. CRC Press, Boca Raton, FL
Alphey L (1997) DNA Sequencing. Springer, New York
Archer SL and Rusch NJ (eds) (2001) Potassium Channels in Cardiovascular Biology. Kluwer, New York
Arrigo AP and Müler WEG (eds) (2002) Small Stress Proteins. Springer, New York
Ashcroft FM (2000) Ion Channels and Disease. Academic Press, San Diego, CA
Ashley RH (1966) Ion Channels: A Practical Approach. Oxford University Press, New York
Balkvill F (ed) (2000) The Cytokine Network. Oxford University Press, New York
Bass BL (ed) (2001) RNA Editing. Oxford University Press, New York
Bates AD and Maxwell A (2005) DNA Topology. Oxford University Press, New York
Bauerle PA (ed) (1995) Inducible Gene Expression. Hormonal Signals. Birkhäuser, Cambridge, MA
Beckerle MC (ed) (2002) Cell Adhesion. Oxford University Press, New York
Bernot A (2004) Genome Transcriptome and Protein Analysis. Wiley, Hoboken, NJ
Beugelsdijk TJ (ed) (1997) Automation Technologies in Genome Characterization. Wiley, New York
Bishop JE and Waldholz M (1990) Genome. Simon and Schuster, New York
Blow J (ed) (1996) Eukaryotic DNA Replication. IRL/Oxford University Press, New York
Bonas C and Revillard J-P (2001) Cytokines and Cytokine Receptors. Taylor & Francis, New York
Buxbaum E (2007) Fundamentals of protein structure and function. Springer, D
Buzdin A and Lukyanov S (eds) (2007) Nucleic Acids Hybridization. Modern Applications. Springer, Berlin, D
Calladine CR et al (2004) Understanding DNA: The Molecule and How It Works. Academic Press, San Diego, CA
Callards R and Gearing A (1994) The Cytokine Factsbook. Academic Press, San Diego, CA
Campbell J and Modrich P (2006) DNA Repair. Elsevier, Philadelphia, PA
Carey M and Smale ST (eds) (2000) Transcriptional Regulation in Eukaryotes: Concepts, Strategies, and Techniques. Cold Spring Harbor Laboratory Press, Cold Spring Harbor, NY
Cesarini G et al (eds) (2005) Protein Domains. Wiley, Hoboken, NJ
Clapp JP (ed) (1995) Species Diagnostics Protocols, PCR and Other Nucleic Acid Methods. Humana Press, Totova, NJ
Collado-Vides J and Hofestadt R (2002) Gene Regulation and Metabolism. MIT Press, Cambridge, MA
Conley EC (1995–1996) The Ion Channel Factsbook, vols 1–4. Academic Press, San Diego, CA
Cortese R (ed) (1995) Combinatorial Libraries. Synthesis, Screening and Application Potentials. De Gruyter, New York
Curiel DT and Douglas J (eds) (2002) Vector Targeting for Therapeutic Gene Delivery. Wiley-Liss, Hoboken, NJ
Dale JW and von Schantz M (2002) From Genes to Genomes. Concepts and Applications of DNA Technology. Wiley, Hoboken, NJ
Demeunynck M et al (eds) (2003) DNA and RNA Binders. Wiley, Hoboken, NJ
De Murcia G and Shall S (2000) From DNA Damage and Stress Signaling to Cell Death: Poly ADP-Ribosylation Reactions. Oxford University Press, Oxford, UK
DePamphilis ML (ed) (1999) Concepts in Eukaryotic DNA Replication. Cold Spring Harbor Laboratory Press, Cold Spring Harbor, NY
DePamphilis ML (ed) (2006) DNA Replication and Human Disease. Cold Spring Harbor Laboratory Press, Cold Spring Harbor, NY
Derynck R (ed) (2008) The TGF-β Family. Cold Spring Harbor Laboratory Press, Cold Spring Harbor, NY
Docherty K (ed) (1996) Gene Transcription: DNA Binding Proteins. Wiley, New York
Docherty K (ed) (1996) Gene Transcription: RNA analysis. Wiley, New York
Doerfler W (2000) Foreign DNA in Mammalian Systems. Wiley-VCH, New York
Doolittle RF (ed) (1996) Computer Methods for Macromolecular Sequence Analysis. Methods in Enzymology 266. Academic Press, San Diego, CA
Eckstein F and Lilley DMJ (1997) Catalytic RNA. Springer, New York
Eggleston DS et al (eds) (1998) The Many Faces of RNA. Academic Press, San Diego, CA
Ellis RJ (ed) (1966) The Chaperonins. Academic Press, San Diego, CA
Epstein RJ (2003) Human Molecular Biology. Cambridge University Press, New York
Eun H-M (1996) Enzymology Primer for Recombinant DNA Technology. Academic Press, San Diego, CA
Farrel RE and Leppertt G (1997) rDNA Manual. Springer, New York
Ferry G (2007) Max Perutz and the Secret of Life. Chatto & Windus/Cold Spring Harbor Laboratory Press, Cold Spring Harbor, NY
Fitzgerald KA et al (2001) The Cytokine Factsbook/Cytokine Webfacts. Academic Press, San Diego, CA
Ganten D and Ruckpaul K (eds) (2006) Encyclopedic Reference of Genomics and Proteomics in Molecular Medicine. Springer, Berlin, D
Gellisen G (2005) Production of Recombinant Proteins. Novel Microbial and Eukaryotic Expression Systems. Wiley, Hoboken, NJ
Gelvin SB and Schilperoort RA (1994) Plant Molecular Biology Manual. Kluwer, Norwell, MA
Gesteland RF et al (2005) The RNA World. Cold Spring Harbor Laboratory Press, Cold Spring Harbor, NY
Gomperts B et al (2002) Signal Transduction. Academic Press, San Diego, CA
Goodbourn S (ed) (1996) Eukaryotic Gene Transcription. Oxford University Press, New York
Goodfellow JM (ed) (1995) Computer Modeling in Molecular Biology. VCH, New York
Goodsell DS (2004) Bionanotechnology: Lessons from Nature. Wiley-Liss, Hoboken, NJ
Göringer HU (ed) (2008) RNA Editing. Springer, Berlin, D

Grandi G (ed) (2004) Genomics, Proteomics and Vaccines. Wiley, Chichester, UK
Gribskov M and Devereux J (1991) Sequence Analysis Primer. Freeman, New York
Griffin AM and Griffin HG (eds) (1994) Computer Analysis of Sequence Data. Humana Press, Totowa, NJ
Grubin DG (2000) Cell Polarity. Oxford University Press, Oxford, UK
Gutkind JS (ed) (2000) Signaling Networks and Cell Cycle Control: The Molecular Basis of Cancer and Other Diseases. Humana Press, Totowa, NJ
Gutman Y and Lazarovici P (1997) Toxins and Signal Transduction. Harwood Academic, Langhorne, PA
Hannon G (ed) (2003) RNAi: A Guide to Gene Silencing. Cold Spring Harbor Laboratory Press, Cold Spring Harbor, NY
Harris DA (1997) Molecular and Cellular Biology. Horizon, Wymondham, UK
Harris DA (ed) (1997) Prions: Molecular and Cellular Biology. Horizon, Wymondham, UK
Hawkins JD (1996) Gene Structure and Expression. Cambridge University Press, New York
Hecker M and Müllner S (eds) (2003) Proteomics of Microorganisms. Fundamental Aspects and Application. Springer, New York
Heilmeyer L and Friedrich P (eds) (2001) Protein Modules in Cellular Signaling. IOS Press, Amsterdam, the Netherlands
Hershey JWB et al (eds) (1996) Translational Control. Cold Spring Harbor Laboratory Press, Cold Spring Harbor, NY
Hertzberg EL (ed) (2000) Gap Junctions. JAI Press, Stamford, CT
Hideaki H et al (eds) (1996) Intracellular Signal Transduction. Academic Press, San Diego, CA
Hille B (2001) Ion Channels of Excitable Membranes. Sinauer, Sunderland, MA
Hills D and Moritz C (eds) (1996) Molecular Systematics. Sinauer, Sunderland, MA
Horton RM and Tait RC (ed) (1998) Genetic Engineering with PCR. Horizon, Wymondham
Houdebine LM (ed) (1997) Transgenic Animals. Generation and Use. Harwood, Newark, NJ
Howard J (2001) Mechanics of Motor Proteins and the Cytoskeleton. Sinauer, Sunderland, MA
Hoy MA (1994) Insect Molecular Genetics. An Introduction to Principles and Applications. Academic Press, San Diego, CA
Huang L et al (1999) Non-Viral Vectors in Gene Therapy. Academic Press, San Diego, CA
Hughes MA (1996) Plant Molecular Genetics. Addison Wesley, Boston, MA
Humphrey-Smith I and Hecker M (eds) (2006) Microbial Proteomics. Functional Biology of Whole Organisms. Wiley, Hoboken, NJ
Innis MA et al (eds) (1995) PCR Strategies. Academic Press, San Diego, CA
Isenberg G (ed) (2002) Encyclopedic Reference to Cell Biology. Springer, New York
Jost J-P and Saluz H-P (eds) (1993) DNA Methylation: Molecular Biology and Biological Significance. Birkhäuser, New York
Kannicht C (ed) (2002) Posttranslational Modification Reactions: Tools for Functional Proteomics. Humana Press, Totowa, NJ
Kaplitt MG and Loewy AD (eds) (1995) Viral Vectors. Gene Therapy and Neuroscience Applications. Academic Press, San Diego, CA
Kazazian HH (ed) (2002) Wiley Encyclopedia of Molecular Medicine. Wiley, Hoboken, NJ
Kirby LT (1990) DNA Fingerprinting. An Introduction. Stockton Press, New York
Kleanthous C (ed) (2001) Protein-Protein Recognition. Oxford University Press, Oxford, UK
Kneale GG (ed) (1994) DNA-Protein Interactions. Principles and Protocols. Humana Press, Totowa, NJ
Korn B et al (1998) Positional Cloning by Exon Trapping and cDNA Selection. Wiley, New York
Krainer AR (ed) (1997) Eukaryotic mRNA Processing. Oxford University Press, New York
Lapointe J and Brakier-Gingras L (eds) (2003) Translation Mechanisms. Kluwer/Plenum, New York
Latchman DS (1999) Eukaryotic Transcription Factors. Academic Press, San Diego, CA
Lesk AM (2004) Introduction to Protein Science: Architecture, Function, and Genomics. Oxford University Press, New York
Lesk AM (2005) Database Annotation. Wiley, Hoboken, NJ
Levy SB (ed) (1997) Antibiotic Resistance. Wiley, New York
Lilley DMJ (ed) (1995) DNA–Protein. Structural Interactions. IRL Press, New York
Litvak S (1996) Retroviral Reverse Transcriptases. Chapman and Hall, New York
Locker J (ed) (2001) Transcription Factors. Academic Press, San Diego, CA
Lund P (ed) (2001) Molecular Chaperones in the Cell. Oxford University Press, New York
Mantovani A (ed) (1999) Chemokines. Karger, New York
Marsh M (ed) (2001) Endocytosis. Oxford University Press, New York
Marshak DR et al (eds) (2001) Stem Cell Biology. Cold Spring Harbor Laboratory Press, Cold Spring Harbor, NY
Masters JR et al (eds) (2007) Embryonic Stem Cells. Springer, D
Mayer RJ et al (eds) (2007) Cell Biology of the Ubiquitin-Proteasome System. Wiley, Hoboken, NJ
McDonald CJ (ed) (1997) Enzymes in Molecular Biology. Essential Data. Wiley, New York
Meager T (1998) The Molecular Biology of Cytokines. Wiley, New York
Meijer L et al (eds) (2000) Progress in Cell Cycle Research. Kluwer Academic, New York
Meyers RA (ed) (2005) Encyclopedia of Molecular Biology and Molecular Medicine. Wiley, Hoboken, NJ
Meyers RA (ed) (2007) Proteins. From Analytics to Structural Geneomics. Wiley, Hoboken, NJ
Meyers RA (ed) (2007) Genomics and Genetics. From Molecular Details to Analysis and Techniques. Wiley, Hoboken, NJ
Mishra NC (2002) Nucleases. Molecular Biology and Applications. Wiley-Interscience, Hoboken, NJ
Moley JF and Kim SH (1994) Molecular Genetics of Surgical Oncology. CRC Press, Boca Raton, FL
Morimoto RI et al (eds) (1994) The Biology of Heat Shock Proteins and Molecular Chaperones. Cold Spring Harbor Laboratory Press, Cold Spring Harbor, NY
Mullis KB et al (eds) (1994) The Polymerase Chain Reaction. Birkhäuser, Boston, MA

Nagai K and Mattaj IW (eds) (1995) RNA—Protein Interactions, IRL Press, New York
Newton C and Graham A (1997) PCR. Springer, New York
Nickoloff JA and Hoekstra MF (eds) (2001) DNA Damage and Repair. Humana Press, Totowa, NJ
Nicola NA (ed) (1995) Guidebook to Cytokines and Their Receptors. Oxford University Press, New York
Nierhaus KH and Wilson DN (eds) (2004) Protein Synthesis and Ribosome Structure. Wiley, Hoboken, NJ
North RA (ed) (1994) Ligand- and Voltage-Gated Ion Channels. CRC Press, Boca Raton, FL
Parker P and Pawson T (eds) (1996) Cell Signalling. Cold Spring Harbor Laboratory Press, Cold Spring Harbor, NY
Pena SDJ (ed) (1993) DNA Fingerprinting: State of the Science. Birkhäuser, Basel, Switzerland
Pennington SR and Dunn MJ (eds) (2001) Proteomics: From Protein Sequence to Function. Springer, New York
Perdew GH et al (2006) Regulation of Gene Expression. Molecular Mechanisms. Humana Press, Totowa, NJ
Petsko GA and Ringe D (2004) Protein Structure and Function. Sinauer, Sutherland, MA
Privalsky ML (ed) (2001) Transcriptional Corepressors: Mediators of Eukaryotic Gene Repression. Springer, New York
Ptashne M (1992) Genetic Switch: Phage Lambda and Higher Organisms. Blackwell, Cambridge, MA
Ptashne M (2004) A Genetic Switch. Phage Lambda Revisited. Cold Spring Harbor Laboratory Press, Woodbury, NY
Ptashne M and Gann A (2002) Genes & Signals. Cold Spring Harbor Laboratory Press, Cold Spring Harbor, NY
Rapley R and Harbron S (eds) (2005) Molecular Analysis and Genome Discovery. Wiley, Hoboken, NJ
Ravid K and Licht JD (2001) Transcription Factors: Normal and Malignant Development of Blood Cells. Wiley, New York
Richter D (ed) (2001) Cell Polarity and Subcellular Localization. Springer, New York
Richter JD (ed) (1997) mRNA Formation and Function. Academic Press, New York
Rush M and D'Eustachio P (eds) (2002) The Small GTPase Ran. Kluwer, Norwell, MA
Saluz HP and Wiebauer K (1995) DNA and Nucleoprotein Structure In Vivo. CRC Press, Boca Raton, FL
Schenkel J (1997) RNP Particles. Springer, New York
Schleef M (ed) (2001) Plasmids for Therapy and Vaccination. Wiley-VCH, New York
Schleef M (ed) (2005) DNA-Pharmaceuticals. Formulation and Delivery in Gene Therapy, DNA Vaccination and Immunotherapy. Wiley, Hoboken, NJ
Schuck P (ed) (2007) Protein Interactions. Springer, Berlin, D
Sealfon SC (1994) Receptor Molecular Biology. Academic Press, San Diego, CA
Sechi S (ed) (2006) Quantitative Proteomics. Humana Press, Totowa, NJ
Sedivy JM and Joyner AL (2000) Gene Targeting. Oxford University Press, New York
Setlow JK and Hollaender A (eds) (1979) Genetic Engineering: Principles and Methods. Plenum, New York
Shinitzky M (ed) (1995) Biomembranes. Signal Transduction Across Membranes. VCH, New York
Sinden RR (1994) DNA Structure and Function. Academic Press, San Diego, CA
Skalka AM and Goff SP (eds) (1993) Reverse Transcriptase. Cold Spring Harbor Laboratory Press, Cold Spring Harbor, NY
Sohail M (ed) (2005) Gene Silencing by RNA Interference. Technology and Application. CRC Press, Boca Raton, FL
Söll D et al (eds) (2001) RNA. Elsevier, Oxford, UK
Söll D and Rajbahandary UL (eds) (1994) tRNA. Structure, Biosynthesis and Function. AMS Press, Washington, DC
Sonenberg N et al (eds) (2006) Translational Control in Biology and Medicine. Cold Spring Harbor Laboratory Press, Cold Spring Harbor, NY (See also Hershey JWB et al 1996)
Stillman B and Stewart D (2007) Regulatory RNA. Cold Spring Harbor Laboratory Press, Woodbury, NY
Tait RC (1997) An Introduction to Molecular Biology. Horizon, Wymondham, UK
Thomson AW and Lotzes MT (eds) (2003) The Cytokine Handbook. Academic Press, San Diego, CA
Tomiuk J et al (1996) Transgenic Organisms. Springer, New York
Vaddi K et al (1997) The Chemokine Factsbook. Academic Press, San Diego, CA
Vega MA (ed) (1995) Gene Targeting. CRC Press, Boca Raton, FL
Wagner R (2000) Transcription Regulation in Prokaryotes. Oxford University Press, New York
Watson JD et al (2008) Molecular Biology of the Gene. Cold Spring Harbor Laboratory Press, Cold Spring Harbor, NY
Watson JD et al (2007) Recombinant DNA. Sinauer/Freeman, Sunderland, MA
Watters D and Lavin M (eds) (1999) Signaling Pathways in Apoptosis. OPA, Amsterdam, The Netherlands
Weis K (ed) (2002) Nuclear Transport. Springer, New York
White RJ (2001) Gene Transcription: Mechanisms and Control. Blackwell, Oxford, UK
Wilkins MR et al (eds) (1997) Proteome Research: New Frontiers in Functional Genomics. Springer, New York
Wilks AE and Harpur AG (1996) Intracellular Signal Transduction: The Jak-STAT Pathway. Chapman and Hall, New York
Wingender E (1993) Gene Regulation in Eukaryotes. VHC, New York
Woodgett JR (ed) (1995) Protein Kinases. Oxford University Press, New York

Mutation, Toxicology ►*radiation*

Amdur MO et al (eds) (1991) Casaretts and Doull's Toxicology. The Basic Sciences of Poisons. Pergamon Press, Elmsford, NY
Benson WH and Di Giulio RT (eds) (2007) Genomic Approaches for Cross-Species Extrapolation in Toxicology. CRC Press, Boca Raton, FL

Borlak J (ed) (2005) Handbook of Toxicogenomics. A Strategic View of Current Research and Applications. Wiley, Hoboken, NJ
Braman J (ed) (2001) In Vitro Mutagenesis Protocols. Humana Press, Totowa, NJ
Brusick D (1987) Principles of Genetic Toxicology. Plenum, New York
Choy WN (ed) (2001) Genetic Toxicology and Cancer Risk Assesment. Marcel Dekker, New York
Cooper DN and Krawczak M (1993) Human Gene Mutation. Bios Scientific, Oxford, UK
Cotton RGH et al (1998) Mutation Detection: A Practical Approach. IRL/Oxford University Press, New York
Davis G and Kayser KJ (eds) Chromosomal Mutagenesis. Humana Press, Totowa, NJ
Feil R and Metzger D (eds) (2007) Conditional Mutagenesis: An Approach to Disease Models. Springer, Berlin, D
Friedberg EC and Walker GC, Siede W (1995) DNA Repair and Mutagenesis. American Society for Microbiology Press, Washington, DC
Gold LS and Zeiger E (eds) (1996) Handbook of Carcinogenic Potency and Genotoxicity Databases. CRC Press, Boca Raton, FL
Halliwell B and Gutteridge JMC (1999) Free Radicals in Biology and Medicine. Oxford University Press, New York
Hamadeh HK and Afshari CA (eds) (2004) Toxicogenomics. Principles and Applicatgions. Wiley, Hoboken, NJ
Hodgson E (2004) A Textbook of Modern Toxicology. Wiley-Inhterscience. Hoboken, NJ
Hollaender A (ed) (1971) Chemical Mutagens. Plenum, New York
Klaassen CD (ed) (1996) Casarett and Doull's Toxicology. The Basic Science of Poisons. McGraw-Hill, New York
Keohavong P and Grant SG (eds) (2005) Molecular Toxicology Protocols. Humana Press, Totowa, NJ
Landegren U (ed) (1996) Laboratory Protocols for Mutation Detection. Oxford University Press, New York
Lankenau D-H (ed) (2007) Genome Integrity. Facets and Perspectives. Springer, Berlin
Li AP and Heflich RH (eds) (1991) Genetic Toxicology. CRC Press, Boca Raton, FL
Loveless A (1966) Genetic and Allied Effects of Alkylating Agents, Pennsylvania State University Press, University Park, PA
Martinez-Vidal JL and Garrido A (eds) (2005) Pesticide Protocols. Humana Press, Totowa, NJ
McPherson MJ (1991) Directed Mutagenesis. IRL Press, New York
Mozayani A (ed) (2004) Handbook of Drug Interactions. A Clinical and Forensic Guide. Humana Press, Totowa, NJ
Pfeifer GP (ed) (1966) Technologies for Detection of DNA Damage and Mutation. Plenum, New York
Ross WCJ (1962) Biological Alkylating Agents. Butterworth, London, UK
Roy MJ (ed) (2004) Physician's Guide to Terrorist Attack. Humana Press, Totowa, NJ
Sankaranarayanan K (ed) (2000) Protocols in Mutagenesis. Elsevier Science, New York
Schardein JL (2000) Chemically Induced Birth Defects. Marcel Dekker, New York
Singer B and Basrtsch H (2000) Exocylic DNA Adducts in Mutagenesis and Carcinogenesis. Oxford University Press, New York
Taylor GR (ed) (1997) Laboratory Methods for the Detection of Mutations and Polymorphisms in DNA. CRC Press, Boca Raton, FL
Taylor GR and Day IN (eds) (2005) Guide to Mutation Detection. Wiley, Hoboken, NJ
Theophilus BDM and Rapley R (2002) PCR Mutation Detection Protocols. Humana Press, Totowa, NJ
Trestrail JH III (2007) Criminal Poisoning. An Investigational Guide for Law Enforcement, Toxicologists, Forensic Scientists, and Attorneys. Humana Press, Totowa, NJ
Vogel F, Röhrborn G (eds) (1970) Chemical Mutagenesis in Mammals and Man. Springer, New York

Neurobiology ►*molecular biology*, ►*behavior*

Ancill RJ et al (eds) (1995) Schizophrenia. Exploring the Spectrum of Psychosis. Wiley, New York
Andreasen NC (2001) Brave New Brain: Conquering Mental Illness in the Era of the Genome. Oxford University Press, New York
Beal MF et al (eds) (2006) Neurodegenerative Diseases: Neurobiology, Pathogenesis and Therapeutics. Cambridge University Press, Cambridge, UK
Bellen H (ed) (1999) Neurotransmitter Release: Frontiers in Molecular Biology. Oxford University Press, London, UK
Bermudez-Rattoni F (ed) (2007) Neural Plasticity and Memory. From Genes to Brain Imaging. CRC Press, Boca Raton, FL
Bliss TVP et al (2004) Long-Term Potentiation: Enhancing Neuroscience for 30 Years. Oxford University Press, Oxford, UK
Blum K and Noble EP (eds) (1997) Handbook of Psychiatric Genetics. CRC Press, Boca Raton, FL
Bogousslavsky J and Boller F (2005) Neurological Disorders in Famous Artists. Karger, Basel, Switzerland
Bohlen VO and Dermitzel HO (2002) Neurotransmitters and Neuromodulators. Handbook of Receptors and Biological Effect. Wiley, New York
Bourtchouladze R (2002) Memories Are Made of This. Columbia University Press, New York
Brandt T et al (eds) (2002) Neurological Disorders. Academic Press, San Diego, CA
Broderick PA et al (eds) (2005) Bioimaging in Neurodegeneration. Humana Press, Totowa, NJ
Budnik V and Gramates LS (eds) (1999) Neuromuscular Junctions in *Drosophila*. Academic Press, San Diego, CA
Burrows M (1996) The Neurobiology of the Insect Brain. Oxford University Press, New York
Butler AB and Hodos W (1996) Comparative Vertebrate Neuroanatomy. Evolution and Adaptation. Wiley, New York

Buzsáki G (2006) Rhythms of the Brain. Oxford University Press, Oxford, UK
Byrne JH and Roberts JL (eds) (2004) From Molecules to Networks. An Introduction to Cellular and Molecular Neuroscience. Elsevier, Amsterdam, The Netherlands
Carnevale NT and Hines ML (2006) The Neuron Book. Cambridge University Press, New York
Cervós-Navarro J and Ulrich H (1995) Metabolic and Degenerative Diseases of the Nervous System. Pathology, Biochemistry and Genetics. Academic Press, San Diego, CA
Charney DS et al (1999) Neurobiology and Mental Illness. Oxford University Press, New York
Chin HR and Moldin SO (eds) (2002) Methods in Genomic Neuroscience. CRC Press, Boca Raton, FL
Chiocca EA and Breakfield XO (eds) (1998) Gene Therapy for Neurological Disorders and Brain Tumors. Humana Press, Totowa, NJ
Christensen TA (ed) (2005) Methods in Insect Sensory Neuroscience. CRC Press, Boca Raton, FL
Cowan WM et al (eds) (2001) Synapses. Johns Hopkins University Press, Baltimore, MD
Crick F (1994) The Astonishing Hypothesis: The Scientific Search for the Soul. Simon and Schuster, Old Tappan, NJ
Dominiczak AF et al (eds) (1999) Molecular Genetics of Hypertension. Academic Press, San Diego, CA
Evers-Kiebomas G et al (eds) (2002) Prenatal Testing for Late-Onset Neurogenetic Diseases. Bios Scientific, Oxford
Finger S (2001) Origins of Neuroscience. A History of Explorations into Brain Function. Oxford University Press, New York
Fisch GS (ed) (2003) Genetics and Genomics of Neurobehavioral Disorders. Humana Press, Totowa, NJ
Fitzgerald M (2003) Autism and Creativity: Is There a Link Between Autism in Men and Exceptional Ability? Brunner-Routledge. London, UK
Frackowiak RSJ et al (eds) (2004) Human Brain Function. Elsevier. San Diego, CA
Friedman JM et al (eds) (1999) Neurofibromatosis: Phenotype, Natural History and Pathogenesis. Johns Hopkins Press, Baltimore, MD
Frith U and Hill E (eds) (2003) Autism: Mind and Brain. Oxford University Press, New York
Galaburda AM (ed) (1993) Dyslexia and Development. Neurobiological Aspects of Extra-ordinary Brains. Harvard University Press, Cambridge, MA
Gershon ES and Cloninger CR (eds) (1994) Genetic Approcahes to Mental Disorders. American Psychiatric Press, Washington, DC
Geschwind DH and Gregg JP (2002) Microarrays for the Neurosciences: An Essential Guide. MIT Press, Cambridge, MA
Gozal D and Molfese DL (eds) (2005) Attention Deficit Hyperactivity Disorder. From Genes to Patients. Humana Press, Totowa, NJ
Gundelman HE et al (eds) (2005) The Neurology of AIDS. Oxford University Press, New York
Hammond C (2008) Cellular and Molecular Neurophysiology. Elsevier, New York
Harper PS and Perutz M (eds) (2001) Glutamine Repeats and Neurodegenerative Diseases: Molecular Aspects. Oxford University Press, New York
Harris JC (2006) Intellectual Disability: Understanding Its Development, Causes, Classification, Evaluation, and Treatment. Oxford University Press, New York
Hemmen JL van and Swejnowski TJ (eds) (2006) 23 Problems in Systems Neuroscience. Oxford University Press, New York
Higgins SJ (ed) (1999) Molecular Biology of the Brain. Princeton University Press, Princeton, NY
Honavar V and Uhr L (eds) (1994) Artificial Intelligence and Neural Networks. Steps Toward Principled Integration. Academic Press, San Diego, CA
Hörnlimann B et al (eds) (2001) Prionen und Prionkrankheiten, De Gruyter, Berlin, D
Ingoglia NA and Murray M (eds) (2001) Axonal Regeneration in the Central Nervous System. Marcel Dekker, New York
Janigro D (ed) (2006) The Cell Cycle in the Central Nervous System. Humana Press, Totowa, NJ
Jeste DV and Friedman JH (eds) (2005) Psychiatry for Neurologists. Humana Press, Totowa, NJ
Jones BC and Mormède P (eds) (2007) Neurobehavioral Genetics. Methods and Applications. CRC Press, Boca Raton, FL
Jones EG (2006) The Thalamus. Cambridge University Press, New York
Johnston D and Wu SM-S (1994) Foundations of Cellular Neurophysiology. MIT Press, Cambridge, MA
Kandel ER (2006) In Search of Memory: The Emergence of a New Science of Mind. WW Norton, New York
Kasper S and Hirscfeld RMA (2006) Handbook of Bipolar Disorder: Diagnosis and Therapeutic Approaches. Taylor & Francis, Boca Raton, FL
Keynes RD and Aidley DJ (2001) Nerve and Muscle. Cambridge University Press, New York
Koslow SH (2005) Databasing the Brain. From Data to Knowledge. Wiley, Hoboken, NJ
Latchman D (ed) (1995) Genetic Manipulation of the Nervous System. Academic Press, San Diego, CA
Leboyer M and Bellevier F (eds) (2002) Psychiatric Genetics. Methods and Reviews. Humana Press, Totowa, NJ
Lyon G et al (2006) Neurology of Hereditary Metabolic Diseases of Children. McGraw-Hill, New York
Mai J et al (2008) Atlas of the human brain. Elsevier, San Diego, CA
Mason T (ed) (2006) Forensic Psychiatry. Influences of Evil. Humana Press, Totowa, NJ
Meyer G (2007) Genetic Control of Neuronal Migrations in Human Cortical Development. Springer, Berlin, D
McGaugh JL (ed) (1995) Brain and Memory. Modulation and Mediation of Neuroplasticity. Oxford University Press, New York
Merchant KM (1996) Pharmacological Regulation of Gene Expression in the CNS. Towards an Understanding of Basal Ganglion Functions. CRC Press, Boca Raton, FL
Methods in Neurosciences, continuous volumes. Academic Press, San Diego, CA
Micevych PE and Hammer RP Jr (eds) (1995) Neurobiological Effects of Sex Steroid Hormones. Cambridge University Press, New York

Minegar A and Alexander JS (eds) (2005) Inflammatory Disorders of the Nervous System. Humana Press, Totowa, NJ

Nicholls JG et al (eds) (2000) From Neuron to Brain. Sinauer, Sunderland, MA

Nieuwenhuys R et al (1997) The Central Nervous System of Vertebrates. Springer, Berlin, New York

Nishino S and Sakurai T (eds) (2005) The Orexin/Hypocretin System. Physiology and Pathophysiology. Humana Press, Totowa, NJ

Noback CR et al (2005) The Human Nervous System. Structure and Function. Humana Press, Totowa, NJ

North G and Greenspan RJ (eds) (2007) Invertebrate Neurobiology. Cold Spring Harbor Laboratory Press, Woodbury, NY

Oehmichen M et al (2006) Forensic Neuropathology and Neurology. Springer-Vlg. Berlin, Germany

Onaivi ES (ed) (2005) Marijuana and Cannabinoid Research. Methods and Protocols. Humana Press, Totowa, NJ

O'Rahilly R and Muller F (1999) The Embryonic Human Brain. An Atlas of Developmental Stages. Wiley-Liss, New York

Pasterkamp RJ (ed) (2007) Semaphorins. Springer, Berlin, D

Pfaff D (1999) Drive: Neurobiological and Molecular Mechanisms of Sexual Motivation. MIT Press, Cambridge, MA

Pfaff DW et al (eds) (1999) Genetic Influences on Neural and Behavioral Functions. CRC Press, Boca Raton, FL

Phillips MI and Evans D (eds) (1994) Neuroimmunology. Academic Press, San Diego, CA

Popko B (ed) (1999) Mouse Models in the Study of Genetic Neurological Disorders. Kluwer Academic/Plenum, New York

Potter NT (2003) Neurogenetics. Methods and Protocols. Humana Press, Totowa, NJ

Pulst S-M (ed) (1999) Neurogenetics. Oxford University Press, New York

Purves D et al (eds) (2007) Neuroscience. Sinauer, Sunderland, MA

Rao MS (2001) Stem Cells and CNS Development. Humana Press, Totowa, NJ

Rao MS (ed) (2005) Neural Development and Stem Cells. Humana Press, Totowa, NJ

Revest P and Longstaff A (1997) Molecular Neuroscience. Springer, New York

Richter RW and Richter BZ (eds) (2004) Alzheimer Disease. A Physician's Guide to Practical Management. Humana Press, Totowa, NJ

Rosenzweig MR et al (2004) Biological Psychology: An Introduction to Behavioral and Cognitive Neuroscience. Sinauer, Sunderland, MA

Rosner R (ed) (2003) Principles and Practice of Forensic Psychiatry. Arnold, London, UK

Rumbaugh DM, Washburn DA (2003) Intelligence of Apes and Other Rational Beings. Yale University Press, New Haven, CN

Sanberg PR and Davis CD (2006) Cell Therapy, Stem Cells and Brain Repair. Humana Press, Totowa, NJ

Sanes DH et al (2000) Development of the Nervous System. Academic Press, San Diego, CA

Schachter DL (1996) Searching for Memory: The Brain, the Mind, and the Past. Basic Books, New York

Schmajuk NA (1997) Animal Learning and Cognition. Cambridge University Press, New York

Schmidt WJ and Reith MEA (eds) (2005) Dopamine and Glutamate in Psychiatric Disorders. Humana Press, Totowa, NJ

Schousboe A and Bräuner-Osborne H (eds) (2004) Molecular Neuropharmacology. Strategies and Methods. Humana Press, Totowa, NJ

Siegel G et al (2006) Basic Neurochemistry. Elsevier, Philadelphia, PA

Smith CUM (1996) Elements of Molecular Neurobiology. Wiley, New York

Snowling MJ (2000) Dyslexia. Blackwell, Oxford, UK

Smythies J (2002) The Dynamic Neuron: A Comparative Survey of the Neurochemical Basis of Synaptic Plasticity. MIT Press, Cambridge, MA

Stone WS et al (eds) (2004) Early Clinical Intervention in Schizophrenia. Humana Press, Totowa, NJ

Striedter GF (2004) Principles of Brain Evolution. Sinauer, Sunderland, MA

Tatlisumak T and Fisher M (eds) (2006) Handbook of Experimental Neurology. Methods and Techniques in Animal Research. Cambridge University Press, New York

Thiel G (ed) (2006) Transcription Factors in the Nervous System. Development, Brain Function and Diseases. Wiley, Hoboken, NJ

Tolnay M and Probst A (eds) (2001) Neuropathology and Genetics of Dementia. Kluwer, Dordrecht, the Netherlands

Tuszynski MH and Kordower J (eds) (1998) CNS Regeneration. Academic Press, San Diego, CA

Ursano RJ, Norwood AE, Fullerton CS (eds) (2004) Bioterrorism: Psychological and Health Interventions. Cambridge University Press, New York

Vogel F (2000) Genetics and the Electroencephalogram. Springer, New York

Waxman SG (ed) (2005) Multiple Sclerosis As a Neuronal Disease. Elsevier, San Diego, CA

Wells RD and Ashizawa T (eds) (2006) Genetic Instabilities and Hereditary Neurological Diseases. Elsevier, Philadelphia, PA

Wess J (1995) Molecular Mechanism of Muscarinic Acetylcholine Receptor Function. Springer, New York

Wheal H and Thomson A (eds) (1995) Excitatory Amino Acids and Synaptic Transmission. Academic Press, San Diego, CA

Wiley RG (2005) Molecular Neurosurgery and Targeted Toxins. Humana Press, Totowa, NJ

Yuste R et al (eds) (1999) Imaging Neurons: A Laboratory Manual. Cold Spring Harbor Laboratory Press, Cold Spring Harbor, NY

Yuste R and Konnerth A (eds) (2004) Imaging in Neuroscience and Development: A Laboaratory Manual. Cold Spring Harbor Laboratory Press, Woodbury, NY

Zigmond MJ et al (eds) (1999) Fundamental Neuroscience. Academic Press, San Diego, CA

Zigova T et al (eds) (2002) Neural Stem Cells for Brain and Spinal Cord Repair. Humana Press, Totowa, NJ

Organelles ▶*genetics genomics*

Argyroudi-Akoyunoglu JH and Senger H (eds) (1999) The Chloroplast: From Molecular Biology to Biotechnology. Kluwer Academic, Dordrecht, the Netherlands

Attardi GM and Chomyn A (eds) (1995) Mitochondrial Biogenesis and Genetics. Methods in Enzymology, vol 260. Academic Press, San Diego, CA

Beale G and Knowles J (1978) Extranuclear Genetics. Arnold, London

Biswal UC et al (2003) Chloroplast Biogenesis. Kluwer, Dordrecht, N

Bock R (ed) (2007) Cell and Molecular Biology of Plastids. Springer. Berlin, D

Brown GC et al (1999) Mitochondria and Cell Death. Princeton University Press, Princeton, NJ

Copeland WC (2002) Mitochondrial DNA. Methods and Protocols. Humana Press, Totowa, NJ

Darley-Usmar V and Schapira AHV (eds) (1994) Mitochondria, DNA, Proteins and Disease. Portland Press, Chapel Hill, NC

Di Mauro S and Wallace DC (eds) (1992) Mitochondrial DNA in Human Pathology. Raven Press, New York

DiMauro S et al (eds) (2006) Mitochondrial Medicine. Informa Healthcare. Abingdon, UK

Grun P (1976) Cytoplasmic Genetics and Evolution. Columbia University Press, New York

Gillham NW (1994) Organelle Genes and Genomes. Oxford University Press, New York

Herrmann RG (ed) (1992) Cell Organelles. Springer, New York

Kirk JTO and Tilney Bassett RAE (1978) The Plastids. Elsevier, Amsterdam, the Netherlands

Leister D and Herrmann J (eds) (2007) Mitochondria. Practical Protocols. Humana Press, Totowa, NJ

Lestienne P (1999) Mitochondrial Diseases. Springer, New York

Levings CS III and Vasil IK (eds) (1995) Molecular Biology of Plant Mitochondria. Kluwer Academic, Dordrecht, the Netherlands

Margulis L (1970) Origin of Eukaryotic Cells. Yale University Press, New Haven, CN

Martin WF and Müller M (eds) (2007) Origin of Mitochondria and Hydrogenosomes. Springer, Berlin, D

Möller JM (eds) (1998) Plant Mitochondria from Gene to Function. Backhuys, Leiden, the Netherlands

Ostrer H (1998) Non-Mendelian Genetics in Humans. Oxford University Press, New York

Rochaix J-D and Merchant S (eds) (1998) Molecular Biology of Chloroplasts and Mitochondria in *Chlamydomonas*. Kluwer Academic, Dordrecht, the Netherlands

Sager R (1972) Cytoplasmic Genes and Organelles. Academic Press, New York

Schaffer SW and Suleiman MS (eds) (2007) Mitochondria. Springer, Berlin, D

Scheffler IE (1999) Mitochondria. Wiley-Liss, New York

Singh KK (ed) (1998) Mitochondrial DNA, Mutations in Aging, Disease and Cancer. Springer, New York

Wise RR and Hoober JK (eds) (2007) Structure and Function of Plastids. Springer, Berlin, D

Philosophy ▶*social and ethical issues,* ▶*biotechnology,* ▶*evolution*

Adami C (1998) Introduction to Artificial Life. Telos (Springer), New York

Avise JC (1998) The Genetic Gods. Evolution and Belief in Human Affairs. Harvard University Press, Cambridge, MA

Baker C (2006) The Evolution Dialogues: Science, Christianity, and the Quest for Understanding. AAAS Press, Washington, DC

Ball P (2006) The Devil's Doctor: Paracelsus and the World of Renaissance Magic and Science. Heinemann/Farrar, Straus, Giroux, London/New York

Bayertz K (ed) (1995) GenEthics: Technological Intervention in Human Reproduction As a Philosophical Problem. Cambridge University Press, New York

Bishop JM (2003) How to Win the Nobel Prize: An Unexpected Life in Science. Harvard University Press, Cambridge, MA

Burdyuzha V (ed) (2007) The Future of Life and the Future of Our Civilization. Springer, Berlin, D

Campbell JA (ed) (2004) Darwinism, and Public Education. Michigan State University Press, East Lansing, MI

Collins F (2006) The Language of God: A Scientist Presents Evidence for Belief. Free Press, New York

Danchin A (2003) The Delphic Boat: What Genomes Tell Us. Harvard University Press, Cambridge, MA

Dawkins R (2006) The God Delusion. Bantam/Houghton Miffin, Boston, MA

Forrest B, Gross PR (2004) Creationism's Trojan Horse. The Wedge of Intelligent Design. Oxford University Press, New York

Fukuyama F (2002) Our Posthuman Future. Consequences of the Biotechnology Revolution. Farrar, Straus and Giroux, New York

Gibbons A (2006) The First Human: The Race to Discover Our Earliest Ancestors. Doubleday, New York

Gottschall J and Wilson DS (2005) The Literary Animal: Evolution of the Nature of Narrative. Northwestern University Press, Evanston, IL

Gould SJ (1999) Rocks of Ages: Science and Religion in the Fullness of Life. Ballantine Press, New York

Gould SJ (2003) The Hedgehog, the Fox and the Magister's Pox: Mending the Gap Between Science and the Humanities. Harmony Books, Nevada City, CA

Gratzer W (2000) The Undergrowth of Science: Delusion, Self-Deception and Human Frailty. Oxford University Press, New York

Holton G (1993) Science and Antiscience. Harvard University Press, Cambridge, MA

Horgan J (1996) The End of Science. Facing the Limits of Knowledge in the Twighlight of the Scientific Age. Helix (Addison-Wesley), New York

Horrobin D (2001) The Madness of Adam and Eve: How Schizophrenia Shaped Humanity. Bantam, New York

Jolly A (1999) Lucy's Legacy: Sex and Intelligence in Human Evolution. Harvard University Press, Cambridge, MA
Kilner JF et al (eds) (2002) Cutting-Edge Bioethics. A Christian Exploration of Technologies and Trends. W.B. Eerdmans Publishing Co., Grand Rapids, MI
Kitcher P (1996) The Lives to Come: The Genetic Revolution and Human Possibilities. Penguin, New York
Kitcher P (2003) In Mendel's Mirror. Philosophical Reflections on Biology. Oxford University Press, New York
Knoppers BM (ed) (2003) Populations and Genetics. Legal and Socio-Ethical Perspectives. Martinus Nijhof/Brill Academic, Leiden, the Netherlands
Lane N (2005) Power, Sex, Suicide: Mitochondria and the Meaning of Life. Oxford University Press, Oxford, UK
Lockshin RA (2008) The Joy of Science. Springer, Berlin, D
Mayr E (2004) What Makes Biology Unique? Considerations on the Autonomy of a Scientific Discipline. Cambridge University Press, New York
McGrath A (2004) Dawkin's God. Genes, Memes, and the Meaning of Life. Blackwell. Oxford, UK
Morange M (2001) The Misunderstood Gene. Harvard University Press, Cambridge, MA
Morrism SC (2003) Life's Solution: Inevitable Humans in a Lonely Universe. Cambridge University Press, New York
Moss L (2002) What Genes Can't Do. MIT Press, Cambridge, MA
Murphy MP and O'Neill LAJ (eds) (1997) What is Life: The Next 50 Years. Speculations on the Future of Biology. Cambridge University Press, New York
Park RL (2000) Voodoo Science: The Road from Foolishness to Fraud. Oxford University Press, New York
Pennock RT (ed) (2001) Intelligent Design Creationism and Its Critics. Philosophical, Theological and Scientific Perspectives. MIT Press, Cambridge, MA
Popper KR (1959) The Logic of Scientific Discovery. Basic Books, New York
Popper KR (1983) Realism and the Aim of Science. Rowman and Littlefield, Totowas, NJ
Rees M (2003) Our Final Hour: A scientist's Warning: How Terror, Error, and Environmental Disaster Threatens Humankind's Future in This Century—On Earth and Beyond/Our Final Century: Will the Human Race Survive the Twenty-First Century? Basic Books, New York
Ruse M (2001) Can a Darwinian be a Christian? The Relationship between Science and Religion. Cambridge University Press, New York
Ruse M (2005) The Evolution–Creation Struggle. Harvard University Press, Cambridge, MA
Shanks N (2004) God, the Devil and Darwin. Oxford University Press, New York
Shreeve J (2004) The Genome War: How Craig Venter Tried to Capture the Code of Life and Save the World. Knopf, New York
Stent G (1969) The Coming of the Golden Age: A View of the End of Progress. Natural History Press, Garden City, NY
Sterelny K and Griffith PE (1999) Sex and Death. An Introduction to Philosophy of Biology. University Chicago Press, Chicago, IL
Stock G (2002) Redesigning Humans. Houghton Miflin, Boston, MA
Taper ML and Lele SR (2004) The Nature of Scientific Evidence. University Chicago Press, Chicago, IL
Thomson K (2005) Before Darwin. Reconciling God and Nature. Yale University Press, New Haven, CN
Turney J (1998) Frankenstein's Footsteps: Science, Genetics and Popular Culture. Yale University Press, New Haven, CN
Wade N (2001) Life Script. How the Human Genome Discoveries Will Transform Medicine and Enhance Your Health. Simon & Schuster, New York
Wade N (2006) Before the Dawn: Recovering the Lost History of Our Ancestors. Penguin, New York
Walters L and Palmer JG (1997) The Ethics of Human Gene Therapy. Oxford University Press, New York
Watson JD (2007) Avoid Boring People. Lessons From a Life in Science/and Other Lessons from a Life in Science. Knopf, New York
Wolpert L (2006) Six Impossible Things Before Breakfast: The Evolutionary Origins of Belief. Faber and Faber, UK/ W.W. Norton, New York

Plant and Animal Breeding ►*biometry*, ►*biotechnology*, ►*cytology and cytogenetics*, ►*population genetics*, ►*plant pathology*

Acquaah G (2006) Principles of Plant Genetics and Breeding. Futura Malden, MA
Anderson NO (ed) (2007) Flower Breeding and Genetics. Issues, Challenges and Opportunities for the 21st Century. Springer, Berlin, D
Axford RFE et al (eds) (2000) Breeding for Disease Resistance in Farm Animals. Oxford University Press, New York
Bahl PN (1996) Genetics, Cytogenetics and Breeding of Crop Plants. Science Pubs., Bio-Oxford & IBH, Lebanon, NH
Bos I and Caligari P (2008) Selection Methods in Plant Breeding. Springer, Berlin, D
Buck HT et al (eds) (2007) Wheat Production in Stressed Environment. Springer, Berlin, D
Caligari P and Brown J (2006) An Introduction to Plant Breeding. Blackwell Publishing, Futura Malden, MA
Campbell C (2005) The Botanist and the Vintner. How Wine Was Saved for the World. Algonquin Books, New York
Chowdhary B (ed) (2003) Animal Genomics. Karger, Farmington, CT
Clark AJ (ed) (1998) Animal Breeding. Technology for the 21st Century. Gordon & Breach, Newark, NJ
Clement SL and Quisenberry SS (eds) (1998) Global Plant Genetic Resources for Insect-Resistant Crops. CRC Press, Boca Raton, FL
Clutton-Brock J (1999) A Natural History of Domesticated Mammals. Cambridge University Press, Cambridge, UK
Cooper HD et al (eds) (2001) Broadening the Genetic Base of Crop Production. Wiley, New York
Coors JG and Pandey S (eds) (1999) The Genetics and Exploitation of Heterosis in Crops. ASA,-CSSA-SSSA, Madison, WI

Crabb R (1992) The Hybrid Corn-Makers. West Chicago Publishing Co., Chicago, IL

Dudits D (ed) (2006) A Búza Nemesítésének Tudománya (The Science of Wheat Breeding). MTA Szegedi Biológiai Központ—Winter Fair KFT, Szeged, Hungary

Ellstrand NC (2003) Dangerous Liasons? When Cultivated Plants Mate With Their Wild Relatives. Johns Hopkins Press, Baltimore, MD

Elsevier's Dictionary of Plant Genetic Resources (1995) Elsevier Science, New York

Engels JMM et al (eds) (2002) Managing Plant Genetic Diversity. CABI, Wallingford, Oxon, UK

Evans LT (1996) Crop Evolution, Adaptation and Yield. Cambridge University Press, New York

Evenson RE et al (eds) (1998) Agricultural Value of Plant Genetic Resources. CAB International, Wallingford, UK

Fedoroff N and Brown NM (2004) Mendel in the Kitchen: A Scientist's View of Genetically Modified Foods. Joseph Henry Press, Washington, DC

Finkeldey R and Hattemer HH (2007) Tropical Forest Genetics. Springer, Berlin, D

Fries R and Ruvinsky A (eds) (1999) The Genetics of Cattle. CAB International, Wallingford, UK

George EF et al (eds) (2007) Plant Propagation by Tissue Culture. Springer, Berlin, D

Hammond J et al (eds) (1999) Plant Biotechnology. New Products and Applications. Springer, New York

Hamnilton RW (2004) The Art of Rice. Spirit and Sustenance in Asia. UCLA Foster Museum of Cultural History, Los Angeles, CA

Hayward MD et al (eds) (1993) Plant Breeding. Principles and Prospects. Chapman and Hall, London, UK

Kang MS and Gauch HG Jr (1996) Genotype-by-Environment Interaction. CRC Press, Boca Raton, FL

Kole C (ed) (2007) Oilseeds. Genome Mapping and Molecular Breeding in Plants. Springer, New York

Lamb BC (2000) The Applied Genetics of Plants, Animals, Humans and Fungi. Imperial College Press, London, UK

Le Roy HL (1966) Elemente der Tierzucht: Genetik, Mathematik, Populationsgenetik. Bayerischer Landwirthschafts-Vlg., München, Germany

Lörz H and Wenzel G (eds) (2008) Molecular Marker Systems in Plant Breeding and Crop Improvement. Springer, Berlin, D

Maluszinszki M et al (2003) Doubled Haploid Production in Crop Plants. A Manual. Kluwer Academic Dordrecht, N

Mason IL (1996) A World Dictionary of Livestock Breeds. CAB International/Oxford University Press, New York

Mrode RA (1996) Linear Models for the Production of Animal Breeding Values. CAB International/Oxford University Press, New York

Muir WM and Aggrey SE (eds) (2003) Poultry Genetics, Breeding and Technology. CABI, Wallinngford, UK

Murray JD et al (eds) (1999) Transgenic Animals in Agriculture. CAB International, Oxon

Nanda JS (ed) (2000) Rice Breeding and Genetics. Science Publishers, Enfeld, NH

Nevo E et al (2002) Evolution of Wild Emmer and Wheat Breeding. Population Genetics, Genetic Resources and Genome Organization of Wheat Progenitor, *Triticum dicoccoides*. Springer, New York

Pechan P and de Vries G (2005) Genes on the Menu. Facts for Knowledge-Based Decisions. Springer, New York

Poehlman JM and Sleper DA (1995) Breeding Field Crops. Iowa State University Press, Ames, IA

Pua EC and Davey MR (eds) (2007) Transgenic Crops. Springer, Berlin, D

Purdom CE (1993) Genetics and Fish Breeding. Chapman and Hall, New York

Renaville R and Burny A (2001) Biotechnology in Animal Husbandry. Kluwer, Dordrecht, the Netherlands

Rothschild MF and Ruvinsky A (eds) (1998) The Genetics of the Pig. CAB International, Oxon, UK

Simmonds NW (1979) Principles of Crop Improvement. Longman. Harlow, Essex, UK

Smith R and Frazer G (2005) Bovine Theriogenology, an Issue of Veterinary Clinics: Food Animal Practice. Elsevier, Philadelphia, PA

Sobral BW (1966) The Impact of Plant Molecular Genetics. Birkhäuser, Boston, MA

Stoskopf NC et al (1993) Plant Breeding. Theory and practice. Westview, Boulder, CO

Sybenga J (1992) Cytogenetics in Plant Breeding. Springer, Berlin, Germany

Thomson JA (2002) Genes for Africa: Genetically Modified Crops in the Developing World. University of Cape Town Press, Cape Town, South Africa

Thomson JA (2007) Seeds for the Future. Impact of Genetically Modified Crops. Cornell University Press, Ithaca, NY

Tsuchya T and Gupta PK (eds) (1991) Chromosome Engineering in Plants. Genetics, Breeding, Evolution. Elsevier Science, New York

Upadhyaya NM (ed) (2007) Rice Functional Genomics. Challenges, Progress and Prospects. Springer, Berlin, D

van Harten AM (1998) Mutation Breeding. Theory and Practical Applications. Cambridge University Press, New York

Van Vleck LD et al (1987) Genetics for Animal Sciences. Freeman, New York

Varshney R and Tuberosa R (eds) (2007) Genomics-Assisted Crop Improvement. Springer, Berlin, D

Weller JI (2001) Quantitative Trait Loci Analysis in Animals. Oxford University Press, New York

Verpoorte R et al (eds) (2007) Applications of Plant Metabolic Engineering. Springer, Berlin, D

White TL and Adams WT (2001) Forest Genetics. Wiley, New York

Plant Pathology ►*plant physiology*, ►*molecular biology*

Agrios GN (1997) Plant Pathology. Academic Press, San Diego, CA

Boller T and Meins F Jr (eds) (1992) Plant Gene Research: Genes Involved in Plant Defense. Springer, New York

Bradbury JF and Saddler GS (2002) A Guide to Plant Pathogenic Bacteria. Oxford University Press, New York

Brown TM (ed) (1996) Molecular Genetics and Evolution of Pesticide Resistance. Oxford University Press, New York

Cooke BM et al (eds) (2006) The Epidemiology of Plant Diseases. Springer, Berlin, D
Cooper JI et al (eds) (2006) Virus Diseases and Crop Biosecurity. Springer, Berlin, D
Crute IR et al (eds) (1997) The Gene-for-Gene Relationship in Plant-Parasite Interactions. CAB International/Oxford University Press, New York
Daniel M and Purkayastha RP (eds) (1994) Handbook of Phytoalexin Metabolism and Action. Dekker, New York
Datta SK and Muthukrishnan S (1999) Pathogenesis-related proteins in plants. CRC Press, Boca Raton, FL
Entwistle PF et al (eds) (1994) *Bacillus thüringiensis*. An Environmental Biopesticide. Theory and Practice. Wiley, New York
George EF et al (eds) (2008) Plant Propagation by Tissue Culture. Springer, Berlin, D
Gnanamanickam SS (ed) (2002) Biological Control of Crop Diseases. Marcel Dekker, New York
Goodman RN and Novacky AJ (1994) The Hypersensitive Reaction in Plants to Pathogens. APS Press, The American Phytopathological Society, St. Paul, MN
Gurr SJ et al (1992) Molecular Plant Pathology. IRL Press, New York
Heine H (ed) (2008) Innate Immunity of Plants, Animals and Humans. Springer, Berlin, D
Holliday P (2001) A Dictionary of Plant Pathology. Cambridge University Press, New York
Horst RK (ed) (2007) Westkott's Plant Disease Handbook. Springer, Berlin, D
Khetan SK (2001) Microbial Pest Control. Marcel Dekker, New York
Kranz J (2002) Comparative Epidemiology of Plant Diseases. Springer, New York
Leigh GJ (2004) The World's Greatest Fix: A History of Nitrogen in Agriculture. Oxford University Press, New York
Loyola-Vargas VM and Vázquez-Flota F (2005) Plant Cell Culture Protocols. Humana Press, Totowa, NJ
Makkar HPS (ed) (2007) Plant Secondary Metabolites. Humana Press, Totowa, NJ
Maloy OC and Murray TD (eds) (2000) Encyclopedia of Plant Pathology. Wiley, New York
Mancuso S and Shabala S (eds) (2007) Rhythms in Plants. Phenomenology, Mechanisms, and Adaptive Significance. Springer, Berlin, D
Matthews R and Burnie JM (1995) Heatschock Proteins in Fungal Infections. Springer, New York
Narayanasamy P (2002) Microbial Plant Pathogens and Crop Disease Management. Science Publishers, Enfield, NH
Nicolau BJ and Wurtele ES (eds) (2007) Concepts in Plant Metabolomics. Springer, Berlin, D
Nobel PS (1999) Physicochemical and Environmental Plant Physiology. Academic Press, San Diego, CA
Pedigo L (2002) Entomology and Pest Management. Prentice Hall, New York
Prell HH and Day PR (2001) Plant-Fungal Pathogen Interaction. Springer, New York
Regnault-Roger C et al (2007) Biopesticides of Plant Origin. Springer, Berlin, D
Ronald PC (ed) (2007) Plant-Pathogen Interactions. Springer, Berlin, D
Scandalios JG (ed) (1997) Oxidative Stress and Molecular Biology of Antioxidant Defenses. Cold Spring Harbor Laboratory Press, Plainview, NY
Scheffer RP (1997) The Nature of Disease in Plants. Cambridge University Press, New York
Scholthoff K-BG et al (eds) (1999) Tobacco Mosaic Virus. One Hundred Years of Contributions to Virology. American Phytopathology Society Press, St. Paul, MN
Sigee DC (1993) Bacterial Plant Pathology. Cell and Molecular Aspects. Cambridge University Press, New York
Singh US and Singh RP (1995) Molecular Methods in Plant Pathology. CRC Press, Boca Raton, FL
Strange RN (2003) Plant Pathology. Wiley, New York
Waigmann E and Heinlein M (eds) (2007) Viral Transport in Plants. Springer, Berlin, D
Whitby SM (2002) Biological Warfare Against Crops. Palgrave Macmillan, Houndmills, Basingstoke, Hamphire, UK

Plant Physiology ►*biochemistry*, ►*molecular biology*, ►*biotechnology*

Andersson B et al (eds) (1996) Molecular Genetics of Photosynthesis. Cambridge University Press, New York
Annual Review of Plant Biology
Bassett CL (ed) (2007) Regulation of Gene Expression in Plants. Springer, Berlin, D
Basra AS (ed) (1994) Stress-Induced Gene Expression in Plants. Harwood, Langhorne, PA
Beck CB (2005) An Introduction to Plant Structure and Development. Cambridge University Press, New York
Böger P et al (eds) (2002) Herbicide Classes in Development. Springer, New York
Buchanan R et al (eds) (2002) Biochemistry and Molecular Biology of Plant. Wiley, New York
Carpentier R (ed) (2005) Photosynthesis Research Protocols. Humana Press, Totowa, NJ
Clement C et al (eds) (1999) Anther and Pollen. Springer, New York
Dashek WV (ed) (1997) Methods in Plant Biochemistry and Molecular Biology. CRC Press, Boca Raton, FL
Dey PM and Harborne JB (eds) (1997) Plant Biochemistry. Academic Press, San Diego, CA
Epstein E and Bloom AJ (2004) Mineral Nutrition of Plants: Principles and Perspectives. Sinauer, Sunderland, MA
Fosket DE (1994) Plant Growth and Development. A Molecular Approach. Academic Press, San Diego, CA
Galun E (2007) Plant Patterning. Structural and Molecular Genetic Aspects. World Scientific, Hackensack, NJ
Gilmartin PM and Bowler C (eds) (2002) Molecular Plant Biology. Oxford University Press, New York
Golbeck JH (ed) (2007) Photosystem I. The Light-Driven Plastocyanin: Ferredoxin Oxido-Reductase. Springer, Berlin, D
Gould K and Lee D (2003) Anthocyanins in Leaves. Elsevier, Philadelphia, PA

Grimm B et al (eds) (2006) Chlorophylls and Bacterio-Chlorophylls. Biochemistry, Biophysics, Functions and Applications. Springer, Berlin, D
Gurevitch J et al (2006) The Ecology of Plants. Sinauer, Sunderland, MA
Hall DO and Rao KK (1994) Photosynthesis. Cambridge University Press, New York
Harberd N (2006) Seed to Seed: The Secret Life of Plants. Bloomsbury, London
Hayat S and Ahmad A (eds) (2006) Salicylic Acid - A Plant Hormone. Springer, Berlin, D
Hooykaas PJJ et al (eds) (1999) Biochemistry and Molecular Biology of Plant Hormones. Elsevier Science, Amsterdam, the Netherlands
Howell SH (1998) Molecular Genetics of Plant Development. Cambridge University Press, New York
Khripach VA et al (1999) Brassinosteroids. A New Class of Plant Hormones. Academic Press, San Diego, CA
Kruger NJ et al (eds) (1999) Regulation of Primary Metabolic Pathways in Plants. Kluwer, Dordrecht, the Netherlands
Lea PJ and Leegood RC (eds) (1999) Plant Biochemistry and Molecular Biology. Wiley, New York
Li PH and Palva ET (eds) (2002) Plant Cold Hardiness. Gene Regulation and Genetic Engineering. Kluwer/Plenum, New York
Malhó R (ed) (2006) The Pollen Tube. A Cellular and Molecular Perspective. Springer. New York
Mancuso S and Shabala S (eds) (2007) Rhythms in Plants. Springer, Berlin, D
Mohr H and Schopfer P (1995) Plant Physiology. Springer, New York
Mujib A and Šamaj J (eds) (2006) Somatic Embryogenesis. Springer, New York
Nelson LS et al (2007) Handbook of Poisonous and Injurious Plants. Springer, Berlin, D
Nikolau BJ and Wurtele ES (eds) (2007) Concepts in Plant Metabolomics. Spinger, Berlin, D
Olsen O-A (ed) (2007) Endosperm. Springer, Berlin, D
Öpik H and Rolfe SA (2005) The Physiology of Flowering Plants. Cambridge University Press, New York
Pirson A (ed) (1993) Encyclopedia of Plant Physiology. Springer, New York
Progress in Botany. Springer, New York
Raghavendra AS (ed) (2000) Photosynthesis: Comprehensive treatise. Cambridge University Press, New York
Ramina A et al (eds) (2007) Advances in Plant Ethylene Research. Springer, Berlin, D
Reynolds PHS (ed) (1999) Inducible Gene Expression in Plants. CABI Publishing, New York
Romeo JT and Dixon RA (eds) (2002) Phytochemistry in the Genomics and Post-Genomics Eras. Pergamon, New York
Ross IA (2005) Medicinal Plant of the World. Chemical Constituents, Traditional and Medical Uses. Humana Press, Totowa, NJ
Samaj J and Thelen JJ (eds) (2007) Plant Proteomics. Springer, Berlin, D
Scheel D, Wasternack C (2002) Plant Signal Transduction. Oxford University Press, New York
Srivastava LM (2002) Plant Growth and Development. Academic Press, San Diego, CA
Verbelen J-P and Vissenberg K (eds) (2007) The Expanding Cell. Springer, Berlin, D
Vermerris W and Nicholson R (eds) (2006) Phenolic Compound Biochemistry. Springer, Berlin, D
Verpoorte R et al (eds) (2007) Applications of Plant Metabolic Engineering. Springer, Berlin, D
Wada M et al (eds) (2005) Light Sensing in Plants. Springer, New York,
Willey N (ed) (2007) Phytoremediation. Methods and Reviews. Springer, Berlin, D
Wise RR and Hoober JK (eds) (2006) The Structure and Function of Plastids. Springer, New York
Taiz L and Zeiger E (eds) (2006) Plant Physiology. Sinauer, Sunderland, MA

Population Genetics ►*biometry*, ►*evolution*, ►*cytology and cytogenetics*

Baumberg S et al (eds) (1995) Population Genetics of Bacteria. Cambridge University Press, New York
Bonné-Tamir B and Adam A (eds) (1992) Genetic Diversity Among Jews: Diseases and Markers at the DNA Level. Oxford University Press, New York
Cavalli-Sforza LL and Bodmer WF (1971) The Genetics of Human Populations. Freeman, San Francisco, CA
Cavalli-Sforza LL et al (1994) The History and Geography of Human Genes. Princeton University Press, Princeton, NJ
Chadwick DJ and Cardew G (eds) (1996) Variation in the Human Genome. Wiley, New York
Christiansen FB (2000) Population Genetics of Multiple Loci. Wiley, New York
Crow JF and Kimura M (1970) An Introduction to Population Genetics Theory. Harper and Row, New York
Fisher RA (1958) The Genetical Theory of Natural Selection. Dover Publications, New York
Fisher RA (1965) The Theory of Inbreeding. Academic Press, New York
Goodman RM and Motulsky AG (eds) (1979) Genetic Diseases among Ashkenazi Jews. Raven, New York
Haldane JBS (1935) The Causes of Evolution. Longmans Green, New York
Hartl DL and Clark AG (2007) Principles of Population Genetics. Sinauer, Sunderland, MA
Henry RJ (ed) (2001) Plant Genotyping. The DNA Finger-printing of Plants. Oxford University Press, New York
Hoelzel AR (ed) (1998) Molecular Genetic Analysis of Populations. Oxford University Press, New York
Kimura M (1983) The Neutral Theory of Molecular Evolution. Cambridge University Press, Cambridge, UK
Kimura M et al (eds) (1994) Population Genetics, Molecular Evolution and the Neutral Theory. University of Chicago Press, Chicago, IL
Lerner IM (1954) Genetic Homeostasis. Oliver & Boyd, Edinburgh, UK
Levin SA (ed) (1994) Frontiers of Mathematical Biology. Springer, New York

Li CC (1976) First Course in Population Genetics. Boxwood Press, Pacific Grove, CA
Magurran AE and May RM (eds) (1999) Evolution of Biological Diversity. Oxford University Press, Oxford
Mettler LE et al (1988) Population Genetics and Evolution. Prentice Hall, Englewood Cliffs, NJ
Mitton JB (2000) Selection in Natural Population. Oxford University Press, New York
Mourant AE et al (1978) Genetics of the Jews. Clarendon Press, Oxford
Nei M (1975) Molecular Population Genetics and Evolution. American Elsevier, New York
Papiha SS et al (eds) (1999) Genomic Diversity. Applications in Human Population Genetics. Kluwer/Plenum, New York
Real LA (ed) (1994) Ecological Genetics. Princeton University Press, Princeton, NJ
Roychoudhury AK and Nei M (1988) Human Polymorphic Genes: World Distribution. Oxford University Press, New York
Spiess EB (1977) Genes in Populations. Wiley, New York
Stone L and Lurquin P (2005) A Genetic and Cultural Odyssey: The Life and Work of L. Luca Cavalli-Forza. Columbia University Press, New York
Templeton AR (2006) Population Genetics and Microevolutionary Theory. Wiley, Hobken, NJ
Weir B (ed) (1995) Human Identification. The Use of DNA Markers. Kluwer Academic, Boston, MA
Weiss K (ed) (1996) Variation in the Human Genome. Wiley, New York
Wilmsen TN (ed) (1993) The Natural History of Inbreeding and Outbreeding. Theoretical and Empirical Perspectives. University of Chicago Press, Chicago, IL
Wright S (1968–1976) Evolution and the Genetics of Populations. University of Chicago Press, Chicago, IL

Radiation ▸*mutation*

Alpen EL (1998) Radiation Biophysics. Academic Press, San Diego, CA
Benson WH and Di Giulio RT (eds) (2007) Genomic Approaches for Cross-Species Extrapolation in Toxicology. CRC Press, Boca Raton, FL
Dunlap RR et al (eds) (1993) Public Reactions to Nuclear Waste. Duke University Press, Durham, NC
Fuciarelli AE and Zimbrick JD (eds) (1995) Radiation Damage in DNA. Battelle, Columbus, OH
Heald MA and Marion JB (1995) Classical Electromagnetic Radiation. Saunders, Philadelphia, PA
Hendry JH and Lord BL (1995) Radiation Toxicology. Bone Marrow and Leukemia. Taylor & Francis, London, UK
Lindee S (1995) Suffering Made Real: American Science and the Survivors at Hiroshima. University Chicago Press, Chicago, IL
Mycio M (2005) Wormwood Forest: A Natural History of Chernobyl. Joseph Henry Press, Washington, DC
National Research Council, Committee on the Biological Effect of Ionizing Radiation. (1990) Health Effects of Exposure to Low Levels of Ionizing Radiation. National Academy of Sciences & National Research Council, Washington, DC
Peterson LE and Abrahamson S (eds) (1998) Effects of Ionizing Radiation. Atomic Bomb Survivors and Their Children. Joseph Henry Press (National Academy of Sciences USA), Washington, DC
Petryna A (2002) Life Exposed. Biological Citizens of Chernobyl. Princeton University Press, NJ.
Prasad KN (1995) Handbook of Radiobiology. CRC Press, Boca Raton, FL
Shigematsu I et al (eds) (1995) Effects of A-Bomb Radiation on the Human Body. Harwood, Lanhorne, PA (translation)
Shull WJ (1995) Effects of Atomic Radiation. Wiley–Liss, New York
Sources and Effects of Ionizing Radiation. United Nations Scientific Committee on the Effects of Atomic Radiation. UNSCEAR (2000) Report to the General Assembly, with scientific annexes, vol II: Effects. United Nations. New York
Turner JE (1995) Atoms, Radiation, and Eadiation Protection. Wiley, New York
Wilkening GM (1991) Ionizing Radiation, In: Clayton GD, Clayton FE Patty's Industrial Hygiene and Toxicology. Wiley, New York, pp 599–655
Wilson MA (ed) (1998) Textbook of Nuclear Medicine. Lippincott-Raven, Philadelphia, PA

Recombination *(see also mapping)*

Aguilera A and Rothstein R (eds) (2007) Molecular Genetics of Recombination. Springer, Berlin, D
Bailey NTJ (1961) Introduction to the Mathematical Theory of Genetic Linkage. Clarendon Press, Oxford, UK
Bishop M (1998) Guide to Human Genome Mapping. Academic Press, San Diego, CA
Broach JR et al (eds) (1991) Recombination in Yeast. Cold Spring Harbor Laboratory Press, Cold Spring Harbor, NY
Leach DRF (1996) Genetic Recombination. Blackwell Science, Cambridge, MA
Mather K (1957) The Measurement of Linkage in Heredity. Methuen, London
Nelson DL and Brownstein BH (eds) (1993) YAC Libraries. Oxford University Press, New York
Ott J (1999) Analysis of Human Genetic Linkage. Johns Hopkins University Press, Baltimore, MD
Paszkowski J (ed) (1994) Homologous Recombination and Gene Silencing in Plants. Kluwer Academic, Norwell, MA
Smith PJ and Jones CJ (eds) (2000) DNA Recombination and Repair. Oxford University Press, New York
Stahl FW (1979) Genetic Recombination. Thinking About It in Phage and Fungi. Freeman, San Francisco, CA
Terwilliger JD and Ott J (1994) Handbook of Human Genetic Linkage. Johns Hopkins University Press, Baltimore, MD

Waldman AS (ed) (2004) Genetic Recombination. Reviews and Protocols. Humana Press, Totowa, NJ
Whitehouse HLK (1982) Genetic Recombination. Understanding the Mechanisms. Wiley, New York
Wolff S (ed) (1982) Sister Chromatid Exchange. Wiley, New York

Social, Legal and Ethical Issues ►*philosophy*

Alper JS et al (eds) (2002) The Double-Edged Helix: Social Implications of Genetics in a Diverse Society. Johns Hopkins. University Press, Baltimore, MD
Andrews LB (2001) Future Perfect: Confronting Decisions About Genetics. Columbia University Press, New York
Anker S and Nelkin D (2004) The Molecular Gaze: Art in the Genetic Age. Cold Spring Harbor Laboratory Press, Cold Spring Harbor, NY
Annas GJ and Elias S (eds) (1992) Gene Mapping. Using Law and ethics As Guides. Oxford University Press, New York
Bainbridge D (2003) The X in Sex: How the X Chromosome Controls Our Lives. Harvard University Press, Cambridge, MA
Ball E (2007) The Genetic Strand: Exploring a Family History Through DNA. Simon & Schuster. New York
Ballantyne J et al (eds) (1989) DNA Technology and Forensic Science. Cold Spring Harbor Laboratory Press, Cold Spring Harbor, NY
Beckwith J (2002) Making Genes, Making Waves: A Social Activist in Science. Harvard University Press, Cambridge, MA
Benecke M (2005) Murderous Methods: Using Forensic Science to Solve Lethal Crimes. Columbia University Press, New York
Bok D (2003) Universities at the Marketplace. The Commercialization of Higher Education. Princeton University Press, Princeton, NJ
Bonnicksen AL (2002) Crafting a Cloning Policy: From Dolly to Stem Cells. Georgetown University Press, Washington, DC
Braben DW (2004) Pioneering Research: A Risk Worth Taking. Wiley, Hoboken, NJ
Bryant J et al (2005) Introduction to Bioethics. Wiley. Hoboken, NJ
Bryant JA et al (eds) (2002) Bioethics for Scientists. Wiley, New York
Bryant MS (2006) Confronting the "Good Death": Nazi Euthanasia on Trial, 1945–1953. University Press Colorado. Boulder, CA
Carlson EA (2006) Times of Triumph, Times of Doubt. Science and the Battle for Public Trust. Cold Spring Harbor Laboratory Press, Cold Spring Harbor, NY
Carracedo A (ed) (2005) Forensic DNA Typing Protocols. Humana Press, Totowa, NJ
Chadwick RF et al (eds) (1999) The Ethics of Genetic Screening. Kluwer, Dordrecht, the Netherlands
Chapman AR and Frankel MS (eds) (2003) Designing Our Descendants: The Promise and Perils of Genetic Modifications. Johns Hopkins University Press, Baltimore, MD
Childress JF et al (eds) (2005) Belmont Revisited: Ethical Principles for Research with Human Subjects. Georgetown University Press, Washington, DC
Claude RP (2002) Science in the Service of Human Rights. University Pennsylvania Press, Philadelphia, PA
Doherty P (2006) The Beginner's Guide to Winning the Nobel Prize. A Life in Science. Columbia University Press, New York
DeSallle R and Yudell M (2004) Welcome to the Genome: A User Guide to the Genetic Past, Present and Future, Wiley. Hooken, NJ
Dickenson DL (ed) (2002) Ethical Issues in Maternal-Fetal Medicine. Cambridge University Press, New York
Evans JH (2002) Playing God? Human Genetic Engineering and the Rationalization of Public Bioethical Debate. University Chicago Press, Chicago, IL
Frankel M and Chapman AR (2000) Human Inheritable Genetic Modifications: Assessing Scientific, Religious, and Politicy Issues. American Association of Advance Science, Washington, DC
Frankel MS and Teich AH (eds) (1994) The Genetic Frontier: Ethics, Law and Policy. American Association for the Advancement of Science, Washington, DC
Freeman M and Lewis ADE (2000) Law and Medicine. Current Legal Issues 2000. Oxford University Press, New York
Gaskell G and Bauer MW (2001) Biotechnology 1996–2000 the Years of Controversy. National Museum of Science and Industry Ltd., London, UK
Gazzaniga MS (2005) The Ethical Brain. Dana Press, Washington, DC
Gennard D (2007) Forensic Entomology. An Introduction. Wiley, Hoboken, NJ
Gilbert SF et al (2005) Bioethics and the New Embryology: Springboards for Research. Sinauer, Sunderland, MA
Gilbert-Barness E and Debich-Spicer DE (2005) Handbook of Pediatric Autopsy Pathology. Humana Press, Totowa, NJ
Green RM (2001) The Human Embryo Research Debates: Bioethics in the Vortex of Controversy. Oxford University Press, New York
Henig RM (2006) Pandoras Baby. How the First Test Tube Babies Sparked the Reproductive Revolution. Cold Spring Harbor Laboratory Press, Cold Spring Harbor, NY
Hindmarsh R and Lawrence G (eds) (2001) Altered Genes II. The Future? Scribe Publications, Melbourne, Australia
Holland S et al (eds) (2001) The Human Embryonic Stem Cell Debate: Science, Ethics, and Public Policy. Cambridge University Press, New York
Human Cloning and Human Dignity (2002) The Report of the President's Council on Bioethics. Perseus, New York
Jackson KH (2001) Patent Strategy for Researchers and Research Managers. Wiley, New York
Johnstone M-J (2004) Bioethics. A Nursing Perspective. Elsevier, Philadelphia, PA
Judson HF (2004) The Great Betrayal: Fraud in Science. Harcourt, New York
Kass LR (2002) Life, Liberty and the Defense of Dignity. The Challenges of Bioethics. Encounter, San Francisco, CA

Kassirer JP (2004) On the Take. How Medicine's Complicity with Big Business Can Endanger Your Health. Oxford University Press, New York

Kitcher P (1996) The Lives to Come. Simon and Schuster, New York

Klotzko AJ (2004) A Clone of Your Own? The Science and Ethics of Cloning. Oxford University Press, New York

Kuntz D (ed) (2004) Deadly Medicine: Creating the Master Race. University of North Carolina Press, Chapel Hill, NC

Kush G (ed) (2004) Handbook of Bioethics: Taking Stock of the Field from a Philosophical Perspective. Kluwer, Dordrecht, the Netherlands

Ltazer D (ed) (2004) DNA and the Criminal Justice System. The Technology of Justice. MIT Press, Cambridge, MA

Lindee S (2005) Moments of Truth in Genetic Medicine. Johns Hopkins University Press, Baltimore MD

Little P (2002) Genetic Destinies. Oxford University Press, New York

Levy GR (2006) Baby at Risk. The Uncertain Legacies of Medical Miracles for Babies, Families, and Society. Capital Books, Sterling, VA

Lurquin PF (2002) High Tech Harvest. Understanding Genetically Modified Food Plants. Perseus, Boulder, CA

Lynn R (2001) Eugenics: A Reassesment. Praeger, Westport, CN

Macrina FL (2005) Scientific Integrity: Text and Cases in Responsible Conduct of Research. ASM Press, Washington, DC

Maienschein J (2003) Whose View of Life? Embryos, Cloning and Stem Cells. Harvard University Press, Cambridge, MA

Marty A and Greenberg M (2006) Biological Weapons and Bioterrorism, an Issue of Clinics in Laboratory Medicine. Elsevier, Philadelphia, PA

Mazur DJ (2007) Evaluating the Science and Ethics of Research on Humans: A Guide for IRB Members. Johns Hopkins University Press, Baltimore, MD

McSherry C (2001) Who Owns Academic Work? Battling for Control of Intellectual Property. Harvard University Press, Cambridge, MA

Mehlman MJ and Veatch RM (eds) (2003) Wondergenes: Genetic Enhancement and the Future of Society. Indiana University Press, Bloomington, IN

Miller FG et al (eds) (2003) The Nature and Prospect of Bioethics. Humana Press, Totowa, NJ

Milunsky A (2001) Your Genetic Destiny: Know Your Genes, Secure Your Health, Save Your Life. Perseus, New York

Quinn S (2001) Human Trials: Scientists, Investors and Patients in the Quest for a Cure. Perseus, Cambridge, MA

Parry B (2004) Trading the Genome: Investigating the Commodification of Bioinformation. Columbia University Press, New York

Parson AB (2004) The Proteus Effect: Stem Cells and Their Promise for Medicine. Joseph Henry Press, Washington, DC

Payne-James J et al (eds) (2005) Encyclopedia of Forensic and Legal Medicine. Elsevier/Academic Press, Oxford

Radetzki M et al (2003) Genes and Insurance. Ethical, Legal and Economic Issues. Cambridge University Press, New York

Rapley R and Whitehouse D (eds) (2007) Molecular Forensics. Wiley, Hoboken, NJ

Reilly P (1977) Genetics, Law and Social Policy. Harvard University Press, Cambridge, MA

Reilly PR (2000) Abraham Lincoln's DNA and Other Adventures in Genetics. Cold Spring Harbor Laboratory Press, Cold Spring Harbor, NY

Reilly PR (2006) The Strongest Boy in the World. How Genetic Information is Reshaping Our World. Cold Spring Harbor Laboratory Press, Cold Spring Harbor, NY

Reiss MJ and Straughan R (1996) Improving Nature? Cambridge University Press, New York

Resnik DB (1998) The Ethics of Science: An Introduction. Routledge, New York

Ridley M (2003) Nature Via Nurture: Genes, Experience & What Makes Us Humans. Harper Collins, New York

Roll-Hansen N (2005) The Lysenko Effect: The Politics of Science. Humanity Books. Buffalo, NY

Rothstein MA (ed) (2003) Pharmacogenomics. Social, Ethical, and Clinical Dimensions. Wiley, Hoboken, NJ

Rothstein MA (ed) (2004) Genetics and Life Insurance: Medical Underwriting and Social Policy. MIT Press, Cambridge, MA

Schmitt A et al (eds) (2006) Forensic Anthropology and Medicine. Humana Press, Totowa, NJ

Silver LM (2006) Challenging Nature. The Clash of Science and Spirituality at the New Frontiers of Life. Ecco/Harper Collins, New York

Spielman B (ed) (2006) Bioethics in Law. Humana Press, Totowa, NJ

Stansfield WD (2000) Death of a Rat. Understandings and Appreciations of Science. Prometheus, Amherst, NY

Stark MM (ed) (2005) Clinical Forensic Medicine. Humana Press, Totowa, NJ

Stephens T and Brynner R (2001) Dark Remedy: The Impact of Thalidomide and Its Revival as a Vital Medicine. Perseus, New York

Stern AM (2005) Eugenic Nation: Faults and Frontiers of Better Breeding in Modern America. University of California Press, Berkely, CA

Stock G and Campbell J (2000) Engineering the Human Germline: An Exploration of the Science and Ethics of Altering the Genes We Pass on to Our Children. Oxford University Press, New York

Taverne D (2005) The March of Unreason: Science, Democracy, and the New Fundamentalism. Oxford University Press, New York

Thompson P (ed) (1994) Issues in Evolutionary Ethics. State University of New York Press, Alabany, NY

Tsokos M (ed) (2004–2006) Forensic Pathology Reviews. Humana Press, Totowa, NJ

Venter JC (2007) My Genome: My Life. Viking, New York

Wailoo K and Pemberton S (2006) The Troubled Dream of Genetic Medicine. Ethnicity and Innovation in Tay-Sachs, Cystic Fibrosis, and Sickle Cell Disease, Johns Hopkins University Press, Baltimore, MD

Walker J (2004) Leaps in the Dark: The Making of Scientific Reputations. Oxford University Press, Oxford, UK

Walters L and Palmer JG (1996) The Ethics of Human Gene Therapy. Oxford University Press, New York

Watson JD (2000) A Passion for DNA. Genes, Genomes, and Society. Cold Spring Harbor Laboratory Press, Cold Spring Harbor, NY

Wilson DS (2002) Darwin's Cathedral. Evolution, Religion, and the Nature of Society. University of Chicago Press, Chicago, IL

Winter G (1998) Die Prüfung der Freisetzung von gentechnisch veränderten Organismen. Erich Schmidt, Berlin

Transposable and Mobile Elements ►*genetics genomics*

Arkhipova IR et al (1995) *Drosophila* Retrotransposons. CRC Press, Boca Raton, FL

Berg DE and Howe MM (eds) (1989) Mobile DNA. American Society for Microbiology, Washington, DC

Burt A and Trivers R (2006) Genes in Conflict: The Biology of Selfish Genetic Elements. Belknap Press, Harvard University Press, Cambridge, MA

Bushman F (2002) Lateral DNA Transfer: Mechanisms and Consequences. Cold Spring Harbor Laboratory Press, Cold Spring Harbor, NY

Capy P (1998) Evolution and Impact of Transposable Elements. Kluwer, Dordrecht, N

Craig NL et al (eds) (2002) Mobile DNA II. American Society for Microbiology Press, Washington, DC

Kaper JB and Hacker J (eds) (1999) Pathogenicity Islands and Other Mobile Virulence Elements. American Society for Microbiology, Washington, DC

McClintock B (1988) The Discovery and Characterization of Transposable Elements. The Collected Papers of Barbara McClintock. Garland, New York

McDonald JF (ed) (2000) Transposable Elements and Genome Evolution. Kluwer, Dordrecht, the Netherlands

Miller WJ and Capy P (eds) (2004) Mobile Genetic Elements. Humana Press, Totowa, NJ

Saedler H and Gierl A (eds) (1996) Transposable Elements. Springer, Berlin, Germany

Sherratt DJ (ed) (1995) Mobile Genetic Elements. IRL Press, New York

Volff J-N (ed) (2005) Retrotransposable Elements and Genome Evolution. Karger, Basel, Switzerland

Viruses ►*microbiology*, ►*molecular genetics*

Abraham T (2005) The Twenty-First Century Plague: The Story of SARS. Johns Hopkins University Press, Baltimore, MD

Adolph KW (ed) (1995) Viral Gene Techniques. Academic Press, San Diego, CA

Advances in Virus Research (Serial). Elsevier, Philadelphia, PA

Arvin A and Gershon AA (eds) (2000) Varicella-Zooster Virus: Virology and Clinical Management. Cambridge University Press, New York

Blumberg BS (2002) Hepatitis B: The Hunt for the Killer Virus. Princeton University Press, Princeton, NJ

Boshoff C and Weiss RA (eds) (2007) Kaposi Sarcoma Herpesvirus: New Perspectives. Springer, Berlin, D

Brown SM (ed) (1997) Herpes Simplex Virus Protocols. Humana Press, Totowa, NJ

Brunt A et al (eds) (1997) Viruses of Plants. Descriptions and Lists from VIDE Databases. CAB International. Willingford, UK

Calendar R (ed) (1988) The Bacteriophages. Plenum, New York

Cann AJ (2001) Principles of Molecular Virology. Academic Press, San Diego, CA

Cann AJ (ed) (2000) RNA Viruses: A Practical Approach. Oxford University Press, Oxford, UK

Cann AJ (ed) (2001) DNA Virus Replication. Oxford University Press, New York

Carter J et al (2007) Virology. Principles and Applications. Wiley, Hoboken, NJ

Coffin JM et al (eds) (1997) Retroviruses. Cold Spring Harbor Laboratory Press, Cold Spring Harbor, NY

Crawford DH (2000) The Invisible Enemy: A Natural History of Viruses. Oxford University Press, New York

Cullen B (ed) (1993) Human Retroviruses. IRL/Oxford University Press, New York

Dalgleish AG and Weiss RA (eds) (1999) HIV and the New Viruses. Academic Press, San Diego, CA

Davy C (ed) (2005) Human Papilloma Viruses. Methods and Protocols. Humana Press, Totowa, NJ

De Clercq EDA (ed) (2001) Antiretroviral Therapy. ASM Press, Washington, DC

Domingo E et al (eds) (1999) Origin and Evolution of Viruses. Academic Press, San Diego, CA

Driscoll JS (2004) Antiviral Drugs. Wiley, Hoboken, NJ

Flint SJ et al (eds) (1999) Principles of Virology: Molecular Biology, Pathogenesis, and Control. American Society for Microbiology Press, Washington, DC

Garcea RL and DiMaio D (eds) (2007) The Papillomaviruses. Springer, Berlin, D

Granoff A and Webster RG (eds) (1999) Encyclopedia of Virology. Academic Press, San Diego, CA

Gray J and Desselberger U (2000) Rotaviruses. Methods and Protocols. Humana Press, Totowa, NJ

Greger M (2006) Bird Flu: A Virus of Our Own Hatching. Lantern Books, New York

Hamatake RK and Lau JYN (eds) (2004) Hepatitis B and D Protocols. Vol I. Detection, Genotypes and Characterization. Vol II. Immunology, Model Systems and Clinical Studies. Humana Press, Totowa, NJ

Hendrix RW et al (eds) (1983) Lambda II. Cold Spring Harbor Laboratory Press, Cold Spring Harbor, NY

Hiscox JA (ed) (2006) Viruses and the Nucleus. Wiley, Hoboken, NJ

Isaacs SN (ed) (2004) Vaccinia Virus and Poxvirology. Methods and Protocols. Humana Press, Totowa, NJ

Jirillo E (ed) (2008) Hepatitis C Virus Disease. Immunobiology and Clinical Applications. Springer, Berlin, D
Kaplitt G and Loewy AD (eds) (1995) Viral Vectors. Gene Therapy and Neuroscience Applications. Academic Press, San Diego, CA
Kartikeyan S et al (2007) HIV and AIDS. Basic Elements and Priorities. Springer, Berlin, D
Khalili K and Stoner GL (2001) Human Polyomaviruses: Molecular and Clinical Perspectives. Wiley, New York
Knipe DM and Howley PM (eds) (2001) Fundamental Virology. Lippincott Williams & Wilkins, Philadelphia, PA
Koplow DA (2002) Smallpox. The Fight to Eradicate a Global Scourge. University of California Press, Berkeley, CA
Levy JA (ed) (1995) The Retroviridae. Plenum, New York
Levy JA et al (1994) Virology. Prentice-Hall, Englewood Cliffs, NJ
Lieberman PM (ed) (2005) DNA Viruses. Methods and Protocols. Humana Press, Totowa, NJ
Mahy B (1996) A Dictionary of Virology. Academic Press, San Diego, CA
Mandahar CL (ed) (1999) Molecular Biology of Plant Viruses. Kluwer, Boston, MA
Mercer A et al (eds) (2007) Poxviruses. Birkhäuser Vlg. Basel, Switzerland
Morse SS (ed) (1993) Emerging Viruses. Oxford University Press, New York
Michael N and Kim JH (eds) (1999) HIV Protocols. Humana Press, Totowa, NJ
Minarovits J et al (eds) (2007) Latency Strategies of Herpesviruses. Springer, Berlin, D
Murphy FA et al (1995) Virus Taxonomy. Springer, New York
Murphy FA et al (1999) Veterinary Virology. Academic Press, San Diego, CA
Reeves JD and Derdeyn CA (eds) (2007) Entry Inhibitors in HIIV Therapy. Birkhäuser, Basel, Switzerland
Regenmortel van OHV et al (eds) (2000) Virus Taxonomy. Classification and Nomenclature of Viruses. Academic Press, San Diego, CA
Richman D (ed) (2002) Clinical Virology, AMS Press, Washington, DC
Robertson ES (ed) (2005) Epstein-Barr Virus. Caister Academic Press, Wymondham, UK
Semler BL and Wimmer E (2002) Molecular Biology of Picornaviruses. ASM Press, Washington, DC.
Sinclair J (ed) (1999) Cytomegalovirus Protocols. Humana Press, Totowa, NJ
Stephenson JR and Warnes A (eds) (1998) Diagnostic Virology Protocols. Humana Press, Totowa, NJ
Sterling JC and Stephen KT (eds) (2001) Human Papillomaviruses. Clinical and Scientific Advances. Arnold/Cambridge University Press, New York
Strauss JH and Strauss EG (2007) Viruses and Human Disease. Academic Press, San Diego, CA
Sugrue RJ (ed) (2007) Glycovirology Protocols. Humana Press, Totowa, NJ
Tidona CA and Darai G (2001) The Springer Index of Viruses. Springer, New York
Torrence PF (ed) (2005) Antiviral Drug Discovery for Emerging Diseases and Bioterrorism Threats. Wiley, Hoboken, NJ
Webster RG and Granoff A (eds) (1994) Encyclopedia of Virology. Academic Press, San Diego, CA
White DO and Fenner FJ (1994) Medical Virology. Academic Press, San Diego, CA
Wimmer E (ed) (1994) Cellular Receptors for Animal Viruses. Cold Spring Harbor Laboratory Press, Cold Spring Harbor, NY
Wold WSM and Tollefson AE (eds) (2007) Adenovirus Methods and Protocols. Humana Press, Totowa, NJ
Zhu T (ed) (2005) Human Retrovirus Protocols. Humana Press, Totowa, NJ

Historical notes

"…in pursuing discovery the scientist finds no time to look behind him or besides him.
He can consider only the nearest connections and accept only the quickest answers.
Not for him to dawdle over remote consequences and unborn generations. Or to
Exercise superfluous imagination for the sake of agreement and applause he will not live to hear."

C.D. Darlington (1964) Genetics and Man, Schocken Books, New York, p. 16

Mark Twain on knowledge:

"It ain't what you don't know that gets you into trouble. It's what you know for sure that just ain't so."

Appendix

Gregor Mendel among fellow monks in Alt Brünn *circa* 1862. In front of him sits Abbot FC Napp who accepted him to the Augustinian Order in 1843. In 1868, Mendel became Abbot and Anselm Rambousek (standing second from left) followed him in that position in 1884 after his death. This photograph sheds some light on the characters of his fellows. While some look bored or argumentative, Mendel studies a fuchsia plant. Fortunately, he did not continue with this ornamental plant with chromosome numbers 2n = 22, 55, 66 and 77, that frustrated others before and after his time. He had perhaps good luck but his genius was needed for recognizing his luck. (The photograph is the courtesy of Dr. V Orel.)

Dr. Rölreuter. Řehoř Mendel Ch. Darwin. Carl Friedrich Gärtner.
A. Kölliker August Weismann Th. Boveri. Ch. Naudin
E. Strasburger
Albrecht Kossel Francis Galton W. Johannsen Karl Pearson. hochachtungsvoll Dr. W.O. Focke
W. E. Castle
W. Bateson 1900
Edmund B. Wilson Archibald E. Garrod. Sincerely,
James F. Crow
H. Nilsson R. C. Punnett C. Correns. Ted Puck
Thomas Hunt Morgan Cordially yours R. A. Fisher
Calvin B. Bridges Theophilus S. Painter Charles Yanofsky
A. H. Sturtevant Curt Stern O. Renner. A. F. Blakeslee
Csaba
E. G. Anderson John Belling
R. C. Punnett
H. J. Muller Erwin Baur
Geo. H. Shull Sincerely,
R. Goldschmidt
Philip R. White Seymour Fogel
M. Delbrück Herschel Roman
T. M. Sonneborn
Very truly yours,
Sincerely,
Barbara
Barbara McClintock
with the best compliments from N. Timoféeff-Ressovsky
H G Khorana
A. Kornberg
Keep up the good work.
Sincerely yours,
I. Michael Lerner
Professor of Genetics
Arthur Kornberg
Boris Ephrussi
With best personal regards,
J. D. Watson
James D. Watson
Cordially yours,
G. Ledyard Stebbins
George B.
Beadle
Sincerely
E. B. Lewis
Sincerely yours
Schell
Rollin
Rollin D. Hotchkiss
Ruth Sager
Norman
Norman H. Giles

Reviews of Author's Books

NATURE (Lond) *302*:169 "...sections dealing with molecular genetics ... are remarkably clear and up to date." "...this one is one of the best textbooks of general genetics". "The book is extremely well illustrated... "

THEORETICAL AND APPLIED GENETICS *66*:38 "...a balanced treatment of almost all genetics disciplines"

HEREDITY *33*:123 "...unusually good value."

QUARTERLY REVIEW OF BIOLOGY *74*:74 "The author clearly met his goal: to facilitate communication and understanding across all biological sciences". "...this book should prove very useful as a quick desk reference for students, professionals and non-professionals."

AMERICAN JOURNAL OF MEDICAL GENETICS *83*:145 "He undauntedly covers the entire field of contemporary genetics including not only the classical areas of formal genetics and cytogenetics but also the latest global molecular advances and their applications". "I found the contents very useful... "

ACTA PÆDIATRICA *87*:1211 "The strength of the book lies in its brief explanations ... it is invaluable for anyone interested in this rapidly evolving area, including scientists, teachers, students, and laymen."

ANNALS OF INTERNAL MEDICINE *130*:168 "This book is a valuable reference tool and will be of interest to a wide audience. Other books, including glossaries and dictionaries of genetics are available but are more limited in coverage".

ANNALS OF THE ENTOMOLOGICAL SOCIETY OF AMERICA *91*:894 "The preparation of a book like this is a massive task. If it is done well, it can provide a valuable central resource to information that might otherwise take a lengthy search to uncover". "In this case ... the overall effort has been very successful."

HortScience *33*:1274 "The book is attractive and user-friendly." "This manual is by far the best I have used".

MUTATION RESEARCH FORUM 1999 May "... this book is a true information warehouse". "The words in this book came through as if someone were talking directly to me." "The investment ... is worth it."

JOURNAL OF PHYTOPATHOLOGY *147*:623 "This giant volume is recommended to those who are in contact with the genetic aspects of questions in their field of research or practice." "The most important feature of this book is that all concepts are explained not just defined." "The style is clear even for the beginner".

ACTA AGRONOMICA HUNGARICA *47*(2):237 "...although a number of dictionaries or glossaries of genetics already exist, this ... is quite different from the others in-as-much as it is much more comprehensive, and not only defines the concepts but explains them in concise plain language"... "fills a void".

CHOICE 1998, *36*–1563 "...outstanding compendium of genetics"... "Highly recommended for biologists".

GENOMICS & PROTEOMICS 2003 (Nov./Dec):56. "useful resource...50% more information than in the first edition".

GENOMICS REVIEWS (2004, Internet). "Excellent resource."

Book News (2004, Internet). " By authoring the work single-handedly (clearly a labor of love) Redei (emeritus, U. of Missouri) could maintain a consistency of approach and avoid redundancy and yet still attain impressive breadth and depth in a compact reference."

GENOMICS & PROTEOMICS (2003, Nov/Dec):56. "...useful resource... 50% more information than in the first edition".

NÖVÉNYTERMELÉS (2003, *52*:724). "...excellent tool...I recommend it to the attention of everybody". (translated from Hungarian)

CHOICE (2004, 882:41–2549). "...an expanded version of the highly acclaimed Genetics Manual... reflects recent developments in genetics, molecular and cell biology, genomics and proteomics." ..."valuable... Recommended".

BOOKWORKZ (2004, Internet). "...incomparable reference...thorough coverage of modern genetics, genomics, and proteomics".

E-STREAMS (2004, *7*[1]:2969). "This is a highly usable item and provides effective and efficient access to relevant topics."

THE SCIENCE ADVISORY BOARD (2004, Internet). "... the best got even better." "This book is indispensable, as a reference for research universities, for high schools, for public libraries, and because of the relatively modest price ...it is affordable for individuals too."

BIOspectrum (2004, *10*:653). "it is truly wonderful". "Most students, scientists, all geneticists, including human geneticists, the general public interested in natural sciences, teachers and journalists can benefit from it to a great degree. It is a must for larger biological libraries!" (translated from German).

CLINICAL CHEMISTRY (2004, *50*:1719–1720). "... the book has an encyclopedic quality, far surpassing a dictionary." "...many readers will find themselves engrossed in this fascinating book." ..."This is a wonderful book..."

PROTEOMICS (2004, *4*:544–545). "Once you start reading this book it is difficult to put down. It is a valuable source of such a breadth of information that would surely be difficult to find elsewhere." "...a highly recommended one-stop resource for medics, grad students and researchers."

GENETIC ENGINEERING NEWS (2004, *24*(6), March 15): "...genomics and post-genomics terms, including concepts, theories, and applications, explained concisely for novices and comprehensive enough for those with expertise."

REFERENCE REVIEWS (2004, 18(62004):43–44): "excellent volume"..."is a readable, accessible, well thought out and extremely useful reference book."

MICROBE (Amer. Soc. Microbiol.) (2004, May): "It is impressive that a single author can put together a dictionary of this size and make it useful to the general scientific reader."..."Rédei's definitions are accurate"... :[he] is a clear winner."

THE CHEMICAL EDUCATOR (2005, 10(6):56–57) "Rédei has eminently succeeded in attaining his goal." "Numerous cross-references link the short entries into a most encyclopedic, up-to-date text."..."gold mine of valuable...information."

MEDICAL REFERENCE SERVICES QUARTERLY (2006, 2[(1]:104) "...high quality reference book" "this is the most comprehensive..."

About the Author

George P. Rédei is Professor Emeritus at the University of Missouri, Columbia. Born in Vienna, Austria he completed his formal education in Hungary. Rédei is affiliated with the University of Missouri since 1957 where he has taught formal courses in basic genetics, analytical genetics, history of genetics and genetic engineering. He has been visiting professor at several leading institutions in Germany, Hungary and China.

Rédei has introduced Arabidopsis to genetic research in the USA and published seminal research papers on classical, molecular, population genetics and history of genetics in Science, Nature, Cell, Genetics, Canadian Journal of Genetics and Cytology, Journal of Heredity, Genome, Experientia, EMBO Journal, Proceedings of the National Academy of Sciences USA, Genes & Development, Molecular and General Genetics, Development, Mutation Research, Annual Review of Genetics, Advances in Genetics, Bibliographia Genetica, Biochemical Genetics, Theoretical and Applied Genetics, Biologisches Zentralblatt, Protoplasma, Plant Physiology and other international journals. More than 50,000 students have used his textbook, Genetics (translated also into many foreign languages). Preceding editions and translation of the present volume have been bestsellers.

Rédei is a foreign member of the Hungarian National Academy of Sciences since 1990 and in 2004 the University of Missouri named a section of the new Life Sciences Building in his honor.